FUNDAMENTAL AND PHYSICAL CONSTANTS

quantity	symbol	English	SI
Charge			
electron	e		-1.6022×10^{-19} C
proton	p		$+1.6021 \times 10^{-19}$ C
Density			
air [STP] [32°F, (0°C)]		0.0805 lbm/ft^3	1.29 kg/m^3
air [70°F, (20°C), 1 atm]		0.0749 lbm/ft^3	1.20 kg/m^3
earth [mean]		345 lbm/ft^3	5520 kg/m^3
mercury		849 lbm/ft^3	1.360×10^4 kg/m^3
sea water		64.0 lbm/ft^3	1025 kg/m^3
water [mean]		62.4 lbm/ft^3	1000 kg/m^3
Distance [mean]			
earth radius		2.09×10^7 ft	6.370×10^6 m
earth-moon separation		1.26×10^9 ft	3.84×10^8 m
earth-sun separation		4.89×10^{11} ft	1.49×10^{11} m
moon radius		5.71×10^6 ft	1.74×10^6 m
sun radius		2.28×10^9 ft	6.96×10^8 m
first Bohr radius	a_0	1.736×10^{-10} ft	5.292×10^{-11} m
Gravitational Acceleration			
earth [mean]	g	32.174 (32.2) ft/sec^2	9.8067 (9.81) m/s^2
moon [mean]		5.47 ft/sec^2	1.67 m/s^2
Mass			
atomic mass unit	u	3.66×10^{-27} lbm	1.6606×10^{-27} kg
earth		4.11×10^{23} slugs	6.00×10^{24} kg
earth [customary U.S.]		1.32×10^{25} lbm	n.a.
electron [rest]	m_e	2.008×10^{-30} lbm	9.109×10^{-31} kg
moon		1.623×10^{23} lbm	7.36×10^{22} kg
neutron [rest]	m_n	3.693×10^{-27} lbm	1.675×10^{-27} kg
proton [rest]	m_p	3.688×10^{-27} lbm	1.673×10^{-27} kg
sun		4.387×10^{30} lbm	1.99×10^{30} kg
Pressure, atmospheric		14.696 (14.7) lbf/in^2	1.0133×10^5 Pa
Temperature, standard		32°F (492°R)	0°C (273K)
Velocity			
earth escape		3.67×10^4 ft/sec	1.12×10^4 m/s
light [vacuum]	c	9.84×10^8 ft/sec	2.9979 (3.00) $\times 10^8$ m/s
sound [air, STP]	a	1090 ft/sec	331 m/s
[air, 70°F (20°C)]		1130 ft/sec	344 m/s
Volume, molal ideal gas [STP]		359 ft^3/lbmol	22.41 m^3/kmol
Fundamental Constants			
Avogadro's number	N_A		6.022×10^{23} mol^{-1}
Bohr magneton	μ_B		9.2732×10^{-24} J/T
Boltzmann constant	k	5.65×10^{-24} ft-lbf/°R	1.3807×10^{-23} J/K
Faraday constant	F		96 487 C/mol
gravitational constant	g_c	32.174 lbm-ft/lbf-sec^2	
gravitational constant	G	3.44×10^{-8} ft^4/lbf-sec^4	6.672×10^{-11} N·m^2/kg^2
nuclear magneton	μ_N		5.050×10^{-27} J/T
permeability of a vacuum	μ_0		1.2566×10^{-6} N/A^2 (H/m)
permittivity of a vacuum	ϵ_0		8.854×10^{-12} C^2/N·m^2 (F/m)
Planck's constant	h		6.6256×10^{-34} J·s
Rydberg constant	R_∞		1.097×10^7 m^{-1}
specific gas constant, air	R	53.3 ft-lbf/lbm-°R	287 J/kg·K
Stefan-Boltzmann constant		1.71×10^{-9} Btu/ft^2-hr-°R^4	5.670×10^{-8} W/m^2·K^4
triple point, water		32.02°F, 0.0888 psia	0.01109°C, 0.6123 kPa
universal gas constant	R^*	1545 ft-lbf/lbmol-°R	8314 J/kmol·K
	R^*	1.986 Btu/lbmol-°R	

Mathematics
Machine Design
Fluids
Dynamics and Vibrations
Thermodynamics
Control Systems
Power Cycles
Plant Engineering
Heat Transfer
Economics
HVAC
Law and Ethics
Statics
Support Material
Materials

C. MCCLAIN

Mechanical Engineering Reference Manual

for the PE Exam

Tenth Edition

Michael R. Lindeburg, PE

Professional Publications, Inc. • Belmont, CA

Production Manager: Aline S. Magee
Acquisitions Editor: Gerald R. Galbo
Permissions Editors: Catie Berkenfield and Carol Thein Moore
Copy Editor: Mia Laurence
Book Designer: Charles P. Oey
Typesetter: Kate Hayes
Illustrator: Yvonne M. Sartain
Proofreaders: Jessica R. Whitney-Holden and Margaret S. Yoon
Cover Designer: Charles P. Oey

MECHANICAL ENGINEERING REFERENCE MANUAL
Tenth Edition

Printed in the United States of America

Professional Publications, Inc.
1250 Fifth Avenue, Belmont, CA 94002
(650) 593-9119
www.ppi2pass.com

Current printing of this edition: 3

Library of Congress Cataloging-in-Publication Data
Lindeburg, Michael R.
    Mechanical engineering reference manual : for the PE exam /
    Michael R. Lindeburg. -- 10th ed.
        p.   cm. -- (Engineering reference manual series)
    Includes index.
    ISBN 1-888577-13-4
    1. Mechanical engineering--Problems, exercises, etc.
    2. Mechanical engineering--Handbooks, manuals, etc.   I. Title.
    II. Series.
    TJ159.L5   1997
    621--DC21   97-7741
                                                         CIP

Table of Contents

Preface to the Tenth Edition v
Acknowledgments for the Tenth Edition vii
Introduction ix

Systems of Units 1-1
Engineering Drawing Practice 2-1
Algebra 3-1
Linear Algebra 4-1
Vectors 5-1
Trigonometry 6-1
Analytic Geometry 7-1
Differential Calculus 8-1
Integral Calculus 9-1
Differential Equations 10-1
Probability and Statistical Analysis
    of Data 11-1
Numbering Systems 12-1
Numerical Analysis 13-1
Fluid Properties 14-1
Fluid Statics 15-1
Fluid Flow Parameters 16-1
Fluid Dynamics 17-1
Hydraulic Machines 18-1
Fluid Power 19-1
Fans and Ductwork 20-1
Inorganic Chemistry 21-1
Fuels and Combustion 22-1
Energy, Work, and Power 23-1
Thermodynamic Properties
    of Substances 24-1
Changes in Thermodynamic Properties 25-1
Compressible Fluid Dynamics 26-1
Vapor Power Equipment 27-1
Vapor Power Cycles 28-1
Combustion Power Cycles 29-1
Nuclear Power Cycles 30-1
Advanced and Alternative
    Power-Generating Systems 31-1
Gas Compression Cycles 32-1
Refrigeration Cycles 33-1
Fundamental Heat Transfer 34-1
Natural Convection, Evaporation,
    and Condensation 35-1
Forced Convection and Heat Exchangers 36-1
Radiation and Combined Heat Transfer 37-1
Psychrometrics 38-1
Ventilation 39-1
Heating Load 40-1
Cooling Load 41-1
Air Conditioning Systems and Controls 42-1
Determinate Statics 43-1

Indeterminate Statics 44-1
Engineering Materials 45-1
Material Properties and Testing 46-1
Thermal Treatment of Metals 47-1
Properties of Areas 48-1
Strength of Materials 49-1
Failure Theories 50-1
Basic Machine Design 51-1
Advanced Machine Design 52-1
Pressure Vessels 53-1
Properties of Solid Bodies 54-1
Kinematics 55-1
Kinetics 56-1
Mechanisms and Power Transmission
    Systems 57-1
Vibrating Systems 58-1
Modeling of Engineering Systems 59-1
Analysis of Engineering Systems 60-1
Management Science 61-1
Instrumentation and Measurements 62-1
Manufacturing Processes 63-1
Material Handling and Processing 64-1
Fire Protection Systems 65-1
Environmental Engineering 66-1
Electricity and Electrical Equipment 67-1
Illumination and Sound 68-1
Engineering Economic Analysis 69-1
Engineering Law 70-1
Engineering Ethics 71-1
Engineering Registration in the
    United States 72-1

Appendices A-1
Index I-1
Index of Selected Figures, Tables,
    Appendices, and Compiled Data I-43

Appendices Table of Contents

1.A   Conversion Factors A-1
1.B   Common SI Unit Conversion Factors A-3
7.A   Mensuration of Two-Dimensional Areas A-7
7.B   Mensuration of Three-Dimensional Volumes A-9
9.A   Abbreviated Table of Indefinite Integrals A-10
10.A Laplace Transforms A-11
11.A Areas Under the Standard Normal Curve A-12
14.A Properties of Water at Atmospheric Pressure
       (English units) A-13
14.B Properties of Water at Atmospheric Pressure (SI units) A-13
14.C Viscosity of Water in Other Units A-14
14.D Properties of Air at Atmospheric Pressure
       (English units) A-14
14.E Properties of Air at Atmospheric Pressure (SI units) . A-14
14.F Properties of Common Liquids A-15
14.G Properties of Uncommon Fluids A-16
14.H Vapor Pressures of Various Hydrocarbons and Water . A-17
14.I Specific Gravity of Hydrocarbons A-18
14.J Viscosity Conversion Chart A-19
14.K Viscosity Index Chart: 0–100 V.I. A-20
16.A Area, Wetted Perimeter, and Hydraulic Radius
       of Partially Filled Circular Pipes A-21
16.B Dimensions of Welded and Seamless Steel
       Pipe (selected sizes) (English units) A-22
16.C Dimensions of Welded and Seamless Steel Pipe
       (schedules 40 and 80) (SI units) A-25
16.D Dimensions of Copper Water Tubing (English units) . A-26
16.E Dimensions of Brass and Copper Water Tubing
       (English units) A-27
16.F Dimensions of Seamless Steel Boiler (BWG)
       Tubing (English units) A-28
16.G Dimensions of PVC Pipe (English units) A-29
16.H Standard ANSI Piping Symbols A-30
17.A Specific Roughness and Hazen-Williams
       Constants for Various Pipe Materials A-31
17.B Darcy Friction Factors A-32
17.C Flow of Water Through Schedule-40 Steel Pipe . . . A-36
17.D Equivalent Length of Straight Pipe for Various
       (Generic) Fittings A-37
19.A Symbols for Fluid Power Equipment A-38
21.A Atomic Numbers and Weights of the Elements A-39
21.B Periodic Table of the Elements A-40
21.C Water Chemistry $CaCO_3$ Equivalents A-41
22.A Heats of Combustion for Common Compounds A-43
24.A Properties of Saturated Steam by Temperature . . . A-44
24.B Properties of Saturated Steam by Pressure A-46
24.C Properties of Superheated Steam A-47
24.D Properties of Compressed Water A-49
24.E Enthalpy-Entropy (Mollier) Diagram for Steam . . . A-50
24.F Properties of Low-Pressure Air A-51
24.G Properties of Saturated Refrigerant-12
       by Temperature A-53
24.H Properties of Saturated Refrigerant-12 by Pressure . A-54
24.I Properties of Superheated Refrigerant-12 . . . . . . A-55
24.J Pressure-Enthalpy Diagram for Refrigerant-12 . . . A-57
24.K Properties of Saturated Refrigerant-22
       by Temperature A-58
24.L Pressure-Enthalpy Diagram for Refrigerant-22 . . . A-60
24.M Pressure-Enthalpy Diagram for Refrigerant HFC-134a A-61
24.N Properties of Saturated Steam by Temperature . . . A-62
24.O Properties of Saturated Steam by Pressure A-64
24.P Properties of Superheated Steam A-65
24.Q Properties of Compressed Water A-66
24.R Enthalpy-Entropy (Mollier) Diagram for Steam . . . A-67
24.S Properties of Low-Pressure Air A-68
24.T Properties of Saturated Refrigerant-12 by
       Temperature A-70
24.U Pressure-Enthalpy Diagram for Refrigerant-12 . . . A-72
24.V Properties of Saturated Refrigerant-22
       by Temperature A-73
24.W Pressure-Enthalpy Diagram for Refrigerant-22 . . . A-75
24.X Pressure-Enthalpy Diagram for Refrigerant HFC-134a A-76
24.Y Physical Properties of Selected Solids A-77
24.Z Generalized Compressibility Charts A-78
26.A Isentropic Flow Factors A-79
26.B Isentropic Flow and Normal Shock Parameters . . . A-80
26.C Fanno Flow Factors A-81
26.D Rayleigh Flow Factors A-82

26.E International Standard Atmosphere A-83
34.A Representative Thermal Conductivity A-84
34.B Properties of Metals and Alloys A-85
34.C Properties of Nonmetals A-86
34.D Transient Heat Flow Charts
       (solid spheres of radius r_o) A-87
34.E Transient Heat Flow Charts
       (infinite solid circular cylinders of r_o) A-88
34.F Transient Heat Flow Charts
       (infinite flat slabs of thickness $2L$) A-89
34.G Heisler Transient Heat Flow Chart (temperature
       at center of a sphere of radius r_o) A-90
34.H Heisler Transient Heat Flow Chart (temperature
       at center of infinite cylinder of radius r_o) A-91
34.I Heisler Transient Heat Flow Chart (temperature
       at center of infinite slab of thickness $2L$) A-92
35.A Properties of Saturated Water (English units) A-93
35.B Properties of Saturated Water (SI units) A-93
35.C Properties of Atmospheric Air (English units) A-94
35.D Properties of Atmospheric Air (SI units) A-94
35.E Properties of Saturated Steam at
       One Atmosphere (English units) A-95
35.F Properties of Saturated Steam at
       One Atmosphere (SI units) A-95
36.A Correction Factor, F_c, for the Logarithmic Mean
       Temperature Difference (one shell pass, even
       number of tube passes) A-96
36.B Correction Factor, F_c, for the Logarithmic Mean
       Temperature Difference (two shell passes,
       multiple of four tube passes) A-96
36.C Characteristics of Birmingham Wire Gage
       (BWG) Size Tubing A-97
36.D Heat Exchanger Effectiveness A-98
37.A Emissivities of Various Surfaces A-100
38.A ASHRAE Psychrometric Chart No. 1, Normal
       Temperature—Sea Level (32–120°)
       (customary U.S. units) A-101
38.B ASHRAE Psychrometric Chart No. 1, Normal
       Temperature—Sea Level (0–50°) (SI units) . . . A-102
38.C ASHRAE Psychrometric Chart No. 2,
       Low Temperature—Sea Level (−40–50°)
       (customary U.S. units) A-103
38.D ASHRAE Psychrometric Chart No. 3, High
       Temperature—Sea Level (60–250°)
       (customary U.S. units) A-104
40.A Representative Insulating Properties of
       Selected Building Materials A-105
46.A Typical Mechanical Properties of
       Representative Metals A-107
46.B Typical Mechanical Properties of Thermoplastic
       Resins and Composites A-110
48.A Centroids and Area Moments of Inertia for
       Basic Shapes A-111
49.A Elastic Beam Deflection Equations A-112
49.B Stress Concentration Factors A-115
51.A Properties of Weld Groups A-116
52.A Spring Wire Diameters and Sheet Metal Gauges . . . A-117
52.B Journal Bearing Correlations A-118
54.A Mass Moments of Inertia A-119
57.A Standard Epicyclic Gear Train Ratios A-120
61.A Percentile Values of Student's t-Distribution . . . A-121
61.B Single Sampling Plan Table for Various
       Producer's and Consumer's Risks A-122
61.C Cumulative Probability Chart A-123
62.A Thermoelectric Constants for Thermocouples A-124
65.A Friction Loss in Schedule-40 Steel Pipe
       (customary U.S. units) A-125
65.B Friction Loss in Schedule-40 Steel Pipe (SI units) . A-126
65.C Equivalent Lengths of Valves and Fittings for
       Fire Protection Systems A-127
65.D Standardized Sprinkler System Worksheet A-128
67.A Polyphase Motor Classifications and Characteristics . A-129
67.B DC and Single-Phase Motor Classifications and
       Characteristics A-130
68.A Noise Reduction and Absorption Coefficients A-131
68.B Transmission Loss Through Common Materials . . . A-132
69.A Standard Cash Flow Factors A-133
69.B Factor Tables A-134

Preface to the Tenth Edition

It has been approximately 13 years since this book experienced its last significant revision, and it has been 20 years since the first edition was printed.

During those 20 years, I have authored more than 20 books, taught more than a thousand mechanical engineers in classroom courses, read hundreds of multidisciplinary engineering books and magazines, and reviewed thousands of questionnaires from more than 50 mail surveys conducted to determine what mechanical engineers want and need. I have also received, read, and responded to hundreds of letters from readers who sent me their suggestions.

As the user of this tenth edition of the *Mechanical Engineering Reference Manual*, you are the beneficiary of all these activities.

This book is intended to provide a broad review of mechanical engineering design and analysis principles, as well as a compendium of data to support those principles. In choosing subjects for inclusion in the tenth edition, I was greatly influenced by four sources of information. In decreasing order of importance, these were: (1) unsolicited and survey comments received from literally thousands of mechanical engineers who have purchased previous editions of this book, (2) task analyses of the mechanical engineering profession published by the National Council Examiners for Engineering and Surveying, (3) mechanical engineering curricula at leading colleges and universities, and (4) current literature in the field of mechanical engineering.

Because of its emphasis on undergraduate engineering subjects (the only prerequisite course of study required of most state licensing examinees), this book is an efficient method of preparing for the national Professional Engineering (PE) examination in Mechanical Engineering. Tens of thousands of engineers have used the previous nine editions to do just that.

In addition to state licensing examinees, however, the intended audience for this book includes mechanical engineering majors needing a general, all-inclusive reference book covering their course of study, as well as practicing engineers desiring a convenient handbook of methods and data from familiar but perhaps faintly remembered subjects.

An important requirement for passing the PE examination is adequate exposure to the areas of questions typically found on the examination. While I have not included any actual examination problems, this book contains numerous practice problems in all PE subject areas. Using original problems maintains the validity of the examination process while affording you the practice you need.

I know from first-hand experience that the easiest type of engineering book to write is a simple collection of solved problems. The effort required to compile typical engineering problems pales in comparison to the effort required to present the engineering knowledge needed to solve those typical problems.

I have authored my share of problem collections. *101 Solved Mechanical Engineering Problems* is a case in point. But, the intent of this book is to provide you with a resource for your daily work long after you have taken an exam.

I have organized this book such that each topic is presented in its own stand-alone section. In most cases, each successive chapter builds on previous chapters, and each successive section builds on the information in previous sections.

There are many differences between this edition and the previous edition. In fact, this edition is essentially a brand new book. Here are some of the major differences.

- Many new subjects have been added, and coverage of old subjects is greatly expanded.

- SI units, equations, data, and examples are presented throughout in parallel with their customary U.S. counterparts.

- Many new tables of data have been added, and old tables have been updated.

- The original 23 chapters of the old edition have been expanded into 72 chapters covering smaller amounts of engineering knowledge.

- Sections within each chapter have been organized in a logical, progressively dependent order.

- All appendices have been relocated to the back of the book.

- The index has been expanded, and most entries have been entered in several different ways.

In this tenth edition, I have been even more rigorous in citing the sources used to assemble data tables, even when the original sources are in the public domain.

The end-of-chapter practice problems that most closely parallel exam problems in level of difficulty are identified with the phrase *Time limit: one hour.*

The presentation has been enhanced by a new page design, increased use of figures, and modernized typographic techniques. New typesetting has eliminated the ragged-looking appearance of the previous editions (that was created on the world's first commercial laser printer, incidentally). Most of the illustrations were drawn on computer. Other innovations in typesetting and illustrating have also been used to produce this book, but the details are pretty boring unless you are a publisher.

The scope of this book exceeds that of the PE exam, and some of the practice problems at the end of each chapter are more difficult than typical exam problems. As I wrote the book, I kept the phrase "engineers should know something about everything" in my head. I included the subjects that I thought a mechanical engineer should know something about.

If you want even more practice with problems representative of the level of difficulty of PE exam problems, I suggest you obtain either the *Mechanical Engineering Sample Examination* (which contains 20 problems) or *101 Solved Mechanical Engineering Problems*, both authored by me and published by Professional Publications.

The choice of subjects and data to be included in this book, though greatly influenced by thousands of engineers, was ultimately mine, as was the choice of degree and depth of coverage of these subjects. After all these years, I think that I have a pretty good feel for the needs of a mechanical engineer. You can help improve that feeling by letting me hear from you. Thousands of engineers have done just that, and this book is the result. The next edition will be the result of your input also.

Michael R. Lindeburg, PE

Acknowledgments for the Tenth Edition

It is hard to believe that 20 years ago, I wrote, typed, edited, and illustrated the first edition of (what was to become the basis of) the *Mechanical Engineering Reference Manual*. Like the bumblebee that doesn't know the aerodynamic laws forbidding the possibility of its flight, no one told me producing a book alone was impossible.

The team that collectively produced this book is substantial in number and talent. I'll begin by mentioning the staff at Professional Publications. (I hope they never tire of hearing me thank them.) Dean Suzuki, Chief Operating Officer, took most of my workload to give me the writing time I needed. Aline Sullivan, Production Department Manager, oversaw the printing of this book. She pitched in and proofread some of the more imposing tables of data, as well. Catie Berkenfield, Acquisitions Editor, and Carol Moore, Customer Service Department Manager, obtained the permissions (almost a hundred) that were needed to use figures and data from other publishers' books. Finding some of the original sources was the proverbial weasel in a smokestack.

Within Professional Publications' Production Department, Yvonne Sartain rendered my illustrations. As this book has been converted from the traditional ink-on-paper format to electronic format, she had to redo every single illustration in the old edition as well as create the new illustrations.

Kate Hayes typeset the book. She let her fingers do the walking. And talking. And everything else, for that matter. What is really humbling is the number of characters she keyboarded. There is tremendous "overhead" with technical typesetting. Every Greek letter and every symbol I wrote required her to do more than I care to think about.

These days, there are very few high school or college graduates who can properly check spelling, punctuation, and grammar. Mia Laurence copyedited the manuscript, and she is quite competent in these areas. But, being a good copyeditor is much, much more than that. A copyeditor gives a book personality. A copyeditor gives a book style. And consistency. To be a great copyeditor, you have to be in tune with the subject material. Mia does that, and her work has brought engineering review publications to a new level of professionalism.

Jessica Whitney-Holden proofread the finished illustrations and text, and she did it more times than I did. (And, more times than I even know about.) After trying to read my mind to figure out what I really wanted "it" to look like, she had to remember how "it" was done in previous chapters, and then reconcile the copyedited manuscript against what was typeset. Proofreading is a very nonlinear, intuitive process. To be a great proofreader, you have to do more than compare two things side by side. She is, and she does.

As the second and newest staff proofreader on this book's production team, Margaret Yoon has given new meaning to the term "quick study." After joining Professional Publications more than halfway into the project, she has been a huge contributor to the ultimate quality of this book. She has even been checking my calculations on her own calculator.

There is no question in my mind that this production team is absolutely the most competent, dedicated, driven, talented, and forgiving group of individuals that I have ever had the good fortune to work on my books.

This is the first edition of the *Mechanical Engineering Reference Manual* that was written to the suggestions of a Mechanical Engineering Advisory Committee. The following individuals served on that committee at one time or the other. They reviewed the previous (8th and 9th) editions and made many suggestions that were incorporated into this edition. Some of these reviews have spanned almost five years.

Robert H. Allen, PhD, PE, Consulting Engineer, Baltimore, MD (Statics, Dynamics)

Steve Andrew, PE, Failure Analysis Associates, Menlo Park, CA (all subjects)

Jesse D. Boak, PE, New Castle, PA (Engineering Economic Analysis, Hydraulic Machines, Machine Design)

Arlin Fynaardt, PE, Design Engineer, Richmond, VA (Mathematics, Engineering Economic Analysis, Machine Design)

R. Bruce Hopkins, PE, Consulting Engineer, Cedar Falls, IA (Mechanics of Materials and Machine Design)

Robert L. Huddleston, PE, Churchville, MD (Materials Testing, Mechanics of Materials, Nondestructive Testing)

John L. Mathieson, PhD, State University of New York, Bronx, NY (all subjects)

Joseph J. Misuraca, PE, Excell Engineering, Inc., Piscataway, NJ (all subjects)

Otto J. Nussbaum, PE, (deceased), Newtown, PA (Heat Transfer, HVAC, Fans and Ductwork)

Edward P. O'Brien, PE, Cleveland State University, Westlake, OH (Combustion, HVAC, Combustion)

Gilberto Russo, PhD, Massachusetts Institute of Technology, Cambridge, MA (various subjects)

Irvin P. Vatz, PE, University of Alabama at Huntsville (Fans and Ductwork, Heat Transfer, HVAC)

The following individuals reviewed pieces of the manuscript for the new tenth edition as they were written.

Mahesh Aggarwal, PhD, Gannon University, Erie, PA (Vapor Power Cycle Equipment, Combustion Power Cycles, and Heat Transfer)

Kevin M. Bailey, PE, Spiral Heat Exchanger Product Manager, Alfa Laval Thermal, Inc., Richmond, VA (Forced Convection and Heat Exchangers)

Robert R. Bolmarcich, PhD, Department of Mechanical Engineering, McNeese State University, Lake Charles, LA (HVAC)

John R. Cochran, PE, Air Quality Control Section Supervisor, Black & Veatch, Kansas City, MO (Environmental Engineering)

E. L. Gerber, PhD, PE, Professor, Electrical and Computer Engineering Department, Drexel University, Philadelphia, PA (Instrumentation and Measurements)

Victor Gerez, PhD, PE, Professor and Department Head, Electrical Engineering Department, Montana State University, Bozeman, MT (Electricity and Electrical Equipment)

Jerry Hamelink, PE (Fluid Power Transmission and Fans and Ductwork)

C. Lynn Kiaer, PhD, Assistant Professor, Rose-Hulman Institute of Technology, Mathematics Department, Terre Haute, IN (Management Science)

Glenn D. Nasman, Consulting Engineer, Manager of Vessel Engineers, Raytheon Engineers and Constructors, Cambridge, MA (Pressure Vessels)

Michel A. Saad, PhD, Professor, Department of Mechanical Engineering, Santa Clara University, Santa Clara, CA (Combustion)

I was fortunate to have acquired from Dr. Allan Kraus, PE, formerly on the staff at the Naval Postgraduate School (Monterey CA), the reprint rights to his *Studying to Pass the Professional Engineers Licensing Examination*, from which I have drawn inspiration and ideas for several examples and practice problems. This was a landmark licensing review book. Thirty years ago, it established a standard for professional presentation that most review books still cannot meet. The code-related material in this grand old book is now out of date, but some parts of it could still serve as an effective exam review today.

Finally, I have some words to say about my family. The acknowledgments in most of my books end in some comment to the effect that my family lost me and "suffered" accordingly during the long writing process. This time, I tried valiantly not to let my writing interfere too much with family and personal time. While writing, I still played racquetball and soccer. I still had lunch with my wife, Elizabeth. I still helped my daughters, Jenny and Katie, with their homework. I coached my youth soccer team to a state championship and served as the goalkeeping coach at my older daughter's high school. So, instead of professing regret to my family for the time lost, this time I'll say, "It was fun being around."

What else can I say? Thanks to you all! Let's do it again in a few years!

Michael R. Lindeburg

Introduction

Part 1: How You Can Use This Book

Part 2: Everything You Ever Wanted to Know About the PE Exam

Part 3: How to Prepare for and Pass the PE Exam in Mechanical Engineering

Part 1: How You Can Use This Book

1. QUICKSTART

If you never read the material at the front of your books anyway, and if you are in a hurry to begin and you only want to read one paragraph, here it is.

> Most chapters in this book are independent. Start with any one and look through it. Decide if you are going to work problems in that subject. Solve all the problems you have time for. Use the index extensively. Don't stop studying until the exam. Good luck.

However, if you want to begin a thorough review, you will probably try to find out everything there is to know about the PE exam. The rest of this introduction is for you.

2. IF YOU ARE A PRACTICING ENGINEER

If you are a practicing engineer or an engineering major and have purchased this book as a general reference handbook, it will probably sit in your bookcase until you have a specific need.

However, if you are a practicing engineer preparing for the PE examination in mechanical engineering, a few suggestions may help.

- You should become intimately familiar with this book. This includes knowing the order of the chapters, the approximate locations of important figures and tables, the contents of the appendices, and so on.

- You should use the accompanying solutions manual to help you solve end-of-chapter practice problems.

- You should use the subject title tabs along the side of each page, which roughly correspond to PE exam subjects.

- Know which subjects in this book are not covered in the PE exam. Several chapters in this book are supportive and do not cover exam topics (see Sec. 5). These chapters provide background for the other chapters and are your "insurance" against rogue questions from the fringe of mechanical engineering.

- It is most efficient to skim through a chapter and familiarize yourself with the subjects before starting the practice problems. Then, you will know the location of each subject if you need a quick review.

- It isn't necessary to solve every end-of-chapter practice problem. The number of practice problems you solve will depend on how much time you have and how skilled you are in each area.

- If you decide to work in customary U.S. (English) units, you will find many equations in which the quantity g/g_c appears. For calculations at standard gravity, the numerical value of this fraction is 1.00. Therefore, it is necessary to incorporate this quantity only in problems with a nonstandard gravity, or when you are being meticulous with units.

- To minimize time spent in searching for often-used formulas and data, you should prepare a one-page summary of all the important formulas and information in each subject area.[1] You can then use these summaries during the examination instead of searching for the correct page in this book.

- This book was meant to be used with its subject index. Every significant term, law, theorem, and concept presented in this book has been indexed in every conceivable way—backwards and forwards. If you don't recognize a term used in a problem statement, look for it in the index. More likely than not, you'll find the subject listed.

- Some subjects appear in more than one chapter. You should use the index liberally to learn all there is about a particular subject.

---

[1] *Quick Reference for the Mechanical Engineering PE Exam*, published by Professional Publications, is a summary of equations, methods, and data needed during the exam.

3. IF YOU ARE AN INSTRUCTOR

The earliest editions of this book started as a series of handouts for a PE review course. It originally was intended as a reference for all of the long formulas, illustrations, and tables of data that I did not have time to put on the chalkboard.

If you are teaching a review course for the PE examination without the benefit of recent, first-hand exam experience, you can use the material in this book as a guide to prepare your lectures. You should emphasize the subjects in each chapter and avoid subjects omitted. You can feel confident that subjects omitted from this book have rarely, if ever, appeared on the PE exam.

It has always been my goal to overprepare my own students. For that reason, the examples and practice problems are often more difficult and more varietal than actual examination problems. Also, you will appreciate that it is more efficient to cover several procedural steps in one practice problem than to ask simple "one-liners." That is the reason there are no multiple-choice examples in this book.

There are many practice problems for each major examination subject. All problems are assigned in my review courses. To do all the problems requires approximately 15 to 20 hours of preparation per week for approximately 14 weeks.

Capacity assignment is the goal of my courses. If you assign 15 hours of homework, and a student can only put in 10 hours of preparation that week, that student will have worked to his or her capacity. After the PE examination, your students will honestly say that they could not have prepared any more than they did in your course.

Homework assignments in my courses are not individually graded. Instead, the students are given the accompanying *Solutions Manual* containing solutions to all practice problems in advance. However, each student must turn in a completed set of problems for credit each week. I address special needs or questions written on the assignments by exception.

I have found that a 14-week format works well for a PE review course. Each week, there is a three-hour lecture with a short break. The following table outlines the basic course format that has worked well for me. If you can add more course time, your students will appreciate it. However, I don't think you can cover the full breadth of material in much less time or in many fewer weeks.

I have tried to order the subjects in a logical, progressive manner. For example, heat transfer and HVAC are dependent on thermodynamic principles, so they come after. Also, machine design comes after statics, engineering materials, and mechanics of materials.

Typical PE Review Course Format

meeting	subject covered	chapters
1.	introduction to the examination	
2.	thermodynamics	23–25
3.	power and refrigeration cycles	27–33
4.	compressible fluid flow	26
5.	heat transfer	34–37
6.	fluids and hydraulic machines	14–19
7.	fans and ductwork	20
8.	HVAC	38–42
9.	combustion	22
10.	engineering materials and statics	43–47
11.	mechanics of materials	48–50
12.	machine design	51–53
13.	kinematics of machinery	54–58
14.	engineering economic analysis	69

Lecture coverage of some examination subjects is necessarily brief; other subjects are not covered at all. These omissions are intentional; they are not the result of scheduling omissions. Why? First, time is not on our side in a review course. Second, some subjects rarely contribute to the examination. Third, some subjects are not well received by the students. For example, I have found that very few people study modeling and systems analysis, fire protection systems, material handling, and manufacturing methods. Most mechanical engineers stick with the basics: fluids, thermodynamics, heat transfer, and machine design. Unless you have two quarters in which to teach your PE review, your students' time can be better spent covering other subjects.

All the skipped chapters and their end-of-chapter practice problems are assigned as floating assignments to be made up in the students' "free time."

I strongly believe in exposing my students to a realistic sample examination, but I no longer administer an in-class mock exam. Since the review course usually ends only a few days before the real PE examination, I hesitate to make students sit for several hours in the late evening to take a "final exam." Rather, I distribute and assign a take-home sample examination at the first meeting of the review course.

If a practice test is to be used as an indication of preparedness, caution your students not to even look at the sample exam prior to taking it. Looking at the sample examination or otherwise using it to direct their review will produce unwarranted specialization in subjects contained in the sample examination.

There are many ways to organize a PE review course depending on your available time, budget, intended audience, facilities, and enthusiasm. However, all good course formats have the same result: The students struggle with the workload during the course, and then they breeze through the examination after the course.

Part 2: Everything You Ever Wanted to Know About the PE Exam

The solutions booklet is essentially a bound book containing sheets of printed graph paper. Each sheet has a place for your personal exam identification number, the problem number, the page number, and room for the grader's eventual score. You must place all of the work that you want to be graded on these pages.

4. WHAT IS THE FORMAT OF THE PE EXAM?

The NCEES Professional Engineering examination in Mechanical Engineering consists of two four-hour sessions separated by a one-hour lunch period. Both the morning and afternoon sessions contain ten problems. There are no required problems.

The four-hour morning session contains ten problems in a free-response (essay) format, while the four-hour afternoon session contains ten problems in a multiple-choice format. You must work four problems in each session to receive full credit.

For each session, you will be given an exam booklet that contains problems only for mechanical engineers.

5. WHAT SUBJECTS ARE ON THE PE EXAM?

NCEES has published a breakdown of subjects on the examination. This breakdown has changed semantically several times in the past, but the nature of the problems has not shifted dramatically. A few changes have been made (e.g., the addition of fire protection problems and the deletion of engineering economics), but essentially, the "bread-and-butter" type of problems have not changed.

Irrespective of the published examination structure, the exact number of problems that will appear in each subject area cannot be predicted reliably. There is no guarantee that any single subject will occur in any quantity. One of the reasons for this is that some of the problems span several disciplines. You might consider a steam flow problem to be a thermodynamics problem, while someone else might categorize it as a fluids or compressible flow problem.

6. WHAT IS THE TYPICAL PROBLEM FORMAT?

Problems in both the morning and afternoon sessions start out with a statement of the "Situation." The situation describes a scenario that you might encounter as a professional engineer. All required data should appear in the situation statement.

The subsequent "Requirements" statement lists all of the responses that are needed to obtain full credit on the exam. These requirements are satisfied by writing the calculations, procedures, and conclusions in the solutions booklet.

7. HOW MUCH MATHEMATICS IS NEEDED FOR THE EXAM?

Generally, only simple algebra, trigonometry, and geometry are needed for the PE exam. You will be able to use the trigonometric, logarithm, square root, and similar buttons on your calculator. There is no need to use any other method for these functions.

***Subjects on the Mechanical Engineering PE Exam**[a]*

subject	number of problems[b]
1. machine design (fasteners, gears, brakes, belts, clutches, wire rope, bearings, conveyors)	2
2. stress analysis/structural design (stress analysis of machines, tools, and structures, including theories of failure and fatigue analysis)	2
3. kinematics and dynamics (motion, forces, and energy of machines and vehicles)	1
4. power plants systems (power plant cycles, compressors, and turbines; thermal and mechanical efficiency; cogeneration)	2
5. power plant processes (fuels and combustion, combustion stoichiometry, products of combustion; efficiency)	1
6. power plant components (boilers and pressure vessels, expansion tanks, piping for gases and liquids)	1
7. HVAC/R systems (heating, cooling, and ventilation load calculations; psychrometrics, evaporative cooling, refrigeration specifications)	2
8. HVAC/R components (components used in the operation and control of heating, ventilating, cooling, and refrigeration equipment)	1
9. control systems (analysis and performance of general mechanical, fluid, and thermal first- and second-order systems; stability using root locus and other criteria; control diagram components)	1

10. instrumentation and measurements (specifications of measuring systems, static and dynamic measurement of temperature, pressure, and flow) 1

11. vibrations (one and two degree of freedom systems, forced vibrations, transmissibility, isolation) 1

12. heat transfer (conduction, convection, and radiation in practical applications) 1

13. thermodynamics (work; energy; compressible gases in various devices, such as compressors, engines, and nozzles) 1

14. hydraulics and pneumatics (hydraulic equipment, hydraulic power and control diagrams, pumps and piping head loss calculations) 1

15. management (estimation of production for various products, such as sheet metal parts; selection of alternative manufacturing methods based on economics and cycle studies) 1

16. fire protection (system specifications, sprinkler systems, mobile systems, control valves, flow measurement and control; NFPA 13, 14, 20, and 29) 1

TOTAL 20

[a] As of October 1991.
[b] Morning essay problems and afternoon multiple-choice questions, combined.

There are no pure mathematics (algebra, geometry, trigonometry, etc.) problems on the exam. You will need to apply these subjects to engineering problems.

Except for simple quadratic equations, you will rarely need to find the roots of a high-order equation.

There is little or no use of calculus on the exam. Occasionally, you may need to take a derivative to find a maximum or minimum of some function. Even more rare is the need to integrate to find an average.

There is essentially no need to solve differential equations. Applications of differential equations, including vibrations, systems, and fluid mixing problems are common. However, these applications can usually be handled without having to solve a differential equation.

There is essentially no probability or statistics.

The PE exam is concerned with numerical answers, not with proofs or derivations. You will not be asked to prove or derive formulas.

Problems requiring iterative solutions are common. Generally, there is no need to complete more than two iterations, as by then, you have shown that you understand the basic concept. You will not need to program your calculator to obtain an "exact" answer.

8. ARE THERE ENGINEERING ECONOMICS PROBLEMS?

For most of the early years of engineering licensing, questions on engineering economics appeared routinely on the examinations. This is no longer the case. However, in its outline of exam subjects, NCEES notes, "Some problems may include economic aspects."

What this means is that engineering economics may be incorporated into other problems. For example, you might need to compare the economics of buying and operating two alternative pumping stations whose costs must be calculated from performance criteria. While the degree of engineering economics knowledge may have decreased somewhat, the basic economic concepts (time value of money, present worth, non-annual compounding, comparison of alternatives, etc.) are still valid test subjects.

9. ARE THERE PROFESSIONALISM AND ETHICS QUESTIONS?

For many years, NCEES has discussed adding professionalism and ethics questions to the PE exam. However, these subjects are still not part of the test outline, and there has yet to be a question on them in the exam.

10. WHAT IS THE FORMAT OF OBJECTIVELY SCORED PROBLEMS?

All problems in the afternoon session are *objectively scored*. This means that each of the ten afternoon problems consists of ten multiple-choice questions which are graded by computer. There is no penalty for guessing or wrong answers, but there is also no partial credit for correct methods and assumptions. There are no objectively scored problems in the morning session.

Examinees generally report that problems in the multiple-choice afternoon session are more difficult and time consuming than the morning essay questions. One of the more disquieting aspects of these problems is that the answer choices are seldom exact. Answer choices generally have three significant digits. Questions ask, "Which answer choice is closest to the correct value?" or instruct you to complete the sentence "The value is approximately..." A lot of self-confidence is required to move on to the next question when you don't find an exact match for the answer you calculated.

NCEES tries hard to make sure the multiple-choice problems are not greatly interrelated. Most problems are independent or start anew with given data. A mistake on one of the ten questions won't cause you to get the following question wrong, as well.

Multiple-choice questions are not necessarily in logical or progressive order. It may be necessary to solve these problems from the "bottom up," determining answers to subsequent questions before answering the first few questions.

11. HOW LONG DOES IT TAKE TO DO THE PROBLEMS?

At one time, NCEES specified that the average time required for a registered engineer with "average experience" was 20 minutes per problem. However, none of the examinees will be registered engineers, so this time estimate will apply only in the rare situation where you find a problem just like the ones you do every day.

NCEES now claims that the average time for a minimally qualified "average examinee" with little or no experience is 45 minutes per problem. This time estimate is more representative of what is experienced by well-prepared examinees. That time can be increased greatly if you don't know how to find things in your books.

12. IS THE EXAM TRICKY?

The PE exam is not "tricky." The exam does not overtly try to get you to fail. Examinees manage to fail on a regular basis with perfectly straight-forward problems. The exam problems are difficult in their own right. NCEES does not need to provide misleading or conflicting statements. However, with the multiple choice problems, you might find that commonly made mistakes are represented in the alternative answer choices. Thus, the alternative answers (known as "distracters") will be logical.

Problems are generally practical, dealing with common and plausible situations that you might experience in your job. For example, you will not be asked to determine the heat transfer from the side of a spacecraft after it has landed on a Jovian moon with a methane atmosphere.

13. WHAT MAKES THE PROBLEMS DIFFICULT?

Some problems are difficult because the pertinent theory is not obvious. Many machine design problems are of this variety. There is one acceptable way to design gear teeth, and nothing else is acceptable.

Some problems are difficult because the data needed are hard to find. Solving some HVAC problems depends on having climatological data for a specific location and performance characteristics of specific construction types.

Some problems are difficult because they defy the imagination. Problems involving epicyclic gear trains can be like this. If you cannot visualize the operation of the mechanism, you probably cannot solve it.

Some problems are difficult because they take a long time. Convective heat transfer and HVAC problems fall into this category.

Some problems are difficult because the computational burden is high. Thermodynamic cycle problems can be this way, particularly if you don't have detailed tables of properties and you end up doing double interpolations for every point on the cycle.

Some problems are difficult because the terminology is obscure, and you just don't know what the terms mean. This can happen in almost any subject.

14. DOES THE PE EXAM USE SI UNITS?

The PE exam in mechanical engineering is written essentially all in customary U.S. units. Use of SI units is minimal. Most problems use pounds, feet, seconds, gallons, degrees Fahrenheit, and British thermal units.

The exam apparently does not differentiate between lbf and lbm (pounds-force and pounds-mass) in the problem "Situation" (i.e., problem statement). However, the few official solutions that NCEES has published for public review do make the distinction.

15. WHY DOES NCEES REUSE SOME PROBLEMS?

NCEES reuses some of the more reliable problems from each exam. The percentage of repeat problems isn't high—approximately 25% of the exam. NCEES repeats problems in order to be able to equate the performance of one group of examinees with the performance of another group.

Once in a while, a new problem appears on the exam that few of the examinees choose. Usually, the reason for this is that the subject is too obscure or difficult. Problems in control systems, fire protection, and some of the engineering management subjects (e.g., linear programming) often fall into this category. Also, there have been cases where a low percentage of the examinees get a passing score because the problem was inadvertently stated in a poor or confusing manner.

NCEES tracks the use and "success" of each of the exam problems and does not repeat such rogue problems without modification. This is one of the reasons that historical analysis of oddball problem types should not be used as the basis of your review.

16. WHAT REFERENCE MATERIAL IS PERMITTED IN THE EXAM?

The PE examination is open book. Most states do not limit the number and types of books you can use. Personal notes in a three-ring binder and other semipermanent covers can usually be used. Loose papers, pads of scratch paper, and tablets (including Post-it® Notes) are forbidden. An exception to this restriction is made for oversize charts, graphs, and tables (Mollier diagrams, refrigerant charts, psychrometric charts, etc.) that are commonly needed for particular types of problems.

The references you bring with you into the examination room in the morning do not have to be the same as the references used in the afternoon. You cannot share books with other examinees.

A few states do not permit collections of solved problems such as *Schaum's Outline Series* and solutions manuals to review books. A few states ban all review books.

Strictly speaking, loose paper and scratch pads are not permitted in the examination. Certain types of pre-printed graphs, including psychrometric charts (which are often needed in HVAC problems) and logarithmically scaled graph papers (which are almost never needed) should be three-hole punched and brought in a three-ring binder. Such pages, if used in a solution, should be taped or stapled to the appropriate page in your answer book.

17. HOW MANY BOOKS SHOULD YOU BRING?

It is unnecessary to bring a large quantity of books to the examination. This book and three to five other references of your choice will be sufficient for most of the problems you solve. The examination is very fast paced. You will not have time to use books with which you are not thoroughly familiar.

Many examinees will show up with boxes and boxes of books. The exam doesn't require you to know obscure solution methods or to use difficult-to-find data. Without a doubt, there are things that you will need that are not in this book. But, there are not so many that you need to bring your entire company's library.

18. WHAT ABOUT CALCULATORS?

It goes without saying that you need a good calculator for the exam. However, it may not be obvious that you should bring a spare calculator with you to the examination. It is always unfortunate when an examinee is not able to finish because his or her calculator was dropped, stolen, or stopped working for some unknown reason.

The exam has not been "optimized" for any particular brand or type of calculator. In fact, for most problems, a $15 scientific calculator will produce results as satisfactory as a $200 calculator.

In most states, any calculator may be used as long as it has the following characteristics: is silent, is non-printing, and is without any significant word-processing functions. In most states, there are no restrictions on programmable, preprogrammed, or business/financial

calculators. Similarly, nomographs and specialty slide rules are permitted.[2] However, you cannot share calculators with other examinees.

It is essential that a calculator used for the PE examination has the following functions.

- trigonometric functions
- inverse trigonometric functions
- hyperbolic functions
- π
- square root and x^2
- both common and natural logarithms
- y^x and e^x

For maximum speed, your calculator should also have or be programmed for the following functions.

- interpolation
- extracting roots of quadratic equations
- converting between polar and rectangular vectors
- finding standard deviations and variances
- calculating determinants of matrices
- linear regression
- calculating factors for economic analysis problems

For the mechanical engineering exam, a calculator with built-in thermodynamic properties of air, steam, and water is a significant advantage.

You may not use a laptop computer, cellular telephone, or other communications device during the exam.

Be sure to take your calculator with you whenever you leave the examination room for any length of time.

19. HOW IS THE EXAM GRADED AND SCORED?

The maximum number of points you can earn on the PE exam is 80 points.[3] The minimum number of points for passing (referred to by NCEES as the *cut score*) is 48 points. This means that an average score of 6 points per problem is enough to pass. Poor performance on one problem can be compensated for by good performance on another, as long as a total score of 48 or higher is achieved.

Each essay (free response) problem in the morning session is worth ten points. Each problem is graded by a technical expert (whom NCEES calls a *subject matter expert*) working against a standardized scoring plan. NCEES has named this method the *criterion-referenced grading method*. The scoring plan consists of the specific features, calculations, assumptions, or characteristics that NCEES is looking for in the solution. Each feature must be present in your solution in order to get the two points associated with it. If a feature is missing, or if multiple mathematical errors are made, points

---

[2]Using quick-solution methods and preprogrammed calculators doesn't show much about the depth of your knowledge. Such aids should be used only to check your work.
[3]The score may be scaled upward and reported as 100% when you receive your results.

will be deducted from the maximum score of ten. (It is generally possible to make a single minor mathematical error and still receive eight or ten points.) In developing the grading plan, particular attention is given to the responses necessary to get six points.

Since multiples of two points are subtracted when one of the criteria are missing, the only possible scores on essay problems are ten, eight, six, four, two, and zero points.

As an example of a criterion-graded problem, consider a heat-transfer problem involving a hot duct passing through a cold room. Five of the criteria might require you to:

- Recognize that convection and radiation are both present and operate in parallel.

- Calculate the energy lost from the mass flow rate and temperature change and equate that to the heat transfer losses.

- As is characteristic of the HVAC industry, use the straight temperature difference instead of the logarithmic mean temperature difference.

- Obtain numerical values of film coefficients that deviate no more than ±10% from the scoring plan.

- Set up all equations correctly, though not necessarily complete the problem.

Getting full credit on an essay problem requires solving it completely and correctly, with no significant mathematical errors in the solution.

Only work that you have entered into the solutions booklet on the right-hand pages will be graded. Nothing that you have written on the left-hand pages or in the exam booklet will be graded.

Each of the objectively scored problems in the afternoon session is also worth ten points (one point for each of the ten multiple-choice questions). Grading is straightforward. The computer grades your score sheet. You either get the question right or you don't. There is no deduction for incorrect answers, so guessing is encouraged. However, if you mark two or more answers, no credit is given for the question.

You will receive the results of your examination from your state board (not from NCEES) by mail. Allow at least four months for notification. Your score may or may not be revealed to you, depending on your state's procedure. Even when the score is reported to you, it may have been scaled or normalized to 100%.

20. WHAT IS THE HISTORICAL PASSING RATE?

The passing rate can vary from exam to exam. It might be close to 70% for one exam and then be 40% for the next one. This essentially means that some exams contain more "easy" problems than others. Because it's

impossible to predict how many of the problems you will consider easy, historical passing rates may be irrelevant for your exam.

On the average, the passing rate for first-time examinees seems to hover around 60% or slightly below. The passing rate for repeat examinees is lower—about 50%.

21. CHEATING AND EXAM SUBVERSION

There aren't very many ways of cheating on an open-book test. The proctors are well versed on the few ways that do exist. It goes without saying that you should not talk to other examinees in the room, nor should you pass notes back and forth. The number of people who are released to use the restroom may be limited to prevent private discussions.

When time is up, time is up. Put your pencil down when your proctor tells you to do so. If you continue working, your exam may be confiscated and all of your efforts invalidated.

NCEES regularly reuses good problems that have appeared on previous exams. Therefore, examination security is a serious issue with NCEES, which goes to great lengths to make sure nobody copies the problems. You may not keep your exam booklet, enter text of problems into your calculator, or copy problems into your own material.

The proctors are concerned about exam subversion, which generally means activity that might invalidate the examination or the examination process. Exam subversion usually involves trying to make copies of exam problems.

Part 3: How to Prepare for and Pass the PE Exam in Mechanical Engineering

22. WHAT SHOULD YOU STUDY?

There is a lot of material in this book. Strictly speaking, you don't have to study every subject on the exam in order to pass. However, the more subjects you study, the better will be your chances of passing the examination. You should decide early in the preparation process which subjects you are going to study. The strategy you select will depend on your background. Following are the four most common strategies.

- A broad approach has been successful for examinees who have recently completed academic studies. Their strategy has been to review the fundamentals in a broad range of undergraduate mechanical engineering subjects. (This means studying all or most of the chapters in this book.) The examination includes enough fundamentals problems to make this strategy worthwhile. This is the best strategy.

- Engineers who have little time in which to prepare tend to concentrate on the subjects in which they will find the most problems. By studying the list of examination subjects, some have been able to choose those that will give them the highest probability of finding enough problems that they can solve. This strategy works as long as the examination "cooperates" and has enough of the types of problems they need. Too often, though, examinees who "pick and choose" subjects to review can't find enough problems to complete the exam.

- Engineers who have been away from classroom work for a long time tend to concentrate on the subjects in which they have had extensive experience and hope that problems in those subjects will be on the exam. This method is seldom successful.

- Some engineers plan on modeling their solutions from similar problems they have found in their textbooks. These engineers often spend a lot of time indexing the example and sample problem types in all of their books. This is not a legitimate preparation method, and it is almost never successful.

23. DO YOU NEED A CLASSROOM REVIEW COURSE?

Many first-time examinees take a classroom review course.[4] Review courses are useful for a number of reasons. Courses provide several significant advantages, some of which may apply to you.

1. A course structures and paces your review. It keeps you going forward without getting bogged down in one subject.

2. A course focuses you on a limited amount of material. Without a course, you may not know what subjects to study.

3. A course provides you with the problems you need to solve. You won't have to spend time looking for problems to solve.

4. The course "spoon feeds" you the material. You may not need to read the book!

5. The course instructor can answer your questions when you get stuck.

You probably already know if any of these advantages would be helpful to you. A review course will be of less value if you are thorough and disciplined.

24. HOW LONG SHOULD YOU STUDY?

We've all heard stories of the person who didn't crack a book until the week before the exam and still passed it with flying colors. Yes, these people really exist. But, I'm not one of them, and you probably aren't either. In fact, after having taught thousands of engineers in my own classes, I'm convinced that these people are as rare as the ones who have taken exam five times and still can't pass it.

A thorough review takes approximately 300 hours. Most of this time is spent solving problems. Some of it may be spent in class. Some people spread this time over a year. Others cram it all into two months. Most review courses last for three months or more. When you start studying will depend on how much time you can spend per week.

25. WHAT ITEMS WILL YOU NEED?

There are many references and resources that you should begin to assemble for your review and for use in the examination. Some of the items (particularly anything in loose sheet form) may not be permitted in the examination but will still be valuable during your studies.

[ ] graph paper (a few sheets of each)
  - 10 squares to the inch × 10 squares to the inch
  - semi-log (3 cycles × 10 squares to the inch)
  - log-log (3 cycles × 3 cycles)

[ ] psychrometric charts, at least 10 for normal temperature and pressure, and several each for low-pressure, low-temperature, and high-temperature problems

[ ] long, flexible, clear plastic ruler marked in tenths of an inch or in centimeters and millimeters

[ ] detailed steam tables in both English and SI units (*Steam Tables* by Keenan and Keyes)

[ ] large Mollier diagrams in both English and SI units[5] (as contained in *Steam Tables* by Keenan and Keyes)

[ ] detailed air tables in both English and SI units (e.g., *Gas Tables* by Keenan and Kaye)

---

[4]For the names and phone numbers of review courses close to you, call Professional Publications' referral service.

[5]A large reprint of *ASME's Mollier Diagram* in English units is available from Professional Publications.

[ ] tables of isentropic flow and normal shock factors for various ratios of specific heats[6]

[ ] *Mark's Standard Handbook for Mechanical Engineers*

[ ] *Machinery's Handbook*

[ ] heat transfer book with the following charts, figures, or tables

- charts for solving transient heat flow problems (simple solids other than spheres)

- radiation arrangement factors (F_a)

- correction factors (F_c) for multiple-pass heat exchangers

[ ] HVAC book with the following charts, figures, or tables of data

- outside design conditions versus geographic location (including winter design temperature, winter degree days, summer design temperature, summer degree days, average temperature swing, and wind velocity)

- k or U values for various wall constructions

- infiltration coefficients

- heat loss coefficient for the slab-edge method

- equivalent temperature differences, and related support tables

- cooling load factors (various)

[ ] machine design book

[ ] *Formulas for Stress and Strain* (Roark & Young)

[ ] fluids data book such as Crane's *Flow of Fluids Through Valves, Fittings, and Pipe (Manual 410)*, Ingersoll-Rand's *Cameron Hydraulic Data*, or Colt Industries' *Hydraulic Handbook*

[ ] National Fire Protection Association Standards NFPA 13 (*Installation of Sprinkler Systems*) and NFPA 20 (*Centrifugal Fire Pumps*), or suitable substitute[7]

[ ] scientific dictionary

[ ] standard English dictionary

---

[6] *Consolidated Gas Dynamics Tables*, published by Professional Publications, provides tables of data for isentropic flow, normal shock waves, Fanno flow, and Rayleigh flow for gases with ratios of specific heats of 1.1, 1.2, 1.3, 1.4, and 1.67.
[7] *Fire and Explosion Protection Systems*, published by Professional Publications, is a complete tutorial for the fire protection question on the exam. It summarizes the key concepts in NFPA-13 and NFPA-20.

Professional Publications, Inc., can provide you with the following references and study aids.

- *ASME Mollier diagram*

- *Fire and Explosion Protection Systems*

- *Mechanical Engineering Sample Examination*

- *Quick Reference for the Mechanical Engineering PE Exam*

- *Consolidated Gas Dynamics Tables*

- *Engineering Unit Conversions*

- *101 Solved Mechanical Engineering Problems*

- *Mechanical Engineering Review Course on Cassettes*

26. WHAT YOU WON'T NEED

Generally, people bring too many things to the examination. One general rule is that you shouldn't bring books that you never looked at during your review. If you didn't need a book while doing the practice problems in this book, you won't need it during the exam.

There are some other things that you won't need.

- books on basic and introductory subjects:
  You won't need books that cover trigonometry, geometry, or calculus. You probably won't even need books that cover basic engineering subjects such as statics, mechanics of materials, and so on.

- books on mathematical analysis, or extensive mathematics tabulations

- books on non-exam subjects:
  Such subjects as electricity and electronics, materials science, surveying, drafting, atomic theory and nuclear engineering, the English language, and geography are not part of the exam.

- extensive collections of material properties:
  You are not expected to know the mechanical properties of obscure or exotic alloys. Most mechanical properties are provided as part of the problem statement.

- building codes:
  There is nothing related to steel, concrete, timber, masonry, plumbing, earthquakes, and so on, on the exam. (However, the NFPA fire protection codes may be needed for the fire protection problem, and rudimentary knowledge of the ASME boiler code may be needed.)

- obscure books and materials:
  Books that are in a different language, doctoral theses, and papers presented at technical societies won't be needed on the exam.

- old textbooks or obsolete, rare, and ancient books:
  NCEES exam problem writers are aware of which textbooks are in use. Material that is available only in out-of-print publications and old editions won't be used.

- handbooks in other disciplines:
  You won't need a civil, electrical, or industrial engineering handbook. You won't need to bring the *Handbook of Chemistry and Physics*. (However, *Machinery's Handbook* has occasionally been needed for manufacturing-oriented data.)

- computer science book:
  You won't need to bring books covering computer logic, programming, algorithms, program design, or subroutines for FORTRAN, C, PASCAL, or any other language.

- crafts- and trades-oriented books:
  You are not expected to have detailed knowledge of manufacturing operations (e.g., foundry, metal turning, sheet metal forming, and designing jigs and fixtures).

- manufacturer's literature and catalogs:
  No part of the exam requires you to be familiar with products that are proprietary to any manufacturer.

- government publications:
  With rare exceptions, few government publications are used in engineering design. None are needed in the PE exam.

- your state's laws:
  The NCEES PE exam is a national exam. Nothing unique to your state will appear on it.

27. SHOULD YOU LOOK FOR OLD EXAMS?

The traditional approach for preparing for standardized tests includes working old tests. However, NCEES does not release old tests or problems. Therefore, there are no official problems or tests in circulation.

In addition to the representative problems in this book (marked *Time limit: one hour*), additional practice problems representative of the exam are available from Professional Publications.[8]

---

[8]Two publications are available that have numerous examples of exam-level problems: *Mechanical Engineering Sample Examination* (Lindeburg) and *101 Solved Mechanical Engineering Problems* (Lindeburg).

28. DO YOU NEED A SCHEDULE?

It is important that you develop and adhere to a review outline and schedule. Once you have decided which subjects you are going to study, you can allocate the available time to those subjects in a manner that makes sense to you. If you are not taking a classroom review course (where the order of preparation is determined by the lectures), you should prepare an *Outline of Subjects for Self-Study* to schedule your preparation.

29. HOW YOU CAN MAKE YOUR REVIEW REALISTIC

In the exam, you must be able to quickly recall solution procedures, formulas, and important data. You must remain sharp for eight hours or more. When you played a sport back in school, your coach tried to put you in game-related situations. Preparing for the PE exam isn't much different than preparing for a big game. Some part of your preparation should be realistic and representative of the examination environment.

There are several things that you can do to make your review more realistic. For example, if you gather most of your review resources (i.e., books) in advance and try to use them exclusively during your review, you will become more familiar with them. (Of course, you can add to or change your references if you find inadequacies.)

Another good suggestion is to work all of your homework problems on graph paper. During the actual exam, you will write your answers on graph paper with lines drawn six to the inch. Some examinees find this lined paper difficult to get used to.

Gaining the ability to use your time wisely is one of the most important lessons you can learn during your review. You will undoubtedly encounter review problems that end up taking much longer than you expected. In some instances, you will cause your own delays by spending too much time looking through books for things you need (or, just by looking for the books themselves!). Other times, the problems will just entail too much work. Learn to recognize these situations so that you can make an intelligent decision about such problems during the exam.

Some of the end-of-chapter problems in this book are marked *Time limit: one hour*. These problems are representative of the types of questions, level of difficulty, and duration that you will encounter on the actual exam. Watch the clock while you are solving these problems. Although the spread in problem complexities and lengths is fairly wide, you should be able to complete most of them in less than an hour.

30. WHAT TO DO A FEW DAYS BEFORE THE EXAM

There are a few things that you should do a week or so before the examination date. For example, visit the exam site to find the building, parking areas, examination room, and restrooms. You should also make arrangements for child care and transportation as necessary. Since the examination does not always start or end at the designated time, make sure that such arrangements are flexible.

If your spare calculator is not identical to your primary calculator, spend a few minutes familiarizing yourself with its functioning.

Second in importance to your scholastic preparation is the preparation of your two examination kits. The first kit consists of a bag or box containing items to be brought with you into the examination room.

[ ] letter admitting you to the examination

[ ] photographic identification (e.g., driver's license)

[ ] this book

[ ] other references

[ ] regular dictionary (if English is not your native language)

[ ] scientific/engineering dictionary

[ ] review course notes in a three-ring binder

[ ] cardboard box to use as a bookcase

[ ] main calculator

[ ] spare calculator

[ ] instruction booklets for your calculators

[ ] extra calculator batteries

[ ] mechanical pencils with HB lead or several sharpened #2 pencils

[ ] erasers

[ ] straight-edge and rulers

[ ] compass

[ ] protractor

[ ] scissors

[ ] stapler

[ ] transparent tape

[ ] psychrometric charts and miscellaneous graph paper (punched and in a binder)

[ ] unobtrusive snacks or candies[9]

[ ] travel pack of Kleenex™ (keep in your pocket)

[ ] handkerchief

[ ] headache remedy

[ ] personal medication

[ ] $2.00 in miscellaneous change

[ ] light, comfortable sweater

[ ] loose shoes or slippers

[ ] cushion for your chair

[ ] earplugs

[ ] wristwatch with alarm

[ ] wire coat hanger (doubles as an emergency car door opener)

[ ] extra set of car keys

The second kit consists of the following items that should be left in a separate bag or box in your car in case they are needed.

[ ] copy of your application

[ ] proof of delivery of your original application

[ ] light lunch

[ ] beverages

[ ] sunglasses

[ ] extra pair of prescription glasses

[ ] raincoat, boots, gloves, hat, and umbrella

[ ] street map of the examination area

[ ] note to the parking patrol for your windshield explaining where you are, what you are doing, and why your time may have expired

[ ] battery-powered desk lamp

The following items cannot be used during the examination and should be left at home.

[ ] fountain pens

[ ] radio or tape player

---

[9]Food and drink are not permitted in the examination room.

[ ] battery charger

[ ] extension cords

[ ] scratch paper/loose pages

[ ] note pads

[ ] Unless you have a good reason (e.g., impending childbirth), leave your pager at home.

31. WHAT TO DO THE DAY BEFORE THE EXAM

Take the day before the examination off from work to relax. Do not cram the last night. A good night's sleep is the best way to start the examination. If you live a considerable distance from the examination site, consider getting a hotel room in which to spend the night.

Make sure your exam kits are packed and ready to go.

32. WHAT TO DO THE MORNING OF THE EXAM

You should arrive at least 30 minutes before the examination starts. This will allow time for finding a convenient parking place, bringing your materials to the examination room, and making room and seating changes that may be necessary. Be aware, though, that the examination room may not be open or ready at the designated time.

33. WHAT TO DO DURING THE EXAM

All of the procedures typically associated with timed, proctored, computer-graded performance testing will be in effect when you take the PE examination.

The proctors will distribute the examination booklets and answer sheets (or blank solution booklets) if they are not already on the desks. However, you should not open them until instructed to do so. You may read the information on the front and back covers, and you should write your name in the upper right corner of the front cover. Listen carefully to everything the proctors say.

Do not ask your proctors technical questions. Even if they are knowledgeable in engineering, they will not be permitted to answer your questions.

For the morning essay (free response) problems, record your examinee number on every page in the solutions booklet. Also, you must list the problem numbers you want graded on the outside of your solutions booklet. You may work more than four problems, but only the four that you list on the front cover will be graded.

Record your answers to the afternoon multiple-choice problems on the answer sheet contained in the test booklet. The proctors will guide you through the process of putting your name and other biographical information on this sheet when the time comes, which will take approximately 15 minutes. You will be given the full four hours to work problems. Time to complete the answer sheet is not part of your four hours.

The common suggestions to "use #2 pencils, completely fill the bubbles, and erase completely" apply here. Actually, mechanical pencils with HB lead can also be used. Use of ballpoint pens and felt-tip markers is discouraged.

If you finish the exam early, and there is still more than 30 minutes remaining, you will be permitted to leave the room. If you finish less than 30 minutes before the end of the exam, you must remain until the end. This is done to be considerate of the people who are still working.

When you leave, you must return your exam booklet. You may not keep the exam booklet for later review.

34. HOW TO CHOOSE YOUR PROBLEMS

Read through all of the problems before starting to work. You shouldn't do the first problem you find that you think you can solve. To save you from rereading and reevaluating each of the problems later in the session, you should classify each problem at the beginning of the session. The following categories have been used by many successful examinees.

- problems you can do easily

- problems you can do with effort

- problems on which you think you might get partial credit

- problems you cannot do

All of the problems on the exam are worth the same number of points, so do the problems in order of increasing difficulty. There is nothing to be gained by attempting the difficult or long problems, even if they are in your area of expertise, if easier or shorter problems are available. There is no penalty for selecting all of the problems in a particular subject (e.g., all of the HVAC problems).

35. HOW TO SOLVE THE ESSAY PROBLEMS

Follow these guidelines when solving the morning essay (free response) problems.

- Use only pencil.

- Be neat. Print all text.

- Do not rewrite the problem statement or redraw any of the figures. That only takes time and will not get you any more points.

- List all of your assumptions (perfect gas, massless spring, zero friction, adiabatic or isentropic process, etc.).

- Draw a box around each completed answer so the grader knows where to find it.

- Label each answer with a symbol (e.g., "$F = 5$ lbf").

- List the units for each answer.

- Write on one side (the right side) of the page only. The left side is to be used as scratch paper.

- Use one page per problem. Don't solve more than one problem per page.

- Go through all of the calculations a second time and check for mathematical errors, or solve the problem by an alternate method.

36. HOW TO SOLVE MULTIPLE-CHOICE PROBLEMS

When you begin the afternoon session with its multiple-choice problems, observe the following suggestions.

- Use only pencil.

- Do not spend more than one hour per problem. If you have not finished a problem in that amount of time, make a note of it and move on.

- Set your wristwatch alarm for five minutes before the end of the four-hour multiple-choice session and use that remaining time to guess at all of the remaining unsolved multiple-choice problems. You will be successful with about 25% of your guesses, and these points will more than make up for the few points that you might earn by working during the last five minutes.

- Make mental notes about any problems for which you cannot find a correct response, that have two correct responses, or for which you believe have some technical flaw. Errors in the exam are rare, but they do occur. Being able to point out an error might be important later. Since such problems are almost always discovered during the scoring process and discounted from the examination, it is not necessary to tell your proctor, but be sure to mark the one best answer before moving on.

- Make sure all of your responses on the answer sheet are dark and completely fill in the bubbles.

37. SHOULD YOU CITE YOUR SOURCES?

Solving problems on the PE exam is more like running a race than presenting a formal paper with tight references. Generally, it will not be necessary to give the source of your solution method or data. The technical experts grading your problem will be familiar with all traditional variations in procedures used to solve the problems they are grading.

It certainly is not necessary to cite your sources for common information, such as economic analysis factors, sizes of schedule-40 steel pipe, fluid properties, and standard formulas that any engineer should recognize. However, when you use an obscure solution method (e.g., something your own company has developed) or data that even you had a hard time finding, you might want to list the source. That will lend credibility to something that might, otherwise, seem like hocus-pocus to the grader.

There is no prohibition about citing this book as a reference in the exam. In fact, engineers have been doing so with previous editions for years. Engineering is engineering. As long as you use them correctly, the data, formulas, and principles are the same regardless of where you got them.

38. HOW TO SOLVE PROBLEMS CAREFULLY

Many points are lost to carelessness. Keep the following items in mind when you are solving the practice problems. Hopefully, these suggestions will be automatic in the exam.

[ ] Reread the question. Did you answer all parts of it?

[ ] Reread the question. Did you use all of the given data?

[ ] Reread the question. Did you use all of the given figures?

[ ] Did you use all of the efficiencies given?

[ ] Did you list all of your assumptions (perfect gas, massless spring, zero friction, adiabatic or isentropic process, etc.)?

[ ] Did you use absolute temperatures where required?

[ ] Did you convert from gage to absolute pressures?

[ ] Did you recheck all data obtained from other sources, tables, and figures?

- Did you enter the air or steam table with absolute pressure and absolute temperature?

- In finding the friction factor, did you enter the Moody diagram at the correct Reynolds number?

- Did you use the table for the right substance, or did you accidentally find air properties from the water table?

[ ] Did you convert between radius and diameter?

[ ] Did you use any symbols in your solution that need to be defined?

[ ] Does your answer have units?

[ ] Are your figures clearly labeled?

[ ] Did you recheck all of your mathematical equations?

[ ] Do the units cancel out in your calculations?

[ ] Did you list the sources of any obscure data or methods?

[ ] Did you tag any problem that you could not do or that you did not finish?

1 Systems of Units

1. Introduction 1-1
2. Common Units of Mass 1-1
3. Mass and Weight 1-1
4. Acceleration of Gravity 1-2
5. Consistent Systems of Units 1-2
6. The English Engineering System 1-2
7. Other Formulas Affected by Inconsistency . 1-3
8. Weight and Weight Density 1-3
9. The English Gravitational System 1-3
10. The Absolute English System 1-4
11. Metric Systems of Units 1-4
12. The cgs System 1-4
13. SI Units (The mks System) 1-5
14. Rules for Using SI Units 1-5
15. Primary Dimensions 1-7
16. Dimensionless Groups 1-7
17. Lineal and Board Foot Measurements 1-8
    Practice Problems 1-8

1. INTRODUCTION

The purpose of this chapter is to eliminate some of the confusion regarding the many units available for each engineering variable. In particular, an effort has been made to clarify the use of the so-called English systems, which for years have used the *pound* unit both for force and mass—a practice that has resulted in confusion for even those familiar with it.

2. COMMON UNITS OF MASS

The choice of a mass unit is the major factor in determining which system of units will be used in solving a problem. It is obvious that one will not easily end up with a force in pounds if the rest of the problem is stated in meters and kilograms. Actually, the choice of a mass unit determines more than whether a conversion factor will be necessary to convert from one system to another (e.g., between SI and English units). An inappropriate choice of a mass unit may actually require a conversion factor *within* the system of units.

The common units of mass are the gram, pound, kilogram, and slug.[1] There is nothing mysterious about these units. All represent different quantities of matter,

as Fig. 1.1 illustrates. In particular, note that the pound and slug do not represent the same quantity of matter.[2]

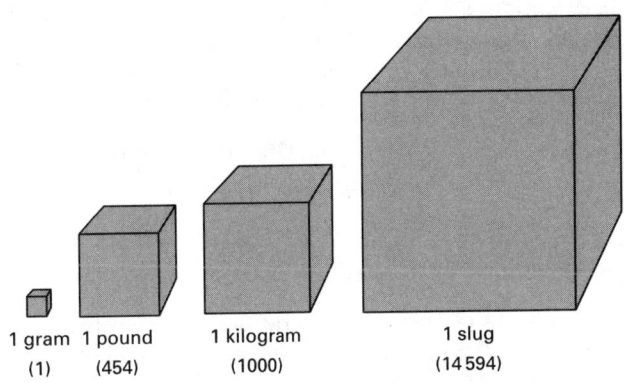

1 gram   1 pound   1 kilogram   1 slug
  (1)   (454)   (1000)   (14 594)

Figure 1.1 *Common Units of Mass*

3. MASS AND WEIGHT

In SI, *kilograms* are used for mass and *newtons* for weight (force). The units are different, and there is no confusion between the variables. However, for years, the term *pound* has been used for both mass and weight. This usage has obscured the distinction between the two: mass is a constant property of an object; weight varies with the gravitational field. Even the conventional use of the abbreviations *lbm* and *lbf* (to distinguish between pounds-mass and pounds-force) has not helped eliminate the confusion.

It is true that an object with a mass of one pound will have an earthly weight of one pound, but this is true only on the earth. The weight of the same object will be much less on the moon. Therefore, care must be taken when working with mass and force in the same problem.

The relationship that converts mass to weight is familiar to every engineering student:

$$W = mg \qquad \qquad 1.1$$

Equation 1.1 illustrates that an object's weight will depend on the local acceleration of gravity as well as the object's mass. The mass will be constant, but gravity will depend on location. Mass and weight are not the same.

---

[1] Normally, one does not distinguish between a unit and a multiple of that unit, as is done here with the gram and the kilogram. However, these two units actually are bases for different consistent systems.

[2] A slug is equal to 32.1740 pounds-mass.

4. ACCELERATION OF GRAVITY

Gravitational acceleration on the earth's surface is usually taken as 32.2 ft/sec² or 9.81 m/s². These values are rounded from the more exact standard values of 32.1740 ft/sec² and 9.8066 m/s². However, the need for greater accuracy must be evaluated on a problem-by-problem basis. Usually, three significant digits are adequate, since gravitational acceleration is not constant anyway but is affected by location (primarily latitude and altitude) and major geographical features.

The term *standard gravity*, g_0, is derived from the acceleration at essentially any point at sea level and approximately 45° N latitude. If additional accuracy is needed, the gravitational acceleration can be calculated from Eq. 1.2. This equation neglects the effects of large land and water masses. ϕ is the latitude in degrees.

$$g_{\text{surface}} = g'[1 + (5.305 \times 10^{-3}) \sin^2 \phi$$
$$- (5.9 \times 10^{-6}) \sin^2 2\phi] \qquad 1.2$$
$$g' = 32.0881 \text{ ft/sec}^2$$
$$= 9.78045 \text{ m/s}^2$$

If the effects of the earth's rotation are neglected, the gravitational acceleration at an altitude h above the earth's surface is given by Eq. 1.3. R_e is the earth's radius.

$$g_h = g_{\text{surface}} \left(\frac{R_e}{R_e + h} \right)^2 \qquad 1.3$$
$$R_e = 3960 \text{ mi}$$
$$= 6.37 \times 10^6 \text{ m}$$

5. CONSISTENT SYSTEMS OF UNITS

A set of units used in a calculation is said to be *consistent* if no conversion factors are needed.[3] For example, a moment is calculated as the product of a force and a lever arm length.

$$M = F \times d \qquad 1.4$$

A calculation using Eq. 1.4 would be consistent if M was in newton-meters, F was in newtons, and d was in meters. The calculation would be inconsistent if M was in ft-kips, F was in kips, and d was in inches (because a conversion factor of 1/12 would be required).

The concept of a consistent calculation can be extended to a system of units. A *consistent system of units* is one in which no conversion factors are needed for any calculation. For example, Newton's second law of motion can be written without conversion factors. Newton's second law simply states that the force required to accelerate an object is proportional to the acceleration of the object. The constant of proportionality is the object's mass.

$$F = ma \qquad 1.5$$

[3]The terms *homogeneous* and *coherent* are also used to describe a consistent set of units.

Notice that Eq. 1.5 is $F = ma$, not $F = Wa/g$ or $F = ma/g_c$. Equation 1.5 is consistent: it requires no conversion factors. This means that in a consistent system where conversion factors are not used, once the units of m and a have been selected, the units of F are fixed. This has the effect of establishing units of work and energy, power, fluid properties, and so on.

It should be mentioned that the decision to work with a consistent set of units is desirable but unnecessary, depending on tradition and environment. Problems in fluid flow and thermodynamics are routinely solved in the United States with inconsistent units. This causes no more of a problem than working with inches and feet when calculating a moment. It is necessary only to use the proper conversion factors.

6. THE ENGLISH ENGINEERING SYSTEM

Through common and widespread use, pounds-mass (lbm) and pounds-force (lbf) have become the standard units for mass and force in the *English Engineering System*. (The customary U.S. solutions presented in this book use this system.)

There are subjects in the United States where the practice of using pounds for mass is firmly entrenched. For example, most thermodynamics, fluid flow, and heat transfer problems have traditionally been solved using the units of lbm/ft³ for density, Btu/lbm for enthalpy, and Btu/lbm-°F for specific heat. Unfortunately, some equations contain both lbm-related and lbf-related variables, as does the steady flow conservation of energy equation, which combines enthalpy in Btu/lbm with pressure in lbf/ft².

The units of pounds-mass and pounds-force are as different as the units of gallons and feet, and they cannot be canceled. A mass conversion factor, g_c, is needed to make the equations containing lbf and lbm dimensionally consistent. This factor is known as the *gravitational constant* and has a value of 32.1740 lbm-ft/lbf-sec². The numerical value is the same as the standard acceleration of gravity, but g_c is not the local gravitational acceleration, g.[4] g_c is a conversion constant, just as 12.0 is the conversion factor between feet and inches.

The English Engineering System is an inconsistent system as defined according to Newton's second law. $F = ma$ cannot be written if lbf, lbm, and ft/sec² are the units used. The g_c term must be included.

$$F \text{ in lbf} = \frac{(m \text{ in lbm}) \left(a \text{ in } \frac{\text{ft}}{\text{sec}^2} \right)}{g_c \text{ in } \frac{\text{lbm-ft}}{\text{lbf-sec}^2}} \qquad 1.6$$

It is important to note in Eq. 1.6 that g_c does more than "fix the units." Since g_c has a numerical value of 32.174, it actually changes the calculation numerically.

[4]It is acceptable (and recommended) that g_c be rounded to the same number of significant digits as g. Therefore, a value of 32.2 for g_c would typically be used.

A force of 1.0 pound will not accelerate a 1.0-pound mass at the rate of 1.0 ft/sec^2.

In the English Engineering System, work and energy are typically measured in ft-lbf (mechanical systems) or in British thermal units, Btu (thermal and fluid systems). One Btu is equal to 778.26 ft-lbf.

Example 1.1

Calculate the weight in lbf of a 1.0 lbm object in a gravitational field of 27.5 ft/sec^2.

Solution

From Eq. 1.6,

$$F = \frac{ma}{g_c} = \frac{(1\text{ lbm})\left(27.5\ \dfrac{\text{ft}}{\text{sec}^2}\right)}{32.2\ \dfrac{\text{lbm-ft}}{\text{lbf-sec}^2}}$$

$$= 0.854\text{ lbf}$$

7. OTHER FORMULAS AFFECTED BY INCONSISTENCY

It is not a significant burden to include g_c in a calculation, but it may be difficult to remember when g_c should be used. Knowing when to include the gravitational constant can be learned through repeated exposure to the formulas in which it is needed, but it is safer to carry the units along in every calculation.

The following is a representative (but not exhaustive) listing of formulas that require the g_c term. In all cases, it is assumed that the standard English Engineering System units will be used.

- kinetic energy

$$E = \frac{m\text{v}^2}{2g_c} \quad [\text{in ft-lbf}] \qquad \qquad 1.7$$

- potential energy

$$E = \frac{mgz}{g_c} \quad [\text{in ft-lbf}] \qquad \qquad 1.8$$

- pressure at a depth

$$p = \frac{\rho g h}{g_c} \quad [\text{in lbf/ft}^2] \qquad \qquad 1.9$$

Example 1.2

A rocket with a mass of 4000 lbm travels at 27,000 ft/sec. What is its kinetic energy in ft-lbf?

Solution

From Eq. 1.7,

$$E_k = \frac{m\text{v}^2}{2g_c} = \frac{(4000\text{ lbm})\left(27{,}000\ \dfrac{\text{ft}}{\text{sec}}\right)^2}{(2)\left(32.2\ \dfrac{\text{lbm-ft}}{\text{lbf-sec}^2}\right)}$$

$$= 4.53 \times 10^{10}\text{ ft-lbf}$$

8. WEIGHT AND WEIGHT DENSITY

Weight is a force exerted on an object due to its placement in a gravitational field. If a consistent set of units is used, Eq. 1.1 can be used to calculate the weight of a mass. In the English Engineering System, however, Eq. 1.10 must be used.

$$W = \frac{mg}{g_c} \qquad \qquad 1.10$$

Both sides of Eq. 1.10 can be divided by the volume of an object to derive the *weight density*, γ, of the object. Equation 1.11 illustrates that the weight density (in lbf/ft^3) can also be calculated by multiplying the mass density (in lbm/ft^3) by g/g_c. Since g and g_c usually have the same numerical values, the only effect of Eq. 1.12 is to change the units of density.

$$\frac{W}{V} = \left(\frac{m}{V}\right)\left(\frac{g}{g_c}\right) \qquad \qquad 1.11$$

$$\gamma = \frac{W}{V} = \left(\frac{m}{V}\right)\left(\frac{g}{g_c}\right) = \frac{\rho g}{g_c} \qquad \qquad 1.12$$

Weight does not occupy volume. Only mass has volume. The concept of weight density has evolved to simplify certain calculations, particularly fluid calculations. For example, pressure at a depth is calculated from Eq. 1.13. (Compare this with Eq. 1.9.)

$$p = \gamma h \qquad \qquad 1.13$$

9. THE ENGLISH GRAVITATIONAL SYSTEM

Not all English systems are inconsistent. Pounds can still be used as the unit of force as long as pounds are not used as the unit of mass. Such is the case with the consistent *English Gravitational System*.

If acceleration is given in ft/sec^2, the units of mass for a consistent system of units can be determined from Newton's second law. The combination of units in Eq. 1.14 is known as a *slug*. g_c is not needed at all since this system is consistent. It would be needed only to convert slugs to another mass unit.

$$\text{units of } m = \frac{\text{units of } F}{\text{units of } a}$$

$$= \frac{\text{lbf}}{\dfrac{\text{ft}}{\text{sec}^2}} = \frac{\text{lbf-sec}^2}{\text{ft}} \qquad \qquad 1.14$$

Slugs and pounds-mass are not the same, as Fig. 1.1 illustrates. However, both are units for the same quantity: mass. Equation 1.15 will convert between slugs and pounds-mass.

$$\text{no. of slugs} = \frac{\text{no. of lbm}}{g_c} \qquad 1.15$$

It is important to recognize that the number of slugs is not derived by dividing the number of pounds-mass by the local gravity. g_c is used regardless of the local gravity. The conversion between feet and inches is not dependent on local gravity; neither is the conversion between slugs and pounds-mass.

Since the English Gravitational System is consistent, Eq. 1.16 can be used to calculate weight. Notice that the local gravitational acceleration is used.

$$W \text{ in lbf} = (m \text{ in slugs})\left(g \text{ in } \frac{\text{ft}}{\text{sec}^2}\right) \qquad 1.16$$

10. THE ABSOLUTE ENGLISH SYSTEM

The obscure *Absolute English System* takes the approach that mass must have units of pounds-mass (lbm) and the units of force can be derived from Newton's second law. The units for F cannot be simplified any more than they are in Eq. 1.17. This particular combination of units is known as a *poundal*.[5] A poundal is not the same as a pound.

$$\text{units of } F = (\text{units of } m)(\text{units of } a)$$
$$= (\text{lbm})\left(\frac{\text{ft}}{\text{sec}^2}\right)$$
$$= \frac{\text{lbm-ft}}{\text{sec}^2} \qquad 1.17$$

Poundals have not seen widespread use in the United States. The English Gravitational System (using slugs for mass) has greatly eclipsed the Absolute English System in popularity. Both are consistent systems, but there seems little need for poundals in modern engineering.

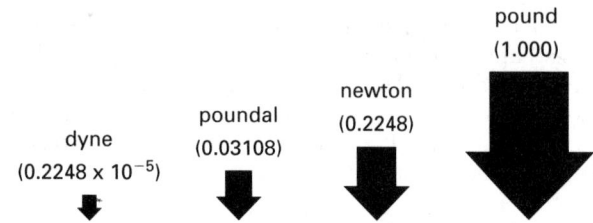

Figure 1.2 *Common Force Units*

[5]A poundal is equal to 0.03108 pounds-force.

11. METRIC SYSTEMS OF UNITS

Strictly speaking, a *metric system* is any system of units that is based on meters or parts of meters. This broad definition includes *mks systems* (based on meters, kilograms, and seconds) as well as *cgs systems* (based on centimeters, grams, and seconds).

Metric systems avoid the pounds-mass versus pounds-force ambiguity in two ways. First, a unit of weight is not established at all. All quantities of matter are specified as mass. Second, force and mass units do not share a common name.

The term *metric system* is not explicit enough to define which units are to be used for any given variable. For example, within the cgs system there is variation in how certain electrical and magnetic quantities are represented (resulting in the ESU and EMU systems). Also, within the mks system, it is common practice in some industries to use kilocalories as the unit of thermal energy, while the SI unit for thermal energy is the joule. Thus, there is a lack of uniformity even within the metricated engineering community.[6]

The "metric" parts of this book use SI, which is the most developed and codified of the so-called metric systems.[7] There will be occasional variances with local engineering custom, but it is difficult to anticipate such variances within a book that must itself be consistent.[8]

12. THE cgs SYSTEM

The *cgs system* is used widely by chemists and physicists. It is named for the three primary units used to construct its derived variables: the centimeter, the gram, and the second.

When Newton's second law is written in the cgs system, the following combination of units results.

$$\text{units of force} = (m \text{ in g})\left(a \text{ in } \frac{\text{cm}}{\text{s}^2}\right)$$
$$= \text{g·cm/s}^2 \qquad 1.18$$

This combination of units for force is known as a *dyne*. Energy variables in the cgs system have units of dyne·cm or, equivalently, g·cm^2/s^2. This combination is known as an *erg*. There is no uniformly accepted unit of power in the cgs system, although calories per second is frequently used.

[6]In the "field test" of the metric system conducted over the past 200 years, other conventions are to use kilograms-force (kgf) instead of newtons and kgf/cm^2 for pressure (instead of pascals).
[7]SI units are an outgrowth of the *General Conference of Weights and Measures*, an international treaty organization that established the *Système International d'Unites (International System of Units)* in 1960. The United States subscribed to this treaty in 1975.
[8]Conversion to pure SI units is essentially complete in Australia, Canada, New Zealand, and South Africa. The use of nonstandard metric units is more common among European civil engineers, who have had little need to deal with the inertial properties of mass. However, even the American Society of Civil Engineers declared its support for SI units in 1985.

The fundamental volume unit in the cgs system is the cubic centimeter (cc). Since this is the same volume as one thousandth of a liter, units of milliliters (ml) are also used.

13. SI UNITS (THE mks SYSTEM)

SI units comprise an *mks system* (so named because it uses the meter, kilogram, and second as base units). All other units are derived from the base units, which are completely listed in Table 1.1. This system is fully consistent, and there is only one recognized unit for each physical quantity (variable).

Table 1.1 SI Base Units

quantity	name	symbol
length	meter	m
mass	kilogram	kg
time	second	s
electric current	ampere	A
temperature	kelvin	K
amount of substance	mole	mol
luminous intensity	candela	cd

Three types of units are used: base units, supplementary units, and derived units. The *base units* are dependent only on accepted standards or reproducible phenomena. The *supplementary units* (Table 1.2) have not been classified as being base units or derived units. The *derived units* (Tables 1.3 and 1.4) are made up of combinations of base and supplementary units.

Table 1.2 SI Supplementary Units

quantity	name	symbol
plane angle	radian	rad
solid angle	steradian	sr

In addition, there is a set of non-SI units that may be used. This concession is primarily due to the significance and widespread acceptance of these units. Use of the non-SI units listed in Table 1.5 will usually create an inconsistent expression requiring conversion factors.

The SI unit of force can be derived from Newton's second law. This combination of units for force is known as a *newton*.

$$\text{units of force} = (m \text{ in kg})\left(a \text{ in } \frac{\text{m}}{\text{s}^2}\right)$$
$$= \text{kg·m/s}^2 \qquad 1.19$$

Energy variables in SI units have units of N·m, or equivalently, kg·m^2/s^2. Both of these combinations are known as a *joule*. The units of power are joules per second, equivalent to a *watt*.

Table 1.3 Some SI Derived Units with Special Names

quantity	name	symbol	expressed in terms of other units
frequency	hertz	Hz	1/s
force	newton	N	kg·m/s^2
pressure, stress	pascal	Pa	N/m^2
energy, work, quantity of heat	joule	J	N·m
power, radiant flux	watt	W	J/s
quantity of electricity, electric charge	coulomb	C	
electric potential, potential difference, electromotive force	volt	V	W/A
electric capacitance	farad	F	C/V
electric resistance	ohm	Ω	V/A
electric conductance	siemen	S	A/V
magnetic flux	weber	Wb	V·s
magnetic flux density	tesla	T	Wb/m^2
inductance	henry	H	Wb/A
luminous flux	lumen	lm	
illuminance	lux	lx	lm/m^2

Example 1.3

A 10 kg block hangs from a cable. What is the tension in the cable? (Standard gravity equals 9.81 m/s^2.)

Solution

$$F = mg = (10 \text{ kg})\left(9.81 \frac{\text{m}}{\text{s}^2}\right)$$
$$= 98.1 \text{ kg·m/s}^2 \ (98.1 \text{ N})$$

Example 1.4

A 10 kg block is raised vertically 3 m. What is the change in potential energy?

Solution

$$\Delta E_p = mg\Delta h = (10 \text{ kg})\left(9.81 \frac{\text{m}}{\text{s}^2}\right)(3 \text{ m})$$
$$= 294 \text{ kg·m}^2/\text{s}^2 \ (294 \text{ J})$$

14. RULES FOR USING SI UNITS

In addition to having standardized units, the set of SI units also has rigid syntax rules for writing the units and combinations of units. Each unit is abbreviated with a specific symbol. The following rules for writing and combining these symbols should be adhered to.

Table 1.4 *Some SI Derived Units*

quantity	description	symbol
area	square meter	m^2
volume	cubic meter	m^3
speed—linear	meter per second	m/s
—angular	radian per second	rad/s
acceleration—linear	meter per second squared	m/s^2
—angular	radian per second squared	rad/s^2
density, mass density	kilogram per cubic meter	kg/m^3
concentration (of amount of substance)	mole per cubic meter	mol/m^3
specific volume	cubic meter per kilogram	m^3/kg
luminance	candela per square meter	cd/m^2
absolute viscosity	pascal second	Pa·s
kinematic viscosity	square meters per second	m^2/s
moment of force	newton meter	N·m
surface tension	newton per meter	N/m
heat flux density, irradiance	watt per square meter	W/m^2
heat capacity, entropy	joule per kelvin	J/K
specific heat capacity, specific entropy	joule per kilogram kelvin	J/kg·K
specific energy	joule per kilogram	J/kg
thermal conductivity	watt per meter kelvin	W/m·K
energy density	joule per cubic meter	J/m^3
electric field strength	volt per meter	V/m
electric charge density	coulomb per cubic meter	C/m^3
surface density of charge, flux density	coulomb per square meter	C/m^2
permittivity	farad per meter	F/m
current density	ampere per square meter	A/m^2
magnetic field strength	ampere per meter	A/m
permeability	henry per meter	H/m
molar energy	joule per mole	J/mol
molar entropy, molar heat capacity	joule per mole kelvin	J/mol·K
radiant intensity	watt per steradian	W/sr

- The expressions for derived units in symbolic form are obtained by using the mathematical signs of multiplication and division. For example, units of velocity are m/s. Units of torque are N·m (not N-m or Nm).

- Scaling of most units is done in multiples of 1000.

- The symbols are always printed in roman type, regardless of the type used in the rest of the text. The only exception to this is in the use of the symbol for liter, where the use of the lower case "el" (l) may be confused with the numeral one (1). In this case, "liter" should be written out in full, or the script ℓ or L used.

Table 1.5 *Acceptable Non-SI Units*

quantity	unit name	symbol or abbreviation	relationship to SI unit
area	hectare	ha	$1 \text{ ha} = 10\,000 \text{ m}^2$
energy	kilowatt-hour	kW·h	$1 \text{ kW·h} = 3.6 \text{ MJ}$
mass	metric ton[a]	t	$1 \text{ t} = 1000 \text{ kg}$
plane angle	degree (of arc)	°	$1° = 0.017\,453 \text{ rad}$
speed of rotation	revolution per minute	r/min	$1 \text{ r/min} = 2\pi/60 \text{ rad/s}$
temperature interval	degree Celsius	°C	$1°C = 1K$
time	minute	min	$1 \text{ min} = 60 \text{ s}$
	hour	h	$1 \text{ h} = 3600 \text{ s}$
	day (mean solar)	d	$1 \text{ d} = 86\,400 \text{ s}$
	year (calendar)	a	$1 \text{ a} = 31\,536\,000 \text{ s}$
velocity	kilometer per hour	km/h	$1 \text{ km/h} = 0.278 \text{ m/s}$
volume	liter[b]	ℓ	$1 \text{ ℓ} = 0.001 \text{ m}^3$

[a]The international name for metric ton is *tonne*. The metric ton is equal to the *megagram* (Mg).
[b]The international symbol for liter is the lowercase l, which can be easily confused with the numeral 1. Several English-speaking countries have adopted the script ℓ or uppercase L (as does this book) as a symbol for liter in order to avoid any misinterpretation.

- Symbols are not pluralized: 1 kg, 45 kg (not 45 kgs).

- A period after a symbol is not used, except when the symbol occurs at the end of a sentence.

- When symbols consist of letters, there is always a full space between the quantity and the symbols: 45 kg (not 45kg). However, when the first character of a symbol is not a letter, no space is left: 32°C (not 32° C or 32 °C); or 42° 12′ 45″ (not 42 ° 12 ′ 45 ″).

- All symbols are written in lowercase, except when the unit is derived from a proper name: m for meter; s for second; A for ampere, Wb for weber, N for newton, W for watt.

- Prefixes are printed without spacing between the prefix and the unit symbol (e.g., km is the symbol for kilometer).

- In text, symbols should be used when associated with a number. However, when no number is involved, the unit should be spelled out. Examples: The area of the carpet is 16 m^2, not 16 square meters. Carpet is sold by the square meter, not by the m^2.

- Where a decimal fraction of a unit is used, a zero should always be placed before the decimal marker: 0.45 kg (not .45 kg). This practice draws attention to the decimal marker and helps avoid errors of scale.

Table 1.6 SI Prefixes[a]

prefix	symbol	value
exa	E	10^{18}
peta	P	10^{15}
tera	T	10^{12}
giga	G	10^{9}
mega	M	10^{6}
kilo	k	10^{3}
hecto	h	10^{2}
deka	da	10^{1}
deci	d	10^{-1}
centi	c	10^{-2}
milli	m	10^{-3}
micro	μ	10^{-6}
nano	n	10^{-9}
pico	p	10^{-12}
femto	f	10^{-15}
atto	a	10^{-18}

[a]There is no "B" (billion) prefix. In fact, the word "billion" means 10^9 in the United States but 10^{12} in most other countries. This unfortunate ambiguity is handled by avoiding the use of the term billion.

- A practice in some countries is to use a comma as a decimal marker, while the practice in North America, the United Kingdom, and some other countries is to use a period (or dot) as the decimal marker. Furthermore, in some countries that use the decimal comma, a dot is frequently used to divide long numbers into groups of three. Because of these differing practices, spaces must be used instead of commas to separate long lines of digits into easily readable blocks of three digits with respect to the decimal marker: 32 453.246 072 5. A space (half-space preferred) is optional with a four-digit number: 1 234 or 1234.

- Some confusion may arise with the word "tonne" (1000 kg). When this word occurs in French text of Canadian origin, the meaning may be a ton of 2000 pounds.

15. PRIMARY DIMENSIONS

Regardless of the system of units chosen, each variable representing a physical quantity will have the same *primary dimensions*. For example, velocity may be expressed in miles per hour (mph) or meters per second (m/s), but both units have dimensions of length per unit time. Length and time are two of the primary dimensions, as neither can be broken down into more basic dimensions. The concept of primary dimensions is useful when converting little-used variables between different systems of units, as well as in correlating experimental results (i.e., dimensional analysis).

There are three different sets of primary dimensions in use.[9] In the $ML\theta T$ system, the primary dimensions are mass (M), length (L), time (θ), and temperature (T). Notice that all symbols are uppercase. In order to avoid confusion between time and temperature, the Greek letter theta is used for time.[10]

All other physical quantities can be derived from these primary dimensions.[11] For example, work in SI units has units of N·m. Since a newton is a kg·m/s², the primary dimensions of work are ML^2/θ^2. The primary dimensions for many important engineering variables are shown in Table 1.7. If it is more convenient to stay with traditional English units, it may be more desirable to work in the $FML\theta TQ$ system (sometimes called the *engineering dimensional system*). This system adds the primary dimensions of force (F) and heat (Q). Thus, work (ft-lbf in the English system) has the primary dimensions of FL. (Compare this with the primary dimensions for work in the $ML\theta T$ system.) Thermodynamic variables are similarly simplified.

Dimensional analysis will be more conveniently carried out when one of the four-dimension systems ($ML\theta T$ or $FL\theta T$) is used. Whether the $ML\theta T$, $FL\theta T$, or $FML\theta TQ$ system is used depends on what is being derived and who will be using it, and whether or not a consistent set of variables is desired. Conversion constants such as g_c and J will almost certainly be required if the $ML\theta T$ system is used to generate variables for use in the English systems. It is also much more convenient to use the $FML\theta TQ$ system when working in the fields of thermodynamics, fluid flow, heat transfer, and so on.

16. DIMENSIONLESS GROUPS

A *dimensionless group* is derived as a ratio of two forces or other quantities. Considerable use of dimensionless groups is made in certain subjects, notably fluid mechanics and heat transfer. For example, the Reynolds number, Mach number, and Froude number are used to distinguish between distinctly different flow regimes in pipe flow, compressible flow, and open channel flow, respectively.

Table 1.8 contains information about the most common dimensionless groups used in fluid mechanics and heat transfer.

---

[9]One of these, the $FL\theta T$ system, is not discussed here but appears in Table 1.7.
[10]This is the most common usage. There is a lack of consistency in the engineering world about the symbols for the primary dimensions in dimensional analysis. Some writers use t for time instead of θ. Some use H for heat instead of Q. And, in the worst mix-up of all, some have reversed the use of T and θ.
[11]A *primary dimension* is the same as a *base unit* in the SI set of units. The SI units add several other base units, as shown in Table 1.1, to deal with variables that are difficult to derive in terms of the four primary base units.

Table 1.7 *Dimensions of Common Variables*

variable	dimensional system		
	$ML\theta T$	$FL\theta T$	$FML\theta TQ$
mass (m)	M	$F\theta^2/L$	M
force (F)	ML/θ^2	F	F
length (L)	L	L	L
time (θ)	θ	θ	θ
temperature (T)	T	T	T
work (W)	ML^2/θ^2	FL	FL
heat (Q)	ML^2/θ^2	FL	Q
acceleration (a)	L/θ^2	L/θ^2	L/θ^2
frequency (N)	$1/\theta$	$1/\theta$	$1/\theta$
area (A)	L^2	L^2	L^2
coefficient of thermal expansion (β)	$1/T$	$1/T$	$1/T$
density (ρ)	M/L^3	$F\theta^2/L^4$	M/L^3
dimensional constant (g_c)	1.0	1.0	$ML/\theta^2 F$
specific heat at constant pressure (c_p); at constant volume (c_v)	$L^2/\theta^2 T$	$L^2/\theta^2 T$	Q/MT
heat transfer coefficient (h); overall (U)	$M/\theta^3 T$	$F/\theta LT$	$Q/\theta L^2 T$
power (P)	ML^2/θ^3	FL/θ	FL/θ
heat flow rate ($\dot{Q}$)	ML^2/θ^3	FL/θ	Q/θ
kinematic viscosity (ν)	L^2/θ	L^2/θ	L^2/θ
mass flow rate ($\dot{m}$)	M/θ	$F\theta/L$	M/θ
mechanical equivalent of heat (J)			FL/Q
pressure (p)	$M/L\theta^2$	F/L^2	F/L^2
surface tension (σ)	M/θ^2	F/L	F/L
angular velocity (ω)	$1/\theta$	$1/\theta$	$1/\theta$
volumetric flow rate ($\dot{m}/\rho = \dot{V}$)	L^3/θ	L^3/θ	L^3/θ
conductivity (k)	$ML/\theta^3 T$	$F/\theta T$	$Q/L\theta T$
thermal diffusivity (α)	L^2/θ	L^2/θ	L^2/θ
velocity (v)	L/θ	L/θ	L/θ
viscosity, absolute (μ)	$M/L\theta$	$F\theta/L^2$	$F\theta/L^2$
volume (V)	L^3	L^3	L^3

17. LINEAL AND BOARD FOOT MEASUREMENTS

The term *lineal* is often mistaken as a typographical error for *linear*. Although "lineal" has its own specific meaning slightly different from "linear," the two are often used interchangeably by engineers.[12] The adjective *lineal* is often encountered in the building trade (e.g., 12 lineal feet of lumber), where the term is used to distinguish it from board feet measurement.

A *board foot* (abbreviated bd-ft) is not a measure of length. Rather, it is a measure of volume used with lumber. Specifically, a board foot is equal to 144 in^3 (2.36×10^{-3} m^3). The name is derived from the volume of a board 1 foot square and 1 inch thick. In that sense, it is parallel in concept to the acre-foot. Since lumber cost is directly related to lumber weight and volume, the board foot unit is used in determining the overall lumber cost.

PRACTICE PROBLEMS

1. Convert 250 degrees Fahrenheit to degrees Celsius.

2. Convert the Stefan-Boltzmann constant from English to SI units.

3. How many U.S. tons of 13,000 Btu/lbm coal must be burned to produce as much energy as from a complete nuclear conversion of 1 g of its mass?

---

[12] *Lineal* is best used when discussing a line of succession (e.g., a lineal descendant of a particular person). *Linear* is best used when discussing length (e.g., a linear dimension of a room).

Table 1.8 *Common Dimensionless Groups*

name	symbol	formula	interpretation
Biot number	Bi	$\dfrac{hL}{k_s}$	$\dfrac{\text{surface conductance}}{\text{internal conduction of solid}}$
Cauchy number	Ca	$\dfrac{\text{v}^2}{\dfrac{B_s}{\rho}} = \dfrac{\text{v}^2}{a^2}$	$\dfrac{\text{inertia force}}{\text{compressive force}} = \text{Mach number}^2$
Eckert number	Ec	$\dfrac{\text{v}^2}{2c_p\Delta T}$	$\dfrac{\text{temperature rise due to energy conversion}}{\text{temperature difference}}$
Euler number	Eu	$\dfrac{\Delta p}{\rho V^2}$	$\dfrac{\text{pressure force}}{\text{inertia force}}$
Fourier number	Fo	$\dfrac{kt}{\rho c_p L^2} = \dfrac{\alpha t}{L^2}$	$\dfrac{\text{rate of conduction of heat}}{\text{rate of storage of energy}}$
Froude number[a]	Fr	$\dfrac{\text{v}^2}{gL}$	$\dfrac{\text{inertia force}}{\text{gravity force}}$
Graetz number[a]	Gz	$\left(\dfrac{D}{L}\right)\left(\dfrac{\text{v}\rho c_p D}{k}\right)$	$\dfrac{(\text{Re})(\text{Pr})}{L/D}$ $\dfrac{\text{heat transfer by convection in entrance region}}{\text{heat transfer by conduction}}$
Grashof number[a]	Gr	$\dfrac{g\beta\Delta T L^3}{\nu^2}$	$\dfrac{\text{buoyancy force}}{\text{viscous force}}$
Knudsen number	Kn	$\dfrac{\lambda}{L}$	$\dfrac{\text{mean free path of molecules}}{\text{characteristic length of object}}$
Lewis number[a]	Le	$\dfrac{\alpha}{D_c}$	$\dfrac{\text{thermal diffusivity}}{\text{molecular diffusivity}}$
Mach number	M	$\dfrac{\text{v}}{a}$	$\dfrac{\text{macroscopic velocity}}{\text{speed of sound}}$
Nusselt number	Nu	$\dfrac{hL}{k}$	$\dfrac{\text{temperature gradient at wall}}{\text{overall temperature difference}}$
Péclet number	Pé	$\dfrac{\text{v}\rho c_p D}{k}$	$(\text{Re})(\text{Pr})$ $\dfrac{\text{heat transfer by convection}}{\text{heat transfer by conduction}}$
Prandtl number	Pr	$\dfrac{\mu c_p}{k} = \dfrac{\nu}{\alpha}$	$\dfrac{\text{diffusion of momentum}}{\text{diffusion of heat}}$
Reynolds number	Re	$\dfrac{\rho \text{v} L}{\mu} = \dfrac{\text{v}L}{\nu}$	$\dfrac{\text{inertia force}}{\text{viscous force}}$
Schmidt number	Sc	$\dfrac{\mu}{\rho D_c} = \dfrac{\nu}{D_c}$	$\dfrac{\text{diffusion of momentum}}{\text{diffusion of mass}}$
Sherwood number[a]	Sh	$\dfrac{k_D L}{D_c}$	$\dfrac{\text{mass diffusivity}}{\text{molecular diffusivity}}$
Stanton number	St	$\dfrac{h}{\text{v}\rho c_p} = \dfrac{h}{c_p G}$	$\dfrac{\text{heat transfer at wall}}{\text{energy transported by stream}}$
Stokes number	Sk	$\dfrac{\Delta p L}{\mu \text{v}}$	$\dfrac{\text{pressure force}}{\text{viscous force}}$
Strouhal number[a]	Sl	$\dfrac{L}{t\text{v}} = \dfrac{L\omega}{\text{v}}$	$\dfrac{\text{frequency of vibration}}{\text{characteristic frequency}}$
Weber number	We	$\dfrac{\rho \text{v}^2 L}{\sigma}$	$\dfrac{\text{inertia force}}{\text{surface tension force}}$

[a]Multiple definitions exist.

2 Engineering Drawing Practice[1]

1. Normal Views of Lines and Planes 2-1
2. Intersecting and Perpendicular Lines 2-1
3. Types of Views 2-1
4. Principal (Orthographic) Views 2-2
5. Auxiliary (Orthographic) Views 2-2
6. Oblique (Orthographic) Views 2-2
7. Axonometric (Orthographic Oblique) Views 2-3
8. Perspective Views 2-3
9. Sections 2-3
10. Tolerances 2-4
11. Surface Finish 2-4

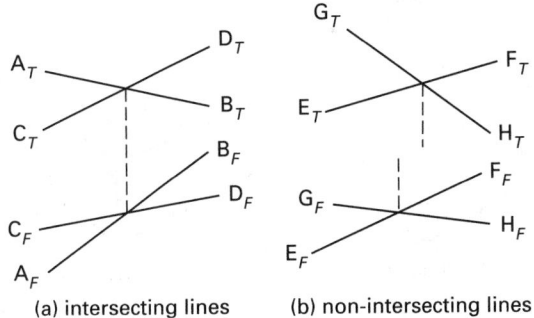

Figure 2.1 *Intersecting and Non-Intersecting Lines*

1. NORMAL VIEWS OF LINES AND PLANES

A *normal view* of a line is a perpendicular projection of the line onto a viewing plane parallel to the line. In the normal view, all points of the line are equidistant from the observer. Therefore, the true length of a line is viewed and can be measured.

Generally, however, a line will be viewed from an oblique position and will appear shorter than it actually is. The normal view can be constructed by drawing an auxiliary view (see Sec. 5) from the orthographic view.[2]

Similarly, a normal view of a plane figure is a perpendicular projection of the figure onto a viewing plane parallel to the plane of the figure. All points of the plane are equidistant from the observer. Therefore, the true size and shape of any figure in the plane can be determined.

2. INTERSECTING AND PERPENDICULAR LINES

A single orthographic view is not sufficient to determine whether two lines intersect. However, if two or more views show the lines as having the same common point (i.e., crossing at the same position in space), then the lines intersect. In Fig. 2.1, the subscripts F and T refer to front and top views, respectively.

According to the *perpendicular line principle*, two perpendicular lines appear perpendicular only in a normal view of either one or both of the lines. Conversely, if two lines appear perpendicular in any view, the lines are perpendicular only if the view is a normal view of one or both of the lines.

3. TYPES OF VIEWS

Objects can be illustrated in several different ways depending on the number of views, the angle of observation, and the degree of artistic latitude taken for the purpose of simplifying the drawing process.[3] Table 2.1 categorizes the types of views.

Table 2.1 *Types of Views of Objects*

orthographic views
  principal views
  auxiliary views
  oblique views
    cavalier projection
    cabinet projection
    clinographic projection
  axonometric views
    isometric
    dimetric
    trimetric
perspective views
  parallel perspective
  angular perspective

---

[1]This chapter is not meant to show "how to do it" as much as it is to present the conventions and symbols of engineering drawing.
[2]The technique for constructing a normal view is covered in engineering drafting texts.

[3]The omission of perspective from a drawing is an example of a step taken to simplify the drawing process.

The different types of views are easily distinguished by their *projectors* (i.e., projections of parallel lines on the object). For a cube, there are three sets of projectors corresponding to the three perpendicular axes. In an *orthographic (orthogonal) view*, the projectors are parallel. In a *perspective (central) view*, some or all of the projectors converge to a point.

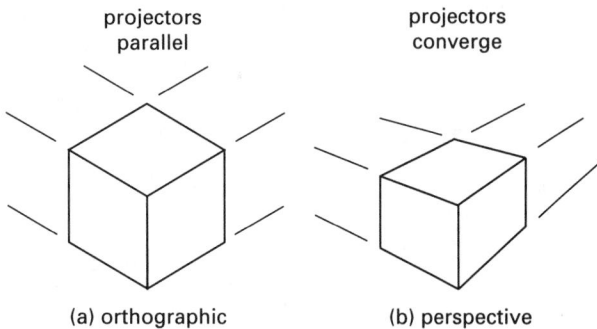

Figure 2.2 *Orthographic and Perspective Views of a Block*

4. PRINCIPAL (ORTHOGRAPHIC) VIEWS

In a *principal view* (also known as a *planar view*), one of the sets of projectors is normal to the view. That is, one of the planes of the object is seen in a normal view. The other two sets of projectors are orthogonal and are usually oriented horizontally and vertically on the paper. Because background details of an object may not be visible in a principal view, it is necessary to have at least three principal views to completely illustrate a symmetrical object. At most, six principal views will be needed to illustrate complex objects.

The relative positions of the six views have been standardized and are shown in Fig. 2.3, which also defines the *width* (also known as *depth*), *height*, and *length* of the object. The views that are not needed to illustrate features or provide dimensions (i.e., redundant views) can be omitted. The usual combination selected consists of the top, front, and right side views.

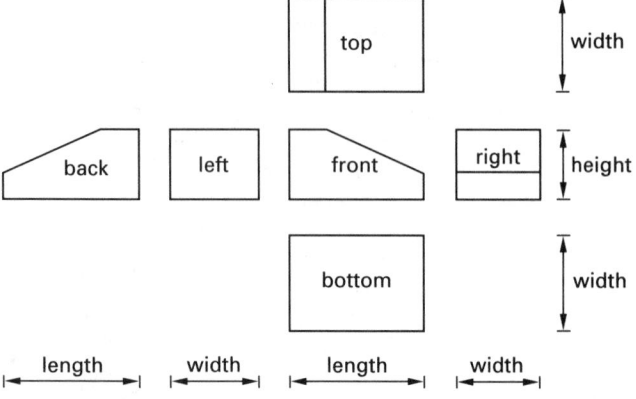

Figure 2.3 *Positions of Standard Orthographic Views*

It is common to refer to the front, side, and back views as *elevations* and to the top and bottom views as *plan views*. These terms are not absolute since any plane can be selected as the front.

5. AUXILIARY (ORTHOGRAPHIC) VIEWS

An *auxiliary view* is needed when an object has an inclined plane or curved feature or when there are more details than can be shown in the six principal views. As with the other orthographic views, the auxiliary view is a normal (face-on) view of the inclined plane.

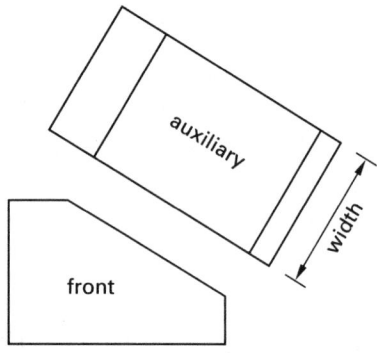

Figure 2.4 *Auxiliary View*

The projectors in an auxiliary view are perpendicular to only one of the directions in which a principal view is observed. Accordingly, only one of the three dimensions of width, height, and depth can be measured (scaled). In a *profile auxiliary view*, the object's width can be measured. In a *horizontal auxiliary view (auxiliary elevation)*, the object's height can be measured. In a *frontal auxiliary view*, the depth of the object can be measured.

6. OBLIQUE (ORTHOGRAPHIC) VIEWS

If the object is turned so that three principal planes are visible, it can be completely illustrated by a single *oblique view*.[4] In an oblique view, the direction from which the object is observed is not (necessarily) parallel to any of the directions from which principal and auxiliary views are observed.

In two common methods of oblique illustration, one of the view planes coincides with an orthographic view plane. Two of the drawing axes are at right angles to each other; one of which is vertical, and the other (the *oblique axis*) which is oriented at 30° or 45° (originally chosen to coincide with standard drawing triangles). The ratio of scales used for the horizontal, vertical, and oblique axes can be 1:1:1 or 1:1:$^{1}/_{2}$. The latter ratio helps to overcome the visual distortion due to the absence of perspective in the oblique direction.

---

[4]Oblique views are not unique in this capability—perspective drawings share it. Oblique and perspective drawings are known as *pictorial drawings* because they give depth to the object by illustrating it in three dimensions.

Cavalier (45° oblique axis and 1:1:1 scale ratio) and *cabinet* (45° oblique axis and 1:1:½ scale ratio) *projections* are the two common types of oblique views that incorporate one of the orthographic views. If an angle of 9.5° is used (as in illustrating crystalline lattice structures), the technique is known as *clinographic projection*.

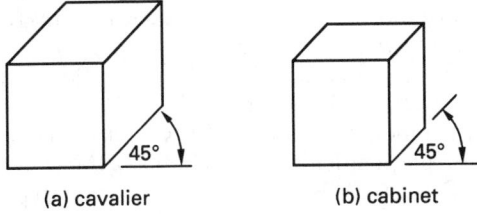

(a) cavalier   (b) cabinet

Figure 2.5 *Cavalier and Cabinet Oblique Drawings*

7. AXONOMETRIC (ORTHOGRAPHIC OBLIQUE) VIEWS

In axonometric views, the view plane is not parallel to any of the principal orthographic planes. Axonometric views and axonometric drawings are not the same. In a *view (projection)*, one or more of the face lengths is foreshortened. In a *drawing*, the lengths are drawn full length, resulting in a distorted illustration. Table 2.2 lists the proper ratios that should be observed.

Table 2.2 *Axonometric Foreshortening*

view	projector intersection angles	proper ratio of sides
isometric	120°, 120°, 120°	0.82:0.82:0.82
dimetric	131°25′, 131°25′, 97°10′	1:1:½
	103°38′, 103°38′, 152°44′	¾:¾:1
trimetric	102°28′, 144°16′, 113°16′	1:⅔:⅞
	138°14′, 114°46′, 107°	1:¾:⅞

In an *isometric view*, the three projectors intersect at equal angles (120°) with the plane. This simplifies construction with standard 30° drawing triangles. All of the faces are foreshortened an equal amount, to $\sqrt{2/3}$, or approximately 81.6% of the true length. In a *dimetric view*, two of the projectors intersect at equal angles, and only two of the faces are equally reduced in length. In a *trimetric view*, all three intersection angles are different, and all three faces are reduced different amounts.

8. PERSPECTIVE VIEWS

In a *perspective view*, one or more sets of projectors converge to a fixed point known as the *center of vision*. In the *parallel perspective*, all vertical lines remain vertical in the picture; all horizontal frontal lines remain horizontal. Therefore, one face is parallel to the observer and only one set of projectors converges. In the *angular perspective*, two sets of projectors converge. In the little-used *oblique perspective*, all three sets of projectors converge.

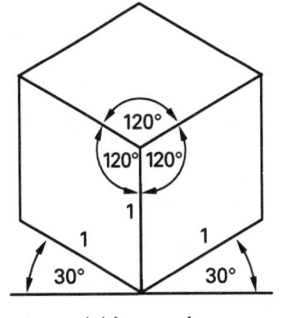

(a) isometric

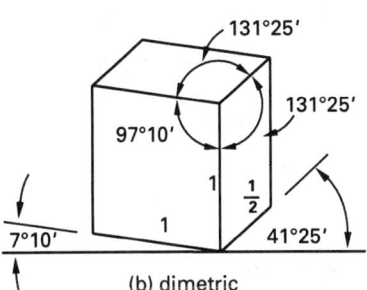

(b) dimetric

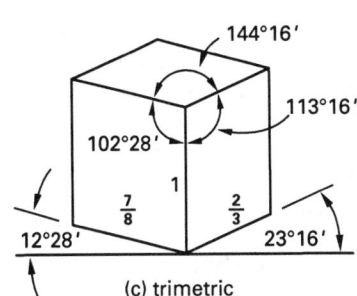

(c) trimetric

Figure 2.6 *Types of Axonometric Views*

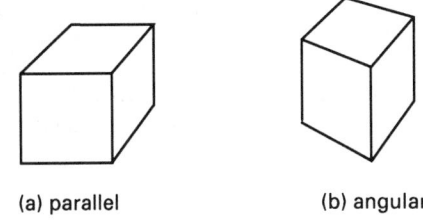

(a) parallel   (b) angular

Figure 2.7 *Types of Perspective Views*

9. SECTIONS

The term *section* is an imaginary cut taken through an object to reveal the shape or interior construction.[5] Figure 2.8 illustrates the standard symbol for a *sectioning cut* and the resulting sectional view. Section arrows are perpendicular to the cutting plane and indicate the viewing direction.

---

[5]The term *section* is also used to mean a *cross section*—a slice of finite but negligible thickness that is taken from an object to show the cross section or interior construction at the plane of the slice.

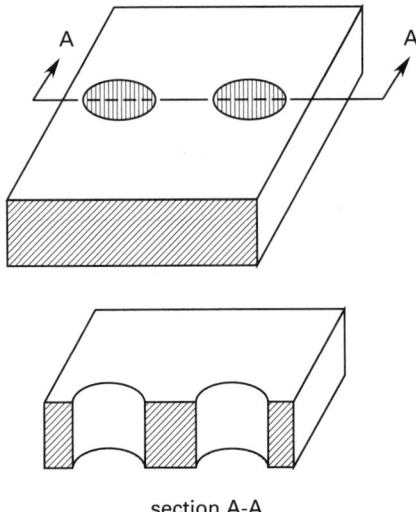

Figure 2.8 *Sectioning Cut Symbol and Sectional View*

10. TOLERANCES

The *tolerance* for a dimension is the total permissible variation or difference between the acceptable limits. The tolerance for a dimension can be specified in two ways: either as a general rule in the title block (e.g., ± 0.001 inch unless otherwise specified) or as specific limits given with each dimension (e.g., 2.575 inches ± 0.005 inch).

11. SURFACE FINISH

Surface finish is specified by a combination of parameters.[6] The basic symbol for designating these factors is shown in Fig. 2.9. In the symbol, A is the maximum *roughness height index*, B is the optional minimum roughness height, C is the peak-to-valley *waviness height*, D is the optional peak-to-valley *wavinessspacing (width)* rating, E is the optional *roughness width*

cutoff (roughness sampling length), F is the *lay*, and G is the *roughness width*. Unless minimums are specified, all parameters are maximum allowable values, and all lesser values are permitted.

Since the roughness varies, the waviness height is an arithmetic average within a sampled square, and the designation A is known as the *roughness weight*, R_a.[7] Values are normally given in micrometers (μm) or microinches (μin) in SI or customary U.S. units, respectively. A value for the roughness width cutoff of 0.80 mm (32 μin) is assumed when D is not specified. The lay symbol, F, can be $=$ (parallel to indicated surface), $\perp$ (perpendicular), C (circular), M (multidirectional), P (pitted), R (radial), or X (crosshatch).

If a small circle is placed at the A position, no machining is allowed and only cast, forged, die-cast, injection-molded, etc., surfaces are acceptable.

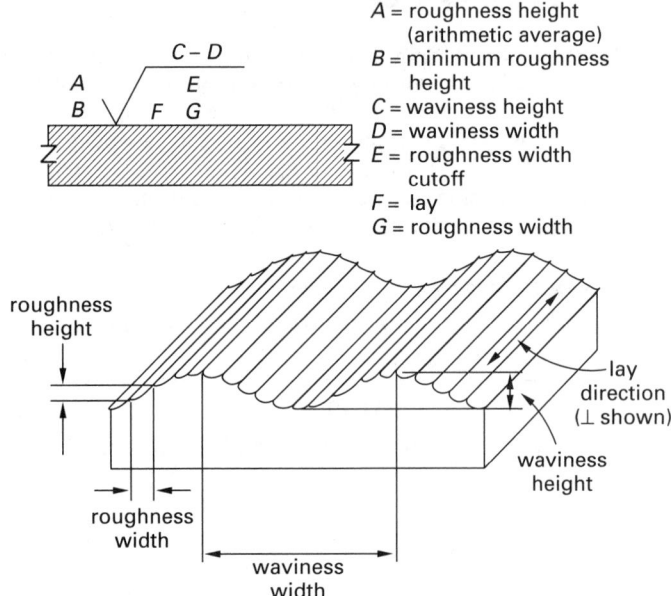

A = roughness height (arithmetic average)
B = minimum roughness height
C = waviness height
D = waviness width
E = roughness width cutoff
F = lay
G = roughness width

Figure 2.9 *Surface Finish Designations*

---

[6]Specification does not indicate appearance (i.e., color, luster) or performance (i.e., hardness, corrosion resistance, microstructure).

[7]The symbol R_a is the same as the AA (arithmetic average) and CLA (centerline average) terms used in other (and earlier) standards.

3 Algebra

1. Introduction 3-1
2. Symbols Used in This Book 3-1
3. Greek Alphabet 3-1
4. Types of Numbers 3-1
5. Significant Digits 3-1
6. Equations 3-2
7. Fundamental Algebraic Laws 3-3
8. Polynomials 3-3
9. Roots of Quadratic Equations 3-3
10. Roots of General Polynomials 3-3
11. Extraneous Roots 3-4
12. Descartes' Rule of Signs 3-4
13. Rules for Exponents and Radicals 3-5
14. Logarithms 3-5
15. Logarithm Identities 3-5
16. Partial Fractions 3-6
17. Simultaneous Linear Equations 3-7
18. Complex Numbers 3-7
19. Operations on Complex Numbers 3-8
20. Limits 3-9
21. Sequences and Progressions 3-10
22. Standard Sequences 3-11
23. Series 3-11
24. Tests for Series Convergence 3-11
25. Series of Alternating Sign 3-12
    Practice Problems 3-12

1. INTRODUCTION

Engineers working in design and analysis encounter mathematical problems on a daily basis. Although algebra and simple trigonometry are often sufficient for routine calculations, there are many instances when certain advanced subjects are needed. This chapter and the following ten, in addition to supporting the calculations used in other chapters, consolidate the mathematical concepts most often needed by engineers.

2. SYMBOLS USED IN THIS BOOK

Many symbols, letters, and Greek characters are used to represent variables in the formulas used throughout this book. These symbols and characters are defined in the nomenclature section of each chapter. However, some of the other symbols in this book are listed in Table 3.2.

3. GREEK ALPHABET

Table 3.1 The Greek Alphabet

A	α	alpha	N	ν	nu
B	β	beta	Ξ	ξ	xi
Γ	γ	gamma	O	o	omicron
Δ	δ	delta	Π	π	pi
E	ϵ	epsilon	P	ρ	rho
Z	ζ	zeta	Σ	σ	sigma
H	η	eta	T	τ	tau
Θ	θ	theta	Υ	υ	upsilon
I	ι	iota	Φ	ϕ	phi
K	κ	kappa	X	χ	chi
Λ	λ	lambda	Ψ	ψ	psi
M	μ	mu	Ω	ω	omega

4. TYPES OF NUMBERS

The *numbering system* consists of three types of numbers: real, imaginary, and complex. *Real numbers*, in turn, consist of rational numbers and irrational numbers. *Rational real numbers* are numbers that can be written as the ratio of two integers (e.g., 4, $^2/_5$, and $^1/_3$).[1] *Irrational real numbers* are nonterminating, nonrepeating numbers that cannot be expressed as the ratio of two integers (e.g., π and $\sqrt{2}$). Real numbers can be positive or negative.

Imaginary numbers are square roots of negative numbers. The symbols i and j are both used to represent the square root of -1.[2] For example, $\sqrt{-5} = \sqrt{5}\sqrt{-1} = \sqrt{5}i$. *Complex numbers* consist of combinations of real and imaginary numbers (e.g., $3 - 7i$).

5. SIGNIFICANT DIGITS

The significant digits in a number include the leftmost, nonzero digits to the rightmost digit written. Final answers from computations should be rounded off to the number of decimal places justified by the data. The answer can be no more accurate than the least accurate number in the data. Of course, rounding should be done on final calculation results only. It should not be done on interim results.

---

[1] Notice that 0.3333333 is a nonterminating number, but as it can be expressed as a ratio of two integers (i.e., $^1/_3$), it is a rational number.

[2] The symbol j is used to represent the square root of -1 in electrical calculations to avoid confusion with the current variable, i.

Table 3.2 *Symbols Used in This Book*

symbol	name	use	example
$\sum$	sigma	series summation	$\sum_{i=1}^{3} x_i = x_1 + x_2 + x_3$
π	pi	$3.1415927\ldots$	$p = \pi D$
e	base of natural logs	$2.71828\ldots$	
$\prod$	pi	series multiplication	$\prod_{i=1}^{3} x_i = x_1 x_2 x_3$
Δ	delta	change in quantity	$\Delta h = h_2 - h_1$
$-$	over bar	average value	$\overline{x}$
$\cdot$	over dot	per unit time	$\dot{m} =$ mass flowing per second
!	factorial[a]		$x! = x(x-1)(x-2)\cdots(2)(1)$
$\mid \ \mid$	absolute value[b]		$\mid -3 \mid = +3$
$\approx$	approximately equal to		$x \approx 1.5$
$\propto$	proportional to		$x \propto y$
∞	infinity		$x \to \infty$
log	base 10 logarithm		$\log(5.74)$
ln	natural logarithm		$\ln(5.74)$
exp	exponential power		$\exp(x) = e^x$
rms	root-mean-square	$\sqrt{\dfrac{1}{n}\sum_{i=1}^{n} x_i^2}$	V_{rms}
$\angle$	phasor or angle		$\angle 53°$

[a] *Zero factorial* $(0!)$ is frequently encountered in the form of $(n-n)!$ when calculating permutations and combinations. Zero factorial is defined as 1.
[b] The notation $\text{abs}(x)$ is also used to indicate the absolute value.

Table 3.3 *Examples of Significant Digits*

number as written	number of significant digits	implied range
341	3	340.5 to 341.5
34.1	3	34.05 to 34.15
0.00341	3	0.003405 to 0.003415
341×10^7	3	340.5×10^7 to 341.5×10^7
3.41×10^{-2}	3	3.405×10^{-2} to 3.415×10^{-2}
3410	4	3409.5 to 3410.5
341.0	4	340.95 to 341.05

There are two ways that significant digits can affect calculations. For the operations of multiplication and division, the final answer is rounded to the number of significant digits in the least significant multiplicand, divisor, or dividend. So, $2.0 \times 13.2 = 26$ since the first multiplicand (2.0) has two significant digits only.

For the operations of addition and subtraction, the final answer is rounded to the position of the least significant digit in the addenda, minuend, or subtrahend. So, $2.0 + 13.2 = 15.2$ because both addenda are significant to the tenths' position; but $2 + 13.4 = 15$ since the 2 is significant only in the ones' position.

The multiplication rule should not be used for addition or subtraction, as this can result in strange answers. For example, it would be incorrect to round $1700 + 0.1$ to 2000 simply because 0.1 has only one significant digit.

6. EQUATIONS

An *equation* is a mathematical statement of equality such as $5 = 3 + 2$. *Algebraic equations* are written in terms of *variables*. In the equation $y = x^2 + 3$, the value of variable y depends on the value of variable x. Therefore, y is the *dependent variable*, and x is the *independent variable*. The dependency of y on x is clearer when the equation is written in *functional form*: $y = f(x)$.

A *parametric equation* uses one or more independent variables (*parameters*) to describe a function.[3] For example, the parameter θ can be used to write the parametric equations of a unit circle.

$$x = \cos\theta \qquad \qquad 3.1$$

$$y = \sin\theta \qquad \qquad 3.2$$

---

[3] As used in this section, there is no difference between a parameter and an independent variable. However, the term *parameter* is also used as a descriptive measurement that determines or characterizes the form, size, or content of a function. For example, the radius is a parameter of a circle, and mean and variance are parameters of a probability distribution. Once these parameters are specified, the function is completely defined.

A unit circle can also be described by a *non-parametric equation*.[4]

$$x^2 + y^2 = 1 \qquad 3.3$$

7. FUNDAMENTAL ALGEBRAIC LAWS

Algebra provides the rules that allow complex mathematical relationships to be expanded or condensed. Algebraic laws may be applied to complex numbers, variables, and real numbers. The general rules for changing the form of a mathematical relationship are given as follows:

- commutative law for addition:

$$A + B = B + A \qquad 3.4$$

- commutative law for multiplication:

$$AB = BA \qquad 3.5$$

- associative law for addition:

$$A + (B + C) = (A + B) + C \qquad 3.6$$

- associative law for multiplication:

$$A(BC) = (AB)C \qquad 3.7$$

- distributive law:

$$A(B + C) = AB + AC \qquad 3.8$$

8. POLYNOMIALS

A *polynomial* is a rational expression—usually the sum of several variable terms known as *monomials*—that does not involve division. The *degree of the polynomial* is the highest power to which a variable in the expression is raised. The following *standard polynomial forms* are useful when trying to find the roots of an equation.

$$(a + b)(a - b) = a^2 - b^2 \qquad 3.9$$

$$(a \pm b)^2 = a^2 \pm 2ab + b^2 \qquad 3.10$$

$$(a \pm b)^3 = a^3 \pm 3a^2b + 3ab^2 \pm b^3 \qquad 3.11$$

$$(a^3 \pm b^3) = (a \pm b)(a^2 \mp ab + b^2) \qquad 3.12$$

$$(a^n - b^n) = (a - b)(a^{n-1} + a^{n-2}b + a^{n-3}b^2 + \cdots$$
$$+ b^{n-1}) \quad [n \text{ is any positive integer}] \quad 3.13$$

$$(a^n + b^n) = (a + b)(a^{n-1} - a^{n-2}b + a^{n-3}b^2 - \cdots$$
$$+ b^{n-1}) \quad [n \text{ is any positive odd integer}]$$
$$3.14$$

The *binomial theorem* defines a polynomial of the form $(a + b)^n$.

$$(a + b)^n = a^n + na^{n-1}b + C_2 a^{n-2}b^2 + \cdots$$
$$\underset{[i=0]}{} \quad \underset{[i=1]}{} \quad \underset{[i=2]}{}$$
$$+ C_i a^{n-i}b^i + \cdots + nab^{n-1} + b^n \qquad 3.15$$

$$C_i = \frac{n!}{i!(n-i)!} \qquad [i = 0,1,2,\ldots n] \qquad 3.16$$

---

[4]Since only the coordinate variables are used, this equation is also said to be in *Cartesian equation form*.

The coefficients of the expansion can be determined quickly from *Pascal's triangle*, Fig. 3.1. Notice that each entry is the sum of the two entries directly above it.

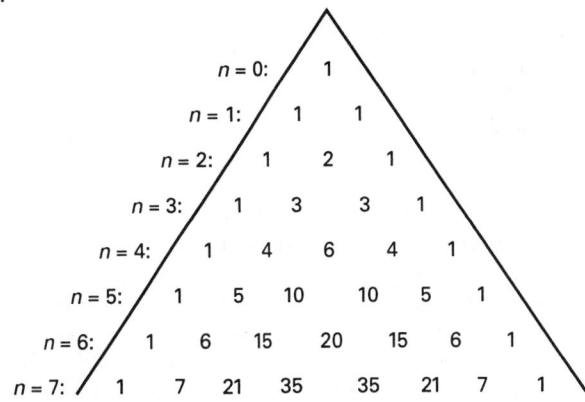

Figure 3.1 *Pascal's Triangle*

The values $r_1, r_2, \ldots, r_n$ of the independent variable x that satisfy a polynomial equation $f(x) = 0$ are known as *roots* or *zeros* of the polynomial. A polynomial of degree n with real coefficients will have at most n real roots, although they need not all be distinctly different.

9. ROOTS OF QUADRATIC EQUATIONS

A *quadratic equation* is an equation of the general form $ax^2 + bx + c = 0 \, [a \neq 0]$. The *roots*, x_1 and x_2, of the equation are the two values of x that satisfy it.

$$x_1, x_2 = \frac{-b \pm \sqrt{b^2 - 4ac}}{2a} \qquad 3.17$$

$$x_1 + x_2 = -\frac{b}{a} \qquad 3.18$$

$$x_1 x_2 = \frac{c}{a} \qquad 3.19$$

The types of roots of the equation can be determined from the *discriminant* (i.e., the quantity under the radical in Eq. 3.17).

- If $(b^2 - 4ac) > 0$, the roots are real and unequal.

- If $(b^2 - 4ac) = 0$, the roots are real and equal. This is known as a *double root*.

- If $(b^2 - 4ac) < 0$, the roots are complex and unequal.

10. ROOTS OF GENERAL POLYNOMIALS

It is more difficult to find roots of cubic and higher-degree polynomials because few general techniques exist.

- *inspection*: Finding roots by inspection is equivalent to making reasonable guesses about the roots and substituting into the polynomial.

- *graphing*: If the value of a polynomial $f(x)$ is calculated and plotted for different values of x, an approximate value of a root can be determined as the value of x at which the plot crosses the x-axis.

- *numerical methods*: If an approximate value of a root is known, numerical methods (bisection method, Newton's method, etc.) can be used to refine the value. The more efficient techniques are too complex to be performed by hand.

- *factoring*: If at least one root (say, $x = r$) of a polynomial $f(x)$ is known, the quantity $(x - r)$ can be factored out of $f(x)$ by long division. The resulting quotient will be lower by one degree, and the remaining roots may be easier to determine. This method is particularly applicable if the polynomial is in one of the standard forms presented in Sec. 8.

- *special cases*: Certain polynomial forms can be simplified by substitution or solved by standard formulas if they are recognized as being special cases. (The standard solution to the quadratic equation is such a special case.) For example, $ax^4 + bx^2 + c = 0$ can be reduced to a polynomial of degree 2 if the substitution $u = x^2$ is made.

11. EXTRANEOUS ROOTS

With simple equalities, it may appear possible to derive roots by basic algebraic manipulations.[5] However, multiplying each side of an equality by a power of a variable may introduce *extraneous roots*. Such roots do not satisfy the original equation even though they are derived according to the rules of algebra. Checking a calculated root is always a good idea, but is particularly necessary if the equation has been multiplied by one of its own variables.

Example 3.1

Use algebraic operations to determine a value that satisfies the following equation. Determine if the value is a valid or extraneous root.
$$\sqrt{x - 2} = \sqrt{x} + 2$$

Solution

Square both sides.
$$x - 2 = x + 4\sqrt{x} + 4$$

Subtract x from each side, and combine the constants.
$$4\sqrt{x} = -6$$

[5]In this sentence, *equality* means a combination of two expressions containing an equal sign. Any two expressions can be linked in this manner, even those that are not actually equal. For example, the expressions for two non-intersecting ellipses can be equated even though there is no intersection point. Finding extraneous roots is more likely when the underlying equality is false to begin with.

Solve for x.
$$x = \left(\frac{-6}{4}\right)^2 = \frac{9}{4}$$

Substitute $x = 9/4$ into the original equation.
$$\sqrt{\frac{9}{4} - 2} = \sqrt{\frac{9}{4}} + 2$$
$$\frac{1}{2} = \frac{7}{2}$$

Since the equality is not established, $x = 9/4$ is an extraneous root.

12. DESCARTES' RULE OF SIGNS

Descartes' rule of signs determines the maximum number of positive (and negative) real roots that a polynomial will have by counting the number of sign reversals (i.e., changes in sign from one term to the next) in the polynomial. The polynomial $f(x)$ must have real coefficients and must be arranged in terms of descending powers of x.

- The number of positive roots of the polynomial equation $f(x) = 0$ will not exceed the number of sign reversals.

- The difference between the number of sign reversals and the number of positive roots is an even number.

- The number of negative roots of the polynomial equation $f(x) = 0$ will not exceed the number of sign reversals in the polynomial $f(-x)$.

- The difference between the number of sign reversals in $f(-x)$ and the number of negative roots is an even number.

Example 3.2

Determine the possible numbers of positive and negative roots that satisfy the following polynomial equation.
$$4x^5 - 5x^4 + 3x^3 - 8x^2 - 2x + 3 = 0$$

Solution

There are four sign reversals, so up to four positive roots exist. To keep the difference between the number of positive roots and the number of sign reversals an even number, the number of positive real roots is limited to zero, two, and four.

Substituting $-x$ for x in the polynomial results in
$$-4x^5 - 5x^4 - 3x^3 - 8x^2 + 2x + 3 = 0$$

There is only one sign reversal, so the number of negative roots cannot exceed one. There must be exactly one negative real root in order to keep the difference to an even number (zero in this case).

13. RULES FOR EXPONENTS AND RADICALS

In the expression $b^n = a$, b is known as the *base* and n is the *exponent* or *power*. In Eqs. 3.20 through 3.33, a, b, m, and n are any real numbers with limitations listed.

$$b^0 = 1 \qquad [b \neq 0] \qquad \text{3.20}$$

$$b^1 = b \qquad \text{3.21}$$

$$b^{-n} = \frac{1}{b^n} = \left(\frac{1}{b}\right)^n \qquad [b \neq 0] \qquad \text{3.22}$$

$$\left(\frac{a}{b}\right)^n = \frac{a^n}{b^n} \qquad [b \neq 0] \qquad \text{3.23}$$

$$(ab)^n = a^n b^n \qquad \text{3.24}$$

$$b^{\frac{m}{n}} = \sqrt[n]{b^m} = \left(\sqrt[n]{b}\right)^m \qquad \text{3.25}$$

$$(b^n)^m = b^{nm} \qquad \text{3.26}$$

$$b^m b^n = b^{m+n} \qquad \text{3.27}$$

$$\frac{b^m}{b^n} = b^{m-n} \qquad [b \neq 0] \qquad \text{3.28}$$

$$\sqrt[n]{b} = b^{\frac{1}{n}} \qquad \text{3.29}$$

$$\left(\sqrt[n]{b}\right)^n = \left(b^{\frac{1}{n}}\right)^n = b \qquad \text{3.30}$$

$$\sqrt[n]{ab} = \sqrt[n]{a}\,\sqrt[n]{b} = a^{\frac{1}{n}} b^{\frac{1}{n}} = (ab)^{\frac{1}{n}} \qquad \text{3.31}$$

$$\sqrt[n]{\frac{a}{b}} = \frac{\sqrt[n]{a}}{\sqrt[n]{b}} = \left(\frac{a}{b}\right)^{\frac{1}{n}} \qquad [b \neq 0] \qquad \text{3.32}$$

$$\sqrt[m]{\sqrt[n]{b}} = \sqrt[mn]{b} = b^{\frac{1}{mn}} \qquad \text{3.33}$$

14. LOGARITHMS

Logarithms can be considered to be exponents. For example, the exponent n in the expression $b^n = a$ is the logarithm of a to the base b. Therefore, the two expressions $\log_b a = n$ and $b^n = a$ are equivalent.

The base for *common logs* is 10. Usually, "log" will be written when common logs are desired, although "$\log_{10}$" appears occasionally. The base for *natural (Napierian) logs* is 2.71828..., a number which is given the symbol e. When natural logs are desired, usually ln will be written, although $\log_e$ is also used.

Most logarithms will contain an integer part (the *characteristic*) and a decimal part (the *mantissa*). The logarithm of any number less than one is negative. If the number is greater than one, its logarithm is positive. Although the logarithm may be negative, the mantissa is always positive. For negative logarithms, the characteristic is found by expressing the logarithm as the sum of a negative characteristic and a positive mantissa.

For common logarithms of numbers greater than one, the characteristics will be positive and equal to one less than the number of digits in front of the decimal. If

the number is less than one, the characteristic will be negative and equal to one more than the number of zeros immediately following the decimal point.

If a negative logarithm is to be used in a calculation, it must first be converted to *operational form* by adding the characteristic and mantissa. The operational form should be used in all calculations and is the form displayed by scientific calculators.

The logarithm of a negative number is a complex number.

Example 3.3

Use logarithm tables to determine the operational form of $\log_{10}(0.05)$.

Solution

Since the number is less than one and there is one leading zero, the characteristic is found by observation to be -2. From a book of logarithm tables, the mantissa of 5.0 is 0.699. Two ways of combining the mantissa and characteristic are possible:

$$method\ 1: \quad \overline{2}.699$$

$$method\ 2: \quad 8.699 - 10$$

The operational form of this logarithm is $-2 + 0.699 = -1.301$.

15. LOGARITHM IDENTITIES

Prior to the widespread availability of calculating devices, logarithm identities were used to solve complex calculations by reducing the solution method to table look-up, addition, and subtraction. Logarithm identities are still useful in simplifying expressions containing exponentials and other logarithms. In Eqs. 3.34 through 3.45, $a \neq 1$, $b \neq 1$, $x > 0$, and $y > 0$.

$$\log_b(b) = 1 \qquad \text{3.34}$$

$$\log_b(1) = 0 \qquad \text{3.35}$$

$$\log_b(b^n) = n \qquad \text{3.36}$$

$$\log(x^a) = a \log(x) \qquad \text{3.37}$$

$$\log\left(\sqrt[n]{x}\right) = \log\left(x^{\frac{1}{n}}\right) = \frac{\log(x)}{n} \qquad \text{3.38}$$

$$b^{n \log_b(x)} = x^n = \text{antilog}[n \log_b(x)] \qquad \text{3.39}$$

$$b^{\frac{\log_b(x)}{n}} = x^{\frac{1}{n}} \qquad \text{3.40}$$

$$\log(xy) = \log(x) + \log(y) \qquad \text{3.41}$$

$$\log\left(\frac{x}{y}\right) = \log(x) - \log(y) \qquad \text{3.42}$$

$$\log_a(x) = \log_b(x) \log_a(b) \qquad \text{3.43}$$

$$\ln(x) = \log_{10}(x) \ln(10) \approx 2.3026 \log_{10}(x) \qquad \text{3.44}$$

$$\log_{10}(x) = \ln(x) \log_{10}(e) \approx 0.4343 \ln(x) \qquad \text{3.45}$$

Example 3.4

The surviving fraction, x, of a radioactive isotope is given by $x = e^{-0.005t}$. For what value of t will the surviving percentage be 7%?

Solution

$$x = 0.07 = e^{-0.005t}$$

Taking the natural log of both sides,

$$\ln(0.07) = \ln\left(e^{-0.005t}\right)$$

From 3.36, $\ln e^x = x$. Therefore,

$$-2.66 = -0.005t$$

$$t = 532$$

16. PARTIAL FRACTIONS

The method of *partial fractions* is used to transform a proper polynomial fraction of two polynomials into a sum of simpler expressions, a procedure known as *resolution*.[6,7] The technique can be considered to be the act of "unadding" a sum to obtain all of the addends.

Suppose $H(x)$ is a proper polynomial fraction of the form $P(x)/Q(x)$. The object of the resolution is to determine the partial fractions u_1/v_1, u_2/v_2, etc., such that

$$H(x) = \frac{P(x)}{Q(x)} = \frac{u_1}{v_1} + \frac{u_2}{v_2} + \frac{u_3}{v_3} + \cdots \qquad \textbf{3.46}$$

The form of the denominator polynomial $Q(x)$ will be the main factor in determining the form of the partial fractions. The task of finding the u_i and v_i is simplified by categorizing the possible forms of $Q(x)$.

case 1: $Q(x)$ factors into n different linear terms. That is,

$$Q(x) = (x - a_1)(x - a_2)\cdots(x - a_n) \qquad \textbf{3.47}$$

Then,

$$H(x) = \sum_{i=1}^{n} \frac{A_i}{x - a_i} \qquad \textbf{3.48}$$

case 2: $Q(x)$ factors into n identical linear terms. That is,

$$Q(x) = (x - a)(x - a)\cdots(x - a) \qquad \textbf{3.49}$$

Then,

$$H(x) = \sum_{i=1}^{n} \frac{A_i}{(x - a)^i} \qquad \textbf{3.50}$$

---

[6]To be a *proper polynomial fraction*, the degree of the numerator must be less than the degree of the denominator. If the polynomial fraction is improper, the denominator can be divided into the numerator to obtain whole and fractional polynomials. The method of partial fractions can then be used to reduce the fractional polynomial.

[7]This technique is particularly useful for calculating integrals and inverse Laplace transforms in subsequent chapters.

case 3: $Q(x)$ factors into n different quadratic terms, $x^2 + p_i x + q_i$. Then,

$$H(x) = \sum_{i=1}^{n} \frac{A_i x + B_i}{x^2 + p_i x + q_i} \qquad \textbf{3.51}$$

case 4: $Q(x)$ factors into n identical quadratic terms, $x^2 + px + q$. Then,

$$H(x) = \sum_{i=1}^{n} \frac{A_i x + B_i}{(x^2 + px + q)^i} \qquad \textbf{3.52}$$

Once the general forms of the partial fractions have been determined from inspection, the *method of undetermined coefficients* is used. The partial fractions are all cross-multiplied to obtain $Q(x)$ as the denominator, and the coefficients are found by equating $P(x)$ and the cross-multiplied numerator.

Example 3.5

Resolve $H(x)$ into partial fractions.

$$H(x) = \frac{x^2 + 2x + 3}{x^4 + x^3 + 2x^2}$$

Solution

Here, $Q(x) = x^4 + x^3 + 2x^2$, which factors into $x^2(x^2 + x + 2)$. This is a combination of cases 2 and 3.

$$H(x) = \frac{A_1}{x} + \frac{A_2}{x^2} + \frac{A_3 + A_4 x}{x^2 + x + 2}$$

Cross-multiplying to obtain a common denominator yields

$$\frac{(A_1 + A_4)x^3 + (A_1 + A_2 + A_3)x^2 + (2A_1 + A_2)x + 2A_2}{x^4 + x^3 + 2x^2}$$

Since the original numerator is known, the following simultaneous equations result:

$$A_1 + A_4 = 0$$
$$A_1 + A_2 + A_3 = 1$$
$$2A_1 + A_2 = 2$$
$$2A_2 = 3$$

The solutions are $A_1 = 0.25$, $A_2 = 1.5$, $A_3 = -0.75$, and $A_4 = -0.25$.

$$H(x) = \frac{1}{4x} + \frac{3}{2x^2} - \frac{x + 3}{(4)(x^2 + x + 2)}$$

17. SIMULTANEOUS LINEAR EQUATIONS

A *linear equation* with n variables is a polynomial of degree 1 describing a geometric shape in n-space. A *homogeneous linear equation* is one that has no constant term, and a *nonhomogeneous linear equation* has a constant term.

A solution to a set of simultaneous linear equations represents the intersection point of the geometric shapes in n-space. For example, if the equations are limited to two variables (e.g., $y = 4x - 5$), they describe straight lines. The solution to two simultaneous linear equations in 2-space is the point where the two lines intersect. The set of the two equations is said to be a *consistent system* when there is such an intersection.[8]

Simultaneous equations do not always have unique solutions, and some have none at all. In addition to crossing in 2-space, lines can be parallel or they can be the same line expressed in a different equation format (i.e., dependent equations). In some cases, parallelism and dependency can be determined by inspection. In most cases, however, matrix and other advanced methods must be used to determine whether a solution exists. A set of linear equations with no simultaneous solution is known as an *inconsistent system*.

Several methods exist for solving linear equations simultaneously by hand.[9]

- *graphing*: The equations are plotted and the intersection point is read from the graph. This method is possible only with two-dimensional problems.

- *substitution*: An equation is rearranged so that one variable is expressed as a combination of the other variables. The expression is then substituted into the remaining equations wherever the selected variable appears.

- *reduction*: All terms in the equations are multiplied by constants chosen to eliminate one or more variables when the equations are added or subtracted. The remaining sum can then be solved for the other variables. This method is also known as *eliminating the unknowns*.

- *Cramer's rule*: This method is covered in Chap. 4.

Example 3.6

Solve the following set of linear equations by (a) substitution and (b) reduction.

$$2x + 3y = 12 \qquad [\text{Eq. 1}]$$
$$3x + 4y = 8 \qquad [\text{Eq. 2}]$$

[8]A homogeneous system always has at least one solution: the *trivial solution*, in which all variables have a value of zero.
[9]Other methods exist, but they require a computer.

Solution

(a) From Eq. 1, solve for variable x.

$$x = 6 - 1.5y \qquad [\text{Eq. 3}]$$

Substitute $6 - 1.5y$ into Eq. 2 wherever x appears.

$$(3)(6 - 1.5y) + 4y = 8$$
$$18 - 4.5y + 4y = 8$$
$$y = 20$$

Substitute 20 for y in Eq. 3.

$$x = 6 - (1.5)(20) = -24$$

The solution $(-24, 20)$ should be checked to verify that it satisfies both original equations.

(b) Eliminate variable x by multiplying Eq. 1 by 3 and Eq. 2 by 2.

$$3 \times \text{Eq. 1: } 6x + 9y = 36 \qquad [\text{Eq. 1}']$$
$$2 \times \text{Eq. 2: } 6x + 8y = 16 \qquad [\text{Eq. 2}']$$

Subtract Eq. 2' from Eq. 1'.

$$y = 20 \quad [\text{Eq. 1}' - \text{Eq. 2}']$$

Substitute $y = 20$ into Eq. 1'.

$$6x + (9)(20) = 36$$
$$x = -24$$

The solution $(-24, 20)$ should be checked to verify that it satisfies both original equations.

18. COMPLEX NUMBERS

A *complex number*, **Z**, is a combination of real and imaginary numbers. When expressed as a sum (e.g., $a + bi$), the complex number is said to be in *rectangular* or *trigonometric form*. The complex number can be plotted on the real-imaginary coordinate system known as the *complex plane*, as illustrated in Fig. 3.2.

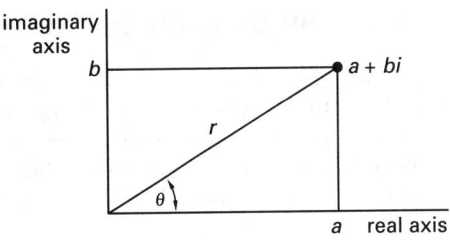

Figure 3.2 *A Complex Number in the Complex Plane*

The complex number $\mathbf{Z} = (a+bi)$ can also be expressed in *exponential form*.[10] The quantity r is known as the *modulus* of $\mathbf{Z}$; θ is the *argument*.

$$a + bi = re^{i\theta} \qquad 3.53$$

$$r = \text{mod}(\mathbf{Z}) = \sqrt{a^2 + b^2} \qquad 3.54$$

$$\theta = \arg(\mathbf{Z}) = \arctan\left(\frac{b}{a}\right) \qquad 3.55$$

Similarly, the *phasor form* (also known as the *polar form*) is

$$\mathbf{Z} = r\angle\theta \qquad 3.56$$

The *rectangular form* can be determined from r and θ.

$$a = r\cos\theta \qquad 3.57$$

$$b = r\sin\theta \qquad 3.58$$

$$\mathbf{Z} = a + bi = r\cos\theta + ir\sin\theta$$
$$= r(\cos\theta + i\sin\theta) \qquad 3.59$$

The *cis form* is a shorthand method of writing a complex number in rectangular (trigonometric) form.

$$a + bi = r(\cos\theta + i\sin\theta) = r\,\text{cis}\,\theta \qquad 3.60$$

Euler's equation (Eq. 3.61) expresses the equality of complex numbers in exponential and trigonometric form.

$$e^{i\theta} = \cos\theta + i\sin\theta \qquad 3.61$$

Related expressions are

$$e^{-i\theta} = \cos\theta - i\sin\theta \qquad 3.62$$

$$\cos\theta = \frac{e^{i\theta} + e^{-i\theta}}{2} \qquad 3.63$$

$$\sin\theta = \frac{e^{i\theta} - e^{-i\theta}}{2i} \qquad 3.64$$

Example 3.7

What is the exponential form of the complex number $\mathbf{Z} = (3 + 4i)$?

Solution

$$r = \sqrt{a^2 + b^2} = \sqrt{3^2 + 4^2} = \sqrt{25} = 5$$

$$\theta = \arctan\left(\frac{b}{a}\right) = \arctan\left(\frac{4}{3}\right) = 0.927 \text{ rad}$$

$$\mathbf{Z} = re^{i\theta} = 5e^{i(0.927)}$$

19. OPERATIONS ON COMPLEX NUMBERS

Most algebraic operations (addition, multiplication, exponentiation, etc.) work with complex numbers, but notable exceptions are the inequality operators. The concept of one complex number being less than or greater than another complex number is meaningless.

[10]The terms *polar form*, *phasor form*, and *exponential form* are all used somewhat interchangeably.

When adding two complex numbers, real parts are added to real parts and imaginary parts are added to imaginary parts.

$$(a_1 + ib_1) + (a_2 + ib_2) = (a_1 + a_2) + i(b_1 + b_2) \qquad 3.65$$

$$(a_1 + ib_1) - (a_2 + ib_2) = (a_1 - a_2) + i(b_1 - b_2) \qquad 3.66$$

Multiplication of two complex numbers in rectangular form is accomplished by the use of the algebraic distributive law, remembering that $i^2 = -1$.

Division of complex numbers in rectangular form requires use of the *complex conjugate*. The complex conjugate of the complex number $(a + bi)$ is $(a - bi)$. By multiplying the numerator and the denominator by the complex conjugate, the denominator will be converted to the real number $a^2 + b^2$. This technique is known as *rationalizing* the denominator and is illustrated in Ex. 3.8(c).

Multiplication and division are often more convenient when the complex numbers are in exponential or phasor forms, as Eqs. 3.67 and 3.68 show.

$$(r_1 e^{i\theta_1})(r_2 e^{i\theta_2}) = r_1 r_2 e^{i(\theta_1 + \theta_2)} \qquad 3.67$$

$$\frac{r_1 e^{i\theta_1}}{r_2 e^{i\theta_2}} = \left(\frac{r_1}{r_2}\right) e^{i(\theta_1 - \theta_2)} \qquad 3.68$$

Taking powers and roots of complex numbers requires *de Moivre's theorem*, Eqs. 3.69 and 3.70.

$$\mathbf{Z}^n = \left(re^{i\theta}\right)^n = r^n e^{in\theta} \qquad 3.69$$

$$\sqrt[n]{\mathbf{Z}} = \left(re^{i\theta}\right)^{\frac{1}{n}} = \sqrt[n]{r}e^{i\left(\frac{\theta + k(360)}{n}\right)}$$
$$[k = 0, 1, 2, \ldots n - 1] \qquad 3.70$$

Example 3.8

Perform the following complex arithmetic.
(a) $(3 + 4i) + (2 + i)$
(b) $(7 + 2i)(5 - 3i)$
(c) $\dfrac{2 + 3i}{4 - 5i}$

Solution

(a)
$$(3 + 4i) + (2 + i) = (3 + 2) + (4 + 1)i$$
$$= 5 + 5i$$

(b)
$$(7 + 2i)(5 - 3i) = (7)(5) - (7)(3i) + (2i)(5)$$
$$- (2i)(3i)$$
$$= 35 - 21i + 10i - 6i^2$$
$$= 35 - 21i + 10i - (6)(-1)$$
$$= 41 - 11i$$

(c) Multiply the numerator and denominator by the complex conjugate of the denominator.

$$\frac{2+3i}{4-5i} = \frac{(2+3i)(4+5i)}{(4-5i)(4+5i)}$$

$$= \frac{-7+22i}{(4)^2+(5)^2}$$

$$= \left(\frac{-7}{41}\right) + i\left(\frac{22}{41}\right)$$

20. LIMITS

A *limit* (*limiting value*) is the value a function approaches when an independent variable approaches a target value. For example, suppose the value of $y = x^2$ is desired as x approaches 5. This could be written as

$$\lim_{x\to5} x^2 = 25 \qquad\qquad 3.71$$

The power of limit theory is wasted on simple calculations such as this but is appreciated when the function is undefined at the target value. The object of limit theory is to determine the limit without having to evaluate the function at the target. The general case of a limit evaluated as x approaches the target value a is written as

$$\lim_{x\to a} f(x) \qquad\qquad 3.72$$

It is not necessary for the actual value $f(a)$ to exist for the limit to be calculated. The function $f(x)$ may be undefined at point a. However, it is necessary that $f(x)$ be defined on both sides of point a for the limit to exist. If $f(x)$ is undefined on one side, or if $f(x)$ is discontinuous at $x = a$ (as in Fig. 3.3(c) and 3.3(d)), the limit does not exist.

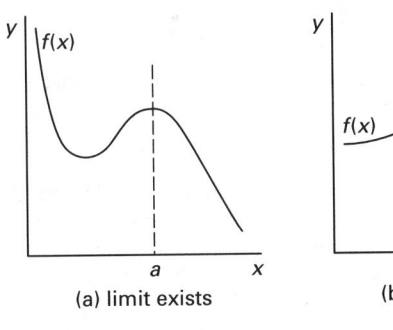

(a) limit exists   (b) limit exists

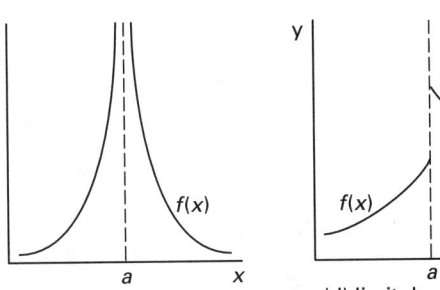

(c) limit does not exist   (d) limit does not exist

Figure 3.3 *Existence of Limits*

The following theorems can be used to simplify expressions when calculating limits.

$$\lim_{x\to a} x = a \qquad\qquad 3.73$$

$$\lim_{x\to a} (mx + b) = ma + b \qquad\qquad 3.74$$

$$\lim_{x\to a} b = b \qquad\qquad 3.75$$

$$\lim_{x\to a} (kF(x)) = k \lim_{x\to a} F(x) \qquad\qquad 3.76$$

$$\lim_{x\to a}\left(F_1(x) \left\{\begin{matrix} + \\ - \\ \times \\ \div \end{matrix}\right\} F_2(x)\right)$$

$$= \lim_{x\to a}(F_1(x)) \left\{\begin{matrix} + \\ - \\ \times \\ \div \end{matrix}\right\} \lim_{x\to a}(F_2(x)) \qquad 3.77$$

The following identities can be used to simplify limits of trigonometric expressions.

$$\lim_{x\to0} \sin x = 0 \qquad\qquad 3.78$$

$$\lim_{x\to0}\left(\frac{\sin x}{x}\right) = 1 \qquad\qquad 3.79$$

$$\lim_{x\to0} \cos x = 1 \qquad\qquad 3.80$$

The following standard methods (tricks) can be used to determine limits.

- If the limit is taken to infinity, all terms can be divided by the largest power of x in the expression. This will leave at least one constant. Any quantity divided by a power of x vanishes as x approaches infinity.

- If the expression is a quotient of two expressions, any common factors should be eliminated from the numerator and denominator.

- *L'Hôpital's rule*, Eq. 3.81, should be used when the numerator and denominator of the expression both approach zero or both approach infinity.[11] $P^k(x)$ and $Q^k(x)$ are the kth derivatives of the functions $P(x)$ and $Q(x)$, respectively.[12] (L'Hôpital's rule can be applied repeatedly as required.)

$$\lim_{x\to a}\left(\frac{P(x)}{Q(x)}\right) = \lim_{x\to a}\left(\frac{P^k(x)}{Q^k(x)}\right) \qquad 3.81$$

---

[11] L'Hôpital's rule should not be used when only the denominator approaches zero. In that case, the limit approaches infinity regardless of the numerator.

[12] The calculation of derivatives is covered in Chap. 8.

Example 3.9

Evaluate the following limits.

(a) $\lim\limits_{x \to 3} \left(\dfrac{x^3 - 27}{x^2 - 9} \right)$

(b) $\lim\limits_{x \to \infty} \left(\dfrac{3x - 2}{4x + 3} \right)$

(c) $\lim\limits_{x \to 2} \left(\dfrac{x^2 + x - 6}{x^2 - 3x + 2} \right)$

Solution

(a) Factor the numerator and denominator. (L'Hôpital's rule can also be used.)

$$\lim_{x \to 3} \left(\frac{x^3 - 27}{x^2 - 9} \right) = \lim_{x \to 3} \left(\frac{(x-3)(x^2 + 3x + 9)}{(x-3)(x+3)} \right)$$

$$= \lim_{x \to 3} \left(\frac{x^2 + 3x + 9}{x + 3} \right) = \frac{(3)^2 + (3)(3) + 9}{3 + 3}$$

$$= \frac{9}{2}$$

(b) Divide through by the largest power of x. (L'Hôpital's rule can also be used.)

$$\lim_{x \to \infty} \left(\frac{3x - 2}{4x + 3} \right) = \lim_{x \to \infty} \left(\frac{3 - \dfrac{2}{x}}{4 + \dfrac{3}{x}} \right)$$

$$= \frac{3 - \dfrac{2}{\infty}}{4 + \dfrac{3}{\infty}}$$

$$= \frac{3 - 0}{4 + 0} = \frac{3}{4}$$

(c) Use L'Hôpital's rule. (Factoring can also be used.) Take the first derivative of the numerator and denominator.

$$\lim_{x \to 2} \left(\frac{x^2 + x - 6}{x^2 - 3x + 2} \right) = \lim_{x \to 2} \left(\frac{2x + 1}{2x - 3} \right)$$

$$= \frac{(2)(2) + 1}{(2)(2) - 3}$$

$$= \frac{5}{1} = 5$$

21. SEQUENCES AND PROGRESSIONS

A *sequence* is an ordered *progression* of numbers, a_i, such as $1, 4, 9, 16, 25, \ldots$ The *terms* in a sequence can be all positive, negative, or of alternating signs. a_n is known as the *general term* of the sequence.

$$\{A\} = \{a_1, a_2, a_3, \ldots a_n\} \qquad 3.82$$

A sequence is said to *diverge* (i.e., be *divergent*) if the terms approach infinity or if the terms fail to approach any finite value, and is said to *converge* (i.e.,

be *convergent*) if the terms approach any finite value (including zero). That is, the sequence converges if the limit defined by Eq. 3.83 exists.

$$\lim_{n \to \infty} a_n \begin{cases} \text{converges if } L \text{ is finite} \\ \text{diverges if } L \text{ is infinite} \\ \text{or does not exist} \end{cases} \qquad 3.83$$

The main task associated with a sequence is determining the next (or the general) term. If several terms of a sequence are known, the next (unknown) term must usually be found by intuitively determining the pattern of the sequence. In some cases, though, the method of *Rth-order differences* can be used to determine the next term. This method consists of subtracting each term from the following term to obtain a set of differences. If the differences are not all equal, the next order of differences can be calculated.

Example 3.10

What is the general term of the sequence A?

$$\{A\} = \left\{ 3, \frac{9}{2}, \frac{27}{6}, \frac{81}{24}, \ldots \right\}$$

Solution

The solution is purely intuitive. The numerator is recognized as a power series based on the number 3. The denominator is recognized as the factorial sequence. The general term is

$$a_n = \frac{3^n}{n!}$$

Example 3.11

Find the sixth term in the sequence $\{7, 16, 29, 46, 67, a_6\}$.

Solution

The sixth term is not intuitively obvious, so the method of *Rth-order differences* is tried. The pattern is not obvious from the first order differences, but the second order differences are all 4.

$$\delta_5 - 21 = 4$$

$$\delta_5 = 25$$

$$a_6 - 67 = \delta_5 = 25$$

$$a_6 = 92$$

Example 3.12

Does the sequence with general term e^n/n converge or diverge?

Solution

See if the limit exists.

$$\lim_{n\to\infty}\left(\frac{e^n}{n}\right)=\frac{\infty}{\infty}$$

Since ∞/∞ is inconclusive, apply L'Hôpital's rule. Take the derivative of both the numerator and the denominator with respect to n.

$$\lim_{n\to\infty}\left(\frac{e^n}{1}\right)=\frac{\infty}{1}$$
$$=\infty$$

The sequence diverges.

22. STANDARD SEQUENCES

There are four standard sequences: the geometric, arithmetic, harmonic, and p sequences.

- *geometric sequence*: The geometric sequence converges for $-1<r\le 1$ and diverges otherwise. a is known as the *first term*; r is known as the *common ratio*.

$$a_n=ar^{n-1}\qquad\left[\begin{array}{l}a\text{ is a constant}\\n=1,2,3,\dots,\infty\end{array}\right]\qquad 3.84$$

  example: $\{1,2,4,8,16,32\}\quad(a=1,r=2)$

- *arithmetic sequence*: The arithmetic sequence always diverges.

$$a_n=a+(n-1)d\qquad\left[\begin{array}{l}a\text{ and }d\text{ are constants}\\n=1,2,3,\dots,\infty\end{array}\right]\quad 3.85$$

  example: $\{2,7,12,17,22,27\}\quad(a=2,d=5)$

- *harmonic sequence*: The harmonic sequence always converges.

$$a_n=\frac{1}{n}\qquad[n=1,2,3,\dots,\infty]\qquad 3.86$$

  example: $\{1,1/2,1/3,1/4,1/5,1/6\}$

- *p-sequence*: The p sequence converges if $p\ge 0$ and diverges if $p<0$. (Notice that this is different than the p series.)

$$a_n=\frac{1}{n^p}\qquad[n=1,2,3,\dots,\infty]\qquad 3.87$$

  example: $\{1,1/4,1/9,1/16,1/25,1/36\}\qquad[p=2]$

23. SERIES

A *series* is the sum of terms in a sequence. There are two types of series. A *finite series* has a finite number of terms, and an *infinite series* has an infinite number of terms.[13] The main tasks associated with series are determining the sum of the terms and whether the series converges. A series is said to *converge* (be *convergent*) if the sum, S_n, of its term exists.[14] A finite series is always convergent.

The performance of a series based on standard sequences (defined in Sec. 21) is well known.

- *geometric series*:

$$S_n=\sum_{i=1}^{n}ar^{i-1}=\frac{a(1-r^n)}{1-r}\qquad\text{[finite]}\qquad 3.88$$

$$S_n=\sum_{i=1}^{\infty}ar^{i-1}=\frac{a}{1-r}\qquad\left[\begin{array}{c}\text{infinite}\\-1<r<1\end{array}\right]\;3.89$$

- *arithmetic series*: The infinite series diverges unless $a=d=0$.

$$S_n=\sum_{i=1}^{n}(a+(i-1)d)=\frac{n(2a+(n-1)d)}{2}\\ \text{[finite]}\qquad 3.90$$

- *harmonic series*: The infinite series diverges.

- *p-series*: The infinite series diverges if $p\le 1$. The infinite series converges if $p>1$. (Notice that this is different than the p sequence.)

24. TESTS FOR SERIES CONVERGENCE

It is obvious that all *finite series* (i.e., series having a finite number of terms) converge. That is, the sum, S_n, defined by Eq. 3.91 exists.

$$S_n=\sum_{i=1}^{n}a_i\qquad 3.91$$

Convergence of an infinite series can be determined by taking the limit of the sum. If the limit exists, the series converges; otherwise, it diverges.

$$\lim_{n\to\infty}S_n=\lim_{n\to\infty}\sum_{i=1}^{n}a_i\qquad 3.92$$

In most cases, the expression for the general term a_n will be known, but there will be no simple expression for the sum S_n. Therefore, Eq. 3.92 cannot be used to

---

[13]The term *infinite series* does not imply the sum is infinite.
[14]This is different from the definition of convergence for a sequence where only the last term was evaluated.

determine convergence. It is helpful, but not conclusive, to look at the limit of the general term. If L, as defined in Eq. 3.93, is nonzero, the series diverges. If L equals zero, the series may either converge or diverge. Additional testing is needed in that case.

$$\lim_{n\to\infty} a_n \begin{cases} =0 & \text{inconclusive} \\ \neq 0 & \text{diverges} \end{cases} \qquad 3.93$$

Two tests can be used independently or after Eq. 3.93 has proven inconclusive: the ratio and comparison tests. The *ratio test* calculates the limit of the ratio of two consecutive terms.

$$\lim_{n\to\infty} \frac{a_{n+1}}{a_n} \begin{cases} <1 & \text{converges} \\ =1 & \text{inconclusive} \\ >1 & \text{diverges} \end{cases} \qquad 3.94$$

The *comparison test* is an indirect method of determining convergence of an unknown series. It compares a standard series (geometric and p series are commonly used) against the unknown series. If all terms in a positive standard series are smaller than the terms in the unknown series and the standard series diverges, the unknown series must also diverge. Similarly, if all terms in the standard series are larger than the terms in the unknown series and the standard series converges, then the unknown series also converges.

In mathematical terms, if A and B are both series of positive terms such that $a_n < b_n$ for all values of n, then (1) B diverges if A diverges, and (2) A converges if B converges.

Example 3.13

Does the infinite series A converge or diverge?

$$A = 3 + \frac{9}{2} + \frac{27}{6} + \frac{81}{24} + \cdots$$

Solution

The general term was found in Ex. 3.9 to be

$$a_n = \frac{3^n}{n!}$$

Since limits of factorials are not easily determined, use the ratio test.

$$\lim_{n\to\infty} \left(\frac{a_{n+1}}{a_n}\right) = \lim_{n\to\infty} \left(\frac{\frac{3^{n+1}}{(n+1)!}}{\frac{3^n}{n!}}\right) = \lim_{n\to\infty} \left(\frac{3}{n+1}\right)$$
$$= \frac{3}{\infty} = 0$$

Since the limit is less than 1, the infinite series converges.

Example 3.14

Does the infinite series A converge or diverge?

$$A = 2 + \frac{3}{4} + \frac{4}{9} + \frac{5}{16} + \cdots$$

Solution

By observation, the general term is

$$a_n = \frac{1+n}{n^2}$$

The general term can be expanded by partial fractions to

$$a_n = \frac{1}{n} + \frac{1}{n^2}$$

However, $1/n$ is the harmonic series. Since the harmonic series is divergent and this series is larger than the harmonic series (by the term $1/n^2$), this series also diverges.

25. SERIES OF ALTERNATING SIGN[15]

Some series contain both positive and negative terms. The ratio and comparison tests can both be used to determine if a series with alternating signs converges. If a series containing all positive terms converges, then the same series with some negative terms also converges. Therefore, the all-positive series should be tested for convergence. If the all-positive series converges, the original series is said to be *absolutely convergent*. (If the all-positive series diverges and the original series converges, the original series is said to be *conditionally convergent*.)

Alternatively, the ratio test can be used with the absolute value of the ratio. The same criteria apply.

$$\lim_{n\to\infty} \left|\frac{a_{n+1}}{a_n}\right| \begin{cases} <1 & \text{converges} \\ =1 & \text{inconclusive} \\ >1 & \text{diverges} \end{cases} \qquad 3.95$$

PRACTICE PROBLEMS

1. Calculate the following sum.

$$\sum_{j=1}^{5} [(j+1)^2 - 1]$$

2. If every 0.1 sec a quantity increases by 0.1% of its current value, calculate the doubling time.

---

[15]This terminology is commonly used even though it is not necessary that the signs strictly alternate.

4 Linear Algebra

1. Matrices 4-1
2. Special Types of Matrices 4-1
3. Row Equivalent Matrices 4-2
4. Minors and Cofactors 4-2
5. Determinants 4-3
6. Matrix Algebra 4-3
7. Matrix Addition and Subtraction 4-4
8. Matrix Multiplication 4-4
9. Transpose 4-4
10. Singularity and Rank 4-5
11. Classical Adjoint 4-5
12. Inverse 4-5
13. Writing Simultaneous Linear Equations in
    Matrix Form 4-6
14. Solving Simultaneous Linear Equations . . . 4-6
15. Eigenvalues and Eigenvectors 4-7
    Practice Problems 4-8

1. MATRICES

A *matrix* is an ordered set of *entries* (*elements*) arranged rectangularly and set off by brackets.[1] The entries can be variables or numbers. A matrix by itself has no particular value—it is merely a convenient method of representing a set of numbers.

The size of a matrix is given by the number of rows and columns, and the nomenclature $m \times n$ is used for a matrix with m rows and n columns. For a *square matrix*, the number of rows and columns will be the same, a quantity known as the *order* of the matrix.

Bold uppercase letters are used to represent matrices, while lowercase letters represent the entries. For example, a_{23} would be the entry in the second row and third column of matrix $\mathbf{A}$.

$$\mathbf{A} = \begin{bmatrix} a_{11} & a_{12} & a_{13} \\ a_{21} & a_{22} & a_{23} \\ a_{31} & a_{32} & a_{33} \end{bmatrix}$$

A *submatrix* is the matrix that remains when selected rows or columns are removed from the original matrix.[2] For example, for matrix $\mathbf{A}$, the submatrix remaining after the second row and second column have been removed is

$$\begin{bmatrix} a_{11} & a_{13} \\ a_{31} & a_{33} \end{bmatrix}$$

An *augmented matrix* results when the original matrix is extended by repeating one or more of its rows or columns or by adding rows and columns from another matrix. For example, for the matrix $\mathbf{A}$, the augmented matrix created by repeating the first and second columns is

$$\begin{bmatrix} a_{11} & a_{12} & a_{13} & \vdots & a_{11} & a_{12} \\ a_{21} & a_{22} & a_{23} & \vdots & a_{21} & a_{22} \\ a_{31} & a_{32} & a_{33} & \vdots & a_{31} & a_{32} \end{bmatrix}$$

2. SPECIAL TYPES OF MATRICES

Certain types of matrices are given special designations.

- *cofactor matrix*: the matrix that is formed when every entry is replaced by the cofactor (Sec. 4) of that entry

- *column matrix*: a matrix with only one column

- *complex matrix*: a matrix with complex number entries

- *diagonal matrix*: a square matrix with all zero entries except for the a_{ij} for which $i = j$

- *echelon matrix*: a matrix in which the number of zeros preceding the first nonzero entry of a row increases row by row until only zero rows remain. A *row-reduced echelon matrix* is an echelon matrix in which the first nonzero entry in each row is a 1 and all other entries in the columns are zero.

- *identity matrix*: a diagonal (square) matrix with all nonzero entries equal to 1, usually designated as $\mathbf{I}$, having the property that $\mathbf{AI} = \mathbf{IA} = \mathbf{A}$

- *null matrix*: the same as a zero matrix

- *row matrix*: a matrix with only one row

- *scalar matrix*:[3] a diagonal (square) matrix with all diagonal entries equal to some scalar k

- *singular matrix*: a matrix whose determinant is zero (See Sec. 10.)

---

[1]The term *array* is synonymous with *matrix*, although the former is more likely to be used in computer applications.
[2]By definition, a matrix is a submatrix of itself.

[3]Although the term *complex matrix* means a matrix with complex entries, the term *scalar matrix* means more than a matrix with scalar entries.

- *skew symmetric matrix*: a square matrix whose transpose (Sec. 9) is equal to the negative of itself (i.e., $\mathbf{A} = -\mathbf{A}^t$)

- *square matrix*: a matrix with the same number of rows and columns (i.e., $m = n$)

- *symmetric(al) matrix*: a square matrix whose transpose is equal to itself (i.e., $\mathbf{A}^t = \mathbf{A}$), which occurs only when $a_{ij} = a_{ji}$

- *triangular matrix*: a square matrix with zeros in all positions above or below the diagonal

- *unit matrix*: the same as the identity matrix

- *zero matrix*: a matrix with all zero entries

$$\begin{bmatrix} 9 & 0 & 0 & 0 \\ 0 & -6 & 0 & 0 \\ 0 & 0 & 1 & 0 \\ 0 & 0 & 0 & 5 \end{bmatrix} \quad \begin{bmatrix} 2 & 18 & 2 & 18 \\ 0 & 0 & 1 & 9 \\ 0 & 0 & 0 & 9 \\ 0 & 0 & 0 & 0 \end{bmatrix} \quad \begin{bmatrix} 1 & 9 & 0 & 0 \\ 0 & 0 & 1 & 0 \\ 0 & 0 & 0 & 1 \\ 0 & 0 & 0 & 0 \end{bmatrix}$$

(a) diagonal   (b) echelon   (c) row-reduced echelon

$$\begin{bmatrix} 1 & 0 & 0 & 0 \\ 0 & 1 & 0 & 0 \\ 0 & 0 & 1 & 0 \\ 0 & 0 & 0 & 1 \end{bmatrix} \quad \begin{bmatrix} 2 & 0 & 0 & 0 \\ 7 & 6 & 0 & 0 \\ 9 & 1 & 1 & 0 \\ 8 & 0 & 4 & 5 \end{bmatrix} \quad \begin{bmatrix} 3 & 0 & 0 & 0 \\ 0 & 3 & 0 & 0 \\ 0 & 0 & 3 & 0 \\ 0 & 0 & 0 & 3 \end{bmatrix}$$

(d) identity   (e) triangular   (f) scalar

Figure 4.1 *Examples of Special Matrices*

3. ROW EQUIVALENT MATRICES

A matrix $\mathbf{B}$ is said to be *row equivalent* to a matrix $\mathbf{A}$ if it is obtained by a finite sequence of *elementary row operations* on $\mathbf{A}$:

- interchanging the ith and jth rows

- multiplying the ith row by a nonzero scalar

- replacing the ith row by the sum of the original ith row and k times the jth row

However, two matrices that are row equivalent as defined do not have the same determinants. (See Sec. 5.)

Gauss-Jordan elimination is the process of using these elementary row operations to row-reduce a matrix to echelon or row-reduced echelon forms, as illustrated in Ex. 4.8. When a matrix has been converted to a row-reduced echelon matrix, it is said to be in *row canonical form*. Thus, the terms *row-reduced echelon form* and *row canonical form* are synonymous.

4. MINORS AND COFACTORS

Minors and cofactors are determinants of submatrices associated with particular entries in the original square matrix. The *minor* of entry a_{ij} is the determinant of a submatrix resulting from the elimination of the single row i and the single column j. For example, the minor corresponding to entry a_{12} in a 3×3 matrix $\mathbf{A}$ is the determinant of the matrix created by eliminating row 1 and column 2.

$$\text{minor of } a_{12} = \begin{vmatrix} a_{21} & a_{23} \\ a_{31} & a_{33} \end{vmatrix} \qquad 4.1$$

The *cofactor* of entry a_{ij} is the minor of a_{ij} multiplied by either $+1$ or -1, depending on the position of the entry. (That is, the cofactor either exactly equals the minor or it differs only in sign.) The sign is determined according to the following positional matrix (for 3×3 matrices).[4]

$$\begin{bmatrix} +1 & -1 & +1 & \cdots \\ -1 & +1 & -1 & \cdots \\ +1 & -1 & +1 & \cdots \\ \vdots & \vdots & \vdots & \end{bmatrix}$$

For example, the cofactor of entry a_{12} in matrix $\mathbf{A}$ (described in Sec. 4) is

$$\text{cofactor of } a_{12} = -\begin{vmatrix} a_{21} & a_{23} \\ a_{31} & a_{33} \end{vmatrix} \qquad 4.2$$

Example 4.1

What is the cofactor corresponding to the -3 entry in the following matrix?

$$\mathbf{A} = \begin{bmatrix} 2 & 9 & 1 \\ -3 & 4 & 0 \\ 7 & 5 & 9 \end{bmatrix}$$

Solution

The minor's submatrix is created by eliminating the row and column of the -3 entry.

$$\mathbf{M} = \begin{bmatrix} 9 & 1 \\ 5 & 9 \end{bmatrix}$$

The minor is the determinant of $\mathbf{M}$.

$$|\mathbf{M}| = (9)(9) - (5)(1) = 76$$

The sign corresponding to the -3 position is negative. Therefore, the cofactor is -76.

---

[4]The sign of the cofactor a_{ij} is positive if $(i + j)$ is even and is negative if $(i + j)$ is odd.

5. DETERMINANTS

A *determinant* is a scalar calculated from a square matrix. The determinant of matrix $\mathbf{A}$ can be represented as $D\{\mathbf{A}\}$, Det $(\mathbf{A})$, $\Delta\mathbf{A}$, or $|\mathbf{A}|$.[5] The following rules can be used to simplify the calculation of determinants.

- If $\mathbf{A}$ has a row or column of zeros, the determinant is zero.

- If $\mathbf{A}$ has two identical rows or columns, the determinant is zero.

- If $\mathbf{B}$ is obtained from $\mathbf{A}$ by adding a multiple of a row (column) to another row (column) in $\mathbf{A}$, then $|\mathbf{B}| = |\mathbf{A}|$.

- If $\mathbf{A}$ is triangular, the determinant is equal to the product of the diagonal entries.

- If $\mathbf{B}$ is obtained from $\mathbf{A}$ by multiplying one row or column in $\mathbf{A}$ by a scalar k, then $|\mathbf{B}| = k|\mathbf{A}|$.

- If $\mathbf{B}$ is obtained from the $n \times n$ matrix $\mathbf{A}$ by multiplying by the scalar matrix k, then $|\mathbf{k} \times \mathbf{A}| = k^n|\mathbf{A}|$.

- If $\mathbf{B}$ is obtained from $\mathbf{A}$ by switching two rows or columns in $\mathbf{A}$, then $|\mathbf{B}| = -|\mathbf{A}|$.

Calculation of determinants is laborious for all but the smallest or simplest of matrices. For a 2×2 matrix, the formula used to calculate the determinant is easy to remember.

$$\mathbf{A} = \begin{bmatrix} a & b \\ c & d \end{bmatrix}$$

$$|\mathbf{A}| = \begin{vmatrix} a & b \\ c & d \end{vmatrix} = ad - bc \qquad 4.3$$

Two methods are commonly used for calculating the determinant of 3×3 matrices by hand. The first uses an augmented matrix constructed from the original matrix and the first two columns (as presented in Sec. 1).[6] The determinant is calculated as the sum of the products in the left-to-right downward diagonals less the sum of the products in the left-to-right upward diagonals.

$$\mathbf{A} = \begin{bmatrix} a & b & c \\ d & e & f \\ g & h & i \end{bmatrix}$$

augmented $\mathbf{A} = \begin{bmatrix} a & b & c & a & b \\ d & e & f & d & e \\ g & h & i & g & h \end{bmatrix}$ $\qquad 4.4$

$$|\mathbf{A}| = aei + bfg + cdh - gec - hfa - idb \qquad 4.5$$

---

[5]The vertical bars should not be confused with the square brackets used to set off a matrix, nor with absolute value.
[6]It is not actually necessary to construct the augmented matrix, but doing so helps avoid errors.

The second method of calculating the determinant is somewhat slower than the first for a 3×3 matrix but illustrates the method that must be used to calculate determinants of 4×4 and larger matrices. This method is known as *expansion by cofactors*. One row (column) is selected as the base row (column). The selection is arbitrary, but the number of calculations required to obtain the determinant can be minimized by choosing the row (column) with the most zeros. The determinant is equal to the sum of the products of the entries in the base row (column) and their corresponding cofactors.

$$\mathbf{A} = \begin{bmatrix} a & b & c \\ d & e & f \\ g & h & i \end{bmatrix}$$

$$|\mathbf{A}| = a\begin{vmatrix} e & f \\ h & i \end{vmatrix} - d\begin{vmatrix} b & c \\ h & i \end{vmatrix} + g\begin{vmatrix} b & c \\ e & f \end{vmatrix} \qquad 4.6$$

Example 4.2

Calculate the determinant of matrix $\mathbf{A}$ (a) by cofactor expansion, and (b) by the augmented matrix method.

$$\mathbf{A} = \begin{bmatrix} 2 & 3 & -4 \\ 3 & -1 & -2 \\ 4 & -7 & -6 \end{bmatrix}$$

Solution

(a) Since there are no zero entries, it does not matter which row or column is chosen as the base. Choose the first column as the base.

$$|\mathbf{A}| = 2\begin{vmatrix} -1 & -2 \\ -7 & -6 \end{vmatrix} - 3\begin{vmatrix} 3 & -4 \\ -7 & -6 \end{vmatrix} + 4\begin{vmatrix} 3 & -4 \\ -1 & -2 \end{vmatrix}$$

$$= (2)(6 - 14) - (3)(-18 - 28) + (4)(-6 - 4)$$

$$= 82$$

$$|\mathbf{A}| = 72 - (-10) = 82$$

6. MATRIX ALGEBRA[7]

Matrix algebra differs somewhat from standard algebra.

- *equality*: Two matrices, $\mathbf{A}$ and $\mathbf{B}$, are equal only if they have the same numbers of rows and columns *and* if all corresponding entries are equal.

- *inequality*: The $>$ and $<$ operators are not used in matrix algebra.

---

[7]Since matrices are used to simplify the presentation and solution of sets of linear equations, matrix algebra is also known as *linear algebra*.

- *commutative law of addition*:

$$\mathbf{A} + \mathbf{B} = \mathbf{B} + \mathbf{A} \qquad 4.7$$

- *associative law of addition*:

$$\mathbf{A} + (\mathbf{B} + \mathbf{C}) = (\mathbf{A} + \mathbf{B}) + \mathbf{C} \qquad 4.8$$

- *associative law of multiplication*:

$$(\mathbf{AB})\mathbf{C} = \mathbf{A}(\mathbf{BC}) \qquad 4.9$$

- *left distributive law*:

$$\mathbf{A}(\mathbf{B} + \mathbf{C}) = \mathbf{AB} + \mathbf{AC} \qquad 4.10$$

- *right distributive law*:

$$(\mathbf{B} + \mathbf{C})\mathbf{A} = \mathbf{BA} + \mathbf{CA} \qquad 4.11$$

- *scalar multiplication*:

$$k(\mathbf{AB}) = (k\mathbf{A})\mathbf{B} = \mathbf{A}(k\mathbf{B}) \qquad 4.12$$

It is important to recognize that, except for trivial and special cases, matrix multiplication is not commutative. That is,

$$\mathbf{AB} \neq \mathbf{BA}$$

7. MATRIX ADDITION AND SUBTRACTION

Addition and subtraction of two matrices is possible only if both matrices have the same numbers of rows and columns (i.e., order). They are accomplished by adding or subtracting the corresponding entries of the two matrices.

8. MATRIX MULTIPLICATION

A matrix can be multiplied by a scalar, an operation known as *scalar multiplication*, in which case all entries of the matrix are multiplied by that scalar. For example, for the 2×2 matrix $\mathbf{A}$,

$$k\mathbf{A} = \begin{bmatrix} ka_{11} & ka_{12} \\ ka_{21} & ka_{22} \end{bmatrix}$$

A matrix can be multiplied by another matrix, but only if the left-hand matrix has the same number of columns as the right-hand matrix has rows. *Matrix multiplication* occurs by multiplying the elements in each left-hand matrix row by the entries in each right-hand matrix column, adding the products, and placing the sum at the intersection point of the participating row and column.

Matrix division can only be accomplished by multiplying by the inverse of the denominator matrix. There is no specific division operation in matrix algebra.

Example 4.3

Determine the product matrix $\mathbf{C}$.

$$\mathbf{C} = \begin{bmatrix} 1 & 4 & 3 \\ 5 & 2 & 6 \end{bmatrix} \begin{bmatrix} 7 & 12 \\ 11 & 8 \\ 9 & 10 \end{bmatrix}$$

Solution

The left-hand matrix has three columns, and the right-hand matrix has three rows. Therefore, the two matrices can be multiplied.

The first row of the left-hand matrix and the first column of the right-hand matrix are worked with first. The corresponding entries are multiplied and the products are summed.

$$c_{11} = (1)(7) + (4)(11) + (3)(9) = 78$$

The intersection of the top row and left column is the entry in the upper left-hand corner of the matrix $\mathbf{C}$.

The remaining entries are calculated similarly.

$$c_{12} = (1)(12) + (4)(8) + (3)(10) = 74$$
$$c_{21} = (5)(7) + (2)(11) + (6)(9) = 111$$
$$c_{22} = (5)(12) + (2)(8) + (6)(10) = 136$$

The product matrix is

$$\mathbf{C} = \begin{bmatrix} 78 & 74 \\ 111 & 136 \end{bmatrix}$$

9. TRANSPOSE

The *transpose*, $\mathbf{A}^t$, of an $m \times n$ matrix $\mathbf{A}$ is an $n \times m$ matrix constructed by taking the ith row and making it the ith column. The diagonal is unchanged. For example,

$$\mathbf{A} = \begin{bmatrix} 1 & 6 & 9 \\ 2 & 3 & 4 \\ 7 & 1 & 5 \end{bmatrix}$$

$$\mathbf{A}^t = \begin{bmatrix} 1 & 2 & 7 \\ 6 & 3 & 1 \\ 9 & 4 & 5 \end{bmatrix}$$

Transpose operations have the following characteristics.

$$(\mathbf{A}^t)^t = \mathbf{A} \qquad 4.13$$
$$(k\mathbf{A})^t = k(\mathbf{A}^t) \qquad 4.14$$
$$\mathbf{I}^t = \mathbf{I} \qquad 4.15$$
$$(\mathbf{A} \times \mathbf{B})^t = \mathbf{B}^t \times \mathbf{A}^t \qquad 4.16$$
$$(\mathbf{A} + \mathbf{B})^t = \mathbf{A}^t + \mathbf{B}^t \qquad 4.17$$
$$|\mathbf{A}^t| = |\mathbf{A}| \qquad 4.18$$

10. SINGULARITY AND RANK

A *singular matrix* is one whose determinant is zero. Similarly, a *nonsingular* matrix is one whose determinant is nonzero.

The *rank* of a matrix is the maximum number of linearly independent row or column vectors.[8] A matrix has rank r if it has at least one nonsingular square submatrix of order r but has no nonsingular square submatrix of order more than r. While the submatrix must be square (in order to calculate the determinant), the original matrix need not be.

The rank of an $m \times n$ matrix will be, at most, the smaller of m and n. The rank of a null matrix is zero. The ranks of a matrix and its transpose are the same. If a matrix is in echelon form, the rank will be equal to the number of rows containing at least one nonzero entry. For a 3×3 matrix, the rank can either be 3 (if it is nonsingular), 2 (if any one of its 2×2 submatrices is nonsingular), 1 (if it and all 2×2 submatrices are singular), or 0 (if it is null).

The determination of rank is laborious if done by hand. Either the matrix is reduced to echelon form by using elementary row operations, or exhaustive enumeration is used to create the submatrices and many determinants are calculated. If a matrix has more rows than columns and row-reduction is used, the work required to put the matrix in echelon form can be reduced by working with the transpose of the original matrix.

Example 4.4

What is the rank of matrix $\mathbf{A}$?

$$\mathbf{A} = \begin{bmatrix} 1 & -2 & -1 \\ -3 & 3 & 0 \\ 2 & 2 & 4 \end{bmatrix}$$

Solution

Matrix $\mathbf{A}$ is singular because $|\mathbf{A}| = 0$. However, there is at least one 2×2 nonsingular submatrix:

$$\begin{vmatrix} 1 & -2 \\ -3 & 3 \end{vmatrix} = (1)(3) - (-3)(-2) = -3$$

Therefore, the rank is 2.

Example 4.5

Determine the rank of matrix $\mathbf{A}$ by reducing it to echelon form.

$$\mathbf{A} = \begin{bmatrix} 7 & 4 & 9 & 1 \\ 0 & 2 & -5 & 3 \\ 0 & 4 & -10 & 6 \end{bmatrix}$$

[8]The *row rank* and *column rank* are the same.

Solution

By inspection, the matrix can be row-reduced by subtracting two times the second row from the third row. The matrix cannot be further reduced. Since there are two nonzero rows, the rank is 2.

$$\begin{bmatrix} 7 & 4 & 9 & 1 \\ 0 & 2 & -5 & 3 \\ 0 & 0 & 0 & 0 \end{bmatrix}$$

11. CLASSICAL ADJOINT

The *classical adjoint* is the transpose of the cofactor matrix. The resulting matrix can be designated as $\mathbf{A}_{adj}$, $\text{adj}\{\mathbf{A}\}$ or $\mathbf{A}^{adj}$.

Example 4.6

What is the classical adjoint of matrix $\mathbf{A}$?

$$\mathbf{A} = \begin{bmatrix} 2 & 3 & -4 \\ 0 & -4 & 2 \\ 1 & -1 & 5 \end{bmatrix}$$

Solution

The matrix of cofactors is

$$\begin{bmatrix} -18 & 2 & 4 \\ -11 & 14 & 5 \\ -10 & -4 & -8 \end{bmatrix}$$

The transpose of the matrix of cofactors is

$$\mathbf{A}_{adj} = \begin{bmatrix} -18 & -11 & -10 \\ 2 & 14 & -4 \\ 4 & 5 & -8 \end{bmatrix}$$

12. INVERSE

The product of a matrix $\mathbf{A}$ and its inverse, $\mathbf{A}^{-1}$, is the identity matrix, $\mathbf{I}$. Only square matrices have inverses, but not all square matrices are invertible. A matrix has an inverse if and only if it is nonsingular (i.e., its determinant is nonzero).

$$\mathbf{A} \times \mathbf{A}^{-1} = \mathbf{A}^{-1} \times \mathbf{A} = \mathbf{I} \qquad 4.19$$

$$(\mathbf{A} \times \mathbf{B})^{-1} = \mathbf{B}^{-1} \times \mathbf{A}^{-1} \qquad 4.20$$

The inverse of a 2×2 matrix is easily determined by formula.

$$\mathbf{A} = \begin{bmatrix} a & b \\ c & d \end{bmatrix}$$

$$\mathbf{A}^{-1} = \frac{\begin{bmatrix} d & -b \\ -c & a \end{bmatrix}}{|\mathbf{A}|} \qquad 4.21$$

For a 3×3 or larger matrix, the inverse is determined by dividing every entry in the classical adjoint by the determinant of the original matrix.

$$\mathbf{A}^{-1} = \frac{\mathbf{A}_{adj}}{|\mathbf{A}|} \qquad 4.22$$

Example 4.7

What is the inverse of matrix $\mathbf{A}$?

$$\mathbf{A} = \begin{bmatrix} 4 & 5 \\ 2 & 3 \end{bmatrix}$$

Solution

The determinant is calculated as

$$|\mathbf{A}| = (4)(3) - (2)(5) = 2$$

Using Eq. 4.22, the inverse is

$$\mathbf{A}^{-1} = \frac{\begin{bmatrix} 3 & -5 \\ -2 & 4 \end{bmatrix}}{2} = \begin{bmatrix} \frac{3}{2} & -\frac{5}{2} \\ -1 & 2 \end{bmatrix}$$

Check.

$$\mathbf{A} \times \mathbf{A}^{-1} = \begin{bmatrix} 4 & 5 \\ 2 & 3 \end{bmatrix} \begin{bmatrix} \frac{3}{2} & -\frac{5}{2} \\ -1 & 2 \end{bmatrix} = \begin{bmatrix} 6-5 & -10+10 \\ 3-3 & -5+6 \end{bmatrix}$$

$$= \begin{bmatrix} 1 & 0 \\ 0 & 1 \end{bmatrix} = \mathbf{I} \qquad [ok]$$

13. WRITING SIMULTANEOUS LINEAR EQUATIONS IN MATRIX FORM

Matrices are used to simplify the presentation and solution of sets of simultaneous linear equations. For example, the following three methods of presenting simultaneous linear equations are equivalent:

$$a_{11}x_1 + a_{12}x_2 = b_1$$
$$a_{21}x_1 + a_{22}x_2 = b_2$$

$$\begin{bmatrix} a_{11} & a_{12} \\ a_{21} & a_{22} \end{bmatrix} \begin{bmatrix} x_1 \\ x_2 \end{bmatrix} = \begin{bmatrix} b_1 \\ b_2 \end{bmatrix}$$

$$\mathbf{A} \times \mathbf{X} = \mathbf{B}$$

In the second and third representations, $\mathbf{A}$ is known as the *coefficient matrix*, $\mathbf{X}$ as the *variable matrix*, and $\mathbf{B}$ as the *constant matrix*.

Not all systems of simultaneous equations have solutions, and those that do may not have unique solutions.

The existence of a solution can be determined by calculating the determinant of the coefficient matrix. These rules are summarized in Table 4.1.

- If the system of linear equations is homogeneous (i.e., $\mathbf{B}$ is a zero matrix) and $|\mathbf{A}|$ is zero, there are an infinite number of solutions.

- If the system is homogeneous and $|\mathbf{A}|$ is nonzero, only the trivial solution exists.

- If the system of linear equations is nonhomogeneous (i.e., $\mathbf{B}$ is not a zero matrix) and $|\mathbf{A}|$ is nonzero, there is a unique solution to the set of simultaneous equations.

- If $|\mathbf{A}|$ is zero, a nonhomogeneous system of simultaneous equations may still have a solution. The requirement is that the determinants of all substitutional matrices (see Sec. 14) are zero, in which case there will be an infinite number of solutions. Otherwise, no solution exists.

Table 4.1 *Solution Existence Rules for Simultaneous Equations*

	$\mathbf{B} = 0$	$\mathbf{B} \neq 0$
$|\mathbf{A}| = 0$	infinite number of solutions (linearly dependent equations)	either an infinite number of solutions or no solution at all
$|\mathbf{A}| \neq 0$	trivial solution only $(x_i = 0)$	unique nonzero solution

14. SOLVING SIMULTANEOUS LINEAR EQUATIONS

Gauss-Jordan elimination can be used to obtain the solution to a set of simultaneous linear equations. The coefficient matrix is augmented by the constant matrix. Then, elementary row operations are used to reduce the coefficient matrix to canonical form. All of the operations performed on the coefficient matrix are performed on the constant matrix. The variable values that satisfy the simultaneous equations will be the entries in the constant matrix when the coefficient matrix is in canonical form.

Determinants are used to calculate the solution to linear simultaneous equations through a procedure known as *Cramer's rule*.

The procedure is to calculate determinants of the original coefficient matrix $\mathbf{A}$ and of the n matrices resulting from the systematic replacement of a column in $\mathbf{A}$ by the constant matrix $\mathbf{B}$. For a system of three equations in three unknowns, there are three substitutional matrices, $\mathbf{A}_1$, $\mathbf{A}_2$, and $\mathbf{A}_3$, as well as the original coefficient matrix, for a total of four matrices whose determinants must be calculated.

The values of the unknowns that simultaneously satisfy all of the linear equations are

$$x_1 = \frac{|\mathbf{A}_1|}{|\mathbf{A}|} \qquad 4.23$$

$$x_2 = \frac{|\mathbf{A}_2|}{|\mathbf{A}|} \qquad 4.24$$

$$x_3 = \frac{|\mathbf{A}_3|}{|\mathbf{A}|} \qquad 4.25$$

Example 4.8

Use Gauss-Jordan elimination to solve the following system of simultaneous equations.

$$2x + 3y - 4z = 1$$
$$3x - y - 2z = 4$$
$$4x - 7y - 6z = -7$$

Solution

The augmented matrix is created by appending the constant matrix to the coefficient matrix.

$$\begin{bmatrix} 2 & 3 & -4 & \vdots & 1 \\ 3 & -1 & -2 & \vdots & 4 \\ 4 & -7 & -6 & \vdots & -7 \end{bmatrix}$$

Elementary row operations are used to reduce the coefficient matrix to canonical form. For example, two times the first row is subtracted from the third row. This step obtains the 0 needed in the a_{31} position.

$$\begin{bmatrix} 2 & 3 & -4 & \vdots & 1 \\ 3 & -1 & -2 & \vdots & 4 \\ 0 & -13 & 2 & \vdots & -9 \end{bmatrix}$$

This process continues until the following form is obtained.

$$\begin{bmatrix} 1 & 0 & 0 & \vdots & 3 \\ 0 & 1 & 0 & \vdots & 1 \\ 0 & 0 & 1 & \vdots & 2 \end{bmatrix}$$

$x = 3$, $y = 1$, and $z = 2$ satisfy this system of equations.

Example 4.9

Use Cramer's rule to solve the following system of simultaneous equations.

$$2x + 3y - 4z = 1$$
$$3x - y - 2z = 4$$
$$4x - 7y - 6z = -7$$

Solution

The determinant of the coefficient matrix is

$$|\mathbf{A}| = \begin{vmatrix} 2 & 3 & -4 \\ 3 & -1 & -2 \\ 4 & -7 & -6 \end{vmatrix} = 82$$

The determinants of the substitutional matrices are

$$|\mathbf{A}_1| = \begin{vmatrix} 1 & 3 & -4 \\ 4 & -1 & -2 \\ -7 & -7 & -6 \end{vmatrix} = 246$$

$$|\mathbf{A}_2| = \begin{vmatrix} 2 & 1 & -4 \\ 3 & 4 & -2 \\ 4 & -7 & -6 \end{vmatrix} = 82$$

$$|\mathbf{A}_3| = \begin{vmatrix} 2 & 3 & 1 \\ 3 & -1 & 4 \\ 4 & -7 & -7 \end{vmatrix} = 164$$

The values of x, y, and z that satisfy the linear equations are

$$x = \frac{246}{82} = 3$$

$$y = \frac{82}{82} = 1$$

$$z = \frac{164}{82} = 2$$

15. EIGENVALUES AND EIGENVECTORS

Eigenvalues and eigenvectors (also known as *characteristic values* and *characteristic vectors*) of a square matrix $\mathbf{A}$ are the scalars k and matrices $\mathbf{X}$ such that

$$\mathbf{A} \times \mathbf{X} = k\mathbf{X} \qquad 4.26$$

The scalar k is an eigenvalue of $\mathbf{A}$ if and only if the matrix $(k\mathbf{I} - \mathbf{A})$ is singular; that is, if $|k\mathbf{I} - \mathbf{A}| = 0$. This equation is called the *characteristic equation* of the matrix $\mathbf{A}$. When expanded, the determinant is called the *characteristic polynomial*. The method of using the characteristic polynomial to find eigenvalues and eigenvectors is illustrated in Ex. 4.10.

If all of the eigenvalues are unique (i.e., nonrepeating), then Eq. 4.27 is valid.

$$[k\mathbf{I} - \mathbf{A}]\mathbf{X} = 0 \qquad 4.27$$

Example 4.10

Find the eigenvalues and nonzero eigenvectors of the matrix $\mathbf{A}$.

$$\mathbf{A} = \begin{bmatrix} 2 & 4 \\ 6 & 4 \end{bmatrix}$$

Solution

$$k\mathbf{I} - \mathbf{A} = \begin{bmatrix} k & 0 \\ 0 & k \end{bmatrix} - \begin{bmatrix} 2 & 4 \\ 6 & 4 \end{bmatrix} = \begin{bmatrix} k-2 & -4 \\ -6 & k-4 \end{bmatrix}$$

The characteristic polynomial is found by setting the determinant $|k\mathbf{I} - \mathbf{A}|$ equal to zero.

$$(k-2)(k-4) - (-6)(-4) = 0$$
$$k^2 - 6k - 16 = (k-8)(k+2) = 0$$

The roots of the characteristic polynomial are $k = +8$ and $k = -2$. These are the eigenvalues of $\mathbf{A}$.

Substituting $k = 8$,

$$k\mathbf{I} - \mathbf{A} = \begin{bmatrix} 8-2 & -4 \\ -6 & 8-4 \end{bmatrix} = \begin{bmatrix} 6 & -4 \\ -6 & 4 \end{bmatrix}$$

This can be interpreted as the linear equation $6x_1 - 4x_2 = 0$. The values of x that satisfy this equation define the eigenvector. An eigenvector $\mathbf{X}$ associated with the eigenvalue $+8$ is

$$\mathbf{X} = \begin{bmatrix} x_1 \\ x_2 \end{bmatrix} = \begin{bmatrix} 4 \\ 6 \end{bmatrix}$$

All other eigenvectors for this eigenvalue are multiples of $\mathbf{X}$. Normally $\mathbf{X}$ is reduced to smallest integers.

$$\mathbf{X} = \begin{bmatrix} 2 \\ 3 \end{bmatrix}$$

Similarly, the eigenvector associated with the eigenvalue -2 is

$$\mathbf{X} = \begin{bmatrix} x_1 \\ x_2 \end{bmatrix} = \begin{bmatrix} +4 \\ -4 \end{bmatrix}$$

Reducing this to smallest integers,

$$\mathbf{X} = \begin{bmatrix} +1 \\ -1 \end{bmatrix}$$

PRACTICE PROBLEMS

1. Use Cramer's rule to solve for the values of x, y, and z that simultaneously satisfy the following equations.

$$x + y = -4$$
$$x + z - 1 = 0$$
$$2z - y + 3x = 4$$

5 Vectors

1. Introduction 5-1
2. Vectors in n-Space 5-1
3. Unit Vectors 5-2
4. Vector Representation 5-2
5. Conversion Between Systems 5-2
6. Vector Addition 5-3
7. Multiplication by a Scalar 5-3
8. Vector Dot Product 5-3
9. Vector Cross Product 5-4
10. Mixed Triple Product 5-4
11. Vector Triple Product 5-5
12. Vector Functions 5-5

1. INTRODUCTION

A physical property or quantity can be a scalar, vector, or tensor. A *scalar* has only magnitude. Knowing its value is sufficient to define a scalar. Mass, enthalpy, density, and speed are examples of scalars.

Force, momentum, displacement, and velocity are examples of vectors. A *vector* is a directed straight line with a specific magnitude. Thus, a vector is specified completely by its direction (consisting of the vector's *angular orientation* and its *sense*) and magnitude. A vector's *point of application* (*terminal point*) is not needed to define the vector.[1] Two vectors with the same direction and magnitude are said to be *equal vectors* even though their *lines of action* may be different.[2]

A vector can be designated by a boldface variable (as in this book) or as a combination of the variable and some other symbol. For example, the notations $\mathbf{V}$, $\bar{V}$, $\hat{V}$, $\vec{V}$, and $\underline{V}$ are used by different authorities to represent vectors. The magnitude of a vector can be designated by either $|\mathbf{V}|$ or V (not bold).

Stress, dielectric constant, and magnetic susceptibility are examples of tensors. A *tensor* has magnitude in a specific direction but the direction is not unique. Tensors are frequently associated with *anisotropic materials* that have different properties in different directions. A tensor in three-dimensional space is defined by nine components, compared with the three that are required

to define vectors. These components are written in matrix form. Stress, σ, at a point, for example, would be defined by the following tensor matrix.

$$\sigma = \begin{pmatrix} \sigma_{xx} & \sigma_{xy} & \sigma_{xz} \\ \sigma_{yx} & \sigma_{yy} & \sigma_{yz} \\ \sigma_{zx} & \sigma_{zy} & \sigma_{zz} \end{pmatrix}$$

2. VECTORS IN n-SPACE

In some cases, a vector, $\mathbf{V}$, will be designated by its two endpoints in n-dimensional vector space. A common example is three-dimensional force-space. Usually, one of the points will be the origin, in which case the vector is said to be "based at the origin," "origin-based," or "zero-based."[3] If one of the endpoints is the origin, specifying a terminal point P would represent a force directed from the origin to point P.

If a coordinate system is superimposed on the vector space, a vector can be specified in terms of the n coordinates of its two endpoints. The magnitude of the vector $\mathbf{V}$ is the distance in vector space between the two points, as given by Eq. 5.1. Similarly, the direction is defined by the angle the vector makes with one of the axes.

$$|\mathbf{V}| = \sqrt{(x_2 - x_1)^2 + (y_2 - y_1)^2} \qquad 5.1$$

$$\phi = \arctan\left(\frac{y_2 - y_1}{x_2 - x_1}\right) \qquad 5.2$$

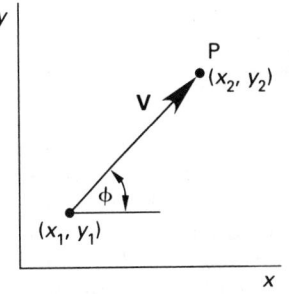

Figure 5.1 *Vector in Two-Dimensional Space*

---

[1] A vector that is constrained to act at or through a certain point is a *bound vector* (*fixed vector*). A *sliding vector* (*transmissible vector*) can be applied anywhere along its line of action. A *free vector* is not constrained and can be applied at any point in space.
[2] A distinction is sometimes made between equal vectors and equivalent vectors. *Equivalent vectors* produce the same effect but are not necessarily equal.

[3] Any vector directed from P$_1$ to P$_2$ can be transformed into a zero-based vector by subtracting the coordinates of point P$_1$ from the coordinates of terminal point P$_2$. The transformed vector will be equivalent to the original vector.

The *components* of a vector are the projections of the vector on the coordinate axes. (For a zero-based vector, the components and the coordinates of the endpoint are the same.) Simple trigonometric principles are used to resolve a vector into its components. A vector constructed from its components is known as a *resultant vector*.

$$V_x = |\mathbf{V}| \cos \phi_x \qquad 5.3$$
$$V_y = |\mathbf{V}| \cos \phi_y \qquad 5.4$$
$$V_z = |\mathbf{V}| \cos \phi_z \qquad 5.5$$
$$|\mathbf{V}| = \sqrt{V_x^2 + V_y^2 + V_z^2} \qquad 5.6$$

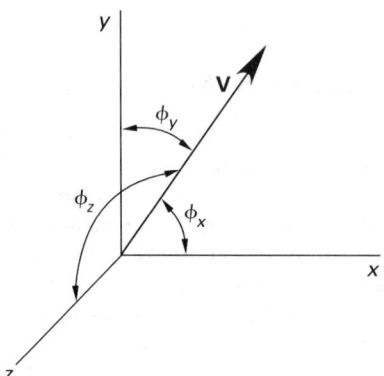

Figure 5.2 *Direction Angles of a Vector*

In Eqs. 5.3 through 5.5, ϕ_x, ϕ_y, and ϕ_z are the *direction angles*—the angles between the vector and the x-, y-, and z-axis, respectively. The cosines of these angles are known as *direction cosines*. The sum of the squares of the direction cosines is equal to 1.

$$\cos^2 \phi_x + \cos^2 \phi_y + \cos^2 \phi_z = 1 \qquad 5.7$$

3. UNIT VECTORS

Unit vectors are vectors with unit magnitudes (i.e., magnitudes of 1). They are represented in the same notation as other vectors. Although they can have any direction, the standard unit vectors (the *Cartesian unit vectors* $\mathbf{i}$, $\mathbf{j}$, and $\mathbf{k}$) have the directions of the x-, y-, and z-coordinate axes and constitute the *Cartesian triad*.

A vector $\mathbf{V}$ can be written in terms of unit vectors and its components.

$$\mathbf{V} = |\mathbf{V}|\mathbf{a} = V_x\mathbf{i} + V_y\mathbf{j} + V_z\mathbf{k} \qquad 5.8$$

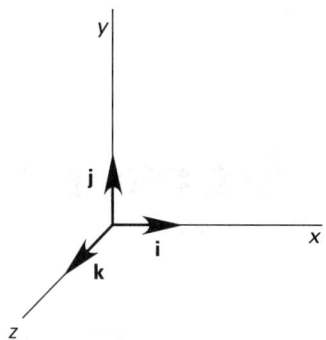

Figure 5.3 *Cartesian Unit Vectors*

The unit vector, $\mathbf{a}$, has the same direction as the vector $\mathbf{V}$ but has a length of 1. This unit vector is calculated by dividing the original vector, $\mathbf{V}$, by its magnitude, $|\mathbf{V}|$.

$$\mathbf{a} = \frac{\mathbf{V}}{|\mathbf{V}|} = \frac{V_x\mathbf{i} + V_y\mathbf{j} + V_z\mathbf{k}}{\sqrt{V_x^2 + V_y^2 + V_z^2}} \qquad 5.9$$

4. VECTOR REPRESENTATION

The most common method of representing a vector is by writing it in *rectangular form*—a vector sum of its orthogonal components. In rectangular form, each of the orthogonal components has the same units as the resultant vector.

$$\mathbf{A} \equiv A_x\mathbf{i} + A_y\mathbf{j} + A_z\mathbf{k} \qquad \text{[three dimensions]}$$

However, the vector is also completely defined by its magnitude and associated angle. These two quantities can be written together in *phasor form*, sometimes referred to as *polar form*.

$$\mathbf{A} \equiv |\mathbf{A}|\angle\phi = A\angle\phi$$

5. CONVERSION BETWEEN SYSTEMS

The choice of the $\mathbf{ijk}$ triad may be convenient but is arbitrary. A vector can be expressed in terms of another set of unit vectors, $\mathbf{uvw}$.

$$\mathbf{V} = V_x\mathbf{i} + V_y\mathbf{j} + V_z\mathbf{k} = V_x'\mathbf{u} + V_y'\mathbf{v} + V_z'\mathbf{w} \qquad 5.10$$

The two representations are related.

$$V_x' = \mathbf{V} \cdot \mathbf{u} = (\mathbf{i} \cdot \mathbf{u})V_x + (\mathbf{j} \cdot \mathbf{u})V_y + (\mathbf{k} \cdot \mathbf{u})V_z \qquad 5.11$$
$$V_y' = \mathbf{V} \cdot \mathbf{v} = (\mathbf{i} \cdot \mathbf{v})V_x + (\mathbf{j} \cdot \mathbf{v})V_y + (\mathbf{k} \cdot \mathbf{v})V_z \qquad 5.12$$
$$V_z' = \mathbf{V} \cdot \mathbf{w} = (\mathbf{i} \cdot \mathbf{w})V_x + (\mathbf{j} \cdot \mathbf{w})V_y + (\mathbf{k} \cdot \mathbf{w})V_z \qquad 5.13$$

Equations 5.11 through 5.13 can be expressed in matrix form. The dot products are known as the *coefficients of transformation*, and the matrix containing them is the *transformation matrix*.

$$\begin{pmatrix} V'_x \\ V'_y \\ V'_z \end{pmatrix} = \begin{pmatrix} \mathbf{i} \cdot \mathbf{u} & \mathbf{j} \cdot \mathbf{u} & \mathbf{k} \cdot \mathbf{u} \\ \mathbf{i} \cdot \mathbf{v} & \mathbf{j} \cdot \mathbf{v} & \mathbf{k} \cdot \mathbf{v} \\ \mathbf{i} \cdot \mathbf{w} & \mathbf{j} \cdot \mathbf{w} & \mathbf{k} \cdot \mathbf{w} \end{pmatrix} \begin{pmatrix} V_x \\ V_y \\ V_z \end{pmatrix} \quad \textit{5.14}$$

6. VECTOR ADDITION

Addition of two vectors by the *polygon method* is accomplished by placing the tail of the second vector at the head (tip) of the first. The sum (i.e., the *resultant vector*) is a vector extending from the tail of the first vector to the head of the second. Alternatively, the two vectors can be considered as two of the sides of a parallelogram, while the sum represents the diagonal. This is known as addition by the *parallelogram method*.

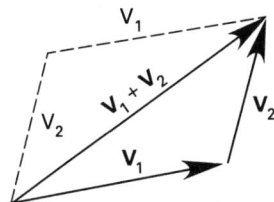

Figure 5.4 *Addition of Two Vectors*

The components of the resultant vector are the sums of the components of the added vectors ($V_{1x} + V_{2x}$, $V_{1y} + V_{2y}$, $V_{1z} + V_{2z}$).

Vector addition is both commutative and associative.

$$\mathbf{V}_1 + \mathbf{V}_2 = \mathbf{V}_2 + \mathbf{V}_1 \quad \textit{5.15}$$
$$\mathbf{V}_1 + (\mathbf{V}_2 + \mathbf{V}_3) = (\mathbf{V}_1 + \mathbf{V}_2) + \mathbf{V}_3 \quad \textit{5.16}$$

7. MULTIPLICATION BY A SCALAR

A vector, $\mathbf{V}$, can be multiplied by a scalar, c. If the original vector is represented by its components, each of the components is multiplied by c.

$$c\mathbf{V} = c|\mathbf{V}|\,\mathbf{a} = cV_x\mathbf{i} + cV_y\mathbf{j} + cV_z\mathbf{k} \quad \textit{5.17}$$

Scalar multiplication is distributive.

$$c(\mathbf{V}_1 + \mathbf{V}_2) = c\mathbf{V}_1 + c\mathbf{V}_2 \quad \textit{5.18}$$

8. VECTOR DOT PRODUCT

The *dot product* (*scalar product*), $\mathbf{V}_1 \cdot \mathbf{V}_2$, of two vectors is a scalar that is proportional to the length of the projection of the first vector onto the second vector.[4]

---

[4]The dot product is also written in parentheses without a dot, that is, $(\mathbf{V}_1\mathbf{V}_2)$.

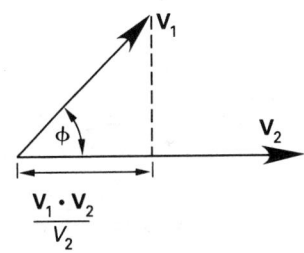

Figure 5.5 *Vector Dot Product*

The dot product is commutative and distributive.

$$\mathbf{V}_1 \cdot \mathbf{V}_2 = \mathbf{V}_2 \cdot \mathbf{V}_1 \quad \textit{5.19}$$
$$\mathbf{V}_1 \cdot (\mathbf{V}_2 + \mathbf{V}_3) = \mathbf{V}_1 \cdot \mathbf{V}_2 + \mathbf{V}_1 \cdot \mathbf{V}_3 \quad \textit{5.20}$$

The dot product can be calculated in two ways, as Eq. 5.21 indicates. ϕ is limited to 180° and is the angle between the two vectors.

$$\mathbf{V}_1 \cdot \mathbf{V}_2 = |\mathbf{V}_1|\,|\mathbf{V}_2| \cos\phi$$
$$= V_{1x}V_{2x} + V_{1y}V_{2y} + V_{1z}V_{2z} \quad \textit{5.21}$$

When Eq. 5.21 is solved for the angle between the two vectors, ϕ, it is known as the *Cauchy-Schwartz theorem*.

$$\cos\phi = \frac{V_{1x}V_{2x} + V_{1y}V_{2y} + V_{1z}V_{2z}}{|\mathbf{V}_1|\,|\mathbf{V}_2|} \quad \textit{5.22}$$

The dot product can be used to determine whether a vector is a unit vector and to show that two vectors are orthogonal (perpendicular). For two non-null orthogonal vectors,

$$\mathbf{V}_1 \cdot \mathbf{V}_2 = 0 \quad \textit{5.23}$$

For any unit vector, $\mathbf{u}$,

$$\mathbf{u} \cdot \mathbf{u} = 1 \quad \textit{5.24}$$

Equations 5.23 and 5.24 can be extended to the Cartesian unit vectors.

$$\mathbf{i} \cdot \mathbf{i} = 1 \quad \textit{5.25}$$
$$\mathbf{j} \cdot \mathbf{j} = 1 \quad \textit{5.26}$$
$$\mathbf{k} \cdot \mathbf{k} = 1 \quad \textit{5.27}$$
$$\mathbf{i} \cdot \mathbf{j} = 0 \quad \textit{5.28}$$
$$\mathbf{i} \cdot \mathbf{k} = 0 \quad \textit{5.29}$$
$$\mathbf{j} \cdot \mathbf{k} = 0 \quad \textit{5.30}$$

Example 5.1

What is the angle between the zero-based vectors $\mathbf{V}_1 = (-\sqrt{3}, 1)$ and $\mathbf{V}_2 = (2\sqrt{3}, 2)$?

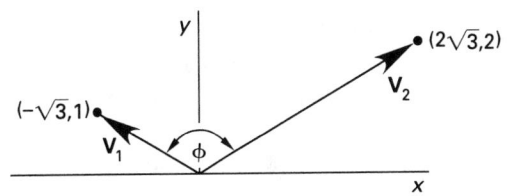

Solution

From Eq. 5.22,

$$\cos\phi = \frac{V_{1x}V_{2x} + V_{1y}V_{2y}}{|\mathbf{V}_1||\mathbf{V}_2|} = \frac{V_{1x}V_{2x} + V_{1y}V_{2y}}{\sqrt{V_{1x}^2 + V_{1y}^2}\sqrt{V_{2x}^2 + V_{2y}^2}}$$

$$= \frac{(-\sqrt{3})(2\sqrt{3}) + (1)(2)}{\sqrt{(-\sqrt{3})^2 + (1)^2}\sqrt{(2\sqrt{3})^2 + (2)^2}}$$

$$= \frac{-4}{8} = -\frac{1}{2}$$

$$\phi = \arccos\left(-\frac{1}{2}\right) = 120°$$

9. VECTOR CROSS PRODUCT

The *cross product (vector product)*, $\mathbf{V}_1 \times \mathbf{V}_2$, of two vectors is a vector that is orthogonal (perpendicular) to the plane of the two vectors.[5] The unit vector representation of the cross product can be calculated as a third-order determinant.

$$\mathbf{V}_1 \times \mathbf{V}_2 = \begin{vmatrix} \mathbf{i} & V_{1x} & V_{2x} \\ \mathbf{j} & V_{1y} & V_{2y} \\ \mathbf{k} & V_{1z} & V_{2z} \end{vmatrix} \qquad 5.31$$

The direction of the cross-product vector corresponds to the direction a right-hand screw would progress if vectors $\mathbf{V}_1$ and $\mathbf{V}_2$ are placed tail-to-tail in the plane they define and $\mathbf{V}_1$ is rotated into $\mathbf{V}_2$. The direction can also be found from the *right-hand rule*. (See Sec. 43.6.)

The magnitude of the cross product can be determined from Eq. 5.32, in which ϕ is the angle between the two vectors and is limited to 180°. The magnitude corresponds to the area of a parallelogram that has $\mathbf{V}_1$ and $\mathbf{V}_2$ as two of its sides.

$$|\mathbf{V}_1 \times \mathbf{V}_2| = |\mathbf{V}_1||\mathbf{V}_2|\sin\phi \qquad 5.32$$

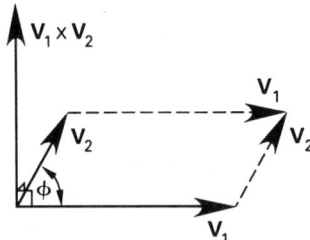

Figure 5.6 *Vector Cross Product*

Vector cross multiplication is distributive but not commutative.

$$\mathbf{V}_1 \times \mathbf{V}_2 = -(\mathbf{V}_2 \times \mathbf{V}_1) \qquad 5.33$$

$$c(\mathbf{V}_1 \times \mathbf{V}_2) = (c\mathbf{V}_1) \times \mathbf{V}_2 = \mathbf{V}_1 \times (c\mathbf{V}_2) \qquad 5.34$$

$$\mathbf{V}_1 \times (\mathbf{V}_2 + \mathbf{V}_3) = \mathbf{V}_1 \times \mathbf{V}_2 + \mathbf{V}_1 \times \mathbf{V}_3 \qquad 5.35$$

---

[5]The cross product is also written in square brackets without a cross, that is, $[\mathbf{V}_1\mathbf{V}_2]$.

If the two vectors are parallel, their cross product will be zero.

$$\mathbf{i} \times \mathbf{i} = \mathbf{j} \times \mathbf{j} = \mathbf{k} \times \mathbf{k} = 0 \qquad 5.36$$

Equation 5.36 can be extended to the unit vectors.

$$\mathbf{i} \times \mathbf{j} = -\mathbf{j} \times \mathbf{i} = \mathbf{k} \qquad 5.37$$

$$\mathbf{j} \times \mathbf{k} = -\mathbf{k} \times \mathbf{j} = \mathbf{i} \qquad 5.38$$

$$\mathbf{k} \times \mathbf{i} = -\mathbf{i} \times \mathbf{k} = \mathbf{j} \qquad 5.39$$

Example 5.2

Find a unit vector orthogonal to $\mathbf{V}_1 = \mathbf{i} - \mathbf{j} + 2\mathbf{k}$ and $\mathbf{V}_2 = 3\mathbf{j} - \mathbf{k}$.

Solution

The cross product is a vector orthogonal to $\mathbf{V}_1$ and $\mathbf{V}_2$.

$$\mathbf{V}_1 \times \mathbf{V}_2 = \begin{vmatrix} \mathbf{i} & 1 & 0 \\ \mathbf{j} & -1 & 3 \\ \mathbf{k} & 2 & -1 \end{vmatrix}$$

$$= -5\mathbf{i} + \mathbf{j} + 3\mathbf{k}$$

Check to see whether this is a unit vector.

$$|\mathbf{V}_1 \times \mathbf{V}_2| = \sqrt{(-5)^2 + (1)^2 + (3)^2} = \sqrt{35}$$

Since its length is $\sqrt{35}$, the vector must be divided by $\sqrt{35}$ to obtain a unit vector.

$$\mathbf{a} = \frac{-5\mathbf{i} + \mathbf{j} + 3\mathbf{k}}{\sqrt{35}}$$

10. MIXED TRIPLE PRODUCT

The *mixed triple product (triple scalar product* or just *triple product)* of three vectors is a scalar quantity representing the volume of a parallelepiped with the three vectors making up the sides. It is calculated as a determinant. Since Eq. 5.40 can be negative, the absolute value must be used to obtain the volume in that case.

$$\mathbf{V}_1 \cdot (\mathbf{V}_2 \times \mathbf{V}_3) = \begin{vmatrix} V_{1x} & V_{1y} & V_{1z} \\ V_{2x} & V_{2y} & V_{2z} \\ V_{3x} & V_{3y} & V_{3z} \end{vmatrix} \qquad 5.40$$

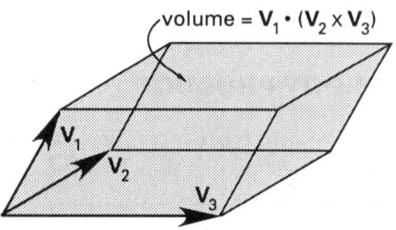

Figure 5.7 *Vector Mixed Triple Product*

The mixed triple product has the property of *circular permutation*, as defined by Eq. 5.41.

$$\mathbf{V}_1 \cdot (\mathbf{V}_2 \times \mathbf{V}_3) = (\mathbf{V}_1 \times \mathbf{V}_2) \cdot \mathbf{V}_3 \qquad 5.41$$

11. VECTOR TRIPLE PRODUCT

The *vector triple product* is a vector defined by Eq. 5.42. The quantities in parentheses on the right-hand side are scalars.

$$\mathbf{V}_1 \times (\mathbf{V}_2 \times \mathbf{V}_3) = (\mathbf{V}_1 \cdot \mathbf{V}_3)\mathbf{V}_2 - (\mathbf{V}_1 \cdot \mathbf{V}_2)\mathbf{V}_3 \qquad 5.42$$

12. VECTOR FUNCTIONS

A vector can be a function of another parameter. For example, a vector $\mathbf{V}$ is a function of variable t when its V_x, V_y, and V_z are functions of t.

$$\mathbf{V}(t) = (2t - 3)\mathbf{i} + (t^2 + 1)\mathbf{j} + (-7t + 5)\mathbf{k} \qquad 5.43$$

When the functions of t are differentiated (or integrated) with respect to t, the vector itself is differentiated (integrated).[6] (Chapters 8 and 9 cover differentiation and integration.)

$$\frac{d\mathbf{V}(t)}{dt} = \left(\frac{dV_x}{dt}\right)\mathbf{i} + \left(\frac{dV_y}{dt}\right)\mathbf{j} + \left(\frac{dV_z}{dt}\right)\mathbf{k} \qquad 5.44$$

Similarly, the integral of the vector is

$$\int \mathbf{V}(t)\,dt = \mathbf{i}\int V_x\,dt + \mathbf{j}\int V_y\,dt + \mathbf{k}\int V_z\,dt \qquad 5.45$$

---

[6]This is particularly valuable when converting among position, velocity, and acceleration vectors.

6 Trigonometry

1. Degrees and Radians 6-1
2. Plane Angles 6-1
3. Triangles 6-2
4. Right Triangles 6-2
5. Circular Transcendental Functions 6-2
6. Small Angle Approximations 6-3
7. Graphs of the Functions 6-3
8. Signs of the Functions 6-3
9. Functions of Related Angles 6-3
10. Trigonometric Identities 6-3
11. Inverse Trigonometric Functions 6-4
12. Hyperbolic Transcendental Functions 6-4
13. Hyperbolic Identities 6-5
14. General Triangles 6-5
15. Spherical Trigonometry 6-5
16. Solid Angles 6-6
    Practice Problems 6-6

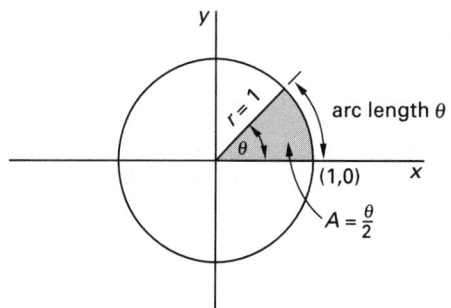

Figure 6.1 *Radians and Area of Unit Circle*

2. PLANE ANGLES

A *plane angle* (usually referred to as just an *angle*) consists of two intersecting lines and an intersection point known as the *vertex*. The angle can be referred to by a capital letter representing the vertex (e.g., B in Fig. 6.2), a Greek letter representing the angular measure (e.g., β), or by three capital letters, where the middle letter is the vertex and the other two letters are two points on different lines, and either the symbol $\angle$ or $\measuredangle$ (e.g., $\measuredangle ABC$).

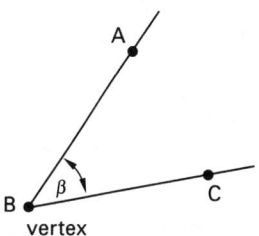

Figure 6.2 *Angle*

The angle between two intersecting lines generally is understood to be the smaller angle created.[2] Angles have been classified as follows.

- *acute angle*: an angle less than 90° ($\pi/2$ rad)

- *obtuse angle*: an angle more than 90° ($\pi/2$ rad) but less than 180° (π rad)

- *reflex angle*: an angle more than 180° (π rad) but less than 360° (2π rad)

1. DEGREES AND RADIANS

Degrees and *radians* are two units for measuring angles. One complete circle is divided into 360 degrees (written 360°) or 2π radians (abbreviated *rad*).[1] The conversions between degrees and radians are

multiply	by	to obtain
radians	$\dfrac{180}{\pi}$	degrees
degrees	$\dfrac{\pi}{180}$	radians

The number of radians in an angle θ corresponds to the area within a circular sector with arc length θ and a radius of one. Alternatively, the area of a sector with central angle θ radians is $\theta/2$ for a *unit circle* (i.e., a circle with a radius of one unit).

---

[1]The abbreviation *rad* is also used to represent *radiation absorbed dose*, a measure of radiation exposure.

[2]In books on geometry, the term *ray* is used instead of *line*.

- *related angle*: an angle that differs from another by some multiple of 90° ($\pi/2$ rad)

- *right angle*: an angle equal to 90° ($\pi/2$ rad)

- *straight angle*: an angle equal to 180° (π rad), that is, a straight line

Complementary angles are two angles whose sum is 90° ($\pi/2$ rad). *Supplementary angles* are two angles whose sum is 180° (π rad). *Adjacent angles* share a common vertex and one (the interior) side. Adjacent angles are supplementary only if their exterior sides form a straight line.

Vertical angles are the two angles with a common vertex and with sides made up by two intersecting straight lines. Vertical angles are equal.

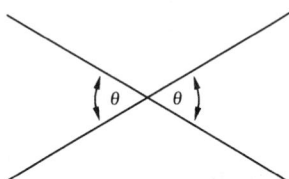

Figure 6.3 *Vertical Angles*

Angle of elevation and *angle of depression* are surveying terms referring to the angle above and below the horizontal plane of the observer, respectively.

3. TRIANGLES

A *triangle* is a three-sided closed polygon with three angles whose sum is 180° (π rad). Triangles are identified by their vertices and the symbol Δ (e.g., ΔABC in Fig. 6.4). A side is designated by its two endpoints (e.g., AB in Fig. 6.4) or by a lowercase letter corresponding to the capital letter of the opposite vertex (e.g., c).

In *similar triangles*, the corresponding angles are equal and the corresponding sides are in proportion. (Since there are only two independent angles in a triangle, showing that two angles of one triangle are equal to two angles of the other triangle is sufficient to show similarity.) The symbol for similarity is $\sim$. In Fig. 6.4, $\Delta ABC \sim \Delta DEF$ (i.e., ΔABC is similar to ΔDEF).

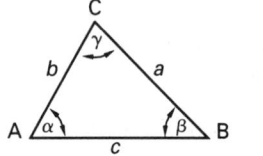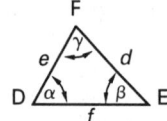

Figure 6.4 *Similar Triangles*

4. RIGHT TRIANGLES

A *right triangle* is a triangle in which one of the angles is 90° ($\pi/2$ rad). The remaining two angles are complementary. If one of the acute angles is chosen as the reference, the sides forming the right angle are known as the *adjacent side*, x, and the *opposite side*, y. The longest side is known as the *hypotenuse*, r. The *Pythagorean theorem* relates the lengths of these sides.

$$x^2 + y^2 = r^2 \qquad 6.1$$

In certain cases, the lengths of unknown sides of right triangles can be determined by inspection.[3] This occurs when the lengths of the sides are in the ratios of 3:4:5, 1:1:$\sqrt{2}$, 1:$\sqrt{3}$:2, and 5:12:13.

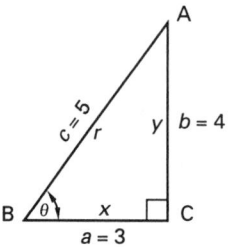

Figure 6.5 *3:4:5 Right Triangle*

5. CIRCULAR TRANSCENDENTAL FUNCTIONS

The *circular transcendental functions* (usually referred to as the *transcendental functions*, *trigonometric functions*, or *functions of an angle*) are calculated from the sides of a right triangle. Equations 6.2 through 6.7 refer to Fig. 6.5.

$$\text{sine: } \sin\theta = \frac{y}{r} = \frac{\text{opposite}}{\text{hypotenuse}} \qquad 6.2$$

$$\text{cosine: } \cos\theta = \frac{x}{r} = \frac{\text{adjacent}}{\text{hypotenuse}} \qquad 6.3$$

$$\text{tangent: } \tan\theta = \frac{y}{x} = \frac{\text{opposite}}{\text{adjacent}} \qquad 6.4$$

$$\text{cotangent: } \cot\theta = \frac{x}{y} = \frac{\text{adjacent}}{\text{opposite}} \qquad 6.5$$

$$\text{secant: } \sec\theta = \frac{r}{x} = \frac{\text{hypotenuse}}{\text{adjacent}} \qquad 6.6$$

$$\text{cosecant: } \csc\theta = \frac{r}{y} = \frac{\text{hypotenuse}}{\text{opposite}} \qquad 6.7$$

[3]These cases are almost always contrived examples. There is nothing intrinsic in nature to cause the formation of triangles with these proportions.

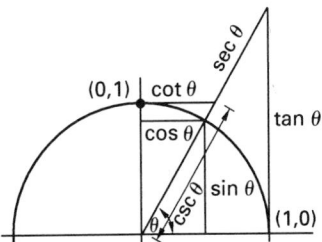

Figure 6.6 *Trigonometric Functions in a Unit Circle*

Three of the transcendental functions are reciprocals of the others. Notice that, while the tangent and cotangent functions are reciprocals of each other, the sine and cosine functions are not.

$$\cot \theta = \frac{1}{\tan \theta} \qquad \textit{6.8}$$

$$\sec \theta = \frac{1}{\cos \theta} \qquad \textit{6.9}$$

$$\csc \theta = \frac{1}{\sin \theta} \qquad \textit{6.10}$$

The trigonometric functions correspond to the lengths of various line segments in a right triangle with unit hypotenuse. Figure 6.6 shows such a triangle inscribed in a unit circle.

6. SMALL ANGLE APPROXIMATIONS

When an angle is very small, the hypotenuse and adjacent sides are essentially equal in length and certain approximations can be made. (The angle θ must be expressed in radians in Eqs. 6.11 and 6.12.)

$$\sin \theta \approx \tan \theta \approx \theta \Big|_{\theta < 10° \ (0.175 \, \text{rad})} \qquad \textit{6.11}$$

$$\cos \theta \approx 1 \Big|_{\theta < 5° \ (0.0873 \, \text{rad})} \qquad \textit{6.12}$$

7. GRAPHS OF THE FUNCTIONS

Figure 6.7 illustrates the periodicity of the sine, cosine, and tangent functions.[4]

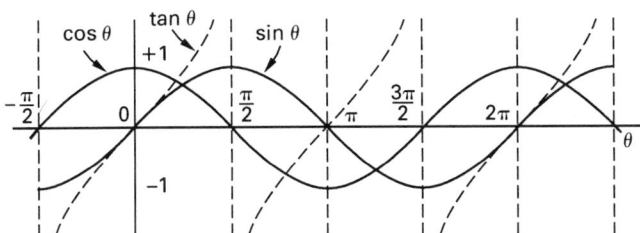

Figure 6.7 *Graphs of Sine, Cosine, and Tangent Functions*

---

[4]The remaining functions, being reciprocals of these three functions, are also periodic.

8. SIGNS OF THE FUNCTIONS

Table 6.1 shows how the sine, cosine, and tangent functions vary in sign with different values of θ. All three functions are positive for angles $0° \leq \theta \leq 90°$ ($0 \leq \theta \leq \pi/2$ rad), but only the sine is positive for angles $90° < \theta \leq 180°$ ($\pi/2$ rad $< \theta \leq \pi$ rad). The concept of quadrants is used to summarize the signs of the functions: angles up to $90°$ ($\pi/2$ rad) are in quadrant I, between $90°$ and $180°$ ($\pi/2$ and π rad) are in quadrant II, and so on.

Table 6.1 *Signs of the Functions by Quadrant*

	quadrant			
function	I	II	III	IV
sine	+	+	−	−
cosine	+	−	−	+
tangent	+	−	+	−

9. FUNCTIONS OF RELATED ANGLES

Figure 6.7 shows that the sine, cosine, and tangent curves are symmetric with respect to the horizontal axis. Furthermore, portions of the curves are symmetric with respect to a vertical axis. The values of the sine and cosine functions repeat every $360°$ (2π rad), and the absolute values repeat every $180°$ (π rad). This can be written as

$$\sin (\theta + 180°) = -\sin \theta \qquad \textit{6.13}$$

Similarly, the tangent function repeats every $180°$ (π rad), and its absolute value repeats every $90°$ ($\pi/2$ rad).

Table 6.2 summarizes the functions of the related angles.

Table 6.2 *Functions of Related Angles*

$f(\theta)$	$-\theta$	$90° - \theta$	$90° + \theta$	$180° - \theta$	$180° + \theta$
sin	$-\sin \theta$	$\cos \theta$	$\cos \theta$	$\sin \theta$	$-\sin \theta$
cos	$\cos \theta$	$\sin \theta$	$-\sin \theta$	$-\cos \theta$	$-\cos \theta$
tan	$-\tan \theta$	$\cot \theta$	$-\cot \theta$	$-\tan \theta$	$\tan \theta$

10. TRIGONOMETRIC IDENTITIES

There are many relationships between trigonometric functions. For example, Eqs. 6.14 through 6.16 are well known.

$$\sin^2 \theta + \cos^2 \theta = 1 \qquad \textit{6.14}$$

$$1 + \tan^2 \theta = \sec^2 \theta \qquad \textit{6.15}$$

$$1 + \cot^2 \theta = \csc^2 \theta \qquad \textit{6.16}$$

Other relatively common identities are listed as follows.[5]

- *double-angle formulas*:

$$\sin 2\theta = 2\sin\theta\cos\theta = \frac{2\tan\theta}{1+\tan^2\theta} \qquad 6.17$$

$$\cos 2\theta = \cos^2\theta - \sin^2\theta = 1 - 2\sin^2\theta$$
$$= 2\cos^2\theta - 1 = \frac{1-\tan^2\theta}{1+\tan^2\theta} \qquad 6.18$$

$$\tan 2\theta = \frac{2\tan\theta}{1-\tan^2\theta} \qquad 6.19$$

$$\cot 2\theta = \frac{\cot^2\theta - 1}{2\cot\theta} \qquad 6.20$$

- *two-angle formulas*:

$$\sin(\theta \pm \phi) = \sin\theta\cos\phi \pm \cos\theta\sin\phi \qquad 6.21$$

$$\cos(\theta \pm \phi) = \cos\theta\cos\phi \mp \sin\theta\sin\phi \qquad 6.22$$

$$\tan(\theta \pm \phi) = \frac{\tan\theta \pm \tan\phi}{1 \mp \tan\theta\tan\phi} \qquad 6.23$$

$$\cot(\theta \pm \phi) = \frac{\cot\phi\cot\theta \mp 1}{\cot\phi \pm \cot\theta} \qquad 6.24$$

- *half-angle formulas* ($\theta < 180°$):

$$\sin\frac{\theta}{2} = \sqrt{\frac{1-\cos\theta}{2}} \qquad 6.25$$

$$\cos\frac{\theta}{2} = \sqrt{\frac{1+\cos\theta}{2}} \qquad 6.26$$

$$\tan\frac{\theta}{2} = \sqrt{\frac{1-\cos\theta}{1+\cos\theta}} = \frac{\sin\theta}{1+\cos\theta} = \frac{1-\cos\theta}{\sin\theta} \qquad 6.27$$

- *miscellaneous formulas* ($\theta < 90°$):

$$\sin\theta = 2\sin\left(\frac{\theta}{2}\right)\cos\left(\frac{\theta}{2}\right) \qquad 6.28$$

$$\sin\theta = \sqrt{\frac{1-\cos 2\theta}{2}} \qquad 6.29$$

$$\cos\theta = \cos^2\left(\frac{\theta}{2}\right) - \sin^2\left(\frac{\theta}{2}\right) \qquad 6.30$$

$$\cos\theta = \sqrt{\frac{1+\cos 2\theta}{2}} \qquad 6.31$$

$$\tan\theta = \frac{2\tan\left(\dfrac{\theta}{2}\right)}{1-\tan^2\left(\dfrac{\theta}{2}\right)}$$
$$= \frac{2\sin\left(\dfrac{\theta}{2}\right)\cos\left(\dfrac{\theta}{2}\right)}{\cos^2\left(\dfrac{\theta}{2}\right) - \sin^2\left(\dfrac{\theta}{2}\right)} \qquad 6.32$$

$$\tan\theta = \sqrt{\frac{1-\cos 2\theta}{1+\cos 2\theta}}$$
$$= \frac{\sin 2\theta}{1+\cos 2\theta} = \frac{1-\cos 2\theta}{\sin 2\theta} \qquad 6.33$$

$$\cot\theta = \frac{\cot^2\left(\dfrac{\theta}{2}\right) - 1}{2\cot\left(\dfrac{\theta}{2}\right)}$$
$$= \frac{\cos^2\left(\dfrac{\theta}{2}\right) - \sin^2\left(\dfrac{\theta}{2}\right)}{2\sin\left(\dfrac{\theta}{2}\right)\cos\left(\dfrac{\theta}{2}\right)} \qquad 6.34$$

$$\cot\theta = \sqrt{\frac{1+\cos 2\theta}{1-\cos 2\theta}}$$
$$= \frac{1+\cos 2\theta}{\sin 2\theta} = \frac{\sin 2\theta}{1-\cos 2\theta} \qquad 6.35$$

11. INVERSE TRIGONOMETRIC FUNCTIONS

Finding an angle from a known trigonometric function is a common operation known as an *inverse trigonometric operation*. The inverse function can be designated by adding "inverse," "arc-," or the superscript -1 to the name of the function. For example,

$$\text{inverse }\sin(0.5) = \arcsin(0.5) = \sin^{-1}(0.5) = 30°$$

12. HYPERBOLIC TRANSCENDENTAL FUNCTIONS

Hyperbolic transcendental functions (normally referred to as *hyperbolic functions*) are specific equations containing combinations of the terms e^θ and $e^{-\theta}$. These combinations appear regularly in certain types of problems (e.g., analysis of cables and heat transfer from fins) and are given specific names and symbols to simplify presentation.[6]

$$\text{hyperbolic sine: }\sinh\theta = \frac{e^\theta - e^{-\theta}}{2} \qquad 6.36$$

$$\text{hyperbolic cosine: }\cosh\theta = \frac{e^\theta + e^{-\theta}}{2} \qquad 6.37$$

---

[5]It is an idiosyncrasy that these formulas are conventionally referred to as *formulas*, not *identities*.

[6]The hyperbolic sine and cosine functions are pronounced (by some) as "cinch" and "cosh," respectively.

hyperbolic tangent: $\tanh \theta = \dfrac{e^\theta - e^{-\theta}}{e^\theta + e^{-\theta}} = \dfrac{\sinh \theta}{\cosh \theta}$ *6.38*

hyperbolic cotangent: $\coth \theta = \dfrac{e^\theta + e^{-\theta}}{e^\theta - e^{-\theta}} = \dfrac{\cosh \theta}{\sinh \theta}$ *6.39*

hyperbolic secant: $\operatorname{sech} \theta = \dfrac{2}{e^\theta + e^{-\theta}} = \dfrac{1}{\cosh \theta}$ *6.40*

hyperbolic cosecant: $\operatorname{csch} \theta = \dfrac{2}{e^\theta - e^{-\theta}} = \dfrac{1}{\sinh \theta}$ *6.41*

Hyperbolic functions cannot be related to a right triangle, but they are related to a rectangular (equilateral) hyperbola, as shown in Fig. 6.8. The shaded area has a value of $\theta/2$ and is sometimes given the units of *hyperbolic radians*.

$$\sinh \theta = \frac{y}{a} \qquad \textit{6.42}$$

$$\cosh \theta = \frac{x}{a} \qquad \textit{6.43}$$

$$\tanh \theta = \frac{y}{x} \qquad \textit{6.44}$$

$$\coth \theta = \frac{x}{y} \qquad \textit{6.45}$$

$$\operatorname{sech} \theta = \frac{a}{x} \qquad \textit{6.46}$$

$$\operatorname{csch} \theta = \frac{a}{y} \qquad \textit{6.47}$$

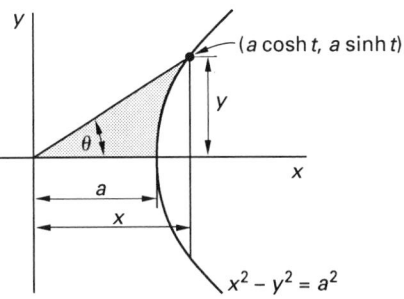

Figure 6.8 *Equilateral Hyperbola and Hyperbolic Functions*

13. HYPERBOLIC IDENTITIES

The hyperbolic identities are different from the standard trigonometric identities. Some of the most important identities are presented as follows.

$$\cosh^2 \theta - \sinh^2 \theta = 1 \qquad \textit{6.48}$$

$$1 - \tanh^2 \theta = \operatorname{sech}^2 \theta \qquad \textit{6.49}$$

$$1 - \coth^2 \theta = -\operatorname{csch}^2 \theta \qquad \textit{6.50}$$

$$\cosh \theta + \sinh \theta = e^\theta \qquad \textit{6.51}$$

$$\cosh \theta - \sinh \theta = e^{-\theta} \qquad \textit{6.52}$$

$$\sinh (\theta \pm \phi) = \sinh \theta \cosh \phi \pm \cosh \theta \sinh \phi \qquad \textit{6.53}$$

$$\cosh (\theta \pm \phi) = \cosh \theta \cosh \phi \pm \sinh \theta \sinh \phi \qquad \textit{6.54}$$

$$\tanh (\theta \pm \phi) = \frac{\tanh \theta \pm \tanh \phi}{1 \pm \tanh \theta \tanh \phi} \qquad \textit{6.55}$$

14. GENERAL TRIANGLES

A *general triangle* (also known as an *oblique triangle*) is one that is not specifically a right triangle, as shown in Fig. 6.9. Equation 6.56 calculates the area of a general triangle.[7]

$$\text{area} = \tfrac{1}{2}ab \sin C = \tfrac{1}{2}bc \sin A = \tfrac{1}{2}ca \sin B \qquad \textit{6.56}$$

The *law of sines* (Eq. 6.57) relates the sines of the angles.

$$\frac{\sin A}{a} = \frac{\sin B}{b} = \frac{\sin C}{c} \qquad \textit{6.57}$$

The *law of cosines* relates the cosine of an angle to an opposite side. (Equation 6.58 can be extended to the two remaining sides.)

$$a^2 = b^2 + c^2 - 2bc \cos A \qquad \textit{6.58}$$

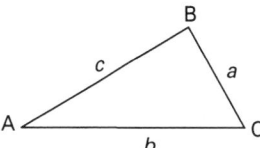

Figure 6.9 *General Triangle*

The *law of tangents* relates the sum and difference of two sides. (Equation 6.59 can be extended to the two remaining sides.)

$$\frac{a-b}{a+b} = \frac{\tan \left(\dfrac{A-B}{2} \right)}{\tan \left(\dfrac{A+B}{2} \right)} \qquad \textit{6.59}$$

15. SPHERICAL TRIGONOMETRY

A *spherical triangle* is a triangle that has been drawn on the surface of a sphere, as shown in Fig. 6.10. The *trihedral angle* O–ABC is formed when the vertices A, B, and C are joined to the center of the sphere. The *face angles* (BOC, COA, and AOB in Fig. 6.10) are used to measure the sides (a, b, and c in Fig. 6.10). The *vertex angles* are A, B, and C. Thus, angles are used to measure both vertex angles and sides.

---

[7]Other methods of calculating the area of a general triangle are given in Chap. 7.

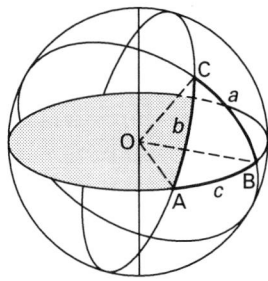

Figure 6.10 *Spherical Triangle ABC*

The following rules are valid for spherical triangles for which each side and angle is less than 180°.

- The sum of the three vertex angles is greater than 180° and less than 540°.

$$180° < A + B + C < 540° \qquad \textit{6.60}$$

- The sum of any two sides is greater than the third side.

- The sum of the three sides is less than 360°.

$$0° < a + b + c < 360° \qquad \textit{6.61}$$

- If the two sides are equal, the corresponding angles opposite are equal, and conversely.

- If two sides are unequal, the corresponding angles opposite are unequal. The greater angle is opposite the greater side.

The *spherical excess*, ϵ, is the amount by which the sum of the vertex angles exceeds 180°. The *spherical defect*, d, is the amount by which the sum of the sides differs from 360°.

$$\epsilon = A + B + C - 180° \qquad \textit{6.62}$$
$$d = 360° - (a + b + c) \qquad \textit{6.63}$$

There are many trigonometric identities that define the relationships between angles in a spherical triangle. Some of the more common identities are presented as follows.

- *law of sines*:

$$\frac{\sin A}{\sin a} = \frac{\sin B}{\sin b} = \frac{\sin C}{\sin c} \qquad \textit{6.64}$$

- *first law of cosines*:

$$\cos a = \cos b \cos c + \sin b \sin c \cos A \qquad \textit{6.65}$$

- *second law of cosines*:

$$\cos A = -\cos B \cos C + \sin B \sin C \cos a \qquad \textit{6.66}$$

16. SOLID ANGLES

A *solid angle*, ω, is a measure of the angle subtended at the vertex of a cone. The solid angle has units of *steradians* (abbreviated *sr*). A steradian is the solid angle subtended at the center of a unit sphere (i.e., a sphere with a radius of one) by a unit area on its surface. Since the surface area of a sphere of radius r is r^2 times the surface area of a unit sphere, the solid angle is equal to the area cut out by the cone divided by r^2. (The surface area of a spherical segment is given in App. 7.B.)

$$\omega = \frac{\text{surface area}}{r^2} \qquad \textit{6.67}$$

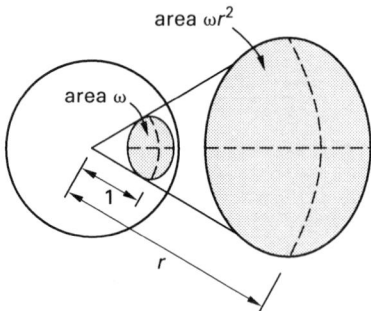

Figure 6.11 *Solid Angle*

PRACTICE PROBLEMS

1. A 5 lbm (5 kg) block sits on a 20° incline without slipping. Draw the freebody with respect to axes parallel and perpendicular to the surface of the incline. Determine the magnitude of all forces on the block.

7 Analytic Geometry

1. Mensuration of Regular Shapes 7-1
2. Areas with Irregular Boundaries 7-1
3. Geometric Definitions 7-2
4. Concave Curves 7-2
5. Convex Regions 7-3
6. Congruency 7-3
7. Coordinate Systems 7-3
8. Curves 7-3
9. Symmetry of Curves 7-4
10. Straight Lines 7-4
11. Direction Numbers, Angles, and Cosines . . 7-5
12. Intersection of Two Lines 7-6
13. Planes 7-6
14. Distances Between Geometric Figures 7-7
15. Angles Between Geometric Figures 7-8
16. Conic Sections 7-8
17. Circle 7-9
18. Parabola 7-10
19. Ellipse 7-11
20. Hyperbola 7-11
21. Sphere 7-12
22. Helix 7-12
    Practice Problems 7-12

Solution

Points O, B, and C constitute a circular segment and are used to find the central angle of the circular segment.

$$\left(\tfrac{1}{2}\right)(\angle \text{BOC}) = \arccos\left(\tfrac{1}{3}\right) = 70.53°$$

$$\phi = \angle \text{BOC} = (2)(70.53°) = 141.06° = 2.462 \text{ rad}$$

From App. 7.A, the area in flow and arc length are

$$A = \tfrac{1}{2}r^2(\phi - \sin\phi)$$
$$= \left(\tfrac{1}{2}\right)(3\,\text{in})^2[2.462\,\text{rad} - \sin(2.462\,\text{rad})]$$
$$= 8.251\,\text{in}^2$$

$$s = r\phi$$
$$= (3\,\text{in})(2.462\,\text{rad}) = 7.386\,\text{in}$$

The hydraulic radius is

$$r_h = \frac{A}{S}$$
$$= \frac{8.251\,\text{in}^2}{7.386\,\text{in}} = 1.12\,\text{in}$$

1. MENSURATION OF REGULAR SHAPES

The dimensions, perimeter, area, and other geometric properties constitute the *mensuration* (i.e., the measurements) of a geometric shape. Appendices 7.A and 7.B contain formulas and tables used to calculate these properties.

Example 7.1

In the study of open channel fluid flow, the hydraulic radius is defined as the ratio of flow area to wetted perimeter. What is the hydraulic radius of a 6 in diameter pipe filled to a depth of 2 in?

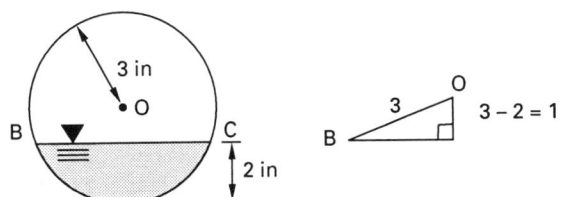

2. AREAS WITH IRREGULAR BOUNDARIES

Areas of sections with irregular boundaries (such as creek banks) cannot be determined precisely, and approximation methods must be used. If the irregular side can be divided into a series of cells of equal width, either the trapezoidal rule or Simpson's rule can be used.

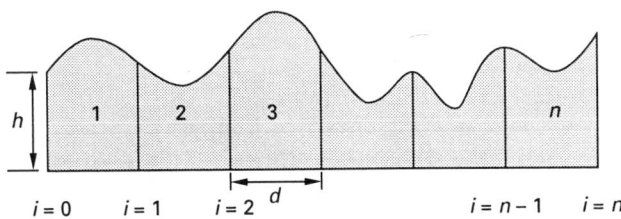

Figure 7.1 *Irregular Areas*

If the irregular side of each cell is fairly straight, the *trapezoidal rule* is appropriate.

$$A = \left(\frac{d}{2}\right)\left(h_0 + h_n + 2\sum_{i=1}^{n-1} h_i\right) \qquad 7.1$$

If the irregular side of each cell is curved (parabolic), *Simpson's rule* should be used. (n must be even to use Simpson's rule.)

$$A = \left(\frac{d}{3}\right)\left(h_0 + h_n + 4\sum_{\substack{i \text{ odd} \\ i=1}}^{n-1} h_i + 2\sum_{\substack{i \text{ even} \\ i=2}}^{n-2} h_i\right) \quad 7.2$$

3. GEOMETRIC DEFINITIONS

The following terms are used in this book to describe the relationship or orientation of one geometric figure to another.

- *abscissa*: the horizontal coordinate, typically designated as x in a rectangular coordinate system

- *asymptote*: a straight line that is approached but not intersected by a curved line

- *asymptotic*: approaching the slope of another line; attaining the slope of another line in the limit

- *center*: a point equidistant from all other points

- *collinear*: falling on the same line

- *concave*: curved inward (in the direction indicated)[1]

- *convex*: curved outward (in the direction indicated)

- *convex hull*: a closed figure whose surface is convex everywhere

- *coplanar*: falling on the same plane

- *inflection point*: a point where the second derivative changes sign or the curve changes from concave to convex. (Also known as a *point of contraflexure*.)

- *locus of points*: a set or collection of points having some common property and being so infinitely close together as to be indistinguishable from a line

- *node*: a point on a line from which other lines enter or leave

- *normal*: rotated 90°; being at right angles

- *ordinate*: the vertical coordinate, typically designated as y in a rectangular coordinate system

- *orthogonal*: rotated 90°; being at right angles

---
[1]This is easily remembered since one must go inside to explore a cave.

- *saddle point*: a point in three-dimensional space where all adjacent points are higher in one direction (the direction of the saddle) and lower in an orthogonal direction (the direction of the sides)

- *tangent*: having equal slopes at a common point

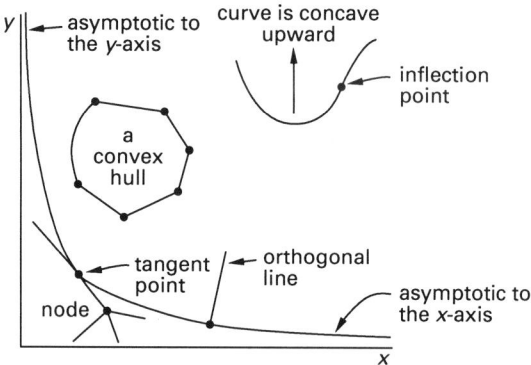

Figure 7.2 *Geometric Definitions*

4. CONCAVE CURVES

Concavity is a term that is applied to curved lines. A *concave up curve* is one whose function's first derivative increases continuously from negative to positive values. Straight lines drawn tangent to concave up curves are all below the curve. The graph of such a function may be thought of as being able to "hold water."

The first derivative of a *concave down curve* decreases continuously from positive to negative. A graph of a concave down function may be thought of as "spilling water." See Fig. 7.3.

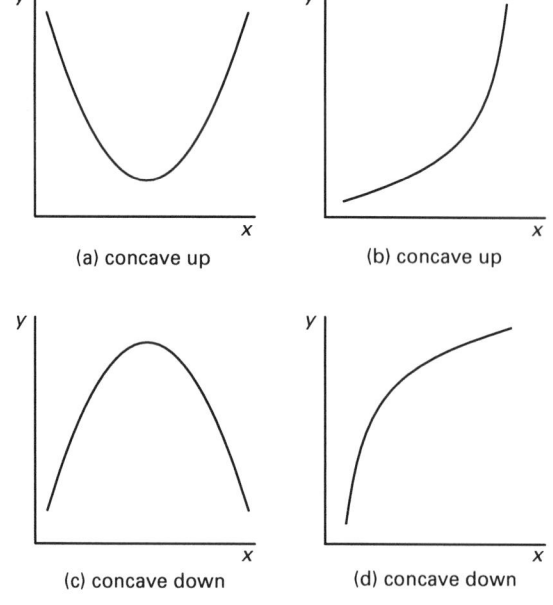

Figure 7.3 *Concave Curves*

5. CONVEX REGIONS

Convexity is a term that is applied to sets and regions.[2] It plays an important role in many mathematics subjects. A set or multidimensional region is *convex* if it contains the line segment joining any two of its points; that is, if a straight line is drawn connecting any two points in a convex region, that line will lie entirely within the region. For example, the interior of a parabola is a convex region, as is a solid sphere. The *void* or *null region* (i.e., an empty set of points), single points, and straight lines are convex sets. A convex region bounded by separate, connected line segments is known as a *convex hull*. (See Fig. 7.4.)

Within a convex region, a local maximum is also the global maximum. Similarly, a local minimum is also the global minimum. (See Sec. 8.3.) The intersection of two convex regions is also convex.

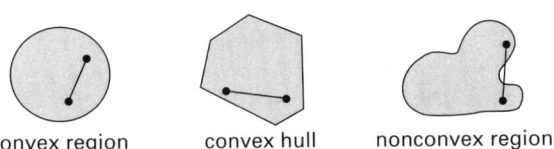

convex region   convex hull   nonconvex region

Figure 7.4 *Convexity*

6. CONGRUENCY

Congruence in geometric figures is analogous to *equality* in algebraic expressions. Congruent line segments are segments that have the same length. Congruent angles have the same angular measure. Congruent triangles have the same vertex angles and side lengths.

In general, *congruency*, indicated by the symbol ≅, means that there is one-to-one correspondence between all points on two objects. This correspondence is defined by the *mapping function* or *isometry*, which can be a translation, rotation, or reflection. Since the identity function is a valid mapping function, every geometric shape is congruent to itself.

Two congruent objects can be in different spaces. For example, a triangular area in three-dimensional space can be mapped into a triangle in two-dimensional space.

---

[2]It is tempting to define regions that fail the convexity test as being "concave." However, it is more proper to define such regions as "nonconvex." In any case, it is important to recognize that convexity depends on the reference point: an observer within a sphere will see the spherical boundary as convex; an observer outside the sphere may see the boundary as nonconvex.

7. COORDINATE SYSTEMS

The manner in which a geometric figure is described depends on the coordinate system that is used. The three-dimensional *rectangular coordinate system* (also known as the *Cartesian coordinate system*) with its x-, y-, and z-coordinates is the most commonly used in engineering. Table 7.1 summarizes the components needed to specify a point in the various coordinate systems. Figure 7.5 illustrates the use of and conversion between the coordinate systems.

Table 7.1 *Components of Coordinate Systems*

name	dimensions	components
rectangular	2	x, y
rectangular	3	x, y, z
polar	2	r, θ
cylindrical	3	r, θ, z
spherical	3	r, θ, ϕ

8. CURVES

A *curve* (commonly called a *line*) is a function over a finite or infinite range of the independent variable. When a curve is drawn in two- or three-dimensional space, it is known as a *graph of the curve*. It may or may not be possible to describe the curve mathematically. The *degree of a curve* is the highest exponent in the function. For example, Eq. 7.3 is a fourth-degree curve.

$$f(x) = 2x^4 + 7x^3 + 6x^2 + 3x + 9 = 0 \qquad 7.3$$

An *ordinary cycloid* (Fig. 7.6) is a curve traced out by a point on the rim of a wheel that rolls without slipping. It is described in parametric form by Eqs. 7.4 and 7.5 and by Eq. 7.6. In Eqs. 7.4 through 7.6, using the minus sign results in a cusp at the origin; using the plus sign results in a vertex (trough) at the origin.

$$x = r(\theta \pm \sin\theta) \qquad 7.4$$
$$y = r(1 \pm \cos\theta) \qquad 7.5$$
$$x = r \arccos\left(\frac{r-y}{r}\right) \pm \sqrt{2ry - y^2} \qquad 7.6$$

An *epicycloid* is a curve generated by a point on the rim of a wheel that rolls on the outside of a circle. A *hypocycloid* is a curve generated by a point on the rim of a wheel that rolls on the inside of a circle. The equation of a hypocycloid of four cusps is

$$x^{\frac{2}{3}} + y^{\frac{2}{3}} = r^{\frac{2}{3}} \qquad \text{[4 cusps]} \qquad 7.7$$

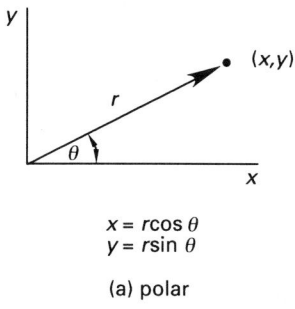

$$x = r\cos\theta$$
$$y = r\sin\theta$$

(a) polar

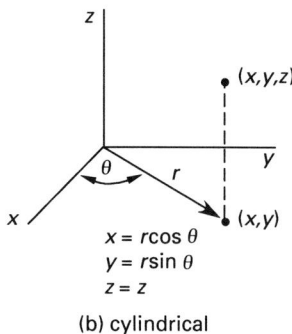

$$x = r\cos\theta$$
$$y = r\sin\theta$$
$$z = z$$

(b) cylindrical

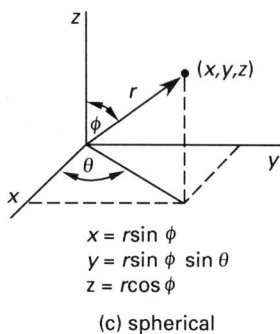

$$x = r\sin\phi$$
$$y = r\sin\phi\,\sin\theta$$
$$z = r\cos\phi$$

(c) spherical

Figure 7.5 *Different Coordinate Systems*

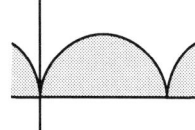

Figure 7.6 *Cycloid (cusp at origin)*

9. SYMMETRY OF CURVES

Two points, P and Q, are symmetrical with respect to a line if the line is a perpendicular bisector of the line segment PQ. If the graph of a curve is unchanged when y is replaced with $-y$, the curve is symmetrical with respect to the x-axis. If the curve is unchanged when x is replaced with $-x$, the curve is symmetrical with respect to the y-axis.

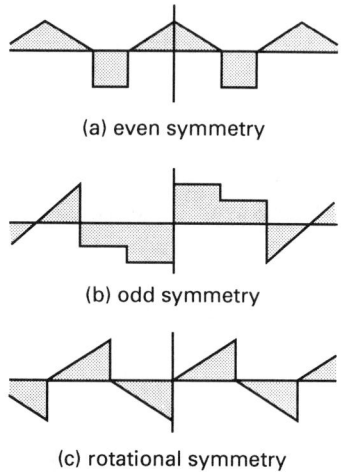

(a) even symmetry

(b) odd symmetry

(c) rotational symmetry

Figure 7.7 *Waveform Symmetry*

Repeating waveforms can be symmetrical with respect to the y-axis. A curve $f(x)$ is said to have *even symmetry* if $f(x) = f(-x)$. (Alternatively, $f(x)$ is said to be a *symmetrical function*.) With even symmetry, the function to the left of $x = 0$ is a reflection of the function to the right of $x = 0$. (In effect, the y-axis is a mirror.) The cosine curve is an example of a curve with even symmetry.

A curve is said to have *odd symmetry* if $f(x) = -f(-x)$. (Alternatively, $f(x)$ is said to be an *asymmetrical function*.)[3] The sine curve is an example of a curve with odd symmetry.

A curve is said to have *rotational symmetry (half-wave symmetry)* if $f(x) = -f(x + \pi)$.[4] Curves of this type are identical except for a sign reversal on alternate half-cycles.

Table 7.2 describes the type of function resulting from the combination of two functions.

Table 7.2 *Combinations of Functions*

	operation			
	$+$	$-$	$\times$	$\div$
$f_1(x)$ even, $f_2(x)$ even	even	even	even	even
$f_1(x)$ odd, $f_2(x)$ odd	odd	odd	even	even
$f_1(x)$ even, $f_2(x)$ odd	neither	neither	odd	odd

10. STRAIGHT LINES

Figure 7.8 illustrates a straight line in two-dimensional space. The *slope* of the line is m, the *y-intercept* is b, and the *x-intercept* is a. The equation of the line

---

[3]The semantics of "even symmetry" and "asymmetrical function" are contradictory.
[4]The symbol π represents half of a full cycle of the waveform, not the value $3.141\ldots$.

can be represented in several forms, and the procedure for finding the equation depends on the form chosen to represent the line. In general, the procedure involves substituting one or more known points on the line into the equation in order to determine the coefficients.

- *general form*:

$$Ax + By + C = 0 \qquad 7.8$$

$$A = -mB \qquad 7.9$$

$$B = \frac{-C}{b} \qquad 7.10$$

$$C = -aA = -bB \qquad 7.11$$

- *slope-intercept form*:

$$y = mx + b \qquad 7.12$$

$$m = \frac{-A}{B} = \tan\theta = \frac{y_2 - y_1}{x_2 - x_1} \qquad 7.13$$

$$b = \frac{-C}{B} \qquad 7.14$$

$$a = \frac{-C}{A} \qquad 7.15$$

- *point-slope form*:

$$y - y_1 = m(x - x_1) \qquad 7.16$$

- *intercept form*:

$$\frac{x}{a} + \frac{y}{b} = 1 \qquad 7.17$$

- *two-point form*:

$$\frac{y - y_1}{x - x_1} = \frac{y_2 - y_1}{x_2 - x_1} \qquad 7.18$$

- *normal form*:

$$x(\cos\beta) + y(\sin\beta) - d = 0 \qquad 7.19$$

(d and β are constants; x and y are variables.)

- *polar form*:

$$r = \frac{d}{\cos(\beta - \alpha)} \qquad 7.20$$

(d and β are constants; r and α are variables.)

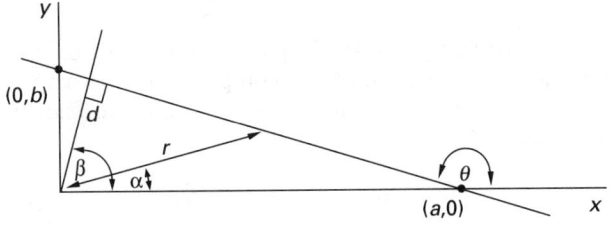

Figure 7.8 *Straight Line*

11. DIRECTION NUMBERS, ANGLES, AND COSINES

Given a directed line from (x_1, y_1, z_1) to (x_2, y_2, z_2), the *direction numbers* are

$$L = x_2 - x_1 \qquad 7.21$$

$$M = y_2 - y_1 \qquad 7.22$$

$$N = z_2 - z_1 \qquad 7.23$$

The distance between two points is

$$d = \sqrt{L^2 + M^2 + N^2} \qquad 7.24$$

The *direction cosines* are

$$\cos\alpha = \frac{L}{d} \qquad 7.25$$

$$\cos\beta = \frac{M}{d} \qquad 7.26$$

$$\cos\gamma = \frac{N}{d} \qquad 7.27$$

Note that

$$\cos^2\alpha + \cos^2\beta + \cos^2\gamma = 1 \qquad 7.28$$

The *direction angles* are the angles between the axes and the lines. They are found from the inverse functions of the direction cosines.

$$\alpha = \arccos\left(\frac{L}{d}\right) \qquad 7.29$$

$$\beta = \arccos\left(\frac{M}{d}\right) \qquad 7.30$$

$$\gamma = \arccos\left(\frac{N}{d}\right) \qquad 7.31$$

The direction cosines can be used to write the equation of the straight line in terms of the unit vectors. The line **R** would be defined as

$$\mathbf{R} = \mathbf{i}\cos\alpha + \mathbf{j}\cos\beta + \mathbf{k}\cos\gamma \qquad 7.32$$

Similarly, the line may be written in terms of its direction numbers.

$$\mathbf{R} = L\mathbf{i} + M\mathbf{j} + N\mathbf{k} \qquad 7.33$$

Example 7.2

A line passes through the points $(4, 7, 9)$ and $(0, 1, 6)$. Write the equation of the line in terms of its (a) direction numbers and (b) direction cosines.

Solution

(a) The direction numbers are

$$L = 4 - 0 = 4$$
$$M = 7 - 1 = 6$$
$$N = 9 - 6 = 3$$

Using Eq. 7.33,

$$\mathbf{R} = 4\mathbf{i} + 6\mathbf{j} + 3\mathbf{k}$$

(b) The distance between the two points is

$$d = \sqrt{(4)^2 + (6)^2 + (3)^2} = 7.81$$

The line in terms of its direction cosines is

$$\mathbf{R} = \frac{4\mathbf{i} + 6\mathbf{j} + 3\mathbf{k}}{7.81}$$

$$= 0.512\mathbf{i} + 0.768\mathbf{j} + 0.384\mathbf{k}$$

12. INTERSECTION OF TWO LINES

The intersection of two lines is a point. The location of the intersection point can be determined by setting the two equations equal and solving them in terms of a common variable. Alternatively, Eqs. 7.34 and 7.35 can be used to calculate the coordinates of the intersection point.

$$x = \frac{B_2 C_1 - B_1 C_2}{A_2 B_1 - A_1 B_2} \qquad 7.34$$

$$y = \frac{A_1 C_2 - A_2 C_1}{A_2 B_1 - A_1 B_2} \qquad 7.35$$

13. PLANES

A *plane* in three-dimensional space is completely determined by one of the following:

- three noncollinear points

- two nonparallel vectors $\mathbf{V}_1$ and $\mathbf{V}_2$ and their intersection point P_0

- a point P_0 and a vector, $\mathbf{N}$, normal to the plane (i.e., the *normal vector*)

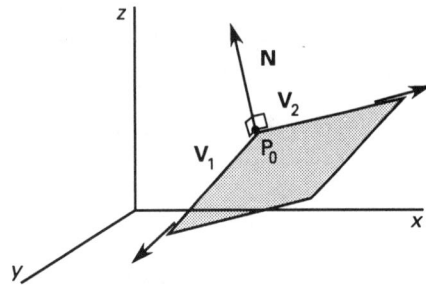

Figure 7.9 *Plane in Three-Dimensional Space*

The plane can be specified mathematically in one of two ways: as a rectangular or as a parametric equation. The general form is

$$A(x - x_0) + B(y - y_0) + C(z - z_0) = 0 \qquad 7.36$$

x_0, y_0, and z_0 are the coordinates of the intersection point of any two vectors in the plane. The coefficients A, B, and C are the same as the coefficients of the normal vector, $\mathbf{N}$.

$$\mathbf{N} = \mathbf{V}_1 \times \mathbf{V}_2 = A\mathbf{i} + B\mathbf{j} + C\mathbf{k} \qquad 7.37$$

Equation 7.36 can be simplified:

$$Ax + By + Cz + D = 0 \qquad 7.38$$
$$D = -(Ax_0 + By_0 + Cz_0) \qquad 7.39$$

The following procedure can be used to determine the equation of a plane from three noncollinear points, P_1, P_2, and P_3, or from a normal vector and a single point.

step 1:   (If the normal vector is known, go to step 3.) Determine the equations of the vectors $\mathbf{V}_1$ and $\mathbf{V}_2$ from two pairs of the points. For example, determine $\mathbf{V}_1$ from points P_1 and P_2, and determine $\mathbf{V}_2$ from P_1 and P_3. Express the vectors in the form $A\mathbf{i} + B\mathbf{j} + C\mathbf{k}$.

$$\mathbf{V}_1 = (x_2 - x_1)\mathbf{i} + (y_2 - y_1)\mathbf{j}$$
$$+ (z_2 - z_1)\mathbf{k} \qquad 7.40$$
$$\mathbf{V}_2 = (x_3 - x_1)\mathbf{i} + (y_3 - y_1)\mathbf{j}$$
$$+ (z_3 - z_1)\mathbf{k} \qquad 7.41$$

step 2:   Find the normal vector, $\mathbf{N}$, as the cross product of the two vectors.

$$\mathbf{N} = \mathbf{V}_1 \times \mathbf{V}_2$$
$$= \begin{vmatrix} \mathbf{i} & (x_2 - x_1) & (x_3 - x_1) \\ \mathbf{j} & (y_2 - y_1) & (y_3 - y_1) \\ \mathbf{k} & (z_2 - z_1) & (z_3 - z_1) \end{vmatrix} \qquad 7.42$$

step 3:   Write the general equation of the plane in rectangular form (Eq. 7.36) using the coefficients A, B, and C from the normal vector and any one of the three points as P_0.

The parametric equations of a plane can be written also as a linear combination of the components of two vectors in the plane. Referring to Fig. 7.9, the two known vectors are

$$\mathbf{V}_1 = V_{1x}\mathbf{i} + V_{1y}\mathbf{j} + V_{1z}\mathbf{k} \qquad 7.43$$
$$\mathbf{V}_2 = V_{2x}\mathbf{i} + V_{2y}\mathbf{j} + V_{2z}\mathbf{k} \qquad 7.44$$

If s and t are scalars, the coordinates of each point in the plane can be written as Eqs. 7.45 through 7.47. These are the parametric equations of the plane.

$$x = x_0 + sV_{1x} + tV_{2x} \qquad \textit{7.45}$$
$$y = y_0 + sV_{1y} + tV_{2y} \qquad \textit{7.46}$$
$$z = z_0 + sV_{1z} + tV_{2z} \qquad \textit{7.47}$$

Example 7.3

The following points are coplanar.

$$P_1 = (2, 1, -4)$$
$$P_2 = (4, -2, -3)$$
$$P_3 = (2, 3, -8)$$

Determine the equation of the plane in (a) general form and (b) parametric form.

Solution

(a) Use the first two points to find a vector, $\mathbf{V}_1$.

$$\begin{aligned}
\mathbf{V}_1 &= (x_2 - x_1)\mathbf{i} + (y_2 - y_1)\mathbf{j} + (z_2 - z_1)\mathbf{k} \\
&= (4 - 2)\mathbf{i} + (-2 - 1)\mathbf{j} + (-3 - (-4))\mathbf{k} \\
&= 2\mathbf{i} - 3\mathbf{j} + 1\mathbf{k}
\end{aligned}$$

Similarly, use the first and third points to find $\mathbf{V}_2$.

$$\begin{aligned}
\mathbf{V}_2 &= (x_3 - x_1)\mathbf{i} + (y_3 - y_1)\mathbf{j} + (z_3 - z_1)\mathbf{k} \\
&= (2 - 2)\mathbf{i} + (3 - 1)\mathbf{j} + (-8 - (-4))\mathbf{k} \\
&= 0\mathbf{i} + 2\mathbf{j} - 4\mathbf{k}
\end{aligned}$$

From Eq. 7.42, determine the normal vector as the determinant.

$$\mathbf{N} = \begin{vmatrix} \mathbf{i} & 2 & 0 \\ \mathbf{j} & -3 & 2 \\ \mathbf{k} & 1 & -4 \end{vmatrix}$$

Expand the determinant across the top row.

$$\begin{aligned}
\mathbf{N} &= \mathbf{i}(12 - 2) - 2(-4\mathbf{j} - 2\mathbf{k}) \\
&= 10\mathbf{i} + 8\mathbf{j} + 4\mathbf{k}
\end{aligned}$$

The rectangular form of the equation of the plane uses the same constants as in the normal vector. Use the first point and write the equation of the plane in the form of Eq. 7.36.

$$10(x - 2) + 8(y - 1) + 4(z + 4) = 0$$

The three constant terms can be combined by using Eq. 7.39.

$$D = -[(10)(2) + (8)(1) + (4)(-4)] = -12$$

The equation of the plane is

$$10x + 8y + 4z - 12 = 0$$

(b) The parametric equations based on the first point and for any values of s and t are

$$x = 2 + 2s + 0t$$
$$y = 1 - 3s + 2t$$
$$z = -4 + 1s - 4t$$

The scalars s and t are not unique. Two of the three coordinates can also be chosen as the parameters. Dividing the rectangular form of the plane's equation by 4 to isolate z results in an alternate set of parametric equations.

$$x = x$$
$$y = y$$
$$z = 3 - 2.5x - 2y$$

14. DISTANCES BETWEEN GEOMETRIC FIGURES

The smallest distance, d, between various geometric figures is given by the following equations.

- between two points in (x, y, z) format:
$$d = \sqrt{(x_2 - x_1)^2 + (y_2 - y_1)^2 + (z_2 - z_1)^2} \quad \textit{7.48}$$

- between a point (x_0, y_0) and a line $Ax + By + C = 0$:
$$d = \frac{|Ax_0 + By_0 + C|}{\sqrt{A^2 + B^2}} \qquad \textit{7.49}$$

- between a point (x_0, y_0, z_0) and a plane $Ax + By + Cz + D = 0$:
$$d = \frac{|Ax_0 + By_0 + Cz_0 + D|}{\sqrt{A^2 + B^2 + C^2}} \qquad \textit{7.50}$$

- between two parallel lines $Ax + By + C = 0$:
$$d = \left| \frac{|C_2|}{\sqrt{A_2^2 + B_2^2}} - \frac{|C_1|}{\sqrt{A_1^2 + B_1^2}} \right| \qquad \textit{7.51}$$

Example 7.4

What is the minimum distance between the line $y = 2x + 3$ and the origin $(0,0)$?

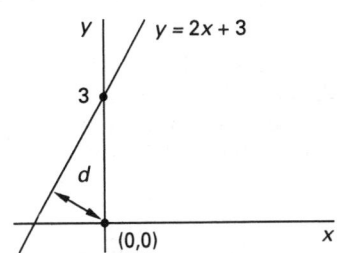

Solution

Put the equation in general form.

$$Ax + By + C = 2x - y + 3 = 0$$

Use Eq. 7.49 with $(x, y) = (0, 0)$.

$$d = \frac{|Ax + By + C|}{\sqrt{A^2 + B^2}} = \frac{(2)(0) + (-1)(0) + 3}{\sqrt{(2)^2 + (-1)^2}}$$

$$= \frac{3}{\sqrt{5}}$$

15. ANGLES BETWEEN GEOMETRIC FIGURES

The angle, ϕ, between various geometric figures is given by the following equations.

- between two lines in $Ax + By + C = 0$, $y = mx + b$, or direction angle formats,

$$\phi = \arctan\left(\frac{A_1 B_2 - A_2 B_1}{A_1 A_2 + B_1 B_2}\right) \qquad 7.52$$

$$= \arctan\left(\frac{m_2 - m_1}{1 + m_1 m_2}\right) \qquad 7.53$$

$$= |\arctan(m_1) - \arctan(m_2)| \qquad 7.54$$

$$= \arccos\left(\frac{L_1 L_2 + M_1 M_2 + N_1 N_2}{d_1 d_2}\right) \qquad 7.55$$

$$= \arccos(\cos\alpha_1 \cos\alpha_2 + \cos\beta_1 \cos\beta_2 + \cos\gamma_1 \cos\gamma_2) \qquad 7.56$$

If the lines are parallel, then $\phi = 0$.

$$\frac{A_1}{A_2} = \frac{B_1}{B_2} \qquad 7.57$$

$$m_1 = m_2 \qquad 7.58$$

$$\alpha_1 = \alpha_2;\ \beta_1 = \beta_2;\ \gamma_1 = \gamma_2 \qquad 7.59$$

If the lines are perpendicular, then $\phi = 90°$.

$$A_1 A_2 = -B_1 B_2 \qquad 7.60$$

$$m_1 = \frac{-1}{m_2} \qquad 7.61$$

$$\alpha_1 + \alpha_2 = \beta_1 + \beta_2 = \gamma_1 + \gamma_2 = 90° \qquad 7.62$$

- between two planes in $A\mathbf{i} + B\mathbf{j} + C\mathbf{k} = 0$ format, the coefficients A, B, and C are the same as the coefficients for the normal vector. (See Eq. 7.37.) ϕ is equal to the angle between the two normal vectors.

$$\cos\phi = \frac{|A_1 A_2 + B_1 B_2 + C_1 C_2|}{\sqrt{A_1^2 + B_1^2 + C_1^2}\sqrt{A_2^2 + B_2^2 + C_2^2}} \qquad 7.63$$

Example 7.5

Use Eqs. 7.52, 7.53, and 7.54 to find the angle between the lines.

$$y = -0.577x + 2$$

$$y = +0.577x - 5$$

Solution

(a) Write both equations in general form.

$$-0.577x - y + 2 = 0$$

$$0.577x - y - 5 = 0$$

(b) Use Eq. 7.53.

$$\phi = \arctan\left(\frac{m_2 - m_1}{1 + m_1 m_2}\right)$$

$$= \arctan\left(\frac{0.577 - (-0.577)}{1 + (0.577)(-0.577)}\right) = 60°$$

From Eq. 7.52,

$$\phi = \arctan\left[\frac{A_1 B_2 - A_2 B_1}{A_1 A_2 + B_1 B_2}\right]$$

$$= \arctan\left[\frac{(-0.577)(-1) - (0.577)(-1)}{(-0.577)(0.577) + (-1)(-1)}\right] = 60°$$

(c) Use Eq. 7.54.

$$\phi = |\arctan(m_1) - \arctan(m_2)|$$

$$= |\arctan(-0.577) - \arctan(0.577)|$$

$$= |-30° - 30°| = 60°$$

16. CONIC SECTIONS

A *conic section* is any one of several curves produced by passing a plane through a cone as shown in Fig. 7.10. If α is the angle between the vertical axis and the cutting plane and β is the cone generating angle, Eq. 7.64 gives the *eccentricity*, ϵ, of the conic section. Values of the eccentricity are given in Fig. 7.10.

$$\epsilon = \frac{\cos\alpha}{\cos\beta} \qquad 7.64$$

All conic sections are described by second-degree polynomials (i.e., are *quadratic equations*) of the form[5]

$$Ax^2 + Bxy + Cy^2 + Dx + Ey + F = 0 \qquad 7.65$$

---

[5]One or more straight lines are produced when the cutting plane passes through the cone's vertex. Straight lines can be considered to be quadratic functions without second-degree terms.

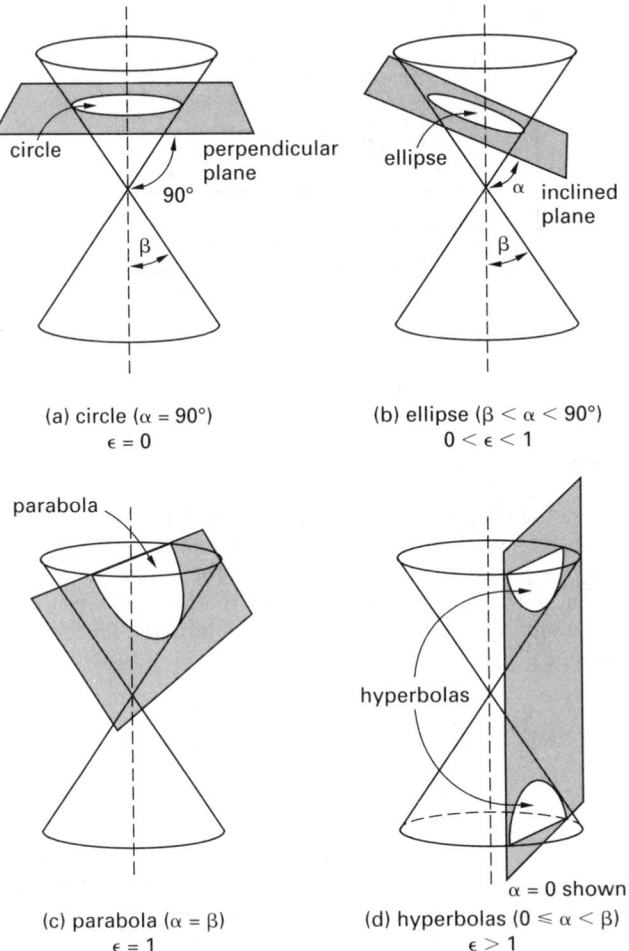

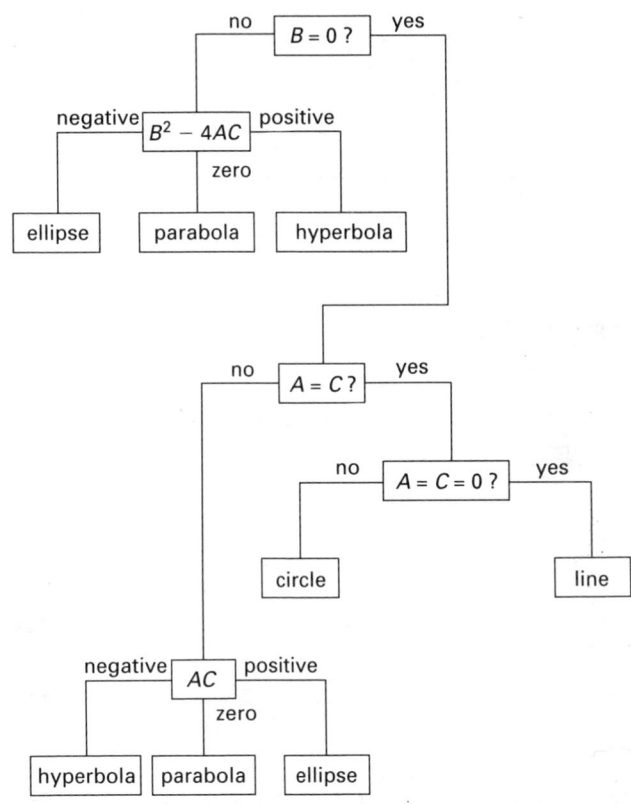

(a) circle ($\alpha = 90°$)
$\epsilon = 0$

(b) ellipse ($\beta < \alpha < 90°$)
$0 < \epsilon < 1$

(c) parabola ($\alpha = \beta$)
$\epsilon = 1$

(d) hyperbolas ($0 \leq \alpha < \beta$)
$\epsilon > 1$

Figure 7.10 *Conic Sections Produced by Cutting Planes*

Figure 7.11 *Determining Conic Sections from Quadratic Equations*

This is the *general form*, which allows the figure axes to be at any angle relative to the coordinate axes. The *standard forms* presented in the following sections pertain to figures whose axes coincide with the coordinate axes, thereby eliminating certain terms of the general equation.

Figure 7.11 can be used to determine which conic section is described by the quadratic function. The quantity $B^2 - 4AC$ is known as the *discriminant*. Figure 7.11 determines only the type of conic section; it does not determine whether the conic section is degenerate (e.g., a circle with a negative radius).

Example 7.6

What geometric figures are described by the following equations?

(a) $4y^2 - 12y + 16x + 41 = 0$

(b) $x^2 - 10xy + y^2 + x + y + 1 = 0$

(c) $x^2 + 4y^2 + 2x - 8y + 1 = 0$

(d) $x^2 + y^2 - 6x + 8y + 20 = 0$

Solution

(a) Referring to Fig. 7.11, $B = 0$ since there is no xy term. $A = 0$ since there is no x^2 term, and $AC = (0)(4) = 0$. This is a parabola.

(b) $B \neq 0$; $B^2 - 4AC = (-10)^2 - (4)(1)(1) = +96$. This is a hyperbola.

(c) $B = 0$; $A \neq C$; $AC = (1)(4) = +4$. This is an ellipse.

(d) $B = 0$; $A = C$; $A = C = 1 \, (\neq 0)$. This is a circle.

17. CIRCLE

The general form of the equation of a circle is

$$Ax^2 + Ay^2 + Dx + Ey + F = 0 \qquad 7.66$$

The *center-radius form* of the equation of a circle with radius r and center at (h, k) is

$$(x - h)^2 + (y - k)^2 = r^2 \qquad 7.67$$

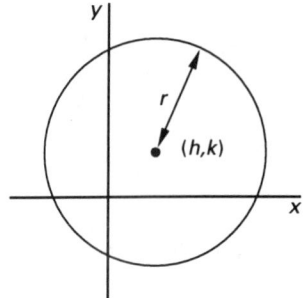

Figure 7.12 Circle

The two forms can be converted by use of Eqs. 7.68 through 7.70.

$$h = \frac{-D}{2A} \qquad 7.68$$

$$k = \frac{-E}{2A} \qquad 7.69$$

$$r^2 = \frac{D^2 + E^2 - 4AF}{4A^2} \qquad 7.70$$

If the right-hand side of Eq. 7.70 is positive, the figure is a circle. If it is zero, the circle shrinks to a point. If the right-hand side is negative, the figure is imaginary. A *degenerate circle* is one in which the right-hand side is less than or equal to zero.

18. PARABOLA

A *parabola* is the locus of points equidistant from the *focus* (point F in Fig. 7.13) and a line called the *directrix*. A parabola is symmetric with respect to its *parabolic axis*. The line normal to the parabolic axis and passing through the focus is known as the *latus rectum*. The eccentricity of a parabola is 1.

There are two common types of parabola in the Cartesian plane—those that open right and left, and those that open up and down. Equation 7.65 is the general form of the equation of a parabola. With Eq. 7.71, the parabola points to the right if $CD > 0$ and to the left if $CD < 0$. With Eq. 7.72, the parabola points up if $AE > 0$ and down if $AE < 0$.

$$Cy^2 + Dx + Ey + F = 0 \Big|_{\substack{C,D \neq 0 \\ \text{opens horizontally}}} \qquad 7.71$$

$$Ax^2 + Dx + Ey + F = 0 \Big|_{\substack{A,E \neq 0 \\ \text{opens vertically}}} \qquad 7.72$$

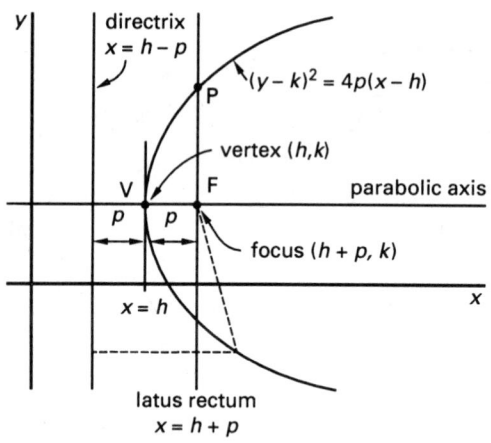

Figure 7.13 Parabola

The *standard form* of the equation of a parabola with vertex at (h, k), focus at $(h + p, k)$, and directrix at $x = h - p$ and that opens to the right or left is given by Eq. 7.73. The parabola opens to the right (points to the left) if $p > 0$ and opens to the left (points to the right) if $p < 0$.

$$(y - k)^2 = 4p(x - h) \Big|_{\text{opens horizontally}} \qquad 7.73$$

$$y^2 = 4px \Big|_{\substack{\text{vertex at origin} \\ h = k = 0}} \qquad 7.74$$

The *standard form* of the equation of a parabola with vertex at (h, k), focus at $(h, k + p)$, and directrix at $y = k - p$ and that opens up or down is given by Eq. 7.75. The parabola opens up (points down) if $p > 0$ and opens down (points up) if $p < 0$.

$$(x - h)^2 = 4p(y - k) \Big|_{\text{opens vertically}} \qquad 7.75$$

$$x^2 = 4py \Big|_{\text{vertex at origin}} \qquad 7.76$$

The general and vertex forms of the equations can be reconciled with Eqs. 7.77 through 7.79. Whether the first or second forms of these equations are used depends on whether the parabola opens to the sides or vertically (i.e., whether $A = 0$ or $C = 0$), respectively.

$$h = \begin{cases} \dfrac{E^2 - 4CF}{4CD} & \text{[opens horizontally]} \\[2mm] \dfrac{-D}{2A} & \text{[opens vertically]} \end{cases} \qquad 7.77$$

$$k = \begin{cases} \dfrac{-E}{2C} & \text{[opens horizontally]} \\[2mm] \dfrac{D^2 - 4AF}{4AE} & \text{[opens vertically]} \end{cases} \qquad 7.78$$

$$p = \begin{cases} \dfrac{-D}{4C} & \text{[opens horizontally]} \\[2mm] \dfrac{-E}{4A} & \text{[opens vertically]} \end{cases} \qquad 7.79$$

19. ELLIPSE

An *ellipse* has two foci separated along the *major axis* by a distance $2c$. The line perpendicular to the major axis passing through the center of the ellipse is the *minor axis*. The lines perpendicular to the major axis passing through the foci are the *latus recta*. The distance between the two vertices is $2a$. The ellipse is the locus of points such that the sum of the distances from the two foci is $2a$. Referring to Fig. 7.14,

$$F_1P + PF_2 = 2a \qquad 7.80$$

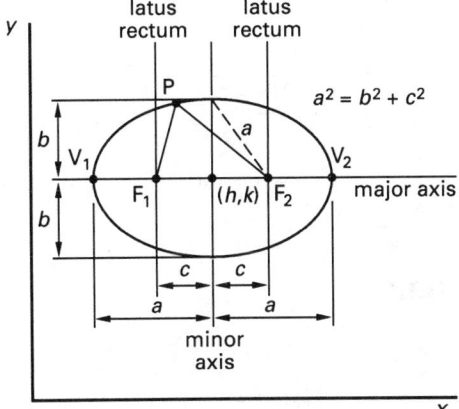

Figure 7.14 *Ellipse*

Equation 7.81 is the standard equation for an ellipse with axes parallel to the coordinate axes. (Equation 7.65 is the general form.) F is not independent of A, C, D, and E for the ellipse.

$$Ax^2 + Cy^2 + Dx + Ey + F = 0\Big|_{\substack{AC > 0 \\ A \neq C}} \qquad 7.81$$

Equation 7.82 gives the standard form of the equation of an ellipse centered at (h, k). Distances a and b are known as the *semimajor distance* and *semiminor distance*, respectively.

$$\frac{(x-h)^2}{a^2} + \frac{(y-k)^2}{b^2} = 1 \qquad 7.82$$

The distance between the two foci is $2c$.

$$2c = 2\sqrt{a^2 - b^2} \qquad 7.83$$

The *aspect ratio* of the ellipse is

$$\text{aspect ratio} = \frac{a}{b} \qquad 7.84$$

The *eccentricity*, ϵ, of the ellipse is always less than 1. If the eccentricity is zero, the figure is a circle (another form of a *degenerative ellipse*).

$$\epsilon = \frac{\sqrt{a^2 - b^2}}{a} < 1 \qquad 7.85$$

The standard and center forms of the equations of an ellipse can be reconciled by using Eqs. 7.86 through 7.89.

$$h = \frac{-D}{2A} \qquad 7.86$$

$$k = \frac{-E}{2C} \qquad 7.87$$

$$a = \sqrt{C} \qquad 7.88$$

$$b = \sqrt{A} \qquad 7.89$$

20. HYPERBOLA

A *hyperbola* has two foci separated along the *transverse axis* by a distance $2c$. Lines perpendicular to the transverse axis passing through the foci are the *conjugate axes*. The distance between the two vertices is $2a$, and the distance along a conjugate axis between two points on the hyperbola is $2b$. The hyperbola is the locus of points such that the difference in distances from the two foci is $2a$. Referring to Fig. 7.15,

$$F_2P - PF_1 = 2a \qquad 7.90$$

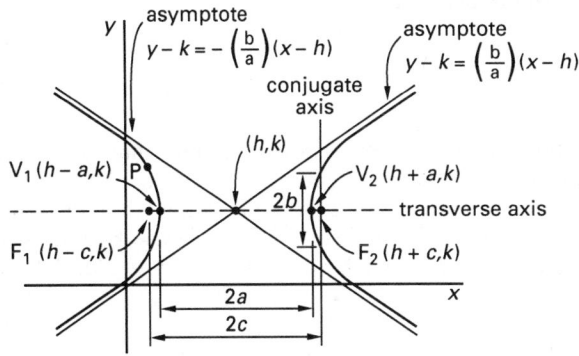

Figure 7.15 *Hyperbola*

Equation 7.91 is the standard equation of a hyperbola. Coefficients A and C must have opposite signs.

$$Ax^2 + Cy^2 + Dx + Ey + F = 0\big|_{AC < 0} \qquad 7.91$$

Equation 7.92 gives the standard form of the equation of a hyperbola centered at (h, k) and opening to the left and right.

$$\frac{(x-h)^2}{a^2} - \frac{(y-k)^2}{b^2} = 1\Big|_{\text{opens horizontally}} \qquad 7.92$$

Equation 7.93 gives the standard form of the equation of a hyperbola centered at (h, k) and opening up and down.

$$\frac{(y-k)^2}{a^2} - \frac{(x-h)^2}{b^2} = 1\Big|_{\text{opens vertically}} \qquad 7.93$$

The distance between the two foci is $2c$.

$$2c = 2\sqrt{a^2 + b^2} \qquad 7.94$$

The *eccentricity*, ϵ, of the hyperbola is calculated from Eq. 7.95 and is always greater than 1.

$$\epsilon = \frac{c}{a} = \frac{\sqrt{a^2 + b^2}}{a} > 1 \qquad 7.95$$

The hyperbola is asymptotic to the lines given by Eqs. 7.96 and 7.97.

$$y = \pm \frac{b}{a}(x - h) + k \bigg|_{\text{opens horizontally}} \qquad 7.96$$

$$y = \pm \frac{a}{b}(x - h) + k \bigg|_{\text{opens vertically}} \qquad 7.97$$

For a *rectangular (equilateral) hyperbola*, the asymptotes are perpendicular, $a = b$, $c = \sqrt{2}a$, and the eccentricity is $\epsilon = \sqrt{2}$. If the hyperbola is centered at the origin (i.e., $h = k = 0$), then the equations are $x^2 - y^2 = a^2$ (opens horizontally) and $y^2 - x^2 = a^2$ (opens vertically).

If the asymptotes are the x- and y-axes, the equation of the hyperbola is simply

$$xy = \pm \frac{a^2}{2} \qquad 7.98$$

The general and center forms of the equations of a hyperbola can be reconciled by using Eqs. 7.99 through 7.103. Whether the hyperbola opens left and right or up and down depends on whether M/A or M/C is positive, respectively, where M is defined by Eq. 7.99.

$$M = \frac{D^2}{4A} + \frac{E^2}{4C} - F \qquad 7.99$$

$$h = \frac{-D}{2A} \qquad 7.100$$

$$k = \frac{-E}{2C} \qquad 7.101$$

$$a = \begin{cases} \sqrt{-C} & \text{[opens horizontally]} \\ \sqrt{-A} & \text{[opens vertically]} \end{cases} \qquad 7.102$$

$$b = \begin{cases} \sqrt{A} & \text{[opens horizontally]} \\ \sqrt{C} & \text{[opens vertically]} \end{cases} \qquad 7.103$$

21. SPHERE

Equation 7.104 is the general equation of a sphere. The coefficient A cannot be zero.

$$Ax^2 + Ay^2 + Az^2 + Bx + Cy + Dz + E = 0 \qquad 7.104$$

Equation 7.105 gives the standard form of the equation of a sphere centered at (h, k, l) with radius r.

$$(x - h)^2 + (y - k)^2 + (z - l)^2 = r^2 \qquad 7.105$$

The general and center forms of the equations of a sphere can be reconciled by using Eqs. 7.106 through 7.109.

$$h = \frac{-B}{2A} \qquad 7.106$$

$$k = \frac{-C}{2A} \qquad 7.107$$

$$l = \frac{-D}{2A} \qquad 7.108$$

$$r = \sqrt{\frac{B^2 + C^2 + D^2}{4A^2} - \frac{E}{A}} \qquad 7.109$$

Planes tangent to spheres and other solids are covered in Chap. 8, Sec. 7.

22. HELIX

A *helix* is a curve generated by a point moving on, around, and along a cylinder such that the distance the point moves parallel to the cylindrical axis is proportional to the angle of rotation about that axis. For a cylinder of radius r, Eqs. 7.110 through 7.112 define the three-dimensional positions of points along the helix. The quantity $2\pi k$ is the *pitch* of the helix.

$$x = r\cos\theta \qquad 7.110$$
$$y = r\sin\theta \qquad 7.111$$
$$z = k\theta \qquad 7.112$$

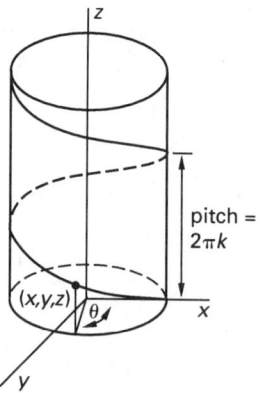

Figure 7.16 *Helix*

PRACTICE PROBLEMS

1. The diameter of a sphere and the base of a cone are equal. What percentage of that diameter must the cone's height be so that both volumes are equal?

8 Differential Calculus

1. Derivative of a Function 8-1
2. Elementary Derivative Operations 8-1
3. Critical Points 8-2
4. Derivatives of Parametric Equations 8-3
5. Partial Differentiation 8-4
6. Implicit Differentiation 8-4
7. Tangent Plane Function 8-5
8. Gradient Vector 8-5
9. Directional Derivative 8-6
10. Normal Line Vector 8-6
11. Divergence of a Vector Field 8-7
12. Curl of a Vector Field 8-7
13. Taylor's Formula 8-7
14. Common Series Approximations 8-8
    Practice Problems 8-8

1. DERIVATIVE OF A FUNCTION

In most cases, it is possible to transform a continuous function, $f(x_1, x_2, x_3, \ldots)$, of one or more independent variables into a derivative function.[1] In simple cases, the *derivative* can be interpreted as the slope (tangent or rate of change) of the curve described by the original function. Since the slope of the curve depends on x, the derivative function will also depend on x. The derivative, $f'(x)$, of a function $f(x)$ is defined mathematically by Eq. 8.1. However, limit theory is seldom needed to actually calculate derivatives.

$$f'(x) = \lim_{\triangle x \to 0} \frac{\triangle f(x)}{\triangle x} \qquad 8.1$$

The derivative of a function $f(x)$, also known as the *first derivative*, is written in various ways, including

$$f'(x), \frac{df(x)}{dx}, \frac{df}{dx}, \mathbf{D}f(x), \mathbf{D}_x f(x), \dot{f}(x), sf(x)$$

A *second derivative* may exist if the derivative operation is performed on the first derivative—that is, a derivative is taken of a derivative function. This is written as

$$f''(x), \frac{d^2 f(x)}{dx^2}, \frac{d^2 f}{dx^2}, \mathbf{D}^2 f(x), \mathbf{D}_x^2 f(x), \ddot{f}(x), s^2 f(x)$$

---

[1] A function, $f(x)$, of one independent variable, x, is used in this section to simplify the discussion. Although the derivative is taken with respect to x, the independent variable can be anything.

A *regular* (*analytic* or *holomorphic*) *function* possesses a derivative. A point at which a function's derivative is undefined is called a *singular point*, as Fig. 8.1 illustrates.

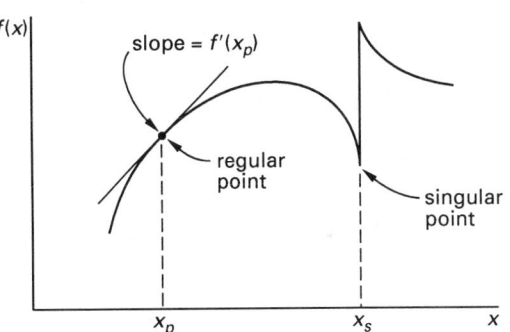

Figure 8.1 *Derivatives and Singular Points*

2. ELEMENTARY DERIVATIVE OPERATIONS

Equations 8.2 through 8.5 summarize the elementary derivative operations on polynomials and exponentials. Equations 8.2 and 8.3 are particularly useful. (a, n, and k represent constants. $f(x)$ and $g(x)$ are functions of x.)

$$\mathbf{D}k = 0 \qquad 8.2$$

$$\mathbf{D}x^n = nx^{n-1} \qquad 8.3$$

$$\mathbf{D}\ln x = \frac{1}{x} \qquad 8.4$$

$$\mathbf{D}e^{ax} = ae^{ax} \qquad 8.5$$

Equations 8.6 through 8.17 summarize the elementary derivative operations on transcendental (trigonometric) functions.

$$\mathbf{D}\sin x = \cos x \qquad 8.6$$

$$\mathbf{D}\cos x = -\sin x \qquad 8.7$$

$$\mathbf{D}\tan x = \sec^2 x \qquad 8.8$$

$$\mathbf{D}\cot x = -\csc^2 x \qquad 8.9$$

$$\mathbf{D}\sec x = \sec x \tan x \qquad 8.10$$

$$\mathbf{D}\csc x = -\csc x \cot x \qquad 8.11$$

$$\mathbf{D}\arcsin x = \frac{1}{\sqrt{1 - x^2}} \qquad 8.12$$

$$\mathbf{D}\arccos x = -\mathbf{D}\arcsin x \qquad 8.13$$

$$\mathbf{D}\arctan x = \frac{1}{1+x^2} \qquad 8.14$$

$$\mathbf{D}\text{arccot } x = -\mathbf{D}\arctan x \qquad 8.15$$

$$\mathbf{D}\text{arcsec } x = \frac{1}{x\sqrt{x^2-1}} \qquad 8.16$$

$$\mathbf{D}\text{arccsc } x = -\mathbf{D}\text{arcsec } x \qquad 8.17$$

Equations 8.18 through 8.23 summarize the elementary derivative operations on hyperbolic transcendental functions. Derivatives of hyperbolic functions are not completely analogous to those of the regular transcendental functions.

$$\mathbf{D}\sinh x = \cosh x \qquad 8.18$$

$$\mathbf{D}\cosh x = \sinh x \qquad 8.19$$

$$\mathbf{D}\tanh x = \text{sech}^2 x \qquad 8.20$$

$$\mathbf{D}\coth x = -\text{csch}^2 x \qquad 8.21$$

$$\mathbf{D}\text{sech } x = -\text{sech } x \tanh x \qquad 8.22$$

$$\mathbf{D}\text{csch } x = -\text{csch } x \coth x \qquad 8.23$$

Equations 8.24 through 8.29 summarize the elementary derivative operations on functions and combinations of functions.

$$\mathbf{D}kf(x) = k\mathbf{D}f(x) \qquad 8.24$$

$$\mathbf{D}(f(x) \pm g(x)) = \mathbf{D}f(x) \pm \mathbf{D}g(x) \qquad 8.25$$

$$\mathbf{D}(f(x)\cdot g(x)) = f(x)\mathbf{D}g(x) + g(x)\mathbf{D}f(x) \qquad 8.26$$

$$\mathbf{D}\left(\frac{f(x)}{g(x)}\right) = \frac{g(x)\mathbf{D}f(x) - f(x)\mathbf{D}g(x)}{(g(x))^2} \qquad 8.27$$

$$\mathbf{D}(f(x))^n = n(f(x))^{n-1}\mathbf{D}f(x) \qquad 8.28$$

$$\mathbf{D}f(g(x)) = \mathbf{D}_g f(g)\mathbf{D}_x g(x) \qquad 8.29$$

Example 8.1

What is the slope at $x = 3$ of the curve $f(x) = x^3 - 2x$?

Solution

The derivative function found from Eq. 8.3 determines the slope.

$$f'(x) = 3x^2 - 2$$

The slope at $x = 3$ is

$$f'(3) = (3)(3)^2 - 2 = 27 - 2 = 25$$

Example 8.2

What are the derivatives of the following functions?

(a) $f(x) = 5\sqrt[3]{x^5}$

(b) $f(x) = (\sin x)(\cos^2 x)$

(c) $f(x) = ln(\cos(e^x))$

Solution

(a) Using Eqs. 8.3 and 8.24,

$$f'(x) = 5\mathbf{D}\sqrt[3]{x^5} = 5\mathbf{D}\left[(x^5)^{\frac{1}{3}}\right]$$

$$= (5)\left(\tfrac{1}{3}\right)(x^5)^{-\frac{2}{3}}\mathbf{D}x^5$$

$$= (5)\left(\tfrac{1}{3}\right)(x^5)^{-\frac{2}{3}}(5)(x^4)$$

$$= \frac{25\,x^{\frac{2}{3}}}{3}$$

(b) Using Eq. 8.26,

$$f'(x) = \sin x\mathbf{D}\cos^2 x + \cos^2 x\mathbf{D}\sin x$$

$$= (\sin x)(2\cos x)\mathbf{D}\cos x + \cos^2 x \cos x$$

$$= (\sin x)(2\cos x)(-\sin x) + \cos^2 x \cos x$$

$$= -2\sin^2 x \cos x + \cos^3 x$$

(c) Using Eq. 8.29,

$$f'(x) = \left(\frac{1}{\cos(e^x)}\right)\mathbf{D}\cos(e^x)$$

$$= \left(\frac{1}{\cos(e^x)}\right)(-\sin(e^x))\mathbf{D}e^x$$

$$= \left(\frac{-\sin(e^x)}{\cos(e^x)}\right)e^x$$

$$= -e^x\tan(e^x)$$

3. CRITICAL POINTS

Derivatives are used to locate the local *critical points* of functions of one variable—that is, *extreme points* (also known as *maximum* and *minimum* points) as well as the *inflection points* (*points of contraflexure*). The plurals *extrema*, *maxima*, and *minima* are used without the word "points." These points are illustrated in Fig. 8.2. There is usually an inflection point between two adjacent local extrema.

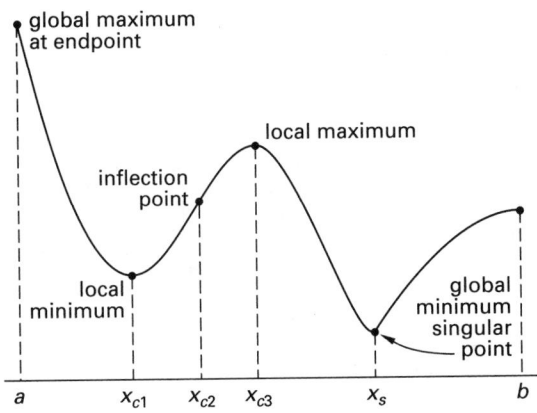

Figure 8.2 *Extreme and Inflection Points*

The first derivative is calculated to determine where the critical points are. The second derivative is calculated to determine whether a critical point is a local maximum, minimum, or inflection point, according to the following conditions. With this method, no distinction is made between local and global extrema. Therefore, the extrema should be compared with the function values at the endpoints of the interval, as illustrated in Ex. 8.3.[2] Note that $f'(x) \neq 0$ at an inflection point.

$$f'(x_c) = 0 \text{ at any extreme point, } x_c \qquad 8.30$$

$$f''(x_c) < 0 \text{ at a maximum point} \qquad 8.31$$

$$f''(x_c) > 0 \text{ at a minimum point} \qquad 8.32$$

$$f''(x_c) = 0 \text{ at an inflection point} \qquad 8.33$$

Example 8.3

Find the global extrema of the function $f(x)$ on the interval $[-2, +2]$.

$$f(x) = x^3 + x^2 - x + 1$$

Solution

The first derivative is

$$f'(x) = 3x^2 + 2x - 1$$

Since the first derivative is zero at extreme points, set $f'(x)$ equal to zero and solve for the roots of the quadratic equation.

$$3x^2 + 2x - 1 = (3x - 1)(x + 1) = 0$$

The roots are $x_1 = {}^1/_3$, $x_2 = -1$. These are the locations of the two extrema.

The second derivative is

$$f''(x) = 6x + 2$$

---

[2]It is also necessary to check the values of the function at singular points (i.e., points where the derivative does not exist).

Substituting x_1 and x_2 into $f''(x)$,

$$f''(x_1) = (6)\left(\tfrac{1}{3}\right) + 2 = 4$$

$$f''(x_2) = (6)(-1) + 2 = -4$$

Therefore, x_1 is a local minimum point (because $f''(x)$ is positive), and x_2 is a local maximum point (because $f''(x)$ is negative). The inflection point between these two extrema is found by setting $f''(x)$ equal to zero.

$$f''(x) = 6x + 2 = 0 \text{ or } x = -\tfrac{1}{3}$$

Since the question asked for the global extreme points, it is necessary to compare the values of $f(x)$ at the local extrema with the values at the endpoints.

$$f(-2) = -1$$

$$f(-1) = 2$$

$$f\left(\tfrac{1}{3}\right) = 22/27$$

$$f(2) = +11$$

Therefore, the actual global extrema are the endpoints.

4. DERIVATIVES OF PARAMETRIC EQUATIONS

The derivative of a function $f(x_1, x_2, \ldots x_n)$ can be calculated from the derivatives of the parametric equations $f_1(s), f_2(s), \ldots f_n(s)$. The derivative will be expressed in terms of the parameter, s, unless the derivatives of the parametric equations can be expressed explicitly in terms of the independent variables.

Example 8.4

A circle is expressed parametrically by the equations

$$x = 5\cos\theta$$

$$y = 5\sin\theta$$

Express the derivative dy/dx (a) as a function of the parameter θ and (b) as a function of x and y.

Solution

(a) Taking the derivative of each parametric equation with respect to θ,

$$\frac{dx}{d\theta} = -5\sin\theta$$

$$\frac{dy}{d\theta} = 5\cos\theta$$

Then,

$$\frac{dy}{dx} = \frac{\dfrac{dy}{d\theta}}{\dfrac{dx}{d\theta}} = \frac{5\cos\theta}{-5\sin\theta} = -\cot\theta$$

(b) The derivatives of the parametric equations are closely related to the original parametric equations.

$$\frac{dx}{d\theta} = -5\sin\theta = -y$$

$$\frac{dy}{d\theta} = 5\cos\theta = x$$

$$\frac{dy}{dx} = \frac{\frac{dy}{d\theta}}{\frac{dx}{d\theta}} = \frac{-x}{y}$$

5. PARTIAL DIFFERENTIATION

Derivatives can be taken with respect to only one independent variable at a time. For example, $f'(x)$ is the derivative of $f(x)$ and is taken with respect to the independent variable x. If a function, $f(x_1, x_2, x_3, \ldots)$, has more than one independent variable, a *partial derivative* can be found, but only with respect to one of the independent variables. All other variables are treated as constants. Symbols for a partial derivative of f taken with respect to variable x are $\partial f/\partial x$ and $f_x(x, y)$.

The geometric interpretation of a partial derivative $\partial f/\partial x$ is the slope of a line tangent to the surface (a sphere, ellipsoid, etc.) described by the function when all variables except x are held constant. In three-dimensional space with a function described by $z = f(x, y)$, the partial derivative $\partial f/\partial x$ (equivalent to $\partial z/\partial x$) is the slope of the line tangent to the surface in a plane of constant y. Similarly, the partial derivative $\partial f/\partial y$ (equivalent to $\partial z/\partial y$) is the slope of the line tangent to the surface in a plane of constant x.

Example 8.5

What is the partial derivative $\partial z/\partial x$ of the following function?

$$z = 3x^2 - 6y^2 + xy + 5y - 9$$

Solution

The partial derivative with respect to x is found by considering all variables other than x to be constants.

$$\frac{\partial z}{\partial x} = 6x - 0 + y + 0 - 0 = 6x + y$$

Example 8.6

A surface has the equation $x^2 + y^2 + z^2 - 9 = 0$. What is the slope of a line that lies in a plane of constant y and is tangent to the surface at $(x, y, z) = (1, 2, 2)$?[3]

---
[3]Although only implied, it is required that the point actually be on the surface (i.e., it must satisfy the equation $f(x, y, z) = 0$).

Solution

Solve for the dependent variable. Then, consider variable y to be a constant.

$$z = \sqrt{9 - x^2 - y^2}$$

$$\frac{\partial z}{\partial x} = \frac{\partial(9 - x^2 - y^2)^{\frac{1}{2}}}{\partial x}$$

$$= \left(\tfrac{1}{2}\right)(9 - x^2 - y^2)^{-\frac{1}{2}}\left[\frac{\partial(9 - x^2 - y^2)}{\partial x}\right]$$

$$= \left(\tfrac{1}{2}\right)(9 - x^2 - y^2)^{-\frac{1}{2}}(-2x)$$

$$= \frac{-x}{\sqrt{9 - x^2 - y^2}}$$

At the point $(1, 2, 2)$, $x = 1$ and $y = 2$.

$$\left.\frac{\partial z}{\partial x}\right|_{(1,2,2)} = \frac{-1}{\sqrt{9 - (1)^2 - (2)^2}} = -1/2$$

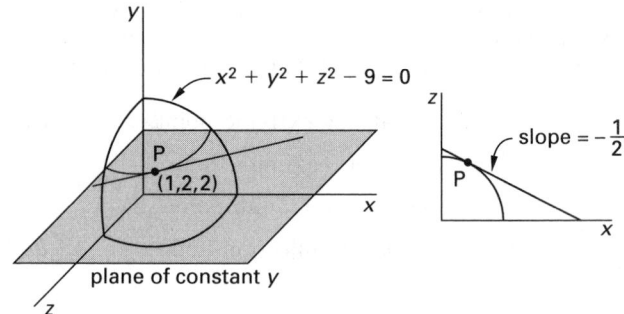

6. IMPLICIT DIFFERENTIATION

When a relationship between n variables cannot be manipulated to yield an explicit function of $n - 1$ independent variables, that relationship implicitly defines the nth variable. Finding the derivative of the implicit variable with respect to any other independent variable is known as *implicit differentiation*.

An implicit derivative is the quotient of two partial derivatives. The two partial derivatives are chosen so that dividing one by the other eliminates a common differential. For example, if z cannot be explicitly extracted from $f(x, y, z) = 0$, the partial derivatives $\partial z/\partial x$ and $\partial z/\partial y$ can still be found as follows.

$$\frac{\partial z}{\partial x} = \frac{-\dfrac{\partial f}{\partial x}}{\dfrac{\partial f}{\partial z}} \qquad 8.34$$

$$\frac{\partial z}{\partial y} = \frac{-\dfrac{\partial f}{\partial y}}{\dfrac{\partial f}{\partial z}} \qquad 8.35$$

Example 8.7

Find the derivative dy/dx of

$$f(x, y) = x^2 + xy + y^3$$

Solution

Implicit differentiation is required because x cannot be extracted from $f(x, y)$.

$$\frac{\partial f}{\partial x} = 2x + y$$

$$\frac{\partial f}{\partial y} = x + 3y^2$$

$$\frac{dy}{dx} = \frac{-\dfrac{\partial f}{\partial x}}{\dfrac{\partial f}{\partial y}} = \frac{-(2x + y)}{x + 3y^2}$$

Example 8.8

Solve Ex. 8.6 using implicit differentiation.

Solution

$$f(x, y, z) = x^2 + y^2 + z^2 - 9 = 0$$

$$\frac{\partial f}{\partial x} = 2x$$

$$\frac{\partial f}{\partial z} = 2z$$

$$\frac{\partial z}{\partial x} = \frac{-\dfrac{\partial f}{\partial x}}{\dfrac{\partial f}{\partial z}} = \frac{-2x}{2z} = -\frac{x}{z}$$

At the point $(1, 2, 2)$,

$$\frac{\partial z}{\partial x} = -1/2$$

7. TANGENT PLANE FUNCTION

Partial derivatives can be used to find the equation of a plane tangent to a three-dimensional surface defined by $f(x, y, z) = 0$ at some point, P_0.

$$T(x_0, y_0, z_0) = (x - x_0)\frac{\partial f(x, y, z)}{\partial x}\Big|_{P_0}$$
$$+ (y - y_0)\frac{\partial f(x, y, z)}{\partial y}\Big|_{P_0}$$
$$+ (z - z_0)\frac{\partial f(x, y, z)}{\partial z}\Big|_{P_0}$$
$$= 0 \qquad \textit{8.36}$$

The coefficients of x, y, and z are the same as the coefficients of $\mathbf{i}$, $\mathbf{j}$, and $\mathbf{k}$ of the normal vector at point P_0. (See Secs. 7.13 and 8.10.)

Example 8.9

What is the equation of the plane that is tangent to the surface defined by $f(x, y, z) = 4x^2 + y^2 - 16z = 0$ at the point $(2, 4, 2)$?

Solution

First, calculate the partial derivatives and substitute the coordinates of the point.

$$\frac{\partial f(x, y, z)}{\partial x}\Big|_{P_0} = 8x|_{(2,4,2)} = (8)(2) = 16$$

$$\frac{\partial f(x, y, z)}{\partial y}\Big|_{P_0} = 2y|_{(2,4,2)} = (2)(4) = 8$$

$$\frac{\partial f(x, y, z)}{\partial z}\Big|_{P_0} = -16|_{(2,4,2)} = -16$$

Next, substitute into Eq. 8.36.

$$T(2, 4, 2) = (16)(x - 2) + (8)(y - 4) - (16)(z - 2)$$
$$= 2x + y - 2z - 4$$

8. GRADIENT VECTOR

The slope of a function is the change in one variable with respect to a distance in a chosen direction. Usually, the direction is parallel to a coordinate axis. However, the maximum slope at a point on a surface may not be in a direction parallel to one of the coordinate axes.

The *gradient vector function* $\nabla f(x, y, z)$ (pronounced "del f") gives the maximum rate of change of the function $f(x, y, z)$.

$$\nabla f(x, y, z) = \left(\frac{\partial f(x, y, z)}{\partial x}\right)\mathbf{i} + \left(\frac{\partial f(x, y, z)}{\partial y}\right)\mathbf{j}$$
$$+ \left(\frac{\partial f(x, y, z)}{\partial z}\right)\mathbf{k} \qquad \textit{8.37}$$

Example 8.10

A two-dimensional function is defined as

$$f(x, y) = 2x^2 - y^2 + 3x - y$$

(a) What is the gradient vector for this function? (b) What is the direction of the line passing through the point $(1, -2)$ that has a maximum slope? (c) What is the maximum slope at the point $(1, -2)$?

Solution

(a) It is necessary to calculate two partial derivatives in order to use Eq. 8.37.

$$\frac{\partial f(x,y)}{\partial x} = 4x + 3$$

$$\frac{\partial f(x,y)}{\partial y} = -2y - 1$$

$$\nabla f(x,y) = (4x+3)\mathbf{i} + (-2y-1)\mathbf{j}$$

(b) The direction of the line passing through $(1, -2)$ with maximum slope is found by inserting $x = 1$ and $y = -2$ into the gradient vector function.

$$\mathbf{V} = [(4)(1) + 3]\,\mathbf{i} + [(-2)(-2) - 1]\,\mathbf{j}$$
$$= 7\mathbf{i} + 3\mathbf{j}$$

(c) The magnitude of the slope is

$$|\mathbf{V}| = \sqrt{(7)^2 + (3)^2} = \sqrt{58} = 7.62$$

9. DIRECTIONAL DERIVATIVE

Unlike the gradient vector (covered in Sec. 8), which calculates the maximum rate of change of a function, the *directional derivative*, indicated by $\nabla_u f(x, y, z)$, $D_u f(x, y, z)$, or $f'_u(x, y, z)$, gives the rate of change in the direction of a given vector, $\mathbf{u}$ or $\mathbf{U}$. The subscript u implies that the direction vector is a unit vector, but it does not need to be, as only the direction cosines are calculated from it.

$$\nabla_u f(x,y,z) = \left(\frac{\partial f(x,y,z)}{\partial x}\right)\cos\alpha + \left(\frac{\partial f(x,y,z)}{\partial y}\right)\cos\beta$$
$$+ \left(\frac{\partial f(x,y,z)}{\partial z}\right)\cos\gamma \qquad \textit{8.38}$$

$$\mathbf{U} = U_x\mathbf{i} + U_y\mathbf{j} + U_z\mathbf{k} \qquad \textit{8.39}$$

$$\cos\alpha = \frac{U_x}{|\mathbf{U}|} = \frac{U_x}{\sqrt{U_x^2 + U_y^2 + U_z^2}} \qquad \textit{8.40}$$

$$\cos\beta = \frac{U_y}{|\mathbf{U}|} \qquad \textit{8.41}$$

$$\cos\gamma = \frac{U_z}{|\mathbf{U}|} \qquad \textit{8.42}$$

Example 8.11

What is the rate of change of $f(x,y) = 3x^2 + xy - 2y^2$ at the point $(1, -2)$ in the direction $4\mathbf{i} + 3\mathbf{j}$?

Solution

The direction cosines are given by Eqs. 8.40 and 8.41.

$$\cos\alpha = \frac{U_x}{|\mathbf{U}|} = \frac{4}{\sqrt{(4)^2 + (3)^2}} = 4/5$$

$$\cos\beta = \frac{U_y}{|\mathbf{U}|} = 3/5$$

The partial derivatives are

$$\frac{\partial f(x,y)}{\partial x} = 6x + y$$

$$\frac{\partial f(x,y)}{\partial y} = x - 4y$$

The directional derivative is given by Eq. 8.38.

$$\nabla_u f(x,y) = \left(\frac{4}{5}\right)(6x+y) + \left(\frac{3}{5}\right)(x - 4y)$$

Substituting the given values of $x = 1$ and $y = -2$,

$$\nabla_u f(1,-2) = \left(\frac{4}{5}\right)[(6)(1) - 2] + \left(\frac{3}{5}\right)[1 - (4)(-2)]$$

$$= \frac{43}{5} = 8.6$$

10. NORMAL LINE VECTOR

Partial derivatives can be used to find the vector normal to a three-dimensional surface defined by $f(x, y, z) = 0$ at some point P_0. Notice that the coefficients of $\mathbf{i}$, $\mathbf{j}$, and $\mathbf{k}$ are the same as the coefficients of x, y, and z calculated for the equation of the tangent plane at point P_0. (See Secs. 7.13 and 8.7.)

$$\mathbf{N} = \left.\frac{\partial f(x,y,z)}{\partial x}\right|_{P_0}\mathbf{i} + \left.\frac{\partial f(x,y,z)}{\partial y}\right|_{P_0}\mathbf{j}$$
$$+ \left.\frac{\partial f(x,y,z)}{\partial z}\right|_{P_0}\mathbf{k} \qquad \textit{8.43}$$

Example 8.12

What is the vector normal to the surface of $f(x, y, z) = 4x^2 + y^2 - 16z = 0$ at the point $(2, 4, 2)$?

Solution

The equation of the tangent plane at this point was calculated in Ex. 8.9 to be

$$T(2,4,2) = 2x + y - 2z - 4 = 0$$

A vector normal to the tangent plane through this point is

$$\mathbf{N} = 2\mathbf{i} + \mathbf{j} - 2\mathbf{k}$$

11. DIVERGENCE OF A VECTOR FIELD

The *divergence*, div $\mathbf{F}$, of a vector field $\mathbf{F}(x, y, z)$ is a scalar function defined by Eqs. 8.44 through 8.46.[4] The divergence of $\mathbf{F}$ can be interpreted as the *accumulation* of flux (i.e., a flowing substance) in a small region (i.e., at a point). One of the uses of the divergence is to determine whether flow (represented in direction and magnitude by $\mathbf{F}$) is compressible. Flow is incompressible if div $\mathbf{F} = 0$, since the substance is not accumulating.

$$\mathbf{F} = P(x, y, z)\mathbf{i} + Q(x, y, z)\mathbf{j} + R(x, y, z)\mathbf{k} \quad \text{8.44}$$

$$\text{div}\,\mathbf{F} = \frac{\partial P}{\partial x} + \frac{\partial Q}{\partial y} + \frac{\partial R}{\partial z} \quad \text{8.45}$$

It may be easier to calculate the divergence from Eq. 8.46.

$$\text{div}\,\mathbf{F} = \boldsymbol{\nabla} \cdot \mathbf{F} \quad \text{8.46}$$

The vector del operator, $\boldsymbol{\nabla}$, is defined as

$$\boldsymbol{\nabla} = \frac{\partial}{\partial x}\mathbf{i} + \frac{\partial}{\partial y}\mathbf{j} + \frac{\partial}{\partial z}\mathbf{k} \quad \text{8.47}$$

If there is no divergence, then the dot product calculated in Eq. 8.46 is zero.

Example 8.13

Calculate the divergence of the following vector function.

$$\mathbf{F}(x, y, z) = xz\mathbf{i} + e^x y\mathbf{j} + 7x^3 y\mathbf{k}$$

Solution

From Eq. 8.45,

$$\text{div}\,\mathbf{F} = \frac{\partial}{\partial x}(xz) + \frac{\partial}{\partial y}(e^x y) + \frac{\partial}{\partial z}(7x^3 y)$$

$$= z + e^x + 0 = z + e^x$$

12. CURL OF A VECTOR FIELD

The *curl*, curl $\mathbf{F}$, of a vector field $\mathbf{F}(x, y, z)$ is a vector field defined by Eqs. 8.51 and 8.52. The curl $\mathbf{F}$ can be interpreted as the *vorticity* per unit area of flux (i.e., a flowing substance) in a small region (i.e., at a point). One of the uses of the curl is to determine whether flow (represented in direction and magnitude by $\mathbf{F}$) is rotational. Flow is irrotational if curl $\mathbf{F} = 0$.[5]

$$\mathbf{F} = P(x, y, z)\mathbf{i} + Q(x, y, z)\mathbf{j} + R(x, y, z)\mathbf{k} \quad \text{8.48}$$

---

[4]Notice that a bold letter, $\mathbf{F}$, is used to indicate that the vector is a function of x, y, and z.

[5]If the velocity vector is $\mathbf{V}$, then the *vorticity* is

$$\boldsymbol{\omega} = \boldsymbol{\nabla} \times \mathbf{V} = \omega_x \mathbf{i} + \omega_y \mathbf{j} + \omega_z \mathbf{k} \quad \text{8.49}$$

The *circulation* (defined in Chap. 17) is the line integral of the velocity $\mathbf{V}$ along a closed curve.

$$\Gamma = \oint \mathbf{V} \cdot d\mathbf{s} = \oint \boldsymbol{\omega} \cdot d\mathbf{A} \quad \text{8.50}$$

$$\text{curl}\,\mathbf{F} = \left(\frac{\partial R}{\partial y} - \frac{\partial Q}{\partial z}\right)\mathbf{i} + \left(\frac{\partial P}{\partial z} - \frac{\partial R}{\partial x}\right)\mathbf{j} + \left(\frac{\partial Q}{\partial x} - \frac{\partial P}{\partial y}\right)\mathbf{k} \quad \text{8.51}$$

It may be easier to calculate the curl from Eq. 8.52. (The vector del operator, $\boldsymbol{\nabla}$, was defined in Eq. 8.47.)

$$\text{curl}\,\mathbf{F} = \boldsymbol{\nabla} \times \mathbf{F}$$

$$= \begin{vmatrix} \mathbf{i} & \mathbf{j} & \mathbf{k} \\ \dfrac{\partial}{\partial x} & \dfrac{\partial}{\partial y} & \dfrac{\partial}{\partial z} \\ P(x, y, z) & Q(x, y, z) & R(x, y, z) \end{vmatrix} \quad \text{8.52}$$

Example 8.14

Calculate the curl of the following vector function.

$$\mathbf{F}(x, y, z) = 3x^2 \mathbf{i} + 7e^x y\,\mathbf{j}$$

Solution

Using Eq. 8.52,

$$\text{curl}\,\mathbf{F} = \begin{vmatrix} \mathbf{i} & \mathbf{j} & \mathbf{k} \\ \dfrac{\partial}{\partial x} & \dfrac{\partial}{\partial y} & \dfrac{\partial}{\partial z} \\ 3x^2 & 7e^x y & 0 \end{vmatrix}$$

Expand the determinant across the top row.

$$\mathbf{i}\left[\frac{\partial}{\partial y}(0) - \frac{\partial}{\partial z}(7e^x y)\right] - \mathbf{j}\left[\frac{\partial}{\partial x}(0) - \frac{\partial}{\partial z}(3x^2)\right]$$

$$+ \mathbf{k}\left[\frac{\partial}{\partial x}(7e^x y) - \frac{\partial}{\partial y}(3x^2)\right]$$

$$= \mathbf{i}(0 - 0) - \mathbf{j}(0 - 0) + \mathbf{k}(7e^x y - 0) = 7e^x y\mathbf{k}$$

13. TAYLOR'S FORMULA

Taylor's formula (series) can be used to expand a function around a point (i.e., approximate the function at one point based on the function's value at another point). The approximation consists of a series, each term composed of a derivative of the original function and a polynomial. Using Taylor's formula requires that the original function be continuous in the interval $[a, b]$ and have the required number of derivatives. To expand a function, $f(x)$, around a point, a, in order to obtain $f(b)$, Taylor's formula is[6]

$$f(b) = f(a) + \frac{f'(a)}{1!}(b - a) + \frac{f''(a)}{2!}(b - a)^2 + \cdots$$

$$+ \frac{f^n(a)}{n!}(b - a)^n + R_n(b) \quad \text{8.53}$$

---

[6]If $a = 0$, Eq. 8.53 is known as the *Maclaurin series*.

In Eq. 8.53, the expression f^n designates the nth derivative of the function $f(x)$. To be a useful approximation, point a must satisfy two requirements: it must be relatively close to point b, and the function and its derivatives must be known or easy to calculate. The last term, $R_n(b)$, is the uncalculated remainder after n derivatives. It is the difference between the exact and approximate values. By using enough terms, the remainder can be made arbitrarily small. That is, $R_n(b)$ approaches zero as n approaches infinity.

It can be shown that the remainder term can be calculated from Eq. 8.54, where c is some number in the interval $[a, b]$. With certain functions, the constant c can be completely determined. In most cases, however, it is possible only to calculate an upper bound on the remainder from Eq. 8.55. M_n is the maximum (positive) value of $f^{(n+1)}(x)$ on the interval $[a, b]$.

$$R_n(b) = \frac{f^{n+1}(c)}{(n+1)!}(b-a)^{n+1} \qquad 8.54$$

$$|R_n(b)| \leq M_n \frac{\left|(b-a)^{n+1}\right|}{(n+1)!} \qquad 8.55$$

14. COMMON SERIES APPROXIMATIONS

Taylor's formulas can be used (by expanding about $a = 0$) to derive the following series approximations:

$$\sin x \approx x - \frac{x^3}{3!} + \frac{x^5}{5!} - \frac{x^7}{7!} + \cdots$$
$$+ (-1)^n \frac{x^{2n+1}}{(2n+1)!} \qquad 8.56$$

$$\cos x \approx 1 - \frac{x^2}{2!} + \frac{x^4}{4!} - \frac{x^6}{6!} + \cdots + (-1)^n \frac{x^{2n}}{(2n)!} \qquad 8.57$$

$$\sinh x \approx x + \frac{x^3}{3!} + \frac{x^5}{5!} + \frac{x^7}{7!} + \cdots + \frac{x^{2n+1}}{(2n+1)!} \qquad 8.58$$

$$\cosh x \approx 1 + \frac{x^2}{2!} + \frac{x^4}{4!} + \frac{x^6}{6!} + \cdots + \frac{x^{2n}}{(2n)!} \qquad 8.59$$

$$e^x \approx 1 + x + \frac{x^2}{2!} + \frac{x^3}{3!} + \cdots + \frac{x^n}{n!} \qquad 8.60$$

$$\ln(1 + x) \approx x - \frac{x^2}{2} + \frac{x^3}{3} - \frac{x^4}{4} + \cdots + (-1)^{n+1} \frac{x^n}{n} \qquad 8.61$$

$$\frac{1}{1-x} \approx 1 + x + x^2 + x^3 + \cdots + x^n \qquad 8.62$$

PRACTICE PROBLEMS

1. What are the values of $a, b,$ and c in the following expression such that $n(\infty) = 100, n(0) = 10,$ and $dn(0)/dt = 0.5$?

$$n(t) = \frac{a}{1 + be^{ct}}$$

2. Find all minima, maxima, and inflection points for

$$y = x^3 - 9x^2 - 3$$

9 Integral Calculus

1. Integration 9-1
2. Elementary Operations 9-1
3. Integration by Parts 9-2
4. Separation of Terms 9-2
5. Double and Higher-Order Integrals 9-3
6. Initial Values 9-3
7. Definite Integrals 9-4
8. Average Value 9-4
9. Area 9-4
10. Arc Length 9-5
11. Pappus' Theorems 9-5
12. Surface of Revolution 9-5
13. Volume of Revolution 9-5
14. Moments of a Function 9-6
15. Fourier Series 9-6
16. Fast Fourier Transforms 9-7
17. Integral Functions 9-8

1. INTEGRATION

Integration is the inverse operation of differentiation. For that reason, *indefinite integrals* are sometimes referred to as *antiderivatives*.[1] Although expressions can be functions of several variables, integrals can only be taken with respect to one variable at a time. The *differential term* (dx in Eq. 9.1) indicates that variable. In Eq. 9.1, the function $f'(x)$ is the *integrand*, and x is the variable of integration.

$$\int f'(x)(dx) = f(x) + C \qquad 9.1$$

While most of a function, $f(x)$, can be "recovered" through integration of its derivative, $f'(x)$, a constant term will be lost. This is because the derivative of a constant term vanishes (i.e., is zero), leaving nothing to recover from. A *constant of integration*, C, is added to the integral to recognize the possibility of such a term.

2. ELEMENTARY OPERATIONS

Equations 9.2 through 9.8 summarize the elementary integration operations on polynomials and exponentials.[2]

[1]The difference between an indefinite and definite integral (covered in Sec. 7) is simple: an *indefinite integral* is a function, while a *definite integral* is a number.
[2]More extensive listings, known as *tables of integrals*, are widely available.

Equations 9.2 and 9.3 are particularly useful. (C and k represent constants. $f(x)$ and $g(x)$ are functions of x.)

$$\int k\,dx = kx + C \qquad 9.2$$

$$\int x^m\,dx = \frac{x^{m+1}}{m+1} + C \qquad [m \neq -1] \qquad 9.3$$

$$\int \frac{1}{x}\,dx = \ln|x| + C \qquad 9.4$$

$$\int e^{kx}\,dx = \frac{e^{kx}}{k} + C \qquad 9.5$$

$$\int xe^{kx}\,dx = \frac{e^{kx}(kx-1)}{k^2} + C \qquad 9.6$$

$$\int k^{ax}\,dx = \frac{k^{ax}}{a\ln k} + C \qquad 9.7$$

$$\int \ln x\,dx = x\ln x - x + C \qquad 9.8$$

Equations 9.9 through 9.20 summarize the elementary integration operations on transcendental functions.

$$\int \sin x\,dx = -\cos x + C \qquad 9.9$$

$$\int \cos x\,dx = \sin x + C \qquad 9.10$$

$$\int \tan x\,dx = \ln|\sec x| + C \qquad 9.11$$

$$\int \cot x\,dx = \ln|\sin x| + C \qquad 9.12$$

$$\int \sec x\,dx = \ln|(\sec x + \tan x)| + C \qquad 9.13$$

$$\int \csc x\,dx = \ln|(\csc x - \cot x)| + C \qquad 9.14$$

$$\int \frac{dx}{k^2 + x^2} = \frac{1}{k}\arctan\frac{x}{k} + C \qquad 9.15$$

$$\int \frac{dx}{\sqrt{k^2 - x^2}} = \arcsin\frac{x}{k} + C \qquad [k^2 > x^2] \qquad 9.16$$

$$\int \frac{dx}{x\sqrt{x^2 - k^2}} = \frac{1}{k}\operatorname{arcsec}\frac{x}{k} + C \qquad [x^2 > k^2] \qquad 9.17$$

$$\int \sin^2 x\,dx = \tfrac{1}{2}x - \tfrac{1}{4}\sin 2x + C \qquad 9.18$$

$$\int \cos^2 x\,dx = \tfrac{1}{2}x + \tfrac{1}{4}\sin 2x + C \qquad 9.19$$

$$\int \tan^2 x\,dx = \tan x - x + C \qquad 9.20$$

Equations 9.21 through 9.26 summarize the elementary integration operations on hyperbolic transcendental functions. Integrals of hyperbolic functions are not completely analogous to those of the regular transcendental functions.

$$\int \sinh x \, dx = \cosh x + C \qquad 9.21$$

$$\int \cosh x \, dx = \sinh x + C \qquad 9.22$$

$$\int \tanh x \, dx = \ln|(\cosh x)| + C \qquad 9.23$$

$$\int \coth x \, dx = \ln|(\sinh x)| + C \qquad 9.24$$

$$\int \operatorname{sech} x \, dx = \arctan(\sinh x) + C \qquad 9.25$$

$$\int \operatorname{csch} x \, dx = \ln\left|\tanh\left(\frac{x}{2}\right)\right| + C \qquad 9.26$$

Equations 9.27 through 9.31 summarize the elementary integration operations on functions and combinations of functions.

$$\int k f(x) \, dx = k \int f(x) \, dx \qquad 9.27$$

$$\int (f(x) + g(x)) \, dx = \int f(x) \, dx + \int g(x) \, dx \qquad 9.28$$

$$\int \frac{f'(x)}{f(x)} \, dx = \ln|f(x)| + C \qquad 9.29$$

$$\int f(x) \, dg(x) = f(x)\int dg(x) - \int g(x) \, df(x) + C$$
$$= f(x)g(x) - \int g(x) \, df(x) + C \qquad 9.30$$

Example 9.1

Find the integral with respect to x of

$$3x^2 + \tfrac{1}{3}x - 7 = 0$$

Solution

This is a polynomial function, and Eq. 9.3 can be applied to each of the three terms.

$$\int \left(3x^2 + \tfrac{1}{3}x - 7\right) dx = x^3 + \tfrac{1}{6}x^2 - 7x + C$$

3. INTEGRATION BY PARTS

Equation 9.30, repeated here, is known as *integration by parts*. $f(x)$ and $g(x)$ are functions. The use of this method is illustrated by Ex. 9.2.

$$\int f(x) \, dg(x) = f(x)g(x) - \int g(x) \, df(x) + C \qquad 9.31$$

Example 9.2

Find the integral

$$\int x^2 e^x \, dx$$

Solution

$x^2 e^x$ is factored into two parts so that integration by parts can be used.

$$f(x) = x^2$$
$$dg(x) = e^x \, dx$$
$$df(x) = 2x \, dx$$
$$g(x) = \int dg(x) = \int e^x \, dx = e^x$$

From Eq. 9.31, disregarding the constant of integration (which cannot be evaluated),

$$\int f(x) \, dg(x) = f(x)g(x) - \int g(x) \, df(x)$$
$$\int x^2 e^x \, dx = x^2 e^x - \int e^x (2x) \, dx$$

The second term is also factored into two parts, and integration by parts is used again. This time,

$$f(x) = x$$
$$dg(x) = e^x \, dx$$
$$df(x) = dx$$
$$g(x) = \int dg(x) = \int e^x \, dx = e^x$$

From Eq. 9.31,

$$\int 2x e^x \, dx = 2 \int x e^x \, dx$$
$$= (2)\left(x e^x - \int e^x \, dx\right)$$
$$= (2)(x e^x - e^x)$$

Then, the complete integral is

$$\int x^2 e^x \, dx = x^2 e^x - (2)(x e^x - e^x) + C$$
$$= e^x (x^2 - 2x + 2) + C$$

4. SEPARATION OF TERMS

Equation 9.28 shows that the integral of a sum of terms is equal to a sum of integrals. This technique is known as *separation of terms*. In many cases, terms are easily separated. In other cases, the technique of *partial fractions* (see Sec. 3.16) can be used to obtain individual terms. These techniques are illustrated by Exs. 9.3 and 9.4.

Example 9.3

Find the integral

$$\int \frac{(2x^2 + 3)^2}{x} dx$$

Solution

$$\int \frac{(2x^2 + 3)^2}{x} dx = \int \frac{4x^4 + 12x^2 + 9}{x} dx$$
$$= \int \left(4x^3 + 12x + \frac{9}{x} \right) dx$$
$$= x^4 + 6x^2 + 9 \ln|x| + C$$

Example 9.4

Find the integral

$$\int \frac{3x + 2}{3x - 2} dx$$

Solution

The integrand is larger than 1, so use long division to simplify it.

$$\begin{array}{r} 1 \text{ rem } \frac{4}{3x-2} \\ 3x-2 \overline{\smash{)} 3x+2} \\ \underline{3x-2} \\ 4 \text{ remainder} \end{array}$$

$$\int \frac{3x + 2}{3x - 2} dx = \int \left(1 + \frac{4}{3x - 2} \right) dx$$
$$= \int dx + \int \frac{4}{3x - 2} dx$$
$$= x + \frac{4}{3} \ln|(3x - 2)| + C$$

5. DOUBLE AND HIGHER-ORDER INTEGRALS

A function can be successively integrated. (This is analogous to successive differentiation.) A function that is integrated twice is known as a *double integral*; if integrated three times, it is a *triple integral*; and so on. Double and triple integrals are used to calculate areas and volumes, respectively.

The successive integrations do not need to be with respect to the same variable. Variables not included in the integration are treated as constants.

There are several notations used for a multiple integral, particularly when the product of length differentials represents a differential area or volume. A double

integral (i.e., two successive integrations) can be represented by one of the following notations.

$$\iint f(x, y) dx \, dy, \quad \int_{R^2} f(x, y) dx \, dy,$$

$$\text{or} \quad \iint_{R^2} f(x, y) dA$$

A triple integral can be represented by one of the following notations.

$$\iiint f(x, y, z) dx \, dy \, dz,$$

$$\int_{R^3} f(x, y, z) dx \, dy \, dz,$$

$$\text{or} \quad \iiint_{R^3} f(x, y, z) dV$$

Example 9.5

Find the integral

$$\iint (x^2 + y^3 x) dx \, dy$$

Solution

$$\int (x^2 + y^3 x) dx = \tfrac{1}{3} x^3 + \tfrac{1}{2} y^3 x^2 + C_1$$

$$\int \left(\tfrac{1}{3} x^3 + \tfrac{1}{2} y^3 x^2 + C_1 \right) dy = \tfrac{1}{3} y x^3 + \tfrac{1}{8} y^4 x^2 + C_1 y + C_2$$

So,

$$\iint (x^2 + y^3 x) dx \, dy = \tfrac{1}{3} y x^3 + \tfrac{1}{8} y^4 x^2 + C_1 y + C_2$$

6. INITIAL VALUES

The constant of integration, C, can be found only if the value of the function $f(x)$ is known for some value of x_0. The value $f(x_0)$ is known as an *initial value*. To completely define a function, as many initial values, $f(x_0), f'(x_0), f''(x_0)$, and so on, as there are integrations are needed.

Example 9.6

It is known that $f(x) = 4$ when $x = 2$ (i.e., the initial condition is $f(2) = 4$). Find the original function.

$$\int (3x^3 - 7x) dx$$

Solution

The function is

$$f(x) = \int (3x^3 - 7x)dx = \frac{3}{4}x^4 - \frac{7}{2}x^2 + C$$

Substituting the initial value determines that C equals 6.

$$4 = \left(\frac{3}{4}\right)(2)^4 - \left(\frac{7}{2}\right)(2)^2 + C$$
$$4 = 12 - 14 + C$$
$$C = 6$$

The function is

$$f(x) = \frac{3}{4}x^4 - \frac{7}{2}x^2 + 6$$

7. DEFINITE INTEGRALS

A *definite integral* is restricted to a specific range of the independent variable. (Unrestricted integrals of the types shown in all preceding examples are known as *indefinite integrals*.) A definite integral restricted to the region bounded by *lower* and *upper limits* (also known as *bounds*) x_1 and x_2 is written as

$$\int_{x_1}^{x_2} f(x)dx$$

Equation 9.32 indicates how definite integrals are evaluated. It is known as the *fundamental theorem of calculus*.

$$\int_{x_1}^{x_2} f'(x)dx = f(x)\Big|_{x_1}^{x_2} = f(x_2) - f(x_1) \qquad 9.32$$

A common use of a definite integral is the calculation of work performed by a force, F, that moves from position x_1 to x_2.

$$W = \int_{x_1}^{x_2} F\,dx \qquad 9.33$$

Example 9.7

Evaluate the definite integral

$$\int_{\frac{\pi}{4}}^{\frac{\pi}{3}} \sin x\,dx$$

Solution

From Eq. 9.32,

$$\int_{\frac{\pi}{4}}^{\frac{\pi}{3}} \sin x\,dx = \Big[-\cos x\Big]_{\frac{\pi}{4}}^{\frac{\pi}{3}}$$
$$= -\cos\left(\frac{\pi}{3}\right) - \left(-\cos\left(\frac{\pi}{4}\right)\right)$$
$$= -0.5 - (-0.707) = 0.207$$

8. AVERAGE VALUE

The average value of a function $f(x)$ that is integrable over the interval $[a, b]$ is

$$\text{average value} = \frac{1}{b-a}\int_a^b f(x)dx \qquad 9.34$$

9. AREA

Equation 9.35 calculates the area, A, bounded by $x = a$, $x = b$, $f_1(x)$ above and $f_2(x)$ below. ($f_2(x) = 0$ if the area is bounded by the x-axis.) This is illustrated in Fig. 9.1.

$$A = \int_a^b (f_1(x) - f_2(x))dx \qquad 9.35$$

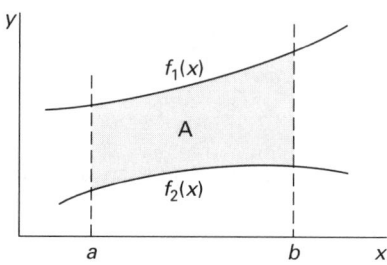

Figure 9.1 *Area Between Two Curves*

Example 9.8

Find the area between the x-axis and the parabola $y = x^2$ in the interval $[0,4]$.

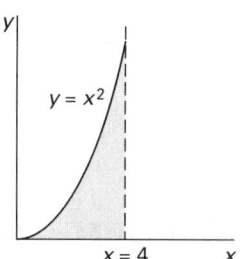

Solution

Referring to Eq. 9.35,

$$f_1(x) = x^2$$
$$f_2(x) = 0$$
$$A = \int_a^b (f_1(x) - f_2(x))dx = \int_0^4 x^2 dx$$
$$= \left[\frac{x^3}{3}\right]_0^4 = 64/3$$

10. ARC LENGTH

Equation 9.36 gives the length of a curve defined by $f(x)$ whose derivative exists in the interval $[a, b]$.

$$\text{length} = \int_a^b \sqrt{1 + (f'(x))^2}\, dx \qquad 9.36$$

11. PAPPUS' THEOREMS[3]

The first and second theorems of Pappus are:[4]

- *First Theorem*: Given a curve, C, that does not intersect the y-axis, the area of the *surface of revolution* generated by revolving C around the y-axis is equal to the product of the length of the curve and the circumference of the circle traced by the centroid of curve C.

$$A = \text{length} \times \text{circumference} = \text{length} \times 2\pi \times \text{radius} \qquad 9.37$$

- *Second Theorem*: Given a plane region, R, that does not intersect the y-axis, the *volume of revolution* generated by revolving R around the y-axis is equal to the product of the area and the circumference of the circle traced by the centroid of area R.

$$V = \text{area} \times \text{circumference} = \text{area} \times 2\pi \times \text{radius} \qquad 9.38$$

12. SURFACE OF REVOLUTION

The surface area obtained by rotating $f(x)$ about the x-axis is

$$A = 2\pi \int_{x=a}^{x=b} f(x)\sqrt{1 + (f'(x))^2}\, dx \qquad 9.39$$

The surface area obtained by rotating $f(y)$ about the y-axis is

$$A = 2\pi \int_{y=c}^{y=d} f(y)\sqrt{1 + (f'(y))^2}\, dy \qquad 9.40$$

Example 9.9

The curve $f(x) = {}^1\!/_2 x$ over the region $x = [0, 4]$ is rotated about the x-axis. What is the surface of revolution?

---

[3]This section is an introduction to surfaces and volumes of revolution but does not involve integration.
[4]Some authorities call the first theorem the second and vice versa.

Solution

The surface of revolution is

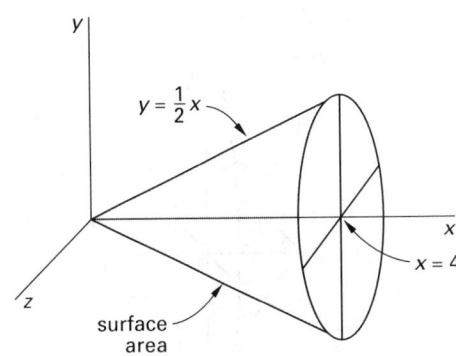

Since $f(x) = {}^1\!/_2 x$, $f'(x) = {}^1\!/_2$. From Eq. 9.39, the area is

$$A = 2\pi \int_{x=a}^{x=b} f(x)\sqrt{1 + (f'(x))^2}\, dx$$

$$= 2\pi \int_0^4 \frac{1}{2} x \sqrt{1 + \left(\frac{1}{2}\right)^2}\, dx$$

$$= \frac{\sqrt{5}}{2}\pi \int_0^4 x\, dx$$

$$= \frac{\sqrt{5}}{2}\pi \left[\frac{x^2}{2}\right]_0^4$$

$$= \frac{\sqrt{5}}{2}\pi \left(\frac{(4)^2 - (0)^2}{2}\right)$$

$$= 4\sqrt{5}\pi$$

13. VOLUME OF REVOLUTION

The volume obtained by rotating $f(x)$ about the x-axis is

$$V = \pi \int_{x=a}^{x=b} f^2(x)\, dx \qquad 9.41$$

The volume obtained by rotating $f(x)$ about the y-axis is

$$V = 2\pi \int_{x=a}^{x=b} x f(x)\, dx \qquad 9.42$$

Example 9.10

The curve $f(x) = x^2$ over the region $x = [0, 4]$ is rotated about the x-axis. What is the volume of revolution?

Solution

The volume of revolution is

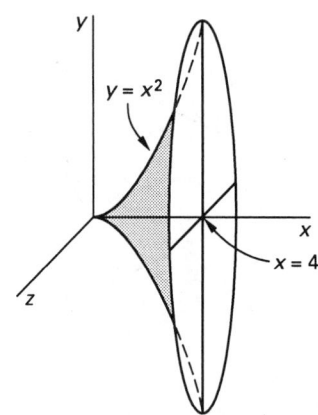

$$V = \pi \int_a^b f^2(x)dx = \pi \int_0^4 (x^2)^2 dx$$

$$= \pi \left[\frac{x^5}{5}\right]_0^4 = \pi \left(\frac{1024}{5} - 0\right) = 204.8\pi$$

14. MOMENTS OF A FUNCTION

The *first moment of a function* is a concept used in finding centroids and centers of gravity. Equations 9.43 and 9.44 are for one- and two-dimensional problems, respectively. It is the exponent of x (1 in this case) that gives the moment its name.

$$\text{first moment} = \int x f(x)dx \qquad 9.43$$

$$\text{first moment} = \iint x f(x, y)dx\, dy \qquad 9.44$$

The *second moment of a function* is a concept used in finding moments of inertia with respect to an axis. Equations 9.45 and 9.46 are for two- and three-dimensional problems, respectively. Second moments with respect to other axes are analogous.

$$(\text{second moment})_x = \iint y^2 f(x, y)dy\, dx \qquad 9.45$$

$$(\text{second moment})_x = \iiint (y^2 + z^2) f(x, y, z)dy\, dz\, dx$$

$$9.46$$

15. FOURIER SERIES

Any periodic waveform can be written as the sum of an infinite number of sinusoidal terms, known as *harmonic terms* (i.e., an infinite series). Such a sum of terms is known as a *Fourier series*, and the process of finding the terms is *Fourier analysis*. (Extracting the original waveform from the series is known as *Fourier inversion*.) Since most series converge rapidly, it is possible to obtain a good approximation to the original waveform with a limited number of sinusoidal terms.

Fourier's theorem is Eq. 9.47.[5] The object of a Fourier analysis is to determine the coefficients a_n and b_n. The constant a_0 can often be determined by inspection since it is the average value of the waveform.

$$f(t) = a_0 + a_1 \cos \omega t + a_2 \cos 2\omega t + \cdots$$
$$+ b_1 \sin \omega t + b_2 \sin 2\omega t + \cdots \qquad 9.47$$

ω is the *natural (fundamental) frequency* of the waveform. It depends on the actual waveform period, T.

$$\omega = \frac{2\pi}{T} \qquad 9.48$$

To simplify the analysis, the time domain can be normalized to the radian scale. The normalized scale is obtained by dividing all frequencies by ω. Then the Fourier series becomes

$$f(t) = a_0 + a_1 \cos t + a_2 \cos 2t + \cdots$$
$$+ b_1 \sin t + b_2 \sin 2t + \cdots \qquad 9.49$$

The coefficients a_n and b_n are found from the following relationships.

$$a_0 = \frac{1}{2\pi} \int_0^{2\pi} f(t)dt$$
$$= \frac{1}{T} \int_0^T f(t)dt \qquad 9.50$$

$$a_n = \frac{1}{\pi} \int_0^{2\pi} f(t)\cos nt\, dt$$
$$= \frac{2}{T} \int_0^T f(t)\cos nt\, dt \qquad [n \geq 1] \qquad 9.51$$

$$b_n = \frac{1}{\pi} \int_0^{2\pi} f(t)\sin nt\, dt$$
$$= \frac{2}{T} \int_0^T f(t)\sin nt\, dt \qquad [n \geq 1] \qquad 9.52$$

While Eqs. 9.51 and 9.52 are always valid, the work of integration and finding a_n and b_n can be greatly simplified if the waveform is recognized as being symmetrical. Table 9.1 summarizes the simplifications.

---

[5]The independent variable used in this section is t, since Fourier analysis is most frequently used in the time domain.

Table 9.1 *Fourier Analysis Simplifications for Symmetrical Waveforms*

	even symmetry $f(-t) = f(t)$	odd symmetry $f(-t) = -f(t)$
	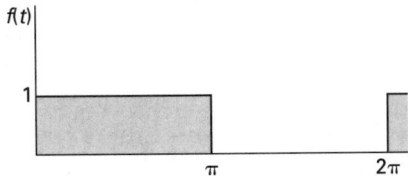	
full-wave symmetry* $f(t + 2\pi) = f(t)$ $\|A_2\| = \|A_1\|$ $\|A_{total}\| = \|A_1\|$ *any repeating wave form	$b_n = 0$ [all n] $a_n = \frac{1}{\pi}\int_0^{2\pi} f(t) \cos nt\, dt$ [all n]	$a_0 = 0$ $a_n = 0$ [all n] $b_n = \frac{1}{\pi}\int_0^{2\pi} f(t) \sin nt\, dt$ [all n]
half-wave symmetry* $f(t + \pi) = -f(t)$ $\|A_2\| = \|A_1\|$ $\|A_{total}\| = 2\|A_1\|$ *same as rotational symmetry	$a_n = 0$ [even n] $b_n = 0$ [all n] $a_n = \frac{2}{\pi}\int_0^{\pi} f(t) \cos nt\, dt$ [odd n]	$a_0 = 0$ $a_n = 0$ [all n] $b_n = 0$ [even n] $b_n = \frac{2}{\pi}\int_0^{\pi} f(t) \sin nt\, dt$ [odd n]
quarter-wave symmetry $f(t + \pi) = -f(t)$ $\|A_2\| = \|A_1\|$ $\|A_{total}\| = 4\|A_1\|$	$a_0 = 0$ $a_n = 0$ [even n] $b_n = 0$ [all n] $a_n = \frac{4}{\pi}\int_0^{\frac{\pi}{2}} f(t) \cos nt\, dt$ [odd n]	$a_0 = 0$ $a_n = 0$ [all n] $b_n = 0$ [even n] $b_n = \frac{4}{\pi}\int_0^{\frac{\pi}{2}} f(t) \sin nt\, dt$ [odd n]

Example 9.11

Find the first four terms of a Fourier series that approximates the repetitive step function illustrated.

$$f(t) = \begin{cases} 1 & 0 < t < \pi \\ 0 & \pi < t < 2\pi \end{cases}$$

Solution

From Eq. 9.50,

$$a_0 = \frac{1}{2\pi}\int_0^{\pi} (1)\, dt + \frac{1}{2\pi}\int_{\pi}^{2\pi} (0)\, dt = 1/2$$

This value of $1/2$ corresponds to the average value of $f(t)$. It could have been found by observation.

$$a_1 = \frac{1}{\pi}\int_0^{\pi} (1)\cos t\, dt + \frac{1}{\pi}\int_{\pi}^{2\pi} (0)\cos t\, dt$$
$$= \frac{1}{\pi}\Big[\sin t\Big]_0^{\pi} + 0 = 0$$

In general,

$$a_n = \frac{1}{\pi}\left[\frac{\sin nt}{n}\right]_0^{\pi} = 0$$

$$b_1 = \frac{1}{\pi}\int_0^{\pi} (1)\sin t\, dt + \frac{1}{\pi}\int_{\pi}^{2\pi} (0)\sin t\, dt$$
$$= \frac{1}{\pi}\Big[-\cos t\Big]_0^{\pi} = \frac{2}{\pi}$$

In general,

$$b_n = \frac{1}{\pi}\left[\frac{-\cos nt}{n}\right]_0^{\pi} = \begin{cases} 0 & \text{for } n \text{ even} \\ \dfrac{2}{\pi n} & \text{for } n \text{ odd} \end{cases}$$

The series is

$$f(t) = \tfrac{1}{2} + \frac{2}{\pi}\left[\sin t + \tfrac{1}{3}\sin 3t + \tfrac{1}{5}\sin 5t + \cdots\right]$$

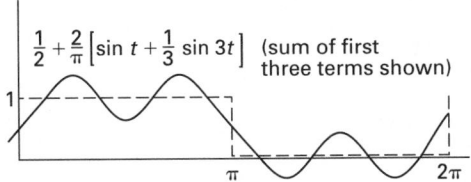

$\frac{1}{2} + \frac{2}{\pi}\left[\sin t + \frac{1}{3}\sin 3t\right]$ (sum of first three terms shown)

16. FAST FOURIER TRANSFORMS

Many mathematical operations are needed to implement a true Fourier transform. While the terms of a Fourier series might be slowly derived by integration, a faster method is needed to analyze real-time data. The *fast Fourier transform* (FFT) is a computer algorithm implemented in *spectrum analyzers (signal analyzers* or *FFT analyzers)* and replaces integration and multiplication operations with table look-ups and additions.[6]

---

[6] *Spectrum analysis*, also known as *frequency analysis, signature analysis*, and *time-series analysis*, develops a relationship (usually graphical) between some property (e.g., amplitude or phase shift) versus frequency.

Since the complexity of the transform is reduced, the transformation occurs more quickly, enabling efficient analysis of waveforms with little or no periodicity.[7]

Using a spectrum analyzer requires choosing the frequency band (e.g., 0 to 20 kHz) to be monitored. (This step automatically selects the sampling period. The lower the frequencies sampled, the longer the sampling period.) If they are not fixed by the analyzer, the numbers of time-dependent input variable samples (e.g., 1024) and frequency-dependent output variable values (e.g., 400) are chosen.[8] There are half as many frequency lines as data points because each line contains two pieces of information—real (amplitude) and imaginary (phase). The *resolution* of the resulting frequency analysis is

$$\text{resolution} = \frac{\text{frequency bandwidth}}{\text{no. of output variable values}} \qquad \textit{9.53}$$

17. INTEGRAL FUNCTIONS

Integrals that cannot be evaluated as finite combinations of elementary functions are called *integral functions*. These functions are evaluated by series expansion. Some of the more common functions are listed as follows.[9,10]

- *integral sine function*:

$$Si(x) = \int_0^x \frac{\sin x}{x} dx$$

$$= x - \frac{x^3}{3 \cdot 3!} + \frac{x^5}{5 \cdot 5!} - \frac{x^7}{7 \cdot 7!} + \cdots \qquad \textit{9.54}$$

- *integral cosine function*:

$$Ci(x) = \int_{-\infty}^x \frac{\cos x}{x} dx = -\int_x^\infty \frac{\cos x}{x} dx$$

$$= C_E + \ln x - \frac{x^2}{2 \cdot 2!} + \frac{x^4}{4 \cdot 4!} - \cdots \qquad \textit{9.55}$$

- *integral exponential function*:

$$Ei(x) = \int_{-\infty}^x \frac{e^x}{x} dx = -\int_{-x}^\infty \frac{e^{-x}}{x} dx$$

$$= C_E + \ln x + x + \frac{x^2}{2 \cdot 2!} + \frac{x^3}{3 \cdot 3!} + \cdots \qquad \textit{9.56}$$

- *error function*:

$$\text{erf}(x) = \frac{2}{\sqrt{\pi}} \int_0^x e^{-x^2} dx$$

$$= \left(\frac{2}{\sqrt{\pi}}\right)\left(\frac{x}{1 \cdot 0!} - \frac{x^3}{3 \cdot 1!} + \frac{x^5}{5 \cdot 2!} - \frac{x^7}{7 \cdot 3!} + \cdots\right) \qquad \textit{9.57}$$

---

[7]Hours and days of manual computations are compressed into milliseconds.

[8]Two samples per time-dependent cycle (at the maximum frequency) is the lower theoretical limit for sampling, but the practical minimum rate is approximately 2.5 samples per cycle. This will ensure that *alias components* (i.e., low-level frequency signals) do not show up in the frequency band of interest.

[9]Other integral functions include the Fresnel integral, gamma function, and elliptic integral.

[10]C_E in Eqs. 9.55 and 9.56 is *Euler's constant*.

$$C_E = \int_{+\infty}^0 e^{-x} \ln x \, dx$$

$$= \lim_{m \to \infty} \left(1 + \tfrac{1}{2} + \tfrac{1}{3} + \cdots + \frac{1}{m} - \ln m\right)$$

$$= 0.577215665$$

10 Differential Equations

1. Types of Differential Equations 10-1
2. Homogeneous, First-Order Linear
   Differential Equations with
   Constant Coefficients 10-2
3. First-Order Linear Differential Equations . . 10-2
4. First-Order Separable
   Differential Equations 10-3
5. First-Order Exact Differential Equations . . 10-3
6. Homogeneous, Second-Order Linear
   Differential Equations with
   Constant Coefficients 10-3
7. Nonhomogeneous Differential
   Equations 10-4
8. Named Differential Equations 10-5
9. Laplace Transforms 10-5
10. Step and Impulse Functions 10-6
11. Algebra of Laplace Transforms 10-6
12. Convolution Integral 10-7
13. Using Laplace Transforms 10-7
14. Third- and Higher-Order Linear
    Differential Equations with
    Constant Coefficients 10-8
15. Application: Engineering Systems 10-8
16. Application: Mixing 10-8
17. Application: Exponential Growth
    and Decay 10-9
18. Application: Epidemics 10-10
19. Application: Surface Temperature 10-10
20. Application: Evaporation 10-10
    Practice Problems 10-10

1. TYPES OF DIFFERENTIAL EQUATIONS

A *differential equation* is a mathematical expression combining a function (e.g., $y = f(x)$) and one or more of its derivatives. The *order* of a differential equation is the highest derivative in it. *First-order differential equations* contain only first derivatives of the function, *second-order differential equations* contain second derivatives (and may contain first derivatives as well), and so on.

A *linear differential equation* can be written as a sum of products of multipliers of the function and its derivatives. If the multipliers are scalars, the differential equation is said to have *constant coefficients*. If the function or one of its derivatives is raised to some power (other than one) or is embedded in another function (e.g., y embedded in $\sin y$ or e^y), the equation is said to be *nonlinear*.

Each term of a *homogeneous differential equation* contains either the function (y) or one of its derivatives—that is, the sum of derivative terms is equal to zero. In a *nonhomogeneous differential equation*, the sum of derivative terms is equal to a nonzero *forcing function* of the independent variable (e.g., $g(x)$). In order to solve a nonhomogeneous equation, it is often necessary to solve the homogeneous equation first. The homogeneous equation corresponding to a nonhomogeneous equation is known as a *reduced equation* or *complementary equation*.

The following examples illustrate the types of differential equations.

- $y' - 7y = 0$ homogeneous, first-order linear, with constant coefficients

- $y'' - 2y' + 8y = \sin 2x$ nonhomogeneous, second-order linear, with constant coefficients

- $y'' - (x^2 - 1)y^2 = \sin 4x$ nonhomogeneous, second-order, nonlinear

An *auxiliary equation* (also called the *characteristic equation*) can be written for a homogeneous linear differential equation with constant coefficients, regardless of order. This auxiliary equation is simply the polynomial formed by replacing all derivatives with variables raised to the power of their respective derivatives.

The purpose of solving a differential equation is to derive an expression for the function in terms of the independent variable. The expression does not need to be explicit in the function, but there can be no derivatives in the expression. Since, in the simplest cases, solving a differential equation is equivalent to finding an indefinite integral, it is not surprising that *constants of integration* must be evaluated from knowledge of how the system behaves. Additional data are known as *initial values*, and any problem that includes them is known as an *initial value problem*.[1]

Most differential equations require lengthy solutions and are not efficiently solved by hand. However, several types are fairly simple and are presented in this chapter.

Example 10.1

Write the complementary differential equation for the following nonhomogeneous differential equation.

$$y'' + 6y' + 9y = e^{-14x}\sin 5x$$

Solution

The complementary equation is found by eliminating the forcing function, $e^{-14x}\sin 5x$.

$$y'' + 6y' + 9y = 0$$

Example 10.2

Write the auxiliary equation to the following differential equation.

$$y'' + 4y' + y = 0$$

Solution

Replacing each derivative with a polynomial term whose degree equals the original order, the auxiliary equation is

$$r^2 + 4r + 1 = 0$$

2. HOMOGENEOUS, FIRST-ORDER LINEAR DIFFERENTIAL EQUATIONS WITH CONSTANT COEFFICIENTS

A homogeneous, first-order linear differential equation with constant coefficients has the general form of Eq. 10.1.

$$y' + ky = 0 \qquad 10.1$$

The auxiliary equation is $r + k = 0$ and has a root of $r = -k$. Equation 10.2 is the solution.

$$y = Ae^{rx} = Ae^{-kx} \qquad 10.2$$

---

[1]The term *initial* implies that time is the independent variable. While this may explain the origin of the term, initial value problems are not limited to the time domain. A *boundary value problem* is similar, except that the data come from different points. For example, additional data in the form $y(x_0)$ and $y'(x_0)$ or $y(x_0)$ and $y'(x_1)$ that need to be simultaneously satisfied constitute an initial value problem. Data of the form $y(x_0)$ and $y(x_1)$ constitute a boundary value problem. Until solved, it is difficult to know whether a boundary value problem has no, one, or multiple solutions.

If the initial condition is known to be $y(0) = y_0$, the solution is

$$y = y_0 e^{-kx} \qquad 10.3$$

3. FIRST-ORDER LINEAR DIFFERENTIAL EQUATIONS

A first-order linear differential equation has the general form of Eq. 10.4. $p(x)$ and $g(x)$ can be constants or any function of x (but not of y). However, if $p(x)$ is a constant and $g(x)$ is zero, it is easier to use Sec. 2 to solve the equation.

$$y' + p(x)y = g(x) \qquad 10.4$$

The *integrating factor* (which is actually a function) to this differential equation is

$$u(x) = \exp\left[\int p(x)dx\right] \qquad 10.5$$

The closed-form solution to Eq. 10.4 is

$$y = \frac{1}{u(x)}\left[\int u(x)g(x)dx + C\right] \qquad 10.6$$

For the special case where $p(x)$ and $g(x)$ are both constants, Eq. 10.4 becomes

$$y' + ay = b \qquad 10.7$$

If the initial condition is $y(0) = y_0$, the solution to Eq. 10.7 is

$$y = \left(\frac{b}{a}\right)(1 - e^{-ax}) + y_0(e^{-ax}) \qquad 10.8$$

Example 10.3

Find a solution to the following differential equation.

$$y' - y = 2xe^{2x} \qquad y(0) = 1$$

Solution

This is a first-order linear equation with $p(x) = -1$ and $g(x) = 2xe^{2x}$. The integrating factor is

$$u(x) = \exp\left[\int p(x)dx\right] = \exp\left[\int -1dx\right] = e^{-x}$$

The solution is given by Eq. 10.6.

$$y = \frac{1}{u(x)} \left[\int u(x)g(x)dx + C \right]$$

$$= \frac{1}{e^{-x}} \left[\int e^{-x} 2xe^{2x}dx + C \right]$$

$$= e^x \left[2 \int xe^x dx + C \right]$$

$$= e^x [2xe^x - 2e^x + C]$$

$$= e^x [2e^x(x-1) + C]$$

From the initial condition,

$$y(0) = 1$$

$$e^0 [(2)(e^0)(0-1) + C] = 1$$

$$1[(2)(1)(-1) + C] = 1$$

Therefore, $C = 3$. The complete solution is

$$y = e^x [2e^x(x-1) + 3]$$

4. FIRST-ORDER SEPARABLE DIFFERENTIAL EQUATIONS

First-order separable differential equations can be placed in the form of Eq. 10.9. For clarity, y' is written as dy/dx.

$$m(x) + n(y)\frac{dy}{dx} = 0 \qquad \text{10.9}$$

Equation 10.9 can be placed in the form of Eq. 10.10, both sides of which are easily integrated. An initial value will establish the constant of integration.

$$m(x)dx = -n(y)dy \qquad \text{10.10}$$

5. FIRST-ORDER EXACT DIFFERENTIAL EQUATIONS

A *first-order exact differential equation* has the form

$$f_x(x,y) + f_y(x,y)y' = 0 \qquad \text{10.11}$$

Notice that $f_x(x,y)$ is the exact derivative of $f(x,y)$ with respect to x, and $f_y(x,y)$ is the exact derivative of $f(x,y)$ with respect to y. The solution is

$$f(x,y) - C = 0 \qquad \text{10.12}$$

6. HOMOGENEOUS, SECOND-ORDER LINEAR DIFFERENTIAL EQUATIONS WITH CONSTANT COEFFICIENTS

Homogeneous second-order linear differential equations with constant coefficients have the form of Eq. 10.13. They are most easily solved by finding the two roots of the auxiliary equation (Eq. 10.14).

$$y'' + k_1y' + k_2y = 0 \qquad \text{10.13}$$

$$r^2 + k_1r + k_2 = 0 \qquad \text{10.14}$$

There are three cases. If the two roots of Eq. 10.14 are real and different, the solution is

$$y = A_1e^{r_1x} + A_2e^{r_2x} \qquad \text{10.15}$$

If the two roots are real and the same, the solution is

$$y = A_1e^{rx} + A_2xe^{rx} \qquad \text{10.16}$$

$$r = \frac{-k_1}{2} \qquad \text{10.17}$$

If the two roots are imaginary, they will be of the form of $(\alpha + i\omega)$ and $(\alpha - i\omega)$, and the solution is

$$y = A_1e^{\alpha x}\cos\omega x + A_2e^{\alpha x}\sin\omega x \qquad \text{10.18}$$

In all three cases, A_1 and A_2 must be found from the two initial conditions.

Example 10.4

Solve the following differential equation.

$$y'' + 6y' + 9y = 0$$

$$y(0) = 0 \qquad y'(0) = 1$$

Solution

The auxiliary equation is

$$r^2 + 6r + 9 = 0$$

$$(r+3)(r+3) = 0$$

The roots to the auxiliary equation are $r_1 = r_2 = -3$. Therefore, the solution has the form of Eq. 10.16.

$$y = A_1e^{-3x} + A_2xe^{-3x}$$

The first initial condition is

$$y(0) = 0$$

$$A_1e^0 + A_2(0)e^0 = 0$$

$$A_1 + 0 = 0$$

$$A_1 = 0$$

To use the second initial condition, the derivative of the equation is needed. Making use of the known fact that $A_1 = 0$,

$$y' = \frac{d}{dx}\left(A_2xe^{-3x}\right) = -3A_2xe^{-3x} + A_2e^{-3x}$$

Using the second initial condition,

$$y'(0) = 1$$

$$-3A_2(0)e^0 + A_2e^0 = 1$$

$$0 + A_2 = 1$$

$$A_2 = 1$$

The solution is

$$y = xe^{-3x}$$

7. NONHOMOGENEOUS DIFFERENTIAL EQUATIONS

A nonhomogeneous equation has the form of Eq. 10.19. $f(x)$ is known as the *forcing function*.

$$y'' + p(x)y' + q(x)y = f(x) \qquad \textbf{10.19}$$

The solution to Eq. 10.19 is the sum of two equations. The *complementary solution*, y_c, solves the complementary (i.e., homogeneous) problem. The *particular solution*, y_p, is any specific solution to the nonhomogeneous Eq. 10.19 that is known or can be found. Initial values used to evaluate any unknown coefficients in the complementary solution *after* y_c and y_p have been combined. (The particular solution will not have any unknown coefficients.)

$$y = y_c + y_p \qquad \textbf{10.20}$$

Two methods are available for finding a particular solution. The *method of undetermined coefficients*, as presented here, can be used only when $p(x)$ and $q(x)$ are constant coefficients and $f(x)$ takes on one of the forms in Table 10.1.

The particular solution can be read from Table 10.1 if the forcing function is of one of the forms given. Of course, the coefficients A_i and B_i are not known—these are the *undetermined coefficients*. The exponent s is the smallest nonnegative number (and will be 0, 1, or 2), which ensures that no term in the particular solution, y_p, is also a solution to the complementary equation, y_c. s must be determined prior to proceeding with the solution procedure.

Table 10.1 *Particular Solutions**

form of $f(x)$	form of y_p
$P_n(x) = a_0x^n + a_1x^{n-1}$ $+ \cdots + a_n$	$x^s(A_0x^n + A_1x^{n-1} + \cdots + A_n)$
$P_n(x)e^{\alpha x}$	$x^s(A_0x^n + A_1x^{n-1} + \cdots + A_n)e^{\alpha x}$
$P_n(x)e^{\alpha x}\begin{Bmatrix} \sin \omega x \\ \cos \omega x \end{Bmatrix}$	$x^s[(A_0x^n + A_1x^{n-1} + \cdots + A_n)e^{\alpha x}\cos \omega x$ $+(B_0x^n + B_1x^{n-1} + \cdots + B_n)e^{\alpha x}\sin \omega x]$

*$P_n(x)$ is a polynomial of degree n.

Once y_p (including s) is known, it is differentiated to obtain y_p' and y_p'', and all three functions are substituted into the original nonhomogeneous equation. The resulting equation is rearranged to match the forcing function, $f(x)$, and the unknown coefficients are determined, usually by solving simultaneous equations.

If the forcing function, $f(x)$, is more complex than the forms shown in Table 10.1, or if either $p(x)$ or $q(x)$ is a function of x, the method of *variation of parameters* should be used. This complex and time-consuming method is not covered in this book.

Example 10.5

Solve the following nonhomogeneous differential equation.

$$y'' + 2y' + y = e^x\cos x$$

Solution

step 1: Find the solution to the complementary (homogeneous) differential equation.

$$y'' + 2y' + y = 0$$

Since this is a differential equation with constant coefficients, write the auxiliary equation.

$$r^2 + 2r + 1 = 0$$

The auxiliary equation factors in $(r+1)^2 = 0$ with two identical roots at $r = -1$. Therefore, the solution to the homogeneous differential equation is

$$y_c(x) = C_1e^{-x} + C_2xe^{-x}$$

step 2: Use Table 10.1 to determine the form of a particular solution. Since the forcing function has the form $P_n(x)e^{\alpha x}\cos \omega x$ with $P_n(x) = 1$ (equivalent to $n = 0$), $\alpha = 1$, and $\omega = 1$, the particular solution has the form

$$y_p(x) = x^s(Ae^x\cos x + Be^x\sin x)$$

step 3: Determine the value of s. Check to see if any of the terms in $y_p(x)$ will themselves solve the homogeneous equation. Try $Ae^x\cos x$ first.

$$\frac{d}{dx}(Ae^x\cos x) = Ae^x\cos x - Ae^x\sin x$$

$$\frac{d^2}{dx^2}(Ae^x\cos x) = -2Ae^x\sin x$$

Substitute these quantities into the homogeneous equation.

$$y'' + 2y' + y = 0$$

$$-2Ae^x\sin x + 2Ae^x\cos x$$

$$-2Ae^x\sin x + Ae^x\cos x = 0$$

$$3Ae^x\cos x - 4Ae^x\sin x = 0$$

Disregarding the trivial ($A = 0$) solution, $Ae^x\cos x$ does not solve the homogeneous equation.

Next, try $Be^x\sin x$.

$$\frac{d}{dx}(Be^x\sin x) = Be^x\cos x + Be^x\sin x$$

$$\frac{d^2}{dx^2}(Be^x\sin x) = 2Be^x\cos x$$

Substitute these quantities into the homogeneous equation.

$$y'' + 2y' + y = 0$$
$$2Be^x\cos x + 2Be^x\cos x$$
$$+ 2Be^x\sin x + Be^x\sin x = 0$$
$$3Be^x\sin x + 4Be^x\cos x = 0$$

Disregarding the trivial ($B = 0$) case, $Be^x\sin x$ does not solve the homogeneous equation.

Since none of the terms in $y_p(x)$ solve the homogeneous equation, $s = 0$, and a particular solution has the form

$$y_p(x) = Ae^x\cos x + Be^x\sin x$$

step 4: Use the method of unknown coefficients to determine A and B in the particular solution. Drawing on the previous steps, substitute the quantities derived from the particular solution into the nonhomogeneous equation.

$$y'' + 2y' + y = e^x\cos x$$
$$-2Ae^x\sin x + 2Be^x\cos x$$
$$+ 2Ae^x\cos x - 2Ae^x\sin x$$
$$+ 2Be^x\cos x + 2Be^x\sin x$$
$$+ Ae^x\cos x + Be^x\sin x = e^x\cos x$$

Combining terms,

$$(-4A + 3B)e^x\sin x + (3A + 4B)e^x\cos x = e^x\cos x$$

Equating the coefficients of like terms on either side of the equal sign results in the following simultaneous equations.

$$-4A + 3B = 0$$
$$3A + 4B = 1$$

The solution to these equations is

$$A = 3/25$$
$$B = 4/25$$

A particular solution is

$$y_p(x) = \left(\frac{3}{25}\right)(e^x\cos x) + \left(\frac{4}{25}\right)(e^x\sin x)$$

step 5: Write the general solution.

$$y(x) = y_c(x) + y_p(x)$$
$$= C_1e^{-x} + C_2xe^{-x} + \left(\frac{3}{25}\right)(e^x\cos x)$$
$$+ \left(\frac{4}{25}\right)(e^x\sin x)$$

The values of C_1 and C_2 would be determined at this time if initial conditions were known.

8. NAMED DIFFERENTIAL EQUATIONS

Some differential equations with specific forms are named after the individuals who developed solution techniques for them.

- *Bessel equation of order ν:*

$$x^2y'' + xy' + (x^2 - \nu^2)y = 0 \qquad 10.21$$

- *Cauchy equation:*

$$a_0x^n\frac{d^ny}{dx^n} + a_1x^{n-1}\frac{d^{n-1}y}{dx^{n-1}} + \cdots$$
$$+ a_{n-1}x\frac{dy}{dx} + a_ny = f(x) \quad 10.22$$

- *Euler equation:*

$$x^2y'' + \alpha xy' + \beta y = 0 \qquad 10.23$$

- *Gauss' hypergeometric equation:*

$$x(1-x)y'' + [c-(a+b+1)x]y' - aby = 0 \qquad 10.24$$

- *Legendre equation of order λ:*

$$(1-x^2)y'' - 2xy' + \lambda(\lambda+1)y = 0 \quad [-1 < x < 1]$$
$$10.25$$

9. LAPLACE TRANSFORMS

Traditional methods of solving nonhomogeneous differential equations by hand are usually difficult and/or time consuming. *Laplace transforms* can be used to reduce many solution procedures to simple algebra.

Every mathematical function, $f(t)$, for which Eq. 10.26 exists has a Laplace transform, written as $\mathcal{L}(f)$ or $F(s)$. The transform is written in the s-domain, regardless of the independent variable in the original function.[2] (The variable s is equivalent to a derivative operator, although it may be handled in the equations as a simple variable.) Equation 10.26 converts a function into a Laplace transform.

$$\mathcal{L}(f(t)) = F(s) = \int_0^\infty e^{-st} f(t) dt \qquad 10.26$$

Equation 10.26 is not often needed because tables of transforms are readily available. (Appendix 10.A contains some of the most common transforms.)

Extracting a function from its transform is the *inverse Laplace transform* operation. Although other methods exist, this operation is almost always done by finding the transform in a set of tables.[3]

$$f(t) = \mathcal{L}^{-1}(F(s)) \qquad 10.27$$

Example 10.6

Find the Laplace transform of the following function.

$$f(t) = e^{at} \qquad [s > a]$$

Solution

Applying Eq. 10.26,

$$\mathcal{L}(e^{at}) = \int_0^\infty e^{-st} e^{at} dt = \int_0^\infty e^{-(s-a)t} dt$$

$$= -\left[\frac{e^{-(s-a)t}}{s-a}\right]_0^\infty = \frac{1}{s-a} \qquad [s > a]$$

10. STEP AND IMPULSE FUNCTIONS

Many forcing functions are sinusoidal or exponential in nature; others, however, can only be represented by a step or impulse function. A *unit step function*, u_t, is a function describing the disturbance of magnitude 1 that is not present before time t but is suddenly there after time t. A step of magnitude 5 at time $t = 3$ would be represented as $5u_3$. (The notation $5u(t-3)$ is used in some books.)

The *unit impulse function*, δ_t, is a function describing a disturbance of magnitude 1 that is applied and removed so quickly as to be instantaneous. An impulse of magnitude 5 at time 3 would be represented by $5\delta_3$. (The notation $5\delta(t-3)$ is used in some books.)

---

[2]It is traditional to write the original function as a function of the independent variable t rather than x. However, Laplace transforms are not limited to functions of time.

[3]Other methods include integration in the complex plane, convolution, and simplification by partial fractions.

Example 10.7

What is the notation for a forcing function of magnitude 6 that is applied at $t = 2$ and completely removed at $t = 7$?

Solution

The notation is $f(t) = 6(u_2 - u_7)$.

Example 10.8

Find the Laplace transform of u_0, a unit step at $t = 0$.

$$f(t) = 0 \text{ for } t < 0$$
$$f(t) = 1 \text{ for } t \geq 0$$

Solution

Since the Laplace transform is an integral that starts at $t = 0$, the value of $f(t)$ prior to $t = 0$ is irrelevant.

$$\mathcal{L}(u_0) = \int_0^\infty e^{-st}(1) dt = -\left[\frac{e^{-st}}{s}\right]_0^\infty$$

$$= 0 - \left(\frac{-1}{s}\right) = \frac{1}{s}$$

11. ALGEBRA OF LAPLACE TRANSFORMS

Equations containing Laplace transforms can be simplified by applying the following principles.

- *linearity theorem*: (c is a constant.)

$$\mathcal{L}(cf(t)) = c\mathcal{L}(f(t)) = cF(s) \qquad 10.28$$

- *superposition theorem*: ($f(t)$ and $g(t)$ are different functions.)

$$\mathcal{L}(f(t)) \pm g(t)) = \mathcal{L}(f(t)) \pm \mathcal{L}(g(t))$$
$$= F(s) \pm G(s) \qquad 10.29$$

- *time-shifting theorem (delay theorem)*:

$$\mathcal{L}(f(t-b)u_b) = e^{-bs} F(s) \qquad 10.30$$

- *Laplace transform of a derivative*:

$$\mathcal{L}(f^n(t)) = -f^{n-1}(0) - sf^{n-2}(0) - \cdots$$
$$- s^{n-1} f(0) + s^n F(s) \qquad 10.31$$

- *other properties*:

$$\mathcal{L}\left(\int_0^t f(u) du\right) = \frac{1}{s} F(s) \qquad 10.32$$

$$\mathcal{L}(t f(t)) = -\frac{dF}{ds} \qquad 10.33$$

$$\mathcal{L}\left(\frac{1}{t} f(t)\right) = \int_s^\infty F(u) du \qquad 10.34$$

Mathematics

12. CONVOLUTION INTEGRAL

A complex Laplace transform, $F(s)$, will often be recognized as the product of two other transforms, $F_1(s)$ and $F_2(s)$, whose corresponding functions $f_1(t)$ and $f_2(t)$ are known. Unfortunately, Laplace transforms cannot be commuted with ordinary multiplication. That is, $f(t) \neq f_1(t)f_2(t)$ even though $F(s) = F_1(s)F_2(s)$.

However, it is possible to extract $f(t)$ from its *convolution*, $h(t)$, as calculated from either of the *convolution integrals* in Eq. 10.35. This process is demonstrated in Ex. 10.9. χ is a dummy variable.

$$f(t) = \mathcal{L}^{-1}(F_1(s)F_2(s))$$
$$= \int_0^t f_1(t-\chi)f_2(\chi)d\chi$$
$$= \int_0^t f_1(\chi)f_2(t-\chi)d\chi \qquad 10.35$$

Example 10.9

Use the convolution integral to find the inverse transform of

$$F(s) = \frac{3}{s^2(s^2+9)}$$

Solution

$F(s)$ can be factored as

$$F_1(s)F_2(s) = \left(\frac{1}{s^2}\right)\left(\frac{3}{s^2+9}\right)$$

Since the inverse transforms of $F_1(s)$ and $F_2(s)$ are $f_1(t) = t$ and $f_2(t) = \sin 3t$, respectively, the convolution integral from Eq. 10.35 is

$$f(t) = \int_0^t (t-\chi)\sin 3\chi \, d\chi$$
$$= \int_0^t (t\sin 3\chi - \chi \sin 3\chi)d\chi$$
$$= t\int_0^t \sin 3\chi \, d\chi - \int_0^t \chi \sin 3\chi \, d\chi$$

Expanding using integration by parts,

$$f(t) = \frac{3t - \sin 3t}{9}$$

13. USING LAPLACE TRANSFORMS

Any nonhomogeneous linear differential equation with constant coefficients can be solved with the following procedure, which reduces the solution to simple algebra. A complete table of transforms simplifies or eliminates step 5.

step 1: Put the differential equation in standard form (i.e., isolate the y'' term).

$$y'' + k_1 y' + k_2 y = f(t) \qquad 10.36$$

step 2: Take the Laplace transform of both sides. Use the linearity and superposition theorems, Eqs. 10.28 and 10.29.

$$\mathcal{L}(y'') + k_1\mathcal{L}(y') + k_2\mathcal{L}(y) = \mathcal{L}(f(t)) \qquad 10.37$$

step 3: Use Eqs. 10.38 and 10.39 to expand the equation. (These are specific forms of Eq. 10.31.) Use a table to evaluate the transform of the forcing function.

$$\mathcal{L}(y'') = s^2\mathcal{L}(y) - sy(0) - y'(0) \qquad 10.38$$
$$\mathcal{L}(y') = s\mathcal{L}(y) - y(0) \qquad 10.39$$

step 4: Use algebra to solve for $\mathcal{L}(y)$.

step 5: If needed, use partial fractions (see Sec. 3.16) to simplify the expression for $\mathcal{L}(y)$.

step 6: Take the inverse transform to find $y(t)$.

$$y(t) = \mathcal{L}^{-1}(\mathcal{L}(y)) \qquad 10.40$$

Example 10.10

Find $y(t)$ for the following differential equation.

$$y'' + 2y' + 2y = \cos t$$
$$y(0) = 1, \qquad y'(0) = 0$$

Solution

step 1: The equation is already in standard form.

step 2: $\mathcal{L}(y'') + 2\mathcal{L}(y') + 2\mathcal{L}(y) = \mathcal{L}(\cos t)$

step 3: Use Eqs. 10.38 and 10.39. Use App. 10.A to find the transform of $\cos t$.

$$s^2\mathcal{L}(y) - sy(0) - y'(0) + 2s\mathcal{L}(y) - 2y(0) + 2\mathcal{L}(y) = \frac{s}{s^2+1}$$

But, $y(0) = 1$ and $y'(0) = 0$.

$$s^2\mathcal{L}(y) - s + 2s\mathcal{L}(y) - 2 + 2\mathcal{L}(y) = \frac{s}{s^2+1}$$

step 4: Combine terms and solve for $\mathcal{L}(y)$.

$$\mathcal{L}(y)(s^2 + 2s + 2) - s - 2 = \frac{s}{s^2+1}$$

$$\mathcal{L}(y) = \frac{\dfrac{s}{s^2+1} + s + 2}{s^2 + 2s + 2}$$

$$= \frac{s^3 + 2s^2 + 2s + 2}{(s^2+1)(s^2+2s+2)}$$

step 5: Expand the expression for $\mathcal{L}(y)$ by partial fractions. (Refer to Sec. 3.16.)

$$\mathcal{L}(y) = \frac{s^3 + 2s^2 + 2s + 2}{(s^2 + 1)(s^2 + 2s + 2)} = \frac{A_1 s + B_1}{s^2 + 1} + \frac{A_2 s + B_2}{s^2 + 2s + 2}$$

$$= \frac{\begin{bmatrix} s^3(A_1 + A_2) + s^2(2A_1 + B_1 + B_2) \\ + s(2A_1 + 2B_1 + A_2) + (2B_1 + B_2) \end{bmatrix}}{(s^2 + 1)(s^2 + 2s + 2)}$$

The following simultaneous equations result.

$$\begin{array}{rcl}
A_1 + A_2 & = 1 \\
2A_1 \quad\quad + B_1 + B_2 & = 2 \\
2A_1 + A_2 + 2B_1 & = 2 \\
2B_1 + B_2 & = 2
\end{array}$$

These equations have the solutions $A_1 = {}^1/_5$, $A_2 = {}^4/_5$, $B_1 = {}^2/_5$, and $B_2 = {}^6/_5$.

step 6: Refer to App. 10.A and take the inverse transforms.

$$y = \mathcal{L}^{-1}(\mathcal{L}(y))$$

$$= \mathcal{L}^{-1}\left(\frac{\left(\frac{1}{5}\right)(s+2)}{s^2 + 1} + \frac{\left(\frac{1}{5}\right)(4s+6)}{s^2 + 2s + 2} \right)$$

$$= \left(\tfrac{1}{5}\right)\left(\mathcal{L}^{-1}\left(\frac{s}{s^2+1}\right) + 2\mathcal{L}^{-1}\left(\frac{1}{s^2+1}\right) \right.$$

$$+ 4\mathcal{L}^{-1}\left(\frac{s - (-1)}{[s - (-1)]^2 + 1} \right)$$

$$\left. + 2\mathcal{L}^{-1}\left(\frac{1}{[s - (-1)]^2 + 1} \right) \right)$$

$$= \left(\tfrac{1}{5}\right)\left(\cos t + 2\sin t + 4e^{-t}\cos t + 2e^{-t}\sin t \right)$$

14. THIRD- AND HIGHER-ORDER LINEAR DIFFERENTIAL EQUATIONS WITH CONSTANT COEFFICIENTS

The solutions of third- and higher-order linear differential equations with constant coefficients are extensions of the solutions for second-order equations of this type. Specifically, if an equation is homogeneous, the auxiliary equation is written and its roots are found. If the equation is nonhomogeneous, Laplace transforms can be used to simplify the solution.

Consider the following homogeneous differential equation with constant coefficients.

$$y^n + k_1 y^{n-1} + \cdots + k_{n-1} y' + k_n y = 0 \qquad 10.41$$

The auxiliary equation to Eq. 10.41 is

$$r^n + k_1 r^{n-1} + \cdots + k_{n-1} r + k_n = 0 \qquad 10.42$$

For each real and distinct root r, the solution contains the term

$$y = Ae^{rx} \qquad 10.43$$

For each real root r that repeats m times, the solution contains the term

$$y = \left(A_1 + A_2 x + A_3 x^2 + \cdots + A_m x^{m-1} \right) e^{rx} \qquad 10.44$$

For each pair of complex roots of the form $r = \alpha \pm i\omega$, the solution contains the terms

$$y = e^{\alpha x}(A_1 \sin \omega x + A_2 \cos \omega x) \qquad 10.45$$

15. APPLICATION: ENGINEERING SYSTEMS

There is a wide variety of engineering systems (mechanical, electrical, fluid flow, heat transfer, and so on) whose behavior is described by linear differential equations with constant coefficients. Mechanical systems are covered in greater detail in Chaps. 59 and 60. Oscillatory and decaying exponential motion are characteristic of these systems and are covered in Chap. 58.

16. APPLICATION: MIXING

A typical mixing problem involves a liquid-filled tank. The liquid may initially be pure or contain some solute. Liquid (either pure or as a solution) enters the tank at a known rate. A drain may be present to remove thoroughly mixed liquid. The concentration of the solution (or, equivalently, the amount of solute in the tank) at some given time is generally unknown.

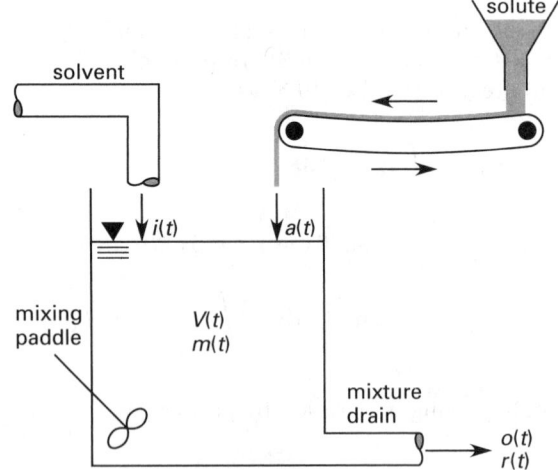

Figure 10.1 *Fluid Mixture Problem*

If $m(t)$ is the mass of solute in the tank at time t, the rate of solute change will be $m'(t)$. If the solute is being added at the rate of $a(t)$ and being removed at the rate of $r(t)$, the rate of change is

$$m'(t) = \text{rate of addition} - \text{rate of removal}$$

$$= a(t) - r(t) \qquad 10.46$$

The rate of solute addition $a(t)$ must be known and, in fact, may be constant. However, $r(t)$ depends on the concentration, $c(t)$, of the mixture and volumetric flow rate at time t. If $o(t)$ is the volumetric flow rate out of the tank, then

$$r(t) = c(t) \times o(t) \qquad 10.47$$

However, the concentration depends on the mass of solute in the tank at time t. Recognizing that the volume, $V(t)$, of the liquid in the tank may be changing with time,

$$c(t) = \frac{m(t)}{V(t)} \qquad 10.48$$

The differential equation describing this problem is

$$m'(t) = a(t) - \frac{m(t)o(t)}{V(t)} \qquad 10.49$$

Example 10.11

A tank contains 100 gal of pure water at the beginning of an experiment. Pure water flows into the tank at the rate of 1 gal/min. Brine containing $1/4$ lbm of salt per gallon enters the tank from a second source at the rate of 1 gal/min. A perfectly mixed solution drains from the tank at the rate of 2 gal/min. How much salt is in the tank 8 min after the experiment begins?

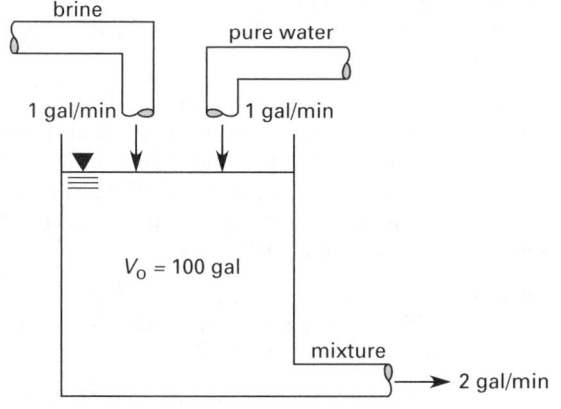

Solution

Let $m(t)$ be the mass of salt in the tank at time t. 0.25 lbm of salt enters the tank per minute (i.e., $a(t) = 0.25$ lbm/min). The salt removal rate depends on the concentration in the tank. That is,

$$r(t) = o(t) \times c(t)$$
$$= \left(2 \; \frac{\text{gal}}{\text{min}} \right) \left(\frac{m(t)}{100 \; \text{gal}} \right) = \left(0.02 \; \frac{1}{\text{min}} \right) m(t)$$

From Eq. 10.46, the rate of change of salt in the tank is

$$m'(t) = a(t) - r(t)$$
$$= 0.25 \; \frac{\text{lbm}}{\text{min}} - \left(0.02 \; \frac{1}{\text{min}} \right) m(t)$$

$$m'(t) + \left(0.02 \; \frac{1}{\text{min}} \right) m(t) = 0.25 \; \text{lbm/min}$$

This is a first-order linear differential equation of the form of Eq. 10.7. Since the initial condition is $m(0) = 0$, the solution is

$$m(t) = \left(\frac{0.25 \; \dfrac{\text{lbm}}{\text{min}}}{0.02 \; \dfrac{1}{\text{min}}} \right) \left(1 - e^{\left(0.02 \frac{1}{\text{min}} \right) t} \right)$$
$$= (12.5 \; \text{lbm}) \left(1 - e^{\left(-0.02 \frac{1}{\text{min}} \right) t} \right)$$

At $t = 8$,

$$m(t) = \left(12.5 \; \frac{1}{\text{min}} \right) \left(1 - e^{-\left(0.02 \frac{1}{\text{min}} \right)(8 \; \text{min})} \right)$$
$$= (12.5 \; \text{lbm})(1 - 0.852)$$
$$= 1.85 \; \text{lbm}$$

17. APPLICATION: EXPONENTIAL GROWTH AND DECAY

Equation 10.50 describes the behavior of a substance whose quantity, $m(t)$, changes at a rate proportional to the quantity present. The constant of proportionality, k, will be negative for decay (e.g., radioactive decay) and positive for growth (e.g., compound interest).

$$m'(t) = km(t) \qquad 10.50$$
$$m'(t) - km(t) = 0 \qquad 10.51$$

If the initial quantity of substance is $m(0) = m_0$, Eq. 10.51 has the solution

$$m(t) = m_0 e^{kt} \qquad 10.52$$

If $m(t)$ is known for some time t, the constant of proportionality is

$$k = \left(\frac{1}{t} \right) \ln \left(\frac{m(t)}{m_0} \right) \qquad 10.53$$

For the case of a decay, the *half-life*, $t_{1/2}$, is the time at which only half of the substance remains. The relationship between k and $t_{1/2}$ is

$$kt_{1/2} = \ln \left(\tfrac{1}{2} \right) = -0.693 \qquad 10.54$$

18. APPLICATION: EPIDEMICS

During an epidemic in a population of n people, the density of sick (contaminated, contagious, affected, etc.) individuals is $\rho_s(t) = s(t)/n$, where $s(t)$ is the number of sick individuals at time t. Similarly, the density of well (uncontaminated, unaffected, susceptible, etc.) individuals is $\rho_\omega(t) = \omega(t)/n$, where $\omega(t)$ is the number of well individuals. Assuming there is no quarantine, the population size is constant, individuals move about freely, and sickness does not limit the activities of individuals, the rate of contagion, $\rho_s'(t)$, will be $k\rho_s(t)\rho_\omega(t)$, where k is a proportionality constant.

$$\rho_s'(t) = k\rho_s(t)\rho_\omega(t) = k\rho_s(t)(1 - \rho_s(t)) \qquad 10.55$$

This is a separable differential equation with the solution

$$\rho_s(t) = \frac{\rho_s(0)}{\rho_s(0) + (1 - \rho_s(0))e^{-kt}} \qquad 10.56$$

19. APPLICATION: SURFACE TEMPERATURE

Newton's law of cooling states that the surface temperature, T, of a cooling object changes at a rate proportional to the difference between the surface and ambient temperatures. The constant k is a positive number.

$$T'(t) = -k(T(t) - T_{\text{ambient}}) \quad [k>0] \qquad 10.57$$

$$T'(t) + kT(t) - kT_{\text{ambient}} = 0 \quad [k>0] \qquad 10.58$$

This first-order linear differential equation with constant coefficients has the following solution (from Eq. 10.8).

$$T(t) = T_{\text{ambient}} + (T(0) - T_{\text{ambient}})e^{-kt} \qquad 10.59$$

If the temperature is known at some time t, the constant k can be found from Eq. 10.60.

$$k = \left(\frac{-1}{t}\right) \ln\left(\frac{T(t) - T_{\text{ambient}}}{T(0) - T_{\text{ambient}}}\right) \qquad 10.60$$

20. APPLICATION: EVAPORATION

The mass of liquid evaporated from a liquid surface is proportional to the exposed surface area. Since quantity, mass, and remaining volume are all proportional, the differential equation is

$$\frac{dV}{dt} = -kA \qquad 10.61$$

For a spherical drop of radius r, Eq. 10.61 reduces to

$$\frac{dr}{dt} = -k \qquad 10.62$$

$$r(t) = r(0) - kt \qquad 10.63$$

For a cube with sides of length s, Eq. 10.61 reduces to

$$\frac{ds}{dt} = -2k \qquad 10.64$$

$$s(t) = s(0) - 2kt \qquad 10.65$$

PRACTICE PROBLEMS

1. Solve the following differential equation for y.

$$y'' - 4y' - 12y = 0$$

2. Solve the following differential equation for y.

$$y' - y = 2xe^{2x} \qquad y(0) = 1$$

3. The oscillation exhibited by the top story of a certain building in free motion is given by the following differential equation.

$$x'' + 2x' + 2x = 0 \qquad x(0) = 0; x'(0) = 1$$

(a) What is x as a function of time? (b) What is the building's fundamental natural frequency of vibration? (c) What is the amplitude of oscillation? (d) What is x as a function of time if a lateral wind load is applied with a form of $\sin t$?

4. (*Time limit: one hour*) A 90 lbm (40 kg) bag of a chemical is accidentally dropped in an aerating lagoon. The chemical is water soluble and nonreacting. The lagoon is 120 ft (35 m) in diameter and filled to a depth of 10 ft (3 m). The aerators circulate and distribute the chemical evenly throughout the lagoon.

Water enters the lagoon at the rate of 30 gal/min (115 L/min). Fully mixed water is pumped into a reservoir at the rate of 30 gal/min (115 L/min).

The established safe concentration of this chemical is 1 ppb (part per billion). How many days will it take for the concentration of the discharge water to reach this level?

11 Probability and Statistical Analysis of Data

1. Set Theory 11-1
2. Combinations of Elements 11-2
3. Permutations 11-2
4. Probability Theory 11-2
5. Joint Probability 11-3
6. Complementary Probabilities 11-4
7. Conditional Probability 11-4
8. Probability Density Functions 11-4
9. Binomial Distribution 11-4
10. Hypergeometric Distribution 11-5
11. Multiple Hypergeometric Distribution 11-5
12. Poisson Distribution 11-5
13. Continuous Distribution Functions 11-5
14. Exponential Distribution 11-6
15. Normal Distribution 11-6
16. Application: Reliability 11-7
17. Analysis of Experimental Data 11-8
18. Measures of Central Tendency 11-10
19. Measures of Dispersion 11-10
20. Central Limit Theorem 11-11
21. Confidence Level 11-11
22. Application: Confidence Limits 11-11
23. Application: Basic Hypothesis Testing . . 11-12
24. Application: Statistical Process Control . 11-12
25. Measures of Experimental Adequacy . . . 11-13
26. Linear Regression 11-13
    Practice Problems 11-15

1. SET THEORY

A *set* (usually designated by a capital letter) is a population or collection of individual items known as *elements* or *members*. The *null set*, ϕ, is empty (i.e., contains no members). If A and B are two sets, A is a *subset* of B if every member in A is also in B. A is a *proper subset* of B if B consists of more than the elements in A. These relationships are denoted

$$A \subseteq B \qquad \text{[subset]}$$
$$A \subset B \qquad \text{[proper subset]}$$

The *universal set*, U, is one from which other sets draw their members. If A is a subset of U, then A' (also designated as $A^{-1}, \tilde{A}, -A$, and $\bar{A}$) is the *complement* of A and consists of all elements in U that are not in A. This is illustrated in a *Venn diagram* in Fig. 11.1(a).

The *union of two sets*, denoted by $A \cup B$ and shown in Fig. 11.1(b), is the set of all elements that are either in A or B or both. The *intersection of two sets*, denoted

by $A \cap B$ and shown in Fig. 11.1(c), is the set of all elements that belong to both A and B. If $A \cap B = \phi$, A and B are said to be *disjoint sets*.

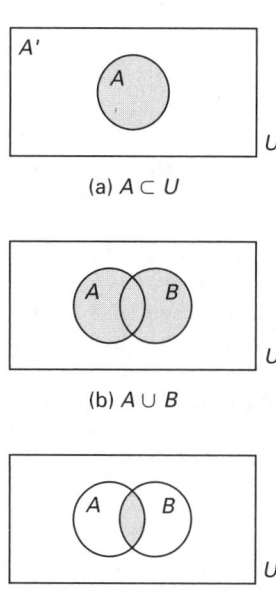

(a) $A \subset U$

(b) $A \cup B$

(c) $A \cap B$

Figure 11.1 *Venn Diagrams*

If A, B, and C are subsets of the universal set, the following laws apply.

- *identity laws*:

$$A \cup \phi = A \qquad \qquad 11.1$$
$$A \cup U = U \qquad \qquad 11.2$$
$$A \cap \phi = \phi \qquad \qquad 11.3$$
$$A \cap U = A \qquad \qquad 11.4$$

- *idempotent laws*:

$$A \cup A = A \qquad \qquad 11.5$$
$$A \cap A = A \qquad \qquad 11.6$$

- *complement laws*:

$$A \cup A' = U \qquad \qquad 11.7$$
$$(A')' = A \qquad \qquad 11.8$$
$$A \cap A' = \phi \qquad \qquad 11.9$$
$$U' = \phi \qquad \qquad 11.10$$

- *commutative laws*:

$$A \cup B = B \cup A \qquad 11.11$$
$$A \cap B = B \cap A \qquad 11.12$$

- *associative laws*:

$$(A \cup B) \cup C = A \cup (B \cup C) \qquad 11.13$$
$$(A \cap B) \cap C = A \cap (B \cap C) \qquad 11.14$$

- *distributive laws*:

$$A \cup (B \cap C) = (A \cup B) \cap (A \cup C) \qquad 11.15$$
$$A \cap (B \cup C) = (A \cap B) \cup (A \cap C) \qquad 11.16$$

- *de Morgan's laws*:

$$(A \cup B)' = A' \cap B' \qquad 11.17$$
$$(A \cap B)' = A' \cup B' \qquad 11.18$$

2. COMBINATIONS OF ELEMENTS

There are a finite number of ways in which n elements can be combined into distinctly different groups of r items. For example, suppose a farmer has a chicken, a rooster, a duck, and a cage that holds only two birds. The possible *combinations* of three birds taken two at a time are (chicken, rooster), (chicken, duck), and (rooster, duck). The birds in the cage will not remain stationary, so the combination (rooster, chicken) is not distinctly different from (chicken, rooster). That is, the groups are not *order conscious*.

The number of combinations of n items taken r at a time is written $C(n,r)$, C_r^n, $_nC_r$, or $\binom{n}{r}$ (pronounced "n choose r") and given by Eq. 11.19. It is sometimes referred to as the *binomial coefficient*.

$$\binom{n}{r} = C(n,r) = \frac{n!}{(n-r)!r!} \qquad \text{[for } r \leq n] \quad 11.19$$

Example 11.1

Six people are on a sinking yacht. There are four life jackets. How many combinations of survivors are there?

Solution

The groups are not order conscious. From Eq. 11.19,

$$C(6,4) = \frac{6!}{(6-4)!4!} = \frac{6 \cdot 5 \cdot 4 \cdot 3 \cdot 2 \cdot 1}{(2 \cdot 1)(4 \cdot 3 \cdot 2 \cdot 1)}$$
$$= 15$$

3. PERMUTATIONS

An order-conscious subset of r items taken from a set of n items is the *permutation* $P(n,r)$, also written P_r^n and $_nP_r$. The permutation is order conscious because the arrangement of two items (say a_i and b_i) as $a_i b_i$

is different from the arrangement $b_i a_i$. The number of permutations is

$$P(n,r) = \frac{n!}{(n-r)!} \qquad \text{[for } r \leq n] \quad 11.20$$

If groups of the entire set of n items are being enumerated, the number of permutations of n items taken n at a time is

$$P(n,n) = \frac{n!}{(n-n)!} = \frac{n!}{0!} = n! \qquad 11.21$$

A *ring permutation* is a special case of n items taken n at a time. There is no identifiable beginning or end, and the number of permutations is divided by n.

$$P_{\text{ring}}(n,n) = \frac{P(n,n)}{n} = (n-1)! \qquad 11.22$$

Example 11.2

A pianist knows four pieces but will have enough stage time to play only three of them. Pieces played in a different order constitute a different program. How many different programs can be arranged?

Solution

The groups are order conscious. From Eq. 11.20,

$$P(4,3) = \frac{4!}{(4-3)!} = \frac{4 \cdot 3 \cdot 2 \cdot 1}{1} = 24$$

Example 11.3

Seven diplomats from different countries enter a circular room. The only furnishings are seven chairs arranged around a circular table. How many ways are there of arranging the diplomats?

Solution

All seven diplomats must be seated, so the groups are permutations of seven objects taken seven at a time. Since there is no head chair, the groups are ring permutations. From Eq. 11.22,

$$P_{\text{ring}}(7,7) = (7-1)! = 6 \cdot 5 \cdot 4 \cdot 3 \cdot 2 \cdot 1 = 720$$

4. PROBABILITY THEORY

The act of conducting an experiment (trial) or taking a measurement is known as *sampling*. *Probability theory* determines the relative likelihood that a particular event will occur. An *event*, e, is one of the possible outcomes of the *trial*. Taken together, all of the possible events constitute a finite *sample space*, $E = [e_1, e_2, \ldots, e_n]$. The trial is drawn from the *population* or *universe*. Populations can be finite or infinite in size.

Events can be numerical or nonnumerical, discrete or continuous, and dependent or independent. An example of a nonnumerical event is getting tails on a coin toss. The roll of a die is a discrete numerical event. The measured diameter of a bolt produced from an automatic screw machine is a numerical event. Since the diameter can (within reasonable limits) take on any value, its measured value is a continuous numerical event.

An event is *independent* if its outcome is unaffected by previous outcomes (i.e., previous runs of the experiment) and *dependent* otherwise. Whether or not an event is independent depends on the population size and how the sampling is conducted. Sampling (a trial) from an infinite population is implicitly independent. When the population is finite, *sampling with replacement* produces independent events, while *sampling without replacement* changes the population and produces dependent events.

The terms *success* and *failure* are loosely used in probability theory to designate obtaining and not obtaining, respectively, the tested-for condition. "Failure" is not the same as a *null event* (i.e., one that has a zero probability of occurrence).

The *probability* of event e_1 occurring is designated as $p\{e_1\}$ and is calculated as the ratio of the total number of ways the event can occur to the total number of outcomes in the sample space.

Example 11.4

There are 380 students in a rural school—200 girls and 180 boys. One student is chosen at random and is checked for gender and height. (a) Define and categorize the population. (b) Define and categorize the sample space. (c) Define the trials. (d) Define and categorize the events. (e) In determining the probability that the student chosen is a boy, define success and failure. (f) What is the probability that the student is a boy?

Solution

(a) The population consists of 380 students and is finite.

(b) In determining the gender of the student, the sample space consists of the two outcomes $E = $ [girl, boy]. This sample space is nonnumerical and discrete. In determining the height, the sample space consists of a range of values and is numerical and continuous.

(c) The trial is the actual sampling (i.e., the determination of gender and height).

(d) The events are the outcomes of the trials (i.e., the gender and height of the student). These events are independent if each student returns to the population prior to the random selection of the next student; otherwise, the events are dependent.

(e) The event is a success if the student is a boy and is a failure otherwise.

(f) From the definition of probability,

$$p\{\text{boy}\} = \frac{\text{no. of boys}}{\text{no. of students}} = \frac{180}{380} = \frac{9}{19} = 0.47$$

5. JOINT PROBABILITY

Joint probability rules specify the probability of a combination of events. If n mutually exclusive events from the set E have probabilities $p\{e_i\}$, the probability of any one of these events occurring in a given trial is the sum of the individual probabilities. Notice that the events in Eq. 11.23 come from a single sample space and are linked by the word *or*.

$$p\{e_1 \text{ or } e_2 \text{ or } \ldots \text{ or } e_k\} = p\{e_1\} + p\{e_2\} + \ldots + p\{e_k\}$$
$$\text{11.23}$$

Given two independent sets of events, E and G, Eq. 11.24 gives the probability that either event e_i or g_i, or both, will occur. Notice that the events in Eq. 11.24 come from two different sample spaces and are linked by the word *or*.

$$p\{e_i \text{ or } g_i\} = p\{e_i\} + p\{g_i\} - p\{e_i\}p\{g_i\} \qquad \text{11.24}$$

Given two independent sets of events, E and G, Eq. 11.25 gives the probability that events e_i and g_i will both occur. Notice that the events in Eq. 11.25 come from two different sample spaces and are linked by the word *and*.

$$p\{e_i \text{ and } g_i\} = p\{e_i\}p\{g_i\} \qquad \text{11.25}$$

Example 11.5

A bowl contains five white balls, two red balls, and three green balls. What is the probability of getting either a white ball or a red ball in one draw from the bowl?

Solution

The two possible events are mutually exclusive and come from the same sample space, so Eq. 11.23 can be used.

$$p\{\text{white or red}\} = p\{\text{white}\} + p\{\text{red}\} = \frac{5}{10} + \frac{2}{10} = \frac{7}{10}$$

Example 11.6

One bowl contains five white balls, two red balls, and three green balls. Another bowl contains three yellow balls and seven black balls. What is the probability of getting a red ball from the first bowl and a yellow ball from the second bowl in one draw from each bowl?

Solution

The two trials are independent, so Eq. 11.25 can be used.

$$p\{\text{red and yellow}\} = p\{\text{red}\}p\{\text{yellow}\}$$
$$= \left(\frac{2}{10}\right)\left(\frac{3}{10}\right) = \frac{6}{100}$$

6. COMPLEMENTARY PROBABILITIES

The probability of an event occurring is equal to one minus the probability of the event not occurring. This is known as a *complementary probability*.

$$p\{e_i\} = 1 - p\{\text{not } e_i\} \qquad \textit{11.26}$$

Equation 11.26 can be used to simplify some probability calculations. Specifically, calculation of the probability of numerical events being "greater than" or "less than" or quantities being "at least" a certain number can often be simplified by calculating the probability of a complementary event.

Example 11.7

A fair coin is tossed five times.[1] What is the probability of getting at least one tail?

Solution

The probability of getting at least one tail in five tosses could be calculated as

$$p\{\text{at least 1 tail}\} = p\{1 \text{ tail}\} + p\{2 \text{ tails}\}$$
$$+ p\{3 \text{ tails}\} + p\{4 \text{ tails}\}$$
$$+ p\{5 \text{ tails}\}$$

However, it is easier to calculate the complementary probability of getting no tails (i.e., getting all heads).

$$p\{\text{at least 1 tail}\} = 1 - p\{0 \text{ tails}\}$$
$$= 1 - (0.5)^5 = 0.96875$$

7. CONDITIONAL PROBABILITY

Given two dependent sets of events, E and G, the probability that event e_k will occur given the fact that the dependent event g has already occurred is written as $p\{e_k|g\}$ and given by *Bayes' theorem*, Eq. 11.27.

$$p\{e_k|g\} = \frac{p\{e_k \text{ and } g\}}{p\{g\}} = \frac{p\{g|e_k\}p\{e_k\}}{\sum\limits_{i=1}^{n} p\{g|e_i\}p\{e_i\}} \qquad \textit{11.27}$$

---

[1]It makes no difference whether one coin is tossed five times or five coins are each tossed once.

8. PROBABILITY DENSITY FUNCTIONS

A *density function* is a nonnegative function whose integral taken over the entire range of the independent variable is unity. A *probability density function* (PDF) is a mathematical formula that gives the probability of a discrete numerical event. A *numerical event* is an occurrence that can be described (usually) by an integer. For example, 27 cars passing through a bridge toll booth in an hour is a discrete numerical event.

A probability density function, $f(x)$, gives the probability that discrete event x will occur. That is, $p\{x\} = f(x)$. Important discrete probability density functions are the binomial, hypergeometric, and Poisson distributions.

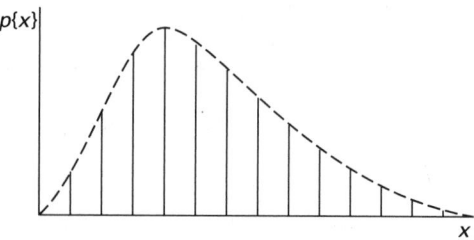

Figure 11.2 *Probability Density Function*

9. BINOMIAL DISTRIBUTION

The *binomial probability density function* is used when all outcomes can be categorized as either successes or failures. The probability of success in a single trial is $\hat{p}$, and the probability of failure is the complement, $\hat{q} = 1 - \hat{p}$. The population is assumed to be infinite in size so that sampling does not change the values of $\hat{p}$ and $\hat{q}$. (The binomial distribution can also be used with finite populations when sampling with replacement.)

Equation 11.28 gives the probability of x successes in n independent *successive trials*. The quantity $\binom{n}{x}$ is the binomial coefficient, identical to the number of combinations of n items taken x at a time. (See Eq. 11.19.)

$$p\{x\} = f(x) = \binom{n}{x} \hat{p}^x \hat{q}^{n-x} \qquad \textit{11.28}$$
$$\binom{n}{x} = \frac{n!}{(n-x)!x!} \qquad \textit{11.29}$$

Equation 11.28 is a true (discrete) distribution, taking on values for each integer value up to n. The mean, μ, and variance, σ^2 (see Secs. 18 and 19), of this distribution are

$$\mu = n\hat{p} \qquad \textit{11.30}$$
$$\sigma^2 = n\hat{p}\hat{q} \qquad \textit{11.31}$$

Example 11.8

Five percent of a large batch of high-strength steel bolts purchased for bridge construction are defective. (a) If seven bolts are randomly sampled, what is the probability that exactly three will be defective? (b) What is the probability that two or more bolts will be defective?

Solution

(a) The bolts are either defective or not, so the binomial distribution can be applied.

$$\hat{p} = 0.05$$

$$\hat{q} = 1 - 0.05 = 0.95$$

From Eq. 11.28,

$$p\{3\} = f(3) = \binom{7}{3}\hat{p}^3\hat{q}^{7-3}$$

$$= \left(\frac{7 \cdot 6 \cdot 5 \cdot 4 \cdot 3 \cdot 2 \cdot 1}{4 \cdot 3 \cdot 2 \cdot 1 \cdot 3 \cdot 2 \cdot 1}\right)(0.05)^3(0.95)^4$$

$$= 0.00356$$

(b) The probability that two or more bolts will be defective could be calculated as

$$p\{x \geq 2\} = p\{2\} + p\{3\} + p\{4\} + p\{5\} + p\{6\} + p\{7\}$$

This method would require six probability calculations. It is easier to use the complement of the desired probability.

$$p\{x \geq 2\} = 1 - p\{x \leq 1\} = 1 - [p\{0\} + p\{1\}]$$

$$p\{0\} = \binom{7}{0}(0.05)^0(0.95)^7 = (0.95)^7$$

$$p\{1\} = \binom{7}{1}(0.05)^1(0.95)^6 = (7)(0.05)(0.95)^6$$

$$p\{x \geq 2\} = 1 - [(0.95)^7 + (7)(0.05)(0.95)^6]$$

$$= 1 - [0.6983 + 0.2573]$$

$$= 0.0444$$

10. HYPERGEOMETRIC DISTRIBUTION

Probabilities associated with sampling from a finite population without replacement are calculated from the *hypergeometric distribution*. If a population of finite size M contains K items with a given characteristic (e.g., red color, defective construction), then the probability of finding x items with that characteristic in a sample of n items is

$$p\{x\} = f(x) = \frac{\binom{K}{x}\binom{M-K}{n-x}}{\binom{M}{n}} \qquad [\text{for } x \leq n] \quad \textit{11.32}$$

11. MULTIPLE HYPERGEOMETRIC DISTRIBUTION

Sampling without replacement from finite populations containing several different types of items is handled by the *multiple hypergeometric distribution*. If a population of finite size M contains K_i items of type i (such that $\sum K_i = M$), the probability of finding x_1 items of type 1, x_2 items of type 2, and so on, in a sample size n (such that $\sum x_i = n$) is

$$p\{x_1, x_2, x_3, \ldots\} = \frac{\binom{K_1}{x_1}\binom{K_2}{x_2}\binom{K_3}{x_3}\cdots}{\binom{M}{n}} \quad \textit{11.33}$$

12. POISSON DISTRIBUTION

Certain events occur relatively infrequently but at a regular rate. The probability of such an event occurring is given by the *Poisson distribution*. Suppose an event occurs, on the average, λ times per period. The probability that the event will occur x times per period is

$$p\{x\} = f(x) = \frac{e^{-\lambda}\lambda^x}{x!} \qquad [\lambda > 0] \quad \textit{11.34}$$

λ is both the mean and the variance of the Poisson distribution.

Example 11.9

The number of customers arriving at a hamburger stand in some period is a Poisson distribution having a mean of eight. What is the probability that exactly six customers will arrive in the next period?

Solution

$\lambda = 8$ and $x = 6$. From Eq. 11.34,

$$p\{6\} = \frac{e^{-\lambda}\lambda^x}{x!} = \frac{e^{-8}(8)^6}{6!} = 0.122$$

13. CONTINUOUS DISTRIBUTION FUNCTIONS

Most numerical events are *continuously distributed* and are not constrained to discrete or integer values. For example, the resistance of a 10% 1 Ω resistor may be any value between 0.9 and 1.1 Ω. The probability of an exact numerical event is zero for continuously distributed variables. That is, there is no chance that a numerical event will be *exactly* x.[2] It is possible to determine only the probability that a numerical event will be less than x, greater than x, or between the values of x_1 and x_2, but not exactly equal to x.

---

[2]It is important to understand the rationale behind this statement. Since the variable can take on any value and has an infinite number of significant digits, we can infinitely continue to increase the precision of the value. For example, the probability is zero that a resistance will be exactly 1 Ω because the resistance is really 1.03 or 1.0260008 or 1.02600080005, and so on.

While an expression, $f(x)$, for a probability density function can be written, it is used to derive the *continuous distribution function* (CDF), $F(x_0)$, which gives the probability of numerical event x_0 or less occurring.

$$p\{X < x_0\} = F(x_0) = \int_0^{x_0} f(x)dx \qquad 11.35$$

$$f(x) = \frac{dF(x)}{dx} \qquad 11.36$$

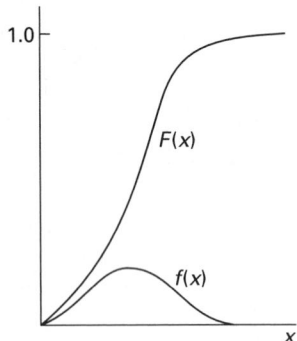

Figure 11.3 *Continuous Distribution Function*

14. EXPONENTIAL DISTRIBUTION

The continuous *exponential distribution* is given by its probability density and continuous distribution functions.

$$f(x) = \lambda e^{-\lambda x} \qquad 11.37$$

$$p\{X < x\} = F(x) = 1 - e^{-\lambda x} \qquad 11.38$$

The mean and variance of the exponential distribution are

$$\mu = \frac{1}{\lambda} \qquad 11.39$$

$$\sigma^2 = \frac{1}{\lambda^2} \qquad 11.40$$

15. NORMAL DISTRIBUTION

The *normal distribution (Gaussian distribution)* is a symmetrical distribution commonly referred to as the *bell-shaped curve*, which represents the distribution of outcomes of many experiments, processes, and phenomena. The probability density and continuous distribution functions for the normal distribution with mean μ and variance σ^2 are

$$f(x) = \frac{e^{-\frac{1}{2}\left(\frac{x-\mu}{\sigma}\right)^2}}{\sigma\sqrt{2\pi}} \qquad [-\infty < x < +\infty] \qquad 11.41$$

$$p\{\mu < X < x_0\} = F(x_0)$$
$$= \frac{1}{\sigma\sqrt{2\pi}} \int_0^{x_0} e^{-\frac{1}{2}\left(\frac{x-\mu}{\sigma}\right)^2} dx \qquad 11.42$$

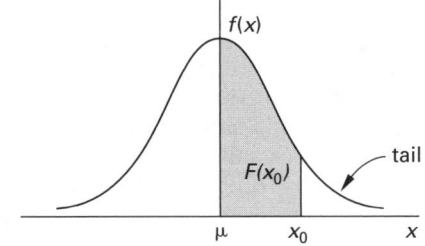

Figure 11.4 *Normal Distribution*

Since $f(x)$ is difficult to integrate, Eq. 11.41 is seldom used directly, and a *standard normal table* (see App. 11.A) is used instead. The standard normal table is based on a normal distribution with a mean of zero and a standard deviation of 1. Since the range of values from an experiment or phenomenon will not generally correspond to the standard normal table, a value, x_0, must be converted to a *standard normal value*, z. In Eq. 11.43, μ and σ are the mean and standard deviation, respectively, of the distribution from which x_0 comes.

$$z = \frac{x_0 - \mu}{\sigma} \qquad 11.43$$

Numbers in the standard normal table given by App. 11.A are the probabilities of the normalized x being between zero and z and represent the areas under the curve up to point z. When x is less than μ, z will be negative. However, the curve is symmetrical, so the table value corresponding to positive z can be used. The probability of x being greater than z is the complement of the table value. The curve area past point z is known as the *tail of the curve*.

The *error function*, $\mathrm{erf}(x)$, and its complement, $\mathrm{erfc}(x)$, are defined by Eqs. 11.44 and 11.45. The error function is used to determine the probable error of a measurement.

$$\mathrm{erf}(x_0) = \frac{2}{\sqrt{\pi}} \int_0^{x_0} e^{-x^2} dx \qquad 11.44$$

$$\mathrm{erfc}(x_0) = 1 - \mathrm{erf}(x_0) \qquad 11.45$$

Example 11.10

The mass, m, of a particular hand-laid fiberglass (Fibreglas™) part is normally distributed with a mean of 66 kg and a standard deviation of 5 kg. (a) What percent of the parts will have a mass less than 72 kg? (b) What percent of the parts will have a mass in excess of 72 kg? (c) What percent of the parts will have a mass between 61 and 72 kg?

Solution

(a) The value of 72 kg must be normalized by using Eq. 11.43. The standard normal variable is

$$z = \frac{x - \mu}{\sigma} = \frac{72 \text{ kg} - 66 \text{ kg}}{5 \text{ kg}} = 1.2$$

Reading from App. 11.A, the area under the normal curve is 0.3849. This represents the probability of the mass, m, being between 66 kg and 72 kg (i.e., z being between 0 and 1.2). However, the probability of the mass being less than 66 kg is also needed. Since the curve is symmetrical, this probability is 0.5. Therefore,

$$p\{m < 72 \text{ kg}\} = p\{z < 1.2\} = 0.5 + 0.3849 = 0.8849$$

(b) The probability of the mass exceeding 72 kg is the area under the tail past point z.

$$p\{m > 72 \text{ kg}\} = p\{z > 1.2\} = 0.5 - 0.3849 = 0.1151$$

(c) The standard normal variable corresponding to $m = 61$ kg is

$$z = \frac{x - \mu}{\sigma} = \frac{61 \text{ kg} - 66 \text{ kg}}{5 \text{ kg}} = -1$$

Since the two masses are on opposite sides of the mean, the probability will have to be determined in two parts.

$$\begin{aligned} p\{61 < m < 72\} &= p\{61 < m < 66\} + p\{66 < m < 72\} \\ &= p\{-1 < z < 0\} + p\{0 < z < 1.2\} \\ &= 0.3413 + 0.3849 = 0.7262 \end{aligned}$$

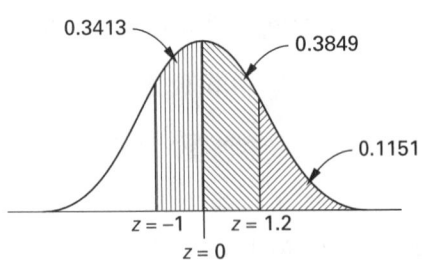

16. APPLICATION: RELIABILITY

Introduction

Reliability, $R\{t\}$, is the probability that an item will continue to operate satisfactorily up to time t. The *bathtub distribution*, Fig. 11.5, is often used to model the probability of failure of an item (or, the number of failures from a large population of items) as a function of time. Items initially fail at a high rate, a phenomenon known as *infant mortality*. For the majority of the operating time, known as the *steady-state operation*, the

failure rate is constant (i.e., is due to random causes). After a long period of time, the items begin to deteriorate and the failure rate increases. (No mathematical distribution describes all three of these phases simultaneously.)

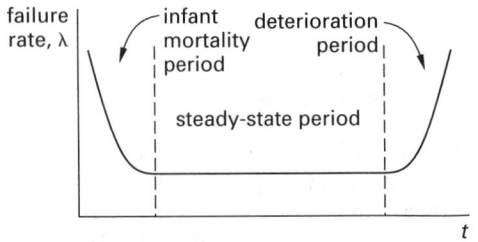

Figure 11.5 *Bathtub Reliability Curve*

The *hazard function*, $z\{t\}$, represents the *conditional probability of failure*—the probability of failure in the next time interval, given that no failure has occurred thus far.[3]

$$z\{t\} = \frac{f(t)}{R(t)} = \frac{\dfrac{dF(t)}{dt}}{1 - F(t)} \qquad 11.46$$

Exponential Reliability

Steady-state reliability is often described by the *negative exponential distribution*. This assumption is appropriate whenever an item fails only by random causes and does not experience deterioration during its life. The parameter λ is related to the *mean time to failure* (MTTF) of the item.[4]

$$R\{t\} = e^{-\lambda t} = e^{\frac{-t}{\text{MTTF}}} \qquad 11.47$$

$$\lambda = \frac{1}{\text{MTTF}} \qquad 11.48$$

Equation 11.47 and the exponential continuous distribution function, Eq. 11.38, are complementary.

$$R\{t\} = 1 - F(t) = 1 - (1 - e^{-\lambda t}) = e^{-\lambda t} \qquad 11.49$$

The hazard function for the negative exponential distribution is

$$z\{t\} = \lambda \qquad 11.50$$

Thus, the hazard function for exponential reliability is constant and does not depend on t (i.e., on the age of the item). In other words, the expected future life of an item is independent of the previous history (length of operation). This lack of memory is consistent with the assumption that only random causes contribute to failure during steady-state operations. And since random causes are unlikely discrete events, their probability of

---

[3]The symbol $z\{t\}$ is traditionally used for the hazard function and is not related to the standard normal variable.
[4]The term "mean time *between* failures" is improper. However, the term *mean time before failure* (MTBF) is acceptable.

occurrence can be represented by a Poisson distribution with mean λ. That is, the probability of having x failures in any given period is

$$p\{x\} = \frac{e^{-\lambda}\lambda^x}{x!} \qquad 11.51$$

Serial System Reliability

In the analysis of system reliability, the binary variable X_i is defined as one if item i operates satisfactorily and zero otherwise. Similarly, the binary variable Φ is one only if the entire system operates satisfactorily. Thus, Φ will depend on a *performance function* containing the X_i.

A *serial system* is one for which all items must operate correctly for the system to operate. Each item has its own reliability, R_i. For a serial system of n items, the performance function is

$$\Phi = X_1 X_2 X_3 \cdots X_n = \min(X_i) \qquad 11.52$$

The probability of a serial system operating correctly is

$$p\{\Phi = 1\} = R_{\text{serial system}} = R_1 R_2 R_3 \cdots R_n \qquad 11.53$$

Parallel System Reliability

A *parallel system* with n items will fail only if all n items fail. Such a system is said to be *redundant* to the nth degree. Using redundancy, a highly reliable system can be produced from components with relatively low individual reliabilities.

The performance function of a redundant system is

$$\begin{aligned} \Phi &= 1 - (1 - X_1)(1 - X_2)(1 - X_3)\cdots(1 - X_n) \\ &= \max(X_i) \end{aligned} \qquad 11.54$$

The reliability of the parallel system is

$$R = p\{\Phi = 1\} = 1 - (1-R_1)(1-R_2)(1-R_3) \cdots (1-R_n) \qquad 11.55$$

Example 11.11

The reliability of an item is exponentially distributed with mean time to failure (MTTF) of 1000 hr. What is the probability that the item will not have failed after 1200 hr of operation?

Solution

The probability of not having failed before time t is the reliability. From Eqs. 11.48 and 11.49,

$$\lambda = \frac{1}{\text{MTTF}} = \frac{1}{1000\,\text{hr}} = 0.001$$

$$R\{1200\} = e^{-\lambda t} = e^{(-0.001)(1200\ \text{hr})} = 0.3$$

Example 11.12

What are the reliabilities of the following systems?

(a)

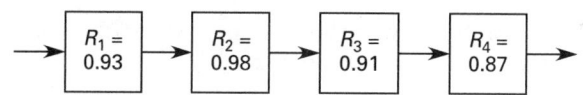

(b)

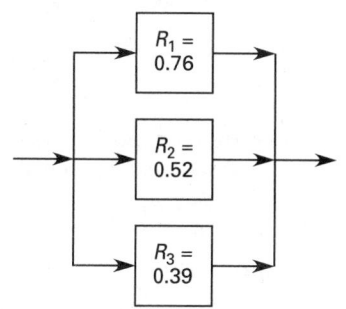

Solution

(a) This is a serial system. From Eq. 11.53,

$$R = R_1 R_2 R_3 R_4 = (0.93)(0.98)(0.91)(0.87) = 0.72$$

(b) This is a parallel system. From Eq. 11.55,

$$\begin{aligned} R &= 1 - (1 - R_1)(1 - R_2)(1 - R_3) \\ &= 1 - (1 - 0.76)(1 - 0.52)(1 - 0.39) = 0.93 \end{aligned}$$

17. ANALYSIS OF EXPERIMENTAL DATA

Experiments can take on many forms. An experiment might consist of measuring the mass of one cubic foot of concrete, or measuring the speed of a car on a roadway. Generally, such experiments are performed more than once to increase the precision and accuracy of the results.

Both systematic and random variations in the process being measured will cause the observations to vary, and the experiment would not be expected to yield the same

(b)

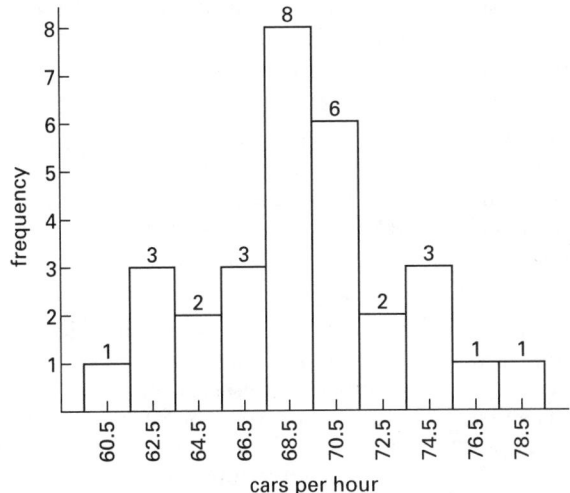

result each time it was performed. Eventually, a collection of experimental outcomes (observations) will be available for analysis.

The *frequency distribution* is a systematic method for ordering the observations from small to large, according to some convenient numerical characteristic. The *step interval* should be chosen so that the data are presented in a meaningful manner. If there are too many intervals, many of them will have zero frequencies; if there are too few intervals, the frequency distribution will have little value. Generally, 10 to 15 intervals are used.

Once the frequency distribution is complete, it can be represented graphically as a *histogram*. The procedure in drawing a histogram is to mark off the interval limits (also known as *class limits*) on a number line and then draw contiguous bars with lengths that are proportional to the frequencies in the intervals and that are centered on the midpoints of their respective intervals. The continuous nature of the data can be depicted by a *frequency polygon*. The number or percentage of observations that occur up to and including some value can be shown in a *cumulative frequency table*.

(c)

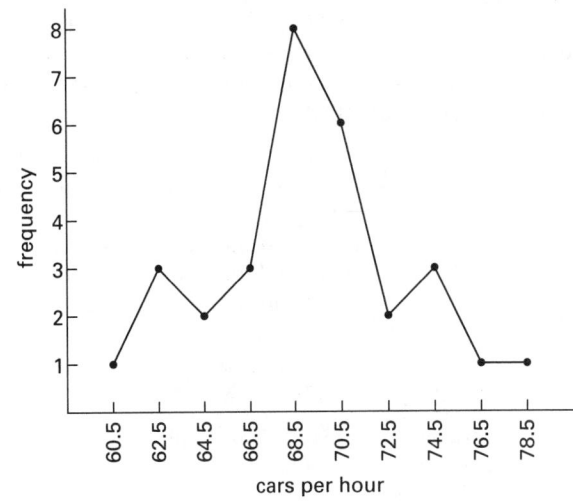

Example 11.13

The number of cars that travel through an intersection between 12 noon and 1 p.m. is measured for 30 consecutive working days. The results of the 30 observations are

79, 66, 72, 70, 68, 66, 68, 76, 73, 71, 74, 70, 71, 69, 67, 74, 70, 68, 69, 64, 75, 70, 68, 69, 64, 69, 62, 63, 63, 61

(a) What are the frequency and cumulative distributions? (Use a distribution interval of two cars per hour.) (b) Draw the histogram. (Use a cell size of two cars per hour.) (c) Draw the frequency polygon. (d) Graph the cumulative frequency distribution.

Solution

(d)

(a)

cars per hour	frequency	cumulative frequency	cumulative percent
60–61	1	1	3
62–63	3	4	13
64–65	2	6	20
66–67	3	9	30
68–69	8	17	57
70–71	6	23	77
72–73	2	25	83
74–75	3	28	93
76–77	1	29	97
78–79	1	30	100

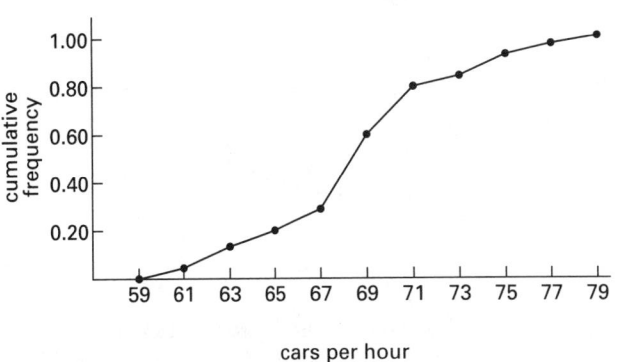

18. MEASURES OF CENTRAL TENDENCY

It is often unnecessary to present the experimental data in their entirety, either in tabular or graphical form. In such cases, the data and distribution can be represented by various parameters. One type of parameter is a measure of *central tendency*. Mode, median, and mean are measures of central tendency.

The *mode* is the observed value that occurs most frequently. The mode may vary greatly between series of observations. Therefore, its main use is as a quick measure of the central value since little or no computation is required to find it. Beyond this, the usefulness of the mode is limited.

The *median* is the point in the distribution that partitions the total set of observations into two parts containing equal numbers of observations. It is not influenced by the extremity of scores on either side of the distribution. The median is found by counting up (from either end of the frequency distribution) until half of the observations have been accounted for.

Similar in concept to the median are *percentile ranks, quartiles,* and *deciles.* The median could also have been called the *50th percentile* observation. Similarly, the 80th percentile would be the observed value (e.g., the number of cars per hour) for which the cumulative frequency was 80%. The quartile and decile points on the distribution divide the observations or distribution into segments of 25% and 10%, respectively.

The *arithmetic mean* is the arithmetic average of the observations. The sample mean, $\bar{x}$, can be used as an unbiased estimator of the population mean, μ. The *mean* may be found without ordering the data (as was necessary to find the mode and median). The mean can be found from the following formula.

$$\bar{x} = \left(\frac{1}{n}\right)(x_1 + x_2 + \cdots + x_n) = \frac{\sum x_i}{n} \qquad \textit{11.56}$$

The *geometric mean* is used occasionally when it is necessary to average ratios. The geometric mean is calculated as

$$\text{geometric mean} = \sqrt[n]{x_1 x_2 x_3 \cdots x_n} \quad [x_i > 0] \quad \textit{11.57}$$

The *harmonic mean* is defined as

$$\text{harmonic mean} = \frac{n}{\dfrac{1}{x_1} + \dfrac{1}{x_2} + \cdots + \dfrac{1}{x_n}} \qquad \textit{11.58}$$

The *root-mean-squared* (rms) *value* of a series of observations is defined as

$$x_{\text{rms}} = \sqrt{\frac{\sum x_i^2}{n}} \qquad \textit{11.59}$$

Example 11.14

Find the mode, median, and arithmetic mean of the distribution represented by the data given in Ex. 11.13.

Solution

The mode is the interval 68–69, since this interval has the highest frequency. If 68.5 is taken as the interval center, then 68.5 would be the mode.

Since there are 30 observations, the median is the value that separates the observations into two groups of 15. From Ex. 11.13, the median occurs someplace within the 68–69 interval. Up through interval 66–67, there are nine observations, so six more are needed to make 15. Interval 68–69 has eight observations, so the median is found to be 6/8 or 3/4 of the way through the interval. Since the real limits of the interval are 67.5 and 69.5, the median is located at

$$67.5 + \left(\tfrac{3}{4}\right)(69.5 - 67.5) = 69$$

The mean can be found from the raw data or from the grouped data using the interval center as the assumed observation value. Using the raw data,

$$\bar{x} = \frac{\sum x}{n} = \frac{2069}{30} = 68.97$$

19. MEASURES OF DISPERSION

The simplest statistical parameter that describes the variability in observed data is the *range*. The range is found by subtracting the smallest value from the largest. Since the range is influenced by extreme (low probability) observations, its use as a measure of variability is limited.

The *standard deviation* is a better estimate of variability because it considers every observation. That is, N in Eq. 11.60 is the total population size, not the sample size, n.

$$\sigma = \sqrt{\frac{\sum (x_i - \mu)^2}{N}} = \sqrt{\frac{\sum x_i^2}{N} - \mu^2} \qquad \textit{11.60}$$

The standard deviation of a sample (particularly a small sample) is a biased (i.e., is not a good) estimator of the population standard deviation. An *unbiased estimator* of the population standard deviation is the *sample standard deviation, s.*[5]

$$s = \sqrt{\frac{\sum (x_i - \bar{x})^2}{n-1}} = \sqrt{\frac{\sum x_i^2 - \dfrac{(\sum x_i)^2}{n}}{n-1}} \qquad \textit{11.61}$$

---

[5]There is a subtle yet significant difference between *standard deviation of the sample,* σ (obtained from Eq. 11.60 for a finite sample drawn from a larger population) and the *sample standard deviation, s* (obtained from Eq. 11.61). While σ can be calculated, it has no significance or use as an estimator. It is true that the difference between σ and s approaches zero when the sample size, n, is large, but this convergence does nothing to legitimize the use of σ as an estimator of the true standard deviation. (Some people say "large" is 30, others say 50 or 100.)

If the sample standard deviation, s, is known, the standard deviation of the sample, σ_{sample}, can be calculated.

$$\sigma_{\text{sample}} = s\sqrt{\frac{n-1}{n}} \qquad 11.62$$

The *variance* is the square of the standard deviation. Since there are two standard deviations, there are two variances. The *variance of the sample* is σ^2, and the *sample variance* is s^2.

The *relative dispersion* is defined as a measure of dispersion divided by a measure of central tendency. The *coefficient of variation* is a relative dispersion calculated from the sample standard deviation and the mean.

$$\text{coefficient of variation} = \frac{s}{\bar{x}} \qquad 11.63$$

Skewness is a measure of a frequency distribution's lack of symmetry.

$$\text{skewness} = \frac{\bar{x} - \text{mode}}{s}$$
$$\approx \frac{(3)(\bar{x} - \text{median})}{s} \qquad 11.64$$

Example 11.15

For the data given in Ex. 11.13, calculate (a) the sample range, (b) the standard deviation of the sample, (c) an unbiased estimator of the population standard deviation, (d) the variance of the sample, and (e) the sample variance.

Solution

$$\sum x_i = 2069$$
$$\left(\sum x_i\right)^2 = (2069)^2 = 4{,}280{,}761$$
$$\sum x_i^2 = 143{,}225$$
$$n = 30$$
$$\bar{x} = \frac{2069}{30} = 68.967$$

(a) $R = x_{\text{max}} - x_{\text{min}} = 79 - 61 = 18$

(b) $\sigma = \sqrt{\dfrac{\sum x_i^2}{n} - (\bar{x})^2} = \sqrt{\dfrac{143{,}225}{30} - \left(\dfrac{2069}{30}\right)^2}$

$\quad = \sqrt{17.766} = 4.215$

(c) $s = \sqrt{\dfrac{\sum x_i^2 - \dfrac{(\sum x_i)^2}{n}}{n-1}} = \sqrt{\dfrac{143{,}225 - \dfrac{4{,}280{,}761}{30}}{29}}$

$\quad = \sqrt{18.378} = 4.287$

(d) $\sigma^2 = 17.77$

(e) $s^2 = 18.38$

20. CENTRAL LIMIT THEOREM

Measuring a sample of n items from a population with mean μ and standard deviation σ is the general concept of an experiment. The sample mean, $\bar{x}$, is one of the parameters that can be derived from the experiment. This experiment can be repeated k times, yielding a set of averages $(\bar{x}_1, \bar{x}_2, \ldots, \bar{x}_k)$. The k numbers in the set themselves represent samples from distributions of averages. The average of averages, $\bar{\bar{x}}$, and sample standard deviation of averages, $s_{\bar{x}}$ (known as the *standard error of the mean*), can be calculated.

The *central limit theorem* defines the mean of the sample averages. The theorem can be stated in several ways, but the essential elements are the following points.

(1) The averages, $\bar{x}_i$, are normally distributed variables, even if the original data from which they are calculated are not normally distributed.

(2) The grand average, $\bar{\bar{x}}$ (i.e., the average of the averages), approaches and is an unbiased estimator of μ.

$$\mu \approx \bar{\bar{x}} \qquad 11.65$$

The standard deviation of the original distribution, σ, is much larger than the standard error of the mean.

$$\sigma \approx \sqrt{n}\,s_{\bar{x}} \qquad 11.66$$

21. CONFIDENCE LEVEL

The results of experiments are seldom correct 100% of the time. Recognizing this, researchers accept a certain probability of being wrong. In order to minimize this probability, the experiment is repeated several times. The number of repetitions depends on the desired level of confidence in the results.

If the results have a 5% probability of being wrong, the *confidence level*, C, is 95% that the results are correct, in which case the results are said to be *significant*. If the results have only a 1% probability of being wrong, the confidence level is 99%, and the results are said to be *highly significant*. Other confidence levels (90%, 99.5%, etc.) are used as appropriate.

22. APPLICATION: CONFIDENCE LIMITS

As a consequence of the central limit theorem, sample means of n items taken from a normal distribution with mean μ and standard deviation σ will be normally distributed with mean μ and variance σ^2/n. Thus, the probability that any given average, $\bar{x}$, exceeds some value, L, is

$$p\{\bar{x} > L\} = p\left\{z > \left|\frac{L-\mu}{\frac{\sigma}{\sqrt{n}}}\right|\right\} \qquad 11.67$$

L is the *confidence limit* for the confidence level $1 - p\{\bar{x} > L\}$ (expressed as a percent). Values of z are read directly from the standard normal table. As an example, $z = 1.645$ for a 95% confidence level since only 5% of the curve is above that z in the upper tail. Similar values are given in Table 11.1. This is known as a *one-tail confidence limit* because all of the probability is given to one side of the variation.

Table 11.1 *Values of z for Various Confidence Levels*

confidence level, C	one-tail limit z	two-tail limit z
90%	1.28	1.645
95%	1.645	1.96
97.5%	1.96	2.17
99%	2.33	2.575
99.5%	2.575	2.81
99.75%	2.81	3.00

With *two-tail confidence limits*, the probability is split between the two sides of variation. There will be upper and lower confidence limits, UCL and LCL, respectively.

$$p\{\text{LCL} < \bar{x} < \text{UCL}\} = p\left\{ \frac{\text{LCL} - \mu}{\frac{\sigma}{\sqrt{n}}} < z < \frac{\text{UCL} - \mu}{\frac{\sigma}{\sqrt{n}}} \right\} \qquad 11.68$$

23. APPLICATION: BASIC HYPOTHESIS TESTING

A *hypothesis test* is a procedure that answers the question, "Did these data come from [a particular type of] distribution?" There are many types of tests, depending on the distribution and parameter being evaluated. The simplest hypothesis test determines whether an average value obtained from n repetitions of an experiment could have come from a population with known mean, μ, and standard deviation, σ. A practical application of this question is whether a manufacturing process has changed from what it used to be or should be. Of course, the answer (i.e., "yes" or "no") cannot be given with absolute certainty—there will be a confidence level associated with the answer.

The following procedure is used to determine whether the average of n measurements can be assumed (with a given confidence level) to have come from a known population.

step 1: Assume random sampling from a normal population.

step 2: Choose the desired confidence level, C.

step 3: Decide on a one-tail or two-tail test. If the hypothesis being tested is that the average has or has not *increased* or *decreased*, choose a one-tail test. If the hypothesis being tested

is that the average has or has not *changed*, choose a two-tail test.

step 4: Use Table 11.1 or the standard normal table to determine the z-value corresponding to the confidence level and number of tails.

step 5: Calculate the actual standard normal variable, z'.

$$z' = \frac{\bar{x} - \mu}{\frac{\sigma}{\sqrt{n}}} \qquad 11.69$$

step 6: If $z' \geq z$, the average can be assumed (with confidence level C) to have come from a different distribution.

Example 11.16

When it is operating properly, a chemical plant has a daily production rate that is normally distributed with a mean of 880 tons/day and a standard deviation of 21 tons/day. During an analysis period, the output is measured on 50 consecutive days, and the mean output is found to be 871 tons/day. With a 95% confidence level, determine whether the plant is operating properly.

Solution

step 1: Given.

step 2: $C = 0.95$ is given.

step 3: Since a specific direction in the variation is not given (i.e., the example does not ask whether the average has decreased), use a two-tail hypothesis test.

step 4: From Table 11.1, $z = 1.96$.

step 5: From Eq. 11.69,

$$z' = \left| \frac{\bar{x} - \mu}{\frac{\sigma}{\sqrt{n}}} \right| = \left| \frac{871 - 880}{\frac{21}{\sqrt{50}}} \right| = 3.03$$

Since $3.03 > 1.96$, the distributions are not the same. There is at least a 95% probability that the plant is not operating correctly.

24. APPLICATION: STATISTICAL PROCESS CONTROL

All manufacturing processes contain variation due to random and nonrandom causes. Random variation cannot be eliminated. *Statistical process control* (SPC) is the act of monitoring and adjusting the performance of a process to detect and eliminate nonrandom variation.

Statistical process control is based on taking regular (hourly, daily, etc.) samples of n items and calculating the mean, $\bar{x}$, and range, R, of the sample. To simplify the calculations, the range is used as a measure of the dispersion. These two parameters are graphed on their respective x-bar and R-*control charts*.[6] Confidence limits are drawn at $\pm 3\sigma/\sqrt{n}$. From a statistical standpoint, the control chart tests a hypothesis each time a point is plotted. When a point falls outside these limits, there is a 99.75% probability that the process is out of control. Until a point exceeds the control limits, no action is taken.[7]

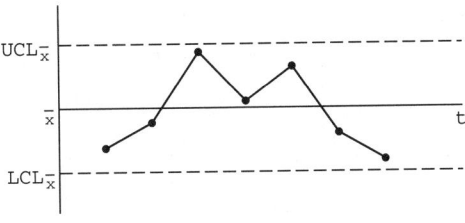

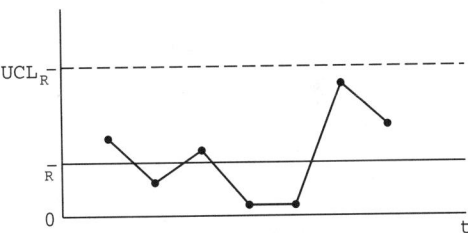

Figure 11.6 *Typical Statistical Process Control Charts*

25. MEASURES OF EXPERIMENTAL ADEQUACY

An experiment is said to be *accurate* if it is unaffected by experimental error. In this case, *error* is not synonymous with *mistake*, but rather includes all variations not within the experimenter's control.

For example, suppose a gun is aimed at a point on a target and five shots are fired. The mean distance from the point of impact to the sight in point is a measure of the alignment accuracy between the barrel and sights. The difference between the actual value and the experimental value is known as *bias*.

Precision is not synonymous with accuracy. Precision is concerned with the repeatability of the experimental results. If an experiment is repeated with identical results, the experiment is said to be precise.

The average distance of each impact from the centroid of the impact group is a measure of the precision of the experiment. Thus, it is possible to have a highly precise experiment with a large bias.

Most of the techniques applied to experiments in order to improve the accuracy of the experimental results (e.g., repeating the experiment, refining the experimental methods, or reducing variability) actually increase the precision.

Sometimes the word *reliability* is used with regard to the precision of an experiment. Thus, a "reliable estimate" is used in the same sense as a "precise estimate."

Stability and *insensitivity* are synonymous terms. A stable experiment will be insensitive to minor changes in the experimental parameters. For example, suppose the centroid of a bullet group is 2.1 in from the point of aim at 65°F and 2.3 in away at 80°F. The sensitivity of the experiment to temperature change would be

$$\text{sensitivity} = \frac{\Delta x}{\Delta T} = \frac{2.3 \text{ in} - 2.1 \text{ in}}{80°\text{F} - 65°\text{F}} = 0.0133 \text{ in}/°\text{F}$$

26. LINEAR REGRESSION

If it is necessary to draw a straight line $(y = mx + b)$ through n data points $(x_1, y_1), (x_2, y_2), \ldots, (x_n, y_n)$, the following method based on the *method of least squares* can be used.

step 1: Calculate the following nine quantities.

$$\sum x_i, \quad \sum x_i^2, \quad \left(\sum x_i\right)^2, \quad \bar{x} = \frac{\sum x_i}{n}, \quad \sum x_i y_i,$$

$$\sum y_i, \quad \sum y_i^2, \quad \left(\sum y_i\right)^2, \quad \bar{y} = \frac{\sum y_i}{n}$$

step 2: Calculate the slope, m, of the line.

$$m = \frac{n\sum(x_i y_i) - (\sum x_i)(\sum y_i)}{n\sum x_i^2 - (\sum x_i)^2} \qquad 11.70$$

step 3: Calculate the y-intercept, b.

$$b = \bar{y} - m\bar{x} \qquad 11.71$$

step 4: To determine the goodness of fit, calculate the correlation coefficient, r.

$$r = \frac{n\sum(x_i y_i) - (\sum x_i)(\sum y_i)}{\sqrt{[n\sum x_i^2 - (\sum x_i)^2][n\sum y_i^2 - (\sum y_i)^2]}} \qquad 11.72$$

If m is positive, r will be positive; if m is negative, r will be negative. As a general rule, if the absolute value of r exceeds 0.85, the fit is good; otherwise, the fit is poor. r equals 1.0 if the fit is a perfect straight line.

---

[6]Other charts (e.g., the σ-*chart*, p-*chart*, and c-*chart*) are less common but are used as required.

[7]Other indications that a correction may be required are seven measurements on one side of the average and seven consecutively increasing measurements. Rules such as these detect shifts and trends.

A low value of r does not eliminate the possibility of a nonlinear relationship existing between x and y. It is possible that the data describe a parabolic, logarithmic, or other nonlinear relationship. (Usually this will be apparent if the data are graphed.) It may be necessary to convert one or both variables to new variables by taking squares, square roots, cubes, or logarithms, to name a few of the possibilities, in order to obtain a linear relationship. The apparent shape of the line through the data will give a clue to the type of variable transformation that is required. The curves in Fig. 11.7 may be used as guides to some of the simpler variable transformations.

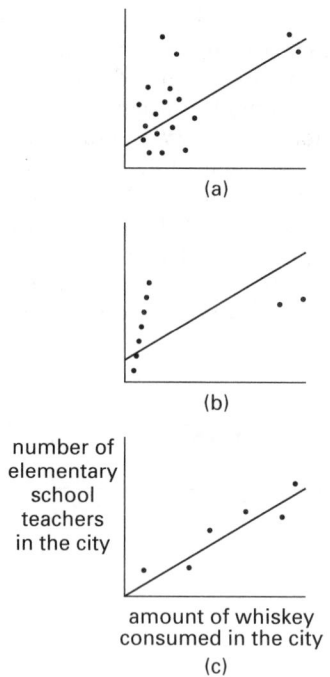

Figure 11.8 *Common Regression Difficulties*

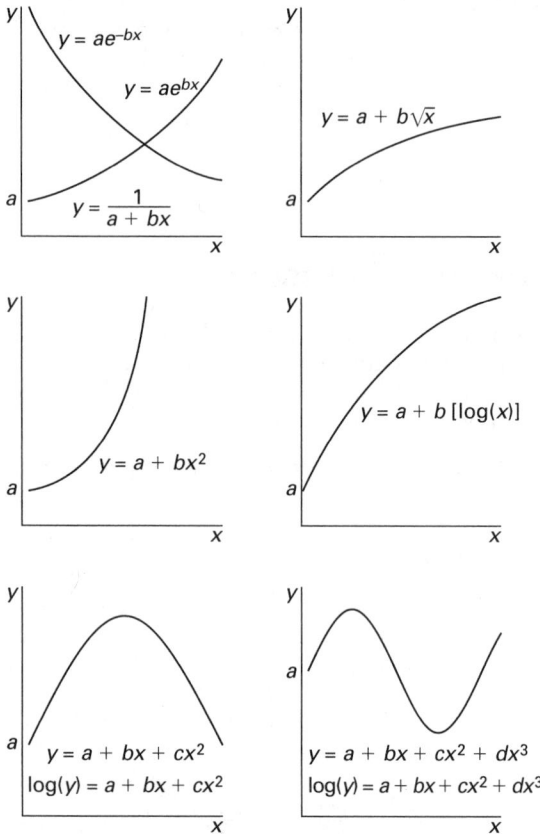

Figure 11.7 *Nonlinear Data Curves*

Figure 11.8 illustrates several common problems encountered in trying to fit and evaluate curves from experimental data. Figure 11.8(a) shows a graph of clustered data with several extreme points. There will be moderate correlation due to the weighting of the extreme points, although there is little actual correlation at low values of the variables. The extreme data should be excluded, or the range should be extended by obtaining more data.

Figure 11.8(b) shows that good correlation exists in general, but extreme points are missed and the overall correlation is moderate. If the results within the small linear range can be used, the extreme points should be excluded. Otherwise, additional data points are needed, and curvilinear relationships should be investigated.

Figure 11.8(c) illustrates the problem of drawing conclusions of cause and effect. There may be a predictable relationship between variables, but that does not imply a cause and effect relationship. In the case shown, both variables are functions of a third variable, the city population. But, there is no direct relationship between the plotted variables.

Example 11.17

An experiment is performed in which the dependent variable y is measured against the independent variable x. The results are as follows.

x	y
1.2	0.602
4.7	5.107
8.3	6.984
20.9	10.031

(a) What is the least squares straight line equation that represents this data? (b) What is the correlation coefficient?

Solution

(a)

$$\sum x_i = 35.1$$

$$\sum y_i = 22.72$$

$$\sum x_i^2 = 529.23$$

$$\sum y_i^2 = 175.84$$

$$\left(\sum x_i\right)^2 = 1232.01$$

$$\left(\sum y_i\right)^2 = 516.38$$

$$\bar{x} = 8.775$$

$$\bar{y} = 5.681$$

$$\sum x_i y_i = 292.34$$

$$n = 4$$

From Eq. 11.70, the slope is

$$m = \frac{(4)(292.34) - (35.1)(22.72)}{(4)(529.23) - (35.1)^2} = 0.42$$

From Eq. 11.71, the y-intercept is

$$b = 5.681 - (0.42)(8.775) = 2.0$$

The equation of the line is

$$y = 0.42x + 2.0$$

(b) From Eq. 11.72, the correlation coefficient is

$$r = \frac{(4)(292.34) - (35.1)(22.72)}{\sqrt{\begin{array}{c}[(4)(529.23) - 1232.01] \\ \times [(4)(175.84) - 516.38]\end{array}}} = 0.914$$

Example 11.18

Repeat Ex. 11.17 assuming the relationship between the variables is nonlinear.

Solution

The first step is to graph the data. Since the graph has the appearance of the fourth case, it can be assumed that the relationship between the variables has the form of $y = a + b[\log(x)]$. Therefore, the variable change $z = \log(x)$ is made, resulting in the following set of data.

z	y
0.0792	0.602
0.672	5.107
0.919	6.984
1.32	10.031

If the regression analysis is performed on this set of data, the resulting equation and correlation coefficient are

$$y = -0.036 + 7.65z$$

$$r = 0.999$$

This is a very good fit. The relationship between the variable x and y is approximately

$$y = -0.036 + 7.65\big(\log(x)\big)$$

PRACTICE PROBLEMS

1. The time taken by a toll taker to collect the toll from vehicles crossing a bridge is an exponential distribution with mean of 23 sec. What is the probability that a random vehicle will be processed in 25 sec or more (i.e., will take longer than 25 sec)?

2. (*Time limit: one hour*) A survey field crew measures one leg of a traverse four times. The following results are obtained.

repetition	measurement	direction
1	1249.529	forward
2	1249.494	backward
3	1249.384	forward
4	1249.348	backward

The crew chief is under orders to obtain readings with confidence limits of 90%.

(a) Which readings are acceptable? (b) Which readings are not acceptable? (c) Explain how to determine which readings are not acceptable. (d) What is the most probable value of the distance? (e) What is the error in the most probable value (at 90% confidence)? (f) If the distance is one side of a square traverse whose sides are all equal, what is the most probable closure error? (g) What is the probable error of part (f) expressed as a fraction? (h) Is this error of closure within the second order of accuracy? (i) Define accuracy and distinguish it from precision. (j) Give an example of systematic error.

3. California law requires a statistical analysis of the average speed driven by motorists on a road prior to the use of radar speed control. The following speeds (all in mi/hr) were observed in a random sample of 40 cars.

44, 48, 26, 25, 20, 43, 40, 42, 29, 39, 23, 26, 24, 47, 45, 28, 29, 41, 38, 36, 27, 44, 42, 43, 29, 37, 34, 31, 33, 30, 42, 43, 28, 41, 29, 36, 35, 30, 32, 31

(a) Tabulate the frequency distribution of the data. (b) Draw the frequency histogram. (c) Draw the frequency polygon. (d) Tabulate the cumulative frequency distribution. (e) Draw the cumulative frequency graph. (f) What is the upper quartile speed? (g) What are the mode, median, and mean speeds? (h) What is the standard deviation of the sample data? (i) What is the sample standard deviation? (j) What is the sample variance?

4. The number of cars entering a toll plaza on a bridge during the hour after midnight follows a Poisson distribution with a mean of 20.

(a) What is the probability that 17 cars will pass through the toll plaza during that hour on any given night? (b) What is the probability that 3 or fewer cars will pass through the toll plaza at that hour on any given night?

5. 100 bearings were tested to failure. The average life was 1520 hr, and the standard deviation was 120 hr. The manufacturer advertises a 1600 hr life. Should the company change its claim? Evaluate using confidence limits of 95% and 99%.

6. (a) Find the best equation for a line passing through the points given. (b) Find the correlation coefficient.

x	y
400	370
800	780
1250	1210
1600	1560
2000	1980
2500	2450
4000	3950

7. Find the best equation for a line passing through the points given.

s	t
20	43
18	141
16	385
14	1099

8. The number of vehicles lining up behind a flashing railroad crossing has been observed for five trains of different lengths, as given. What is the mathematical formula that relates the two variables?

no. of cars in train	no. of vehicles
2	14.8
5	18.0
8	20.4
12	23.0
27	29.9

9. A mechanical component exhibits a negative exponential failure distribution with a mean time to failure of 1000 hr. What is the maximum operating time such that the reliability remains above 99%?

12 Numbering Systems

1. Positional Numbering Systems 12-1
2. Converting Base-b Numbers to Base-10 . . . 12-1
3. Converting Base-10 Numbers to Base-b . . . 12-1
4. Binary Number System 12-1
5. Octal Number System 12-2
6. Hexadecimal Number System 12-2
7. Conversions Among Binary, Octal, and
   Hexadecimal Numbers 12-3
8. Complement of a Number 12-3
9. Application of Complements to Computer
   Arithmetic 12-3
10. Computer Representation of Negative
    Numbers 12-4

1. POSITIONAL NUMBERING SYSTEMS

A *base-b number*, N_b, is made up of individual *digits*. In a *positional numbering system*, the position of a digit in the number determines that digit's contribution to the total value of the number. Specifically, the position of the digit determines the power to which the *base* (also known as the *radix*), b, is raised. For *decimal numbers*, the radix is 10, hence the description *base-10 numbers*.

$$(a_n a_{n-1} \cdots a_2 a_1 a_0)_b = a_n b^n + a_{n-1} b^{n-1} + \cdots \\ + a_2 b^2 + a_1 b + a_0 \quad \textbf{12.1}$$

The leftmost digit, a_n, contributes the greatest to the number's magnitude and is known as the *most significant digit* (MSD). The rightmost digit, a_0, contributes the least and is known as the *least significant digit* (LSD).

2. CONVERTING BASE-b NUMBERS TO BASE-10

Equation 12.1 converts base-b numbers to base-10 numbers. The calculation of the right-hand side of Eq. 12.1 is performed in the base-10 arithmetic and is known as the *expansion method*.[1]

Converting base-b fractions to base-10 is similar to converting whole numbers and is accomplished by Eq. 12.2.

$$(0.a_1 a_2 \cdots a_m)_b = a_1 b^{-1} + a_2 b^{-2} + \cdots + a_m b^{-m} \quad \textbf{12.2}$$

[1]Equation 12.1 works with any base number. The *double-dabble (double and add) method* is a specialized method of converting from base-2 to base-10 numbers.

3. CONVERTING BASE-10 NUMBERS TO BASE-b

The *remainder method* is used to convert base-10 numbers to base-b numbers. This method consists of successive divisions by the base, b, until the quotient is zero. The base-b number is found by taking the remainders in the reverse order from which they were found. This method is illustrated in Exs. 12.1 and 12.3.

Converting a base-10 fraction to base-b requires multiplication of the base-10 fraction and subsequent fractional parts by the base. The base-b fraction is formed from the integer parts of the products taken in the same order in which they were determined. This is illustrated in Ex. 12.2(d).

4. BINARY NUMBER SYSTEM

There are only two *binary digits (bits)* in the *binary number system*: zero and one.[2] Thus, all binary numbers consist of strings of bits (i.e., zeros and ones). The leftmost bit is known as the *most significant bit* (MSB), and rightmost bit is the *least significant bit* (LSB).

As with digits from other numbering systems, bits can be added, subtracted, multiplied, and divided, although only digits 0 and 1 are allowed in the results. The rules of bit addition are

$$0 + 0 = 0$$
$$0 + 1 = 1$$
$$1 + 0 = 1$$
$$1 + 1 = 0 \text{ carry } 1$$

Example 12.1

(a) Convert $(1011)_2$ to base-10. (b) Convert $(75)_{10}$ to base-2.

Solution

(a) Using Eq. 12.1 with $b = 2$,

$$(1)(2)^3 + (0)(2)^2 + (1)(2)^1 + 1 = 11$$

[2]Alternatively, the binary states may be called *true* and *false*, *on* and *off*, *high* and *low*, or *positive* and *negative*.

(b) Use the remainder method. (See Sec. 3.)

$$75 \div 2 = 37 \text{ remainder } 1$$
$$37 \div 2 = 18 \text{ remainder } 1$$
$$18 \div 2 = 9 \text{ remainder } 0$$
$$9 \div 2 = 4 \text{ remainder } 1$$
$$4 \div 2 = 2 \text{ remainder } 0$$
$$2 \div 2 = 1 \text{ remainder } 0$$
$$1 \div 2 = 0 \text{ remainder } 1$$

The binary representation of $(75)_{10}$ is $(1001011)_2$.

5. OCTAL NUMBER SYSTEM

The *octal (base-8) system* is one of the alternatives to working with long binary numbers. Only the digits 0 through 7 are used. The rules for addition in the octal system are the same as for the decimal system except that the digits 8 and 9 do not exist. For example,

$$7 + 1 = 6 + 2 = 5 + 3 = (10)_8$$
$$7 + 2 = 6 + 3 = 5 + 4 = (11)_8$$
$$7 + 3 = 6 + 4 = 5 + 5 = (12)_8$$

Example 12.2

Perform the following operations.

(a) $(2)_8 + (5)_8$

(b) $(7)_8 + (6)_8$

(c) Convert $(75)_{10}$ to base-8.

(d) Convert $(0.14)_{10}$ to base-8.

(e) Convert $(13)_8$ to base-10.

(f) Convert $(27.52)_8$ to base-10.

Solution

(a) The sum of 2 and 5 in base-10 is 7, which is less than 8 and, therefore, is a valid number in the octal system. The answer is $(7)_8$.

(b) The sum of 7 and 6 in base-10 is 13, which is greater than 8 (and, therefore, needs to be converted). Using the remainder method (Sec. 3),

$$13 \div 8 = 1 \text{ remainder } 5$$
$$1 \div 8 = 0 \text{ remainder } 1$$

The answer is $(15)_8$.

(c) Use the remainder method (Sec. 3),

$$75 \div 8 = 9 \text{ remainder } 3$$
$$9 \div 8 = 1 \text{ remainder } 1$$
$$1 \div 8 = 0 \text{ remainder } 1$$

The answer is $(113)_8$.

(d) Refer to Sec. 3.

$$0.14 \times 8 = 1.12$$
$$0.12 \times 8 = 0.96$$
$$0.96 \times 8 = 7.68$$
$$0.68 \times 8 = 5.44$$
$$0.44 \times 8 = \text{etc.}$$

The answer, $(0.1075 \cdots)_8$, is constructed from the integer parts of the products.

(e) Use Eq. 12.1.

$$(1)(8) + 3 = (11)_{10}$$

(f) Use Eqs. 12.1 and 12.2.

$$(2)(8)^1 + (7)(8)^0 + (5)(8)^{-1} + (2)(8)^{-2} = 16 + 7 + \frac{5}{8} + \frac{2}{64}$$
$$= (23.656)_{10}$$

6. HEXADECIMAL NUMBER SYSTEM

The *hexadecimal (base-16) system* is a shorthand method of representing the value of four binary digits at a time.[3] Since 16 distinctly different characters are needed, the capital letters A through F are used to represent the decimal numbers 10 through 15. The progression of hexadecimal numbers is illustrated in Table 12.1.

Example 12.3

(a) Convert $(4D3)_{16}$ to base-10. (b) Convert $(1475)_{10}$ to base-16. (c) Convert $(0.8)_{10}$ to base-16.

Solution

(a) The hexadecimal number D is 13 in base-10. Using Eq. 12.1,

$$(4)(16)^2 + (13)(16)^1 + 3 = (1235)_{10}$$

(b) Use the remainder method. (Sec. 3.)

$$1475 \div 16 = 92 \text{ remainder } 3$$
$$92 \div 16 = 5 \text{ remainder } 12$$
$$5 \div 16 = 0 \text{ remainder } 5$$

Since $(12)_{10}$ is $(C)_{16}$, (or hex C), the answer is $(5C3)_{16}$.

(c) Refer to Sec. 3.

$$0.8 \times 16 = 12.8$$
$$0.8 \times 16 = 12.8$$
$$0.8 \times 16 = \text{etc.}$$

Since $(12)_{10} = (C)_{16}$, the answer is $(0.CCCCC\cdots)_{16}$.

---

[3]The term *hex number* is often heard.

7. CONVERSIONS AMONG BINARY, OCTAL, AND HEXADECIMAL NUMBERS

The octal system is closely related to the binary system since $(2)^3 = 8$. Conversion from a binary to an octal number is accomplished directly by starting at the LSB (right-hand bit) and grouping the bits in threes. Each group of three bits corresponds to an octal digit. Similarly, each digit in an octal number generates three bits in the equivalent binary number.

Conversion from a binary to a hexadecimal number starts by grouping the bits (starting at the LSB) into fours. Each group of four bits corresponds to a hexadecimal digit. Similarly, each digit in a hexadecimal number generates four bits in the equivalent binary number.

Table 12.1 *Binary, Octal, Decimal, and Hexadecimal Equivalents*

binary	octal	decimal	hexadecimal
0	0	0	0
1	1	1	1
10	2	2	2
11	3	3	3
100	4	4	4
101	5	5	5
110	6	6	6
111	7	7	7
1000	10	8	8
1001	11	9	9
1010	12	10	A
1011	13	11	B
1100	14	12	C
1101	15	13	D
1110	16	14	E
1111	17	15	F
10000	20	16	10

Conversion between octal and hexadecimal numbers is easiest when the number is first converted to a binary number.

Example 12.4

(a) Convert $(5431)_8$ to base-2. (b) Convert $(1001011)_2$ to base-8. (c) Convert $(1011111101111001)_2$ to base-16.

Solution

(a) Convert each octal digit to binary digits.

$$(5)_8 = (101)_2$$
$$(4)_8 = (100)_2$$
$$(3)_8 = (011)_2$$
$$(1)_8 = (001)_2$$

The answer is $(101100011001)_2$.

(b) Group the bits into threes starting at the LSB.

$$1 \quad 001 \quad 011$$

Convert these groups into their octal equivalents.

$$(1)_2 = (1)_8$$
$$(001)_2 = (1)_8$$
$$(011)_2 = (3)_8$$

The answer is $(113)_8$.

(c) Group the bits into fours starting at the LSB.

$$1011 \quad 1111 \quad 0111 \quad 1001$$

Convert these groups into their hexadecimal equivalents.

$$(1011)_2 = (B)_{16}$$
$$(1111)_2 = (F)_{16}$$
$$(0111)_2 = (7)_{16}$$
$$(1001)_2 = (9)_{16}$$

The answer is $(BF79)_{16}$.

8. COMPLEMENT OF A NUMBER

The *complement*, N^*, of a number, N, depends on the machine (computer, calculator, etc.) being used. Assuming that the machine has a maximum number, n, of digits per integer number stored, the b's and $(b-1)$'s complements are

$$N_b^* = b^n - N \hspace{2cm} 12.3$$
$$N_{b-1}^* = N_b^* - 1 \hspace{2cm} 12.4$$

For a machine that works in base-10 arithmetic, the *tens* and *nines complements* are

$$N_{10}^* = 10^n - N \hspace{2cm} 12.5$$
$$N_9^* = N_{10}^* - 1 \hspace{2cm} 12.6$$

For a machine that works in base-2 arithmetic, the *twos* and *ones complements* are

$$N_2^* = 2^n - N \hspace{2cm} 12.7$$
$$N_1^* = N_2^* - 1 \hspace{2cm} 12.8$$

9. APPLICATION OF COMPLEMENTS TO COMPUTER ARITHMETIC

Equations 12.9 and 12.10 are the practical applications of complements to computer arithmetic.

$$(N^*)^* = N \hspace{2cm} 12.9$$
$$M - N = M + N^* \hspace{2cm} 12.10$$

The binary ones complement is easily found by switching all of the ones and zeros to zeros and ones, respectively. It can be combined with a technique known as *end-around carry* to perform subtraction. End-around carry is the addition of the *overflow bit* to the sum of N and its ones complement.

Example 12.5

(a) Simulate the operation of a base-10 machine with a capacity of four digits per number and calculate the difference $(18)_{10} - (6)_{10}$ with tens complements.

(b) Simulate the operation of a base-2 machine with a capacity of five digits per number and calculate the difference $(01101)_2 - (01010)_2$ with twos complements.

(c) Solve part (b) with a ones complement and end-around carry.

Solution

(a) The tens complement of 6 is

$$(6)_{10}^* = (10)^4 - 6 = 10,000 - 6 = 9994$$

Using Eq. 12.10,

$$18 - 6 = 18 + (6)_{10}^* = 18 + 9994 = 10,012$$

However, the machine has a maximum capacity of four digits. Therefore, the leading 1 is dropped, leaving 0012 as the answer.

(b) The twos complement of $(01010)_2$ is

$$\begin{aligned} N_2^* &= (2)^5 - N = (32)_{10} - N \\ &= (100000)_2 - (01010)_2 \\ &= (10110)_2 \end{aligned}$$

From Eq. 12.10,

$$\begin{aligned} (01101)_2 - (01010)_2 &= (01101)_2 + (10110)_2 \\ &= (100011)_2 \end{aligned}$$

Since the machine has a capacity of only five bits, the leftmost bit is dropped, leaving $(00011)_2$ as the difference.

(c) The ones complement is found by reversing all the digits.

$$(01010)_1^* = (10101)_2$$

Adding the ones complement,

$$(01101)_2 + (10101)_2 = (100010)_2$$

The leading bit is the overflow bit, which is removed and added to give the difference.

$$(00010)_2 + (1)_2 = (00011)_2$$

10. COMPUTER REPRESENTATION OF NEGATIVE NUMBERS

On paper, a minus sign indicates a negative number. This representation is not possible in a machine. Hence, one of the n digits, usually the MSB, is reserved for sign representation. (This reduces the machine's capacity to represent numbers to $n - 1$ bits per number.) It is arbitrary whether the sign bit is 1 or 0 for negative numbers as long as the MSB is different for positive and negative numbers.

The ones complement is ideal for forming a negative number since it automatically reverses the MSB. For example, $(00011)_2$ is a five-bit representation of decimal 3. The ones complement is $(11100)_2$, which is recognized as a negative number because the MSB is 1.

Example 12.6

Simulate the operation of a six-digit binary machine that uses ones complements for negative numbers.

(a) What is the machine representation of $(-27)_{10}$? (b) What is the decimal equivalent of the twos complement of $(-27)_{10}$? (c) What is the decimal equivalent of the ones complement of $(-27)_{10}$?

Solution

(a) $(27)_{10} = (011011)_2$. The negative of this number is the same as the ones complement: $(100100)_2$.

(b) The twos complement is one more than the ones complement. (See Eqs. 12.7 and 12.8.) Therefore, the twos complement is

$$(100100)_2 + 1 = (100101)_2$$

This represents $(-26)_{10}$.

(c) From Eq. 12.9, the complement of a complement of a number is the original number. Therefore, the decimal equivalent is -27.

13 Numerical Analysis

1. Numerical Methods 13-1
2. Finding Roots: Bisection Method 13-1
3. Finding Roots: Newton's Method 13-2
4. Nonlinear Interpolation: Lagrangian
   Interpolating Polynomial 13-2
5. Nonlinear Interpolation:
   Newton's Interpolating Polynomial 13-3
   Practice Problems 13-4

1. NUMERICAL METHODS

Although the roots of second-degree polynomials are easily found by a variety of methods (by factoring, completing the square, or using the quadratic equation), easy methods of solving cubic and higher-order equations exist only for specialized cases. However, cubic and higher-order equations occur frequently in engineering, and they are difficult to factor. Trial and error solutions, including graphing, are usually satisfactory for finding only the general region in which the root occurs.

Numerical analysis is a general subject that covers, among other things, iterative methods for evaluating roots to equations. The most efficient numerical methods are too complex to present and, in any case, work by hand. However, some of the simpler methods are presented here. Except in critical problems that must be solved in real time, a few extra calculator or computer iterations will make no difference.[1]

2. FINDING ROOTS: BISECTION METHOD

The *bisection method* is an iterative method that "brackets" (also known as "straddles") an interval containing the *root* or *zero* of a particular equation.[2] The size of the interval is halved after each iteration. As the method's name suggests, the best estimate of the root after any iteration is the midpoint of the interval. The maximum error is half the interval length. The procedure continues until the size of the maximum error is "acceptable."[3]

---

[1]Most advanced hand-held calculators now have "root finder" functions that use numerical methods to iteratively solve equations.

[2]The equation does not have to be a pure polynomial. The bisection method requires only that the equation be defined and determinable at all points in the interval.

[3]The bisection method is not a closed method. Unless the root actually falls on the midpoint of one iteration's interval, the method continues indefinitely. Eventually, the magnitude of the maximum error is small enough as to not matter.

The disadvantages of the bisection method are (a) the slowness in converging to the root, (b) the need to know the interval containing the root before starting, and (c) the inability to determine the existence of or find other real roots in the starting interval.

The bisection method starts with two values of the independent variable, $x = L_0$ and $x = R_0$, which straddle a root. Since the function passes through zero at a root, $f(L_0)$ and $f(R_0)$ will have opposite signs. The following algorithm describes the remainder of the bisection method.

Let n be the iteration number. Then, for $n = 0, 1, 2, \ldots$, perform the following steps until sufficient accuracy is attained.

step 1: Set $m = \left(\frac{1}{2}\right)(L_n + R_n)$.

step 2: Calculate $f(m)$.

step 3: If $f(L_n)f(m) \leq 0$, set $L_{n+1} = L_n$ and $R_{n+1} = m$. Otherwise, set $L_{n+1} = m$ and $R_{n+1} = R_n$.

step 4: $f(x)$ has at least one root in the interval (L_{n+1}, R_{n+1}). The estimated value of that root, x^*, is

$$x^* \approx \left(\tfrac{1}{2}\right)(L_{n+1} + R_{n+1})$$

The maximum error is $(^1/_2)(R_{n+1} - L_{n+1})$.

Example 13.1

Use two iterations of the bisection method to find a root of

$$f(x) = x^3 - 2x - 7$$

Solution

The first step is to find L_0 and R_0, which are the values of x that straddle a root and have opposite signs. A table can be made and values of $f(x)$ calculated for random values of x.

x	-2	-1	0	$+1$	$+2$	$+3$
$f(x)$	-11	-6	-7	-8	-3	$+14$

Since $f(x)$ changes sign between $x = 2$ and $x = 3$, $L_0 = 2$ and $R_0 = 3$.

first iteration, $n = 0$:

$$m = \left(\tfrac{1}{2}\right)(2 + 3) = 2.5$$
$$f(2.5) = (2.5)^3 - (2)(2.5) - 7 = 3.625$$

Since $f(2.5)$ is positive, a root must exist in the interval $(2, 2.5)$. Therefore, $L_1 = 2$ and $R_1 = 2.5$. At this point, the best estimate of the root is

$$x^* \approx \left(\tfrac{1}{2}\right)(2 + 2.5) = 2.25$$

The maximum error is $(^1\!/_2)(2.5 - 2) = 0.25$.

second iteration, $n = 1$:

$$m = \left(\tfrac{1}{2}\right)(2 + 2.5) = 2.25$$
$$f(2.25) = (2.25)^3 - (2)(2.25) - 7 = -0.1094$$

Since $f(2.25)$ is negative, a root must exist in the interval $(2.25, 2.5)$. Therefore, $L_2 = 2.25$ and $R_2 = 2.5$. The best estimate of the root is

$$x^* \approx \left(\tfrac{1}{2}\right)(2.25 + 2.5) = 2.375$$

The maximum error is $(^1\!/_2)(2.5 - 2.25) = 0.125$.

3. FINDING ROOTS: NEWTON'S METHOD

Many other methods have been developed to overcome one or more of the disadvantages of the bisection method. These methods have their own disadvantages.[4]

Newton's method is a particular form of *fixed-point iteration*. In this sense, "fixed point" is often used as a synonym for "root" or "zero." However, fixed-point iterations get their name from functions with the characteristic property $x = g(x)$ such that the limit of $g(x)$ is the fixed point (i.e., is the root).

All fixed-point techniques require a starting point. Preferably, the starting point will be close to the actual root.[5] And, while Newton's method converges quickly, it requires the function to be continuously differentiable.

Newton's method algorithm is simple. At each iteration ($n = 0, 1, 2$, etc.), Eq. 13.1 estimates the root. The maximum error is determined by looking at how much the estimate changes after each iteration. If the change between the previous and current estimates (representing the magnitude of error in the estimate) is too large, the current estimate is used as the independent variable for the subsequent iteration.[6]

$$x_{n+1} = g(x_n) = x_n - \frac{f(x_n)}{f'(x_n)} \qquad \textbf{13.1}$$

---

[4]The *regula falsi (false position) method* converges faster than the bisection method but is unable to specify a small interval containing the root. The *secant method* is prone to round-off errors and gives no indication of the remaining distance to the root.

[5]Theoretically, the only penalty for choosing a starting point too far away from the root will be a slower convergence to the root.

[6]Actually, the theory defining the maximum error is more definite than this. For example, for a large enough value of n, the error decreases approximately linearly. Therefore, the consecutive values of x_n converge linearly to the root as well.

Example 13.2

Solve Ex. 13.2 using two iterations of Newton's method. Use $x_0 = 2$.

Solution

The function and its first derivative are

$$f(x) = x^3 - 2x - 7$$
$$f'(x) = 3x^2 - 2$$

first iteration, $n = 0$:

$$x_0 = 2$$
$$f(x_0) = f(2) = (2)^3 - (2)(2) - 7 = -3$$
$$f'(x_0) = f'(2) = (3)(2)^2 - 2 = 10$$
$$x_1 = x_0 - \frac{f(x_0)}{f'(x_0)} = 2 - \frac{-3}{10} = 2.3$$

second iteration, $n = 1$:

$$x_1 = 2.3$$
$$f(x_1) = (2.3)^3 - (2)(2.3) - 7 = 0.567$$
$$f'(x_1) = (3)(2.3)^2 - 2 = 13.87$$
$$x_2 = x_1 - \frac{f(x_1)}{f'(x_1)} = 2.3 - \frac{0.567}{13.87} = 2.259$$

4. NONLINEAR INTERPOLATION: LAGRANGIAN INTERPOLATING POLYNOMIAL

Interpolating between two points of known data is common in engineering. Primarily due to its simplicity and speed, straight-line interpolation is used most often. Even if more than two points on the curve are explicitly known, they are not used. Since straight-line interpolation ignores all but two of the points on the curve, it ignores the effects of curvature.

A more powerful technique that accounts for the curvature is the *Lagrangian interpolating polynomial*. This method uses an nth degree parabola (polynomial) as the interpolating curve.[7] This method requires that $f(x)$ be continuous and real-valued on the interval $[x_0, x_n]$ and that $n + 1$ values of $f(x)$ are known corresponding to $x_0, x_1, x_2, \ldots, x_n$.

The procedure for calculating $f(x)$ at some intermediate point x^* starts by calculating the Lagrangian interpolating polynomial for each known point.

$$L_k(x^*) = \prod_{\substack{i=0 \\ i \neq k}}^{n} \frac{x^* - x_i}{x_k - x_i} \qquad \textbf{13.2}$$

---

[7]The Lagrangian interpolating polynomial reduces to straight-line interpolation if only two points are used.

The value of $f(x)$ at x^* is calculated from Eq. 13.3.

$$f(x^*) = \sum_{k=0}^{n} f(x_k) L_k(x^*) \qquad \textit{13.3}$$

The Lagrangian interpolating polynomial has two primary disadvantages. The first is that a large number of additions and multiplications are needed.[8] The second is that the method does not indicate how many interpolating points should be (or should have been) used. Other interpolating methods have been developed that overcome these disadvantages.[9]

Example 13.3

A real-valued function has the following values:

$$f(1) = 3.5709$$
$$f(4) = 3.5727$$
$$f(6) = 3.5751$$

Use the Lagrangian interpolating polynomial to determine the value of the function at 3.5.

Solution

The procedure for applying Eq. 13.2, the Lagrangian interpolating polynomial, is illustrated in tabular form. Notice that the term corresponding to $i = k$ is omitted from the product.

$$
\begin{array}{ccc}
 & i=0 \quad i=1 \quad i=2 \\
k=0: \; L_0(3.5) = \left(\frac{3.5-1}{1-1}\right)\left(\frac{3.5-4}{1-4}\right)\left(\frac{3.5-6}{1-6}\right) = 0.08333 \\
k=1: \; L_1(3.5) = \left(\frac{3.5-1}{4-1}\right)\left(\frac{3.5-4}{4-4}\right)\left(\frac{3.5-6}{4-6}\right) = 1.04167 \\
k=2: \; L_2(3.5) = \left(\frac{3.5-1}{6-1}\right)\left(\frac{3.5-4}{6-4}\right)\left(\frac{3.5-6}{6-6}\right) = -0.12500 \\
\end{array}
$$

Equation 13.3 is used to calculate the estimate.

$$
\begin{aligned}
f(3.5) &= (3.5709)(0.08333) + (3.5727)(1.04167) \\
&\quad + (3.5751)(-0.12500) \\
&= 3.57225
\end{aligned}
$$

---

[8] As with the numerical methods for finding roots previously discussed, the number of calculations probably will not be an issue if the work is performed by a calculator or computer.
[9] Other common methods for performing interpolation include the *Newton form* and *divided difference table*.

5. NONLINEAR INTERPOLATION: NEWTON'S INTERPOLATING POLYNOMIAL

Newton's form of the interpolating polynomial is more efficient than the Lagrangian method of interpolating between known points.[10] Given $n + 1$ known points for $f(x)$, the *Newton form of the interpolating polynomial* is

$$f(x^*) = \sum_{i=0}^{n} \left(f[x_0, x_1, \ldots, x_i] \prod_{j=0}^{i-1} (x^* - x_j) \right) \qquad \textit{13.4}$$

$f[x_0, x_1, \ldots, x_i]$ is known as the *i*th *divided difference*.

$$f[x_0, x_1, \ldots, x_i] = \sum_{k=0}^{i} \left(\frac{f(x_k)}{\substack{(x_k - x_0) \cdots (x_k - x_{k-1}) \\ \times (x_k - x_{k+1}) \cdots (x_k - x_i)}} \right) \qquad \textit{13.5}$$

It is necessary to define the following two terms.

$$f[x_0] = f(x_0) \qquad \textit{13.6}$$

$$\prod (x^* - x_j) = 1 \qquad [i = 0] \qquad \textit{13.7}$$

Example 13.4

Repeat Ex. 13.3 using Newton's form of the interpolating polynomial.

Solution

Since there are $n + 1 = 3$ data points, $n = 2$. Evaluate the terms for $i = 0$ to 2.

$i = 0$:

$$f[x_0] \prod_{j=0}^{-1} (x^* - x_j) = f(x_0)(1) = f(x_0)$$

$i = 1$:

$$f[x_0, x_1] \prod_{j=0}^{0} (x^* - x_j) = f[x_0, x_1](x^* - x_0)$$

$$f[x_0, x_1] = \frac{f(x_0)}{x_0 - x_1} + \frac{f(x_1)}{x_1 - x_0}$$

$i = 2$:

$$f[x_0, x_1, x_0] \prod_{j=0}^{1} (x^* - x_j) = f[x_0, x_1, x_0](x^* - x_0)$$
$$\times (x^* - x_1)$$

$$
\begin{aligned}
f[x_0, x_1, x_2] &= \frac{f(x_0)}{(x_0 - x_1)(x_0 - x_2)} \\
&\quad + \frac{f(x_1)}{(x_1 - x_0)(x_1 - x_2)} \\
&\quad + \frac{f(x_2)}{(x_2 - x_0)(x_2 - x_1)}
\end{aligned}
$$

---

[10] In this case, "efficiency" relates to the ease in adding new known points without having to repeat all previous calculations.

Substituting known values,

$$f(3.5) = 3.5709 + \left[\frac{3.5709}{1-4} + \frac{3.5727}{4-1}\right](3.5-1)$$

$$+ \left[\frac{3.5709}{(1-4)(1-6)} + \frac{3.5727}{(4-1)(4-6)}\right.$$

$$\left. + \frac{3.5751}{(6-1)(6-4)}\right]$$

$$\times (3.5-1)(3.5-4)$$

$$= 3.57225$$

This is the same answer as was determined in Ex. 13.3.

PRACTICE PROBLEMS

1. A function is given as $y = 3x^{0.93} + 4.2$. What is the percent error if the value of y at $x = 2.7$ is found by using straight-line interpolation between $x = 2$ and $x = 3$?

2. Given the following data points, find y by straight-line interpolation for $x = 2.75$.

x	y
1	4
2	6
3	2
4	−14

14 Fluid Properties

1. Characteristics of a Fluid 14-1
2. Types of Fluids 14-2
3. Fluid Pressure and Vacuum 14-2
4. Density 14-3
5. Specific Volume 14-4
6. Specific Gravity 14-4
7. Specific Weight 14-5
8. Mole Fraction 14-6
9. Viscosity 14-6
10. Kinematic Viscosity 14-8
11. Viscosity Conversions 14-8
12. Viscosity Index 14-9
13. Vapor Pressure 14-9
14. Osmotic Pressure 14-9
15. Surface Tension 14-10
16. Capillary Action 14-11
17. Compressibility 14-12
18. Bulk Modulus 14-13
19. Speed of Sound 14-14
20. Properties of Solutions 14-15
    Practice Problems 14-15

Nomenclature

a	speed of sound	ft/sec	m/s
A	area	ft^2	m^2
d	diameter	ft	m
E	bulk modulus	lbf/ft^2	Pa
F	force	lbf	N
g	gravitational acceleration	ft/sec^2	m/s^2
g_c	gravitational conversion constant (32.174)	lbm-ft/lbf-sec^2	n.a.
h	height	ft	m
k	ratio of specific heats	–	–
L	length	ft	m
M	molar concentration	lbmol/ft^3	kmol/m^3
M	Mach number	–	–
MW	molecular weight	lbm/lbmol	kg/kmol
n	number of moles	–	–
p	pressure	lbf/ft^2	Pa
r	radius	ft	m
R	specific gas constant	ft-lbf/lbm-°R	J/kg·K
R^*	universal gas constant	1545 ft-lbf/ lbm-°R	8314 J/kmol·K
SG	specific gravity	–	–
T	absolute temperature	°R	K
v	velocity	ft/sec	m/s
V	volume	ft^3	m^3
VI	viscosity index	various	various
x	mole fraction	–	–
y	distance	ft	m
Z	compressibility factor	–	–

Symbols

β	compressibility	ft^2/lbf	Pa^{-1}
β	contact angle	deg	deg
γ	specific weight	lbf/ft^3	n.a.
μ	absolute viscosity	lbf-sec/ft^2	Pa·s
ν	kinematic viscosity	ft^2/sec	m^2/s
π	osmotic pressure	lbf/ft^2	Pa
ρ	density	lbm/ft^3	kg/m^3
σ	surface tension	lbf/ft	N/m
τ	shear stress	lbf/ft^2	Pa
v	specific volume	ft^3/lbm	m^3/kg

Subscripts

c	critical
o	original
p	constant pressure
T	constant temperature

1. CHARACTERISTICS OF A FLUID

Liquids and gases can both be categorized as fluids, although this chapter is primarily concerned with incompressible liquids. There are certain characteristics shared by all fluids, and these characteristics can be used, if necessary, to distinguish between liquids and gases.[1]

- *Compressibility*: Liquids are only slightly compressible and are assumed to be incompressible for most purposes. Gases are highly compressible.

- *Shear resistance*: Liquids and gases cannot support shear, and they deform continuously to minimize applied shear forces.

- *Shape and volume*: As a consequence of their inability to support shear forces, liquids and gases take on the shapes of their containers. Only liquids have free surfaces. Liquids have fixed volumes, regardless of their container volumes, and these volumes are not significantly affected by temperature and pressure. Gases take on the volumes

---

[1]The differences between liquids and gases become smaller as temperature and pressure are increased. Gas and liquid properties become the same at the critical temperature and pressure.

Fluids

of their containers. If allowed to do so, gas volumes will change as temperature and pressure are varied.

- *Resistance to motion*: Due to viscosity, liquids resist instantaneous changes in velocity, but the resistance stops when liquid motion stops. Gases have very low viscosities.

- *Molecular spacing*: Molecules in liquids are relatively close together and are held together with strong forces of attraction. Liquid molecules have low kinetic energy. The distance each liquid molecule travels between collisions is small. In gases, the molecules are relatively far apart, and the attractive forces are weak. Kinetic energy of the molecules is high. Gas molecules travel larger distances between collisions.

- *Pressure*: The pressure at a point in a fluid is the same in all directions. Pressure exerted by a fluid on a solid surface (e.g., container wall) is always normal to that surface.

2. TYPES OF FLUIDS

For computational convenience, fluids are generally divided into two categories: ideal fluids and real fluids. *Ideal fluids* are assumed to have no viscosity (and hence, no resistance to shear), be incompressible, and have uniform velocity distributions when flowing. In an ideal fluid, there is no friction between moving layers of fluid, and there are no eddy currents or turbulence.

Real fluids exhibit finite viscosities and non-uniform velocity distributions, are compressible, and experience friction and turbulence in flow. Real fluids are further divided into *Newtonian fluids* and *non-Newtonian fluids*, depending on their viscous behavior. The differences between Newtonian and the various types of non-Newtonian fluids are described in Sec. 9.

For convenience, most fluid problems assume real fluids with Newtonian characteristics. This is an appropriate assumption for water, air, gases, steam, and other simple fluids (alcohol, gasoline, acid solutions, etc.). However, slurries, pastes, gels, suspensions, and polymer/electrolyte solutions may not behave according to simple fluid relationships.

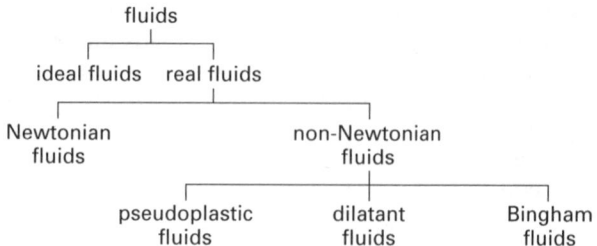

Figure 14.1 *Types of Fluids*

3. FLUID PRESSURE AND VACUUM

In the English system, fluid pressure is measured in pounds per square inch (lbf/in^2 or psi) and pounds per square foot (lbf/ft^2 or psf), although tons (2000 pounds) per square foot (tsf) are occasionally used. In SI units, pressure is measured in pascals (Pa). Because a pascal is very small, kilopascals (kPa) are usually used. Other units of pressure are bars; millibars; atmospheres; inches and feet of water; millimeters, centimeters, and inches of mercury; and torrs.

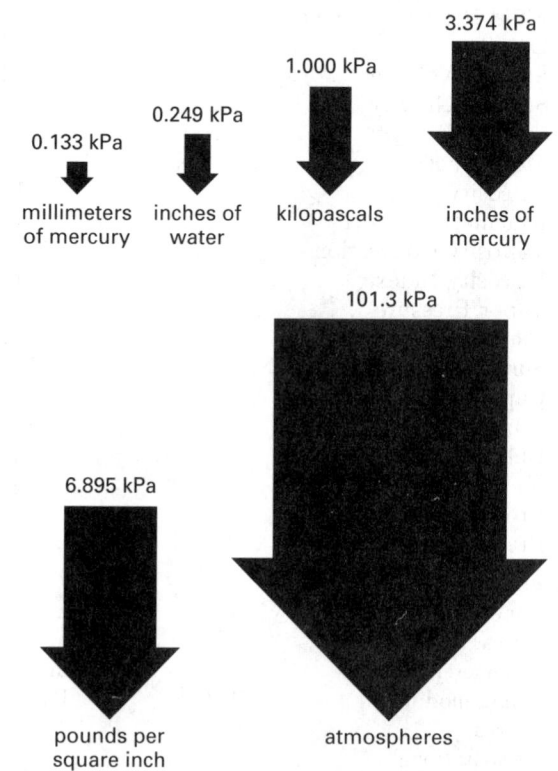

Figure 14.2 *Relative Sizes of Pressure Units*

Fluid pressures are measured with respect to two pressure references: zero pressure and atmospheric pressure. Pressures measured with respect to a true zero pressure reference are known as *absolute pressures*. Pressures measured with respect to atmospheric pressure are known as *gage pressures*.[2] Most pressure gauges read the excess of the test pressure over atmospheric pressure (i.e., the gage pressure). To distinguish between these two pressure measurements, the letters "a" and "g" are traditionally added to the the unit symbols in the English unit system (e.g., 14.7 psia and 4015 psfg). For SI units, the actual words "gauge" and "absolute" can be added to the measurement (e.g., 25.1 kPa absolute). Alternatively, the pressure is assumed to be absolute unless the "g" is used (e.g., 15 Pag).

---

[2]The spelling *gage* persists even though pressures are measured with *gauges*. In some countries, the term *meter pressure* is used instead of gage pressure.

Absolute and gage pressures are related by Eq. 14.1. It should be mentioned that $p_{\text{atmospheric}}$ in Eq. 14.1 is the actual atmospheric pressure existing when the gage measurement is taken. It is not standard atmospheric pressure, unless that pressure is implicitly or explicitly applicable. Also, since a barometer measures atmospheric pressure, *barometric pressure* is synonymous with atmospheric pressure. Table 14.1 lists standard atmospheric pressure in various units.

$$p_{\text{absolute}} = p_{\text{gage}} + p_{\text{atmospheric}} \qquad 14.1$$

Table 14.1 Standard Atmospheric Pressure

1.000 atm	(atmosphere)
14.696 psia	(pounds per square inch absolute)
2116.2 psfa	(pounds per square foot absolute)
407.1 in wg	(inches of water, inches water gage)
33.93 ft wg	(feet of water, feet water gage)
29.921 in Hg	(inches of mercury)
760.0 mm Hg	(millimeters of mercury)
760.0 torr	
1.013 bars	
1013 millibars	
1.013×10^5 Pa	(pascals)
101.3 kPa	(kilopascals)

A *vacuum* measurement is implicitly a pressure below atmospheric (i.e., a negative gage pressure). It must be assumed that any measured quantity given as a vacuum is a quantity to be subtracted from the atmospheric pressure. Thus, when a condenser is operating with a vacuum of 4.0 in Hg (4 in of mercury), the absolute pressure is approximately 29.92 in Hg − 4.0 in Hg = 25.92 in Hg. Vacuums are always stated as positive numbers.

$$p_{\text{absolute}} = p_{\text{atmospheric}} - p_{\text{vacuum}} \qquad 14.2$$

4. DENSITY

The *density*, ρ, of a fluid is its mass per unit volume.[3] In SI units, density is measured in kg/m^3. In a consistent English system, density is measured in slugs/ft^3, even though fluid density is traditionally reported in lbm/ft^3.

$$\rho = \text{ fluid density} \qquad 14.3$$

The density of a fluid in a liquid form is usually given, known in advance, or easily obtained from tables in any one of a number of sources. Most English fluid data are

---

[3]Mass is an absolute property of a substance. Weight is not absolute, since it depends on the local gravity. Some fluids books continue to use γ as the symbol for weight density. The equations using γ that result (such as Bernoulli's equation) cannot be used with SI data, since the equations are not consistent. Thus, engineers end up with two different equations for the same thing.

reported on a per pound basis, and the data included in this book follow that tradition. To make the conversion from pounds to slugs, divide by g_c as an implied step whenever using pound-basis fluid data.

$$\rho_{\text{slugs}} = \frac{\rho_{\text{lbm}}}{g_c} \qquad 14.4$$

The density of an ideal gas can be found from the specific gas constant and the ideal gas law.

$$\rho = \frac{p}{RT} \qquad 14.5$$

Table 14.2 Approximate Room-Temperature Densities of Common Fluids

fluid	lbm/ft^3	kg/m^3
air (STP)	0.0807	1.29
air (70°F, 1 atm)	0.075	1.20
alcohol	49.3	790
ammonia	38	602
gasoline	44.9	720
glycerin	78.8	1260
mercury	848	13 600
water	62.4	1000

(Multiply lbm/ft^3 by 16.0 to obtain kg/m^3.)

Example 14.1

The density of water is typically taken as 62.4 lbm/ft^3 for engineering problems where greater accuracy is not required. What is the value in (a) slugs/ft^3 and (b) kg/m^3?

Solution

(a) Equation 14.4 can be used to calculate the slug-density of water.

$$\rho = \frac{\rho_{\text{lbm}}}{g_c} = \frac{62.4 \frac{\text{lbm}}{\text{ft}^3}}{32.2 \frac{\text{lbm-ft}}{\text{lbf-sec}^2}} = 1.94 \text{ lbf-sec}^2/\text{ft-ft}^3$$
$$= 1.94 \text{ slugs/ft}^3$$

(b) The conversion between lbm/ft^3 and kg/m^3 is approximately 16.0, derived as follows.

$$\rho = \left(62.4 \frac{\text{lbm}}{\text{ft}^3}\right) \left(\frac{35.31 \frac{\text{ft}^3}{\text{m}^3}}{2.205 \frac{\text{lbm}}{\text{kg}}}\right)$$
$$= \left(62.4 \frac{\text{lbm}}{\text{ft}^3}\right)\left(16.01 \frac{\text{kg-ft}^3}{\text{m}^3\text{-lbm}}\right) = 999 \text{ kg/m}^3$$

In SI problems, it is common to take the density of water as 1000 kg/m^3.

5. SPECIFIC VOLUME

Specific volume, v, is the volume occupied by a unit mass of fluid.[4] Since specific volume is the reciprocal of density, typical units will be ft^3/lbm, ft^3/lbmole, or m^3/kg.[5]

$$v = \frac{1}{\rho} \qquad 14.6$$

6. SPECIFIC GRAVITY

Specific gravity (SG) is a dimensionless ratio of a fluid's density to some standard reference density.[6] For liquids and solids, the reference is the density of pure water. There is some variation in this reference density, however, since the temperature at which the water density is evaluated is not standardized. Temperatures of 39.2°F (4°C), 60°F, and 70°F have been reported.[7]

Fortunately, the density of water is the same to three significant digits over the normal ambient temperature range: 62.4 lbm/ft^3 or 1000 kg/m^3. However, to be precise, the temperature of both the fluid and water should be specified (e.g., "... the specific gravity of the 20°C fluid is 1.05 referred to 4°C water ...").

$$SG_{\text{liquid}} = \frac{\rho_{\text{liquid}}}{\rho_{\text{water}}} \qquad 14.7$$

Since the SI density of water is very nearly 1.000 g/cm^3 (1000 kg/m^3), the numerical values of density in g/cm^3 and specific gravity are the same. Such is not the case with English units.

The standard reference used to calculate the specific gravity of gases is the density of air. Since the density of a gas depends on temperature and pressure, both must be specified for the gas and air (i.e., two temperatures and two pressures must be specified). While STP (standard temperature and pressure) conditions are commonly specified, they are not universal.[8] Table 14.3 lists several common sets of standard conditions.

$$SG_{\text{gas}} = \frac{\rho_{\text{gas}}}{\rho_{\text{air}}} \qquad 14.8$$

Table 14.3 *Commonly Quoted Values of Standard Temperature and Pressure*

system	temperature	pressure
SI	273.15K	101.325 kPa
scientific	0.0°C	760 mm Hg
U.S. engineering	32°F	14.696 psia
natural gas industry (U.S.)	60°F	14.65, 14.73, or 15.025 psia
natural gas industry (Canada)	60°F	14.696 psia

If it is known or implied that the temperature and pressure of the air and gas are the same, the specific gravity of the gas will be equal to the ratio of molecular weights and the inverse ratio of specific gas constants. The density of air evaluated at STP is listed in Table 14.2. At 70°F (21.1°C) and 1.0 atm, the density is approximately 0.075 lbm/ft^3 (1.20 kg/m^3).

$$\begin{aligned} SG_{\text{gas}} &= \frac{MW_{\text{gas}}}{MW_{\text{air}}} = \frac{MW_{\text{gas}}}{29.0} \\ &= \frac{R_{\text{air}}}{R_{\text{gas}}} = \frac{53.3}{R_{\text{gas}}} \end{aligned} \qquad 14.9$$

Specific gravities of petroleum liquids and aqueous solutions (of acid, antifreeze, salts, etc.) can be determined by use of a *hydrometer*. In its simplest form, a hydrometer is constructed as a graduated scale weighted at one end so it will float vertically. The height at which the hydrometer floats depends on the density of the fluid, and the graduated scale can be calibrated directly in specific gravity.[9]

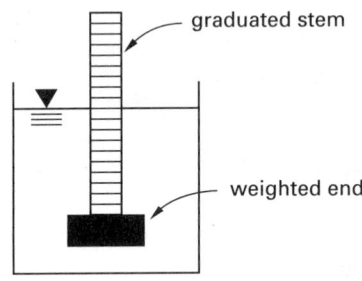

Figure 14.3 *Hydrometer*

There are two standardized hydrometer scales (i.e., methods for calibrating the hydrometer stem.)[10] Both state specific gravity in degrees, although temperature is not being measured. The *American Petroleum Institute* (API) scale (°API) may be used with all liquids, not only with oils or other hydrocarbons. For the specific gravity value, a standard reference temperature of 60°F (15.6°C) is implied for both the liquid and the water.

---

[4]Care must be taken to distinguish between the symbol upsilon, v, used for specific volume, and italic roman "vee," v, used for velocity in many engineering textbooks.

[5]Units of ft^3/slug are also possible, but this combination of units is almost never encountered.

[6]The symbols S.G., sp.gr., S, and G are also used. In fact, petroleum engineers in the United States use γ, a symbol that civil engineers use for density. There is no standard engineering symbol for specific gravity.

[7]Density of liquids is sufficiently independent of pressure to make consideration of pressure in specific gravity calculations unnecessary.

[8]The abbreviation "SC" (standard conditions) is interchangeable with "STP."

[9]This is a direct result of the buoyancy principle of Archimedes.

[10]In addition to °Be and °API mentioned in this chapter, the *Twaddell scale* (°Tw) is used in chemical processing, the *Brix* and *Balling's scales* are used in the sugar industry, and the *Salometer scale* is used to measure salt (NaCl and CaCl$_2$) solutions.

$$°\text{API} = \frac{141.5}{\text{SG}} - 131.5 \qquad \textit{14.10}$$

$$\text{SG} = \frac{141.5}{°\text{API} + 131.5} \qquad \textit{14.11}$$

The *Baumé scale* (°Be) has been used in the past. It is somewhat confusing because there are actually two Baumé scales—one for liquids heavier than water and another for liquids lighter than water. (There is also a discontinuity in the scales at SG = 1.00.) As with the API scale, the specific gravity value assumes 60°F is the standard temperature for both scales.

$$\text{SG} = \frac{140.0}{130.0 + °\text{Be}} \qquad [\text{SG} \leq 1.00] \qquad \textit{14.12}$$

$$\text{SG} = \frac{145.0}{145.0 - °\text{Be}} \qquad [\text{SG} \geq 1.00] \qquad \textit{14.13}$$

Example 14.2

Determine the specific gravity of carbon dioxide gas (molecular weight = 44) at 150°F (66°C) and 20 psia (138 kPa) using STP air as a reference.

SI Solution

$$R^* = 8314 \, \text{J/kmol·K}$$

$$R = \frac{R^*}{\text{MW}} = \frac{8314 \, \dfrac{\text{J}}{\text{kmol·K}}}{44 \, \dfrac{\text{kg}}{\text{kmol}}} = 189.0 \, \text{J/kg·K}$$

$$\rho = \frac{p}{RT} = \frac{1.38 \times 10^5 \, \text{Pa}}{\left(189.0 \, \dfrac{\text{J}}{\text{kg·K}}\right)(66 + 273)\text{K}}$$

$$= 2.15 \, \text{kg/m}^3$$

$$\text{SG} = \frac{2.15 \, \dfrac{\text{kg}}{\text{m}^3}}{1.29 \, \dfrac{\text{kg}}{\text{m}^3}} = 1.67$$

Customary U.S. Solution

Since the conditions of the carbon dioxide and air are different, Eq. 14.8 cannot be used. It is necessary to calculate the density of the carbon dioxide from Eq. 14.5. The specific gas constant of carbon dioxide is 35.1 ft-lbf/lbm-°R. The density is

$$\rho = \frac{p}{RT} = \frac{\left(20 \, \dfrac{\text{lbf}}{\text{in}^2}\right)\left(144 \, \dfrac{\text{in}^2}{\text{ft}^2}\right)}{\left(35.1 \, \dfrac{\text{ft-lbf}}{\text{lbm-°R}}\right)(150 + 460)\text{°R}}$$

$$= 0.135 \, \text{lbm/ft}^3$$

From Table 14.2, the density of STP air is 0.0807 lbm/ft³. From Eq. 14.8, the specific gravity of carbon dioxide at the conditions given is

$$\text{SG} = \frac{0.135 \, \dfrac{\text{lbm}}{\text{ft}^3}}{0.0807 \, \dfrac{\text{lbm}}{\text{ft}^3}} = 1.67$$

7. SPECIFIC WEIGHT

Specific weight, γ, is the weight of fluid per unit volume. The use of specific weight is most often encountered in civil engineering works from the United States, where it is commonly called "density." Mechanical and chemical engineers seldom encounter the term. The usual units of specific weight are lbf/ft³.[11] Specific weight is not an absolute property of a fluid, since it depends not only on the fluid but on the local gravitational field as well.

$$\gamma = g\rho \qquad [\text{SI}] \qquad \textit{14.14(a)}$$

$$\gamma = \rho \times \frac{g}{g_c} \qquad [\text{U.S.}] \qquad \textit{14.14(b)}$$

If the gravitational acceleration is 32.2 ft/sec², as it is almost everywhere on the earth, the specific weight in lbf/ft³ will be numerically equal to the density in lbm/ft³. This is illustrated in Ex. 14.3.

Example 14.3

What is the sea level ($g = 32.2$ ft/sec²) specific weight (in lbf/ft³) of liquids with densities of (a) 1.95 slug/ft³, and (b) 58.3 lbm/ft³?

Solution

(a) From Eq. 14.14(a),

$$\gamma = g\rho = \left(32.2 \, \frac{\text{ft}}{\text{sec}^2}\right)\left(1.95 \, \frac{\text{slug}}{\text{ft}^3}\right)$$

$$= \left(32.2 \, \frac{\text{ft}}{\text{sec}^2}\right)\left(1.95 \, \frac{\text{lbf-sec}^2}{\text{ft-ft}^3}\right) = 62.8 \, \text{lbf/ft}^3$$

(b) From Eq. 14.14(b),

$$\gamma = \frac{\gamma \times \dfrac{g}{g_c}}{g}$$

$$= \left(58.3 \, \frac{\text{lbm}}{\text{ft}^3}\right)\left(\frac{32.2 \, \dfrac{\text{ft}}{\text{sec}^2}}{32.2 \, \dfrac{\text{lbm-ft}}{\text{lbf-sec}^2}}\right) = 58.3 \, \text{lbf/ft}^3$$

---

[11]Notice that the units are lbf/ft³, not lbm/ft³. Pound-mass (lbm) is a mass unit, not a weight (force) unit.

8. MOLE FRACTION

Mole fraction is an important parameter in many practical engineering problems, particularly in chemistry and chemical engineering. The composition of a fluid consisting of two or more distinctly different substances, A, B, C, and so on, can be described by the mole fractions, x_A, x_B, x_C, and so on, of each substance.[12] The mole fraction of component A is the number of moles of that component, n_A, divided by the total number of moles in the combined fluid.

$$x_A = \frac{n_A}{n_A + n_B + n_C + \cdots} \qquad 14.15$$

Mole fraction is a number between 0 and 1.000. *Mole percent* is the mole fraction multiplied by 100, expressed in percent.

9. VISCOSITY

The *viscosity* of a fluid is a measure of that fluid's resistance to flow when acted upon by an external force such as a pressure differential or gravity. Some fluids, such as heavy oils, jellies, and syrups, are very viscous. Other fluids, such as water, lighter hydrocarbons, and gases, are not as viscous.

Most viscous liquids will flow more easily when their temperatures are raised. However, the behavior of a fluid when temperature, pressure, or stress is varied will depend on the type of fluid. The different types of fluids can be determined with a *sliding plate viscometer* test.[13]

---

[12]There are other methods of specifying the composition. The subjects of moles, mole fraction, gravimetric fraction, and volumetric fraction are covered in Chap. 21.

[13]This test is conceptually simple but is not always practical, since the liquid leaks out between the plates. In research work with liquids, it is common to determine viscosity with a *concentric cylinder viscometer*, also known as a *cup-and-bob viscometer*. Viscosities of perfect gases can be predicted by the kinetic theory of gases.

Viscosity can also be measured by a *Saybolt viscometer*, which is essentially a container that allows a given quantity of fluid to leak out through one of two different-sized orifices. The more viscous the fluid, the more time will be required for the fluid to leak out. *Saybolt Seconds Universal* (SSU) and *Saybolt Seconds Furol* (SSF) are scales of such viscosity measurement based on the smaller and larger orifices, respectively. Seconds can be converted (empirically) to viscosity in other units.

The following relations are approximate conversions between SSU, stokes, and poise.

SSU < 100 sec:
$$\nu_{\text{stokes}} = (0.00226)(\text{SSU}) - \frac{1.95}{\text{SSU}}$$

$$\mu_{\text{poise}} = (\text{SG})(\nu_{\text{stokes}})$$

SSU > 100 sec:
$$\nu_{\text{stokes}} = (0.00220)(\text{SSU}) - \frac{1.35}{\text{SSU}}$$

$$\mu_{\text{poise}} = (\text{SG})(\nu_{\text{stokes}})$$

Consider two plates of area A separated by a fluid with thickness y_0, as shown in Fig. 14.4. The bottom plate is fixed, and the top plate is kept in motion at a constant velocity, v_0, by a force, F.

Experiments with Newtonian fluids have shown that the force, F, required to maintain the velocity, v_0, is proportional to the velocity and the area and is inversely proportional to the separation of the plates. That is,

$$\frac{F}{A} \propto \frac{dv}{dy} \qquad 14.16$$

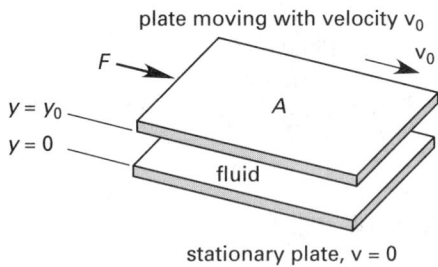

plate moving with velocity v_0

Figure 14.4 *Sliding Plate Viscometer*

The constant of proportionality needed to make Eq. 14.16 an equality is the *absolute viscosity*, μ, also known as the *coefficient of viscosity*.[14] The reciprocal of absolute viscosity, $1/\mu$, is known as the *fluidity*.

$$\frac{F}{A} = \mu \frac{dv}{dy} \qquad 14.17$$

F/A is the *fluid shear stress*, τ. The quantity dv/dy (v_0/y_0) is known by various names, including *rate of strain, shear rate, velocity gradient*, and *rate of shear formation*. Equation 14.17 is known as *Newton's law of viscosity*, from which Newtonian fluids get their name.[15] Equation 14.18 is simply the equation of a straight line.

$$\tau = \mu \frac{dv}{dy} \qquad 14.18$$

Not all fluids are Newtonian (although most common fluids are), and Eq. 14.17 is not universally applicable. Figure 14.5 illustrates how differences in fluid shear stress behavior (at constant temperature and pressure) can be used to define Bingham, pseudoplastic, and dilatant fluids, as well as Newtonian fluids.

Gases, water, alcohol, and benzene are examples of *Newtonian fluids*. In fact, all liquids with a simple chemical formula are Newtonian. Also, most solutions of simple compounds, such as sugar and salt, are Newtonian. For

---

[14]Another name for absolute viscosity is *dynamic viscosity*. The name *absolute viscosity* is preferred, if for no other reason than to avoid confusion with *kinematic viscosity*.

[15]Sometimes Eq. 14.18 is written with a minus sign to compare viscous behavior with other behavior. However, the direction of positive shear stress is arbitrary.

a highly viscous fluid, the straight line (see Fig. 14.5) will be closer to the τ axis. For low-viscosity fluids, the straight line will be closer to the dv/dy axis.

Pseudoplastic fluids (muds, motor oils, polymer solutions, natural gums, and most slurries) exhibit viscosities that decrease with an increasing velocity gradient. Such fluids present no serious pumping problems.

Bingham fluids (Bingham plastics), typified by toothpaste, jellies, bread dough, and some slurries, are capable of indefinitely resisting a small shear stress, but move easily when the stress becomes large—that is, Bingham fluids become pseudoplastic when the stress increases.

Dilatant fluids are rare but include clay slurries, various starches, some paints, milk-chocolate with nuts, and other candy compounds. They exhibit viscosities that increase with increasing agitation (i.e., with increasing velocity gradients), but they return rapidly to their normal viscosity after the agitation ceases. Pump selection is critical for dilatant fluids because these fluids can become almost solid if the shear rate is high enough.

Plastic materials, such as tomato catsup, behave similarly to pseudoplastic fluids once movement begins—their viscosities decrease with agitation. However, a finite force must be applied before any fluid movement occurs.

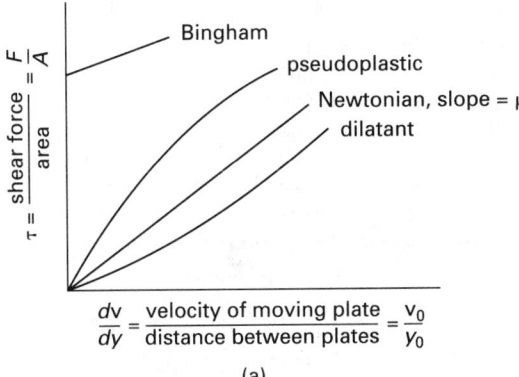

(a)

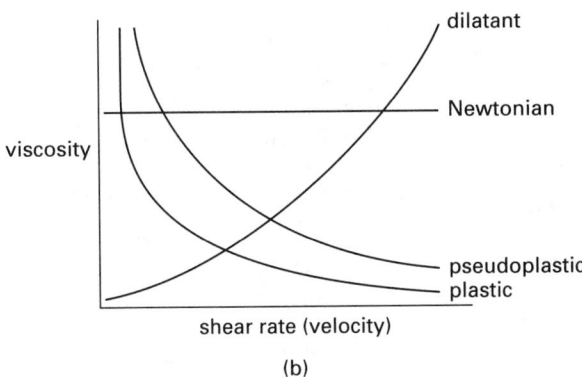

(b)

Figure 14.5 *Shear Stress Behavior for Different Types of Fluids*

Viscosity can also change with time (all other conditions being constant). If viscosity decreases with time, the fluid is said to be a *thixotropic fluid*. If viscosity increases (usually up to a finite value) with time, the fluid is a *rheopectic fluid*. Viscosity does not change in time-independent fluids. *Colloidal materials*, such as gelatinous compounds, lotions, shampoos, and low-temperature solutions of soaps in water and oil, behave like *thixotropic liquids*—their viscosities decrease as the rate of shear is increased. However, viscosity does not return to its original state after the agitation ceases.

Molecular cohesion is the dominating cause of viscosity in liquids. As the temperature of a liquid increases, these cohesive forces decrease, resulting in a decrease in viscosity.

In gases, the dominant cause of viscosity is random collisions between gas molecules. This molecular agitation increases with increases in temperature. Therefore, viscosity in gases increases with temperature.

Although viscosity of liquids increases slightly with pressure, the increase is insignificant over moderate pressure ranges. Therefore, the absolute viscosity of both gases and liquids is usually considered to be essentially independent of pressure.[16]

The units of absolute viscosity, as derived from Eq. 14.18, are lbf-sec/ft^2. Such units are actually used in the English engineering system.[17] Another common unit used throughout the world is the *poise* (abbreviated P), equal to a dyne·s/cm^2. These dimensions are the same primary dimensions as in the English system, $F\theta/L^2$ or $M/L\theta$, and functionally the same as a g/cm·s. Since the poise is a large unit, the *centipoise* (abbreviated cP) scale is generally used. By coincidence, the viscosity of pure water at room temperature is approximately 1 cP.

Absolute viscosity is measured in pascal-seconds (Pa·s) in SI units.

Example 14.4

A liquid ($\mu = 5.2 \times 10^{-5}$ lbf-sec/ft^2) is flowing in a rectangular duct. The equation of the symmetrical velocity (in ft/sec) is approximately v $= 3y^{0.7}$ ft/sec, where y is in inches. (a) What is the velocity gradient at $y = 3.0$ in from the duct wall? (b) What is the shear stress in the fluid at that point?

Solution

(a) The velocity is not a linear function of y, so dv/dy must be calculated as a derivative.

$$\frac{dv}{dy} = \frac{d}{dy}\left(3y^{0.7}\right)$$
$$= (3)\left(0.7y^{-0.3}\right) = 2.1y^{-0.3}$$

[16]This is not true for kinematic viscosity, however.
[17]Units of lbm/ft-sec are also used for absolute viscosity in the English system, although it is difficult to see how such units are derived from the sliding-plate test. These units are obtained by dividing lbf-sec/ft^2 units by g_c.

At $y = 3$ in,

$$\frac{dv}{dy} = (2.1)\left(3^{-0.3}\right) = 1.51 \text{ ft/sec-in}$$

(b) From Eq. 14.18, the shear stress is

$$\tau = \mu \frac{dv}{dy}$$

$$= \left(5.2 \times 10^{-5} \frac{\text{lbf-sec}}{\text{ft}^2}\right)\left(1.51 \frac{\text{ft}}{\text{sec-in}}\right)\left(12 \frac{\text{in}}{\text{ft}}\right)$$

$$= 9.42 \times 10^{-4} \text{ lbf/ft}^2$$

10. KINEMATIC VISCOSITY

Another quantity with the name *viscosity* is the ratio of absolute viscosity to mass density. This combination of variables, known as *kinematic viscosity*, ν, appears sufficiently often in fluids and other problems as to warrant its own symbol and name. Thus, kinematic viscosity is the name given to a frequently occurring combination of variables.

$$\nu = \frac{\mu}{\rho} \qquad \text{[SI]} \qquad \textit{14.19(a)}$$

$$\nu = \frac{\mu g_c}{\rho} \qquad \text{[U.S.]} \qquad \textit{14.19(b)}$$

The primary dimensions of kinematic viscosity are L^2/θ. Typical units are ft^2/sec and cm^2/s (the *stoke*, St). It is also common to give kinematic viscosity in *centistokes*, cSt. The SI units of kinematic viscosity are m^2/s.

It is essential that consistent units be used with Eq. 14.19(a). The following sets of units are consistent:

$$\text{ft}^2/\text{sec} = \frac{\text{lbf-sec}/\text{ft}^2}{\text{slugs}/\text{ft}^3}$$

$$\text{m}^2/\text{s} = \frac{\text{Pa}\cdot\text{s}}{\text{kg}/\text{m}^3}$$

$$\text{s (stoke)} = \frac{\text{P (poise)}}{\text{g}/\text{cm}^3}$$

$$\text{cSt (centistokes)} = \frac{\text{cP (centipoise)}}{\text{g}/\text{cm}^3}$$

Unlike absolute viscosity, kinematic viscosity is greatly dependent on both temperature and pressure, since these variables affect the density of the fluid. Referring to Eq. 14.19, even if absolute viscosity is independent of temperature or pressure, the change in density will change the kinematic viscosity.

11. VISCOSITY CONVERSIONS

The most common units of absolute and kinematic viscosity are listed in Table 14.4.

Table 14.4 *Common Viscosity Units*

	absolute (μ)	kinematic (ν)
English	lbf-sec/ft^2 (slug/ft-sec)	ft^2/sec
conventional metric	dyne·s/cm^2 (poise)	cm^2/s (stoke)
SI	Pa·s (N·s/m^2)	m^2/s

Table 14.5 contains conversions between the various viscosity units.

Table 14.5 *Viscosity Conversions*[a]

multiply	by	to obtain
absolute viscosity, μ		
dyne·s/cm^2	0.10	Pa·s
lbf-sec/ft^2	478.8	P
lbf-sec/ft^2	47,880	cP
lbf-sec/ft^2	47.88	Pa·s
slug/ft-sec	47.88	Pa·s
lbm/ft-sec	1.488	Pa·s
cP	1.0197×10^{-4}	kgf·s/m^2
cP	2.0885×10^{-5}	lbf-sec/ft^2
cP	0.001	Pa·s
Pa·s	0.020885	lbf-sec/ft^2
Pa·s	1000	cP
kinematic viscosity, ν		
ft^2/sec	92,903	cSt
ft^2/sec	0.092903	m^2/s
m^2/s	10.7639	ft^2/sec
m^2/s	1×10^6	cSt
cSt	1×10^{-6}	m^2/sec
cSt	1.0764×10^{-5}	ft^2/sec
absolute viscosity to kinematic viscosity		
cP	$1/\rho$ in g/cm^3	cSt
cP	$6.7195 \times 10^{-4}/\rho$ in lbm/ft^3	ft^2/sec
lbf-sec/ft^2	$32.174/\rho$ in lbm/ft^3	ft^2/sec
kgf·s/m^2	$9.807/\rho$ in kg/m^3	m^2/s
Pa·s	$1000/\rho$ in g/cm^3	cSt
kinematic viscosity to absolute viscosity		
cSt	ρ in g/cm^3	cP
cSt	$0.001 \times \rho$ in lbm/ft^3	Pa·s
m^2/s	$0.10197 \times \rho$ in kg/m^3	kgf·s/m^2
m^2/s	$1000 \times \rho$ in g/cm^3	Pa·s
ft^2/sec	$0.031081 \times \rho$ in lbm/ft^3	lbf-sec/ft^2
ft^2/sec	$1488.2 \times \rho$ in lbm/ft^3	cP

[a]cP: centipoise; cSt: centistoke; P: poise

Example 14.5

Water at 60°F has a specific gravity of 0.999 and a kinematic viscosity of 1.12 cs. What is the absolute viscosity in lbf-sec/ft^2?

Solution

The density of a liquid expressed in g/cm^3 is numerically equal to its specific gravity.

$$\rho = 0.999 \, \text{g/cm}^3$$

The centistoke (cSt) is a measure of kinematic viscosity. Kinematic viscosity is converted first to the absolute viscosity units of centipoise. From Table 14.5,

$$\mu_{\text{cP}} = \nu_{\text{cSt}} \, \rho_{\text{g/cm}^3}$$

$$= (1.12)(0.999) = 1.119 \, \text{cP}$$

Next, centipoises are converted to lbf-sec/ft^2.

$$\mu_{\text{lbf-sec/ft}^2} = \mu_{\text{cP}} \left(2.0885 \times 10^{-5}\right)$$

$$= (1.119) \left(2.0885 \times 10^{-5}\right)$$

$$= 2.34 \times 10^{-5} \, \text{lbf-sec/ft}^2$$

12. VISCOSITY INDEX

Viscosity index (VI) is a measure of a fluid's sensitivity to changes in viscosity with changes in temperature. It has traditionally been applied to crude and refined oils through use of a 100-point scale.[18] The viscosity is measured at two temperatures: 100°F and 210°F (38°C and 99°C). Appendix 14.K can be used to convert these temperatures into a viscosity index in accordance with standard ASTM D2270.

13. VAPOR PRESSURE

Molecular activity in a liquid will allow some of the molecules to escape the liquid surface. Strictly speaking, a small portion of the liquid vaporizes. Molecules of the vapor also condense back into the liquid. The vaporization and condensation at constant temperature are equilibrium processes. The equilibrium pressure exerted by these free molecules is known as the *vapor pressure* or *saturation pressure*. (Vapor pressure does not include the pressure of other substances in the mixture.)

Some liquids, such as propane, butane, ammonia, and Freon, have significant vapor pressures at normal temperatures. Liquids near their boiling point or that vaporize easily are said to be *volatile liquids*.[19] Other liquids, such as mercury, have insignificant vapor pressures at the same temperature. Liquids with low vapor pressures are used in accurate barometers.

The tendency toward vaporization is dependent on the temperature of the liquid. *Boiling* occurs when the liquid temperature is increased to the point that the vapor pressure is equal to the local ambient pressure. Thus, a liquid's boiling temperature depends on the local ambient pressure as well as on the liquid's tendency to vaporize.

Vapor pressure is usually considered to be a nonlinear function of temperature only. It is possible to derive correlations between vapor pressure and temperature, and such correlations usually involve a logarithmic transformation of vapor pressure.[20] Vapor pressure can also be graphed against temperature in a (logarithmic) *Cox chart* (such as App. 14.H) when values are needed over larger temperature extremes. Although there is also some variation with external pressure, the external pressure effect is negligible under normal conditions.

Typical values of vapor pressure are given in Table 14.6.

Table 14.6 Typical Vapor Pressures

fluid	lbf/ft^2, 68°F	kPa, 20°C
mercury	0.00362	0.000173
turpentine	1.115	0.0534
water	48.9	2.34
ethyl alcohol	122.4	5.86
ether	1231	58.9
butane	4550	218
Freon-12	12,200	584
propane	17,900	855
ammonia	18,550	888

14. OSMOTIC PRESSURE

Osmosis is a special case of diffusion in which molecules of the *solvent* move from one fluid to another (i.e., from the *solvent* to the *solution*) in one direction only, usually through a *semipermeable membrane*.[21] Osmosis continues until sufficient solvent has passed through the membrane to make the activity (or solvent pressure) of the solution equal to that of the solvent.[22] The pressure at equilibrium is known as the *osmotic pressure*, π.

---

[18]Use of the *viscosity index* has been adopted by other parts of the chemical process industry (CPI), including in the manufacture of solvents, polymers, and other synthetics. The 100-point scale may be exceeded (on both ends) for these uses. Refer to standard ASTM D2270 for calculating extreme values of the viscosity index.

[19]Because a liquid that vaporizes easily has an aroma, the term *aromatic liquid* is also occasionally used.
[20]The *Clausius-Clapeyron equation* and *Antoine equation* are two such logarithmic correlations of vapor pressure with temperature.
[21]A semipermeable membrane will be impermeable to the solute but permeable for the solvent.
[22]Two solutions in equilibrium (i.e., whose activities are equal) are said to be in *isopiestic equilibrium*.

Figure 14.6 illustrates an *osmotic pressure apparatus*. The fluid column can be interpreted as the result of an osmotic pressure that has developed through diffusion into the solution. The fluid column will continue to increase in height until equilibrium is reached. Alternatively, the fluid column can be adjusted so that the solution pressure just equals the osmotic pressure that would develop otherwise, in order to prevent the flow of solvent. For the arrangement in Fig. 14.6, the osmotic pressure can be calculated from the difference in fluid level heights.

$$\pi = \rho g h \quad \text{[SI]} \qquad 14.20(a)$$

$$\pi = \frac{\rho g h}{g_c} \quad \text{[U.S.]} \qquad 14.20(b)$$

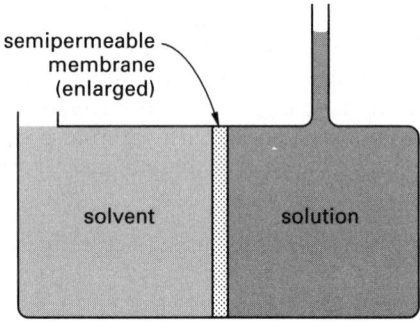

Figure 14.6 *Osmotic Pressure Apparatus*

In dilute solutions, osmotic pressure obeys the ideal gas law. The solute acts like a gas in exerting pressure against the membrane. The solvent exerts no pressure since it can pass through. In Eq. 14.21, M is the molarity (concentration). Consistent units must be used.

$$\pi = MR^*T \qquad 14.21$$

Example 14.6

An aqueous solution is in isopiestic equilibrium with a 0.1 molarity sucrose solution at 22°C. What is the osmotic pressure?

Solution

Referring to Eq. 14.21,

$M = 0.1$ mol/L of solvent

$R^* = 0.0821$ atm·L/mol·K

$T = 22 + 273 = 295\text{K}$

$$\pi = \left(0.1 \ \frac{\text{mol}}{\text{L}}\right) \left(0.0821 \ \frac{\text{atm·L}}{\text{mol·K}}\right) (295\text{K}) = 2.42 \text{ atm}$$

15. SURFACE TENSION

The membrane or "skin" that seems to form on the free surface of a fluid is due to the intermolecular cohesive forces and is known as *surface tension*, σ. Surface tension is the reason that insects are able to sit on water and a needle is able to float on it. Surface tension also causes bubbles and droplets to take on a spherical shape, since any other shape would have more surface area per unit volume.

Data on the surface tension of liquids is important in determining the performance of heat-, mass-, and momentum-transfer equipment, including heat transfer devices.[23] Surface tension data is needed to calculate the nucleate boiling point (i.e., the initiation of boiling) of liquids in a pool (using the *Rohsenow equation*) and the maximum heat flux of boiling liquids in a pool (using the *Zuber equation*).

Surface tension can be interpreted as the tension between two points a unit distance apart on the surface or as the amount of work required to form a new unit of surface area in an apparatus similar to that shown in Fig. 14.7. Typical units of surface tension are lbf/ft (ft-lbf/ft^2), dyne/cm, and N/m.

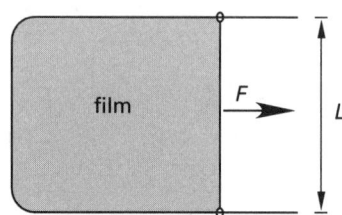

Figure 14.7 *Wire Frame for Stretching a Film*

The apparatus shown in Fig. 14.7 consists of a wire frame with a sliding side that has been dipped in a liquid to form a film. Surface tension is determined by measuring the force necessary to keep the sliding side stationary, against the surface tension pull of the film.[24] (The film does not act like a spring, since the force, F, does not increase as the film is stretched.) Since the film has two surfaces (i.e., two surface tensions), the surface tension is

$$\sigma = \frac{F}{2L} \qquad 14.22$$

Alternatively, surface tension can also be determined by measuring the force required to pull a wire ring out of the liquid, as shown in Fig. 14.8.[25] Since the ring's inner and outer sides are in contact with the liquid, the wetted perimeter is twice the circumference. The surface tension is

$$\sigma = \frac{F}{4\pi r} \qquad 14.23$$

---

[23]Surface tension plays a role in processes involving dispersion, emulsion, flocculation, foaming, and solubilization. It is not surprising that surface data tension is particularly important in determining the performance of equipment in the chemical process industry (CPI), such as distillation columns, packed towers, wetted-wall columns, strippers, and phase-separation equipment.
[24]The force includes the weight of the sliding side wire if the frame is oriented vertically, with gravity acting on the sliding side wire to stretch the film.
[25]This apparatus is known as a *Du Novy torsion balance*. The ring is made of platinum with a diameter of 4.00 cm.

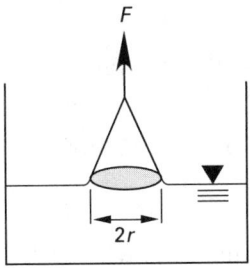

Figure 14.8 *Du Novy Ring Surface Tension Apparatus*

Surface tension depends slightly on the gas in contact with the free surface. Surface tension values are usually quoted for air contact. Typical values of surface tension are listed in Table 14.7.

Table 14.7 *Approximate Values of Surface Tension (air contact)*

fluid	lbf/ft, 68°F	N/m, 20°C
n-octane	0.00149	0.0217
ethyl alcohol	0.00156	0.0227
acetone	0.00162	0.0236
kerosene	0.00178	0.0260
carbon tetrachloride	0.00185	0.0270
turpentine	0.00186	0.0271
toluene	0.00195	0.0285
benzene	0.00198	0.0289
olive oil	0.0023	0.034
glycerin	0.00432	0.0631
water	0.00499	0.0728
mercury	0.0356	0.519

(Multiply lbf/ft by 14.59 to obtain N/m.)
(Multiply dyne/cm by 0.001 to obtain N/m.)

At temperatures below freezing, the substance will be a solid, so surface tension is a moot point. As the temperature of a liquid is raised, the surface tension decreases because the cohesive forces decrease. Surface tension is zero at a substance's critical temperature. If a substance's critical temperature is known, the *Othmer correlation*, Eq. 14.24, can be used to determine the surface tension at one temperature from the surface tension at another temperature.[26]

$$\sigma_2 = \sigma_1 \left[\frac{T_c - T_2}{T_c - T_1} \right]^{\frac{11}{9}} \qquad \text{14.24}$$

Surface tension is the reason that the pressure on the inside of bubbles and droplets is greater than on the outside. Equation 14.25 gives the relationship between the surface tension in a hollow bubble surrounded by a

gas and the difference between the inside and outside pressures. For a spherical droplet or a bubble in a liquid, where in both cases there is only one surface in tension, the surface tension is twice as large. (r is the radius of the bubble or droplet.)

$$\sigma_{\text{bubble}} = \frac{r(p_{\text{inside}} - p_{\text{outside}})}{4} \qquad \text{14.25}$$

$$\sigma_{\text{droplet}} = \frac{r(p_{\text{inside}} - p_{\text{outside}})}{2} \qquad \text{14.26}$$

16. CAPILLARY ACTION

Capillary action (capillarity) is the name given to the behavior of a liquid in a thin-bore tube. Capillary action is caused by surface tension between the liquid and a vertical solid surface.[27] In water, the adhesive forces between the liquid molecules and the surface are greater than (i.e., dominate) the cohesive forces between the water molecules themselves.[28] The adhesive forces cause the water to attach itself to and climb a solid vertical surface. It can be said that the water "reaches up and tries to wet as much of the interior surface as it can." In so doing, the water rises above the general water surface level. This is illustrated in Fig. 14.9.

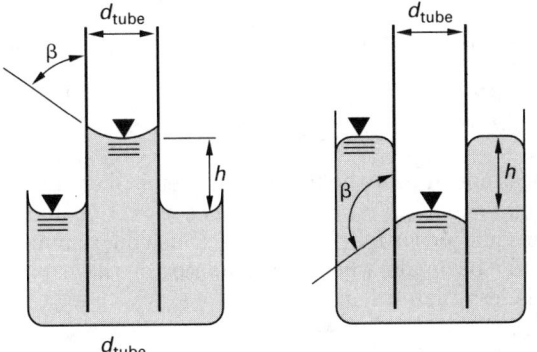

(a) adhesive force dominates   (b) cohesive force dominates

Figure 14.9 *Capillarity of Liquids*

Figure 14.9 also illustrates that the same surface tension forces that keep a droplet spherical are at work on the surface of the liquid in the tube. The curved liquid surface, known as the *meniscus*, can be considered to be an incomplete droplet. If the tube inside diameter is less than approximately 0.1 in (2.5 mm), the meniscus is essentially hemispherical, and $r_{\text{meniscus}} = r_{\text{tube}}$.

For a few other liquids, such as mercury, the molecules have a strong affinity for each other (i.e., the cohesive forces dominate). The liquid avoids contact with the tube surface. In such liquids, the meniscus will be below the general surface level.

---

[26] An extensive listing of critical temperatures and surface tension data was presented in "633 Organic Chemicals: Surface Tension Data," *Chemical Engineering Magazine*, March 1991, pp. 140–150.

[27] In fact, observing the rise of liquid in a capillary tube is another method of determining the surface tension of a liquid.
[28] *Adhesion* is the attractive force between molecules of different substances. *Cohesion* is the attractive force between molecules of the same substance.

Table 14.8 Contact Angles, β

materials	angle
mercury–glass	140°
water–paraffin	107°
water–silver	90°
kerosene–glass	26°
glycerin–glass	19°
water–glass	0°
ethyl alcohol–glass	0°

The *angle of contact*, β, is an indication of whether adhesive or cohesive forces dominate. For contact angles less than 90°, adhesive forces dominate. For contact angles greater than 90°, cohesive forces dominate.

Equation 14.27 can be used to predict the capillary rise in a small-bore tube. Surface tension and contact angles can be obtained from Tables 14.7 and 14.8, respectively.

$$h = \frac{4\,\sigma\cos\beta}{\rho d_{\text{tube}}\,g} \qquad \text{[SI]} \quad 14.27(a)$$

$$h = \left(\frac{4\,\sigma\cos\beta}{\rho d_{\text{tube}}}\right)\left(\frac{g_c}{g}\right) \qquad \text{[U.S.]} \quad 14.27(b)$$

$$\sigma = \frac{h\rho d_{\text{tube}}\,g}{4\cos\beta} \qquad \text{[SI]} \quad 14.28(a)$$

$$\sigma = \left(\frac{h\rho d_{\text{tube}}}{4\cos\beta}\right)\left(\frac{g}{g_c}\right) \qquad \text{[U.S.]} \quad 14.28(b)$$

$$r_{\text{meniscus}} = \frac{d_{\text{tube}}}{2\cos\beta} \qquad\qquad 14.29$$

If it is assumed that the meniscus is hemispherical, then $r_{\text{meniscus}} = r_{\text{tube}}$, $\beta = 0°$, and $\cos\beta = 1.0$, and the above equations can be simplified. (Such an assumption can only be made when the diameter of the capillary tube is less than 0.1 in.)

Example 14.7

To what height will 68°F (20°C) ethyl alcohol rise in a 0.005 in (0.127 mm) internal diameter glass capillary tube? The density of the alcohol is 49 lbm/ft^3 (790 kg/m^3).

SI Solution

$$\sigma = 0.0227 \text{ N/m}$$

$$\beta = 0°$$

The acceleration due to gravity is 9.81 m/s^2. From Eq. 14.27, the height is

$$h = \frac{4\sigma\cos\beta}{\rho d_{\text{tube}}\,g}$$

$$= \frac{(4)\left(0.0227\,\dfrac{\text{N}}{\text{m}}\right)(1.0)\left(1000\,\dfrac{\text{mm}}{\text{m}}\right)}{\left(790\,\dfrac{\text{kg}}{\text{m}^3}\right)(0.127\text{ mm})\left(9.81\,\dfrac{\text{m}}{\text{sec}^2}\right)}$$

$$= 0.0923 \text{ m}$$

Customary U.S. Solution

From Tables 14.7 and 14.8, respectively, the surface tension and contact angle are

$$\sigma = 0.00156 \text{ lbf/ft}$$

$$\beta = 0°$$

The acceleration due to gravity is 32.2 ft/sec^2. From Eq. 14.27, the height is

$$h = \frac{4\sigma\cos\beta g_c}{\rho d_{\text{tube}}\,g}$$

$$= \frac{(4)\left(0.00156\,\dfrac{\text{lbf}}{\text{ft}}\right)(1.0)\left(32.2\,\dfrac{\text{lbm-ft}}{\text{lbf-sec}^2}\right)\left(12\,\dfrac{\text{in}}{\text{ft}}\right)}{\left(49\,\dfrac{\text{lbm}}{\text{ft}^3}\right)(0.005\text{ in})\left(32.2\,\dfrac{\text{ft}}{\text{sec}^2}\right)}$$

$$= 0.306 \text{ ft}$$

17. COMPRESSIBILITY[29]

Compressibility (also known as the *coefficient of compressibility*), β, is the fractional change in the volume of a fluid per unit change in pressure in a constant-temperature process.[30] Typical units are 1/psi, 1/psf, 1/atm, and 1/kPa. It is the reciprocal of the bulk modulus, a quantity that is more commonly tabulated than compressibility.

$$\beta = \frac{-\dfrac{\Delta V}{V_0}}{\Delta p} = \frac{1}{E} \qquad\qquad 14.30$$

Compressibility can also be written in terms of partial derivatives.

$$\beta = \left(\frac{-1}{V_0}\right)\left(\frac{\partial V}{\partial p}\right)_T = \left(\frac{-1}{\rho_0}\right)\left(\frac{\partial\rho}{\partial p}\right)_T \qquad 14.31$$

Compressibility changes only slightly with temperature. The small compressibility of liquids is typically considered to be insignificant, giving rise to the common understanding that liquids are incompressible.

The density of a compressible fluid depends on the fluid's pressure. For small changes in pressure, the density at one pressure can be calculated from the density at another pressure from Eq. 14.32.

$$\rho_2 \approx \rho_1\left[1 + \beta(p_2 - p_1)\right] \qquad 14.32$$

---

[29]Compressibility should not be confused with the *thermal coefficient of expansion*, $(1/V_0)\,(\partial V/\partial T)_p$, which is the fractional change in volume per unit temperature change in a constant-pressure process (with units of 1/°F or 1/°C), or the dimensionless *compressibility factor*, Z, used with the ideal gas law.

[30]Other symbols used for compressibility are c, C, and K. Equation 14.30 is written with a negative sign to show that volume decreases as pressure increases.

Gases, of course, are easily compressed. The compressibility of an ideal gas depends on its pressure, p, its ratio of specific heats, k, and the nature of the process.[31] Depending on the process, the compressibility may be known as *isothermal compressibility* or *(adiabatic) isentropic compressibility*. Of course, compressibility is zero for constant-volume processes and is infinite (or undefined) for constant-pressure processes.

$$\beta_T = \frac{1}{p} \quad \text{[isothermal ideal gas processes]} \quad \textit{14.33}$$

$$\beta_s = \frac{1}{kp} \quad \text{[adiabatic ideal gas processes]} \quad \textit{14.34}$$

Table 14.9 Approximate Compressibility of Common Liquids at 1 atm

liquid	temperature	β, 1/psi	β, 1/atm
mercury	32°F	0.027×10^{-5}	0.39×10^{-5}
glycerin	60°F	0.16×10^{-5}	2.4×10^{-5}
water	60°F	0.33×10^{-5}	4.9×10^{-5}
ethyl alcohol	32°F	0.68×10^{-5}	10×10^{-5}
chloroform	32°F	0.68×10^{-5}	10×10^{-5}
gasoline	60°F	1.0×10^{-5}	15×10^{-5}
hydrogen	20K	11×10^{-5}	160×10^{-5}
helium	2.1K	48×10^{-5}	700×10^{-5}

(Multiply 1/psi by 0.14504 to obtain 1/kPa.)
(Multiply 1/psi by 14.696 to obtain 1/atm.)

Example 14.8

Water at 68°F (20°C) and 1 atm has a density of 62.3 lbm/ft^3 (997 kg/m^3). What is the new density if the pressure is isothermally increased from 14.7 psi (100 kPa) to 400 psi (2760 kPa)? Assume that the bulk modulus has a constant value of 320,000 psi (2.2×10^6 kPa).

SI Solution

Compressibility is the reciprocal of the bulk modulus.

$$\beta = \frac{1}{E} = \frac{1}{2.2 \times 10^6 \text{ kPa}} = 4.55 \times 10^{-7} \text{ 1/kPa}$$

All other information needed to use Eq. 14.32 is provided.

$$\begin{aligned} \rho_2 &= \rho_1[1 + \beta(p_2 - p_1)] \\ &= \left(997 \frac{\text{kg}}{\text{m}^3}\right)\left[1 + \left(4.55 \times 10^{-7} \frac{1}{\text{kPa}}\right)\right. \\ &\quad \left. \times (2760 \text{ kPa} - 100 \text{ kPa})\right] \\ &= 998.2 \text{ kg/m}^3 \end{aligned}$$

[31]For air, $k = 1.4$.

Customary U.S. Solution

Compressibility is the reciprocal of the bulk modulus.

$$\beta = \frac{1}{E} = \frac{1}{320,000 \frac{\text{lbf}}{\text{in}^2}} = 0.3125 \times 10^{-5} \text{ in}^2/\text{lbf}$$

All other information needed to use Eq. 14.32 is provided.

$$\begin{aligned} \rho_2 &= \rho_1 \left[1 + \beta(p_2 - p_1)\right] \\ &= \left(62.3 \frac{\text{lbm}}{\text{ft}^3}\right)\left[1 + \left(0.3125 \times 10^{-5} \frac{\text{in}^2}{\text{lbf}}\right)\right. \\ &\quad \left. \times \left(400 \frac{\text{lbf}}{\text{in}^2} - 14.7 \frac{\text{lbf}}{\text{in}^2}\right)\right] \\ &= 62.38 \text{ lbm/ft}^3 \end{aligned}$$

18. BULK MODULUS

The *bulk modulus*, E, of a fluid is analogous to the modulus of elasticity of a solid. Typical units are psi, atm, and kPa. The term Δp in Eq. 14.35 represents an increase in stress. The term $\Delta V/V_0$ is a *volumetric strain*. Analogous to Hooke's law describing elastic formation, the *bulk modulus* of a fluid (liquid or gas) is given by Eq. 14.35.

$$E = \frac{\text{stress}}{\text{strain}} = \frac{-\Delta p}{\dfrac{\Delta V}{V_0}} \quad \textit{14.35(a)}$$

$$E = -V_0 \left(\frac{\partial p}{\partial V}\right)_T \quad \textit{14.35(b)}$$

The term *secant bulk modulus* is associated with Eq. 14.35(a) (using finite differences), while the terms *tangent bulk modulus* and *point bulk modulus* are associated with Eq. 14.35(b) (using partial derivatives).

The bulk modulus is the reciprocal of compressibility.

$$E = \frac{1}{\beta} \quad \textit{14.36}$$

The bulk modulus changes only slightly with temperature. Water's bulk modulus is usually taken as 300,000 psi (2.1×10^6 kPa) unless greater accuracy is required, in which case, Table 14.10 or App. 14.A can be used.

Table 14.10 Approximate Bulk Modulus of Water

pressure (psi)	32°F	68°F	120°F	200°F	300°F
	(thousands of psi)				
15	292	320	332	308	
1500	300	330	340	319	218
4500	317	348	362	338	271
15,000	380	410	420	405	350

(Multiply psi by 6.8948 to obtain kPa.)
Reprinted with permission from Victor L. Streeter, *Handbook of Fluid Dynamics*, copyright © 1961, McGraw-Hill Book Company.

19. SPEED OF SOUND

The *speed of sound* (*acoustical velocity* or *sonic velocity*), a, in a fluid is a function of its bulk modulus (or, equivalently, of its compressibility).[32] Equation 14.37 gives the speed of sound through a liquid.

$$a = \sqrt{\frac{E}{\rho}} = \sqrt{\frac{1}{\beta\rho}} \quad \text{[SI]} \quad \textit{14.37(a)}$$

$$a = \sqrt{\frac{Eg_c}{\rho}} = \sqrt{\frac{g_c}{\beta\rho}} \quad \text{[U.S.]} \quad \textit{14.37(b)}$$

Equation 14.38 gives the speed of sound in an ideal gas. The temperature, T, must be in degrees absolute (i.e., °R or K). For air, the ratio of specific heats is $k = 1.40$, the molecular weight is 29.0, and the universal gas constant is $R^* = 1545.3$ ft-lbf/lbmol-°R (8314 J/kmol·K).

$$a = \sqrt{\frac{E}{\rho}} = \sqrt{\frac{kp}{\rho}}$$
$$= \sqrt{kRT} = \sqrt{\frac{kR^*T}{\text{MW}}} \quad \text{[SI]} \quad \textit{14.38(a)}$$

$$a = \sqrt{\frac{Eg_c}{\rho}} = \sqrt{\frac{kg_cp}{\rho}}$$
$$= \sqrt{kg_cRT} = \sqrt{\frac{kg_cR^*T}{\text{MW}}} \quad \text{[U.S.]} \quad \textit{14.38(b)}$$

Since k and R are constant for an ideal gas, the speed of sound is a function of temperature only. Equation 14.39 can be used to calculate the new speed of sound when temperature is varied.

$$\frac{a_1}{a_2} = \sqrt{\frac{T_1}{T_2}} \quad \textit{14.39}$$

The *Mach number* of an object is the ratio of the object's speed to the speed of sound in the medium through which it is traveling.

$$\text{M} = \frac{\text{v}}{a} \quad \textit{14.40}$$

The term *subsonic travel* implies M < 1.[33] Similarly, *supersonic travel* implies M > 1, but usually M < 5. Travel above M = 5 is known as *hypersonic travel*. Travel in the transition region between subsonic and supersonic (i.e., 0.8 < M < 1.2) is known as *transonic travel*. A *sonic boom* (a shock-wave phenomenon) occurs when an object travels at supersonic speed.

Example 14.9

What is the velocity of sound in 150°F water? The density is 61.2 lbm/ft³ (980 kg/m³), and the bulk modulus is 328,000 psi (2.26×10^6 kPa).

[32]The symbol c is also used for the sonic velocity.
[33]In the language of compressible fluid flow, this is known as the *subsonic flow regime*.

SI Solution

$$a = \sqrt{\frac{(2.26 \times 10^6 \text{ kPa})\left(1000 \frac{\text{Pa}}{\text{kPa}}\right)}{980 \frac{\text{kg}}{\text{m}^3}}} = 1519 \text{ m/s}$$

Customary U.S. Solution

From Eq. 14.37,

$$a = \sqrt{\frac{Eg_c}{\rho}}$$

$$= \sqrt{\frac{\left(328{,}000 \frac{\text{lbf}}{\text{in}^2}\right)\left(144 \frac{\frac{\text{lbf}}{\text{ft}^2}}{\frac{\text{lbf}}{\text{in}^2}}\right)\left(32.2 \frac{\text{lbm-ft}}{\text{lbf-sec}^2}\right)}{61.2 \frac{\text{lbm}}{\text{ft}^3}}}$$

$$= 4985 \text{ ft/sec}$$

Example 14.10

What is the velocity of sound in 150°F (66°C) air at standard atmospheric pressure?

SI Solution

$$R = \frac{8314 \frac{\text{J}}{\text{kmol·K}}}{29 \frac{\text{kg}}{\text{kmol}}} = 286.7 \text{ J/kg·K}$$

$$T = 66°C + 273 = 339\text{K}$$

$$a = \sqrt{kg_cRT} = \sqrt{(1.4)\left(286.7 \frac{\text{J}}{\text{kg·K}}\right)(339\text{K})}$$

$$= 369 \text{ m/s}$$

Customary U.S. Solution

The specific gas constant, R, for air is

$$R = \frac{R^*}{\text{MW}} = \frac{1545.3 \frac{\text{ft-lbf}}{\text{lbmol-°R}}}{29.0 \frac{\text{lbm}}{\text{lbmol}}}$$

$$= 53.3 \text{ ft-lbf/lbm-°R}$$

The absolute temperature is

$$T = 150°F + 460 = 610°R$$

From Eq. 14.38(b),

$$a = \sqrt{kg_cRT}$$

$$= \sqrt{(1.4)\left(32.2\,\frac{\text{ft-lbm}}{\text{lbf-sec}^2}\right)\left(53.3\,\frac{\text{ft-lbf}}{\text{lbm-}^\circ\text{R}}\right)(610^\circ\text{R})}$$

$$= 1211\text{ ft/sec}$$

20. PROPERTIES OF SOLUTIONS

There are few convenient ways of predicting the properties of nonreacting, nonvolatile organic and aqueous solutions (acids, brines, alcohol mixtures, coolants, etc.) and mixtures from the individual properties of the components.

Volumes of two combining organic liquids (e.g., acetone and chloroform) are essentially additive. The volume change upon mixing will seldom be more than a few tenths of a percent. The volume change in aqueous solutions is often slightly greater, but is still limited to a few percent (e.g., 3% for some solutions of methanol and water).

Thus, the specific gravity (density, specific weight, etc.) can be considered to be a volumetric weighting of the individual specific gravities. Most times, however, the specific gravity of a known solution must be calculated from known data regarding one of the various density scales (see Sec. 6) or determined through research.

Most other important fluid properties of aqueous solutions, such as viscosity, compressibility, surface tension, and vapor pressure, have been measured and are usually determined through research.[34] It is important to be aware of the operating conditions of the solution. Data for one concentration or condition should not be used for another concentration or condition.

PRACTICE PROBLEMS

(Use $g = 32.2$ ft/sec^2 or 9.81 m/s^2 unless told to do otherwise in the problem.)

1. What is the absolute pressure if a gauge reads 8.7 psi (60 kPa) vacuum?

2. Calculate the kinematic viscosity of air at 80°F (27°C) and 70 psia (480 kPa).

---

[34]There is no substitute for a complete fluid properties data book.

15

Fluid Statics

1. Pressure-Measuring Devices 15-1
2. Manometers 15-3
3. Hydrostatic Pressure 15-4
4. Fluid Height Equivalent to Pressure 15-4
5. Multifluid Barometers 15-5
6. Pressure on a Horizontal Plane Surface . . . 15-6
7. Pressure on a Vertical Plane Surface 15-6
8. Pressure on an Inclined Plane Surface 15-7
9. Pressure on a General Plane Surface 15-8
10. Special Cases: Vertical Surfaces 15-9
11. Forces on Curved and Compound Surfaces 15-10
12. Torque on a Gate 15-11
13. Hydrostatic Forces on a Dam 15-11
14. Pressure Due to Several Immiscible Liquids 15-12
15. Pressure from Compressible Fluids 15-12
16. Externally Pressurized Liquids 15-14
17. Hydraulic Ram 15-14
18. Buoyancy 15-14
19. Intact Stability:
     Stability of Floating Objects 15-16
20. Fluid Masses Under
     External Acceleration 15-19
     Practice Problems 15-19

Nomenclature

a	acceleration	ft/sec^2	m/s^2
A	area	ft^2	m^2
b	base length	ft	m
d	diameter	ft	m
e	eccentricity	ft	m
F	force	lbf	N
FS	factor of safety	–	–
g	gravitational acceleration	ft/sec^2	m/s^2
g_c	gravitational conversion constant (32.2)	lbm-ft/lbf-sec^2	n.a.
h	height	ft	m
I	moment of inertia	ft^4	m^4
J	polar moment of inertia	ft^4	m^4
k	radius of gyration	ft	m
k	ratio of specific heats	–	–
L	length	ft	m
m	mass	lbm	kg
M	mechanical advantage	–	–
M	moment	ft-lbf	N·m
n	polytropic exponent	–	–
N	normal force	lbf	N
p	pressure	lbf/ft^2	Pa
r	radius	ft	m
R	resultant force	lbf	N
R	specific gas constant	ft-lbf/lbm-°R	J/kg·K
SG	specific gravity	–	–
t	wall thickness	ft	m
T	temperature	°R	K
v	velocity	ft/sec	m/s
V	volume	ft^3	m^3
W	weight	lbf	N
x	distance	ft	m
x	fraction	–	–
y	distance	ft	m

Symbols

γ	specific weight	lbf/ft^3	n.a.
η	efficiency	–	–
θ	angle	deg	deg
μ	coefficient of friction	–	–
ρ	density	lbm/ft^3	kg/m^3
σ	stress	lbf/ft^2	N/m^2
υ	specific volume	ft^3/lbm	m^3/kg
ω	angular velocity	rad/sec	rad/s

Subscripts

a	atmospheric
b	buoyant
bg	between CB and CG
c	centroidal
f	frictional
F	force
h	hoop
l	lever or longitudinal
m	manometer fluid, mercury, or metacentric
p	plunger
r	ram
R	resultant
t	tank
v	vapor or vertical
w	water

1. PRESSURE-MEASURING DEVICES

There are many devices for measuring and indicating fluid pressure. Some devices measure gage pressure; others measure absolute pressure. The effects of non-standard atmospheric pressure and nonstandard gravitational acceleration must be determined, particularly for devices relying on columns of liquid to indicate pressure. Table 15.1 lists the common types of devices and the ranges of pressure appropriate for each.

Table 15.1 *Common Pressure-Measuring Devices*

device	approximate range (in atmospheres)
water manometer	0–0.1
mercury barometer	0–1
mercury manometer	0.001–1
metallic diaphragm	0.01–200
transducer	0.001–700
Bourdon pressure gauge	1–3000
Bourdon vacuum gauge	0.1–1

The *Bourdon pressure gauge* is the most common pressure-indicating device. This mechanical device consists of a coiled hollow tube that tends to straighten out (i.e., unwind) when the tube is subjected to an internal pressure. The degree to which the coiled tube unwinds depends on the difference between the internal and external pressures. A Bourdon gauge directly indicates *gage pressure*. Extreme accuracy is generally not a characteristic of Bourdon gauges.

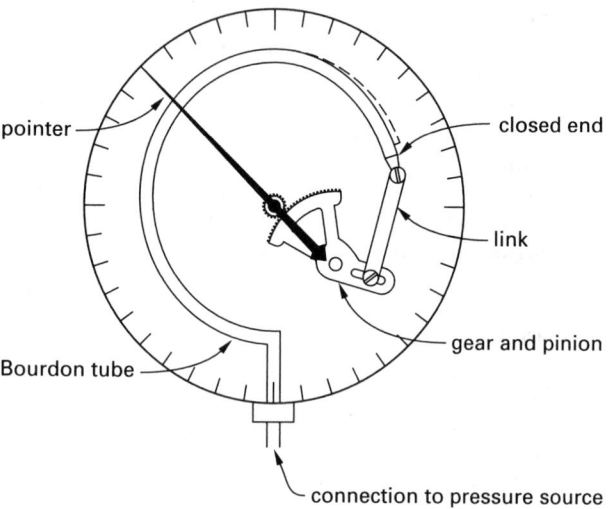

Figure 15.1 *C-Bourdon Pressure Gauge*

In non-SI installations, gauges are always calibrated in psi, unless the dial is marked "altitude" (measuring in feet of water) or "vacuum" (measuring in inches of mercury) on its face. To avoid confusion, the gauge dial will be clearly marked if other units are indicated.

The *barometer* is a common device for measuring the absolute pressure of the atmosphere.[1] It is constructed by filling a long tube open at one end with mercury (or alcohol, or some other liquid) and inverting the tube such that the open end is below the level of a mercury-filled container. If the vapor pressure of the mercury in the tube is neglected, the fluid column will be supported only by the atmospheric pressure transmitted through the container fluid at the lower, open end.

---

[1]A barometer can be used to measure the pressure inside any vessel. However, the barometer must be completely enclosed in the vessel, which may not be possible. Also, it is difficult to read a barometer enclosed within a tank.

Strain gauges, diaphragm gauges, quartz-crystal transducers and other devices using the *piezoelectric effect* are also used to measure stress and pressure, particularly when pressure fluctuates quickly (e.g., as in a rocket combustion chamber). With these devices, calibration is required to interpret pressure from voltage generation or changes in resistance, capacitance, or inductance. These devices are generally unaffected by atmospheric pressure or gravitational acceleration.

Manometers (U-tube manometers) can also be used to indicate small pressure differences, and for this purpose, they provide great accuracy. (Manometers are not suitable for measuring pressures much larger than 10 psi (70 kPa), however.) A difference in manometer fluid surface heights is converted into a pressure difference. If one end of a manometer is open to the atmosphere, the manometer indicates gage pressure. It is theoretically possible, but impractical, to have a manometer indicate absolute pressure, since one end of the manometer would have to be exposed to a perfect vacuum.

A *static pressure tube (piezometer tube)* is a variation of the manometer. It is a simple method of determining the static pressure in a pipe or other vessel, regardless of fluid motion in the pipe. A vertical transparent tube is connected to a hole in the pipe wall.[2] (None of the tube projects into the pipe.) The static pressure will force the contents of the pipe up into the tube. The height of the contents will be an indication of gage pressure in the pipe.

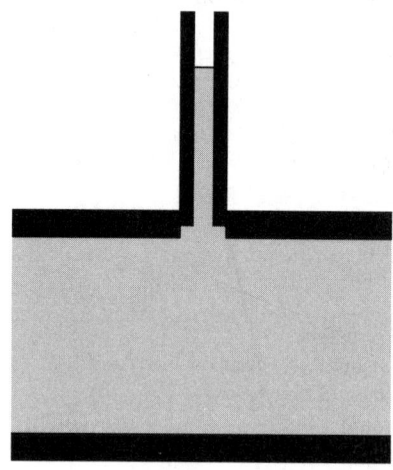

Figure 15.2 *Static Pressure Tube*

The device used to measure the pressure should not be confused with the method used to obtain exposure to the pressure. For example, a static pressure *tap* in a pipe is merely a hole in the pipe wall. A Bourdon gauge, manometer, or transducer can then be used with the tap to indicate pressure.

---

[2]Where greater accuracy is required, multiple holes may be drilled around the circumference of the pipe and connected through a manifold (*piezometer ring*) to the pressure-measuring device.

Tap holes are generally $\frac{1}{8}$ in to $\frac{1}{4}$ in in diameter, drilled at right angles to the wall, and smooth and flush with the pipe wall. No part of the gauge or connection projects into the pipe. The tap holes will be at least 5 to 10 pipe diameters downstream from any source of turbulence (e.g., a bend, fitting, or valve).

2. MANOMETERS

Figure 15.3 illustrates a simple U-tube manometer used to measure the difference in pressure between two vessels. When both ends of the manometer are connected to pressure sources, the name *differential manometer* is used. If one end of the manometer is open to the atmosphere, the name *open manometer* is used.[3] The open manometer implicitly measures gage pressures.

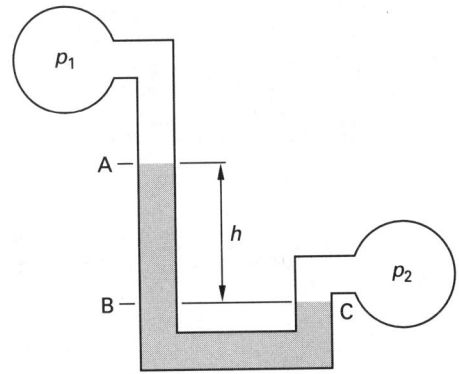

Figure 15.3 *Simple U-Tube Manometer*

Since the pressure at point B in Fig. 15.3 is the same as at point C, the pressure differential produces the vertical fluid column of height h. In Eq. 15.2, A is the area of the tube. Equations 15.2 and 15.3 assume consistent density units. In the absence of any capillary action, the inside diameters of the manometer tubes are irrelevant.

$$F_{\text{net}} = \text{weight of fluid column} \qquad \textbf{15.1}$$

$$(p_2 - p_1) \times A = \rho_m g h \times A \qquad \textbf{15.2}$$

$$p_2 - p_1 = \rho_m g h \qquad \text{[SI]} \qquad \textbf{15.3}$$

In countries that do not use SI units, densities are commonly quoted in pounds per cubic foot. In that case, Eq. 15.3 can be written as

$$p_2 - p_1 = \rho_m \times \frac{g}{g_c} \times h$$
$$= \gamma_m h \qquad \text{[U.S.]} \qquad \textbf{15.4}$$

The quantity g/g_c has a value of 1.0 lbf/lbm in almost all cases, and thus γ_m is numerically equal to ρ_m with units of lbf/ft^3.

---

[3]If one of the manometer legs is inclined, the term *inclined manometer* or *draft gauge* is used. Although only the vertical distance between the manometer fluid surfaces should be used to calculate the pressure difference, with small pressure differences it may be more accurate to read the inclined distance (which is larger than the vertical distance) and compute the vertical distance from the angle of inclination.

Equations 15.3 and 15.4 assume that the manometer fluid height is small, or that only low-density gases fill the tubes above the manometer fluid. If a high-density fluid (such as water) is present above the measuring fluid, or if the columns h_1 or h_2 are very long, corrections will be necessary.

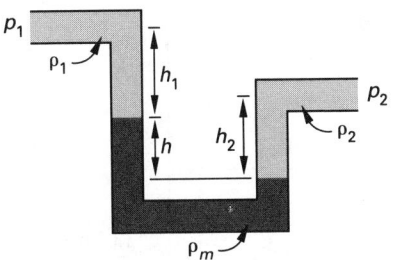

Figure 15.4 *Manometer Requiring Corrections*

Fluid column h_2 "sits on top" of the manometer fluid, forcing the manometer fluid to the left. This increase must be subtracted out. Similarly, the column h_1 restricts the movement of the manometer fluid. The observed measurement must be increased to correct for this restriction.

$$p_2 - p_1 = g(\rho_m h + \rho_1 h_1 - \rho_2 h_2) \qquad \text{[SI]} \qquad \textbf{15.5(a)}$$

$$p_2 - p_1 = \frac{g}{g_c} \times (\rho_m h + \rho_1 h_1 - \rho_2 h_2)$$
$$= \gamma_m h + \gamma_1 h_1 - \gamma_2 h_2 \qquad \text{[U.S.]} \; \textbf{15.5(b)}$$

When a manometer is used to measure the pressure difference across an orifice or other fitting (as in Fig. 15.5), it is not necessary to correct the manometer reading for all of the liquid present above the manometer fluid. This is because parts of the correction for both sides of the manometer are the same. Thus, the distance y in Fig. 15.5 is an irrelevant distance.

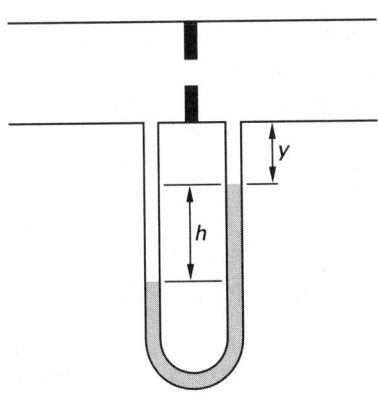

Figure 15.5 *Irrelevant Distance*

Manometer tubes are generally large enough in diameter to avoid significant capillary effects. Corrections for capillarity are seldom necessary.

Example 15.1

The pressure at the bottom of a tank of water ($\rho = 62.4$ lbm/ft^3; 998 kg/m^3) is measured with a mercury manometer. (The density of mercury is 848 lbm/ft^3; 13 575 kg/m^3.) What is the gage pressure at the bottom of the water tank?

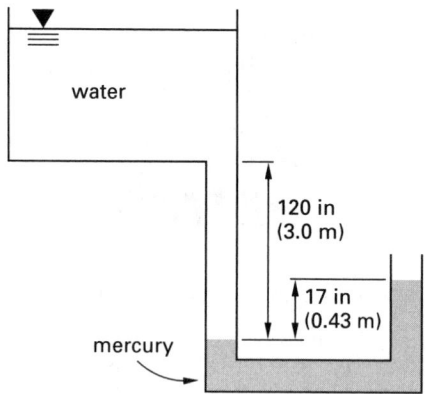

SI Solution

$$\Delta p = \gamma_m h_m - \gamma_w h_w$$

$$= \left(9.81 \, \frac{\text{m}}{\text{s}^2}\right) \left[\left(13\,575 \, \frac{\text{kg}}{\text{m}^3}\right) (0.43 \, \text{m}) \right.$$

$$\left. - \left(998 \, \frac{\text{kg}}{\text{m}^3}\right) (3.0 \, \text{m}) \right]$$

$$= 27\,892 \text{ Pa } (27.9 \text{ kPa}) \text{ gage}$$

Customary U.S. Solution

From Eq. 15.5(b),

$$\Delta p = \gamma_m h_m - \gamma_w h_w$$

$$= \frac{\left(848 \, \frac{\text{lbf}}{\text{ft}^3}\right)(17 \text{ in}) - \left(62.4 \, \frac{\text{lbf}}{\text{ft}^3}\right)(120 \text{ in})}{\left(12 \, \frac{\text{in}}{\text{ft}}\right)^3}$$

$$= 4.01 \text{ lbf/in}^2 \text{ (psig)}$$

3. HYDROSTATIC PRESSURE

Hydrostatic pressure is the pressure a fluid exerts on an immersed object or container walls.[4] Pressure is equal to the force per unit area of surface.

$$p = \frac{F}{A} \qquad \text{15.6}$$

---

[4]The term *hydrostatic* is used with all fluids, not only water.

Hydrostatic pressure in a stationary, incompressible fluid behaves according to the following characteristics.

- Pressure is a function of vertical depth (and density) only. The pressure will be the same at two points with identical depths.

- Pressure varies linearly with (vertical) depth.

- Pressure is independent of an object's area and size and the weight (mass) of water above the object. Figure 15.6 illustrates the *hydrostatic paradox*. The pressures at depth h are the same in all four columns because pressure depends on depth, not volume.

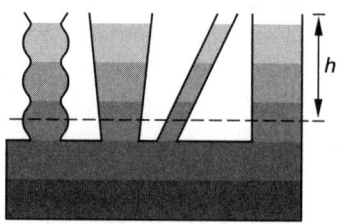

Figure 15.6 *Hydrostatic Paradox*

- Pressure at a point has the same magnitude in all directions (*Pascal's law*). Thus, pressure is a scalar quantity.

- Pressure is always normal to a surface, regardless of the surface's shape or orientation. (This is a result of the fluid's inability to support shear stress.)

- The resultant of the pressure distribution acts through the *center of pressure*.

4. FLUID HEIGHT EQUIVALENT TO PRESSURE

Pressure varies linearly with depth. The relationship between pressure and depth for an incompressible fluid is given by Eq. 15.7.

$$p = \rho g h \qquad \text{[SI]} \qquad \text{15.7(a)}$$

$$p = \frac{\rho g h}{g_c} = \gamma h \qquad \text{[U.S.]} \qquad \text{15.7(b)}$$

Since ρ and g are constants, Eq. 15.7 shows that p and h are linearly related. Knowing one determines the other.[5] For example, the height of a fluid column needed to produce a pressure is

$$h = \frac{p}{\rho g} \qquad \text{[SI]} \qquad \text{15.8(a)}$$

$$h = \frac{p g_c}{\rho g} = \frac{p}{\gamma} \qquad \text{[U.S.]} \qquad \text{15.8(b)}$$

---

[5]In fact, pressure and height of a fluid column can be used interchangeably. The height of a fluid column is known as *head*. For example: "The fan developed a static head of 3 inches of water," or "The pressure head at the base of the water tank was 8 meters." When the term "head" is used, it is essential to specify the fluid.

Table 15.2 lists six important fluid height equivalents that many engineers commit to memory.[6]

Table 15.2 Approximate Fluid Height Equivalents at 68°F (20°C)

liquid	height equivalents
water	0.0361 psi/in
water	62.4 psf/ft
water	9.8 kPa/m
water	2.31 ft/psi
mercury	0.491 psi/in
mercury	133.4 kPa/m

A barometer is an example of the measurement of pressure by the height of a fluid column. If the vapor pressure of the barometer liquid is neglected, the atmospheric pressure will be given by Eq. 15.9.

$$p_a = \rho g h \qquad \text{[SI]} \quad \textit{15.9(a)}$$

$$p_a = \frac{\rho g h}{g_c} = \gamma h \qquad \text{[U.S.]} \quad \textit{15.9(b)}$$

If the vapor pressure of the barometer liquid is significant (as it would be with alcohol or water), the vapor pressure effectively reduces the height of the fluid column, as Eq. 15.10 illustrates.

$$p_a - p_v = \rho g h \qquad \text{[SI]} \quad \textit{15.10(a)}$$

$$p_a - p_v = \frac{\rho g h}{g_c} = \gamma h \qquad \text{[U.S.]} \quad \textit{15.10(b)}$$

Example 15.2

A vacuum pump is used to drain a flooded mine shaft of 68°F (20°C) water.[7] The vapor pressure of water at this temperature is 0.34 psi (2.34 kPa). The pump is incapable of lifting the water higher than 400 in (10.16 m). What is the atmospheric pressure?

SI Solution

$$p_a = p_v + \rho g h$$

$$= 2.34 \text{ kPa} + \frac{\left(998 \frac{\text{kg}}{\text{m}^3}\right)\left(9.81 \frac{\text{m}}{\text{s}^2}\right)(10.16 \text{ m})}{1000 \frac{\text{Pa}}{\text{kPa}}}$$

$$= 101.8 \text{ kPa}$$

(alternate SI solution, using Table 15.2)

$$p_a = p_v + \rho g h$$

$$= 2.34 \text{ kPa} + \left(9.8 \frac{\text{kPa}}{\text{m}}\right)(10.16 \text{ m})$$

$$= 101.9 \text{ kPa}$$

Customary U.S. Solution

From Table 15.2, the height equivalent of water is approximately 0.0361 psi/in. Notice that psi/in is the same as lbf/in³, the units of γ. From Eq. 15.10, the atmospheric pressure is

$$p_a = p_v + \rho g h = p_v + \gamma h$$

$$= 0.34 \frac{\text{lbf}}{\text{in}^2} + \left(0.0361 \frac{\frac{\text{lbf}}{\text{in}^2}}{\text{in}}\right)(400 \text{ in})$$

$$= 14.78 \text{ lbf/in}^2 \text{ (psia)}$$

5. MULTIFLUID BAROMETERS

It is theoretically possible to fill a barometer tube with several different immiscible fluids.[8] Upon inversion, the fluids will separate, leaving the most dense fluid at the bottom and the least dense fluid at the top. All of the fluids will contribute, by superposition, to the balance between the external atmospheric pressure and the weight of the fluid column.

$$p_a - p_v = g \sum \rho_i h_i \qquad \text{[SI]} \quad \textit{15.11(a)}$$

$$p_a - p_v = \frac{g}{g_c} \sum \rho_i h_i = \sum \gamma_i h_i \qquad \text{[U.S.]} \quad \textit{15.11(b)}$$

The pressure at any intermediate point within the fluid column is found by starting at a location where the pressure is known, and then adding or subtracting $\rho g h$ to get to the point where the pressure is needed. Usually, the known pressure will be the atmospheric pressure located in the barometer barrel at the level (elevation) of the fluid outside of the barometer.

Example 15.3

Neglecting vapor pressure, what is the pressure of the air (at point E) in the container shown? The external pressure is 1.0 atm.

---

[6]Of course, these values are recognized to be the approximate specific weights of the liquids.
[7]A reciprocating or other direct-displacement pump would be a better choice to drain a mine.

[8]In practice, barometers are never constructed this way. This theory is more applicable to a category of problems dealing with up-ended containers, as illustrated in Ex. 15.3.

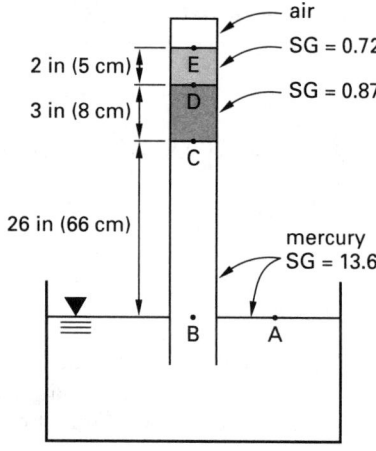

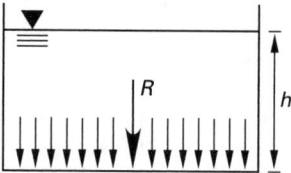

SI Solution

$$p_E = p_{atm} - g\rho_{water} \sum (SG_i)h_i$$

$$= 101\,300\text{ Pa} - \left(9.81\ \frac{\text{m}}{\text{s}^2}\right)\left(1000\ \frac{\text{kg}}{\text{m}^3}\right)$$

$$\times\ [(0.66\text{ m})(13.6) +\ (0.08\text{ m})(0.87)$$

$$+\ (0.05\text{ m})(0.72)]$$

$$= 12\,210\text{ Pa }(12.2\text{ kPa})$$

Customary U.S. Solution

The pressure at point B is the same as the pressure at point A—1.0 atm. The density of mercury is 13.6 × 0.0361 = 0.491 lbm/in³. The pressure at point C is

$$p_C = 14.7\text{ psia} - (26\text{ in})\left(0.491\ \frac{\frac{\text{lbf}}{\text{in}^2}}{\text{in}}\right)$$

$$= 1.93\text{ lbf/in}^2\text{ (psia)}$$

Similarly, the pressure at point E (and anywhere within the captive air) is

$$p_E = 14.7\ \frac{\text{lbf}}{\text{in}^2} - (26\text{ in})\left(0.491\ \frac{\text{lbf}}{\text{in}^3}\right)$$

$$-\ (3\text{ in})(0.87)\left(0.0361\ \frac{\text{lbf}}{\text{in}^3}\right)$$

$$-\ (2\text{ in})(0.72)\left(0.0361\ \frac{\text{lbf}}{\text{in}^3}\right)$$

$$= 1.79\text{ lbf/in}^2\text{ (psia)}$$

6. PRESSURE ON A HORIZONTAL PLANE SURFACE

The pressure on a horizontal plane surface is uniform over the surface because the depth of the fluid is uniform. The resultant of the pressure distribution acts

through the center of pressure of the surface, which corresponds to the centroid of the surface.

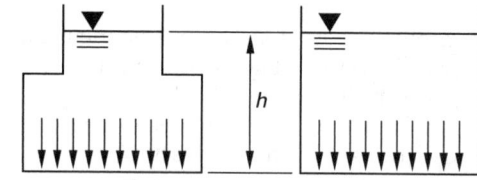

Figure 15.7 *Hydrostatic Pressure on a Horizontal Plane Surface*

The uniform pressure at depth h is given by Eq. 15.12.[9]

$$p = \rho gh \qquad \text{[SI]} \qquad \textit{15.12(a)}$$

$$p = \frac{\rho gh}{g_c} = \gamma h \qquad \text{[U.S.]} \qquad \textit{15.12(b)}$$

The total vertical force on the horizontal plane of area A is given by Eq. 15.13.

$$R = pA \qquad\qquad \textit{15.13}$$

It is tempting, but not always correct, to calculate the vertical force on a submerged surface as the weight of the fluid above it. Such an approach works only when there is no change in the cross-sectional area of the fluid above the surface. This is a direct result of the *hydrostatic paradox*. Figure 15.8 illustrates two containers with the same pressure distribution (force) on their bottom surfaces.

Figure 15.8 *Two Containers with the Same Pressure Distribution*

7. PRESSURE ON A VERTICAL PLANE SURFACE

The pressure on a vertical rectangular plane surface increases linearly with depth. The pressure distribution will be triangular, as in Fig. 15.9(a), if the plane surface extends to the surface; otherwise, the distribution will be trapezoidal, as in Fig. 15.9(b).

---

[9]The phrase *pressure at a depth* is universally understood to mean the *gage pressure*, as given by Eq. 15.12.

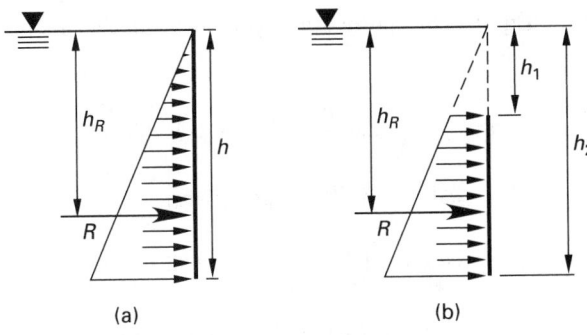

Figure 15.9 *Hydrostatic Pressure on a Vertical Plane Surface*

The resultant force is calculated from the *average pressure*.

$$\bar{p} = \tfrac{1}{2}(p_1 + p_2) \qquad \text{15.14}$$

$$= \tfrac{1}{2}\rho g(h_1 + h_2) \qquad \text{[SI]} \quad \text{15.15(a)}$$

$$\bar{p} = \frac{\tfrac{1}{2}\rho g(h_1 + h_2)}{g_c} = \tfrac{1}{2}\gamma(h_1 + h_2) \quad \text{[U.S.]} \quad \text{15.15(b)}$$

$$R = \bar{p}A \qquad \text{15.16}$$

Although the resultant is calculated from the average depth, the resultant does not act at the average depth. The resultant of the pressure distribution passes through the centroid of the pressure distribution. For the triangular distribution of Fig. 15.9(a), the resultant is located at a depth of $h_R = {}^2/_3h$. For the more general case of Fig. 15.9(b), the resultant is located from Eq. 15.17.

$$h_R = \left(\tfrac{2}{3}\right)\left(h_1 + h_2 - \frac{h_1 h_2}{h_1 + h_2}\right) \qquad \text{15.17}$$

8. PRESSURE ON AN INCLINED PLANE SURFACE

The average pressure and resultant force on an inclined rectangular plane surface are calculated in much the same fashion as for the vertical plane surface. The pressure varies linearly with depth. The resultant is calculated from the average pressure, which, in turn, depends on the average depth.

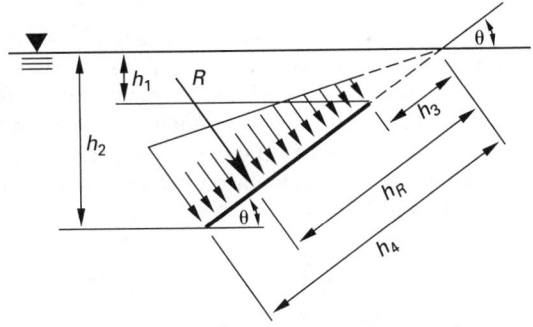

Figure 15.10 *Hydrostatic Pressure on an Inclined Rectangular Plane Surface*

The average pressure and resultant are given by Eqs. 15.18 through 15.21.

$$\bar{p} = \tfrac{1}{2}(p_1 + p_2) \qquad \text{15.18}$$

$$\bar{p} = \tfrac{1}{2}\rho g(h_1 + h_2) \qquad \text{15.19}$$

$$\bar{p} = \tfrac{1}{2}\rho g(h_3 + h_4)\sin\theta \qquad \text{[SI]} \quad \text{15.20(a)}$$

$$\bar{p} = \frac{\tfrac{1}{2}\rho g(h_1 + h_2)}{g_c} = \frac{\tfrac{1}{2}\rho g(h_3 + h_4)\sin\theta}{g_c}$$

$$= \tfrac{1}{2}\gamma(h_1 + h_2) = \tfrac{1}{2}\gamma(h_3 + h_4)\sin\theta \quad \text{[U.S.]} \quad \text{15.20(b)}$$

$$R = \bar{p}A \qquad \text{15.21}$$

As with the vertical plane surface, the resultant acts at the centroid of the pressure distribution, not at the average depth. Equation 15.17 is rewritten in terms of inclined depths.[10]

$$h_R = \left(\frac{\tfrac{2}{3}}{\sin\theta}\right)\left(h_1 + h_2 - \frac{h_1 h_2}{h_1 + h_2}\right) \qquad \text{15.22}$$

$$h_R = \left(\tfrac{2}{3}\right)\left(h_3 + h_4 - \frac{h_3 h_4}{h_3 + h_4}\right) \qquad \text{15.23}$$

Example 15.4

The tank shown is filled with water ($\rho = 62.4$ lbm/ft^3; $\rho = 1000$ kg/m^3). (a) What is the force on a 1 ft (1 m) length of the inclined portion of the wall?[11] (b) At what depth (vertical distance) is the resultant located?

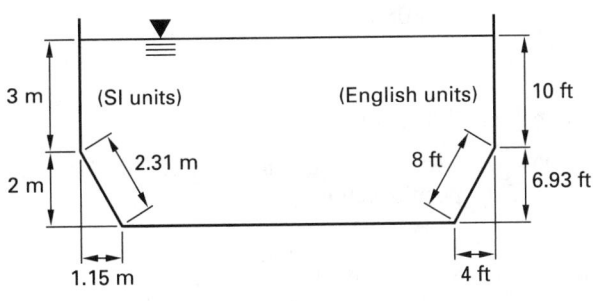

---

[10]Notice that h_R is an inclined distance. If a vertical distance is wanted, it must usually be calculated from h_R and $\sin\theta$. Equation 15.22 can be derived simply by dividing Eq. 15.17 by $\sin\theta$.

[11]Since the width of the tank (the distance into and out of the illustration) is unknown, it is common to calculate the pressure or force per unit width of tank wall. This is the same as calculating the pressure on a 1 ft (1 m) wide section of wall.

SI Solution

(a) The depth of the tank bottom is

$$h_2 = 3 \text{ m} + 2 \text{ m} = 5 \text{ m}$$

From Eq. 15.20, the average pressure on the inclined section is

$$\bar{p} = \left(\tfrac{1}{2}\right)\left(1000 \ \frac{\text{kg}}{\text{m}^3}\right)\left(9.81 \ \frac{\text{m}}{\text{s}^2}\right)(3 \text{ m} + 5 \text{ m})$$
$$= 39\,240 \text{ Pa} \quad \text{(gage)}$$

The total force on a 1 m section of wall is

$$R = \bar{p}A = (39\,240 \text{ Pa})(2.31 \text{ m})(1 \text{ m})$$
$$= 90\,644 \text{ N} \quad (90.6 \text{ kN})$$

(b) θ must be known to determine h_R.

$$\theta = \arctan\left(\frac{2 \text{ m}}{1.15 \text{ m}}\right) = 60°$$

The location of the resultant can be calculated from Eq. 15.23 once h_3 and h_4 are known.

$$h_3 = \frac{3 \text{ m}}{\sin 60°} = 3.464 \text{ m}$$

$$h_4 = \frac{5 \text{ m}}{\sin 60°} = 5.774 \text{ m}$$

$$h_R = \left(\tfrac{2}{3}\right)\left[3.464 + 5.774 - \frac{(3.464)(5.774)}{3.464 + 5.774}\right]$$
$$= 4.715 \text{ m} \quad [\text{inclined}]$$

The vertical depth at which the resultant acts is

$$h = h_R \sin \theta = (4.715 \text{ m})(\sin 60°)$$
$$= 4.08 \text{ m} \quad [\text{vertical}]$$

Customary U.S. Solution

(a) The water density is given in traditional U.S. mass units. The specific weight, γ, is

$$\gamma = \frac{\rho g}{g_c} = \frac{\left(62.4 \ \frac{\text{lbm}}{\text{ft}^3}\right)\left(32.2 \ \frac{\text{ft}}{\text{sec}^2}\right)}{32.2 \ \frac{\text{lbm-ft}}{\text{lbf-sec}^2}}$$

$$= 62.4 \text{ lbf/ft}^3$$

The depth of the tank bottom is

$$h_2 = 10 \text{ ft} + 6.93 \text{ ft} = 16.93 \text{ ft}$$

From Eq. 15.20, the average pressure on the inclined section is

$$\bar{p} = \left(\tfrac{1}{2}\right)\left(62.4 \ \frac{\text{lbf}}{\text{ft}^3}\right)(10 \text{ ft} + 16.93 \text{ ft})$$

$$= 840.2 \text{ lbf/ft}^2 \text{ (gage) (psfg)}$$

The total force on a 1 ft section of wall is

$$R = \bar{p}A = \left(840.2 \ \frac{\text{lbf}}{\text{ft}^2}\right)(8 \text{ ft})(1 \text{ ft})$$

$$= 6722 \text{ lbf}$$

(b) θ must be known to determine h_R.

$$\theta = \arctan\left(\frac{6.93 \text{ ft}}{4 \text{ ft}}\right) = 60°$$

The location of the resultant can be calculated from Eq. 15.23 once h_3 and h_4 are known.

$$h_3 = \frac{10 \text{ ft}}{\sin 60°} = 11.55 \text{ ft}$$

$$h_4 = \frac{16.93 \text{ ft}}{\sin 60°} = 19.55 \text{ ft}$$

From Eq. 15.23,

$$h_R = \left(\tfrac{2}{3}\right)\left[11.55 + 19.55 - \frac{(11.55)(19.55)}{11.55 + 19.55}\right]$$

$$= 15.89 \text{ ft} \quad [\text{inclined}]$$

The vertical depth at which the resultant acts is

$$h = h_R \sin \theta = (15.89 \text{ ft})(\sin 60°)$$
$$= 13.76 \text{ ft} \quad [\text{vertical}]$$

9. PRESSURE ON A GENERAL PLANE SURFACE

Figure 15.11 illustrates a nonrectangular plane surface that may or may not extend to the liquid surface and that may or may not be inclined. As with other regular surfaces, the resultant force depends on the average pressure and acts through the center of pressure (CP). The average pressure is calculated from the depth of the surface's centroid (CG).

$$\bar{p} = \rho g h_c \sin \theta \qquad \text{[SI]} \qquad \textit{15.24(a)}$$

$$\bar{p} = \frac{\rho g h_c \sin \theta}{g_c} = \gamma h_c \sin \theta \qquad \text{[U.S.]} \qquad \textit{15.24(b)}$$

$$R = \bar{p}A \qquad \qquad \textit{15.25}$$

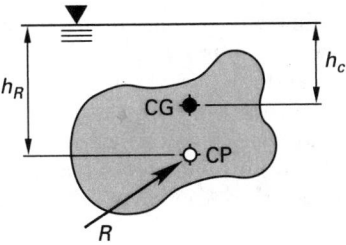

Figure 15.11 *General Plane Surface*

The resultant force acts at depth h_R normal to the plane surface. I_c in Eq. 15.26 is the centroidal area moment of inertia, with dimensions of L^4 (length^4) about an axis parallel to the surface. (Appendix 48.A contains equations for the moment of inertia of common shapes.) Both h_c and h_R are measured parallel to the plane surface. That is, if the plane surface is inclined, h_c and h_R are inclined distances.

$$h_R = h_c + \frac{I_c}{A\, h_c} \qquad 15.26$$

Example 15.5

The top edge of a vertical circular observation window in a submarine is located 4.0 ft (1.25 m) below the surface of the water. The window is 1.0 ft (0.3 m) in diameter. The density of the water is 62.4 lbm/ft^3 (1000 kg/m^3). Neglect the salinity of the water. (a) What is the resultant force on the window? (b) At what depth does the resultant force act?

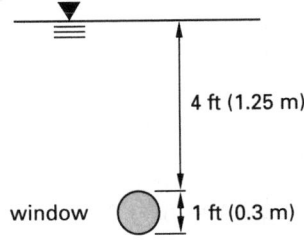

4 ft (1.25 m)

window   1 ft (0.3 m)

SI Solution

(a) The radius of the window is

$$r = \frac{0.3\text{ m}}{2} = 0.15\text{ m}$$

$$h_c = 1.25\text{ m} + 0.15\text{ m} = 1.4\text{ m}$$

$$A = \pi(0.15\text{ m})^2 = 0.0707\text{ m}^2$$

$$\bar{p} = \left(1000\ \frac{\text{kg}}{\text{m}^3}\right)\left(9.81\ \frac{\text{m}}{\text{s}^2}\right)(1.4\text{ m})$$

$$= 13\,734\text{ Pa \ (gage)}$$

$$R = (13\,734\text{ Pa})(0.0707\text{ m}^2) = 971\text{ N}$$

(b) $$I_c = \left(\frac{\pi}{4}\right)(0.15)^4 = 3.976 \times 10^{-4}\text{ m}^4$$

$$h_R = 1.4\text{ m} + \frac{3.976 \times 10^{-4}\text{ m}^4}{(0.0707\text{ m}^2)(1.4\text{ m})}$$

$$= 1.404\text{ m}$$

Customary U.S. Solution

(a) The depth at which the centroid of the circular window is located is

$$h_c = 4.0\text{ ft} + 0.5\text{ ft} = 4.5\text{ ft}$$

The area of the circular window is

$$A = \pi r^2 = \pi(0.5\text{ ft})^2$$
$$= 0.7854\text{ ft}^2$$

As in Ex. 15.4, the resultant force will be calculated using γ as the weight density.

The average pressure is

$$\bar{p} = \gamma h_c = \left(62.4\ \frac{\text{lbf}}{\text{ft}^3}\right)(4.5\text{ ft})$$

$$= 280.8\text{ lbf/ft}^2 \text{ (gage) (psfg)}$$

The resultant is calculated from Eq. 15.21.

$$R = \bar{p}A = \left(280.8\ \frac{\text{lbf}}{\text{ft}^2}\right)(0.7854\text{ ft}^2)$$

$$= 220.5\text{ lbf}$$

(b) The centroidal area moment of inertia of a circle is

$$I_c = \frac{\pi}{4}r^4 = \left(\frac{\pi}{4}\right)(0.5\text{ ft})^4$$

$$= 0.049\text{ ft}^4$$

From Eq. 15.26, the depth at which the resultant force acts is

$$h_R = 4.5\text{ ft} + \frac{0.049\text{ ft}^4}{(0.7854\text{ ft}^2)(4.5\text{ ft})}$$

$$= 4.514\text{ ft}$$

10. SPECIAL CASES: VERTICAL SURFACES

Several simple wall shapes and configurations recur frequently. Figure 15.12 indicates the location of their hydrostatic pressure resultants (*centers of pressure*). In all cases, the surfaces are vertical and extend to the liquid's surface.

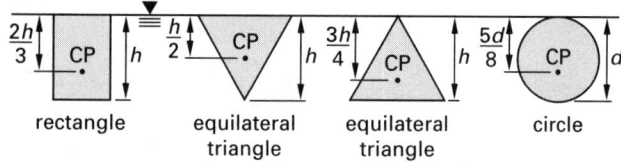

Figure 15.12 *Centers of Pressure for Common Configurations*

11. FORCES ON CURVED AND COMPOUND SURFACES

Figure 15.13 illustrates a curved surface. The resultant force acting on such a curved surface is not difficult to determine, although the x- and y-components of the resultant usually must be calculated first. The magnitude and direction of the resultant are found by conventional methods.

$$R = \sqrt{R_x^2 + R_y^2} \qquad \text{15.27}$$

$$\theta = \arctan\left(\frac{R_y}{R_x}\right) \qquad \text{15.28}$$

The horizontal component of the resultant hydrostatic force is found in the same manner as for a vertical plane surface.

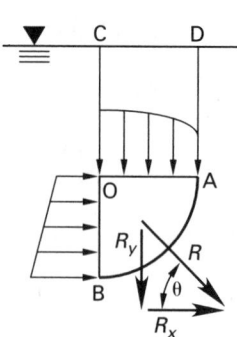

Figure 15.13 *Pressure Distributions on a Curved Surface*

The fact that the surface is curved does not affect the calculation of the horizontal force. In Fig. 15.13, the horizontal pressure distribution on curved surface BA is the same as the horizontal pressure distribution on imaginary projected surface BO.

The vertical component of force on the curved surface is most easily calculated as the weight of the liquid above it.[12] In Fig. 15.13, the vertical component of force on the curved surface BA is the weight of liquid within the area ABCD, with a vertical line of action passing through the centroid of the area ABCD.

---

[12]Calculating the vertical force component is not in conflict with the hydrostatic paradox as long as the cross-sectional area of liquid above the curved surface does not decrease between the curved surface and the liquid's free surface. If there is a change in the cross-sectional area, the vertical component of force is equal to the weight of fluid in an unchanged cross-sectional area (i.e., the equivalent area).

Figure 15.14 illustrates a curved surface with no liquid above it. However, it is not difficult to show that the resultant force acting upward on the curved surface HG is equal in magnitude (and opposite in direction) to the force that would be acting downward due to the missing area EFGH. Such an imaginary area used to calculate hydrostatic pressure is known as an *equivalent area*.

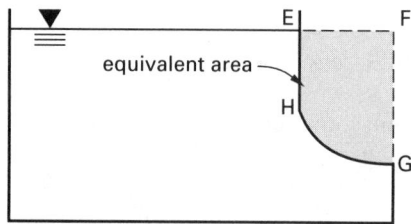

Figure 15.14 *Equivalent Area*

Example 15.6

What is the total force on a 1 ft section of the wall in Ex. 15.4?

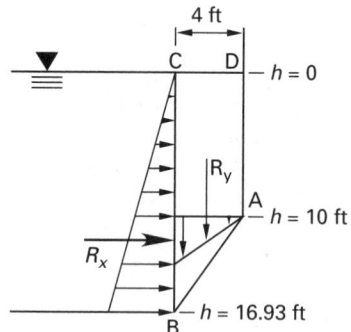

Solution

The average depth is

$$\overline{h} = \left(\tfrac{1}{2}\right)(0 + 16.93 \text{ ft}) = 8.465 \text{ ft}$$

The average pressure and horizontal component of the resultant on a 1 ft section of wall are

$$\overline{p} = \gamma\overline{h} = \left(62.4 \ \frac{\text{lbf}}{\text{ft}^3}\right)(8.465 \text{ ft})$$

$$= 528.2 \text{ lbf/ft}^2 \text{ (gage) (psfg)}$$

$$R_x = \overline{p}A = \left(528.2 \ \frac{\text{lbf}}{\text{ft}^2}\right)(16.93 \text{ ft})(1 \text{ ft})$$

$$= 8942 \text{ lbf}$$

The volume of a 1 ft section of area ABCD is

$$V_{\text{ABCD}} = (1 \text{ ft})\left[(4 \text{ ft})(10 \text{ ft}) + \left(\tfrac{1}{2}\right)(4 \text{ ft})(6.93 \text{ ft})\right]$$

$$= 53.86 \text{ ft}^3$$

The vertical component is

$$R_y = \gamma V = \left(62.4 \ \frac{\text{lbf}}{\text{ft}^3}\right)(53.86 \ \text{ft}^3)$$

$$= 3361 \ \text{lbf}$$

The total resultant force is

$$R = \sqrt{(8942 \ \text{lbf})^2 + (3361 \ \text{lbf})^2}$$

$$= 9553 \ \text{lbf}$$

12. TORQUE ON A GATE

When an openable gate or door is submerged in such a manner as to have unequal depths of liquid on either of its sides, or when there is no liquid present on one side of a gate or door, the hydrostatic pressure will act to either open or close the door. If the gate does not open, this pressure is resisted, usually by a latching mechanism on the gate itself.[13] The magnitude of the resisting latch force can be determined from the *hydrostatic torque (hydrostatic moment)* acting on the gate. The moment is almost always taken with respect to the gate hinges.

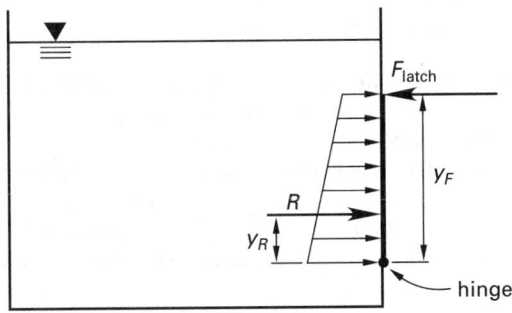

Figure 15.15 *Torque on a Hinge*

The applied moment is calculated as the product of the resultant force on the gate and the distance from the hinge to the resultant on the gate. This applied moment is balanced by the resisting moment, calculated as the latch force times the separation of the latch and hinge.

$$M_{\text{applied}} = M_{\text{resisting}} \qquad 15.29$$

$$R \times y_R = F_{\text{latch}} \times y_F \qquad 15.30$$

13. HYDROSTATIC FORCES ON A DAM

The techniques presented in the preceding sections are applicable to dams. That is, the horizontal force on the dam face can be as in Ex. 15.6, regardless of inclination

---

[13]Any contribution to resisting force from stiff hinges or other sources of friction is typically neglected.

or curvature of the dam face. The vertical force on the dam face is calculated as the weight of the water above the dam face. Of course, the vertical force is zero if the dam face is vertical.

Figure 15.16 illustrates a typical dam, defining its *heel, toe,* and *crest.*

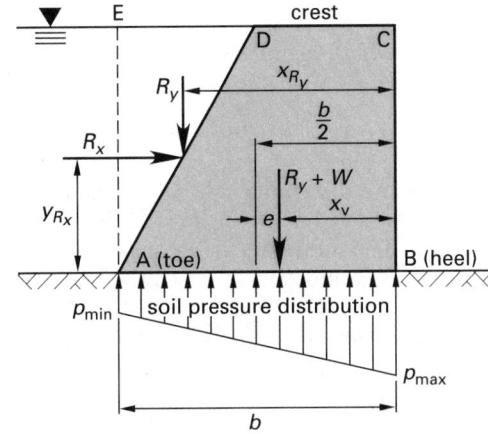

Figure 15.16 *Dam*

However, there are other considerations for gravity dams.[14] Most notably, the dam must not tip over or slide away due to the hydrostatic pressure. Furthermore, the pressure distribution within the soil under the dam is not uniform, and soil loading may be excessive.

The *overturning moment* is a measure of the horizontal pressure's tendency to tip the dam over, pivoting it about the heel of the dam (point B in Fig. 15.16). (Usually, moments are calculated with respect to the pivot point.) The overturning moment is calculated as the product of the horizontal component of hydrostatic pressure (i.e., the x-component of the resultant) and the vertical distance between the heel and the line of action of the force.

$$M_{\text{overturning}} = R_x \times y_{R_x} \qquad 15.31$$

In most configurations, the overturning is resisted jointly by moments from the dam's own weight, W, and the vertical component of the resultant, R_y (i.e., the weight of area EAD in Fig. 15.16).[15]

$$M_{\text{resisting}} = (R_y \times x_{R_y}) + (W \times x_{\text{CG}}) \qquad 15.32$$

The *factor of safety against overturning* is

$$(\text{FS})_{\text{overturning}} = \frac{M_{\text{resisting}}}{M_{\text{overturning}}} \qquad 15.33$$

---

[14]A *gravity dam* is one that is held in place and orientation by its own mass (weight) and the friction between its base and the ground.
[15]The density of concrete or masonry is usually taken to be approximately 150 lbm/ft^3 (2400 kg/m^3).

In addition to causing the dam to tip over, the horizontal component of hydrostatic force will also cause the dam to tend to slide along the ground. This tendency is resisted by the frictional force between the dam bottom and soil. The frictional force, F_f, is calculated as the product of the *normal force*, N, and the *coefficient of static friction*, μ.

$$F_f = \mu_{\text{static}} \, N = \mu_{\text{static}}(W + R_y) \qquad 15.34$$

The *factor of safety against sliding* is

$$(\text{FS})_{\text{sliding}} = \frac{F_f}{R_x} \qquad 15.35$$

The soil pressure distribution beneath the dam is usually assumed to vary linearly from a minimum to a maximum value. (The minimum value must be greater than zero, since soil cannot be in a state of tension. The maximum pressure should not exceed the allowable soil pressure.) Equation 15.36 predicts the minimum and maximum soil pressures.

$$p_{\text{max}}, p_{\text{min}} = \left(\frac{R_y + W}{b}\right)\left(1 \pm \frac{6e}{b}\right) \qquad 15.36$$

The *eccentricity*, e, in Eq. 15.36 is the distance between the mid-length of the dam and the line of action of the total vertical force, $W + R_y$. (The eccentricity must be less than $b/6$ for the entire base to be in compression.) Notice that distances x_v and x_{CG} are different.

$$e = \frac{b}{2} - x_v \qquad 15.37$$

$$x_v = \frac{M_{\text{resisting}}}{R_y + W} \qquad 15.38$$

14. PRESSURE DUE TO SEVERAL IMMISCIBLE LIQUIDS

Figure 15.17 illustrates the non-uniform pressure distribution due to two immiscible liquids (e.g., oil on top and water below).

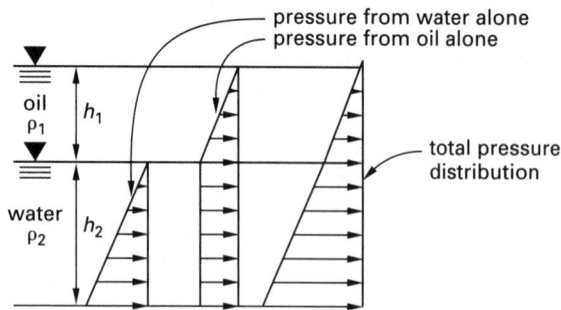

Figure 15.17 Pressure Distribution from Two Immiscible Liquids

The pressure due to the upper liquid (oil), once calculated, serves as a *surcharge* to the liquid below (water).

The pressure at the tank bottom is given by Eq. 15.39. (The principle can be extended to three or more immiscible liquids as well.)

$$p_{\text{bottom}} = \rho_1 g h_1 + \rho_2 g h_2 \qquad [\text{SI}] \qquad 15.39(a)$$

$$p_{\text{bottom}} = \frac{\rho_1 g h_1}{g_c} + \frac{\rho_2 g h_2}{g_c}$$
$$= \gamma_1 h_1 + \gamma_2 h_2 \qquad [\text{U.S.}] \qquad 15.39(b)$$

15. PRESSURE FROM COMPRESSIBLE FLUIDS

Fluid density, thus far, has been assumed to be independent of pressure. In reality, even "incompressible" liquids are slightly compressible. Sometimes, the effect of this compressibility cannot be neglected.

The familiar $p = \rho g h$ equation is a special case of Eq. 15.40. (It is assumed that $h_2 > h_1$. The minus sign in Eq. 15.40 indicates that pressure decreases as elevation (height) increases.)

$$\int_{p_1}^{p_2} \frac{dp}{\rho g} = -(h_2 - h_1) \qquad [\text{SI}] \qquad 15.40(a)$$

$$\int_{p_1}^{p_2} \frac{g_c \, dp}{\rho g} = -(h_2 - h_1) \qquad [\text{U.S.}] \qquad 15.40(b)$$

If the fluid is a perfect gas, and if compression is an isothermal (i.e., constant temperature) process, the relationship between pressure and density is given by Eq. 15.41. The isothermal assumption is appropriate, for example, for the earth's *stratosphere* (i.e., above 35,000 ft or 11 000 m), where the temperature is assumed to be constant at approximately $-67°\text{F}$ $(-55°\text{C})$.

$$pv = \frac{p}{\rho} = RT = \text{constant} \qquad 15.41$$

In the isothermal case, Eq. 15.41 can be rewritten as Eq. 15.42. (For air, $R = 53.3$ ft-lbf/lbm-°R; $R = 286.7$ J/kg·K.) Of course, the temperature, T, must be in degrees absolute (i.e., in °R or K). Equation 15.42 is known as the *barometric height relationship*[16] because knowledge of atmospheric temperature and the pressures at two points is sufficient to determine the height difference between the two points.

$$h_2 - h_1 = \left(\frac{RT}{g}\right) \ln\left(\frac{p_1}{p_2}\right) \qquad [\text{SI}] \qquad 15.42(a)$$

$$h_2 - h_1 = \left(\frac{g_c RT}{g}\right) \ln\left(\frac{p_1}{p_2}\right) \qquad [\text{U.S.}] \qquad 15.42(b)$$

---

[16]You may recognize this as being equivalent to the work done in an isothermal compression process. The elevation (height) difference $h_2 - h_1$ (with units of feet) can be interpreted as the work done per unit mass during compression (with units of ft-lbf/lbm).

The pressure at an elevation (height) h_2 in a layer of perfect gas that has been isothermally compressed is given by Eq. 15.43.

$$p_2 = p_1 e^{\frac{g(h_1 - h_2)}{RT}} \qquad \text{[SI]} \qquad \textbf{15.43(a)}$$

$$p_2 = p_1 e^{\frac{g(h_1 - h_2)}{g_c RT}} \qquad \text{[U.S.]} \qquad \textbf{15.43(b)}$$

If the fluid is a perfect gas, and if compression is an *adiabatic process*, the relationship between pressure and density is given by Eq. 15.44[17]. k is the *ratio of specific heats*, a property of the gas. ($k = 1.4$ for air, hydrogen, oxygen, and carbon monoxide, among others.)

$$pv^k = p\left(\frac{1}{\rho}\right)^k = \text{constant} \qquad \textbf{15.44}$$

The following three equations apply to adiabatic compression of an ideal gas.

$$h_2 - h_1 = \left(\frac{k}{k-1}\right)\left(\frac{RT_1}{g}\right)\left[1 - \left(\frac{p_2}{p_1}\right)^{\frac{k-1}{k}}\right]$$
$$\text{[SI]} \qquad \textbf{15.45(a)}$$

$$h_2 - h_1 = \left(\frac{k}{k-1}\right)\left(\frac{g_c}{g}\right)RT_1\left[1 - \left(\frac{p_2}{p_1}\right)^{\frac{k-1}{k}}\right]$$
$$\text{[U.S.]} \qquad \textbf{15.45(b)}$$

$$p_2 = p_1\left[1 - \left(\frac{k-1}{k}\right)\left(\frac{g}{RT_1}\right)(h_2 - h_1)\right]^{\frac{k}{k-1}}$$
$$\text{[SI]} \qquad \textbf{15.46(a)}$$

$$p_2 = p_1\left[1 - \left(\frac{k-1}{k}\right)\left(\frac{g}{g_c}\right)\left(\frac{h_2 - h_1}{RT_1}\right)\right]^{\frac{k}{k-1}}$$
$$\text{[U.S.]} \qquad \textbf{15.46(b)}$$

$$T_2 = T_1\left[1 - \left(\frac{k-1}{k}\right)\left(\frac{g}{RT_1}\right)(h_2 - h_1)\right]$$
$$\text{[SI]} \qquad \textbf{15.47(a)}$$

$$T_2 = T_1\left[1 - \left(\frac{k-1}{k}\right)\left(\frac{g}{g_c}\right)\left(\frac{h_2 - h_1}{RT_1}\right)\right]$$
$$\text{[U.S.]} \qquad \textbf{15.47(b)}$$

The three adiabatic compression equations can be used for the more general *polytropic compression* case simply by substituting the *polytropic exponent*, n, for

[17]There is no heat or energy transfer to or from the ideal gas in an adiabatic process. However, this is not the same as an isothermal process.

k.[18] Unlike the ratio of specific heats, the polytropic exponent is a function of the process, not the gas. The polytropic compression assumption is appropriate for the earth's *troposphere*.[19] Assuming a linear decrease in temperature with altitude of $-0.00356°F/ft$ ($-0.00649°C/m$), a polytropic exponent of $n = 1.235$ can be derived.

Example 15.7

The air pressure and temperature at sea level are 1.0 standard atmosphere and $68°F$ ($20°C$), respectively. Assume isentropic compression with $n = 1.235$. What is the pressure at an altitude of 5000 ft (1525 m)?

SI Solution

The absolute temperature of the air is $20°C + 273 = 293K$. From Eq. 15.46, the pressure at 1525 m altitude is

$$p_2 = (1.0 \text{ atm})\left(1 - \left(\frac{1.235 - 1}{1.235}\right)\left(9.81\frac{\text{m}}{\text{s}^2}\right)\right.$$
$$\left.\times\left[\frac{1525 \text{ m}}{\left(286.7\frac{\text{J}}{\text{kg}\cdot\text{K}}\right)(293\text{K})}\right]\right)^{\frac{1.235}{1.235-1}}$$
$$= 0.834 \text{ atm}$$

Customary U.S. Solution

The absolute temperature of the air is $68°F + 460 = 528°R$. From Eq. 15.46 (substituting $n = 1.235$ for polytropic compression), the pressure at 5000 ft altitude is

$$p_2 = (1.0 \text{ atm})\left(1 - \left(\frac{1.235 - 1}{1.235}\right)\left(\frac{32.2\frac{\text{ft}}{\text{sec}^2}}{32.2\frac{\text{lbm-ft}}{\text{lbf-sec}^2}}\right)\right.$$
$$\left.\times\left[\frac{5000 \text{ ft}}{\left(53.3\frac{\text{ft-lbf}}{\text{lbm-}°R}\right)(528°R)}\right]\right)^{\frac{1.235}{1.235-1}}$$
$$= 0.835 \text{ atm}$$

[18]Actually, polytropic compression is the general process. Isothermal compression is a special case ($n = 1$) of the polytropic process, as is adiabatic compression ($n = k$).
[19]The *troposphere* is the part of the earth's atmosphere we live in and where most atmospheric disturbances occur. The *stratosphere*, starting at approximately 35,000 ft (11 000 m), is cold, clear, dry, and still. Between the troposphere and the stratosphere is the *tropopause*, a transition layer that contains most of the atmosphere's dust and moisture. Temperature actually increases slightly in the tropopause.

16. EXTERNALLY PRESSURIZED LIQUIDS

If the gas above a liquid in a closed tank is pressurized to a gage pressure of p_t, this pressure will add to the hydrostatic pressure anywhere in the fluid. The pressure at the tank bottom illustrated in Fig. 15.18 is given by Eq. 15.48.

$$p_{\text{bottom}} = p_t + \rho g h \qquad \text{[SI]} \quad 15.48(a)$$

$$p_{\text{bottom}} = p_t + \frac{\rho g h}{g_c} = p_t + \gamma h \qquad \text{[U.S.]} \quad 15.48(b)$$

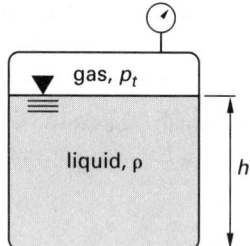

Figure 15.18 *Externally Pressurized Liquid*

17. HYDRAULIC RAM

A *hydraulic ram (hydraulic jack, hydraulic press, fluid press*, etc.) is illustrated in Fig. 15.19. This is a force-multiplying device. A force, F_p, is applied to the *plunger*, and a useful force, F_r, appears at the *ram*. Even though the pressure in the hydraulic fluid is the same everywhere, the forces at the two cylinders will be proportional to their respective cross-sectional areas.[20]

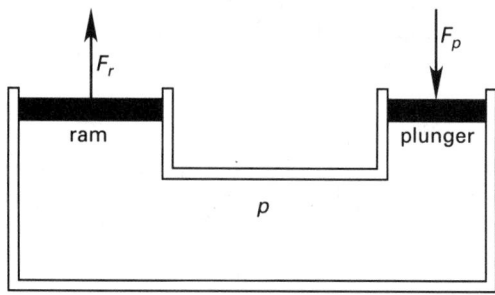

Figure 15.19 *Hydraulic Ram*

Since the pressure is the same everywhere, Eq. 15.49 can be solved for p for the ram and plunger.

$$F = pA = p(\pi r^2) \qquad 15.49$$

$$p_p = p_r \qquad 15.50$$

$$\frac{F_p}{A_p} = \frac{F_r}{A_r} \qquad 15.51$$

$$\frac{F_p}{d_p^2} = \frac{F_r}{d_r^2} \qquad 15.52$$

---

[20] *Pascal's law*: Pressure in a liquid is the same in all directions.

Small, manually actuated hydraulic rams usually have a lever handle to increase the mechanical advantage of the ram from 1.0 to M, as illustrated in Fig. 15.20. In most cases, the pivot and connection mechanism will not be frictionless, and some of the applied force will be used to overcome the friction. This friction loss is accounted for by a *lever efficiency* or *lever effectiveness*, η.

$$M = \frac{L_1}{L_2} \qquad 15.53$$

$$F_p = \eta\, M F_l \qquad 15.54$$

$$\frac{\eta\, M F_l}{A_p} = \frac{F_r}{A_r} \qquad 15.55$$

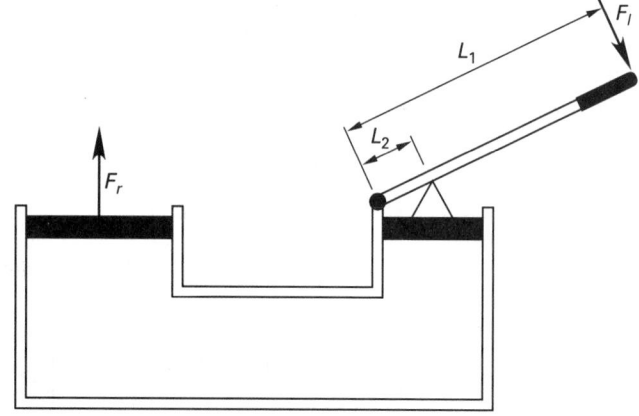

Figure 15.20 *Hydraulic Ram with Mechanical Advantage*

18. BUOYANCY

Buoyant force is an upward force that acts on all objects that are partially or completely submerged in a fluid. The fluid can be a liquid, as in the case of a ship floating at sea, or the fluid can be a gas, as in a balloon floating in the atmosphere.

There is a buoyant force on all submerged objects, not just those that are stationary or ascending. There will be, for example, a buoyant force on a rock sitting at the bottom of a pond. There will also be a buoyant force on a rock sitting exposed on the ground, since the rock is "submerged" in air. A buoyant force due to displaced air also exists, although it may be insignificant, in the case of partially exposed floating objects such as icebergs.

Buoyant force always acts to cancel the object's weight (i.e., buoyancy acts against gravity). The magnitude of the buoyant force is predicted from *Archimedes' principle (the buoyancy theorem)*: the buoyant force on a submerged object is equal to the weight of the displaced fluid.[21] An equivalent statement of Archimedes' principle is: a floating object displaces liquid equal in weight to its own weight.

---

[21] The volume term in Eq. 15.56 is the total volume of the object only in the case of complete submergence.

$$F_{\text{buoyant}} = \rho g V_{\text{displaced}} \qquad \text{[SI]} \quad \textit{15.56(a)}$$

$$F_{\text{buoyant}} = \frac{\rho g V_{\text{displaced}}}{g_c} = \gamma V_{\text{displaced}} \quad \text{[U.S.]} \ \textit{15.56(b)}$$

In the case of stationary (i.e., not moving vertically) floating or submerged objects, the buoyant force and object weight are in equilibrium. If the forces are not in equilibrium, the object will rise or fall until equilibrium is reached. That is, the object will sink until its remaining weight is supported by the bottom, or it will rise until the weight of displaced liquid is reduced by breaking the surface.[22]

The specific gravity (SG) of an object submerged in water can be determined from its dry and submerged weights. Neglecting the buoyancy of any surrounding gases,

$$\text{SG} = \frac{W_{\text{dry}}}{W_{\text{dry}} - W_{\text{submerged}}} \qquad \textit{15.57}$$

Figure 15.21 illustrates an object floating partially exposed in a liquid. Neglecting the insignificant buoyant force from the displaced air (or other gas), the fractions of volume exposed and submerged are easily determined.

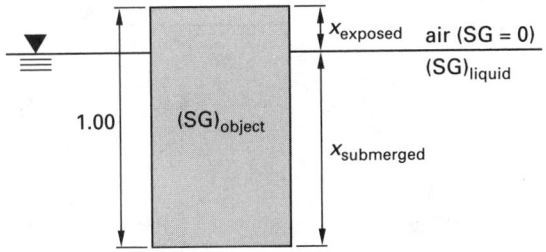

Figure 15.21 *Partially Submerged Object*

$$x_{\text{submerged}} = \frac{\rho_{\text{object}}}{\rho_{\text{liquid}}} = \frac{(\text{SG})_{\text{object}}}{(\text{SG})_{\text{liquid}}} \qquad \textit{15.58}$$

$$x_{\text{exposed}} = 1 - x_{\text{submerged}} \qquad \textit{15.59}$$

Figure 15.22 illustrates a somewhat more complicated situation—that of an object floating at the interface between two liquids of different densities. The fractions of immersion in each liquid are given by the following equations.

$$x_1 = \frac{(\text{SG})_2 - (\text{SG})_{\text{object}}}{(\text{SG})_2 - (\text{SG})_1} \qquad \textit{15.60}$$

$$x_2 = 1 - x_1 = \frac{(\text{SG})_{\text{object}} - (\text{SG})_1}{(\text{SG})_2 - (\text{SG})_1} \qquad \textit{15.61}$$

---

[22]An object can also stop rising or falling due to a change in the fluid's density. The buoyant force will increase with increasing depth in the ocean due to an increase in density at great depths. The buoyant force will decrease with increasing altitude in the atmosphere due to a decrease in density at great heights.

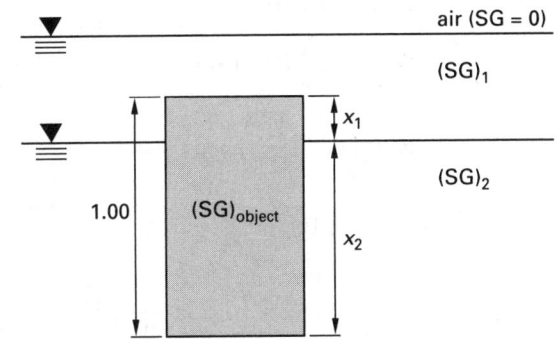

Figure 15.22 *Object Floating in Two Liquids*

A more general case of a floating object is shown in Fig. 15.23. Problems of this type are easily solved by equating the object's weight with the sum of the buoyant forces.

In the case of Fig. 15.23 (with two liquids), the following relationships apply.

$$(\text{SG})_{\text{object}} = x_1(\text{SG})_1 + x_2(\text{SG})_2 \qquad \textit{15.62}$$

$$(\text{SG})_{\text{object}} = (1 - x_0 - x_2)(\text{SG})_1 + x_2(\text{SG})_2 \quad \textit{15.63}$$

$$(\text{SG})_{\text{object}} = x_1(\text{SG})_1 + (1 - x_0 - x_1)(\text{SG})_2 \quad \textit{15.64}$$

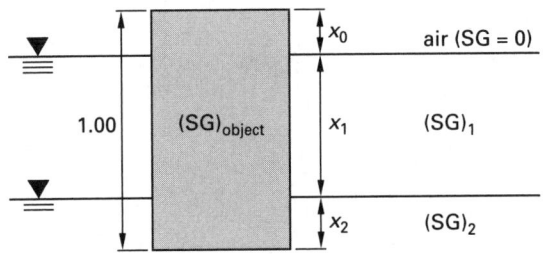

Figure 15.23 *General Two-Liquid Buoyancy Problem*

Example 15.8

An empty polyethylene telemetry balloon and payload have a mass of 500 lbm (225 kg). The balloon is charged with helium when the atmospheric conditions are 60°F (15.6°C) and 14.8 psia (102 kPa). The specific gas constant of helium is 386.3 ft-lbf/lbm-°R (2079 J/kg·K). What volume of helium is required for lift-off from a sea-level platform?

SI Solution

$$\rho_{\text{air}} = \frac{p}{RT} = \frac{1.02 \times 10^5 \text{ Pa}}{\left(286.7 \ \dfrac{\text{J}}{\text{kg·K}}\right)(15.6 + 273)\text{ K}}$$

$$= 1.233 \text{ kg/m}^3$$

$$\rho_{\text{helium}} = \frac{1.02 \times 10^5 \text{ Pa}}{\left(2079 \ \dfrac{\text{J}}{\text{kg} \cdot \text{K}}\right)(288.6\text{K})} = 0.17 \text{ kg/m}^3$$

$$m = 225 \text{ kg} + \left(0.17 \ \frac{\text{kg}}{\text{m}^3}\right) V_{\text{He}}$$

$$m_b = \left(1.233 \ \frac{\text{kg}}{\text{m}^3}\right) V_{\text{He}}$$

$$225 \text{ kg} + \left(0.17 \ \frac{\text{kg}}{\text{m}^3}\right) V_{\text{He}} = \left(1.233 \ \frac{\text{kg}}{\text{m}^3}\right) V_{\text{He}}$$

$$V_{\text{He}} = 211.7 \text{ m}^3$$

Customary U.S. Solution

The gas densities are

$$\rho_{\text{air}} = \frac{p}{RT} = \frac{\left(14.8 \ \dfrac{\text{lbf}}{\text{in}^2}\right)\left(144 \ \dfrac{\text{in}^2}{\text{ft}^2}\right)}{\left(53.3 \ \dfrac{\text{ft-lbf}}{\text{lbm-}^\circ\text{R}}\right)(60 + 460)^\circ\text{R}}$$

$$= 0.07689 \text{ lbm/ft}^3$$

$$\gamma_{\text{air}} = \rho \times \frac{g}{g_c} = 0.07689 \text{ lbf/ft}^3$$

$$\rho_{\text{helium}} = \frac{\left(14.8 \ \dfrac{\text{lbf}}{\text{in}^2}\right)\left(144 \ \dfrac{\text{in}^2}{\text{ft}^2}\right)}{\left(386.3 \ \dfrac{\text{ft-lbf}}{\text{lbm-}^\circ\text{R}}\right)(520^\circ\text{R})}$$

$$= 0.01061 \text{ lbm/ft}^3$$

$$\gamma_{\text{helium}} = 0.01061 \text{ lbf/ft}^3$$

The total weight of the balloon, payload, and helium is

$$W = 500 \text{ lbf} + \left(0.01061 \ \frac{\text{lbf}}{\text{ft}^3}\right) V_{\text{He}}$$

The buoyant force is the weight of the displaced air. Neglecting the payload volume, the displaced air volume is the same as the helium volume.

$$F_b = \left(0.07689 \ \frac{\text{lbf}}{\text{ft}^3}\right) V_{\text{He}}$$

At lift-off, the weight of the balloon is just equal to the buoyant force.

$$W = F_b$$

$$500 \text{ lbf} + \left(0.01061 \ \frac{\text{lbf}}{\text{ft}^3}\right) V_{\text{He}} = \left(0.07689 \ \frac{\text{lbf}}{\text{ft}^3}\right) V_{\text{He}}$$

$$V_{\text{He}} = 7544 \text{ ft}^3$$

19. INTACT STABILITY: STABILITY OF FLOATING OBJECTS

A stationary object is said to be in *static equilibrium*. However, an object in static equilibrium is not necessarily stable. For example, a coin balanced on edge is in static equilibrium, but it will not return to the balanced position if it is disturbed. An object is said to be *stable* (i.e., in *stable equilibrium*) if it tends to return to the equilibrium position when slightly displaced.

Stability of floating and submerged objects is known as *intact stability*.[23] There are two forces acting on a stationary floating object: the buoyant force and the object's weight. The buoyant force acts upward through the centroid of the displaced volume (not the object's volume). This centroid is known as the *center of buoyancy*. The gravitational force on the object (i.e., the object's weight) acts downward through the entire object's center of gravity.

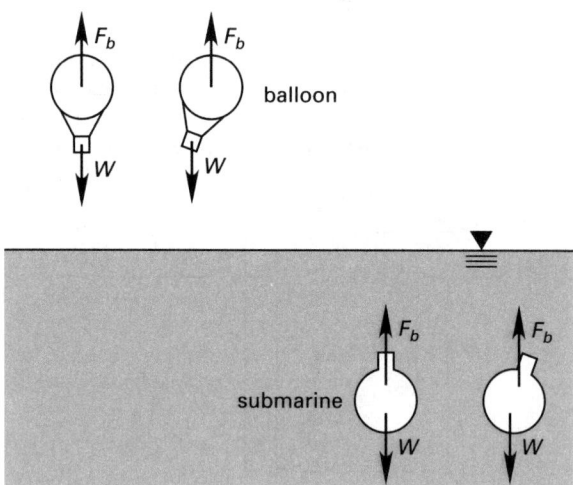

Figure 15.24 *Stability of a Submerged Object*

For a totally submerged object (e.g., a balloon or submarine) to be stable, the center of buoyancy must be above the center of gravity. The object will be stable because a *righting moment* will be created if the object tips over, since the center of buoyancy will move outward from the center of gravity.

The stability criterion is different for partially submerged objects (e.g., surface ships). For partially submerged objects to be stable, the *metacenter* must be above the center of gravity. If the vessel shown in Fig. 15.25 heels over, the location of the center of gravity of the object does not change. However, the center of buoyancy shifts to the centroid of the new submerged section 123. The centers of buoyancy and gravity are no longer in line. The righting couple resists further overturning.

---

[23]The subject of intact stability, being a part of naval architecture curriculum, is not covered extensively in most fluids books. However, it is covered extensively in basic ship design and naval architecture books.

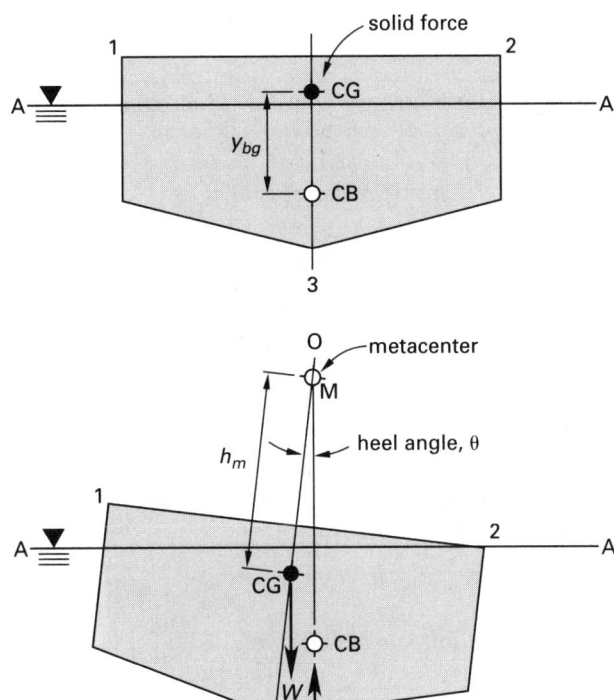

Figure 15.25 *Stability of a Partially Submerged Floating Object*

Eq. 15.65. Variable I is the centroidal area moment of inertia of the original waterline (free surface) cross section about a longitudinal (fore and aft) waterline axis; V is the displaced volume.

If the distance, y_{bg}, separating the centers of buoyancy and gravity is known, Eq. 15.65 can be solved for the metacentric height. y_{bg} is positive when the center of gravity is above the center of buoyancy. This is the normal case. Otherwise, y_{bg} is negative.

$$y_{bg} + h_m = \frac{I}{V} \qquad \text{15.65}$$

The righting moment (also known as the *restoring moment*) is the stabilizing moment exerted when the ship rolls. Values of the righting moment are typically specified with units of foot-tons (MN·m).

$$M_{\text{righting}} = h_m \gamma_w V_{\text{displaced}} \sin\theta \qquad \text{15.66}$$

The transverse (roll) and longitudinal (pitch) *roll periods* also depend on the metacentric height. The roll characteristics are found from the differential equation formed by equating the righting moment to the product of the ship's transverse mass moment of inertia and the angular acceleration. Larger metacentric heights result in lower roll periods. If k is the radius of gyration about the roll axis, the roll period is

$$T_{\text{roll}} = \frac{2\pi k}{\sqrt{gh_m}} \qquad \text{15.67}$$

The roll and pitch periods must be adjusted for the appropriate level of crew and passenger comfort. A "beamy" ice-breaking ship will have a metacentric height much larger than normally required for intact stability, resulting in a low, nauseating, roll period. The designer of a passenger ship, however, would have to decrease the intact stability (i.e., decrease the metacentric height) in order to achieve an acceptable ride characteristic. This requires a metacentric height that is less than approximately 6% of the beam length.

This righting couple exists when the extension of the buoyant force, F_b, intersects line O-O above the center of gravity at M, the *metacenter*. If M lies below the center of gravity, an overturning couple will exist. The distance between the center of gravity and the metacenter is called the *metacentric height*, and it is reasonably constant for heel angles less than 10°. Also, for angles less than 10°, the center of buoyancy follows a locus for which the metacenter is the instantaneous center.

The metacentric height is one of the most important and basic parameters in ship design. It determines the ship's ability to remain upright as well as the ship's roll and pitch characteristics.[24]

"Acceptable" minimum values of the metacentric height have been established from experience, and these depend on the ship type and class. For example, many submarines are required to have a metacentric height of 1 ft (0.3 m) when surfaced. This will increase to approximately 3.5 ft (1.2 m) for some of the largest surface ships. If an acceptable metacentric height is not achieved initially, the center of gravity must be lowered or the keel depth increased. The beam width can also be increased slightly to increase the waterplane moment of inertia.

For a surface vessel rolling through an angle less than approximately 10°, the distance between the vertical center of gravity and the metacenter can be found from

Example 15.9

A 600,000 lbm (280 000 kg) rectangular barge has external dimensions of 24 ft width, 98 ft length, and 12 ft height (7 m × 30 m × 3.6 m). It floats in seawater ($\gamma_w = 64.0$ lbf/ft³; $\rho_w = 1024$ kg/m³). The center of gravity is 7.8 ft (2.4 m) from the top of the barge as loaded. Find (a) the location of the center of buoyancy when the barge is floating on an even keel, and (b) the approximate location of the metacenter when the barge experiences a 5° list.

SI Solution

(a) Refer to the following diagram. Let dimension y represent the depth of the submerged barge.

---

[24]Other verbs synonymous with *roll* are *list* and *heel*.

Fluids

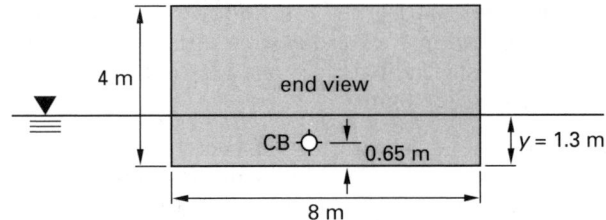

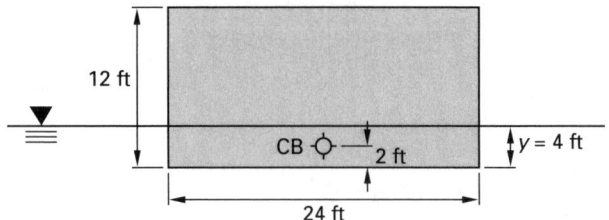

Customary U.S. Solution

(a) Refer to the following diagram. Let dimension y represent the depth of the submerged barge.

From Archimedes' principle, the buoyant force equals the weight of the barge. This, in turn, equals the weight of the displaced seawater.

$$F_b = W = V\gamma_w$$

$$(280\,000 \text{ kg})\left(9.81 \frac{\text{m}}{\text{s}^2}\right) = y(7 \text{ m})(30 \text{ m})\left(1024 \frac{\text{kg}}{\text{m}^3}\right)$$

$$\times \left(9.81 \frac{\text{m}}{\text{s}^2}\right)$$

$$y = 1.30 \text{ m}$$

The center of buoyancy is located at the centroid of the submerged cross section. When floating on an even keel, the submerged cross section is rectangular with a height of 1.30 m. The height of the center of buoyancy above the keel is

$$\frac{1.30 \text{ m}}{2} = 0.65 \text{ m}$$

(b) While the location of the new center of buoyancy could be determined, the location of the metacenter does not change significantly for small angles of list. Therefore, for approximate calculations, the angle of list is not significant.

The area moment of inertia of the longitudinal waterline cross section is

$$I = \frac{Lw^3}{12} = \frac{(30 \text{ m})(7 \text{ m})^3}{12}$$

$$= 858 \text{ m}^4$$

The submerged volume is

$$V = (1.3 \text{ m})(7 \text{ m})(30 \text{ m}) = 273 \text{ m}^3$$

The distance between the center of gravity and the center of buoyancy is

$$y_{bg} = 3.6 \text{ m} - 2.4 \text{ m} - 0.65 \text{ m} = 0.55 \text{ m}$$

The metacentric height measured above the center of gravity is

$$h_m = \frac{I}{V} - y_{bg}$$

$$= \frac{858 \text{ m}^4}{273 \text{ m}^3} - 0.55 \text{ m} = 2.6 \text{ m}$$

From Archimedes' principle, the buoyant force equals the weight of the barge. This, in turn, equals the weight of the displaced seawater.

$$F_b = W = V\gamma_w$$

$$600{,}000 \text{ lbf} = y(24 \text{ ft})(98 \text{ ft})\left(64 \frac{\text{lbf}}{\text{ft}^3}\right)$$

$$y = 4.00 \text{ ft}$$

The center of buoyancy is located at the centroid of the submerged cross section. When floating on an even keel, the submerged cross section is rectangular with a height of 4.00 ft. The height of the center of buoyancy above the keel is

$$\frac{4.00 \text{ ft}}{2} = 2.00 \text{ ft}$$

(b) While the location of the new center of buoyancy could be determined, the location of the metacenter does not change significantly for small angles of the list. Therefore, for approximate calculations, the angle of list is not significant.

The area moment of inertia of the longitudinal waterline cross section is

$$I = \frac{Lw^3}{12} = \frac{(98 \text{ ft})(24 \text{ ft})^3}{12}$$

$$= 112{,}900 \text{ ft}^4$$

The submerged volume is

$$V = (4 \text{ ft})(24 \text{ ft})(98 \text{ ft}) = 9408 \text{ ft}^3$$

The distance between the center of gravity and the center of buoyancy is

$$y_{bg} = 12 \text{ ft} - 7.8 \text{ ft} - 2.0 \text{ ft} = 2.2 \text{ ft}$$

The metacentric height measured above the center of gravity is

$$h_m = \frac{I}{V} - y_{bg}$$

$$= \frac{112{,}900 \text{ ft}^4}{9408 \text{ ft}^3} - 2.2 \text{ ft}$$

$$= 9.8 \text{ ft}$$

20. FLUID MASSES UNDER EXTERNAL ACCELERATION

Up to this point, fluid masses have been stationary, and gravity has been the only force acting on them. If a fluid mass is subjected to an external acceleration (moved sideways, rotated, etc.), an additional force will be introduced. This force will change the equilibrium position of the fluid surface as well as the hydrostatic pressure distribution.

Figure 15.26 illustrates a liquid mass subjected to constant accelerations in the vertical and/or horizontal directions. (a_y is negative if the acceleration is downward.) The surface is inclined at the angle predicted by Eq. 15.68. The planes of equal hydrostatic pressure beneath the surface are also inclined at the same angle.[25]

$$\theta = \arctan\left(\frac{a_x}{a_y + g}\right) \qquad 15.68$$

$$p = \rho g h\left(1 + \frac{a_y}{g}\right) \qquad \text{[SI]} \quad 15.69(a)$$

$$p = \left(\frac{\rho g h}{g_c}\right)\left(1 + \frac{a_y}{g}\right) = \gamma h\left(1 + \frac{a_y}{g}\right)$$
$$\text{[U.S.]} \quad 15.69(b)$$

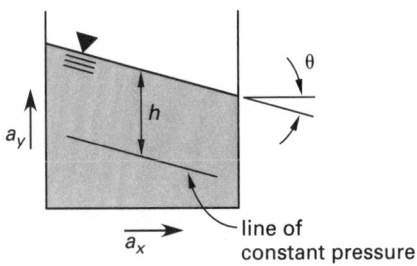

Figure 15.26 *Fluid Mass Under Constant Linear Acceleration*

Figure 15.27 illustrates a fluid mass rotating at constant angular velocity, ω, in rad/sec.[26] The resulting surface is parabolic in shape. The elevation of the fluid surface at point A at distance r from the axis of rotation is given by Eq. 15.71. The distance h in Fig. 15.27 is measured from the lowest fluid elevation during rotation. h is not measured from the original elevation of the stationary fluid.

$$\theta = \arctan\left(\frac{\omega^2 r}{g}\right) \qquad 15.70$$

$$h = \frac{(\omega r)^2}{2g} = \frac{\text{v}^2}{2g} \qquad 15.71$$

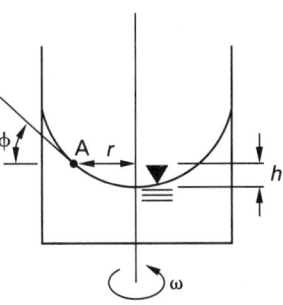

Figure 15.27 *Rotating Fluid Mass*

PRACTICE PROBLEMS

(Use $g = 32.2$ ft/sec^2 or 9.81 m/s^2 unless told to do otherwise in the problem.)

1. A blimp contains 10,000 lbm (4500 kg) of hydrogen at 56°F (13°C) and 30.2 in Hg (770 mm Hg). What is its lift if the hydrogen and air are in thermal and pressure equilibrium?

2. A hollow 6 ft (1.8 m) diameter sphere floats half-submerged in seawater. What mass of concrete is required as an external anchor to just submerge the sphere completely?

---

[25]Once the orientation of the surface is known, the pressure distribution can be determined without considering the acceleration. The hydrostatic pressure at a point depends only on the height of the liquid above that point. The acceleration affects that height but does not change the $p = \rho g h$ relationship.

[26]Even though the rotational speed is not increasing, the fluid mass experiences a constant *centripetal acceleration* radially outward from the axis of rotation.

16 Fluid Flow Parameters

1. Introduction to Fluid Energy Units 16-1
2. Kinetic Energy 16-2
3. Potential Energy 16-2
4. Pressure Energy 16-2
5. Bernoulli Equation 16-2
6. Pitot Tube 16-4
7. Impact Energy 16-4
8. Hydraulic Radius 16-6
9. Equivalent Diameter 16-6
10. Reynolds Number 16-7
11. Laminar Flow 16-7
12. Turbulent Flow 16-8
13. Critical Flow 16-8
14. Fluid Velocity Distribution in Pipes 16-8
15. Energy Grade Line 16-9
16. Specific Energy 16-9
17. Pipe Materials and Sizes 16-9
18. Types of Valves 16-11
    Practice Problems 16-12

Nomenclature

A	area	ft^2	m^2
C	correction factor	–	–
d	depth	ft	m
D	diameter	ft	m
E	specific energy	ft-lbf/lbm	J/kg
g	gravitational acceleration	ft/sec^2	m/s^2
g_c	gravitational conversion constant (32.2)	lbm-ft/lbf-sec^2	n.a.
G	mass flow rate per unit area	lbm/ft^2-sec	kg/m^2·s
h	height, or head	ft	m
L	length	ft	m
p	pressure	lbf/ft^2	Pa
r	radius	ft	m
Re	Reynolds number	–	–
s	wetted perimeter	ft	m
v	velocity	ft/sec	m/s
$\dot{V}$	volumetric flow rate	ft^3/sec	m^3/s
y	distance	ft	m
z	elevation	ft	m

Symbols

α	angle	rad	rad
μ	absolute viscosity	lbf-sec/ft^2	Pa·s
ν	kinematic viscosity	ft^2/sec	m^2/s
ρ	density	lbm/ft^3	kg/m^3
ϕ	angle	rad	rad

Subscripts

e	equivalent
h	hydraulic
i	impact, or inner
o	outer
p	pressure
r	radius
s	static
t	total
v	velocity
z	potential

1. INTRODUCTION TO FLUID ENERGY UNITS

Several important fluids and thermodynamics equations, such as Bernoulli's equation and the steady-flow energy equation, are special applications of the *conservation of energy* concept. However, it is not always obvious how some formulations of these equations can be termed "energy." For example, elevation (z), with units of feet, is often called *gravitational energy*.[1]

Since every fluids problem is different, energy must be expressed per unit mass (i.e., *specific energy*). With SI formulations, the choice of units is unambiguous: J/kg is the only choice.

$$\frac{\text{J}}{\text{kg}} = \frac{\text{N·m}}{\text{kg}} = \frac{\text{m}^2}{\text{s}^2} \qquad 16.1$$

With problems formulated in English units, specific energy units will be ft-lbf/lbm.[2] (If the consistent set of units ft-lbf/slug is chosen, the same equations can be used with SI and consistent English units, since the primary dimensions are the same, that is, L^2/θ^2.)

$$\frac{\text{ft-lbf}}{\text{slug}} = \frac{\text{ft}^2}{\text{sec}^2} \qquad 16.2$$

The gravitational conversion constant, g_c, must be used if the units ft-lbf/lbm are used for specific fluid energy.

In many cases, the ratio g/g_c appears in equations. Since g and g_c have the same numerical value in most

---

[1]Foot and ft-lbf/lbm may be thought of as one and the same. Certainly, the set of units ft-lbf/lbm represents energy per unit mass. Unfortunately, lbf and lbm do not really cancel out to yield ft.

[2]Btu could be used for energy, as is common in thermodynamics problems, instead of ft-lbf. However, this is almost never done in fluids problems.

instances, this ratio affects the units without affecting the calculation. Because of this, some engineers omit the term g/g_c entirely. While it is easy to justify such a practice, the resulting equations are not dimensionally consistent.[3]

2. KINETIC ENERGY

Energy is required to accelerate a stationary body. Thus, a moving mass of fluid possesses more energy than an identical, stationary mass. The energy difference is the *kinetic energy* of the fluid.[4] If the kinetic energy is evaluated per unit mass, the term *specific kinetic energy* is used. Equation 16.3 gives the specific kinetic energy corresponding to a velocity, v.

$$E_v = \frac{v^2}{2} \qquad \text{[SI]} \qquad 16.3(a)$$

$$E_v = \frac{v^2}{2g_c} \qquad \text{[U.S.]} \qquad 16.3(b)$$

The units of specific kinetic energy in consistent units are clearly m^2/s^2 or ft^2/sec^2, which coincide with Eqs. 16.1 and 16.2 as representing energy per consistent mass unit. The units in traditional English units are

$$\frac{\left(\frac{ft}{sec}\right)^2}{\frac{lbm\text{-}ft}{lbf\text{-}sec^2}} = \frac{ft\text{-}lbf}{lbm} \qquad 16.4$$

3. POTENTIAL ENERGY

Work is performed in elevating a body. Thus, a mass of fluid at a high elevation will have more energy than an identical mass of fluid at a lower elevation. The energy difference is the *potential energy* of the fluid.[5] Like kinetic energy, potential energy is usually expressed per unit mass. Equation 16.5 gives the potential energy of fluid at an elevation z.

$$E_z = zg \qquad \text{[SI]} \qquad 16.5(a)$$

$$E_z = \frac{zg}{g_c} \qquad \text{[U.S.]}[6] \qquad 16.5(b)$$

The units of potential energy in a consistent system are again m^2/s^2 or ft^2/sec^2. The units in a traditional English system are ft-lbf/lbm.

z is the elevation of the fluid. The reference point (i.e., zero elevation point) is entirely arbitrary and can be chosen for convenience. This is because potential energy always appears in a difference equation (i.e., ΔE_z), and the reference point cancels out.

4. PRESSURE ENERGY

Work is performed, and energy is expended, when a substance is compressed. Thus, a mass of fluid at a high pressure will have more energy than an identical mass of fluid at a lower pressure. The energy difference is the *pressure energy* of the fluid.[7] Pressure energy is usually found in equations along with kinetic and potential energies, and is expressed as energy per unit mass. Equation 16.6 gives the pressure energy of fluid at pressure p.

$$E_p = \frac{p}{\rho} \qquad 16.6$$

The consistent SI units of pressure energy are

$$\frac{Pa \cdot m^3}{kg} = \frac{N \cdot m}{kg} = \frac{J}{kg} \qquad 16.7$$

The units of pressure energy in the consistent English system are

$$\frac{lbf\text{-}ft^3}{ft^2\text{-}slug} = \frac{ft\text{-}lbf}{slug} \qquad 16.8$$

The units of pressure energy in the traditional English system are

$$\frac{lbf\text{-}ft^3}{ft^2\text{-}lbm} = \frac{ft\text{-}lbf}{lbm} \qquad 16.9$$

5. BERNOULLI EQUATION

The *Bernoulli equation* is an energy conservation equation based on several reasonable assumptions. The equation assumes the following.

- The fluid is incompressible.

- There is no fluid friction.

- Changes in thermal energy are negligible.[8]

The Bernoulli equation states that the *total energy* of a fluid flowing without friction losses in a pipe is constant.[9] The total energy possessed by the fluid is the

---

[3]More than being dimensionally inconsistent, the resulting equations will be incorrect in any nonstandard (other than one gravity) gravitational field.

[4]The terms *velocity energy* and *dynamic energy* are used less often.

[5]The term *gravitational energy* is also used.

[6]Since $g = g_c$ (numerically), it is tempting to write $E_z = z$. In fact, many engineers in the United States do just that.

[7]The terms *static energy* and *flow energy* are also used. The name *flow energy* results from the need to push (pressurize) a fluid to get it to flow through a pipe. However, flow energy and kinetic energy are not the same.

[8]In thermodynamics, the fluid flow is said to be *adiabatic*.

[9]Strictly speaking, this is the *total specific energy*, since the energy is per unit mass. However, the word "specific," being understood, is seldom used. Of course, "the total energy of the system" means something else and requires knowing the fluid mass in the system.

sum of its pressure, kinetic, and potential energies. Drawing on Eqs. 16.3, 16.5, and 16.6, the Bernoulli equation is written as

$$E_t = E_p + E_v + E_z \qquad \textbf{16.10}$$

$$E_t = \frac{p}{\rho} + \frac{v^2}{2} + zg \qquad \text{[SI]} \qquad \textbf{16.11(a)}$$

$$E_t = \frac{p}{\rho} + \frac{v^2}{2g_c} + \frac{zg}{g_c} \qquad \text{[U.S.]} \qquad \textbf{16.11(b)}$$

Equation 16.11 is valid for both laminar and turbulent flows. It can also be used for gases and vapors if the incompressibility assumption is valid.[10]

The quantities known as *total head*, h_t, and *total pressure*, p_t, can be calculated from total energy.

$$h_t = \frac{E_t}{g} \qquad \text{[SI]} \qquad \textbf{16.12(a)}$$

$$h_t = E_t \times \frac{g_c}{g} \qquad \text{[U.S.]} \qquad \textbf{16.12(b)}$$

$$p_t = \rho g h_t \qquad \text{[SI]} \qquad \textbf{16.13(a)}$$

$$p_t = \rho h_t \times \frac{g}{g_c} \qquad \text{[U.S.]} \qquad \textbf{16.13(b)}$$

Example 16.1

A pipe draws water from the bottom of a reservoir and discharges it freely at point C, 100 ft (30 m) below the surface. The flow is frictionless. (a) What is the total specific energy at an elevation 50 ft (15 m) below (i.e., point B) the water surface? (b) What is the velocity at point C?

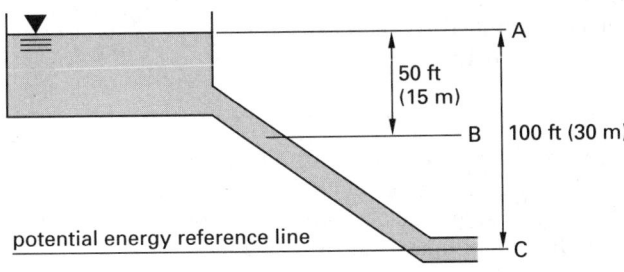

SI Solution

(a) At point A, the velocity and gage pressure are both zero. Therefore, the total energy consists only of potential energy. Point C is chosen as the reference ($z = 0$) elevation.

$$E_A = z_A g = (30 \text{ m})\left(9.81 \frac{\text{m}}{\text{s}^2}\right)$$
$$= 294.3 \text{ m}^2/\text{s}^2 \quad (\text{J/kg})$$

---

[10]A gas or vapor can be considered to be incompressible as long as its pressure does not change by more than 10% between points 1 and 2, and its velocity is less than Mach 0.3 everywhere.

At point B, the fluid is moving and possesses kinetic energy. The fluid is also under hydrostatic pressure and possesses pressure energy. These energy forms have come at the expense of potential energy. (This is a direct result of the Bernoulli equation.) Also, the flow is frictionless. Thus, there is no net change in the total energy between points A and B.

$$E_B = E_A = 294.3 \text{ m}^2/\text{s}^2 \quad (\text{J/kg})$$

(b) At point C, the gage pressure and pressure energy are again zero, since the discharge is at atmospheric pressure. The potential energy is zero, since $z = 0$. The total energy of the system has been converted to kinetic energy. From Eq. 16.11,

$$E_t = 294.3 \frac{\text{m}^2}{\text{s}^2} = 0 + \frac{v^2}{2} + 0$$
$$v = 24.3 \text{ m/s}$$

Customary U.S. Solution

(a)
$$E_A = \frac{z_A g}{g_c} = \frac{(100 \text{ ft})\left(32.2 \dfrac{\text{ft}}{\text{sec}^2}\right)}{32.2 \dfrac{\text{lbm-ft}}{\text{lbf-sec}^2}}$$
$$= 100 \text{ ft-lbf/lbm}$$
$$E_B = E_A = 100 \text{ ft-lbf/lbm}$$

(b)
$$E_t = 100 \frac{\text{ft-lbf}}{\text{lbm}} = 0 + \frac{v^2}{2g_c} + 0$$
$$v^2 = (2)\left(32.2 \frac{\text{lbm-ft}}{\text{lbf-sec}^2}\right)\left(100 \frac{\text{ft-lbf}}{\text{lbm}}\right)$$
$$= 6440 \text{ ft}^2/\text{sec}^2$$
$$v = 80.2 \text{ ft/sec}$$

Example 16.2

Water (62.4 lbm/ft³; 1000 kg/m³) is pumped up a hillside into a reservoir. The pump discharges water at the rate of 6 ft/sec (2 m/s) and pressure of 150 psig (1000 kPa). Disregarding friction, what is the maximum elevation (above the centerline of the pump's discharge) of the reservoir's water surface?

SI Solution

At the centerline of the pump's discharge, the potential energy is zero. The pressure and velocity energies are

$$E_p = \frac{p}{\rho} = \frac{(1000 \text{ kPa})\left(1000 \dfrac{\text{Pa}}{\text{kPa}}\right)}{1000 \dfrac{\text{kg}}{\text{m}^3}} = 1000 \text{ J/kg}$$

$$E_v = \frac{v^2}{2} = \frac{\left(2 \dfrac{\text{m}}{\text{s}}\right)^2}{2} = 2 \text{ J/kg}$$

The total energy at the pump's discharge is

$$E_{t,1} = E_p + E_v$$
$$= 1000 \ \frac{J}{kg} + 2 \ \frac{J}{kg} = 1002 \ J/kg$$

Since the flow is frictionless, the same energy is possessed by the water at the reservoir's surface. Since the velocity and gage pressure at the surface are zero, all of the available energy has been converted to potential energy.

$$E_{t,2} = E_{t,1}$$
$$z_2 g = 1002 \ J/kg$$
$$z_2 = \frac{1002 \ \dfrac{J}{kg}}{9.81 \ \dfrac{m}{s^2}} = 102.1 \ m$$

Notice that the volumetric flow rate of the water is not relevant since the water velocity was known. Similarly, the pipe size is not needed.

Customary U.S. Solution

$$E_p = \frac{p}{\rho} = \frac{\left(150 \ \dfrac{lbf}{in^2}\right)\left(144 \ \dfrac{in^2}{ft^2}\right)}{62.4 \ \dfrac{lbm}{ft^3}}$$
$$= 346.15 \ ft\text{-}lbf/lbm$$

$$E_v = \frac{v^2}{2g_c} = \frac{\left(6 \ \dfrac{ft}{sec}\right)^2}{(2)\left(32.2 \ \dfrac{lbm\text{-}ft}{lbf\text{-}sec^2}\right)}$$
$$= 0.56 \ ft\text{-}lbf/lbm$$

$$E_{t,1} = E_p + E_v$$
$$= 346.15 \ \frac{ft\text{-}lbf}{lbm} + 0.56 \ \frac{ft\text{-}lbf}{lbm} = 346.71 \ ft\text{-}lbf/lbm$$

$$E_{t,2} = E_{t,1}$$
$$\frac{z_2 g}{g_c} = 346.71 \ ft\text{-}lbf/lbm$$

$$z_2 = \frac{\left(346.71 \ \dfrac{ft\text{-}lbf}{lbm}\right)\left(32.2 \ \dfrac{lbm\text{-}ft}{lbf\text{-}sec^2}\right)}{32.2 \ \dfrac{ft}{sec^2}} = 346.7 \ ft$$

6. PITOT TUBE

A *pitot tube* (also known as an *impact tube* or *stagnation tube*) is simply a hollow tube that is placed longitudinally in the direction of fluid flow, allowing the flow to enter one end at the fluid's *velocity of approach*. It is used to measure velocity of flow and finds uses in both subsonic and supersonic applications.

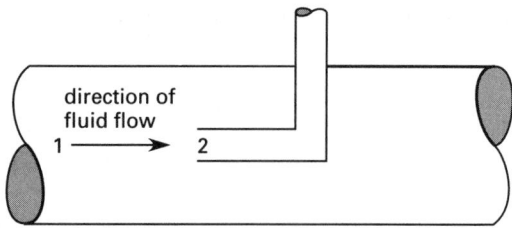

Figure 16.1 *Pitot Tube*

When the fluid enters the pitot tube, it is forced to come to a stop (at the *stagnation point*), and the velocity energy is transformed into pressure energy. If the fluid is a low-velocity gas, the stagnation is assumed to occur without compression heating of the gas. If there is no friction (the common assumption), the process is said to be adiabatic.

Bernoulli's equation can be used to predict the static pressure at the stagnation point. Since the velocity of the fluid within the pitot tube is zero, the upstream velocity can be calculated if the static and stagnation pressures are known.

$$\frac{p_1}{\rho} + \frac{v_1^2}{2} = \frac{p_2}{\rho} \qquad \text{16.14}$$

$$v_1 = \sqrt{\frac{(2)(p_2 - p_1)}{\rho}} \qquad \text{[SI]} \quad \text{16.15(a)}$$

$$v_1 = \sqrt{\frac{2g_c(p_2 - p_1)}{\rho}} \qquad \text{[U.S.]} \quad \text{16.15(b)}$$

In reality, both friction and heating occur, and the fluid may be compressible. These errors are taken care of by a correction factor known as the *impact factor*, C_i, which is applied to the derived velocity. C_i is usually very close to 1.00 (e.g., 0.99 or 0.995).

$$v_{actual} = C_i \ v_{indicated}$$

Since accurate measurements of fluid velocity are dependent on one-dimensional fluid flow, it is essential that any obstructions or pipe bends be more than ten pipe diameters upstream from the pitot tube.

7. IMPACT ENERGY

Impact energy, E_i (also known as *stagnation energy* and *total energy*), is the sum of the kinetic and pressure energy terms.[11] Equation 16.17 is applicable to fluids

---

[11]It is confusing to label Eq. 16.16 *total* when the gravitational energy term has been omitted. However, the reference point for gravitational energy is arbitrary, and in this application, the reference coincides with the centerline of the fluid flow. In truth, the effective pressure developed in a fluid which has been brought to rest adiabatically does not depend on the elevation or altitude of the fluid. This situation is seldom ambiguous. The application will determine which definition of total head or total energy is intended.

and gases flowing with velocities less than approximately Mach 0.3.

$$E_i = E_p + E_v \qquad 16.16$$

$$E_i = \frac{p}{\rho} + \frac{v^2}{2} \qquad \text{[SI]} \qquad 16.17(a)$$

$$E_i = \frac{p}{\rho} + \frac{v^2}{2g_c} \qquad \text{[U.S.]} \qquad 16.17(b)$$

Impact head, h_i, is calculated from the impact energy in a manner analogous to Eq. 16.12. Impact head represents the height the liquid will rise in a piezometer-pitot tube when the fluid has been brought to rest (i.e., stagnated) in an adiabatic manner. Such a case is illustrated in Fig. 16.2. If a gas or high-velocity, high-pressure fluid is flowing, it will be necessary to use a mercury manometer or pressure gauge to measure stagnation head.

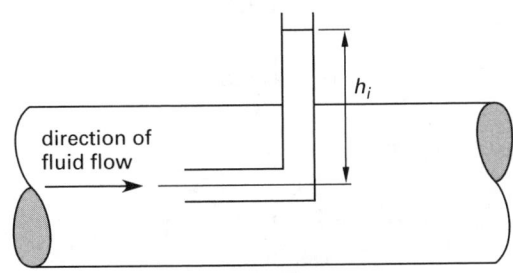

Figure 16.2 *Pitot Tube-Piezometer Apparatus*

Example 16.3

The static pressure of air (0.075 lbm/ft^3; 1.2 kg/m^3) flowing in a pipe is measured by a precision gauge to be 10.00 psig (68.95 kPa). A pitot tube-manometer indicates 20.6 in (0.52 m) of mercury. Losses are insignificant. What is the velocity of the air in the pipe?

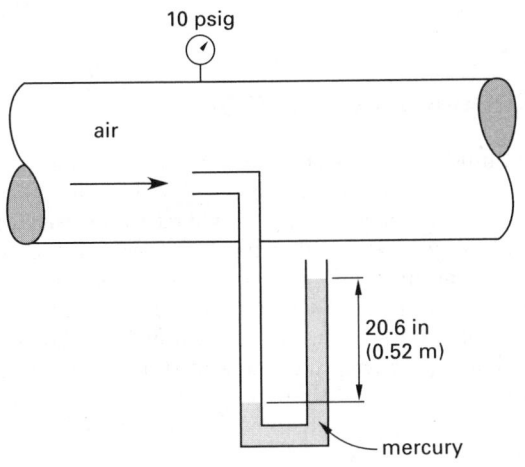

SI Solution

From Table 14.2, the density of mercury is $13\,600$ kg/m^3. The impact pressure is

$$p_i = \rho g h$$

$$= \left(13\,600\ \frac{\text{kg}}{\text{m}^3}\right)\left(9.81\ \frac{\text{m}}{\text{s}^2}\right)\left(\frac{0.52\ \text{m}}{1000\ \dfrac{\text{Pa}}{\text{kPa}}}\right)$$

$$= 69.38\ \text{kPa}$$

From Eq. 16.15, the velocity is

$$v = \sqrt{\frac{(2)(p_i - p_s)}{\rho}}$$

$$= \sqrt{\frac{(2)(69.38\ \text{kPa} - 68.95\ \text{kPa})\left(1000\ \dfrac{\text{Pa}}{\text{kPa}}\right)}{1.2\ \dfrac{\text{kg}}{\text{m}^3}}}$$

$$= 26.8\ \text{m/s}$$

Customary U.S. Solution

The pitot tube measures impact (stagnation) pressure. From Table 15.2, the fluid height equivalent (specific weight) of mercury is 0.491 psi/in. Therefore, the impact pressure is

$$p_i = (20.6\ \text{in})\left(0.491\ \frac{\text{psi}}{\text{in}}\right) = 10.11\ \text{psig}$$

Since impact pressure is the sum of the static and kinetic (velocity) pressures, the kinetic pressure is

$$p_v = p_i - p_s$$
$$= 10.11\ \text{psig} - 10.00\ \text{psig} = 0.11\ \text{psig}$$

The velocity is calculated from Eq. 16.15.

$$v = \sqrt{\frac{2g_c(p_i - p_s)}{\rho}}$$

$$= \sqrt{\frac{(2)\left(32.2\ \dfrac{\text{lbm-ft}}{\text{lbf-sec}^2}\right)\left(0.11\ \dfrac{\text{lbf}}{\text{in}^2}\right)\left(144\ \dfrac{\text{in}^2}{\text{lbf}}\right)}{0.075\ \dfrac{\text{lbm}}{\text{ft}^3}}}$$

$$= 117\ \text{ft/sec}$$

8. HYDRAULIC RADIUS

The *hydraulic radius* is defined as the area in flow divided by the *wetted perimeter*.[12] (The hydraulic radius is not the same as the radius of a pipe.) The area in flow is the cross-sectional area of the fluid flowing. When a fluid is flowing under pressure in a pipe (i.e., *pressure flow*), the area in flow will be the internal area of the pipe. However, the fluid may not completely fill the pipe and may flow simply because of a sloped surface (i.e., *gravity flow* or *open channel flow*).

The wetted perimeter is the length of the line representing the interface between the fluid and the pipe or channel. It does not include the *free surface* length (i.e., the interface between fluid and atmosphere).

$$r_h = \frac{\text{area in flow}}{\text{wetted perimeter}} \qquad 16.18$$

Consider a circular pipe flowing completely full. The area in flow is πr^2. The wetted perimeter is the entire circumference, $2\pi r$. The hydraulic radius is

$$r_{h,\text{pipe}} = \frac{\pi r^2}{2\pi r} = \frac{r}{2} \qquad 16.19$$

The hydraulic radius of a pipe flowing half full is also $r/2$, since the flow area and wetted perimeter are both halved. However, it is time-consuming to calculate the hydraulic radius for pipe flow at any intermediate depth, due to the difficulty in evaluating the flow area and wetted perimeter. Appendix 16.A greatly simplifies such calculations.

Example 16.4

A pipe (internal diameter = 6 units) carries water with a depth of 2 units flowing under the influence of gravity. (a) Calculate the hydraulic radius analytically. (b) Verify the result by using App. 16.A.

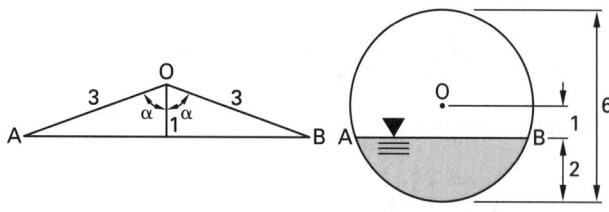

Solution

(a) The equations for a circular segment must be used. The radius is $6/2 = 3$.

---

[12]The hydraulic radius can also be calculated as one-fourth of the equivalent diameter of the pipe or channel, as will be subsequently shown. That is, $r_h = \frac{1}{4} D_e$.

Points A, O, and B are used to find the central angle of the circular segment.

$$\phi = 2\alpha = (2)\left[\arccos\left(\tfrac{1}{3}\right)\right]$$
$$= (2)(70.53°) = 141.06°$$

ϕ must be expressed in radians.

$$\phi = 2\pi \left(\frac{141.06°}{360°}\right) = 2.46 \text{ rad}$$

The area of the circular segment (i.e., the area in flow) is

$$A = \tfrac{1}{2} r^2 (\phi - \sin \phi) \quad [\phi \text{ in radians}]$$
$$= (0.5)(3)^2 (2.46 \text{ rad} - \sin(2.46 \text{ rad}))$$
$$= 8.235 \text{ units}^2$$

The arc length (i.e., the wetted perimeter) is

$$s = r\phi = (3)(2.46 \text{ rad}) = 7.38 \text{ units}$$

The hydraulic radius is

$$r_h = \frac{A}{s} = \frac{8.235 \text{ units}^2}{7.38 \text{ units}} = 1.12 \text{ units}$$

(b) The ratio d/D is needed to use App. 16.A.

$$\frac{d}{D} = \frac{2 \text{ units}}{6 \text{ units}} = 0.333$$

From App. 16.A,

$$\frac{r_h}{D} \approx 0.186$$
$$r_h = (0.186)(6 \text{ units}) = 1.12 \text{ units}$$

9. EQUIVALENT DIAMETER

Many fluid, thermodynamic, and heat transfer processes are dependent on the physical length of an object. The general name for this controlling variable is *characteristic dimension*. The characteristic dimension in evaluating fluid flow is the *equivalent diameter* (also known as the *hydraulic diameter*). The equivalent diameter for a full-flowing pipe is simply its inside diameter. The equivalent diameters of other cross sections in flow are given in Table 16.1. If the hydraulic radius is known, it can be used to calculate the equivalent diameter.

$$D_e = 4r_h \qquad 16.20$$

Table 16.1 *Equivalent Diameters for Common Conduit Shapes*

conduit cross section	D_e
flowing full	
annulus (outer diameter D_o, inner diameter D_i)	$D_o - D_i$
square (side L)	L
rectangle (sides L_1 and L_2)	$\dfrac{2L_1L_2}{L_1 + L_2}$
flowing partially full	
half-filled circle (diameter D)	D
rectangle (h deep, L wide)	$\dfrac{4hL}{L + 2h}$
wide, shallow stream (h deep)	$4h$
triangle (h deep, L broad, s side)	$\dfrac{hL}{s}$
trapezoid (h deep, a wide at top, b wide at bottom, s side)	$\dfrac{2h(a + b)}{b + 2s}$

Example 16.5

Determine the equivalent diameter and hydraulic radius for the open trapezoidal channel shown.

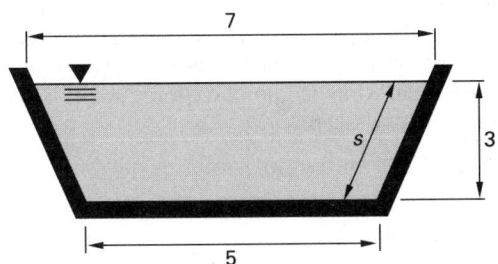

Solution

$$s = \sqrt{(3)^2 + (1)^2} = 3.16$$

Using Table 16.1,

$$D_e = \frac{2h(a + b)}{b + 2s} = \frac{(2)(3)(7 + 5)}{5 + (2)(3.16)} = 6.36$$

From Eq. 16.20,

$$r_h = \frac{D_e}{4} = \frac{6.36}{4} = 1.59$$

10. REYNOLDS NUMBER

The *Reynolds number*, Re, is a dimensionless number interpreted as the ratio of inertial forces to viscous forces in the fluid.[13]

$$\text{Re} = \frac{\text{inertial forces}}{\text{viscous forces}} \qquad \textit{16.21}$$

The inertial forces are proportional to the flow diameter, velocity, and fluid density. (Increasing these variables will increase the momentum of the fluid in flow.) The viscous force is represented by the fluid's absolute viscosity, μ. Thus, the Reynolds number is calculated as

$$\text{Re} = \frac{D_e \text{v} \rho}{\mu} \qquad \text{[SI]} \qquad \textit{16.22(a)}$$

$$\text{Re} = \frac{D_e \text{v} \rho}{g_c \mu} \qquad \text{[U.S.]} \qquad \textit{16.22(b)}$$

Since μ/ρ is defined as the *kinematic viscosity*, ν, Eq. 16.22 can be simplified.[14]

$$\text{Re} = \frac{D_e \text{v}}{\nu} \qquad \textit{16.23}$$

Occasionally, the *mass flow rate per unit area*, $G = \rho \text{v}$, will be known. This variable expresses the quantity of fluid flowing in kg/m$^2 \cdot$s or lbm/ft^2-sec.

$$\text{Re} = \frac{D_e G}{\mu} \qquad \text{[SI]} \qquad \textit{16.24(a)}$$

$$\text{Re} = \frac{D_e G}{g_c \mu} \qquad \text{[U.S.]} \qquad \textit{16.24(b)}$$

11. LAMINAR FLOW

Laminar flow gets its name from the word *laminae* (layers). If all of the fluid particles move in paths parallel to the overall flow direction (i.e., in layers), the flow is said to be *laminar*. (The terms *viscous flow* and *streamline flow* are also used.) This occurs in pipeline flow when the Reynolds number is less than (approximately) 2100. Laminar flow is typical when the flow channel is small, the velocity is low, and the fluid is viscous. Viscous forces are dominant in laminar flow.

In laminar flow, a stream of dye inserted in the flow will continue from the source in a continuous, unbroken line with very little mixing of the dye and surrounding liquid. The fluid particle paths coincide with imaginary *streamlines*. (Streamlines and velocity vectors are always tangent to each other.) A "bundle" of these streamlines (i.e., a *streamtube*) constitutes a complete fluid flow.

---

[13]Engineering authors are not in agreement about the symbol for the Reynolds number. In addition to Re (used in this book), engineers commonly use **Re**, R, $\Re$, N_{Re}, and N_R.
[14]This simplification implies a caveat as well. If the viscosity is known or is given in a problem, the units must be used to determine if this viscosity is μ or ν.

12. TURBULENT FLOW

A fluid is said to be in *turbulent flow* if the Reynolds number is greater than (approximately) 4000. (This is the most common situation.) Turbulent flow is characterized by a three-dimensional movement of the fluid particles superimposed on the overall direction of motion. A stream of dye injected into a turbulent flow will quickly disperse and uniformly mix with the surrounding flow. Inertial forces dominate in turbulent flow. At very high Reynolds numbers, the flow is said to be *fully turbulent*.

13. CRITICAL FLOW

The flow is said to be in a *critical zone* or *transition region* when the Reynolds number is between 2100 and 4000. These numbers are known as the lower and upper *critical Reynolds numbers* for fluid flow, respectively. (Critical Reynolds numbers for other processes are different.) It is difficult to design for the transition region, since fluid behavior is not consistent and few processes operate in the critical zone. In the event a critical zone design is required, the conservative assumption of turbulent flow will result in the greatest value of friction loss.

14. FLUID VELOCITY DISTRIBUTION IN PIPES

With laminar flow, the viscous effects make some fluid particles adhere to the pipe wall. The closer a particle is to the pipe wall, the greater the tendency will be for the fluid to adhere to the pipe wall. The following statements characterize laminar flow.

- The velocity distribution is parabolic.

- The velocity is zero at the pipe wall.

- The velocity is maximum at the center and equal to twice the average velocity.

$$v_{ave} = \frac{\dot{V}}{A} = \frac{v_{max}}{2} \qquad 16.25$$

With turbulent flow, there is generally no distinction made between the velocities of particles near the pipe wall and particles at the pipe centerline.[15] All of the fluid particles are assumed to have the same velocity. This velocity is known as the *average* or *bulk velocity*. It can be calculated from the volume flowing.

$$v_{ave} = \frac{\dot{V}}{A} \qquad 16.26$$

---

[15]This disregards the *boundary layer*, a thin layer near the pipe wall, where the velocity goes from zero to v_{ave}.

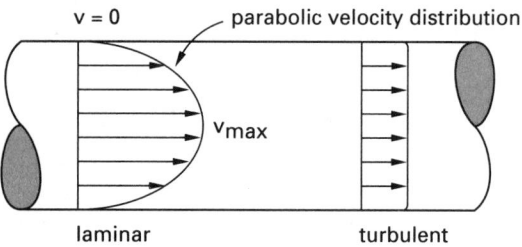

Figure 16.3 *Laminar and Turbulent Velocity Distributions*

In actuality, no flow is completely turbulent, and there is a difference between the *centerline velocity* and the average velocity. The error decreases as the Reynolds number increases. The ratio of v_{ave}/v_{max} starts at approximately 0.75 for Re = 4000 and increases to approximately 0.86 at Re = 10^6. Most problems ignore the difference between v_{ave} and v_{max}, but care should be taken when a centerline measurement (as from a pitot tube) is used to evaluate the average velocity.

For turbulent flow (Re ≈ 10^5) in a smooth, circular pipe of radius r_o, the velocity at a radial distance r from the centerline is given by the $1/7$-*power law*:

$$v_r = v_{max} \left(\frac{r_o - r}{r_o} \right)^{\frac{1}{7}} \qquad [\text{turbulent flow}] \qquad 16.27$$

The fluid's *velocity profile*, given by Eq. 16.27, is valid in smooth pipes up to a Reynolds number of approximately 100,000. Above that, up to a Reynolds number of approximately 400,000, an exponent of $1/8$ fits experimental data better. For rough pipes, the exponent is larger (e.g., $1/5$).

The ratio of the average velocity of maximum velocity is known as the *pipe coefficient* or *pipe factor*. Considering all the other coefficients used in pipe flow, these names are somewhat vague and ambiguous. Therefore, they are not in widespread use.

Equation 16.27 can be integrated to determine the average velocity.

$$v_{ave} = \left(\frac{49}{60} \right) v_{max} = 0.817\, v_{max} \quad [\text{turbulent flow}]$$
$$16.28$$

When the flow is laminar, the velocity profile within a pipe will be parabolic and of the form of Eq. 16.29. (Equation 16.27 is for turbulent flow and does not describe a parabolic velocity profile.) The velocity at a radial distance r from the centerline in a pipe of radius r_o is

$$v_r = v_{max} \left(\frac{r_o - r}{r_o} \right)^2 \qquad [\text{laminar flow}] \qquad 16.29$$

When the velocity profile is parabolic, the flow rate and pressure drop can easily be determined. The average velocity is half of the maximum velocity given in the velocity profile equation.

$$v_{ave} = \frac{1}{2} v_{max} \qquad [\text{laminar flow}] \qquad 16.30$$

The average velocity is used to determine the flow quantity and friction loss. The friction loss is determined by traditional means.

$$\dot{V} = A v_{\text{ave}} \qquad 16.31$$

The kinetic energy of laminar flow can be found by integrating the velocity profile equation, resulting in Eq. 16.32.

$$E_v = v_{\text{ave}}^2 \qquad \text{[laminar flow]} \quad \text{[SI]} \quad 16.32(a)$$

$$E_v = \frac{v_{\text{ave}}^2}{g_c} \qquad \text{[laminar flow]} \quad \text{[U.S.]} \quad 16.32(b)$$

15. ENERGY GRADE LINE

The *energy grade line* (EGL) is a graph of the total energy (total specific energy) along a length of pipe.[16] In a frictionless pipe without pumps or turbines, the total specific energy is constant, and the EGL will be horizontal. (This is a restatement of the Bernoulli equation.)

$$\text{elevation of EGL} = h_p + h_v + h_z \qquad 16.33$$

The *hydraulic grade line* (HGL) is the graph of the sum of the pressure and gravitational heads, plotted as a position along the pipeline. Since the pressure head can increase at the expense of the velocity head, the HGL can increase in elevation if the flow area is increased.

$$\text{elevation of HGL} = h_p + h_z \qquad 16.34$$

The difference between the EGL and the HGL is the velocity head, h_v, of the fluid.

$$h_v = \text{elevation of EGL} - \text{elevation of HGL} \qquad 16.35$$

The following rules apply to these grade lines in a frictionless environment without pumps or turbines.

- The EGL is always horizontal.

- The HGL is always equal to or below the EGL.

- For still ($v = 0$) fluid at a free surface, EGL = HGL (i.e., the EGL coincides with the fluid surface in a reservoir).

- If flow velocity is constant (i.e., flow in a constant-area pipe), the HGL will be horizontal and parallel to the EGL, regardless of pipe orientation or elevation.

- When the flow area decreases, the HGL decreases.

- When the flow area increases, the HGL increases.

- In a free jet (i.e., a stream of water from a hose), the HGL coincides with the jet elevation, following a parabolic path.

[16]The term *energy line* (EL) is also used.

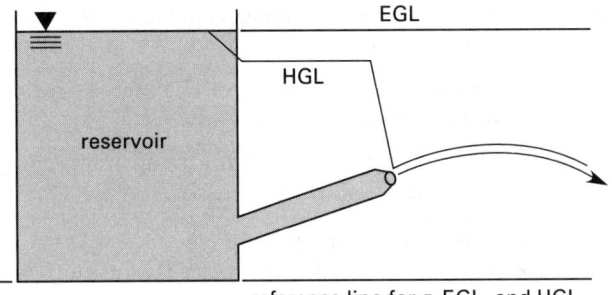

Figure 16.4 *Energy and Hydraulic Grade Lines Without Friction*

16. SPECIFIC ENERGY

Specific energy is a term that is used primarily with open channel flow. It is the total energy with respect to the channel bottom, consisting of pressure and velocity energy terms only.

$$E_{\text{specific}} = E_p + E_v \qquad 16.36$$

Since the channel bottom is chosen as the reference elevation ($z = 0$) for gravitational energy, there is no contribution by gravitational energy to specific energy.

$$E_{\text{specific}} = \frac{p}{\rho} + \frac{v^2}{2} \qquad \text{[SI]} \quad 16.37(a)$$

$$E_{\text{specific}} = \frac{p}{\rho} + \frac{v^2}{2g_c} \qquad \text{[U.S.]} \quad 16.37(b)$$

However, since p is the hydrostatic pressure at the channel bottom due to a fluid depth, d, p/ρ can be interpreted as the depth of the fluid.

$$h_{\text{specific}} = d + \frac{v^2}{2g} \qquad 16.38$$

Specific energy is constant when the flow depth and width are constant (i.e., *uniform flow*). A change in channel width will cause a change in flow depth, and since width is not part of the equation for specific energy, there will be a corresponding change in specific energy. There are other ways that specific energy can decrease, also.[17]

17. PIPE MATERIALS AND SIZES

Many materials are used for pipes. The material used depends on the application. Water supply distribution, waste-water collection, and air conditioning refrigerant lines all place different demands on pipe material performance. Pipe materials are chosen on the basis of strength to withstand internal pressures, strength to withstand external loads from backfill and traffic,

[17]Specific energy changes dramatically in a *hydraulic jump* or *hydraulic drop*.

Fluids

smoothness, corrosion resistance, chemical inertness, cost, and other factors.

The following are characteristics of the major types of commercial pipe materials that are in use.

- *ductile cast iron*: long lived, strong, impervious, heavy, scour resistant, but costly

- *asbestos cement*: immune to electrolysis and corrosion, light in weight, but weak structurally; environmentally limited

- *concrete*: durable, water tight, low maintenance, smooth interior

- *vitrified clay*: resistant to corrosion, acids (e.g., hydrogen sulfide from septic sewage), scour, and erosion

- *steel*: high strength, ductile, resistant to shock, very smooth interior, but susceptible to corrosion

- *plastic* (PVC and ABS):[18] chemically inert, resistant to corrosion, very smooth, lightweight, low cost

- *copper and brass*: used primarily for water, condensate, and refrigerant lines; in some cases, easily bent by hand, good thermal conductivity

Table 16.2 lists recommendations for pipe materials in several common applications.

Table 16.2 *Recommended Pipe Materials by Application*

service	pipe material
most refrigerants (suction, liquid, and hot gas lines)	hard copper tubing (type L)[a]; standard wall steel pipe, lap welded or seamless
chilled water	hard copper tubing; plain (black) or galvanized steel pipe[b]
condenser or make-up water	hard copper tubing; plain or galvanized steel pipe[b]
steam or condensate	hard copper tubing; steel pipe[b]
hot water	hard copper tubing; steel pipe

[a]Soft copper may be used for $1/4$ in (6.3 mm) and $3/8$ in (9.5 mm) (outside diameter) with wall thicknesses of 0.30 in (7.6 mm) and 0.32 in (8.1 mm), respectively. Soft copper refrigeration lines are commonly used up to $1\frac{3}{8}$ in (35 mm) (outside diameter). Mechanical joints should not be used with soft copper tubing larger than $7/8$ in (22 mm).

[b]Standard wall steel pipe or type M hard copper tubing are usually satisfactory for air conditioning applications. However, the pressure rating of the pipe material should be checked at the design temperature.

[18]PVC: polyvinyl chloride; ABS: acrylonitrile-butadiene-styrene.

The required wall thickness of a pipe is proportional to the pressure the pipe must carry. However, not all pipes operate at high pressures. Therefore, pipes and tubing may be available in different wall thicknesses (*schedules*, *series*, or *types*). Steel pipe, for example, is available in schedules 40, 80, and others.[19]

For initial estimates, the approximate schedule of steel pipe has traditionally been calculated from Eq. 16.39. p is the operating pressure in psig; S is the allowable stress in the pipe material; and E is the *joint efficiency*, also known as the *joint quality factor* (typically 1.00 for seamless pipe, 0.85 for electric resistance-welded pipe, 0.80 for electric fusion-welded pipe, and 0.60 for furnace butt-welded pipe). For seamless carbon steel (A53) pipe used below 650°F (340°C), the allowable stress is approximately 12,000 to 15,000 psi. So, with butt-welded joints, a value of 6500 psi is often used for the product SE.

$$\text{schedule} \approx \frac{1000p}{SE} \qquad \qquad 16.39$$

Steel pipe is available in black (i.e., plain) and galvanized (inside, outside, or both) varieties. Steel pipe is manufactured in plain-carbon and stainless varieties. AISI 316 stainless is particularly corrosion-resistant.

The actual dimensions of some pipes (concrete, clay, some cast iron, etc.) coincide with their *nominal dimensions*. For example, a 12 in concrete pipe has an inside diameter of 12 in, and no further refinement is needed. However, some pipes and tubing (e.g., steel pipe, copper and brass tubing, and some cast iron) are called out by a nominal diameter that has nothing to do with the internal diameter of the pipe. For example, a 16 in schedule-40 steel pipe has an actual inside diameter of 15 in. In some cases, the nominal size does not coincide with the external diameter, either.

The potential for confusion exists when copper is specified as a pipe material, as there are two different sets of dimensions for copper pipe. Copper pipe in the K, L, and M categories is available in both annealed rolls (referred to as *copper tubing*) and as hardened lengths (referred to as *copper pipe*). Dimensions for copper pipe, ambiguously referred to as *copper water tubing*, are listed in App. 16.D. Copper and brass pipe can also have the dimensions given in App. 16.E.

Since the term "copper pipe" is ambiguous, the designation and/or application may be used to determine the correct dimensions. Type L copper tubing in straight lengths is used principally in domestic and commercial plumbing because of its lower cost and the availability of solder fittings. However, brass piping may be used with high-temperature and corrosive fluids.

[19]Other schedules of steel pipe, such as 30, 60, 120, etc., also exist, but in limited sizes, as Table 16.3 indicates. Schedule-40 pipe roughly corresponds to the standard weight (S) designation used in the past. Schedule-80 roughly corresponds to the extra-strong (X) designation. There is no uniform replacement for double-extra-strong (XX) pipe.

Table 16.3 *Dimensions of Commercial Steel Pipe[a] (English Units)*

nominal diameter	outside diameter	10	20	30	40	60	80	100	120	140	160
						wall thickness (in)					
$\frac{1}{2}$	0.840	. . . .	. . . .	. . . .	0.109	. . . .	0.147	. . . .	. . . .	. . . .	0.187
$\frac{3}{4}$	1.05	. . . .	. . . .	. . . .	0.113	. . . .	0.154	. . . .	. . . .	. . . .	0.218
1	1.315	. . . .	. . . .	. . . .	0.133	. . . .	0.179	. . . .	. . . .	. . . .	0.250
$1\frac{1}{4}$	1.660	. . . .	. . . .	. . . .	0.140	. . . .	0.191	. . . .	. . . .	. . . .	0.250
$1\frac{1}{2}$	1.900	. . . .	. . . .	. . . .	0.145	. . . .	0.200	. . . .	. . . .	. . . .	0.281
2	2.375	. . . .	. . . .	. . . .	0.154	. . . .	0.218	. . . .	. . . .	. . . .	0.343
$2\frac{1}{2}$	2.875	. . . .	. . . .	. . . .	0.203	. . . .	0.276	. . . .	. . . .	. . . .	0.375
3	3.500	. . . .	. . . .	. . . .	0.216	. . . .	0.300	. . . .	. . . .	. . . .	0.437
$3\frac{1}{2}$	4.000	. . . .	. . . .	. . . .	0.226	. . . .	0.318	. . . .	. . . .	. . . .	. . . .
4	4.500	. . . .	. . . .	. . . .	0.237	. . . .	0.337	. . . .	0.437	. . . .	0.531
5	5.563	. . . .	. . . .	. . . .	0.258	. . . .	0.375	. . . .	0.500	. . . .	0.625
6	6.625	. . . .	. . . .	. . . .	0.280	. . . .	0.432	. . . .	0.562	. . . .	0.718
8	8.625	. . . .	0.250	0.277	0.322	0.406	0.500	0.593	0.718	0.812	0.906
10	10.75	. . . .	0.250	0.307	0.365	0.500	0.593	0.718	0.843	1.000	1.125
12	12.75	. . . .	0.250	0.330	0.406	0.562	0.687	0.843	1.000	1.125	1.312
14	14.00	0.250	0.312	0.375	0.437	0.593	0.750	0.937	1.062	1.250	1.406
16	16.00	0.250	0.312	0.375	0.500	0.656	0.843	1.031	1.218	1.437	1.562
18	18.00	0.250	0.312	0.437	0.562	0.718	0.937	1.156	1.343	1.562	1.750
20	20.00	0.250	0.375	0.500	0.593	0.812	1.031	1.250	1.500	1.750	1.937
24	24.00	0.250	0.375	0.562	0.687	0.937	1.218	1.500	1.750	2.062	2.312

[a]Also, see Apps. 16.B and 16.C.

It is essential that tables of pipe sizes, such as App. 16.A, be used when working problems involving steel and copper pipes, since there is no other way to obtain the inside diameters of such pipes.[20]

18. TYPES OF VALVES

Valves used for *shutoff service* (e.g., gate, plug, ball, and butterfly valves) are used fully open or fully closed. *Gate valves* offer minimum resistance to flow. They are used in clean fluid and slurry services when valve operation is infrequent. Many turns of the handwheels are required to raise or lower their gates. *Plug valves* provide for tight shutoff. A 90° turn of their handles is sufficient to rotate the plugs fully open or closed. *Ball valves* offer an unobstructed flow path and tight shutoff. They are often used with slurries and viscous fluids, as well as with cryogenic fluids. A 90° turn of their handles rotates the balls fully open or closed. *Butterfly valves* (when specially designed with appropriate seats) can be used for shutoff operation. They are particularly applicable to large flows of low-pressure (vacuum up to 150 psig (1 MPa)) gases or liquids. Their straight-through, open-disk design results in minimal solids build-up and low pressure drops.

Other valve types (e.g., globe, needle, Y-, angle, and butterfly valves) are more suitable for *throttling service*. *Globe valves* provide positive shutoff and precise metering on clean fluids. However, since the seat is parallel to the direction of flow and the fluid makes two right-angle turns, there is substantial resistance and pressure drop through them, as well as relatively fast erosion of the seat. Globe valves are intended for infrequent operation. *Needle valves* are similar to globe valves, except that the plug is a tapered, needle-like cone. Needle valves provide accurate metering of small flows of clean fluids. Needle valves are applicable to cryogenic fluids. *Y-valves* are similar to globe valves in operation, but their seats are inclined to the direction of flow, offering more of a straight-through passage and unobstructed flow than the globe valve. *Angle valves* are essentially globe valves where the fluid makes a 90° turn. They can be used for throttling and shut-off of clean or viscous fluids and slurries. *Butterfly valves* are often used for

---

[20]It is a characteristic of standard steel pipes that the schedule number does not affect the outside diameter of the pipe. An 8 in schedule-40 pipe has the same exterior dimensions as an 8 in schedule-80 pipe. However, the interior flow area will be less for the schedule-80 pipe.

throttling services with the same limitations and benefits as those listed for shutoff use.

Other valves are of the *check (nonreverse-flow)* variety. These react automatically to changes in pressure to prevent reversals of flow. Special check valves can also prevent excess flow. Figure 16.5 illustrates *swing, lift,* and *angle lift check valves*.

Figure 16.5 *Types of Valves*

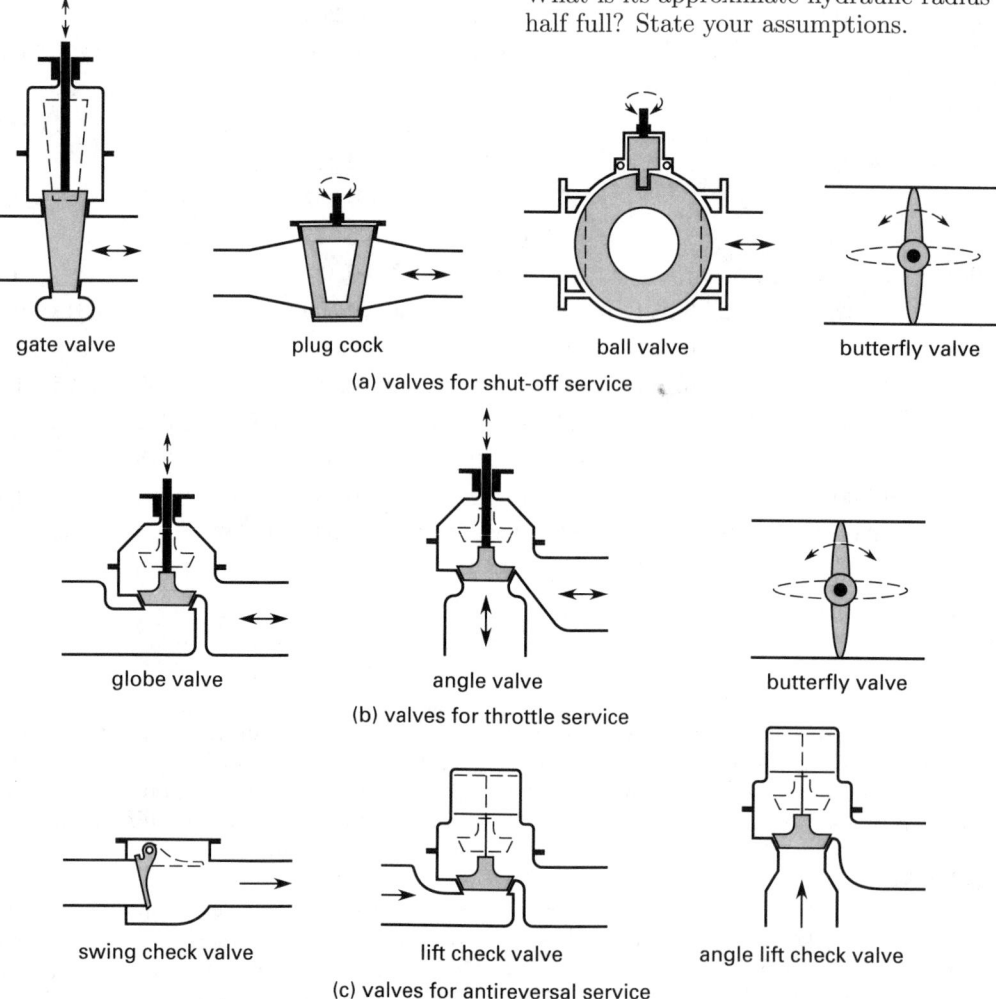

(a) valves for shut-off service

gate valve plug cock ball valve butterfly valve

(b) valves for throttle service

globe valve angle valve butterfly valve

(c) valves for antireversal service

swing check valve lift check valve angle lift check valve

PRACTICE PROBLEMS

Use the following values unless told to do otherwise in the problem:

$$g = 32.2 \text{ ft/sec}^2 \ (9.81 \text{ m/s}^2)$$

$$\rho_{\text{water}} = 62.4 \text{ lbm/ft}^3 \ (1000 \text{ kg/m}^3)$$

$$p_{\text{atmospheric}} = 14.7 \text{ psia} \ (101.3 \text{ kPa})$$

1. A 10 in (25 cm) composition pipe is compressed by a tree root until its inside height is only 7.2 in (18 cm). What is its approximate hydraulic radius when flowing half full? State your assumptions.

Table 16.4 *Characteristics of Common Valve Types*

valve types	fluid properties	switching frequency	pressure drop (fully open)	control response	maximum pressure, atm	maximum temperature, °C
butterfly	clean	low	low	poor	50	400
ball	clean	low	low	very poor	160	300
plug	clean	low	low	very poor	160	300
gate	clean	low	low	very poor	50	400
globe	clean	high	medium to high	very good	80	300
diaphragm	clean to slurried	very high	low to medium	very good	16	150

17 Fluid Dynamics

1.	Hydraulics and Hydrodynamics	17-2
2.	Conservation of Mass	17-2
3.	Maximum Velocities in Pipes	17-3
4.	Stream Potential and Stream Function	17-3
5.	Head Loss Due to Friction	17-4
6.	Relative Roughness	17-4
7.	Friction Factor	17-5
8.	Energy Loss Due to Friction: Laminar Flow	17-7
9.	Energy Loss Due to Friction: Turbulent Flow	17-7
10.	Friction Loss for Water Flow in Steel Pipes	17-8
11.	Friction Loss in Noncircular Ducts	17-9
12.	Friction Loss for Steam and Gases	17-9
13.	Effect of Viscosity on Head Loss	17-10
14.	Friction Loss with Slurries and Non-Newtonian Fluids	17-11
15.	Minor Losses	17-11
16.	Valve Flow Coefficients	17-13
17.	Shear Stress in Circular Pipes	17-13
18.	Introduction to Pumps and Turbines	17-14
19.	Extended Bernoulli Equation	17-14
20.	Energy and Hydraulic Grade Lines with Friction	17-15
21.	Discharge from Tanks	17-16
22.	Discharge from Pressurized Tanks	17-17
23.	Coordinates of a Fluid Stream	17-17
24.	Time to Empty a Tank	17-18
25.	Discharge from Large Orifices	17-18
26.	Culverts	17-19
27.	Siphons	17-19
28.	Series Pipe Systems	17-19
29.	Parallel Pipe Systems	17-20
30.	Other Pipe Branching Problems	17-20
31.	Flow Measuring Devices	17-20
32.	Pitot-Static Gauge	17-22
33.	Venturi Meter	17-23
34.	Orifice Meter	17-24
35.	Flow Nozzle	17-25
36.	Flow Measurements of Compressible Fluids	17-26
37.	Impulse-Momentum Principle	17-26
38.	Jet Propulsion	17-27
39.	Open Jet on a Vertical Flat Plate	17-28
40.	Open Jet on a Horizontal Flat Plate	17-28
41.	Open Jet on an Inclined Plate	17-28
42.	Open Jet on a Single Stationary Blade	17-29
43.	Open Jet on a Single Moving Blade	17-29
44.	Open Jet on a Multiple-Bladed Wheel	17-29
45.	Impulse Turbine Power	17-29
46.	Confined Streams in Pipe Bends	17-30
47.	Water Hammer	17-31
48.	Lift	17-32
49.	Circulation	17-33
50.	Lift from Rotating Cylinders	17-33
51.	Drag	17-34
52.	Drag on Spheres and Disks	17-35
53.	Terminal Velocity	17-35
54.	Nonspherical Particles	17-36
55.	Flow Around a Cylinder	17-36
56.	Flow Over a Parallel Flat Plate	17-37
57.	Similarity	17-38
58.	Viscous and Inertial Forces Dominate	17-38
59.	Inertial and Gravitational Forces Dominate	17-40
60.	Surface Tension Force Dominates	17-40
	Practice Problems	17-41

Nomenclature

a	length	ft	m
a	speed of sound	ft/sec	m/s
A	area	ft^2	m^2
C	Hazen-Williams coefficient	–	–
C	coefficient	–	–
d	diameter	in	cm
D	diameter	ft	m
E	specific energy	ft-lbf/lbm	J/kg
E	bulk modulus	lbf/ft^2	Pa
f	Darcy friction factor	–	–
f	fraction split	–	–
F	force	lbf	N
Fr	Froude number	–	–
g	gravitational acceleration	ft/sec^2	m/s^2
g_c	gravitational conversion constant (32.2)	lbm-ft/lbf-sec^2	n.a.
G	mass flow rate per unit area	lbm/ft^2-sec	kg/m$^2\cdot$s
h	height or head	ft	m
HP	horsepower	hp	n.a.
I	impulse	lbf-sec	N·s
k	ratio of specific heats	–	–
K	minor loss coefficient	–	–
l	length	ft	m
L	length	ft	m
m	mass	lbm	kg
$\dot{m}$	mass flow rate	lbm/sec	kg/s
MW	molecular weight	lbm/lbmol	kg/kmol
n	Manning roughness constant	–	–
p	pressure	lbf/ft^2	Pa
P	power	ft-lbf/sec	W

Fluids

P	momentum	lbm-ft/sec	kg·m/s
Q	flow rate	gal/min	n.a.
r	radius	ft	m
rpm	rotational speed	rev/min	rev/min
R	resultant force	lbf	N
R^*	universal gas constant	ft-lbf/lbmol-°R	J/kmol·K
Re	Reynolds number	–	–
SG	specific gravity	–	–
t	time	sec	s
t	thickness	ft	m
T	absolute temperature	°R	K
u	x-component of velocity	ft/sec	m/s
v	y-component of velocity	ft/sec	m/s
v	velocity	ft/sec	m/s
V	volume	ft³	m³
$\dot{V}$	volumetric flow rate	ft³/sec	m³/s
W	work	ft-lbf	J
We	Weber number	–	–
WHP	water horsepower	hp	n.a.
x	x-coordinate of position	ft	m
y	y-coordinate of position	ft	m
Y	expansion factor	–	–
z	elevation	ft	m

Symbols

β	diameter ratio	–	–
Γ	circulation	ft²/sec	m²/s
ϵ	specific roughness	ft	m
η	efficiency	–	–
η	non-Newtonian viscosity	lbf-sec/ft²	Pa·s
θ	angle	degrees	degrees
μ	absolute viscosity	lbf-sec/ft²	Pa·s
ν	kinematic viscosity	ft²/sec	m²/s
ρ	density	lbm/ft³	kg/m³
σ	surface tension	lbf/ft	N/m
τ	shear stress	lbf/ft²	Pa
υ	specific volume	ft³/lbm	m³/kg
ϕ	angle	degrees	degrees
Φ	stream potential	–	–
Ψ	stream function	–	–
ω	angular velocity	rad/sec	rad/s

Subscripts

A	added (by pump)
b	blade or buoyant
c	contraction
d	discharge
D	drag
e	equivalent
E	extracted (by turbine)
f	friction or flow
i	inside
I	instrument
L	lift

m	minor, model, or manometer fluid
o	orifice or outside
p	pressure or prototype
r	ratio
s	static
t	total, tank, or theoretical
v	velocity
va	velocity of approach
z	potential

1. HYDRAULICS AND HYDRODYNAMICS

This chapter investigates fluid moving through pipes, measurements with venturis and orifices, and other motion-related topics such as model theory, lift and drag, and pumps. In a strict interpretation, any fluid-related phenomenon that isn't hydro*statics* should be hydro*dynamics*. However, tradition has separated the study of moving fluids into the fields of hydraulics and hydrodynamics.

In a general sense, *hydraulics* is the study of the practical laws of fluid flow and resistance in pipes and open channels. Hydraulic formulas are often developed from experimentation, empirical factors, and curve fitting, without an attempt to justify why the fluid behaves the way it does.

On the other hand, *hydrodynamics* is the study of fluid behavior based on theoretical considerations. Hydrodynamicists start with Newton's laws of motion and try to develop models of fluid behavior. Models developed in this manner are complicated greatly by the inclusion of viscous friction and compressibility. Therefore, hydrodynamic models assume a perfect fluid with constant density and zero viscosity. The conclusions reached by hydrodynamicists can differ greatly from those reached by hydraulicians.[1]

2. CONSERVATION OF MASS

Fluid mass is always conserved in fluid systems, regardless of the pipeline complexity, orientation of the flow, or which fluid is flowing. This single concept is often sufficient to solve simple fluid problems.

$$\dot{m}_1 = \dot{m}_2 \qquad 17.1$$

When applied to fluid flow, the conservation of mass law is known as the *continuity equation*.

$$\rho_1 A_1 v_1 = \rho_2 A_2 v_2 \qquad 17.2$$

If the fluid is incompressible, then $\rho_1 = \rho_2$.

$$A_1 v_1 = A_2 v_2 \qquad 17.3$$

$$\dot{V}_1 = \dot{V}_2 \qquad 17.4$$

---

[1]Perhaps the most disparate conclusion is *D'Alembert's paradox*. In 1744, D'Alembert derived theoretical results "proving" that there is no resistance to bodies moving through an ideal (non-viscous) fluid.

Various units and symbols are used for *volumetric flow rate*. (Though this book uses $\dot{V}$, the symbol Q is often used when the flow rate is expressed in gallons.) MGD (millions of gallons per day) and mgpcd (millions of gallons per capita day) are units commonly used in municipal water works problems. MMSCFD (millions of standard cubic feet per day) may be used to express gas flows.

Calculation of flow rates is often complicated by the interdependence between flow rate and friction loss. Each affects the other. Hence, many pipe flow problems must be solved iteratively. Usually, a reasonable friction factor is assumed and is used to calculate an initial flow rate. The flow rate establishes the flow velocity, from which a revised friction factor can be determined.

3. MAXIMUM VELOCITIES IN PIPES

Fluid friction in pipes is kept at acceptable levels by maintaining reasonable fluid velocities. Table 17.1 lists typical maximum fluid velocities.

Table 17.1 *Typical Maximum Fluid Velocities*

| | maximum velocity | |
fluid and application	ft/sec	m/s
water: city service	2–5	0.6–1.5
water: boiler feed	10	3
air: compressor suction	75–200	23–60
air: compressor discharge	100–250	30–75
refrigerant: suction	15–35	4.5–11
refrigerant: discharge	35–60	11–18
steam, saturated: heating	65–100	20–30
steam, saturated: miscellaneous	100–200	30–60
steam, superheated: turbine feed	160–250	50–75

4. STREAM POTENTIAL AND STREAM FUNCTION

An application of hydrodynamic theory is the derivation of the stream function from stream potential. The *stream potential function (velocity potential function)*, Φ, is the algebraic sum of the component velocity potential functions.[2]

$$\Phi = \Phi_x(x,y) + \Phi_y(x,y) \qquad 17.5$$

The velocity component of the resultant in the x-direction is

$$u = \frac{\partial \Phi}{\partial x} \qquad 17.6$$

The velocity component of the resultant in the y-direction is

$$v = \frac{\partial \Phi}{\partial y} \qquad 17.7$$

The total derivative of the stream potential function is

$$d\Phi = \frac{\partial \Phi}{\partial x}dx + \frac{\partial \Phi}{\partial y}dy$$
$$= u\,dx + v\,dy \qquad 17.8$$

An *equipotential line* is a line along which the function Φ is constant (i.e., $d\Phi = 0$). The slope of the equipotential line is derived from Eq. 17.8.

$$\frac{dy}{dx}\bigg|_{\text{equipotential}} = -\frac{u}{v} \qquad 17.9$$

For flow through a porous, permeable medium, pressure will be constant along equipotential lines (i.e., along lines of constant Φ). However, for an ideal, non-viscous fluid flowing in a frictionless environment, Φ has no physical significance. Even though Φ has a theoretical basis, it does not coincide with any measurable physical quantity.

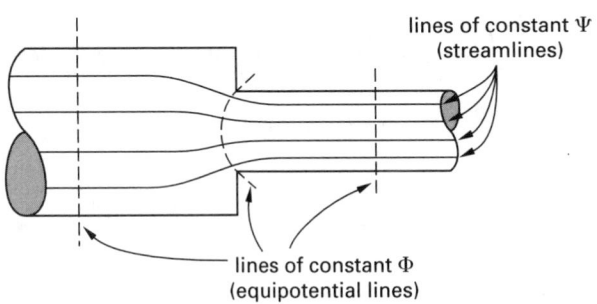

Figure 17.1 *Equipotential Lines and Streamlines*

The *stream function (Lagrange stream function)*, $\Psi(x,y)$, defines the direction of flow at a point.

$$u = \frac{\partial \Psi}{\partial y} \qquad 17.10$$

$$v = -\frac{\partial \Psi}{\partial x} \qquad 17.11$$

The stream function can also be written in total derivative form.

$$d\Psi = \frac{\partial \Psi}{\partial x}dx + \frac{\partial \Psi}{\partial y}dy$$
$$= -v\,dx + u\,dy \qquad 17.12$$

The stream function, $\Psi(x,y)$, satisfies Eq. 17.12. For a given streamline, $d\Psi = 0$, and each streamline is a line representing a constant value of Ψ. A streamline is perpendicular to an equipotential line.

$$\frac{dy}{dx}\bigg|_{\text{streamline}} = \frac{v}{u} \qquad 17.13$$

[2]The two-dimensional derivation of the stream function can be extended to three dimensions, if necessary. The stream function can also be expressed in the cylindrical coordinate system.

Example 17.1

The stream potential function for water flowing through a particular valve is

$$\Phi = 3xy - 2y$$

What is the stream function, Ψ?

Solution

First, work with Φ to obtain u and v.

$$u = \frac{\partial \Phi}{\partial x} = \frac{\partial (3xy - 2y)}{\partial x} = 3y$$

$$v = \frac{\partial \Phi}{\partial y} = 3x - 2$$

u and v are also related to the stream function, Ψ. From Eq. 17.10,

$$u = \frac{\partial \Psi}{\partial y}$$

$$\partial \Psi = u \, \partial y$$

$$\Psi = \int 3y \, dy = \frac{3}{2}y^2 + \text{ some function of } x + C_1$$

Similarly, from Eq. 17.11,

$$v = -\frac{\partial \Psi}{\partial x}$$

$$\partial \Psi = -v \, \partial x$$

$$\Psi = -\int (3x - 2) \, dx$$

$$= 2x - \frac{3}{2}x^2 + \text{ some function of } y + C_2$$

Ψ is found by superposition of these two results.

$$\Psi = \frac{3}{2}y^2 + 2x - \frac{3}{2}x^2 + C$$

5. HEAD LOSS DUE TO FRICTION

The original Bernoulli equation was based on an assumption of frictionless flow. In actual practice, friction occurs during fluid flow. This friction "robs" the fluid of energy, so that the fluid at the end of a pipe section has less energy than it does at the beginning.[3]

$$E_1 > E_2 \qquad\qquad 17.14$$

---

[3]The friction generates minute amounts of heat. The heat is lost to the surroundings.

Most formulas for calculating friction loss use the symbol h_f to represent the *head loss due to friction*.[4] This loss is added into the original Bernoulli equation to restore the equality. Of course, the units of h_f must be the same as the units for the other terms in the Bernoulli equation. If the Bernoulli equation is written in terms of energy, the units will be ft-lbf/lbm or J/kg.

$$E_1 = E_2 + E_f \qquad\qquad 17.15$$

Consider the constant-diameter, horizontal pipe in Fig. 17.2. An incompressible fluid is flowing at a steady rate. Since the elevation of the pipe does not change, the potential energy is constant. Since the pipe has a constant area, the kinetic energy (velocity) is constant. Therefore, the friction energy loss must show up as a decrease in pressure energy. Since the fluid is incompressible, this can only occur if the pressure decreases in the direction of flow.

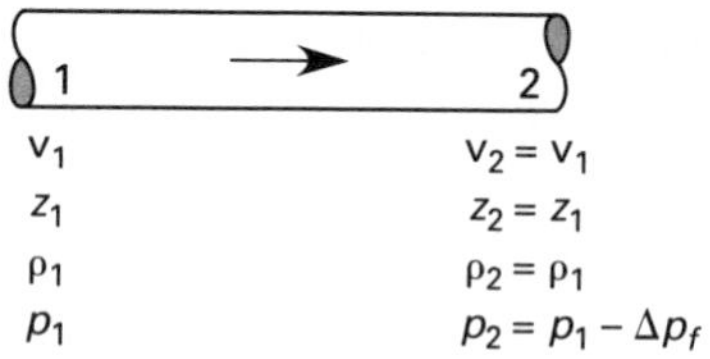

Figure 17.2 *Pressure Drop in a Pipe*

6. RELATIVE ROUGHNESS

It is intuitive that pipes with rough inside surfaces will experience greater friction losses than smooth pipes.[5] *Specific roughness*, ϵ, is a parameter that measures the average size of imperfections inside the pipe. Table 17.2 lists values of ϵ for common pipe materials. (Also, see App. 17.A.)

Table 17.2 *Values of Specific Roughness for Common Pipe Materials*

material	ϵ	
	ft	m
plastic (PVC, ABS)	0.000005	1.5×10^{-6}
copper and brass	0.000005	1.5×10^{-6}
steel	0.0002	6.0×10^{-5}
plain cast iron	0.0008	2.4×10^{-3}
concrete	0.004	1.2×10^{-3}

---

[4]Other names and symbols for this friction loss are *friction head loss* (h_L), *lost work* (LW), *friction heating* ($\mathcal{F}$), *skin friction loss* (F_f), and *pressure drop due to friction* (Δp_f). All terms and symbols mean basically the same thing, although the units may need to be converted.

[5]Surprisingly, this intuitive statement is valid only for turbulent flow. The roughness does not (ideally) affect the friction loss for laminar flow.

However, an imperfection the size of a sand grain will have much more effect in a small-diameter hydraulic line than in a large-diameter sewer. Therefore, the *relative roughness*, ϵ/D, is a better indicator of pipe roughness. Both ϵ and D have units of length (e.g., feet or meters), and the relative roughness is dimensionless.

7. FRICTION FACTOR

The *Darcy friction factor*, f, is one of the parameters used to calculate the friction loss.[6] The friction factor is not constant, but decreases as the Reynolds number (fluid velocity) increases, up to a certain point, known as *fully turbulent flow* (or *rough-pipe flow*). Once the flow is fully turbulent, the friction factor remains constant and depends only on the relative roughness and not on the Reynolds number. For very smooth pipes, fully turbulent flow is achieved only at very high Reynolds numbers.

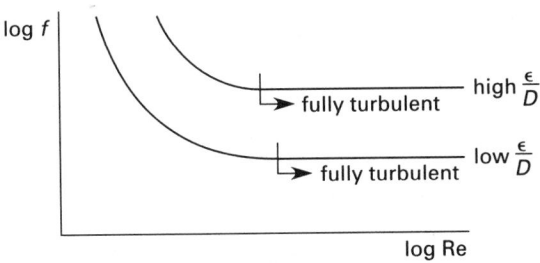

Figure 17.3 *Friction Factor as a Function of Reynolds Number*

The friction factor is not dependent on the material of the pipe but is affected by the roughness. For example, for a given Reynolds number, the friction factor will be the same for any smooth pipe material (glass, plastic, smooth brass and copper, etc.).

The friction factor is determined from the relative roughness, ϵ/D, and the Reynolds number, Re, by various methods. These methods include explicit and implicit equations, the Moody diagram, and tables. The values obtained are based on experimentation, primarily the work of J. Nikuradse in the early 1930s.

When a moving fluid initially encounters a parallel surface (as when a moving gas encounters a flat plate or when a fluid first enters the mouth of a pipe), the flow will generally not be turbulent, even for very rough surfaces. The flow will be laminar for a certain *critical distance* before becoming turbulent. This phenomenon, as it relates to heat transfer, is discussed in greater detail in Chap. 36.

---

[6]There are actually two friction factors: the Darcy friction factor and the *Fanning friction factor*, f_{Fanning}, also known as the *skin friction coefficient* and *wall shear stress factor*. Both factors are in widespread use, sharing the same symbol, f. Civil and (most) mechanical engineers use the Darcy friction factor. The Fanning friction factor is encountered more often in the chemical industry. One can be derived from the other: $f_{\text{Darcy}} = 4f_{\text{Fanning}}$.

Friction Factors for Laminar Flow

The easiest method of obtaining the friction factor for laminar flow (Re < 2100) is to calculate it. Equation 17.16, known as the *Darcy equation*, illustrates that roughness is not a factor in determining the frictional loss in laminar flow.

$$f = \frac{64}{\text{Re}} \qquad 17.16$$

Friction Factors for Turbulent Flow: by Formula

One of the earliest attempts to predict the friction factor for turbulent flow in smooth pipes resulted in the *Blasius equation* (claimed "valid" for 3000 < Re < 100,000).

$$f = \frac{0.316}{\text{Re}^{0.25}} \qquad 17.17$$

The *Nikuradse equation* can also be used to determine the friction factor for smooth pipes (i.e., when $\epsilon/D = 0$). Unfortunately, this equation is implicit in f and must be solved iteratively.

$$\frac{1}{\sqrt{f}} = 2.0 \log_{10}\left(\text{Re}\sqrt{f} - 0.80\right) \qquad 17.18$$

The *Karman-Nikuradse equation* predicts the fully turbulent friction factor (i.e., when Re is very large).

$$\frac{1}{\sqrt{f}} = 1.74 - 2\log_{10}\left(\frac{2\epsilon}{D}\right) \qquad 17.19$$

The most widely known method of calculating the friction factor for any pipe roughness and Reynolds number is another implicit formula, the *Colebrook equation*. Most other equations are variations of this equation. (Notice that the relative roughness, ϵ/D, is used to calculate f.)

$$\frac{1}{\sqrt{f}} = -2\log_{10}\left(\frac{\frac{\epsilon}{D}}{3.7} + \frac{2.51}{\text{Re}\sqrt{f}}\right) \qquad 17.20$$

A suitable approximation would appear to be the Swamee-Jain equation, which claims to have less than 1% error (as measured against the Colebrook equation) for relative roughnesses between 0.000001 and 0.01, and for Reynolds numbers between 5000 and 100,000,000.[7] Even with a 1% error, this equation produces more accurate results than can be read from the Moody diagram.

$$f = \frac{0.25}{\left[\log_{10}\left(\frac{\frac{\epsilon}{D}}{3.7} + \frac{5.74}{\text{Re}^{0.9}}\right)\right]^2} \qquad 17.21$$

---

[7]*ASCE Hydraulic Division Journal*, Vol. 102, May 1976, p. 657. This is not the only explicit approximation to the Colebrook equation in existence.

Friction Factors for Turbulent Flow: by Moody Chart

The *Moody friction factor chart*, Fig. 17.4, presents the friction factor graphically as a function of Reynolds number and relative roughness. There are different lines for selected discrete values of relative roughness. Due to the complexity of this graph, it is easy to mislocate the Reynolds number or use the wrong curve. Nevertheless, the Moody chart remains the most common method of obtaining the friction factor.

Friction Factors for Turbulent Flow: by Table

Appendix 17.B (based on the Colebrook equation), or a similar table, will usually be the most convenient method of obtaining friction factors for turbulent flow.

Example 17.2

Determine the friction factor for a Reynolds number of Re = 400,000 and a relative roughness of $\epsilon/D = 0.004$ using (a) the Moody diagram, (b) App. 17.B, and (c) the Swamee-Jain approximation. (d) Check the table value of f with the Colebrook equation.

Solution

(a) From Fig. 17.4, the friction factor is approximately 0.028.

(b) Appendix 17.B lists the friction factor as 0.0287.

(c) From Eq. 17.21,

$$f = \frac{0.25}{\left[\log_{10}\left(\dfrac{0.004}{3.7} + \dfrac{5.74}{(400{,}000)^{0.9}}\right)\right]^2}$$

$$= 0.0288$$

(d) From Eq. 17.20,

$$\frac{1}{\sqrt{0.0287}} = -2\log_{10}\left(\frac{0.004}{3.7} + \frac{2.51}{(400{,}000)\sqrt{0.0287}}\right)$$

$$5.903 = 5.903$$

Figure 17.4 *Moody Friction Factor Chart*

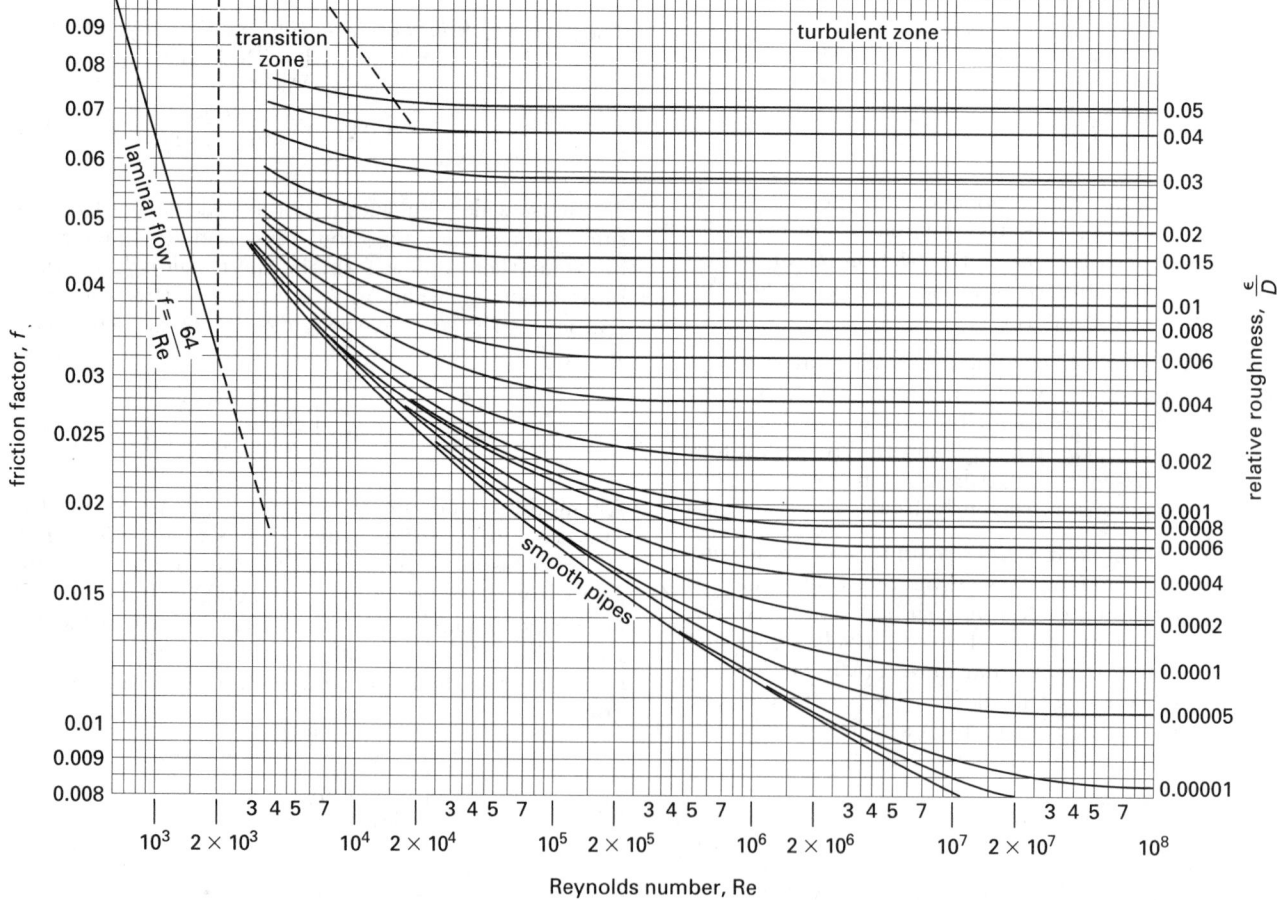

Reproduced from "Friction Factor for Pipe Flow" by L.F. Moody, *Trans. ASME*, Vol. 66, with permission of The American Society of Mechanical Engineers.

8. ENERGY LOSS DUE TO FRICTION: LAMINAR FLOW

Two methods are available for calculating the frictional energy loss for fluids experiencing laminar flow. The most common is the *Darcy equation* (also known as the *Weisbach equation* or *Darcy-Weisbach equation*), which can be used for both laminar and turbulent flow.[8] One of the advantages to using the Darcy equation is that the assumption of laminar flow does not need to be confirmed if f is known.

$$h_f = \frac{fL\text{v}^2}{2Dg} \qquad\qquad 17.22$$

$$E_f = h_f g = \frac{fL\text{v}^2}{2D} \qquad [\text{SI}] \qquad 17.23(a)$$

$$E_f = h_f \times \left(\frac{g}{g_c}\right) = \frac{fL\text{v}^2}{2Dg_c} \qquad [\text{U.S.}] \qquad 17.23(b)$$

If the flow is truly laminar and the fluid is flowing in a circular pipe, then the *Hagen-Poiseuille equation* can be used.

$$E_f = \frac{32\mu\text{v}L}{D^2\rho} \qquad [\text{SI}] \qquad 17.24(a)$$

$$E_f = \frac{32\mu\text{v}Lg_c}{D^2\rho} \qquad [\text{U.S.}] \qquad 17.24(b)$$

An alternate form of the Hagen-Poiseuille equation substitutes $\dot{V}/A$ for v.

$$E_f = \frac{128\mu\dot{V}L}{\pi D^4\rho} \qquad [\text{SI}] \qquad 17.25(a)$$

$$E_f = \frac{128\mu\dot{V}Lg_c}{\pi D^4\rho} \qquad [\text{U.S.}] \qquad 17.25(b)$$

If necessary, h_f can be converted to an actual pressure drop in psi or Pa by multiplying by the fluid density.

$$\Delta p = h_f \times \rho g \qquad [\text{SI}] \qquad 17.26(a)$$

$$\Delta p = h_f \times \rho\left(\frac{g}{g_c}\right) \qquad [\text{U.S.}] \qquad 17.26(b)$$

Values of the Darcy friction factor, f, are often quoted for new, clean pipe. The friction head losses and pumping power requirements calculated from these values are minimal values. Depending on the nature of the service, scale and impurity buildup within pipes may decrease the pipe diameters over time. Since the frictional loss is proportional to the fifth power of the diameter, such diameter decreases can produce dramatic increases in the friction loss.

$$\frac{h_{f,\text{scaled}}}{h_{f,\text{new}}} = \left(\frac{D_{\text{new}}}{D_{\text{scaled}}}\right)^5 \qquad\qquad 17.27$$

Equation 17.27 accounts only for the decrease in diameter. Any increase in roughness (i.e., friction factor) will produce a proportional increase.

Because the "new, clean" condition is transitory in most applications, an uprating factor of 10 to 30% is often applied to either the friction factor, f, or the head loss, h_f. Of course, even larger increases should be considered when extreme fouling is expected.

Another approach eliminates the need to estimate the scaled pipe diameter. This simplistic approach multiplies the initial friction loss by a factor based on the age of the pipe. For example, for schedule-40 pipe 4 to 10 in (10 to 25 cm) in diameter, the multipliers of 1.4, 2.2, and 5.0 have been proposed for pipe ages of 5, 10, and 20 years, respectively. For larger pipes, the corresponding multipliers are 1.3, 1.6, and 2.0. Obviously, use of these values should be based on a clear understanding of the method's limitations.

9. ENERGY LOSS DUE TO FRICTION: TURBULENT FLOW

The *Darcy equation* is used almost exclusively to calculate the head loss due to friction for turbulent flow.

$$h_f = \frac{fL\text{v}^2}{2Dg} \qquad\qquad 17.28$$

The head loss can be converted to pressure drop.

$$\Delta p = h_f \times \rho g \qquad [\text{SI}] \qquad 17.29(a)$$

$$\Delta p = h_f \times \rho\left(\frac{g}{g_c}\right) \qquad [\text{U.S.}] \qquad 17.29(b)$$

In problems where the pipe size is unknown, it will be impossible to obtain an accurate value of the friction factor, f (since f depends on velocity). In such problems, an iterative solution will be necessary.

It is not uncommon for civil engineers to use the *Hazen-Williams equation* to calculate head loss. This method requires knowledge of the Hazen-Williams *roughness coefficient*, C, values of which are widely tabulated.[9] (See App. 17.A.) The advantage of using this equation is that C does not depend on the Reynolds number.

$$h_f = \frac{(3.022)(\text{v})^{1.85}L}{(C)^{1.85}(D)^{1.165}} \qquad [\text{U.S.}] \qquad 17.30$$

Or, in terms of other units,

$$h_f = \frac{(10.44)(L)(\dot{V}_{\text{gpm}})^{1.85}}{(C)^{1.85}(d_{\text{inches}})^{4.8655}} \qquad [\text{U.S.}] \qquad 17.31$$

The Hazen-Williams equation is empirical and is not dimensionally homogeneous. It is taken as a matter of faith that the units of h_f are feet.

---

[8]The difference is that the friction factor can be derived by hydrodynamics: $f = 64/\text{Re}$. For turbulent flow, f is empirical.

[9]An approximate value of $C = 140$ is often chosen for initial calculations for new pipe. $C = 100$ is more appropriate for pipe that has been in service for some time.

The Hazen-Williams equation should be used only for turbulent flow. It gives good results for liquids that have kinematic viscosities around 1.2×10^{-5} ft^2/sec (1.1×10^{-6} m^2/s), which corresponds to the viscosity of 60°F (16°C) water. At extremely high and low temperatures, the Hazen-Williams equation can be 20% or more in error for water.

Example 17.3

50°F water is pumped through 1000 ft of 4 in, schedule-40 welded steel pipe at the rate of 300 gpm. What friction loss (in ft-lbf/lbm) is predicted by the Darcy equation?

Solution

First, it is necessary to collect data on the pipe and water. The fluid viscosity, pipe dimensions, and other parameters can be found from the appendices.

$$\nu = 1.41 \times 10^{-5} \text{ ft}^2/\text{sec}$$

$$\epsilon = 0.0002 \text{ ft}$$

$$D = 0.3355 \text{ ft}$$

$$A = 0.0884 \text{ ft}^2$$

The flow quantity is converted from gallons per minute to cubic feet per second.

$$\dot{V} = (300 \text{ gpm}) \left(0.002228 \; \frac{\frac{\text{ft}^3}{\text{sec}}}{\text{gpm}} \right) = 0.6684 \text{ ft}^3/\text{sec}$$

The velocity is

$$\text{v} = \frac{\dot{V}}{A} = \frac{0.6684 \; \frac{\text{ft}^3}{\text{sec}}}{0.0884 \text{ ft}^2} = 7.56 \text{ ft/sec}$$

The Reynolds number is

$$\text{Re} = \frac{D\text{v}}{\nu} = \frac{(0.3355 \text{ ft}) \left(7.56 \; \frac{\text{ft}}{\text{sec}} \right)}{1.41 \times 10^{-5} \; \frac{\text{ft}^2}{\text{sec}}}$$

$$= 1.8 \times 10^5$$

The relative roughness is

$$\frac{\epsilon}{D} = \frac{0.0002}{0.3355} = 0.0006$$

From the friction factor table (or the Moody friction factor chart), $f = 0.0195$.

Equation 17.23(b) is used to calculate the friction loss.

$$E_f = h_f \times \left(\frac{g}{g_c} \right) = \frac{fL\text{v}^2}{2Dg_c}$$

$$= \frac{(0.0195)(1000 \text{ ft}) \left(7.56 \; \frac{\text{ft}}{\text{sec}} \right)^2}{(2)(0.3355 \text{ ft}) \left(32.2 \; \frac{\text{lbm-ft}}{\text{lbf-sec}^2} \right)}$$

$$= 51.6 \text{ ft-lbf/lbm}$$

Example 17.4

Calculate the head loss due to friction for the pipe in Ex. 17.3 using the Hazen-Williams formula. Assume $C = 100$.

Solution

Substituting the parameters derived in Ex. 17.3 into Eq. 17.30,

$$h_f = \frac{(3.022) \left(7.56 \; \frac{\text{ft}}{\text{sec}} \right)^{1.85} (1000 \text{ ft})}{(100)^{1.85} (0.3355 \text{ ft})^{1.165}} = 90.8 \text{ ft}$$

Alternatively, the given data can be substituted directly into Eq. 17.31.

$$h_f = \frac{(10.44)(1000 \text{ ft})(300 \text{ gpm})^{1.85}}{(100)^{1.85}(4.026 \text{ in})^{4.8655}} = 90.9 \text{ ft}$$

10. FRICTION LOSS FOR WATER FLOW IN STEEL PIPES

Friction loss and velocity for water flowing through steel pipe (as well as for other liquids and other pipe materials) in table and chart form are in widespread use. (Appendix 17.C is an example of such a table.) These tables and charts are unable to compensate for the effects of fluid temperature and different pipe roughness. Unfortunately, the assumptions made in developing the tables and charts are seldom listed. Another disadvantage is that the values can be read to only a few significant figures.

Since water's specific volume is essentially constant within the normal temperature range, tables and charts can be used to determine water velocity. Friction loss data, however, should be considered accurate to only ±20%. Alternatively, a 20% safety margin should be established in choosing pumps and motors.

Tables and charts almost always give the friction loss per 100 ft or 10 m of pipe. The pressure drop is proportional to the length, so the value read can be scaled for other pipe lengths. Flow velocity is independent of the pipe length.

11. FRICTION LOSS IN NONCIRCULAR DUCTS

The frictional energy loss by a fluid flowing in a rectangular, annular, or other noncircular duct can be calculated from the Darcy equation by using the *equivalent diameter (hydraulic diameter)*, D_e, in place of the diameter variable, D.[10] The friction factor, f, is determined in any of the conventional manners.

12. FRICTION LOSS FOR STEAM AND GASES

The Darcy equation can be used to calculate the frictional energy loss for all incompressible liquids, not just for water. Alcohol, gasoline, fuel oil, and refrigerants, for example, are all handled well, since the effect of viscosity is considered in determining the friction factor, f.[11]

In fact, the Darcy equation is commonly used with noncondensing vapors and compressed gases, such as air, nitrogen, and steam.[12] In such cases, reasonable accuracy will be achieved as long as the fluid is not moving too fast (i.e., less than Mach 0.3) and is incompressible. The fluid is assumed to be incompressible if the pressure (or density) change along the section of interest is less than 10% of the starting pressure.

If possible, it is preferred to base all calculations on the average properties of the fluid.[13] Specifically, the fluid velocity would normally be calculated as

$$v = \frac{\dot{m}}{\rho_{ave} A} \qquad 17.32$$

However, the average density of a gas depends on the average pressure, which is unknown at the start of a problem. The solution is to write the Reynolds number and Darcy equation in terms of the constant mass flow rate per unit area, G, instead of velocity, v, which varies.

$$G = v_{ave}\rho_{ave} \qquad 17.33$$

$$Re = \frac{DG}{\mu} \qquad \text{[SI]} \qquad 17.34(a)$$

$$Re = \frac{DG}{g_c\mu} \qquad \text{[U.S.]} \qquad 17.34(b)$$

$$\Delta p_f = p_1 - p_2 = \rho_{ave}h_f = \frac{fLG^2}{2D\rho_{ave}} \qquad 17.35$$

---

[10]Although it is used for both, this approach is better suited for turbulent flow than for laminar flow. Also, the accuracy of this method decreases as the flow area becomes more noncircular. The friction drop in long, narrow slits is poorly predicted, for example. However, there is no other convenient method of predicting friction drop. Experimentation should be used with a particular flow geometry if extreme accuracy is required.
[11]Since viscosity is not an explicit factor in the formula, it should be obvious that the Hazen-Williams equation is primarily used for water.
[12]Use of the Darcy equation is limited only by the availability of the viscosity data needed to calculate the Reynolds number.
[13]Of course, the entrance (or exit) conditions can be used if great accuracy is not needed.

Assuming a perfect gas with a molecular weight of MW, the ideal gas law can be used to calculate ρ_{ave} from the absolute temperature, T, and $p_{ave} = (p_1 + p_2)/2$.

$$p_1^2 - p_2^2 = \frac{fLG^2R^*T}{D(\text{MW})} \qquad \text{[SI]} \qquad 17.36(a)$$

$$p_1^2 - p_2^2 = \frac{fLG^2R^*T}{Dg_c(\text{MW})} \qquad \text{[U.S.]} \qquad 17.36(b)$$

To summarize, use the following guidelines when working with compressible gases or vapors flowing in a pipe or duct: (1) If the pressure drop, based on the entrance pressure, is less than 10%, the fluid can be assumed to be incompressible, and the gas properties can be evaluated at any point known along the pipe. (2) If the pressure drop is between 10% and 40%, use of the midpoint properties will yield reasonably accurate friction losses. If the pressure drop is greater than 40%, the pipe can be divided into shorter sections and the losses calculated for each section, or exact calculations based on compressible flow theory must be made.

Calculating a friction loss for steam flow can be frustrating if steam viscosity data are unavailable. Generally, the steam viscosities listed in compilations of heat transfer data (see Apps. 35.A and 35.B) are sufficiently accurate. Various empirical methods are also in use. For example, the *Babcock formula* (Eq. 17.37) for pressure drop when steam with a specific volume of v flows in a pipe of diameter d is

$$\Delta p_{psi} = (0.470)\left(\frac{d_{in} + 3.6}{d_{in}^6}\right)(\dot{m}_{lbm/sec})^2 L_{ft}v$$

$$17.37$$

Use of empirical formulas is not limited to steam. Theoretical formulas (e.g., the *complete isothermal flow equation*) and specialized empirical formulas (e.g., the *Weymouth*, *Panhandle*, and *Spitzglass formulas*) have been developed, particularly by the gas pipeline industry. Each of these provides reasonable accuracy within their operating limits. However, none should be used without knowing the assumptions and operational limitations that were used in their derivations.

Example 17.5

0.0011 kg/s of 25°C nitrogen gas flows isothermally through a 175 m section of smooth tubing (inside diameter = 0.012 m). The viscosity of the nitrogen is 1.8×10^{-5} Pa·s. The pressure of the nitrogen is 200 kPa originally. At what pressure is the nitrogen delivered?

Solution

The flow area of the pipe is

$$A = \frac{\pi}{4}D^2 = \frac{\pi(0.012\text{ m})^2}{4} = 1.131 \times 10^{-4}\text{ m}^2$$

The mass flow rate per unit area is

$$G = \frac{\dot{m}}{A} = \frac{0.0011 \, \frac{\text{kg}}{\text{s}}}{1.131 \times 10^{-4} \, \text{m}^2} = 9.73 \, \text{kg/m}^2 \cdot \text{s}$$

The Reynolds number is

$$\text{Re} = \frac{DG}{\mu} = \frac{(0.012 \, \text{m}) \left(9.73 \, \frac{\text{kg}}{\text{m}^2 \cdot \text{s}} \right)}{1.8 \times 10^{-5} \, \text{Pa} \cdot \text{s}} = 6487$$

The flow is turbulent, and the pipe is said to be smooth. Therefore, the friction factor is interpolated (from App. 17.B) as 0.0347.

The molecular weight of nitrogen is 28.0 kg/kmol. The temperature must be in degrees absolute: $T = 25°\text{C} + 273 = 298\text{K}$. The universal gas constant is 8314.3 J/kmol·K.

From Eq. 17.36, the final pressure is

$$p_2^2 = p_1^2 - \frac{fLG^2 R^* T}{D(\text{MW})}$$

$$= (200{,}000 \, \text{Pa})^2$$

$$- \frac{(0.0347)(175 \, \text{m}) \left(9.73 \, \frac{\text{kg}}{\text{m}^2 \cdot \text{s}} \right)^2 \times \left(8314.3 \, \frac{\text{J}}{\text{kmol} \cdot \text{K}} \right) (298\text{K})}{(0.012 \, \text{m}) \left(28 \, \frac{\text{kg}}{\text{kmol}} \right)}$$

$$= (4 \times 10^{10}) - (4.24 \times 10^9) = 3.576 \times 10^{10} \, (\text{Pa})^2$$

$$p_2 = \sqrt{3.576 \times 10^{10}} = 1.89 \times 10^5 \, \text{Pa} \quad (189 \, \text{kPa})$$

The percentage drop in pressure should not be more than 10%:

$$\frac{200 - 189}{200} = 0.055 \, (5.5\%) \quad [\text{o.k.}]$$

Example 17.6

Superheated steam at 140 psi and 500°F enters a 200 ft long steel pipe with an internal diameter of 3.826 in. The pipe is insulated so that there is no heat loss. (a) Use the Babcock formula to determine the maximum velocity such that the steam does not experience more than a 10% drop in pressure. (b) Verify the velocity by calculating the pressure drop with the Darcy equation.

Solution

(a) From superheat tables, the specific volume of the steam is 3.954 ft³/lbm. The maximum pressure drop is 10% of 140 psi or 14 psi.

From Eq. 17.37,

$$\Delta p_{\text{psi}} = (0.470) \left(\frac{d_{\text{in}} + 3.6}{d_{\text{in}}^6} \right) (\dot{m}_{\text{lbm/sec}})^2 L_{\text{ft}} v$$

$$14 \, \text{psi} = (0.470) \left[\frac{3.826 \, \text{in} + 3.6}{(3.826 \, \text{in})^6} \right] \dot{m}^2$$

$$\times (200 \, \text{ft}) \left(3.954 \, \frac{\text{ft}^3}{\text{lbm}} \right)$$

$$\dot{m} = 3.99 \, \text{lbm/sec} \quad (4 \, \text{lbm/sec})$$

(b) Assume a Darcy friction factor of 0.02 (typical for turbulent flow in steel pipe). The steam flow velocity is

$$\text{v} = \frac{Q}{A} = \frac{\dot{m}}{\rho A} = \frac{\dot{m} v}{A}$$

$$= \frac{\left(4 \, \frac{\text{lbm}}{\text{sec}} \right) \left(3.954 \, \frac{\text{ft}^3}{\text{lbm}} \right)}{\left(\frac{\pi}{4} \right) \left(\frac{3.826 \, \text{in}}{12 \, \frac{\text{in}}{\text{ft}}} \right)^2}$$

$$= 198 \, \text{ft/sec}$$

$$h_f = \frac{f L \text{v}^2}{2Dg}$$

$$= \frac{(0.02)(200 \, \text{ft}) \left(198 \, \frac{\text{ft}}{\text{sec}} \right)^2}{(2) \left(\frac{3.826 \, \text{in}}{12 \, \frac{\text{in}}{\text{ft}}} \right) \left(32.2 \, \frac{\text{ft}}{\text{sec}^2} \right)}$$

$$= 7637 \, \text{ft of steam}$$

$$\Delta p = \rho h_f \times \left(\frac{g}{g_c} \right) = \left(\frac{h_f}{v} \right) \times \left(\frac{g}{g_c} \right)$$

$$= \left[\frac{7637 \, \text{ft}}{\left(3.954 \, \frac{\text{ft}^3}{\text{lbm}} \right) \left(144 \, \frac{\text{in}^2}{\text{ft}^2} \right)} \right]$$

$$\times \left(\frac{32.2 \, \frac{\text{ft}}{\text{sec}^2}}{32.2 \, \frac{\text{ft-lbm}}{\text{lbf-sec}^2}} \right)$$

$$= 13.4 \, \text{lbf/in}^2 \quad (\text{psi})$$

13. EFFECT OF VISCOSITY ON HEAD LOSS

Friction loss in a pipe is affected by the fluid viscosity. For both laminar and turbulent flow, viscosity is considered when the Reynolds number is calculated. When viscosities substantially increase without a corresponding decrease in flow rate, two things usually happen: (1) the friction loss greatly increases, and (2) the flow becomes laminar.

It is sometimes necessary to estimate head loss for a new fluid viscosity based on head loss at an old fluid viscosity. The estimation procedure used depends on the flow regimes for the new and old fluids.

For laminar flow, the friction factor is directly proportional to the viscosity. If the flow is laminar for both fluids, the ratio of new-to-old head losses will be equal to the ratio of new-to-old viscosities. Thus, if a flow is already known to be laminar at one viscosity and the fluid viscosity increases, a simple ratio will define the new friction loss.

If both flows are fully turbulent, the friction factor will not change. If flow is fully turbulent and the viscosity decreases, the Reynolds number will increase. Theoretically, this will have no effect on the friction loss.

There are no analytical ways of estimating the change in friction loss when the flow regime changes between laminar and turbulent or between semiturbulent and fully turbulent. Various graphical methods are used, particularly by the pump industry, for calculating power requirements.

14. FRICTION LOSS WITH SLURRIES AND NON-NEWTONIAN FLUIDS

A *slurry* is a mixture of a liquid (usually water) and a solid (e.g., coal, paper pulp, foodstuffs). The liquid is generally used as the transport mechanism (i.e., the *carrier*) for the solid.

Friction loss calculations for slurries vary in sophistication depending on what information is available. In many cases, only the slurry's specific gravity is known. In that case, use is made of the fact that friction loss can be reasonably predicted by multiplying the friction loss based on the pure carrier (e.g., water) by the specific gravity of the slurry.

Another approach is possible if the density and viscosity are known in the operating range. The traditional Darcy equation (Eq. 17.28) and Reynolds number can be used for thin slurries as long as the flow velocity is high enough to keep solids from settling. (Settling is more of a concern for laminar flow. With turbulent flow, the direction of velocity components fluctuates, assisting the solids to remain in suspension.)

The most analytical approach to slurries or other non-Newtonian fluids requires laboratory-derived rheological data. Non-Newtonian viscosity (η, in Pa·s) is fitted to data of the shear rate (dv/dy, in s^{-1}) according to two common models: the power-law model and the Bingham-plastic model. (These two models are applicable to both laminar and turbulent flow, although each has its advantages and disadvantages.)

The *power-law model* has two empirical constants, m and n, that must be determined.

$$\eta = m \left(\frac{dv}{dy} \right)^{n-1} \qquad 17.38$$

The *Bingham-plastic model* also requires finding two empirical constants: the yield stress τ_0 (in Pa) and the Bingham-plastic limiting viscosity μ_∞ (in Pa·s).

$$\eta = \frac{\tau_0}{\dfrac{dv}{dy}} + \mu_\infty \qquad 17.39$$

Once m and n (or τ_0 and μ_∞) have been determined, the friction factor is determined from one of various models (e.g., Buckingham-Reiner, Dodge-Metzner, Metzner-Reed, Hanks-Ricks, Darby, or Hanks-Dadia). Specialized texts and articles cover these models in greater detail. The friction loss is calculated from the traditional Moody equation.

15. MINOR LOSSES

In addition to the frictional energy lost due to viscous effects, friction losses also result from fittings in the line, changes in direction, and changes in flow area. These losses are known as *minor losses*, since they are usually much smaller in magnitude than the pipe wall frictional loss.[14] Two methods are used to calculate minor losses: equivalent lengths and loss coefficients.

With the *method of equivalent lengths*, each fitting or other flow variation is assumed to produce friction equal to the pipe wall friction from an *equivalent length* of pipe. For example, a 2 in globe valve may produce the same amount of friction as 54 ft (its equivalent length) of 2 in pipe. The equivalent lengths for all minor losses are added to the pipe length term, L, in the Darcy equation. This method can be used with all liquids, but it is generally limited to turbulent flow.

$$L_t = L + \sum L_e \qquad 17.40$$

Equivalent lengths are simple to use, but the method depends on having a table of equivalent length values. The actual value for a fitting will depend on the fitting manufacturer, as well as the fitting material (e.g., brass, cast iron, or steel) and the method of attachment (e.g.,

---

[14]Example and practice problems often include the instruction to "Ignore minor losses." In some industries, valves are considered to be "components," not fittings. In such cases, instructions to "Ignore minor losses in fittings" would be ambiguous, since minor losses in valves would be included in the calculations. However, this interpretation is rare in examples and practice problems.

weld, thread, or flange).[15] Because of these many variations, it may be necessary to use a "generic table" of equivalent lengths during the initial design stages. (See Table 17.3 and App. 17.C.)

Table 17.3 *Typical Equivalent Lengths*
(schedule-40, screwed steel fittings)

	pipe size		
	1"	2"	4"
fitting type	equivalent length, ft		
regular 90° elbow	5.2	8.5	13.0
long radius 90° elbow	2.7	3.6	4.6
regular 45° elbow	1.3	2.7	5.5
tee, flow through line (run)	3.2	7.7	17.0
tee, flow through stem	6.6	12.0	21.0
180° return bend	5.2	8.5	13.0
globe valve	29.0	54.0	110.0
gate valve	0.84	1.5	2.5
angle valve	17.0	18.0	18.0
swing check valve	11.0	19.0	38.0
coupling or union	0.29	0.45	0.65

An alternative method of calculating the minor loss for a fitting is to use the *method of loss coefficients*. Each fitting has a *loss coefficient*, K, associated with it, which, when multiplied by the kinetic energy, gives the loss. Thus, a loss coefficient is the minor loss expressed in fractions (or multiples) of the velocity head.

$$h_m = K h_{\mathrm{v}} \qquad 17.41$$

The loss coefficient for any minor loss can be calculated if the equivalent length is known. However, there is no advantage to using one method over the other, except for consistency in calculations.

$$K = \frac{f L_e}{D} \qquad 17.42$$

Exact friction loss coefficients for bends, fittings, and valves are unique to each manufacturer. Furthermore, except for contractions, enlargements, exits, and entrances, the coefficients decrease fairly significantly (according to the fourth power of the diameter ratio) with increases in valve size. Therefore, a single K value is seldom applicable to an entire family of valves. Nevertheless, generic tables and charts have been developed. These compilations can be used for initial estimates as long as the general nature of the data is recognized.

Table 17.4 *Typical Loss Coefficients*[a]

device	K
angle valve	5
bend, close return	2.2
butterfly valve[b]	
2 in to 8 in	45 f_t
10 in to 14 in	35 f_t
16 in to 24 in	25 f_t
check valve, swing,	
fully open	2.3
corrugated bends	1.3 to 1.6 times value for smooth bend
elbow, 90°, standard	0.9
elbow, 90°, long radius	0.6
elbow, 45°	0.42
gate valve	
fully open	0.19
$1/4$ closed	1.15
$1/2$ closed	5.6
$3/4$ closed	24
globe valve	10
meter	
disk or wobble	3.4 to 10
rotary (star or cog-wheel piston)	10
reciprocating piston	15
turbine wheel (double flow)	5 to 7.5
tee, standard	1.8

[a]The actual loss coefficient will usually depend on the size of the valve. Average values are given.
[b]Loss coefficients for butterfly valves are calculated from the friction factors for the pipes with complete turbulent flow.

Loss coefficients for specific fittings and valves must be known in order to be used. They cannot be derived theoretically. However, the loss coefficients for certain changes in flow area can be calculated from the following equations.[16]

- *sudden enlargements:* (D_1 is the smaller of the two diameters)

$$K = \left[1 - \left(\frac{D_1}{D_2}\right)^2\right]^2 \qquad 17.43$$

- *sudden contractions:* (D_1 is the smaller of the two diameters)

$$K = \left(\tfrac{1}{2}\right)\left[1 - \left(\frac{D_1}{D_2}\right)^2\right] \qquad 17.44$$

- *pipe exit:* (projecting exit, sharp-edged or rounded)

$$K = 1.0 \qquad 17.45$$

---

[15]In the language of pipe fittings, a *threaded fitting* is known as a *screwed fitting*, even though no screws are used.

[16]No attempt is made to imply great accuracy with these equations. Correlation between actual and theoretical losses is fair.

- *pipe entrance:*

reentrant: $K = 0.78$

sharp-edged: $K = 0.50$

rounded:

$\frac{r}{D}$	K
0.02	0.28
0.04	0.24
0.06	0.15
0.10	0.09
0.15	0.04

- *tapered diameter changes:*

$$\beta = \frac{\text{small diameter}}{\text{large diameter}} = \frac{D_1}{D_2}$$

ϕ = wall-to-horizontal angle

enlargement, $\phi \leq 22°$:

$$K = 2.6 \, \sin\phi \, (1 - \beta^2)^2 \qquad \textit{17.46}$$

enlargement, $\phi > 22°$:

$$K = (1 - \beta^2)^2 \qquad \textit{17.47}$$

contraction, $\phi \leq 22°$:

$$K = 0.8 \, \sin\phi \, (1 - \beta^2) \qquad \textit{17.48}$$

contraction, $\phi > 22°$:

$$K = 0.5 \, \sqrt{\sin\phi} \, (1 - \beta^2) \qquad \textit{17.49}$$

Example 17.7

Determine the total equivalent length of the piping system shown. The pipeline contains one gate valve, five regular 90° elbows, one tee (flow through the run), and 228 ft of straight pipe. All fittings are 1 in screwed steel pipe. Disregard entrance and exit losses.

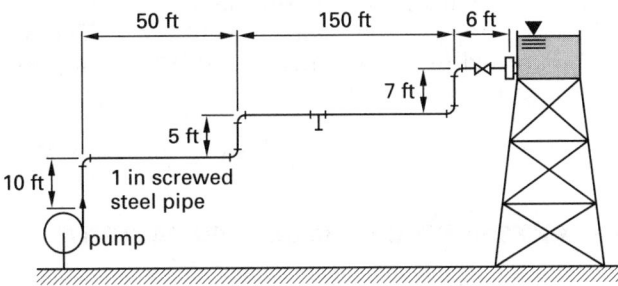

Solution

From Table 17.3, the equivalent lengths are

1	gate valve	1×0.84	=	0.84
5	regular elbows	5×5.2	=	26.0
1	tee run	1×3.2	=	3.2
	straight pipe			228.0
			total L_e =	258.0 ft

16. VALVE FLOW COEFFICIENTS

Valve flow capacities depend on the geometry of the inside of the valve. The *flow coefficient*, C_v, for a valve (particularly a control valve) relates the flow quantity (in gallons per minute) of a fluid with specific gravity to the pressure drop (in pounds per square inch). (The flow coefficient for a valve is not the same as the coefficient of flow for an orifice or venturi meter.) As Eq. 17.50 shows, the flow coefficient is not dimensionally homogeneous and is specifically limited to English units.

$$Q_{\text{gpm}} = C_v \sqrt{\frac{\Delta p_{\text{psi}}}{\text{SG}}} \qquad \textit{17.50}$$

When selecting a control valve for a particular application, the value of C_v is first calculated. Depending on the application and installation, C_v may be further modified by dividing by *piping geometry* and *Reynolds number factors*. (These additional procedures are often specified by the valve manufacturer.) Then, a valve with the required value of C_v is selected.

Although the flow coefficient concept is generally limited to control valves, its use can be extended to all fittings and valves. The relationship between C_v and the loss coefficient, K, is

$$C_v = \frac{(29.9)(d_{\text{in}})^2}{\sqrt{K}} \qquad \textit{17.51}$$

17. SHEAR STRESS IN CIRCULAR PIPES

Shear stress in fluid always acts to oppose the motion of the fluid. (That is the reason the term *frictional force* is occasionally used.) Shear stress for a fluid in laminar flow can always be calculated from the basic definition of absolute viscosity.

$$\tau = \mu \, \frac{d\text{v}}{dy} \qquad \textit{17.52}$$

In the case of the flow in a circular pipe, dr can be substituted for dy in the expression for *shear rate (velocity gradient)*, $d\text{v}/dy$.

$$\tau = \mu \, \frac{d\text{v}}{dr} \qquad \textit{17.53}$$

Equation 17.54 calculates the shear stress between fluid layers a distance r from the pipe centerline from the pressure drop across a length, L, of the pipe.[17] Equation 17.54 is valid for both laminar and turbulent flows.

$$\tau = \frac{(p_1 - p_2)r}{2L} \qquad [r \leq \tfrac{D}{2}] \qquad \textit{17.54}$$

---

[17]In highly turbulent flow, shear stress is not caused by viscous effects but rather by momentum effects. Equation 17.54 is derived from a shell momentum balance. Such an analysis requires the concept of *momentum flux*. In a circular pipe with laminar flow, momentum flux is maximum at the pipe wall, zero at the flow centerline, and varies linearly in between.

The quantity $(p_1 - p_2)$ can be calculated from the Darcy equation (Eq. 17.28). If v is the average flow velocity, the shear stress at the wall (where $r = D/2$) is

$$\tau_{\text{wall}} = \frac{f\rho v^2}{8} \qquad \text{[SI]} \qquad 17.55(a)$$

$$\tau_{\text{wall}} = \frac{f\rho v^2}{8g_c} \qquad \text{[U.S.]} \qquad 17.55(b)$$

Equation 17.54 can be rearranged somewhat to give the relationship between the pressure gradient and the shear stress at the wall.

$$\frac{dp}{dL} = \frac{4\tau_{\text{wall}}}{D} \qquad\qquad 17.56$$

Equation 17.55 can be combined with the Hagen-Poiseuille equation (Eq. 17.24) if the flow is laminar. (v in Eq. 17.57 is the average velocity of fluid flow.)

$$\tau = \frac{16\mu vr}{D^2} \qquad \left[r \leq \frac{D}{2}\right] \qquad 17.57$$

At the pipe wall, $r = D/2$, and the shear stress is maximum. Therefore,

$$\tau_{\text{wall}} = \frac{8\mu v}{D} \qquad\qquad 17.58$$

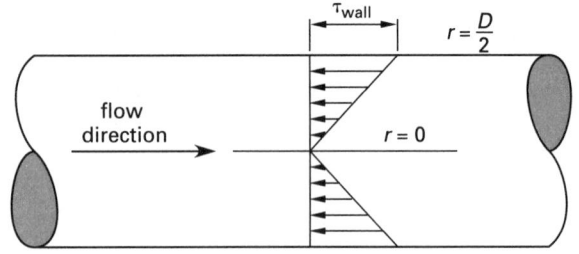

Figure 17.5 *Shear Stress Distribution in a Circular Pipe*

18. INTRODUCTION TO PUMPS AND TURBINES[18]

A *pump* adds energy to the fluid flowing through it. The amount of energy that a pump puts into the fluid stream can be determined by the difference between the total energy on either side of the pump. In most situations, a pump will add primarily pressure energy. The specific energy added (a positive number) on a per-unit mass basis (i.e., J/kg or ft-lbf/lbm) is given by Eq. 17.59.

$$E_A = E_{t,2} - E_{t,1} \qquad\qquad 17.59$$

---

[18]Pumps and turbines are covered in greater detail in Chap. 18.

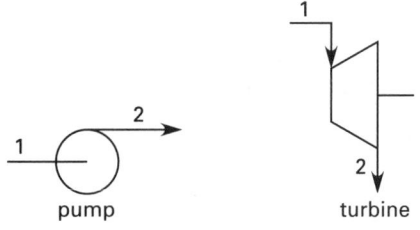

Figure 17.6 *Pump and Turbine Representation*

The *head added* by a pump is

$$h_A = \frac{E_A}{g} \qquad \text{[SI]} \qquad 17.60(a)$$

$$h_A = \frac{E_A g_c}{g} \qquad \text{[U.S.]} \qquad 17.60(b)$$

The specific energy added by a pump can also be calculated from the input power, HP_{input} or kW_{input}, if the mass flow rate is known. The input power to the pump will be the output power of the electric motor or engine driving the pump.

$$E_A = \frac{\left(1000\,\dfrac{\text{W}}{\text{kW}}\right)(\text{kW}_{\text{input}})(\eta_{\text{pump}})}{\dot{m}} \qquad \text{[SI]} \qquad 17.61(a)$$

$$E_A = \frac{\left(550\,\dfrac{\text{ft-lbf}}{\text{sec-hp}}\right)(\text{HP}_{\text{input}})(\eta_{\text{pump}})}{\dot{m}} \qquad \text{[U.S.]} \qquad 17.61(b)$$

The *water horsepower* (WHP, also known as the *hydraulic horsepower* and *theoretical horsepower*) is the amount of power actually entering the fluid.

$$\text{WHP} = (\text{HP}_{\text{input}})\,\eta_{\text{pump}} \qquad\qquad 17.62$$

A *turbine* extracts energy from the fluid flowing through it. As with a pump, the energy extraction can be obtained by evaluating the Bernoulli equation on both sides of the turbine and taking the difference. The energy extracted (a positive number) on a per-unit mass basis is given by Eq. 17.63.

$$E_E = E_{t,1} - E_{t,2} \qquad\qquad 17.63$$

19. EXTENDED BERNOULLI EQUATION

The original Bernoulli equation assumes frictionless flow and does not consider the effects of pumps and turbines. When friction is present, and when there are minor losses such as fittings and other energy-related devices in a pipeline, the energy balance is affected. The *extended Bernoulli equation* takes these additional factors into account.

$$(E_p + E_v + E_z)_1 + E_A$$
$$= (E_p + E_v + E_z)_2 + E_E + E_f + E_m \qquad 17.64$$

$$\frac{p_1}{\rho} + \frac{v_1^2}{2} + z_1 g + E_A$$

$$= \frac{p_2}{\rho} + \frac{v_2^2}{2} + z_2 g + E_E + E_f + E_m \quad [\text{SI}] \quad \textit{17.65(a)}$$

$$\frac{p_1}{\rho} + \frac{v_1^2}{2g_c} + \frac{z_1 g}{g_c} + E_A$$

$$= \frac{p_2}{\rho} + \frac{v_2^2}{2g_c} + \frac{z_2 g}{g_c} + E_E + E_f + E_m \quad [\text{U.S.}] \quad \textit{17.65(b)}$$

As defined, E_A, E_E, and E_f are all positive terms. None of the terms in Eq. 17.64 is negative.

The concepts of sources and sinks can be used to decide whether the friction, pump, and turbine terms appear on the left or right side of the Bernoulli equation. An *energy source* puts energy into the system. The incoming fluid and a pump contribute energy to the system. An *energy sink* removes energy from the system. The leaving fluid, friction, and a turbine remove energy from the system. In an energy balance, all energy must be accounted for, and the energy sources just equal the energy sinks.

$$\sum E_{\text{sources}} = \sum E_{\text{sinks}} \qquad \textit{17.66}$$

Therefore, the energy added by a pump always appears on the entrance side of the Bernoulli equation. Similarly, the frictional energy loss always appears on the discharge side.

20. ENERGY AND HYDRAULIC GRADE LINES WITH FRICTION[19]

The *energy grade line* (EGL, also known as *total energy line*) is a graph of the total energy versus position in a pipeline. Since a pitot tube measures total (stagnation) energy, EGL will always coincide with the elevation of a pitot-piezometer fluid column. When friction is present, the EGL will always slope down, in the direction of flow. Figure 17.7 illustrates the EGL for a complex pipe network. The difference between EGL$_{\text{frictionless}}$ and EGL$_{\text{with friction}}$ is the energy loss due to friction.

Notice that the EGL line in Figure 17.7 is discontinuous at point 2, since the friction in pipe section B-C cannot be portrayed without disturbing the spatial correlation of points in the figure. Since the friction loss is proportional to v^2, the slope is steeper when the fluid velocity increases (i.e., when the pipe decreases in flow area), as it does in section D-E. Disregarding air friction, the EGL becomes horizontal at point 6 when the fluid becomes a free jet.

The *hydraulic grade line* (HGL) is a graph of the sum of pressure and potential energies versus position in the pipeline. (That is, the EGL and HGL differ by the kinetic energy.) The HGL will always coincide with

---

[19]The energy grade line and hydraulic grade line were covered in Sec. 16.15 for the frictionless case.

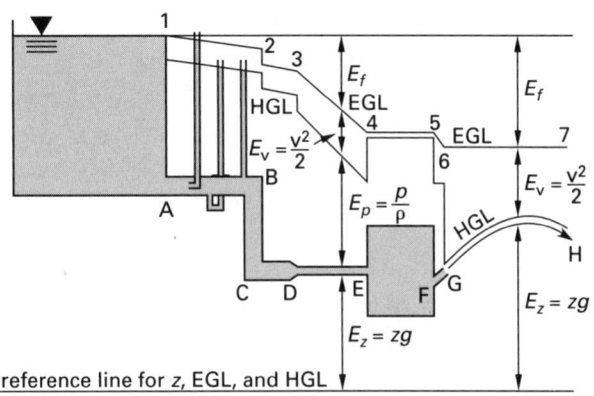

Figure 17.7 *Energy and Hydraulic Grade Lines*

the height of the fluid column in a static piezometer tube. The reference point for elevation is arbitrary, and the pressure energy is usually referenced to atmospheric pressure. Thus, the pressure energy, E_p, for a free jet will be zero, and the HGL will consist only of the potential energy, as shown in section G-H.

The easiest way to draw the energy and hydraulic grade lines is to start with the EGL. The EGL can be drawn simply by recognizing that the rate of divergence from the horizontal EGL$_{\text{frictionless}}$ line is proportional to v^2. Then, since EGL and HGL differ by the velocity head, the HGL can be drawn parallel to the EGL when the pipe diameter is constant. The larger the pipe diameter, the closer the two lines will be.

The EGL for a pump will increase in elevation by E_A across the pump. (The actual energy "path" taken by the fluid is unknown, and a dotted line is used to indicate a lack of knowledge about what really happens in the pump.) The placement of the HGL for a pump will depend on whether the pump increases the fluid velocity and elevation as well as the fluid pressure. In most cases, only the pressure will be increased. Figure 17.8 illustrates the HGL for the case of a pressure increase only.

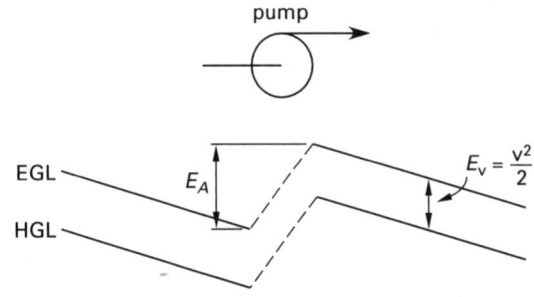

Figure 17.8 *EGL and HGL for a Pump*

The EGL and HGL for minor losses (fittings, contractions, expansions, etc.) are shown in Fig. 17.9.

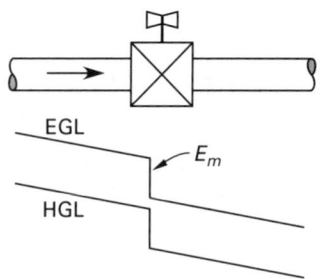

(a) valve, fitting, or obstruction

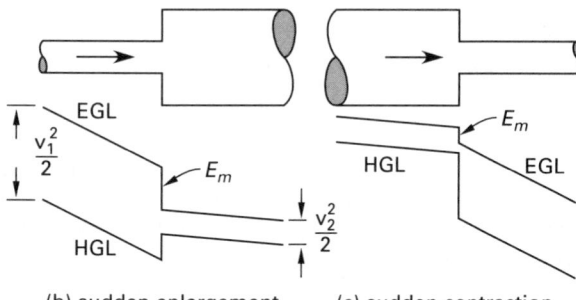

(b) sudden enlargement　　(c) sudden contraction

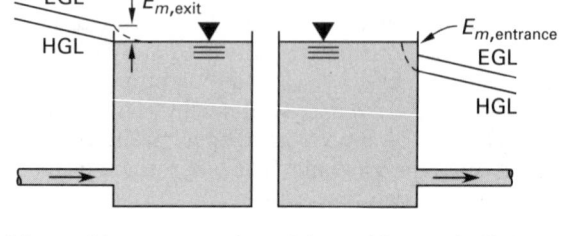

(d) transition to reservoir　　(e) transition to pipeline

Figure 17.9 EGL and HGL for Minor Losses

21. DISCHARGE FROM TANKS

The velocity of a jet issuing from an orifice in a tank can be determined by comparing the total energies at the free fluid surface and the jet itself. At the fluid surface, $p_1 = 0$ (atmospheric) and $v_1 = 0$. (v_1 is known as the *velocity of approach*.) The only energy the fluid has is potential energy. At the jet, $p_2 = 0$. All of the potential energy difference ($z_1 - z_2$) has been converted to kinetic energy. The theoretical velocity of the jet can be derived from the Bernoulli equation. Equation 17.67 is known as the equation for *Torricelli's speed of efflux*.

$$v_t = \sqrt{2gh} \qquad 17.67$$

$$h = z_1 - z_2 \qquad 17.68$$

The actual jet velocity is affected by the orifice geometry. The *coefficient of velocity*, C_v, is an empirical factor that accounts for the friction and turbulence at the orifice. Typical values of C_v are given in Table 17.5.

$$v_o = C_v \sqrt{2gh} \qquad 17.69$$

$$C_v = \frac{\text{actual velocity}}{\text{theoretical velocity}} = \frac{v_o}{v_t} \qquad 17.70$$

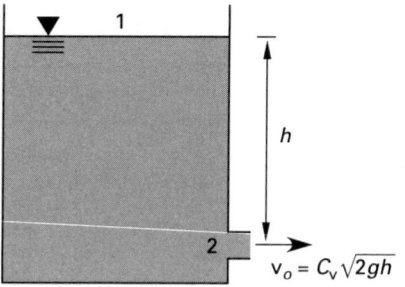

Figure 17.10 Discharge from a Tank

The specific energy loss due to turbulence and friction at the orifice is calculated as a multiple of the jet's kinetic energy.

$$E_f = \left(\frac{1}{C_v^2} - 1\right)\left(\frac{v_o^2}{2}\right) = \left(1 - C_v^2\right) gh \quad \text{[SI]} \qquad 17.71(a)$$

$$E_f = \left(\frac{1}{C_v^2} - 1\right)\left(\frac{v_o^2}{2g_c}\right) = \left(1 - C_v^2\right) h \times \left(\frac{g}{g_c}\right)$$
$$\text{[U.S.]} \qquad 17.71(b)$$

The total head producing discharge (*effective head*) is the difference in elevations that would produce the same velocity from a frictionless orifice.

$$h_{\text{effective}} = C_v^2 h \qquad 17.72$$

The orifice guides quiescent water from the tank into the jet geometry. Unless the orifice is very smooth and the transition is gradual, momentum effects will continue to cause the jet to contract after it has passed through. The velocity calculated from Eq. 17.69 is usually assumed to be the velocity at the *vena contracta*, the section of smallest cross-sectional area.

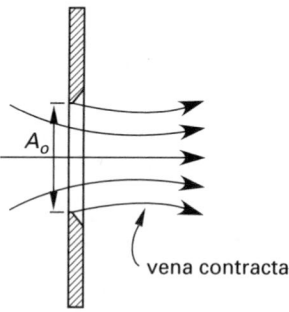

Figure 17.11 Vena Contracta of a Fluid Jet

For a thin plate or sharp-edged orifice, the vena contracta is often assumed to be located approximately one-half an orifice diameter past the orifice, although the distance can vary from $0.3D_o$ to $0.8D_o$. The area of the vena contracta can be calculated from the orifice area and the *coefficient of contraction*, C_c. For water flowing with a high Reynolds number through a small sharp-edged orifice, the contracted area is approximately 61 to 63% of the orifice area.

Table 17.5 *Approximate Orifice Coefficients for Turbulent Water*

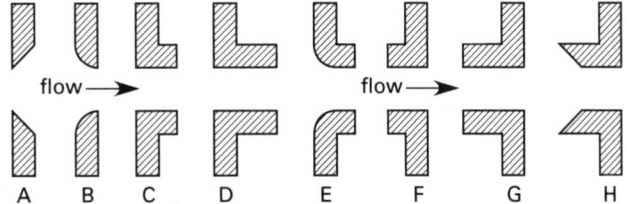

illustration	description	C_d	C_c	C_v
A	sharp-edged	0.62	0.63	0.98
B	round-edged	0.98	1.00	0.98
C	short tube (fluid separates from walls)	0.61	1.00	0.61
D	sharp tube (no separation)	0.82	1.00	0.82
E	sharp tube with rounded entrance	0.97	0.99	0.98
F	reentrant tube, length less than one-half of pipe diameter	0.54	0.55	0.99
G	reentrant tube, length 2 to 3 pipe diameters	0.72	1.00	0.72
H	Borda	0.51	0.52	0.98
(none)	smooth, well-tapered nozzle	0.98	0.99	0.99

$$A_{\text{vena contracta}} = C_c A_o \qquad 17.73$$

$$C_c = \frac{\text{area of vena contracta}}{\text{orifice area}} \qquad 17.74$$

The theoretical discharge rate from a tank is $\dot{V} = A_o\sqrt{2gh}$. However, this relationship needs to be corrected for friction and contraction by multiplying by C_v and C_c. The *coefficient of discharge*, C_d, is the product of the coefficients of velocity and contraction.

$$\dot{V} = C_d v_t A_o = C_d A_o \sqrt{2gh} \qquad 17.75$$

$$\begin{aligned} C_d &= C_v C_c \\ &= \frac{\text{actual discharge}}{\text{theoretical discharge}} \qquad 17.76 \end{aligned}$$

22. DISCHARGE FROM PRESSURIZED TANKS

If the gas or vapor above the liquid in a tank is at gage pressure p, and the discharge is to atmospheric pressure, the head causing discharge will be

$$h = z_1 - z_2 + \frac{p}{\rho g} \qquad \text{[SI]} \qquad 17.77(a)$$

$$h = z_1 - z_2 + \left(\frac{p}{\rho}\right) \times \left(\frac{g_c}{g}\right) \qquad \text{[U.S.]} \qquad 17.77(b)$$

The discharge velocity can be calculated from Eq. 17.69 using the increased discharge head.

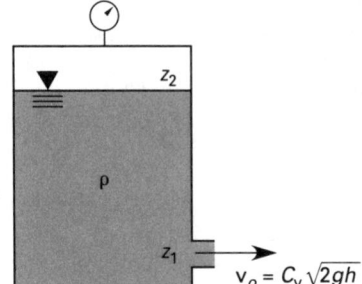

Figure 17.12 *Discharge from a Pressurized Tank*

23. COORDINATES OF A FLUID STREAM

Fluid discharged from an orifice in a tank gets its initial velocity from the conversion of potential energy. After discharge, no additional energy conversion occurs, and all subsequent velocity changes are due to external forces.

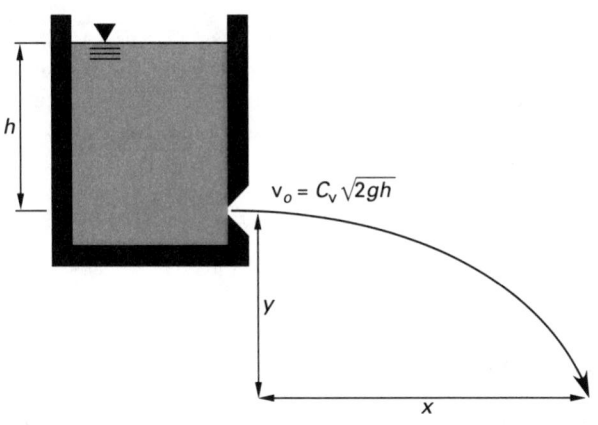

Figure 17.13 *Coordinates of a Fluid Stream*

In the absence of air friction (drag), there are no retarding or accelerating forces in the x-direction on the fluid stream. The x-component of velocity is constant.

$$v_x = v_o \qquad 17.78$$

$$x = v_o t = v_o \sqrt{\frac{2y}{g}} = 2C_v \sqrt{hy} \qquad 17.79$$

After discharge, the fluid stream is acted upon by a constant gravitational acceleration. The y-component of velocity is zero at discharge but increases linearly with time.

$$v_y = gt \qquad 17.80$$

$$y = \frac{gt^2}{2} = \frac{gx^2}{2v_o^2} = \frac{x^2}{4hC_v^2} \qquad 17.81$$

24. TIME TO EMPTY A TANK

If the fluid in an open or vented tank is not replenished at the rate of discharge, the static head forcing discharge through the orifice will decrease with time. If the tank has a varying cross section, A_t, Eq. 17.82 specifies the basic relationship between the change in elevation and elapsed time. (The negative sign indicates that z decreases as t increases.)

$$\dot{V} dt = -A_t dz \qquad 17.82$$

If A_t can be expressed as a function of h, Eq. 17.83 can be used to determine the time to lower the fluid elevation from z_1 to z_2.

$$t = \int_{z_1}^{z_2} \frac{-A_t dz}{C_d A_o \sqrt{2gz}} \qquad 17.83$$

For a tank with a constant cross-sectional area, A_t, the time required to lower the fluid elevation is

$$t = \frac{2A_t(\sqrt{z_1} - \sqrt{z_2})}{C_d A_o \sqrt{2g}} \qquad 17.84$$

If a tank is replenished at a rate of $\dot{V}_{in}$, Eq. 17.85 can be used to calculate the discharge time. If the tank is replenished at a rate greater than the discharge rate, t in Eq. 17.85 will represent the time to raise the fluid level from z_1 to z_2.

$$t = \int_{z_1}^{z_2} \frac{A_t dz}{(C_d A_o \sqrt{2gz}) - \dot{V}_{in}} \qquad 17.85$$

Example 17.8

A tank 15 ft in diameter discharges 150°F water ($\rho = 61.20$ lbm/ft^3) through a sharp-edged 1.0 in diameter orifice ($C_d = 0.62$) in the bottom. The original water depth is 12 ft. The tank is continually pressurized to 50 psig. What is the time to empty the tank?

Solution

The area of the orifice is

$$A_o = \frac{\pi D^2}{4} = \frac{\pi (1 \text{ in})^2}{(4)\left(12 \, \dfrac{\text{in}}{\text{ft}}\right)^2} = 0.00545 \text{ ft}^2$$

The tank area constant with respect to z is

$$A_t = \frac{\pi D^2}{4} = \frac{\pi (15 \text{ ft})^2}{4} = 176.7 \text{ ft}^2$$

The total initial head includes the effect of the pressurization.

$$z_1 = 12 \text{ ft} + \left[\frac{\left(50 \, \dfrac{\text{lbf}}{\text{in}^2}\right)\left(144 \, \dfrac{\text{in}^2}{\text{ft}^2}\right)}{61.2 \, \dfrac{\text{lbm}}{\text{ft}^3}}\right]\left(\frac{32.2 \, \dfrac{\text{lbm-ft}}{\text{lbf-sec}^2}}{32.2 \, \dfrac{\text{ft}}{\text{sec}^2}}\right)$$

$$= 12 \text{ ft} + 117.6 \text{ ft} = 129.6 \text{ ft}$$

When the fluid has reached the level of the orifice, the fluid potential head will be zero, but the pressurization will remain.

$$z_2 = 117.6 \text{ ft}$$

The time to empty the tank is given by Eq. 17.84.

$$t = \frac{(2)(176.7 \text{ ft}^2)\left(\sqrt{129.6 \text{ ft}} - \sqrt{117.6 \text{ ft}}\right)}{(0.62)(0.00545 \text{ ft}^2)\sqrt{(2)\left(32.2 \, \dfrac{\text{ft}}{\text{sec}^2}\right)}}$$

$$= 7036 \text{ sec}$$

25. DISCHARGE FROM LARGE ORIFICES

When an orifice diameter is large compared with the discharge head, the jet velocity at the top edge of the orifice will be less than the velocity at the bottom edge. Since the velocity is related to the square root of the head, the distance used to calculate the effective jet velocity should be measured from the fluid surface to a point above the centerline of the orifice.

This correction is generally neglected, however, since it is small for heads of more than twice the orifice diameter. Furthermore, if an orifice is intended to work regularly with small heads, the orifice should be calibrated in place. The discrepancy can then be absorbed into the discharge coefficient, C_d.

26. CULVERTS

A *culvert* is a water path (usually a large diameter pipe) used to channel water around or through an obstructing feature. In most instances, a culvert is used to restore a natural water path obstructed by a manufactured feature. For example, when a road is built across (perpendicular to) a natural ravine or arroyo, a culvert can be used to channel water under the road.

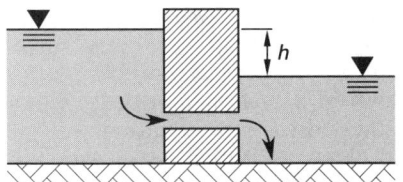

Figure 17.14 *Simple Pipe Culvert*

Because culverts usually operate only partially full and with low heads, Torricelli's equation does not apply. Therefore, most culvert designs are empirical. However, if the entrance and exit of a culvert are both submerged, the culvert will flow full, and the discharge will be independent of the barrel slope. Equation 17.86 can be used to calculate the discharge.

$$\dot{V} = C_d A \sqrt{2gh} \qquad \text{17.86}$$

If the culvert is long (more than 60 ft or 20 m), or if the entrance is not gradual, the available energy will be divided between friction and velocity heads. The effective head used in Eq. 17.86 should be

$$h_{\text{effective}} = h - h_{f,\text{barrel}} - h_{m,\text{entrance}} \qquad \text{17.87}$$

The friction loss in the barrel can be found in the usual manner, from either the Darcy equation or the Hazen-Williams equation. The entrance loss is calculated using the standard method of loss coefficients. Representative values of the loss coefficient, K, are given in Table 17.6. Since the fluid velocity is not initially known but is needed to find the friction factor, a trial-and-error solution will be necessary.

Table 17.6 *Representative Loss Coefficients for Culvert Entrances*

entrance	K
smooth and gradual transition	0.08
flush vee or bell shape	0.10
projecting vee or bell shape	0.15
flush, square-edged	0.50
projecting, square-edged	0.90

27. SIPHONS

A *siphon* is a bent or curved tube that carries fluid from a container at a high elevation to another container at a lower elevation. Normally, it would not seem difficult to have a fluid flow to a lower elevation. However, the fluid seems to flow "uphill" in a siphon. Figure 17.15 illustrates a siphon.

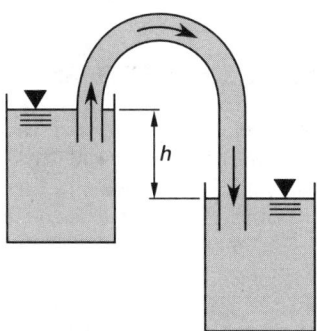

Figure 17.15 *Siphon*

Starting a siphon requires the tube to be completely filled with liquid. Then, since the fluid weight is greater in the longer arm than in the shorter arm, the fluid in the longer arm "falls" out of the siphon, "pulling" more liquid into the shorter arm and over the bend.

Operation of a siphon is essentially independent of atmospheric pressure. The theoretical discharge is the same as predicted by the Torricelli equation. A correction for discharge is necessary, but little data is available on typical values of C_d. Therefore, siphons should be tested and calibrated in place.

$$\dot{V} = C_d A \text{v} = C_d A \sqrt{2gh} \qquad \text{17.88}$$

28. SERIES PIPE SYSTEMS

A system of pipes in series consists of two or more lengths of different-diameter pipes connected together. In the case of the series pipe from a reservoir discharging to the atmosphere shown in Fig. 17.16, the available head will be split between the velocity head and the friction loss.

$$h = h_{\text{v}} + h_f \qquad \text{17.89}$$

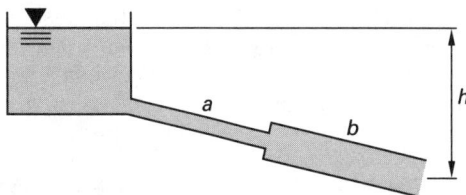

Figure 17.16 *Series Pipe System*

If the flow rate or velocity in any part of the system is known, the friction loss can easily be found as the sum of the friction losses in the individual sections. The solution is somewhat more simple than first appears, since the velocity of all sections can be written in terms of only one velocity.

$$h_{f,t} = h_{f,a} + h_{f,b} \qquad \text{17.90}$$
$$A_a \text{v}_a = A_b \text{v}_b \qquad \text{17.91}$$

If neither the velocity nor the flow quantity are known, a trial-and-error solution will be required, since a friction factor must be known to calculate h_f. A good starting point is to assume fully turbulent flow.[20]

29. PARALLEL PIPE SYSTEMS

A *pipe loop* is a set of two pipes placed in parallel, both originating and terminating at the same junction. Adding a second pipe in parallel to a first is a standard method of increasing the capacity of a line.

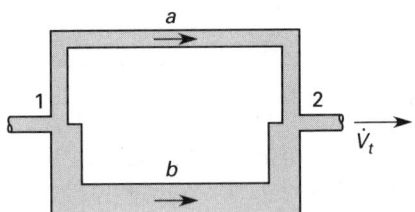

Figure 17.17 *Parallel Pipe System*

There are three principles that govern the distribution of flow between the two branches:

- The flow divides in such a manner as to make the head loss in each branch the same.

$$h_{f,a} = h_{f,b} \qquad 17.92$$

- The head loss between the two junctions is the same as the head loss in each branch.

$$h_{f,1-2} = h_{f,a} \qquad 17.93$$

- The total flow rate is the sum of the flow rates in the two branches.

$$\dot{V}_t = \dot{V}_a + \dot{V}_b \qquad 17.94$$

If the pipe diameters are known, Eq. 17.92 through Eq. 17.94 can be solved simultaneously for the branch velocities. In such problems, it is common to neglect minor losses, the velocity head, and the variation in the friction factor, f, with velocity.

30. OTHER PIPE BRANCHING PROBLEMS

In the *three-reservoir problem*, there are many possible choices for the unknown quantity (pipe length, diameter, head, flow rate, etc.). In all but the simplest cases, the solution technique is by trial and error based on conservation of mass and energy.[21]

---

[20]If the friction loss is calculated by the Hazen-Williams formula, the solution will be considerably more difficult.
[21]Systematic solution methods for the many variations of this problem are given in the *Civil Engineering Reference Manual* by Michael R. Lindeburg (Professional Publications, Inc.).

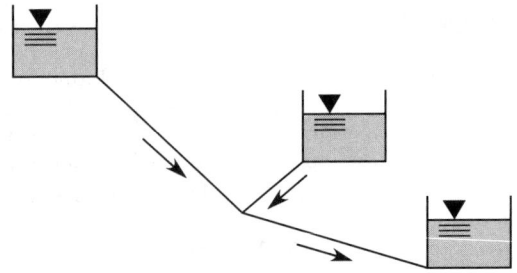

Figure 17.18 *Three-Reservoir System*

Network flows in a *multiloop system* cannot be determined by any closed-form equation. Most real-world problems involving multiloop systems are analyzed on a computer. Computer programs are based on the *Hardy-Cross method*, which can also be performed manually when there are only a few loops. In this method, flows in all of the branches are first assumed, and adjustments are made in consecutive iterations to the assumed flow. The Hardy-Cross method is presented in greater detail in other sources.[22]

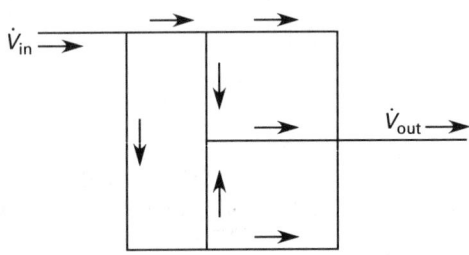

Figure 17.19 *Multiloop System*

31. FLOW MEASURING DEVICES

A device that measures flow can be calibrated to indicate either velocity or volumetric flow rate. There are many methods available to obtain the flow rate. Some are complicated, requiring the use of transducers and solid-state electronics, and others can be evaluated using the Bernoulli equation. Some are more appropriate for one variety of fluid than others, and some are limited to specific ranges of temperature and pressure.

Table 17.7 categorizes a few common flow measurement methods. Many other methods and variations thereof exist, particularly for specialized industries. Some of the methods listed are so basic that only a passing mention will be made of them. Others, particularly those that can be analyzed with energy and mass conservation laws, will be covered in greater detail in subsequent sections.

---

[22]Ibid.

The utility meters used to measure gas and water usage are examples of *displacement meters*. Such devices are cyclical, fixed-volume devices with counters to record the numbers of cycles. Displacement devices are generally unpowered, drawing on only the pressure energy to overcome mechanical friction. Most configurations for positive-displacement pumps (e.g., reciprocating piston, helical screw, and nutating disk) have also been converted to measurement devices.

The venturi nozzle, orifice plate, and flow nozzle are examples of *obstruction meters*. These devices rely on a decrease in static pressure to measure the flow velocity. One disadvantage of these devices is that the pressure drop is proportional to the square of the velocity, limiting the range over which any particular device can be used.

An obstruction meter that somewhat overcomes the velocity range limitation is the *variable-area meter*, also known as a *rotameter*, illustrated in Fig. 17.20.[23] This device consists of a float (which is actually more dense than the fluid) and a transparent sight tube. With proper design, the effects of fluid density and viscosity can be minimized. The sight glass can be directly calibrated in volumetric flow rate, or the height of the float above the zero position can be used in a volumetric calculation.

Table 17.7 *Flow Measuring Devices*

I.   direct (primary) measurements
        positive-displacement meters
        weight and mass scales
        volume tanks

II. indirect (secondary) measurements
        obstruction meters
            – venturi meters
            – orifice plate meters
            – flow nozzles
            – variable-area meters
        velocity probes
            – static pressure probes
            – direction sensing probes
            – pitot tubes
            – pitot-static meters
        miscellaneous methods
            – turbine and propeller meters
            – hot-wire meters
            – magnetic flow meters
            – sonic flow meters
            – mass flow meters

---

[23]The rotameter has its own disadvantages, however. It must be installed vertically; the fluid cannot be opaque; and it is more difficult to manufacture for use with high-temperature, high-pressure fluids.

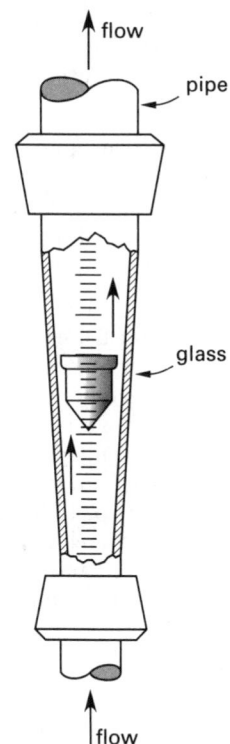

Figure 17.20 *Variable-Area Rotameter*

It is necessary to be able to measure static pressures in order to use obstruction meters and pitot-static tubes. In some cases, a *static pressure probe* is used. Figure 17.21 illustrates a simplified static pressure probe. In practice, such probes are sensitive to burrs and irregularities in the tap openings, orientation to the flow (i.e., *yaw*), and interaction with the pipe walls and other probes. A *direction-sensing probe* overcomes some of these problems.

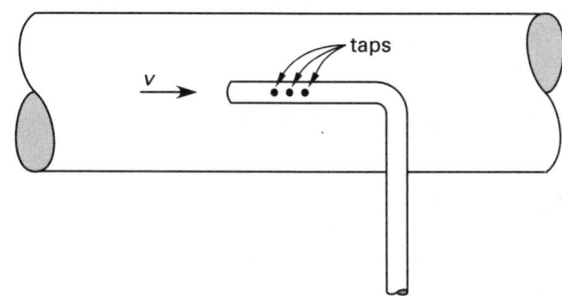

Figure 17.21 *Simple Static Pressure Probe*

A weather station *anemometer* used to measure wind velocity is an example of a simple *turbine meter*. Similar devices are used to measure the speed of a stream or river, in which case the name *current meter* may be used. Turbine meters are further divided into cup-type meters and propeller-type meters, depending on the orientation of the turbine axis relative to the flow direction. (The turbine axis and flow direction are parallel for propeller-type meters; they are perpendicular for cup-type meters.) Since the wheel motion is proportional to the flow velocity, the velocity is determined by counting the number of revolutions made by the wheel per unit time.

A more sophisticated turbine flowmeter uses a reluctance-type pickup coil to detect wheel motion. The permeability of a magnetic circuit changes each time a wheel blade passes the pole of a permanent magnet in the meter body. This change is detected to indicate velocity or flow rate.

A *hot-wire anemometer* measures velocity by determining the cooling effect of fluid (usually a gas) flowing over an electrically heated tungsten, platinum, or nickel wire. Cooling is primarily by convection; radiation and conduction are neglected. Circuitry can be used either to keep the current constant (in which case, the changing resistance is measured) or to keep the temperature constant (in which case, the changing current is measured). Additional circuitry can be used to compensate for thermal lag if the velocity changes rapidly.

A voltage proportional to the velocity will be generated when a conductor passes through a magnetic field.[24] This characteristic can be used to measure flow velocity if the fluid is electrically conductive. *Magnetic flowmeters* are ideal for measuring the flow of liquid metals, but variations of the device can also be used when the fluid is only slightly conductive. In some cases, precise quantities of conductive ions can be added to the fluid to permit measurement by this method.

In an *ultrasonic flowmeter*, two electric or magnetic transducers are placed a short distance apart on the outside of the pipe. One transducer serves as a transmitter of ultrasonic waves; the other transducer is a receiver. As an ultrasonic wave travels from the transmitter to the receiver, its velocity will be increased (or decreased) by the relative motion of the fluid. The phase shift between the fluid-carried waves and the waves passing through a stationary medium can be measured and converted to fluid velocity.

32. PITOT-STATIC GAUGE

Measurements from pitot tubes are used to determine total (stagnation) energy. Piezometer tubes and wall taps are used to measure static pressure energy. The difference between the total and static energies is the kinetic energy of the flow. Figure 17.22 illustrates a comparative method of directly measuring the velocity head for an incompressible fluid.

$$\frac{v^2}{2} = \frac{p_t - p_s}{\rho} = hg \qquad \text{[SI]} \qquad 17.95(a)$$

$$\frac{v^2}{2g_c} = \frac{p_t - p_s}{\rho} = h \times \left(\frac{g}{g_c}\right) \quad \text{[U.S.]} \quad 17.95(b)$$

$$v = \sqrt{2gh} \qquad\qquad 17.96$$

---

[24]The magnitude of this induced voltage is predicted by *Faraday's law*.

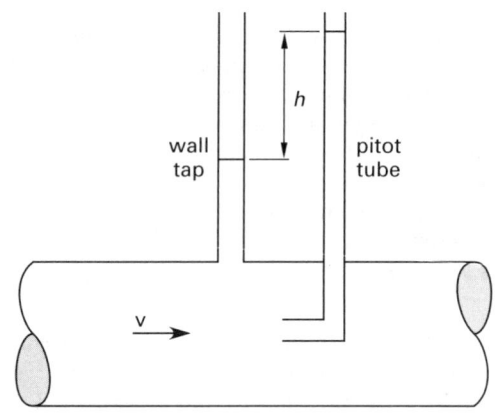

Figure 17.22 *Comparative Velocity Head Measurement*

The pitot tube and static pressure tap shown in Fig. 17.22 can be combined into a *pitot-static gauge*. In a pitot-static gauge, one end of the manometer is acted upon by the static pressure, also referred to as *transverse pressure*. The other end of the manometer experiences the total pressure. The difference in elevations of the manometer fluid columns is the velocity head. This distance must be corrected if the density of the flowing fluid is significant.

$$\frac{v^2}{2} = \frac{p_t - p_s}{\rho} = \frac{h(\rho_m - \rho)g}{\rho} \qquad \text{[SI]} \qquad 17.97(a)$$

$$\frac{v^2}{2g_c} = \frac{p_t - p_s}{\rho} = \frac{h(\rho_m - \rho)}{\rho} \times \left(\frac{g}{g_c}\right) \quad \text{[U.S.]} \quad 17.97(b)$$

$$v = \sqrt{\frac{2gh(\rho_m - \rho)}{\rho}} \qquad\qquad 17.98$$

Another correction, which is seldom made, is to multiply the velocity calculated from Eq. 17.98 by C_I, the *coefficient of the instrument*. Since the flow past the pitot-static tube is slightly faster than the free-fluid velocity, the static pressure measured will be slightly lower than the true value. This makes the indicated velocity slightly higher than the true value. C_I, a number close to but less than 1.0, corrects for this.

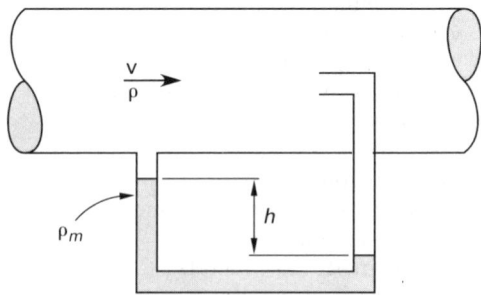

Figure 17.23 *Pitot-Static Gauge*

A pitot-static tube indicates the velocity at only one point in a pipe. If the flow is laminar, and if the pitot-static tube is in the center of the pipe, v_{max} will be determined. The average velocity, however, will be only half the maximum value.

Pitot tube measurements are sensitive to the condition of the opening and errors in installation alignment. The *yaw angle* (i.e., the acute angle between the pitot tube axis and the flow streamline) should be zero.

Example 17.9

Water ($\rho = 1000$ kg/m^3; 62.4 lbm/ft^3) is flowing through a pipe. A pitot-static gage registers 3.0 in (0.076 m) of mercury. What is the velocity of the water in the pipe?

SI Solution

The density of mercury is $\rho = 13\,580$ kg/m^3. The velocity can be calculated directly from Eq. 17.98.

$$v = \sqrt{\frac{(2)\left(9.81\ \frac{m}{s^2}\right)(0.076\ m)\left(13\,580 - 1000\ \frac{kg}{m^3}\right)}{1000\ \frac{kg}{m^3}}}$$

$$= 4.33\ m/s$$

Customary U.S. Solution

The density of mercury is $\rho = 848.6$ lbm/ft^3.

From Eq. 17.98,

$$v = \sqrt{\frac{(2)\left(32.2\ \frac{ft}{sec^2}\right)\left(\frac{3}{12}\ ft\right)\left(848.6 - 62.4\ \frac{lbm}{ft^3}\right)}{62.4\ \frac{lbm}{ft^3}}}$$

$$= 14.24\ ft/sec$$

33. VENTURI METER

Figure 17.24 illustrates a simple *venturi*. (Sometimes the venturi is called a *converging-diverging nozzle*.) This flow measuring device can be inserted directly into a pipeline. Since the diameter changes are gradual, there is very little friction loss.[25] Static pressure measurements are taken at the throat and upstream of the diameter change. These measurements are traditionally made by manometer.

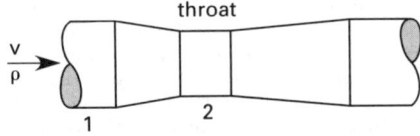

Figure 17.24 *Venturi Meter*

---

[25]The actual friction loss is approximately 10% of the pressure difference $p_1 - p_2$.

The analysis of *venturi meter* performance is relatively simple. The traditional derivation of upstream velocity starts by assuming a horizontal orientation and frictionless, incompressible, and turbulent flow. Then, the Bernoulli equation is written for points 1 and 2. Equation 17.99 shows that the static pressure decreases as the velocity increases. This is known as the *venturi effect*.

$$\frac{v_1^2}{2} + \frac{p_1}{\rho} = \frac{v_2^2}{2} + \frac{p_2}{\rho} \qquad \text{[SI]} \quad \textit{17.99(a)}$$

$$\frac{v_1^2}{2g_c} + \frac{p_1}{\rho} = \frac{v_2^2}{2g_c} + \frac{p_2}{\rho} \qquad \text{[U.S.]} \quad \textit{17.99(b)}$$

The two velocities are related by the continuity equation.

$$A_1 v_1 = A_2 v_2 \qquad \textit{17.100}$$

Combining Eqs. 17.99 and 17.100 and eliminating the unknown v_1 produces an expression for the throat velocity. Also, a *coefficient of velocity* is used to account for the small effect of friction. (C_v is very close to 1.0, usually 0.98 or 0.99.)

$$v_2 = C_v v_{2,\text{ideal}}$$

$$= \left[\frac{C_v}{\sqrt{1 - \left(\frac{A_2}{A_1}\right)^2}}\right]\sqrt{\frac{2(p_1 - p_2)}{\rho}} \qquad \text{[SI]} \quad \textit{17.101(a)}$$

$$v_2 = \left[\frac{C_v}{\sqrt{1 - \left(\frac{A_2}{A_1}\right)^2}}\right]\sqrt{\frac{2g_c(p_1 - p_2)}{\rho}} \qquad \text{[U.S.]} \quad \textit{17.101(b)}$$

The *velocity of approach factor*, F_{va}, also known as the *meter constant*, is the reciprocal of the denominator of Eq. 17.101. The *beta ratio* can be incorporated into the formula for F_{va}.

$$\beta = \frac{D_2}{D_1} \qquad \textit{17.102}$$

$$F_{va} = \frac{1}{\sqrt{1 - \left(\frac{A_2}{A_1}\right)^2}} = \frac{1}{\sqrt{1 - \beta^4}} \qquad \textit{17.103}$$

If a manometer is used to measure the pressure difference directly, Eq. 17.101 can be rewritten in terms of the manometer fluid reading.

$$v_2 = \left(\frac{C_v}{\sqrt{1 - \beta^4}}\right)\sqrt{\frac{2g(\rho_m - \rho)h}{\rho}}$$

$$= C_v F_{va}\sqrt{\frac{2g(\rho_m - \rho)h}{\rho}} \qquad \textit{17.104}$$

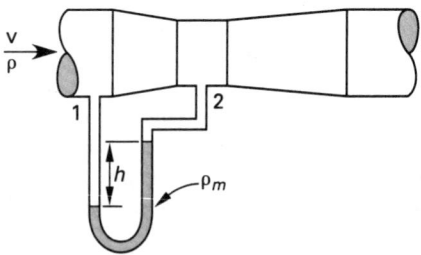

Figure 17.25 *Venturi Meter with Manometer*

The flow rate through a venturi meter can be calculated from the throat area. There is an insignificant amount of contraction of the flow as it passes through the throat, and the *coefficient of contraction* is seldom encountered in venturi meter work. However, the *coefficient of discharge* ($C_d = C_c C_v$) is quoted, nevertheless. Values of C_d range from slightly less than 0.90 to over 0.99, depending on the Reynolds number. C_d is seldom less than 0.95 for turbulent flow.

$$\dot{V} = C_d A_2 \text{v}_{2,\text{ideal}} \qquad 17.105$$

The product $C_d F_{\text{va}}$ is known as the *coefficient of flow* or *flow coefficient*, not to be confused with the coefficient of discharge.[26] This factor is used for convenience, since it combines the losses with the meter constant.

$$C_f = C_d F_{\text{va}} = \frac{C_d}{\sqrt{1 - \beta^4}} \qquad 17.106$$

$$\dot{V} = C_f A_2 \sqrt{\frac{2g(\rho_m - \rho)h}{\rho}} \qquad 17.107$$

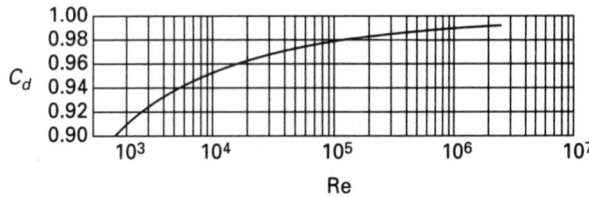

Figure 17.26 *Typical Venturi Meter Discharge Coefficients*

34. ORIFICE METER

The *orifice meter* (or *orifice plate*) is used more frequently than the venturi meter to measure flow rates in small pipes. It consists of a thin or sharp-edged plate with a central, round hole through which the fluid flows. Such a plate is easily clamped between two flanges in an existing pipeline.

While (for small pipes) the orifice meter may consist of a thin plate without significant thickness, various types of bevels and rounded edges are also used with thicker

---

[26]Some writers use the symbol K for the flow coefficient.

plates. There is no significant difference in the analysis procedure between "flat plate," "sharp-edged," or "square-edged" orifice meters. Any effect that the orifice edges have is accounted for in the discharge and flow coefficient correlations. Similarly, the direction of the bevel will affect the coefficients, but not the analysis method.

As with the venturi meter, pressure taps are used to obtain the static pressure upstream of the orifice plate and at the *vena contracta* (i.e., at the point of minimum pressure).[27] A differential manometer connected to the two taps conveniently indicates the difference in static pressures.

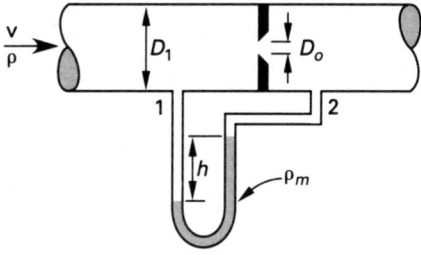

Figure 17.27 *Orifice Meter with Differential Manometer*

The derivation of the governing equations for an orifice meter is similar to that of the venturi meter. (The obvious falsity of assuming frictionless flow through the orifice is corrected by the coefficient of discharge.) The major difference is that the coefficient of contraction is taken into consideration in writing the mass continuity equation, since the pressure is measured at the vena contracta, not at the orifice.

$$A_2 = C_c A_o \qquad 17.108$$

$$\text{v}_o = \left[\frac{C_{\text{v}}}{\sqrt{1 - \left(\frac{C_c A_o}{A_1}\right)^2}} \right] \sqrt{\frac{2(p_1 - p_2)}{\rho}} \quad \text{[SI]} \quad 17.109(a)$$

---

[27]Calibration of the orifice meter is sensitive to tap placement. Upstream taps are placed between one-half and two pipe diameters upstream from the orifice. (An upstream distance of one pipe diameter is often quoted and used.) There are three tap-placement options: flange, vena contracta, and standardized. Flange taps are used with prefabricated orifice meters that are inserted (by flange bolting) in pipes. If the location of the vena contracta is known, a tap can be placed there. However, the location of the vena contracta depends on the diameter ratio $\beta = D_o/D$ and varies from approximately 0.4 to 0.7 pipe diameters downstream. Due to the difficulty of locating the vena contracta, the standardized $1D$-$^1/_2 D$ configuration is often used. The upstream tap is one diameter before the orifice; the downstream tap is one-half diameter after the orifice. Since approaching flow should be stable and uniform, care must be taken not to install the orifice meter less than approximately five diameters after a bend or elbow.

$$v_o = \left[\frac{C_v}{\sqrt{1 - \left(\dfrac{C_c A_o}{A_1} \right)^2}} \right] \sqrt{\frac{2g_c(p_1 - p_2)}{\rho}} \quad \text{[U.S.]} \quad \textit{17.109(b)}$$

If a manometer is used to indicate the differential pressure $p_1 - p_2$, the velocity at the vena contracta can be calculated from Eq. 17.110.

$$v_o = \left[\frac{C_v}{\sqrt{1 - \left(\dfrac{C_c A_o}{A_1} \right)^2}} \right] \sqrt{\frac{2g(\rho_m - \rho)h}{\rho}} \quad \textit{17.110}$$

Although the orifice meter is simpler and less expensive than a venturi meter, its discharge coefficient is much less than that of a venturi meter. C_d usually ranges from 0.55 to 0.75, with values of 0.60 and 0.61 often being quoted. (The coefficient of contraction has a large effect, since $C_d = C_v C_c$.) Also, its pressure recovery is poor (i.e., there is a permanent pressure reduction), and it is susceptible to inaccuracies from wear and abrasion.[28]

The velocity of approach factor, F_{va}, for an orifice meter is defined differently than for a venturi meter, since it takes into consideration the contraction of the flow. However, the velocity of approach factor is still combined with the coefficient of discharge into the flow coefficient, C_f. Figure 17.28 illustrates how the flow coefficient varies with the area ratio and the Reynolds number.

$$F_{va} = \frac{1}{\sqrt{1 - \left(\dfrac{C_c A_o}{A_1} \right)^2}} \quad \textit{17.111}$$

$$C_f = C_d F_{va} \quad \textit{17.112}$$

The flow rate through an orifice meter is given by Eq. 17.113.

$$\dot{V} = C_f A_o \sqrt{\frac{2g(\rho_m - \rho)h}{\rho}} \quad \textit{17.113}$$

---

[28]The actual permanent pressure loss varies from 40 to 90% of the differential pressure. The loss depends on the diameter ratio $\beta = D_o/D_1$, and for turbulent flow is not particularly sensitive to the Reynolds number. For $\beta = 0.5$, the loss is 73% of the measured pressure difference, $p_1 - p_2$. This decreases to approximately 56% of the pressure difference when $\beta = 0.65$ and to 38% when $\beta = 0.8$. For any diameter ratio, the pressure drop coefficient, K, in multiples of the orifice velocity head is

$$K = \frac{1 - \beta^2}{(C_f)^2}$$

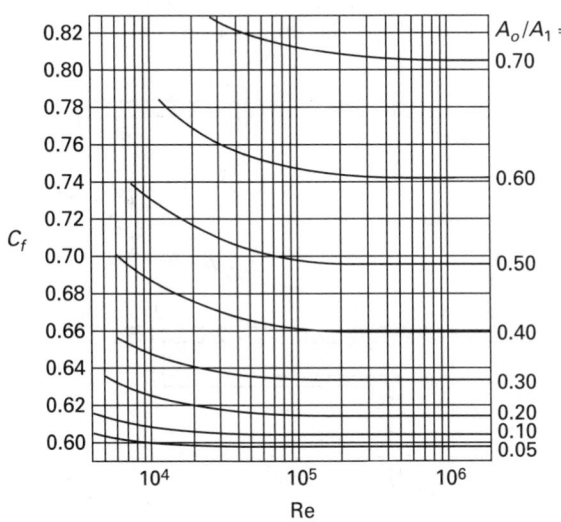

Figure 17.28 *Typical Flow Coefficients for Orifice Plates*

Example 17.10

150°F water ($\rho = 61.2 \, \text{lbm/ft}^3$) flows in an 8 in schedule-40 steel pipe at the rate of 2.23 ft³/sec. A sharp-edged orifice with a 7 in diameter hole is placed in the line. A mercury differential manometer is used to record the pressure difference. If the orifice has a flow coefficient, C_f, of 0.62, what deflection (in inches) of mercury is observed? (Mercury has a density of 848.6 lbm/ft³.)

Solution

The orifice area is

$$A_o = \frac{\pi}{4} D_o^2 = \left(\frac{\pi}{4} \right) \left(\frac{7}{12} \right)^2$$

$$= 0.2673 \, \text{ft}^2$$

Equation 17.113 is solved for h.

$$h = \frac{\dot{V}^2 \rho}{2g C_f^2 A_o^2 (\rho_m - \rho)}$$

$$= \frac{\left(2.23 \, \dfrac{\text{ft}^3}{\text{sec}} \right)^2 \left(61.2 \, \dfrac{\text{lbm}}{\text{ft}^3} \right) \left(12 \, \dfrac{\text{in}}{\text{ft}} \right)}{(2) \left(32.2 \, \dfrac{\text{ft}}{\text{sec}^2} \right) (0.62)^2} \times (0.2673 \, \text{ft}^2)^2 \left(848.6 - 61.2 \, \dfrac{\text{lbm}}{\text{ft}^3} \right)}$$

$$= 2.62 \, \text{in}$$

35. FLOW NOZZLE

A typical flow nozzle is illustrated in Fig. 17.29. This device consists only of a converging section. It is somewhat between an orifice plate and a venturi meter in performance, possessing some of the advantages and disadvantages of each. The venturi performance equations can be used for the flow nozzle.

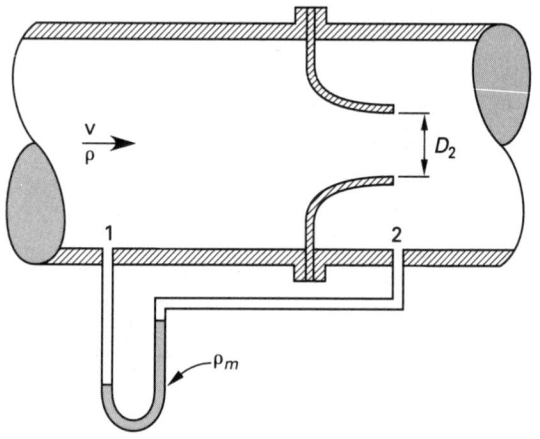

Figure 17.29 *Flow Nozzle*

The geometry of the nozzle entrance is chosen to prevent separation of the fluid from the wall. The converging portion and the subsequent parallel section keep the coefficients of velocity and contraction close to 1.0. However, the absence of a diffuser section disrupts the orderly return of fluid to its original condition. The permanent pressure drop is more similar to that of the orifice meter than the venturi meter.

Since the nozzle geometry greatly affects the performance, values of C_d and C_f have been established for only a limited number of specific proprietary nozzles.[29]

36. FLOW MEASUREMENTS OF COMPRESSIBLE FLUIDS

Volume measurements of compressible fluids (i.e., gases) are not very meaningful. The volume of a gas will depend on its temperature and pressure. For that reason, flow quantities of gases discharged should be stated as mass flow rates.

$$\dot{m} = \rho_2 A_2 v_2 \qquad 17.114$$

Equation 17.114 requires that the velocity and area be measured at the same point. Even more important is that the density must be measured at that point, as well. However, it is common practice in flow measurement work to use the density of the upstream fluid at position 1. (Note that this is not the stagnation density.)

The significant error introduced by this simplification is corrected by the use of an *expansion factor*, Y. For venturi meters and flow nozzles, values of the expansion factor are generally calculated theoretical values.

---

[29]Some of the proprietary nozzles for which detailed performance data exist are the ASME long-radius nozzle (low-β and high-β series) and the International Standards Association (ISA) nozzle (German standard nozzle).

Values of Y are determined experimentally for orifice plates.

$$\dot{m} = Y \rho_1 A_2 v_2 \qquad 17.115$$

Derivation of the theoretical formula for the expansion factor for venturi meters and flow nozzles is based on thermodynamic principles and an assumption of adiabatic flow.

$$Y = \sqrt{\frac{(1-\beta^4)\left[\left(\dfrac{p_2}{p_1}\right)^{\frac{2}{k}} - \left(\dfrac{p_2}{p_1}\right)^{\frac{k+1}{k}}\right]}{\left(\dfrac{k-1}{k}\right)\left(1 - \dfrac{p_2}{p_1}\right)\left[1 - \beta^4\left(\dfrac{p_2}{p_1}\right)^{\frac{2}{k}}\right]}} \qquad 17.116$$

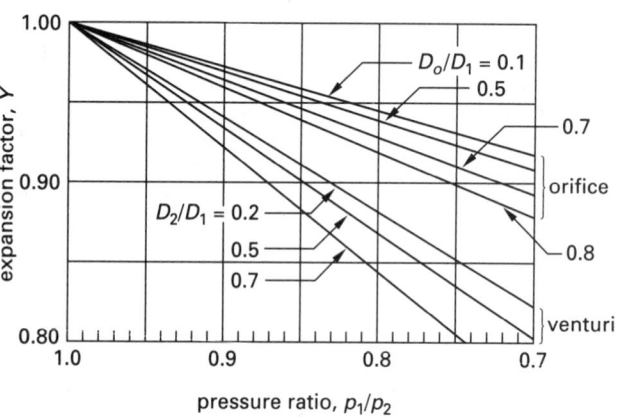

Reprinted from *Transport Processes: Momentum, Heat, and Mass*, by Christie J. Geankoplis with permission of Prentice Hall, Inc., ©1983.

Figure 17.30 *Approximate Expansion Factors*

Once the expansion factor is known, it can be used with the standard flow rate ($\dot{V}$) equations for venturi meters, flow nozzles, and orifice meters. For example, for a venturi meter, the mass flow rate would be calculated from Eq. 17.117.

$$\dot{m} = \left(\frac{C_d A_2 Y}{\sqrt{1-\beta^4}}\right)\sqrt{2\rho_1(p_1 - p_2)} \qquad \text{[SI]} \qquad 17.117(a)$$

$$\dot{m} = \left(\frac{C_d A_2 Y}{\sqrt{1-\beta^4}}\right)\sqrt{2g_c\rho_1(p_1 - p_2)} \qquad \text{[U.S.]} \qquad 17.117(b)$$

37. IMPULSE-MOMENTUM PRINCIPLE

(The convention of this section is to make F and x positive when they are directed toward the right. F and y are positive when directed upward. Also, the fluid is assumed to flow horizontally from left to right, and it has no initial y-component of velocity.)

The *momentum*, **P** (also known as *linear momentum* to distinguish it from *angular momentum*, which is not considered here), of a moving object is a vector quantity defined as the product of the object's mass and velocity.[30]

$$\mathbf{P} = m\mathbf{v} \qquad \text{[SI]} \qquad 17.118(a)$$

$$\mathbf{P} = \frac{m\mathbf{v}}{g_c} \qquad \text{[U.S.]} \qquad 17.118(b)$$

The *impulse*, I, of a constant force is calculated as the product of the force's magnitude and the length of time the force is applied.

$$\mathbf{I} = \mathbf{F}\Delta t \qquad 17.119$$

The *impulse-momentum principle* states that the impulse applied to a body is equal to the change in momentum. (This is also known as the *law of conservation of momentum*, even though fluid momentum is not always conserved.) Equation 17.120 is one way of stating Newton's second law.

$$\mathbf{I} = \Delta \mathbf{P} \qquad 17.120$$

$$F\Delta t = m\Delta v = m(v_2 - v_1) \qquad \text{[SI]} \qquad 17.121(a)$$

$$F\Delta t = \frac{m\Delta v}{g_c} = \frac{m(v_2 - v_1)}{g_c} \qquad \text{[U.S.]} \qquad 17.121(b)$$

For fluid flow, there is a mass flow rate, $\dot{m}$, but no mass per se. Since $\dot{m} = m/\Delta t$, the impulse-momentum equation can be rewritten as follows.

$$F = \dot{m}\Delta v \qquad \text{[SI]} \qquad 17.122(a)$$

$$F = \frac{\dot{m}\Delta v}{g_c} \qquad \text{[U.S.]} \qquad 17.122(b)$$

Equation 17.122 calculates the constant force required to accelerate or retard a fluid stream. This would occur when fluid enters a reduced or enlarged flow area. If the flow area decreases, for example, the fluid will be accelerated by a wall force up to the new velocity. Ultimately, this force must be resisted by the pipe supports.

As Eq. 17.122 illustrates, fluid momentum is not always conserved, since it is generated by the external force, F. Examples of external forces are gravity (considered zero for horizontal pipes), gage pressure, friction, and turning forces from walls and vanes. Only if these external forces are absent is fluid momentum conserved.

Since force is a vector, it can be resolved into its x- and y-components of force.

$$F_x = \dot{m}\Delta v_x \qquad 17.123$$

$$F_y = \dot{m}\Delta v_y \qquad 17.124$$

---

[30]The symbol B is also used for momentum. In many texts, however, momentum is given no symbol at all.

If the flow is initially at velocity v but is directed through an angle θ with respect to the original direction, the x- and y-components of velocity can be calculated from Eqs. 17.125 and 17.126.

$$\Delta v_x = v(\cos\theta - 1) \qquad 17.125$$

$$\Delta v_y = v\sin\theta \qquad 17.126$$

Since F and v are vector quantities and Δt and m are scalars, F must have the same direction as $v_2 - v_1$. This provides an intuitive method of determining the direction in which the force acts. Essentially, one needs to ask, "In which direction must the force act in order to push the fluid stream into its new direction?" (The force, F, is the force on the fluid. The force on the pipe walls or pipe supports has the same magnitude but is opposite in direction.)

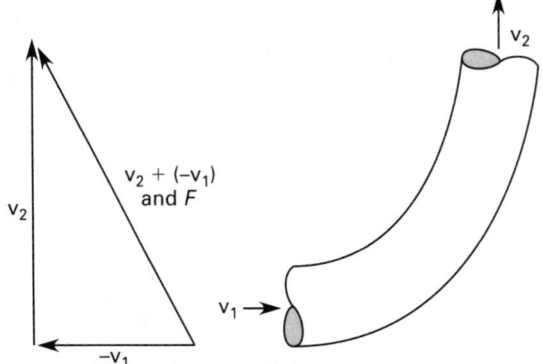

Figure 17.31 *Force on a Confined Fluid Stream*

If a jet is completely stopped by a flat plate placed perpendicular to its flow, then $\theta = 90°$ and $\Delta v_x = -v$. If a jet is turned around so that it ends up returning to where it originated, then $\theta = 180°$ and $\Delta v_x = -2v$. Notice that a positive Δv indicates an increase in velocity. A negative Δv indicates a decrease in velocity.

38. JET PROPULSION

A basic application of the impulse-momentum principle is the analysis of jet propulsion. Air enters a jet engine and is mixed with a small amount of jet fuel. The air and fuel mixture is compressed and ignited, and the exhaust products leave the engine at a greater velocity than was possessed by the original air. The change in momentum of the air produces a force on the engine.

The governing equation for a jet engine is Eq. 17.127. In the special case of VTOL (vertical takeoff and landing) aircraft, there will also be a y-component of force. The mass of the jet fuel is small compared with the air mass, and the fuel mass is commonly disregarded.

$$F_x = \dot{m}\,(v_2 - v_1) \qquad 17.127$$

$$F_x = \dot{V}_2\rho_2 v_2 - \dot{V}_1\rho_1 v_1 \qquad \text{[SI]} \qquad 17.128(a)$$

$$F_x = \frac{\dot{V}_2\rho_2 v_2 - \dot{V}_1\rho_1 v_1}{g_c} \qquad \text{[U.S.]} \qquad 17.128(b)$$

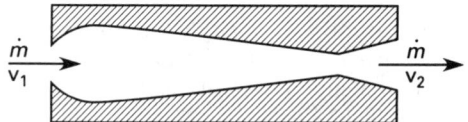

Figure 17.32 *Jet Engine*

39. OPEN JET ON A VERTICAL FLAT PLATE

Figure 17.33 illustrates an open jet on a vertical flat plate. The fluid approaches the plate with no vertical component of velocity; it leaves the plate with no horizontal component of velocity. (This is another way of saying there is no splash-back.) Thus, all of the velocity in the x-direction is canceled. (The minus sign in Eq. 17.129 indicates that the force is opposite the initial velocity direction.)

$$\Delta \mathrm{v} = -\mathrm{v} \qquad \qquad 17.129$$

$$F_x = -\dot{m}\mathrm{v} \qquad \text{[SI]} \quad 17.130(a)$$

$$F_x = \frac{-\dot{m}\mathrm{v}}{g_c} \qquad \text{[U.S.]} \quad 17.130(b)$$

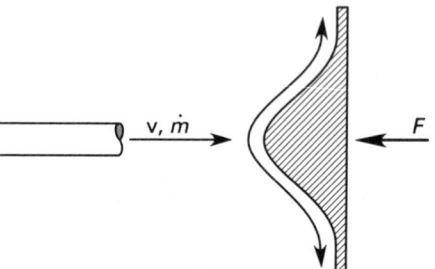

Figure 17.33 *Jet on a Vertical Plate*

Since the flow is divided, half going up and half going down, the net velocity change in the y-direction is zero. There is no force in the y-direction on the fluid.

40. OPEN JET ON A HORIZONTAL FLAT PLATE

If a jet of fluid is directed upward, its velocity will decrease due to the effect of gravity. The force exerted on the fluid by the plate will depend on the fluid velocity at the plate surface, v_y, not the original jet velocity, v_o. All of this velocity is canceled. Since the flow divides evenly in both horizontal directions ($\Delta \mathrm{v}_x = 0$), there is no force component in the x-direction.

$$\mathrm{v}_y = \sqrt{\mathrm{v}_o^2 - 2gh} \qquad \qquad 17.131$$

$$\Delta \mathrm{v}_y = -\sqrt{\mathrm{v}_o^2 - 2gh} \qquad \qquad 17.132$$

$$F_y = -\dot{m}\sqrt{\mathrm{v}_o^2 - 2gh} \qquad \text{[SI]} \quad 17.133(a)$$

$$F_y = \frac{-\dot{m}\sqrt{\mathrm{v}_o^2 - 2gh}}{g_c} \qquad \text{[U.S.]} \quad 17.133(b)$$

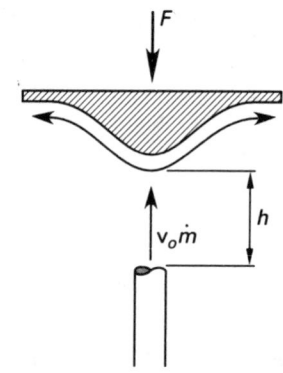

Figure 17.34 *Open Jet on a Horizontal Plate*

41. OPEN JET ON AN INCLINED PLATE

An open jet will be diverted both up and down a stationary, inclined plate, as shown in Fig. 17.35. In the absence of friction, the velocity in each diverted flow will be v, the same as in the approaching jet. The fractions f_1 and f_2 of the jet that are diverted up and down can be found from Eq. 17.134 through Eq. 17.137.

$$f_1 = \frac{1 + \cos\theta}{2} \qquad \qquad 17.134$$

$$f_2 = \frac{1 - \cos\theta}{2} \qquad \qquad 17.135$$

$$f_1 - f_2 = \cos\theta \qquad \qquad 17.136$$

$$f_1 + f_2 = 1.0 \qquad \qquad 17.137$$

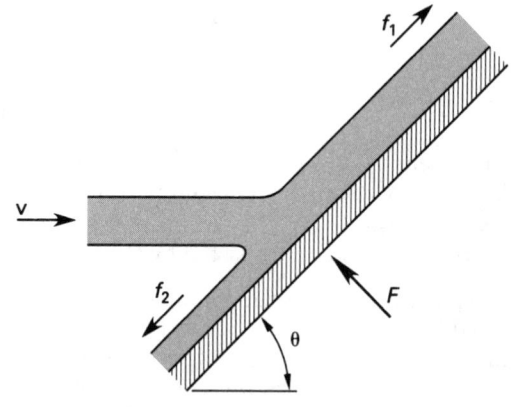

Figure 17.35 *Open Jet on an Inclined Plate*

If the flow along the plate is frictionless, there will be no force component parallel to the plate. The force perpendicular to the plate is given by Eq. 17.138.

$$F = \dot{m}V\sin\theta \qquad \text{[SI]} \quad 17.138(a)$$

$$F = \frac{\dot{m}V\sin\theta}{g_c} \qquad \text{[U.S.]} \quad 17.138(b)$$

42. OPEN JET ON A SINGLE STATIONARY BLADE

Figure 17.36 illustrates a fluid jet being turned through an angle θ by a stationary blade (also called a *vane*). It is common to assume that $|v_2| = |v_1|$, although this will not be strictly true if friction between the blade and fluid is considered. Since the fluid is both retarded (in the x-direction) and accelerated (in the y-direction), there will be two components of force on the fluid.

$$\Delta v_x = v_2 \cos\theta - v_1 \qquad 17.139$$

$$\Delta v_y = v_2 \sin\theta \qquad 17.140$$

$$F_x = \dot{m}(v_2 \cos\theta - v_1) \qquad \text{[SI]} \quad 17.141(a)$$

$$F_x = \frac{\dot{m}(v_2 \cos\theta - v_1)}{g_c} \qquad \text{[U.S.]} \quad 17.141(b)$$

$$F_y = \dot{m}v_2 \sin\theta \qquad \text{[SI]} \quad 17.142(a)$$

$$F_y = \frac{\dot{m}v_2 \sin\theta}{g_c} \qquad \text{[U.S.]} \quad 17.142(b)$$

$$F = \sqrt{F_x^2 + F_y^2} \qquad 17.143$$

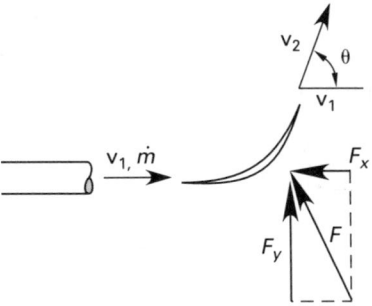

Figure 17.36 *Open Jet on a Stationary Blade*

43. OPEN JET ON A SINGLE MOVING BLADE

If a blade is moving away at velocity v_b from the source of the fluid jet, only the *relative velocity difference* between the jet and blade produces a momentum change. Furthermore, not all of the fluid jet overtakes the moving blade. The equations used for the single stationary blade can be used by substituting $(v - v_b)$ for v and by using the effective mass flow rate, m_{eff}.

$$\Delta v_x = (v - v_b)(\cos\theta - 1) \qquad 17.144$$

$$\Delta v_y = (v - v_b)\sin\theta \qquad 17.145$$

$$\dot{m}_{eff} = \left(\frac{v - v_b}{v}\right)\dot{m} \qquad 17.146$$

$$F_x = \dot{m}_{eff}(v - v_b)(\cos\theta - 1) \qquad \text{[SI]} \quad 17.147(a)$$

$$F_x = \frac{\dot{m}_{eff}(v - v_b)(\cos\theta - 1)}{g_c} \qquad \text{[U.S.]} \quad 17.147(b)$$

$$F_y = \dot{m}_{eff}(v - v_b)\sin\theta \qquad \text{[SI]} \quad 17.148(a)$$

$$F_y = \frac{\dot{m}_{eff}(v - v_b)\sin\theta}{g_c} \qquad \text{[U.S.]} \quad 17.148(b)$$

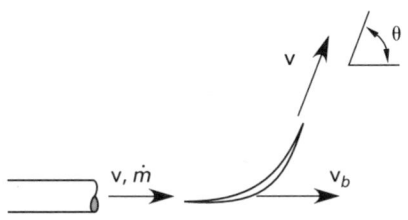

Figure 17.37 *Open Jet on a Moving Blade*

44. OPEN JET ON A MULTIPLE-BLADED WHEEL

An *impulse turbine* consists of a series of blades (buckets or vanes) mounted around a wheel. The tangential velocity of the blades is approximately parallel to the jet. The effective mass flow rate, $\dot{m}_{eff}$, used in calculating the reaction force is the full discharge rate, since when one blade moves away from the jet, other blades will have moved into position. Thus, all of the fluid discharged is captured by the blades. Equations 17.147 and 17.148 are applicable if the total flow rate is used.

$$v_b = \frac{\text{rpm} \times 2\pi r}{60} = \omega r \qquad 17.149$$

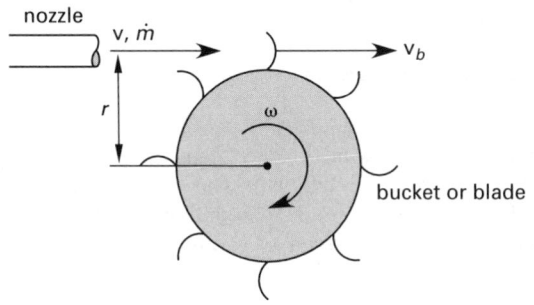

Figure 17.38 *Impulse Turbine*

45. IMPULSE TURBINE POWER

The total power potential of a fluid jet can be calculated from the kinetic energy of the jet and the mass flow rate.[31] (This neglects the pressure energy, which is small by comparison.)

$$P_{jet} = \frac{\dot{m}v^2}{2} \qquad \text{[SI]} \quad 17.150(a)$$

$$P_{jet} = \frac{\dot{m}v^2}{2g_c} \qquad \text{[U.S.]} \quad 17.150(b)$$

The power transferred from a fluid jet to the blades of a turbine is calculated from the x-component of force on the blades. The y-component of force does no work.

[31]The full jet discharge is used in this section. If only a single blade is involved, the effective mass flow rate, $\dot{m}_{eff}$, must be used.

$$P = F_x v_b \qquad\qquad 17.151$$

$$P = \dot{m} v_b \, (v - v_b)(1 - \cos\theta) \qquad \text{[SI]} \quad 17.152(a)$$

$$P = \frac{\dot{m} v_b \, (v - v_b)(1 - \cos\theta)}{g_c} \qquad \text{[U.S.]} \; 17.152(b)$$

The maximum theoretical blade velocity is the velocity of the jet: $v_b = v$. This is known as the *runaway speed* and can only occur when the turbine is unloaded. If Eq. 17.152 is maximized with respect to v_b, however, the maximum power will be found to occur when the blade is traveling at half of the jet velocity: $v_b = v/2$. The power (force) is also affected by the deflection angle of the blade. Power is maximized when $\theta = 180°$. Figure 17.39 illustrates the relationship between power and the variables θ and v_b.

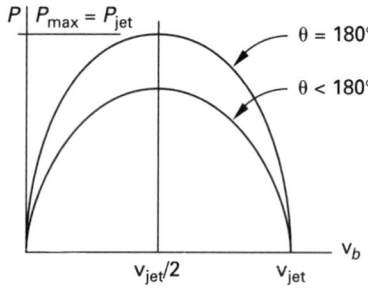

Figure 17.39 Turbine Power

Putting $\theta = 180°$ and $v_b = v/2$ into Eq. 17.152 results in $P_{\max} = \dot{m} v^2/2$, which is the same as P_{jet} in Eq. 17.146. If the machine is 100% efficient, 100% of the jet power can be transferred to the machine.

46. CONFINED STREAMS IN PIPE BENDS

As presented in Sec. 37, momentum can also be changed by pressure forces. Such is the case when fluid enters a pipe fitting or bend. Since the fluid is confined, the forces due to static pressure must be included in the analysis. (The effects of gravity and friction are neglected.)

$$F_x = p_2 A_2 \, \cos\theta - p_1 A_1 + \dot{m}(v_2 \, \cos\theta - v_1)$$
$$\text{[SI]} \quad 17.153(a)$$

$$F_x = p_2 A_2 \, \cos\theta - p_1 A_1 + \frac{\dot{m}(v_2 \, \cos\theta - v_1)}{g_c}$$
$$\text{[U.S.]} \; 17.153(b)$$

$$F_y = (p_2 A_2 + \dot{m} v_2) \, \sin\theta \qquad \text{[SI]} \quad 17.154(a)$$

$$F_y = \left(p_2 A_2 + \frac{\dot{m} v_2}{g_c} \right) \sin\theta \qquad \text{[U.S.]} \; 17.154(b)$$

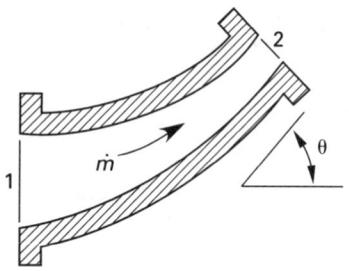

Figure 17.40 Pipe Bend

Example 17.11

60°F water ($\rho = 62.4 \; \text{lbm/ft}^3$) at 40 psig enters a 12 in × 8 in reducing elbow at 8 ft/sec and is turned through an angle of 30°. (a) What is the resultant force exerted on the water by the elbow? (b) What other forces should be considered in the design of supports for the fitting?

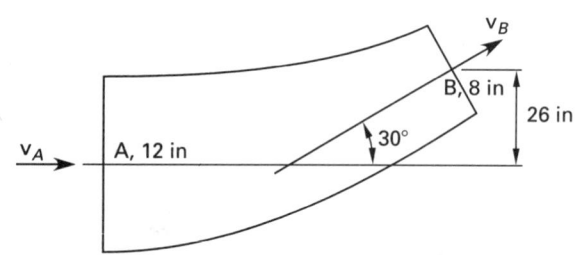

Solution

(a) The velocity and pressure at point B are both needed. The velocity is easily calculated from the continuity equation.

$$A_A = \frac{\pi}{4} D_A^2 = \left(\frac{\pi}{4}\right) \left(\frac{12 \text{ in}}{12 \, \frac{\text{in}}{\text{ft}}}\right)^2 = 0.7854 \text{ ft}^2$$

$$A_B = \left(\frac{\pi}{4}\right) \left(\frac{8}{12}\right)^2 = 0.3491 \text{ ft}^2$$

$$v_B = \frac{v_A A_A}{A_B}$$

$$= \left(8 \, \frac{\text{ft}}{\text{sec}}\right) \left(\frac{0.7854 \text{ ft}^2}{0.3491 \text{ ft}^2}\right) = 18 \text{ ft/sec}$$

$$p_A = \left(40 \, \frac{\text{lbf}}{\text{in}^2}\right) \left(144 \, \frac{\text{in}^2}{\text{ft}^2}\right) = 5760 \text{ lbf/ft}^2$$

The Bernoulli equation is used to calculate p_B. (Notice that gage pressures are used. Absolute pressures could also be used, but the addition of p_{atm}/ρ to both sides of the Bernoulli equation would not affect p_B.)

$$\frac{5760 \; \frac{\text{lbf}}{\text{ft}^2}}{62.4 \; \frac{\text{lbm}}{\text{ft}^3}} + \frac{\left(8 \; \frac{\text{ft}}{\text{sec}}\right)^2}{(2)\left(32.2 \; \frac{\text{lbm-ft}}{\text{lbf-sec}^2}\right)}$$

$$= \frac{p_B}{62.4} + \frac{(18)^2}{(2)(32.2)} + \left(\frac{26 \; \text{in}}{12 \; \frac{\text{in}}{\text{ft}}}\right) \times \left(\frac{g}{g_c}\right)$$

$$p_B = 5373 \; \text{lbf/ft}^2$$

The mass flow rate is

$$\dot{m} = \dot{V}\rho = \text{v}A\rho$$

$$= \left(8 \; \frac{\text{ft}}{\text{sec}}\right)(0.7854 \; \text{ft}^2)\left(62.4 \; \frac{\text{lbm}}{\text{ft}^3}\right)$$

$$= 392.1 \; \text{lbm/sec}$$

From Eq. 17.153,

$$F_x = \left(5373 \; \frac{\text{lbf}}{\text{ft}^2}\right)(0.3491 \; \text{ft}^2)\cos 30°$$

$$- \left(5760 \; \frac{\text{lbf}}{\text{ft}^2}\right)(0.7854 \; \text{ft}^2)$$

$$+ \frac{\left(392.1 \; \frac{\text{lbm}}{\text{sec}}\right)\left[\left(18 \; \frac{\text{ft}}{\text{sec}}\right)\cos 30° - 8 \; \frac{\text{ft}}{\text{sec}}\right]}{32.2 \; \frac{\text{lbm-ft}}{\text{lbf-sec}^2}}$$

$$= -2755 \; \text{lbf}$$

From Eq. 17.154,

$$F_y = \left[(5373)(0.3491) + \frac{(392.1)(18)}{32.2}\right]\sin 30°$$

$$= 1047 \; \text{lbf}$$

The resultant force on the water is

$$R = \sqrt{F_x^2 + F_y^2} = \sqrt{(-2755 \; \text{lbf})^2 + (1047 \; \text{lbf})^2}$$

$$= 2947 \; \text{lbf}$$

(b) In addition to counteracting the resultant force, R, the support should be designed to carry the weight of the elbow and the water in it. Also, the support must carry a part of the pipe and water weight tributary to the elbow.

47. WATER HAMMER

Water hammer in a long pipe is an increase in fluid pressure caused by a sudden velocity decrease. The sudden velocity decrease will usually be caused by a valve closing. Analysis of the water hammer phenomenon can take two approaches, depending on whether or not the pipe material is assumed to be elastic.

If the pipe material is assumed to be inelastic (i.e., rigid pipe), the time required for the water hammer shock wave to travel from the suddenly closed valve to a point of interest depends only on the velocity of sound in the fluid (a) and the distance (L) between the two points. This is also the time required to bring all of the fluid in the pipe to rest.

$$t = \frac{L}{a} \qquad\qquad 17.155$$

When the water hammer shock wave reaches the original source of water, the pressure wave will dissipate. A rarefaction wave (at the pressure of the water source) will return at velocity a to the valve. The time for the compression shock wave to travel to the source and the rarefaction wave to return to the valve is given by Eq. 17.156. This is also the length of time that the pressure is constant at the valve.

$$t = \frac{2L}{a} \qquad\qquad 17.156$$

The fluid pressure increase resulting from the shock wave is calculated by equating the kinetic energy change of the fluid with the average pressure during the compression process. The pressure increase is independent of the length of pipe. If the velocity is decreased by an amount Δv instantaneously, the increase in pressure will be

$$\Delta p = \rho a \Delta\text{v} \qquad \text{[SI]} \quad 17.157(a)$$

$$\Delta p = \frac{\rho a \Delta\text{v}}{g_c} \qquad \text{[U.S.]} \quad 17.157(b)$$

It is interesting that the pressure increase at the valve depends on Δv but not on the actual length of time it takes to close the valve. Therefore, there is no difference in pressure buildups at the valve for an "instantaneous closure," "rapid closure," or "sudden closure."[32] It is only necessary for the closure to occur rapidly.

Having a very long pipe is equivalent to assuming an instantaneous closure. When the pipe is long, the time for the shock wave to travel round-trip is much longer than the time to close the valve. Thus, the valve will be closed when the rarefaction wave returns to the valve.

If the pipe is short, it will be difficult to close the valve before the rarefaction wave returns to the valve. With a short pipe, the pressure buildup will be less than is predicted by Eq. 17.157. (Having a short pipe is equivalent to the case of "slow closure.") The actual pressure history is complex, and no simple method exists for calculating the pressure buildup in short pipes.

---

[32]The pressure elsewhere along the pipe, however, will be lower for slow closures than for instantaneous closures.

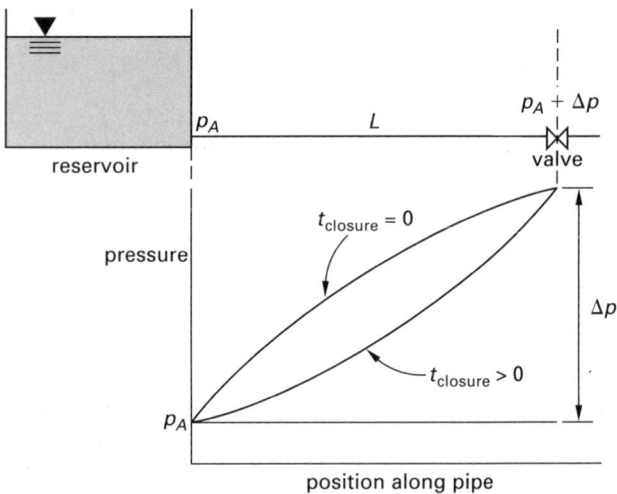

Figure 17.41 *Water Hammer*

Installing a *surge tank, accumulator, slow-closing valve,* or *pressure-relief valve* in the line will protect against water hammer damage. The surge tank (or *surge chamber*) is an open tank or reservoir. Since the water is unconfined, large pressure buildups do not occur. An accumulator is a closed tank that is partially filled with air. Since the air is much more compressible than the water, it will be compressed by the water hammer shock wave. The energy of the shock wave is dissipated when the air is compressed.

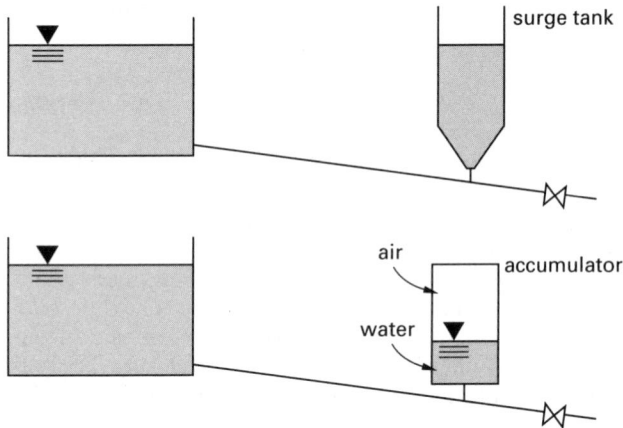

Figure 17.42 *Water Hammer Protective Devices*

If the pipe material is elastic (the typical assumption for steel and plastic), the previous analysis of water hammer effects (t and Δp) is still valid. However, the calculation of the speed of sound in water must account for the elasticity of the pipe material. This is accomplished by using Eq. 17.158 for the modulus of elasticity (bulk modulus) when calculating the speed of sound, a. (In Eq. 17.158, t is the pipe wall thickness, and D is the pipe inside diameter.)

$$E = \frac{E_{water}\, t_{pipe}\, E_{pipe}}{t_{pipe}\, E_{pipe} + D_{pipe} E_{water}} \qquad 17.158$$

Example 17.12

Water ($\rho = 1000$ kg/m^3, $E = 2 \times 10^9$ Pa) is flowing at 4 m/s through a long length of 4 in schedule-40 steel pipe ($D_i = 0.102$ m, $t = 0.00602$ m, $E = 2 \times 10^{11}$ Pa) when a valve suddenly closes completely. What is the theoretical increase in pressure?

Solution

From Eq. 17.158, the modulus of elasticity to be used in calculating the speed of sound is

$$E = \frac{(2 \times 10^9 \text{ Pa})(0.00602 \text{ m})(2 \times 10^{11} \text{ Pa})}{\substack{(0.00602 \text{ m})(2 \times 10^{11} \text{ Pa}) \\ +(0.102 \text{ m})(2 \times 10^9 \text{ Pa})}}$$

$$= 1.71 \times 10^9 \text{ Pa}$$

The speed of sound in the pipe is

$$a = \sqrt{\frac{E}{\rho}} = \sqrt{\frac{1.71 \times 10^9 \text{ Pa}}{1000 \, \dfrac{\text{kg}}{\text{m}^2}}}$$

$$= 1308 \text{ m/s}$$

From Eq. 17.155, the pressure increase is

$$\Delta p = \rho a \Delta \text{v} = \left(1000 \, \frac{\text{kg}}{\text{m}^3}\right)\left(1308 \, \frac{\text{m}}{\text{s}}\right)\left(4 \, \frac{\text{m}}{\text{s}}\right)$$

$$= 5.23 \times 10^6 \text{ Pa}$$

48. LIFT

Lift is an upward force that is exerted on an object (flat plate, airfoil, rotating cylinder, etc.) as the object passes through a fluid. Lift combines with drag to form the resultant force on the object, as shown in Fig. 17.43.

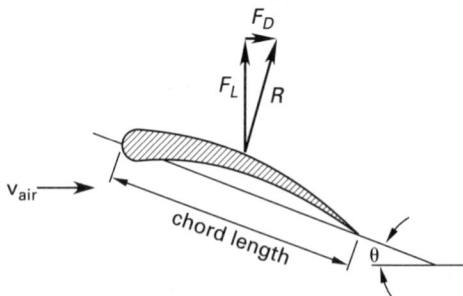

Figure 17.43 *Lift and Drag on an Airfoil*

The generation of lift from air flowing over an airfoil is predicted by Bernoulli's equation. Air molecules must travel a longer distance over the top surface of the airfoil than over the lower surface, and, therefore, they travel faster over the top surface. Since the total energy of the air is constant, the increase in kinetic energy comes at the expense of pressure energy. The static pressure on the top of the airfoil is reduced, and a net upward force is produced.

Within practical limits, the lift produced can be increased at lower speeds by increasing the curvature of the wing. This increased curvature is achieved by the use of *flaps*. When a plane is traveling slowly (e.g., during take-off or landing), its flaps are extended to create the lift needed.

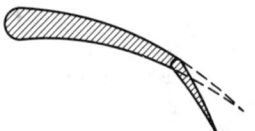

Figure 17.44 *Use of Flaps in an Airfoil*

The lift produced can be calculated from Eq. 17.159, whose use is not limited to airfoils.

$$F_L = \frac{C_L A \rho \mathrm{v}^2}{2} \qquad \text{[SI]} \qquad 17.159(a)$$

$$F_L = \frac{C_L A \rho \mathrm{v}^2}{2g_c} \qquad \text{[U.S.]} \qquad 17.159(b)$$

The dimensions of an airfoil or wing are frequently given in terms of chord length and aspect ratio. The *chord length* is the front-to-back dimension of the airfoil. The *aspect ratio* is the ratio of the *span* (wing length) to chord length. The area, A, in Eq. 17.159 is the airfoil's area projected onto the plane of the chord. Thus, for a rectangular airfoil, $A = \text{chord} \times \text{span}$.

The dimensionless *coefficient of lift*, C_L, is used to measure the effectiveness of the airfoil. The coefficient of lift depends on the shape of the airfoil and the Reynolds number. No simple relationship can be given for calculating the coefficient of lift for airfoils, but the theoretical coefficient of lift for a thin plate in two-dimensional flow at a low angle of attack θ is given by Eq. 17.160. Actual airfoils are able to achieve only 80 to 90% of this theoretical value.

$$C_L = 2\pi \sin \theta \qquad 17.160$$

The coefficient of lift for an airfoil cannot be increased without limit merely by increasing θ. Eventually, the *stall angle* is reached, at which point the coefficient of lift decreases dramatically.

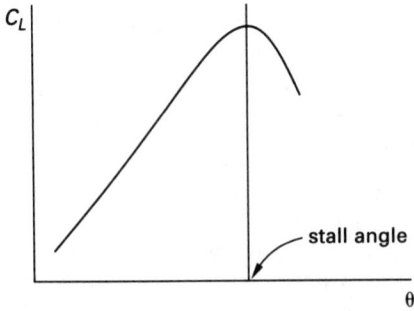

Figure 17.45 *Typical Plot of Lift Coefficient*

49. CIRCULATION

A theoretical concept for calculating the lift generated by an object (an airfoil, propeller, turbine blade, etc.) is *circulation*. Circulation, Γ, is defined by Eq. 17.161.[33] Its units are length²/time.

$$\Gamma = \oint \mathrm{v} \cos \theta \, dl \qquad 17.161$$

Figure 17.46 illustrates an arbitrary closed curve drawn around a point (or body) in steady flow. The tangential components of velocity, v, at all points around the curve are $\mathrm{v} \cos \theta$. It is a fundamental theorem that circulation has the same value for every closed curve that can be drawn around a body.

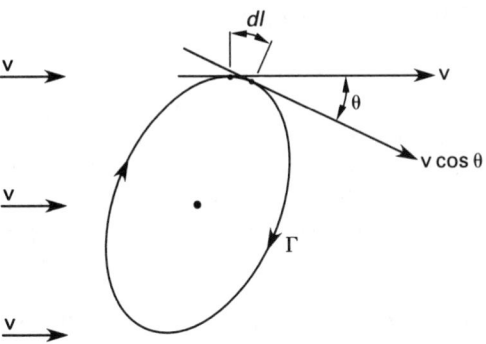

Figure 17.46 *Circulation Around a Point*

Lift on a body traveling with relative velocity v through a fluid of density ρ can be calculated from the circulation by using Eq. 17.162.[34]

$$F_L = \rho \, \mathrm{v}\Gamma \times \text{chord length} \qquad 17.162$$

There is no actual circulation of air "around" an airfoil, but this mathematical concept can be used, nevertheless. However, since the flow of air "around" an airfoil is not symmetrical in path or velocity, experimental determination of C_L is favored over theoretical calculations of circulation.

50. LIFT FROM ROTATING CYLINDERS

When a cylinder is placed transversely to an airflow traveling at velocity v_∞, the velocity at a point on the surface of the cylinder is $2\mathrm{v}_\infty \sin \theta$. Since the flow is symmetrical, however, no lift is produced.

---

[33]Equation 17.161 is analogous to the calculation of work being done by a constant force moving around a curve. If the force makes an angle of θ with the direction of motion, the work done as the force moves a distance dl around a curve is $W = \oint F \cos \theta \, dl$. In calculating circulation, velocity takes the place of force.

[34]U is the traditional symbol of velocity in circulation studies.

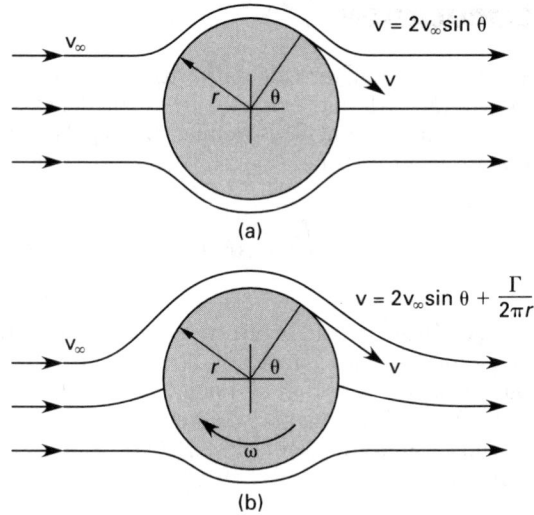

Figure 17.47 *Flow Over a Cylinder*

If the cylinder with radius r is rotating at ω rad/sec while it moves with a relative velocity v_∞ through the air, the *Kutta-Joukowsky result* (theorem) can be used to calculate the lift per unit length of cylinder.[35] This is known as the *Magnus effect*.

$$F_L \text{ (per unit length)} = \rho v_\infty \Gamma \quad \text{[SI]} \quad 17.163(a)$$

$$F_L = \frac{\rho v_\infty \Gamma}{g_c} \quad \text{[U.S.]} \quad 17.163(b)$$

$$\Gamma = 2\pi r^2 \omega \quad 17.164$$

Equation 17.163 assumes that there is no slip (i.e., that the air drawn around the cylinder by rotation moves at ω), and in that ideal case, the maximum coefficient of lift is 4π. Practical rotating devices, however, seldom achieve a coefficient of lift in excess of 9 or 10, and even then, the power expenditure is excessive.

51. DRAG

Drag is a frictional force that acts parallel but opposite to the direction of motion. It combines with the lift (acting perpendicular to the direction of motion) to produce a resultant force on the object. The total drag force is made up of *skin friction* and *pressure drag* (also known as *form drag*). These components, in turn, can be subdivided and categorized into *wake drag*, *induced drag*, and *profile drag*.

---

[35] A similar analysis can be used to explain why a pitched baseball curves. The rotation of the ball produces a force that changes the path of the ball as it travels.

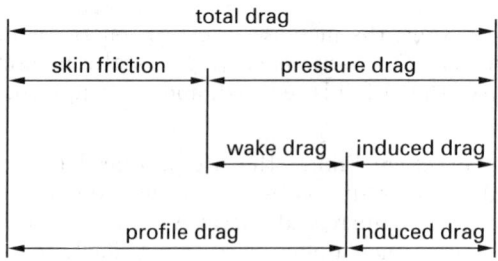

Figure 17.48 *Components of Total Drag*

Most aeronautical engineering books contain descriptions of these drag terms. However, the difference between the situations where either skin friction drag or pressure drag predominates is illustrated in Fig. 17.49.

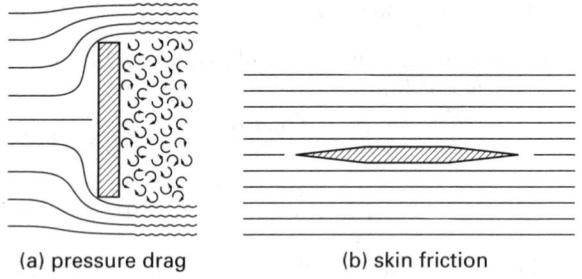

Figure 17.49 *Extreme Cases of Pressure Drag and Skin Friction*

Total drag is most easily calculated from the dimensionless *drag coefficient*, C_D. It can be shown by dimensional analysis that the drag coefficient depends only on the Reynolds number.

$$F_D = \frac{C_D A \rho v^2}{2} \quad \text{[SI]} \quad 17.165(a)$$

$$F_D = \frac{C_D A \rho v^2}{2g_c} \quad \text{[U.S.]} \quad 17.165(b)$$

In most cases, the area, A, in Eq. 17.165 is the projected area (i.e., the *frontal area*) normal to the stream. This is appropriate for spheres, cylinders, and automobiles. In a few cases (e.g., for airfoils and flat plates), the area is a projection of the object onto a plane parallel to the stream.

Typical drag coefficients for production cars vary from approximately 0.35 to approximately 0.55, with most modern cars being nearer the lower end. By comparison, other low-speed drag coefficients are approximately 0.05 (aircraft wing), 0.10 (sphere in turbulent flow), and 1.2 (flat plate).

Aero horsepower is a term used by automobile manufacturers to designate the power required to move a car horizontally at 50 mi/hr (80.5 km/h) against the drag force. Aero horsepower varies from approximately 7 hp (5.2 kW) for a streamlined subcompact car to approximately 100 hp (75 kW) for a box-shaped truck.

52. DRAG ON SPHERES AND DISKS

The drag coefficient varies linearly with the Reynolds number for laminar flow around a sphere or disk. In this region, the drag is almost entirely due to skin friction. For Reynolds numbers below approximately 0.4, experiments have shown that the drag coefficient can be calculated from Eq. 17.164.[36] In calculating the Reynolds number, the sphere or disk diameter should be used as the characteristic dimension.

$$C_D = \frac{24}{\mathrm{Re}} \qquad 17.166$$

Substituting this value of C_D into Eq. 17.165 results in *Stokes' law*, which is applicable to slow motion (ascent or descent) of spherical particles and bubbles traveling at velocity v through a fluid. Stokes' law is based on the assumptions that (1) flow is laminar, (2) Newton's law of viscosity is valid, and (3) all higher-order velocity terms (v^2, etc.) are negligible.

$$F_D = 3\pi \mu \mathrm{v} D \qquad 17.167$$

The drag coefficients for disks and spheres operating outside the region covered by Stokes' law have been determined experimentally. In the turbulent region, pressure drag is predominant. Figure 17.50 can be used to obtain approximate values for C_D.

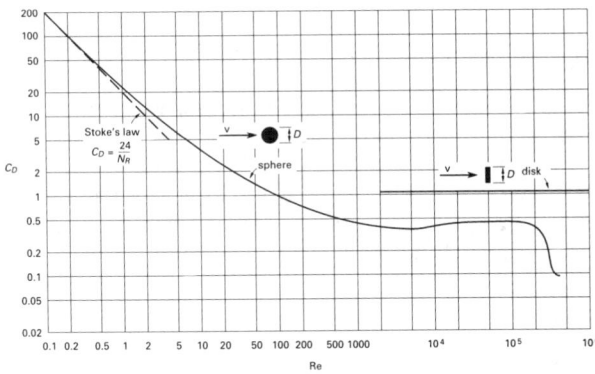

Figure 17.50 *Drag Coefficients for Spheres and Circular Flat Disks*

Figure 17.50 shows that there is a dramatic drop in the drag coefficient around Re = 10^5. The explanation for this is that the point of separation of the boundary layer shifts, decreasing the width of the wake. Since the drag force is primarily pressure drag at higher Reynolds numbers, a reduction in the wake reduces the pressure drag. Thus, anything that can be done to a sphere (scuffing or wetting a baseball, dimpling a golf ball, etc.) to induce a smaller wake will reduce the drag. There can be no shift in the boundary layer separation point for a thin disk, since the disk has no depth in the direction of

---

[36]Some sources report that the region in which Stokes' law applies extends to Re = 1.0.

---

flow. Therefore, the drag coefficient remains the same at all turbulent Reynolds numbers.

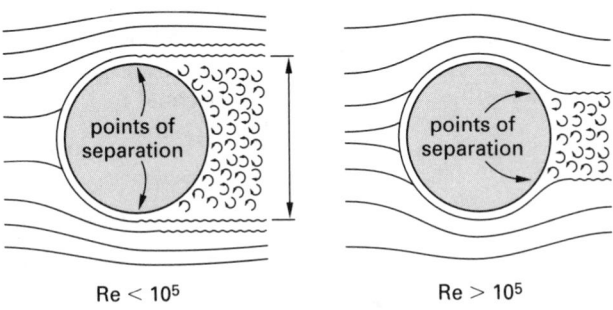

Figure 17.51 *Turbulent Flow Around a Sphere at Various Reynolds Numbers*

53. TERMINAL VELOCITY

The velocity of an object falling through a fluid will continue to increase until the drag force equals the net downward force (i.e., the weight less the buoyant force). The maximum velocity attained is known as the *terminal velocity*. At terminal velocity,

$$F_D = mg - F_b \qquad \text{[SI]} \qquad 17.168(a)$$

$$F_D = m \times \left(\frac{g}{g_c}\right) - F_b \qquad \text{[U.S.]} \qquad 17.168(b)$$

If the drag coefficient is known, the terminal velocity can be calculated from Eq. 17.169. For small, heavy objects falling in air, the buoyant force can be neglected.

$$\mathrm{v} = \sqrt{\frac{2(mg - F_b)}{C_D A \rho_{\mathrm{fluid}}}} = \sqrt{\frac{2Vg(\rho_{\mathrm{object}} - \rho_{\mathrm{fluid}})}{C_D A \; \rho_{\mathrm{fluid}}}} \qquad 17.169$$

For a sphere of diameter D, the terminal velocity is

$$\mathrm{v} = \sqrt{\frac{4Dg(\rho_{\mathrm{sphere}} - \rho_{\mathrm{fluid}})}{3C_D \rho_{\mathrm{fluid}}}} \qquad 17.170$$

If the spherical particle is very small, Stokes' law may apply. In that case, the terminal velocity can be calculated from Eq. 17.171.

$$\mathrm{v} = \frac{D^2 g(\rho_{\mathrm{sphere}} - \rho_{\mathrm{fluid}})}{18 \, \mu} \qquad \text{[SI]} \qquad 17.171(a)$$

$$\mathrm{v} = \frac{D^2(\rho_{\mathrm{sphere}} - \rho_{\mathrm{fluid}})}{18 \, \mu} \times \left(\frac{g}{g_c}\right) \qquad \text{[U.S.]} \qquad 17.171(b)$$

54. NONSPHERICAL PARTICLES

Only the most simple bodies can be modeled as spheres. One method of overcoming the complexity of dealing with the flow of real particles is to correlate performance with *sphericity*. For a particle and a sphere with the same volume, the sphericity is defined by Eq. 17.172. Sphericity will always be less than or equal to 1.0.

$$\text{sphericity} = \frac{A_\text{sphere}}{A_\text{particle}} \qquad 17.172$$

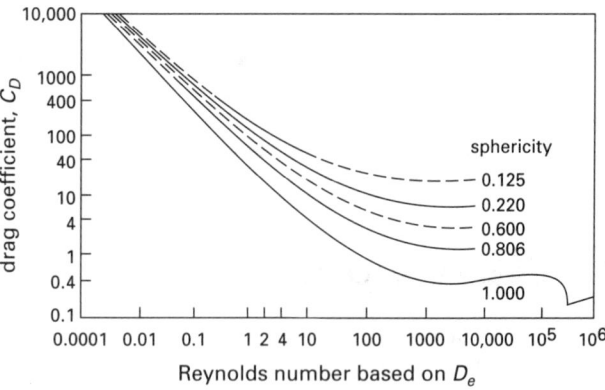

Used with permission from *Unit Operations*, by George Granger Brown, et. al., John Wiley & Sons, Inc., 1950.

Figure 17.52 *Drag Coefficients for Nonspherical Particles*

Another parameter that can be used to describe the deviation from ideal spherical behavior is the ratio of equivalent to average diameters, D_e/D_ave. The *average diameter* can be determined by screening a sample of the particles and evaluating the size distribution. The *equivalent diameter* is the diameter of a sphere having the same volume as the particle.

55. FLOW AROUND A CYLINDER

The characteristic drag coefficient plot for cylinders placed normal to the fluid flow is similar to the plot for spheres. The plot shown in Fig. 17.53 is for infinitely long cylinders, since there is additional wake drag at the cylinder ends. In calculating the Reynolds number, the cylinder diameter should be used as the characteristic dimension.

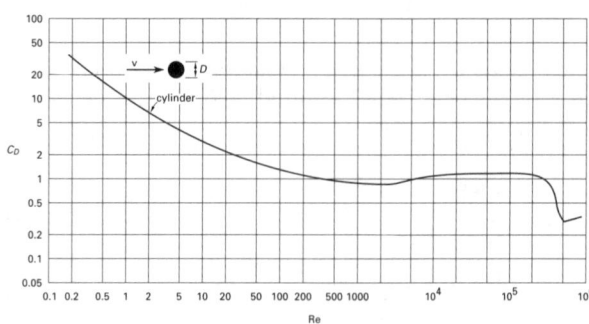

Figure 17.53 *Drag Coefficient for a Cylinder*

Table 17.8 *Sphericity and D_e/D_ave Ratios for Nonspherical Particles*

shape	sphericity	D_e/D_ave
sphere	1.00	1.00
octahedron	0.817	0.965
cube	0.806	1.24
prisms		
$\quad a \times a \times 2a$	0.767	1.564
$\quad a \times 2a \times 2a$	0.761	0.985
$\quad a \times 2a \times 3a$	0.725	1.127
cylinders		
$\quad h = 2r$	0.874	1.135
$\quad h = 3r$	0.860	1.31
$\quad h = 10r$	0.691	1.96
$\quad h = 20r$	0.580	2.592
disks		
$\quad h = 1.33r$	0.858	1.00
$\quad h = r$	0.827	0.909
$\quad h = r/3$	0.594	0.630
$\quad h = r/10$	0.323	0.422
$\quad h = r/15$	0.254	0.368

Used with permission from *Unit Operations*, by George Granger Brown, et. al., John Wiley & Sons, Inc., 1950.

Example 17.13

A 50 ft (15 m) high flagpole is constructed of a uniformly smooth cylinder 10 in (25 cm) in diameter. The surrounding air is at 40°F (0°C) and 14.6 psia (100 kPa). What is the total drag force on the flagpole in a 30 mi/hr (50 km/h) gust? (Neglect variations in the wind speed with height above the ground.)

SI Solution

At 0°C, the absolute viscosity of air is

$$\mu_{p,0°\text{C}} = 1.709 \times 10^{-5} \text{ Pa·s}$$

The density of the air is

$$\rho_p = \frac{p}{RT} = \frac{(100 \text{ kPa})\left(1000 \dfrac{\text{Pa}}{\text{kPa}}\right)}{\left(287 \dfrac{\text{J}}{\text{kg·K}}\right)(0°\text{C} + 273)}$$

$$= 1.276 \text{ kg/m}^3$$

The kinematic viscosity is

$$\nu_p = \frac{\mu}{\rho} = \frac{1.709 \times 10^{-5} \text{ Pa·s}}{1.276 \dfrac{\text{kg}}{\text{m}^3}}$$

$$= 1.339 \times 10^{-5} \text{ m}^2/\text{s}$$

The wind speed is

$$\text{v} = \frac{\left(50 \frac{\text{km}}{\text{h}}\right)\left(1000 \frac{\text{m}}{\text{km}}\right)}{3600 \frac{\text{sec}}{\text{h}}} = 13.89 \text{ m/s}$$

The characteristic dimension of the flagpole is its diameter. The Reynolds number is

$$\text{Re} = \frac{L\text{v}}{\nu} = \frac{\left(\frac{25 \text{ cm}}{100 \frac{\text{cm}}{\text{m}}}\right)\left(13.89 \frac{\text{m}}{\text{s}}\right)}{1.339 \times 10^{-5} \frac{\text{m}^2}{\text{s}}}$$

$$= 2.59 \times 10^5$$

From Fig. 17.53 for this Reynolds number, the drag coefficient is approximately 1.2.

The frontal area of the flagpole is

$$A = DL = \frac{(25 \text{ cm})(15 \text{ m})}{100 \frac{\text{cm}}{\text{m}}}$$

$$= 3.75 \text{ m}^2$$

From Eq. 17.165(b), the drag on the flagpole is

$$F_D = \frac{C_D A \rho \text{v}^2}{2}$$

$$= \frac{(1.2)(3.75 \text{ m}^2)\left(1.276 \frac{\text{kg}}{\text{m}^3}\right)\left(13.89 \frac{\text{m}}{\text{s}}\right)^2}{2}$$

$$= 554 \text{ N}$$

Customary U.S. Solution

At 40°F, the absolute viscosity of air is

$$\mu_{p,40°F} = 3.62 \times 10^{-7} \text{ lbf-sec/ft}^2$$

The density of the air is

$$\rho_p = \frac{p}{RT} = \frac{\left(14.6 \frac{\text{lbf}}{\text{in}^2}\right)\left(144 \frac{\text{in}^2}{\text{ft}^2}\right)}{\left(53.3 \frac{\text{ft-lbf}}{\text{lbm-°R}}\right)(40°F + 460)}$$

$$= 0.07889 \text{ lbm/ft}^3$$

The kinematic viscosity is

$$\nu_p = \frac{\mu g_c}{\rho} = \frac{\left(3.62 \times 10^{-7} \frac{\text{lbf-sec}}{\text{ft}^2}\right)\left(32.2 \frac{\text{ft-lbm}}{\text{lbf-sec}^2}\right)}{0.07889 \frac{\text{lbm}}{\text{ft}^3}}$$

$$= 1.478 \times 10^{-4} \text{ ft}^2/\text{sec}$$

The wind speed is

$$\text{v} = \frac{\left(30 \frac{\text{mi}}{\text{hr}}\right)\left(5280 \frac{\text{ft}}{\text{mi}}\right)}{3600 \frac{\text{sec}}{\text{hr}}} = 44 \text{ ft/sec}$$

The characteristic dimension of the flagpole is its diameter. The Reynolds number is

$$\text{Re} = \frac{L\text{v}}{\nu} = \frac{\left(\frac{10 \text{ in}}{12 \frac{\text{in}}{\text{ft}}}\right)\left(44 \frac{\text{ft}}{\text{sec}}\right)}{1.478 \times 10^{-4} \frac{\text{ft}^2}{\text{sec}}}$$

$$= 2.48 \times 10^5$$

From Fig. 17.53 for this Reynolds number, the drag coefficient is approximately 1.2.

The frontal area of the flagpole is

$$A = DL = \frac{(10 \text{ in})(50 \text{ ft})}{12 \frac{\text{in}}{\text{ft}}} = 41.67 \text{ ft}^2$$

From Eq. 17.165(a), the drag on the flagpole is

$$F_D = \frac{C_D A \rho \text{v}^2}{2g_c}$$

$$= \frac{(1.2)(41.67 \text{ ft}^2)\left(0.07889 \frac{\text{lbm}}{\text{ft}^3}\right)\left(44 \frac{\text{ft}}{\text{sec}}\right)^2}{(2)\left(32.2 \frac{\text{ft-lbm}}{\text{lbf-sec}^2}\right)}$$

$$= 118.6 \text{ lbf}$$

56. FLOW OVER A PARALLEL FLAT PLATE

The drag experienced by a flat plate oriented parallel to the direction of flow is almost totally skin friction drag. Prandtl's *boundary layer theory* can be used to evaluate the frictional effects. Such an analysis shows that the shape of the boundary layer is a function of the Reynolds number. The characteristic dimension used in calculating the Reynolds number is the chord length, L (i.e., the dimension of the plate parallel to the flow).

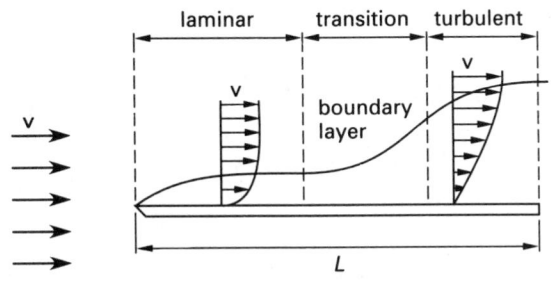

Figure 17.54 *Flow Over a Parallel Flat Plate*
(one side)

When skin friction predominates, it is common to use the symbol C_f (i.e., the *skin friction coefficient*) for the drag coefficient. For laminar flow over a smooth, flat plate, the drag coefficient based on the boundary layer theory is given by Eq. 17.173, which is known as the *Blasius solution*.[37] The critical Reynolds number for laminar flow is often reported to be 530,000. However, the transition region between laminar flow and turbulent flow actually occupies the Reynolds number range of 100,000 to 1,000,000.

$$C_f = \frac{1.328}{\sqrt{\text{Re}}} \qquad \textit{17.173}$$

Prandtl reported that the skin friction coefficient for turbulent flow is

$$C_f = \frac{0.455}{(\log \text{Re})^{2.58}} \qquad \textit{17.174}$$

The drag force is calculated from Eq. 17.175. The factor 2 is used because there is friction on two sides of the flat plate.

$$F_D = 2 \times \frac{C_f A \rho v^2}{2} = C_f A \rho v^2 \qquad \text{[SI]} \quad \textit{17.175(a)}$$

$$F_D = 2 \times \frac{C_f A \rho v^2}{2g_c} = \frac{C_f A \rho v^2}{g_c} \qquad \text{[U.S.]} \quad \textit{17.175(b)}$$

57. SIMILARITY

Similarity considerations between a *model* (subscript m) and a full-sized object (subscript p, for *prototype*) imply that the model can be used to predict the performance of the prototype. Such a model is said to be *mechanically similar* to the prototype.

Complete *mechanical similarity* requires both geometric and dynamic similarity.[38] *Geometric similarity* means that the model is true to scale in length, area, and volume. The *model scale (length ratio)* is defined as

$$L_r = \frac{\text{size of model}}{\text{size of prototype}} \qquad \textit{17.176}$$

The area and volume ratios are based on the model scale.

$$\frac{A_m}{A_p} = (L_r)^2 \qquad \textit{17.177}$$

$$\frac{V_m}{V_p} = (L_r)^3 \qquad \textit{17.178}$$

---

[37] Other correlations substitute the coefficient 1.44, the original value calculated by Prandtl, for 1.328 in Eq. 17.173. The Blasius solution is considered to be more accurate.

[38] Complete mechanical similarity also requires kinematic and thermal similarity, which are not discussed in this book.

Dynamic similarity means that the ratios of all types of forces are equal for the model and the prototype. These forces result from inertia, gravity, viscosity, elasticity (i.e., fluid compressibility), surface tension, and pressure.

The number of possible ratios of forces is large. For example, the ratios of viscosity/inertia, inertia/gravity, and inertia/surface tension are only three of the ratios of forces that must match for every corresponding point on the model and prototype. Fortunately, some force ratios can be neglected because the forces are negligible or are self-canceling.

In some cases, the geometric scale may be deliberately distorted. For example, with scale models of harbors and rivers, the water depth might be only a fraction of an inch if the scale is followed loyally. Not only will the surface tension be excessive, but the shifting of the harbor or river bed may not be properly observed. Therefore, the vertical scale is chosen to be different from the horizontal scale. Such models are known as *distorted models*. Experience is needed to interpret observations made of distorted models.

The following sections deal with the most common similarity problems, but not all. For example, similarity of steady laminar flow in a horizontal pipe requires the Stokes number, similarity of high-speed (near-sonic) aircraft requires the Mach number, and similarity of capillary rise in a tube requires the *Eötvös number* ($g\rho L^2/\sigma$).

58. VISCOUS AND INERTIAL FORCES DOMINATE

Consider the testing of a completely submerged object, such as the items listed in Table 17.9. Surface tension will be negligible. The fluid can be assumed to be incompressible for low velocity. Gravity does not change the path of the fluid particles significantly during the passage of the object.

Table 17.9 *Cases with Reynolds Number Similarity*

> submarines
> torpedoes
> aircraft (subsonic)
> airfoils (subsonic)
> fans
> pumps
> turbines
> drainage through tank orifices
> closed-pipe flow (turbulent)
> flow meters
> open channel flow (without wave action)

Only inertial, viscous, and pressure forces are significant. Because they are the only forces that are acting, these three forces are in equilibrium. Since they are in equilibrium, knowing any two forces will define the third force. This third force is dependent and can be omitted from the similarity analysis. For submerged objects, pressure is traditionally chosen as the dependent force.

The dimensionless ratio of the inertial forces to the viscous forces is the Reynolds number. Equating the model's and prototype's Reynolds numbers will ensure similarity.[39]

$$\text{Re}_m = \text{Re}_p \qquad 17.179$$

$$\frac{L_m v_m}{\nu_m} = \frac{L_p v_p}{\nu_p} \qquad 17.180$$

If the model is tested in the same fluid and at the same temperature in which the prototype is expected to operate, setting the Reynolds numbers equal is equivalent to setting $L_m v_m = L_p v_p$.

Example 17.14

A 1/30th size scale model of a helicopter fuselage is tested in a wind tunnel at 120 mi/hr (190 km/h). The conditions in the wind tunnel are 50 psia and 100°F (350 kPa and 50°C). What is the corresponding speed of a prototype traveling in 14.0 psia, 40°F still air (100 kPa and 0°C)?

SI Solution

From App. 14.E,

$$\mu_{p,0°C} = 1.709 \times 10^{-5} \text{ Pa·s}$$

$$\mu_{m,50°C} = 1.951 \times 10^{-5} \text{ Pa·s}$$

The densities of air at the two conditions are

$$\rho_p = \frac{p}{RT} = \frac{(100 \text{ kPa})\left(1000 \dfrac{\text{Pa}}{\text{kPa}}\right)}{\left(287 \dfrac{\text{J}}{\text{kg·K}}\right)(0°C + 273)}$$

$$= 1.276 \text{ kg/m}^3$$

$$\rho_m = \frac{p}{RT} = \frac{(350 \text{ kPa})\left(1000 \dfrac{\text{Pa}}{\text{kPa}}\right)}{\left(287 \dfrac{\text{J}}{\text{kg·K}}\right)(50°C + 273)}$$

$$= 3.776 \text{ kg/m}^3$$

---

[39]An implied assumption is that the drag coefficients are the same for the model and the prototype. In the case of pipe flow, it is assumed that flow will be in the turbulent region with the same relative roughness.

The kinematic viscosities are

$$\nu_p = \frac{\mu}{\rho} = \frac{1.709 \times 10^{-5} \text{ Pa·s}}{1.276 \dfrac{\text{kg}}{\text{m}^3}}$$

$$= 1.339 \times 10^{-5} \text{ m}^2/\text{s}$$

$$\nu_m = \frac{\mu}{\rho} = \frac{1.951 \times 10^{-5} \text{ Pa·s}}{3.776 \dfrac{\text{kg}}{\text{m}^3}}$$

$$= 5.167 \times 10^{-6} \text{ m}^2/\text{s}$$

From Eq. 17.180,

$$v_p = v_m \left(\frac{L_m}{L_p}\right)\left(\frac{\nu_p}{\nu_m}\right)$$

$$= \frac{\left(190 \dfrac{\text{km}}{\text{h}}\right)\left(\dfrac{1}{30}\right)\left(1.339 \times 10^{-5} \dfrac{\text{m}^2}{\text{s}}\right)}{5.167 \times 10^{-6} \dfrac{\text{m}^2}{\text{s}}}$$

$$= 16.4 \text{ km/h}$$

Customary U.S. Solution

Since surface tension and gravitational forces on the air particles are insignificant and since the flow velocities are low, viscous and inertial forces dominate. The Reynolds numbers of the model and prototype are equated.

The kinematic viscosity of air must be evaluated at the respective temperatures and pressures. As kinematic viscosity tables for air are not readily available, the viscosity must be calculated. Although kinematic viscosity depends on the temperature and pressure, absolute viscosity is essentially independent of pressure. From App. 14.D,

$$\mu_{p,40°F} = 3.62 \times 10^{-7} \text{ lbf-sec/ft}^2$$

$$\mu_{m,100°F} = 3.96 \times 10^{-7} \text{ lbf-sec/ft}^2$$

The densities of air at the two conditions are

$$\rho_p = \frac{p}{RT} = \frac{\left(14.0 \dfrac{\text{lbf}}{\text{in}^2}\right)\left(144 \dfrac{\text{in}^2}{\text{ft}^2}\right)}{\left(53.3 \dfrac{\text{ft-lbf}}{\text{lbm-°R}}\right)(40°F + 460)}$$

$$= 0.0756 \text{ lbm/ft}^3$$

$$\rho_m = \frac{p}{RT} = \frac{\left(50.0 \dfrac{\text{lbf}}{\text{in}^2}\right)\left(144 \dfrac{\text{in}^2}{\text{ft}^2}\right)}{\left(53.3 \dfrac{\text{ft-lbf}}{\text{lbm-°R}}\right)(100°F + 460)}$$

$$= 0.2412 \text{ lbm/ft}^3$$

The kinematic viscosities are

$$\nu_p = \frac{\mu g_c}{\rho} = \frac{\left(3.62 \times 10^{-7} \dfrac{\text{lbf-sec}}{\text{ft}^2}\right) g_c}{0.0756 \dfrac{\text{lbm}}{\text{ft}^3}}$$

$$= 4.79 \times 10^{-6} \times g_c \text{ ft}^2/\text{sec}$$

$$\nu_m = \frac{\mu g_c}{\rho} = \frac{\left(3.96 \times 10^{-7} \dfrac{\text{lbf-sec}}{\text{ft}^2}\right) g_c}{0.2412 \dfrac{\text{lbm}}{\text{ft}^3}}$$

$$= 1.64 \times 10^{-6} \times g_c \text{ ft}^2/\text{sec}$$

From Eq. 17.180,

$$v_p = v_m \left(\frac{L_m}{L_p}\right) \left(\frac{\nu_p}{\nu_m}\right)$$

$$= \frac{\left(120 \dfrac{\text{mi}}{\text{hr}}\right)\left(\dfrac{1}{30}\right)\left(4.79 \times 10^{-6} \times g_c \dfrac{\text{ft}^2}{\text{sec}}\right)}{1.64 \times 10^{-6} \times g_c \dfrac{\text{ft}^2}{\text{sec}}}$$

$$= 11.7 \text{ mi/hr}$$

59. INERTIAL AND GRAVITATIONAL FORCES DOMINATE

Table 17.10 lists the cases when elasticity and surface tension forces can be neglected but gravitational forces cannot. Omitting these two forces from the similarity calculations leaves pressure, inertia, viscosity, and gravity forces, which are in equilibrium. Pressure is chosen as the dependent force and is omitted from the analysis.

Table 17.10 *Cases with Froude Number Similarity*

> surface ships
> seaplane hulls
> surface wave action
> surge and flood waves
> oscillatory wave action
> bow waves from ships
> flow over weirs
> flow over spillways
> open channel flow with varying surface levels
> motion of a fluid jet

There are only two possible combinations of the remaining three forces. The ratio of inertial to viscous forces is the Reynolds number. The ratio of the inertial forces to the gravitational forces is the *Froude number*, Fr.[40]

---

[40]There are two definitions of the Froude number. Dimensional analysis determines the Froude number as Eq. 17.181 (v^2/Lg), a form which is used in model similitude. However, in open channel flow studies performed by civil engineers, the Froude number is taken as the square root of Eq. 17.181. Whether the derived form or its square root is used can sometimes be determined from the application. If the Froude number is squared (e.g., as in $dE/dx = 1 - \text{Fr}^2$), the square root form is probably needed. In similarity problems, it doesn't make any difference which definition is used.

The Froude number is used when gravitational forces are significant, such as in wave motion produced by a ship or seaplane hull.

$$\text{Fr} = \frac{v^2}{Lg} \qquad \textit{17.181}$$

Thus, similarity is ensured when Eqs. 17.182 and 17.183 are satisfied.

$$\text{Re}_m = \text{Re}_p \qquad \textit{17.182}$$
$$\text{Fr}_m = \text{Fr}_p \qquad \textit{17.183}$$

As an alternative, Eqs. 17.182 and 17.183 can be solved simultaneously. This results in the following requirement for similarity, which indicates that it is necessary to test the model in a manufactured liquid with a specific viscosity.

$$\frac{\nu_m}{\nu_p} = \left(\frac{L_m}{L_p}\right)^{\frac{3}{2}} = (L_r)^{\frac{3}{2}} \qquad \textit{17.184}$$

Sometimes it is not possible to satisfy Eqs. 17.182 and 17.183. This occurs when a model fluid viscosity is called for that is not available. If only one of the equations is satisfied, the model is said to be *partially similar*. In such a case, corrections based on other factors are used.

Another problem with trying to achieve similarity in open channel flow problems is the need to scale surface drag. It can be shown that the ratio of Manning's roughness constants is given by Eq. 17.185. In some cases, it may not be possible to create a surface smooth enough to satisfy this requirement.

$$n_r = (L_r)^{\frac{1}{6}} \qquad \textit{17.185}$$

60. SURFACE TENSION FORCE DOMINATES

Table 17.11 lists some of the cases where surface tension is the predominant force. Such cases can be handled by equating the Weber numbers, We, of the model and prototype.[41] (The *Weber number* is the ratio of inertial force to surface tension.)

$$\text{We} = \frac{v^2 L \rho}{\sigma} \qquad \textit{17.186}$$
$$\text{We}_m = \text{We}_p \qquad \textit{17.187}$$

---

[41]There are two definitions of the Weber number. The alternate definition is the square root of Eq. 17.184. In similarity problems, it does not make any difference which definition of the Weber number is used.

Table 17.11 *Cases with Weber Number Similarity*

        waves
        droplets
        bubbles
        air entrainment

PRACTICE PROBLEMS

(Use $g = 32.2$ ft/sec^2 (9.81 m/s^2) and 60°F (16°C) water unless told to do otherwise.)

1. 1.5 ft^3/sec (40 L/s) of 70°F (20°C) water flows through 1200 ft (355 m) of 6 in (nominal) diameter new schedule-40 pipe. What is the friction loss?

2. 500 gal/min (30 L/s) of 100°F (40°C) water flows through 300 ft (90 m) of 6 in, schedule-40 pipe. The pipe contains two 6 in flanged, steel elbows, two full-open gate valves, a full-open 90° angle valve, and a backflow limiter. The discharge is located 20 ft (6 m) higher than the entrance. What is the pressure difference between the two ends of the pipe?

3. 70°F (20°C) air is flowing at 60 ft/sec (18 m/s) through 300 ft (90 m) of 6 in, schedule-40 pipe. The pipe contains two 6 in flanged, steel elbows, two full-open gate valves, a full-open 90° angle valve, and a backflow limiter. The discharge is located 20 ft (6 m) higher than the entrance. What is the pressure difference between the two ends of the pipe?

4. 5.0 ft^3/sec (130 L/s) of water flows through a pipe that changes gradually in diameter from 6 in (150 mm) at point A to 18 in (460 mm) at point B. Point B is 15 ft (4.6 m) higher than point A. The respective pressures at points A and B are 10 psia (69 kPa) and 7 psia (48.3 kPa). Is the direction of flow from A to B or from B to A?

5. The velocity of discharge from a fire hose is 50 ft/sec (15 m/s). The hose is oriented 45° from the horizontal. Disregarding air friction, what is the maximum range of the discharge?

6. 60°F (15°C) benzene (specific gravity at 60°F (15°C) of 0.885) flows through an 8 in/3^1/$_2$ in (200 mm/90 mm) venturi meter whose coefficient of discharge is 0.99. A mercury manometer indicates a 4 in difference in the heights of the mercury columns. What is the volumetric flow rate of the benzene?

7. A mercury manometer is used to measure a pressure difference across an orifice meter in a water line. The difference in mercury levels is 7 in (17.8 cm). What is the pressure differential?

8. A sharp-edged ISA orifice is used in a 12 in (300 mm) water line. (Figure 17.28 is applicable.) The water temperature is 70°F (20°C), and the flow rate is 10 ft^3/sec (250 L/s). The head loss across the orifice must not exceed 25 ft (7.5 m). What is the smallest orifice that can be used?

9. 2000 gal/min (125 L/s) of brine with a specific gravity of 1.2 pass through an 85% efficient pump. The centerlines of the pump's 12 in (300 mm) inlet and 8 in (200 mm) outlet are at the same elevation. The inlet suction gage indicates 6 in (150 mm) of mercury below atmospheric. The discharge pressure gage is located 4 ft (1.2 m) above the centerline of the pump's outlet and indicates 20 psig (138 kPa). What is the input power to the pump?

10. 100 ft^3/sec (2.6 kL/s) of water pass through a horizontal turbine. The water's pressure is reduced from 30 psig (210 kPa) to 5 psi (35 kPa) vacuum. Ignoring friction, velocity, and other factors, what power is generated?

11. A 2 in (50 mm) diameter horizontal water jet has an absolute velocity (with respect to a stationary point) of 40 ft/sec (12 m/s) as it strikes a curved blade. The blade is moving horizontally away with an absolute velocity of 15 ft/sec (4.5 m/s). Water is deflected 60° from the horizontal. What is the force on the blade?

12. (*Time limit: one hour*) A cast iron pipe has an inside diameter of 24 in (600 mm) and a wall thickness of 0.75 in (20 mm). The pipe's modulus of elasticity is 20×10^6 psi (140 GPa). The pipeline is 500 ft (150 m) long. 70°F (20°C) water is flowing at 6 ft/sec (2 m/s) when a valve in the pipeline is suddenly closed. (a) What will be the maximum pressure in the pipeline? (b) Over what length of time would the valve have to be closed in order to cause the same pressure as occurs with the instantaneous closure?

13. (*Time limit: one hour*) A car traveling through 70°F (20°C) air has the following characteristics:

frontal area	28 ft^2 (2.6 m^2)
mass	3300 lbm (1500 kg)
drag coefficient	0.42
rolling resistance	1% of weight
engine thermal efficiency	28%
fuel heating value	115,000 Btu/gal (460 MJ/L)

(a) Considering only the drag and rolling resistance, what is the fuel consumption when the car is traveling at 55 mi/hr (90 km/h)? (b) What is the percentage increase in fuel consumption at 65 mi/hr (105 km/h) (compared to at 55 mi/hr (90 km/h))?

14. A 1/20th airplane model is tested in a wind tunnel at full velocity and temperature. What is the ratio of the wind tunnel pressure to normal ambient pressure?

15. 68°F (20°C) castor oil (kinematic viscosity at 68°F (20°C) of 0.0111 ft^2/sec (0.00103 m^2/s)) flows through a pump whose impeller turns at 1000 rpm. A similar pump twice the first pump's size is tested with 68°F (20°C) air. What should be the speed of the second pump to ensure similarity?

16. 8 MGD (millions of gallons per day) (350 L/s) of
70°F (20°C) water flow into the smooth schedule-40
steel pipe network shown. Minor losses are insignificant. (a) What is the quantity of water flowing in each
branch? (b) What is the head loss between the inlet
and the outlet?

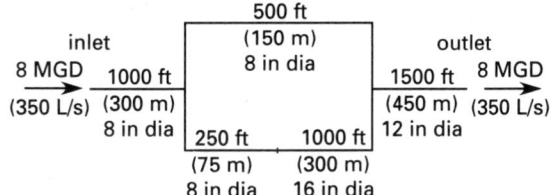

18 Hydraulic Machines

.... 18-2
.... 18-2
.... 18-2
.... 18-3
.... 18-3
.... 18-4
.... 18-4
.... 18-6
.... 18-7
.... 18-8
.... 18-8
.... 18-10
.... 18-11
.... 18-13
.... 18-13
.... 18-14
17. Cavitation Coefficient 18-14
18. Suction Specific Speed 18-15
19. Pump Performance Curves 18-15
20. System Curves 18-15
21. Operating Point 18-16
22. Pumps in Parallel 18-17
23. Pumps in Series 18-17
24. Affinity Laws 18-17
25. Pump Similarity 18-18
26. Pumping Liquids Other Than Cold Water 18-19
27. Turbine Specific Speed 18-20
28. Types of Turbines 18-20
29. Hydroelectric Generating Plants 18-20
30. Impulse Turbines 18-21
31. Reaction Turbines 18-22
32. Axial-Flow Turbines 18-23
    Practice Problems 18-23

Nomenclature

C	coefficient	–	–
D	diameter	ft	m
E	specific energy	ft-lbf/lbm	J/kg
f	Darcy friction factor	–	–
f	frequency of voltage	Hz	Hz
g	gravitational acceleration	ft/sec^2	m/s^2
g_c	gravitational constant	lbm-ft/ lbf-sec^2	n.a.
h	height or head	ft	m
h_{ac}	acceleration head	ft	m
K	dimensionless factor	–	–
L	length	ft	m
m	mass	lbm	kg
$\dot{m}$	mass flow rate	lbm/sec	kg/s

n	dimensionless exponent	–	–
n	rotational speed	rpm	rpm
NPSHA	net positive suction head available	ft	m
NPSHR	net positive suction head required	ft	m
p	pressure	lbf/ft^2	Pa
P	power	ft-lbf/sec	W
Q	volumetric flow rate	gal/min	L/s
SA	suction specific speed available	rpm	rpm
SG	specific gravity	–	–
t	time	sec	s
T	transmitted torque	ft-lbf	N·m
v	velocity	ft/sec	m/s
$\dot{V}$	volumetric flow rate	ft^3/sec	m^3/s
W	work	ft-lbf	kW·h
WHP	water horsepower	hp	n.a.
WkW	water kilowatts	n.a.	kW
z	elevation	ft	m

Symbols

γ	specific weight	lbf/ft^3	n.a.
η	efficiency	–	–
θ	angle	deg	deg
ν	kinematic viscosity	ft^2/sec	m^2/s
ρ	density	lbm/ft^3	kg/m^3
σ	cavitation number	–	–
ω	angular velocity	rad/sec	rad/s

Subscripts

atm	atmospheric
A	added (by pump)
b	blade
cr	critical
d	discharge
f	friction
i	inlet
j	jet
m	motor
n	nozzle
o	outlet
p	pressure or pump
s	suction or specific
ss	suction specific
t	total or tangential
th	theoretical
v	velocity or volumetric
vp	vapor pressure
z	potential

Fluids

1. HYDRAULIC MACHINES

Pumps and turbines are the two basic types of hydraulic machines discussed in this chapter. Pumps convert mechanical energy into fluid energy, increasing the energy possessed by the fluid. Turbines convert fluid energy into mechanical energy, extracting energy from the fluid.

2. TYPES OF PUMPS

Pumps can be classified according to the method by which pumping energy is transferred to the fluid. This classification separates pumps into positive displacement pumps and kinetic pumps.

The most common types of *positive displacement pumps* are *reciprocating action pumps* (which use pistons, plungers, diaphragms, or bellows) and *rotary action pumps* (using vanes, screws, lobes, or progressing cavities). Such pumps discharge a fixed volume for each stroke or revolution. Energy is added intermittently to the fluid.

Kinetic pumps transform fluid kinetic energy into fluid static pressure energy. The pump imparts the kinetic energy; the pump mechanism or housing is constructed in a manner that causes the transformation. *Jet pumps* and *ejector pumps* fall into the kinetic pump category, but centrifugal pumps are the primary examples.

In the operation of a *centrifugal pump*, liquid flowing into the *suction side* (the *inlet*) is captured by the *impeller* and thrown to the outside of the pump casing. Within the casing, the velocity imparted to the fluid by the impeller is converted into pressure energy. The fluid leaves the pump through the *discharge line* (the *exit*). It is a characteristic of most centrifugal pumps that the fluid is turned approximately 90° from the original flow direction.

Table 18.1 *Generalized Characteristics of Positive Displacement and Kinetic Pumps*

characteristic	positive displacement pumps	kinetic pumps
flow rate	low	high
pressure rise per stage	high	low
constant quantity over operating range	flow rate	pressure rise
self-priming	yes	no
discharge stream	pulsing	steady
works with high viscosity fluids	yes	no

3. POSITIVE DISPLACEMENT RECIPROCATING PUMPS

Reciprocating positive displacement (PD) pumps are used with viscous fluids and slurries (up to about 8000 SSU), when the fluid is sensitive to shear, and when a high discharge pressure is required.[1] By entrapping a volume of liquid in the cylinder, reciprocating pumps provide a fixed-displacement volume per cycle. They are self-priming and inherently leak-free. Within the pressure limits of the line and pressure relief valve and the current capacity of the motor circuit, reciprocating pumps can provide an infinite discharge pressure.[2]

There are three main types of reciprocating pumps: power, direct-acting, and diaphragm. A *power pump* is a *cylinder-operated pump*. It can be single-acting or double-acting. A *single-acting pump* discharges liquid (or takes suction) only on one side of the piston, and there is only one transfer operation per crankshaft revolution. A *double-acting pump* discharges from both sides, and there are two transfers per revolution of the crank.

Traditional reciprocating pumps with pistons and rods can be either single-acting or double-acting and are suitable up to approximately 2000 psi (14 MPa). *Plunger pumps* are only single-acting and are suitable up to approximately 10,000 psi (70 MPa).

Simplex pumps have one cylinder, *duplex pumps* have two cylinders, *triplex pumps* have three cylinders, and so forth. *Direct-acting pumps* (sometimes referred to as *steam pumps*) are always double-acting. They use steam or unburned fuel gas as a motive fluid.

PD pumps are limited by both their NPSHR characteristics, acceleration head, and (for rotary pumps) slip.[3] Because the flow is unsteady, a certain amount of energy, the *acceleration head* (h_{ac}), is required to accelerate the fluid flow each stroke or cycle. If the acceleration head is too large, the NPSHR requirements may not be attainable. Acceleration head can be reduced by increasing the pipe diameter, shortening the suction piping, decreasing the pump speed, or placing a *pulsation damper* (*stabilizer*) in the suction line.[4]

Generally, friction losses with pulsating flow are calculated based on the maximum velocity attained by the fluid. Since this is difficult to determine, the maximum velocity can be approximated by multiplying the average velocity (calculated from the rated capacity) by the factors in Table 18.2.

When the suction line is "short," the acceleration head can be calculated from the length of the suction line, the average velocity in the line, and the rotational speed.[5]

---

[1] For SSU > 240, multiply SSU viscosity by 0.216 to get viscosity in centistokes.
[2] For this reason, a relief valve should be included in every installation of positive displacement pumps. Rotary pumps typically have integral relief valves, but external relief valves are often installed to provide easier adjusting, cleaning, and checking.
[3] Manufacturers of PD pumps prefer the term *net positive inlet pressure* (NPIP) to NPSH. NPIPA corresponds to NPSHA; NPIPR corresponds to NPSHR. Pressure and head are related by $p = \gamma h$.
[4] Pulsation dampers are not needed with rotary-action PD pumps, as the discharge is essentially constant.
[5] With a properly designed pulsation damper, the effective length of the suction line is reduced to approximately 10 pipe diameters.

In Eq. 18.1, C and K are dimensionless factors. K represents the relative compressibility of the liquid. (Typical values are 1.4 for hot water; 1.5 for amine, glycol, and cold water; and 2.5 for hot oil.) Values of C are given in Table 18.3.

$$h_{ac} = \left(\frac{C}{K}\right)\left(\frac{L_{\text{suction}} v_{\text{ave}} n_{\text{rpm}}}{g}\right) \qquad \textit{18.1}$$

Table 18.2 Typical v_{max}/v_{ave} Velocity Ratios[a,b]

pump type	single-acting	double-acting
simplex	3.2	2.0
duplex	1.6	1.3
triplex	1.1	1.1
quadriplex	1.1	1.1
quintuplex and up	1.05	1.05

[a]Without stabilization. With properly sized stabilizers, use 1.05 to 1.1 for all cases.
[b]Multiply the values by 1.3 for metering pumps where lost fluid motion is relied on for capacity control.

Table 18.3 Typical Acceleration Head C-Values[a]

pump type	single-acting	double-acting
simplex	0.4	0.2
duplex	0.2	0.115
triplex	0.066	0.066
quadriplex	0.040	0.040
quintuplex and up	0.028	0.028

[a]Typical values for common connecting rod lengths and crank radii.

4. ROTARY PUMPS

Rotary pumps are *positive displacement* (PD) pumps that move fluid by means of screws, progressing cavities, gears, lobes, or vanes turning within a fixed casing (the *stator*). Rotary pumps are useful for high viscosities (up to 4×10^6 SSU for screw pumps). The rotation creates a cavity of fixed volume near the pump input; atmospheric or external pressure forces the liquid into that cavity. Near the outlet, the cavity is collapsed, forcing the liquid out. Figure 18.1 illustrates the external circumferential piston rotary pump.

Discharge from rotary pumps is relatively smooth. Acceleration head is negligible. Pulsation dampers and suction stabilizers are not required.

Slip in rotary pumps is the amount (sometimes expressed as a percentage) of each rotational fluid volume that "leaks" back to the suction line on each revolution. Slip reduces pump capacity. It is a function of clearance, differential pressure, and viscosity. Slip is proportional to the third power of the clearance between the rotating element and the casing. Slip decreases with increases in

viscosity; it increases linearly with increases in differential pressure. Slip is not affected by rotational speed. The *volumetric efficiency* is defined by Eq. 18.2. Figure 18.2 illustrates the relationship between flow rate, speed, slip, and differential pressure.

$$\eta_{v} = \frac{Q_{\text{actual}}}{Q_{\text{ideal}}} = \frac{Q_{\text{ideal}} - Q_{\text{slip}}}{Q_{\text{ideal}}} \qquad \textit{18.2}$$

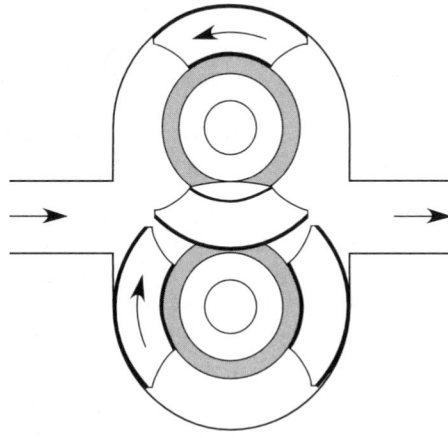

Figure 18.1 External Circumferential Piston Rotary Pump

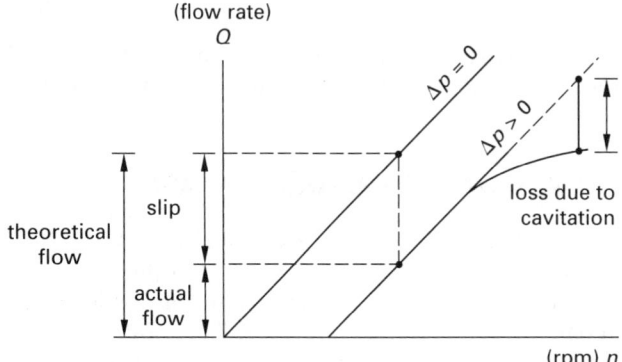

Figure 18.2 Slip in Rotary Pumps

Except for screw pumps, rotary pumps are generally not used for handling abrasive fluids or materials with suspended solids. Few rotary pumps are suitable when variable flow is required.

5. DIAPHRAGM PUMPS

Hydraulically operated *diaphragm pumps* have a diaphragm that completely separates the pumped fluid from the rest of the pump. A reciprocating plunger pressurizes and moves a hydraulic fluid that, in turn, flexes the diaphragm. Single-ball check valves in the suction and discharge lines determine the direction of flow during both phases of the diaphragm action.

Metering is a common application of diaphragm pumps. They have no packing and are essentially leak-proof. This makes them ideal when fugitive emissions are undesirable. Diaphragm pumps are suitable for pumping a wide range of materials, from liquefied gases to coal slurries, though the upper viscosity limit is approximately 3500 SSU. Within the limits of their reactivities, hazardous and reactive materials can also be handled.

Diaphragm pumps are limited by capacity, suction pressure, and discharge pressure and temperature. Because of their construction and size, most diaphragm pumps are limited to discharge pressures of 5000 psi (35 MPa) or less, and most high-capacity pumps are limited to 2000 psi (14 MPa). Suction pressures are similarly limited to 5000 psi (35 MPa). A minimum pressure of 3 to 9 psi (20 to 60 kPa) is often quoted as the minimum liquid-side pressure for metering applications.

The discharge is inherently pulsating, and the dampers or stabilizers are often used. (The acceleration head term is required when calculating NPSHR.) The discharge can be smoothed out somewhat by using two or three (i.e., duplex or triplex) plungers.

Diaphragms are commonly manufactured from stainless steel (type 316) and polytetrafluorethylene (PTFE) or other elastomers. PTFE diaphragms are suitable in the range of $-50°F$ to $300°F$ ($-45°C$ to $150°C$), while metal diaphragms (and some ketone resin diaphragms) are used up to approximately $400°F$ ($200°C$), with life expectancy being reduced at higher temperatures. Although most diaphragm pumps usually operate below 200 spm (strokes per minute), diaphragm life will be improved by limiting the maximum speed to 100 spm.

6. TYPES OF CENTRIFUGAL PUMPS

Centrifugal pumps can be classified according to the way their impellers impart energy to the fluid. Each category of pump is suitable for a different application and (specific) speed range. Figure 18.3 illustrates a typical centrifugal pump and its schematic symbol.

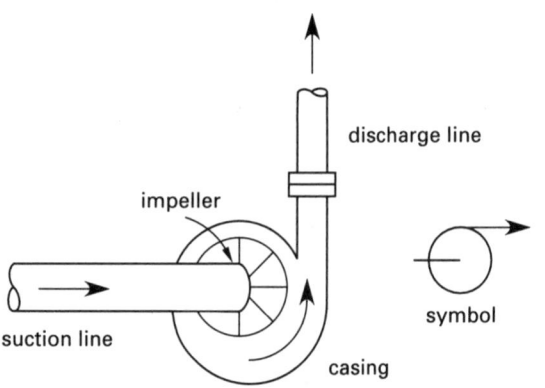

Figure 18.3 *Centrifugal Pump and Symbol*

Radial-flow impellers impart energy primarily by centrifugal force. Liquid enters the impeller at the hub and flows radially to the outside of the casing. Radial-flow pumps are suitable for adding high pressure at low fluid flow rates. *Axial-flow impellers* impart energy to the fluid by acting as compressors. Fluid enters and exits along the axis of rotation. Axial-flow pumps are suitable for adding low pressures at high fluid flow rates.[6]

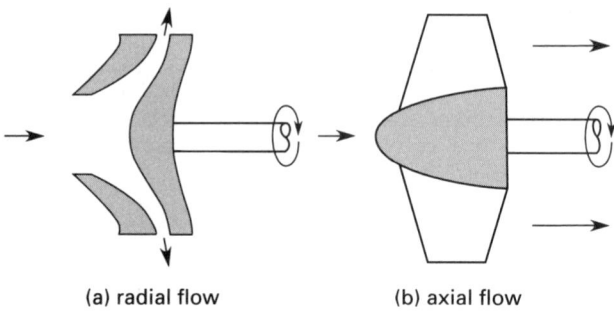

Figure 18.4 *Radial and Axial Flow Impellers*

Radial-flow pumps can be designed for either single- or double-suction operation. In a *single-suction pump*, fluid enters from only one side of the impeller. In a *double-suction pump*, fluid enters from both sides of the impeller.[7] (That is, the impeller is two-sided.) Operation is similar to having two single-suction pumps in parallel.

A *multiple-stage pump* consists of two or more impellers within a single casing. The discharge of one stage feeds the input of the next stage, and operation is similar to having several pumps in series. In this manner, higher heads are achieved than would be possible with a single impeller.

The *circular blade pitch* is the impeller's circumference divided by the number of impeller vanes. The impeller *tip speed* (not to be confused with the specific and suction specific speeds) is easily calculated from the impeller diameter and rotational speed. The impeller "tip speed" is actually the tangential velocity at the periphery. Tip speed is typically somewhat less than 1000 ft/sec (300 m/s).

$$v_{tip} = \frac{\pi D n}{60 \frac{\sec}{\min}} = \frac{D\omega}{2} \qquad 18.3$$

7. TERMINOLOGY OF HYDRAULIC MACHINES

A pump will always have an inlet (designated the *suction*) and an outlet (designated the *discharge*). The subscripts s and d refer to the inlet and outlet of the pump, not the pipeline.

---

[6]There is a third category of centrifugal pumps known as *mixed flow pumps*. Mixed flow pumps have operational characteristics in between those of radial flow and axial flow pumps.

[7]The double-suction pump can handle a greater fluid flow rate than a single-suction pump with the same specific speed. Also, the double-suction pump will have a lower NPSHR.

All of the terms that are discussed in this section are *head* terms, and as such, have the units of length. In work with hydraulic machines, it is common to hear such phrases as "...a pressure head of 50 feet..." and "...a static discharge head of 85 meters..." The term *head* is often substituted for pressure or pressure drop. Any head term (*pressure head, atmospheric head, vapor pressure head,* etc.) can be calculated from pressure by using Eq. 18.4.[8]

$$h = \frac{p}{\gamma} \qquad 18.4$$

$$h = \frac{p}{g\rho} \qquad [\text{SI}] \qquad 18.5(a)$$

$$h = \left(\frac{p}{\rho}\right) \times \left(\frac{g_c}{g}\right) \qquad [\text{U.S.}] \qquad 18.5(b)$$

Some of the terms used in the description of pipelines appear to be similar and may be initially confusing (e.g., suction head and total suction head). The following general rules will help to clarify the meanings:[9]

Rule 18.1: The word *suction* or *discharge* limits the quantity to the suction line or discharge line, respectively. The absence of either word implies that both the suction and discharge lines are included. Example: discharge head.

Rule 18.2: The word *static* means that static head only is included (not velocity head, friction head, etc.). Example: static suction head.

Rule 18.3: The word *total* means that static head, velocity head, and friction head are all included. (Note that total does not mean the combination of suction and discharge.) Example: total suction head.

The following terms are commonly encountered.

- *friction head* (h_f): The head required to overcome resistance to flow in the pipe, fittings, valves, entrances, and exits.

$$h_f = \frac{fL\text{v}^2}{2Dg} \qquad 18.6$$

- *velocity head* (h_v): The specific kinetic energy of the fluid. Also known as *dynamic head.*

$$h_\text{v} = \frac{\text{v}^2}{2g} \qquad 18.7$$

[8]Equation 18.5 can be used to define *pressure head, atmospheric head,* and *vapor pressure head,* whose meanings and derivations should be obvious.
[9]The term *dynamic* is not as consistently applied as are the terms described in the rules. In particular, it is not clear whether friction head is to be included with the velocity head.

- *static suction head* ($h_{z(s)}$): The vertical distance above the centerline of the pump inlet to the free level of the fluid source. If the free level of the fluid is below the pump inlet, h_z will be negative and is known as *static suction lift.*

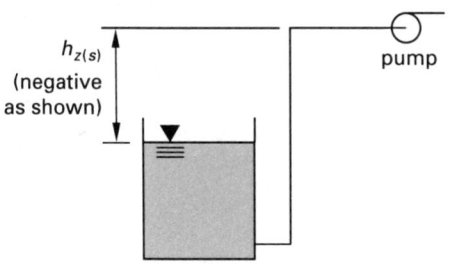

Figure 18.5 *Static Suction Lift*

- *static discharge head* ($h_{z(d)}$): The vertical distance above the centerline of the pump inlet to the point of free discharge or surface level of the discharge tank.

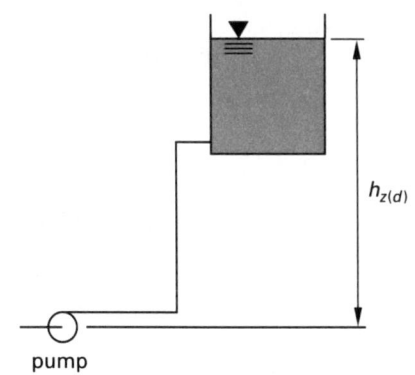

Figure 18.6 *Static Discharge Head*

Example 18.1

Write the symbolic equations for the following terms: (a) total suction head, (b) total discharge head, and (c) total head.

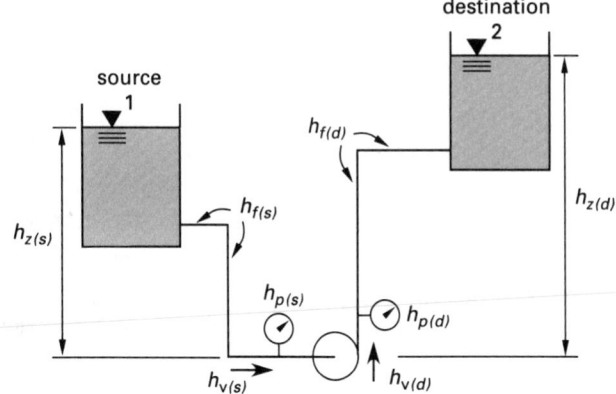

Solution

(a) The total suction head at the pump inlet is the sum of static (pressure) head and velocity head at the pump suction.

$$h_{t(s)} = h_{p(s)} + h_{v(s)}$$

Total suction head can also be calculated from the conditions existing at the source (1), in which case suction line friction would also be considered. (With an open reservoir, $h_{p(1)}$ will be zero if gage pressures are used, and $h_{v(1)}$ will be zero if the source is large.)

$$h_{t(s)} = h_{p(1)} + h_{z(s)} + h_{v(1)} - h_{f(s)}$$

(b) The total discharge head at the pump outlet is the sum of the static (pressure) and velocity heads at the pump outlet. Friction head is zero since the fluid has not yet traveled through any length of pipe when it is discharged.

$$h_{t(d)} = h_{p(d)} + h_{v(d)}$$

The total discharge head can also be evaluated at the destination (2) if the friction head, $h_{f(d)}$, between the discharge and the destination is known. (With an open reservoir, $h_{p(2)}$ will be zero if gage pressures are used, and $h_{v(2)}$ will be zero if the destination is large.)

$$h_{t(d)} = h_{p(2)} + h_{z(d)} + h_{v(2)} + h_{f(d)}$$

(c) The total head added by the pump is the total discharge head less the total suction head. Assuming suction from and discharge to reservoirs exposed to the atmosphere and assuming negligible reservoir velocities,

$$h_t = h_{t(d)} - h_{t(s)} \approx h_{z(d)} - h_{z(s)} + h_{f(d)} + h_{f(s)}$$

8. PUMPING POWER

The energy (head) added by a pump can be determined from the difference in total energy on either side of the pump. Writing the Bernoulli equation for the discharge and suction conditions produces Eq. 18.8, an equation for the *total dynamic head*.

$$h_A = h_t = h_{t(d)} - h_{t(s)} \qquad \text{18.8}$$

$$h_A = \frac{p_d - p_s}{\rho g} + \frac{v_d^2 - v_s^2}{2g} + (z_d - z_s) \qquad \text{[SI]} \quad \textbf{18.9(a)}$$

$$h_A = \frac{(p_d - p_s)g_c}{\rho g} + \frac{v_d^2 - v_s^2}{2g} + (z_d - z_s) \qquad \text{[U.S.]} \quad \textbf{18.9(b)}$$

In most applications, the change in velocity and potential heads is either zero or small in comparison to the increase in pressure head. Equation 18.9 then reduces to Eq. 18.10.

$$h_A = \frac{p_d - p_s}{\rho g} \qquad \text{[SI]} \quad \textbf{18.10(a)}$$

$$h_A = \left(\frac{p_d - p_s}{\rho}\right) \times \left(\frac{g_c}{g}\right) \qquad \text{[U.S.]} \quad \textbf{18.10(b)}$$

The head added by the pump can also be calculated from the impeller and fluid speeds. Equation 18.11 is useful for radial- and mixed-flow pumps for which the incoming fluid has little or no rotational velocity component (i.e., up to a specific speed of approximately 2000 U.S. or 40 SI). In Eq. 18.11, v_{impeller} is the tangential impeller velocity at the radius being considered, and v_{fluid} is the average tangential velocity imparted to the fluid by the impeller. The impeller efficiency, η_{impeller}, is typically 0.85 to 0.95. This is much higher than the total pump efficiency (Sec. 9) because it does not include mechanical and fluid friction losses.

$$h_A = \frac{\eta_{\text{impeller}} v_{\text{impeller}} v_{\text{fluid}}}{g} \qquad \textbf{18.11}$$

The pumping power depends on the head added (h_A) and the mass flow rate. For example, the product $\dot{m}\,h_A$ has the units of foot-pounds per second (in customary U.S. units), which can be easily converted to horsepower. Pump output power is known as *hydraulic power* or *water power*. Hydraulic power is the net power actually transferred to the fluid.

Horsepower is the unit of power in the United States and other non-SI countries, which gives rise to the terms *hydraulic horsepower* and *water horsepower*, WHP. Various relationships for finding the hydraulic horsepower are given in Table 18.4.

The unit of power in SI units is the watt (kilowatt). Table 18.5 can be used to determine *hydraulic kilowatts*, WkW.

Table 18.4 Hydraulic Horsepower Equations[a]

	Q in $\dfrac{\text{gal}}{\text{min}}$	$\dot{m}$ in $\dfrac{\text{lbm}}{\text{sec}}$	$\dot{V}$ in $\dfrac{\text{ft}^3}{\text{sec}}$
h_A in feet	$\dfrac{h_A Q(\text{SG})}{3956}$	$\left(\dfrac{h_A \dot{m}}{550}\right)\times\left(\dfrac{g}{g_c}\right)$	$\dfrac{h_A \dot{V}(\text{SG})}{8.814}$
Δp in psi	$\dfrac{\Delta p Q}{1714}$	$\left[\dfrac{\Delta p \dot{m}}{(238.3)(\text{SG})}\right]\times\left(\dfrac{g}{g_c}\right)$	$\dfrac{\Delta p \dot{V}}{3.819}$
Δp in $\dfrac{\text{lbf}}{\text{ft}^2}$	$\dfrac{\Delta p Q}{2.468 \times 10^5}$	$\left[\dfrac{\Delta p \dot{m}}{(34{,}320)(\text{SG})}\right]\times\left(\dfrac{g}{g_c}\right)$	$\dfrac{\Delta p \dot{V}}{550}$
W in $\dfrac{\text{ft-lbf}}{\text{lbm}}$	$\dfrac{W Q(\text{SG})}{3956}$	$\dfrac{W \dot{m}}{550}$	$\dfrac{W \dot{V}(\text{SG})}{8.814}$

(Multiply horsepower by 0.7457 to obtain kilowatts.)
[a] Table 18.4 is based on $\rho_{\text{water}} = 62.4 \text{ lbm/ft}^3$ and $g = 32.2 \text{ ft/sec}^2$.

Table 18.5 Hydraulic Kilowatt Equations[a]

	Q in $\dfrac{\text{L}}{\text{s}}$	$\dot{m}$ in $\dfrac{\text{kg}}{\text{s}}$	$\dot{V}$ in $\dfrac{\text{m}^3}{\text{s}}$
h_A in meters	$\dfrac{(9.81)h_A\,Q(\text{SG})}{1000}$	$\dfrac{(9.81)h_A\,\dot{m}}{1000}$	$(9.81)h_A\,\dot{V}(\text{SG})$
Δp in kPa	$\dfrac{\Delta p\,Q}{1000}$	$\dfrac{\Delta p\,\dot{m}}{1000(\text{SG})}$	$\Delta p\,\dot{V}$
W in $\dfrac{\text{J}}{\text{kg}}$	$\dfrac{W\,Q(\text{SG})}{1000}$	$\dfrac{W\,\dot{m}}{1000}$	$W\,\dot{V}(\text{SG})$

(Multiply kilowatts by 1.341 to obtain horsepower.)
[a]Table 18.5 is based on $\rho_{\text{water}} = 1000$ kg/m^3 and $g = 9.81$ m/s^2.

Example 18.2

A pump adds 550 ft of pressure head to 100 lbm/sec of water. (a) Complete the following table of performance data. (b) What is the hydraulic power in horsepower and kilowatts? (Assume $\rho = 62.4$ lbm/ft^3 or 1000 kg/m^3, and $g = 9.81$ m/s^2.)

item	customary U.S.	SI
$\dot{m}$	100 lbm/sec	— kg/s
h	550 ft	— m
Δp	— lbf/ft^2	— kPa
$\dot{V}$	— ft^3/sec	— m^3/s
W	— ft-lbf/lbm	— J/kg
P	— hp	— kW

Solution

(a) Work initially with the customary U.S. data.

$$\Delta p = \rho h \times \left(\frac{g}{g_c}\right) = \left(62.4\,\frac{\text{lbm}}{\text{ft}^3}\right)(550\text{ ft}) \times \left(\frac{g}{g_c}\right)$$
$$= 34{,}320 \text{ lbf/ft}^2$$

$$\dot{V} = \frac{\dot{m}}{\rho} = \frac{100\,\dfrac{\text{lbm}}{\text{sec}}}{62.4\,\dfrac{\text{lbm}}{\text{ft}^3}} = 1.603 \text{ ft}^3/\text{sec}$$

$$W = h \times \left(\frac{g}{g_c}\right) = 550 \text{ ft-lbf/lbm}$$

Now, convert to SI units.

$$\dot{m} = \frac{100\,\dfrac{\text{lbm}}{\text{sec}}}{2.201\,\dfrac{\text{lbm}}{\text{kg}}} = 45.43 \text{ kg/s}$$

$$h = \frac{550\text{ ft}}{3.281\,\dfrac{\text{ft}}{\text{m}}} = 167.6 \text{ m}$$

$$\Delta p = \left(34{,}320\,\frac{\text{lbf}}{\text{ft}^2}\right)\left(\frac{1}{144\,\dfrac{\text{in}^2}{\text{ft}^2}}\right)\left(6.895\,\frac{\text{kPa}}{\dfrac{\text{lbf}}{\text{in}^2}}\right)$$
$$= 1643 \text{ kPa}$$

$$\dot{V} = \left(1.603\,\frac{\text{ft}^3}{\text{sec}}\right)\left(0.0283\,\frac{\text{m}^3}{\text{ft}^3}\right) = 0.0454 \text{ m}^3/\text{s}$$

$$W = \left(550\,\frac{\text{ft-lbf}}{\text{lbm}}\right)\left(1.356\,\frac{\text{J}}{\text{ft-lbf}}\right)\left(2.201\,\frac{\text{lbm}}{\text{kg}}\right)$$
$$= 1642 \text{ J/kg}$$

(b) From Table 18.4, the hydraulic horsepower is

$$\text{WHP} = \frac{\dot{m}h_A}{550} \times \left(\frac{g}{g_c}\right)$$
$$= \left[\frac{\left(100\,\dfrac{\text{lbm}}{\text{sec}}\right)(550\text{ ft})}{550\,\dfrac{\text{ft-lbf}}{\text{hp-sec}}}\right] \times \left(\frac{g}{g_c}\right)$$
$$= 100 \text{ hp}$$

From Table 18.5, the power is

$$\text{WkW} = \frac{\dot{m}\Delta p}{(1000)(\text{SG})} = \frac{\left(45.43\,\dfrac{\text{kg}}{\text{s}}\right)(1643\text{ kPa})}{\left(1000\,\dfrac{\text{W}}{\text{kW}}\right)(1.0)}$$
$$= 74.6 \text{ kW}$$

9. PUMPING EFFICIENCY

Hydraulic power is the net energy actually transferred to the fluid per unit time. The input power delivered by the motor to the pump is known as the *brake pump power*. (For example, the term *brake horsepower*, bhp, is commonly used.) Due to frictional losses between the fluid and the pump and mechanical losses in the pump itself, the brake pump power will be greater than the hydraulic power.

The difference between brake and hydraulic power is accounted for by the pump efficiency, η_p. (Figure 18.7 gives typical pump efficiencies as a function of the

Fluids

pump's specific speed. The difference between the brake and hydraulic powers is known as the *friction power* (or *friction horsepower*).

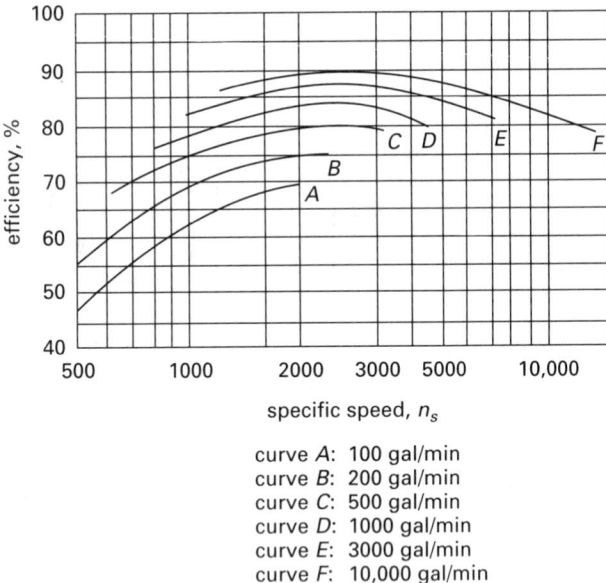

curve A: 100 gal/min
curve B: 200 gal/min
curve C: 500 gal/min
curve D: 1000 gal/min
curve E: 3000 gal/min
curve F: 10,000 gal/min

Figure 18.7 *Average Pump Efficiency Versus Specific Speed*

$$\text{brake pump power} = \frac{\text{hydraulic power}}{\eta_p} \qquad 18.12$$

$$\begin{matrix}\text{friction}\\\text{power}\end{matrix} = \begin{matrix}\text{brake}\\\text{pump}\\\text{power}\end{matrix} - \begin{matrix}\text{hydraulic}\\\text{power}\end{matrix} \qquad 18.13$$

Pumping efficiency is not constant for any specific pump; rather, it depends on the operating point (Sec. 21) and the speed of the pump. A specific pump's characteristic efficiency curves will be published by its manufacturer.

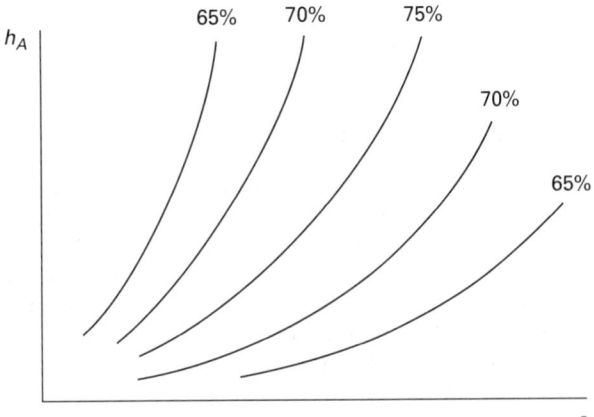

Figure 18.8 *Typical Centrifugal Pump Efficiency Curves*

Unfortunately, efficiency curves published by a manufacturer may not be representative of the actual installed efficiency. The manufacturer's efficiency may not include such losses in the suction elbow, discharge diffuser, couplings, bearing frame, seals, or pillow blocks. Up to 15% of the motor horsepower may be lost to these factors. Therefore, the manufacturer should be requested to provide the pump's installed *wire-to-water efficiency* (i.e., the fraction of the electrical power drawn that is converted to hydraulic power).

The pump must be driven by an engine or motor.[10] The power delivered to the motor is greater than the power delivered to the pump, as accounted for by the motor efficiency, η_m. If the pump motor is electrical, its input power requirements will be stated in kilowatts.[11]

$$\begin{matrix}\text{motor}\\\text{input}\\\text{power}\end{matrix} = \frac{\text{brake pump power}}{\eta_m} \qquad 18.14$$

$$\begin{matrix}(\text{motor}\\\text{input}\\\text{power})_{\text{kW}}\end{matrix} = \frac{0.7457 \times (\text{brake pump power})_{\text{hp}}}{\eta_m} \qquad 18.15$$

The *overall efficiency* of the pump installation is the product of the pump and motor efficiencies.

$$\eta = \eta_p \eta_m = \frac{\text{hydraulic horsepower}}{\text{motor horsepower}} \qquad 18.16$$

10. COST OF ELECTRICITY

The power utilization of pump motors is usually measured in kilowatts. The kilowatt usage represents the rate that energy is transferred by the pump motor. The total amount of work, W, done by the pump motor is found by multiplying the rate of energy usage by the length of time the pump is in operation.

$$W = Pt \qquad 18.17$$

Although the units of horsepower-hours are occasionally encountered, it is more common to measure electrical work in *kilowatt-hours* (kW-hr). Accordingly, the cost of electrical energy is stated per kW-hr (e.g., $0.10 per kW-hr).

$$\text{cost} = \frac{W_{\text{kW-hr}} \times \text{cost per kW-hr}}{\eta_m} \qquad 18.18$$

11. STANDARD MOTOR SIZES AND SPEEDS

An effort should be made to specify standard motor sizes when selecting the source of pumping power. Table 18.6 lists NEMA (National Electrical Manufacturers

---

[10]The source of power is sometimes called the *prime mover*.
[11]A *watt* is a joule per second.

Association) standard motor sizes by horsepower *nameplate rating*.[12] The rated horsepower is the maximum power the motor can provide without incurring damage. Other size motors may also be available by special order.

Table 18.6 *NEMA Standard Motor Sizes*
(brake horsepower)

$\frac{1}{8}$,	$\frac{1}{6}$,	$\frac{1}{4}$,	$\frac{1}{3}$,				
0.5,	0.75,	1,	1.5,	2,	3,	5,	7.5
10,	15,	20,	25,	30,	40,	50,	60
75,	100,	125,	150,	200,	250,	300,	350
400,	450,	500,	600,	700,	800,	900,	1000
1250,	1500,	1750,	2000,	2250,	2500,	2750,	3000
3500,	4000,	4500,	5000				

Larger horsepower motors are usually three-phase induction motors. The *synchronous speed*, n in rpm, of such motors is the speed of the rotating field, which depends on the number of poles per stator phase and the frequency, f. The number of poles must be an even number. The frequency is typically 60 Hz (hertz, previously cycles per second), as in the United States, or 50 Hz, as in European countries. Table 18.7 lists common synchronous speeds.

$$n = \frac{120 \times f}{\text{no. of poles}} \quad (\text{rpm}) \qquad \textit{18.19}$$

Table 18.7 *Common Synchronous Speeds*

number	n (rpm)	
of poles	60 Hz	50 Hz
2	3600	3000
4	1800	1500
6	1200	1000
8	900	750
10	720	600
12	600	500
14	514	428
18	400	333
24	300	250
48	150	125

Induction motors do not run at their synchronous speeds when loaded. Rather, they run at slightly less than synchronous speed. The deviation is known as the *slip*. Slip is typically around 4% and is seldom greater than 10% at full load for motors in the 1 to 75 hp range.

---

[12]The nameplate rating gets its name from the information stamped on the motor's identification plate. Besides the horsepower rating, other nameplate data used to classify the motor are the service class, voltage, full-load current, speed, number of phases, frequency of the current, and maximum ambient temperature (or, for older motors, the motor temperature rise).

$$\begin{array}{l} \text{slip} \\ \text{(in rpm)} \end{array} = \text{synchronous speed} - \text{actual speed} \qquad \textit{18.20}$$

$$\begin{array}{l} \text{percent} \\ \text{slip} \end{array} = 100\% \times \frac{\begin{array}{c}\text{synchronous speed}\\ -\text{ actual speed}\end{array}}{\text{synchronous speed}} \qquad \textit{18.21}$$

Induction motors may also be specified in terms of their kVA (kilovolt-amp) ratings. The kVA rating is not the same as the power in kilowatts, although one can be derived from the other if the motor's power factor is known. Such power factors typically range from 0.8 to 0.9, depending on the installation and motor size.

$$\text{kVA rating} = \frac{\text{motor power (in kW)}}{\text{power factor}} \qquad \textit{18.22}$$

Example 18.3

A pump driven by an electrical motor moves 25 gal/min of water from reservoir A to reservoir B, lifting the water a total of 245 ft. The efficiencies of the pump and motor are 64% and 84%, respectively. Electricity costs \$0.08/kW-hr. Neglect velocity head, friction, and minor losses. (a) What size motor is required? (b) How much does it cost to operate the pump for 6 hr?

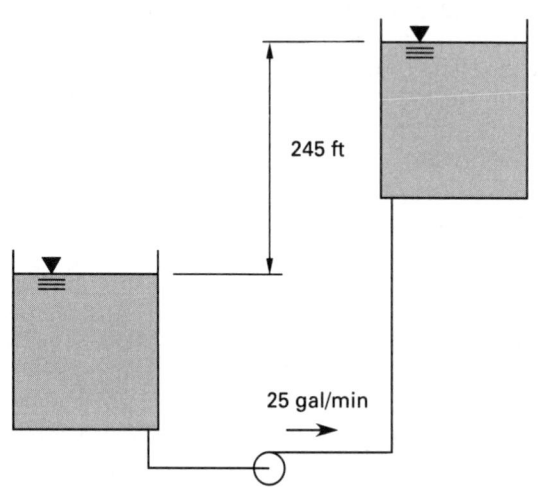

Solution

(a) The head added is 245 ft.

From Table 18.4, the motor power required is

$$\begin{aligned} P = \frac{h_A\, Q(\text{SG})}{3956\eta_p} &= \frac{(245 \text{ ft})\left(25\ \dfrac{\text{gal}}{\text{min}}\right)(1.0)}{(3956)(0.64)} \\ &= 2.42 \text{ hp} \end{aligned}$$

From Table 18.6, select a 3 hp motor.

(b) From Eqs. 18.17 and 18.18,

$$\text{cost} = \text{cost per kW-hr} \times \left(\frac{Pt}{\eta_m}\right)$$

$$= \frac{\left(0.08 \ \frac{\$}{\text{kW-hr}}\right)\left(0.7457 \ \frac{\text{kW}}{\text{hp}}\right)(2.42 \text{ hp})(6 \text{ hr})}{0.84}$$

$$= \$1.03$$

12. PUMP SHAFT LOADING

The torque on a pump or motor shaft, brake power, and speed are all related. The power used in Eqs. 18.23 through 18.25 is the brake (shaft) power, not the hydraulic power.

$$T_{\text{in-lbf}} = \frac{63{,}025 \times P_{\text{hp}}}{n} \qquad \textbf{18.23}$$

$$T_{\text{ft-lbf}} = \frac{5252 \times P_{\text{hp}}}{n} \qquad \textbf{18.24}$$

$$T_{\text{N}\cdot\text{m}} = \frac{9549 \times P_{\text{kW}}}{n} \qquad \textbf{18.25}$$

The torque can also be calculated from the change in the momentum of the fluid flow. For radial impellers, the fluid enters through the eye and is turned 90°. The direction change is related to the increase in momentum and the shaft torque. When fluid is introduced axially through the eye of the impeller, the tangential velocity at the inlet (eye), $v_{t,i}$, is zero.[13]

$$T = \left(\frac{\dot{m}}{g_c}\right)\left(v_{t(d)}r_{\text{impeller}} - v_{t(i)}r_{\text{eye}}\right) \qquad \textbf{18.26}$$

Centrifugal pumps may be driven directly from a motor, or a speed changer may be used. Rotary pumps generally require a speed reduction. *Gear motors* have integral speed reducers. V-belt drives are widely used because of their initial low cost, although timing belts and chains can be used in some applications.

When a belt or chain is used, the pump's and motor's maximum overhung loads must be checked. This is particularly important for high-power, low-speed applications (such as rotary pumps). *Overhung load* is the side load (force) put on shafts and bearings. The overhung load is calculated from Eq. 18.27. The empirical factor K is 1.0 for chain drives, 1.25 for timing belts, and 1.5 for V-belts.

$$\text{overhung load} = \frac{2KT}{D_{\text{sheave}}} \qquad \textbf{18.27}$$

---

[13]The tangential component of fluid velocity is sometimes referred to as the *velocity of whirl*.

If a direct-drive cannot be used and the overhung load is excessive, the installation can incorporate a jack shaft or outboard bearing.

Example 18.4

A centrifugal pump delivers 275 lbm/sec (125 kg/s) of water while turning at 850 rpm. The impeller has straight radial vanes and an outside diameter of 10 in (25.4 cm). Water enters the impeller through the eye. The driving motor develops 30 hp (22 kW). What are the (a) theoretical torque, (b) pump efficiency, and (c) total dynamic head?

SI Solution

(a) From Eq. 18.3, the impeller's tangential velocity is

$$v_t = \frac{\pi D n}{60 \ \frac{\text{s}}{\text{min}}} = \frac{\pi(25.4 \text{ cm})(850 \text{ rpm})}{\left(60 \ \frac{\text{s}}{\text{min}}\right)\left(100 \ \frac{\text{cm}}{\text{m}}\right)}$$

$$= 11.3 \text{ m/s}$$

Since water enters axially, the incoming water has no tangential component. From Eq. 18.26,

$$T = \dot{m}v_{t(d)}r_{\text{impeller}}$$

$$= \frac{\left(125 \ \frac{\text{kg}}{\text{s}}\right)\left(11.3 \ \frac{\text{m}}{\text{s}}\right)\left(\frac{25.4 \text{ cm}}{2}\right)}{100 \ \frac{\text{cm}}{\text{m}}}$$

$$= 179.4 \text{ N}\cdot\text{m}$$

(b) From Eq. 18.25, the theoretical power is

$$P_{\text{kW}} = \frac{nT_{\text{N}\cdot\text{m}}}{9549} = \frac{(850 \text{ rpm})(179.4 \text{ N}\cdot\text{m})}{9549}$$

$$= 15.97 \text{ kW}$$

The pump efficiency is

$$\eta_p = \frac{P_{\text{theoretical}}}{P_{\text{actual}}}$$

$$= \frac{15.97 \text{ kW}}{22 \text{ kW}} = 0.726 \quad (72.6\%)$$

(c) From Table 18.5, the total dynamic head is

$$h_A = \frac{P_{\text{kW}}\left(1000 \ \frac{\text{W}}{\text{kW}}\right)}{\dot{m}\left(9.81 \ \frac{\text{m}}{\text{s}^2}\right)}$$

$$= \frac{(15.97 \text{ kW})\left(1000 \ \frac{\text{W}}{\text{kW}}\right)}{\left(125 \ \frac{\text{kg}}{\text{s}}\right)\left(9.81 \ \frac{\text{m}}{\text{s}^2}\right)}$$

$$= 13.0 \text{ m}$$

Customary U.S. Solution

(a) From Eq. 18.3, the impeller's tangential velocity is

$$v_t = \frac{\pi D n}{60 \frac{\text{sec}}{\text{min}}} = \frac{\pi (10 \text{ in})(850 \text{ rpm})}{\left(60 \frac{\text{sec}}{\text{min}}\right)\left(12 \frac{\text{in}}{\text{ft}}\right)}$$

$$= 37.08 \text{ ft/sec}$$

Since water enters axially, the incoming water has no tangential component. From Eq. 18.26,

$$T = \frac{\dot{m} v_{t(d)} r_{\text{impeller}}}{g_c}$$

$$= \frac{\left(275 \frac{\text{lbm}}{\text{sec}}\right)\left(37.08 \frac{\text{ft}}{\text{sec}}\right)\left(\frac{10 \text{ in}}{2}\right)}{\left(32.2 \frac{\text{lbm-ft}}{\text{lbf-sec}^2}\right)\left(12 \frac{\text{in}}{\text{ft}}\right)}$$

$$= 131.9 \text{ ft-lbf}$$

(b) From Eq. 18.24, the theoretical power is

$$P_{\text{hp}} = \frac{T_{\text{ft-lbf}} n}{5252}$$

$$= \frac{(131.9 \text{ ft-lbf})(850 \text{ rpm})}{5252} = 21.35 \text{ hp}$$

The pump efficiency is

$$\eta_p = \frac{P_{\text{theoretical}}}{P_{\text{actual}}}$$

$$= \frac{21.35 \text{ hp}}{30 \text{ hp}} = 0.712 \quad (71.2\%)$$

(c) From Table 18.4, the total dynamic head is

$$h_A = \left[\frac{\left(550 \frac{\text{ft-lbf}}{\text{hp-sec}}\right)(P_{\text{hp}})}{\dot{m}}\right] \times \left(\frac{g_c}{g}\right)$$

$$= \left[\frac{\left(550 \frac{\text{ft-lbf}}{\text{hp-sec}}\right)(21.35 \text{ hp})}{275 \frac{\text{lbm}}{\text{sec}}}\right]$$

$$\times \left(\frac{32.2 \frac{\text{lbm-ft}}{\text{lbf-sec}^2}}{32.2 \frac{\text{ft}}{\text{sec}^2}}\right)$$

$$= 42.7 \text{ ft}$$

13. SPECIFIC SPEED

The capacity and efficiency of a centrifugal pump are partially governed by the impeller design. For a desired flow rate and added head, there will be one optimum impeller design. The quantitative index used to optimize the impeller design is known as *specific speed*, n_s, also known as *impeller specific speed*. Table 18.8 lists the impeller designs that are appropriate for different specific speeds.[14]

Table 18.8 Specific Speed versus Impeller Design

approximate range of specific speed (rpm)		
customary U.S. units	SI units	impeller type
500–1000	10–20	radial vane
2000–3000	40–60	Francis (mixed) vane
4000–7000	80–140	mixed flow
9000 and above	180 and above	axial flow

Highest heads per stage are developed at low specific speeds. However, for best efficiency, specific speed should be greater than 650 (13 in SI units). If the specific speed for a given set of conditions drops below 650 (13), a multiple-stage pump should be selected.[15]

Specific speed is a function of a pump's capacity, head, and rotational speed at peak efficiency, as shown in Eq. 18.28. For a given pump and impeller configuration, the specific speed remains essentially constant over a range of flow rates and heads.

While specific speed is not dimensionless, the units are meaningless. Specific speed may be assigned units of rpm, but most often it is expressed simply as a pure number. (Q or $\dot{V}$ in Eq. 18.28 is half of the full flow rate for double-suction pumps.)

$$n_s = \frac{n\sqrt{\dot{V} \text{ in } \frac{\text{m}^3}{\text{s}}}}{(h_A \text{ in m})^{0.75}} \qquad \text{[SI]} \qquad \textit{18.28(a)}$$

$$n_s = \frac{n\sqrt{Q \text{ in } \frac{\text{gal}}{\text{min}}}}{(h_A \text{ in ft})^{0.75}} \qquad \text{[U.S.]} \qquad \textit{18.28(b)}$$

---

[14]Specific speed is useful for more than just selecting an impeller type. Maximum suction lift, pump efficiency, and net positive suction head required (NPSHR) can be correlated with specific speed.

[15]Advances in *partial emission, forced vortex centrifugal pumps* allow operation down to specific speeds of 150 (3 in SI). Such pumps have been used for low-flow, high-head applications, such as high-pressure petrochemical cracking processes.

The numerical range of acceptable performance for each impeller type is redefined when SI units are used. The SI specific speed is obtained by dividing the customary U.S. specific speed by 51.66.

A common definition of specific speed is the speed (in rpm) at which a *homologous pump* would have to turn in order to deliver one gallon per minute at one foot total added head.[16] This definition is implicit to Eq. 18.28 but is not very useful otherwise.

Specific speed can be used to determine the type of impeller needed. Once a pump is selected, its specific speed and Eq. 18.28 can be used to determine other operational parameters (e.g., maximum rotational speed). Specific speed can be used with Fig. 18.7 to obtain an approximate pump efficiency.

Example 18.5

A centrifugal pump powered by a direct-drive induction motor is needed to discharge 150 gal/min against a 300 ft total head when turning at the fully loaded speed of 3500 rpm. What type of pump should be selected?

Solution

From Eq. 18.28, the specific speed is

$$n_s = \frac{n\sqrt{Q}}{(h_A)^{0.75}} = \frac{3500\sqrt{150}}{(300)^{0.75}} = 595$$

From Table 18.8, the pump should be a radial vane type. However, pumps achieve their highest efficiencies when specific speed exceeds 650. (See Fig. 18.7.) To increase the specific speed, the rotational speed can be increased, or the total added head can be decreased. Since the pump is direct-driven and 3600 rpm is the maximum speed for induction motors (see Table 18.7), the total added head should be divided evenly between two stages, or two pumps should be used in series.

In a two-stage system, the specific speed would be

$$n_s = \frac{3500\sqrt{150}}{(150)^{0.75}} = 1000$$

This is satisfactory for a radial vane pump.

Example 18.6

An induction motor turning at 1200 rpm is to be selected to drive a single-stage, single-suction centrifugal water pump through a direct drive. The total dynamic head added by the pump is 26 ft. The flow rate is 900 gpm. What size motor should be selected?

---

[16] *Homologous pumps* are geometrically similar. This means that each pump is a scaled up or down version of the others. Such pumps are said to belong to a *homologous family*.

Solution

The specific speed is

$$n_s = \frac{n\sqrt{Q}}{(h_A)^{0.75}} = \frac{1200\sqrt{900}}{(26)^{0.75}}$$

$$= 3127$$

From Fig. 18.7, the pump efficiency will be approximately 82%.

From Table 18.4, the minimum motor horsepower is

$$P_{hp} = \frac{h_A Q(\text{SG})}{\left(3956 \ \frac{\text{ft-gal}}{\text{hp-min}}\right)\eta_p}$$

$$= \frac{(26 \text{ ft})\left(900 \ \frac{\text{gal}}{\text{min}}\right)(1.0)}{\left(3956 \ \frac{\text{ft-gal}}{\text{hp-min}}\right)(0.82)}$$

$$= 7.2 \text{ hp}$$

Example 18.7

A single-stage pump driven by a 3600 rpm motor is currently delivering 150 gal/min. The total dynamic head is 430 ft. What would be the approximate increase in efficiency if the single-stage pump is replaced by a double-stage pump?

Solution

The specific speed is

$$n_s = \frac{n\sqrt{Q}}{(h_A)^{0.75}} = \frac{3600\sqrt{150}}{(430)^{0.75}}$$

$$= 467$$

From Fig. 18.7, the approximate efficiency is 45%.

In a two-stage pump, each stage adds half of the head. The specific speed per stage would be

$$n_s = \frac{n\sqrt{Q}}{(h_A)^{0.75}} = \frac{3600\sqrt{150}}{\left(\dfrac{430}{2}\right)^{0.75}}$$

$$= 785$$

From Fig. 18.7, the efficiency for this configuration is approximately 60%.

The increase in efficiency is 60% − 45% = 15%. Whether or not the cost of multistaging is worthwhile in this low-volume application would have to be determined.

14. CAVITATION

Cavitation is the spontaneous vaporization of the fluid, resulting in a degradation of pump performance. If the fluid pressure is less than the vapor pressure, small pockets of vapor will form. These pockets usually form only within the pump itself, although cavitation slightly upstream within the suction line is also possible. As the vapor pockets reach the surface of the impeller, the local high fluid pressure collapses them. Noise, vibration, impeller pitting, and structural damage to the pump casing are manifestations of cavitation.

Cavitation can be caused by any of the following conditions:

- discharge head far below the pump head at peak efficiency

- high suction lift or low suction head

- excessive pump speed

- high liquid temperature (i.e., high vapor pressure)

15. NET POSITIVE SUCTION HEAD

The occurrence of cavitation is predictable. Cavitation will occur when the net pressure in the fluid drops below the vapor pressure. This criterion is commonly stated in terms of head: cavitation occurs when the available head is less than the required head for satisfactory operation.

$$\text{available head} < \text{required head} \qquad \begin{bmatrix} \text{criterion for} \\ \text{cavitation} \end{bmatrix}$$

The minimum fluid energy required at the pump inlet for satisfactory operation (i.e., the required head) is known as the *net positive suction head required*, NPSHR.[17] NPSHR is a function of the pump and will be given by the pump manufacturer as part of the pump performance data.[18] NPSHR is dependent on the flow rate. However, if NPSHR is known for one flow rate, it can be determined for another flow rate from Eq. 18.29.

$$\frac{\text{NPSHR}_2}{\text{NPSHR}_1} = \left(\frac{Q_2}{Q_1} \right)^2 \qquad 18.29$$

Net positive suction head available, NPSHA, is the actual fluid energy at the inlet. There are two different methods for calculating NPSHA, both of which are correct and yield identical answers. Equation 18.30(a) is

based on the conditions at the fluid surface at the top of the fluid source (e.g., tank or reservoir). There is a potential energy term but no kinetic energy term. Equation 18.30(b) is based on the conditions at the immediate entrance (suction, subscript s) to the pump. At that point, some of the potential head has been converted to velocity head. Frictional losses are implicitly part of the reduced pressure head, as is the atmospheric pressure head. The effect of higher altitudes is explicit in Eq. 18.30(a) and implicit in Eq. 18.30(b).

$$\text{NPSHA} = h_{\text{atm}} + h_{z(s)} - h_{f(s)} - h_{\text{vp}} \qquad 18.30(a)$$

$$\text{NPSHA} = h_{p(s)} + h_{v(s)} - h_{\text{vp}} \qquad 18.30(b)$$

The net positive suction head available (NPSHA) for most positive displacement pumps includes a term for acceleration head.[19]

$$\text{NPSHA} = h_{\text{atm}} + h_{z(s)} - h_{f(s)} - h_{\text{vp}} - h_{\text{ac}} \qquad 18.31$$

$$\text{NPSHA} = h_{p(s)} + h_{v(s)} - h_{\text{vp}} - h_{\text{ac}} \qquad 18.32$$

If NPSHA is less than NPSHR, the fluid will cavitate. The criterion for cavitation is given by Eq. 18.33. (In practice, it is desirable to have a safety margin.)

$$\text{NPSHA} < \text{NPSHR} \qquad \text{[criterion for cavitation]} \quad 18.33$$

Example 18.8

2.0 ft³/sec (56 L/s) of 60°F (16°C) water are pumped from an elevated feed tank to an open reservoir through 6 in (15.2 cm), schedule-40 steel pipe, as shown. The friction loss for the piping and fittings in the suction line is 2.6 ft (0.9 m). The friction loss for the piping and fittings in the discharge line is 13 ft (4.3 m). The atmospheric pressure is 14.7 psia (101 kPa). What is the NPSHA?

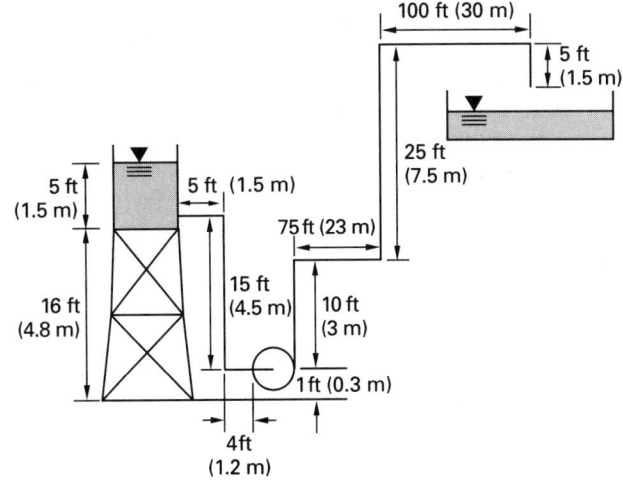

---

[17]If NPSHR (a head term) is multiplied by the fluid specific weight, it is known as the *net inlet pressure required*, NIPR. Similarly, NPSHA can be converted to NIPA.

[18]It is also possible to calculate NPSHR from other information, such as suction specific speed. However, this still depends on information provided by the manufacturer.

---

[19]The friction loss and the acceleration are both maximum values, but they do not occur in phase. Combining them is conservative.

SI Solution

The density of water is approximately 1000 kg/m^3. The atmospheric head is

$$h_{\text{atm}} = \frac{p}{\rho g} = \frac{(101 \text{ kPa})\left(1000 \frac{\text{Pa}}{\text{kPa}}\right)}{\left(1000 \frac{\text{kg}}{\text{m}^3}\right)\left(9.81 \frac{\text{m}}{\text{s}^2}\right)}$$
$$= 10.3 \text{ m}$$

The vapor pressure is the same as the saturation pressure corresponding to the water temperature. From steam tables for 16°C water, the saturation pressure is approximately 0.01818 bars. The vapor pressure head is

$$h_{\text{vp}} = \frac{p}{\rho g} = \frac{(0.01818 \text{ bars})\left(1 \times 10^5 \frac{\text{Pa}}{\text{bar}}\right)}{\left(1000 \frac{\text{kg}}{\text{m}^3}\right)\left(9.81 \frac{\text{m}}{\text{s}^2}\right)}$$
$$= 0.2 \text{ m}$$

From Eq. 18.30(a), the NPSHA is

$$\text{NPSHA} = h_{\text{atm}} + h_{z(s)} - h_{f(s)} - h_{\text{vp}}$$
$$= 10.3 \text{ m} + (1.5 \text{ m} + 4.8 \text{ m} - 0.3 \text{ m})$$
$$- 0.9 \text{ m} - 0.2 \text{ m}$$
$$= 15.2 \text{ m}$$

Customary U.S. Solution

The specific weight of water is approximately 62.4 lbf/ft^3. The atmospheric head is

$$h_{\text{atm}} = \frac{p}{\gamma} = \frac{\left(14.7 \frac{\text{lbf}}{\text{in}^2}\right)\left(144 \frac{\text{in}^2}{\text{ft}^2}\right)}{62.4 \frac{\text{lbf}}{\text{ft}^3}}$$
$$= 33.9 \text{ ft}$$

The vapor pressure is the same as the saturation pressure corresponding to the water temperature. From steam tables for 60°F water, the saturation pressure is approximately 0.256 lbf/in^2. The vapor pressure head is

$$h_{\text{vp}} = \frac{p}{\gamma} = \frac{\left(0.256 \frac{\text{lbf}}{\text{in}^2}\right)\left(144 \frac{\text{in}^2}{\text{ft}^2}\right)}{62.4 \frac{\text{lbf}}{\text{ft}^3}}$$
$$= 0.6 \text{ ft}$$

From Eq. 18.30(a), the NPSHA is

$$\text{NPSHA} = h_{\text{atm}} + h_{z(s)} - h_{f(s)} - h_{\text{vp}}$$
$$= 33.9 \text{ ft} + (5 \text{ ft} + 16 \text{ ft} - 1 \text{ ft}) - 2.6 \text{ ft} - 0.6 \text{ ft}$$
$$= 50.7 \text{ ft}$$

16. PREVENTING CAVITATION

Cavitation is eliminated by increasing NPSHA or decreasing NPSHR. NPSHA can be increased by:

- increasing the height of the fluid source
- lowering the pump
- reducing friction and minor losses by shortening the suction line or using a larger pipe size
- reducing the temperature of the fluid at the pump entrance
- pressurizing the fluid supply tank
- reducing the flow rate or velocity (i.e., reducing the pump speed)

NPSHR can be reduced by:

- placing a throttling valve or restriction in the discharge line[20]
- using an oversized pump
- using a double-suction pump
- using an impeller with a larger eye
- using an inducer

High NPSHR applications, such as boiler feed pumps needing 150 to 250 ft (50 to 80 m), should use one or more booster pumps in front of each high-NPSHR pump. Such booster pumps are typically single-stage, double-suction pumps running at low speed. Their NPSHR can be 25 ft (8 m) or less.

Note that throttling the input line to a pump and venting or evacuating the receiving tank both increase cavitation. Throttling the input line increases the friction head and decreases NPSHA. Evacuating the receiving tank increases the flow rate, increasing NPSHR while simultaneously increasing the friction head and reducing NPSHA.

17. CAVITATION COEFFICIENT

The *cavitation coefficient* (or *cavitation number*) is a dimensionless number that can be used in modeling and extrapolating experimental results. The actual cavitation coefficient is compared with the critical cavitation number obtained experimentally. If the actual cavitation number is less than the critical cavitation number, cavitation will occur. Absolute pressure must be used in calculating σ.

---

[20]This will increase the total head (h_A) added by the pump, thereby reducing the pump's output and driving the pump's operating point into a region of lower NPSHR.

$$\sigma = \frac{2(p - p_{vp})}{\rho v^2} = \frac{NPSHA}{h_A} \qquad \text{[SI]} \qquad \textbf{18.34(a)}$$

$$\sigma = \frac{2g_c(p - p_{vp})}{\rho v^2} = \frac{NPSHA}{h_A} \qquad \text{[U.S.]} \qquad \textbf{18.34(b)}$$

$$\sigma < \sigma_{cr} \qquad \text{[criterion for cavitation]} \qquad \textbf{18.35}$$

The two forms of Eq. 18.34 yield slightly different results. The first form is essentially the ratio of the net pressure available for collapsing a vapor bubble to the velocity pressure creating the vapor. It is useful in model experiments. The second form is applicable to tests of production model pumps.

18. SUCTION SPECIFIC SPEED

The formula for *suction specific speed*, n_{ss}, can be derived by substituting NPSHR for total head in the expression for specific speed. Q is halved for double-suction pumps.

$$n_{ss} = \frac{n\sqrt{\dot{V} \text{ in } \frac{m^3}{s}}}{(NPSHR \text{ in m})^{0.75}} \qquad \text{[SI]} \qquad \textbf{18.36(a)}$$

$$n_{ss} = \frac{n\sqrt{Q \text{ in } \frac{gal}{min}}}{(NPSHR)^{0.75}} \qquad \text{[U.S.]} \qquad \textbf{18.36(b)}$$

Suction specific speed is an index of the suction characteristics of the impeller. Ideally, it should be approximately 8500 (165 in SI) for both single- and double-suction pumps. This assumes the pump is operating at or near its point of optimum efficiency.

Suction specific speed can be used to determine the maximum recommended operating speed by substituting 8500 (165 in SI) for n_{ss} in Eq. 18.37 and solving for n.

If the suction specific speed is known, it can be used to determine the NPSHR. If the pump is known to be operating at or near its optimum efficiency, an approximate NPSHR value can be found by substituting 8500 (165 in SI) for n_{ss} in Eq. 18.36 and solving for NPSHR.

Suction specific speed available, SA, is obtained when NPSHA is substituted for total head in the expression for specific speed. The suction specific speed available must exceed the suction specific speed required to prevent cavitation.[21]

19. PUMP PERFORMANCE CURVES

For a given impeller diameter and constant speed, the head added will decrease as the flow rate increases. This can be shown graphically on the *pump performance*

curve (pump curve) supplied by the pump manufacturer. Other operating characteristics (e.g., power requirement, NPSHR, and efficiency) also vary with flow rate, and these are usually plotted on a common graph, as shown in Fig. 18.9.[22] Manufacturers' pump curves show performance over a limited number of calibration speeds. If an operating point is outside the range of published curves, the affinity laws can be used to estimate a speed at which the pump gives the required performance.

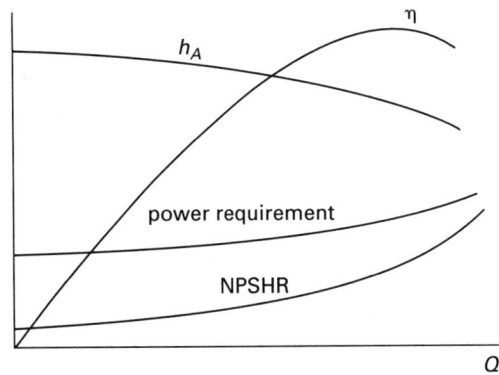

Figure 18.9 *Pump Performance Curves*

Figure 18.9 is for a pump with a fixed impeller diameter and rotational speed. The characteristics of a pump operated over a range of speed or for different impeller diameters are illustrated in Fig. 18.10.

20. SYSTEM CURVES

A *system curve* (or *system performance curve*) is a plot of the static and friction energy losses experienced by the fluid for different flow rates. Unlike the pump curve, which depends only on the pump, the system curve depends only on the configuration of the suction and discharge lines. (The following equations assume equal pressures at the fluid source and destination surfaces, which is the case for pumping from one atmospheric reservoir to another. The velocity head is insignificant and is disregarded.)

$$h_A = h_z + h_f \qquad \textbf{18.37}$$
$$h_z = h_{z(d)} - h_{z(s)} \qquad \textbf{18.38}$$
$$h_f = h_{f(s)} + h_{f(d)} \qquad \textbf{18.39}$$

If the fluid reservoirs are large, or if the fluid reservoir levels are continually replenished, the net static suction head $(h_{z(1)} - h_{z(2)})$ will be constant for all flow rates. The friction loss, h_f, varies with v^2 (and hence with Q)

---

[21]Since speed and flow rate are constants, this is another way of saying NPSHA must equal or exceed NPSHR.

[22]The term *pump curve* is commonly used to designate the h_A versus Q characteristics, whereas *pump characteristics curve* implies all of the pump data.

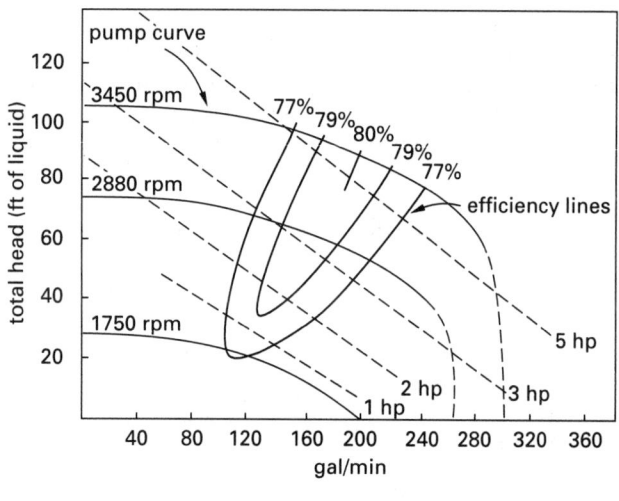

(a) variable speed

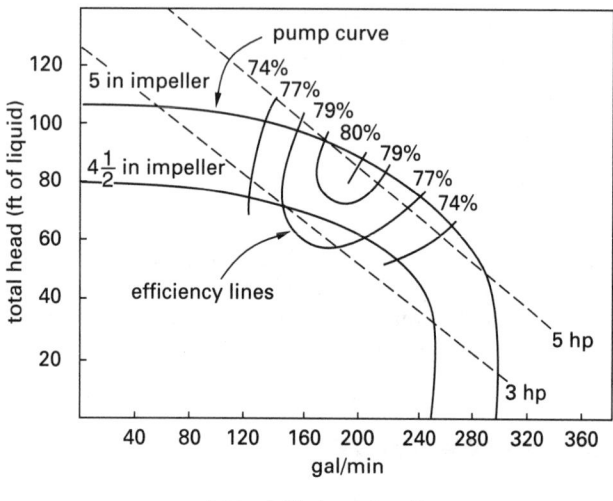

(b) variable impeller diameter

Figure 18.10 *Centrifugal Pump Characteristics Curves*

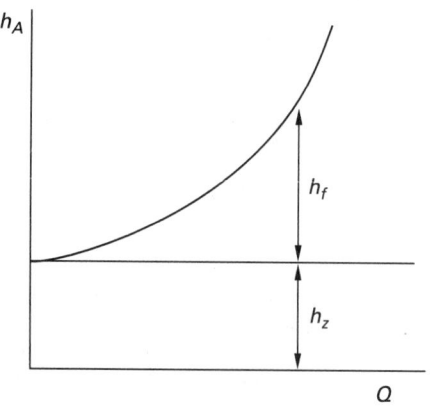

Figure 18.11 *System Curve*

21. OPERATING POINT

The intersection of the pump curve and the system curve determines the *operating point*, as shown in Fig. 18.12. The operating point defines the system head and system flow rate.

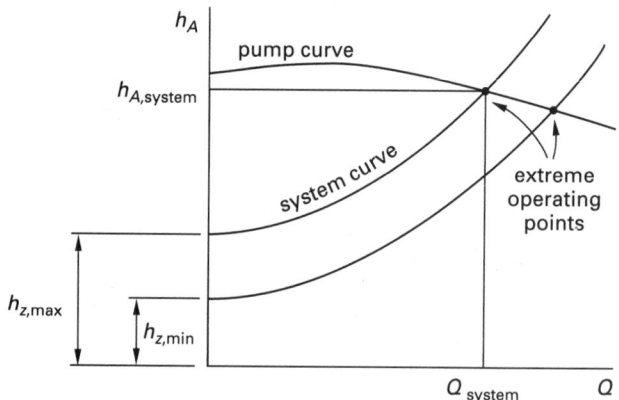

Figure 18.12 *Extreme Operating Points*

When selecting a pump, the system curve is plotted on manufacturers' pump curves for different speeds and/or impeller diameters (i.e., Fig. 18.10). There will be several possible operating points corresponding to the various pump curves shown. Generally, the design operating point should be close to the highest pump efficiency. This, in turn, will determine speed and impeller diameter.

In many systems, the static head will vary as the source reservoir is drained or as the destination reservoir fills. The system head is then defined by a pair of matching system friction curves intersecting the pump curve. The two intersection points are the *extreme operating points*—the maximum and minimum capacity requirements.

After a pump is installed, it may be desired to change the operating point. This can be done without replacing the pump by placing a throttling valve in the discharge

in the Darcy friction formula. This makes it easy to find friction losses for other flow rates once one friction loss is known.[23]

$$\frac{h_{f,1}}{h_{f,2}} = \left(\frac{Q_1}{Q_2}\right)^2 \qquad \text{18.40}$$

Figure 18.11 illustrates a system curve with a positive static head. In the case of a negative static head (i.e., a fluid source above the fluid destination), the system curve would be depressed below the Q-axis and would cross it at some positive value of Q. Of course, the curve could also start at the origin.

---

[23]Equation 18.40 implicitly assumes that the friction factor, f, is constant. This may be true over a limited range of flow rates, but it is not true over large ranges unless the Hazen-Williams friction loss equation is being used. Nevertheless, Eq. 18.40 is often used to quickly construct preliminary versions of the system curve.

line. The operating point can then be moved along the pump curve by opening or closing the valve, as illustrated in Fig. 18.13. (A throttling valve should never be placed in the suction line since that would reduce NPSHA.)

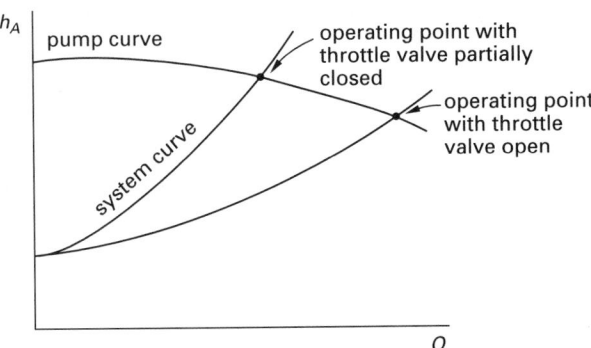

Figure 18.13 *Throttling the Discharge*

22. PUMPS IN PARALLEL

Parallel operation is obtained by having two pumps discharging into a common header. This type of connection is advantageous when the system demand varies greatly and when high reliability is required. A single pump providing total flow would have to operate far from its optimum efficiency at one point or another. With two pumps in parallel, one can be shut down during low demand. This allows the remaining pump to operate close to its optimum efficiency point.

Figure 18.14 illustrates that parallel operation increases the capacity of the system while maintaining the same total head.

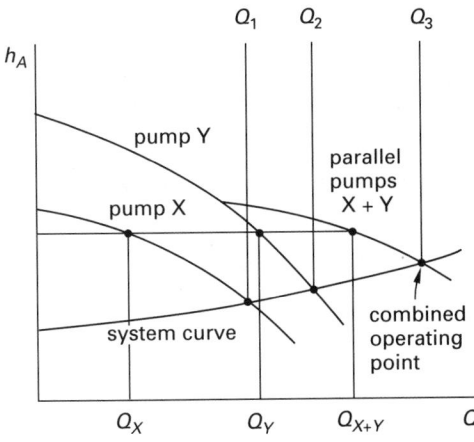

Figure 18.14 *Pumps Operating in Parallel*

The performance curve for a set of pumps in parallel can be plotted by adding the capacities of the two pumps at various heads. Capacity does not increase at heads

above the maximum head of the smaller pump. Furthermore, a second pump will operate only when its discharge head is greater than the discharge head of the pump already running.

When the parallel performance curve is plotted with the system head curve, the operating point is the intersection of the system curve with the X + Y curve. With pump X operating alone, the capacity is given by Q_1. When pump Y is added, the capacity increases to Q_3 with a slight increase in total head.

23. PUMPS IN SERIES

Series operation is achieved by having one pump discharge into the suction of the next. This arrangement is used primarily to increase the discharge head, although a small increase in capacity also results.

The performance curve for a set of pumps in series can be plotted by adding the heads of the two pumps at various capacities.

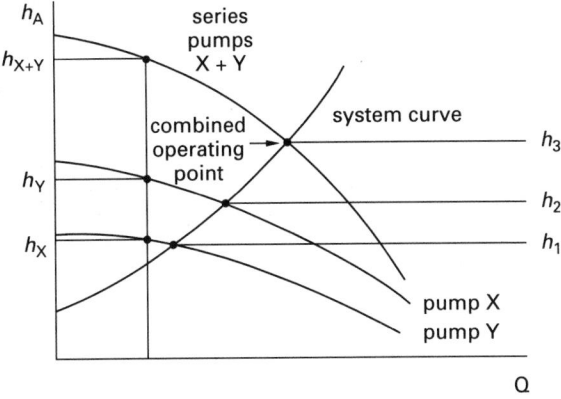

Figure 18.15 *Pumps Operating in Series*

24. AFFINITY LAWS

Most parameters (impeller diameter, speed, and flow rate) determining a pump's performance can vary. If the impeller diameter is held constant and the speed is varied, the following ratios are maintained with no change of efficiency.

$$\frac{Q_2}{Q_1} = \frac{n_2}{n_1} \qquad 18.41$$

$$\frac{h_2}{h_1} = \left(\frac{n_2}{n_1}\right)^2 = \left(\frac{Q_2}{Q_1}\right)^2 \qquad 18.42$$

$$\frac{P_2}{P_1} = \left(\frac{n_2}{n_1}\right)^3 = \left(\frac{Q_2}{Q_1}\right)^3 \qquad 18.43$$

If the speed is held constant and the impeller size is varied,

$$\frac{Q_2}{Q_1} = \frac{D_2}{D_1} \qquad 18.44$$

$$\frac{h_2}{h_1} = \left(\frac{D_2}{D_1}\right)^2 \qquad\qquad 18.45$$

$$\frac{P_2}{P_1} = \left(\frac{D_2}{D_1}\right)^3 \qquad\qquad 18.46$$

The affinity laws are based on the assumption that the efficiencies of the two pumps are the same. In reality, larger pumps are somewhat more efficient than smaller pumps. Therefore, extrapolations to greatly different sized pumps should be avoided. Equation 18.47 can be used to estimate the efficiency of a different sized pump. The dimensionless exponent varies from 0 to approximately 0.26, with 0.2 being a typical value.

$$\frac{1 - \eta_{\text{smaller}}}{1 - \eta_{\text{larger}}} = \left(\frac{D_{\text{larger}}}{D_{\text{smaller}}}\right)^n \qquad\qquad 18.47$$

Example 18.9

A pump operating at 1770 rpm delivers 500 gal/min against a total head of 200 ft. Changes in the piping system have increased the total head to 375 ft. At what rpm should this pump be operated to achieve this new head at the same efficiency?

Solution

From Eq. 18.42,

$$n_2 = n_1 \sqrt{\frac{h_2}{h_1}} = 1770 \text{ rpm} \sqrt{\frac{375 \text{ ft}}{200 \text{ ft}}}$$
$$= 2424 \text{ rpm}$$

Example 18.10

A pump is required to pump 500 gal/min while providing a total dynamic head of 425 ft. The hydraulic system has no static head change. Only the 1750 rpm performance curve is known for the pump. At what speed must the pump be turned to achieve the desired performance with no change in efficiency or impeller size?

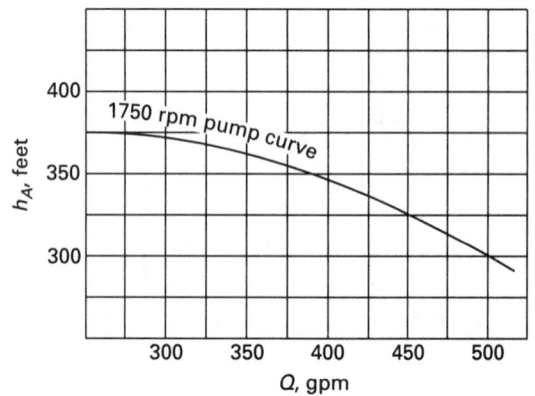

Solution

A flow of 500 gal/min with a head of 425 ft does not correspond to any point on the 1750 rpm curve.

From Eq. 18.42, the quantity h/Q^2 is constant.

$$\frac{h}{Q^2} = \frac{425 \text{ ft}}{\left(500 \ \dfrac{\text{gal}}{\text{min}}\right)^2} = 1.7 \times 10^{-3} \quad \text{[mixed units]}$$

In order to use the affinity laws, the operating point on the 1750 rpm curve must be determined. Random values of Q are chosen and the corresponding values of h are determined such that the ratio h/Q^2 is unchanged.

Q	h
475	383
450	344
425	307
400	272

These points are plotted and connected to draw the system curve. The intersection of the system and 1750 rpm pump curve at 440 gal/min defines the operating point at that speed. From Eq. 18.41,

$$n_2 = \frac{n_1 Q_2}{Q_1} = \frac{(1750 \text{ rpm})\left(500 \ \dfrac{\text{gal}}{\text{min}}\right)}{440 \ \dfrac{\text{gal}}{\text{min}}}$$
$$= 1989 \text{ rpm}$$

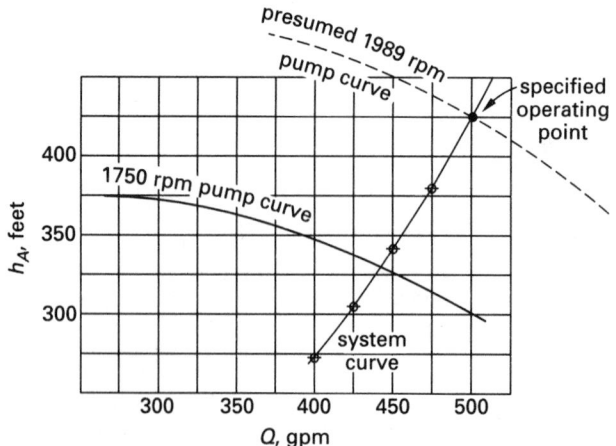

25. PUMP SIMILARITY

The performance of one pump can be used to predict the performance of a *dynamically similar (homologous) pump*. This can be done by using Eqs. 18.48 through 18.53.

$$\frac{n_1 D_1}{\sqrt{h_1}} = \frac{n_2 D_2}{\sqrt{h_2}} \qquad 18.48$$

$$\frac{Q_1}{D_1^2 \sqrt{h_1}} = \frac{Q_2}{D_2^2 \sqrt{h_2}} \qquad 18.49$$

$$\frac{P_1}{\rho_1 D_1^2 h_1^{1.5}} = \frac{P_2}{\rho_2 D_2^2 h_2^{1.5}} \qquad 18.50$$

$$\frac{Q_1}{n_1 D_1^3} = \frac{Q_2}{n_2 D_2^3} \qquad 18.51$$

$$\frac{P_1}{\rho_1 n_1^3 D_1^5} = \frac{P_2}{\rho_2 n_2^3 D_2^5} \qquad 18.52$$

$$\frac{n_1 \sqrt{Q_1}}{(h_1)^{0.75}} = \frac{n_2 \sqrt{Q_2}}{(h_2)^{0.75}} \qquad 18.53$$

These so-called *similarity laws* assume that both pumps:

- operate in the turbulent region

- have the same pump efficiency

- operate at the same percentage of wide-open flow

Similar pumps also will have the same specific speed and cavitation number.

As with the affinity laws, these relationships assume that the efficiencies of the larger and smaller pumps are the same. In reality, larger pumps will be more efficient than smaller pumps. Therefore, extrapolations to much larger or much smaller sizes should be avoided.

Example 18.11

A 6 in pump operating at 1770 rpm discharges 1500 gal/min of cold water (SG = 1.0) against an 80 ft head at 85% efficiency. A homologous 8 in pump operating at 1170 rpm is being considered as a replacement. (a) What total head and capacity can be expected from the new pump? (b) What would be the new horsepower requirement?

Solution

(a) From Eq. 18.48,

$$h_2 = \left(\frac{D_2 n_2}{D_1 n_1}\right)^2 h_1 = \left[\frac{(8 \text{ in})(1170 \text{ rpm})}{(6 \text{ in})(1770 \text{ rpm})}\right]^2 (80 \text{ ft})$$

$$= 62.14 \text{ ft}$$

From Eq. 18.51,

$$Q_2 = \left(\frac{n_2 D_2^3}{n_1 D_1^3}\right) Q_1 = \left[\frac{(1170 \text{ rpm})(8 \text{ in})^3}{(1770 \text{ rpm})(6 \text{ in})^3}\right] \left(1500 \frac{\text{gal}}{\text{min}}\right)$$

$$= 2350.3 \text{ gal/min}$$

(b) From Table 18.4, the hydraulic horsepower is

$$\text{WHP}_2 = \frac{h_2 Q_2 (\text{SG})}{3956} = \frac{(62.14 \text{ ft})\left(2350.3 \frac{\text{gal}}{\text{min}}\right)(1.0)}{3956 \frac{\text{ft-gal}}{\text{hp-min}}}$$

$$= 36.92 \text{ hp}$$

From Eq. 18.48,

$$\eta_{\text{larger}} = 1 - \frac{1 - \eta_{\text{smaller}}}{\left(\dfrac{D_{\text{larger}}}{D_{\text{smaller}}}\right)^{0.2}}$$

$$= 1 - \frac{1 - 0.85}{\left(\dfrac{8 \text{ in}}{6 \text{ in}}\right)^{0.2}} = 0.858$$

$$\text{BHP}_2 = \frac{\text{WHP}_2}{\eta_p} = \frac{36.92 \text{ hp}}{0.858} = 43.0 \text{ hp}$$

26. PUMPING LIQUIDS OTHER THAN COLD WATER

Many pump parameters are determined from tests with cold, clear water at 85°F (29°C). The following guidelines can be used when pumping water at other temperatures or when pumping other fluids.

- Head developed is independent of the liquid's specific gravity. Pump performance curves from tests with water can be used with other Newtonian fluids (e.g., gasoline, alcohol, and aqueous solutions) having similar viscosities.

- Head, flow rate, and efficiency are all reduced when pumping highly viscous non-Newtonian fluids. No exact method exists for determining the reduction factors, other than actual tests of an installation using both fluids. Some sources have published charts of correction factors based on tests over limited viscosity and size ranges.[24]

- The hydraulic horsepower depends on the specific gravity of the fluid. If the pump characteristic curve is used to find the operating point, multiply the horsepower reading by the specific gravity. Tables 18.4 and 18.5 incorporate the specific gravity term in the calculation of hydraulic power where required.

- Efficiency is not affected by changes in temperature that cause only the specific gravity to change.

- Efficiency is nominally affected by changes in temperature that cause the viscosity to change.

---

[24]The chart published by the Hydraulics Institute is widely distributed.

Equation 18.54 is an approximate relationship suggested by the Hydraulics Institute when extrapolating the efficiency (in decimal form) from cold water to hot water. n is an experimental exponent established by the pump manufacturer, generally in the range of 0.05 to 0.1.

$$\eta_{\text{hot}} = 1 - (1 - \eta_{\text{cold}}) \left(\frac{\nu_{\text{hot}}}{\nu_{\text{cold}}} \right)^n \qquad 18.54$$

- NPSHR is not significantly affected by minor variations in the water temperature.

- When hydrocarbons are pumped, the NPSHR determined from cold water can usually be reduced. This reduction is apparently due to the slow vapor release of complex organic liquids. If the hydrocarbon's vapor pressure at the pumping temperature is known, Fig. 18.16 will give the percentage of the cold-water NPSHR.

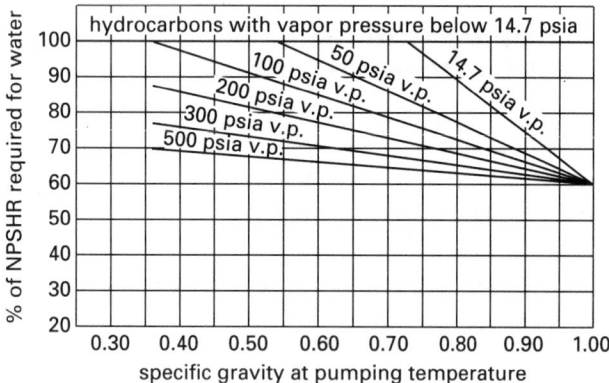

Figure 18.16 *Hydrocarbon NPSHR Correction Factor*

- Pumping many fluids requires expertise that goes far beyond simply extrapolating parameters in proportion to the fluid's specific gravity. Such special cases include pumping liquids containing abrasives, liquids that solidify, highly corrosive liquids, liquids with vapor or gas, highly viscous fluids, paper stock, and hazardous fluids.

Example 18.12

A centrifugal pump has an NPSHR of 12 psi based on cold water. What NPSHR should be used with 10°F liquid isobutane?

Solution

From App. 14.H, the vapor pressure of 10°F isobutane is approximately 15 psia. From App. 14.I, the specific gravity at 10°F is approximately 0.60. From Fig. 18.16, the intersection of a specific gravity of 0.60 and 15 psia is above the horizontal 100% line. The full NPSHR of 12 psia should be used.

27. TURBINE SPECIFIC SPEED

Like centrifugal pumps, turbines are classified according to the manner in which the impeller extracts energy from the fluid flow. This is measured by the turbine-specific speed equation, which is different from the equation used to calculate specific speed for pumps.

$$n_s = \frac{n\sqrt{P \text{ in kW}}}{(h_t \text{ in meters})^{1.25}} \qquad \text{[SI]} \quad 18.55(a)$$

$$n_s = \frac{n\sqrt{P \text{ in hp}}}{(h_t \text{ in feet})^{1.25}} \qquad \text{[U.S.]} \quad 18.55(b)$$

28. TYPES OF TURBINES

Each of the three types of turbines is associated with a range of specific speeds.

- Axial-flow turbines are used for low heads, high rotational speeds, and large flow rates. These propeller turbines operate with specific speeds in the 80 to 200 range (300 to 760 in SI). Their best efficiencies, however, are produced with specific speeds between 120 and 160 (460 and 610 in SI).

- For reaction turbines, the specific speed varies from 10 to 100 (38 and 380 in SI). Best efficiencies are found in the 40 to 60 (150 and 230 in SI) range with heads below 600 to 800 ft (180 to 240 m).

- Radial-flow (impulse) turbines have the lowest specific speeds but are used when heads are high. These turbines have specific speeds below 5 (19 in SI).

29. HYDROELECTRIC GENERATING PLANTS

In a typical hydroelectric generating plant using reaction turbines, the turbine is generally housed in a *powerhouse*, with water conducted to the turbine through the *penstock* piping. Water originates in a reservoir, dam, or *forebay* (in the instance where the reservoir is a long distance from the turbine).

After the water passes through the turbine, it is discharged through the draft tube to the receiving reservoir, known as the *tailwater*. The *draft tube* is used to keep the turbine up to 15 ft (5 m) above the tailwater surface, while still being able to extract the total available head. If a draft tube is not employed, water may be returned to the tailwater by way of a channel known as the *tail race*. The turbine, draft tube, and all related parts comprise what is known as the *setting*.

When a forebay is not part of the generating plant's design, it will be desirable to provide a *surge chamber* in order to relieve the effects of rapid changes in flow rate. In the case of a sudden power demand, the surge chamber would provide an immediate source of water, without waiting for a contribution from the feeder reservoir.

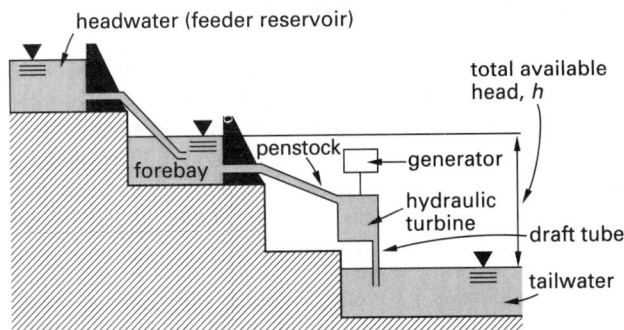

Figure 18.17 *Typical Hydroelectric Plant*

Similarly, in the case of a sudden decrease in discharge through the turbine, the excess water would surge back into the surge chamber.

30. IMPULSE TURBINES

An *impulse turbine* consists of a rotating shaft (called a *turbine runner*) on which buckets or blades are mounted. (This is commonly called a *Pelton wheel*.)[25] A jet of water (or other fluid) hits the buckets and causes the turbine to rotate. The kinetic energy of the jet is converted into rotational kinetic energy. The jet is essentially at atmospheric pressure.

Impulse turbines are generally employed where the available head is very high, above 800 to 1600 ft (250 to 500 m). (There is no exact value for the critical head, hence the range. What is important is that impulse turbines are *high-head turbines*.) Efficiencies are in the range of 80 to 90%, with the higher efficiencies being associated with turbines having two or more jets per runner.

The total available head in this installation is h_t, but not all of this energy can be extracted. Some of the energy is lost to friction in the penstock. Minor losses also occur, but these small losses are usually disregarded. In the penstock, immediately before entering the nozzle, the remaining head is divided between the pressure head and the velocity head.[26]

$$h' = h_t - h_f = (h_p + h_v)_{\text{penstock}} \qquad \textbf{18.56}$$

Another loss, h_n, occurs in the nozzle itself. The head remaining to turn the turbine is

$$\begin{aligned} h'' &= h' - h_n \\ &= h_t - h_f - h_n \end{aligned} \qquad \textbf{18.57}$$

---

[25] In a Pelton wheel turbine, the spoon-shaped buckets are divided into two halves, with a ridge between the halves. Half of the water is thrown to each side of the bucket. A Pelton wheel is known as a *tangential turbine (tangential wheel)* because the centerline of the jet is directed at the centers of the buckets.

[26] Care must be taken to distinguish between the conditions existing in the penstock, the nozzle throat, and the jet itself. The velocity in the nozzle throat and jet will be the same, but this is different from the penstock velocity. Similarly, the pressure in the jet is zero, although it is nonzero in the penstock.

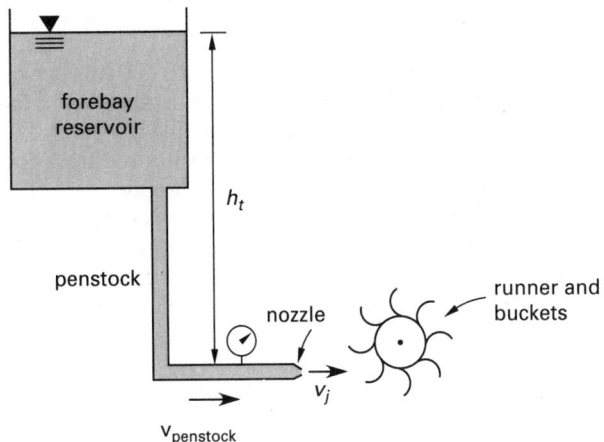

Figure 18.18 *Impulse Turbine Installation*

h_f is calculated from either the Darcy or the Hazen-Williams equation. The nozzle loss is calculated from the *nozzle coefficient*, C_v.

$$h_n = h'(1 - C_v^2) \qquad \textbf{18.58}$$

The total head at the nozzle exit is converted into velocity head according to the Torricelli equation.

$$v_j = \sqrt{2gh''} = C_v\sqrt{2gh'} \qquad \textbf{18.59}$$

In order to maximize power output, buckets are usually designed to partially reverse the direction of the water jet flow. The forces on the turbine buckets can be found from the impulse-momentum equations. If the water is turned through an angle θ and the wheel's *tangential velocity* is v_b, the energy transmitted by each unit mass of water to the turbine runner is[27]

$$E = v_b(v_j - v_b)(1 - \cos\theta) \qquad \text{[SI]} \qquad \textbf{18.60(a)}$$

$$E = \left[\frac{v_b(v_j - v_b)}{g_c}\right](1 - \cos\theta) \quad \text{[U.S.]} \qquad \textbf{18.60(b)}$$

$$v_b = \frac{(\text{rpm})\, 2\pi r}{60} = \omega r \qquad \textbf{18.61}$$

The theoretical *turbine power* is found by multiplying Eq. 18.60 by the mass flow rate. The actual power will be less than the theoretical output. Typical turbine efficiencies range between 80 and 90%. For a mass flow rate in kg/s, the theoretical power (in kilowatts) will be

$$P_{\text{th}} = \frac{\dot{m}E}{1000} \text{ in kW} \qquad \text{[SI]} \qquad \textbf{18.62(a)}$$

For a mass flow rate in lbm/sec, the theoretical horsepower will be

$$P_{\text{th}} = \frac{\dot{m}E}{550} \text{ in hp} \qquad \text{[U.S.]} \qquad \textbf{18.62(b)}$$

---

[27] $\theta = 180°$ would be ideal. However, the actual angle is limited to approximately 165° to keep the deflected jet out of the way of the incoming jet.

Example 18.13

A Pelton wheel impulse turbine develops 100 hp (brake) while turning at 500 rpm. The water is supplied from a penstock with an internal area of 0.3474 ft^2. The water subsequently enters a nozzle with a reduced flow area. The total head is 200 ft before nozzle loss. The turbine efficiency is 80%, and the nozzle coefficient, C_v, is 0.95. Disregard penstock friction losses. What are the (a) flow rate (in ft^3/sec), (b) area of the jet, and (c) pressure head in the penstock just before the nozzle?

Solution

(a) From Eq. 18.59 with $h' = h_t = 200$ ft, the jet velocity is

$$v_j = C_v\sqrt{2gh'} = 0.95\sqrt{(2)\left(32.2\ \frac{\text{ft}}{\text{sec}^2}\right)(200\ \text{ft})}$$
$$= 107.8\ \text{ft/sec}$$

From Eq. 18.58, the nozzle loss is

$$h_n = h'(1 - C_v^2) = (200\ \text{ft})[1 - (0.95)^2]$$
$$= 19.5\ \text{ft}$$

From Table 18.4, the flow rate is

$$\dot{V} = \frac{(8.814)(\text{hp})}{h_A(\text{SG})\eta} = \frac{(8.814)(100\ \text{hp})}{(200\ \text{ft} - 19.5\ \text{ft})(1)(0.8)}$$
$$= 6.104\ \text{ft}^3/\text{sec}$$

(b) The jet area is

$$A_j = \frac{\dot{V}}{v_j} = \frac{6.104\ \dfrac{\text{ft}^3}{\text{sec}}}{107.8\ \dfrac{\text{ft}}{\text{sec}}} = 0.0566\ \text{ft}^2$$

(c) The velocity in the penstock is

$$v_{\text{penstock}} = \frac{\dot{V}}{A} = \frac{6.104\ \dfrac{\text{ft}^3}{\text{sec}}}{0.3474\ \text{ft}^2}$$
$$= 17.57\ \text{ft/sec}$$

The pressure head in the penstock is

$$h_p = h' - h_v = h' - \frac{v^2}{2g}$$
$$= 200\ \text{ft} - \frac{\left(17.57\ \dfrac{\text{ft}}{\text{sec}}\right)^2}{(2)\left(32.2\ \dfrac{\text{ft}}{\text{sec}^2}\right)}$$
$$= 195.2\ \text{ft}$$

31. REACTION TURBINES

Reaction turbines (also known as *Francis turbines* or *radial-flow turbines*) are essentially centrifugal pumps operating in reverse. They are used when the total available head is small, typically below 600 to 800 ft. However, their energy conversion efficiency is higher than that of impulse turbines, typically in the 85 to 95% range.

In a reaction turbine, water enters the turbine housing with a pressure greater than atmospheric pressure. The water completely surrounds the turbine runner (impeller) and continues through the draft tube. There is no vacuum or air pocket between the turbine and the tailwater.

All of the power, affinity, and similarity relationships used with centrifugal pumps can be used with reaction turbines.

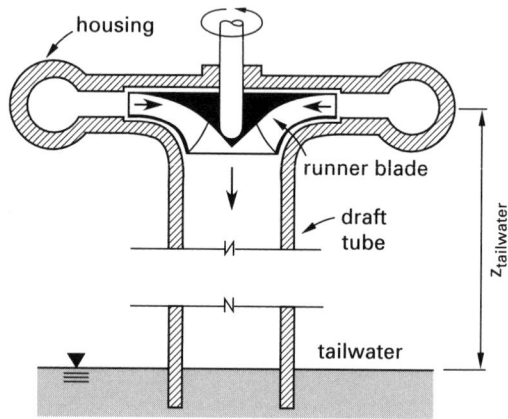

Figure 18.19 Reaction Turbine

Example 18.14

A reaction turbine with a draft tube develops 500 hp (brake) when 50 ft^3/sec of water flow through it. Water enters the turbine at 20 ft/sec with a 100 ft pressure head. The elevation of the turbine above the tailwater level is 10 ft. Disregarding friction, what are the (a) total available head and (b) turbine efficiency?

Solution

(a) The available head is the difference between the forebay and tailwater elevations. The tailwater depression is known, but the height of the forebay above the turbine is not known. At the turbine entrance, this unknown potential energy has been converted to pressure and velocity head. Therefore, the total available head (exclusive of friction) is

$$h_t = z_{\text{forebay}} - z_{\text{tailwater}}$$
$$= h_p + h_v - z_{\text{tailwater}}$$
$$= 100 \text{ ft} + \frac{\left(20 \dfrac{\text{ft}}{\text{sec}}\right)^2}{(2)\left(32.2 \dfrac{\text{ft}}{\text{sec}^2}\right)} - (-10 \text{ ft})$$
$$= 116.2 \text{ ft}$$

(b) From Table 18.4, the theoretical hydraulic horsepower is

$$P_{\text{th}} = \frac{h_A \dot{V}(\text{SG})}{8.814} = \frac{(116.2 \text{ ft})\left(50 \dfrac{\text{ft}^3}{\text{sec}}\right)(1.0)}{8.814}$$
$$= 659.2 \text{ hp}$$

The efficiency of the turbine is

$$\eta = \frac{P_{\text{brake}}}{P_{\text{th}}} = \frac{500 \text{ hp}}{659.2 \text{ hp}} = 0.758 \quad (75.8\%)$$

32. AXIAL-FLOW TURBINES

Figure 18.20 illustrates an *axial-flow turbine*, also known as a *propeller turbine*. Analysis of performance is similar to that of reaction and impulse turbines.

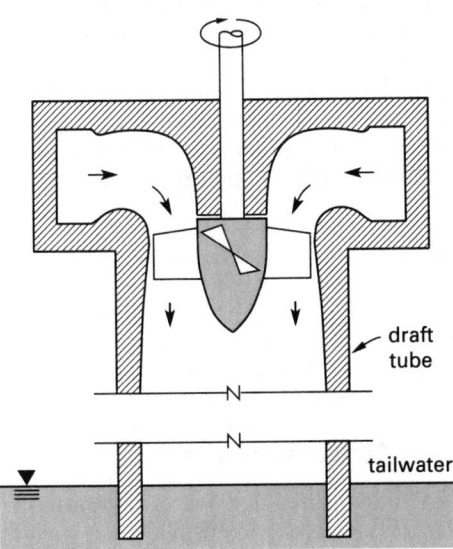

Figure 18.20 *Axial-Flow Turbine*

PRACTICE PROBLEMS

Pumping Power

1. 37 gal/min (65 L/s) of 80°F (27°C) SAE 40 oil are increased in pressure from 1 atm to 40 psig (275 kPa). What hydraulic power is required?

2. 1.25 ft³/sec (35 L/s) of 70°F (21°C) water are pumped from the bottom of a tank through 700 ft (230 m) of 4 in (10.2 cm), schedule-40 steel pipe. The line includes a 50 ft (15 m) rise in elevation, two right-angle elbows, a wide-open gate valve, and a swing check valve. All fittings and valves are regular screwed. The inlet pressure is 50 psig (345 kPa), and a working pressure of 20 psig (140 kPa) is needed at the end of the pipe. What is the hydraulic power for this pumping application?

3. 80 gal/min (5 L/s) of 80°F (27°C) water are lifted 12 ft (4 m) vertically by a pump through 50 ft (15 m) of a 2 in (5.1 cm) diameter rubber hose. The discharge end of the hose is submerged in 8 ft (2.5 m) of water as shown. What head is added by the pump?

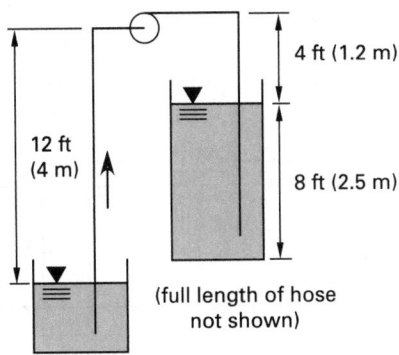

4. A 20 hp motor drives a centrifugal pump. The pump discharges 60°F (16°C) water at 12 ft/sec (4 m/s) into a 6 in (15.2 cm) steel, schedule-40 line. The inlet is 8 in (20.3 cm) steel, schedule-40 pipe. The pump suction is at 5 psig (35 kPa) below standard atmospheric pressure. The friction head loss in the system is 10 ft (3.3 m). The pump efficiency is 70%. What is the maximum height above the pump inlet that water is available at standard atmospheric pressure?

5. (*Time limit: one hour*) A pump station is used to fill a tank on a hill above from a lake below. The flow rate is 10,000 gal/hr (10.5 L/s) of 60°F (16°C) water. The atmospheric pressure is 14.7 psia (101 kPa). The pump is 12 ft (4 m) above the lake, and the tank surface level is 350 ft (115 m) above the pump. The discharge line is 4 in (10.2 cm) diameter schedule-40 steel pipe. The equivalent length of the inlet line between the lake and the pump is 300 ft (100 m). The total equivalent length between the lake and the tank is 7000 ft (2300 m), including all fittings, bends, screens, and valves. The cost of electricity is $0.04 per kW-hr. The overall efficiency of the pump and motor set is 70%. (a) What does it cost to operate the pump for one hour? (b) What motor power is required? (c) What is the NPSHA for this application?

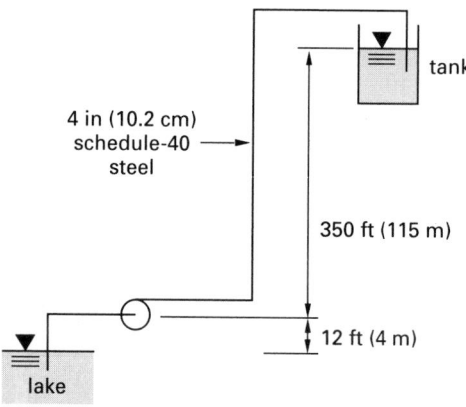

6. (*Time limit: one hour*) A three-zone heating system uses hot water passing through the network shown. The heater increases the water temperature by 20°F (10°C). All pipes are copper, type L. The pump efficiency is 45%. The pipe equivalent lengths, pipe sizes, and flow rates through each part of the network are known. (Flow-balancing valves are not shown.)

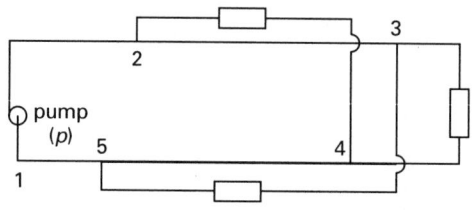

circuit	equivalent length, ft (m)	diameter, in (cm)	flow rate, gpm (L/s)
5–1–p–2	40 (13)	$2\frac{1}{2}$ (6.35)	60 (3.9)
2–4	70 (23)	$1\frac{1}{2}$ (3.81)	20 (1.3)
2–3	55 (18)	2 (5.08)	40 (2.6)
3–4	65 (22)	$1\frac{1}{2}$ (3.81)	20 (1.3)
3–5	60 (20)	$1\frac{1}{2}$ (3.81)	20 (1.3)
4–5	50 (17)	2 (5.08)	40 (2.6)

(a) What is the total head added by the pump? (b) What size motor should be chosen to drive the pump? (c) How much heat is added to the water each hour?

Pumping Other Fluids

7. (*Time limit: one hour*) Gasoline with a specific gravity of 0.7 and viscosity of 6×10^{-6} ft²/sec (5.6×10^{-7} m²/s) is transferred from a tanker to a storage tank. The interior of the storage tank is maintained at atmospheric pressure by a vapor-recovery system. The free surface in the storage tank is 60 ft (20 m) above the tanker's free surface. The pipe consists of 500 ft (170 m) of 3 in (7.62 cm) schedule-40 steel pipe with six flanged elbows and two wide-open gate valves. The pump and

motor both have individual efficiencies of 88%. Electricity costs $0.045 per kW-hr. The pump's performance data (based on cold, clear water) is known. (a) What is the transfer rate? (b) What is the total cost of operating the pump for one hour?

flow rate, gpm (L/s)	head, ft (m)
0 (0)	127 (42)
100 (6.3)	124 (41)
200 (12)	117 (39)
300 (18)	108 (36)
400 (24)	96 (32)
500 (30)	80 (27)
600 (36)	55 (18)

Specific Speed

8. A double-suction water pump moving 300 gal/sec (1.1 kL/s) turns at 900 rpm. The pump adds 20 ft (7 m) of head to the water. What is the specific speed?

9. A two-stage centrifugal pump draws water from an inlet 10 ft (3 m) below its eye. Each stage of the pump adds 150 ft (50 m) of head. What is the approximate maximum suggested speed for this application?

Cavitation

10. 100 gal/min (6.3 L/s) of hot water at 281°F and 80 psia (138°C and 550 kPa) discharges through 30 ft (10 m) of 1.5 in (3.81 cm) schedule-40 steel pipe into a 2 psig (14 kPa) tank. The inlet and outlet are both 20 ft (6 m) below the surface of the water when the tank is full. The inlet line contains two wide-open gate valves and two long-radius elbows. All components are regular screwed. The pump's NPSHR is 10 ft (3 m) for this application. Will the pump cavitate?

11. The velocity of the tip of a marine propeller is 4.2 times the boat velocity. The propeller is located 8 ft (3 m) below the surface. The temperature of the seawater is 68°F (20°C). The density is approximately 64.0 lbm/ft³ (1024 kg/m³), and the salt content is 2.5% by weight. What is the practical maximum boat velocity, as limited strictly by cavitation?

Pump and System Curves

12. The inlet of a centrifugal water pump is 7 ft (2.3 m) above the free surface from which it draws. The inlet consists of 12 ft (4 m) of 2 in (5.08 cm) schedule-40 steel pipe and contains one long-radius elbow and one check valve. The discharge line is 2 in (5.08 cm) schedule-40 steel pipe and includes two long-radius elbows and an 80 ft (27 m) run. The discharge is 20 ft (6.3 m) above the free surface. All components are regular screwed. The water temperature is 70°F (21°C). The following pump curve data is applicable. (a) What is the flow rate? (b) What can be said about the use of this pump in this installation?

flow rate, gpm (L/s)	head, ft (m)
0 (0)	110 (37)
10 (0.6)	108 (36)
20 (1.2)	105 (35)
30 (1.8)	102 (34)
40 (2.4)	98 (33)
50 (3.2)	93 (31)
60 (3.6)	87 (29)
70 (4.4)	79 (26)
80 (4.8)	66 (22)
90 (5.7)	50 (17)

13. (*Time limit: one hour*) A heating system uses 100 radiative heaters with a rated capacity of 10,000 Btu/hr each (3 kW). Circulating water enters each heater at 200°F (90°C) and leaves at 180°F (80°C). The heaters are connected in a reverse-return circuit, making all pipe lengths equal. The equivalent line length is 420 ft (140 m), which includes all losses for valves, fittings, and bends. The building specification requires that the system piping be sized such that the pressure drop per equivalent foot (meter) of pipe is 0.25 to 0.65 in of water (0.021 to 0.054 m of water). Three different pumps are available to supply the heating system. (a) Which pump should be used for a 1,000,000 Btu/hr (257 kW) system? (b) What pipe size should be used for a 1,000,000 Btu/hr (257 kW) system?

flow rate[a] (gpm)	head[b] (ft)		
	pump 1	pump 2	pump 3
10	5.3	8.8	13.3
20	5.5	9.0	13.4
30	5.5	9.0	13.4
40	5.4	8.9	13.3
50	5.2	8.7	13.1
60	4.9	8.4	12.7
70	4.5	8.0	12.3
80	3.9	7.4	11.8
90	3.3	6.8	11.1
100	2.5	5.8	10.0
110	1.4	4.3	8.3

[a]Multiply gpm by 0.0631 to obtain L/s.
[b]Multiply ft by 0.305 to obtain m.

Affinity Laws

14. A pump was intended to run at 1750 rpm when driven by a 0.5 hp (0.37 kW) motor. What is the required power rating of a motor that will turn the pump at 2000 rpm?

15. (*Time limit: one hour*) A centrifugal pump running at 1400 rpm has the curve shown. The pump will be installed in an existing pipeline with known head requirements given by the formula $H = 30 + 2Q^2$. H is the system head in feet of water. Q is the flow rate in ft³/sec. (a) What is the flow rate if the pump is turned at 1400 rpm? (b) What power is required to drive the pump? (c) What is the flow rate if the pump is turned at 1200 rpm?

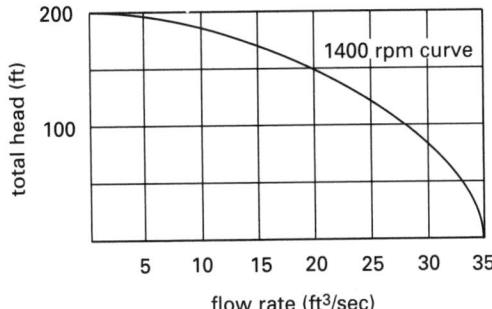

Turbines

16. Water at 500 psig and 60°F (3.5 MPa and 16°C) drives a 250 hp (185 kW) turbine at 1750 rpm against a back pressure of 30 psig (210 kPa). The water discharges through a 4 in (100 mm) diameter nozzle at 35 ft/sec (10.5 m/s). The water is deflected 80° by a single blade moving directly away at 10 ft/sec (3 m/s). (a) What type of turbine is appropriate for this application? (b) What is the total force acting on the blade?

17. (*Time limit: one hour*) A Francis-design hydraulic reaction turbine with 22 in (560 mm) diameter blades runs at 610 rpm. The turbine develops 250 hp (185 kW) when 25 ft³/sec (700 L/s) of water flow through it. The pressure head at the turbine entrance is 92.5 ft (30.8 m). The elevation of the turbine above the tailwater level is 5.26 ft (1.75 m). The inlet and outlet velocity are both 12 ft/sec (3.6 m/s). (a) What is the effective head? (b) What is the overall turbine efficiency? (c) What will be the turbine speed (in rpm) if the effective head is 225 ft (75 m)? (d) What horsepower is developed if the effective head is 225 ft (75 m)? (e) What is the flow rate if the effective head is 225 ft (75 m)?

19 Fluid Power

1. Introduction to Fluid Power 19-1
2. Fluid Power Symbols 19-1
3. Hydraulic Fluids 19-2
4. Valves 19-2
5. Tubing, Hose, and Pipe for Fluid Power . . 19-2
6. Pressure Rating of Pipe and Tubing 19-3
7. Burst Pressure 19-3
8. Pressure Drop 19-4
9. Fluid Power Pumps 19-4
10. Electric Motors 19-5
11. Strainers and Filters 19-5
12. Accumulators 19-5
13. Linear Actuators and Cylinders 19-6

Symbols

η	efficiency	percent	percent
ν	kinematic viscosity[3]	SSU	cS
ρ	mass density	lbm/ft^3	kg/m^3

Subscripts

ut   ultimate tensile
y yield

Nomenclature[1]

A	area[2]	in^2	cm^2
C	corrosion allowance	in	cm
C_v	valve flow factor	–	–
d	inside diameter	in	cm
D	outside diameter	in	cm
f	friction factor	–	–
F	force	lbf	N
I	effective (rms) line current	A	A
k	ratio of specific heats	–	–
l	length	in	cm
L	length	ft	m
n	polytropic exponent	–	–
n	rotational speed	rpm	rpm
p	pressure	lbf/in^2	Pa
pf	power factor	–	–
P	power	hp	kW
Q	flow rate (liquids)	gal/min	L/min
Re	Reynolds number	–	–
S	maximum allowable tensile stress	lbf/in^2	kPa
S	strength (with subscript)	lbf/in^2	kPa
SG	specific gravity	–	–
t	nominal thickness	in	cm
t'	design thickness	in	cm
T	temperature	°F	K
T	torque	in-lbf	N·m
v	linear velocity	ft/sec	m/s
V	capacity or volume	in^3	L
V	effective (rms) line voltage	V	V
y	temperature derating factor	–	–

[1]Symbols and units consistent with the fluid power industry are used in this chapter.
[2]The symbol S is used in some references to designate surface area.

1. INTRODUCTION TO FLUID POWER

Fluid power (hydraulic power) equipment is hydraulically operating equipment that generates hydraulic pressure at one point in order to perform useful tasks at another. The equipment typically consists of a power source (i.e., an electric motor or internal combustion engine), pump, actuator cylinders or rotary fluid motors, control valves, high-pressure tubing or hose, fluid reservoir, and hydraulic fluid. During operation, the power source pressurizes the hydraulic fluid, and the control valves direct the fluid to the cylinders or hydraulic motors.

2. FLUID POWER SYMBOLS

Most symbols for fluid power equipment are simplified representations of their physical counterparts. Symbols have been standardized by the ANSI and ISO.[4] Appendix 19.A lists some of the more common symbols.

[3]The following conversions may be used between Saybolt Seconds Universal (SSU) and centistokes (cS):

$$\nu_{\text{cS}} = 0.2253\nu_{\text{SSU}} - \frac{194.4}{\nu_{\text{SSU}}} \qquad 32 < \text{SSU} < 100$$

$$\nu_{\text{cS}} = 0.2193\nu_{\text{SSU}} - \frac{134.6}{\nu_{\text{SSU}}} \qquad 100 < \text{SSU} < 240$$

$$\nu_{\text{cS}} = \frac{\nu_{\text{SSU}}}{4.635} \qquad \text{SSU} > 240$$

[4]Refer to ANSI Y32.10 and ISO 1219.

Fluids

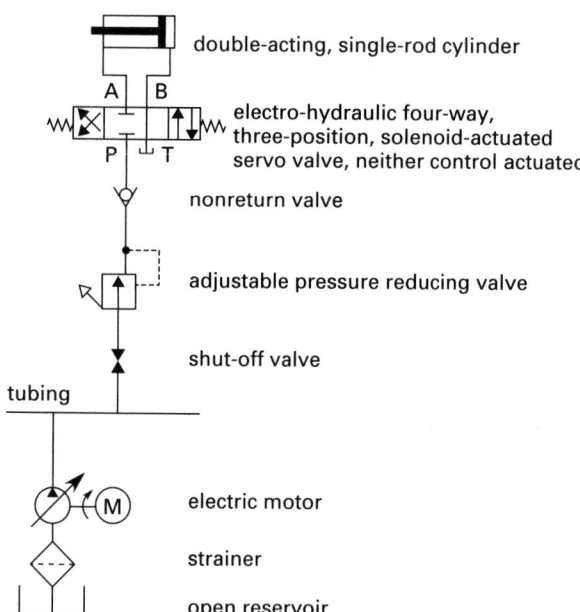

Figure 19.1 *Typical Fluid Power Circuit*

3. HYDRAULIC FLUIDS

Petroleum-based oils and fire-resistant fluids are the two general categories of *hydraulic fluids*. Petroleum oils are enhanced with additives intended to inhibit or prevent rust, foam, wear, and oxidation. The largest drawback to petroleum oils, however, is flammability. *Fire-resistant fluids* are categorized as water and oil emulsions,[5] water-glycol mixtures, or straight synthetic fluids (e.g., silicone or phosphate esters, ester blends, and chlorinated hydrocarbon-based fluids).

In addition to cost, the properties most relevant in selecting hydraulic fluids are lubricity (i.e., the ability to reduce friction and prevent wear), viscosity, viscosity index,[6] pour point, flash point, rust resistance, oxidation resistance, and foaming resistance. Relative to petroleum oils, fire resistance is usually achieved to the detriment of the other properties.

Water-oil emulsions are the lowest-cost fire-resistant fluids. They generally perform as well as or better than most petroleum fluids.

Due to their similarity to standard antifreeze solutions, water-glycol mixtures are a good choice for low-temperature use. Periodic checking is required to monitor alkalinity and water evaporation. The water content should not be allowed to drop below approximately 35 to 50%.

Synthetic hydraulic fluids are the most costly of the fire-resistant fluids. They have high lubricity. Special formulations may be needed for low-temperature use, and their viscosity indexes are generally lower than those of petroleum oils. A significant factor is that they are not chemically consistent with the seal materials in use for petroleum oils.[7] Therefore, synthetics cannot be used in all existing systems.

The temperature of the hydraulic fluid entering the pump is typically 100 to 120°F (38 to 49°C). The temperature is commonly limited by most specifications to 120°F (49°C) for water-based fluids and 130°F (54°C) for all other fluids.

4. VALVES

Control valves are used to direct the flow of hydraulic fluid. *Pressure reduction valves* are used to reduce the pressure to components that cannot tolerate or do not need the full system pressure. *Pressure relief valves* are used to protect the system from dangerously high pressures.

Although seated poppet (globe, gate, and plunger) valves can be used, the *spool-type valve* is most common for channel selection and directional control.

Control valves are described according to their number of ports, their normal configuration, and their number of positions. The "way" of a control valve is equal to the number of ports. Thus, a three-way valve will have one port for pressurized fluid and two possible discharge ports. A four-way valve will have two ports for pressurized fluid and two discharge ports. If a valve prevents through-flow when de-energized (i.e., when off), it is designated as "normally closed" or "NC." If the valve permits through-flow when de-energized, it is designated as "normally open" or "NO."

In some valves, the fluid can be infinitely split between two discharge ports. In others, there are two distinct *positions*. Thus, a four-way, two-position valve (designated as a "4/2 valve") could switch the discharges completely.

5. TUBING, HOSE, AND PIPE FOR FLUID POWER

Special steel fluid power tubing is available, particularly in the smaller sizes (e.g., up to approximately 2 in in diameter).[8] Fluid power tubing is available in two different pressure ratings: 0 to 1000 psi (0 to 6.89 MPa) and 1000 to 2500 psi (6.89 to 17.2 MPa).

---

[5] An emulsion of water in oil (as opposed to oil in water) is known as an *invert emulsion*.

[6] *Viscosity index* is the relative rate of change in viscosity with temperature. The viscosities of fluids with high viscosity indexes change less with variations in temperature than those of fluids with low viscosity indexes.

[7] Some of the seal materials that are suitable for use with synthetics include butyl rubber, ethylene-propylene rubber, silicone, Teflon™, and nylon.

[8] Copper pipe may react with some hydraulic fluids and is rarely used.

Fluids

Types S (seamless) and F (furnace butt-welded) varieties of A53 and A106 grade B black steel pipe can be used when larger diameter pipes are required.[9] The dimensions are the same as for normal steel pipe of the same schedule.

There are numerous types of flexible hose. Most are reinforced with steel wire, steel braid, or other high-strength fiber. Fluid compatibility, bend radius, and operating pressure are selection criteria. The internal diameter of flexible hose is the hose's nominal size. When pressurized to more than 250 psi (1.7 MPa), the working pressure of hose is taken as 25% of the burst pressure.

6. PRESSURE RATING OF PIPE AND TUBING

Allowable working pressure (*pressure rating*) for pipe in fluid power systems is calculated in the same manner as that for pressure piping in other types of power piping systems.[10] The maximum working pressure is given by Eq. 19.1.

$$p_{max} = \frac{2S_{psi}(t'_{in} - C)}{D - 2y(t'_{in} - C)} \quad \text{[U.S. only]} \qquad \textbf{19.1}$$

In Eq. 19.1, S is the maximum allowable stress. The maximum allowable stress depends on the metal composition and temperature, but since most fluid power systems run at approximately 100°F (38°C), the maximum stress corresponding to that temperature should be used. By convention, the maximum allowable stress is calculated as the ultimate tensile strength divided by a factor of safety (referred to as a *design factor*) of 3 or, occasionally, 4.[11] Values of 12,500 psi (86.1 MPa) and 17,000 psi (117 MPa) are commonly used for initial studies for A285 carbon steel pipes and tubes, and these values correspond to approximate factors of safety over the ultimate strength of 4 and 3, respectively.

In Eq. 19.1, t' is the *design thickness* and is calculated as 87.5% of the nominal wall thickness listed in the pipe tables. C is an allowance for corrosion, threading, and variations in mechanical strength. A value of $C = 0.05$ in applies for threaded steel tubes up to $^3/_8$ in diameter. For larger threaded steel tubes, C is equal to the depth of the thread. $C = 0.05$ in for unthreaded steel tubes up to 0.5 in. $C = 0.065$ in for unthreaded steel tubes

up to $1^1/_4$ in and larger. For both ferritic and austenitic steels, y is 0.4 for operation at temperatures common to fluid power systems.[12]

Equation 19.1 is a code-based approach to calculating the working pressure. Three other theoretical methods are also used, primarily with steel hydraulic tubing connected with flared fittings, to calculate the working pressure.[13] Dimensions used in Eqs. 19.2, 19.3, and 19.4 are nominal, tabulated values. The *Barlow formula* is the standard thin-wall cylinder formula.

$$p = \frac{2St}{D} \qquad \textbf{19.2}$$

The *Boardman formula* is

$$p = \frac{2St}{D - 0.8t} \qquad \textbf{19.3}$$

The *Lamé formula* is

$$p = \frac{S(D^2 - d^2)}{D^2 + d^2}. \qquad \textbf{19.4}$$

Since power fluid systems are subject to rapid valve closures, the pressures calculated from Eqs. 19.1 through 19.4 must be reduced for the effect of water hammer. The amount of reduction for water hammer is calculated from the *water hammer factor*, the ratio of pressure (in psi) to flow rate (in gpm).[14] The derating in working stress due to water hammer is

$$\Delta p_{psi} = \text{water hammer factor} \times Q_{gpm} \quad \text{[U.S. only]} \quad \textbf{19.5}$$

The working pressure after water hammer may be further reduced for connections and fittings. The amount of the reduction is approximately 25%.

7. BURST PRESSURE

The *burst pressure* is calculated from Eq. 19.6, which is derived from the thin-walled cylinder theory. S_{ut} is the ultimate tensile strength.

$$p_{burst} = \frac{2S_{ut}t_{in}}{D_{in}} \qquad \textbf{19.6}$$

Example 19.1

A 2 in (nominal) schedule-80 A53 grade B (ultimate tensile strength of 60 ksi) steel tube is used in a fluid power system. The pipe carries 100 gpm of 100°F

---

[9]The "grade" of a steel is usually its tensile strength in ksi. For example, the tensile strength of A516 grade 60 steel is 60 ksi. The materials, grades, specifications, classes, and types of steel pipes and tubes are easily confused. A285, A515, and A516 are common designations for the carbon steel used in pipes and tubes. The grade of the material may relate to its tensile strength, ductility, or other property. Additional specifications may apply to the manufacturer of the pipe. A285 pipe, for example, is often specified according to the additional specifications A53 and A106. The type (e.g., F or S) relates to how seams are formed.
[10]ASA B31.1.1
[11]The abbreviations SMYS and SMTS stand for "standard minimum yield strength" and "standard minimum tensile strength," respectively.

[12]Values of C and y are specified by Part 2 of ASME's Code for Power Piping (ASME B31.1).
[13]Tubing must conform to SAE J524, J525, and J356.
[14]Including an allowance for water hammer should be based on the type of fluid system, not on the material used for the pipe or tube. Some sources suggest that the allowance for water hammer should be included only with cast-iron pipes. While it is true that most cast irons experience brittle (not ductile) failure, omitting the pressure increase with ductile pipes denies that water hammer actually occurs.

hydraulic fluid. The water hammer factor for this pipe is 6.52 psi/gpm. Considering a 25% reduction for connections and fittings, what is the working pressure?

Solution

From App. 16.B, the dimensions of schedule-80 steel pipe are

$$D = 2.375 \text{ in}$$

$$t_{\text{nominal}} = 0.218 \text{ in}$$

The design wall thickness is

$$t_{\text{design}} = 0.875 t_{\text{nominal}} = (0.875)(0.218 \text{ in})$$

$$= 0.191 \text{ in}$$

The allowable stress is

$$S = \frac{S_{\text{ut}}}{3} = \frac{60{,}000 \ \frac{\text{lbf}}{\text{in}^2}}{3}$$

$$= 20{,}000 \ \text{lbf/in}^2$$

From Eq. 19.1 using $S = 20{,}000 \ \text{lbf/in}^2$, $C = 0.05$ in, and $y = 0.4$,

$$p_{100°\text{F}} = \frac{2 S_{\text{psi}}(t_{\text{in}} - C)}{D - 2y(t_{\text{in}} - C)}$$

$$= \frac{(2)\left(20{,}000 \ \frac{\text{lbf}}{\text{in}^2}\right)(0.191 \text{ in} - 0.05 \text{ in})}{2.375 \text{ in} - (2)(0.4)(0.191 \text{ in} - 0.05 \text{ in})}$$

$$= 2493 \ \text{lbf/in}^2$$

The derating due to water hammer is

$$\Delta p_{\text{psi}} = \text{water hammer factor} \times Q_{\text{gpm}}$$

$$= \left(6.52 \ \frac{\frac{\text{lbf}}{\text{in}^2}}{\frac{\text{gal}}{\text{min}}}\right)\left(100 \ \frac{\text{gal}}{\text{min}}\right)$$

$$= 652 \ \text{lbf/in}^2$$

The working pressure before fitting allowance is

$$2493 \ \frac{\text{lbf}}{\text{in}^2} - 652 \ \frac{\text{lbf}}{\text{in}^2} = 1841 \ \text{lbf/in}^2$$

The fitting derating is 25%. The working pressure is

$$p = (1 - 0.25)\left(1841 \ \frac{\text{lbf}}{\text{in}^2}\right) = 1381 \ \text{lbf/in}^2$$

8. PRESSURE DROP

Flows below 15 ft/sec (4.5 m/s) in small-diameter (i.e., less than 1 in (2.5 cm) nominal) tubing are almost always laminar or transitional.[15] Using traditional units,

---

[15]Since the friction factor is larger for laminar flow than for transitional and turbulent flows, laminar flow is the conservative assumption.

the Reynolds number and friction factor are given by Eqs. 19.7 and 19.8. Tabulations of friction losses for hydraulic oil often assume a viscosity of 33.1 cS (corresponding to 155 SSU) and a density of 62.4 lbm/ft³ (1000 kg/m³).

$$\text{Re} = \frac{(3162)Q_{\text{gpm}}}{d_{\text{in}}\nu_{\text{cS}}} \qquad \text{19.7}$$

$$f = \frac{64}{\text{Re}} \quad [\text{laminar flow}] \qquad \text{19.8}$$

The friction loss is given by Eq. 19.9. For an adequate factor of safety, the value calculated should be multiplied by 2 or 3. This will account for inaccuracies based on an incorrect choice of density or viscosity.

$$\Delta p_{f,\text{psi}} = \frac{2.15 \times 10^{-4} f \rho_{\text{lbm/ft}^3} L_{\text{ft}}(Q_{\text{gpm}})^2}{(d_{\text{in}})^5} \qquad \text{19.9}$$

The pressure drop through valves and fittings can be determined in the conventional manner if the equivalent length of the valve or fitting is known. If the valve flow factor, C_{v}, is known, Eq. 19.10 can be used.

$$\Delta p_{\text{v,psi}} = \frac{(Q_{\text{gpm}})^2(\text{SG})}{(C_{\text{v}})^2} \qquad \text{19.10}$$

9. FLUID POWER PUMPS

Most fluid power pumps are positive displacement pumps. Rotary and reciprocating designs are both used. All of the standard formulas for pump performance (horsepower, torque, etc.) from Chap. 18 apply. For example, the horsepower needed to drive the pump is given by Eq. 19.11. η_{pump} is typically taken as 0.85.

$$P_{\text{hp}} = \frac{p_{\text{psi}} Q_{\text{gpm}}}{1714 \eta_{\text{pump}}} \qquad \text{19.11}$$

The torque on the pump shaft is

$$T_{\text{in-lbf}} = p_{\text{psi}} \times \frac{\text{displacement in} \ \frac{\text{in}^3}{\text{rev}}}{2\pi}$$

$$= \frac{P_{\text{hp}}(63{,}025)}{n_{\text{rpm}}}$$

$$= \frac{(36.77)Q_{\text{gpm}} p_{\text{psi}}}{n_{\text{rpm}}} \qquad \text{19.12}$$

The flow rate in pumps is related to the displacement per revolution by Eq. 19.13.

$$Q_{\text{gpm}} = \frac{n_{\text{rpm}}\left(\text{displacement in} \ \frac{\text{in}^3}{\text{rev}}\right)}{231} \qquad \text{19.13}$$

To prevent cavitation, the flow velocity in suction lines is generally limited by specification to 1 to 5 ft/sec (0.3

to 1.5 m/s). The flow velocity in discharge lines is limited to 10 to 15 ft/sec (3 to 4.5 m/s).[16] The actual fluid velocity can easily be calculated from Eq. 19.14.

$$v_{ft/sec} = \frac{0.3208 Q_{gpm}}{A_{in^2}} \qquad 19.14$$

When starting a hydraulic pump, the fluid viscosity should be 4000 SSU (870 cS) or less. During steady operation at higher temperaures, viscosity should be above 70 SSU (13 cS).

Pump life is a term that refers to bearing life in hours. Most motors have a *rated life* at some specific speed and pressure. The rated (bearing) life can be modified for other speeds and pressures with Eq. 19.15.

$$\text{life} = (\text{rated life}) \left(\frac{\text{rated speed}}{\text{actual speed}} \right)$$
$$\times \left(\frac{\text{rated pressure}}{\text{actual pressure}} \right)^3 \qquad 19.15$$

10. ELECTRIC MOTORS

The power needed to drive a hydraulic pump is given by Eq. 19.11. Since motors are rated by their developed power, Eq. 19.11 also specifies the minimum motor size. Most fixed (i.e., not mobile) fluid power pumps are driven by single- or three-phase induction motors. Table 19.1 can be used to solve for typical electrical parameters describing the performance of an induction motor.[17]

Table 19.1 *Electric Motor Variables*[a]

find	given	formula single-phase	formula three-phase
I_{amps}	P_{hp}	$\dfrac{P_{hp}(746)}{V\eta(pf)}$	$\dfrac{P_{hp}(746)}{\sqrt{3}V\eta(pf)}$
I_{amps}	P_{kW}	$\dfrac{P_{kW}(1000)}{V(pf)}$	$\dfrac{P_{kW}(1000)}{\sqrt{3}V(pf)}$
I_{amps}	P_{kVA}	$\dfrac{P_{kVA}(1000)}{V}$	$\dfrac{P_{kVA}(1000)}{\sqrt{3}V}$
P_{kW}		$\dfrac{IV(pf)}{1000}$	$\dfrac{\sqrt{3}IV(pf)}{1000}$
P_{kVA}		$\dfrac{IV}{1000}$	$\dfrac{\sqrt{3}IV}{1000}$
P_{hp}		$\dfrac{IV\eta(pf)}{746}$	$\dfrac{\sqrt{3}IV\eta(pf)}{746}$

[a] η is the motor efficiency.

[16]Velocities could be faster—up to 25 ft/sec (7.5 m/s)—but are limited to the lower values in order to prevent excessive friction and noise.
[17]"Line voltage" and "line-to-line voltage" are synonymous terms. There is also a "phase voltage" in three-phase systems. Phase and line voltage are related by $V_{line} = \sqrt{3}V_{phase}$.

11. STRAINERS AND FILTERS

Although fluid power systems are theoretically closed, they are never free from dirt, grit, and metal particles. Power systems in daily use should be protected by a 1 μm filter. Backup and occasional-use systems may be able to use a 25 μm filter.

A filter installed in the suction line will protect all components but will contribute to suction pressure loss. For that reason, it is common practice to install the filter after the pump, protecting all components except the pump. The pump is protected by a coarse screen in the suction line.

The maximum permissible pressure drop across a new and clean suction strainer or filter installed below the fluid level (i.e., submerged) is commonly limited by specification to 0.25 psi (1.7 kPa) for fire-resistant fluids and 0.50 psi (3.4 kPa) for all others.

12. ACCUMULATORS

Accumulators store potential energy in the form of pressurized hydraulic fluid. They are commonly used with intermittent duty cycles or to provide emergency power.[18] However, they can also be used to compensate for leakage, act as shock absorbers, and dampen pulsations.

The three basic types of accumulators are weight loaded, mechanical spring loaded, and gas loaded (i.e., hydro-pneumatic). *Hydro-pneumatic accumulators*, where a piston, diaphragm, or bladder separates the hydraulic fluid from the gas, are (by far) the most common.[19,20] Most accumulators are high-pressure tanks, and as such, should conform to ASME's Code for Unfired Pressure Vessels.

The volume of hydraulic fluid released by or captured in an accumulator is equal to the change in volume of the compressed gas. Compression and expansion of the gas in an accumulator is governed by standard thermodynamic principles. However, calculation of volumetric changes is complicated by the speed of the process, since the gas may heat or cool during the volume change.

The expression for polytropic processes, Eq. 19.16, is the most useful. (In Eq. 19.16, pressures should be strictly absolute pressures. However, the accumulator pressures are so high that use of gage pressures is a common practice.) $V_{charged}$ is the volume of the gas in the accumulator when it is charged with hydraulic fluid; $V_{discharged}$ is the gas volume after the fluid discharge.

$$p_{charged} V_{charged}^n = p_{discharged} V_{discharged}^n \qquad 19.16$$

$$V_{discharged} = V_{charged} + \text{discharge} \qquad 19.17$$

[18]Use of accumulators is becoming less common.
[19]Reactive gases, such as hydrogen and oxygen, should never be used as accumulator gases.
[20]Spring and weight-loaded accumulators are much rarer.

Fluids

Table 19.2 illustrates the variation in the polytropic exponent, n, on the charge and discharge rates. If the discharge is rapid, the process will be adiabatic, and $n = k$ (the ratio of specific heats). If it is very slow, then the process will be isothermal, and $n = 1$.

Table 19.2 *Typical Polytropic Exponents for Accumulator Sizing*

time for discharge or charging	diatomic gas[a]	monatomic gas[b]
1 min or less	1.4	1.7
2	1.3	1.5
3	1.15	1.25
more than 3 min	1.0	1.0

[a] Nitrogen, air, etc.
[b] Helium, argon, etc.

The *precharge gas pressure* is the pressure in the accumulator when it is completely filled with gas and is empty of hydraulic fluid. Ideally, the precharge pressure would be the minimum system pressure. However, rapid discharge of the gas is an adiabatic (or semi-adiabatic) process, and the gas will cool to below its original system temperature. Some of the accumulator fluid will not discharge. Therefore, the precharge pressure must be higher than the minimum system pressure.

After discharge, the accumulator will slowly warm back up to the system temperature in a constant-volume process.

$$\frac{p_{\text{discharged,cold}}}{T_{\text{discharged,cold}}} = \frac{p_{\text{discharged,warm}}}{T_{\text{discharged,warm}}} \qquad 19.18$$

The slow increase in temperature can also be considered to be an isothermal compression from the charged condition.

$$p_{\text{charged,warm}} V_{\text{charged,warm}}$$
$$= p_{\text{discharged,warm}} V_{\text{discharged,warm}} \qquad 19.19$$

Since the accumulator has a rigid body, the discharged volume is equal to the total accumulator volume, regardless of the temperature or pressure in the discharged accumulator. Equations 19.17 and 19.19 can be combined to give a direct expression for the precharge pressure.

$$p_{\text{precharge}} = p_{\text{discharged,warm}}$$
$$= \frac{p_{\text{charged,warm}} V_{\text{charged,warm}}}{V_{\text{accumulator}}}$$
$$= \frac{p_{\text{charged,warm}} (V_{\text{accumulator}} - \text{discharge})}{V_{\text{accumulator}}}$$
$$\qquad 19.20$$

Example 19.2

An accumulator must supply 250 in³ of fluid at 2000 psi minimum pressure. When charged, the maximum accumulator pressure cannot exceed 3000 psi. Discharge is

assumed to be adiabatic. What are (a) the accumulator size, and (b) the required precharge pressure?

Solution

(a) From Eq. 19.17, the charged accumulator gas volume is

$$V_{\text{charged}} = V_{\text{discharged}} - 250 \text{ in}^3$$

The gas starts out compressed at 3000 psi with volume $V_{\text{warm,charged}}$. It expands to volume $V_{\text{cold,discharged}}$ in an adiabatic process. The pressure after the discharge is 2000 psi. Use Eqs. 19.16 and 19.17.

$$p_{\text{charged}} V_{\text{charged}}^n = p_{\text{discharged}} V_{\text{discharged}}^n$$

$$V_{\text{discharged}} = (V_{\text{discharged}} - 250 \text{ in}^3) \left(\frac{3000 \frac{\text{lbf}}{\text{in}^2}}{2000 \frac{\text{lbf}}{\text{in}^2}} \right)^{\frac{1}{1.4}}$$

Rearranging terms and solving, the total accumulator volume is

$$V_{\text{accumulator}} = V_{\text{discharged}} = 994 \text{ in}^3$$

(b) Use Eq. 19.20 to calculate the precharge pressure.

$$p_{\text{precharge}} = \frac{p_{\text{charged,warm}} (V_{\text{accumulator}} - \text{discharge})}{V_{\text{accumulator}}}$$
$$= \frac{\left(3000 \frac{\text{lbf}}{\text{in}^2} \right) (994 \text{ in}^3 - 250 \text{ in}^3)}{994 \text{ in}^3}$$
$$= 2245 \text{ lbf/in}^2$$

13. LINEAR ACTUATORS AND CYLINDERS

Linear actuators (e.g., *hydraulic rams* and *hydraulic cylinders*) are the most common type of fluid power actuators.[21] The force exerted by a hydraulic cylinder is easily calculated. In Eq. 19.21, η_{cylinder} accounts for cylinder friction and is typically taken as 0.85. Hydraulic rams are sized in tons (2000 lbf/ton).

$$F = \eta_{\text{cylinder}} pA \qquad 19.21$$

Actuating speed of a cylinder rod depends on the fluid flow rate, Q, into the cylinder and the effective cylinder area, A. In most cases, retraction speed is less than the actuating speed, since the fluid acts on the piston area less the rod area.

$$v_{\text{ft/sec}} = 0.3208 \times \frac{Q_{\text{gpm}}}{A_{\text{in}^2}} \qquad 19.22$$

When fully extended, cylinder rods are like long columns and are subject to buckling failure. Therefore, piston rod sizing should include a check for column loading.

---

[21] Strictly speaking, a hydraulic motor is a rotary actuator or "linear motor."

20 Fans and Ductwork

1. Static Pressure 20-1
2. Velocity Pressure 20-2
3. Total Pressure 20-2
4. Standard and Actual Flow Rates 20-2
5. Variable Flow Rates 20-3
6. Axial Fans 20-3
7. Centrifugal Fans 20-3
8. Fan Specific Speed 20-4
9. Fan Power 20-4
10. Temperature Increase Across the Fan 20-5
11. Fan Curves 20-5
12. Multirating Tables 20-5
13. System Curve 20-7
14. System Effect 20-7
15. Operating Point 20-7
16. Affinity Laws 20-8
17. Fan Similarity 20-9
18. Operation at Nonstandard Conditions . . 20-10
19. Friction Losses in Round Ducts 20-11
20. Rectangular Ducts 20-14
21. Friction Losses in Fittings 20-14
22. Coefficient of Entry 20-15
23. Static Regain 20-15
24. Divided-Flow Fittings 20-16
25. Duct Design Principles 20-17
26. Leakage 20-18
27. Collapse of Ducts 20-18
28. Dampers 20-18
29. Velocity-Reduction Method 20-19
30. Equal-Friction Method 20-19
31. Combination Method 20-21
32. Static Regain Method 20-21
33. Total Pressure Design Method 20-24
34. Air Distribution 20-24
35. Exhaust Duct Systems 20-25
    Practice Problems 20-25

Nomenclature[1]

A	area	ft^2	m^2
ACFM	actual flow rate	cfm	L/s
BHP	brake horsepower	hp	–
c_p	specific heat at constant pressure	Btu/ lbm-°F	kJ/kg·K
C	coefficient	–	–
d	diameter	in	mm
D	diameter	ft	m
E	energy	ft-lbf	J
FHP	friction horsepower	hp	–
FP	friction pressure	in wg	Pa
g	acceleration of gravity	ft/sec^2	m/s^2
h	head or height	ft	m
K	factor	–	–
L	length	ft	m
$\dot{m}$	mass flow rate	lbm/sec	kg/s
ME	mechanical efficiency	–	–
n	fan speed	rpm	rpm
p	pressure	in wg	Pa
P	power	hp	kW
Q	volumetric flow rate	cfm	L/s
r	radius	ft	m
R	aspect ratio	–	–
R	regain coefficient	–	–
SCFM	standard flow rate	cfm	L/s
SE	static efficiency	–	–
SP	static pressure	in wg	Pa
SR	static regain	in wg	Pa
T	absolute temperature	°R	K
TP	total pressure	in wg	Pa
v	velocity	fpm	m/s
VP	velocity pressure	in wg	Pa

Symbols

γ	specific weight	lbf/ft^3	–
η	efficiency	–	–
ρ	density	lbm/ft^3	kg/m^3

Subscripts

br	branch
d	density, or discharge
down	downstream
e	entry, or equivalent
f	friction
k	kinetic
m	motor
s	specific
std	standard
up	upstream
v	velocity

---

[1]There is only marginal consistency in the symbols used by this industry. For example, the symbol for total pressure can be p_t, p_T, TP, P_t, T_p, h_t, and many other variations. The symbol for fitting loss coefficient is almost universally K in industry; ASHRAE uses C.

1. STATIC PRESSURE

The force of moving or stationary air perpendicular to a duct wall is known as the *static pressure*, SP. Static pressure can be measured in the field by a static tube or static tap. It is usually reported in inches of water, abbreviated in wg (or in. w.g.) for "inches of water gage," or in pascals. The pressure, height of a fluid column, and specific weight of the fluid are related by Eq. 20.1. The specific weight, γ, of water is approximately 0.0361 lbf/in^3 (1000 kg/m^3).

$$h = \frac{p}{\gamma} \qquad 20.1$$

$$\mathrm{SP}_{\mathrm{in\ wg}} = \frac{p_{\mathrm{psig}}}{0.0361\ \dfrac{\mathrm{lbf}}{\mathrm{in}^3}} \qquad 20.2$$

2. VELOCITY PRESSURE

The *velocity pressure*, VP, is the kinetic energy of the air expressed in inches of water. Velocity pressure is measured with a pitot tube, velometer, hot wire or rotating vane anemometer, or calibrated orifice or nozzle. The velocity head of a mass of moving air is

$$h_{\mathrm{V}} = \frac{v^2}{2g} \qquad 20.3$$

Air velocities are typically measured in ft/min (fpm) in the United States and in m/s in SI countries. As expressed, Eq. 20.3 calculates the kinetic energy as the height of an air column, not a height of a water column. The specific weights of air and water are approximately 0.075 lbf/ft^3 (1.2 kg/m^3) and 62.4 lbf/ft^3 (1000 kg/m^3), respectively. Therefore, the velocity pressure in inches of water is[2]

$$\mathrm{VP}_{\mathrm{in\ wg}} = \left[\frac{\left(\dfrac{v}{60\ \frac{\mathrm{sec}}{\mathrm{min}}} \right)^2 \left(12\ \dfrac{\mathrm{in}}{\mathrm{ft}} \right)}{(2)\left(32.2\ \dfrac{\mathrm{ft}}{\mathrm{sec}^2} \right)} \right] \left(\frac{0.075\ \dfrac{\mathrm{lbf}}{\mathrm{ft}^3}}{62.4\ \dfrac{\mathrm{lbf}}{\mathrm{ft}^3}} \right)$$

$$= \left(\frac{v_{\mathrm{fpm}}}{4005} \right)^2 \qquad 20.4$$

In SI units, with pressure in pascals, the velocity pressure is

$$\mathrm{VP}_{\mathrm{Pa}} = \frac{\rho(v_{\mathrm{m/s}})^2}{2} = \frac{\left(1.2\ \dfrac{\mathrm{kg}}{\mathrm{m}^3} \right)(v_{\mathrm{m/s}})^2}{2}$$

$$= 0.6(v_{\mathrm{m/s}})^2 \qquad 20.5$$

3. TOTAL PRESSURE

The *total pressure*, TP, is the sum of the velocity and static pressures. Contributions from potential energy are insignificant in virtually all duct design problems.

The total pressure decreases in the direction of flow. (However, the static pressure can increase with diameter increases.)

$$\mathrm{TP} = \mathrm{SP} + \mathrm{VP} \qquad 20.6$$

The change in total pressure is the algebraic sum of the changes in static and velocity pressures. In straight ducts with no branches or diameter changes, the change in total pressure is the same as the friction loss. ΔTP is positive when total pressure decreases.

$$\Delta\mathrm{TP} = \Delta\mathrm{SP} + \Delta\mathrm{VP} \qquad 20.7$$

4. STANDARD AND ACTUAL FLOW RATES

Air flow through fans is typically measured in units of cubic feet per minute, cfm (L/s). When the flow is at the *standard conditions* of 70°F (21°C) and 14.7 psia (101 kPa), the air flow is designated as SCFM (*standard cubic feet per minute*). Air flow at any other condition is designated as ACFM (*actual cubic feet per minute*). The two quantities are related by the *density factor*, K_d.[3] In Eq. 20.9, absolute temperature must be used.[4]

$$\mathrm{SCFM} = \frac{\mathrm{ACFM}}{K_d} \qquad 20.8$$

$$K_d = \frac{\rho_{\mathrm{std}}}{\rho_{\mathrm{actual}}}$$

$$= \left(\frac{p_{\mathrm{std}}}{p_{\mathrm{actual}}} \right) \left(\frac{T_{\mathrm{actual}}}{T_{\mathrm{std}}} \right) \qquad 20.9$$

Table 20.1 *Pressure at Altitudes*

altitude[a] ft (m)	ratio of $p_{\mathrm{actual}}/p_{\mathrm{std}}$
sea level (0)	1.00
1000 (305)	0.965
2000 (610)	0.930
3000 (915)	0.896
4000 (1220)	0.864
5000 (1525)	0.832
6000 (1830)	0.801
7000 (2135)	0.772

[a]Multiply ft by 0.305 to obtain m.

---

[2]The constant 4005 is reported at 4004 and 4004.4 in some references. 4005 is the most common.

[3]Some sources use an *air density ratio* that is the reciprocal of the density factor, K_d defined by Eq. 20.9. In some confusing cases, the same name (i.e., density factor) is used with the reciprocal value.

[4]The temperature correction should be based on the temperature and pressure of the air through the duct system. Though atmospheric pressure and temperature both decrease with higher altitudes, air entering any occupied space will generally be heated to normal temperatures. Therefore, the temperature correction will not generally be used unless the duct system carries air for process heating or cooling.

5. VARIABLE FLOW RATES

Ventilation and air conditioning rates usually vary with time. It is essential in modern, large systems to be able to vary the flow rate, as large amounts of energy are saved. *Variable flow rates* (i.e., *capacity control, flow rate modulation*) can be achieved through use of system dampers, fan speed control, variable blade pitch, and inlet vanes.

System dampers downstream of the fan are rarely used for capacity control. They increase friction loss, are noisy, and are nonlinear in their response. Speed control of the fan through fluid or magnetic coupling has low noise levels, but a high initial cost. For that reason, it also is seldom used.

With *blade pitch control* (*controllable pitch*), all blades are connected and simultaneously controlled while the fan is operating. Changing the blade angle of attack is efficient, quiet, and linear in response.

Inlet vanes are the most commonly used device for automatic control of centrifugal and in-line fans. Inlet vanes pre-spin and throttle the air prior to its entry into the wheel. Inlet vanes are relatively inefficient, noisy, and nonlinear in response.

Though a direct drive with flexible coupling can be used when flow is steady, most fans are run by v-belts.[5] In some applications requiring variable volume, either *variable-pitch pulleys* or variable (multispeed) motors can be used.

6. AXIAL FANS

Axial-flow fans are essentially propellers mounted with small tip clearances in ducts. They develop static pressure by changing the air-flow velocity. Axial flow fans are usually used when it is necessary to move large quantities of air (i.e., greater than 500,000 cfm; 235 000 L/s) against low static pressures (i.e., less than 12 in of water, 3 kPa), although the pressures and flow rates are much lower at most installations.

Compared with centrifugal fans, axial flow fans are more compact and less expensive. However, they run faster than centrifugals, draw more power, are less efficient, and are noisier. Axial flow fans are capable of higher velocities than centrifugal fans. In addition, overloading is less likely due to the flatter power curve. (See Fig. 20.1(a).) Fan noise is lowest at maximum efficiencies.

Axial fans can be further categorized into propeller, tubeaxial, and vaneaxial varieties. *Propeller fans* (such as the popular ceiling-mounted fans) are usually used only for exhaust and make-up duty, where the system static pressure is not more than $1/2$ in wg (125 Pa). Since they generally don't have housings, they are not

capable of generating static pressures in excess of about 1 in wg (250 Pa). Though they are light and inexpensive, they are the least efficient (about 50%) and the most noisy axial fans.

Tubeaxial fans, also known as *duct fans*, generally move air against less than 3 in of water (750 Pa). They have four to eight blades, and the clearance between the blade tips and surrounding duct is low. Fan efficiency is approximately 75 to 80%. Tubeaxials can be recognized by their hub diameters, which are less than 50% of the tip-to-tip diameter.

Vaneaxial fans can be distinguished from tubeaxial fans by their hub diameters, which are greater than 50% of the tip-to-tip diameter. Furthermore, the fan assembly will usually have vanes downstream from the fan to straighten the air flow and recover the rotational kinetic energy that would otherwise be lost. Vaneaxials typically have as many as 24 blades, and the blades may have cross sections similar to airfoils. Because they recover the rotational energy, vaneaxials are capable of moving air against pressures of up to 12 in of water (3.8 kPa). Their efficiencies are typically 85 to 90%.

7. CENTRIFUGAL FANS

Centrifugal fans are used in installations moving less than 1×10^6 cfm (470 000 L/s) and pressures less than 60 in of water (15 kPa). Like centrifugal pumps, they develop static pressure by imparting a centrifugal force on the rotating air. Depending on the blade curvature, kinetic energy can be made greater (forward-curved blades) or less (backward-curved blades) than the tangential velocity of the impeller blades.

Forward-curved centrifugals (also called *squirrel cage fans*) are the most widely used centrifugals for general ventilation and packaged units. They operate at relatively low speeds, about half that of backward-curved fans. This makes them useful in high-temperature applications where stress due to rotation is a factor. Compared with backward-curved centrifugals, forward-curved blade fans have a greater capacity (due to their higher velocities) but require larger scrolls. However, since the fan blades are "cupped," they cannot be used when the air contains particles or contaminants. Efficiencies are the lowest of all centrifugals—70 to 75%.

Motors driving centrifugal fans with forward-curved blades can be overloaded if the duct losses are not calculated correctly. The power drawn increases rapidly with increases in the delivery rate. The motors are usually sized with some safety factor to compensate for the possibility that the actual system pressure will be less than the design pressure. For forward-curved blades, the maximum efficiency occurs near the point of maximum static pressure. Since their tip speeds are low, they are quiet. The fan noise is lowest at maximum pressure.

---

[5]When selecting v-belts for fans, use a load factor of 1.4 (i.e., a power of 1.4 times the motor power).

Radial fans (also called *straight-blade fans*, *paddle wheel fans*, and *shaving wheel fans*) have blades that are neither forward- nor backward-inclined. They are the workhorse of most industrial exhaust applications and can be used in material-handling and conveying systems where large amounts of bulk material pass through them. Such fans are low-volume, high-pressure (up to 60 in wg; 15 kPa), high-noise, high-temperature, and low-efficiency (65 to 70%) units. *Radial tip fans* constitute a subcategory of radial fans. Their performance characteristics are between those of forward-curved and conventional radial fans.

Backward-curved centrifugals are quiet, medium-to-high volume and pressure, and high-efficiency units. They can be used in most applications with clean air below 1000°F (540°C) and up to about 40 in wg (10 kPa). They are available in three styles: flat, curved, and airfoil. Airfoil fans have the highest efficiency (up to 90%), while the other types have efficiencies between 80 and 90%. Because of these high efficiencies, power savings easily compensate for higher installation or replacement costs.

Motor overloading is less likely with backward-curved blades than with forward-curved blades, and for that reason may be referred to as *non-overloading fans*. These fans are normally equipped with motors sized to the peak power requirement so that the motors will not overload at any other operating condition.

Such fans operate over a great range of flows without encountering unstable air. Though their efficiencies are greater, they are noisier than forward-curved fans. The fan noise is lowest at the highest efficiencies. For the same operating speed, backward-curved blade fans develop more pressure than forward-curved fans.

8. FAN SPECIFIC SPEED

The fan specific speed is calculated from Eq. 20.10. Specific speed will be 10,000 to 20,000 (110 to 220) for radial centrifugals, 12,000 to 50,000 (130 to 550) for centrifugals, 40,000 to 170,000 (440 to 1900) for vaneaxials, 100,000 to 200,000 (1100 to 2200) for tubeaxials, and 120,000 to 300,000 (1300 to 3300) for propeller fans.

$$n_s = \frac{n_{\rm rpm}\sqrt{Q_{\rm L/s}}}{(\rm SP_{\rm Pa})^{0.75}} \qquad \text{[SI]} \qquad 20.10(a)$$

$$n_s = \frac{n_{\rm rpm}\sqrt{Q_{\rm cfm}}}{(\rm SP_{\rm in\ wg})^{0.75}} \qquad \text{[U.S.]} \qquad 20.10(b)$$

9. FAN POWER

The *air horsepower* (*blower horsepower*), AHP, or *air kilowatts*, AkW, is the power required to move the air.

$$\rm AkW = \frac{Q_{\rm L/s}(TP_{\rm Pa})}{10^6} \qquad \text{[SI]} \qquad 20.11(a)$$

$$\rm AHP = \frac{Q_{\rm cfm}(TP_{\rm in\ wg})}{6356} \qquad \text{[U.S.]} \qquad 20.11(b)$$

The actual power delivered to a fan from its motor is the *brake horsepower*, BHP, or *brake kilowatts*, BkW. Centrifugal fan efficiencies are in the range of 50 to 65%, although values as high as 80% are possible. The mechanical efficiency, ME, of a fan is normally read from the fan curves once the operating point is known, but can be calculated from Eq. 20.12.

$$\rm ME = \frac{AHP}{BHP} = \frac{AkW}{BkW} \qquad 20.12$$

The electrical power delivered to the motor will include the friction, windage, and other electrical losses in the motor. The electrical power is

$$\rm EHP = \frac{BHP}{\eta_m} = \frac{AHP}{\eta_m(ME)} \qquad 20.13$$

The *static efficiency*, SE, of a fan is defined as

$$\rm SE = (ME)\left(\frac{SP}{TP}\right) \qquad 20.14$$

Example 20.1

A fan moves 27,000 cfm (12 700 L/s) of air at 1800 fpm (9.2 m/s) against a static pressure of 2 in wg (500 Pa). The electrical motor driving the fan delivers 13.12 hp (9.77 kW) to it. What are the (a) total efficiency and (b) static efficiency?

SI Solution

(a) From Eq. 20.5, the velocity pressure is

$$\rm VP = (0.6)(v_{m/s})^2$$
$$= (0.6)\left(9.2\ \frac{m}{s}\right)^2 = 51\ Pa$$

The total pressure is

$$\rm TP = SP + VP = 500\ Pa + 51\ Pa$$
$$= 551\ Pa$$

From Eq. 20.11, the air kilowatts are

$$\rm AkW = \frac{Q_{\rm L/s}(TP_{\rm Pa})}{10^6}$$
$$= \frac{\left(12\,700\ \dfrac{L}{s}\right)(551\ Pa)}{10^6}$$
$$= 7\ kW$$

The fan efficiency is

$$\rm ME = \frac{AHP}{BHP} = \frac{7\ kW}{9.77\ kW}$$
$$= 0.72\ (72\%)$$

(b) From Eq. 20.14, the static efficiency is

$$\text{SE} = (\text{ME}) \left(\frac{\text{SP}}{\text{TP}} \right)$$

$$= (0.72) \left(\frac{500 \text{ Pa}}{551 \text{ Pa}} \right)$$

$$= 0.65 \ (65\%)$$

Customary U.S. Solution

(a) From Eq. 20.4, the velocity pressure is

$$\text{VP} = \left(\frac{\text{v}}{4005} \right)^2$$

$$= \left(\frac{1800 \text{ fpm}}{4005} \right)^2 = 0.2 \text{ in wg}$$

The total pressure is

$$\text{TP} = \text{SP} + \text{VP} = 2.0 \text{ in wg} + 0.2 \text{ in wg}$$

$$= 2.2 \text{ in wg}$$

From Eq. 20.11, the air horsepower is

$$\text{AHP} = \frac{Q(\text{TP})}{6356}$$

$$= \frac{(27{,}000 \text{ cfm})(2.2 \text{ in wg})}{6356}$$

$$= 9.35 \text{ hp}$$

The fan efficiency is

$$\text{ME} = \frac{\text{AHP}}{\text{BHP}} = \frac{9.35 \text{ hp}}{13.12 \text{ hp}}$$

$$= 0.71 \ (71\%)$$

(b) From Eq. 20.14, the static efficiency is

$$\text{SE} = (\text{ME}) \left(\frac{\text{SP}}{\text{TP}} \right)$$

$$= (0.71) \left(\frac{2 \text{ in wg}}{2.2 \text{ in wg}} \right)$$

$$= 0.65 \ (65\%)$$

10. TEMPERATURE INCREASE ACROSS THE FAN

The difference between the brake horsepower and air horsepower represents the power lost in the fan. This *friction horsepower* heats the air passing through the fan. If the fan motor is also in the air stream, it will also contribute to the heating effect to the extent that the motor's efficiency is not 100%.[6]

$$\text{FHP} = \text{BHP} - \text{AHP} = \text{BHP}(1 - \text{ME}) \qquad \textit{20.15}$$

The temperature increase across the fan is given by Eq. 20.16. Consistent units must be used.

$$\Delta T = \frac{\text{FHP}}{\dot{m} \, c_p} \qquad \textit{20.16}$$

If traditional units are used and the air is essentially at standard conditions, Eqs. 20.17 and 20.18 give the relationship between temperature change and the sensible heating or cooling effect.

$$\Delta T_{°\text{F}} = \frac{(3160)(\text{heating or cooling effect in kW})}{Q_{\text{cfm}}}$$

$$= \frac{(0.926) \left(\text{heating or cooling effect in } \dfrac{\text{Btu}}{\text{hr}} \right)}{Q_{\text{cfm}}}$$

$$= \frac{(2356)(\text{heating or cooling effect in hp})}{Q_{\text{cfm}}} \qquad \textit{20.17}$$

$$\Delta T_{°\text{C}} = \frac{(829)(\text{heating or cooling effect in kW})}{Q_{\text{L/sec}}}$$

$$= \frac{(1760)(\text{heating or cooling effect in kW})}{Q_{\text{cfm}}} \qquad \textit{20.18}$$

11. FAN CURVES

The operational parameters of fans are usually presented graphically by fan manufacturers. Total pressure, power, and efficiency are typically plotted on *fan characteristic curves*. Figure 20.1 contains typical curves for the three main types of fans. The dip in total pressure for axial flow and forward-curved centrifugal fans is characteristic.

12. MULTIRATING TABLES

Some manufacturers provide *fan rating tables* similar to Table 20.2. These tables, known as *multirating tables*, give the fan curve data in tabular, rather than in graphical, format. The highest mechanical efficiency for each pressure range will be in the middle third of the flow rate (Q) range. Manufacturers often indicate (by underlining or shading) points of operation that are within

---

[6]If the electrical input power and air horsepower are known, it isn't necessary to determine where the friction losses occur. The bearings, pulleys, and belts may all contribute to friction. However, the heating depends only on the difference between the input power and the power contributing to pressure and velocity.

Figure 20.1 *Typical Fan Curves*

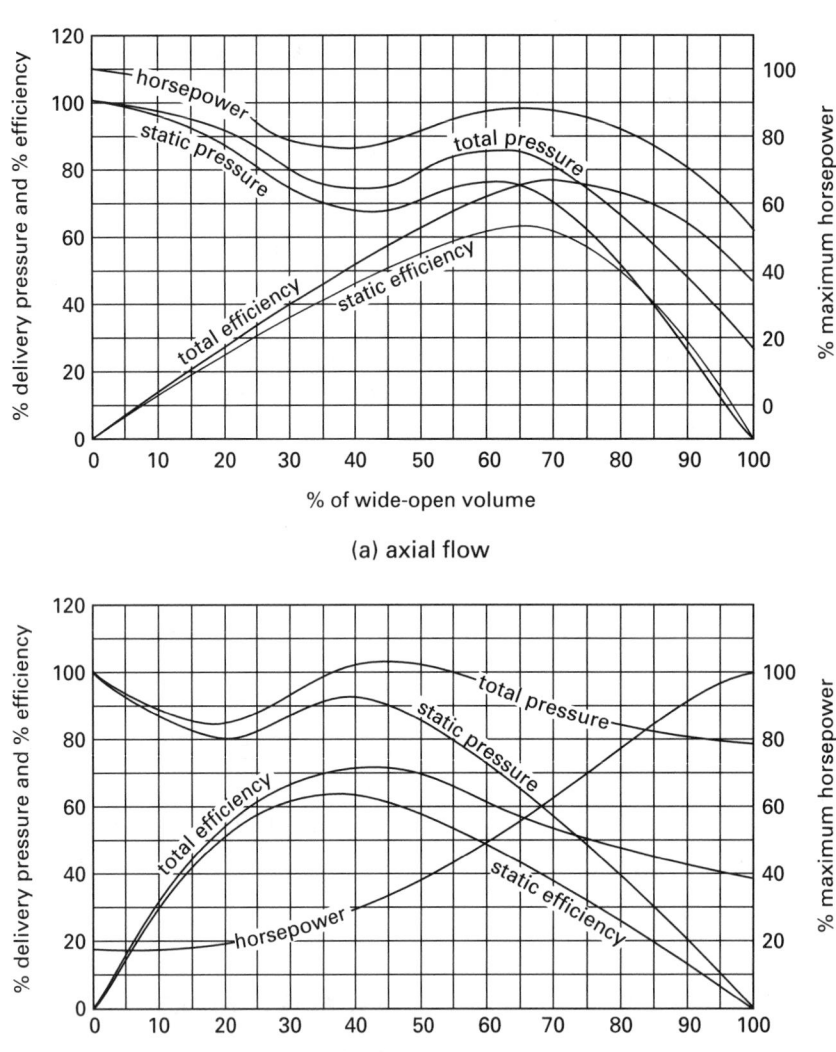

(a) axial flow

(b) forward-curved centrifugal

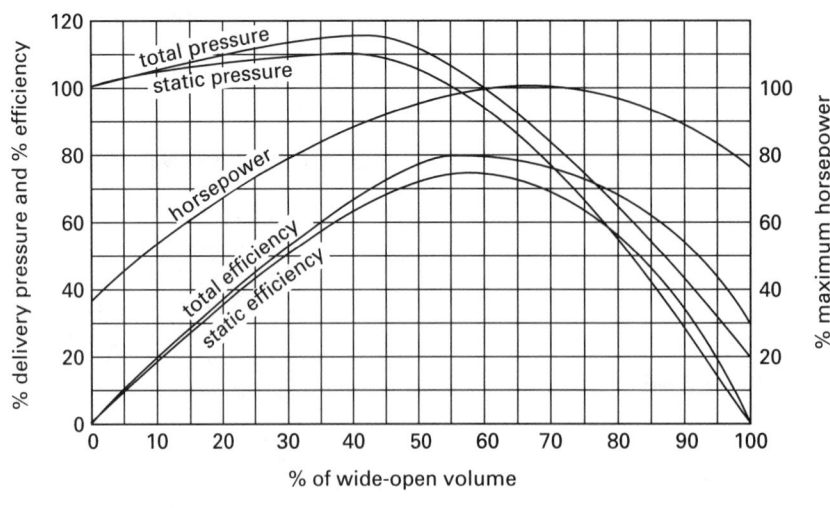

(c) backward-curved centrifugal

2% of the peak efficiency. If the peak efficiency point is not indicated, the actual efficiency can be calculated for each point using Eq. 20.12. Otherwise, selections can be limited to the middle third of the column.

Table 20.2 *Typical Fan Rating Table (portion)*

	SP = 1 in wg		SP = 2 in wg	
Q (cfm)	n (rpm)	P (bhp)	n (rpm)	P (bhp)
5000	440	1.20	617	2.67
10,000	492	2.18	626	4.20
15,000	600	4.06	706	6.45
20,000	816	9.59	830	10.83

13. SYSTEM CURVE

As with liquid flow in pipes, the pressure loss due to friction of air flowing in ducts varies with the square of the velocity. And, since $Q = Av$, the pressure loss varies with the square of Q. The graph of the friction loss versus the flow rate is the *system characteristic curve* (*system curve*).

If one point on the curve is known, the remainder of the curve can be found or generated from Eq. 20.19.

$$\frac{p_2}{p_1} = \left(\frac{Q_2}{Q_1}\right)^2 \qquad 20.19$$

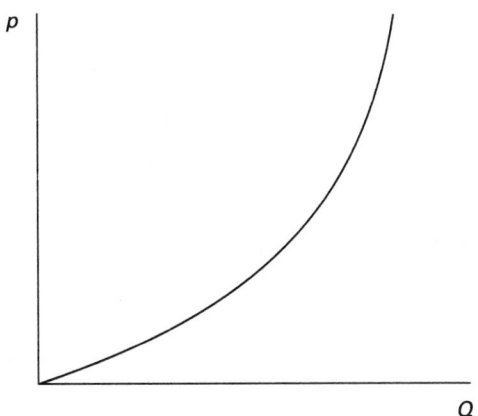

Figure 20.2 *Typical System Curve*

14. SYSTEM EFFECT

Almost all fans are rated under ideal laboratory conditions. Not only is the air at standard conditions, but also many of the physical features that would normally cause turbulence (bearings, diffusers, plenums, duct corners, etc.) are not present when the fan is tested. For that reason, rated performance is rarely achieved in practice.

Most fans are tested without being attached to a duct system. Merely connecting a duct system to a fan will produce a degradation in fan performance from rated values. This degradation, known as the *system effect*, is in addition to the duct friction and other losses. The *system effect factor* is the additional pressure (in in wg) that must be added to the calculated duct friction. The system effect factor depends on the flow rate (velocity) through the fan and the type of fan. For that reason, it must be based on information provided by the fan manufacturer.

15. OPERATING POINT

The intersection of the fan and system curves defines the *operating point* (*point of operation*).[7] If a fan is to be chosen by plotting the system curve on various fan curves, the following guidelines should be observed.

- To minimize the required motor power, the operating point should be as close as possible to the peak efficiency.

- For fans whose pressure characteristics have a dip (e.g., forward-curved centrifugals and axial flow fans), the operating point should be to the right of the peak fan pressure. This will avoid the noise and uneven motor loading that accompany pressure and volume fluctuations. (See Fig. 20.3.)

- A fan with a steep pressure curve should be chosen to avoid large variations in flow rate with changes in duct friction.

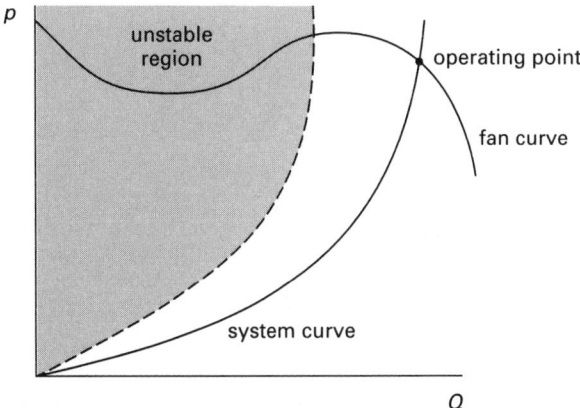

Figure 20.3 *Operating Point and Unstable Region*

System configurations (length of runs, bends, equipment, etc.) are usually less flexible than fans when it comes to changing performance. If the operating point for a specific fan does not provide the required air flow

---

[7]The *rating point* (*point of rating*) is the one single point on the fan curve that corresponds to stated (often the rated) performance. The *duty point* (*point of duty*) is one single point on the system curve where a fan is to operate.

(efficiency, power, etc.), there are several different steps that can be taken.

- Use a different fan.

- Change the fan speed.

- Change the fan size.

- Use two fans in parallel. The combined flow at a particular pressure will be the sum of the individual fan flows corresponding to that pressure.

- Use two fans in series. The combined pressure at a particular flow rate will be the sum of the individual fan pressures corresponding to that flow rate.

Example 20.2

The pressure loss due to friction in the system shown is 1.5 in wg (375 Pa) when the flow rate is 3500 cfm (1650 L/s). Velocity head and outlet pressure are negligible. What will be the flow rate if a fan with the characteristics shown is used?

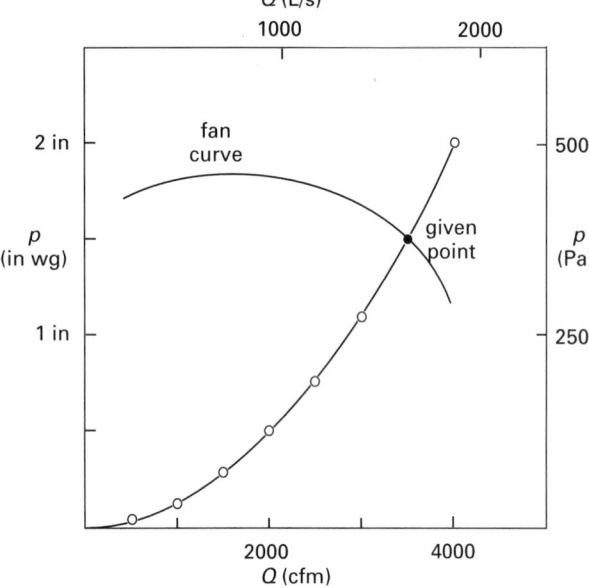

SI Solution

The fan curve is given. One point on the system curve is known. Use Eq. 20.19 to derive the remainder of the system curve. For 1500 L/s, the pressure drop would be

$$p_2 = p_1 \left(\frac{Q_2}{Q_1} \right)^2$$

$$= (375 \text{ Pa}) \left(\frac{1500 \frac{\text{L}}{\text{s}}}{1650 \frac{\text{L}}{\text{s}}} \right)^2$$

$$= 310 \text{ Pa}$$

The remainder of the system curve can be determined in the same manner. The fan and system curves intersect at approximately 1550 L/s.

Q (L/s)	p (Pa)
500	35
750	80
1000	140
1250	215
1500	310
1650	375
1750	420

Customary U.S. Solution

The fan curve is given. One point on the system curve is known. Use Eq. 20.19 to derive the remainder of the system curve. For 3000 cfm, the pressure drop would be

$$p_2 = p_1 \left(\frac{Q_2}{Q_1} \right)^2$$

$$= (1.5 \text{ in wg}) \left(\frac{3000 \text{ cfm}}{3500 \text{ cfm}} \right)^2$$

$$= 1.1 \text{ in wg}$$

The remainder of the system curve can be determined in the same manner. The fan and system curves intersect at approximately 3300 cfm.

Q (cfm)	p (in wg)
500	0.03
1000	0.12
1500	0.28
2000	0.49
2500	0.77
3000	1.1
3500	1.5
4000	2.0

16. AFFINITY LAWS

Within reasonable limits, the speed of v-belt driven fans can be easily changed by changing pulleys. The following *affinity laws (fan laws)* can be used to predict performance of a particular fan at different speeds. These fan laws assume the fan size, fan efficiency, and air density are the same.[8,9]

$$\frac{Q_2}{Q_1} = \frac{n_2}{n_1} \qquad \qquad 20.20$$

---

[8]These fan laws are simplifications of the similarity laws presented in the next section. The similarity laws must be used if the density changes.

[9]For any given efficiency, the locus of equal-efficiency points on the pressure-capacity (p-Q) diagram is a parabola starting at the origin and crossing the different fan curves corresponding to different speeds. The intersection points of the fan curves and the parabolic equal-efficiency curve are known as *corresponding points*. Theoretically, the fan laws can only be used at these points.

$$\frac{p_2}{p_1} = \left(\frac{n_2}{n_1}\right)^2 \qquad \textit{20.21}$$

$$\frac{AHP_2}{AHP_1} = \left(\frac{n_2}{n_1}\right)^3 \qquad \textit{20.22}$$

Since the efficiency at the two speeds is assumed to be the same, Eq. 20.22 can be rewritten in terms of power drawn.

$$\frac{BHP_2}{BHP_1} = \left(\frac{n_2}{n_1}\right)^3 \qquad \textit{20.23}$$

A change in the fan speed cannot be used to put the operating point into a stable region. Changing the fan speed does not change the relative position of the operating point with respect to the dip present in some fan curves. The locus of peak points follows the Q^2 rule also. So, if an operating point is to the left of the peak point (i.e., is in an unstable region) at one fan speed, the new operating point will be to the left of the peak point on the new fan curve corresponding to the new speed.

17. FAN SIMILARITY

The performance of one fan can be used to predict the performance of a *dynamically similar (homologous) fan*. This can be done by using Eqs. 20.24 through 20.28.

$$\frac{Q_A}{Q_B} = \left(\frac{D_A}{D_B}\right)^3 \left(\frac{n_A}{n_B}\right)$$
$$= \left(\frac{D_A}{D_B}\right)^2 \sqrt{\frac{p_A}{p_B}} \sqrt{\frac{\gamma_B}{\gamma_A}} \qquad \textit{20.24}$$

$$\frac{p_A}{p_B} = \left(\frac{D_A}{D_B}\right)^2 \left(\frac{n_A}{n_B}\right)^2 \left(\frac{\gamma_A}{\gamma_B}\right) \qquad \textit{20.25}$$

$$\frac{AHP_A}{AHP_B} = \left(\frac{D_A}{D_B}\right)^5 \left(\frac{n_A}{n_B}\right)^3 \left(\frac{\gamma_A}{\gamma_B}\right)$$
$$= \left(\frac{D_A}{D_B}\right)^2 \left(\frac{p_A}{p_B}\right)^{\frac{3}{2}} \sqrt{\frac{\gamma_B}{\gamma_A}}$$
$$= \left(\frac{Q_A}{Q_B}\right) \left(\frac{p_A}{p_B}\right) \qquad \textit{20.26}$$

$$\frac{n_A}{n_B} = \left(\frac{D_B}{D_A}\right) \sqrt{\frac{p_A}{p_B}} \sqrt{\frac{\gamma_B}{\gamma_A}}$$
$$= \sqrt{\frac{Q_B}{Q_A}} \left(\frac{p_A}{p_B}\right)^{\frac{3}{4}} \left(\frac{\gamma_B}{\gamma_A}\right)^{\frac{3}{4}} \qquad \textit{20.27}$$

$$\frac{D_A}{D_B} = \sqrt{\frac{Q_A}{Q_B}} \left(\frac{p_B}{p_A}\right)^{\frac{1}{4}} \left(\frac{\gamma_A}{\gamma_B}\right)^{\frac{1}{4}} \qquad \textit{20.28}$$

Similarity laws may be used to predict the performance of a larger fan from a smaller fan's performance, since the efficiency of the larger fan can be expected to exceed

that of the smaller fan. Larger fans should not be used to predict the performance of smaller fans. Even extrapolations to larger fans should be viewed cautiously when there is a significant decrease in air density or when the ratio of the larger-to-smaller fan diameters, the ratio of speed, or the product of the diameter and speed ratios exceed 3.0.

Example 20.3

A fan turning at 800 rpm draws 6.2 hp (4.6 kW) while delivering 10,000 cfm (4700 L/s) against a static pressure of 2.25 in wg (560 Pa). (a) If the fan is driven at 1400 rpm, what will be the power drawn? (b) If duct length is increased such that the static pressure loss is 2.85 in wg (713 Pa), what speed will be required?

SI Solution

(a) Use Eq. 20.23.

$$AkW_2 = AkW_1 \left(\frac{n_2}{n_1}\right)^3$$
$$= (4.6 \text{ kW}) \left(\frac{1400 \text{ rpm}}{800 \text{ rpm}}\right)^3$$
$$= 24.7 \text{ kW}$$

(It is unlikely that the same motor will be able to provide this increased power.)

(b) Use Eq. 20.21.

$$n_2 = n_1 \sqrt{\frac{p_2}{p_1}}$$
$$= (800 \text{ rpm}) \sqrt{\frac{713 \text{ Pa}}{560 \text{ Pa}}}$$
$$= 903 \text{ rpm}$$

Customary U.S. Solution

(a) Use Eq. 20.23.

$$BHP_2 = BHP_1 \left(\frac{n_2}{n_1}\right)^3$$
$$= (6.2 \text{ hp}) \left(\frac{1400 \text{ rpm}}{800 \text{ rpm}}\right)^3$$
$$= 33.2 \text{ hp}$$

(It is unlikely that the same motor will be able to provide this increased power.)

(b) Use Eq. 20.21.

$$n_2 = n_1 \sqrt{\frac{p_2}{p_1}}$$
$$= (800 \text{ rpm}) \sqrt{\frac{2.85 \text{ in wg}}{2.25 \text{ in wg}}}$$
$$= 900 \text{ rpm}$$

Fluids

18. OPERATION AT NONSTANDARD CONDITIONS

Fan tests used to develop curves and rating tables are based on air at standard conditions—70°F and 14.7 psia (21°C and 101 kPa). Small variations in density due to normal temperature and humidity fluctuations can be disregarded. However, if the system operates at extremely elevated temperatures or reduced atmospheric pressures, corrections will be necessary.

Fans are constant-volume devices. They deliver the same volume of air (and at the same duct speed) regardless of temperature, pressure, and humidity ratio. Therefore, the actual flow rate, ACFM, should be used to select a fan from rating tables. The speed can be read directly from the fan table. The standard (i.e., table) values of power and pressure (static, velocity, and total) should be modified by the *density factor*, K_d[10] (Eq. 20.9), with the density.

$$K_d = \frac{\text{SP}_{\text{table}}}{\text{SP}_{\text{actual}}} = \frac{\text{VP}_{\text{table}}}{\text{VP}_{\text{actual}}} = \frac{\text{TP}_{\text{table}}}{\text{TP}_{\text{actual}}}$$

$$= \frac{\text{BHP}_{\text{table}}}{\text{BHP}_{\text{actual}}} \qquad \text{20.29}$$

Example 20.4

A fan is chosen to move 18,000 SCFM (8500 L/s) of air against a static pressure of 0.85 in wg (210 Pa). The fan draws 4.2 hp (3.1 kW) when moving standard air in that configuration. However, the fan is used to provide 150°F (66°C) air for drying. What will be the (a) required power, and (b) friction loss?

SI Solution

(a) Equation 20.9 gives the density factor. The pressure is unchanged.

$$K_d = \frac{T_{\text{actual}}}{T_{\text{std}}}$$

$$= \frac{66°\text{C} + 273}{21°\text{C} + 273} = 1.15$$

From Eq. 20.29,

$$\text{AkW}_{\text{actual}} = \frac{\text{AkW}_{\text{std}}}{K_d} = \frac{3.1\text{ kW}}{1.15}$$

$$= 2.7\text{ kW}$$

(b) $$\qquad p_{f,\text{actual}} = \frac{p_{f,\text{std}}}{K_d} = \frac{210\text{ Pa}}{1.15}$$

$$= 183\text{ Pa}$$

---

[10]There is also a correction for viscosity. However, the viscosity change is so insignificant that it is disregarded.

Customary U.S. Solution

(a) Equation 20.9 gives the density factor. The pressure is unchanged.

$$K_d = \frac{T_{\text{actual}}}{T_{\text{std}}}$$

$$= \frac{150°\text{F} + 460}{70°\text{F} + 460} = 1.15$$

From Eq. 20.29,

$$\text{BHP}_{\text{actual}} = \frac{\text{BHP}_{\text{std}}}{K_d} = \frac{4.2\text{ hp}}{1.15}$$

$$= 3.7\text{ hp}$$

(b) $$\qquad p_{f,\text{actual}} = \frac{p_{f,\text{std}}}{K_d} = \frac{0.85\text{ in wg}}{1.15}$$

$$= 0.74\text{ in wg}$$

Example 20.5

70°F (21°C) air in a duct located at an altitude of 5000 ft (1525 m) moves at 1500 fpm (7.7 m/s). The actual flow rate is 39,000 cfm (18 300 L/s). The duct resistance at that altitude is 1.5 in wg (375 Pa). (a) What duct resistance should be used with fan rating tables? (b) What input power to the fan is required if the fan efficiency is 75%?

SI Solution

(a) From Table 20.1, the atmospheric pressure ratio at 1525 m altitude is 0.832. The density is

$$p_{\text{actual}} = (0.832)(101.3\text{ kPa}) = 84.3\text{ kPa}$$

The density factor is

$$K_d = \frac{p_{\text{std}}}{p_{\text{actual}}}$$

$$= \frac{101.3\text{ kPa}}{84.3\text{ kPa}} = 1.2$$

From Eq. 20.29, the standard friction pressure loss is

$$\text{FP}_{\text{table}} = K_d(\text{FP}_{\text{actual}})$$

$$= (1.2)(375\text{ Pa}) = 450\text{ Pa}$$

(b) Since the volumetric flow rate does not change, the duct speed is also unchanged. The fan supplies both velocity and static pressure. From Eq. 20.4, the velocity pressure is

$$\text{VP} = (0.6)(v_{\text{m/s}})^2 = (0.6)\left(7.7\ \frac{\text{m}}{\text{s}}\right)^2$$

$$= 36\text{ Pa}$$

The total pressure energy supplied by the fan is

$$TP = SP + VP = 375 \text{ Pa} + 36 \text{ Pa}$$
$$= 411 \text{ Pa}$$

From Eq. 20.11, the original power drawn is

$$AkW = \frac{Q_{L/s}(TP_{Pa})}{10^6}$$
$$= \frac{\left(18\,300 \frac{L}{s}\right)(411 \text{ Pa})}{10^6}$$
$$= 7.52 \text{ kW}$$

From Eqs. 20.12 and 20.29, the standardized brake kilowatts are

$$AkW_{table} = \frac{K_d(AkW_{actual})}{ME}$$
$$= \frac{(1.2)(7.52 \text{ kW})}{0.75} = 12 \text{ kW}$$

Customary U.S. Solution

(a) From Table 20.1, the atmospheric pressure ratio at 5000 ft altitude is 0.832. The density is

$$p_{actual} = (0.832)(14.7 \text{ psia}) = 12.2 \text{ psia}$$

The density factor is

$$K_d = \frac{p_{std}}{p_{actual}}$$
$$= \frac{14.7 \text{ psia}}{12.2 \text{ psia}} = 1.2$$

From Eq. 20.29, the standard friction pressure loss is

$$FP_{table} = K_d(FP_{actual})$$
$$= (1.2)(1.5 \text{ in wg}) = 1.8 \text{ in wg}$$

(b) Since the volumetric flow rate does not change, the duct speed is also unchanged. The fan supplies both velocity and static pressure. From Eq. 20.4, the velocity pressure is

$$VP = \left(\frac{v}{4005}\right)^2$$
$$= \left(\frac{1500 \text{ fpm}}{4005}\right)^2 = 0.14 \text{ in wg}$$

The total pressure energy supplied by the fan is

$$TP = SP + VP$$
$$= 1.5 \text{ in wg} + 0.14 \text{ in wg} = 1.64 \text{ in wg}$$

From Eq. 20.11, the original power drawn is

$$AHP = \frac{Q(TP)}{6356}$$
$$= \frac{(39,000 \text{ cfm})(1.64 \text{ in wg})}{6356}$$
$$= 10.06 \text{ hp}$$

From Eqs. 20.12 and 20.29, the standardized brake horsepower is

$$BHP_{table} = \frac{K_d(BHP_{actual})}{ME}$$
$$= \frac{(1.2)(10.06 \text{ hp})}{0.75} = 16.1 \text{ hp}$$

19. FRICTION LOSSES IN ROUND DUCTS

Friction loss (*friction pressure*, FP) can be calculated from the standard Moody equation. Equation 20.30 expresses the Moody equation in typical air-moving units for standard air moving in clean galvanized commercial ducts for a specific roughness (ϵ) of 0.0005 ft (0.15 mm).[11]

$$FP_{in \text{ wg}} = \frac{(3.9 \times 10^{-9})(v_{fpm})^{2.43}L_{ft}}{(Q_{cfm})^{0.61}} \qquad 20.30$$

However, Eq. 20.30 is almost never used in the HVAC industry. Rather, friction losses are typically determined from graphs. Figures 20.4 and 20.5 assume clean commercial-quality round ducts with a normal number of joints, and conditions close to standard air. For smooth ducts with no joints, the friction loss is 60 to 95% of the value determined from Figs. 20.4 and 20.5.[12]

Duct flow areas are calculated from their nominal dimensions. Any size duct can be manufactured. However, there are standard sizes of premanufactured round duct, and these sizes should be chosen to minimize cost. Generally, commercial duct manufacturers produce every whole-inch size up to at least 20 in in diameter (510 mm). After that, ducts are available in 2 in (50 mm) increments. Odd-number sizes may be available with premium pricing.

---

[11]Specific roughness is the reciprocal of the number of duct diameters required to cause a static pressure loss of one velocity pressure. Typical values for galvanized duct are 0.0003 to 0.0005 ft (0.09 to 0.15 mm).

[12]For corrugated ductwork, the friction loss is approximately twice that shown in Figs. 20.4 and 20.5. However, the analysis is not precise enough to make most corrections, including those for different number of joints or duct material. Corrections should be limited to extreme cases when the ductwork deviates significantly from commercial (e.g., air flow through brickwork or corrugated duct).

Figure 20.4 *Standard Friction Loss in Standard Duct[a] (inches of water per 100 ft of duct; temperatures of 50 to 90°F)*
(Recommended operating points shown as shaded region.)

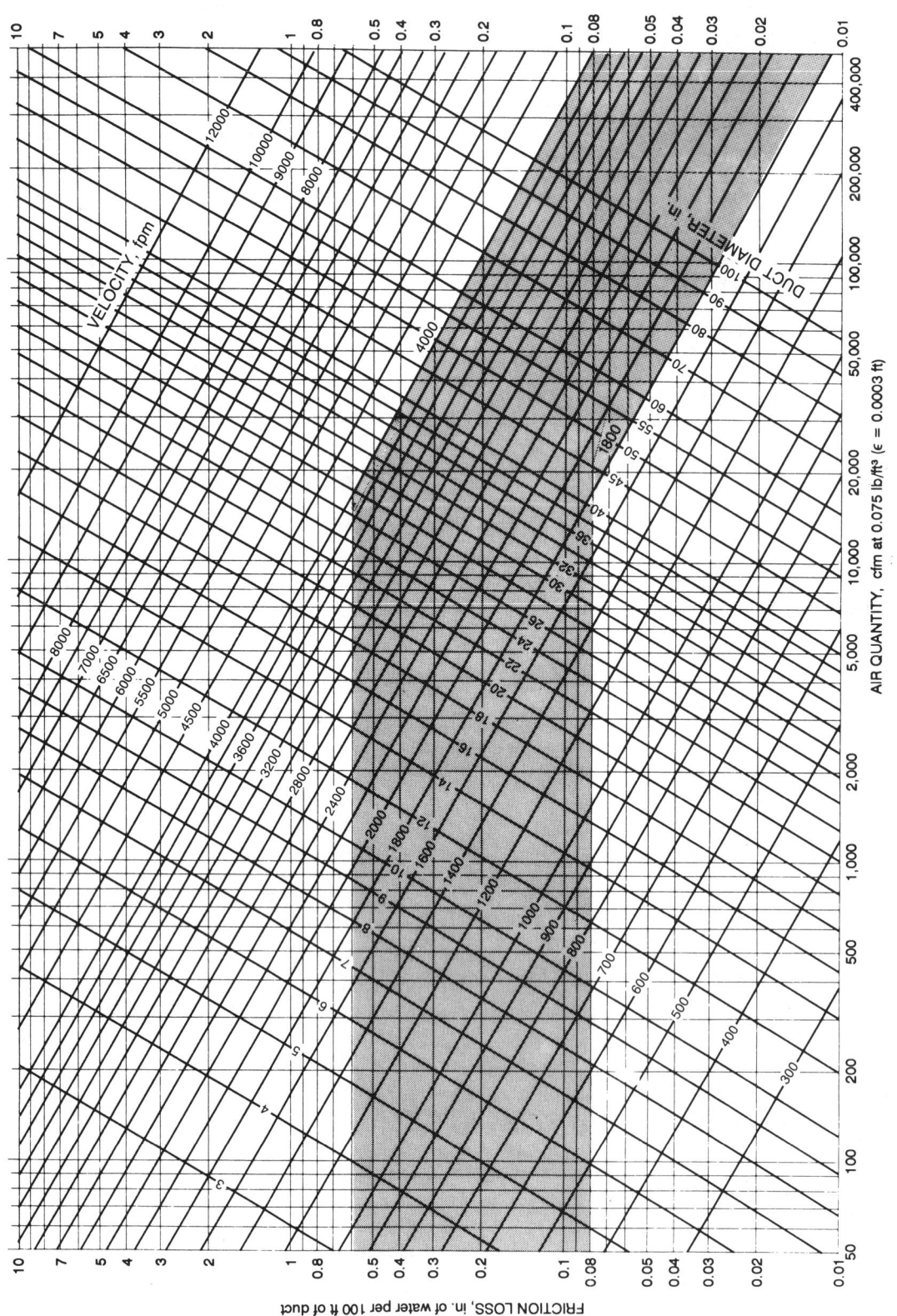

[a] Clean, round galvanized duct with a specific roughness of 0.0005 ft (0.15 mm) and approximately 25 beaded slip-couplings (joints) per 100 ft (30 m). Can also be used for smooth commercial spiral duct with about 10 joints per 100 ft (30 m).

Figure 20.5 *Standard Friction Loss in Standard Duct*[a] *(pascals per meter of duct; temperatures of 10 to 30°C)*
(Recommended operating points shown as shaded region.)

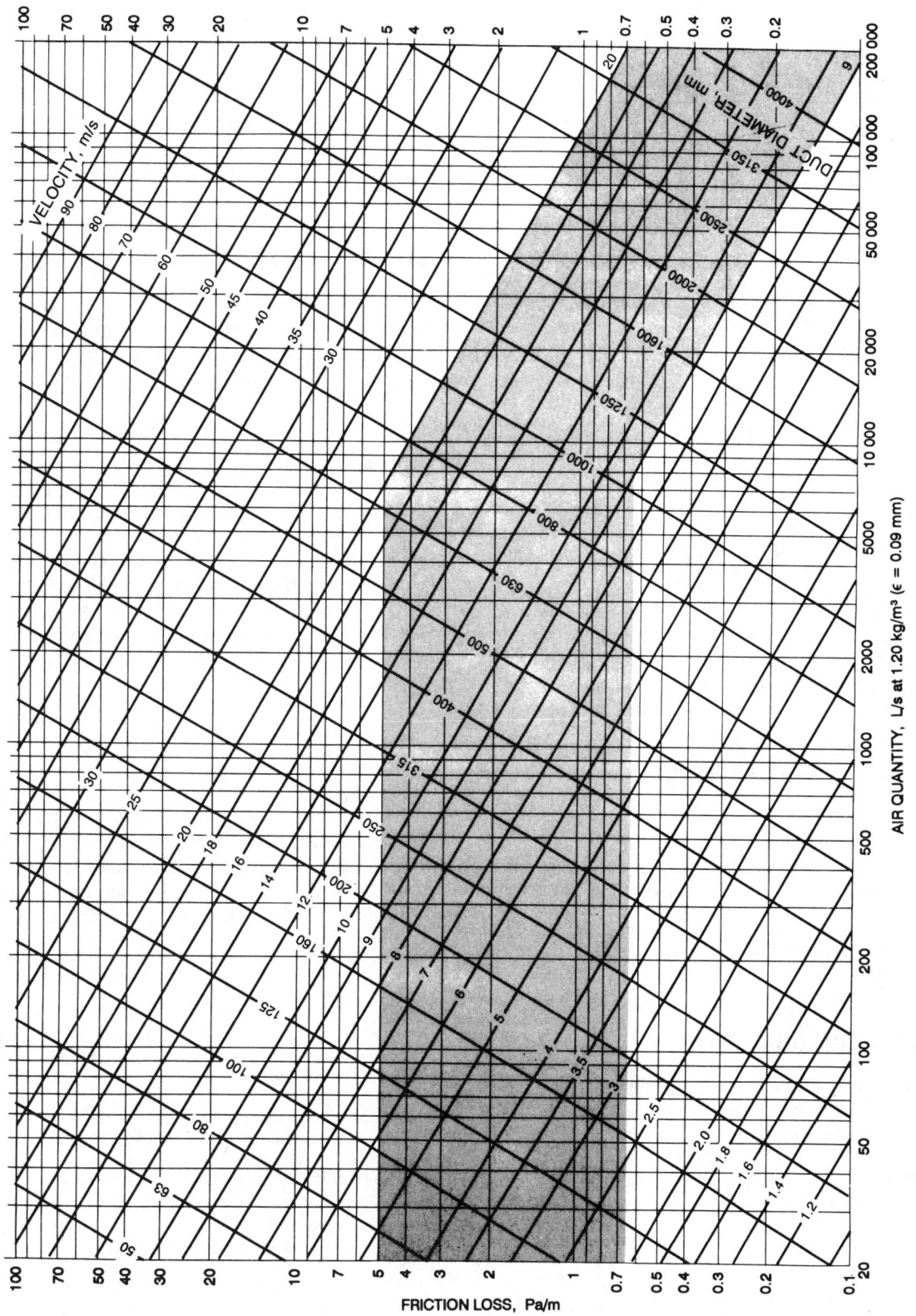

[a] Clean, round galvanized duct with a specific roughness of 0.0005 ft to 0.15 mm and approximately 1 beaded slip-coupling (joint) per meter. Can also be used for smooth commercial spiral duct with about 1 joint per 3 m.

Reprinted with permission from *1993 ASHRAE Handbook: Fundamentals*, SI Edition. American Society of Heating, Refrigeration, and Air Conditioning Engineers, Inc., Atlanta, GA, ©1993.

Example 20.6

2000 cfm (1000 L/s) of air flow in a 13 in (315 mm) diameter duct. What are the (a) velocity and (b) friction loss per 100 ft (per meter) of duct?

SI Solution

(a) Use Fig. 20.5. Locate the intersection of the 1000 L/s and 315 mm lines. The velocity is approximately 13 m/s.

(b) Move to the left to the vertical scale. The friction loss is approximately 6 Pa/m of duct.

Customary U.S. Solution

(a) Use Fig. 20.4. Locate the intersection of the 2000 cfm and 13 in lines. The velocity is approximately 2200 fpm. (This answer could also be calculated from v = Q/A.)

(b) Drop straight down to the horizontal scale. The friction loss is approximately 0.5 in wg per 100 ft of duct.

Example 20.7

2000 cfm (1000 L/s) of standard air move through a duct with a velocity of 1600 fpm (8 m/s). (a) What size duct is required? (b) What is the friction loss?

SI Solution

(a) Locate the intersection of 1000 L/s and 8 m/s. The required duct diameter is 400 mm.

(b) Move horizontally to the left to the vertical scale. The friction loss is approximately 1.7 Pa/m.

Customary U.S. Solution

(a) Locate the intersection of 2000 cfm and 1600 fpm. The required duct diameter is approximately 15 in.

(b) Drop straight down to the horizontal scale. The friction loss is approximately 0.23 in wg per 100 ft of duct.

20. RECTANGULAR DUCTS

Duct systems are initially designed for round ducts. Then, conversions to rectangular ducts are made as required. A round duct with diameter D can be converted to a rectangular duct with equal friction per unit length if the desired aspect ratio is known. The *aspect ratio*, R, of a rectangular duct should be kept below 8 for ease of manufacture.

$$\text{short side} = \frac{D(1+R)^{\frac{1}{4}}}{1.3R^{\frac{5}{8}}} \qquad 20.31$$

$$R = \frac{\text{long side}}{\text{short side}} \qquad 20.32$$

The *equivalent diameter* of a rectangular duct with an aspect ratio less than 8 is given by Eq. 20.33. The round duct will have the same friction and capacity as the rectangular duct. Figures 20.4 and 20.5 can be used with D_e and the actual flow rate to find the friction loss.[13] The velocity indicated by the chart will be incorrect but can be calculated as Q/A.

$$D_e = \frac{(1.3)(\text{short side} \times \text{long side})^{\frac{5}{8}}}{(\text{short side} + \text{long side})^{\frac{1}{4}}} \qquad 20.33$$

21. FRICTION LOSSES IN FITTINGS

Friction losses (*dynamic losses*) due to bends, fittings, enlargements, contractions, and obstructions are calculated in the same ways as for liquid friction losses—either by loss coefficient or equivalent length methods.[14,15]

With the *loss coefficient method*, the friction loss is calculated as a multiple of the velocity pressure. Though reported values vary widely, typical values of the loss coefficient, K, for common features are given in Table 20.3.[16] Loss coefficients are usually based on the upstream velocity pressure. However, there are some cases (i.e., where plenum air with negligible velocity enters a duct) where the downstream velocity is used by convention. The coefficient should always be used with the velocity at the point corresponding to the coefficient's subscript.

$$FP = K(VP) = K\left(\frac{v}{4005}\right)^2 \qquad 20.34$$

In a straight duct of constant diameter, the velocity pressure is unchanged. The change in static pressure is the friction loss. Since there is no change in the velocity pressure, the friction loss will produce an equivalent decrease in total pressure. In other words, the friction loss is the change in total pressure. The same loss coefficient is used for calculating the change in static pressure and the change in total pressure.

$$\begin{aligned} FP &= \Delta TP = \Delta SP \\ &= K(VP)_{up} \quad \begin{bmatrix} \text{constant area duct} \\ \text{with no branches} \end{bmatrix} \end{aligned} \qquad 20.35$$

---

[13] This is the formula used by ASHRAE and most other authorities for the *equivalent* diameter. Some sources merely equate formulas for round and rectangular areas and use the *hydraulic diameter*, which is $\sqrt{(4 \times \text{short side} \times \text{long side})/\pi}$. When rounded to the nearest whole duct size, the difference is often insignificant. Although the two ducts may have the same cross-sectional area, they will not have the same capacity or friction.

[14] Unlike losses for liquid flows, however, fitting losses for duct systems are significant. They are not "minor" losses.

[15] Any fitting or feature that causes a static pressure loss of 0.75 in wg (200 Pa) or higher is a potential source of unwanted noise.

[16] Static pressure losses due to equipment (e.g., filters, coils, and heat exchangers) are determined from manufacturers' literature.

Table 20.3 *Typical Fitting Loss Coefficients*[a,b]

feature		K_{up}	K_{down}
abrupt expansion from A_{up} to A_{down}	$\dfrac{A_{up}}{A_{down}} = 0$	1.00	
	0.2	0.64	
	0.4	0.36	
	0.6	0.16	
	0.8	0.04	
abrupt contraction from A_{up} to A_{down}	$\dfrac{A_{down}}{A_{up}} = 0.2$		0.32
	0.4		0.25
	0.6		0.16
	0.8		0.06
round pipe of diameter D_1 across duct of diameter D_2	$\dfrac{D_1}{D_2} = 0.1$	0.2	
	0.25	0.55	
	0.5	2.0	
tapered reducing section[c]			
taper angle[d] =	20°	0.012	
	40°	0.032	
	60°	0.07	
bell-mouthed entrance[e]			0.04
90° round elbows[f,g]			
smooth die-stamped			
	$r/D = 1.25$	0.55	
	1.5	0.39	
	1.75	0.32	
	2.0	0.27	
	2.25	0.26	
	2.5	0.22	
	2.75	0.26	
mitered with turning vanes		0.5	
mitered with throat gore		0.9	
mitered with heel gore		1.0	
straight miter (two piece)		1.2	
straight miter (three piece)		0.34	
straight miter (four piece)		0.22	
straight miter (five piece)		0.2	

[a]Subscripts "up" and "down" refer to upstream and downstream, respectively.
[b]In multiples of velocity pressure.
[c]The total energy loss is small for all taper angles. The advantage of a very long taper is insignificant.
[d]The *taper angle* is the angle one side makes with the straight wall. The *included angle* refers to twice the taper angle.
[e]When stationary air is drawn into a bell-mouthed opening, the fan must supply the velocity pressure (1.0) as well as overcome the entrance friction (0.04). Because of that, some sources report this value as (1.04).
[f] Also, see Table 20.4.
[g]For 60° elbows, multiply by $^2/_3$. For 45° elbows, multiply by $^1/_2$. For 30° elbows, multiply by $^1/_3$.

Loss coefficients are *zero-length losses*. That is, they only include the dynamic effects. If a large fitting has a specific length, that length must be included in the run-of-duct length when calculating the duct friction.

The fitting loss can be assumed to be the result of an equivalent length of duct. These lengths are given in multiples of duct diameter in Table 20.4. This is the *equivalent length method*.

Table 20.4 *Typical Equivalent Lengths*[a]

			L_e
90° smooth round elbow of radius r and diameter D	$\dfrac{r}{D} =$	0.5	$45D$
		0.75	$23D$
		1.0	$17D$
		1.5	$12D$
		2.0	$10D$
90°F miter elbow			$65D$

[a]In terms of inside duct diameter, D.

22. COEFFICIENT OF ENTRY

The *coefficient of entry*, C_e, is the ratio of the actual to ideal velocities as stationary air is drawn into an inlet.[17] The coefficient of entry is not the same as the loss coefficient, K_e, which is the entrance friction expressed as a multiple of the velocity pressure. Typical values of the coefficient of entry are given in Table 20.5. Equation 20.38 is the relationship between the coefficient of entry and the loss coefficient.

$$C_e = \sqrt{\frac{VP_{duct}}{SP_{duct}}} \qquad 20.36$$

$$Q = v_{actual}A = C_e v_{ideal}A$$
$$= (4005)C_e A \times \sqrt{SP_{duct}} \qquad 20.37$$

$$K_e = \frac{1 - C_e^2}{C_e^2} \qquad 20.38$$

Table 20.5 *Typical Coefficients of Entry*

entrance	C_e
plain opening (round, rectangular, square)	0.72
flanged openings (round, rectangular, square)	0.82
bell-mouthed	0.98
tapered square[a]	0.93
conical[a]	0.96

[a]For included taper angle of 30 to 60°. Values of other angles are close.

23. STATIC REGAIN

Disregarding friction loss, energy is constant along the run of a duct. If the velocity pressure decreases due to an increase in duct area or a branch takeoff, the static

---

[17]This is analogous to the coefficient of velocity, C_v, used in liquid flow measurement devices.

Fluids

pressure will increase. This increase is known as the *static regain*, SR.[18]

Ideally, the regain would be exactly equal to the decrease in velocity pressure. Actually, 10 to 25% of the energy is lost due to friction, turbulence, and other factors. The portion of the theoretical regain that is realized is given by the *static regain coefficient, R. R* has typical values of 0.75 to 0.90 for well-designed ducts without reducing sections. (When $v_{down} > v_{up}$, the regain will be a static pressure loss. Use $R = 1.1$ in that case.)

$$R = \frac{SR_{actual}}{SR_{ideal}} = \frac{SR_{actual,Pa}}{(0.6)\left(v_{up}^2 - v_{down}^2\right)} \quad \text{[SI]} \quad \textit{20.39(a)}$$

$$R = \frac{SR_{actual}}{SR_{ideal}} = \frac{SR_{actual,in\ wg}}{\dfrac{v_{up}^2 - v_{down}^2}{(4005)^2}} \quad \text{[U.S.]} \quad \textit{20.39(b)}$$

24. DIVIDED-FLOW FITTINGS

Figure 20.6 illustrates typical commercial *divided-flow* (i.e., *branch takeoff*) *fittings* and the terminology that describes them. Air enters upstream, from the left. After the air reduction at the branch takeoff, the downstream velocity will (generally) be less than the upstream velocity. The change in static pressure due to the change in velocity is

$$\Delta SP_{up\text{-}down} = \Delta TP_{up\text{-}down} - \Delta VP_{up\text{-}down}$$
$$= R\Delta VP_{up\text{-}down}$$
$$= R(VP_{up} - VP_{down}) \quad \text{[positive for decreases]}$$
$$\textit{20.40}$$

The total pressure change from upstream to downstream is

$$\Delta TP_{up\text{-}down} = \Delta SP_{up\text{-}down} + \Delta VP_{up\text{-}down}$$
$$= (R+1)\Delta VP_{up\text{-}down}$$
$$= (R+1)(VP_{up} - VP_{down})$$
$$\text{[positive for decreases]} \quad \textit{20.41}$$

The total pressure change from upstream through the branch is

$$\Delta TP_{up\text{-}br} = \Delta SP_{up\text{-}br} + \Delta VP_{up\text{-}br}$$
$$= K_{br}(VP_{up}) \quad \textit{20.42}$$

The change in static pressure from upstream through the branch is

$$\Delta SP_{up\text{-}br} = \Delta TP_{up\text{-}br} - \Delta VP_{up\text{-}br}$$
$$= K_{br}(VP_{up}) - \Delta VP_{up\text{-}br}$$
$$= (K_{br} - 1)(VP_{up}) + VP_{br}$$
$$\text{[positive for decreases]} \quad \textit{20.43}$$

Manufacturers of commercial fittings provide graphs of the loss coefficient as a function of the ratio of branch-to-upstream velocities.[19] Typical values of the *branch loss coefficient*, K_{br}, are given in Table 20.6.

Table 20.6 *Typical Branch Loss Coefficient (K_{br}) Values*[a]

ratio of v_{br}/v_{up}	angle of takeoff		
	90°	60°	45°
0.5	1.1	0.8	0.5
1.0	1.5	0.8	0.5
1.5	2.2	1.1	0.9
2.0	3.0	2.9	2.8
2.5	4.3	3.3	3.2
3.0	5.6	5.2	4.9

[a]Round ducts only.

Example 20.8

The velocity in a main duct before a branch is 3200 fpm (16 m/s). The velocity in the branch duct is 2560 fpm (12.8 m/s). What is the change in static pressure from the main duct through the branch?

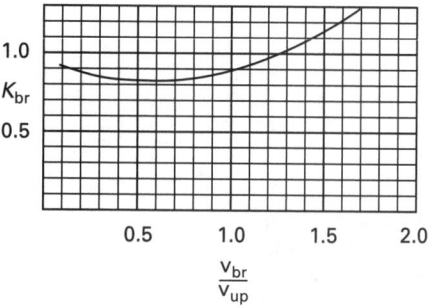

SI Solution

The ratio of the branch-to-upstream velocity is

$$\frac{v_{br}}{v_{up}} = \frac{12.8\ \dfrac{m}{s}}{16\ \dfrac{m}{s}} = 0.8$$

From the graph, the loss coefficient is $K_{br} = 0.85$.

The upstream and branch velocity pressures are

$$VP_{up} = (0.6)(v_{m/s})^2 = (0.6)\left(16\ \frac{m}{s}\right)^2$$
$$= 154\ \text{Pa}$$

$$VP_{br} = (0.6)\left(12.8\ \frac{m}{s}\right)^2$$
$$= 98\ \text{Pa}$$

---

[18]If a downstream or branch velocity is low enough, it is possible for the regain to actually exceed the dynamic losses due to fitting turbulence. This may hide the true inefficiency of the fitting.

[19]Some manufacturers also provide direct-reading charts that give the friction loss directly in terms of in wg.

Figure 20.6 *Types of Commercial Divided-Flow Fittings*

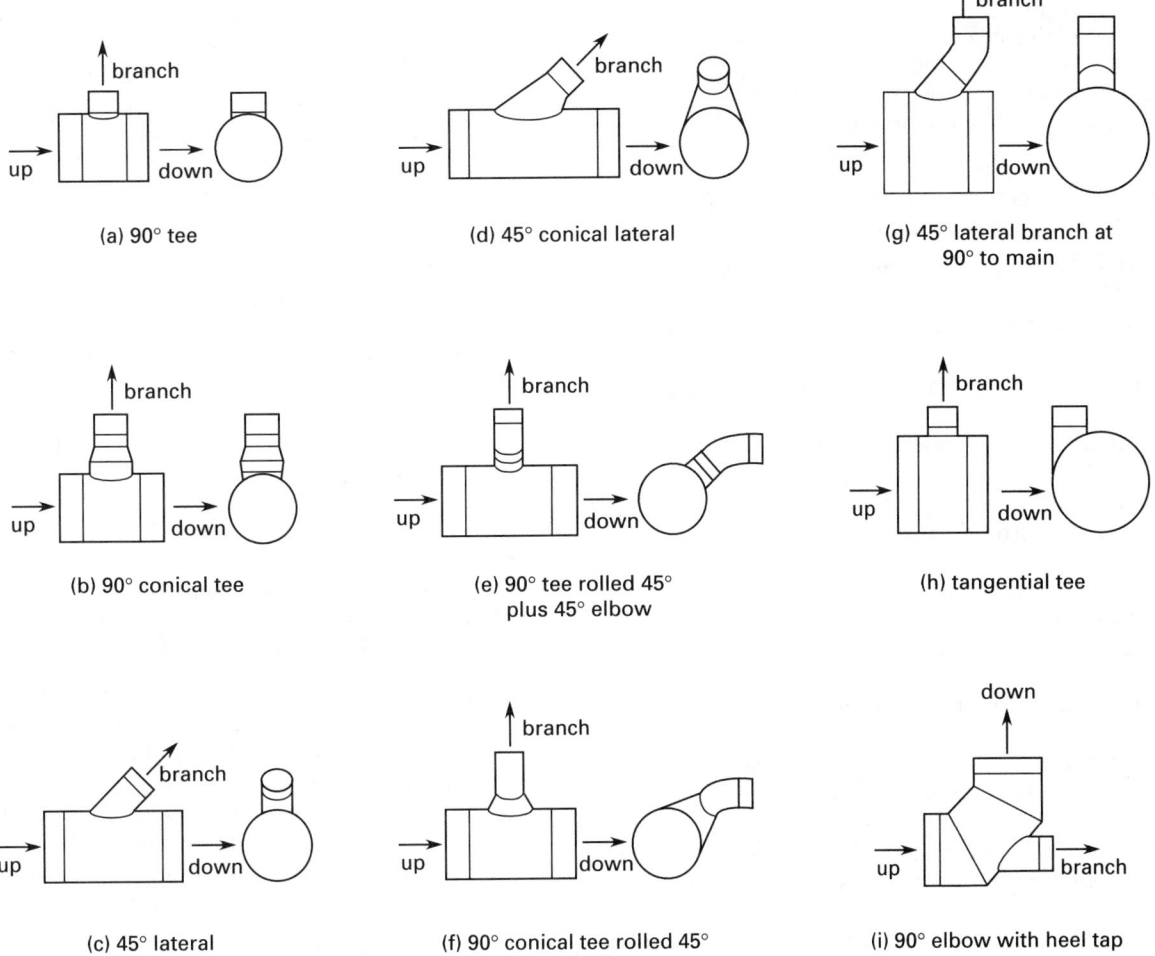

(a) 90° tee

(d) 45° conical lateral

(g) 45° lateral branch at 90° to main

(b) 90° conical tee

(e) 90° tee rolled 45° plus 45° elbow

(h) tangential tee

(c) 45° lateral

(f) 90° conical tee rolled 45° plus 45° elbow

(i) 90° elbow with heel tap

From Eq. 20.43, the change in static pressure is

$$\Delta SP = (K_{br} - 1)(VP_{up}) + VP_{br}$$
$$= (0.85 - 1)(154 \text{ Pa}) + 98 \text{ Pa}$$
$$= 75 \text{ Pa} \quad [\text{decrease}]$$

Customary U.S. Solution

The ratio of the branch-to-upstream velocity is

$$\frac{v_{br}}{v_{up}} = \frac{2560 \text{ fpm}}{3200 \text{ fpm}} = 0.8$$

From the graph, the loss coefficient is $K_{br} = 0.85$. The upstream and branch velocity pressures are

$$VP_{up} = \left(\frac{v_{up}}{4005}\right)^2 = \left(\frac{3200 \text{ fpm}}{4005}\right)^2$$
$$= 0.64 \text{ in wg}$$

$$VP_{br} = \left(\frac{2560 \text{ fpm}}{4005}\right)^2$$
$$= 0.41 \text{ in wg}$$

From Eq. 20.43, the change in static pressure is

$$\Delta SP = (K_{br} - 1)(VP_{up}) + VP_{br}$$
$$= (0.85 - 1)(0.64 \text{ in wg}) + 0.41 \text{ in wg}$$
$$= 0.31 \text{ in wg} \quad [\text{decrease}]$$

25. DUCT DESIGN PRINCIPLES

Duct systems are categorized as low-velocity (up to 2000 to 2500 fpm or 10.2 to 12.8 m/s) and high-velocity (above 2500 fpm or 12.8 m/s). *Low-velocity systems*, also known as *conventional systems*, are usually designed with the velocity-reduction and equal-friction methods. *High-velocity systems* are able to take advantage of benefits associated with the static regain method.

Duct systems are categorized according to the static pressure at the fan: low-pressure (0 to 2 in wg; 0 to 500 Pa), medium-pressure (2 to 6 in wg; 500 to 1500 Pa), and high-pressure (6 to 10 in wg; 1500 to 2500 Pa).

Supply duct systems take air from the fan and bring it to the ventilated space. *Exhaust duct systems* carry air from ventilated space back to the fan.

Compared with conventional systems, high-pressure and high-velocity systems require less space, cost less for ductwork, and provide better control of the conditioned space. However, they are noisier, require a more precise duct design, and require larger (more expensive) fans.

The following general recommendations apply to all duct designs and design methods.

- Make duct routes as direct as possible.

- Avoid sudden changes in direction and diameter.

- Use radius-to-diameter ratios of 1.5 or higher.

- Eliminate obstructions in and through the ducts.

- Use radiused elbows whenever possible, and when not, use turning vanes.

- Make rectangular ducts as square as possible. Avoid aspect ratios greater than 8:1, and use 4:1 or less whenever space permits.

- Use smooth metal construction whenever possible.

- Maintain an incremental size difference of at least 2 in in adjacent duct sections.

- Include a small volume allowance above the sum of all the outlet volumes to account for leakage.[20]

- Size the fan with excess capacity to compensate for inaccuracies in the design.

- Install balancing dampers in all branches, even when the static regain method is used for the design. In order to minimize noise, install dampers as close as possible to the main duct.

- Use the lowest possible duct velocities in order to minimize fan power and noise.

26. LEAKAGE

Ducts are not intended to be leak-free. However, the volumetric leakage should be less than 1% for well-sealed ducts and 2 to 5% for unsealed ducts.[21] Ducts are not pressure vessels and are not intended to be tested by sealing and pressurization. (The term "airtight" should be avoided.) Leakage should be tested volumetrically with the air in motion.

---

[20]Some sources say to include up to 10% excess air to account for leaks. While this may sound nominal, the fan laws show that increasing the fan speed 10% to obtain the extra flow will increase the horsepower 30%. It is unlikely that a motor would be able to provide 30% more power. Therefore, more reliance should be placed on tight ductwork than on excess air.

[21]ASHRAE and SMACNA recommendations.

27. COLLAPSE OF DUCTS

Under certain conditions, ducts may collapse inward. This may happen in medium- and high-velocity systems when a fire damper or blast gate suddenly closes, but can also occur in long, large-diameter air return systems. The negative pressure created between a closed damper and the retreating mass of air may collapse the duct. The negative pressure required to collapse a duct depends on the duct construction and must be specified by the duct manufacturer.

To present collapse in air return systems, increased metal gage and/or angle rings may be used. To prevent collapse due to sudden closures, a negative-pressure relief valve can be installed immediately downstream of each fire damper.

28. DAMPERS

Since friction loss is proportional to the distance from the fan to the outlet, duct runs to close outlets will have lower friction losses than the main duct run. If not constrained, most of the air flow will escape out of the lower-friction runs.

Dampers are included in all runs to keep the pressure drop the same in all duct runs, even those that are short. *Balancing* is the act of closing down the dampers to equalize the friction losses. It is a good idea to install dampers even when sophisticated design methods are used. Dampers can be manually operated (for balancing or other occasional use), motorized (for zone control and variable volume), or gravity operated.

Figure 20.7 illustrates three general types of dampers. *Volume dampers* should be used only in branch ducts when a splitter damper cannot be used.[22] *Splitter dampers* should be used at the junction of the main and branch ducts. *Automatic dampers* are usually chosen with parallel blades for applications with two distinct positions. Dampers with opposed blades are chosen when air flow is to be controlled over a wide range. *Gravity dampers* are self-closing and are intended to prevent backflow. *Fire dampers* close automatically to prevent the spread of smoke throughout the system.

Balancing/volume-adjusting dampers should be installed close to the main supply, as far away as possible from the outlets. However, outlets can be designed to act as dampers. Such decorative *grilles* may be *fixed pattern dampers* (i.e., *perforated plate* or *fixed-bar grilles*) or *adjustable bar grilles* with a manual control lever. The disadvantage of combining terminal air distribution with dampering is that the noise created by the friction is projected directly into the room. In order to limit noise, the maximum flow velocities and flow rates are specified with standard commercial grilles by the manufacturers.

---

[22]In the ventilation industry, dampers used only to adjust volume are also known as *blast gates*.

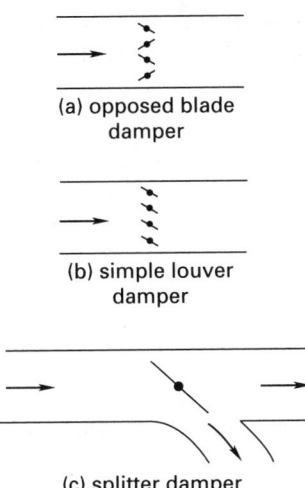

(a) opposed blade damper

(b) simple louver damper

(c) splitter damper

Figure 20.7 *Types of Dampers*

29. VELOCITY-REDUCTION METHOD

In the *velocity-reduction method*, the fan discharge velocity is selected by judgment. Arbitrary reductions in velocity are made down the run, usually at branch takeoffs. This method requires expertise on the part of the designer. It is used primarily for estimating simple layouts. The following steps constitute the velocity-reduction method.

step 1: Select the velocity leaving the fan from judgment. As the duct branches off, use judgment to select a reduced velocity for each branch. Table 20.7 lists typical maximum values of duct velocities for conventional low-velocity systems.

step 2: Determine the air flow requirement, Q, for each outlet.

step 3: Calculate the duct size from $A = Q/\mathrm{v}$.

step 4: By inspection, find the highest-resistance (i.e., longest) duct run. Calculate the static pressure drop in the longest run.

step 5: Specify dampers in all of the runs for balancing.

30. EQUAL-FRICTION METHOD

The *equal-friction method* is applicable to simple low-velocity systems. The method gets its name from the procedure that arbitrarily keeps the friction loss per unit length the same in all duct runs. Velocity pressure and regain are disregarded. The system will require extensive dampering, as no attempt is made to equalize pressure drops in the branches.

Table 20.7 *Typical Maximum Duct Velocities (in fpm)[a]*

application	large supply ducts	small supply ducts	return ducts
residences	800	600	600
apartments/hotel bedrooms	1500	1100	1000
theaters	1600	1200	1200
deluxe offices		1100	800
average offices		1300	1000
general offices	2200	1400	1200
restaurants	1800	1400	1200
small shops		1500	1200
department stores			
lower floors	2100	1600	1200
upper floors	1800	1400	1200

[a]Multiply fpm by 0.00508 to obtain m/s.

A common assumption in preliminary design studies is a friction loss of 0.08 in wg per 100 ft (0.65 Pa/m) for virtually all situations except offices (0.10 in wg per 100 ft; 0.82 Pa/m) and industrial uses (0.15 in wg per 100 ft; 1.2 Pa/m). These values represent a good compromise between duct installation and operating costs.

step 1: Select the main duct velocity from Table 20.7, contract specifications, or judgment.

step 2: From the velocity and flow rate in the main duct, find the friction loss per unit length from Fig. 20.4 or 20.5.

step 3: After each branch, reduce the main duct flow rate by the branch flow. Find the new velocity and duct size to keep the same friction loss per unit length. (This means that all points will be along a vertical line on Fig. 20.4 or a horizontal line on Fig. 20.5.)

step 4: Determine the static pressure drop in the highest-resistance duct. The fan must supply this static pressure drop plus the desired outlet pressure.

step 5: Compare the actual system pressure with the design pressure, if known. If they are significantly different, repeat all the steps with a different main duct velocity.

step 6: Size branch runs the same way—keeping the same friction loss per unit length. Use dampers to equalize the pressure drops.

Example 20.9

A theater duct system is shown. All bends have a radius-to-diameter ratio of 1.5. The branch takeoff between sections A and B has a branch loss coefficient of $K_{br} = 1.5$. The design pressure at each outlet is 0.15 in wg (38 Pa). Use the equal-friction method to size the system.

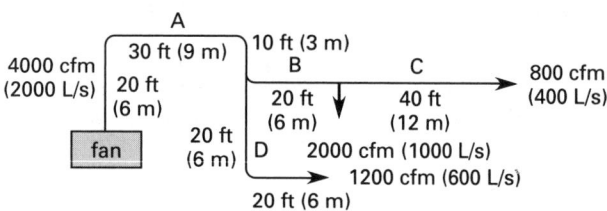

SI Solution

step 1: From Table 20.7, choose the main duct velocity (section A) as 1600 fpm. From the table footnote, the SI velocity is

$$v_{main} = (1600 \text{ fpm})(0.00508) = 8.1 \text{ m/s}$$

step 2: The total air flow from the fan is 2000 L/s. From Fig. 20.5, the main duct diameter (section A) is approximately 560 mm. (This may not correspond to a standard duct size.) The friction loss is 1.2 Pa/m.

step 3: After the first takeoff, the flow rate in section B is

$$2000 \; \frac{L}{s} - 600 \; \frac{L}{s} = 1400 \text{ L/s}$$

From Fig. 20.5 for 1400 L/s and 1.2 Pa/m, the diameter is 490 mm and the velocity is 7.4 m/s.

step 4: By inspection, the longest run is ABC. From Table 20.4, the equivalent length of each bend is 12D.

$$L_{e,bend} = 12D = \frac{(12)(560 \text{ mm})}{1000 \; \frac{mm}{m}}$$
$$= 6.72 \text{ m} \quad [\text{say, 7 m}]$$

The equivalent length of the entire run is

6 m (from fan to first bend)
7 m (equivalent length of first bend)
9 m (first bend to second bend)
7 m (equivalent length of second bend)
3 m (jog between sections A and B)
6 m (section B)
<u>12 m</u> (section C)
total: 50 m

The straight-through friction loss in the longest run is

$$(50 \text{ m})\left(1.2 \; \frac{Pa}{m}\right) = 60 \text{ Pa}$$

Use Eq. 20.42 to find the friction loss in the branch takeoff between sections A and B.

$$\Delta TP_{A-B} = K_{br}(VP_{up}) = K_{br}(0.6)(v_{up})^2$$
$$= (1.5)(0.6)\left(8.1 \; \frac{m}{s}\right)^2$$
$$= 59 \text{ Pa}$$

The fan must be able to supply a static pressure of

$$SP_{fan} = 60 \text{ Pa} + 59 \text{ Pa} + 38 \text{ Pa}$$
$$= 157 \text{ Pa}$$

The total pressure supplied by the fan is

$$TP_{fan} = 157 \text{ Pa} + (0.6)\left(8.1 \; \frac{m}{s}\right)^2$$
$$= 196 \text{ Pa}$$

Customary U.S. Solution

step 1: From Table 20.7, choose the main duct velocity (section A) as 1600 fpm.

step 2: The total air flow from the fan is 4000 cfm. From Fig. 20.4, the main duct diameter (section A) is 21 in. The friction loss is 0.15 in wg per 100 ft.

step 3: After the first takeoff, the flow rate in section B is

$$4000 \text{ cfm} - 1200 \text{ cfm} = 2800 \text{ cfm}$$

From Fig. 20.4 for 2800 cfm and 0.15 in wg per 100 ft, the diameter is 18 in and the velocity is 1500 fpm. Similarly, the diameters at sections C and D are 11.5 in (say 12 in) and 13 in, respectively. The velocity at section C is 1000 fpm.

step 4: By inspection, the longest run is ABC. From Table 20.4, the equivalent length of each bend is 12D.

$$L_{e,bend} = 12D = \frac{(12)(21 \text{ in})}{12 \; \frac{in}{ft}} = 21 \text{ ft}$$

The equivalent length of the entire run is

20 ft (from fan to first bend)
21 ft (equivalent length of first bend)
30 ft (first bend to second bend)
21 ft (equivalent length of second bend)
10 ft (jog between sections A and B)
20 ft (section B)
<u>40 ft</u> (section C)
total: 162 ft

The straight-through friction loss in the longest run is

$$\left(\frac{162 \text{ ft}}{100 \text{ ft}}\right)(0.15 \text{ in wg per } 100 \text{ ft}) = 0.24 \text{ in wg}$$

Use Eq. 20.42 to find the friction loss in the branch takeoff between sections A and B.

$$\Delta \text{TP}_{A-B} = K_{br}(\text{VP}_{up})$$
$$= (1.5)\left(\frac{1600 \text{ fpm}}{4005}\right)^2 = 0.24 \text{ in wg}$$

The fan must be able to supply a static pressure of

$$\text{SP}_{fan} = 0.24 \text{ in wg} + 0.24 \text{ in wg} + 0.15 \text{ in wg}$$
$$= 0.63 \text{ in wg}$$

The total pressure supplied by the fan is

$$\text{TP}_{fan} = 0.63 \text{ in wg} + \left(\frac{1600 \text{ fpm}}{4005}\right)^2$$
$$= 0.79 \text{ in wg}$$

31. COMBINATION METHOD

A *combination method* is sometimes used. The main duct is sized by the equal-friction method. The branch runs are sized so as to dissipate the remaining friction (compared with the main duct run) and to (theoretically) eliminate the need for dampers.

The desired outlet pressure is subtracted from the pressure at the main duct branch takeoff to get the pressure that must be dissipated in the branch run. This pressure is divided by the estimated equivalent length to find the pressure drop per unit length. Figures 20.4 and 20.5 can be used to find the duct size and velocity.[23]

Example 20.10

An air supply system consists of a long run and two branches. The total friction loss in the long run is 0.15 in wg (38 Pa). The longest duct was sized with the equal-friction method using a pressure drop of 0.2 in wg per 100 ft (1.6 Pa/m). The equivalent length of the branch takeoff at A is 12 ft (3.6 m). The equivalent length of the elbow in duct A is 18 ft (5.4 m). Rather than use a damper in duct A to equal the pressure drop, duct A will be sized small enough to equalize the losses through increased velocity. Use the combination method to size duct A.

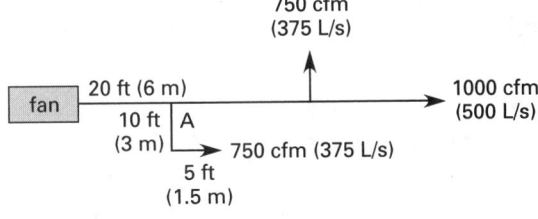

SI Solution

Subtracting the pressure drop from the fan to the branch takeoff, the pressure left to be dissipated in duct A is

$$38 \text{ Pa} - (6 \text{ m})\left(1.6 \frac{\text{Pa}}{\text{m}}\right) = 28 \text{ Pa}$$

The required loss per 100 ft in duct A is

$$\frac{28 \text{ Pa}}{3 \text{ m} + 1.5 \text{ m} + 3.6 \text{ m} + 5.4 \text{ m}} = 2.07 \text{ Pa/m}$$

Use Fig. 20.5. With 375 L/s and 2.07 Pa/m, the velocity is approximately 6.9 m/s, and the diameter is approximately 260 mm.

Customary U.S. Solution

Subtracting the pressure drop from the fan to the branch takeoff, the pressure left to be dissipated in duct A is

$$0.15 \text{ in wg} - \left(\frac{20 \text{ ft}}{100 \text{ ft}}\right)(0.2 \text{ in wg per } 100 \text{ ft}) = 0.11 \text{ in wg}$$

The required loss per 100 ft in duct A is

$$\frac{(0.11 \text{ in wg})(100 \text{ ft})}{10 \text{ ft} + 5 \text{ ft} + 12 \text{ ft} + 18 \text{ ft}} = 0.24 \text{ in wg per } 100 \text{ ft}$$

Use Fig. 20.4. With 750 cfm and 0.24 in wg, the velocity is 1240 fpm and the diameter is 10 in.

32. STATIC REGAIN METHOD

In the *static regain method*, the diameter of each branch is reduced in order to increase the static pressure. The reduction is such that the static regain offsets the friction loss in the succeeding section. (This method can also be used to size branch ducts as long as the duct sizes are reasonable.) In Eq. 20.44, point A is before the branch takeoff, point B is immediately after the branch takeoff, and point C is just prior to the next branch takeoff.

$$\Delta \text{SP}_{A-B} = \text{FP}_{B-C} \qquad \qquad 20.44$$

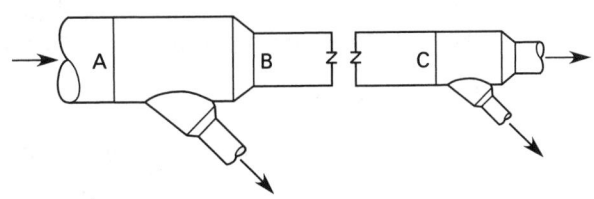

Figure 20.8 *Duct Runs for Static Regain*

---

[23]To limit noise, very high velocities should not be selected.

The analytical relationship between the velocities in the sections is given by Eq. 20.45. C is equal to 0.0832 for a static regain coefficient (R) of 0.75. Values of C for other values of R can be calculated from Eq. 20.46.[24]

$$v_A^2 - v_B^2 = C(v_B)^{2.43}\left(\frac{L_{BC}}{(Q_B)^{0.61}}\right) \quad \text{[U.S. only]} \quad \textit{20.45}$$

$$C = \frac{6.256 \times 10^{-2}}{R} \quad \text{[U.S. only]} \quad \textit{20.46}$$

The velocity v_B appears on both sides, making Eq. 20.45 difficult to use. For that reason, duct diameters are often chosen by trial and error. However, graphical aids (e.g., Fig. 20.9) can be used to determine the unknown velocity without extensive trial and error iterations. To use Fig. 20.9, the quantity $L_{AB}/Q_B^{0.61}$ is calculated. The intersection of the L/Q line and the v_A line defines v_B. In most cases, it is assumed that the regain will equal the friction loss in the following section. In that case, v_B is read directly from the horizontal scale. However, Fig. 20.9 can also be used to determine a velocity that will increase or decrease the static pressure by some given amount. If a loss in static pressure is required, move to the right of the intersection point until the vertical separation between the two curves equals the desired loss. (Use the vertical scale on the right edge to determine the vertical separation.) Then, drop down and read the velocity from the horizontal scale. A gain is handled similarly—by moving to the left.

The following steps constitute a simplified static regain method.

step 1: Use Table 20.7 to choose a velocity in the main run.

step 2: Size the main run using $A = Q/v$.

step 3: Find the equivalent length of the main duct from the fan to the first branch takeoff. Assume any unknown bend radii.

step 4: Use Fig. 20.4 or 20.5 to find the friction loss, FP_{main}, in the main run up to the branch takeoff.

step 5: Determine the fan pressure. Assuming that all subsequent friction after the first branch takeoff will be recovered with static regain, the static pressure supplied by the fan will be

$$\text{SP}_{\text{fan}} = \text{FP}_{\text{main}} + \text{grille discharge pressure} \quad \textit{20.47}$$

---

[24]Equation 20.45 is based on a friction factor of 0.0270.

Equation 20.47 assumes that the fan discharge velocity and velocity in the main duct run are the same. If the velocities are different, Eq. 20.39 is used to determine a static regain that reduces the pressure supplied by the fan.

step 6: Calculate the flow rate in the duct after the branch takeoff.

$$Q = Q_{\text{main}} - \text{branch flow} \quad \textit{20.48}$$

step 7: Knowing the flow rate and length of the next section, determine the velocity in that section from Fig. 20.9. For other values of R, or when a regain chart is not available, the duct size must be found by trial and error. The duct size is varied until the friction loss equals the regain.

step 8: Solve for the duct size from $A = Q/v$.

Example 20.11

The fan in the duct system shown moves a total of 1500 cfm. The velocity of the air in the fan is 1700 fpm. Bends have equivalent lengths of 15 ft. The required outlet grille pressure is 0.25 in wg. Use the static regain method with a static regain coefficient of 0.75 to size the main duct run fan-A-F.

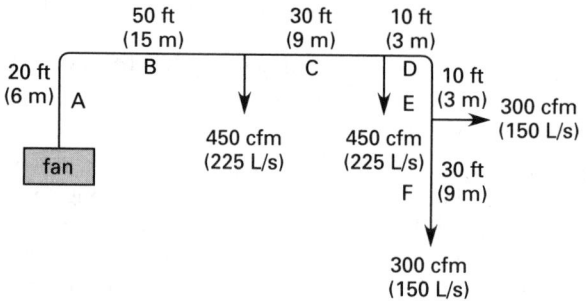

Solution

step 1: Choose 1500 fpm as the main duct velocity.

step 2: The area and diameter of the main duct are

$$A = Q/v = \frac{1500 \text{ cfm}}{1500 \text{ fpm}} = 1 \text{ ft}^2$$

$$D = \sqrt{\frac{4A}{\pi}} = \sqrt{\frac{(4)(1 \text{ ft}^2)}{\pi}}\left(12 \frac{\text{in}}{\text{ft}}\right)$$

$$= 13.5 \text{ in} \quad \text{[say 14 in]}$$

Figure 20.9 Static Regain Chart (R = 0.75)

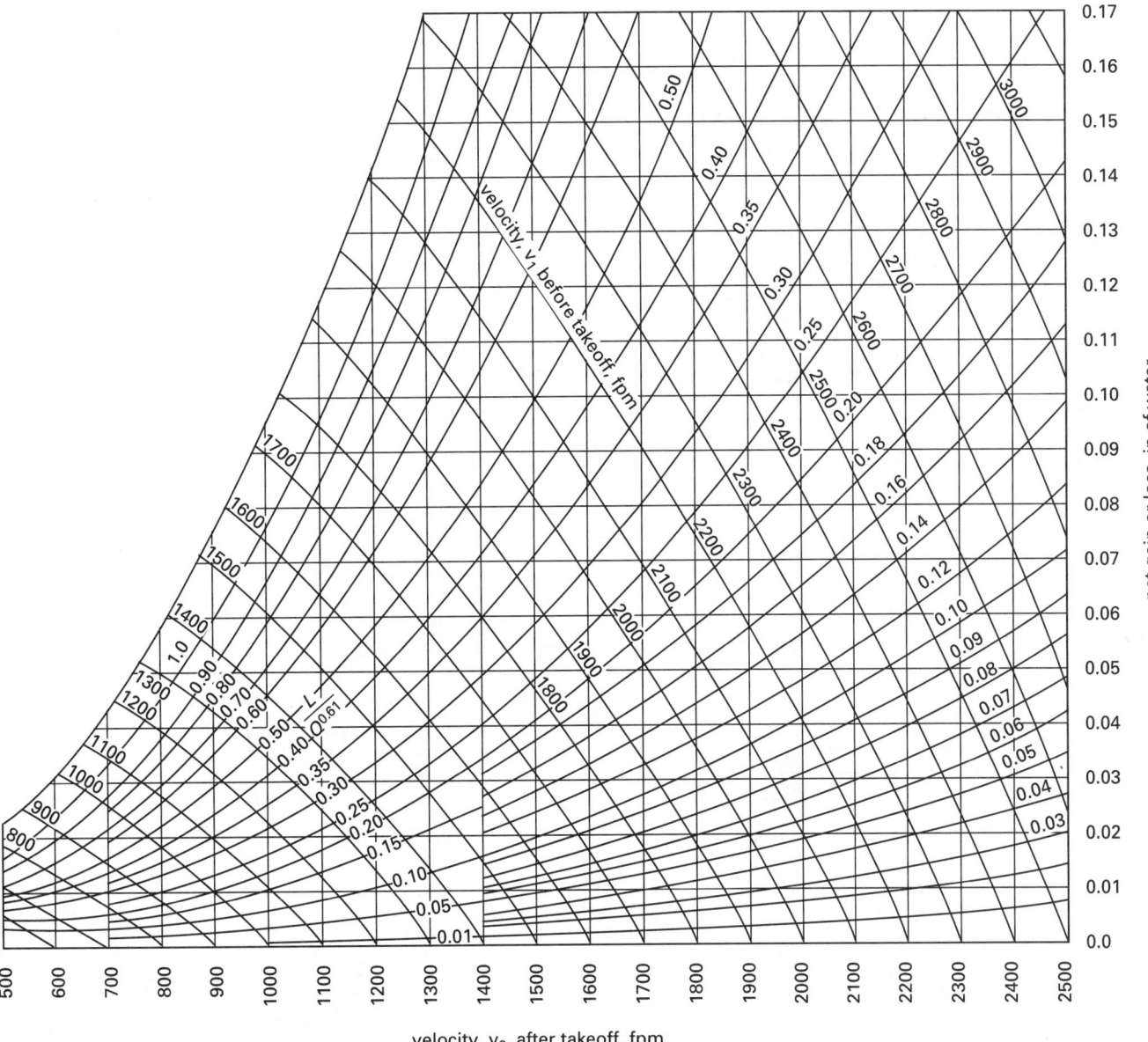

velocity, v_2, after takeoff, fpm

step 3: The equivalent length of the main duct from the fan to the first takeoff and bend is

$$L = 20 \text{ ft} + 15 \text{ ft} + 50 \text{ ft} = 85 \text{ ft}$$

step 4: From Fig. 20.4, the friction loss in the main run up to the branch takeoff is approximately 0.20 in wg per 100 ft. The actual friction loss is

$$\text{FP}_{\text{main}} = (0.20 \text{ in wg per 100 ft}) \left(\frac{85 \text{ ft}}{100 \text{ ft}} \right)$$

$$= 0.17 \text{ in wg}$$

step 5: Since the main duct velocity is lower than the fan discharge velocity, there will be static regain from the fan. From Eq. 20.39,

$$\text{SR}_{\text{fan}} = \frac{R(v_{\text{fan}}^2 - v_{\text{main}}^2)}{(4005)^2}$$

$$= (0.75) \left(\frac{(1700 \text{ fpm})^2 - (1500 \text{ fpm})^2}{(4005)^2} \right)$$

$$= 0.03 \text{ in wg}$$

$$\text{SP}_{\text{fan}} = \text{FP}_{\text{main}} + \text{grille pressure} - \text{SR}_{\text{fan}}$$

$$= 0.17 \text{ in wg} + 0.25 \text{ in wg} - 0.03 \text{ in wg}$$

$$= 0.39 \text{ in wg}$$

step 6: The flow rates, equivalent lengths, and $L/Q^{0.61}$ ratios for each section are

section	L	Q	$\dfrac{L}{Q^{0.61}}$
C	30	1050	0.43
D to E	$10 + 15 + 10 = 35$	600	0.71
F	30	300	0.92

step 7: From Fig. 20.9, the velocities are

section	v
C	1130 fpm
D to E	800 fpm
F	560 fpm

step 8: Solve for the duct size from $A = Q/v$.

$$D = \sqrt{\frac{4A}{\pi}} = \sqrt{\frac{4Q}{\pi v}}$$

$$D_{\text{C}} = \sqrt{\frac{(4)(1050 \text{ cfm})}{\pi(1130 \text{ fpm})}} \left(12 \, \frac{\text{in}}{\text{ft}}\right)$$
$$= 13 \text{ in}$$

$$D_{\text{D/F}} = \sqrt{\frac{(4)(600 \text{ cfm})}{\pi(800 \text{ fpm})}} \left(12 \, \frac{\text{in}}{\text{ft}}\right)$$
$$= 12 \text{ in}$$

$$D_{\text{F}} = \sqrt{\frac{(4)(300 \text{ cfm})}{\pi(560 \text{ fpm})}} \left(12 \, \frac{\text{in}}{\text{ft}}\right)$$
$$= 10 \text{ in}$$

33. TOTAL PRESSURE DESIGN METHOD

The previous duct design methods focus on static pressure. The static regain method actually compensates for the inefficiency of a fitting by increasing subsequent duct sizes. None of the duct design methods mentioned attempt to minimize the friction loss. Nevertheless, energy is lost due to friction and turbulence at fittings.

The true measure of the loss in a fitting is represented by the change in total pressure it causes. Unlike static pressure, total pressure along a duct run will always decrease, never increase. Features (i.e., fittings) where the total pressure drops by a significant amount represent inefficient features (i.e., the "wrong" fitting for that location). These should be replaced with fittings with lower loss coefficients. This is the basic premise of the *total pressure design method*.

34. AIR DISTRIBUTION

An *outlet* is any opening through which air enters the ventilated space. A *grille* is a decorative covering for an outlet. A *register* is a grill with an internal damper. A *diffuser* is a grille used to guide the direction of projected air.

For a grille (outlet, register, etc.) to work properly, the air must have a minimum static pressure (typically 0.1 to 0.3 in wg; 25 to 75 Pa) at the grille outlet. This grille pressure is added to the static and velocity pressures of the air when the fan is sized.

The *gross area* or *core area* of the grille is its total cross-sectional area. The total area of the opening in the grille is the *free area* or *daylight area*. The outlet velocity can be found from the core velocity and outlet's coefficient of discharge, C_d, which is typically between 0.7 and 0.9.

$$v_{\text{outlet}} = \frac{v_{\text{core}} A_{\text{core}}}{C_d A_{\text{free}}} \qquad 20.49$$

The *throw* (also known as the *blow*) is the distance from the outlet to the *distribution point*. (See Fig. 20.10.) When an outlet has been properly selected, the average *terminal velocity* at the distribution point should be approximately 50 fpm (0.25 m/s) for sedentary occupants up to 75 fpm (0.38 m/s) for slightly active occupants.[25,26] Therefore, the throw is roughly the distance from the outlet where the average air velocity is 50 fpm (0.25 m/s). As it emerges from the outlet, duct air will entrain room air. The increase in air flow width is known as the *rise*, and the absolute width of the air flow is the *spread*. (Even straight outlets have air flows that diverge with a total included angle of up to 20°.)

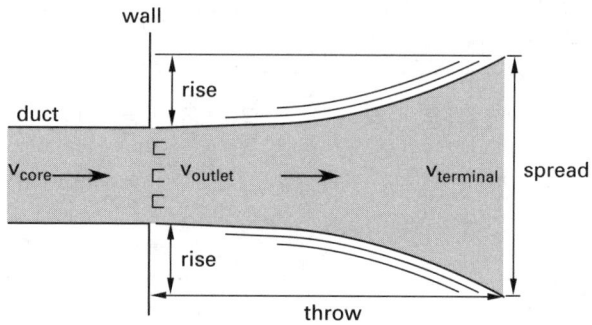

Figure 20.10 *Air Distribution Terminology*

The conservation of momentum law predicts the amount of entrained air. The velocity of the room air is initially zero. The *induction ratio*, IR, is the ratio of combined to outlet air masses.

$$m_{\text{outlet}} v_{\text{outlet}} = (m_{\text{outlet}} + m_{\text{entrained}}) v_{\text{combined}} \qquad 20.50$$

---

[25]For industrial work, the velocity may be as high as 300 fpm (1.5 m/s).
[26]Some sources say the minimum air movement should be 20 fpm (0.1 m/s) or above.

$$IR = \frac{v_{outlet}}{v_{combined}} = \frac{m_{outlet} + m_{entrained}}{m_{outlet}} \qquad 20.51$$

The centerline velocity can be predicted for a distance x from the outlet. In Eq. 20.52, K is the *outlet constant* supplied by the outlet manufacturer. The outlet velocity should be in the range of 300 to 500 fpm (1.5 to 2.5 m/s) for ultra-quiet areas, 500 to 750 fpm (2.5 to 3.8 m/s) for residences, theaters, and libraries, and 600 to 1000 fpm (3.0 to 5.1 m/s) for offices and service areas. In noisy industrial areas, velocities as high as 2000 fpm (10.2 m/s) may be tolerable.

$$v_{centerline} = \frac{K Q_{outlet}}{x \sqrt{\dfrac{C_d A_{core} A_{free}}{A_{gross}}}} \qquad 20.52$$

The centerline velocity of the air emerging from a duct is approximately twice that of the average velocity across the duct face. Since the throw is roughly the distance at which the average distribution velocity is 50 fpm (0.25 m/s), the throw is

$$throw = \frac{K Q_{outlet}}{(100\ \text{fpm}) \sqrt{\dfrac{C_d A_{core} A_{free}}{A_{gross}}}} \quad \begin{bmatrix} \text{consistent} \\ \text{units} \end{bmatrix}$$

$$20.53$$

Generally, the throw should be 75% of the distance from the outlet face to the opposing normal surface. For example, for a ceiling-mounted outlet and a 12 ft (3.6 m) ceiling height, the throw would be approximately 9 ft (2.7 m). The throw should be increased 25 to 50% when the air is released along a wall or near the ceiling in order to compensate for the friction between the air and that surface.

35. EXHAUST DUCT SYSTEMS

Exhaust (return air) duct systems are designed somewhat differently from supply systems.[27] Low-velocity designs often use the equal-friction method. Converging-flow fittings behave differently than do divided-flow fittings. Since the duct operates under a negative pressure, collapse is always a consideration. The negative suction rating of the fan should also not be exceeded.

---

[27]Some buildings (e.g., those with high-velocity supply systems and where space is limited) don't have return air systems. Even when there is a return air system, some rooms (e.g., those generating odors) may not have return air inlets.

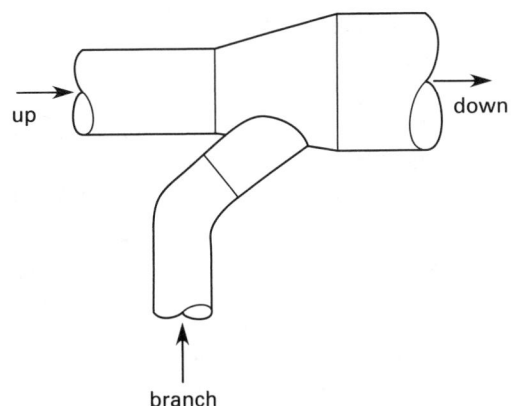

Figure 20.11 *Converging-Flow Fitting*

Figure 20.11 shows a typical converging-flow fitting. The branch loss coefficient (branch-to-downstream) for the fitting is defined by Eq. 20.54.

$$K_{br} = \frac{\Delta TP_{br\text{-}down}}{VP_{down}} \qquad 20.54$$

The *main loss coefficient* (upstream-to-downstream) is

$$K_{main} = \frac{\Delta TP_{up\text{-}down}}{VP_{down}} \qquad 20.55$$

The change in static pressure from the branch to downstream is

$$\Delta SP_{br\text{-}down} = (1 + K_{br})VP_{down} - VP_{br} \qquad 20.56$$

The change in static pressure from upstream to downstream is

$$\Delta SP_{up\text{-}down} = (1 + K_{main})VP_{down} - VP_{br} \qquad 20.57$$

PRACTICE PROBLEMS

(Note: Round all duct dimensions to the next larger whole inch or multiples of 25 mm.)

1. A round 18 in (457 mm) duct is to be replaced by a rectangular duct with aspect ratio of 4:1. What should be the dimensions of the rectangular duct?

2. Draw the system curve for a fan that moves 10,000 SFCM (4700 L/s) against a pressure of 4 in wg (1 kPa).

3. A $^1/_8$ scale model fan is tested at 300 rpm. At standard air conditions and at that speed, the model fan moves 40 cfm (19 L/s) against 0.5 in wg (125 Pa). What is the air power if a full-sized fan operates at 5000 ft (1500 m) altitude at half the model's speed?

4. A duct system consists of 750 ft (230 m) of galvanized 20 in (508 mm) ducts, four round elbows (radius-to-diameter ratio of 1.5). Two 2 in (51 mm) diameter pipes pass perpendicularly through the duct at two locations.

The flow rate is 6000 SCFM (4200 L/s). What is the friction loss?

5. 1500 SCFM (700 L/s) of air flows in an 18 in (457 mm) round duct. After a branch reduction of 300 SCFM (140 L/s), the fitting reduces to 14 in (356 mm) in the through direction. What is the static regain in the through direction?

6. (*Time limit: one hour*) A fan in a small theater delivers 1500 cfm (700 L/s) through the system shown. The duct is rectangular with an aspect ratio of 1.5:1. All elbows have a radius-to-diameter ratio of 1.5. The terminal pressure at each outlet must be 0.25 in wg (63 Pa) or higher. Takeoff fitting losses are to be disregarded. (a) Use the equal-friction method to size the system. (b) What is the static pressure at the fan? (c) What is the total pressure at the fan?

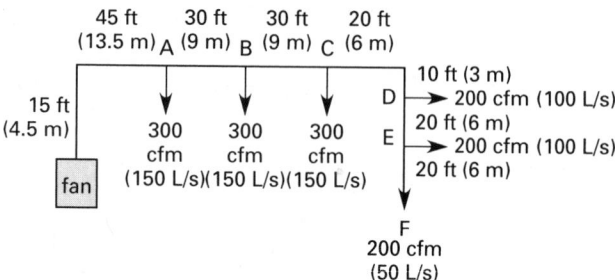

7. (*Time limit: one hour*) A fan in a theater delivers 300 cfm (140 L/s) of air through round ducts to each of the twelve outlets shown. The minimum terminal pressure at the outlets is 0.15 in wg (38 Pa). All elbows have a radius-to-diameter ratio of 1.25. Branch takeoff fitting losses are to be disregarded. (a) Use the equal-friction method to size the system. (b) What is the static pressure at the fan? (c) What is the total pressure at the fan?

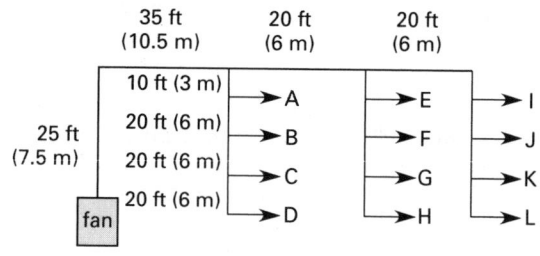

8. (*Time limit: one hour*) Size the longest run in Prob. 7 using the static regain method with a regain coefficient of 0.75. (Customary U.S. solution only required.)

9. (*Time limit: one hour*) 3000 cfm (1400 L/s) of air with a density of 0.075 lbm/ft³ (1.2 kg/m³) enters a 12 in (305 mm) diameter duct at section A. The four-piece 90° elbows have a radius-to-diameter ratio of 1.5 and an equivalent length of 14 diameters. The Darcy friction factor is 0.02 everywhere in the system. The static regain coefficient for this system is 0.65. (a) Use the static regain method to calculate the diameters at sections B and C. (b) Specify if dampers should be placed in sections A, B, and/or C. (c) What is the regain in sections B and C?

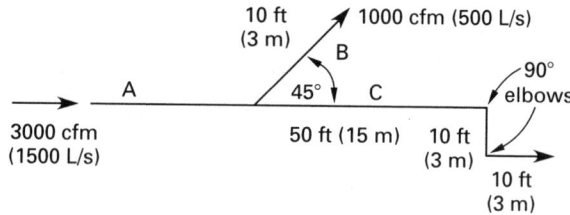

21 Inorganic Chemistry

1. Atomic Structure 21-1
2. Isotopes 21-2
3. The Periodic Table 21-2
4. Oxidation Number 21-3
5. Compounds 21-3
6. Formation of Compounds 21-3
7. Moles 21-4
8. Formula and Molecular Weights 21-4
9. Equivalent Weight 21-4
10. Gravimetric Fraction 21-5
11. Empirical Formula Development 21-5
12. Chemical Reactions 21-5
13. Balancing Chemical Equations 21-6
14. Stoichiometric Reactions 21-6
15. Nonstoichiometric Reactions 21-7
16. Solutions of Gases in Liquids 21-7
17. Solutions of Solids in Liquids 21-7
18. Units of Concentration 21-7
19. pH and pOH 21-8
20. Buffers 21-9
21. Neutralization 21-9
22. Enthalpy of Formation 21-9
23. Enthalpy of Reaction 21-9
24. Corrosion 21-10
25. Uniform Attack Corrosion 21-10
26. Intergranular Corrosion 21-10
27. Pitting 21-11
28. Concentration-Cell Corrosion 21-11
29. Erosion Corrosion 21-11
30. Selective Leaching 21-11
31. Hydrogen Embrittlement 21-11
32. Monitoring of Boiler Feedwater Quality . . 21-11
33. Galvanic Action 21-11
34. Stress Corrosion 21-12
35. Fretting Corrosion 21-12
36. Cavitation Corrosion 21-12
37. Water Supply Chemistry 21-13
38. Acidity and Alkalinity in Water 21-13
39. Water Hardness 21-13
40. Comparison of Alkalinity and Hardness . . 21-14
41. Water Softening with Lime 21-14
42. Water Softening by Ion Exchange 21-14
43. Characteristics of Boiler Feedwater 21-15
44. Production and Regeneration of
    Boiler Feedwater 21-16
45. Regeneration of Ion Exchange Resins . . . 21-16
    Practice Problems 21-17

Nomenclature

A	atomic weight	lbm/lbmol	kg/kmol
EW	equivalent weight	lbm/lbmol	kg/kmol
F	formality	n.a.	FW/L
FW	formula weight	lbm/lbmol	kg/kmol
H	enthalpy	Btu/lbmol	kcal/mol
H	Henry's law constant	1/atm	1/atm
m	mass	lbm	kg
m	molality	n.a.	mol/1000 g
M	molarity	n.a.	mol/L
MW	molecular weight	lbm/lbmol	kg/kmol
n	number of moles	–	–
N	normality	n.a.	GEW/L
N_A	Avogadro's number	–	1/mol
p	pressure	lbf/ft^2	Pa
V	volume	ft^3	m^3
x	gravimetric fraction	–	–
x	mole fraction	–	–
x	relative abundance	–	–
X	fraction ionized	–	–
Z	atomic number	–	–

Subscripts

A	Avogadro
f	formation
r	reaction
t	total

1. ATOMIC STRUCTURE

An *element* is a substance that cannot be decomposed into simpler substances during ordinary chemical reactions.[1] An *atom* is the smallest subdivision of an element that can take part in a chemical reaction. A *molecule* is the smallest subdivision of an element or compound that can exist in a natural state.

The atomic nucleus consists of neutrons and protons, known as *nucleons*. The masses of neutrons and protons are essentially the same—one *atomic mass unit*, amu. (One amu is exactly 1/12 of the mass of an atom of carbon-12, approximately equal to 1.66×10^{-27} kg.[2]) The *relative atomic weight* or just *atomic weight*, A, of an atom is approximately equal to the number of

---

[1] Atoms of an element can be decomposed into subatomic particles in nuclear reactions.
[2] Until 1961, the atomic mass unit was defined as 1/16 of the mass of one atom of oxygen-16. The carbon-12 reference has now been universally adopted.

Thermodynamics

protons and neutrons in the nucleus.[3] The *atomic number*, Z, of an atom is equal to the number of protons in the nucleus.

The atomic number and atomic weight of an element E are written in symbolic form as $_Z\mathrm{E}^A$, E_Z^A, or $_Z^A\mathrm{E}$. For example, carbon is the sixth element; radioactive carbon has an atomic mass of 14. Therefore, the symbol for carbon-14 is C_6^{14}. Since the atomic number is superfluous if the chemical symbol is given, the atomic number can be omitted (e.g., C^{14}).

2. ISOTOPES

Although an element can have only a single atomic number, atoms of that element can have different atomic weights. Many elements possess *isotopes*. The nuclei of isotopes differ from one another only in the number of neutrons. Isotopes behave the same way chemically.[4] Therefore, isotope separation must be done physically (e.g., by centrifuging or gaseous diffusion) rather than chemically.

Hydrogen has three isotopes. H_1^1 is *normal hydrogen* with a single proton nucleus. H_1^2 is known as *deuterium* (*heavy hydrogen*), with a nucleus of a proton and neutron. (This nucleus is known as a *deuteron*.) Finally, H_1^3 (*tritium*) has two neutrons in the nucleus. While normal hydrogen and deuterium are stable, tritium is radioactive. Many elements have more than one stable isotope. Tin, for example, has ten.

The *relative abundance*, x_i, of an isotope i is equal to the fraction of that isotope in a naturally occurring sample of the element. The *chemical atomic weight* is the weighted average of the isotope weights.

$$A_{\mathrm{average}} = x_1 A_1 + x_2 A_2 + \cdots \qquad 21.1$$

3. THE PERIODIC TABLE

The *periodic table* (App. 21.B) is organized around the *periodic law*: The properties of the elements depend on the atomic structure and vary with the atomic number in a systematic way. Elements are arranged in order of increasing atomic numbers from left to right. Adjacent elements in horizontal rows differ decidedly in both physical and chemical properties. However, elements in the same column have similar properties. Graduations in properties, both physical and chemical, are most pronounced in the *periods* (i.e., the horizontal rows).

The vertical columns are known as *groups*, numbered in Roman numerals. Elements in a group are called *cogeners*. Each vertical group except 0 and VIII has A

and B subgroups (*families*). The elements of a family resemble each other more than they resemble elements in the other family of the same group. Graduations in properties are definite but less pronounced in vertical families. The trend in any family is toward more *metallic properties* as the atomic weight increases.

Metals (elements at the left end of the periodic chart) have low electron affinities, are reducing agents, form positive ions, and have positive oxidation numbers. They have high electrical conductivities, luster, generally high melting points, ductility, and malleability.

Nonmetals (elements at the right end of the periodic chart) have high electron affinities, are oxidizing agents, form negative ions, and have negative oxidation numbers. *Nonmetals* are poor electrical conductors, have little or no luster, and form brittle solids.

The *metalloids* (e.g., boron, silicon, germanium, arsenic, antimony, tellurium, and polonium) have characteristics of both metals and nonmetals. Electrically, they are semiconductors.

Elements in the periodic table are often categorized into the following groups.

- *actinons:* elements 90 to 103[5]

- *actinides:* same as actinons

- *alkali metals:* group IA

- *alkaline earth metals:* group IIA

- *halogens:* group VIIA

- *heavy metals:* metals near the center of the chart

- *inner transition elements:* same as transition metals

- *lanthanides:* same as lanthanons

- *lanthanons:* elements 58 to 71[6]

- *light metals:* elements in the first two groups

- *metals:* everything except the nonmetals

- *metalloids:* elements along the dark line in the chart separating metals and nonmetals

- *noble gases:* group 0

- *nonmetals:* elements 2, 5 to 10, 14 to 18, 33 to 36, 52 to 54, 85, and 86

- *rare earths:* same as lanthanons

---

[3]The term *weight* is used even though all chemical calculations involve mass. The atomic weight of an atom includes the mass of the electrons. Published *chemical atomic weights* of elements are averages of all the atomic weights of stable isotopes, taking into consideration the relative abundances of the isotopes.

[4]There are slight differences, known as *isotope effects*, in the chemical behavior of isotopes. These effects usually influence only the rate of reaction, not the kind of reaction.

---

[5]The *actinons* resemble element 89, *actinium*. Therefore, element 89 is sometimes included as an actinon.

[6]The *lanthanons* resemble element 57, *lanthanum*. Therefore element 57 is sometimes included as a lanthanon.

- *transition elements:* same as transition metals

- *transition metals:* all B families and group VIII[7]

4. OXIDATION NUMBER

The *oxidation number (oxidation state)* is an electrical charge assigned by a set of prescribed rules. It is actually the charge assuming all bonding is ionic. The sum of the oxidation numbers equals the net charge. For monoatomic ions, the oxidation number is equal to the charge. The oxidation numbers of some common ions and radicals are given in Table 21.1.

For atoms in a free-state molecule, the oxidation number is zero. Hydrogen gas is a diatomic molecule, H_2. Thus, the oxidation number of the hydrogen molecule, H_2, is zero. The same is true for the atoms in O_2, N_2, Cl_2, and so on. Also, the sum of all the oxidation numbers of atoms in a neutral molecule is zero.

For a charged *radical* (a group of atoms that combine as a single unit), the net oxidation number is equal to the charge on the radical.

5. COMPOUNDS

Combinations of elements are known as *compounds*. *Binary compounds* contain two elements; *ternary (tertiary) compounds* contain three elements. A *chemical formula* is a representation of the relative numbers of each element in the compound. For example, the formula $CaCl_2$ shows that there are one calcium atom and two chlorine atoms in one molecule of calcium chloride.

Generally, the numbers of atoms are reduced to their lowest terms. However, there are exceptions. For example, acetylene is C_2H_2, and hydrogen peroxide is H_2O_2.

6. FORMATION OF COMPOUNDS

Compounds form according to the *law of definite (constant) proportions*: A pure compound is always composed of the same elements combined in a definite proportion by mass. For example, common table salt is always NaCl. It is not sometimes NaCl and other times Na_2Cl or $NaCl_3$ (which do not exist, in any case).

Furthermore, compounds form according to the *law of (simple) multiple proportions*: When two elements combine to form more than one compound, the masses of one element that combine with the same mass of the other are in the ratios of small integers.

In order to evaluate whether a compound formula is valid, it is necessary to know the *oxidation numbers* of the interacting atoms. Although some atoms have more than one possible oxidation number, most do not.

---

[7]The *transition metals* are elements whose electrons occupy the d sublevel. They can have various oxidation numbers, including +2, +3, +4, +6, and +7.

Table 21.1 Oxidation Numbers of Atoms and Charge Numbers of Radicals

name	symbol	oxidation or charge number
acetate	$C_2H_3O_2$	−1
aluminum	Al	+3
ammonium	NH_4	+1
barium	Ba	+2
borate	BO_3	−3
boron	B	+3
bromine	Br	−1
calcium	Ca	+2
carbon	C	+4, −4
carbonate	CO_3	−2
chlorate	ClO_3	−1
chlorine	Cl	−1
chlorite	ClO_2	−1
chromate	CrO_4	−2
chromium	Cr	+2, +3, +6
copper	Cu	+1, +2
cyanide	CN	−1
dichromate	Cr_2O_7	−2
fluorine	F	−1
gold	Au	+1, +3
hydrogen	H	+1
hydroxide	OH	−1
hypochlorite	ClO	−1
iron	Fe	+2, +3
lead	Pb	+2, +4
lithium	Li	+1
magnesium	Mg	+2
mercury	Hg	+1, +2
nickel	Ni	+2, +3
nitrate	NO_3	−1
nitrite	NO_2	−1
nitrogen	N	−3, +1, +2, +3, +4, +5
oxygen	O	−2
perchlorate	ClO_4	−1
permanganate	MnO_4	−1
phosphate	PO_4	−3
phosphorus	P	−3, +3, +5
potassium	K	+1
silicon	Si	+4, −4
silver	Ag	+1
sodium	Na	+1
sulfate	SO_4	−2
sulfite	SO_3	−2
sulfur	S	−2, +4, +6
tin	Sn	+2, +4
zinc	Zn	+2

The sum of the oxidation numbers must be zero if a neutral compound is to form. For example, H_2O is a valid compound since the two hydrogen atoms have a total positive oxidation number of $2 \times 1 = +2$. The oxygen ion has an oxidation number of -2. These oxidation numbers sum to zero.

On the other hand, $NaCO_3$ is not a valid compound formula. The sodium (Na) ion has an oxidation number of $+1$. However, the carbonate radical has a *charge number* of -2. The correct sodium carbonate molecule is Na_2CO_3.

7. MOLES

The *mole* is a measure of the quantity of an element or compound. Specifically, a mole of an element will have a mass equal to the element's atomic (or molecular) weight. The three main types of moles are based on mass being measured in grams, kilograms, and pounds.[8] Obviously, a gram-based mole of carbon (12.0 grams) is not the same quantity as a pound-based mole of carbon (12.0 pounds). Although "mol" is understood in SI countries to mean a gram-mole, the term *mole* is ambiguous, and the units mol (gmol), kmol (kgmol), or lbmol must be specified, or the type of mole must be spelled out.[9]

One gram-mole of any substance has a number of particles (atoms, molecules, ions, electrons, etc.) equal to 6.022×10^{23}, *Avogadro's number*, N_A. A pound-mole contains approximately 454 times the number of particles in a gram-mole.

Avogadro's law (hypothesis) holds that equal volumes of all gases at the same temperature and pressure contain equal numbers of gas molecules. Specifically, at standard scientific conditions (1.0 atm and 0°C), one gram-mole of any gas contains 6.022×10^{23} molecules and occupies 22.4 L. Of course, a pound-mole occupies 454 times that volume (359 ft^3).

"Molar" is used as an adjective when describing properties of a mole. For example, a *molar volume* is the volume of a mole.

Example 21.1

How many moles are in 0.01 g of gold ($A = 196.97$; $Z = 79$)?

Solution

The number of gram-moles of gold present is

$$n = \frac{m}{A} = \frac{0.01 \text{ g}}{196.97 \, \frac{\text{g}}{\text{mol}}} = 5.077 \times 10^{-5} \text{ mol}$$

---

[8]Theoretically, a slug-mole could be defined, but it is not used.
[9]There are also variations on the presentation of these units, such as g mol, gmole, g-mole, kmole, kg-mol, lb-mole, pound-mole, and p-mole. In most cases, the intent is clear.

8. FORMULA AND MOLECULAR WEIGHTS

The *formula weight*, FW, of a molecule (compound) is the sum of the atomic weights of all elements in the molecule. The *molecular weight*, MW, is generally the same as the formula weight. The units of molecular weight are actually g/mol, kg/kmol, or lb/lbmol. However, units are sometimes omitted because weights are relative. For example,

$$CaCO_3: \text{FW} = \text{MW} = 40.1 + 12 + 3 \times 16 = 100.1$$

An *ultimate analysis* (which determines how much of each element is present in a compound) will not determine the molecular formula. It will determine only the relative proportions of each element. Therefore, except for hydrated molecules and other linked structures, the molecular weight will be an integer multiple of the formula weight.

For example, an ultimate analysis of hydrogen peroxide (H_2O_2) will show that the compound has one oxygen atom for each hydrogen atom. In this case, the formula would be assumed to be HO and the formula weight would be approximately 17, although the actual molecular weight is 34.

For *hydrated molecules* (e.g., $FeSO_4 \cdot 7H_2O$), the mass of the *water of hydration* (also known as the *water of crystallization*) is included in the formula and in the molecular weight.

9. EQUIVALENT WEIGHT

The *equivalent weight* (i.e., an *equivalent*) is the amount of substance (in grams) that supplies one gram-mole (i.e., 6.022×10^{23}) of reacting units. For acid-base reactions, an acid equivalent supplies one gram-mole of H^+ ions. A base equivalent supplies one gram-mole of OH^- ions. In oxidation-reduction reactions, an equivalent of a substance gains or loses a gram-mole of electrons. Similarly, in electrolysis reactions, an equivalent weight is the weight of substance that either receives or donates one gram-mole of electrons at an electrode.

The equivalent weight can be calculated as the molecular weight divided by the change in oxidation number experienced in a chemical reaction. A substance can have several equivalent weights.

$$\text{EW} = \frac{\text{MW}}{\Delta \text{ oxidation number}} \qquad 21.2$$

Example 21.2

What are the equivalent weights of the following compounds?

(a) Al in the reaction

$$Al^{+++} + 3e^- \longrightarrow Al$$

(b) H_2SO_4 in the reaction

$$H_2SO_4 + H_2O \longrightarrow 2H^+ + SO_4^{--} + H_2O$$

(c) NaOH in the reaction

$$NaOH + H_2O \longrightarrow Na^+ + OH^- + H_2O$$

Solution

(a) The atomic weight of aluminum is approximately 27. Since the change in the oxidation number is 3, the equivalent weight is $27/3 = 9$.

(b) The molecular weight of sulfuric acid is approximately 98. Since the acid changes from a neutral molecule to ions with two charges each, the equivalent weight is $98/2 = 49$.

(c) Sodium hydroxide has a molecular weight of approximately 40. The originally neutral molecule goes to a singly charged state. Therefore, the equivalent weight is $40/1 = 40$.

10. GRAVIMETRIC FRACTION

The *gravimetric fraction*, x_i, of an element i in a compound is the fraction by weight of that element in the compound. The gravimetric fraction is found from an *ultimate analysis* (also known as a *gravimetric analysis*) of the compound.

$$x_i = \frac{m_i}{m_1 + m_2 + \cdots + m_i + \cdots + m_n} = \frac{m_i}{m_t} \quad 21.3$$

The *percentage composition* is the gravimetric fraction converted to percentage.

$$\% \text{ composition} = x_i \times 100\% \quad 21.4$$

If the gravimetric fractions are known for all elements in a compound, the *combining weights* of each element can be calculated. (The term *weight* is used even though mass is the traditional unit of measurement.)

$$m_i = x_i m_t \quad 21.5$$

11. EMPIRICAL FORMULA DEVELOPMENT

It is relatively simple to determine the *empirical formula* of a compound from the atomic and combining weights of elements in the compound. The empirical formula gives the relative number of atoms (i.e., the formula weight is calculated from the empirical formula).

step 1: Divide the gravimetric fractions (or percentage compositions) by the atomic weight of each respective element.

step 2: Determine the smallest ratio from step 1.

step 3: Divide all of the ratios from step 1 by the smallest ratio.

step 4: Write the chemical formula using the results from step 3 as the numbers of atoms. Multiply through as required to obtain all integer numbers of atoms.

Example 21.3

A clear liquid is analyzed, and the following percentage compositions are recorded: carbon, 37.5%; hydrogen, 12.5%; oxygen, 50%. What is the chemical formula for the liquid?

Solution

step 1: Divide the percentage compositions by the atomic weight.

$$C: \quad \frac{37.5}{12} = 3.125$$

$$H: \quad \frac{12.5}{1} = 12.5$$

$$O: \quad \frac{50}{16} = 3.125$$

steps 2 and 3: The smallest ratio is 3.125. Divide all ratios by 3.125.

$$C: \quad \frac{3.125}{3.125} = 1$$

$$H: \quad \frac{12.5}{3.125} = 4$$

$$O: \quad \frac{3.125}{3.125} = 1$$

step 4: The empirical formula is CH_4O.

If it had been known that the liquid behaved chemically as though it had a hydroxyl (OH) radical, the formula would have been written as CH_3OH. This is recognized as methyl alcohol.

12. CHEMICAL REACTIONS

During chemical reactions, bonds between atoms are broken and new bonds are usually formed. The starting substances are known as *reactants*; the ending substances are known as *products*. In a chemical reaction, reactants are either converted to simpler products or synthesized into more complex compounds. There are four common types of reactions.

- *direct combination* (or *synthesis*): This is the simplest type of reaction where two elements or compounds combine directly to form a compound.

$$2H_2 + O_2 \longrightarrow 2H_2O$$

$$SO_2 + H_2O \longrightarrow H_2SO_3$$

- *decomposition* (or *analysis*): Bonds within a compound are disrupted by heat or other energy to produce simpler compounds or elements.

$$2HgO \longrightarrow 2Hg + O_2$$

$$H_2CO_3 \longrightarrow H_2O + CO_2$$

- *single displacement* (or *replacement*): This type of reaction has one element and one compound as reactants.

$$2Na + 2H_2O \longrightarrow 2NaOH + H_2$$

$$2KI + Cl_2 \longrightarrow 2KCl + I_2$$

- *double displacement* (or *replacement*):[10] These are reactions with two compounds as reactants and two compounds as products.

$$AgNO_3 + NaCl \longrightarrow AgCl + NaNO_3$$

$$H_2SO_4 + ZnS \longrightarrow H_2S + ZnSO_4$$

13. BALANCING CHEMICAL EQUATIONS

The coefficients in front of element and compound symbols in chemical reaction equations are the numbers of molecules or moles taking part in the reaction. (For gaseous reactants and products, the coefficients also represent the numbers of volumes. This is a direct result of Avogadro's hypothesis that equal numbers of molecules in the gas phase occupy equal volumes at the same conditions.)[11]

Since matter cannot be destroyed in a normal chemical reaction (i.e., mass is conserved), the numbers of each element must match on both sides of the equation. When the numbers of each element match, the equation is said to be "balanced." The total atomic weights on both sides of the equation will be equal when the equation is balanced.

Balancing simple chemical equations is largely a matter of deductive trial and error. More complex reactions require use of oxidation numbers.

Example 21.4

Balance the following reaction equation.

$$Al + H_2SO_4 \longrightarrow Al_2(SO_4)_3 + H_2$$

---

[10]Another name for replacement is *metathesis*.
[11]When water is part of the reaction, the interpretation that the coefficients are volumes is valid only if the reaction takes place at a high enough temperature to vaporize the water.

Solution

As written, the reaction is not balanced. For example, there is one aluminum on the left, but there are two on the right. The starting element in the balancing procedure is chosen somewhat arbitrarily.

step 1: Since there are two aluminums on the right, multiply Al by 2.

$$2Al + H_2SO_4 \longrightarrow Al_2(SO_4)_3 + H_2$$

step 2: Since there are three sulfate radicals (SO_4) on the right, multiply H_2SO_4 by 3.

$$2Al + 3H_2SO_4 \longrightarrow Al_2(SO_4)_3 + H_2$$

step 3: Now there are six hydrogens on the left, so multiply H_2 by 3 to balance the equation.

$$2Al + 3H_2SO_4 \longrightarrow Al_2(SO_4)_3 + 3H_2$$

14. STOICHIOMETRIC REACTIONS

Stoichiometry is the study of the proportions in which elements and compounds react and are formed. A *stoichiometric reaction* (also known as a *perfect reaction* or an *ideal reaction*) is one in which just the right amounts of reactants are present. After the reaction stops, there are no unused reactants.

Stoichiometric problems are known as *weight and proportion problems* because their solutions use simple ratios to determine the masses of reactants required to produce given masses of products, or vice versa. The procedure for solving these problems is essentially the same regardless of the reaction.

step 1: Write and balance the chemical equation.

step 2: Determine the atomic (molecular) weight of each element (compound) in the equation.

step 3: Multiply the atomic (molecular) weights by their respective coefficients and write the products under the formulas.

step 4: Write the given mass data under the weights determined in step 3.

step 5: Fill in the missing information by calculating simple ratios.

Example 21.5

Caustic soda (NaOH) is made from sodium carbonate (Na_2CO_3) and slaked lime ($Ca(OH)_2$) according to the given reaction. How many kilograms of caustic soda can be made from 2000 kg of sodium carbonate?

Solution

$$Na_2CO_3 + Ca(OH)_2 \longrightarrow 2NaOH + CaCO_3$$

molecular weights	106	74	2×40	100
given data	2000 kg		X kg	

The simple ratio used is

$$\frac{NaOH}{Na_2CO_3} = \frac{80}{106} = \frac{X}{2000}$$

Solving for the unknown mass, $X = 1509$ kg.

15. NONSTOICHIOMETRIC REACTIONS

In many cases, it is not realistic to assume a stoichiometric reaction because an excess of one or more reactants is necessary to assure that all of the remaining reactants take part in the reaction. Combustion is an example where the stoichiometric assumption is, more often than not, invalid. Excess air is generally needed to ensure that all of the fuel is burned.

With nonstoichiometric reactions, the reactant that is used up first is called the *limiting reactant*. The amount of product will be dependent on (limited by) the limiting reactant.

The *theoretical yield* or *ideal yield* of a product is the maximum amount of product per unit amount of limiting reactant that can be obtained from a given reaction if the reaction goes to completion. The *percentage yield* is a measure of the efficiency of the actual reaction.

$$\text{percentage yield} = \frac{\text{actual yield} \times 100\%}{\text{theoretical yield}} \qquad 21.6$$

16. SOLUTIONS OF GASES IN LIQUIDS

Henry's law states that the amount (i.e., mole fraction) of a slightly soluble gas dissolved in a liquid is proportional to the partial pressure of the gas. This law applies separately to each gas to which the liquid is exposed, as if each gas was present alone. The algebraic form of Henry's law is given by Eq. 21.7, in which H is the *Henry's law constant* in mole fractions/atmosphere.

$$x_i = H \times p_i \qquad 21.7$$

Generally, the solubility of gases in liquids decreases with increasing temperature.

The volume of gas absorbed at a partial pressure of 1 atm and 0°C is known as the *absorption coefficient*. Typical absorption coefficients for solutions in water are: H_2, 0.017 1/L; He, 0.009 1/L; N_2, 0.015 1/L; O_2, 0.028 1/L; CO, 0.025 1/L; and CO_2, 0.88 1/L.

Example 21.6

At 20°C and 1 atm, 1 L of water will absorb 0.043 g of oxygen or 0.019 g of nitrogen. Atmospheric air is 20.9% oxygen by volume, and the remainder is assumed to be nitrogen. What masses of oxygen and nitrogen will be absorbed by 1 L of water exposed to 20°C air at 1 atm?

Solution

Since partial pressure is volumetrically weighted,

$$m_{\text{oxygen}} = (0.209)\left(0.043 \, \frac{g}{L}\right) = 0.009 \text{ g/L}$$

$$m_{\text{nitrogen}} = (1.000 - 0.209)\left(0.019 \, \frac{g}{L}\right) = 0.015 \text{ g/L}$$

17. SOLUTIONS OF SOLIDS IN LIQUIDS

When a solid is added to a liquid, the solid is known as the *solute* and the liquid is known as the *solvent*.[12] If the dispersion of the solute throughout the solvent is at the molecular level, the mixture is known as a *solution*. If the solute particles are larger than molecules, the mixture is known as a *suspension*.[13]

In some solutions, the solvent and solute molecules bond loosely together. This loose bonding is known as *solvation*. If water is the solvent, the bonding process is also known as *aquation* or *hydration*.

The solubility of most solids in liquid solvents usually increases with increasing temperature. Pressure has very little effect on the solubility of solids in liquids.

When the solvent has dissolved as much solute as it can, it is a *saturated solution*.[14] Adding more solute to an already saturated solution will cause the excess solute to settle to the bottom of the container, a process known as *precipitation*. Other changes (in temperature, concentration, etc.) can be made to cause precipitation from saturated and unsaturated solutions.

18. UNITS OF CONCENTRATION

There are many units of concentration to express solution strengths.

F— *formality:* The number of gram formula weights (i.e., molecular weights in grams) per liter of solution.

m— *molality:* The number of gram-moles of solute per 1000 grams of solvent. A "molal" solution contains 1 gram-mole per 1000 grams of solvent.

---

[12]The term *solvent* is often associated with volatile liquids, but the term is more general than that. (A *volatile liquid* evaporates rapidly and readily at normal temperatures.) Water is the solvent in aqueous solutions.
[13]An *emulsion* is not a mixture of a solid in a liquid. It is a mixture of two immiscible liquids.
[14]Under certain circumstances, a *supersaturated solution* can exist for a limited amount of time.

M— *molarity:* The number of gram-moles of solute per liter of solution. A "molar" (i.e., 1 M) solution contains 1 gram-mole per liter of solution. Molarity is related to normality: $N = M \times \Delta$ oxidation number.

N— *normality:* The number of gram equivalent weights of solute per liter of solution. A solution is "normal" (i.e., 1 N) if there is exactly one gram equivalent weight per liter of solution.

x— *mole fraction:* The number of moles of solute divided by the number of moles of solvent and all solutes.

meq/L— *milligram equivalent weights* of solute per liter of solution: calculated by multiplying normality by 1000.

mg/L— *milligrams per liter:* The number of milligrams of solute per liter of solution. Same as ppm for solutions of water.

ppm— *parts per million:* The number of pounds (or grams) of solute per million pounds (or grams) of solution. Same as mg/L for solutions of water. May be abbreviated as ppmw for "parts per million by weight."

For compounds whose molecules do not dissociate in solution (e.g., table sugar), there is no difference between molarity and formality. There is a difference, however, for compounds that dissociate into ions (e.g., table salt). Consider a solution derived from 1 gmol of magnesium nitrate $Mg(NO_3)_2$ in enough water to bring the volume to 1 L. The formality is 1.0 (i.e., the solution is 1.0 formal). However, 3 moles of ions will be produced—1 mole of Mg^{++} ions and 2 moles of NO_3^- ions. Therefore, molarity is 1.0 for the magnesium ion and 2.0 for the nitrate ion.

The use of formality avoids the ambiguity in specifying concentrations for ionic solutions. Also, the use of formality avoids the problem of determining a molecular weight when there are no discernible molecules (e.g., as in a crystalline solid such as NaCl). However, in their quest for uniformity in nomenclature, most modern chemists do not make the distinction between molarity and formality, and molarity is used as if it were formality.

Example 21.7

A solution is made by dissolving 0.353 g of $Al_2(SO_4)_3$ in 730 g of water. Assuming 100% ionization, what is the concentration expressed as normality, molarity, and mg/L?

Solution

The molecular weight of $Al_2(SO_4)_3$ is

$$MW = (2)(26.98) + (3)[32.06 + (4)(16)] = 342.14$$

The equivalent weight is

$$EW = \frac{342.14}{6} = 57.02$$

The number of gram equivalent weights used is

$$\frac{0.353}{57.02} = 6.19 \times 10^{-3} \text{ GEW}$$

The number of liters of solution (same as the solvent volume if the small amount of solute is neglected) is 0.73.

The normality is

$$N = \frac{6.19 \times 10^{-3} \text{ GEW}}{0.73 \text{ L}} = 8.48 \times 10^{-3}$$

The number of moles of solute used is 0.353 g/342.14 g/mol = 1.03 $\times 10^{-3}$ mol.

The molarity is

$$M = \frac{1.03 \times 10^{-3} \text{ mol}}{0.73 \text{ L}} = 1.41 \times 10^{-3}$$

The number of milligrams is

$$\frac{0.353 \text{ g}}{0.001 \frac{\text{g}}{\text{mg}}} = 353 \text{ mg}$$

$$\text{mg/L} = \frac{353 \text{ mg}}{0.73 \text{ L}} = 483.6$$

19. pH AND pOH

A measure of the strength of an acid or base is the number of hydrogen or hydroxide ions in a liter of solution. Since these are very small numbers, a logarithmic scale is used.

$$pH = -\log_{10}[H^+] = \log_{10}\left(\frac{1}{[H^+]}\right) \qquad \textbf{21.8}$$

$$pOH = -\log_{10}[OH^-] = \log_{10}\left(\frac{1}{[OH^-]}\right) \qquad \textbf{21.9}$$

The quantities $[H^+]$ and $[OH^-]$ in square brackets are the *ionic concentrations* in moles of ions per liter. The number of moles can be calculated from Avogadro's law by dividing the actual number of ions per liter by 6.022×10^{23}. Alternatively, for a partially ionized compound, X, in a solution of known molarity M, the ionic concentration is

$$[X] = (\text{fraction ionized}) \times M \qquad \textbf{21.10}$$

A *neutral solution* has a pH of 7.[15] Solutions with a pH below 7 are acidic; the smaller the pH, the more acidic the solution. Solutions with a pH above 7 are basic.

---

[15]The pH of a neutral solution depends on the temperature. At 25°C, the pH is 7. When the temperature is higher (lower) than 25°C, the pH will be less (greater) than 7.

The relationship between pH and pOH is

$$pH + pOH = 14 \qquad 21.11$$

Example 21.8

A 4.2% ionized 0.01M ammonia solution is prepared from ammonium hydroxide (NH_4OH). Calculate the pH, pOH, and concentrations of H^+ and OH^-.

Solution

From Eq. 21.9,

$$[OH^-] = (\text{fraction ionized}) \times M = (0.042)(0.01)$$
$$= 4.2 \times 10^{-4} \text{ mol/L}$$

From Eq. 21.8,

$$pOH = -\log[OH^-] = -\log(4.2 \times 10^{-4})$$
$$= 3.38$$

From Eq. 21.7,

$$pH = 14 - pOH = 14 - 3.38$$
$$= 10.62$$

The $[H^+]$ ionic concentration can be extracted from the definition of pH.

$$[H^+] = 10^{-pH} = 10^{-10.62}$$
$$= 2.4 \times 10^{-11} \text{ mol/L}$$

20. BUFFERS

A *buffer solution* resists changes in acidity and maintains a relatively constant pH when a small amount of an acid or base is added to it. Buffers are usually combinations of weak acids and their salts. A buffer is most effective when the acid and salt concentrations are equal.

21. NEUTRALIZATION

Acids and bases neutralize each other to form water.

$$H^+ + OH^- \longrightarrow H_2O$$

Assuming 100% ionization of the solute, the volumes, V, required for complete neutralization can be calculated from the normalities, N, or the molarities, M.

$$V_{\text{base}} N_{\text{base}} = V_{\text{acid}} N_{\text{acid}} \qquad 21.12$$

$$V_{\text{base}} M_{\text{base}} \Delta_{\text{base charge}} = V_{\text{acid}} M_{\text{acid}} \Delta_{\text{acid charge}} \qquad 21.13$$

22. ENTHALPY OF FORMATION

Enthalpy, H, is the potential energy that a substance possesses by virtue of its temperature, pressure, and phase.[16] The *enthalpy of formation* (*heat of formation*), ΔH_f, of a compound is the energy absorbed during the formation of 1 gmol of the compound from the elements.[17] The enthalpy of formation is assigned a value of zero for elements in their free states at 25°C and 1 atm. This is the so-called *standard state* for enthalpies of formation.

Table 21.2 contains enthalpies of formation for some common elements and compounds. The enthalpy of formation depends on the temperature and phase of the compound. A standard temperature of 25°C is used in tables.[18] Compounds are solid (s) unless indicated to be gaseous (g) or liquid (l). Some (aq) values are also encountered, referring to an aqueous solution.

23. ENTHALPY OF REACTION

The *enthalpy of reaction* (*heat of reaction*), ΔH_r, is the energy absorbed during a chemical reaction under constant volume conditions. It is found by summing the enthalpies of formation of all products and subtracting the sum of enthalpies of formation of all reactants. This is essentially a restatement of the energy conservation principle and is known as *Hess' law of energy summation*.

$$\Delta H_r = \sum \Delta H_{f,\text{products}} - \sum \Delta H_{f,\text{reactants}} \qquad 21.14$$

Reactions that give off energy (i.e., have negative enthalpies of reaction) are known as *exothermic reactions*. Many (but not all) exothermic reactions begin spontaneously. On the other hand, *endothermic reactions* absorb energy and require heat or electrical energy to begin.

Example 21.9

Using enthalpies of formation, calculate the heat of stoichiometric combustion (standardized to 25°C) of gaseous methane (CH_4) and oxygen.

Solution

The balanced chemical equation for the stoichiometric combustion of methane is

$$CH_4 + 2O_2 \longrightarrow 2H_2O + CO_2$$

---

[16]The older term *heat* is rarely encountered today.
[17]The symbol H is used to denote molar enthalpies. The symbol h is used for specific enthalpies (i.e., energy per kilogram or per pound).
[18]It is possible to correct the enthalpies of formation to other reaction temperatures.

Table 21.2 *Standard Enthalpies of Formation (kcal/mol at 25°C)[a]*

element/compound	ΔH_f
Al (s)	0.00
Al_2O_3	−399.09
C (graphite)	0.00
C (diamond)	0.45
C (g)	171.70
CO (g)	−26.42
CO_2 (g)	−94.05
CH_4 (g)	−17.90
C_2H_2 (g)	54.19
C_2H_4 (g)	12.50
C_2H_6 (g)	−20.24
CCl_4 (g)	−25.5
$CHCl_4$ (g)	−24
CH_2Cl_2 (g)	−21
CH_3Cl (g)	−19.6
CS_2 (g)	27.55
COS (g)	−32.80
$(CH_3)_2S$ (g)	−8.98
CH_3OH (g)	−48.08
C_2H_5OH (g)	−56.63
$(CH_3)_2O$ (g)	−44.3
C_3H_6 (g)	9.0
C_6H_{12} (g)	−29.98
C_6C_{10} (g)	−1.39
C_6H_6 (g)	19.82
Fe (s)	0.00
Fe (g)	99.5
Fe_2O_3 (s)	−196.8
Fe_3O_4 (s)	−267.8
H_2 (g)	0.00
H_2O (g)	−57.80
H_2O (l)	−68.32
H_2O_2 (g)	−31.83
H_2S (g)	−4.82
N_2 (g)	0.00
NO (g)	21.60
NO_2 (g)	8.09
NO_3 (g)	13
NH_3 (g)	−11.04
O_2 (g)	0.00
O_3 (g)	34.0
S (s)	0.00
SO_2 (g)	−70.96
SO_3 (g)	−94.45

[a]Multiply kcal/mol by 1800/MW to obtain Btu/lbm.

Using the footnote to Table 21.2, this value can be converted to Btu/lbm. The molecular weight of methane is

$$MW_{CH_4} = 12 + (4)(1) = 16$$

$$\text{higher heating value} = \frac{\left(191.75 \frac{\text{kcal}}{\text{mol}}\right)(1800)}{16}$$

$$= 21{,}570 \text{ Btu/lbm}$$

Using enthalpies of formation from Table 21.2 in Eq. 21.13, the enthalpy of reaction per mole of methane is

$$\Delta H_r = 2\Delta H_{f,H_2O} + \Delta H_{f,CO_2} - \Delta H_{f,CH_4}$$
$$- 2\Delta H_{f,O_2}$$
$$= (2)\left(-57.80 \frac{\text{kcal}}{\text{mol}}\right) + \left(-94.05 \frac{\text{kcal}}{\text{mol}}\right)$$
$$- \left(-17.90 \frac{\text{kcal}}{\text{mol}}\right) - (2)(0)$$
$$= -191.75 \text{ kcal/mol } CH_4 \quad [\text{exothermic}]$$

Notice that the enthalpy of formation of oxygen gas (its free-state configuration) is zero.

24. CORROSION

Corrosion is an undesirable degradation of a material resulting from a chemical or physical reaction with the environment. Conditions within the crystalline structure can accentuate or retard corrosion. The main types of corrosion are listed in subsequent sections. Corrosion rates are reported in units of mils per year (mpy) and micrometers per year (μm/y).

25. UNIFORM ATTACK CORROSION

Uniform rusting of steel and oxidation of aluminum over entire exposed surfaces are examples of *uniform attack corrosion*. Uniform attack is usually prevented by the use of paint, plating, and other protective coatings.

26. INTERGRANULAR CORROSION

Some metals are particularly sensitive to *intergranular corrosion*, IGC—selective or localized attack at metal-grain boundaries. For example, the Cr_2O_3 oxide film on stainless steel contains numerous imperfections at grain boundaries, and these boundaries can be attacked and enlarged by chlorides.

Intergranular corrosion may occur after a metal has been heated, in which case it may be known as *weld decay*. In the case of type 304 austenitic stainless steels, heating to 930 to 1300°F (500 to 700°C) in a welding process causes chromium carbides to precipitate out, reducing the corrosion resistance.[19] Reheating to 1830 to 2010°F (1000 to 1100°C) followed by rapid cooling will redissolve the chromium carbides and restore corrosion resistance.

27. PITTING

Pitting is a localized perforation on the surface. It can occur even when where there is little or no other visible damage. Chlorides and other halogens (e.g., HF and HCl) in the presence of water foster pitting in passive alloys, especially in stainless steels and aluminum alloys.

28. CONCENTRATION-CELL CORROSION

Concentration-cell corrosion (also known as *crevice corrosion* and *intergranular attack*, IGA) occurs when a metal is in contact with different electrolyte concentrations. It usually occurs in crevices, between two assembled parts, under riveted joints, or where there are scale and surface deposits that create stagnant areas in a corrosive medium.

29. EROSION CORROSION

Erosion corrosion is the deterioration of metals buffeted by the entrained solids in a corrosive medium.

30. SELECTIVE LEACHING

Selective leaching is the dealloying process in which one of the alloy ingredients is removed from the solid solution. This occurs because the lost ingredient has a lower corrosion resistance than the remaining ingredient. *Dezincification* is the classic case where zinc is selectively destroyed in brass. Other examples are the dealloying of nickel from copper-nickel alloys, iron from steel, and aluminum from copper-aluminum alloys.

31. HYDROGEN EMBRITTLEMENT

Hydrogen damage occurs when hydrogen gas diffuses through and decarburizes steel (i.e., reacts with carbon to form methane). *Hydrogen embrittlement* (also known as *caustic embrittlement*) is hydrogen damage where the source of hydrogen is the products of caustic corrosion.

Though riveted joints in early boilers were highly susceptible to hydrogen embrittlement, modern boilers are forged and welded, eliminating most places where corrosive solutions can concentrate. In high-pressure boilers, however, hydrogen can still be released in areas of major heat flux (i.e., boiler tubes and drums, economizers, and superheaters) under existing scale deposits. Sodium nitrate is used to combat hydrogen embrittlement.

---

[19]*Austenitic stainless steels* are the 300 series. They consist of chromium nickel alloys with up to 8% nickel. They are not hardenable by heat treatment, are nonmagnetic, and offer the greatest resistance to corrosion. *Martensitic stainless steels* are hardenable and magnetic. *Ferritic stainless steels* are magnetic and not hardenable.

32. MONITORING OF BOILER FEEDWATER QUALITY

Water chemistry is monitored through continuous *in-line sampling* and periodic *grab sampling*. Water impurities are expressed in milligrams per liter (mg/L), parts per million (ppm), parts per billion (ppb), unit equivalents per million (epm), and milliequivalents per milliliter (meq/mL).[20,21] ppm and ppb can be "as $CaCO_3$" equivalents or "as substance." Electrical conductivity (a measure of total dissolved solids) is measured in micromhos (μS) or ppm.

Water quality is maintained by monitoring pH, electrical resistivity silica, and hardness (calcium, magnesium, and bicarbonates), the primary corrosion species (sulfates, chlorides, and sodium), dissolved gases (primarily oxygen), and the concentrations of buffering chemicals (e.g., phosphate, hydrazine, or AVT).

Continuous monitoring of ionic concentrations is complicated by a process known as *hideout*, in which a chemical species (e.g., *crevice salts*) disappears by precipitation or adsorption during low-flow and high heat transfer. Upon cooling (as during a shut-down), the hideout chemicals reappear.

33. GALVANIC ACTION

Galvanic action (*galvanic corrosion* or *two-metal corrosion*) results from the difference in oxidation potentials of metallic ions. The greater the difference in oxidation potentials, the greater will be the galvanic corrosion. If two metals with different oxidation potentials are placed in an electrolytic medium (e.g., seawater), a galvanic cell will be created. The metal with the higher potential (i.e., the more "noble" metal) will act as an anode and will corrode. The metal with the lower potential, being the cathode, will be unchanged. In one extreme type of intergranular corrosion known as *exfoliation*, open endgrains separate into layers.

Metals are often classified according to their position in the *galvanic series* listed in Table 21.3. As would be expected, the metals in this series are in approximately the same order as their half-cell potentials. However, alloys and proprietary metals are also included in the series.

---

[20]Milligrams per liter (mg/L) is essentially the same as parts per million (ppm). The older units of grains per gallon (gpm) are still occasionally encountered. 1 grain equals 1/7000th of a pound. Multiply grains per gallon (gpg) by 17.1 to get ppm. Unit equivalents per million are derived by dividing the concentration in ppm by the equivalent weight. Equivalent weight is the molecular weight divided by the valence or oxidation number.

[21]Use the following relationship to convert meq/mL to ppm.

$$\text{ppm} = (1000) \left(\frac{\text{meq}}{\text{mL}} \right) (\text{equivalent weight})$$

$$= \frac{(1000) \left(\frac{\text{meq}}{\text{mL}} \right) (\text{formula weight})}{\text{valence}}$$

Table 21.3 *The Galvanic Series in Seawater (anodic to cathodic)*

magnesium
zinc
Alclad 3S
cadmium
2024 aluminum alloy
low-carbon steel
cast iron
stainless steel (active)
    No. 410
    No. 430
    No. 404
    No. 316
Hastelloy A
lead
lead-tin alloys
tin
nickel
brass, copper-zinc
copper
bronze, copper-tin
90/10 copper-nickel
70/30 copper-nickel
Inconel
silver solder
silver
stainless steels (passive)
Monel metal
Hastelloy C
titanium
graphite
gold

Precautionary measures can be taken to inhibit or eliminate galvanic action when use of dissimilar metals is unavoidable.

- Use dissimilar metals that are close neighbors in the galvanic series.

- Use *sacrificial anodes*. In marine saltwater applications, sacrificial zinc plates can be used.

- Use protective coatings, oxides, platings, or inert spacers to reduce or eliminate the access of corrosive environments to the metals.[22]

---

[22]While cadmium, nickel, chromium, and zinc are often used as protective deposits on steel, porosities in the surfaces can act as small galvanic cells, resulting in invisible subsurface corrosion.

34. STRESS CORROSION

When subjected to sustained surface tensile stresses (including low residual stresses from manufacturing) in corrosive environments, certain metals exhibit catastrophic *stress corrosion* cracking, SCC. When the stresses are cyclic, this type of corrosion is called *corrosion fatigue*, which can lead to fatigue failures well below normal yield stresses.

Stress corrosion occurs because the more highly stressed grains (at the crack tip) are slightly more anodic than neighboring grains with lower stresses. Although intergranular cracking (at grain boundaries) is more common, corrosion cracking may be *intergranular* (between the grains), *transgranular* (through the grains), or a combination of the two, depending on the alloy. Cracks propagate, often with extensive branching, until failure occurs.

The precautionary measures that can be taken to inhibit or eliminate stress corrosion are as follows.

- Avoid using metals that are susceptible to stress corrosion. These include austenitic stainless steels without heat treatment in seawater, certain tempers of the aluminum alloys 2124, 2219, 7049, and 7075 in seawater, and copper alloys exposed to ammonia.

- Protect open-grain surfaces from the environment. For example, press-fitted parts in drilled holes can be assembled with wet zinc chromate paste. Also, weldable aluminum can be "buttered" with pure aluminum rod.

- Stress-relieve by annealing heat treatment after welding or cold working.

35. FRETTING CORROSION

Fretting corrosion occurs when two highly loaded members have a common surface at which rubbing and sliding take place. The phenomenon is a combination of wear and chemical corrosion. Metals that depend on a film of surface oxide for protection, such as aluminum and stainless steel, are especially susceptible.

Fretting corrosion can be reduced by the following methods.

- Lubricate the rubbing surfaces.

- Seal the surfaces.

- Reduce vibration and movement.

36. CAVITATION CORROSION

Cavitation is the formation and collapse of minute bubbles of vapor in liquids. It is caused by a combination of reduced pressure and increased velocity in the fluid. In effect, very small amounts of the fluid vaporize (i.e., boil) and almost immediately condense. The repeated collapse of the bubbles hammers and work-hardens the surface.

When the surface work-hardens, it becomes brittle. Small amounts of the surface flake away, and the surface becomes pitted. This is known as *cavitation corrosion*. Eventually, the entire piece may work-harden and become brittle, leading to structural failure.

37. WATER SUPPLY CHEMISTRY

Most water supply composition data is not given in units of molarity, normality, molality, and so on. Rather, the most common measure of solution strength is the *$CaCO_3$ equivalent* measurement. With this method, substances are reported in milligrams per liter (mg/L, same as parts per million, ppm) "as $CaCO_3$," even when $CaCO_3$ is unrelated to the substance or reaction that produced the substance.

Actual gravimetric amounts of a substance can be converted to amounts as $CaCO_3$ by use of the conversion factors in App. 21.C. These factors are easily derived from stoichiometric principles.

The reason for converting all substance quantities to amounts as $CaCO_3$ is that equal $CaCO_3$ amounts constitute stoichiometric reaction quantities. For example, 100 mg/L as $CaCO_3$ of sodium ion (Na^+) react with 100 mg/L as $CaCO_3$ of chloride ion (Cl^-) to produce 100 mg/L as $CaCO_3$ of salt (NaCl), even though the gravimetric quantities differ and $CaCO_3$ is not part of the reaction.

Example 21.10

Lime is added to water to remove carbon dioxide gas.

$$CO_2 + Ca(OH)_2 \longrightarrow CaCO_3 \downarrow + H_2O$$

If water contains 5 mg/L of CO_2, how much lime is required for its removal?

Solution

From App. 21.C, the factor that converts CO_2 as substance to CO_2 as $CaCO_3$ is 2.27.

$$CO_2 \text{ as } CaCO_3 \text{ equivalent} = (2.27)\left(5\,\frac{\text{mg}}{\text{L}}\right)$$
$$= 11.35 \text{ mg/L as } CaCO_3$$

Therefore, the $CaCO_3$ equivalent of lime required will also be 11.35 mg/L.

From App. 21.C again, the factor that converts lime as $CaCO_3$ to lime as substance is (1/1.35).

$$Ca(OH)_2 \text{ substance} = \frac{11.35\frac{\text{mg}}{\text{L}}}{1.35}$$
$$= 8.41 \text{ mg/L as substance}$$

38. ACIDITY AND ALKALINITY IN WATER

Acidity is a measure of acids in solutions. Acidity in surface water (e.g., lakes and streams) is caused by formation of *carbonic acid* (H_2CO_3) from carbon dioxide in the air.[23] Acidity in water is typically given in terms of the $CaCO_3$ equivalent that would neutralize the acid.

$$CO_2 + H_2O \longrightarrow H_2CO_3 \qquad \textit{21.15}$$

$$H_2CO_3 + H_2O \longrightarrow HCO_3^- + H_3O^+ \quad [\text{pH} > 4.5]$$
$$\textit{21.16}$$

$$HCO_3^- + H_2O \longrightarrow CO_3^{--} + H_3O^+ \quad [\text{pH} > 8.3]$$
$$\textit{21.17}$$

Alkalinity is a measure of the amount of negative (basic) ions in the water. Specifically, OH^-, CO_3^{--}, and HCO_3^- all contribute to alkalinity.[24] The measure of alkalinity is the sum of concentrations of each of the substances measured as $CaCO_3$.

Alkalinity and acidity of a titrated sample is determined from color changes in indicators added to the titrant.

Example 21.11

Water from a city well is analyzed and is found to contain 20 mg/L as substance of HCO_3^- and 40 mg/L as substance of CO_3^{--}. What is the alkalinity of this water?

Solution

From App. 21.C, the factors converting HCO_3^- and CO_3^{--} ions to $CaCO_3$ equivalents are 0.82 and 1.67, respectively.

$$\text{alkalinity} = (0.82)\left(20\,\frac{\text{mg}}{\text{L}}\right) + (1.67)\left(40\,\frac{\text{mg}}{\text{L}}\right)$$
$$= 83.2 \text{ mg/L as } CaCO_3$$

39. WATER HARDNESS

Water hardness is caused by multivalent (doubly charged, triply charged, etc., but not singly charged) positive metallic ions such as calcium, magnesium, iron, and manganese. (Iron and manganese are not as common, however.) Hardness reacts with soap to reduce its cleansing effectiveness and to form scum on the water surface and a ring around the bathtub.

---

[23]Carbonic acid is very aggressive and must be neutralized to eliminate the cause of water pipe corrosion. If the pH of water is greater than 4.5, carbonic acid ionizes to form bicarbonate (Eq. 21.15). If the pH is greater than 8.3, carbonate ions form that cause water hardness by combining with calcium. (See Eq. 21.16.)
[24]Other ions, such as NO_3^-, also contribute to alkalinity, but their presence is rare. If detected, they should be included in the calculation of alkalinity.

Water containing bicarbonate (HCO_3^-) ions can be heated to precipitate a carbonate molecule.[25] This hardness is known as *temporary hardness* or *carbonate hardness*.[26]

$$Ca^{++} + 2HCO_3^- + heat \longrightarrow CaCO_3 \downarrow + CO_2 + H_2O$$

$$21.18$$

$$Mg^{++} + 2HCO_3^- + heat \longrightarrow MgCO_3 \downarrow + CO_2 + H_2O$$

$$21.19$$

Remaining hardness due to sulfates, chlorides, and nitrates is known as *permanent hardness* or *noncarbonate hardness* because it cannot be removed by heating. The amount of permanent hardness can be determined numerically by causing precipitation, drying, and then weighing the precipitate.

$$Ca^{++} + SO_4^{--} + Na_2CO_3 \longrightarrow 2Na^+ + SO_4^{--} + CaCO_3 \downarrow$$

$$21.20$$

$$Mg^{++} + 2Cl^- + 2NaOH \longrightarrow 2Na^+ 2 + Cl^- + Mg(OH)_2 \downarrow$$

$$21.21$$

Total hardness is the sum of temporary and permanent hardnesses, both expressed in mg/L as $CaCO_3$.

40. COMPARISON OF ALKALINITY AND HARDNESS

Hardness measures the presence of positive, multivalent ions in the water supply. Alkalinity measures the presence of negative (basic) ions such as hydrates, carbonates, and bicarbonates. Since positive and negative ions coexist, an alkaline water can also be hard.

If certain assumptions are made, then it is possible to draw conclusions about the water composition from the hardness and alkalinity. For example, if the effects of Fe^{+2} and OH^- are neglected, the following rules apply. (All concentrations are measured as $CaCO_3$.)

- *hardness = alkalinity*: There is no noncarbonate hardness. There are no SO_4^{-2}, Cl^-, or NO_3^- ions present.

- *hardness > alkalinity*: Noncarbonate hardness is present.

- *hardness < alkalinity*: All hardness is carbonate hardness. The extra HCO_3^- comes from other sources (e.g., $NaHCO_3$).

---

[25] Hard water forms scale when heated. This scale, if it forms in pipes, eventually restricts water flow. Even in small quantities, the scale insulates boiler tubes. Therefore, water used in steam-producing equipment must be essentially hardness-free.
[26] The hardness is known as *carbonate* hardness even though it is caused by *bicarbonate* ions, not carbonate ions.

Titration with indicator solutions is used to determine the alkalinity. The *phenolphthalein alkalinity* (or "P reading" in mg/L as $CaCO_3$) measures hydrate alkalinity and half of the carbonate alkalinity. The *methyl orange alkalinity* (or "M reading" in mg/L as $CaCO_3$) measures the total alkalinity (including the phenolphthalein alkalinity). The following table can be used to interpret these tests.[27]

Table 21.4 *Interpretation of Alkalinity Tests*

case	hydrate as $CaCO_3$	carbonate as $CaCO_3$	bicarbonate as $CaCO_3$
$P = 0$	0	0	M
$0 < P < \dfrac{M}{2}$	0	2P	M − 2P
$P = \dfrac{M}{2}$	0	2P	0
$\dfrac{M}{2} < P < M$	2P − M	2(M − P)	0
$P = M$	M	0	0

41. WATER SOFTENING WITH LIME

Water softening can be accomplished with lime and soda ash to precipitate calcium and magnesium ions from the solution. Lime treatment has the added benefits of disinfection, iron removal, and clarification. Practical limits of *precipitation softening* are 30 mg/L of $CaCO_3$ and 10 mg/L of $Mg(OH)_2$ (as $CaCO_3$) because of intrinsic solubilities. Water treated by this method usually leaves the softening apparatus with a hardness of between 50 and 80 mg/L as $CaCO_3$.

42. WATER SOFTENING BY ION EXCHANGE

In the *ion exchange process* (also known as *zeolite process* or *base exchange method*), water is passed through a filter bed of exchange material. This exchange material is known as *zeolite*. Ions in the insoluble exchange material are displaced by ions in the water.

The processed water will have a zero hardness. However, if there is no need for water with zero hardness (as in municipal water supply systems), some water can be bypassed around the unit.

There are three types of ion exchange materials. *Green sand (glauconite)* is a natural substance that is mined and treated with manganese dioxide. *Siliceous-gel zeolite* is an artificial solid used in small volume deionizer columns. *Polystyrene resins* are also synthetic and dominate the softening field.

---

[27] The titration may be affected by the presence of silica and phosphates, which also contribute to alkalinity. The effect is small, but the titration may not be a completely accurate measure of carbonates and bicarbonates.

The earliest synthetic zeolites were gelular *ion exchange resins* using a three-dimensional copolymer (e.g., styrene-divinyl benzene). Porosity through the continuous-phase gel was near zero, and dry contact surface areas of 500 ft^2/lbm (0.1 m^2/g) or less were common.

Macroreticular synthetic resins are discontinuous, three-dimensional copolymer beads in a rigid-sponge type formation. Each bead is made up of thousands of microspheres of the gel resin. Porosity is increased, and dry contact surface areas are approximately 270,000 to 320,000 ft^2/lbm (55 to 65 m^2/g).

During operation, the calcium and magnesium ions are removed according to the following reaction in which R is the zeolite anion.

$$\begin{Bmatrix} Ca \\ Mg \end{Bmatrix} \begin{Bmatrix} (HCO_3)_2 \\ SO_4 \\ Cl_2 \end{Bmatrix} + Na_2R$$

$$\longrightarrow Na_2 \begin{Bmatrix} (HCO_3)_2 \\ SO_4 \\ Cl_2 \end{Bmatrix} + \begin{Bmatrix} Ca \\ Mg \end{Bmatrix} R \qquad \textit{21.22}$$

The resulting sodium compounds are soluble.

Typical saturation capacities of synthetic resins are 1.0 to 1.5 meq/mL for anion exchange resins and 1.7 to 1.9 meq/mL for cation exchange resins. However, working capacities are more realistic measures. Working capacities are approximately 10 to 15 kilograins/ft^3 (23 to 35 kg/m^3) before regeneration.

Flow rates through the bed are typically 1 to 6 gpm/ft^3 (2 to 13 L/s·m^3) of resin volume. The flow rate in terms of gpm/ft^2 (L/s·m^2) across the exposed surface will depend on the geometry of the bed, but values of 3 to 15 gpm/ft^2 (2 to 10 L/s·m^2) are typical.[28]

Example 21.12

A municipal plant processes water with a total initial hardness of 200 mg/L. The designed discharge hardness is 50 mg/L. If an ion exchange unit is used, what is the bypass factor?

Solution

The water passing through the ion exchange unit is reduced to zero hardness. If x is the water fraction bypassed around the zeolite bed,

$$(1-x)\left(0 \ \frac{mg}{L}\right) + x\left(200 \ \frac{mg}{L}\right) = 50 \ mg/L$$

$$x = 0.25$$

43. CHARACTERISTICS OF BOILER FEEDWATER

Water that replaces liquid that has been converted to steam in a boiler is known as *feedwater*. Water that has been added to replace losses is known as *make-up water*.

The purity of the feedwater returned to the boiler (after condensing) depends on the purity of the make-up water, since impurities continually build up. *Blowdown* is the intentional periodic release of some of the feedwater in order to remove chemicals whose concentrations have built up over time.

Water impurities cause scaling (which reduces fluid flow and heat transfer rates) and corrosion (which reduces strength). Deposits from calcium, magnesium, and silica compounds are particularly troublesome.[29] As feedwater purity increases, deposits from copper, iron, and nickel oxides (corrosion products from pipe and equipment) become more problematic.

Water can also contain dissolved oxygen, nitrogen, and carbon dioxide. The nitrogen is inert and does not need to be considered. However, both the oxygen and carbon dioxide need to be removed. High-temperature dissolved oxygen readily attacks pipe and boiler metal. Most of the oxygen in make-up water is removed by heating the water. Although the solubility of oxygen in water decreases with temperature, water at high pressures can hold large amounts of oxygen. Hydrazine (N_2H_4) and sodium sulfite (Na_2SO_3) are used for *oxygen scavenging*.[30] Because sodium sulfite forms sodium sulfate at high pressures, only hydrazine should be used above 1500 to 1800 psi (10.3 to 12.4 MPa).

Carbon dioxide combines with water to form *carbonic acid*. In power plants, this is more likely to occur in condenser return lines than elsewhere. Acidity of the condensate is reduced by *neutralizing amines* such as morpholine, cyclohexylamine, ethanolamine (the tetrasodium salt of ethylenediamine tetra-acetic acid, also known as Na$_4$EDTA), and diethylaminoethanol.

Most types of corrosion are greatly reduced when the pH is within the range of 9 to 10. Below this range, corrosion and deposits of sulfates, carbonates, and silicates become a major problem. For this reason, boilers are often shut down when the pH drops below 8.

Filming amines (*polar amines*) such as octadecylamine do not neutralize acidity or raise pH. They form a non-wettable layer which protects surfaces from corrosive compounds. Filming and neutralizing amines are often used in conjunction, though they are injected at different locations in the system.

Other purity requirements depend on the boiler equipment, and in particular, the steam pressure. Also, specific locations within the system can tolerate different concentrations.[31] The following general guidelines

---

[28] Much higher values, up to 15 to 20 gpm/ft^3 or 40 to 50 gpm/ft^2 (33 to 45 L/s·m^2 or 27 to 34 L/s·m^2), may occur in certain types of units and at certain times (e.g., start-up and leak conditions).

[29] Silica is most troubling because silicate deposits cannot be removed by chemical means. They must be removed mechanically.
[30] 8 ppm of sodium nitrate react with 1 ppm of oxygen. To achieve a complete reaction, an excess of 2 to 3 ppm of sodium nitrate is required. Hydrazine reacts on a one-to-one basis and is commonly available as a 35% solution. Therefore, approximately 3 ppm of solution is required per ppm of oxygen.
[31] For example, silica in the boiler feedwater may be limited to 1 to 5 ppm, but the silica content in steam should be limited to 0.01 to 0.03 ppm.

apply to boilers operating in the 900 to 2500 psig (6.2 to 17.3 MPa) range.[32]

- All water feedwater entering the boiler should be free from dissolved oxygen, carbon dioxide, suspended solids, and hardness.

- pH of water entering the boiler should be in the range of 8.5 to 9.0. pH of water in the boiler should be in the range of 10.8 to 11.5.

- Silica in boiler water should be limited to 5 ppm (as $CaCO_3$) at 900 psig (6.2 MPa), with a gradual reduction to 1 ppm (as $CaCO_3$) at 2500 psig (17.3 MPa).

44. PRODUCTION AND REGENERATION OF BOILER FEEDWATER

The processes used to treat raw water for use in power-generating plants depend on the incoming water quality. Depending on need, filters may be used to remove solids, softeners or exchange units may remove permanent (sulfate) hardness, and activated carbon may remove organics.

Bicarbonate hardness is converted to sludge when the water is boiled, is removed during blow-down, and does not need to be chemically removed. However, a strong *dispersant* (*sludge conditioner*) must be added to prevent scale formation.[33] Alternatively, the calcium salts can be converted to sludges of phosphate salts by adding phosphates. Prior to the 1970s, *caustic phosphate treatments* using sodium phosphate, SP, in the form of Na_3PO_4, $NaHPO_4$, and NaH_2PO_4, was the most common method used to treat low-pressure boiler feedwater with high solids. Sodium phosphate converts impurities and buffers pH.

However, magnesium phosphate is a very sticky sludge, and using phosphates may cause more problems than not using them. Also, with higher pressures and temperatures, *wasting* becomes problematic.[34] Therefore, many modern plants use an *all-volatile treatment*, AVT, also known as a *zero-solids treatment*. Early AVTs used ammonia (NH_3) for pH control, but ammonia causes *denting* in systems that operate without blowdown.[35]

Chelant AVTs do not add any solid chemicals to the boiler. They work by keeping calcium and magnesium in solution. The compounds are removed through continuous blowdown.[36] Chelant treatments can be used in boilers up to approximately 1500 psi (10.3 MPa) but require high-purity (low-solids) feedwater.[37]

Subsequent advances in corrosion protection have been based on use of stainless steel or titanium-tubed condensers, condensate polishing, and even better AVT formulations. Supplementing ammonia-AVTs are volatile amines, chelants, and boric acid treatments. Boric acid effectively combats denting, intergranular attack, and intergranular stress corrosion cracking.

Water that is processed through an ion-exchange unit is known as *deionized water*. Deionized water is "hungry" for minerals and picks up contamination as it passes through the power plant. When the operating pressure is 1500 psig (10.3 MPa) or above, condensed steam cannot be reused without additional demineralization.[38] *Demineralization* is not generally needed for lower-pressure units, but some *condensate polishing* is still necessary.

In low-pressure plants, condensate polishing may consist of passing the water through a filter (of sand, anthracite, or pre-coat cellulose filters), a strong-acid cation unit, and a mixed-bed unit. The filter removes gross particles (referred to as *crud*), the strong-acid unit removes the dissolved iron, and the mixed-bed unit polishes the condensate.[39]

In modern high-pressure plants, the filtration step is often omitted, and crud removal occurs in the strong-acid unit. Furthermore, if the mixed-bed unit is made large enough or the cation component is increased, the strong-acid unit can also be omitted. A mixed-bed unit operating by itself is known as a *naked mixed-bed unit*. Demineralization of condensed steam is commonly accomplished with naked mixed beds.

45. REGENERATION OF ION EXCHANGE RESINS

Ion exchange material has a finite capacity for ion removal. When the zeolite is saturated or has reached some other prescribed limit, it must be regenerated (rejuvenated).

Standard ion exchange units are regenerated when the alkalinity of their effluent increases to the *set point*. Most condensate polishing units that also collect crud are operated to a *pressure-drop endpoint*. The pressure

---

[32]Tolerable concentrations at cold start-up can be as much as 5 to 100 times higher.

[33]Typical sludge conditioner dispersants are natural organics (lignins, tannins, and starches) and synthetics (sodium polyacrylate, sodium polymethacrylate, sulfonated polystyrene, and maleic anhydride).

[34]*Wasting* is a term used in the power-generating industry to describe the process of general pipe wall thinning.

[35]*Denting* is the constricting of the intersection between tubes and support plates in boilers due to a buildup of corrosion.

[36]AVTs work in a completely different way than phosphates. It is not necessary to use phosphates and AVTs simultaneously.

[37]If the feedwater becomes contaminated, a supplementary phosphate treatment will be required.

[38]The term *deionization* refers to the process that produces make-up water. *Demineralization* refers to the process used to prepare condensed steam for reuse.

[39]*Crud* is primarily iron-corrosion products ranging from dissolved to particulate matter.

drop through the ion exchange unit is primarily dependent on the amount of crud collected. (See Ftn. 39.) When the pressure drop reaches a set point, the resin is regenerated.

Regeneration of synthetic ion exchange resins is accomplished by passing a *regenerating solution* over/through the resin. Although regeneration can occur in the ion exchange unit itself, external regeneration is becoming common. This involves removing the bed contents hydraulically, backwashing to separate the components (for mixed beds), regenerating the bed components separately, washing, then recombining and transferring the bed components back into service.

Common regeneration compounds are NaCl (for water hardness removal units), H_2SO_4 (for cation exchange resins), and NaOH (for anion exchange resins). The amount of regeneration solution depends on the resin's degree of saturation. A rule of thumb is to expect to use 6 to 10 lbm of regeneration compound per cubic foot of resin (100 to 160 kg per cubic meter). Alternatively, dosage of the regeneration compound may be specified in terms of hardness removed (e.g., 0.4 lbm of salt per 1000 grains of hardness removed). These rates are applicable to deionization plants for boiler make-up water. For condensate polishing, saturation levels of 10 to 25 lbm/ft^3 (160 to 400 kg/m^3) are used.

PRACTICE PROBLEMS

1. The gravimetric analysis of a compound is 40% carbon, 6.7% hydrogen, and 53.3% oxygen. What is the simplest formula for the compound?

2. A municipal water supply has the ionic concentrations listed.

Al^{+++}	0.5 mg/L
Ca^{++}	80.2 mg/L
Cl^-	85.9 mg/L
CO_2	19 mg/L
Fe^{++}	1.0 mg/L
HCO_3^-	185 mg/L
Mg^{++}	24.3 mg/L
Na^+	46.0 mg/L
SO_4^{--}	125 mg/L

(a) What is the total water hardness in mg/L? (b) How much lime ($Ca(OH)_2$) is required to remove the carbonate hardness?

3. The following concentrations of inorganic compounds are found during a routine analysis of a municipal water supply. Some of the water is treated with lime, and some is passed through a zeolite process. The zeolite has an exchange capacity of 10,000 grains of hardness per cubic foot (22.9 kg/m^3). 0.5 lbm of salt per 1000 grains of $CaCO_3$ hardness (3.5 kg of salt per kg of $CaCO_3$ hardness) are used to rejuvenate the zeolite.

$Ca(HCO_3)_2$	137 mg/L (as $CaCO_3$)
CO_2	0 mg/L
$MgSO_4$	72 mg/L (as $CaCO_3$)

(a) How much lime ($Ca(OH)_2$) is required to soften 1 million gallons (4 million liters) of this water to 100 mg/L as $CaCO_3$ if an excess of 30 mg/L is required for a complete reaction? (b) How much salt is required to rejuvenate the zeolite if 1 million gallons (4 million liters) are softened in that process?

22 Fuels and Combustion

1.	Hydrocarbons	22-1
2.	Cracking of Hydrocarbons	22-2
3.	Fuel Analysis	22-2
4.	Weighting of Thermodynamic Properties	22-2
5.	Standard Conditions	22-2
6.	Moisture	22-3
7.	Ash and Mineral Matter	22-3
8.	Sulfur	22-3
9.	Wood	22-3
10.	Waste Fuels	22-4
11.	Incineration	22-4
12.	Coal	22-4
13.	Low-Sulfur Coal	22-4
14.	Clean Coal Technologies	22-5
15.	Coke	22-5
16.	Liquid Fuels	22-5
17.	Fuel Oils	22-5
18.	Gasoline	22-6
19.	Oxygenated Gasoline	22-6
20.	Diesel Fuel	22-7
21.	Alcohol	22-7
22.	Gaseous Fuels	22-7
23.	Ignition Temperature	22-8
24.	Atmospheric Air	22-8
25.	Combustion Reactions	22-8
26.	Stoichiometric Reactions	22-9
27.	Stoichiometric Air	22-9
28.	Incomplete Combustion	22-11
29.	Flue Gas Analysis	22-11
30.	Actual and Excess Air	22-12
31.	Calculations Based on Flue Gas Analysis	22-12
32.	Temperature of Flue Gas	22-13
33.	Smoke	22-13
34.	Dew Point of Flue Gas Moisture	22-13
35.	Heat of Combustion	22-14
36.	Maximum Theoretical Combustion (Flame) Temperature	22-15
37.	Combustion Losses	22-15
38.	Combustion Efficiency	22-16
39.	Draft	22-16
40.	Stack Effect	22-16
41.	Stack Friction	22-18
	Practice Problems	22-18

Nomenclature

A	area	ft^2	m^2
B	volumetric fraction	–	–
c	specific heat	Btu/lbm-°F	kJ/kg·°C
d	diameter	in	cm
D	diameter	ft	m
D	draft	in wg	kPa
g	acceleration of gravity	ft/sec^2	m/s^2
g_c	gravitational constant (32.2)	$ft\text{-}lbm/lbf\text{-}sec^2$	–
G	gravimetric fraction	–	–
h	enthalpy	Btu/lbm	kJ/kg
h	head	ft	m
H	height	ft	m
HHV	higher heating value	Btu/lbm	kJ/kg
HV	heating value	Btu/lbm	kJ/kg
K	constant	–	–
L	length	ft	m
LHV	lower heating value	Btu/lbm	kJ/kg
m	mass	lbm	kg
M	fraction moisture	–	–
ON	octane number	–	–
p	pressure	lbf/ft^2	Pa
P	power	Btu/sec	kW
PN	performance number	–	–
q	heat loss	Btu/lbm	kJ/kg
Q	flow rate	ft^3/sec	m^3/s
R	ratio	lbm/lbm	kg/kg
R	specific gas constant	ft-lbf/lbm-°R	kJ/kg·K
T	temperature	°F	°C
v	velocity	ft/sec	m/s

Symbols

γ	specific weight	lbf/ft^3	–
η	efficiency	–	–
ω	humidity ratio	–	–

Subscripts

a/f	air/fuel
fg	vaporization
g	gas
i	initial
p	constant pressure
SE	stack effect

1. HYDROCARBONS

With the exception of sulfur and related compounds, most fuels are hydrocarbons. Hydrocarbons are further categorized into subfamilies such as *alkynes* (C_nH_{2n-2}, such as acetylene C_2H_2), *alkenes* (C_nH_{2n}, such as ethylene C_2H_4), and *alkanes* (C_nH_{2n+2}, such as octane C_8H_{18}). The alkynes and alkenes are referred to as *unsaturated hydrocarbons*, while the alkanes are referred to as *saturated hydrocarbons*. The alkanes are also known

Table 22.1 *Approximate Specific Heats (at constant pressure) of Gases* (c_p *in Btu/lbm-°R; at 1 atm*)

gas	temperature (°R)							
	500	1000	1500	2000	2500	3000	4000	5000
air	0.240	0.249	0.264	0.277	0.286	0.294	0.302	—
carbon dioxide	0.196	0.251	0.282	0.302	0.314	0.322	0.332	0.339
carbon monoxide	0.248	0.257	0.274	0.288	0.298	0.304	0.312	0.316
hydrogen	3.39	3.47	3.52	3.63	3.77	3.91	4.14	4.30
nitrogen	0.248	0.255	0.270	0.284	0.294	0.301	0.310	0.315
oxygen	0.218	0.236	0.253	0.264	0.271	0.276	0.286	0.294
sulfur dioxide	0.15	0.16	0.18	0.19	0.20	0.21	0.23	—
water vapor	0.444	0.475	0.519	0.566	0.609	0.645	0.696	0.729

(Multiply Btu/lbm-°R by 4.187 to obtain kJ/kg·K.)

as the *paraffin series* and *methane series*. The alkenes are subdivided into the chain-structured *olefin series* and the ring-structured *napthalene series*. *Aromatic hydrocarbons* (C_nH_{2n-6}, such as benzene C_6H_6) constitute another subfamily. Names for common hydrocarbon compounds are listed in App. 22.A.

2. CRACKING OF HYDROCARBONS

Cracking is the process of splitting hydrocarbon molecules into smaller molecules. For example, alkane molecules crack into a smaller member of the alkane subfamily and a member of the alkene subfamily. Cracking is used to obtain lighter hydrocarbons (such as those used in gasoline) from heavy hydrocarbons (e.g., crude oil).

Cracking can proceed under the influence of high temperatures (*thermal cracking*) or catalysts (*catalytic cracking* or "cat cracking"). Since (from Le Châtelier's principle) cracking at high pressure favors recombination, catalytic cracking is performed at pressures near atmospheric. Catalytic cracking also produces gasolines with better antiknock properties than does thermal cracking.

3. FUEL ANALYSIS

Fuels analyses are reported as either percentages by weight (for liquid and solid fuels) or percentages by volume (for gaseous fuels). Percentages by weight are known as *gravimetric analyses*, while percentages by volume are known as *volumetric analyses*. An *ultimate analysis* is a special type of gravimetric analysis in which the constituents are reported by atomic species rather than by compound. In an ultimate analysis, combined hydrogen from moisture in the fuel is added to hydrogen from the combustive compounds. (See Sec. 6.)

A *proximate analysis* (not "approximate") gives the gravimetric fraction of moisture, volatile matter, fixed carbon, and ash. Sulfur may be combined with the ash or may be specified separately.

Gas analyses are typically specified as volumetric fractions. For a gas in a mixture, its *volumetric fraction* (i.e., its volumetric percentage) is the same as its *mole fraction* and *partial pressure fraction*.

A volumetric fraction can be converted to a gravimetric fraction by multiplying by the molecular weight and then dividing by the sum of the products of all the volumetric fractions and molecular weights. Conversions between volumetric and gravimetric analyses are covered in Chap. 24.

4. WEIGHTING OF THERMODYNAMIC PROPERTIES

Many gaseous fuels (and all gaseous combustion products) are mixtures of different compounds. Some thermodynamic properties of mixtures are gravimetrically weighted, while others are volumetrically weighted. Specific heat, specific gas constant, enthalpy, internal energy, and entropy are gravimetrically weighted. For gases, molecular weight, density, and all molar properties are volumetrically weighted.[1]

When a compound experiences a large temperature change, the thermodynamic properties should be evaluated at the average temperature. Table 22.1 can be used to find the specific heat of gases at various temperatures.

5. STANDARD CONDITIONS

Though "standard conditions" usually means 70°F (21°C) and 1 atm pressure, *standard temperature and pressure*, STP, for manufactured fuel gases is 60°F

---

[1] For gases, molar properties include molar specific heats, enthalpy per mole, and internal energy per mole.

(16°C) and 1 atm pressure. Since this convention is not well standardized, the actual temperature and pressure should be stated.[2]

Some combustion equipment (e.g., particulate collectors) operates within a narrow range of temperatures and pressures. These conditions are referred to as *normal temperature and pressure*, NTP.

6. MOISTURE

If an ultimate analysis of a fuel is given, all of the oxygen is assumed to be in the form of free water.[3] The amount of hydrogen combined as free water is assumed to be one-eighth of the oxygen weight.[4] All remaining hydrogen, known as the *available hydrogen*, is assumed to be combustible.

$$G_{\text{H,combined}} = \frac{G_O}{8} \qquad 22.1$$

$$G_{\text{H,available}} = G_{\text{H,total}} - \frac{G_O}{8} \qquad 22.2$$

For coal, the *"bed" moisture level* refers to the moisture level when the coal is mined. The terms *dry* and *as fired* are often used in commercial coal specifications. The "as fired" condition corresponds to a specific moisture content when placed in the furnace. The "as fired" heating value should be used, since the moisture actually decreases the combustion heat. The relationship between the two heating values is given by Eq. 22.3, where M is the moisture content from a proximate analysis.

$$\text{HV}_{\text{as fired}} = \text{HV}_{\text{dry}}(1 - M) \qquad 22.3$$

Moisture in fuel is undesirable because it increases fuel weight (transportation costs) and decreases available combustion heat.[5]

7. ASH AND MINERAL MATTER

Mineral matter is the noncombustible material in a fuel. *Ash* is the residue remaining after combustion. Ash may contain some combustible carbon. The two terms ("mineral matter" and "ash") are often used interchangeably when reporting fuel analyses.

Ash may also be categorized according to where it is recovered. Dry and wet *bottom ashes* are recovered from *ash pits*. (As little as 10% of the total ash content may be recovered in the ash pit.) *Fly ash* is carried out of the boiler by the flue gas. Fly ash can be deposited on walls and heat transfer surfaces. It will be discharged

from the stack if not captured. *Economizer ash* and *air heater ash* are recovered from the devices the ash is named after.

The finely powdered ash that covers combustion grates protects them from high temperatures.[6] If the ash has a low (i.e., below 2200°F; 1200°C) fusion temperature (melting point), it may form *clinkers* in the furnace and/or *slag* in other high-temperature areas. In extreme cases, it can adhere to the surfaces. Ashes with high melting (fusion) temperatures (i.e., above 2600°F; 1430°C) are known as *refractory ashes*. The T_{250} *temperature* is used as an index of slagging tendencies of an ash. This is the temperature at which the slag becomes molten with a viscosity of 250 poise. Slagging will be experienced when the T_{250} temperature is exceeded.

The actual melting point depends on the ash composition. Ash is primarily a mixture of silica (SiO_2), alumina (Al_2O_3), and ferric oxide (Fe_2O_3).[7] The relative proportions of each will determine the melting point, with lower melting points resulting from high amounts of ferric oxide and calcium oxide. The melting points of pure alumina and pure silica are in the 2700 to 2800°F (1480 to 1540°C) range.

Coal ash is either of a bituminous type or lignite type. Bituminous-type ash (from midwestern and eastern coals) contains more ferric oxide than lime and magnesia. Lignite-type ash (from western coals) contains more lime and magnesia than ferric oxide.

8. SULFUR

Several forms of sulfur are present in coal and fuel oils. *Pyritic sulfur* (FeS_2) is the primary form. *Organic sulfur* is combined with hydrogen and carbon in other compounds. *Sulfate sulfur* is iron sulfate and gypsum ($CaSO_4 \cdot 2H_2O$). Sulfur in elemental, organic, and pyritic forms oxidizes to sulfur dioxide. *Sulfur trioxide* can be formed under certain conditions. Sulfur trioxide combines with water to form sulfuric acid and is a major source of boiler/stack corrosion and pollution.

$$SO_3 + H_2O \longrightarrow H_2SO_4$$

9. WOOD

Wood is not an industrial fuel, though it may be used in small quantities in developing countries. Most woods have heating values around 8300 Btu/lbm (19 MJ/kg), with specific values depending on the species and moisture content. Variation in wood properties are so great that generalized properties are meaningless.

---

[2]Both of these are different from the standard temperature and pressure used for scientific work. The standard temperature in that case is 32°F (0°C).

[3]This assumes that none of the oxygen is in the form of carbonates.

[4]The value of $^1/_8$ follows directly from the combustion reaction of hydrogen and oxygen.

[5]A moisture content up to 5% is reported to be beneficial in some mechanically fired boilers.

[6]Some boiler manufacturers rely on the thermal protection the ash provides. For example, coal burned in cyclone boilers should have a minimum ash content of 7% to cover and protect the cyclone barrel tubes. Boiler wear and ash carryover will increase with lower ash contents.

[7]Calcium oxide (CaO), magnesium oxide ("magnesia," MgO), titanium oxide ("titania," TiO_2), ferrous oxide (FeO), and alkalies (Na_2O and K_2O) may be present in smaller amounts.

10. WASTE FUELS

Waste fuels are increasingly being burned or incinerated in industrial boilers and furnaces. Such fuels include digester and landfill gases, waste process gases, flammable waste liquids, and volatile organic compounds (VOCs) such as benzene, toluene, xylene, ethanol, and methane. Other waste fuels include oil shale, tar sands, green wood, seed and rice hulls, biomass refuse, peat, tire shreddings, and shingle/roofing waste.

The term *refuse-derived fuels*, RDF, is used to describe fuel produced from municipal waste. After separation (removal of glass, plastics, metals, corrugated cardboard, etc.), the waste is merely pushed into the combustion chamber. If the waste is to be used elsewhere, it is compressed and baled.

The heating value of RDF depends on the moisture content and fraction of combustible material. For RDFs derived from typical municipal wastes, the heating value will range from 3000 to 6000 Btu/lbm (7 to 14 MJ/kg), though higher ranges 7500 to 8500 Btu/lbm (17.5 to 19.8 MJ/kg) can be obtained by careful selection of ingredients. Pelletized RDF (containing some coal and a limestone binder) with heating values around 8000 Btu/lbm (18.6 MJ/kg) can be used as a supplemental fuel in coal-fired units.

Scrap tires are an attractive fuel source because of their high heating values—12,000 to 16,000 Btu/lbm (28 to 37 MJ/kg). To be compatible with existing coal-loading equipment, tires are chipped or shredded to 1-inch size (25 mm). Tires in this form are known as *tire-derived fuel*, TDF. Metal (from tire reinforcement) may or may not be present.

TDF has been shown capable of supplying up to 90% of a steam-generating plant's total Btu input without any deterioration in particulate emissions, pollutants, and stack opacity. In fact, compared with some low-quality coals (e.g., lignite), TDF is far superior: about 2.5 times the heating value and about 2.5 times less sulfur per Btu.

11. INCINERATION

Many toxic wastes are incinerated rather than "burned." Incineration and combustion are not the same. *Incineration* is the term used to describe a disposal process that uses combustion to render wastes ineffective (nonharmful, nontoxic, etc.). Wastes and combustible fuel are combined in a furnace, and the heat of combustion destroys the waste.[8] Wastes may themselves be combustible, though they may not be self-sustaining if the moisture content is too high.

Incinerated wastes are categorized into seven types. Type 0 is *trash* (highly combustible paper and wood, with 10% or less moisture); type 1 is *rubbish* (combustible waste with up to 25% moisture); type 2 is *refuse* (a mixture of rubbish and garbage, with up to 50% moisture); type 3 is *garbage* (residential waste with up to 70% moisture); type 4 is animal solids and pathological wastes (85% moisture); type 5 is industrial process wastes in gaseous, liquid, and semiliquid form; and, type 6 is industrial process wastes in solid and semisolid form requiring incineration in hearth, retort, or grate burning equipment.

12. COAL

Coal consists of volatile matter, fixed carbon, moisture, noncombustible mineral matter ("ash"), and sulfur. *Volatile matter* is driven off as a vapor when the coal is heated, and it is directly related to flame size. *Fixed carbon* is the combustible portion of the solid remaining after the volatile matter is driven off. Moisture is present in the coal as free water and (for some mineral compounds) as water of hydration. Sulfur, an undesirable component, contributes to heat content.

Coals are categorized into anthracitic, bituminous, and lignitic types. *Anthracite coal* is clean, dense, and hard. It is comparatively difficult to ignite but burns uniformly and smokelessly with a short flame. *Bituminous coal* varies in composition, but generally has a higher volatile content than anthracite, starts easily, and burns freely with a long flame. Smoke and soot are possible if bituminous coal is improperly fired. *Lignite coal* is a coal of woody structure, very high in moisture and with a low heating value. It normally ignites slowly due to its moisture, breaks apart when burning, and burns with little smoke or soot.

Coal is burned efficiently in a particular furnace only if it is uniform in size. Screen sizes are used to grade coal, but descriptive terms can also be used.[9] *Run-of-mine coal*, ROM, is coal as mined. *Lump coal* is in the 1 to 6 in (25 to 150 mm) range. *Nut coal* is smaller, followed by even smaller *pea coal screenings*, and *fines* (dust).

13. LOW-SULFUR COAL

Switching to low-sulfur coal is one way of meeting strict sulfur emission standards. Western and eastern low-sulfur coals have different properties.[10,11] Eastern low-sulfur coals are generally low-impact coals. (That is, few changes need to be made to the power plant when switching to them.) Western coals are generally high-impact coals. Properties of typical high- and low-sulfur fuels are shown in Table 22.2.

---

[8]Rotary kilns can accept waste in many forms. They are "workhorse" incinerators.

[9]The problem with descriptive terms is that one company's "pea coal" may be as small as $1/4$ in (6 mm), while another's may start at $1/2$ in (13 mm).

[10]In the United States, low-sulfur coals predominantly come from the western United States ("western subbituminous"), although some come from the east ("eastern bituminous").

[11]Some of the impact considerations are coal preparation, firing rate, ash volume and handling, slagging, corrosion rates, dust collection and suppression, and fire and explosion prevention.

The lower sulfur content results in less boiler corrosion. However, of all the coal variables, the different ash characteristics are the most significant with regard to the steam generator components. The slagging and fouling tendencies are prime concerns.

Table 22.2 *Typical Properties of High- and Low-Sulfur Coals*[a]

property	high-sulfur	low-sulfur eastern	low-sulfur western
higher heating value,			
Btu/lbm	10,500	13,400	8000
(MJ/kg)	(24.4)	(31.2)	(18.6)
moisture content, percent	11.7	6.9	30.4
ash content, percent	11.8	4.5	6.4
sulfur content, percent	3.2	0.7	0.5
slag melting			
temperature, °F	2400	2900	2900
(°C)	(1320)	(1590)	(1590)

(Multiply Btu/lbm by 2.326 to obtain kJ/kg.)
[a] All properties are "as received."

14. CLEAN COAL TECHNOLOGIES

A lot of effort has been put into developing technologies that will reduce acid rain, pollution and air toxics (NO_x and SO_2), and global warming. These technologies are loosely labeled as *clean coal technologies*, CCTs. Whether or not these technologies can be retrofitted into an existing plant or designed into a new plant depends on the economics of the process.

With *coal cleaning*, coal is ground to ultrafine sizes to remove sulfur and ash-bearing minerals.[12] However, finely ground coal creates problems in handling, storage, and dust production. The risk of fire and explosion increases. Different approaches to reducing the problems associated with transporting and storing finely ground coal include the use of dust suppression chemicals, pelletizing, transportation of coal in liquid slurry form, and pelletizing followed by reslurrying. Some of these technologies may not be suitable for retrofit into existing installations.

With *coal upgrading*, moisture is thermally removed from low-rank coal (e.g., lignite or subbituminous coal). With some technologies, sulfur and ash are also removed when the coal is upgraded.

Reduction in sulfur dioxide emissions is the goal of *SO_2 control* technologies. These technologies include conventional use of lime and limestone in *flue gas desulfurization* (FGD) systems, *furnace sorbent-injection* (FSI) and *duct sorbent-injection*. *Advanced scrubbing* is included in FGD technologies.

Redesigned burners and injectors and adjustment of the flame zone are typical types of *NO_x control*. Use of secondary air, injection of ammonia or urea, and selective

catalytic reduction (SCR) are also effective in NO_x reduction.

Fluidized-bed combustion (FBC) reduces NO_x emissions by reducing combustion temperatures to around 1500°F (815°C). FBC is also effective in removing up to 90% of the SO_2. *Atmospheric FBC* operates at atmospheric pressure, but higher thermal efficiencies are achieved in *pressurized FBC* units operating at pressures up to 10 atm.

Integrated gasification/combined cycle (IGCC) processes are able to remove 99% of all sulfur while reducing NO_x to well below current emission standards. *Synthetic gas* (syngas) is derived from coal. Syngas has a lower heating value than natural gas, but it can be used to drive gas turbines in combined cycles or as a reactant in the production of other liquid fuels.

15. COKE

Coke, typically used in blast furnaces, is produced by heating coal in the absence of oxygen. The heavy hydrocarbons crack (i.e., the hydrogen is driven off), leaving only a carbonaceous residue containing ash and sulfur. Coke burns smokelessly. *Breeze* is coke smaller than $5/8$ in (16 mm). It is not suitable for use in blast furnaces, but steam boilers can be adapted to use it. *Char* is produced from coal in a 900°F (500°C) carbonization process. The volatile matter is removed, but there is little cracking. The process is used to solidify tars, bitumens, and some gases.

16. LIQUID FUELS

Liquid fuels are lighter hydrocarbon products refined from crude petroleum oil. They include liquefied petroleum gas (LPG), gasoline, kerosene, jet fuel, diesel fuels, and heating oils. Important characteristics of a liquid fuel are its composition, ignition temperature, flash point,[13] viscosity, and heating value.

17. FUEL OILS

In the United States, fuel oils are categorized into grades 1 through 6 according to their viscosities.[14] Viscosity is the major factor in determining firing rate and the need for preheating for pumping or atomizing prior to burning. Grades 1 and 2 can be easily pumped at ambient temperatures. In the United States, the heaviest fuel oil used is grade 6, also known as *Bunker C oil*.[15, 16]

---

[12] 80% or more of *micronized coal* is 44 microns or less in size.

[13] This is different from the *flash point* that is the temperature at which fuel oils generate enough vapor to sustain ignition in the presence of spark or flame.
[14] Grade 3 became obsolete in 1948. Grade 5 is also subdivided into light and heavy categories.
[15] 120°F (48°C) is the optimum temperature for pumping no. 6 fuel oil. At that temperature, no. 6 oil has a viscosity of approximately 3000 SSU. Further heating is necessary to lower the viscosity to 150 to 350 SSU for atomizing.
[16] To avoid *coking* of oil, heating coils in contact with oil should not be hotter than 240°F (116°C).

Table 22.3 *Typical Properties of Common Commercial Fuels*

	butane	no. 1 diesel	no. 2 diesel	ethanol	gasoline	JP-4	methanol	propane
chemical formula	C_4H_{10}	–	–	C_2H_5OH	–	–	CH_3OH	C_3H_8
molecular weight	58.12	≈ 170	≈ 184	46.07	≈ 126			44.09
heating value								
higher Btu/lbm	21,240	19,240	19,110	12,800	20,260			21,646
lower Btu/lbm	19,620	18,250	18,000	11,500	18,900	18,400	9078	19,916
lower Btu/gal	102,400	133,332	138,110	76,152	116,485	123,400	60,050	81,855
latent heat of								
vaporization Btu/lbm		115	105	361	142			147
specific gravity[a]	2.01	0.876	0.920	0.794	0.68–0.74	0.8017	0.793	1.55

(Multiply Btu/lbm by 2.326 to obtain kJ/kg.)
(Multiply Btu/gal by 0.2786 to obtain MJ/m^3.)
[a]Specific gravities of propane and butane are with respect to air.

Fuel oils are also classified according to their viscosities as *distillate oils* (lighter) and *residual fuel oils* (heavier).

Like coal, fuel oils contain sulfur and ash that may cause pollution, slagging on the hot end of the boiler, and corrosion in the cold end. Table 22.4 lists typical properties of fuel oils.

Table 22.4 *Typical Properties of Fuel Oils*[a]

grade	specific gravity	heating value MBtu/gal[b]	heating value GJ/m^3
1	0.805	134	37.3
2	0.850	139	38.6
4	0.903	145	40.4
5	0.933	148	41.2
6	0.965	151	41.9

(Multiply MBtu/gal by 0.2786 to obtain GJ/m^3.)
[a]Actual values will vary depending on composition.
[b]MBtu is thousands of Btus.

18. GASOLINE

Gasoline is not a pure compound. It is a blended mixture of various hydrocarbons that has the desired flammability, volatility, heating value, and octane rating. There are numerous blends that can be used to produce gasoline.

Gasoline's heating value depends only slightly on composition. Within a variation of $1^1/_2\%$, the heating value can be taken as 20,200 Btu/lbm (47.0 MJ/kg) for regular gasoline and as 20,300 Btu/lbm (47.2 MJ/kg) for high-octane aviation fuel.

Since gasoline is a mixture of hydrocarbons, different fractions will evaporate at different temperatures. The *volatility* is the percentage of the fuel that has evaporated by a given temperature. Typical volatility specifications call for 10% at 167°F (75°C), 50% at 221°F (105°C), and 90% at 275°F (135°C). Low volatility causes difficulty starting and poor engine performance at low temperatures.

The *octane number*, ON, is a measure of knock resistance. It is based on comparison, performed in a standardized one-cylinder engine, with the burning of isooctane and *n*-heptane. *n*-heptane, C_7H_{16}, is rated zero and produces violent knocking. Isooctane, C_8H_{18}, is rated 100 and produces relatively knock-free operation. The percentage blend by volume of these fuels that matches the performance of the gasoline is the octane rating. The *research octane number* (RON) is a measure of the fuel's antiknock characteristics while idling; the *motor octane number* (MON) applies to high-speed, high-acceleration operation. The octane rating reported for commercial gasoline is an average of the two.

The *performance number* (PN) of gasoline containing antiknock compounds (e.g., tetraethyl lead, TEL) is related to the octane number.

$$ON = 100 + \frac{PN - 100}{3} \qquad 22.4$$

19. OXYGENATED GASOLINE

In parts of the United States, gasoline is "oxygenated" during the cold winter months. The addition of *oxygenates* raises the combustion temperature, reducing carbon monoxide and unburned hydrocarbons.[17] Common oxygenates used in *reformulated gasoline* (RFG) include methyl tertiary-butyl ether (MTBE) and ethanol. Methanol, ethyl tertiary-butyl ether (ETBE) tertiary-amyl methyl ether (TAME) and tertiary-amyl ethyl ether (TAEE) may also be used. Oxygenates are added to bring the minimum oxygen level to 2 to 3% by weight.[18]

---

[17]Oxygenation may not be successful in reducing carbon dioxide. Since the heating value of the oxygenates is lower, fuel consumption of oxygenated fuels is higher. On a per-gallon (per-liter) basis, oxygenation reduces carbon dioxide. On a per-mile (per-kilometer) basis, however, oxygenation appears to increase carbon dioxide.
[18]Other specifications on gasoline during the winter months intended to reduce pollution may include maximum percentages of benzene and total aromatics, and limits on Reid vapor pressure.

Table 22.5 Typical Properties of Common Oxygenates

	MTBE[c]	TAME	ETBE	TAEE
specific gravity	0.744	0.740	0.770	0.791
octane	110	112	105	100
heating value (MBtu/gal)[a]	93.6			
Reid vapor pressure (psig)[b]	8	4	3	2
percent oxygen by weight	18.2	15.7	15.7	13.8
volumetric percent needed to achieve gasoline				
2.7% oxygen by weight	15.1	17.2	17.2	19.4
2.0% oxygen by weight	11.0	12.4	12.7	13.0

(Multiply MBtu/gal by 0.2786 to obtain MJ/m^3.)
[a]MBtu is thousands of Btus.
[b]The Reid vapor pressure is the vapor pressure when heated to 100°F (38°C).
[c]MTBE is water soluble and does not degrade. As a suspected carcinogen, it is proving to be a serious threat to water supplies.

20. DIESEL FUEL

Properties and specifications for various grades of diesel fuel oil are similar to specifications for fuel oils. Grade 1-D ("D" for diesel) is a light distillate oil for high-speed engines in service requiring frequent speed and load changes. Grade 2-D is a distillate of lower volatility for engines in industrial and heavy mobile service. Grade 4-D is for use in medium speed engines under sustained loads.

Diesel oils are specified by a *cetane number*, which is a measure of the ignition quality (ignition delay) of a fuel. Like the octane number for gasoline, the cetane number is determined by comparison with standard fuels. Cetane, $C_{16}H_{34}$, has a cetane number of 100. *n*-methyl-napthalene, $C_{11}H_{10}$, has a cetane number of zero. A cetane number of approximately 30 is required for satisfactory operation of low-speed diesel engines. High-speed engines, such as those used in cars, require a cetane number of 45 or more. The cetane number can be increased by use of such additives as amyl nitrate, ethyl nitrate, and ether.

A diesel fuel's *pour point* number refers to its viscosity. A fuel with a pour point of 10°F (−12°C) will flow freely above that temperature. A fuel with a high pour point will thicken in cold temperatures.

The *cloud point* refers to the temperature at which wax crystals cloud the fuel at lower temperatures. The cloud point should be 20°F (−7°C) or higher. Below that temperature, the engine will not run well.

21. ALCOHOL

Both methanol and ethanol can be used in internal combustion engines. *Methanol* (*methyl alcohol*) is produced from natural gas and coal, although it can also be produced from wood and organic debris. *Ethanol* (*ethyl alcohol, grain alcohol*) is distilled from grain, sugar cane, potatoes, and other agricultural products containing various amounts of sugars, starches, and cellulose.

Although methanol generally works as well as ethanol, only ethanol can be produced in large quantities from inexpensive agricultural products and by-products.

Alcohol is water-soluble. The concentration of alcohol is measured by its *proof*, where 200 proof is pure alcohol. (180 proof is 90% alcohol and 10% water.)

Gasohol is a mixture of approximately 90% gasoline and 10% alcohol (generally ethanol).[19] Alcohol's heating value is less than gasoline's, so fuel consumption (per distance traveled) is higher with gasohol. Also, since alcohol absorbs moisture more readily than gasoline, corrosion of fuel tanks becomes problematic. In some engines, significantly higher percentages of alcohol may require such modifications as including larger carburetor jets, timing advances, heaters for preheating fuel in cold weather, tank lining to prevent rusting, and alcohol-resistant gaskets.

Mixtures of gasoline and alcohol can be designated by the first letter and the fraction of the alcohol. E10 is a mixture of 10% ethanol in gasoline. M85 is a blend of 85% methanol and 15% gasoline.

Alcohol is a poor substitute for diesel fuel because alcohol's cetane number is low—from −20 to +8. Straight injection of alcohol results in poor performance and heavy knock.

22. GASEOUS FUELS

Various gaseous fuels are used as energy sources, but most applications are limited to natural gas and *liquefied petroleum gases*, LPGs (i.e., propane, butane, and mixtures of the two).[20,21] Natural gas is a mixture of methane (55 to 95%), higher hydrocarbons

---

[19]In fact, oxygenated gasoline may use more than 10% alcohol.
[20]A number of *manufactured gases* are of practical (and historical) interest in specific industries, including *coke-oven gas, blast-furnace gas, water gas, producer gas*, and (in the UK) *town gas*. However, these gases are not in widespread use.
[21]At atmospheric pressure, propane boils at −40°F (−40°C), while butane boils at 32°F (0°C).

(primarily ethane), and other noncombustible gases. Typical heating values for natural gas range from 950 to 1100 Btu/ft^3 (35 to 41 MJ/m^3).

The production of *synthetic gas* through coal gasification may be applicable to large power generating plants. The cost of gasification, though justifiable to reduce sulfur and other pollutants, is too high for syngas to become a widespread substitute for natural gas.

23. IGNITION TEMPERATURE

The *ignition temperature* (*autoignition temperature*) is the minimum temperature at which combustion can be sustained. It is the temperature at which more heat is generated by the combustion reaction than is lost to the surroundings, after which combustion becomes self-sustaining. For coal, the minimum ignition temperature varies from around 800°F (425°C) for bituminous varieties to 900 to 1100°F (480 to 590°C) for anthracite. For sulfur and charcoal, the ignition temperatures are approximately 470°F (240°C) and 650°F (340°C), respectively.

For gaseous fuels, the ignition temperature depends on the air/fuel ratio, temperature, pressure, and length of time the source of heat is applied. Ignition can be instantaneous or with a lag, depending on the temperature. Generalizations can be made for any gas, but the generalized temperatures will be meaningless without specifying all of these factors.

24. ATMOSPHERIC AIR

It is important to make a distinction between "air" and "oxygen." Atmospheric air is a mixture of oxygen, nitrogen, and small amounts of carbon dioxide, water vapor, argon, and other inert ("rare") gases. For the purpose of combustion calculations, all constituents except oxygen are grouped with nitrogen. It is necessary to supply 4.32 (i.e., 1/0.2315) masses of air to obtain one mass of oxygen. Similarly, it is necessary to supply 4.773 volumes of air to obtain one volume of oxygen. The average molecular weight of air is 28.97. The specific gas constant is 53.35 ft-lbf/lbm-°R (287 kJ/kg·K).

Table 22.6 Composition of Dry Air ^{a}

component	percent by weight	percent by volume
oxygen	23.15	20.95
nitrogen/inerts	76.85	79.05
ratio of nitrogen^a to oxygen	3.320	3.773^b
ratio of air to oxygen	4.320	4.773

^{a}Inert gases and N_2 included as N_2.
^{b}The value is also reported by various sources as 3.76, 3.78, and 3.784.

25. COMBUSTION REACTIONS

A limited number of elements appear in combustion reactions. Carbon, hydrogen, sulfur, hydrocarbons, and oxygen are the reactants. Carbon dioxide and water vapor are the main products, with carbon monoxide, sulfur dioxide, and sulfur trioxide occurring in lesser amounts. Nitrogen and excess oxygen emerge essentially unchanged from the stack.

Combustion reactions occur according to the normal chemical reaction principles. Balancing combustion reactions is usually easiest if carbon is balanced first, followed by hydrogen and then oxygen. When a gaseous fuel has several combustible gases, the volumetric fuel composition can be used as coefficients in the chemical equation.

Table 22.7 lists ideal combustion reactions. These reactions do not include any nitrogen or water vapor that are present in the combustion air.

Example 22.1

A gaseous fuel is 20% hydrogen and 80% methane by volume. What volume of oxygen is required to burn 120 volumes of fuel at the same conditions?

Solution

Write the unbalanced combustion reaction.

$$H_2 + CH_4 + O_2 \longrightarrow CO_2 + H_2O$$

Use the volumetric analysis as coefficients of the fuel.

$$0.2H_2 + 0.8CH_4 + O_2 \longrightarrow CO_2 + H_2O$$

Balance the carbons.

$$0.2H_2 + 0.8CH_4 + O_2 \longrightarrow 0.8CO_2 + H_2O$$

Balance the hydrogens.

$$0.2H_2 + 0.8CH_4 + O_2 \longrightarrow 0.8CO_2 + 1.8H_2O$$

Balance the oxygens.

$$0.2H_2 + 0.8CH_4 + 1.7O_2 \longrightarrow 0.8CO_2 + 1.8H_2O$$

For gaseous components, the coefficients correspond to the volumes. Since one (0.2 + 0.8) volume of fuel requires 1.7 volumes of oxygen, the required oxygen is

$$(1.7)(120 \text{ volumes of fuel}) = 204 \text{ volumes of oxygen}$$

Table 22.7 *Ideal Combustion Reactions*

fuel	formula	reaction equation (excluding nitrogen)
carbon (to CO)	C	$2C + O_2 \longrightarrow 2CO$
carbon (to CO_2)	C	$2C + 2O_2 \longrightarrow 2CO_2$
sulfur (to SO_2)	S	$S + O_2 \longrightarrow SO_2$
sulfur (to SO_3)	S	$2S + 3O_2 \longrightarrow 2SO_3$
carbon monoxide	CO	$2CO + O_2 \longrightarrow 2CO_2$
methane	CH_4	$CH_4 + 2O_2 \longrightarrow CO_2 + 2H_2O$
acetylene	C_2H_2	$2C_2H_2 + 5O_2 \longrightarrow 4CO_2 + 2H_2O$
ethylene	C_2H_4	$C_2H_4 + 3O_2 \longrightarrow 2CO_2 + 2H_2O$
ethane	C_2H_6	$2C_2H_6 + 7O_2 \longrightarrow 4CO_2 + 6H_2O$
hydrogen	H_2	$2H_2 + O_2 \longrightarrow 2H_2O$
hydrogen sulfide	H_2S	$2H_2S + 3O_2 \longrightarrow 2H_2O + 2SO_2$
propane	C_3H_8	$C_3H_8 + 5O_2 \longrightarrow 3CO_2 + 4H_2O$
n-butane	C_4H_{10}	$2C_4H_{10} + 13O_2 \longrightarrow 8CO_2 + 10H_2O$
octane	C_8H_{18}	$2C_8H_{18} + 25O_2 \longrightarrow 16CO_2 + 18H_2O$
olefin series	C_nH_{2n}	$2C_nH_{2n} + 3nO_2 \longrightarrow 2nCO_2 + 2nH_2O$
paraffin series	C_nH_{2n+2}	$2C_nH_{2n+2} + (3n + 1)O_2 \longrightarrow 2nCO_2 + (2n + 2)H_2O$

(Multiply oxygen volumes by 3.773 to get nitrogen volumes.)

26. STOICHIOMETRIC REACTIONS

Stoichiometric quantities (*ideal quantities*) are the exact quantities of reactants that are needed to complete a combustion reaction with no reactants left over. Table 22.7 contains some of the more common chemical reactions. Stoichiometric volumes and masses can always be determined from the balanced chemical reaction equation. Table 22.8 can be used to quickly determine stoichiometric amounts for some fuels.

27. STOICHIOMETRIC AIR

Stoichiometric air (*ideal air*) is the air necessary to provide the exact amount of oxygen for complete combustion of a fuel. Stoichiometric air includes atmospheric nitrogen. For each volume of oxygen, 3.773 volumes of nitrogen pass unchanged through the reaction.[22]

---

[22]The only major change in the nitrogen gas is its increase in temperature. Dissociation of nitrogen and formation of nitrogen compounds can occur but are essentially insignificant.

Stoichiometric air can be stated in units of mass (pounds or kilograms of air) for solid and liquid fuels, and in units of volume (cubic feet or cubic meters of air) for gaseous fuels. When stated in terms of mass, the stoichiometric ratio of air to fuel masses is known as the ideal *air/fuel ratio*, $R_{a/f}$.

$$R_{a/f,\text{ideal}} = \frac{m_{\text{air,ideal}}}{m_{\text{fuel}}} \qquad 22.5$$

The ideal air/fuel ratio can always be determined from the combustion reaction equation. It can also be determined by adding the oxygen and nitrogen amounts from Table 22.8.

For fuels whose ultimate analysis is known, the approximate stoichiometric air (oxygen and nitrogen) requirement in pounds of air per pound of fuel (kilograms of air per kilogram of fuel) can be quickly calculated from Eq. 22.6.[23] All oxygen in the fuel is assumed to be free moisture. All of the reported oxygen is assumed to be locked up in the form of water. Any free oxygen (i.e., oxygen dissolved in liquid fuels) is subtracted from the oxygen requirements.

$$R_{a/f,\text{ideal}} = (34.5)\left(\frac{G_C}{3} + G_H - \frac{G_O}{8} + \frac{G_S}{8}\right)$$
$$\text{[solid fuels]} \qquad 22.6$$

For fuels consisting of a mixture of gases, Eq. 22.7 and the constants J_i from Table 22.9 can be used to quickly determine the stoichiometric air requirements.

$$R_{a/f,\text{ideal}} = \sum J_i G_i \quad \text{[gaseous fuels]} \qquad 22.7$$

For fuels consisting of a mixture of gases, the air/fuel ratio can also be expressed in volumes of air per volume of fuel.

$$\text{volumetric air/fuel ratio} = \sum K_i B_i \quad \text{[gaseous fuels]}$$
$$22.8$$

---

[23]This is a "compromise" equation. Variations in the atomic weights will affect the coefficients slightly. The coefficient 34.5 is reported as 34.43 in some older books. 34.5 is the exact value needed for carbon and hydrogen, which constitute the bulk of the fuel. 34.43 is the correct value for sulfur, but the error is small and is disregarded in this equation.

Thermodynamics

Table 22.8 Consolidated Combustion Data[a,b,c,d]

fuel	for 1 mole of fuel — air O₂	N₂	other products CO₂	H₂O	SO₂	for 1 ft³ of fuel — air O₂	N₂	other products CO₂	H₂O	for 1 lbm of fuel — air O₂	N₂	other products CO₂	H₂O	SO₂	units of fuel
C carbon	1.0 379.5 32.0	3.773 1432 106	1.0 379.5 44.0							0.0833 31.63 2.667	0.3143 119.3 8.883	0.0833 31.63 3.667			moles ft³ lbm
H₂ hydrogen	0.5 189.8 16.0	1.887 716.1 53.0		1.0 379.5 18.0		0.001317 0.5 0.04216	0.004969 1.887 0.1397		0.002635 1.0 0.04747	0.248 94.12 7.936	0.9357 355.1 26.29		0.496 188.25 8.936		moles ft³ lbm
S sulfur	1.0 379.5 32.0	3.773 1432 106.0			1.0 379.5 64.06					0.03119 11.84 0.998	0.1177 44.67 3.306			0.03119 11.84 1.998	moles ft³ lbm
CO carbon monoxide	0.5 189.8 16.0	1.887 716.1 53.0	1.0 379.5 44.01			0.001317 0.5 0.04216	0.004969 1.887 0.1397	0.002635 1.0 0.1160		0.01785 6.774 0.5712	0.06735 25.56 1.892	0.03570 13.55 1.572			moles ft³ lbm
CH₄ methane	2.0 759 64.0	7.546 2864 212.0	1.0 379.5 44.01	2.0 758 36.03		0.00527 2.0 0.1686	0.01988 7.546 0.5586	0.002635 1.0 0.1160	0.00527 2.0 0.0949	0.1247 47.31 3.989	0.4705 178.5 13.21	0.06233 23.66 2.743	0.1247 47.31 2.246		moles ft³ lbm
C₂H₂ acetylene	2.5 948.8 80.0	9.433 3580 265.0	2.0 758 88.02	1.0 379.5 18.02		0.006588 2.5 0.2108	0.02486 9.443 0.6983	0.00527 2.0 0.2319	0.002635 1.0 0.04747	0.09601 36.44 3.072	0.3622 137.5 10.18	0.07681 29.15 3.380	0.03841 14.57 0.6919		moles ft³ lbm
C₂H₄ ethylene	3.0 1139 96.0	11.32 4297 318.0	2.0 758 88.02	2.0 758 36.03		0.007905 3.0 0.2951	0.02983 11.32 0.8380	0.00527 2.0 0.2319	0.00527 2.0 0.0949	0.1069 40.58 3.422	0.4033 153.1 11.34	0.07129 27.05 3.137	0.07129 27.05 1.284		moles ft³ lbm
C₂H₆ ethane	3.5 1328 112.0	13.21 5010 371.0	2.0 758 88.02	3.0 1139 54.05		0.009223 3.5 0.2951	0.03480 13.21 0.9776	0.00527 2.0 0.2319	0.007905 3.0 0.1424	0.1164 44.17 3.724	0.4392 166.7 12.34	0.06651 25.24 2.927	0.09977 37.86 1.797		moles ft³ lbm

(Multiply lbm/ft³ by 0.06243 to obtain kg/m³.)

[a] Rounding of molecular weights and air composition may introduce slight inconsistencies in the table values. This table is based on atomic weights with at least four significant digits, a ratio of 3.773 volumes of nitrogen per volume of oxygen, and 379.5 ft³ per mole at 1 atm and 60°F.

[b] Volumes per unit mass are at 1 atm and 60°F (16°C). To obtain volumes at other temperatures, multiply by (T°F + 460)/520 or (T°C + 283)/289.

[c] The volume of water applies only when the combustion products are at such high temperatures that all of the water is in vapor form.

[d] This table can be used to directly determine some SI ratios. For kg/kg ratios, the values are the same as lbm/lbm. For L/L or m³/m³, the values are the same as ft³/ft³. For mol/mol, use mole/mole. For mixed units (e.g., ft³/lbm), conversions are required.

Table 22.9 *Approximate Air/Fuel Ratio Coefficients for Components of Natural Gas[a]*

fuel component	J (gravimetric)	K (volumetric)
acetylene, C_2H_2	13.25	11.945
butane, C_4H_{10}	15.43	31.06
carbon monoxide, CO	2.463	2.389
ethane, C_2H_6	16.06	16.723
ethylene, C_2H_4	14.76	14.33
hydrogen, H_2	34.23	2.389
hydrogen sulfide, H_2S	6.074	7.167
methane, CH_4	17.20	9.556
oxygen, O_2	-4.320	-4.773
propane, C_3H_8	15.65	23.89

[a]Rounding of molecular weights and air composition may introduce slight inconsistencies in the table values. This table is based on atomic weights with at least four significant digits and a ratio of 3.773 volumes of nitrogen per volume of oxygen.

Example 22.2

Use Table 22.8 to determine the theoretical volume of 90°F (32°C) air required to burn 1 volume of 60°F (16°C) carbon monoxide to carbon dioxide.

Solution

From Table 22.8, 0.5 volumes of oxygen are required to burn 1 volume of carbon monoxide to carbon dioxide. 1.887 volumes of nitrogen accompany the oxygen. The total amount of air at the temperature of the fuel is 0.5 + 1.887 = 2.387 volumes.

This volume will expand at the higher temperature. The volume at the higher temperature is

$$\frac{(90°F + 460)(2.387 \text{ volumes})}{60°F + 460} = 2.53 \text{ volumes}$$

Example 22.3

How much air is required for the ideal combustion of (a) coal with an ultimate analysis of 93.5% carbon, 2.6% hydrogen, 2.3% oxygen, 0.9% nitrogen, and 0.7% sulfur, (b) fuel oil with a gravimetric analysis of 84% carbon, 15.3% hydrogen, 0.4% nitrogen, and 0.3% sulfur, and (c) natural gas with a volumetric analysis of 86.92% methane, 7.95% ethane, 2.81% nitrogen, 2.16% propane, and 0.16% butane?

Solution

(a) Use Eq. 22.6.

$$R_{a/f,\text{ideal}} = (34.5)\left(\frac{G_C}{3} + G_H - \frac{G_O}{8} + \frac{G_S}{8}\right)$$
$$= (34.5)\left(\frac{0.935}{3} + 0.026 - \frac{0.023}{8} + \frac{0.007}{8}\right)$$
$$= 11.58 \text{ lbm/lbm (kg/kg)}$$

(b) Use Eq. 22.6.

$$R_{a/f,\text{ideal}} = (34.5)\left(\frac{G_C}{3} + G_H + \frac{G_S}{8}\right)$$
$$= (34.5)\left(\frac{0.84}{3} + 0.153 + \frac{0.003}{8}\right)$$
$$= 14.95 \text{ lbm/lbm (kg/kg)}$$

(c) Use Eq. 22.8 and the coefficients from Table 22.9.

$$\frac{\text{volumetric}}{\text{air/fuel ratio}} = \sum K_i B_i$$
$$= (0.8692)(9.556) + (0.0795)(16.723)$$
$$+ (0.0216)(23.89) + (0.0016)(31.06)$$
$$= 10.20 \text{ ft}^3/\text{ft}^3 \text{ (m}^3/\text{m}^3)$$

28. INCOMPLETE COMBUSTION

Incomplete combustion occurs when there is insufficient oxygen to burn all of the hydrogen, carbon, and sulfur in the fuel. Without enough available oxygen, carbon burns to carbon monoxide.[24] Carbon monoxide in the flue gas indicates incomplete and inefficient combustion. Incomplete combustion is caused by cold furnaces, low combustion temperatures, poor air supply, smothering from improperly vented stacks, and insufficient mixing of air and fuel.

29. FLUE GAS ANALYSIS

Combustion products that pass through a furnace's exhaust system are known as *flue gases* (*stack gases*). Flue gases are almost all nitrogen.[25] (Nitrogen oxides are not present in large enough amounts to be included separately in combustion reactions.)

The actual composition of flue gases can be obtained in a number of ways, including by electronic detectors, less expensive "length-of-stain" detectors, and by direct sampling with an Orsat apparatus.

The *Orsat apparatus* determines the volumetric percentages of CO_2, CO, O_2, and N_2 in a flue gas. The sampled flue gas passes through a series of chemical compounds. The first compound absorbs only CO_2, the next only O_2, and the third only CO. The unabsorbed gas is assumed to be N_2 and is found by subtracting the volumetric percentages of all other components from 100%. An Orsat analysis is a dry analysis; the percentage of water vapor is not usually determined. A wet volumetric analysis (needed to compute the dew-point temperature) can be derived if the volume of water vapor is added to the Orsat volumes.

---

[24]Toxic alcohols, ketones, and aldehydes may also be formed during incomplete combustion.
[25]This assumption is helpful in making quick determinations of the thermodynamic properties of flue gases.

Modern electronic detectors can determine free oxygen (and other gases) independently of the other gases. Because the relationship between oxygen and excess air is relatively insensitive to fuel composition, oxygen measurements are replacing standard carbon dioxide measurements in determining combustion efficiency.

30. ACTUAL AND EXCESS AIR

Complete combustion occurs when all of the fuel is burned. Usually, *excess air* is required to achieve complete combustion. Excess air is expressed as a percentage of the theoretical air requirements. Different fuel types burn more efficiently with different amounts of excess air. Coal-fired boilers need approximately 30 to 35% excess air, oil-based units need about 15%, and natural gas burners need about 10%.

The actual air/fuel ratio can be estimated from the flue gas analysis and the fraction of carbon in the fuel.

$$R_{a/f,\text{actual}} = \frac{m_{\text{air,actual}}}{m_{\text{fuel}}}$$

$$= \frac{3.04 B_{\text{N}_2} G_{\text{C}}}{B_{\text{CO}_2} + B_{\text{CO}}} \qquad 22.9$$

Too much free oxygen or too little carbon dioxide in the flue gas is indicative of excess air. The relationship between excess air and oxygen in the flue gases is not highly affected by fuel composition.[26] The relationship between excess air and the volumetric fraction of oxygen in the flue gas is given in Table 22.10.

Reducing the air/fuel ratio will have several results: (1) The furnace temperature will increase due to a reduction in cooling air. (2) The flue gas will decrease in quantity. (3) The heat loss will decrease. (4) The furnace efficiency will increase. (5) Pollutants will (usually) decrease.

With a properly adjusted furnace and good mixing, the flue gas will contain no carbon monoxide, and the amount of carbon dioxide will be maximum. The stoichiometric amount of carbon dioxide in the flue gas is known as the *ultimate CO_2*. The air/fuel mixture should be adjusted until the maximum level of carbon dioxide is attained.

Table 22.10 Approximate Volumetric Percentage of Oxygen in Stack Gas

fuel[a]	excess air							
	0%	1%	5%	10%	20%	50%	100%	200%
fuel oils,								
no. 2–6	0	0.22	1.06	2.02	3.69	7.29	10.8	14.2
natural gas	0	0.25	1.18	2.23	4.04	7.83	11.4	14.7
propane	0	0.23	1.08	2.06	3.75	7.38	10.9	14.3

[a]Values for coal are only marginally lower than the values for fuel oils.

[26]The relationship between excess air and CO_2 is much more dependent on fuel type and composition.

Example 22.4

Propane (C_3H_8) is burned completely with 20% excess air. What is the volumetric fraction of carbon dioxide in the flue gas?

Solution

The balanced chemical reaction equation is

$$\text{C}_3\text{H}_8 + 5\text{O}_2 \longrightarrow 3\text{CO}_2 + 4\text{H}_2\text{O}$$

With 20% excess air, the oxygen volume is $(1.2)(5) = 6$.

$$\text{C}_3\text{H}_8 + 6\text{O}_2 \longrightarrow 3\text{CO}_2 + 4\text{H}_2\text{O} + \text{O}_2$$

From Table 22.6, there are 3.773 volumes of nitrogen for every volume of oxygen.

$$(6)(3.773) = 22.6$$

$$\text{C}_3\text{H}_8 + 6\text{O}_2 + 22.6\text{N}_2 \longrightarrow 3\text{CO}_2 + 4\text{H}_2\text{O} + \text{O}_2 + 22.6\text{N}_2$$

For gases, the coefficients can be interpreted as volumes. The volumetric fraction of carbon dioxide is

$$B_{\text{CO}_2} = \frac{3}{3 + 4 + 1 + 22.6}$$

$$= 0.0980 \ (9.8\%)$$

31. CALCULATIONS BASED ON FLUE GAS ANALYSIS

Equation 22.10 gives the approximate percentage (by volume) of actual excess air.

$$\text{actual excess air,} \atop \% \text{ by volume} = \frac{(100\%)(B_{\text{O}_2} - 0.5 B_{\text{CO}})}{0.264 B_{\text{N}_2} - B_{\text{O}_2} + 0.5 B_{\text{CO}}} \qquad 22.10$$

The ultimate CO_2 (i.e., the maximum theoretical carbon dioxide) can be determined from Eq. 22.11.

$$\text{ultimate CO}_2, \atop \% \text{ by volume} = \frac{(100\%) B_{\text{CO}_2,\text{actual}}}{1 - 4.773 B_{\text{O}_2,\text{actual}}} \qquad 22.11$$

The mass ratio of dry flue gases to solid fuel is given by Eq. 22.12.

$$\frac{\text{mass of flue gas}}{\text{mass of solid fuel}} = \frac{\left[11 B_{\text{CO}_2} + 8 B_{\text{O}_2} + (7)(B_{\text{CO}} + B_{\text{N}_2})\right]\left(G_{\text{C}} + \dfrac{G_{\text{S}}}{1.833}\right)}{(3)(B_{\text{CO}_2} + B_{\text{CO}})} \qquad 22.12$$

Example 22.5

A coal has a proximate analysis of 75% carbon. The volumetric analysis of the flue gas is 80.2% nitrogen, 12.6% carbon dioxide, 6.2% oxygen, and 1.0% carbon monoxide. Find the (a) actual air/fuel ratio, (b) percentage excess air, (c) ultimate carbon dioxide, and (d) mass of flue gas per mass of fuel.

Solution

(a) Use Eq. 22.9.

$$R_{a/f,\text{actual}} = \frac{3.04 B_{N_2} G_C}{B_{CO_2} + B_{CO}}$$

$$= \frac{(3.04)(0.802)(0.75)}{0.126 + 0.01}$$

$$= 13.4 \text{ lbm air/lbm fuel (kg air/kg fuel)}$$

(b) Use Eq. 22.10.

$$\frac{\text{actual}}{\text{excess air}} = \frac{(100\%)(B_{O_2} - 0.5 B_{CO})}{0.264 B_{N_2} - B_{O_2} + 0.5 B_{CO}}$$

$$= \frac{(100\%)[(0.062 - (0.5)(0.01)]}{(0.264)(0.802) - 0.062 + (0.5)(0.01)}$$

$$= 36.8\% \text{ by volume}$$

(c) Use Eq. 22.11.

$$\text{ultimate CO}_2 = \frac{(100\%) B_{CO_2,\text{actual}}}{1 - 4.773 B_{O_2,\text{actual}}}$$

$$= \frac{(100\%)(0.126)}{1 - (4.773)(0.062)}$$

$$= 17.9\% \text{ by volume}$$

(d) Use Eq. 22.12.

$$\frac{\text{mass of flue gas}}{\text{mass of solid fuel}} = \frac{\begin{bmatrix} 11 B_{CO_2} + 8 B_{O_2} \\ + (7)(B_{CO} + B_{N_2}) \end{bmatrix}\left(G_C + \dfrac{G_S}{1.833}\right)}{(3)(B_{CO_2} + B_{CO})}$$

$$= \frac{\begin{bmatrix} (11)(0.126) + (8)(0.062) \\ + (7)(0.01 + 0.802)\end{bmatrix}(0.75)}{(3)(0.126 + 0.01)}$$

$$= 13.9 \text{ lbm flue gas/lbm fuel} \\ \text{(kg flue gas/kg fuel)}$$

32. TEMPERATURE OF FLUE GAS

The temperature of the gas at the furnace outlet—before the gas reaches any other equipment—should be approximately 550°F (300°C). Overly low temperatures mean there is too much excess air. Overly high temperatures—above 750°F (400°C)—mean that heat is being wasted to the atmosphere and indicate other problems (ineffective heat transfer surfaces, overfiring, defective combustion chamber, etc.).

The *net stack temperature* is the difference between the stack and local environment temperatures. The net stack temperature should be as low as possible without causing corrosion of the low end.

33. SMOKE

The amount of smoke can be used as an indicator of combustion cleanliness. Smoky combustion may indicate improper air/fuel ratio, insufficient draft, leaks, insufficient preheat, or misadjustment of the fuel system.

Smoke measurements are made in a variety of ways, with the standards depending on the equipment used. Photoelectric sensors in the stack may be used to continuously monitor smoke. The *smoke spot number* (SSN) and ASTM smoke scale are used with continuous stack monitors. For coal-fired furnaces, the maximum desirable smoke number is SSN 4. For grade 2 fuel oil, the SSN should be less than 1; for grade 4, SSN 4; for grades 5L, 5H, and low-sulfur residual fuels, SSN 3; for grade 6, SSN 4.

The *Ringelman scale* is a subjective method in which the smoke density is visually compared to five standardized white-black grids. Ringelman chart no. 0 is solid white; chart no. 5 is solid black. Ringelman chart no. 1, which is 20% black, is the preferred operating point for most power plants.

34. DEW POINT OF FLUE GAS MOISTURE

The *dew point* is the temperature at which the water vapor in the flue gas begins to condense in a constant pressure process. To avoid condensation and corrosion in the stack, the temperature of the flue gases must be above the dew point.

Dalton's law predicts the dew point of moisture in the flue gas. The partial pressure of the water vapor depends on the mole fraction (i.e., the volumetric fraction) of water vapor. The higher the water vapor pressure, the higher the dew point. Air entering a furnace can also contribute to moisture in the flue gas. This moisture should be added to the water vapor from combustion when calculating the mole fraction.

$$\text{partial pressure} = (\text{water vapor mole fraction})$$
$$\times (\text{flue gas pressure}) \qquad 22.13$$

Once the water vapor's partial pressure is known, the dew point can be found from the steam tables as the saturation temperature corresponding to the partial pressure.

When there is no sulfur in the fuel, the dew point is typically around 100°F (40°C). The presence of sulfur in virtually any quantity increases the actual dew point to approximately 300°F (150°C).[27]

---

[27] The theoretical dew point is even higher—up to 350 to 400°F (175 to 200°C). For complex reasons, the theoretical value is not attained.

35. HEAT OF COMBUSTION

The *heating value* of a fuel can be determined experimentally in a *bomb calorimeter*, or it can be estimated from the fuel's chemical analysis. The *higher heating value*, HHV (or *gross heating value*), of a fuel includes the heat of vaporization (condensation) of the water vapor formed from hydrogen during combustion. The *lower heating value*, LHV (or *net heating value*), assumes that all the products of combustion remain gaseous. The LHV is generally the value to use in calculations of thermal energy generated, since the heat of vaporization is not recovered in the stack system.

Traditionally, heating values have been reported on an HHV basis for coal-fired systems but on an LHV basis for natural gas-fired combustion turbines. There is an 11% difference between HHV and LHV thermal efficiencies for gas-fired systems and a 4% difference for coal-fired systems.

The HHV can be calculated from the LHV if the enthalpy of vaporization, h_{fg}, is known at the pressure of the water vapor.[28] In Eq. 22.14, m_{water} is the mass of water produced per unit (pound, mole, m³, etc.) of fuel.

$$HHV = LHV + m_{water}h_{fg} \qquad 22.14$$

Only the hydrogen that is not locked up with oxygen in the form of water is combustible. This is known as the *available hydrogen*. The correct percentage of combustible hydrogen, $G_{H,available}$, is calculated from the hydrogen and oxygen fraction. Equation 22.15 assumes that all of the oxygen is present in the form of water.

$$G_{H,available} = G_{H,total} - \frac{G_O}{8} \qquad 22.15$$

Dulong's formula calculates the higher heating value of coals and coke with a 2 to 3% accuracy for moisture contents below approximately 10%.[29] The gravimetric or volumetric analysis percentages for each combustible element (including sulfur) are multiplied by the heating value per unit (mass or volume) from App. 22.A and summed.

$$HHV_{MJ/kg} = 32.78G_C + (141.8)\left(G_H - \frac{G_O}{8}\right) + 9.264G_S$$
$$\text{[SI]} \qquad 22.16(a)$$

$$HHV_{Btu/lbm} = 14{,}093G_C + (60{,}958)\left(G_H - \frac{G_O}{8}\right) + 3983G_S$$
$$\text{[U.S.]} \qquad 22.16(b)$$

[28]For the purpose of initial studies, the heat of vaporization is usually assumed to be 1040 Btu/lbm (2.42 kJ/kg). This corresponds to a partial pressure of approximately 1 psia (7 kPa) and a dew point of 100°F (38°C).
[29]The coefficients in Eq. 22.16 are slightly different from the coefficients originally proposed by Dulong. Equation 22.16 reflects currently accepted heating values that were unavailable when Dulong developed his formula. Equation 22.16 makes the following assumptions: (1) None of the oxygen is in carbonate form. (2) There is no free oxygen. (3) The hydrogen and carbon are not combined as hydrocarbons. (4) Carbon is amorphous, not graphitic. (5) Sulfur is not in sulfate form. (6) Sulfur burns to sulfur dioxide.

The higher heating value of gasoline can be approximated from the Baumé specific gravity.

$$HHV_{gasoline,MJ/kg} = 42.61 + (0.093)(°Baumé - 10)$$
$$\text{[SI]} \qquad 22.17(a)$$

$$HHV_{gasoline,Btu/lbm} = 18{,}320 + (40)(°Baumé - 10)$$
$$\text{[U.S.]} \qquad 22.17(b)$$

The heating value of petroleum oils (including diesel fuel) can also be approximately determined from the oil's specific gravity. The values derived from Eq. 22.18 may not exactly agree with values for specific oils because the equation does not account for refining methods and sulfur content. Equation 22.18 was originally intended for combustion at constant volume, as in a gasoline engine. However, the effect is very small, and Eq. 22.18 is widely used as an approximation for all types of combustion, including constant pressure combustion in industrial boilers.

$$HHV_{fuel oil,MJ/kg} = 51.92 - (8.792)(SG)^2$$
$$\text{[SI]} \qquad 22.18(a)$$

$$HHV_{fuel oil,Btu/lbm} = 22{,}320 - (3780)(SG)^2$$
$$\text{[U.S.]} \qquad 22.18(b)$$

Example 22.6

A coal has an ultimate analysis of 93.9% carbon, 2.1% hydrogen, 2.3% oxygen, 0.3% nitrogen, and 1.4% ash. What are its (a) higher and (b) lower heating values in Btu/lbm?

Solution

(a) The noncombustible ash, oxygen, and nitrogen do not contribute to heating value. Some of the hydrogen is in the form of water. From Eq. 22.2, the available hydrogen is

$$G_{H,available} = G_{H,total} - \frac{G_O}{8}$$

$$= 2.1\% - \frac{2.3\%}{8} = 1.8\%$$

From App. 22.A, the higher heating values of carbon and hydrogen are 14,093 Btu/lbm and 60,958 Btu/lbm, respectively. From Eq. 22.16, the total heating value per pound of coal is

$$HHV = (0.939)\left(14{,}093\ \frac{Btu}{lbm}\right)$$

$$+ (0.018)\left(60{,}958\ \frac{Btu}{lbm}\right)$$

$$= 14{,}331\ Btu/lbm$$

(b) All of the combustible hydrogen forms water vapor. The mass of water produced is the hydrogen mass and eight times the oxygen mass.

$$m_{\text{water}} = m_{\text{H,available}} + m_{\text{oxygen}}$$
$$= 0.018 + (8)(0.018)$$
$$= 0.162 \text{ lbm water/lbm coal}$$

Assume that the partial pressure of the water vapor is approximately 1 psia (7 kPa). Then, the heat of condensation will be 1040 Btu/lbm. From Eq. 22.14, the lower heating value is approximately

$$\text{LHV} = \text{HHV} - m_{\text{water}} h_{fg}$$
$$= 14{,}331 \, \frac{\text{Btu}}{\text{lbm}} - \left(0.162 \, \frac{\text{lbm}}{\text{lbm}}\right)\left(1040 \, \frac{\text{Btu}}{\text{lbm}}\right)$$
$$= 14{,}163 \text{ Btu/lbm}$$

Alternatively, the lower heating value of the coal can be calculated from Eq. 22.16 by substituting the lower (net) hydrogen heating value from App. 22.A, 51,623 Btu/lbm, for the gross heating value of 60,598 Btu/lbm. This yields 14,160 Btu/lbm.

36. MAXIMUM THEORETICAL COMBUSTION (FLAME) TEMPERATURE

It can be assumed that the maximum theoretical increase in flue gas temperature will occur if all of the combustion energy is absorbed adiabatically by the combustion products. This provides a method of estimating the *maximum theoretical combustion temperature*, also sometimes called the *maximum flame temperature* or *adiabatic flame temperature*.[30]

In Eq. 22.19, the mass of the products is the sum of the fuel, oxygen, and nitrogen masses for stoichiometric combustion. The mean specific heat is a gravimetrically weighted average of the values of c_p for all combustion gases. (Since nitrogen comprises the majority of the combustion gases, the mixture's specific heat will be approximately that of nitrogen.) The heat of combustion can be found either from the lower heating value, LHV, or from a difference in air enthalpies across the furnace.

$$T_{\text{max}} = T_i + \frac{\text{heat of combustion}}{m_{\text{products}} c_{p,\text{mean}}} \qquad \text{22.19}$$

Actual flame temperatures are always lower than the theoretical temperature. Most fuels produce flame temperatures in the range of 3350 to 3800°F (1850 to 2100°C).

37. COMBUSTION LOSSES

A portion of the combustion energy is lost in heating the dry flue gases, dfg.[31] This is known as *dry flue gas*

---

[30]Flame temperature is limited by the dissociation of common reaction products (CO_2, N_2, etc.). At high enough temperatures (3400 to 3800°F; 1880 to 2090°C), the endothermic dissociation process reabsorbs combustion heat and the temperature stops increasing. The temperature at which this occurs is known as the *maximum flame temperature*. This definition of flame temperature is not a function of heating values and flow rates.

[31]The abbreviation "dfg" for dry flue gas is peculiar to the combustion industry. It may not be recognized outside of that field.

loss. In Eq. 22.20, $m_{\text{flue gas}}$ is the mass of dry flue gas per pound of fuel. It can be estimated from Eq. 22.9. Although the full temperature difference is used, the specific heat should be evaluated at the average temperature of the flue gas. For quick estimates, the dry flue gas can be assumed to be pure nitrogen.

$$q_1 = m_{\text{flue gas}} c_p (T_{\text{flue gas}} - T_{\text{incoming air}}) \qquad \text{22.20}$$

Heat is lost in the vapor formed during the combustion of hydrogen. In Eq. 22.21, m_{vapor} is the mass of vapor per pound of fuel. G_H is the gravimetric fraction of hydrogen in the fuel. h_g is the enthalpy of superheated steam at the flue gas temperature and the partial pressure of the water vapor. h_f is the enthalpy of saturated liquid at the air's entrance temperature.

$$q_2 = m_{\text{vapor}}(h_g - h_f) = 8.94 G_H (h_g - h_f) \qquad \text{22.21}$$

Heat is lost when it is absorbed by moisture originally in the combustion air (and by free moisture in the fuel, if any). In Eq. 22.22, $m_{\text{combustion air}}$ is the mass of combustion air per pound of fuel. ω is the humidity ratio. h'_g is the enthalpy of superheated steam at the air's entrance temperature and partial pressure of the water vapor.

$$q_3 = m_{\text{atmospheric water vapor}}(h_g - h'_g)$$
$$= \omega m_{\text{combustion air}}(h_g - h'_g) \qquad \text{22.22}$$

When carbon monoxide appears in the flue gas, potential energy is lost in incomplete combustion. The difference in the two heating values in Eq. 22.23 is 9746 Btu/lbm (22.67 MJ/kg).[32]

$$q_4 = \frac{(\text{HHV}_C - \text{HHV}_{CO}) G_C B_{CO}}{B_{CO_2} + B_{CO}} \qquad \text{22.23}$$

For solid fuels, potential energy is lost in unburned carbon in the ash. (Some carbon may be carried away in the flue gas, as well.) This is known as *combustible loss* or *unburned fuel loss*. In Eq. 22.24, m_{ash} is the mass of ash produced per pound of fuel consumed. $G_{C,\text{ash}}$ is the gravimetric fraction of carbon in the ash. The heating value of carbon is 14,093 Btu/lbm (32.8 MJ/kg).

$$q_5 = \text{HHV}_C m_{\text{ash}} G_{C,\text{ash}} \qquad \text{22.24}$$

Energy is also lost through radiation from boiler surfaces. This can be calculated if enough information is known. The *radiation loss* is fairly insensitive to different firing rates, and once calculated, it can be considered constant for different conditions.

Other conditions where energy can be lost include air leaks, poor pulverizer operation, excessive blowdown, steam leaks, missing or loose insulation, and excessive soot-blower operation. Losses due to these sources must be evaluated on a case-by-case basis.

---

[32]Obsolete values of 10,150 Btu/lbm (23.61 MJ/kg) and 10,190 Btu/lbm (23.70 MJ/kg) are still encountered for the difference in heating values between carbon and carbon monoxide.

38. COMBUSTION EFFICIENCY

The *combustion efficiency* (also referred to as *boiler efficiency*, *furnace efficiency*, and *thermal efficiency*) is the overall thermal efficiency of the combustion reaction. Coal and fuel oil furnaces with air heaters and economizers have 75 to 85% efficiencies. Gas furnaces have 75 to 80% efficiencies.

In Eq. 22.25, m_{steam} is the mass of steam produced per pound of fuel burned. The useful heat may also be determined from the boiler rating. One *boiler horsepower* is equal to approximately 33,475 Btu/hr (9.808 kW).[33]

$$
\eta = \frac{\text{useful heat extracted}}{\text{heating value}}
$$

$$
= \frac{m_{steam}(h_{steam} - h_{feedwater})}{\text{LHV}} \qquad 22.25
$$

Calculating the efficiency by subtracting all known losses is known as the *loss method*. Minor sources of thermal energy such as the entering air and feedwater are essentially disregarded.

$$
\eta = \frac{\text{LHV} - q_1 - q_2 - q_3 - q_4 - q_5 - \text{radiation}}{\text{LHV}} \qquad 22.26
$$

Combustion efficiency can be improved by decreasing either the temperature or the volume of the flue gas, or both. Since the latent heat of moisture is a loss, and since the amount of moisture generated corresponds to the hydrogen content of the fuel, a minimum efficiency loss due to moisture formation cannot be eliminated. This minimum loss is approximately 13% for natural gas, 8% for oil, and 6% for coal.

39. DRAFT

The amount of air that flows through a furnace is determined by the furnace draft. *Draft* is the difference in static pressures that causes the flue gases to flow.[34] It is usually stated in inches of water (kPa). *Natural draft* (ND) furnaces rely on the *stack effect* (*chimney draft*) to draw off combustion gases. Air flows through the furnace and out the stack due to the pressure differential caused by reduced densities in the stack.[35]

Forced draft (FD) fans located before the furnace are used to supply air for burning. Combustion occurs under pressure, hence the descriptive term "pressure-fired unit" for such fans. FD fans are run at relatively high speeds (1200 to 1800 rpm) with direct-drive motors. Two or more fans are used in parallel to provide for efficient operation at low furnace demand.

FD fans create a positive pressure (e.g., 2 to 10 in wg; 0.5 to 2.5 kPa). This pressure is reduced to a very small negative pressure after passing through the air heater, ducts, and windbox system. The negative pressure in the furnace serves to keep combustion gases from leaking out into the furnace and boiler areas. The pressure continues to drop as it passes though the boiler, economizer, air heater, and pollution-control equipment.

Whereas FD fans force air through the system, *induced draft* (ID) fans are used to draw combustion products through the furnace bed, stack, and pollution control system by injecting air into the stack after combustion.

The term "suction units" is used with ID fans. ID fans are located after dust collectors and precipitators (often at the base of the stack) in order to reduce the abrasive effects of fly ash. They are run at slower speeds than forced draft fans in order to reduce the abrasive effects even further. Unlike FD fans, ID fans are usually very large and powerful because they have to handle all of the combustion gases, not just combustion air.

Some stack systems operate in a condition of *balanced draft*. Balanced draft is the term used when the static pressure is equal to atmospheric pressure. This requires the use of both ID and FD fans. In order to keep combustion products inside the combustion chamber and stack system, balanced draft systems may actually operate with a slight negative pressure.

Modern welded stacks are essentially "air-tight" and can operate at pressures above atmospheric, often eliminating the need for ID fans.

The *draft loss* is the static pressure drop due to friction through the boiler and stack. The *available draft* is the difference between the theoretical draft (stack effect) and the draft loss. The available draft is zero in a balanced system.

$$
D_{available} = D_{theoretical} - D_{friction} \qquad 22.27
$$

The *net rating* or *fan boost* of the fan is the total pressure (the difference of draft losses and draft gains) supplied by the fan at maximum operating conditions. The fan power for ID and FD fans is calculated as with any fan application—from the net rating and the flow rate. (See Chap. 20.) It is customary in sizing ID and FD draft fans to include increases of approximately 15% for flow rate, 30% for pressure, and 20°F (10°C) for temperature.

40. STACK EFFECT

Stack effect (*chimney action* or *natural draft*) is a pressure difference caused by the difference in air and flue gas densities. It can be relied on to draw combustion products at least partially up through the stack. The stack effect is determined with no flow of flue gas. With flow, some of the stack effect is converted to velocity head, and the remainder is used to overcome friction.

---

[33]Boiler horsepower is sometimes equated with a gross heating rate of 44,633 Btu/hr (13.08 kW). However, this is the total incoming heating value assuming a standardized 75% combustion efficiency.

[34]The British term "draught" is synonymous with "draft."

[35]Chimneys that rely on natural draft are sometimes referred to as *gravity chimneys*.

Generally, the higher the chimney, the greater the stack effect. However, the stack effect in modern plants is greatly reduced by the friction and cooling effects of economizers, air heaters, and precipitators. Fans supply the needed pressure differential. Therefore, the primary function of modern stacks is to carry the combustion products a sufficient distance upward to dilute the combustion products. Modern stacks are seldom built higher than 200 ft (60 m), and most are less than half that value.

The theoretical stack effect, $D_{\text{theoretical}}$, is calculated from the densities of the flue gas and atmospheric air. In Eq. 22.28, H_{stack} is the total height of the stack. The average temperature is used to determine the average density in Eq. 22.28.[36]

$$D_{\text{theoretical}} = H_{\text{stack}}(\gamma_{\text{air}} - \gamma_{\text{flue gas,ave}}) \qquad 22.28$$

If a few assumptions are made, the stack effect can be calculated from the average temperature of the flue gas (i.e., the temperature at the average stack elevation).[37] The average flue gas temperature is the temperature halfway up the stack. H_{stack} is the total height of the stack.

$$D_{\text{theoretical}} = \left(\frac{p_{\text{air}}H_{\text{stack}}g}{R_{\text{air}}}\right)\left(\frac{1}{T_{\text{air}}} - \frac{1}{T_{\text{flue gas,ave}}}\right)$$
$$\text{[SI]} \qquad 22.29(a)$$

$$D_{\text{theoretical}} = \left(\frac{p_{\text{air}}H_{\text{stack}}g}{R_{\text{air}}g_c}\right)\left(\frac{1}{T_{\text{air}}} - \frac{1}{T_{\text{flue gas,ave}}}\right)$$
$$\text{[U.S.]} \qquad 22.29(b)$$

The *stack effect head*, h_{SE}, is used to calculate the ideal stack velocity. The coefficient of velocity, C_v, is approximately 0.30 to 0.50.

$$h_{\text{SE}} = \frac{D_{\text{theoretical}}}{\gamma_{\text{flue gas,ave}}} \qquad 22.30$$

$$\text{v} = C_v\sqrt{2gh_{\text{SE}}} \qquad 22.31$$

The required stack area is

$$A_{\text{stack}} = \frac{Q_{\text{flue gas}}}{\text{v}} \qquad 22.32$$

---

[36]The actual stack effect will be less than the value calculated in Eq. 22.28. For realistic problems, the achievable stack effect probably should be considered to be 80% of the ideal.
[37]One assumption is that the pressure in the stack is equal to atmospheric pressure. Then, the density difference will be due to only the temperature difference. Also, the flue gas is assumed to be air. This makes it unnecessary to know the flue gas composition, molecular weight, and so on.

Example 22.7

The air surrounding an 80 ft (24 m) tall vertical stack is at 70°F (21°C) and one standard atmospheric pressure. The temperature of the flue gases at the average stack elevation is 400°F (200°C). What is the theoretical natural draft?

SI Solution

The absolute temperatures are

$$T_{\text{air}} = 21°C + 273 = 294\text{K}$$
$$T_{\text{flue gases}} = 200°C + 273 = 473\text{K}$$

Calculate the stack effect from Eq. 22.29.

$$D_{\text{theoretical}} = \left(\frac{p_{\text{air}}H_{\text{stack}}g}{R_{\text{air}}}\right)\left(\frac{1}{T_{\text{air}}} - \frac{1}{T_{\text{flue gas}}}\right)$$
$$= \left[\frac{(101.3\,\text{kPa})(24\,\text{m})\left(9.81\,\dfrac{\text{m}}{\text{s}^2}\right)}{287\,\dfrac{\text{J}}{\text{kg·K}}}\right]$$
$$\times \left(\frac{1}{294\text{K}} - \frac{1}{473\text{K}}\right)$$
$$= 0.107\,\text{kPa}$$

Customary U.S. Solution

The absolute temperatures are

$$T_{\text{air}} = 70°F + 460 = 530°R$$
$$T_{\text{flue gases}} = 400°F + 460 = 860°R$$

Calculate the stack effect from Eq. 22.29.

$$D_{\text{theoretical}} = \left(\frac{p_{\text{air}}h_{\text{stack}}g}{R_{\text{air}}g_c}\right)\left(\frac{1}{T_{\text{air}}} - \frac{1}{T_{\text{flue gas}}}\right)$$
$$= \left[\frac{\left(14.7\,\dfrac{\text{lbf}}{\text{in}^2}\right)\left(144\,\dfrac{\text{in}^2}{\text{ft}^2}\right)(80\,\text{ft})}{53.3\,\dfrac{\text{ft-lbf}}{\text{lbm-}°R}}\right]$$
$$\times \left(\frac{1}{530°R} - \frac{1}{860°R}\right)\left(\frac{32.2\,\dfrac{\text{ft}}{\text{sec}^2}}{32.2\,\dfrac{\text{ft-lbm}}{\text{lbf-sec}^2}}\right)$$
$$= 2.3\,\text{lbf/ft}^2$$

Convert this to inches of water.

$$D_{\text{theoretical}} = \frac{\left(2.3\,\dfrac{\text{lbf}}{\text{ft}^2}\right)\left(12\,\dfrac{\text{in}}{\text{ft}}\right)}{62.4\,\dfrac{\text{lbf}}{\text{ft}^3}}$$
$$= 0.44\,\text{in wg}$$

41. STACK FRICTION

Flue gas velocities of 15 to 40 ft/sec (4.5 to 12 m/s) are typical. As with most fluids, the friction pressure drop through stack "piping" is proportional to the velocity head.[38]

$$D_{\text{friction}} = \frac{Kv^2}{2g} \qquad 22.33$$

Values of the friction coefficient, K, for chimney fittings are similar to those used in ventilating ductwork (Chap. 20). Values for specific pieces of equipment (burners, pressure regulators, vents, etc.) must be provided by the manufacturers.

The approximate friction loss coefficient for a circular section of ductwork or chimney of length L and diameter d is given by Eq. 22.34. Equation 22.34 assumes a Darcy friction factor of 0.0233. For other friction factors, the constant terms can be scaled up or down proportionately. For noncircular ducts, the diameter can be replaced by four times the hydraulic radius (i.e., four times the duct area divided by the duct perimeter).

$$D_{\text{friction,kPa}} = \frac{(0.119)L_m \rho_{\text{kg/m}^3}(Q_{\text{L/s}}^2)}{d_{\text{cm}}^5}$$

$$[\text{SI}] \qquad 22.34(a)$$

$$D_{\text{friction,in wg}} = \frac{(48)L_{\text{ft}} \gamma_{\text{lbf/ft}^3}(Q_{\text{cfs}}^2)}{d_{\text{inches}}^5}$$

$$[\text{U.S.}] \qquad 22.34(b)$$

PRACTICE PROBLEMS

1. Methane with a heating value of 24,000 Btu/lbm (55.8 MJ/kg) is burned in a furnace with a 50% efficiency. How much water can be heated from 60 to 200°F (15 to 95°C) when 7 ft^3 (200 L) at gas industry standard conditions are burned?

2. 15 lbm/hr (6.8 kg/h) of propane (C_3H_8) are burned stoichiometrically in air. What volume of dry carbon dioxide (CO_2) is formed after cooling to 70°F (21°C) and 14.7 psia (101 kPa)?

3. In a particular installation, 30% excess air at 15 psia (103 kPa) and 100°F (40°C) is needed for the combustion of methane. How much nitrogen passes through the furnace if methane is burned at the rate of 4000 ft^3/hr (31 L/s)?

4. How much air is required to completely burn one unit mass of a fuel that is 84% carbon, 15.3% hydrogen, 0.3% sulfur, and 0.4% nitrogen by weight?

5. Propane (C_3H_8) is burned with 20% excess air. What is the gravimetric percentage of carbon dioxide in the flue gas?

6. The ultimate analysis of a coal is 80% carbon, 4% hydrogen, 2% oxygen, and the rest ash. The flue gases

are 60°F and 14.7 psia (15.6°C and 101.3 kPa) when sampled, and are 80% nitrogen, 12% carbon dioxide, 7% oxygen, and 1% carbon monoxide by volume. How much air is required to burn 1 lbm (1 kg) of coal under these conditions?

7. What is the approximate heating value of an oil with a specific gravity of 40° API?

8. The ultimate analysis of a coal is 75% carbon, 5% hydrogen, 3% oxygen, 2% nitrogen, and the rest ash. Atmospheric air is 60°F (16°C) and standard pressure. (a) What is the theoretical temperature of the combustion products? (b) Estimate the actual temperature of the combustion products. State your assumptions.

9. A fuel oil has the following ultimate analysis: 85.43% carbon, 11.31% hydrogen, 2.7% oxygen, 0.34% sulfur, and 0.22% nitrogen. The oil is burned with 60% excess air. Evaluate flue gas volumes at 600°F (320°C). (a) What volume of wet flue gases will be produced? (b) What volume of dry flue gases will be produced? (c) What will be the volumetric fraction of carbon dioxide?

10. The ultimate analysis of a coal is 51.45% carbon, 16.69% ash, 15.71% moisture, 7.28% oxygen, 4.02% hydrogen, 3.92% sulfur, and 0.93% nitrogen. 15,395 lbm (6923 kg) of the coal are burned, and 2816 lbm (1267 kg) of ash containing 20.9% carbon (by weight) are recovered. 13.3 lbm (kg) of dry gases are produced per pound (kilogram) of fuel burned. How much air was used per unit mass of fuel?

11. A coal is 65% carbon by weight. During combustion, 3% of the coal is lost in the ash pit. Combustion uses 9.87 lbm (kg) of air per pound (kg) of fuel. The flue gas analysis is 81.5% nitrogen, 9.5% carbon dioxide, and 9% oxygen. What is the percentage of excess air by mass?

12. A coal has an ultimate analysis of 67.34% carbon, 4.91% oxygen, 4.43% hydrogen, 4.28% sulfur, 1.08% nitrogen, and the rest ash. 3% of the carbon is lost during combustion. The flue gases are 81.9% nitrogen, 15.5% carbon dioxide, 1.6% carbon monoxide, and 1% oxygen by volume. What is the heat loss due to the formation of carbon monoxide?

13. A natural gas is 93% methane, 3.73% nitrogen, 1.82% hydrogen, 0.45% carbon monoxide, 0.35% oxygen, 0.25% ethylene, 0.22% carbon dioxide, and 0.18% hydrogen sulfide by volume. The gas is burned with 40% excess air. Atmospheric air is 60°F and standard atmospheric pressure. (a) What is the gas density? (b) What are the theoretical air requirements? (c) What is the wet flue gas analysis? (d) What is the dry flue gas analysis?

14. Coal enters a steam generator at 73°F (23°C). The coal has an ultimate analysis of 78.42% carbon, 8.25% oxygen, 5.68% ash, 5.56% hydrogen, 1.09% nitrogen, and 1.0% sulfur. The coal's heating value is 14,000 Btu/lbm (32.6 MJ/kg). 7.03% of the coal's weight is lost in the ash pit. The ash contains 31.5% carbon.

---

[38]The term *piping* is used to mean the stack passages.

Air at 67°F (19°C) wet bulb and 73°F (23°C) dry bulb is supplied. The flue gases consist of 80.08% nitrogen, 14.0% carbon dioxide, 5.5% oxygen, and 0.42% carbon monoxide. The flue gases are at a temperature of 575°F (300°C). Saturated water at 212°F (100°C) and 1.0 atm enters the steam generator. 11.12 lbm (kg) of water are evaporated in the boiler per pound (kilogram) of dry coal consumed. What is the complete heat balance, enumerating all combustion losses in the furnace?

15. Coal has a gravimetric analysis of 83% carbon, 5% hydrogen, 5% oxygen, and 7% noncombustible matter. 10% of the fired coal mass, including all of the noncombustible matter, is recovered in the ash pit. 26 lbm (26 kg) of air at 70°F (21°C) and 1 atmosphere are used per lbm (kg) of coal burned. When loaded into the furnace, the coal temperature is 60°F (15.6°C). The stack gas from coal combustion has a temperature of 550°F (290°C). What percentage of the combustion heat is carried away by the stack gases?

16. (*Time limit: one hour*) A utility boiler burns coal with an ultimate analysis of 76.56% carbon, 7.7% oxygen, 6.1% silicon, 5.5% hydrogen, 2.44% sulfur, and 1.7% nitrogen. 410 lbm/hr of refuse are removed with a composition of 30% carbon and 0% sulfur. All sulfur and the remaining carbon is burned. The power plant has the following characteristics.

- coal feed rate: 15,300 lbm/hr

- electric power rating: 17 MW

- generator efficiency: 95%

- steam generator efficiency: 86%

- cooling water rate: 225 ft³/sec

(a) What is the emission rate of solid particulates in lbm/hr? (b) How much sulfur dioxide is produced per hour? (c) What is the temperature rise of the cooling water? (d) What efficiency must the flue gas particulate collectors have in order to meet a limit of 0.1 lbm of particulates per million Btu per hour (0.155 kg/MW)?

17. (*Time limit: one hour*) 250 SCFM (118 L/s) of propane are mixed with an oxidizer consisting of 60% oxygen and 40% nitrogen by volume in a proportion allowing 40% excess oxygen by weight. The maximum velocity for the two reactants when combined is 400 ft/min (2 m/s) at 14.7 psia (101 kPa) and 80°F (27°C). Maximum velocity for the products is 800 ft/min

(4 m/s) at 8 psia (55 kPa) and 460°F (240°C). (a) Size the inlet pipe. (b) Size the stack. (c) What is the actual flow of oxygen? (d) What is the volume of flue gases? (e) What is the dew point of the flue gases?

18. (*Time limit: one hour*) An industrial process uses hot gas at 3600°R (1980°C) and 14.7 psia (101 kPa). It is proposed that propane be burned stoichiometrically in a mixture of nitrogen and oxygen. After passing through the process, gas will be exhausted through a duct, being cooled slowly to 100°F (38°C) and 14.7 psia (101 kPa) before discharge. The following data are available.

- The enthalpies of formation (at the standard reference temperature) are

$$C_3H_8(g) \quad \Delta H_f = +28,800 \text{ Btu/lbmol}$$
$$(+67.0 \text{ GJ/kmol})$$

$$CO_2(g) \quad \Delta H_f = -169,300 \text{ Btu/lbmol}$$
$$(-393.8 \text{ GJ/kmol})$$

$$H_2O(g) \quad \Delta H_f = -104,040 \text{ Btu/lbmol}$$
$$(-242 \text{ GJ/kmol})$$

- The enthalpy increases from the standard reference temperature to 3600°R (1980°C) are

$$CO_2 : \quad 39,791 \text{ Btu/lbmol} \quad (92.6 \text{ GJ/kmol})$$
$$H_2O : \quad 31,658 \text{ Btu/lbmol} \quad (73.6 \text{ GJ/kmol})$$
$$N_2 : \quad 24,471 \text{ Btu/lbmol} \quad (56.9 \text{ GJ/kmol})$$

(a) What are the proportions by weight of the oxygen and nitrogen? (b) What is the amount of liquid water, if any, present in the stack gas? (c) How much water is removed from the stack gas?

19. (*Time limit: one hour*) An internal combustion engine is normally fuel-injected with gasoline (C_8H_{18}; lower heating value of 23,200 Btu/lbm; 54 MJ/kg). It is desired to switch to alcohol (C_2H_5OH; lower heating value of 11,930 Btu/lbm; 27.7 MJ/kg). Minimal changes to the engine are to be made. The indicated and mechanical efficiencies are unchanged. Stoichiometric air/fuel ratios are being used for initial calculations. (a) If the power output is to be unchanged, what must be the percentage change in specific fuel consumption? (b) If the fuel-injection velocity is to be unchanged, what is the percentage change in jet size? (c) If no changes are made to the engine, what will be the percentage change in power output?

23 Energy, Work, and Power

1. Energy of a Mass 23-1
2. Law of Conservation of Energy 23-1
3. Work 23-1
4. Potential Energy of a Mass 23-2
5. Kinetic Energy of a Mass 23-3
6. Spring Energy 23-3
7. Pressure Energy of a Mass 23-3
8. Internal Energy of a Mass 23-3
9. Work-Energy Principle 23-4
10. Conversion Between Energy Forms 23-4
11. Power 23-5
12. Efficiency 23-6

Nomenclature

E	energy	ft-lbf	J
F	force	lbf	N
g	gravitational acceleration	ft/sec^2	m/s^2
g_c	gravitational constant	lbm-ft/lbf-sec^2	n.a.
h	height	ft	m
I	mass moment of inertia	lbm-ft^2	kg·m^2
J	Joule's constant	ft-lbf/Btu	n.a.
k	spring constant	lbf/ft	N/m
m	mass	lbm	kg
p	pressure	lbf/ft^2	Pa
P	power	ft-lbf/sec	W
q	heat	Btu/lbm	J/kg
r	radius	ft	m
s	distance	ft	m
t	time	sec	s
T	torque	ft-lbf	N·m
u	specific energy	ft-lbf/lbm	J/kg
U	internal energy	Btu	J
v	velocity	ft/sec	m/s
W	work	ft-lbf	J
x	displacement	ft	m

Symbols

δ	displacement	ft	m
η	efficiency	–	–
θ	angular position	rad	rad
ρ	mass density	lbm/ft^3	kg/m^3
v	specific volume	ft^3/lbm	m^3/kg
ϕ	angle	deg	deg
ω	angular velocity	rad/sec	rad/s

Subscripts

0	initial
f	frictional
O	center

1. ENERGY OF A MASS

The *energy* of a mass represents the capacity of the mass to do work. Such energy can be stored and released. There are many forms that it can take, including mechanical, thermal, electrical, and magnetic energies. Energy is a positive, scalar quantity (although the change in energy can be either positive or negative).

The total energy of a body can be calculated from its mass, m, and its *specific energy*, U (i.e., the energy per unit mass).[1]

$$E = mU \qquad 23.1$$

Typical units of mechanical energy are foot-pounds and joules. (A joule is equivalent to the units of N·m and kg·m^2/s^2.) In traditional English-unit countries, the *British thermal unit* (Btu) is used for thermal energy, whereas the kilocalorie (kcal) is still used in some applications in SI countries. *Joule's constant* or the *Joule equivalent* (778.26 ft-lbf/Btu, usually shortened to 778, three significant digits) is used to convert between English mechanical and thermal energy units.

$$\text{energy in Btu} = \frac{\text{energy in ft-lbf}}{J} \qquad 23.2$$

Two other units of large amounts of energy are the therm and the quad. A *therm* is 10^5 Btu (1.055×10^8 J). A *quad* is equal to a quadrillion (10^{15}) Btu. This is 1.055×10^{18} J, roughly the energy contained in 200 million barrels of oil.

2. LAW OF CONSERVATION OF ENERGY

The *law of conservation of energy* says that energy cannot be created or destroyed. However, energy can be converted into different forms. Therefore, the sum of all energy forms is constant.

$$\sum E = \text{constant} \qquad 23.3$$

3. WORK

Work, W, is the act of changing the energy of a particle, body, or system. For a mechanical system, *external work* is work done by an external force, whereas *internal work* is done by an internal force. Work is a

---

[1]The use of symbols E and U is not consistent in the engineering field.

signed, scalar quantity. Typical units are inch-pounds, foot-pounds, and joules. Mechanical work is seldom expressed in British thermal units or kilocalories.

For a mechanical system, work is positive when a force acts in the direction of motion and helps a body move from one location to another. Work is negative when a force acts to oppose motion. (Friction, for example, always opposes the direction of motion and can do only negative work.) The work done on a body by more than one force can be found by superposition.

From a thermodynamic standpoint, work is positive if a particle or body does work on its surroundings. Work is negative if the surroundings do work on the object. (Thus, blowing up a balloon represents negative work to the balloon.) Although this may be a difficult concept, it is consistent with the conservation of energy, since the sum of negative work and the positive energy increase is zero (i.e., no net energy change in the system).[2]

The work performed by a variable force or torque is calculated from the dot products of Eqs. 23.4 and 23.5,

$$W_{\text{variable force}} = \int \mathbf{F} \cdot d\mathbf{s} \quad \text{[linear systems]} \qquad 23.4$$

$$W_{\text{variable torque}} = \int \mathbf{T} \cdot d\boldsymbol{\theta} \quad \text{[rotational systems]} \quad 23.5$$

The work done by a force or torque of constant magnitude is

$$W_{\text{constant force}} = \mathbf{F} \cdot \mathbf{s} = Fs\cos\phi \quad \text{[linear systems]} \quad 23.6$$

$$W_{\text{constant torque}} = \mathbf{T} \cdot \boldsymbol{\theta} = Fr\theta\cos\phi$$
$$\text{[rotational systems]} \qquad 23.7$$

The nonvector forms of Eqs. 23.6 and 23.7 illustrate that only the component of force or torque in the direction of motion contributes to work.

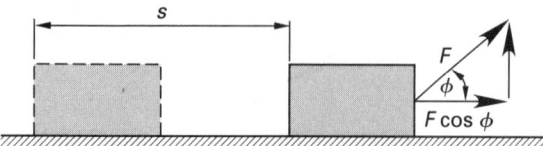

Figure 23.1 *Work of a Constant Force*

Common applications of the work done by a constant force are frictional work and gravitational work. The work to move an object a distance s against a frictional force of F_f is

$$W_{\text{friction}} = F_f s \qquad 23.8$$

The work done by gravity when a mass m changes in elevation from h_1 to h_2 is

$$W_{\text{gravity}} = mg(h_2 - h_1) \qquad \text{[SI]} \qquad 23.9(a)$$

$$W_{\text{gravity}} = \left(\frac{mg}{g_c}\right)(h_2 - h_1) \quad \text{[U.S.]} \quad 23.9(b)$$

The work done by or on a *linear spring* whose length or deflection changes from δ_1 to δ_2 is given by Eq. 23.10.[3] It does not make any difference whether the spring is a compression spring or an extension spring.

$$W_{\text{spring}} = \tfrac{1}{2}k\left(\delta_2^2 - \delta_1^2\right) \qquad 23.10$$

Example 23.1

A lawn mower engine is started by pulling a cord wrapped around a sheave. The sheave radius is 8.0 cm. The cord is wrapped around the sheave two times. If a constant tension of 90 N is maintained in the cord during starting, what work is done?

Solution

The starting torque on the engine is

$$T = Fr = (90\text{ N})\left(\frac{8\text{ cm}}{100\ \frac{\text{cm}}{\text{m}}}\right)$$
$$= 7.2\text{ N·m}$$

The cord wraps around the sheave $(2)(2\pi) = 12.6$ rad. From Eq. 23.7, the work done by a constant torque is

$$W = T\theta = (7.2\text{ N·m})(12.6\text{ rad})$$
$$= 90.7\text{ J}$$

Example 23.2

A 200 lbm crate is pushed 25 ft at constant velocity across a warehouse floor. There is a frictional force of 60 lbf between the crate and floor. What work is done by the frictional force on the crate?

Solution

From Eq. 23.8,

$$W_{\text{friction}} = F_f s = (60\text{ lbf})(25\text{ ft})$$
$$= 1500\text{ ft-lbf}$$

4. POTENTIAL ENERGY OF A MASS

Potential energy (gravitational energy) is a form of mechanical energy possessed by a body due to its relative position in a gravitational field. Potential energy is lost when the elevation of a body decreases. The lost potential energy usually is converted to kinetic energy or heat.

$$E_{\text{potential}} = mgh \qquad \text{[SI]} \qquad 23.11(a)$$

$$E_{\text{potential}} = \frac{mgh}{g_c} \qquad \text{[U.S.]} \qquad 23.11(b)$$

---

[2]This is just a partial statement of the *first law of thermodynamics*.

[3]A *linear spring* is one for which the linear relationship $F = kx$ is valid.

In the absence of friction and other nonconservative forces, the change in potential energy of a body is equal to the work required to change the elevation of the body.

$$W = \Delta E_{\text{potential}} \qquad 23.12$$

5. KINETIC ENERGY OF A MASS

Kinetic energy is a form of mechanical energy associated with a moving or rotating body. The kinetic energy of a body moving with instantaneous linear velocity v is

$$E_{\text{kinetic}} = \tfrac{1}{2}m\text{v}^2 \qquad \text{[SI]} \qquad 23.13(a)$$

$$E_{\text{kinetic}} = \frac{m\text{v}^2}{2g_c} \qquad \text{[U.S.]} \qquad 23.13(b)$$

According to the *work-energy principle* (see Section 23.9), the kinetic energy is equal to the work necessary to initially accelerate a stationary body, or to bring a moving body to rest.

$$W = \Delta E_{\text{kinetic}} \qquad 23.14$$

A body can also have rotational kinetic energy.

$$E_{\text{rotational}} = \tfrac{1}{2}I\omega^2 \qquad \text{[SI]} \qquad 23.15(a)$$

$$E_{\text{rotational}} = \frac{I\omega^2}{2g_c} \qquad \text{[U.S.]} \qquad 23.15(b)$$

Example 23.3

A solid disk flywheel ($I = 200 \text{ kg·m}^2$) is rotating with a speed of 900 rpm. What is its rotational kinetic energy?

Solution

The angular rotational velocity is

$$\omega = \frac{(900 \text{ rpm})(2\pi)}{60 \, \dfrac{\text{s}}{\text{min}}} = 94.25 \text{ rad/s}$$

From Eq. 23.15, the rotational kinetic energy is

$$E = \tfrac{1}{2}I\omega^2 = \left(\tfrac{1}{2}\right)(200 \text{ kg·m}^2)\left(94.25 \, \frac{\text{rad}}{\text{s}}\right)^2$$

$$= 888 \times 10^3 \text{ J} \quad (888 \text{ kJ})$$

6. SPRING ENERGY

A spring is an energy storage device, since the spring has the ability to perform work. In a perfect spring, the amount of energy stored is equal to the work required to compress the spring initially. The stored spring energy does not depend on the mass of the spring. Given a spring with spring constant (stiffness) k, the *spring energy* is

$$E_{\text{spring}} = \tfrac{1}{2}k\delta^2 \qquad 23.16$$

Example 23.4

A body of mass m falls from height h onto a massless, simply supported beam. The mass adheres to the beam. If the beam has a lateral stiffness k, what will be the deflection, δ, of the beam?

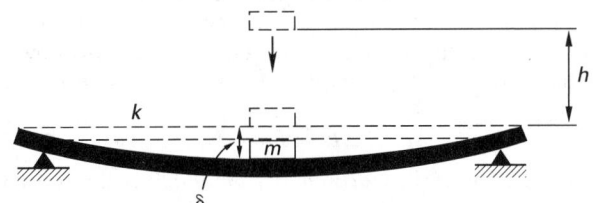

Solution

The initial energy of the system consists of only the potential energy of the body. Using consistent units, the change in potential energy is

$$E = mg(h + \delta) \qquad \text{[consistent units]}$$

All of this energy is stored in the spring. Therefore,

$$\tfrac{1}{2}k\delta^2 = mg(h + \delta)$$

Solving for the deflection,

$$\delta = \frac{mg + \sqrt{mg(2hk + mg)}}{k}$$

7. PRESSURE ENERGY OF A MASS

Since work is done in increasing the pressure of a system (i.e., it takes work to blow up a balloon), mechanical energy can be stored in pressure form. This is known as *pressure energy, static energy, flow energy, flow work,* and *p-V work (energy)*. For a system of pressurized mass m, the pressure energy is

$$E_{\text{flow}} = \frac{mp}{\rho} = mpv \qquad 23.17$$

8. INTERNAL ENERGY OF A MASS

The total internal energy, usually given the symbol U, of a body increases when the body's temperature increases.[4] In the absence of any work done on or by the

---

[4]The *thermal energy*, represented by the body's enthalpy, is the sum of internal and pressure energies. A more detailed discussion of thermal energy appears in the following chapters.

Thermodynamics

body, the change in internal energy is equal to the heat flow, Q, into the body. Q is positive if the heat flow is into the body and negative otherwise.

$$U_2 - U_1 = Q \qquad 23.18$$

The property of internal energy is encountered primarily in thermodynamics problems. Typical units are British thermal units, joules, and kilocalories.

9. WORK-ENERGY PRINCIPLE

Since energy can neither be created nor destroyed, external work performed on a conservative system goes into changing the system's total energy. This is known as the *work-energy principle* (or *principle of work and energy*).

$$W = \Delta E = E_2 - E_1 \qquad 23.19$$

Generally, the term *work-energy principle* is limited to use with mechanical energy problems (i.e., conversion of work into kinetic or potential energies). When energy is limited to kinetic energy, the work-energy principle is a direct consequence of Newton's second law but is valid for only inertial reference systems.

By directly relating forces, displacements, and velocities, the work-energy principle introduces some simplifications into many mechanical problems.

- It is not necessary to calculate or know the acceleration of a body to calculate the work performed on it.

- Forces that do not contribute to work (e.g., are normal to the direction of motion) are irrelevant.

- Only scalar quantities are involved.

- It is not necessary to individually analyze the particles or component parts in a complex system.

Example 23.5

A 4000 kg elevator starts from rest, accelerates uniformly to a constant speed of 2.0 m/s, and decelerates uniformly to a stop 20 m above its initial position. Neglecting friction and other losses, what work was done on the elevator?

Solution

By the work-energy principle, the work done on the elevator is equal to the change in the elevator's energy. Since the initial and final kinetic energies are zero, the only mechanical energy change is the potential energy change.

Taking the initial elevation of the elevator as the reference (i.e., $h_1 = 0$),

$$W = E_{2,\text{potential}} - E_{1,\text{potential}} = mg(h_2 - h_1)$$

$$= (4000 \text{ kg})\left(9.81 \frac{\text{m}}{\text{s}^2}\right)(20 \text{ m}) = 785 \times 10^3 \text{ J} \quad (785 \text{ kJ})$$

10. CONVERSION BETWEEN ENERGY FORMS

Conversion of one form of energy into another does not violate the conservation of energy law. However, most problems involving conversion of energy are really just special cases of the work-energy principle. For example, consider a falling body that is acted upon by a gravitational force. The conversion of potential energy into kinetic energy can be interpreted as equating the work done by the constant gravitational force to the change in kinetic energy.

In general terms, *Joule's law* states that one energy form can be converted without loss into another. There are two specific formulations of *Joule's Law*. As related to electricity, $P = I^2R = V^2/R$ is the common formulation of Joule's law. As related to thermodynamics and ideal gases, Joule's law states that "the change in internal energy of an ideal gas is a function of the temperature change, not of the volume." This latter form can also be stated more formally as "at constant temperature, the internal energy of a gas approaches a finite value that is independent of the volume as the pressure goes to zero."

Example 23.6

A 2.0 lbm projectile is launched straight up with an initial velocity of 700 ft/sec. Neglecting air friction, calculate the (a) kinetic energy immediately after launch, (b) kinetic energy at maximum height, (c) potential energy at maximum height, (d) total energy at an elevation where the velocity has dropped to 300 ft/sec, and (e) maximum height attained.

Solution

(a) From Eq. 23.13, the kinetic energy is

$$E_{\text{kinetic}} = \frac{mv^2}{2g_c} = \frac{(2 \text{ lbm})\left(700 \frac{\text{ft}}{\text{sec}}\right)^2}{(2)\left(32.2 \frac{\text{lbm-ft}}{\text{sec}^2\text{-lbf}}\right)}$$

$$= 15{,}217 \text{ ft-lbf}$$

(b) The velocity is zero at the maximum height. Therefore, the kinetic energy is zero.

(c) At the maximum height, all of the kinetic energy has been converted into potential energy. Therefore, the potential energy is 15,217 ft-lbf.

(d) Although some of the kinetic energy has been transformed into potential energy, the total energy is still 15,217 ft-lbf.

(e) Since all of the kinetic energy has been converted into potential energy, the maximum height can be found from Eq. 23.11.

$$E_{\text{potential}} = \frac{mgh}{g_c}$$

$$15{,}217 \text{ ft-lbf} = \frac{(2 \text{ lbm}) \left(32.2 \dfrac{\text{ft}}{\text{sec}^2}\right)(h)}{32.2 \dfrac{\text{lbm-ft}}{\text{sec}^2\text{-lbf}}}$$

$$h = 7609 \text{ ft}$$

Example 23.7

A 4500 kg ore car rolls down an incline and passes point A traveling at 1.2 m/s. The ore car is stopped by a spring bumper that compresses 0.6 m. A constant friction force of 220 N acts on the ore car at all times. What spring constant is required?

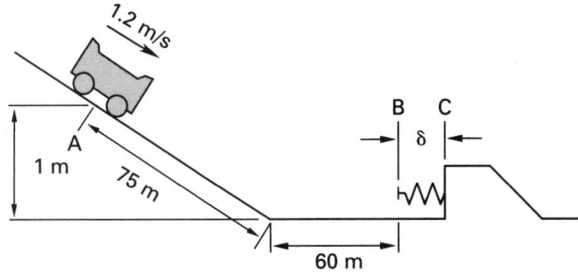

Solution

The car's total energy at point A is the sum of the kinetic and potential energies.

$$E_{\text{total,A}} = E_{\text{kinetic}} + E_{\text{potential}}$$
$$= \tfrac{1}{2}mv^2 + mgh$$
$$= \left(\tfrac{1}{2}\right)(4500 \text{ kg})\left(1.2 \,\frac{\text{m}}{\text{s}}\right)^2$$
$$\quad + (4500 \text{ kg})\left(9.81 \,\frac{\text{m}}{\text{s}^2}\right)(1 \text{ m})$$
$$= 47\,385 \text{ J}$$

At point B, the potential energy has been converted into additional kinetic energy. However, except for friction, the total energy is the same as at point A. Since the frictional force does negative work, the total energy remaining at point B is

$$E_{\text{total,B}} = E_{\text{total,A}} - W_{\text{friction}}$$
$$= 47\,385 \text{ J} - (220 \text{ N})(75 \text{ m} + 60 \text{ m})$$
$$= 17\,685 \text{ J}$$

At point C, the maximum compression point, the remaining energy has gone into compressing the spring a distance $\delta = 0.6$ m and performing a small amount of frictional work.

$$E_{\text{total,B}} = E_{\text{total,C}} = W_{\text{spring}} + W_{\text{friction}}$$
$$= \tfrac{1}{2}k\delta^2 + F_f\delta$$
$$17\,685 \text{ J} = \left(\tfrac{1}{2}\right)(k)(0.6 \text{ m})^2 + (220 \text{ N})(0.6 \text{ m})$$

The spring constant can be determined directly.

$$k = 97\,520 \text{ N/m} \quad (97.5 \text{ kN/m})$$

11. POWER

Power is the amount of work done per unit time. It is a scalar quantity. (Although power is calculated from two vectors, the vector dot-product operation is seldom needed.)

$$P = \frac{W}{\Delta t} \qquad\qquad 23.20$$

For a body acted upon by a force or torque, the instantaneous power can be calculated from the velocity.

$$P = Fv \qquad \text{[linear systems]} \qquad 23.21$$
$$P = T\omega \qquad \text{[rotational systems]} \qquad 23.22$$

Typical basic units of power are ft-lbf/sec and watts (J/s), although *horsepower* is widely used. Table 23.1 can be used to convert units of power.

Table 23.1 *Useful Power Conversion Formulas*

$$
\begin{aligned}
1 \text{ hp} &= 550 \text{ ft-lbf/sec} \\
&= 33{,}000 \text{ ft-lbf/min} \\
&= 0.7457 \text{ kW} \\
&= 0.7068 \text{ Btu/sec} \\[6pt]
1 \text{ kW} &= 737.6 \text{ ft-lbf/sec} \\
&= 44{,}250 \text{ ft-lbf/min} \\
&= 1.341 \text{ hp} \\
&= 0.9483 \text{ Btu/sec} \\[6pt]
1 \text{ Btu/sec} &= 778.26 \text{ ft-lbf/sec} \\
&= 46{,}680 \text{ ft-lbf/min} \\
&= 1.415 \text{ hp}
\end{aligned}
$$

Example 23.8

When traveling at 100 kph, a car supplies a constant horizontal force of 50 N to the hitch of a trailer. What tractive power (in horsepower) is required for the trailer alone?

Solution

From Eq. 23.21, the power being generated is

$$P = Fv = \frac{(50 \text{ N}) \left(100 \, \frac{\text{km}}{\text{h}}\right) \left(1000 \, \frac{\text{m}}{\text{km}}\right)}{\left(60 \, \frac{\text{s}}{\text{min}}\right) \left(60 \, \frac{\text{min}}{\text{h}}\right) \left(1000 \, \frac{\text{W}}{\text{kW}}\right)}$$

$$= 1.389 \text{ kW}$$

Using a conversion from Table 23.1, the horsepower is

$$P = \left(1.341 \, \frac{\text{hp}}{\text{kW}}\right) (1.389 \text{ kW})$$

$$= 1.86 \text{ hp}$$

12. EFFICIENCY

For energy-using systems (such as cars, electrical motors, elevators, etc.), the *energy-use efficiency*, η, of a system is the ratio of an ideal property to an actual property. The property used is commonly work, power, or for thermodynamic problems, heat. When the rate of work is constant, either work or power can be used to calculate the efficiency. Otherwise, power should be used. Except in rare instances, the numerator and denominator of the ratio must have the same units.[5]

$$\eta = \frac{P_{\text{ideal}}}{P_{\text{actual}}} \qquad [P_{\text{actual}} \geq P_{\text{ideal}}] \qquad 23.23$$

For energy-producing systems (such as electrical generators, prime movers, and hydroelectric plants), the *energy-production efficiency* is

$$\eta = \frac{P_{\text{actual}}}{P_{\text{ideal}}} \qquad [P_{\text{ideal}} \geq P_{\text{actual}}] \qquad 23.24$$

The efficiency of an *ideal machine* is 1.0 (100%). However, all *real machines* have efficiencies of less than 1.0.

---

[5]The *energy-efficiency ratio* used to evaluate refrigerators, air conditioners, and heat pumps, for example, has units of Btu per watt-hour (Btu/W-hr).

24 Thermodynamic Properties of Substances

1. Phases of a Pure Substance 24-2
2. Determining Phase 24-3
3. Properties of a Substance 24-4
4. Mass: m 24-4
5. Temperature: T 24-4
6. Pressure: p 24-5
7. Density: ρ 24-5
8. Specific Volume: v and V 24-5
9. Internal Energy: u and U 24-5
10. Enthalpy: h and H 24-5
11. Entropy: s and S 24-6
12. Specific Heat: c and C 24-7
13. Ratio of Specific Heats: k 24-8
14. Triple Point Properties 24-8
15. Critical Properties 24-8
16. Latent Heats 24-8
17. Quality: x 24-9
18. Compressibility 24-9
19. Using the Mollier Diagram 24-9
20. Using Saturation Tables 24-10
21. Using Superheat Tables 24-10
22. Using Compressed Liquid Tables 24-10
23. Using Gas Tables 24-10
24. Pressure-Enthalpy Charts 24-11
25. Properties of Solids 24-11
26. Properties of Subcooled Liquids 24-11
27. Properties of Saturated Liquids 24-12
28. Properties of Saturated Vapors 24-12
29. Properties of Liquid-Vapor Mixtures . . . 24-12
30. Properties of Superheated Vapors 24-13
31. Equation of State for Ideal Gases 24-13
32. Properties of Ideal Gases 24-15
33. Specific Heats of Ideal Gases 24-15
34. Kinetic Gas Theory 24-15
35. Gravimetric, Volumetric, and
     Mole Fractions 24-17
36. Partial Pressure and Partial Volume of
     Gas Mixtures 24-17
37. Properties of Nonreacting Ideal Gas
     Mixtures 24-17
38. Equation of State for Real Gases 24-19
39. Specific Heats of Real Gases 24-20
     Practice Problems 24-20

Nomenclature

a	van der Waals factor	atm-ft^6/lbmol	Pa·m^6/kmol
AW	atomic weight	lbm/lbmol	kg/kmol
b	van der Waals factor	ft^3/lbmol	m^3/kmol
B	volumetric fraction	–	–
c	specific heat	Btu/lbm-°F	kJ/kg·K
C	molar specific heat	Btu/lbmol-°F	kJ/kmol·K
E	energy	Btu	kJ
G	gravimetric fraction	–	–
h	enthalpy	Btu/lbm	kJ/kg
H	molar enthalpy	Btu/lbmol	kJ/kmol
J	Joule's constant (778.17)	ft-lbf/Btu	n.a.
k	ratio of specific heats	–	–
κ	Boltzmann constant	n.a.	kJ/molecule·K
m	mass	lbm	kg
MW	molecular weight	lbm/lbmol	kg/kmol
n	number of moles	–	–
N	number of molecules	–	–
N_A	Avogadro's number	molecules/lbmol	molecules/kmol
p	absolute pressure	lbf/ft^2	Pa
q	heat	Btu/lbm	kJ/kg
Q	molar heat	Btu/lbmol	kJ/kmol
Q	total heat	Btu	kJ
R	specific gas constant	ft-lbf/lbm-°R	kJ/kg·K
R^*	universal gas constant	ft-lbf/lbmol-°R	kJ/kmol·K
s	entropy	Btu/lbm-°R	kJ/kg·K
S	molar entropy	Btu/lbmol-°R	kJ/kmol·K
T	temperature	°F	°C
T	absolute temperature	°R	K
u	internal energy	Btu/lbm	kJ/kg
U	molar internal energy	Btu/lbmol	kJ/kmol
v	velocity	ft/sec	m/s
V	molar specific volume	ft^3/lbmol	m^3/kmol
V	volume	ft^3	m^3
x	mole fraction	–	–
x	quality	–	–
Z	compressibility factor	–	–

Symbols

α	isentropic compressibility	1/°R	1/K
β	isobaric compressibility	1/°R	1/K
κ	isothermal compressibility	1/°R	1/K
ρ	density	lbm/ft^3	kg/m^3
v	specific volume	ft^3/lbm	m^3/kg
ϕ	entropy function	Btu/lbm-°R	kJ/kg·K

Subscripts

A	Avogadro
c	critical
f	fluid (liquid)
fg	liquid-to-gas (vaporization)
g	gas (vapor)
i	ice

Thermodynamics

k kinetic
l latent
m mean
o outside (environment)
p constant pressure, or most probable
r ratio, or reduced
rms   root-mean-square
s sensible, or solid
sat   saturated
t total
v constant volume

1. PHASES OF A PURE SUBSTANCE

Thermodynamics is the study of a substance's energy-related properties. The properties of a substance and the procedures used to determine those properties depend on the state and the phase of the substance. The *thermodynamic state* of a substance is defined by two or more independent thermodynamic properties. For example, the temperature and pressure of a substance are two properties commonly used to define the state of a superheated vapor.

The common *phases* of a substance are solid, liquid, and gas. However, because substances behave according to different rules, it is convenient to categorize them into more than these three phases.[1]

solid: A solid does not take on the shape or volume of its container.

subcooled liquid: If a liquid is not saturated (i.e., the liquid is not at its boiling point), it is said to be subcooled. Water at 1 atm and room temperature is subcooled, as the addition of a small amount of heat will not cause vaporization.

saturated liquid: A saturated liquid has absorbed as much heat energy as it can without vaporizing. Liquid water at standard atmospheric pressure and 212°F (100°C) is an example of a saturated liquid.

liquid-vapor mixture: A liquid and vapor of the same substance can coexist at the same temperature and pressure. This is called a two-phase, liquid-vapor mixture.

saturated vapor: A vapor (e.g., steam at standard atmospheric pressure and 212°F (100°C)) that is on the verge of condensing is said to be saturated.

superheated vapor: A superheated vapor is one that has absorbed more heat than is needed merely to vaporize it. A superheated vapor will not condense when small amounts of heat are removed.

ideal gas: A gas is a highly superheated vapor. If the gas behaves according to the ideal gas laws, it is called an ideal gas.

real gas: A real gas does not behave according to the ideal gas laws.

gas mixtures: Most gases mix together freely. Two or more pure gases together constitute a gas mixture.

vapor/gas mixtures: Atmospheric air is an example of a mixture of several gases and water vapor.

These phases and subphases can be illustrated with a pure substance in the piston/cylinder arrangement shown in Fig. 24.1. The pressure in this system is constant and is determined by the weight of the piston, which moves freely to permit volume changes.

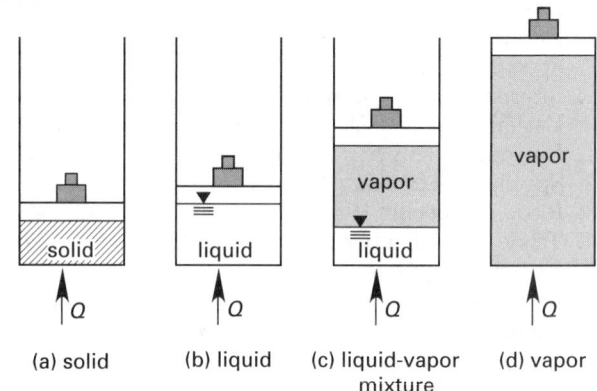

Figure 24.1 *Phase Changes at Constant Pressure*

In Fig. 24.1(a) the volume is minimum. This is the solid phase. The temperature will rise as heat, Q, is added to the solid. This increase in temperature is accompanied by a small increase in volume. The temperature increases until the melting point is reached.

The solid will begin to melt as heat is added to it at the melting point. The temperature will not increase until all of the solid has been turned into liquid. The liquid phase, with its small increase in volume, is illustrated by Fig. 24.1(b).

If the subcooled liquid continues to receive heat, its temperature will rise. This temperature increase continues until evaporation is imminent. The liquid at this point is said to be saturated. Any increase in heat energy will cause a portion of the liquid to vaporize. This is shown in Fig. 24.1(c), in which a liquid-vapor mixture exists.

As with melting, evaporation occurs at constant temperature and pressure but with a very large increase in volume. The temperature cannot increase until the last drop of liquid has been evaporated, at which point the vapor is said to be saturated. This is shown in Fig. 24.1(d).

Additional heat will result in high-temperature *superheated vapor*. This vapor may or may not behave according to the ideal gas laws.

---

[1]Plasma, *cryogenic fluids* (*cryogens*) that boil at temperatures less than approximately 200°R (110K), and solids near absolute zero are not discussed in this chapter.

2. DETERMINING PHASE

It is possible to develop a three-dimensional surface that predicts the substance's phase based on the properties of pressure, temperature, and specific volume. Such an *equilibrium solid* is illustrated in Fig. 24.2.[2]

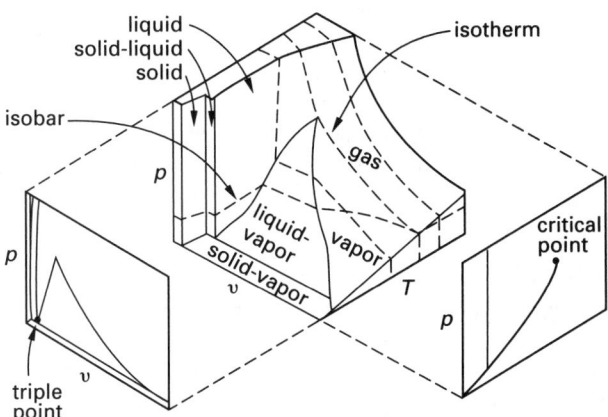

Figure 24.2 *Equilibrium Solid*

If one property is held constant during a process, a two-dimensional projection of the equilibrium solid can be used. This projection is known as an *equilibrium diagram* or a *phase diagram*, of which Fig. 24.3 is an example.

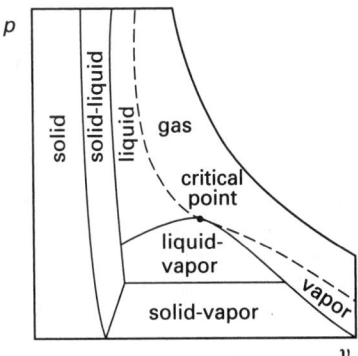

Figure 24.3 *Constant Temperature Phase Diagram*

The most important part of a phase diagram is limited to the liquid-vapor region. A general phase diagram showing this region and the bell-shaped dividing line (known as the *vapor dome*) is shown in Fig. 24.4.

---

[2]Figure 24.2 is applicable for most substances, excluding water. In Fig. 24.2, the specific volume "steps in" (i.e., decreases) when the liquid turns to a solid as the temperature drops below freezing. However, water is unique in that it expands upon freezing. Thus, the pVT diagram for water shows a "step out" instead of a "step in" upon freezing. The remainder of the pVT diagram for water is the same as Fig. 24.2.

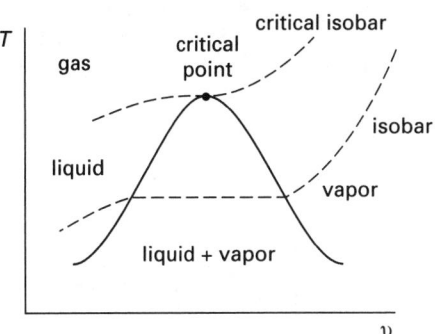

Figure 24.4 *Vapor Dome with Isobars*

The vapor dome region can be drawn with many variables for the axes. For example, either temperature or pressure can be used for the vertical axis. Energy, specific volume, or entropy can be chosen for the horizontal axis. The principles presented here apply to all combinations.

The left-hand part of the vapor dome separates the liquid phase from the liquid-vapor phase. This part of the line is known as the *saturated liquid line*. Similarly, the right-hand part of the line separates the liquid-vapor phase from the vapor phase. This line is called the *saturated vapor line*.

Lines of constant pressure (*isobars*) can be superimposed on the vapor dome. Each isobar is horizontal as it passes through the two-phase region, indicating that both temperature and pressure remain unchanged as a liquid vaporizes.

Notice that there is no real dividing line between liquid and vapor at the top of the vapor dome. Far above the vapor dome, there is no distinction between liquids and gases, as their properties are identical. The phase is assumed to be a gas.

The implied dividing line between liquid and gas is the isobar that passes through the topmost part of the vapor dome. This is known as the *critical isobar*, and the highest point of the vapor dome is known as the *critical point*. This critical isobar also provides a way to distinguish between a vapor and a gas. A substance below the critical isobar (but to the right of the vapor dome) is a vapor. Above the critical isobar, it is a gas.

Figure 24.5 illustrates a vapor dome for which pressure has been chosen as the vertical axis and enthalpy (h) has been chosen as the horizontal axis. The shape of the dome is essentially the same, but the lines of constant temperature (*isotherms*) have a different direction than isobars.

Figure 24.5 also illustrates the subscripting convention used to identify points on the saturation line. The subscript f (fluid) is used to indicate a saturated liquid.

The subscript g (gas) is used to indicate a saturated vapor.[3] The subscript fg is used to indicate the difference in saturation properties.

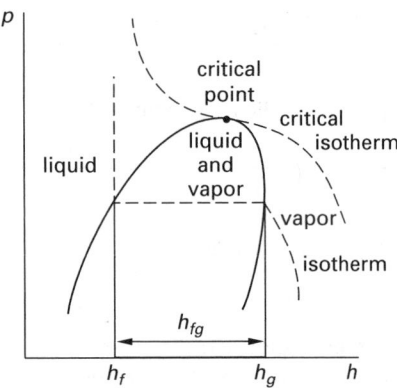

Figure 24.5 *Vapor Dome with Isotherms*

The vapor dome is a good tool for illustration, but it cannot be used to determine a substance's phase. Such a determination must be made based on the substance's pressure and temperature.

For example, consider water at a pressure of 1 atm. Its boiling temperature (the *saturation temperature*) at this pressure is 212°F (100°C). If the water has a lower temperature, for example, 85°F (29°C), the water will be liquid (i.e., will be subcooled). On the other hand, if the water's temperature (at 1 atm) is 270°F (132°C), the water must be in vapor form (i.e., must be superheated).

This example is valid only for water at atmospheric pressure. Water at other pressures will have other boiling temperatures. (The lower the pressure, the lower the boiling temperature.) However, the rules given here follow directly from the previous example and will become more meaningful as this chapter progresses.

Rule 1: A substance is a subcooled liquid if its temperature is less than the saturation temperature corresponding to its pressure.

Rule 2: A substance is in the liquid-vapor region if its temperature is equal to the saturation temperature corresponding to its pressure.

Rule 3: A substance is a superheated vapor if its temperature is greater than the saturation temperature corresponding to its pressure.

The rules that follow can be stated using pressure as the determining variable.

Rule 4: A substance is a subcooled liquid if its pressure is greater than the saturation pressure corresponding to its temperature.

Rule 5: A substance is in the liquid-vapor region if its pressure is equal to the saturation pressure corresponding to its temperature.

Rule 6: A substance is a superheated vapor if its pressure is less than the saturation pressure corresponding to its temperature.

3. PROPERTIES OF A SUBSTANCE

The thermodynamic *state* or condition of a substance is determined by its properties. *Intensive properties* are independent of the amount of substance present. Temperature, pressure, and stress are examples of intensive properties. *Extensive properties* are dependent on the amount of substance present. Examples are volume, strain, charge, and mass.

In this chapter, and in most books on thermodynamics, both lowercase and uppercase forms of the same characters are used to represent property variables. The two forms are used to distinguish between the units of mass. For example, lowercase h represents *specific enthalpy* (usually just called "enthalpy") in units of Btu/lbm or kJ/kg. Uppercase H is used to represent the *molar enthalpy* in units of Btu/lbmol or kJ/kmol.

4. MASS: m

The mass of a substance is a measure of its quantity. Mass is independent of location and gravitational field strength. In thermodynamics, the customary U.S. and SI units of mass (m) are pound-mass (lbm) and kilogram (kg), respectively.

5. TEMPERATURE: T

Temperature is a thermodynamic property of a substance that depends on energy content. Heat energy entering a substance will increase the temperature of that substance. Normally, heat energy will flow only from a hot object to a cold object. If two objects are in *thermal equilibrium* (are at the same temperature), no heat will flow between them.

If two systems are in thermal equilibrium, they must be at the same temperature. If both systems are in equilibrium with a third, then all three are at the same temperature. This concept is known as the *Zeroth Law of Thermodynamics.*

---

[3]Although this book makes it a rule never to call a vapor a gas, this convention is not adhered to in the field of thermodynamics. The subscript g is standard for a saturated vapor.

The scales most commonly used for measuring temperature are the Fahrenheit and Celsius scales.[4] The relationship between these two scales is

$$T_{\circ F} = 32^\circ + \left(\frac{9}{5}\right) T_{\circ C} \qquad 24.1$$

The *absolute temperature scale* defines temperature independently of the properties of any particular substance. This is unlike the Celsius and Fahrenheit scales, which are based on the freezing point of water. The absolute temperature scale should be used for all calculations.

In the customary U.S. system, the absolute scale is the *Rankine scale.*[5]

$$T_{\circ R} = T_{\circ F} + 459.67^\circ \qquad 24.2$$
$$\Delta T_{\circ R} = \Delta T_{\circ F} \qquad 24.3$$

The absolute temperature scale in the SI system is the *Kelvin scale.*[6]

$$T_K = T_{\circ C} + 273.15^\circ \qquad 24.4$$
$$\Delta T_K = \Delta T_{\circ C} \qquad 24.5$$

The relationships between temperature differences in the customary U.S. and SI systems are independent of the freezing point of water.

$$\Delta T_{\circ C} = \left(\frac{5}{9}\right) \Delta T_{\circ F} \qquad 24.6$$

$$\Delta T_K = \left(\frac{5}{9}\right) \Delta T_{\circ R} \qquad 24.7$$

Table 24.1 *Temperature Scales*

	Kelvin	Celsius	Rankine	Fahrenheit
normal boiling point of water	373.15K	100.00°C	671.67°R	212.00°F
triple point of water (see Sec. 24.14)	273.16	0.01	491.69	32.02
	273.15	0.00	491.67	32.00 ice point
absolute zero	0	−273.15	0	−459.67

---

[4]The term *centigrade* was replaced by the term *Celsius* in 1948.
[5]Normally, three significant temperature digits (i.e., 460°) are sufficient.
[6]Normally, three significant temperature digits (i.e., 273°) are sufficient.

6. PRESSURE: p

Customary U.S. pressure units are pounds per square inch (psi). Standard SI pressure units are kPa or MPa, although bars are also used in tabulations of thermodynamic data. The values of a *standard atmosphere* in various units are given in Chap. 14.

7. DENSITY: ρ

Customary U.S. density units in tabulations of thermodynamic data are pounds per cubic foot (lbm/ft^3). Standard SI density units are kilograms per cubic meter (kg/m^3). Density is the reciprocal of specific volume.

$$\rho = \frac{1}{v} \qquad 24.8$$

8. SPECIFIC VOLUME: v AND V

Specific volume, v, is the volume occupied by one unit mass of a substance. Customary U.S. units in tabulations of thermodynamic data are cubic feet per pound (ft^3/lbm). Standard SI specific volume units are cubic meters per kilogram (m^3/kg). *Molar specific volume*, V, has units of ft^3/lbmol (m^3/kmol) and is seldom encountered. Specific volume is the reciprocal of density.

$$v = \frac{1}{\rho} \qquad 24.9$$
$$V = (MW) \times v \qquad 24.10$$

9. INTERNAL ENERGY: u AND U

Internal energy includes all of the potential and kinetic energies of the atoms or molecules in a substance. Energies in the translational, rotational, and vibrational modes are included. Since this movement increases as the temperature increases, internal energy is a function of temperature. It does not depend on the process or path taken to reach a particular temperature.

In the United States, the *British thermal unit*, Btu, is used to measure all forms of thermodynamic energy. (One Btu is approximately the energy given off by burning one wooden match.) Standard units of *specific internal energy* (u) are Btu/lbm and kJ/kg. The units of *molar internal energy* (U) are Btu/lbmol and kJ/kmol. Equation 24.11 gives the relationship between the specific and molar quantities.

$$U = (MW) \times u \qquad 24.11$$

10. ENTHALPY: h AND H

Enthalpy (also known at various times in history as *total heat* and *heat content*) represents the total useful energy of a substance. Useful energy consists of two parts—the

Thermodynamics

internal energy, u, and the *flow energy* (also known as *flow work* and *p-V work*), pV. Therefore, enthalpy has the same units as internal energy.

$$h = u + pv \qquad \text{24.12}$$

$$H = U + pV \qquad \text{24.13}$$

$$H = (\text{MW}) \times h \qquad \text{24.14}$$

Enthalpy is defined as useful energy because, ideally, all of it can be used to perform useful tasks. It takes energy to increase the temperature of a substance. If that internal energy is recovered, it can be used to heat something else (e.g., to vaporize water in a boiler). Also, it takes energy to increase pressure and volume (as in blowing up a balloon). If pressure and volume are decreased, useful energy is given up.

The customary U.S. units of Eqs. 24.12 and 24.13 are not consistent, since flow work (as written) has units of ft-lbf/lbm, not Btu/lbm. (There is also a consistency problem if pressure is defined in lbf/ft^2 and given in lbf/in^2.) Strictly, Eq. 24.12 should be written as

$$h = u + \frac{pv}{J} \qquad [\text{U.S.}] \qquad \text{24.15}$$

The conversion factor, J, in Eq. 24.15 is known as *Joule's constant*. It has a value of 778.17 ft-lbf/Btu. (In SI units, Joule's constant has a value of 1.0 N·m/J and is unnecessary.) As in Eqs. 24.13 and 24.14, Joule's constant is often omitted from the statement of generic thermodynamic equations, but it is always needed with customary U.S. units for dimensional consistency.

11. ENTROPY: s AND S

Absolute entropy is a measure of the energy that is no longer available to perform useful work within the current environment. Other definitions used (the "disorder of the system," the "randomness of the system," etc.) are frequently quoted. Although these alternate definitions cannot be used in calculations, they are consistent with the *third law of thermodynamics* (also known as the *Nernst theorem*). This law states that the absolute entropy of a perfect crystalline solid in thermodynamic equilibrium is (approaches) zero when the temperature is (approaches) absolute zero.[7] Equation 24.16 expresses the third law mathematically.

$$\lim_{T \to 0\text{K}} s = 0 \qquad \text{24.16}$$

An increase in entropy is known as *entropy production*. The total absolute entropy in a system is equal to the summation of all absolute entropy productions that have occurred over the life of the system.

$$s = \sum \Delta s \qquad \text{24.17}$$

---

[7]A molecule with zero entropy exists in only one quantum state. The energy state is known precisely, without *uncertainty*.

For an isothermal process taking place at a constant temperature T_o, the entropy production depends on the amount of energy transfer.

$$\Delta s = \frac{q}{T_o} \qquad \text{24.18}$$

$$\Delta S = \frac{Q}{T_o} \qquad \text{24.19}$$

$$\Delta S = (\text{MW}) \times \Delta s \qquad \text{24.20}$$

For processes that occur over a varying temperature, the entropy production must be found by integration.

$$\Delta s = \int ds = \int \frac{dq}{T} \qquad \text{24.21}$$

From Eqs. 24.18 and 24.21, it is apparent that the units of specific absolute entropy are Btu/lbm-°R and kJ/kg·K. For molar entropy, the units are Btu/lbmol-°R and kJ/kmol·K.

Unlike absolute entropy, *standardized entropy*, usually just called "entropy," is not referenced to absolute zero conditions, but is measured with respect to some other convenient thermodynamic state. For water, the reference condition is the liquid phase at the triple point. (See Sec. 14.)

Example 24.1

Three planets of identical size and mass are oriented in space such that radiant energy transfers can occur. The average temperatures of planets A, B, and C are 530°R, 520°R, and 510°R (294K, 289K, and 283K), respectively. All three planets are massive enough that small energy losses or gains can be considered to be isothermal processes (i.e., they will not change the average temperature).

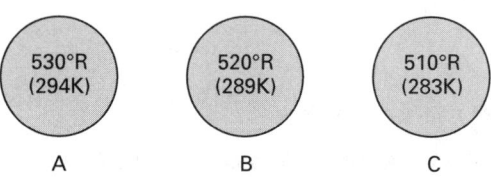

(a) Can a radiation energy transfer occur spontaneously from planet B to planet C? (b) What are the entropy productions for planets B and C if an energy transfer of 1000 Btu/lbm (2330 kJ/kg) occurs by radiation? (c) What is the overall entropy change as a result of the energy transfer in (b)? (d) Does entropy always increase? (e) Can planet B ever be returned to its original condition?

Solution

(a) A radiation transfer can occur spontaneously because planet B is hotter than planet C. Energy will flow spontaneously from a hot object to a cold object.

(b) In SI units, the entropy productions are

$$\Delta s_{\rm B} = \frac{q}{T_o} = \frac{-2330 \ \dfrac{\rm kJ}{\rm kg}}{289 {\rm K}} = -8.062 \ {\rm kJ/kg\cdot K}$$

$$\Delta s_{\rm C} = \frac{2330 \ \dfrac{\rm kJ}{\rm kg}}{283 {\rm K}} = 8.233 \ {\rm kJ/kg\cdot K}$$

In customary U.S. units, the entropy productions are

$$\Delta s_{\rm B} = \frac{q}{T_o} = \frac{-1000 \ \dfrac{\rm Btu}{\rm lbm}}{520^\circ {\rm R}} = -1.923 \ {\rm Btu/lbm\text{-}{}^\circ R}$$

$$\Delta s_{\rm C} = \frac{1000 \ \dfrac{\rm Btu}{\rm lbm}}{510^\circ {\rm R}} = 1.961 \ {\rm Btu/lbm\text{-}{}^\circ R}$$

(c) The entropy change is not the same for the two planets. Entropy is not conserved in an energy transfer process. The overall entropy production is

$$\Delta s = \Delta s_{\rm B} + \Delta s_{\rm C} = -8.062 \ \frac{\rm kJ}{\rm kg\cdot K} + 8.233 \ \frac{\rm kJ}{\rm kg\cdot K}$$

$$= 0.171 \ {\rm kJ/kg\cdot K}$$

In customary U.S. units,

$$\Delta s = \Delta s_{\rm B} + \Delta s_{\rm C}$$

$$= -1.923 \ \frac{\rm Btu}{\rm lbm\text{-}{}^\circ R} + 1.961 \ \frac{\rm Btu}{\rm lbm\text{-}{}^\circ R}$$

$$= 0.038 \ {\rm Btu/lbm\text{-}{}^\circ R}$$

(d) Local entropy can decrease, as shown by planet B's negative entropy production. However, overall entropy always increases when the total universe is considered, as shown in part (c).

(e) Planet B can be brought back to its original condition if 1000 Btu/lbm (2330 kJ/kg) of energy is transferred from planet A to planet B. (Heat will not flow spontaneously from planet C to planet B, as heat will not flow spontaneously from a cold object to a hot object.)[8]

12. SPECIFIC HEAT: c AND C

An increase in internal energy is needed to cause a rise in temperature. Different substances differ in the quantity of heat needed to produce a given temperature increase. The ratio of heat, Q, required to change the temperature of a mass, m, by an amount ΔT is called the *specific heat* (*heat capacity*) of the substance, c.

---

[8]This is one way of stating the *Second Law of Thermodynamics*.

Because specific heats of solids and liquids are slightly temperature dependent, the mean specific heats are used when evaluating processes covering a large temperature range.

$$Q = mc \, \Delta T \qquad\qquad \textit{24.22}$$

$$c = \frac{Q}{m \Delta T} \qquad\qquad \textit{24.23}$$

The lowercase c implies that the units are Btu/lbm-°F or J/kg·°C. Typical values of specific heat are given in Table 24.2. The *molar specific heat*, designated by the symbol C, has units of Btu/lbmol-°F or J/kmol·°C.

$$C = ({\rm MW}) \times c \qquad\qquad \textit{24.24}$$

For gases, the specific heat depends on the type of process during which the heat exchange occurs. Specific heats for constant-volume and constant-pressure processes are designated by c_v and c_p, respectively.

$$Q = mc_v \, \Delta T \qquad {\rm [constant\text{-}volume \ process]} \qquad \textit{24.25}$$

$$Q = mc_p \, \Delta T \qquad {\rm [constant\text{-}pressure \ process]} \qquad \textit{24.26}$$

Approximate values of c_p and c_v for common gases are given in Table 24.7. c_v and c_p for solids and liquids are essentially the same. However, the designation c_p is often encountered for solids and liquids.

The law of *Dulong and Petit* predicts the approximate molar specific heat (in cal/mol·°C) at high temperatures from the atomic weight.[9] This law is valid for solid elements having atomic weights greater than 40 and for most metallic elements. It is not valid at room temperature for carbon, silicon, phosphorus, and sulfur. 6.3 cal/mol·°C is known as the *Dulong and Petit value*.

$$c \times ({\rm AW}) = 6.3 \pm 0.1 \qquad\qquad \textit{24.27}$$

Example 24.2

Compare the value of specific heat of pure iron calculated from Dulong and Petit's law with the value from Table 24.2.

Solution

The atomic weight of iron is 55.8. From Eq. 24.27,

$$c = \frac{6.3}{\rm AW} = \frac{6.3 \ \dfrac{\rm cal}{\rm mol\cdot{}^\circ C}}{55.8 \ \dfrac{\rm g}{\rm mol}} = 0.11 \ {\rm cal/g\cdot{}^\circ C}$$

This is the same value as is given in Table 24.1.

---

[9]Dulong and Petit's law becomes valid at different temperatures for different substances, and a more specific definition of "high temperature" is impossible. For lead, the law is valid at 200K. For copper, it is not valid until above 400K.

Table 24.2 Approximate Specific Heats
of Selected Liquids and Solids[a]

substance	c_p Btu/lbm-°F	kJ/kg·K
aluminum, pure	0.23	0.960
aluminum, 2024-T4	0.2	0.840
ammonia	1.16	4.860
asbestos	0.20	0.840
benzene	0.41	1.720
brass, red	0.093	0.390
bronze	0.082	0.340
concrete	0.21	0.880
copper, pure	0.094	0.390
Freon-12	0.24	1.000
gasoline	0.53	2.200
glass	0.18	0.750
gold, pure	0.031	0.130
ice	0.49	2.050
iron, pure	0.11	0.460
iron, cast (4% C)	0.10	0.420
lead, pure	0.031	0.130
magnesium, pure	0.24	1.000
mercury	0.033	0.140
oil, light hydrocarbon	0.5	2.090
silver, pure	0.06	0.250
steel, 1010	0.10	0.420
steel, stainless 301	0.11	0.460
tin, pure	0.055	0.230
titanium, pure	0.13	0.540
tungsten, pure	0.032	0.130
water	1.0	4.190
wood (typical)	0.6	2.500
zinc, pure	0.088	0.370

(Multiply Btu/lbm-°F by 4.1868 to obtain kJ/kg·K.)
[a]Values in cal/g·°C are the same as Btu/lbm-°F.

Table 24.3 Approximate Triple Points

substance	pressure (atm)	temperature °R	K
ammonia	0.060	352	196
argon	0.676	151	84
carbon dioxide	5.10	390	217
helium	0.0508	4	2
hydrogen	0.0676	26	14
nitrogen	0.127	14	7.8
oxygen	0.00265	99	55
water	0.00592	492.02	273.34

Table 24.4 Approximate Critical Properties

substance	pressure (atm)	temperature °R	K
air	37.2	235.8	131.0
ammonia	111.5	730.1	405.6
argon	48.0	272.2	151.2
carbon dioxide	72.9	547.8	304.3
carbon monoxide	34.6	242.2	134.6
chlorine	75.9	751.0	417.2
ethane	48.8	549.8	305.4
ethylene	50.7	509.5	283.1
helium	2.3	10.0	5.56
hydrogen	12.8	60.5	33.6
mercury	180.0	2109.0	1171.7
methane	45.8	343.9	191.1
neon	25.7	79.0	43.9
nitrogen	33.5	227.2	126.2
oxygen	49.7	278.1	154.5
propane	42.0	666.3	370.2
sulfur dioxide	77.6	775.0	430.6
water vapor	218.2	1165.4	647.4
xenon	58.2	521.9	289.9

13. RATIO OF SPECIFIC HEATS: k

For gases, the *ratio of specific heats*, k, is defined by Eq. 24.28. Typical values are given in Table 24.7.

$$k = \frac{c_p}{c_v} \qquad 24.28$$

14. TRIPLE POINT PROPERTIES

The *triple point* of a substance is a unique state at which solid, liquid, and gaseous phases can coexist. Table 24.3 lists values of the triple point for several common substances.

15. CRITICAL PROPERTIES

If the temperature and pressure of a liquid are increased, a state will eventually be reached at which the liquid and gas phases are indistinguishable. This state is known as the *critical point*, and the properties (generally temperature, pressure, and specific volume) at that point are known as the *critical properties*. At the critical point, the heat of vaporization, h_{fg}, becomes zero. Above the critical temperature, the substance will be a gas no matter how high the pressure. Critical properties are listed in Table 24.4.

16. LATENT HEATS

The total energy (*total heat*, Q_t) entering a substance is the sum of the energy that changes the phase of the substance (*latent heat*, Q_l) and energy that changes the temperature of the substance (*sensible heat*, Q_s). During a phase change (solid to liquid, liquid to vapor, etc.), energy will be transferred to or from the substance without a change in temperature.[10]

$$Q_t = Q_s + Q_l \qquad 24.29$$

[10]Changes in crystalline form are also latent changes.

Examples of latent energies are the *latent heat of fusion* (i.e., change from solid to liquid), h_{sl}, *latent heat of vaporization*, h_{fg}, and *latent heat of sublimation* (i.e., direct change from solid to vapor without becoming liquid), h_{ig}.[11,12] The energy required for these latent changes to occur in water is given in Table 24.5.

Table 24.5 *Latent Heats for Water at One Atmosphere*

effect	Btu/lbm	kJ/kg	cal/g
fusion	143.4	333.5	79.7
vaporization	970.3	2256.7	539.1
sublimation	1220	2838	677.8

(Multiply Btu/lbm by 2.326 to obtain kJ/kg.)
(Multiply Btu/lbm by 5/9 to obtain cal/g.)

Example 24.3

How much energy is required to just vaporize 1.0 lbm (0.45 kg) of water that is originally at 75°F (24°C) and 1 atm?

SI Solution

$$Q_s = mc\,(T_2 - T_1)$$
$$= (0.45\text{ kg})\left(4.190\,\frac{\text{kJ}}{\text{kg·°C}}\right)(100°\text{C} - 24°\text{C})$$
$$= 143.3\text{ kJ}$$

$$Q_l = mh_{fg} = (0.45\text{ kg})\left(2256.7\,\frac{\text{kJ}}{\text{kg}}\right) = 1015.5\text{ kJ}$$

$$Q_t = Q_s + Q_l = 143.3\text{ kJ} + 1015.5\text{ kJ}$$
$$= 1158.8\text{ kJ}$$

Customary U.S. Solution

The sensible heat required to raise the temperature of the water from 75°F to 212°F is given by Eq. 24.22.

$$Q_s = mc\,(T_2 - T_1)$$
$$= (1\text{ lbm})\left(1.0\,\frac{\text{Btu}}{\text{lbm-°F}}\right)(212°\text{F} - 75°\text{F})$$
$$= 137.0\text{ Btu}$$

From Table 24.5, the latent heat required to vaporize the water is

$$Q_l = mh_{fg} = (1\text{ lbm})\left(970.3\,\frac{\text{Btu}}{\text{lbm}}\right) = 970.3\text{ Btu}$$

The total heat required is

$$Q_t = Q_s + Q_l = 137.0\text{ Btu} + 970.3\text{ Btu}$$
$$= 1107.3\text{ Btu}$$

---

[11]The subscript s (for "solid") is sometimes used in place of i (for "ice").
[12]*Sublimation* can only occur below the triple point, where it is too cold for the liquid phase to exist at all.

17. QUALITY: x

Within the vapor dome, water is at its saturation pressure and temperature. There are an infinite number of thermodynamic states in which the water can simultaneously exist in liquid and vapor phases. The *quality* is the fraction by weight of the total mass that is vapor.

$$x = \frac{m_{\text{vapor}}}{m_{\text{vapor}} + m_{\text{liquid}}} \qquad 24.30$$

18. COMPRESSIBILITY

Three different compressibilities are distinguished. The *isobaric compressibility*, β, is defined as

$$\beta = \left(\frac{1}{v}\right)\left(\frac{\partial v}{\partial T}\right)_p \qquad 24.31$$

The *isothermal compressibility*, κ, is

$$\kappa = \left(-\frac{1}{v}\right)\left(\frac{\partial v}{\partial p}\right)_T \qquad 24.32$$

The *isentropic compressibility*, α, is

$$\alpha = \left(-\frac{1}{v}\right)\left(\frac{\partial v}{\partial p}\right)_s \qquad 24.33$$

19. USING THE MOLLIER DIAGRAM

The *Mollier diagram (enthalpy-entropy diagram)* is a graph of enthalpy versus entropy for steam. It is particularly suitable for determining property changes between the superheated vapor and the liquid-vapor regions. For this reason, the Mollier diagram covers only a limited region.

The Mollier diagram plots the enthalpy for a unit mass of steam as the ordinate and plots the entropy as the abscissa. Lines of constant pressure (isobars) slope upward from left to right. Below the saturation line, curves of *constant moisture content* (the complement of quality) slope down from left to right. Above the saturation line are lines of constant temperature and lines of constant superheat.

Example 24.4

Find the following properties using the Mollier diagram: (a) enthalpy and entropy of steam at 700 psia and 1000°F, (b) enthalpy of steam at 1 psia and 80% quality, and (c) final temperature of steam throttled from 700 psia and 1000°F to 450 psia.

Solution

(a) Reading directly from the Mollier diagram (App. 24.E), $h = 1510$ Btu/lbm, and $s = 1.7$ Btu/lbm-°R. (b) 80% quality is the same as 20% moisture. Reading at the intersection of 20% moisture and 1 psia, $h = 900$ Btu/lbm.

(c) By definition, a throttling process does not change the enthalpy. This process is represented by a horizontal line to the right on the Mollier diagram. Starting at the intersection of 700 psia and 1000°F and moving horizontally to the right until 450 psia is reached defines the endpoint of the process. The final temperature is interpolated as approximately 990°F.

20. USING SATURATION TABLES

The information presented graphically on an enthalpy-entropy diagram can be obtained with greater accuracy from *saturation tables*, also known as *property tables* and (in the case of water) *steam tables*. These tables represent extensive tabulations of data for liquid and vapor phases of a substance.

Saturation tables contain values of enthalpy (h), entropy (s), internal energy (u), and specific volume (v). Within the vapor dome, these properties are functions of temperature. Therefore, Apps. 24.A and 24.N (for water) are organized in this manner.

However, as shown in Fig. 24.4, there is a unique pressure associated with each temperature (i.e., there is only one horizontal isobar for each temperature). Since the pressure does not vary in even increments when temperature is changed, the second column of App. 24.A (24.N) varies irregularly. A second property table, App. 24.B (24.O), is set up so that the pressure increments are uniform.

In App. 24.A (24.N), the first column after temperature gives the corresponding saturation pressure. The next two columns give specific volume. The first of these gives the specific volume of the saturated liquid, v_f; the second column gives the specific volume of the saturated vapor, v_g.

The relationship between v_f, v_{fg}, and v_g is given by Eq. 24.34.

$$v_g = v_f + v_{fg} \qquad 24.34$$

The subsequent columns list the same data for enthalpy and entropy.

$$h_g = h_f + h_{fg} \qquad 24.35$$
$$s_g = s_f + s_{fg} \qquad 24.36$$

In App. 24.B (24.O), the first column after the pressure gives the corresponding saturation temperature. The next two columns give specific volume in a manner similar to that of App. 24.A (24.N).

21. USING SUPERHEAT TABLES

In the superheated region, pressure and temperature are independent properties. Therefore, for each pressure, a large number of temperatures is possible. Appendix 24.C (24.P) is a superheated steam table that gives the properties of specific volume, enthalpy, and entropy for various combinations of temperature and pressure. The *degrees of superheat* (e.g., "100°F of superheat") represents the difference between actual and saturation temperatures.

22. USING COMPRESSED LIQUID TABLES

Water is only slightly compressible. For most thermodynamic problems, changes in properties for a liquid are negligible. In problems where the exact values are needed, tables such as App. 24.D (24.Q) of compressed liquid properties are available. Such tables may present the properties directly or may give the properties at the saturation state and the corrections to those properties for various pressures.

Example 24.5

A piston compresses 1.0 lbm of 300°F (140°C) saturated water to 1000 psia (7.5 MPa). What is the final specific volume?

SI Solution

From App. 24.Q, the specific volume of the water at 7.5 MPa and 140°C is read directly as 1.0752 cm^3/g.

Customary U.S. Solution

From App. 24.D, the specific volume of the water in its original saturated state is 0.017449 ft^3/lbm.

The difference in specific volumes for the saturated and compressed water is given by App. 24.D. (To simplify the presentation, each entry in this table has been multiplied by 10^5.)

$$v - v_f = -6.9 \times 10^{-5} \text{ ft}^3/\text{lbm}$$

The specific volume of the compressed water is

$$0.017449 \, \frac{\text{ft}^3}{\text{lbm}} - \left(6.9 \times 10^{-5} \, \frac{\text{ft}^3}{\text{lbm}} \right) = 0.01738 \text{ ft}^3/\text{lbm}$$

23. USING GAS TABLES

Gas tables are essentially superheat tables for which an assumption has been made about the pressure. For example, App. 24.F (24.S) is a gas table for air at low pressures. "Low pressure" means less than several hundred psi pressure. However, reasonably good results can be expected even if pressures are higher.

Gas tables are indexed by temperature. That is, implicit in their use is the assumption that properties are functions of temperature only. Gas tables are not arranged in the same way as other property tables.

The *volume ratio* (v_r) and the *pressure ratio* (p_r) columns can be used when gases take part in an isentropic process. These terms should not be confused with the reduced variables introduced in Sec. 38. These two columns are not the specific volume or pressure. They are ratios with arbitrary references that make analysis of isentropic processes easier. Their use is illustrated in Ex. 24.7 and is based on Eqs. 24.37 and 24.38.

$$\frac{v_{r,1}}{v_{r,2}} = \frac{V_1}{V_2} \qquad [\Delta s = 0] \qquad \textit{24.37}$$

$$\frac{p_{r,1}}{p_{r,2}} = \frac{p_1}{p_2} \qquad [\Delta s = 0] \qquad \textit{24.38}$$

Entropy is not listed at all in App. 24.F (24.S). Instead, a column of *entropy function* (ϕ) with the same units as entropy is given. The entropy function is not the same as specific entropy, but it can be used to calculate the change in entropy as the gas goes through a process. This entropy change is calculated with Eq. 24.39.

$$s_2 - s_1 = \phi_2 - \phi_1 - R \ln\left(\frac{p_2}{p_1}\right) \qquad \textit{24.39}$$

Example 24.6

What is the enthalpy of air at 100°F (310K) and 50 psia (345 kPa)?

SI Solution

Since the pressure is low (less than 20 atm), App. 24.S can be used. From the $T = 310$K line, the enthalpy is read directly as 310.24 kJ/kg.

Customary U.S. Solution

Since the pressure is low (less than 300 psia), App. 24.F can be used. 100°F = 560°. From the $T = 560°$ line, the enthalpy is read directly as 133.86 Btu/lbm.

Example 24.7

Air is originally at 60°F and 14.7 psia. It is compressed isentropically to 86.5 psia. What are its new temperature and enthalpy?

Solution

From App. 24.F for 520°R, $p_{r,1} = 1.2147$. Using Eq. 24.38,

$$p_{r,2} = \frac{p_{r,1}p_2}{p_1} = \frac{(1.2147)(86.5 \text{ psia})}{14.7 \text{ psia}}$$

$$= 7.148$$

Searching the p_r column of App. 24.F results in $T = 860°$R and $h = 206.46$ Btu/lbm.

24. PRESSURE-ENTHALPY CHARTS

For convenience (and by tradition), properties of refrigerants are typically shown graphically in a *pressure-enthalpy (p-h) diagram*. Figure 24.6 shows a skeleton

p-h diagram. The vapor dome and saturation lines from Fig. 24.5 are recognizable. Lines of constant specific volume, constant entropy, and constant temperature are added.

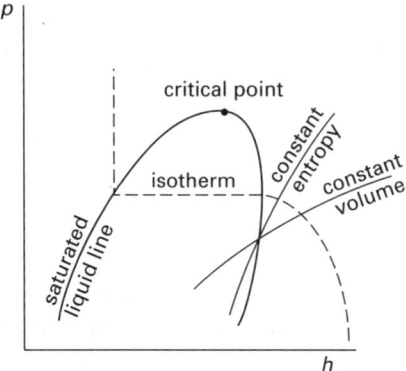

Figure 24.6 *Pressure-Enthalpy Diagram*

Isotherms are horizontal within the vapor dome, corresponding to the constant saturation pressure. In the subcooled-liquid region to the left of the vapor dome, isotherms are essentially vertical, since the temperature (not the pressure) of a subcooled liquid determines enthalpy. In the superheated region, the isotherms gradually approach vertical as the gaseous refrigerant becomes more like an ideal gas.

The enthalpy of superheated refrigerant is easily determined from the pressure and temperature. Isentropic compression follows a line of constant entropy. Throttling follows a line of constant enthalpy (i.e., a vertical line).

25. PROPERTIES OF SOLIDS

There are few mathematical relationships that predict the thermodynamic properties of solids. Properties such as temperature, specific heat, and density, usually are known. If the properties are not known, they must be found from tables.

The reference point for properties of solids is usually absolute zero temperature. That is, properties such as enthalpy and entropy are defined to be zero at 0°R (0K). This is an arbitrary convention. The choice of reference point does not affect the *change* in properties between two temperatures.

26. PROPERTIES OF SUBCOOLED LIQUIDS

A subcooled liquid is at a temperature less than the saturation temperature corresponding to its pressure. Unless the pressure of the liquid is very high, the various thermodynamic properties can be considered to be functions of only the liquid's temperature.

Example 24.8

What is the enthalpy of water at 30 psia (2.0 bars, 0.2 MPa) and 240°F (110°C, 383K)?

SI Solution

The logic is the same as presented in the customary U.S. solution.

$$T_{sat,0.2 \text{ MPa}} = 120.2°C$$

$$T_{actual} < T_{sat} \qquad \text{[liquid phase]}$$

$$h_{f,110°C} = 461.3 \text{ kJ/kg}$$

Customary U.S. Solution

Although the substance is water, its phase is unknown. It could be liquid, vapor, or a combination of the two. From App. 24.B, the saturation (boiling) temperature for 30 psia water is approximately 250°F. Since the actual water temperature is less than the saturation temperature, the water is liquid.

As a liquid, the properties are essentially functions of temperature only. From App. 24.A for 240°F, $h = 208.4$ Btu/lbm.

27. PROPERTIES OF SATURATED LIQUIDS

Either the temperature or the pressure of a saturated liquid must be known in order to identify its thermodynamic state. One defines the other, since there is a one-to-one relationship between saturation pressure and saturation temperature. The first two columns of Apps. 24.A and 24.B (24.N and 24.O) can be used to determine saturation temperatures and pressures.

Since the saturation tables are set up specifically for saturated substances, enthalpy, entropy, internal energy, and specific volume can be read directly from the h_f, s_f, u_f, and v_f columns, respectively. Density can be calculated as the reciprocal of the specific volume. The liquid's vapor pressure is the same as the saturation pressure listed in the table.

28. PROPERTIES OF SATURATED VAPORS

Properties of saturated vapors can be read directly from the saturation tables. The vapor's pressure or temperature can be used to define its thermodynamic state. Enthalpy, entropy, internal energy, and specific volume can be read directly as h_g, s_g, u_g, and v_g, respectively.

29. PROPERTIES OF LIQUID-VAPOR MIXTURES

When the thermodynamic state of a substance is within the vapor dome, there is a one-to-one correspondence between the saturation temperature and saturation

pressure. One determines the other. The thermodynamic state is uniquely defined by any two independent properties (temperature and quality, pressure and enthalpy, entropy and quality, etc.).

If the quality of a liquid-vapor mixture is known, it can be used to calculate all of the primary thermodynamic properties. If a thermodynamic property has a value between the saturated liquid and saturated vapor values (i.e., h is between h_f and h_g), any of Eqs. 24.40 through 24.43 can be solved for the quality.

$$h = h_f + x h_{fg} \qquad \text{24.40}$$

$$s = s_f + x s_{fg} \qquad \text{24.41}$$

$$u = u_f + x u_{fg} \qquad \text{24.42}$$

$$v = v_f + x v_{fg} \qquad \text{24.43}$$

Example 24.9

What is the final enthalpy of superheated steam that is expanded isentropically (i.e., with no change in entropy) from 100 psia (700 kPa) and 500°F (250°C) to 3 psia (20 kPa)?

SI Solution

From App. 24.P, the entropy, s_1, of the superheated steam at 700 kPa and 250°C is 7.1053 kJ/kg·K. Since the expansion is isentropic, this is also the final entropy, s_2.

From App. 24.O for 20 kPa (0.20 bars) vapor, the entropy of a saturated liquid, s_f, is 0.8320 kJ/kg·K; and the entropy of a saturated vapor, s_g, is 7.9085 kJ/kg·K. Since $s_2 < s_g$, the expanded steam is in the liquid-vapor region. The quality of the mixture is given by Eq. 24.41.

$$x = \frac{s - s_f}{s_{fg}} = \frac{s - s_f}{s_f - s_g}$$

$$= \frac{7.1053 \, \dfrac{\text{kJ}}{\text{kg·K}} - 0.8320 \, \dfrac{\text{kJ}}{\text{kg·K}}}{7.9085 \, \dfrac{\text{kJ}}{\text{kg·K}} - 0.8320 \, \dfrac{\text{kJ}}{\text{kg·K}}}$$

$$= 0.8865$$

From App. 24.O, the enthalpy of saturated liquid, h_f, at 20 kPa is 251.4 kJ/kg. The heat of vaporization, h_{fg}, is 2358.3 kJ/kg. The final enthalpy is given by Eq. 24.40.

$$h = h_f + x h_{fg}$$

$$= 251.4 \, \frac{\text{kJ}}{\text{kg}} + (0.8865)\left(2358.3 \, \frac{\text{kJ}}{\text{kg}}\right)$$

$$= 2342.0 \text{ kJ/kg}$$

Customary U.S. Solution

From App. 24.C, the entropy, s_1, of the superheated steam at 100 psia and 500°F is 1.7085 Btu/lbm-°R.

Since the expansion is isentropic, this is also the final entropy, s_2.

From App. 24.B for 3 psia vapor, the entropy of a saturated liquid, s_f, is 0.2009 Btu/lbm-°R; the entropy of vaporization, s_{fg}, is 1.6852 Btu/lbm-°R; and the entropy of a saturated vapor, s_g, is 1.8861 Btu/lbm-°R. Since $s_2 < s_g$, the expanded steam is in the liquid-vapor region. The quality of the mixture is given by Eq. 24.41.

$$x = \frac{s - s_f}{s_{fg}}$$

$$= \frac{1.7085 \, \frac{\text{Btu}}{\text{lbm-°R}} - 0.2009 \, \frac{\text{Btu}}{\text{lbm-°R}}}{1.6852 \, \frac{\text{Btu}}{\text{lbm-°R}}}$$

$$= 0.8946$$

From App. 24.B, the enthalpy of saturated liquid, h_f, at 3 psia is 109.39 Btu/lbm. The heat of vaporization, h_{fg}, is 1013.1 Btu/lbm. The final enthalpy is given by Eq. 24.40.

$$h = h_f + x h_{fg}$$

$$= 109.39 \, \frac{\text{Btu}}{\text{lbm}} + (0.8946) \left(1013.1 \, \frac{\text{Btu}}{\text{lbm}} \right)$$

$$= 1015.7 \, \text{Btu/lbm}$$

Example 24.10

What is the enthalpy of 200°F (90°C, 363K) steam with a quality of 90%?

SI Solution

Using App. 24.N and Eq. 24.40,

$$h = h_f + x h_{fg}$$

$$= 376.92 \, \frac{\text{kJ}}{\text{kg}} + (0.9) \left(2283.2 \, \frac{\text{kJ}}{\text{kg}} \right) = 2431.8 \, \text{kJ/kg}$$

Customary U.S. Solution

Using App. 24.A and Eq. 24.40,

$$h = h_f + x h_{fg}$$

$$= 168.07 \, \frac{\text{Btu}}{\text{lbm}} + (0.9) \left(977.9 \, \frac{\text{Btu}}{\text{lbm}} \right)$$

$$= 1048.2 \, \text{Btu/lbm}$$

30. PROPERTIES OF SUPERHEATED VAPORS

Unless a vapor is highly superheated, its properties should be found from a superheat table, such as App. 24.C or 24.P, for water vapor. Since temperature and pressure are independent for a superheated vapor, both must be known in order to define the thermodynamic state.

If the vapor's temperature and pressure do not correspond to the superheat table entries, single or double interpolation will be required. Such interpolation can be avoided by using more complete tables, but where required, linear interpolation is standard practice.

Example 24.11

What is the enthalpy of water at 300 psia (20 bars, 2.0 MPa) and 900°F (500°C, 773K)?

SI Solution

The enthalpy can be read directly from App. 24.P as 3467.6 kJ/kg.

Customary U.S. Solution

It may not be obvious what phase the water is in. From App. 24.B, the saturation (boiling) temperature for 300 psia water is 417.43°F. Since the actual water temperature is higher than the saturation temperature, the water exists as a vapor.

From App. 24.C, the enthalpy can be read directly as 1473.6 Btu/lbm.

31. EQUATION OF STATE FOR IDEAL GASES

An *equation of state* is a relationship that predicts the state (a property such as pressure, temperature, volume, etc.) from a set of two other independent properties.

Avogadro's law states that equal volumes of different gases at the same temperature and pressure contain equal numbers of molecules. For one mole of any gas, Avogadro's law can be stated as the *equation of state for ideal gases*. (Temperature, T, in Eq. 24.44 must be in degrees absolute.)

$$\frac{pV}{T} = R^* \qquad \qquad 24.44$$

In Eq. 24.44, R^* is known as the *universal gas constant*. It is "universal" (within a system of units) because the same value can be used with any gas. Its value depends on the units used for pressure, temperature, and volume, as well as on the units of mass.

The ideal gas equation of state can be modified for more than one mole of gas. If there are n moles,

$$pV = nR^*T \qquad \qquad 24.45$$

The number of moles can be calculated from the substance's mass and molecular weight.

$$n = \frac{m}{\text{MW}} \qquad \qquad 24.46$$

Table 24.6 *Values of the Universal Gas Constant, R**

units in SI and other metric systems

8.3143 kJ/kmol·K

8314.3 J/kmol·K

0.08206 atm·L/mol·K

1.986 cal/mol·K

8.314 J/mol·K

82.06 atm·cm³/mol·K

0.08206 atm·m³/kmol·K

8314.3 kg·m²/s² kmol·K

8314.3 m³·Pa/kmol·K

8.314×10^7 erg/mol·K

units in English systems

1545.33 ft-lbf/lbmol-°R

1.986 Btu/lbmol-°R

0.7302 atm-ft³/lbmol-°R

10.73 ft³-lbf/in²-lbmol-°R

Equations 24.45 and 24.46 can be combined. R is the *specific gas constant*. It is specific because it is valid only for a gas with a molecular weight of MW.

$$pV = \frac{mR^*T}{MW} = m\left(\frac{R^*}{MW}\right)T = mRT \qquad \textbf{24.47}$$

$$R = \frac{R^*}{MW} \qquad \textbf{24.48}$$

Values of the specific gas constant and the molecular weights of several common gases are given in Table 24.7.

Example 24.12

What mass of nitrogen is contained in a 2000 ft³ (57 m³) tank if the pressure and temperature are 1 atm and 70°F (21°C), respectively?

SI Solution

First, convert to absolute temperature.

$$T = 21°C + 273 = 294K$$

From Table 24.6, $R = 297$ J/kg·K. From Eq. 24.47,

$$m = \frac{pV}{RT}$$

$$= \frac{(1\ \text{atm})\left(1.013 \times 10^5\ \dfrac{\text{Pa}}{\text{atm}}\right)(57\ \text{m}^3)}{\left(297\ \dfrac{\text{J}}{\text{kg·K}}\right)(294K)}$$

$$= 66.1\ \text{kg}$$

Customary U.S. Solution

$$T = 70°F + 460 = 530°$$

From Table 24.7, $R = 55.16$ ft-lbf/lbm-°R. From Eq. 24.47,

$$m = \frac{pV}{RT}$$

$$= \frac{(1\ \text{atm})\left(14.7\ \dfrac{\text{lbf}}{\text{in}^2\text{-atm}}\right)\left(144\ \dfrac{\text{in}^2}{\text{ft}^2}\right)(2000\ \text{ft}^3)}{\left(55.16\ \dfrac{\text{ft-lbf}}{\text{lbm-°R}}\right)(530°\text{R})}$$

$$= 144.8\ \text{lbm}$$

Example 24.13

A 25 ft³ (0.71 m³) tank contains 10 lbm (4.5 kg) of an ideal gas. The gas has a molecular weight of 44 and is at 70°F (21°C). What is the pressure of the gas?

SI Solution

From Eq. 24.48, the specific gas constant is

$$R = \frac{R^*}{MW} = \frac{8314.3\ \dfrac{\text{J}}{\text{kmol·K}}}{44\ \dfrac{\text{kg}}{\text{kmol}}}$$

$$= 189\ \text{J/kg·K}$$

The absolute temperature is $T = 21°C + 273 = 294K$. From Eq. 24.47, the pressure is

$$p = \frac{mRT}{V} = \frac{(4.5\ \text{kg})\left(189\ \dfrac{\text{J}}{\text{kg·K}}\right)(294K)}{(0.71\ \text{m}^3)\left(1000\ \dfrac{\text{Pa}}{\text{kPa}}\right)}$$

$$= 352.2\ \text{kPa}$$

Customary U.S. Solution

$$R = \frac{R^*}{MW} = \frac{1545.33\ \dfrac{\text{ft-lbf}}{\text{lbmol-°R}}}{44\ \dfrac{\text{lbm}}{\text{lbmol}}}$$

$$= 35.12\ \text{ft-lbf/lbm-°R}$$

$$T = 70°F + 460 = 530°\text{R}$$

$$p = \frac{mRT}{V} = \frac{(10\ \text{lbm})\left(35.12\ \dfrac{\text{ft-lbf}}{\text{lbm-°R}}\right)(530°\text{R})}{25\ \text{ft}^3}$$

$$= 7445\ \text{lbf/ft}^2$$

Table 24.7 *Approximate Properties of Selected Gases*

gas	symbol	temp °F	MW	customary U.S. units			SI units			k
				R ft-lbf/lbm-°R	c_p Btu/lbm-°R	c_v Btu/lbm-°R	R J/kg·K	c_p J/kg·K	c_v J/kg·K	
acetylene	C_2H_2	68	26.038	59.35	0.350	0.274	319.32	1465	1146	1.279
air		100	28.967	53.35	0.240	0.171	287.03	1005	718	1.400
ammonia	NH_3	68	17.032	90.73	0.523	0.406	488.16	2190	1702	1.287
argon	Ar	68	39.944	38.69	0.124	0.074	208.15	519	311	1.669
butane (n)	C_4H_{10}	68	58.124	26.59	0.395	0.361	143.04	1654	1511	1.095
carbon dioxide	CO_2	100	44.011	35.11	0.207	0.162	188.92	867	678	1.279
carbon monoxide	CO	100	28.011	55.17	0.249	0.178	296.82	1043	746	1.398
chlorine	Cl_2	100	70.910	21.79	0.115	0.087	117.25	481	364	1.322
ethane	C_2H_6	68	30.070	51.39	0.386	0.320	276.50	1616	1340	1.206
ethylene	C_2H_4	68	28.054	55.08	0.400	0.329	296.37	1675	1378	1.215
Freon (R-12)[a]	CCl_2F_2	200	120.925	12.78	0.159	0.143	68.76	666	597	1.115
helium	He	100	4.003	386.04	1.240	0.744	2077.03	5192	3115	1.667
hydrogen	H_2	100	2.016	766.53	3.420	2.435	4124.18	14 319	10 195	1.405
hydrogen sulfide	H_2S	68	34.082	45.34	0.243	0.185	243.95	1017	773	1.315
krypton	Kr		83.800	18.44	0.059	0.035	99.22	247	148	1.671
methane	CH_4	68	16.043	96.32	0.593	0.469	518.25	2483	1965	1.264
neon	Ne	68	20.183	76.57	0.248	0.150	411.94	1038	626	1.658
nitrogen	N_2	100	28.016	55.16	0.249	0.178	296.77	1043	746	1.398
nitric oxide	NO	68	30.008	51.50	0.231	0.165	277.07	967	690	1.402
nitrous oxide	NO_2	68	44.01	35.11	0.221	0.176	188.92	925	736	1.257
octane vapor	C_8H_{18}		114.232	13.53	0.407	0.390	72.78	1704	1631	1.045
oxygen	O_2	100	32.000	48.29	0.220	0.158	259.82	921	661	1.393
propane	C_3H_8	68	44.097	35.04	0.393	0.348	188.55	1645	1457	1.129
sulfur dioxide	SO_2	100	64.066	24.12	0.149	0.118	129.78	624	494	1.263
water vapor[a]	H_2O	212	18.016	85.78	0.445	0.335	461.50	1863	1402	1.329
xenon	Xe		131.300	11.77	0.038	0.023	63.32	159	96	1.661

(Multiply Btu/lbm-°F by 4186.8 to obtain J/kg·K.)
(Multiply ft-lbf/lbm-°R by 5.3803 to obtain J/kg·K.)
[a] Values for steam and Freon are approximate and should be used only for low pressures and high temperatures.

32. PROPERTIES OF IDEAL GASES

A gas can be considered to behave ideally if its pressure is very low and the temperature is much higher than its critical temperature. (Otherwise, the substance is in vapor form.) Under these conditions, the molecule size is insignificant compared with the distance between molecules, and molecules do not come into contact. By definition, an *ideal gas* behaves according to the various ideal gas laws.

Values of h, u, and v for gases usually are read from gas tables. Since it contains a pv term, enthalpy can be related to the equation of state. Depending on the units chosen, a conversion factor may be needed.

$$h = u + pv = u + RT \qquad 24.49$$

Furthermore, density is the reciprocal of specific volume, v. Therefore, the density of an ideal gas can be derived from the equation of state by setting $m = 1$ and solving for the reciprocal of volume.

$$\rho = \frac{1}{v} = \frac{p}{RT} \qquad 24.50$$

33. SPECIFIC HEATS OF IDEAL GASES

The specific heats of an ideal gas can be calculated from its specific gas constant. Depending on the units chosen, a conversion factor may or may not be needed in Eqs. 24.51 through 24.54.

$$c_p - c_v = R \qquad 24.51$$
$$C_p - C_v = R^* \qquad 24.52$$
$$c_p = \frac{Rk}{k - 1} \qquad 24.53$$
$$C_p = \frac{R^*k}{k - 1} \qquad 24.54$$

34. KINETIC GAS THEORY

The *kinetic gas theory* predicts the velocity distribution of gas molecules as a function of temperature. This theory makes the following assumptions.

- Gas molecules do not attract one another.

- The volume of the gas molecules is negligible compared with the volume of the gas.

- The molecules behave like hard spheres.

- The container volume is large enough that interactions with the wall are not a predominant activity of the molecules.

Equation 24.55 is the *Maxwell-Boltzmann distribution* of gas molecule velocities. The variable κ is the *Boltzmann constant*, which has a value of 1.3803×10^{-23} J/molecule·K. (These units are the same as kg · m²/s² · molecule · K.) Kinetic gas theory calculations are traditionally performed in SI units. English units can be used, however, if all constants and units are consistent.

$$\frac{dN}{dv} = \left(\frac{4N}{\sqrt{\pi}}\right)(v^2)\left(\frac{m}{2\kappa T}\right)^{\frac{3}{2}} e^{\frac{-mv^2}{2\kappa T}} \qquad 24.55$$

$$\kappa = \frac{R^*}{N_A} \qquad 24.56$$

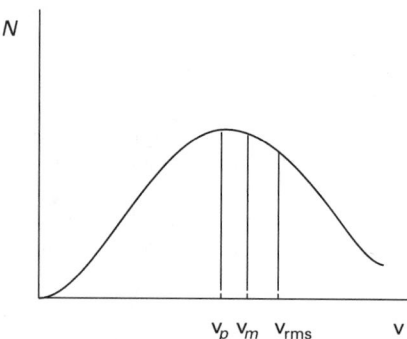

Figure 24.7 *Maxwell-Boltzmann Velocity Distribution*

The *most probable speed* of a molecule with mass m is

$$v_p = \sqrt{\frac{2\kappa T}{m}} \qquad 24.57$$

The *mean speed* of a molecule with mass m is

$$v_m = 2\sqrt{\frac{2\kappa T}{\pi m}} \qquad 24.58$$

The *root-mean-square speed* of a molecule with mass m is

$$v_{\text{rms}} = \sqrt{\frac{3\kappa T}{m}} \qquad 24.59$$

The three velocities are illustrated in Fig. 24.7 and are related.

$$\frac{v_m}{v_p} = 1.128 \qquad 24.60$$

$$\frac{v_{\text{rms}}}{v_p} = 1.225 \qquad 24.61$$

Temperature has a molecular interpretation derived from the kinetic gas theory. It can be shown that the root-mean-square velocity is related to the absolute temperature. That is, absolute temperature is proportional to the square of the rms velocity.

$$T = \left(\frac{m}{3\kappa}\right)v_{\text{rms}}^2 \qquad 24.62$$

Since $\frac{1}{2}mv^2$ is the definition of kinetic energy, the mean translational kinetic energy of the molecule is proportional to the mean absolute temperature.

$$E_k = \tfrac{1}{2}mv_{\text{rms}}^2 = \tfrac{3}{2}\kappa T \qquad 24.63$$

The pressure of a gas can be calculated as the total change in momentum of all the gas molecules bouncing off the walls of the container.

$$p = \frac{\rho v_{\text{rms}}^2}{3} = \frac{Nmv_{\text{rms}}^2}{3V} \qquad 24.64$$

A small-enough particle suspended in a fluid will exhibit small random movements due to the statistical (i.e., random) collisions of fluid molecules on the particle's surface. Such motion is known as *Brownian movement*.

Example 24.14

What are the kinetic energy and rms velocity of 275K argon molecules (MW = 39.9)?

Solution

From Eq. 24.63,

$$E_k = \tfrac{3}{2}\kappa T$$
$$= (1.5)\left(1.3803 \times 10^{-23}\,\frac{\text{J}}{\text{molecule·K}}\right)(275\text{K})$$
$$= 5.69 \times 10^{-21}\ \text{J/molecule}$$

The molecular mass of argon is its mass per mole divided by the number of molecules in a mole.

$$m = \frac{\text{MW}}{N_A} = \frac{39.9\,\dfrac{\text{kg}}{\text{kmol}}}{\left(6.023 \times 10^{23}\,\dfrac{\text{molecules}}{\text{mol}}\right)\left(1000\,\dfrac{\text{mol}}{\text{kmol}}\right)}$$
$$= 6.62 \times 10^{-26}\ \text{kg/molecule}$$

From Eq. 24.59, the rms velocity is

$$v_{\text{rms}} = \sqrt{\frac{3\kappa T}{m}}$$
$$= \sqrt{\frac{(3)\left(1.3803 \times 10^{-23}\,\dfrac{\text{J}}{\text{molecule·K}}\right)(275\text{K})}{6.62 \times 10^{-26}\,\dfrac{\text{kg}}{\text{molecule}}}}$$
$$= 414.7\ \text{m/s}$$

35. GRAVIMETRIC, VOLUMETRIC, AND MOLE FRACTIONS

The *gravimetric fraction*, G_A (also known as the *mass fraction*), of a component A in a mixture of components A, B, C, and so on, is the ratio of the component's mass to the total mixture mass.

$$G_A = \frac{m_A}{m} = \frac{m_A}{m_A + m_B + m_C} \qquad 24.65$$

The *volumetric fraction*, B_A, of a component A is the ratio of the component's partial volume to the overall mixture volume.

$$B_A = \frac{V_A}{V} = \frac{V_A}{V_A + V_B + V_C} \qquad 24.66$$

It is possible to convert between gravimetric and volumetric fractions.

$$G_A = \frac{B_A(\text{MW})_A}{B_A(\text{MW})_A + B_B(\text{MW})_B + B_C(\text{MW})_C} \qquad 24.67$$

$$B_A = \frac{\dfrac{G_A}{(\text{MW})_A}}{\dfrac{G_A}{(\text{MW})_A} + \dfrac{G_B}{(\text{MW})_B} + \dfrac{G_C}{(\text{MW})_C}} \qquad 24.68$$

The *mole fraction*, x_A, of a component A is the ratio of the number of moles of substance A to the total number of moles of all substances.

$$x_A = \frac{n_A}{n} = \frac{n_A}{n_A + n_B + n_C} \qquad 24.69$$

For nonreacting mixtures of ideal gases, the mole fraction and volumetric fraction (and partial pressure ratio) are the same.

$$x_A = B_A \Big|_{\text{ideal gases}} \qquad 24.70$$

36. PARTIAL PRESSURE AND PARTIAL VOLUME OF GAS MIXTURES

A *gas mixture* consists of an aggregation of molecules of each gas component, the molecules of any single component being distributed uniformly and moving as if they alone occupied the space. The *partial volume*, V_A, of a gas A in a mixture of nonreacting gases A, B, C, and so on, is the volume that gas A alone would occupy at the temperature and pressure of the mixture.

The partial volume can be calculated from the volumetric fraction and total volume.

$$V_A = B_A V \qquad 24.71$$

Amagat's law (also known as *Amagat-Leduc's rule*) states that the total volume of a mixture of nonreacting gases is equal to the sum of the partial volumes.

$$V = V_A + V_B + V_C \qquad 24.72$$

The *partial pressure*, p_A, of a gas A in a mixture of nonreacting gases A, B, C, and so on, is the pressure gas A alone would exert in the total volume at the temperature of the mixture.

$$p_A = \frac{m_A R_A T}{V} = \frac{n_A R^* T}{V} \qquad 24.73$$

The partial pressure can also be calculated from the mole fraction and the total pressure. However, for ideal gases, the partial pressure ratio, mole fraction, and volumetric fraction are the same.

$$\frac{p_A}{p} = x_A = B_A \qquad 24.74$$

If the average specific gas constant, $\overline{R}$, for the gas mixture is known, it can be used with the gravimetric fraction to calculate the partial pressure.

$$p_A = \frac{G_A R_A p}{\overline{R}} \qquad 24.75$$

According to *Dalton's law of partial pressures*, the *total pressure* of a gas mixture is the sum of the partial pressures.

$$p = p_A + p_B + p_C \qquad 24.76$$

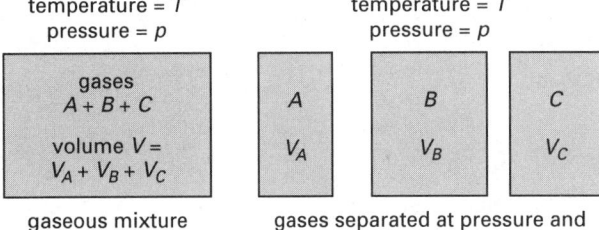

Figure 24.8 *Mixture of Ideal Gases*

37. PROPERTIES OF NONREACTING IDEAL GAS MIXTURES

A mixture's average molecular weight and density are volumetrically weighted averages of its components' values.

$$\text{MW} = \sum B_i (\text{MW})_i \qquad 24.77$$

$$\rho = \sum B_i \rho_i \qquad 24.78$$

On the other hand, a mixture's internal energy, enthalpy, specific heats, and specific gas constant are equal to the sum of the values of its individual components (i.e., the mixture average is gravimetrically weighted).

$$u = \frac{\sum m_i u_i}{m} = \sum_i G_i u_i \qquad 24.79$$

$$h = \frac{\sum m_i h_i}{m} = \sum_i G_i h_i \qquad 24.80$$

$$c_v = \frac{\sum m_i c_{v,i}}{m} = \sum_i G_i c_{v,i} \qquad \text{24.81}$$

$$c_p = \frac{\sum m_i c_{p,i}}{m} = \sum_i G_i c_{p,i} \qquad \text{24.82}$$

$$R = \frac{\sum m_i R_i}{m} = \sum_i G_i R_i \qquad \text{24.83}$$

If the mixing is reversible and adiabatic, the entropy will also be equal to the sum of the individual entropies. However, each individual entropy, s_i, must be evaluated at the temperature and volume of the mixture and at the individual partial pressure, p_i.

$$s = \frac{\sum m_i s_i}{m} = \sum_i G_i s_i \qquad \text{24.84}$$

Equations 24.79, 24.80, and 24.84 are mathematical formulations of *Gibbs theorem* (also known as *Gibbs rule*). This theorem states that the total property (e.g., U, H, or S) of a mixture of ideal gases is the sum of the properties that the individual gases would have if each occupied the total mixture volume alone at the same temperature.

Table 24.8 summarizes the composite gas properties. Notice that the ratio of specific heats, k, is not a composite gas property.

Table 24.8 *Summary of Composite Ideal Gas Properties*

gravimetrically (mass) weighted	volumetrically (mole fraction) weighted
u	U
h	H
c_p	C_p
c_v	C_v
R	MW
s	S
	ρ

Example 24.15

0.14 lbm (0.064 kg) of octane vapor (MW = 114) is mixed with 2.0 lbm (0.91 kg) of air (MW = 29.0) in the manifold of an engine. The total pressure in the manifold is 12.5 psia (86.1 kPa), and the temperature is 520° (290K). Assume octane behaves ideally. (a) What is the total volume of this mixture? (b) What is the partial pressure of the air in the mixture?

SI Solution

(a) The number of moles of octane and air are

$$n_{\text{octane}} = \frac{m}{MW} = \frac{0.064 \text{ kg}}{114 \dfrac{\text{kg}}{\text{kmol}}} = 5.61 \times 10^{-4} \text{ kmol}$$

$$n_{\text{air}} = \frac{0.91 \text{ kg}}{29 \dfrac{\text{kg}}{\text{kmol}}} = 0.0314 \text{ kmol}$$

From Eq. 24.45, the total volume of any gas is

$$V = \frac{nR^*T}{p}$$

$$= (5.61 \times 10^{-4} \text{ kmol} + 0.0314 \text{ kmol})$$

$$\times \left[\frac{\left(8314.3 \dfrac{\text{J}}{\text{kmol·K}} \right)(290\text{K})}{86.1 \times 10^3 \text{ Pa}} \right]$$

$$= 0.895 \text{ m}^3$$

(b) The mole fraction of the air is

$$x_{\text{air}} = \frac{n_{\text{air}}}{n_{\text{total}}} = \frac{0.0314 \text{ kmol}}{5.61 \times 10^{-4} \text{ kmol} + 0.0314 \text{ kmol}}$$

$$= 0.982$$

The partial pressure of air is

$$p_{\text{air}} = x_{\text{air}}p = (0.982)(86.1 \text{ kPa})$$

$$= 84.6 \text{ kPa}$$

Customary U.S. Solution

(a) The number of moles of octane and air are

$$n_{\text{octane}} = \frac{m}{MW} = \frac{0.14 \text{ lbm}}{114 \dfrac{\text{lbm}}{\text{lbmol}}} = 0.001228 \text{ lbmol}$$

$$n_{\text{air}} = \frac{2.0 \text{ lbm}}{29.0 \dfrac{\text{lbm}}{\text{lbmol}}} = 0.068966 \text{ lbmol}$$

From Eq. 24.45, the total volume is

$$V = \frac{nR^*T}{p}$$

$$= (0.001228 \text{ lbmol} + 0.068966 \text{ lbmol})$$

$$\times \left[\frac{\left(1545.33 \dfrac{\text{ft-lbf}}{\text{lbmol-°R}} \right)(520\text{°R})}{\left(12.5 \dfrac{\text{lbf}}{\text{in}^2} \right)\left(144 \dfrac{\text{in}^2}{\text{ft}^2} \right)} \right]$$

$$= 31.34 \text{ ft}^3$$

(b) The mole fraction of the air is

$$x_{\text{air}} = \frac{n_{\text{air}}}{n} = \frac{0.068966 \text{ lbmol}}{0.001228 \text{ lbmol} + 0.068966 \text{ lbmol}}$$

$$= 0.9825$$

The partial air pressure is

$$p_{\text{air}} = x_{\text{air}}p = (0.9825)\left(12.5 \dfrac{\text{lbf}}{\text{in}^2} \right)$$

$$= 12.3 \text{ psia}$$

38. EQUATION OF STATE FOR REAL GASES

Real gases do not meet the basic assumptions defining an ideal gas. Specifically, the molecules of a real gas occupy a volume that is not negligible in comparison with the total volume of the gas. (This is especially true for gases at low temperatures.) Furthermore, real gases are subject to *van der Waals' forces*, which are attractive forces between gas molecules.

There are two methods of accounting for real gas behavior. The first method is to modify the ideal gas equation of state with various empirical correction factors. Since the modifications are empirical, the resulting equations of state are known as *correlations*. One well-known correlation is *van der Waals' equation of state*.

$$\left(p + \frac{a}{V^2}\right)(V - b) = nR^*T \qquad 24.85$$

The van der Waals corrections usually need to be made only when a gas is below its critical temperature. For an ideal gas, the a and b terms are zero. When the spacing between molecules is close, as it would be at low temperatures, the molecules attract each other and reduce the pressure exerted by the gas. The pressure is then corrected by the a/V^2 term. b is a constant that accounts for the molecular volume in a dense state.

Other correlations of this type include the *Clausius, Bertholet, Dieterici,* and *Beattie-Bridgeman equations of state*. However, the most accurate empirical correlation is the *virial equation of state*, which has the form shown in Eqs. 24.86 and 24.87. The constants B, C, D, and so on, are called the *virial coefficients*.

$$pV = nR^*T\left(1 + \frac{B}{V} + \frac{C}{V^2} + \frac{D}{V^3} + \cdots\right) \qquad 24.86$$

$$pV = nR^*T\left(1 + B'p + C'p^2 + D'p^3 + \cdots\right) \qquad 24.87$$

The *principle (law) of corresponding states* provides a second method of correcting for real gas behavior. This law states that the *reduced properties* of all gases are identical (i.e., all gases behave similarly when their reduced variables are used). Specifically, there is one property, the *compressibility factor*, Z, that has the same value for all gases when evaluated at the same values of the *reduced variables*.[13] (The reduced variables are not the same as the ratios defined in Sec. 23.)

$$Z = f(T_r, p_r, v_r) \qquad 24.88$$

$$T_r = \frac{T}{T_c} \qquad 24.89$$

$$p_r = \frac{p}{p_c} \qquad 24.90$$

$$v_r = \frac{v}{v_c} \qquad 24.91$$

---

[13]Some engineering disciplines (e.g., petroleum engineering) use the symbol z for compressibility factor.

Compressibility factors are almost always read from *generalized compressibility charts* such as App. 24.Z. The compressibility factor can then be used to correct the ideal gas equation of state.

$$pv = ZRT \qquad 24.92$$

$$pV = mZRT \qquad 24.93$$

Example 24.16

What is the specific volume of carbon dioxide at 2680 psia (182 atm) and 300°F (150°C)?

SI Solution

The absolute temperature is

$$T = 150°C + 273 = 423K$$

From Table 24.4, the critical temperature and pressure of carbon dioxide are 304.3K and 72.9 atm, respectively. The reduced variables are

$$T_r = \frac{T}{T_c} = \frac{423K}{304.3K} = 1.39$$

$$p_r = \frac{p}{p_c} = \frac{182 \text{ atm}}{72.9 \text{ atm}} = 2.5$$

From App. 24.Z, Z is read as 0.75. R is read from Table 24.6. Solving Eq. 24.92 for v,

$$v = \frac{ZRT}{p} = \frac{(0.75)\left(189\ \dfrac{\text{J}}{\text{kg·K}}\right)(423K)}{(182 \text{ atm})\left(101.3 \times 10^3\ \dfrac{\text{Pa}}{\text{atm}}\right)}$$

$$= 3.25 \times 10^{-3} \text{ m}^3/\text{kg}$$

Customary U.S. Solution

The absolute temperature is

$$T = 300°F + 460 = 760°R$$

From Table 24.4, the critical temperature and pressure of carbon dioxide are 547.8°R and 72.9 atm, respectively. The reduced variables are

$$T_r = \frac{T}{T_c} = \frac{760°R}{547.8°R} = 1.39$$

$$p_r = \frac{p}{p_c} = \frac{2680 \text{ psia}}{(72.9 \text{ atm})\left(14.7\ \dfrac{\text{psia}}{\text{atm}}\right)} = 2.5$$

From App. 24.Z, Z is read as 0.75. R is read from Table 24.6. Solving Eq. 24.92 for v,

$$v = \frac{ZRT}{p} = \frac{(0.75)\left(35.11\ \dfrac{\text{ft-lbf}}{\text{lbm-°R}}\right)(760°R)}{\left(2680\ \dfrac{\text{lbf}}{\text{in}^2}\right)\left(144\ \dfrac{\text{in}^2}{\text{ft}^2}\right)}$$

$$= 0.0519 \text{ ft}^3/\text{lbm}$$

39. SPECIFIC HEATS OF REAL GASES

By definition, the specific heat of an ideal gas is independent of temperature. However, the specific heat of a real gas varies with temperature and (slightly) with pressure. There are several ways to find the specific heat of a gas at different temperatures: tables, graphs, and correlations.[14] Table 22.1 is a typical tabulation of specific heats versus temperature.

Equation 24.94 is a typical temperature correlation. The variation with pressure, being small at low pressures, is disregarded. The coefficients A, B, C, and D will depend on the units, the temperature range over which the correlation is to be used, and the desired accuracy. Typical coefficients are given in Table 24.9.

$$c_p = A + BT + CT^2 + \frac{D}{\sqrt{T}} \qquad 24.94$$

$$c_v = c_p - R \qquad \text{[SI]} \qquad 24.95(a)$$

$$c_v = c_p - \frac{R}{J} \qquad \text{[U.S.]} \qquad 24.95(b)$$

PRACTICE PROBLEMS

1. What is the molar enthalpy of 250°F (120°C) steam with a quality of 92%?

2. What is the ratio of specific heats for air at 600°F (300°C)?

3. What is the density of helium at 600°F (300°C) and one standard atmosphere?

Table 24.9 Correlation Coefficients for Calculating Specific Heat[a] (Btu/lbm-°R)

gas	temperature range °R	K	A	B	C	D
air	400 to 1200	220 to 670	0.2405	-1.186×10^{-5}	20.1×10^{-9}	0
	1200 to 4000	670 to 2220	0.2459	3.22×10^{-5}	-3.74×10^{-9}	-0.833
CH_4	400 to 1000	220 to 560	0.453	0.62×10^{-5}	268.8×10^{-9}	0
	1000 to 4000	560 to 2220	1.152	32.58×10^{-5}	-41.29×10^{-9}	-22.42
CO	400 to 1200	220 to 670	0.2534	-2.35×10^{-5}	26.88×10^{-9}	0
	1200 to 4000	670 to 2220	0.2763	3.04×10^{-5}	-3.89×10^{-9}	-1.5
CO_2	400 to 4000	220 to 2220	0.328	3.2×10^{-5}	-4.4×10^{-9}	-3.33
H_2	400 to 1000	220 to 560	2.853	145×10^{-5}	-883×10^{-9}	0
	1000 to 2500	560 to 1390	3.447	-4.7×10^{-5}	70.3×10^{-9}	0
	2500 to 4000	1390 to 2220	2.841	45×10^{-5}	-31.2×10^{-9}	0
H_2O	400 to 1800	220 to 1000	0.4267	2.425×10^{-5}	23.85×10^{-9}	0
	1800 to 4000	1000 to 2220	0.3275	14.67×10^{-5}	-13.59×10^{-9}	0
N_2	400 to 1200	220 to 670	0.2510	-1.63×10^{-5}	20.4×10^{-9}	0
	1200 to 4000	670 to 2220	0.2192	4.38×10^{-5}	-5.14×10^{-9}	-0.124
O_2	400 to 1200	220 to 670	0.213	0.188×10^{-5}	20.3×10^{-9}	0
	1200 to 4000	670 to 2220	0.340	-0.36×10^{-5}	0.616×10^{-9}	-3.19

[a]Multiply values in this table by 2.326 to obtain coefficients for specific heats in kJ/kg·K.
Adapted from a table published in *Engineering Thermodynamics*, C. O. Mackay, W. N. Barnard, and F. O. Ellenwood (John Wiley, New York, 1957).

[14]It is also possible to calculate the specific heat over a small temperature range if the enthalpies are known (e.g., from an air table) from $c_p = \Delta h/\Delta T$. However, if the enthalpies are known, it is unlikely that the specific heat will be needed.

25 Changes in Thermodynamic Properties

1. Systems 25-1
2. Types of Processes 25-2
3. Polytropic Processes 25-2
4. Throttling Processes 25-2
5. Reversible Processes 25-3
6. Finding Work and Heat Graphically 25-4
7. Sign Convention 25-4
8. First Law of Thermodynamics for
   Closed Systems 25-4
9. Thermal Equilibrium 25-5
10. First Law of Thermodynamics for
    Open Systems 25-5
11. Basic Thermodynamic Relations 25-6
12. Property Changes in Ideal Gases 25-7
13. Property Changes in
    Incompressible Fluids and Solids 25-10
14. Heat Reservoirs 25-10
15. Cycles 25-10
16. Thermal Efficiency 25-10
17. Second Law of Thermodynamics 25-11
18. Availability 25-11
    Practice Problems 25-12

Nomenclature

c	specific heat	Btu/lbm-°F	kJ/kg·K
C	molar specific heat	Btu/lbmol-°F	kJ/kmol·K
E	energy	Btu	kJ
g	acceleration of gravity	ft/sec^2	m/s^2
g_c	gravitational constant	lbm-ft/lbf-sec^2	n.a.
h	enthalpy	Btu/lbm	kJ/kg
H	molar enthalpy	Btu/lbmol	kJ/kmol
J	Joule's constant (778.17)	ft-lbf/Btu	n.a.
k	ratio of specific heats	–	–
m	mass	lbm	kg
n	number of moles	lbmol	kmol
n	polytropic exponent	–	–
p	pressure	lbf/ft^2	kPa
P	power	Btu/sec	kW
q	heat	Btu/lbm	kJ/kg
Q	molar heat	Btu/lbmol	kJ/kmol
Q	total heat	Btu	kJ
R	specific gas constant	ft-lbf/lbm-°R	kJ/kg·K
R^*	universal gas constant	ft-lbf/lbmol-°R	kJ/kmol·K
s	entropy	Btu/lbm-°F	kJ/kg·K
S	molar entropy	Btu/lbmol-°F	kJ/kmol·K
T	temperature	°F	°C
T	absolute temperature	°R	K
u	internal energy	Btu/lbm	kJ/kg
U	molar internal energy	Btu/lbmol	kJ/kmol
v	velocity	ft/sec	m/s
V	molar specific volume	ft^3/lbmol	m^3/kmol
V	volume	ft^3	m^3
W	specific work	ft-lbf/lbm	kJ/kg
W	total work	ft-lbf	kJ
z	elevation	ft	m

Symbols

β	isobaric compressibility	1/°R	1/K
η	efficiency	–	–
μ_J	Joule-Thomson coefficient	°F-ft^2/lbf	K/Pa
v	specific volume	ft^3/lbm	m^3/kg
Φ	availability function	Btu/lbm	kJ/kg

Subscripts

al	aluminum
H	hot (high-temperature)
k	kinetic
L	cold (low-temperature)
n	polytropic
o	outside (environment)
p	constant pressure, or potential
st	steel
t	total
th	thermal
v	constant volume
w	water

1. SYSTEMS

A thermodynamic *system* is defined as the matter enclosed within an arbitrary but precisely defined *control volume*. Everything external to the system is defined as the *surroundings*, *environment*, or *universe*. The environment and system are separated by the *system boundaries*. The surface of the control volume is known as the *control surface*. The control surface can be real (e.g., piston and cylinder walls) or imaginary.

If mass flows through the system across system boundaries, the system is an *open system*. Pumps, heat exchangers, and jet engines are examples of open systems. An important type of open system is the *steady-flow open system* in which matter enters and exits at the same rate. Pumps, turbines, heat exchangers, and boilers are all steady-flow open systems.

If no mass crosses the system boundaries, the system is said to be a *closed system*. The matter in a closed system may be referred to as a *control mass*. Closed systems can have variable volumes. The gas compressed by a piston in a cylinder is an example of a closed system with a variable control volume.

In most cases, energy in the form of heat, work, or electrical energy can enter or exit any open or closed system. Systems closed to both matter and energy transfer are known as *isolated systems*.

2. TYPES OF PROCESSES

Changes in thermodynamic properties of a system often depend on the type of process experienced. This is particularly true of gaseous systems. Several common types of processes are listed as follows, along with their heat, energy, and work relationships.[1]

- *adiabatic process*—a process in which no heat or other energy crosses the system boundary.[2] Adiabatic processes include isentropic and throttling processes.

$$Q = 0 \qquad \qquad 25.1$$
$$\Delta U = -W \qquad \qquad 25.2$$

- *constant pressure process*—also known as an *isobaric process*

$$\Delta p = 0 \qquad \qquad 25.3$$
$$Q = \Delta H \qquad \qquad 25.4$$

- *constant temperature process*—also known as an *isothermal process*

$$\Delta T = 0 \qquad \qquad 25.5$$
$$Q = W \qquad \qquad 25.6$$

- *constant volume process*—also known as an *isochoric* or *isometric process*

$$\Delta V = 0 \qquad \qquad 25.7$$
$$Q = \Delta U \qquad \qquad 25.8$$
$$W = 0 \qquad \qquad 25.9$$

- *isentropic process*—an adiabatic process in which there is no change in system entropy (i.e., it is reversible)

$$\Delta S = 0 \qquad \qquad 25.10$$
$$Q = 0 \qquad \qquad 25.11$$

- *throttling process*—an adiabatic process in which there is no change in system enthalpy but for which there is a significant pressure drop[3]

$$\Delta H = 0 \qquad \qquad 25.12$$
$$p_2 < p_1 \qquad \qquad 25.13$$

---

[1] The heat, energy, and work relationships are easily derived from the *first law of thermodynamics*.
[2] An adiabatic process is not the same as a constant temperature process, however.
[3] The classic throttling process is the expansion of a gas in a pipe through a porous plug that offers significant resistance to flow.

A system that is in equilibrium at the start and finish of a process may or may not be in equilibrium during the process. A *quasistatic process* (*quasiequilibrium process*) is one that can be divided into a series of infinitesimal deviations (steps) from equilibrium. During each step, the property changes are small, and all intensive properties are uniform throughout the system. The interim equilibrium at each step is known as a *quasiequilibrium*.

Transport processes are generally more complex than can be analyzed by the simple thermodynamic relationships in this chapter. In a *transport process*, there is a transfer of some quantity. Drying, distillation, and evaporation are examples of heat transfer processes. Fluid flow, mixing, and sedimentation are examples of momentum transfer. Distillation, absorption, extraction, and leaching are examples of mass transfer processes.

3. POLYTROPIC PROCESSES

A *polytropic process* is one that obeys Eq. 25.14, the *polytropic equation of state*. Gases always constitute the system in polytropic processes.

$$p_1(V_1)^n = p_2(V_2)^n \qquad \qquad 25.14$$

n is the *polytropic exponent*, a property of the equipment, not of the gas. For efficient air compressors, n is typically between 1.25 and 1.30.

Depending on the value of the polytropic exponent, the polytropic equation of state can also be used for other processes.

$$
\begin{aligned}
n &= 0: &&\text{constant pressure process} \\
n &= 1: &&\text{constant temperature process} \\
n &= k: &&\text{isentropic process} \\
n &= \infty: &&\text{constant volume process}
\end{aligned}
$$

The *polytropic specific heat*, c_n, is defined as

$$c_n = \frac{n-k}{n-1} \times c_v \qquad \qquad 25.15$$
$$Q = mc_n \Delta T \qquad \qquad 25.16$$

4. THROTTLING PROCESSES

For ideal gases, throttling processes are constant temperature processes. Some real gases at normal temperatures, however, decrease in temperature when throttled.[4] Others increase in temperature. Whether or not an increase or a decrease in temperature occurs depends on the temperature and pressure at which the throttling occurs.

For any given set of initial conditions, there is one temperature at which no temperature change occurs when

---

[4] This tendency can be used to liquefy gases and vapors (e.g., refrigerants) by passing them through an expansion valve.

a real gas is throttled. This is called the *inversion temperature* or *inversion point*. The inversion temperature is dependent on the initial gas conditions.

Table 25.1 *Approximate Maximum Inversion Temperatures*

substance	°R	K
air	1085	603
argon	1301	723
carbon dioxide	≈ 2700	≈ 1500
helium	≈ 72	≈ 40
hydrogen	364	202
nitrogen	1118	621

(Multiply K by 1.8 to obtain °R.)

The *Joule-Thomson coefficient*, μ_J (also known as the *Joule-Kelvin coefficient*), is defined as the ratio of the change in temperature to the change in pressure when a real gas is throttled. The Joule-Thomson coefficient is zero for an ideal gas.

$$\mu_J = \left(\frac{\partial T}{\partial p}\right)_h \qquad 25.17$$

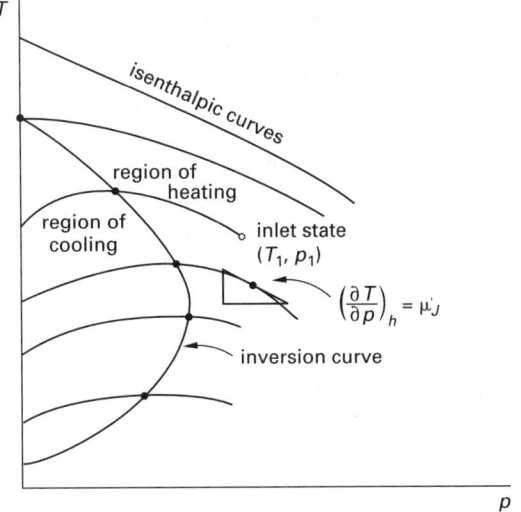

Figure 25.1 *Inversion Curve*

The Joule-Thomson coefficient can be calculated from the specific heat, c_p, and other properties.

$$\mu_J = \left(\frac{1}{c_p}\right)\left[T\left(\frac{\partial V}{\partial T}\right)_p - v\right] = \left(\frac{v}{c_p}\right)(T\beta - 1)$$

$$= \left(\frac{-1}{c_p}\right)\left(\frac{\partial h}{\partial p}\right)_T \qquad 25.18$$

It is convenient to plot lines of constant enthalpy (*isenthalpic curves*) on a T-p diagram. The various curves in Fig. 25.1 represent different sets of starting conditions (p_1 and T_1). The slope of a curve represents the Joule-Thomson coefficient, which can be zero, positive, or negative at that point. The points where the slope is zero are known as *inversion points*. The locus of inversion points is known as the *inversion curve*.

States to the right of an inversion point will have negative Joule-Thomson coefficients. States to the left of an inversion point will have positive Joule-Thomson coefficients. Throttling totally within the region bounded by the temperature axis and the inversion curve will always result in a decrease in temperature. Similarly, throttling totally outside of the inversion curve will always produce an increase in temperature. Throttling across the inversion curve may produce a higher or lower final temperature, depending on initial and final conditions.

5. REVERSIBLE PROCESSES

Processes can be categorized as reversible and irreversible. A *reversible process* is one that is performed in such a way that at the conclusion of the process, *both* the system and the local surroundings can be restored to their initial states. A process that does not meet these requirements is an *irreversible process*. A reversible adiabatic process is implicitly isentropic.

Although all real-world processes result in an overall increase in entropy, it is possible to conceptualize processes that have zero entropy change. For a reversible process,

$$\Delta s = 0 \Big|_{\text{reversible}} \qquad 25.19$$

Processes that contain friction are never reversible. Other processes that are irreversible are listed as follows.

- stirring a viscous fluid

- slowing down a moving fluid

- unrestrained expansion of gas

- throttling

- changes of phase (freezing, condensation, etc.)

- chemical reaction

- diffusion

- current flow through electrical resistance

- electrical polarization

- magnetization with hysteresis

- releasing a stretched spring

- inelastic deformation

- heat conduction

A system that has experienced an irreversible process can be returned to its original state. An example of this is the water in a closed boiler-turbine installation. The entropy increases when the water is vaporized in the boiler (an irreversible process). The entropy decreases when the steam is expanded through the turbine. When the water returns to the boiler, its entropy is increased to its original value. However, the environment cannot be returned to its original condition; hence the cycle overall is irreversible.

6. FINDING WORK AND HEAT GRAPHICALLY

A process between two thermodynamic states can be shown graphically. The line representing the locus of quasiequilibrium states between the initial and final states is known as the *path* of the process.

It is sometimes convenient to see what happens to the pressure and volume of a system by plotting the path on a *p-V* diagram. In addition, the work done by or on the system can be determined from the graph. This is possible because the integral calculating *p-V work* represents area under a curve in the *p-V* plane.

$$W = \int_{V_1}^{V_2} p \, dV \qquad 25.20$$

Similarly, the amount of heat absorbed or released from a system can be determined as the area under the path on the *T-s* diagram.

$$Q = \int_{s_1}^{s_2} T \, ds \qquad 25.21$$

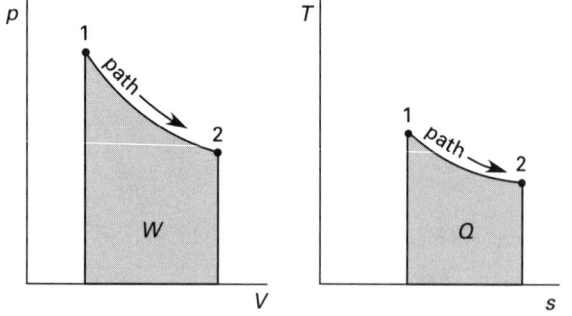

Figure 25.2 Process Work and Heat

The variables $p, V, T,$ and s are *point functions* because their values are independent of the path taken to arrive at the thermodynamic state. Work and heat (W and Q), however, are *path functions* because they depend on the path taken.

7. SIGN CONVENTION

A definite sign convention is used in calculating work, heat, and other property changes in systems.[5] This sign convention takes the system (not the environment) as the reference. (For example, a net heat gain would mean the system gained energy and the environment lost energy.)

- Heat, Q, is positive if heat flows into the system.

- Work, W, is positive if the system does work on the surroundings.

- Changes in enthalpy, entropy, and internal energy ($\Delta H, \Delta S,$ and ΔU) are positive if these properties increase within the system.

8. FIRST LAW OF THERMODYNAMICS FOR CLOSED SYSTEMS

There is a basic principle that underlies all property changes as a system experiences a process: all energy must be accounted for. Energy that enters a system must either leave the system or be stored in some manner, and energy cannot be created or destroyed. These statements are the primary manifestations of the *first law of thermodynamics*: the work done in an adiabatic process depends solely on the system's endpoint conditions, not on the nature of the process.

For closed systems, the first law can be written in differential form.[6,7] Since Q and W are not properties, infinitesimal changes are designated as δ.

$$\delta Q = dU + \delta W \qquad 25.22$$

Most simple thermodynamic problems can be solved without resorting to differential calculus. The first law can then be written in finite terms.

$$Q = \Delta U + W \qquad 25.23$$

Equation 25.23 says that the heat, Q, entering a closed system can either increase the temperature (i.e., increase U) or be used to perform work (increase W) on the surroundings. In a non-adiabatic closed system, heat energy entering the system also can leak to the surroundings. However, in Eq. 25.23, the Q term is understood to be the net heat entering the system, exclusive of the loss.

---

[5]This sign convention is automatically followed if the formulas in this chapter are used.

[6]This formulation of the conservation of energy implicitly assumes that kinetic and potential energies are negligible.

[7]Q, U, and W must all have the same units. This is less of a problem with SI units than with English units. For example, if Q and U are in Btu/lbm and W is in ft-lbf/lbm, Joule's constant must be incorporated into the first law.

$$Q = \Delta U + W/J \quad [\text{U.S.}]$$

In accordance with the standard sign convention given in Sec. 7, Q will be negative if the net heat exchange to the system is a loss. ΔU will be negative if the internal energy of the system decreases. W will be negative if the surroundings do work on the system (e.g., a piston compressing gas in a cylinder).

9. THERMAL EQUILIBRIUM

A simple application of the first law is the calculation of a thermal equilibrium point for a nonreacting system. *Thermal equilibrium* is reached when all parts of the system are at the same temperature.

The drive toward thermal equilibrium occurs spontaneously, without the addition of external work, whenever masses (two liquids, a solid and a liquid, etc.) with different temperatures are mixed. Therefore, the work term, W, in the first law is zero.

The energy comes from the heat given off by the cooling mass (i.e., mass A). Energy is stored by increasing the temperature of the warmed mass (mass B). These changes are equal, but opposite, since the net change is zero.

$$Q_{\text{net}} = Q_{\text{in},A} - Q_{\text{out},B} = 0 \qquad 25.24$$

The form of the equation for Q depends on the phases of the substances and the nature of the system. If the substances are either solid or liquid, Eq. 25.25 can be used. (For use with open systems, the m in Eq. 25.25 should be replaced with $\dot{m}$.)

$$m_A c_{p,A}(T_{1,A} - T_{2,A}) = m_B c_{p,B}(T_{2,B} - T_{1,B}) \qquad 25.25$$

Example 25.1

A block of steel is removed from a furnace and quenched in an insulated aluminum tank filled with water. The water and aluminum tank are initially in equilibrium at 75°F (24°C), and the final equilibrium temperature after quenching is 100°F (38°C). What is the initial temperature of the steel?

steel block mass:		
m_{st}	2.0 lbm	0.9 kg
steel specific heat:		
$c_{p,\text{st}}$	0.11 Btu/lbm-°F	0.460 kJ/kg·K
aluminum tank mass:		
m_{al}	5.0 lbm	2.25 kg
aluminum specific heat:		
$c_{p,\text{al}}$	0.21 Btu/lbm-°F	0.880 kJ/kg·K
water mass:		
m_w	12.0 lbm	5.4 kg
water specific heat:		
$c_{p,w}$	1.0 Btu/lbm-°F	4.190 kJ/kg·K

SI Solution

The heat lost by the steel is equal to the heat gained by the tank and water.

$$m_{\text{st}}c_{p,\text{st}}(T_{1,\text{st}} - T_{\text{eq}})$$
$$= (m_{\text{al}}c_{p,\text{al}} + m_w c_{p,w})(T_{\text{eq}} - T_{1,w})$$
$$(0.9 \text{ kg})\left(0.460 \ \frac{\text{kJ}}{\text{kg·K}}\right)(T_{1,\text{st}} - 38°\text{C})$$
$$= \left[(2.25 \text{ kg})\left(0.880 \ \frac{\text{kJ}}{\text{kg·K}}\right)\right.$$
$$\left. + (5.4 \text{ kg})\left(4.190 \ \frac{\text{kJ}}{\text{kg·K}}\right)\right](38°\text{C} - 24°\text{C})$$
$$T_{1,\text{st}} = 870°\text{C}$$

Customary U.S. Solution

Proceeding as in the SI solution,

$$(2.0 \text{ lbm})\left(0.11 \ \frac{\text{Btu}}{\text{lbm-}°\text{F}}\right)(T_{1,\text{st}} - 100°\text{F})$$
$$= \left[(5.0 \text{ lbm})\left(0.21 \ \frac{\text{Btu}}{\text{lbm-}°\text{F}}\right)\right.$$
$$\left. + (12.0 \text{ lbm})\left(1 \ \frac{\text{Btu}}{\text{lbm-}°\text{F}}\right)\right](100°\text{F} - 75°\text{F})$$
$$T_{1,\text{st}} = 1583°\text{F}$$

10. FIRST LAW OF THERMODYNAMICS FOR OPEN SYSTEMS

The first law of thermodynamics can also be written for open systems, but more terms are required to account for the many energy forms. The first law formulation is essentially the Bernoulli energy conservation equation extended to non-adiabatic processes.

If the mass flow rate is constant, the system is a *steady-flow system*, and the first law is known as the *steady-flow energy equation*, SFEE, Eq. 25.26. It is customary to express the SFEE in terms of energy per unit mass or per mole.

$$Q = \Delta U + \Delta E_p + \Delta E_k + W_{\text{flow}} + W_{\text{shaft}} \qquad 25.26$$

Some of the terms in Eq. 25.26 are illustrated in Fig. 25.3. Q is the net heat flow into the system, inclusive of any losses. It can be supplied from furnace flame, electrical heating, nuclear reaction, or other sources. If the system is adiabatic, Q is zero.

ΔE_p and ΔE_k are the fluid's potential and kinetic energy changes. Generally, these terms are insignificant compared to the thermal energy transfers.

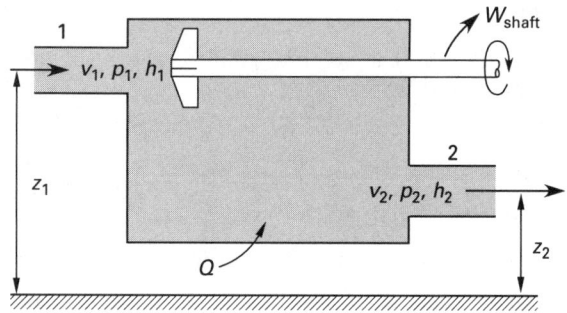

Figure 25.3 Steady Flow Device

W_{shaft} is the *shaft work*—work that the steady-flow device does on the surroundings. Its name is derived from the output shaft that serves to transmit energy out of the system. For example, turbines and internal combustion engines have output shafts. W_{shaft} can be negative, as in the case of a pump or compressor.

W_{flow} is the *p-V work* (*flow energy, flow work*, etc.). There is a pressure (p_2) at the exit of the steady-flow device in Fig. 25.3. This exit pressure opposes the entrance of the fluid. Therefore, the flow work term represents the work required to cause the flow into the system against the exit pressure. The flow work can be calculated from Eq. 25.27. (Consistent units must be used.)

$$W_{\text{flow}} = p_2 v_2 - p_1 v_1 \qquad \textbf{25.27}$$

ΔU is the change in the system's internal energy. Since the combination of internal energy and flow work constitutes enthalpy, it will seldom be necessary to work with either internal energy or flow work.

$$h = u + pv \qquad \textbf{25.28}$$
$$\Delta h = \Delta u + W_{\text{flow}} \qquad \textbf{25.29}$$

Equation 25.30 is a useful formulation of the SFEE. Units of power can be obtained by multiplying both sides by $\dot{m}$.

$$q = h_2 - h_1 + \frac{v_2^2 - v_1^2}{2} + (z_2 - z_1)g + W_{\text{shaft}}$$
$$\text{[SI]} \quad \textbf{25.30(a)}$$

$$q = h_2 - h_1 + \frac{v_2^2 - v_1^2}{2g_c J} + \frac{(z_2 - z_1)g}{g_c J} + \frac{W_{\text{shaft}}}{J}$$
$$\text{[U.S.]} \quad \textbf{25.30(b)}$$

Example 25.2

4 lbm/sec (1.8 kg/s) of steam enter a turbine with a velocity of 65 ft/sec (20 m/s) and an enthalpy of 1350 Btu/lbm (3140 kJ/kg). The steam enters the condenser after being expanded to 1075 Btu/lbm (2500 kJ/kg) at 125 ft/sec (38 m/s). There is a total heat loss from the turbine casing of 50 Btu/sec (53 kJ/s). Potential energy changes are insignificant. What power is generated at the turbine shaft?

SI Solution

Equation 25.30 can be solved for the shaft work.

$$P_{\text{shaft}} = \dot{m} W_{\text{shaft}} = \dot{Q}_t + \dot{m}\left(h_1 - h_2 + \frac{v_1^2 - v_2^2}{2}\right)$$

$$= -53\,000\ \frac{\text{J}}{\text{s}} + \left(1.8\ \frac{\text{kg}}{\text{s}}\right)$$

$$\times \left[3140 \times 10^3\ \frac{\text{J}}{\text{kg}} - 2500 \times 10^3\ \frac{\text{J}}{\text{kg}} \right.$$

$$\left. + \frac{\left(20\ \frac{\text{m}}{\text{s}}\right)^2 - \left(38\ \frac{\text{m}}{\text{s}}\right)^2}{2} \right]$$

$$= 1.098 \times 10^6\ \text{W} \quad (1.1\ \text{MW})$$

Customary U.S. Solution

Proceeding as in the SI solution,

$$P_{\text{shaft}} = \dot{m} W_{\text{shaft}} = \dot{Q}_t + \dot{m}\left(h_1 - h_2 + \frac{v_1^2 - v_2^2}{2g_c J}\right)$$

$$= -50\ \frac{\text{Btu}}{\text{sec}} + \left(4.0\ \frac{\text{lbm}}{\text{sec}}\right)\left[1350\ \frac{\text{Btu}}{\text{lbm}} - 1075\ \frac{\text{Btu}}{\text{lbm}} \right.$$

$$\left. + \frac{\left(65\ \frac{\text{ft}}{\text{sec}}\right)^2 - \left(125\ \frac{\text{ft}}{\text{sec}}\right)^2}{(2)\left(32.2\ \frac{\text{lbm-ft}}{\text{lbf-sec}^2}\right)\left(778\ \frac{\text{ft-lbf}}{\text{Btu}}\right)} \right]$$

$$= 1049\ \text{Btu/sec}$$

11. BASIC THERMODYNAMIC RELATIONS

The following relations for a simple compressible substance (though not necessarily an ideal gas) can be derived from the first law of thermodynamics and other basic definitions.[8]

$$du = T\,ds - p\,dv \qquad \textbf{25.31}$$
$$dh = T\,ds + v\,dp \qquad \textbf{25.32}$$

---

[8]It is impractical to list every relationship for property changes. Most formulas can be written in several forms. For example, all equations can be written either for a unit mass (i.e., per lbm or kg) or for a mole (i.e., per lbmol or kmol). For example, Eq. 25.31 on a molar basis would be $dU = T\,dS - p\,dV$.

12. PROPERTY CHANGES IN IDEAL GASES

For real solids, liquids, and vapors, changes in most properties can be determined only by subtracting the initial property value from the final property value. For simple compressible substances (i.e., ideal gases), however, many changes in properties can be found directly, without knowing the initial and final property values.

Some of the relations for determining property changes do not depend on the type of process. For example, a general relationship that applies to any ideal gas experiencing any process is easily derived from the equation of state.

$$\frac{p_1 V_1}{T_1} = \frac{p_2 V_2}{T_2} \qquad 25.33$$

When temperature is held constant, Eq. 25.33 reduces to *Boyle's law*.

$$p_1 V_1 = p_2 V_2 \qquad 25.34$$

When pressure is held constant, Eq. 25.33 reduces to *Charles' law*.

$$\frac{V_1}{T_1} = \frac{V_2}{T_2} \qquad 25.35$$

Similarly, the changes in enthalpy, internal energy, and entropy are independent of the process.

$$\Delta h = c_p \Delta T \qquad 25.36$$

$$\Delta u = c_v \Delta T \qquad 25.37$$

$$\Delta s = c_p \ln\left(\frac{T_2}{T_1}\right) - R \ln\left(\frac{p_2}{p_1}\right)$$

$$= c_v \ln\left(\frac{T_2}{T_1}\right) + R \ln\left(\frac{v_2}{v_1}\right) \qquad 25.38$$

Enthalpy and internal energy are *point functions*, so Eqs. 25.36 and 25.37 are valid for all processes, even though the specific heats at constant pressure and volume are part of the calculation. These equations should not be confused with the heat transfer relations, which have a similar form. Heat, Q, is a path function and depends on the path taken.

$$q = c_p \Delta T \Big|_p \qquad 25.39$$

$$q = c_v \Delta T \Big|_v \qquad 25.40$$

Relationships between the properties follow for ideal gases experiencing specific processes. (The standard thermodynamic sign convention is automatically followed.) All relations can be written in slightly different forms, including per-unit mass and per-mole bases. For compactness, only the per-unit mass equations are listed. However, all equations can be converted to a molar basis by substituting H for h, V for v, R^* for R, and so on. Other forms can be derived by substituting the equation of state where appropriate.[9]

---

[9]For example, RT can be substituted anywhere pv appears.

Constant Pressure, Closed Systems

$$p_2 = p_1 \qquad 25.41$$

$$T_2 = T_1\left(\frac{v_2}{v_1}\right) \qquad 25.42$$

$$v_2 = v_1\left(\frac{T_2}{T_1}\right) \qquad 25.43$$

$$q = h_2 - h_1 \qquad 25.44$$

$$= c_p\left(T_2 - T_1\right) \qquad 25.45$$

$$= c_v\left(T_2 - T_1\right) + p(v_2 - v_1) \qquad 25.46$$

$$u_2 - u_1 = c_v\left(T_2 - T_1\right) \qquad 25.47$$

$$= \frac{c_v p(v_2 - v_1)}{R} \qquad 25.48$$

$$= \frac{p(v_2 - v_1)}{k - 1} \qquad 25.49$$

$$W = p(v_2 - v_1) \qquad 25.50$$

$$= R(T_2 - T_1) \qquad 25.51$$

$$s_2 - s_1 = c_p \ln\left(\frac{T_2}{T_1}\right) \qquad 25.52$$

$$= c_p \ln\left(\frac{v_2}{v_1}\right) \qquad 25.53$$

$$h_2 - h_1 = q \qquad 25.54$$

$$= c_p(T_2 - T_1) \qquad 25.55$$

$$= \frac{kp(v_2 - v_1)}{k - 1} \qquad 25.56$$

Constant Volume, Closed Systems

$$p_2 = p_1\left(\frac{T_2}{T_1}\right) \qquad 25.57$$

$$T_2 = T_1\left(\frac{p_2}{p_1}\right) \qquad 25.58$$

$$v_2 = v_1 \qquad 25.59$$

$$q = u_2 - u_1 \qquad 25.60$$

$$= c_v\left(T_2 - T_1\right) \qquad 25.61$$

$$u_2 - u_1 = q \qquad 25.62$$

$$= c_v\left(T_2 - T_1\right) \qquad 25.63$$

$$= \frac{c_v v(p_2 - p_1)}{R} \qquad 25.64$$

$$= \frac{v(p_2 - p_1)}{k - 1} \qquad 25.65$$

$$W = 0 \qquad 25.66$$

$$s_2 - s_1 = c_v \ln\left(\frac{T_2}{T_1}\right) \qquad 25.67$$

$$= c_v \ln\left(\frac{p_2}{p_1}\right) \qquad 25.68$$

$$h_2 - h_1 = c_p(T_2 - T_1) \qquad 25.69$$

$$= \frac{kv(p_2 - p_1)}{k - 1} \qquad 25.70$$

Thermodynamics

Constant Temperature, Closed Systems

$$p_2 = p_1 \left(\frac{v_1}{v_2} \right) \tag{25.71}$$

$$T_2 = T_1 \tag{25.72}$$

$$v_2 = v_1 \left(\frac{p_1}{p_2} \right) \tag{25.73}$$

$$q = W \tag{25.74}$$

$$= T(s_2 - s_1) \tag{25.75}$$

$$= p_1 v_1 \ln \left(\frac{v_2}{v_1} \right) \tag{25.76}$$

$$= p_1 v_1 \ln \left(\frac{p_1}{p_2} \right) \tag{25.77}$$

$$= RT \ln \left(\frac{v_2}{v_1} \right) \tag{25.78}$$

$$= RT \ln \left(\frac{p_1}{p_2} \right) \tag{25.79}$$

$$u_2 - u_1 = 0 \tag{25.80}$$

$$W = q \tag{25.81}$$

$$= p_1 v_1 \ln \left(\frac{v_2}{v_1} \right) \tag{25.82}$$

$$= p_1 v_1 \ln \left(\frac{p_1}{p_2} \right) \tag{25.83}$$

$$= RT \ln \left(\frac{v_2}{v_1} \right) \tag{25.84}$$

$$= RT \ln \left(\frac{p_1}{p_2} \right) \tag{25.85}$$

$$s_2 - s_1 = \frac{q}{T} \tag{25.86}$$

$$= R \ln \left(\frac{v_2}{v_1} \right) \tag{25.87}$$

$$= R \ln \left(\frac{p_1}{p_2} \right) \tag{25.88}$$

$$h_2 - h_1 = 0 \tag{25.89}$$

Isentropic, Closed Systems (Reversible Adiabatic)

$$p_2 = p_1 \left(\frac{v_1}{v_2} \right)^k \tag{25.90}$$

$$= p_1 \left(\frac{T_2}{T_1} \right)^{\frac{k}{k-1}} \tag{25.91}$$

$$T_2 = T_1 \left(\frac{v_1}{v_2} \right)^{k-1} \tag{25.92}$$

$$= T_1 \left(\frac{p_2}{p_1} \right)^{\frac{k-1}{k}} \tag{25.93}$$

$$v_2 = v_1 \left(\frac{p_1}{p_2} \right)^{\frac{1}{k}} \tag{25.94}$$

$$= v_1 \left(\frac{T_1}{T_2} \right)^{\frac{1}{k-1}} \tag{25.95}$$

$$q = 0 \tag{25.96}$$

$$u_2 - u_1 = -W \tag{25.97}$$

$$= c_v (T_2 - T_1) \tag{25.98}$$

$$= \frac{c_v (p_2 v_2 - p_1 v_1)}{R} \tag{25.99}$$

$$= \frac{p_2 v_2 - p_1 v_1}{k - 1} \tag{25.100}$$

$$W = u_1 - u_2 \tag{25.101}$$

$$= c_v (T_1 - T_2) \tag{25.102}$$

$$= \frac{p_1 v_1 - p_2 v_2}{k - 1} \tag{25.103}$$

$$= \left(\frac{p_1 v_1}{k - 1} \right) \left[1 - \left(\frac{p_2}{p_1} \right)^{\frac{k-1}{k}} \right] \tag{25.104}$$

$$s_2 - s_1 = 0 \tag{25.105}$$

$$h_2 - h_1 = c_p (T_2 - T_1) \tag{25.106}$$

$$= \frac{k(p_2 v_2 - p_1 v_1)}{k - 1} \tag{25.107}$$

Polytropic, Closed Systems

$$p_2 = p_1 \left(\frac{v_1}{v_2} \right)^n \tag{25.108}$$

$$= p_1 \left(\frac{T_2}{T_1} \right)^{\frac{n}{n-1}} \tag{25.109}$$

$$T_2 = T_1 \left(\frac{v_1}{v_2} \right)^{n-1} \tag{25.110}$$

$$= T_1 \left(\frac{p_2}{p_1} \right)^{\frac{n-1}{n}} \tag{25.111}$$

$$v_2 = v_1 \left(\frac{p_1}{p_2} \right)^{\frac{1}{n}} \tag{25.112}$$

$$= v_1 \left(\frac{T_1}{T_2} \right)^{\frac{1}{n-1}} \tag{25.113}$$

$$q = \frac{c_v (n - k)(T_2 - T_1)}{n - 1} \tag{25.114}$$

$$u_2 - u_1 = c_v (T_2 - T_1) \tag{25.115}$$

$$= q - W \tag{25.116}$$

$$W = \frac{R(T_1 - T_2)}{n - 1} \tag{25.117}$$

$$= \frac{p_1 v_1 - p_2 v_2}{n - 1} \tag{25.118}$$

$$= \left(\frac{p_1 v_1}{n - 1} \right) \left[1 - \left(\frac{p_2}{p_1} \right)^{\frac{n-1}{n}} \right] \tag{25.119}$$

$$s_2 - s_1 = \left[\frac{c_v(n-k)}{n-1}\right]\left[\ln\left(\frac{T_2}{T_1}\right)\right] \qquad 25.120$$

$$h_2 - h_1 = c_p(T_2 - T_1) \qquad 25.121$$

$$= \frac{n(p_2 v_2 - p_1 v_1)}{n-1} \qquad 25.122$$

Isentropic, Steady-Flow Systems

$p_2, v_2,$ and T_2 are the same as for isentropic, closed systems.

$$q = 0 \qquad 25.123$$

$$W = h_1 - h_2 \qquad 25.124$$

$$= \left(\frac{kp_1 v_1}{k-1}\right)\left[1 - \left(\frac{p_2}{p_1}\right)^{\frac{k-1}{k}}\right]$$

$$= c_p T_1\left[1 - \left(\frac{p_2}{p_1}\right)^{\frac{k-1}{k}}\right] \qquad 25.125$$

$$u_2 - u_1 = c_v(T_2 - T_1) \qquad 25.126$$

$$h_2 - h_1 = -W \qquad 25.127$$

$$= c_p(T_2 - T_1) \qquad 25.128$$

$$= \frac{k(p_2 v_2 - p_1 v_1)}{k-1} \qquad 25.129$$

$$s_2 - s_1 = 0 \qquad 25.130$$

Polytropic, Steady-Flow Systems

$p_2, v_2,$ and T_2 are the same as for polytropic, closed systems.

$$q = \frac{c_v(n-k)(T_2 - T_1)}{n-1} \qquad 25.131$$

$$W = h_1 - h_2 \qquad 25.132$$

$$= \left[\frac{nc_v(n-k)T_1}{n-1}\right]\left[1 - \left(\frac{p_2}{p_1}\right)^{\frac{n-1}{n}}\right] \qquad 25.133$$

$$u_2 - u_1 = c_v(T_2 - T_1) \qquad 25.134$$

$$h_2 - h_1 = -W \qquad 25.135$$

$$= c_p(T_2 - T_1) \qquad 25.136$$

$$= \frac{n(p_2 v_2 - p_1 v_1)}{n-1} \qquad 25.137$$

$$s_2 - s_1 = \left[\frac{c_v(n-k)}{n-1}\right]\left[\ln\left(\frac{T_2}{T_1}\right)\right] \qquad 25.138$$

Throttling, Steady-Flow Systems

$$p_1 v_1 = p_2 v_2 \qquad 25.139$$

$$p_2 < p_1 \qquad 25.140$$

$$v_2 > v_1 \qquad 25.141$$

$$T_2 = T_1 \qquad 25.142$$

$$q = 0 \qquad 25.143$$

$$W = 0 \qquad 25.144$$

$$u_2 - u_1 = 0 \qquad 25.145$$

$$h_2 - h_1 = 0 \qquad 25.146$$

$$s_2 - s_1 = R\ln\left(\frac{p_1}{p_2}\right) \qquad 25.147$$

$$= R\ln\left(\frac{v_2}{v_1}\right) \qquad 25.148$$

Example 25.3

2.0 lbm (0.9 kg) of hydrogen are cooled from 760°F to 660°F (400°C to 350°C) in a constant volume process. The specific heat at constant volume, c_v, is 2.435 Btu/lbm-°F (10.2 kJ/kg·K). How much heat is removed?

SI Solution

Use Eq. 25.61 on a per unit mass basis. The total heat transfer for m kilograms is

$$Q = mc_v(T_2 - T_1)$$

$$= (0.9\text{ kg})\left(10.2\frac{\text{kJ}}{\text{kg·K}}\right)(350°C - 400°C)$$

$$= -459\text{ kJ}$$

The minus sign is consistent with the convention that a heat loss is negative.

Customary U.S. Solution

The total heat transfer for m lbm is

$$Q = mc_v(T_2 - T_1)$$

$$= (2.0\text{ lbm})\left(2.435\frac{\text{Btu}}{\text{lbm-°F}}\right)(660°F - 760°F)$$

$$= -487\text{ Btu}$$

Example 25.4

4 lbmol (4 kmol) of air initially at 1 atm and 530°R (295K) are compressed isothermally to 8 atm. How much total heat is removed during the compression?

SI Solution

Equation 25.85 applies to constant temperature compression. R^* is used in place of R to convert the calculations to a molar basis.

$$Q = nR^*T\ln\left(\frac{p_1}{p_2}\right)$$

$$= (4.0\text{ kmol})\left(8.314\frac{\text{kJ}}{\text{kmol·K}}\right)(295K)\ln\left(\frac{1\text{ atm}}{8\text{ atm}}\right)$$

$$= -20\,400\text{ kJ}$$

Customary U.S. Solution

Writing Eq. 25.85 on a molar basis,

$$Q = nR^*T \ln\left(\frac{p_1}{p_2}\right)$$

$$= (4.0 \text{ lbmol})\left(1545 \ \frac{\text{ft-lbf}}{\text{lbmol-}°\text{R}}\right)(530°\text{R}) \ln\left(\frac{1 \text{ atm}}{8 \text{ atm}}\right)$$

$$= -6.81 \times 10^6 \text{ ft-lbf}$$

13. PROPERTY CHANGES IN INCOMPRESSIBLE FLUIDS AND SOLIDS

In order to simplify the solution to practical problems, enthalpy, entropy, internal energy, and specific volume in liquids and solids are often considered to be functions of temperature only. The effect of pressure is disregarded.

There are times, however, when it is necessary to evaluate the property changes in a solid or liquid system, no matter how small they may be. Changes for liquids can be evaluated by using compressed liquid tables. If certain assumptions are made, some property changes can also be calculated without knowing the initial and final values. The main assumptions are incompressibility and constant specific heats.

$$c_p = c_v = c \qquad \text{25.149}$$

$$dv = 0 \qquad \text{25.150}$$

$$du = c \, dT \qquad \text{25.151}$$

$$ds = \frac{du}{T} \qquad \text{25.152}$$

$$dh = c \, dT + v \, dp \qquad \text{25.153}$$

If the specific heat is constant, Eqs. 25.154 through 25.157 are valid.

$$v_2 - v_1 = 0 \qquad \text{25.154}$$

$$u_2 - u_1 = c \, (T_2 - T_1) \qquad \text{25.155}$$

$$s_2 - s_1 = c \ln\left(\frac{T_2}{T_1}\right) \qquad \text{25.156}$$

$$h_2 - h_1 = c \, (T_2 - T_1) + v \, (p_2 - p_1) \qquad \text{25.157}$$

14. HEAT RESERVOIRS

It is convenient to show a source of energy as an infinite constant-temperature *reservoir*. Figure 25.4 illustrates a source of energy (known as a *high-temperature reservoir* or *source reservoir*). By convention, the reservoir temperature is designated T_H, and the heat transfer from it is Q_H. The energy derived from such a theoretical source might actually be supplied by combustion, electrical heating, or nuclear reaction.

Similarly, energy is released to a *low-temperature reservoir* known as a *sink reservoir* or *energy sink*. The most common practical sink is the local environment. T_L and

Q_L are used to represent the reservoir temperature and energy absorbed. It is common to refer to Q_L as the rejected energy or energy rejected to the environment.

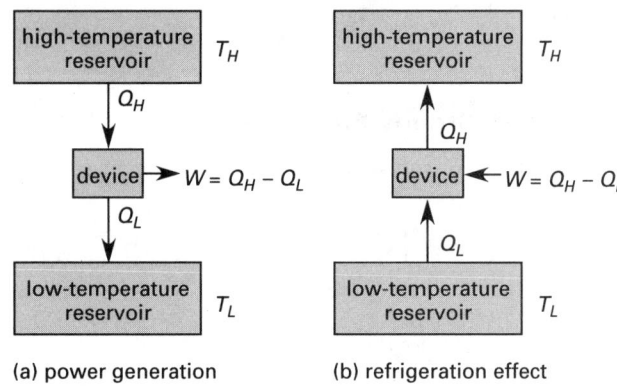

(a) power generation   (b) refrigeration effect

Figure 25.4 *Energy Reservoirs*

15. CYCLES

Although heat can be extracted and work performed in a single process, a cycle is necessary to obtain work in a useful quantity and duration. A *cycle* is a series of processes that eventually brings the system back to its original condition. Most cycles are continually repeated.

A cycle is completely defined by the working substance, the high- and low-temperature reservoirs, the means of doing work on the system, and the means of removing energy from the system.[10]

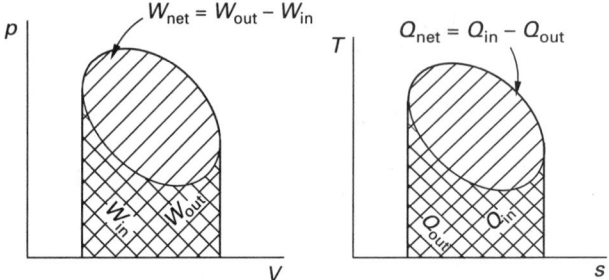

Figure 25.5 *Net Work and Net Heat*

A cycle will appear as a closed curve when plotted on p-V and T-s diagrams. The area within the p-V or T-s curve represents the net work or net heat, respectively.

16. THERMAL EFFICIENCY

The *thermal efficiency* of a *power cycle* is defined as the ratio of useful work output to the supplied input energy.[11] W in Eq. 25.158 is the *net work*, since some

---

[10]The *Carnot cycle* depends only on the source and sink temperatures, not on the working fluid. However, most practical cycles depend on the working fluid.

[11]The effectiveness of refrigeration and compression cycles is measured by other parameters.

of the gross output work may be used to run certain parts of the cycle. For example, a small amount of turbine output power may run boiler feed pumps.

$$\eta_{th} = \frac{\text{net work output}}{\text{energy input}} = \frac{W_{net}}{Q_{in}}$$

$$= \frac{W_{out} - W_{in}}{Q_{in}} \qquad 25.158$$

The first law can be written as

$$Q_{in} = Q_{out} + W_{net} \qquad 25.159$$

Equations 25.158 and 25.159 can be combined to define the thermal efficiency in terms of heat variables alone.

$$\eta_{th} = \frac{Q_{in} - Q_{out}}{Q_{in}} \qquad 25.160$$

Equation 25.160 shows that obtaining the maximum efficiency requires minimizing the Q_{out} term. The most efficient power cycle possible is the *Carnot cycle*.

17. SECOND LAW OF THERMODYNAMICS

The *second law of thermodynamics* can be stated in several ways. Equation 25.161 is the mathematical relation defining the second law. The equality holds for reversible processes; the inequality holds for irreversible processes.

$$\Delta s \geq \int_{T_1}^{T_2} \frac{dq}{T} \qquad 25.161$$

Equation 25.161 effectively states that net entropy must always increase in practical (irreversible) cyclical processes.

> A natural process that starts in one equilibrium state and ends in another will go in the direction that causes the entropy of the system and the environment to increase.

The *Kelvin-Planck statement* of the second law effectively says that it is impossible to build a cyclical engine that will have a thermal efficiency of 100%.

> It is impossible to operate an engine operating in a cycle that will have no other effect than to extract heat from a reservoir and turn it into an equivalent amount of work.

This formulation is not a contradiction of the first law of thermodynamics. The first law does not preclude the possibility of converting heat entirely into work—it only denies the possibility of creating or destroying energy. The second law says that if some heat is converted entirely into work, some other energy must be rejected to a low-temperature sink (i.e., lost to the surroundings).

Figure 25.6 illustrates violations of the second law.

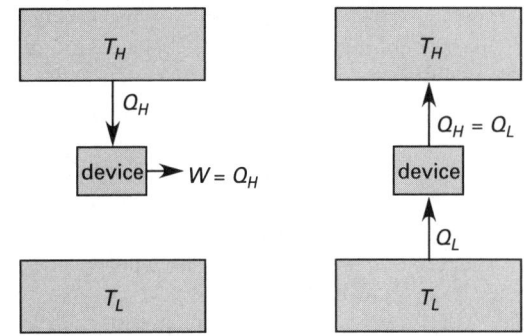

Figure 25.6 *Second Law Violations*

18. AVAILABILITY

The maximum possible work that can be obtained from a cycle is known as the *availability*. Availability is independent of the device but is dependent on the temperature of the local environment. Both the first and second law must be applied to determine availability.

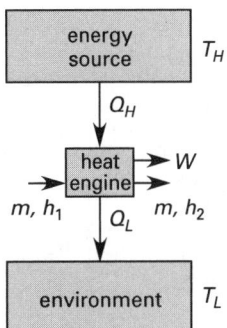

Figure 25.7 *Typical Heat Engine*

In Fig. 25.7, the *heat engine* is surrounded by an environment at absolute temperature T_L. The net heat available for conversion to useful work is $Q = Q_H - Q_L$. Work, W, is performed at a constant rate on the environment by the engine.[12] Assuming steady flow, and neglecting kinetic and potential energies, the first law can be written as

$$mh_1 + Q = mh_2 + W \qquad 25.162$$

Entropy is not a part of the first law. The leaving entropy, however, can be calculated from the entering entropy and the entropy production.

$$ms_2 = ms_1 + \frac{Q}{T_L} \qquad 25.163$$

Since Eqs. 25.162 and 25.163 both contain Q, they can be combined.

$$W = m(h_1 - T_L s_1 - h_2 + T_L s_2) \qquad 25.164$$

---

[12]This is the traditional development of the availability concept. Actually, the symbols $\dot{Q}, \dot{m}$, and P (power) should be used to represent the time rate of heat, mass, and work.

This equation can be simplified by introducing the *steady-flow availability function*, Φ. The maximum work output (availability) is

$$W_{max} = \Phi_1 - \Phi_2 \qquad 25.165$$

$$\Phi = h - T_L s \qquad 25.166$$

If the equality in Eq. 25.165 holds, both the process within the control volume and the energy transfers between the system and the environment must be reversible. Maximum work output, therefore, will be obtained in a reversible process. The difference between the maximum and the actual work output is known as the *process irreversibility*.

Example 25.5

What is the maximum useful work that can be produced per pound (per kilogram) of steam that enters a steady flow system at 800 psia (5.0 MPa) saturated and leaves in equilibrium with the environment at 70°F and 14.7 psia (20°C and 100 kPa)?

SI Solution

From the saturated steam table, $h_1 = 2794$ kJ/kg and $s_1 = 5.9726$ kJ/kg·K.

The final properties are obtained from the saturated steam table for water at 20°C. $h_2 = 83.9$ kJ/kg and $s_2 = 0.2965$ kJ/kg·K.

The availability is calculated from Eq. 25.164 using $T_L = 20 + 273 = 293$K.

$$
\begin{aligned}
W_{max} &= m\left[h_1 - h_2 - T_L(s_1 - s_2)\right] \\
&= (1 \text{ kg})\left[2794\,\frac{\text{kJ}}{\text{kg}} - 83.9\,\frac{\text{kJ}}{\text{kg}} \right. \\
&\quad \left. - (293\text{K})\left(5.9726\,\frac{\text{kJ}}{\text{kg·K}} - 0.2965\,\frac{\text{kJ}}{\text{kg·K}}\right)\right] \\
&= 1047 \text{ kJ}
\end{aligned}
$$

Customary U.S. Solution

From the saturated steam table, $h_1 = 1199.3$ Btu/lbm and $s_1 = 1.4160$ Btu/lbm-°R.

The final properties are obtained from the saturated steam table for water at 70°F. $h_2 = 38.09$ Btu/lbm and $s_2 = 0.07463$ Btu/lbm-°R.

The availability is calculated from Eq. 25.164 using $T_L = 70 + 460 = 530$°R.

$$
\begin{aligned}
W_{max} &= m[h_1 - h_2 - T_L(s_1 - s_2)] \\
&= (1 \text{ lbm})\left[1199.3\,\frac{\text{Btu}}{\text{lbm}} - 38.09\,\frac{\text{Btu}}{\text{lbm}}\right. \\
&\quad \left. - (530°\text{R})\left(1.4160\,\frac{\text{Btu}}{\text{lbm-°R}} - 0.07463\,\frac{\text{Btu}}{\text{lbm-°R}}\right)\right] \\
&= 450.3 \text{ Btu}
\end{aligned}
$$

PRACTICE PROBLEMS

1. Cast iron is heated from 80°F to 780°F (27°C to 416°C). What heat is required per unit mass?

2. The ventilation rate in a building is 3×10^5 ft³/hr (2.4 m³/s). The air is heated from 35°F to 75°F (2°C to 24°C) by water whose temperature decreases from 180°F to 150°F (82°C to 66°C). What is the water flow rate in gal/min (L/s)?

3. 8.0 ft³ (0.25 m³) of 180°F, 14.7 psia (82°C, 101.3 kPa) air are cooled to 100°F (38°C) in a constant-pressure process. What work is done?

4. What is the availability of an isentropic process using steam with a quality of 95% and operating between 300 psia and 50 psia (2 MPa and 0.35 MPa)?

5. (*Time limit: one hour*) A closed air heater receives 540°F, 100 psia (280°C, 700 kPa) air and heats it to 1540°F (840°C). The outside temperature is 100°F (40°C). The pressure of the air drops 20 psi (150 kPa) as it passes through the heater. What is the percentage loss in available energy due to the pressure drop?

6. (*Time limit: one hour*) Xenon gas at 20 psia and 70°F (150 kPa and 21°C) is compressed to 3800 psia and 70°F (25 MPa and 21°C) by a compressor/heat exchanger combination. The compressed gas is stored in a 100 ft³ (3 m³) rigid tank initially charged with xenon gas at 20 psia (150 kPa). (a) What is the mass of the xenon gas initially in the tank? (b) What is the average mass flow rate of xenon gas into the tank if the compressor fills the tank in exactly one hour? (c) If filling takes exactly one hour and electricity costs $0.045 per kW-hr, what is the cost of filling the tank?

7. (*Time limit: one hour*) The mass of a 20 ft³ (0.6 m³) steel tank is 40 lbm (20 kg). The steel has a specific heat of 0.11 Btu/lbm-°R (0.46 kJ/kg·K). The tank is placed in a room where the surrounding air is 70°F and 14.7 psia (21°C and 101.3 kPa). After the tank is evacuated to 1 psia and 70°F (7 kPa and 21°C), a valve is suddenly opened, allowing the tank to fill with room air. The air enters the tank in a well-mixed, turbulent condition. Find the air temperature inside the tank immediately after filling and before any heat loss to the tank or room occurs.

26 Compressible Fluid Dynamics

1. Introduction 26-1
2. Steady-Flow Energy Equation 26-2
3. Isentropic Flow 26-3
4. Isentropic Flow Factors 26-3
5. Ratio of Specific Heats 26-3
6. Critical Constants 26-4
7. Choked Flow 26-4
8. Changing Duct Area 26-4
9. Convergent-Divergent Nozzles 26-5
10. Design of Supersonic Nozzles 26-5
11. Nozzle Performance Characteristics 26-8
12. Shock Waves 26-9
13. Oblique Shock Waves 26-11
14. Critical Back-Pressure Ratio 26-12
15. Compressible Flow Through Orifices . . . 26-15
16. Adiabatic Flow with Friction
    in Constant-Area Ducts 26-15
17. Isothermal Flow with Friction 26-17
18. Frictionless Flow with Heating or Cooling 26-19
19. Steam Flow Through Nozzles 26-20
20. Pitot Tubes 26-21
    Practice Problems 26-21

Nomenclature

a	speed of sound	ft/sec	m/s
A	area	ft^2	m^2
c	specific heat	Btu/lbm-°F	kJ/kg·K
C	coefficient	–	–
D	diameter	ft	m
f	Fanning friction factor	–	–
F	thrust force	lbf	N
g	acceleration of gravity	ft/sec^2	m/s^2
g_c	gravitational constant	lbm-ft/lbf-sec^2	n.a.
h	enthalpy	Btu/lbm	kJ/kg
h	head	ft	m
I	impulse	lbf-sec	N·s
I_{sp}	specific impulse	sec	s
J	Joule's constant (778.17)	ft-lbf/Btu	n.a.
k	ratio of specific heats	–	–
L	distance	ft	m
L	length	ft	m
m	mass	lbm	kg
M	Mach number	–	–
MW	molecular weight	lbm/lbmol	kg/kmol
p	pressure	lbf/ft^2	Pa
q	heat per unit mass	Btu/lbm	kJ/kg
R	specific gas constant	ft-lbf/lbm-°R	kJ/kg·K
R^*	universal gas constant	ft-lbf/lbmol-°R	kJ/kmol·K
R_{cp}	critical pressure ratio	–	–
s	entropy	Btu/lbm-°F	kJ/kg·K
t	time	sec	s
T	absolute temperature	°R	K
v	velocity	ft/sec	m/s
V	volume	ft^3	m^3
z	elevation	ft	m

Symbols

γ	specific weight	lbf/ft^3	–
δ	semivertex angle	degrees	degrees
η	efficiency	–	–
θ	shock semivertex angle	degrees	degrees
ρ	mass density	lbm/ft^3	kg/m^3
υ	specific volume	ft^3/lbm	m^3/kg

Subscripts

0	total (stagnation)
a	ambient
cp	critical pressure
d	discharge
e	exit
eff	effective
f	friction
F	thrust
h	hydraulic
i	inlet
p	constant pressure
sp	specific
t	throat
v	velocity
x	before the shock wave
y	after the shock wave

1. INTRODUCTION

A *high-velocity gas* is defined as a gas moving with a velocity in excess of approximately 300 ft/sec (100 m/s). A high gas velocity is often achieved at the expense of internal energy. A drop in internal energy, u, is seen as a drop in enthalpy, h, since $h = u + pv$. Since the Bernoulli equation does not account for this conversion, it cannot be used to predict the thermodynamic properties of the gas. Furthermore, density changes and shock waves complicate the use of traditional evaluation tools such as energy and momentum conservation equations.

Thermodynamics

2. STEADY-FLOW ENERGY EQUATION

Consider a gas or vapor flowing from a high-pressure reservoir through a duct, as shown in Fig. 26.1. The properties in the source reservoir, designated by the subscript zero (0), are known alternatively as the *total properties*, *stagnation properties*, or *chamber properties*.

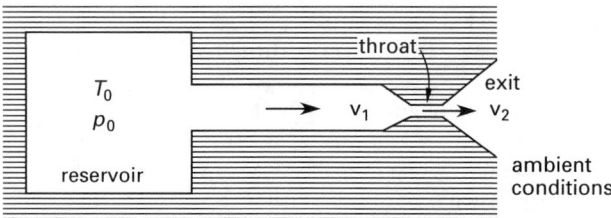

Figure 26.1 *High-Velocity Flow Locations*

The flow can be assumed to be one dimensional as long as the flow cross section varies slowly along the path length. The gas possesses energy in static, kinetic, and internal (thermal) forms. Since the flow is fast, there is no time for significant heat transfer to occur, so an assumption of adiabatic flow is appropriate. If the duct run is short, there will be little or no friction. Equation 26.1 is an energy balance of an adiabatic open-flow system for one unit mass of the gas. This is the *steady-flow energy equation*, SFEE, for an adiabatic system.[1]

$$h_1 + \frac{v_1^2}{2} + z_1 g = h_2 + \frac{v_2^2}{2} + z_2 g \quad \text{[SI]} \qquad 26.1(a)$$

$$J h_1 + \frac{v_1^2}{2g_c} + z_1 \left(\frac{g}{g_c}\right) = J h_2 + \frac{v_2^2}{2g_c} + z_2 \left(\frac{g}{g_c}\right)$$
$$\text{[U.S.]} \qquad 26.1(b)$$

It is often necessary to write the velocity at a point in terms of the total properties. The velocity is zero in the source reservoir. Since the gas density is small, the potential energy terms are always disregarded. Equation 26.2 can be used to calculate the discharge velocity for high-pressure steam and other substances for which the ideal gas assumption would be inappropriate. The *theoretical maximum velocity* is achieved when all internal and pressure energies are converted to kinetic energy (i.e., when $h_2 = 0$). This is never achieved in practice, however.

$$v_2 = \sqrt{2(h_0 - h_2)} \qquad \text{[SI]} \qquad 26.2(a)$$

$$v_2 = \sqrt{2g_c J(h_0 - h_2)} = 223.8\sqrt{h_0 - h_2} \quad \text{[U.S.]} \qquad 26.2(b)$$

---

[1]The *steady-flow energy equation*, SFEE, is also known as the *general flow equation*, GFE.

If it is appropriate to assume constant values of specific heat or specific gas constant, ideal gas relationships can be substituted into the SFEE.

$$v_2 = \sqrt{2c_p(T_0 - T_2)} = \sqrt{\frac{2Rk(T_0 - T_2)}{k-1}}$$

$$= \sqrt{\left(\frac{2k}{k-1}\right)\left(\frac{p_0}{\rho_0} - \frac{p_2}{\rho_2}\right)} \qquad \text{[SI]} \qquad 26.3(a)$$

$$v_2 = \sqrt{2g_c J c_p(T_0 - T_2)} = \sqrt{\frac{2g_c Rk(T_0 - T_2)}{k-1}}$$

$$= \sqrt{\left(\frac{2g_c k}{k-1}\right)\left(\frac{p_0}{\rho_0} - \frac{p_2}{\rho_2}\right)} \qquad \text{[U.S.]} \qquad 26.3(b)$$

Example 26.1

Steam at 200 psia (1.4 MPa) and 500°F (250°C) enters an insulated nozzle with negligible velocity and is expanded to 20 psia (150 kPa) and 98% quality. Find the steam's exit velocity.

SI Solution

The initial enthalpy is found from the superheat tables or a Mollier diagram. The final enthalpy is found from property tables or from a Mollier diagram.

$$h_0 = 2927 \text{ kJ/kg}$$
$$h_2' = 2649 \text{ kJ/kg}$$

From Eq. 26.2, the exit velocity is

$$v_2 = \sqrt{(2)(h_0 - h_2')}$$

$$= \sqrt{(2)\left(2927 - 2649 \frac{\text{kJ}}{\text{kg}}\right)\left(1000 \frac{\text{J}}{\text{kJ}}\right)} = 746 \text{ m/s}$$

Customary U.S. Solution

The initial enthalpy is found from the superheat tables or a Mollier diagram. The final enthalpy is found from property tables or from a Mollier diagram.

$$h_0 = 1268.8 \text{ Btu/lbm}$$
$$h_2' = 1137.2 \text{ Btu/lbm}$$

From Eq. 26.2, the exit velocity is

$$v_2 = \sqrt{2g_c J(h_0 - h_2')}$$

$$= \sqrt{\begin{array}{c}(2)\left(32.2 \frac{\text{lbm-ft}}{\text{lbf-sec}^2}\right)\left(778 \frac{\text{ft-lbf}}{\text{Btu}}\right) \\ \times \left(1268.8 \frac{\text{Btu}}{\text{lbm}} - 1137.2 \frac{\text{Btu}}{\text{lbm}}\right)\end{array}}$$

$$= 2568 \text{ ft/sec}$$

3. ISENTROPIC FLOW

If the gas flow is adiabatic and frictionless (that is, reversible), the entropy change is zero and the flow is known as *isentropic flow*. As a practical matter, completely isentropic flow does not exist. Some high-velocity, steady-state flow processes, however, proceed with little increase in entropy and are considered to be isentropic. The irreversible effects are accounted for by various correction factors, such as nozzle and discharge coefficients.

4. ISENTROPIC FLOW FACTORS

In isentropic flow, total pressure, total temperature, and total density remain constant, regardless of the flow area and velocity. The instantaneous properties, known as *static properties*, do change along the flow path, however.[2] Equations 26.5 through 26.7 predict these static properties as functions of the Mach number, M.[3]

$$M = \frac{v}{a} = \frac{v}{\sqrt{kRT}} = \frac{v}{\sqrt{\frac{kR^*T}{(MW)}}} \qquad \text{[SI]} \qquad \textbf{26.4(a)}$$

$$M = \frac{v}{a} = \frac{v}{\sqrt{kg_cRT}} = \frac{v}{\sqrt{\frac{kg_cR^*T}{(MW)}}} \qquad \text{[U.S.]} \qquad \textbf{26.4(b)}$$

$$\left[\frac{T_0}{T}\right] = \left(\tfrac{1}{2}\right)(k-1)M^2 + 1 \qquad \textbf{26.5}$$

$$\left[\frac{p_0}{p}\right] = \left[\left(\tfrac{1}{2}\right)(k-1)M^2 + 1\right]^{\frac{k}{k-1}} = \left[\frac{T_0}{T}\right]^{\frac{k}{k-1}} \qquad \textbf{26.6}$$

$$\left[\frac{\rho_0}{\rho}\right] = \left[\left(\tfrac{1}{2}\right)(k-1)M^2 + 1\right]^{\frac{1}{k-1}} = \left[\frac{T_0}{T}\right]^{\frac{1}{k-1}} \qquad \textbf{26.7}$$

The isentropic flow ratios given by Eqs. 26.5 through 26.7 are functions only of the Mach number, M, and ratio of specific heats, k. Therefore, the ratios can be easily tabulated, as has been done in App. 26.A. The numbers in such tables are known as *isentropic flow factors*.

Example 26.2

Air ($k = 1.4$, MW = 29.0) flows isentropically from a large tank at 530°R (294K) through a convergent-divergent nozzle and is expanded to supersonic velocities. At a point where the Mach number is 2.5, what are the (a) gas temperature and (b) actual velocity?

SI Solution

(a) The temperature isentropic flow factor, T_0/T, can be calculated from Eq. 26.5 or its inverse read directly from the M = 2.5 line in App. 26.B.

---

[2]The *static properties* are not the same as the *stagnation properties*.

[3]For example, Eq. 26.5 can be derived from Eq. 26.1 by dividing both sides by the speed of sound.

$$\left[\frac{T_0}{T}\right] = \left(\tfrac{1}{2}\right)(k-1)M^2 + 1$$

$$= \left(\tfrac{1}{2}\right)(1.4-1)(2.5)^2 + 1 = 2.25$$

The temperature is

$$T = \frac{T_0}{\left[\dfrac{T_0}{T}\right]} = \frac{294\text{K}}{2.25} = 130.7\text{K}$$

(b) Since the Mach number is 2.5, the gas velocity is 2.5 times the speed of sound. This speed of sound is calculated from the static temperature, not from the total temperature. From Eq. 26.4,

$$v = Ma = M\sqrt{\frac{kR^*T}{(MW)}}$$

$$= 2.5\sqrt{\frac{(1.4)\left(8314\,\dfrac{\text{J}}{\text{kmol}\cdot\text{K}}\right)(130.7\text{K})}{29.0\,\dfrac{\text{kg}}{\text{kmol}}}} = 573\text{ m/s}$$

Customary U.S. Solution

(a) The temperature isentropic flow factor, T/T_0, can be read from the M = 2.5 line in App. 26.B.

$$\left[\frac{T}{T_0}\right] = 0.4444$$

$$T = T_0\left[\frac{T}{T_0}\right] = (530°\text{R})(0.4444) = 235.5°\text{R}$$

(b) Since the Mach number is 2.5, the gas velocity is

$$v = M\sqrt{\frac{kg_cR^*T}{(MW)}}$$

$$= 2.5\sqrt{\frac{(1.4)\left(32.2\,\dfrac{\text{lbm-ft}}{\text{lbf-sec}^2}\right)\left(1545\,\dfrac{\text{ft-lbf}}{\text{lbmol-°R}}\right)(235.5°\text{R})}{29.0\,\dfrac{\text{lbm}}{\text{lbmol}}}}$$

$$= 1880\text{ ft/sec}$$

5. RATIO OF SPECIFIC HEATS

The ratio of specific heats, $k = c_p/c_v$, is remarkably similar for gases with similar structures. For monatomic gases (He, Ar, Ne, Kr, etc.), it is approximately 1.67. For diatomic gases (N_2, O_2, H_2, CO, NO, and air), it is approximately 1.4. For triatomic gases (H_2O and CO_2), it is approximately 1.3. The value of k is less than 1.3 for more complex gases.

In most problems, the ratio of specific heats is assumed to be constant over all temperatures and pressures encountered. However, when the gas experiences a large temperature change, the ratio of specific heats should be evaluated at the average temperature.

Rather than using mathematical relationships to calculate exact solutions, it is common to use factors from tables appropriate for that ratio of specific heat. Since tables are typically available only for $k = 1.0$, 1.1, 1.2, 1.3, 1.4, and 1.67, a solution will not be exact when the ratio of specific heats is some intermediate value.

6. CRITICAL CONSTANTS

The point where sonic velocity has been achieved (i.e., $M = 1$) is known as a *critical point*. Sonic properties are designated by an asterisk, $*$. The insentropic flow factors at that point are known as *critical ratios* or *critical constants*. For example, the *critical pressure ratio* can be calculated from Eq. 26.8.

$$R_{cp} = \left[\frac{p^*}{p_0}\right] = \left(\frac{2}{k+1}\right)^{\frac{k}{k-1}} \qquad 26.8$$

The *critical temperature ratio* and *critical density ratio* are also easily derived.

$$\left[\frac{T^*}{T_0}\right] = \frac{2}{k+1} \qquad 26.9$$

$$\left[\frac{\rho^*}{\rho_0}\right] = \left(\frac{2}{k+1}\right)^{\frac{1}{k-1}} \qquad 26.10$$

Example 26.3

What is the critical pressure ratio for air ($k = 1.4$)?

Solution

From Eq. 26.8,

$$R_{cp} = \left(\frac{2}{k+1}\right)^{\frac{k}{k-1}} = \left(\frac{2}{1.4+1}\right)^{\frac{1.4}{1.4-1}} = 0.5283$$

7. CHOKED FLOW

Equation 26.6 seems to indicate that the gas velocity (and hence, the mass flow rate) in a discharge duct can be increased by increasing the ratio of source reservoir to ambient pressures (or by decreasing the ratio of ambient to source reservoir pressures). This is a logical conclusion, but valid only up to a certain gas velocity in the duct—the *sonic velocity* (i.e., speed of sound).

Changes in ambient pressure required to change the gas velocity in the duct travel upstream at sonic velocity. Therefore, if the gas in the duct is already traveling at sonic velocity, the fact that the ambient pressure has been changed cannot be transmitted back to the source

reservoir. Sonic velocity will occur when the ratio of ambient to source pressures drops to the *critical pressure ratio* (0.5283 for air). Once sonic velocity has been achieved in a duct or nozzle throat, the mass flow rate will be at its maximum and will remain constant for all subsequent decreases in ambient pressure.[4] This condition is called *choked flow*.

8. CHANGING DUCT AREA

Equation 26.11 defines the relationship between the change in flow area and the initial gas velocity for isentropic flow.

$$\frac{dA}{A} = \frac{dp}{\rho v^2}(1 - M^2) \qquad 26.11$$

Table 26.1 is based on Eq. 26.11 and summarizes the effects of changing flow area on the various thermodynamic properties for a gas in isentropic flow.

Table 26.1 *Effect of Changing Duct Area on Isentropic Flow*

	initial Mach no.	
	M< 1	M> 1
area decreasing		
	total properties constant	total properties constant
	velocity increases	velocity decreases
	Mach no. increases	Mach no. decreases
	pressure decreases	pressure increases
	density decreases	density increases
	temperature decreases	temperature increases
	enthalpy decreases	enthalpy increases
	internal energy decreases	internal energy increases
	entropy is constant	entropy is constant
area increasing		
	total properties constant	total properties constant
	velocity decreases	velocity increases
	Mach no. decreases	Mach no. increases
	pressure increases	pressure decreases
	density increases	density decreases
	temperature increases	temperature decreases
	enthalpy increases	enthalpy decreases
	internal energy increases	internal energy decreases
	entropy is constant	entropy is constant

---

[4]When the flow is already choked and the back pressure is lowered, neither the throat velocity nor the mass flow rate will increase. When the flow is already choked and the source pressure is increased, the throat velocity will remain constant. However, the increased density in the source will cause the mass flow rate to increase.

9. CONVERGENT-DIVERGENT NOZZLES

Sonic velocity from a reservoir can be achieved with almost any configuration of reservoir orifice. All that is necessary to achieve sonic velocity in an orifice is to lower the ambient pressure until the pressure ratio equals the critical pressure ratio. Achieving supersonic flow, however, specifically requires a *convergent-divergent nozzle* (also known as a *C-D nozzle* or *venturi nozzle*) in addition to the proper pressure conditions.

The properties at the point of flow constriction (i.e., the *throat*, subscript *t*) in a *C-D* nozzle are known as the *throat properties*. If sonic velocity is achieved in the throat, the properties are known as *critical properties*. Similarly, the *exit properties* (subscript *e*) represent the gas properties at the discharge point.

The rules for property changes given in Table 26.1, as well as the following statements, apply to a *C-D* nozzle.

- If sonic velocity occurs anywhere in the nozzle, it occurs in the throat.
- If sonic velocity occurs in the throat, the velocity may or may not be supersonic in the diverging section.
- If supersonic velocity occurs anywhere in the nozzle, it occurs in the diverging section.[5]
- If supersonic velocity occurs in the diverging section of the nozzle, the velocity is sonic in the throat.

From these four statements, it is clear that having sonic velocity in the throat is a necessary but not a sufficient condition for achieving supersonic velocity in the diverging section. In order to ensure supersonic velocity existing everywhere in the diverging section, there is an additional necessary condition. Specifically, the nozzle must be designed to expand the gases to the *design pressure ratio*, equal to $p_{ambient}/p_0$.

10. DESIGN OF SUPERSONIC NOZZLES

In order to design a nozzle capable of expanding a gas to some given velocity or Mach number, it is sufficient to have an expression for the flow area versus Mach number. Since the reservoir cross-sectional area is an unrelated variable, it is not possible to use the stagnation area as a reference area and to develop the ratio $[A_0/A]$ as was done for temperature, pressure, and density. The usual choice for a reference area is the *critical area*—the area at which the gas velocity is (or could be) sonic. This area is designated as A^*.

$$\frac{A}{A^*} = \frac{1}{M}\left[\frac{\left(\frac{1}{2}\right)(k-1)M^2 + 1}{\left(\frac{1}{2}\right)(k-1) + 1}\right]^{\frac{k+1}{(2)(k-1)}} \qquad 26.12$$

---

[5]This is a valid statement only if the flow in the converging section is subsonic. If a *C-D* nozzle is placed in an existing supersonic flow, there will be supersonic flow in the converging section of the nozzle.

Figure 26.2 is a plot of Eq. 26.12 versus Mach number. As long as the Mach number is less than 1.0, the area must decrease in order for the velocity to increase. However, if the Mach number is greater than 1.0, the area must increase in order for the velocity to increase. This is consistent with Table 26.1.

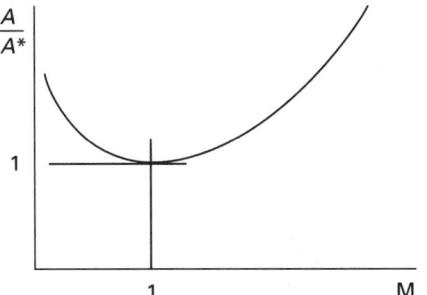

Figure 26.2 A/A^* *versus* M

It is not possible to change the Mach number to any desired value over any arbitrary distance along the axis of a converging-diverging nozzle. If the rate of change, dA/dx, is too great, the assumptions of one-dimensional flow become invalid. Usually, the converging section has a steeper angle (known as the *convergent angle*) than the diverging section. If the diverging angle is too great, a normal shock wave may form in that part of the nozzle.

Equation 26.13 gives the ratio of mass flow rate to the critical area. This equation is useful when the flow rate is known and the critical area is needed.

$$\frac{\dot{m}}{A^*} = \frac{\rho_0\sqrt{kRT_0}}{\left[\left(\frac{1}{2}\right)(k-1)+1\right]^{\frac{k+1}{(2)(k-1)}}} \qquad \text{[SI]} \qquad 26.13(a)$$

$$\frac{\dot{m}}{A^*} = \frac{\rho_0\sqrt{kg_cRT_0}}{\left[\left(\frac{1}{2}\right)(k-1)+1\right]^{\frac{k+1}{(2)(k-1)}}} \qquad \text{[U.S.]} \qquad 26.13(b)$$

Equation 26.14 gives the ratio of v/a^*—the ratio of gas velocity at some point to the sonic velocity in the throat. This formula is useful if the total properties and velocity at some point are known and the Mach number is needed. It is possible to solve for the static temperature and speed of sound, but this is more tedious than is first apparent.[6]

$$\frac{v}{a^*} = \sqrt{\frac{\left(\frac{1}{2}\right)(k+1)M^2}{\left(\frac{1}{2}\right)(k-1)M^2 + 1}} \qquad 26.14$$

By now, it should be apparent that the algebraic burden of solving gas dynamics problems is tremendous. The easiest way to solve such problems is with isentropic flow tables, such as Apps. 26.A and 26.B. The notation

---

[6]The speed of sound in a sonic throat, a^*, depends on the throat temperature, not on the total temperature.

and symbols in these appendices, or in any other table that might be available, should be noted carefully.

Example 26.4

The Mach number is 0.8 at a point in a nozzle where the area is 1.5 square units. Air ($k = 1.4$) is flowing. (a) What would the throat area have to be in order to achieve sonic velocity there? (b) What is the area at a point where the Mach number is 0.4?

Solution

(a) From the isentropic flow table for M = 0.8, $[A/A^*]$ = 1.0382. Therefore, the throat area would have to be

$$A_t = \frac{A_{M=0.8}}{\left[\dfrac{A}{A^*}\right]_{M=0.8}} = \frac{1.5 \text{ units}^2}{1.0382} = 1.445 \text{ units}^2$$

(b) In this case, the critical point will be used as a reference point, even though sonic velocity may not actually be achieved. For M = 0.4, $[A/A^*]$ = 1.5901.

$$A_{M=0.4} = \left[\frac{A}{A^*}\right]_{M=0.4} \times A_{M=1}$$

$$= \frac{A_{M=0.8} \times \left[\dfrac{A}{A^*}\right]_{M=0.4}}{\left[\dfrac{A}{A^*}\right]_{M=0.8}}$$

$$= \frac{(1.5 \text{ units}^2)(1.5901)}{1.0382} = 2.30 \text{ units}^2$$

Example 26.5

A frictionless, adiabatic nozzle receives 10 lbm/sec (5.0 kg/s) of air from a reservoir whose stagnation properties are 200°F and 30 psia (100°C and 200 kPa). At a particular point in the nozzle, a velocity of 1400 ft/sec (450 m/s) is attained. What are the (a) Mach number, (b) pressure, and (c) cross-sectional area at that point?

SI Solution

The absolute total temperature is

$$T_0 = 100°\text{C} + 273 = 373\text{K}$$

The total density in the reservoir is

$$\rho_0 = \frac{p_0}{RT_0}$$

$$= \frac{(200 \text{ kPa})\left(1000 \dfrac{\text{Pa}}{\text{kPa}}\right)}{\left(287 \dfrac{\text{J}}{\text{kg·K}}\right)(373\text{K})}$$

$$= 1.868 \text{ kg/m}^3$$

Read the property ratios for Mach 1 from the isentropic flow table: $[T/T_0] = 0.8333$, and $[\rho/\rho_0] = 0.6339$.

The sonic properties at the throat are

$$T^* = \left[\frac{T}{T_0}\right] T_0 = (0.8333)(373°\text{R}) = 310.8\text{K}$$

$$\rho^* = \left[\frac{\rho}{\rho_0}\right] \rho_0$$

$$= (0.6339)\left(1.868 \frac{\text{kg}}{\text{m}^3}\right) = 1.184 \text{ kg/m}^3$$

The sonic velocity at the throat is

$$a^* = \sqrt{kRT^*}$$

$$= \sqrt{(1.4)\left(287 \frac{\text{J}}{\text{kg·K}}\right)(310.8\text{K})}$$

$$= 353.4 \text{ m/s}$$

The throat area is

$$A^* = \frac{\dot{m}}{\rho^* a^*}$$

$$= \frac{5.0 \dfrac{\text{kg}}{\text{s}}}{\left(1.184 \dfrac{\text{kg}}{\text{m}^3}\right)\left(353.4 \dfrac{\text{m}}{\text{s}}\right)}$$

$$= 0.01195 \text{ m}^2$$

The ratio of actual-to-sonic throat velocities is

$$\left[\frac{\text{v}}{a^*}\right] = \frac{450 \dfrac{\text{m}}{\text{s}}}{353.4 \dfrac{\text{m}}{\text{s}}} = 1.273$$

Searching the $[\text{v}/a^*]$ column of the isentropic flow tables for this value gives M = 1.36. The corresponding ratios are $[p/p_0] = 0.3323$ and $[A/A^*] = 1.094$. At the point where the velocity is 450 m/s, the static pressure and area are

$$p_{450} = \left[\frac{p}{p_0}\right] p_0 = (0.3323)(200 \text{ kPa})$$

$$= 66.46 \text{ kPa}$$

$$A_{450} = \left[\frac{A}{A^*}\right] A^* = (1.094)(0.01195 \text{ m}^2)$$

$$= 0.01307 \text{ m}^2$$

Customary U.S. Solution

The absolute total temperature is

$$T_0 = 200°\text{F} + 460 = 660°\text{R}$$

The total density in the reservoir is

$$\rho_0 = \frac{p_0}{RT_0}$$

$$= \frac{\left(30 \, \frac{lbf}{in^2}\right)\left(144 \, \frac{in^2}{ft^2}\right)}{\left(53.3 \, \frac{ft\text{-}lbf}{lbm\text{-}°R}\right)(660°R)}$$

$$= 0.1228 \, lbm/ft^3$$

Read the property ratios for Mach 1 from the isentropic flow table: $[T/T_0] = 0.8333$, and $[\rho/\rho_0] = 0.6339$.

The sonic properties at the throat are

$$T^* = \left[\frac{T}{T_0}\right]T_0 = (0.8333)(660°R) = 550°R$$

$$\rho^* = \left[\frac{\rho}{\rho_0}\right]\rho_0$$

$$= (0.6339)\left(0.1228 \, \frac{lbm}{ft^3}\right) = 0.07784 \, lbm/ft^3$$

The sonic velocity at the throat is

$$a^* = \sqrt{kg_c R T^*}$$

$$= \sqrt{(1.4)\left(32.2 \, \frac{ft\text{-}lbm}{lbf\text{-}sec^2}\right)\left(53.3 \, \frac{ft\text{-}lbf}{lbm\text{-}°R}\right)(550°R)}$$

$$= 1149.6 \, ft/sec$$

The throat area is

$$A^* = \frac{\dot{m}}{\rho^* a^*}$$

$$= \frac{10 \, \frac{lbm}{sec}}{\left(0.07784 \, \frac{lbm}{ft^3}\right)\left(1149.6 \, \frac{ft}{sec}\right)}$$

$$= 0.1118 \, ft^2$$

The ratio of actual-to-sonic throat velocities is

$$\left[\frac{v}{a^*}\right] = \frac{1400 \, \frac{ft}{sec}}{1149.6 \, \frac{ft}{sec}} = 1.218$$

Searching the $[v/a^*]$ column of the isentropic flow tables for this value gives M = 1.28. The corresponding ratios are $[p/p_0] = 0.3708$ and $[A/A^*] = 1.058$. At the point where the velocity is 1400 ft/sec, the static pressure and area are

$$p_{1400} = \left[\frac{p}{p_0}\right]p_0 = (0.3708)\left(30 \, \frac{lbf}{in^2}\right)$$

$$= 11.12 \, lbf/in^2$$

$$A_{1400} = \left[\frac{A}{A^*}\right]A^* = (1.058)(0.1118 \, ft^2)$$

$$= 0.1183 \, ft^2$$

Example 26.6

An attitude-adjustment jet in a satellite uses high-pressure gas with a ratio of specific heats of 1.4 and a molecular weight of 21.0 lbm/mol (kg/kmol). The chamber conditions are 450 psia and 4700°R (3.2 MPa and 2600K). The gas is expanded supersonically to 2.97 psia (20.5 kPa) at the nozzle exit. What are the (a) sonic velocity in the throat, (b) required exit-to-throat area ratio, and (c) exit velocity?

SI Solution

(a) The specific gas constant is

$$R = \frac{R^*}{(MW)} = \frac{8314.3 \, \frac{J}{kmol\cdot K}}{21 \, \frac{kg}{kmol}}$$

$$= 395.9 \, J/kg\cdot K$$

Use the chamber properties as the total properties. Since the exit velocity is supersonic, sonic flow is achieved in the throat. From the isentropic flow tables at Mach 1, $[T/T_0] = 0.8333$. The temperature at the throat is

$$T_t = \left[\frac{T}{T_0}\right]T_0 = (0.8333)(2600K)$$

$$= 2167K$$

The velocity is sonic at the throat.

$$a^* = \sqrt{kRT}$$

$$= \sqrt{(1.4)\left(395.9 \, \frac{J}{kg\cdot K}\right)(2167K)}$$

$$= 1096 \, m/s$$

(b) The static pressure ratio at the exit is

$$\frac{p}{p_0} = \frac{20.5 \, kPa}{(3.2 \, MPa)\left(1000 \, \frac{kPa}{MPa}\right)}$$

$$= 0.006406$$

Locate this pressure ratio in the supersonic region of the isentropic flow table. The corresponding Mach number is approximately 4.0. Read $[A/A^*] = 10.7187$.

(c) At Mach 4, $[T/T_0] = 0.2381$. The exit temperature is

$$T_e = \left[\frac{T}{T_0}\right]T_0 = (0.2381)(2600K)$$

$$= 619.1K$$

The exit velocity is

$$v_e = Ma_e = 4\sqrt{kRT_e}$$

$$= 4\sqrt{(1.4)\left(395.9 \, \frac{J}{kg\cdot K}\right)(619.1K)}$$

$$= 2343 \, m/s$$

Customary U.S. Solution

(a) The specific gas constant is

$$R = \frac{R^*}{(MW)} = \frac{1545 \, \frac{\text{ft-lbf}}{\text{lbmol-}^\circ\text{R}}}{21 \, \frac{\text{lbm}}{\text{lbmol}}}$$

$$= 73.57 \text{ ft-lbf/lbm-}^\circ\text{R}$$

Use the chamber properties as the total properties. Since the exit velocity is supersonic, sonic flow is achieved in the throat. From the isentropic flow tables at Mach 1, $[T/T_0] = 0.8333$. The temperature at the throat is

$$T_t = \left[\frac{T}{T_0}\right] T_0 = (0.8333)(4700^\circ\text{R})$$

$$= 3917^\circ\text{R}$$

The velocity is sonic at the throat.

$$a^* = \sqrt{kg_c RT}$$

$$= \sqrt{(1.4)\left(32.2 \, \frac{\text{ft-lbm}}{\text{lbf-sec}^2}\right)\left(73.57 \, \frac{\text{ft-lbf}}{\text{lbm-}^\circ\text{R}}\right)(3917^\circ\text{R})}$$

$$= 3604 \text{ ft/sec}$$

(b) The static pressure ratio at the exit is

$$\frac{p}{p_0} = \frac{2.97 \, \frac{\text{lbf}}{\text{in}^2}}{450 \, \frac{\text{lbf}}{\text{in}^2}} = 0.0066$$

Locate this pressure ratio in the supersonic region of the isentropic flow table. The corresponding Mach number is 4.0. Read $[A/A^*] = 10.7187$.

(c) At Mach 4, $[T/T_0] = 0.2381$. The exit temperature is

$$T_e = \left[\frac{T}{T_0}\right] T_0 = (0.2381)(4700^\circ\text{R})$$

$$= 1119^\circ\text{R}$$

The exit velocity is

$$v_e = \text{Ma}_e = 4\sqrt{kg_c RT_e}$$

$$= 4\sqrt{\frac{(1.4)\left(32.2 \, \frac{\text{ft-lbm}}{\text{lbf-sec}^2}\right)\left(73.57 \, \frac{\text{ft-lbf}}{\text{lbm-}^\circ\text{R}}\right)}{\times(1119^\circ\text{R})}}$$

$$= 7706 \text{ ft/sec}$$

11. NOZZLE PERFORMANCE CHARACTERISTICS

The *nozzle velocity coefficient* is the ratio of actual to ideal (isentropic) exit velocities.

$$C_v = \frac{v_{\text{actual}}}{v_{\text{ideal}}} \qquad 26.15$$

The *nozzle discharge coefficient*, C_d, is the ratio of actual to ideal (isentropic) mass flow rates. C_d is typically high for well-designed nozzles, in the 0.97 to 0.98 range.

$$C_d = \frac{\dot{m}_{\text{actual}}}{\dot{m}_{\text{ideal}}} \qquad 26.16$$

$$\dot{m}_{\text{actual}} = \rho_e v_e A_e \qquad 26.17$$

The *nozzle efficiency*, η, is defined as the ratio of the actual to ideal (isentropic) energies extracted from the flowing gas. From Eq. 26.18, this is equal to the ratio of the square of actual to ideal velocities.

$$\eta = \frac{\Delta h_{\text{actual}}}{\Delta h_{\text{ideal}}} = \left(\frac{v_{\text{actual}}}{v_{\text{ideal}}}\right)^2 = C_v^2 \qquad 26.18$$

The *effective exhaust velocity*, v_{eff}, is determined from the thrust, F, and the actual mass flow rate.[7]

$$v_{\text{eff}} = \frac{F}{\dot{m}_{\text{actual}}} \qquad \text{[SI]} \qquad 26.19(a)$$

$$v_{\text{eff}} = \frac{F g_c}{\dot{m}_{\text{actual}}} \qquad \text{[U.S.]} \qquad 26.19(b)$$

Equation 26.20 gives the *thrust*, F, developed by the nozzle. (If the nozzle is operating at its design pressure ratio, then the exit and ambient pressures will be equal.)

$$F = \dot{m}_{\text{actual}}v_e + A_e(p_e - p_a)$$

$$= \dot{m}_{\text{actual}}v_{\text{eff}} \qquad \text{[SI]} \qquad 26.20(a)$$

$$F = \frac{\dot{m}_{\text{actual}}v_e}{g_c} + A_e(p_e - p_a)$$

$$= \frac{\dot{m}_{\text{actual}} \, v_{\text{eff}}}{g_c} \qquad \text{[U.S.]} \qquad 26.20(b)$$

The *coefficient of thrust*, C_F, is defined as

$$C_F = \frac{F}{p_0 A_t} \qquad 26.21$$

The *specific impulse*, I_{sp}, has units of seconds and is the ratio of thrust to rate of fuel consumption.

$$I_{\text{sp}} = \frac{F}{\dot{m}_{\text{actual}} \, g} = \frac{v_{\text{eff}}}{g} \qquad \text{[SI]} \qquad 26.22(a)$$

$$I_{\text{sp}} = \frac{F}{\dot{m}_{\text{actual}}} \times \frac{g_c}{g} = \frac{v_{\text{eff}}}{g} \qquad \text{[U.S.]} \qquad 26.22(b)$$

For constant-thrust engines, the *total impulse*, I_{total}, is the product of thrust and effective length of time, t_{eff}, the thrust is produced. For variable-thrust engines, the

---

[7]Symbols used in this section have been chosen to be consistent with other symbols in this chapter. However, it is traditional in the field of propulsion system design to use the following symbols: c, effective exhaust velocity; c^*, characteristic exhaust velocity; I_t, total impulse; and p_o, ambient (outside) pressure.

total impulse is the integral of thrust with respect to time.

$$I_{\text{total}} = \int F(t)\, dt$$

$$= F_{\text{ave}} t_{\text{eff}} = I_{\text{sp}} \times \text{total fuel weight} \quad \textit{26.23}$$

The *effective time*, t_{eff}, is

$$t_{\text{eff}} = \frac{I_{\text{total}}}{F_{\text{ave}}} = \frac{I_{\text{sp}} \times \text{total fuel weight}}{F_{\text{ave}}} \quad \textit{26.24}$$

The *characteristic exhaust velocity*, v^*, is defined as

$$\text{v}^* = \frac{\text{v}_{\text{eff}}}{C_F} = \frac{gI_{\text{sp}}}{C_F} = \frac{p_0 A_t}{\dot{m}_{\text{actual}}} \quad [\text{SI}] \quad \textit{26.25(a)}$$

$$\text{v}^* = \frac{\text{v}_{\text{eff}}}{C_F} = \frac{gI_{\text{sp}}}{C_F} = \frac{g_c p_0 A_t}{\dot{m}_{\text{actual}}} \quad [\text{U.S.}] \quad \textit{26.25(b)}$$

The *characteristic length*, L^*, of the combustion chamber is

$$L^* = \frac{V_{\text{chamber}}}{A_t} \quad \textit{26.26}$$

Example 26.7

What are the (a) nozzle efficiency and (b) coefficient of velocity for the steam nozzle in Ex. 26.1?

SI Solution

(a) If the expansion through the nozzle had been isentropic (i.e., straight down on the Mollier diagram), the exit enthalpy would have been approximately 2510 kJ/kg. The nozzle efficiency is defined by Eq. 26.18.

$$\eta_{\text{nozzle}} = \frac{h_0 - h_2'}{h_0 - h_2} = \frac{2927\,\dfrac{\text{kJ}}{\text{kg}} - 2649\,\dfrac{\text{kJ}}{\text{kg}}}{2927\,\dfrac{\text{kJ}}{\text{kg}} - 2510\,\dfrac{\text{kJ}}{\text{kg}}} = 0.66667$$

(b) The coefficient of velocity is defined by Eq. 26.18.

$$C_{\text{v}} = \sqrt{\eta_{\text{nozzle}}} = \sqrt{0.66667} = 0.816$$

Customary U.S. Solution

(a) If the expansion through the nozzle had been isentropic (i.e., straight down on the Mollier diagram), the exit enthalpy would have been approximately 1082.2 Btu/lbm. The nozzle efficiency is defined by Eq. 26.18.

$$\eta_{\text{nozzle}} = \frac{h_0 - h_2'}{h_0 - h_2} = \frac{1268.8\,\dfrac{\text{Btu}}{\text{lbm}} - 1137.2\,\dfrac{\text{Btu}}{\text{lbm}}}{1268.8\,\dfrac{\text{Btu}}{\text{lbm}} - 1082.2\,\dfrac{\text{Btu}}{\text{lbm}}} = 0.70525$$

(b) The coefficient of velocity is defined by Eq. 26.18.

$$C_{\text{v}} = \sqrt{\eta_{\text{nozzle}}} = \sqrt{0.70525} = 0.840$$

12. SHOCK WAVES

Section 9 stated that a nozzle must be designed to the design pressure ratio in order to keep the flow supersonic in the diverging section of the nozzle. It is possible, though, to have supersonic velocity only in part of the diverging section. Once the flow is supersonic in a part of the diverging section, however, it cannot become subsonic by an isentropic process.

Therefore, the gas experiences a *shock wave* as the velocity drops from supersonic to subsonic. Shock waves are very thin (several molecules thick) and separate areas of radically different thermodynamic properties. Since the shock wave forms normal to the flow direction, it is known as a *normal shock wave*. The strength of a shock wave is measured by the change in Mach number across it.

The velocity always changes from supersonic to subsonic across a shock wave. Since there is no loss of heat energy, a shock wave is an adiabatic process, and total temperature is constant. However, the process is not isentropic, and total pressure decreases. Momentum is also conserved. Table 26.2 lists the property changes across a shock wave.

Table 26.2 *Property Changes Across a Normal Shock Wave*

property	change
total temperature	is constant
total pressure	decreases
total density	decreases
velocity	decreases
Mach number	decreases
pressure	increases
density	increases
temperature	increases
entropy	increases
internal energy	increases
enthalpy	is constant
momentum	is constant

The properties before a shock wave are given the subscript x. The subscript y is used for the properties after a shock wave. These subscripts may be combined with the subscript 0 to represent a total (stagnation) property. (For example, $p_{0,x} < p_{0,y}$.)

Although it is possible to calculate the ratio of a property before and after a shock wave, the relationships are complex, and it is more convenient to use *normal shock factors* from a *normal shock table* such as App. 26.B. These factors are dependent only on the ratio of specific heats, the Mach number, M_x, immediately before the shock wave.

Example 26.8

A shock wave in air ($k = 1.4$) occurs at a point where the Mach number is 2.4. If the static pressure before the shock wave is 5.0 atm, what are the (a) Mach number after the shock, (b) static pressure after the shock, and (c) total pressure after the shock?

Solution

All of the factors are read from the normal shock table (App. 26.B) for $M_x = 2.4$.

(a) The Mach number after the shock is read directly as $M_y = 0.5231$.

(b) The ratio of static pressures is read as $[p_y/p_x] = 6.553$. The static pressure after the shock is

$$p_y = \left[\frac{p_y}{p_x}\right] \times p_x = (6.553)(5.0) = 32.8 \text{ atm}$$

(c) The ratio of static pressure before the shock to total pressure after the shock is read as $[p_x/p_{0,y}] = 0.1266$. The total pressure after the shock is

$$p_{0,y} = \frac{p_x}{\left[\frac{p_x}{p_{0,y}}\right]} = \frac{5.0 \text{ atm}}{0.1266} = 39.5 \text{ atm}$$

Example 26.9

The temperature and pressure in a desert prior to an explosive test are 70°F and 14.7 psia (21°C and 101.3 kPa), respectively. At the center of the blast, the pressure is raised to 2 atm. (a) What is the wave velocity? What are the (b) pressure, (c) temperature, and (d) velocity behind the wave?

SI Solution

(a) Even though the expansion is spherical, the values from the normal shock tables can be used. Let the x-subscript correspond to the still air into which the shock wave is expanding. Let the y-subscript correspond to the gas inside the expanding high-pressure region.

The absolute temperature of the desert air is

$$T_x = 21°C + 273 = 294K$$

The speed of sound in the desert air is

$$a = \sqrt{kRT}$$
$$= \sqrt{(1.4)\left(287 \ \frac{J}{\text{kg·K}}\right)(294K)}$$
$$= 343.7 \text{ m/s}$$

From the normal shock tables for $[p_y/p_x] = 2$, the Mach number is 1.36. The velocity of the expanding shock wave is

$$v = (1.36)\left(343.7 \ \frac{m}{s}\right) = 467.4 \text{ m/s}$$

(b) Behind the wave, the pressure is

$$(2)(101.3 \text{ kPa}) = 202.6 \text{ kPa}$$

(c) From the isentropic flow tables at $M = 1.36$, $[T_y/T_x] = 1.229$. The temperature behind the shock wave is

$$T = \left[\frac{T_y}{T_x}\right]T_x = (1.229)(294K)$$
$$= 361.3K$$

(d) The velocity of sound behind the shock wave is

$$a = \sqrt{kRT}$$
$$= \sqrt{(1.4)\left(287 \ \frac{J}{\text{kg · K}}\right)(361.3K)}$$
$$= 381.0 \text{ m/s}$$

From the isentropic flow tables at $M = 1.36$, $M_y = 0.75718$. The velocity of the air behind the shock wave is

$$v_y = M_y a = (0.75718)\left(381.0 \ \frac{m}{s}\right)$$
$$= 288.5 \text{ m/s}$$

Customary U.S. Solution

(a) Use the normal shock tables. Let the x-subscript correspond to the still air into which the shock wave is expanding. Let the y-subscript correspond to the gas inside the expanding high-pressure region.

The absolute temperature of the desert air is

$$T_x = 70°F + 460 = 530°R$$

The speed of sound in the desert air is

$$a = \sqrt{kg_c RT}$$
$$= \sqrt{\frac{(1.4)\left(32.2 \ \frac{\text{ft-lbm}}{\text{lbf-sec}^2}\right)\left(53.3 \ \frac{\text{ft-lbf}}{\text{lbm-°R}}\right)}{\times (530°R)}}$$
$$= 1128 \text{ ft/sec}$$

From the normal shock tables for $[p_y/p_x] = 2$, the Mach number is 1.36. The velocity of the expanding shock wave is

$$v = (1.36)\left(1128 \ \frac{\text{ft}}{\text{sec}}\right) = 1534 \text{ ft/sec}$$

(b) Behind the wave, the pressure is

$$(2)\left(14.7 \ \frac{\text{lbf}}{\text{in}^2}\right) = 29.4 \text{ lbf/in}^2$$

(c) From the isentropic flow tables at $M = 1.36$, $[T_y/T_x] = 1.229$. The temperature behind the shock wave is

$$T = \left[\frac{T_y}{T_x}\right]T_x = (1.229)(530°R)$$
$$= 651.4°R$$

(d) The velocity of sound behind the shock wave is

$$a = \sqrt{kg_cRT}$$

$$= \sqrt{\begin{array}{c}(1.4)\left(32.2 \dfrac{\text{ft-lbm}}{\text{lbf-sec}^2}\right)\left(53.3 \dfrac{\text{ft-lbf}}{\text{lbm-}°\text{R}}\right)\\ \times(651.4°\text{R})\end{array}}$$

$$= 1251 \text{ ft/sec}$$

From the isentropic flow tables at M = 1.36, M_y = 0.75718. The velocity of the air behind the shock wave is

$$v_y = M_y a = (0.75718)\left(1251 \dfrac{\text{ft}}{\text{sec}}\right)$$

$$= 947.2 \text{ ft/sec}$$

13. OBLIQUE SHOCK WAVES

The shock waves described in Sec. 12 are called *normal shock waves* because they are normal to the direction of flow. Supersonic flow past wedges and cones creates *oblique shock waves* that are inclined from the direction of flow. For a wedge, the shock wave is in the form of two inclined planes. For a conical projectile, the shock wave is in the form of a conical envelope. The shock wave may be attached or may form in front of the object.

There is no convenient formula for determining the *shock angle* θ, although it can be obtained graphically from Fig. 26.3.

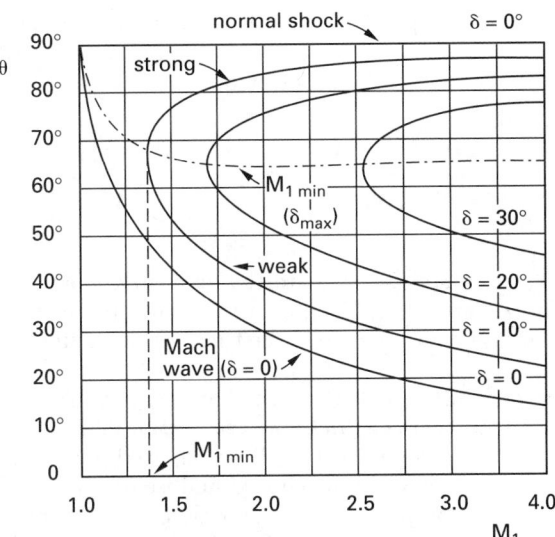

(This figure cannot be used for cones.)

Figure 26.3 *Shock Angle for Wedges (k = 1.4)*

Figure 26.3 illustrates that, for higher Mach numbers, there are two angles corresponding to each Mach number. Below the broken line on the chart, a *weak shock*

wave (more like a Mach wave than a shock wave) exists.[8] Above the broken line, a strong oblique shock wave (more like a normal shock wave) exists. Though both can theoretically exist simultaneously, it is believed that only the weaker wave can exist attached to the body.

If the *semivertex angle* (also known as the *deviation angle*), δ, is small enough, the shock wave will be attached. Attached shock waves are more likely to occur with wedges and narrow cones. Detached shock waves are more likely to form on thick wedges and objects with blunt leading edges.[9]

For attached shock waves, the conditions (static pressure, static temperature, etc.) after the shock wave can be found from the normal shock tables if $M_1 \sin \theta$ is used in place of M_x. If the shock wave is not attached, the shock tables cannot be used.

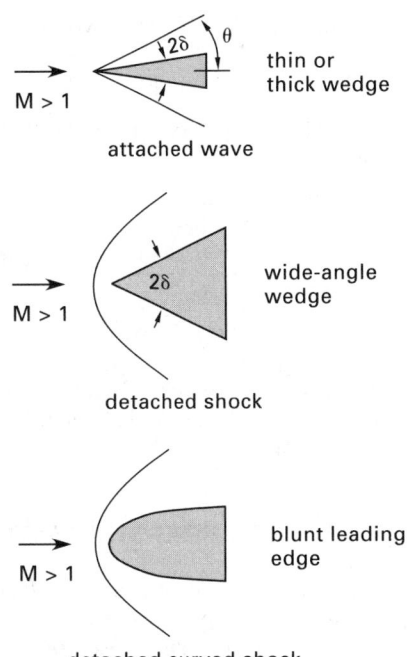

Figure 26.4 *Attached and Detached Oblique Shock Waves*

Unlike normal shock waves, the velocity after an oblique shock wave will not necessarily be subsonic. The Mach number for the normal component of velocity after an oblique shock wave will be less than 1, but the Mach

---

[8]A *Mach wave* is an infinitesimal oblique shock wave whose normal component of the Mach number is 1.0. That is, $M_1 \sin \theta$ is 1.0, even though $M_1 > 1$. Changes in flow properties across a Mach wave are negligible. A Mach wave is not a discontinuity in the flow, and flow through it is isentropic. Therefore, Mach waves can exist anywhere in supersonic flow. The inclination of a Mach wave to the initial direction of flow is known as the *Mach angle* and is given by $\theta = \sin^{-1}(1/M_1)$.

[9]For wedges in air, the shock wave will be attached if the Mach number is greater than the minimum value indicated by Fig. 26.3 for the wedge semivertex angle. The shock wave will be detached if the Mach number is less than this value, or if the wedge semivertex angle is greater than $\sin^{-1}(1/k)$, which is 45.58° for air. For cones, the semivertex angle must be less than 57.5° for attached shock waves.

number of the resultant velocity may exceed 1. The Mach number after the shock is

$$M_2 = \frac{M_y \text{ from shock table}}{\sin (\theta - \delta)} \qquad 26.27$$

Example 26.10

A wedge with a semivertex angle of $10°$ is traveling through air at Mach 2. An attached oblique shock wave forms. What are the (a) shock angle, (b) Mach number after the shock wave, and (c) ratio of static pressures before and after the shock wave?

Solution

(a) From Fig. 26.3, the semivertex shock angle will be approximately $40°$.

(b) Use the normal shock table with $M_1 \sin \theta$ in place of M_x.

$$M_1 \sin \theta = 2 \sin 40° = 1.29$$

For $M_x = 1.29$, $M_y = 0.79108$. From Eq. 26.27,

$$\begin{aligned} M_2 &= \frac{M_y}{\sin (\theta - \delta)} \\ &= \frac{0.79108}{\sin (40° - 10°)} \\ &= 1.58 \end{aligned}$$

(c) The ratio of static pressures before and after the shock wave is read from the normal shock table for Mach 1.29: $[p_y/p_x] = 1.7748$.

14. CRITICAL BACK-PRESSURE RATIO

Figure 26.5(a) illustrates a converging-diverging nozzle separating high- and low-pressure reservoirs. The total pressure in the high-pressure reservoir is constant. The pressure in the low-pressure reservoir (i.e., the *back pressure*) is variable. The nozzle geometry, particularly the throat and exit areas, A_t and A_e, is assumed to be known. Figure 26.5(b) illustrates the ratio of static pressure at the corresponding point to the total pressure (p/p_0) in the first reservoir. The following eight cases (*flow regimes*) represent decidedly different performances.

case A: If $p_2 = p_0$, there will be no flow.

case B: When p_2 is lowered to just slightly less than p_0, flow will be initiated. As long as the pressure ratio p_t/p_0 is less than the first critical pressure ratio, R_{cp} (0.5283 for air), flow in the divergent section will be subsonic, the exit pressure, p_e, will be equal to p_2. Flow will be isentropic, and the nozzle acts like a venturi. The conditions at any point in the nozzle where the area is known can be determined in the following manner.

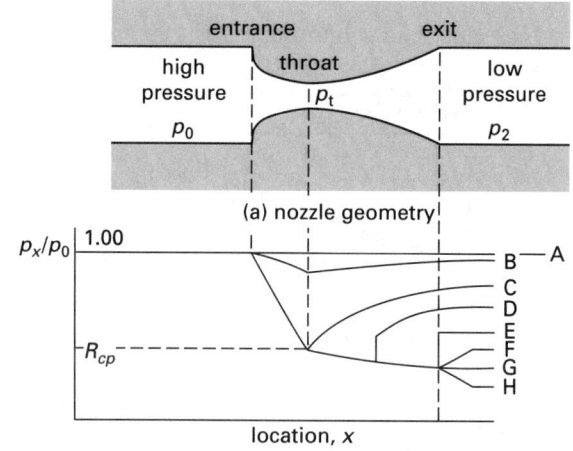

(a) nozzle geometry

(b) pressure ratio

Figure 26.5 *Flow Regimes in a C-D Nozzle*

step 1: Calculate the exit pressure ratio. (For any other point, substitute the known pressure for p_2.)

$$\frac{p_e}{p_0} = \frac{p_2}{p_0} \qquad 26.28$$

step 2: Locate the pressure ratio calculated in step 1 in the isentropic flow table. Read the exit Mach number, M_e, and the critical area ratio $[A/A^*]_e$. (Since sonic velocity is not actually achieved, the theoretical critical area A^* does not exist in this nozzle.)

step 3: Calculate the ratio of actual throat area to the hypothetical critical throat area.

$$\frac{A_t}{A^*} = \frac{A_t \left[\dfrac{A}{A^*} \right]_e}{A_e} \qquad 26.29$$

step 4: Locate the area ratio calculated in step 3 in the isentropic flow table. Read the throat Mach number, M_t.

case C: If p_2 is lowered so that p_t/p_0 equals the *first critical (back) pressure ratio* (0.5283 for air), sonic velocity will be achieved in the throat. Flow will be choked and isentropic. The gas properties in the nozzle can be found from the procedure given in case B. The exit velocity will be subsonic, and the exit pressure will equal the back pressure ($p_e = p_2$).

case D: If p_2 is lowered slightly below that of case C (but not below the pressure ratio corresponding to the area ratio A_e/A^*), the mass flow rate will not increase above that of case C because the flow will be choked. However, since the back pressure p_2 is lower, the velocity will be higher. Supersonic velocity will be achieved in some parts of the nozzle. Flow will drop back to subsonic via a normal shock wave. The exit velocity will be subsonic, and the exit pressure will equal the back pressure ($p_e = p_2$). The location (i.e., the area) in the divergent section where the shock wave occurs can be found from the following trial-and-error procedure.

step 1: Assume a Mach number, M_x, at which the shock wave occurs.

step 2: For M_x, read the ratio of total pressures $[p_{0,y}/p_{0,x}]$ from the normal shock table.

step 3: Calculate the ratio of static exit pressure to total pressure after the shock wave. $p_{0,x}$ is the total pressure in the source reservoir.

$$\frac{p_e}{p_{0,y}} = \frac{p_e \left[\dfrac{p_{0,y}}{p_{0,x}} \right]}{p_{0,x}} \qquad 26.30$$

step 4: Determine the exit Mach number, M_e, by locating the ratio calculated in step 3 in the $[p/p_0]$ column of the normal shock tables. Read the area ratio $[A/A^*]$.

step 5: Calculate the area ratio.

$$\frac{A_e \left[\dfrac{p_{0,y}}{p_{0,x}} \right]}{A_t} \qquad 26.31$$

step 6: Compare the area ratio from steps 4 and 5. If the values differ, repeat from step 1. (The area ratio from step 5 decreases when the assumed Mach number increases.) If the values are equal, the shock wave occurs at the assumed Mach number, M_x.

step 7: Calculate the area at which the shock wave occurs.

$$A = A_t \left[\frac{A}{A^*} \right] \qquad 26.32$$

case E: If the back pressure is equal to the *second critical (back) pressure ratio* (i.e., the pressure ratio corresponding to the area ratio A_e/A^*), the shock wave will stand at the exit. The mass flow rate is the same as in case C.

case F: If the back pressure ratio is less than the second critical back pressure ratio but greater than the third, the condition is known as *over-expansion*. The exit area is too large, and gas will be discharged at a pressure less than the ambient conditions. The gas will "pop" back up to the ambient pressure through an *oblique compression shock wave* outside the nozzle, but this will not change the conditions inside the nozzle. The mass flow rate is the same as in case C.

case G: If the back pressure ratio is equal to the *third critical (back) pressure ratio*, there will be no shock wave. Pressure p_2 is known as the *design pressure* or the *isentropic pressure* for the nozzle. Flow will be isentropic. The mass flow rate is the same as in case C.

case H: If the pressure ratio is less than the third critical back pressure ratio, the condition is known as *under-expansion*. The exit area is too small for the back pressure. Expansion will be incomplete in the nozzle, and the gas will be discharged at a pressure above the local ambient pressure. Expansion continues outside the nozzle, and the pressure drops down to the ambient pressure by way of a *rarefaction shock wave* (also known as an *expansion wave* or an *oblique expansion shock wave*) outside the nozzle. The mass flow rate is the same as in case C.

Both over- and under-expansion reduce the efficiency of the nozzle. The effect of over-expansion is to reduce the gas exit velocity. In the case of rocket propulsion nozzles, the decreased exit velocity produces a proportional decrease in thrust (see Eq. 26.20). In the case of a fixed steam nozzle, the energy available to the steam turbine is reduced.

A given percentage of over-expansion (based on the ratio of actual-to-theoretical exit areas) can reduce the available energy as much as ten times the reduction for the same percentage of under-expansion. For that reason, steam nozzles feeding turbines are sometimes designed 10 to 20% too small to ensure under-expansion under light or partial loads.

Example 26.11

A supersonic wind tunnel is constructed so that the back pressure can be varied during the testing of a nozzle. The wind tunnel is fed from a reservoir with a total pressure of 1000 psia (7 MPa) and a total temperature of 1000°R (560K). A nozzle in the wind tunnel has a throat area of 1.5 in² (9.68 cm²) and an exit area of

3.457 m^2 (22.26 cm^2). What is the highest back pressure for which the flow will remain supersonic throughout the entire length of the nozzle?

SI Solution

The desired operating conditions (i.e., that of no shock wave in the nozzle) can be satisfied by case E, where there is a shock wave at (or just outside of) the exit.[10] Since the velocity is supersonic in the diverging section, the velocity is sonic in the throat. The ratio of exit-to-throat areas, corresponding to $[A/A^*]$ since Mach 1 has been achieved, is

$$\frac{A_e}{A_t} = \frac{22.26 \text{ cm}^2}{9.68 \text{ cm}^2} = 2.3$$

Searching the isentropic flow tables for this value, the exit Mach number is found to be 2.35. For that point, $[p/p_0] = 0.07396$. The exit pressure before the shock wave is

$$p_{x,e} = \left[\frac{p}{p_0}\right] p_0 = (0.07396)(7.0 \text{ MPa})$$
$$= 0.5177 \text{ MPa}$$

This pressure is the design pressure for the nozzle and corresponds to case G. However, if there is a shock wave at the exit, the pressure after the shock wave will be higher. From the normal shock table for M$_x$ = 2.35, $[p_y/p_x] = 6.276$. The maximum back pressure is

$$p_y = \left[\frac{p_y}{p_x}\right] p_{x,e} = (6.276)(0.5177 \text{ MPa})$$
$$= 3.249 \text{ MPa}$$

Customary U.S. Solution

The desired operating conditions (i.e., that of no shock wave in the nozzle) can be satisfied by case E, where there is a shock wave at (or just outside of) the exit.[11] Since the velocity is supersonic in the diverging section, the velocity is sonic in the throat. The ratio of exit-to-throat areas, corresponding to $[A/A^*]$ since Mach 1 has been achieved, is

$$\frac{A_e}{A_t} = \frac{3.457 \text{ in}^2}{1.5 \text{ in}^2} = 2.3$$

Searching the isentropic flow tables for this value, the exit Mach number is found to be 2.35. For that point, $[p/p_0] = 0.07396$. The exit pressure before the shock wave is

$$p_{x,e} = \left[\frac{p}{p_0}\right] p_0 = (0.07396)\left(1000 \frac{\text{lbf}}{\text{in}^2}\right)$$
$$= 73.96 \text{ lbf/in}^2$$

This pressure is the design pressure for the nozzle and corresponds to case G. However, if there is a shock wave at the exit, the pressure after the shock wave will be higher. From the normal shock table for M$_x$ = 2.35, $[p_y/p_x] = 6.276$. The maximum back pressure is

$$p_y = \left[\frac{p_y}{p_x}\right] p_{x,e} = (6.276)\left(73.96 \frac{\text{lbf}}{\text{in}^2}\right)$$
$$= 464.2 \text{ lbf/in}^2$$

Example 26.12

A nozzle is fed from a large reservoir containing 100 psia (0.7 MPa) air. The back pressure is adjustable. A shock wave stands at the exit of the nozzle. The Mach number just before the shock wave is 3.0. For which values of the back pressure will the shock wave at the exit disappear completely?

SI Solution

There are two ways the shock wave can disappear. If the back pressure is decreased to the design pressure (case G), the shock wave will move out of the nozzle and dissipate. At Mach 3, $[p/p_0] = 0.02722$. Therefore, the design pressure for this nozzle is

$$p_{\text{design}} = \left[\frac{p}{p_0}\right] p_0 = (0.02722)(0.7 \text{ MPa})\left(1000 \frac{\text{kPa}}{\text{MPa}}\right)$$
$$= 19.05 \text{ kPa}$$

If the back pressure is increased, the shock wave will move upstream and vanish at the throat (case C). From the supersonic portion of the isentropic flow tables at Mach 3, $[A/A^*] = 4.235$. The same value occurs in the subsonic part of the isentropic flow table. The Mach number corresponding to this value in the subsonic part of the table is 0.138. At this Mach number, the shock wave will vanish and the flow will be subsonic everywhere in the nozzle. At Mach 0.138, $[p/p_0] = 0.987$. The back pressure is

$$p_{\text{back}} = \left[\frac{p}{p_0}\right] p_0 = (0.987)(0.7 \text{ MPa})$$
$$= 0.6909 \text{ MPa}$$

Customary U.S. Solution

Shock wave at the exit (case G): At Mach 3, $[p/p_0] = 0.02722$. Therefore, the design pressure is

$$p_{\text{design}} = \left[\frac{p}{p_0}\right] p_0 = (0.02722)\left(100 \frac{\text{lbf}}{\text{in}^2}\right)$$
$$= 2.722 \text{ lbf/in}^2$$

---

[10]Therefore, this example is *not* correctly solved by assuming case G and finding the design pressure.
[11]See Ftn. 10.

Shock wave at the throat (case C): At Mach 3, $[A/A^*] = 4.235$. The Mach number corresponding to this value in the subsonic part of the table is 0.138. At Mach 0.138, $[p/p_0] = 0.987$. The back pressure is

$$p_{back} = \left[\frac{p}{p_0} \right] p_0 = (0.987) \left(100 \ \frac{lbf}{in^2} \right)$$
$$= 98.7 \ lbf/in^2$$

15. COMPRESSIBLE FLOW THROUGH ORIFICES

Compressible flow through orifices (simple holes, perforations, re-entrant tubes, etc.) is similar to flow through nozzles, cases A, B, and C. If the ratio of ambient-to-source pressures is less than the first critical pressure ratio, the velocity will be sonic in the orifice. However, since there is no diverging section, the flow can never become supersonic. Due to the high turbulence at the orifice, the coefficient of velocity and nozzle efficiency may be relatively low. This will be manifested in a lower-than-ideal mass flow rate.

16. ADIABATIC FLOW WITH FRICTION IN CONSTANT-AREA DUCTS

Although flow through nozzles is assumed to be frictionless (because nozzles are short), friction cannot be disregarded in long constant-area ducts. If the velocity is high or if the duct is insulated, flow may still be adiabatic, however. Adiabatic flow with friction is known as *Fanno flow*. In Fanno flow, total pressure always decreases in the direction of flow. Total temperature, however, is constant, regardless of the velocity of the entering gas.

The *Fanno line* is a plot of temperature (or enthalpy) versus entropy for a specific flow rate. The line illustrates how the velocity tends toward Mach 1, regardless of the entering velocity.

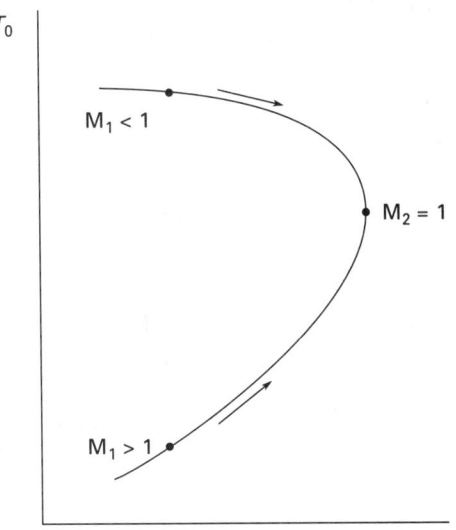

Figure 26.6 *Fanno Line*

If the velocity of the gas entering the duct is subsonic, the pressure and temperature will drop in the direction of flow. Since the density also drops, the velocity will increase to maintain the same mass flow rate. Eventually, the velocity in the duct becomes sonic, the flow is choked, and no further increase in mass flow rate can occur.

If the velocity of the gas entering the duct is supersonic, friction causes the pressure and temperature to increase in the direction of flow. Since the density also increases, the velocity will decrease to maintain the same mass flow rate. Eventually, the velocity in the duct becomes sonic.

When the flow is Mach 1 at the entrance, friction causes the pressure, density, and temperature to increase and the velocity to decrease. The Mach number is 1 at the exit.

Since the flow is adiabatic, the total temperature is constant. Equation 26.5 can be used to find the temperature at any point along the duct. The ratio of static pressures for two points along the duct is given by Eq. 26.33.

$$\frac{p_2}{p_1} = \left(\frac{\rho_1}{\rho_2} \right) \left(1 + \left[\left(\tfrac{1}{2} \right) (k-1) M_1^2 \right] \left[1 - \left(\frac{\rho_1}{\rho_2} \right)^2 \right] \right) \quad \text{26.33}$$

The duct length required for the flow to reach Mach 1 is found by solving Eq. 26.34 for L_{max}. For round ducts, the *hydraulic diameter*, D_h, is equal to the inside diameter. $f_{Fanning}$ is the *Fanning friction factor*, also known as the *coefficient of friction*, C_f.[12]

$$\frac{4 f_{Fanning} L_{max}}{D_h} = \left(\frac{1 - M^2}{k M^2} \right)$$
$$+ \left(\frac{1+k}{2k} \right) \ln \left(\frac{(1+k) M^2}{2 \left[1 + \left(\tfrac{1}{2} \right) (k-1) M^2 \right]} \right) \quad \text{26.34}$$

The length of duct between two points with known Mach numbers can be found from Eq. 26.35. The quantity $4 f_{Fanning} L_{max}/D_h$ is found from Eq. 26.34, tables, or graphs.

$$x = \left(\frac{D_h}{4f} \right) \left[\left(\frac{4 f_{Fanning} x_{max}}{D_h} \right)_1 - \left(\frac{4 f_{Fanning} x_{max}}{D_h} \right)_2 \right] \quad \text{26.35}$$

The static gas properties at any point can be related to the sonic properties farther downstream.

$$\frac{p}{p^*} = \frac{1}{M} \sqrt{ \frac{1+k}{(2) \left[1 + \left(\tfrac{1}{2} \right) (k-1) M^2 \right]} } \quad \text{26.36}$$

---

[12]Multiplying the Fanning friction factor by 4, as is done in Eq. 26.34, converts it to the Darcy friction factor.

$$\frac{T}{T^*} = \frac{1+k}{(2)\left[1 + \left(\frac{1}{2}\right)(k-1)M^2\right]} \qquad 26.37$$

$$\frac{p_0}{p_0^*} = \left(\frac{1}{M}\right)\left(\left(\frac{2}{1+k}\right)\left[1 + \left(\frac{1}{2}\right)(k-1)M^2\right]\right)^{\frac{1+k}{(2)(k-1)}} \qquad 26.38$$

Example 26.13

Air enters an adiabatic round duct with an inside diameter of 0.3 ft (0.09 m). The Fanning friction factor is 0.003. The entrance air pressure is 12 psia (80 kPa), and the entrance Mach number is 0.3. What are the (a) Mach number and (b) pressure 50 ft (15 m) downstream?

SI Solution

(a) From App. 26.C for M = 0.3,

$$\frac{4f_{\text{Fanning}}x_{\max}}{D_h} = 5.299$$

The value 15 m farther into the duct is

$$\frac{4f_{\text{Fanning}}x}{D_h} = \frac{(4)(0.003)(15\text{ m})}{0.09\text{ m}}$$
$$= 2.000$$

From Eq. 26.35,

$$\left(\frac{4f_{\text{Fanning}}x_{\max}}{D_h}\right)_2 = 5.299 - 2.000$$
$$= 3.299$$

Interpolating in App. 26.C, M_2 is approximately 0.356.

(b) At the entrance, for Mach 0.3, the ratio $[p/p^*]$ is 3.619. Therefore, the sonic pressure is

$$p^* = \frac{p}{\left[\dfrac{p}{p^*}\right]} = \frac{80\text{ kPa}}{3.619} = 22.11\text{ kPa}$$

15 m downstream, at Mach 0.35, the ratio $[p/p^*]$ is 3.092. p^* is unchanged.

$$p_2 = \left[\frac{p}{p^*}\right]p^* = (3.092)(22.11\text{ kPa})$$
$$= 68.36\text{ kPa}$$

Customary U.S. Solution

(a) From App. 26.C for M = 0.3,

$$\frac{4f_{\text{Fanning}}x_{\max}}{D_h} = 5.299$$

The value 50 ft farther into the duct is

$$\frac{4f_{\text{Fanning}}x}{D_h} = \frac{(4)(0.003)(50\text{ ft})}{0.3\text{ ft}}$$
$$= 2.000$$

From Eq. 26.35,

$$\left(\frac{4f_{\text{Fanning}}x_{\max}}{D_h}\right)_2 = 5.299 - 2.000$$
$$= 3.299$$

Interpolating in App. 26.C, M_2 is approximately 0.356.

(b) At the entrance, for Mach 0.3, the ratio $[p/p^*]$ is 3.619. Therefore, the sonic pressure is

$$p^* = \frac{p}{\left[\dfrac{p}{p^*}\right]} = \frac{12\text{ psia}}{3.619} = 3.316\text{ psia}$$

50 ft downstream, at Mach 0.35, the ratio $[p/p^*]$ is 3.092. p^* is unchanged.

$$p_2 = \left[\frac{p}{p^*}\right]p^* = (3.092)(3.316\text{ psia})$$
$$= 10.25\text{ psia}$$

Example 26.14

Air enters a round duct at Mach 2. The inside diameter of the duct is 2.0 in (5 cm), and the Fanning friction factor is 0.005. The total pressure and total temperature at the duct entrance are 78.25 psia and 1080°R (540 kPa and 600K), respectively. What are the (a) temperature and (b) pressure at the entrance? (c) What is the static pressure at a point where the Mach number is 1.75? (d) What is the distance between the points where the Mach numbers are 2.0 and 1.75?

SI Solution

(a) From the isentropic flow table at Mach 2, $[T/T_0] = 0.5556$.

$$T = \left[\frac{T}{T_0}\right]T_0 = (0.5556)(600\text{K})$$
$$= 333.4\text{K}$$

(b) From the isentropic flow table at Mach 2, $[p/p_0] = 0.1278$.

$$p = \left[\frac{p}{p_0}\right]p_0 = (0.1278)(540\text{ kPa})$$
$$= 69.0\text{ kPa}$$

(c) It is not known if the duct is long enough for choked flow (M = 1) to occur. However, the conditions at the hypothetical choke point can be used as a reference for other conditions. The following values are read from a Fanno flow table.

From App. 26.C, at Mach 2.0, $[p/p^*] = 0.408$. At Mach 1.75, $[p/p^*] = 0.493$.

$$p_{M=1.75} = \frac{\left[\dfrac{p}{p^*}\right]_{M=1.75} p_{M=2.0}}{\left[\dfrac{p}{p^*}\right]_{M=2.0}}$$

$$= \frac{(0.493)(69.0 \text{ kPa})}{0.408}$$

$$= 83.4 \text{ kPa}$$

(d) At Mach 2.0, $[4fL/D] = 0.305$. At Mach 1.75, $[4fL/D] = 0.225$. At Mach 2, the distance to reach Mach 1 is

$$L_{M=2.0} = \frac{\left[\dfrac{4fL}{D}\right] D}{4f} = \frac{(0.305)(5 \text{ cm})}{(4)(0.005)}$$

$$= 76.25 \text{ cm}$$

At Mach 1.75, the distance to reach Mach 1 is

$$L_{M=1.75} = \frac{\left[\dfrac{4fL}{D}\right] D}{4f} = \frac{(0.225)(5 \text{ cm})}{(4)(0.005)}$$

$$= 56.25 \text{ cm}$$

The distance between the two points is

$$L = L_{M=2.0} - L_{M=1.75}$$

$$= 76.25 \text{ cm} - 56.25 \text{ cm} = 20 \text{ cm}$$

Customary U.S. Solution

(a) From the isentropic flow table at Mach 2, $[T/T_0] = 0.5556$.

$$T = \left[\frac{T}{T_0}\right] T_0 = (0.5556)(1080°\text{R})$$

$$= 600°\text{R}$$

(b) From the isentropic flow table at Mach 2, $[p/p_0] = 0.1278$.

$$p = \left[\frac{p}{p_0}\right] p_0 = (0.1278)\left(78.25 \ \frac{\text{lbf}}{\text{in}^2}\right)$$

$$= 10 \text{ lbf/in}^2$$

(c) From App. 26.C, at Mach 2.0, $[p/p^*] = 0.408$. At Mach 1.75, $[p/p^*] = 0.493$.

$$p_{M=1.75} = \frac{\left[\dfrac{p}{p^*}\right]_{M=1.75} p_{M=2.0}}{\left[\dfrac{p}{p^*}\right]_{M=2.0}}$$

$$= \frac{(0.493)\left(10 \ \dfrac{\text{lbf}}{\text{in}^2}\right)}{0.408}$$

$$= 12.1 \text{ lbf/in}^2$$

(d) At Mach 2.0, $[4fL/D] = 0.305$. At Mach 1.75, $[4fL/D] = 0.225$. At Mach 2, the distance to reach Mach 1 is

$$L_{M=2.0} = \frac{\left[\dfrac{4fL}{D}\right] D}{4f} = \frac{(0.305)(2 \text{ in})}{(4)(0.005)}$$

$$= 30.5 \text{ in}$$

At Mach 1.75, the distance to reach Mach 1 is

$$L_{M=1.75} = \frac{\left[\dfrac{4fL}{D}\right] D}{4f} = \frac{(0.225)(2 \text{ in})}{(4)(0.005)}$$

$$= 22.5 \text{ in}$$

The distance between the two points is

$$L = L_{M=2.0} - L_{M=1.75}$$

$$= 30.5 \text{ in} - 22.5 \text{ in} = 8 \text{ in}$$

17. ISOTHERMAL FLOW WITH FRICTION

As the Mach number approaches 1.0, an infinite heat transfer rate would be required to keep the flow isothermal. For that reason, gas flow in most systems is more of an adiabatic process than an isothermal process. A notable exception is encountered with long-distance gas pipelines.[13] The earth acts like a large heat reservoir and supplies the energy needed to keep the flow isothermal.

For flow that is initially subsonic with a Mach number less than $1/\sqrt{k}$, pressure, density, and total pressure decrease in the direction of flow. The velocity, Mach number, and total temperature all increase. The Mach number approaches $1/\sqrt{k}$.

For flow that is initially subsonic with a Mach number greater than $1/\sqrt{k}$, and for flow that is initially supersonic, pressure and density increase, and velocity, Mach number, and total temperature decrease. The total pressure increases for Mach numbers less than $\sqrt{2}/(1+k)$ and decreases otherwise. The Mach number approaches $1/\sqrt{k}$.

The Darcy equation can be used to calculate the friction loss, and the thermodynamic relationships for isothermal processes are all applicable. For pipelines with diameters greater than 24 in (60 cm) experiencing turbulent flow, for any given pressure drop between two points a distance L apart, the mass flow rate is given by the *Weymouth equation*.

$$\dot{m} = \sqrt{\frac{(p_1^2 - p_2^2) D^5 g_c \left(\dfrac{\pi}{4}\right)^2}{4 f_{\text{Fanning}} LRT}} \qquad 26.39$$

---

[13]The low velocities (typically less than 20 ft/sec (6 m/s)) typical of natural gas pipelines seem out of place in this chapter. However, the gas experiences a density change along the length of pipeline, qualifying it for coverage in this chapter.

The *Fanning friction factor*, $f_{Fanning}$, in the Weymouth equation is

$$f_{Fanning} = \frac{0.00349}{(D_{ft})^{0.333}} \qquad 26.40$$

The *transmission factor* is defined as

$$\text{transmission factor} = \sqrt{\frac{1}{f_{Fanning}}} \qquad 26.41$$

Example 26.15

A 40 in (100 cm) inside diameter pipeline carries natural gas (essentially methane with a molecular weight of 16) between pumping stations 75 mi (120 km) apart. The gas leaves the upstream pumping station at 650 psia (4.5 MPa). The pressure has decreased to 450 psia (3.1 MPa) by the time it reaches the next station. The gas temperature is 40°F (4°C) along the entire pipeline. (a) Calculate the mass flow rate from the Weymouth equation. (b) Calculate the mass flow rate from the Bernoulli equation using the average gas properties along the pipeline.

Solution

The pipeline diameter is

$$D = \frac{40 \text{ in}}{12 \frac{\text{in}}{\text{ft}}} = 3.333 \text{ ft}$$

The flow area of the pipeline is

$$A = \frac{\pi D^2}{4} = \frac{\pi (3.333 \text{ ft})^2}{4}$$
$$= 8.725 \text{ ft}^2$$

The absolute temperature of the natural gas is

$$T = 40°F + 460 = 500°R$$

Convert the pressures.

$$p_1 = \left(650 \frac{\text{lbf}}{\text{in}^2}\right)\left(144 \frac{\text{in}^2}{\text{ft}^2}\right)$$
$$= 9.36 \times 10^4 \text{ lbf/ft}^2$$

$$p_2 = \left(450 \frac{\text{lbf}}{\text{in}^2}\right)\left(144 \frac{\text{in}^2}{\text{ft}^2}\right)$$
$$= 6.48 \times 10^4 \text{ lbf/ft}^2$$

The specific gas constant for methane is

$$R = \frac{R^*}{(MW)} = \frac{1545 \frac{\text{ft-lbf}}{\text{lbmol-°R}}}{16 \frac{\text{lbm}}{\text{lbmol}}}$$
$$= 96.56 \text{ ft-lbf/lbm-°R}$$

(a) The Fanning friction factor is given by Eq. 26.40.

$$f_{Fanning} = \frac{0.00349}{(D_{ft})^{0.333}}$$
$$= \frac{0.0349}{(3.333 \text{ ft})^{0.333}} = 0.00234$$

From Eq. 26.39, the mass flow rate is

$$\dot{m} = \sqrt{\frac{(p_1^2 - p_2^2)D^5 g_c \left(\frac{\pi}{4}\right)^2}{4 f_{Fanning} LRT}}$$

$$= \sqrt{\frac{\left[\left(9.36 \times 10^4 \frac{\text{lbf}}{\text{ft}^2}\right)^2 - \left(6.48 \times 10^4 \frac{\text{lbf}}{\text{ft}^2}\right)^2\right] \times (3.333 \text{ ft})^5 \left(32.2 \frac{\text{ft-lbm}}{\text{lbf-sec}^2}\right)\left(\frac{\pi}{4}\right)^2}{(4)(0.00234)(75 \text{ mi})\left(5280 \frac{\text{ft}}{\text{mi}}\right) \times \left(96.56 \frac{\text{ft-lbf}}{\text{lbm-°R}}\right)(500°R)}}$$

$$= 456 \text{ lbm/sec}$$

(b) Since the pressure drops approximately 30% along the pipeline, use the average pressure to calculate the average density.

$$\rho_{ave} = \frac{p}{RT}$$
$$= \frac{\left(9.36 \times 10^4 \frac{\text{lbf}}{\text{ft}^2}\right) + \left(6.48 \times 10^4 \frac{\text{lbf}}{\text{ft}^2}\right)}{(2)\left(96.56 \frac{\text{ft-lbf}}{\text{lbm-°R}}\right)(500°R)}$$
$$= 1.640 \text{ lbm/ft}^3$$

The entire pressure drop is due to friction. Neglecting the low kinetic energy and unknown changes in potential energy, the Bernoulli equation can be written for the two ends of the pipeline. (The specific weight and density are numerically equal.)

$$\frac{p_1}{\gamma} = \frac{p_2}{\gamma} + h_f$$

$$\frac{9.36 \times 10^4 \frac{\text{lbf}}{\text{ft}^2}}{1.640 \frac{\text{lbm}}{\text{ft}^3}} = \frac{6.48 \times 10^4 \frac{\text{lbf}}{\text{ft}^2}}{1.640 \frac{\text{lbm}}{\text{ft}^3}} + h_f$$

$$h_f = 17{,}560 \text{ ft}$$

Use the same friction factor as in part (a).

$$f_{Darcy} = 4 f_{Fanning} = (4)(0.00234) = 0.00936$$

Solve for the average flow velocity from the Darcy equation.

$$h_f = \frac{fLv^2}{2Dg}$$

$$17{,}560 \text{ ft} = \frac{(0.00936)(75 \text{ mi})\left(5280 \,\frac{\text{ft}}{\text{mi}}\right)v^2}{(2)(3.333 \text{ ft})\left(32.2 \,\frac{\text{ft}}{\text{sec}^2}\right)}$$

$$v_{ave} = 31.89 \text{ ft/sec}$$

The mass flow rate is

$$\begin{aligned}
\dot{m} &= \rho_{ave} v_{ave} A \\
&= \left(1.64 \,\frac{\text{lbm}}{\text{ft}^3}\right)\left(31.89 \,\frac{\text{ft}}{\text{sec}}\right)(8.725 \text{ ft}^2) \\
&= 456 \text{ lbm/sec}
\end{aligned}$$

18. FRICTIONLESS FLOW WITH HEATING OR COOLING

If heat is added to or removed from a gas flowing through a frictionless constant-area duct, the flow is said to be *diabatic flow* (flow *with* heat transfer) or *Rayleigh flow*. Heating or cooling can be caused either by chemical reactions taking place in the gas, phase changes, electrical sources, or by heating or refrigeration acting through the duct walls. (Heating due to friction is not Rayleigh flow.) The heat transfer rate per unit mass is

$$\dot{q} = c_p(T_{0,2} - T_{0,1}) \qquad \textbf{26.42}$$

The *Rayleigh line* is a plot of temperature (or enthalpy) versus entropy for a specific flow rate. It illustrates how the velocity tends toward Mach 1 for both subsonic and supersonic entrances. Entropy is maximum at the "nose" corresponding to Mach 1. The maximum static temperature occurs at $M = 1/\sqrt{k}$.

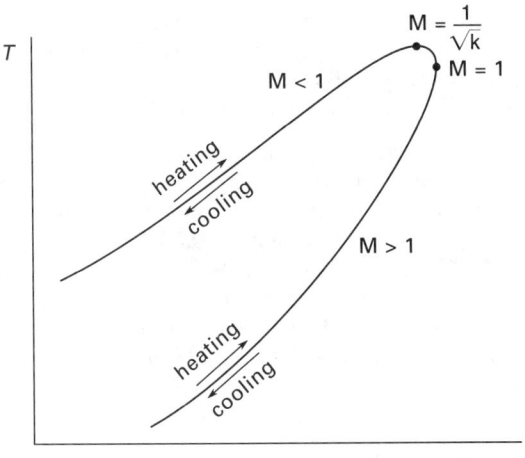

Figure 26.7 *Rayleigh Line*

Equations 26.43 through 26.46 give the ratios of static-to-sonic parameters. Since it is difficult to extract the Mach number from these expressions, graphs and tables are often used to solve Rayleigh flow problems.

$$\frac{T}{T^*} = \left[\frac{M(1+k)}{1+kM^2}\right]^2 \qquad \textbf{26.43}$$

$$\frac{T_0}{T_0^*} = \frac{(2)(1+k)M^2\left[1+\left(\frac{1}{2}\right)(k-1)M^2\right]}{(1+kM^2)^2} \qquad \textbf{26.44}$$

$$\frac{p}{p^*} = \frac{1+k}{1+kM^2} \qquad \textbf{26.45}$$

$$\frac{p_0}{p_0^*} = \left(\frac{2}{1+k}\right)^{\frac{k}{k-1}}\left(\frac{1+k}{1+kM^2}\right)$$

$$\times \left[1+\left(\tfrac{1}{2}\right)(k-1)M^2\right]^{\frac{k}{k-1}} \qquad \textbf{26.46}$$

Example 26.16

Air with a specific heat of 0.24 Btu/lbm-°R (1 kJ/kg·K) enters a constant area duct at 50°F (10°C). While passing through the duct, the air's enthalpy is increased by 150 Btu/lbm (350 kJ/kg). If the inlet Mach number is 0.3, what are the (a) exit temperature, (b) exit Mach number, and (c) exit stagnation temperature?

SI Solution

From the isentropic flow table (not a Rayleigh table) for Mach 0.3, $[T/T_0] = 0.9825$. The total temperature at the inlet is

$$\begin{aligned}
T_{0,i} &= \frac{T}{\left[\dfrac{T}{T_0}\right]} \\
&= \frac{10°C + 273}{0.9825} = 288 \text{K}
\end{aligned}$$

Sonic velocity may not be achieved if the duct is not long enough. The sonic temperature is determined as a reference point. From App. 26.D at Mach 0.3, the ratio of total temperatures at the inlet and sonic points is $[T_0/T_0^*] = 0.3468$. The total temperature at the sonic point is

$$\begin{aligned}
T_0^* &= \frac{T_{0,i}}{\left[\dfrac{T_0}{T_0^*}\right]} \\
&= \frac{288 \text{K}}{0.3468} = 830.4 \text{K}
\end{aligned}$$

From App. 26.D, at Mach 0.3, the ratio of static temperatures at the inlet and sonic points is $[T/T^*] = 0.409$. The static temperature at the sonic point is

$$\begin{aligned}
T^* &= \frac{T_i}{\left[\dfrac{T}{T^*}\right]} \\
&= \frac{288 \text{K}}{0.409} = 704.2 \text{K}
\end{aligned}$$

The total (stagnation) temperature at the exit is

$$T_{0,e} = T_{0,i} + \frac{q}{c_p}$$

$$= 288\text{K} + \left(\frac{350 \, \dfrac{\text{kJ}}{\text{kg}}}{1 \, \dfrac{\text{kJ}}{\text{kg·K}}} \right)$$

$$= 638\text{K}$$

The ratio of total temperature at the exit and sonic points is

$$\frac{T_{0,e}}{T_0^*} = \frac{638\text{K}}{830.3\text{K}} = 0.7684$$

Interpolating from App. 26.D, the Mach number corresponding to this ratio of total temperatures is $\text{M}_e = 0.557$ (use 0.56). At that Mach number, $[T/T^*] = 0.872$. The temperature at the exit is

$$T_e = \left[\frac{T}{T^*} \right] T^* = (0.872)(704.2\text{K})$$

$$= 614\text{K}$$

Customary U.S. Solution

From the isentropic flow table for Mach 0.3, $[T/T_0] = 0.9825$. The total temperature at the inlet is

$$T_{0,i} = \frac{T}{\left[\dfrac{T}{T_0} \right]}$$

$$= \frac{50°\text{F} + 460}{0.9825} = 519.1°\text{R}$$

From App. 26.D at Mach 0.3, the ratio of total temperatures at the inlet and sonic points is $[T_0/T_0^*] = 0.3468$. The total temperature at the sonic point is

$$T_0^* = \frac{T_{0,i}}{\left[\dfrac{T_0}{T_0^*} \right]}$$

$$= \frac{519.1°\text{R}}{0.3468} = 1497°\text{R}$$

From App. 26.D, at Mach 0.3, the ratio of static temperatures at the inlet and sonic points is $[T/T^*] = 0.409$. The static temperature at the sonic point is

$$T^* = \frac{T_i}{\left[\dfrac{T}{T^*} \right]}$$

$$= \frac{519.1°\text{R}}{0.409} = 1269°\text{R}$$

The total (stagnation) temperature at the exit is

$$T_{0,e} = T_{0,i} + \frac{q}{c_p}$$

$$= 519.1°\text{R} + \frac{150 \, \dfrac{\text{Btu}}{\text{lbm}}}{0.24 \, \dfrac{\text{Btu}}{\text{lbm}}}$$

$$= 1144°\text{R}$$

The ratio of total temperature at the exit and sonic points is

$$\frac{T_{0,e}}{T_0^*} = \frac{1144°\text{R}}{1497°\text{R}} = 0.7642$$

Interpolating from App. 26.D, the Mach number corresponding to this ratio of total temperatures is $\text{M}_e = 0.554$ (use 0.56). At that Mach number, $[T/T^*] = 0.872$. The temperature at the exit is

$$T_e = \left[\frac{T}{T^*} \right] T^* = (0.872)(1269°\text{R})$$

$$= 1107°\text{R}$$

19. STEAM FLOW THROUGH NOZZLES

If steam is assumed to be an ideal gas, then all of the relationships presented in this chapter can be used. For low temperatures, steam's ratio of specific heats is approximately 1.33. However, the value is lower at the temperatures normally encountered in steam nozzles.

The actual value of the ratio of specific heats can be readily found. Since $\Delta h = c_p \Delta T$, the specific heat, c_p, can be calculated from the superheat tables as $\Delta h / \Delta T$ for states in close proximity to the actual steam condition. Then, c_v and k can be calculated from the ideal gas relationships. For steam, the specific gas constant is approximately 85.78 ft-lbf/lbf-°R (0.4615 kJ/kg·K).

Equation 26.1 and (if the entrance velocity is small) Eq. 26.2 are always applicable to the flow of steam through nozzles. However, since steam expansion is generally not isentropic, the exit velocity will be less than calculated, and Eqs. 26.15 through 26.18 should be used. Isentropic efficiencies of 85 to 95% are typical, as are coefficients of velocity, C_v, of 0.95 to 0.99.

The critical pressure ratio can be calculated from Eq. 26.8 using an average ratio of specific heats. However, for typical steam inlet pressures, the critical pressure ratio is often taken as 0.5457 (corresponding to $k = 1.3$) for highly superheated steam and 0.57 (corresponding to $k = 1.18$) for moderately superheated steam, and 0.577 to 0.58 (corresponding to $k = 1.13$) for saturated (i.e., "wet") steam.[14] Thus, sonic velocity at the throat can be achieved if the back pressure is 54 to 57% of the inlet pressure.

---

[14]Different values of k should be used to design the converging and diverging sections if the steam is dry in the converging section and wet in the diverging section.

It is often necessary to calculate the throat area that will produce sonic velocity for a given mass flow rate. Treating steam as an ideal gas, the ratio of mass flow to throat area for sonic flow is

$$\frac{\dot{m}}{A^*} = \rho_t a = \rho_t \sqrt{kRT_t} \quad \text{[SI]} \quad \textit{26.47(a)}$$

$$\frac{\dot{m}}{A^*} = \rho_t a = \rho_t \sqrt{kg_c RT_t} \quad \text{[U.S.]} \quad \textit{26.47(b)}$$

Generally, only the inlet conditions are known, not the throat conditions. The ratios calculated from Eqs. 26.4 and 26.5 (or from a table for the appropriate value of k, if available) are used to write the throat temperature and pressure in terms of the inlet temperature and pressure. Using 1.3 for the ratio of specific heats, 85.8 ft-lbf/lbm-°R (0.4615 kJ/kg·K) as the specific gas constant, 0.6276 as the ratio of throat-to-inlet densities, and 0.8696 as the ratio of throat-to-inlet temperatures, the approximate ratio of mass flow to throat area is given by Eq. 26.48.

$$\frac{\dot{m}}{A^*} \approx 14.34 \rho_i \sqrt{T_i} \quad \text{[SI]} \quad \textit{26.48(a)}$$

$$\frac{\dot{m}}{A^*} \approx 35.1 \rho_i \sqrt{T_i} \quad \text{[U.S.]} \quad \textit{26.48(b)}$$

The exit area can be found from the mass flow rate and the exit conditions. Modern turbines are multistaged, and the condenser pressure is the back pressure for the last stage only. For other stages, the back pressure must be specified.

20. PITOT TUBES

When the initial velocity of a gas is subsonic, air flow is brought to rest in a pitot tube through an isentropic process. Eqs. 26.49 and 26.50 can be used to calculate the velocity and Mach number, respectively.

$$v = \sqrt{\left(\frac{2kRT}{k-1}\right)\left[\left(\frac{p_0}{p}\right)^{\frac{k-1}{k}} - 1\right]} \quad \text{[SI]} \quad \textit{26.49(a)}$$

$$v = \sqrt{\left(\frac{2kg_c RT}{k-1}\right)\left[\left(\frac{p_0}{p}\right)^{\frac{k-1}{k}} - 1\right]} \quad \text{[U.S.]} \quad \textit{26.49(b)}$$

$$M = \sqrt{\frac{(2)\left[\left(\frac{p_0}{p}\right)^{\frac{k-1}{k}} - 1\right]}{k-1}} \quad \textit{26.50}$$

When the initial velocity is supersonic, the gas cannot be brought to rest isentropically at the pitot tube inlet. (An isentropic stagnation process would require a converging-diverging nozzle at the pitot tube inlet.) Rather, a shock wave forms in front of the pitot tube.

The pressure measured by the pitot tube is the downstream stagnation pressure, not the upstream stagnation pressure. Normal shock tables tabulate the value of $[p_x/p_{0,y}]$ so that the Mach number can be read directly.

Example 26.17

A pitot tube measures an air pressure of 140 kPa. The corresponding static pressure is 30 kPa. (a) Is the flow subsonic or supersonic? (b) What is the Mach number?

Solution

(a) The ratio of static-to-total pressure is

$$\frac{p}{p_0} = \frac{30 \text{ kPa}}{140 \text{ kPa}}$$

$$= 0.214$$

Searching the isentropic low table for this value, the Mach number appears to be approximately 1.66. However, since this is greater than 1.0, flow is supersonic, and there is a normal shock wave before the pitot tube.

(b) Searching the $[p_x/p_{0,y}]$ column in the normal shock tables, M = 1.80.

PRACTICE PROBLEMS

1. 150°F, 10 psia (65°C, 70 kPa) air flows at 750 ft/sec (225 m/s) through a converging section into a chamber whose back pressure is 5.5 psia (40 kPa). What are the (a) Mach number, (b) pressure, and (c) total temperature at the throat?

2. A round-nosed bullet travels at 2000 ft/sec (600 m/s) through 32°F, 14.7 psia (0°C, 101.3 kPa) air. (a) Is the bullet speed subsonic or supersonic? What are the stagnation values of (b) pressure, (c) temperature, and (d) enthalpy at the bullet face?

3. 4.5 lbm/sec (2 kg/s) of air with total properties of 160 psia and 240°F (1.1 MPa and 120°C) expands through a converging-diverging nozzle to 20 psia (140 kPa). What are the (a) throat Mach number, (b) throat area, (c) exit Mach number, and (d) exit area?

4. A wedge-shaped leading edge with a semivertex angle of 20° travels at 2700 ft/sec (800 m/s) through 60°F, 14.7 psia (16°C, 101.3 kPa) air. What is the maximum semivertex shock angle?

5. A large tank contains 100 psia, 80°F (0.7 MPa, 27°C) air. The tank feeds a converging-diverging nozzle with a throat area of 1 in² (6.45 × 10⁻⁴ m²). At a particular point in the nozzle, the Mach number is 2. What are the (a) temperature, (b) area, and (c) mass flow rate at that point?

6. The Mach number before a shock wave is 2. The temperature before the shock wave is 500°R (280K). What is the air velocity behind the shock wave?

7. Air with a total pressure of 100 psia (0.7 MPa) and a total temperature of 70°F (21°C) flows through a converging-diverging nozzle. At a particular point, the ratio of flow area to the critical throat area is 1.555. What are the (a) Mach number, (b) temperature, and (c) pressure at that point?

8. 20 lbm/sec (9 kg/s) of air with a static pressure of 10 psia (70 kPa) and a total temperature of 40°F (4°C) flow through a 1 ft^2 (0.09 m^2) round duct. What is the smallest area to which the duct can be reduced?

9. The static pressure in a supersonic wind tunnel is 1.38 psia (9.51 kPa). A pitot tube records a total pressure of 20 psia (140 kPa) behind the shock wave that forms at its entrance. What is the Mach number in the tunnel?

10. A boiler produces 100 psia, 800°F (0.7 MPa, 425°C) steam. An isentropic nozzle expands the steam to 60 psia (400 kPa). What is the steam velocity?

11. Air flows through a converging section. At a particular point where the flow area is 0.1 ft^2 (0.0009 m^2), the static pressure is 50 psia (350 kPa), the static temperature is 1000°R (560K), and the velocity is 600 ft/sec (180 m/s). What are the (a) Mach number, (b) total temperature, (c) total pressure, (d) critical exit area, (e) critical pressure, and (f) critical temperature?

12. 3 lbm/sec (1.4 kg/s) of steam ($k = 1.31$) enter an 85% efficient transonic nozzle at 300 ft/sec (90 m/s). The steam expands from 200 psia and 600°F (1.5 MPa and 300°C) to 80 psia (500 kPa). What is the throat area?

13. (*Time limit: one hour*) 35,200 lbm/hr (4.4 kg/s) of steam flow ($k = 1.30$) through 95 ft (30 m) of 3 in (7.6 cm) inside diameter pipe. The pipe is insulated to prevent heat loss. The Fanning friction factor is 0.012. The average steam pressure along the pipe length is 200 psia (1.5 MPa). At the exit, the steam is at 115 psia and 540°F (0.7 MPa and 300°C). 30 ft (9 m) of pipe are added to the end of the pipe. Will the steam flow be maintained?

14. (*Time limit: one hour*) 3600 lbm/hr (0.45 kg/s) of air flow through an adiabatic converging-diverging nozzle. The air enters with total properties of 160 psia and 660°R (1.1 MPa and 370K). The exit pressure is 14.7 psia (101.3 kPa). The ratio of specific heats for the air is 1.4 throughout the nozzle. The nozzle has a coefficient of discharge of 0.90. The total diverging angle is 6°. The length of the converging section is 5% of the diverging section. (a) Draw and dimension the nozzle. (b) What is the throat area?

27 Vapor Power Equipment

1. Typical System Integration 27-1
2. Furnaces 27-3
3. Steam Generators 27-4
4. Blowdown 27-5
5. Pumps 27-6
6. Pump Efficiency 27-6
7. Temperature Increase in
     Pressurized Liquids 27-7
8. Turbines 27-8
9. Turbine Work, Power, and Efficiency 27-8
10. Two-Stage Expansion 27-9
11. Heat Exchangers 27-10
12. Condensers 27-11
13. Condenser Fouling 27-12
14. Nozzles, Orifices, and Valves 27-12
15. Superheaters 27-12
16. Reheaters 27-13
17. Desuperheaters 27-13
18. Feedwater Heaters 27-13
19. Evaporators 27-14
20. Deaerators 27-14
21. Economizers 27-15
22. Air Heaters 27-15
23. Electrical Generators 27-15
24. Load and Capacity 27-15
25. Control and Regulation of Turbines 27-16
      Practice Problems 27-16

Nomenclature

c_p	specific heat	Btu/lbm-°R	kJ/kg·K
C	concentration	ppm	ppm
g_c	gravitational constant (32.2)	ft-lbf/lbm-°R	n.a.
h	enthalpy	Btu/lbm	kJ/kg
HWD	hot well depression	°F	°C
HV	heating value	Btu/lbm	kJ/kg
J	Joule's constant (778)	ft-lbf/Btu	n.a.
m	mass	lbm	kg
$\dot{m}$	mass flow rate	lbm/sec	kg/s
p	pressure	lbf/ft^2	kPa
P	power	Btu/sec	kW
Q	heat	Btu	kJ
Q	heat flow rate	Btu/sec	kW
t	time (duration)	sec	s
T	temperature	°R	K
TTD	terminal temperature difference	°F	°C
v	velocity	ft/sec	m/s
W	work	ft-lbf	kW
y	bleed fraction	–	–

Symbols

η	efficiency	–	–
v	specific volume	ft^3/lbm	m^3/kg

Subscripts

f	saturated fluid
fg	fluid-to-gas (vaporization)
m	mechanical or intermediate
p	at pressure p
s	isentropic
sat	saturated

1. TYPICAL SYSTEM INTEGRATION

This chapter takes a quick look at some of the equipment necessary to implement a typical utility power-generating plant. It is impractical to show all of the lines, valves, tanks, and redundant elements in even a simple power-generating system. There are numerous ways of combining the devices described in this chapter into a working power-generating system. However, Fig. 27.1(a) illustrates the main elements and interconnections for *unit operation* (i.e., in a typical drum boiler unit).

The integration of elements in high-pressure *once-through boilers* is somewhat different, as Fig. 27.1(b) illustrates. (The term "once-through" does not mean the steam is discarded after use. It means that the steam is heated along one long progressive transfer path.) Once-through boilers do not have steam drums. The steam flow rate is controlled by the boilerfeed pump. At start-up and at low load, the steam generator produces more steam than the turbine requires. At low loads, super-heated steam is passed to a flash tank where it becomes available for feedwater heating, or it is passed directly to the condenser and returned to the condenser. This lets the flash tank act like a drum in a drum boiler unit. Valving and valve sequencing is more complex with once-through operation.

Power plant output may be qualified by the terms "thermal," "electrical," "gross," or "net." For example, a 1,000,000 Btu/hr *thermal* (i.e, "1 MBth" or "MBt") power plant indicates the energy transfer to the feed-water. 1000 MW *electrical* (i.e., "1000 MWe") power plant refers to the generator output. Similarly, "MWt" refers to thermal power in megawatts.

Some of the steam and some of the generated electrical power are used to drive auxiliary devices (motors, fans, soot-blowers, pumps, etc.). The *gross electrical output* is the power before the auxiliary loads have been removed; the *net electrical output* is after.

Power Cycles

Figure 27.1. *Typical Integration of Power-Generating Elements*

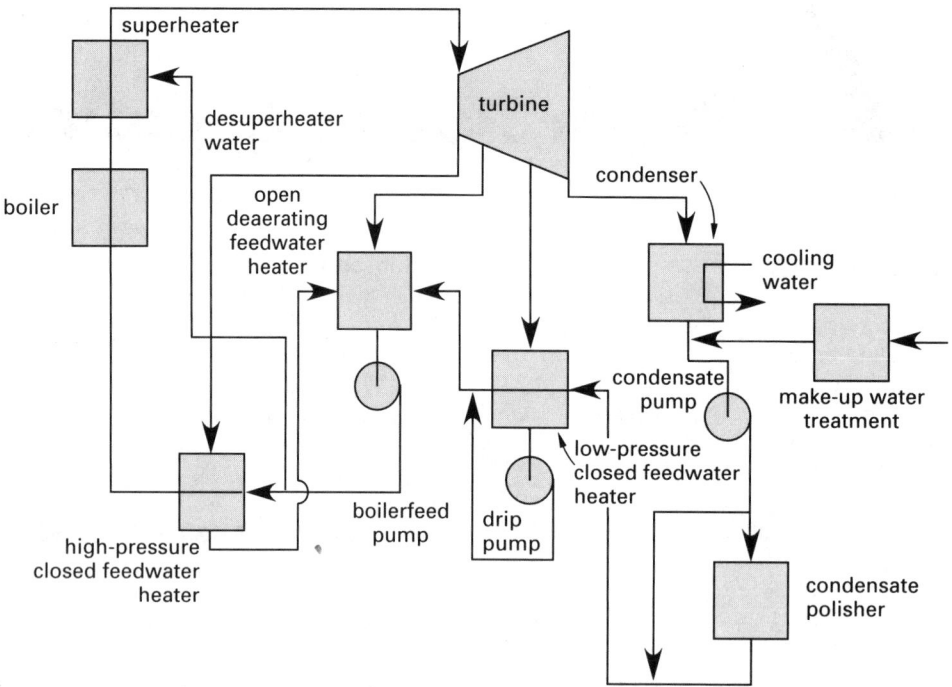

(a) typical unit operation

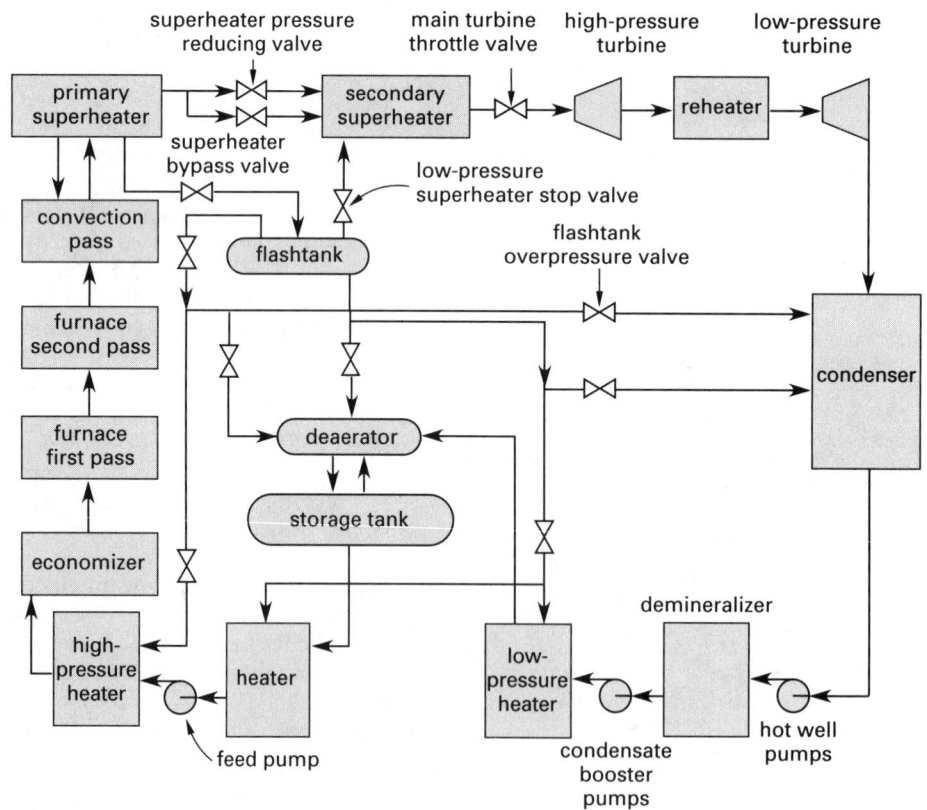

(b) typical once-through operation
(bleeds not shown)

2. FURNACES

Most modern power-generating plants burn coal; some burn natural gas, and a few burn oil. In the past, solid fuel furnaces have been dominated by "pile burning" on grates, and more recently, by "suspension burning" of pulverized coal. Current fluidized-bed combustion offers additional benefits (particularly environmental) over both traditional methods.

The amount of air introduced into the combustion chamber affects the completeness of combustion, pollutant production, and furnace efficiency. For efficient combustion, air may be introduced at more than one point along the combustion path. *Primary air* is introduced first, followed by *secondary air* and *tertiary air*.

A high turndown ratio is desirable for furnaces that are frequently started cold.[1] For gaseous fuels, the *turndown ratio* is the ratio of the maximum-to-minimum fuel-to-air ratios over which the burner will operate satisfactorily. The maximum turndown ratio is limited by the flame velocity. *Flame blow-off* results when a fuel-air mixture is injected too fast. *Flash-back* occurs when the flame velocity exceeds the mixture velocity.

In older lump-coal furnaces, coal was introduced to the furnace by a *stoker*. Stoking is the act of adding and distributing coal. In low-volume home furnaces and early fire-tube boilers, stoking was accomplished by hand shoveling. Stoking in lump-coal plants and in modern refuse-burning plants is performed mechanically and automatically.

Figure 27.2 illustrates several types of ram stokers for lump coal and refuse. Fuel can also be brought into the furnace by screws and sprinklers. With *overfeed stoking*, coal is placed above the air flow. With *underfeed stoking (retort stoking)*, coal is pushed into position under the air flow by plungers. Air enters the combustion chamber through ports in the furnace walls known as *tuyères*. The combustion area holding burning coal is known as the *retort*.

Coal in all modern power-generating plants is pulverized to some extent prior to use. In some cases, coal is ground into a fine or microfine powder.[2] The finely ground coal is suspended in a gaseous atmosphere while burning. There are two general ways of feeding *pulverized-coal furnaces*. The *bin system (central*

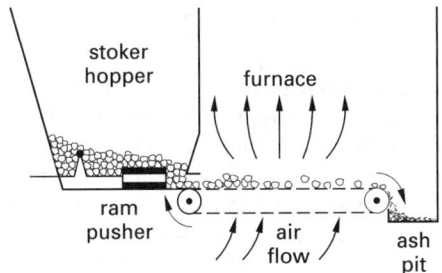

(a) overfeed traveling bed grate
with ram stoker

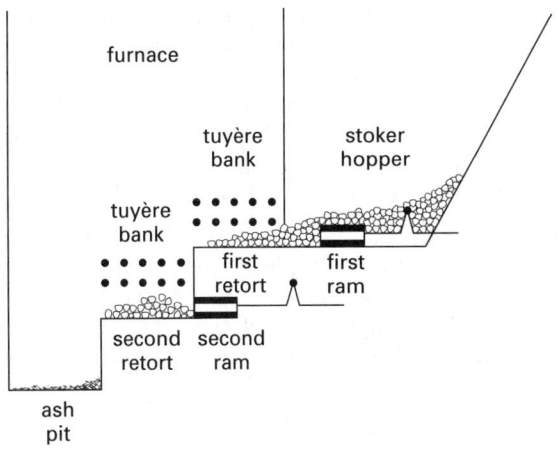

(b) underfeed retort furnace
with ram stoker

Figure 27.2 *Stokers*

system) stores dried and pulverized coal for later use.[3] Air transport, pressure pulse, and screw conveyor systems can be used to transfer pulverized coal to the furnace as needed.[4] *Direct-firing systems (unit systems)* pulverize coal on demand. Whether the bin or direct-firing system is used depends on the type of coal and the reliability of the pulverizing and feed equipment.[5]

High-sulfur, high-ash coals, "scrap" fuels, and toxic materials can be burned in *fluidized-bed furnaces*. Solid fuel is turned into a turbulent, fluid-like mass by mixing it with a bed material and blowing air through the mixture at a controlled rate. The bed material can be any solid substance that is not consumed. Sand and

---

[1]The turndown ratio is not of much interest for continuously fired furnaces.

[2]Generally, approximately 65 to 70% of the pulverized coal will pass through a 200-mesh (74-micron) sieve, which has 200 openings per inch. The nominal aperture for 200 mesh is 0.0029 in (0.074 mm), which is like talcum powder. Anthracitic coals can be ground so that up to 90% passes through a 200 mesh sieve. *Micronized coal* is even finer: 80 to 90% will pass through a 325-mesh sieve (43 microns).

[3]An *eductor* is a special type of jet pump that uses air at 20 to 80 psig (140 to 550 kPa) to withdraw pulverized coal from an unpressurized storage hopper, transport it in dense phase through a small-diameter feed line, and inject it into the burner. The coal flow rate is controlled by the air pressure. A control valve is not needed.

[4]Ground coal can also be carried for longer distances in slurry form.

[5]Reliability of pulverizers is not the problem it was when pulverizing was first introduced. Pulverizing on demand is now the norm.

limestone are the most common options. The mixture of fuel and bed material becomes fluidized and assumes free-flowing properties when the air flow is at the fluidizing velocity.

Fluidized-bed combustion, FBC, using limestone as the bed material considerably reduces some pollutants, eliminating the need for expensive air-pollution control equipment. Approximately 90% of the sulfur is absorbed by the limestone, forming calcium sulfate. Combustion takes place rapidly at 1500 to 1750°F (815 to 950°C), about 400°F (220°C) lower than conventional boilers. This is below the point at which nitrogen oxides are formed.

3. STEAM GENERATORS

A *steam generator* is a combination of a furnace and a boiler. The *boiler* is constructed so that the combustion heat is transferred to feedwater. Modern boilers are *water-tube boilers* (i.e., water passes through tubes surrounded by combustion products).[6] Some of the tubes may be surface mounted or embedded in the walls of the furnace (known as the *setting* or "brickwork").

Other tubes may be placed across the path of the combustion gases. In the typical *cross-drum boiler*, the combustion gases flow perpendicular to the water tubes. Tubes may be straight or bent (i.e., *straight-tube boilers* or *bent-tube boilers*). Steam accumulates in one or more headers at the top of the boiler (known as the *steam drums*).[7]

Circulation of water through the boiler can occur in one of four ways. For pressures below approximately 2800 psi (19.3 MPa), *natural-circulation boilers* can be used. Density is the force driving circulation up the *riser* to the steam drum and back down the *downcomer* to the water-supply header or mud drum. With *controlled-circulation boilers*, common between 2400 and 2800 psi (16.6 and 19.3 MPa), a pump keeps the water circulating between the steam drum and water-supply header. In *once-through boilers* (with moisture separators but no steam drums) and *supercritical-pressure boilers* (with neither steam drums nor separators) that operate around 3500 psi (24.1 MPa) and 1000°F (540°C), water circulates by use of the boilerfeed pump and must be ultrapure to prevent solids build-up.[8]

Capacities of steam generators are given in mass of steam per hour. This is not an exact determination of

---

[6] *Fire-tube boilers*, in which the combustion products pass through small-diameter tubes and transfer heat to a surrounding water jacket, are seldom used today because they are limited to low-pressure steam. The maximum operating pressure for fire-tube boilers is approximately 150 psi (1 MPa) for riveted construction and 400 psia (3 MPa) for welded construction.

[7] A drum at the bottom of the boiler for collecting sediment is called the *mud drum*.

[8] Though demonstration units have operated as high as 1200°F (650°C), commercial operating temperatures have been restricted to about 1050°F (570°C) by metallurgical considerations. Similarly, commercial pressures are limited to about 3500 psig (24.1 MPa).

the thermodynamic output, of course, unless the conditions of the entering water and leaving steam are also specified. The net energy input to a steam generator, known as the *heat absorption*, is given by Eq. 27.1.[9]

$$Q_{in,net} = \dot{m}_{steam}(h_{steam} - h_{feed}) \qquad 27.1$$

Not all of the combustion energy released in a steam generator will be transferred to the water. The *boiler efficiency*, which ranges from 75 to 90%, is calculated from Eq. 27.2.

$$\eta_{boiler} = \frac{Q_{in,net}}{Q_{in,gross}}$$
$$= \frac{\dot{m}_{steam}(h_{steam} - h_{feed})}{\dot{m}_{fuel}HV} \qquad 27.2$$

The steam mass flow rate in Eq. 27.2 will be as high as it can be at *maximum capacity*. At *normal capacity*, the boiler efficiency will be as high as it can be.

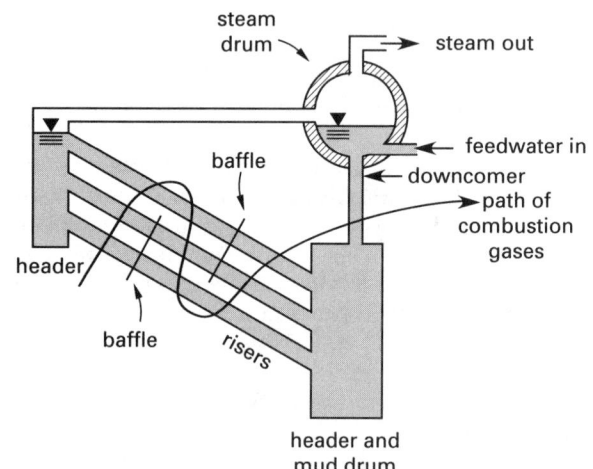

Figure 27.3 *Cross-Drum, Straight-Tube, Water-Tube Steam Generator*

There are several obsolete units that are occasionally encountered. These units compare energy and power on the basis of steam produced at one atmospheric pressure. One *boiler horsepower* is equal to 33,479 Btu/hr (9.81 kW). The *factor of evaporation* is defined by Eq. 27.3. $h_{fg, 1\,atm}$ has a value of 970.3 Btu/lbm (2257 MJ/kg).

$$\text{factor of evaporization} = \frac{h_{steam} - h_{feed}}{h_{fg,\,1\,atm}} \qquad 27.3$$

---

[9] In the United States, various confusing and ambiguous abbreviations for units of heat absorption have been used, including kB, kBtu, kBH, and kB/hr (1000 Btus per hour); and, mB, MB, mBtu, MBtu, MMBH, and MB/hr (1,000,000 Btus per hour). In some particularly unfortunate cases, "MB" is used to mean 1000 Btus (no hour rate), and "MMB" is used to mean 1,000,000 Btus (no hour rate). The actual meaning often has to be determined from the context.

The *equivalent evaporation* is

$$\text{equivalent evaporation} = \frac{\dot{m}_{\text{steam}}(h_{\text{steam}} - h_{\text{feed}})}{h_{fg,\ 1\ \text{atm}}} \quad 27.4$$

For an electrical generating system, the *heat rate* (*station rate, plant heat rate*, etc.) in Btu/kW-hr (kJ/kW·s) is defined as the total energy input to the steam generator divided by the electrical energy output. Typical full-throttle values in modern, large coal-fired plants are around 7500 to 8500 Btu/kW-hr (2.2 to 2.5 kJ/kW·s), with the lower values being more efficient. Values for partial throttle operation, and values for older plants, are higher. If the output is mechanical, the heat rate is the total energy input divided by the horsepower output.

Water passing through the boiler experiences a small pressure drop due to the friction and expansion. This pressure drop slightly increases the power to pump the water. Calculations are typically based on conditions at the boiler outlet. The error due to neglecting the pressure drop through the boiler is negligible.

4. BLOWDOWN

As water is evaporated in the boiler, dissolved and suspended solids are left behind. To prevent these solids from accumulating and causing fouling, constriction, and corrosion, some of the boiler water is bled off, a process known as *blowdown* (or *blowoff*). Blowdown may be intermittent or continuous. The blowdown rate is easily determined from the circulation rate (steam or water) and actual and permitted concentrations. With intermittent blowdown, the mass of water to be released after a period Δt of accumulation is given by Eq. 27.5. Concentrations of total solids, C, are typically given in ppm or mg/L.

$$m_{\text{blowdown}} = \left(\frac{C_{\text{feedwater}}}{C_{\text{maximum limit}} - C_{\text{feedwater}}}\right)\Delta t \dot{m}_{\text{steam}}$$
$$27.5$$

For continuous blowdown, the blowdown rate is

$$\dot{m}_{\text{blowdown}} = \left(\frac{C_{\text{feedwater}}}{C_{\text{maximum limit}} - C_{\text{feedwater}}}\right)\dot{m}_{\text{steam}}$$
$$27.6$$

Automatic (unattended) blowdown systems work in conjunction with total dissolved solids (TDS) monitors that continuously evaluate the ionic conductivity of the water. Sodium is completely soluble and is the most common ion in boiler water. Automatic blowdown systems operate by maintaining a preselected conductivity set-point in the range of 2400 to 2800 μmhos (μS).

Example 27.1

The total solids concentration in a boiler may not exceed 2500 mg/L. The concentration of total solids in the make-up water is 175 mg/L. The concentration of total solids in the condensate return is 25 mg/L. The boiler produces steam at the rate of 20,000 lbm/hr (2.5 kg/s), of which 5000 lbm/hr (0.63 kg/s) are used for process heating elsewhere. What should be the rate of continuous blowdown?

SI Solution

Assume a blowdown rate of 0.06 kg/s. Then, the mass of the make-up water will be

$$\dot{m}_{\text{make-up water}} = 0.63\ \frac{\text{kg}}{\text{s}} + 0.06\ \frac{\text{kg}}{\text{s}} = 0.69\ \text{kg/s}$$

The rate of condensate return is

$$\dot{m}_{\text{condensate return}} = 2.5\ \frac{\text{kg}}{\text{s}} - 0.63\ \frac{\text{kg}}{\text{s}} = 1.87\ \text{kg/s}$$

The concentration of total solids in the boiler feedwater is a weighted average of the concentration of the make-up water and the condensate return.

$$C_{\text{feedwater}} = \frac{\begin{array}{c}\dot{m}_{\text{make-up water}}C_{\text{make-up water}} \\ + \dot{m}_{\text{condensate return}}C_{\text{condensate return}}\end{array}}{\dot{m}_{\text{make-up water}} + \dot{m}_{\text{condensate return}}}$$

$$= \frac{\begin{array}{c}\left(0.69\ \dfrac{\text{kg}}{\text{s}}\right)\left(175\ \dfrac{\text{mg}}{\text{L}}\right) \\ + \left(1.87\ \dfrac{\text{kg}}{\text{s}}\right)\left(25\ \dfrac{\text{mg}}{\text{L}}\right)\end{array}}{0.69\ \dfrac{\text{kg}}{\text{s}} + 1.87\ \dfrac{\text{kg}}{\text{s}}}$$

$$= 65.4\ \text{mg/L}$$

From Eq. 27.6,

$$\dot{m}_{\text{blowdown}} = \left(\frac{C_{\text{feedwater}}}{C_{\text{maximum limit}} - C_{\text{feedwater}}}\right)\dot{m}_{\text{steam}}$$

$$= \left(\frac{65.4\ \dfrac{\text{mg}}{\text{L}}}{2500\ \dfrac{\text{mg}}{\text{L}} - 65.4\ \dfrac{\text{mg}}{\text{L}}}\right)\left(2.5\ \frac{\text{kg}}{\text{s}}\right)$$

$$= 0.067\ \text{kg/s}$$

If necessary, perform a second iteration with this new blowdown rate to derive a more precise value.

Customary U.S. Solution

Assume a blowdown rate of 500 lbm/hr. Then, the mass of the make-up water will be

$$\dot{m}_{\text{make-up water}} = 5000\ \frac{\text{lbm}}{\text{hr}} + 500\ \frac{\text{lbm}}{\text{hr}} = 5500\ \text{lbm/hr}$$

The rate of condensate return is

$$\dot{m}_{\text{condensate return}} = 20{,}000\ \frac{\text{lbm}}{\text{hr}} - 5000\ \frac{\text{lbm}}{\text{hr}}$$
$$= 15{,}000\ \text{lbm/hr}$$

The concentration of total solids in the boiler feedwater is a weighted average of the concentration of the make-up water and the condensate return.

Power Cycles

$$C_{\text{feedwater}} = \cfrac{\dot{m}_{\text{make-up water}}C_{\text{make-up water}} + \dot{m}_{\text{condensate return}}C_{\text{condensate return}}}{\dot{m}_{\text{make-up water}} + \dot{m}_{\text{condensate return}}}$$

$$= \cfrac{\left(5500\,\dfrac{\text{lbm}}{\text{hr}}\right)\left(175\,\dfrac{\text{mg}}{\text{L}}\right) + \left(15{,}000\,\dfrac{\text{lbm}}{\text{hr}}\right)\left(25\,\dfrac{\text{mg}}{\text{L}}\right)}{5500\,\dfrac{\text{lbm}}{\text{hr}} + 15{,}000\,\dfrac{\text{lbm}}{\text{hr}}}$$

$$= 65.2\ \text{mg/L}$$

From Eq. 27.6,

$$\dot{m}_{\text{blowdown}} = \left(\frac{C_{\text{feedwater}}}{C_{\text{maximum limit}} - C_{\text{feedwater}}}\right)\dot{m}_{\text{steam}}$$

$$= \left(\frac{65.2\,\dfrac{\text{mg}}{\text{L}}}{2500\,\dfrac{\text{mg}}{\text{L}} - 65.2\,\dfrac{\text{mg}}{\text{L}}}\right)\left(20{,}000\,\frac{\text{lbm}}{\text{hr}}\right)$$

$$= 536\ \text{lbm/hr}$$

If necessary, perform a second iteration with this new blowdown rate to derive a more precise value.

5. PUMPS

Centrifugal pumps are used in power-generation plants. Positive displacement pumps have become nearly obsolete for this application.[10]

The purpose of a pump is to increase the total energy content of the fluid flowing through it. Pumps can be considered adiabatic devices because the fluid gains (or loses) very little heat during the short time it passes through them. If the inlet and outlet are the same size and at the same elevation, the kinetic and potential energy terms can be neglected.[11] Then, the steady flow energy equation, SFEE, reduces to Eq. 27.7.

$$W_{\text{pump}} = m(h_2 - h_1) = mv(p_2 - p_1) \qquad 27.7$$
$$P_{\text{pump}} = \dot{m}(h_2 - h_1) \qquad 27.8$$

Equation 27.7 assumes that the pump is capable of isentropic compression. However, due to inefficiencies, some of the input energy is converted to heat. This heat energy raises the temperature (and hence, the internal energy portion of enthalpy) without increasing the pressure. The actual exit enthalpy, h_2', takes the inefficiency into consideration.

$$h_2' = h_1 + \frac{h_2 - h_1}{\eta_s} \qquad 27.9$$

---

[10]Obsolescence is due to the difficulty in controlling reciprocating pumps during start-up and partial-load. With modern variable-frequency drives, however, this may no longer be the case.

[11]Even if the pump inlet and outlet are different sizes and at different elevations, the kinetic and potential energy changes are small compared to the pressure energy increase.

Combining Eqs. 27.8 and 27.9, the actual input pump power is

$$P_{\text{pump}}' = \dot{m}(h_2' - h_1)$$

$$= \frac{\dot{m}(h_2 - h_1)}{\eta_s \eta_m} \qquad 27.10$$

The ideal exit enthalpy, h_2, can be calculated from the specific volume and the exit pressure, p_2. The calculation is simplified if the fluid is assumed to be incompressible. (Equation 27.11 cannot be used for the compression of gases and vapors.)

$$h_2 = h_1 + v(p_2 - p_1)\Big|_{\substack{\text{isentropic}\\\text{incompressible}}} \qquad \text{[SI]} \quad 27.11(a)$$

$$h_2 = h_1 + \frac{v(p_2 - p_1)}{J}\Big|_{\substack{\text{isentropic}\\\text{incompressible}}} \qquad \text{[U.S.]} \quad 27.11(b)$$

6. PUMP EFFICIENCY

Line 1-to-2 in Fig. 27.4 illustrates the ideal condition line for a pump. Since the compression is isentropic, the line is directed vertically upward. The actual work, however, results in an increase in entropy. Line 1-to-2' is the actual condition line for non-isentropic compression.

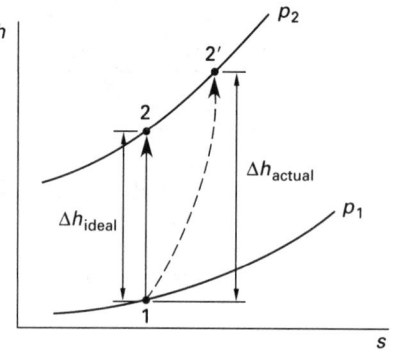

Figure 27.4 *Compression by a Pump*

The ideal and actual work for a pump are

$$W_{\text{ideal}} = h_2 - h_1 \qquad 27.12$$
$$W_{\text{actual}} = h_2' - h_1 \qquad 27.13$$

The definition of *isentropic efficiency*, η_s, for a pump is the inverse of what it is for a turbine.

$$\eta_s = \frac{W_{\text{ideal}}}{W_{\text{actual}}}$$

$$= \frac{h_2 - h_1}{h_2' - h_1} \qquad 27.14$$

Equation 27.15 defines the pump efficiency. Since W_{ideal} is small—less than 5 Btu/lbm (10 kJ/kg)—and does not affect the thermal efficiency greatly, the pump mechanical efficiency is often ignored, and η_{pump} is taken as η_s.

$$\eta_{\text{pump}} = \eta_s \eta_m \qquad 27.15$$

7. TEMPERATURE INCREASE IN PRESSURIZED LIQUIDS

The temperature of water (or any other liquid) increases when it is pressurized (i.e., passes through a pump). However, the increase in temperature is very small—far less than what would occur with a gas or vapor.

The increase in temperature can be thought of as being caused by two processes: an isentropic compression and an irreversible compression (if the pump is not 100% efficient). The increase due to isentropic compression is so small as to be negligible. For example, the increase in temperature of 100°F (38°C) water compressed isentropically from saturation to 1000 psi (6.9 MPa) is approximately 0.3°F (0.17°C).

The irreversible portion of the process incorporates the additional energy that must be put into the water to overcome friction and turbulence. These factors produce a profound heating effect (as compared to the isentropic portion). It is proper to consider only this irreversible work when calculating the increase in water temperature.

Example 27.2

A boilerfeed pump increases the pressure of 90°F (30°C) water from 1 atm to 150 psia (1000 kPa). The pump's isentropic efficiency is 80%. The water's specific heat is 1.0 Btu/lbm-°F (4.187 kJ/kg·K). What is the temperature of the water leaving the pump?

SI Solution

Since the properties of a liquid are essentially independent of pressure, the properties of 30°C water can be read from a saturated steam table (App. 24.O).

$$h_{30°C} = 125.79 \text{ kJ/kg}$$

$$v_{30°C} = 1.0043 \text{ cm}^3/\text{g} = 0.0010043 \text{ m}^3/\text{kg}$$

The ideal enthalpy of the feedwater entering the boiler (point 2 on Fig. 27.4) is equal to the enthalpy at point 1 plus the energy put into the water by the pump. Assuming the water is incompressible, the specific volumes at points 1 and 2 are the same. From Eq. 27.11,

$$h_2 = h_1 + v_1(p_2 - p_1)$$

$$= 125.79 \text{ kJ/kg}$$

$$+ \frac{\left(0.0010043 \, \frac{\text{m}^3}{\text{kg}}\right)(1000 - 101.3 \text{ kPa})\left(1000 \, \frac{\text{Pa}}{\text{kPa}}\right)}{1000 \, \frac{\text{J}}{\text{kJ}}}$$

$$= 125.79 \, \frac{\text{kJ}}{\text{kg}} + 0.903 \, \frac{\text{kJ}}{\text{kg}} = 126.69 \text{ kJ/kg}$$

This calculation assumes that the pump is capable of isentropic compression. Because of the pump's inefficiency, not all of the 0.903 kJ/kg goes into raising the pressure. Some of the energy goes into raising the temperature. (Since $h = u + pv$, both energy contributions increase the enthalpy.) Therefore, to get to 1000 kPa, more than 0.903 kJ/kg must be added to the water. The actual enthalpy at point 2′ is

$$h_2' = h_1 + \frac{h_2 - h_1}{\eta_s}$$

$$= 125.79 \, \frac{\text{kJ}}{\text{kg}} + \frac{0.903 \, \frac{\text{kJ}}{\text{kg}}}{0.80} = 126.9 \text{ kJ/kg}$$

The enthalpy was increased by 1.13 kJ/kg. Since the specific heat is 4.187 kJ/kg·K, the final temperature is

$$T_2' = T_1 + \frac{\Delta h}{c_p} = 30°C + \frac{1.13 \, \frac{\text{kJ}}{\text{kg}}}{4.187 \, \frac{\text{kJ}}{\text{kg·K}}}$$

$$= 30.27°C$$

Customary U.S. Solution

Since the properties of a liquid are essentially independent of pressure, the properties of 90°F water can be read from a saturated steam table (App. 24.A).

$$h_{90°F} = 58.07 \text{ Btu/lbm}$$

$$v_{90°F} = 0.01610 \text{ ft}^3/\text{lbm}$$

The ideal enthalpy of the feedwater entering the boiler (point 2 on Fig. 27.4) is equal to the enthalpy at point 1 plus the energy put into the water by the pump. Assuming the water is incompressible, the specific volumes at points 1 and 2 are the same. From Eq. 27.11,

$$h_2 = h_1 + \frac{v_1(p_2 - p_1)}{J}$$

$$= 58.07 \text{ Btu/lbm}$$

$$+ \frac{\left(0.01610 \, \frac{\text{ft}^3}{\text{lbm}}\right)\left(150 \, \frac{\text{lbf}}{\text{in}^2} - 14.7 \, \frac{\text{lbf}}{\text{in}^2}\right)\left(144 \, \frac{\text{in}^2}{\text{ft}^2}\right)}{778.2 \, \frac{\text{ft-lbf}}{\text{Btu}}}$$

$$= 58.07 \, \frac{\text{Btu}}{\text{lbm}} + 0.403 \, \frac{\text{Btu}}{\text{lbm}} = 58.47 \text{ Btu/lbm}$$

This calculation assumes that the pump is capable of isentropic compression. Because of the pump's inefficiency, not all of the 0.403 Btu/lbm goes into raising the pressure. Some of the energy goes into raising the temperature. (Since $h = u + pv$, both energy contributions increase the enthalpy.) Therefore, to get to 150 psia, more than 0.403 Btu/lbm must be added to the water. The actual enthalpy at point 2 is

$$h_2' = h_1 + \frac{h_2 - h_1}{\eta_s}$$

$$= 58.07 \, \frac{\text{Btu}}{\text{lbm}} + \frac{0.403 \, \frac{\text{Btu}}{\text{lbm}}}{0.80} = 58.57 \text{ Btu/lbm}$$

The enthalpy was increased by 0.5 Btu/lbm. Since the specific heat is 1.0 Btu/lbm-°F, the corresponding temperature increase is 0.5°F. The final temperature is $90°F + 0.5°F = 90.5°F$.

8. TURBINES

There are two general categories of steam turbine operation. A *reaction turbine* consists of a rotating drum with small nozzles (reaction jets) located around the drum's periphery. Steam is discharged from the nozzles, and the drum turns in reaction to the steam's action. An *impulse turbine* is characterized by stationary jets discharging against vanes mounted on the periphery of a wheel. Modern steam turbines use both principles to extract energy from the steam.

There are several designations given to turbines. A *back pressure turbine* (*topping turbine* or *superposed turbine*) exhausts either to a high-pressure industrial process or to a second turbine operating in a lower-pressure range.[12] In a two-turbine installation, the first is designated as the *high-pressure (HP) turbine*, and the second is known as the *low-pressure (LP) turbine*. (If there are three turbines in series, the middle one is known as an *intermediate-pressure (IP) turbine*.) Turbines that operate at supercritical conditions are known as *supercritical-pressure (SP) turbines*.

Some of the steam entering a turbine may be removed before it has expanded to the turbine exit pressure. This removal is known as a *bleed*, and the steam is known as *bleed steam*. The *extraction rate* or *bleed rate* is the rate, usually expressed in Btu/hr (MJ/h), at which the partially expanded steam is bled off. A turbine with one or more bleeds is known as an *extraction turbine*.

In a *reheating turbine*, steam is removed from the turbine, returned to its original temperature in a reheater, and returned to a lower-pressure portion of the same turbine.

9. TURBINE WORK, POWER, AND EFFICIENCY

Turbines can generally be thought of as pumps operating in reverse. A turbine extracts energy from the fluid which, in turn, decreases in temperature and pressure. The process is essentially adiabatic because the fluid loses very little heat during the short time it passes through the turbine.

Figure 27.5 diagrams the expansion of steam in a turbine from a high pressure, p_1, to a low pressure, p_2. The broken line is the *condition line*—the locus of all states of steam during the expansion process. The ideal condition line is isentropic, which is represented by a vertical line downward on the Mollier (h versus s) diagram.

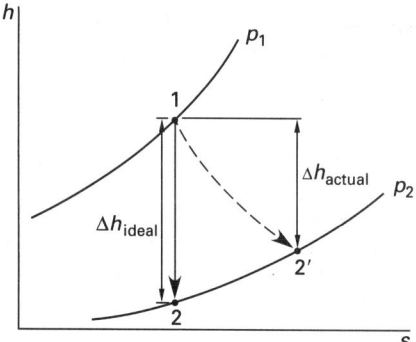

Figure 27.5 *Single-Stage Turbine Expansion*

Changes in potential and kinetic energies are small and can be neglected. The ideal steady flow energy equation, SFEE, reduces to Eq. 27.16 and predicts the work output of the turbine. The work is known as *shaft work* because it is transmitted from the turbine to a generator through a shaft.[13]

$$W_{\text{ideal}} = m(h_1 - h_2) \qquad 27.16$$

Equation 27.16 assumes that the turbine is capable of isentropic expansion. However, not all of the available fluid energy can be extracted. Friction within the turbine increases the enthalpy without decreasing the pressure. The actual exit enthalpy, h_2', takes this inefficiency into consideration. If the expansion process is not isentropic, entropy will increase and the actual final enthalpy, h_2', will be higher than the ideal final enthalpy, h_2.

The *isentropic efficiency* (*adiabatic efficiency*), η_s, of a turbine is the ratio of actual to ideal energy extractions.[14] Actual isentropic efficiencies vary from approximately 65% for 1 MW unit to over 80% for 100 MW units.

$$\eta_s = \frac{W_{\text{actual}}}{W_{\text{ideal}}} = \frac{h_1 - h_2'}{h_1 - h_2} \qquad 27.17$$

The actual energy extracted is

$$W_{\text{actual}} = m(h_1 - h_2') = m\eta_s(h_1 - h_2) \qquad 27.18$$
$$h_2' = h_1 - \eta_s(h_1 - h_2) \qquad 27.19$$

In addition to thermodynamic friction losses, there are additional friction losses in bearings and other moving mechanical parts. The net turbine work is

$$\begin{aligned} W_{\text{turbine}} &= W_{\text{actual}} - W_{\text{friction}} \\ &= \eta_m W_{\text{actual}} \qquad 27.20 \end{aligned}$$

---

[12]Topping turbines are added when older, low-pressure plants are "repowered." The furnace and steam generator are replaced so that high-pressure steam is produced, and a topping turbine is added. The topping turbine uses the high pressure steam. The original low-pressure turbine uses the exhaust steam from the topping turbine. Most of the remaining equipment in the cycle is unchanged.

[13]Other names for *shaft work* are *brake work*, *useful work*, and *net work*.

[14]The terminology "adiabatic efficiency" is ambiguous because an "adiabatic system" does not require a 100% efficient turbine.

Most steam turbines have high mechanical efficiencies—in the order of 98%. Inasmuch as the friction losses are very small, the isentropic and turbine efficiencies are essentially identical. However, the *overall turbine efficiency* incorporates the mechanical friction losses.

$$\eta_{\text{turbine}} = \eta_s \eta_m \qquad 27.21$$

The isentropic efficiency of a device affects both the flow rate and the thermodynamic properties of the substance flowing through the device. The mechanical efficiency affects the flow rate, but it does not affect the thermodynamic properties and should not be used to determine final enthalpies. Both the isentropic and mechanical efficiencies reduce the amount of useful energy that can be generated in a turbine, and they affect the amount of substance flowing through the turbine. The actual power generated in the turbine is

$$P_{\text{turbine}} = \dot{m}_{\text{actual}} \eta_{\text{turbine}} (h_1 - h_2) \qquad 27.22$$

$$\dot{m}_{\text{actual}} = \frac{\dot{m}_{\text{ideal}}}{\eta_{\text{turbine}}}$$

$$= \frac{\dot{m}_{\text{ideal}}}{\eta_s \eta_m} \qquad 27.23$$

10. TWO-STAGE EXPANSION

If expansion through the turbine is multiple stage, or if the turbine has a bleed at an intermediate pressure, p_m, the expansion process for each stage or bleed will be as illustrated by Fig. 27.6. In this figure, a fraction y of the original steam mass is removed after expanding to pressure p_m. The remaining fraction $1 - y$ expands to pressure p_2.

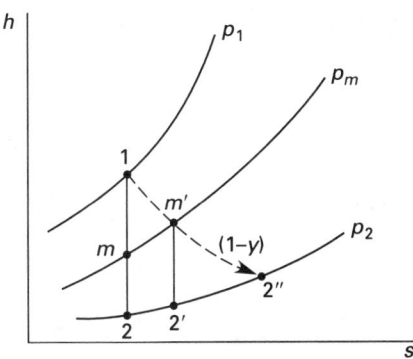

Figure 27.6 *Two-Stage Turbine Expansion*

The ideal and actual work generated in the first stage are

$$W_{\text{ideal},1} = m(h_1 - h_m) \qquad 27.24$$

$$W_{\text{actual},1} = m(h_1 - h'_m)$$

$$= \eta_{s,1} m(h_1 - h_m) \qquad 27.25$$

The ideal and actual work generated by the remaining fraction, $1 - y$, in the second stage are

$$W_{\text{ideal},2} = (1 - y)m(h'_m - h'_2) \qquad 27.26$$

$$W_{\text{actual},2} = (1 - y)m(h'_m - h''_2)$$

$$= (1 - y)m\eta_{s,2}(h'_m - h'_2) \qquad 27.27$$

The total work done per unit mass of the steam is

$$W_{\text{turbine}} = W_{\text{actual},1} + W_{\text{actual},2}$$

$$= m[h_1 - h'_m + (1 - y)(h'_m - h''_2)]$$

$$= m[\eta_{s,1}(h_1 - h_m) + (1 - y)\eta_{s,2}$$

$$\times (h'_m - h'_2)] \qquad 27.28$$

Example 27.3

Steam is expanded from 700°F (360°C) and 200 psia (1500 kPa) to 5 psia (50 kPa) in an 87% efficient turbine. What is the final enthalpy of the steam?

SI Solution

Refer to Fig. 27.5. From the superheated steam tables,

$$h_1 = 3169.2 \text{ kJ/kg}$$

$$s_1 = 7.1363 \text{ kJ/kg·K}$$

At this point, proceed as if the turbine is 100% efficient. From the saturated steam tables for 50 kPa,

$$s_f = 1.0910 \text{ kJ/kg·K}$$

$$s_g = 7.5939 \text{ kJ/kg·K}$$

$$h_f = 340.49 \text{ kJ/kg}$$

$$h_{fg} = 2305.4 \text{ kJ/kg}$$

Since at this point it is assumed that the expansion is isentropic (i.e., 100% efficient), $s_2 = s_1$. The quality at point 2 can be found as

$$x = \frac{s_2 - s_f}{s_g - s_f} = \frac{7.1363 \frac{\text{kJ}}{\text{kg·K}} - 1.0910 \frac{\text{kJ}}{\text{kg·K}}}{7.5939 \frac{\text{kJ}}{\text{kg·K}} - 1.0910 \frac{\text{kJ}}{\text{kg·K}}}$$

$$= 0.9296$$

The ideal final enthalpy can be found from the quality.

$$h_2 = h_f + x h_{fg}$$

$$= 340.49 \frac{\text{kJ}}{\text{kg}} + (0.9296)\left(2305.4 \frac{\text{kJ}}{\text{kg}}\right)$$

$$= 2483.6 \text{ kJ/kg}$$

However, this value of h_2 assumes the expansion through the turbine is isentropic. Since the turbine is capable of extracting only 87% of the ideal energy, the actual final enthalpy is

$$h_2' = h_1 - \eta_s(h_1 - h_2)$$

$$= 3169.2 \ \frac{\text{kJ}}{\text{kg}} - (0.87)\left(3169.2 \ \frac{\text{kJ}}{\text{kg}} - 2483.6 \ \frac{\text{kJ}}{\text{kg}}\right)$$

$$= 2572.7 \ \text{kJ/kg}$$

Customary U.S. Solution

Refer to Fig. 27.5. From the superheated steam tables,

$$h_1 = 1373.8 \ \text{Btu/lbm}$$

$$s_1 = 1.7234 \ \text{Btu/lbm-}°\text{F}$$

At this point, proceed as though the turbine is 100% efficient. From the saturated steam tables for 5 psia,

$$s_f = 0.2349 \ \text{Btu/lbm-}°\text{F}$$

$$s_{fg} = 1.6093 \ \text{Btu/lbm-}°\text{F}$$

$$h_f = 130.17 \ \text{Btu/lbm}$$

$$h_{fg} = 1000.9 \ \text{Btu/lbm}$$

Since it is assumed that the expansion is isentropic (i.e., 100% efficient), $s_2 = s_1$. The quality at point 2 can be found as

$$x = \frac{s_2 - s_f}{s_{fg}} = \frac{1.7234 \ \dfrac{\text{Btu}}{\text{lbm-}°\text{F}} - 0.2349 \ \dfrac{\text{Btu}}{\text{lbm-}°\text{F}}}{1.6093 \ \dfrac{\text{Btu}}{\text{lbm-}°\text{F}}}$$

$$= 0.9249$$

The ideal final enthalpy can be found from the quality.

$$h_2 = h_f + x h_{fg} = 130.17 \ \frac{\text{Btu}}{\text{lbm}} + (0.9249)\left(1000.9 \ \frac{\text{Btu}}{\text{lbm}}\right)$$

$$= 1055.9 \ \text{Btu/lbm}$$

However, this value of h_2 assumes the expansion through the turbine is isentropic. Since the turbine is capable of extracting only 87% of the ideal energy, the actual final enthalpy is

$$h_2' = h_1 - \eta_s(h_1 - h_2)$$

$$= 1373.8 \ \frac{\text{Btu}}{\text{lbm}} - (0.87)\left(1373.8 \ \frac{\text{Btu}}{\text{lbm}} - 1055.9 \ \frac{\text{Btu}}{\text{lbm}}\right)$$

$$= 1097.2 \ \text{Btu/lbm}$$

Example 27.4

Repeat Ex. 27.3 using the Mollier (h versus s) diagram.

SI Solution

h_1 is read directly from the Mollier diagram at the intersection of 360°C and 1500 kPa. (Greater accuracy is possible with a large Mollier diagram.)

$$h_1 \approx 3170 \ \text{kJ/kg}$$

The ideal final enthalpy, h_2, is found by dropping straight down to the 50 kPa line. h_2 is read as approximately 2485 kJ/kg. Since the turbine is capable of extracting only 87% of the ideal energy, the actual final enthalpy is calculated from Eq. 27.19.

$$h_2' = h_1 - \eta_s(h_1 - h_2)$$

$$= 3170 \ \frac{\text{kJ}}{\text{kg}} - (0.87)\left(3170 \ \frac{\text{kJ}}{\text{kg}} - 2485 \ \frac{\text{kJ}}{\text{kg}}\right)$$

$$= 2574 \ \text{kJ/kg}$$

Customary U.S. Solution

h_1 is read directly from the Mollier diagram at the intersection of 700°F and 200 psia. (Greater accuracy is possible with a large Mollier diagram.)

$$h_1 \approx 1374 \ \text{Btu/lbm}$$

The ideal final enthalpy, h_2, is found by dropping straight down to the 5 psia line. h_2 is read as approximately 1055 Btu/lbm. Since the turbine is capable of extracting only 87% of the ideal energy, the actual final enthalpy is calculated from Eq. 27.19.

$$h_2' = h_1 - \eta_s(h_1 - h_2)$$

$$= 1374 \ \frac{\text{Btu}}{\text{lbm}} - (0.87)\left(1374 \ \frac{\text{Btu}}{\text{lbm}} - 1055 \ \frac{\text{Btu}}{\text{lbm}}\right)$$

$$= 1096 \ \text{Btu/lbm}$$

11. HEAT EXCHANGERS

A *heat exchanger* transfers energy from one fluid to another through a wall separating them. If the heat transfer is assumed to be adiabatic, the total energy of both input streams must be the same as the total energy of both output streams. No work is done within a heat exchanger, and the potential and kinetic energies of the fluids can be ignored.

$$\dot{m}_A h_{A,\text{in}} + \dot{m}_B h_{B,\text{in}} = \dot{m}_A h_{A,\text{out}} + \dot{m}_B h_{B,\text{out}} \quad \textit{27.29}$$

12. CONDENSERS

Condensers are special-purpose heat exchangers that remove the heat of vaporization from steam.[15] Steam passes over tubes in which cooling water (or, occasionally, air) passes. Liquid water falls to the lower portion of the condenser known as the *hot well* or *condensate well*. Heat is transferred through the tubing walls to the cooling water and is then rejected to the environment. Since the condensation takes place on a cold surface, the term *surface condenser* is occasionally encountered.[16]

The heat flow, $\dot{Q}$, out of a condenser is referred to as the *heat load* and the *condenser duty*. Since heat flows out of the condenser, $\dot{Q}$ and the quantity $(h_2 - h_1)$ will be negative.

$$\dot{Q} = \dot{m}_{\text{steam}}(h_2 - h_1) \qquad 27.30$$

Equation 27.30 uses the steam properties to calculate the heat load. The heat load can also be calculated from the temperature increase in the cooling water. The *range* is the initial difference between the incoming cooling water temperature and the saturation temperature corresponding to the condenser pressure. The *terminal temperature difference* is the difference in temperatures of the cooling water and the condensed liquid leaving the condenser. It is seldom less than $5°F$ ($2.7°C$). A terminal temperature difference of $10°F$ ($5.6°C$) is usually assumed in initial performance estimates. The *rise* is the increase in the cooling water temperature as it passes through the condenser. The rise seldom exceeds $20°F$ ($11°C$).

$$\dot{Q} = \dot{m}_{\text{cooling water}} c_p (T_{\text{out}} - T_{\text{in}}) \qquad 27.31$$

The *hot well depression*, HWD, also known as the *condensate depression*, *condenser hot well subcooling*, and *degrees of freedom*, is the difference in temperature between the saturation temperature corresponding to the condenser pressure and the steam condensate in the condenser's hot well. It typically varies between 5 and $10°F$ (2.7 and $5.4°C$). Large hot well depressions are undesirable because energy is wasted and because the solubility of gases in the condensate is higher at lower temperatures. To keep the temperature high, some steam is bypassed to the hot well.

$$\text{HWD} = T_{\text{sat},p} - T_{\text{condensate,out}} \qquad 27.32$$

The *cleanliness factor* is the ratio (expressed as a percentage) of the actual to the ideal (new, clean, etc.) heat transfer coefficient. For a given installation, it is the ratio of the actual to the ideal heat loads.

Condensation is assumed to occur at constant pressure. It is also assumed that the water will leave the condenser as a saturated liquid (corresponding to the condenser pressure).[17] In real power-generating systems, subcooling represents wastage of energy, as nothing is gained by subcooling the water.

Due to the volume decrease that occurs during condensation, the condenser pressure can be low and is usually far below atmospheric. The *back pressure* is the absolute pressure in the condenser, usually expressed in inches of mercury (bars). When the back pressure is lowered, more energy can be extracted from the steam.

Utility condensers operate at a high vacuum. It is inevitable that some *noncondensing gases* (primarily oxygen and other air components) will enter the condenser. These gases, commonly referred to as "air" regardless of composition, enter dissolved in the steam and through air leaks.[18] At high temperatures, these gases are highly corrosive, particularly to the copper used in condenser tubing. In addition, the effect of these gases is to increase the back pressure. As Fig. 27.7 shows, increasing the back pressure reduces the energy available to the power cycle. Noncondensing gases are removed from condensers with steam-jet air ejectors. A general rule of thumb is that the air-removal rate should be less than 1 SCFM (0.5 L/s) per 100 MW thermal.

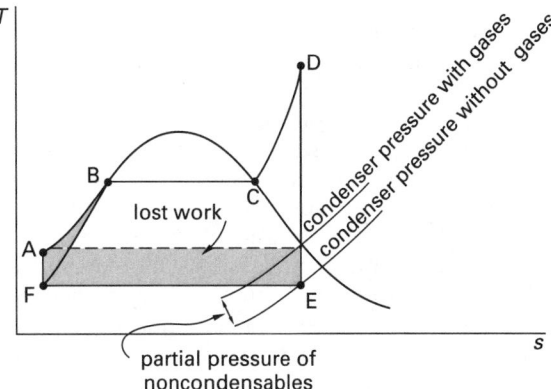

Figure 27.7 *Effect of Noncondensing Gases on Power Cycle Performance*

Turbine output cannot be increased indefinitely by decreasing the back pressure, however. As back pressure is lowered below what produces sonic velocity in the

---

[15] *Dump condensers* are separate condensers used to condense steam when the turbine is not in service (i.e., are tripped).

[16] Besides surface condensers, *jet condensers* constitute the other major condenser category. While surface condensers both produce a vacuum and recover the condensate, jet condensers produce only a vacuum. The condensate escapes with the jet. Since the condensate is lost, jet condensers are not used extensively in power-generating systems.

[17] The term "straight condensing" is sometimes used to mean converting saturated steam to saturated water at the same pressure and temperature. However, this may be confused with the *straight condensing cycle*. Therefore, "pure condensing" is probably a better choice. Besides pure condensing, other modes of condenser operation are desuperheating with condensing, condensing with subcooling, and desuperheating with condensing and subcooling.

[18] The most common *leak points* are cracks in expansion joints between the turbine and condenser, cracks in the condenser shell at condensate-return lines, low-pressure heater vents, and condensate pump seals.

Power Cycles

last turbine stage, condensate temperature is lowered without any increase in turbine output. This results in a loss similar to that of subcooling the condensate.

13. CONDENSER FOULING

As Fig. 27.7 illustrates, an increase in condenser pressure will decrease the energy extracted from the steam. Besides from an accumulation of noncondensing gases, the pressure increase can also occur when the condenser is unable to remove the heat fast enough due to fouling. Within the vapor dome, reductions in enthalpy, temperature, and pressure go hand-in-hand. When the enthalpy is not reduced, the pressure is not reduced.

The most common causes of fouling on the water side of water-cooled, steam surface condensers are microbiological growth, scale formation, and tube pluggage by debris and aquatic organisms. These problems are more intense in once-through condensers that draw water from lakes, rivers, and oceans.

Since condensers are warm, they are prone to microbiological growth (slime and biofilm). This *microfouling* also provides a base for most *microbiologically influenced corrosion*, MIC. Microfouling is easily detected by a gradual increase in the terminal temperature difference. Growth of biological organisms is controlled along the coolant intake by continuous injection and *shock treatment* (sudden injection) of *biocides*.

Scaling is the formation of mineral compounds, most notably calcium carbonate and (to a lesser extent) calcium sulfate and phosphate, manganese compounds, and silicates, on the condenser surfaces. Sometimes adding sulfuric acid to the coolant will keep the pH low enough to prevent calcium carbonate deposits. Phosphates, phosphonates, and polymers can also be used. As with make-up water treatment, though, no generic treatment works for all installations. Once formed, scale must be mechanically removed or treated chemically. Calcium carbonate dissolves in acid, but other deposits require more exotic treatments.

Tube pluggage from leaves, vegetation, and anything else that can pass through intake screens will reduce coolant flow. This can be detected by weekly pressure-drop checks or sudden changes in terminal temperature difference. *Macrofouling* is a term referring to infestation by Asian clams, saltwater barnacles, marine blue mussels, zebra mussels, or similar.

Zebra mussels, accidentally introduced in the United States in 1986, are particularly troublesome for several reasons. First, young mussels are microscopic and easily pass through intake screens. Second, they attach to anything, even other mussels, to produce thick colonies. Third, adult mussels quickly sense some biocides, most notably those that are halogen-based (including chlorine), and quickly close and remain closed for days or weeks. An ongoing biocide program aimed at pre-adult mussels, combined with slippery polymer-based surface coatings, is probably the best approach at prevention.

Once a condenser is colonized, mechanical removal by scraping or water blasting is the only practical option.

Chlorine is a common treatment for once-through systems. However, chlorine reacts with organic precursors to produce trihalorganics, including trihalomethanes, which are suspected carcinogenic compounds, and its use may be prohibited in the United States. Or, when chlorine is used, a *dehalogenation agent* (e.g., sodium bisulfite) may be needed downstream of the condenser. In addition to specialized *molluscicides*, other more expensive substitutes for chlorine include bromine, chlorine dioxide, ozone, nonoxidizing biocides, copper, cyanuric acid, hydrogen peroxide, polymers, potassium permanganate, and sodium hypochlorite.

Heat treatment (*thermal backflushing*) is another alternative for zebra mussel prevention. Exposure to 95 to 105°F (35 to 41°C) water results in 100% mortality within approximately an hour. A temperature of 89°F (32°C) for six hours or more also appears to be effective. Water can be heated to these temperatures by recirculating condenser discharge, or for more distant points, by steam sparging.

14. NOZZLES, ORIFICES, AND VALVES

Since a flowing fluid is in contact with nozzle, orifice, and valve walls for only a very short period of time, flow through them can be considered adiabatic. No work is done on the fluid as it passes through. If the potential energy changes are neglected, the SFEE reduces to Eq. 27.33.

$$h_1 + \frac{v_1^2}{2} = h_2 + \frac{v_2^2}{2} \quad \text{[SI]} \quad 27.33(a)$$

$$h_1 + \frac{v_1^2}{2g_cJ} = h_2 + \frac{v_2^2}{2g_cJ} \quad \text{[U.S.]} \quad 27.33(b)$$

If the kinetic energy changes are neglected, $h_1 = h_2$. A constant enthalpy pressure drop is characteristic of a *throttling valve*. Pressure-reduction and superheater-bypass valves used to control turbine output in sliding-pressure operation are typically of the "tortuous-path" variety. Steam makes numerous abrupt direction changes, and pressure is reduced in a throttling process.

15. SUPERHEATERS

A *superheater* is essentially a heat exchanger used to increase the energy of the steam. A *radiant superheater* is exposed directly to combustion flame and receives energy by radiation. A *convection superheater* is typically exposed to the first pass of stack gases and is screened from direct view of the combustion flame by rows of boiler tubes. Alternatively, a superheater can be a separately fired, independent unit.

Since the overall heat transfer coefficient increases with increases in combustion product flow, the steam temperature increases with convection superheaters when the combustion rate increases. The opposite is true for radiant superheaters (since furnace temperature does not increase rapidly enough with increasing steam flow). In order to maintain a constant superheat, modern superheaters use both radiant and convection sections in series. Damper control and tiltable burners may also be used to maintain constant superheat.

A superheater does not increase steam pressure; it only increases temperature. In fact, steam pressure will decrease a slight amount (e.g., 2 to 5%) by virtue of the superheater's resistance to flow.

16. REHEATERS

The name *reheater* (*resuperheater*) is given to specific parts of a furnace or to separately fired superheaters that add energy to steam after it has already passed at least once through a turbine. This is done to reduce the moisture content of steam that, after partial expansion, may have already entered the vapor dome. Reheaters do not differ much from convection-type superheaters, except that the volume of steam handled is greater (since the pressure and density are less). In fact, reheaters are typically placed after the primary superheater tubing in the stack and before the secondary superheater tubing, if any. Radiant-type reheaters are also occasionally used.

The *reheat temperature* is usually the same as the original steam temperature (known as the *throttle temperature*), usually 1000 to 1050°F (540 to 570°C) in traditional drum boiler units and 1000 to 1100°F (540 to 590°C) or more in supercritical boilers. Optimum reheat pressure is approximately 25% of the throttle pressure. *Multiple reheat*, double or triple, is applicable only with supercritical throttle pressures (i.e., above 3206 psi (22.1 MPa)). Otherwise, multiple reheat will result in superheated exhaust.

17. DESUPERHEATERS

Some power plant auxiliary equipment is designed to use saturated (as opposed to superheated) steam. It is common to bleed steam from the superheater and run it through a *desuperheater* (*attemperator*) to decrease the steam enthalpy. This can be done by looping the steam back through cooler water in the steam drum or a feedwater heater, or more commonly, by injecting cooling water (i.e., a *fixed-orifice desuperheater*).[19] Since the water and steam leave at the same properties, the heat balance equation for water injection is

$$\dot{m}_{\text{cold water}}(h_{\text{out,steam}} - h_{\text{in,water}})$$
$$= \dot{m}_{\text{steam}}(h_{\text{in,steam}} - h_{\text{out,steam}}) \qquad 27.34$$

---

[19]Other types of superheaters are surface-water absorption, steam atomizing, ejector-recycle, venturi, and variable orifice.

The *turndown ratio* (*capacity turndown*) is the ratio of maximum-to-minimum mass steam flow rates at which the temperature can be accurately controlled by the desuperheater. Minimum flow rates are influenced by the lowest velocity at which water droplets can be held in suspension.

Desuperheaters are usually needed to run auxiliary equipment (e.g., turbine-driven boilerfeed pumps); they should also be installed across high-pressure turbines. If a high-pressure turbine trips or is taken out of service, the high-pressure steam can be desuperheated and routed to the low-pressure turbine.

18. FEEDWATER HEATERS

A *feedwater heater* uses steam to increase the temperature of water entering the steam generator, thus increasing the boiler efficiency. Each 10°F (5.6°C) increase in water temperature increases a corresponding increase in boiler efficiency of approximately 1%. In a normal condensing cycle, the water being heated comes from the condenser. Steam for heating is bled off from an *extraction turbine*. Eight or more stages of feedwater heating may be used in modern plants.

Open heaters (also known as *direct-contact heaters* and *mixing heaters*) physically mix the steam and water, as shown in Fig. 27.8. Because they are essentially large boxes under internal steam pressure, open heaters are usually closest to the condenser (i.e., on the suction side of the boilerfeed pumps), where the pressure is low. Each open heater requires its own feed pump. Because open heaters are not highly pressurized, the temperature of the leaving water is seldom over 220°F (104°C) and is usually near 212°F (100°C).

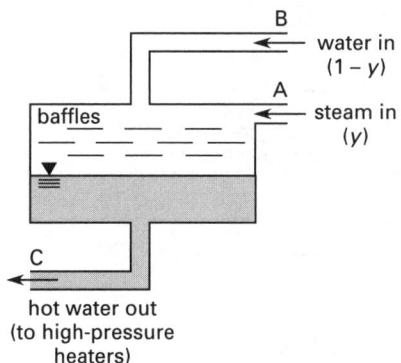

Figure 27.8 *Open Feedwater Heater*

Open heaters may be simply vented to the atmosphere, or dissolved gases that come out of solution may be extracted by a vent condenser or a vacuum pump.

When saturated or nearly saturated water is reduced in pressure, some of it may "flash" into steam. The reduction in pressure can occur in condensate return lines due to fluid friction or from pressure losses through valves and traps. When flashing occurs, the steam will impede liquid flow, a condition known as "binding."[20] To prevent boilerfeed pumps handling water near the boiling temperature from racing due to lack of feed liquid, open feedwater heaters are mounted quite high (i.e., 12 to 20 ft (3.6 to 6 m) up or higher). This provides the necessary NPSHR for the pumps below.

The adiabatic energy balance around the open feedwater heater shown in Fig. 27.8 is

$$(1 - y)h_B + yh_A = h_C \qquad 27.35$$

If the three enthalpies are known, the *bleed fraction, y,* can be determined.

$$y = \frac{h_C - h_B}{h_A - h_B} \qquad 27.36$$

A *closed feedwater heater* is a heat exchanger that can operate at either high or low pressures. There is no mixing of the water and steam in the feedwater heater. The cooled steam (known as the *drips*) leaves the feedwater heater in liquid form.

There are various methods of disposal of the drips. The condensate may be combined with the feedwater (as in Fig. 27.9) after passing through a *drip pump* to raise its pressure. (Since the drip pump work is small, it can be omitted for first approximations.) Alternatively, the drips can be returned to the condenser *hot well.* For the closed feedwater shown in Fig. 27.9 receiving bleed fraction y of steam, the adiabatic heat balance is

$$h_C = yh_A + (1 - y)h_D + W_{\text{drip pump per pound}} \qquad 27.37$$

For a feedwater heater, the *terminal temperature difference*, TTD (also known as the *approach*), is defined as the difference in saturation temperature corresponding to the steam pressure and the temperature of the leaving water. It typically varies between 5 and 20°F (2.8 and 11°C).

$$\text{TTD} = T_{\text{sat},p,A} - T_C = T_B - T_C \qquad 27.38$$

In the United States, *tubesheet feedwater heaters* have been used almost exclusively. Since tubesheet heaters are limited to temperature changes of about 10°F/min (5.6°C/min), they are susceptible to thermal stresses when used with cycling plants. With *header-type feedwater heaters*, all of the tubes are welded to headers. This type of design tolerates temperature change rates up to 30°F/min (17°C/min) and is more suitable for cycling loads.

19. EVAPORATORS

An *evaporator* is a closed shell-and-tube heat exchanger that is used to produce distilled boiler feedwater. Steam bled off from other points is the source of heating. Steam passes through the tubes. As the water is evaporated, the distilled water steam is routed to a holding tank, an open feedwater heater, or a special condenser.

Single-effect (single-stage) evaporators produce distilled water from the steam of a single evaporator. *Multiple-effect evaporators* are more common and use the steam produced in one evaporator as the heat source for the next evaporator. In that way, the heat of evaporization is used several times. Three and four effects in series are typical for power-generating plants.

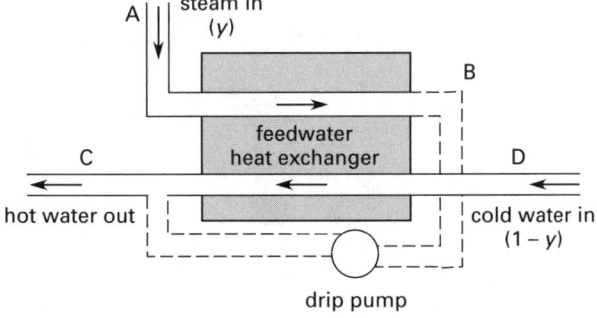

Figure 27.9 *Closed Feedwater Heater*

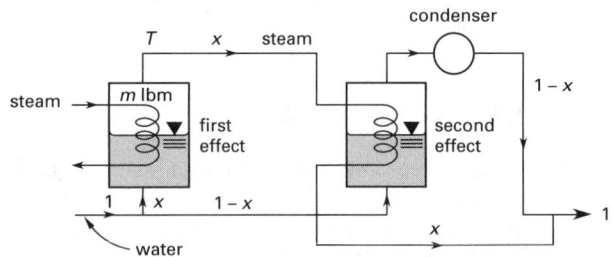

Figure 27.10 *Double-Effect Evaporator*

---

[20]Flashing is intentional in a *flash evaporator* (*flash tank, flash chamber*, etc.). Since the temperature of the flash steam produced depends on the pressure maintained in the flash tank, precise temperature control can be maintained where needed. The enthalpy of the flash steam is the same as live steam at that pressure and temperature. The fraction of condensate that will flash when dropped from pressure p_1 to p_2 is

$$\text{flash fraction} = \frac{h_{f,p_1} - h_{f,p_2}}{h_{fg,p_2}}$$

20. DEAERATORS

Deaerators (*deaerating heaters*) are a special category of open feedwater heaters. (Although the pressure is higher than in normal open feedwater heaters, deaerators are still direct-contact heaters.) They remove dissolved gases (e.g., oxygen and carbon dioxide) from feedwater. This is done in a baffled chamber in which the water is broken into droplets and heated by exposure to high-temperature steam. Gas solubility is essentially zero at the saturation temperature.[21] Gases are removed by a vent condenser.

21. ECONOMIZERS

An *economizer* is a water-tube heat exchanger consisting of tubes heated by the last pass of the combustion gases.[22] It increases the temperature of boiler feedwater entering the steam generator and reduces the required combustion energy input by utilizing energy that would otherwise be wasted. The heat transfer takes place in the stack, not in the boiler. Because an economizer can add significant resistance to the flow of stack gases, a forced draft fan is usually required.

The overall coefficient of heat transfer is in the range of approximately 2 to 3 Btu/hr-ft^2-°F (11 to 17 W/m^2·°C) for air passing at 2000 lbm/hr-ft^2 (2.7 kg/s·m^2) to approximately 5 to 7 Btu/hr-ft^2-°F (28 to 40 W/m^2·°C) for air passing at 6000 lbm/hr-ft^2 (8.1 kg/s·m^2).

External corrosion of the economizer heat transfer surface is avoided by keeping the temperature high enough (i.e., above the dew point of the stack gases) to prevent acid formation. This is generally 275 to 350°F (135 to 175°C), depending on the amount of sulfur dioxide in the stack gases. Internal corrosion is avoided by maintaining proper feedwater chemistry.

22. AIR HEATERS

Increasing the temperature of combustion air increases combustion efficiency. While the temperature of incoming air for stoker furnaces is limited to approximately 250 to 350°F (120 to 175°C) by mechanical cooling considerations, at least some portion of combustion air used with pulverized coal must be heated to approximately 600°F (315°C) or more.

An *air heater* (*air preheater*) recovers energy from the stack gases and transfers it to incoming combustion air. *Convection preheaters* (also known as *recuperative heaters*) are conventional heat exchangers that transfer energy through tubes or flat plates. *Regenerative preheaters* use a slowly rotating drum with honeycomb-like passageways that are progressively exposed to incoming air and leaving stack gases. The overall heat transfer coefficient for both types is generally low, varying from approximately 1.5 Btu/hr-ft^2-°F (8.5 W/m^2·°C) for air passing at 2000 lbm/hr-ft^2 (2.7 kg/s·m^2) to approximately 3.0 Btu/hr-ft^2-°F (17 W/m^2·°C) for air passing at 6000 lbm/hr-ft^2 (8.1 kg/s·m^2).

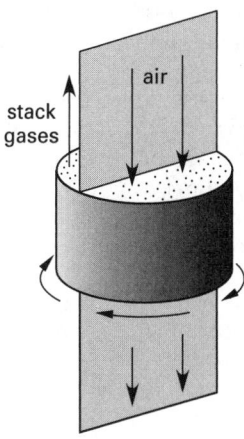

Figure 27.11 *Regenerative Air Preheater*

23. ELECTRICAL GENERATORS

Electrical generators convert energy extracted from the steam into electrical energy. Generator efficiencies are high—around 95% for 1 MW unit, 96% for 10 MW units, to 98% for 100 MW units and above. The *steam rate* (*water rate*) is defined as the steam mass flow rate divided by the generator output in kilowatts.[23] Typical units for steam rate are lbm/kW-hr (kg/kW·h). Values depend greatly on the throttle steam conditions and condenser pressure. For 1000°F (540°C) steam and condenser pressures of 2 to 4 in of mercury, the steam rate is approximately 5.6 to 5.9 lbm/kW-hr (7.1 to 7.4 g/kW·s), with the lower values corresponding to the lower condenser pressures.

24. LOAD AND CAPACITY

Electrical demand varies with time of day, month, and season. The *load curve* is a curve of fluctuating instantaneous load (electrical demand in MW or steam requirements) versus time of day for an average day in a particular season. The *load duration curve* is a curve of load (in MW) versus the number of hours (in a year) the load was experienced. Large power demand variations are met by taking entire power-generating plants on the electrical grid online or offline. Smaller variations are met at the plant level by taking individual boilers and turbines online or offline. Capacity at the plant and regional levels is categorized into *firm capacity* (always available, even in emergencies), *spinning reserve* (floating capacity on the electrical bus), *hot reserve* (in operation but not in service), and *cold reserve* (not in operation but operational).

The *load factor* is the ratio of the peak-to-average loads. The *capacity factor* (*plant factor*) of a unit is the ratio of the average load to rated capacity. The *demand*

---

[21]After heating, the residual oxygen concentration will be approximately 0.005 mg/L. Vacuum deaerators using steam-jet extractors or vacuum pumps without heating leave water with residual concentrations of approximately 0.2 mg/L of oxygen and 2 to 10 mg/L of carbon dioxide.

[22]Originally, economizer tubes have been cast iron to protect them from corrosion. Now, with better furnace oxygen control, cast-iron tubes have been essentially replaced by steel tubes.

[23]For turbines acting as prime movers, the steam flow rate is divided by the horsepower output, with units of lbm/hp-hr.

factor is the ratio of the maximum demand to the total maximum possible connected load (assuming everybody turned everything on at once). The *output factor* (*use factor*) accounts for downtime and is the ratio of actual energy output during a particular period to the output that would have occurred during that period if the plant had been operating at full load rating. The *operation factor* is a ratio of operational time to total time (including downtime).

A *base-load unit* is a unit or plant intended to satisfy a constant fixed demand by running continuously. A base-load unit will usually have a capacity factor of 50% or more. A *peaking unit* is intended to satisfy only peak loads.

25. CONTROL AND REGULATION OF TURBINES

Traditionally, turbines in the United States have been designed to operate at *constant throttle* pressure. Base-load units were not intended to drop below the minimum load they were designed for, typically 25 to 30% of capacity. However, daily cycling of base-load units (down to as low as 10%) has now become common.

Sliding load units are turbines that can infinitely adjust their output, perhaps down to as low as 10% of the rated capacities. With the commonly used *sliding pressure operation*, also known as *variable throttle pressure operation*, steam generator pressure is varied to change the turbine output. Since temperature and pressure are related, this subjects the turbine and other equipment to thermal-stress damage due to rapid and repeated temperature swings.

An alternative to using sliding pressure operation is using control valves to redirect the steam flow through different components.[24] The amount of steam entering the turbine is controlled by *boiler throttle valves* or *turbine (steam admission) throttle valves*. For example, at low loads, steam may be routed around the second superheater; at intermediate loads, superheated steam may pass through a pressure reduction valve; and at high loads, the turbine may receive fully superheated steam at full pressure. The process of implementing this control method in older plants is known as *cycle redesign*, *backfitting*, and *retrofitting*.

Different methods of governing can be used to maintain the turbine operating point. With *throttle control* (also known as *cut-off governing* and *full-arc admission*), the pressure and temperature of the steam admitted to the turbine are constant and all turbine nozzles are operational, but valves control the steam flow rate through each nozzle.[25] With *nozzle control*, also known as *partial-arc admission*, the steam flow rate to the turbine is controlled by "cutting out" some of the steam nozzles. With *bypass governing*, steam pressure and temperature are constant, but the stage at which steam is introduced to the turbine(s) is changed. With *blast governing*, the steam is not admitted to the turbine continuously. Rather, the steam flow is repeatedly turned on and off in a ratio that achieves the desired duty cycle.

At the steam level, variations in output are achieved by varying the firing (fueling) rate, by varying the input air and/or draft, and by using separately fired superheaters, desuperheaters, movable burner heads, and other techniques.

At the individual turbine, steam accumulators can supply short-term peak loads.[26] This method is commonly used in once-through boilers, but may also be used with sliding pressure operation to supply all low-load steam.

PRACTICE PROBLEMS

1. What is the isentropic efficiency of a process that expands dry steam from 100 psia to 3 psia (700 kPa to 20 kPa) and 90% quality?

2. A 5000 kW steam turbine uses 200 psia (1.5 MPa) steam with 100°F (50°C) of superheat. The condenser is at 1 in Hg (3.4 kPa) absolute. (a) What is the water rate at full load? (b) If the actual load is only 2500 kW and the steam is throttled to reduce the availability, what is the loss in available energy per unit steam mass?

3. A 10,000 kW steam turbine operates on 400 psia (3.0 MPa), 750°F (420°C) dry steam, expanding to 2 in Hg (6.8 kPa) absolute. What is the adiabatic heat drop available for power production?

4. A 750 kW steam turbine has a water rate of 20 lbm/kW-hr (2.5×10^{-3} kg/kW·s). Steam with 50°F (30°C) of superheat is expanded from 165 psia (1.0 MPa absolute) to 26 in Hg (90 kPa) absolute. 65°F (18°C) cooling water is available. The terminal temperature difference is zero. Find the quantity of cooling water required.

5. 332,000 lbm/hr (41.8 kg/s) of 81°F (27°C) water enters a two-pass, counterflow, closed feedwater heater. The feedwater heater is constructed of 1850 ft² (170 m²) of ⅝ in (15.8 mm) copper tubing. Saturated steam is bled from a turbine at 4.45 psia (30 kPa) and condenses to saturated liquid. The heated water leaves at

---

[24] Actually, most modern plants use a combination of sliding pressure and backfitting, a system known as *hybrid throttle pressure operation*.

[25] The terminology is confusing. A *throttle valve* controls the mass flow rate, but a *throttling valve* changes the pressure. Actually, partly closing any valve will result in reductions in both the mass flow rate and the pressure. Phrases such as "...at 85% throttle..." usually refer to the capacity ratio, not the mass flow rate.

[26] A volume of saturated water in the accumulator at the system pressure is kept heated by boiler steam. When the load peaks and the system pressure drops (below the saturation pressure), some of the accumulator water flashes into steam, supplying the majority of the increase in demand.

150°F (65°C) with an enthalpy of 1100 Btu/lbm (2.56 MJ/kg). (a) What is the overall heat transfer coefficient? (b) What is the steam extraction rate?

6. A two-pass surface condenser constructed of 1 in (25.4 mm) BWG tubing receives 82,000 lbm/hr (10.3 kg/s) of steam from a turbine. Steam enters the condenser with an enthalpy of 980 Btu/lbm (2.280 MJ/kg). The condenser operates at a pressure of 1 in Hg (3.4 kPa) absolute. Water is circulated at 8 fps (2.4 m/s) through an equivalent length of 120 ft (36 m) of extra strong 30 in (76.2 cm) steel pipe. An additional head loss of 6 in wg (1.5 kPa) is incurred in the intake screens. (a) What is the head added by the circulating water pump? (b) If the water temperature increases 10°F (5.6°C), what is the circulation rate of cooling water?

7. 100 lbm/hr (0.013 kg/s) of 60°F (16°C) water are turned into 14.7 psia (101.3 kPa) saturated steam in an electric boiler. Radiation losses are 35%. What is the cost if electricity is $0.04 per kW-hr?

8. A gas burner produces 250 lbm/hr (0.032 kg/s) of 98% dry steam at 40 psia (300 kPa) from 60°F (16°C) feedwater. The fuel gas enters at 80°F (26°C) and 4 in Hg (13.6 kPa) and has a heating value of 550 Btu/ft³ (20.5 MJ/m³) at standard industrial conditions. The barometric pressure is 30.2 in Hg (102.4 kPa). 13.5 ft³/min (6.4 L/s) of fuel gas are consumed. What is the efficiency of the boiler?

9. A boiler evaporates 8.23 lbm (8.23 kg) of 120°F (50°C) water per pound (per kilogram) of coal fired, producing 100 psia (700 kPa) saturated steam. The coal is 2% moisture by weight as fired, and dry coal is 5% ash. 12% of the coal is removed from the ash pit. (Ash has the same composition as unfired, dry coal.) Coal is initially at 60°F (16°C), and combustion occurs at 14.7 psia (101.3 kPa). The combustion products leave at 600°F (315°C). What is the efficiency of the boiler?

10. 500 psia (3.5 MPa) steam is superheated to 1000°F (500°C) before expanding through a 75% efficient turbine to 5 psia (30 kPa). No subcooling occurs. The pump work is negligible compared to the 200 MW generated. (a) What quantity of steam is required? (b) What heat is removed by the condenser?

11. 191,000 lbm/hr (24 kg/s) of 635°F (335°C) combustion gases flow through a 20 ft (6 m) wide boiler stack whose front and back plates are 5 ft-10 in (1.78 m) apart. An integral economizer is being designed to heat water from 212°F to 285°F (100°C to 140°C) by dropping the stack gas temperature to 470°F (240°C).

Layers of 24 tubes with dimensions 0.957 in (24.3 mm) I.D., 1.315 in (33.4 mm) O.D., and 20 ft (6 m) length will be placed on a 2.315 in (58.8 mm) pitch in horizontal banks. The overall coefficient of heat transfer for the tubes is 10 Btu/hr-ft²-°F (57 W/m²·°C). How many 24-tube layers are required?

12. (*Time limit: one hour*) Water is used in an adiabatic steam desuperheater. 1000 lbm/hr (0.13 kg/s) of 200 psia (1.5 MPa), 600°F (300°C) steam enters with negligible velocity. 50 lbm/hr (0.0063 kg/s) of 82°F (28°C) water enters with negligible velocity. 100 psia (700 kPa) steam leaves the desuperheater at 2000 ft/sec (600 m/s). (a) What is the temperature of the leaving steam? (b) What is the quality of the leaving steam?

13. (*Time limit: one hour*) (See diagram.) Waste steam at 400°F (200°C) and 100 psia (700 kPa) was originally used only for heating cold water. Cold water entered at 70°F (21°C) and 60 psia (400 kPa). 2000 lbm/hr (0.25 kg/s) of hot water at 180°F (80°C) and 20 psia (150 kPa) were produced. Now, the same quantity of steam is to be expanded through a low-pressure turbine. The low-pressure turbine has an isentropic efficiency of 60% and a mechanical efficiency of 96%. The steam will then flow through a mixing heater. A pressure drop of 5 psi (30 kPa) occurs through the heater. The heater output must remain at 180°F (80°C) and 20 psia (350 kPa), but the output at point F may decrease. (a) What power is developed in the turbine? (b) What is the flow rate at point F?

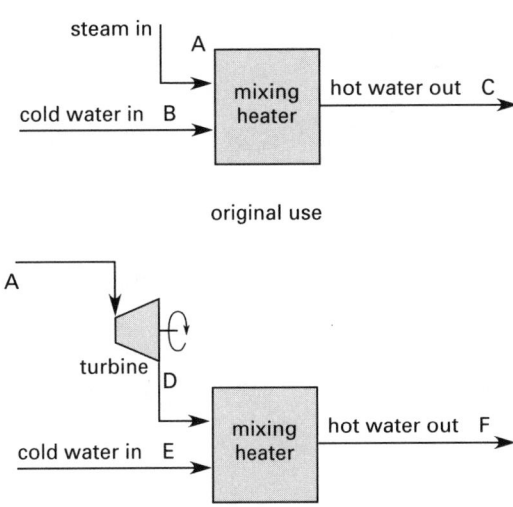

original use

proposed use of waste steam

28 Vapor Power Cycles

1. General Vapor Power Cycles 28-1
2. Sign Convention 28-2
3. Thermal Efficiencies 28-2
4. Cycle Designations 28-2
5. Carnot Cycle 28-2
6. Basic Rankine Cycle 28-3
7. Rankine Cycle with Superheat 28-4
8. Rankine Cycle with Superheat and Reheat . 28-5
9. Rankine Cycle with Regeneration
    (Regenerative Cycle) 28-7
10. Supercritical Cycle 28-7
11. Binary Cycle 28-7
    Practice Problems 28-8

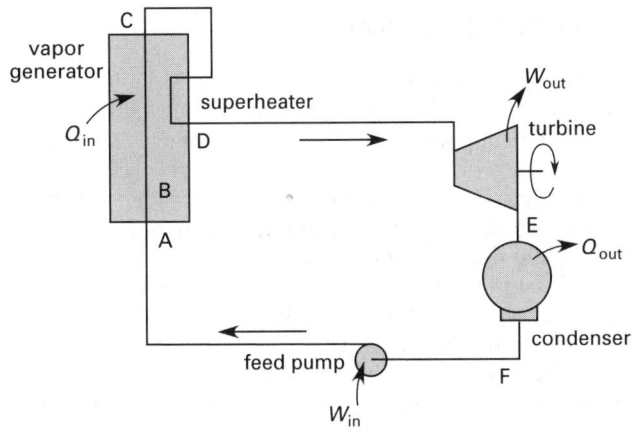

Figure 28.1 *Simplified Vapor Power Cycle*

Nomenclature

h	enthalpy	Btu/lbm	kJ/kg
HR	heat rate	Btu/kW-hr	MJ/kW·h
p	pressure	lbf/ft^2	Pa
Q	heat	Btu	kJ
RF	reheat factor	–	–
s	entropy	Btu/lbm-°R	kJ/kg·K
T	temperature	°R	K
W	work	Btu	kJ
x	quality	–	–
y	bleed fraction	–	–

Symbols

η	efficiency	–	–
v	specific volume	ft^3/lbm	m^3/kg

Subscripts

f	fluid (liquid)
fg	fluid-to-gas (vaporization)
s	isentropic
th	thermal

1. GENERAL VAPOR POWER CYCLES

The most general form of a vapor (almost always steam) power cycle incorporates a vapor generator, turbine, condenser, and feed pump as illustrated in Fig. 28.1.

The following thermodynamic processes take place in a typical steam vapor power cycle.

A to B: Subcooled water is heated to the saturation temperature corresponding to the pressure in the steam generator.

B to C: Saturated water is vaporized in the steam generator, producing saturated vapor.

C to D: An optional superheating process increases the steam temperature and enthalpy.

D to E: Steam expands in the turbine and does work as it decreases in temperature, pressure, and quality.

E to F: Vapor is returned to a liquid phase in the condenser.

F to A: The pressure of the liquid water is brought up to the steam generator pressure by the boilerfeed pump.

After making a full cycle, the steam is brought back to the exact same thermodynamic conditions it started the cycle with. While the entropy of the steam returns to its original value after each full cycle, the entropy of the environment increases due to the heat transfers (losses, etc.) from the steam generator and condenser. Thus, even if the pump compression and turbine expansion are isentropic processes, the cycle does irreversible "damage" to the entropy of the environment.

2. SIGN CONVENTION

The sign convention normally adhered to in textbooks for changes in thermodynamic properties is not followed for pump work. Since the pump does work on the water, the pump work would be negative according to the traditional sign convention. However, since the power industry customarily refers to pump work as a positive quantity, that convention is followed in this chapter.[1]

3. THERMAL EFFICIENCIES

The *thermal efficiency*, η_{th}, of a power cycle is the ratio of net work out (or net heat in) divided by the heat input.[2] The upper limit of thermal efficiency is the thermal efficiency of the theoretical Carnot cycle (Eq. 28.8). Typical actual values of thermal efficiency depend to a great extent on the operating temperature and pressure and on the degree of cycle sophistication. Cycles with high-pressure superheat, reheat, and multiple stages of regeneration will be far more efficient in converting thermal energy to electricity than the simple low-pressure cycles that prevailed before 1960. However, the best single-cycle plants seldom have thermal efficiencies in excess of 40%.

$$\eta_{th} = \frac{W_{out} - W_{in}}{Q_{in}}$$
$$= \frac{Q_{in} - Q_{out}}{Q_{in}} \qquad \textit{28.1}$$

The *heat rate*, HR (or *station rate*), is the ratio of the total energy input divided by the net electrical output. This is not merely the reciprocal of the thermal efficiency because of the units for heat rate. Heat rate has units of Btu/kW-hr (MJ/kW·h). It can be calculated from the thermal efficiency and the appropriate conversion factor.[3]

$$\mathrm{HR} = \frac{Q_{in}}{W_{out} - W_{in}} = \frac{3.6 \; \frac{\mathrm{MJ}}{\mathrm{kW \cdot h}}}{\eta_{th}} \qquad [\mathrm{SI}] \quad \textit{28.2(a)}$$

$$\mathrm{HR} = \frac{Q_{in}}{W_{out} - W_{in}} = \frac{3413 \; \frac{\mathrm{Btu}}{\mathrm{kW\text{-}hr}}}{\eta_{th}} \qquad [\mathrm{U.S.}] \quad \textit{28.2(b)}$$

For heat rates expressed in Btu/hp-hr, the relationship is

$$\mathrm{HR} = \frac{2543 \; \frac{\mathrm{Btu}}{\mathrm{hp\text{-}hr}}}{\eta_{th}} \qquad \textit{28.3}$$

---

[1]Nobody says "The pump work is negative four Btus per pound."
[2]Only a *cycle* (i.e., an integrated system of devices) can have a *thermal* efficiency. A single device (e.g., a pump) can have mechanical and isentropic efficiencies, but it cannot have a thermal efficiency.
[3]The conversion between kW and Btu/hr is given in some sources as 3412. The conversion between hp and Btu/hr is given in some sources as 2544.

4. CYCLE DESIGNATIONS

There are numerous descriptive terms used to describe categories of cycles. If a condenser is not part of the power-generating process, the steam cannot expand below atmospheric pressure. A cycle without a condenser is known as a *noncondensing cycle*. The steam may be used for process heating or feedwater heating, or may simply be discharged to the atmosphere. Steam may condense elsewhere (in a feedwater heater, for example). It is the absence of a condenser that identifies a noncondensing cycle, not the absence of condensation.

If there is a condenser, the cycle is known as a *condensing cycle*. If there are no bleeds (i.e., no extraction of steam prior to steam entering the condenser), the cycle is known as *straight condensing*. In an *extraction cycle*, steam is withdrawn from the turbine, usually for feedwater heating, although the bleed steam can also be used for other process heating.

5. CARNOT CYCLE

The Carnot cycle is an ideal power cycle that is impractical to implement. However, its theoretical work output sets the maximum attainable from any heat engine, as evidenced by the isentropic (reversible) processes between states (D and A) and (B and C) in Fig. 28.2. The working fluid in a Carnot cycle is irrelevant.

The processes involved are:

A to B: isothermal expansion of saturated liquid to saturated vapor

B to C: isentropic expansion of vapor

C to D: isothermal compression of vapor

D to A: isentropic compression

The properties at the various states can be found from the following solution methods. The letters f and g stand for saturated fluid and saturated gas, respectively. They do not correspond to states F and G on any cycle diagram. The Carnot cycle can be evaluated by working around, finding T, p, x, h, and s at each node.

At A: From the property table for T_{high}, read p_A, h_A, and s_A for a saturated fluid.

At B: $T_B = T_A$; $p_B = p_A$; $x = 1$; h_B is read from the table as h_g; s_B read as s_g.

At C: Either p_C or T_C must be known. Read p_C from the T_C line on the property table or vice versa; $x_C = (s_B - s_f)/s_{fg}$; $h_C = h_f + x_C h_{fg}$; $s_C = s_B$.

At D: $T_D = T_C$; $p_D = p_C$; $x_D = (s_A - s_f)/s_{fg}$; $h_D = h_f + x_D h_{fg}$; $s_D = s_A$.

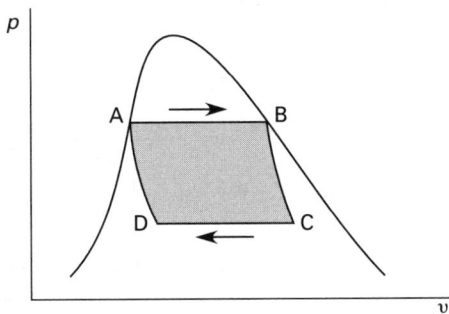

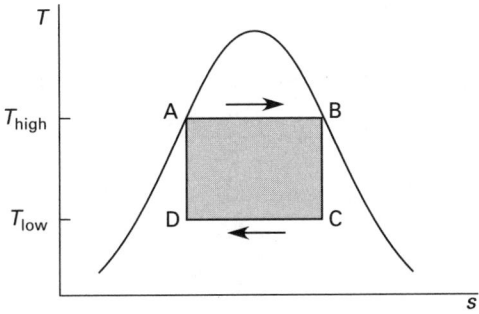

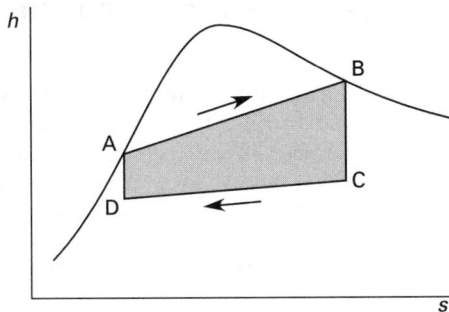

Figure 28.2 *Carnot Cycle*

The turbine and pump work terms per unit mass are

$$W_{\text{turbine}} = h_B - h_C \qquad \textbf{28.4}$$

$$W_{\text{pump}} = h_A - h_D \qquad \textbf{28.5}$$

The heat flows per unit mass into and out of the system are

$$Q_{\text{in}} = T_{\text{high}}(s_B - s_A) = h_B - h_A \qquad \textbf{28.6}$$

$$Q_{\text{out}} = T_{\text{low}}(s_C - s_D) = T_{\text{low}}(s_B - s_A) = h_C - h_D \quad \textbf{28.7}$$

The thermal efficiency of the entire cycle is

$$
\eta_{\text{th}} = \frac{Q_{\text{in}} - Q_{\text{out}}}{Q_{\text{in}}} = \frac{W_{s,\text{turbine}} - W_{s,\text{pump}}}{Q_{\text{in}}}
$$
$$
= \frac{(h_B - h_C) - (h_A - h_D)}{h_B - h_A} = \frac{T_{\text{high}} - T_{\text{low}}}{T_{\text{high}}} \qquad \textbf{28.8}
$$

If isentropic efficiencies for the pump and turbine are known, proceed as follows. Calculate all properties assuming that the efficiencies are 100%. Then, modify h_C and h_A as given in Eqs. 28.9 and 28.10. Use the new values to find the actual thermal efficiency of the cycle.

$$h'_C = h_B - \eta_{s,\text{turbine}}(h_B - h_C) \qquad \textbf{28.9}$$

$$h'_A = h_D + \frac{h_A - h_D}{\eta_{s,\text{pump}}} \qquad \textbf{28.10}$$

$$W'_{\text{turbine}} = h_B - h'_C \qquad \textbf{28.11}$$

$$W'_{\text{pump}} = h'_A - h_D \qquad \textbf{28.12}$$

6. BASIC RANKINE CYCLE

The basic Rankine cycle is similar to the Carnot cycle except that the compression process occurs in the liquid region. The Rankine cycle is closely approximated in steam turbine plants. The efficiency of the Rankine cycle is lower than that of a Carnot cycle operating between the same temperature limits because the mean temperature at which heat is added to the system is lower than T_{high}.

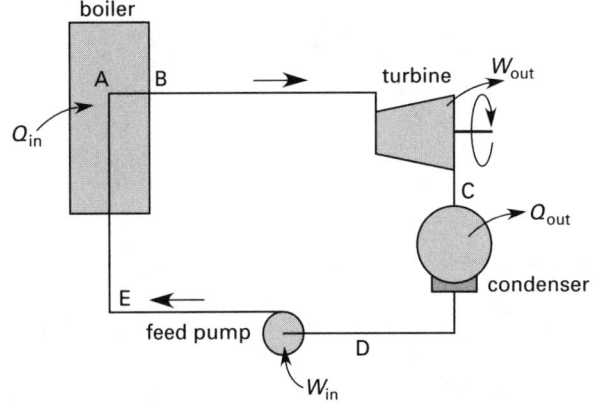

Figure 28.3 *Basic Rankine Cycle*

The processes used in the basic Rankine cycle are:

A to B: vaporization in the boiler

B to C: adiabatic expansion in the turbine

C to D: condensation

D to E: adiabatic compression to boiler pressure

E to A: heating liquid to saturation temperature

The properties at each point can be found from the following procedure. The letters f and g refer to saturated fluid and saturated gas, respectively. They do not correspond to locations F and G on any diagram. Usually T_{high} and T_{low} are known. The procedure is to work around the cycle, finding $T, p, x, h,$ and s at each node.

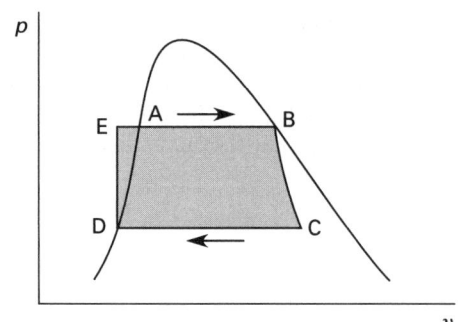

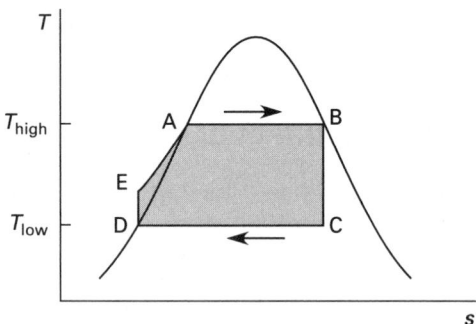

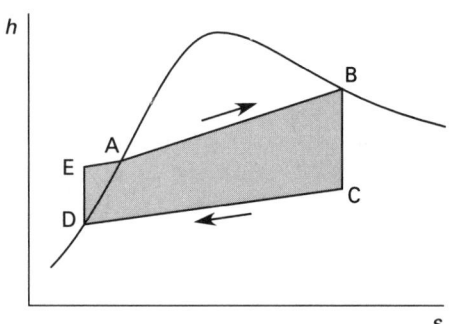

Figure 28.4 *Basic Rankine Cycle*

At A: From the property table for T_{high}, read p_A, h_A, and s_A for a saturated liquid.

At B: $T_B = T_A$; $p_B = p_A$; $x = 1$; h_B is read from the table as h_g; s_B read as s_g.

At C: Either p_C or T_C must be known. Read p_C from the T_C line on the property table or vice versa; $x_C = (s_B - s_f)/s_{fg}$; $h_C = h_f + x_C h_{fg}$; $s_C = s_B$.

At D: $T_D = T_C$; $p_D = p_C$; $x_D = 0$; h_D is read as h_f; s_D read as s_f; v_D read as v_f.

At E: $p_E = p_A$; $s_E = s_D$; $h_E = h_D + W_{pump} = h_D + v_D(p_E - p_D)$ (watch units); T_E is found as the saturation temperature for a liquid with enthalpy equal to h_E.

The per unit mass work and heat flow terms are

$$W_{turbine} = h_B - h_C \qquad 28.13$$

$$W_{pump} = h_E - h_D \approx v_D(p_E - p_D)$$
$$\text{[watch units]} \quad 28.14$$

$$Q_{in} = h_B - h_E \qquad 28.15$$

$$Q_{out} = h_C - h_D \qquad 28.16$$

The thermal efficiency of the entire cycle is

$$\eta_{th} = \frac{Q_{in} - Q_{out}}{Q_{in}} = \frac{W_{turbine} - W_{pump}}{Q_{in}}$$
$$= \frac{(h_B - h_C) - (h_E - h_D)}{h_B - h_E} \qquad 28.17$$

If isentropic efficiencies for the pump and the turbine are known, calculate all properties as if these efficiencies were 100%. Then use the following relationships to modify h_C and h_E. Use the new values to recalculate the thermal efficiency.

$$h'_C = h_B - \eta_{s,turbine}(h_B - h_C) \qquad 28.18$$

$$h'_E = h_D + \frac{h_E - h_D}{\eta_{s,pump}} \qquad 28.19$$

$$W'_{turbine} = h_B - h'_C \qquad 28.20$$

$$W'_{pump} = h'_E - h_D \qquad 28.21$$

$$Q'_{in} = h_B - h'_E \qquad 28.22$$

7. RANKINE CYCLE WITH SUPERHEAT

Superheating occurs when heat in excess of that required to produce saturated vapor is added to the water. Superheat is used to raise the vapor above the critical temperature, to raise the mean effective temperature at which heat is added, and to keep the expansion primarily in the vapor region to reduce wear on the turbine blades. A maximum practical metallurgical limit on superheat is approximately 1150°F (625°C).

The processes in the Rankine cycle with superheat are similar to the basic Rankine cycle.

A to B: heating water to the saturation temperature in the boiler

B to C: vaporization of water in the boiler

C to D: superheating steam in the superheater region of the boiler

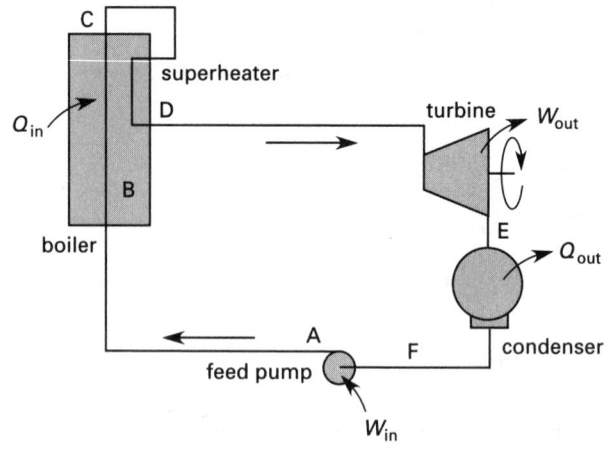

Figure 28.5 *Rankine Cycle with Superheat*

D to E: adiabatic expansion in the turbine

E to F: condensation

F to A: adiabatic compression to boiler pressure

The properties at each point can be found from the following procedure. The subscripts f and g refer to saturated fluid and saturated gas, respectively. They do not correspond to any points on the property plot diagram.

At A: Point A is covered below. Start the analysis at point B.

At B: From the property table for T_B or p_B, read h_B and s_B for a saturated fluid.

At C: $T_C = T_B$; $p_C = p_B$; $x = 1$; h_C is read as h_g; s_C read as s_g.

At D: T_D usually is known; $p_D = p_C$; h_D is read from superheat tables with T_D and p_D known. Same for s_D and v_D.

At E: T_E usually is known; p_E is read from the saturated table for T_E; $x_E = (s_E - s_f)/s_{fg}$; $h_E = h_f + x_E h_{fg}$; $s_E = s_D$. If point E is in the superheated region, the Mollier diagram should be used to find the properties at point E.

At F: $T_F = T_E$; $p_F = p_E$; $x = 0$; h_F, s_F, and v_F are read as h_f, s_f, and v_f from the saturated table.

At A: $p_A = p_B$; $h_A = h_F + v_F(p_A - p_F)$ (watch units); $s_A = s_F$. T_A is equal to the saturation temperature for a liquid with enthalpy equal to h_A.

The per unit mass work and heat flow terms are

$$W_{\text{turbine}} = h_D - h_E \qquad 28.23$$

$$W_{\text{pump}} = h_A - h_F = v_F(p_A - p_F) \qquad 28.24$$
$$[\text{watch units}]$$

$$Q_{\text{in}} = h_D - h_A \qquad 28.25$$

$$Q_{\text{out}} = h_E - h_F \qquad 28.26$$

The thermal efficiency of the entire cycle is

$$\eta_{\text{th}} = \frac{Q_{\text{in}} - Q_{\text{out}}}{Q_{\text{in}}} = \frac{W_{\text{turbine}} - W_{\text{pump}}}{Q_{\text{in}}}$$
$$= \frac{(h_D - h_A) - (h_E - h_F)}{h_D - h_A} \qquad 28.27$$

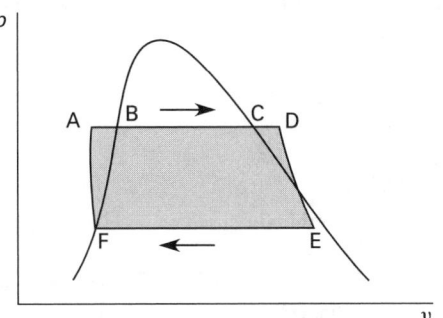

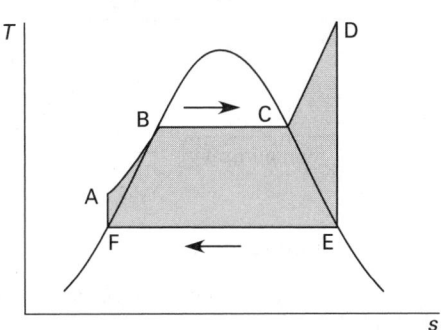

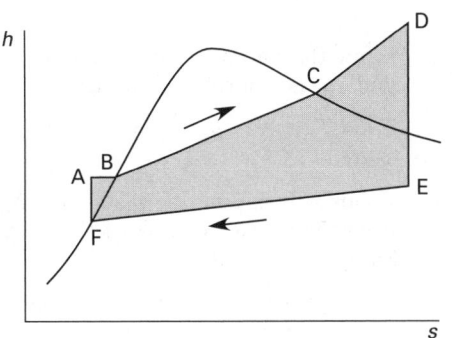

Figure 28.6 *Rankine Cycle with Superheat*

If pump and turbine efficiencies are known, calculate all quantities as if those efficiencies were 100%. Then modify h_E and h_A prior to recalculating the thermal efficiency.

$$h'_E = h_D - \eta_{s,\text{turbine}}(h_D - h_E) \qquad 28.28$$

$$h'_A = h_F + \frac{h_A - h_F}{\eta_{s,\text{pump}}} \qquad 28.29$$

$$W'_{\text{turbine}} = h_D - h'_E \qquad 28.30$$

$$W'_{\text{pump}} = h'_A - h_F \qquad 28.31$$

$$Q'_{\text{in}} = h_D - h'_A \qquad 28.32$$

8. RANKINE CYCLE WITH SUPERHEAT AND REHEAT

Reheat is used to increase the mean effective temperature at which heat is added without producing significant expansion in the liquid-vapor region. The analysis given assumes that $T_D = T_F$, as is usually the case. It is possible, however, that the two temperatures will be different.

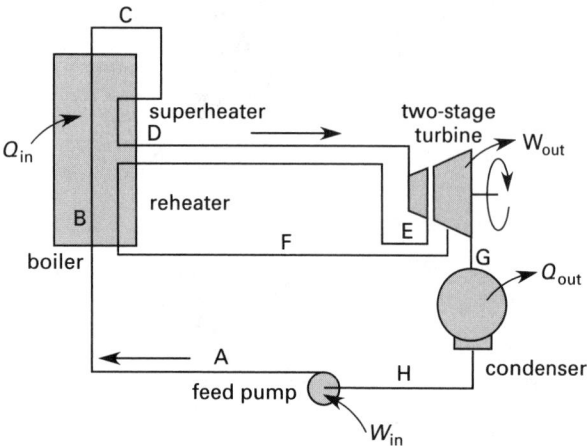

Figure 28.7 Reheat Cycle

The properties at each point can be found from the following procedure.

At A: Point A is covered below. Start the analysis at point B.

At B: From the vapor table for T_B, read p_B, h_B, and s_B as a saturated fluid.

At C: $T_C = T_B$; $p_C = p_B$; $x_C = 1$; h_C is read as h_g; s_C read as s_g.

At D: T_D usually is known; $p_D = p_C$; h_D is read from superheat tables with T_D and p_D known. Same for s_D and v_D.

At E: p_E usually is known; T_E is read from the property table for p_E; $s_E = s_D$; $x_E = (s_E - s_f)/s_{fg}$; $h_E = h_f + x_E h_{fg}$. (Use the Mollier diagram if superheated.)

At F: T_F usually is known; $p_F = p_E$; h_F is read from superheat tables with T_F and p_F known. Same for s_F and v_F.

At G: T_G usually is known; p_G is read from property tables for T_G; $s_G = s_F$; $x_G = (s_G - s_f)/s_{fg}$; $h_G = h_F + x_G h_{fg}$. (Use the Mollier diagram if superheated.)

At H: $T_H = T_G$; $p_H = p_G$; $x_H = 0$; h_H, s_H and v_H are read as h_f, s_f, and v_f.

At A: $p_A = p_B$; $h_A = h_H + v_H(p_A - p_H)$ (watch units); $s_A = s_H$. T_A is the saturation temperature for a liquid with enthalpy equal to h_A.

The per unit mass work and heat flow terms are

$$W_{\text{turbine}} = (h_D - h_E) + (h_F - h_G) \qquad 28.33$$

$$W_{\text{pump}} = (h_A - h_H) = v_H(p_A - p_H) \qquad 28.34$$

$$Q_{\text{in}} = (h_D - h_A) + (h_F - h_E) \qquad 28.35$$

$$Q_{\text{out}} = h_G - h_H \qquad 28.36$$

The thermal efficiency of the entire cycle is

$$\eta_{\text{th}} = \frac{W_{\text{turbine}} - W_{\text{pump}}}{Q_{\text{in}}} = \frac{Q_{\text{in}} - Q_{\text{out}}}{Q_{\text{in}}}$$

$$= \frac{(h_D - h_A) + (h_F - h_E) - (h_G - h_H)}{(h_D - h_A) + (h_F - h_E)} \qquad 28.37$$

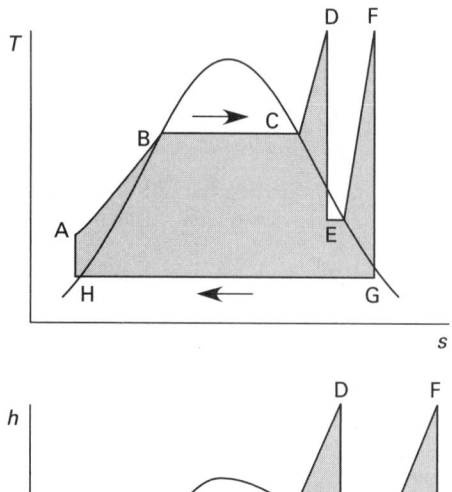

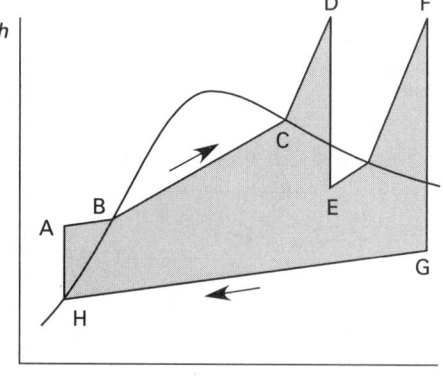

Figure 28.8 Reheat Cycle

If the reheat cycle is specified with isentropic efficiencies for the pump and the turbine, calculate all of the preceding quantities as if the efficiencies were 100%. Then modify h_E, h_G, and h_A before recalculating the thermal efficiency.

$$h'_E = h_D - \eta_{s,\text{turbine}}(h_D - h_E) \qquad 28.38$$

$$h'_G = h_F - \eta_{s,\text{turbine}}(h_F - h_G) \qquad 28.39$$

$$h'_A = h_H + \frac{h_A - h_H}{\eta_{s,\text{pump}}} \qquad 28.40$$

$$W'_{\text{turbine}} = (h_D - h'_E) + (h_F - h'_G) \qquad 28.41$$

$$W'_{\text{pump}} = h'_A - h_H \qquad 28.42$$

$$Q'_{\text{in}} = (h_D - h'_A) + (h_F - h'_E) \qquad 28.43$$

The *reheat factor*, RF, is the ratio of the actual turbine work (with all of the reheat stages) in a multi-stage expansion to the ideal turbine work assuming a one-stage, isentropic expansion from the same entering conditions to the same condenser pressure. The fractional improvement due to the reheat is RF−1. The reheat factor depends greatly on the steam properties and complexity of the cycle. Values are typically in the range of 1.05 to 1.10.

9. RANKINE CYCLE WITH REGENERATION (REGENERATIVE CYCLE)

If the mean effective temperature at which heat is added can be increased, the overall thermal efficiency of the cycle will be improved. This can be accomplished by raising the temperature at which the condensed fluid enters the boiler.

In the regenerative cycle, portions of the steam in the turbine are withdrawn at various points. Heat is transferred from this bleed stream to the feedwater coming from the condenser. Although only two bleeds are used in the following analysis, seven or more exchange locations can be used in a large installation. The regenerative cycle always involves superheating, although conceptually it does not need to.

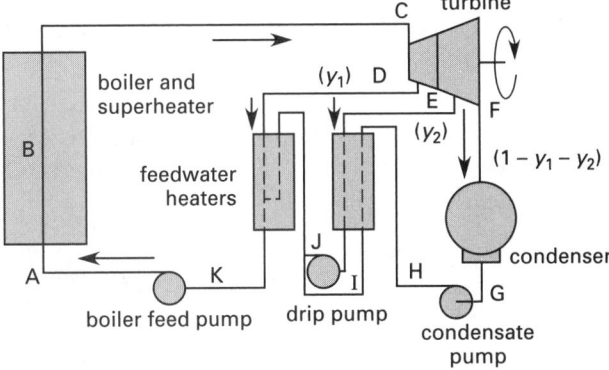

Figure 28.9 *Regenerative Cycle with Two Feedwater Heaters*

In the following analysis, y_1 is the first bleed fraction, and y_2 is the second bleed fraction.

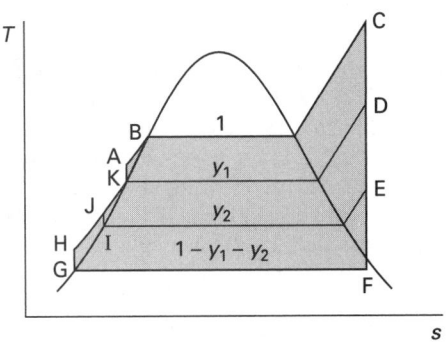

Figure 28.10 *Regenerative Cycle Property Plot*

The ideal per unit mass work and heat flow terms for the regenerative cycle are

$$W_{\text{turbine}} = (h_C - h_D) + (1 - y_1)(h_D - h_E)$$
$$+ (1 - y_1 - y_2)(h_E - h_F) \qquad 28.44$$

$$W_{\text{pumps}} = (h_A - h_K) + y_2(h_J - h_I)$$
$$+ (1 - y_1 - y_2)(h_H - h_G) \qquad 28.45$$

$$Q_{\text{in}} = h_C - h_A \qquad 28.46$$

$$Q_{\text{out}} = (1 - y_1 - y_2)(h_F - h_G) \qquad 28.47$$

The thermal efficiency of the entire cycle is

$$\eta_{\text{th}} = \frac{Q_{\text{in}} - Q_{\text{out}}}{Q_{\text{in}}}$$
$$= \frac{(h_C - h_A) - (1 - y_1 - y_2)(h_F - h_G)}{h_C - h_A} \qquad 28.48$$

10. SUPERCRITICAL CYCLE

The thermal efficiency is increased by raising the average temperature at which heat is added. This is evident from Eq. 28.8 that shows the ideal thermal efficiency depends on the highest temperature achieved in the cycle. Therefore, some modern plants are designed to achieve supercritical temperatures. The operating temperatures are limited only by metallurgical considerations.

Figure 28.11 shows that no heat is added at constant temperature. Reheating (resuperheating) is used to keep the steam quality high when it expands from point B.

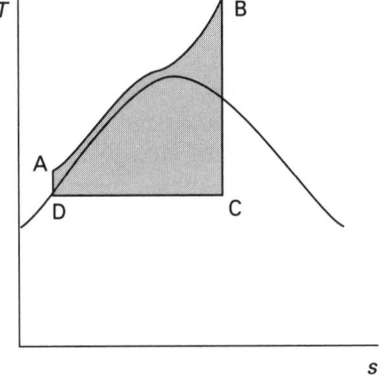

Figure 28.11 *Supercritical Rankine Cycle Property Plot*

11. BINARY CYCLE

The binary cycle utilizes two different fluids, such as mercury and water, achieving conditions unobtainable with a single working fluid.[4] The binary cycle is essentially two Rankine cycles, and Rankine procedures should be used to evaluate it.

---

[4]The mercury binary cycle is of academic interest. However, there are no commercial installations using this design.

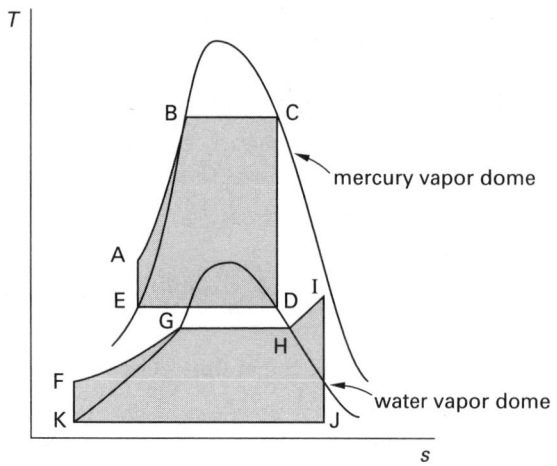

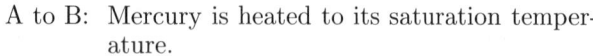

Figure 28.12 *Binary Cycle*

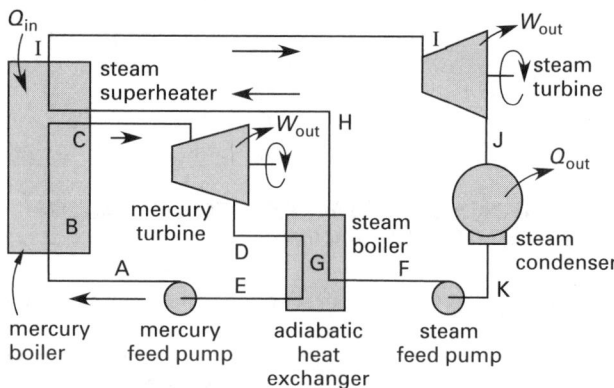

Figure 28.13 *Binary Cycle*

A to B: Mercury is heated to its saturation temperature.

B to C: Mercury is vaporized in the boiler.

C to D: Mercury expands adiabatically in the turbine.

D to E: Mercury condenses in the combination condenser-boiler.

E to A: Mercury is compressed adiabatically.

F to G: Water is heated to its saturation temperature.

G to H: Water vaporizes in a boiler.

H to I: Steam is superheated.

I to J: Steam expands adiabatically in the turbine.

J to K: Water condenses.

K to F: Water is compressed adiabatically.

If m mass units of mercury flow for every mass unit of steam and the heat transfer between the mercury and steam is adiabatic, the thermal efficiency of the binary cycle is

$$\eta_{th} = \frac{W_{turbines} - W_{pumps}}{Q_{in}}$$

$$= \frac{\begin{array}{c} m(h_C - h_D) + (h_I - h_J) \\ - m(h_A - h_E) - (h_F - h_K) \end{array}}{m(h_C - h_A) + (h_I - h_H)} \qquad 28.49$$

PRACTICE PROBLEMS

1. A steam cycle operates between 650°F and 100°F (340°C and 38°C). What is the maximum possible thermal efficiency?

2. A steam Carnot cycle operates between 650°F and 100°F (340°C and 38°C). The turbine and compressor (pump) isentropic efficiencies are 90% and 80%, respectively. What is the thermal efficiency?

3. A steam turbine cycle produces 600 MWe of energy. The condenser load is 3.07×10^9 Btu/hr (900 MW). What is the thermal efficiency?

4. A steam Rankine cycle operates with 100 psia (700 kPa) saturated steam that is reduced to 1 atm through expansion in a turbine with an isentropic efficiency of 80%. Water is at 80°F (27°C) and 1 atm when it enters the boilerfeed pump. The pump's isentropic efficiency is 60%. What is the cycle's thermal efficiency?

5. A turbine and the boilerfeed pumps in a reheat cycle have isentropic efficiencies of 88% and 96%, respectively. The cycle starts with water at 60°F (16°C) and produces 600°F (300°C), 600 psia (4 MPa) steam. The steam is reheated when its pressure drops during the first expansion to 20 psia (150 kPa). What is the thermal efficiency of the cycle?

6. The cycle in Prob. 5 is modified to include a bleed of 270°F (130°C) steam (within the second expansion) for feedwater heating in a closed feedwater heater. The condenser pressure is unchanged. Steam leaves the feedwater heater as saturated liquid. The terminal temperature difference in the heater is 6°F (3°C). Condensate and drip pump work are both negligible. What is the thermal efficiency of the cycle?

7. (*Time limit: one hour*) A reheat steam cycle operates as shown. The pump work between points F and G is 0.15 Btu/lbm (0.3 kJ/kg).

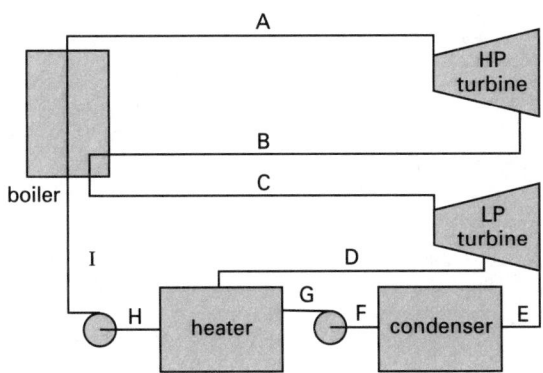

At A: 900 psia (6.2 MPa)
       800°F (420°C)

At B: 200 psia (1.5 MPa)
       1270 Btu/lbm (2960 kJ/kg)

At C: 190 psia (1.4 MPa)
       800°F (420°C)

At D: 50 psia (350 kPa)
       1280 Btu/lbm (2980 kJ/kg)

At E: 2 in Hg absolute
       1075 Btu/lbm (2500 kJ/kg)

At F: 69.73 Btu/lbm (162.5 kJ/kg)

At H: 250.2 Btu/lbm (583.0 kJ/kg)
       0.0173 ft^3/lbm

At I: 253.1 Btu/lbm (589.7 kJ/kg)

(a) What is the isentropic efficiency of the high-pressure turbine? (b) What is the thermal efficiency of the cycle?

8. (*Time limit: one hour*) A precision air turbine is used to drive a small dentist's drill. 140°F (60°C) air enters the turbine at the rate of 15 lbm/hr (1.9 g/s). The output of the turbine is 0.25 hp (0.19 kW). The turbine exhausts to 15 psia (103.5 kPa). The flow is steady and the process is adiabatic. The isentropic efficiency of the expansion process is 60%. (a) What is the exhaust temperature? (b) What is the inlet air pressure? (c) What is the change in entropy through the turbine?

Power Cycles

29 Combustion Power Cycles

1. Introduction to Air-Standard Cycles 29-1
2. Engine Terminology 29-2
3. Four- and Two-Stroke Engines 29-2
4. Carburetion and Fuel Injection 29-2
5. Operating Characteristics of
   Combustion Engines 29-3
6. Brake and Indicated Properties 29-3
7. Engine Power and Torque 29-3
8. Air-Standard Carnot Cycle 29-4
9. Air-Standard Otto Cycle 29-5
10. PLAN Formula 29-8
11. Introduction to Diesel Engines 29-10
12. Stationary and Marine Gas Diesels 29-10
13. Air-Standard Diesel Cycle 29-10
14. Air-Standard Dual Cycle 29-11
15. Introduction to Stirling Engines 29-12
16. Stirling Cycle 29-12
17. Ericsson Cycle 29-13
18. Effect of Altitude on Output Power 29-13
19. Gas Turbines 29-14
20. Steam Injection 29-14
21. Turbine Fuels 29-14
22. Indirect Coal-Fired Turbines 29-15
23. Brayton Gas Turbine Cycle 29-15
24. Air-Standard Brayton Cycle
    with Regeneration 29-18
25. Brayton Cycle with Regeneration,
    Intercooling, and Reheating 29-19
26. Heat Recovery Steam Generators 29-20
27. Cogeneration Cycles 29-21
28. Combined Cycles 29-22
29. High-Performance Power Systems 29-22
30. Repowering and Life-Extension 29-22
    Practice Problems 29-23

Nomenclature

A	bore area	in^2	n.a.
AFR	air-fuel ratio	lbm/lbm	kg/kg
BHP	brake horsepower	hp	n.a.
BkW	brake kilowatts	n.a.	kW
BSFC	brake specific fuel consumption	lbm/hp-hr	kg/kW·h
BWR	back-work ratio	–	–
c	clearance	–	–
c	specific heat	Btu/lbm-°F	kJ/kg·K
F	force	lbf	N
FAR	fuel-air ratio	lbm/lbm	kg/kg
FHP	friction (horse) power	hp	kW
FU	fuel utilization	–	–
h	enthalpy	Btu/lbm	kJ/kg
HV	heating value	Btu/lbm	kJ/kg
IHP	indicated (horse) power	hp	kW
k	ratio of specific heats	–	–
L	stroke length	ft	n.a.
$\dot{m}$	mass flow rate	lbm/sec	kg/s
MEP	mean effective pressure	lbf/ft^2	kPa
n	engine speed (rpm)	rev/min	r/min
N	number of power strokes per minute	min^{-1}	n.a.
p	mean effective pressure	psig	n.a.
p	pressure	lbf/ft^2	kPa
P	power	hp	kW
PTR	power-to-heat ratio	–	–
q	heat	Btu/lbm	kJ/kg
Q	heat	Btu	kJ
r	moment arm or radius	ft	m
r	ratio	–	–
R	specific gas constant	ft-lbf/lbm-°R	kJ/kg·K
s	entropy	Btu/lbm-°F	kJ/kg·K
T	temperature	°R	K
T	torque	ft-lbf	N·m
u	internal energy	Btu/lbm	kJ/kg
U	overall coefficient of heat transfer	Btu/ft²-hr-°F	W/m²·K
V	volume	ft^3	m^3
$\dot{V}$	volumetric flow rate	ft^3/sec	m^3/s
W	work per unit mass	Btu/lbm	kJ/kg
W	work	Btu	kJ

Symbols

η	efficiency	–	–
ρ	mass density	lbm/ft^3	kg/m^3
v	specific volume	ft^3/lbm	m^3/kg

Subscripts

a	air
c	cut-off
f	fuel
m	mechanical
p	pressure or constant pressure
r	relative
s	isentropic
th	thermal
v	volume or constant volume

1. INTRODUCTION TO AIR-STANDARD CYCLES

Combustion power cycles differ from vapor power cycles in that the combustion products cannot be returned to their initial conditions for reuse. Due to the computational difficulties of working with mixtures of fuel vapor, combustion products, and air, combustion power cycles are often analyzed as air-standard cycles.

An *air-standard cycle* is a closed system using a fixed amount of ideal air as the working fluid. In contrast to a combustion process, the heat of combustion is included in the calculations without consideration of the heat source or delivery mechanism. (That is, the combustion process is replaced by a process of instantaneous heat transfer from high-temperature surroundings.) Similarly, the cycle ends with an instantaneous transfer of waste heat to the surroundings. All processes are considered to be internally reversible. Because the air is ideal, it has a constant specific heat.[1]

Actual engine efficiencies for internal combustion engine cycles may be up to 50% lower than the efficiencies calculated from air-standard analyses. Empirical corrections must be applied to theoretical calculations based on the characteristics of the engine. The large amount of excess air used in turbine combustion cycles results in better agreement between actual and ideal performance than for reciprocating engines.

2. ENGINE TERMINOLOGY

Internal combustion (IC) engines can be categorized into *spark ignition* (SI) and *compression ignition* (CI) categories. SI engines (i.e., typical gasoline engines) use a spark to ignite the air-fuel mixture, while the heat of compression ignites the air-fuel mixture in CI engines (i.e., typical diesel engines).

The diameter of the circular cylinder is the *bore*. The maximum distance traveled by the piston is the *stroke*. An engine with a bore of diameter D and stroke of length L is sometimes referred to as a $D \times L$ engine. The product of the cylinder area and stroke is the *swept volume*. The fractional *clearance* is the ratio of *clearance (clear) volume* to the *swept volume*, usually expressed as a percentage. At *top-dead-center* (TDC), the piston is at maximum reach from the crankshaft. At *bottom-dead-center*, the piston is closest to the crankshaft.

The expanding combustion gases act on one end of the piston in *single-acting engines* and on both ends of the piston in *double-acting engines*. Double-acting internal combustion engines are essentially nonexistent due to difficulties in sealing and power transmission. However, some double-acting reciprocating steam engines are used for stationary applications, and double-acting Stirling machines are common.[2] The total power generated in a double-acting engine is the sum of the power generated by the *head end* and *crank end*.

3. FOUR- AND TWO-STROKE ENGINES

Engines are categorized as either four-stroke or two-stroke. In a *four-stroke cycle*, four separate piston stroke. In a *four-stroke cycle*, four separate piston

movements (*strokes*) and two complete crankshaft revolutions are required to accomplish all of the processes: *intake, compression, power,* and *exhaust strokes*.[3]

Two-stroke cycles are used in small gasoline engines (such as in lawn mowers and other garden equipment, outboards, and motorcycles) as well as in some diesel engines. In a two-stroke cycle, only two piston movements and one crankshaft revolution occur per power stroke. The air-fuel mixture is drawn in through the intake port, and exhaust gases expand out through the exhaust port, near the end of the power stroke. The intake, exhaust, and power strokes overlap. The term *scavenging* is used to describe the act of blowing the exhaust products out with the air-fuel mixture.

Because power strokes occur twice as often, a two-stroke engine produces more power (for a given engine weight and displacement) than a four-stroke engine. However, the increase in power is only approximately 70 to 90% (not 100%) due to various inefficiencies, including incomplete mixing and scavenging.

In the typical commercial two-stroke engine, some of the exhaust gases dilute the air-fuel mixture. Also, oil is mixed with the fuel to provide engine lubrication. Both of these practices lead to poorer fuel economy and higher emissions.

With the proper modifications (e.g., supercharging, indirect fuel injection, scavenging with pure air, lean-burning *stratified charge combustion*, and *exhaust gas dilution*), however, two-stroke gasoline engines may someday be an attractive alternative to traditional automobile engines. The primary benefits (as perceived by automobile engine manufacturers) of two-stroke engines are reduced engine vibration, more uniform torque and power, and the ability to achieve reduced air pollution emissions, particularly nitrogen oxides.[4] This theoretically eliminates the need for a nitrogen-reducing catalyst.[5]

4. CARBURETION AND FUEL INJECTION

In a normally aspirated engine, fuel is mixed with air outside of the cylinder in the *carburetor*. Mixing continues in the manifold. A butterfly valve in the carburetor throttles the flow of air. The air-fuel mixture enters through ports closed by flat-headed valves.[6]

In *fuel-injected* (FI) *engines*, pressurized fuel is injected directly into the cylinder at just the right moment in the

---

[1]The term *cold air-standard cycle* is sometimes used to describe a cycle where the specific heats are considered to be constant at their room-temperature values.
[2]Stirling engines are not very common in the first place. But among Stirling engines, double-action is not unheard of.

[3]The intake and exhaust strokes do not affect the thermodynamics of the cycle. These strokes do not appear on the cycle diagrams. The four strokes do not correspond to the four processes in the Otto and diesel cycles.
[4]Another benefit, that of fewer parts and similar design, is sacrificed in commercial vehicle designs because of the need for strict emissions control.
[5]*Two-way catalytic converters* (using platinum and palladium) are still required for hydrocarbon and carbon monoxide reduction. However, these catalysts are less expensive than the *three-way converters* that also use rhodium for nitrogen control.
[6]The term *poppet valve* (i.e., a valve that rises and returns to its seat) is almost never heard anymore.

cycle. The timing may be controlled electronically or mechanically (by distributor disk, camshaft lobes, etc.). Pressurization may be supplied by a fuel pump or by injector pistons (either spring-loaded or cam-actuated). With *direct injection*, fuel is injected directly into the combustion chamber, typically into the crown at the top of the piston. With *indirect injection*, fuel is injected into a *precombustion chamber (prechamber)* where it is mixed with air prior to moving into the combustion chamber. Diesel engines using indirect injection are quieter and less polluting, though they are slightly less fuel-efficient.

5. OPERATING CHARACTERISTICS OF COMBUSTION ENGINES

The *specific fuel consumption* (SFC) is the fuel usage rate divided by the power generated. Typical units are lbm/hp-hr and kg/kW·h.

The *air-fuel ratio* (AFR) is the ratio of the air mass that enters the engine to each mass of fuel burned. The *fuel-air ratio* (FAR) is the reciprocal of the air-fuel ratio.

$$\text{AFR} = \frac{\dot{m}_{\text{air}}}{\dot{m}_{\text{fuel}}} = \frac{1}{\text{FAR}} \qquad 29.1$$

For Otto and diesel cycles, the *compression ratio*, r_v, is the ratio of two volumes.[7]

$$r_v = \frac{V_{\max}}{V_{\min}} \qquad 29.2$$

The *thermal efficiency*, η_{th}, of a combustion engine cycle is the ratio of net output power to input energy, both expressed in the same units.

$$\eta_{\text{th}} = \frac{W_{\text{out}} - W_{\text{in}}}{Q_{\text{in}}} = \frac{Q_{\text{in}} - Q_{\text{out}}}{Q_{\text{in}}} \qquad 29.3$$

The *mechanical efficiency*, η_m, of a combustion engine is

$$\eta_m = \frac{\text{BHP}}{\text{IHP}} = \frac{\text{actual power developed}}{\text{ideal power developed}} \qquad 29.4$$

In reciprocating engines, the *volumetric efficiency*, η_v, is the ratio of the actual to ideal volumes of entering gases.

$$\eta_v = \frac{\dot{V}_{\text{actual}}}{\dot{V}_{\text{ideal}}} \qquad 29.5$$

The *relative efficiency*, η_r, is the ratio of the actual and ideal thermal efficiencies. The ideal efficiency, η_i, can usually be calculated from other cycle properties (e.g., the compression ratio).

$$\eta_r = \frac{\eta_{\text{th,actual}}}{\eta_{\text{th,ideal}}} \qquad 29.6$$

---
[7]For compressors and compression cycles, the term *compression ratio* is also used as the ratio of pressures.

6. BRAKE AND INDICATED PROPERTIES

The performance characteristics (e.g., horsepower) of internal combustion engines can be reported with or without the effect of power-reducing friction and other losses. A value of a property that includes the effect of friction is known as a *brake value*. If the effect of friction is removed, the property is known as an *indicated value*.[8]

Common brake properties are *brake horsepower* (BHP), *brake specific fuel consumption* (BSFC), and *brake mean effective pressure* (BMEP). Common indicated properties are *indicated horsepower* (IHP), *indicated specific fuel consumption* (ISFC), and *indicated mean effective pressure* (IMEP).

The brake and indicated horsepowers differ by the *friction horsepower* (FHP).

$$\text{FHP} = \text{IHP} - \text{BHP} \qquad 29.7$$

Except for Eq. 29.7, indicated and brake properties are usually not combined in calculations since they refer to two different operating conditions. For example, the calculation of the actual fuel mass flow rate requires two brake parameters.

$$\dot{m}_f = (\text{BSFC})(\text{BHP}) \qquad 29.8$$

7. ENGINE POWER AND TORQUE

The power and torque curves are graphs of maximum power and maximum torque that the engine can develop over its speed range. For any given speed, the maximum power that the engine can develop can be determined by loading the engine with progressively higher loads until stalling occurs. This is done on a *dynamometer* or with a *prony brake*, which applies a frictional resistance to a rotating drum attached to the engine's power takeoff, as shown in Fig. 29.1. Since the prony brake measures net available power (inclusive of frictional losses), the operating characteristics derived are brake values (i.e., brake torque and brake horsepower).

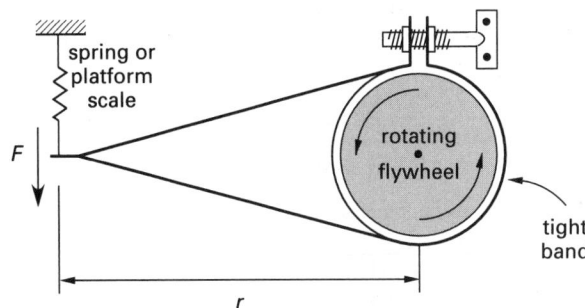

Figure 29.1 *Idealized Prony Brake*

---
[8]It may be helpful to think of the *i* in "indicated" as meaning "ideal."

The *brake torque* developed is

$$T = rF \qquad \text{29.9}$$

The brake torque can also be calculated from the brake horsepower (brake kilowatts) and rotational speed, n, in rpm.

$$T = \frac{(60\,000)(\text{BkW})}{2\pi n} = \frac{(9549)(\text{BkW})}{n} \quad \text{[SI]} \qquad \text{29.10(a)}$$

$$T = \frac{(33,000)(\text{BHP})}{2\pi n} = \frac{(5252)(\text{BHP})}{n} \quad \text{[U.S.]} \qquad \text{29.10(b)}$$

The *brake horsepower (brake kilowatts)* can be calculated from the brake torque or directly from the results of the prony brake test. Notice that rF is the brake torque.

$$\text{BkW} = \frac{2\pi r F n}{60\,000} = \frac{rFn}{9549} \quad \text{[SI]} \qquad \text{29.11(a)}$$

$$\text{BHP} = \frac{2\pi r F n}{33,000} = \frac{rFn}{5252} \quad \text{[U.S.]} \qquad \text{29.11(b)}$$

Power ratings listed for most commercial engines correspond to the "standard conditions" of 500 ft (150 m) altitude, a dry barometric pressure of 29.00 in of mercury, water vapor pressure of 0.38 in of mercury, and temperature of 85°F (29°C).[9],[10]

The *continuous duty rating* is the rated power that the manufacturer claims the engine is able to provide on a continuous (governed or steady rpm) basis without incurring damage. The *intermittent rating* represents the peak power that can be produced on an occasional basis.

8. AIR-STANDARD CARNOT CYCLE

Theoretically, the *air-standard Carnot combustion cycle* can be implemented either as a reciprocating or steady-flow device. However, like the Carnot vapor cycle, the air-standard Carnot combustion cycle is not a practical engine cycle.[11] The value of the cycle is in establishing a maximum thermal efficiency against which all other cycles can be compared. The Carnot cycle consists of the following processes.

A to B: isentropic compression

B to C: isothermal expansion power stroke

C to D: isentropic expansion

D to A: isothermal compression

---

[9]This "yet another" definition of standard conditions is specified in SAE Standard J816b.

[10]Some manufacturers rate their engines at sea level and 60°F (15.6°C). This increases the reported power by approximately 4%.

[11]In particular, it is not possible to design equipment that will transfer heat to a working fluid at a constant temperature in a reversible process over a reasonably finite time.

The T-s diagram shown in Fig. 29.2 is the same for any Carnot cycle. However, the isothermal expansion and compression do not occur within a vapor dome as they do in a vapor power cycle and, therefore, do not occur at constant pressure. Thus, the p-v diagram is different than for a vapor cycle.

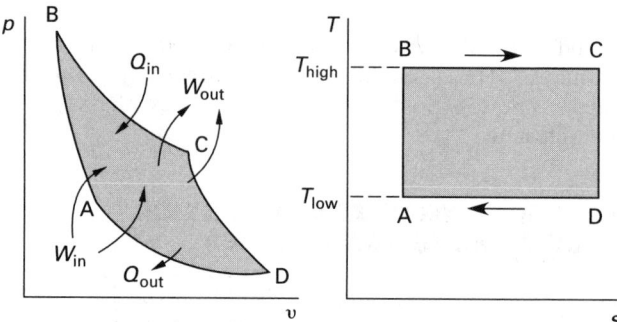

Figure 29.2 *Air-Standard Carnot Cycle*

The *isentropic pressure ratio* is a ratio of pressures.

$$r_{p,s} = \frac{p_B}{p_A} = \frac{p_C}{p_D} = \left(\frac{T_D}{T_C}\right)^{\frac{k}{1-k}} \qquad \text{29.12}$$

The *isentropic compression ratio* for reciprocating equipment is a ratio of volumes.

$$r_{v,s} = \frac{v_A}{v_B} = \frac{v_D}{v_C} = \left(\frac{T_D}{T_C}\right)^{\frac{1}{1-k}} \qquad \text{29.13}$$

Pressure, temperature, and volume can be calculated from the isentropic and isothermal relationships for ideal gases.

At A:

$$p_A = p_D\left(\frac{v_D}{v_A}\right) = p_B\left(\frac{v_B}{v_A}\right)^k = p_B\left(\frac{T_A}{T_B}\right)^{\frac{k}{k-1}} \qquad \text{29.14}$$

$$T_A = T_D = T_B\left(\frac{v_B}{v_A}\right)^{k-1} = T_B\left(\frac{p_A}{p_B}\right)^{\frac{k-1}{k}} \qquad \text{29.15}$$

$$v_A = v_D\left(\frac{p_D}{p_A}\right) = v_B\left(\frac{p_B}{p_A}\right)^{\frac{1}{k}} = v_B\left(\frac{T_B}{T_A}\right)^{\frac{1}{k-1}} \qquad \text{29.16}$$

At B:

$$p_B = p_A\left(\frac{v_A}{v_B}\right)^k = p_A\left(\frac{T_B}{T_A}\right)^{\frac{k}{k-1}} = p_C\left(\frac{v_C}{v_B}\right) \qquad \text{29.17}$$

$$T_B = T_A\left(\frac{v_A}{v_B}\right)^{k-1} = T_A\left(\frac{p_B}{p_A}\right)^{\frac{k-1}{k}} = T_C \qquad \text{29.18}$$

$$v_B = v_A\left(\frac{p_A}{p_B}\right)^{\frac{1}{k}} = v_A\left(\frac{T_A}{T_B}\right)^{\frac{1}{k-1}} = v_C\left(\frac{p_C}{p_B}\right) \qquad \text{29.19}$$

At C:

$$p_C = p_B \left(\frac{v_B}{v_C} \right) = p_D \left(\frac{v_D}{v_C} \right)^k = p_D \left(\frac{T_C}{T_D} \right)^{\frac{k}{k-1}} \quad 29.20$$

$$T_C = T_B = T_D \left(\frac{v_D}{v_C} \right)^{k-1} = T_D \left(\frac{p_C}{p_D} \right)^{\frac{k-1}{k}} \quad 29.21$$

$$v_C = v_B \left(\frac{p_B}{p_C} \right) = v_D \left(\frac{p_D}{p_C} \right)^{\frac{1}{k}} = v_D \left(\frac{T_D}{T_C} \right)^{\frac{1}{k-1}} \quad 29.22$$

At D:

$$p_D = p_C \left(\frac{v_C}{v_D} \right)^k = p_C \left(\frac{T_D}{T_C} \right)^{\frac{k}{k-1}} = p_A \left(\frac{v_A}{v_D} \right) \quad 29.23$$

$$T_D = T_C \left(\frac{v_C}{v_D} \right)^{k-1} = T_C \left(\frac{p_D}{p_C} \right)^{\frac{k-1}{k}} = T_A \quad 29.24$$

$$v_D = v_C \left(\frac{p_C}{p_D} \right)^{\frac{1}{k}} = v_C \left(\frac{T_C}{T_D} \right)^{\frac{1}{k-1}} = v_A \left(\frac{p_A}{p_D} \right) \quad 29.25$$

The work and heat flow terms are

$$W_{out} = c_v(T_C - T_D) + T_B(s_C - s_B) \quad 29.26$$

$$W_{in} = |c_v(T_A - T_B)| + T_D(s_D - s_A) \quad 29.27$$

$$q_{out,D-A} = |T_{low}(s_A - s_D)| = \left| p_D v_D \ln \left(\frac{v_A}{v_D} \right) \right| \quad 29.28$$

$$q_{in,B-C} = T_{high}(s_C - s_B) = p_B v_B \ln \left(\frac{v_C}{v_B} \right) \quad 29.29$$

The thermal efficiency of the air-standard Carnot cycle can be calculated from the heat and work terms, but it is easier to work with the two temperature extremes. The temperature forms of the thermal efficiency equation are easy to derive since $q = T\Delta s$.

$$\eta_{th} = \frac{W_{out} - W_{in}}{Q_{in}} = \frac{Q_{in} - Q_{out}}{Q_{in}}$$
$$= \frac{T_B - T_A}{T_B} = \frac{T_C - T_D}{T_C} \quad 29.30$$

The ideal thermal efficiency can also be calculated from the compression ratio.

$$\eta_{th} = 1 - r_{v,s}^{1-k} = 1 - r_{p,s}^{\frac{1-k}{k}} \quad 29.31$$

9. AIR-STANDARD OTTO CYCLE

The *air-standard Otto cycle* consists of the following processes and is illustrated in Fig. 29.3. The Otto cycle is a *four-stroke cycle*.

A to B: isentropic compression

B to C: constant volume heat addition

C to D: isentropic expansion

D to A: constant volume heat rejection

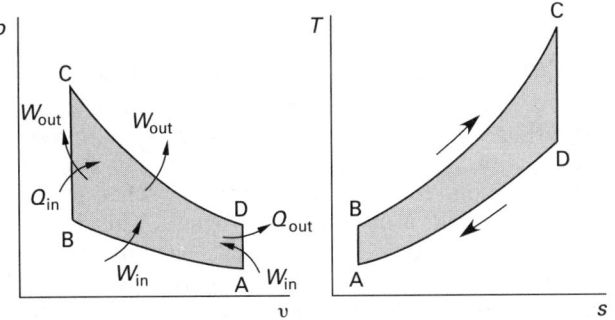

Figure 29.3 *Air-Standard Otto Cycle*

Although it is possible to define a pressure ratio for the Otto cycle, usually only the volumetric *compression ratio*, r_v, is used.

$$r_v = \frac{V_A}{V_B} = \frac{V_D}{V_C} \quad 29.32$$

Depending on the octane, compression ratios are typically in the range of 5:1 to 10.5:1. Most modern automobile engines designed to run on unleaded gasoline have compression ratios of approximately 8:1 to 8.5:1.

The ratios of absolute temperatures are equal because the piston travels a fixed distance, regardless of which stroke the cycle is on.

$$\frac{T_D}{T_C} = \frac{T_A}{T_B} \quad 29.33$$

Pressure, volume, and temperature relationships for the isentropic and constant volume ideal gas processes can be evaluated using air tables or the ideal gas equations. Using the ideal gas equations makes the implicit assumption that the specific heats are constant during the four processes, while the specific heats actually vary greatly over the wide temperature extremes encountered. By using the air tables this source of error is eliminated, but because the working fluid is not actually pure air, the additional refinement is probably unwarranted.

Assuming an ideal gas, the work and heat flow terms are

$$q_{in,B-C} = c_v(T_C - T_B) \quad 29.34$$

$$q_{out,D-A} = |c_v(T_A - T_D)| \quad 29.35$$

$$W_{in,A-B} = c_v(T_B - T_A) = \frac{p_B v_B - p_A v_A}{k-1} \quad 29.36$$

$$W_{out,C-D} = |c_v(T_D - T_C)| = \left| \frac{p_D v_D - p_C v_C}{k-1} \right| \quad 29.37$$

The thermal efficiency for the Otto cycle can be calculated in a number of ways.

$$\eta_{th} = \frac{Q_{in} - Q_{out}}{Q_{in}} = \frac{W_{out} - W_{in}}{Q_{in}}$$
$$= \frac{T_C - T_D}{T_C} = \frac{T_B - T_A}{T_B} \quad 29.38$$

The ideal thermal efficiency can be calculated from the compression ratio.

$$\eta_{th} = 1 - r_v^{1-k} \qquad \textbf{29.39}$$

From Eq. 29.39, it can be inferred that the ideal thermal efficiency of an Otto cycle can be increased by increasing either the ratio of specific heats or the compression ratio or both. Increasing the ratio of specific heats can only be accomplished by substituting another gas (e.g., helium) for the atmospheric nitrogen. This is not a practical alternative. Increases in the compression ratio are limited to approximately 10:1 by fuel detonation.[12] (Cylinder sealing is a problem only for compression ratios above approximately 16:1.)

The air-standard Otto cycle is less efficient than the Carnot cycle operating between the same temperature limits. However, Eqs. 29.31 and 29.39 show that equal efficiencies are obtained if the compression ratios are made equal.

Example 29.1

An internal combustion engine is to be evaluated on the basis of an air-standard Otto cycle. The pressure and temperature at the intake are 14.7 psia and 600°R (101.3 kPa and 330K), respectively. The maximum pressure and temperature in the cycle are 340 psia and 2610°R (2.3 MPa and 1450K), respectively. Consider air to be an ideal gas. What are the (a) pressure at the end of the compression stroke, (b) temperature at the end of the compression stroke, and (c) cycle efficiency?

SI Solution

Refer to Fig. 29.3. The pressure at point B can be related to the temperatures at points A and B.

$$p_B = p_A \left(\frac{T_B}{T_A}\right)^{\frac{k}{k-1}}$$

$$= (101.3 \text{ kPa}) \left(\frac{T_B}{330\text{K}}\right)^{\frac{1.4}{1.4-1}}$$

$$= (0.1013 \text{ MPa}) \left(\frac{T_B}{330\text{K}}\right)^{3.5}$$

The temperature at point B can be related to the pressures at points B and C.

$$T_B = \frac{T_C p_B}{p_C} = \frac{(1450\text{K})(p_B)}{2.3 \text{ MPa}}$$

$$= 630.4 p_B$$

---

[12]*Detonation* is a premature autoignition of the fuel. After the spark ignites one portion of the air-fuel mixture, the advancing flame front compresses the remainder of the mixture. If the compression ratio is too high, the compressed remainder may detonate.

Solve these two equations simultaneously.

$$p_B = 1.01 \text{ MPa}$$
$$T_B = 637\text{K}$$

Use Eq. 29.38.

$$\eta_{th} = \frac{T_B - T_A}{T_B} = \frac{637\text{K} - 330\text{K}}{637\text{K}}$$

$$= 0.482 \ (48.2\%)$$

Customary U.S. Solution

Refer to Fig. 29.3. The pressure at point B can be related to the temperatures at points A and B.

$$p_B = p_A \left(\frac{T_B}{T_A}\right)^{\frac{k}{k-1}}$$

$$= (14.7 \text{ psia}) \left(\frac{T_B}{600°\text{R}}\right)^{\frac{1.4}{1.4-1}}$$

$$= (14.7 \text{ psia}) \left(\frac{T_B}{600°\text{R}}\right)^{3.5}$$

The temperature at point B can be related to the pressures at points B and C.

$$T_B = \frac{T_C p_B}{p_C} = \frac{(2610°\text{R})(p_B)}{340 \text{ psia}}$$

$$= 7.676 p_B$$

Solve these two equations simultaneously.

$$p_B = 152.5 \text{ psia}$$
$$T_B = 1171°\text{R}$$

Use Eq. 29.38.

$$\eta_{th} = \frac{T_B - T_A}{T_B} = \frac{1171°\text{R} - 600°\text{R}}{1171°\text{R}}$$

$$= 0.488 \ (48.8\%)$$

Example 29.2

An Otto engine has 15% clearance. Air enters the engine at 14.0 psia and 580°R (96.6 kPa and 320K). The air-fuel ratio is 16.67. The heating value of gasoline is 19,500 Btu/lbm (45.5 MJ/kg). What are the temperatures and pressures at all points in the cycle?

SI Solution

(This solution takes a cold air-standard approach. The customary U.S. solution uses an air table.)

Refer to Fig. 29.3.

At A:

$$V_A = V_{swept} + V_{clearance} = (1 + c)V_{swept}$$
$$= (1 + 0.15)V_{swept} = 1.15V_{swept}$$

$p_A = 96.6$ kPa   [given]

$T_A = 320$K   [given]

At B:

$$V_B = 0.15V_{swept}$$

The compression ratio is

$$r_v = \frac{V_A}{V_B} = \frac{1.15V_{swept}}{0.15V_{swept}}$$
$$= 7.67$$

The compression from B to A is isentropic.

$$T_B = T_A\left(\frac{V_A}{V_B}\right)^{k-1} = (320K)(7.67)^{1.4-1}$$
$$= 722.9K$$

$$p_B = p_A\left(\frac{V_A}{V_B}\right)^{k} = (96.6 \text{ kPa})(7.67)^{1.4}$$
$$= 1674 \text{ kPa (1.674 MPa)}$$

At C: The heat added by the fuel is

$$q = \frac{HV}{AFR} = \frac{45.5 \dfrac{MJ}{\text{kg fuel}}}{16.67 \dfrac{\text{kg air}}{\text{kg fuel}}}$$
$$= 2.7295 \text{ MJ/kg air}$$

The combustion heat increases the internal energy of the air. For any process, $\Delta u = c_v \Delta T$. The cold specific heat (0.718 kJ/kg·K) of air is used.

$$T_C = T_B + \frac{\Delta u}{c_v} = T_B + \frac{q}{c_v}$$
$$= 722.9K + \frac{\left(2.7295 \dfrac{MJ}{kg}\right)\left(1000 \dfrac{kJ}{MJ}\right)}{0.718 \dfrac{kJ}{kg \cdot K}}$$
$$= 4524K$$

$$p_C = \frac{p_B T_C}{T_B} = \frac{(1.674 \text{ MPa})(4524K)}{722.9K}$$
$$= 10.48 \text{ MPa}$$

At D: The process from C to D is isentropic. Since the piston travels the same distance going from A to B as it does from C to D, the C-D expansion ratio is the same as the A-B compression ratio.

$$T_D = T_C\left(\frac{V_C}{V_D}\right)^{k-1} = (4524K)\left(\frac{1}{7.67}\right)^{1.4-1}$$
$$= 2003K$$

$$p_D = p_C\left(\frac{V_C}{V_D}\right)^{k} = (10.48 \text{ MPa})\left(\frac{1}{7.67}\right)^{1.4}$$
$$= 0.605 \text{ MPa}$$

Customary U.S. Solution

(This solution uses an air table. The SI solution takes a cold air-standard cycle approach.)

Refer to Fig. 29.3.

At A:

$$V_A = V_{swept} + V_{clearance} = (1 + c)V_{swept}$$
$$= (1 + 0.15)V_{swept} = 1.15V_{swept}$$

$p_A = 14.0$ psia   [given]

$T_A = 580°$R   [given]

From the air table at 580°R, $v_{r,A} = 120.7$ and $p_{r,A} = 1.78$.

At B:

$$V_B = 0.15V_{swept}$$

The compression ratio is

$$r_v = \frac{V_A}{V_B} = \frac{1.15V_{swept}}{0.15V_{swept}}$$
$$= 7.67$$

Consider the compression from B to A to be isentropic. Then,

$$v_{r,B} = \frac{v_{r,A}}{r_v}$$
$$= \frac{120.7}{7.67} = 15.74$$

The temperature corresponding to this volume ratio in the air table is 1273°R.

$$T_B = 1273°R$$
$$u_B = 222.72 \text{ Btu/lbm}$$
$$p_{r,B} = 29.93$$
$$p_B = \frac{p_A p_{r,B}}{p_{r,A}} = \frac{(14.0 \text{ psia})(29.93)}{1.78}$$
$$= 235.4 \text{ psia}$$

At C: The heat added by the fuel per pound of air is

$$q = \frac{HV}{AFR} = \frac{19{,}500 \dfrac{\text{Btu}}{\text{lbm fuel}}}{16.67 \dfrac{\text{lbm air}}{\text{lbm fuel}}}$$
$$= 1170 \text{ Btu/lbm air}$$

Enthalpy is the sum of internal energy and pV energy. Even though the enthalpy at point B could be found from the air table, the pressure at point C is not yet known. Therefore, this heat addition cannot merely be added to the enthalpy at point B. However, internal energy does not include the pV term. The internal energy at the end of the heat addition is

$$u_C = u_B + q = 222.72 \frac{\text{Btu}}{\text{lbm}} + 1170 \frac{\text{Btu}}{\text{lbm}}$$
$$= 1392.72 \text{ Btu/lbm}$$

Locate this value of internal energy in the air table.

$$T_C = 6420°R$$

$$v_C = 0.08685$$

$$p_{r,C} = 27{,}381$$

Use the ideal gas law to find the pressure at C.

$$p_C = \frac{p_B T_C}{T_B} = \frac{(235.4 \text{ psia})(6420°R)}{1273°R}$$

$$= 1187.2 \text{ psia}$$

At D: The process from C to D is isentropic.

$$v_{r,D} = v_{r,C} r_v$$

$$= (0.08685)(7.67) = 0.6661$$

Locate this volume ratio in the air table.

$$T_D = 3565°R$$

$$p_{r,D} = 1981.9$$

$$p_D = \frac{p_C p_{r,D}}{p_{r,C}} = \frac{(1187.2 \text{ psia})(1981.9)}{27{,}381}$$

$$= 85.9 \text{ psia}$$

10. PLAN FORMULA

The performance of an internal combustion engine operating on the Otto cycle can be predicted from the PLAN formula, Eq. 29.40.[13] It is necessary to know the engine bore area (A in square inches), stroke (L in feet), number of engine power strokes per minute (N), and mean effective pressure (p in psig). It is essential to use the units as defined.[14] The *brake horsepower* (BHP) will be calculated if the *brake mean effective pressure* (BMEP) is used. The *indicated horsepower* (IHP) will be calculated if the *indicated mean effective pressure* (IMEP) is used.

$$\text{kW} = \frac{pLAN}{60} \qquad \text{[SI]} \qquad 29.40(a)$$

$$\text{hp} = \frac{pLAN}{33{,}000} \qquad \text{[U.S.]} \qquad 29.40(b)$$

Equation 29.41 gives the number of engine power strokes per minute for two- and four-stroke engines.

$$N = \frac{(2n)(\text{no. cylinders})}{\text{no. strokes per cycle}} \qquad 29.41$$

The indicated *mean effective pressure*, p, is determined graphically from an actual p-v plot (known as an *indicator drawing, indicator diagram,* or *indicator card*)

of the actual cycle. This is done by taking the overall area of the indicator drawing and dividing it by the total width of the drawing. The resulting number is then multiplied by the indicator spring constant (scale).

The mean effective pressure is a theoretical average pressure that would produce the same amount of net work during one stroke as the varying pressure does in the entire cycle. The mean effective pressure can be calculated if the cycle performance is known. The mean effective pressure in nonsupercharged engines is typically limited to approximately 100 psi (700 kPa).[15]

$$\text{MEP} = \frac{W_{\text{net}}}{V_A - V_B} \qquad 29.42$$

In some cases, the power generated by an engine is correlated as CND^2, where C is a constant. This is the same as Eq. 29.40, with the coefficient C incorporating all of the constant terms.

Figure 29.4 illustrates how the actual Otto cycle deviates from the ideal cycle (shown by the broken line) and the causes of the deviations. Due to heat transfers during the A-to-B and C-to-D processes, the lines are not true adiabats. Rather, these two processes are polytropic, with a polytropic exponent of approximately 1.3. The net effect of all of these deviations is to reduce the actual efficiency to approximately half of the ideal efficiency.

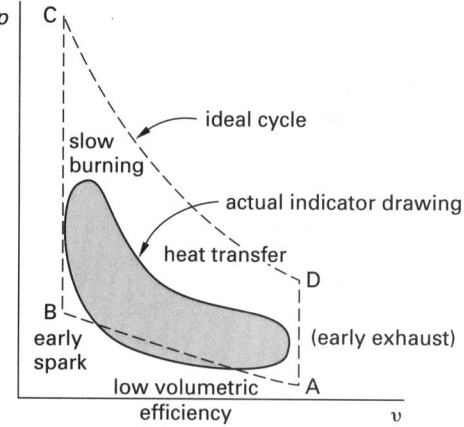

***Figure 29.4** Otto Cycle Indicator Drawing*

Example 29.3

A single-cylinder four-stroke engine has a 10 in (25.4 cm) bore and an 18 in (45.7 cm) stroke. The engine is run at 200 rpm while being tested on a prony brake. The gross weight on the brake is 140 lbf (620 N), the tare is 25 lbf (110 N), and the arm length is 66 in (1.7 m). The indicator card shows an area of 1.20 in² (7.7 cm²) with an overall length of 3 in (7.6 cm).

---

[13]The PLAN formula is also applicable to reciprocating steam engines.
[14]The units are strictly traditional English units.

[15]With the advent of supercharging, *peak firing pressure*, also known as *peak cylinder pressure* (PCP), has become a benchmark for engine performance.

The spring scale used to draw the indicator card is 200 psi/in (540 kPa/cm). What are the (a) indicated mean effective pressure, (b) indicated horsepower, (c) brake horsepower, and (d) mechanical efficiency?

SI Solution

(a) The indicated mean effective pressure is

$$p = \frac{(\text{diagram area}) \,(\text{pressure scale factor})}{\text{diagram length}}$$

$$= \frac{(7.7 \text{ cm}^2) \left(540 \, \dfrac{\text{kPa}}{\text{cm}}\right)}{7.6 \text{ cm}}$$

$$= 547 \text{ kPa}$$

(b) The stroke is

$$L = \frac{45.7 \text{ cm}}{100 \, \dfrac{\text{cm}}{\text{m}}} = 0.457 \text{ m}$$

The cylinder area is

$$A = \frac{\pi D^2}{4} = \frac{\pi (25.4 \text{ cm})^2}{4}$$

$$= 506.7 \text{ cm}^2 \quad (0.05067 \text{ m}^2)$$

The number of power strokes per minute is

$$N = \frac{(2n)(\text{no. cylinders})}{\text{no. strokes per cycle}}$$

$$= \frac{\left(2 \, \dfrac{\text{strokes}}{\text{rev}}\right) \left(200 \, \dfrac{\text{rev}}{\text{min}}\right)(1 \text{ cylinder})}{4 \text{ strokes per power stroke}}$$

$$= 100 \text{ power strokes/min}$$

Equation 29.40 gives the indicated power. The conversion to horsepower is not needed.

$$\begin{array}{l}\text{indicated} \\ \text{power}\end{array} = pLAN$$

$$= \frac{\begin{array}{c}(547 \text{ kPa}) \left(1000 \, \dfrac{\text{Pa}}{\text{kPa}}\right)(0.457 \text{ m}) \\[2mm] \times (0.05067 \text{ m}^2) \left(100 \, \dfrac{\text{power strokes}}{\text{min}}\right)\end{array}}{\left(60 \, \dfrac{\text{s}}{\text{min}}\right) \left(1000 \, \dfrac{\text{W}}{\text{kW}}\right)}$$

$$= 21.1 \text{ kW}$$

(c) The brake horsepower is found from the brake information. The brake arm length is

$$r = 1.7 \text{ m}$$

The force on the brake caused by the engine excludes the tare weight.

$$F = 620 \text{ N} - 110 \text{ N} = 510 \text{ N}$$

The brake power is

$$\text{brake power} = 2\pi r F n$$

$$= \frac{(2)\pi(1.7 \text{ m})(510 \text{ N}) \left(200 \, \dfrac{\text{rev}}{\text{min}}\right)}{\left(60 \, \dfrac{\text{s}}{\text{min}}\right) \left(1000 \, \dfrac{\text{W}}{\text{kW}}\right)}$$

$$= 18.16 \text{ kW}$$

(d) The mechanical efficiency is

$$\eta_m = \frac{\text{brake power}}{\text{indicated power}} = \frac{18.16 \text{ kW}}{21.1 \text{ kW}}$$

$$= 0.861 \ (86.1\%)$$

Customary U.S. Solution

(a) The indicated mean effective pressure is

$$p = \frac{(\text{diagram area})(\text{pressure scale factor})}{\text{diagram length}}$$

$$= \frac{(1.2 \text{ in}^2) \left(200 \, \dfrac{\text{lbf}}{\text{in}^2}\right)}{3 \text{ in}}$$

$$= 80 \text{ lbf/in}^2$$

(b) The stroke is

$$L = \frac{18 \text{ in}}{12 \, \dfrac{\text{in}}{\text{ft}}} = 1.5 \text{ ft}$$

The cylinder area is

$$A = \frac{\pi D^2}{4} = \frac{\pi (10 \text{ in})^2}{4}$$

$$= 78.54 \text{ in}^2$$

The number of power strokes per minute is

$$N = \frac{(2n)(\text{no. cylinders})}{\text{no. strokes per cycle}}$$

$$= \frac{\left(2 \, \dfrac{\text{strokes}}{\text{rev}}\right) \left(200 \, \dfrac{\text{rev}}{\text{min}}\right)(1 \text{ cylinder})}{4 \text{ strokes per power stroke}}$$

$$= 100 \text{ power strokes/min}$$

From Eq. 29.40, the indicated horsepower is

$$\text{IHP} = \frac{pLAN}{33{,}000}$$

$$= \frac{\begin{array}{c}\left(80 \, \dfrac{\text{lbf}}{\text{in}^2}\right)(1.5 \text{ ft})(78.54 \text{ in}^2) \\[2mm] \times \left(100 \, \dfrac{\text{power strokes}}{\text{min}}\right)\end{array}}{33{,}000 \, \dfrac{\text{ft-lbf}}{\text{hp-min}}}$$

$$= 28.56 \text{ hp}$$

(c) The brake horsepower is found from the brake information. The brake arm length is

$$r = \frac{66 \text{ in}}{12 \frac{\text{in}}{\text{ft}}} = 5.5 \text{ ft}$$

The force on the brake caused by the engine excludes the tare weight.

$$F = 140 \text{ lbf} - 25 \text{ lbf} = 115 \text{ lbf}$$

From Eq. 29.11, the brake horsepower is

$$
\begin{aligned}
\text{BHP} &= \frac{2\pi r F n}{33,000} \\
&= \frac{(2)\pi(5.5 \text{ ft})(115 \text{ lbf})\left(200 \frac{\text{rev}}{\text{min}}\right)}{33,000 \frac{\text{ft-lbf}}{\text{hp-min}}} \\
&= 24.09 \text{ hp}
\end{aligned}
$$

(d) The mechanical efficiency is

$$
\begin{aligned}
\eta_m &= \frac{\text{BHP}}{\text{IHP}} = \frac{24.09 \text{ hp}}{28.56 \text{ hp}} \\
&= 0.843 \quad (84.3\%)
\end{aligned}
$$

11. INTRODUCTION TO DIESEL ENGINES

A diesel engine is a compression-ignition, internal combustion engine. Spark plugs are not used for ignition.[16] Rather, high compression ratios produce autoignition of the air-fuel mixture. The higher the compression ratio, the higher the thermal efficiency. Compression ratios for diesel engines are ratios of volumes. The compression ratio varies from about 13.5:1 to 17.5:1, depending on whether or not the engine is turbocharged. Diesel engines can burn a variety of fuels including No. 6 heavy fuel oil, natural gas, and light distillate fuel oils.

Four-stroke diesels offer many significant advantages over two-stroke diesels. (1) With two-stroke engines, a significant amount of energy is expended by a blower to supply air for scavenging. When the load on the engine drops, the blower continues to run at its peak load, reducing engine efficiency even more. (2) Two-cycle engines often require premium fuels, compared to No. 2 oil that most four-stroke engines can use. (3) Valve, piston, and ring burning are more common in two-stroke engines since there is less time for cooling during the cycle. Four-stroke engines can use aluminum pistons, while heavy cast-iron pistons are needed for heat dissipation in two-stroke engines. (4) In a two-stroke engine, piston rings pass over the intake ports on every stroke,

leading to increased ring wear. (5) Since the valves and injectors operate every cycle with two-stroke engines, overhauls are required more frequently.

Supercharging compresses and increases the amount of air that enters the cylinder per stroke. (This increases the volumetric efficiency above 100%.) *Turbocharging* is a form of supercharging in which the exhaust gases drive the supercharger. With more air, more fuel can be burned, increasing the power per stroke. Supercharging also helps deliver air for combustion at higher altitudes. This results in better fuel economy than normally aspirated diesels. Supercharging also reduces smoke, particularly during lugging.

The temperature of compressed air from a turbocharger can be cooled by passing it through an aftercooler. An *aftercooler* is a closed heat exchanger that transfers heat from compressed air to cooler air. In vehicles, cool air flow is found in the manifold.

Emissions from diesel engines include nitrogen oxides, unburned hydrocarbons (*polycyclic aromatic hydrocarbons*—PAH), and carbon monoxide, much the same as from gasoline engines. Particulate emission (smoke or soot) and noise are also characteristic diesel pollutants.

12. STATIONARY AND MARINE GAS DIESELS

Large stationary engines operating on the diesel cycle can be built to run on natural gas.[17] There are two types of gas-burning diesels: gas-diesel and lean-burn engines. Natural gas under pressure has poor ignition and combustion characteristics. Therefore, in the *gas-diesel engine*, a small amount (about 3%) of oil is burned in addition to the natural gas. The oil is injected into the regular chamber or a precombustion chamber. Compression ignites the oil, and the resulting flame "torch" ignites the natural gas.

Since most gas-diesel engines produce less than 10 MW of power, they can be used for small electrical generation plants, gas pipeline compression, refrigeration chillers, and so on.[18]

In the *lean-burn engine*, a spark ignites a very lean mixture in a precombustion chamber. The resulting flame jet issuing from the precombustion chamber ignites the main fuel mixture. Most commercial lean-burn engines produce power in the range of 400 to 3000 kW.

13. AIR-STANDARD DIESEL CYCLE

The processes in an air-standard diesel cycle are illustrated in Fig. 29.5.

---

[16]A *glow plug* may be used to improve cold-weather starting.

[17]Large diesels are those producing up to approximately 10 MW. This is large for diesels but small for gas turbines. In a *combined diesel or gas turbine system* (CODOG) for marine propulsion, diesels are used for low power and cruise operation; the turbine takes over when high speeds are needed.

[18]This is an awkward size for gas turbines. Power turbines are not very practical in this size range.

A to B: isentropic compression

B to C: constant pressure heating

C to D: isentropic expansion

D to A: constant volume cooling

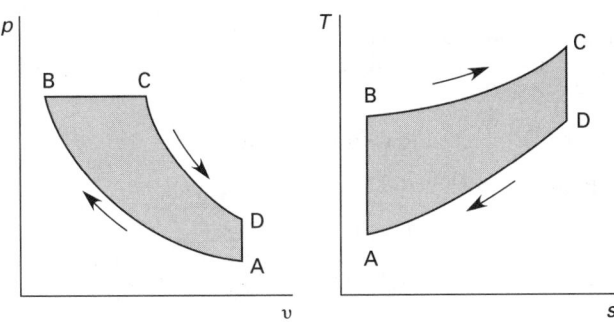

Figure 29.5 *Air-Standard Diesel Cycle*

Diesel engines can be either four-stroke or two-stroke devices. The four-stroke and two-stroke engines are analyzed the same way, since the four processes on the p-v and T-s diagrams do not correspond to the four strokes. The main difference is the number of power strokes per revolution—one for the two-stroke engine and one-half for the four-stroke engine.

The diesel cycle *compression ratio*, r_v, is

$$r_v = \frac{V_A}{V_B} \qquad 29.43$$

The *cut-off ratio*, $r_{\text{cut-off}}$, is defined by Eq. 29.44. The volume V_C is known as the *cut-off volume*.

$$r_{\text{cut-off}} = \frac{V_C}{V_B} = \frac{T_C}{T_B} \qquad 29.44$$

Pressure, volume, and temperature at each point in the cycle can be evaluated with ideal gas equations. The work and heat flow terms are evaluated with the following equations. (Equation 29.45 uses c_p because B-to-C is a constant pressure process.)

$$q_{\text{in}} = c_p(T_C - T_B) \qquad 29.45$$
$$q_{\text{out}} = |c_v(T_A - T_D)| \qquad 29.46$$
$$W_{\text{in}} = c_v(T_B - T_A) \qquad 29.47$$
$$W_{\text{out}} = |c_v(T_D - T_C) + (c_p - c_v)(T_C - T_B)| \qquad 29.48$$

The air-standard diesel cycle is always less efficient than the air-standard Otto cycle for equal compression ratios. However, for a specific maximum cylinder pressure, the diesel cycle is more efficient than the Otto cycle. The thermal efficiency can be calculated in a number of ways.

$$\eta_{\text{th}} = \frac{Q_{\text{in}} - Q_{\text{out}}}{Q_{\text{in}}} = \frac{W_{\text{out}} - W_{\text{in}}}{Q_{\text{in}}}$$
$$= 1 - \frac{T_D - T_A}{k(T_C - T_B)} \qquad 26.49$$

In standard diesels (i.e., those constructed until about 1980), thermal efficiency was not much higher than about 35%. Most manufacturers of stationary diesels are now reporting full-load thermal efficiencies of about 45%, and some as high as 50% (based on lower heating value). This is the highest thermal efficiency of any type of prime mover (including turbines).

14. AIR-STANDARD DUAL CYCLE

The *air-standard dual cycle* shown in Fig. 29.6 is a combination of the diesel and Otto cycles. It more accurately predicts the performance of a spark-ignited internal combustion engine, since the combustion energy is added partly at constant volume and partly at constant pressure.

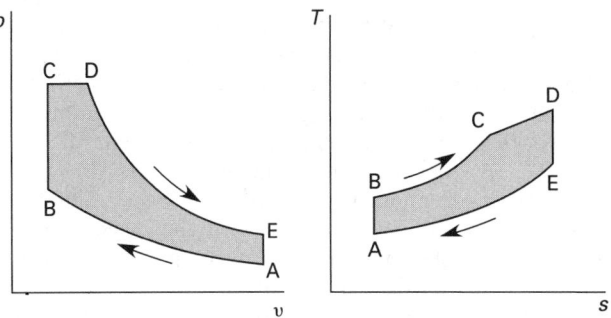

Figure 29.6 *Air-Standard Dual Cycle*

The processes in the dual cycle are

A to B: isentropic compression

B to C: constant volume heating

C to D: constant pressure heating

D to E: isentropic expansion

E to A: constant volume cooling

As with the Otto and diesel cycles, the *compression ratio*, r_v, is defined as a ratio of volumes.

$$r_v = \frac{V_A}{V_B} \qquad 29.50$$

The *pressure ratio*, r_p, is

$$r_p = \frac{p_C}{p_B} \qquad 29.51$$

The *cut-off ratio*, $r_{\text{cut-off}}$, is

$$r_{\text{cut-off}} = \frac{V_D}{V_C} \qquad 29.52$$

Pressure, temperature, and volume at various points in the cycle can be evaluated with ideal gas relationships. The work and heat terms are defined as follows.

$$q_{in} = c_v(T_C - T_B) + c_p(T_D - T_C) \qquad 29.53$$

$$q_{out} = |c_v(T_A - T_E)| \qquad 29.54$$

$$W_{in} = c_v(T_B - T_A) \qquad 29.55$$

$$W_{out} = |p_C(v_D - v_C) + c_v(T_D - T_E)|$$
$$= |(c_p - c_v)(T_D - T_C) + c_v(T_D - T_E)| \quad 29.56$$

The efficiency of a dual cycle is between those of the Otto and diesel cycles.

$$\eta_{th} = \frac{Q_{in} - Q_{out}}{Q_{in}} = \frac{W_{out} - W_{in}}{Q_{in}}$$
$$= 1 - \frac{T_E - T_A}{(T_C - T_B) + k(T_D - T_C)} \qquad 29.57$$

15. INTRODUCTION TO STIRLING ENGINES

During a typical Stirling cycle, the gaseous working fluid (e.g., helium or hydrogen) is shuttled through a heat exchanger circuit consisting of a heat acceptor, a regenerator, and a heat rejector. A piston compresses the working fluid in a compression space near the *heat rejector*. A *displacer* then shuttles the gas through a heat exchanger circuit to an expansion space. Along the way, the gas absorbs heat stored in the *regenerator*.[19] The piston is then moved so that the gas is expanded. During the expansion, heat is absorbed through the *acceptor* walls. The displacer then shuttles the gas back through the heat exchanger circuit at a constant volume, heating the regenerator along the way. This returns the Stirling machine to its original condition.

Heating of the working fluid occurs externally. The heat source can be from a combustion, nuclear (typically radioisotope), or solar process. Cooling is usually by a water-glycol mixture. Most Stirling machines draw power from combustion in a separate *combustor*. Since the combustion process is external, Stirling engines are able to use a variety of fuels and burn them at lower temperatures, resulting in lower emissions.

There are two methods of configuring Stirling machines. In a *kinematic Stirling engine*, double-acting pistons function separately as the compressor and displacer. Sealing of kinematic engines is a major challenge. *Free-piston engines* avoid leakage problems by eliminating the mechanical connection with the piston. Magnetic coupling (magnetic flux linkage) between the piston (and anything moving with it) and the surroundings limit the output of free-piston machines to electrical power generation.

Though Stirling engines are unlikely to be widely used in commercial vehicles in the near future, they are used to limited extents in small cryogenic applications. Since

---

[19]A *regenerator* is a device that captures heat energy and transfers it to the working fluid. In the case of the Stirling cycle, the heated gas passes through and heats a wire or ceramic mesh. When cool gas is passed back over the mesh, the heat transfer is reversed, and the gas is heated.

solar energy can be used as the heating source, Stirling engines are attractive for generating small amounts of electricity (i.e., 20 kW or less per high-efficiency engine/generator).

16. STIRLING CYCLE

The *Stirling cycle* has a thermal efficiency that can equal that of the Carnot cycle. The processes are shown in Fig. 29.7.

A to B: constant volume heating

B to C: isothermal heating and expansion

C to D: constant volume cooling

D to A: isothermal cooling and compression

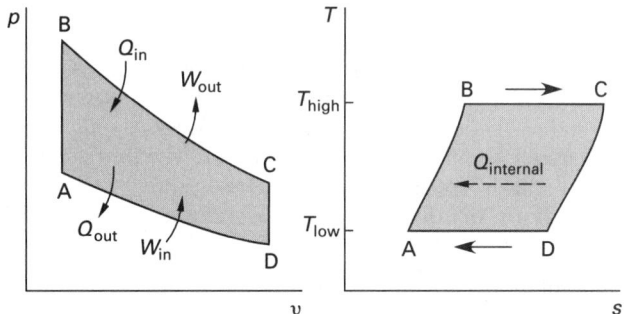

Figure 29.7 *Stirling Cycle*

Pressure, temperature, and volume for the various points on the cycle can be evaluated from ideal gas relationships.

The work and heat terms are given by Eqs. 29.58 through 29.61. There is also a heat transfer to the working fluid in the A-to-B process. However, this heat transfer takes place in a reversible regenerator, and the heat transferred is from the working fluid's C-to-D process. Since the heat remains within the system boundary, it is not included in the Q_{in} or Q_{out} terms.

$$q_{in,B-C} = T_{high}(s_C - s_B) = RT_{high} \ln\left(\frac{v_C}{v_B}\right)$$
$$= RT_{high} \ln\left(\frac{p_B}{p_C}\right) \qquad 29.58$$

$$q_{out,D-A} = |T_{low}(s_A - s_D)| = \left|RT_{low} \ln\left(\frac{v_A}{v_D}\right)\right|$$
$$= \left|RT_{low} \ln\left(\frac{p_D}{p_A}\right)\right| \qquad 29.59$$

$$W_{in,D-A} = Q_{out,D-A} \qquad 29.60$$

$$W_{out,B-C} = Q_{in,B-C} \qquad 29.61$$

The thermal efficiency of the Stirling cycle is equal to the Carnot cycle efficiency if the regenerator used to transfer heat from the C-to-D process to the A-to-B process is reversible. With a reversible regenerator, the thermal efficiency is

$$\eta_{th} = \frac{Q_{in} - Q_{out}}{Q_{in}} = \frac{W_{out} - W_{in}}{Q_{in}}$$

$$= \frac{T_{high} - T_{low}}{T_{high}} \qquad \text{29.62}$$

17. ERICSSON CYCLE

The processes of the Ericsson cycle are shown in Fig. 29.8. This cycle offers the best chance of achieving a thermal efficiency approaching that of the Carnot cycle.[20]

A to B: isothermal compression

B to C: constant pressure heating

C to D: isothermal expansion

D to A: constant pressure cooling

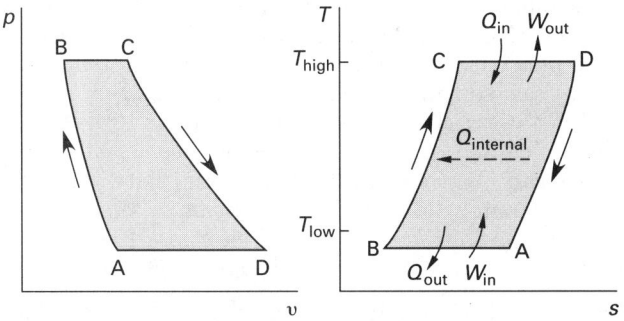

Figure 29.8 Ericsson Cycle

Pressure, temperature, and volume for the various points on the cycle can be evaluated from ideal gas relationships.

The work and heat terms are given by Eqs. 29.63 through 29.66. There is also a heat transfer to the working fluid in the B-to-C process. However, if a reversible regenerator is used to capture heat from the D-to-A process, no external heat will be required.

$$q_{in,C-D} = T_{high}(s_D - s_C) = RT_{high} \ln\left(\frac{v_D}{v_C}\right)$$

$$= RT_{high} \ln\left(\frac{p_C}{p_D}\right) \qquad \text{29.63}$$

$$q_{out,A-B} = |T_{low}(s_B - s_A)| = \left| RT_{low} \ln\left(\frac{v_B}{v_A}\right) \right|$$

$$= \left| RT_{low} \ln\left(\frac{p_A}{p_B}\right) \right| \qquad \text{29.64}$$

$$W_{in,A-B} = Q_{out,A-B} \qquad \text{29.65}$$

$$W_{out,C-D} = Q_{in,C-D} \qquad \text{29.66}$$

---

[20]The Ericsson cycle is approximated if the Brayton gas turbine cycle (covered in Sec. 23) is modified to include regenerative heat exchange, intercooling, and reheating.

With a reversible regenerator, the thermal efficiency of the Ericsson cycle is equal to that of the Carnot cycle.

$$\eta_{th} = \frac{Q_{in} - Q_{out}}{Q_{in}} = \frac{W_{out} - W_{in}}{Q_{in}}$$

$$= \frac{T_{high} - T_{low}}{T_{high}} \qquad \text{29.67}$$

18. EFFECT OF ALTITUDE ON OUTPUT POWER

Since a lower atmospheric pressure decreases atmospheric density, the oxygen per intake stroke available to engines operating on the Otto, diesel, and dual cycles decreases with altitude. The following steps constitute a procedure for determining the variation in power when the altitude is changed. It is assumed that the engine speed is constant.

step 1: Let 1 and 2 be the lower and higher altitudes, respectively.

step 2: Calculate the frictionless power.

$$IHP_1 = \frac{BHP_1}{\eta_{m,1}} \qquad \text{29.68}$$

step 3: Calculate the friction power, which is assumed to be constant at constant speed.

$$FHP = IHP_1 - BHP_1 \qquad \text{29.69}$$

step 4: Calculate the air densities ρ_{a1} and ρ_{a2} from App. 26.E.

step 5: Calculate the new frictionless power.

$$IHP_2 = IHP_1\left(\frac{\rho_{a2}}{\rho_{a1}}\right) \qquad \text{29.70}$$

step 6: Calculate the new net power.

$$BHP_2 = IHP_2 - FHP \qquad \text{29.71}$$

step 7: Calculate the new mechanical efficiency.

$$\eta_{m,2} = \frac{BHP_2}{IHP_2} \qquad \text{29.72}$$

step 8: The volumetric air flow rates are the same.

$$\dot{V}_{a2} = \dot{V}_{a1} \qquad \text{29.73}$$

step 9: The original air and fuel rates are

$$\dot{m}_{f1} = (BSFC_1)(BHP_1) \qquad \text{29.74}$$

$$\dot{m}_{a1} = (AFR)(\dot{m}_{f1}) \qquad \text{29.75}$$

$$\dot{V}_{a1} = \frac{\dot{m}_{a1}}{\rho_{a1}} \qquad \text{29.76}$$

step 10: The new air mass flow rate is

$$\dot{m}_{a2} = \dot{V}_{a2}\rho_{a2} \qquad \text{[see step 8]} \qquad 29.77$$

step 11: For engines with metered injection, $\dot{m}_{f2} = \dot{m}_{f1}$. For engines with carburetors,

$$\dot{m}_{f2} = \frac{\dot{m}_{a2}}{\text{AFR}} \qquad 29.78$$

step 12: The new fuel consumption is

$$\text{BSFC}_2 = \frac{\dot{m}_{f2}}{\text{BHP}_2} \qquad 29.79$$

19. GAS TURBINES

Combustion turbines (CTs), or "gas" turbines (GTs), are the preferred combustion engines in applications much above 10 MW. Large units regularly operate in the 100 to 200 MW range (up to approximately 230 MW).[21] Some smaller CTs—typically less than 40,000 hp (30 MW)—for such applications as marine propulsion and pipeline compression are rated in standard horsepower.

There are two general CT categories. The traditional heavy-duty, industrial CT, and the smaller, lighter aeroderivative CT. Heavy-duty turbines typically have a single shaft. The rotor is supported on two bearings, and the thrust bearing is on the compressor end. The generator is direct-driven from the compressor (i.e, *cold-end drive*). Small individual combustors surround the hot end radially. The *axial exhaust* duct goes directly into the steam generator in cogeneration and combined cycle plants.

The *aeroderivative combustion turbine* is basically a jet engine that exhausts into a turbine generator. Output is less than 50 MW per unit, and most aeroderivative CTs produce less than 40 MW. Split shafts are common in this range. The power turbine and generator are usually mounted at the "hot end" of the gas generator. The generator is run from a gearbox, allowing the turbine to run at higher, more efficient speeds.

The compression ratio (based on pressures) in the compression stage is typically 11:1 to 16:1, with most heavy-duty turbines in the 14:1 to 15:1 range. Aeroderivative turbines have higher compression ratios—typically 19:1 to 21:1, even as high as 30:1. Most heavy-duty combustion turbines have 16 to 18 compression stages. However, the Mach number at the tip of the of the first-stage rotor has risen from approximately 0.9 to approximately 1.4, reducing the number of compression stages to approximately 12 to 15 in modern transsonic turbines.

The temperature of the gas entering the expander section is typically 2200 to 2350°F (1200 to 1290°C).[22] The exhaust temperature is typically 1000 to 1100°F (540 to 590°C), which makes the exhaust an ideal heat source for combined cycles. Most combustion turbines have three to four expander stages. The exhaust flow rate in modern heavy-duty turbines per 100 MW is approximately 525 to 550 lbm/sec (240 to 250 kg/s).

Combustors vary widely in design. Large, single chambers (i.e., "cans") have large residence times and allow heavy fuels to be burned completely. Smaller, multiple chambers and annular burners perform better for gaseous and distillate fuels. Film cooling is no longer adequate for wall cooling. Intensive convection cooling and thermal barrier coatings, including ceramic tiles, are needed.

20. STEAM INJECTION

Steam and water can be injected into the combustor to lower the exhaust temperature, inhibiting the formation of nitrogen oxides. An injection rate on the order of one-half pound of water per pound of fuel is sufficient to keep NOx emissions below older limits in the 75 to 150 ppm range. More water can be used to reduce the emissions somewhat below those values. However, to reach the strictest limits (e.g., less than 10 ppm), selective catalytic reduction (SCR) is needed. NOx emissions can also be controlled to intermediate values (to approximately 25 ppm) without steam in "*dry combustors*" (*low-NOx burners*) based on *staged lean combustion*, also known as *sequential combustion*.

Steam is also injected into the expansion section (sometimes referred to as the *expander*) of CTs, a process known as *power-boosting*. Power-boost steam injection is particularly applicable with aeroderivative turbines where steam pressures are consistent with higher (20:1 or more) compression ratios. Steam injection is popular in cogeneration plants where process steam use is variable, and excess steam can be routed back to the turbine. NOx reduction occurs, but this is not the primary purpose. Steam, which absorbs heat better than air, is also widely used for cooling gas turbines.

21. TURBINE FUELS

Modern turbines can burn a wide range of gaseous and distillate fuels, and they can switch from one fuel to another over the entire load range. Natural gas is the most economical and, therefore, the most common fuel in combustion turbines used for electrical power generation. Propane, No. 2 oil, and kerosene are used as backup fuels.

Turbines can also be partially fueled by *synthetic gas* ("syngas") generated from coal. The *gasifier* may be a fixed-bed, fluid-bed, or entrained flow type. Both

---

[21] In the past, gas turbines were plagued by poor reliability, availability, and maintainability (RAM). The RAM record of modern turbines, particularly aeroderivative types, is excellent.

[22] New materials, coatings, and other devices in *advanced turbine systems* (ATS) may someday push the upper limit to the 2600°F (1425°C) mark.

air-blown and oxygen-blown gasification processes can be used, although the oxygen-blown process requires a separate oxygen plant. The heating value of syngas is low, approximately 240 Btu/scf (8.9 MJ/m^3). So, combustion turbines cannot make full load on syngas alone.

In the *gasification process*, coal-water slurry is pumped into the gasifier. Oxygen or air is added, forming a hot, partially burned gas consisting of carbon monoxide, hydrogen sulfide, and carbonyl sulfide. Most of the noncarbon material in the coal melts and flows out of the gasifier as *slag*. Hot-gas clean-up equipment removes particulates, sulfur, and other impurities from the gas. Gasification is relatively insensitive to coal feedstock.

In future advances, finely ground coal may be used in *direct coal-fueled turbines* (DCFTs). In the past, attempts to inject coal directly into the turbine have been plagued by severe erosion of turbine blades by tiny ash particles, pluggage of the gas flow passages by ash deposits, and corrosion of high-temperature metallic surfaces by alkali compounds. Other considerations are increased emissions and cost.

Traditional gas turbines burn fuel within the envelope of the engine. However, the in-line combustor does not provide adequate residence time to burn coal. Therefore, an "external combustor" must be used with DCFTs.[23] The shortcomings associated with burning ground coal may be overcome by some form of staged combustion, sometimes referred to as "rich-quench-lean" (RQL) firing.

With RQL external combustion, finely ground coal in powder or slurry form (known as *coal-water mixture* or CWM) is burned in a fuel-rich, oxygen-starved first-stage combustor at about 3000°F (1650°C). The low oxygen level inhibits NOx formation from fuel-bound nitrogen. The fuel-rich gas is quenched with water to approximately 2000°F (1100°C), inhibiting thermal NOx formation, and solidifying coal ash so that it can be removed in a cyclone separator, inertial slag separator, or ceramic filter.[24] Clean gas is then sent to the fuel-lean second-stage combustor where additional air is injected, and the temperature increases to approximately 2800°F (1540°C) as the carbon monoxide and hydrogen components burn. Sulfur emissions are controlled by injection of calcium-based sorbents in either the first- or second-stage combustors.

22. INDIRECT COAL-FIRED TURBINES

With *indirect coal-fired turbines* (ICFTs), the coal combustion products never enter the turbine expander. Rather, heat from combustion is transferred through a

[23]The external combuster is located "off base" on a separate combustion "island."
[24]There is also promise in *pressurized slagging combustors* where slag is removed in liquid form.

closed heat exchanger to the compressed air. Only clean air flows through the expander. Since the heat transfer occurs at over 2000°F (1100°C), well over the 1600°F (870°C) limit of traditional metallic heat exchangers, a special ceramic heat exchanger must be used. When used in a combined cycle, the term *externally fired combined cycle* (EFCC) is used to describe this process.

Theoretically, the thermal efficiency of the cycle can be increased by closing the process (i.e., making it a *closed turbine cycle*) where the exhaust from the expander is returned to the compressor for reuse.

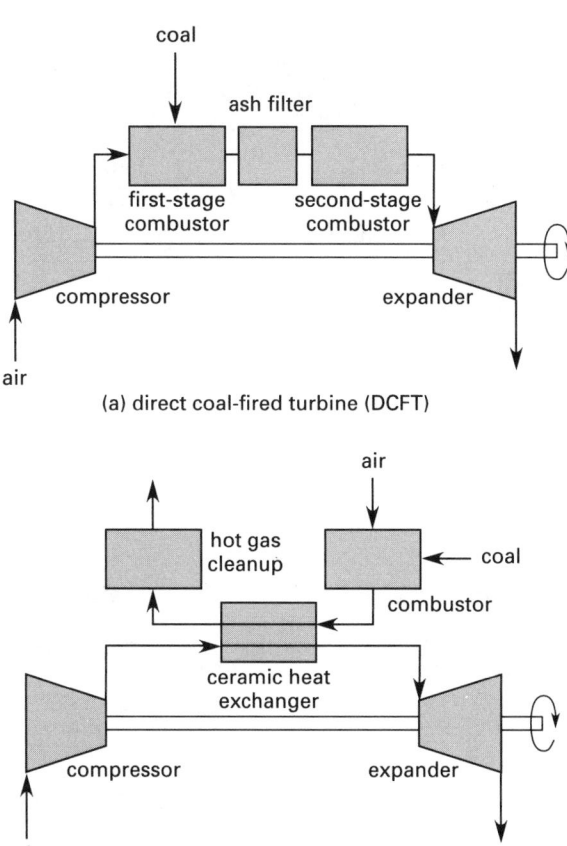

Figure 29.9 *Direct and Indirect Coal-Fired Turbines*

23. BRAYTON GAS TURBINE CYCLE

Strictly speaking, the *Brayton gas turbine cycle* (also known as the *Joule cycle*) is an internal combustion cycle. It differs from the previous cycles in that each process is carried out in a different location, air flow and fuel injection are steady, and air-standard calculations are realistic since a large air-fuel ratio is used to keep combustion temperatures below metallurgical limits.

Figure 29.10 illustrates the physical arrangement of components used to achieve the Brayton cycle. Almost all installations drive the compressor from the turbine. (Approximately 50 to 75% of the turbine power is required to drive the high-efficiency compressor.) The actual arrangement differs from the air-standard property plot shown in Fig. 29.11 in that the exhaust products exiting at point D are not cooled and returned to the compressor at point A. The processes in the air-standard Brayton cycle follow.

A to B: isentropic compression in the compressor

B to C: constant pressure heat addition in the combustor

C to D: isentropic expansion in the turbine

D to A: constant pressure cooling

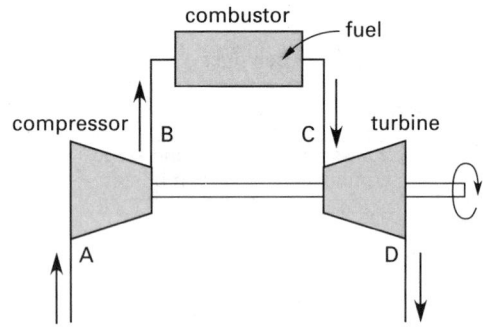

Figure 29.10 *Gas Turbine*

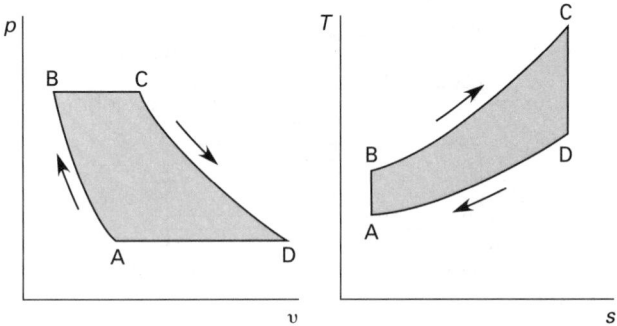

Figure 29.11 *Brayton Gas Turbine Cycle*

Combustors are said to be *open* if the incoming air and combustion products flow to the turbine. A high percentage, 30 to 60%, of excess air is needed to cool the gases. Depending on the turbine construction details, the temperature of the air entering the turbine will be between 1200°F and 1800°F (650°C and 1000°C). Turbojet and turboprop engines typically use *open combustors*.

A *closed combustor* is a heat exchanger. Air flowing to the turbine is not combined with the combustion gases. This allows any type of fuel to be used, but the bulk of the heat exchanger limits closed combustors to stationary power plants and pipeline pumping stations.

Ideal gas relationships for steady-flow systems can be used to evaluate the p, V, and T properties at points A, B, C, and D. Constant values of k, c_p, and c_v can be

assumed if air is considered ideal.[25] An air table also can be used if the isentropic efficiencies are known. Usually, it is assumed that p_B and p_C are equal.

At A:

$$p_A = p_D \text{ [usually atmospheric]} = p_B \left(\frac{v_B}{v_A}\right)^k$$

$$= p_B \left(\frac{T_A}{T_B}\right)^{\frac{k}{k-1}} \qquad 29.80$$

$$T_A = T_D \left(\frac{v_A}{v_D}\right) = T_B \left(\frac{v_B}{v_A}\right)^{k-1}$$

$$= T_B \left(\frac{p_A}{p_B}\right)^{\frac{k-1}{k}} \qquad 29.81$$

$$v_A = v_D \left(\frac{T_A}{T_D}\right) = v_B \left(\frac{p_B}{p_A}\right)^{\frac{1}{k}}$$

$$= v_B \left(\frac{T_B}{T_A}\right)^{\frac{1}{k-1}} \qquad 29.82$$

At B:

$$p_B = p_C = p_A \left(\frac{v_A}{v_B}\right)^k = p_A \left(\frac{T_B}{T_A}\right)^{\frac{k}{k-1}} \qquad 29.83$$

$$T_B = T_A \left(\frac{v_A}{v_B}\right)^{k-1} = T_A \left(\frac{p_B}{p_A}\right)^{\frac{k-1}{k}}$$

$$= T_C \left(\frac{v_B}{v_C}\right) \qquad 29.84$$

$$v_B = v_A \left(\frac{p_A}{p_B}\right)^{\frac{1}{k}} = v_A \left(\frac{T_A}{T_B}\right)^{\frac{1}{k-1}}$$

$$= v_C \left(\frac{T_B}{T_C}\right) \qquad 29.85$$

At C:

$$p_C = p_B = p_D \left(\frac{v_D}{v_C}\right)^k = p_D \left(\frac{T_C}{T_D}\right)^{\frac{k}{k-1}} \qquad 29.86$$

$$T_C = T_B \left(\frac{v_C}{v_B}\right) = T_D \left(\frac{v_D}{v_C}\right)^{k-1}$$

$$= T_D \left(\frac{p_C}{p_D}\right)^{\frac{k-1}{k}} \qquad 29.87$$

$$v_C = v_B \left(\frac{T_C}{T_B}\right) = v_D \left(\frac{p_D}{p_C}\right)^{\frac{1}{k}}$$

$$= v_D \left(\frac{T_D}{T_C}\right)^{\frac{1}{k-1}} \qquad 29.88$$

[25]Another typical assumption, justified on the basis of large amounts of excess air, is that the gas is all nitrogen.

At D:

$$p_D = p_C \left(\frac{v_C}{v_D}\right)^k = p_C \left(\frac{T_D}{T_C}\right)^{\frac{k}{k-1}} \qquad \textit{29.89}$$

$$T_D = T_C \left(\frac{v_C}{v_D}\right)^{k-1} = T_C \left(\frac{p_D}{p_C}\right)^{\frac{k-1}{k}}$$

$$= T_A \left(\frac{v_D}{v_A}\right) \qquad \textit{29.90}$$

$$v_D = v_C \left(\frac{p_C}{p_D}\right)^{\frac{1}{k}} = v_C \left(\frac{T_C}{T_D}\right)^{\frac{1}{k-1}}$$

$$= v_A \left(\frac{T_D}{T_A}\right) \qquad \textit{29.91}$$

The work and heat flow terms are

$$q_{in} = c_p(T_C - T_B) = h_C - h_B \qquad \textit{29.92}$$

$$q_{out} = |c_p(T_A - T_D)| = |h_A - h_D| \qquad \textit{29.93}$$

$$W_{turbine} = |c_p(T_D - T_C)| = |h_D - h_C| \qquad \textit{29.94}$$

$$W_{compressor} = c_p(T_B - T_A) = h_B - h_A \qquad \textit{29.95}$$

$$\eta_{th} = \frac{Q_{in} - Q_{out}}{Q_{in}} = \frac{W_{turbine} - W_{compressor}}{Q_{in}}$$

$$= \frac{(h_C - h_B) - (h_D - h_A)}{h_C - h_B} \qquad \textit{29.96}$$

If the gas is ideal so that c_p is constant, then

$$\eta_{th} = \frac{(T_C - T_B) - (T_D - T_A)}{T_C - T_B} \qquad \textit{26.97}$$

If the isentropic efficiency is less than 100% for either or both the compressor and turbine, the actual enthalpies or temperatures must be used to calculate the work, heat, and thermal efficiency.

$$h'_B = h_A + \frac{h_B - h_A}{\eta_{s,compressor}} \qquad \textit{29.98}$$

$$T'_B = T_A + \frac{T_B - T_A}{\eta_{s,compressor}} \qquad \textit{29.99}$$

$$h'_D = h_C - \eta_{s,turbine}(h_C - h_D) \qquad \textit{29.100}$$

$$T'_D = T_C - \eta_{s,turbine}(T_C - T_D) \qquad \textit{29.101}$$

Simple cycle (SC) operation means that the turbine is not part of a cogeneration or combined cycle system. The full-load thermal efficiency of existing heavy-duty combustion turbines in simple cycles is approximately 34 to 36%, while new turbines on the cutting edge of technology (i.e., *advanced turbine systems*, ATS) are able to achieve 38 to 38.5%. Aeroderivative turbines commonly achieve efficiencies up to 42%. The heat rate can be found from the cycle efficiency.

The *back work ratio* (BWR), which is approximately 50 to 75%, is the ratio of the compressor work to the turbine expansion work.

Example 29.4

Air enters the compressor of a gas turbine at 14.7 psia and 540°R (101.3 kPa and 300K). The pressure ratio is 4.5:1. The conditions at the turbine inlet are 64 psia and 2200°R (440 kPa and 1220K). The turbine's expansion pressure ratio is 1:4, and exhaust is to the atmosphere. The isentropic efficiency of the turbine is 85%. What is the thermal efficiency of the cycle?

SI Solution

Assume an ideal gas. (The customary U.S. solution solves this example using air tables.)

Refer to Fig. 29.11.

At A:

$$T_A = 300K \quad \text{[given]}$$

$$p_A = 101.3 \text{ kPa} \quad \text{[given]}$$

At B:

$$T_B = T_A \left(\frac{p_B}{p_A}\right)^{\frac{k-1}{k}}$$

$$= (300K)(4.5)^{\frac{1.4-1}{1.4}} = 461K$$

$$p_B = (4.5)(101.3 \text{ kPa}) = 456 \text{ kPa}$$

At C:

$$T_C = 1220K \quad \text{[given]}$$

$$p_C = 440 \text{ kPa} \quad \text{[given]}$$

At D:

$$p_D = \frac{440 \text{ kPa}}{4} = 110 \text{ kPa}$$

If the expansion had been isentropic, the temperature would have been

$$T_D = T_C \left(\frac{p_D}{p_C}\right)^{\frac{k-1}{k}}$$

$$= (1220K) \left(\frac{1}{4}\right)^{\frac{1.4-1}{1.4}}$$

$$= 821K$$

For ideal gases, the specific heats are constant. Therefore, the change in internal energy (and enthalpy, approximately) is proportional to the change in temperature. The actual temperature is

$$T'_D = T_C - \eta_{s,turbine}(T_C - T_D)$$

$$= 1220K - (0.85)(1220K - 821K)$$

$$= 881K$$

The thermal efficiency is given by Eq. 29.97.

$$\eta_{th} = \frac{(T_C - T_B) - (T_D - T_A)}{T_C - T_B}$$

$$= \frac{(1220\text{K} - 461\text{K}) - (881\text{K} - 300\text{K})}{1220\text{K} - 461\text{K}}$$

$$= 0.235 \quad (23.5\%)$$

Customary U.S. Solution

(Use an air table. The SI solution assumes an ideal gas.)

At A:

$$T_A = 540°\text{R} \quad \text{[given]}$$

$$p_A = 14.7 \text{ psia} \quad \text{[given]}$$

$$h_A = 129.06 \text{ Btu/lbm}$$

$$p_{r,A} = 1.3860$$

At B: The A-to-B process is isentropic.

$$p_{r,B} = p_{r,A}\left(\frac{p_B}{p_A}\right) = (1.3860)(4.5)$$

$$= 6.237$$

Locate this pressure ratio in the air table.

$$T_B = 827.6°\text{R}$$

$$h_B = 198.5 \text{ Btu/lbm}$$

$$p_B = (4.5)(14.7 \text{ psia}) = 66.15 \text{ psia}$$

At C:

$$T_C = 2200°\text{R} \quad \text{[given]}$$

$$p_C = 64 \text{ psia} \quad \text{[given]}$$

$$h_C = 560.59 \text{ Btu/lbm}$$

$$p_{r,C} = 256.6$$

At D:

$$p_{r,D} = p_{r,C}\left(\frac{p_D}{p_C}\right)$$

$$= \frac{256.6}{4} = 64.15$$

Locate this pressure ratio in the air table.

$$T_D = 1554.6°\text{R}$$

$$h_D = 383.66 \text{ Btu/lbm}$$

Since the efficiency of the expansion process is 85%,

$$h'_D = h_C - \eta_{s,\text{turbine}}(h_C - h_D)$$

$$= 560.59 \frac{\text{Btu}}{\text{lbm}}$$

$$\quad - (0.85)\left(560.59 \frac{\text{Btu}}{\text{lbm}} - 383.66 \frac{\text{Btu}}{\text{lbm}}\right)$$

$$= 410.2 \text{ Btu/lbm}$$

The thermal efficiency is

$$\eta_{th} = \frac{(h_C - h_B) - (h_D - h_A)}{h_C - h_B}$$

$$= \frac{\left(560.59 \dfrac{\text{Btu}}{\text{lbm}} - 198.5 \dfrac{\text{Btu}}{\text{lbm}}\right) - \left(410.2 \dfrac{\text{Btu}}{\text{lbm}} - 129.06 \dfrac{\text{Btu}}{\text{lbm}}\right)}{560.59 \dfrac{\text{Btu}}{\text{lbm}} - 198.5 \dfrac{\text{Btu}}{\text{lbm}}}$$

$$= 0.224 \quad (22.4\%)$$

24. AIR-STANDARD BRAYTON CYCLE WITH REGENERATION

Regeneration is used to improve the efficiency of the Brayton cycle. Regeneration involves transferring some of the heat from the exhaust products to the air in the compressor. The transfer occurs in a *regenerator*, which is a crossflow heat exchanger. There is no effect on turbine work, compressor work, or net output. However, the cycle is more efficient since less heat is added. Of course, T_B cannot be greater than T_F. Similarly, T_C cannot be greater than T_E.

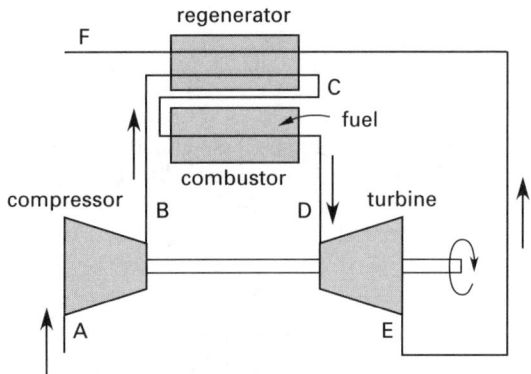

Figure 29.12 *Gas Turbine with Regeneration*

The processes involved are:

A to B: isentropic compression

B to C: constant pressure heat addition in regenerator

C to D: constant pressure heat addition in combustor

D to E: isentropic expansion

E to F: constant pressure heat removal in the regenerator

F to A: constant pressure heat removal in the sink

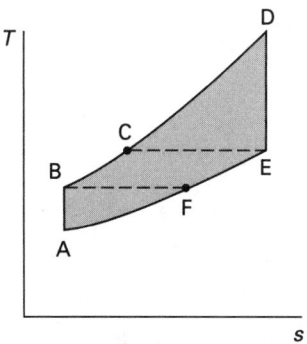

Figure 29.13 *Brayton Cycle with 100% Efficient Regeneration*

If $T_C = T_E$, the regenerator is said to be 100% efficient. Otherwise, the *regenerator efficiency*, also known as *regeneration effectiveness*, is calculated from Eq. 29.102. Actual regeneration efficiency rarely exceeds 75%.

$$\eta_{\text{regenerator}} = \frac{h_C - h_B}{h_E - h_B} \qquad 29.102$$

The thermal efficiency of the air-standard Brayton cycle with regeneration is

$$\eta_{\text{th}} = \frac{W_{\text{out}} - W_{\text{in}}}{Q_{\text{in}}} = \frac{(h_D - h_E) - (h_B - h_A)}{h_D - h_C} \qquad 29.103$$

If air is considered to be an ideal gas, temperatures can be substituted for enthalpies in Eqs. 29.102 and 29.103.

25. BRAYTON CYCLE WITH REGENERATION, INTERCOOLING, AND REHEATING

Multiple stages of compression and expansion can be used to improve the efficiency of the Brayton cycle. Physical limitations usually preclude more than two stages of intercooling and reheat. (This section assumes only one stage of each.) The physical arrangement is shown in Fig. 29.14.

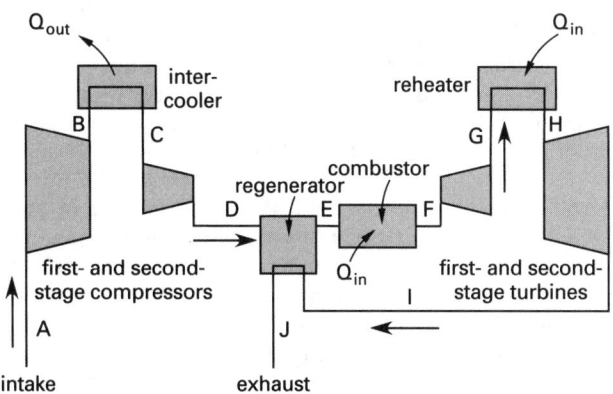

Figure 29.14 *Augmented Gas Turbine*

The processes are:

A to B: isentropic compression

B to C: cooling at constant pressure (usually back to T_A)

C to D: isentropic compression

D to F: constant pressure heat addition

F to G: isentropic expansion

G to H: reheating at constant pressure in combustor or reheater (usually back to T_F)

H to I: isentropic expansion

I to A: constant pressure heat rejection

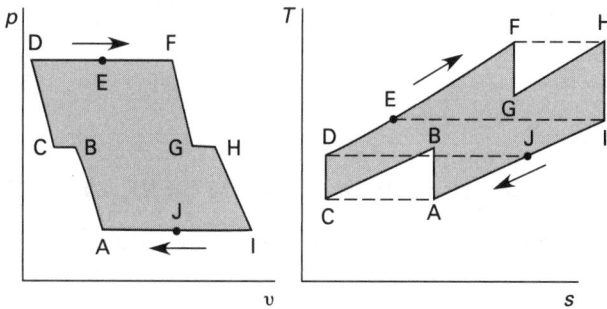

Figure 29.15 *Augmented Brayton Cycle*

Calculation of the work and heat flow terms and of thermal efficiency is similar to that in the previous cycles, except that there are two W_{turbine}, $W_{\text{compressor}}$, and Q_{in} terms. If efficiencies for the compressor, turbine, and regenerator are given, the following relationships are required.

$$p'_B = p_B \qquad 29.104$$

$$h'_B = h_A + \frac{h_B - h_A}{\eta_{s,\text{compressor}}} \qquad 29.105$$

$$p'_D = p_D \qquad 29.106$$

$$h'_D = h_C + \frac{h_D - h_C}{\eta_{s,\text{compressor}}} \qquad 29.107$$

$$p'_E = p_E \qquad 29.108$$

$$h'_E = h_D + \eta_{\text{regenerator}} \, (h_I - h_D) \qquad 29.109$$

$$p'_G = p_G \qquad 29.110$$

$$h'_G = h_F - \eta_{s,\text{turbine}} \, (h_F - h_G) \qquad 29.111$$

$$p'_I = p_I \qquad 29.112$$

$$h'_I = h_H - \eta_{s,\text{turbine}} \, (h_H - h_I) \qquad 29.113$$

Optimum performance improvement with multiple staging is achieved when the pressure ratio across each turbine stage (either compression stages or expansion stages) is the same. Referring to Figs. 29.14 and 29.15, for the compressors, $p_B = p_C = \sqrt{p_A p_D}$. For the turbine, $p_G = p_H = \sqrt{p_F p_I}$.

Figure 29.16 *Typical Heat Recovery Steam Generator*

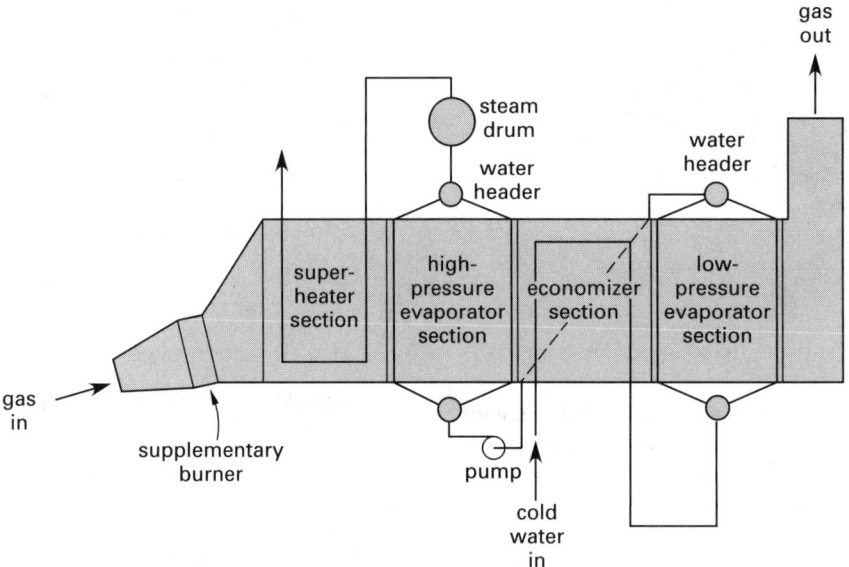

26. HEAT RECOVERY STEAM GENERATORS

All simple power cycles end up discarding most of the incoming heat energy. For example, a thermal efficiency of 40% means that 60% of the heat energy is lost. Combustion cycles (gas turbine cycles in particular) produce high-temperature, 800 to 1100°F (430 to 590°C) exhaust streams.[26] Some of this energy can be recovered.

All cogeneration and combined cycles make use of some type of *heat-recovery steam generator* (HRSG), also known as a *waste-heat boiler*. Although fire-tube designs are applicable to low gas flow rates—less than 100,000 lbm/hr (12.6 kg/s)—due to the large mass flow rates of exhaust gases typically encountered and the high-pressure 1000 to 2800 psi (7 to 19 MPa) steam generated, HRSGs are typically of the water-tube design. When turbines are fueled by natural gas, the exhaust is clean and does not pose a corrosion problem for HRSGs.[27]

The exhaust gases pass sequentially through superheater, high-pressure evaporator, economizer, and low-pressure evaporator sections. (A selective catalytic conversion (SCR) section may also be used, though no heat transfer occurs in it.) To extract more heat energy, extended surfaces (i.e., fins) may be used on the

tubes. The maximum fin temperature occurs at the tip, and fin material limits the operating temperature of the HRSG. Since the tube-side heat transfer coefficient is high, on the order of 1000 to 3000 Btu/hr-ft²-°F (1.7 to 5.2 kW/m·K) in economizers and LP evaporators, the *fin density* is also high—on the order of 2 to 5 fins per inch. The fin density is lower in the superheater section.

As in traditional steam generators, steam drums in HRSGs can be either natural or forced circulation in design. The *circulation ratio* is the ratio of the mass of circulating steam-water (in the risers and downcomers) to the mass of generated steam. It varies from 10 to 40 for natural circulation and from 3 to 10 for forced circulation.

The temperature difference between the gas side and steam sides varies along the run of the HRSG and with variations in gas flow, inlet temperature, and extent of supplemental firing. The *pinch point* is the minimum temperature difference. The *approach point* is the difference between the saturation temperature and the temperature of the leaving water (in the economizer). Trial and error solutions are usually required to determine the entering and leaving temperatures at each section. For trial and error solutions, the pinch and approach points may initially be assumed to be 20°F and 15°F (11°C and 8°C), respectively.

HRSGs may be unfired, supplementary fired, or furnace fired. (Fired HRSGs are more common with combined cycles than with cogeneration.) *Unfired HRSGs* are usually one- or two-pass designs. Since the turbine exhaust contains approximately 16% oxygen by volume, additional fuel can be sprayed into the exhaust stream, usually after the superheater portion of the HRSG. This option is known as a *supplementary-fired*

---

[26] The exhaust from municipal waste incinerators is approximately 1800°F (980°C), and some chemical waste incinerators produce 2000 to 2400°F (1090 to 1320°C) combustion gases.

[27] Combustion gas from municipal incinerators contains particulates that can cause slagging, and such gas is corrosive at both high and low temperatures. Above 800°F (430°C), hydrogen chloride (HCl) formed from the combustion of waste plastics, is very corrosive. Chlorine, formed when incinerating chlorinated wastes, is even more corrosive than HCl on carbon steel above 400 to 450°F (200 to 230°C). Therefore, HRSGs associated with the burning of such wastes must operate in a narrow temperature band below 400°F (200°C) and above the acid-vapor dew point.

Figure 29.17 *Pinch and Approach Points*

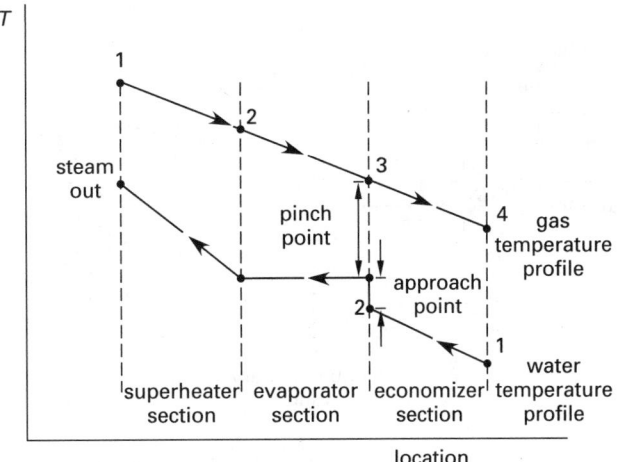

HRSG. Temperatures are limited to approximately 1700°F (930°C) by the liner material. Other than that, supplementary-fired units are similar to unfired versions.

Temperatures in excess of 1700°F (930°C) can be achieved with *furnace-fired HRSGs.* Temperatures up to about 2300°F (1260°C) can be achieved with a duct burner and water-cooled membrane walls; higher temperatures require special register burners with their own air chambers.

Referring to Fig. 29.17, the heat balance in the superheater and evaporator sections is

$$\dot{m}_{\text{gas}} c_{p,\text{gas}} (T_{\text{gas},1} - T_{\text{gas},3})$$
$$= \dot{m}_{\text{steam}} (h_{\text{steam,out}} - h_{\text{water},2}) \qquad 29.114$$

The heat balance for the complete HRSG is

$$\dot{m}_{\text{gas}} c_{p,\text{gas}} (T_{\text{gas},1} - T_{\text{gas},4})$$
$$= \dot{m}_{\text{steam}} (h_{\text{steam,out}} - h_{\text{water},1}) \qquad 29.115$$

The *transferred duty* (or just "duty") of the HRSG is the total heat transfer rate for all components (superheater, evaporator, economizer, etc.).

HRSGs often operate far from their design point. Equation 29.116 can be used to predict the performance of each section of the HRSG at one mass flow rate based on known performance at another mass flow rate.

$$\frac{U_1 A_1}{U_2 A_2} = \frac{\dfrac{\dot{Q}_1}{T_1}}{\dfrac{\dot{Q}_2}{T_2}} \approx \left(\frac{\dot{m}_1}{\dot{m}_2} \right)^{0.65} \qquad 29.116$$

27. COGENERATION CYCLES

In *cogeneration* plants, HRSGs convert waste heat (generally from turbine exhaust) into low-pressure high-quality or saturated process steam with pressures of 10 to 300 psig (70 to 210 kPa).[28] *Process steam* is steam that is used for some other process, such as *space heating* (also known as *district heating*), other than electrical power generation.[29]

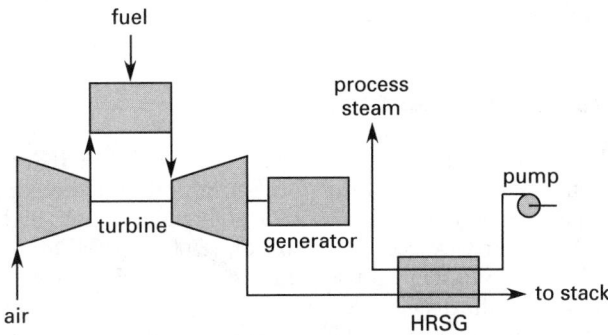

Figure 29.18 *Simple Cogeneration Process*

Cogeneration systems can be configured as topping or bottoming cycles. In a *topping cycle*, the primary fuel produces electricity first. The turbine exhaust (either steam or gas) is captured to make steam. In a *bottoming cycle*, the waste heat from an industrial process is captured to produce steam first for the process and then to power a turbine. Bottoming cycles, being less efficient and requiring supplementary firing, are less common.

---

[28]Not all combined cycles utilize exhaust from gas turbines. High-temperature diesel engine exhaust is rich in oxygen—ideal for use as preheated combustion air in smaller plants.

[29]Theoretically, the steam could be used for cooling in an absorption system. This probably is never done in practice, however.

Figure 29.19 *Combined Cycle Process*

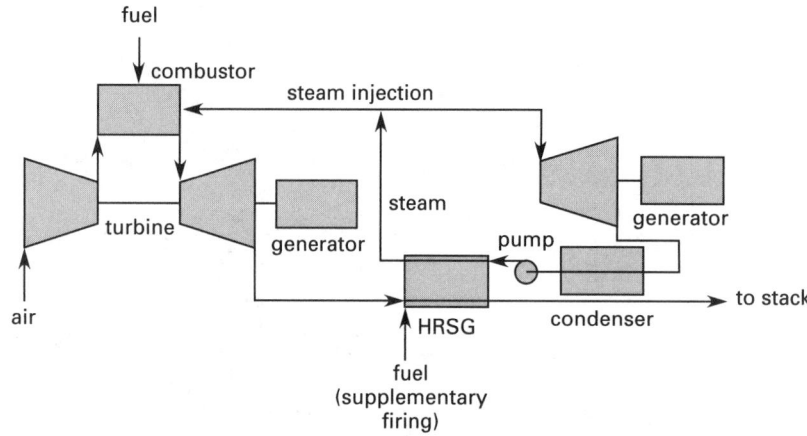

The *power-to-heat ratio* (PTR) is

$$\text{PTR} = \frac{P_{\text{turbine}}}{\dot{Q}_{\text{recovered}}} \qquad 29.117$$

The *fuel utilization* (FU) is not the same as thermal efficiency. Fuel utilizations for different units should be compared only for equal PTRs. Highly efficient cycles have fuel utilizations above 80%.

$$\text{FU} = \frac{P_{\text{turbine}} + \dot{Q}_{\text{recovered}}}{\dot{Q}_{\text{in}}} \qquad 29.118$$

28. COMBINED CYCLES

In a *combined cycle*, the heat recovered in an HRSG is used to vaporize water for use in a Rankine steam cycle. Steam from the HRSG is high pressure, high temperature, exceeding 750 psi and 700°F (5 MPa and 370°C). Supplementary firing in the HRSG may be used. Combined cycle efficiencies are much higher than simple cycles, typically 52 to 54%. Some state-of-the-art prototype ("proof of concept") installations are able to achieve combined efficiencies of 56 to 58%. A combined efficiency of 60% appears to be achievable.[30]

A plethora of confusing acronyms identify combined cycle (CC) variations, including GFCC (gas-fired turbine combined cycle), GTCC (gas turbine combined cycles), GCC (gasification combined cycle), CGCC (coal gasification combined cycle), DCCC (diesel coal combined cycle), EFCC (externally fired combined cycle), and IGCC (integrated gasification combined cycle). Some of these terms are synonymous.

---

[30]The practical upper limit for combined cycle efficiencies is considered to be approximately 70%. The Carnot efficiency sets the ideal upper limit. This is 81% for cycles operating between 70°F (21°C) and 2350°F (1290°C).

Figure 29.20 illustrates an IGCC cycle. "Integrated" refers to the fact that the gasifier uses part of the compressor discharge rather than operating as an independent source. IGCC is an inherently low-emissions process because ash is removed in the gasification process and sulfur is removed from the fuel gas, not the flue gas. Overall efficiencies with IGCC are approximately 42%—lower than some combined cycles, but still much higher than the traditional coal-fired plant with emission controls.

29. HIGH-PERFORMANCE POWER SYSTEMS

It may be possible to combine all of the advanced technologies into a combined cycle and make coal combustion as efficient as natural gas.[31] In the prototypical *high-performance power system* (HIPPS), a *high-temperature advanced furnace* (HITAF, which integrates the combustion of coal, heat transfer, and emissions control into a single unit) provides most of the heat for a Brayton topping cycle. Supplementary firing with clean fuel (such as natural gas) boosts the air temperature to achieve optimum gas turbine performance. Energy in the HITAF can also be used for superheating and reheating in the conventional Rankine bottoming cycle. Gas turbine exhaust is routed to the HITAF as combustion air.

30. REPOWERING AND LIFE-EXTENSION

The average useful life of most coal-fired utility power plants is 30 to 40 years. However, because of increasing demand, utilities are unlikely to retire any significant portion of their old capacity. Rather, the aging plants are improved. *Repowering* is a popular term used by utilities to mean upgrading an existing plant. In a sense,

---

[31]However, it may not be economical. Time will tell.

Figure 29.20 *Integrated Gasification Combined Cycle*

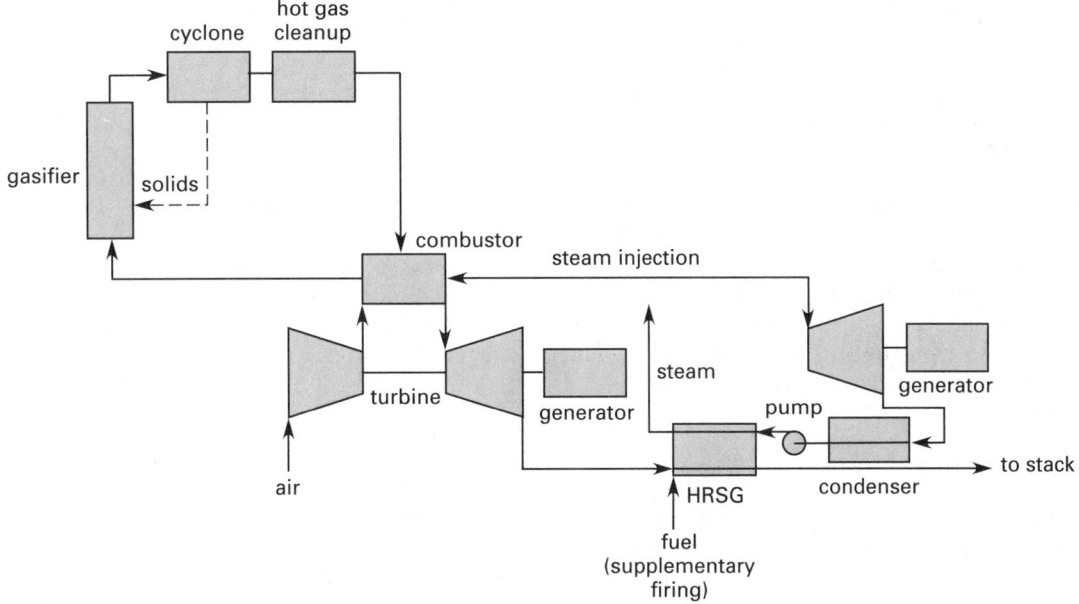

repowering is a variation of *life extension*.[32] However, life extension also includes programs to increase reliability, availability, and maintainability (RAM) without changing equipment. For example, many programs concentrate on weak areas, such as boiler tube failure (BTF, the leading cause of forced outages).

There are many existing electrical generating plants in the United States that don't even have reheat boilers. Repowering concentrates on replacing the furnace and boiler with clean-burning, more efficient coal combustion (as in fluidized-bed combustion) or gasification while retaining the remainder of the existing plant (i.e., the coal handling, steam cycle, and generating equipment).

In the face of massive boiler repairs or replacement, a utility may choose to repower by replacing boilers with gas turbines and HRSGs, using the existing steam turbine in a topping cycle. This is a good option if the existing steam cycle is below that of the HRSG output, generally, less than 1450 psig (10 MPa). There are three general ways the gas turbine can be integrated into the existing plant. (1) In the *hot windbox system*, some or all of the turbine exhaust is routed to the furnace as combustion air. (2) Turbine exhaust can be used for feedwater heating. (3) Turbine exhaust can be used as the heat source in a *supplementary boiler* or separate superheater.

Though often overlooked, the fuel flexibility and efficiency of large diesel engines make diesel combined cycles (DCC) ideal candidates in repowering options in smaller power plants.

PRACTICE PROBLEMS

1. Air expands isentropically at the rate of 10 ft^3/sec (280 L/s) from 200 psia and 1500°F (1.4 MPa and 820°C) to 50 psia (350 kPa). What are the air's (a) final temperature, (b) final volume, and (c) enthalpy change?

2. A 10 in (250 mm) bore, 18 in (460 mm) stroke, two-cylinder, four-stroke internal combustion engine operates with a mean effective pressure of 95 psig (660 kPa) at 200 rpm. The actual torque developed is 600 ft-lbf (820 N·m). What is the friction horsepower?

3. A four-cycle internal combustion engine has a bore of 3.1 in (80 mm) and a stroke of 3.8 in (97 mm). When the engine is running at 4000 rpm, the air-fuel mixture enters the cylinders during the intake stroke with a velocity of 100 ft/sec (30 m/s). The intake valve opens at TDC and closes at 40° past BDC. The volumetric efficiency is 65%. What is the effective area of the intake valve?

4. An engine runs on the Otto cycle. The compression ratio is 10:1. The total intake volume of the cylinders is 11 ft^3 (0.3 m^3). Air enters at 14.2 psia and 80°F (98 kPa and 27°C). After the compression stroke, 160 Btu (179 kJ) of heat are added. (a) What is the temperature after the heat addition? (b) What is the thermal efficiency of the cycle?

[32]Utilities in the United States are adverse to using the term *life extension*. The reason lies in the 1970 Clean Air Act. Power plants built prior to 1971 were exempted from emissions control requirements. These "grandfathered" plants lose their exemption if they are significantly modified. "Refurbishment" and "repair," maybe; "life extension," no.

5. A 4.25 in × 6 in (110 mm × 150 mm), six-cylinder diesel engine runs at 1200 rpm while consuming 28 lbm of fuel per hour (0.0035 kg/s). The actual volumetric fraction of carbon dioxide in the exhaust is 9% (dry basis). At another throttle setting, when the air-fuel ratio is 15:1, there is 13.7% (dry basis) carbon dioxide in the exhaust. The atmospheric conditions are 14.7 psia and 70°F (101.3 kPa and 21°C). What is the volumetric efficiency at 1200 rpm?

6. At standard atmospheric conditions, a fully loaded diesel engine runs at 2000 rpm and develops 1000 bhp (750 kW). The air-fuel ratio is 23:1. The brake specific fuel consumption is 0.45 lbm/bhp-hr (76 kg/GJ). The mechanical efficiency is 80%, independent of the altitude. What will be the (a) brake horsepower and (b) brake specific fuel consumption at an altitude of 5000 ft (1500 m)?

7. In an air-standard gas turbine, air at 14.7 psia and 60°F (101.3 kPa and 16°C) enters a compressor and is compressed through a volume ratio of 5:1. The compressor efficiency is 83%. Air enters the turbine at 1500°F (820°C) and expands to 14.7 psia (101.3 kPa). The turbine efficiency is 92%. What is the thermal efficiency of the cycle?

8. A 65% efficient regenerator is added to the gas turbine described in Prob. 7. Assume the specific heat remains constant. What is the new thermal efficiency?

9. (*Time limit: one hour*) A gasoline-fueled internal combustion engine runs at 4600 rpm. The engine is four-stroke and V-8 in configuration with a displacement of 265 in^3 (4.3 L). The work required to compress the air-fuel mixture is 1200 ft-lbf (1.6 kJ) per piston per cycle. The work done by the exhaust gases in expansion is 1500 ft-lbf (2.0 kJ) per piston per cycle. The input energy from fuel combustion is 1.27 Btu (1.33 kJ) per piston per cycle. Atmospheric air is at 14.7 psia and 70°F (101.3 kPa and 21°C). The air-fuel ratio is 20:1. The heating value of gasoline is 18,900 Btu/lbm (44 MJ/kg). Neglect the effects of friction. What are the (a) indicated horsepower, (b) thermal efficiency, (c) mass per hour of gasoline consumed, and (d) specific fuel consumption?

10. (*Time limit: one hour*) A mixture of carbon dioxide and helium in an engine undergoes the following processes in a cycle.

A to B: compression and heat removal

B to C: constant volume heating

C to A: isentropic expansion

point	temperature	pressure
A	520°R (290K)	14.7 psia (101.3 kPa)
B	1240°R (690K)	unknown
C	1600°R (890K)	568.6 psia (3.920 MPa)

Both gases are ideal gases. (a) What are the gravimetric fractions of the carbon dioxide and helium in the mixture? (b) What work is done during the isentropic expansion process? (c) Draw the temperature-entropy and pressure-volume diagrams for the cycle.

11. (*Time limit: one hour*) When the atmospheric conditions are 14.7 psia and 80°F (101.3 kPa and 27°C), a diesel engine with metered fuel injection has the following operating characteristics.

brake horsepower:	200 bhp (150 kW)
brake specific fuel consumption:	0.48 lbm/hp-hr (81 kg/GJ)
air-fuel ratio:	22:1
mechanical efficiency:	86%

The engine is moved to an altitude where the atmospheric conditions are 12.2 psia and 60°F (84 kPa and 16°C). The running speed is unchanged. What are the corresponding operating characteristics?

12. (*Time limit: one hour*) A gas turbine operating on the Brayton cycle with an 8:1 pressure ratio is located at 7000 ft (2100 m) altitude. The conditions at that altitude are 12 psia and 35°F (82 kPa and 2°C). While consuming 0.609 lbm/hp-hr (100 kg/GJ) of fuel and 50,000 cfm (23 500 L/s) of air, the turbine develops 6000 bhp (4.5 MW). The turbine efficiency is 80%, and the compressor efficiency is 85%. The fuel has a lower heating value of 19,000 Btu/lbm (44 MJ/kg). The fuel mass is small compared to the air mass. The turbine receives combustor gases at 1800°F (980°C). The air inlet filter area is 254 ft^2 (22.9 m^2). The turbine is moved to sea level where the conditions are 14.7 psia and 70°F (101.3 kPa and 21°C). The combustion efficiency and combustor temperature remain the same. What are the new (a) brake horsepower and (b) brake specific fuel consumption?

30 Nuclear Power Cycles

1. Introduction 30-1
2. Reactor Types 30-1
3. Pressurized Water Reactors 30-1
4. Boiling Water Reactors 30-2
5. Standardized Reactors 30-2
6. Corrosion and Embrittlement 30-2
7. Externalities 30-2

1. INTRODUCTION

A *nuclear power cycle* is a Rankine vapor power cycle in which a nuclear reactor core replaces the traditional furnace. Because there is no combustion, nuclear plants do not emit carbon dioxide, sulfur dioxide, or nitrogen oxides. However, they continue to be plagued by technical problems, poor economics, licensing difficulties, and poor public acceptance.

In the United States, approximately 18% of all electricity is generated by nuclear plants.[1] Commercial nuclear reactors produce 600 to 1500 MW of power, though most operate in the 800 to 1000 MW range. Limitations associated with the method of heat extraction have kept pressures and temperatures lower than what is achievable in simple Rankine cycles. This limits the maximum attainable thermal efficiency to (approximately) 30%.

2. REACTOR TYPES

Nuclear reactor systems are characterized by their neutron moderators and methods used to transfer heat from the reactor to the steam. Although there are a number of different reactor designs, only *light-water (moderated) reactors* (LWRs), which includes pressured water reactors and boiling water reactors, have been commercialized in any significant numbers.[2,3]

---

[1] This figure is low compared to many other countries that don't have the large coal reserves that the United States has. In France, for example, the percentage is above 70%.

[2] Other successful but noncommercial designs include heavy water reactors (HWR), gas-cooled reactors (GCR) and high-temperature gas-cooled reactors (HTGR), organic liquid-cooled reactors (OCR), and liquid metal-cooled reactors (LMR). The only commercial gas-cooled reactor to operate in the United States—the 330 MW, helium-cooled Ft. St. Vrain run by the Colorado Public Service Company—was permanently closed in 1989. The U.S. Department of Energy (DOE) *liquid metal fast breeder reactor* (LMFBR) program was canceled in 1983.

[3] Slightly more than half of the nuclear power plants in the world are pressurized water reactors, while most of the remainder are boiling water reactors. A handful of breeder reactors are in various stages of design, development, and use throughout the world.

3. PRESSURIZED WATER REACTORS

In a *pressurized water reactor* (PWR), heated liquid water in the *primary loop* travels from the reactor through a heat exchanger, which serves as the *steam generator*. Steam flowing in the *secondary loop* absorbs heat from the primary loop water. Since water in the primary loop cannot be permitted to vaporize, the primary loop is limited to approximately 2000 psia (13.8 MPa), which corresponds to 640°F (340°C). Water at approximately 1850 psi (12.7 MPa) and 550°F (290°C) enters the steam generator and leaves as 625°F (330°C) steam. State-of-the-art systems are generating steam at approximately 1040 psig (7.2 MPa).

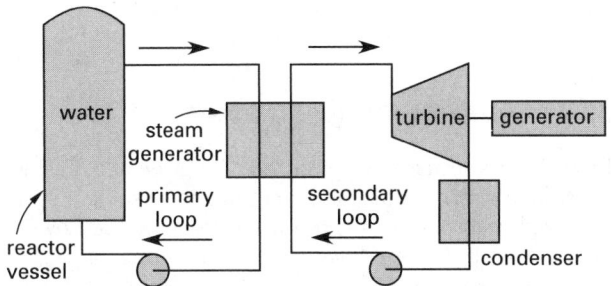

Figure 30.1 *Pressurized Water Reactor*

Steam produced in the secondary loop is passed through a *moisture separator* before entering the turbine. These are typically "passive swirl" (i.e., centrifugal cyclone) in nature and produce steam with a quality of 99.5% or higher.

If the steam were to expand directly to the condenser pressure, the final steam quality would be approximately 75 to 80%. To avoid turbine problems associated with a high moisture content, steam is treated between the HP and LP turbine elements, when the pressure drops to approximately 10 to 25% of the throttle pressure.[4] The steam passes through an external *moisture separator reheater*. Reheating to approximately 25°F (14°C) below the saturation temperature occurs in a shell-and-tube exchanger. Steam from the throttle (*single-stage reheat*) or extraction steam is the source of reheat.

---

[4] The moisture separation-reheating process using steam-to-steam results in a loss (or, at best, a very small gain) in thermal efficiency. Its function is primarily moisture separation, not efficiency improvement.

Power Cycles

4. BOILING WATER REACTORS

In a *boiling water reactor* (BWR), steam is generated directly within the nuclear core. It is passed through a series of separators and driers (integral to the reactor vessel) before entering the turbine. The pressure in a typical BWR is approximately 1000 psia (6.9 MPa). With this configuration, the turbine and all the equipment that processes water and steam eventually become radioactive. Radioactivity in steam is primarily N-16, an isotope with a half-life of only 7 sec. Therefore, radioactivity in steam exists only during power generation.

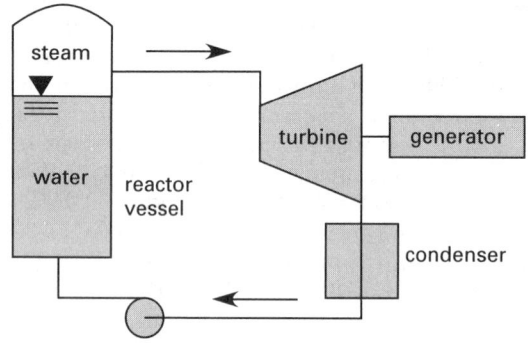

Figure 30.2 *Boiling Water Reactor*

5. STANDARDIZED REACTORS

In order to lower cost, shorten the licensing period, shorten construction time, and increase safety, standardized reactors are being designed, built, and evaluated.[5] Manufacturers have coined the names *advanced light water reactor* (ALWR), *advanced boiling water reactor* (ABWR), *advanced water-cooled reactor* (AWCR), and *safe integral reactor* (SIR). These designs are also collectively referred to as *simplified boiling water reactor* (SBWR) designs.

SBWRs use prefabricated construction and standardized components. The numbers of pumps, valves, transducers, and instruments are reduced. Control rooms are highly computerized. Multiplexed cabling or fiber optics provides communication between devices. Related equipment is located in a single building. Safety systems are inherent and redundant.

SBWRs can be categorized as evolutionary or passive.[6] *Evolutionary designs* take the best, proven design features from existing technology. They rely on pumps for cooling during normal operation. *Passive designs* rely on natural forces (e.g., gravity and natural convection) for core and reactor cooling during normal operation.

These are significant departures from traditional reactor designs.

Standardized reactors have been designed for intrinsic safety. In the event of an accident, the safety systems operate without the need for pumps, traditional heat exchangers, electricity, or even operator intervention—typically for up to three days. For example, in a *loss-of-coolant accident* (LOCA), cooling water from tanks above the reactor would be automatically released, flowing downward by gravity. Cooling air would flow upward by natural convection. Flooding by borated water would ensure reactor shutdown.

6. CORROSION AND EMBRITTLEMENT

The two major causes of reactor outages are refueling and repair or replacement of the steam generator. Most nuclear plants eventually suffer from various corrosion-related problems in their primary loops. This includes *intergranular stress corrosion cracking* (IGSCC) and *neutron embrittlement.*

Neutron irradiation of reactor vessels reduces steel's toughness. The problem is most severe at the reactor's *beltline* where the neutron flux is highest. Beltline welds are considered to be the limiting features. Ongoing embrittlement can be reduced by *fuel management* (i.e., loading new fuel rods in the center of the core and moving spent fuel rods to the outside) and by the addition of internal shielding. Existing embrittlement can be eliminated by in situ thermal annealing of the reactor vessel.[7]

7. EXTERNALITIES

Some states have mandated that the effects of externalities be included when new power plants are being economically evaluated. *Externalities* (also known as *Pace values*) are difficult-to-evaluate economic factors.[8] In engineering economics problems, these might be known as "nonquantifiables." A typical externality for a fossil-fueled plant is the effect of carbon dioxide on global warming.

When the value of externalities associated with fossil-fuel plants is included, nuclear power becomes economically viable. Of course, nuclear power has its own set of unevaluated externalities. These include low-level waste disposal, radon emissions, poor public acceptance, and severe accidents.

---

[5]Although standardization has been used for years in other countries (e.g., France), the first advanced standardized nuclear plant was GE's 1356 MW ABWR unit in Tokyo.

[6]The term "natural forces" has been proposed as a replacement for "passive," which connotes a "we don't care" attitude to the general public.

[7]Thermal annealing is an emerging technology. The embrittled area is heated to approximately 850°F (450°C) and is held at that temperature for as long as is necessary to anneal the metal.

[8]Pace values are named after the Pace University Center for Environmental Studies.

31 Advanced and Alternative Power-Generating Systems

1. Introduction 31-1
2. Solar Thermal Energy 31-2
3. Solar Power Cycles 31-3
4. Photovoltaic Energy Conversion 31-4
5. Wind Power 31-4
6. Ocean Thermal Energy Conversion 31-6
7. Tidal and Wave Power Plants 31-7
8. Geothermal Energy 31-7
9. Fuel Cells 31-7
10. Magnetohydrodynamics 31-8
11. Thermoelectric and
 Thermionic Generators 31-9
12. Methods of Energy Storage 31-9
13. Batteries 31-10
 Practice Problems 31-10

Nomenclature

a	interference factor	–	–
A	area	ft^2	m^2
B	magnetic flux density	–	T
c_p	specific heat	Btu/lbm-°F	kJ/kg·K
C_P	power factor	–	–
E	electric field	–	V/m
F_L	loss factor	–	–
F_R	removal factor	–	–
F_s	shading factor	–	–
g	acceleration of gravity	ft/sec^2	m/s^2
I	incident energy	Btu/ft^2-hr	W/m^2
k	ratio of specific heats	–	–
k	thermal conductivity	Btu-ft/hr-ft^2-°F	W/m·K
K	loading factor	–	–
KE	kinetic energy	ft-lbf/lbm	J/kg
$\dot{m}$	mass flow rate	lbm/sec	kg/s
p	pressure	lbf/ft^2	Pa
P	power	ft-lbf/sec	W
q	heat transfer	Btu/hr	W
Q	volumetric flow rate	ft^3/sec	m^3/s
r	radius	ft	m
S	Seebeck coefficient	–	V/K
T	temperature	°F	K
U	overall coefficient of heat transfer	Btu/hr-ft^2-°F	W/m^2·K
v	velocity	ft/sec	m/s
Z	figure of merit	1/°R	1/K

Symbols

α	absorptance	–	–
η	efficiency	–	–
ρ	density	lbm/ft^3	kg/m^3
ρ	electrical resistivity	–	Ω·m
τ	transmittance	–	–

Subscripts

s isentropic

1. INTRODUCTION

Many advanced energy technologies are mature but uneconomical, are limited to specific applications, or are still in various stages of development. The various *renewables* (e.g., *renewable energy sources* such as wind and solar energy) are in the first category. Radioisotope sources (used primarily for space probes) are in the second. Emerging energy systems (e.g., fusion and magnetohydrodynamics) are in the third category.[1]

The 1973 energy crisis and oil embargo illuminated the need for alternative energy sources. However, with the large fossil fuel reserves in the United States, wide geographic distribution of these reserves, and the high efficiencies being achieved in combined cycle plants, it is likely that coal and natural gas will continue to be the most economic source of baseload electricity well into the future. Of all the other renewable energy sources, wind energy comes closest in price. Even so, wind power is still 25 to 100% more expensive than coal/gas power.

Renewables are limited by localization and capacity factor problems. Not only are most renewables confined to specific locations, but even then, their capacity factors are often below 30%. (By comparison, new coal plants have capacity factors in excess of 85%.)

---

[1]Land-based hydroelectric power is a big exception. This mature renewable energy source is limited by site availability, not by economics.

2. SOLAR THERMAL ENERGY

Solar energy arrives at the outside of the earth's atmosphere at an average rate of 429.2 Btu/ft^2-hr (1.354 kW/m^2), a value known as the *solar constant*. 40 to 70% of this energy survives absorption and reflection in the atmosphere and reaches the earth's surface. The actual incident energy, I, sometimes referred to as *insolation*, depends on many factors, including geographic location, tilt and orientation of the receiving surface, calendar day, time of day, and weather conditions. Average values for clear days are given in maps and tabulations.[2] Some of this energy can be captured in *active solar systems* and used for space heating, domestic hot water (DHW) generation, and cooling (using heat pumps and absorption chillers).[3]

Solar energy is captured in *solar collectors*. The sun's energy enters the collector and, because of the *greenhouse effect*, is trapped inside.[4] Heat is absorbed by *heat transfer fluid* pumped through tubes mounted on the *absorber plate*.[5] The tubes and absorber plate can be left uncoated, painted black, or treated with a *selective surface* coating that absorbs more energy than it reradiates.[6] Water at 100 to 150°F (38 to 66°C) can easily be generated in this manner. Thermal energy in the heat transfer fluid can be used directly (as in swimming pool heaters), or it can be transferred to and stored in a tank of water or rock pebble beds.

Most collectors in simple heating systems are *flat-plate collectors*. These are essentially wide, flat boxes with clear plastic or glass coverings known as the *glazing*. *Concentrating (focusing) collectors* use mirrors and/or lenses to focus the sun's energy on a small absorber area. Except for some parabolic mirror designs, focusing collectors use a controller and motor to track the sun across the sky. *Evacuated-tube collectors* are more complex, but their efficiencies are higher. A U-shaped tube carries the transfer fluid through an air-filled transparent cylinder, which itself is enclosed in a transparent

vacuum cylinder. Evacuated collectors are useful when extremely hot transfer fluid is needed and are generally limited to commercial projects.

Ideally, collectors should point due south and be tilted (from the horizontal) an angle equal to the latitude. For winter use, the tilt should be somewhat higher (so that the maximum energy is received on the coldest days). The variations from south-pointing and latitude tilt may be ±10 to 15° without significant degradation in performance.

The heat absorbed by the solar collector can be calculated from the incident energy, the *absorptance*, α, of the absorber, and the *transmittance*, τ, of the cover plate. The *shading factor*, F_s, in Eq. 31.1 has a value of approximately 0.95 to 0.97 and accounts for dirt on the cover plates and shading from the glazing supports. It is neglected in most initial studies.

$$q_{absorbed} = F_s A \alpha \tau I \qquad 31.1$$

The heat absorbed by the transfer fluid is

$$q_{fluid} = \dot{m} c_p (T_{out} - T_{in}) \qquad 31.2$$

The difference between the heat absorbed by the collector and transfer fluid constitutes the conduction and convection losses.

$$q_{loss} = q_{absorbed} - q_{fluid}$$
$$= U_L A (T_{plate} - T_{air}) \qquad 31.3$$

The average plate temperature, T_{plate}, in Eq. 31.3 is seldom known. However, the incoming fluid temperature is usually known and is used by convention in place of the plate temperature. The collector *heat removal efficiency factor*, F_R, is used to correct for the substitution in variables. For liquid collectors, F_R ranges from 0.8 to 0.95 and is usually specified by the collector manufacturer.

$$q_{fluid} = q_{absorbed} - q_{loss}$$
$$= F_R F_s A \alpha \tau I - F_R U_L A (T_{in} - T_{air}) \qquad 31.4$$

The *collector efficiency* is the ratio of energy absorbed by the transfer fluid to the original incident energy striking the collector.

$$\eta = \frac{q_{fluid}}{IA} \qquad 31.5$$

Since convective and radiation losses increase at higher collector temperatures, the collector efficiency decreases as the difference between ambient air and average plate (or inlet) temperatures increases. When the efficiency is plotted against the ratio of inlet-ambient temperature difference to incident energy, $(T_{in} - T_{ambient})/I$, the falloff rate is approximately linear with a slope of $-F_R U_L$. Typical values of $F_R U_L$ for flat-plate collectors range from 0.6 to 1.1 Btu/hr-ft^2-°F (3.4 to 6.2 W/m^2·K). When the temperature difference is zero,

[2]Some isolation maps use units of langleys/day. A *langley* is equal to 1 cal/cm^2, 3.69 Btu/ft^2, and 41 840 J/m^2.
[3]*Passive solar systems*, which include strategically oriented buildings, walls, and thermal collectors, and which rely on natural convection and conduction for storage and heat transfer, are not included in this chapter.
[4]As received, solar radiation has wavelengths of 0.2 to 3.0 μm. Thermal radiation reradiated from the collector plate has a wavelength of approximately 3 μm. Good covering materials have high-shortwave transmittance (85 to 95%) and low-shortwave transmittance (less than 2%). White crystal glass, low-iron tempered and sheet glasses, and tempered float glass satisfy these requirements. Polycarbonates, acrylics, and fiberglass also perform well but suffer from weathering and durability problems.
[5]Water can be used as the heat transfer medium, but it is subject to freezing, boiling, and chemical breakdown; and the system is subject to corrosion. To counteract these problems, ethylene glycol-water and glycerine-water mixtures are often used. Ethylene glycol, however, is toxic, and a heat exchanger must be used to keep the heat transfer fluid separate from domestic water. "State-of-the-art" fluids include silicone or hydrocarbon (aromatic and paraffinic) oils.
[6]Common selective surface coatings with low-to-moderate costs include copper oxide, black nickel, black chrome, lead oxide, and aluminum conversion.

Table 31.1 *Properties of Transfer Fluids*

fluid	specific gravity	viscosity[a] cP	specific heat Btu/lbm-°F	specific heat kJ/kg·K	freezing point °F	freezing point °C
water	1.00	0.5–0.9	1.00	4.19	+32	0
50% water-50% ethylene glycol	1.05	1.2–4.4	0.83	3.5	−33	−36
50% water-50% propylene glycol	1.02	1.4–7.0	0.85	3.6	−28	−33
paraffinic oil	0.82	12–30	0.51	2.1	+15	−9
aromatic oil[b]	0.85	0.6–0.8	0.45	1.9	−100	−73
silicone oil[b]	0.94	10–20	0.38	1.6	−120	−84

[a]Viscosity in the range of 80 to 140°F (27 to 60°C).
[b]Typical values given.

q_{loss} will also be zero, and the y-intercept will be the theoretical maximum efficiency, $F_R \tau \alpha$. Typical values of $F_R \tau \alpha$ range from 0.50 to 0.75.

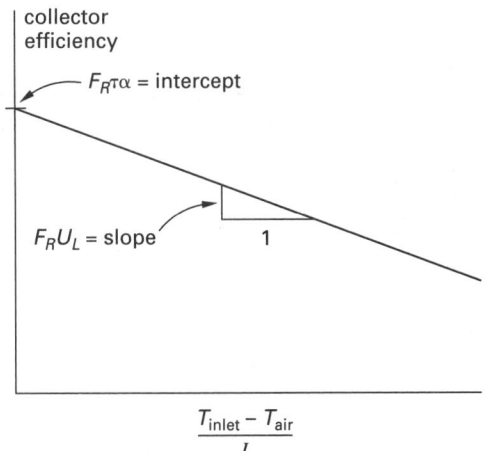

Figure 31.1 *Solar Collector Efficiency Curve*

3. SOLAR POWER CYCLES

Solar energy generating systems (SEGS) use solar energy to generate electricity in traditional power cycles. Three main approaches are taken: *distributed collector systems* (DCS), also known as "trough-electric systems"; *central receiver systems* (CRS), also known as "power-tower systems"; and *dish/Stirling* (D/S) systems.[7] In a *trough-electric system*, parabolic tracking trough concentrators focus sunlight on evacuated glass tubes that run along the collectors' focal lines. Synthetic oil in tubes in the glass cylinders is heated to 575 to 750°F (300 to 400°C). Heat is transferred in a heat exchanger to steam used in a traditional Rankine cycle. Trough-electric technology is relatively mature, but due to the low temperatures, average annual thermal efficiencies are only 10 to 15%.

In the emerging *power-tower system*, a field of *heliostats* (tracking mirrors) concentrates solar energy onto a receiver on a central tower. Since directly generated steam would be subject to variations in solar flux from passing clouds, heat generation and use are decoupled. Heat can be stored in molten eutectic salts such as sodium nitrate, which has a melting point of approximately 430°F (220°C). In one scenario, molten salt would be heated to approximately 1050°F (560°C) and cooled to approximately 550°F (290°C) in the heat exchanger. With these higher temperatures, typical thermal efficiencies of 15 to 20% are possible.

In a *dish-engine system*, the heat engine and electrical generator are located at the focus of a parabolic dish. Most current research centers on Stirling engines (kinematic and free-piston designs) since they are externally heated. Heat can be transferred to the Stirling cycle through direct irradiation, heat pipes, or pool boiling.[8] Due to size and wind loading, 25 to 50 kW is about the highest expected output per dish. However, since the units are modular, they can be grouped in a field to produce any desired output. With operating temperatures up to 1400°F (800°C), thermal efficiencies of 24 to 28% are already being realized, and over 30% has been achieved in some prototypes.

---

[7]This discussion excludes the ultra-high concentrations and temperatures that are available with high-tech optical, two-stage collectors. These systems produce extremely high temperatures on extremely small targets. The discussion also omits the *solar chimney* concept in which collectors establish a 54 to 72°F (30 to 40°C) thermal gradient in the air within tall towers. The stack effect produces air movement, which drives fan-driven generators.

[8]One dish-engine design uses a *reflux pool-boiling receiver*. The receiver uses a pool of molten liquid metal (e.g, sodium, potassium, or a mixture thereof) to transfer heat from the exposed face of the receiver to the helium-filled heater tubes of the Stirling engine. As the liquid metal boils, it vaporizes. It gives up heat when it condenses on the heater tubes. Condensed liquid drips back to the pool by gravity (hence the term "reflux").

Power Cycles

In addition to technical and economic problems, SEGS require large areas. Modern trough systems would require approximately 5 to 6 km^2 to generate 1000 MWe. Even with this space, practical and economic issues limit trough-electric systems to about 200 MWe and tower systems to approximately 100 to 300 MWe.

4. PHOTOVOLTAIC ENERGY CONVERSION

A *photovoltaic cell* (*PV cell* or *solar cell*) generates a voltage from incident light, usually light in the visible region. Traditional PV cells are sliced from monolithic silicon crystals and (when cost is unimportant) from gallium arsenide (GaAs) in crystalline form. Thin vacuum-deposited semiconductor films (known as *amorphous films*), including silicon, copper indium diselenide (CIS), and cadmium telluride, are less costly but have much lower efficiencies. Spherical silicon particles on a metallic substrate constitute another emerging technology.

Each silicon cell produces the same current and a voltage of approximately $^1/_2$V. To obtain larger voltages or currents, cells are connected into *modules*. Large collections of modules are known as *arrays*.

Efficiency is measured by electrical energy output as a fraction of solar energy input. Although maximum efficiencies of single-crystal and thin-film cells are approximately 35% and 15%, respectively, module efficiencies are much less than single-cell efficiencies due to electrical and optical losses. Most commercial silicon modules have efficiencies of 12 to 13%.

Several approaches are taken to increase module efficiencies: (1) optical concentrating of solar energy onto existing technology cells, (2) production of more efficient solar cells. Module efficiencies as high as 30% have been achieved in the lab using optical concentration. Since PV cells are wavelength sensitive, higher efficiencies can be achieved in layered *triple junction cells*.[9]

Experience with existing utility PV "plants" (most of which are less than 1 MW) have capacity factors of approximately 20 to 35%. (Desert sites and two-axis tracking might increase capacity factors to 35%.) Availability is high, 90 to 97%. PV cell output is direct current. Output must be inverted (i.e., converted to alternating current) when connecting to the electric grid.

While PV power cannot compete economically with baseload power generation, it might someday be competitive with peak power alternatives. It is significant that peak PV output coincides with some peak power uses (e.g., air conditioning peaks on hot days). The main problems facing commercialization of PV technologies are high cost, limited lifetimes due to deterioration in performance, and storage of generated energy.

---

[9]For example, blue and green light could be captured in amorphous silicon layers, while a silicon-germanium layer could intercept infrared light.

5. WIND POWER

Wind energy conversion systems (WECS), also known as "wind turbines," usually consist of a conventional induction generator driven through a drivetrain by a large rotor. Most rotors consist of a hub and two or three blades. The main shaft, gearbox, and generator are protected from the elements by an enclosure known as a *nacelle*. The WECS is mounted on a tower 80 to 160 ft (25 to 50 m) tall. Most WECS feeding the electrical grid turn at constant speed.

WECS are classified by the orientation of their rotational axes.[10] With *upwind machines (head-on, horizontal-axis rotors* or *wind-axis rotors)*, the axis of rotation is parallel to the windstream and the blades face the direction from which the wind is coming. With *downwind machines*, the blades face away from the source of the wind. With *vertical-axis rotors (Darrieus rotors* or *"eggbeater windmills")*, the axis is perpendicular to both the surface of the earth and the windstream.

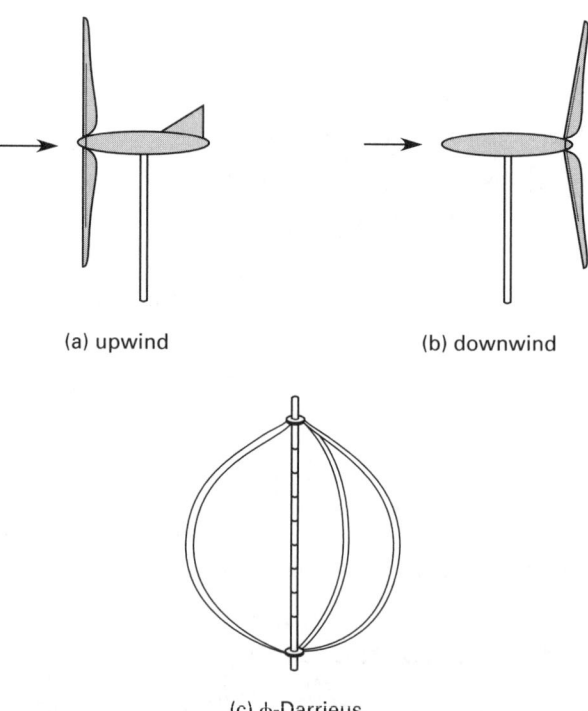

(a) upwind   (b) downwind

(c) φ-Darrieus

Figure 31.2 *Types of Rotors*

Most WECS are upwind machines, even though they require yaw drives to keep them facing into the wind. However, the propeller turbulence (*wash*) can put a strain on the tower. Downwind machines suffer from *tower shadow*. The tower eclipses some of the windstream, and the forces on the up- and down-blades are

---

[10]*Cross-wind horizontal-axis rotors*, including *Davonius rotor* machines, where the axis is horizontal and perpendicular to the windstream, are very uncommon.

different.[11] The resulting vibration can be severe. However, downwind machines can yaw freely without a yaw drive, and blade stresses are less.

Vertical-axis rotors do not have to be turned to face the wind and require no support tower. They have low starting torques, but may not be self-starting. Their high tip/wind speed ratio produces relatively high power outputs. Since the rotor shape is the shape that a flexible cable would take if spun, the bending stresses are low, no matter how fast the rotor turns. The drivetrain and generator are located on the ground, providing for easy maintenance. However, most often they are not installed high enough above the ground to catch the highest velocity wind.

Because of frictional and inertial effects, WECS will not operate much below 5 mph, and generally wind speeds of 10 mph or more are needed. Fifteen mph is generally considered an ideal average speed. Since power increases with the cube of the velocity, WECS are particularly efficient at high velocities. However, forces on WECS become unacceptably high much above 30 mph. Various methods are used to limit speed. Most commercial designs are either *stall-limited* (stall-controlled) or *pitch-limited* so that the blades are automatically feathered at higher speeds. Other options include use of blade ailerons, yawing the rotor out of the wind, mechanical braking of the shaft, blade tip brakes, and electrical dynamic braking.

Most sites have a positive *wind shear*—the wind velocity increases exponentially (for a distance) with altitude. Because of this, WECS manufacturers use tower heights up to 200 to 230 ft (60 to 70 m) to generate the increased power available at those heights.

Graphs of the number of hours per year that the wind reaches each hourly mean velocity are known as *annual average velocity duration (AAVD) curves*. Curves showing the distribution of annual average wind power per unit subtended area as a function of wind speed are called *annual average power density distribution (AAPD) curves*. The annual average wind energy density distribution is equal to the annual average power density distribution times the number of hours per year the corresponding wind speeds occur.

The total ideal power available in a windstream is found by multiplying the kinetic energy per unit mass by the flow rate.

$$P_{\text{ideal}} = \dot{m}(\text{KE}) = \dot{Q}\rho(\text{KE})$$
$$= Av\rho\left(\frac{v^2}{2}\right) = \frac{A\rho v^3}{2}$$
$$= \frac{\pi(r_{\text{rotor}})^2\rho v^3}{2} \qquad \text{[SI]} \qquad 31.6(a)$$

$$P_{\text{ideal}} = \dot{m}(\text{KE}) = \dot{Q}\rho(\text{KE})$$
$$= Av\rho\left(\frac{v^2}{2g_c}\right) = \frac{A\rho v^3}{2g_c}$$
$$= \frac{\pi(r_{\text{rotor}})^2\rho v^3}{2g_c} \qquad \text{[U.S.]} \qquad 31.6(b)$$

WECS are unable to extract the ideal power from the airstream. To do so would require decelerating the air flow to zero velocity (i.e., removing all of the kinetic energy). The actual power can be determined if the actual pressures and velocities before and after the WECS are known.

$$P_{\text{actual}} = Av\rho\left(\frac{p_2 - p_1}{\rho} + \frac{v_2^2 - v_1^2}{2}\right)$$
$$\text{[SI]} \qquad 31.7(a)$$

$$P_{\text{actual}} = Av\rho\left(\frac{p_2 - p_1}{\rho} + \frac{v_2^2 - v_1^2}{2g_c}\right)$$
$$\text{[U.S.]} \qquad 31.7(b)$$

The actual power generated will depend on the *power coefficient*, C_P. The power coefficient of an ideal wind machine rotor with a propeller rotor varies with the ratio of blade tip speed to free-flow windstream speed and approaches the maximum value of 0.593:1 (known as the *Betz coefficient*) when this ratio reaches a value of 5:1 or 6:1. The maximum power coefficients for the best two-blade rotors is approximately 0.47, and for Darrieus vertical-axis rotors, the maximum value is approximately 0.35. Figure 31.3 illustrates the variation in power coefficient for rotors.

$$P_{\text{actual}} = C_P P_{\text{ideal}} \qquad 31.8$$

The velocity in the immediate wake of the rotor depends on the *interference factor*, a, and is given by $v_2 = v_1(1 - a)$. If some simplifying assumptions are made, the power coefficient can be calculated from the interference factor.[12]

$$C_P = 4a(1 - a)^2 \qquad 31.9$$

Commercial WECS come in all sizes. Units for residential use are seldom larger than 2.5 kW. Modern units connected to utility grids can be up to 100 to 500 kW in size (as limited by economics and mechanical stresses),

---

[11]This happens to a certain extent in all horizontal-axis machines because the wind speed is lower, closer to the ground.

[12]The radial pressure gradient and the kinetic energy of the swirl velocity component in the rotor's wake are neglected.

Figure 31.3 *Power Coefficients*

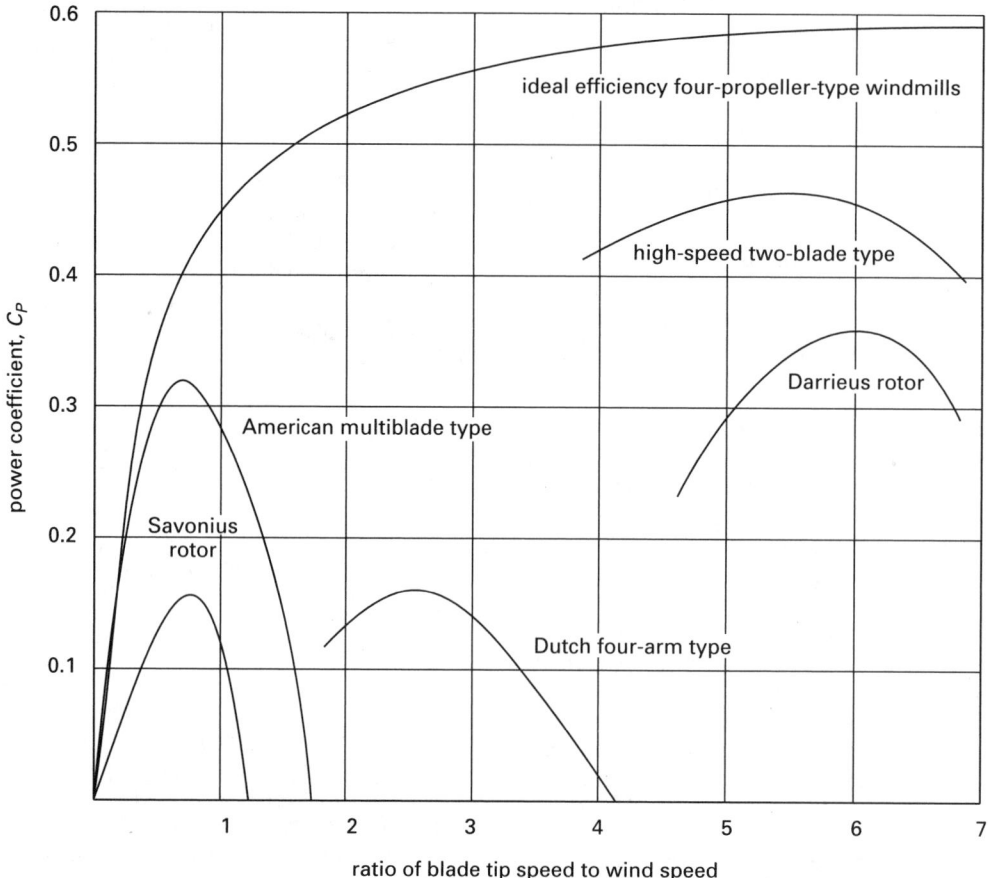

(National Technical Information Service)

though some mammoth 1 to 3 MW units have been built over the years. Individual units can be combined into large "wind farms" to produce any amount of power needed.

Like solar energy, WECS produce little or no environmental impact.[13] However, they require vast amounts of space. Structural corrosion from air pollutants and fatigue failures from normal operation and random gust loads are prevalent. During normal operation, WECS are plagued by blade pitting and soiling by insects and air pollutants. Many first-generation units built in the late 1970s and early 1980s (the heyday of U.S. construction) have failed because of underestimated operating conditions and forces.

Research into *teetering* (*hinged* or *pivoted*) *two-blade machines* is expected to produce lower stresses in the future. Blades are also being redesigned to eliminate areas of laminar flow at all speeds. Only laminar flow is sensitive to blade roughness due to pitting and soiling.

6. OCEAN THERMAL ENERGY CONVERSION

During the summer, the temperature at the surface of tropical oceans is almost constant at 77°F (25°C). 2000 to 3000 ft (600 to 900 m) below, however, the temperature can be as low as 38 to 40°F (4 to 5°C). An *ocean thermal energy conversion* (OTEC) system uses the difference in temperatures to drive a vapor cycle.

In a *closed-cycle plant*, warm water from the surface supplies heat to a boiler; cold water from the depths provides cooling in the condenser. A traditional low-pressure vapor cycle operating with ammonia, propylene, or other refrigerant drives a generator.

In an *open-cycle plant*, the warm seawater is injected into a near vacuum. The water vaporizes and the resulting steam expands through a low-pressure turbogenerator before being condensed with the cold seawater.[14]

The net-to-gross power ratio, typically around 0.20, is primarily a function of the power needed to pump water

---

[13]Many birds are killed by spinning rotors.

---

[14]The closed process was originally proposed in 1881 by French engineer Jacques-Arsene D'Arsonval. The open process is sometimes referred to as the *Claude cycle*, named after Georges Claude, a student of D'Arsonval, who demonstrated the practicality (at a net power loss) of the open process off the coast of Cuba in 1929.

from the depths. In addition to low efficiencies due to the small temperature differential, other problems associated with OTEC systems include high component and material costs, corrosion, biofouling, and long-distance power transmission. Production of electrically intensive products (e.g., hydrogen from electrolysis of seawater or even desalinated fresh water) has been proposed as a solution to the latter problem. The major environmental effect would be a slight cooling at the surface of the ocean and a warming of the depths.

7. TIDAL AND WAVE POWER PLANTS

Electric power can be generated by capturing water brought into a lagoon or bay by tidal action, and then releasing the water through traditional hydroelectric generators when the tide recedes. The technology is neither new nor complex. However, only about a hundred sites in the world have large enough rises and falls for tidal power to be practical.

With *wave power plants*, advancing and receding ocean waves repeatedly compress and force air through air turbines. (The turbines spin in the same direction regardless of the wave direction.) The turbines drive electrical generators. Wave machines do not suffer from the power transmission problem of OTEC units, since they can be located close to shore.

8. GEOTHERMAL ENERGY

There are three general types of geothermal energy sources: vapor-dominated, liquid-dominated, and hot rock sources.[15] The 750 MW "Geysers" in northern California is an example of a *vapor-dominated reservoir* driving a *direct steam cycle*. Multiple wells provide steam at 100 to 120 psi (690 to 820 kPa) and 400°F (205°C) which is collected, separated to remove liquid, filtered to remove abrasive particles, and passed through turbines. Evaporative air cooling towers are used to cool condensing water. Water lost in evaporation is replaced by condensed water, and approximately 80% of the steam is evaporated in this manner. The remaining 20%, containing minerals, is reinjected into the ground through deep injection wells. Substantial noncondensible gases (primarily hydrogen sulfide and ammonia, but also including carbon monoxide, hydrogen, methane, and nitrogen) constitute up to 5% of the steam volume and are removed by steam ejectors, vacuum pumps, or compressors and released through the cooling tower.[16] Thermal efficiency is approximately 22%.

Most geothermal sites are *liquid-dominated reservoirs* and produce mixtures of hot water (i.e., brine) and

hot steam. In *hot water systems*, the hot water is discarded and only the steam is used. If the hot water temperature is approximately 330°F (165°C) or higher, a *flash steam cycle* can be used. Hot water and steam are pumped to an evacuated *flash tank*. The resulting high-pressure flash steam drives the turbine. In the *dual-flash cycle*, low-pressure hot water remaining in the first flash tank is routed to a second, low-pressure flash tank. Low-pressure steam is used in the LP turbine section. Liquid-dominated reservoirs typically do not produce large quantities of noncondensible gases.

If the temperature of the hot water is between approximately 330°F and 250°F (165°C and 120°C), a *binary cycle* using a separate heat transfer fluid, such as pentane or isobutane, is required. Hot water passes through a closed heat exchanger, vaporizing the heat transfer fluid. Cooled water is discarded in injection wells. The vapor drives a binary turbine generator in a traditional Rankine cycle. An air-cooled condenser maintains the turbine back pressure.

Fewer than ten naturally occurring vapor-dominated reservoirs are known to exist in the world. With drilling, however, hot geological rock can be found anywhere. The temperature of the earth's crust increases by 30°F (16.7°C) for every kilometer of depth. Anomalous temperatures of 570 to 1300°F (300 to 700°C) can be found within 20,000 ft (6000 m) of the surface at some locations. In *hot rock systems*, water is injected through injection wells into artificially made fractured rock beds 0.6 to 3.6 mi (1 to 6 km) below the surface.[17] Pressurized hot water at approximately 400°F (200°C) would be removed from an adjacent production well. Steam flashing from the hot water drives turbines located at the surface.

Difficulties associated with natural geothermal energy include the scarcity of natural sites, method of cooling, corrosion and fouling by mineral-laden water, and high moisture contents during steam expansion. Environmental problems include disposal of high-mineral waste water, pollution of the air by noncondensing noxious gases, drift, waste heat release, noise from steam venting, and ground subsidence.[18] Hot rock systems, still in their infancy, have many technical difficulties to overcome, including drilling and controlled fracturing techniques and loss of water.

9. FUEL CELLS

A *fuel cell* converts chemical energy directly into electrical energy. With two electrodes separated by an electrolyte, a fuel cell is similar to a continuously fueled battery. Unlike a battery, however, the chemical process continues indefinitely as the spent reactants are

---

[15]*Hydrothermal sources* (steam and water), *hot dry rock sources* (steam only), and *geopressured sources* (high-pressure liquid) are three slightly different categorizations of geothermal sources.
[16]Because of the high noncondensible component of steam at the Geysers, removal of noncondensibles is the largest single use of steam—up to 6%.

[17]Nuclear blasts are no longer considered a viable method of producing a cavity deep within the earth.
[18]The salt content of geothermal waters varies from 2% to about 20%. By comparison, the salt content of seawater is approximately 3.3%. Hydrogen sulfide is soluble in water and escapes by evaporation during the cooling process.

Power Cycles

replaced. Fuel (e.g., hydrogen or methane) is supplied to the anode (the negative terminal), and oxidizer (e.g., oxygen) is supplied to the cathode (the positive terminal). The anode and cathode are both porous to allow diffusion of the fuel and oxygen through them. Fuel reacts with the electrolyte at the cathode; oxygen reacts at the anode. Liquid water is produced, along with heat and electricity, in a process that is essentially the reverse of electrolysis. (Combustion does not occur in the fuel cell.)

When free hydrogen gas is not available, hydrogen-rich gas can be generated from fossil fuels in a *reformer (reformer reactor)*. Reforming ("front-end reforming") can take place within the fuel cell if the temperature is high enough. This is referred to as the *direct fuel cycle* or *internal fuel reforming*. If the cell temperature is too low or if the fuel is not gaseous, a separate reformer is needed. Steam reforming of natural gas, adiabatic reforming of heavy distillates, and coal gasification are all applicable processes.

The major factor limiting the amount of power generated is the speed of the diffusion processes through the electrolyte between the electrodes. Because of this, fuel cells are named for the electrolyte used.

Commercial units combine individual cells, each of which is about 0.5 cm thick and generates approximately 700 mV, into *stacks*. Stacks containing hundreds of cells are used to generate up to 200 to 250 kW of power.[19] Multiple stacks can be used to produce any amount of power desired.

Since fuel cells are not heat engines, they are not limited by Carnot efficiencies. Theoretical fuel cell efficiencies are high, ideally on the order of 80% for hydrogen, 90% for methane, and up to 98% for more complex hydrocarbons.[20] However, not all of the heat generated can be recovered. This and other factors reduce the practical stack efficiency to 35 to 45% based on higher heating values (HHV).

Small hydrogen-oxygen fuel cells have been used in spacecraft for years. First-generation commercial utility-sized *phosphoric acid fuel cells* (PAFCs) operating at 400°F (200°C) are a mature technology. Efficiencies are typically around 40%.

Second-generation *molten carbonate fuel cells* (MCFCs), operating at 1200°F (650°C), offer higher efficiencies. By reforming coal, MCFCs could theoretically replace gas turbines in ICGCC plants. HHV efficiencies of carbonate fuel cell stacks are projected to be approximately 50 to 55%.

Third-generation monolithic *solid oxide fuel cells* (MSOFC or SOFC) operating at 1800°F (1100°C) are being developed. A separate reformer reactor is not needed; the high temperature with or without a catalyst can reform the fuel gas (natural gas or gasified coal)

internally. The high temperatures also make SOFCs ideal candidates for cogeneration and combined cycle plants, where overall efficiencies may eventually reach 50 to 70%.

Ongoing research continues with *alkaline fuel cells* (which power the U.S. space shuttle) and *proton exchange membrane* (PEM) cells (originally called *solid polymer fuel cells*).

Fuel cells are attractive in the transportation sector because they produce clean energy from readily available reactants. Fuel can be supplied by hydrogen-rich liquids such as methanol, and oxygen is supplied by the air. Temperatures are low enough to keep NOx from being a problem, carbon dioxide is one third of that from hydrocarbon combustion, and all other polluting emissions are substantially reduced. However, because of size, mass, poor start-up and transient responses, and expensive catalysts, fuel cells have been incorporated into only a few bus-sized experimental vehicles.

10. MAGNETOHYDRODYNAMICS

In a *magnetohydrodynamic* (MHD) *generator*, a high-temperature, ionized plasma flows through a supersonic nozzle.[21] The high-velocity plasma in the expansion channel passes through a magnetic field generated by toroidal coil or plates. A constant electric field is generated perpendicularly to the magnetic field (in the direction of the moving gas) by the moving ions.[22] Direct current is generated, so inverters are needed to produce grid-quality power.

High magnetic fields (5 T) and high temperatures—4600°F (2800K) at the generator entrance and 4400°F (2700K) at the exit—are required to create the plasma. The high temperatures are achieved by combustion of gasified coal (with or without oxygen enrichment) in a high-temperature air heater prior to the additional combustion of fossil fuels (coal or natural gas) in a subsequent combustor. Designs for a commercial high-temperature air heater are the subject of ongoing research.

The combustion products are doped (seeded) with salts of easily ionized elements, such as potassium or cesium, to obtain the necessary (1%) ionization. When supplied at approximately 150% of the stoichiometric rate, these alkali metal ions also combine with virtually all sulfur in the fuel. The resulting compounds (potassium sulfate) can be recovered in electrostatic precipitators for reuse as seed. With all sulfur removed, it may be possible to allow the stack gas to cool to as low as 395°F (200°C).

---

[19]For example, a stack of 470 1 m^2 cells will generate approximately 750 kW.

[20]All fuel cell efficiencies are temperature-dependent.

---

[21]MHD machines are not limited to plasmas. In MHD marine (submarine) propulsion systems, the cycle is reversed. Seawater flowing through the duct is acted upon by magnetic and electrical fields. The fields force the water through the duct, propelling the vessel forward.

[22]Generation of this electric field is known as the *Hall effect*.

NOx generation was once thought to be problematic. Large amounts (10,000 to 12,000 ppm) of NOx are produced at the high plasma temperatures. However, various modern control methods, including staged fuel-rich combustion, and decomposition in the radiant furnace apparently reduce NOx to acceptable levels.

MHD generators are attractive because there are no highly stressed rotating parts and because the higher temperatures are compatible with bottoming cycles. Most of the heat remaining in the combustion gases can be recovered in a high-temperature air heater (HTAH) for use in cogeneration or combined bottoming cycles.

From a thermodynamic standpoint, the MHD cycle is the same as the Brayton gas turbine cycle, except that the work output during the expansion is electrical, not mechanical. Since the movement of ionized particles through a magnetic field results in a conversion of mechanical energy to electrical energy, the efficiency of that process is not limited to the Carnot maximum. The mechanical-electrical conversion efficiency, known as the *loading factor*, K, of this process is high: 0.80 to 0.90 (i.e., 80 to 90%). For an isentropic process, $K = 1$. For an isothermal process, $K = 0$.

$$K = \frac{E}{vB} \qquad 31.10$$

The temperature change along the expansion channel is

$$\frac{T_2}{T_1} = \left(\frac{p_2}{p_1}\right)^{\frac{K(k-1)}{k}} \qquad 31.11$$

In a constant-velocity generator, the electrical energy removed decreases the enthalpy between the entrance and exit of the generator. Assuming a constant velocity generator and an ideal gas, the isentropic efficiency of the expansion process is

$$\eta_s = \frac{\Delta h_{actual}}{\Delta h_{ideal}} = \frac{\Delta T_{actual}}{\Delta T_{ideal}} \qquad 31.12$$

While the mechanical-electrical conversion efficiency is not limited to the Carnot efficiency, the thermal-electrical conversion efficiency is. Prototypes have achieved thermal efficiencies of approximately 20% with MHD channels alone. When MHD generation is integrated with cogeneration and combined cycles, thermal efficiencies of 40 to 50% seem reasonable.

11. THERMOELECTRIC AND THERMIONIC GENERATORS

A *thermocouple* generates electric potential directly from heat and is an example of a *thermoelectric generator*. Depending on the temperature, certain semiconductors also exhibit thermoelectric effects: bismuth telluride and selenide at room temperature, lead telluride alloys at 390 to 930°F (200 to 500°C), and silicon germanium alloys at 750 to 1830°F (400 to 1000°C).

The *figure of merit*, Z, for a thermoelectric material is an indicator of the effectiveness of the thermoelectric conversion process. Typical values are 3×10^{-3} K^{-1} for bismuth telluride alloys to 1×10^{-3} K^{-1}. In Eq. 31.13, S is the *Seebeck coefficient*, and k is the thermal conductivity.

$$Z = \frac{S^2}{\rho k} \qquad 31.13$$

Current *radioisotope thermoelectric generators* (RTGs) used in space probes contain radiatively coupled *unicouples* using plutonium dioxide as the heat source.[23] Typical output per unicouple is approximately 2.5 W at 3.5 V. Theoretically, any number of unicouples can be combined into a *multicouple*, and multicouples can be stacked to obtain power in any amount. Most RTG stacks for space missions produce less than 1 kWe.

Thermionic generators, also known as *thermionic energy converters* (TECs), consist of two closely spaced tungsten plates (an emitter cathode and a collector anode) known as *shoes*. The hot shoe is maintained at approximately 2600 to 3150°F (1700 to 2000K), and the cold shoe is maintained at approximately 1350°F (1000K). Current is generated when electrons boil off the hot shoe by thermionic emission and flow to the cold shoe. Since the plates are separated by a low-pressure ionized gas (usually cesium vapor), the device is sometimes referred to as a *plasma diode*.

Thermoelectric and thermionic power generators are power cycles that convert heat into work. As such, their efficiencies are limited to the Carnot efficiency. Most RTGs have efficiencies of 10%, and TECs have efficiencies around 15%. For space missions, these low efficiencies are offset by other desirable operating characteristics.

12. METHODS OF ENERGY STORAGE

Thermal energy is stored in heated beds. In simple home solar installations, pebble and rock beds or tanks filled with water and/or other high-density materials can be used. In power utility applications, large tanks containing oil and rock have been proposed. In high-tech applications, molten salt and liquid metals can be used. Molten salt is particularly attractive in two-stage solar electric plants.

Electrical energy can be stored electrically, mechanically, or chemically. Large energy storage capacitor banks and (theoretically) superconducting coils can store electrical energy in electric and magnetic field forms. Mechanical storage involves converting electrical energy into potential, gravitational, or kinetic energy. Tanks ("accumulators") or caverns of compressed air, gas, or liquid store energy in pressure.[24] Elevated water ("pumped hydro" storage) stores gravitational energy.

---

[23]Some of the Pioneer and Voyager space probes used RTGs.
[24]CAES is the acronym used in the electrical power industry for *compressed-air energy storage*.

Flywheels store rotational kinetic energy. Electrical energy can be used to charge batteries, or it can be used to produce hydrogen through electrolysis for later use in fuel cells.

13. BATTERIES

Batteries store and produce electrical energy electrochemically. The common flashlight cell and mercury-oxide "button" cells don't have enough energy or power for *electric vehicles* (EVs), however. In addition to traditional lead-acid batteries, promising EV technologies include nickel-cadmium, nickel-metal hydride, and sodium-sulfur. All other exotic batteries appear to be too expensive, toxic, or limited for commercial use.

Indicators of battery performance are their *specific power* (in W/kg); *specific energy*, also known as *energy density* (in W·h/kg); and *specific capacity* (in A·h/kg). Generally, specific energy decreases as specific power increases. Thus, batteries can have either high specific power or high specific energy. The best high-tech batteries (e.g., NiMH) operate with specific powers of 150 to 200 W/kg and specific energies in excess of 80 W/kg.[25]

The specific capacity of lead-acid batteries, the most commercially mature technology, is theoretically around 120 A·h/kg; the typical energy density is around 30 W·h/kg. Drawbacks to using lead-acid batteries in EVs are many: high mass, limited recharging cycles, low range, long recharge time, and life reduction if fully discharged.

Nickel-metal hydride (NiMH) batteries appear to be the most advanced EV-capable batteries. They have long life spans and high power and energy densities. They recharge fully in approximately an hour.

Fiber-nickel-cadmium (FNC) batteries are already in limited use, particularly in military and civilian aircraft. FNC batteries can be recharged several thousand times, and they operate over a wide temperature range. Cadmium constitutes a disposal problem.

Sodium-sulfur batteries are attractive because they have three to four times the storage capacity of lead-acid batteries. Their major drawback is that they must be operated at elevated temperatures of 662 to 716°F (350 to 380°C) in order to keep the sulfur electrode molten. They also need a liquid cooling system.

Polymer-electrolyte technology, such as that used in flat cells for powering instant film packs and smart credit cards, hold promise for the future.

PRACTICE PROBLEMS

1. An ocean thermal energy conversion plant draws 40°F (4.4°C) water from a depth of 1200 ft (360 m). The water temperature at the surface is 82°F (27.8°C). What is the maximum achievable thermal efficiency?

---

[25]For EVs, goals of 400 W/kg and 200 W·h/kg have been established.

32 Gas Compression Cycles

1. Types of Compressors 32-1
2. Boosters, Intensifiers, and Amplifiers 32-2
3. Compressor Control 32-2
4. Storage Tanks 32-2
5. Compressor Capacity 32-2
6. Compression Processes 32-2
7. Clearance and Clearance Volume 32-3
8. Reciprocating, Single-Stage Compression . . 32-3
9. Polytropic Head 32-5
10. Polytropic Efficiency 32-5
11. Cooling and Intercooling 32-7
12. Multistage Compression 32-7
13. Dynamic Compressors 32-8
14. Compressor Characteristic Curves 32-8
15. Operation at Changed Inlet Conditions . . . 32-8
    Practice Problems 32-9

Nomenclature

c	clearance	percent	percent
c_p	specific heat at constant pressure	Btu/lbm-°F	kJ/kg·K
h	enthalpy	Btu/lbm	kJ/kg
H	head	ft	m
k	ratio of specific heats	–	–
m	mass	lbm	kg
$\dot{m}$	mass flow rate	lbm/sec	kg/s
MW	molecular weight	lbm/lbmol	kg/kmol
n	polytropic exponent	–	–
p	pressure	lbf/ft^2	Pa
P	power	Btu/lbm-sec	W/kg
r	ratio	–	–
R	specific gas constant	ft-lbf/lbm-°R	kJ/kg·K
R^*	universal gas constant	ft-lbf/lbmol-°R	kJ/kmol·K
s	entropy	Btu/lbm-°F	kJ/kg·K
T	temperature	°R	K
V	volume	ft^3	m^3
W	work	Btu/lbm	kJ/kg
Z	compressibility factor	–	–

Symbols

η	efficiency	–	–
v	specific volume	ft^3/lbm	m^3/kg

Subscripts

c	compression
m	mechanical
n	polytropic
p	pressure
s	isentropic
v	volumetric

1. TYPES OF COMPRESSORS

There are two main types of gas compressors: reciprocating and centrifugal.[1] *Reciprocating compressors* are appropriate for high-pressure, low-volume applications, such as air conditioning systems. *Rotating compressors* (also known as *dynamic compressors* and *blowers*) are used in low- to moderate-pressure, high-volume applications such as gas turbines and turbojets.[2,3]

Since any compressor will be limited to operating at a finite *pressure ratio*, higher pressures are achieved by routing the output of one compression stage to the input of a subsequent compression stage. Compressors are categorized by the number of *compression stages* (e.g., a two-stage compressor). Subsequent compressions usually occur in different parts of a multistage compressor and are seldom carried out in separate compressors.

Like pumps and fans, dynamic compressors can be radial or axial in design.[4] In radial compressors (i.e., centrifugal compressors), air enters through the eye, is accelerated by the impeller blades, and leaves 90° from the original inlet direction. By mounting two or more impellers on a single shaft, centrifugal compressors can easily be constructed as multistage machines. In *axial-flow compressors*, air flows essentially straight through as it is compressed by blades in the rotor. Axial-flow compressors are explicitly multistage machines, since each row of blades represents a single stage.

In comparison, radial (centrifugal) machines compress small volumes through large pressure ratios, and axial machines compress large volumes through small pressure ratios. Combined *axial-centrifugal compressors* are used when large volumes need to be compressed through

---

[1]Other types of *rotary positive-displacement compressors*, such as screw, globe, water-ring, and sliding vane designs, are not covered in this chapter.

[2]There is much overlap in the operating characteristics of all types of compressors. The decision to use one type in favor of another will also be influenced by cost, speed, and available driving mechanisms.

[3]The terms *turbocompressor* and *turboblower* are also used to describe these dynamic machines. The "turbo" prefix does not require that the input power comes from engine exhaust or combustion products (i.e., a turbocompressor is not a *turbocharger*). Input power can be from any shaft drive.

[4]Traditional radial (centrifugal) compressors are used for compression of air and other nitrogen, oxygen, carbon dioxide, etc. Axial compressors are traditionally found in the steel industry to produce blast furnace air and in plants with large air-separation processes.

large pressure ratios. The axial stages first reduce the volume; the subsequent radial stages (on the same shaft) complete the pressure increase.

2. BOOSTERS, INTENSIFIERS, AND AMPLIFIERS

Most compressed air equipment (pneumatic tools, painting, instrumentation, etc.) operates in the range of 90 to 125 psig (620 to 860 kPa). Some higher-pressure processes (e.g., plastic molding, metal-working, and pressure testing machines) may require up to 500 psig (3.5 MPa). This ultra-high pressure plant air can be obtained by dedicated multistage compressors, but a power savings can be achieved by starting with the lower-pressure plant air.

A *pressure amplifier* (*pressure intensifier*) is one method of producing high-pressure air from a lower-pressure source. A large piston is driven by plant air at low pressures. This piston, in turn, drives a smaller piston, producing a smaller quantity of higher-pressure air. Output is seldom higher than 25 cfm (12 L/s). The output pressure is determined by the input pressure and the ratio of piston areas. Amplifiers do not require any input electrical power. However, they exhaust substantial amounts of compressed plant air, and they require complex valving.

A *pressure booster* is essentially another electrically driven single-stage compressor (usually air-cooled) drawing on compressed plant air. Pressure ratios are typically 2:1 to 7:1. Flow rates are typically up to 500 cfm (240 L/s).

3. COMPRESSOR CONTROL

Reciprocating compressors used to supply plant air can either run continuously or intermittently. There is essentially no "control" when running continuously, although pressure will be limited to the relief valve setting, usually just above the highest working pressure required. For reciprocating compressors operating intermittently, the maximum pressure will be the cutoff pressure, usually 15 to 20 psi (100 to 140 kPa) higher than the highest pressure needed.

There are three general ways of controlling dynamic compressors. The most effective method is *speed control*, which is applicable to drivers (i.e., prime movers) such as steam or combustion turbines whose speeds can be varied. Figure 32.5 illustrates the variation in performance with changes in rotational speed.

When the driving speed is fixed, as it is with electrical motors, guide vane control and suction throttling can be used. With *guide vane control*, vanes in the compressor impart rotation to the inlet stream. By changing the components of inlet velocity, flow rate and pressure are changed. By imparting either a rotation or

counter-rotation to the inlet stream, guide-vane control can either decrease or increase the flow and pressure, respectively. With the inefficient *suction throttling* control method, a butterfly valve in the suction line is partially closed. This decreases the entering suction pressure and flow rate.

4. STORAGE TANKS

Compressed-air storage tanks (*receiver tanks*) are used to reduce the duty cycle of plant compressors. The compressor starts when the tank pressure is reduced to the *cut-in pressure* and runs until the tank pressure reaches the *cut-out pressure*. Larger tanks result in longer off-periods for the compressor and fewer compressor on-cycles. However, they increase the running time per cycle.

5. COMPRESSOR CAPACITY

In the United States, the capacity of a compressor can be expressed in any of several ways.

- mass flow rate: $\dot{m}$

- volumetric flow rate referred to standard conditions (usually 14.7 psia and 60°F): SCFM (standard cubic feet per minute); SCFH (standard cubic feet per hour); MMSCFD (million standard cubic feet per 24-hour day).[5]

- volumetric flow rate referred to inlet conditions: ICFS (inlet cubic feet per second); ICFM (inlet cubic feet per minute); ICFH (inlet cubic feet per hour).

6. COMPRESSION PROCESSES

The thermodynamic behavior of a gas in a *reciprocating compressor* is controlled by the movement of the piston and heat transfer to compressor surfaces. Such compression and expansion are polytropic processes. The work done per cycle depends on the polytropic exponent, n, which is a function of the compressor. Efficient air compressors have polytropic exponents between 1.25 and 1.30, but values up to 1.35 are not uncommon. If the compression is known to be isentropic, the gas' ratio of specific heats, k, should be used in place of the polytropic exponent, n.

Ideal compression in an uncooled *dynamic compressor* is usually considered to be adiabatic (isentropic), with deviations from ideal performance accounted for by the

---

[5]The ASME Power Test Code defines "standard air" as air at a temperature of 68°F (20°C), a total pressure of 14.7 psia (101.3 kPa), and a relative humidity of 36%. The term *free air* is also used by manufacturers to describe the air demands of their equipment. Free air is air at atmospheric pressure and ambient temperature.

adiabatic (isentropic) compression efficiency.[6] With intercooling (see Sec. 11), compression in dynamic compressors can also approach an isothermal process. Deviations from ideal isothermal performance are accounted for by the *isothermal compression efficiency.*

$$\eta_{\text{adiabatic}} = \frac{W_{\text{adiabatic}}}{W_{\text{actual}}} \qquad 32.1$$

$$\eta_{\text{isothermal}} = \frac{W_{\text{isothermal}}}{W_{\text{actual}}} \qquad 32.2$$

Figure 32.1 shows that the work of an isothermal compression is less than the work of an adiabatic compression.[7] The work of a polytropic compression will be somewhere between the two.

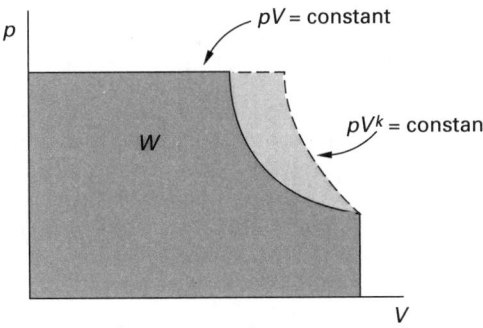

Figure 32.1 *Comparison of Isothermal and Adiabatic Work (zero clearance)*

7. CLEARANCE AND CLEARANCE VOLUME

Reciprocating compressors are characterized by their *clearance volume.*[8] This is the volume (volume V_D in Fig. 32.3) at the head of the cylinder when the piston is at its most extended position in its stroke. (This position is known as *top-dead-center* or TDC.)[9] The gases remaining in the clearance volume after the discharge valve closes at top-dead-center are known as the *residual gases.*

---

[6]Eq. 32.1 is the definition of isentropic efficiency. For some reason, in the study of gas compression, the term *adiabatic* is used rather than the terms *reversible adiabatic* or *isentropic.*
[7]Note this confusing point: The work done *in* an adiabatic (isentropic) *process* is less than the work done in an isothermal *process*, but (for the same pressure limits) the work done *by* an isentropic *compressor* is more than the work done by an isothermal *compressor.*
[8]Although there is clearance between the casing and blades of a dynamic compressor, the terms *clearance volume* and *clearance percent* always refer to reciprocating compressors.
[9]The piston is said to be at *bottom-dead-center* (BDC) when it is at its most retracted position in the stroke.

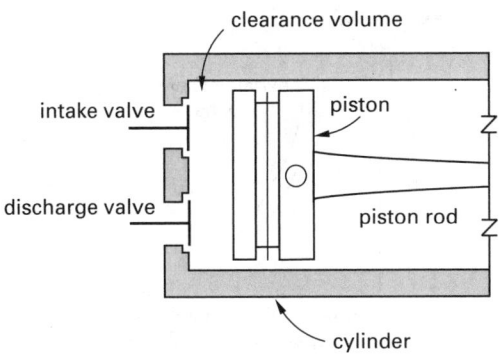

Figure 32.2 *Clearance Volume*

The *percent clearance* (also known as *clearance*), c, is calculated from the *swept volume (piston displacement)*, $V_B - V_D$ in Eq. 32.3. Swept volume is not the total volume of the free gas.

$$c = 100\% \times \frac{\text{clearance volume}}{\text{swept volume}} = 100\% \times \frac{V_D}{V_B - V_D} \qquad 32.3$$

The residual gases expand along with the next intake of gas during the intake stroke to reduce the volumetric capacity per stroke. However, clearance affects only the volumetric efficiency; it does not affect the required input power. Two compressors with the same gas flow rate but with different clearances will require the same input power because the expanding residual gases give back their work of compression when they expand. The power requirement depends on only the mass of the gas passing through the compressor.

During steady-state compressor operation, the mass of gas entering the cylinder equals the mass of gas discharged, and clearance is not considered.

8. RECIPROCATING, SINGLE-STAGE COMPRESSION

The following processes describe a single stage of compression in a reciprocating compressor.

A to B: constant pressure suction (intake valve open, discharge valve closed)

B to C: polytropic compression (both valves closed)

C to D: constant pressure delivery (exhaust valve open, intake valve closed)

D to A: polytropic expansion (both valves closed)

The processes describing the compression do not constitute a true cycle because the gas mass in the B-to-C process is not the same as the gas mass in the C-to-D process. For that reason, the broken lines used in Fig. 32.3 show the cylinder volume, not the gas conditions.

Power Cycles

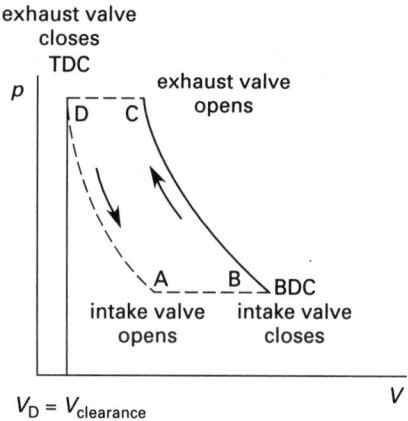

Figure 32.3 *Single-Stage Compression*

The main parameters affecting performance are the polytropic exponent, n; compression ratio, r_p; and volumetric efficiency, η_v. For convenience, the polytropic exponents for the expansion and compression processes are assumed to be the same. (However, it is not difficult to consider different values.)

The *compression ratio* for reciprocating and rotating compressors, defined by Eq. 32.4, is the ratio of pressures (not the ratio of volumes as it is for internal combustion engines).[10]

$$r_p = \frac{p_D}{p_A} = \frac{p_C}{p_B} \qquad 32.4$$

The *volumetric efficiency*, η_v, is the ratio of the actual mass of gas compressed to the mass of gas in the swept volume. Volumetric efficiency and capacity are evaluated at the inlet conditions unless otherwise noted.

$$\eta_v = \left.\frac{\text{actual mass of gas compressed}}{\text{mass of gas in swept volume}}\right|_{\substack{\text{inlet} \\ \text{conditions}}} \qquad 32.5$$

$$= \frac{V_B - V_A}{V_B - V_D} = 1 - \left(r_p^{\frac{1}{n}} - 1\right)\left(\frac{c}{100}\right) \qquad 32.6$$

The gas mass flow rate through the compressor can be determined from the mass per stroke and the number of strokes per minute. Assuming an ideal gas and evaluating the gas at the compressor's inlet conditions, the ideal and actual gas masses per stroke are

$$m_{\text{stroke,ideal}} = \frac{p_A(V_B - V_D)}{R T_A} \qquad 32.7$$

$$m_{\text{stroke,actual}} = \eta_v \, m_{\text{stroke,ideal}} \qquad 32.8$$

The mass flow rate is

$$\dot{m} = m_{\text{stroke,actual}} \times \text{rotational speed} \qquad 32.9$$

---

[10]The term *ratio of compression*, r_c, is often used to distinguish the ratio of pressures from the *ratio of volumes*.

The relationships between the pressures and volumes are predicted by the ideal gas laws for polytropic processes.

$$\frac{V_A}{V_D} = \left(\frac{p_D}{p_A}\right)^{\frac{1}{n}} = r_p^{\frac{1}{n}} \qquad 32.10$$

$$\frac{V_B}{V_C} = \left(\frac{p_C}{p_B}\right)^{\frac{1}{n}} = r_p^{\frac{1}{n}} \qquad 32.11$$

Equations 32.12 through 32.14 give the ideal work of compression for a polytropic *steady-flow* process. (From a thermodynamic standpoint, work is negative if the surroundings compress the system. The work of compression in these equations is expressed as a positive number.)

$$W_{BC} = h_C - h_B = c_p(T_C - T_A) \qquad 32.12$$

$$= \left(\frac{n}{1-n}\right)(p_C v_C - p_B v_B) \qquad 32.13$$

$$= \left(\frac{n p_B v_B}{1-n}\right)\left[\left(\frac{p_C}{p_B}\right)^{\frac{n-1}{n}} - 1\right] \qquad 32.14$$

During the D-to-A recovery stroke, the residual gases expand and do work on the piston. The net work per cycle is

$$W_{\text{net}} = W_{BC} - W_{DA} = h_B - h_A$$
$$= c_p(T_B - T_A) \qquad 32.15$$

"Work done by compressor" almost always means the change in energy imparted to the air. It does not mean the work supplied to the compressor (i.e., it does not include the compression efficiencies).

In addition to the work of compression, the compressor motor supplies power to overcome friction, discharge the gas against the receiver pressure, and move the piston during noncompression parts of the cycle. Equation 32.16 gives the *brake work*, which is expressed in brake horsepower (BHP) and brake kilowatts (BkW).

$$\text{brake work} = \frac{W_{\text{net}}}{\eta_m \eta_s} \qquad 32.16$$

The power requirement is

$$P = \dot{m} \times \text{brake work} \qquad 32.17$$

Typical single-stage compressors for common plant processes produce approximately 4 to 5 cfm of 90 to 125 psig (2 to 2.5 L/s of 620 to 860 kPa) air per brake horsepower delivered. Single-stage pressure boosters starting with compressed plant air are highly efficient and can produce as much as 13 cfm (6.1 L/s) per brake horsepower.

Example 32.1

Air enters a reciprocating compressor at 14.7 psia and 70°F (101.3 kPa and 21°C). 100 ft³/min (50 L/s) of free air is compressed isentropically to 55 psia (379 kPa). The clearance is 6%. What is the volumetric efficiency?

SI Solution

Solve by using Eq. 32.6. (The customary U.S. solution takes a different approach.) Since the compression is isentropic, the polytropic exponent, n, is equal to the ratio of specific heats, k.

$$\eta_v = 1 - \left(r_p^{\frac{1}{n}} - 1\right)\left(\frac{c}{100}\right)$$

$$= 1 - \left[\left(\frac{379 \text{ kPa}}{101.3 \text{ kPa}}\right)^{\frac{1}{1.4}} - 1\right]\left(\frac{6\%}{100}\right)$$

$$= 0.906 \quad (90.6\%)$$

Customary U.S. Solution

(The SI solution takes a different approach.) Refer to Fig. 32.3. Calculate the volumes per minute. Since the compression is isentropic, the polytropic exponent, n, is equal to the ratio of specific heats, k.

$$\frac{V_A}{V_D} = \left(\frac{p_D}{p_A}\right)^{\frac{1}{k}}$$

$$= \left(\frac{55 \frac{\text{lbf}}{\text{in}^2}}{14.7 \frac{\text{lbf}}{\text{in}^2}}\right)^{\frac{1}{1.4}} = 2.566$$

$$V_A = 2.566 V_D$$

Since $p_C = p_D$ and $p_B = p_A$, $V_B/V_C = 2.566$ as well.

The actual volumetric flow rate entering the compressor each minute is 100 ft^3 of free air.

$$V_B - V_A = 100 \text{ ft}^3$$

$$V_B = 100 \text{ ft}^3 + V_A$$

$$= 100 \text{ ft}^3 + 2.566 V_D$$

The swept volume (piston displacement) is

$$\text{swept volume} = V_B - V_D$$

V_D is the clearance volume, which is 6% of the swept volume.

$$V_D = c(\text{swept volume}) = (0.06)(V_B - V_D)$$

Solving for V_B,

$$V_B = 17.667 V_D$$

Combine the two equations for V_B.

$$17.667 V_D = 100 \text{ ft}^3 + 2.566 V_D$$

$$V_D = 6.622 \text{ ft}^3$$

$$V_B = 17.667 V_D = (17.667)(6.622 \text{ ft}^3)$$

$$= 117.0 \text{ ft}^3$$

The volumetric efficiency (calculated from volumes per minute) is

$$\eta_v = \frac{V_B - V_A}{V_B - V_D}$$

$$= \frac{100 \text{ ft}^3}{117.0 \text{ ft}^3 - 6.622 \text{ ft}^3}$$

$$= 0.906 \quad (90.6\%)$$

9. POLYTROPIC HEAD

When the net work calculated in Eq. 32.15 is expressed in feet (numerically equivalent to ft-lbf/lbm) or meters, it may be referred to as the *polytropic head* or (if the process is adiabatic) as the *adiabatic head*. While the discharge pressure depends on the gas, the head does not. The head, H, can be calculated from Eq. 32.18. Z_{ave} is the compressibility factor averaged over the inlet and discharge conditions.

$$H = \left(\frac{n}{n-1}\right) Z_{\text{ave}} R T_{\text{inlet}} \left(r_p^{\frac{n-1}{n}} - 1\right) \qquad \textit{32.18}$$

10. POLYTROPIC EFFICIENCY

The *polytropic efficiency*, η_n, is the ratio of ideal polytropic compression work to actual compression work. (For isentropic processes, the polytropic efficiency is 100%.)

$$\eta_n = \frac{W_{\text{ideal}}}{W_{\text{actual}}} = \frac{\dot{m}H}{P_{\text{actual}}} \qquad \textit{32.19}$$

The polytropic efficiency can also be calculated from the polytropic exponent, n.

$$\frac{n}{n-1} = \frac{\eta_n k}{k-1} \qquad \textit{32.20}$$

Example 32.2

A dynamic compressor receives 50,000 cfm (23.5 kL/s) of a gas at 15 psia (104 kPa) and 70°F (21°C). The pressure is increased by the compressor to 85 psia (590 kPa). The gas has a molecular weight of 31 and a ratio of specific heats of 1.17. The compressibility factor at the inlet is 0.99, and the average compressibility factor over the compression process is 0.98. The actual shaft power is 9225 hp (6880 kW). 183 hp (136 kW) are required to overcome friction and windage losses. What are the (a) theoretical compression power, (b) polytropic head, and (c) polytropic efficiency?

SI Solution

The absolute inlet temperature is

$$T_{\text{inlet}} = 21°C + 273 = 294 \text{K}$$

The compression ratio is

$$r_p = \frac{p_{\text{discharge}}}{p_{\text{inlet}}}$$

$$= \frac{590 \text{ kPa}}{104 \text{ kPa}} = 5.673$$

The specific gas constant is

$$R = \frac{R^*}{\text{MW}} = \frac{8314 \frac{\text{J}}{\text{kmol·K}}}{31 \frac{\text{kg}}{\text{kmol}}}$$

$$= 268.2 \frac{\text{J}}{\text{kg} \cdot \text{K}}$$

The inlet density is

$$\rho_{\text{inlet}} = \frac{p_{\text{inlet}}}{ZRT_{\text{inlet}}}$$

$$= \frac{(104 \text{ kPa})\left(1000 \frac{\text{Pa}}{\text{kPa}}\right)}{(0.99)\left(268.2 \frac{\text{J}}{\text{kg·K}}\right)(294\text{K})}$$

$$= 1.332 \text{ kg/m}^3$$

The mass flow rate is

$$\dot{m} = \dot{V}\rho$$

$$= \left(23.5 \frac{\text{kL}}{\text{s}}\right)\left(1 \frac{\text{kL}}{\text{m}^3}\right)\left(1.332 \frac{\text{kg}}{\text{m}^3}\right)$$

$$= 31.3 \text{ kg/s}$$

This process is not known to be adiabatic or isentropic. The polytropic exponent is not known. An iterative process is required. Assume the polytropic compression efficiency, η_n, is 80%. Use Eqs. 32.18 and 32.20 to determine the theoretical polytropic head for this iteration using the assumed efficiency.

$$\frac{n}{n-1} = \frac{\eta_n k}{k-1}$$

$$= \frac{(0.80)(1.17)}{1.17 - 1} = 5.506$$

$$H = \left(\frac{n}{n-1}\right) Z_{\text{ave}} R T_{\text{inlet}} \left(r_p^{\frac{n-1}{n}} - 1\right)$$

$$= (5.506)(0.98)\left(268.2 \frac{\text{J}}{\text{kg·K}}\right)(294\text{K})$$

$$\times \left[(5.673)^{\frac{1}{5.506}} - 1\right]$$

$$= 1.577 \times 10^5 \text{ J/kg}$$

The theoretical power for this iteration is

$$P_{\text{theoretical}} = \frac{P_{\text{ideal}}}{\eta_n}$$

$$= \frac{\dot{m}H}{\eta_n}$$

$$= \frac{\left(31.3 \frac{\text{kg}}{\text{s}}\right)\left(1.577 \times 10^5 \frac{\text{J}}{\text{kg}}\right)}{(0.80)\left(1000 \frac{\text{W}}{\text{kW}}\right)}$$

$$= 6170 \text{ kW}$$

The actual compression power is known to be

$$P_{\text{actual}} = P_{\text{shaft}} - P_{\text{friction}}$$

$$= 6880 \text{ kW} - 136 \text{ kW} = 6744 \text{ kW}$$

Since the actual power is higher, the assumed efficiency is too high.

For the next iteration, use an assumed efficiency of

$$\eta_{n,2} = \eta_{n,1}\left(\frac{P_{\text{theoretical}}}{P_{\text{actual}}}\right)$$

$$= (0.80)\left(\frac{6170 \text{ kW}}{6744 \text{ kW}}\right) = 0.732$$

Customary U.S. Solution

The absolute inlet temperature is

$$T_{\text{inlet}} = 70°\text{F} + 460 = 530°\text{R}$$

The compression ratio is

$$r_p = \frac{p_{\text{discharge}}}{p_{\text{inlet}}}$$

$$= \frac{85 \text{ psia}}{15 \text{ psia}} = 5.667$$

The specific gas constant is

$$R = \frac{R^*}{\text{MW}} = \frac{1545 \frac{\text{ft-lbf}}{\text{lbmol-}°\text{R}}}{31 \frac{\text{lbm}}{\text{lbmol}}}$$

$$= 49.84 \text{ ft-lbf/lbm-}°\text{R}$$

The inlet density is

$$\rho_{\text{inlet}} = \frac{p_{\text{inlet}}}{ZRT_{\text{inlet}}}$$

$$= \frac{\left(15 \frac{\text{lbf}}{\text{in}^2}\right)\left(144 \frac{\text{in}^2}{\text{ft}^2}\right)}{(0.99)\left(49.84 \frac{\text{ft-lbf}}{\text{lbm-}°\text{R}}\right)(530°\text{R})}$$

$$= 0.0826 \text{ lbm/ft}^3$$

The mass flow rate is

$$\dot{m} = \dot{V}\rho$$

$$= \left(50{,}000 \; \frac{\text{ft}^3}{\text{min}}\right)\left(0.0826 \; \frac{\text{lbm}}{\text{ft}^3}\right)$$

$$= 4130 \; \text{lbm/min}$$

This process is not known to be adiabatic or isentropic. The polytropic exponent is not known. An iterative process is required. Assume the polytropic compression efficiency, η_n, is 80%. Use Eqs. 32.18 and 32.20 to determine the theoretical polytropic head for this iteration using the assumed efficiency.

$$\frac{n}{n-1} = \frac{\eta_n k}{k-1}$$

$$= \frac{(0.80)(1.17)}{1.17-1} = 5.506$$

$$H = \left(\frac{n}{n-1}\right)Z_{\text{ave}}RT_{\text{inlet}}\left(r_p^{\frac{n-1}{n}}-1\right)$$

$$= (5.506)(0.98)\left(49.84 \; \frac{\text{ft-lbf}}{\text{lbm-}^\circ\text{R}}\right)(530^\circ\text{R})$$

$$\times \left[(5.667)^{\frac{1}{5.506}}-1\right]$$

$$= 52{,}784 \; \text{ft-lbf/lbm}$$

The theoretical horsepower for this iteration is

$$P_{\text{theoretical}} = \frac{P_{\text{ideal}}}{\eta_n}$$

$$= \frac{\dot{m}H}{\eta_n}$$

$$= \frac{\left(4130 \; \frac{\text{lbm}}{\text{min}}\right)\left(52{,}784 \; \frac{\text{ft-lbf}}{\text{lbm}}\right)}{(0.80)\left(33{,}000 \; \frac{\text{ft-lbf}}{\text{hp-min}}\right)}$$

$$= 8257 \; \text{hp}$$

The actual compression power is known to be

$$P_{\text{actual}} = P_{\text{shaft}} - P_{\text{friction}}$$

$$= 9225 \; \text{hp} - 183 \; \text{hp} = 9042 \; \text{hp}$$

Since the actual power is higher, the assumed efficiency is too high.

For the next iteration, use an assumed efficiency of

$$\eta_{n,2} = \eta_{n,1}\left(\frac{P_{\text{theoretical}}}{P_{\text{actual}}}\right)$$

$$= (0.80)\left(\frac{8257 \; \text{hp}}{9042 \; \text{hp}}\right) = 0.731$$

After two additional iterations, the values are essentially stable.

$$P_{\text{theoretical}} = 9000 \; \text{hp} \quad [9042 \; \text{hp ideally}]$$

$$H = 53{,}460 \; \text{ft-lbf/lbm}$$

$$\eta_n = 0.741$$

11. COOLING AND INTERCOOLING

For a given compression ratio, the work for isothermal compression is less than the work for adiabatic compression.[11] Furthermore, standard hydrocarbon lubricants used in air compressors are limited to approximately 365°F (185°C). Therefore, cooling is often used in compressors whose compression ratio exceeds 4.0. (A third benefit from cooling is a reduction in ring and valve loading.) For reciprocating compressors, the cooling is accomplished by surrounding the cylinder with a water jacket or by incorporating finned heat radiators in the jacket design.

When a partially compressed gas is withdrawn, cooled, and compressed further, the term *intercooling* is used. Intercoolers are typically used with centrifugal compressors. The term *perfect intercooling* refers to the case where the gas is cooled to the original inlet temperature (i.e., $T_B = T_D$ and $p_C = p_B$ in Fig. 32.3).[12]

If necessary, an *aftercooler* removes heat (and reduces the pressure) from the compressed gas after the compression process is complete.

Compressed air for plant use usually leaves the aftercooler at 25 to 75°F (15 to 45°C) above the ambient temperature.

12. MULTISTAGE COMPRESSION

Figure 32.4 shows the path of a gas experiencing two-stage compression and intercooling. For a reciprocating compressor, the curved solid lines represent polytropic compressions, and the broken lines represent the cylinder volume. For a centrifugal compressor, the curved solid lines represent isentropic (or isothermal) compressions and the broken lines have no meaning.

Analysis of a multistage compressor is similar to that of a single-stage compressor. The minimum work occurs when the compression ratios of all stages are the same. This is known as *optimum staging*. In that case,

$$r_p = \frac{p_C}{p_B} = \frac{p_E}{p_D} \qquad \textit{32.21}$$

Since $p_C = p_D$,

$$p_C^2 = p_E p_B \qquad \textit{32.22}$$

---

[11]In practice, the additional cost of the intercooler apparatus must be compared with the power savings.
[12]The discharge from an intercooler will usually be approximately 20°F (10°C) higher than the jacket water temperature.

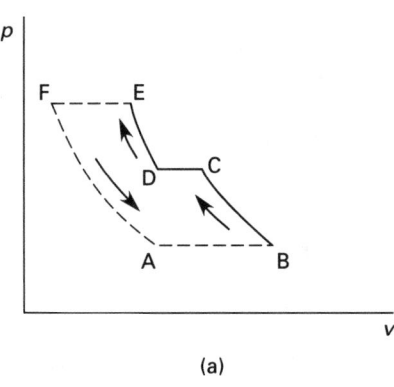

(a)

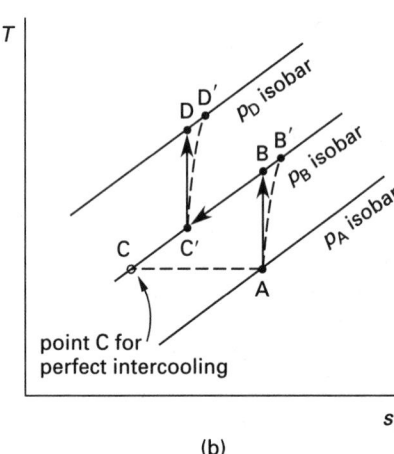

point C for perfect intercooling

(b)

Figure 32.4 *Two-Stage Compression with Intercooling*

Typical water-cooled two-stage compressors for common plant processes produce approximately 2 to 3 cfm of 200 to 500 psig (1 to 1.5 L/s of 1.4 to 3.5 MPa) air per brake horsepower delivered.

13. DYNAMIC COMPRESSORS

The ideal gas relationships for adiabatic and isothermal processes can be used to analyze the performance of dynamic gas compressors. In fact, most of the equations in Sec. 8 can be used for isentropic processes if the ratio of specific heats, k, is used in place of the polytropic exponent, n.

14. COMPRESSOR CHARACTERISTIC CURVES

Performance of dynamic compressors, like that of pumps and fans, is described by characteristic curves. Compressor manufacturers usually provide curves of discharge pressure versus inlet volumetric flow rate and brake power versus flow rate. Alternatively, head versus flow rate or pressure ratio versus flow rate may be substituted for the discharge pressure curve.

Characteristic curves are different for reciprocating and centrifugal compressors. Reciprocating compressors are essentially constant-volume, variable-pressure devices. Since their capacities are essentially fixed, the characteristic curve is quite steep. This is shown in Fig. 32.6.

Figure 32.5 illustrates the characteristic curve for a dynamic compressor. It is more similar to curves for pumps and fans. The *surge limit* is the leftmost area on the curve. In surge, flow becomes unstable and the compressor pressure intermittently drops below the system pressure, resulting in backflow.[13] Figure 32.5 also illustrates that, even with decreasing pressure, the flow rate does not increase indefinitely. Flow approaches a maximum flow rate at the *choke* or *stone wall point*.

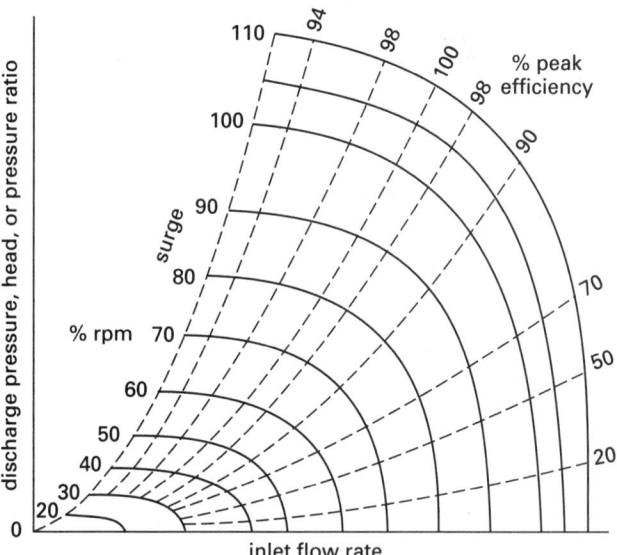

Figure 32.5 *Dynamic Compressor Head-Capacity Curves*

The system curve (not shown) will intersect the characteristic curve at the operating point, as it does with pumps and fans. Ideally, the *operating point* should coincide with the maximum efficiency line. Since each stage of compression will have its own characteristic curve, there will be as many operating points as there are stages. In a properly matched multistage compressor, all of the stages will be operating at high efficiencies. If the stages are not well matched or if the flow rate or inlet pressure change, one or more of the stages may operate "off" its design point. In the worst cases, a stage may be in a surge or choke condition.

15. OPERATION AT CHANGED INLET CONDITIONS

Discharge pressure and power curves can only be used for the gas and inlet conditions (typically, "standard" inlet conditions) intended. Those curves cannot be used if the gas or any of the inlet conditions are changed. In

---

[13]Dynamic compressors should have antisurge controls to open a recirculation valve. These increase the flow sufficiently to keep operation out of surge.

most cases, curves applicable for the new conditions are not available. In those cases, a curve for head versus inlet flow rate can be laboriously generated for the new conditions.[14]

Ideally, energy transfer (head) per unit mass is a function of only the velocity of the impeller. Head is the same for all gases, regardless of density. Therefore, while the gas properties affect the discharge pressure, they do not affect the head. Moving a compressor to another altitude will affect the discharge pressure, mass flow rate, and power requirement. It will not affect the ideal compression ratio or the volumetric flow rate.

In reality, compressors (particularly piston compressors) do not deliver as much flow at higher altitudes as they do at sea level. As the air viscosity decreases, leakage through pump clearances increases. The effect is known as *droop*. The disparity between sea-level and high-altitude performance becomes more pronounced at higher discharge pressures.

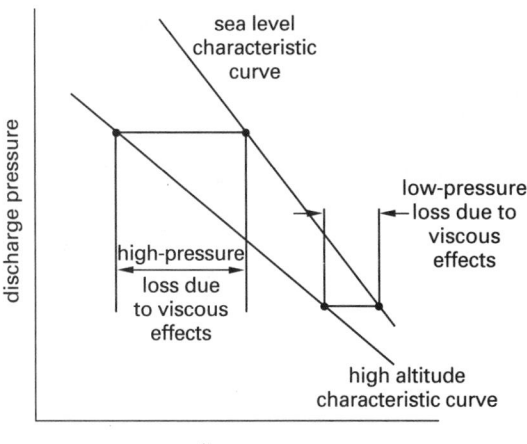

Figure 32.6 *Reciprocating Compressor Droop*

PRACTICE PROBLEMS

1. What is the isentropic efficiency of a compressor that takes in air at 8 psia and $-10°F$ (55 kPa and $-20°C$) and discharges it at 40 psia and 315°F (275 kPa and 160°C)?

2. A reciprocating air compressor has a 7% clearance and discharges 65 psia (450 kPa) air at the rate of 48 lbm/min (0.36 kg/s). The polytropic exponent is 1.33. What mass of air is compressed each minute?

3. Air at 14.7 psia and 500°F (101.3 kPa and 260°C) is compressed in a centrifugal compressor to 6 atm. The

[14]Generating the head-flow rate curve is not as easy as applying Eq. 32.18. The polytropic exponent, n, must simultaneously satisfy Eqs. 32.18, 32.19, and 32.20. The actual compression power is taken from the existing power-flow rate curve provided by the manufacturer. Therefore, calculating even a single point on the head-flow rate curve is an iterative process. This is illustrated in Ex. 32.2.

isentropic efficiency of the compression process is 65%. What are the (a) compression work, (b) final temperature, and (c) increase in entropy?

4. (*Time limit: one hour*) Compressors A and B both discharge into a common tank. The storage tank contains 100 psia, 90°F air (700 kPa, 32°C). Both compressors receive air at 14.7 psia and 80°F (101.3 kPa and 27°C). The flow rate through compressor A is 600 cfm (300 L/s). Process C uses 100 cfm of 80 psia, 85°F (50 L/s of 550 kPa, 29°C) air. Process D uses 120 cfm of 85 psia, 80°F (60 L/s of 590 kPa, 27°C) air. Process E uses 8 lbm/min (0.06 kg/s) of 85°F (29°C) air. Air is an ideal gas, and compressibility effects are to be disregarded. Calculate the flow rate through compressor B.

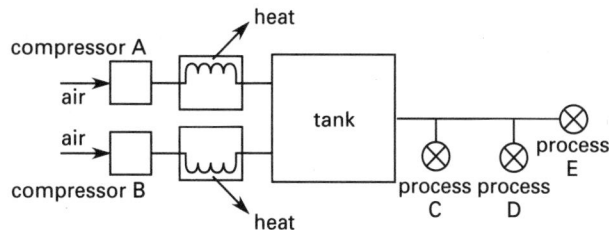

5. (*Time limit: one hour*) 300 cfm (150 L/s) of air at 14.7 psia and 90°F (101.3 kPa and 32°C) enter a compressor. The compressor discharges air into a water-cooled heat exchanger. The compressed air is stored at 300 psig and 90°F (2.07 MPa and 32°C) in a 1000 ft³ (27 m³) tank. The tank feeds three air-driven tools with the flow rates and properties listed. The pressures to the air-driven tools are regulated and remain constant as the tank pressure changes. Air is an ideal gas, and compressibility effects are to be disregarded. How long can the system run?

	tool 1	tool 2	tool 3
flow rate			
(cfm)	40	15	unknown
(L/s)	19	7	unknown
flow rate			
(lbm/min)	unknown	unknown	6
(kg/s)	unknown	unknown	0.045
minimum pressure			
(psig)	90	50	80
(kPa)	620	350	550
temperature			
(°F)	90	85	80
(°C)	32	29	27

Power Cycles

33 Refrigeration Cycles

1. Introduction to Refrigeration 33-1
2. Refrigerants 33-2
3. Heat Pumps 33-2
4. Coefficient of Performance 33-2
5. Energy Efficiency Ratio 33-2
6. Refrigeration Capacity 33-3
7. Carnot Refrigeration Cycle 33-3
8. Vapor Compression Cycle 33-3
9. Air Refrigeration Cycle 33-6
10. Heat-Driven Refrigeration Cycles 33-6
11. Absorption Cycle 33-6
12. High-Side Equipment 33-7
13. Low-Side Equipment 33-8
14. Refrigerant Line Sizing 33-9
    Practice Problems 33-9

Nomenclature

A	area	ft^2	m^2
c_p	specific heat	Btu/lbm-°F	kJ/kg·K
COP	coefficient of performance	–	–
D	diameter	ft	m
EER	energy efficiency ratio	Btu/W-hr	n.a.
h	enthalpy	Btu/lbm	kJ/kg
k	ratio of specific heats	–	–
L	stroke	ft	m
$\dot{m}$	mass flow rate	lbm/sec	kg/s
n	rotational speed	rpm	rpm
p	pressure	lbf/ft^2	kPa
q	heat	Btu/lbm	kJ/kg
Q	heat	Btu	kJ
r	ratio	–	–
s	entropy	Btu/lbm-°F	kJ/kg·K
T	temperature	°R	K
$\dot{V}$	volumetric flow rate	ft^3/min	L/s
W	work	Btu/lbm	kJ/kg
x	quality	–	–

Symbols

η	efficiency	–	–
v	specific volume	ft^3/lbm	m^3/kg

Subscripts

f	saturated liquid
g	saturated vapor
p	pressure or constant pressure
s	isentropic
v	volumetric

1. INTRODUCTION TO REFRIGERATION

Refrigeration is the process of transferring heat from a low-temperature area to a high-temperature area. Since heat flows spontaneously only from high- to low-temperature areas (according to the second law of thermodynamics), refrigeration needs an external energy source to force the heat transfer to occur. This energy source is a pump or compressor that does work in compressing the refrigerant. It is necessary to perform this work on the refrigerant in order to get it to discharge energy to the high-temperature area.

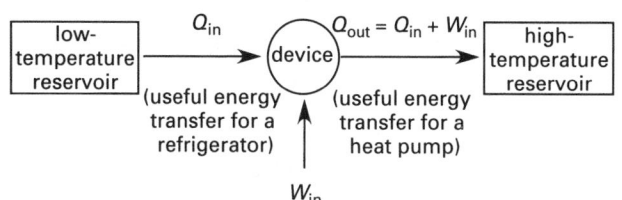

Figure 33.1 *Device Operating on a Refrigeration Cycle*

In a power cycle, heat from combustion is the input and work is the desired effect. Refrigeration cycles, though, are power cycles in reverse; work is the input and cooling is the desired effect. (For every power cycle, there is a corresponding refrigeration cycle.) In a refrigerator, the heat is absorbed from a low-temperature area and is rejected to a high-temperature area.[1] The pump work is also rejected to the high-temperature area.

General refrigeration devices consist of a coil (the *evaporator*) that absorbs heat, a *condenser* that rejects heat, a compressor, and a pressure-reduction device (the *expansion valve* or *throttling valve*).[2]

In operation, liquid refrigerant passes through the evaporator where it picks up heat from the low-temperature area and vaporizes, becoming slightly superheated. The vaporized refrigerant, that is, the "suction gas," is compressed by the compressor and, in so doing, increases even more in temperature. The high-pressure, high-temperature refrigerant passes through the condenser

---

[1]It is common thermodynamic jargon to refer to the low- and high-temperature areas as *environments* or *thermal reservoirs*. In particular, a low-temperature area is called a *source*, and a high-temperature area is referred to as a *sink*.

[2]In home refrigerators, the *expansion valve* takes the form of a long capillary tube.

coils, and, being hotter than the high-temperature environment, loses energy. Finally, the pressure is reduced in the expansion valve, where some of the liquid refrigerant also flashes into a vapor.

If the low-temperature area from which the heat is being removed is air from occupied space (that is, air is being cooled), the device is known as an *air conditioner*; if the heat is being removed from water, the device is known as a *chiller*. An air conditioner produces cold air; a chiller produces cold water.

In small refrigeration systems, such as those used for in-window home air conditioning, the compressor, condenser, evaporator (cooling) coil, and fan are combined into a single housing. These are referred to as *unit* (or *unitary*) *air conditioners*. In large commercial systems, the compressor is separate from the evaporation coil.

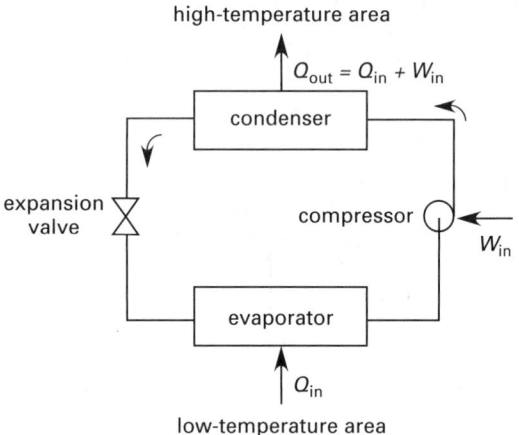

Figure 33.2 *Refrigeration Device*

2. REFRIGERANTS

Prior to the Montreal Protocol, refrigerants R-12 and R-22, in pure form or in blends (e.g., R-502), were the traditional choices for systems using reciprocating compressors.[3] R-22 is the refrigerant "of choice" in unitary air conditioners. Now, R-123 is replacing R-11 in new equipment. (R-22 is a hydrochlorofluorocarbon and does not yet need a replacement.) Refrigerant R-11 has been used with centrifugal compressors in all capacities (up to and above 1000 tons). It is being replaced by R-134a.

However, simply substituting R-123 for R-11 or R-134a for R-12 results in a 5 to 30% decrease in cooling capacity, depending on the operating conditions. If the reduced capacity is not acceptable, it can be restored by (1) replacing the compressor impeller with one more efficient, (2) increasing the compressor rotating speed, or (3) replacing the heat exchanger and tubing.[4]

---

[3]For more information on the Montreal Protocol, see Chap. 66.
[4]Power varies with the cube of the rotational speed. Increasing the compressor speed can increase the power significantly.

3. HEAT PUMPS

Heat pumps also operate on refrigeration cycles. Like standard refrigerators, they transfer heat from low-temperature areas to high-temperature areas. The device shown in Fig. 33.1 could represent either a heat pump or a refrigerator. There is no significant difference in the mechanisms or construction of heat pumps and refrigerators. The only difference is their purpose.

A refrigerator's main function is to cool the low-temperature area. The *useful energy transfer* for a refrigerator is the heat removed from the cold area. A heat pump's main function is to warm the high-temperature area. The useful energy transfer is the heat rejected to the high-temperature area. A heat pump is almost always used when *space heating* of an occupied area is needed. The attraction of *heat pump devices* is that they can provide both heating in the winter and cooling in the summer. During the summer, however, they run refrigeration cycles and are technically refrigerators. Strictly speaking, *heat pump cycles* are used only for space heating.

4. COEFFICIENT OF PERFORMANCE

The concept of thermal efficiency is not used with devices operating on refrigeration cycles. Rather, the *coefficient of performance* (COP) is defined as the ratio of useful energy transfer (as defined in Sec. 3) to the work input. The higher the coefficient of performance, the greater will be the effect for a given work input. Since the useful energy transfer is different for refrigerators and heat pumps, the coefficients of performance will also be different.

In calculating coefficients of performance, the refrigerant is considered to be the system. Therefore, Q_{in} is the energy that enters the refrigerant. (Q_{in} is not the energy that enters the high-temperature area because that area is not the system.)

$$\text{COP}_{\text{refrigerator}} = \frac{Q_{in}}{W_{in}} = \frac{Q_{in}}{Q_{out} - Q_{in}}$$
$$= \text{COP}_{\text{heat pump}} - 1 \qquad 33.1$$

The coefficient of performance for a heat pump includes the desired heating effect of the pump work input.

$$\text{COP}_{\text{heat pump}} = \frac{Q_{in} + W_{in}}{W_{in}} = \frac{Q_{in} + W_{in}}{Q_{out} - Q_{in}}$$
$$= \text{COP}_{\text{refrigerator}} + 1 \qquad 33.2$$

5. ENERGY EFFICIENCY RATIO

In the United States, the *energy efficiency ratio* (EER) is defined as the useful energy transfer in Btu/hr divided by input power in watts. This is just the coefficient of performance expressed in mixed units.

$$\text{EER} = 3.41 \times \text{COP} \qquad 33.3$$

As with the coefficient of performance, the definition of useful energy transfer depends on whether the device is being used as a refrigerator or as a heat pump.

$$\text{EER}_{\text{refrigerator}} = \frac{\dot{Q}_{\text{in,Btu/hr}}}{\dot{W}_{\text{in,watts}}}$$

$$= \text{EER}_{\text{heat pump}} - 1 \qquad 33.4$$

$$\text{EER}_{\text{heat pump}} = \frac{\left(\dot{Q}_{\text{in}} + \dot{W}_{\text{in}}\right)_{\text{Btu/hr}}}{\dot{W}_{\text{in,watts}}}$$

$$= \text{EER}_{\text{refrigerator}} + 1 \qquad 33.5$$

6. REFRIGERATION CAPACITY

The rate of energy removal from the low-temperature area is known as the *refrigeration capacity* or *refrigeration effect*. While the kilowatt is the appropriate SI unit, in the United States capacity is measured in *refrigeration tons*, where one ton is 200 Btu/min or 12,000 Btu/hr (3.517 kW) of heat removal. The ton is derived from the heat flow required to melt one ton of ice in 24 hours.

Since refrigeration capacity has the units of power (e.g., Btu/hr), it can be combined with the coefficient of performance to calculate the required pump horsepower.

$$\frac{\text{pump}}{\text{horsepower}} = \frac{4.715\,\dot{Q}_{\text{in,tons}}}{\text{COP}} = \frac{4.715\,\dot{Q}_{\text{in,Btu/hr}}}{(12,000)\,(\text{COP})} \qquad 33.6$$

The rate of energy removal can be used to calculate the refrigerant mass flow rate.

$$\dot{m} = \frac{\dot{Q}_{\text{in}}}{\Delta h_{\text{evaporator}}} \qquad 33.7$$

7. CARNOT REFRIGERATION CYCLE

The *Carnot refrigeration cycle* is a Carnot power cycle running in reverse. Because it is reversible, the Carnot refrigeration cycle has the highest coefficient of performance for any given temperature limits of all the refrigeration cycles. As shown in Fig. 33.3, the processes all occur within the vapor dome.

A to B: isentropic expansion

B to C: isothermal heating (vaporization)

C to D: isentropic compression

D to A: isothermal cooling (condensation)

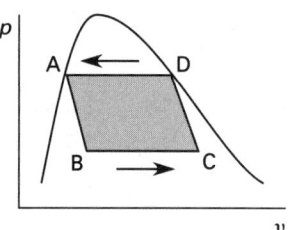

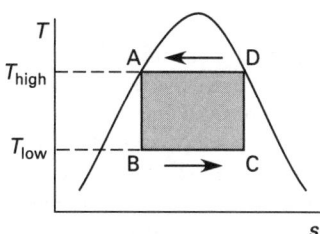

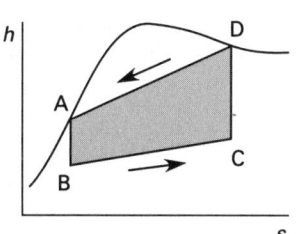

Figure 33.3 *Carnot Refrigeration Cycle*

The analysis procedure for the Carnot refrigeration cycle is reversed but otherwise identical to the procedure for analyzing the Carnot power cycle. The coefficients of performance are

$$\text{COP}_{\text{refrigerator}} = \frac{Q_{\text{in}}}{W_{\text{in}}} = \frac{Q_{\text{in}}}{Q_{\text{out}} - Q_{\text{in}}}$$

$$= \frac{T_{\text{low}}}{T_{\text{high}} - T_{\text{low}}} = \text{COP}_{\text{heat pump}} - 1 \quad 33.8$$

$$\text{COP}_{\text{heat pump}} = \frac{Q_{\text{in}} + W_{\text{in}}}{W_{\text{in}}} = \frac{Q_{\text{out}}}{Q_{\text{out}} - Q_{\text{in}}}$$

$$= \frac{T_{\text{high}}}{T_{\text{high}} - T_{\text{low}}} = \text{COP}_{\text{refrigerator}} + 1 \quad 33.9$$

8. VAPOR COMPRESSION CYCLE

The *vapor compression cycle* is essentially a reversed Rankine vapor cycle.[5] It is the most common type of refrigeration cycle, finding application in household refrigerators, air conditioners for cars and houses, chillers, and so on. The processes are illustrated in Fig. 33.4.

A to B: isenthalpic expansion

B to C: constant pressure heating (vaporization)

C to D: isentropic compression

D to A: constant pressure cooling (condensation)

---

[5]An irreversible expansion through a throttling valve takes the place of the boiler.

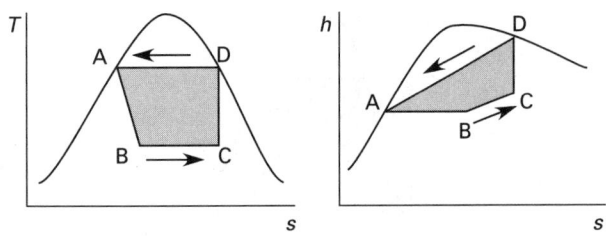

Figure 33.4 *Wet Vapor Compression Cycle*

If the refrigerant leaves the evaporator (point C in Fig. 33.4) with a quality of less than 1.0, the cycle is known as a *wet vapor compression cycle*. *Wet compression* is undesirable because of compressor wear and performance problems. For that reason, refrigerators are designed so that the refrigerant leaves the evaporator either saturated or slightly superheated, as shown by point C in Fig. 33.5. Compression of saturated or superheated vapor is said to be *dry compression*, and the cycle is known as a *dry vapor compression cycle*.

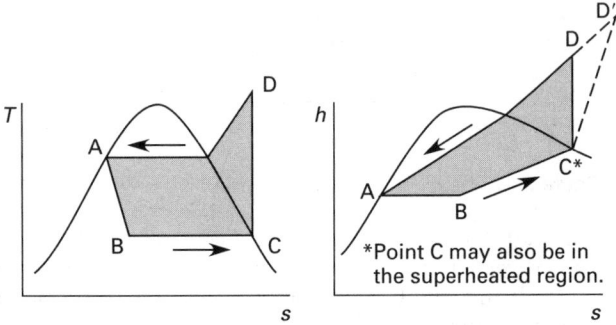

*Point C may also be in the superheated region.

Figure 33.5 *Dry Vapor Compression Cycle*

The refrigerant used depends on the temperatures of the low- and high-temperature areas, as well as on the power of the compressor. Although there are more than a hundred commercial refrigerants commonly available, *fluorinated hydrocarbons* (e.g., Freon™) are currently used (at least where they are not banned) for most residential and commercial applications. Where even lower temperatures are required (as in commercial freezing equipment), ammonia is used as the refrigerant.

If the refrigerant is saturated when it leaves the evaporator, the following solution method can be used. (The subscripts f and g refer to saturated fluid and vapor properties read from a refrigerant table. They do not correspond to any point on the figures shown in this chapter.)

At A: $x = 0$ (saturated liquid); either p_A or T_A must be known, or they can be found as the saturation temperature and pressure having an enthalpy of h_A. $h_A = h_B$; $p_A = p_D$.

At B: $T_B = T_C$; $h_B = h_A$ (because A-to-B is isenthalpic).

At C: $x = 1$ (saturated vapor); $T_C = T_B$; $p_C = p_{sat}$ for temperature T_C; h_C and s_C read as h_g and s_g from the refrigerant table.

At D: $p_D = p_A$; if s_D and either p_D or T_D are known, h_D can be found by searching the refrigerant superheat table. $s_D = s_C$ (because C-to-D is isentropic).

The refrigerant mass flow rate is

$$\dot{m} = \frac{\dot{Q}_{in}}{h_C - h_B} \qquad 33.10$$

The compressor power is

$$\dot{W}_{in} = \dot{m}(h_D - h_C) \qquad 33.11$$

The refrigeration effect is

$$\dot{Q}_{in} = \dot{m}(h_C - h_B) \qquad 33.12$$

The coefficient of performance as a refrigerator is

$$\text{COP}_{refrigerator} = \frac{\dot{Q}_{in}}{\dot{W}_{in}} = \frac{h_C - h_B}{h_D - h_C} \qquad 33.13$$

If the compression is not isentropic (i.e., an isentropic compressor efficiency is known), h_D will be affected. This, in turn, will determine T_D. (p_D is not affected.)

$$h'_D = h_C + \frac{h_D - h_C}{\eta_{s,compressor}} \qquad 33.14$$

Example 33.1

A refrigerator using HFC-134 has a cooling effect of 10,000 Btu/hr (2.9 kW). Refrigerant leaves the evaporator saturated at 0°F (−18°C). The pressure of the refrigerant entering the condenser is 120 psia (830 kPa; 8.3 bars). The refrigerant leaves the condenser as a saturated liquid. The compressor's isentropic efficiency is 80%. Find (a) the temperature of the refrigerant immediately after compression, (b) the refrigerant flow rate, (c) the compressor input shaft work, and (d) the refrigerator's coefficient of performance.

SI Solution

Refer to Fig. 33.5 and App. 24.X. Values will vary somewhat with the precision to which the pressure-enthalpy chart can be read.

At point C: Locate the intersection of the horizontal 0°F line and saturated vapor line.

$$T_C = -18°C \quad [\text{given}]$$
$$p_C = 1.6 \text{ bars}$$
$$h_C = 385 \text{ kJ/kg}$$

At point D: Follow a line of constant entropy upward and to the right to the horizontal 8.3 bars line.

$$T_D = 40°C$$

$$p_D = 8.3 \text{ bars} \quad [\text{given}]$$

$$h_D = 425 \text{ kJ/kg}$$

$$h'_D = h_C + \frac{h_D - h_C}{\eta_{s,\text{compressor}}}$$

$$= 385 \frac{\text{kJ}}{\text{kg}} + \frac{425 \frac{\text{kJ}}{\text{kg}} - 385 \frac{\text{kJ}}{\text{kg}}}{0.80}$$

$$= 435 \text{ kJ/kg}$$

At point A: Move horizontally to the saturated liquid line.

$$p_A = p_D = 8.3 \text{ bars}$$

$$T_A = 32°C$$

$$h_A = 244 \text{ kJ/kg}$$

At point B: Throttling processes are constant-enthalpy processes. Follow a vertical line downward to the horizontal 1.6 bar line.

$$p_B = p_C = 1.6 \text{ bar}$$

$$h_B = h_A = 244 \text{ kJ/kg} \quad [\text{throttling process}]$$

(a) Use the chart to find the temperature corresponding to 8.3 bars and 435 kJ/kg. The temperature immediately after the compression process is $T'_D \approx 50°C$.

(b) The required refrigerant flow is found from the total heat load and the enthalpy change per pound.

$$\dot{m} = \frac{Q_{\text{in}}}{h_C - h_B}$$

$$= \frac{(2.9 \text{ kW})\left(1.0 \frac{\text{kJ}}{\text{kW·s}}\right)}{385 \frac{\text{kJ}}{\text{kg}} - 244 \frac{\text{kJ}}{\text{kg}}}$$

$$= 0.0206 \text{ kg/s}$$

(c) The compressor work is

$$W_{\text{in}} = \dot{m}(h'_D - h_C)$$

$$= \left(0.0206 \frac{\text{kg}}{\text{s}}\right)\left(435 \frac{\text{kJ}}{\text{kg}} - 385 \frac{\text{kJ}}{\text{kg}}\right)\left(1.0 \frac{\text{kW·s}}{\text{kJ}}\right)$$

$$= 1.03 \text{ kW}$$

(d) The coefficient of performance as a refrigerator is

$$\text{COP} = \frac{Q_{\text{in}}}{W_{\text{in}}} = \frac{2.9 \text{ kW}}{1.03 \text{ kW}}$$

$$= 2.8$$

Customary U.S. Solution

Refer to Fig. 33.5 and App. 24.M. Values will vary somewhat with the precision to which the pressure-enthalpy chart can be read.

At point C: Locate the intersection of the horizontal 0°F line and saturated vapor line.

$$T_C = 0°F \quad [\text{given}]$$

$$p_C = 21 \text{ psia}$$

$$h_C = 102 \text{ Btu/lbm}$$

At point D: Follow a line of constant entropy upward and to the right to the horizontal 120 psia line.

$$T_D = 103°F$$

$$p_D = 120 \text{ psia} \quad [\text{given}]$$

$$h_D = 119 \text{ Btu/lbm}$$

$$h'_D = h_C + \frac{h_D - h_C}{\eta_{s,\text{compressor}}}$$

$$= 102 \frac{\text{Btu}}{\text{lbm}} + \frac{119 \frac{\text{Btu}}{\text{lbm}} - 102 \frac{\text{Btu}}{\text{lbm}}}{0.80}$$

$$= 123 \text{ Btu/lbm}$$

At point A: Move horizontally to the saturated liquid line.

$$p_A = p_D = 120 \text{ psia}$$

$$T_A = 91°F$$

$$h_A = 41 \text{ Btu/lbm}$$

At point B: Throttling processes are constant-enthalpy process. Follow a vertical line downward to the horizontal 21 psia line.

$$p_B = p_C = 21 \text{ psia}$$

$$h_B = h_A = 41 \text{ Btu/lbm} \quad [\text{throttling process}]$$

(a) Use the chart to find the temperature corresponding to 120 psia and 123 Btu/lbm. The temperature immediately after the compression process is $T'_D \approx 117°F$.

(b) The required refrigerant flow is found from the total heat load and the enthalpy change per pound.

$$\dot{m} = \frac{Q_{\text{in}}}{h_C - h_B}$$

$$= \frac{10,000 \frac{\text{Btu}}{\text{hr}}}{102 \frac{\text{Btu}}{\text{lbm}} - 41 \frac{\text{Btu}}{\text{lbm}}}$$

$$= 164 \text{ lbm/hr}$$

(c) The compressor work is

$$W_{\text{in}} = \dot{m}(h'_D - h_C)$$

$$= \left(164 \frac{\text{lbm}}{\text{hr}}\right)\left(123 \frac{\text{Btu}}{\text{lbm}} - 102 \frac{\text{Btu}}{\text{lbm}}\right)$$

$$= 3444 \text{ Btu/hr}$$

Power Cycles

(d) The coefficient of performance as a refrigerator is

$$\text{COP} = \frac{Q_{\text{in}}}{W_{\text{in}}} = \frac{10{,}000 \ \dfrac{\text{Btu}}{\text{hr}}}{3444 \ \dfrac{\text{Btu}}{\text{hr}}}$$

$$= 2.9$$

9. AIR REFRIGERATION CYCLE

Any gas will cool when expanded. This is the principle behind the air refrigeration cycle. The *air refrigeration cycle* (also known as a *Brayton cooling cycle*) is essentially a reversed Brayton turbine cycle. It is not common because of its high power consumption. However, air is nonflammable, readily available, and nontoxic. Therefore, the air refrigeration cycle is often used in aircraft air conditioning and gas liquefaction applications where a refrigerant with such characteristics is needed.

Compressed-air air conditioners, now essentially nonexistent, were common prior to the development of Freon and other chlorofluorocarbon refrigerants. These units consisted of a compressor, heat exchanger, and expansion turbine. The heat increase due to compression was dissipated by the heat exchanger. The expanded, cooler air was discharged directly into the room.

A to B: isentropic expansion

B to C: constant pressure heating

C to D: isentropic compression

D to A: constant pressure cooling

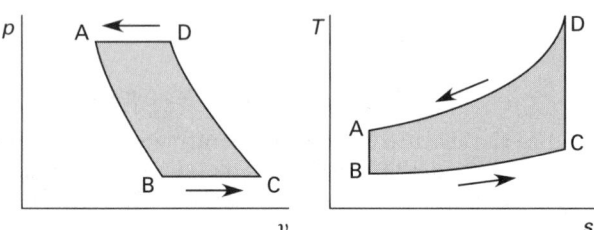

Figure 33.6 *Air Refrigeration Cycle*

If air is considered to be an ideal gas, the ideal gas equations can be used to find the properties at each point.

$$\frac{T_{\text{A}}}{T_{\text{B}}} = \frac{T_{\text{D}}}{T_{\text{C}}} = \left(\frac{p_{\text{high}}}{p_{\text{low}}}\right)^{\frac{k-1}{k}} \qquad 33.15$$

Since $\Delta h = c_p \Delta T$ for an ideal gas, the coefficient of performance can be calculated from Eq. 33.16.

$$\text{COP}_{\text{refrigerator}} = \frac{T_{\text{C}} - T_{\text{B}}}{(T_{\text{D}} - T_{\text{A}}) - (T_{\text{C}} - T_{\text{B}})} \qquad 33.16$$

If air is not considered to be an ideal gas, an air table must be used to find the coefficient of performance.

$$\text{COP}_{\text{refrigerator}} = \frac{h_{\text{C}} - h_{\text{B}}}{(h_{\text{D}} - h_{\text{A}}) - (h_{\text{C}} - h_{\text{B}})} \qquad 33.17$$

The coefficient of performance can also be calculated from the pressure ratio, r_p. (This formula cannot be easily corrected for nonisentropic expansion or compression.)

$$\text{COP}_{\text{refrigerator}} = \frac{1}{r_p^{\frac{k-1}{k}} - 1} \qquad 33.18$$

$$r_p = \frac{p_{\text{high}}}{p_{\text{low}}} \qquad 33.19$$

10. HEAT-DRIVEN REFRIGERATION CYCLES

A *heat-driven refrigeration cycle*, also known as a *heat-activated refrigeration cycle*, is practical when large quantities of waste or inexpensive heat energy are available. In addition to using waste heat, combustion of natural gas and LPG, solar collectors, geothermal sources, and waste steam can provide the energy to drive a refrigeration cycle.

One method of obtaining refrigeration is to use the available heat to drive a Stirling heat engine. The work from the engine drives the refrigerator's compressor. Thus, the apparatus consists of combined power-generating and refrigeration units, as shown in Fig. 33.7.

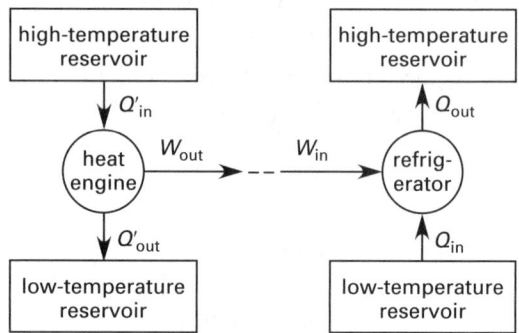

Figure 33.7 *Heat-Driven Cooling Cycle*

11. ABSORPTION CYCLE

The *absorption cycle* is similar to the vapor compression cycle, with one major change—there is no compressor, and no external work is used to compress the refrigerant.[6] Rather, a generator-absorber-recuperator apparatus produces a solution of refrigerant in another liquid, and external heat superheats the refrigerant.

---

[6] A very small amount of work may be used by the liquid return pump between the absorber and generator shown in Fig. 33.8. However, even this pump can be replaced by a *thermosiphon* (i.e., using an inert gas that circulates the liquid by expanding and contracting in response to heat).

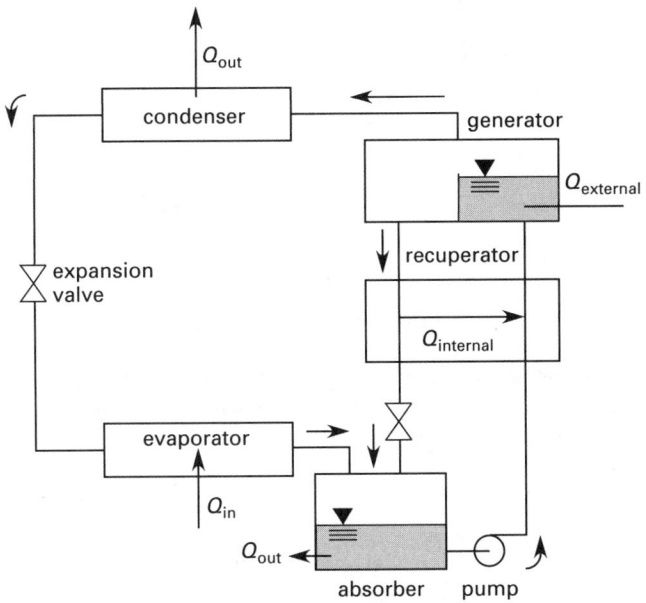

Figure 33.8 *Absorption Cycle Apparatus*

Two working fluids are required in the absorption cycle.[7] As in the vapor-compression cycle, the cycle starts with the refrigerant passing through the evaporator, removing heat from the low-temperature area. The refrigerant (in saturated vapor condition) then enters an *absorber* where it is absorbed by the liquid *absorbent*.

The mixture of liquid refrigerant and absorbent is next pumped into the higher pressure of the *generator*. (This operation requires very little work because the fluid is liquid.) Heat from an external source (e.g., solar collectors, geothermal generators, or combustion of natural gas) drives off the refrigerant in a superheated (though usually subatmospheric) condition. The absorbent left behind is returned to the absorber.

An optional heat exchanger (known as the *recuperator*) may be used to transfer heat from the returning absorbent to the mixture entering the generator. This heat transfer helps improve the system efficiency since the absorbent can carry more refrigerant at a lower temperature. If a recuperator is not used, it may be necessary to cool the liquid in the absorber by some other means (e.g., passing cooling water from another source through the absorber).

An optional *rectifier* (not shown) between the generator and condenser is needed in ammonia-water systems to remove any remaining traces of absorbent from the refrigerant.

The absorption cycle is not very efficient (e.g., coefficients of performance less than 1.0) and the initial equipment costs are higher than for the vapor compression cycle, but when cheap or free energy is available,

---

[7]Common pairs are ammonia (the refrigerant) and water (the absorbent), and water (the refrigerant) and lithium bromide (the absorbent).

the cycle can be economical. The cycle is also bulky and involves toxic fluids; hence it is unsuitable for home and auto cooling.

In Fig. 33.8, the coefficient of performance is

$$\text{COP}_{\text{refrigerator}} = \frac{Q_{\text{in}}}{Q_{\text{external}}} = \frac{Q_{\text{in}}}{Q_{\text{out}} - Q_{\text{in}}} \quad \textbf{33.20}$$

The *generator-absorber-heat exchanger* (GAX) *cycle* is an advanced variation of the absorption cycle. With an ammonia-water absorption cycle, the lower range of the high-temperature absorption process overlaps the higher range of the low-temperature heat input process. With the proper equipment, a portion of the rejected absorber heat can be used to supply heat to the generator.

12. HIGH-SIDE EQUIPMENT

High-side equipment, primarily the compressor and condenser, is the equipment that operates at the high pressure in the refrigerator. Reciprocating piston compressors are used for small- and medium-size refrigerators. For the smallest units, for example, those used in kitchen refrigerators and other unitary machines, hermetically sealed single-stage motor and compressor packages have traditionally been used. Sealed units eliminate the leakage through mechanical seals that is common in large commercial units. Compressor operators, power requirements, volumetric efficiency, and other thermodynamic considerations are similar to those for other compressors. (See Chap. 32.) Centrifugal compressors are used in the largest machines (usually, 200 tons or larger).

Scroll compressors, a product of computer-aided manufacturing and precision machining, were introduced commercially in the late 1980s as replacements for reciprocating compressors in small residential air conditioners. They are approximately 10% more efficient than piston compressors and have higher reliabilities.

Condensers used in small- and medium-sized (i.e., up to approximately 100 tons) refrigerators are typically air-cooled heat exchangers. For efficient operation, the condensing temperature should not be lower than 10°F (5°C) and not more than 30°F (17°C) above the initial air temperature. For larger capacities, water-cooled condensers are used.

Example 33.2

Refrigerant HFC-134a is used in a single-acting, two-cylinder compressor running at 300 rpm. The bore diameter is 6 in (152 mm), and the stroke length is 7 in (178 mm). Saturated vapor enters the compressor and is compressed isentropically to 150 psia (1.0 MPa; 10 bars). Saturated liquid enters the expansion valve. Evaporation occurs at 50 psia (350 kPa; 3.5 bars). The cooling effect is 17 tons (60.4 kW). What is the volumetric efficiency?

SI Solution

The swept volume per stroke is

$$V = AL = \frac{\pi D^2 L}{4}$$

$$= \frac{\pi (152 \text{ mm})^2 \left(178 \frac{\text{mm}}{\text{stroke-cylinder}}\right)}{(4)\left(1000 \frac{\text{mm}}{\text{m}}\right)^3}$$

$$= 0.00323 \text{ m}^3/\text{stroke-cylinder}$$

Since the compressor is single-acting, each cylinder contributes one compression stroke each revolution. The ideal volumetric flow rate is

$$\dot{V}_{\text{ideal}} = Vn$$

$$= \frac{\left(0.00323 \frac{\text{m}^3}{\text{stroke-cylinder}}\right)}{60 \frac{\text{s}}{\text{min}}}$$

$$= 0.0323 \text{ m}^3/\text{s}$$

Refer to Fig. 33.5. At point C, the pressure is 3.5 bars. From App. 24.X, the enthalpy of saturated 3.5 bar vapor is approximately 400 kJ/kg. The specific volume is approximately 0.060 m³/kg. At point A, the pressure is 10 bars. The enthalpy of saturated 10 bar liquid is approximately 255 kJ/kg. This is also the enthalpy at point B.

The actual volumetric flow rate is

$$\dot{V}_{\text{actual}} = \dot{m}_C v_C = \frac{Q v_C}{h_C - h_B}$$

$$= \frac{(60.4 \text{ kW})\left(1.0 \frac{\text{kJ}}{\text{kW·s}}\right)\left(0.060 \frac{\text{m}^3}{\text{kg}}\right)}{400 \frac{\text{kJ}}{\text{kg}} - 255 \frac{\text{kJ}}{\text{kg}}}$$

$$= 0.0250 \text{ m}^3/\text{s}$$

The volumetric efficiency is

$$\eta_v = \frac{\dot{V}_{\text{actual}}}{\dot{V}_{\text{ideal}}}$$

$$= \frac{0.0250 \frac{\text{m}^3}{\text{s}}}{0.0323 \frac{\text{m}^3}{\text{s}}}$$

$$= 0.774 \ (77.4\%)$$

Customary U.S. Solution

The swept volume per stroke is

$$V = AL = \frac{\pi D^2 L}{4}$$

$$= \frac{\pi (6 \text{ in})^2 \left(7 \frac{\text{in}}{\text{stroke-cylinder}}\right)}{(4)\left(12 \frac{\text{in}}{\text{ft}}\right)^3}$$

$$= 0.1145 \text{ ft}^3/\text{stroke-cylinder}$$

Since the compressor is single-acting, each cylinder contributes one compression stroke each revolution. The ideal volumetric flow rate is

$$\dot{V}_{\text{ideal}} = Vn$$

$$= \left(0.1145 \frac{\text{ft}^3}{\text{stroke-cylinder}}\right)(300 \text{ rpm})$$
$$\times (2 \text{ cylinders})$$

$$= 68.7 \text{ ft}^3/\text{min}$$

Refer to Fig. 33.5. At point C, the pressure is 50 psia. From App. 24.M, the enthalpy of saturated 50 psia vapor is approximately 107 Btu/lbm. The specific volume is approximately 0.94 ft³/lbm. At point A, the pressure is 150 psia. The enthalpy of saturated 150 psia liquid is approximately 46 Btu/lbm. This is also the enthalpy at point B.

The actual volumetric flow rate is

$$\dot{V}_{\text{actual}} = \dot{m}_C v_C = \frac{Q v_C}{h_C - h_B}$$

$$= \frac{(17 \text{ tons})\left(200 \frac{\text{Btu}}{\text{min-ton}}\right)\left(0.94 \frac{\text{ft}^3}{\text{lbm}}\right)}{107 \frac{\text{Btu}}{\text{lbm}} - 46 \frac{\text{Btu}}{\text{lbm}}}$$

$$= 52.4 \text{ ft}^3/\text{min}$$

The volumetric efficiency is

$$\eta_v = \frac{\dot{V}_{\text{actual}}}{\dot{V}_{\text{ideal}}}$$

$$= \frac{52.4 \frac{\text{ft}^3}{\text{min}}}{68.7 \frac{\text{ft}^3}{\text{min}}}$$

$$= 0.763 \ (76.3\%)$$

13. LOW-SIDE EQUIPMENT

Low-side equipment, primarily the evaporator, is the equipment that operates at the low pressure in the refrigerator. *Evaporators* for cooling air are usually externally finned and are known as *cooling coils*. Evaporators for cooling water and other liquids are mostly shell-and-coil and shell-and-tube heat exchangers known as *coolers* and *chillers*.[8] Tubes are generally smooth inside.[9]

Coolers and chillers for water generally operate with an average temperature difference of 6 to 20°F (3 to 11°C), and optimally with a 10 to 14°F (5 to 8°C) difference. To avoid freezing problems, entering refrigerant should be above 28°F (−2°C).

14. REFRIGERANT LINE SIZING

The size of the refrigerant return line (from the condenser) is not a critical design parameter. The pressure drop in the throttling valve is far greater than the pressure drop from the fluid flow.

A small suction line, however, can greatly decrease compressor capacity. The suction line between the evaporator and the compressor is typically sized from tables based on the allowing pressure drop, refrigeration capacity, and length. Tables are presented in different formats depending on the source, and different tables are needed for each different refrigerant. Maximum allowable pressure drops of 3 psi (20 kPa) on high-temperature units and 1.5 psi (10 kPa) on low-temperature units are the established rules of thumb, and tables are given for this range of pressure drops.

Suction lines should not be sized too large, as a reasonable velocity is needed to carry oil from the evaporator back to the compressor. For horizontal suction lines, 750 ft/min (3.8 m/s) is a recommended minimum velocity. For vertical suction lines, the velocity should be 1200 to 1400 ft/min (6.1 to 7.1 m/s).

PRACTICE PROBLEMS

1. A heat pump operates on the Carnot cycle between 40°F and 700°F (4°C and 370°C). What is the coefficient of performance?

2. A heat pump using refrigerant R-12 operates on the Carnot cycle. The refrigerant evaporates and is compressed at 35.7 psia and 172.4 psia (246 kPa and 1190 kPa), respectively. What is the coefficient of performance?

3. A refrigerator uses refrigerant R-12. The input power is 585 W. Heat absorbed from the cooled space is 450 Btu/hr (0.13 kW). What is the coefficient of performance?

Power Cycles

---

[8]In industrial settings, the shell-and-coil design is less desirable because the coil cannot be mechanically cleaned.
[9]*Microfins* inside the evaporator tube increase the heat transfer efficiency.

34 Fundamental Heat Transfer

1. Introduction to Conductive Heat Transfer . 34-2
2. Other Modes of Heat Transfer 34-2
3. Simplifying Assumptions 34-2
4. Conductivity 34-2
5. Varying Conductivity 34-2
6. Thermal Resistance 34-3
7. R-Value 34-3
8. Specific Heat 34-3
9. Characteristic Dimension 34-3
10. Biot Number 34-4
11. Thermal Diffusivity 34-4
12. Fourier Number 34-4
13. Modified Fourier Number 34-4
14. Temperature 34-4
15. Temperature Profile 34-5
16. Heat Transfer 34-5
17. Fourier's Law 34-5
18. Electrical Analogy 34-5
19. Sandwiched Planes 34-6
20. Heat Transfer Through a Film 34-6
21. Overall Coefficient of Heat Transfer 34-6
22. Temperature at a Point 34-7
23. Complex Wall 34-7
24. Logarithmic Mean Area 34-8
25. Radial Conduction Through a
    Hollow Cylinder 34-8
26. Radial Conduction Through a
    Composite Cylinder 34-8
27. Pipe Insulation 34-9
28. Critical Insulation Thickness 34-10
29. Economic Insulation Thickness 34-10
30. Insulation Thickness to Prevent
    Freezing of Water Pipe 34-10
31. Insulation Thickness to Prevent Sweating . 34-10
32. Radial Conduction Through a
    Spherical Shell 34-10
33. Conduction Through a Cube 34-11
34. Transient Conduction 34-11
35. Newton's Method 34-11
36. Lumped Parameter Method 34-12
37. Total Amount of Heat Transferred 34-12
38. Lumped Parameter Electrical Analogy . . 34-12
39. Graphical Solutions to
    Transient Heat Transfer 34-13
40. Internal Heat Generation 34-14
41. Flat Plate with Heat Generation 34-14
42. Cylinder with Heat Generation 34-14
43. Fins 34-16
44. Infinite Cylindrical Fin 34-17
45. Finite Cylindrical Fin with
    Adiabatic Tip 34-17
46. Finite Cylindrical Fin with
    Convective Tip 34-18
47. Finite Rectangular Fin with
    Adiabatic Tip 34-18
48. Finite Rectangular Fin with
    Convective Tip 34-18
49. Circular Fin 34-18
50. Shape Factors 34-20
51. Steam and Electric Heat Tracing 34-20
52. Heat Pipes 34-20
    Practice Problems 34-22

Nomenclature

A	area	ft^2	m^2
Bi	Biot number	–	–
c_p	specific heat	Btu/lbm-°F	J/kg·K
C	thermal capacitance	Btu/°F	J/K
d	diameter	ft	m
E	effectiveness	–	–
E	voltage	V	V
Fo	Fourier number	–	–
G	heat generation rate	Btu/hr-ft^3	W/m^3
h	film coefficient	Btu/hr-ft^2-°F	W/m^2·K
I	current	A	A
k	thermal conductivity	Btu-ft/hr-ft^2-°F	W/m·K
L	length	ft	m
m	mass	lbm	kg
m	$\sqrt{hP/kA}$	1/ft	1/m
P	perimeter	ft	m
P	power	W	W
q	heat transfer per unit area	Btu/hr-ft^2	W/m^2
Q	heat transfer rate	Btu/hr	W
r	radius	ft	m
r	empirical rate constant	1/hr	1/s
R	electrical resistance	Ω	Ω
R	thermal resistance	hr-°F/Btu	K/W
S	shape factor	ft	m
t	time	hr	s
t	thickness	ft	m
T	temperature	°F	K
U	internal energy	Btu	J
U	overall heat transfer coefficient	Btu/hr-ft^2-°F	W/m^2·K
V	volume	ft^3	m^3
w	width	ft	m
x	distance	ft	m
z	depth	ft	m

Symbols

α	diffusivity	ft^2/hr	m^2/s
γ	empirical constant	1/°F	1/K
η	efficiency	–	–
ρ	mass density	lbm/ft^3	kg/m^3
ρ	electrical resistivity	Ω-in	Ω·cm
τ	thermal time constant	hr	s

Heat Transfer

Subscripts

0	initial
b	base
c	characteristic
corr	corrected
e	equivalent
f	fin
h	film
i	inner or ith layer
j	jth film
L	per unit length
m	logarithmic mean
o	outer
r	radiation, radius, or at radius r
ref	reference
s	surface
t	at time t
th	thermal
T	at temperature T
x	at distance x
∞	at infinity

1. INTRODUCTION TO CONDUCTIVE HEAT TRANSFER

Conduction is the flow of heat through solids or stationary fluids. Thermal conductance in metallic solids is due to molecular vibrations within the metallic crystalline lattice and movement of free valence electrons through the lattice. Insulating solids, which have fewer free electrons, conduct heat primarily by the agitation of adjacent atoms vibrating about their equilibrium positions. This vibrational mode of heat transfer is several orders of magnitude less efficient than conduction by free electrons.

In stationary liquids, heat is transmitted by longitudinal vibrations, similar to sound waves. The *net transport theory* explains heat transfer through gases. Hot molecules move faster than cold molecules. Hot molecules travel to cold areas with greater frequency than cold molecules travel to hot areas.

2. OTHER MODES OF HEAT TRANSFER

The other primary modes of heat transfer are convection (the transfer of heat in a moving fluid) and radiation (the transfer of energy between isolated bodies by electromagnetic waves through a vacuum or transparent gas or liquid).[1] Natural convection is covered in Chap. 35; forced convection is covered in Chap. 36; and radiation is covered in Chap. 37.

---

[1] There are two additional, but similar, methods of removing heat energy: change of phase and ablation. Heat transfer can result from the evaporation of liquids and the condensation of vapors. This is the principle used in heat pipes. (See Sec. 52.) In other cases, such as during high-speed reentry into the earth's atmosphere, the heat flux may be very high (e.g., up to 10,000 Btu/ft²-sec (114 MW/m²) and the time of application very short (e.g., 1 or 2 min). It may be necessary to allow a portion of the material to melt or vaporize in order to remove the heat. This is known as *ablation*.

3. SIMPLIFYING ASSUMPTIONS

Determining heat transfer by conduction can be an easy task if sufficient simplifying assumptions are made. Major discrepancies can arise, however, when the simplifying assumptions are not met. The following assumptions are commonly made in simple problems.

- The heat transfer is steady-state.

- The heat path is one-dimensional. (Objects are infinite in one or more directions and do not have any end effects.)

- The heat path has a constant area.

- The heat path consists of a homogeneous material with constant conductivity.

- The heat path consists of anisotropic material.[2]

- There is no internal heat generation.

Many real heat transfer cases violate one or more of these assumptions. Unfortunately, problems with closed-form solutions (suitable for working by hand) are in the minority. More complex problems must be solved by appropriate iterative, graphical, or numerical methods.[3]

4. CONDUCTIVITY

The *thermal conductivity* (also known as the *thermal conductance*) is a measure of the rate a substance transfers thermal energy through a unit thickness.[4] Units of thermal conductivity are Btu-ft/hr-ft²-°F or Btu-in/hr-ft²-°F (W/m·K or W·cm/m²·K). The units of Btu-ft/hr-ft²-°F are the same as Btu/hr-ft-°F. These units are not the same as Btu-in/hr-ft, however, which are also used.[5] Conductivity of a substance should not be confused with the *overall conductivity*, U, of an object. (See Sec. 21.) Appendices 34.A and 34.B list representative thermal conductivities for commonly encountered substances.

5. VARYING CONDUCTIVITY

Solids exhibit conductivities that vary with temperature.[6] Over limited ranges, thermal conductivity in common solids is assumed to vary linearly with temperature, as indicated in Eq. 34.1. k_{ref} is the conductivity at the reference temperature, usually 0°F (−18°C).

---

[2] Examples of *anisotropic materials*, materials whose heat transfer properties depend on the direction of heat flow, are crystals, plywood and other laminated sheets, and the core elements of some electrical transformers.

[3] Finite-difference methods are commonly used.

[4] Another (lesser-encountered) meaning for *conductivity* is the reciprocal of thermal resistance (i.e., conductivity = kA/L).

[5] Temperature units of °R can also be used in place of °F, since conductivity is always multiplied by ΔT, and $\Delta T°_F = \Delta T°_R$.

[6] Conductivity at absolute zero is zero because there is no atomic/molecular motion. For the first few degrees about absolute zero, conductivity increases with increases in temperature.

Table 34.1 *Typical Ranges of Conductivity*
(See also Apps. 34.A and 34.B.)

material	conductivity range	
	Btu-ft/hr-ft²-°F	W/m·K
gases (at 1 atm)	0.004–0.10	0.007–0.17
insulators	0.02–0.12	0.03–0.21
nonmetallic liquids	0.05–0.40	0.09–0.70
nonmetallic solids	0.02–1.5	0.03–2.6
liquid metals	5.0–45	8.7–78
metallic alloys	8.0–70	14–120
pure metals	30–240	52–420

(Multiply Btu-ft/hr-ft²-°F by 12 to get Btu-in/hr-ft²-°F.)
(Multiply Btu-ft/hr-ft²-°F by 1.73073 to get W/m·K.)
(Multiply Btu-ft/hr-ft²-°F by 4.1365 ×10⁻³ to get cal·cm/
s·cm²·°C.)

Values of γ and γ' are not common, as graphs and tabulations of conductivity versus temperature are more readily available.

$$k_T = k_{\text{ref}}(1 + \gamma T)$$
$$= k_{\text{ref}} + \gamma' T \qquad 34.1$$

Conductivity decreases with temperature for pure metals and increases with temperature for most alloys. For insulating materials, it increases with temperature. Conductivity in water and aqueous solutions increases with increases in temperature up to approximately 250°F (120°C) and then gradually decreases. Conductivity decreases with increases in concentrations of aqueous solutions, as it does with most other liquids. Conductivity increases with increases in pressure. Of the nonmetallic liquids, water is the best thermal conductor. Conductivity in gases increases almost linearly with temperature but is fairly independent of pressure in common ranges.

Thermal conductivity is often assumed to be constant over the entire length of the transmission path. In most calculations, either the average thermal conductivity or the conductivity at the arithmetic mean temperature is used.[7] When k varies linearly (or nearly so) with temperature, k is evaluated at the average temperature $(1/2)(T_1 + T_2)$.

In rare cases where k does not vary linearly with temperature, the solution must proceed iteratively. An initial estimate of the average conductivity yields a heat transfer which, in turn, is used to solve for the surface temperatures. These temperatures are used to determine a new average conductivity, and so on.

---

[7]The mean thermal conductivity is not the same as the thermal conductivity at the mean temperature. However, it is very nearly so, and this simplification is widely used.

6. THERMAL RESISTANCE

The *thermal resistance* is defined by Eq. 34.2. Typical units are °F-hr/Btu (K/W).

$$R_{\text{th}} = \frac{T_1 - T_2}{Q} \qquad 34.2$$

For a plate, the thermal resistance depends on the thickness (path length), L, and is

$$R_{\text{th}} = \frac{L}{kA} \qquad 34.3$$

For a film, the thermal resistance is

$$R_{\text{th,film}} = \frac{1}{hA} \qquad 34.4$$

For a curved layer (i.e., a layer of insulation on a pipe), the thermal resistance is

$$R_{\text{th}} = \frac{\ln\left(\dfrac{r_o}{r_i}\right)}{2\pi kL} \qquad 34.5$$

7. *R*-VALUE

The *R-value* of a substance is the thermal resistance on a unit area basis. Typical units are °F-ft²-hr/Btu (K·m²/W). The *R*-value concept is usually encountered in the construction industry as a means of comparing insulating materials. It is not the same as the thermal resistance.

$$R\text{-value} = \frac{T_1 - T_2}{\dfrac{Q}{A}} = R_{\text{th}}A \qquad 34.6$$

8. SPECIFIC HEAT

The *specific heat*, c_p, is the energy required to change the temperature of a unit mass of a body one degree. Values of specific heat are given in Apps. 34.B and 34.C.

$$c_p = \frac{Q}{m\Delta T} \qquad 34.7$$

9. CHARACTERISTIC DIMENSION

The *characteristic dimension* (*characteristic length*), L_c, of an object is the ratio of its volume to its surface area.

$$L_c = \frac{V}{A_s} \qquad 34.8$$

Heat Transfer

For a long cylinder of radius r, the characteristic dimension is

$$L_{c,\text{cylinder}} = \frac{V}{A_s} = \frac{\pi r^2 L}{2\pi r L}$$
$$= \frac{r}{2} \qquad 34.9$$

For a sphere, the characteristic dimension is one-third of the radius (i.e., $L_{c,\text{sphere}} = r/3$). For an infinite slab or a long square rod, the characteristic dimension is one-half of the slab/rod thickness.

10. BIOT NUMBER

The *Biot number*, Bi (also known as the *Biot modulus* and *transient modulus*), is a comparison of the internal thermal resistance to the external resistance of a body.[8] If the Biot number is small (i.e., less than 0.1), the internal thermal resistance will be small, and the body temperature will be essentially uniform throughout during heating or cooling. The length used to calculate the Biot number is the characteristic length, not an external body dimension.

$$\text{Bi} = \frac{hL_c}{k} \qquad 34.10$$

11. THERMAL DIFFUSIVITY

The *thermal diffusivity*, α, of a substance is a measure of the speed of propagation of a specific temperature into a solid. The higher the diffusivity, the faster will be the penetration of a specific temperature into the solid.

$$\alpha = \frac{k}{\rho c_p} \qquad 34.11$$

12. FOURIER NUMBER

The *Fourier number*, Fo, also known as *relative time*, is the ratio of the rate of heat transferred by conduction to the rate of energy stored.[9] The Fourier number is useful in working transient heat transfer problems. The larger the Fourier number, the larger the amount of heat conducted through the solid as compared to the amount of heat stored. This is manifested as a deeper penetration of a specific temperature into the solid over a given period of time and a faster temperature change at a given depth in the object.

$$\text{Fo} = \frac{kt}{\rho c_p L_c^2} = \frac{\alpha t}{L_c^2} \qquad 34.12$$

---

[8]Biot is pronounced "bee'-oe."
[9]The Fourier number is sometimes given the symbol τ.

Table 34.2 *Representative Thermal Diffusivities (at 32°F (0°C) unless specified otherwise)*

material	ft^2/hr	m^2/s
aluminum, pure	3.33	8.59×10^{-5}
aluminum, 2024	1.76	4.54×10^{-5}
asbestos	0.010	2.58×10^{-7}
brass (70% Cu, 30% Zn, 68°F)	1.27	3.28×10^{-5}
brick, fire clay (400°F)	0.020	5.15×10^{-7}
copper	4.42	11.4×10^{-5}
cork	0.006	1.5×10^{-7}
glass, Pyrex™	0.023	5.93×10^{-7}
glass, window	0.013	3.35×10^{-7}
gold (68°F)	4.68	12.1×10^{-5}
ice	0.046	11.9×10^{-7}
iron, pure	0.70	1.81×10^{-5}
iron, cast (4% C, 68°F)	0.66	1.70×10^{-5}
lead (70°F)	0.95	2.45×10^{-5}
magnesium (60°F)	3.68	9.49×10^{-5}
mercury	0.172	44.4×10^{-7}
nickel	0.60	1.55×10^{-5}
rubber, soft	0.003	0.77×10^{-7}
steel (1% C)	0.48	1.24×10^{-5}
silver	6.60	17.0×10^{-5}
soil, dry	≈ 0.01	3×10^{-7}
soil, moist	≈ 0.03	8×10^{-7}
steel, stainless (68°F)	0.17	0.44×10^{-5}
steel (1% C, 68°F)	0.45	1.2×10^{-5}
tin	1.57	4.05×10^{-5}
tungsten	2.39	6.17×10^{-5}
water	0.005	1.3×10^{-7}
zinc	1.60	4.13×10^{-5}

(Multiply ft^2/hr by 2.58×10^{-5} to obtain m^2/s.)
(Multiply ft^2/sec by 0.092903 to obtain m^2/s.)
(Multiply ft^2/sec by 3600 to obtain ft^2/hr.)

13. MODIFIED FOURIER NUMBER

A modified Fourier number is used with graphical solutions to transient heat transfer problems. The modified Fourier number is similar in format to the Fourier number except that the length, L, is defined differently. For example, the length of a sphere used in calculating the modified Fourier number is the sphere's radius, r, not the characteristic length of $r/3$.

$$\text{modified Fourier number} = \frac{\alpha t}{L^2} \qquad 34.13$$

14. TEMPERATURE

The *ambient temperature* is the same as *environment temperature*, *far-field temperature*, *local temperature*, or (occasionally) *air temperature*. However, due to the thermal resistance of a film, it is not the same as the *surface temperature*. The *bulk temperature* is the temperature of a thoroughly mixed liquid or gas. The term "bulk temperature" is occasionally used to represent the ambient temperature.

In most heat transfer calculations, the temperature variable shows up as a change in temperature. It is convenient to recognize that $\Delta T_{\circ F} = \Delta T_{\circ R}$ and $\Delta T_{\circ C} = \Delta T_K$.

15. TEMPERATURE PROFILE

A *temperature profile* is a graph of the temperature versus location within an object. A profile will often be drawn on a representation of the cross section of the object through which the energy transfer occurs. In uniform materials, the profile will consist of straight lines, with steeper lines representing materials with lower thermal conductivities. The temperature scale is omitted and replaced with values of temperature at the interface points.

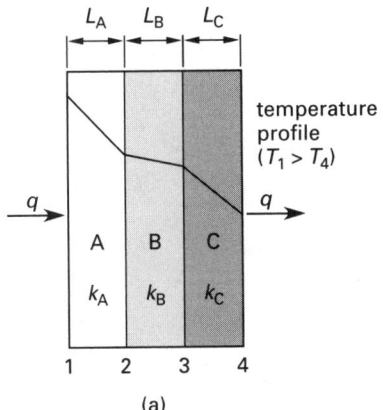

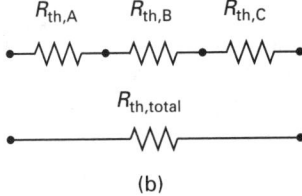

Figure 34.1 *Typical Temperature Profile*

16. HEAT TRANSFER

The term *heat transfer* is used in three different contexts. The intended interpretation must be determined from the needs of the investigation. The most common usage has units of energy per unit of time (e.g., Btu/hr-ft²).[10] Strictly speaking, this should be called a *heat transfer rate* and should be written $\dot{q}$. However,

[10]In the United States, the units of Btu/hr are often written as Btuh. Similarly, MBh generally means millions of Btus per hr, but can mean thousands of Btus per hour in some industries (e.g., HVAC).

common usage omits the rate symbol. A second usage, referred to in this book as *total heat transfer*, Q, is the energy transfer taken over the entire body area or mass.

$$Q = qA \qquad 34.14$$
$$Q = mc_p\Delta T \qquad 34.15$$

The term "heat transfer" is also used when meaning the total energy change (in Btu) over some period of time. In this book, symbol U (internal energy) is used for this purpose.

17. FOURIER'S LAW

If the assumptions listed in Sec. 3 are valid, *Fourier's law*, Eq. 34.16, is applicable. On its own, heat always flows from a higher temperature to a lower temperature. The heat transfer from high-temperature point 1 to lower-temperature point 2 through an infinite plane of thickness L and homogeneous conductivity, k, is

$$q_{1-2} = \frac{-k(T_2 - T_1)}{L}$$
$$= \frac{k(T_1 - T_2)}{L} \qquad 34.16$$
$$Q_{1-2} = q_{1-2}A = \frac{kA(T_1 - T_2)}{L} \qquad 34.17$$

The temperature difference $T_2 - T_1$ is the *temperature gradient* or *thermal gradient*. Heat transfer is always positive. The minus sign in Eq. 34.16 indicates that the heat flow direction is opposite that of the thermal gradient. The direction of heat flow is obvious in most problems. Therefore, the minus sign is usually omitted.

18. ELECTRICAL ANALOGY

Ohm's law can be solved for current and written in terms of the electrical potential gradient, $E_1 - E_2$.

$$I = \frac{E_1 - E_2}{R} \qquad 34.18$$

Drawing on the concept of thermal resistance, Eqs. 34.18 and 34.19 are analogous.

$$Q = \frac{T_1 - T_2}{R_{th}} \qquad 34.19$$

The electrical analogy is particularly valuable in analyzing heat transfer through multiple layers (e.g., a layered wall or cylinder with several layers of different insulation). Each layer is considered a series resistance. The sum of the individual thermal resistances is the total thermal resistance. (See Fig. 34.1(b).)

$$R_{th,total} = \Sigma R_{th,i} \qquad 34.20$$

19. SANDWICHED PLANES

Figure 34.1(a) illustrates the cross section of a *sandwiched plane* with three homogeneous, but different, materials. This type of construction is sometimes known as a *composite wall*. The heat flow due to conduction through a series ($i = 1$ to n) of plane surfaces is

$$Q = \frac{T_1 - T_2}{R_{th,total}} = \frac{A(T_1 - T_2)}{\sum\limits_{i=1}^{n} \dfrac{L_i}{k_i}} \qquad 34.21$$

$$R_{th,total} = \sum_{i=1}^{n} \frac{L_i}{k_i A} \qquad 34.22$$

20. HEAT TRANSFER THROUGH A FILM

Heat conducted through solids is often removed by a physical transport process at an exposed fluid surface. For example, heat transmitted through a heat exchanger wall is removed by a moving coolant. Unless the fluid is extremely turbulent, the fluid molecules immediately adjacent to the exposed surface move much slower than molecules farther away. Molecules immediately adjacent to the wall may be stationary altogether. The fluid molecules that are affected by the exposed surface constitute a layer known as a film.[11] The film has a thermal resistance just like any other sandwiched plane.

Because the film thickness is not easily determined, the thermal resistance of a film is given by a *film coefficient* (*convective heat transfer coefficient* or *unit conductance*), h, with units of Btu/hr-ft²-°F (W/m²·K). If the film coefficient is not constant over the entire surface, the symbol for average film coefficient, $\bar{h}$ is used. Subscripts are used (e.g., h_o or h_i) to indicate whether the film is located on the outside or inside of the object. The heat flow through a film is

$$Q = hA(T_1 - T_2) \qquad 34.23$$

$$R_{th} = \frac{1}{hA} \qquad 34.24$$

If the fluid is extremely turbulent, the thermal resistance will be small (i.e., h will be very large). In such cases, the wall temperature is essentially the same as the fluid temperature.

21. OVERALL COEFFICIENT OF HEAT TRANSFER

When conduction is the only mode of heat transfer, the *overall coefficient of heat transfer*, U, also known as the *overall conductivity, overall heat transmittance*, or simply *U-factor*, is defined by Eq. 34.25.

---

[11]The film coefficient concept can also be used to quantify the thermal resistance of imperfect bonding between two planes (i.e., *contact resistance*), a dust layer, or scale buildup.

$$U = \frac{1}{A R_{th,total}}$$
$$= \frac{1}{\sum\limits_{i} \dfrac{L_i}{k_i} + \sum\limits_{j} \dfrac{1}{h_j}} \qquad 34.25$$

The overall coefficient is usually used in conjunction with the outside (exposed) surface area (because it is easier to measure), but not always. Therefore, it will generally be written with a subscript (e.g., U_o or U_i) indicating which area it is based on. The heat transfer is

$$Q = U_o A_o (T_1 - T_2) = U_i A_i (T_1 - T_2) \qquad 34.26$$

When heat transfer occurs by two or three modes, the overall coefficient of heat transfer takes all active modes—conduction, convection, and radiation—into consideration. In that sense, the overall coefficient, U, is defined by Eq. 34.27, rather than being used in it.

$$U = \frac{Q}{A(T_1 - T_2)} \qquad 34.27$$

Example 34.1

A 100 ft² (9.3 m²) wall consists of 4 in (10 cm) of red brick ($k = 0.38$ Btu-ft/hr-ft²-°F; 0.66 W/m·K), 1 in (2.5 cm) of pine ($k = 0.06$ Btu-ft/hr-ft²-°F; 0.10 W/m·K), and ¹/₂ in (1.2 cm) of plasterboard ($k = 0.30$ Btu-ft/hr-ft²-°F; 0.52 W/m·K). The internal and external film coefficients are 1.65 Btu/hr-ft²-°F and 6.00 Btu/hr-ft²-°F (9.38 W/m²·K and 34.1 W/m²·K), respectively. The inside and outside air temperatures are 72°F (22°C) and 30°F (−1°C), respectively. Determine the heat transfer.

SI Solution

Convert the thicknesses from centimeters to meters.

$$L_{brick} = \frac{10 \text{ cm}}{100 \ \dfrac{\text{cm}}{\text{m}}} = 0.10 \text{ m}$$

$$L_{pine} = \frac{2.5 \text{ cm}}{100 \ \dfrac{\text{cm}}{\text{m}}} = 0.025 \text{ m}$$

$$L_{plasterboard} = \frac{1.2 \text{ cm}}{100 \ \dfrac{\text{cm}}{\text{m}}} = 0.012 \text{ m}$$

From Eq. 34.25, the overall coefficient of heat transfer is

$$\frac{1}{U} = \frac{0.10 \text{ m}}{0.66 \dfrac{\text{W}}{\text{m·K}}} + \frac{0.025 \text{ m}}{0.10 \dfrac{\text{W}}{\text{m·K}}} + \frac{0.012 \text{ m}}{0.52 \dfrac{\text{W}}{\text{m·K}}}$$

$$+ \frac{1}{9.38 \dfrac{\text{W}}{\text{m}^2\text{·K}}} + \frac{1}{34.1 \dfrac{\text{W}}{\text{m}^2\text{·K}}}$$

$$= 0.561 \frac{\text{m}^2\text{·K}}{\text{W}}$$

$$U = \frac{1}{0.561 \dfrac{\text{m}^2\text{·K}}{\text{W}}}$$

$$= 1.78 \text{ W/m}^2\text{·K}$$

From Eq. 34.26 (and recognizing that $\Delta T_{°\text{C}} = \Delta T_{\text{K}}$), the heat transfer is

$$Q = UA\Delta T = \left(1.78 \frac{\text{W}}{\text{m}^2\text{·K}}\right)(9.3 \text{ m}^2)$$

$$\times [22°\text{C} - (-1°\text{C})]$$

$$= 380.7 \text{ W}$$

Customary U.S. Solution

Convert the thicknesses from inches to feet.

$$L_{\text{brick}} = \frac{4 \text{ in}}{12 \dfrac{\text{in}}{\text{ft}}} = 0.333 \text{ ft}$$

$$L_{\text{pine}} = \frac{1 \text{ in}}{12 \dfrac{\text{in}}{\text{ft}}} = 0.083 \text{ ft}$$

$$L_{\text{plasterboard}} = \frac{0.5 \text{ in}}{12 \dfrac{\text{in}}{\text{ft}}} = 0.042 \text{ ft}$$

From Eq. 34.25, the overall coefficient of heat transfer is

$$\frac{1}{U} = \frac{0.333 \text{ ft}}{0.38 \dfrac{\text{Btu-ft}}{\text{hr-ft}^2\text{-°F}}} + \frac{0.083 \text{ ft}}{0.06 \dfrac{\text{Btu-ft}}{\text{hr-ft}^2\text{-°F}}} + \frac{0.042 \text{ ft}}{0.30 \dfrac{\text{Btu-ft}}{\text{hf-ft}^2\text{-°F}}}$$

$$+ \frac{1}{1.65 \dfrac{\text{Btu}}{\text{hr-ft}^2\text{-°F}}} + \frac{1}{6.00 \dfrac{\text{Btu}}{\text{hr-ft}^2\text{-°F}}}$$

$$= 3.17 \text{ hr-ft}^2\text{-°F/Btu}$$

$$U = \frac{1}{3.17 \dfrac{\text{hr-ft}^2\text{-°F}}{\text{Btu}}}$$

$$= 0.315 \text{ Btu/hr-ft}^2\text{-°F}$$

From Eq. 34.26, the heat transfer is

$$Q = UA\Delta T$$

$$= \left(0.315 \frac{\text{Btu}}{\text{hr-ft}^2\text{-°F}}\right)(100 \text{ ft}^2)(72°\text{F} - 30°\text{F})$$

$$= 1323 \text{ Btu/hr}$$

22. TEMPERATURE AT A POINT

The temperature at any point on or within a simple or complex wall can be found if the heat transfer, q or Q, is known. The procedure is to calculate the thermal resistance up to the point of unknown temperature and then to solve Eq. 34.19 for the temperature difference. Since one of the temperatures is known, the unknown temperature is found from the temperature difference.

23. COMPLEX WALL

Figure 34.2 illustrates a complex wall. Heat transfer through such a wall is intrinsically a two-dimensional problem. Since the temperature profile (Sec. 15) will not be the same for each heat transfer path, there will be adjacent areas with different temperatures. Heat will naturally flow from the hotter to the cooler areas.

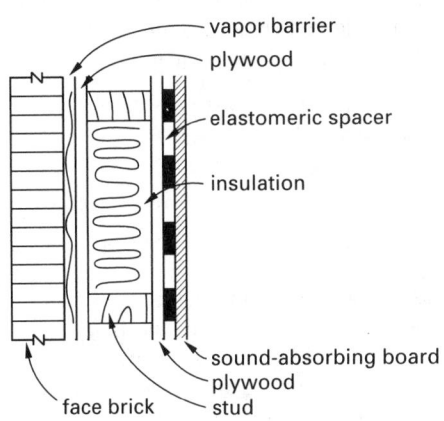

Figure 34.2 Complex Wall Construction

There is no easy way to analytically obtain an exact value of heat transfer through a complex wall. A lower bound on the heat transfer can be obtained by considering the wall as having multiple one-dimensional transmission paths (i.e., ignoring the interaction of adjacent paths. Similarly, an upper bound can be obtained by assuming the entire transmission path consists of a homogeneous material whose conductivity is obtained by weighting each material's conductivity by its volume within the wall.

Thermal conductivities and resistances for common wall constructions are widely available in heating, ventilating, and air conditioning (HVAC) textbooks. Values in HVAC tabulations usually include the inside and outside film coefficients and are based on some time of year or on wind speed.

24. LOGARITHMIC MEAN AREA

Cylinders and spheres are examples of objects whose heat transfer paths increase in area from the inside to outside. In such instances, the *logarithmic mean area*, A_m, should be used.[12]

$$Q = UA_m(T_1 - T_2) \qquad 34.28$$

$$A_m = \frac{A_o - A_i}{\ln\left(\dfrac{A_o}{A_i}\right)} \qquad 34.29$$

25. RADIAL CONDUCTION THROUGH A HOLLOW CYLINDER

The logarithmic mean area (excluding the ends) for a hollow cylinder of length L and wall thickness $r_o - r_i$ is

$$A_m = \frac{2\pi L(r_o - r_i)}{\ln\left(\dfrac{r_o}{r_i}\right)} \qquad 34.30$$

The overall radial heat transfer through an uninsulated hollow cylinder without films is given by Eq. 34.31. This equation disregards heat transfer from the ends and assumes that the length is sufficiently large so that the heat transfer is radial at all locations.

$$\begin{aligned}
Q = qA_m &= \frac{kA_m(T_1 - T_2)}{r_o - r_i} \\
&= \frac{2\pi kL(T_1 - T_2)}{\ln\dfrac{r_o}{r_i}} \qquad 34.31
\end{aligned}$$

The temperature at a point within a layer a distance r from the center is given by Eq. 34.32. T_i is the temperature at the inside of the layer. For a pipe wall, the inside wall temperature is not necessarily the same as the temperature of the pipe's contents.

$$T_r = T_i - (T_i - T_o)\left[\frac{\ln\left(\dfrac{r}{r_i}\right)}{\ln\left(\dfrac{r_o}{r_i}\right)}\right] \qquad 34.32$$

26. RADIAL CONDUCTION THROUGH A COMPOSITE CYLINDER

A *composite cylinder* consists of two or more concentric, cylindrical layers. This configuration is typically encountered as an *insulated pipe*. The logarithmic mean area and heat transfer are

$$Q = qA_m = \frac{2\pi L(T_1 - T_2)}{\dfrac{\ln\left(\dfrac{r_b}{r_a}\right)}{k_{\text{pipe}}} + \dfrac{\ln\left(\dfrac{r_c}{r_b}\right)}{k_{\text{insulation}}}} \qquad 34.33$$

When films are present, Eq. 34.34 can be expanded. h_b is a film coefficient representing contact or interface resistance, if any.

$$Q = \frac{2\pi L(T_i - T_o)}{\dfrac{1}{r_a h_a} + \dfrac{\ln\left(\dfrac{r_b}{r_a}\right)}{k_{\text{pipe}}} + \dfrac{1}{r_b h_b} + \dfrac{\ln\left(\dfrac{r_c}{r_b}\right)}{k_{\text{insulation}}} + \dfrac{1}{r_c h_c}} \qquad 34.34$$

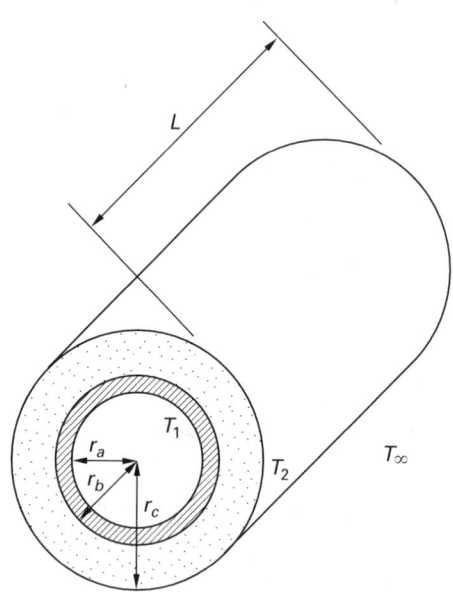

Figure 34.3 *Composite Cylinder (Insulated Pipe)*

Compared to the thermal resistance of the insulation, the thermal resistance of metal pipe walls is negligible. Little accuracy will be lost if the pipe wall thickness term is omitted from the denominator.[13] An equivalent assumption is that there is no temperature change across the thickness of the pipe wall.

Example 34.2

Liquid oxygen at $-290°F$ ($-180°C$) is stored in a cylindrical tank with thermal conductance of 28.0 Btu-ft/hr-ft²-°F (48.0 W/m·K), 5 ft (1.5 m) inside diameter, and 20 ft (6.0 m) long. The wall thickness is ³/₈ in (1 cm). The tank is insulated with 1.0 ft (30 cm) of powdered diatomaceous silica with an average thermal conductivity of 0.022 Btu-ft/hr-ft²-°F (0.038 W/m·K).

---

[12]If $A_o/A_i \leq 2$, using the arithmetic mean $(^1\!/_2)(A_o + A_i)$ in place of the logaritmic mean area will result in a maximum error of 4%, with the heat transfer being too high.

[13]Since the thermal conductivity values may be in error by as much as 20%, it makes little sense to strive for perfection by including the thermal resistance of a thin-walled metal pipe.

The surrounding air temperature is 70°F (21°C), and the outside film coefficient is 6.0 Btu/ hr-ft²-°F (34 W/m²·K). The inside film coefficient is assumed to be infinite. Disregard heat transfer through the tank ends. Calculate the radial heat gain to the liquid oxygen.

SI Solution

The wall thickness is

$$t = \frac{10 \text{ cm}}{100 \frac{\text{cm}}{\text{m}}} = 0.01 \text{ m}$$

In Fig. 34.3, the corresponding radii are

$$r_a = \frac{1.5 \text{ m}}{2} = 0.75 \text{ m}$$

$$r_b = r_a + t = 0.75 \text{ m} + 0.01 \text{ m} = 0.76 \text{ m}$$

$$r_c = r_b + t_{\text{insulation}} = 0.76 \text{ m} + 0.30 \text{ m} = 1.06 \text{ m}$$

The heat transfer is calculated directly from Eq. 34.34.

$$Q = \frac{2\pi L(T_1 - T_2)}{\dfrac{1}{r_a h_a} + \dfrac{\ln\left(\dfrac{r_b}{r_a}\right)}{k_{\text{pipe}}} + \dfrac{\ln\left(\dfrac{r_c}{r_b}\right)}{k_{\text{insulation}}} + \dfrac{1}{r_c h_c}}$$

$$= \frac{(2\pi)(6 \text{ m})[21°C - (-180°C)]}{\dfrac{1}{(0.75 \text{ m})(\infty)} + \dfrac{\ln\left(\dfrac{0.76 \text{ m}}{0.75 \text{ m}}\right)}{48.0 \dfrac{\text{W}}{\text{m·K}}} + \dfrac{\ln\left(\dfrac{1.06 \text{ m}}{0.76 \text{ m}}\right)}{0.038 \dfrac{\text{W}}{\text{m·K}}}}$$
$$\qquad + \dfrac{1}{(1.06 \text{ m})\left(34 \dfrac{\text{W}}{\text{m}^2\text{·K}}\right)}$$

$$= 862.7 \text{ W}$$

Customary U.S. Solution

The wall thickness is

$$t = \frac{\frac{3}{8} \text{ in}}{12 \dfrac{\text{in}}{\text{ft}}}$$

$$= 0.031 \text{ ft}$$

In Fig. 34.3, the corresponding radii are

$$r_a = \frac{5 \text{ ft}}{2} = 2.5 \text{ ft}$$

$$r_b = r_a + t = 2.5 \text{ ft} + 0.031 \text{ ft} = 2.53 \text{ ft}$$

$$r_c = r_b + t_{\text{insulation}} = 2.53 \text{ ft} + 1 \text{ ft} = 3.53 \text{ ft}$$

The heat transfer is calculated directly from Eq. 34.34.

$$Q = \frac{2\pi L(T_1 - T_2)}{\dfrac{1}{r_a h_a} + \dfrac{\ln\left(\dfrac{r_b}{r_a}\right)}{k_{\text{pipe}}} + \dfrac{\ln\left(\dfrac{r_c}{r_b}\right)}{k_{\text{insulation}}} + \dfrac{1}{r_c h_c}}$$

$$= \frac{(2\pi)(20 \text{ ft})[70°F - (-290°F)]}{\dfrac{1}{(25 \text{ ft})(\infty)} + \dfrac{\ln\left(\dfrac{2.53 \text{ ft}}{2.5 \text{ ft}}\right)}{28.0 \dfrac{\text{Btu-ft}}{\text{hr-ft}^2\text{-}°F}} + \dfrac{\ln\left(\dfrac{3.53 \text{ ft}}{2.53 \text{ ft}}\right)}{0.022 \dfrac{\text{Btu-ft}}{\text{hr-ft}^2\text{-}°F}}}$$
$$\qquad + \dfrac{1}{(3.53 \text{ ft})\left(6.0 \dfrac{\text{Btu}}{\text{hr-ft}^2\text{-}°F}\right)}$$

$$= 2979 \text{ Btu/hr}$$

27. PIPE INSULATION

Commercial pipe insulation is usually fiberglass or calcium silicate.[14] Fiberglass has a high insulation value, is low in cost and low in weight, and is easy to install. However, conventional fiberglass is limited to uses below approximately 850°F (450°C).[15] Mineral wool (mineral-fiber), calcium silicate, and composite materials are used for high-temperature (i.e., above approximately 1000°F (540°C)) installations. Traditional calcium silicate, known as *cal sil*, can withstand more mechanical abuse than other insulating materials. However, it requires a saw for cutting, while fiberglass and mineral wool can easily be cut with a knife. An outer jacket (cover) of plastic (white Kraft all-service jacket, ASJ), aluminum, or stainless steel will protect insulation from dirt and moisture.[16,17]

Insulation products are available as pipe wrap, blankets, and boards. Integral protective jacketing of pipe wrap can be provided, with sealing tape closing the hinged wrap around the pipe. Insulating boards are designed for large flat surfaces, or, when manufactured with multiple strip hinges, for large curved surfaces such as cylindrical tanks. Compressible blankets are highly flexible and are used for irregular surfaces.

---

[14] Asbestos insulation, commonly used in the past, has fallen from favor. This includes *85% magnesia* insulation consisting of 85% magnesium carbonate and 15% asbestos fiber.
[15] High-temperature resin binders can increase the useful range of fiberglass insulation to approximately 1000°F (540°C).
[16] Metallic jacketing may represent a safety hazard for plant personnel since the surface is generally conductive and reflective. Jacket temperature should be less than 130°F (54°C) in areas where contact by personnel is possible.
[17] An exterior jacket should always be removed when adding a second layer of insulation. Retaining the original jacket may produce a condensation site or damage the jacket material or adhesive.

Table 34.3 *Representative Conductivities of Pipe Insulation (Btu-ft/hr-ft^2-$°$F)*

	insulation temperature					
	100°F (38°C)	200°F (93°C)	300°F (149°C)	400°F (204°C)	500°F (260°C)	600°F (316°C)
calcium silicate	0.033	0.037	0.041	0.046	0.057	0.060
cellular glass	0.039	0.047	0.055	0.064	0.074	0.085
fiberglass	0.026	0.030	0.034			
magnesia, 85%	0.034	0.037	0.041	0.044		
polyurethane	0.016	0.016	0.016			

(Multiply Btu-ft/hr-ft^2-$°$F by 12 to get Btu in/hr-ft^2-$°$F.)
(Multiply Btu-ft/hr-ft^2-$°$F by 1.7307 to get W/m·K.)
(Multiply Btu-ft/hr-ft^2-$°$F by 4.1365 $\times 10^{-3}$ to get cal·cm/s·cm^2·$°$C.)

28. CRITICAL INSULATION THICKNESS

The addition of insulation to a bare pipe or wire increases the surface area. Adding insulation to a small-diameter pipe may actually increase the heat loss above bare-pipe levels. Adding insulation past the *critical thickness* will decrease heat loss.[18] The *critical radius* is usually very small (e.g., a few millimeters), and it is most relevant in the case of insulating thin wires. The critical radius, measured from the center of the pipe or wire, is

$$r_{\text{critical}} = \frac{k_{\text{insulation}}}{h} \qquad 34.35$$

29. ECONOMIC INSULATION THICKNESS

Optimizing an insulation installation usually requires choosing an insulating material and then selecting its thickness. Fiberglass and mineral wool insulations are commonly less expensive than traditional calcium silicate. However, other factors are included in the economic analysis, including cost of installation labor, thermal efficiency, current fuel cost, and useful life.[19]

Adding insulation conserves more energy, but the costs of material and installation are also greater. Insulation thickness is optimized by balancing the heat losses against the cost of insulating. The *economic insulation thickness* is the thickness that minimizes the annual cost of ownership and operation. The economic thickness will vary with pipe diameter.

---

[18]There is another, less commonly used, meaning for the term *critical thickness*: the thickest required insulation. In situations where the required insulation thickness is different for energy conservation, condensation control, personnel protection, and process temperature control, the "critical" thickness is the thickness that controls the design.
[19]The Thermal Insulation Manufacturer's Association (TIMA) can provide manual and computerized methods of evaluating the economics of different insulations.

30. INSULATION THICKNESS TO PREVENT FREEZING OF WATER PIPE

Freezing of water flowing in a pipe will be prevented if the exit water temperature is kept from dropping below 32°F. The maximum allowable heat loss from the water is

$$Q_{\text{maximum}} = \dot{m}c_p(T_{\text{entrance,water}} - 32°\text{F}) \qquad 34.36$$

If there is no water flow, a pipe exposed for long periods to subfreezing temperatures cannot be prevented from freezing.

31. INSULATION THICKNESS TO PREVENT SWEATING

Sweating is the condensation of moisture from the surrounding air on the surface of the pipe insulation. Sweating is prevented by keeping the surface temperature above the air's dew-point temperature.[20]

32. RADIAL CONDUCTION THROUGH A SPHERICAL SHELL

The logarithmic mean area and heat transfer are

$$A_m = \sqrt{A_o A_i} \qquad 34.37$$

$$\begin{aligned} Q &= qA_m \\ &= k\sqrt{A_o A_i}\left(\frac{T_i - T_o}{r_o - r_i}\right) \\ &= \frac{4\pi k(r_o r_i)(T_i - T_o)}{r_o - r_i} \qquad 34.38 \end{aligned}$$

---

[20]The minimum thickness is found by equating the heat transfer through the insulation by conduction to the heat transfer from the surface by convection and radiation. As a first approximation, a total heat transfer coefficient, h_{total}, for convection and radiation of 0.65 Btu/hr-ft^2-$°$F (3.7 W/m^2·K) can be used.

The thermal resistance of each layer is

$$R_{\text{th}} = \left(\frac{1}{4\pi k}\right)\left(\frac{r_o - r_i}{r_o r_i}\right) \qquad 34.39$$

Spheres have the smallest surface area-to-volume ratio, so spherical tanks are used where heat transfer is to be minimized. Unlike for plane surfaces and cylindrical tanks, steady-state heat transfer from spheres cannot be reduced indefinitely by the addition of increasing amounts of insulation. For a sphere, steady-state conduction can never be less than the minimum value given by Eq. 34.40, no matter how thick the wall is. If r_o is infinite, Eq. 34.38 becomes

$$Q_{\text{minimum}} = 4\pi k r_i (T_i - T_o) \qquad 34.40$$

33. CONDUCTION THROUGH A CUBE

Conduction through a roughly cubic shell consisting of a small cavity surrounded by thick walls can be calculated from Eq. 34.41. The constant 0.725 is a semi-empirical correction factor.[21]

$$Q \approx 0.725 k \sqrt{A_o A_i} \left(\frac{T_i - T_o}{r_o - r_i}\right) \quad \left[\frac{A_o}{A_i} \geq 2\right] \qquad 34.41$$

34. TRANSIENT CONDUCTION

Matter changes temperature when it gains or loses thermal energy. A change in the temperature gradient over time invalidates the use of Eq. 34.14 except for the calculation of an instantaneous heat transfer. As the temperature gradient decreases, the heat transfer also decreases. *Transient heat transfer* (also known as *unsteady-state heat transfer*) problems with constant-temperature surroundings can be solved by a variety of methods, including Newton's method (for simplified problems), the lumped parameter method, and graphical methods.

35. NEWTON'S METHOD

If the temperature difference between a body and the surroundings is not too large, the rate of heat transfer by all modes is approximately proportional to the temperature difference. This empirical relationship is valid for all modes of heat transfer and is known as *Newton's law of cooling* (or *heating*) and the *Newtonian method*.[22,23] The temperature at time t (starting at zero) is

$$T_t = T_\infty + (T_0 - T_\infty)e^{-rt} \qquad 34.42$$

---

[21]Studies describing more accurate methods are available.
[22]*Newton's law of cooling* applies to the combined heat transfer mechanisms of conduction, convection, and radiation.
[23]Newton's law of cooling is essentially the same as the lumped parameter method (Sec. 36). The choice of the two methods depends on what information is known. If information about the temperature history is known, Newton's law of cooling is applicable. If material properties are known, the lumped parameter method should be used.

For each configuration, r is a rate constant that is usually found from other known information. Either the rate constant or the time for a change in temperature from T_0 to T_t can be found from Eq. 34.43.

$$rt = -\ln\left(\frac{T_t - T_\infty}{T_0 - T_\infty}\right) \qquad 34.43$$

Example 34.3

Water in a 5 in (12.7 cm) diameter, 10 ft (3.0 m) long tank cools from 200°F (93°C) to 190°F (88°C) in 1.25 hr. The ambient temperature is 75°F (24°C). Approximately how long will it take for the water to cool from 200°F to 125°F (93°C to 52°C)?

SI Solution

From Eq. 34.43,

$$\begin{aligned}
r &= -\left(\frac{1}{t}\right)\ln\left(\frac{T_t - T_\infty}{T_0 - T_\infty}\right) \\
&= -\left(\frac{1}{1.25\text{ h}}\right)\ln\left(\frac{88°\text{C} - 24°\text{C}}{93°\text{C} - 24°\text{C}}\right) \\
&= 0.0602 \text{ 1/h}
\end{aligned}$$

Now, solve Eq. 34.43 for the unknown time.

$$\begin{aligned}
t &= -\left(\frac{1}{r}\right)\ln\left(\frac{T_t - T_\infty}{T_0 - T_\infty}\right) \\
&= -\left(\frac{1}{0.0602 \frac{1}{\text{h}}}\right)\ln\left(\frac{52°\text{C} - 24°\text{C}}{93°\text{C} - 24°\text{C}}\right) \\
&= 14.98 \text{ h}
\end{aligned}$$

Customary U.S. Solution

From Eq. 34.43,

$$\begin{aligned}
r &= -\left(\frac{1}{t}\right)\ln\left(\frac{T_t - T_\infty}{T_0 - T_\infty}\right) \\
&= -\left(\frac{1}{1.25\text{ hr}}\right)\ln\left(\frac{190°\text{F} - 75°\text{F}}{200°\text{F} - 75°\text{F}}\right) \\
&= 0.0667 \text{ 1/hr}
\end{aligned}$$

Now, solve Eq. 34.43 for the unknown time.

$$\begin{aligned}
t &= -\left(\frac{1}{r}\right)\ln\left(\frac{T_t - T_\infty}{T_0 - T_\infty}\right) \\
&= -\left(\frac{1}{0.0667 \frac{1}{\text{hr}}}\right)\ln\left(\frac{125°\text{F} - 75°\text{F}}{200°\text{F} - 75°\text{F}}\right) \\
&= 13.74 \text{ hr}
\end{aligned}$$

Heat Transfer

36. LUMPED PARAMETER METHOD

If the Biot number is less than approximately 0.10, the internal thermal resistance of a body will be smaller in comparison to the external thermal resistance. In that case, the *lumped parameter method* can be used to approximate the transient (time dependent) heat flow.[24,25] The temperature at time t (starting at zero) is

$$T_t = T_\infty + (T_0 - T_\infty)e^{-\text{BiFo}} \qquad 34.44$$

The instantaneous heat transfer at time t is

$$Q_t = hA_s(T_t - T_\infty) = hA_s(T_0 - T_\infty)e^{-\text{BiFo}} \qquad 34.45$$

37. TOTAL AMOUNT OF HEAT TRANSFERRED

Since the heat transfer at any time t is proportional to the temperature difference, the cumulative thermal energy, U_{total}, transferred up to time, t, is proportional to the remaining temperature gradient.

$$U_{\text{total},t} = \int Q_t dt$$
$$= \left(\frac{T_t - T_\infty}{T_0 - T_\infty}\right) U_{\text{total},0} \qquad 34.46$$

$$U_{\text{total},0} = mc_p(T_0 - T_\infty) = V\rho c_p(T_0 - T_\infty) \qquad 34.47$$

38. LUMPED PARAMETER ELECTRICAL ANALOGY

The analogy of a capacitor discharging through a resistor is often used to illustrate lumped parameter performance.[26] This analogy has given rise to the concepts of thermal capacitance and thermal resistance. The equivalent *thermal capacitance* (also known as the *capacity* of the system) and equivalent *thermal resistance* are

$$C_e = c_p\rho V \qquad 34.48$$
$$R_e = \frac{1}{hA_s} \qquad 34.49$$

These equivalent variables can be used to calculate the performance of the system.

$$T_t = T_\infty + (T_0 - T_\infty)e^{\frac{-t}{R_e C_e}} \qquad 34.50$$

The instantaneous heat transfer at time t is

$$Q_t = hA_s(T_t - T_\infty) = hA_s(T_0 - T_\infty)e^{\frac{-t}{R_e C_e}} \qquad 34.51$$

---

[24]Because the assumptions are the same, this method is also referred to as *Newton's method.*
[25]The maximum error will be less than 7%.
[26]The analogy is often made, but there is no significant advantage to solving problems in this manner.

The thermal *time constant*, τ, so named because it has units of time, is a measure of the rate of decay of the thermal gradient. The heat content (i.e., temperature gradient, $T_t - T_\infty$) of the source will be reduced by 63.2% in one time constant (i.e., will be reduced to 36.8% of its original value, $T_0 - T_\infty$).

$$\tau = \frac{c_p\rho L_c}{h} = \frac{c_p\rho V}{hA} \qquad 34.52$$

Example 34.4

A 0.03125 in (0.8 mm) diameter copper wire is heated by a short circuit to 300°F (150°C) before a slow-blow fuse burns out and all heating ceases. The ambient temperature is 100°F (38°C), and the film coefficient on the wire is 1.65 Btu/hr-ft²-°F (9.37 W/m²·K). Use the following copper properties to determine how long it will take for the wire temperature to drop to 120°F (49°C).

conductivity (k):	224 Btu-ft/hr-ft²-°F	(388 W/m·K)
specific heat (c_p):	0.091 Btu/lbm-°F	(380 J/kg·K)
density (ρ):	558 lbm/ft³	(8940 kg/m³)

SI Solution

The characteristic length of the wire is half of the radius.

$$L_c = \frac{r}{2} = \frac{d}{4}$$
$$= \frac{0.0008 \text{ m}}{4} = 0.0002 \text{ m}$$

The Biot number is

$$\text{Bi} = \frac{hL_c}{k}$$
$$= \frac{\left(9.37 \dfrac{\text{W}}{\text{m}^2\cdot\text{K}}\right)(0.0002 \text{ m})}{388 \dfrac{\text{W}}{\text{m}\cdot\text{K}}}$$
$$= 4.83 \times 10^{-6} \quad [< 0.1]$$

The Fourier number is

$$\text{Fo} = \frac{kt}{\rho c_p L_c^2}$$
$$= \frac{\left(388 \dfrac{\text{W}}{\text{m}\cdot\text{K}}\right)t}{\left(8940 \dfrac{\text{kg}}{\text{m}^3}\right)\left(380 \dfrac{\text{J}}{\text{kg}\cdot\text{K}}\right)(0.0002 \text{ m})^2}$$
$$= 2855t \quad [t \text{ in seconds}]$$

From Eq. 34.44,

$$T_t = T_\infty + (T_0 - T_\infty)e^{-\text{BiFo}}$$
$$49°C = 38°C + (150°C - 38°C)e^{-(4.83\times10^{-6})(2855t)}$$
$$0.0982 = e^{-0.0138t}$$
$$t = \left(\frac{-1}{0.0138}\right)\ln(0.0982) = 168 \text{ s} \quad (0.047 \text{ h})$$

Customary U.S. Solution

From Sec. 9, the characteristic length of the wire is half of the radius.

$$L_c = \frac{r}{2} = \frac{d}{4}$$

$$= \frac{0.03125 \text{ in}}{(4)\left(12\ \dfrac{\text{in}}{\text{ft}}\right)} = 0.00065 \text{ ft}$$

The Biot number is

$$\text{Bi} = \frac{hL_c}{k}$$

$$= \frac{\left(1.65\ \dfrac{\text{Btu}}{\text{hr-ft}^2\text{-}^\circ\text{F}}\right)(0.00065 \text{ ft})}{224\ \dfrac{\text{Btu-ft}}{\text{hr-ft}^2\text{-}^\circ\text{F}}}$$

$$= 4.79 \times 10^{-6} \quad [< 0.1]$$

The Fourier number is

$$\text{Fo} = \frac{kt}{\rho c_p L_c^2}$$

$$= \frac{\left(224\ \dfrac{\text{Btu-ft}}{\text{hr-ft}^2\text{-}^\circ\text{F}}\right)t}{\left(558\ \dfrac{\text{lbm}}{\text{ft}^3}\right)\left(0.091\ \dfrac{\text{Btu}}{\text{lbm-}^\circ\text{F}}\right)(0.00065 \text{ ft})^2}$$

$$= 1.044 \times 10^7 t \quad [t \text{ in hours}]$$

From Eq. 34.44,

$$T_t = T_\infty + (T_0 - T_\infty)e^{-\text{BiFo}}$$

$$120^\circ\text{F} = 100^\circ\text{F} + (300^\circ\text{F} - 100^\circ\text{F})e^{-(4.79\times10^{-6})(1.044\times10^7)t}$$

$$0.10 = e^{-50t}$$

$$t = \left(\frac{-1}{50}\right)\ln(0.1) = 0.046 \text{ hr} \quad (166 \text{ sec})$$

39. GRAPHICAL SOLUTIONS TO TRANSIENT HEAT TRANSFER

If the Biot number is greater than approximately 0.1, the internal resistance of the object cannot be disregarded. If Fo $\geq$ 0.2 and the geometry is sufficiently simple, a graphical approach can be used. Graphs for transient heat flow problems are available for simple shapes, including *semi-infinite solids* (i.e., an object that is infinite in one direction only), spheres, large (infinite) slabs, and long (infinite) cylinders.

Graphs for solving transient heat transfer problems are known as *temperature-time charts*. (See Apps. 34.D, 34.E, and 34.F.) *Heisler charts* exhibit similar information for a particular point in the solid, but use a logarithmic scale to cover a greater range of values (in particular, longer time periods).[27] (See Apps. 34.G, 34.H, and 34.I.)

Values in temperature-time charts are plotted as functions of the modified Biot and modified Fourier numbers. (See Sec. 13.) Temperature-time charts may be general enough to determine the temperature history at any point in the solid, or they may be unique to some point within the shape (e.g., the center of a sphere or the centerline of a slab).[28] Example 34.5 illustrates the use of temperature-time charts.

Example 34.5

Whole peaches at 80°F (27°C) are to be cooled in 14.7 psia, 10°F (101 kPa, −12°C) air by natural convection prior to being completely frozen. The average film coefficient on the peaches is 5.8 Btu/hr-ft²-°F (33 W/m²·K). The peaches are placed on trays and are spaced far enough apart so that they do not affect one another. The peaches are considered to be homogeneous spheres with 3 in (7.6 cm) diameters. Use the properties listed to determine how long it will take for the peach centers to cool to 40°F (4°C).

conductivity (k): 0.4 Btu-ft/hr-ft²-°F (0.7 W/m·K)
specific heat (c_p): 1.0 Btu/lbm-°F (4.2 kJ/kg·K)
thermal
   diffusivity (α): 0.04 ft²/hr (1.0 ×10⁻⁶ m²/s)

SI Solution

The characteristic length of a sphere is

$$L_{c,\text{sphere}} = \frac{r}{3} = \frac{d}{6}$$

$$= \frac{0.076 \text{ m}}{6} = 0.0127 \text{ m}$$

The Biot number is

$$\text{Bi} = \frac{hL}{k}$$

$$= \frac{\left(33\ \dfrac{\text{W}}{\text{m}^2\text{·K}}\right)(0.0127 \text{ m})}{0.7\ \dfrac{\text{W}}{\text{m·K}}}$$

$$= 0.599$$

---

[27]The charts are named after H. P. Heisler, who published his original graphical methods in the *ASME Transactions* in 1947.
[28]Auxiliary *position-correction charts* have also been developed by researchers and are available to correct the center-line or surface temperature to other locations within the body. These are available in most heat transfer textbooks.

Since Bi > 0.1, the lumped parameter method cannot be used. Calculate the parameters needed to use App. 34.D.

$$\frac{T_t - T_\infty}{T_0 - T_\infty} = \frac{4°C - (-12°C)}{27°C - (-12°C)}$$
$$= 0.410$$

$$r_o = \frac{d}{2} = \frac{0.076 \text{ m}}{2} = 0.038 \text{ m}$$

$$\frac{k}{hr_o} = \frac{0.7 \dfrac{W}{m \cdot K}}{\left(33 \dfrac{W}{m^2 \cdot K}\right)(0.038 \text{ m})}$$
$$= 0.56$$

Since the desired temperature corresponds to the center of the peach, use App. 34.D. The value of $\alpha t/r_o^2$ is found from the chart to be approximately 0.3.

$$\frac{\alpha t}{r_o^2} = 0.3$$

$$t = \frac{0.3 r_o^2}{a} = \frac{(0.3)(0.038 \text{ m})^2}{1.0 \times 10^{-6} \dfrac{m^2}{s}}$$
$$= 433 \text{ s} \quad (0.117 \text{ h})$$

Customary U.S. Solution

The characteristic length of a sphere is

$$L_{c,\text{sphere}} = \frac{r}{3} = \frac{d}{6}$$
$$= \frac{3 \text{ in}}{(6)\left(12 \dfrac{in}{ft}\right)} = 0.0417 \text{ ft}$$

The Biot number is

$$\text{Bi} = \frac{hL}{k}$$
$$= \frac{\left(5.8 \dfrac{Btu}{hr\text{-}ft^2\text{-}°F}\right)(0.0417 \text{ ft})}{0.4 \dfrac{Btu\text{-}ft}{hr\text{-}ft^2\text{-}°F}}$$
$$= 0.605$$

Since Bi > 0.1, the lumped parameter method cannot be used. Calculate the parameters needed to use App. 34.D.

$$\frac{T_t - T_\infty}{T_0 - T_\infty} = \frac{40°F - 10°F}{80°F - 10°F}$$
$$= 0.43$$

$$r_o = \frac{d}{2} = \frac{3 \text{ in}}{(2)\left(12 \dfrac{in}{ft}\right)} = 0.125 \text{ ft}$$

$$\frac{k}{hr_o} = \frac{0.4 \dfrac{Btu\text{-}ft}{hr\text{-}ft^2\text{-}°F}}{\left(5.8 \dfrac{Btu}{hr\text{-}ft^2\text{-}°F}\right)(0.125 \text{ ft})}$$
$$= 0.55$$

Since the desired temperature corresponds to the center of the peach, use App. 34.D. The value of $\alpha t/r_o^2$ is found from the chart to be approximately 0.3.

$$\frac{\alpha t}{r_o^2} = 0.3$$

$$t = \frac{0.3 r_o^2}{a} = \frac{(0.3)(0.125 \text{ ft})^2}{0.04 \dfrac{ft^2}{hr}}$$
$$= 0.117 \text{ hr} \quad (422 \text{ sec})$$

40. INTERNAL HEAT GENERATION

Some objects generate their own heat internally, for example, electrical heating elements, nuclear fuel rods, chemical reaction beds, beds of fine coal on furnace grates, and composting heaps. The relationship between the heat generation rate, G (in Btu/hr-ft^3 or W/m^3), and distribution of temperature within the object can be evaluated for simple steady-state cases involving homogeneous material and uniform heat generation.

41. FLAT PLATE WITH HEAT GENERATION

An infinite flat plate with thickness $t = 2L$ will dissipate internally generated heat equally and uniformly from its two surfaces.[29] The temperature a distance x from the closest surface is given by Eq. 34.53.

$$T_x = T_s + \frac{GLx}{k} - \frac{Gx^2}{2k} \quad [x \le L] \qquad 34.53$$

$$T_s = T_{\text{center}} - \frac{GL^2}{2k} \qquad 34.54$$

42. CYLINDER WITH HEAT GENERATION

An infinitely long cylinder with outside radius r_o will dissipate internally generated heat uniformly along its entire length. The temperature at distance r from the center is given by Eq. 34.55.

$$T_r = T_s + \left(\frac{Gr_o^2}{4k}\right)\left[1 - \left(\frac{r}{r_o}\right)^2\right]$$
$$= T_{\text{center}} - \frac{Gr^2}{4k} \qquad 34.55$$

$$T_s = T_{\text{center}} - \frac{Gr_o^2}{4k} \qquad 34.56$$

---

[29]Assuming the plate to be infinite in two out of its three dimensions eliminates the need to determine the heat transfer from its ends.

Example 34.6

Both surfaces of a large 0.25 in thick (6.4 mm) copper plate are kept at 80°F (27°C) by a circulating coolant. The thermal conductivity of the copper is 224 Btu-ft/hr-ft^2-°F (388 W/m·K). The electrical resistivity of the copper is $0.68 \times 10^{-6}\,\Omega$-in ($1.7 \times 10^{-6}\,\Omega$·cm). A DC electrical potential of 0.01 V is applied across the two surfaces. What is the temperature at the geometric midpoint of the plate?

SI Solution

The thickness of the copper plate is

$$t = 2L = 0.0064 \text{ m}$$

The heat generation rate, G, is per unit volume (i.e., per cubic meter) of the copper plate. Therefore, it is convenient to work with a square section of the plate having a volume of 1 m^3. The length of the sides of a square of the copper plate is

$$V = \text{thickness} \times w^2$$
$$1 \text{ m}^3 = (0.0064 \text{ m})w^2$$
$$w = 12.5 \text{ m}$$

From the definition of resistivity, the electrical resistance of the plate is

$$
\begin{aligned}
R_{\text{electrical}} &= \frac{\rho(\text{thickness})}{A} \\
&= \frac{(1.7 \times 10^{-6}\ \Omega\text{·cm})(0.0064 \text{ m})}{(12.5 \text{ m})^2 \left(100\ \dfrac{\text{cm}}{\text{m}}\right)} \\
&= 6.96 \times 10^{-13}\ \Omega
\end{aligned}
$$

The thermal power generation, G, in the section is the same as the electrical power dissipated. For a DC voltage, $P = IE = E^2/R = I^2R$.

$$
\begin{aligned}
G &= \frac{E^2}{R_{\text{electrical}}} = \frac{(0.01 \text{ V})^2}{6.96 \times 10^{-13}\ \Omega} \\
&= 1.44 \times 10^8 \text{ W/m}^3
\end{aligned}
$$

Equation 34.54 is used to calculate the center temperature. By definition, L is half of the plate thickness.

$$L = \frac{0.0064 \text{ m}}{2} = 0.0032 \text{ m}$$

$$
\begin{aligned}
T_{\text{center}} &= T_s + \frac{GL^2}{2k} \\
&= 27°\text{C} + \frac{\left(1.44 \times 10^8\ \dfrac{\text{W}}{\text{m}^3}\right)(0.0032 \text{ m})^2}{(2)\left(388\ \dfrac{\text{W}}{\text{m·K}}\right)} \\
&= 28.9°\text{C}
\end{aligned}
$$

Customary U.S. Solution

The thickness of the copper plate is

$$t = 2L = \frac{0.25 \text{ in}}{12\ \dfrac{\text{in}}{\text{ft}}} = 0.02083 \text{ ft}$$

The heat generation rate, G, is per unit volume (i.e., per cubic foot) of the copper plate. Therefore, it is convenient to work with a square section of the plate having a volume of 1 ft^3. The length of the sides of a square of the copper plate is

$$V = (\text{thickness})w^2$$
$$1 \text{ ft}^3 = (0.02083 \text{ ft})w^2$$
$$w = 6.93 \text{ ft}$$

From the definition of resistivity, the electrical resistance of the plate is

$$
\begin{aligned}
R_{\text{electrical}} &= \frac{\rho(\text{thickness})}{A} \\
&= \frac{(0.68 \times 10^{-6}\ \Omega\text{-in})(0.02083 \text{ ft})}{(6.93 \text{ ft})^2 \left(12\ \dfrac{\text{in}}{\text{ft}}\right)} \\
&= 2.46 \times 10^{-11}\ \Omega
\end{aligned}
$$

The thermal power generation, G, in the section is the same as the electrical power dissipated. For a DC voltage, $P = IE = E^2/R = I^2R$.

$$
\begin{aligned}
G &= \frac{E^2}{R_{\text{electrical}}} = \frac{(0.01 \text{ V})^2}{2.46 \times 10^{-11}\ \Omega} \\
&= 4.07 \times 10^6 \text{ W/ft}^3
\end{aligned}
$$

Convert from watts to Btu/hr.

$$
\begin{aligned}
G &= \left(4.07 \times 10^6\ \frac{\text{W}}{\text{ft}^3}\right)\left(3.413\ \frac{\text{Btu}}{\text{W-hr}}\right) \\
&= 1.39 \times 10^7 \text{ Btu/hr-ft}^3
\end{aligned}
$$

Equation 34.55 is used to calculate the center temperature. By definition, L is half of the plate thickness.

$$L = \frac{0.25 \text{ in}}{(2)\left(12\ \dfrac{\text{in}}{\text{ft}}\right)} = 0.01042 \text{ ft}$$

$$
\begin{aligned}
T_{\text{center}} &= T_s + \frac{GL^2}{2k} \\
&= 80°\text{F} + \frac{\left(1.39 \times 10^7\ \dfrac{\text{Btu}}{\text{hr-ft}^3}\right)(0.01042 \text{ ft})^2}{(2)\left(224\ \dfrac{\text{Btu-ft}}{\text{hr-ft}^2\text{-°F}}\right)} \\
&= 83.4°\text{F}
\end{aligned}
$$

43. FINS

Fins (extended surfaces) are objects that receive and move thermal energy by conduction along their length and width, prior to (in most cases) convective and radiative heat removal. Radiators include simple fins, fin tubes, finned channels, and heat pipes. Some simple configurations can be considered and evaluated as fins even though that is not their intended function.[30]

External fins are attached at their base to a source of thermal energy at temperature T_b. It is assumed that the temperature is constant across the face of the fin at any point along its extent.

The *radiator efficiency (fin efficiency)*, η, is the ratio of the actual to ideal heat transfers assuming the entire fin is at the base temperature, T_b.

$$\eta_f = \frac{Q_{\text{actual}}}{Q_{\text{ideal}}}$$
$$= \frac{Q_{\text{actual}}}{hA_f(T_b - T_\infty)} \qquad \textit{34.57}$$

The fin efficiency can also be calculated by Eq. 34.58.[31,32] P is the perimeter length of the exposed face.

$$\eta_f = \frac{\tanh(Lm)}{Lm} \qquad \textit{34.58}$$

$$m = \sqrt{\frac{hP}{kA_b}} \qquad \textit{34.59}$$

The area, A_b, in Eq. 34.59 is the cross-sectional area of the fin. In rectangular and cylindrical fins, this area is uniform along the fin length. For a pipe, it is the solid annular area.

Not all heat radiators increase the heat transfer. In some cases, the fin can act as insulation. The functionality of a radiator depends on its size (specifically, its face perimeter, P). If the quantity hA/Pk is less than 1.0, the perimeter will be large enough and the radiator will increase the heat transfer.[33] If hA/Pk is greater than 1.0, the radiator will insulate. This may be the case where the film coefficient, h, is large, as it is with boiling or high-velocity (turbulent) liquids.

---

[30] As an example, a wire being soldered at one end can be considered as a rod-shaped fin. Two pipes being butt-welded together end-to-end can be considered as two infinite fins, each dissipating half of the welding energy.

[31] The arguments of hyperbolic functions are in radians.

[32] Graphs of fin efficiency are available for rectangular, triangular, parabolic, and annual fins.

[33] In practice, the extra material, installation, and maintenance expense associated with fins is rarely justified unless $hA/Pk < 0.25$.

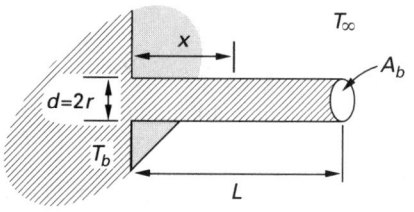

(a) finite cylindrical fin

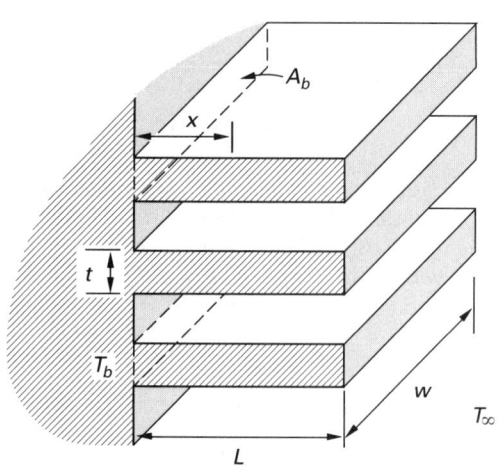

(b) finite rectangular fin

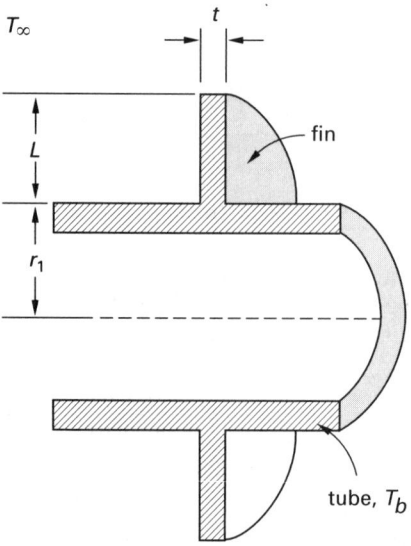

(c) annular (circular) fin

Figure 34.4 *Finned Heat Radiators*

The *radiator effectiveness (fin effectiveness)* is the ratio of the actual heat transfer with a fin to the heat transfer that would have occurred without a fin. This is not the same as the radiator efficiency. Fin efficiencies are less than 100%; effectiveness is generally greater than 100%.

$$E = \frac{Q_{\text{actual}}}{Q_{\text{no fin}}}$$

$$= \frac{Q_{\text{actual}}}{hA_b(T_b - T_\infty)} \qquad \textit{34.60}$$

The film coefficient, h, should be evaluated at the average of the fin surface and environmental temperatures. Since the fin surface temperature varies along its length, the midpoint surface temperature is used in the calculation.

$$T_h = \left(\tfrac{1}{2}\right)\left[\left(\tfrac{1}{2}\right)(T_b + T_{x=L}) + T_\infty\right] \qquad \textit{34.61}$$

Radiation may be significant along all or a part of the fin. A *combined* (*total, overall,* etc.) *film coefficient*, h_{total}, that accounts for both convective and radiative heat transfers should be used.

$$h_{\text{total}} = h + h_r \qquad \textit{34.62}$$

Unless stated otherwise, the formulas for temperature of and heat transfer from a fin do not consider the heat transfer from the exposed end. For that reason, the fin is said to possess an *adiabatic tip* or *insulated tip*. If the tip area is significant or if it is desired to include the heat transfer from the tip for other reasons, the film coefficient for the fin can be used as an approximation.[34] Closed form solutions for most common non-adiabatic cases are available in heat transfer textbooks.

Fins should not be used inside fluid-carrying pipes. Aside from the difficulty of manufacturing, spiral or transverse fins within a pipe would trap fluid and substantially increase the internal film coefficient. Longitudinal fins would "laminarize" the flow, also increasing the internal film coefficient.

44. INFINITE CYLINDRICAL FIN

The temperature at a distance x from the base of an infinite cylindrical (rod) fin is given by Eq. 34.63. The variable m is given by Eq. 34.59.

$$T_x = T_\infty + (T_b - T_\infty)e^{-mx} \qquad \textit{34.63}$$

The heat transfer from the barrel of the cylinder is given by Eq. 34.64. The quantity under the square root is not the same as the variable defined by Eq. 34.59.

$$Q = \sqrt{hPkA_b}\,(T_b - T_\infty) \qquad \textit{34.64}$$

$$A_b = \pi r^2 \qquad \textit{34.65}$$

$$P = \pi d = 2\pi r \qquad \textit{34.66}$$

---

[34]Even this is an approximation, as the additional heat transfer will change the temperature distribution along the fin length. However, the effect is probably minimal.

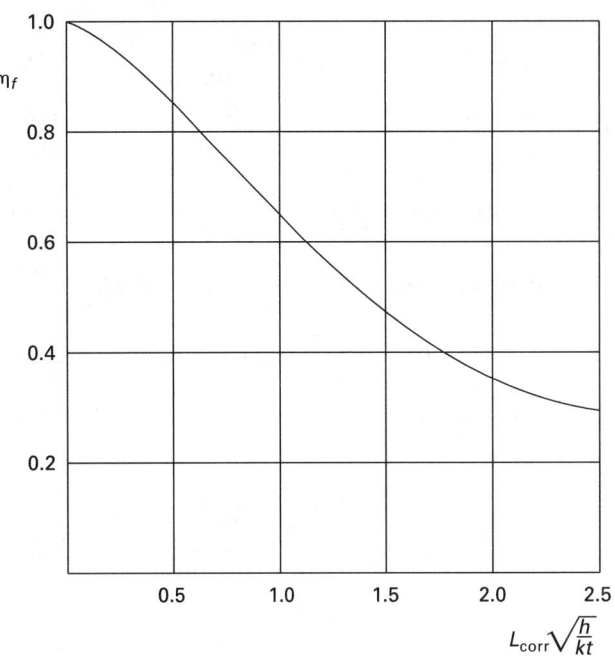

(a) rectangular (straight) fins

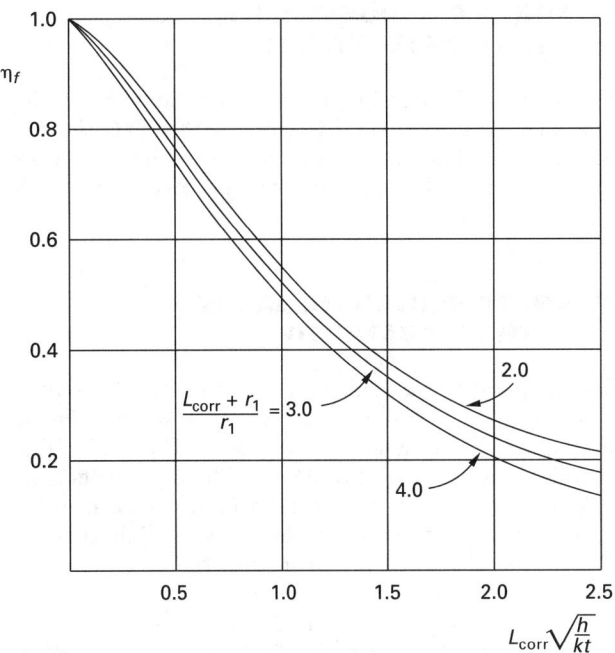

(b) circular (transverse) fins

Reproduced from *Transport Processes: Momentum, Heat and Mass*, by Christie J. Geankoplis, Allyn and Bacon, Inc., © 1983, with permission from Prentice Hall, Inc.

Figure 34.5 *Fin Efficiency*
(Refer to Fig. 34.4 for the dimensions of fins.)

45. FINITE CYLINDRICAL FIN WITH ADIABATIC TIP

The temperature at a distance x from the base of a finite cylindrical (rod) fin, also known as a *pin fin*, is given by Eq. 34.67.

$$T_x = T_\infty + (T_b - T_\infty) \left[\frac{\cosh\left(m(L-x)\right)}{\cosh\left(mL\right)} \right] \quad 34.67$$

$$P = \pi d = 2\pi r \quad 34.68$$

$$A_b = \pi r^2 \quad 34.69$$

$$m = \sqrt{\frac{hP}{kA_b}} = \sqrt{\frac{2h}{kr}} \quad 34.70$$

The heat transfer from the barrel of the cylinder is

$$Q = \sqrt{hPkA_b}(T_b - T_\infty)\tanh\left(mL\right) \quad 34.71$$

The fin efficiency is

$$\eta_f = \frac{\tanh\left(\sqrt{\frac{2hL^2}{kr}}\right)}{\frac{2hL^2}{kr}} \quad 34.72$$

46. FINITE CYLINDRICAL FIN WITH CONVECTIVE TIP

Within a small margin of error (less than 8%), the equations for finite cylindrical fins with adiabatic tips can be used when the tip contributes to heat transfer if the length, L, is replaced with the corrected length $L_{\text{corr}} = L + 1/4d$.

47. FINITE RECTANGULAR FIN WITH ADIABATIC TIP

The temperature at a distance x from the base of a finite rectangular fin (also known as a *straight fin* or *longitudinal fin*) of length L is given by Eq. 34.73. The variable m is given by Eq. 34.59. The film coefficient, h, is an average value that can be used with the entire exposed surface of the fin (excluding the adiabatic tip). The product Lt is the *profile area* of the fin.

$$T_x = T_\infty + (T_b - T_\infty) \left[\frac{\cosh\left(m(L-x)\right)}{\cosh\left(mL\right)} \right] \quad 34.73$$

The heat transfer from the body of the fin is

$$Q = \sqrt{hPkA_b}(T_b - T_\infty)\tanh\left(mL\right) \quad 34.74$$

$$P = (2)(w + t) \quad 34.75$$

$$A_b = wt \quad 34.76$$

$$mL = L\sqrt{\frac{hP}{kA}}$$

$$\approx L^{1.5}\sqrt{\frac{2h}{kLt}} \quad [w \gg t] \quad 34.77$$

$$\eta_f = \frac{\tanh\left(mL\right)}{mL} \quad 34.78$$

48. FINITE RECTANGULAR FIN WITH CONVECTIVE TIP

$$T_x = T_\infty + (T_b - T_\infty)$$
$$\times \left[\frac{\cosh\left(m(L-x)\right) + \left(\frac{h}{mk}\right)\sinh\left(m(L-x)\right)}{\cosh\left(mL\right) + \left(\frac{h}{mk}\right)\sinh\left(mL\right)} \right]$$
$$34.79$$

$$P = (2)(w + t) \quad 34.80$$

$$A_b = wt \quad 34.81$$

The heat transfer from the body of the fin is

$$Q = \sqrt{hPkA}(T_b - T_\infty)$$
$$\times \left[\frac{\sinh\left(mL\right) + \left(\frac{h}{mk}\right)\cosh\left(mL\right)}{\cosh\left(mL\right) + \left(\frac{h}{mk}\right)\sinh\left(mL\right)} \right] \quad 34.82$$

The approximate efficiency of the fin can be calculated with a margin of error of less than 8% if the length in Eq. 34.78 is replaced with the *corrected fin length*, $L_{\text{corr}} = L + t/2$.

$$\eta_f = \frac{\tanh\left(m\left(L + \frac{t}{2}\right)\right)}{m\left(L + \frac{t}{2}\right)} \quad 34.83$$

49. CIRCULAR FIN

Figure 34.4(c) shows a *circular fin*, also known as a *transverse fin*, installed on the exterior of a pipe. The exterior transverse area (corresponding to the "tip" area on straight fins) is substantial and is probably never insulated. Therefore, the corrected length should be used in the fin equations.

$$L_{\text{corr}} = L + \frac{t}{2} \quad 34.84$$

The heat transfer from the two surfaces is

$$Q = \eta_f h A_f (T_b - T_\infty) \quad 34.85$$

$$A_f = 2\pi[(L_{\text{corr}} + r_1)^2 - (r_1)^2] \quad 34.86$$

The fin efficiency can be found from Eq. 34.58 or read from Fig. 34.5(b).

Example 34.7

The base of a 0.5 in $\times$ 0.5 in $\times$ 10 in (1.2 cm $\times$ 1.2 cm $\times$ 25 cm) long rectangular rod is maintained at 300°F (150°C) by an electrical heating element. The conductivity of the rod is 80 Btu-ft/hr-ft^2-°F (140 W/m·K). The ambient air temperature is 80°F (27°C), and the average film coefficient is 1.65 Btu/hr-ft^2-°F (9.4 W/m^2·K). (a) What is the temperature of the rod 3 in (7.6 cm) from the base? (b) What energy input in kW is required to maintain the base temperature?

SI Solution

(a) Since the tip area is approximately 1% of the total surface, the rod is assumed to have an adiabatic tip.

The perimeter length is

$$P = (2)(w + t) = (2)(0.012 \text{ m} + 0.012 \text{ m}) = 0.048 \text{ m}$$

The cross-sectional area of the fin at its base is

$$A_b = wt = (0.012 \text{ m})(0.012 \text{ m})$$
$$= 0.000144 \text{ m}^2$$

From Eq. 34.59,

$$m = \sqrt{\frac{hP}{kA_b}}$$
$$= \sqrt{\frac{\left(9.4 \dfrac{\text{W}}{\text{m}^2 \cdot \text{K}}\right)(0.048 \text{ m})}{\left(140 \dfrac{\text{W}}{\text{m} \cdot \text{K}}\right)(0.000144 \text{ m}^2)}}$$
$$= 4.73 \text{ 1/m}$$

From Eq. 34.73, the temperature 7.6 cm from the base is

$$T_x = T_\infty + (T_b - T_\infty)\left[\frac{\cosh\left(m(L-x)\right)}{\cosh\left(mL\right)}\right]$$
$$= 27°\text{C} + (150°\text{C} - 27°\text{C})$$
$$\times \left[\frac{\cosh\left(\left(4.73 \dfrac{1}{\text{m}}\right)(0.25 \text{ m} - 0.076 \text{ m})\right)}{\cosh\left(\left(4.73 \dfrac{1}{\text{m}}\right)(0.25 \text{ m})\right)}\right]$$
$$= 120.6°\text{C}$$

(b) The total heat loss is

$$Q = \sqrt{hPkA_b}(T_b - T_\infty)\tanh\left(mL\right)$$
$$= \left[\sqrt{\begin{array}{l}\left(9.4 \dfrac{\text{W}}{\text{m}^2 \cdot \text{K}}\right)(0.048 \text{ m}) \\ \times \left(140 \dfrac{\text{W}}{\text{m} \cdot \text{K}}\right)(0.000144 \text{ m}^2) \end{array}} \right.$$
$$\times (150°\text{C} - 27°\text{C})\tanh\left(\left(4.73 \dfrac{1}{\text{m}}\right)(0.25 \text{ m})\right)$$
$$= 9.72 \text{ W}$$

Customary U.S. Solution

(a) Since the tip area is approximately 1% of the total surface, the rod is assumed to have an adiabatic tip.

The perimeter length is

$$P = (2)(w + t) = \frac{(2)(0.5 \text{ in} + 0.5 \text{ in})}{12 \dfrac{\text{in}}{\text{ft}}} = 0.167 \text{ ft}$$

The cross-sectional area of the fin at its base is

$$A_b = wt = \frac{(0.5 \text{ in})(0.5 \text{ in})}{\left(12 \dfrac{\text{in}}{\text{ft}}\right)^2}$$
$$= 0.00174 \text{ ft}^2$$

From Eq. 34.59,

$$m = \sqrt{\frac{hP}{kA_b}}$$
$$= \sqrt{\frac{\left(1.65 \dfrac{\text{Btu}}{\text{hr-ft}^2\text{-°F}}\right)(0.167 \text{ ft})}{\left(80 \dfrac{\text{Btu-ft}}{\text{hr-ft}^2\text{-°F}}\right)(0.00174 \text{ ft}^2)}}$$
$$= 1.41 \text{ 1/ft}$$

From Eq. 34.73, the temperature 3 in from the base is

$$T_x = T_\infty + (T_b - T_\infty)\left[\frac{\cosh\left(m(L-x)\right)}{\cosh\left(mL\right)}\right]$$
$$= 80°\text{F} + (300°\text{F} - 80°\text{F})$$
$$\times \left[\frac{\cosh\left(\dfrac{\left(1.41 \dfrac{1}{\text{ft}}\right)(10 \text{ in} - 3 \text{ in})}{12 \dfrac{\text{in}}{\text{ft}}}\right)}{\cosh\left(\dfrac{\left(1.41 \dfrac{1}{\text{ft}}\right)(10 \text{ in})}{12 \dfrac{\text{in}}{\text{ft}}}\right)}\right]$$
$$= 248.4°\text{F}$$

(b) The total heat loss is

$$Q = \sqrt{hPkA_b}(T_b - T_\infty)\tanh\left(mL\right)$$
$$= \sqrt{\begin{array}{l}\left(1.65 \dfrac{\text{Btu}}{\text{hr-ft}^2\text{-°F}}\right)(0.167 \text{ ft}) \\ \times \left(80 \dfrac{\text{Btu-ft}}{\text{hr-ft}^2\text{-°F}}\right)(0.00174 \text{ ft}^2) \end{array}}$$
$$\times (300°\text{F} - 80°\text{F})\tanh\left(\dfrac{\left(1.41 \dfrac{1}{\text{ft}}\right)(10 \text{ in})}{12 \dfrac{\text{in}}{\text{ft}}}\right)$$
$$= 35.6 \text{ Btu/hr}$$

Heat Transfer

50. SHAPE FACTORS

A horizontal pipe buried in the ground is a case of an object experiencing multidimensional heat transfer. Heat transfer will be by conduction, but the heat flow path will not be radial to the pipe everywhere. The complexity of this and similar problems is handled by use of a *shape factor*, S. Shape factors for several common configurations are given in Table 34.4.[35] The heat transfer is given by Eq. 34.87.

$$Q = Sk(T_1 - T_2) \qquad 34.87$$

51. STEAM AND ELECTRIC HEAT TRACING

Viscous fluids become difficult to pump if their temperatures drop significantly. This is a concern in lines that cannot be flushed, blown down, or drained while a process is on-line. Pipes exposed to low temperatures can be heated to keep their contents flowing by wrapping with *heat tracing*.

Electric heat tracing, also called *heat tape*, consists of long, flexible heating elements. There are two basic types: self-regulating and constant-watt products. *Self-regulating heat tape* is more common. The lower the ambient temperature, the greater the current.[36] *Constant-watt heat tape* puts out a constant heat (e.g., 8, 12, or 15 W/ft) at all times. The current drawn from a constant-watt heat tape can be calculated from Eq. 34.88, where P_L is the rated power per unit length.

$$I = \frac{P_L L}{E} \qquad 34.88$$

Design and installation of steam tracing is more of an art than a science. *Steam tracing* typically consists of $^3/_8$ in or $^1/_2$ in (9.5 mm or 12.7 mm) types K and L copper or stainless steel tubing through which low-pressure, 15 to 125 psig (100 to 860 kPa), steam runs.[37] A *steam header* (*manifold*) will feed up to 20 or so tracings. Each tracing should have its own isolation/control valve. Each tracing runs approximately 75 to 150 ft (20 to 45 m) between its supply valve and its terminating steam trap. Temperature-sensing (thermostatic or thermodynamic) *steam traps* should be spaced regularly, including at the ends of each tracing run and the bottom of vertical runs, to remove condensate.[38] Vacuum breakers and strainers are included in most designs.

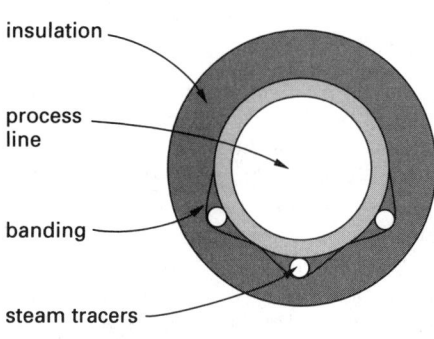

Figure 34.6 *Steam Tracing*

Tracing should be located inside the pipe insulation, along the bottom of the pipe.[39] Tracing is attached by bands or graphite-based heat transfer cement. However, a union should be inserted at pipe flange points, located outside of the pipe insulation. It is common also to install an expansion loop wherever a union is installed.

52. HEAT PIPES

A *heat pipe* is a cylindrical device with a high thermal conductance used to transfer large amounts of thermal energy on a continuous basis without moving parts or additional energy input. Heat pipes can be used on a small scale (as in satellites to remove heat from electronics) or on a large scale (as in a steam condenser or furnace economizer).

Operation of a heat pipe is based on the latent heat of vaporization of a working fluid. Heat is absorbed at the evaporator end of the heat pipe. The working fluid vaporizes and moves to the condenser end. The vapor condenses at the cooler end, releasing the heat of vaporization. The condensed working fluid returns to the evaporator end where the cycle is repeated on a continuous basis. The heat transfer occurs without any external power other than the thermal energy transferred.

There are two types of heat pipes. In the *thermo-siphon heat pipe*, gravity, buoyancy, and vapor pressure are the forces that move the different phases of the working fluid. While the thermo-siphon heat pipe does not have to be vertical, the heat source (i.e., the evaporator) must be below the heat sink (i.e., the condenser).[40] The vapor generated travels upward because of its buoyancy; condensed fluid moves downward under the influence of gravity.

---

[35] The case of two parallel cylinders is covered in most heat transfer textbooks.

[36] At very low temperatures, this can result in a high initial current and can trip circuit breakers.

[37] Low-pressure steam is not only cheaper than high-pressure steam, but also it liberates more heat during condensation.

[38] If the pressure drops to zero at the end of the tracing, temperature-triggered steam traps are required.

[39] Two other methods (used when the heating load is high) of using steam to heat process piping are running the steam line inside the process pipe or jacketing the entire process pipe with steam. These two methods essentially create shell-in-tube heat exchangers.

[40] Depending on the heat pipe, an orientation of as little as $10°$ from the horizontal may be satisfactory for common thermo-siphon heat pipes. With a separate condensate-collection tube returning condensate directly to the evaporator end, completely horizontal operation is possible.

Table 34.4 *Shape Factors*

configuration	illustration	shape factor, S
sphere, buried $(d = 2r)$	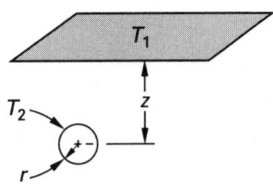	$S = \dfrac{2\pi d}{1 - \left(\dfrac{d}{4z}\right)}$
finite horizontal isothermal cylinder, buried $(z \ll L)$ $(d = 2r)$	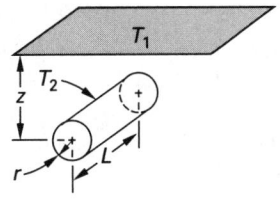	$S = \dfrac{2\pi L}{\text{arccosh}\left(\dfrac{2z}{d}\right)}$
infinite horizontal isothermal cylinder, buried (per unit length) $(d = 2r)$		$S/L = \dfrac{2\pi}{\text{arccosh}\left(\dfrac{2}{\frac{z}{d}}\right)}$
finite vertical isothermal cylinder $(d \ll L)$	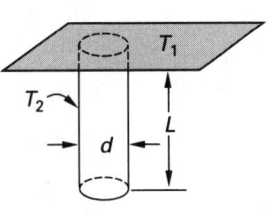	$S = \dfrac{2\pi L}{\ln\left(\dfrac{4L}{d}\right)}$
intersecting planes of thickness t $(a > t/5)$ $(b > t/5)$	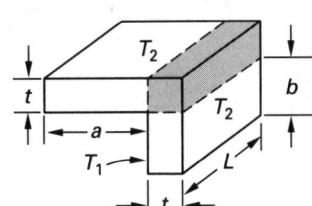	$S = 0.54L$

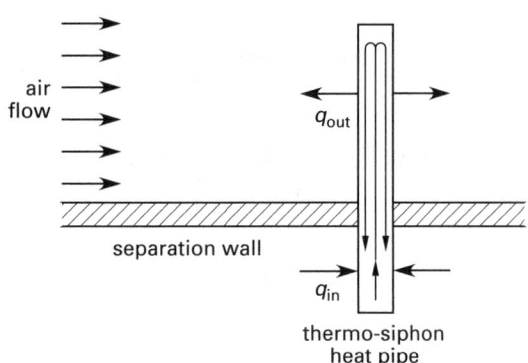

Figure 34.7 *Thermo-Siphon Heat Pipe*

A *capillary-action heat pipe* contains a wick constructed of gauze, wire mesh, or other material throughout its length. The condensate is returned to the evaporator end through this wick. There are no restrictions on orientation, and because gravity is not required, capillary action heat pipes can be used in zero-gravity.

PRACTICE PROBLEMS

Thermal Conductivity

1. Experiments have shown that the thermal conductivity, k, of a particular material varies with temperature, T, according to the following relationship.

$$k_T = (0.030)(1 + 0.0015T)$$

What is the value of k that should be used for a transfer of heat through the material if the hot-side temperature is 350° (°F or °C) and the cold-side temperature is 150° (°F or °C)?

Conduction

2. What is the heat flow through insulating brick, 1.0 ft (30 cm) thick, in an oven wall with a thermal conductivity of 0.038 Btu-ft/hr-ft²-°F (0.066 W/m·K)? The thermal gradient is 350°F (195°C).

3. A composite wall is made up of 3.0 in (7.6 cm) of material A exposed to 1000°F (540°C), 5.0 in (13 cm) of material B, and 6.0 in (15 cm) of material C exposed to 200°F (90°C). The mean thermal conductivities for materials A, B, and C are 0.06, 0.5, and 0.8 Btu-ft/hr-ft²-°F (0.1, 0.9, and 1.4 W/m·K), respectively. What are the temperatures at the A-B and B-C material interfaces?

Unsteady Heat Flow

4. The heat supply of a large building is turned off at 5:00 P.M. when the interior temperature is 70°F (21°C). The outdoor temperature is constant at 40°F (4°C). The thermal capacity of the building and its contents is 100,000 Btu/°F (60 MJ/K), and the conductance is 6500 Btu/hr-°F (1.1 kW/K). What is the interior temperature at 1:00 A.M.?

5. Steel ball bearings varying in diameter from 1/4 in to 1 1/2 in (6.35 to 38.1 mm) are quenched from 1800°F (980°C) in an oil bath that remains at 110°F (43°C). The ball bearings are removed when their internal (center) temperature reaches 250°F (120°C). The average film coefficient is 56 Btu/hr-ft²-°F (320 W/m²·K). (a) What is the time constant for the cooling process? (b) Derive a linear equation for the time the ball bearings remain in the oil bath as a function of diameter.

Internal Heat Generation

6. A 0.4 in (1.0 cm) diameter uranium dioxide fuel rod with a thermal conductivity of 1.1 Btu-ft/hr-ft²-°F (1.9 W/m·K) is clad with 0.020 in (0.5 mm) of stainless steel. The fuel rod generates 4×10^7 Btu/hr-ft³ (4.1×10^8 W/m³) internally. A coolant at 500°F (260°C) circulates around the clad rod. The outside film coefficient is 10,000 Btu/hr-ft²-°F (57 kW/m²·K). What is the temperature at the longitudinal centerline of the rod?

Finned Radiators

7. Two long pieces of 1/16 in (1.6 mm) copper wire are connected end-to-end with a hot soldering iron. The minimum melting temperature of the solder is 450°F (230°C). The surrounding air temperature is 80°F (27°C). The unit film coefficient is 3 Btu/hr-ft²-°F (17 W/m²·K). What is the minimum rate of heat application to keep the solder molten?

35 Natural Convection, Evaporation, and Condensation

1.	Introduction	35-2
2.	Heat Transfer by Natural Convection	35-2
3.	Film Coefficients	35-2
4.	Nusselt Number	35-2
5.	Prandtl Number	35-2
6.	Grashof Number	35-2
7.	Rayleigh Number	35-3
8.	Reynolds Number	35-3
9.	Correlations	35-3
10.	Film Temperature	35-3
11.	Nusselt Equation	35-3
12.	Film Coefficients for Air	35-4
13.	Film Coefficients for Air at Other Pressures	35-4
14.	Film Coefficients for Water	35-4
15.	Film Coefficients for Air on Heated Flat Plates	35-6
16.	Film Coefficients for the Outside of a Coil	35-7
17.	Film Coefficients for Thin Wires	35-7
18.	Film Coefficient with Enclosed Air Spaces	35-7
19.	Condensing Vapor	35-7
20.	Condensation on Vertical and Inclined Surfaces	35-8
21.	Turbulent Condensation on Vertical Plates and Tubes	35-9
22.	Condensation on the Outside of a Sphere	35-9
23.	Condensation Inside Tubes	35-9
24.	Evaporation from Horizontal Tubes	35-10
25.	Evaporation from Flat Plates	35-11
26.	Simplified Vaporization Equations for Water	35-11
27.	Effects of Radiation on Evaporation Film Coefficients	35-11
	Practice Problems	35-11

Nomenclature

A	area	ft^2	m^2
B	separation distance	ft	m
c_p	specific heat	Btu/lbm-°F	J/kg·K
C	constant	–	–
d	diameter	ft	m
g	gravitational acceleration[a]	ft/hr^2	m/s^2
Gr	Grashof number	–	–
h	film coefficient	Btu/hr-ft^2-°F	W/m^2·K
h_{fg}	latent heat of vaporization	Btu/lbm	J/kg
k	thermal conductivity	Btu-ft/hr-ft^2-°F	W/m·K
L	characteristic length	ft	m
m	exponent	–	–
m	mass	lbm	kg
$\dot{m}$	mass condensation rate	lbm/hr	kg/s
n	exponent	–	–
N	number of tube layers	–	–
Nu	Nusselt number	–	–
p	pressure	lbf/ft^2	kPa
P	wetted perimeter	ft	m
Pr	Prandtl number	–	–
q	heat transfer per unit area	Btu/hr-ft^2	W/m^2
Q	heat transfer rate	Btu/hr	W
r	radius	ft	m
Ra	Rayleigh number	–	–
Re	Reynolds number	–	–
s	side length	ft	m
T	temperature	°F	K
v	velocity	ft/hr	m/s

Symbols

α	thermal diffusivity	ft^2/sec	m^2/s
β	volumetric coefficient of expansion	1/°R	1/K
Γ	mass flow rate per unit area	lbm/hr-ft^2	kg/s·m^2
θ	angle (measured from horizontal)	degrees	degrees
μ	viscosity[b,c,d]	lbm/hr-ft	kg/s·m
ν	kinematic viscosity	ft^2/sec	m^2/s
ρ	mass density	lbm/ft^3	kg/m^3

Subscripts

0	initial, or zero gage pressure
a	atmospheric
b	boiling
B	based on separation distance
c	condensing
f	fluid
h	film
l	liquid
p	pressure
r	radiation
s	surface
sat	saturated
v	vapor
∞	at infinity

[a] g has a value of 4.17×10^8 ft/hr^2 (1.27×10^8 m/h^2).
[b] The use of mass units in viscosity values is typical in the subject of convective heat transfer.
[c] Most data compilations give fluid viscosity in units of seconds. In the United States, heat transfer is traditionally given on a per hour basis. Therefore, a conversion factor of 3600 is needed when calculating dimensionless numbers from table data.
[d] The combination of units kg/s·m is the same as a N·s/m^2 or Pa·s.

Heat Transfer

1. INTRODUCTION

Natural convection (also known as *free convection*) is the removal of heat from a surface by a fluid that moves upward under the influence of a density gradient. As the fluid warms, it becomes lighter and rises from the heating surface. The fluid is acted upon by buoyant and gravitational forces. The fluid does not have a component of motion parallel to the surface.[1]

Natural convection is attractive from an engineering design standpoint because no motors, fans, pumps, or other equipment with moving parts are required. However, the transfer surface must be much larger than it would be with forced convection.[2]

2. HEAT TRANSFER BY NATURAL CONVECTION

Equation 35.1 is the basic equation used to calculate the steady-state heat transfer by natural convection in both heating and cooling configurations. The *film coefficient* (*heat transfer coefficient*), h, is seldom known to great accuracy.[3] The average film coefficient, $\bar{h}$, is used where there are variations over the heat transfer surface.[4]

$$Q = qA = hA(T_s - T_\infty) \qquad 35.1$$

3. FILM COEFFICIENTS

Typical values of film coefficients for natural convection are listed in Table 35.1.

4. NUSSELT NUMBER

The *Nusselt number*, Nu, is defined by Eq. 35.2. The Nusselt number is sometimes written with a subscript (e.g., Nu_h or Nu_f) to indicate that the fluid properties are evaluated at the film temperature. (See Eq. 35.11.)

$$Nu = \frac{hd}{k} \qquad 35.2$$

5. PRANDTL NUMBER

The dimensionless *Prandtl number*, Pr, is defined by Eq. 35.3. It represents the ratio of momentum diffusion to thermal diffusion. The values used are for the

fluid, not for the surface material. For gases, the values used in calculating the Prandtl number, and hence the Prandtl number itself, do not vary significantly with temperature.

$$Pr = \frac{c_p \mu}{k} = \frac{c_p \nu \rho}{k} = \frac{\nu}{\alpha} \qquad 35.3$$

Table 35.1 Typical[a] Film Coefficients for Natural Convection

	Btu/hr-ft²-°F	W/m²·K
no change in phase:		
air, still	0.8–4.4	5.0–25.0
condensing:		
steam		
horizontal surface	1700–4300	9600–24 400
vertical	700–2000	4000–11 300
organic solvents	150–500	850–2800
ammonia	500–1000	2800–5700
evaporating:		
water	800–2000	4500–11 300
organic solvents	100–300	550–1700
ammonia	200–400	1100–2300

(Multiply Btu/hr-ft²-°F by 5.6783 to obtain W/m²·K.)
[a]Values outside of these ranges have been observed. However, these ranges are typical of those encountered in industrial processes.

6. GRASHOF NUMBER

The dimensionless *Grashof number*, Gr, is the ratio of buoyant to viscous forces. Dynamic similarity in free convection problems is assured by equating the Grashof numbers.

The *characteristic length*, L, is defined in Table 35.2 for various configurations.[5] The coefficient of volumetric expansion, β, for ideal gases is the reciprocal of the absolute film temperature. Gravity, g, and viscosity, μ, must have the same unit of time in order to make Gr dimensionless. The quantity $g\beta\rho^2/\mu^2$ is tabulated in Apps. 35.A through 35.F, so the component values generally do not need to be evaluated individually.

$$Gr = \frac{L^3 g\beta\rho^2(T_s - T_\infty)}{\mu^2}$$
$$= \frac{L^3 g\beta(T_s - T_\infty)}{\nu^2} \qquad 35.4$$

The Grashof number may be written with a subscript indicating which dimension is to be used as the characteristic length. For example, the symbol Gr_d or Gr_D could be used to represent the Grashof number in which diameter is the characteristic length.

For air, the critical Grashof number for laminar flow is approximately 10^9. Below 10^9, the air flow will be laminar; above 10^9, it will be turbulent.

---

[1]Rotating spheres and cylinders are special categories of convective heat transfer where the fluid has a component of relative motion parallel to the heat transfer surface.
[2]Natural convection requires approximately 2 to 10 times more surface area than does forced convection.
[3]An error of up to 25% can be expected.
[4]Though $\bar{h}$ has historically been used in books on the subject of heat transfer, most modern books and this book use the symbol h. The fact that the film coefficient is an inaccurate, average value is implicit.

[5]The length of the side of a square, the mean length of a rectangle, and 90% of the diameter of a circle have historically been used as the *characteristic length*. However, the ratio of surface area to perimeter gives better agreement with experimental data.

7. RAYLEIGH NUMBER

The *Rayleigh number*, Ra, is the product of the Grashof and Prandtl numbers.

$$\text{Ra} = \text{GrPr} = \frac{L^3 g \beta \rho^2 (T_s - T_\infty) c_p}{k \mu} \qquad 35.5$$

The quantity $g\beta\rho^2 c_p / k\mu$ is tabulated in some books and given the symbol a.

$$\text{Ra} = aL^3(T_s - T_\infty) \qquad 35.6$$

$$a = \frac{g\beta\rho^2 c_p}{k\mu} \qquad 35.7$$

8. REYNOLDS NUMBER

The free-stream velocity is always zero with natural convection, so the traditional Reynolds number is also always zero. The Grashof and Rayleigh numbers take the place of determining whether flow is laminar or turbulent.[6] The film Reynolds number (Sec. 21) is used to determine if condensation is turbulent.

9. CORRELATIONS

Equation 35.1 is simple to use. The main difficulty is finding the film coefficient, h. Various theoretical, empirical, and semi-empirical correlations have been developed using dimensional analysis and experimentation. These correlations are of several forms.

Theoretical correlations are developed completely from dimensional analysis and theoretical considerations.

Empirical correlations are determined by fitting a curve through observed data points. The film coefficient has traditionally been correlated with either the *heat flux*, $q = Q/A$ (in Btu/hr-ft^2 or kW/m^2), or with the difference in temperature between the heated surface and the liquid. For example, the convective film coefficient for boiling water can be predicted approximately by Eqs. 35.8 and 35.9.[7]

$$\log(h_b) \approx -2.05 + 2.5 \log(T_s - T_{\text{bulk}}) + 0.014 T_{\text{sat}}$$
$$\text{[U.S. only]} \qquad 35.8$$

$$h_b \approx 190 + 0.043q \qquad \text{[U.S. only]} \qquad 35.9$$

Semi-empirical correlations, derived from dimensional analysis with exponents and constants determined from experimentation, are the form of Eq. 35.10.[8] Exponents m and n are often sufficiently close so that a common value can be used.[9]

---

[6]Rising air nevertheless has a velocity. The critical Reynolds number for laminar flow of air is approximately 550 (corresponding to a Grashof number of 10^9).

[7]Equations 35.8 and 35.9 actually yield some "pretty good" initial estimates.

[8]Equation 35.10 is known as a *Nusselt-type correlation*.

[9]The implication of exponents m and n being identical is that the temperature differences and fluid velocities are small.

$$\text{Nu} = C \times (\text{Pr}^m \text{Gr}^n)$$
$$\approx C \times (\text{PrGr})^n = C \times (\text{Ra})^n \qquad 35.10$$

Each correlation can only be used in particular configurations (i.e., a correlation for horizontal cylinders cannot generally be used for vertical cylinders), and even then, only within a particular range of parameters (e.g., Prandtl or Grashof numbers).

The usefulness of a correlation depends on how well it predicts actual performance. Though correlations should always be accompanied by a parameter range, the percentage accuracy of the correlation is generally not stated. Considering that correlations are often accurate to only $\pm 20\%$, a value derived from a correlation near the end of its applicable range should be considered "ballpark."

Often, two or more parameter ranges will be given for a particular configuration. (See the GrPr ranges in Table 35.2.) The lower range of parameters correspond to laminar air flow, while the higher parameter ranges correspond to turbulent air flow. Correlations are less reliable near the transition region between the two regimes.

10. FILM TEMPERATURE

Film properties are evaluated at the average of the surface temperature, T_s, and the *bulk temperature*, T_∞. When there is a variation of the surface temperature, as there could be along the length of a long tube used for heat transfer, the surface temperature is assumed to be the temperature at midlength along the tube.[10]

$$T_h = \left(\tfrac{1}{2}\right)(T_s + T_\infty) \qquad 35.11$$

11. NUSSELT EQUATION

The *Nusselt equation* and equations of its form are often used to find the film coefficient for natural convective heating and cooling. The thermal conductivity, k, in Eq. 35.12 is for the transfer fluid, not for the surface wall, and is evaluated at the film temperature, T_h.

$$\frac{hL}{k} = C(\text{GrPr})^n \qquad \text{[SI and U.S.]} \qquad 35.12$$

For laminar convection ($1000 < \text{GrPr} < 10^9$), n has a value of approximately $1/4$. For turbulent convection ($\text{GrPr} > 10^9$), n is approximately $1/3$. For sublaminar convection ($\text{GrPr} < 1000$), n is less than $1/4$ (typically taken as $1/5$), and graphical solutions are commonly used.

The values of the dimensionless empirical constants C and n in Eq. 35.12 and given in Table 35.2 can be used with all fluids and any consistent systems of units. Table 35.2 is limited in application to single heat transfer surfaces (i.e., a single tube or a single plate).

---

[10]Variations in the surface temperature with time, however, cannot be so easily handled.

Heat Transfer

Table 35.2 *Parameters for the Nusselt Equation (any substance; isothermal surfaces, U.S. or SI units)*

configuration	L	GrPr	C	n
vertical plate or vertical cylinder[b]	height[a]	$< 10^4$	1.36	0.20
		10^4 to 10^9	0.59	$1/4$
		10^9 to 10^{12}	0.13	$1/3$
inclined plate (θ measured from horizontal)	Use vertical plate constants, subsituting $\sin\theta\, \mathrm{Gr}$ for Gr.			
horizontal cylinder[c]	outside diameter	10^3 to 10^9	0.53	$1/4$
		10^9 to 10^{12}	0.13	$1/3$
thin horizontal wire	diameter	$< 10^{-5}$	0.49	0
		10^{-5} to 10^{-3}	0.71	0.04
		10^{-3} to 1	1.09	0.10
		1 to 10^4	1.09	0.20
		10^4 to 10^9		
	(See also Eq. 35.18 and Fig. 35.1.)			
horizontal plate[d]: hot surface facing up or cold surface facing down	$(1/2)(s_1 + s_2)$ or $0.9d$	10^5 to 2×10^7	0.54	$1/4$
		2×10^7 to 3×10^{10}	0.14	$1/3$
horizontal plate[d]: hot surface facing down or cold surface facing up	$(1/2)(s_1 + s_2)$ or $0.9d$	3×10^5 to 3×10^{10}	0.27	$1/4$
sphere[e]	radius	10^3 to 10^9	0.63	$1/4$
		$> 10^9$	0.15	$1/3$

[a]For short vertical plates, the characteristic length is approximately (height $\times$ width)/(height + width).
[b]A vertical cylinder can be considered a vertical plate as long as $d/L \geq 35/(\mathrm{Gr}_L)^{\frac{1}{4}}$.
[c]The values for the laminar range can also be used for heat transfer to liquid metals.
[d]For a circular flat disc, the characteristic length is 90% of the disc diameter.
[e]Ranges and values reported by different researchers show significant variation. Values can also be used for short cylinders and blocks with a characteristic length of (height $\times$ width)/(height + width).

Horizontal pipe diameters greater than approximately 8 in (20 cm) and plate heights greater than approximately 2 ft (0.6 m) have little effect on film coefficients.[11] Therefore, characteristic lengths for tall plates and large-diameter pipes should be limited to 2 ft (0.6 m) and 0.67 ft (0.2 m), respectively, when calculating the film coefficients for free convection.[12]

12. FILM COEFFICIENTS FOR AIR

Table 35.3 contains simplified equations for calculating film coefficients for air at standard atmospheric pressure. These equations are derived from the Nusselt equation using simplifying assumptions.[13]

---

[11]Some researchers report 3 ft (0.9 m) as the limiting value.
[12]These limiting values for characteristic length are taken into consideration automatically in the simplified equations given in Table 35.3. The characteristic length, L, does not appear in the turbulent flow equations.
[13]The main assumption is the film temperature.

13. FILM COEFFICIENTS FOR AIR AT OTHER PRESSURES

The film coefficients derived from Table 35.3 can be multiplied by Eqs. 35.13 or 35.14 to correct for pressures other than atmospheric.

$$\text{correction factor} = \left(\frac{p_{\text{actual}}}{p_{\text{std. atmosphere}}} \right)^{\frac{1}{2}}$$

$$\text{[laminar]} \quad 35.13$$

$$\text{correction factor} = \left(\frac{p_{\text{actual}}}{p_{\text{std. atmosphere}}} \right)^{\frac{2}{3}}$$

$$\text{[turbulent]} \quad 35.14$$

14. FILM COEFFICIENTS FOR WATER

Equation 35.15 is a simplified equation for calculating the film coefficient for water on vertical planes and

Table 35.3 Natural Convection Film Coefficients:
Simplified Equations for Air
(isothermal surfaces, 1 atm)

configuration	GrPr	simplified equation	
vertical plate or vertical cylinder[a]	10^4 to 10^9	$h = (1.37)\left(\dfrac{T_s - T_\infty}{L}\right)^{\frac{1}{4}}$	[SI]
		$h = (0.29)\left(\dfrac{T_s - T_\infty}{L}\right)^{\frac{1}{4}}$	[U.S.]
	10^9 to 10^{12}	$h = (1.24)(T_s - T_\infty)^{\frac{1}{3}}$	[SI]
		$h = (0.19)(T_s - T_\infty)^{\frac{1}{3}}$	[U.S.]
horizontal cylinder	10^3 to 10^9	$h = (1.32)\left(\dfrac{T_s - T_\infty}{d}\right)^{\frac{1}{4}}$	[SI]
		$h = (0.27)\left(\dfrac{T_s - T_\infty}{d}\right)^{\frac{1}{4}}$	[U.S.]
	10^9 to 10^{12}	$h = (1.24)(T_s - T_\infty)^{\frac{1}{3}}$	[SI]
		$h = (0.18)(T_s - T_\infty)^{\frac{1}{3}}$	[U.S.]
horizontal plate: hot surface facing up or cold surface facing down (square)[b]	10^5 to 2×10^7	$h = (1.32)\left(\dfrac{T_s - T_\infty}{d}\right)^{\frac{1}{4}}$	[SI]
		$h = (0.27)\left(\dfrac{T_s - T_\infty}{L}\right)^{\frac{1}{4}}$	[U.S.]
	2×10^7 to 3×10^{10}	$h = (1.52)(T_s - T_\infty)^{\frac{1}{3}}$	[SI]
		$h = (0.22)(T_s - T_\infty)^{\frac{1}{3}}$	[U.S.]
horizontal plate: cold surface facing up or hot surface facing down (square)[b]	3×10^5 to 3×10^{10}	$h = (0.59)\left(\dfrac{T_s - T_\infty}{d}\right)^{\frac{1}{4}}$	[SI]
		$h = (0.12)\left(\dfrac{T_s - T_\infty}{L}\right)^{\frac{1}{4}}$	[U.S.]

[a]A vertical cylinder can be considered a vertical plate as long as $d/L \geq 35/(\mathrm{Gr}_L)^{\frac{1}{4}}$.
[b]For horizontal circular disc, use $L = 0.9 \times$ disc diameter.

cylinders at room temperature (i.e., 70°F (21°C)) and in the range $10^4 < \mathrm{GrPr} < 10^9$.

$$h = (127)\left(\frac{T_s - T_\infty}{L}\right)^{\frac{1}{4}} \quad \text{[SI]} \quad \textbf{35.15(a)}$$

$$h = (26)\left(\frac{T_s - T_\infty}{L}\right)^{\frac{1}{4}} \quad \text{[U.S.]} \quad \textbf{35.15(b)}$$

Example 35.1

A horizontal 4.0 in (10 cm) diameter (nominal) pipe carries steam through a 20 ft (6 m) long room. The temperature of the exterior of the pipe surface is 300°F(150°C).

The temperature of the air in the room is 100°F (36°C). What is the natural convective heat loss from the exterior of the pipe?

SI Solution

The characteristic length is the pipe outside diameter.

$$L = 0.10 \text{ m}$$

The temperature gradient is

$$T_s - T_\infty = 150°C - 36°C = 114°C$$

The air properties are evaluated at the film temperature.

$$T_h = \left(\tfrac{1}{2}\right)(T_s + T_\infty) = \left(\tfrac{1}{2}\right)(150°C + 36°C)$$
$$= 93°C$$

The air properties at 93°C are found from App. 35.D.

$$k = 0.03115 \text{ W/m·K}$$
$$\rho = 0.964 \text{ kg/m}^3$$
$$\mu = 2.15 \times 10^{-5} \text{ kg/s·m}$$
$$\beta = 2.74 \times 10^{-3} \text{ 1/K}$$
$$\text{Pr} = 0.694$$

The Grashof number is

$$\text{Gr} = \frac{L^3 g \beta \rho^2 (T_s - T_\infty)}{\mu^2}$$

$$= \frac{(0.10 \text{ m})^3 \left(9.81 \,\dfrac{\text{m}}{\text{s}^2}\right)\left(2.74 \times 10^{-3} \,\dfrac{1}{\text{K}}\right)}{\left(2.15 \times 10^{-5} \,\dfrac{\text{kg}}{\text{s·m}}\right)^2}$$
$$ \times \left(0.964 \,\dfrac{\text{kg}}{\text{m}^3}\right)^2 (114 \text{ K})$$

$$= 6.16 \times 10^6$$

$$\text{GrPr} = (6.16 \times 10^6)(0.694) = 4.28 \times 10^6$$

Using Eq. 35.12 and values from Table 35.2,

$$h = \frac{kC(\text{GrPr})^n}{L}$$

$$= \frac{\left(0.03115 \,\dfrac{\text{W}}{\text{m·K}}\right)(0.53)(4.28 \times 10^6)^{\frac{1}{4}}}{0.10 \text{ m}}$$

$$= 7.51 \text{ W/m}^2\text{·K}$$

The heat transfer from the pipe is

$$Q = qA = \pi d L h (T_s - T_\infty)$$

$$= \pi (0.10 \text{ m})(6 \text{ m})\left(7.51 \,\dfrac{\text{W}}{\text{m}^2\text{·K}}\right)(150°C - 36°C)$$

$$= 1614 \text{ W}$$

Customary U.S. Solution

The characteristic length is the pipe outside diameter.

$$L = \frac{4 \text{ in}}{12 \,\dfrac{\text{in}}{\text{ft}}} = 0.333 \text{ ft}$$

The temperature gradient is

$$T_s - T_\infty = 300°F - 100°F = 200°F$$

The air properties are evaluated at the film temperature.

$$T_h = \left(\tfrac{1}{2}\right)(T_s + T_\infty) = \left(\tfrac{1}{2}\right)(300°F + 100°F)$$
$$= 200°F$$

The air properties at 200°F are found from App. 35.C.

$$k = 0.0174 \text{ Btu-ft/hr-ft}^2\text{-}°F$$
$$\rho = 0.060 \text{ lbm/ft}^3$$
$$\mu = 1.44 \times 10^{-5} \text{ lbm/ft-sec}$$
$$\beta = 1.52 \times 10^{-3} \text{ 1/}°F$$
$$\text{Pr} = 0.72$$

The Grashof number is

$$\text{Gr} = \frac{L^3 g \beta \rho^2 (T_s - T_\infty)}{\mu^2}$$

$$= \frac{(0.333 \text{ ft})^3 \left(32.2 \,\dfrac{\text{ft}}{\text{sec}^2}\right)\left(1.52 \times 10^{-3} \,\dfrac{1}{°F}\right)}{\left(1.44 \times 10^{-5} \,\dfrac{\text{lbm}}{\text{ft-sec}}\right)^2}$$
$$ \times \left(0.060 \,\dfrac{\text{lbm}}{\text{ft}^3}\right)^2 (200°F)$$

$$= 6.28 \times 10^6$$

Using Eq. 35.12 and values from Table 35.2,

$$h = \frac{kC(\text{GrPr})^n}{L}$$

$$= \frac{\left(0.0174 \,\dfrac{\text{Btu-ft}}{\text{hr-ft}^2\text{-}°F}\right)(0.53)(4.52 \times 10^6)^{\frac{1}{4}}}{0.333 \text{ ft}}$$

$$= 1.28 \text{ Btu/hr-ft}^2\text{-}°F$$

The heat transfer from the pipe is

$$Q = qA = \pi d L h (T_s - T_\infty)$$

$$= \pi (0.333 \text{ ft})(20 \text{ ft})\left(1.28 \,\dfrac{\text{Btu}}{\text{hr-ft}^2\text{-}°F}\right)$$
$$ \times (300°F - 100°F)$$

$$= 5356 \text{ Btu/hr}$$

15. FILM COEFFICIENTS FOR AIR ON HEATED FLAT PLATES

If the film coefficient for air heated on a flat plate is known for either the vertical or horizontal configuration, an approximate film coefficient for the corresponding configuration can be determined from Eqs. 35.16 and 35.17.

$$h_{\text{horizontal,facing upward}} \approx 1.27 h_{\text{vertical}} \qquad 35.16$$
$$h_{\text{horizontal,facing downward}} \approx 0.67 h_{\text{vertical}} \qquad 35.17$$

16. FILM COEFFICIENTS FOR THE OUTSIDE OF A COIL

For air, the film coefficient on the outside of a coil of tubing is approximately the same as the film coefficient for a single horizontal tube.

17. FILM COEFFICIENTS FOR THIN WIRES

Since the boundary layer (i.e., the film) thickness is not small compared to the diameter, correlations derived from the traditional Nusselt equation may not be adequate. Equation 35.18 is a theoretical relationship for calculating the film coefficient for horizontal thin wires with laminar convection.[14] Table 35.2 and Fig. 35.1 can also be used.

$$h = \cfrac{2k}{d \times \ln\left(1 + \cfrac{2}{1 + \cfrac{2}{(0.4)(Gr)^{\frac{1}{4}}}}\right)} \qquad 35.18$$

$$[GrPr < 10^3]$$

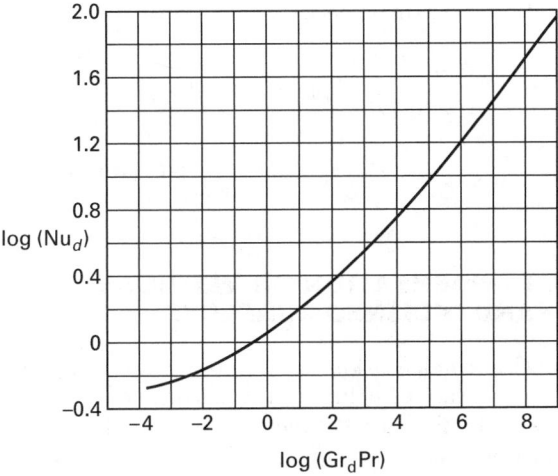

Figure 35.1 *Correlation for Horizontal Small Tubes and Thin Wires*

18. FILM COEFFICIENT WITH ENCLOSED AIR SPACES

A double-paned window is an example of a configuration where heat is transferred across an enclosed air space. The heat flow mechanism is largely conduction across the air layer at low Grashof numbers. At higher Grashof numbers, convection becomes more of a factor.

---

[14]The assumptions used to derive this theoretical equation include (1) the film thickness is much smaller than the diameter, and (2) the diameter is 2.5 times the height of an equivalent vertical surface.

Equations 35.19 through 35.21 can be used to calculate the film coefficient for gases enclosed between two vertical plates (or inside a vertical annulus) and for which the ratio of height to separation distance, L/B, is greater than 3. The temperature difference used in calculating the Grashof number is the difference in surface temperatures for the two plates. The characteristic length used in calculating the Nusselt and Grashof numbers is the separation distance (clear spacing), B, between the two plates.

$$Nu_B = 1.0 \quad [Gr_B Pr < 2 \times 10^3] \qquad 35.19$$

$$Nu_B = \frac{(0.20)(Gr_B Pr)^{\frac{1}{4}}}{\left(\frac{L}{B}\right)^{\frac{1}{9}}} \quad \begin{bmatrix} 6 \times 10^3 < Gr_B Pr < 2 \times 10^5 \\ 0.5 < Pr < 2.0 \\ 11 < \frac{L}{B} < 42 \end{bmatrix}$$

$$35.20$$

$$Nu_B = \frac{(0.073)(Gr_B Pr)^{\frac{1}{3}}}{\left(\frac{L}{B}\right)^{\frac{1}{9}}} \quad \begin{bmatrix} 2 \times 10^5 < Gr_B Pr < 2 \times 10^7 \\ 0.5 < Pr < 2.0 \\ 11 < \frac{L}{B} < 42 \end{bmatrix}$$

$$35.21$$

Equations 35.22 and 35.23 can be used for liquids between two vertical plates (or inside a vertical annulus).

$$Nu_B = 1.0 \quad [Gr_B Pr < 1 \times 10^3] \qquad 35.22$$

$$Nu_B = \frac{(0.28)(Gr_B Pr)^{\frac{1}{4}}}{\left(\frac{L}{B}\right)^{\frac{1}{4}}} \quad [1 \times 10^3 < Gr_B Pr < 1 \times 10^7]$$

$$35.23$$

Equations 35.24 and 35.25 can be used for gases between horizontal plates with the lower plate hotter than the upper.

$$Nu_B = (0.21)(Gr_B Pr)^{\frac{1}{4}} \quad [7 \times 10^3 < Gr_B Pr < 3 \times 10^5]$$

$$35.24$$

$$Nu_B = (0.061)(Gr_B Pr)^{\frac{1}{3}} \quad [Gr_B Pr > 3 \times 10^5] \qquad 35.25$$

Equation 35.26 can be used for liquids between horizontal plates with the lower plate hotter than the upper.

$$Nu_B = (0.069)(Gr_B Pr)^{\frac{1}{3}}(Pr)^{0.074}$$
$$[1.5 \times 10^5 < Gr_B Pr < 1 \times 10^9] \quad 35.26$$

19. CONDENSING VAPOR

When a vapor condenses on a cooler surface, the condensate forms a thin layer on the surface. This layer insulates the surface and creates a thermal resistance. However, if the condensate falls or flows from the

Heat Transfer

surface (as it would from a horizontal tube), the condensate also removes thermal energy from the surface. Film coefficients are relatively high (e.g., in the order of 2000 to 4000 Btu/hr-ft²-°F (11.5 to 22.7 kW/m²·K)).[15]

Filmwise condensation occurs when the condensing surface is smooth and free from impurities.[16] A continuous film of condensate covers the entire surface. The film flows smoothly down over the surface under the action of gravity and eventually falls off. However, if the surface contains impurities or irregularities that prevent complete wetting, the film will be discontinuous, a condition known as *dropwise condensation*.[17]

Equation 35.27, based on Nusselt's theoretical work, predicts film coefficients for the laminar filmwise condensation of a pure saturated vapor on the outside of a horizontal tube with a diameter between 1 and 3 in (2.5 and 7.6 cm).[18] Equation 35.27 is in fair agreement with experimental data, with calculated values generally being low.

The actual surface temperature is often unknown in initial studies. However, for steam, condensation frequently occurs with a temperature difference of $T_{\text{sat},v} - T_s$ between 5°F and 40°F (3°C and 22°C). For first approximations, the vapor density, ρ_v, can be taken as zero. Proper units must be observed in order to keep the argument of exponentiation unitless.[19]

$$h_c = (0.725)\left[\frac{\rho_l(\rho_l - \rho_v)gh'_{fg}(k_l)^3}{d\mu_l(T_{\text{sat},v} - T_s)}\right]^{\frac{1}{4}}$$

[SI and U.S.]   *35.27*

The fluid (subscript l) properties are evaluated at the film temperature.

$$T_h = \left(\tfrac{1}{2}\right)(T_{\text{sat},v} + T_s)$$ *35.28*

The latent heat of condensation, h_{fg}, corresponding to the steam pressure (not the film temperature), is used for the effective heat of condensation, h'_{fg}, unless there is significant subcooling of the condensate as it flows

over the tube (i.e., the tube is much colder than the saturation temperature). In that case, the effective heat of condensation is given by Eq. 35.29.[20]

$$h'_{fg} = h_{fg} + 0.68c_p(T_{\text{sat},v} - T_s)$$ *35.29*

Superheated vapor has essentially the same film coefficient as saturated vapor. If the vapor is condensing from a superheated condition, or if the quality of the vapor surrounding the cooling surface is less than 100%, Eq. 35.27 is still used with the vapor's saturation temperature corresponding to the system pressure. The vapor's superheated temperature is not used and the effect of superheat is disregarded.

The temperature difference used to calculate the heat transfer is the difference between the vapor's saturation temperature and the surface temperature.

$$Q = h_c A(T_{\text{sat},v} - T_s)$$ *35.30*

In the case of a tube bank with N layers of horizontal tubes arranged vertically over one another, the condensate will drop from one layer to another. A conservative estimate of the film coefficient is given by Eq. 35.31, which assumes that the increased thickness of the film on the lower tubes due to the accumulation of condensate will be partially offset by the increase in liquid agitation (i.e., turbulence).

$$h_c = (0.725)\left[\frac{\rho_l(\rho_l - \rho_v)gh'_{fg}(k_l)^3}{Nd\mu_l(T_{\text{sat},v} - T_s)}\right]^{\frac{1}{4}}$$

[SI and U.S.]   *35.31*

20. CONDENSATION ON VERTICAL AND INCLINED SURFACES

Filmwise condensation of pure saturated vapors on vertical surfaces (including the insides and outsides of tubes) or on flat surfaces inclined at an angle θ from the horizontal is predicted by Eq. 35.32.[21,22,23,24] As with

---

[15]A film coefficient of 2000 Btu/hr-ft²-°F (11.5 kW/m²·K) is routinely assumed as a first estimate for condensation of steam on the outside of tubes.
[16]Filmwise condensation can always be expected with clean steel and aluminum tubes under ordinary conditions, as well as with heavily contaminated tubes. Dropwise condensation generally requires smooth surfaces with minute amounts of contamination, rather than rough surfaces. Since dropwise condensation can be expected only under carefully controlled conditions, the assumption of filmwise condensation is generally warranted.
[17]Film coefficients for dropwise condensation can be 4 to 8 times larger than for filmwise condensation because the film is thinner and the thermal resistance is smaller.
[18]When a noncondensing gas is present simultaneously with the condensing vapor, the gas adds to the thermal resistance. Even small amounts of air (e.g., less than 5% by volume) can reduce the film coefficient significantly (e.g., reductions up to 80%). Noncondensable gases should be avoided if the highest rates of condensation are required.
[19]In particular, μ should be in units of lbm/ft-hr (kg/s·m) and g should be 4.17×10^8 ft/hr² (9.81 m/s²).

[20]Equation 35.29 was derived by Nusselt with a constant of $3/8$. For $\text{Pr} > 0.5$ and $c_p(T_{\text{sat},v} - T_s) < h_{fg}$, the constant 0.68 yields values that are in better agreement with experimental data.
[21]Equation 35.32 cannot be used for condensation on inclined tubes. The film flow is not parallel with the longitudinal axis of an inclined tube, resulting in an effective inclination angle that varies with location along the tube.
[22]Equation 35.32 was derived by Nusselt with a coefficient of 0.943. However, ripples in the laminar film appear at condensation Reynolds numbers (see Sec. 21) as low as 30 or 40. Experimental data show actual film coefficients are approximately 20% higher than the theoretical. A coefficient of 1.13 in place of the 0.943 reflects this increase. Retaining the 0.943 value, however, yields a conservative value.
[23]Some researchers measure their angles from the vertical, in which case the cosine function will be used. It should be noted that $\cos\theta_{\text{vertical}} = \sin\theta_{\text{horizontal}}$.
[24]Equation 35.32 can be used to find the condensing film coefficient on a vertical tube when $\dot{m}/\mu_d < 1020$, where $\dot{m}$ is the total condensation in pounds per hour.

condensation on horizontal surfaces, the latent heat of condensation is evaluated at the vapor temperature, while the remaining fluid properties are evaluated at the film temperature. The characteristic length, L, in Eq. 35.32 is the surface length.

$$h_c = (0.943)\left[\frac{\rho_l(\rho_l - \rho_v)gh'_{fg}(k_l)^3 \sin\theta}{L\mu_l(T_{\text{sat},v} - T_s)}\right]^{\frac{1}{4}}$$

[SI and U.S.] *35.32*

Nusselt showed that the film coefficient for condensation on a vertical flat surface of height L is the same as for a tube of diameter $L/2.86$. This explains the similarity between Eqs. 35.27 and 35.32.

21. TURBULENT CONDENSATION ON VERTICAL PLATES AND TUBES

On a tall vertical surface or on a surface with a large amount of condensate, the condensate on the lower portion of the surface may flow turbulently. This will increase the heat transfer rate.

The condensation will remain laminar as long as the *film Reynolds number* (also known as the *condensation Reynolds number*), Re_h, is less than approximately 1800 for a vertical tube.[25] The variable P in Eq. 35.33 is the wetted perimeter, equal to the width, w, for a vertical plate and πd for a vertical tube with diameter d. The maximum value of Re_h occurs at the lower edge of the condensing surface.

$$\text{Re}_h = \frac{4\dot{m}}{P\mu_h} \qquad 35.33$$

The condensation Reynolds number is sometimes expressed in terms of mass flow per unit depth of plate, Γ, so that[26]

$$\text{Re}_h = \frac{4\Gamma}{\mu_h} \qquad 35.34$$

$$\Gamma = \frac{\dot{m}}{P} \qquad 35.35$$

From a practical standpoint, the condensate generation rate, $\dot{m}$, cannot be calculated until h is known, resulting in an iterative solution procedure.

$$\dot{m} = \rho_l A v = \frac{Q}{h_{fg}}$$

$$= \frac{hA(T_{\text{sat},v} - T_s)}{h_{fg}} \qquad 35.36$$

---

[25] The critical film Reynolds number for horizontal tubes is 3600 because the film flows down two sides. Turbulent condensing flow inside a tube requires a large-diameter tube, and rarely occurs otherwise.

[26] Some references define Re_f as Γ/μ, in which case the critical Reynolds number would be 450 instead of 1800.

The approximate average heat film coefficient for turbulent condensation with $\text{Re}_h > 1800$ is given by Eq. 35.37.

$$h_c = (0.0076)(\text{Re}_h)^{\frac{2}{5}}\left[\frac{\rho_l(\rho_l - \rho_v)g(k_l)^3}{(\mu_l)^2}\right]^{\frac{1}{3}}$$

[SI and U.S.] *35.37*

Generally, the assumption of a laminar film should be checked with Eq. 35.37 for every condensation problem.

22. CONDENSATION ON THE OUTSIDE OF A SPHERE

Equation 35.38 can be used to calculate the film coefficient on the exterior of an isothermal sphere.

$$h_c = (0.815)\left[\frac{\rho_l(\rho_l - \rho_v)gh'_{fg}(k_l)^3}{d\mu_l(T_{\text{sat},v} - T_s)}\right]^{\frac{1}{4}}$$

[SI and U.S.] *35.38*

23. CONDENSATION INSIDE TUBES

Equation 35.32 (or 35.37 in the turbulent case) can be used for condensation inside vertical tubes. Condensation inside horizontal tubes is complicated by the issue of condensate removal. Since the condensate specific volume is much less than the specific volume of the vapor, the accumulation will not occlude much of the wall surface. With a large enough diameter pipe and/or the assumption of vapor traps to remove the condensate, use of Eq. 35.32 is justified.[27]

If condensate is forced rapidly through the tube by a pump, then the film coefficient will be essentially the same as for forced convection. For low vapor velocities inside tubes, Eq. 35.32 can be used with a leading coefficient of 0.612 for condensing steam (the *Kern correlation*) and 0.555 for condensing refrigerants.

Example 35.2

A brass tube with 1 in (2.54 cm) outside and $^7/_8$ in (2.22 cm) inside diameters is surrounded by 7.5 psia (50 kPa). The wall temperature is $60°F$ ($15°C$). What is the film coefficient on the outside of the tube?

SI Solution

The saturation temperature for 50 kPa steam is $81°C$.

$$T_{\text{sat},v} = 81°C$$

The film properties are evaluated at the average of the wall and saturation temperatures.

$$T_h = \left(\tfrac{1}{2}\right)(T_{\text{sat},v} + T_s)$$
$$= \left(\tfrac{1}{2}\right)(81°C + 15°C) = 48°C \quad [\text{say } 50°C]$$

---

[27] A value of 1200 Btu/hr-ft²-°F (6800 W/m²·K) is routinely used for condensation of steam inside radiators and fan coils.

Film properties are obtained (by interpolation) from App. 35.B for liquid water.

$$k_{50°C} = 0.6435 \text{ W/m·K}$$

$$\mu_{50°C} = 5.72 \times 10^{-4} \text{ kg/s·m}$$

$$\rho_{l,50°C} = 989.1 \text{ kg/m}^3$$

$$\rho_{v,50°C} \approx 0$$

$$h_{fg,50 \text{ kPa}} = 2388.8 \text{ kJ/kg} = 2.3888 \times 10^6 \text{ J/kg}$$

$$d = 0.0254 \text{ m}$$

$$g = 9.81 \text{ m/s}^2$$

From Eq. 35.27, the film coefficient is

$$h_c = (0.725) \left[\frac{\rho_l(\rho_l - \rho_v)gh'_{fg}(k_l)^3}{d\mu_l(T_{\text{sat},v} - T_s)} \right]^{\frac{1}{4}}$$

$$= (0.725) \left[\frac{ \begin{array}{c} \left(989.1 \frac{\text{kg}}{\text{m}^3}\right)\left(989.1 \frac{\text{kg}}{\text{m}^3} - 0\right) \\ \times \left(9.81 \frac{\text{m}}{\text{s}^2}\right)\left(2.3888 \times 10^6 \frac{\text{J}}{\text{kg}}\right) \\ \times \left(0.6435 \frac{\text{W}}{\text{m·K}}\right)^3 \end{array} }{ \begin{array}{c} (0.0254 \text{ m})\left(5.72 \times 10^{-4} \frac{\text{kg}}{\text{s·m}}\right) \\ \times (81°C - 15°C) \end{array} } \right]^{\frac{1}{4}}$$

$$= 6477 \text{ W/m}^2\text{·K}$$

Customary U.S. Solution

The saturation temperature for 7.5 psia steam is 180°F.

$$T_{\text{sat},v} = 180°F$$

The film properties are evaluated at the average of the wall and saturation temperatures.

$$T_h = \left(\tfrac{1}{2}\right)(T_{\text{sat},v} + T_s)$$
$$= \left(\tfrac{1}{2}\right)(180°F + 60°F) = 120°F$$

Film properties are obtained from App. 35.A for liquid water.

$$k_{120°F} = 0.372 \text{ Btu/hr-ft-°F}$$

$$\mu_{120°F} = \left(0.392 \times 10^{-3} \frac{\text{lbm}}{\text{sec-ft}}\right)\left(3600 \frac{\text{sec}}{\text{hr}}\right)$$
$$= 1.41 \text{ lbm/hr-ft}$$

$$\rho_{l,120°F} = \frac{1}{v_{1,120°F}} = \frac{1}{0.01620 \frac{\text{ft}^3}{\text{lbm}}}$$
$$= 61.73 \text{ lbm/ft}^3$$

$$\rho_{v,120°F} = \frac{1}{v_{v,120°F}} = \frac{1}{203.27 \frac{\text{ft}^3}{\text{lbm}}}$$
$$\approx 0$$

$$h_{fg,7.5 \text{ psia}} = 990.2 \text{ Btu/lbm}$$

$$d = \frac{1 \text{ in}}{12 \frac{\text{in}}{\text{ft}}} = 0.0833 \text{ ft}$$

$$g = \left(32.2 \frac{\text{ft}}{\text{sec}^2}\right)\left(3600 \frac{\text{sec}}{\text{hr}^2}\right)^2$$
$$= 4.17 \times 10^8 \text{ ft/hr}^2$$

From Eq. 35.27, the film coefficient is

$$h_c = (0.725) \left[\frac{\rho_l(\rho_l - \rho_v)gh'_{fg}(k_l)^3}{d\mu_l(T_{\text{sat},v} - T_s)} \right]^{\frac{1}{4}}$$

$$= (0.725) \left[\frac{ \begin{array}{c} \left(61.73 \frac{\text{lbm}}{\text{ft}^3}\right)\left(61.73 \frac{\text{lbm}}{\text{ft}^3} - 0\right) \\ \times \left(4.17 \times 10^8 \frac{\text{ft}}{\text{hr}^2}\right)\left(990.2 \frac{\text{Btu}}{\text{lbm}}\right) \\ \times \left(0.372 \frac{\text{Btu}}{\text{hr-ft-°F}}\right)^3 \end{array} }{ \begin{array}{c} (0.0833 \text{ ft})\left(1.41 \frac{\text{lbm}}{\text{hr-ft}}\right) \\ \times (180°F - 60°F) \end{array} } \right]^{\frac{1}{4}}$$

$$= 1123 \text{ Btu/hr-ft}^2\text{-°F}$$

24. EVAPORATION FROM HORIZONTAL TUBES

If a liquid is vaporizing from a heated surface, the change of phase can occur in three distinctly different ways: pool, nucleate, and film boiling. *Pool boiling* occurs when the temperature of the heating element is near the vaporization (boiling or saturation) temperature. There will be little or no bubble formation and liquid agitation (rolling). Heat transfer will be essentially convective in nature.

As the temperature of the heating element exceeds the vaporization temperature, vapor bubbles begin to form, a mechanism known as *nucleate boiling*. Equations for determining the film coefficient for nucleate boiling are complex, and correlations typically depend on the heat flux, $q = Q/A$.

When the surface temperature is much greater (i.e., approximately 200°F (93°C) or more) than the vaporization temperature, a film of vapor will cover the surface of the heating element. This is known as *film boiling*.

The *Bromley equation*, Eq. 35.39, can be used for laminar film boiling around horizontal tubes with diameters up to approximately $\frac{1}{2}$ in (1.2 cm).[28] This makes the equation useful for liquids that are heated by commercial electrical heating rods. No forced movement of the fluid across the heating element is permitted. Agitation is solely the result of boiling. All properties should be evaluated at the saturation temperature corresponding to the vapor pressure.

$$h_b = (0.62) \left[\frac{\rho_v(\rho_l - \rho_v)g[h_{fg} + 0.4c_{p,v} \times (T_s - T_{\text{sat},v})](k_v)^3}{d\mu_v(T_s - T_{\text{sat},v})} \right]^{\frac{1}{4}}$$

[SI and U.S.] *35.39*

25. EVAPORATION FROM FLAT PLATES

Most methods for calculating the film coefficient for evaporation from flat plates are unwieldy, and graphical methods are often used. (See Fig. 35.2.)

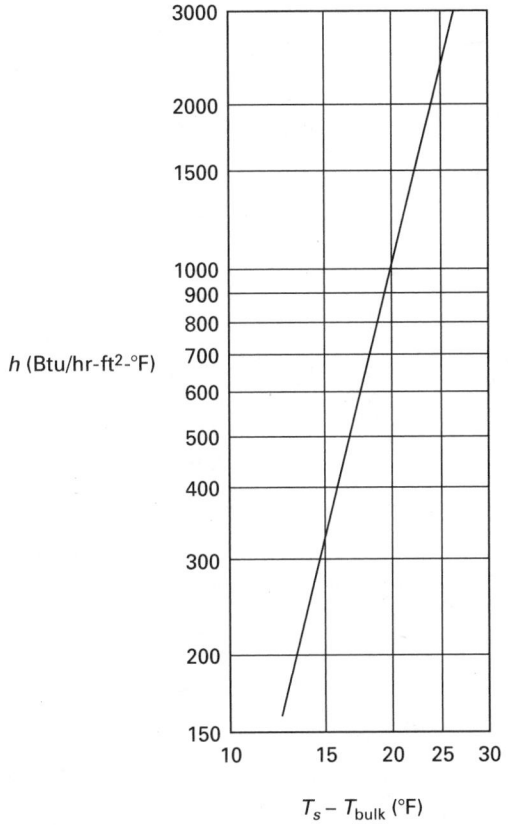

h (Btu/hr-ft²-°F)

$T_s - T_{\text{bulk}}$ (°F)

Figure 35.2 *Film Coefficient for Evaporating Water on Horizontal and Vertical Plates (atmospheric pressure)*

---

[28]The flow is assumed to be laminar when the flow path around the tube is short.

26. SIMPLIFIED VAPORIZATION EQUATIONS FOR WATER

Many researchers have developed empirical equations for calculating the film coefficient for boiling water. These are of the form of Eq. 35.40 and are correlated with heat flux, $q = Q/A$. Values of the coefficient C and exponent n for water boiling at atmospheric pressure are listed in Table 35.4. Equation 35.41 is used to correct the film coefficients derived from Eq. 35.40 for other pressures.

$$h_0 = C(T_s - T_{\text{sat},v})^n \quad \text{[SI]} \qquad 35.40$$

$$h_p = h_0 \left(\frac{p_{\text{actual}}}{p_{\text{std. atmosphere}}} \right)^{\frac{2}{5}} \qquad 35.41$$

Table 35.4 *Simplified Evaporation Constants for Water (atmospheric pressure; SI units only)*

orientation	Q/A (kW/m²)	C	n
horizontal surface (in wide vessel)	< 16	1042	$\frac{1}{3}$
	16 to 240	5.56	3
vertical surface (in wide vessel)	< 3	537	$\frac{1}{7}$
	3 to 63	7.96	3

(Multiply kW/m² by 317 to obtain Btu/hr-ft².)
(Multiply Btu/hr-ft²-°F by 5.6783 to obtain W/m²·K.)

27. EFFECTS OF RADIATION ON EVAPORATION FILM COEFFICIENTS

Since the temperature of a heating element is likely to be high, the effects of radiation must be considered. The combined heat transfer is more than the sum of the convective and radiative components. Radiation increases the film thickness, reducing the values of the film coefficient and convective heat transfer. If the radiation film coefficient, h_r, is known, the total film coefficient can be found by solving Eq. 35.42 iteratively.

$$h_{\text{total}} = h_b \left(\frac{h_b}{h_{\text{total}}} \right)^{\frac{1}{3}} + h_r \qquad 35.42$$

The total heat transfer in film boiling is

$$Q = qA = h_{\text{total}} A(T_s - T_{\text{sat},v}) \qquad 35.43$$

PRACTICE PROBLEMS

1. What is the density of 87% wet steam at 50 psia (350 kPa)?

2. What is the viscosity of 100°F (38°C) water in units of lbm/hr-ft (kg/s·m)?

3. A fluid in a tank is maintained at 85°F (29°C) by a copper tube carrying hot water. The water decreases in temperature from 190°F (88°C) to 160°F (71°C) as it flows through the tube. At what temperature should the fluid's film coefficient be evaluated?

4. A bare, horizontal conductor with circular cross section with an outside diameter of 0.6 in (1.5 cm) dissipates 8.0 W/ft (25 W/m). The conductor is cooled by free convection, and the surrounding air temperature is 60°F (15°C). Assume the film temperature is 100°F (38°C). What is the conductor's surface temperature?

36

Forced Convection and Heat Exchangers

1. Introduction	36-2
2. Heat Transfer by Forced Convection	36-2
3. Dimensionless Numbers	36-2
4. Dimensionless Number Ranges	36-2
5. Bulk Temperature	36-3
6. Film Temperature	36-3
7. Flow Over Flat Plates	36-3
8. Laminar Flow Over Isothermal Flat Plates	36-4
9. Laminar Flow Over Flat Plates with Constant Heat Flux	36-4
10. Turbulent Flow Over Isothermal Flat Plates	36-4
11. Laminar Flow Inside Tubes with Constant Wall Temperature	36-5
12. Laminar Flow Inside Tubes with Constant Heat Flux	36-5
13. Turbulent Flow Inside Straight Tubes	36-5
14. Turbulent Flow Inside Coiled Tubes	36-6
15. Turbulent Air Flow in Tubes	36-7
16. Turbulent Water Flow in Tubes	36-7
17. Turbulent Oil Flow in Tubes	36-7
18. Turbulent Liquid Metal Flow in Tubes	36-7
19. Flow Through Noncircular Ducts	36-8
20. Crossflow Over a Single Cylinder	36-8
21. Crossflow Over Tube Bundles	36-9
22. Pressure Drop Across Tube Bundles	36-9
23. Flow Over Spheres	36-10
24. Change of Phase in Tubes with Forced Convection	36-10
25. Heat Transfer in Packed Beds	36-10
26. Heat Exchangers	36-10
27. Heat Exchanger Designations	36-11
28. Commercial Heat Exchangers	36-11
29. Transverse Baffles in Heat Exchangers	36-13
30. Temperature Difference Terminology	36-13
31. Temperature Cross	36-14
32. Logarithmic Temperature Difference	36-14
33. Rules of Thumb for Cooling Water	36-14
34. Heat Transfer in Heat Exchangers	36-14
35. Overall Heat Transfer Coefficient	36-15
36. Tube Length Required	36-15
37. Fouling	36-17
38. Cooling Superheated Steam	36-20
39. NTU Method	36-20
40. Heat Transfer Fluids	36-22
41. Control of Process Heat Exchange Operations	36-23
42. Cooling Electronic Enclosures	36-23
Practice Problems	36-24

Nomenclature

A	area	ft^2	m^2
c_p	specific heat	Btu/lbm-°F	J/kg·K
C	a constant or coefficient	–	–
C	thermal capacity rate	Btu/hr-°F	W/K
d	diameter	ft	m
f	friction factor	–	–
F	factor	–	–
G	mass flow rate per unit area	lbm/hr-ft^2	kg/s·m^2
Gz	Graetz number	–	–
h	film coefficient	Btu/hr-ft^2-°F	W/m^2·K
HTEF	heat transfer efficiency factor	–	–
k	thermal conductivity	Btu-ft/hr-ft^2-°F	W/m·K
L	length	ft	m
m	exponent	–	–
m	mass	lbm	kg
$\dot{m}$	mass flow rate	lbm/hr	kg/s
M	number of tube rows per layer	–	–
n	exponent	–	–
N	number of tube layers	–	–
Nu	Nusselt number	–	–
p	pressure	lbf/ft^2	kPa
P	power	hp	kW
Pe	Peclet number	–	–
Pr	Prandtl number	–	–
q	heat transfer per unit area	Btu/hr-ft^2	W/m^2
Q	heat transfer rate	Btu/hr	W
r	radius	ft	m
R	thermal resistance	hr-ft^2-°F/Btu	m^2·K/W
Ra	Rayleigh number	–	–
Re	Reynolds number	–	–
s	spacing or side length	ft	m
St	Stanton number	–	–
T	temperature[a,b]	°F	K
U	overall coefficient of heat transfer	Btu/hr-ft^2-°F	W/m^2·K
v	velocity	ft/hr	m/s
$\dot{V}$	volumetric flow rate	ft^3/hr	m^3/s
x	distance x	ft	m

Symbols

α	thermal diffusivity	ft^2/sec	m^2/s
$\mathcal{E}$	effectiveness	–	–
η	efficiency	–	–
μ	viscosity[c,d]	lbm/hr-ft	kg/s·m
ν	kinematic viscosity	ft^2/sec	m^2/s
ρ	mass density	lbm/ft^3	kg/m^3

Heat Transfer

Subscripts

0	initial
a	atmospheric
A	at end A
ave	average
b	bulk
B	at end B
c	correction
d	based on diameter
D	drag
f	fouling or friction
G	constant mass flow rate
H	hydraulic
i	inside
l	liquid or longitudinal
L	over length L
lm	log mean
m	mean
max	maximum
min	minimum
o	outside
s	surface
t	transverse or at time t
T	temperature
V	constant volumetric flow rate
x	at point x
∞	free-stream (far field) or at time $= \infty$

[a] The symbol θ is used for temperature in some books.

[b] It is common in heat exchanger literature to use lowercase t as the cold side temperature. This eliminates the requirement for "cold" and "hot" designations.

[c] The use of mass units in viscosity values is typical in the subject of convective heat transfer.

[d] Most data compilations give fluid viscosity in units of seconds. In the United States, heat transfer is traditionally stated on a per hour basis. Therefore, a conversion factor of 3600 sec/hr is needed when calculating dimensionless numbers from table data. The combination of units kg/s·m is the same as a Pa·s and N/m²·s.

1. INTRODUCTION

As with natural convection, *forced convection* depends on the movement of a fluid to remove heat from a surface. With forced convection, a fan, a pump, or relative motion causes the fluid motion. If the flow is over a flat surface, the fluid particles near the surface will flow more slowly due to friction with the surface. The *boundary layer* of slow-moving particles comprises the major thermal resistance. The thermal resistance of the tube and other heat exchanger components is often disregarded.

2. HEAT TRANSFER BY FORCED CONVECTION

Newton's law of convection, Eq. 36.1, gives the heat transfer for Newtonian fluids in forced convection over exterior surfaces.[1,2] The film coefficient, h, is also known as the *coefficient of forced convection*. T_∞ is the *free-stream temperature*.

---

[1] Newton's law of convection is the same for natural and forced convection. Only the methods used to evaluate the film coefficient are different.

[2] The results of this chapter do not generally apply to non-Newtonian fluids.

$$Q = qA = hA(T_s - T_\infty) \qquad 36.1$$

For flow within a tube, the more easily determined *bulk temperature* (Sec. 5) is used in place of the free-stream temperature.

$$Q = qA = hA(T_s - T_b) \qquad 36.2$$

3. DIMENSIONLESS NUMBERS

The Nusselt number, Nu, Prandtl number, Pr, and Reynolds number, Re, are

$$\text{Nu} = \frac{hd}{k} \qquad 36.3$$

$$\text{Pr} = \frac{c_p \mu}{k} = \frac{\nu}{\alpha} \qquad 36.4$$

$$\text{Re} = \frac{\text{v}d}{\nu} = \frac{dG}{\mu} \qquad 36.5$$

$$\text{G} = v_\infty \rho_\infty \qquad 36.6$$

The density and velocity used in Eq. 36.6 must correspond to the same point. In the most common case, both are free-stream values. It would be incorrect to use the free-stream velocity with the density evaluated at the film temperature.

The *Peclet number*, Pe, is the product of the Reynolds and Prandtl numbers.

$$\text{Pe} = \text{RePr} = \frac{d\text{v}\rho c_p}{k} \qquad 36.7$$

The *Graetz number*, Gz, is used in the reporting of empirical data for laminar flow in tubes.

$$\text{Gz} = \text{Re}_d \text{Pr} \left(\frac{d}{L} \right) \qquad 36.8$$

The *Stanton number*, St, is encountered in correlations of fluid friction and heat transfer.

$$\text{St} = \frac{\text{Nu}}{\text{RePr}} = \frac{h}{\text{v}\rho c_p} \qquad 36.9$$

4. DIMENSIONLESS NUMBER RANGES

All heat transfer correlations have associated ranges of the Prandtl and Reynolds numbers, whether stated or not. The endpoints of these ranges are indistinct and depend on the fluid properties. For example, Eq. 36.33 can be used with a Reynolds number as low as 2100 as long as the Prandtl number is less than 10 (i.e., it cannot be used at that lower limit for fluids with viscosities more than twice that of water). Therefore, the lower limit is established as 10,000 instead of 2100, and the equation is deemed applicable to all fluids. This explains why researchers report different ranges for the same correlation.

5. BULK TEMPERATURE

The *bulk temperature*, T_b, also known as the *mixing temperature*, is the energy-average fluid temperature. The bulk temperature concept is usually encountered with tube flow where there is no free-stream temperature. The centerline temperature is a candidate for theoretical considerations, but it cannot be easily measured. Therefore, the bulk temperature is used to calculate the heat transfer for flow in tubes.

The bulk temperature used to calculate local properties is evaluated over the tube cross-sectional area at the point along the tube length being considered. The bulk temperature used in the calculation of an average film coefficient over the entire length of a tube or heat exchanger is evaluated as the average of the entrance and exit temperatures. For this reason, it is often referred to as the *mean bulk temperature*.

$$T_b = \left(\tfrac{1}{2}\right)\left(T_{\text{in}} + T_{\text{out}}\right) \qquad \textit{36.10}$$

The bulk temperature should be used to calculate the film coefficient of a fluid flowing in a tube or heat exchanger unless the properties change a lot (i.e., as they do with high-viscosity fluids).[3] It may be necessary to solve heat transfer equations iteratively in order to determine the bulk temperature, since the film coefficient (based on the bulk temperature) is needed in order to determine the outlet temperature.

The mass flow rate in a tube is constant everywhere. Where there are large variations in temperature along the length of a tube, the density, velocity, and temperature must all be consistent. It would be incorrect to use a density evaluated at the bulk temperature with an entrance velocity.

6. FILM TEMPERATURE

As with natural convection, the free-field temperature, T_∞, is used to calculate the film temperature in external flow configurations.

$$T_{\text{film}} = \left(\tfrac{1}{2}\right)\left(T_s + T_\infty\right) \qquad \textit{36.11}$$

Film properties inside tubes are evaluated at the average of the surface temperature, T_s, and the bulk temperature, T_b. When there is a variation of the surface temperature, as there could be along the length of a tube used for heat transfer, the surface temperature is assumed to be the temperature at midlength along the tube.

$$T_{\text{film}} = \left(\tfrac{1}{2}\right)\left(T_s + T_b\right) \qquad \textit{36.12}$$

7. FLOW OVER FLAT PLATES

The boundary layer of a fluid flowing over a flat plate is assumed to have a parabolic velocity distribution.[4] The layer has three distinct regions: laminar, transition, and turbulent. From the leading edge, the layer is laminar and the thickness increases gradually until the transition region where the thickness increases dramatically. Thereafter, the boundary layer is turbulent. The laminar region is always present, though its length decreases as velocity increases. Turbulent flow may not develop at all with short plates.

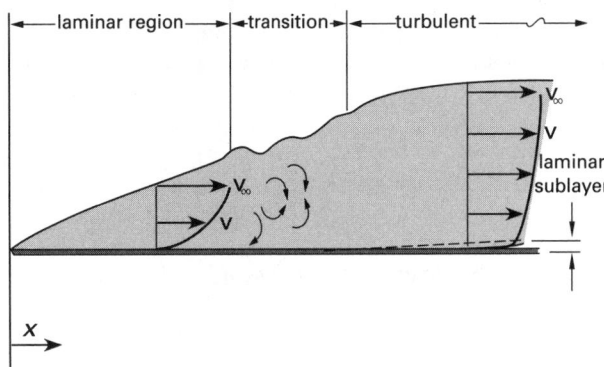

Figure 36.1 *Flow Over a Flat Plate*

The Reynolds number is used to determine which of the three flow regimes is applicable. Laminar flow on smooth flat plates occurs for Reynolds numbers up to approximately 2×10^5; turbulent flow exists for Reynolds numbers greater than approximately 3×10^6.[5] Transition flow is in between. The distance from the leading edge at which turbulent flow is initially experienced is determined from the *critical Reynolds number*, commonly taken as $\text{Re} = 5 \times 10^5$ for smooth flat plates, though the actual value is highly dependent on surface roughness. Distance, x, is measured from the leading edge.

$$\text{Re}_x = \frac{v_\infty x \rho}{\mu} = \frac{v_\infty x}{\nu} \qquad \textit{36.13}$$

The heat transfer from a flat plate is

$$Q = h_{\text{ave}} A \left(T_s - T_\infty\right) \qquad \textit{36.14}$$

---

[3]In that case, evaluate the film coefficient at the inlet and outlet and take the logarithmic mean of the two. This requires calculating the film coefficients twice.

[4]The velocity distribution does not have to be parabolic. In *Couette flow*, there are two closely spaced parallel surfaces, one which is stationary and the other moving with constant velocity. The velocity gradient is assumed to be linear between the plates.
[5]Turbulent flow can begin at Reynolds numbers less than 3×10^5 if the plate is rough. This discussion assumes the plate is smooth.

8. LAMINAR FLOW OVER ISOTHERMAL FLAT PLATES

The *Pohlhausen solution* is theoretically exact and can be used to determine the local thermal film coefficient at a distance x from the leading edge of an isothermal flat plate in laminar flow.[6] Fluid properties are evaluated at the film temperature.

$$\text{Nu}_{x,\text{local}} = (0.332)(\text{Re}_x)^{\frac{1}{2}}\text{Pr}^{\frac{1}{3}}$$
$$[0.6 < \text{Pr} < 1; \text{Re}_x < 2 \times 10^5]^{7} \quad \textit{36.15}$$

The *Blasius solution* is exact and can be derived from the Pohlhausen correction by setting $(\text{Pr})^{\frac{1}{3}} = 1.0$.[8] The Blasius solution is useful with water and gases.[9] The fluid properties are evaluated at the film temperature.

$$\text{Nu}_{x,\text{local}} = (0.332)(\text{Re}_x)^{\frac{1}{2}}$$
$$[0.6 < \text{Pr} < 50; \text{Re}_x < 2 \times 10^5]^{10} \quad \textit{36.16}$$

The local *skin-friction coefficient*, $C_{f,x}$, for laminar flow a distance x from the leading edge is[11,12]

$$C_{f,x} = \frac{0.664}{(\text{Re}_x)^{\frac{1}{2}}} \quad [\text{Re}_x < 2 \times 10^5]^{13} \quad \textit{36.17}$$

---

[6]The x in the subscripts does not mean "at point x." Rather, it means that the value of x is used as the characteristic length in the dimensionless number.

[7]Some authorities: $[\text{Re} < 5 \times 10^5]$.

[8]The Blasius solution can also be derived from dimensional analysis.

[9]When $\text{Pr} = 1$, the velocity profile (i.e., the hydrodynamic boundary layer) of the boundary layer for laminar fluid flow is the same as the temperature profile (i.e., the thermal boundary layer) for laminar convective heat transfer. This is essentially true for water, and approximately true for most gases that have Prandtl numbers in the 0.6 to 1.0 range. (Most gases have Prandtl numbers in the 0.65 to 0.84 range. These numbers raised to fractional powers result in a value very close to 1.0.) For liquids, however, the Prandtl number ranges from very small (liquid metals) to very large (viscous oils).

[10]See Ftn. 7.

[11]The *skin-friction coefficient* is also known as the *Fanning friction factor*. The Darcy friction factor, more commonly used in fluid flow problems, is

$$f_{\text{Darcy}} = 4f_{\text{Fanning}}$$

[12]The *drag coefficient* for submerged bodies is the sum of the *skin-friction coefficient* and the *profile drag coefficient*. For flat plates, there is no profile drag. Therefore, the drag coefficient is the same as the skin-friction coefficient, and the drag per unit area is

$$\frac{\text{drag}}{A} = \frac{C_{f,x}\rho(\text{v}_\infty)^2}{2g_c}$$
$$\text{drag} = A_{\text{surface}}C_{f,\text{ave}}$$

[13]See Ftn. 7.

The average film coefficient over a distance L (in the direction of flow) from the leading edge can be calculated from Eq. 36.18. Fluid properties are evaluated at the film temperature.

$$\text{Nu}_{L,\text{ave}} = (0.664)(\text{Re}_L)^{\frac{1}{2}}\text{Pr}^{\frac{1}{3}}$$
$$[\text{Pr} > 0.6; \text{Re}_L < 2 \times 10^5]^{14,15,16} \quad \textit{36.18}$$

$$\text{Re}_L = \frac{\text{v}_\infty L}{\nu} \quad \textit{36.19}$$

The average skin-friction coefficient is

$$C_{f,\text{ave}} = \frac{1.328}{(\text{Re}_L)^{\frac{1}{2}}} \quad \textit{36.20}$$

9. LAMINAR FLOW OVER FLAT PLATES WITH CONSTANT HEAT FLUX

Equation 36.21 can be used to determine the local thermal film coefficient at a distance x from the leading edge for a fluid in laminar flow over a flat plate with constant heat flux (i.e., constant heat transfer). Fluid properties are evaluated at the film temperature.

$$\text{Nu}_{x,\text{local}} = (0.453)(\text{Re}_x)^{\frac{1}{2}}\text{Pr}^{\frac{1}{3}}$$
$$[0.6 < \text{Pr} < 50; \text{Re}_x < 5 \times 10^5] \quad \textit{36.21}$$

10. TURBULENT FLOW OVER ISOTHERMAL FLAT PLATES

For completely turbulent flow ($\text{Pr} > 0.7$; $3 \times 10^6 < \text{Re} < 10^8$), the local and average film coefficients are given by Eqs. 36.22 and 36.23, respectively. All of the fluid properties are evaluated at the film temperature.

$$\text{Nu}_{x,\text{local}} = (0.0288)(\text{Re}_x)^{0.8}\text{Pr}^{\frac{1}{3}}$$
$$[0.6 < \text{Pr} < 60; 3 \times 10^6 < \text{Re}_x < 10^8]^{17} \quad \textit{36.22}$$

Equation 36.23 is applicable for the entire plate and includes the heat transfer contribution of the leading laminar portion of flow. This is sometimes called a *laminar-turbulent average*.

$$\text{Nu}_{L,\text{ave}} = (0.0366\text{Re}_L - 871)^{0.8}\text{Pr}^{\frac{1}{3}}$$
$$\left[\begin{array}{c} 0.6 < \text{Pr} < 60; 5 \times 10^5 \\ < \text{Re}_x < 10^7; \text{Re}_L < 10^8 \end{array}\right]^{18,19,20,21} \quad \textit{36.23}$$

---

[14]See Ftn. 7.

[15]Some authorities: $[\text{Re}_L < 3 \times 10^5]$.

[16]Some authorities: $[\text{Pr} > 0.7]$.

[17]Some authorities report an empirical coefficient of 0.0296 instead of the theoretical 0.0288.

[18]See Ftn. 15.

[19]See Ftn. 16.

[20]Some authorities report the coefficient as 0.036 or 0.037.

[21]Some authorities report the 871 correction as 850 or 23,200, or they omit it all together.

The local and average skin-friction coefficients for turbulent flow are

$$C_{f,x} = \frac{0.0576}{(\text{Re}_x)^{\frac{1}{5}}} \qquad \qquad 36.24$$

$$C_{f,\text{ave}} = \frac{0.072}{(\text{Re}_L)^{\frac{1}{5}}} \qquad \qquad 36.25$$

Equations 36.24 and 36.25 do not include the skin-friction for the laminar section. The total friction from the leading edge to point x is obtained by subtracting the turbulent drag for the critical length and adding the laminar drag for the entire section (including the laminar section).

$$C_{f,\text{total}} = \frac{0.072}{(\text{Re}_L)^{\frac{1}{5}}} - (0.072)\left(\frac{x}{L}\right)(\text{Re}_{x,\text{critical}})^{\frac{1}{5}}$$
$$+ \frac{(1.328)\left(\dfrac{x_{\text{critical}}}{L}\right)}{(\text{Re})^{\frac{1}{2}}} \qquad 36.26$$

For a critical Reynolds number of 5×10^5, this simplifies to

$$C_{f,\text{total}} = \frac{0.072}{(\text{Re}_L)^{\frac{1}{5}}} - (0.00334)\left(\frac{x_{\text{critical}}}{L}\right) \qquad 36.27$$

11. LAMINAR FLOW INSIDE TUBES WITH CONSTANT WALL TEMPERATURE

Laminar flow in smooth tubes occurs at Reynolds numbers less than 2000. As with flow over a flat plate, the velocity distribution is parabolic, but the extent of the parabola is limited to the tube radius. In the *entrance region*, the parabola does not extend to the centerline. Further on, a point is reached where the parabolic distribution is complete, and the flow is said to be *fully developed* laminar flow.[22] At that point, the average velocity is one-half of the maximum (centerline) velocity.

For laminar flow inside a tube with constant wall temperature, the film coefficient decreases with distance from the entrance and approaches the fully developed laminar value given by Eq. 36.28. (All fluid properties are evaluated at the bulk temperature.)

$$\text{Nu}_d = 3.658 \quad [\text{Re} < 2000; \text{Pr} > 0.6]^{23} \qquad 36.28$$

The tube length required to establish fully developed laminar film is known as the *thermal entry length*.

$$x_{\text{thermal entry}} = 0.05 d_H \text{Re}_{d_H} \text{Pr} \qquad 36.29$$

The *Seider-Tate correlation* predicts the average film coefficient along the entire length of laminar flow. All of the fluid properties are evaluated at the bulk temperature except μ_s, which is evaluated at the surface (wall) temperature.

$$\text{Nu}_{d,\text{ave}} = (1.86)\left[\text{Re}_d \text{Pr}\left(\frac{d}{L}\right)\right]^{\frac{1}{3}}\left(\frac{\mu}{\mu_s}\right)^{0.14}$$

$$\left[\begin{array}{l} \text{Pr} > 0.48;\ \text{Re}_d < 2000;\ 0.0044 < \dfrac{\mu}{\mu_s} \\[2mm] < 9.75;\ \left[\text{Re}_d\text{Pr}\left(\dfrac{d}{L}\right)\right]^{\frac{1}{3}}\left(\dfrac{\mu}{\mu_s}\right)^{0.14} > 2 \end{array}\right]$$
$$36.30$$

12. LAMINAR FLOW INSIDE TUBES WITH CONSTANT HEAT FLUX

Some tubes experience a constant *heat flux* (i.e., a constant heat transfer) per unit length (area).[24] The film coefficient decreases with distance from the tube entrance. The laminar flow is *fully developed* when the difference between the surface (wall) and mean fluid temperature is constant. This essentially occurs for when $(x/r_i)/\text{Re}_x\text{Pr} > 0.100$. As Table 36.1 indicates, the fully developed laminar Nusselt number is 4.364.

$$\text{Nu}_d = 4.364 \quad [\text{Re}_d < 2000;\ \text{Pr} > 0.6] \qquad 36.31$$

Table 36.1 Nusselt Numbers for Laminar Flow Inside Tubes with Constant Heat Flux

$\dfrac{x/r}{\text{Re}_d\text{Pr}}$	Nu_d
0	∞
0.002	12.00
0.004	9.93
0.010	7.49
0.020	6.14
0.040	5.19
0.100	4.51
∞	4.364

13. TURBULENT FLOW INSIDE STRAIGHT TUBES

The theoretical *Nusselt equation* can be used to find the inside film coefficient for turbulent flow inside round, horizontal tubes.[25] All fluid properties except specific

---

[22]The term *fully developed* is also used when referring to full turbulence. In this section it is understood that the flow is laminar.
[23]Some authorities report the value as 3.656.

[24]An example of a tube with constant heat flux is a pipe that has been uniformly wrapped with an electric heat strip along the pipe length.
[25]Equation 36.32 can also be used to obtain conservative values for turbulent flow in vertical tubes.

Heat Transfer

heat are evaluated at the film temperature. Specific heat is evaluated at the bulk temperature.

$$\mathrm{Nu} = (0.0225)(\mathrm{Re}^{0.8}\mathrm{Pr}^{\frac{1}{3}})$$
$$\left[0.6 < \mathrm{Pr} < 160; \ \mathrm{Re} > 10^4; \ \frac{L}{d} > 60\right]^{26}$$

36.32

Equation 36.32 is difficult to use in design work because the film temperature is an inconvenient concept with tube flow. With tube flow, it is more common to base all fluid properties on the bulk temperature. The *Dittus-Boelter equation*, as modified by W. H. McAdams, evaluates all fluid properties at the bulk temperature.

$$\mathrm{Nu} = (0.023)(\mathrm{Re}^{0.8}\mathrm{Pr}^n)$$
$$\left[0.7 < \mathrm{Pr} < 120; \ \mathrm{Re} > 10^4; \ \frac{L}{d} > 60\right]^{27}$$

36.33

The exponent, n, in Eq. 36.33 has a value of 0.3 when the surface (wall) temperature is less than the bulk fluid temperature, and n is 0.4 when the surface (wall) temperature is greater than the bulk fluid temperature.

Within the normal range of most gases, $\mathrm{Pr}^n \approx 1.0$, resulting in Eq. 36.34.

$$\mathrm{Nu} = (0.023)(\mathrm{Re}^{0.8}) \qquad \textit{36.34}$$

If there is a large change in viscosity during the heat transfer process, as there would be with oils and other viscous fluids heated in a long tube, Eq. 36.34 is modified into the *Seider-Tate equation* for turbulent flow. All fluid properties in Eq. 36.35 are evaluated at the bulk temperature except for μ_s, which is evaluated at the surface temperature.

$$\mathrm{Nu} = (0.023)\left[(\mathrm{Re})^{0.8}\mathrm{Pr}^{\frac{1}{3}}\right]\left(\frac{\mu}{\mu_s}\right)^{0.14}$$
$$\left[0.7 < \mathrm{Pr} < 160; \ \mathrm{Re} > 10^4; \ \frac{L}{d} > 60\right]^{28}$$

36.35

For L/d ratios less than 60 in pipes with sharp leading edges, the righthand side of Eq. 36.35 can be multiplied by either Eq. 36.36 or 36.37.

$$1 + \left(\frac{d}{L}\right)^{0.7} \qquad \left[2 < \frac{L}{d} < 20\right] \qquad \textit{36.36}$$

$$1 + \left(\frac{6d}{L}\right) \qquad \left[20 < \frac{L}{d} < 60\right] \qquad \textit{36.37}$$

---

[26]Some authorities: $[0.5 < \mathrm{Pr} < 100]$.
[27]Some authorities: $[\mathrm{Pr} < 100]$.
[28]Some authorities report the coefficient as 0.027 instead of 0.023. Some authorities: $[\mathrm{Pr} > 0.6]$. The upper Prandtl number limit is also reported as 700, 16,700, and 17,000.

14. TURBULENT FLOW INSIDE COILED TUBES

For flow inside helically coiled tubes and Reynolds numbers above 10^4, the film coefficient derived for a straight pipe should be multiplied by

$$1 + \frac{3.5d_{\mathrm{tube}}}{d_{\mathrm{coil}}} \qquad \textit{36.38}$$

Example 36.1

Water flows at 10 ft/sec (3 m/s) through the inside of a 2.00 in (51 mm) inside diameter, 2.125 in (54 mm) outside diameter tube. The tube wall temperature is 170°F (75°C) along its entire length. The water enters at 70°F (20°C) and is heated to 130°F (56°C). What is the inside film coefficient?

SI Solution

The film properties evaluated at the bulk temperature are

$$T_b = \left(\tfrac{1}{2}\right)(T_{\mathrm{in}} + T_{\mathrm{out}})$$
$$= \left(\tfrac{1}{2}\right)(20°\mathrm{C} + 56°\mathrm{C}) = 38°\mathrm{C}$$

From App. 35.B,

$$\rho_{38°\mathrm{C}} = 994.7 \ \mathrm{kg/m^3}$$
$$c_{p,38°\mathrm{C}} = 4.183 \ \mathrm{kJ/kg \cdot K}$$
$$\mu_{38°\mathrm{C}} = 0.682 \times 10^{-3} \ \mathrm{kg/m \cdot s}$$
$$k_{38°\mathrm{C}} = 0.6283 \ \mathrm{W/m \cdot K}$$
$$\mathrm{Pr} = 4.51$$

The Reynolds number is

$$\mathrm{Re} = \frac{\mathrm{v}D}{\nu} = \frac{\mathrm{v}D\rho}{\mu}$$
$$= \frac{\left(3 \ \frac{\mathrm{m}}{\mathrm{s}}\right)(0.051 \ \mathrm{m})\left(994.7 \ \frac{\mathrm{kg}}{\mathrm{m^3}}\right)}{0.682 \times 10^{-3} \ \frac{\mathrm{kg}}{\mathrm{m \cdot s}}}$$
$$= 2.23 \times 10^5 \quad [\text{turbulent}]$$

From Eq. 36.33, the film coefficient is

$$h = (0.023)(\mathrm{Re}^{0.8}\mathrm{Pr}^n)\left(\frac{k}{d}\right)$$
$$= \frac{(0.023)(2.23 \times 10^5)^{0.8}(4.51)^{0.4}\left(0.6283 \ \frac{\mathrm{W}}{\mathrm{m \cdot K}}\right)}{0.051 \ \mathrm{m}}$$
$$= 9832 \ \mathrm{W/M^2 \cdot K}$$

Customary U.S. Solution

The film properties evaluated at the bulk temperature are

$$T_b = \left(\tfrac{1}{2}\right)(T_{\text{in}} + T_{\text{out}})$$
$$= \left(\tfrac{1}{2}\right)(70°F + 130°F) = 100°F$$

From App. 35.A,

$$\rho_{100°F} = 62.0 \text{ lbm/ft}^3$$

$$c_{p,100°F} = 0.998 \text{ Btu/lbm-°F}$$

$$\nu_{100°F} = 0.74 \times 10^{-5} \text{ ft}^2/\text{sec}$$

$$k_{100°F} = 0.364 \text{ Btu-ft/hr-ft}^2\text{-°F}$$

$$\text{Pr} = 4.52$$

The Reynolds number is

$$\text{Re} = \frac{\text{v}D}{\nu}$$

$$= \frac{\left(10 \dfrac{\text{ft}}{\text{sec}}\right)(2.00 \text{ in})}{\left(0.74 \times 10^{-5} \dfrac{\text{ft}^2}{\text{sec}}\right)\left(12 \dfrac{\text{in}}{\text{ft}}\right)}$$

$$= 2.25 \times 10^5 \quad [\text{turbulent}]$$

From Eq. 36.33, the film coefficient is

$$h = (0.023)(\text{Re}^{0.8}\text{Pr}^n)\left(\frac{k}{d}\right)$$

$$= \frac{(0.023)(2.25 \times 10^5)^{0.8}(4.52)^{0.4}\left(0.364 \dfrac{\text{Btu-ft}}{\text{hr-ft}^2\text{-°F}}\right)}{\dfrac{2 \text{ in}}{12 \dfrac{\text{in}}{\text{ft}}}}$$

$$= 1757 \text{ Btu/hr-ft}^2\text{-°F}$$

15. TURBULENT AIR FLOW IN TUBES

A reasonably accurate approximation for turbulent air at one atmospheric pressure and between 0°F and 240°F (-17°C and 116°C) flowing in a circular tube is given by Eq. 36.39.

$$h \approx (0.00351 + 0.000001583T_{°F})\left[\frac{(G_{\text{lbm/hr-ft}^2})^{0.8}}{(d_{\text{ft}})^{0.2}}\right]$$
$$[\text{U.S.}] \quad \textit{36.39}$$

Equations 36.40 and 36.41 are approximations that do not depend on the air temperature.

$$h \approx \frac{(3.52)(\text{v}_{\text{m/s}})^{0.8}}{(d_{\text{meters}})^{0.2}} \quad [\text{SI}] \quad \textit{36.40}$$

$$h \approx \frac{(0.5)(\text{v}_{\text{ft/sec}})^{0.8}}{(d_{\text{inches}})^{0.2}} \quad [\text{U.S.}] \quad \textit{36.41}$$

16. TURBULENT WATER FLOW IN TUBES

A reasonably accurate approximation for turbulent water between 40°F and 220°F (4°C and 105°C) flowing in a round tube is

$$h \approx \frac{(1429)(1 + 0.0146T_{b,°C})\text{v}^{0.8}}{(d_{\text{meters}})^{0.2}}$$
$$[\text{SI}] \quad \textit{36.42}$$

$$h \approx \frac{(150)(1 + 0.011T_{b,°F})\text{v}^{0.8}}{(d_{\text{inches}})^{0.2}}$$
$$[\text{U.S.}] \quad \textit{36.43}$$

17. TURBULENT OIL FLOW IN TUBES

A reasonably accurate approximation for oil heated in pipes is

$$h \approx \frac{0.034\text{v}}{(\mu_b)^{0.63}} \quad [\text{U.S.}] \quad \textit{36.44}$$

18. TURBULENT LIQUID METAL FLOW IN TUBES

When the surface (wall) temperature is constant, Eq. 36.45 can be used to calculate the average film coefficient for liquid metals (e.g., mercury, sodium, and lead-bismuth alloys) experiencing fully developed turbulent flow inside tubes.[29] Fluid properties are evaluated at the mean bulk temperature.

$$\text{Nu}_d = 5.0 + (0.025)(\text{Re}_d\text{Pr})^{0.8}$$
$$\left[\text{Re}_d\text{Pr} > 100; \frac{L}{d} > 60\right] \quad \textit{36.45}$$

With a constant heat flux, Eqs. 36.46 through 36.48 can be used to calculate the average film coefficient for liquid metals with fully developed turbulent flow inside tubes. Equation 36.48 is known as the *Lubarsky-Kaufman correlation*.

$$\text{Nu}_d = 4.82 + (0.0185)(\text{Re}_d\text{Pr})^{0.827}$$
$$\left[\begin{array}{c} 3.6 \times 10^3 < \text{Re}_d < 9.05 \times 10^5; \\ 100 < \text{Re}_d\text{Pr} < 10,000 \end{array}\right] \quad \textit{36.46}$$

$$\text{Nu}_d = 7.0 + (0.025)(\text{Re}_d\text{Pr})^{0.8}$$
$$\left[\text{Re}_d\text{Pr} > 100; \frac{L}{d} > 60\right] \quad \textit{36.47}$$

$$\text{Nu}_d = (0.625)(\text{Re}_d\text{Pr})^{0.4}$$
$$\left[100 < \text{Re}_d\text{Pr} < 10,000; \frac{L}{d} > 60\right] \quad \textit{36.48}$$

---

[29]The term *fully developed* is also used when referring to full laminar flow. In this section it is understood that the flow is turbulent.

Heat Transfer

19. FLOW THROUGH NONCIRCULAR DUCTS

Dimensional analysis shows that a *characteristic length* is required in the Nusselt number, but it does not identify the length to be used. It has been common practice to correlate empirical pressure drop and heat transfer data with the *hydraulic diameter*, d_H, of non-circular (e.g., rectangular, square, elliptical, polygonal) ducts.[30,31] The Nusselt number for laminar and turbulent flow through noncircular ducts is given by Eq. 36.49.

$$\text{Nu} = \frac{h d_H}{k} \qquad 36.49$$

$$d_H = (4)\left(\frac{\text{area in flow}}{\text{wetted perimeter}}\right) \qquad 36.50$$

Annular flow is the flow of fluid through an annulus. Fluid flow is annular in simple tube-in-tube heat exchangers of the type shown in Fig. 36.6.[32] For an annulus, Eq. 36.51 gives the hydraulic diameter.[33]

$$d_H = d_{i,\text{shell}} - d_{o,\text{tube}} \qquad 36.51$$

The film coefficient for fully developed laminar flow through noncircular ducts can be calculated from the Nusselt numbers in Table 36.2.

20. CROSSFLOW OVER A SINGLE CYLINDER

Figure 36.2 illustrates a cylinder (e.g., a tube or wire) in crossflow. Equation 36.52 can be used with any fluid to calculate the film coefficient.[34] The fluid properties are evaluated at the film temperature. The entire surface area of the tube is used when calculating the heat transfer. Flow is laminar up to a Reynolds number of approximately 5×10^5.

$$\text{Nu} = C_1 (\text{Re}_d)^n \text{Pr}^{\frac{1}{3}} \quad [\text{Pr} > 0.7] \qquad 36.52$$

---

[30] A *duct* is any closed channel through which a fluid flows. Tubes and pipes are examples of round ducts. "Ducts" are not limited to air conditioning ducts.

[31] The use of the hydraulic diameter as the characteristic length is convenient and logical. Though an approximation, empirical data support using the hydraulic diameter in most cases. Notable exceptions are flow through ducts with narrow angles (e.g., an equilateral triangle with a narrow angle) and flow *parallel* to banks of tubes.

[32] Flow is not annular through more complex shell-and-tube heat exchangers.

[33] Though the hydraulic diameter is widely used as the characteristic length in calculating the Nusselt number for annuli, it is not a universal choice. Some researchers recommend using an equivalent diameter defined as

$$D_{\text{equivalent}} = \frac{(4)(\text{area in flow})}{\text{heated perimeter}}$$

$$= \frac{(d_{i,\text{shell}})^2 - (d_{o,\text{tube}})^2}{d_{o,\text{tube}}}$$

[34] There are more sophisticated correlations.

Table 36.2 *Nusselt Numbers for Fully Developed Laminar Flow Ducts*

	Nusselt number	
configuration	constant wall temperature	constant heat flux
circular	3.658	4.364
square	2.98	3.63
rectangular, aspect ratio		
1:$\sqrt{2}$	–	3.78
1:2	3.39	4.11
1:3	–	4.77
1:4	4.44	5.35
1:8	5.95	6.60
parallel plates	7.54	8.235
triangular, isosceles	2.35	3.00

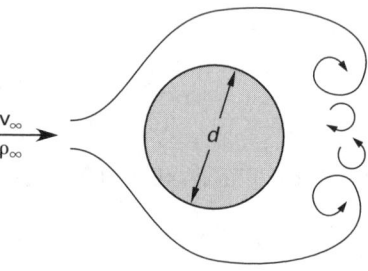

Figure 36.2 *Single Cylinder in Crossflow*

Equation 36.52 can be simplified for air since $(\text{Pr})^{\frac{1}{3}} \approx 1.00$. Equation 36.53 is sometimes referred to as the *Hilbert-Morgan equation*.

$$\text{Nu} = C_2 (\text{Re}_d)^n \qquad 36.53$$

$$C_2 \approx 1.1 C_1 \qquad 36.54$$

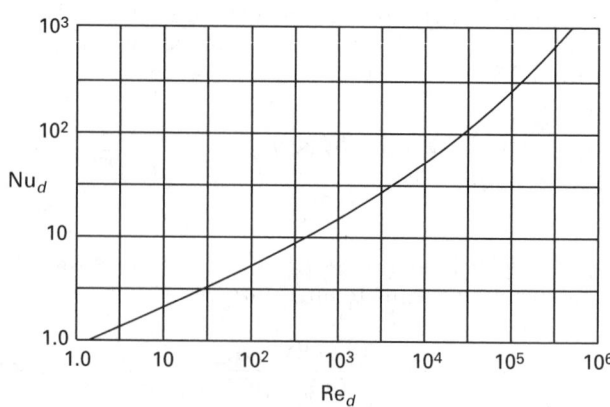

Figure 36.3 *Average Nusselt Number for Cylinder in Crossflow (air)*

Table 36.3 Constants for Tubes in Crossflow
(air and other gases)

Re	C_1	C_2	n
0.4^a–4	0.989	0.891	0.330
4–40	0.911	0.821	0.385
40–4000	0.683	0.615	0.466
4000–40,000	0.193	0.174	0.618
40,000–400,000^b	0.0266	0.0239	0.805

^{a}Some sources give the lower limit as 1.0.
^{b}Some sources give the upper limit as 250,000.

21. CROSSFLOW OVER TUBE BUNDLES

Industrial heat exchangers contain a large number of tubes arranged in layers to increase the heat transfer surface. Two or more layers are considered a tube bundle.[35] The tubes can be arranged in several ways, including a square (i.e., in-line) or rotated square (i.e., staggered), as shown in Fig. 36.4(a) and (b). Figure 36.4 also defines the *longitudinal pitch* (spacing), s_l, and *transverse pitch* (spacing), s_t. The number of transverse rows, M, is the number of tubes in the first layer encountered by the flow. N is the number of layers.

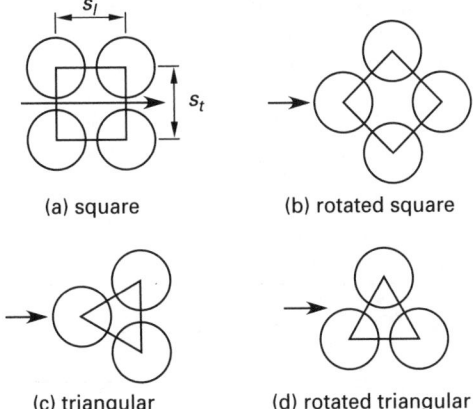

(a) square   (b) rotated square

(c) triangular   (d) rotated triangular

Figure 36.4 Tube Bundles in Crossflow

For the square and rotated square configurations, the maximum fluid velocity based on the minimum area is used to calculate the Reynolds number.[36] L is the length of each tube.

$$v_{max} = \frac{\dot{V}}{A_{min}} \qquad 36.55$$

$$A_{min} = LM(s_t - d_o) \quad \text{[in-line]} \qquad 36.56$$

$$A_{min} = LM \times \text{minimum} \left\{ \begin{array}{l} 2s_t - d_o \\ \sqrt{s_t{}^2 + s_l{}^2} - d_o \end{array} \right\}$$
$$\text{[staggered]} \qquad 36.57$$

[35]The tubes in a single layer perform as single tubes in crossflow. See Sec. 20.
[36]Equation 36.56 is an example of where common sense needs to be used. If there are M tubes in the layer, there are only $M - 1$ openings. However, there is also space on the outside of the last tubes in the layer. The actual configuration will determine the true nature of the calculation.

The *Colburn equation*, Eq. 36.58, is used to determine the film coefficient for turbulent flow in heat exchangers with 10 or more transverse rows (i.e., $M \geq 10$) in the square or rotated square configurations.[37,38,39] Fluid properties are evaluated at the film and bulk temperatures, according to their subscripts.

$$Nu_d = \frac{h d_o}{k_{film}} = (0.26)(Re_{max})^{0.6}(Pr_{film})^{0.3}$$
$$\text{[in-line; Re} > 5000]^{40} \qquad 36.58$$

$$Nu_d = \frac{h d_o}{k_{film}} = (0.33)(Re_{max})^{0.6}(Pr_{film})^{0.3}$$
$$\text{[staggered; Re} > 5000]^{41} \qquad 36.59$$

$$Re_{max} = \frac{G_{max} d_o}{\mu_{film}} = \frac{v_{max} \rho d_o}{\mu_{film}} \qquad 36.60$$

Most commercial heat exchangers have 10 or more transverse rows. However, if there are fewer than 10 transverse rows, the film coefficient calculated from Eq. 36.59 should be multiplied by the correction factor from Table 36.4.

Table 36.4 Tube Bundle Correction Factor

number of transverse rows	correction factor	
	in-line	staggered
1	0.64	0.68
2	0.80	0.75
3	0.87	0.83
4	0.90	0.89
5	0.92	0.92
6	0.94	0.95
7	0.96	0.97
8	0.98	0.98
9	0.99	0.99
10	1.00	1.00

22. PRESSURE DROP ACROSS TUBE BUNDLES

Empirical relationships for the pressure drop for flow across tube bundles are available in most heat transfer books. Though these relationships are of theoretical interest, the shellside pressure drop through a commercial heat exchanger with baffles is much more difficult to calculate. Though some correlations exist for specific configurations, most predictions are derived from the performance of similar units.

[37]A theoretical derivation shows that the exponent is $1/3$, not 0.3. However, the 0.3 is consistent with experimental data.
[38]Each tube in the first transverse row essentially has the same heat transfer as a single tube.
[39]Other researchers omit the Prandtl number term and report slightly different coefficients. Such a formulation makes the assumption that $(Pr)^{0.3} \approx 1$.
[40]Some authorities: [Re > 6000].
[41]See Ftn. 40.

23. FLOW OVER SPHERES

The general Nusselt correlation for fluid flow over a sphere is given by Eq. 36.61. Flow is laminar up to a Reynolds number of approximately 3×10^5. Fluid properties should be evaluated at the film temperature.

$$\text{Nu}_d = (0.97 + 0.68\sqrt{\text{Re}_d})\text{Pr}^{\frac{1}{3}}$$
$$[1 < \text{Re}_d < 2000] \qquad \textbf{36.61}$$

An empirical correlation with a higher Reynolds number limit is

$$\text{Nu}_d = 2.0 + (0.6)(\text{Re}_d)^{\frac{1}{2}}\text{Pr}^{\frac{1}{3}}$$
$$[1 < \text{Re}_d < 70,000] \qquad \textbf{36.62}$$

For gases, Eq. 36.63 can be used.

$$\text{Nu}_d = (0.37)(\text{Re}_d)^{0.6}(\text{Pr})^{\frac{1}{3}}$$
$$[25 < \text{Re}_d < 150,000]^{42} \qquad \textbf{36.63}$$

For air, Eq. 36.64 can be used.

$$\text{Nu}_d = (0.33)(\text{Re}_d)^{0.6} \quad [20 < \text{Re}_d < 150,000] \qquad \textbf{36.64}$$

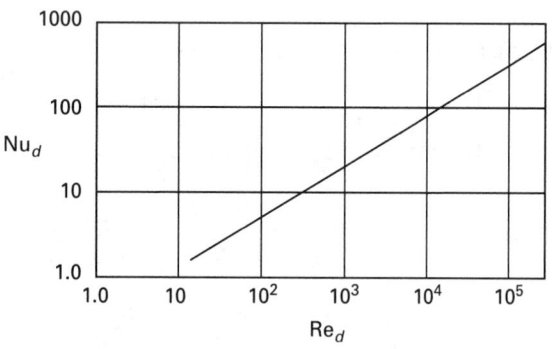

Figure 36.5 *Average Nusselt Number for Flow Over a Sphere*

24. CHANGE OF PHASE IN TUBES WITH FORCED CONVECTION

The processes of vaporization and condensation inside tubes experiencing forced convection are complex. Limited empirical relationships have been developed, but all are very specific in their derivations and limited in application. The ability to extrapolate these results to different conditions is questionable.

25. HEAT TRANSFER IN PACKED BEDS

Catalytic reactors, pebble-bed heat exchangers, and fluidized-bed furnaces are examples of *packed beds*. The heat transfer is based on the bed volume, not on surface area. The film coefficient per unit volume of packed bed, h_V, is a function of the solid particle surface area and the empty cross-sectional area of the bed per unit volume.

$$Q = h_V V_{\text{bed}}(T_{s,\text{bed particle}} - T_{b,\text{gas}}) \qquad \textbf{36.65}$$

---

[42]Some authorities: $[25 < \text{Re}_d < 100,000]$ and $[17 < \text{Re}_d < 70,000]$.

26. HEAT EXCHANGERS

Two fluids flow through or over a heat exchanger.[43] Heat from the hot fluid passes through the exchanger walls to the cold fluid.[44] The heat transfer mechanism is essentially completely forced convection.

Heat exchangers are categorized into simple *tube-in-tube heat exchangers* (also known as *jacketed pipe heat exchangers*), single-pass shell-and-tube heat exchangers, multiple-pass shell-and-tube heat exchangers, and crossflow heat exchangers.[45] *Shell-and-tube heat exchangers*, also known as *sathes* and *S & T heat exchangers*, consist of a large housing, the *shell*, with many smaller tubes running through it. The *tube fluid* passes through the tubes, while the *shell fluid* passes through the shell and around tubes.[46]

In a *single-pass heat exchanger*, each fluid is exposed to the other fluid only once. Operation is known as *parallel flow* (same as *cocurrent flow*) if both fluids flow in the same direction along the longitudinal axis of the exchanger and *counterflow* (same as *countercurrent flow*) if the fluids flow in opposite directions.[47,48] Counterflow is more efficient, and the heat transfer area required is less than that with parallel flow since the temperature gradient is more constant.

For increased efficiency, most exchangers are *multiple-pass heat exchangers*. The tubes pass through the shell more than once, and the shell fluid is routed around baffles.

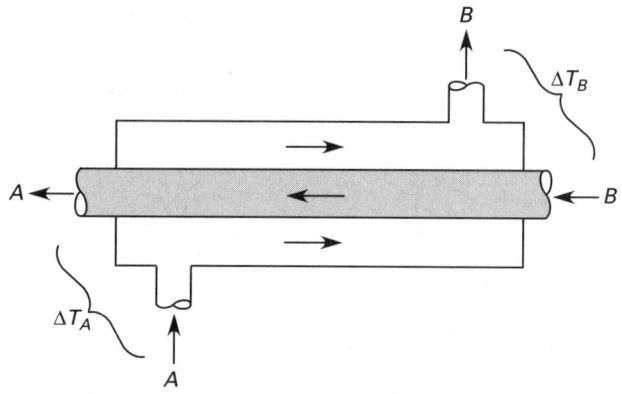

Figure 36.6 *Simple Heat Exchanger*
(single-pass, counterflow, tube-in-tube)

---

[43]These fluids do not have to be liquids. *Air-cooled exchangers* are gaining in popularity because they reduce water consumption in traditional cooling applications.

[44]A *recuperative heat exchanger*, typified by the traditional shell-and-tube exchanger, maintains separate flow channels for each of the fluids. A *regenerative heat exchanger* has only one flow path, to which the two fluids are exposed on an alternating basis.

[45]Fin coil heat exchangers are a special case of crossflow heat exchangers.

[46]Tubular heat exchangers are also known as *shell-and-tube heat exchangers*.

[47]Flow through shell-and-tube heat exchangers is neither purely parallel nor purely counterflow. Thus, these exchangers are sometimes designated as *parallel counterflow exchangers*.

[48]The designation *cocurrent* is not an abbreviation for *countercurrent*.

In a *crossflow heat exchanger*, one fluid flow is normal to the other.[49] Crossflow exchangers can operate with both fluids unmixed (typical when fluids are constrained to move through tubes and passageways), or one or both fluids may be mixed within the heat exchanger by forcing the fluids around tubes, baffles, or passages. If the fluid is mixed, its temperature is essentially uniform across the outlet. In TEMA X shells (see Sec. 27) experiencing pure crossflow and air-cooled exchangers (see Sec. 28), the fluids are generally unmixed.

One of the fluids in a crossflow heat exchanger can have multiple passes through the other fluid. Since the flow cannot be parallel in a crossflow heat exchanger, the designations used are *counter-crossflow* and *cocurrent-crossflow*. The distinction between mixed and unmixed fluids is further complicated by whether the fluids are mixed or unmixed between passes.

27. HEAT EXCHANGER DESIGNATIONS

A heat exchanger with X shell passes and Y tube passes is designated as an *X-Y heat exchanger*. In addition, most manufacturers follow the TEMA standards for design, fabrication, and material selection.[50,51,52] Heat exchanger types can be described by a three-character TEMA designation. For example, a one-two TEMA E shell and tube heat exchanger would be a shell-and-tube heat exchanger with one shell pass and two tube passes.

28. COMMERCIAL HEAT EXCHANGERS

Fixed tubesheet and U-tube bundles are the two most common commercial heat exchanger designs.[53] Fixed tubesheet heat exchangers (e.g., TEMA types BEM, AEM, and NEN) use straight tubes and offer the lowest cost heat transfer surface. A series of straight tubes are sealed between flat, perforated metal tubesheets.

The straight tubes are replaceable and easily cleaned on the inside. Since the tube bundle cannot be removed, the shellside of the tubes can only be cleaned chemically. Therefore, the shellside fluid must be nonfouling. There are no gaskets on the shell side, and the two fluids cannot accidentally mix. The no-gasket, closed-shell design is applicable to high vacuum and high pressure work as well as to potentially toxic fluids.

---

[49]An automobile radiator is an example of a crossflow exchanger.
[50]The Tubular Exchangers Manufacturers Association (TEMA) 25 N. Broadway, Tarrytown, NY, 10591, publishes the definitive standards for shell-and-tube heat exchanger construction and performance.
[51]Other applicable standards are published by the American Society of Mechanical Engineers (ASME) and the American Petroleum Institute (API).
[52]Similar to ASME's Pressure Vessel Code, TEMA standards B, C, and R are applicable to shell-and-tube heat exchangers with shell diameters not exceeding 60 in (152 cm), pressures not exceeding 3000 psi (20.1 MPa), and product of shell diameter and pressure not exceeding 60,000 lbf/in (10.5 MN/m).
[53]Other variations and commercial designs include *packed floating head, bayonet, thimble, jacketed pipe,* and *platecoil (flat)* heat exchangers.

Table 36.5 *TEMA Heat Exchanger Designations*

front-end head[a] type

A	channel and removable head
B	bonnet (integral, removable head)
C	channel (integral with tubesheet; removable plate cover)
D	special, high-pressure closure[b]

shell type

E	one-pass shell
F	two-pass shell with longitudinal baffle[b]
F	split flow, one-pass tube[b]
G	split flow, two-pass tube[b]
H	double split flow[b]
J	divided flow, one-pass tube[b]
K	kettle type reboiler[b]

rear-end head type

L	fixed tubesheet (like "A" head)
M	fixed tubesheet (like "B" head)
N	fixed tubesheet (like "C" head)
P	outside packed floating head
S	floating head with backing device (including clamp ring)
T	pull-through floating head
U	U-tube bundle
W	packed floating tubesheet with lantern ring
X	crossflow heat exchanger

[a]The term *head* is synonymous with *cover*.
[b]Specialty exchangers such as reboilers, steam heaters, vapor condensers, and feedwater heaters.

Removable-bundle heat exchangers differ in the types of removable heads. *Floating-head exchangers* (TEMA types AEW and BEW) have straight-through tubes with one tubesheet that is fixed to the shell and another that is fixed only to the tubes and "floats" within the shell. Floating-head exchangers are able to compensate for thermal expansion and contraction. Since they are separated by only a gasket, both fluids must be nonvolatile and nontoxic, and operation must be below approximately 300°F (150°C) and 300 psig (2.1 MPa).

With *outside-packed, floating-head exchangers* (TEMA types BEP and AEP), only the shellside fluid is exposed to the packing (i.e., the gasket). Corrosive gases, vapors, and liquids can be circulated in the tubes.

Internal clampring, floating-head exchangers (TEMA types AES and BES) are useful for applications with high-fouling fluids that require frequent inspection and cleaning or with high temperature differentials between the two fluids. Multipass arrangements are possible.

Pull-through floating-head exchangers (TEMA types AET and BET) have a floating head that is bolted directly to the floating tubesheet. The bundle can be

Heat Transfer

Figure 36.7 *Fixed Tubesheet Heat Exchanger (one shell pass; two tube passes)*

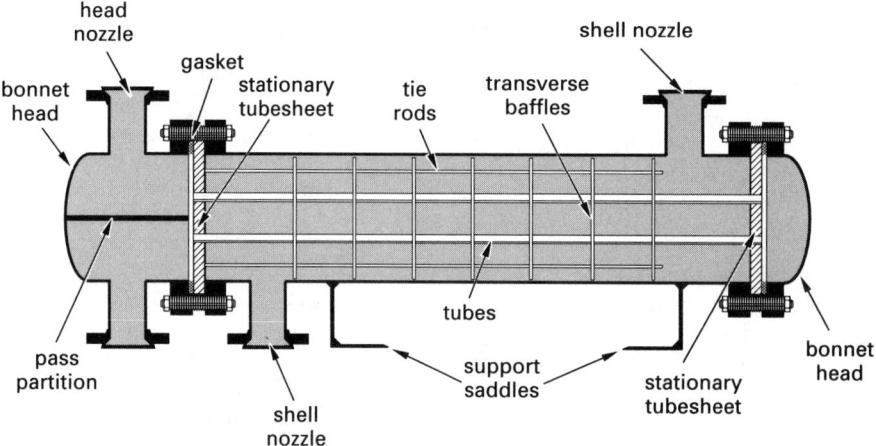

Figure 36.8 *U-Tube Bundle Heat Exchanger (one shell pass; two tube passes)*

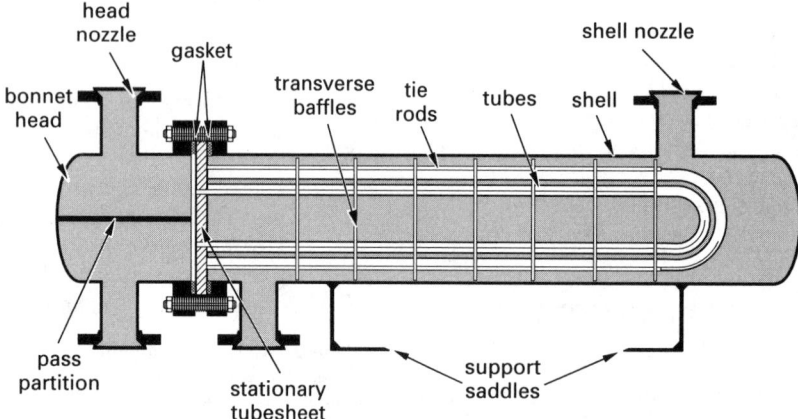

removed without removing the shell or floating head covers. The design accommodates fewer tubes for a given shell diameter and offers less surface area than other removable-bundle exchangers. Although expensive, this design permits frequent cleaning.

U-tube heat exchangers (TEMA types AEU and BEU) have a bundle of tubes, each bent in a series of concentrically tighter U-shapes. They are lower in initial cost since there is only one tube sheet. The tube bundle can be removed for inspection and cleaning. The individual tubes automatically compensate for thermal expansion and contraction. These exchangers are ideal for intermittent service or where thermal shock is expected. However, the insides of the tubes cannot be mechanically cleaned along their entire length, and the tubes cannot be replaced.[54] The design cannot be single-pass on the tube side, and true countercurrent flow is not possible. *Plate-type heat exchangers* (often called *panel coil exchangers*, *welded-plate exchangers*, or simply *plate exchangers*) are used as immersion heaters in tanks for the heating or cooling of solutions and as evaporator

surfaces where liquids cascade down their sides.[55] They are constructed from two sheets welded and embossed or expanded to form a series of passes through which one of the fluids flow.

Plate-and-frame heat exchangers combine several thin-gauge plates with embossed flow paths that are separated by an elastomeric or asbestos fiber gasket.[56] Several two-plate units are clamped together into a compact unit that is suitable for large temperature crosses or small temperature approaches. (See Sec. 30).

---

[55]"Platecoil" and "Temp-Plate" are trade names for the plate-type heat exchanger.

[56]The choice of gasket material depends on the temperatures, pressures, and fluids encountered. The temperature limits of popular gasket materials are 230 to 275°F (110 to 135°C) for nitrile, 300°F (150°C) for resin-cured butyl rubber, 320 to 350°F (160 to 175°C) for ethylene-propylene diene monomer (EDPM), and 212 to 350°F (100 to 175°C) for fluorocarbon rubber base. Nitrile is a general service material suitable for water applications. EDPM is suitable for steam and higher temperature aqueous solutions. All of the foregoing materials are limited to approximately 400 psig (2.8 MPa). Compressed asbestos gaskets can be used up to approximately 600°F (320°C), but they are limited to approximately 250 psig(1.7MPa).

---

[54]A few of the outer bends can be replaced. However, a complete retubing is usually necessary.

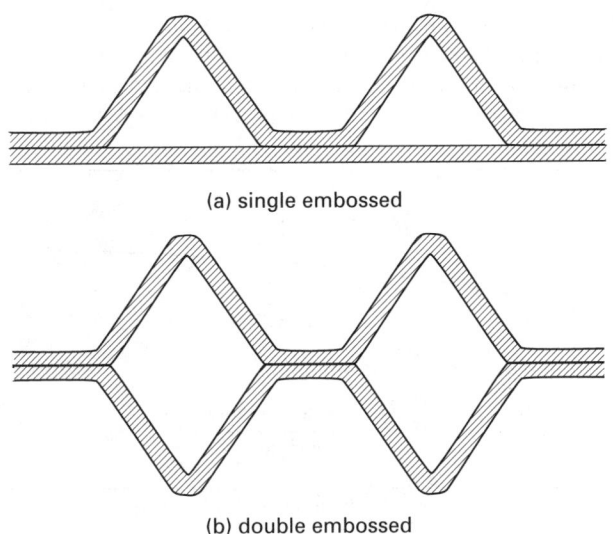

Figure 36.9 *Cross-Section of Plate-Type Heat Exchanger*

In addition to a high heat transfer rate due to their thin plate material, other advantages of plate and plate-and-frame heat exchangers include compactness, accessibility to both sides of each plate for maintenance and cleaning, and flexibility of design (since the number of plates in the frame can be varied).

For a given heat transfer coefficient, plate heat exchangers have a lower pressure drop. However, the narrow passageways produce high pressure drops in high-volume, low-pressure gas applications. These exchangers are limited to approximately 300 psig (2.1 MPa) and compatible gasket materials.

Air-cooled exchangers include a motor and fan assembly that forces air over a series of (typically coiled) tubes. To increase the heat transfer, the tubes are usually finned, hence the names *fin-tube exchanger* and *heavy-duty coil*.

29. TRANSVERSE BAFFLES IN HEAT EXCHANGERS

Several types of *baffles* are used inside the shells of commercial heat exchangers to increase the velocity (and, accordingly, the film coefficient) on the shellside. The most common type is the *segmental baffle*, also known as the *crossflow baffle*.

30. TEMPERATURE DIFFERENCE TERMINOLOGY

There are several specialized temperature difference terms used in the analysis of heat transfer.

The difference in hot and cold fluid temperatures is seldom constant in a heat exchanger. The *approach* (*temperature approach* or *approach temperature*) is the smallest difference in temperature between the two fluids. For a double-pipe, single-pass counterflow heat exchanger, the temperature approach is defined by Eq. 36.66.

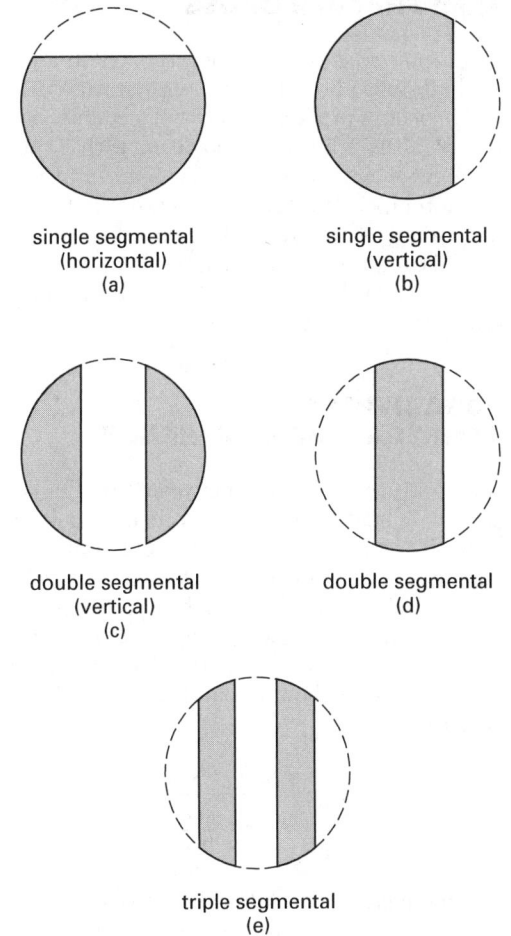

Figure 36.10 *Types of Baffles*

$$\Delta T_{\text{approach}} = T_{\text{hot,in}} - T_{\text{cold,out}} \qquad 36.66$$

Traditional shell-and-tube heat exchangers seldom have temperature approaches less than 10°F (6°C). Exceptions are some refrigeration systems that work with temperature approaches of 5 to 9°F (3 to 5°C), and plate-and-frame heat exchangers (Sec. 28) that can work well with as little as a 2°F (1°C) temperature approach. In combustion air preheaters using flue gas as the heating source, the temperature approach should be approximately 36°F (20°C).

The *extreme temperature difference* for heat exchangers is defined as

$$\Delta T_{\text{extreme}} = T_{\text{hot,in}} - T_{\text{cold,in}} \qquad 36.67$$

The ratio of the cold fluid change to the extreme temperature difference, a form of "temperature efficiency," is[57]

$$\eta_T = \frac{T_{\text{cold,out}} - T_{\text{cold,in}}}{T_{\text{hot,in}} - T_{\text{cold,in}}} \qquad 36.68$$

---

[57]The symbol S is also used for this quantity by some authors.

31. TEMPERATURE CROSS

A *temperature cross* occurs when the exit temperature of the cold fluid is above the exit temperature of the hot fluid. This occurs predominately with counterflow heat exchangers, although it can also occur with a shell-and-tube exchanger with one shell pass and multiple tube passes. A temperature cross indicates that there is a relatively small temperature difference between the two fluids. This requires either a large heat transfer area or a relative high fluid velocity (to increase the overall heat transfer coefficient).

32. LOGARITHMIC TEMPERATURE DIFFERENCE

The temperature difference between two fluids is not constant in a heat exchanger. When calculating the heat transfer for a tube whose temperature difference changes along its length, the *logarithmic mean temperature difference*, ΔT_{lm} or LMTD, is used.[58,59,60] In Eq. 36.69, ΔT_A and ΔT_B are the temperature differences at ends A and B, respectively, regardless of whether the fluid flow is parallel or counterflow, as shown in Fig. 36.6.[61,62]

$$\Delta T_{lm} = \frac{\Delta T_A - \Delta T_B}{\ln\left(\dfrac{\Delta T_A}{\Delta T_B}\right)} \qquad 36.69$$

For multiple-pass and crossflow heat exchangers, a multiplicative correction factor, F_c, is required for ΔT_{lm}. When one of the fluids does not change temperature, as in a feedwater heater or other condensation/evaporation environment, the correction factor is 1.00.[63] The

---

[58]An exception occurs in HVAC calculations where ΔT at mid-length has traditionally been used to calculate the heat transfer in air conditioning ducts. Considering the imprecise nature of HVAC calculations, the added sophistication of using the logarithmic mean temperature difference is probably unwarranted.

[59]The symbol ΔT_m is also widely used for the log-mean temperature difference. However, this can also be interpreted as the arithmetic mean temperature.

[60]The logarithmic temperature difference is used even with change of phase (e.g., boiling liquid or condensing vapor) and the temperature is constant in one tube.

[61]It doesn't make any difference which end is A and which is B. If the numerator in Eq. 36.69 is negative, the denominator will also be negative.

[62]Using Eq. 36.69 presents many difficulties, particularly with computer-based heat transfer analysis. As ΔT_A and ΔT_B become equal, Eq. 36.69 becomes indeterminate, even though the correct relationship is $\Delta T_{lm} = \Delta T_A = \Delta T_B$. Also, the first derivative of Eq. 36.69, used in some calculations, is undefined when ΔT_A and ΔT_B are equal, even though the correct value is 0.5. A replacement expression that avoids these difficulties with (generally) less than a 0.3% error is

$$\Delta T_{lm} \approx \frac{(\Delta T_A)^{\frac{1}{3}} + (\Delta T_B)^{\frac{1}{3}}}{2}$$

[63]A general rule for good designs of boiling/evaporative systems is that the logarithmic mean temperature difference should be kept less than 110°F (60°C).

---

procedure in all other cases is to calculate ΔT_{lm} as if the fluids were in counterflow. The correction factor depends on the type of heat exchanger and is almost always given graphically.[64] (See Apps. 36.A and 36.B.)

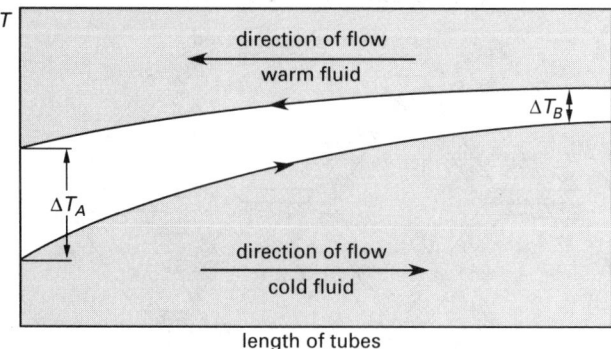

Figure 36.11 *Temperature Profile for LMTD Calculation (single-pass, counterflow exchanger, no change of phase)*

33. RULES OF THUMB FOR COOLING WATER

The following rules of thumb will result in more efficient heat transfer operations when water is used for cooling.

- For shell-and-tube heat exchangers, the water temperature should not increase by more than 20°F (10°C) when $\Delta T_{lm} < 70°F$ (40°C) and should not increase by more than 35°F (20°C) when $\Delta T_{lm} > 70°F$ (40°C).

- The water's exit temperature should not exceed 122°F (50°C) due to the increased potential for solids precipitation and fouling.

- The water's entrance temperature should be at least 10°F (5°C) above the freezing temperature of the liquid being cooled.

34. HEAT TRANSFER IN HEAT EXCHANGERS

Equation 36.70 calculates the steady-state heat transfer (also known as the *heat duty* and *heat load*) in a heat exchanger or feedwater heater.[65,66,67] The *overall*

---

[64]Calculating the F_c factor is preferred over reading it from a chart. The error of the F_c factor read from a chart can be as great as 3 to 5%.

[65]*Closed feedwater heaters* are heat exchangers whose purpose is to heat water with condensing steam.

[66]There are three heat loads referred to in heat exchanger specifications: the *specific heat load*, which is the design heat transfer; the heat released by the hot fluid; and the heat absorbed by the cold fluid. All three would be the same if operation was adiabatic, but due to practical losses, they are not. If they differ by more than 10%, the cause of the discrepancy should be evaluated.

[67]Transient (i.e., startup) performance of heat exchangers is poorly covered in most textbooks. An excellent article on this subject is Chester A. Plant's "Evaluate Heat-Exchanger Performance," *Chemical Engineering Magazine*, p. 104, July 1992.

heat transfer coefficient, U, also known as the *overall conductance* and the *overall coefficient of heat transfer*, can be specified for use with either the outside or inside tube areas.[68] The heat transfer is independent of whether the outside or inside area is used.

$$Q = U_o A_o F_c \Delta T_{lm} = U_i A_i F_c \Delta T_{lm} \qquad 36.70$$

35. OVERALL HEAT TRANSFER COEFFICIENT

The overall heat transfer coefficient, U, is calculated from the film coefficients and the tube material conductivities. The overall heat transfer coefficient, based on outside and inside areas and exclusive of a fouling factor, is

$$\frac{1}{U_o} = \frac{1}{h_o} + \left(\frac{r_o}{k_{\text{tube}}}\right) \ln\left(\frac{r_o}{r_i}\right) + \frac{r_o}{r_i h_i} \qquad 36.71$$

$$\frac{1}{U_i} = \frac{1}{h_i} + \left(\frac{r_i}{k_{\text{tube}}}\right) \ln\left(\frac{r_o}{r_i}\right) + \frac{r_i}{r_o h_o} \qquad 36.72$$

The second term in Eqs. 36.71 and 36.72 is sometimes approximated by the term t/k when the tube diameters are "large." However, the thermal resistance of the tube is very small and is often omitted entirely. If the tube resistance term is kept at all, it is not very difficult to use Eqs. 36.71 and 36.72 as written.

Table 36.6 *Typical Values of Overall Coefficient of Heat Transfer (U-Values)[a] (Btu/hr-ft²-°F)*

		clean surface	with normal fouling
heating applications			
hot side	*cold side*		
steam	aqueous solution	300–550	150–275
steam	light oils	110–140	60–110
steam	medium lube oils	110–130	50–100
steam	Bunker C or no. 6 oil	70–90	60–80
steam	air or gases	5–10	4–8
hot water	aqueous solution	200–250	110–160
cooling applications			
cold side	*hot side*		
water	aqueous solution	200–250	105–155
water	medium lube oil	20–30	10–20
water	air or gases	5–10	4–8
freon/ammonia	aqueous solution	60–90	40–60
calcium or sodium brine	aqueous solution	175–200	80–125

(Multiply Btu/hr-ft²-°F by 5.6783 to obtain W/m²·K.)
[a]Overall heat transfer coefficient values are strongly dependent on the type of heat exchanger, as well as on the hot- and cold-side fluids.

---

[68]It is more common (and preferred) to use the outside tube area because the tube outside diameter is more easily measured.

If the tube thermal resistance term is omitted entirely and if the tube is thin-walled, Eqs. 36.73 and 36.74 can be used as approximations.

$$\frac{1}{U_o} \approx \frac{1}{h_o} + \frac{r_o}{r_i h_i} \approx \frac{1}{h_o} + \frac{1}{h_i} \qquad 36.73$$

$$\frac{1}{U_i} \approx \frac{1}{h_i} + \frac{r_i}{r_o h_o} \approx \frac{1}{h_i} + \frac{1}{h_o} \qquad 36.74$$

In reality, it is very difficult to predict the heat transfer coefficient for most types of commercial heat exchangers. Values can be predicted by comparison with similar units, or "tried-and-true" rules of thumb can be used. One such rule of thumb for baffled shell-and-tube heat exchangers predicts the clean, average heat transfer coefficient as 60% of the value for the same arrangement of tubes in pure crossflow.

36. TUBE LENGTH REQUIRED

For a simple, tube-in-tube counterflow heat exchanger as shown in Fig. 36.6, the length of tube required for a fluid to change temperature from T_{in} to T_{out} is given by Eq. 36.75. Velocity, v, must have units of ft/hr. The maximum normal length for straight tubes is 20 ft (0.6 m). Multiple-pass heat exchangers are needed for longer lengths.

$$L = \frac{\rho \text{v}(d_i)^2 c_p (T_{\text{out}} - T_{\text{in}})}{4 U_o d_o \Delta T_{lm}} \qquad 36.75$$

Example 36.2

All thermal resistance other than the internal film for the tube in Ex. 36.1 can be disregarded. What tube length is required?

SI Solution

The two "end" temperature differences are

$$\Delta T_A = 75°C - 20°C = 55°C$$
$$\Delta T_B = 75°C - 56°C = 19°C$$

The logarithmic temperature difference is

$$\Delta T_{lm} = \frac{\Delta T_A - \Delta T_B}{\ln\left(\frac{\Delta T_A}{\Delta T_B}\right)} = \frac{55°C - 19°C}{\ln\left(\frac{55°C}{19°C}\right)}$$

$$= 33.9°C$$

Since $U_o A_o = U_i A_i$ and $U_i = h_i$,

$$U_o = \frac{h_i A_i}{A_o} = h_i \left(\frac{d_i}{d_o}\right)^2$$

$$= \frac{\left(9832 \ \frac{\text{W}}{\text{m}^2\cdot\text{K}}\right)(51 \text{ mm})^2}{(54 \text{ mm})^2}$$

$$= 8770 \text{ W/m}^2\cdot\text{K}$$

Table 36.7 *Approximate Thermal Conductivity of Common Heat Exchanger Materials (Btu-ft/hr-ft²-°F)*

	temperature, °F													
	200	300	400	500	600	700	800	900	1000	1100	1200	1300	1400	1500
aluminum (annealed)														
type 1100-0	126	124	123	122	121	120	118							
type 3003-0	111	111	111	111	111	111	111							
type 3004-0	97	98	99	100	102	103	104							
type 6061-0	102	103	104	105	106	106	106							
aluminum (tempered)														
type 1100 (all tempers)	123	122	121	120	118	118	118							
type 3003 (all tempers)	96	97	98	99	100	102	104							
type 3004 (all tempers)	97	98	99	100	102	103	104							
type 6061—T4 & T6	95	96	97	98	99	100	102							
type 6063—T5 & T6	116	116	116	116	116	115	114							
type 6063—T42	111	111	111	111	111	111	111							
cast iron	31	31	30	29	28	27	26	25						
carbon steel	29.2	28.4	27.6	26.6	25.6	24.6	23.5	22.5	21.4	20.2	19.0	17.6	16.2	15.6
carbon moly (1/2%) steel	25.2	25.1	24.8	24.3	23.7	23.0	22.2	21.4	20.4	19.5	18.4	16.7	15.3	15.0
chrome moly steels														
1% Cr, 1/2% Mo	21.9	22.0	21.9	21.7	21.3	20.8	20.2	19.7	19.1	18.5	17.7	16.5	15.0	14.8
2 1/4% Cr, 1% Mo	21.3	21.5	21.5	21.4	21.1	20.7	20.2	19.7	19.1	18.5	18.0	17.2	15.6	15.3
5% Cr, 1/2% Mo	18.1	18.7	19.1	19.2	19.2	19.0	18.7	18.4	18.0	17.6	17.1	16.6	16.0	15.8
12% Cr	14	15	15	15	16	16	16	16	17	17	17	18		
austenitic stainless steels														
18% Cr, 8% Ni	9.3	9.8	10	11	11	12	12	13	13	14	14	14	15	15
25% Cr, 20% Ni	7.8	8.4	8.9	9.5	10	11	11	12	12	13	14	14	15	15
admiralty	70	75	79	84	89									
naval brass	71	74	77	80	83									
copper	225	225	224	224	223									
copper and nickel alloys														
90% Cu, 10% Ni	30	31	34	37	42	47	49	51	53					
80% Cu, 20% Ni	22	23	25	27	29	31	34	37	40					
70% Cu, 30% Ni	18	19	21	23	25	27	30	33	37					
30% Cu, 70% Ni Alloy 400	15	15	16	16	17	18	18	19	20	20				

(Multiply Btu-ft/hr-ft²-°F by 1.731 to obtain W/m·K.)
© 1988 by Tubular Exchanger Manufacturers Association.

The tube length required is

$$L = \frac{\rho v (d_i)^2 c_p (T_{\text{out}} - T_{\text{in}})}{4 U_o d_o \Delta T_m}$$

$$= \left(994.7 \; \frac{\text{kg}}{\text{m}^3}\right) \left(3 \; \frac{\text{m}}{\text{s}}\right) (0.051 \; \text{m})^2$$

$$\times \frac{\left(4.183 \; \frac{\text{kJ}}{\text{kg·K}}\right) \left(1000 \; \frac{\text{J}}{\text{kJ}}\right) (56°\text{C} - 20°\text{C})}{(4) \left(8770 \; \frac{\text{W}}{\text{m}^2·\text{K}}\right) (0.054 \; \text{m})(33.9°\text{C})}$$

$$= 18.2 \; \text{m}$$

Customary U.S. Solution

The two "end" temperature differences are

$$\Delta T_A = 170°\text{F} - 70°\text{F} = 100°\text{F}$$

$$\Delta T_B = 170°\text{F} - 130°\text{F} = 40°\text{F}$$

The logarithmic temperature difference is

$$\Delta T_{lm} = \frac{\Delta T_A - \Delta T_B}{\ln\left(\frac{\Delta T_A}{\Delta T_B}\right)} = \frac{100°\text{F} - 40°\text{F}}{\ln\left(\frac{100°\text{F}}{40°\text{F}}\right)}$$

$$= 65.5°\text{F}$$

Since $U_o A_o = U_i A_i$ and $U_i = h_i$,

$$U_o = \frac{h_i A_i}{A_o} = h_i \left(\frac{d_i}{d_o}\right)^2$$

$$= \frac{\left(1757 \; \frac{\text{Btu}}{\text{hr-ft}^2\text{-}°\text{F}}\right) (2.00 \; \text{in})^2}{(2.125 \; \text{in})^2}$$

$$= 1556 \; \text{Btu/hr-ft}^2\text{-}°\text{F}$$

The tube length required is

$$L = \frac{\rho v (d_i)^2 c_p (T_{\text{out}} - T_{\text{in}})}{4 U_o d_o \Delta T_m}$$

$$= \left(62.0 \; \frac{\text{lbm}}{\text{ft}^3}\right) \left(10 \; \frac{\text{ft}}{\text{sec}}\right) \left(3600 \; \frac{\text{sec}}{\text{hr}}\right) (2.00 \; \text{in})^2$$

$$\times \frac{\left(0.998 \; \frac{\text{Btu}}{\text{lbm-}°\text{F}}\right) (130°\text{F} - 70°\text{F})}{(4) \left(1556 \; \frac{\text{Btu}}{\text{hr-ft}^2\text{-}°\text{F}}\right)}$$

$$\times (2.125 \; \text{in})(65.5°\text{F}) \left(12 \; \frac{\text{in}}{\text{ft}}\right)$$

$$= 51.4 \; \text{ft}$$

37. FOULING

Fouling is corrosion, precipitation of compounds in solution (i.e., *scaling*), settling of suspended particulate solids, and biological activity (i.e., growth of algae and other life forms) that adhere to a heat transfer. Fouling of heat transfer surfaces decreases the heat transfer and increases pumping power.[69,70]

Unless the tube walls are known to be "new and clean," a *fouling factor*, R_f, should be incorporated into the expression for the overall heat transfer coefficient.[71,72] On a macroscopic basis, where inside and outside coefficients and fouling factors are combined, Eq. 36.76 can be used.

$$R_f = \frac{1}{U_{\text{fouled}}} - \frac{1}{U_{\text{clean}}} \qquad \textbf{36.76}$$

The tubeside and shellside fouling factors can be individually combined to determine the fouled overall heat transfer coefficient. Values of the fouling factor for normal operation are given in Table 36.8.

$$\frac{1}{U_{o,\text{fouled}}} = \frac{1}{h_o} + \left(\frac{r_o}{k_{\text{tube}}}\right) \ln\left(\frac{r_o}{r_i}\right) + R_{f,o}$$

$$+ \left(\frac{r_o}{r_i}\right) \left(R_{f,i} + \frac{1}{h_i}\right)$$

$$\approx \frac{1}{h_o} + \frac{1}{h_i} + R_{f,o} + R_{f,i} \qquad \textbf{36.77}$$

Table 36.8 *Typical Fouling Factors for Shell-and-Tube Heat Exchangers*[a] *(hr-ft²-°F/Btu)*

fluid	below 125°F (50°C)	above 125°F (50°C)
air	0.002	0.002
water		
city/well	0.001	0.002
brine, salt, or		
seawater	0.0005	0.001
hard water	0.003	0.005
distilled	0.0005	0.0005
treated boiler		
feedwater	0.001	0.001
untreated cooling		
tower	0.002	0.002
steam		
clean	0.0005	0.0005
oil-bearing	0.001	0.001
diesel exhaust		

(Multiply hr-ft²-°F/Btu by 0.17611 to obtain m²·K/W.)
[a] Fouling factors depend on the heat exchanger type. Lower values apply to heat exchangers with lower fouling tendencies.

---

[69] Fouling is known in the heat exchanger industry as the "silent thief."

[70] Condenser tube fouling accounts for up to 50% of the total thermal resistance in steam power plants. For refrigeration condensers, each 0.0001 hr-ft²-°F/Btu (1.75×10^{-5} K·m²/W increase in the fouling factor increases the centrifugal refrigeration compressor power by approximately 1.1%.

[71] In a shell-and-tube heat exchanger, it is preferable to run a fouling liquid through straight tubes, since the interior of straight tubes is accessible for inspection and cleaning. This is not the case for the exteriors of tubes and U-tube bundles.

[72] Fouled tubes can be cleaned chemically or mechanically to return them to almost "new" performance. Mechanical cleaning that occurs during normal operation is known as *online cleaning*, while *offline* mechanical cleaning requires the unit to be taken out of service. The two most common online mechanical cleaning systems are *brush-and-basket* and *sponge rubber ball* types. Brushes, rubber, scrapers, cutters, vibrators, and water lances are common offline cleaning systems.

Fouling is intrinsically a changing characteristic. Equation 36.77 calculates the fouled overall heat transfer coefficient but does not predict when that value will be reached. Fouling as a function of time can be modeled as *linear*, *falling-rate*, or *asymptotic* (approaching $R_{f,\infty}$). The values of coefficients $a, b, R_{f,\infty}$, and t' must be known or determined from empirical data or other knowledge.

$$R_{f,t} = at \quad \text{[linear]} \qquad\qquad \textbf{36.78}$$

$$(R_{f,t})^2 = (R_{f,0})^2 + bt \quad \text{[falling-rate]} \quad \textbf{36.79}$$

$$R_{f,t} = R_{f,\infty}\left(1 - e^{-t/t'}\right) \quad \text{[asymptotic]} \quad \textbf{36.80}$$

Example 36.3

A single-pass shell-in-tube heat exchanger is constructed of $1\frac{1}{4}$ in schedule-40 pipe for the tube and $2\frac{1}{2}$ in schedule-40 pipe for the shell. The thermal conductivity of the pipe material is 35 Btu-ft/hr-ft²-°F (61 W/m·K). Flow is counterflow.

The heat exchanger cools 10 gal/min (0.6 L/s) of water flowing in the shell from 55°F (13°C) to 45°F (7°C). The water film coefficient is 246 Btu/hr-ft²-°F (1400 W/m²·K). The cooling solution is 10 gal/min (0.6 L/s) of brine initially at 10°F (−12°C). The brine has a specific heat of 0.68 Btu/lbm-°F (2.85 kJ/kg·K) and a density of 77.0 lbm/ft³ (1233 kg/m³). The brine film coefficient is 265 Btu/hr-ft²-°F (1500 W/m²·K).

How long should the heat exchanger be in a typical fouled condition?

SI Solution

The dimensions of the schedule-40 tube and shell pipes are

	tube	shell
inside diameter	35.05 mm	62.71 mm
outside diameter	42.16 mm	73.03 mm

The water's bulk temperature is

$$T_{b,\text{water}} = \left(\tfrac{1}{2}\right)(7°C + 13°C) = 10°C$$

The fluid properties at 10°C are interpolated from App. 35.B.

$$\rho_{10°C} = 999 \text{ kg/m}^3$$

$$c_{p,10°C} = 4.20 \text{ kJ/kg·K}$$

The mass flow rates are

$$\dot{m}_{\text{water}} = \dot{V}\rho = \frac{\left(0.6 \dfrac{\text{L}}{\text{s}}\right)\left(999 \dfrac{\text{kg}}{\text{m}^3}\right)}{1000 \dfrac{\text{L}}{\text{m}^3}}$$

$$= 0.599 \text{ kg/s}$$

$$\dot{m}_{\text{brine}} = \dot{V}\rho = \frac{\left(0.6 \dfrac{\text{L}}{\text{s}}\right)\left(1233 \dfrac{\text{kg}}{\text{m}^3}\right)}{1000 \dfrac{\text{L}}{\text{m}^3}}$$

$$= 0.740 \text{ kg/s}$$

The heat transfer is found from the temperature gain of the water. The logarithmic mean temperature difference is not used here, even though this is a heat exchanger.

$$Q = \dot{m}c_p\Delta T$$

$$= \left(0.599 \frac{\text{kg}}{\text{s}}\right)\left(4.20 \frac{\text{kJ}}{\text{kg·K}}\right)\left(1000 \frac{\text{J}}{\text{kJ}}\right)$$

$$\times (13°C - 7°C)$$

$$= 1.509 \times 10^4 \text{ W}$$

The brine's final temperature is

$$T_{\text{brine,out}} = T_{\text{brine,in}} + \frac{Q}{\dot{m}_{\text{brine}}c_p}$$

$$= -12°C + \frac{1.509 \times 10^4 \text{ W}}{\left(0.740 \dfrac{\text{kg}}{\text{s}}\right)}$$

$$\times \left(2.85 \frac{\text{kJ}}{\text{kg·K}}\right)\left(1000 \frac{\text{J}}{\text{kJ}}\right)$$

$$= -4.84°C$$

The logarithmic mean temperature difference is

$$\Delta T_{lm} = \frac{\Delta T_A - \Delta T_B}{\ln\left(\dfrac{\Delta T_A}{\Delta T_B}\right)}$$

$$= \frac{(13°C - (-4.84°C)) - (7°C - (-12°C))}{\ln\left(\dfrac{13°C - (-4.84°C)}{7°C - (-12°C)}\right)}$$

$$= 18.41°C$$

Fouling factors are converted from the values given in Table 36.8.

$$R_{f,o} = \left(0.001 \frac{\text{hr-ft}^2\text{-°F}}{\text{Btu}}\right)\left(0.17611 \frac{\text{m}^2\text{·K·Btu}}{\text{W-hr-ft}^2\text{-°F}}\right)$$

$$= 1.8 \times 10^{-4} \text{ m}^2\text{·K/W} \quad \text{[brine]}$$

$$R_{f,i} = \left(0.0005 \frac{\text{hr-ft}^2\text{-°F}}{\text{Btu}}\right)\left(0.17611 \frac{\text{m}^2\text{·K·Btu}}{\text{W-hr-ft}^2\text{-°F}}\right)$$

$$= 8.8 \times 10^{-5} \text{ m}^2\text{·K/W} \quad \text{[water below 125°F]}$$

The overall heat transfer coefficient can be calculated from Eq. 36.77. (It is not necessary to convert ratios of dimensions from inches to feet.)

$$\frac{1}{U_o} = \frac{1}{h_o} + \left(\frac{r_o}{k_{\text{tube}}}\right)\ln\left(\frac{r_o}{r_i}\right)$$
$$+ R_{f,o} + \left(\frac{r_o}{r_i}\right)\left[R_{f,i} + \left(\frac{1}{h_i}\right)\right]$$
$$= \frac{1}{1400 \dfrac{\text{W}}{\text{m}^2\cdot\text{K}}} + \left[\frac{0.04216 \text{ m}}{(2)\left(61 \dfrac{\text{W}}{\text{m}\cdot\text{K}}\right)}\right]$$
$$\times \ln\left(\frac{42.16 \text{ mm}}{35.05 \text{ mm}}\right)$$
$$+ \left(1.8 \times 10^{-4} \frac{\text{m}^2\cdot\text{K}}{\text{W}}\right) + \left(\frac{42.16 \text{ mm}}{35.05 \text{ mm}}\right)$$
$$\times \left(8.8 \times 10^{-5} \frac{\text{m}^2\cdot\text{K}}{\text{W}} + \frac{1}{1500 \dfrac{\text{W}}{\text{m}^2\cdot\text{K}}}\right)$$
$$= 1.866 \times 10^{-3} \text{ m}^2\cdot\text{K/W}$$
$$U_o = \frac{1}{\left(1.866 \times 10^{-3} \dfrac{\text{m}^2\cdot\text{K}}{\text{W}}\right)}$$
$$= 536 \text{ W/m}^2\cdot\text{K}$$

The heat transfer is known. Therefore, the length can be calculated from Eq. 36.70.

$$Q = U_o A_o F_c \Delta T_{lm} = U_o \pi D_o L F_c \Delta T_{lm}$$
$$L = \frac{Q}{U_o \pi D_o F_c \Delta T_{lm}}$$
$$= \frac{1.509 \times 10^4 \text{ W}}{\left(536 \dfrac{\text{W}}{\text{m}^2\cdot\text{K}}\right)\pi(0.04216 \text{ m})(1)(18.41°\text{C})}$$
$$= 11.54 \text{ m}$$

Customary U.S. Solution

The dimensions of the schedule-40 tube and shell pipes are

	tube	shell
inside diameter	1.38 in	2.47 in
outside diameter	1.66 in	2.88 in

The water's bulk temperature is

$$T_{b,\text{water}} = \left(\tfrac{1}{2}\right)(45°\text{F} + 55°\text{F}) = 50°\text{F}$$

The fluid properties at 50°F are obtained from App. 35.A.

$$\rho_{50°\text{C}} = 62.4 \text{ lbm/ft}^3$$
$$c_{p,50°\text{C}} = 1.00 \text{ Btu/lbm-°F}$$

The mass flow rates are

$$\dot{m}_{\text{water}} = \dot{V}\rho$$
$$= \frac{\left(10 \dfrac{\text{gal}}{\text{min}}\right)\left(60 \dfrac{\text{min}}{\text{hr}}\right)\left(62.4 \dfrac{\text{lbm}}{\text{ft}^3}\right)}{7.48 \dfrac{\text{gal}}{\text{ft}^3}}$$
$$= 5005 \text{ lbm/hr}$$
$$\dot{m}_{\text{brine}} = \dot{V}\rho$$
$$= \frac{\left(10 \dfrac{\text{gal}}{\text{min}}\right)\left(60 \dfrac{\text{min}}{\text{hr}}\right)\left(77.0 \dfrac{\text{lbm}}{\text{ft}^3}\right)}{7.48 \dfrac{\text{gal}}{\text{ft}^3}}$$
$$= 6176 \text{ lbm/hr}$$

The heat transfer is found from the temperature gain of the water. The logarithmic mean temperature difference is not used here, even though this is a heat exchanger.

$$Q = \dot{m}c_p\Delta T$$
$$= \left(5005 \dfrac{\text{lbm}}{\text{hr}}\right)\left(1.0 \dfrac{\text{Btu}}{\text{lbm-°F}}\right)(55°\text{F} - 45°\text{F})$$
$$= 50{,}050 \text{ Btu/hr}$$

The brine's final temperature is

$$T_{\text{brine,out}} = T_{\text{brine,in}} + \frac{Q}{\dot{m}_{\text{brine}}c_p}$$
$$= 10°\text{F} + \frac{50{,}050 \dfrac{\text{Btu}}{\text{hr}}}{\left(6176 \dfrac{\text{lbm}}{\text{hr}}\right)\left(0.68 \dfrac{\text{Btu}}{\text{lbm-°F}}\right)}$$
$$= 21.9°\text{F}$$

The logarithmic mean temperature difference is

$$\Delta T_{lm} = \frac{\Delta T_A - \Delta T_B}{\ln\left(\dfrac{\Delta T_A}{\Delta T_B}\right)}$$
$$= \frac{(45°\text{F} - 10°\text{F}) - (55°\text{F} - 21.9°\text{F})}{\ln\left(\dfrac{45°\text{F} - 10°\text{F}}{55°\text{F} - 21.9°\text{F}}\right)}$$
$$= 34.0°\text{F}$$

Fouling factors are given in Table 36.8.

$$R_{f,i} = 0.0005 \text{ hr-ft}^2\text{-°F/Btu} \quad [\text{brine}]$$
$$R_{f,o} = 0.001 \text{ hr-ft}^2\text{-°F/Btu} \quad [\text{water below } 125°\text{F}]$$

The overall heat transfer coefficient can be calculated from Eq. 36.77. (It is not necessary to convert ratios of dimensions from inches to feet.)

$$\frac{1}{U_o} = \frac{1}{h_o} + \left(\frac{r_o}{k_{\text{tube}}}\right)\ln\left(\frac{r_o}{r_i}\right)$$
$$+ R_{f,o} + \left(\frac{r_o}{r_i}\right)\left(R_{f,i} + \frac{1}{h_i}\right)$$

$$= \frac{1}{246\ \dfrac{\text{Btu}}{\text{hr-ft}^2\text{-}°\text{F}}}$$

$$+ \left[\frac{1.66\ \text{in}}{(2)\left(12\ \dfrac{\text{in}}{\text{ft}}\right)\left(35\ \dfrac{\text{Btu-ft}}{\text{hr-ft}^2\text{-}°\text{F}}\right)}\right]$$

$$\times \ln\left(\frac{1.66\ \text{in}}{1.38\ \text{in}}\right) + 0.001\ \frac{\text{hr-ft}^2\text{-}°\text{F}}{\text{Btu}} + \left(\frac{1.66\ \text{in}}{1.38\ \text{in}}\right)$$

$$\times \left(0.0005\ \frac{\text{hr-ft}^2\text{-}°\text{F}}{\text{Btu}} + \frac{1}{265\ \dfrac{\text{Btu}}{\text{hr-ft}^2\text{-}°\text{F}}}\right)$$

$$= 0.01057\ \text{hr-ft}^2\text{-}°\text{F/Btu}$$

$$U_o = \frac{1}{0.01057\ \dfrac{\text{hr-ft}^2\text{-}°\text{F}}{\text{Btu}}}$$

$$= 94.6\ \text{Btu/hr-ft}^2\text{-}°\text{F}$$

The heat transfer is known. Therefore, the length can be calculated from Eq. 36.70.

$$Q = U_o A_o F_c \Delta T_{lm} = U_o \pi D_o L F_c \Delta T_{lm}$$

$$L = \frac{Q}{U_o \pi D_o F_c \Delta T_{lm}}$$

$$= \frac{\left(50{,}050\ \dfrac{\text{Btu}}{\text{hr}}\right)\left(12\ \dfrac{\text{in}}{\text{ft}}\right)}{\left(94.6\ \dfrac{\text{Btu}}{\text{hr-ft}^2\text{-}°\text{F}}\right)\pi(1.66\ \text{in})(1)(34.0°\text{F})}$$

$$= 35.8\ \text{ft}$$

38. COOLING SUPERHEATED STEAM

It is commonly believed that highly superheated steam renders heat transfer cooling surfaces inefficient due to the steam's low heat-transfer properties. However, the controlling factor is the cooling surface temperature, not the degree of superheat. If the cooling surface temperature is maintained below the cooling fluid below the condensation temperature, performance will be the same for superheated steam as it is for saturated steam. If the superheated steam flow rate and/or the superheat is sufficiently high, the steam will heat the cooling surface to above the condensation temperature, and condensation will slow or cease. In the latter case, a desuperheater will be required.

39. NTU METHOD

Use of Eq. 36.70 is known as the *F-method* or the *LMTD method*. The F-method is suitable for design (i.e., the calculation of the required heat transfer area).

Some heat exchanger analysis problems, such as where both outlet temperatures are unknown, appear to be unsolvable or solvable only by trial and error using the F-method.[73] However, the *number of transfer units (NTU) method* (also known as the *efficiency method* and *effectiveness method*) can be used to handle these problems more easily.[74] The steps in the NTU method depend on whether or not both exit temperatures are known.

The first step in the NTU method is to calculate the *thermal capacity rates*, C, for the two fluids. The smaller capacity rate is designated $C_{\min}$. The larger is designated $C_{\max}$.

$$C = \dot{m}c_p \qquad\qquad 36.81$$

The *heat exchanger effectiveness*, $\mathcal{E}$, is defined as the ratio of the actual heat transfer to the maximum possible heat transfer.[75,76] This ratio is generally not known in advance.

$$\mathcal{E} = \frac{Q_{\text{actual}}}{Q_{\text{ideal}}} \qquad\qquad 36.82$$

If the cold fluid has the minimum capacity rate (i.e., $C_{\min} = C_{\text{cold}}$), the effectiveness is given by Eq. 36.83.

$$\mathcal{E} = \frac{C_{\text{hot}}(T_{\text{hot,in}} - T_{\text{hot,out}})}{C_{\min}(T_{\text{hot,in}} - T_{\text{cold,in}})} \qquad\qquad 36.83$$

If the hot fluid has the minimum capacity rate (i.e., $C_{\min} = C_{\text{hot}}$), the effectiveness is given by Eq. 36.84.

$$\mathcal{E} = \frac{C_{\text{cold}}(T_{\text{cold,out}} - T_{\text{cold,in}})}{C_{\min}(T_{\text{hot,in}} - T_{\text{cold,in}})} \qquad\qquad 36.84$$

If the effectiveness is known, the heat transfer can be calculated from Eq. 36.85. Notice that the temperature difference is the difference of two entering temperatures, which are generally both known.

$$Q = \mathcal{E}C_{\min}(T_{\text{hot,in}} - T_{\text{cold,in}}) \qquad\qquad 36.85$$

The outlet temperatures are generally not known, and Eqs. 36.83 and 36.84 cannot be used. In this case, the NTU method starts by calculating the ratio $C_{\min}/C_{\max}$ and then finding the number of transfer units, NTU, from Eq. 36.86.

$$\text{NTU} = \frac{U_o A_o}{C_{\min}} = \frac{U_i A_i}{C_{\min}} \qquad\qquad 36.86$$

---

[73] Actually, any shell-and-tube heat exchanger with an even number of tube passes has a closed-form analytical solution for the outlet temperature. However, the mathematics are laborious and the form of the solution varies with the type of flow and heat exchanger design.

[74] The names *number of thermal units* (NTU), *heat transfer units* (HTU), and *temperature ratio* (TR) are synonymous with *number of transfer units* (NTU).

[75] The maximum possible transfer can occur only if the heat exchanger has an infinite length.

[76] Other names used in the literature to define the effectiveness are *efficiency, thermodynamic efficiency, temperature efficiency,* and *performance parameter.* The symbol P and ϵ are also used in place of $\mathcal{E}$.

Next, the effectiveness is determined from Eqs. 36.87 or 36.88 or from traditional charts such as those in App. 36.D. For a single-pass heat exchanger operating in counterflow, the effectiveness is

$$\mathcal{E} = \frac{1 - \exp\left(-\text{NTU}\left(1 - \frac{C_{\min}}{C_{\max}}\right)\right)}{1 - \left(\frac{C_{\min}}{C_{\max}}\right)\exp\left(-\text{NTU}\left(1 - \frac{C_{\min}}{C_{\max}}\right)\right)} \quad 36.87$$

For a single-pass heat exchanger operating in parallel flow operation, the effectiveness is

$$\mathcal{E} = \frac{1 - \exp\left(-\text{NTU}\left(1 + \frac{C_{\min}}{C_{\max}}\right)\right)}{1 + \frac{C_{\min}}{C_{\max}}} \quad 36.88$$

Equations 36.87 and 36.88 and similar (though more complex) relationships for the remaining exchanger configurations, given in heat transfer textbooks, were used to develop the charts in App. 36.D.

Example 36.4

40,000 lbm/hr (5.0 kg/s) of lubrication oil are cooled in a single-pass countercurrent heat exchanger by 10,000 lbm/hr (1.3 kg/s) of water. The oil enters at 225°F (105°C). The water enters at 95°F (35°C). The oil's specific heat is 0.45 Btu/lbm-°F (1.9 kJ/kg·K). The overall coefficient of heat transfer for the heat exchanger is 50 Btu/hr-ft^2-°F (280 W/m^2·K) based on an effective heat transfer area of 150 ft^2 (14 m^2) (a) What is the heat transfer? (b) What is the water's exit temperature?

SI Solution

(a) Assume an initial water exit temperature of 93°C to obtain the water's specific heat directly from App. 35.B.

$$c_{p,93°C,\text{water}} = 4.229 \text{ kJ/kg·K}$$

The thermal capacity rates are

$$C_{\text{water}} = \dot{m}c_p = \left(1.3 \frac{\text{kg}}{\text{s}}\right)\left(4.229 \frac{\text{kJ}}{\text{kg·K}}\right)$$
$$= 5.5 \text{ kW/K}$$

$$C_{\text{oil}} = \dot{m}c_p = \left(5.0 \frac{\text{kg}}{\text{s}}\right)\left(1.9 \frac{\text{kJ}}{\text{kg·K}}\right)$$
$$= 9.5 \text{ kW/K}$$

$$C_{\min} = C_{\text{water}} = 5.5 \text{ kW/K}$$
$$C_{\max} = C_{\text{oil}} = 9.5 \text{ kW/K}$$

$$\frac{C_{\min}}{C_{\max}} = \frac{5.5 \frac{\text{kW}}{\text{K}}}{9.5 \frac{\text{kW}}{\text{K}}}$$
$$= 0.579$$

The number of transfer units is

$$\text{NTU} = \frac{UA}{C_{\min}} = \frac{\left(280 \frac{\text{W}}{\text{m}^2\text{·K}}\right)(14 \text{ m}^2)}{\left(5.5 \frac{\text{kW}}{\text{K}}\right)\left(1000 \frac{\text{J}}{\text{kJ}}\right)}$$
$$= 0.71$$

From App. 36.D(a) for a single-pass counterflow heat exchanger with NTU = 0.71 and $C_{\min}/C_{\max} = 0.579$, the effectiveness is $\mathcal{E} \approx 0.45$. (The actual value calculated from Eq. 36.87 is 0.453.)

The heat transfer is

$$Q = \mathcal{E}C_{\min}(T_{\text{hot,in}} - T_{\text{cold,in}})$$
$$= (0.45)\left(5.5 \frac{\text{kW}}{\text{K}}\right)\left(1000 \frac{\text{W}}{\text{kW}}\right)(105°C - 35°C)$$
$$= 1.74 \times 10^5 \text{ W}$$

(b) The water's exit temperature is

$$T_{\text{cold,out}} = T_{\text{cold,in}} + \frac{Q}{\dot{m}c_p}$$
$$= 35°C + \frac{1.74 \times 10^5 \text{ W}}{\left(1.3 \frac{\text{kg}}{\text{s}}\right)\left(4.229 \frac{\text{kJ}}{\text{kg·K}}\right)\left(1000 \frac{\text{J}}{\text{kJ}}\right)}$$
$$= 66.6°C$$

Customary U.S. Solution

(a) Assume an initial water exit temperature of approximately 200°F in order to obtain the water's specific heat. (The value is not sensitive to temperature.) From App. 35.A,

$$c_{p,200°F,\text{water}} = 1.00 \text{ Btu/lbm-°F}$$

The thermal capacity rates are

$$C_{\text{water}} = \dot{m}c_p = \left(10,000 \frac{\text{lbm}}{\text{hr}}\right)\left(1.00 \frac{\text{Btu}}{\text{lbm-°F}}\right)$$
$$= 10,000 \text{ Btu/hr-°F}$$

$$C_{\text{oil}} = \dot{m}c_p = \left(40,000 \frac{\text{lbm}}{\text{hr}}\right)\left(0.45 \frac{\text{Btu}}{\text{lbm-°F}}\right)$$
$$= 18,000 \text{ Btu/hr-°F}$$

$$C_{\min} = C_{\text{water}} = 10,000 \text{ Btu/hr-°F}$$
$$C_{\max} = C_{\text{oil}} = 18,000 \text{ Btu/hr-°F}$$

$$\frac{C_{\min}}{C_{\max}} = \frac{10,000 \frac{\text{Btu}}{\text{hr-°F}}}{18,000 \frac{\text{Btu}}{\text{hr-°F}}}$$
$$= 0.556$$

The number of transfer units is

$$\text{NTU} = \frac{UA}{C_{\min}} = \frac{\left(50 \, \dfrac{\text{Btu}}{\text{hr-ft}^2\text{-}°\text{F}}\right)(150 \text{ ft}^2)}{10{,}000 \, \dfrac{\text{Btu}}{\text{hr-}°\text{F}}}$$

$$= 0.75$$

From App. 36.D(a) for a single-pass counterflow heat exchanger with NTU $= 0.75$ and $C_{\min}/C_{\max} = 0.556$, the effectiveness is $\mathcal{E} \approx 0.47$. (The actual value calculated from Eq. 36.87 is 0.471.)

The heat transfer is

$$Q = \mathcal{E}C_{\min}(T_{\text{hot,in}} - T_{\text{cold,in}})$$

$$= (0.47)\left(10{,}000 \, \frac{\text{Btu}}{\text{hr-}°\text{F}}\right)(225°\text{F} - 95°\text{F})$$

$$= 6.11 \times 10^5 \text{ Btu/hr}$$

(b) The water's exit temperature is

$$T_{\text{cold,out}} = T_{\text{cold,in}} + \frac{Q}{\dot{m}c_p}$$

$$= 95°\text{F} + \frac{6.11 \times 10^5 \, \dfrac{\text{Btu}}{\text{hr}}}{\left(10{,}000 \, \dfrac{\text{lbm}}{\text{hr}}\right)\left(1.00 \, \dfrac{\text{Btu}}{\text{lbm-}°\text{F}}\right)}$$

$$= 156°\text{F}$$

40. HEAT TRANSFER FLUIDS

Indirect heating (cooling) occurs when a process fluid is heated (cooled) by an intermediate fluid. Indirect heat transfer is used when the process fluid cannot be directly exposed to sources of heat or cooling, as when the process fluid is sensitive to hot spots or requires uniform heating.

A *heat transfer fluid* is any substance that is used for indirect heat transfer. Water, brines, mineral oils and other inorganic liquids, gases, molten inorganic salts, and liquid metals (sodium and mercury) are all used in indirect heat transfer.[77] The term "heat transfer fluid," however, is generally used to refer to commercial inorganic liquids.

Water is often the best heat transfer fluid, though it is limited in unpressurized applications to approximately 35 to 190°F (2 to 90°C). Adding ethylene glycol extends the range to approximately −50 to 250°F (−45 to 120°C), but heat transfer properties are impaired.

Commercial heat transfer fluids occupy specific low- and high-temperature niches: indirect heat transfer operations below 32°F (0°C) and above (approximately) 350°F (175°C) up to 600 to 800°F (315 to 425°C). Heat transfer fluids may be used in liquid-phase heat transfer or vapor-phase heat transfer.

A heat transfer fluid can be selected in one of three ways, depending on whether the process is being designed from the ground up, the fluid is being retrofitted into an existing installation, or the pumping energy is to be minimized. From Eq. 36.35, the film coefficient for turbulent flow in a tube is

$$h = (0.023)\left(\frac{k}{d}\right)\left(\frac{dv\rho}{\mu}\right)^{0.8}\left(\frac{c_p\mu}{k}\right)^n \qquad 36.89$$

For a constant mass flow rate, G,

$$h = 0.023G^{0.8}d^{-0.2}K_G \qquad 36.90$$

$$K_G = (c_p)^n k^{1-n} \mu^{n-0.8} \qquad 36.91$$

For a constant volumetric flow rate, $\dot{V}$,

$$h = 0.023\dot{V}^{0.8}d^{-0.2}K_V \qquad 36.92$$

$$K_V = \rho^{0.8}(c_p)^n k^{1-n} \mu^{n-0.8} \qquad 36.93$$

The two *heat transfer factors*, K_G and K_V, contain only physical properties of the fluid. The design engineer keeps the mass flow rate constant and compares K_G values; the plant engineer keeps the volumetric flow rate constant and compares K_V values.[78]

The *Fried heat transfer efficiency factor*, HTEF, is a third factor, usually intermediate between K_G and K_V. It represents the ratio of the film coefficient to the frictional energy expended in pumping the fluid.

$$\text{HTEF} = \frac{19.75\rho^{0.57}(c_p)^n k^{1-n}}{\mu^{0.52}} \qquad 36.94$$

Choice of heat transfer fluid also depends on cost, thermal (heat transfer) properties, fire safety, environmental and toxicological issues, containment issues, chemical stability over the required temperature range, and compatibility with packing, gasket, and wall materials.

Film coefficients for heat transfer fluids are calculated in the same manner as for other fluids (i.e., with Eq. 36.35 for tubeside flow). Fluid properties are generally provided by the manufacturer.[79] Equation 36.95 is a very simplified relationship that can be used for initial estimates with organic liquids.

$$h \approx (423)\left[\frac{(\text{v}_{\text{m/s}})^{0.8}}{(d_{\text{meters}})^{0.2}}\right] \qquad \text{[SI]} \quad 36.95(a)$$

$$h \approx (60)\left[\frac{(\text{v}_{\text{ft/sec}})^{0.8}}{(d_{\text{inches}})^{0.2}}\right] \qquad \text{[U.S.]} \quad 36.95(b)$$

---

[77]Liquids used for commerical heat transfer fluids include alkylated benzenes, alkylated biphenyls, alkylated napthalenes, hydrogenated and unhydrogenated polyphenyls, benzylated aromatics, diphenyl/diphenyl oxide eutectics, aromatic-ether-based fluids, polyalkylene glycols, dicarboloxylic acid esters, polymethyl siloxanes, mineral oils, and inorganic nitrate salts.

[78]Some commercial literature uses the variables "Const. G" and "Const. V" in place of K_G and K_V.

[79]Commercial heat transfer fluids often have the word "therm" in their names, for example, "Dowtherm" (Dow Chemical), "Therminol" (Monsanto), "Mobiltherm" (Mobil Oil), and so on.

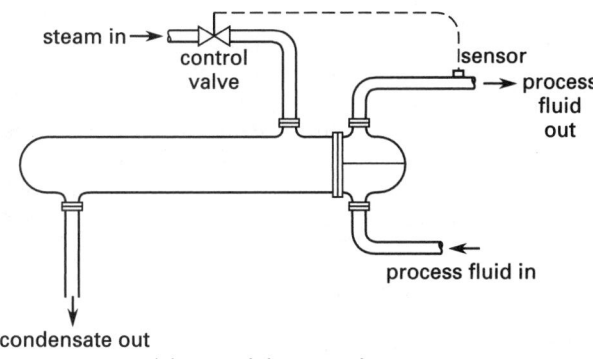

(a) steam inlet control

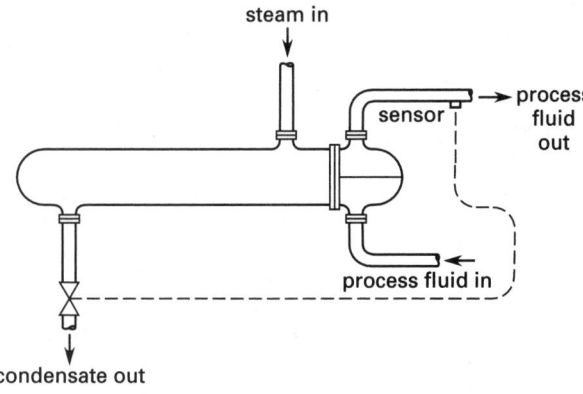

(b) condensate flow control

Figure 36.12 *Control Methods for Steam-Heated Exchangers*

41. CONTROL OF PROCESS HEAT EXCHANGE OPERATIONS

Correct operation of many chemical and other processes requires consistent outlet temperatures, regardless of flow rates. Traditional systems control the inlet flow rate of one fluid by monitoring the outlet temperature of the other. This is generally satisfactory.

Control of steam-heated exchangers at high downturn is more problematic. (*Downturn* is operation with one fluid's flow rate substantially reduced.) Control of steam-heated exchangers has traditionally been with simple steam-inlet control and condensate removal. However, condensate-flow is more successful at high downturn.[80]

42. COOLING ELECTRONIC ENCLOSURES

Electronic equipment generates significant amounts of heat. In some cases, the components generate enough heat to damage themselves. The heat can be calculated from Eq. 36.96, where the current and voltage are

effective values, also known as rms (room-mean-square) values, measured at the power cord.

$$P = \text{current} \times \text{voltage} \qquad 36.96$$

Equipment cooling is accomplished with forced air (fans or blowers), heat exchangers, and air conditioners.[81] Component cooling is usually accomplished by mounting a heat radiator (i.e., fins) on the component.

Fans occupy minimal space and can move large volumes of air against low static heads. Blowers operate (and are most efficient when operating) against higher static heads.[82] Cabinets should always be pressurized by introducing filtered air, rather than drawing the air in. (A vacuum inside the cabinet will attract unfiltered air through the cracks around the panels and door.)

When the cabinet is sealed, or when the ambient temperature near the enclosure is near the maximum permissible operating temperature of the equipment, a heat exchanger (air-to-air or air-to-water) or air conditioner is needed. Air conditioners are the only alternative when the ambient temperature is above the maximum permissible operating temperature.[83]

Rules of thumb regarding forced air cooling are

- The effective area of an air intake will be approximately 65% of the grill area.

- To prevent choking, the flow area inside the cabinet should equal the effective intake area.

- The flow path should not be short circuited. The exhaust area should be downstream, beyond all of the heat-producing components.

- A single excessively hot component can be effectively protected with a baffle (to direct a small amount of high-velocity air over it) or with a dedicated fan.

- Using a booster fan in a two-stage configuration can eliminate the need for a larger, noisier fan.

- Static pressure drop through an enclosure is usually 0.25 to 0.50 in wg (inches of water gage), and is seldom greater than 1.0 in wg.

- Depending on the type of unit and mounting arrangement, the fan/blower motor power may or may not need to be added to the cooled equipment's power.

---

[80]There are problems with condensate-return control, as well. Water hammer and repeated thermal expansion and contraction due to alternating exposure to steam and condensate shorten the exchanger's life. Condensate leaving the exchanger may be well below the saturation temperature, and the condensate will readily absorb uncondensable gases, becoming aggressively corrosive.

[81]Air travels parallel to the rotational axis of a *fan* (i.e., radial flow), air travels normal to the rotational axis of a (centrifugal) *blower*.
[82]Multistage blowers can reach pressures up to 100 in of water.
[83]Environmental heat gain must be added to the equipment power dissipation. It is common to include the surface area of all four sides but to omit half of the area of the top and all of the area of the base when calculating the heat gain. Insulation should be used liberally to limit environmental heat gain.

The airflow required to dissipate equipment power P with an airflow temperature rise of ΔT is given by Eqs. 36.97 and 36.98, which assume a constant specific heat of 0.24 Btu/lbm-°F (1010 J/kg·K) and an air density of 0.075 lbm/ft^3 (1.2 kg/m^3). A 25% increase in air flow is usually added to the calculated air flow as a safety factor.

$$\dot{V}_{\text{cfm}} = \frac{3.16 P_{\text{watts}}}{\Delta T_{°F}}$$

$$= \frac{1.76 P_{\text{watts}}}{\Delta T_{°C}} \qquad 36.97$$

$$\dot{V}_{\text{L/s}} = \frac{0.0278 P_{\text{watts}}}{\Delta T_{°C}} \qquad 36.98$$

Maximum temperature occurs at the inside top of a cabinet. This is a relevant factor only in critical components. Usually, the average cabinet temperature is used in designing the cooling system.

PRACTICE PROBLEMS

Logarithmic Mean Temperature Difference

1. Water is heated from 55°F to 87°F (15°C to 30°C) by stack gases that are cooled from 350°F to 270°F (175°C to 130°C). What is the logarithmic mean temperature difference?

2. A fluid in a tank is maintained at 85°F (30°C) by an immersed hot water coil. The hot water enters at 190°F (90°C) and leaves at 160°F (70°C). What is the logarithmic mean temperature difference?

Film Temperature

3. What film temperature should be used for the heated fluid in Prob. 2?

4. If the fluid in Prob. 2 is gradually heated from 85°F (30°C) to 110°F (45°C), what initial film temperature should be used?

Reynolds Number

5. Light No. 10 oil is heated from 95°F to 105°F (35°C to 41°C) in a 0.6 in (1.52 cm) inside diameter tube. The average velocity of the oil is 2.0 ft/sec (0.6 m/s). The viscosity of the oil at 100°F (38°C) is 45 centistokes. What is the Reynolds number?

Convective Heat Transfer

6. A white, uninsulated, rectangular duct passes through a 50 ft (15 m) wide room. The duct is 18 in (45 cm) wide and 12 in (30 cm) high. The room and its contents are at 70°F (21°C). Air at 100°F (40°C) enters the duct flowing at 800 ft/min (4.0 m/s). The combined convection and radiation film coefficient is 2.0 Btu/hr-ft^2-°F (11 W/m^2·K). (a) What is the heat transfer to the room? (b) What is the temperature of the air after it has traveled in the duct the full 50 ft (15 m)? (c) What is the pressure drop in in (cm) of water due to friction in the duct?

Laminar Flow in Tubes

7. A steel pipe carrying 350°F (175°C) air is 100 ft (30 m) long. The outside and inside diameters are 4.00 in and 3.50 in (10 cm and 9.0 cm), respectively. The pipe is covered with 2.0 in (5.0 cm) of insulation with a thermal conductivity of 0.05 Btu-ft/hr-ft^2-°F (0.086 W/m·K). The pipe passes through a 50°F (10°C) basement. Flow is laminar and fully developed. What is the heat loss?

Turbulent Flow in Tubes

8. An uninsulated horizontal pipe with 4.00 in (10 cm) outside diameter carries saturated 300 psia (2.1 MPa) steam through a 70°F (21°C) room. The steam flow rate is 5000 lbm/hr (0.63 kg/s). What decrease in quality will occur in the first 50 ft (15 m) of length?

Heat Exchangers

9. A tubular feedwater heater is being designed to heat 2940 lbm/hr (0.368 kg/s) of water from 70°F (21°C) to 190°F (90°C). Saturated steam at 134 psi (923 kPa) is condensing on the outside of the tubes. The tubes are a copper alloy. Each tube has a 1 in (2.54 cm) outside diameter and a 0.9 in (2.29 cm) inside diameter. The water velocity inside the tubes is 3 ft/sec (0.9 m/s). What outside tube surface area is required?

10. A two-pass surface feedwater heater is being designed to heat 500,000 lbm/hr (60 kg/s) of water from 200°F to 390°F (100°C to 200°C). The water flows at 5 ft/sec (1.5 m/s) through the tubes. Dry, saturated steam at 400°F (205°C) is to be used as the heating medium. The heater is to operate straight condensing (i.e., the condensed steam will not be mixed with the heated water). Saturated water at 400°F (205°C) is removed from the heater. The tubes in the heater are 7/8 in (2.2 cm) outside diameter with 1/16 in (1.6 mm) walls. The overall heat transfer coefficient is estimated as 700 Btu/hr-ft^2-°F. (a) How many tubes are required? (b) What should be the tube length?

11. A single-pass heat exchanger is tested in a clean condition and is found to heat 100 gal/min (6.3 L/s) of 70°F (21°C) water to 140°F (60°C). The hot side uses 230°F (110°C) steam. The tube's inner surface area is 50 ft^2 (4.7 m^2). After being used in the field for several months, the exchanger heats 100 gal/min (6.3 L/s) of 70°F (21°C water) to 122°F (50°C). What is the fouling factor?

Tubes in Crossflow

12. A glass thermometer has an outside diameter of 0.35 in (8.9 mm). Its uniform temperature is 100°F (38°C). The thermometer is inserted perpendicularly into a 100 ft/sec (30 m/s) airflow. The air temperature is 150°F (66°C). What is the film coefficient on the outside of the thermometer?

37 Radiation and Combined Heat Transfer

1. Thermal Radiation 37-1
2. Emissive Power 37-1
3. Black, Real, and Gray Bodies 37-1
4. Radiation from a Body 37-2
5. Black Body Shape Factor 37-2
6. Gray Body Shape Factor 37-3
7. Net Radiation Heat Transfer 37-4
8. Reciprocity Theorem 37-4
9. Radiation with Reflection/Reradiation . . . 37-4
10. Combined Heat Transfer 37-4
11. Equilibrium Condition with
    Combined Heat Transfer 37-6
12. Solar Radiation 37-7
13. Nocturnal Radiation 37-7
    Practice Problems 37-7

Nomenclature

A	area	ft^2	m^2
E	emissive power	Btu/hr-ft^2	W/m^2
F	factor	–	–
$\mathfrak{F}$	gray body shape factor	–	–
G	geometric flux	ft^2	m^2
h	film coefficient	Btu/hr-ft^2-°F	W/m^2·K
q	unit heat transfer	Btu/hr-ft^2	W/m^2
Q	heat transfer rate	Btu/hr	W
T	temperature	°R	K

Symbols

α	absorptivity	–	–
ϵ	emissivity	–	–
ρ	reflectivity	–	–
σ	Stefan-Boltzmann constant	Btu/hr-ft^2-°R^4	W/m^2·K^4
τ	transmissivity	–	–

Subscripts

12	from body 1 to body 2
21	from body 2 to body 1
a	arrangement
e	emissivity
o	outer
i	inner

1. THERMAL RADIATION

Thermal radiation is electromagnetic radiation with wavelengths in the 0.1 to 100 μm range. All bodies, even "cold" ones, radiate thermal radiation.

Thermal radiation hitting a body can be absorbed, reflected, or transmitted. The *radiation conservation law* is[1]

$$\alpha + \rho + \tau = 1 \qquad \textbf{37.1}$$

2. EMISSIVE POWER

The rate of thermal radiation emitted per unit area of a body is the *emissive power*, E.

$$E = \frac{Q_{\text{radiation}}}{A} \qquad \textbf{37.2}$$

Kirchhoff's radiation law states that for any two bodies in thermal equilibrium, the ratios of emissive power to absorptivity are equal. (Bodies at the same temperature are said to be in *thermal equilibrium.*)

$$\frac{E_1}{\alpha_1} = \frac{E_2}{\alpha_2} \qquad \textbf{37.3}$$

3. BLACK, REAL, AND GRAY BODIES

Since absorptivity, α, cannot exceed 1.0, Kirchhoff's law places an upper limit on emissive power. Bodies that radiate at this upper limit (i.e., $\alpha = 1$) are known as *black bodies* or *ideal radiators*. A black body emits the maximum possible radiation for its temperature and absorbs all incident energy.[2]

Real bodies do not radiate at the ideal level. The ratio of actual to ideal emissive powers is the *emissivity*, ϵ.

$$\epsilon = \frac{E_{\text{actual}}}{E_{\text{black}}} \qquad \textbf{37.4}$$

Emissivity generally has the following characteristics.

- Emissivity varies widely with the surface condition of a material.

- Emissivity is low with highly polished metals.

- Emissivity is high with most nonmetals.

- Emissivity increases with increases in temperature.

---

[1]Notice that emissivity, ϵ, does not appear in the conservation law.
[2]Black body performance can be approximated but not achieved in practice.

The emissivity (and hence the emissive power) usually depends on the temperature of the body.[3] A body that emits at constant emissivity, regardless of wavelength, is known as a *gray body*.

For a black body, both emissivity and absorptivity are 1.0. However, emissivity also equals absorptivity for any body in thermal equilibrium.[4,5]

$$\epsilon = \alpha \quad \text{[thermal equilibrium]} \qquad 37.5$$

For a gray body, the reflectivity is constant and

$$\rho + \epsilon = 1 \qquad 37.6$$

4. RADIATION FROM A BODY

The *Stefan-Boltzmann law*, also known as the *fourth-power law*, gives the total emissive power (E) from a black body. The temperature (T) is expressed in degrees Rankine or in Kelvins. σ is the *Stefan-Boltzmann constant*.

$$E_{\text{black}} = \sigma T^4 \qquad 37.7$$
$$\sigma = 0.1713 \times 10^{-8} \text{ Btu/hr-ft}^2\text{-}^\circ\text{R}^4$$
$$\sigma = 5.67 \times 10^{-8} \text{ W/m}^2\text{·K}^4$$

The radiation from a gray body follows directly from the definition of emissivity.

$$E_{\text{gray}} = \epsilon E_{\text{black}} = \epsilon \sigma T^4 \qquad 37.8$$

5. BLACK BODY SHAPE FACTOR

For two black bodies radiating to each other, a *shape factor*, F_{12}, accounts for the spatial arrangement of the two bodies. (Generally, the smaller body is designated as body 1.) The shape factor is the fraction of the total radiation leaving body 1 that will travel directly to body 2. The shape factor for two black bodies, therefore, is often referred to as the *arrangement factor, geometric factor, geometrical factor, geometric shape factor, angle factor, view factor, interaction factor,* and *configuration factor*, as well as the *black body shape factor*.

The shape factor is 1.0 for two infinite, parallel planes, two infinite coaxial cylinders, and two concentric spheres, since all emitted radiation is absorbed. It is more difficult to evaluate the black body shape factor for more complex arrangements of bodies and surfaces. However, many of the simpler cases have been solved, and their graphical solutions are available.

---

[3] This is equivalent to saying the emissivity depends on the wavelength of the radiation.

[4] "Steady-state operation" would be a better term here since the term "equilibrium" implies temperature equality with another body. In fact, the phrase "the bodies are in thermal equilibrium" means that the body temperatures are equal. In this case, the term "equilibrium" means that the body's temperature is constant.

[5] Equation 37.5 follows directly from Kirchhoff's law (Eq. 37.3).

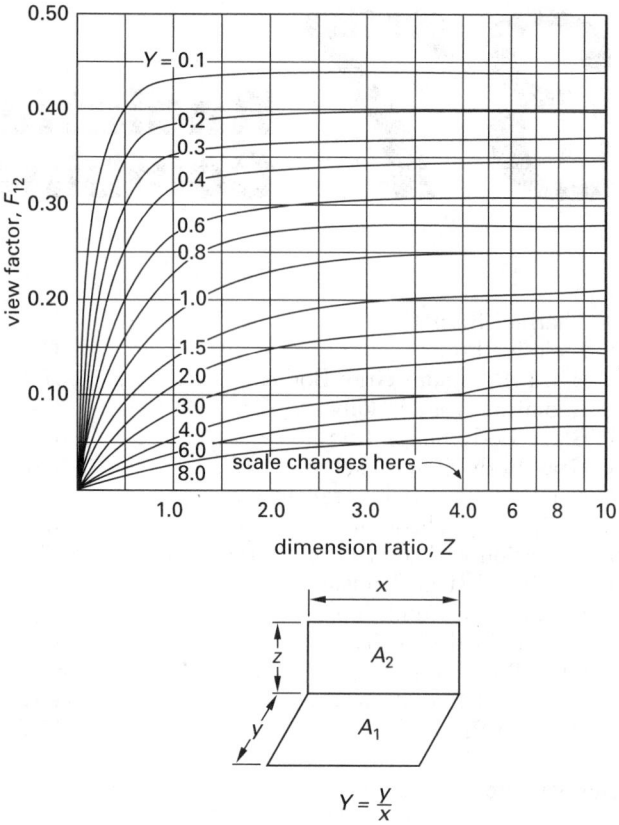

View factors for perpendicular rectangles with a common side. From H. C. Hottel, "Radiant Heat Transmission," *Mechanical Engineering Magazine 52* (1930). By permission of The American Society of Mechanical Engineers.

Figure 37.1 *Black Body Shape Factor*
Adjacent Perpendicular Rectangles

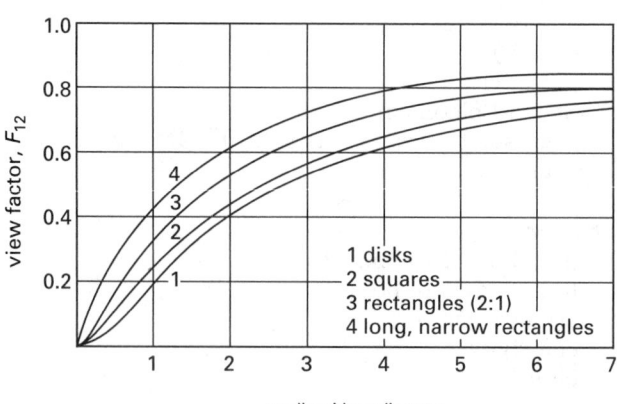

View factors for parallel squares, rectangles, and disks. From H. C. Hottel, "Radiant Heat Transmission," *Mechanical Engineering Magazine 52* (1930). By permission of The American Society of Mechanical Engineers.

Figure 37.2 *Black Body Shape Factor*
(for directly opposed, parallel, finite surfaces)

Table 37.1 *Arrangement and Emissivity Factors*

arrangement	area	F_a	F_e
infinite parallel planes	A_1 or A_2	1	$\dfrac{1}{\dfrac{1}{\epsilon_1} + \dfrac{1}{\epsilon_2} - 1}$
completely enclosed body; small compared with enclosure[a]	A_1	1	ϵ_1
completely enclosed body; large compared with enclosure[a]	A_1	1	$\dfrac{1}{\dfrac{1}{\epsilon_1} + \dfrac{1}{\epsilon_2} - 1}$
concentric spheres or infinite cylinders with diffuse radiation[a]	A_1	1	$\dfrac{1}{\dfrac{1}{\epsilon_1} + \left(\dfrac{A_1}{A_2}\right)\left(\dfrac{1}{\epsilon_2} - 1\right)}$
concentric spheres or infinite cylinders with specular (mirror-like) radiation[a]	A_1	1	$\dfrac{1}{\dfrac{1}{\epsilon_1} + \dfrac{1}{\epsilon_2} - 1}$
two perpendicular rectangles with a common edge	A_1 or A_2	(See Fig. 37.1)	$\epsilon_1 \epsilon_2$
directly opposed, parallel disks, squares, or rectangles of equal size	A_1 or A_2	(See Fig. 37.2)	$\epsilon_1 \epsilon_2$
directly opposed, parallel disks, squares, or rectangles of equal size, connected by nonconducting, reradiating walls	A_1 or A_2	(See Fig. 37.3)	$\epsilon_1 \epsilon_2$

[a]Object 1 is the smaller, enclosed body.

6. GRAY BODY SHAPE FACTOR

Real bodies deviate from black body behavior. To account for the effect of less than ideal emissivities, the black body shape factor, F_{12}, is replaced by the gray body shape factor, $\Im_{12}$. The *gray body shape factor* accounts for the spatial arrangements of the bodies and their emissivities. The lower limit for the gray body shape factor is $\epsilon_1 \epsilon_2$; the upper limit is 1.0.

The gray body shape factor can be written as the product of the black body shape factor (i.e., now referred to as the *arrangement factor, F_a*) and the *emissivity factor, F_e*. The emissivity factor accounts for the departure of the surface from black-body conditions.

$$\Im_{12} = F_{12}F_e = F_a F_e \qquad \textit{37.9}$$

For two gray bodies that radiate to each other (and to no others), the gray body shape factor can be calculated from the black body shape factor.[6] A_1 in Eq. 37.10 is the smaller area.

[6]Although Eq. 37.10 is limited to two bodies that exchange heat with each other and with no other bodies, not all of each body's radiation has to reach the other body. This is evident in the presence of the black body shape factor, F_{12}.

$$A_1 \Im_{12} = \dfrac{1}{\dfrac{1-\epsilon_1}{\epsilon_1 A_1} + \dfrac{1}{A_1 F_{12}} + \dfrac{1-\epsilon_2}{\epsilon_2 A_2}} \qquad \textit{37.10}$$

A special case is two gray bodies with uniform thermal radiation and $F_{12} = 1$. Examples of this case include infinite parallel gray plates, infinite length concentric gray cylinders, and concentric gray spheres. In such cases, Eq. 37.11 can be used. A_1 is the smaller area.

$$A_1 \Im_{12} = \dfrac{1}{\dfrac{1-\epsilon_1}{\epsilon_1 A_1} + \dfrac{1}{A_1} + \dfrac{1-\epsilon_2}{\epsilon_2 A_2}} \qquad \textit{37.11}$$

With two infinite parallel plates, $A_1 = A_2$. Then, the gray body shape factor is

$$\Im_{12} = \dfrac{1}{\dfrac{1}{\epsilon_1} + \dfrac{1}{\epsilon_2} - 1} \qquad \textit{37.12}$$

For a small gray body enclosed by a black body,

$$\Im_{12} = \epsilon_1 \qquad \textit{37.13}$$

7. NET RADIATION HEAT TRANSFER

The net heat transfer due to radiation between two gray bodies at different temperatures is given by Eq. 37.14. The area of body 1 must be used with $\Im_{12}$, and the area of body 2 must be used with $\Im_{21}$. Whether $\Im_{12}$ or $\Im_{21}$ is used depends on which is easier to evaluate.

$$E_{\text{net},12} = \sigma\Im_{12}[(T_1)^4 - (T_2)^4]$$
$$= \sigma F_a F_e[(T_1)^4 - (T_2)^4] \qquad 37.14$$
$$Q_{\text{net},12} = A_1 E_{\text{net},12} \qquad 37.15$$

8. RECIPROCITY THEOREM

Equation 37.17 is known as the *reciprocity theorem for radiation*.

$$Q_{\text{net}} = \sigma A_1 \Im_{12}[(T_1)^4 - (T_2)^4]$$
$$= \sigma A_2 \Im_{21}[(T_1)^4 - (T_2)^4] \qquad 37.16$$
$$A_1 \Im_{12} = A_2 \Im_{21} \qquad 37.17$$

The product of the area and the shape factor is known as the *geometric flux*, G.

$$G_{12} = A_1 \Im_{12} \qquad 37.18$$

The reciprocity theorem can be written in terms of the geometric flux.

$$G_{12} = G_{21} \qquad 37.19$$

9. RADIATION WITH REFLECTION/RERADIATION

Surfaces that reradiate absorbed thermal radiation are known as *refractory materials* or *refractories*. (Furnace walls that reradiate almost all of the thermal energy they receive from combustion flames back to boiler tubes are an example of refractories.) The shape factor for cases with reradiation is traditionally given the symbol $\overline{F}_{12}$. Equation 37.20 (similar to Eq. 37.10) is used to calculate the gray body shape factor when there are two communicating bodies. (Values of $\overline{F}_{12}$ for two black bodies connected by a nonconducting but reradiating wall are given in Fig. 37.3.)

$$A_1 \Im_{12} = \cfrac{1}{\cfrac{1-\epsilon_1}{\epsilon_1 A_1} + \cfrac{1}{A_1 \overline{F}_{12}} + \cfrac{1-\epsilon_2}{\epsilon_2 A_2}} \qquad 37.20$$

10. COMBINED HEAT TRANSFER

When heat is transferred by both radiation and convection, it is convenient to define the *radiant heat transfer coefficient*, $h_{\text{radiation}}$. T_∞ is the free-stream (far-field)

temperature for convective heat transfer.[7] T_1 and T_2 should both be expressed as absolute temperatures.

$$h_{\text{radiation}} = \frac{Q_{\text{net}}}{A_1(T_1 - T_\infty)}$$
$$= \frac{\sigma F_a F_e[(T_1)^4 - (T_2)^4)]}{T_1 - T_\infty} \qquad 37.21$$

The *combined heat transfer coefficient* is[8]

$$h_{\text{total}} = h_{\text{radiation}} + h_{\text{convective}} \qquad 37.22$$

The combined radiation and convective heat transfer is

$$Q = Q_{\text{radiation}} + Q_{\text{convection}}$$
$$= h_{\text{total}} A_1(T_1 - T_\infty) \qquad 37.23$$

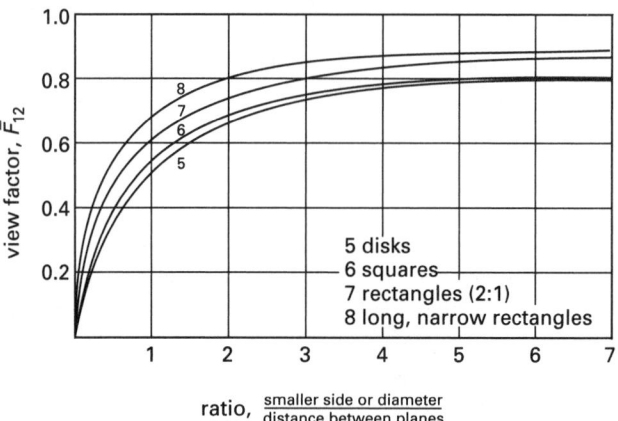

Figure 37.3 *Black Body Shape Factor (for parallel finite squares, rectangles, and disks connected by nonconducting, reradiating wall)*

Example 37.1

A 1.0 ft (30 cm) diameter uninsulated horizontal duct carries hot air through a basement. The duct surface temperature is 200°F (95°C); the duct has an emissivity of 0.8. Air surrounding the duct in the basement is at 40°F (5°C); the basement walls are at 0°F (-20°C). The convective film coefficient on the exterior of the duct is 0.96 Btu/hr-ft²-°F (5.5 W/m²·K). What is the heat loss per unit area of duct?

SI Solution

Since the convective film coefficient for the duct is known, the heat losses from convection and radiation can be calculated independently.

---

[7]T_∞ can be the same as either T_1 or T_2. In that case, the solutions to common types of problems will be greatly simplified since the heat transfer can be calculated separately. If the temperatures are different, trial and error will be necessary to determine the surface temperature.

[8]It is understood that radiation and convection are the combination of heat transfer mechanisms. Conductive heat transfer does not use the film coefficient concept.

The absolute temperatures are

$$T_{\text{duct}} = 95°C + 273 = 368K$$

$$T_\infty = 5°C + 273 = 278K$$

$$T_{\text{walls}} = -20°C + 273 = 253K$$

The convective heat loss is

$$q_{\text{convection}} = \frac{Q}{A} = h(T_{\text{duct}} - T_\infty)$$

$$= \left(5.5 \ \frac{W}{m^2 \cdot K}\right)(368K - 278K)$$

$$= 495 \ W/m^2$$

Since the duct is entirely enclosed by the basement, the arrangement factor is $F_a = 1.0$. The emissivity factor is $F_e = \epsilon_{\text{duct}}$. The radiation heat transfer is

$$E_{\text{net}} = \sigma F_a F_e[(T_{\text{duct}})^4 - (T_{\text{walls}})^4]$$

$$= \left(5.67 \times 10^{-8} \ \frac{W}{m^2 \cdot K^4}\right)(0.8)(1.0)$$

$$\times [(368K)^4 - (253K)^4]$$

$$= 646 \ W/m^2$$

The total heat loss is

$$q_{\text{total}} = q_{\text{convection}} + E_{\text{net}}$$

$$= 495 \ \frac{W}{m^2} + 646 \ \frac{W}{m^2}$$

$$= 1141 \ W/m^2$$

Customary U.S. Solution

Since the convective film coefficient for the duct is known, the heat losses from convection and radiation can be calculated independently.

The absolute temperatures are

$$T_{\text{duct}} = 200°F + 460 = 660°R$$

$$T_\infty = 40°F + 460 = 500°R$$

$$T_{\text{walls}} = 0°F + 460 = 460°R$$

The convective heat loss is

$$q_{\text{convection}} = \frac{Q}{A} = h(T_{\text{duct}} - T_\infty)$$

$$= \left(0.96 \ \frac{Btu}{hr\text{-}ft^2\text{-}°F}\right)(660°R - 500°R)$$

$$= 153.6 \ Btu/hr\text{-}ft^2$$

Since the duct is entirely enclosed by the basement, the arrangement factor is $F_a = 1.0$. The emissivity factor is $F_e = \epsilon_{\text{duct}}$. The radiation heat transfer is

$$E_{\text{net}} = \sigma F_a F_e[(T_{\text{duct}})^4 - (T_{\text{walls}})^4]$$

$$= \left(0.1713 \times 10^{-8} \ \frac{Btu}{hr\text{-}ft^2\text{-}°R^4}\right)(0.8)(1.0)$$

$$\times [(660°R)^4 - (460°R)^4]$$

$$= 198.7 \ Btu/hr\text{-}ft^2$$

The total heat loss is

$$q_{\text{total}} = q_{\text{convection}} + E_{\text{net}}$$

$$= 153.6 \ \frac{Btu}{hr\text{-}ft^2} + 198.7 \ \frac{Btu}{hr\text{-}ft^2}$$

$$= 352.3 \ Btu/hr\text{-}ft^2$$

Example 37.2

The duct in Ex. 37.1 loses heat at the rate of 500 Btu/hr-ft² (1600 W/m²). The duct temperature and film coefficient are unknown. All other values are the same. What is the duct surface temperature?

SI Solution

The unknown surface temperature cannot be extracted directly from the combined heat transfer equation. It is more convenient to solve this problem by trial and error.

Assume laminar convective heat transfer and a duct temperature of 393°R (120°C). From Chap. 35, the convective film coefficient on the outside of the duct is approximately

$$h_{\text{convective}} = (1.32)\left(\frac{T_{\text{duct}} - T_\infty}{L}\right)^{0.25}$$

$$= (1.32)\left(\frac{393K - 278K}{0.3 \ m}\right)^{0.25}$$

$$= 5.84 \ W/m^2 \cdot K$$

The convective heat loss is

$$q_{\text{convection}} = \frac{Q}{A} = h(T_{\text{duct}} - T_\infty)$$

$$= \left(5.84 \ \frac{W}{m^2 \cdot K}\right)(393K - 278K)$$

$$= 672 \ W/m^2$$

The radiation loss is

$$E_{\text{net}} = \sigma F_a F_e[(T_{\text{duct}})^4 - (T_{\text{walls}})^4]$$

$$= \left(5.67 \times 10^{-8} \ \frac{W}{m^2 \cdot K^4}\right)(0.8)(1.0)$$

$$\times [(393K)^4 - (253K)^4]$$

$$= 896 \ W/m^2$$

The total heat loss for this iteration is

$$q_{\text{total}} = q_{\text{convection}} + E_{\text{net}}$$

$$= 672 \ \frac{W}{m^2} + 896 \ \frac{W}{m^2}$$

$$= 1568 \ W/m^2$$

Since the calculated heat loss agrees with the known heat loss, the assumed surface temperature is correct. (Usually, several trial and error iterations would be required to converge on the solution.)

Heat Transfer

Customary U.S. Solution

The unknown surface temperature cannot be extracted directly from the combined heat transfer equation. It is more convenient to solve this problem by trial and error.

Assume laminar convective heat transfer and a duct temperature of 710°R (250°F). From Chap. 35, the convective film coefficient on the outside of the duct is approximately

$$h_{\text{convective}} = (0.27) \left(\frac{T_{\text{duct}} - T_\infty}{L} \right)^{0.25}$$
$$= (0.27) \left(\frac{710°\text{R} - 500°\text{R}}{1 \text{ ft}} \right)^{0.25}$$
$$= 1.03 \text{ Btu/hr-ft}^2\text{-°F}$$

The convective heat loss is

$$q_{\text{convection}} = \frac{Q}{A} = h(T_{\text{duct}} - T_\infty)$$
$$= \left(1.03 \frac{\text{Btu}}{\text{hr-ft}^2\text{-°F}} \right)(710°\text{R} - 500°\text{R})$$
$$= 216.3 \text{ Btu/hr-ft}^2$$

The radiation loss is

$$E_{\text{net}} = \sigma F_a F_e [(T_{\text{duct}})^4 - (T_{\text{walls}})^4]$$
$$= \left(0.1713 \times 10^{-8} \frac{\text{Btu}}{\text{hr-ft}^2\text{-°R}^4} \right)(0.8)(1.0)$$
$$\times [(710°\text{R})^4 - (460°\text{R})^4]$$
$$= 286.9 \text{ Btu/hr-ft}^2$$

The total heat loss for this iteration is

$$q_{\text{total}} = q_{\text{convection}} + E_{\text{net}}$$
$$= 216.3 \frac{\text{Btu}}{\text{hr-ft}^2} + 286.9 \frac{\text{Btu}}{\text{hr-ft}^2}$$
$$= 503.2 \text{ Btu/hr-ft}^2$$

Since the calculated heat loss agrees with the known heat loss, the assumed surface temperature is correct. (Usually, several trial and error iterations would be required to converge on the solution.)

11. EQUILIBRIUM CONDITION WITH COMBINED HEAT TRANSFER

A single body that remains at a constant temperature is said to be in an equilibrium condition.[9] To be in equilibrium, the body must continually lose all of the energy gained. Depending on the situation, a body might radiate all of the energy gained by convection, or it may lose by convection all of the energy gained by radiation.

---

[9]See Ftn. 4.

Example 37.3

A small temperature probe with an emissivity of 0.8 measures the temperature of a gas flowing in a large pipe as 850°F (450°C). The pipe walls are 350°F (180°C). The convective film coefficient for the probe in the gas flow is 27 Btu/hr-ft²-°F (150 W/m²·K). The probe's surface temperature is constant. There is no heat transfer to the probe by conduction. What is the actual gas temperature?

SI Solution

The absolute temperatures are

$$T_{\text{probe}} = 450°\text{C} + 273 = 723\text{K}$$
$$T_{\text{pipe}} = 180°\text{C} + 273 = 453\text{K}$$

Since the probe is entirely enclosed by the large pipe, the arrangement factor is $F_a = 1$, and the emissivity factor is $F_e = 0.8$.

Since the probe's surface temperature is constant, the heat gained by the probe through convection from the gas stream is being lost through radiation to the cooler pipe walls.

$$q_{\text{gain,convection}} = E_{\text{loss}}$$
$$h(T_{\text{gas}} - T_{\text{probe}}) = \sigma F_a F_e [(T_{\text{probe}})^4 - (T_{\text{pipe}})^4]$$
$$\left(150 \frac{\text{W}}{\text{m}^2\text{·K}} \right)(T_{\text{gas}} - 723\text{K}) = \left(5.67 \times 10^{-8} \frac{\text{W}}{\text{m}^2\text{·K}} \right)(0.8)$$
$$\times (1.0)[(723\text{K})^4 - (453\text{K})^4]$$

Solving directly,

$$T_{\text{gas}} = 793\text{K}$$

Customary U.S. Solution

The absolute temperatures are

$$T_{\text{probe}} = 850°\text{F} + 460 = 1310°\text{R}$$
$$T_{\text{pipe}} = 350°\text{F} + 460 = 810°\text{R}$$

Since the probe is entirely enclosed by the large pipe, the arrangement factor is $F_a = 1$, and the emissivity factor is $F_e = 0.8$.

Since the probe's surface temperature is constant, the heat gained by the probe through convection from the gas stream is being lost through radiation to the cooler pipe walls.

$$q_{\text{gain,convection}} = E_{\text{loss}}$$
$$h(T_{\text{gas}} - T_{\text{probe}}) = \sigma F_a F[(T_{\text{probe}})^4 - (T_{\text{pipe}})^4]$$
$$\left(27 \frac{\text{Btu}}{\text{hr-ft}^2\text{-}°\text{F}}\right)$$
$$\times (T_{\text{gas}} - 1310°\text{R}) = \left(0.1713 \times 10^{-8} \frac{\text{Btu}}{\text{hr-ft}^2\text{-}°\text{R}^4}\right)$$
$$\times (0.8)(1.0)$$
$$\times [(1310°\text{R})^4 - (810°\text{R})^4]$$

Solving directly,

$$T_{\text{gas}} = 1438°\text{R}$$

12. SOLAR RADIATION

The average solar energy hitting the outer edge of the earth's atmosphere is approximately 442 Btu/hr-ft^2 (1.41 kW/m^2) and is known as the *solar constant*. The actual instantaneous value depends on the altitude, latitude, time of year, time of day, sky conditions, and orientation angle of the receiving body.[10,11]

13. NOCTURNAL RADIATION

Measurements of the night sky show that the effective temperature of the sky for purposes of radiation is approximately 410°R (210K) on cold clear nights. Since this temperature is below the freezing point of water, it is possible for standing water and tree fruits to freeze even when the air temperature is above the freezing point of water.

Example 37.4

On a cold, clear night, the surface of a pond has a convective film coefficient of 5 Btu/hr-ft^2-°R (28 W/m^2·K). The emissivity of the pond surface is 0.96. The air temperature is 60°F (15°C). The pond temperature remains the same all night long. Evaporative losses are negligible. What is the water temperature?

SI Solution

With an effective sky temperature of 210K, the pond will lose heat through radiation. Since pond temperature is constant, it gains heat from the air through convection. Therefore, the pond temperature is less than 15°C.

The absolute temperatures are

$$T_{\text{sky}} = 210\text{K}$$
$$T_{\text{air}} = 15°\text{C} + 273 = 288\text{K}$$

The equilibrium equation is

$$q_{\text{gain,convection}} = E_{\text{loss}}$$
$$h(T_{\text{air}} - T_{\text{pond}}) = \sigma F_a F_e[(T_{\text{pond}})^4 - (T_{\text{sky}})^4]$$
$$\left(28 \frac{\text{W}}{\text{m}^2\text{·K}}\right)(288\text{K} - T_{\text{pond}}) = \left(5.67 \times 10^{-8} \frac{\text{W}}{\text{m}^2\text{·K}^4}\right)(0.96)$$
$$\times (1.0)[(T_{\text{pond}})^4 - (210°\text{K})^4]$$

Solving by trial and error,

$$T_{\text{pond}} \approx 280\text{K} \ (7°\text{C})$$

Customary U.S. Solution

With an effective sky temperature of 410°R, the pond will lose heat through radiation. Since pond temperature is constant, it gains heat from the air through convection. Therefore, the pond temperature is less than 60°F.

The absolute temperatures are

$$T_{\text{sky}} = 410°\text{R}$$
$$T_{\text{air}} = 60°\text{F} + 460 = 520°\text{R}$$

The equilibrium equation is

$$q_{\text{gain,convection}} = E_{\text{loss}}$$
$$h(T_{\text{air}} - T_{\text{pond}}) = \sigma F_a F_e[(T_{\text{pond}})^4 - (T_{\text{sky}})^4]$$
$$\left(5 \frac{\text{Btu}}{\text{hr-ft}^2\text{-}°\text{R}}\right)$$
$$\times (520°\text{R} - T_{\text{pond}}) = \left(0.1713 \times 10^{-8} \frac{\text{Btu}}{\text{hr-ft}^2\text{-}°\text{R}^4}\right)$$
$$\times (0.96)(1.0)$$
$$\times [(T_{\text{pond}})^4 - (410°\text{R})^4]$$

Solving by trial and error,

$$T_{\text{pond}} \approx 508°\text{R} \ (48°\text{F})$$

PRACTICE PROBLEMS

Arrangement Factors

1. A 6 in (15 cm) thick furnace wall has a 3 in (8 cm) square peephole. The interior of the furnace is at 2200°F (1200°C). The surrounding air temperature is 70°F (20°C). What is the heat loss due to radiation when the peephole is open?

Combined Heat Transfer

2. A 9 in (23 cm) diameter duct is painted with white lacquer. The surface of the duct is at 200°F (95°C). The duct carries hot air through a room whose walls are 70°F (20°C). The air in the room is at 80°F (27°C). What is the unit heat transfer?

---

[10]The value of the solar constant given can be used at altitudes higher than approximately 50,000 ft (15,000 m).
[11]Most heat transfer textbooks detail the method of calculating the exact instantaneous solar heat gain. Books on the subject of HVAC cover this subject particularly well.

3. The walls of a cold storage unit have the cross section shown. (a) What is the heat transfer per unit area of wall? (b) What is the temperature of the aluminum foil?

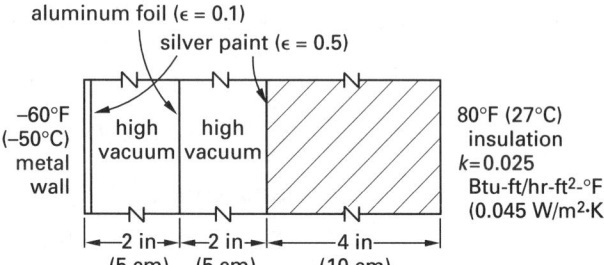

aluminum foil ($\epsilon = 0.1$)

silver paint ($\epsilon = 0.5$)

–60°F (–50°C) metal wall

high vacuum

high vacuum

80°F (27°C) insulation k=0.025 Btu-ft/hr-ft²-°F (0.045 W/m²·K)

2 in (5 cm) — 2 in (5 cm) — 4 in (10 cm)

4. Dry air at 1 atmospheric pressure flows at 500 ft³/min (0.25 m³/s) through 50 ft (15 m) of 12 in (30 cm) diameter uninsulated duct. The emissivity of the duct surface is 0.28. Air enters the duct at 45°F (7°C). The walls, air, and contents of the room through which the duct passes are at 80°F (27°C). An engineer states that the air leaving the duct will be at 50°F (10°C). Consider both convection and radiation to prove or disprove the engineer's statement.

5. A steel pipe is painted on the outside with dull gray (oil-based) paint. The pipe is 35 ft (10 m) long. The pipe is 4.00 in (10.2 cm) inside diameter and 4.25 in (10.8 cm) outside diameter. The pipe carries 200 ft³/min (0.1 m³/s) of 500°F (260°C), 25 psig (170 kPa) air through a 70°F (20°C) room. The conditions of the air at the end of the pipe are 350°F (180°C) and 15 psig (100 kPa). (a) What is the overall coefficient of heat transfer? (b) Using theoretical methods or empirical correlations, what is the calculated overall coefficient of heat transfer? (c) Explain the possible reasons for differences between the actual and calculated overall coefficients of heat transfer.

6. A semiconductor device is modeled as an upright circular cylinder 0.75 in (19 mm) in diameter and 1.5 in (38 mm) high. The device emits 5.0 W and is cooled by a combination of natural convection and radiation. The surface emissivity is 0.65. The base is insulated and transmits no heat. The air and surroundings are at 14.7 psia (101 kPa) and 75°F (24°C). (a) What is the surface temperature of the device? (b) What percentages of heat are lost through convection and radiation?

7. The temperature of a gas in a duct with 600°F (315°C) walls is evaluated with a 0.5 in (13 mm) diameter thermocouple probe. The emissivity of the probe is 0.8. The gas flow rate is 3480 lbm/hr-ft² (4.7 kg/s·m²). The gas velocity is 400 ft/min (2 m/s). The film coefficient on the probe is given empirically as

$$h = \frac{0.024 G^{0.8}}{D^{0.4}}$$

h in Btu/hr-ft²-°F

G in lbm/hr-ft²

D in ft

$$h = \frac{2.9 G^{0.8}}{D^{0.4}}$$

h in W/m²·K

G in kg/s·m²

D in m

(a) If the actual gas temperature is 300°F (150°C), what is the probe's reading? (b) If the probe reading indicates that the gas temperature is 300°F (150°C), what is the actual gas temperature?

38 Psychrometrics

1. Introduction to Psychrometrics 38-1
2. Properties of Atmospheric Air 38-1
3. Vapor Pressure 38-2
4. Energy Content of Air 38-3
5. The Psychrometric Chart 38-3
6. Enthalpy Corrections 38-4
7. Basis of Properties 38-4
8. Lever Rule 38-5
9. Adiabatic Mixing of Two Air Streams 38-5
10. Air Conditioning Processes 38-6
11. Sensible Heat Ratio 38-6
12. Bypass Factor and Coil Efficiency 38-7
13. Adiabatic Saturation Processes 38-7
14. Straight Humidification 38-8
15. Sensible Cooling and Heating 38-8
16. Cooling with Coil Dehumidification 38-8
17. Cooling with Humidification 38-10
18. Cooling with Spray Dehumidification . . . 38-10
19. Heating with Humidification 38-10
20. Heating and Dehumidification 38-11
21. Air Washers 38-11
22. Wet Cooling Towers 38-11
23. Cooling Tower Blowdown 38-13
24. Dry Cooling Towers 38-13
    Practice Problems 38-13

Nomenclature

ADP	apparatus dew point	°F	°C
B	volumetric fraction	–	–
BF	bypass factor	–	–
C	concentration	ppm	mg/L
c_p	specific heat	Btu/lbm-°F	kJ/kg·°C
G	gravimetric fraction	–	–
h	enthalpy	Btu/lbm	kJ/kg
m	mass	lbm	kg
n	number of moles	–	–
p	pressure	lbf/ft^2	kPa
PF	performance factor	–	–
q	heat	Btu/lbm	J/kg
R	specific gas constant	ft-lbf/lbm-°R	kJ/kg·K
RF	rating factor	–	–
SHR	sensible heat ratio	–	–
T	temperature	°F	°C
TDS	total dissolved solids	ppm	mg/L
TU	transfer units	–	–
V	volume	ft^3	m^3
x	mole fraction	–	–

Symbols

η	efficiency	–	–
μ	degree of saturation	–	–
ρ	mass density	lbm/ft^3	kg/m^3
υ	specific volume	ft^3/lbm	m^3/kg
ϕ	relative humidity	–	–
ω	humidity ratio	lbm/lbm	kg/kg

Subscripts

a	dry air
db	dry-bulb
dp	dew-point
fg	vaporization
l	latent
s	sensible
sat	saturation
t	total
unsat	unsaturated
v	vapor
w	water
wb	wet-bulb

1. INTRODUCTION TO PSYCHROMETRICS

Atmospheric air contains small amounts of moisture and can be considered to be a mixture of two ideal gases—dry air and water vapor. All of the thermodynamic rules relating to the behavior of nonreacting gas mixtures apply to atmospheric air. From Dalton's law, for example, the total atmospheric pressure is the sum of the dry air partial pressure and the water vapor pressure.[1]

$$p = p_a + p_w \qquad 38.1$$

The study of the properties and behavior of atmospheric air is known as *psychrometrics*. Properties of atmospheric air are seldom evaluated, however, from theoretical thermodynamic principles. Rather, specialized techniques and charts have been developed for that purpose.

2. PROPERTIES OF ATMOSPHERIC AIR

At first, psychrometrics seems complicated by three different definitions of temperature. These three terms are not interchangeable.

- *dry-bulb temperature*, T_{db}: This is the equilibrium temperature that a regular thermometer measures if exposed to atmospheric air.

---

[1] Equation 38.1 points out a problem in semantics. The term *air* means *dry air*. The term *atmosphere* refers to the combination of dry air and water vapor. It is common to refer to the atmosphere as *moist air*.

HVAC

- *wet-bulb temperature*, T_{wb}: This is the temperature of air that has gone through an adiabatic saturation process. (See Sec. 13.)

- *dew-point temperature*, T_{dp}: This is the dry-bulb temperature at which water starts to condense out when moist air is cooled in a constant pressure process.

For every temperature, there is a unique vapor pressure, p_{sat}, which represents the maximum pressure the water vapor can exert. The actual vapor pressure, p_v, can be less than or equal to, but not greater than, the saturation value. The saturation pressure is found from steam tables as the pressure corresponding to the dry-bulb temperature of the atmospheric air.

$$p_v \leq p_{sat} \qquad 38.2$$

If the vapor pressure equals the saturation pressure, the air is said to be saturated.[2] *Saturated air* is a mixture of dry air and saturated water vapor. When the air is saturated, all three temperatures are equal.

$$T_{db} = T_{wb} = T_{dp} \Big|_{sat} \qquad 38.3$$

Unsaturated air is a mixture of dry air and superheated water vapor.[3] When the air is unsaturated, the dew-point temperature will be less than the wet-bulb temperature. The *wet-bulb depression* is the difference between the dry-bulb and wet-bulb temperatures.

$$T_{dp} < T_{wb} < T_{db} \Big|_{unsat} \qquad 38.4$$

The amount of water vapor in atmospheric air is specified by three different parameters. The *humidity ratio*, ω (also known as the *specific humidity*), is the mass ratio of water vapor to dry air. If both masses are expressed in pounds (kilograms), the units of humidity ratio are lbm/lbm (kg/kg). However, since there is so little water vapor, the water vapor mass is often reported in *grains* of water. (There are 7000 grains per pound.) Accordingly, the humidity ratio will have the units of grains per pound.

$$\omega = \frac{m_w}{m_a} \qquad 38.5$$

Since $m = \rho V$, and since $V_w = V_a$, the humidity ratio can be written as

$$\omega = \frac{\rho_w}{\rho_a} \qquad 38.6$$

---

[2] Actually, the water vapor is saturated, not the air. However, this particular inconsistency in terms is characteristic of psychrometrics.

[3] As strange as it sounds, atmospheric water vapor is almost always superheated. This can be shown by drawing an isotherm passing through the vapor dome on a p-V diagram. The only place where the water vapor pressure is less than the saturation pressure is in the superheated region.

From the equation of state for an ideal gas, $m = pV/RT$. Since $V_w = V_a$ and $T_w = T_a$, the humidity ratio can be written in one additional form.

$$\omega = \frac{R_a p_w}{R_w p_a} = \frac{53.35 \, p_w}{85.78 \, p_a} = (0.622)\left(\frac{p_w}{p_a}\right) \qquad 38.7$$

The *degree of saturation*, μ (also known as the *saturation ratio* and the *percentage humidity*), is the ratio of the actual humidity ratio to the saturated humidity ratio at the same temperature and pressure.

$$\mu = \frac{\omega}{\omega_{sat}} \qquad 38.8$$

A third index of moisture content is the *relative humidity*—the partial pressure of the water vapor divided by the saturation pressure.

$$\phi = \frac{p_w}{p_{sat}} \qquad 38.9$$

From the equation of state for an ideal gas, $\rho = p/RT$, so the relative humidity can be written as

$$\phi = \frac{\rho_w}{\rho_{sat}} \qquad 38.10$$

Combining the definitions of specific and relative humidities,

$$\phi = 1.608 \, \omega \left(\frac{p_a}{p_{sat}}\right) \qquad 38.11$$

3. VAPOR PRESSURE

There are at least five methods in use for determining the partial pressure, p_w, of the water vapor in the air. The first method, derived from Eq. 38.9, is to multiply the relative humidity, ϕ, by the water's saturation pressure. The saturation pressure, in turn, is obtained from steam tables as the pressure corresponding to the air's dry-bulb temperature.

$$p_w = \phi p_{sat,db} \qquad 38.12$$

A more direct method is to read the saturation pressure (from the steam tables) corresponding to the air's dew-point temperature.

$$p_w = p_{sat,dp} \qquad 38.13$$

The third method can be used if water's mole (volumetric) fraction is known.

$$p_w = x_w p_t = B_w p_t \qquad 38.14$$

The fourth method is to calculate the actual vapor pressure from the empirical *Carrier equation*, valid for Customary U.S. units only.[4]

$$p_w = p_{sat,wb} - \frac{(p_t - p_{sat,wb})(T_{db} - T_{wb})}{2830 - 1.44 T_{wb}} \quad \text{[U.S.]} \quad 38.15$$

---

[4] An *air washer* is basically a *spray chamber* through which air passes. When supplied with chilled water from a refrigeration source, the air washer can cool, dehumidify, or humidify the air. Air washers can be used without refrigeration to cool and humidify the air through an evaporative cooling process.

The fifth method is based on the humidity ratio.

$$p_w = \frac{p_t \omega}{0.622 + \omega} \qquad 38.16$$

Example 38.1

Use the methods described in the previous section to determine the partial pressure of water vapor in standard atmospheric air at 60°F (16°C) dry-bulb and 50% relative humidity.

SI Solution

method 1: From the steam tables, the saturation pressure corresponding to 16°C is 0.01818 bars. The partial pressure of the vapor is

$$p_w = \phi p_{\text{sat}} = (0.50)(0.01818 \text{ bars})\left(100 \ \frac{\text{kPa}}{\text{bar}}\right)$$
$$= 0.909 \text{ kPa}$$

method 2: The dew-point temperature (reading straight across on the psychrometric chart) is approximately 5°C. The saturation pressure from the steam table corresponding to 5°C is approximately 0.0087 bars (0.87 kPa).

method 3: The humidity ratio is 0.0056 kg/kg. From Eq. 38.16,

$$p_w = \frac{p_t \omega}{0.622 + \omega}$$
$$= \frac{(101.3 \text{ kPa})\left(0.0056 \ \frac{\text{kg}}{\text{kg}}\right)}{0.622 + 0.0056 \ \frac{\text{kg}}{\text{kg}}}$$
$$= 0.904 \text{ kPa}$$

Customary U.S. Solution

method 1: From the steam tables, the saturation pressure corresponding to 60°F is 0.2563 lbf/in². The partial pressure of the vapor is

$$p_w = \phi p_{\text{sat}} = (0.50)\left(0.2563 \ \frac{\text{lbf}}{\text{in}^2}\right)$$
$$= 0.128 \text{ lbf/in}^2$$

method 2: The dew-point temperature (reading straight across the psychrometric chart) is approximately 41°F. The saturation pressure from the steam table corresponding to 41°F is approximately 0.127 lbf/in².

method 3: Use the Carrier equation. The wet-bulb temperature of the air is approximately 50°F. From the steam tables, the saturation pressure corresponding to that temperature is 0.1780 lbf/in².

$$p_w = p_{\text{sat,wb}} - \frac{(p_t - p_{\text{sat,wb}})(T_{\text{db}} - T_{\text{wb}})}{2830 - 1.44 T_{\text{wb}}}$$

$$= 0.1780 \ \frac{\text{lbf}}{\text{in}^2} - \frac{\left(14.7 \ \frac{\text{lbf}}{\text{in}^2} - 0.1780 \ \frac{\text{lbf}}{\text{in}^2}\right) \times (60°F - 50°F)}{2830 - (1.44)(50°F)}$$

$$= 0.125 \text{ lbf/in}^2$$

4. ENERGY CONTENT OF AIR

Since moist air is a mixture of dry air and water vapor, its total enthalpy, h (i.e., energy content), takes both components into consideration. Total enthalpy is conveniently shown on the diagonal scales of the psychrometric chart, but it can also be calculated. As Eq. 38.18 indicates, the reference temperature for the enthalpy of dry air is 0°F (0°C).

$$h_t = h_a + \omega h_w \qquad 38.17$$
$$h_a = c_{p,\text{air}}T \approx \left(1.005 \ \frac{\text{kJ}}{\text{kg-°C}}\right)T_{°C} \quad \text{[SI]} \quad 38.18(a)$$
$$h_a = c_{p,\text{air}}T \approx \left(0.240 \ \frac{\text{Btu}}{\text{lbm-°F}}\right)T_{°F} \quad \text{[U.S.]} \quad 38.18(b)$$
$$h_w = c_{p,\text{water vapor}}T + h_{\text{fg}}$$
$$\approx \left(1.805 \ \frac{\text{kJ}}{\text{kg-°C}}\right)T_{°C} + 2501 \ \frac{\text{kJ}}{\text{kg}}$$
$$\text{[SI]} \quad 38.19(a)$$
$$h_w = c_{p,\text{water vapor}}T + h_{fg}$$
$$\approx \left(0.444 \ \frac{\text{Btu}}{\text{lbm-°F}}\right)T_{°F} + 1061 \ \frac{\text{Btu}}{\text{lbm}}$$
$$\text{[U.S.]} \quad 38.19(b)$$

5. THE PSYCHROMETRIC CHART

It is possible to develop mathematical relationships for enthalpy and specific volume (the two most useful thermodynamic properties) for atmospheric air. However, these relationships are almost never used. Rather, psychrometric properties can be read directly from *psychrometric charts* ("psych charts," as they are usually referred to), as illustrated in Apps. 38.A and 38.B. There are different psychrometric charts for low, medium, and high temperature ranges, as well as charts for different atmospheric pressures (i.e., elevations).

The usage of several scales varies somewhat from chart to chart. In particular, the use of the enthalpy scale depends on the chart used. Furthermore, not all psychrometric charts contain all scales.

A psychrometric chart is easy to use, despite the multiplicity of scales. The thermodynamic state (i.e., the position on the chart) is defined by specifying the values of any two parameters on intersecting scales (e.g., dry-bulb and wet-bulb temperature, or dry-bulb temperature and relative humidity). Once the state point has been located on the chart, all other properties can be read directly.

6. ENTHALPY CORRECTIONS

Some psychrometric charts have separate lines or scales for dew point and enthalpy. However, the deviation between lines of constant dew point and lines of constant enthalpy is small. Therefore, other psychrometric charts use only one set of diagonal lines for both scales. The error introduced is small, seldom greater than 0.1 to 0.2 Btu/lbm (0.23 to 0.46 kJ/kg). When extreme precision is needed, correction factors from the psychrometric chart can be used.

Example 38.2

Air at 50°F (10°C) dry bulb has a humidity ratio of 0.006 lbm/lbm (0.006 kg/kg). (a) Use the psychrometric chart to determine the enthalpy of the air. (b) Calculate the enthalpy of the air directly. (c) How much heat is needed to heat one unit mass of the air from 50°F to 140°F (10°C to 60°C) without changing the moisture content?

SI Solution

(a) Use the moisture content and dry-bulb temperature scales to locate the point corresponding to the original conditions. From the psychrometric chart, the enthalpy is approximately 25 kJ/kg.

(b) Use Eqs. 38.17 and 38.18.

$$h_a = c_{p,\text{air}} T$$
$$\approx \left(1.005 \ \frac{\text{kJ}}{\text{kg} \cdot {}^\circ\text{C}} \right) T_{{}^\circ\text{C}}$$
$$= \left(1.005 \ \frac{\text{kJ}}{\text{kg} \cdot {}^\circ\text{C}} \right) (10^\circ\text{C}) = 10 \ \text{kJ/kg}$$

$$h_w = c_{p,\text{water vapor}} T + h_{fg}$$
$$\approx \left(1.805 \ \frac{\text{kJ}}{\text{kg} \cdot {}^\circ\text{C}} \right) T_{{}^\circ\text{C}} + 2501 \ \frac{\text{kJ}}{\text{kg}}$$
$$= \left(1.805 \ \frac{\text{kJ}}{\text{kg} \cdot {}^\circ\text{C}} \right) (10^\circ\text{C}) + 2501 \ \frac{\text{kJ}}{\text{kg}}$$
$$= 2519 \ \text{kJ/kg}$$

$$h_t = h_a + \omega h_w$$
$$= 10 \ \frac{\text{kJ}}{\text{kg}} + \left(0.006 \ \frac{\text{kg}}{\text{kg}} \right) \left(2519 \ \frac{\text{kJ}}{\text{kg}} \right)$$
$$= 25.1 \ \text{kJ/kg}$$

(c) The psychrometric chart does not go up to 60°C. Therefore, the energy difference must be calculated mathematically. Although the initial enthalpy could be subtracted from the calculated final enthalpy, it is equivalent merely to calculate the difference based on the variable terms.

$$q = h_{t,2} - h_{t,1}$$
$$= (c_{p,\text{air}} + \omega c_{p,\text{water vapor}})(T_2 - T_1)$$
$$= \left[\left(1.005 \ \frac{\text{kJ}}{\text{kg} \cdot {}^\circ\text{C}} \right) + \left(0.006 \ \frac{\text{kg}}{\text{kg}} \right) \left(1.805 \ \frac{\text{kJ}}{\text{kg} \cdot {}^\circ\text{C}} \right) \right]$$
$$\times (60^\circ\text{C} - 10^\circ\text{C})$$
$$= 50.8 \ \text{kJ/kg}$$

Customary U.S. Solution

(a) Use the moisture content and dry-bulb temperature scales to locate the point corresponding to the original conditions. From the psychrometric chart, the enthalpy is approximately 18.5 Btu/lbm.

(b) Use Eqs. 38.17 and 38.18.

$$h_a = c_{p,\text{air}} T$$
$$\approx \left(0.240 \ \frac{\text{Btu}}{\text{lbm-}{}^\circ\text{F}} \right) T_{{}^\circ\text{F}}$$
$$= \left(0.240 \ \frac{\text{Btu}}{\text{lbm-}{}^\circ\text{F}} \right) (50^\circ\text{F}) = 12 \ \text{Btu/lbm}$$

$$h_w = c_{p,\text{water vapor}} T + h_{fg}$$
$$\approx \left(0.444 \ \frac{\text{Btu}}{\text{lbm-}{}^\circ\text{F}} \right) T_{{}^\circ\text{F}} + 1061 \ \frac{\text{Btu}}{\text{lbm}}$$
$$= \left(0.444 \ \frac{\text{Btu}}{\text{lbm-}{}^\circ\text{F}} \right) (50^\circ\text{F}) + 1061 \ \frac{\text{Btu}}{\text{lbm}}$$
$$= 1083.2 \ \text{Btu/lbm}$$

$$h_t = h_a + \omega h_w$$
$$= 12 \ \frac{\text{Btu}}{\text{lbm}} + \left(0.006 \ \frac{\text{lbm}}{\text{lbm}} \right) \left(1083.2 \ \frac{\text{Btu}}{\text{lbm}} \right)$$
$$= 18.5 \ \text{Btu/lbm}$$

(c) The psychrometric chart does not go up to 140°F. Therefore, the energy difference must be calculated mathematically. Although the initial enthalpy could be subtracted from the calculated final enthalpy, it is equivalent merely to calculate the difference based on the variable terms.

$$q = h_{t,2} - h_{t,1}$$
$$= (c_{p,\text{air}} + \omega c_{p,\text{water vapor}})(T_2 - T_1)$$
$$= \left[\left(0.240 \ \frac{\text{Btu}}{\text{lbm-}{}^\circ\text{F}} \right) + \left(0.006 \ \frac{\text{lbm}}{\text{lbm}} \right) \right.$$
$$\left. \times \left(0.444 \ \frac{\text{Btu}}{\text{lbm-}{}^\circ\text{F}} \right) \right] (140^\circ\text{F} - 50^\circ\text{F})$$
$$= 21.84 \ \text{Btu/lbm}$$

7. BASIS OF PROPERTIES

Several of the properties read from the psychrometric chart (specific volume, enthalpy, etc.) are given "per pound of dry air." This basis does not mean that the water vapor's contribution is absent. For example, if the enthalpy of atmospheric air is 28.0 Btu per pound of dry air, the energy content of the water vapor has been included. However, to get the energy of a mass of moist air, the enthalpy of 28 Btu/lbm would be multiplied by the mass of the dry air (m_a) only, not by the combined air and water masses.

$$h_t = m_a h_{\text{chart}} \qquad \textbf{38.20}$$

Example 38.3

During the summer, air in a room reaches 75°F and 50% relative humidity. Find the air's (a) wet-bulb temperature, (b) humidity ratio, (c) enthalpy, (d) specific volume, (e) dew-point temperature, (f) actual vapor pressure, and (g) degree of saturation.

Solution

Locate the point where the 75°F vertical line intersects the curved 50% humidity line. Read all other values directly from the chart.

(a) Follow the diagonal line up to the left until it intersects the wet-bulb temperature scale. Read $T_{\text{wb}} = 62.6°F$.

(b) Follow the horizontal line to the right until it intersects the humidity ratio scale. Read $\omega = 64.8$ gr (0.0093 lbm) of moisture per pound of dry air.

(c) Finding the enthalpy is different on different charts. Some charts use the same diagonal lines for wet-bulb temperature and humidity. Corrections are required in such cases. Other charts employ two alignment scales to use in conjunction with a straightedge. Read 28.1 Btu per pound of dry air and a correction of approximately −0.09 Btu/lbm.

$$h = 28.1 \frac{\text{Btu}}{\text{lbm}} - 0.09 \frac{\text{Btu}}{\text{lbm}}$$
$$= 28.01 \text{ Btu/lbm dry air}$$

(d) Interpolate between diagonal specific volume lines. Read $v = 13.68$ cubic feet per pound of dry air.

(e) Follow the horizontal line to the left until it intersects the dew-point scale. Read $T_{\text{dp}} = 55.1°F$.

(f) From the steam tables, the saturation pressure corresponding to a dry-bulb temperature of 75°F is approximately 0.43 psia. From Eq. 38.9, the water vapor pressure is

$$p_w = \phi p_{\text{sat}} = (0.50)(0.43 \text{ psia}) = 0.215 \text{ psia}$$

(g) The humidity ratio at 75°F saturated is 131.5 gr (0.0188 lbm) per pound. From Eq. 38.9, the degree of saturation is

$$\mu = \frac{\omega}{\omega_{\text{sat}}} = \frac{64.8 \frac{\text{gr}}{\text{lbm}}}{131.5 \frac{\text{gr}}{\text{lbm}}} = 0.49$$

8. LEVER RULE

With few exceptions (e.g., relative humidity and enthalpy correction), the scales on a psychrometric chart are linear. Because they are linear, any one property can be used as the basis for interpolation or extrapolation for another property on an intersecting linear scale. This applies regardless of orientation of the scales. The scales do not have to be orthogonal.

Furthermore, since psychrometric properties are extensive properties (i.e., they depend on the quantity of air present), the mass of air can be used as the basis for interpolation or extrapolation. This principle, known as the *lever rule* or *inverse lever rule*, is used when determining the properties of a mixture. (See Sec. 9.)

9. ADIABATIC MIXING OF TWO AIR STREAMS

Figure 38.1 shows the mixing of two moist air streams. The state of the mixture can be determined if the flow rates and psychrometric properties of the two component streams are known.

The two input states are located on the psychrometric chart and a straight line is drawn between them. The state of the mixture air will be on the straight line. The lever rule based on air masses is used to locate the mixture point. (Since the water vapor adds little to the mixture mass, the ratio of moist air masses can be approximated by the ratio of dry air masses, which, in turn, can be approximated by the ratio of air flow volumes.)

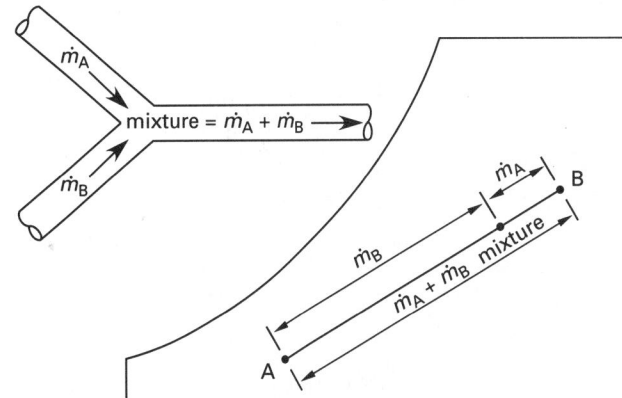

Figure 38.1 *Mixing of Two Air Streams*

The lever rule can be used to find the mixture properties algebraically. For the dry-bulb temperature (or any other property with a linear scale), the mixture temperature is

$$T_{\text{mixture}} = T_A + \left(\frac{\dot{m}_B}{\dot{m}_A + \dot{m}_B}\right)(T_B - T_A)$$
$$\approx T_A + \left(\frac{\dot{V}_B}{\dot{V}_A + \dot{V}_B}\right)(T_B - T_A) \qquad \textbf{38.21}$$

Example 38.4

5000 ft³/min (2.36 m³/s) of air at 40°F (4°C) dry-bulb and 35°F (2°C) wet-bulb are mixed with 15,000 ft³/min (7.08 m³/s) of air at 75°F (24°C) dry-bulb and 50% relative humidity. Find the mixture dry-bulb temperature.

SI Solution

An approximate mixture temperature can be found by disregarding the change in density and taking a weighted average. (The psychrometric chart can be used to determine a more precise value mixture temperature. The more precise approach is used in the Customary U.S. Solution.)

$$
\begin{aligned}
T_{\text{mixture}} &\approx \frac{\dot{V}_A T_A + \dot{V}_B T_B}{\dot{V}_A + \dot{V}} \\
&= \frac{\left(2.36 \ \frac{\text{m}^3}{\text{s}}\right)(4°\text{C}) + \left(7.08 \ \frac{\text{m}^3}{\text{s}}\right)(24°\text{C})}{2.36 \ \frac{\text{m}^3}{\text{s}} + 7.08 \ \frac{\text{m}^3}{\text{s}}} \\
&= 19°\text{C}
\end{aligned}
$$

Customary U.S. Solution

Locate the two points on the psychrometric chart, and draw a line between them. Estimate the specific volumes.

$$
v_A = 12.65 \ \text{ft}^3/\text{lbm}
$$

$$
v_B = 13.68 \ \text{ft}^3/\text{lbm}
$$

Calculate the dry air masses.

$$
\begin{aligned}
\dot{m}_A &= \frac{\dot{V}_A}{v_A} = \frac{5000 \ \dfrac{\text{ft}^3}{\text{min}}}{12.65 \ \dfrac{\text{ft}^3}{\text{lbm}}} \\
&= 395 \ \text{lbm/min}
\end{aligned}
$$

$$
\begin{aligned}
\dot{m}_B &= \frac{\dot{V}_B}{v_B} = \frac{15{,}000 \ \dfrac{\text{ft}^3}{\text{min}}}{13.68 \ \dfrac{\text{ft}^3}{\text{lbm}}} \\
&= 1096 \ \text{lbm/min}
\end{aligned}
$$

Use Eq. 38.21.

$$
\begin{aligned}
T_{\text{mixture}} &= T_A + \left(\frac{\dot{m}_B}{\dot{m}_A + \dot{m}_B}\right)(T_B - T_A) \\
&= 40°\text{F} + \left(\frac{1096 \ \dfrac{\text{lbm}}{\text{min}}}{1096 \ \dfrac{\text{lbm}}{\text{min}} + 395 \ \dfrac{\text{lbm}}{\text{min}}}\right) \\
&\quad \times (75°\text{F} - 40°\text{F}) \\
&= 65.7°\text{F}
\end{aligned}
$$

10. AIR CONDITIONING PROCESSES

The psychrometric chart is particularly useful in analyzing air conditioning processes because the paths of many processes are straight lines. Sensible heating and cooling processes, for example, follow horizontal straight lines. Adiabatic saturation processes follow lines of constant enthalpy (essentially parallel to lines of constant wet-bulb temperature). The paths of pure humidification and dehumidification follow vertical paths. Figure 38.2 summarizes the directions of these paths.

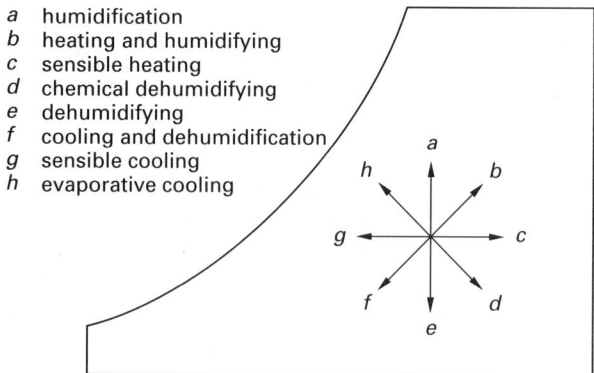

a humidification
b heating and humidifying
c sensible heating
d chemical dehumidifying
e dehumidifying
f cooling and dehumidification
g sensible cooling
h evaporative cooling

Figure 38.2 *Air Conditioning Processes*

11. SENSIBLE HEAT RATIO

In general, the slope of any process line on the psychrometric chart is determined from the *sensible heat ratio*, also known as the *sensible heat factor*, SHF, and *sensible-total ratio*, S/T, scale on the chart. In an air conditioning process, the sensible heat ratio, SHR, is the ratio of sensible heat added (or removed) to total heat added (or removed). (The use of such scales varies from chart to chart. In Apps. 38.A and 38.B, the process slope is determined from the sensible heat factor protractor and then translated (i.e., moved) to the appropriate point on the chart.)

$$
\text{SHR} = \frac{q_s}{q_t} = \frac{q_s}{q_s + q_l} \qquad \textit{38.22}
$$

The sensible heat ratio is always the slope of the line representing the change from the beginning point to the ending point on the psychrometric chart. Different designations are given to the sensible heat ratio, however, depending on where the changes occur.

If the sensible and latent energies change as the air passes through an occupied room, the term *room sensible heat ratio*, RSHR, is used. If the changes occur as the air passes through an air conditioning coil (apparatus), the term *coil* (or *apparatus*) *sensible heat ratio* is used, CSHR. Since the air conditioning apparatus usually removes heat and moisture from both the conditioned room and from outside makeup air, the term *grand sensible heat ratio*, GSHR, can be used in place of the coil sensible heat ratio. The *effective sensible heat ratio*, ESHR, is the slope of the line between the apparatus dew point on the saturation line and the design conditions of the conditioned space.

The sensible heat ratio is a psychrometric slope; it is not a geometric slope.

Example 38.5

During the summer, air from a conditioner enters an occupied space at 55°F (13°C) dry-bulb and 30% relative humidity. The ratio of sensible to latent loads in the space is 0.45:1. The humidity ratio of the air leaving the room is 60 gr/lbm (8.6 g/kg). What is the dry-bulb temperature of the leaving air?

SI Solution

The sensible heat ratio is 0.45. Use the psychrometric chart (App. 38.B) to determine the slope corresponding to this ratio. Draw a temporary line from the center of the protractor to the 0.45 mark on the sensible heat factor (inside) scale.

Locate 13°C dry-bulb and 30% relative humidity on the psychrometric chart. Draw a line through this point parallel to the temporary line which is drawn with a slope of 0.45. The intersection of this line and the horizontal line corresponding to 8.6 g/kg determines the condition of the leaving air. The dry-bulb temperature is approximately 25.2°C.

Customary U.S. Solution

The sensible heat ratio is 0.45. Use the psychrometric chart (App. 38.A) to determine the slope corresponding to this ratio. Draw a temporary line from the center of the protractor to 0.45 on the sensible heat factor (inside) scale.

Locate 55°F dry-bulb and 30% relative humidity on the psychrometric chart. Draw a line through this point parallel to the temporary line which is drawn with a slope of 0.45. The intersection of this line and the horizontal line corresponding to 60 gr/lbm determines the condition of the leaving air. The dry-bulb temperature is approximately 75°F.

12. BYPASS FACTOR AND COIL EFFICIENCY

Conditioning of air is accomplished by passing it through cooling or heating coils. Ideally, all of the air will come into contact with the coil for a long enough time and will leave at the coil temperature. In reality, this does not occur, and the air does not reach the coil temperature. The *bypass factor* can be thought of as the percentage of the air that is not cooled (or heated) by the coil. Under this interpretation, the remaining air (which is cooled or heated by the coil) is assumed to reach the coil temperature. The bypass factor expressed in decimal form is

$$\text{BF} = \frac{T_{\text{db,out}} - T_{\text{coil}}}{T_{\text{db,in}} - T_{\text{coil}}} \qquad 38.23$$

Bypass factors depend largely on the type of coil used. Bypass factors for large commercial units (such as those used in department stores) are small—around 10%. For small residential units, they are approximately 35%.

The *coil efficiency* is the complement of the bypass factor.

$$\eta_{\text{coil}} = 1.0 - \text{BF} \qquad 38.24$$

13. ADIABATIC SATURATION PROCESSES

To measure the wet-bulb temperature, air must experience an *adiabatic saturation process*, also known as *evaporative cooling*. To become saturated, the air must pick up the maximum amount of moisture it can hold at that temperature. This moisture comes from the vaporization of liquid water. For the process to be adiabatic, there can be no external source of energy to vaporize the liquid water needed to saturate the air.

At first analysis, the terms adiabatic and saturation seem contradictory. Adiabatic saturation is possible, however, if the latent heat of vaporization comes from the air itself. If the air gives up sensible heat, that energy can be used to vaporize liquid water. Of course, the air temperature decreases when sensible heat is given up. That is the reason that the wet-bulb temperature is generally less than the dry-bulb temperature. Only when the air is saturated will the two temperatures be equal.

An adiabatic saturation process can be produced with a *sling psychrometer*, which is essentially a regular thermometer with its bulb wrapped in wet cotton or gauze. Rapidly twirling the thermometer through the air at the end of a cord will cause the water in the gauze to evaporate. The latent heat needed to vaporize the water will come from the sensible heat of the air, and the thermometer will measure the wet-bulb temperature.

Since the increase in the water vapor's latent heat content equals the decrease in the air's sensible heat, the total enthalpies before and after adiabatic saturation are the same. Therefore, an adiabatic saturation process follows a line of constant enthalpy on the psychrometric chart. These lines are, for approximation purposes, parallel to lines of constant wet-bulb temperature.

Adiabatic saturation processes occur in cooling towers, air washers, and evaporative coolers ("swamp coolers").

The bypass factor concept is not used with adiabatic saturation processes. Instead, the *saturation efficiency* (*humidification efficiency*) is used. The saturation efficiency of large commercial air washers is typically 90 to 95%.

$$\eta_{\text{sat}} = \frac{T_{\text{db,air,in}} - T_{\text{db,air,out}}}{T_{\text{db,air,in}} - T_w} \qquad 38.25$$

HVAC

14. STRAIGHT HUMIDIFICATION

Straight (pure) *humidification* increases the water content of the air without changing the dry-bulb temperature. This is represented by a vertical condition line on the psychrometric chart. The *humidification load* is the mass of water added to the air per unit time (usually per hour).

15. SENSIBLE COOLING AND HEATING

There is no change in the dew point or moisture content of the air with *sensible heating* and *cooling*. Since the moisture content is constant, these processes are represented by horizontal condition lines on the psychrometric chart (moving right for heating and left for cooling).

The energy change during the process can be calculated from enthalpies read directly from the psychrometric chart or approximated from the dry-bulb temperatures. In Eq. 38.26, $c_{p,\text{air}}$ is usually taken as 0.240 Btu/lbm-°F (1.005 kJ/kg·°C), and $c_{p,\text{moisture}}$ is taken as approximately 0.444 Btu/lbm-°F (1.805 kJ/kg·°C).

$$q = m_a(h_2 - h_1)$$
$$= m_a(c_{p,\text{air}} + \omega c_{p,\text{moisture}})(T_2 - T_1) \quad \textbf{38.26}$$

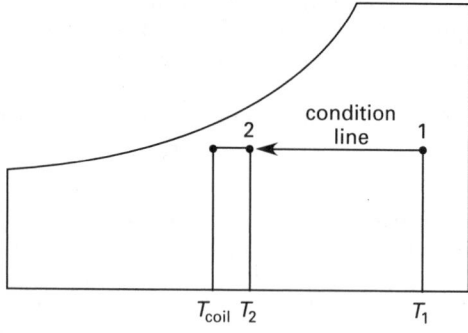

Figure 38.3 *Sensible Cooling*

16. COOLING WITH COIL DEHUMIDIFICATION

If the cooling coil's temperature is below the air's dew point (as is usually the case), moisture will condense on the coil. The effective coil temperature in this instance is referred to as the *apparatus dew point*, ADP, and is determined from the intersection of the condition line and the curved saturation line on the psychrometric chart. The apparatus dew point is the temperature to which the air would be cooled if 100% of it contacted the coil.

The mass of condensing water will be

$$m_w = m_a(\omega_1 - \omega_2) \quad \textbf{38.27}$$

The total energy removed from the air includes both sensible and latent components. The latent heat is calculated from the heat of vaporization evaluated at the pressure of the water vapor.

$$q_t = q_s + q_l = m_a(h_1 - h_2) \quad \textbf{38.28}$$
$$q_l = m_a(\omega_1 - \omega_2)h_{fg} \quad \textbf{38.29}$$

Referring to Fig. 38.4, it is convenient to think of air experiencing sensible cooling from point 1 to point 3, after which the air follows the saturation line down from point 3 to point 4 (the apparatus dew point). Water condenses out between points 3 and 4. For convenience, the condition line is drawn as a straight line between points 1 and 4. The slope of the ADP-2-1 line corresponds to the sensible heat ratio. (See Sec. 11.) Since some of the air does not contact the coil at all, the final condition of the air will actually be at point 2, on the condition line.

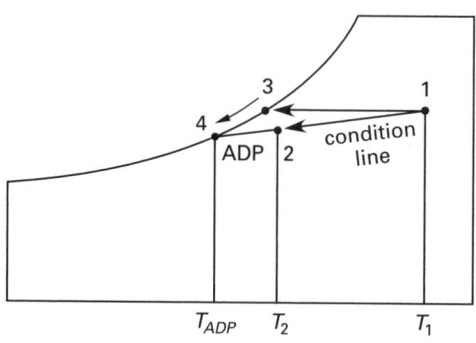

Figure 38.4 *Cooling and Dehumidification*

In practice, point 1 is usually known and either point 2 or point 4 are unknown. If point 2 is known, point 4 (the apparatus dew point) can be found graphically by extending the condition line over to the saturation line. (In some cases, the sensible heat ratio must be used to locate the apparatus dew point.) If point 4 is known, point 2 can be found from the bypass factor.

$$\text{BF} = \frac{T_{2,\text{db}} - \text{ADP}}{T_{1,\text{db}} - \text{ADP}} \quad \textbf{38.30}$$

Water condenses out over the entire temperature range from point 3 to point 4. The temperature of the water being removed is assumed to be the dew-point temperature at point 2.

Example 38.6

A coil has a bypass factor of 20% and an apparatus dew point of 55°F (13°C). Air enters the coil at 85°F (29°C) dry-bulb and 69°F (21°C) wet-bulb. What are the (a) latent heat loss, (b) sensible heat loss, and (c) sensible heat ratio?

SI Solution

(a) Locate the point corresponding to the entering air on the psychrometric chart. The enthalpy and humidity ratio are approximately

$$h_1 = 60.4 \text{ kJ/kg}$$
$$\omega_1 = 0.0123 \text{ kg/kg}$$

Use Eq. 38.30 to calculate the dry-bulb temperature of the air leaving the coil.

$$\begin{aligned} T_{2,db} &= ADP + BF(T_{1,db} - ADP) \\ &= 13°C + (0.20)(29°C - 13°C) \\ &= 16.2°C \end{aligned}$$

Draw a condition line between the entering air and the apparatus dew point on the psychrometric chart. Locate the point corresponding to 16.2°C dry-bulb on the condition line. The leaving enthalpy and humidity ratio are approximately

$$h_2 = 40.8 \text{ kJ/kg}$$
$$\omega_2 = 0.0100 \text{ kg/kg}$$

The total energy loss per kilogram is

$$\begin{aligned} q_t = h_1 - h_2 &= 60.4 \frac{\text{kJ}}{\text{kg}} - 40.8 \frac{\text{kJ}}{\text{kg}} \\ &= 19.6 \text{ kJ/kg of dry air} \end{aligned}$$

Since the partial pressure of the water vapor is unknown, evaluate $h_{fg} \approx 2501$ kJ/kg.

From Eq. 38.29, on a kilogram basis,

$$\begin{aligned} \frac{q_l}{m_a} &= (\omega_1 - \omega_2)h_{fg} \\ &= \left(0.0123 \frac{\text{kg}}{\text{kg}} - 0.0100 \frac{\text{kg}}{\text{kg}}\right)\left(2501 \frac{\text{kJ}}{\text{kg}}\right) \\ &= 5.75 \text{ kJ/kg of dry air} \end{aligned}$$

(b) The sensible heat loss is

$$\begin{aligned} q_s &= q_t - q_l \\ &= 19.6 \frac{\text{kJ}}{\text{kg}} - 5.75 \frac{\text{kJ}}{\text{kg}} = 13.9 \text{ kJ/kg of dry air} \end{aligned}$$

(c) The sensible heat ratio is

$$SHR = \frac{q_s}{q_t} = \frac{11.1 \frac{\text{kJ}}{\text{kg}}}{16.4 \frac{\text{kJ}}{\text{kg}}}$$
$$= 0.68$$

Customary U.S. Solution

(a) Locate the point corresponding to the entering air on the psychrometric chart. The enthalpy and humidity ratio are approximately

$$h_1 = 33.1 \text{ Btu/lbm}$$
$$\omega_1 = 0.0116 \text{ lbm/lbm}$$

Use Eq. 38.30 to calculate the dry-bulb temperature of the air leaving the coil.

$$\begin{aligned} T_{2,db} &= ADP + BF(T_{1,db} - ADP) \\ &= 55°F + (0.20)(85°F - 55°F) \\ &= 61°F \end{aligned}$$

Draw a condition line between the entering air and apparatus dew point on the psychrometric chart. Locate the point corresponding to 61°F dry-bulb on the condition line. The leaving enthalpy and humidity ratio are approximately

$$h_2 = 25.1 \text{ Btu/lbm}$$
$$\omega_2 = 0.0097 \text{ lbm/lbm}$$

The total energy loss per pound is

$$\begin{aligned} q_t = h_1 - h_2 &= 33.1 \frac{\text{Btu}}{\text{lbm}} - 25.1 \frac{\text{Btu}}{\text{lbm}} \\ &= 8.0 \text{ Btu/lbm of dry air} \end{aligned}$$

Since the partial pressure of the water vapor is not known, evaluate $h_{fg} \approx 1060$ Btu/lbm.

From Eq. 38.29, on a pound basis,

$$\begin{aligned} \frac{q_l}{m_a} &= (\omega_1 - \omega_2)h_{fg} \\ &= \left(0.0116 \frac{\text{lbm}}{\text{lbm}} - 0.0097 \frac{\text{lbm}}{\text{lbm}}\right)\left(1060 \frac{\text{Btu}}{\text{lbm}}\right) \\ &= 2.01 \text{ Btu/lbm of dry air} \end{aligned}$$

(b) The sensible heat loss is

$$\begin{aligned} q_s &= q_t - q_l \\ &= 8.0 \frac{\text{Btu}}{\text{lbm}} - 2.0 \frac{\text{Btu}}{\text{lbm}} = 6.0 \text{ Btu/lbm of dry air} \end{aligned}$$

(c) The sensible heat ratio is

$$SHR = \frac{q_s}{q_t} = \frac{6.0 \frac{\text{Btu}}{\text{lbm}}}{8 \frac{\text{Btu}}{\text{lbm}}}$$
$$= 0.75$$

HVAC

17. COOLING WITH HUMIDIFICATION

When air passes through a water spray (as in an *air washer*), an *adiabatic saturation process* known as *evaporative cooling* occurs.[5] The air leaves with a lower temperature and a higher moisture content. This is represented on the psychrometric chart by a condition line parallel to the lines of constant enthalpy (essentially constant wet-bulb temperature).

Adiabatic saturation is a constant-enthalpy process, since any evaporation of the water requires heat to be drawn from the air. Since the removed heat goes into the remaining water, the water temperature increases. When the water spray is continuously recirculated, the water temperature gradually increases to the wet-bulb temperature of the incoming air. The minimum leaving air temperature will be the water temperature (i.e., the wet-bulb temperature of the incoming air).

During steady-state operation, the temperature of the water spray will normally be stable at the air's wet-bulb temperature. However, the water temperature can also be artificially maintained lower than the wet-bulb temperature (but greater than the dew-point temperature). Line 1–3 in Fig. 38.5 illustrates such a process.

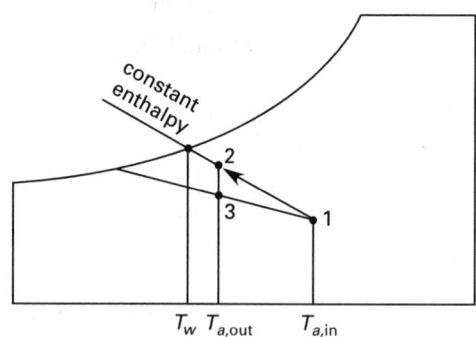

Figure 38.5 *Cooling with Humidification (adiabatic saturation)*

To prevent ice buildup, the cooled air temperature should be kept from dropping below the freezing point of water. The entering wet-bulb temperature should be kept above 35°F (1.7°C).

Example 38.7

Air at 90°F (32°C) dry-bulb and 65°F (18°C) wet-bulb enters an evaporative cooler. The air leaves at 90% relative humidity. The continuously recirculated spray water is stable at 65°F (18°C). What is the dry-bulb temperature of the leaving air?

Solution

Since the spray water is the same temperature as the wet-bulb temperature of the entering air, the cooler has

reached its steady-state operating conditions. Locate the entering point on the psychrometric chart and draw a line of constant enthalpy (or constant 65°F (18°C) wet-bulb temperature) up to the 90% relative humidity curve. Read the dry-bulb temperature as approximately 67°F (19°C).

18. COOLING WITH SPRAY DEHUMIDIFICATION

If air passes through a water spray whose temperature is less than the entering air's wet-bulb temperature, both the dry-bulb and wet-bulb temperatures will decrease.[6] If the leaving water temperature is below the entering air's dew point, dehumidification will occur. As with any evaporative cooling, the air will give up thermal energy to the water. The final water temperature will depend on the thermal energy pickup and water flow rate. All air temperatures decrease, and some moisture condenses. The *performance factor* is defined as

$$\text{PF} = 1 - \frac{T_{\text{air,wb,out}} - T_{w,\text{out}}}{T_{\text{air,wb,in}} - T_{w,\text{in}}} \qquad 38.31$$

19. HEATING WITH HUMIDIFICATION

If air is humidified by injecting steam (*steam humidification*) or by passing the air through a hot water spray, the dry-bulb temperature and enthalpy of the air will increase.[7] The final air enthalpy and/or the required steam enthalpy can be determined from a conservation of energy equation. In Eq. 38.32, the mass of the air used is the dry air mass, which does not change.

$$m_a h_{a,\text{in}} + m_w h_w = m_a h_{a,\text{out}} \qquad 38.32$$

From a conservation of mass for the water,

$$m_a \omega_{\text{in}} + m_w = m_a \omega_{\text{out}} \qquad 38.33$$

Figure 38.6 illustrates that the condition line will be above the line of constant enthalpy that radiates from the point corresponding to the incoming air. However, even though heat is added to the water, the air temperature can either decrease (as in the 1–2 process shown) or increase (as in the 1–3 process shown).

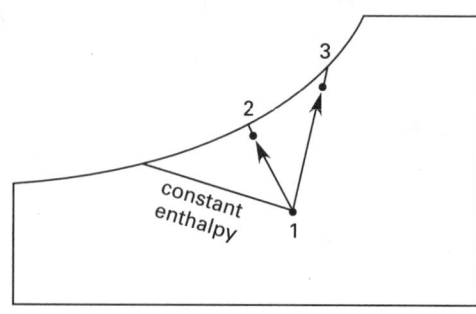

Figure 38.6 *Heating with Steam Humidification*

---

[5]An *air washer* is basically a *spray chamber* through which air passes. When supplied with chilled water from a refrigeration source, the air washer can cool, dehumidify, or humidify the air. Air washers can be used without refrigeration to cool and humidify the air through an evaporative cooling process.

[6]This can unintentionally occur during the start-up of an air washer used for humidification, or the water can be kept intentionally chilled.

[7]When a spray of hot water is used, the water must be continually heated. Unlike a cold water spray, a natural equilibrium water temperature is not achieved.

20. HEATING AND DEHUMIDIFICATION

Air passing through a solid or liquid *adsorbent bed*, such as silica gel or activated alumina, will decrease in humidity. This is sometimes referred to as *chemical dehumidification*, *chemical dehydration*, or "absorbent" dehumidification.[8] If only latent heat was involved, this process would be the reverse of an adiabatic saturation process. However, as moisture is removed, exothermic chemical energy is generated in addition to the heat of vaporization liberated. Since thermal energy is generated, this is not an adiabatic process.

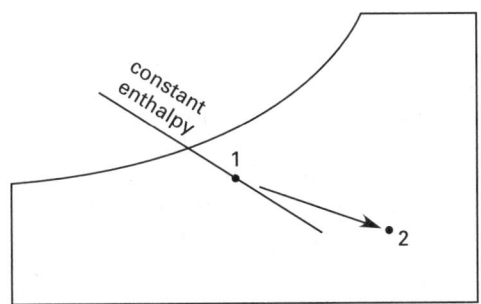

Figure 38.7 *Heating and Dehumidification*

21. AIR WASHERS

An *air washer* is a device that passes air through a dense spray of recirculating water. The water is used to change the properties of the air. Air washers are used in air purifying and cleaning processes (i.e., removal of solids, liquids, gases, vapors, and odors), as well as for evaporative cooling and dew-point control.[9]

The difference between a spray humidifier and spray dehumidifier is the temperature of the spray water. In an *adiabatic air washer*, the spray water is recirculated without being heated or cooled. After equilibrium is reached, the water temperature will be equal to the air's entering wet-bulb temperature. The air will be cooled and humidified, leaving partially or completely saturated at its entering wet-bulb temperature. However, if the spray water is chilled, the air will be cooled and dehumidified. And, if the spray water is heated, the air will be humidified and (possibly) heated.

An air washer's *saturation efficiency*, typically 90 to 95%, is measured by the drop in dry-bulb temperature relative to the entering wet-bulb depression.

$$\eta_{\text{sat}} = \frac{T_{\text{in,db}} - T_{\text{out,db}}}{T_{\text{in,db}} - T_{\text{in,wb}}} \qquad 38.34$$

Air velocity through washers is approximately 500 ft/min (2.6 m/s). Velocities outside the range of 300 to 750 ft/min (1.5 to 3.8 m/s) are probably faulty. The water pressure is typically 20 to 40 psig (140 to 280 kPa). The spray quantity per bank of nozzles is in the range of 1.5 to 5 gal/min per 1000 ft^3 (3.3 to 11 L/s per 1000 m^3) of air. Screen, louvers, and mist eliminator plates will generate a static pressure drop of approximately 0.2 to 0.5 in wg (50 to 125 kPa) at 500 ft/min (2.6 m/s). Other operating parameters used to describe air washer performance include air mass flow rate per unit area (lbm/hr-ft^2 or kg/m^2·s), air and liquid heat transfer coefficients per volume of chamber (Btu/hr-°F-ft^3 or kW/°C·m^3), and the *spray ratio* (the mass of water sprayed to the mass of air passing through the washer per unit time).

22. WET COOLING TOWERS

Conventional wet *cooling towers* cool warm water by exposing it to colder air.[10] They are usually used to provide cold water to power plant and large refrigeration condensers. The air is used to change the properties of the water, which leaves cooler. As it leaves, the saturated (or nearly saturated) warmed air takes sensible and latent heat from the water.

Cooling towers are generally counterflow, crossflow, or a combination. Though natural-draft and atmospheric towers exist, limited space usually requires that cooling towers operate with mechanical draft. Fans are located at the base of *forced draft towers* and blow air into the water cascading down. With *induced mechanical draft*, fans are located at the top of the tower, drawing air upward. Some portion of the exhaust air might reenter the cooling tower. This is known as *recycle air* (*recirculation air*). Recycle air decreases the efficiency of the tower.

During countercurrent operation, warm water is introduced at the top of the tower and is distributed by troughs or spray nozzles. The water passes over staggered slats or interior *fill* (also known as *packing*).[11] Air flows upward, contacting the water on its downward path. A portion of the water evaporates, cooling the remainder of the water. The water temperature cannot decrease below the wet-bulb temperature. The actual final water temperature depends on a number of factors, including the state of the incoming air, the heat load, and the design (and efficiency) of the cooling tower.

---

[8]The correct term for a substance that collects water on its surface is an *adsorbent*. By virtue of their great porosities, adsorbent particles have large surface areas. The attractive forces on the surfaces of these solids cause a thin layer of condensed water to form. Adsorbents are reactivated by heating.

[9]Air washers are generally not used for removing carbonaceous or greasy particles.

[10]Though larger in size, a cooling tower is similar in operation to an air washer. In fact, an air washer can be used to cool water. Since air washer operation is not countercurrent, however, larger air flows are required to obtain the same cooling effect.

[11]Modern *filled towers* use corrugated *cellular fill* to maximize the air-water contact area. Standard polyvinyl chloride (PVC) fill is useful up to about 125°F (52°C). From 125°F to 140°F (52°C to 60°C), chlorinated PVC fill is recommended. Polypropylene fill should be used above 140°F (60°C).

"Fill-less" towers, where the sprayed water merely falls through oncoming air, are used in some industries (food, steel, and paper processes) where a high-product carryover can lead to coating or buildup on the fill material.

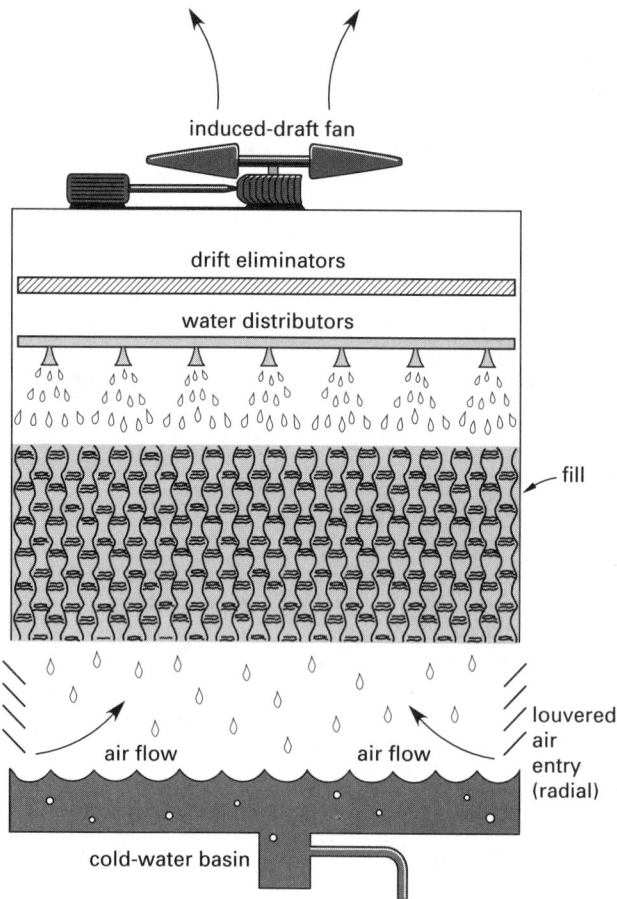

Figure 38.8 *Counterflow Wet Cooling Tower*

There are several environmental issues associated with wet cooling towers. Makeup water, though relatively little is needed, may be difficult to obtain. Moist plume discharges cause shadowing of adjacent areas and fogging and icing on nearby highways. Disposal of blowdown wastewater is also problematic.

Equation 38.35 is a per-unit energy balance that can be used to evaluate cooling tower performance. Each term, including the circulating water flow rate, is per unit mass (e.g., pound or kilogram) of dry air. Since the energy contribution of the makeup water is small, that term can be omitted for a first approximation.

$$m_{w,\text{in}}h_{w,\text{in}} + h_{a,\text{in}} + (\omega_{a,\text{out}} - \omega_{a,\text{in}})h_{\text{makeup}}$$
$$= m_{w,\text{out}}h_{w,\text{out}} + h_{a,\text{out}}$$
$$= (m_{w,\text{in}} + \omega_{a,\text{in}} - \omega_{a,\text{out}})h_{w,\text{out}} + h_{a,\text{out}} \quad \textbf{38.35}$$

If operation is at standard pressure, a psychrometric chart can be used to obtain the air enthalpies. For operation at different altitudes (i.e., different atmospheric pressures), the mathematical psychrometric relationships in Sec. 4 can be used to calculate the enthalpy. From Eq. 38.16, the humidity ratio is

$$\omega = \frac{0.622 p_{\text{water vapor}}}{p_{\text{total}} - p_{\text{water vapor}}} \quad \textbf{38.36}$$

When a cooling tower is used to provide cold water for the condenser of a refrigeration system, the water circulation will be approximately 3 gal/min per ton (0.19 L/s) of refrigeration. Approximately 2 to 4 gal/min of water are distributed per square foot (1.4 to 2.7 L/s per square meter) of tower, and the air velocity should be approximately 700 ft/min (3.6 m/s) through the net free area. Coolants for condensers in reciprocating refrigeration systems usually call for an 85 to 90°F (29 to 32°C) water temperature. (This corresponds to a condensing temperature of approximately 100 to 110°F (38 to 43°C).) Various valves, mixing, louvers, and dampers are used to maintain a constant output water temperature.

The lowest temperature to which water can be cooled by purely evaporative means is the wet-bulb temperature of the entering air. The *cooling efficiency*, η_w, is based on the water temperature. The *water range* (or *range*) is defined as the actual difference between the entering and leaving water temperatures. (For water-cooled refrigeration condensers, this is equal to the water's temperature increase in the condenser.) The *approach* is defined as the difference between the leaving water temperature and the entering air wet-bulb temperatures.[12] Cooling efficiency is typically 50 to 70%.[13] Natural draft towers can cool the water to within 10 to 12°F (5.5 to 6.7°C) of the wet-bulb temperature. Forced draft towers can cool the water to within 5 to 6°F (2.8 to 3.3°C).

$$\eta_w = \frac{\text{range}}{\text{approach} + \text{range}}$$
$$= \frac{T_{w,\text{in}} - T_{w,\text{out}}}{T_{w,\text{in}} - T_{\text{air,wb,in}}} \quad \textbf{38.37}$$

As Eq. 38.37 indicates, the actual wet-bulb temperature of the cooling air is particularly important in determining cooling tower performance. The higher the wet-bulb temperature, the lower the efficiency. In rating their cooling towers, most manufacturers have adopted the practice of using wet-bulb temperatures that will be exceeded only 2.5% of the time or less.

Performance of a cooling tower also depends on the relative humidity of the air. High relative humidities decrease the water evaporation rate, decreasing the efficiency.

The *heat load* (*tower load* or *cooling duty*) is calculated from the range and the water mass flow rate.

$$q = m_w c_p (T_{w,\text{in}} - T_{w,\text{out}})$$
$$= m_w (h_{w,\text{in}} - h_{w,\text{out}}) \quad \textbf{38.38}$$

Cooling towers are sometimes rated in tower units, which are essentially proportional to the tower cost. The number of *tower units* is equal to a rating factor times the flow rate. *Rating factors* define the relative difficulty in cooling, essentially the relative amount

---

[12]Thus, approach for a cooling tower is analogous to the terminal temperature difference in the surface-condenser.
[13]The term "thermal efficiency" is sometimes used here inappropriately.

of contact area or fill volume required. Manufacturers provide charts showing the relationship between rating factor, approach, range, and wet-bulb temperature.

$$TU = RF \times \dot{V}_{gpm} \qquad 38.39$$

23. COOLING TOWER BLOWDOWN

Water losses occur from evaporation, windage, and blowdown. *Evaporation loss* can be calculated from the humidity ratio increase and is approximately 0.1% per °F (0.18% per °C) decrease in water temperature.[14] *Windage loss*, also known as *drift*, is water lost in small droplets and carried away by the air flow. Windage loss is typically in the 0.1 to 0.3% range for mechanical draft towers. Since windage droplets are a mixture (not a thermodynamic solution of two gases), they are not adequately represented by the humidity ratio.

Makeup water must be provided to replace all water losses. As more and more water enters the system, total dissolved solids, TDS (e.g., chlorides), will build up over time. Water can be treated to prevent deposit, and a portion of the water can be periodically or continuously bled off. *Cycles of concentration, C (ratio of concentration)*, is the ratio of total dissolved solids in the recirculating water to the total dissolved solids in the makeup water.[15]

$$
\begin{aligned}
C &= \frac{(\text{TDS})_{\text{recirculating}}}{(\text{TDS})_{\text{makeup}}} \\
&= \frac{m_{\text{evaporation}} + m_{\text{blowdown}} + m_{\text{windage}}}{m_{\text{blowdown}} + m_{\text{windage}}} \qquad 38.40
\end{aligned}
$$

Though windage removes some of the solids, most must be removed by bleeding some of the water off. This is known as *blowdown* or *bleed-off*. If the maximum cycles of concentration are known, the blowdown is

$$m_{\text{blowdown}} = \frac{m_{\text{evaporation}} + (1 - C_{\text{max}})m_{\text{windage}}}{C_{\text{max}} - 1} \qquad 38.41$$

Additives should be used to prevent specific problems encountered such as scale buildup, corrosion, biological growth, foaming, and discoloration.

24. DRY COOLING TOWERS

Dry cooling is used when environmental protection and water conservation are issues. It is used primarily by nonutility generators (e.g., waste-to-energy and cogeneration plants).

There are two types of dry cooling towers. Both use finned-tube heat exchangers. In a *direct-condensing tower*, steam travels through large-diameter "trunks" to a crossflow heat exchanger where it is condensed and cooled by the cooler air.[16] In an *indirect-condensing dry cooling tower*, steam is condensed by cold water jets (surface or jet condenser). The hot condensate is then pumped to crossflow heat exchangers where it is sensibly cooled (no condensation) by the air. Air flow may be mechanical or natural draft. Most U.S. installations are direct-condensing. Worldwide, natural-draft indirect systems are more predominant, particularly for power plants with capacities in excess of 100 MW.

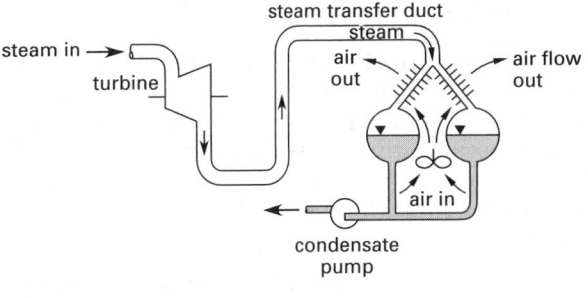

(a) direct

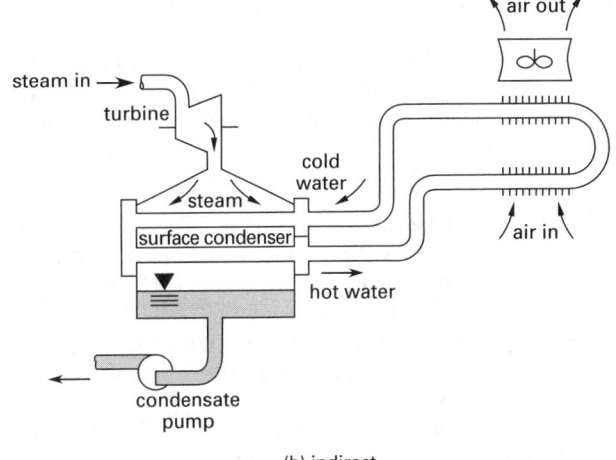

(b) indirect

Figure 38.9 *Dry Cooling Towers*

PRACTICE PROBLEMS

1. A room contains air at 80°F (27°C) dry-bulb and 67°F (19°C) wet-bulb. The total pressure is 1 atm. What are the (a) humidity, (b) enthalpy, and (c) specific heat?

2. If one row of cooling coils bypasses one-third of the air passing through it, what is the theoretical bypass factor for four rows of identical cooling coils in series?

---

[14]This value is approximate and is reported in various ways. Some authorities state "0.1% per degree Fahrenheit"; others say "1% per 10°F"; and yet others, "1% per 10 to 13°F."
[15]Multiply grains/gallon (gr/gal) by 17.1 to obtain parts per million (ppm) or milligrams per liter (mg/L).

[16]The term "direct contact" does not mean that the air and steam are combined in a single vessel.

3. 1000 ft^3/min (0.5 m^3/s) of air at 50°F (10°C) dry-bulb and 95% relative humidity are mixed with 1500 ft^3/min (0.75 m^3/s) of air at 76°F (24°C) and 45% relative humidity. What are the (a) dry-bulb temperature, (b) specific humidity, and (c) dew point of the mixture?

4. Air at 60°F (16°C) dry-bulb and 45°F (7°C) wet-bulb passes through an air washer with a humidifying efficiency of 70%. What are the (a) effective bypass factor of the system and (b) dry-bulb temperature of the air leaving the washer?

5. 95°F (35°C) dry-bulb, 75°F (24°C) wet-bulb air passes through a cooling tower and leaves at 85°F (29°C) dry-bulb and 90% relative humidity. What are the (a) enthalpy change and (b) change in moisture content per cubic foot (meter) of air?

6. An air washer receives 1800 ft^3/min (0.85 m^3/s) of air at 70°F (21°C) and 40% relative humidity and discharges the air at 75% relative humidity. A recirculating water spray with a constant temperature of 50°F (10°C) is used. (a) What mass of makeup water is required per minute? (b) What will be the condition of the discharged air?

7. Repeat Prob. 6 using saturated steam at atmospheric pressure in place of the 50°F (10°C) water spray.

8. During performances, a theater experiences a sensible heat load of 500,000 Btu/hr (150 kW) and a moisture load of 175 lbm/hr (80 kg/h). Air enters the theater at 65°F (18°C) and 55% relative humidity and is removed when it reaches 75°F (24°C) or 60% relative humidity, whichever comes first. (a) What is the ventilation rate in mass of air per hour? (b) What are the conditions of the air leaving the theater?

9. 500 ft^3/min (0.25 m^3/s) of air at 80°F (27°C) dry-bulb and 70% relative humidity are removed from a room. 150 ft^3/min (0.075 m^3/s) pass through an air conditioner and leave saturated at 50°F (10°C). The remaining 350 ft^3/min (0.175 m^3/s) bypass the air conditioner and mix with the conditioned air at 1 atm. What is the mixture's (a) temperature, (b) humidity ratio, and (c) relative humidity? (d) What is the heat load (in tons) of the air conditioner?

10. (*Time limit: one hour*) A dehumidifier takes 5000 ft^3/min (2.36 m^3/s) of air at 95°F (35°C) dry-bulb and 70% relative humidity and discharges it at 60°F (16°C) dry-bulb and 95% relative humidity. The dehumidifier uses an R-12 refrigeration cycle operating between 100°F (saturated) (38°C) and 50°F (10°C). (a) Locate the air entering and leaving points on the psychrometric chart. (b) Find the quantity of water removed from the air. (c) Find the quantity of heat removed from the air. (d) Draw the temperature-entropy and enthalpy-entropy diagrams for the refrigeration cycle. (e) Find the temperature, pressure, enthalpy, entropy, and specific volume for each endpoint of the refrigeration cycle.

11. (*Time limit: one hour*) 1500 ft^3/min (0.71 m^3/s) of saturated 25 psia (170 kPa) air are heated from 200°F to 400°F (93°C to 204°C) in a constant pressure, constant moisture drying process. (a) What is the final relative humidity? (b) What is the final specific humidity? (c) How much heat is required per unit mass of dry air? (d) What is the final dew point?

12. (*Time limit: one hour*) 410 lbm/hr (0.052 kg/s) of dry 800°F (427°C) air pass through a scrubber to reduce particulate emissions. To protect the elastomeric seals in the scrubber, the air temperature is reduced to 350°F (177°C) by passing the air through a spray of 80°F (27°C) water. The pressure in the spray chamber is 20 psia (140 kPa). (a) How much water is evaporated per hour? (b) What will be the relative humidity of the air leaving the spray chamber?

13. (*Time limit: one hour*) An evaporative counterflow air cooling tower removes 1×10^6 Btu/hr (290 kW) from a water flow. The temperature of the water is reduced from 120°F to 110°F (49°C to 43°C). Air enters the cooling tower at 91°F (33°C) and 60% relative humidity, and air leaves at 100°F (38°C) and 82% relative humidity. Calculate the (a) air flow rate and (b) quantity of makeup water.

39 Ventilation

1. Ventilation 39-1
2. Ventilation Standards 39-1
3. Infiltration 39-1
4. Infiltration in Tall Buildings 39-2
5. Indoor Design Condition 39-2
6. Humidification 39-2
7. Oxygen Needs 39-3
8. Carbon Dioxide Buildup 39-3
9. Odor Removal 39-3
10. Sensible and Latent Heat 39-3
11. Ventilation for Heat Removal 39-3
12. Fume Exhaust Hoods 39-4
13. Dilution Ventilation 39-5
14. Recirculation of Cleaned Air 39-5
15. Clean Rooms 39-5
    Practice Problems 39-6

Nomenclature

A	area	ft^2	m^2
c_p	specific heat	Btu/lbm-°F	kJ/kg·°C
B	crack coefficient	ft^3/ft-hr	m^3/m·h
C	concentration	lbm/ft^3	kg/m^3
h	enthalpy	Btu/lbm	kJ/kg
K	mixing factor	–	–
L	length	ft	m
$\dot{m}$	mass flow rate	lbm/min	kg/min
n	exponent	–	–
MW	molecular weight	lbm/lbmol	kg/kmol
p	pressure	lbf/in^2	kPa
$\dot{q}$	heat transfer rate	Btu/min	W
R	contaminant generation rate	lbm/min	kg/min
SG	specific gravity	–	–
t	time	hr	–
T	temperature	°F	°C
TLV	threshold limit value	various	various
v	velocity	mi/hr	–
$\dot{V}$	flow rate	ft^3/min	m^3/min
y	elevation	ft	–

Symbols

η	efficiency	–	–
ρ	density	lbm/ft^3	kg/m^3
ω	humidity ratio	lbm/lbm	kg/kg

Subscripts

ADP	apparatus dew point
fg	vaporization
id	inside design
in	entering the room
l	latent
s	sensible

1. VENTILATION

Ventilation primarily refers to air that is necessary to satisfy the needs of occupants.[1] The term may mean the air that is introduced into an occupied space, or it may refer to the new air that is deliberately drawn in from the outside and mixed with return air. Ventilation, however, does not normally include unintentional infiltration through cracks and openings.

Ventilation air is provided to the occupied space primarily to remove heat and moisture generated in the space. Heat and moisture can both be generated metabolically as well as by equipment and processes. To a lesser extent, ventilation is also used to remove odors, provide oxygen, prevent carbon dioxide buildup, and remove noxious fumes. Generally, however, all of the other needs will be met if removal of body heat is accomplished.

2. VENTILATION STANDARDS

Minimum ventilation requirements are given by local building codes, local ordinances, health regulations, and construction specifications. Actual ventilation rates generally depend on intended usage. A common minimum design standard is 15 ft^3/min (0.42 m^3/min; 7.1 L/s) of new, outside air per person. In areas with heavy smoking (e.g., cocktail and smoking lounges), 30 to 60 ft^3/min (0.84 to 1.68 m^3/min; 14.1 to 28.2 L/s) per person is required. Some nonsmoking areas may require 20 ft^3/min (0.57 m^3/min; 9.4 L/s) anyway to avoid the "sick building syndrome" (as when formaldehyde-emitting furniture and building materials are present).

Some ventilation requirements are specified by the numbers of air changes (i.e., room volumes) required per hour. This is known as the *air change method*. A typical minimum value for toilet rooms, for example, is four air changes per hour. For other uses (automotive, boiler rooms, engine rooms, etc.) the number of air changes can be significantly higher (e.g., 25 to 100 per hour).

3. INFILTRATION

Infiltration refers to the air that unintentionally enters an occupied space through cracks around doors and windows and through openings in the buildings.

---

[1]The term *process air* is the most common designation given to "ventilation" needed for manufacturing processes.

With the *crack length method*, the amount of infiltration, $\dot{V}$, is determined from the *crack coefficient*, B, and the crack length, L. Values of the crack coefficient vary greatly and depend on the type of window or door, wind velocity, orientation, and degree of closure.[2] Alternatively, the infiltration may be found from the plane area. (This method is more common when determining infiltration through entire walls.) As with the crack length method, the crack area coefficient B' depends on many factors.

$$\dot{V} = BL = B'A \qquad 39.1$$

More sophisticated correlations recognize the dependence on the difference in outside and inside pressures. Values of B'' and n must be known or assumed and must be consistent with the units of pressure.

$$\dot{V} = B''A(\Delta p)^n \qquad 39.2$$

Since air entering through cracks on the windward side must leave through cracks on the leeward side, only half of the total crack length is used when all four sides of a building are exposed to wind. However, the amount of crack length used also depends on the building orientation. When only one wall is exposed to wind, that wall's total crack length is used. With two exposed walls, the wall with the larger crack length is used. When three walls are exposed, only two walls contribute to crack length. The crack length used should never be less than half of the total crack length.

The air change method can also be used for infiltration. Infiltration into modern (tight) residential construction may be as low as 0.2 air changes per hour, while older residences in good condition may experience ten times as much. In the past, a rule of thumb used (to size furnaces) in the absence of other information is that infiltration into residences with windows on one, two, or three sides would be one, one and one-half, or two air changes per hour, respectively. Experience is needed to modify these values for use with modern, energy-efficient construction.

4. INFILTRATION IN TALL BUILDINGS

Wind velocity is a major factor in calculating infiltration. However, the wind velocity increases with elevation. The wind velocity profile depends on many factors.

For buildings more than 100 ft (30 m) in height, the stack (chimney) effect must also be considered in the winter. Studies have shown that the combined infiltration from wind velocity and chimney effect is proportional to the square root of the sum of the heads acting on the building. In the past, the *effective wind velocity* as defined by Eq. 39.3 has been used to determine winter infiltration at an elevation y above or below the

building's midheight.[3] y is positive above the midheight and negative below the midheight. v_o is the design wind velocity that would be used if the chimney effect is neglected.

$$v_{\text{effective,mph}} = \sqrt{v_o^2 - 1.75 y_{\text{ft}}} \qquad 39.3$$

5. INDOOR DESIGN CONDITION

The *inside design condition* refers to the thermodynamic state of the air that is removed from an occupied space. The inside design temperature, T_{id}, represents the maximum dry-bulb air temperature—a not-to-be-exceeded limit—that the space will reach.

HVAC books contain tables of recommended inside design conditions and charts of *comfort ranges* that can be used to select suitable combinations of temperature and humidity. Within a comfort range, choice of the actual inside design condition is subjective and requires modification based on experience for the needs of the particular industries, the season, and levels of physical exertion.

Most people feel comfortable when the dry-bulb temperature is kept between 74°F and 77°F (23.3°C and 25°C) and the relative humidity is 30 to 35% (in the winter) or 45 to 50% (in the summer). 75°F (23.9°C) dry-bulb and 50% relative humidity is often selected as an inside design condition for initial studies. This temperature is for the breathing line, 3 to 5 ft (0.9 to 1.5 m) above the floor.[4]

Attention also needs to be given to the *temperature swing*, the difference of the thermostat's on and off settings. For commercial applications, the swing during the summer should be approximately 2 to 4°F (1.1 to 2.2°C) above, and swing during the winter should be approximately 4°F (2.2°C) below.

6. HUMIDIFICATION

Ventilation provides humidification to the occupied space, particularly during the winter. Air should not be completely dry when it enters an occupied space. Air that is too dry will cause discomfort and susceptibility to respiratory ailments. Also, some pathogenic bacteria that survive in low- and high-humidity air will die very quickly in air with midrange humidities.[5]

Some manufacturing and materials handling processes require specific humidity for efficient operation. *Hygroscopic materials*, such as wood, paper, textiles, leather,

---

[2]Typical values of the crack length coefficient are given in most HVAC books. Manufacturers' literature should be used for specific name-brand windows and doors.

[3]This assumes that a "neutral zone" exists halfway up the building.

More sophisticated methods exist, but this equation is still valid for approximations.

[4]The temperature variation with height above the floor is approximately 0.75°F/ft (1.4°C/m).

[5]In particular, airborne type 1 pneumococcus, group C staphylococcus, and staphylococcus are quickly killed in relative humidities of 45 to 55%. Other viruses, including measles, influenza, and encephalomyelitis, survive longer in very dry air than in midrange relative humidities.

and many food and chemical products, readily absorb moisture. A constant humidity level is required to obtain consistent manufacturing conditions with such products.

Dry air prevents static electricity from dissipating (into the air). Therefore, dry air can cause intermittent electrical/electronic failures; affect the handling of static-prone materials such as paper, films, and plastics; and ignite potentially explosive atmospheres of dust and gases.

Humidification can be provided by evaporating water in the occupied space (the *evaporative pan method*) or by injecting water (the *water spray method*) or steam into the duct flow. Most commercial humidification is accomplished by placing one or more steam manifolds in the air distribution duct. *Booster humidification (spot humidification)* from a separate, independent source is required when a higher humidity is needed in a limited area within a larger controlled space. Steam flow is controlled by controlling and high-limit humidistats placed downstream of the steam manifold.[6]

7. OXYGEN NEEDS

The oxygen concentration should not fall below 12% by volume. Providing oxygen is not an issue in traditional buildings. Infiltration alone provides the oxygen needed. However, in closed environments such as mines, tunnels, manholes, and closed tanks, forced ventilation and/or oxygen masks are necessary.

8. CARBON DIOXIDE BUILDUP

Diluting exhaled carbon dioxide, like providing oxygen, is only an issue in completely closed environments. Infiltration alone provides the dilution needed. Healthy individuals can probably tolerate a concentration of 0.5% (by volume), though the air will be noticeably stale.[7] Equation 39.4 is used for finding the approximate time for a 3% buildup of carbon dioxide in a closed area.[8] The carbon dioxide concentration should not exceed 5% under any circumstances.

$$t_{hours} = \frac{(1.4)V_{room,m^3}}{\text{no. of occupants}} \quad \text{[SI]} \quad \textit{39.4(a)}$$

$$t_{hours} = \frac{(0.04)V_{room,ft^3}}{\text{no. of occupants}} \quad \text{[U.S.]} \quad \textit{39.4(b)}$$

[6]Steam flow is turned off when the humidity reaches the high-limit humidistat setting, typically 90% relative humidity. This prevents oversaturation of the air when there is a failure in the air conditioning system or controlling humidistat.

[7]Some submarines have operated at 1% carbon dioxide by volume.

[8]This equation assumes an initial (atmospheric) carbon dioxide content of 0.03% and carbon dioxide production of 0.011 ft³/min (0.00031 m³/min; 5.2 mL/s) per person.

9. ODOR REMOVAL

The air flow required through a room to remove body odors depends on the room size and level of activity. Body odors become more pronounced when the relative humidity is above 55%. Except for very cramped areas, common ventilation standards are normally sufficient for odor removal. Approximately one-third of the air should be new air.

10. SENSIBLE AND LATENT HEAT

Metabolic heat contains both sensible and latent components. *Sensible heat* is pure thermal energy that increases the air's dry-bulb temperature. *Latent heat* is moisture that increases air's humidity ratio. Table 39.1 gives the approximate amounts of metabolic heat in a 75°F (23.9°C) environment. The "adjusted" column refers to a normal mix of men, women, and children. For design purposes, the heat gain for an adult female is approximately 85% of the adult male rate; the heat gain for a child is 75% of the adult male rate.

Table 39.1 *Approximate Heat Generation by Occupants (Btu/hr)*

activity	total adult males	total adjusted	sensible[a] adjusted	latent[a] adjusted
seated, at rest, theater, classroom	390	330	225	105
moderately active office work	475	450	250	200
standing, light work, slowly walking	550	450	250	200
moderate dancing	900	850	305	545
walking 3 mph, moderately heavy work	1000	1000	375	625
heavy work	1500	1450	580	870

(Multiply Btu/hr by 0.293 to obtain watts.)

[a]The sensible-latent split given is for a 75°F (23.9°C) environment. For an 80°F (26.7°C) environment, total heat remains the same, but sensible heat decreases approximately 20% and latent heat increases accordingly.

11. VENTILATION FOR HEAT REMOVAL

Ventilation requirements can be calculated from sensible heat and/or moisture (i.e., latent heat) generation rates. In Eq. 39.5, T_{in} is the dry-bulb temperature of the air entering the room.

$$\dot{q}_s = \dot{m}c_p(T_{id} - T_{in})$$
$$= \dot{V}\rho c_p(T_{id} - T_{in}) \quad \textit{39.5}$$

In ventilation work, volumetric flow rates are traditionally given in ft³/min (cfm), m³/min, or L/s. The constant 1.08 (0.02) in Eq. 39.6 is the product of an air density of 0.075 lbm/ft³ (1.2 kg/m³), a specific heat of 0.24 Btu/lbm-°F (1.0 kJ/kg·°C), and the factor 60 min/hr (60 s/min).

$$\dot{V}_{m^3/min} = \frac{\dot{q}_{s,kW}}{\left(0.02 \ \dfrac{kJ \cdot min}{m^3 \cdot s \cdot °C}\right)(T_{id,°C} - T_{in,°C})}$$

[SI] *39.6(a)*

$$\dot{V}_{cfm} = \frac{\dot{q}_{s,Btu/hr}}{\left(1.08 \ \dfrac{Btu\text{-}min}{ft^3\text{-}hr\text{-}°F}\right)(T_{id,°F} - T_{in,°F})}$$

[U.S.] *39.6(b)*

$$\dot{V}_{L/s} = \frac{\dot{q}_{s,Btu/hr}}{\left(1.20 \ \dfrac{Btu\text{-}sec}{L\text{-}hr\text{-}°F}\right)(T_{id,°F} - T_{in,°F})}$$

[mixed units] *39.6(c)*

The sensible heat loads will usually be more significant than the latent load, and ventilation will be determined solely on that basis. When large moisture sources are present, however, the latent loads may control.

$$\dot{q}_l = \dot{m}_{water}h_{fg} = \dot{m}_{air}\Delta\omega h_{fg}$$
$$= \dot{V}\rho\Delta\omega h_{fg} \qquad 39.7$$

The constant 4775 (49.36) in Eq. 39.8 is the product of the air density of 0.075 lbm/ft³ (1.2 kg/m³), a latent heat of vaporization at the approximate partial pressure of the water vapor in the air of 1061 Btu/lbm (2468 kJ/kg), and the factor 60 min/hr (60 s/min).[9]

$$\dot{V}_{m^3/min} = \frac{\dot{q}_{l,kW}}{\left(49.36 \ \dfrac{kJ \cdot min}{m^3 \cdot s}\right)\Delta\omega_{kg/kg}}$$

[SI] *39.8(a)*

$$\dot{V}_{cfm} = \frac{\dot{q}_{l,Btu/hr}}{\left(4775 \ \dfrac{Btu\text{-}min}{ft^3\text{-}hr}\right)\Delta\omega_{lbm/lbm}}$$

[U.S.] *39.8(b)*

12. FUME EXHAUST HOODS

Fume exhaust hoods are used to provide localized removal of dangerous dusts and vapors. There are two types of hoods: total or partial enclosures (e.g., canopies, booths, and cabinets) and nonenclosure types (e.g., slotted and receiving hoods). *Nonenclosure hoods* are usually placed around the perimeter of the source and are not generally used to capture contaminants that rise. Since a substantial portion of the air can escape

---

[9]There is some variation in these constants depending on what heat of vaporization is used. For example, some sources use 1076 Btu/lbm (2503 kJ/kg), in which case the constant is 4840 Btu-min/ft³-hr (50.06 kJ·min/m³·s).

capture by going "around" a plain hood opening, most *enclosure hoods* include flanges. Flanges need to extend only about 6 in (15 cm) to be effective. They reduce the required air flow by as much as 25%. Canopy hoods are generally used over open tanks. The canopy portion should extend beyond the edge of the tank.

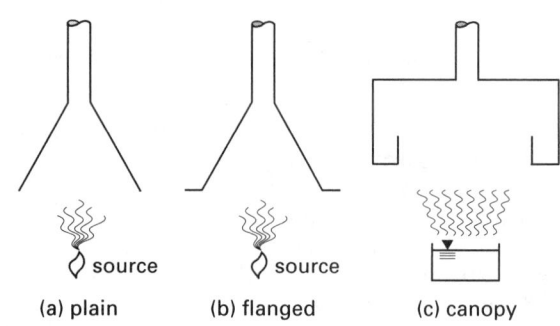

| (a) plain | (b) flanged | (c) canopy |

Figure 39.1 Exhaust Hoods

Some contaminants are released with almost no velocity of their own. This is the case with products of natural evaporation. Other contaminants (paint booth overspray, dust from grinding wheels, etc.) are released at high velocity. With distance, the velocity dissipates and reaches zero at the *null point*. Although capture is easiest at the null point, it is difficult in most situations to determine the distance to the null point. Even when the distance to the null point is known, the directionality may vary. Therefore, a high air intake velocity is needed to capture the moving contaminants near the point of generation. Table 39.2 lists typical minimum *control velocities*.

Table 39.2 Minimum Control Velocities for Enclosure Hoods

process	minimum control velocity	
	ft/min	m/min
evaporation from open tanks	50–100	15–30
paint spraying, welding, plating	100–200	30–60
stone cutting, mixing, conveying	200–500	60–150
grinding, crushing	500–2000	150–600

(Multiply ft/min by 0.3 to obtain m/min.)

Air velocity decreases with increasing distance from the source, varying almost inversely with the square of the distance. Therefore, the hood opening should be as close as possible to the contaminant source.

Since heated air rises, different design principles are needed for high-temperature (e.g., molten metal) processes. The heated air mixes with the surrounding air, and a larger volume of diluted air must be captured.

13. DILUTION VENTILATION

Dilution ventilation (*toxicity dilution*) refers to the dilution of contaminated air in order to reduce its health or explosion hazard.[10] Dilution is less effective than outright removal of hazards by exhaust ventilation. Dilution is generally applicable to organic liquids and solvents whose toxicity is low, when the workers are not too close to the source, and when vapor generation is fairly uniform.[11] The required volume of dilution air must not be so great as to make air velocities unreasonable.

Dilution is achieved by providing enough air to reduce a vapor's concentration to an acceptable level. Various designations are given to acceptable levels, although the *threshold limit value* (TLV) in parts per million (ppm) by volume (mg/m^3 in SI) and *lower explosive limit* in parts per hundred (pph) by volume (mg/m^3 in SI) are the most common.[12] The TLV is assumed to be the concentration that workers may be continuously exposed to during a certain period. TLVs are subject to ever-changing legislation and ongoing research.[13]

Three types of TLVs are used. The *time-weighted average* (TLV-TWA) is the time-weighted average concentration that workers may be exposed to for eight hours per day, day after day, without experiencing adverse effects. The *short-term exposure limit* (TLV-STEL) is the time-weighted average concentration that workers may be exposed to for fifteen minutes, up to four times per eight-hour period.[14] The *ceiling value* (TLV-C) is the concentration that should not be exceeded, even instantaneously. Depending on the substance, one, two, or all three of these limits may be applicable.[15]

Assuming that a contaminant is uniformly distributed throughout the plant air, at equilibrium the contaminant generation rate, R, is equal to the ventilated removal rate.

$$R = C\dot{V} \qquad 39.9$$

To account for irregular vapor evolution, inefficient ventilation, and toxicity, an empirical multiplicative "effectiveness of mixing factor" (the *K factor*) between 3 and 10 is used.

$$KR = C\dot{V} \qquad 39.10$$

---

[10]See Ftn. 2.
[11]Dilution ventilation is not recommended for carbon tetrachloride, chloroform, and gasoline, among others. Dusts are seldom removed successfully by dilution.
[12]Another unit used for dust concentrations in respirable air is *millions of particles per cubic foot* (mppcf) determined by *midget impinger* techniques. The conversion between mppcf and other units is not exact, depending primarily on the particle size and density. However, equivalences of 5.6 and 6.4 mppcf to 1.0 mg/m^3 are quoted. In the absence of any other information, an average value of 6 is recommended.
[13]In the United States, TLVs are updated annually in the Industrial Ventilation, *Manual of Recommended Practice*, published by the American Conference of Governmental Industrial Hygienists.
[14]Other restrictions may apply to TLV-STEL. For example, there may be a sixty-minute waiting period between successive exposures at this level.
[15]For example, irritant gases may be controlled only by the TLV-C value.

The maximum concentration, C, is usually the threshold limit value. The required airflow is

$$\dot{V}_{L/s} = \frac{(4.02 \times 10^8)KR_{kg/min}}{(MW)TLV_{ppm}} \qquad \text{[SI]} \quad \textit{39.11(a)}$$

$$= \frac{(4.02 \times 10^8)K(SG)R_{L/min}}{(MW)TLV_{ppm}}$$

$$\dot{V}_{cfm} = \frac{(3.86 \times 10^8)KR_{lbm/min}}{(MW)TLV_{ppm}} \qquad \text{[U.S.]} \quad \textit{39.11(b)}$$

$$= \frac{(4.03 \times 10^8)K(SG)R_{pints/min}}{(MW)TLV_{ppm}}$$

When two or more hazardous substances that have similar toxicologic effects are simultaneously present (i.e., act on the same organ or metabolic process), the combined effect should be considered. If the air is to be breathed, the air flow rates for each substance must be calculated and the separate flow rates summed. The mixture threshold value is exceeded when

$$\left(\frac{C}{TLV}\right)_1 + \left(\frac{C}{TLV}\right)_2 + \cdots > 1 \qquad \textit{39.12}$$

The additive nature implied by Eq. 39.12 is assumed unless the two substances are known to act independently instead of additively. In that case, the threshold limit is exceeded only when the ratio C/TLV for at least one component in the mixture exceeds unity. The highest ventilation rate calculated for each component independently is the design ventilation rate.

14. RECIRCULATION OF CLEANED AIR

The volume of ventilation air required to remove contaminants will be greatly reduced if some of the air can be cleaned and returned. Dust and particulate matter in air can be removed by two types of air cleaners. *Air filters* are applicable when the concentration is between 0.5 and 50 grains per 1000 cubic feet (1.1 to 110 mg/m^3). *Dust collectors* are used at the higher concentrations normally found in manufacturing processes. Equilibrium will be achieved when the contaminate generation rate equals the rate at which the air filter removes particles from the air.

$$R = \eta_{filter}C\dot{V} \qquad 39.13$$

15. CLEAN ROOMS

Clean rooms are defined by the number of particles (pollen, skin flakes, etc.) above a given size (usually 0.5 microns) in a cubic foot of air. Most semiconductor clean rooms are Class-100 or better, meaning that there will be no more than one hundred 0.5 micron-sized particles per cubic foot.

Clean rooms technology generally relies on high-efficiency prefilters, either *high-efficiency particulate arresters* (HEPAs) or *ultra-low particle arrester* (ULPA) filters in the supply, positive room pressure, fast air movement, and floor grates (i.e., downflow air movement). A positive pressure of approximately 0.1 in of water (25 Pa) is typical. Twenty air changes per hour is a typical minimum, while air flow velocities of 75 to 100 ft/min (0.38 to 0.5 m/s) are used in the best clean rooms.

High-efficiency (60 to 90%) prefilters reduce the load on HEPAs. Usually, HEPAs (99.97% efficiency at the 0.3 micron level) are suitable for Class-100 clean rooms, while ULPAs are needed for Class-10 or better clean rooms. Large centralized equipment may be used, or modular *air handling units* (AHUs) drawing air from the main general supply may be used for individual clean rooms. Adjustable-frequency drives can be used to change the air flow in order to reduce energy usage or change the cleanliness. Stainless steel is the preferred material for ducts and hoods, as it does not have the flaking problem associated with galvanized metals.

Ventilation requirements are similar to regular designs. Makeup air should be 25% of the total air flow and not less than 15 ft^3/min (0.42 m^3/min; 7.1 L/s) per person.

PRACTICE PROBLEMS

1. An office room has floor dimensions of 60 ft by 95 ft (18 m by 29 m) and a ceiling height of 10 ft (3 m). 45 people occupy the office, and half of them smoke. (a) Calculate the ventilation rate based on six air changes per hour. (b) State your assumptions and determine the ventilation rate based on occupancy.

2. 150 ppm of methanol (TLV = 200 ppm; MW = 32.04; SG = 0.792) and 285 ppm of methylene chloride (TLV = 500 ppm; MW = 84.94; SG = 1.336) are found in the air in a plating booth. Two pints of each are evaporated per hour. Use a mixing safety factor (i.e., a K value) of 6. What ventilation rate is required?

3. An auditorium is designed to seat 4500 people. The ventilation rate is 60 ft^3/min (1.68 m^3/min) per person of outside air. The outside temperature is 0°F (−18°C) dry-bulb, and the outside pressure is 14.6 psia (100.6 kPa). Air leaves the auditorium at 70°F (21°C) dry-bulb. There is no recirculation. The furnace has been sized assuming an internal heat gain of 1,250,000 Btu/hr (370 kW) and the outside conditions described. (a) At what temperature should the air enter the auditorium? (b) How much heat should be supplied to the ventilation air? (c) Has the furnace been sized properly?

4. A room is maintained at design conditions of 75°F (23.9°C) dry-bulb and 50% relative humidity. The air outside is at 95°F (35°C) dry-bulb and 75°F (23.9°C) wet-bulb. Conditioned air enters the room and increases 20°F (−6.7°C) in temperature before being removed from the room. The sensible and latent loads are 200,000 Btu/hr (60 kW) and 50,000 Btu/hr (15 kW), respectively. 2000 ft^3/min (57 m^3/min) of outside air are used. What are the (a) apparatus dew point and (b) quantity of air flowing through the air conditioner?

40 Heating Load

1. Introduction 40-1
2. Inside Design Conditions 40-1
3. Outside Design Conditions 40-2
4. Adjacent Space Conditions 40-2
5. Walls and Ceilings 40-2
6. Air Spaces 40-3
7. Ground Slabs 40-3
8. Ventilation and Infiltration Air 40-4
9. Humidification 40-4
10. Internal Heat Sources 40-4
11. Thermal Inertia 40-5
12. Furnace Sizing 40-5
13. Degree Days and Kelvin Days 40-5
14. Fuel Consumption 40-6
15. Conservation Through Thermostat Setback . 40-6
    Practice Problems 40-6

Nomenclature

A	area	ft^2	m^2
C	thermal conductance of air space	Btu/ft^2-hr-°F	W/m^2·°C
DD	degree days	°F-day	°C·day
E	effective emissivity	–	–
F	slab edge coefficient	Btu/ft-hr-°F	W/m·°C
h	enthalpy	Btu/lbm	kJ/kg
h	surface heat transfer coefficient	Btu/ft^2-hr-°F	W/m^2·°C
HV	heating value	various	various
k	thermal conductivity	Btu-ft/ft^2-hr-°F	W·cm/m^2·°C
L	length	ft	cm
m	mass	lbm	kg
M	masonry M factor	–	–
N	number of days in heating season	–	–
p	perimeter length	ft	m
P	power	hp	kW
$\dot{q}$	heat transfer rate	Btu/hr	kW
R	thermal resistance	ft^2-hr-°F/Btu	m^2·°C/W
SF	service factor	–	–
T	temperature	°F	°C
U	overall coefficient of heat transfer	Btu/ft^2-hr-°F	W/m^2·°C
$\dot{V}$	volumetric flow rate	ft^3/min	m^3/s

Symbols

ϵ	emissivity	–	–
η	efficiency	–	–
ρ	density	lbm/ft^3	kg/m^3
ω	humidity ratio	lbm/lbm	kg/kg

Subscripts

a	dry air
fg	vaporization
i	inside design
o	outside design

1. INTRODUCTION

A building's *heating load* is the maximum heat loss (typically expressed in Btu/hr or kW) during the heating season.[1] The *maximum heating load* occurs when the outside temperature is the lowest. The maximum heating load corresponds to the minimum furnace size, even though the lowest temperature occurs only a few times each year. The *average heating load* can be derived from the maximum heating load and it is used to determine the annual fuel requirements.

Heating load consists of heat to make up for transmission and infiltration losses. Determining transmission losses is essentially a heat transfer problem. *Transmission loss* is heat lost through walls, roof, and floor. *Infiltration loss* is heat required to warm ventilation and infiltration air. Though no "credit" for solar heat gain is taken in heating load calculations, reliable sources of internal heating are considered.[2] Modifications for thermal inertia due to high-mass walls and ceilings are generally not made in calculations of heating load. When thermal inertia is considered, the approach taken is simplistic. (See Sec. 11.)

Calculation of heating load is greatly simplified by having access to tabulations of data.[3] Data on climatological conditions are essential, and data on building materials and construction will greatly simplify the task of calculating heat transfer coefficients. Data of this nature is available in a variety of formats. Heat transmission data are available for specific materials as well as for composite walls of specific construction. Both types of data are useful.

Calculations of heating load are based on many assumptions. Because of the intrinsic unreliability of some of the data, an *exposure allowance* of up to 15% may be added to the calculated ideal heating load. This helps to account for unexpected heat losses and severe climatic conditions.

2. INSIDE DESIGN CONDITIONS

For the purposes of initial heating load calculations for residences and office spaces, the *inside design temperature* is generally taken as 70 to 72°F (21.1 to 22.2°C).

---

[1] A *therm* is 100,000 Btu.

[2] The sky is assumed to be overcast during the heating season, so solar heat gain is minimal. Reliable sources of internal heating include permanently mounted equipment and lights.

[3] It is essential that engineers working in this area obtain their own compilations of this type of data.

For industrial spaces, such as factories and warehouses, the inside design temperature is lower: 60 to 65°F (15.6 to 18.3°C). Humidity is typically 30 to 35% relative humidity, but is generally not considered except in certain manufacturing industries (e.g., textiles) where moisture content is critical.

3. OUTSIDE DESIGN CONDITIONS

The outside temperature and average wind speed (for infiltration) are needed to determine heating load. For estimates of annual heating costs, information on the winter degree days is needed (see Sec. 13). These values are almost always obtained from tables of climatological data.

Design conditions are probabilistic in nature. There is always some probability that a temperature will be exceeded. Depending on the nature of the facility, it may be more economical to select an outside design temperature that will be exceeded (say) 5 days out of 100 days. Many tables give design temperatures for 1%, 2.5%, and 5% exceedance probabilities.

4. ADJACENT SPACE CONDITIONS

Since the conductive heat transfer through shared walls depends on the temperatures on both faces, determining the heating load for single rooms and separately heated offices also requires knowing the temperatures in adjacent spaces. For residential calculations, this may require knowing the temperature in attics, large closets, and basements. For attics ventilated by large open louvers, the approximate attic temperature is the average of the inside and outside design temperatures.

5. WALLS AND CEILINGS

Each material used in constructing a wall, ceiling, and so on, contributes resistance to heat flow. This resistance can be specified in a variety of ways. *Total resistance*, R (with units of Btu/ft^2-hr-°F or W/m^2·°C), is the total resistance to heat flow through all of the material. *Unit resistance* (with units of Btu-ft/ft^2-hr-°F or W·cm/m^2·°C) is the resistance per unit thickness of the material.[4] The total resistance is the product of the resistivity and the material thickness. *Conductance* and *conductivity*, k, are the reciprocals of total resistance and unit resistance, respectively.

$$R = \frac{L}{k} \qquad 40.1$$

The heat transfer through walls, doors, windows, and ceilings is calculated from the traditional heat transfer

[4]Unit resistance and conductivity are combined with a length when calculating thermal resistance. There are several sets of units in use for conductivity, depending on how the length is measured. For lengths in feet, Btu/ft-hr-°F and Btu-ft/ft^2-hr-°F are the same. For lengths in inches (centimeters), the units Btu-in/ft^2-hr-°F (W·cm/m^2·hr-°F) must be used.

equation, Eq. 40.2. The *overall coefficient of heat transfer*, U, can be calculated for each transmission path from the conductivities and resistances of the individual components in that path, or it can be obtained from tabulations of typical wall/ceiling construction.

$$\dot{q} = U A (T_i - T_o) \qquad 40.2$$

$$U = \frac{1}{\Sigma R_i}$$

$$= \frac{1}{\dfrac{1}{h_i} + \sum \dfrac{L}{k} + \sum \dfrac{1}{C} + \dfrac{1}{h_o}} \qquad 40.3$$

Table 40.1 *Typical Overall Coefficients of Heat Transfer (Btu/ft^2-hr-°F)*

exterior walls	average U-factors, insulated as shown			
	none	1 in	2 in	4 in
• standard 2 by 4 construction sheathed in wood or insulating board, covered with wood siding, shingles, brick, and sheetrocked or plastered	0.22	0.12	0.09	0.07
12 in concrete blocks	0.49	0.14	—	—
8 in poured concrete walls	0.70	0.16	—	—

interior ceilings	average U-factors, insulated between rafters as shown					
	none	1 in	2 in	3 in	4 in	6 in
• ceiling applied directly to wood rafters with wood sheathing covered by asphalt or cedar shingles	0.64	0.19	0.12	0.09	0.077	0.05
	average U-factors, insulated between joists as shown					
	none	1 in	2 in	3 in	4 in	6 in
• horizontal ceiling under a pitched roof with no flooring on the ceiling						
• rafters covered by wood sheathing and asphalt or cedar shingles	0.32	0.15	0.10	0.09	0.077	0.05

There is no heat loss in ceilings between heated floors. For ceilings over insulated crawl spaces, use $1/2$ of the U-value shown. For ceilings over vented or unheated crawl spaces, use the indicated U-value. For ceilings over unheated basements, use $1/3$ of the indicated U-value.

concrete slab on grade floors	average U-factors, per lineal foot		
	none	1 in × 12 in	1 in × 24 in
	0.81	0.46	0.21

windows	average U-factors
• single glazing	1.31
with storm window	0.45
thermopane	0.61
doors	
• $3/4$ in wood	0.69
$1^5/8$ in wood	0.46
$1^5/8$ in wood with storm door	0.32

(Multiply in by 2.54 to obtain cm.)
(Multiply Btu/ft^2-hr-°F by 5.68 to obtain W/m^2·C.)

Table 40.2 *Representative Thermal Conductances of Planar Air Spaces*[a,b] *(Btu/ft²-hr-°F (W/m²·°C))*

air space orientation	heat flow direction	space thickness in (mm)	effective space emissivity, E			
			0.05	0.20	0.50	0.82
horizontal	up	0.75–4.0 (19–102)	0.41 (2.3)	0.55 (3.1)	0.82 (4.7)	1.11 (6.30)
horizontal	down	0.75 (19)	0.28 (1.6)	0.42 (2.4)	0.69 (3.9)	0.98 (5.6)
		1.5 (38)	0.18 (1.0)	0.31 (1.8)	0.58 (3.3)	0.87 (4.9)
		4.0 (102)	0.11 (0.62)	0.25 (1.4)	0.52 (3.0)	0.81 (4.6)
vertical	horizontal	0.75–4.0 (19–102)	0.28 (1.6)	0.42 (2.4)	0.69 (3.9)	0.99 (5.6)

(Multiply in by 25.4 to obtain mm.)
(Multiply Btu/ft²-hr-°F by 5.68 to obtain W/m²·°C.)
[a]For a 50°F (27.8°C) temperature differential.
[b]Compiled from a variety of sources.

Unlike most heat transfer problems, little effort is expended in calculating surface heat transfer (film) coefficients from theoretical correlations. Tables of typical values are used. When tabulations of overall coefficients of heat transfer are used, it is important to know if an outside film coefficient has been included. If it has, the table's assumption about outside wind speed must be known and compared with actual wind conditions. Multiplicative corrections for other wind speeds generally accompany such tables.

Table 40.3 *Representative Surface Heat Transfer (Film) Coefficients for Air[a] (nonreflecting surfaces)*

surface orientation, heat flow direction	air speed[b]	Btu/ft²-hr-°F	W/m²·°C
horizontal, heat flow up	0	1.63	9.26
	7½ mph	4.00	22.7
	15 mph	6.00	34.1
vertical, heat flow horizontal	0	1.46	8.29
	7½ mph	4.00	22.7
	15 mph	6.00	34.1
horizontal, heat flow down	0	1.08	6.13

(Multiply mph by 1.61 to obtain km/h.)
(Multiply Btu/ft-hr-°F by 5.68 to obtain W/m²·°C.)
[a]Compiled from a variety of sources.
[b]Use 0 indoors. Use 7½ mph in summer. Use 15 mph in winter.

6. AIR SPACES

The thermal conductance of an air space, C, used in Eq. 40.3 depends on the space width, emissivities of both sides, orientation, mean temperature, and temperature differential. Typical values for a 50°F (27.8°C) temperature differential are given in Table 40.2.[5] The effective space emissivity, E, used in the table is a function of the surface emissivity, ϵ, and is defined by Eq. 40.4. Surface emissivities range from approximately 0.05 for bright aluminum foil through 0.25 for bright galvanized steel to 0.90 for typical building materials (wood, sheetrock, masonry, etc.).

$$E = \cfrac{1}{\cfrac{1}{\epsilon_1} + \cfrac{1}{\epsilon_2} - 1}$$

40.4

7. GROUND SLABS

Heat is lost from ground slabs both through the face and from the exposed edges. Since the ground temperature is usually higher than the winter air temperature, a lower temperature difference (typically 5°F (2.8°C)) should be used to find the heat loss through the face.

This loss is generally small, and an overall coefficient of 0.05 Btu/ft²-hr-°F (0.3 W/m³·°C) is typical. The loss is essentially constant throughout the year since the soil temperature under a building does not vary appreciably. The radial loss from the edges can be found

[5]Data in Table 40.2 depend on many factors and are merely representative. Values can vary 20% or more.

HVAC

from the *slab edge coefficient*, F, and from the perimeter of the exposed edge. Thickness of the slab is disregarded. Coefficients for concrete slabs on grade depend on the construction method, the amount of insulation, and weather conditions. The coefficient varies from approximately 0.5 Btu/ft-hr-°F (0.9 W/m·°C) for slabs with insulated edges in warm climates to approximately 2.7 Btu/hr-°F (4.9 W/m·°C) for slabs with no insulation located in cold climates.[6]

$$\dot{q} = pF(T_i - T_o) \qquad 40.5$$

8. VENTILATION AND INFILTRATION AIR

Outside ventilation air must be warmed before being introduced into the occupied space. All infiltrated air (from window sashes, door jams, and other cracks) will also go through the air conditioner, so its sensible heat load is combined with the ventilation air deliberately drawn in. The amount of sensible heat required was covered in Chap. 39.

If the outside design temperature is 32°F (0°C) or less, there will be little or no moisture in the incoming air. However, for air with higher temperatures, heat is required to warm any moisture that enters with outside air.

Although the sensible heating of the dry air and accompanying moisture can be calculated separately and added, it is more expedient to use enthalpy values read from the psychrometric chart.

$$
\begin{aligned}
\dot{q}_{\text{kW}} &= \dot{m}_{a,\text{kg/s}}(h_i - h_o)_{\text{kJ/kg}} \\
&= \dot{V}_{\text{m}^3/\text{s}}\rho_{\text{kg/m}^3}(h_i - h_o)_{\text{kJ/kg}} \\
&= \left(1.2\ \frac{\text{kg}}{\text{m}^3}\right)\dot{V}_{\text{m}^3/\text{s}}(h_i - h_o)_{\text{kJ/kg}} \\
&= \left(0.0012\ \frac{\text{kg}}{\text{L}}\right)\dot{V}_{\text{L/s}}(h_i - h_o)_{\text{kJ/kg}}
\end{aligned}
$$
$$\text{[SI]} \qquad 40.6(a)$$

$$
\begin{aligned}
\dot{q}_{\text{Btu/hr}} &= \dot{m}_{a,\text{lbm/hr}}(h_i - h_o)_{\text{Btu/lbm}} \\
&= \left(60\ \frac{\text{min}}{\text{hr}}\right)\dot{V}_{\text{ft}^3/\text{min}}\rho_{\text{lbm/ft}^3}(h_i - h_o)_{\text{Btu/lbm}} \\
&\approx \left(4.5\ \frac{\text{lbm-min}}{\text{ft}^3\text{-hr}}\right)\dot{V}_{\text{ft}^3/\text{min}}(h_i - h_o)_{\text{Btu/lbm}}
\end{aligned}
$$
$$\text{[U.S.]} \qquad 40.6(b)$$

Special accounting for infiltrated air is necessary when the conditioned space supplies combustion air. This normally is the case for residential fireplaces and space heaters. Each volume of gaseous fuel gas burned uses approximately nine volumes of heated air. Unless the face of the fireplace is blocked off, heated air continues to be lost up the flue, even after combustion stops.

---

[6]The slab edge coefficient may be given in Btu/hr-°F (W/°C) per lineal distance. A *lineal distance* between two points is the linear distance that one would have to walk to go from one point to the other.

9. HUMIDIFICATION

Outside air entering at very low temperatures may need to be humidified. The latent heat required to add moisture to the outside air was covered in Chap. 38.

10. INTERNAL HEAT SOURCES

Internal heat sources are heat sources within the conditioned space.[7] Heat sources may introduce sensible and latent loads. Two general methods are used to estimate internal sources. The load can be based on the equipment nameplate rating. Alternatively, tables of typical values can be used. Such tables are helpful in estimating the fractions of each load that are sensible and latent.[8] The sensible heat supplied by permanent machinery and lighting should be subtracted from the heating load.

Some equipment is rated in horsepower; other equipment is rated by its wattage.[9] The *service factor*, SF, used in Eq. 40.7 is the fraction of the rated power being developed. Service factors greater than 1.00 are possible in overload conditions.

$$
\begin{aligned}
\dot{q}_{\text{kW}} &= \left(0.7457\ \frac{\text{kW}}{\text{hp}}\right)(\text{SF})\left(\frac{P_{\text{hp}}}{\eta}\right) \\
&= (\text{SF})\left(\frac{P_{\text{kW}}}{\eta}\right) \qquad \text{[SI]} \qquad 40.7(a)
\end{aligned}
$$

$$
\begin{aligned}
\dot{q}_{\text{Btu/hr}} &= \left(2545\ \frac{\text{Btu}}{\text{hp-hr}}\right)(\text{SF})\left(\frac{P_{\text{hp}}}{\eta}\right) \\
&= \left(3413\ \frac{\text{Btu}}{\text{kW-hr}}\right)(\text{SF})\left(\frac{P_{\text{kW}}}{\eta}\right)
\end{aligned}
$$
$$\text{[U.S.]} \qquad 40.7(b)$$

The equipment efficiency, η, is included in the denominator in Eq. 40.7. This is appropriate when the motor and equipment being driven are both in the conditioned space. If the motor is not actually in the space but the driven equipment is, then the efficiency should be omitted.[10] (An example of this configuration would be

---

[7]Residential heating sources such as toasters and coffee brewers are sometimes referred to as *domestic heat sources*.

[8]In the absence of tabular data, simple, logical rules of thumb can be used. For example, when cooking under a hood, all moisture is captured, the latent load is zero, and the sensible load transferred due to radiation is taken as some assumed fraction of the nameplate rating. If not cooking under a hood, two-thirds of the load is sensible and one-third is latent. However, tables are usually essential when determining the heat gain for esoteric sources such as doughnut machines, deep-fry kettles, and waffle irons for ice cream sandwiches.

[9]The *watt density* (strictly, the *linear watt density*) for baseboard heaters, heat tracing, and similar devices is the amount of heating generated per lineal foot (meter). It is not based on area.

[10]There is a third, less likely alternative. The conditioned space may contain only the motor, with the driven machinery being someplace else. In that case, only the energy lost in the energy conversion appears in the load. The nameplate power is multiplied by $(1 - \eta)/\eta$.

an in-duct fan that is driven by a motor located outside of the duct.) Typical motor efficiencies are 80% for 1 hp motors and 90% for 10 hp and higher motors.

For fluorescent lights, the rated wattage should be increased by 20 to 25% to account for ballast heating. (No increase is used with incandescent lights.) However, it may be inappropriate to assume that all heat generated enters the conditioned space. If the air space above the lights in a dropped ceiling is not directly conditioned, then only some fraction (e.g., 60%) of the heat enters the occupied space.

Unless the room is reasonably occupied on a permanent basis, the heating load should not be reduced by the metabolic heat of the occupants.

Some buildings have internal sources generating large amounts of energy. Under no circumstance, however, should the theoretical heating load be reduced by these internal heat gains to a point where the inside temperature would be 40°F (4.4°C) or lower in the absence of these internal sources.

11. THERMAL INERTIA

Buildings with massive masonry (including concrete) walls are thermally more stable than those with thin walls. However, the effect of *thermal lag* (*thermal inertia* or *thermal flywheel effect*) is difficult to incorporate in most studies. A simplified approach known as the *M-factor method* has been developed. This method correlates a factor, M, with the degree days and mass per unit wall area (lbm/ft^2 or kg/m^2). The M-factor modifies the overall coefficient of heat transfer, U. The actual heat transfer is

$$\dot{q} = MUA(T_i - T_o) \qquad 40.8$$

As written, Eq. 40.8 is appropriate for analyzing the heat loss through a wall. For design, particularly when wall construction with a maximum heat loss coefficient, U, is specified by the building code, a wall may be designed to have an instantaneous heat loss coefficient of U/M and still have performance that meets code.

12. FURNACE SIZING

A furnace must be capable of keeping a building warm on the coldest days of the heating season. Therefore, a furnace should be sized based on the coldest temperature reasonably expected (i.e., on the outside design temperature). However, not all days in the heating season will have temperatures that cold, and the entire capacity will rarely be utilized.

Pickup load is the furnace capacity needed to bring a cold building up to the inside design temperature in a reasonable time. Since the outside design temperature provides excess capacity most of the year, a pickup allowance may not be necessary. However, churches, auditoriums, and office buildings needing to be "ready" at

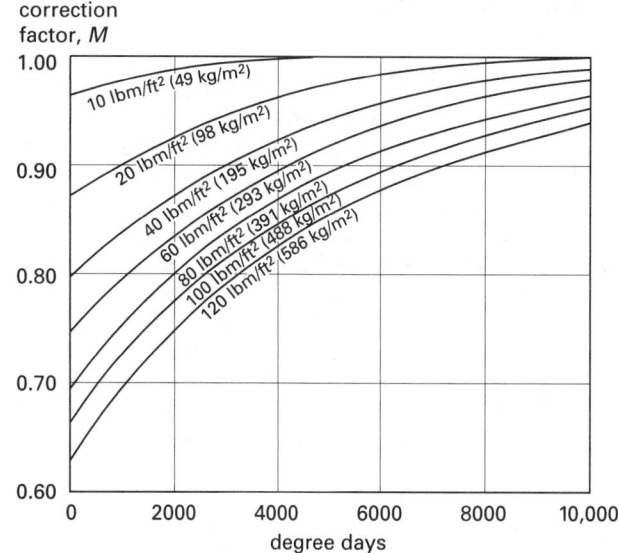

Source: "Mass Masonry Energy," Masonry Industry Committee, ca. 1979.

Figure 40.1 *M-Factor*

specific times need special attention.[11] If a building is heated during the day only, a 10% increase in the rated furnace size is sometimes added as the pickup load to allow for starting up. If a building is left unheated for extended periods, the increase should be 25% or higher.

Some furnaces have nameplate ratings for both fuel input and output heating. The furnace efficiency is incorporated in a rating for "output Btus." Output ratings apply to newly installed furnaces. Output can be expected to decrease with time.

13. DEGREE DAYS AND KELVIN DAYS

The *degree day* (*kelvin day*) concept can be used to determine the average temperature over the entire heating season. Each day whose 24-hour average temperature is less than 65°F (18°C) will accumulate $\overline{T} - 65$ degree days ($\overline{T} - 18$ kelvin days).[12] The sum of these degree day terms over the entire heating season of N days, which is different for each geographical location, is the total *heating degree days*, DD (also known as *winter degree days*) for that location.[13] The average temperature, $\overline{T}$, over the entire heating season can be calculated from the degree days.

$$DD = N(18°C - \overline{\overline{T}}) \qquad \text{[SI]} \qquad 40.9(a)$$

$$DD = N(65°F - \overline{\overline{T}}) \qquad \text{[U.S.]} \qquad 40.9(b)$$

---

[11]There is not much information on this subject. The increase for pickup load is largely a matter of judgment.
[12]Approximate kelvin days can be obtained by multiplying customary U.S. degree days by 5/9. The conversion is not exact because 65°F does not correspond exactly to 18°C.
[13]*Summer degree days (cooling degree days)* are covered in Chap. 41.

14. FUEL CONSUMPTION

The design heating load used to size the furnace is not the average heating load over the heating season. However, when combined with degree days, the design heating load can be used with smaller, simple structures to calculate the total fuel consumption and heating costs during the entire heating season. (This method assumes that there will be no heating when the outside temperature is 65°F (18°C) or higher. This assumption is appropriate, particularly in residential buildings, even though the interior temperature is maintained at 68 to 72°F (20 to 22.2°C) because of the existence of internal heat sources.) The units of fuel consumption in Eq. 40.10 depend on the units of the heating value. Equation 40.10 does not include factors for operating pumps, fans, stokers, and other devices, nor are costs of maintenance, tank insurance, and so on, included.[14] The minimum efficiency, η, of gas- and oil-fired residential furnaces is approximately 70 to 75%.

fuel consumption (in units/heating season)

$$= \frac{\left(86\,400\,\dfrac{\text{s}}{\text{day}}\right)\dot{q}_{\text{kW}}(\text{DD})}{(T_i - T_o)(\text{HV}_{\text{kJ/unit}})\eta_{\text{furnace}}} \quad \text{[SI]} \quad \textit{40.10(a)}$$

fuel consumption (in units/heating season)

$$= \frac{\left(24\,\dfrac{\text{hr}}{\text{day}}\right)\dot{q}_{\text{Btu/hr}}(\text{DD})}{(T_i - T_o)(\text{HV}_{\text{Btu/unit}})\eta_{\text{furnace}}} \quad \text{[U.S.]} \quad \textit{40.10(b)}$$

Equation 40.10 has traditionally been used to estimate fuel consumption. In recent years, various improvements to the "model" have been made. Specifically, Eq. 40.10 is multiplied by an empirical correction, C_D, to correct for the difference between calculated values and actual performance.[15] Values of C_D range from about 0.60 to 0.87 and are correlated with the number of degree-days. Also, the furnace efficiency, η, is replaced with an efficiency factor, k, that includes the effects of rated full-load efficiency, part-load performance, oversizing, and energy conservation devices. A value of 1.0 should be used for k for electric heating. Values of 0.55 and 0.65 are appropriate for older and energy-efficient houses, respectively.

15. CONSERVATION THROUGH THERMOSTAT SETBACK

A variety of *conservation methods* are used to reduce energy consumption during the heating season. These include installing insulation, weatherstripping, and

reducing the thermostat setting for all or part of the day. An extreme case of thermostat setback occurs when the heating system is turned off entirely at night.

If a building is to be maintained at two different temperatures during different parts of the day, the average heating can be found from the duration-weighted average of the two inside temperatures. Alternatively, separate calculations can be made for the periods of different temperatures.

PRACTICE PROBLEMS

1. A conditioned room contains 12,000 W of fluorescent lights and twelve 90% efficient, 10 hp motors operating at 80% of their rated capacities. The lights are pendant-mounted on chains from the ceiling. The motors drive various pieces of machinery located in the conditioned space. What is the internal heat gain?

2. A building is located in the city of New York. There are 4772 (2651 in SI) degree days during the October 15 to May 15 period for this area. The building is heated by fuel oil whose heating value is 153,600 Btu/gal (42 800 MJ/m^3) and which costs $0.15/gal ($0.034/L). The calculated design heat loss is 3.5×10^6 Btu/hr (1 MW) based on 70°F (21.1°C) inside and 0°F (−17.8°C) outside design temperatures. The furnace has an efficiency of 70%. What is the approximate cost for heating this building during the winter (October 15 through May 15)?

3. A flat roof consists of $1^1/_2$ in (38 mm) of insulation installed over 3 in (76 mm) of soft pine sheathing, and a $^3/_4$ in (19 mm) acoustical ceiling suspended 4 in (100 mm) below the pine. The outside wind velocity is 15 mi/hr (24 km/h). The interior design temperature is 80°F (26.7°C). The exterior design temperature is 95°F (35°C). What is the overall coefficient of heat transfer?

4. A 12 ft by 12 ft (3.6 m by 3.6 m) floor is constructed as a concrete slab with two exposed edges. The slab edge heat loss coefficient for these two edges is 0.55 Btu/ft-hr-°F (0.95 W/m·°C). The other two edges form part of a heated basement wall. The inside design temperature is 70°F (21.1°C); the outdoor design temperature is −10°F (−23.3°C). Use the slab edge method to determine the heat loss from the slab.

5. An office in a remodeled historic building has floor dimensions of 100 ft by 40 ft (30 m by 12 m) and a ceiling height of 10 ft (3 m). One of the 40 ft (12 m) walls is shared with an adjacent heated space. The three remaining walls have two 4 ft (1.2 m) wide by 6 ft (1.8 m) high, double glass with $^1/_4$ in (6.4 mm) air space, weatherstripped, double-hung windows per 20 ft (6 m). The crack coefficient for the windows is 32 ft^3/hr-ft (3.0 m^3/h·m). The basement and second floor are heated to 70°F (21.1°C). The wall coefficient is 0.2 Btu/ft^2-hr-°F (1.1 W/m^2·°C), and the outside wind velocity is

---

[14]Equation 40.10 appears to imply that the lower T_o is, the lower the fuel consumption will be. This is obviously untrue. The temperature difference in the denominator actually cancels the same temperature difference used to calculate the heat transfer terms in q. Thus, q is put on a per-degree basis. The average temperature difference used in the calculation of degree days converts the per-degree heat loss to an average heat loss.

[15]It has been shown that Eq. 40.10 overestimates the fuel requirement in most cases. C_D simply reduces the estimate.

15 mi/hr (24 km/h). The inside design temperature is 70°F (21.1°C); the outside design temperature is −10°F (−23.3°C). Including infiltration but disregarding ventilation air, what is the heating load?

6. A building is located in Newark, New Jersey. The heating season lasts for 245 days; the degree days are 5252 (2918 in SI); the outside design temperature is 0°F (−17.8°C). The temperature in the building is maintained at 70°F (21.1°C) between the hours of 8:30 A.M. and 5:30 P.M. During the rest of the day and the night, the temperature is allowed to drop to 50°F (10°C). The building is heated with coal that has a heating value of 13,000 Btu/lbm (30.2 MJ/kg). A heat loss of 650,000 Btu/hr (0.19 MW) has been calculated based on 70°F (21.1°C) inside and 0°F (−17.8°C) outside temperatures. The furnace has an efficiency of 70%. What mass of coal is required each year?

7. The design heat loss of a building is 200,000 Btu/hr (60 kW) based on 70°F (21.1°C) inside and 0°F (−17.8°C) outside design temperatures. At that location, there are 4200 (2333 in SI) degree days over the 210-day heating season. The building is occupied 24 hr/day. What is the percentage reduction in heating fuel if the thermostat is lowered from 70°F to 68°F (21.1°C to 20°C)?

8. (*Time limit: one hour*) A building is located where the annual heating season lasts 21 weeks. The inside design temperature is 70°F (21.1°C). The gas furnace has an efficiency of 75%. Fuel costs $0.25 per therm. The building is occupied only from 8 A.M. until 6 P.M., Monday through Friday. In the past, the building was maintained 70°F (21.1°C) at all times and ventilated with one air change per hour. The thermostat is now being set back 12°F (6.7°C) and the ventilation reduced 50% during the unoccupied times. Infiltration through cracks and humidity changes are negligible. What is the annual savings?

internal volume:	801,000 ft^3 (22 700 m^3)
wall area:	11,040 ft^2 (993 m^2)
wall overall heat transfer coefficient:	0.15 Btu/ft^2-hr-°F (0.85 W/m^2·°C)
window area:	2760 ft^2 (260 m^2)
window overall heat transfer coefficient:	1.13 Btu/ft^2-hr-°F (6.42 W/m^2·°C)
roof area:	26,700 ft^2 (2480 m^2)
roof overall heat transfer coefficient:	0.05 Btu/ft^2-hr-°F (0.3 W/m^2·°C)
slab on grade:	690 lineal ft (210 m) of exposed slab edge
slab edge coefficient:	1.5 Btu/ft-hr-°F (2.6 W/m·°C)

HVAC

41 Cooling Load

1. Introduction 41-1
2. Inside and Outside Design Conditions 41-1
3. Instantaneous Cooling Load
    from Walls and Roofs 41-2
4. Instantaneous Cooling Load
    from Windows 41-2
5. Cooling Load from Internal Heat Sources . . 41-2
6. Ventilation and Infiltration 41-2
7. Heat Gain to Air Conditioning Ducts 41-2
8. Degree Days 41-2
9. Seasonal Cooling Energy 41-3
10. Recirculating Air Bypass 41-3
11. Reheat 41-6
    Practice Problems 41-6

Nomenclature

A	area	ft^2	m^2
BF	bypass factor	–	–
CDD	cooling degree days	°F-day	°C·day
CLF	cooling load factor	–	–
CLTD	cooling load temperature difference	°F	°C
GSHR	grand sensible heat ratio	–	–
h	enthalpy	Btu/lbm	kJ/kg
$\dot{q}$	heat transfer rate	Btu/hr	W
RSHR	room sensible heat ratio	–	–
SC	shading coefficient	–	–
SCL	solar cooling load factor	Btu/hr-ft^2	W/m^2
SEER	seasonable energy efficiency ratio	kW-hr/Btu	–
T	temperature	°F	°C
U	overall coefficient of heat transfer	Btu/ft^2-hr-°C	W/m^2·°C
$\dot{V}$	volumetric flow rate	ft^3/min	L/s

Symbols

v	specific volume	ft^3/lbm	m^3/kg

Subscripts

ADP	apparatus dew point
c	cooling
i	indoor design
in	in (entering the space)
l	latent
o	outdoor design
s	sensible
t	total
te	total equivalent

1. INTRODUCTION

The procedure for finding the *cooling load* (also referred to as the *air conditioning load*) is similar in some respects to the procedure for finding the heating load.[1] The aspects of determining inside and outside design conditions, heat transfer from adjacent spaces, ventilation air requirements, and internal heat gains are the same as for heating load calculations and are not covered in this chapter.

However, the calculation of cooling load is complicated considerably by the thermal lag of the exterior surfaces (i.e., walls and roof). Depending on construction, the solar energy absorbed by exterior surfaces can take hours to appear as an interior cooling load.[2] Further complicating the determination of cooling load are the direct transmission of solar energy through windows and the facts that the delay is different for each surface, the solar energy absorbed changes with time of day, and instantaneous heat gain into the room contributes to instantaneous and delayed cooling loads.

It is important to distinguish between three terms. The *instantaneous heat absorption* is the solar energy that is absorbed at a particular moment. The *instantaneous heat gain* is the energy that enters the conditioned space at that moment. Due to solar lag, the heat gain is a complex combination of heat absorptions from previous hours. The *instantaneous cooling load* is a portion (i.e., is essentially the convective portion) of the instantaneous heat gain.

2. INSIDE AND OUTSIDE DESIGN CONDITIONS

It is impossible to maintain the air passing through a conditioned space at a particular temperature. Cool air enters a space, and warm air leaves. The *inside design temperature*, T_i, is understood to be the temperature of the air removed from the conditioned space. Indoor design temperatures for summer use are generally a few degrees warmer than winter design temperatures—approximately 75°F (23.9°C).

---

[1]The *refrigeration load* or *coil loading* is the cooling load expressed in appropriate units (e.g., tons of refrigeration).
[2]That is, the *instantaneous heat gain* is the heat that enters the conditioned space. Due to thermal lag, it is not the same as the *instantaneous heat absorption* by the building at that same moment. This terminology is not rigidly adhered to.

HVAC

3. INSTANTANEOUS COOLING LOAD FROM WALLS AND ROOFS

Three general methods are used to determine instantaneous cooling load: (1) total equivalent temperature difference method, (2) transfer function method, and (3) cooling load factor and temperature difference methods.

The total equivalent temperature difference (TETD) method determines the instantaneous heat gain. The *total equivalent temperature difference*, ΔT_{te}, depends on the type of construction, geographical location, time of day, and wall orientation. It is read from extensive tabulations. The instantaneous heat gain consists of stored radiant and convective portions. In the TETD/TA (time averaging) method, weighting factors are used to average the radiant portions from current and previous hours. The sum of the convective portions and the weighted average of the series of radiant portions are taken as the cooling load. Computer analysis and considerable judgment are required to use this method.

$$\dot{q}_{\text{heat gain}} = UA\Delta T_{te} \qquad 41.1$$

The *transfer function method* is similar to the total equivalent temperature difference method. A series of weighting factors, known as *room transfer functions*, is applied to cooling load values from the current and previous hours. The transfer functions are related to spatial geometry, configuration, mass, and other characteristics.

The only modern method of calculating the cooling load suitable for quick (manual) analysis is the *cooling load temperature difference method* (CLTD) using the related *solar cooling load factor* (SCL) and *cooling load factor* (CLF) covered in subsequent sections. For exterior surfaces, the cooling load is calculated by Eq. 41.2. Tables of CLTD are needed. Values depend on time of year, location and orientation, type, configuration, and orientation of the surface, as well as other factors. Using the CLTD/SCL/CLF method, the instantaneous cooling load for conduction through opaque walls and roofs is

$$\dot{q}_c = UA(\text{CLTD}) \qquad 41.2$$

4. INSTANTANEOUS COOLING LOAD FROM WINDOWS

Using the CLTD/SCL/CLF method, the cooling load due to solar energy received through windows is calculated in two parts.[3] The first is an immediate conductive part; the second is a delayed radiant part. Appropriate tables are needed to evaluate the *shading coefficient* (SC) and the *solar cooling load factor* (SCL) for the radiant portions.

$$\dot{q}_c = \dot{q}_{\text{conductive}} + \dot{q}_{\text{radiant}}$$
$$= UA(\text{CLTD}) + A(\text{SC})(\text{SCL}) \qquad 41.3$$

---

[3]The term *fenestration* refers to windows or other openings transparent to solar radiation.

5. COOLING LOAD FROM INTERNAL HEAT SOURCES

Latent loads (including metabolic latent loads) are considered instantaneous cooling loads. Only a portion, given by the *cooling load factor* (CLF), of the sensible heat sources show up as instantaneous cooling load. CLF is a function of time and depends on zone type, occupancy period, interior and exterior shading, and other factors. Although tables are usually necessary to evaluate CLF, there are some cases where CLF is assumed to be 1.0. These include when the cooling system is shut down during the night, when there is a high occupant density (as in theaters and auditoriums), and when lights and other sources are operated for 24 hours a day.

$$\dot{q}_c = \dot{q}_l + \dot{q}_s(\text{CLF}) \qquad 41.4$$

6. VENTILATION AND INFILTRATION

Since all air passes through the air conditioner, sensible and latent loads from ventilation and infiltration air are instantaneous cooling loads.

7. HEAT GAIN TO AIR CONDITIONING DUCTS

The calculation of heat absorbed by air conditioning ducts that pass through unconditioned spaces is not sophisticated. The heat transfer is generally estimated from tables or figures of standard configurations (e.g., the heat loss per fixed length of duct per 10 degrees of temperature difference). Extrapolation is used for other duct lengths and temperature differences.

The logarithmic mean temperature difference (generally used when the temperature difference varies along the length) is seldom used in the HVAC industry.[4] Rather, the heat transfer is based on the temperature difference between the environment and the midlength temperature of the duct. If the temperature of the duct at its midlength is not known, one or more iterations will be needed to calculate the temperature drop and heat transfer.

8. DEGREE DAYS

Some sources present tables of *cooling degree days*, CDD, (*summer degree days*).[5] Data in these tables are usually related to a base temperature of 65°F (18.3°C). Cooling is considered to occur only when the temperature is higher than 65°F (18.3°C).

---

[4]This is probably because the accuracy of other data does not warrant a high level of sophistication.
[5]Tables of cooling degree days are far less common than tables of heating degree days.

9. SEASONAL COOLING ENERGY

The approximate total energy used during the cooling season can be calculated from the cooling degree days. SEER is the *seasonal energy efficiency ratio*, which incorporates various equipment and process efficiencies as well as the conversion from Btus to kilowatt-hours. SEER for electrically driven refrigeration is typically in the range of 10 to 12 Btu/W-hr (2.9 to 3.5 W/W). The seasonal cooling cost is determined from the cost per kW-hr.

$$\text{energy}_{kW \cdot h/season}$$

$$= \frac{\left(24 \, \dfrac{h}{day}\right)(\dot{q}_{\text{design cooling,W}})(\text{CDD})}{\left(1000 \, \dfrac{W}{kW}\right)(T_o - T_i)(\text{SEER})}$$

$$\text{[SI]} \quad \textit{41.5(a)}$$

$$\text{energy}_{kW \text{-} hr/season}$$

$$= \frac{\left(24 \, \dfrac{hr}{day}\right)(\dot{q}_{\text{design cooling,Btu/hr}})(\text{CDD})}{\left(1000 \, \dfrac{W}{kW}\right)(T_o - T_i)(\text{SEER})}$$

$$\text{[U.S.]} \quad \textit{41.5(b)}$$

10. RECIRCULATING AIR BYPASS

Some of the return air may be bypassed around the air conditioner through a bypass channel. Figure 41.1 illustrates such a *recirculating air bypass* configuration.[6]

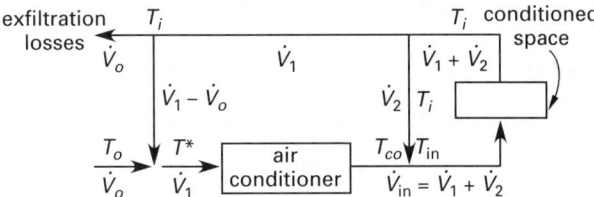

Figure 41.1 *Recirculating Air Bypass*

The inside design temperature, T_i, and the sensible and latent loads are generally known, as are the outside design conditions. The temperature, T_{in}, and flow rate of the air entering the conditioned space are generally not known. The following procedure can be used to determine these unknowns.[7] This procedure can also be used when there is no bypass (i.e., straight recirculation) by setting $\dot{V}_2 = 0$.

---

[6]The separate bypass duct shown in Fig. 41.1 does not actually exist. Bypassed air actually flows unchanged through a separate channel in the air conditioner.

There is a variation of this configuration in which the outdoor air is mixed with the return air before the bypass takeoff. The primary difference between the variations is when the $\dot{V}_o$ term is used.

[7]Slight modifications of the procedure may be necessary, depending on what is known.

Care must be taken in distinguishing between subscripts "i," "in," and "1."

step 1: Locate the indoor (i) and outdoor (o) design conditions on the psychrometric chart. Read h_i, h_o, and v_o.

step 2: Draw a line between the indoor and outdoor points. This line represents all possible ratios of mixing indoor and outdoor air. The ratio of outdoor ventilation air, $\dot{V}_o$, to conditioned recirculating air, $\dot{V}_1$, determines the actual mixture point *. For an initial estimate, assume that the densities of the two air streams are the same. Then, the air masses are proportional to the air volumes. Calculate the temperature T^* from Eq. 41.6.[8]

$$\frac{T^* - T_i}{T_o - T_i} = \frac{\dot{V}_o}{\dot{V}_1} \qquad \textit{41.6}$$

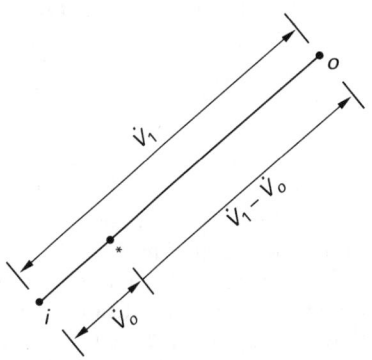

Figure 41.2 *Adiabatic Mixing of Inside and Outside Air*

step 3: If the sensible and latent loads in the conditioned space are known, calculate the *room sensible heat ratio*, RSHR. The latent and sensible loads from outside ventilation air are not included.

$$\text{RSHR} = \frac{\dot{q}_s}{\dot{q}_s + \dot{q}_l} \qquad \textit{41.7}$$

step 4: Draw a line with the slope RSHR (based on the psychrometric chart's sensible heat ratio scale) through point "i." The apparatus dew point (ADP) should be greater than 32°F (0°C), in which case the air will be cooled and dehumidified as it passes through the coil.

Alternatively, if the condition of the air leaving the air conditioner is known, draw a line from that point ("co" for conditioner output) to point "i."

---

[8]Though dry-bulb temperatures are commonly used in Eq. 41.6, they need not be. Since all of the temperature scales are linear, wet-bulb and dew-point temperatures could be used.

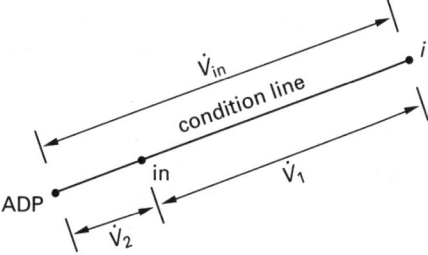

Figure 41.3 *Adiabatic Mixing of Conditioned and Bypass Air*

step 5: The condition of the air entering the conditioned space lies along the line drawn in step 4. The line represents all possible ratios of mixing conditioned and bypassed air. The amounts of conditioned air ($\dot{V}_1$) and bypassed air ($\dot{V}_2$) determines the mixture point "in." For an initial estimate, assume that the densities of the two air streams are the same. Calculate the temperature T_{in}. The ratio of air flows determines the *system bypass factor*.

$$\text{BF}_{\text{system}} = \frac{T_{\text{in}} - T_{\text{ADP}}}{T_i - T_{\text{ADP}}}$$

$$= \frac{\dot{V}_2}{\dot{V}_1 + \dot{V}_2} \qquad 41.8$$

step 6: If neither T_{in} nor $\dot{V}_{\text{in}}$ are known, T_{in} should be chosen such that it is 15 to 20°F (8 to 11°C) less than T_i. The temperature of the air entering the conditioned space and the flow rate through the space are related; one determines the other. The larger the temperature difference $T_i - T_{\text{in}}$ (representing the temperature rise as the air flows through the conditioned space), the lower the flow rate. However, very large temperature differences require extremely efficient mixing within the space, and very low T_{in} temperatures are uncomfortable for occupants near the discharge registers. Therefore, the temperature difference should not exceed 15 to 20°F (8 to 11°C).

step 7: If $\dot{V}_{\text{in}}$ is known, calculate T_{in} from the sensible heating relationship, Eq. 41.9.

$$\dot{V}_{\text{in,L/s}} = \dot{V}_1 + \dot{V}_2$$

$$= \frac{\dot{q}_{s,\text{W}}}{\left(1.20 \, \dfrac{\text{W·s}}{\text{L·°C}}\right)(T_i - T_{\text{in}})} \qquad \text{[SI]} \quad 41.9(a)$$

$$\dot{V}_{\text{in,cfm}} = \dot{V}_1 + \dot{V}_2$$

$$= \frac{\dot{q}_{s,\text{Btu/hr}}}{\left(1.08 \, \dfrac{\text{Btu-min}}{\text{ft}^3\text{-hr-°F}}\right)(T_i - T_{\text{in}})} \qquad \text{[U.S.]} \quad 41.9(b)$$

step 8: Locate point "in" corresponding to T_{in} on the RSHR condition line from step 4.

step 9: Knowing T_{in} establishes the ratio of $\dot{V}_1$ and $\dot{V}_2$ in Eq. 41.8. Calculate $\dot{V}_1$ and $\dot{V}_2$.

$$\dot{V}_2 = (\text{BF}_{\text{system}})\dot{V}_{\text{in}}$$

$$= (\text{BF}_{\text{system}})(\dot{V}_1 + \dot{V}_2) \qquad 41.10$$

$$\dot{V}_1 = \dot{V}_{\text{in}} - \dot{V}_2 \qquad 41.11$$

step 10: Draw a line through the mixture point * and the incoming air point "in." This line represents the process occurring in the air conditioner. Heat from outside air and from within the space are both removed by the air conditioner. Therefore, the slope of this line is the *grand sensible heat ratio*, GSHR, also known as the *coil sensible heat ratio*. If this slope is known in advance, it can be used (with the sensible heat ratio scale on the psychrometric chart) to draw a line through either * or "in," thereby establishing point "in" or *, respectively. If it is not known in advance, it can be determined from the sensible heat ratio scale.

$$\text{GSHR} = \frac{\dot{q}_{s,\text{room}} + \dot{q}_{s,\text{ventilation air}}}{\dot{q}_{t,\text{room}} + \dot{q}_{t,\text{ventilation air}}} \qquad 41.12$$

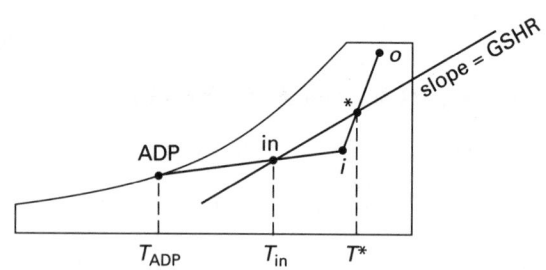

Figure 41.4 *Total Heat Removal Process*

step 11: The required air conditioning capacity is

$$\dot{q}_t = \dot{q}_{t,\text{room}} + \dot{q}_{t,\text{ventilation air}}$$

$$= \dot{q}_{s,\text{room}} + \dot{q}_{l,\text{room}} + \frac{(h_o - h_i)\dot{V}_o}{v_o} \qquad 41.13$$

Equation 41.14 expresses the air conditioning capacity in traditional HVAC units. The constant 4.5 lbm-min/ft³-hr is the product of air density (0.075 lbm/ft³) and 60 min/hr. The constant 1.2 kg/L is the product of air density (1.2 kg/m³) and conversions 1000 W/kW and 0.001 m³/L.

$$\dot{q}_{t,\text{W}} = \dot{q}_{s,\text{room}} + \dot{q}_{l,\text{room}}$$

$$+ \left(1.2 \, \frac{\text{kg}}{\text{L}}\right)(h_o - h_i)\dot{V}_{o,\text{L/s}}$$

$$\text{[SI]} \quad 41.14(a)$$

$$\dot{q}_{t,\text{Btu/hr}} = \dot{q}_{s,\text{room}} + \dot{q}_{l,\text{room}}$$

$$+ \left(4.5 \, \frac{\text{lbm-min}}{\text{ft}^3\text{-hr}}\right)(h_o - h_i)\dot{V}_{o,\text{cfm}}$$

$$\text{[U.S.]} \quad 41.14(b)$$

Example 41.1

The inside design condition for a conditioned space with partial recirculation is 75°F dry-bulb (23.9°C) and 62.5°F (16.9°C) wet-bulb. The outside air is at 94°F dry-bulb (34.4°C) and 78°F (25.6°C) wet-bulb. The sensible space load is 160,320 Btu/hr (46.5 kW). The latent load from occupants and infiltration, but excluding intentional ventilation, is 19,210 Btu/hr (5.6 kW). A total of 1275 ft³/min (600 L/s) of ventilation air is required. The air temperature increases 19°F (10.6°C) as it passes through the conditioned space.

Find the (a) required apparatus dew point, (b) volume of air passing through the space, (c) wet-bulb temperature of the air entering space, (d) system bypass ratio, and (e) dry-bulb temperature of the air entering the conditioner.

SI Solution

(a) The *room sensible heat ratio*, RSHR, is

$$\text{RSHR} = \frac{\dot{q}_s}{\dot{q}_s + \dot{q}_l}$$
$$= \frac{46.5 \text{ kW}}{46.5 \text{ kW} + 5.6 \text{ kW}}$$
$$= 0.89$$

Locate the indoor (i) and outdoor (o) design conditions on the psychrometric chart. Draw a line with the slope 0.89 through point "i." Extend the line to the left to 11.9°C on the saturation line. This is the apparatus dew point (ADP).

(b) The dry-bulb temperature of the air as it enters the conditioned space is

$$T_\text{in} = T_i - 10.6°\text{C} = 23.9°\text{C} - 10.6°\text{C}$$
$$= 13.3°\text{C}$$

Calculate the air flow through the space.

$$\dot{V}_\text{in,L/s} = \frac{\dot{q}_{s,\text{W}}}{\left(1.20 \dfrac{\text{W·s}}{\text{L·°C}}\right)(T_i - T_\text{in})}$$

$$= \frac{(46.5 \text{ kW})\left(1000 \dfrac{\text{W}}{\text{kW}}\right)}{\left(1.2 \dfrac{\text{W·s}}{\text{L·°C}}\right)(23.9°\text{C} - 13.3°\text{C})}$$

$$= 3656 \text{ L/s}$$

(c) Locate point "in" corresponding to T_in on the RSHR condition line from step 4. Read the wet-bulb temperature at this point as 12.5°C.

(d) Calculate the system bypass ratio from Eq. 41.8.

$$\text{BF}_\text{system} = \frac{T_\text{in} - T_\text{ADP}}{T_i - T_\text{ADP}}$$
$$= \frac{13.3°\text{C} - 11.9°\text{C}}{23.9°\text{C} - 11.9°\text{C}}$$
$$= 0.117$$

(e) The flow rates are

$$\dot{V}_2 = (\text{BF}_\text{system})\dot{V}_\text{in} = (0.117)\left(3656 \frac{\text{L}}{\text{s}}\right)$$
$$= 428 \text{ L/s}$$

$$\dot{V}_1 = 3656 \frac{\text{L}}{\text{s}} - 428 \frac{\text{L}}{\text{s}}$$
$$= 3228 \text{ L/s}$$

Use Eq. 41.6 to locate the point corresponding to the air entering the conditioner.

$$\frac{\dot{V}_o}{\dot{V}_1} = \frac{600 \dfrac{\text{L}}{\text{s}}}{3228 \dfrac{\text{L}}{\text{s}}} = 0.186$$

$$\frac{T^* - T_i}{T_o - T_i} = 0.233$$

$$T^* = T_i + (0.233)(T_o - T_i)$$
$$= 23.9°\text{C} + (0.186)(34.4°\text{C} - 23.9°\text{C})$$
$$= 25.9°\text{C}$$

Customary U.S. Solution

(a) The *room sensible heat ratio* is

$$\text{RSHR} = \frac{\dot{q}_s}{\dot{q}_s + \dot{q}_l}$$

$$= \frac{160,320 \dfrac{\text{Btu}}{\text{hr}}}{160,320 \dfrac{\text{Btu}}{\text{hr}} + 19,210 \dfrac{\text{Btu}}{\text{hr}}}$$

$$= 0.89$$

Locate the indoor (i) and outdoor (o) design conditions on the psychrometric chart. Draw a line with the slope 0.89 through point "i." Extend the line to the left to 53.4°F on the saturation line. This is the apparatus dew point (ADP).

(b) The dry-bulb temperature of the air as it enters the conditioned space is

$$T_\text{in} = T_i - 19°\text{F} = 75°\text{F} - 19°\text{F}$$
$$= 56°\text{F}$$

Calculate the air flow through the space.

$$\dot{V}_\text{in,cfm} = \frac{\dot{q}_{s,\text{Btu/hr}}}{\left(1.08 \dfrac{\text{Btu-min}}{\text{ft}^3\text{-hr-°F}}\right)(T_i - T_\text{in})}$$

$$= \frac{160,320 \dfrac{\text{Btu}}{\text{hr}}}{\left(1.08 \dfrac{\text{Btu-min}}{\text{ft}^3\text{-hr-°F}}\right)(75°\text{F} - 56°\text{F})}$$

$$= 7813 \text{ ft}^3/\text{min} \quad (\text{cfm})$$

(c) Locate point "in" corresponding to T_in on the RSHR condition line from step 4. Read the wet-bulb temperature at this point as 54.6°F.

(d) Calculate the system bypass ratio. From Eq. 41.8,

$$\begin{aligned}
\text{BF}_{\text{system}} &= \frac{T_{\text{in}} - T_{\text{ADP}}}{T_i - T_{\text{ADP}}} \\
&= \frac{56°\text{F} - 53.4°\text{F}}{75°\text{F} - 53.4°\text{F}} \\
&= 0.120
\end{aligned}$$

(e) The flow rates are

$$\dot{V}_2 = (\text{BF}_{\text{system}})\dot{V}_{\text{in}} = (0.120)\left(7813\ \frac{\text{ft}^3}{\text{min}}\right)$$

$$= 938\ \text{ft}^3/\text{min}\quad(\text{cfm})$$

$$\dot{V}_1 = 7813\ \frac{\text{ft}^3}{\text{min}} - 938\ \frac{\text{ft}^3}{\text{min}}$$

$$= 6875\ \text{ft}^3/\text{min}\quad(\text{cfm})$$

Use Eq. 41.6 to locate the point corresponding to the air entering the conditioner.

$$\frac{\dot{V}_o}{\dot{V}_1} = \frac{1275\ \dfrac{\text{ft}^3}{\text{min}}}{6875\ \dfrac{\text{ft}^3}{\text{min}}} = 0.185$$

$$\frac{T^* - T_i}{T_o - T_i} = 0.185$$

$$\begin{aligned}
T^* &= T_i + (0.185)(T_o - T_i) \\
&= 75°\text{F} + (0.185)(94°\text{F} - 75°\text{F}) \\
&= 78.5°\text{F}
\end{aligned}$$

11. REHEAT

If the extension of the line containing points "in" and "i" (as drawn in step 4 of Sec. 10) intersects the saturation line below 32°F (0°C) or if it does not intersect the saturation line at all, reheating of the air will be necessary.

As the moisture content does not change, sensible reheating is represented on the psychrometric chart by a horizontal line to the right. Assuming the air leaves the coil saturated, the line will start at the equipment's apparatus dew point on the psychrometric chart's saturation curve. Since energy from heating must be removed by the coil when the air returns through it, the cooling load (refrigeration load) includes the reheat, as shown in Fig. 41.5. The cooling load is the enthalpy difference.

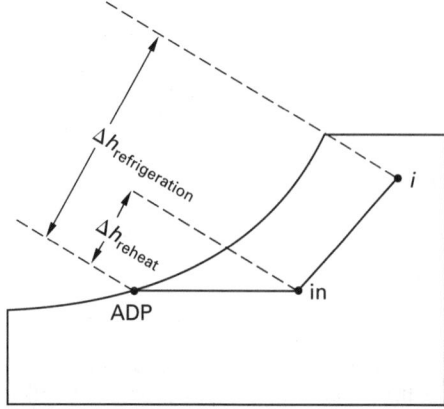

Figure 41.5 *Preventing Coil Icing*

PRACTICE PROBLEMS

1. (*Time limit: one hour*) The bypass air conditioning system shown has the following operating characteristics.

> total air pressure: 14.7 psia (101.3 kPa)
> supply temperature: 58°F (14.4°C) dry-bulb
> sensible load: 200,000 Btu/hr (58.6 kW)
> latent load: 450,000 grains/hr (29 kg/h)
> outside air: 90°F (32.2°C) dry-bulb,
>   76°F (24.4°C) wet-bulb
> make-up air: 2000 ft³/min (940 L/s)
> condition of air leaving washer: saturated

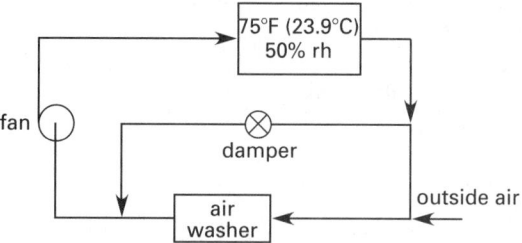

Find the (a) temperature of the air leaving the washer, (b) rate of supply air, and (c) moisture content of the supply air.

2. (*Time limit: one hour*) A building is located at 32°N latitude. The inside design temperature is 78°F (25.6°C). The outside design temperature is 95°F (35°C). The daily temperature range is 22°F (12.2°C). Energy gains through the floor, from lights, and from occupants are insignificant. The building's construction is as follows.

> walls: 1600 ft² (144 m²) facing north
>   1400 ft² (126 m²) facing south
>   1500 ft² (135 m²) facing east
>   1400 ft² (126 m²) facing west
>   4 in (100 mm) brick facing
>   3 in (75 mm) concrete block
>   1 in (25 mm) mineral wool
>   (emissivity = 0.96)
>   2 in (50 mm) furring
>   ³⁄₈ in (9.5 mm) drywall gypsum
>   (emissivity = 0.95)
>   ½ in (12 mm) plaster
>
> roof: 6000 ft² (540 m²)
>   4 in concrete (100 mm)
>   2 in insulation (50 mm)
>   insulation conductivity:
>   0.29 Btu/ft²-hr-°F (1.65 W/m²·°C)
>   single felt layer
>   1 in (25 mm) air gap
>   ½ in (12 mm) acoustical ceiling tile
>
> windows: 100 ft² (9 m²) facing east
>   ¼ in (6.4 mm) thick, single glazing
>   cream-colored Venetian shades
>   no exterior shading

(a) What is the cooling load for mid-July at 4:00 P.M. sun time? (b) Why or why not is this the peak cooling load?

42 Air Conditioning Systems and Controls

1. Types of Air Conditioning Systems 42-1
2. Control Equipment 42-1
3. HVAC Process Control 42-1
4. Typical Controls Integration 42-2
5. Freeze Protection 42-2
6. Feedback and Control 42-2
7. Digital Control 42-5

Nomenclature
R range °F °C
T temperature °F °C

1. TYPES OF AIR CONDITIONING SYSTEMS

Depending on the medium delivered to the conditioned space, air conditioning systems are categorized as all-air, air-and-water, all-water, and unitary. *All-air systems* cool and heat by sending only cold air to the space. Most central units are *single-duct*, which means that the cooling and heating coils are in series. In *dual-duct* units, the heating and cooling coils are in parallel ducts.

In *air-and-water systems*, air and water are both distributed to the conditioned space. In *all-water systems*, the cooling and heating effects are provided solely by water pumped to the conditioned space. With *unitary equipment*, the fan, condenser, and cooling and heating coils are combined in a factory-matched combination, often within the same housing for window and through-the-wall installations.

2. CONTROL EQUIPMENT

Control equipment in HVAC systems consists of sensors, actuators, motors, relays, and controllers. *Sensors* (*transducers*) are used to monitor temperature (*thermostats* or just "*stats*"), enthalpy, and humidity (*humidistats* or *hygrostats*). In complex delivery systems, pressure may also be monitored (i.e., by *pressurestats*). Thermostats are designated as "room," "insertion," or "immersion" according to their placement (i.e., in the occupied space, in a duct, or in a water/steam manifold, respectively).

The signal from a sensor is received by the *controller*, which energizes or deenergizes the appropriate equipment. Since most control systems operate at low voltages (e.g., 24 V AC), a *relay* must be used when the equipment operates at a higher voltage. Relays convert one type (form, or voltage) of energy to another. For example, a room thermostat might energize a relay that, in turn, would provide power to a line-voltage fan motor. Alternatively, the thermostat could activate a pneumatic relay (*electropneumatic switch*).

Actuators (also referred to as *operators* and *motors*) provide the force to open and close valves and dampers. The control signal can be electrical, electronic, or pneumatic. Actuators are designated as *normally open* (N.O.) or *normally closed* (N.C.) depending on their position when deenergized. Most actuators act relatively slowly. *Solenoids*, however, act quickly in response to signals.

A *pneumatic actuator* is essentially a piston/cylinder arrangement or diaphragm/bellows. With pneumatic actuators, a separate compressed-air system is required to supply the force for changing damper settings. With a typical 18 psig (124 kPa) source, the pneumatic signal will be approximately 3 to 15 psig (21 to 100 kPa). Pneumatic actuators are essentially linear. The actuator position is proportional to the air pressure.

3. HVAC PROCESS CONTROL

Control of basic commercial HVAC systems has traditionally meant controlling either the amount of bypass air or the amount of reheat. These two methods are known as face and bypass damper control and reheat control, respectively. *Face-and-bypass damper control* is normally used to control only the dry-bulb temperature. Because of the possibility of bringing in too much moisture (an uncontrolled variable), this method is generally not used with a high percentage of outside air unless the outside air can be dehumidified while bypassing the room air. *Reheat control* is needed when both room temperature and humidity control are needed. Once the proper humidity level is achieved, reheat ensures the proper room temperature.

A third method, *air volume control*, relies on variations in flow rate through the conditioned space. Only one parameter (i.e., dry-bulb temperature) can be adequately controlled in this manner. Volume control is more applicable in the largest systems where the additional cost and complexity can be economically justified. Advances in noise control, monitoring of other comfort parameters, and providing sufficient outside ventilation when volume is low may overcome the criticisms this method has received in past years.

HVAC

4. TYPICAL CONTROLS INTEGRATION

There are numerous variations in equipment layout, mixing sequence, and control methodology. Figure 42.1 schematically illustrates a central station *air handling unit* (AHU) in a multizone system with reheat control. Return air enters from the top; ventilation air enters from the left. Conditioned air is discharged into the distribution ductwork. The cooling effect can be obtained from either the evaporator coils of a vapor-compression refrigeration cycle or from liquid chiller cooling coils. The heating effect may be from hot water or steam coils or from the condenser section of a vapor-compression heat pump. The air flow through various components is controlled by remotely actuated dampers.

Until the fan motor starts, all the dampers are at their deenergized positions. The bypass damper is normally open. The outside air damper is normally closed. (Interlocking the outside air damper to the fan motor prevents induction of cold air and potential coil freeze-up by the stack effect whenever the fan is not running.)

The control sequence begins when the fan motor starts. The fan voltage energizes a relay and/or electric-pneumatic valve (EP), which provides air to the controllers. When the fan starts, damper motor (or damper actuator) DM1 opens the outside damper to a predetermined minimum position, permitting outside air to enter.

Damper motor DM2 is controlled by two sensors: the return air humidistat (humidity controller, HC) and the supply air temperature controller (TC), also known as a *mixing thermostat*. The duplex pressure selector (DS) selects the higher of the two pressure signals from either the HC or TC sensors and positions damper motor DM2 appropriately.[1]

The cooling coil both cools and dehumidifies the air stream. If the latent loads are low, the air is merely cooled to TC's set point temperature. If latent loads are high, more air is passed through the cooling coil to remove the moisture. Reheating is used to prevent overcooling of the space. Reheating is controlled by room thermostats (TR1, TR2, and TR3). When a TR set point is reached, the corresponding steam or hot water valve, V, is opened.

Temperature controller TC also acts as a high-limit controller, preventing the supply air control from increasing above what is required for adequate zone cooling.

When the space has low latent loads, humidification is required. Figure 42.1 does not show the humidification system and controlling humidistats. Small humidification increases can be obtained by increasing the amount of bypass air. Steam or hot water injection is needed for large humidity changes.

---

[1] Motor DM2 controls two dampers to vary the air passing through the bypass and coil. The bypass damper is normally open; the coil damper is normally closed. In some systems, the coil damper is a *face damper*. The face damper is installed immediately before the face of the cooling coil, hence its name.

Also not shown is an outside enthalpy controller. This controller compares the heat content of the return air with that of the outside air. When the refrigeration load can be reduced, the enthalpy controller overrides the temperature controller and increases the outdoor air damper opening. In smaller installations, an *outside air thermostat* can work almost as well.

5. FREEZE PROTECTION

When the outdoor air is at subfreezing temperature, freezing of water in preheat, reheat, and chilled water coils can occur whether or not the coils are in operation. Freezing is caused by direct contact with or incomplete mixing (*stratification*) of the outside air with return air, although reduced warm air flows due to clogged filters can also be a contributing factor. In some systems, freeze-up can be prevented by using antifreeze or by draining the coils when they are not in use.

It is appropriate to protect the coils with thermostats (*freeze stats*). For example, the face dampers upstream of the coils can be closed down when the plenum temperature drops to approximately 35°F (2°C) or when the temperature of the incoming heated water in the heating coils (as determined by an *immersion thermostat*) drops below 120 to 150°F (50 to 65°C). Furthermore, when the fan is not running, the outside air dampers should be closed (as described in Sec. 5), and minimum heat to the heating coils should continue to be provided.

6. FEEDBACK AND CONTROL

Together, sensors and their controllers constitute a traditional feedback loop and control system. Using temperature as the controlled variable, the *set point*, T_{set}, is the temperature that the conditioned space would like to maintain. The *control point* is the actual temperature in the room. The *offset* is the difference between the set and control points.

$$T_{\text{offset}} = T_{\text{set point}} - T_{\text{control point}} \qquad 42.1$$

There are several basic control methods. The most simple is *two-position control*. The controlled device (e.g., a valve or damper) is either fully on or it is fully off. Because of thermal mass and other delays, the temperature will continue to increase for a short time after the heat is turned off. This is known as *temperature overshoot*. Similarly, the temperature will continue to drop for a short time after the heat is turned on. The room temperature oscillates around the set point. The range of temperatures experienced by the room is the *operating differential*, while the difference in on and off set point temperatures is the *control differential*.

With *timed two-position control* (*anticipation control*), a small heater is built into the thermostat. While the room is being heated, the thermostat is also being heated, and this turns the thermostat off sooner than it otherwise would. The overshoot is reduced considerably.

Figure 42.1 *Bypass Air Handling Unit (reheat control configuration)*

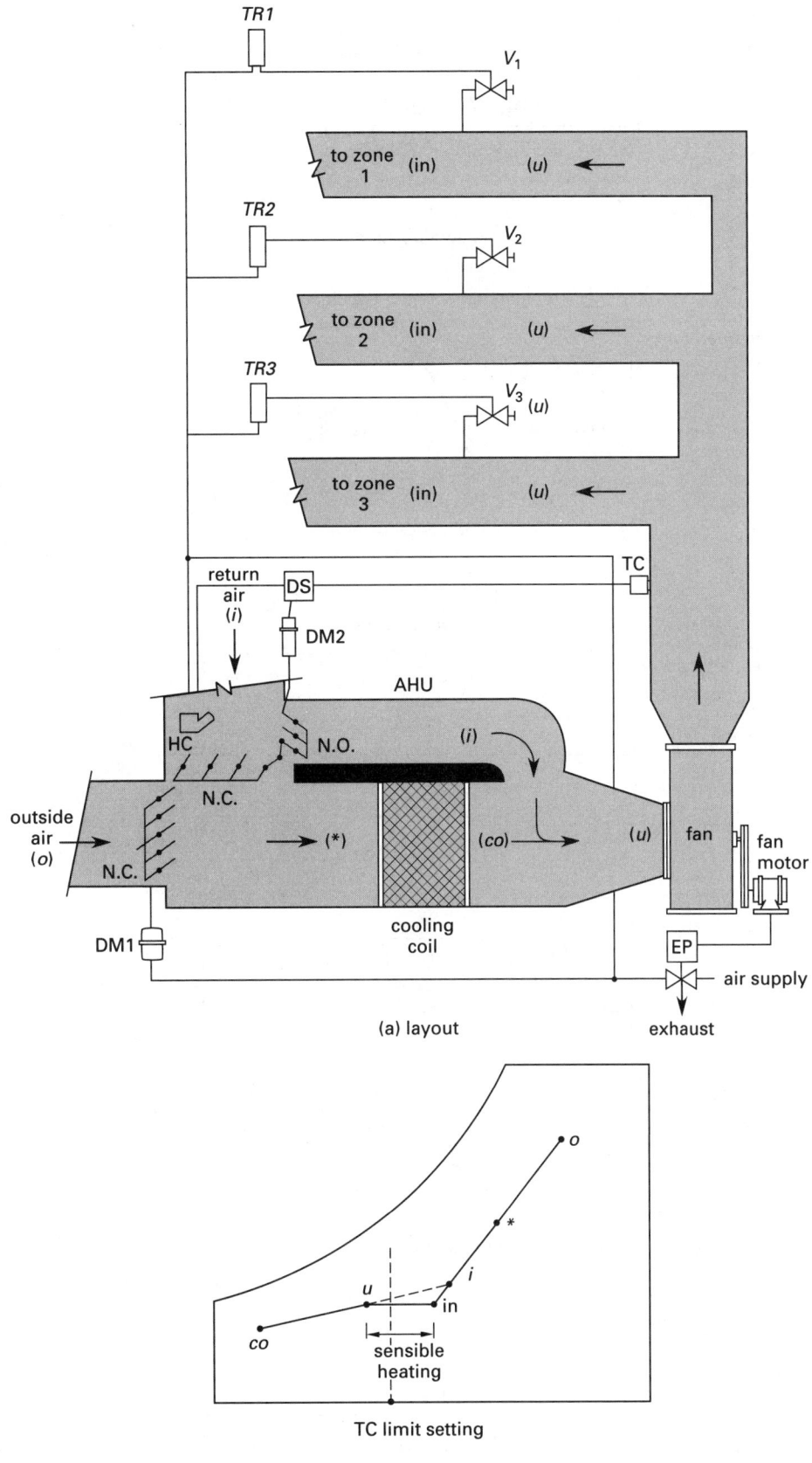

(a) layout

(b) psychrometric plot

HVAC

Figure 42.2 *Response of Temperature Control Methods*

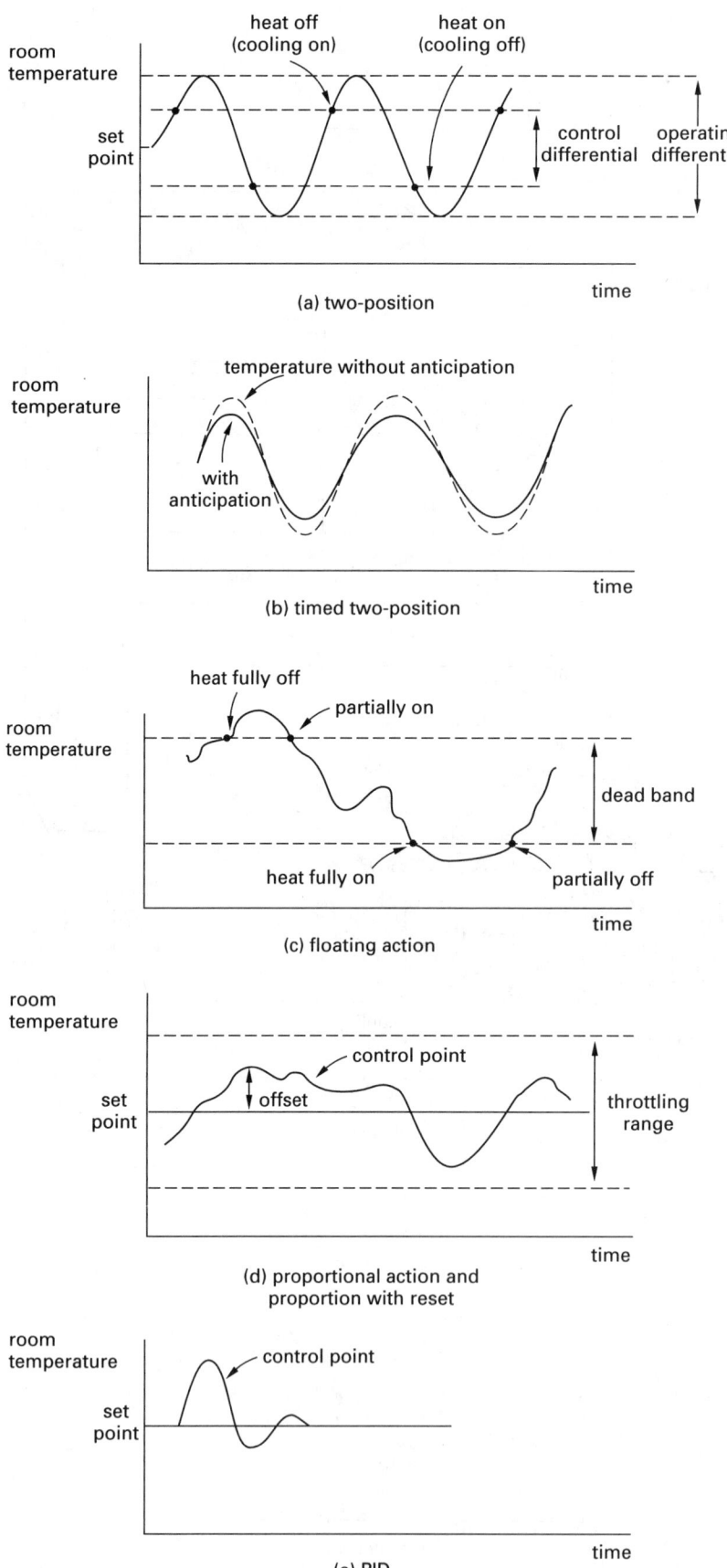

(a) two-position

(b) timed two-position

(c) floating action

(d) proportional action and proportion with reset

(e) PID

With *floating action control*, the controlled device has three positions: fully on, fully off, and a fixed intermediate position. The controller has three corresponding signals: fully on, fully off, and a *neutral position* that is generated while the temperature is within a *dead band* range. Within the dead band, there is an intermediate amount of air heating (or cooling). If the amount of heating (cooling) corresponds to the heat loss (gain), the temperature remains within the dead band. Otherwise, the controller will generate a fully on or fully off signal.

With *proportional action*, the position (e.g., percentage opening) of the damper or valve is proportional to the offset. The *throttling range*, R_{throttle}, is the temperature range over which the damper or valve changes from fully closed to fully open. The throttling range should coincide with the normal range of temperatures encountered. (When the temperature is outside of the throttling range, the system is "out of control.") Since the damper or valve should be 50% open at the set point, within the limits of 0 to 1.00, the fraction open is

$$\text{fraction open} = 0.50 + \frac{T_{\text{offset}}}{R_{\text{throttle}}} \qquad 42.2$$

Proportional action does not provide extra heating (cooling) to compensate for changes; it tends to maintain the existing control point. With proportional action, the settling time is very long.

Proportional action with automatic reset, also known as *proportional plus integral control* (PI), attempts to return the room temperature back to the set point. In effect, the controller "overreacts" and the signal is more than proportional to the offset. The time required for the room temperature to become established at the set point is known as the *settling time*.

With *proportional plus integral plus derivative control* (PID), the control action responds to three different parameters: (1) the magnitude of the offset, (2) the duration of the offset, and (3) the rate at which the temperature is changing. These three aspects correspond to the terms "proportional," "integral," and "derivative," respectively, in the name. Because of the complexity of this algorithm, and since an accurate time base is needed, PID control is implemented through digital control.

7. DIGITAL CONTROL

Direct digital control (DDC) is an alternative to traditional analog control. Devices in analog and digital control systems are analogous. Digital sensors replace analog sensors one for one and are located in the same place. Digital controllers replace analog controllers. Digital actuators replace analog (electric and pneumatic) actuators. Digital devices are connected by simple "twisted pair" control wiring. Each device has its own address, and the digital signal generated by the controller includes the device address.

Digital controllers are essentially *local control computers* (LCC) running algorithms preprogrammed by the manufacturer. PI and (in some cases) PID control are easily implemented. Changes to the dead band, proportional band, set points, low- and high-limits, lockouts, and so on, can be programmed for all the controlled devices after installation.

When installing a digital controller in a system already equipped with analog (pneumatic or electric) control devices, a *digital-to-analog (digital-to-proportional) staging module* is needed. The staging module is essentially an electropneumatic switch that translates digital signals into pneumatic signals.

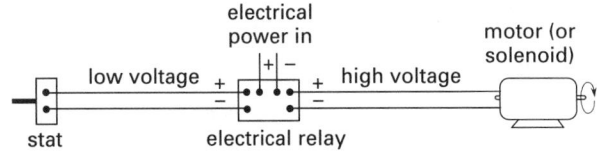

(a) analog all-electric

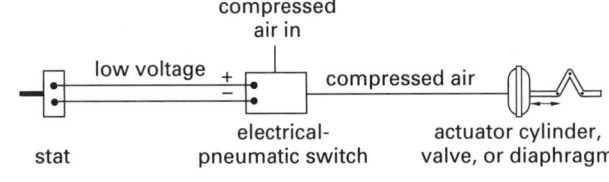

(b) analog pneumatic

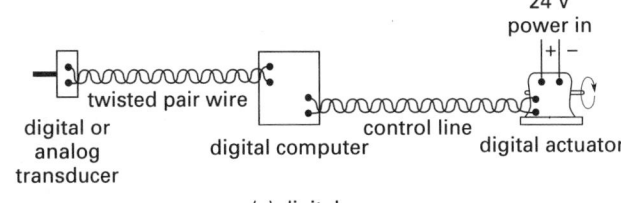

(c) digital

Figure 42.3 *Basic Hook-Up Diagrams*

43 Determinate Statics

1. Introduction to Statics 43-1
2. Internal and External Forces 43-2
3. Unit Vectors 43-2
4. Concentrated Forces 43-2
5. Moments 43-2
6. Moment of a Force About a Point 43-2
7. Varignon's Theorem 43-3
8. Moment of a Force About a Line 43-3
9. Components of a Moment 43-3
10. Couples 43-4
11. Equivalence of Forces and
     Force-Couple Systems 43-4
12. Resultant Force-Couple Systems 43-4
13. Linear Force Systems 43-4
14. Distributed Loads 43-4
15. Moment from a Distributed Load 43-6
16. Types of Force Systems 43-6
17. Conditions of Equilibrium 43-6
18. Two- and Three-Force Members 43-6
19. Reactions 43-6
20. Determinacy 43-7
21. Types of Determinate Beams 43-7
22. Free-Body Diagrams 43-8
23. Finding Reactions in Two Dimensions . . . 43-8
24. Couples and Free Moments 43-9
25. Influence Lines for Reactions 43-10
26. Hinges 43-10
27. Levers 43-10
28. Pulleys 43-11
29. Axial Members 43-11
30. Forces in Axial Members 43-11
31. Trusses 43-12
32. Determinate Trusses 43-13
33. Zero-Force Members 43-14
34. Method of Joints 43-14
35. Cut-and-Sum Method 43-15
36. Method of Sections 43-15
37. Superposition of Loads 43-16
38. Transverse Truss Member Loads 43-16
39. Cables Carrying Concentrated Loads . . . 43-17
40. Parabolic Cables 43-17
41. Cables Carrying Distributed Loads 43-18
42. Catenary Cables 43-19
43. Cables with Ends at Different Elevations . 43-20
44. Two-Dimensional Mechanisms 43-20
45. Equilibrium in Three Dimensions 43-20
46. Tripods 43-22
     Practice Problems 43-23

Nomenclature

a	distance to lowest cable point	ft	m
c	parameter of the catenary	ft	m
d	distance or diameter	ft	m
D	diameter	ft	m
F	force	lbf	N
H	horizontal cable force	lbf	N
L	length	ft	m
M	moment	ft-lbf	N·m
n	number of sheaves	–	–
r	position vector or radius	ft	m
R	reaction force	lbf	N
s	distance along cable	ft	m
S	sag	ft	m
T	tension	lbf	N
w	load per unit length	lbf/ft	N/m
W	weight	lbf	N
x	horizontal distance or position	ft	m
y	vertical distance or position	ft	m
z	distance or position along z-axis	ft	m

Symbols

ϵ	pulley loss factor	–	–
η	pulley efficiency	–	–
θ	angle	deg	rad
ϕ	angle	deg	rad

Subscripts

O	origin
P	point P
R	resultant

1. INTRODUCTION TO STATICS

Statics is a part of the subject known as *engineering mechanics*.[1] It is the study of rigid bodies that are stationary. To be stationary, a rigid body must be in static equilibrium. In the language of statics, a stationary rigid body has no *unbalanced forces* acting on it.

---

[1]Engineering mechanics also includes the subject of dynamics. Interestingly, the subject of mechanics of materials (i.e., strength of materials) is not part of engineering mechanics.

2. INTERNAL AND EXTERNAL FORCES

An *external force* is a force on a rigid body caused by other bodies. The applied force can be due to physical contact (i.e., pushing) or close proximity (e.g., gravitational, magnetic, or electrostatic forces). If unbalanced, an external force will cause motion of the body.

An *internal force* is one that holds parts of the rigid body together. Internal forces are the tensile and compressive forces within parts of the body as found from the product of stress and area. Although internal forces can cause deformation of a body, motion is never caused by internal forces.

3. UNIT VECTORS

A *unit vector* is a vector of unit length directed along a coordinate axis.[2] In the rectangular coordinate system, there are three unit vectors, **i**, **j**, and **k**, corresponding to the three coordinate axes, x, y, and z, respectively.[3] Unit vectors are used in vector equations to indicate direction without affecting magnitude. For example, the vector representation of a 97 N force in the negative x-direction would be written as $\mathbf{F} = -97\mathbf{i}$.

4. CONCENTRATED FORCES

A *force* is a push or pull that one body exerts on another. A *concentrated force*, also known as a *point force*, is a vector having magnitude, direction, and location (i.e., point of application) in three-dimensional space. In this chapter, the symbols **F** and F will be used to represent the vector and its magnitude, respectively.[4]

The vector representation of a three-dimensional force is given by Eq. 43.1. Of course, vector addition is required.

$$\mathbf{F} = F_x\mathbf{i} + F_y\mathbf{j} + F_z\mathbf{k} \qquad 43.1$$

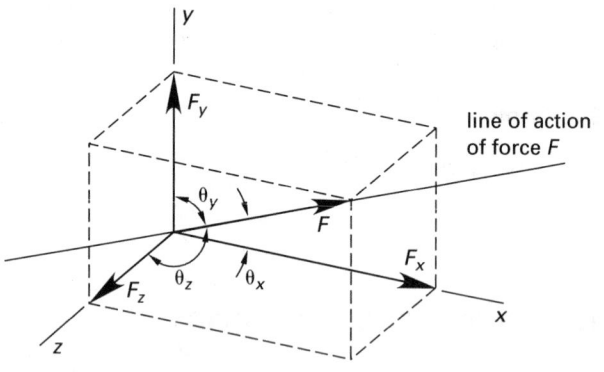

Figure 43.1 *Components and Direction Angles of a Force*

---

[2]Although polar, cylindrical, and spherical coordinate systems can have unit vectors also, this chapter is concerned only with the rectangular coordinate system.

[3]There are other methods of representing vectors, in addition to bold letters. For example, the unit vector **i** is represented in other sources as $\vec{\imath}$ and $\hat{\imath}$.

[4]As with the unit vectors, the symbols **F**, $\overline{F}$, and $\hat{F}$ are used in other sources interchangeably to represent the same vector.

If **u** is a *unit vector* in the direction of the force, the force can be represented as

$$\mathbf{F} = F\mathbf{u} \qquad 43.2$$

The components of the force can be found from the *direction cosines*, the cosines of the true angles made by the force vector with the x-, y-, and z-axes.

$$F_x = F\cos\theta_x \qquad 43.3$$

$$F_y = F\cos\theta_y \qquad 43.4$$

$$F_z = F\cos\theta_z \qquad 43.5$$

$$F = \sqrt{F_x^2 + F_y^2 + F_z^2} \qquad 43.6$$

The *line of action* of a force is the line in the direction of the force extended forward and backward. The force, **F**, and its unit vector, **u**, are along the line of action.

5. MOMENTS

Moment is the name given to the tendency of a force to rotate, turn, or twist a rigid body about an actual or assumed pivot point. (Another name for moment is *torque*, although torque is used mainly with shafts and other power-transmitting machines.) When acted upon by a moment, unrestrained bodies rotate. However, rotation is not required for the moment to exist. When a restrained body is acted upon by a moment, there is no rotation.

An object experiences a moment whenever a force is applied to it.[5] Only when the line of action of the force passes through the center of rotation (i.e., the actual or assumed pivot point) will the moment be zero.

Moments have primary dimensions of length × force. Typical units are foot-pounds, inch-pounds, and newton-meters.[6]

6. MOMENT OF A FORCE ABOUT A POINT

Moments are vectors. The moment vector, $\mathbf{M_O}$, for a force about point O is the *cross product* of the force, **F**, and the vector from point O to the point of application of the force, known as the *position vector*, **r**. The scalar product $|\mathbf{r}|\sin\phi$ is known as the *moment arm*, d.

$$\mathbf{M_O} = \mathbf{r} \times \mathbf{F} \qquad 43.7$$

$$M_O = |\mathbf{M_O}| = |\mathbf{r}||\mathbf{F}|\sin\theta = d|\mathbf{F}| \quad [\theta \le 180°] \qquad 43.8$$

---

[5]The moment may be zero, as when the moment arm length is zero, but there is a (trivial) moment nevertheless.

[6]Units of kilogram-force-meter have also been used in metric countries.

Foot-pounds and newton-meters are also the units of energy. To distinguish between moment and energy, some authors reverse the order of the units. Therefore, pound-feet and meter-newtons become the units of moment. This convention is unnecessary and not universal, since the context is adequate to distinguish between the two.

The line of action of the moment vector is normal to the plane containing the force vector and the position vector. The sense (i.e., the direction) of the moment is determined from the *right-hand rule*.

Right-hand rule: Place the position and force vectors tail to tail. Close your right hand and position it over the pivot point. Rotate the position vector into the force vector, and position your hand such that your fingers curl in the same direction as the position vector rotates. Your extended thumb will coincide with the direction of the moment.[7]

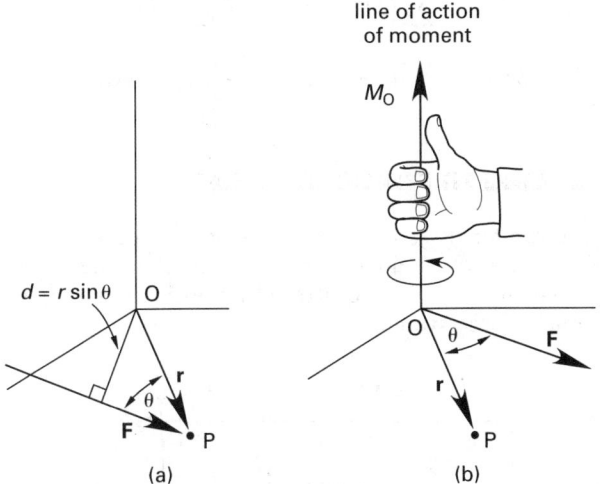

Figure 43.2 Right-Hand Rule

7. VARIGNON'S THEOREM

Varignon's theorem is a statement of how the total moment is derived from a number of forces acting simultaneously at a point.

Varignon's theorem: The sum of individual moments about a point caused by multiple concurrent forces is equal to the moment of the resultant force about the same point.

$$(\mathbf{r} \times \mathbf{F}_1) + (\mathbf{r} \times \mathbf{F}_2) + \cdots = \mathbf{r} \times (\mathbf{F}_1 + \mathbf{F}_2 + \cdots) \quad 43.9$$

8. MOMENT OF A FORCE ABOUT A LINE

Most rotating machines (motors, pumps, flywheels, etc.) have a fixed rotational axis. That is, the machines turn around a line, not around a point. The moment of a force about the rotational axis is not the same as the moment of the force about a point. In particular, the moment about a line is a scalar.[8]

---

[7]The direction of a moment also corresponds to the direction a right-hand screw would progress if it was turned in the direction that rotates **r** into **F**.
[8]Some sources say that the moment of a force about a line can be interpreted as a moment directed along the line. However, this interpretation does not follow from vector operations.

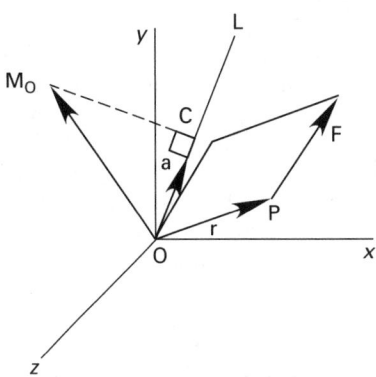

Figure 43.3 Moment of a Force About a Line

The moment M_{OL} of a force **F** about a line OL is the projection OC of the moment $\mathbf{M}_{\mathrm{O}}$ onto the line. Equation 43.10 gives the moment of a force about a line. **a** is the unit vector directed along the line, and a_x, a_y, and a_z are the direction cosines of the axis OL. Notice that Eq. 43.10 is a dot product (i.e., a scalar).

$$M_{\mathrm{OL}} = \mathbf{a} \cdot \mathbf{M}_{\mathrm{O}} = \mathbf{a} \cdot (\mathbf{r} \times \mathbf{F})$$
$$= \begin{bmatrix} a_x & a_y & a_z \\ x_{\mathrm{P}} - x_{\mathrm{O}} & y_{\mathrm{P}} - y_{\mathrm{O}} & z_{\mathrm{P}} - z_{\mathrm{O}} \\ F_x & F_y & F_z \end{bmatrix} \quad 43.10$$

If point O is the origin, then Eq. 43.10 reduces to Eq. 43.11.

$$M_{\mathrm{OL}} = \begin{bmatrix} a_x & a_y & a_z \\ x & y & z \\ F_x & F_y & F_z \end{bmatrix} \quad 43.11$$

9. COMPONENTS OF A MOMENT

The direction cosines of a force can be used to determine the components of the moment about the coordinate axes.

$$M_x = M\cos\theta_x \quad 43.12$$
$$M_y = M\cos\theta_y \quad 43.13$$
$$M_z = M\cos\theta_z \quad 43.14$$

Alternatively, the following three equations can be used to determine the components of the moment from a force applied at point (x, y, z) referenced to an origin at $(0, 0, 0)$.

$$M_x = yF_z - zF_y \quad 43.15$$
$$M_y = zF_x - xF_z \quad 43.16$$
$$M_z = xF_y - yF_x \quad 43.17$$

The resultant moment magnitude can be reconstituted from its components.

$$M = \sqrt{M_x^2 + M_y^2 + M_z^2} \quad 43.18$$

10. COUPLES

Any pair of equal, opposite, and parallel forces constitute a *couple*. A couple is equivalent to a single moment vector. Since the two forces are opposite in sign, the x-, y-, and z-components of the forces cancel out. Therefore, a body is induced to rotate without translation. A couple can be counteracted only by another couple. A couple can be moved to any location without affecting the equilibrium requirements.

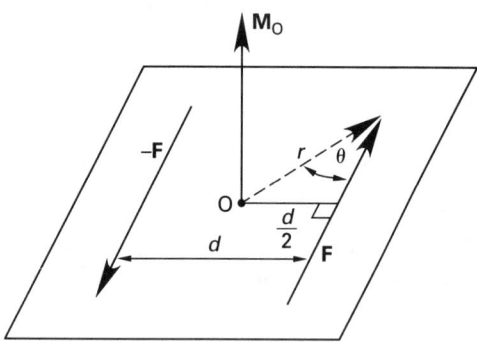

Figure 43.4 *Couple*

In Fig. 43.4, the equal but opposite forces produce a moment vector $\mathbf{M_O}$ of magnitude Fd. The two forces can be replaced by this moment vector, which can be moved to any location on a body. (Such a moment is known as a *free moment, moment of a couple,* or *coupling moment.*)

$$M_O = 2r\,F \sin\theta = Fd \qquad 43.19$$

11. EQUIVALENCE OF FORCES AND FORCE-COUPLE SYSTEMS

If a force, F, is moved a distance d from the original point of application, a couple, M, equal to Fd must be added to counteract the induced couple. The combination of the moved force and the couple is known as a *force-couple system.* Alternatively, a force-couple system can be replaced by a single force located a distance $d = M/F$ away.

12. RESULTANT FORCE-COUPLE SYSTEMS

The equivalence described in the previous section can be extended to three dimensions and multiple forces. Any collection of forces and moments in three-dimensional space is statically equivalent to a single resultant force vector plus a single resultant moment vector. (Either or both of these resultants can be zero.)

The x-, y-, and z-components of the resultant force are the sums of the x-, y-, and z-components of the individual forces, respectively.

$$F_{R,x} = \sum_i (F \cos\theta_x)_i \qquad 43.20$$

$$F_{R,y} = \sum_i (F \cos\theta_y)_i \qquad 43.21$$

$$F_{R,z} = \sum_i (F \cos\theta_z)_i \qquad 43.22$$

The resultant moment vector is more complex. It includes the moments of all system forces around the reference axes plus the components of all system moments.

$$M_{R,x} = \sum_i (yF_z - zF_y)_i + \sum_i (M \cos\theta_x)_i \qquad 43.23$$

$$M_{R,y} = \sum_i (zF_x - xF_z)_i + \sum_i (M \cos\theta_y)_i \qquad 43.24$$

$$M_{R,z} = \sum_i (xF_y - yF_x)_i + \sum_i (M \cos\theta_z)_i \qquad 43.25$$

13. LINEAR FORCE SYSTEMS

A *linear force system* is one in which all forces are parallel and applied along a straight line. A straight beam loaded by several concentrated forces is an example of a linear force system.

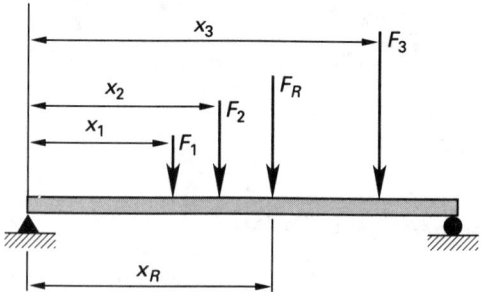

Figure 43.5 *Linear Force System*

For the purposes of statics, all of the forces in a linear force system can be replaced by an *equivalent resultant force*, F_R, equal to the sum of the individual forces. The location of the equivalent force coincides with the location of the centroid of the force group.

$$F_R = \sum_i F_i \qquad 43.26$$

$$x_R = \frac{\sum_i F_i x_i}{\sum_i F_i} \qquad 43.27$$

14. DISTRIBUTED LOADS

If an object is continuously loaded over a portion of its length, it is subject to a *distributed load*. Distributed loads result from *dead load* (i.e., self-weight), hydrostatic pressure, and materials distributed over the object.

If the load per unit length at some point x is $w(x)$, the statically equivalent concentrated load, F_R, can be found from Eq. 43.28. The equivalent load is the area under the loading curve.

$$F_R = \int_{x=0}^{x=L} w(x)\,dx \qquad 43.28$$

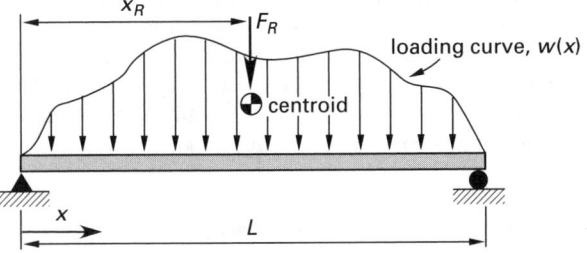

Figure 43.6 *Distributed Loads on a Beam*

The location, x_R, of the equivalent load is calculated from Eq. 43.29. The location coincides with the centroid of the area under the loading curve and is referred to in some problems as the *center of pressure*.

$$x_R = \frac{\int_{x=0}^{x=L} x\,w(x)\,dx}{F_R} \qquad 43.29$$

For a straight beam of length L under a uniform transverse loading of w pounds per foot (newtons per meter),

$$F_R = wL \qquad 43.30$$

$$x_R = \frac{L}{2} \qquad 43.31$$

For a straight beam of length L under a triangular distribution that increases from zero (at $x = 0$) to w (at $x = L$),

$$F_R = \frac{wL}{2} \qquad 43.32$$

$$x_R = \frac{2L}{3} \qquad 43.33$$

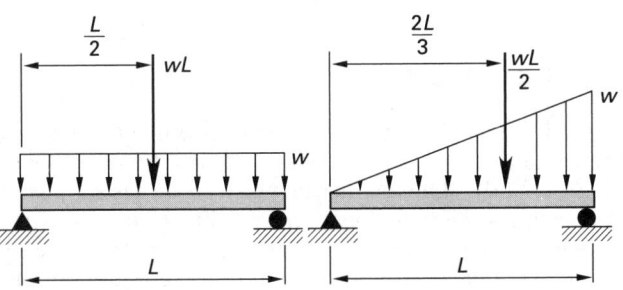

Figure 43.7 *Special Cases of Distributed Loading*

Example 43.1

Find the magnitude and location of the two equivalent forces on the two spans of the beam.

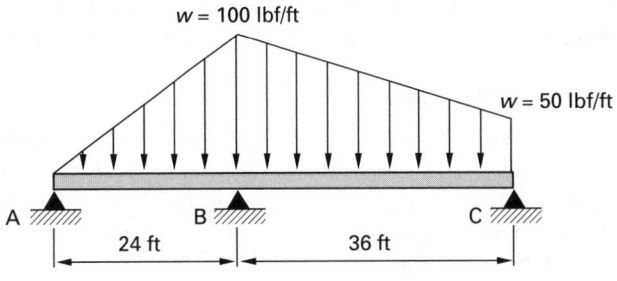

Solution

Span A–B
The area under the triangular loading curve is

$$A = \tfrac{1}{2}bh = \left(\tfrac{1}{2}\right)(24\text{ ft})\left(100\,\frac{\text{lbf}}{\text{ft}}\right)$$
$$= 1200\text{ lbf}$$

The centroid of the loading triangle is located at

$$x_{R,\,\text{A–B}} = \tfrac{2}{3}b = \left(\tfrac{2}{3}\right)(24\text{ ft})$$
$$= 16\text{ ft}$$

Span B–C
The area under the loading curve consists of a uniform load of 50 lbf/ft over the entire span B–C, plus a triangular load that starts at zero at point C and increases to 50 lbf/ft at point B. The area under the loading curve is

$$A = wL + \tfrac{1}{2}bh$$
$$= \left(50\,\frac{\text{lbf}}{\text{ft}}\right)(36\text{ ft}) + \left(\tfrac{1}{2}\right)(36\text{ ft})\left(50\,\frac{\text{lbf}}{\text{ft}}\right)$$
$$= 2700\text{ lbf}$$

The centroid of the trapezoidal loading curve is located at[9]

$$x_{R,\,\text{B–C}} = h\left(\frac{b+2t}{3b+3t}\right)$$
$$= \left(\frac{36\text{ ft}}{3}\right)\left[\frac{50\text{ ft} + (2)(100\text{ ft})}{50\text{ ft} + 100\text{ ft}}\right] = 20\text{ ft}$$

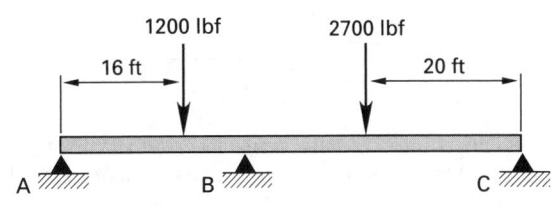

---

[9]The formula for the location of a centroid in a trapezoidal area is found in App. 48.A.

15. MOMENT FROM A DISTRIBUTED LOAD

The total force from a uniformly distributed load w over a distance x is wx. For the purposes of statics, the uniform load can be replaced by a concentrated force of wx located at the centroid of the distributed load, that is, at the midpoint, $x/2$, of the load. Therefore, the moment taken about one end of the distributed load is

$$M_{\text{distributed load}} = \text{force} \times \text{distance}$$
$$= wx\left(\frac{x}{2}\right) = \tfrac{1}{2}wx^2 \qquad 43.34$$

In general, the moment of a distributed load, uniform or otherwise, is the product of the total force and the distance to the centroid of the distributed load.

16. TYPES OF FORCE SYSTEMS

The complexity of methods used to analyze a statics problem depends on the configuration and orientation of the forces. Force systems can be divided into the following categories.

- *concurrent force system:* All of the forces act at the same point.

- *collinear force system:* All of the forces share the same line of action.

- *parallel force system:* All of the forces are parallel (though not necessarily in the same direction).

- *coplanar force system:* All of the forces are in a plane.

- *general three-dimensional system:* All other combinations of nonconcurrent, nonparallel, and noncoplanar forces.

17. CONDITIONS OF EQUILIBRIUM

An object is static when it is stationary. To be stationary, all of the forces on the object must be in equilibrium.[10] For an object to be in equilibrium, the resultant force and moment vectors must both be zero.

$$\mathbf{F}_R = \sum \mathbf{F} = 0 \qquad 43.35$$

$$F_R = \sqrt{F_{R,x}^2 + F_{R,y}^2 + F_{R,z}^2} = 0 \qquad 43.36$$

$$\mathbf{M}_R = \sum \mathbf{M} = 0 \qquad 43.37$$

$$M_R = \sqrt{M_{R,x}^2 + M_{R,y}^2 + M_{R,z}^2} = 0 \qquad 43.38$$

Since the square of any nonzero quantity is positive, Eqs. 43.39 through 43.44 follow directly from Eqs. 43.36 and 43.38.

[10]Thus the term *static equilibrium*, though widely used, is redundant.

$$F_{R,x} = 0 \qquad 43.39$$
$$F_{R,y} = 0 \qquad 43.40$$
$$F_{R,z} = 0 \qquad 43.41$$
$$M_{R,x} = 0 \qquad 43.42$$
$$M_{R,y} = 0 \qquad 43.43$$
$$M_{R,z} = 0 \qquad 43.44$$

Equations 43.39 through 43.44 seem to imply that six simultaneous equations must be solved in order to determine whether a system is in equilibrium. While this is true for general three-dimensional systems, fewer equations are necessary with most problems. Table 43.1 can be used as a guide to determine which equations are most helpful in solving different categories of problems.

Table 43.1 *Number of Equilibrium Conditions Required to Solve Different Force Systems*

type of force system	two-dimensional	three-dimensional
general	4	6
coplanar	3	3
concurrent	2	3
parallel	2	3
coplanar, parallel	2	2
coplanar, concurrent	2	2
collinear	1	1

18. TWO- AND THREE-FORCE MEMBERS

Members limited to loading by two or three forces are special cases of equilibrium. A *two-force member* can be in equilibrium only if the two forces have the same line of action (i.e., are collinear) and are equal but opposite. In most cases, two-force members are loaded axially, and the line of action coincides with the member's longitudinal axis. By choosing the coordinate system so that one axis coincides with the line of action, only one equilibrium equation is needed.

A *three-force member* can be in equilibrium only if the three forces are concurrent or parallel. Stated another way, the force polygon of a three-force member in equilibrium must close on itself.

19. REACTIONS

The first step in solving most statics problems is to determine the reaction forces (i.e., the *reactions*) supporting the body. The manner in which a body is supported determines the type, location, and direction of the reactions. Conventional symbols are often used to define the type of support (such as pinned, roller, etc.). Examples of the symbols are shown in Table 43.2.

Table 43.2 *Types of Two-Dimensional Supports*

type of support	reactions and moments	number of unknowns[a]
simple, roller, rocker, ball, or frictionless surface	reaction normal to surface, no moment	1
cable in tension, or link	reaction in line with cable or link, no moment	1
frictionless guide or collar	reaction normal to rail, no moment	1
built-in, fixed support	two reaction components, one moment	3
frictionless hinge, pin connection, or rough surface	reaction in any direction, no moment	2

[a]The number of unknowns is valid for two-dimensional problems only.

For beams, the two most common types of supports are the roller support and the pinned support. The *roller support*, shown as a cylinder supporting the beam, supports vertical forces only. Rather than support a horizontal force, a roller support simply rolls into a new equilibrium position. Only one equilibrium equation (i.e., the sum of vertical forces) is needed at a roller support. Generally, the terms *simple support* and *simply supported* refer to a roller support.

The *pinned support*, shown as a pin and clevis, supports both vertical and horizontal forces. Two equilibrium equations are needed.

Generally, there will be vertical and horizontal components of a reaction when one body touches another. However, when a body is in contact with a *frictionless surface*, there is no frictional force component parallel to the surface. Therefore, the reaction is normal to the contact surfaces. The assumption of frictionless contact is particularly useful when dealing with systems of spheres and cylinders in contact with rigid supports. Frictionless contact is also assumed for roller and rocker supports.[11]

20. DETERMINACY

When the equations of equilibrium are independent, a rigid body force system is said to be *statically determinate*. A statically determinate system can be solved for all unknowns, which are usually reactions supporting the body.

When the body has more supports than are necessary for equilibrium, the force system is said to be *statically indeterminate*. In a statically indeterminate system, one or more of the supports or members can be removed or reduced in restraint without affecting the equilibrium position.[12] Those supports and members are known as *redundant members*. The number of redundant members is known as the *degree of indeterminacy*. Figure 43.8 illustrates several common indeterminate structures.

A statically indeterminate body requires additional equations to supplement the equilibrium equations. The additional equations typically involve deflections and depend on mechanical properties of the body.

21. TYPES OF DETERMINATE BEAMS

Figure 43.9 illustrates the terms used to describe determinate beam types.

---

[11]Frictionless surface contact, which requires only one equilibrium equation, should not be confused with a frictionless pin connection, which requires two equilibrium equations. A pin connection with friction introduces a moment at the connection, increasing the number of required equilibrium equations to three.

[12]An example of a support reduced in restraint is a pinned joint replaced by a roller joint. The pinned joint restrains the body vertically and horizontally, requiring two equations of equilibrium. The roller joint restrains the body vertically only and requires one equilibrium equation.

Statics

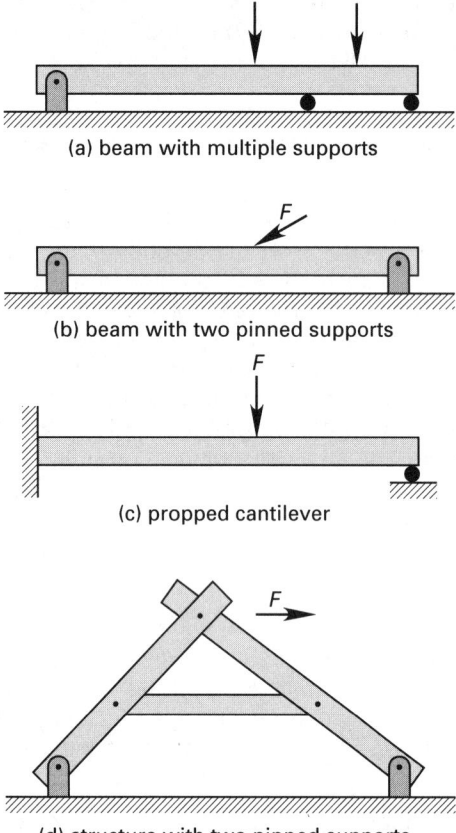

(a) beam with multiple supports

(b) beam with two pinned supports

(c) propped cantilever

(d) structure with two pinned supports

Figure 43.8 *Examples of Indeterminate Systems*

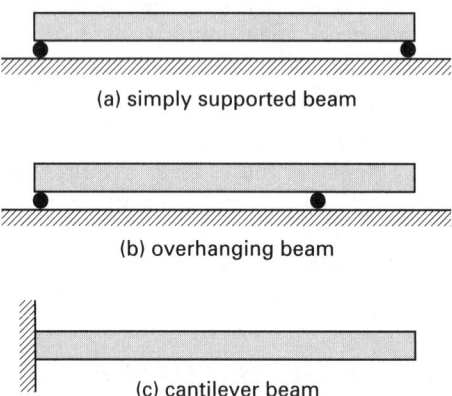

(a) simply supported beam

(b) overhanging beam

(c) cantilever beam

Figure 43.9 *Types of Determinate Beams*

22. FREE-BODY DIAGRAMS

A *free-body diagram* is a representation of a body in equilibrium. It shows all applied forces, moments, and reactions. Free-body diagrams do not consider the internal structure or construction of the body, as Fig. 43.10 illustrates.

Since the body is in equilibrium, the resultants of all forces and moments on the free body are zero. In order to maintain equilibrium, any portions of the body that are removed must be replaced by the forces and moments those portions impart to the body. Typically, the body is isolated from its physical supports in order to help evaluate the reaction forces. In other cases, the body may be sectioned (i.e., cut) in order to determine the forces at the section.

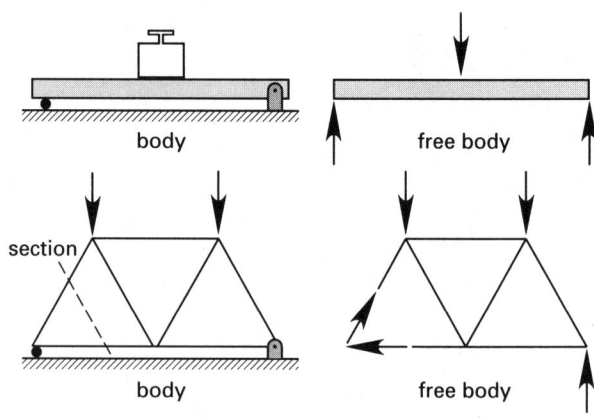

Figure 43.10 *Bodies and Free Bodies*

23. FINDING REACTIONS IN TWO DIMENSIONS

The procedure for finding determinate reactions in two-dimensional problems is straightforward. Determinate structures will have either a roller support and a pinned support or two roller supports.

step 1: Establish a convenient set of coordinate axes. (To simplify the analysis, one of the coordinate directions should coincide with the direction of the forces and reactions.)

step 2: Draw the free-body diagram.

step 3: Resolve the reaction at the pinned support (if any) into components normal and parallel to the coordinate axes.

step 4: Establish a positive direction of rotation (e.g., clockwise) for purposes of taking moments.

step 5: Write the equilibrium equation for moments about the pinned connection. (By choosing the pinned connection as the point about which to take moments, the pinned connection reactions do not enter into the equation.) This will usually determine the vertical reaction at the roller support.

step 6: Write the equilibrium equation for the forces in the vertical direction. Usually, this equation will have two unknown vertical reactions.

step 7: Substitute the known vertical reaction from step 5 into the equilibrium equation from step 6. This will determine the second vertical reaction.

step 8: Write the equilibrium equation for the forces in the horizontal direction. Since there is a maximum of one unknown reaction component in the horizontal direction, this step will determine that component.

step 9: If necessary, combine the vertical and horizontal force components at the pinned connection into a resultant reaction.

Example 43.2

Determine the reactions, R_1 and R_2, on the following beam.

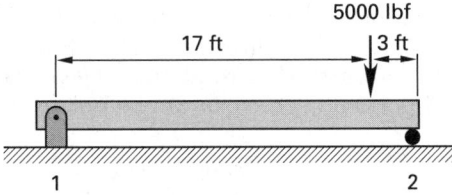

Solution

step 1: The x- and y-axes are established parallel and perpendicular to the beam.

step 2: The free-body diagram is

step 3: R_1 is a pinned support. Therefore, it has two components, $R_{1,x}$ and $R_{1,y}$.

step 4: Assume clockwise moments are positive.

step 5: Take moments about the left end and set them equal to zero.

$$\sum M_{\text{left end}} = (5000 \text{ lbf})(17 \text{ ft}) - R_2(20 \text{ ft}) = 0$$

$$R_2 = 4250 \text{ lbf}$$

step 6: The equilibrium equation for the vertical direction is

$$\sum F_y = R_{1,y} + R_2 - 5000 \text{ lbf} = 0$$

step 7: Substituting R_2 into the vertical equilibrium equation,

$$R_{1,y} + 4250 \text{ lbf} - 5000 \text{ lbf} = 0$$

$$R_{1,y} = 750 \text{ lbf}$$

step 8: There are no applied forces in the horizontal direction. Therefore, the equilibrium equation is

$$\sum F_x = R_{1,x} + 0 = 0$$

$$R_{1,x} = 0$$

24. COUPLES AND FREE MOMENTS

Once a couple on a body is known, the derivation and source of the couple are irrelevant. When the moment on a body is 80 N·m, it does not make any difference whether the force is 40 N with a lever arm of 2 m, or 20 N with a lever arm of 4 m, and so on. Therefore, the point of application of a couple is disregarded when writing the moment equilibrium equation. For this reason, the term *free moment* is used synonymously with *couple*.

Figure 43.11 illustrates two diagrammatic methods of indicating the application of a free moment.

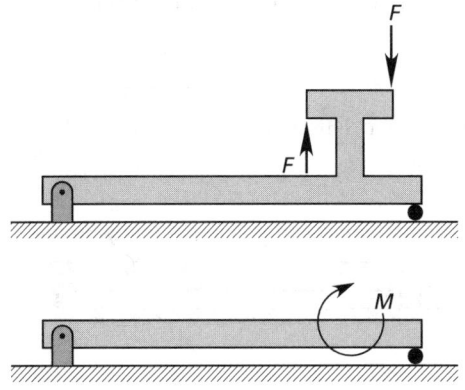

Figure 43.11 *Free Moments*

Example 43.3

What is the reaction R_2 for the beam shown?

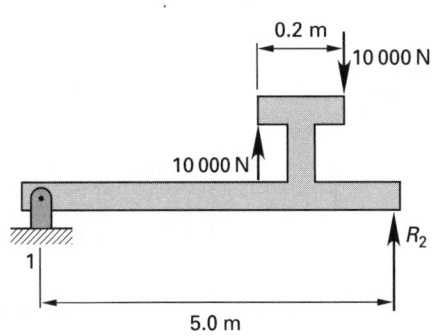

Solution

The two couple forces are equal and cancel each other as they come down the stem of the tee bracket. Therefore, there are no applied vertical forces.

The couple has a value of

$$M = (10\,000 \text{ N})(0.2 \text{ m}) = 2000 \text{ N·m} \qquad \text{[clockwise]}$$

Choose clockwise as the direction for positive moments. Taking moments about the pinned connection,

$$\sum M = 2000 \text{ N·m} - R_2(5 \text{ m}) = 0$$

$$R_2 = 400 \text{ N}$$

25. INFLUENCE LINES FOR REACTIONS

An *influence line* (also known as an *influence graph* and *influence diagram*) is a graph of the magnitude of a reaction as a function of the load placement.[13] The x-axis of the graph corresponds to the location on the body (along the length of a beam). The y-axis corresponds to the magnitude of the reaction.

By convention (and to generalize the graph for use with any load), the load is taken as one force unit. Therefore, for an actual load of F units, the actual reaction, R, is the product of the actual load and the influence line ordinate.

$$R = F \times \text{influence line ordinate} \qquad \textbf{43.45}$$

Example 43.4

Draw the influence line for the left reaction for the beam shown.

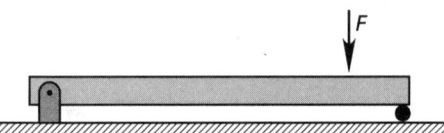

Solution

If the unit load is at the left end, the left reaction will be 1.0. If the unit load is at the right end, it will be supported entirely by the right reaction, so the left reaction will be zero. The influence line for the left reaction varies linearly for intermediate load placement.

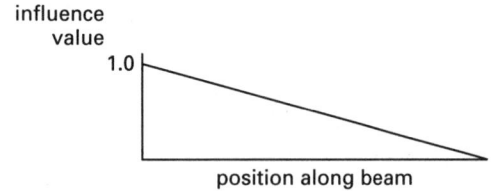

---

[13]Influence diagrams can also be drawn for moments, shears, and deflections.

26. HINGES

Hinges are added to structures to prevent translation while permitting rotation. A frictionless hinge can support a force, but it cannot transmit a moment. Since the moment is zero at a hinge, a structure can be sectioned at the hinge and the remainder of the structure can be replaced by only a force.

Example 43.5

Calculate the reaction R_3 and the hinge force on the two-span beam shown.

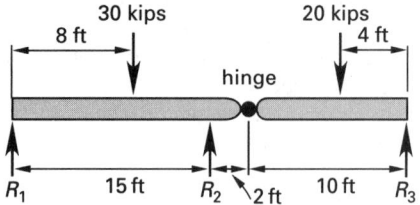

Solution

At first, this beam may appear to be statically indeterminate since it has three supports. However, the moment is known to be zero at the hinge. Therefore, the hinged portion of the span can be isolated.

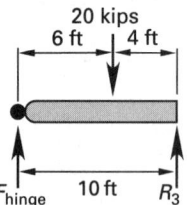

Reaction R_3 is found by taking moments about the hinge. Assume clockwise moments are positive.

$$\sum M_{\text{hinge}} = (20{,}000 \text{ lbf})(6 \text{ ft}) - R_3(10 \text{ ft}) = 0$$

$$R_3 = 12{,}000 \text{ lbf}$$

The hinge force is found by summing vertical forces on the isolated section.

$$\sum F_y = F_{\text{hinge}} + 12{,}000 \text{ lbf} - 20{,}000 \text{ lbf} = 0$$

$$F_{\text{hinge}} = 8000 \text{ lbf}$$

27. LEVERS

A *lever* is a simple mechanical machine with the ability to increase an applied force. The ratio of the load-bearing force to applied force (i.e., the *effort*) is known as the *mechanical advantage* or *force amplification*. As Fig. 43.12 shows, the mechanical advantage is equal to the ratio of lever arms.

$$\text{mechanical advantage} = \frac{F_{\text{load}}}{F_{\text{applied}}} = \frac{\text{applied force lever arm}}{\text{load lever arm}}$$

$$= \frac{\text{distance moved by applied force}}{\text{distance moved by load}} \qquad 43.46$$

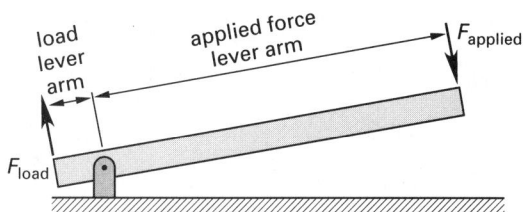

Figure 43.12 Lever

28. PULLEYS

A *pulley* (also known as a *sheave*) is used to change the direction of an applied tensile force. A series of pulleys working together (known as a *block and tackle*) can also provide *pulley advantage* (i.e., mechanical advantage). A *hoist* is any device used to raise or lower an object. A hoist may contain one or more pulleys.

If the pulley is attached by a bracket or cable to a fixed location, it is said to be a *fixed pulley*. If the pulley is attached to a load, or if the pulley is free to move, it is known as a *free pulley*.

Most simple problems disregard friction and assume that all ropes are parallel.[14] In such cases, the pulley advantage is equal to the number of ropes coming to and going from the load-carrying pulley. The diameters of the pulleys are not factors in calculating the pulley advantage.

In other cases, a *loss factor*, ϵ, is used to account for rope rigidity. For most wire ropes and chains with 180° contact, the loss factor at low speeds varies between 1.03 and 1.06. The loss factor is the reciprocal of the *pulley efficiency*, η.

$$\epsilon = \frac{\text{applied force}}{\text{load}} = \frac{1}{\eta} \qquad 43.47$$

29. AXIAL MEMBERS

An *axial member* is capable of supporting axial forces only and is loaded only at its joints (i.e., ends). This type of performance can be achieved through the use of frictionless bearings or smooth pins at the ends. Since the ends are assumed to be pinned (i.e., rotation-free), an axial member cannot support moments. The weight of the member is disregarded or is included in the joint loading.

[14] Although the term *rope* is used here, the principles apply equally well to wire rope, cables, chains, belts, etc.

An axial member can be in either tension or compression. It is common practice to label forces in axial members as (T) or (C) for tension or compression, respectively. Alternatively, tensile forces can be written as positive numbers, while compressive forces are written as negative numbers.

The members in simple trusses are assumed to be axial members. Each member is identified by its endpoints, and the force in a member is designated by the symbol for the two endpoints. For example, the axial force in a member connecting points C and D will be written as **CD**. Similarly, $\mathbf{EF}_y$ is the y-component of the force in the member connecting points E and F.

For equilibrium, the resultant forces at the two joints must be equal, opposite, and collinear. This applies to the total (resultant) force as well as to the x- and y-components at those joints.

Table 43.3 Mechanical Advantages of Rope-Operated Machines

	fixed sheave	free sheave	ordinary pulley block (n sheaves)	differential pulley block
F_{ideal}	W	$\dfrac{W}{2}$	$\dfrac{W}{n}$	$\left(\dfrac{W}{2}\right)\left(1-\dfrac{d}{D}\right)$
F to raise load	ϵW	$\dfrac{\epsilon W}{1+\epsilon}$	$\dfrac{\epsilon^n(\epsilon-1)W}{\epsilon^n-1}$	$\dfrac{\left(\epsilon^2-\dfrac{d}{D}\right)W}{1+\epsilon}$
F to lower load	$\dfrac{W}{\epsilon}$	$\dfrac{W}{1+\epsilon}$	$\dfrac{\left(\dfrac{1}{\epsilon}-1\right)W}{1-\epsilon^n}$	$\left(\dfrac{\epsilon W}{1+\epsilon}\right)\left(\dfrac{1}{\epsilon^2}-\dfrac{d}{D}\right)$
ratio of distance of force to distance of load	1	2	n	$\dfrac{2D}{D-d}$

30. FORCES IN AXIAL MEMBERS

The line of action of a force in an axial member coincides with the longitudinal axis of the member. Depending on the orientation of the coordinate axis system, the direction of the longitudinal axis will have both x- and y-components. Therefore, the force in an axial member will generally have both x- and y-components.

Statics

The following four general principles are helpful in determining the force in an axial member.

- A horizontal member carries only horizontal loads. It cannot carry vertical loads.

- A vertical member carries only vertical loads. It cannot carry horizontal loads.

- The vertical component of an axial member's force is equal to the vertical component of the load applied to the member.

- The total and component forces in an inclined member are proportional to the sides of the triangle outlined by the member and the coordinate axes.[15]

Example 43.6

Member BC is an inclined axial member oriented as shown. A vertical 1000 N force is applied to the top end. What are the x- and y-components of the force in member BC? What is the total force in the member?

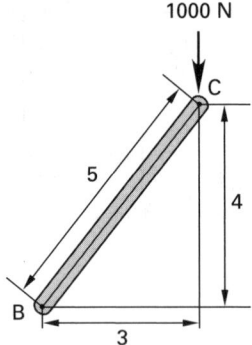

Solution

From the third principle,

$$\mathbf{BC}_y = 1000 \text{ N}$$

From the fourth principle,

$$\mathbf{BC}_x = \left(\frac{3}{4}\right)\mathbf{BC}_y = \left(\frac{3}{4}\right)(1000 \text{ N}) = 750 \text{ N}$$

The total resultant force in member BC can be calculated from the Pythagorean theorem. However, it is easier to use the fourth principle.

$$\mathbf{BC} = \left(\frac{5}{4}\right)\mathbf{BC}_y = \left(\frac{5}{4}\right)(1000 \text{ N}) = 1250 \text{ N}$$

---

[15]This is an application of the principle of similar triangles.

Example 43.7

The 12 ft long axial member FG supports an axial force of 180 lbf. What are the x- and y-components of the applied force?

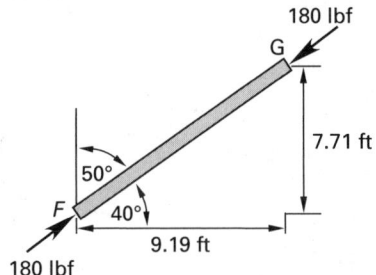

Solution

method 1: direction cosines

The x- and y-direction angles are 40° and 90° − 40° = 50°, respectively.

$$\mathbf{FG}_x = (180 \text{ lbf})\cos 40° = 137.9 \text{ lbf}$$
$$\mathbf{FG}_y = (180 \text{ lbf})\cos 50° = 115.7 \text{ lbf}$$

method 2: similar triangles

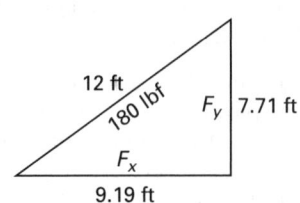

$$\mathbf{FG}_x = \left(\frac{9.19 \text{ ft}}{12 \text{ ft}}\right)(180 \text{ lbf}) = 137.9 \text{ lbf}$$
$$\mathbf{FG}_y = \left(\frac{7.71 \text{ ft}}{12 \text{ ft}}\right)(180 \text{ lbf}) = 115.7 \text{ lbf}$$

31. TRUSSES

A *truss* or *frame* is a set of pin-connected axial *members* (i.e., *two-force members*). The connection points are known as *joints*. Member weights are disregarded, and truss loads are applied only at joints. A *structural cell* consists of all members in a closed loop of members. For the truss to be stable (i.e., to be a *rigid truss*), all of the structural cells must be triangles. Figure 43.13 identifies *chords*, *end posts*, *panels*, and other elements of a typical *bridge truss*.

A *trestle* is a braced structure spanning a ravine, gorge, or other land depression in order to support a road or rail line. Trestles are usually indeterminate, have multiple earth contact points, and are more difficult to evaluate than simple trusses.

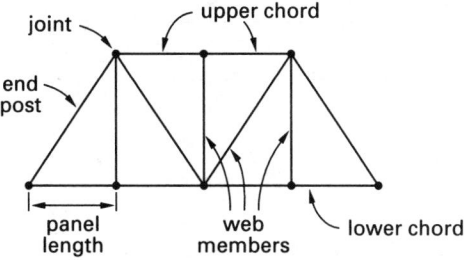

Figure 43.13 *Parts of a Bridge Truss*

Several types of trusses have been given specific names. Some of the more common types of named trusses are shown in Fig. 43.14.

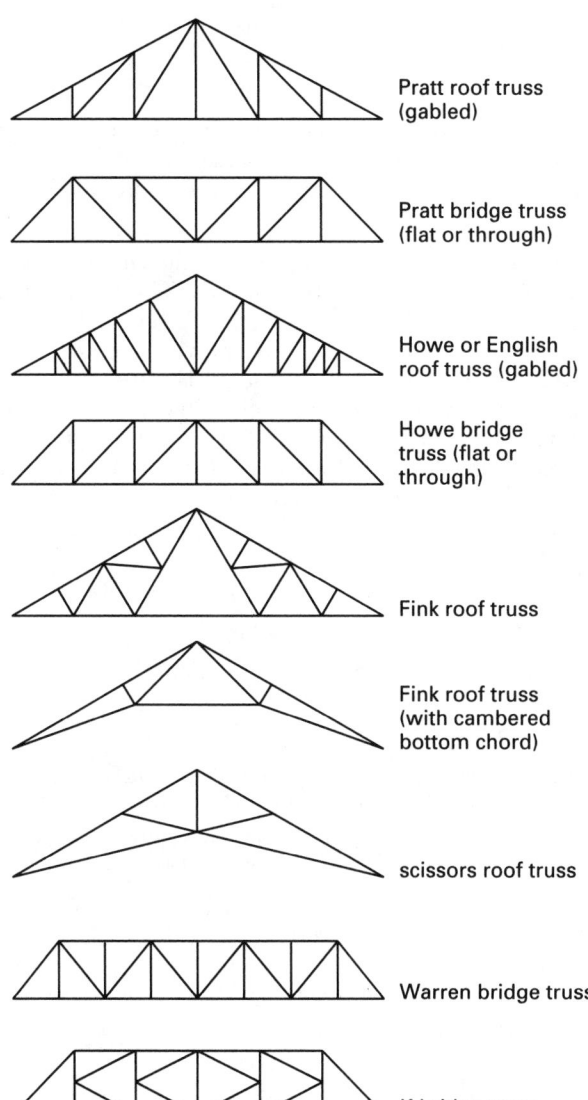

Figure 43.14 *Special Types of Trusses*

Truss loads are considered to act only in the plane of a truss. Therefore, trusses are analyzed as two-dimensional structures. Forces in truss members hold the various truss parts together and are known as *internal forces*. The internal forces are found by drawing free-body diagrams.

Although free-body diagrams of truss members can be drawn, this is not usually done. Instead, free-body diagrams of the pins (i.e., the joints) are drawn. A pin in compression will be shown with force arrows pointing toward the pin, away from the member. (Similarly, a pin in tension will be shown with force arrows pointing away from the pin, toward the member.)[16]

With typical bridge trusses supported at the ends and loaded downward at the joints, the upper chords are almost always in compression, and the end panels and lower chords are almost always in tension.

32. DETERMINATE TRUSSES

A truss will be statically determinate if Eq. 43.48 holds.

$$\text{no. of members} = (2)(\text{no. of joints}) - 3 \qquad \textit{43.48}$$

If the left-hand side is greater than the right-hand side (i.e., there are *redundant members*), the truss is statically indeterminate. If the left-hand side is less than the right-hand side, the truss is unstable and will collapse under certain types of loading.

Equation 43.48 is a special case of the following general criterion.

$$
\begin{aligned}
\text{no. of members} & \\
+ \text{ no. of reactions} & \\
- (2)(\text{no. of joints}) &= 0 \quad \text{[determinate]} \\
&> 0 \quad \text{[indeterminate]} \\
&< 0 \quad \text{[unstable]} \qquad \textit{43.49}
\end{aligned}
$$

For a structure with one roller and one pinned support, there are three reactions (counting the x- and y-components of the pinned support separately). Thus, Eq. 43.48 is for use only with structures supported in this manner.

Furthermore, Eq. 43.48 is a necessary, but not sufficient, condition for truss stability. It is possible to arrange the members in such a manner as to not contribute to truss stability. This will seldom be the case in actual practice, however.

---

[16]The method of showing tension and compression on a truss drawing may appear incorrect. This is because the arrows show the forces on the pins, not on the members.

33. ZERO-FORCE MEMBERS

Forces in truss members can sometimes be determined by inspection. One of these cases is *zero-force members*. A third member framing into a joint already connecting two collinear members carries no internal force unless there is a load applied at that joint. Similarly, both members forming an apex of the truss are zero-force members unless there is a load applied at the apex.

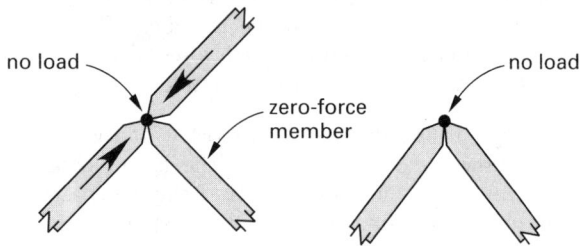

Figure 43.15 *Zero-Force Members*

34. METHOD OF JOINTS

The *method of joints* is one of three methods that can be used to find the internal forces in each truss member. This method is useful when most or all of the truss member forces are to be calculated. Because this method advances from joint to adjacent joint, it is inconvenient when a single isolated member force is to be calculated.

The method of joints is a direct application of the equations of equilibrium in the x- and y-directions. Traditionally, the method starts by finding the reactions supporting the truss. Next, the joint at one of the reactions is evaluated, which determines all the member forces framing into the joint. Then, knowing one or more of the member forces from the previous step, an adjacent joint is analyzed. The process is repeated until all the unknown quantities are determined.

At a joint, there may be up to two unknown member forces, each of which can have dependent x- and y-components.[17] Since there are two equilibrium equations, the two unknown forces can be determined. Even though determinate, however, the sense of a force will often be unknown. If the sense cannot be determined by logic, an arbitrary decision can be made. If the incorrect direction is chosen, the force will be negative.

Example 43.8

Use the method of joints to calculate the force **BD** in the truss shown.

---

[17]Occasionally, there will be three unknown member forces. In that case, an additional equation must be derived from an adjacent joint.

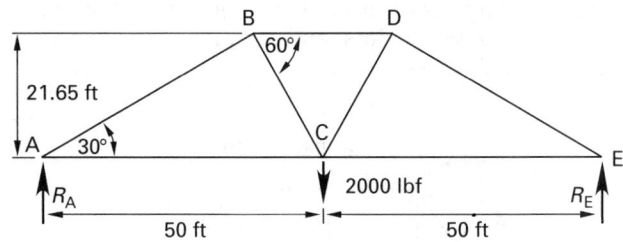

Solution

First, find the reactions. Assume clockwise is positive and take moments about point A.

$$\sum M_A = (2000 \text{ lbf})(50 \text{ ft}) - R_E(50 \text{ ft} + 50 \text{ ft}) = 0$$
$$R_E = 1000 \text{ lbf}$$

Since the sum of forces in the y-direction is also zero,

$$\sum F_y = R_A + 1000 \text{ lbf} - 2000 \text{ lbf} = 0$$
$$R_A = 1000 \text{ lbf}$$

There are three unknowns at joint B (and also D). Therefore, the analysis must start at joint A (or E) where there are only two unknowns (forces **AB** and **AC**).

The free-body diagram of pin A is shown. The direction of R_A is known to be upward. The directions of forces **AB** and **AC** can be assumed, but logic can be used to determine them. Only the vertical component of **AB** can oppose R_A. Therefore, **AB** is directed downward. (This means that member AB is in compression.) Similarly, **AC** must oppose the horizontal component of **AB**. Therefore, **AC** is directed to the right. (This means that member AC is in tension.)

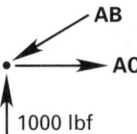

Resolve force **AB** into horizontal and vertical components using trigonometry, direction cosines, or similar triangles. (R_A and **AC** are already parallel to an axis.) Then, use the equilibrium equations to determine the forces.

By inspection, $AB_y = 1000$.

$$AB_y = AB \sin 30°$$
$$1000 \text{ lbf} = 0.5AB$$
$$AB = 2000 \text{ lbf} \ (C)$$
$$AB_x = AB \cos 30° = (2000 \text{ lbf})(0.866)$$
$$= 1732 \text{ lbf}$$

Now, draw the free-body diagram of pin B. (Notice that the direction of force **AB** is toward the pin, just as it was for pin A.) Although the true directions of the forces are unknown, they can be determined logically. The direction of force **BC** is chosen to counteract the vertical component of force **AB**. The direction of force **BD** is chosen to counteract the horizontal components of forces **AB** and **BC**.

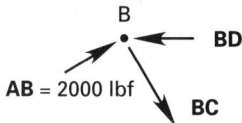

$\mathbf{AB}_x$ and $\mathbf{AB}_y$ are already known. Resolve the force **BC** into horizontal and vertical components.

$$\mathbf{BC}_x = \mathbf{BC} \sin 30° = (0.5)\,\mathbf{BC}$$
$$\mathbf{BC}_y = \mathbf{BC} \cos 30° = (0.866)\,\mathbf{BC}$$

Now, write the equations of equilibrium for point B.

$$\sum F_x = 1732 + (0.5)\,\mathbf{BC} - \mathbf{BD} = 0$$
$$\sum F_y = 1000\ \text{lbf} - (0.866)\,\mathbf{BC} = 0$$

From the second equation, $\mathbf{BC} = 1155$ lbf. Substituting this into the first equation,

$$1732\ \text{lbf} + (0.5)(1155\ \text{lbf}) - \mathbf{BD} = 0$$
$$\mathbf{BD} = 2310\ \text{lbf}\ \ (\text{C})$$

Since **BD** turned out to be positive, its direction was chosen correctly.

The direction of the arrow indicates that the member is compressing the pin. Consequently, the pin is compressing the member. Member BD is in compression.

If the process is continued, all forces can be determined. However, the truss is symmetrical, and it is not necessary to evaluate every joint to calculate all forces.

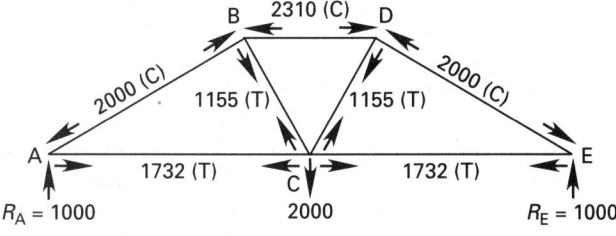

35. CUT-AND-SUM METHOD

The *cut-and-sum method* can be used to find forces in inclined members. This method is strictly an application of the vertical equilibrium condition ($\Sigma F_y = 0$).

The method starts by finding all of the support reactions on a truss. Then a cut is made through the truss in such a way as to pass through one inclined or vertical member only. (At this point, it should be clear that the vertical component of the inclined member must balance all of the external vertical forces.) The equation for vertical equilibrium is written for the free body of the remaining truss portion. This process is illustrated by the following example.

Example 43.9

Find the force in member BC for the truss in Ex. 43.8.

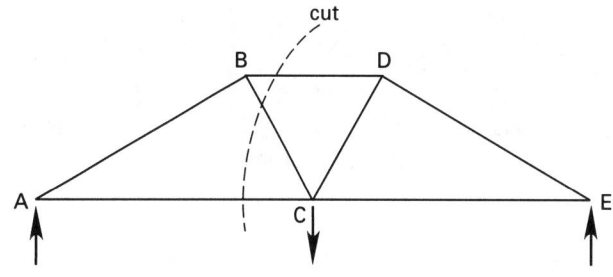

Solution

The reactions were determined in Ex. 43.8. The truss is cut in such a way as to pass through member BC but through no other inclined member. The free body of the remaining portion of the truss is

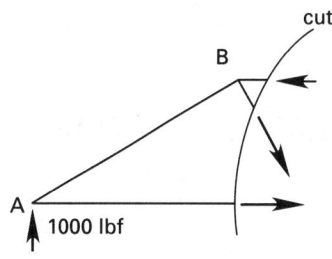

The vertical equilibrium equation is

$$\sum F_y = R_A - \mathbf{BC}_y = 0$$
$$= 1000\ \text{lbf} - (0.866)\mathbf{BC} = 0$$
$$\mathbf{BC} = 1155\ \text{lbf}$$

36. METHOD OF SECTIONS

The *method of sections* is a direct approach to finding forces in any truss member. This method is convenient when only a few truss member forces are unknown.

As with the previous two methods, the first step is to find the support reactions. Then a cut is made through the truss, passing through the unknown member.[18] Finally, all three conditions of equilibrium are applied as needed to the remaining truss portion. (Since there are three equilibrium equations, the cut cannot pass through more than three members in which the forces are unknown.)

Example 43.10

Find the forces in members CD and CE. The support reactions have already been determined.

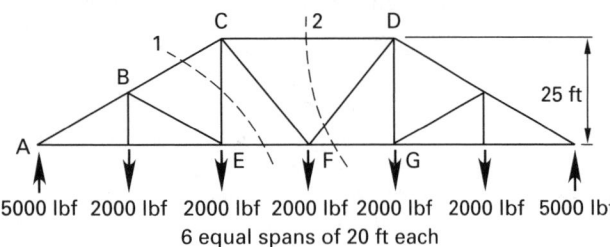

5000 lbf 2000 lbf 2000 lbf 2000 lbf 2000 lbf 2000 lbf 5000 lbf
6 equal spans of 20 ft each

Solution

To find the force **CE**, the truss is cut at section 1.

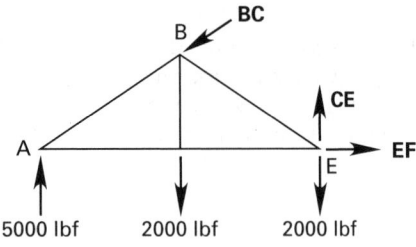

5000 lbf   2000 lbf   2000 lbf

Taking moments about point A will eliminate all of the unknown forces except **CE**. Assume clockwise moments are positive.

$$\sum M_A = (2000 \text{ lbf})(20 \text{ ft}) + (2000 \text{ lbf})(40 \text{ ft}) - (40 \text{ ft})\mathbf{CE}$$
$$= 0$$
$$\mathbf{CE} = 3000 \text{ lbf}$$

To find the force **CD**, the truss is cut at section 2.

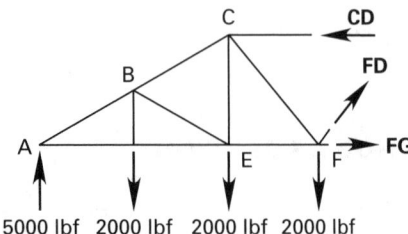

5000 lbf 2000 lbf 2000 lbf 2000 lbf

Taking moments about point F will eliminate all unknowns except **CD**. Assume clockwise moments are positive.

$$\sum M_F = (5000 \text{ lbf})(60 \text{ ft}) - (25)\mathbf{CD}$$
$$- (2000 \text{ lbf})(20 \text{ ft}) - (2000 \text{ lbf})(40 \text{ ft}) = 0$$
$$\mathbf{CD} = 7200 \text{ lbf}$$

37. SUPERPOSITION OF LOADS

Superposition is a term used to describe the process of determining member forces by considering loads one at a time. Suppose, for example, that the force in member FG is unknown and that the truss carries three loads. If the method of superposition is used, the force in member FG (call it **FG**$_1$) is determined with only the first load acting on the truss. **FG**$_2$ and **FG**$_3$ are similarly found. The true member force **FG** is found by adding **FG**$_1$, **FG**$_2$, and **FG**$_3$.

Superposition should be used with discretion since trusses can change shape under load. If a truss deflects such that the load application points are significantly different from those in the undeflected truss, superposition cannot be used for that truss.

In simple truss analysis, change of shape under load is neglected. Superposition, therefore, can be assumed to apply.

38. TRANSVERSE TRUSS MEMBER LOADS

Truss members are usually designed as axial members, not as beams. Trusses are traditionally considered to be loaded at joints only. Figure 43.16, however, illustrates cases of nontraditional *transverse loading* that can actually occur. For example, a truss member's own weight would contribute to a uniform load, as would a severe ice buildup.

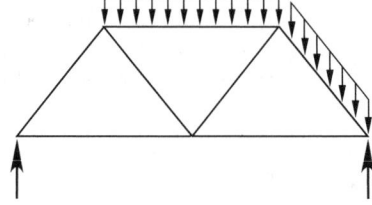

Figure 43.16 *Transverse Truss Member Loads*

Transverse loads add two solution steps to a truss problem. First, the truss member must be individually considered as a beam simply supported at its pinned connections, and the reactions needed to support the transverse loading must be found. These reactions become additional loads applied to the truss joints, and the truss can then be evaluated in the normal manner.

The second step is a check of the structural adequacy (deflection, bending stress, shear stress, buckling, etc.) of the truss member under transverse loading.

---

[18]Knowing where to cut the truss is the key part of this method. Such knowledge is developed only by practice.

39. CABLES CARRYING CONCENTRATED LOADS

An *ideal* cable is assumed to be completely flexible, massless, and incapable of elongation. It therefore acts as an axial two-force tension member between points of concentrated loading. In fact, the term *tension* or *tensile force* is commonly used in place of "member force" when dealing with cables.

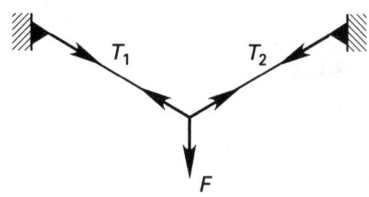

Figure 43.17 *Cable with Concentrated Load*

The methods of joints and sections used in truss analysis can be used to determine the tensions in cables carrying concentrated loads. After separating the reactions into x- and y-components, it is particularly useful to sum moments about one of the reaction points. All cables will be found to be in tension, and (with vertical loads only) the horizontal tension component will be the same in all cable segments. Unlike the case of a rope passing over a series of pulleys, however, the total tension in the cable will not be the same in every cable segment.

Example 43.11

What are the tensions **AB**, **BC**, and **CD**?

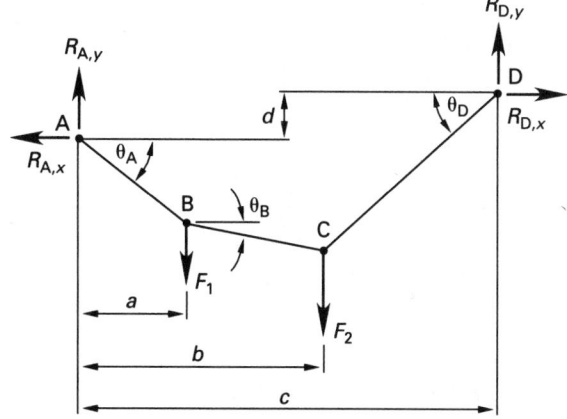

Solution

Separate the two reactions into x- and y-components. (The total reactions R_A and R_D are also the tensions **AB** and **CD**, respectively.)

$$R_{A,x} = -\mathbf{AB}\cos\theta_A$$
$$R_{A,y} = -\mathbf{AB}\sin\theta_A$$
$$R_{D,x} = -\mathbf{CD}\cos\theta_D$$
$$R_{D,y} = -\mathbf{CD}\sin\theta_D$$

Next, take moments about point A to find **CD**. Assume clockwise to be positive.

$$\sum M_A = aF_1 + bF_2 - d\,\mathbf{CD}_x + c\,\mathbf{CD}_y = 0$$

None of the applied loads are in the x-direction. Therefore, the only horizontal loads are the x-components of the reactions. To find tension **AB**, take the entire cable as a free body. Then, sum external forces in the x-direction.

$$\sum F_x = R_{D,x} - R_{A,x} = 0$$
$$\mathbf{CD}\cos\theta_D = \mathbf{AB}\cos\theta_A$$

The x-component of force is the same in all cable segments. To find **BC**, sum x-direction forces at point B.

$$\sum F_x = \mathbf{BC}_x - \mathbf{AB}_x = 0$$
$$\mathbf{BC}\cos\theta_B = \mathbf{AB}\cos\theta_A$$

40. PARABOLIC CABLES

If the distributed load per unit length, w, on a cable is constant with respect to the horizontal axis (as is the load from a bridge floor), the cable will be parabolic in shape.[19] This is illustrated in Fig. 43.18.

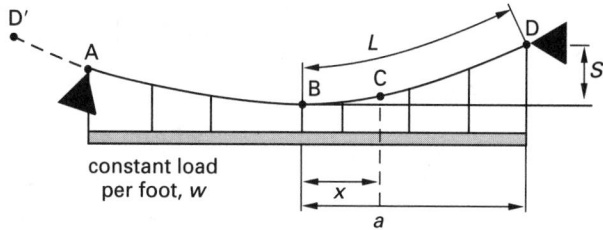

Figure 43.18 *Parabolic Cable*

Assuming the location of the maximum sag (i.e., the lowest cable point) is known, the horizontal component of tension, H, can be found by taking moments about a reaction point. If the cable is cut at the maximum sag point, B, the cable tension on the free body will be horizontal since there is no vertical component to the cable. Cutting the cable in Fig. 43.18 at point B and taking moments about point D will determine the minimum cable tension, H.

$$\sum M_D = wa\left(\frac{a}{2}\right) - HS = 0 \qquad 43.50$$

$$H = \frac{wa^2}{2S} \qquad 43.51$$

---

[19]The parabolic case can also be assumed with cables loaded only by their own weight (e.g., telephone and trolley wires), if both ends are at the same elevations and if the sag is no more than 10% of the distance between supports.

Since the load is vertical everywhere, the horizontal component of tension is constant everywhere in the cable. The tension, T_C, at any point C can be found by applying the equilibrium conditions to the cable segment BC.

$$T_{C,x} = H = \frac{wa^2}{2S} \qquad 43.52$$

$$T_{C,y} = wx \qquad 43.53$$

$$T_C = \sqrt{(T_{C,x})^2 + (T_{C,y})^2}$$
$$= w\sqrt{\left(\frac{a^2}{2S}\right)^2 + x^2} \qquad 43.54$$

The angle of the cable at any point is

$$\tan\theta = \frac{wx}{H} \qquad 43.55$$

The tension and angle are maximum at the supports.

If the lowest sag point, point B, is used as the origin, the shape of the cable is

$$y(x) = \frac{wx^2}{2H} \qquad 43.56$$

The approximate length of the cable from the lowest point to the support (i.e., length BD) is

$$L \approx a\left[1 + \left(\frac{2}{3}\right)\left(\frac{S}{a}\right)^2 - \left(\frac{2}{5}\right)\left(\frac{S}{a}\right)^4\right] \qquad 43.57$$

Example 43.12

A pedestrian foot bridge has two suspension cables and a flexible floor weighing 28 lbf/ft. The span of the bridge is 100 ft. When the bridge is empty, the tension at point C is 1500 lbf. Assuming a parabolic shape, what is the maximum cable sag, S?

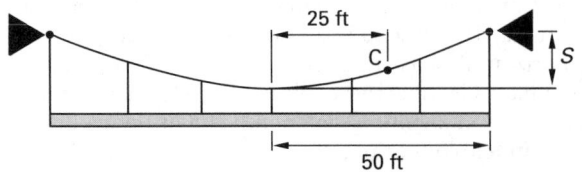

Solution

Since there are two cables, the floor weight per suspension cable is

$$w = \frac{28\,\dfrac{\text{lbf}}{\text{ft}}}{2} = 14\ \text{lbf/ft}$$

From Eq. 43.54,

$$T_C = w\sqrt{\left(\frac{a^2}{2S}\right)^2 + x^2}$$

$$1500\ \text{ft} = 14\ \frac{\text{lbf}}{\text{ft}}\sqrt{\left[\frac{(50\ \text{ft})^2}{2S}\right]^2 + (25\ \text{ft})^2}$$

$$S = 12\ \text{ft}$$

41. CABLES CARRYING DISTRIBUTED LOADS

An idealized tension cable with a distributed load is similar to a linkage made up of a very large number of axial members. The cable is an axial member in the sense that the internal tension acts tangentially to the cable everywhere.

Since the load is vertical everywhere, the horizontal component of cable tension is constant along the cable. The cable is horizontal at the point of lowest sag. There is no vertical tension component, and the cable tension is minimum. By similar reasoning, the cable tension is maximum at the supports.

Figure 43.19 illustrates a general cable with a distributed load. The shape of the cable will depend on the relative distribution of the load. A free-body diagram of segment BC is also shown.

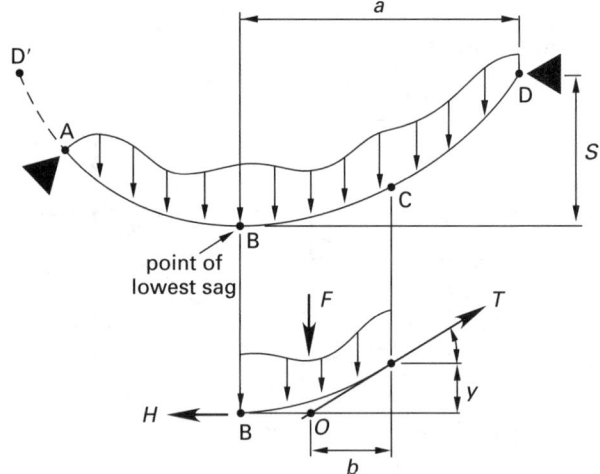

Figure 43.19 *Cable with Distributed Load*

F is the resultant of the distributed load on segment BC, T is the cable tension at point C, and H is the tension at the point of lowest sag (i.e., the point of minimum tension). Since segment BC taken as a free body is a three-force member, the three forces (H, F, and T) must be concurrent to be in equilibrium. The horizontal component of tension can be found by taking moments about point C.

$$\sum M_C = Fb - Hy = 0 \qquad 43.58$$

$$H = \frac{Fb}{y} \qquad 43.59$$

Also, $\tan\theta = y/b$. Therefore,

$$H = \frac{F}{\tan\theta} \qquad 43.60$$

The basic equilibrium conditions can be applied to the free-body cable segment BC to determine the tension in the cable at point C.

$$\sum F_x = T_C\cos\theta - H = 0 \qquad 43.61$$

$$\sum F_y = T_C\sin\theta - F = 0 \qquad 43.62$$

The resultant tension at point C is

$$T_C = \sqrt{H^2 + F^2} \qquad 43.63$$

42. CATENARY CABLES

If the distributed load is constant along the length of the cable, as it is with a loose cable loaded by its own weight, the cable will have the shape of a *catenary*. A vertical axis catenary has a shape determined by Eq. 43.64, where c is a constant and *cosh* is the *hyperbolic cosine*. The quantity x/c is in radians.[20]

$$y(x) = c\cosh\left(\frac{x}{c}\right) \qquad 43.64$$

Referring to Fig. 43.20, the vertical distance, y, to any point C on the catenary is measured from a reference plane located a distance c below the point of greatest sag, point B. The distance c is known as the *parameter of the catenary*. Although the value of c establishes the location of the x-axis, the value of c does not correspond to any physical distance, nor is the reference plane the ground level.

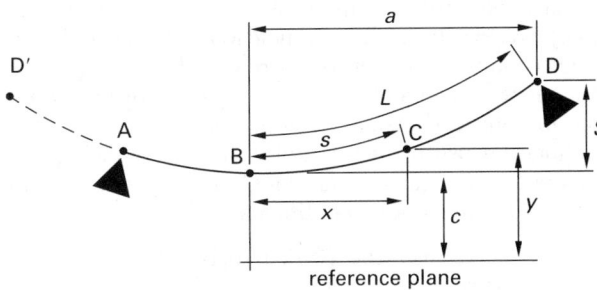

Figure 43.20 *Catenary Cable*

In order to define the cable shape and determine cable tensions, it is necessary to have enough information to calculate c. For example, if a and S are known,

Eq. 43.67 can be solved by trial and error for c.[21] Once c is known, the cable geometry and forces are determined by the remaining equations.

For any point C, the equations most useful in determining the shape of the catenary are

$$y = \sqrt{s^2 + c^2} = c\left(\cosh\left(\frac{x}{c}\right)\right) \qquad 43.65$$

$$s = c\left(\sinh\left(\frac{x}{c}\right)\right) \qquad 43.66$$

$$\text{sag} = S = y_D - c = c\left(\cosh\left(\frac{a}{c}\right) - 1\right) \qquad 43.67$$

$$\tan\theta = \frac{s}{c} \qquad 43.68$$

The equations most useful in determining the cable tensions are

$$H = wc \qquad 43.69$$

$$F = ws \qquad 43.70$$

$$T = wy \qquad 43.71$$

$$\tan\theta = \frac{ws}{H} \qquad 43.72$$

$$\cos\theta = \frac{H}{T} \qquad 43.73$$

Example 43.13

A cable 100 m long is loaded by its own weight. The maximum sag is 25 m and the supports are on the same level. What is the distance between the supports?

Solution

Since the two supports are on the same level, the cable length, L, between the point of maximum sag and support D is half of the total length.

$$L = \frac{100\,\text{m}}{2} = 50\,\text{m}$$

Writing Eqs. 43.65 and 43.67 for point D (with $S = 25$ m),

$$y_D = c + S = \sqrt{L^2 + c^2}$$

$$c + 25\,\text{m} = \sqrt{(50\,\text{m})^2 + c^2}$$

$$c = 37.5\,\text{m}$$

Substituting a for x and $L = 50$ for s in Eq. 43.66,

$$s = c\left(\sinh\left(\frac{x}{c}\right)\right)$$

$$50 = (37.5)\left(\sinh\left(\frac{a}{37.5}\right)\right)$$

$$a = 41.2\,\text{m}$$

The distance between supports is

$$2a = (2)(41.2\,\text{m}) = 82.4\,\text{m}$$

---

[20]In order to use Eqs. 43.64 through 43.67, you must reset your calculator from degrees to radians.

[21]Because obtaining the solution may require trial and error, it will be advantageous to assume a parabolic shape if the cable is taut. (See Ftn. 19.) The error will generally be small.

43. CABLES WITH ENDS AT DIFFERENT ELEVATIONS

A cable will be asymmetrical if its ends are at different elevations. In some cases, as shown in Fig. 43.21, the cable segment will not include the lowest point B. However, if the location of the theoretical lowest point can be derived, the positions and elevations of the cable supports will not affect the analysis. The same procedure is used in proceeding from the theoretical point B to either support. In fact, once the theoretical shape of a cable has been determined, the supports can be relocated anywhere along the cable line without affecting the equilibrium of the supported segment.

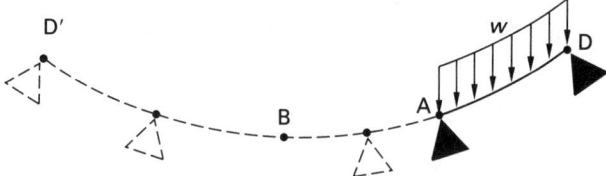

Figure 43.21 *Asymmetrical Segment of Symmetrical Cable*

44. TWO-DIMENSIONAL MECHANISMS

A two-dimensional *mechanism (machine)* is a nonrigid structure. Although parts of the mechanism move, the relationships between forces in the mechanism can be determined by statics. In order to determine an unknown force, one or more of the mechanism components must be considered as a free body. All input forces and reactions must be included on this free body. In general, the resultant force on such a free body will not be in the direction of the member.

Several free bodies may be needed for complicated mechanisms. Sign conventions of acting and reacting forces must be strictly adhered to when determining the effect of one component on another.

Example 43.14

A 70 N·m couple is applied to the mechanism shown. All connections are frictionless hinges. What are the x- and y-components of the reactions at B?

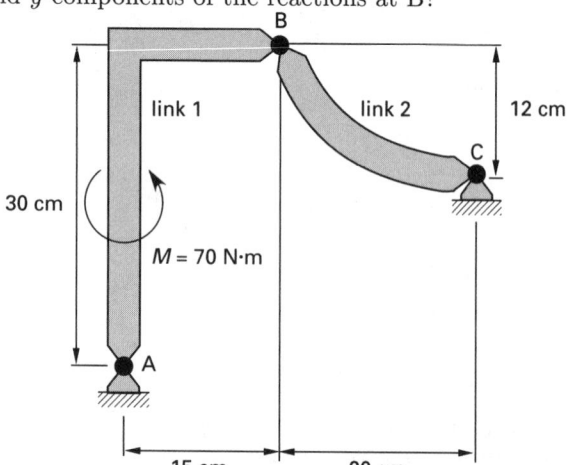

Solution

Isolate links 1 and 2 and draw their free bodies.

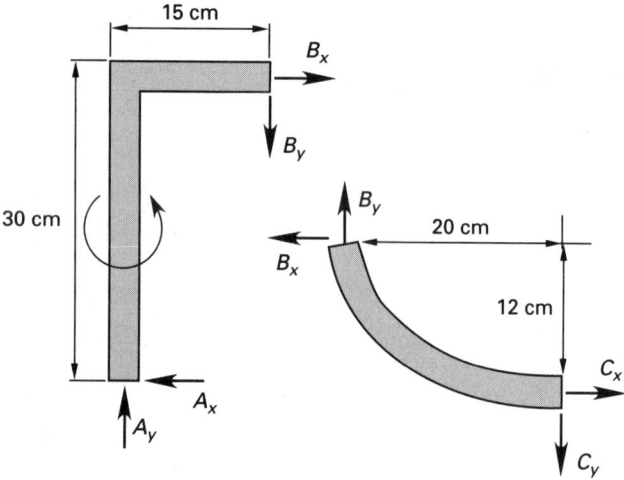

Assume clockwise moments are positive. Take moments about point A on link 1.

$$\sum M_{\rm A} = B_x(0.3\ {\rm m}) + B_y(0.15\ {\rm m}) - 70\ {\rm N{\cdot}m} = 0$$

Assume clockwise moments are positive. Take moments about point C on link 2.

$$\sum M_{\rm C} = B_y(0.20\ {\rm m}) - B_x(0.12\ {\rm m}) = 0$$

Solving these two equations simultaneously determines the force at joint B.

$$B_x = 179\ {\rm N}$$
$$B_y = 108\ {\rm N}$$

45. EQUILIBRIUM IN THREE DIMENSIONS

The basic equilibrium equations can be used with vector algebra to solve a three-dimensional statics problem. When a manual calculation is required, however, it is often more convenient to write the equilibrium equations for one orthogonal direction at a time, thereby avoiding the use of vector notation and reducing the problem to two dimensions. The following method can be used to analyze a three-dimensional structure.

step 1: Establish the $(0,0,0)$ origin for the structure.

step 2: Determine the (x, y, z) coordinates of all load and reaction points.

step 3: Determine the x-, y-, and z-components of all loads and reactions. This is accomplished by using direction cosines calculated from the (x, y, z) coordinates.

step 4: Draw a *coordinate free body* of the structure for each of the three coordinate axes. Include only forces, reactions, and moments that affect the coordinate free body.

step 5: Apply the basic two-dimensional equilibrium equations.

Example 43.15

Beam AC is supported at point A by a frictionless ball joint and at points B and C by cables. A l00 lbf load is applied vertically to point C, and a 180 lbf load is applied horizontally to point B. What are the cable tensions T_1 and T_2?

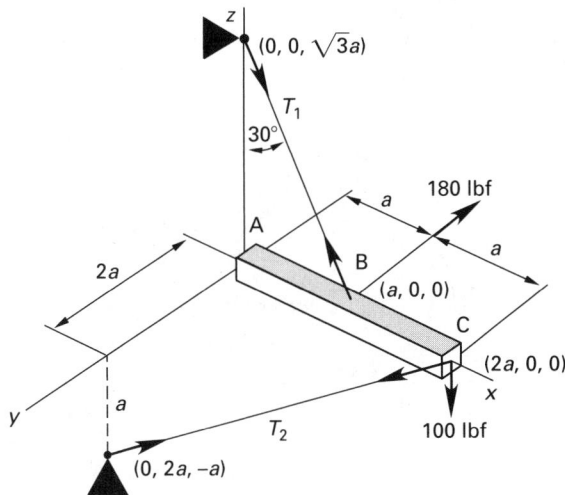

Solution

steps 1 and 2: Point A has already been established as the origin. The locations of all support and load points are shown on the illustration.

step 3: By inspection, for the 180 lbf horizontal load at point B,

$$F_x = 0$$
$$F_y = -180 \text{ lbf}$$
$$F_z = 0$$

By inspection, for the 100 lbf vertical load at point C,

$$F_x = 0$$
$$F_y = 0$$
$$F_z = -100 \text{ lbf}$$

The length of cable 1 is

$$L_1 = \sqrt{(a-0)^2 + (0-0)^2 + (0-\sqrt{3}\,a)^2}$$
$$= 2a$$

The direction cosines of the force from cable 1 at point B are

$$\cos\theta_x = \frac{d_x}{L_1} = \frac{0-a}{2a} = -0.5$$

$$\cos\theta_y = \frac{d_y}{L_1} = \frac{0-0}{2a} = 0$$

$$\cos\theta_z = \frac{d_z}{L_1} = \frac{\sqrt{3}\,a - 0}{2a} = 0.866$$

Therefore, the components of the tension in cable 1 are

$$T_{1,x} = -0.5\,T_1$$
$$T_{1,y} = 0$$
$$T_{1,z} = 0.866\,T_1$$

Similarly, for cable 2,

$$L_2 = \sqrt{(2a-0)^2 + (0-2a)^2 + [0-(-a)]^2}$$
$$= 3a$$

The direction cosines for the force from cable 2 at point C are

$$\cos\theta_x = \frac{d_x}{L_2} = \frac{0-2a}{3a} = -0.667$$

$$\cos\theta_y = \frac{d_y}{L_2} = \frac{2a-0}{3a} = 0.667$$

$$\cos\theta_z = \frac{d_z}{L_2} = \frac{-a-0}{3a} = -0.333$$

Therefore, the components of the tension in cable 2 are

$$T_{2,x} = -0.667\,T_2$$
$$T_{2,y} = 0.667\,T_2$$
$$T_{2,z} = -0.333\,T_2$$

step 4: The three coordinate free-body diagrams are

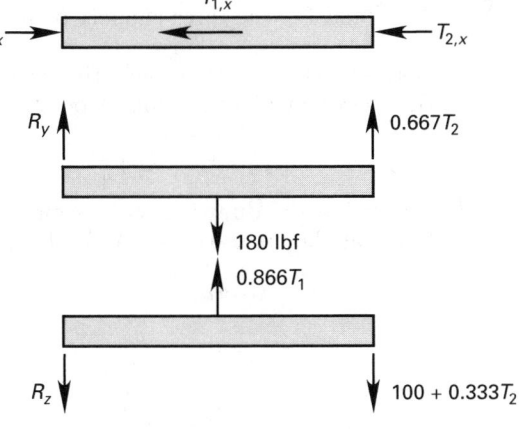

step 5: Tension T_2 can be found by taking moments about point A on the y-coordinate free body.

$$\sum M_A = (0.667 T_2)(2a) - (180 \text{ lbf})(a) = 0$$

$$T_2 = 135 \text{ lbf}$$

Tension T_1 can be found by taking moments about point A on the z-coordinate free body.

$$\sum M_A = (0.866\, T_1)(a) - (0.333)(135 \text{ lbf})(2a)$$
$$- (100 \text{ lbf})(2a) = 0$$

$$T_1 = 335 \text{ lbf}$$

46. TRIPODS

A *tripod* is a simple three-dimensional truss (frame) that consists of three axial members. One end of each member is connected at the *apex* of the tripod, while the other ends are attached to the supports. All connections are assumed to allow free rotation in all directions.

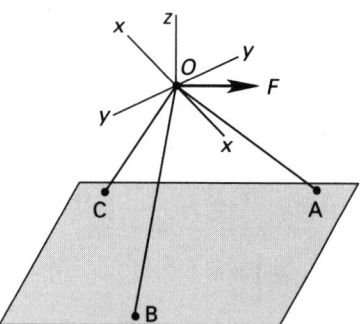

Figure 43.22 *Tripod*

The general solution procedure given in the preceding section can be made more specific for tripods.

step 1: Establish the apex as the origin.

step 2: Determine the x-, y-, and z-components of the force applied to the apex.

step 3: Determine the (x, y, z) coordinates of points A, B, and C—the three support points.

step 4: Determine the length of each tripod leg from the coordinates of the support points.

$$L = \sqrt{x^2 + y^2 + z^2} \qquad 43.74$$

step 5: Determine the direction cosines for the leg forces at the apex. For leg A, for example,

$$\cos \theta_{A,x} = \frac{x_A}{L} \qquad 43.75$$

$$\cos \theta_{A,y} = \frac{y_A}{L} \qquad 43.76$$

$$\cos \theta_{A,z} = \frac{z_A}{L} \qquad 43.77$$

step 6: Write the x-, y-, and z-components of each leg force in terms of the direction cosines. For leg A, for example,

$$F_{A,x} = F_A \cos \theta_{A,x} \qquad 43.78$$

$$F_{A,y} = F_A \cos \theta_{A,y} \qquad 43.79$$

$$F_{A,z} = F_A \cos \theta_{A,z} \qquad 43.80$$

step 7: Write the three sum-of-forces equilibrium equations for the apex.

$$F_{A,x} + F_{B,x} + F_{C,x} + F_x = 0 \qquad 43.81$$

$$F_{A,y} + F_{B,y} + F_{C,y} + F_y = 0 \qquad 43.82$$

$$F_{A,z} + F_{B,z} + F_{C,z} + F_z = 0 \qquad 43.83$$

Example 43.16

Determine the force in each leg of the tripod.

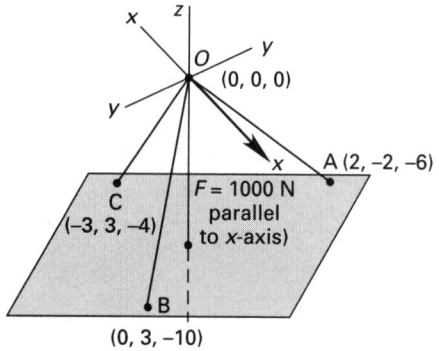

Solution

step 1: The origin is at the apex.

step 2: By inspection, $F_x = +1000$ N. All other components are zero.

steps 3 and 4: The direction cosines for the tripod legs have been calculated and are presented in the following table.

member	x^2	y^2	z^2	L^2	L	$\cos \theta_x$	$\cos \theta_y$	$\cos \theta_z$
OA	4	4	36	44	6.63	0.3015	−0.3015	−0.9046
OB	0	9	100	109	10.44	0.0	0.2874	−0.9579
OC	9	9	16	34	5.83	−0.5146	0.5146	−0.6861

steps 5 and 6: The equilibrium equations are

$$0.3015\, F_A + \qquad 0\, F_B - 0.5146\, F_C + 1000 = 0$$
$$- 0.3015\, F_A + 0.2874\, F_B + 0.5146\, F_C \qquad = 0$$
$$- 0.9046\, F_A - 0.9579\, F_B - 0.6861\, F_C \qquad = 0$$

The solution to these simultaneous equations is

$$F_A = +1531 \text{ N} \quad (\text{T})$$

$$F_B = -3480 \text{ N} \quad (\text{C})$$

$$F_C = +2841 \text{ N} \quad (\text{T})$$

PRACTICE PROBLEMS

1. Two towers are located on level ground 100 ft (30 m) apart. They support a transmission line with a mass of 2 lbm/ft (3 kg/m). The midpoint sag is 10 ft (3 m). What are the (a) midpoint tension and (b) maximum tension in the transmission line? (c) If the maximum tension is 500 lbf (2200 N), what is the sag in the cable?

2. Two legs of a tripod are mounted on a vertical wall. Both legs are horizontal. The apex is 12 distance units from the wall. The right leg is 13.4 units long. The wall mounting points are 10 units apart. A third leg is mounted on the wall 6 units to the left of the right upper leg and and 9 units below the two top legs. A vertical downward load of 200 is supported at the apex. What are the reactions at the three mounting points?

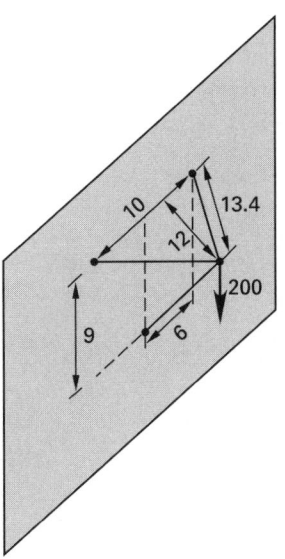

3. The ideal truss shown is supported by a pinned connection at point D and a roller connection at point C. Loads are applied at points A and F. What is the force in member DE?

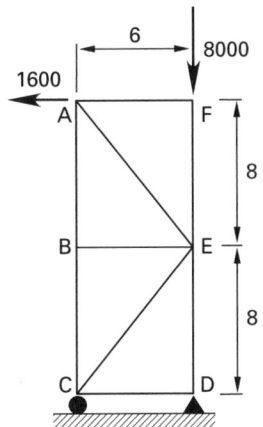

4. A pin-connected tripod is loaded at the apex by a horizontal force of 1200 as shown. Find the magnitudes of the forces in each of the three legs.

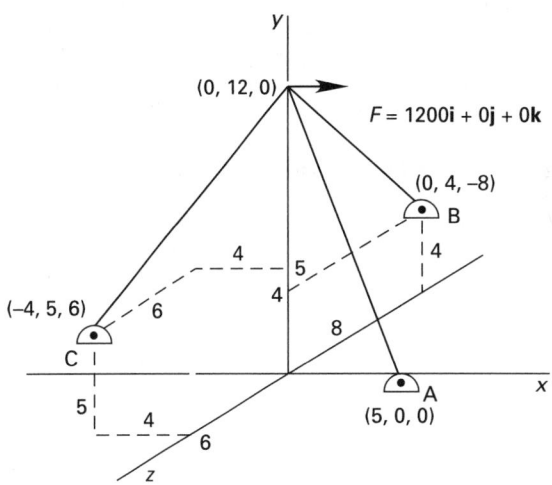

5. A truss is loaded by forces of 4000 at each upper connection point and forces of 60,000 at each lower connection point as shown. What are the forces in members (a) DE and (b) HJ?

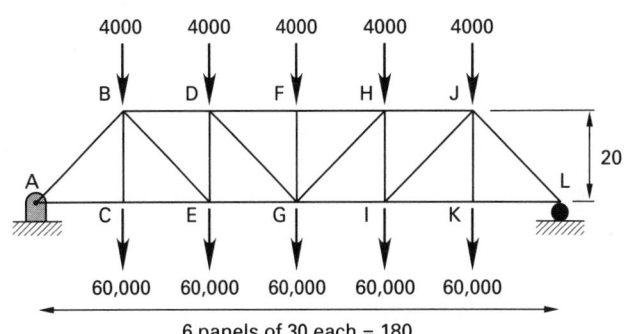

6. The rigid rod AO is supported by guy wires BO and CO, as shown. (Points A, B, and C are all in the same vertical plane.) Vertical and horizontal forces are 12,000 and 6000, respectively, as carried at the end of the rod. What are the x-, y-, and z-components of the reactions at points A, B, and C?

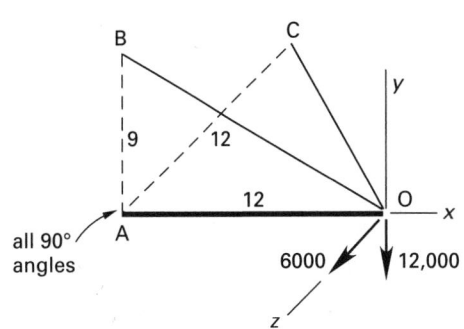

7. When the temperature is 70°F (21.11°C), sections of steel railroad rail are welded end to end to form a continuous, horizontal track exactly 1 mi long (1.6 km). Both ends of the track are constrained by preexisting installed sections of rail. Before the 1 mi section of track can be nailed to the ties, however, the sun warms it to a uniform temperature of 99.14°F (37.30°C). Laborers watch in amazement as the rail pops up in the middle and takes on a parabolic shape. The laborers prop the rail up (so that it doesn't buckle over) while they take souvenir pictures. How high off the ground is the midpoint of the hot rail?

44 Indeterminate Statics

1. Introduction to Indeterminate Statics 44-1
2. Degree of Indeterminacy 44-1
3. Indeterminate Beams 44-1
4. Review of Elastic Deformation 44-1
5. Consistent Deformation Method 44-2
6. Superposition Method 44-4
7. Three-Moment Equation 44-5
8. Fixed-End Moments 44-6
9. Indeterminate Trusses 44-6

Nomenclature

A	area	ft^2	m^2
d	distance	ft	m
E	modulus of elasticity	lbf/ft^2	Pa
F	force	lbf	N
I	area moment of inertia	ft^4	m^4
L	length	ft	m
M	moment	ft-lbf	N·m
R	reaction	lbf	N
S	force	lbf	N
T	temperature	°F	°C
u	force	lbf	N
w	distributed load	lbf/ft	N/m
y'	slope	ft/ft	m/m

Symbols

α	coefficient of thermal expansion	1/°F	1/°C
δ	deformation	ft	m
θ	angle	deg	rad

Subscripts

c	concrete
o	original
st	steel

1. INTRODUCTION TO INDETERMINATE STATICS

A structure that is *statically indeterminate* is one for which the equations of statics are not sufficient to determine all reactions, moments, and internal forces. Additional formulas involving deflection are required to completely determine these variables.

Although there are many configurations of statically indeterminate structures, this chapter is primarily concerned with beams on more than two supports, trusses with more members than are required for rigidity, and miscellaneous composite structures.

2. DEGREE OF INDETERMINACY

The *degree of indeterminacy* (*degree of redundancy*) is equal to the number of reactions or members that would have to be removed in order to make the structure statically determinate. For example, a two-span beam on three simple supports is indeterminate (redundant) to the first degree. The degree of indeterminacy of a pin-connected truss is given by Eq. 44.1.

$$\begin{matrix} \text{degree of} \\ \text{indeterminacy} \end{matrix} = 3 + \begin{matrix} \text{no. of} \\ \text{members} \end{matrix} - \left(2 \times \begin{matrix} \text{no. of} \\ \text{joints} \end{matrix}\right) \quad 44.1$$

3. INDETERMINATE BEAMS

Three common configurations of beams can easily be recognized as being statically indeterminate. These are the *continuous beam*, *propped cantilever beam*, and *fixed-end beam* illustrated in Fig. 44.1.

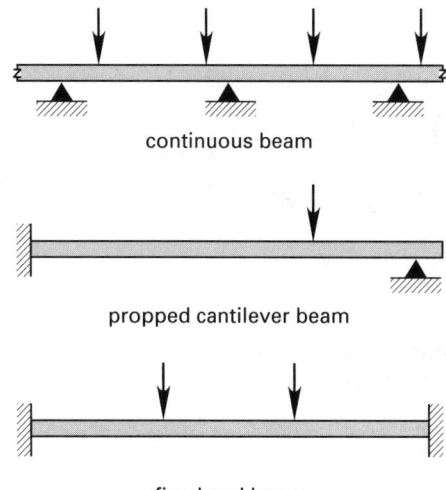

continuous beam

propped cantilever beam

fixed-end beam

Figure 44.1 *Types of Indeterminate Beams*

4. REVIEW OF ELASTIC DEFORMATION[1]

When an axial force, F, acts on an object with length L, cross-sectional area A, and modulus of elasticity E, the deformation[2] is

$$\delta = \frac{FL}{AE} \quad 44.2$$

---

[1]This subject is covered in greater detail in Chap. 49.
[2]The terms *deformation* and *elongation* are often used interchangeably in this context.

When an object with initial length L_o and coefficient of thermal expansion α experiences a temperature change of ΔT degrees, the deformation is

$$\delta = \alpha L_o \Delta T \qquad 44.3$$

5. CONSISTENT DEFORMATION METHOD

The *consistent deformation method*, also known as the *compatibility method*, is one of the methods of solving indeterminate problems. This method is simple to learn and to apply. First, geometry is used to develop a relationship between the deflections of two different members (or for one member at two locations) in the structure. Then, the deflection equations for the two different members at a common point are written and equated, since the deformations must be the same at a common point. This method is illustrated by the following examples.

Example 44.1

A pile is constructed of concrete with a steel jacket. What are the forces in the steel and concrete if a load F is applied? Assume the end caps are rigid and the steel-concrete bond is perfect.

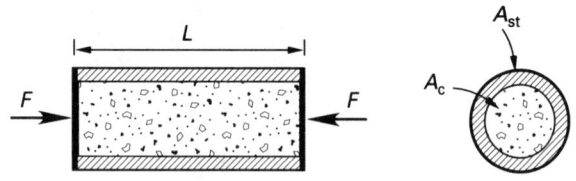

Solution

Let F_c and F_{st} be the loads carried by the concrete and steel, respectively. Then,

$$F_c + F_{st} = F$$

The deformation of the steel is

$$\delta_{st} = \frac{F_{st} L}{A_{st} E_{st}}$$

Similarly, the deflection of the concrete is

$$\delta_c = \frac{F_c L}{A_c E_c}$$

But, $\delta_c = \delta_{st}$ since the bonding is perfect. Therefore,

$$\frac{F_c L}{A_c E_c} - \frac{F_{st} L}{A_{st} E_{st}} = 0$$

The first and last equations are solved simultaneously for F_c and F_{st}.

$$F_c = \frac{F}{1 + \dfrac{A_{st} E_{st}}{A_c E_c}}$$

$$F_{st} = \frac{F}{1 + \dfrac{A_c E_c}{A_{st} E_{st}}}$$

Example 44.2

A uniform bar is clamped at both ends and the axial load applied near one of the supports. What are the reactions?

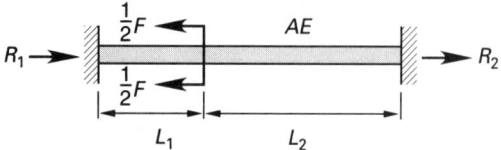

Solution

The first required equation is

$$R_1 + R_2 = F$$

The shortening of section 1 due to the reaction R_1 is

$$\delta_1 = \frac{-R_1 L_1}{AE}$$

The elongation of section 2 due to the reaction R_2 is

$$\delta_2 = \frac{R_2 L_2}{AE}$$

However, the bar is continuous, so $\delta_1 = -\delta_2$. Therefore,

$$R_1 L_1 = R_2 L_2$$

The first and last equations are solved simultaneously to find R_1 and R_2.

$$R_1 = \frac{F}{1 + \dfrac{L_1}{L_2}}$$

$$R_2 = \frac{F}{1 + \dfrac{L_2}{L_1}}$$

Example 44.3

The non-uniform bar shown is clamped at both ends. What are the reactions if a temperature change of ΔT is experienced?

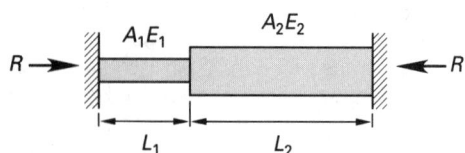

Solution

The thermal deformations of sections 1 and 2 can be calculated directly.

$$\delta_1 = \alpha_1 L_1 \Delta T$$

$$\delta_2 = \alpha_2 L_2 \Delta T$$

The total deformation is $\delta = \delta_1 + \delta_2$. However, the deformation can also be calculated from the principles of mechanics.

$$\delta = \frac{RL_1}{A_1 E_1} + \frac{RL_2}{A_2 E_2}$$

These equations can be solved directly for R.

$$R = \frac{(\alpha_1 L_1 + \alpha_2 L_2)\Delta T}{\dfrac{L_1}{A_1 E_1} + \dfrac{L_2}{A_2 E_2}}$$

Example 44.4

The beam shown is supported by dissimilar members. What are the forces in the members? Assume the bar is rigid and remains horizontal.[3] The beam's mass is insignificant.

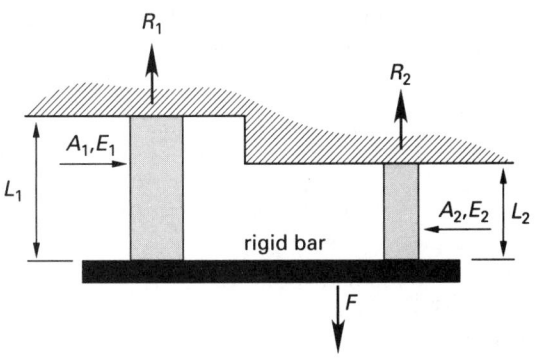

Solution

The required equilibrium condition is

$$R_1 + R_2 = F$$

The elongations of the two tension members are

$$\delta_1 = \frac{R_1 L_1}{A_1 E_1}$$

$$\delta_2 = \frac{R_2 L_2}{A_2 E_2}$$

Since the horizontal bar remains horizontal, $\delta_1 = \delta_2$.

$$\frac{R_1 L_1}{A_1 E_1} = \frac{R_2 L_2}{A_2 E_2}$$

---

[3]This example is easily solved by summing moments about a point on the horizontal beam.

The first and last equations are solved simultaneously to find R_1 and R_2.

$$R_1 = \frac{F}{1 + \dfrac{L_1 A_2 E_2}{L_2 A_1 E_1}}$$

$$R_2 = \frac{F}{1 + \dfrac{L_2 A_1 E_1}{L_1 A_2 E_2}}$$

Example 44.5

The beam shown is supported by dissimilar members. The bar is rigid but is not constrained to remain horizontal. The beam's mass is insignificant. What are the reactions in the vertical members?

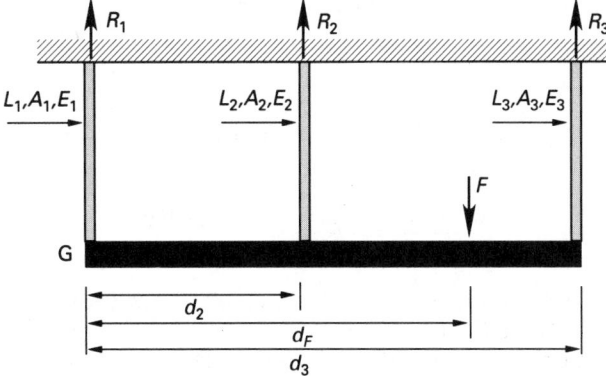

Solution

The forces in the supports are R_1, R_2, and R_3. Any of these may be tensile (positive) or compressive (negative).

$$R_1 + R_2 + R_3 = F$$

The changes in length are

$$\delta_1 = \frac{R_1 L_1}{A_1 E_1}$$

$$\delta_2 = \frac{R_2 L_2}{A_2 E_2}$$

$$\delta_3 = \frac{R_3 L_3}{A_3 E_3}$$

Since the bar is rigid, the deflections will be proportional to the distance from point G.

$$\delta_2 = \delta_1 + \left(\frac{d_2}{d_3}\right)(\delta_3 - \delta_1)$$

Moments can be summed about point G to give a third equation.

$$M_G = R_3 d_3 + R_2 d_2 - F d_F = 0$$

Example 44.6

A load is supported by three tension members. Find the forces in the three members.

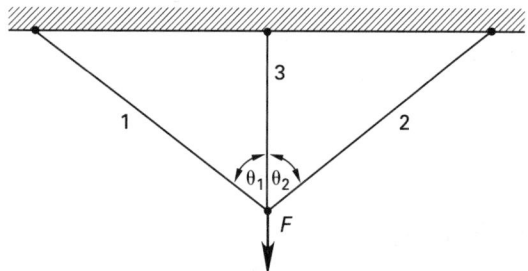

Solution

The equilibrium requirement is

$$F_{1y} + F_3 + F_{2y} = F$$
$$F_1 \cos\theta_1 + F_3 + F_2 \cos\theta_2 = F$$

Assuming the elongations are small compared to the member lengths, the angles θ_1 and θ_2 are unchanged. Then,

$$\frac{F_1 L_1 \cos\theta_1}{A_1 E_1} = \frac{F_3 L_3}{A_3 E_3} = \frac{F_2 L_2 \cos\theta_2}{A_2 E_2}$$

These equations can be solved simultaneously to find F_1, F_2, and F_3. (It may be necessary to work with the x-components of the deflections in order to find a third equation.)

6. SUPERPOSITION METHOD

Two-span (three-support) beams and propped cantilevers are indeterminate to the first degree. Their reactions can be determined from a variation of the consistent deformation procedure known as the *superposition method*.[4] This method requires finding the deflection with one or more supports removed and then satisfying the known conditions.

step 1: Remove enough redundant supports to reduce the structure to a statically determinate condition.

step 2: Calculate the deflections at the previous locations of redundant supports. Use consistent sign conventions.

step 3: Apply each redundant support as a load, and find the deflections at the redundant support points as functions of the redundant support forces.

---

[4]Superposition can also be used with higher-order indeterminate problems. However, the simultaneous equations that must be solved may make superposition unattractive for manual calculations.

step 4: Use superposition to combine (i.e., add) the deflections due to the actual loads and the redundant support loads. The total deflections must agree with the known deflections (usually zero) at the redundant support points.

Example 44.7

A propped cantilever is loaded by a concentrated force at midspan. Determine the reaction, S, at the prop.

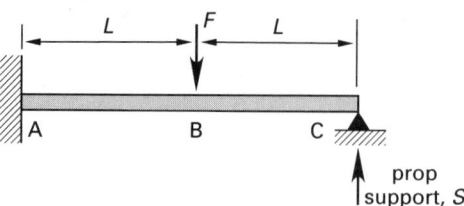

Solution

Start by removing the unknown prop reaction at point C. The cantilever beam is then statically determinate. The deflection and slope at point B can be found or derived from the beam equations.

$$\text{deflection:} \quad \delta_{\text{B}} = \frac{-FL^3}{3EI}$$

$$\text{slope:} \quad y'_{\text{B}} = \frac{-FL^2}{2EI}$$

The slope remains constant to the right of point B. Therefore, the deflection at point C due to the load at point B is

$$\delta_{\text{C},F} = \delta_{\text{B}} + y'_{\text{B}} L$$
$$= \frac{-5FL^3}{6EI}$$

The upward deflection at the cantilever tip due to the prop support, S, alone is

$$\delta_{\text{C},S} = \frac{S(2L)^3}{3EI} = \frac{8SL^3}{3EI}$$

Now, it is known that the actual deflection at point C is zero (the boundary condition). Therefore, the prop support, S, can be determined as a function of the applied load.

$$\delta_{\text{C},S} + \delta_{\text{C},F} = 0$$
$$\frac{8SL^3}{3EI} - \frac{5FL^3}{6EI} = 0$$
$$S = \frac{5F}{16}$$

7. THREE-MOMENT EQUATION

A *continuous beam* has two or more spans (i.e., three or more supports) and is statically indeterminate. The *three-moment equation* is a method of determining the reactions on continuous beams. It relates the moments at any three adjacent supports. The three-moment method can be used with a two-span beam to directly find all three reactions.

When a beam has more than two spans, the equation must be used with three adjacent supports at a time, starting with a support whose moment is known. (The moment is known to be zero at a simply supported end. For a cantilever end, the moment depends only on the loads on the cantilever portion.)

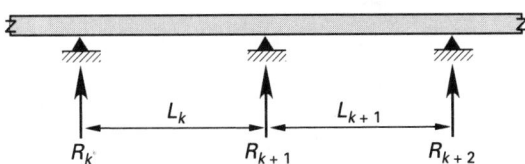

Figure 44.2 *Portion of a Continuous Beam*

In its most general form, the three-moment equation is applicable to beams with non-uniform cross sections. In Eq. 44.4, I_k is the moment of inertia of span k.

$$\frac{M_k L_k}{I_k} + (2M_{k+1})\left(\frac{L_k}{I_k} + \frac{L_{k+1}}{I_{k+1}}\right) + \frac{M_{k+2}L_{k+1}}{I_{k+1}}$$
$$= (-6)\left(\frac{A_k a}{I_k L_k} + \frac{A_{k+1} b}{I_{k+1} L_{k+1}}\right) \quad 44.4$$

Equation 44.4 uses the following special nomenclature.

a distance from the left support to the centroid of the moment diagram on the left span

b distance from the right support to the centroid of the moment diagram on the right span

I_k the moment of inertia of the open span between supports k and $k+1$

L_k length of the span between supports k and $k+1$

M_k bending moment at support k

A_k area of moment diagram between supports k and $k+1$, assuming that the span is simply and independently supported

The products Aa and Ab are known as *first moments of the areas*. It is convenient to derive simplified expressions for Aa and Ab for commonly encountered configurations. Several are presented in Fig. 44.3.

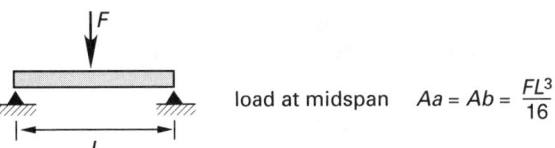

load at midspan $Aa = Ab = \dfrac{FL^3}{16}$

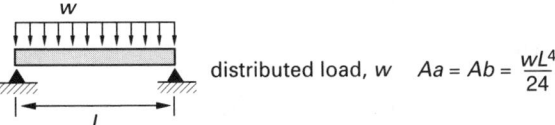

distributed load, w $Aa = Ab = \dfrac{wL^4}{24}$

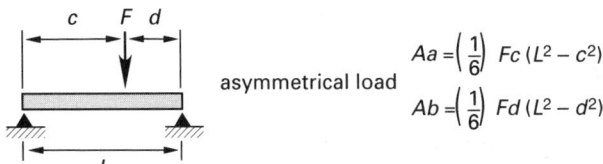

asymmetrical load

$Aa = \left(\dfrac{1}{6}\right) Fc\,(L^2 - c^2)$

$Ab = \left(\dfrac{1}{6}\right) Fd\,(L^2 - d^2)$

Figure 44.3 *Simplified Three-Moment Equation Terms*

For beams with uniform cross sections, the moment of inertia terms can be eliminated.

$$M_k L_k + (2M_{k+1})(L_k + L_{k+1}) + M_{k+2}L_{k+1}$$
$$= (-6)\left(\frac{A_k a}{L_k} + \frac{A_{k+1} b}{L_{k+1}}\right) \quad 44.5$$

Example 44.8

Find the four reactions supporting the beam.

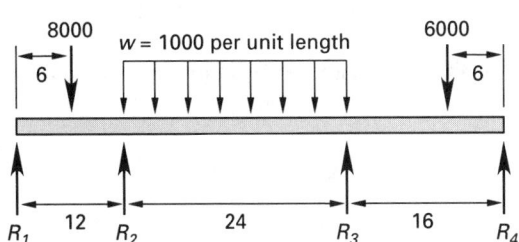

Solution

Spans 1 and 2:

Since the three-moment method can be applied to only two spans at a time, work first with the left and middle spans (spans 1 and 2).

From Fig. 44.3, the quantities $A_1 a$ and $A_2 b$ are

$$A_1 a = \frac{FL^3}{16} = \frac{(8000)(12)^3}{16} = 864{,}000$$

$$A_2 b = \frac{wL^4}{24} = \frac{(1000)(24)^4}{24} = 13{,}824{,}000$$

Statics

Since the left end of the beam is simply supported, M_1 is zero. Therefore, the three-moment equation becomes

$$(2M_2)(L_1 + L_2) + M_3L_2 = (-6)\left(\frac{A_1a}{L_1} + \frac{A_2b}{L_2}\right)$$

$$(2M_2)(12 + 24) + M_3(24) = (-6)\left(\frac{864,000}{12} + \frac{13,824,000}{24}\right)$$

After simplification,

$$3M_2 + M_3 = -162,000$$

Spans 2 and 3:

From the previous calculations,

$$A_2a = A_2b = 13,824,000$$

From Fig. 44.3 for the third span,

$$A_3b = \left(\tfrac{1}{6}\right)Fd(L^2 - d^2)$$
$$= \left(\tfrac{1}{6}\right)(6000)(6)\left[(16)^2 - (6)^2\right]$$
$$= 1,320,000$$

Since the right end is simply supported, $M_4 = 0$, and the three-moment equation is

$$M_2L_2 + (2M_3)(L_2 + L_3) = (-6)\left(\frac{A_2a}{L_2} + \frac{A_3b}{L_3}\right)$$

$$24M_2 + (2M_3)(24 + 16) = (-6)\left(\frac{13,824,000}{24} + \frac{1,320,000}{16}\right)$$

After simplifying,

$$0.3M_2 + M_3 = -49,388$$

There are two equations in two unknowns (M_2 and M_3). A simultaneous solution yields

$$M_2 = -41,708$$
$$M_3 = -36,875$$

Finding reactions:

M_2 can be written in terms of the loads and reactions to the left of support 2. Assuming clockwise moments are positive,

$$M_2 = 12\,R_1 - (6)(8000) = -41,708$$
$$R_1 = 524.3$$

Now that R_1 is known, moments can be taken from support 3 to the left.

$$M_3 = (36)(524.3) + 24R_2 - (30)(8000) - (12)(24,000)$$
$$= -36,875$$
$$R_2 = 19,677.1$$

Similarly, R_4 and R_3 can be determined by working from the right end to the left. Assuming counterclockwise moments are positive,

$$M_3 = 16R_4 - (10)(6000) = -36,875$$
$$R_4 = 1445.3$$
$$M_2 = (40)(1445) + 24R_3 - (34)(6000) - (12)(24,000)$$
$$= -41,708$$
$$R_3 = 16,353.3$$

Check:

It is a good idea to check for equilibrium in the vertical direction.

$$\sum \text{loads} = 8000 + 24,000 + 6000$$
$$= 38,000$$

$$\sum \text{reactions} = 524.3 + 19,677.1 + 1445.3 + 16,353.3$$
$$= 38,000$$

(*Alternate solution:* $R_1 = 4000$; $R_2 = 16,000$; $R_3 = 14,250$; $R_4 = 3750$.)

8. FIXED-END MOMENTS

When the end of a beam is constrained against rotation, it is said to be a *fixed end* (also known as a *built-in end*). The ends of fixed-end beams are constrained to remain horizontal. Cantilever beams have a single fixed end. Some beams, as illustrated in Fig. 44.1, have two fixed ends and are known as *fixed-end beams*.[5]

Fixed-end beams are inherently indeterminate. To reduce the work required to find end moments and reactions, tables and books of fixed-end moments are often used.

9. INDETERMINATE TRUSSES

It is possible to manually calculate the forces in all members of an indeterminate truss. However, due to the time required, it is preferable to limit such manual calculations to trusses that are indeterminate to the first degree. The following *dummy unit load method* can be used to solve trusses with a single redundant member.

step 1: Draw the truss twice. Omit the redundant member on both trusses. (There may be a choice of redundant members.)

step 2: Load the first truss (which is now determinate) with the actual loads.

step 3: Calculate the forces, S, in all of the members. Assign a positive sign to tensile forces.

[5]The definition is loose. The term *fixed-end beam* can also be used to mean any indeterminate beam with at least one built-in end (e.g., a propped cantilever).

step 4: Load the second truss with two unit forces acting collinearly toward each other along the line of the redundant member.

step 5: Calculate the force, u, in each of the members.

step 6: Calculate the force in the redundant member from Eq. 44.6.

$$S_{\text{redundant}} = \frac{-\sum \left(\dfrac{SuL}{AE}\right)}{\sum \left(\dfrac{u^2L}{AE}\right)} \qquad 44.6$$

If AE is the same for all members,

$$S_{\text{redundant}} = \frac{-\sum SuL}{\sum u^2 L} \qquad 44.7$$

The true force in member j of the truss is

$$F_{j,\text{true}} = S_j + (S_{\text{redundant}})\, u_j \qquad 44.8$$

Example 44.9

Find the force in members BC and BD. $AE = 1$ for all members except for CB, which is 2, and AD, which is 1.5.

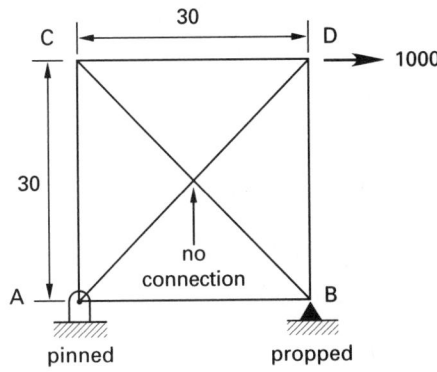

Solution

The two trusses are shown appropriately loaded.

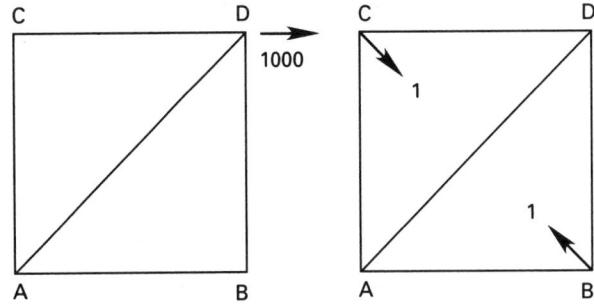

member	L	AE	S	u	$\dfrac{SuL}{AE}$	$\dfrac{u^2L}{AE}$
AB	30	1	0	−0.707 (C)	0	15
BD	30	1	−1000 (C)	−0.707	21,210	15
DC	30	1	0	−0.707	0	15
CA	30	1	0	−0.707	0	15
CB	42.43	2	0	1.0	0	21.22
AD	42.43	1.5	1414 (T)	1.0	39,997	28.29
					61,207	109.51

From Eq. 44.6,

$$S_{\text{BC}} = \frac{-61,207}{109.51}$$
$$= -558.9 \ (\text{C})$$

From Eq. 44.8,

$$F_{\text{BD,true}} = -1000 + (-558.9)(-0.707)$$
$$= -604.9 \ (\text{C})$$

45

Engineering Materials

1. Characteristics of Metals	45-1
2. Unified Numbering System	45-1
3. Blast Furnace Iron Production	45-2
4. Oxygen and Bessemer Processes	45-3
5. Open Hearth Process	45-4
6. Electric Arc Furnace	45-5
7. Advanced Steel-Making Processes	45-5
8. Steel and Alloy Steel Grades	45-5
9. Tool Steel	45-7
10. Stainless Steel	45-7
11. Cast Iron	45-9
12. Wrought Iron	45-10
13. Production of Aluminum	45-11
14. Properties of Aluminum	45-11
15. Aluminum Alloys	45-11
16. Production of Copper	45-12
17. Alloys of Copper	45-12
18. Nickel and Its Alloys	45-13
19. Refractory Metals	45-13
20. Natural Polymers	45-13
21. Degree of Polymerization	45-14
22. Synthetic Polymers	45-14
23. Fluoropolymers	45-15
24. Elastomeric Compounds	45-15
25. Polymer Crystallinity	45-16
26. Thermosetting and Thermoplastic Polymers	45-16
27. Wood	45-16
28. Glass	45-17
29. Ceramics	45-17
30. Abrasives	45-19
31. Modern Composite Materials	45-19
Practice Problems	45-20

Nomenclature

A	area	in^2	m^2
D	diameter	in	m
DP	degree of polymerization	–	–
E	modulus of elasticity	lbf/in^2	Pa
m	mass	lbm	kg
MC	moisture content	%	%
MW	molecular weight	–	–
p	pressure	lbf/in^2	Pa
S_H	circumferential stress	lbf/in^2	Pa
T	temperature	°F	°C

Subscripts

c	concrete

1. CHARACTERISTICS OF METALS

Metals are the most frequently used materials in engineering design. Steel is the most prevalent engineering metal because of the abundance of iron ore, simplicity of production, low cost, and predictable performance. However, other metals play equally important parts in specific products.

Most metals are characterized by the properties in Table 45.1.

Table 45.1 *Properties of Most Metals and Alloys*

high thermal conductivity (low thermal resistance)
high electrical conductivity (low electrical resistance)
high chemical reactivity[a]
high strength
high ductility[b]
high density
high radiation resistance
highly magnetic (ferrous alloys)
optically opaque
electromagnetically opaque

[a]Some alloys, such as stainless steel, are more resistant to chemical attack than pure metals.
[b]Brittle metals, such as some cast irons, are not ductile.

Metallurgy is the subject that encompasses the procurement and production of metals. *Extractive metallurgy* is the subject that covers the refinement of pure metals from their ores.

2. UNIFIED NUMBERING SYSTEM

The Unified Numbering System (UNS) was introduced in the mid-1970s to provide a consistent identification of metals and alloys for use throughout the world. The UNS designation consists of one of seventeen single uppercase letter prefixes followed by five digits. Many of the letters are suggestive of the family of metals, as Table 45.2 indicates.

For each UNS designation, there is a specific percentage range of critical alloying elements. However, the UNS designation is a description, not a specification. Specifications are administered by the American Society of Testing and Materials (ASTM) and similar organizations. The UNS designation refers only to the major alloying and residual elements. It is not an exact specification. One manufacturer may produce an alloy

Table 45.2 UNS Alloy Prefixes

A	aluminum
C	copper
E	rare-earth metals
F	cast irons
G	AISI and SAE carbon and alloy steels
H	AISI and SAE H-steels
J	cast steels (except tool steels)
K	miscellaneous steels and ferrous alloys
L	low-melting metals
M	miscellaneous nonferrous metals
N	nickel
P	precious metals
R	reactive and refractory metals
S	heat- and corrosion-resistant steels (stainless and valve steels and superalloys)
T	tool steels (wrought and cast)
W	welding filler metals
Z	zinc

in the middle of the UNS ranges, while another manufacturer may operate at the low or high end of the ranges.[1] The presence of small amounts of residual elements, directionality due to manufacturing processes (e.g., rolling), and heat treatments are also not part of the specification. Furthermore, the UNS designations are not always sufficiently unique to differentiate between two existing products. Additional specifications or trade names are still required in those instances.

A cross-reference index is generally needed to convert older designations to the UNS designation.[2] However, for many stainless steels, the first three digits are the same as the AISI numbering system. Straight AISI type 304 is written simply as UNS S30400. The last two digits may be used to designate some differentiating characteristic. For example, stainless 304L, with a maximum of 0.03% carbon, is designated as S30403.[3] The UNS designations for other families (e.g., aluminum, copper, and nickel) also incorporate some or all of the common designations in use prior to the UNS.

3. BLAST FURNACE IRON PRODUCTION

Iron (chemical symbol Fe) is obtained from its oxides Fe_2O_3 (*hematite*, 69.9% iron) and, to a lesser extent, Fe_3O_4 (*magnetite*, 72.4% iron).[4] Only about 50% of

---

[1]In addition to economically lowering the carbon content, *argon-oxygen-decarburization* (AOD) during refining has made it possible to control nitrogen and other alloying ingredients precisely. The percentage of expensive alloying ingredients (e.g., molybdenum in stainless steels) will generally be at the low ends of the allowable ranges.

[2]SAE and ASTM jointly publish *Metals and Alloys in the Unified Numbering System*, which includes a cross-reference.

[3]Since straight type 304 has a maximum of 0.08 carbon, some engineers prefer to write S30408 for uniformity.

[4]Since an additional roasting process is needed to remove the sulfur, iron pyrite, FeS_2, is not used in iron production, despite its abundance. Other lower-grade ores, such as $FeCO_3$ (siderite, 48.3% iron) and $Fe_2O_3\cdot n[H_2O]$ (limonite, 60 to 65% iron), are used only in the absence of better ores.

iron ore consists of iron oxides, the remainder being the gangue. *Gangue* is the earth and stone mixed with the iron oxides.

The process used to reduce iron oxides to pure iron takes place in a *blast furnace*. The furnace is charged with alternate layers of iron ore, coke, and limestone in the approximate ratio of 4:2:1, respectively.[5] The limestone serves as a flux for the gangue and the coke ash, enabling the molten gangue and impurities to be drawn off as *slag*.

A traditional blast furnace is shown in Fig. 45.1. The top of the furnace is provided with a pair of conical bells for loading the charge (when open) and limiting the escape of gases (when closed). High-temperature air is injected through nozzles around the periphery of the lower portion of the furnace. These openings are known as *tuyères*.[6]

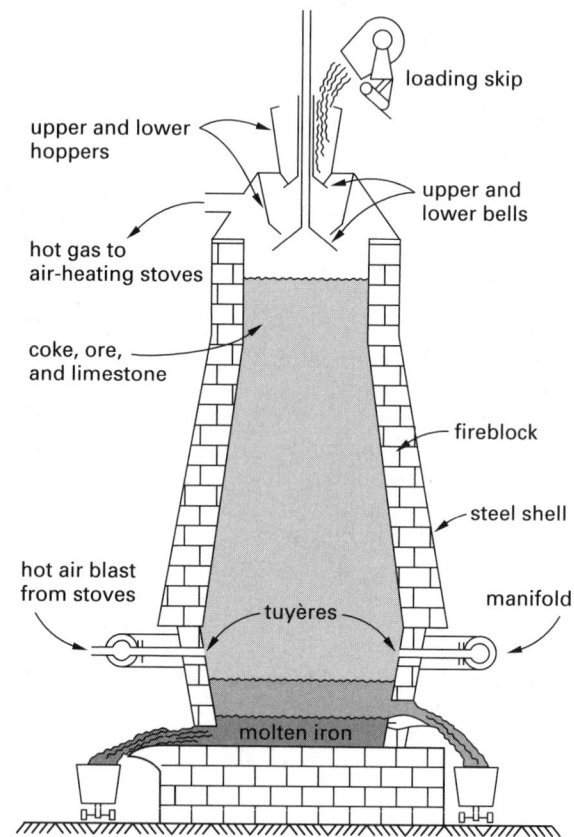

Figure 45.1 Blast Furnace

The hot combustion air is produced in preheaters (stoves) that adjoin the blast furnace. Generally, four stoves are provided for each furnace. Each stove is heated in rotation by burning the carbon monoxide-rich furnace gases. Cold air enters one stove while the

---

[5]*Coke* is coal that has been previously burned in an oxygen-poor environment. The remaining carbonaceous material has a high-combustion energy content. (Clean-air legislation has had a severe impact on coke making.)

[6]*Tuyère* is pronounced twee-yer and too-ur.

remaining stoves are being heated. The air is heated to 1000 to 1300°F (550 to 700°C) before being injected into the blast furnace.

The injected air oxidizes the coke, producing heat and large amounts of carbon monoxide. The carbon monoxide rises to the top of the furnace and, at a temperature of approximately 600°F (300°C), reduces the iron oxide to FeO. The following chemical reactions describe the production of FeO.

$$
\begin{aligned}
C + O_2 &\rightarrow CO_2 \\
CO_2 + C &\rightarrow 2CO \\
2C + O_2 &\rightarrow 2CO \\
3Fe_2O_3 + CO &\rightarrow 2Fe_3O_4 + CO_2 \\
Fe_3O_4 + CO &\rightarrow 3FeO + CO_2
\end{aligned}
$$

As the reduction process continues, the FeO temperature drops down to 1300 to 1500°F (700 to 800°C). The FeO is reduced to a spongy mass of pure iron by the carbon monoxide.

$$FeO + CO \rightarrow Fe + CO_2$$

The molten iron then drops into a region where the temperature is 1500 to 2500°F (800 to 1400°C). The iron becomes saturated with carbides and free carbon. The absorbed carbon lowers the melting point of the iron from approximately 2800°F (1550°C) to approximately 2100°F (1150°C) so that it runs as a liquid to the bottom of the furnace.

The slag melts at approximately the same temperature as the iron but, being less dense, floats on the liquid iron. This allows the slag and iron to be drawn off separately. The iron usually goes in a liquid state to a subsequent refinement process, but may be allowed to cool in molds (forming blocks of iron known as *pigs*). The slag is discarded in a *slag heap*.

Since liquid iron is an excellent solvent, pig iron contains all of the minerals that are not fluxed away by the liquid limestone. The approximate composition of pig iron is 3 to 4% carbon, 1 to 3% silicon, 0.1 to 2% phosphorus, 0.5 to 2% manganese, and 0.01 to 0.1% sulfur. The actual composition will depend on the gangue elements. Calcium, magnesium, and aluminum oxides are fluxed out by the molten limestone and appear in the slag.

The subsequent process that the pig iron undergoes depends on the desired end product (i.e., the desired carbon content). This is shown in Fig. 45.2. The Bessemer, oxygen, open hearth, and electric furnace processes are used to produce steel.

4. OXYGEN AND BESSEMER PROCESSES

Pig iron is saturated with carbon and contains other impurities. The *Bessemer* and *oxygen processes* (also known as the *dissolved oxygen process*, *L-D process*, and the *Linz-Donawitz process*) are used to reduce the

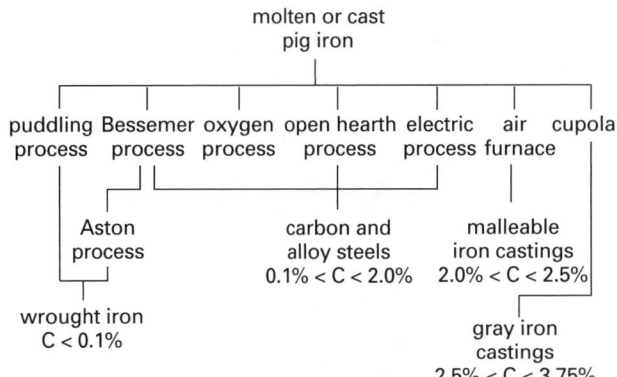

Figure 45.2 *Methods of Refining Pig Iron*

carbon content and purify the iron.[7] The conceptual differences between the oxygen and Bessemer processes shown in Table 45.3 are minor. However, the production and economic advantages of the oxygen process are considerable.

Table 45.3 *Comparison Between Oxygen and Bessemer Processes*

characteristic	oxygen	Bessemer
crucible lining	basic	acidic
oxidizer	pure oxygen	air
oxidizer supply	lance above	nozzles below
reaction time	25 minutes	10–15 minutes
typical charge size	100 English tons	25 English tons
temperature	higher	high
percent scrap steel in charge	up to 30%	10–20%

The chemical refinement takes place in a pear-shaped steel crucible (a *converter*) lined with refractory material. The crucible is filled with a *bath* of molten pig iron at approximately 2200°F (1200°C), steel scrap, and lime. In the oxygen process, a water-cooled oxygen *lance* is lowered to within several feet (approximately a meter) of the bath surface.

High-pressure oxygen flows through the lance at high velocity, pushing aside the molten slag and exposing the molten iron. (In the Bessemer process, air is injected from below.) The silicon and manganese impurities are oxidized first, causing a temperature rise to approximately 3500°F (1900°C). Carbon is oxidized at the higher temperatures. Since the reaction is violent, the bath churns and circulates naturally.

The impurities are completely oxidized in approximately 15 to 25 minutes (see Table 45.3), although loading and unloading extends the cycle time to approximately one hour. Since the refinement also eliminates beneficial elements, measured amounts of carbon, manganese, and other alloying ingredients are subsequently added at the end of the refinement process to obtain the desired steel grade.

---

[7]Linz and Donawitz are the two Austrian towns in which the oxygen process was perfected.

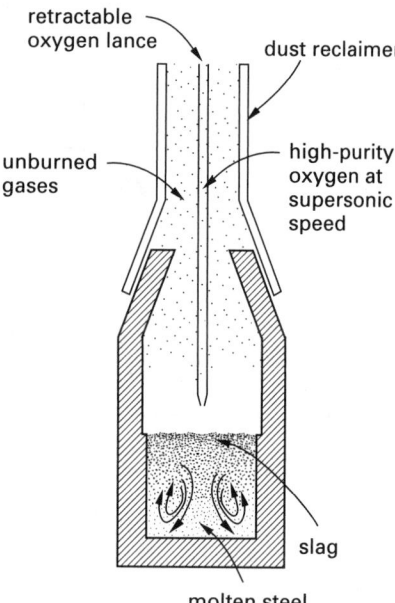

Figure 45.3 *Oxygen Process Crucible*

Because a Bessemer reaction proceeds rapidly and because the process cannot be interrupted for analyses, the Bessemer process is considered to be crude. The progress of refinement is judged by the color and length of flame that issues from the mouth of the crucible. Also, phosphorus and sulfur are not affected by the Bessemer process, making further refinement often necessary. This process, when used at all, produces steel for less critical grades of sheet, wire, and pipe.

Bessemer steel that is subsequently refined in an open hearth process to remove sulfur, phosphorus, and iron oxide is known as *duplexed steel*.

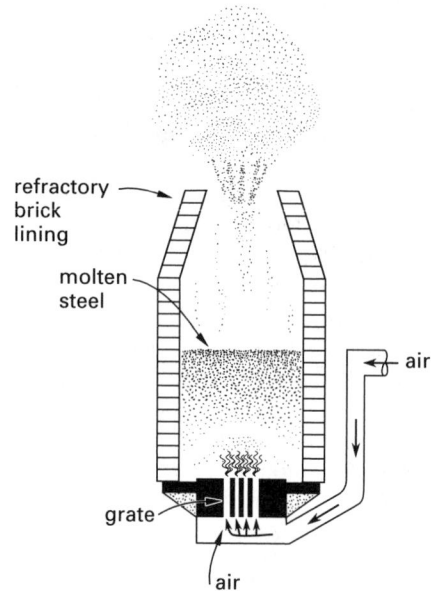

Figure 45.4 *Bessemer Process Converter*

Compared to the open hearth process, smaller batches of steel have traditionally been produced in the Bessemer process. However, larger vessels are now in use with the oxygen process. Also, since the refinement process proceeds faster than in an open hearth, the steel production rates are approximately the same.

5. OPEN HEARTH PROCESS

The major advantages of the open hearth process are control and flexibility in charge composition. A charge of scrap steel and limestone is placed in a fairly shallow furnace receptacle of large area.[8] Heat is provided by burning carbon monoxide (CO) above the charge. When the charge is molten, liquid pig iron from the blast furnace is added. Iron oxide (in the form of mill scale or iron ore) is added as an oxidizer. Gaseous oxygen from a lance is frequently used to accelerate the refinement.

The combustion air and fuel enter the hearth from one side, and a long flame plays over the charge. The combustion gases are withdrawn on the opposite side and pass through brick *checker chambers*, which absorb some of the residual heat. The incoming air is preheated to approximately 1800°F (1000°C) by passing through these chambers. When the chambers cool, the flow direction of air and fuel is reversed.

Since the combustion products contain very little free oxygen, oxidation of the impurities depends on the oxygen in the iron oxide. The carbon burns off as carbon monoxide and carbon dioxide. The molten limestone slag provides a protective covering over the melt (to prevent excessive oxidation and nitrogen absorption) and fluxes away the sulfur, phosphorus, and silicon in the steel.

Open hearth refinement takes eight to twelve hours, with melting and refining each taking approximately half of the time. Continuous monitoring of the steel composition is possible, resulting in a high-grade steel. A typical final composition would be sulfur and phosphorus, less than 0.04% each; manganese, 0.05 to 0.35%; silicon, less than 0.01%; and carbon as desired. The refined molten steel batch is known as a *heat of steel*.

When the steel is sufficiently pure, alloying elements are added to achieve the desired steel grade and properties. For example, manganese is added as necessary to combine with the sulfur remaining in the steel. Iron sulfide weakens the steel, whereas manganese sulfide does not.

Deoxidizers (ferromanganese and ferrosilicon) are added to compensate for the remaining high iron oxide content. If the steel is not deoxidized, the oxide reacts with carbon during solidification and produces large amounts of carbon monoxide gas. This gas is trapped in the steel, producing many voids and leading to the characteristic

---

[8]The charge can be as much as 500 English tons.

appearance of such *rimmed steel*. The gas voids usually do not constitute a defect since they are closed by welding during subsequent hot-working manufacturing processes.

Aluminum can be added to the steel to produce *killed steel*. No gas at all is evolved in killed steel during solidification. *Semikilled steel* has some gas formation, but not as much as rimmed steel.

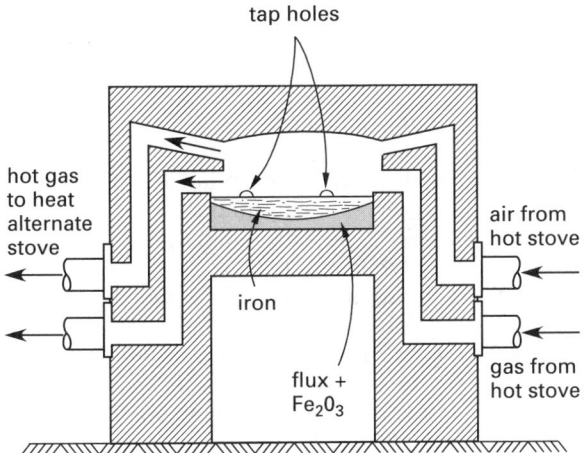

Figure 45.5 *Open Hearth Furnace*

6. ELECTRIC ARC FURNACE

Electric furnaces common to "minimills" utilizing electric arc or induction heating are used to produce tool and special alloy steels.[9] It is possible to produce a high-quality steel because air and gaseous fuels are not required, and the impurities they introduce are eliminated.

To further increase the quality and to reduce the expensive refining time, the charge is usually select scrap or *direct-reduced iron* rather than molten pig iron. As with the open hearth process, iron oxide is added as an oxidizing agent, and the composition is modified following refinement.

In an *electric arc furnace*, heat is generated by electrical arcs from three electrodes (for three-phase current) extending through the furnace wall down into the charge space. Although the electrode voltage is low (approximately 40 V), the current is high (approximately 12,000 A). Coils surround an *induction furnace*, and the heating is created from eddy current flowing within the melt.

The two major problems associated with electric arc furnaces are (1) steel contamination from trace metals present in the charging scrap, and (2) the formation of ionized nitrogen in the arc, a cause of undesirable hardening.

[9]The term "minimill" has become a misnomer. Electric arc furnaces can produce up to 130 tons/hr, not much less than the 200 to 300 tons/hr capacity of a standard basic oxygen furnace.

7. ADVANCED STEEL-MAKING PROCESSES

Air-quality legislation has had a severe impact on the production of coke. New steel-making technologies are being used to reduce or eliminate the need for coke entirely (as in *direct iron-making processes*). Existing blast furnaces can be retrofitted to use pulverized coal (*coal injection*). In new *reduced-coke processes*, coal, iron pellets and fines, and limestone are added to an already molten iron bath. Carbon from the coal combines with oxygen from the ore to produce carbon monoxide and molten iron. Oxygen is injected to burn some of the gas before it leaves the vessel.

8. STEEL AND ALLOY STEEL GRADES

The properties of steel can be adjusted by the addition of alloying ingredients. Some steels are basically mixtures of iron and carbon. Other steels are produced with a variety of ingredients.

The simplest and most common grades of steel belong to the group of *carbon steels*. Carbon is the primary non-iron element, although sulfur, phosphorus, and manganese can also be present. Carbon steel can be subcategorized into *plain carbon steel (nonsulfurized carbon steel), free-machining steel (resulfurized carbon steel)*, and *resulfurized and rephosphorized carbon steel*. Plain carbon steel is subcategorized into *low-carbon steel* (less than 0.30% carbon), *medium-carbon steel* (0.30 to 0.70% carbon), and *high-carbon steel* (0.70 to 1.40% carbon).

Low-carbon steels are used for wire, structural shapes, and screw machine parts. Medium-carbon steels are used for axles, gears, and similar parts requiring medium to high hardness and high strength. High-carbon steels are used for drills, cutting tools, and knives.

Low-alloy steels (containing less than 8.0% total alloying ingredients) include the majority of steel alloys but exclude the high-chromium content *corrosion-resistant (stainless) steels*. Generally, low-alloy steels will have higher strength (e.g., double the yield strength) of plain carbon steel. *Structural steel, high-strength steel*, and *ultrahigh-strength steel* are general types of low-alloy steel.[10]

High-alloy steels contain more than 8.0% total alloying ingredients.

Table 45.4 lists typical alloying ingredients and their effects on steel properties. The percentages represent typical values, not maximum solubilities.

[10]The *ultrahigh-strength steels*, also known as *maraging steels*, are very low-carbon (less than 0.03%) steels with 20 to 30% nickel and small amounts of cobalt, molybdenum, titanium, and aluminum. With precipitation hardening, ultimate tensile strengths up to 400,000 lbf/in² (2.8 GPa), yield strengths up to 250,000 lbf/in² (1.7 GPa), and elongations in excess of 10% are achieved. Maraging steels are used for rocket motor cases, aircraft and missile turbine housings, aircraft landing gears, and other applications requiring high strength, low weight, and toughness.

Materials

Table 45.4 *Steel Alloying Ingredients*

ingredient	range (%)	purpose
aluminum	–	deoxidation
boron	0.001–0.003	increase hardness
carbon	0.1–4.0	increase hardness and strength
chromium	0.5–2	increase hardness and strength
	4–18	increase corrosion resistance
copper	0.1–0.4	increase atmospheric corrosion resistance
iron sulfide	–	increase brittleness
manganese	0.23–0.4	reduce brittleness, combine with sulfur
	> 1.0	increase hardness
manganese sulfide	0.8–0.15 (S)	increase machinability
molybdenum	0.2–5	increase dynamic and high-temperature strength and hardness
nickel	2–5	increase toughness, increase hardness
	12–20	increase corrosion resistance
	> 30	reduce thermal expansion
phosphorus	0.04–0.15	increase hardness and corrosion resistance
silicon	0.2–0.7	increase strength
	2	increase spring steel strength
	1–5	improve magnetic properties
sulfur	–	(see *iron sulfide* and *manganese sulfide*)
titanium	–	fix carbon in inert particles; reduce martensitic hardness
tungsten	–	increase high-temperature hardness
vanadium	0.15	increase strength

Table 45.5 *AISI-SAE Steel Designations*

carbon steels

10XX	nonsulfurized carbon steel (plain-carbon)
11XX	resulfurized carbon steel (free-machining)
12XX	resulfurized and rephosphorized carbon steel

low-alloy steels

13XX	manganese 1.75
23XX	nickel 3.50
25XX	nickel 1.25, chromium 0.65
31XX	nickel 3.50, chromium 1.55
33XX	nickel 3.50, chromium 1.55
40XX	molybdenum 0.25
41XX	chromium 0.50 or 0.95, molybdenum 0.12 or 0.20
43XX	nickel 1.80, chromium 0.50 or 0.80, molybdenum 0.25
46XX	nickel 1.55 or 1.80, molybdenum 0.20 or 0.25
47XX	nickel 1.05, chromium 0.45, molybdenum 0.20
48XX	nickel 3.50, molybdenum 0.25
50XX	chromium 0.38 or 0.40
51XX	chromium 0.80, 0.90, 0.95, 1.00, or 1.05
5XXXX	chromium 0.50, 1.00, or 1.45, carbon 1.00
61XX	chromium 0.60, vanadium 0.10–0.15; or chromium 0.95, vanadium 0.15
86XX	nickel 0.55, chromium 0.50 or 0.65, molybdenum 0.20
87XX	nickel 0.55, chromium 0.50, molybdenum 0.25
92XX	manganese 0.85, silicon 2.00
93XX	nickel 3.25, chromium 1.20, molybdenum 0.12
98XX	nickel 1.00, chromium 0.80, molybdenum 0.25

heat- and corrosion-resistant steels

2XX	chromium-nickel-manganese (nonhardenable, austenitic, nonmagnetic)
3XX	chromium-nickel (nonhardenable, austenitic, nonmagnetic)
4XX	chromium (hardenable, martensitic, magnetic)
4XX	chromium (hardenable, ferritic, magnetic)
5XX	chromium (low-chromium, heat-resisting)

Since steel properties are dependent on composition, steels are designated by composition. Table 45.5 shows the AISI-SAE four-digit designations for typical steels and alloys.[11] The first two digits designate the type of steel; the last two digits designate the percentage of carbon in hundredths of a percent. (For example, AISI steel 1035 is a plain carbon steel with 0.35% carbon. This is also referred to as "35-point carbon" steel.) A number following an alloying ingredient is the nominal percentage of that ingredient.

---

[11]The abbreviations stand for the American Iron and Steel Institute and the Society of Automotive Engineers.

Also, an optional capital letter may be added as a prefix to designate the manufacturing process (A, acid Bessemer; B, basic Bessemer; C, basic open hearth; CB, either B or C at steel mill option; O, basic oxygen).[12]

9. TOOL STEEL

Each grade of tool steel is designed for a specific purpose, and as such, there are few generalizations that can be made about tool steel. Each tool steel exhibits its own blend of the three main performance criteria: toughness, wear resistance, and *hot hardness*.[13]

Some of the few generalizations possible are listed as follows.

- An increase in carbon content increases wear resistance and reduces toughness.

- An increase in wear resistance reduces toughness.

- Hot hardness is independent of toughness.

- Hot hardness is independent of carbon content.

Group A steels are air-hardened, medium-alloy cold-work tool steels. Air-hardening allows the tool to develop a homogeneous hardness throughout, without distortion. This hardness is achieved by large amounts of alloying elements and comes at the expense of wear resistance.

Group D steels are high-carbon, high-chromium tool steels suitable for cold-working applications. These steels are high in abrasion resistance but low in machinability and ductility. Some steels in this group are air hardened, while others are oil quenched. Typical uses are blanking and cold-forming punches.

Group H steels are hot-work tool steels, capable of being used in the 1100 to 2000°F (600 to 1100°C) range. They possess good wear resistance, hot hardness, shock resistance, and resistance to surface cracking. Carbon content is low, between 0.35% and 0.65%. This group is subdivided according to the three primary alloying ingredients: chromium, tungsten, or molybdenum. For example, a particular steel might be designated as a "chromium hot-work tool steel."

Group M steels are molybdenum high-speed steels. Properties are very similar to the group T steels, but group M steels are less expensive since one part molybdenum can replace two parts tungsten. For that reason, most high-speed steel in common use is produced from the M group. Cobalt is added in large percentages (5 to 12%) to increase high-temperature cutting efficiency in heavy-cutting (high-pressure cutting) applications.

Group O steels are oil-hardened, cold-work tool steels. These high-carbon steels use alloying elements to permit oil quenching of large tools and are sometimes referred to as *nondeforming steels*. Chromium, tungsten, and silicon are typical alloying elements.

Group S steels are shock-resistant tool steels. Toughness (not hardness) is the main characteristic, and either water or oil may be used for quenching. Group S steels contain chromium and tungsten as alloying ingredients. Typical uses are hot header dies, shear blades, and chipping chisels.

Group T steels are tungsten high-speed tool steels that maintain a sharp hard cutting edge at temperatures in excess of 1000°F (550°C). The famous 18-4-1 grade T1 (named after the percentages of tungsten, chromium, and vanadium, respectively) is part of this group. Increases in hot hardness are achieved by simultaneous increases in carbon and vanadium (the key ingredient in these tool steels) and special, multiple-step heat treatments.[14]

Group W steels are water-hardened tool steels. These are plain high-carbon steels (modified with small amounts of vanadium or chromium, resulting in high surface hardness but low hardenability). The combination of high surface hardness and ductile core makes group W steels ideal for rock drills, pneumatic tools, and cold header dies. The limitation on this tool steel group is the loss of hardness that begins at temperatures above 300°F (150°C) and is complete at 600°F (300°C).

Example 45.1

The composition of a group M tool steel is being formulated to replace the 18-4-1 group T steel. What are the percentages of alloying ingredients if two-thirds of the tungsten are to be replaced with molybdenum?

Solution

Since one part molybdenum replaces two parts tungsten, the alloy would be designated 6-6-4-1, representing 6% molybdenum, 6% tungsten, 4% chromium, and 1% vanadium.

10. STAINLESS STEEL

Adding chromium improves steel's corrosion resistance. Moderate corrosion resistance is obtained by adding 4 to 6% chromium to low-carbon steel. (Other elements, specifically less than 1% each of silicon and molybdenum, are also usually added.)

For superior corrosion resistance, larger amounts of chromium are needed. At a minimum level of 12%

---

[12]The electric furnace process may be designated by a C, D, or E prefix.

[13]The ability of a steel to resist softening at high temperatures is known as *hot hardness* and *red hardness*.

[14]For example, the 18-4-1 grade is heated to approximately 1050°F (550°C) for two hours, air cooled, and then heated again to the same temperature. The term *double-tempered steel* is used in reference to this process. Most heat treatments are more complex.

chromium, steel is *passivated* (i.e., an inert film of chromic oxide forms over the metal and inhibits further oxidation). The formation of this protective coating is the basis of the corrosion resistance of *stainless steel*.[15]

Passivity is enhanced by oxidizers and aeration but is reduced by abrasion that wears off the protective oxide coating. An increase in temperature may increase or decrease the passivity, depending on the abundance of oxygen.

Stainless steels are generally categorized into ferritic, martensitic (heat-treatable), austenitic, duplex, and high-alloy stainless steels.[16] Table 45.6 categorizes some of the more popular AISI grades of stainless steel.

Ferritic stainless steels, grouped with the AISI 400 series, contain more than 12 to 27% chromium. The body-centered cubic ferrite structure is stable (i.e., does not transform to austenite, a face-centered cubic structure) at all temperatures.

For this reason, ferritic steels cannot be hardened significantly. Since ferritic stainless steels contain no nickel, they are less expensive than the austenitic steels. Turbine blades are typical of the heat-resisting products manufactured from ferritic stainless steels.

The so-called *superferritics*, such as Alloy 2904C (S44735), Sea-Cure (UNS S44660), and Alloy 2903, are highly resistant to chloride pitting and crevice corrosion. Superferritics have been incorporated into marine tubing and heat exchangers for power plant condensers. Like all ferritics, however, superferritics experience embrittlement above 885°F (475°C).

The *martensitic (heat-treatable) stainless steels* (also part of the AISI 400 series) contain no nickel and differ from ferritic stainless steels primarily in higher carbon contents. Cutlery and surgical instruments are typical applications requiring both corrosion resistance and hardness.

The *austenitic stainless steels* are commonly used for general corrosive applications. The stability of the austenite (a face-centered cubic structure) depends primarily on 4 to 22% nickel as an alloying ingredient. The basic composition is approximately 18% chromium and 8% nickel, hence the term "18 to 8 type."

The so-called *superaustenitics*, such as the 317 and 316L series, achieve superior corrosion resistance by adding more molybdenum and nitrogen, respectively. AISI

---

[15]Stainless steels are corrosion resistant in oxidizing environments. In reducing environments (such as with exposure to hydrochloric and other halide acids and salts), the steel will corrode.

[16]There is a fifth category, that of *precipitation-hardened stainless steels*, widely used in the aircraft industry. (Precipitation hardening is also known as *age hardening*.) These steels have not been given specific AISI numbers. UNS designations for precipitation-hardened stainless steels include S13800, S15500, S17400, and S17700.

The *sigma phase* structure that appears at very high chromium levels (e.g., 24 to 50%) is usually undesirable in stainless steels because it reduces corrosion resistance and impact strength. A notable exception is in the manufacture of automobile engine valves.

Table 45.6 *Characteristics of Common Stainless Steels*

	AISI type	application
martensitic (hardenable by heat treatment)	410 420 440C	general purpose
ferritic (more corrosion resistant than martensitic; not hardenable by heat treatment)	405 430 446	hardenable by heat treatment
		hardenable by cold working
austenitic (best corrosion resistance; hardenable only by cold working)	201 202 301 302 302B 304 304L 310 316 321 254SMo (UNS S31254) Alloy 904L (UNS N08904) AL-6XN (UNS N08367)	for elevated-temperature service / modified for welding / superior corrosion resistance
duplex: austenitic-ferritic (tough, weldable, superior corrosion resistance)	Alloy 2205 (UNS S31803) Ferrallium 225 (UNS S32550) 329 (UNS S32900) 7-Mo Plus (S32950) 44LN (UNS S31200) DP-3 (UNS S31260) 2304 (UNS S2304) SAF 2507 (UNS S32750) Code Plus Two	

type 317 (UNS S31700) contains 3 to 4% molybdenum. Variants of type 317L (UNS S317XX) contain up to 5% molybdenum. Variants of type 316L achieve superior corrosion resistance by adding 10 to 14% nitrogen. "6-Mo" superaustenitics with approximately 6% molybdenum, 20% chromium, and 0.10% nitrogen are now well established, particularly in the chemical process industry. "7-Mo" alloys probably represent the ultimate in corrosion resistance while still remaining commercially viable.

Table 45.7 *Typical Compositions of Stainless Steels*

element	ferritic	martensitic	austenitic
carbon	0.08–0.20%	0.15–1.2%	0.03–0.25%
manganese	1–1.5%	1%	2%
silicon	1%	1%	1–2%
chromium	11–27%	11.5–18%	16–26%
nickel	–	–	3.5–22%
phosphorus and sulfur	–	–	normal
molybdenum	–	–	some cases
titanium	–	–	some cases

Austenitic stainless steels are the most weldable of the stainless steels but are nevertheless susceptible to sensitization. They are nonmagnetic and are hardenable only by cold working. They can be polished to a mirror finish, which makes them useful in food-industry applications. Because of the nickel, they are more expensive than ferritic stainless steels.

Welding stainless steels is possible when proper welding rod alloys are used, but is difficult for several reasons:

- Stainless steels, particularly austenitic types, possess relatively low thermal conductivities and high coefficients of thermal expansion. The maintenance of a high temperature gradient (because the heat is not readily dissipated) and a high expansion increases the possibility of *weld bead cracking* (i.e., longitudinal cracking along the weld). Weld bead cracking can be minimized by welding at as low a temperature as possible.

- High temperature sensitizes the steel adjacent to welds, producing local chromium deficits. This phenomenon is known as *sensitization*, and the resulting corrosion is known as *weld decay*.

- Substantial grain growth occurs in ferritic steels, since there is no gamma-alpha transformation to keep grains small. Growth of grains is substantial at normal welding temperatures.

- When stainless steels cool from welding temperatures, martensite forms unless cooling is slowed down. The martensite makes the metal brittle and reduces its ductility.

Because of their nominal costs, austenitic AISI 304 and 316 are the most commonly used corrosion-resistant steels. However, their low strengths make them unsuitable for high-pressure applications. They are also susceptible to wear and galling, and they have limited resistance to localized corrosion and *stress corrosion cracking* (SCC), particularly *chloride stress corrosion cracking* (CSCC) above 130°F (54°C).[17] Alternatives include duplex and high-alloy austenitic stainless steels.

Second-generation *duplex stainless steels* are austenitic-ferritic stainless steels that have the toughness and weldability of austenitic stainless steels and yield strengths and corrosion/wear resistances greater than the 300 series.[18] Alloy 2205 is the most widely used. Types 2304 and 2507 (UNS S32304 and S32507, respectively) are third-generation duplex steels designed to reduce mill costs. With the highest percentages of nitrogen, molybdenum, and nickel of any duplex stainless steel, 2507 has been dubbed a "super-duplex" stainless steel.

---

[17]Chloride stress corrosion cracking is an important issue in heat exchangers for use with ocean and inland water.

[18]The first generation of duplex stainless steels developed in the 1930s, such as AISI type 329 (UNS S32900), lost much of their corrosion resistance after welding unless they were given a post-weld heat treatment. Second-generation stainless steels contain less carbon and 0.15 to 0.30% nitrogen and, when combined with proper welding technique, offer the same level of corrosion resistance as mill-annealed material.

The *high-alloy austenitic stainless steels* containing 22 to 28% chromium, 24 to 32% nickel, and 4 to 6% molybdenum provide superior corrosion resistance at lower cost than the nickel- and titanium-based alloys they replace.

11. CAST IRON

Cast iron is a general name given to a wide range of alloys containing iron, carbon, and silicon, and to a lesser extent, manganese, phosphorus, and sulfur. Generally, the carbon content will exceed 2%. The properties of cast iron depend on the amount of carbon present, as well as the form of the carbon (i.e., graphite or carbide).

Carbon in the form of carbide is stable only at low temperatures. (The carbide is said to be in a *metastable structure*.) At high temperatures, *graphitization* takes place according to the following reaction:

$$Fe_3C \rightarrow 3Fe + C \quad (graphite)$$

The most common type of cast iron is *gray cast iron*. The carbon in gray cast iron is in the form of graphite flakes. Graphite flakes are very soft and constitute points of weakness in the metal, which simultaneously improve machinability and decrease ductility. Gray cast iron is categorized into classes according to its tensile strength, as shown in Table 45.8. Compressive strength is three to five times the tensile strength.

Table 45.8 Classes of Gray Cast Iron[a]

class	minimum tensile strength lbf/in^2	minimum tensile strength MPa	tensile modulus of elasticity lbf/in^2	tensile modulus of elasticity GPa
20	20,000	138	10–14 $\times 10^6$	69–97
25	25,000	172	12–15 $\times 10^6$	83–104
30	30,000	207	13–16.5 $\times 10^6$	90–114
35	35,000	242	14.5–17 $\times 10^6$	100–117
40	40,000	276	16–20 $\times 10^6$	110–138
45	45,000	310	–	–
50	50,000	345	18.8–23 $\times 10^6$	130–159
60	60,000	414	20.4–23.5 $\times 10^6$	141–162

[a]ASTM specification A48

Magnesium and cerium can be added to improve the ductility of gray cast iron. The resulting *nodular cast iron* (also known as *ductile cast iron*) has the best tensile and yield strengths of all the cast irons. It also has good ductility (typically 5%) and machinability. Because of these properties, it is often used for automobile crankshafts.

White cast iron has been cooled quickly from a molten state. No graphite is produced from the cementite, and the carbon remains in the form of a carbide, Fe_3C.[19]

---

[19]White and gray cast irons get their names from the coloration at a fracture.

The carbide is hard and is the reason that white cast iron is difficult to machine. White cast iron is used primarily in the production of malleable cast iron.

Table 45.9 *Common Grades of Ductile Iron*[a]

class/grade	minimum tensile strength		minimum yield strength		elongation
	ksi	MPa	ksi	MPa	%
60-40-18[b]	60	410	40	280	18
65-45-12[c]	65	450	45	310	12
80-55-06[d]	80	550	55	380	6
100-70-03[e]	100	690	70	480	3
120-90-02[f]	120	830	90	620	2

(Multiply ksi by 6.895 to obtain MPa.)
[a] ASTM A-536-70
[b] May be annealed after casting.
[c] An as-cast grade.
[d] An as-cast grade with higher manganese content.
[e] Usually obtained by a normalizing heat treatment.
[f] Oil quenched and tempered to specified hardness.

Malleable cast iron is produced by reheating white cast iron to between 1500°F and 1850°F (800°C and 1000°C) for several days, followed by slow cooling. During this treatment, the carbide is partially converted to nodules of graphitic carbon known as *temper carbon*. The tensile strength is increased to approximately 55,000 lbf/in^2 (380 MPa), and the elongation at fracture increases to approximately 18%.

Mottled cast iron contains both cementite and graphite and is between white and gray cast irons in composition and performance.

Compacted graphitic iron (CGI) is a unique form of cast iron with worm-shaped graphite particles. The shape of the graphite particles gives CGI the best properties of both gray and ductile cast iron: twice the strength of gray cast iron and half the cost of aluminum. The higher strength permits thinner sections. (Some engine blocks are 25% lighter than gray iron castings.) Using computer-controlled refining, volume production of CGI with the consistency needed for commercial applications is now possible.

Silicon is the most important element affecting graphitization. The effects of various elements in cast iron are listed in Table 45.10. Most of the elements that increase hardness do so by promoting the formation of iron carbide.

12. WROUGHT IRON

Wrought iron is low-carbon (less than 0.1%) iron with small amounts (approximately 3%) of slag and gangue in the form of fibrous inclusions. It has good ductility and corrosion resistance. Prior to the use of steel, wrought iron was the most important structural metal.

Table 45.10 *Effects of Elements in Cast Iron*

element	effect
aluminum	deoxidizes molten cast iron
carbon	depending on form, affects machinability, ductility, and shrinkage
manganese	below 0.5%, reduces hardness by combining with sulfur; above 0.5%, increases hardness
phosphorus	increases fluidity and lowers melting temperature
silicon	below 3.25%, softens iron and increases ductility; above 3.25%, hardens iron; above 13%, increases acid and corrosion resistance
sulfur	increases hardness, sulfur is removed by addition of manganese

In the ancient *puddling process*, wrought iron is produced in a *reverberatory furnace* similar to the open hearth furnace.[20] The molten iron floats on a layer of iron oxide that provides the oxygen for removal of almost all of the carbon, sulfur, and manganese. The limestone flux combines with silicon and phosphorus to form slag.

As the iron becomes purer, its melting temperature increases to above the furnace temperature. Spongy masses of congealing iron and slag are collected on the ends of rods inserted into the pool of molten metal. These masses are then removed and forged or hammered to squeeze out most of the slag. The remaining product consists of slag-coated iron particles welded together by the forging processes. The deformed slag particles contribute to the fatigue resistance of wrought iron.

In the modern *Aston process (Byers-Aston process)*, pig metal is melted in a cupola and is then highly purified in a Bessemer converter.[21] Simultaneously and separately, molten slag is prepared in an open hearth furnace and transferred to a mixing ladle. The molten iron is poured into the cooler slag in the mixing ladle (a process known as *shotting*), where the iron rapidly solidifies and releases dissolved gases. The gases fracture the solidifying iron, and molten slag enters the fissures. The excess molten slag is poured off, and the metal mass is pressed and rolled into blooms, billets, and slabs to remove most of the interior slag.[22] The slabs are hot-rolled together to form larger pieces of wrought iron.

---

[20] A *reverberatory furnace* is a cavernous, brick-lined chamber. The metal and ore are melted by flames that play over the top of the melt.
[21] A *cupola* is a tall, open-top vertical stack lined with furnace brick. An air blast is introduced at the base. Heat is produced from the combustion of coke mixed with the ore.
[22] In the rolling mill industry, a *bloom* is a long piece having a cross section greater than approximately 6 in (15 cm) square. A *billet* is smaller than a bloom, with a cross section greater than approximately 1.5 in (4 cm). A *slab* has a minimum thickness of 1.5 in (4 cm) and width between 10 and 15 in (25 and 40 cm). The width of a slab is always at least three times the thickness.

13. PRODUCTION OF ALUMINUM

Aluminum is produced from *bauxite ore*, a mixture of hydroxides of aluminum ($Al_2O_3 \cdot nH_2O$) and oxides or iron, silicon, and titanium. Most of the nonrecycled aluminum produced today is recovered in an electrochemical process known as the *Bayer process*.[23]

In the Bayer process, the ore is crushed and ground into a fine powder. It is then treated with a hot solution of sodium hydroxide, producing a solution of sodium aluminate. The solution is drawn off into a separate tank, leaving the remaining ore constituents (known as *red mud* because of the iron coloration) as a solid deposit to be discarded.

$$Al(OH)_3 + NaOH \rightarrow NaAl(OH)_4$$

As the solution cools, aluminum hydroxide precipitates, leaving a sodium hydroxide solution. The aluminum hydroxide is collected in solid, crystalline form and baked to form *alumina* (aluminum oxide, Al_2O_3). Because of its high melting temperature (3720°F, 2050°C), alumina cannot be economically reduced in a furnace.

Final reduction (*smelting*) is accomplished through an electrolytic process using molten *cryolite* (Na_3AlF_6) as the electrolyte. Large carbon blocks act as anodes, and the carbon-lined steel tank acts as the cathode.

The aluminum oxide dissolves in the cryolite and is separated into molten aluminum and oxygen gas by the electric current. The aluminum collects in the bottom of the tank. Carbon dioxide is released at the anodes. The cryolite recomposes after decomposition and can be reused. (The following reactions disregard the cryolite.)

$$Al_2O_3 \rightarrow 2Al^{+++} + 3O^{--}$$
$$Al^{+++} + 3e^- \rightarrow Al$$
$$C + 2O^{--} \rightarrow CO_2 + 4e^-$$

14. PROPERTIES OF ALUMINUM

Aluminum satisfies applications requiring low weight, corrosion resistance, and good electrical and thermal conductivities. Its corrosion resistance derives from the oxide film that forms over the raw metal, inhibiting further oxidation. The primary disadvantages of aluminum are its cost and low strength.

In pure form, aluminum is soft, ductile, and not very strong. Copper, manganese, magnesium, and silicon can be added to increase its strength, but at the expense of other properties, primarily corrosion resistance.[24] Aluminum is hardened by the *precipitation hardening (age-hardening)* process.

---

[23]refined in a similar electrochemical process are magnesium, copper, zinc, and (to a lesser extent) gold and silver.
[24]One ingenious method of having both corrosion resistance and strength is to produce a composite material. *Alclad* is the name given to aluminum alloy that has a layer of pure aluminum bonded to the surface. The alloy provides the strength, and the pure aluminum provides the corrosion resistance.

The oxide coating that forms readily (particularly at high temperatures) on aluminum complicates welding. However, special processes that perform the welding under a blanket of inert gas (e.g., helium or argon) overcome this complication.[25]

15. ALUMINUM ALLOYS

Except for use in electrical work, most aluminum is alloyed with other elements, primarily copper, magnesium, and silicon.[26] Aluminum alloys are identified by a four-digit number and a letter suffix (e.g., 2014-T4). The number indicates the major alloying ingredient and chemical composition of the alloy, as determined from Table 45.11. The suffix indicates the condition of the alloy, as determined from Tables 45.12 and 45.13.

Table 45.11 Aluminum Designations

designation	major alloying ingredient
1XXX	commercially pure (99+%)
2XXX	copper
3XXX	manganese
4XXX	silicon
5XXX	magnesium
6XXX	magnesium and silicon
7XXX	zinc
8XXX	other

Silicon occurs as a normal impurity in aluminum, and in natural amounts (less than 0.4%), it has little effect on properties. If moderate quantities (above 3%) of silicon are added, the molten aluminum will have high fluidity, making it ideal for castings. Above 12%, silicon improves the hardness and wear resistance of the alloy. When combined with copper and magnesium (as Mg_2Si and $AlCuMgSi$) in the alloy, silicon improves age hardenability. Silicon has negligible effect on the corrosion resistance of aluminum.

Table 45.12 Conditions of Aluminum Alloys

letter suffix	meaning
F	as fabricated
O	soft (after annealing)
H	strain hardened (cold worked) temper
T	heat treated

Copper improves the age hardenability of aluminum, particularly in conjunction with silicon and magnesium. Thus, copper is a primary element in achieving high mechanical strength in aluminum alloys at elevated temperatures. Copper also increases the conductivity of aluminum, but decreases its corrosion resistance.

---

[25]TIG (tungsten-inert gas) and MIG (metal-inert gas) processes are commonly used to weld aluminum.
[26]*EC (electrical-conductor) grade aluminum* consists of approximately 99.45% aluminum.

Magnesium is highly soluble in aluminum and is used to increase strength by improving age hardenability. Magnesium improves corrosion resistance and may be added when exposure to saltwater is anticipated.

Some aluminum alloys can be work-hardened (e.g., 1100, 3003, 5052). The ductility of these alloys decreases as strength is increased through working. Most aluminum alloys (e.g., 2014, 2017, 2024, 6061), however, must be precipitation-hardened.[27] The decrease in ductility with increased strength through heat treatment is small or nonexistent.

The letter suffixes H and T are followed by numbers that provide additional detail about the type of hardening process used to achieve the material properties. Table 45.13 lists the types of treatments associated with the H and T suffix letters.

Table 45.13 *Aluminum Treatment Conditions*

suffix	meaning
H1	strain hardened by working to desired dimensions
H2	strain hardened by cold working, followed by partial annealing
H3	strain hardened and stabilized
T2	annealed (castings only)
T3	solution heat treated, followed by cold working (strain hardening)
T4	solution heat treated, followed by natural aging at room temperature
T5	artificial aging only
T6	solution heat treated, followed by artificial aging
T7	solution heat treated, followed by stabilizing by overaging heat treating
T8	solution heat treated, followed by cold working and subsequent artificial aging
T9	solution heat treated, followed by artificial aging and subsequent cold working

16. PRODUCTION OF COPPER

Copper occurs in the free (metallic) state as well as in ores containing its oxides, sulfides, and carbonates. *Native copper* is recovered by the simple process of heating highly crushed ore. Molten copper flows to the bottom of the furnace.

Oxides and carbonates of copper are reduced in a blast or reverberatory furnace.

Sulfides of copper are heated in air, and the sulfur is replaced by oxygen. The product, which contains both copper and iron oxides, is further reduced in a reverberatory furnace. This step removes the oxygen but leaves some sulfur and iron. The final sulfur-removal process takes place in a furnace similar to a Bessemer converter in which air is injected into the molten copper. The iron oxide combines with a silica furnace liner.

Very low-grade ores are leached with sulfuric acid to recover the copper.

Regardless of the primary recovery method, *electrolysis (electrodeposition)* is generally required to remove the remaining impurities. Thick sheets of impure copper and thin sheets of pure copper are immersed together in an electrolyte of copper sulfate. The pure copper acts as the cathode. A direct current causes copper from the impure sheets to migrate to the pure sheets. Impurities drop to the bottom of the tank as they are released.

17. ALLOYS OF COPPER

Zinc is the most common alloying ingredient in copper. It constitutes a significant part (up to 40% zinc) in brass.[28] (Brazing rod contains even more, approximately 45 to 50% zinc.) Zinc increases copper's hardness and tensile strength. Up to approximately 30%, it increases the percent elongation at fracture. It decreases electrical conductivity considerably. *Dezincification*, a loss of zinc in the presence of certain corrosive media or at high temperatures, is a special problem that occurs in brasses containing more than 15% zinc.

Tin constitutes a major (up to 20%) component in most bronzes. Tin increases fluidity, which improves casting performance. In moderate amounts, corrosion resistance in saltwater is improved. (*Admiralty metal* has approximately 1%; *government bronze* and *phosphorus bronze* have approximately 10% tin.) In moderate amounts (less than 10%), tin increases the alloy's strength without sacrificing ductility. Above 15%, however, the alloy becomes brittle. For this reason, most bronzes contain less than 12% tin. Tin is more expensive than zinc as an alloying ingredient.

Lead is practically insoluble in solid copper. When present in small to moderate amounts, it forms minute soft particles that greatly improve machinability (2 to 3% lead) and wearing (bearing) properties (10% lead).

Silicon increases the mechanical properties of copper by a considerable amount. On a per unit basis, silicon is the most effective alloying ingredient in increasing hardness. *Silicon bronze* (96% copper, 3% silicon, 1% zinc) is used where high strength combined with corrosion resistance is needed (e.g., in boilers).

If aluminum is added in amounts of 9 to 10%, copper becomes extremely hard. Thus, *aluminum bronze* (as an example) trades an increase in brittleness for increased wearing qualities. Aluminum in solution with the copper makes it possible to precipitation harden the alloy.

Beryllium in small amounts (less than 2%) improves the strength and fatigue properties of copper. These properties make precipitation-hardened *copper-beryllium* (beryllium-copper, beryllium bronze, etc.) ideal for small springs. These alloys are also used for producing non-sparking tools.

---

[27]These alloys are all known as *duralumin*.

[28]*Brass* is an alloy of copper and zinc. *Bronze* is an alloy of copper and tin. Unfortunately, brasses are often named for the color of the alloys, leading to some very misleading names. For example, *nickel silver*, *commercial bronze*, and *manganese bronze* are all brasses.

18. NICKEL AND ITS ALLOYS

Like aluminum, nickel is largely hardened by precipitation hardening. Nickel is similar to iron in many of its properties, except that it has higher corrosion resistance and a higher cost. Also, nickel alloys have special electrical and magnetic properties.

Copper and iron are completely miscible with nickel. Copper increases formability. Iron improves electrical and magnetic properties markedly.

Some of the better-known nickel alloys are *monel metal* (30% copper, used hot-rolled where saltwater corrosion resistance is needed), *K-monel metal* (29% copper, 3% aluminum, precipitation-hardened for use in valve stems), *inconel* (14% chromium, 6% iron, used hot-rolled in gas turbine parts), and *inconel-X* (15% chromium, 7% iron, 2.5% titanium, aged after hot rolling for springs and bolts subjected to corrosion). *Hastelloy* (22% chromium) is another well-known nickel alloy.[29]

Nichrome (15 to 20% chromium) has high electrical resistance, high corrosion resistance, and high strength at red heat temperatures, making it useful in resistance heating. *Constantan* (40 to 60% copper, rest nickel) also has high electrical resistance and is used in thermocouples.

Alnico (14% nickel, 8% aluminum, 24% cobalt, 3% copper, rest iron) and *cunife* (20% nickel, 60% copper, rest iron) are two well-known nickel alloys with magnetic properties ideal for permanent magnets. Other magnetic nickel alloys are *permalloy* and *permivar*.

Invar, Nilvar, and *Elinvar* are nickel alloys with low or zero thermal expansion and are used in thermostats, instruments, and surveyors' measuring tapes.

For decades, C-family Alloy C-276 (Alloy 276) has been the nickel-chromium-molybdenum workhorse for piping and reaction vessels in chemical process industries. Newer alloys from the C-family (e.g., Alloy C-22, also designated as Alloy 22, 622, and 5621 hMoW) with more chromium and less tungsten are extremely corrosion resistant, even in extremely aggressive, mixed-acid environments. Alloy 59 (UNS N06059), also referred to as Alloy 5923 hMo, maintains its corrosion resistance in strongly oxidizing environments, including the most severe corrosion conditions in modern pollution-control equipment.

19. REFRACTORY METALS

Reactive and *refractory metals* include alloys based on titanium, tantalum, zirconium, molybdenum, niobium (also known as columbium), and tungsten. These metals are used when superior properties (i.e., corrosion resistance) are needed. They are most often used where high-strength acids are used or manufactured.

20. NATURAL POLYMERS

A *polymer* is a large molecule in the form of a long chain of repeating units. The basic repeating unit is called a *monomer* or just *mer*. (A large molecule with two alternating mers is known as a *copolymer* or *interpolymer*. Vinyl chloride and vinyl acetate form one important family of copolymer plastics.)

Many of the natural organic materials (e.g., rubber and asphalt) are polymers. (Polymers with elastic properties similar to rubber are known as *elastomers*.) Natural rubber is a polymer of the *isoprene latex* mer, (formula $[C_5H_8]_n$, repeating unit of $CH_2=CCH_3-CH=CH_2$, systematic name of 2-methyl-1,3-butadiene). The strength of natural polymers can be increased by causing the polymer chains to cross-link, restricting the motion of the polymers within the solid.

Cross-linking of natural rubber is known as *vulcanization*. Vulcanization is accomplished by heating raw rubber with small amounts of sulfur. The process raises the tensile strength of the material from approximately 300 lbf/in[2] (2.1 MPa) to approximately 3000 lbf/in[2] (21 MPa). The addition of carbon black as a reinforcing *filler* raises this value to approximately 4500 lbf/in[2] (31 MPa) and provides tear resistance and toughness.

The amount of cross-linking between the mers determines the properties of the solid. Figure 45.6 shows how sulfur joins two adjacent isoprene (natural rubber) mers in *complete cross-linking*.[30] If sulfur does not replace both of the double carbon bonds, *partial cross-linking* is said to have occurred.

Example 45.2

What is the approximate fraction (by mass) of sulfur in a completely cross-linked natural rubber polymer?

Solution

Figure 45.6 shows that, for complete cross-linking, each mer of the natural rubber mer requires one sulfur atom.

The atomic weight of sulfur is approximately 32. The molecular weight of the rubber mer is

$$(5)(12) + (8)(1) = 68$$

The fraction of sulfur is

$$\frac{m_{sulfur}}{m_{sulfur} + m_{mer}} = \frac{32\,\frac{g}{mol}}{32\,\frac{g}{mol} + 68\,\frac{g}{mol}}$$
$$= 0.32 \quad (32\%)$$

---

[29]K-monel is one of four special forms of monel metal. There are also H-monel, S-monel, and R-monel.

[30]A tire tread may contain 3 to 4% sulfur. Hard rubber products, which do not require flexibility, may contain as much as 40 to 50% sulfur.

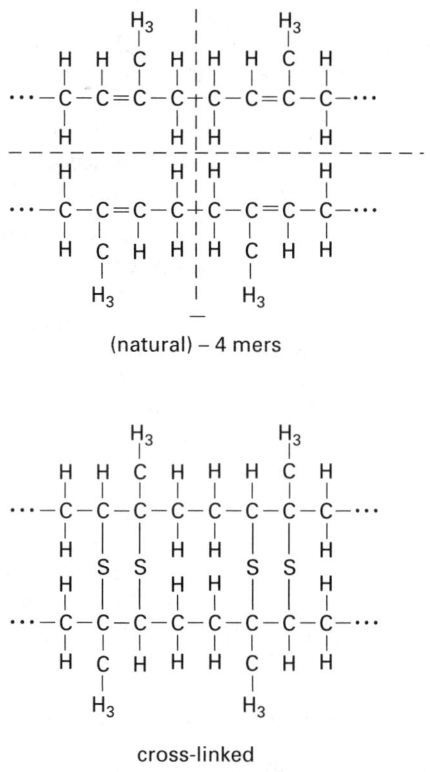

Figure 45.6 *Vulcanization of Natural Rubber*

21. DEGREE OF POLYMERIZATION

The *degree of polymerization*, DP, is the average number of mers in the molecule, typically several hundred to several thousand.[31] (In general, compounds with degrees of less than ten are called *telenomers* or *oligomers*.) The degree of polymerization can be calculated from the mer and polymer molecular weights.

$$DP = \frac{MW_{polymer}}{MW_{mer}} \qquad 45.1$$

A polymer batch usually will contain molecules with different length chains. Therefore, the degree of polymerization will vary from molecule to molecule, and an average degree of polymerization is reported.

The stiffness and hardness of polymers vary with their degrees. Polymers with low degrees are liquids or oils. With increasing degree, they go through waxy to hard resin stages. High-degree polymers have hardness and strength qualities that make them useful for engineering applications. Tensile strength and melting (softening) point also increase with increasing degree of polymerization.

---

[31]Degrees of polymerization for commercial plastics are usually less than 1000.

Example 45.3

A polyvinyl chloride molecule (mer molecular weight of 62.5) is found to contain 860 carbon atoms, 1290 hydrogen atoms, and 430 chlorine atoms. What is the degree of polymerization?

Solution

The approximate atomic weights of carbon, hydrogen, and chlorine are 12, 1, and 35.5, respectively. The molecular weight of the molecule is

$$MW_{polymer} = (860)(12) + (1290)(1) + (430)(35.5)$$
$$= 26{,}875$$

The degree of polymerization is

$$DP = \frac{MW_{polymer}}{MW_{mer}} = \frac{26{,}875}{62.5} = 430$$

Example 45.4

What is the degree of polymerization of a polyvinyl acetate (mer of $C_4H_6O_2$, molecular weight of 86) sample having the following analysis of molecular weights?

range of molecular weights	mole fraction
5001–15,000	0.30
15,001–25,000	0.47
25,001–35,000	0.23

Solution

Using the midpoint of each range, the average polymer molecular weight is

$$MW_{polymer} = (0.30)(10{,}000) + (0.47)(20{,}000)$$
$$+ (0.23)(30{,}000)$$
$$= 19{,}300$$

The degree of polymerization is

$$DP = \frac{MW_{polymer}}{MW_{mer}} = \frac{19{,}300}{86} = 224$$

22. SYNTHETIC POLYMERS

Table 45.14 lists some of the common mers. Polymers are named by adding the prefix "poly" to the name of the basic mer. For example, C_2H_4 is the chemical formula for ethylene. Chains of C_2H_4 are called polyethylene.

Table 45.14 *Names of Common Mers*

name	repeating unit	combined formula
ethylene	CH_2CH_2	C_2H_4
propylene	$CH_2(HCCH_3)$	C_3H_6
styrene	$CH_2CH(C_6H_5)$	C_8H_8
vinyl acetate	$CH_2CH(C_2H_3O_2)$	$C_4H_6O_2$
vinyl chloride	CH_2CHCl	C_2H_3Cl
isobutylene	$CH_2C(CH_3)_2$	C_4H_8
methyl methacrylate	$CH_2C(CH_3)(COOCH_3)$	$C_5H_8O_2$
acrylonitrile	CH_2CHCN	C_3H_3N
epoxide (ethoxylene)	CH_2CH_2O	C_2H_4O
amide (nylon)	$CONH_2$ or $CONH$	$CONH_2$ or $CONH$

Polymers are able to form when double (covalent) bonds break and produce *reaction sites*. The number of bonds in the mer that can be broken open for attachment to other mers is known as the *functionality* of the mer. The ethylene mer in Fig. 45.7 is *bifunctional* since the C=C bond can be broken to form two reaction sites. Other mers are *trifunctional* or *tetrafunctional*. When combining into chains, bifunctional mers form *linear polymers*, whereas trifunctional and tetrafunctional mers form *network polymers*.

Two processes produce reaction sites: addition polymerization and condensation polymerization. With *addition polymerization*, mers simply combine sequentially into chains by breaking double (covalent) bonds. No other products are produced. For example, the formation of polyethylene is given by the following reaction and Fig. 45.7.

$$(2)(CH_2{=}CH_2) \rightarrow -CH_2-CH_2-CH_2-CH_2-$$

Figure 45.7 *Steps in Forming Polyethylene*

Substances called *initiators* are used to start addition polymerization.[32] Initiators break down under heat, light, or other energy and provide free radicals. A free radical disrupts the double bond in a mer and attaches to one side of the mer, opening up the mer's double bond and releasing enough energy to open another mer's double bond.

The process continues (*propagation*) until the initiator is used up, until the mers are used up, or until another free radical terminates the chain. The latter occurrence (i.e., when there is no possibility for additional joining) is known as *saturation*. Since the same radicals that initiated a polymer chain can terminate the chain, initiators are also known as *terminators*.

Condensation polymerization (step reaction polymerization) also requires opening bonds in two molecules to form a larger molecule. This formation is often accompanied by the release of small molecules such as H_2O, CO_2, or N_2. The repeating units derived from the condensation process are not the same as the monomers from which they are formed since portions of the original monomer form the small molecules that are lost.[33] The formation of *phenolic plastics* from formaldehyde and ammonia illustrates condensation polymerization.

$$6HCHO + 4NH_3 \rightarrow (CH_2)_6N_4 + 6H_2O$$

23. FLUOROPOLYMERS

Fluoropolymers (*fluoroplastics*) are a class of paraffinic, thermoplastic polymers in which some or all of the hydrogens have been replaced by fluorine.[34] There are seven major types of fluoropolymers, with overlapping characteristics and applications. They include the fully fluorinated fluorocarbon polymers of Teflon® PTFE (polytetrafluoroethylene), FEP (fluorinated ethylene propylene), and PFA (perfluoroalkoxy), as well as the partially fluorinated polymers of PCTFE (polychlorotrifluoroethylene), ETFE (ethylene tetrafluoroethylene), ECTFE (ethylene chlorotrifluoroethylene), and PVDF (polyvinylidene fluoride).

Fluoropolymers compete with metals, glass, and other polymers in providing corrosion resistance. Choosing the right fluoropolymer depends on the operating environment including temperature, chemical exposure, and mechanical stress.

PTFE, the first available fluoropolymer, is probably the most inert compound known. It has been used extensively for pipe and tank linings, fittings, gaskets, valves, and pump parts. It has the highest operating temperature—approximately 500°F (260°C). Unlike the other fluoropolymers, however, it is not a melt-processed polymer. Like a powdered metallurgy product, PTFE is processed by compression and isostatic molding, followed by sintering. PTFE is also the weakest of all the fluoropolymers.

24. ELASTOMERIC COMPOUNDS

Flexible parts, such as gaskets and O-rings, are manufactured from *elastomeric compounds*. Common elastomerics include natural rubber, butyl rubber, buna-N

---

[32] *Initiators* are usually *organic peroxides* such as *hydrogen peroxide* (H_2O_2) and *benzoyl peroxide* (C_6H_5COO).

[33] The formation of nylon and the famous *Bakelite* material from phenol (C_6H_5OH) are examples of condensation polymerization.
[34] *Fluoroelastomers* are uniquely different from fluoropolymers. They have their own areas of application.

Table 45.15 Characteristics of Fluoropolymers

fluoro-polymer	specific gravity	tensile strength (ksi)	flexural modulus (ksi)	upper service temperature[c] (°C)
PTFE[a]	2.13–2.22	2.0–6.5	70–110	287
FEP	2.12–2.17	2.7–3.1	90	204
PFA	2.12–2.17	4.0–4.5	100	260
PCTFE	2.08–2.2	4.0–6.0	150–260	199
ETFE	1.70	6.5	200	149–182
ECTFE	1.68	6.6–7.8	240	149–179
PVDF[b]	1.76–1.78	3.5–6.2	70–320	150

(Multiply ksi by 6.895 to obtain MPa.)
[a]Properties of PTFE are highly variable, depending on type of resin and method of processing.
[b]Properties of PVDF include those of its copolymers, hence the range of values.
[c]Upper service temperature is also known as *temperature of continuous heat resistance.*

(nitrile rubber), neoprene, ethylene-propylene-diene monomer (EPDM) rubber, chlorosulfonated polyethylene, and various fluoroelastomers. Chemical resistance, temperature, and pressure are the primary factors considered in choosing sealing compounds.

25. POLYMER CRYSTALLINITY

Under favorable conditions, polymer chains align in regular patterns, minimizing volume, and increasing the attractive forces between atoms. This phenomenon is known as *polymer crystallinity*. Crystallinity is often only partial, however, because the chain structure interferes with complete alignment, and only weak van der Waals forces are available to drive the alignment.

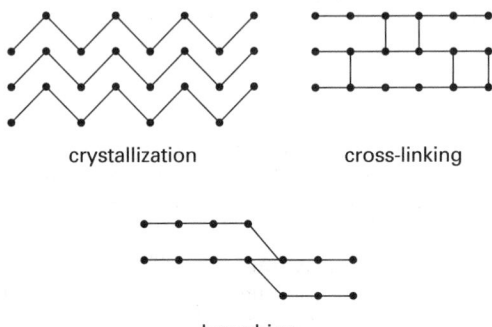

crystallization   cross-linking

branching

Figure 45.8 Polymer Crystallization, Cross-Linking, and Branching

Several factors affect the likelihood of crystallization:

- *Trans mers* have *unsaturated positions* (i.e., linkage points) on opposite sides of the mer, producing less tangling between chains. Trans mers are more likely to crystallize than *cis mers*, which have both unsaturated positions on the same side of the mer.

- When mers join with a high degree of regularity in chain structure and mer orientation, the result is an *isotactic polymer*. Isotactic polymers

favor crystallization over the *syndiotactic polymers* (partial regularity) and *atactic polymers* (no regularity).

- Linear polymers are more likely to crystallize than network polymers.

Crystallization requires a physical rearrangement of polymer chains into regular layers. It should not be confused with *cross-linking* between adjacent polymers or with polymer *branching*.

26. THERMOSETTING AND THERMOPLASTIC POLYMERS

Most polymers can be softened and formed by applying heat and pressure. These are known by various terms including *thermoplastics, thermoplastic resins*, and *thermoplastic polymers*. Polymers that are resistant to heat (and that actually harden or "kick over" through the formation of permanent cross-linking upon heating) are known as *thermosetting plastics*. Table 45.16 lists the common polymers in each category. Thermoplastic polymers retain their chain structures and do not experience any chemical change (i.e., bonding) upon repeated heating and subsequent cooling. Thermoplastics can be formed in a cavity mold, but the mold must be cooled before the product is removed. Thermoplastics are particularly suitable for injection molding. The mold is kept relatively cool, and the polymer solidifies almost instantly.

Thermosetting polymers form complex, three-dimensional networks. Thus, the complexity of the polymer increases dramatically, and a product manufactured from a thermosetting polymer may be thought of as one big molecule. Thermosetting plastics are rarely used with injection molding processes.

27. WOOD

Woods are classified broadly as softwoods or hardwoods, although it is difficult to define these terms exactly. *Softwoods* contain tube-like fibers (*tracheids*) oriented with the longitudinal axis (grain) and cemented together with *lignin*. *Hardwoods* contain more complex structures (e.g., storage cells) in addition to longitudinal fibers. Fibers in hardwoods are also much smaller and shorter than those in softwoods.

The mechanical properties of woods are influenced by moisture content and grain orientation. (Strengths of dry woods are approximately twice those of wet or green woods. Longitudinal strengths may be as much as 40 times higher than cross-grain strengths.) *Moisture content*, MC, is defined by Eq. 45.2.

$$\text{MC} = \frac{\text{wet weight} - \text{oven-dry weight}}{\text{oven-dry weight}} \quad\quad 45.2$$

Table 45.16 *Thermosetting and Thermoplastic Polymers*

thermosetting
- epoxy
- melamine
- natural rubber (polyisoprene)
- phenolic (phenol formaldehyde, Bakelite®)
- polyester (DAP)
- silicone
- urea formaldehyde

thermoplastic
- acetal
- acrylic
- acrylonitrile-butadiene-styrene (ABS)
- cellulosics (e.g., cellophane)
- polymethyl-methacrylate (Plexiglas®, Lucite®)
- polyamide (nylon)
- polyarylate
- polycarbonate
- polyester (PBT and PET)
- polyethylene
- polypropylene
- polystyrene
- polytetrafluoroethylene (Teflon®)
- polyurethane
- polyvinyl chloride (PVC)
- synthetic rubber (Neoprene)
- vinyl

Bakelite® is a trademark of Union Carbide.
Plexiglass® is a trademark of Atohass.
Lucite® is a trademark of ICI Acrylics.
Teflon® is a trademark of Du Pont Company.

Wood is considered to be green if its moisture content is above 19%. Wood is considered to be dry when it has reached its *equilibrium moisture content*, generally between 12% and 15% moisture. Thus, moisture is not totally absent in dry wood.[35]

The approximate mechanical properties of several dry wood varieties are listed in Table 45.17.

Table 45.17 *Approximate Properties of Wood*

variety	flexural strength (lbf/in^2)	tensile strength (lbf/in^2)^a	compressive strength (lbf/in^2)^b	modulus of elasticity (lbf/in^2)
oak (red)	14,400	820	6920	1.81×10^6
mahogany	11,460	740	6780	1.50×10^6
douglas fir	12,200	340	7430	1.95×10^6
cedar	7700	220	5020	1.12×10^6
fir	9800	300	5480	1.49×10^6
spruce	10,200	710	5610	1.57×10^6
yellow pine	12,800	470	7080	1.80×10^6
redwood	10,000	240	6150	1.34×10^6

(Multiply lbf/in^2 by 6.894 to obtain kPa.)
^{a}perpendicular to grain
^{b}parallel to grain

[35]Oven-dry lumber is not used in construction.

28. GLASS

Glass is a term used to designate any material that has a volumetric expansion characteristic similar to Fig. 45.9. Glasses are sometimes considered to be *supercooled liquids* because their crystalline structures solidify in random orientation when cooled below their melting points. It is a direct result of the high liquid viscosities of oxides, silicates, borates, and phosphates that the molecules cannot move sufficiently to form large crystals with cleavage planes.

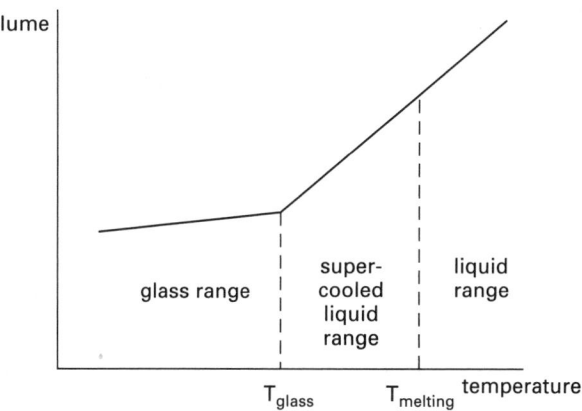

Figure 45.9 *Behavior of a Glass*

As a liquid glass is cooled, its atoms develop more efficient packing arrangements. This leads to a rapid decrease in volume (i.e., a steep slope on the temperature-volume curve). Since no crystallization occurs, the liquid glass simply solidifies without molecular change when cooled below the melting point. (This is known as *vitrification*.) The more efficient packing continues past the point of solidification.

At the *glass transition temperature (fictive temperature)*, the glass viscosity increases suddenly by several orders of magnitude. Since the molecules are more restrained in movement, efficient atomic rearrangement is curtailed, and the volume-temperature curve changes slope. This temperature also divides the region into flexible and brittle regions. At the glass transition temperature, there is a 100-fold to 1000-fold increase in stiffness (modulus of elasticity).

Both organic and inorganic compounds may behave as glasses. *Common glasses* are mixtures of SiO_2, B_2O_3, and various other compounds to modify performance, as shown in Table 45.18.[36]

29. CERAMICS

Ceramics are compounds of metallic and nonmetallic elements. Ceramics form crystalline structures but have no free valence electrons. All electrons are shared ionically or in covalent bonds. Common examples include brick, portland cement, refractories, and abrasives.

[36]This excludes lead-alkali glasses that contain 30 to 60% PbO.

Table 45.18 *Analyses and Properties of Representative Glasses*

type of glass	analysis, percent by weight					softening temperature,[b] °C	coefficient of expansion, $(°C)^{-1}$	characteristics	use
	SiO_2	modifiers[a]	Al_2O_3	B_2O_3	PbO				
fused silica	99.9	–	–	–	–	1667	5.5×10^{-7}	thermal shock resistant	laboratory equipment
96% silica	96.0	–	–	4.0	–	1500	8.0×10^{-7}	thermal shock	laboratory equipment
borosilicate (Pyrex®)	80.5	4.2	2.2	12.9	–	820	32.0×10^{-7}	thermal shock resistant, easy to form	cooking utensils
aluminasilicate	57.7	9.5	25.3	7.4	–	915	42.0×10^{-7}	thermal shock resistant	thermometers
soda-lime silica	73.6	25.4	1.0	–	–	696	92.0×10^{-7}	easy to form	plate, bulbs
lead-alkali	54.0	11.0	–	–	35	630	89.0×10^{-7}	high index of refraction	cut glass
lead-alkali	35.0	7.0	–	–	58	580	91.0×10^{-7}	dielectric	capacitors

[a]Sum of Na_2O, K_2O, Ni_2O, CaO, MgO, and BaO.
[b]The temperature at which glass will sag appreciably under its own weight.
Pyrex® is a trademark of Corning.

Data derived with permission from *Metals, Ceramics and Polymers*, by Oliver H. Wyatt and David Dew-Hughes, Cambridge University Press, © 1974. Reprinted with permission of Cambridge University Press.

(Glass is also considered a ceramic even though it does not crystallize.) Typical ceramic properties are listed in Table 45.19.

Table 45.19 *Properties of Typical Ceramics*

high melting point
high hardness
high compressive strength
high tensile strength (perfect crystals)
low ductility (brittleness)
high shear resistance (low slip)
low electrical conductivity
low thermal conductivity
high corrosion (acid) resistance
low coefficient of thermal expansion

Although perfect ceramic crystals have extremely high tensile strength (e.g., some glass fibers have ultimate strengths of 100,000 lbf/in^2 (700 MPa)), the multiplicity of cracks and other defects in natural crystals reduces their tensile strengths to near-zero levels.

Due to the absence of free electrons, ceramics are typically poor conductors of electrical current, although some (e.g., magnetite, Fe_3O_4) possess semiconductor properties. Other ceramics, such as $BaTiO_3$, SiO_2, and $PbZrO_3$, have *piezoelectric (ferroelectric) qualities* (i.e., generate a voltage when compressed).

Ceramics with similar structures behave similarly. Table 45.20 lists several structure designations (e.g., sodium chloride structure) as used in the study of ceramics.

Table 45.20 *Structural Designations*
A and B are metals; X is a nonmetal

compound	formula	basic form	other examples
sodium chloride	NaCl	AX	FeO, MgO, CaO
cesium chloride	CsCl	AX	
calcium fluoride	CaF_2	AX_2	GeO_2, MgF_2, TO_2
silica	SiO_2	AX_2	
corundum	Al_2O_3	A_2X_3	Cr_2O_3, Fe_2O_3
spinel (ferrites)	$MgAl_2O_4$	AB_2X_4	
Perovskite	$CaTiO_3$	ABX_3	$BaTiO_3$, $PbZrO_3$
zircon	$ZrSiO_4$	ABX_4	

Polymorphs are compounds that have the same chemical formula but have different physical structures. Some ceramics, of which *silica* (SiO_2) is a common example, exhibit *polymorphism*. At room temperature, silica is in the form of *quartz*. At 1607°F (875°C), the structure changes to *tridymite*. A change to a third structure, that of *cristobalite*, occurs at 2678°F (1470°C).

Ferrimagnetic materials (ferrites, spinels, or *ferrispinels)* are ceramics with valuable magnetic qualities. Recent advances in near-room-temperature superconductivity have been based on *lanthanum barium copper oxide* ($La_{2-x}Ba_xCuO_4$), a ceramic oxide, as well as compounds based on yttrium (Y-Ba-Cu-O), bismuth, thallium, and others.

Common ceramics are listed in Table 45.21.

Table 45.21 Common Ceramics

compound	mineral name	use[a]
Al_2O_3	corundum, alumina	abrasives, firebrick
$Al_2Si_2O_5(OH)_4$	kaolinite clay	porcelain paste
$BaTiO_3$	barium titanate	piezoelectricity
BN	boron nitride	refractory
CaF_2	fluorite	flux
CaO		refractory
Fe_2O_3		refractory
Fe_3O_4 or $FeFe_2O_4$	magnetite	thermistors
MgO	periclase	refractory
Mg_2SiO_4	forsterite	refractory
$MnFe_2O_4$		ferrimagnetism
$MgCr_2O_4$	magnesium chromate	piezoelectricity
$MgFe_2O_4$		antiferromagnetism
$NiFe_2O_4$		ferrimagnetism
NaCl	salt	food, chemicals
$PbZrO_3$		piezoelectricity
SiC	silicon carbide	refractory
SiO_2	quartz[b]	refractory
TiC	titanium carbide	refractory
TiO_2	titanium dioxide	refractory
UO_2	uranium dioxide	nuclear fuel
ZrN	zirconium nitride	refractory
$ZnFe_2O_4$		ferrimagnetism

[a]The term *refractory* means the ceramic is used in firebrick, stoneware, and other containers intended for use at high temperatures.

[b]SiO_2 has several temperature-dependent polymorphs, including coesite, cristobalite, and tridymite.

30. ABRASIVES

An *abrasive* is a hard material that can cut other materials. Abrasives are typically ceramic compounds embedded in stiff binders. *Natural abrasives* include *emery* (50 to 60% Al_2O_3, rest iron oxide), corundum, quartz, garnets, and diamonds. *Artificial abrasives* include cemented carbides (e.g., SiC) and artificially made aluminum oxide (Al_2O_3).

Binders for rigid wheel abrasives are kiln-fired vitreous materials derived from clays or feldspars. Other grinding wheels use rubber or synthetic elastomers as the binders. Grains of abrasive are mixed thoroughly with a binder and molded into final form. Diamond wheels are often in the form of thin metal discs with the diamond grains bonded only to the periphery.

Carbides (cemented carbides, sintered carbides) have extreme hardness, wear resistance, and thermal stability.[37] These properties make them useful for high-speed metal cutting. Silicon carbide (SiC, sold under the name of *carborundum*) is probably the best known. Carbides of tungsten, molybdenum, titanium, vanadium, tantalum, and zirconium are also widely used.

---

[37]*Sintering* is the process where a physical mixture of carbide and powdered metal is heated in order to solidify the powder into a single piece. When the metal melts, it acts as the binder for the carbide.

31. MODERN COMPOSITE MATERIALS

There are many types of modern composite material systems including dispersion-strengthened, particle-strengthened, and fiber-strengthened materials. (Steel-reinforced concrete and steel-reinforced wood systems, discussed elsewhere in this book, are also composite systems.)

In *dispersion-strengthened systems* (e.g., aluminum-aluminum oxide systems known as *SAP alloys*, *toughened ceramics*, *metal-metal systems* in which tungsten *whiskers* are blended in a copper alloy matrix, and NiO_2-ThO_2 mixtures known as *TD-nickel*), the matrix is an actual load-bearing element.[38] The discrete particles (*dispersoids*) distributed throughout the matrix occupy less than 15% of the volume, and particle sizes are in the 0.01 to 0.1 μm range.

In *particle-strengthened systems* (e.g., tungsten carbide, known as WC, in a cobalt matrix, and *cermets* produced by sintering), the dispersoid particles are larger than 1.0 μm in size, and they occupy up to 25% of the material volume. The matrix is not the major load-carrying element, but it does contribute to strength.

In *fiber-strengthened systems* (e.g., glass-reinforced epoxies), reinforcing materials vary widely in size, and sizes may range up to several mils.[39] The matrix transmits the loads to the fibers and protects the fibers from chemical attack.

Initial attempts at fiber-reinforced polymers involved the impregnation of natural fibers (cotton, wood, etc.) with Bakelite and phenolic resins. The introduction of various types of glass (e.g., rovings, windings, and woven cloth) as reinforcement was the next development step.[40] Currently, higher-strength epoxy resins have been used as matrices with graphite, boron, beryllium, steel, titanium, aluminum, or magnesium fibers.

Many fiber-reinforced composites, particularly those involving cloth, are highly *anisotropic*. Properties vary with the orientation of lay-ups as well as with weave orientation in the reinforcing materials. In directions transverse to fiber orientation, tensile and compressive strengths are a function of the matrix material. Loads parallel to the fibers are carried by the reinforcement, while flexural strength is limited by the shear bond between the filaments and the matrix material.

Typical reinforcing materials for fiberglass include E- and S-glass.[41] *S-glass* is a silica-alumina-magnesia

---

[38]Ceramics typically fail catastrophically, without warning. To counteract this tendency, *toughened ceramics* incorporate discrete solids such as SiC or TiC whiskers throughout the ceramic matrix. Cracks that start are arrested by the dispersoids. Dispersoids can increase ceramic toughness by as much as 40%.

[39]A *whisker* is a single crystal grown by vapor deposition. Although lengths of several millimeters are typical, diameters are only a few micrometers.

[40]A *roving* consists of a number of parallel strands of fiber. The strands are side by side (not interwoven or twisted together), forming a flat ribbon.

[41]*A-glass* is common soda-lime glass used for windows, bottles, and jars. Other types of glass used for reinforcement include *C-glass* (developed for greater chemical and corrosion resistance), *D-glass* (glass possessing a low dielectric constant), and *M-glass* (glass containing BeO to increase the elastic modulus).

compound with improved tensile properties. It is used mainly in nonwoven, monodirectional, and wound configurations. *E-glass* is a lime-alumina-borosilicate (CaO-Al_2O_3-SiO_2) compound used primarily in woven fabrics.

Table 45.22 *Typical Properties of E- and S-Glass Fibers at Room Temperature*

property	E-glass	S-glass
specific gravity	2.54	2.48
density (lbm/in^3)	0.092	0.090
ultimate tensile strength (lbf/in^2)		
• monofilament	5.0×10^5	6.6×10^5
• 12-end roving	3.7×10^5	5.5×10^5
modulus of elasticity (lbf/in^2)	10.5×10^6	12.5×10^6
coefficient of thermal expansion (1/°F)	2.8×10^{-6}	1.6–2.2×10^{-6}
specific heat (Btu/lbm-°F)	0.192	0.176

(Multiply lbm/in^3 by 27.68 to obtain kg/m^3.)
(Multiply lbf/in^2 by 6.895 to obtain MPa.)
(Multiply 1/°F by 1.8 to obtain 1/°C.)
(Multiply Btu/lbm-°F by 4.19 to obtain kJ/kg-°C.)

Graphite (i.e., carbon) fibers are used where high stiffness and low coefficients of thermal expansion are needed.[42] These advantages are balanced by the disadvantages of brittleness and high cost. The ultimate tensile strength for graphite varies inversely with modulus of elasticity. Graphite fibers range in strength from 180 ksi (1.2 GPa) for yarn configurations to 350 ksi (2.4 GPa) for tow, while the modulus of elasticity varies from 60×10^6 to 20×10^6 lbf/in^2 (410 GPa to 140 GPa).[43]

There are three general categories of carbon fibers: standard, high-modulus, and high-strength. Kevlar aramid fibers provide strength and stiffness that are essentially in between that of glass and carbon fibers.

Resistance to chemical corrosion, rather than strength-to-weight ratio, is the primary factor in selecting *fiber-reinforced plastics* (also referred to as *fiber-reinforced polymers*, both abbreviated FRP) for process tanks, reaction vessels, and pipes.[44] The main resins used for this purpose are vinyl esters, epoxies, polyesters, furans, and phenolics. Vinyl esters, which can handle both acidic and basic fluids as well as strong oxidizers such as chlorine, are the most widely used. Epoxies have better thermal and mechanical properties, but epoxies cannot

be used in highly acidic environments (i.e., pHs less than 3). Polyesters, on the other hand, are acid resistant, but are not alkali resistant (i.e., pHs more than 9). Furans and phenolics are relatively weak and are used in special applications.

A typical FRP tank or pipe consists of three layers: the veil, liner, and structural laminate (from the inside out). The *veil* consists of a thin layer (e.g., 0.25 mm) of glass or polyester fibers saturated with approximately 90% by weight of the resin. The 2.5 mm thick *liner* consists of glass in resin in a 3:1 ratio, respectively. The liner provides chemical resistance. The outermost fiberglass *structural laminate* bears all of the pressure, stresses, and mechanical forces imposed on the tank.[45] Although the structural laminate can be laid up by hand, filament-wound tanks are more popular because manufacturing costs are lower.

Table 45.23 *Typical Properties of Laid-Up Composites (linear lay-up)*

composite	fiber content (%)	specific gravity	modulus (ksi) axial	transverse	shear
graphite-epoxy					
high strength	65	1.58	20,000	1000	650
high modulus	65	1.61	29,000	1000	700
ultrahigh modulus	65	1.69	44,000	1000	950
Kevlar^{TM} 49-epoxy	65	1.39	12,500	800	300
E-glass-epoxy	65	1.99	6000	1500	300
chopped glass-polyester					
sheet molding	30	1.88	2500	2500	100
compound (SMC)	65	1.99	3500	3500	150

(Multiply ksi by 6.895 to obtain MPa.)

PRACTICE PROBLEMS

1. How much (in molecules of HCl per gram of PVC) HCl should be used as an initiator in PVC if the efficiency is 20% and an average molecular weight of 7000g/mol is desired? The final polymer has the following structure.

2. 10 ml of a 0.2% solution (by weight) of hydrogen peroxide is added to 12 g of ethylene to stabilize the polymer. What is the average degree of polymerization if the hydrogen peroxide is completely utilized? Assume that hydrogen breaks down according to

$$H_2O_2 \rightarrow 2(OH^-) + \cdots$$

Assume that the stabilized polymer has the following structure.

---

[42] It is interesting that graphite in *fiber* or *whisker* form is used to provide strength, while graphite in *powder* form is used as a solid lubricant. The *laminar* structure of graphite permits particles to easily slide over one another. This laminar structure does not easily break down, making graphite particularly valuable as a lubricant at high temperatures and pressures—up to at least 3600°F (2000°C). In fact, graphite's coefficient of friction decreases with temperature. Another excellent solid lubricant, *molybdenum sulfide* (MoS$_2$), also has a laminar structure, but its friction coefficient increases sharply above 1600°F (900°C).

[43] *Tow* consists of loose, untwisted fibers.

[44] Other characteristics for which FRP may be used include cost, weight, and ease of manufacture of complex shapes.

[45] ASTM standards D3299 and D4097 both specify the common $pD/2S_H$ formula for wall thickness, where S_H is the allowable circumferential stress that produces a maximum strain of 0.0010.

46 Material Properties and Testing

1. The Tensile Test	46-1
2. Stress-Strain Characteristics of Nonferrous Metals	46-3
3. Stress-Strain Characteristics of Brittle Materials	46-3
4. The Secant Modulus	46-4
5. Poisson's Ratio	46-4
6. Strain Hardening and Necking Down	46-4
7. True Stress and Strain	46-4
8. Ductility	46-5
9. Strain Energy	46-6
10. Resilience	46-6
11. Toughness	46-6
12. Unloading and Reloading	46-6
13. Compressive Strength	46-7
14. The Torsion Test	46-7
15. Relationship Between the Elastic Constants	46-8
16. Fatigue Testing	46-8
17. Testing of Plastics	46-10
18. Nondestructive Testing	46-11
19. Hardness Testing	46-12
20. Toughness Testing	46-13
21. The Creep Test	46-15
22. Effects of Impurities and Strain on Mechanical Properties	46-15
23. Classification of Materials	46-16
Practice Problems	46-16

Nomenclature

A	area	in²	m²
b	width	in	m
B	bulk modulus	lbf/in²	MPa
C	constant	–	–
C_V	impact energy	ft-lbf	J
d	diameter of impression	in	mm
D	diameter	in	m
e	engineering strain	in/in	m/m
E	modulus of elasticity	lbf/in²	MPa
F	force	lbf	N
G	shear modulus	lbf/in²	MPa
J	polar moment of inertia	in⁴	m⁴
k	exponent	1/hr	1/h
K	endurance strength reduction factor	–	–
K	strength coefficient	lbf/in²	MPa
L	length	in	m
LYS	lower yield strength	lbf/in²	MPa
n	exponent	–	–
N	number of cycles	–	–
P	force of impression	lbf	kg
q	fatigue notch sensitivity factor	–	–
r	radius	in	m
s	stress	lbf/in²	MPa
S	strength	lbf/in²	MPa
t	depth (thickness)	in	mm
t	time	hr	h
T	torque	in-lbf	N·m
U_R	modulus of resilience	lbf/in²	MPa
U_T	modulus of toughness	lbf/in²	MPa
UYS	upper yield strength	lbf/in²	MPa

Symbols

β	Andrade's beta	$hr^{-\frac{1}{3}}$	$h^{-\frac{1}{3}}$
δ	elongation	in	m
ϵ	true strain or creep	in/in	m/m
θ	angle of rupture	deg	deg
θ	shear strain	rad	rad
ν	Poisson's ratio	–	–
σ	true stress	lbf/in²	MPa
τ	shear stress	lbf/in²	MPa
ϕ	angle of internal friction	deg	deg

Subscripts

c	compressive
e	endurance
f	fatigue or fracture
o	original
p	particular
s	shear
t	tensile
u	ultimate
y	yield

1. THE TENSILE TEST

Many useful material properties are derived from the results of a standard tensile test. In this test, a prepared material sample (i.e., a *specimen*) is axially loaded in tension, and the resulting elongation, δ, is measured as the load, F, increases. A *load-elongation curve* of tensile test data for a ductile ferrous material (e.g., low-carbon steel or other BCC transition metal) is shown in Fig. 46.1.

When elongation is plotted against the applied load, the graph is applicable only to an object with the same length and area as the test specimen. To generalize the test results, the data are converted to stresses and strains by use of Eqs. 46.1 and 46.2.[1]

---

[1] The most common *test specimen* in the United States has a length of 2.00 in and a diameter of 0.505 in. Since the cross-sectional area of this *0.505 bar* is 0.2 in², the stress in lbf/in² is calculated by multiplying the force in pounds by five.

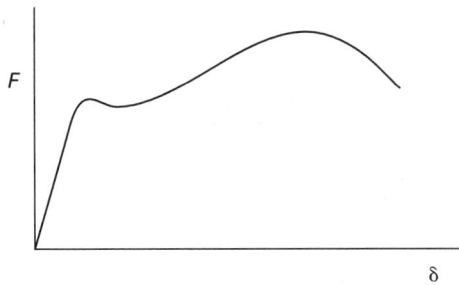

Figure 46.1 *Typical Tensile Test Results for a Ductile Material*

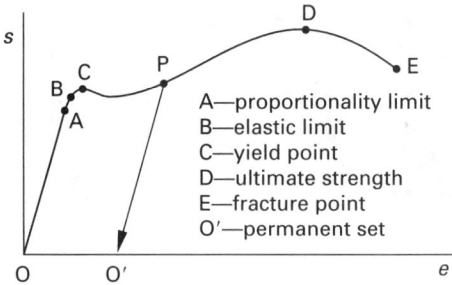

A—proportionality limit
B—elastic limit
C—yield point
D—ultimate strength
E—fracture point
O'—permanent set

Figure 46.2 *Typical Stress-Strain Curve for Steel*

Engineering stress, s (usually called *stress*), is the load per unit original area. Typical engineering stress units are lbf/in^2 and MPa. *Engineering strain*, e (usually called *strain*), is the elongation of the test specimen expressed as a percentage or decimal fraction of the original length. The units in/in and m/m are also used for strain.

$$s = \frac{F}{A_o} \qquad 46.1$$

$$e = \frac{\delta}{L_o} \qquad 46.2$$

If the stress-strain data are plotted, the shape of the resulting line will be essentially the same as the force-elongation curve, although the scales will change.

Segment O-A in Fig. 46.2 is a straight line. The relationship between the stress and strain in this linear region is given by *Hooke's law*, Eq. 46.3. The slope of line segment O-A is the *modulus of elasticity*, E, also known as *Young's modulus*. Table 46.1 lists approximate values of the modulus of elasticity for materials at room temperature. The modulus of elasticity will be lower at higher temperatures.[2]

$$s = Ee \qquad 46.3$$

The stress at point A in Fig. 46.2 is known as the *proportionality limit* (i.e, the maximum stress for which the linear relationship is valid). Strain in the *proportional region* is called *proportional strain*.

The *elastic limit*, point B in Fig. 46.2, is slightly higher than the proportionality limit. As long as the stress is kept below the elastic limit, there will be no *permanent set* (permanent deformation) when the stress is removed. Strain that disappears when the stress is

Table 46.1 *Approximate Modulus of Elasticity of Representative Materials at Room Temperature*

material	lbf/in^2	GPa
aluminum alloys	$10\text{--}11\times10^6$	70–80
brass	$15\text{--}16\times10^6$	100–110
cast iron	$15\text{--}22\times10^6$	100–150
cast iron, ductile	$22\text{--}25\times10^6$	150–170
cast iron, malleable	$26\text{--}27\times10^6$	180–190
copper alloys	$17\text{--}18\times10^6$	110–112
glass	$7\text{--}12\times10^6$	50–80
magnesium alloys	6.5×10^6	45
molybdenum	47×10^6	320
nickel alloys	$26\text{--}30\times10^6$	180–210
steel, hard[a]	30×10^6	210
steel, soft[a]	29×10^6	200
steel, stainless	$28\text{--}30\times10^6$	190–210
titanium	$15\text{--}17\times10^6$	100–110

(Multiply lbf/in^2 by 6.89×10^{-6} to obtain GPa.)
[a]Common values given.

removed is known as *elastic strain*, and the stress is said to be in the *elastic region*. When the applied stress is removed, the *recovery* is 100%, and the material follows the original curve back to the origin.

If the applied stress exceeds the elastic limit, the recovery will be along a line parallel to the straight line portion of the curve, as shown in line segment P-O'. The strain that results (line O-O') is *permanent set* (i.e., a permanent deformation). The terms *plastic strain* and *inelastic strain* are used to distinguish this behavior from elastic strain.

For steel, the *yield point*, point C, is very close to the elastic limit. For all practical purposes, the *yield strength* or *yield stress*, S_y (or S_{yt} to indicate yield in tension), can be taken as the stress that accompanies the beginning of plastic strain. Yield strengths are reported in lbf/in^2, ksi, and MPa.[3]

Figure 46.2 does not show the full complexity of the stress-strain curve near the yield point. Rather than being smooth, the curve is ragged near the yield point. At the upper yield strength, there is a pronounced drop (i.e., "drop of beam") in load-carrying ability to a plateau yield strength after the initial yielding occurs. The plateau value is known as the *lower yield strength* and is commonly reported as the yield strength.

---

[2]For steel at higher temperatures, the modulus of elasticity is reduced as follows.

temperature		% of original value
°F	°C	
70	20	100%
400	200	90%
800	425	75%
1000	540	65%
1200	650	60%

[3]A *kip* is a thousand pounds. *ksi* is the abbreviation for kips per square inch (thousands of lbf/in^2).

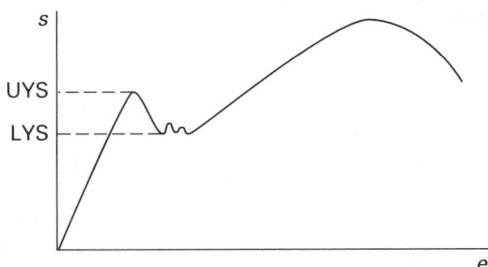

Figure 46.3 *Upper and Lower Yield Strengths*

Table 46.2 *Approximate Yield Strengths of Representative Materials*

	yield strength	
material	lbf/in^2	MPa
iron and steel		
1020	43,000	300
A36	36,000	250
stainless (304)	43,000	300
pure	24,000	160
copper		
beryllium	130,000	900
brass	11,000	75
pure	10,000	70
aluminum		
2024	50,000	345
6061	21,000	145
pure	5000	35
titanium		
alloy 6% Al, 4% V	160,000	1100
pure	20,000	140
nickel		
hastelloy	55,000	380
inconel	40,000	280
monel	35,000	240
pure	20,000	140

(Multiply lbf/in^2 by 6.89×10^{-3} to obtain MPa.)

The *ultimate strength* or *tensile strength*, S_u (or S_{ut} to indicate an ultimate tensile strength), point D in Fig. 46.2, is the maximum stress the material can support without failure. However, since stresses near the ultimate strength are accompanied by large plastic strains, this property is not used for the design of ductile materials.

The *breaking strength* or *fracture strength*, S_f, is the stress at which the material actually fails (point E in Fig. 46.2). For ductile materials, the breaking strength is less than the ultimate strength due to the necking down in the cross-sectional area that accompanies high plastic strains.

2. STRESS-STRAIN CHARACTERISTICS OF NONFERROUS METALS

Most nonferrous materials, such as aluminum, magnesium, copper, and other FCC and HCP metals, do not have well-defined yield points. The stress-strain curve

starts to bend at low stresses, as illustrated by Fig. 46.4. In such cases, the yield strength is commonly defined as the stress that will cause a 0.2% *parallel offset* (i.e., a plastic strain of 0.002).[4] However, the yield strength can also be defined by other offset values (e.g., 0.1% for metals and 1.0% for plastics).

The yield strength is found by extending a line from the offset strain value parallel to the linear portion of the curve until it intersects the curve.

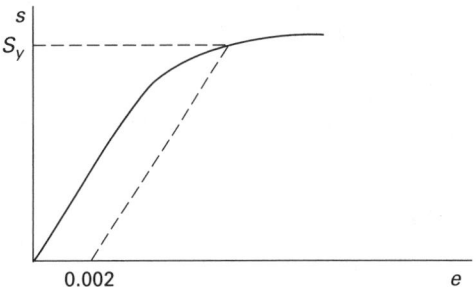

Figure 46.4 *Typical Stress-Strain Curve for a Nonferrous Metal*

With nonferrous metals, the difference between parallel offset and total strain characteristics is important. Sometimes, the yield point will be determined as the stress accompanying *0.5% total strain*.

3. STRESS-STRAIN CHARACTERISTICS OF BRITTLE MATERIALS

Brittle materials, such as glass, cast iron, and ceramics, can support only small stresses before they fail catastrophically (i.e., without warning). As the stress is increased, the elongation is linear and Hooke's law (Eq. 46.3) can be used to predict the strain. Failure occurs within the linear region, and there is very little, if any, necking down. Since the failure occurs at a low strain, brittle materials are not ductile. (The words "brittle" and "ductile" are antonyms.) Figure 46.5 is typical of the stress-strain curve of a brittle material.

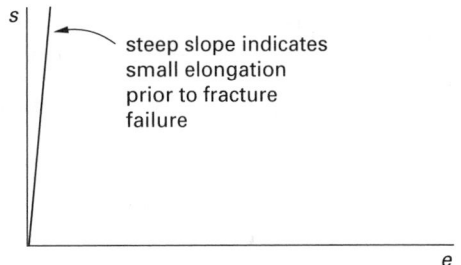

steep slope indicates small elongation prior to fracture failure

Figure 46.5 *Stress-Strain Curve of a Brittle Material*

---

[4]In Great Britain, the 0.2% offset strength is known as the *proof stress*.

4. THE SECANT MODULUS

The modulus of elasticity, E, is usually determined from the steepest portion of the stress-strain curve. (This avoids the difficulty of locating the starting part of the curve.) For materials operating in the nonlinear region, the *secant modulus* gives the average ratio of stress to strain. The secant modulus is the slope of the straight line connecting the origin and the point of operation. Some designs using elastomers or concrete may be based on the secant modulus.

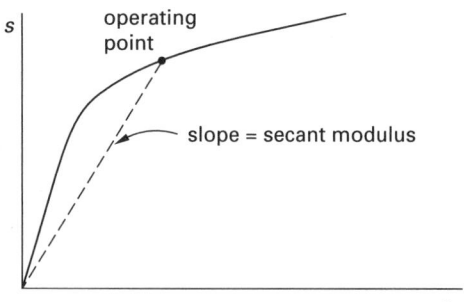

Figure 46.6 *Secant Modulus*

5. POISSON'S RATIO

As a specimen elongates axially during a tensile test, it will also decrease slightly in diameter or breadth. For any specific material, the percentage decrease in diameter, known as the *lateral strain*, will be a fraction of the *axial strain*. The ratio of the lateral strain to the axial strain is known as *Poisson's ratio*, ν, which is taken as approximately 0.3 for most metals.

$$\nu = \frac{e_{\text{lateral}}}{e_{\text{axial}}} = \frac{\dfrac{\Delta D}{D_o}}{\dfrac{\delta}{L_o}} \qquad 46.4$$

Poisson's ratio applies only to elastic strain. When the stress is removed, the lateral strain disappears along with the axial strain.

Table 46.3 *Approximate Values of Poisson's Ratio*

material	Poisson's ratio
liquids	0.50[a]
rubber	0.49
thermosetting plastics	0.40–0.45
aluminum	0.32–0.34 (0.33)[b]
magnesium	0.35
copper	0.33–0.36 (0.33)[b]
titanium	0.34
brass	0.33–0.36
stainless steel	0.30
steel	0.26–0.30 (0.30)[b]
nickel	0.30
beryllium	0.27
cast iron	0.21–0.33 (0.27)[b]
glass (SiO_2)	0.21–0.27 (0.23)[b]
diamond	0.20

[a]Limiting value.
[b]Commonly used for design.

6. STRAIN HARDENING AND NECKING DOWN

When the applied stress exceeds the yield strength, the specimen will experience plastic deformation and will strain harden. (Plastic deformation is primarily due to the shear stress-induced movement of dislocations.) Since the specimen volume is constant (i.e., $A_oL_o = AL$), the cross-sectional area decreases. Initially, the strain hardening more than compensates for the decrease in area, so the material's strength increases, and the engineering stress increases with larger strains.

Eventually, a point is reached when the available strain hardening and increase in strength cannot keep up with the decrease in cross-sectional area. The specimen then begins to neck down (at some local weak point), and all subsequent plastic deformation is concentrated at the neck. The cross-sectional area decreases even more rapidly thereafter since only a small portion of the specimen volume is strain hardening. The engineering stress decreases to failure.

Figure 46.7 shows necking down in two different specimens tested to failure. Very ductile materials pull out to a point, while most moderately ductile materials exhibit a *cup-and-cone failure*. (Failed brittle materials, not shown, do not exhibit any reduction in area.)

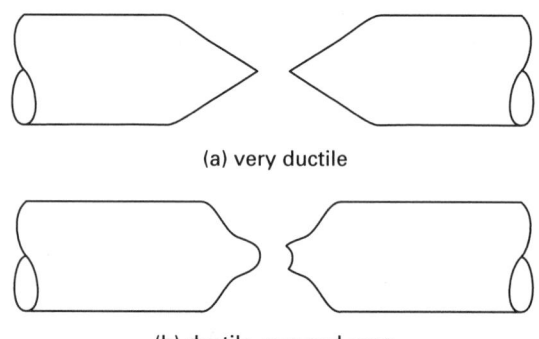

(a) very ductile

(b) ductile, cup and cone

Figure 46.7 *Types of Tensile Ductile Failure*

7. TRUE STRESS AND STRAIN

Engineering stress, given by Eq. 46.1, is calculated for all stress levels from the original cross-sectional area. However, during a tensile test, the area of a specimen decreases as the stress increases. The decrease is only slight in the elastic region but is much more significant after plastic deformation begins.

If the stress is calculated from the instantaneous area, it is known as *true stress* or *physical stress*, σ. (Fractional reduction in area, used in Eq. 46.6, is expressed as a number between 0 and 1.)

$$\sigma = \frac{F}{A}$$
$$= \frac{F}{(1 - \text{fractional reduction in area})\, A_o}$$
$$= s(1 + e) \qquad 46.5$$

Engineering strain, given by Eq. 46.2, is calculated from the original length, although the actual length increases during the tensile test. The *true strain* or *physical strain*, ϵ, is found from Eqs. 46.8 and 46.9.

$$\epsilon = \int_{L_0}^{L} \frac{dL}{L} = \ln\left(\frac{L}{L_o}\right)$$
$$= \ln(1 + e) \quad \text{[prior to necking]} \qquad 46.6$$

Since the plastic deformation occurs through a shearing process, there is essentially no volume decrease during elongation.

$$A_o L_o = AL \qquad 46.7$$

Therefore, true strain can be calculated from the cross-sectional areas and, for a circular specimen, from diameters. If necking down has occurred, true strain must be calculated from the areas or diameters, not the lengths.

$$\epsilon = \ln\left(\frac{A_o}{A}\right) = \ln\left(\frac{D_o}{D}\right)^2 = 2\ln\left(\frac{D_o}{D}\right) \qquad 46.8$$

Figure 46.8 compares engineering and true stresses and strains for a ferrous alloy. A graph of true stress and true strain is known as a *flow curve*. Log σ can also be plotted against log ϵ, resulting in a straight-line relationship.

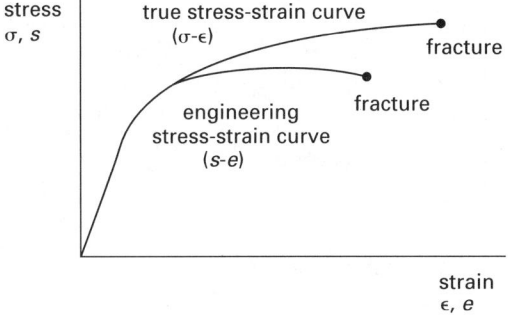

Figure 46.8 *True and Engineering Stresses and Strains for a Ferrous Alloy*

The flow curve of many metals in the plastic region can be expressed by the relationship of Eq. 46.9. K is known as the *strength coefficient*, and n is the *strain-hardening exponent*. Values of both vary greatly with material, composition, and heat treatment. n can vary from 0 (for a perfectly inelastic solid) to 1.0 (for an elastic solid). Typical values are between 0.1 and 0.5.

$$\sigma = K\epsilon^n \qquad 46.9$$

Although true stress and strain are more accurate, almost all engineering work is based on engineering stress and strain, which is justifiable for two reasons: (1) design using ductile materials is limited to the elastic region where engineering and true values differ little, and (2) the reduction in area of most parts at their service stresses is not known; only the original area is known.

Example 46.1

The engineering stress in a solid tension member was 47,000 lbf/in² at failure. The reduction in area was 80%. What were the true stress and strain at failure?

Solution

Since engineering stress, s, is F/A_o, from Eq. 46.5 the true stress is

$$\sigma = \frac{s}{1 - \text{reduction in area}}$$
$$= \frac{47,000 \ \dfrac{\text{lbf}}{\text{in}^2}}{1 - 0.80} = 235,000 \ \text{lbf/in}^2$$

From Eq. 46.6, the true strain is

$$\epsilon = \ln\left(\frac{1}{1 - 0.80}\right) = 1.61 \quad (161\%)$$

8. DUCTILITY

A material that deforms and elongates a great deal before failure is said to be a *ductile material*. (Steel, for example, is a ductile material.) The *percent elongation*, short for *percent elongation at failure*, is the total plastic strain at failure. (Percent elongation does not include the elastic strain, because even at ultimate failure the material snaps back an amount equal to the elastic strain.)

$$\begin{aligned} \text{percent} \atop \text{elongation} &= \frac{L_f - L_o}{L_o} \times 100\% \\ &= e_f \times 100\% \qquad 46.10 \end{aligned}$$

The value of the final strain to be used in Eq. 46.10 is found by extending a line from the failure point downward to the strain axis, parallel to the linear portion of the curve. This is equivalent to putting the two broken specimen pieces together and measuring the total length.

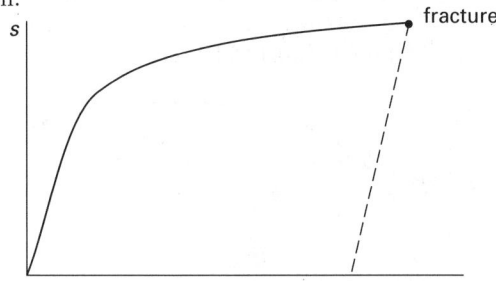

Figure 46.9 *Percent Elongation*

Highly ductile materials exhibit large percent elongations. However, percent elongation is not the same as *ductility*.

$$\text{ductility} = \frac{\text{ultimate failure strain}}{\text{yielding strain}} \qquad 46.11$$

Materials

The *reduction in area* (at the point of failure), expressed as a percentage or decimal fraction, is a third measure of a material's ductility. The reduction in area due to necking down will be 50% or greater for ductile materials and less than 10% for brittle materials.[5]

$$\frac{\text{reduction}}{\text{in area}} = \frac{A_o - A_f}{A_o} \times 100\% \qquad 46.12$$

9. STRAIN ENERGY

Strain energy, also known as *internal work*, is the energy per unit volume stored in a deformed material. The strain energy is equivalent to the work done by the applied tensile force. Simple work is calculated as the product of a force moving through a distance.

$$\text{work} = \text{force} \times \text{distance} = \int F\, dL \qquad 46.13$$

$$\frac{\text{work per}}{\text{unit volume}} = \int \frac{F\, dL}{AL} = \int_0^{\epsilon_{\text{final}}} \sigma\, d\epsilon \qquad 46.14$$

This work per unit volume corresponds to the area under the true stress-strain curve. Units are in-lbf/in^3 (i.e., inch-pounds (a unit of energy) per cubic inch (a unit of volume)), usually shortened to lbf/in^2 (MPa). (Equation 46.16 cannot be simplified further because stress is not proportional to strain for the entire curve.)

10. RESILIENCE

A *resilient material* is able to absorb and release *strain energy* without permanent deformation. *Resilience* is measured by the *modulus of resilience*, also known as the *elastic toughness*, which is the strain energy per unit volume required to reach the yield point. This is represented by the area under the stress-strain curve up to the yield point. Since the stress-strain curve is essentially a straight line up to that point, the area is triangular.

$$U_R = \int_0^{\epsilon_y} \sigma\, d\epsilon = E \int_0^{\epsilon_y} \epsilon\, d\epsilon = \frac{E\epsilon_y^2}{2} = \frac{S_y \epsilon_y}{2} \qquad 46.15$$

The modulus of resilience varies greatly for steel. It can be more than ten times higher for high-carbon spring steel ($U_R = 320$ lbf/in^2, 2.2 MPa) than for low-carbon steel ($U_R = 20$ lbf/in^2, 0.14 MPa).

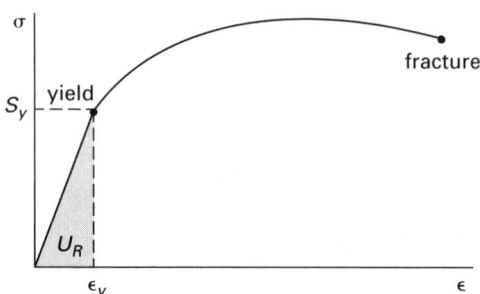

Figure 46.10 *Modulus of Resilience*

[5] *Notch-brittle materials* have reductions in area that are moderate (e.g., 25 to 35%) when tested in the usual manner but close to zero when the test specimen is given a small notch or crack.

11. TOUGHNESS

A *tough material* will be able to withstand occasional high stresses without fracturing. Products subjected to sudden loading, such as chains, crane hooks, railroad couplings, and so on, should be tough. One measure of a material's *toughness* is the *modulus of toughness* (i.e, the strain energy or work per unit volume required to cause fracture). This is the total area under the stress-strain curve, and is given the symbol U_T. Since the area is irregular, the modulus of toughness cannot be exactly calculated by a simple formula. However, the modulus of toughness of ductile materials (with large strains at failure) can be approximately calculated from either Eq. 46.16 or Eq. 46.17.

$$U_T \approx S_u \epsilon_f \quad \text{[ductile]} \qquad 46.16$$

$$U_T \approx \left(\frac{S_y + S_u}{2} \right) \epsilon_f \quad \text{[ductile]} \qquad 46.17$$

For brittle materials, the stress-strain curve may be either linear or parabolic. If the curve is parabolic, Eq. 46.18 approximates the modulus of toughness.

$$U_T \approx \frac{2}{3} S_u \epsilon_f \quad \text{[brittle]} \qquad 46.18$$

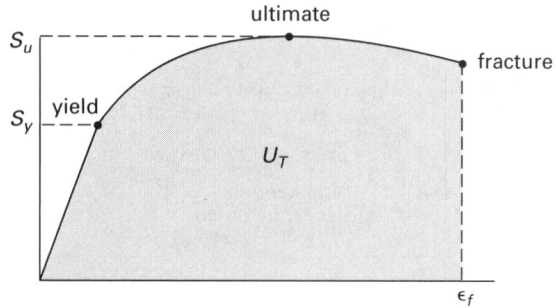

Figure 46.11 *Modulus of Toughness*

12. UNLOADING AND RELOADING

If the load is removed after a specimen is stressed elastically, the material will return to its original state. If the load is removed after a specimen is stressed into the plastic region, the *unloading curve* will follow a sloped path back to zero stress. The slope of the unloading curve will be equal to the original modulus of elasticity, E, illustrated by Fig. 46.12.

If this same material is subsequently reloaded, the *reloading curve* will follow the previous unloading curve up to the continuation of the original stress-strain curve. Therefore, the *apparent yield stress* of the reloaded specimen will be higher. This extra strength is the result of the strain hardening that has occurred.[6] Although the material will have a higher strength, its ductility will have been reduced.

[6] The additional strength is lost if the material is subsequently annealed.

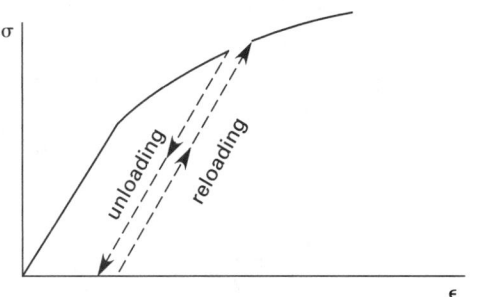

Figure 46.12 *Unloading and Reloading Curves*

13. COMPRESSIVE STRENGTH

Compressive strength, S_{yc} (i.e, ultimate strength in compression), is an important property for brittle materials such as concrete and cast iron that are primarily loaded in compression only.[7] The compressive strengths of these materials are much greater than their tensile strengths, while the compressive strengths for ductile materials, such as steel, are the same as their tensile yield strengths.

Within the linear (elastic) region, Hooke's law is valid for compression of both brittle and ductile materials.

The failure mechanism for ductile materials is plastic deformation alone. Such materials do not rupture in compression. Thus, a ductile material can support a load long after the material is distorted beyond a useful shape.

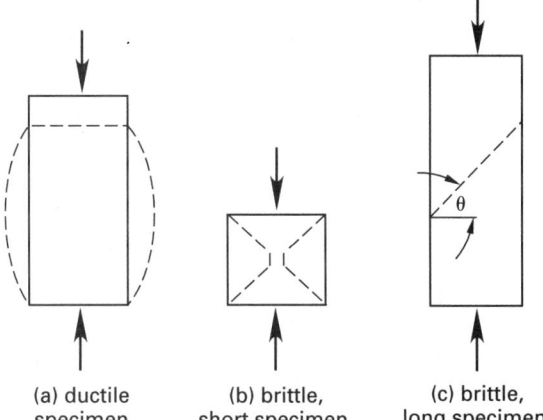

(a) ductile
specimen

(b) brittle,
short specimen

(c) brittle,
long specimen

Figure 46.13 *Compressive Failures*

The failure mechanism for brittle materials is shear along an inclined plane. Figure 46.13 shows the characteristic plane and hourglass failures for brittle materials.[8] Theoretically, only *cohesion* contributes to compressive strength, and the *angle of rupture* (i.e., the incline angle), θ, should be 45°. In real materials, however, internal friction also contributes strength. The angles

---

[7] f_c' is commonly used as the symbol for the compressive strength of concrete.
[8] The *hourglass failure* appears when the material is too short for a complete failure surface to develop.

of rupture for cast iron, concrete, brick, and so on vary roughly between 50° and 60°. If the *angle of internal friction*, ϕ, is known for the material, the angle of rupture can be calculated exactly from *Mohr's theory of rupture*.

$$\theta = 45° + \frac{\phi}{2} \qquad 46.19$$

14. THE TORSION TEST

Figure 46.14 illustrates a simple cube loaded by a shear stress, τ. The volume of the cube does not decrease when loaded, but the shape changes. The *shear strain* is the angle, θ, expressed in radians. The shear strain is proportional to the shear stress, analogous to Hooke's law for tensile loading. G is the *shear modulus*, also known as the *modulus of shear*, *modulus of elasticity in shear*, and *modulus of rigidity*.

$$\tau = G\theta \qquad 46.20$$

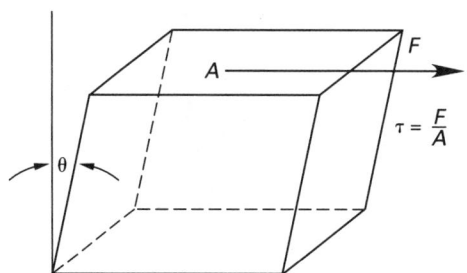

Figure 46.14 *Cube Loaded in Shear*

Table 46.4 *Approximate Values of Shear Modulus*

material	lbf/in^2	GPa
aluminum	3.8×10^6	26
brass	5.5×10^6	38
copper	6.2×10^6	43
cast iron	8.0×10^6	55
magnesium	2.4×10^6	17
steel	11.5×10^6	79
stainless steel	10.6×10^6	73
titanium	6.0×10^6	41
glass	4.2×10^6	29

(Multiply lbf/in^2 by 6.89×10^{-6} to obtain GPa.)

The shear modulus can be calculated from the modulus of elasticity and Poisson's ratio and, therefore, can be derived from the results of a tensile test.

$$G = \frac{E}{(2)(1+\nu)} \qquad 46.21$$

The shear stress can also be calculated from a torsion test, as illustrated by Fig. 46.15. Equation 46.22 relates the angle of twist (in radians) to the shear modulus.

$$\theta = \frac{TL}{JG} = \frac{\tau L}{rG} \quad \text{[radians]} \qquad 46.22$$

Materials

Table 46.5 *Relationships Between Elastic Constants*

elastic constants	in terms of				
	E, ν	E, G	B, ν	B, G	E, B
E	–	–	$(3)(1-2\nu)B$	$\dfrac{9BG}{3B+G}$	–
ν	–	$\dfrac{E}{2G} - 1$	–	$\dfrac{3B-2G}{(2)(3B+G)}$	$\dfrac{3B-E}{6B}$
G	$\dfrac{E}{(2)(1+\nu)}$	–	$\dfrac{(3)(1-2\nu)B}{(2)(1+\nu)}$	–	$\dfrac{3EB}{9B-E}$
B	$\dfrac{E}{(3)(1-2\nu)}$	$\dfrac{GE}{(3)(3G-E)}$	–	–	–

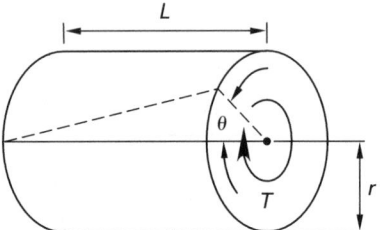

Figure 46.15 *Uniform Bar in Torsion*

The *shear strength*, S_s or S_{ys}, of a material is the maximum shear stress that the material can support without yielding in shear. (The ultimate shear strength property is rarely encountered.) For ductile materials, *maximum shear stress theory* predicts the shear strength as one-half of the tensile yield strength. A more accurate relationship is derived from the *distortion energy theory* (also known as *von Mises theory*).

$$S_{ys} = \frac{S_{yt}}{\sqrt{3}} = 0.577 S_{yt} \qquad 46.23$$

15. RELATIONSHIP BETWEEN THE ELASTIC CONSTANTS

The elastic constants (modulus of elasticity, shear modulus, bulk modulus, and Poisson's ratio) are related in elastic materials. Table 46.5 lists the common relationships.

16. FATIGUE TESTING

A material can fail after repeated stress loadings even if the stress level never exceeds the ultimate strength, a condition known as *fatigue failure*.

The behavior of a material under repeated loadings is evaluated by a fatigue test. A specimen is loaded repeatedly to a specific stress amplitude, s, and the number of applications of that stress required to cause failure, N, is counted. Rotating beam tests that load the specimen in bending are more common than alternating deflection and push-pull tests but are limited to round specimens.[9]

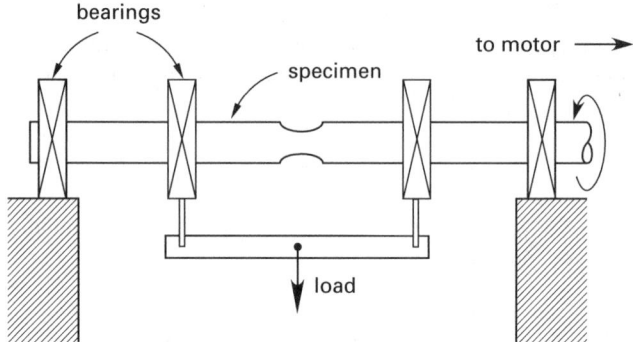

Figure 46.16 *Rotating Beam Test*

This procedure is repeated for different stresses, using eight to fifteen specimens. The results of these tests are graphed, resulting in an *S-N curve* (i.e., stress-number of cycles) shown in Fig. 46.17.

For an alternating stress test, the stress plotted on the *S-N* curve can be the maximum, minimum, or mean value. The choice depends on the method of testing as well as the intended application. The maximum stress should be used in rotating beam tests, since the mean stress is zero. For cyclic, one-dimensional bending, the maximum and mean stresses are commonly used.

---

[9]In the design of ductile steel buildings, the static case is assumed up to 20,000 cycles. However, in critical applications (such as nuclear steam vessels, turbines, and so on) that experience temperature swings, fatigue failure can occur with a smaller number of cycles due to *cyclic strain*, not due to *cyclic stress*.

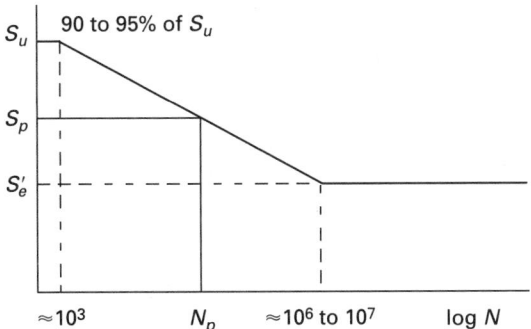

Figure 46.17 *Typical S-N Curve for Steel*

For a specific stress level, say S_p in Fig. 46.17, the number of cycles required to cause failure, N_p, is the *fatigue life*. S_p is the *fatigue strength* corresponding to N_p.

For steel subjected to fewer than approximately 10^3 loadings, the fatigue strength starts at the ultimate strength and drops to 90 to 95% of the ultimate strength at 10^3 cycles. (Although *low-cycle fatigue* theory has its own peculiarities, a part experiencing a small number of cycles can usually be designed or analyzed for static loading.) The curve is linear between 10^3 and 10^6 cycles if a logarithmic N-scale is used. Beyond 10^6 to 10^7 cycles, there is no further decrease in strength.

Therefore, below a certain stress level, called the *endurance limit* or *endurance strength*, S'_e, the material will withstand an almost infinite number of loadings without experiencing failure.[10] This is characteristic of steel and titanium. Therefore, if a dynamically loaded part is to have an infinite life, the stress must be kept below the endurance limit. The ratio S'_e/S_u is known as the *endurance ratio* or *fatigue ratio*. For carbon steel, the endurance ratio is approximately 0.4 for pearlitic, 0.60 for ferritic, and 0.25 for martensitic microstructures. For martensitic alloy steels, it is approximately 0.35.

For steel whose microstructure is unknown, the endurance strength is given approximately by Eq. 46.24.[11]

$$S'_{e,\text{steel}} \begin{cases} = 0.5S_u & [S_u < 200{,}000\,\text{lbf/in}^2] \\ & [S_u < 1.4\,\text{GPa}] \\ = 100{,}000\,\text{lbf/in}^2 & [S_u > 200{,}000\,\text{lbf/in}^2] \\ (700\,\text{MPa}) & [S_u > 1.4\,\text{GPa}] \end{cases}$$

$$46.24$$

For cast iron, the endurance strength-ultimate strength correlation is lower.

$$S'_{e,\text{cast iron}} = 0.4S_u \qquad 46.25$$

Steel and titanium are the most important engineering materials that have well-defined endurance limits.

---

[10]Most endurance tests use some form of sinusoidal loading. However, the fatigue and endurance strengths do not depend much on the shape of the loading curve. Only the maximum amplitude of the stress is relevant. Therefore, the endurance limit can be used with other types of loading (sawtooth, square wave, random, etc.).
[11]The coefficient in Eq. 46.24 actually varies between 0.25 and 0.6. However, 0.5 is commonly quoted.

Many nonferrous metals and alloys, such as aluminum, magnesium, and copper alloys, do not have well-defined endurance limits. The strength continues to decrease with cyclic loading and never levels off. In such cases, the endurance limit is taken as the stress that causes failure at 10^8 or 5×10^8 loadings. Alternatively, the endurance strength is approximated by Eq. 46.26.

$$S'_{e,\text{aluminum}} = \begin{cases} 0.3S_u & [\text{cast}] \\ 0.4S_u & [\text{wrought}] \end{cases} \qquad 46.26$$

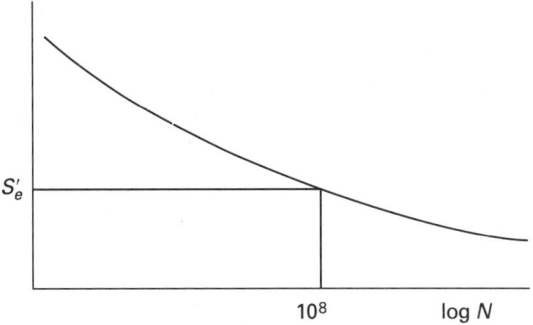

Figure 46.18 *Typical S-N Curve for Aluminum*

The yield strength is an irrelevant factor in cyclic loading. Fatigue failures are fracture failures; they are not yielding failures. They start with microscopic cracks at the material surface. Some of the cracks are present initially; others form when repeated cold working reduces the ductility in strain-hardened areas. These cracks grow minutely with each loading. Since cracks start at the location of surface defects, the endurance limit is increased by proper treatment of the surface. Such treatments include polishing, surface hardening, shot peening, and filleting joints.

The endurance limit is not a true property of the material since the other significant influences, particularly surface finish, are never eliminated. However, representative values of S'_e obtained from ground and polished specimens provide a baseline to which other factors can be applied to account for the effects of surface finish, temperature, stress concentration, notch sensitivity, size, environment, and desired reliability. These other influences are accounted for by reduction factors, K, which are used to calculate a working endurance strength, S_e, for the material.

$$S_e = KS'_e \qquad 46.27$$

Since a rough surface significantly decreases the endurance strength of a specimen, it is not surprising that notches (and other features that produce stress concentration) do so as well. In some cases, the theoretical tensile *stress concentration factor*, K_t, due to notches and other features, can be determined theoretically or experimentally. The ratio of the fatigue strength of a polished specimen to the fatigue strength of a notched

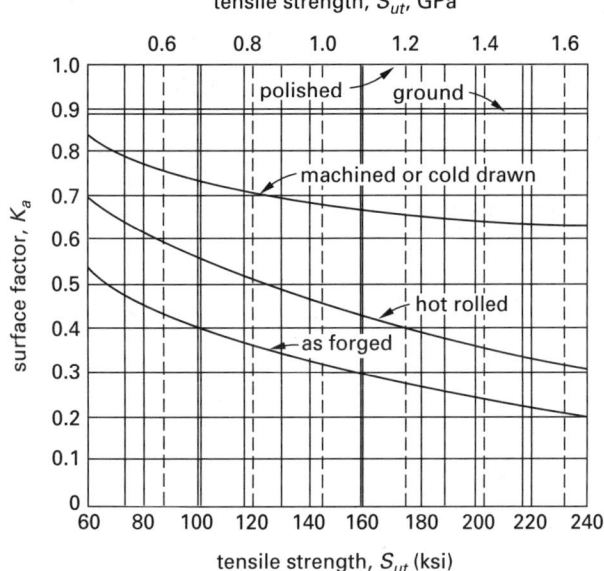

(Reprinted with permission from *Mechanical Engineering Design*, Third Edition, by Joseph Edward Shigley, McGraw-Hill, Inc., © 1977.)

Figure 46.19 *Surface Finish Reduction Factors for Endurance Strength*

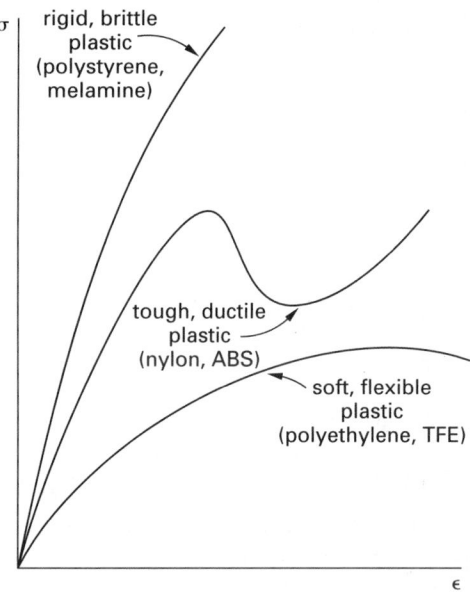

Figure 46.20 *Typical Tensile Test Performance For Plastics (loaded at 2 in/min)*

specimen at the same number of cycles is known as the *fatigue notch factor*, K_f, also known as the *fatigue stress concentration factor*. The *fatigue notch sensitivity*, q, is a measure of the degree of agreement between the stress concentration factor and the fatigue notch factor.

$$q = \frac{K_f - 1}{K_t - 1} \qquad 46.28$$

17. TESTING OF PLASTICS

With reasonable variations, mechanical properties of plastics are evaluated using the same methods as for metals.[12] Although temperature is an important factor in the testing of plastics, tests for tensile strength, endurance, hardness, toughness, and creep rate are similar or the same as for metals. Figure 46.20 illustrates typical tensile test results.[13]

Unlike metals, which follow Hooke's law, plastics are non-Hookean. (They may be Hookean for a short-duration loading.) The modulus of elasticity, for example, changes with stress level, temperature, time, and

chemical environment. A plastic that appears to be satisfactory under one set of conditions can fail quickly under slightly different conditions. Therefore, properties of plastics determined from testing (and from tables) should be used only to compare similar materials, not to predict long-term behavior. Plastic tests are used to determine material specifications, not performance specifications.

On a short-term basis, plastics behave elastically. They distort when loaded and spring back when unloaded. Under prolonged loading, however, creep (cold flow) becomes significant. When loading is removed, there is some instantaneous recovery, some delayed recovery, and some permanent deformation. The recovery might be complete if the load is removed within 10 hours, but if the loading is longer (e.g., 100 hours), recovery may be only partial. Because of this behavior, plastics are subjected to various other tests.

Additional tests used to determine the mechanical properties of plastics include deflection temperature, long-term (e.g., 3000 hours) tensile creep, creep rupture, and *stress-relaxation* (long-duration, constant-strain tensile testing at elevated temperatures).[14] Because some plastics deteriorate when exposed to light, plasma, or chemicals, performance under these conditions can be evaluated, as can the insulating and dielectric properties.

The *creep modulus* (also known as *apparent modulus*), determined from tensile creep testing, is the instantaneous ratio of stress to creep strain. The creep modulus decreases with time. The *deflection temperature* test indicates the dimensional stability of a plastic at high temperatures. A plastic bar is loaded laterally

---

[12]For example, plastic specimens for tensile testing can be produced by injection molding as well as by machining from compression-molded plaques, rather than machining from bar stock.

[13]Plastics are sensitive to the rate of loading. Figure 46.20 illustrates tensile performance based on a 2 in/min loading rate. However, for a fast loading rate (e.g., 2 in/sec), most plastics would exhibit brittle performance. On the other hand, given a slow loading rate (e.g., 2 in/month), most would behave as a soft and flexible plastic. Therefore, with different rates of loading, all three types of stress-strain performance shown in Fig. 46.20 can be obtained from the same plastic.

[14]Plastic pipes have their own special tests (ASTM D1598).

(as a beam) to a known stress level, and the temperature of the bar is gradually increased. The temperature at which the deflection reaches 0.010 in (0.254 mm) is taken as the *thermal deflection temperature* (TDT). The *Vicat softening point*, primarily used with polyethylenes, is the temperature at which a loaded standard needle penetrates 1 mm when the temperature is uniformly increased at a standard rate.

18. NONDESTRUCTIVE TESTING

Nondestructive testing (NDT) or *nondestructive evaluation* (NDE) is used when it is impractical or uneconomical to perform destructive sampling on manufactured products. Typical applications of NDT are inspection of helicopter blades, cast aluminum wheels, and welds in nuclear pressure vessels. Some procedures are particularly useful in providing quality monitoring on a continuous, real-time basis. In addition to visual processes, the main types of nondestructive testing are magnetic particle, eddy current, liquid penetrant, ultrasonic imaging, acoustic emission, and infrared testing, as well as radiography.

The *visual-optical* process differs from normal visual inspection in the use of optical scanning systems, borescopes, magnifiers, and holographic equipment. Flaws are identified as changes in light intensity (reflected, transmitted, or refracted), color changes, polarization changes, or phase changes. This method is limited to the identification of surface flaws or interior flaws in transparent materials.

Magnetic particle testing takes advantage of the attraction of ferromagnetic powders (e.g., the *Magnaflux*™ *process*) and fluorescent particles (e.g., the *Magnaglow*™ *process*) to leakage flux at surface flaws in magnetic materials. The particles accumulate and become visible at such flaws when an intense magnetic field is set up in a part.

This method can locate most surface flaws (such as cracks, laps, and seams) and in some special cases, subsurface flaws. The procedure is fast and simple to interpret. However, parts must be ferromagnetic and clean. Following the test, demagnetization may be required. A high-current power source is required.

Eddy current testing uses alternating current from a test coil to induce eddy currents in electrically conducting, metallic objects. Flaws and other material properties affect the current flow. The change in current is monitored by a detection circuit or on a meter or screen. This method can be used to locate defects of many types, including cracks, voids, inclusions, and weld defects, as well as to find changes in composition, structure, hardness, and porosity.

Intimate contact between the material and the test coil is not required. Operation can be continuous, automatic, and monitored electronically. Sensitivity is easily adjusted. Therefore, this method is ideal for unattended continuous processing. Many variables, however, can affect the current flow, and only electrically conducting materials can be tested with this method.

With *infrared testing*, infrared radiation emitted from objects can be detected and correlated with quality. Any discontinuities that interrupt heat flow, such as flaws, voids, and inclusions, can be detected.

Infrared testing requires access to only one side and is highly sensitive. It is applicable to complex shapes and assemblies of dissimilar components but is relatively slow. The detection can be performed electronically. Results are affected by variations in material size, coatings, and colors, and hot spots can be hidden by cool surface layers.

Liquid penetrant testing is based on a fluorescent dye being drawn by capillary action into surface defects. A developer substance is commonly used to aid in visual inspection. This method can be used with any nonporous material, including metals, plastics, and glazed ceramics. It is capable of finding cracks, porosities, pits, seams, and laps.

Liquid penetrant tests are simple, can be used with complex shapes, and can be performed on site. Parts must be clean and nonporous. However, only small surface defects are detectable.

In *ultrasound imaging testing (ultrasonics)*, mechanical vibrations in the 0.1 to 25 MHz range are induced by pressing a piezoelectric transducer against a workpiece. The transmitted waves are normally reflected back, but the waves are scattered by interior defects. The results are interpreted by reading a screen or meter. The method can be used for metals, plastics, glass, rubber, graphite, and concrete. It is excellent for detecting internal defects such as inclusions, cracks, porosities, laminations, and changes in material structure.

Ultrasound testing is extremely flexible. It can be automated and is very fast. Results can be recorded or interpreted electronically. Penetration through thick steel layers is possible. Direct contact (or immersion in a fluid) is required, but only one surface needs to be accessible. Rough surfaces and complex shapes may cause difficulties, however. A related method, *acoustic emission monitoring*, is used to test pressurized systems.

There are two types of *holographic NDT methods*. *Acoustic holography* is a form of ultrasonic testing that passes an ultrasonic beam through the workpiece (or through a medium such as water surrounding the workpiece) and measures the displacement of the workpiece (or medium). With suitable processing, a three-dimensional hologram is formed that can be visually inspected.

In one form of *optical holography*, a hologram of the unloaded workpiece is imposed on the actual workpiece. If the workpiece is then loaded (stressed), the observed changes (e.g., deflections) from the holographic image will be non-uniform when discontinuities and defects are present.

Materials

Radiography (i.e., *nuclear sensing*) uses neutron, X-ray, gamma-ray (e.g., Ce-137), and isotope (e.g., Co-60) sources. (When neutrons are used, the method is known as *neutron radiography* or *neutron gaging*.) The intensity of emitted radiation is changed when the rays pass through defects, and the intensity changes are monitored on a fluoroscope or recorded on film. This method can be used to detect internal defects, changes in material structure, thickness, and the absence of internal parts. It is also used to check liquid levels in filled containers.

Up to 30 in (0.75 m) of steel can be penetrated by X-ray sources. Gamma sources, which are more portable and lower in cost than X-ray sources, can be used with steel up to 10 in (0.25 m).

Radiography requires access to both sides of the part. Radiography involves some health risk, and there may be government standards associated with its use. Electrical power and cooling water may be required in large installations. Shielding and film processing are also required, making this the most expensive form of nondestructive testing.

19. HARDNESS TESTING

Hardness tests measure the capacity of a surface to resist deformation. The main use of hardness testing is to verify heat treatments, an important factor in product service life. Through empirical correlations, it is also possible to predict the ultimate strength and toughness of some materials.

The *Brinell hardness test* is used primarily with iron and steel castings. The *Brinell hardness number*, BHN, is determined by pressing a hardened steel ball into the surface of a specimen. The diameter of the resulting depression is correlated to the hardness. The standard ball is 10 mm in diameter and loads are 500 kg and 3000 kg for soft and hard materials, respectively.

The Brinell hardness number is the load per unit contact area. If a load, P (in kilograms), is applied through a steel ball of diameter D (in millimeters) and produces a depression of diameter d (in millimeters) and depth t (in millimeters), the Brinell hardness number is calculated from Eq. 46.29.

$$\text{BHN} = \frac{P}{A_{\text{contact}}} = \frac{P}{\pi D t}$$
$$= \frac{2P}{\pi D \left(D - \sqrt{D^2 - d^2}\right)} \qquad 46.29$$

For heat-treated plain-carbon and medium-alloy steels, the ultimate tensile strength in lbf/in^2 can be approximately calculated from the steel's Brinell hardness number.

$$S_u \approx (500)(\text{BHN}) \qquad 46.30$$

The *Rockwell hardness test* is similar to the Brinell test. A steel ball or diamond spheroconical penetrator

(known as a *brale indentor*) is pressed into the material. The machine applies an initial load (10 kg) that sets the penetrator below surface imperfections. Then a significant load is applied. The Rockwell hardness is determined from the depth of penetration and is read directly from a dial.

Although a number of Rockwell scales (A through G) exist, the B and C scales are commonly used for steel. The *Rockwell B scale* is used with a steel ball for mild steel and high-strength aluminum. The *Rockwell C scale* is used with the brale indentor for hard steels having ultimate tensile strengths up to 300 ksi (2 GPa). The *Rockwell A scale* has a wide range and can be used with both soft materials (such as annealed brass) and hard materials (such as cemented carbides).

Other penetration hardness tests include the *Meyer*, *Vickers*, *Meyer-Vickers*, and *Knoop* tests, as described in Table 46.6.

Cutting hardness is a measure of the force per unit area to cut a chip at low speed.

$$\text{cutting hardness} = \frac{F}{bt} \qquad 46.31$$

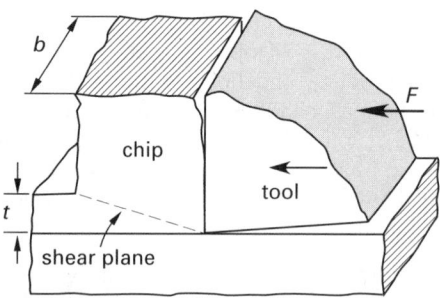

Figure 46.21 *Cutting Hardness*

The *scratch hardness test*, also known as the *Mohs test*, compares the hardness of the material to that of minerals. Minerals of increasing hardness are used to scratch the sample. The resulting *Mohs scale* hardness can be used or correlated to other hardness scales, as in Fig. 46.22.

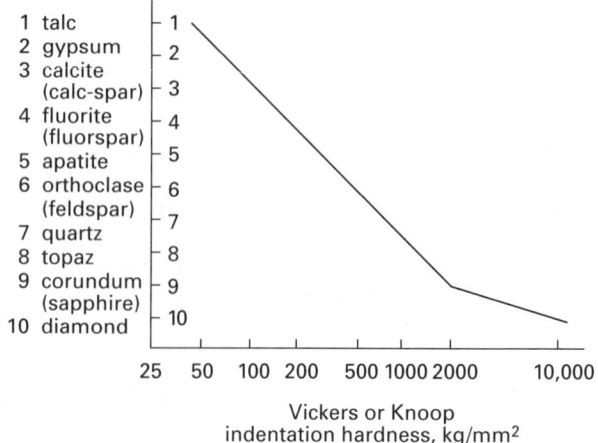

Figure 46.22 *Mohs Hardness Scale*

Table 46.6 *Hardness Penetration Tests*

test	penetrator	diagram	measured dimension	hardness
Brinell	sphere	(a)	diameter, d	$BHN = \dfrac{2P}{\pi D \left(D - \sqrt{D^2 - d^2}\right)}$
Rockwell[a]	sphere or penetrator	(b)	depth, t	$R = C_1 - C_2 t$
Vickers	square pyramid	(b)	mean diagonal, d_1	$VHN = \dfrac{1.854P}{d_1^2}$
Meyer	sphere	(a)	diameter, d	$MHN = \dfrac{4P}{\pi d^2}$
Meyer-Vickers	square pyramid	(b)	mean diagonal, d_1	$M_V = \dfrac{2P}{d_1^2}$
Knoop	asymmetrical pyramid	(c)	long diagonal, L	$K = \dfrac{14.2P}{L^2}$

[a]C_1 and C_2 are constants that depend on the scale.

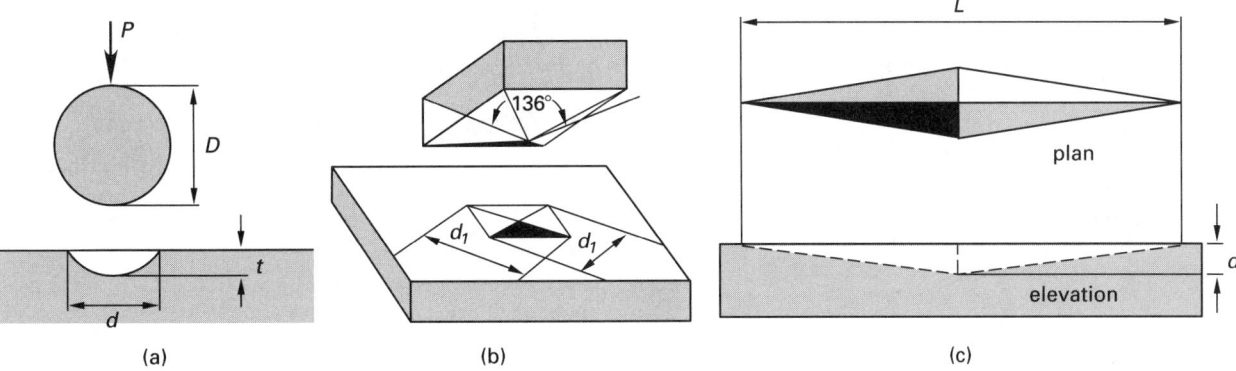

The *file hardness* test is a combination of the cutting and scratch tests. Files of known hardness are drawn across the sample. The file ceases to cut the material when the material and file hardnesses are the same.

All of the preceding hardness tests are *destructive tests* because they mar the material surface. However, *ultrasonic tests* and various *rebound tests* (e.g., the *Shore hardness test* and the *scleroscopic hardness test*) are *nondestructive tests*. In a rebound test, a standard object, usually a diamond-tipped hammer, is dropped from a standard height onto the sample. The height of the rebound is measured and correlated to other hardness scales.

The various hardness tests do not measure identical properties of the material, so correlations between the various scales are not exact. For steel, the Brinell and Vickers hardness numbers are approximately the same below values of 320 Brinell. Also, the Brinell hardness is approximately ten times the Rockwell C hardness (R_c) for $R_c > 20$. Table 46.7 is an accepted correlation between several of the scales for steel. The table should not be used for other materials.

20. TOUGHNESS TESTING

During World War II, the United States experienced spectacular failures in approximately 25% of its Liberty ships and T-2 tankers. The mild steel plates of these ships were connected by welds that lost their ductility and became brittle at winter temperatures. Some of the ships actually broke into two pieces. Such *brittle failures* are most likely to occur when three conditions are met: (1) triaxial stress, (2) low temperature, and (3) rapid loading.

Toughness is a measure of the material's ability to yield and absorb highly localized and rapidly applied stresses. *Notch toughness* is evaluated by measuring the *impact energy* that causes a notched sample to fail.[15]

In the *Charpy test*, popular in the United States, a standardized beam specimen is given a 45° notch. The specimen is then centered on simple supports with the notch

---

[15]Without a notch, the specimen would experience uniaxial stress (tension and compression) at impact. The notch allows triaxial stresses to develop. Most materials become more brittle under triaxial stresses than under uniaxial stresses.

Materials

Table 46.7 *Correlations Between Hardness Scales for Steel*

Brinell number	Vickers number	Rockwell numbers		scleroscope number
		C	B	
780	1150	70	...	106
712	960	66	...	95
653	820	62	...	87
601	717	58	...	81
555	633	55	120	75
514	567	52	119	70
477	515	49	117	65
429	454	45	115	59
401	420	42	113	55
363	375	38	110	51
321	327	34	108	45
293	296	31	106	42
277	279	29	104	39
248	248	24	102	36
235	235	22	99	34
223	223	20	97	32
207	207	16	95	30
197	197	13	93	29
183	183	9	90	27
166	166	4	86	25
153	153	...	82	23
140	140	...	78	21
131	131	...	74	20
121	121	...	70	...
112	112	...	66	...
105	105	...	62	...
99	99	...	59	...
95	95	...	56	...

down. A falling pendulum striker hits the center of the specimen. This test is performed several times with different heights and different specimens until a sample fractures.

The kinetic energy expended at impact, equal to the initial potential energy less the rebound or follow-through height of the pendulum striker, is calculated from measured heights. It is designated C_V and is expressed in either foot-pounds (ft-lbf) or joules (J).[16] The energy required to cause failure is a measure of toughness.

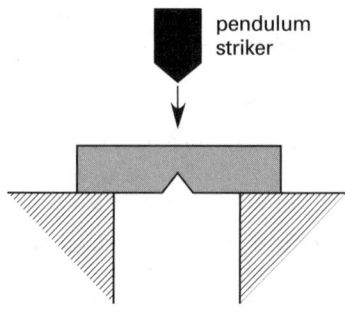

Figure 46.23 *Charpy Test*

---

[16]In Europe, the energy is often expressed per unit cross section of specimen area.

At 70°F (21°C), the energy required to cause failure ranges from 45 ft-lbf (60 J) for carbon steels to approximately 110 ft-lbf (150 J) for chromium-manganese steels. As temperature is reduced, however, the toughness decreases. In BCC metals, such as steel, at a low enough temperature the toughness decreases sharply. The transition from high-energy ductile failures to low-energy brittle failures begins at the *fracture transition plastic (FTP) temperature*.

Since the transition occurs over a wide temperature range, the *transition temperature* (also known as the *ductile transition temperature*) is taken as the temperature at which an impact of 15 ft-lbf (20 J) will cause failure. (15 ft-lbf is used for low-carbon ship steels. Other values may be used with other materials.) This occurs at approximately 30°F (−1°C) for low-carbon steel.

The appearance of the fractured surface is also used to evaluate the transition temperature. The fracture can be fibrous (from shear fracture) or granular (from cleavage fracture), or a mixture of both. The fracture planes are studied and the percentages of ductile failure plotted against temperature. The temperature at which the failure is 50% fibrous and 50% granular is known as the *fracture appearance transition temperature*, FATT.

Table 46.8 *Approximate Ductile Transition Temperatures*

type of steel	transition ductile temperature, °F
carbon steel	30°
high-strength, low-alloy steel	0° to 30°
heat-treated, high-strength, carbon steel	−25°
heat-treated, construction alloy steel	−40° to −80°

Not all materials have a ductile-brittle transition. Aluminum, copper, other FCC metals, and most HCP metals do not lose their toughness abruptly. Figure 46.24 illustrates the failure energy curves for several different types of materials.

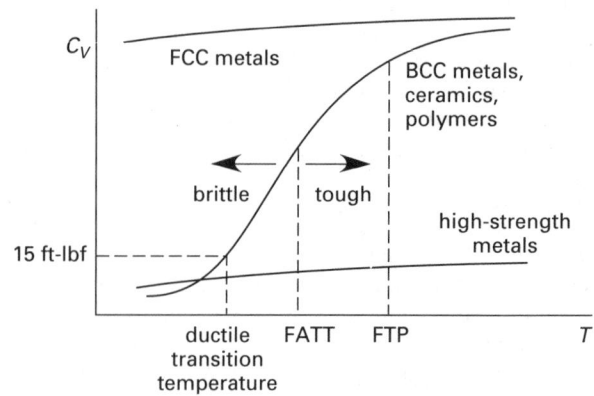

Figure 46.24 *Failure Energy versus Temperature*

Another toughness test is the *Izod test*. This is illustrated in Fig. 46.25 and is similar to the Charpy test in its use of a notched specimen. The height to which a swinging pendulum rebounds and causes the specimen to fail determines the energy of failure.

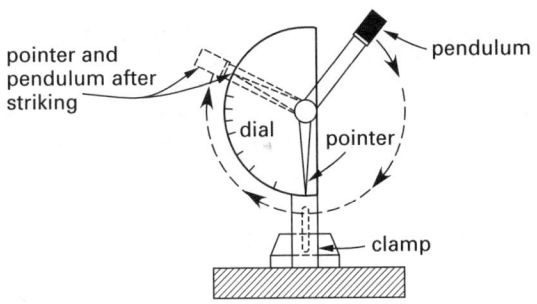

Figure 46.25 *Izod Test*

21. THE CREEP TEST

Creep or *creep strain* is the continuous yielding of a material under constant stress. For metals, creep is negligible at low temperatures (i.e., less than half of the absolute melting temperature), although the usefulness of nonreinforced plastics as structural materials is seriously limited by creep at room temperature.

During a *creep test*, a low tensile load of constant magnitude is applied to a specimen, and the strain is measured as a function of time. The *creep strength* is the stress that results in a specific creep rate, usually 0.001% or 0.0001% per hour. The *rupture strength*, determined from a *stress-rupture test*, is the stress that results in a failure after a given amount of time, usually 100, 1000, or 10,000 hours.

If strain is plotted as a function of time, three different curvatures will be apparent following the initial elastic extension.[17] During the first stage, the *creep rate* ($d\epsilon/dt$) decreases since strain hardening (dislocation generation and interaction with grain boundaries and other barriers) is occurring at a greater rate than annealing (annihilation of dislocations, climb, cross-slip, and some recrystallization). This is known as *primary creep*.

During the second stage, the creep rate is constant, with strain hardening and annealing occurring at the same rate. This is known as *secondary creep* or *cold flow*. During the third stage, the specimen begins to neck down, and rupture eventually occurs. This region is known as *tertiary creep*.

The secondary creep rate is lower than the primary and tertiary creep rates. The secondary creep rate, represented by the slope (on a log-log scale) of the line during the second stage, is temperature and stress dependent. This slope increases at higher temperatures and stresses. The creep rate curve can be represented by the following empirical equation, known as *Andrade's equation*.

[17]In Great Britain, the initial elastic elongation is considered the first stage. Therefore, creep has four stages in British nomenclature.

$$\epsilon = \epsilon_o \left(1 + \beta t^{\frac{1}{3}}\right) e^{kt} \qquad 46.32$$

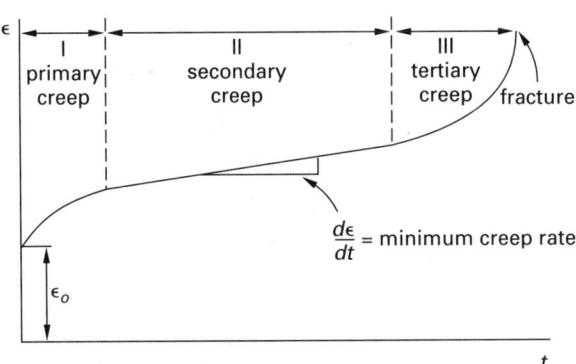

Figure 46.26 *Stages of Creep*

Dislocation climb (glide and creep) is the primary creep mechanism, although diffusion creep and grain boundary sliding also contribute to creep on a microscopic level. On a larger scale, the mechanisms of creep involve slip, subgrain formation, and grain-boundary sliding.

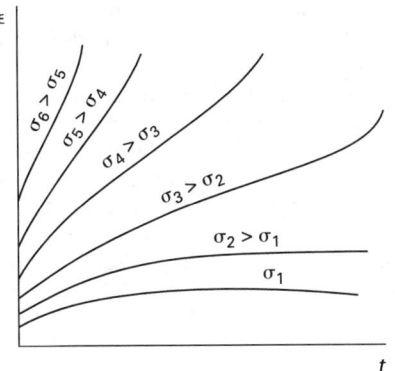

Figure 46.27 *Effect of Stress on Creep Rates*

22. EFFECTS OF IMPURITIES AND STRAIN ON MECHANICAL PROPERTIES

Anything that restricts the movement of dislocations will increase the strength of metals and reduce ductility. Alloying materials, impurity atoms, imperfections, and other dislocations produce stronger materials. This is illustrated in Fig. 46.28.

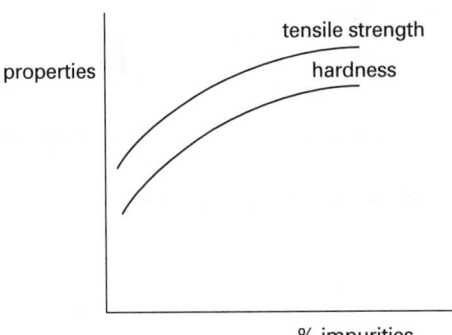

Figure 46.28 *Effect of Impurities on Mechanical Properties*

Additional dislocations are generated by the plastic deformation (i.e., cold working) of metals, and these dislocations can strain-harden the metal. Figure 46.29 shows the effect of strain-hardening on mechanical properties.

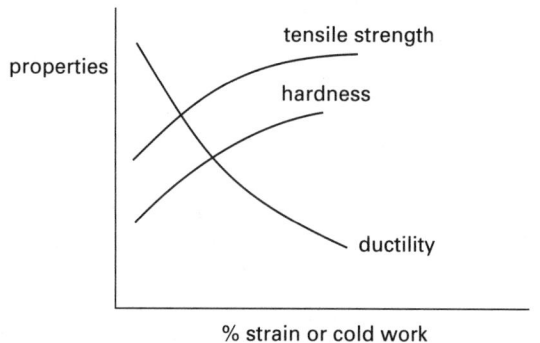

Figure 46.29 *Effect of Strain-Hardening on Mechanical Properties*

23. CLASSIFICATION OF MATERIALS

When used to describe engineering materials, the terms "strong" and "tough" are not synonymous. Similarly, "weak," "soft," and "brittle" have different engineering meanings. A *strong material* has a high ultimate strength, whereas a *weak material* has a low ultimate strength. A *tough material* will yield greatly before breaking, whereas a *brittle material* will not. (A brittle material is one whose strain at fracture is less than approximately 0.5%.) A *hard material* has a high modulus of elasticity, whereas a *soft material* does not. Figure 46.30 illustrates some of the possible combinations of these classifications.

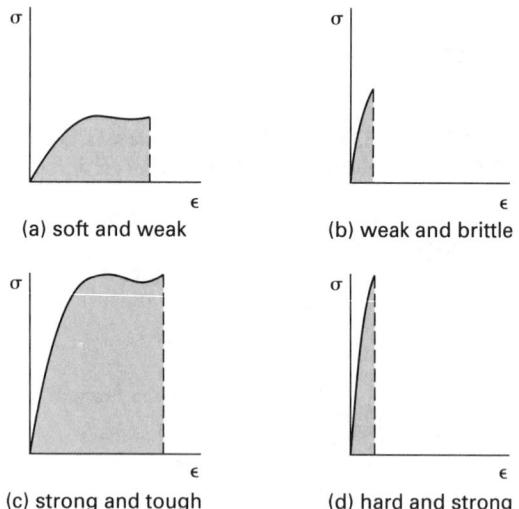

Figure 46.30 *Types of Engineering Materials*

PRACTICE PROBLEMS

1. The engineering stress and engineering strain for a copper specimen are 20,000 lbf/in² (140 MPa) and 0.0200 in/in (0.0200 mm/mm), respectively. Poisson's ratio for the specimen is 0.3. What are the (a) true stress and (b) true strain?

2. A graph of engineering stress-strain is shown. Poisson's ratio for the material is 0.3. Find the (a) 0.5% parallel offset yield strength, (b) elastic modulus, (c) ultimate strength, (d) fracture strength, (e) percentage of elongation at fracture, (f) shear modulus, and (g) toughness.

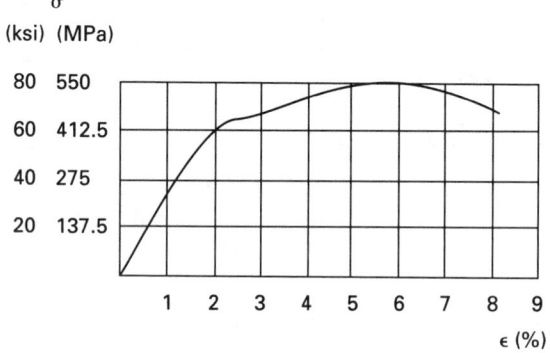

3. A specimen with an unstressed cross-sectional area of 4 in² (25 cm²) necks down to 3.42 in² (22 cm²) before breaking in a standard tensile test. What is the ductility?

4. A constant 15,000 lbf/in² (100 MPa) tensile stress is applied to a specimen. The stress is known to be less than the material's yield strength. The strain is measured at various times. What is the steady-state creep rate for the material?

time (hr)	strain
5	0.018
10	0.022
20	0.026
30	0.031
40	0.035
50	0.040
60	0.046
70	0.058

47 Thermal Treatment of Metals

1. Soluble Alloy Equilibrium Diagrams 47-1
2. The Lever Rule 47-2
3. Eutectic Alloy Equilibrium Diagrams:
   Partial Solubility 47-3
4. Gibbs Phase Rule 47-3
5. Allotropic Changes in Steel 47-4
6. The Iron-Carbon Diagram 47-4
7. Quenching and Rates of Cooling 47-6
8. TTT and CCT Curves 47-6
9. Steel Hardening Processes 47-7
10. Heat Treating Cast Iron 47-8
11. Properties versus Grain Size 47-8
12. Recrystallization 47-8
13. Stress Relief Processes for Steel 47-8
14. Cold Working versus Hot Working 47-9
15. Hardening of Non-Allotropic Alloys 47-9
16. Surface Hardening 47-10
17. Shot-Peening 47-10
    Practice Problems 47-10

Nomenclature

A	atomic fraction	–	–
F	degrees of freedom	–	–
G	gravimetric fraction	–	–
M	martensite transformation temperature	°F	°C
M	molecular weight	lbm/lbmol	kg/kmol
N	number of elements	–	–
P	number of phases	–	–
R_C	Rockwell C hardness	–	–

Subscripts

f	finish
s	start

1. SOLUBLE ALLOY EQUILIBRIUM DIAGRAMS

Most engineering materials are not pure elements but are alloys of two or more elements. Alloys of two elements are known as *binary alloys*. Steel, for example, is an alloy of primarily iron and carbon. Usually, one of the elements is present in a much smaller amount, and this element is known as the *alloying ingredient*. The primary ingredient is known as the *host ingredient*, *base metal*, or *parent ingredient*.

Sometimes, such as with alloys of copper and nickel, the alloying ingredient is 100% soluble in the parent ingredient. Nickel-copper alloy is said to be a *completely miscible alloy* or a *solid-solution alloy*.

The presence of the alloying ingredient changes the thermodynamic properties, notably the freezing (or melting) temperatures of both elements.[1] Usually the freezing temperatures decrease as the percentage of alloying ingredient is increased. Since the freezing points of the two elements are not the same, one of them will start to solidify at a higher temperature than the other. Thus, for any given composition, the alloy might consist of all liquid, all solid, or a combination of solid and liquid, depending on the temperature.

A *phase* of a material at a specific temperature will have a specific composition and crystalline structure and distinct physical, electrical, and thermodynamic properties. (In metallurgy, the word *phase* refers to more than just solid, liquid, and gas phases.)

The regions of an *equilibrium diagram*, also known as a *phase diagram*, illustrate the various alloy phases. The phases are plotted against temperature and composition. (The composition is usually a gravimetric fraction of the alloying ingredient. Only one ingredient's gravimetric fraction needs to be plotted for a binary alloy.)[2]

The equilibrium conditions do not occur instantaneously, and an equilibrium diagram is applicable only to the case of slow cooling.

Figure 47.1 is an equilibrium diagram for copper-nickel alloy. (Most equilibrium diagrams are much more complex.) The *liquidus line* is the boundary above which no solid can exist. The *solidus line* is the boundary below which no liquid can exist. The area between these two lines represents a mixture of solid and liquid phase materials.

---

[1]The term *freezing point* or *melting point* is used depending on whether heat is being removed or added, respectively.

[2]Sometimes, the amount of alloying ingredient is specified in *atomic fraction* or *atomic percent*. The conversions between gravimetric fractions, G_A and G_B, and atomic fractions, A_A and A_B, depend on the ratio of the molecular weights, M_A and M_B.

$$A_A = \frac{M_B G_A}{M_B G_A + M_A G_B}$$

$$A_B = \frac{M_A G_B}{M_B G_A + M_A G_B}$$

$$G_A = \frac{M_A A_A}{M_A A_A + M_B A_B}$$

$$G_B = \frac{M_B A_B}{M_A A_A + M_B A_B}$$

Materials

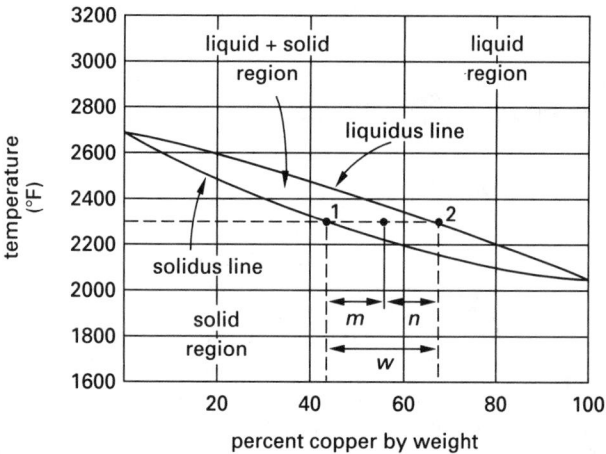

Figure 47.1 *Copper-Nickel Phase Diagram*

Curve (a) in Fig. 47.2 is a *time-temperature* or *temperature-time curve* for a pure metal cooling from liquid to solid state. At a particular point, the temperature remains constant (i.e., there is a *thermal arrest*). This temperature is the *freezing point* of the liquid, indicated on the graph by a horizontal line known as a *shelf* or *plateau*. The metal continues to lose heat energy—its *heat of fusion*—as the phase change from liquid to solid occurs.

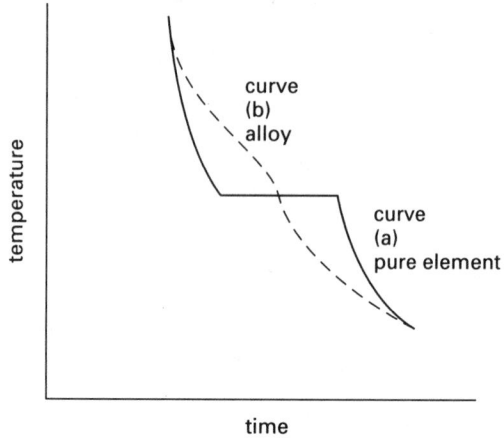

Figure 47.2 *Time-Temperature Curves*

With an alloy of two elements, it is logical to expect two plateaus, since the two elements solidify at different temperatures. However, there is a range of temperatures over which the solidification occurs, and the transformation curve is smooth with an inflection point. This is illustrated by curve (b) in Fig. 47.2.

If the time-temperature curve is plotted for various compositions, the transition temperature will vary with proportions of the constituents. The locus of the curve's inflection points coincides with the liquidus and solidus lines in the equilibrium diagram.

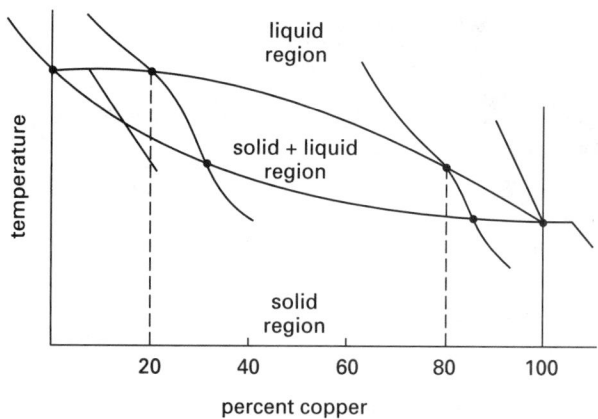

Figure 47.3 *Time-Temperature Curves for a Binary Alloy*

2. THE LEVER RULE

Within the liquid-solid region, the percentage of solid and liquid phases is a function of temperature and composition. Near the liquidus line, there is very little solid phase. Near the solidus line, there is very little liquid phase. The *lever rule* is used to find the relative amounts of solid and liquid phase at any composition. These percentages are given in fraction (or percent) by weight.

Figure 47.1 shows an alloy with an average composition of 55% copper at 2300°F. (A horizontal line representing different conditions at a single temperature is known as a *tie line*.) The liquid composition is defined by point 2, and the solid composition is defined by point 1.

The fractions of solid and liquid phases depend on the distances m, n, and w (equal to $m + n$), which are measured using any convenient scale. (Although the distances can be measured in millimeters or tenths of an inch, it is more convenient to use the percentage alloying ingredient scale. This is illustrated in Ex. 47.1.) Then, the fractions of solid and liquid can be calculated from Eqs. 47.1 and 47.2.

$$\text{fraction solid} = \frac{n}{w} = 1 - \text{fraction liquid} \qquad 47.1$$

$$\text{fraction liquid} = \frac{m}{w} = 1 - \text{fraction solid} \qquad 47.2$$

The lever rule and method of determining the composition of the two components are applicable to any solution or mixture, liquid or solid, in which two phases are present.

Example 47.1

A mixture of 55% copper and 45% nickel exists at 2300°F. What are the fractions of solid and liquid phases and the compositions of each?

Solution

Referring back to Fig. 47.1, the solid portion of the mixture will have the composition at point 1 (44% copper), while the liquid will be at composition 2 (68% copper).

The phase fractions are

$$\text{fraction solid} = \frac{68 - 55}{68 - 44} = 0.54 \quad (54\%)$$

$$\text{fraction liquid} = 1.00 - 0.54 = 0.46 \quad (46\%)$$

3. EUTECTIC ALLOY EQUILIBRIUM DIAGRAMS: PARTIAL SOLUBILITY

Just as only a limited amount of salt can be absorbed by water, there are many instances where a limited amount of the alloying ingredient can be absorbed by the solid mixture. The elements of a binary alloy may be completely soluble in the liquid state but only partially soluble in the solid state.

When the alloying ingredient is present in amounts above the maximum solubility percentage, the alloying ingredient precipitates out. In aqueous solutions, the precipitate falls to the bottom of the container. In metallic alloys, the precipitate remains suspended as pure crystals dispersed throughout the primary metal.

Figure 47.4 is typical of an equilibrium diagram for ingredients displaying a limited solubility.

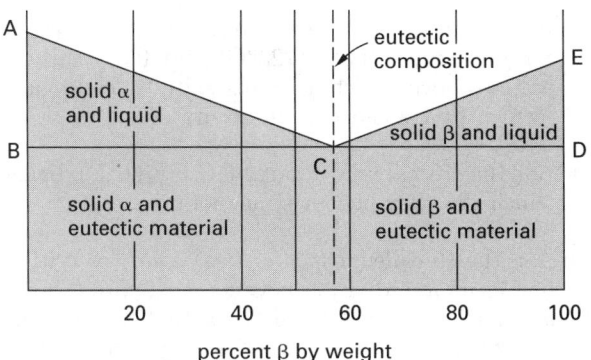

Figure 47.4 Equilibrium Diagram of a Limited Solid Solubility Alloy

In chemistry, a *mixture* is different from a *solution*. Salt in water forms a solution. Sugar crystals mixed with salt crystals form a mixture. A mixture of two solid ingredients with no solubility is known as a *eutectic alloy*.

In Fig. 47.4, the components are perfectly miscible at point C only. This point is known as the *eutectic composition*. The material in the region ABC consists of a mixture of solid component α crystals in a liquid of components α and β. This liquid is known as the *eutectic material*, and it will not solidify until the line B–D (the *eutectic line, eutectic point*, or *eutectic temperature*) is reached, the lowest point at which the eutectic material can exist in liquid form.[3]

Since the two ingredients do not mix, reducing the temperature below the eutectic line results in crystals (layers or plates) of both pure ingredients forming. This

---

[3]The term *point* usually can be interpreted as temperature. Thus, the eutectic point really refers to the eutectic temperature.

is the microstructure of a solid eutectic alloy: alternating pure crystals of the two ingredients. Since two solid substances are produced from a single liquid substance, the process could be written in chemical reaction format as: liquid $\longrightarrow \alpha + \beta$. (Alternatively, upon heating the reaction would be: $\alpha + \beta \longrightarrow$ liquid.) For this reason, the phase change is called a *eutectic reaction*.

There are similar reactions involving other phases and states. Tables 47.1 and 47.2 illustrate these.

Table 47.1 Types of Equilibrium Reactions

reaction name	type of reaction upon cooling
eutectic	liquid → solid α + solid β
peritectic	liquid + solid α → solid β
monotectic	liquid α → liquid β + solid α
eutectoid	solid γ → solid α + solid β
peritectoid	solid α + solid γ → solid β

Table 47.2 Typical Appearance of Equilibrium Diagram at Reaction Points

reaction name	phase reaction	phase diagram
eutectic	$L \rightarrow \alpha(s) + \beta(s)$ cooling	
peritectic	$\alpha(s) + L \rightarrow \beta(s)$ cooling	
eutectoid	$\gamma(s) \rightarrow \alpha(s) + \beta(s)$ cooling	
peritectoid	$\alpha(s) + \gamma(s) \rightarrow \beta(s)$ cooling	

4. GIBBS PHASE RULE

The *Gibbs phase rule* defines the relationship between the number of phases and elements in an equilibrium mixture. For such an equilibrium mixture to exist, the alloy must have been slowly cooled, and thermodynamic equilibrium must have been achieved along the way. At equilibrium, and considering both temperature and pressure to be independent variables, the Gibbs phase rule is

$$P + F = N + 2 \qquad 47.3$$

P is the number of phases existing simultaneously; F is the number of independent variables, known as *degrees of freedom*; and N is the number of elements in the alloy. Composition, temperature, and pressure are examples of degrees of freedom that can be varied.

For example, if water is to be stored in a condition where three phases (solid, liquid, gas) are present simultaneously, then $P = 3$, $N = 1$, and $F = 0$. That is, neither pressure nor temperature can be varied. This state corresponds to the *triple point* of water.

If pressure is constant, then the number of degrees of freedom is reduced by one, and the Gibbs phase rule can be rewritten as

$$P + F = N + 1\big|_{\text{constant pressure}} \qquad 47.4$$

If the Gibbs rule predicts $F = 0$, then an alloy can exist in only one composition.

5. ALLOTROPIC CHANGES IN STEEL

Allotropes have the same composition but different atomic structures (microstructures), volumes, electrical resistance, and magnetic properties. In the case of iron, *allotropic changes* are reversible changes that occur at the *critical points* (i.e., *critical temperatures*).

Iron exists in three primary allotropic forms: alpha-iron, delta-iron, and gamma-iron. The changes are brought about by varying the temperature of the iron. Heating pure iron from room temperature changes its structure from body-centered cubic (BCC) to face-centered cubic (FCC) and then back to body-centered cubic.

Table 47.3 *Allotropic Points for Pure Iron (upon heating)*

1670°F
(910°C): alpha (BCC) to gamma (FCC) transition

2552°F
(1400°C): gamma (FCC) to delta (BCC) transition

2802°F
(1539°C): delta (BCC) to liquid transition

Alpha-iron, also known as *ferrite*, is a BCC structure that exists only below the A_3 line (defined as follows). The maximum carbon solubility is 0.03%, the lowest of all three allotropic forms. Alpha-iron is stable from −460°F to 1670°F (−273°C to 910°C), soft, and strongly magnetic up to approximately 1410°F (766°C). (*Beta-iron* is a nonmagnetic form of BCC alpha-iron that exists between 1418°F and 1670°F (770°C and 910°C). The distinction between alpha- and beta-iron is not usually made.)

Gamma-iron is an FCC arrangement of iron atoms, stable between 1670°F and 2552°F (910°C and 1400°C), and nonmagnetic. The maximum carbon solubility is 1.7%.

Delta-iron is a BCC form of iron existing above 2552°F (1400°C).

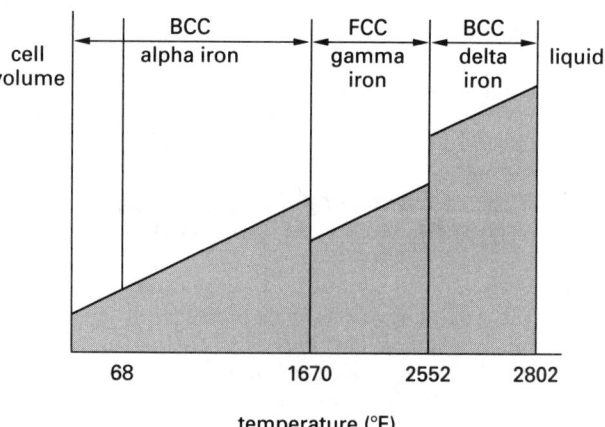

Figure 47.5 *Allotropic Changes of Iron*

Since allotropic changes occur at various temperatures in an iron-carbon mixture, there are *critical lines* but no critical points. Depending on the authority, these critical lines may be labeled A_c, A_r, or just A.[4] Refer to Fig. 47.6.

- A_0: the critical line, about 410°F (210°C), above which cementite becomes nonmagnetic

- A_1: the so-called *lower critical point*, or *eutectoid temperature*, 1333°F (723°C) and 0.8% carbon, a line above which the austenite-to-ferrite and -cementite transformation occurs

- A_2: the critical line, about 1410°F (766°C), below which the alloy becomes magnetic[5]

- A_3: the so-called *upper critical point*, or critical line forming a division between austenite (above) and ferrite (below), with the actual temperature being dependent on composition (see Fig. 47.6)

- A_4: the critical point, 2552°F (1400°C), at which the gamma-delta transformation occurs

In some parts of the iron-carbon diagram, two critical lines coincide. For example, the tie line (1333°F or 723°C) for more than 0.8% carbon is labeled $A_{1,3}$ or A_{13}.

6. THE IRON-CARBON DIAGRAM

The iron-carbon phase diagram is much more complex than idealized equilibrium diagrams due to the existence of many different phases. Each of these phases

---

[4]Labeling the critical lines as A_{c1}, A_{c2}, and so on is a reference to the French word *chauffage* ("heating"). Such critical temperatures are encountered if iron is slowly heated from room temperature. Critical lines labels of A_{r1}, A_{r2}, and so on refer to the French word *refroidissement* ("cooling"), as such critical temperatures are observed upon cooling the iron back to room temperature. As the temperatures (A_{c1} and A_{r1}, etc.) are approximately the same, the distinction is not always made, as it is not in this book.

[5]The A_2 temperature is also known as the *Curie point*.

Figure 47.6 *Iron-Carbon Diagram*

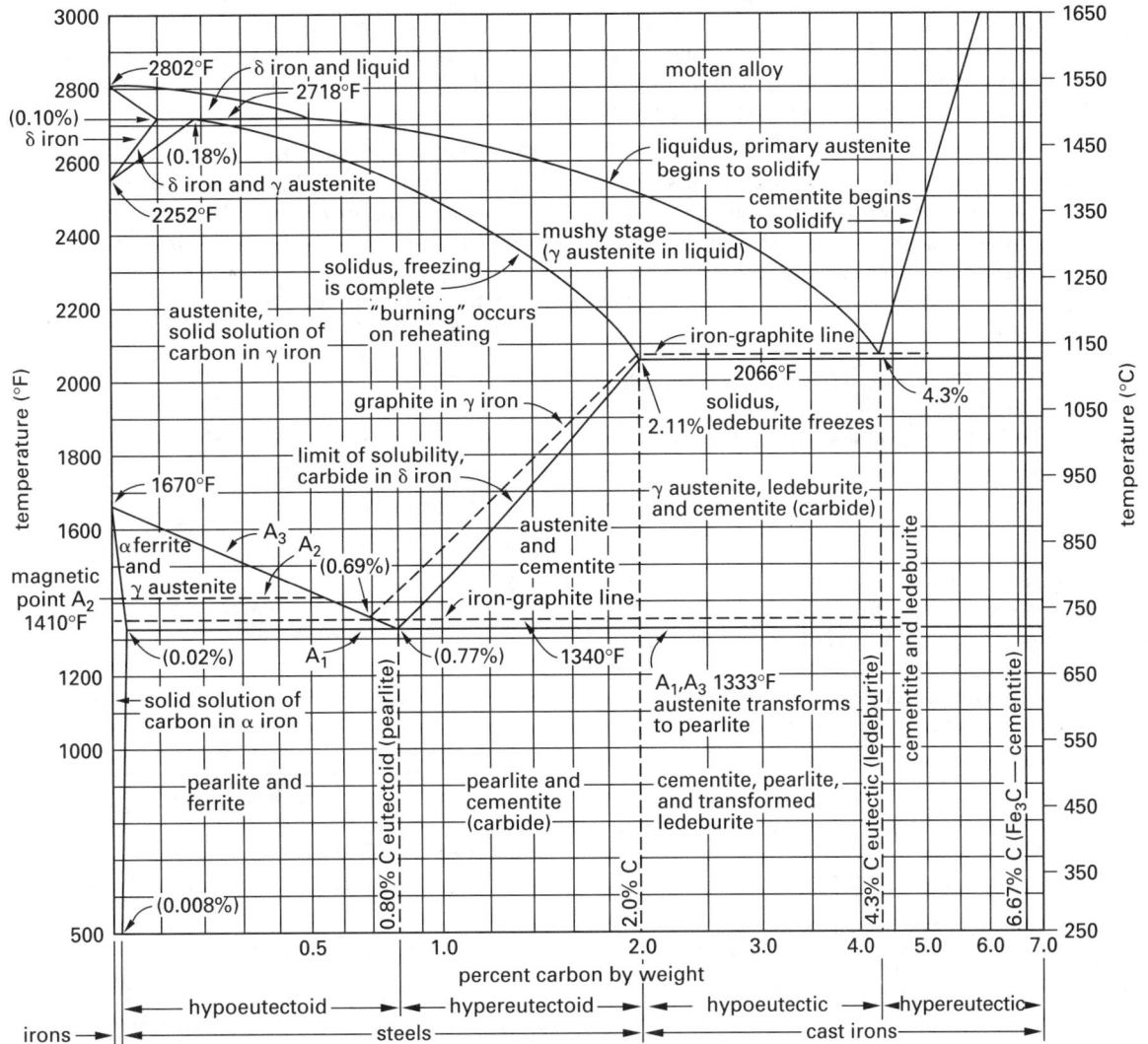

has a different microstructure and therefore different mechanical properties. By treating the steel in a manner that forces the occurrence of particular phases, steel with desired wear and endurance properties can be produced.

Iron-carbon mixtures are categorized into *steel* (less than 2% carbon) and *cast iron* (more than 2% carbon) according to the amounts of carbon in the mixtures. Iron-carbon alloys are further classified as follows.

- *steel*: iron alloy with less than 2.0% carbon
  hypoeutectoid steel: iron alloy with less than 0.8% carbon, consisting of ferrite and pearlite
  eutectoid steel: equilibrium iron alloy with 0.8% carbon, consisting of ferrite and pearlite
  hypereutectoid steel: iron alloy with 0.8 to 2.0% carbon, consisting of cementite and pearlite

- *cast iron*: iron alloy with more than 2% carbon
  hypoeutectic cast iron: iron alloy with 2.0 to 4.3% carbon
  eutectic cast iron: iron alloy with 4.3% carbon
  hypereutectic cast iron: iron alloy with more than 4.3% carbon

The most important eutectic reaction in the iron-carbon system is the formation of a solid mixture of austenite and cementite at approximately 2065°F (1129°C). *Austenite* is a solid solution of carbon in gamma-iron. It is nonmagnetic, decomposes on slow cooling, and does not normally exist below 1333°F (723°C), though it can be partially preserved by extremely rapid cooling.

Cementite (Fe_3C), also known as *carbide* or *iron carbide*, has approximately 6.67% carbon. Cementite is the hardest of all forms of iron, has low tensile strength, and is quite brittle. Cementite ceases to be magnetic above the A_0 line.

Materials

The most important eutectoid reaction in the iron-carbon system is the formation of *pearlite* from the decomposition of austenite at approximately 1333°F (723°C). Pearlite is actually a mixture of two solid components, ferrite and cementite, with the common *lamellar (layered) appearance*. (The name *pearlite* is derived from similarity in appearance to mother-of-pearl.)

Ferrite is essentially pure iron (less than 0.025% carbon) in BCC alpha-iron structure. It is magnetic and has properties complementary to cementite, since it has low hardness, high tensile strength, and high ductility.

Example 47.2

An iron alloy at 1500°F contains 2.0% carbon by weight. (a) What are the percentages of austenite and cementite in the mixture? (b) How much carbon is present in each of the phases?

Solution

Referring to the iron-carbon diagram, Fig. 47.6, compositions of the phases are read from the carbon contents at the intersection of the 1500°F tie line and the phase boundary lines. The austenite will have the composition at point 1 (approximately 1.08% carbon), while the cementite will be at composition 2 (6.67% carbon).

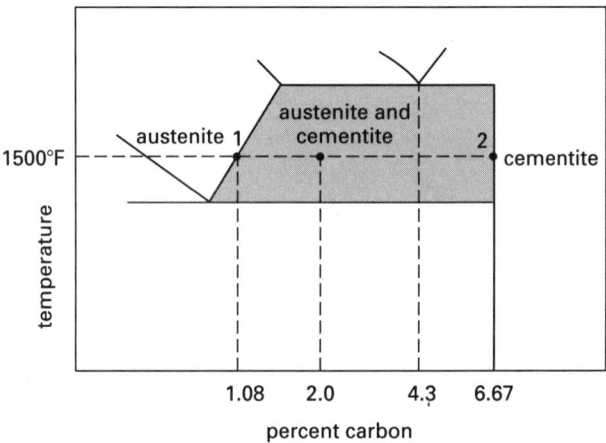

The phase fractions are

$$\text{fraction austenite} = \frac{6.67 - 2.0}{6.67 - 1.08} = 0.835 \quad (83.5\%)$$

$$\text{fraction cementite} = \frac{2.0 - 1.08}{6.67 - 1.08} = 0.165 \quad (16.5\%)$$

7. QUENCHING AND RATES OF COOLING

As steel is heated, the grain sizes remain the same until the A_1 line is reached. Between the A_1 and A_2 lines, the average grain size of austenite in solution decreases.

This characteristic is used in heat treatments wherein steel is heated and then quenched. The quenching can be performed with gases (usually air), oil, water, or brine. Agitation or spraying of these fluids during the quenching process increases the severity of the quenching.

The rate of cooling determines the hardness and ductility. Rapid cooling in water or brine is necessary to quench low- and medium-carbon steels, since steels with small amounts of pearlite are difficult to harden. Oil is used to quench high-carbon and alloy steel or parts with non-uniform cross sections (to prevent warping). Figure 47.7 illustrates the relative rates of cooling for different quenching media.

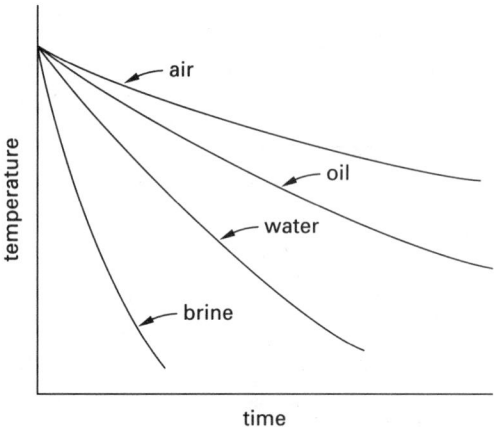

Figure 47.7 *Relative Cooling Rates for Different Quenching Media*

8. TTT AND CCT CURVES

Controlled-cooling-transformation (CCT) *curves* and *time-temperature-transformation* (TTT) *curves* are used to determine how fast an alloy should be cooled to obtain a desired microstructure. Although these curves show different phases, they are not equilibrium diagrams. On the contrary, they show the microstructures that are produced with controlled temperatures or when quenching interrupts the equilibrium process.

TTT curves are determined under ideal, isothermal conditions. For that reason, they are also known as *isothermal transformation diagrams*. CCT curves are experimentally determined under conditions of continuous cooling. Therefore, CCT curves are better suited for designing cooling processes. However, TTT curves are more readily available than CCT curves and are used in lieu of them. Both curves are similar in shape, although the CCT curves are displaced downward and to the right from TTT curves.

Figure 47.8 shows a TTT diagram for a high-carbon (0.80% C or more) steel. Curve 1 represents extremely rapid quenching. The transformation begins at 420°F (216°C) and continues for 8 to 30 seconds, changing all of the austenite to martensite. Such a material is seldom used because martensite has almost no ductility.

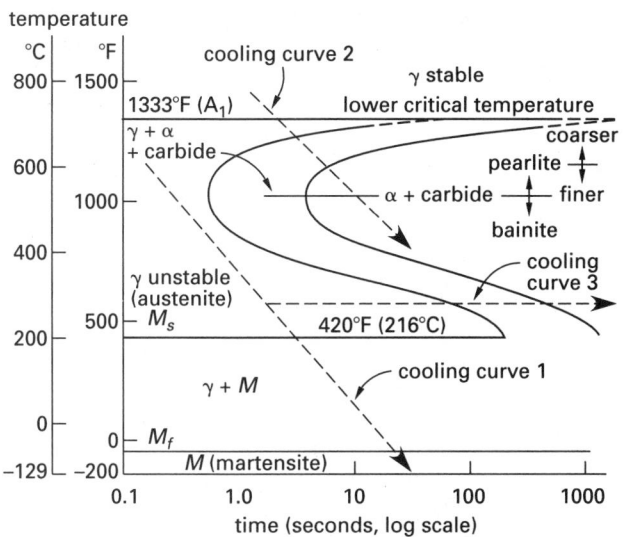

Figure 47.8 *TTT Diagram for High-Carbon Steel*

Curve 2 is a slower quench that converts all of the austenite to fine pearlite. This corresponds to a *normalizing* process.

A horizontal line below the critical temperature is a *tempering* process. If the temperature is decreased rapidly along curve 1 to 520°F (270°C) and is then held constant along cooling curve 3, bainite is produced. This is the principle of *austempering*. Performing the same procedure at 350°F to 400°F (180°C to 200°C) is *martempering*, which produces *tempered martensite*, a soft and tough steel.

The austenite-martensite transformation is extremely important, and the temperatures at which the transition starts and finishes are sometimes referred to as *critical temperatures*.

- M_s: the temperature at which austenite first begins to transform into martensite
- M_f: the temperature at which austenite is fully transformed into martensite

9. STEEL HARDENING PROCESSES

Hard steel resists plastic deformation. Steel is hard if it has a homogeneous, austenitic structure with coarse grains. Some steels (e.g., those that have little carbon) are difficult to harden. The *hardenability* of a steel specimen can be determined in a standard *Jominy end-quench test*.

The basic hardening processes consist of heating to approximately 100°F (50°C) above the A_3 critical line, allowing austenite to form, and then quenching rapidly. Hardened steel consists primarily of martensite or bainite. *Martensite* is a supersaturated solution of carbon in alpha-iron.[6] *Bainite* is not as hard as martensite, but

---

[6]The carbon in martensite distorts the BCC structure of iron. The distorted BCC lattice is known as a *body-centered tetragonal* (BCT) structure.

it does have good impact strength and fairly high hardness. Neither martensite nor bainite are equilibrium substances—they are not found on the iron-carbon equilibrium diagram but are formed during the quenching operation.

The maximum hardness obtained depends on the carbon content. An upper limit of R_C 66–67 is reached with approximately 0.5% carbon. No further increase in hardness is achieved by increasing the carbon content. Since hardening is accompanied by a decrease in toughness, it is usually followed by tempering.

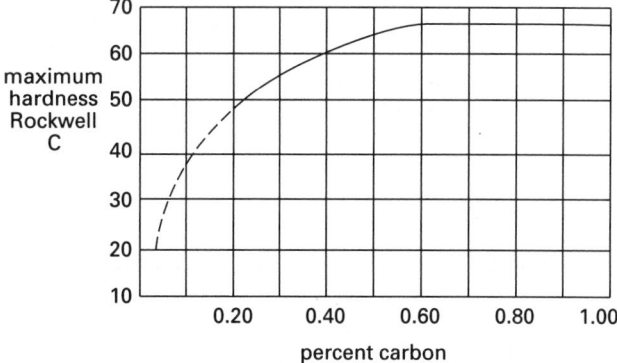

Figure 47.9 *Maximum Hardness of Steel*

There are many steel-hardening processes with special names, listed alphabetically as follows.

- *austempering*: an interrupted quenching process resulting in an austenite-to-bainite transition. Steel is quenched to below approximately 800°F (430°C) but above 400°F (200°C), and is allowed to reach equilibrium. No martensite is formed, and further tempering is not required.

- *austenitizing*: quenching after heating above the A_3 line (for steel with up to 0.8% carbon) or above the A_1 line (for steel with more than 0.8% carbon).

- *martempering*: an interrupted quenching process resulting in an austenite-to-(tempered) martensite transition. Steel is quenched to below 400°F (200°C) and allowed to reach equilibrium. Further tempering is not required.

10. HEAT TREATING CAST IRON

Hardness in low-carbon steels results from the presence of martensite or bainite—supersaturated solutions of carbon in alpha-iron that begin as austenite at higher temperatures and are "frozen" in place by rapid cooling. Cast iron starts as iron carbide (cementite), not austenite, and cast iron is not normally hardened by heating and quenching. Hardness in cast irons is primarily obtained by including alloying ingredients (see Table 45.9) that promote the formation and retention of iron carbide (Fe_3C). Although it is hard, iron carbide is also very brittle.

Materials

Heating can actually reduce hardness in cast iron. Iron carbide dissociates into iron and graphite at high temperatures. Graphite in flake form (*gray cast iron*) and in spheroidal form (*nodular cast iron*) greatly decreases hardness. The presence of silicon in cast iron greatly affects hardness, since silicon promotes the formation of graphite. Cast irons typically contain $1^1/_2\%$ or more silicon, and less than 1% is needed to ensure that the carbon in iron carbide dissociates into iron and graphite. Cast iron with no graphite is known as *white cast iron*. Cast iron that has been heated and slowly cooled to permit graphite to form is known as *malleable cast iron*.

Small castings (e.g., with diameters less than 3 or 4 in) or white cast iron contain less silicon and can be hardened by rapid quenching. This occurs because rapid cooling prevents the dissociation of iron carbide into iron and graphite. However, the interiors of larger castings cannot be cooled fast enough to prevent the formation of weakening graphite.

11. PROPERTIES VERSUS GRAIN SIZE

Many properties are related to grain size, which initially depends on composition but can be changed by heat treatment. Coarse-grained structures have less toughness and ductility but have greater machineability and case hardenability.

As low-carbon steels are heated from room temperature, the grain size remains constant up to the A_1 line. Above the A_1 line, ferrite and pearlite are transformed into austenite, and the grain size decreases. The grain size is minimum at the upper critical line, A_3, and then increases again as the steel is heated above the A_3 line.

Aluminum in small quantities is an important alloying ingredient in steel. As a deoxidizer, it raises the temperature at which rapid grain growth takes place. In steels that have been deoxidized with aluminum (e.g., medium-carbon and alloy steels), no grain growth occurs until the *coarsening temperature* is reached, which is well above the critical temperature.

12. RECRYSTALLIZATION

Recrystallization can be used with all metals to relieve stresses induced during cold working. It involves heating the material in a furnace to a specific temperature (the *recrystallization temperature*) and holding it there for a long time. This induces the formation and growth of strain-free grains within the grains already formed. The resulting microstructure is essentially the microstructure that existed before any cold working but is softer and more ductile than the cold-worked microstructure.

The recrystallization process is more sensitive to temperature than it is to exposure time. Recrystallization will occur naturally over a wide range of temperatures;

however, the reaction rates increase at higher temperatures. Table 47.4 lists approximate temperatures that will produce complete recrystallization in one hour.

Table 47.4 *Approximate Recrystallization Temperatures*

material	recrystallization temperature	
	°F	°C
copper (99.999% pure)	250	120
(5% zinc)	600	315
(5% aluminum)	550	290
(2% beryllium)	700	370
aluminum (99.999% pure)	175	80
(99.0+ % pure)	550	290
(alloys)	600	315
nickel (99.99% pure)	700	370
(99.4% pure)	1100	590
(monel metal)	1100	590
iron (pure)	750	400
(low-carbon steel)	1000	540
magnesium (99.99% pure)	150	65
(alloys)	450	230
zinc	50	10
tin	25	−4
lead	25	−4

13. STRESS RELIEF PROCESSES FOR STEEL

The annealing, normalizing, and tempering processes are used to relieve the internal stresses, refine the grain size, and soften the material (to improve machineability). The high temperatures used in these processes allow some of the carbon to migrate out of the martensite, thereby relieving stresses in the crystalline structure.

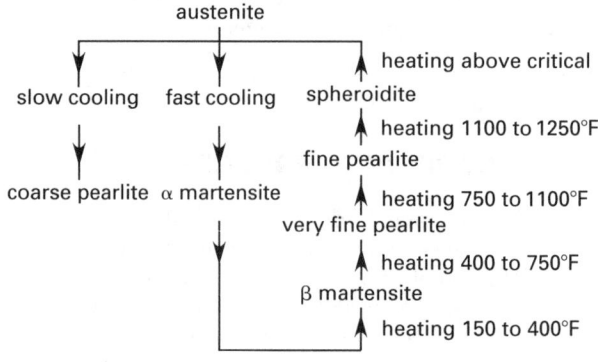

Figure 47.10 *Products of Cooling and Reheating Iron-Carbon Alloys*

The basic *full annealing* process involves heating to approximately 100°F (50°C) above the critical A_3 point, allowing austenite to form fully, and then cooling slowly in a furnace to produce coarse pearlite. Three separate stages constitute annealing. First, the material is stress-relieved by heating in the *recovery stage*. Next, during the *recrystallization stage*, new crystals form within the existing distorted structure. Finally, during the *grain growth stage*, some of the crystals grow in size (by eliminating smaller grains).

The partial *annealing* process, also known as *process annealing*, *spheroidize-annealing*, or just *spheroidizing*, also softens the material and relieves stresses, but heating is to below the A_3 point. The term *spheroidizing* gets its name from the spherical cementite particles that appear in the ferrite matrix, as opposed to the lamellar structure of cementite and ferrite in pearlite.[7]

Normalizing is similar to annealing but is more rapid because it uses air cooling. Heating is to approximately 200°F (100°C) higher than the critical point. Normalizing produces a harder and stronger steel than full annealing.

Tempering, also known as *drawing* or *toughening*, is used with hypoeutectoid steels to change martensite into pearlite. It is used after hardening to produce softer and tougher steel. The steel is heated to below its critical temperature. However, the higher the temperature, the more soft and ductile the steel becomes.

To avoid *blue embrittlement* in steel, tempering should not be done between 450°F and 700°F (230°C and 370°C). Within this range, the *notch toughness* of the steel (as determined from an impact test), is lowered considerably. The reason for this effect is not well understood.

14. COLD WORKING VERSUS HOT WORKING

As a material is worked and the dislocations move, plastic strain builds up. If the strain occurs at a high enough temperature (i.e., above the recrystallization temperature), there will be sufficient thermal energy to anneal out the lattice distortions. Forming operations above the recrystallization temperature are known as *hot working*, since annealing occurs simultaneously with the plastic forming. The material remains ductile.

If the plastic forming takes place at a low temperature (i.e., below the recrystallization temperature), there will be insufficient thermal energy to anneal out the dislocations. The material will become progressively stronger, harder, and more brittle until it eventually fails. This is known as *cold working*.

15. HARDENING OF NON-ALLOTROPIC ALLOYS

The properties of nonferrous substances that do not readily form allotropes cannot be changed by controlled cooling. Such substances are known as *non-allotropic alloys* and include aluminum, copper, and magnesium alloys as well as stainless steels containing nickel.

---

[7]Although the cementite concentrations have changed in shape, the phase is the same: cementite and ferrite.

The primary method of hardening non-allotropic alloys is *solution heat treatment*, which consists of two or three steps: precipitation, quenching, and (optionally) artificial aging. Because of these steps, solution heat treating is also known as *precipitation hardening* and *age hardening*.

Precipitation involves the formation of a new crystalline structure through the application of controlled quenching and tempering. Precipitation disperses hard particles throughout the existing more ductile material. These particles disrupt the long dislocation planes of the material, restricting the movement of dislocations and increasing the strength and stiffness of the alloy. The ultimate strength is raised to the rupture strength of either the particles or the surrounding matrix.

Solution heat treatment culminates in rapid quenching. Quenching speeds must be consistent with the size of the object. Massive specimens may require slower processes that use oil or boiling water.

The final step is to hold the material at a specific temperature for a given amount of time. This is known as *aging* or *artificial aging*. Post-treatment cooling for precipitation hardening is relatively unimportant.

It is important not to over-age aluminum. If the precipitation process goes on too long, the precipitates will not be effective in strengthening the material. Precipitation hardening is optimum at the point where the particles are just starting to form.

Table 47.5 lists some of the more common *temper designations* for 2XXX-, 6XXX-, and 7XXX-series aluminum alloys that can be precipitation hardened. For example, 2024-T4 is a widely used alloy having strength and toughness when hardened and aged.

Table 47.5 *Aluminum Tempers*

temper	description
T2	annealed (castings only)
T3	solution heat-treated, followed by cold working
T4	solution heat-treated, followed by natural aging
T5	artificial aging only
T6	solution heat-treated, followed by artificial aging
T7	solution heat-treated, followed by stabilizing by overaging heat treating
T8	solution heat treated, followed by cold working and subsequent artificial aging

16. SURFACE HARDENING

Often, it is desirable to have a hard (wear-resistant) outer surface with a ductile interior. This combination is needed when the product is subjected to fatigue.

Materials

There are several processes used to *surface harden* (also known as *case harden* and *differential harden*) steel.

- *boron diffusion*: exposure to boron (a powerful hardening ingredient) at low temperatures; slow but distortion-free; suitable for high carbon, spring, and tool steels as well as bonded steel carbides and some age-hardenable alloys.

- *carburizing* (also known as *cementation*): heating for up to 24 hours at approximately 1650°F (900°C) in contact with a carbonaceous material (usually carbon monoxide, CO, gas), followed by rapid cooling. Carburizing is used for steels with less than 0.2% carbon.

- *cyaniding*: heating at 1700°F (925°C) in a cyanide-rich atmosphere or immersion in a cyanide salt bath.

- *flame hardening*: supplying flame heat at the surface in quantities and rates higher than can be conducted into the material's interior, followed by drastic spray quenching. Typically used with steels containing more than 0.4% carbon.

- *induction hardening*: using high-frequency electric currents to heat the metal surface, followed by normal quenching. Typically used with steels containing more than 0.4% carbon.

- *nitriding*: heating at 1000°F (540°C) for up to 100 hours (but usually less than 70 hours) in an ammonia atmosphere, followed by slow cooling (no quenching required).

17. SHOT-PEENING

Shot-peening is the "bombardment" of a metal surface by high-velocity particles (e.g., hard steel shot). As each particle strikes, the target's surface stretches and deforms plastically, creating residual compressive stresses at the surface. The induced compressive stress from shot-peening removes tensile stresses left over from manufacturing operations and offsets the effects of applied tensile operating loads. In gears, the compressive layer improves load-carrying capacity by increasing the bending fatigue strength of the teeth.[8] A 20% improvement in both strength (i.e., endurance limit) and wear is typical for shot-peened parts.

PRACTICE PROBLEMS

1. Refer to the following equilibrium diagram for an alloy of elements A and B. (a) For a 4% A alloy at temperature T_1, what are the compositions of solids α and β? (b) For a 1% A alloy at temperature T_2, how much liquid and how much solid are present?

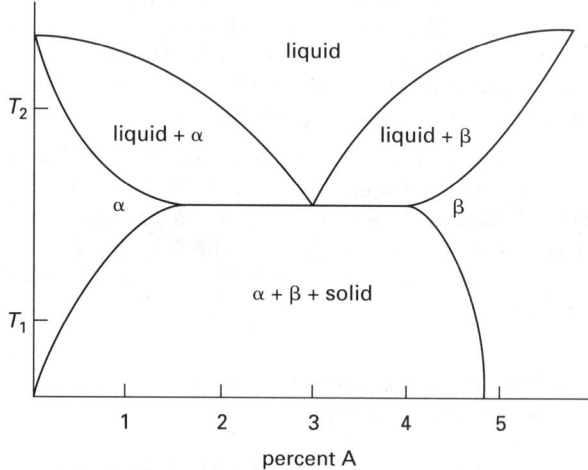

2. Write the procedures used with 2011 aluminum for (a) annealing and (b) precipitation hardening.

---

[8]Fatigue failure never starts in an area under compressive stress.

48 Properties of Areas

1. Centroid of an Area 48-1
2. First Moment of the Area 48-2
3. Centroid of a Line 48-2
4. Theorems of Pappus-Guldinus 48-3
5. Moment of Inertia of an Area 48-3
6. Parallel Axis Theorem 48-4
7. Polar Moment of Inertia 48-5
8. Radius of Gyration 48-6
9. Product of Inertia 48-6
10. Rotation of Axes 48-7
11. Principal Axes 48-7
12. Mohr's Circle 48-7
    Practice Problems 48-8

Nomenclature

A	area	units^2
b	base distance	units
d	separation distance	units
h	height distance	units
I	moment of inertia	units^4
J	polar moment of inertia	units^4
k	radius of gyration	units
L	length	units
P	product of inertia	units^4
Q	first moment of the area	units^3
r	radius, or polar radius of gyration	units
V	volume	units^3
x	distance in the x-direction	units
y	distance in the y-direction	units

Symbols

θ	angle	rad

Subscripts

o	with respect to the origin
c	centroidal

1. CENTROID OF AN AREA

The *centroid* of an area is analogous to the center of gravity of a homogeneous body.[1] The centroid is often described as the point at which a thin homogeneous plate would balance. This definition, however, combines the definitions of centroid and center of gravity and implies that gravity is required to identify the centroid, which is not true.

---
[1] The analogy has been simplified. A three-dimensional body also has a centroid. However, the centroid and center of gravity will not coincide unless the body is homogeneous.

The location of the centroid of an area bounded by the x- and y-axes and the mathematical function $y = f(x)$ can be found by the *integration method* by using Eqs. 48.1 through 48.4. The centroidal location depends only on the geometry of the area and is identified by the coordinates (x_c, y_c). Some references place a bar over the coordinates of the centroid to indicate an average point, such as $(\overline{x}, \overline{y})$.

$$x_c = \frac{\int x \, dA}{A} \qquad \text{48.1}$$

$$y_c = \frac{\int y \, dA}{A} \qquad \text{48.2}$$

$$A = \int f(x) \, dx \qquad \text{48.3}$$

$$dA = f(x) \, dx = g(y) \, dy \qquad \text{48.4}$$

The locations of the centroids of *basic shapes*, such as triangles and rectangles, are well known. The most common basic shapes have been included in App. 48.A. There should be no need to derive centroidal locations for these shapes by the integration method.

The centroid of a complex area can be found from Eqs. 48.5 and 48.6 if the area can be divided into the basic shapes in App. 48.A. This process is simplified when all or most of the subareas adjoin the reference axis. Example 48.1 illustrates this method.

$$x_c = \frac{\sum_i A_i x_{ci}}{\sum_i A_i} \qquad \text{48.5}$$

$$y_c = \frac{\sum_i A_i y_{ci}}{\sum_i A_i} \qquad \text{48.6}$$

Example 48.1

An area is bounded by the x- and y-axes, the line $x = 2$, and the function $y = e^{2x}$. Find the x-component of the centroid.

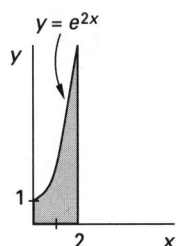

Materials

Solution

First, use Eq. 48.3 to find the area.

$$A = \int f(x)\, dx = \int\limits_{x=0}^{x=2} e^{2x}\, dx$$

$$= \left[\left(\tfrac{1}{2}\right) e^{2x}\right]_0^2 = 27.3 - 0.5 = 26.8 \text{ units}^2$$

Since y is a function of x, dA must be expressed in terms of x. From Eq. 48.4,

$$dA = f(x)\, dx = e^{2x}\, dx$$

Finally, use Eq. 48.1 to find x_c.

$$x_c = \frac{\int x\, dA}{A} = \frac{1}{26.8} \int\limits_{x=0}^{x=2} xe^{2x}\, dx$$

$$= \left(\frac{1}{26.8}\right) \left[\left(\tfrac{1}{2}\right) xe^{2x} - \left(\tfrac{1}{4}\right) e^{2x}\right]_0^2 = 1.54 \text{ units}$$

Example 48.2

Find the y-coordinate of the centroid of the area shown.

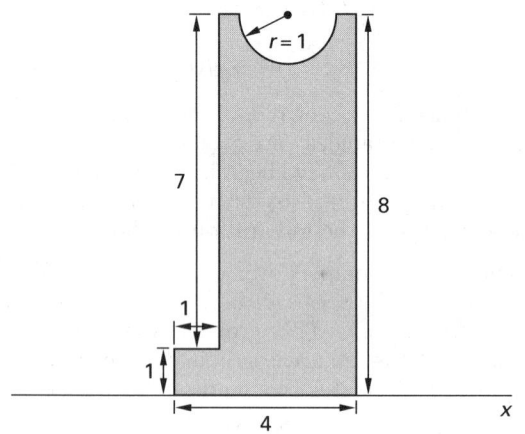

Solution

The x-axis is the reference axis. The area is divided into basic shapes of a 1×1 square, a 3×8 rectangle, and a half-circle of radius 1. (The area could also have been divided into 1×4 and 3×7 rectangles and the half-circle, but the 3×7 rectangle would not then adjoin the x-axis.)

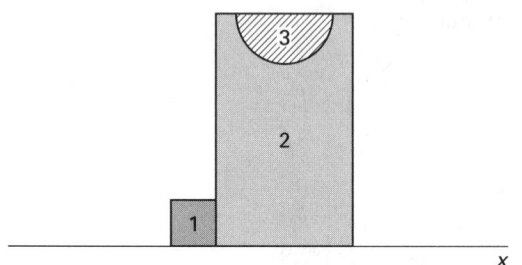

First, calculate the areas of the basic shapes. Notice that the half-circle area is negative since it represents a cutout.

$$A_1 = (1.0)(1.0) = 1.0 \text{ units}^2$$

$$A_2 = (3.0)(8.0) = 24.0 \text{ units}^2$$

$$A_3 = -\left(\tfrac{1}{2}\right) \pi r^2 = \left(-\tfrac{1}{2}\right) \pi (1.0)^2 = -1.57 \text{ units}^2$$

Next, find the y-components of the centroids of the basic shapes. Most are found by inspection, but App. 48.A can be used for the half-circle. Notice that the centroidal location for the half-circle is positive.

$$y_{c1} = 0.5 \text{ units}$$

$$y_{c2} = 4.0 \text{ units}$$

$$y_{c3} = 8.0 - 0.424 = 7.576 \text{ units}$$

Finally, use Eq. 48.6.

$$y_c = \frac{\sum A_i y_{ci}}{\sum A_i} = \frac{(1.0)(0.5)+(24.0)(4.0)+(-1.57)(7.576)}{1.0 + 24.0 - 1.57}$$

$$= \frac{0.5 + 96.0 - 11.9}{23.43} = 3.61 \text{ units}$$

2. FIRST MOMENT OF THE AREA

The quantity $\int x\, dA$ is known as the *first moment of the area* or *first area moment* with respect to the y-axis. Similarly, $\int y\, dA$ is known as the first moment of the area with respect to the x-axis. By rearranging Eqs. 48.1 and 48.2, it is obvious that the first moment of the area can be calculated from the area and centroidal distance.

$$Q_y = \int x\, dA = x_c A \qquad \text{48.7}$$

$$Q_x = \int y\, dA = y_c A \qquad \text{48.8}$$

In basic engineering, the two primary applications of the first moment concept are to determine centroidal locations and shear stress distributions. In the latter application, the first moment of the area is known as the *statical moment*.

3. CENTROID OF A LINE

The location of the *centroid of a line* is defined by Eqs. 48.9 and 48.10, which are analogous to the equations used for centroids of areas.

$$x_c = \frac{\int x\, dL}{L} \qquad \text{48.9}$$

$$y_c = \frac{\int y\, dL}{L} \qquad \text{48.10}$$

Since equations of lines are typically in the form $y = f(x)$, dL must be expressed in terms of x or y.

$$dL = \left(\sqrt{\left(\frac{dy}{dx}\right)^2 + 1} \right) dx \qquad 48.11$$

$$dL = \left(\sqrt{\left(\frac{dx}{dy}\right)^2 + 1} \right) dy \qquad 48.12$$

4. THEOREMS OF PAPPUS-GULDINUS

The *Theorems of Pappus-Guldinus* define the surface and volume of revolution (i.e., the surface area and volume generated by revolving a curve around a fixed axis).

- *Theorem I:* The area of a surface of revolution is equal to the product of the length of the generating curve and the distance traveled by the centroid of the curve while the surface is being generated.

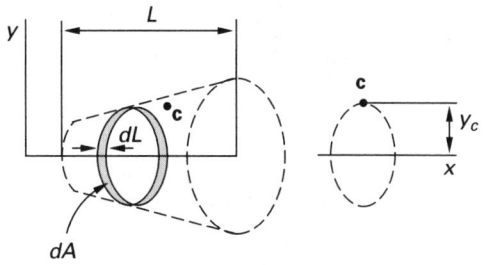

Figure 48.1 *Surface of Revolution*

When a part, dL, of a line L is revolved about the x-axis, a differential ring having surface area dA is generated.

$$dA = 2\pi y \, dL \qquad 48.13$$

$$A = \int dA = 2\pi \int y \, dL = 2\pi y_c L \qquad 48.14$$

- *Theorem II:* The volume of a surface of revolution is equal to the generating area times the distance traveled by the centroid of the area in generating the volume.

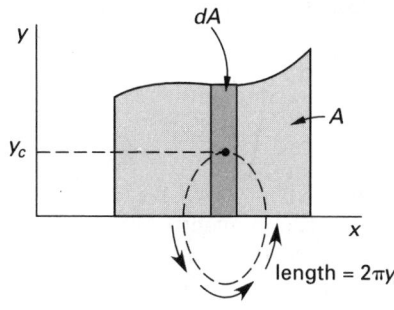

Figure 48.2 *Volume of Revolution*

When a differential plane area, dA, is revolved about the x-axis and does not intersect the y-axis, it generates a ring of volume dV.

$$dV = \pi y^2 dx = \pi y \, dA \qquad 48.15$$

$$V = 2\pi y_c A \qquad 48.16$$

5. MOMENT OF INERTIA OF AN AREA

The *moment of inertia*, I, of an area is needed in mechanics of materials problems. It is convenient to think of the moment of inertia of a beam's cross-sectional area as a measure of the beam's ability to resist bending. Thus, given equal loads, a beam with a small moment of inertia will bend more than a beam with a large moment of inertia.

Since the moment of inertia represents a resistance to bending, it is always positive. Since a beam can be unsymmetrical (e.g., a rectangular beam) and can be stronger in one direction than another, the moment of inertia depends on orientation. Therefore, a reference axis or direction must be specified.

The moment of inertia taken with respect to one of the axes in the rectangular coordinate system is sometimes referred to as the *rectangular moment of inertia*.

The symbol I_x is used to represent a moment of inertia with respect to the x-axis. Similarly, I_y is the moment of inertia with respect to the y-axis. I_x and I_y do not normally combine and are not components of some resultant moment of inertia.

Any axis can be chosen as the reference axis, and the value of the moment of inertia will depend on the reference selected. The moment of inertia taken with respect to an axis passing through the area's centroid is known as the *centroidal moment of inertia*, I_{cx} or I_{cy}. For any set of parallel axes, the centroidal moment of inertia is the smallest possible moment of inertia for the shape.

The *integration method* can be used to calculate the moment of inertia of a function that is bounded by the x- and y-axes and a curve $y = f(x)$. From Eqs. 48.17 and 48.18, it is apparent why the moment of inertia is also known as the *second moment of the area* or *second area moment*.

$$I_x = \int y^2 \, dA \qquad 48.17$$

$$I_y = \int x^2 \, dA \qquad 48.18$$

$$dA = f(x) \, dx = g(y) \, dy \qquad 48.19$$

The moments of inertia of the *basic shapes* are well known and are listed in App. 48.A.

Materials

Example 48.3

What is the centroidal moment of inertia with respect to the x-axis of a rectangle 5.0 units wide and 8.0 units tall?

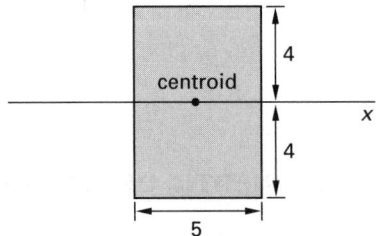

Solution

Since the centroidal moment of inertia is needed, the reference line passes through the centroid. From App. 48.A, the centroidal moment of inertia is

$$I_{cx} = \frac{bh^3}{12} = \frac{(5)(8)^3}{12} = 213.3 \text{ units}^4$$

Example 48.4

What is the moment of inertia with respect to the y-axis of the area bounded by the y-axis, the line $y = 8.0$, and the parabola $y^2 = 8x$?

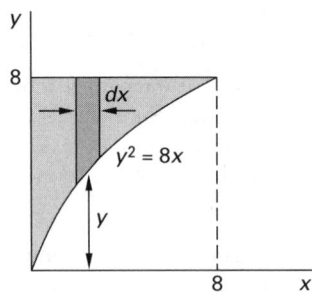

Solution

This problem is more complex than it first appears, since the area is above the curve, bounded not by $y = 0$ but by $y = 8$. In particular, dA must be determined correctly.

$$y = \sqrt{8x}$$
$$dA = (8 - f(x)) \, dx = (8 - y) \, dx$$
$$= (8 - \sqrt{8x}) \, dx$$

Equation 48.18 is used to calculate the moment of inertia with respect to the y-axis.

$$I_y = \int x^2 \, dA = \int_0^8 x^2 \left(8 - \sqrt{8x}\right) \, dx$$
$$= \left[\left(\frac{8}{3}\right) x^3 - \left(\frac{4\sqrt{2}}{7}\right) x^{\frac{7}{2}}\right]_0^8 = 195.0 \text{ units}^4$$

6. PARALLEL AXIS THEOREM

If the moment of inertia is known with respect to one axis, the moment of inertia with respect to another, the parallel axis can be calculated from the *parallel axis theorem*, also known as the *transfer axis theorem*. This theorem is used to evaluate the moment of inertia of areas that are composed of two or more basic shapes. In Eq. 48.20, d is the distance between the centroidal axis and the second, parallel axis.

$$I_{\text{parallel axis}} = I_c + Ad^2 \qquad \textbf{48.20}$$

The second term in Eq. 48.20 is often much larger than the first term. Areas close to the centroidal axis do not affect the moment of inertia considerably. This principle is exploited by structural steel shapes that derive bending resistance from *flanges* located away from the centroidal axis. The *web* does not contribute significantly to the moment of inertia.

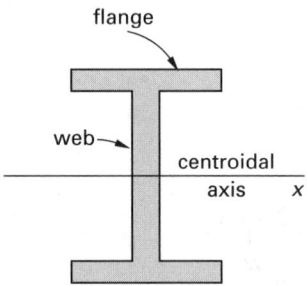

Figure 48.3 *Structural Steel W-Shape*

Example 48.5

Find the moment of inertia about the x-axis for the inverted T-area shown.

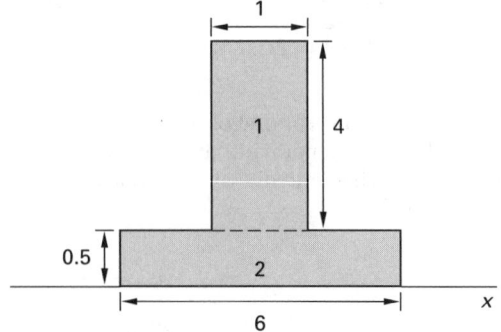

Solution

The area is divided into two basic shapes: 1 and 2. From App. 48.A, the moment of inertia of basic shape 2 with respect to the x-axis is

$$I_{x2} = \frac{bh^3}{3} = \frac{(6.0)(0.5)^3}{3} = 0.25 \text{ units}^4$$

The moment of inertia of basic shape 1 about its own centroid is

$$I_{cx1} = \frac{bh^3}{12} = \frac{(1)(4)^3}{12} = 5.33 \text{ units}^4$$

The x-axis is located 2.5 units from the centroid of basic shape 1. Therefore, from the parallel axis theorem, Eq. 48.20, the moment of inertia of basic shape 1 about the x-axis is

$$I_{x1} = 5.33 + (4)(2.5)^2 = 30.33 \text{ units}^4$$

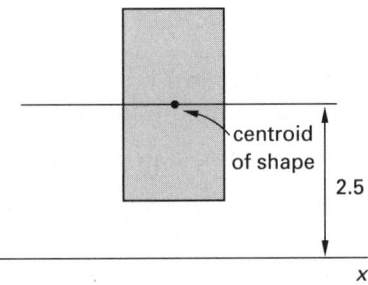

The total moment of inertia of the T-area is

$$I_x = I_{x1} + I_{x2} = 30.33 \text{ units}^4 + 0.25 \text{ units}^4$$
$$= 30.58 \text{ units}^4$$

Example 48.6

Find the moment of inertia about the horizontal centroidal axis for the inverted T-area shown in Ex. 48.5.

Solution

The first step is to find the location of the centroid. The areas and centroidal locations (with respect to the x-axis) of the two basic shapes are

$$A_1 = (4.0)(1.0) = 4.0 \text{ units}^2$$
$$A_2 = (0.5)(6.0) = 3.0 \text{ units}^2$$
$$y_{c1} = 2.5 \text{ units}$$
$$y_{c2} = 0.25 \text{ units}$$

From Eq. 48.6, the composite centroid is located at

$$y_c = \frac{A_1 y_{c1} + A_2 y_{c2}}{A_1 + A_2} = \frac{(4.0)(2.5) + (3.0)(0.25)}{4.0 + 3.0}$$
$$= 1.536 \text{ units}$$

The distances between the centroids of the basic shapes and the composite shape are

$$d_1 = 2.5 - 1.536 = 0.964 \text{ units}$$
$$d_2 = 1.536 - 0.25 = 1.286 \text{ units}$$

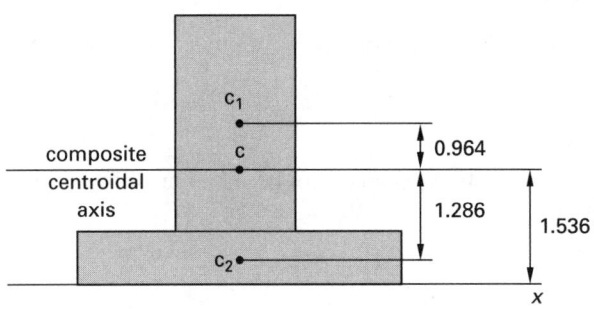

The centroidal moments of inertia of the basic shapes with respect to an axis parallel to the x-axis are

$$I_{cx1} = \frac{bh^3}{12} = \frac{(1)(4)^3}{12} = 5.33 \text{ units}^4$$
$$I_{cx2} = \frac{bh^3}{12} = \frac{(6)(0.5)^3}{12} = 0.0625 \text{ units}^4$$

Using Eq. 48.20, the centroidal moment of inertia of the inverted T-area is

$$\begin{aligned} I_{cx} &= I_{cx1} + I_{cx2} \\ &= I_{cx1} + A_1 d_1^2 + I_{cx2} + A_2 d_2^2 \\ &= 5.33 + (4.0)(0.964)^2 + 0.0625 + (3.0)(1.286)^2 \\ &= 14.07 \text{ units}^4 \end{aligned}$$

7. POLAR MOMENT OF INERTIA

The *polar moment of inertia*, J, is required in torsional shear stress calculations.[2] It can be thought of as a measure of an area's resistance to torsion (twisting). The definition of a polar moment of inertia of a two-dimensional area requires three dimensions because the reference axis for a polar moment of inertia of a plane area is perpendicular to the plane area.

The polar moment of inertia can be derived from Eq. 48.21.

$$J = \int (x^2 + y^2) \, dA \qquad \textbf{48.21}$$

It is often easier to use the *perpendicular axis theorem* to quickly calculate the polar moment of inertia.

- *perpendicular axis theorem:* The moment of inertia of a plane area about an axis normal to the plane is equal to the sum of the moments of inertia about any two mutually perpendicular axes lying in the plane and passing through the given axis.

$$J = I_x + I_y \qquad \textbf{48.22}$$

Since the two perpendicular axes can be chosen arbitrarily, it is most convenient to use the centroidal moments of inertia.

$$J = I_{cx} + I_{cy} \qquad \textbf{48.23}$$

---

[2] The symbols I_z and I_{xy} are also encountered and are more consistent with the nomenclature. However, the symbol J is more common.

Example 48.7

What is the centroidal polar moment of inertia of a circular area of radius r?

Solution

From App. 48.A, the centroidal moment of inertia of a circle with respect to the x-axis is

$$I_{cx} = \frac{\pi r^4}{4}$$

Since the area is symmetrical, I_{cy} and I_{cx} are the same. From Eq. 48.23,

$$J_c = I_{cx} + I_{cy} = \frac{\pi r^4}{4} + \frac{\pi r^4}{4} = \frac{\pi r^4}{2}$$

8. RADIUS OF GYRATION

Every nontrivial area has a centroidal moment of inertia. Usually, some portions of the area are close to the centroidal axis and other portions are farther away. The *radius of gyration*, k, is an imaginary distance from the centroidal axis at which the entire area can be assumed to exist without affecting the moment of inertia. Despite the name "radius," the radius of gyration is not limited to circular shapes or polar axes. This concept is illustrated in Fig. 48.4.

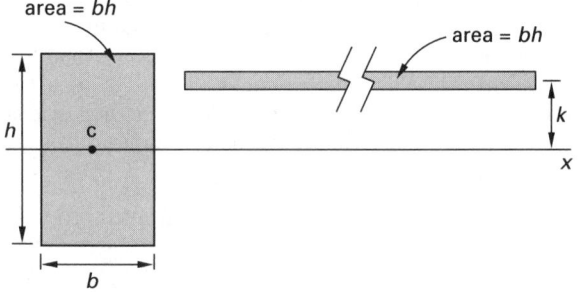

Figure 48.4 *Radius of Gyration*

The method of calculating the radius of gyration is based on the parallel axis theorem. If all of the area is located a distance k from the original centroidal axis, there will be no I_c term in Eq. 48.20. Only the Ad^2 will contribute to the moment of inertia.

$$I = k^2 A \qquad 48.24$$

$$k = \sqrt{\frac{I}{A}} \qquad 48.25$$

The concept of *least radius of gyration* comes up frequently in column design problems. (The column will tend to buckle about an axis that produces the smallest radius of gyration.) Usually, finding the least radius of gyration means solving Eq. 48.25 twice: once with I_x to find k_x, and once with I_y to find k_y. The smallest value of k is the least radius of gyration.

The analogous quantity in the polar system is

$$r = \sqrt{\frac{J}{A}} \qquad 48.26$$

Just as the polar moment of inertia, J, can be calculated from the two rectangular moments of inertia, the polar radius of gyration can be calculated from the two rectangular radii of gyration.

$$r^2 = k_x^2 + k_y^2 \qquad 48.27$$

Example 48.8

What is the radius of gyration of the rectangular shape in Ex. 48.3?

Solution

The area of the rectangle is

$$A = bh = (5)(8) = 40 \text{ units}^2$$

From Eq. 48.25, the radius of gyration is

$$k_x = \sqrt{\frac{I_x}{A}} = \sqrt{\frac{213.3}{40}} = 2.31 \text{ units}$$

2.31 units is the distance from the centroidal x-axis that an infinitely long strip with an area of 40 square units would have to be located in order to have a moment of inertia of 213.3 units^4.

9. PRODUCT OF INERTIA

The *product of inertia*, P_{xy}, of a two-dimensional area is found by multiplying each differential element of area by its x- and y-coordinate and then summing over the entire area.

$$P_{xy} = \int xy \, dA \qquad 48.28$$

The product of inertia is zero when either axis is an axis of symmetry. Since the axes can be chosen arbitrarily, the area may be in one of the negative quadrants, and the product of inertia may be negative.

The parallel axis theorem for products of inertia is given by Eq. 48.29. (Both axes are allowed to move to new positions.) x_c' and y_c' are the coordinates of the centroid in the new coordinate system.

$$P_{x'y'} = P_{c,xy} + x_c' y_c' A \qquad 48.29$$

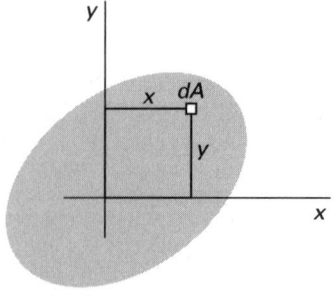

Figure 48.5 *Calculating the Product of Inertia*

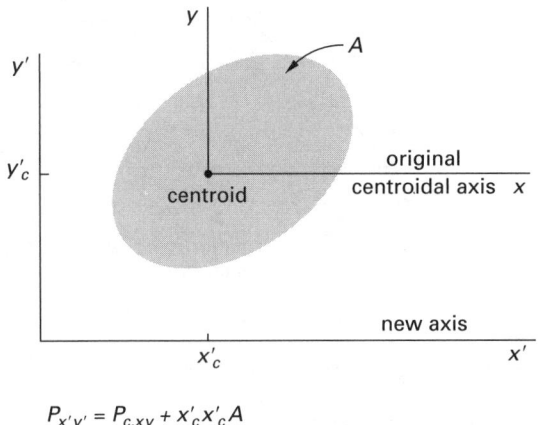

$$P_{x'y'} = P_{c,xy} + x'_c x'_c A$$

Figure 48.6 *Parallel Axis Theorem for Products of Inertia*

10. ROTATION OF AXES

Figure 48.7 shows rotation of the x-y-axes through an angle, θ, into a new set of u-v-axes, without rotating the area. If the moments and product of inertia of the area are known with respect to the old x-y-axes, the new properties can be calculated from Eqs. 48.30 through 48.32.

$$I_u = I_x \cos^2 \theta - 2P_{xy} \sin \theta \cos \theta + I_y \sin^2 \theta$$
$$= \left(\tfrac{1}{2}\right)(I_x + I_y) + \left(\tfrac{1}{2}\right)(I_x - I_y) \cos 2\theta$$
$$- P_{xy} \sin 2\theta \qquad \textit{48.30}$$

$$I_v = I_x \sin^2 \theta + 2P_{xy} \sin \theta \cos \theta + I_y \cos^2 \theta$$
$$= \left(\tfrac{1}{2}\right)(I_x + I_y) - \left(\tfrac{1}{2}\right)(I_x - I_y) \cos 2\theta$$
$$+ P_{xy} \sin 2\theta \qquad \textit{48.31}$$

$$P_{uv} = I_x \sin \theta \cos \theta + P_{xy}(\cos^2 \theta - \sin^2 \theta)$$
$$- I_y \sin \theta \cos \theta$$
$$= \left(\tfrac{1}{2}\right)(I_x - I_y) \sin 2\theta + P_{xy} \cos 2\theta \qquad \textit{48.32}$$

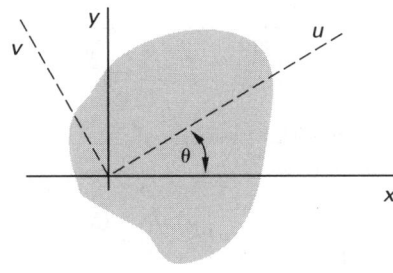

Figure 48.7 *Rotation of Axes*

Since the polar moment of inertia about a fixed axis perpendicular to any two orthogonal axes in the plane is constant, the polar moment of inertia is unchanged by the rotation.

$$J_{xy} = I_x + I_y = I_u + I_v = J_{uv} \qquad \textit{48.33}$$

Example 48.9

What is the centroidal area moment of inertia of a 6×6 square that is rotated $45°$ from its "flat" orientation?

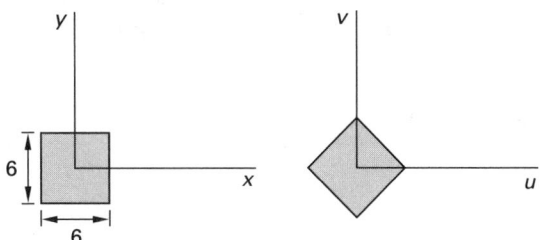

Solution

The centroidal moments of inertia with respect to the x- and y-axes are

$$I_x = I_y = \frac{s^4}{12} = \frac{(6)^4}{12} = 108 \text{ units}^4$$

Since the centroidal x- and y-axes are axes of symmetry, the product of inertia is zero.

Use Eq. 48.30.

$$I_u = I_x \cos^2 \theta - 2P_{xy} \sin \theta \cos \theta + I_y \sin^2 \theta$$
$$= (108)\cos^2 45° - 0 + (108)\sin^2 45°$$
$$= 108 \text{ units}^4$$

The centroidal moment of inertia of a square is the same regardless of rotation angle.

11. PRINCIPAL AXES

Referring to Fig. 48.7, there is one angle, θ, that will maximize the moment of inertia, I_u. This angle can be found from calculus by setting $dI_u/d\theta = 0$. The resulting equation defines two angles, one that maximizes I_u and one that minimizes I_u.

$$\tan 2\theta = \frac{-2P_{xy}}{I_x - I_y} \qquad \textit{48.34}$$

The two angles that satisfy Eq. 48.34 are $90°$ apart. The set of u-v-axes defined by Eq. 48.34 are known as *principal axes*. The moments of inertia about the principal axes are defined by Eq. 48.35 and are known as the *principal moments of inertia*.

$$I_{\text{max,min}} = \left(\tfrac{1}{2}\right)(I_x + I_y) \pm \sqrt{\left(\tfrac{1}{4}\right)(I_x - I_y)^2 + P_{xy}^2} \qquad \textit{48.35}$$

12. MOHR'S CIRCLE

Once I_x, I_y, and P_{xy} are known, *Mohr's circle* can be drawn to graphically determine the moments of inertia about the principal axes. The procedure for drawing Mohr's circle is given as follows.

step 1: Determine I_x, I_y, and P_{xy} for the existing set of axes.

step 2: Draw a set of I-P_{xy}-axes.

step 3: Plot the center of the circle, point **c**, by calculating distance c along the I-axis.

$$c = \left(\tfrac{1}{2}\right)\left(I_x + I_y\right) \qquad 48.36$$

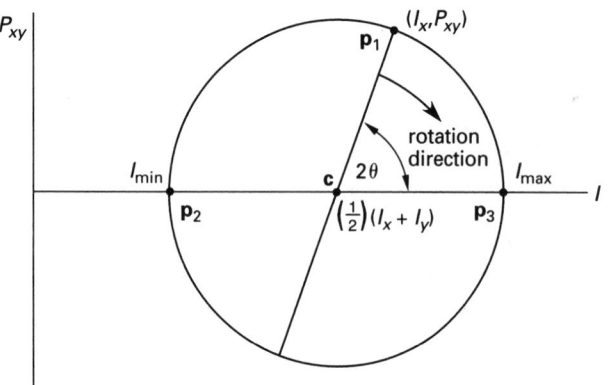

Figure 48.8 *Mohr's Circle*

step 4: Plot the point $\mathbf{p}_1 = (I_x, P_{xy})$.

step 5: Draw a line from point $\mathbf{p}_1$ through center **c** and extend it an equal distance below the I-axis. This is the diameter of the circle.

step 6: Using the center **c** and point $\mathbf{p}_1$, draw the circle. An alternate method of constructing the circle is to draw a circle of radius r.

$$r = \sqrt{\left(\tfrac{1}{4}\right)\left(I_x - I_y\right)^2 + P_{xy}^2} \qquad 48.37$$

step 7: Point $\mathbf{p}_2$ defines $I_{\min}$. Point $\mathbf{p}_3$ defines $I_{\max}$.

step 8: Determine the angle θ as half of the angle 2θ on the circle. This angle corresponds to $I_{\max}$. (The axis giving the minimum moment of inertia is perpendicular to the maximum axis.) The sense of this angle and the sense of the rotation are the same. That is, the direction that the diameter would have to be turned in order to coincide with the $I_{\max}$-axis has the same sense as the rotation of the x-y-axes needed to form the principal u-v-axes.

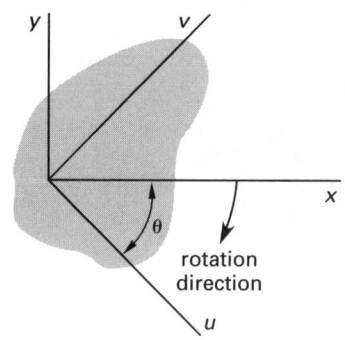

Figure 48.9 *Principal Axes from Mohr's Circle*

PRACTICE PROBLEMS

1. Locate the centroid of the area.

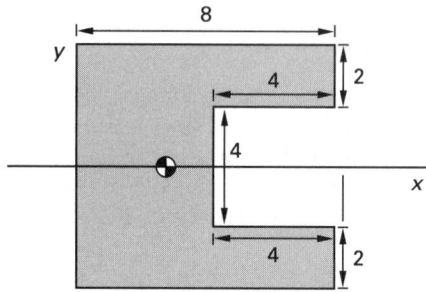

2. Replace the distributed load with three concentrated loads, and indicate the points of application.

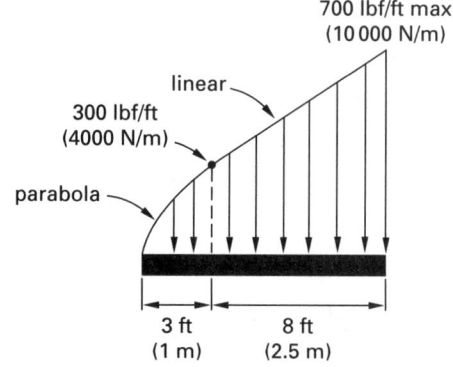

3. Find the centroidal moment of inertia about an axis parallel to the x-axis.

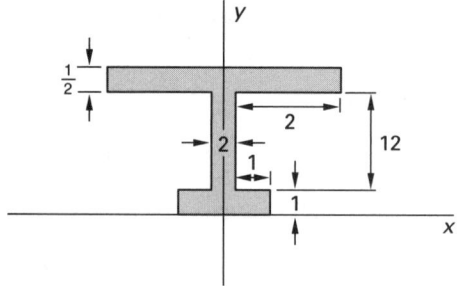

49 Strength of Materials

1. Basic Concepts 49-2
2. Hooke's Law 49-2
3. Elastic Deformation 49-2
4. Total Strain Energy 49-2
5. Stiffness and Rigidity 49-2
6. Thermal Deformation 49-3
7. Stress Concentrations 49-4
8. Combined Stresses (Biaxial Loading) 49-5
9. Mohr's Circle for Stress 49-6
10. Impact Loading 49-7
11. Shear and Moment 49-7
12. Shear and Bending Moment Diagrams . . . 49-8
13. Shear Stress in Beams 49-9
14. Bending Stress in Beams 49-10
15. Strain Energy Due to Bending Moment . . 49-11
16. Eccentric Loading of Axial Members . . . 49-11
17. Beam Deflection:
    Double Integration Method 49-13
18. Beam Deflection:
    Moment Area Method 49-14
19. Beam Deflection:
    Strain Energy Method 49-15
20. Beam Deflection:
    Conjugate Beam Method 49-15
21. Beam Deflection:
    Table Look-Up Method 49-16
22. Beam Deflection: Superposition 49-16
23. Inflection Points 49-16
24. Truss Deflection:
    Strain Energy Method 49-17
25. Truss Deflection: Virtual Work Method . . 49-17
26. Moving Loads on Beams 49-18
27. Modes of Beam Failure 49-18
28. Curved Beams 49-19
29. Composite Structures 49-19
    Practice Problems 49-21

Nomenclature

a	width	in	m
A	area	in^2	m^2
b	width	in	m
c	distance to extreme fiber	in	m
C	constant	–	–
C	couple	in-lbf	N·m
d	distance, depth, or diameter	in	m
e	eccentricity	in	m
E	modulus of elasticity	lbf/in^2	MPa
F	force	lbf	N
G	shear modulus	lbf/in^2	MPa
h	height	in	m
I	moment of inertia	in^4	m^4
J	polar moment of inertia	in^4	m^4
k	spring constant	lbf/in	N/m
K	stress concentration factor	–	–
L	length	in	m
M	moment	in-lbf	N·m
P	force	lbf	N
Q	statical moment	in^3	m^3
r	radius	in	m
R	rigidity	–	–
R	reaction	lbf	N
S	force	lbf	N
t	thickness	in	m
T	temperature	°F	°C
T	torque	in-lbf	N·m
u	unit force	lbf	N
U	energy	in-lbf	N·m
V	vertical shear force	lbf	N
V	volume	in^3	m^3
w	load per unit length	lbf/in	N/m
x	location	in	m
y	location	in	m
Z	section modulus	in^3	m^3

Symbols

α	coefficient of linear thermal expansion	1/°F	1/°C
β	coefficient of volumetric thermal expansion	1/°F	1/°C
γ	coefficient of area thermal expansion	1/°F	1/°C
δ	deformation	in	m
ϵ	strain	–	–
θ	angle	deg	deg
ν	Poisson's ratio	–	–
ρ	radius of curvature	in	m
σ	normal stress	lbf/in^2	MPa
τ	shear stress	lbf/in^2	MPa
ϕ	angle	rad	rad

Subscripts

0	nominal
a	alternating
b	bending
c	centroidal
i	inside
j	jth member
l	left
m	mean
o	original
r	range or right
R	right
th	thermal
w	web

1. BASIC CONCEPTS

Strength of materials (known also as *mechanics of materials*) deals with the elastic behavior of loaded engineering materials.[1] This subject draws heavily on the topics in Chaps. 46 and 48.

Stress is force per unit area, F/A. Typical units of stress are lbf/in^2, ksi (thousands of pounds per square inch), and MPa. Although there are many names given to stress, there are only two primary types, differing in the orientation of the loaded area. With *normal stress*, σ, the area is normal to the force carried. With *shear stress*, τ, the area is parallel to the force.

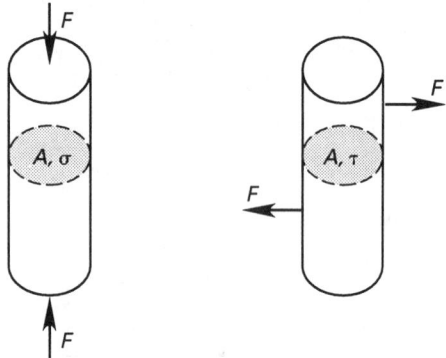

Figure 49.1 *Normal and Shear Stress*

Strain, ϵ, is elongation expressed on a fractional or percentage basis. It may be listed as having units of in/in, mm/mm, and percent, or no units at all. A strain in one direction will be accompanied by strains in orthogonal directions in accordance with Poisson's ratio. *Dilation* is the sum of the strains in the three coordinate directions.

$$\text{dilation} = \epsilon_x + \epsilon_y + \epsilon_z \qquad 49.1$$

2. HOOKE'S LAW

Hooke's law is a simple mathematical statement of the relationship between elastic stress and strain: Stress is proportional to strain. For normal stress, the constant of proportionality is the *modulus of elasticity (Young's modulus)*, E.

$$\sigma = E\epsilon \qquad 49.2$$

For shear stress, the constant of proportionality is the *shear modulus*, G.

$$\tau = G\phi \qquad 49.3$$

---

[1]Plastic behavior and ultimate strength design are not covered in this book.

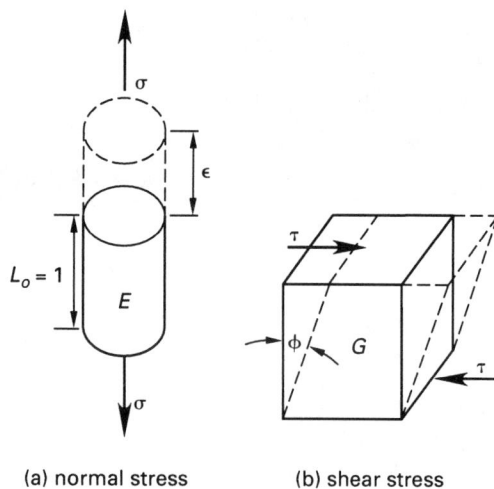

(a) normal stress   (b) shear stress

Figure 49.2 *Application of Hooke's Law*

3. ELASTIC DEFORMATION

Since stress is F/A and strain is δ/L_o, Hooke's law can be rearranged in form to give the elongation of an axially loaded member with a uniform cross section experiencing normal stress. Tension loading is considered positive; compressive loading is negative.

$$\delta = L_o\epsilon = \frac{L_o\sigma}{E} = \frac{L_oF}{EA} \qquad 49.4$$

The actual length of a member under loading is given by Eq. 49.5. The algebraic sign of the deformation must be observed.

$$L = L_o + \delta \qquad 49.5$$

4. TOTAL STRAIN ENERGY

The energy stored in a loaded member is equal to the work required to deform the member. Below the proportionality limit, the total *strain energy* for a member loaded in tension or compression is given by Eq. 49.6.

$$U = \tfrac{1}{2}F\delta = \frac{F^2L_o}{2AE} = \frac{\sigma^2L_oA}{2E} \qquad 49.6$$

5. STIFFNESS AND RIGIDITY

Stiffness is the amount of force required to cause a unit of deformation (displacement) and is often referred to as a *spring constant*. Typical units are pounds per inch and newtons per meter. The stiffness of a spring or other structure can be calculated from the deformation equation by solving for F/δ. Equation 49.7 is valid for tensile and compressive normal stresses. For torsion and bending, the stiffness equation will depend on how the deflection is calculated.

$$k = \frac{F}{\delta} \qquad \text{[general form]} \qquad 49.7(a)$$

$$= \frac{AE}{L_o} \qquad \text{[normal stress form]} \qquad 49.7(b)$$

When more than one spring or resisting member share the load, the relative stiffnesses are known as *rigidities*. Rigidities have no units, and the individual rigidity values have no significance. A ratio of two rigidities, however, indicates how much stronger one member is compared to another. Equation 49.8 is one method of calculating rigidity in a multi-member structure. (Since rigidities are relative numbers, they can be multiplied by the least common denominator to obtain integer values.)

$$R_j = \frac{k_j}{\sum\limits_i k_i} \qquad 49.8$$

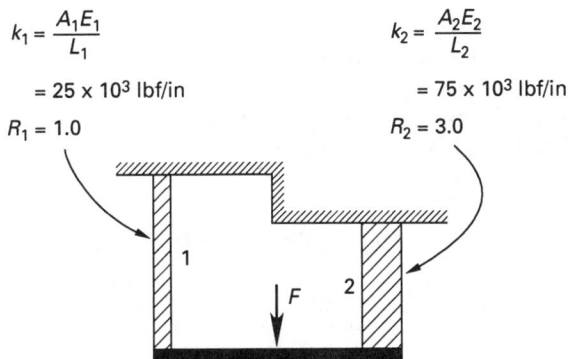

Figure 49.3 *Stiffness and Rigidity*

Rigidity is the reciprocal of deflection. *Flexural rigidity* is the reciprocal of deflection in members that are acted upon by a moment (i.e., are in bending), although that term may also be used to refer to the product, EI, of the modulus of elasticity and the moment of inertia.

6. THERMAL DEFORMATION

If the temperature of an object is changed, the object will experience length, area, and volume changes. The magnitude of these changes will depend on the *coefficient of linear expansion*, α, which is widely tabulated for solids. *The coefficient of volumetric expansion*, β, is encountered less often for solids but is used extensively with liquids and gases.

$$\Delta L = \alpha L_o(T_2 - T_1) \qquad 49.9$$
$$\Delta A = \gamma A_o(T_2 - T_1) \qquad 49.10$$
$$\gamma \approx 2\alpha \qquad 49.11$$
$$\Delta V = \beta V_o(T_2 - T_1) \qquad 49.12$$
$$\beta \approx 3\alpha \qquad 49.13$$

It is a common misconception that a hole in a plate will decrease in size when the plate is heated (because the surrounding material "squeezes in" on the hole). However, changes in temperature affect all dimensions the same way. In this case, the circumference of the hole is a linear dimension that follows Eq. 49.9. As the circumference increases, the hole area also increases.

Table 49.1 *Deflection and Stiffness for Various Systems (due to bending moment alone)*

system	maximum deflection (x)	stiffness (k)
	$\dfrac{Fh}{AE}$	$\dfrac{AE}{h}$
	$\dfrac{Fh^3}{3EI}$	$\dfrac{3EI}{h^3}$
	$\dfrac{Fh^3}{12EI}$	$\dfrac{12EI}{h^3}$
	$\dfrac{wL^4}{8EI}$	$\dfrac{8EI}{L^3}$
	$\dfrac{Fh^3}{12E(I_1 + I_2)}$	$\dfrac{12E(I_1 + I_2)}{h^3}$
	$\dfrac{FL^3}{48EI}$	$\dfrac{48EI}{L^3}$
(w is load per unit length)	$\dfrac{5wL^4}{384EI}$	$\dfrac{384EI}{5L^3}$
	$\dfrac{FL^3}{192EI}$	$\dfrac{192EI}{L^3}$
(w is load per unit length)	$\dfrac{wL^4}{384EI}$	$\dfrac{384EI}{L^3}$

Materials

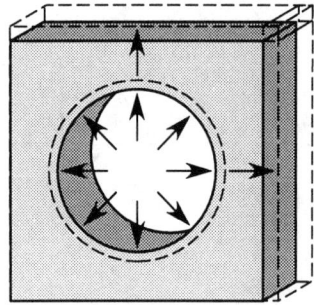

Figure 49.4 *Thermal Expansion of an Area*

If Eq. 49.9 is rearranged, an expression for the *thermal strain* is obtained.

$$\epsilon_{th} = \frac{\delta}{L_o} = \alpha(T_2 - T_1) \qquad \textbf{49.14}$$

Thermal strain is handled in the same manner as strain due to an applied load. For example, if a bar is heated but is not allowed to expand, the stress can be calculated from the thermal strain and Hooke's law.

$$\sigma_{th} = E\epsilon_{th} \qquad \textbf{49.15}$$

Low values of the coefficient of expansion, such as with Pyrex™ glassware, result in low thermally induced stresses and high insensitivity to temperature extremes. Differences in the coefficients of expansion of two materials are used in *bimetallic elements*, such as thermostatic springs and strips.

Table 49.2 *Average Coefficients of Linear Thermal Expansion (multiply all values by 10^{-6})*

substance	1/°F	1/°C
aluminum alloy	12.8	23.0
brass	10.0	18.0
cast iron	5.6	10.1
chromium	3.8	6.8
concrete	6.7	12.0
copper	8.9	16.0
glass (plate)	4.9	8.9
glass (Pyrex™)	1.8	3.2
invar	0.39	0.7
lead	15.6	28.0
magnesium alloy	14.5	26.1
marble	6.5	11.7
platinum	5.0	9.0
quartz, fused	0.2	0.4
steel	6.5	11.7
tin	14.9	26.9
titanium alloy	4.9	8.8
tungsten	2.4	4.4
zinc	14.6	26.3

(Multiply 1/°F by 9/5 to obtain 1/°C.)
(Multiply 1/°C by 5/9 to obtain 1/°F.)

Example 49.1

A replacement steel rail ($L = 20.0$ m, $A = 60\times10^{-4}$ m^2) was installed when its temperature was 5°C. The rail was installed tightly in the line, without an allowance for expansion. If the rail ends are constrained by adjacent rails, and if the spikes prevent buckling, what is the compressive force in the rail at 25°C?

Solution

From Table 49.2, the coefficient of linear expansion for steel is 11.7×10^{-6} 1/°C. From Eq. 49.14, the thermal strain is

$$\epsilon_{th} = \alpha(T_2 - T_1) = \left(11.7 \times 10^{-6} \frac{1}{°C}\right)(25°C - 5°C)$$
$$= 2.34 \times 10^{-4} \text{ m/m}$$

The modulus of elasticity of steel is 20×10^{10} N/m^2 (20×10^4 MPa). The compressive stress is given by Hooke's law.

$$\sigma_{th} = E\epsilon_{th} = \left(20 \times 10^{10} \frac{N}{m^2}\right)\left(2.34 \times 10^{-4} \frac{m}{m}\right)$$
$$= 4.68 \times 10^7 \text{ N/m}^2$$

The compressive force is

$$F = \sigma_{th}A = \left(4.68 \times 10^7 \frac{N}{m^2}\right)(60 \times 10^{-4} \text{ m}^2)$$
$$= 281\,000 \text{ N}$$

7. STRESS CONCENTRATIONS

A *geometric stress concentration* occurs whenever there is a discontinuity or non-uniformity in an object. Examples of non-uniform shapes are stepped shafts, plates with holes and notches, and shafts with keyways. It is convenient to think of stress as streamlines within an object. There will be a stress concentration wherever local geometry forces the streamlines closer together.

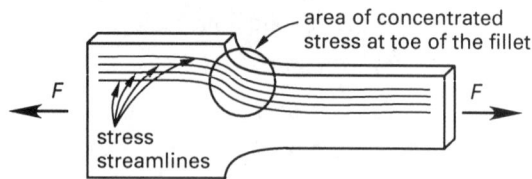

Figure 49.5 *Streamline Analogy to Stress Concentrations*

Stress values determined by simplistic F/A, Mc/I, or Tr/J calculations will be greatly understated. *Stress concentration factors (stress risers)* are correction factors used to account for the non-uniform stress distributions. The symbol K is often used, but this is not universal. The actual stress is determined as the product of the stress concentration factor, K, and the *nominal stress*, σ_0. Values of the stress concentration factor are almost always greater than 1.0 and can run as high as 3.0 and above. The exact value for a given application must be determined from extensive experimentation or from published tabulations of standard configurations such as App. 49.B.

$$\sigma' = K\sigma_0 \qquad \textit{49.16}$$

Stress concentration factors are normally not applied to members with multiple redundancy, for static loading of ductile materials, or where local yielding around the discontinuity reduces the stress. For example, there will be many locations of stress concentration in a lap rivet connection. However, the stresses are kept low by design, and stress concentration is disregarded.

Stress concentration factors are not applicable to every point on an object; they apply only to the point of maximum stress. For example, with filleted shafts, the maximum stress occurs at the toe of the fillet. Therefore, the stress concentration factor should be applied to the stress calculated from the smaller section's properties. For objects with holes or notches, it is important to know if the nominal stress to which the factor is applied is calculated from an area that includes or excludes the holes or notches.

In addition to geometric stress concentrations, there are also *fatigue stress concentrations*. The *fatigue stress concentration* factor is the ratio of the fatigue strength without a stress concentration to the fatigue stress with a stress concentration. Fatigue stress concentration factors depend on the material, material strength, and geometry of the stress concentration (i.e., radius of the notch). Fatigue stress concentration factors can be less than the geometric factors from which they are computed.

8. COMBINED STRESSES (BIAXIAL LOADING)

Loading is rarely confined to a single direction. Many practical cases have different normal and shear stresses on two or more perpendicular planes. Sometimes, one of the stresses may be small enough to be disregarded, reducing the analysis to one dimension. In other cases, however, the shear and normal stresses must be combined to determine the maximum stresses acting on the material.

For any point in a loaded specimen, a plane can be found where the shear stress is zero. The normal stresses associated with this plane are known as the *principal stresses*, which are the maximum and minimum stresses acting at that point in any direction.

For two-dimensional *(biaxial stress)* loading (i.e., two normal stresses combined with a shearing stress), the normal and shear stresses on a plane whose normal line is inclined an angle θ from the horizontal can be found from Eqs. 49.17 and 49.18. Proper sign convention must be adhered to when using the combined stress equations. The positive senses of shear and normal stresses are shown in Fig. 49.6. As is usually the case, tensile

normal stresses are positive; compressive normal stresses are negative. In two dimensions, shear stresses are designated as clockwise (positive) or counterclockwise (negative).[2]

$$\sigma_\theta = \left(\tfrac{1}{2}\right)(\sigma_x + \sigma_y) + \left(\tfrac{1}{2}\right)(\sigma_x - \sigma_y)\cos 2\theta + \tau \sin 2\theta$$
$$\textit{49.17}$$

$$\tau_\theta = \left(-\tfrac{1}{2}\right)(\sigma_x - \sigma_y)\sin 2\theta + \tau \cos 2\theta \qquad \textit{49.18}$$

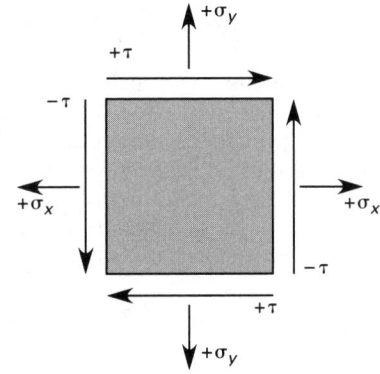

Figure 49.6 *Sign Convention for Combined Stress*

At first glance, the orientation of the shear stresses may seem confusing. However, the arrangement of stresses shown produces equilibrium in the x- and y-directions without causing rotation. Other than a mirror image or a trivial rotation of the arrangement shown in Fig. 49.6, no other arrangement of shear stresses will produce equilibrium.

The maximum and minimum values (as θ is varied) of the normal stress, σ_θ, are the *principal stresses*, which can be found by differentiating Eq. 49.17 with respect to θ, setting the derivative equal to zero, and substituting θ back into Eq. 49.17. Equation 49.19 is derived in this manner. A similar procedure is used to derive the *extreme shear stresses* (i.e., maximum and minimum shear stresses) in Eq. 49.20 from Eq. 49.18. (The term *principal stress* implies a normal stress, never a shear stress.)

$$\sigma_1, \sigma_2 = \left(\tfrac{1}{2}\right)(\sigma_x + \sigma_y) \pm \tau_1 \qquad \textit{49.19}$$

$$\tau_1, \tau_2 = \pm \tfrac{1}{2}\sqrt{(\sigma_x - \sigma_y)^2 + (2\tau)^2} \qquad \textit{49.20}$$

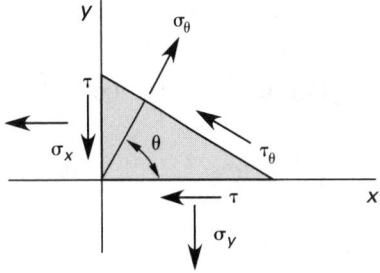

Figure 49.7 *Plane of Principal Stresses*

---

[2]Some sources refer to the shear stress as τ_{xy}; others use the symbol τ_z. When working in two dimensions only, the subscripts xy and z are unnecessary and confusing conventions.

The angles of the planes on which the normal stresses are minimum and maximum are given by Eq. 49.21. θ is measured from the x-axis, clockwise if negative and counterclockwise if positive. Equation 49.21 will yield two angles, 90° apart. These angles can be substituted back into Eqs. 49.17 and 49.18 to determine which angle corresponds to the minimum normal stress and which angle corresponds to the maximum normal stress.[3]

$$\theta_{\sigma_1,\sigma_2} = \left(\tfrac{1}{2}\right) \arctan\left(\frac{2\tau}{\sigma_x - \sigma_y}\right) \qquad \textbf{49.21}$$

The angles of the planes on which the shear stress is minimum and maximum are given by Eq. 49.22. These planes will be 90° apart and will be rotated 45° from the planes of principal normal stresses. As with Eq. 49.21, θ is measured from the x-axis, clockwise if negative and counterclockwise if positive. Generally, the sign of a shear stress on an inclined plane will be unimportant.

$$\theta_{\tau_1,\tau_2} = \left(\tfrac{1}{2}\right) \arctan\left(\frac{\sigma_x - \sigma_y}{-2\tau}\right) \qquad \textbf{49.22}$$

Example 49.2

Find the maximum normal and shear stresses on the object shown. Determine the angle of the plane of principal normal stresses.

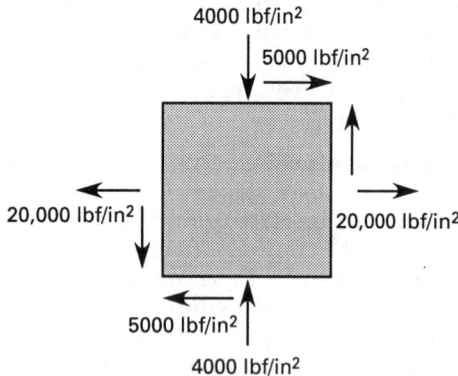

Solution

Find the principal shear stresses first. The applied 4000 lbf/in^2 compressive stress is negative. Equation 49.20 can be used directly.

$$\tau_1 = \tfrac{1}{2}\sqrt{(\sigma_x - \sigma_y)^2 + (2\tau)^2}$$

$$= \tfrac{1}{2}\sqrt{\left[20{,}000\,\frac{\text{lbf}}{\text{in}^2} - \left(-4000\,\frac{\text{lbf}}{\text{in}^2}\right)\right]^2 + \left[(2)\left(5000\,\frac{\text{lbf}}{\text{in}^2}\right)\right]^2}$$

$$= 13{,}000\,\text{lbf/in}^2$$

---

[3]Alternatively, the following procedure can be used to determine the direction of the principal planes. Let σ_x be the algebraically larger of the two given normal stresses. The angle between the direction of σ_x and the direction of σ_1, the algebraically larger principal stress, will always be less than 45°.

From Eq. 49.19, the maximum normal stress is

$$\sigma_1 = \left(\tfrac{1}{2}\right)(\sigma_x + \sigma_y) + \tau_1$$

$$= \left(\tfrac{1}{2}\right)\left[20{,}000\,\frac{\text{lbf}}{\text{in}^2} + \left(-4000\,\frac{\text{lbf}}{\text{in}^2}\right)\right] + 13{,}000\,\frac{\text{lbf}}{\text{in}^2}$$

$$= 21{,}000\,\text{lbf/in}^2 \quad [\text{tension}]$$

The angle of the principal normal stresses is given by Eq. 49.21.

$$\theta = \left(\tfrac{1}{2}\right)\arctan\left(\frac{2\tau}{\sigma_x - \sigma_y}\right)$$

$$= \left(\tfrac{1}{2}\right)\arctan\left(\frac{(2)\left(5000\,\frac{\text{lbf}}{\text{in}^2}\right)}{20{,}000\,\frac{\text{lbf}}{\text{in}^2} - \left(-4000\,\frac{\text{lbf}}{\text{in}^2}\right)}\right)$$

$$= \left(\tfrac{1}{2}\right)(22.6°, 202.6°)$$

$$= 11.3°, 101.3°$$

It is not obvious which angle produces which normal stress. One of the angles can be substituted back into the general equation (Eq. 49.17) for σ_θ.

$$\sigma_{11.3°} = \left(\tfrac{1}{2}\right)\left(20{,}000\,\frac{\text{lbf}}{\text{in}^2} - 4000\,\frac{\text{lbf}}{\text{in}^2}\right)$$

$$+ \left(\tfrac{1}{2}\right)\left[20{,}000\,\frac{\text{lbf}}{\text{in}^2} - \left(-4000\,\frac{\text{lbf}}{\text{in}^2}\right)\right]$$

$$\times \cos(2)(11.3°) + (5000)\sin(2)(11.3°)$$

$$= 21{,}000\,\text{lbf/in}^2$$

Thus, the 11.3° angle corresponds to the maximum normal stress of 21,000 lbf/in^2.

9. MOHR'S CIRCLE FOR STRESS

Mohr's circle can be constructed to graphically determine the principal stresses.

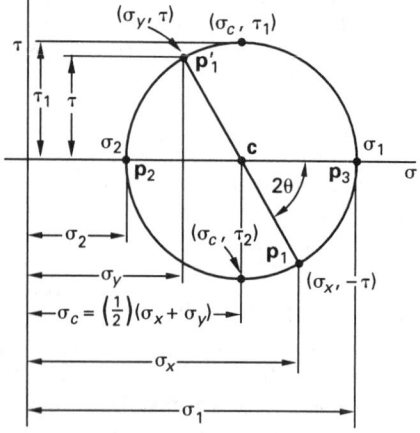

Figure 49.8 *Mohr's Circle for Stress*

step 1: Determine the applied stresses: σ_x, σ_y, and τ. (Tensile normal stresses are positive; compressive normal stresses are negative. Clockwise shear stresses are positive; counterclockwise shear stresses are negative.)

step 2: Draw a set of σ-τ axes.

step 3: Plot the center of the circle, point $\mathbf{c}$, by calculating $\sigma_c = \left(\frac{1}{2}\right)(\sigma_x + \sigma_y)$.

step 4: Plot the point $\mathbf{p}_1 = (\sigma_x, -\tau)$. (Alternatively, plot $\mathbf{p}_1'$ at $(\sigma_y, +\tau)$.)

step 5: Draw a line from point $\mathbf{p}_1$ through center $\mathbf{c}$ and extend it an equal distance below the σ-axis. This is the diameter of the circle.

step 6: Using the center $\mathbf{c}$ and point $\mathbf{p}_1$, draw the circle. An alternative method is to draw a circle of radius r about point $\mathbf{c}$.

$$r = \sqrt{\left(\tfrac{1}{4}\right)(\sigma_x - \sigma_y)^2 + \tau^2} \qquad \textbf{49.23}$$

step 7: Point $\mathbf{p}_2$ defines the smaller principal stress, σ_2. Point $\mathbf{p}_3$ defines the larger principal stress, σ_1.

step 8: Determine the angle θ as half of the angle 2θ on the circle. This angle corresponds to the larger principal stress, σ_1. On Mohr's circle, angle 2θ is measured counterclockwise from the $\mathbf{p}_1$-$\mathbf{p}_1'$ line to the horizontal axis.

Example 49.3

Construct Mohr's circle for Ex. 49.2.

Solution

$$\sigma_c = \left(\tfrac{1}{2}\right)(\sigma_x + \sigma_y)$$
$$= \left(\tfrac{1}{2}\right)\left[20{,}000\ \frac{\text{lbf}}{\text{in}^2} + \left(-4000\ \frac{\text{lbf}}{\text{in}^2}\right)\right]$$
$$= 8000\ \text{lbf/in}^2$$

$$r = \sqrt{\left(\tfrac{1}{4}\right)(\sigma_x - \sigma_y)^2 + \tau^2}$$
$$= \sqrt{\begin{array}{c}\left(\tfrac{1}{4}\right)\left[20{,}000\ \dfrac{\text{lbf}}{\text{in}^2} - \left(-4000\ \dfrac{\text{lbf}}{\text{in}^2}\right)\right]^2 \\[2mm] + \left(5000\ \dfrac{\text{lbf}}{\text{in}^2}\right)^2\end{array}}$$
$$= 13{,}000\ \text{lbf/in}^2$$

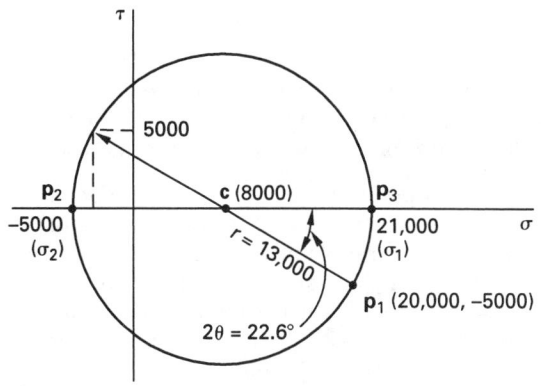

10. IMPACT LOADING

If a load is applied to a structure suddenly, the structure's response will be composed of two parts: a transient response (which decays to zero) and a steady-state response. (These two parts are also known as the *dynamic* and *static responses*.) It is not unusual for the transient loading to be larger than the steady-state response.

Although a *dynamic analysis* of the structure is preferred, the procedure is lengthy and complex. Therefore, arbitrary multiplicative factors may be applied to the steady-state stress to determine the maximum transient. For example, if a load is applied quickly as compared to the natural period of vibration of the structure (e.g., the classic definition of an *impact load*), a dynamic factor of 2.0 might be used. Actual dynamic factors should be determined or validated by testing.

The energy-conservation method (i.e., the work-energy principle) can be used to determine the maximum stress due to a falling mass. The total change in potential energy of the mass from the change in elevation and the deflection δ) is equated to the appropriate expression for total strain energy (see Sec. 4).

11. SHEAR AND MOMENT

Shear at a point is the sum of all vertical forces acting on an object. It has units of pounds, kips, tons, newtons, and so on. Shear is not the same as shear stress, since the area of the object is not considered.

The most typical application is shear at a point on a beam, V, defined as the sum of all vertical forces between the point and one of the ends.[4] The direction (i.e., to the left or right of the point) in which the summation proceeds is not important. Since the values of shear will differ only in sign for summations to the left and right ends, the direction that results in the fewest calculations should be selected.

---

[4]The conditions of equilibrium require that the sum of all vertical forces on a beam be zero. However, the *shear* can be nonzero because only a portion of the beam is included in the analysis. Since that portion extends to the beam end in one direction only, shear is sometimes called *resisting shear* or *one-way shear*.

$$V = \sum_{\substack{\text{point to} \\ \text{one end}}} F_i \qquad\qquad \textbf{49.24}$$

Shear is taken as positive when there is a net upward force to the left of a point and negative when there is a net downward force between the point and the left end.

Moment at a point is the total bending moment acting on an object. In the case of a beam, the moment, M, will be the algebraic sum of all moments and couples located between the investigation point and one of the beam ends. As with shear, the number of calculations required to calculate the moment can be minimized by careful choice of the beam end.[5]

$$M = \sum_{\substack{\text{point to} \\ \text{one end}}} F_i d_i + \sum_{\substack{\text{point to} \\ \text{one end}}} C_i \qquad\qquad \textbf{49.25}$$

Moment is taken as positive when the upper surface of the beam is in compression and the lower surface is in tension (see Fig. 49.12). Since the beam ends will usually be higher than the midpoint, it is commonly said that "a positive moment will make the beam smile."

12. SHEAR AND BENDING MOMENT DIAGRAMS

The value of the shear and moment, V and M, will depend on location along the beam. Both shear and moment can be described mathematically for simple loadings, but the formulas are likely to become discontinuous as the loadings become more complex. It is much more convenient to describe the shear and moment functions graphically. Graphs of shear and moment as functions of position along the beam are known as *shear* and *moment diagrams*. Drawing these diagrams does not require knowing the shape or area of the beam.

The following guidelines and conventions should be observed when constructing a *shear diagram*.

- The shear at any point is equal to the sum of the loads and reactions from the point to the left end.

- The magnitude of the shear at any point is equal to the slope of the moment line at that point.

$$V = \frac{dM}{dx} \qquad\qquad \textbf{49.26}$$

- Loads and reactions acting upward are positive.

- The shear diagram is straight and sloping over uniformly distributed loads.

- The shear diagram is straight and horizontal between concentrated loads.

- The shear is a vertical line and is undefined at points of concentrated loads.

The following guidelines and conventions should be observed when constructing a *bending moment diagram*. By convention, the moment diagram is drawn on the compression side of the beam.

- The moment at any point is equal to the sum of the moments and couples from the point to the left end.[6]

- Clockwise moments about the point are positive.

- The magnitude of the moment at any point is equal to the area under the shear line up to that point. This is equivalent to the integral of the shear function.

$$M = \int V \, dx \qquad\qquad \textbf{49.27}$$

- The *maximum moment* occurs where the shear is zero.

- The moment diagram is straight and sloping between concentrated loads.

- The moment diagram is curved (parabolic upward) over uniformly distributed loads.

These principles are illustrated in Fig. 49.9.

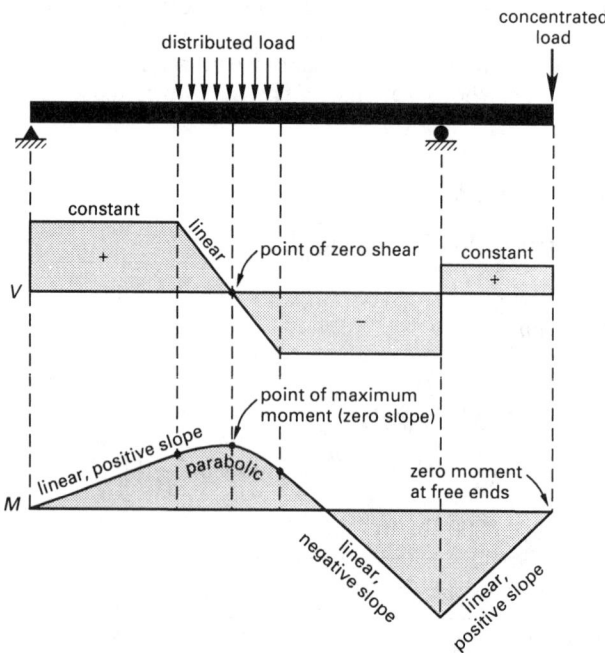

Figure 49.9 *Drawing Shear and Moment Diagrams*

---

[5]The conditions of equilibrium require that the sum of all moments on a beam be zero. However, the *moment* can be nonzero because only a portion of the beam is included in the analysis. Since that portion extends to the beam end in one direction only, moment is sometimes called *bending moment*, *resisting moment*, or *one-way moment*.

[6]If the beam is cantilevered with its built-in end at the left, the fixed-end moment will be unknown. In that case, the moment must be calculated to the right end of the beam.

Example 49.4

Draw the shear and bending moment diagrams for the following beam.

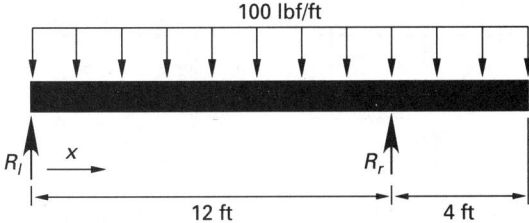

Solution

First, determine the reactions. The uniform load of $100x$ can be assumed to be concentrated at $x/2$.

$$R_r = \frac{\left(\frac{1}{2}\right)(16 \text{ ft})(16 \text{ ft})\left(100 \dfrac{\text{lbf}}{\text{ft}}\right)}{12 \text{ ft}} = 1066.7 \text{ lbf}$$

$$R_l = (16 \text{ ft})\left(100 \frac{\text{lbf}}{\text{ft}}\right) - R_r = 533.3 \text{ lbf}$$

The shear diagram starts at $+533.3$ at the left reaction but decreases linearly at the rate of 100 lbf/ft between the two reactions. Measuring x from the left end, the shear line goes through zero at

$$x = \frac{533.3 \text{ lbf}}{100 \dfrac{\text{lbf}}{\text{ft}}} = 5.333 \text{ ft}$$

The shear just to the left of the right reaction is

$$533.3 \text{ lbf} - (12 \text{ ft})\left(100 \frac{\text{lbf}}{\text{ft}}\right) = -666.7 \text{ lbf}$$

The shear just to the right of the right reaction is

$$-666.7 \text{ lbf} + R_r = +400 \text{ lbf}$$

To the right of the right reaction, the shear diagram decreases to zero at the same constant rate: 100 lbf/ft. This is sufficient information to draw the shear diagram.

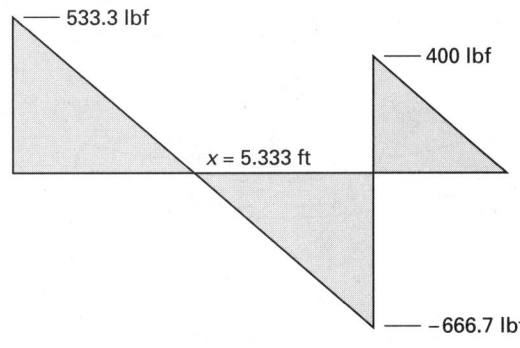

The left end is a free end, so the moment is zero. The bending moment at a distance x to the right of the left end has two parts. The left reaction of 533.3 lbf acts with moment arm x. The moment between the two reactions is

$$M_x = 533.3x - 100x\left(\frac{x}{2}\right)$$

This equation describes a parabolic section (curved upward) with a peak at $x = 5.333$ ft, where the shear is zero. The maximum moment is

$$M_{x=5.333} = (533.3 \text{ lbf})(5.333 \text{ ft}) - \left(50 \frac{\text{lbf}}{\text{ft}}\right)(5.333 \text{ ft})^2$$

$$= 1422.0 \text{ ft-lbf}$$

The moment at the right reaction (where $x = 12$ ft) is

$$M_{x=12} = (533.3 \text{ lbf})(12 \text{ ft}) - \left(50 \frac{\text{lbf}}{\text{ft}}\right)(12 \text{ ft})^2$$

$$= -800 \text{ ft-lbf}$$

The right end is a free end, so the moment is zero. The moment between the right reaction and the right end could be calculated by summing moments to the left end, but it is more convenient to sum moments to the right end. Measuring x from the right end, the moment is derived only from the uniform load.

$$M = 100x\left(\frac{x}{2}\right) = 50x^2$$

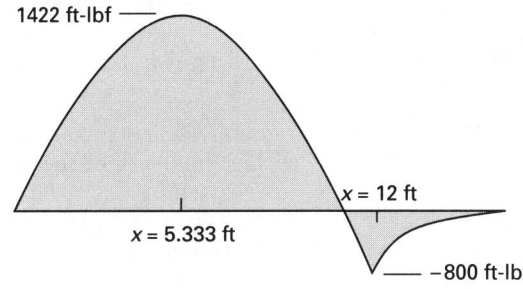

This is sufficient information to draw the moment diagram. Once the maximum moment is located, no attempt is made to determine the exact curvature. The point where $M = 0$ is of limited interest, and no attempt is made to determine the exact location.

Notice that the cross-sectional ɑ needed in this example.

13. SHEAR STRESS IN BEAMS

Shear stress is generally not the limiting factor in most designs. However, it can control (or be limited by code) in wood and concrete beams and in thin tubes.

The average shear stress experienced at a point along the length of a beam depends on the shear, V, at that point and the area, A, of the beam. The shear can be found from the shear diagram.

$$\tau = \frac{V}{A} \qquad 49.28$$

In most cases, the entire area, A, of the beam is used in calculating the average shear stress. However, in flanged beam calculations, it is assumed that only the web carries the average shear stress.[7] The flanges are not included in shear stress calculations.

$$\tau = \frac{V}{t_w d} \qquad 49.29$$

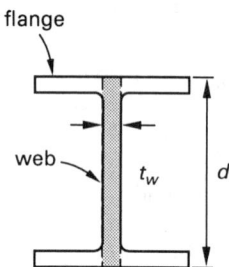

Figure 49.10 Web of a Flanged Beam

Figure 49.6 shows that for biaxial loading, identical shear stresses exist simultaneously in all four directions. One set of parallel shears (a couple) counteracts the rotational moment from the other set of parallel shears. The horizontal shear exists even when the loading is vertical (e.g., when a horizontal beam is loaded by a vertical force). For that reason, the term *horizontal shear* is sometimes used to distinguish it from the applied shear load.

The exact value of the horizontal shear stress is dependent on the location, y_1, within the depth of the beam. The shear stress distribution is given by Eq. 49.30. The shear stress is zero at the top and bottom surfaces of the beam and is maximum at the neutral axis (i.e., the center).

$$\tau_{y_1} = \frac{QV}{Ib} \qquad 49.30$$

In Eq. 49.30, V is the vertical shear at the point along the length of the beam where the shear stress is wanted. I is the beam's centroidal moment of inertia, and b is the width of the beam at the depth y_1 within the beam where the shear stress is wanted. Q is the *statical moment*, as defined by Eq. 49.31.

$$Q = \int_{y_1}^{c} y \, dA \qquad 49.31$$

---

[7]This is more than an assumption; it is a fact. There are several reasons the flanges do not contribute to shear resistance, including a non-uniform shear stress distribution in the flanges. This non-uniformity is too complex to be analyzed by elementary methods.

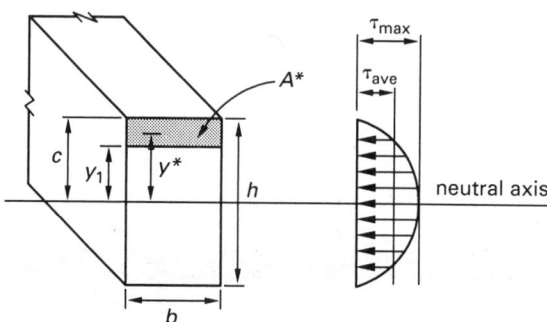

Figure 49.11 Shear Stress Distribution Within a Rectangular Beam

For rectangular beams, $dA = b \, dy$. Then, the statical moment of the area A^* above layer y_1 is equal to the product of the area and the distance from the centroidal axis to the centroid of the area.

$$Q = y^* A^* \qquad 49.32$$

Equation 49.33 calculates the maximum shear stress in a rectangular beam. It is 50% higher than the average shear stress.

$$\tau_{\text{max,rectangular}} = \frac{3V}{2A} = \frac{3V}{2bh} \qquad 49.33$$

For a beam with a circular cross section, the maximum shear stress is

$$\tau_{\text{max,circular}} = \frac{4V}{3A} = \frac{4V}{3\pi r^2} \qquad 49.34$$

For a hollow cylinder used as a beam, the maximum shear stress occurs at the plane of the neutral axis and is

$$\tau_{\text{max,hollow cylinder}} = \frac{2V}{A} \qquad 49.35$$

14. BENDING STRESS IN BEAMS

Normal stress occurs in a bending beam, as shown in Fig. 49.12, where the beam is acted upon by a *transverse force*. Although it is a normal stress, the term *bending stress* or *flexural stress* is used to indicate the source of the stress. The lower surface of the beam experiences tensile stress (which causes lengthening). The upper surface of the beam experiences compressive stress (which causes shortening). There is no normal stress along a horizontal plane passing through the centroid of the cross section, a plane known as the *neutral plane* or the *neutral axis*.

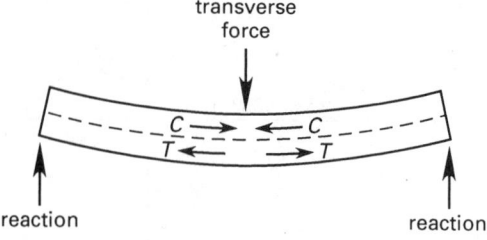

Figure 49.12 Normal Stress Due to Bending

Bending stress varies with location (depth) within the beam. It is zero at the neutral axis, and increases linearly with distance from the neutral axis, as predicted by Eq. 49.36.

$$\sigma_b = \frac{-My}{I_c} \qquad \textbf{49.36}$$

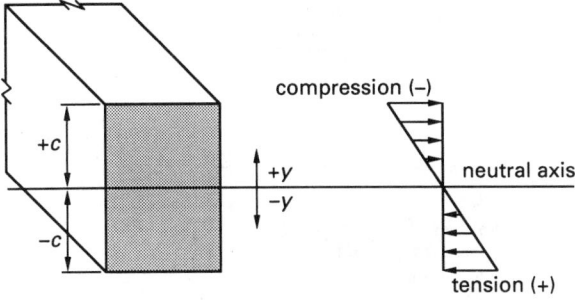

Figure 49.13 *Bending Stress Distribution in a Beam*

The bending moment, M, is used in Eq. 49.36. I_c is the centroidal moment of inertia of the beam's cross section. The negative sign in Eq. 49.36, required by the convention that compression is negative, is commonly omitted.

Since the maximum stress will govern the design, y can be set equal to c to obtain the *extreme fiber stress*. c is the distance from the neutral axis to the *extreme fiber* (i.e., the top or bottom surface most distant from the neutral axis).

$$\sigma_{b,\max} = \frac{Mc}{I_c} \qquad \textbf{49.37}$$

Equation 49.37 shows that the maximum bending stress will occur where the moment is maximum. The region immediately adjacent to the point of maximum bending moment is called the *dangerous section* of the beam. The dangerous section can be found from a bending moment or shear diagram.

For any given beam cross section, I_c and c are fixed. Therefore, these two terms can be combined into the *section modulus*, Z.[8]

$$\sigma_{b,\max} = \frac{M}{Z} \qquad \textbf{49.38}$$

$$Z = \frac{I_c}{c} \qquad \textbf{49.39}$$

Since $c = h/2$, the section modulus of a rectangular $b \times h$ section ($I_c = bh^3/12$) is

$$Z_{\text{rectangular}} = \frac{bh^2}{6} \qquad \textbf{49.40}$$

Example 49.5

The beam in Ex. 49.4 has a 6 in $\times$ 8 in cross section. What are the maximum shear and bending stresses in the beam?

[8]The symbol S is also commonly used for the section modulus.

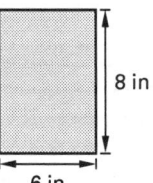

Solution

The maximum shear (from the shear diagram) is 666.7 lbf. (The negative sign is disregarded.) From Eq. 49.33, the maximum shear stress in a rectangular beam is

$$\tau_{\max} = \frac{3V}{2A} = \frac{(3)(666.7\ \text{lbf})}{(2)(6\ \text{in})(8\ \text{in})}$$
$$= 20.8\ \text{lbf/in}^2$$

The centroidal moment of inertia is

$$I_c = \frac{bh^3}{12} = \frac{(6\ \text{in})(8\ \text{in})^3}{12} = 256\ \text{in}^4$$

The maximum bending moment (from the bending moment diagram) is 1422 ft-lbf. From Eq. 49.37, the maximum bending stress is

$$\sigma_{b,\max} = \frac{Mc}{I_c} = \frac{(1422\ \text{ft-lbf})\left(12\ \dfrac{\text{in}}{\text{ft}}\right)(4\ \text{in})}{256\ \text{in}^4}$$
$$= 266.6\ \text{lbf/in}^2$$

15. STRAIN ENERGY DUE TO BENDING MOMENT

The elastic strain energy due to a bending moment stored in a beam is

$$U = \frac{1}{2EI} \int M^2(x)\, dx \qquad \textbf{49.41}$$

The use of Eq. 49.41 is illustrated by Ex. 49.10.

16. ECCENTRIC LOADING OF AXIAL MEMBERS

If a load is applied through the centroid of a tension or compression member's cross section, the loading is said to be *axial loading* or *concentric loading*. *Eccentric loading* occurs when the load is not applied through the centroid.

If an axial member is loaded eccentrically, it will bend and experience bending stress in the same manner as a beam. Since the member experiences both axial stress and bending stress, it is known as a *beam-column*. In Fig. 49.14, e is known as the *eccentricity*.

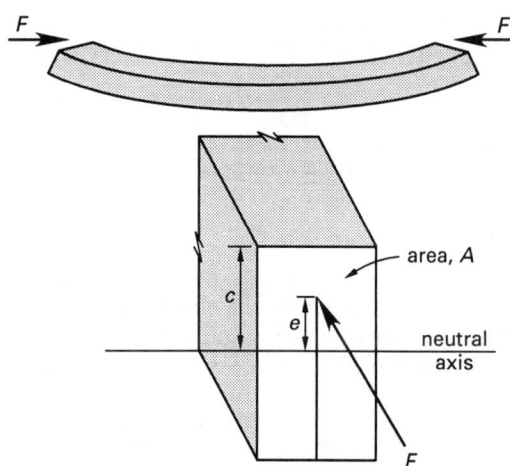

Figure 49.14 *Eccentric Loading of an Axial Member*

Both the axial stress and bending stress are normal stresses oriented in the same direction; therefore, simple addition can be used to combine them. Combined stress theory is not applicable. By convention, F is negative if the force compresses the member (as shown in Fig. 49.14).

$$\sigma_{\text{max,min}} = \frac{F}{A} \pm \frac{Mc}{I_c} \qquad \textit{49.42}$$

$$= \frac{F}{A} \pm \frac{Fec}{I_c} \qquad \textit{49.43}$$

If a pier or column (primarily designed as a compression member) is loaded with an eccentric compressive load, part of the section can still be placed in tension. Tension will exist when the Mc/I_c term in Eq. 49.42 is larger than the F/A term. It is particularly important to eliminate or severely limit tensile stresses in concrete and masonry piers, since these materials cannot support tension.

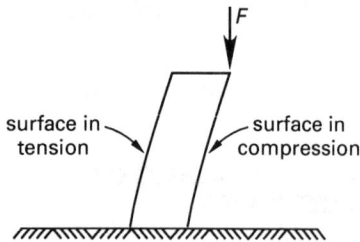

Figure 49.15 *Tension in a Pier*

Regardless of the size of the load, there will be no tension as long as the eccentricity is low. In a rectangular member, the load must be kept within a rhombus-shaped area formed from the middle thirds of the centroidal axes. This area is known as the *core*, *kern*, or *kernel*. Figure 49.16 illustrates the kernel for other cross sections.

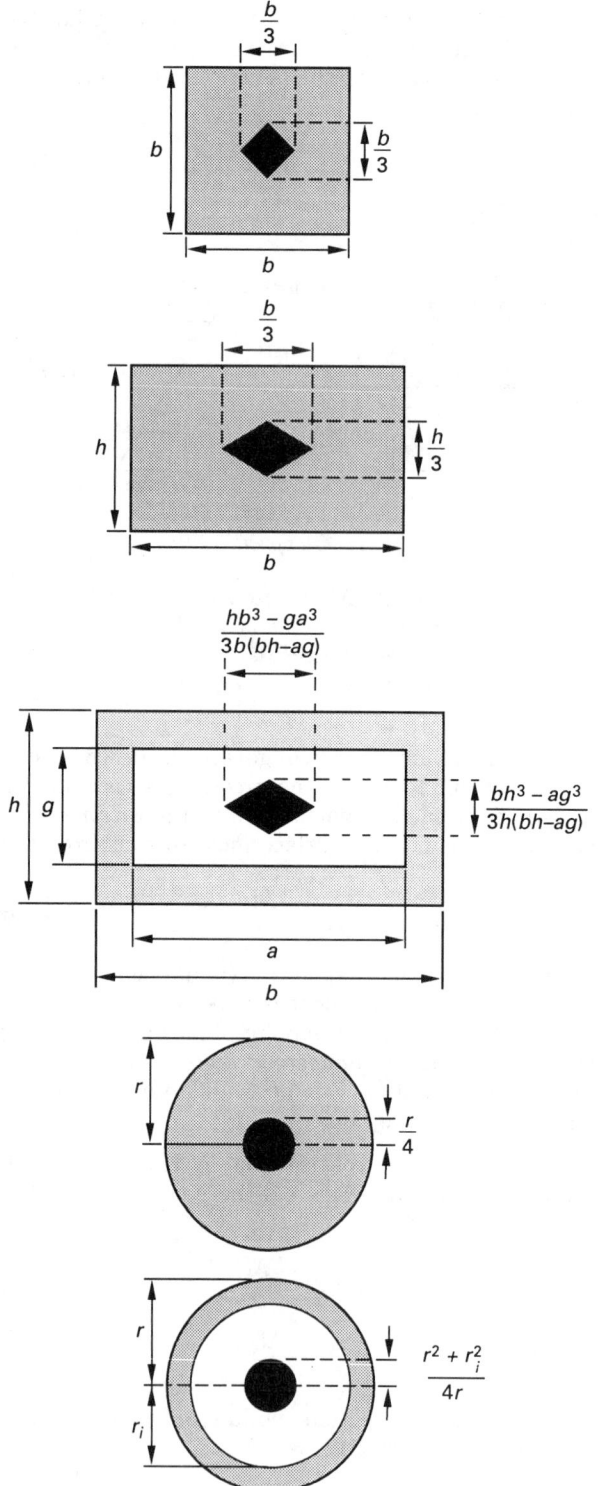

Figure 49.16 *Kerns of Common Cross Sections*

Example 49.6

A built-in hook with a cross section of 1 in × 1 in carries a load of 500 lbf, but the load is not in line with the centroidal axis of the hook's neck. What are the minimum stresses in the neck? Is the neck in tension everywhere?

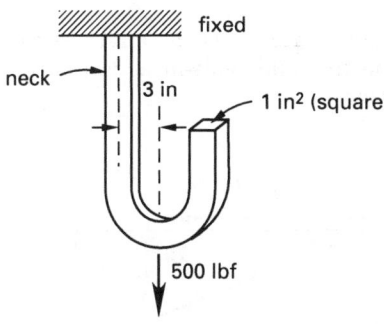

Solution

The centroidal moment of inertia of a 1 in × 1 in section is

$$I_c = \frac{bh^3}{12} = \frac{(1 \text{ in})(1 \text{ in})^3}{12} = 0.0833 \text{ in}^4$$

The hook is eccentrically loaded with an eccentricity of 3 in. From Eq. 49.43, the total stress is the sum of the direct axial tension and the bending stress. As the hook bends to reduce the eccentricity, the inner face of the neck will receive a tensile bending stress. The outer face of the neck will receive a compressive bending stress.

$$
\begin{aligned}
\sigma_{\text{max,min}} &= \frac{F}{A} \pm \frac{Fec}{I} \\
&= \frac{500 \text{ lbf}}{1 \text{ in}^2} \pm \frac{(500 \text{ lbf})(3 \text{ in})(0.5 \text{ in})}{0.0833 \text{ in}^4} \\
&= 500 \frac{\text{lbf}}{\text{in}^2} \pm 9000 \frac{\text{lbf}}{\text{in}^2} \\
&= +9500 \text{ lbf/in}^2, \; -8500 \text{ lbf/in}^2
\end{aligned}
$$

The 500 lbf/in^2 direct stress is tensile, and the inner face experiences a total tensile stress of 9500 lbf/in^2. However, the compressive bending stress of 9000 lbf/in^2 counteracts the direct tensile stress, resulting in an 8500 lbf/in^2 compressive stress at the outer face of the neck.

17. BEAM DEFLECTION: DOUBLE INTEGRATION METHOD

The deflection and the slope of a loaded beam are related to the moment and shear by Eqs. 49.44 through 49.48.

$$y = \text{deflection} \qquad\qquad \textbf{49.44}$$

$$y' = \frac{dy}{dx} = \text{slope} \qquad\qquad \textbf{49.45}$$

$$y'' = \frac{d^2y}{dx^2} = \frac{M(x)}{EI} \qquad\qquad \textbf{49.46}$$

$$y''' = \frac{d^3y}{dx^3} = \frac{V(x)}{EI} \qquad\qquad \textbf{49.47}$$

If the *moment function*, $M(x)$, is known for a section of the beam, the deflection at any point can be found from Eq. 49.48.

$$y = \frac{1}{EI} \int \left[\int M(x)\,dx \right] dx \qquad\qquad \textbf{49.48}$$

In order to find the deflection, constants must be introduced during the integration process. These constants can be found from Table 49.3.

Table 49.3 *Beam Boundary Conditions*

end condition	y	y'	y''	V	M
simple support	0				0
built-in support	0	0			
free end			0	0	0
hinge					0

Example 49.7

Find the tip deflection of the beam shown. EI is 5×10^{10} lbf-in^2 everywhere.

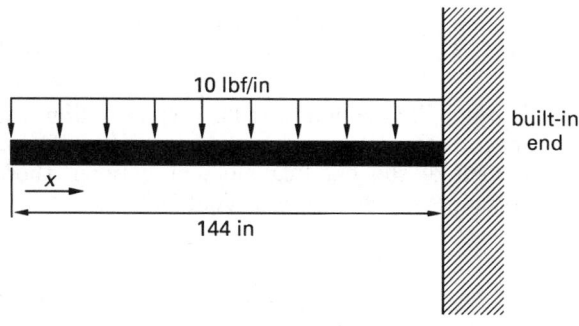

Solution

The moment at any point x from the left end of the beam is

$$M(x) = -10x \left(\tfrac{1}{2}x \right) = -5x^2$$

This is negative by the left-hand rule convention. From Eq. 49.46,

$$y'' = \frac{M(x)}{EI}$$

$$EIy'' = M(x) = -5x^2$$

$$EIy' = \int -5x^2\,dx = -\frac{5}{3}x^3 + C_1$$

Since $y' = 0$ at a built-in support (Table 49.3) and $x = 144$ in at the built-in support,

$$0 = \left(-\frac{5}{3} \right)(144)^3 + C_1$$

$$C_1 = 4.98 \times 10^6$$

$$
\begin{aligned}
EIy &= \int \left(-\frac{5}{3}x^3 + 4.98 \times 10^6 \right) dx \\
&= -\frac{5}{12}x^4 + (4.98 \times 10^6)x + C_2
\end{aligned}
$$

Again, $y = 0$ at $x = 144$, so $C_2 = -5.38 \times 10^8$. Therefore, the deflection as a function of x is

$$y = \left(\frac{1}{EI}\right)\left[\left(-\frac{5}{12}\right)x^4 + (4.98 \times 10^6)x - (5.38 \times 10^8)\right]$$

At the tip $x = 0$, so the deflection is

$$y_{\text{tip}} = \frac{-5.38 \times 10^8}{5 \times 10^{10}} = -0.0108 \text{ in}$$

18. BEAM DEFLECTION: MOMENT AREA METHOD

The moment area method is a semigraphical technique that is applicable whenever slopes of deflection beams are not too great. This method is based on the following two theorems.

- *Theorem I:* The angle between tangents at any two points on the *elastic line* of a beam is equal to the area of the moment diagram between the two points divided by EI. That is,

$$\phi = \int \frac{M(x)\,dx}{EI} \qquad\qquad 49.49$$

- *Theorem II:* One point's deflection away from the tangent of another point is equal to the *statical moment* of the bending moment between those two points divided by EI. That is,

$$y = \int \frac{xM(x)\,dx}{EI} \qquad\qquad 49.50$$

If EI is constant, the statical moment $\int xM(x)\,dx$ can be calculated as the product of the total moment diagram area times the distance from the point whose deflection is wanted to the centroid of the moment diagram.

If the moment diagram has positive and negative parts (areas above and below the zero line), the statical moment should be taken as the sum of two products, one for each part of the moment diagram.

Example 49.8

Find the deflection, y, and the angle, ϕ, at the free end of the cantilever beam shown. Neglect the beam weight.

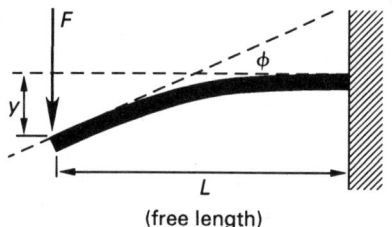

(free length)

Solution

The deflection angle, ϕ, is the angle between the tangents at the free and built-in ends (Theorem I). The moment diagram is

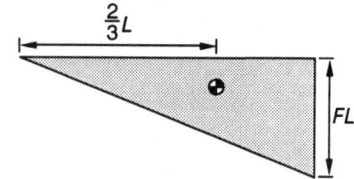

The area of the moment diagram is

$$\left(\tfrac{1}{2}\right)(FL)(L) = \tfrac{1}{2}FL^2$$

From Eq. 49.49,

$$\phi = \frac{FL^2}{2EI}$$

From Eq. 49.50,

$$y = \left(\frac{FL^2}{2EI}\right)\left(\tfrac{2}{3}L\right) = \frac{FL^3}{3EI}$$

Example 49.9

Find the deflection of the free end of the cantilever beam shown. Neglect the beam weight.

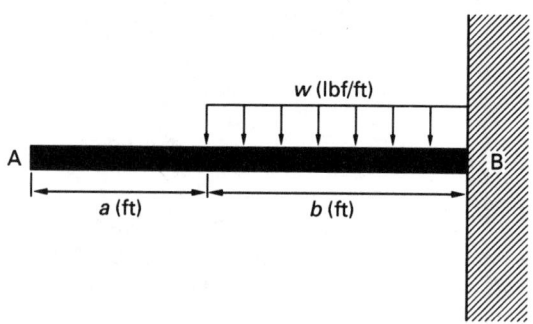

Solution

The distance from point A (where the deflection is wanted) to the centroid is $a + 0.75b$. The area of the moment diagram is $wb^3/6$. From Theorem II,

$$y = \left(\frac{wb^3}{6EI}\right)(a + 0.75b)$$

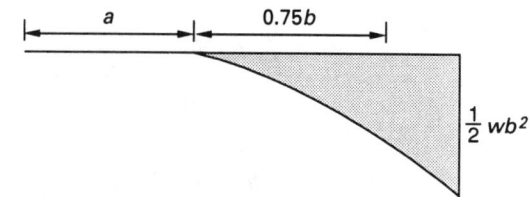

19. BEAM DEFLECTION: STRAIN ENERGY METHOD

The deflection at a point of load application can be found by the strain energy method. This method equates the external work to the total internal strain energy. Since work is a force moving through a distance (which in this case is the deflection), Eq. 49.51 holds true.

$$\tfrac{1}{2}Fy = \sum U \qquad 49.51$$

Example 49.10

Find the deflection at the tip of the stepped beam shown.

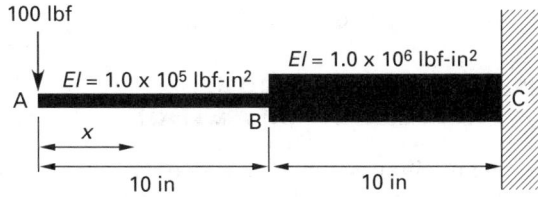

Solution

In section A–B, $M(x) = 100x$ in-lbf.

From Eq. 49.41,

$$U = \frac{1}{2EI} \int M^2(x)\, dx$$

$$= \frac{1}{(2)(1 \times 10^5)} \int_0^{10} (100x)^2 = 16.67 \text{ in-lbf}$$

In section B–C, $M = 100x$.

$$U = \frac{1}{(2)(1 \times 10^6)} \int_{10}^{20} (100x)^2 = 11.67 \text{ in-lbf}$$

Equating the internal work (U) and the external work,

$$\sum U = W$$

$$16.67 \text{ in-lbf} + 11.67 \text{ in-lbf} = \left(\tfrac{1}{2}\right)(100 \text{ lbf})y$$

$$y = 0.567 \text{ in}$$

20. BEAM DEFLECTION: CONJUGATE BEAM METHOD

The *conjugate beam method* changes a deflection problem into one of drawing moment diagrams. The method has the advantage of being able to handle beams of varying cross sections (e.g., stepped shafts) and materials. It has the disadvantage of not easily being able to handle beams with two built-in ends. The following steps constitute the conjugate beam method.

step 1: Draw the moment diagram for the beam as it is actually loaded.

step 2: Construct the M/EI diagram by dividing the value of M at every point along the beam by EI at that point. If the beam is of constant cross section, EI will be constant, and the M/EI diagram will have the same shape as the moment diagram. However, if the beam cross section varies with x, I will change. In that case, the M/EI diagram will not look the same as the moment diagram.

step 3: Draw a conjugate beam of the same length as the original beam. The material and the cross-sectional area of this conjugate beam are not relevant.

   (a) If the actual beam is simply supported at its ends, the conjugate beam will be simply supported at its ends.

   (b) If the actual beam is simply supported away from its ends, the conjugate beam has hinges at the support points.

   (c) If the actual beam has free ends, the conjugate beam has built-in ends.

   (d) If the actual beam has built-in ends, the conjugate beam has free ends.

step 4: Load the conjugate beam with the M/EI diagram. Find the conjugate reactions by methods of statics. Use the superscript * to indicate conjugate parameters.

step 5: Find the conjugate moment at the point where the deflection is wanted. The deflection is numerically equal to the moment as calculated from the conjugate beam forces.

Example 49.11

Find the deflections at the two load points. EI has a constant value of 2.356×10^7 lbf-in^2.

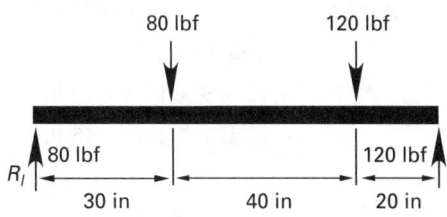

Solution

step 1: The moment diagram for the actual beam is

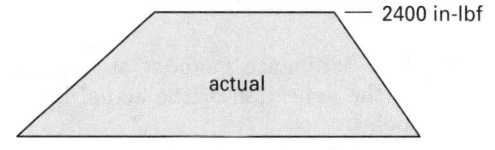

steps 2, 3, and 4: Since the beam cross section is constant, the conjugate load has the same shape as the original moment diagram. The peak load on the conjugate beam is

$$\frac{2400 \text{ in-lbf}}{2.356 \times 10^7 \text{ lbf-in}^2} = 1.019 \times 10^{-4} \text{ 1/in}$$

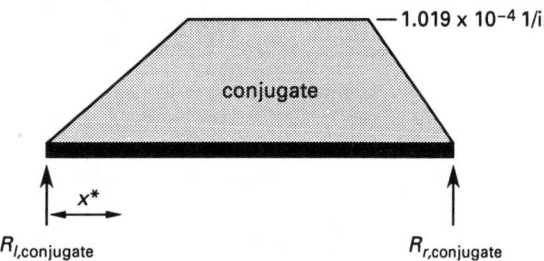

The conjugate reaction, $R_{l,\text{conjugate}}$, is found by the following method. The loading diagram is assumed to be made up of a rectangular load and two negative triangular loads. The area of the rectangular load (which has a centroid at $x_{\text{conjugate}} = 45$ in) is $(90 \text{ in})(1.019 \times 10^{-4} \text{ 1/in}) = 9.171 \times 10^{-3}$.

Similarly, the area of the left triangle (which has a centroid at $x_{\text{conjugate}} = 10$ in) is $(^1/_2)(30 \text{ in})(1.019 \times 10^{-4} \text{ 1/in}) = 1.529 \times 10^{-3}$. The area of the right triangle (which has a centroid at $x_{\text{conjugate}} = 83.33$ in) is $(^1/_2)(20 \text{ in})(1.019 \times 10^{-4} \text{ 1/in}) = 1.019 \times 10^{-3}$.

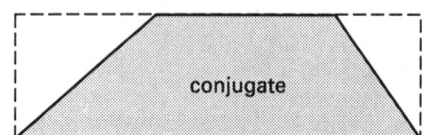

$$\sum M_L^* = (90 \text{ in})R_{r,\text{ conjugate}} + (1.019 \times 10^{-3})$$
$$\times (83.3 \text{ in})$$
$$+ (1.529 \times 10^{-3})(10 \text{ in})$$
$$- (9.171 \times 10^{-3})(45 \text{ in})$$
$$= 0$$
$$R_{r,\text{ conjugate}} = 3.472 \times 10^{-3}$$

Then,

$$R_{l,\text{conjugate}} = (9.171 - 1.019$$
$$- 1.529 - 3.472) \times 10^{-3}$$
$$= 3.151 \times 10^{-3}$$

step 5: The conjugate moment at $x_{\text{conjugate}} = 30$ is the deflection of the actual beam at that point.

$$M_{\text{conjugate}} = (3.151 \times 10^3)(30 \text{ in})$$
$$+ (1.529 \times 10^3)(30 \text{ in} - 10 \text{ in})$$
$$- (9.171 \times 10^{-3})\left(\frac{30 \text{ in}}{90 \text{ in}}\right)(15 \text{ in})$$
$$= 7.926 \times 10^{-2} \text{ in}$$

The conjugate moment (the deflection) at the rightmost load is

$$M_{\text{conjugate}} = (3.472 \times 10^{-3})(20 \text{ in})$$
$$+ (1.019 \times 10^{-3})(13.3 \text{ in})$$
$$- (9.171 \times 10^{-3})\left(\frac{20 \text{ in}}{90 \text{ in}}\right)(10 \text{ in})$$
$$= 6.266 \times 10^{-2} \text{ in}$$

21. BEAM DEFLECTION: TABLE LOOK-UP METHOD

Appendix 49.A is a compilation of the most commonly used beam deflection formulas. These formulas should never need to be derived and should be used whenever possible. They are particularly useful in calculating deflections due to multiple loads using the principle of superposition.

The deflection of very *wide beams* (i.e., those whose widths are larger than 8 or 10 times the thickness) is less than predicted by the equations in App. 49.A for elastic behavior. (This is particularly true for leaf springs.) The large width prevents lateral expansion and contraction of the beam material, reducing the deflection. For wide beams, the calculated deflection should be reduced by multiplying by $(1 - \nu^2)$.

22. BEAM DEFLECTION: SUPERPOSITION

When multiple loads act simultaneously on a beam, all of the loads contribute to deflection. The principle of *superposition* permits the deflections at a point to be calculated as the sum of the deflections from each individual load acting singly.[9] This principle is valid as long as none of the deflections are excessive and all stresses are kept less than the yield point of the beam material.

23. INFLECTION POINTS

The *inflection point* (also known as a *point of contraflexure*) on a horizontal beam in elastic bending occurs where the curvature changes from concave up to concave down, or vice versa. There are three ways of determining the inflection point.

(1) If the elastic deflection equation, $y(x)$, is known, the inflection point can be found consistent with normal calculus methods (i.e., by determining the value of x for which $y''(x) = M(x) = 0$).

---

[9]The principle of superposition is not limited to deflections. It can also be used to calculate the shear and moment at a point and to draw the shear and moment diagrams.

(2) From Eq. 49.46, $y''(x) = M(x)/EI$. y'' is also the reciprocal of the *radius of curvature*, ρ, of the beam. Therefore,

$$y''(x) = \frac{1}{\rho(x)} = \frac{M(x)}{EI} \qquad \textbf{\textit{49.52}}$$

Since the flexural rigidity, EI, is always positive, the radius of curvature, $\rho(x)$, changes sign when the moment equation, $M(x)$, changes sign.

(3) If a shear diagram is known, the inflection point can sometimes be found by noting the point at which the positive and negative shear areas on either side of the point balance.

24. TRUSS DEFLECTION: STRAIN ENERGY METHOD

The deflection of a truss at the point of a single load application can be found by the *strain energy method* if all member forces are known. This method is illustrated by Ex. 49.12.

Example 49.12

Find the vertical deflection of point A under the external load of 707 lbf. $AE = 10^6$ lbf for all members. The internal forces have already been determined.

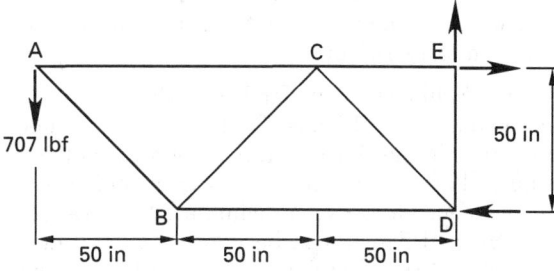

Solution

The length of member AB is $\sqrt{(50 \text{ in})^2 + (50 \text{ in})^2} = 70.7$ in. From Eq. 49.6 the internal strain energy in member AB is

$$U = \frac{F^2 L_o}{2AE} = \frac{(-1000 \text{ lbf})^2 (70.7 \text{ in})}{(2)(10^6 \text{ lbf})} = 35.4 \text{ in-lbf}$$

Similarly, the energy in all members can be determined.

member	L (in)	F (lbf)	U (in-lbf)
AB	70.7	−1000	+35.4
BC	70.7	+1000	+35.4
AC	100	+707	+25.0
BD	100	−1414	+100.0
CD	70.7	−1000	+35.4
CE	50	+2121	+112.5
DE	50	+707	+12.5
			$\overline{356.2}$

The work done by a constant force F moving through a distance y is Fy. In this case, the force increases with y. The average force is $(1/2)F$. The external work is $W_{\text{ext}} = (1/2)(707 \text{ lbf})y$, so

$$\left(\frac{1}{2}\right)(707 \text{ lbf})y = 356.2 \text{ in-lbf}$$

$$y = 1 \text{ in}$$

25. TRUSS DEFLECTION: VIRTUAL WORK METHOD

The *virtual work method* (also known as the *unit load method* and the *Hardy Cross method*) is an extension of the strain energy method.[10] It can be used to determine the deflection of any point on a truss.

step 1: Draw the truss twice.

step 2: On the first truss, place all the actual loads.

step 3: Find the forces, S, due to the actual applied loads in all the members.

step 4: On the second truss, place a dummy one-unit load in the direction of the desired displacement.

step 5: Find the forces, u, due to the one-unit dummy load in all members.

step 6: Find the desired displacement from Eq. 49.53. The summation is over all truss members that have nonzero forces in *both* trusses.

$$\delta = \sum \frac{SuL}{AE} \qquad \textbf{\textit{49.53}}$$

Example 49.13

What is the horizontal deflection of joint F on the truss shown? Use $E = 3 \times 10^7$ lbf/in². Joint A is restrained horizontally.

Solution

steps 1 and 2: Use the truss as drawn.

step 3: The forces in all the truss members are summarized in step 5.

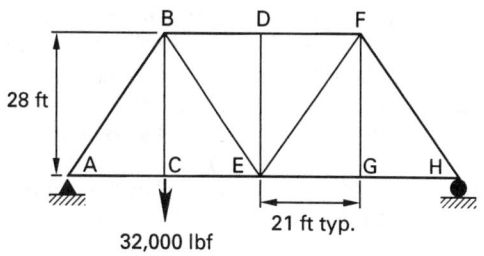

---

[10]Not to be confused with the Hardy-Cross method for determining flows in a piping network by way of iterative calculations.

Materials

step 4: Draw the truss and load it with a unit horizontal force at point F.

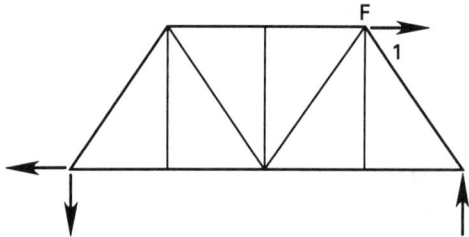

step 5: Find the forces, u, in all members of the second truss. These are summarized in the following table. Notice the sign convention: $+$ for tension and $-$ for compression.

member	S (lbf)	u (lbf)	L (ft)	A (in^2)	$\dfrac{SuL}{AE}$ (ft)
AB	−30,000	5/12	35	17.5	−8.33 × 10^{-4}
CB	32,000	0	28	14	0
EB	−10,000	−5/12	35	17.5	2.75 × 10^{-4}
ED	0	0	28	14	0
EF	10,000	5/12	35	17.5	2.78 × 10^{-4}
GF	0	0	28	14	0
HF	−10,000	−5/12	35	17.5	2.78 × 10^{-4}
BD	−12,000	1/2	21	10.5	−4.00 × 10^{-4}
DF	−12,000	1/2	21	10.5	−4.00 × 10^{-4}
AC	18,000	3/4	21	10.5	9.00 × 10^{-4}
CE	18,000	3/4	21	10.5	9.00 × 10^{-4}
EG	6000	1/4	21	10.5	1.00 × 10^{-4}
GH	6000	1/4	21	10.5	1.00 × 10^{-4}
					12.01 × 10^{-4}

Since 12.01×10^{-4} is positive, the deflection is in the direction of the dummy unit load. In this case, the deflection is to the right.

26. MOVING LOADS ON BEAMS

If a beam supports a single moving load, the maximum bending and shearing stresses at any point can be found by drawing the moment and shear influence diagrams for that point. Once the positions of maximum moment and maximum shear are known, the stresses at the point in question can be calculated.

If a simply supported beam carries a set of moving loads (which remain equidistant as they travel across the beam), the following procedure can be used to find the *dominant load*. The dominant load is the one that occurs directly over the point of maximum moment.

step 1: Calculate and locate the resultant of the load group.

step 2: Assume that one of the loads is dominant. Place the group on the beam such that the distance from one support to the assumed

dominant load is equal to the distance from the other support to the resultant of the load group.

step 3: Check to see that all loads are on the span and that the shear changes sign under the assumed dominant load. If the shear does not change sign under the assumed dominant load, the maximum moment may occur when only some of the load group is on the beam. If it does change sign, calculate the bending moment under the assumed dominant load.

step 4: Repeat steps 2 and 3, assuming that the other loads are dominant.

step 5: Find the maximum shear by placing the load group such that the resultant is a minimum distance from a support.

27. MODES OF BEAM FAILURE

Beams can fail in different ways, including excessive deflection, local buckling, lateral buckling, and rotation.

Excessive deflection occurs when a beam bends more than a permitted amount.[11] The deflection is elastic and no yielding occurs. For this reason, the failure mechanism is sometimes called *elastic failure*. Although the beam does not yield, the excessive deflection may cause cracks in plaster and sheetrock, misalignment of doors and windows, and occupant concern and lack of confidence in the structure.

Local buckling is an overload condition that occurs near large concentrated loads. Such locations include where a column frames into a supporting girder or a reaction point. *Vertical buckling* and *web crippling*, two types of local buckling, can be eliminated by use of *stiffeners*. Such stiffeners can be referred to as *intermediate stiffeners*, *bearing stiffeners*, *web stiffeners*, and *flange stiffeners*, depending on the location and technique of stiffening.

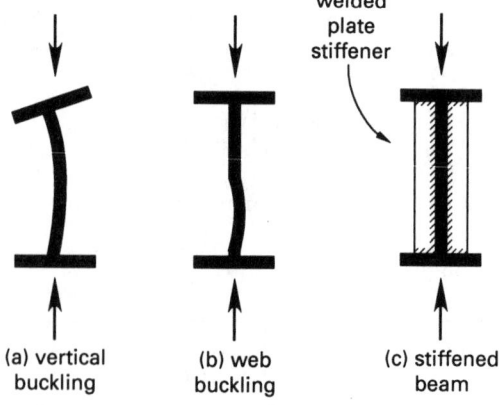

Figure 49.17 *Local Buckling and Stiffeners*

---

[11]The Uniform Building Code, as well as the steel, concrete, and timber codes, specifies maximum permitted deflections in terms of beamlength.

Lateral buckling, such as illustrated in Fig. 49.18, occurs when a long, unsupported member rolls out of its normal plane. To prevent lateral buckling, the beam's compression flange must be supported continuously or at frequent intervals along its length.

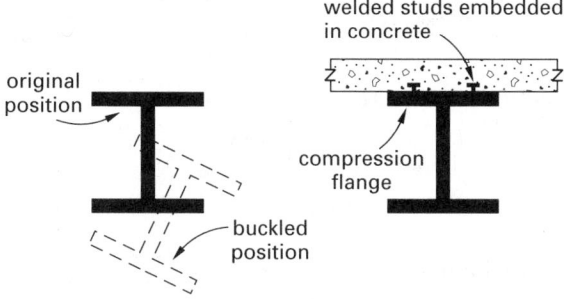

Figure 49.18 *Lateral Buckling and Flange Support*

Rotation is an inelastic (plastic) failure of the beam. When the bending stress at a point exceeds the strength of the beam material, the material yields. As the beam yields, its slope changes. Since the appearance is that of the beam rotating at a hinge at the yield point, the term *plastic hinge* is used to describe the failure mechanism.

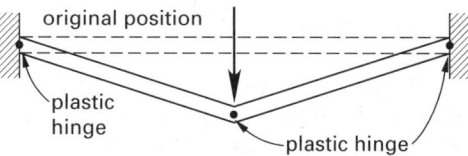

Figure 49.19 *Beam Failure by Rotation*

28. CURVED BEAMS

Many members (e.g., hooks, chain links, clamps, and machine frames) have curved main axes. The distribution of bending stress in a curved beam is nonlinear. Compared to a straight beam, the stress at the inner radius is higher because the inner radius fibers are shorter. Similarly, the stress at the outer radius is lower because the outer radius fibers are longer. Also, the neutral axis is shifted from the center inward toward the center of curvature.

Since the process of finding the neutral axis and calculating the stress amplification is complex, tables and graphs are used for quick estimates and manual computations. The forms of these computational aids vary, but the straight-beam stress is generally multiplied by factors, K, to obtain the stresses at the extreme faces. The factor values depend on the beam cross section and radius of curvature.

$$\sigma_{\text{curved}} = K\sigma_{\text{straight}} = \frac{KMc}{I}$$

Table 49.4 is typical of compilations for round and rectangular beams. Factors K_A and K_B are the multipliers for the inner (high stress) and outer (low stress) faces. The ratio h/r is the fractional distance that the neutral axis shifts inward toward the radius of curvature.

Table 49.4 *Curved Beam Correction Factors*

solid rectangular section	r/c	K_A	K_B	h/r
	1.2	2.89	0.57	0.305
	1.4	2.13	0.63	0.204
	1.6	1.79	0.67	0.149
	1.8	1.63	0.70	0.112
	2.0	1.52	0.73	0.90
	3.0	1.30	0.81	0.041
	4.0	1.20	0.85	0.021
	6.0	1.12	0.90	0.0093
	8.0	1.09	0.92	0.0052
	10.0	1.07	0.94	0.0033

solid circular section	r/c	K_A	K_B	h/r
	1.2	3.41	0.54	0.224
	1.4	2.40	0.60	0.151
	1.6	1.96	0.65	0.108
	1.8	1.75	0.68	0.084
	2.0	1.62	0.71	0.069
	3.0	1.33	0.79	0.03
	4.0	1.23	0.84	0.016
	6.0	1.14	0.89	0.007
	8.0	1.10	0.91	0.0039
	10.0	1.08	0.93	0.0025

29. COMPOSITE STRUCTURES

A *composite structure* is one in which two or more different materials are used. Each material carries part of an applied load. Examples of composite structures include steel-reinforced concrete and steel-plated timber beams.

Most simple composite structures can be analyzed using the *method of consistent deformations*, also known as the *area transformation method*. This method assumes that the strains are the same in both materials at the interface between them. Although the strains are the same, the stresses in the two adjacent materials are not equal, since stresses are proportional to the moduli of elasticity.

The following steps comprise an analysis method based on area transformation.

step 1: Determine the modulus of elasticity for each of the materials used in the structure.

step 2: For each of the materials used, calculate the *modular ratio*, n.

$$n = \frac{E}{E_{\text{weakest}}} \qquad \text{49.54}$$

E_{weakest} is the smallest modulus of elasticity of any of the materials used in the composite structure.

Materials

step 3: For all of the materials except the weakest, multiply the actual material stress area by n. Consider this expanded *(transformed)* area to have the same composition as the weakest material.

step 4: If the structure is a tension or compression member, the distribution or placement of the transformed area is not important. Just assume that the transformed areas carry the axial load. For beams in bending, the transformed area can add to the width of the beam, but it cannot change the depth of the beam or the thickness of the reinforcement.

step 5: For compression or tension numbers, calculate the stresses in the weakest and stronger materials.

$$\sigma_{\text{weakest}} = \frac{F}{A_t} \qquad \textbf{49.55}$$

$$\sigma_{\text{stronger}} = \frac{nF}{A_t} \qquad \textbf{49.56}$$

step 6: For beams in bending, proceed through step 9. Find the centroid of the transformed beam.

step 7: Find the centroidal moment of inertia of the transformed beam $I_{c,t}$.

step 8: Find $V_{\max}$ and $M_{\max}$ by inspection or from the shear and moment diagrams.

step 9: Calculate the stresses in the weakest and stronger materials.

$$\sigma_{\text{weakest}} = \frac{Mc_{\text{weakest}}}{I_{c,t}} \qquad \textbf{49.57}$$

$$\sigma_{\text{stronger}} = \frac{nMc_{\text{stronger}}}{I_{c,t}} \qquad \textbf{49.58}$$

Example 49.14

A short circular steel core is surrounded by a copper tube. The assemblage supports an axial compressive load of 100,000 lbf. The core and tube are well bonded, and the load is applied uniformly. Find the compressive stress in the inner steel core and the outer copper tube.

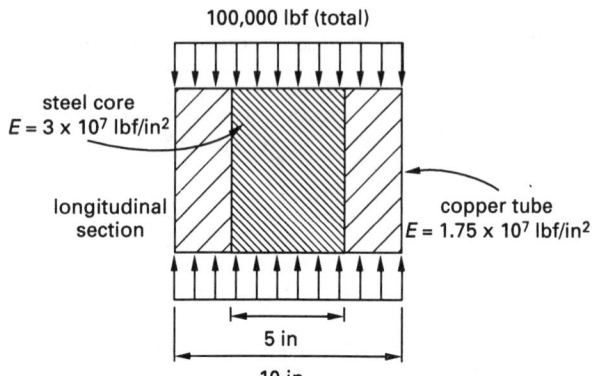

Solution

The moduli of elasticity are given in the illustration. From step 2, the modular ratio is

$$n = \frac{E_{\text{steel}}}{E_{\text{copper}}} = \frac{3 \times 10^7 \,\frac{\text{lbf}}{\text{in}^2}}{1.75 \times 10^7 \,\frac{\text{lbf}}{\text{in}^2}} = 1.714$$

The actual cross-sectional area of the steel is

$$A_{\text{steel}} = \frac{\pi}{4} d^2 = \left(\frac{\pi}{4}\right) (5 \text{ in})^2$$
$$= 19.63 \text{ in}^2$$

The actual cross-sectional area of the copper is

$$A_{\text{copper}} = \left(\frac{\pi}{4}\right) \left(d_o^2 - d_i^2\right) = \left(\frac{\pi}{4}\right) \left[(10 \text{ in})^2 - (5 \text{ in})^2\right]$$
$$= 58.90 \text{ in}^2$$

The steel is the stronger material. Its area must be expanded to an equivalent area of copper. From step 3, the total transformed area is

$$A_t = A_{\text{copper}} + nA_{\text{steel}}$$
$$= 58.90 \text{ in}^2 + (1.714)(19.63 \text{ in}^2) = 92.55 \text{ in}^2$$

Since the two pieces are well bonded and the load is applied uniformly, both pieces experience identical strains. From step 5, the compressive stresses are

$$\sigma_{\text{copper}} = \frac{F}{A_t} = \frac{-100{,}000 \text{ lbf}}{92.55 \text{ in}^2} = -1080 \text{ lbf/in}^2$$
$$\text{[compression]}$$

$$\sigma_{\text{steel}} = \frac{nF}{A_t} = n\sigma_{\text{copper}} = (1.714)\left(-1080 \,\frac{\text{lbf}}{\text{in}^2}\right)$$
$$= -1851 \text{ lbf/in}^2$$

Example 49.15

At a particular point along the length of a steel-reinforced wood beam, the moment is 40,000 ft-lbf. Assume the steel reinforcement is lag bolted to the wood at regular intervals along the beam. What is the maximum bending stress in the wood and steel?

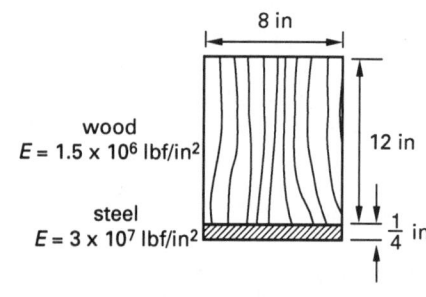

Solution

The moduli of elasticity are given in the illustration. From step 2, the modular ratio is

$$n = \frac{E_{steel}}{E_{wood}} = \frac{3 \times 10^7 \dfrac{lbf}{in^2}}{1.5 \times 10^6 \dfrac{lbf}{in^2}} = 20$$

The actual cross-sectional area of the steel is

$$A_{steel} = (0.25 \text{ in})(8 \text{ in}) = 2 \text{ in}^2$$

The steel is the stronger material. Its area must be expanded to an equivalent area of wood. Since the depth of the beam and reinforcement cannot be increased (step 4), the width must increase. The width of the transformed steel plate is

$$b' = nb = (20)(8 \text{ in}) = 160 \text{ in}$$

The centroid of the transformed section is located 4.45 in from the horizontal axis. The centroidal moment of inertia of the transformed section is $I_{c,t} = 2211.5 \text{ in}^4$. (The calculations for centroidal location and moment of inertia are not presented here.)

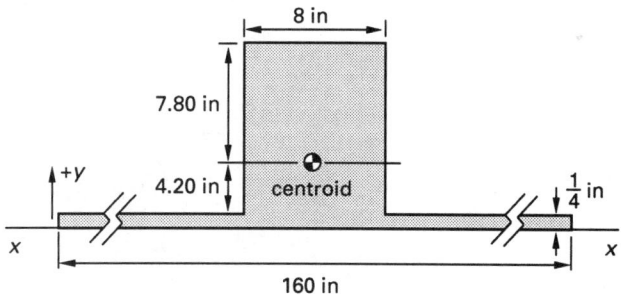

Since the steel plate is bolted to the wood at regular intervals, both pieces experience the same strain. From step 9, the stresses in the wood and steel are

$$\sigma_{max,wood} = \frac{Mc_{wood}}{I}$$

$$= \frac{(40,000 \text{ ft-lbf})\left(12 \dfrac{in}{ft}\right)(7.8 \text{ in})}{2211.5 \text{ in}^4}$$

$$= 1693 \text{ lbf/in}^2$$

$$\sigma_{max,steel} = \frac{nMc_{steel}}{I}$$

$$= \frac{(20)(40,000 \text{ ft-lbf})\left(12 \dfrac{in}{ft}\right)(4.45 \text{ in})}{2211.5 \text{ in}^4}$$

$$= 19,317 \text{ lbf/in}^2$$

PRACTICE PROBLEMS

Note: Unless instructed otherwise in a problem, use the following properties.

steel:
$E = 30 \times 10^6 \text{ lbf/in}^2 \ (20 \times 10^4 \text{ MPa})$
$G = 11.5 \times 10^6 \text{ lbf/in}^2 \ (8.0 \times 10^4 \text{ MPa})$
$\alpha = 6.5 \times 10^{-6} \ 1/°F \ (1.2 \times 10^{-5} \ 1/°C)$
$\nu = 0.3$

aluminum: $E = 10 \times 10^6 \text{ lbf/in}^2 \ (70 \times 10^3 \text{ MPa})$
copper: $E = 17.5 \times 10^6 \text{ lbf/in}^2 \ (12 \times 10^4 \text{ MPa})$

1. A beam 12 ft (3.6 m) long is supported at the left end and 2 ft (0.6 m) from the right end. The beam has a mass of 20 lbm/ft (30 kg/m). A 100 lbf (450 N) load is applied 2 ft (0.6 m) from the left end. An 80 lbf (350 N) load is applied at the right end. What are the (a) maximum moment and (b) maximum shear?

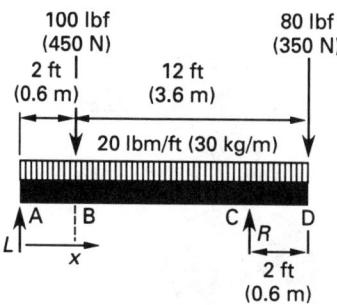

2. A cantilever beam is 6 ft (1.8 m) in length. The cross section is 6 in wide × 4 in high (150 mm wide by 100 mm high). The modulus of elasticity is $1.5 \times 10^6 \text{ lbf/in}^2$ (10 GPa). The beam is loaded by two concentrated forces: 200 lbf (900 N) located 1 ft (0.3 m) from the free end and 120 lbf (530 N) located 2 ft (0.6 m) from the free end. What is the tip deflection?

3. A straight steel beam 200 ft (60 m) long is installed when the temperature is 40°F (4°C). It is supported in such a manner as to allow only 0.5 in (12 mm) longitudinal expansion. If the temperature increases to 110°F (43°C), what will be the compressive stress in the member? Neglect buckling.

4. A 1 in (25 mm) diameter soft steel rod carries a tensile load of 15,000 lbf (67 kN). The elongation is 0.158 in (4 mm). The modulus of elasticity is 2.9×10^7 lbf/in² (200 GPa). What is the total length of the rod?

5. A 3 in (75 mm) diameter horizontal shaft carries a 32 in (80 cm) diameter, 600 lbm (270 kg) pulley on an overhung (cantilever) end. The pulley is 8 in (200 mm) from the face of the outboard bearing. The pulley belt approaches and leaves horizontally. The belt carries upper and lower tensions of 1500 lbf (6.7 kN) and 350 lbf (1.6 kN), respectively. What is the maximum stress in the shaft?

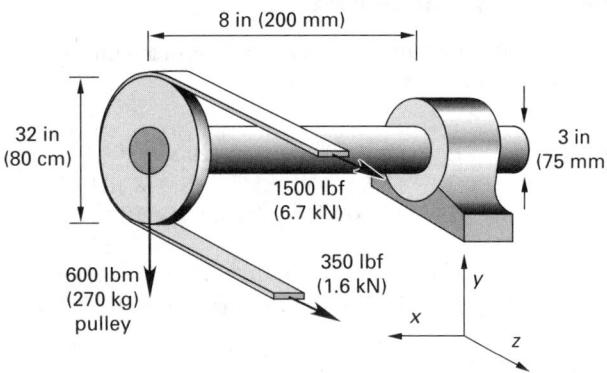

6. A 1.0 in (25 mm) diameter solid rod is held firmly in a chuck. A wrench with a 12 in (300 mm) moment arm applies 60 lbf (270 N) of force 8 in (200 mm) up from the chuck. What are the (a) maximum shear stress and (b) maximum normal stress in the rod?

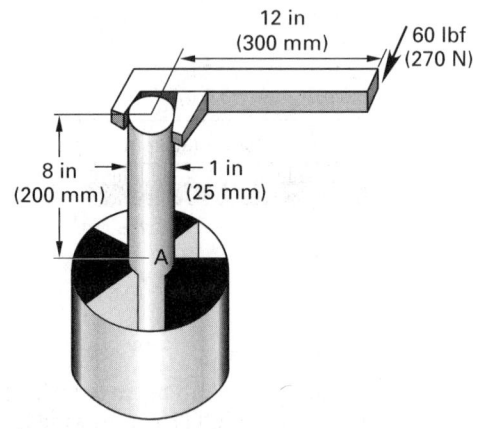

7. A horizontal square bar with a cross-sectional area of 1.5 in^2 (9.7 cm^2) is acted upon by an 18,000 lbf (80 kN) compressive load at each end. The shear stress at a particular point on a plane inclined +30° from the horizontal is 4000 lbf/in^2 (28 MPa). What are the (a) normal stress and (b) shear stress on the inclined plane?

8. (*Time limit: one hour*) The offset wrench handle shown is constructed from a 5/8 in (16 mm) diameter round bar. The bar's modulus of elasticity is 29.6×10^6 lbf/in^2 (204 GPa). The 3 in (75 mm) rise is in the plane of the socket. (a) What are the principal stresses at section A-A? (b) What is the maximum shear at section A-A?

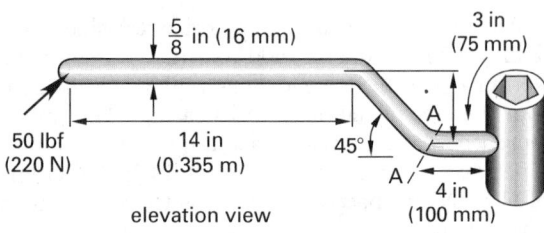

elevation view

9. (*Time limit: one hour*) A brass tube 6 ft long, 2.0 in outside diameter, 1.0 in inside diameter, and with a modulus of elasticity of 1.5×10^7 lbf/in^2 (1.8 m long, 50 mm outside diameter, 25 mm inside diameter, and with a modulus of elasticity of 100 GPa) is used as a cantilever beam. When a concentrated load of 50 lbf (220 N) is applied at the free end, the tip deflection is found to be excessive. To reduce the deflection, it is suggested that a tight-fitting 1 in (25 mm) outside diameter soft steel rod with a modulus of elasticity of 2.9×10^7 lbf/in^2 (200 GPa) be inserted into the entire length of the brass tube. (a) Does the suggestion have merit? (b) What is the percentage change in tip deflection, if any? (Neglect the self-weights.)

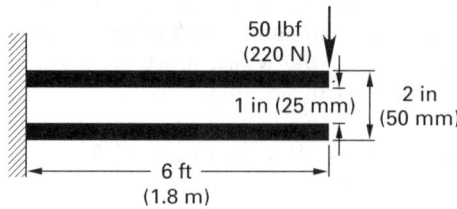

10. A 2500 lbm (1100 kg) stationary flywheel is mounted with a 4 in (100 mm) overhang on a stepped shaft as shown. The step fillet has a 5/16 in (8 mm) radius. Disregarding yielding in the vicinity of the fillet, what is the maximum bending stress in the overhanging section of the shaft?

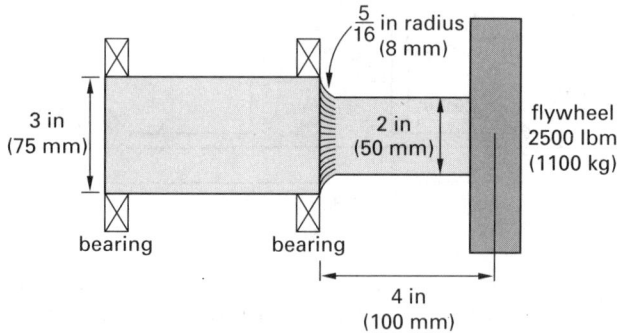

11. A bimetallic spring is constructed of an aluminum strip with a 1/8 in by 1 1/2 in (3.2 mm by 38 mm) cross section bonded on top of a 1/16 in by 1 1/2 in (6.4 mm by 38 mm) steel strip. What is the equivalent, all-aluminum centroidal area moment of inertia of the cross section?

50 Failure Theories

1. Static Loading of Brittle Materials:
   Uniaxial Loading 50-1
2. Static Loading of Brittle Materials:
   Maximum Normal Stress Theory 50-1
3. Static Loading of Brittle Materials:
   Coulomb-Mohr Theory 50-2
4. Static Loading of Brittle Materials:
   Modified Mohr Theory 50-2
5. Static Loading of Ductile Materials:
   Uniaxial Loading 50-3
6. Static Loading of Ductile Materials:
   Maximum Shear Stress Theory 50-3
7. Static Loading of Ductile Materials:
   Strain Energy Theory 50-4
8. Static Loading of Ductile Materials:
   Distortion Energy Theory 50-5
9. Alternating Stress: Soderberg Line 50-6
10. Alternating Stress: Goodman Line 50-6
11. Cumulative Fatigue 50-7
12. Alternating Combined Stresses 50-8
    Practice Problems 50-8

Nomenclature

C	constant	–	–
E	modulus of elasticity	lbf/in²	Pa
FS	factor of safety	–	–
MR	modulus of resilience	in-lbf/in³	J/m³
n	number of cycles	–	–
N	endurance life	–	–
S	strength	lbf/in²	Pa
U	strain energy	in-lbf/in³	J/m³

Symbols

ϵ	strain	in/in	m/m
ν	Poisson's ratio	–	–
σ	normal stress	lbf/in²	Pa
τ	shear stress	lbf/in²	Pa

Subscripts

a	allowable
alt	alternating
c	compressive
e	endurance
eq	equivalent
m	mean
r	range
s	shear
t	tensile
u	ultimate
y	yield

1. STATIC LOADING OF BRITTLE MATERIALS: UNIAXIAL LOADING

Brittle materials such as gray cast iron fail by sudden fracturing, not by yielding. The basic method of designing with brittle materials is to keep the maximum stress below the ultimate strength. Stress concentration factors must be included. The failure criterion is

$$\sigma > S_u \quad \text{[failure criterion]} \qquad 50.1$$

2. STATIC LOADING OF BRITTLE MATERIALS: MAXIMUM NORMAL STRESS THEORY

The *maximum normal stress theory* predicts the failure stress reasonably well for brittle materials under static biaxial loading.[1] Failure is assumed to occur if the largest tensile principal stress, σ_1, is greater than the ultimate tensile strength, or if the largest compressive principal stress, σ_2, is greater than the ultimate compressive strength. Brittle materials generally have much higher compressive strengths than tensile strengths, so both tensile and compressive stresses must be checked.

Figure 50.1 illustrates a standard graphical method of describing the safe operating region. Notice that the axes represent the principal stresses, not the stresses in the x- and y-directions.

Stress concentration factors are applicable to brittle materials under static loading. The *factor of safety*, FS, is the ultimate strength, S_u, divided by the actual stress, σ. (The *margin of safety* is MS = FS − 1.) Where a factor of safety is known in advance, the *allowable stress*, S_a, can be calculated by dividing the ultimate strength by it. The allowable operating region can be constructed from the allowable stresses rather than from the ultimate stresses.

$$\text{FS} = \frac{S_u}{\sigma} \qquad 50.2$$

$$S_a = \frac{S_u}{\text{FS}} \qquad 50.3$$

The failure criterion is

$$\sigma_1, \sigma_2 > \frac{S_u}{\text{FS}} \quad \text{[failure criterion]} \qquad 50.4$$

---

[1] As described in subsequent sections, the maximum normal stress theory is limited to cases where both principal stresses have the same sign (i.e., both are compressive or both are tensile).

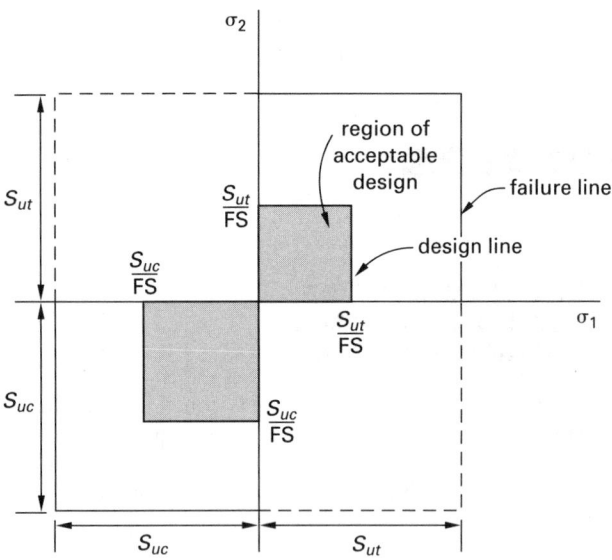

Figure 50.1 *Maximum Normal Stress Theory*

Example 50.1

A lathe bed is made of gray cast iron. The ultimate strengths for the cast iron are 30,000 lbf/in^2 (tension) and 110,000 lbf/in^2 (compression). The bed is subjected to maximum stresses of 17,150 lbf/in^2 in tension and 42,800 lbf/in^2 in compression. What are the factors of safety?

Solution

The factor of safety in tension is

$$
\text{FS} = \frac{S_{ut}}{\sigma_t} = \frac{30{,}000 \; \dfrac{\text{lbf}}{\text{in}^2}}{17{,}150 \; \dfrac{\text{lbf}}{\text{in}^2}}
$$

$$
= 1.75
$$

The factor of safety in compression is

$$
\text{FS} = \frac{S_{uc}}{\sigma_c} = \frac{110{,}000 \; \dfrac{\text{lbf}}{\text{in}^2}}{42{,}800 \; \dfrac{\text{lbf}}{\text{in}^2}}
$$

$$
= 2.57
$$

Tensile failure is the limiting case.

3. STATIC LOADING OF BRITTLE MATERIALS: COULOMB-MOHR THEORY

The maximum normal stress theory is somewhat in conflict with experimental evidence. Reliable operation in the second and fourth quadrants (i.e., when the two principal stresses have opposite signs) has not been observed, even though the stresses are less than the ultimate strengths. The *Coulomb-Mohr theory* is a conservative theory that reduces the acceptable operating

region in the second and fourth quadrants. As with the maximum normal stress theory, the factor of safety is calculated from Eq. 50.2.

The failure criterion is

$$
\frac{(\text{FS})\sigma_1}{S_{ut}} + \frac{(\text{FS})\sigma_2}{S_{uc}} > 1 \quad \text{[failure criterion]} \qquad 50.5
$$

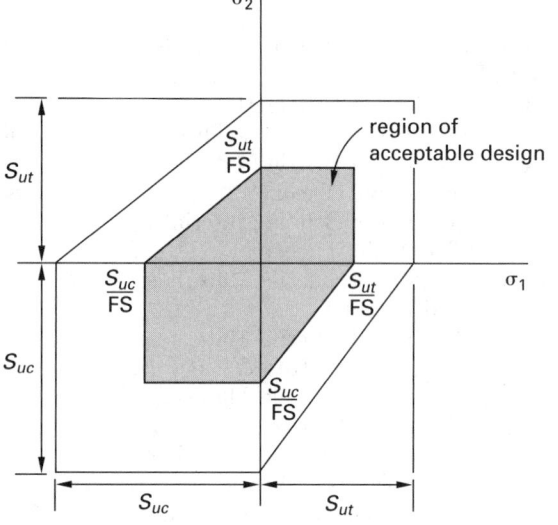

Figure 50.2 *Coulomb-Mohr Theory*

4. STATIC LOADING OF BRITTLE MATERIALS: MODIFIED MOHR THEORY

The Coulomb-Mohr theory is considered conservative because failures typically "miss" the diagonal line in Fig. 50.2 by a considerable margin. The *modified Mohr theory* more closely predicts the envelope of observed failures. Though the failure criterion can be based on deriving equations of the straight lines in the second and fourth quadrants, graphical solutions are more common.

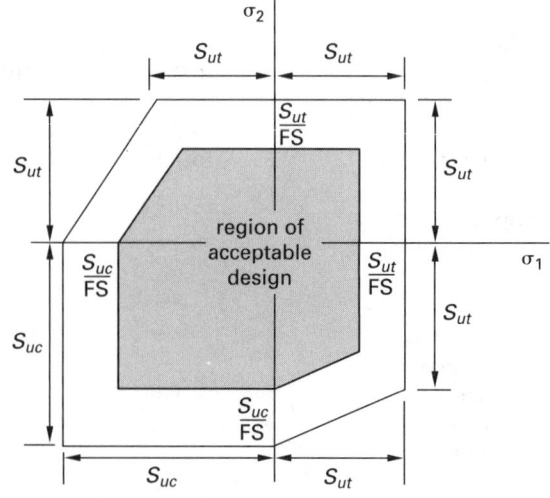

Figure 50.3 *Modified Mohr Theory*

Example 50.2

A 0.5 in diameter dowel is made from cast iron (ultimate tensile strength of 40,000 lbf/in²; ultimate compressive strength of 135,000 lbf/in²). The dowel supports a compressive load of 15,000 lbf and is subjected simultaneously to an unknown torsional shear stress. Use the modified Mohr theory to determine the torsional shear stress that will cause failure.

Solution

The compressive normal stress in the dowel is

$$\sigma_c = \frac{F}{A} = \frac{-15{,}000 \text{ lbf}}{\left(\dfrac{\pi}{4}\right)(0.5 \text{ in})^2}$$

$$= -76{,}400 \text{ lbf/in}^2 \quad \text{[negative because compressive]}$$

To use the modified Mohr theory, the principal stresses must be known. However, the shear stress in this problem is the unknown, so the principal stress cannot be calculated directly. Use a trial and error approach.

Assume the shear stress is 10,000 lbf/in². With $\tau = 10{,}000$ lbf/in², $\sigma_x = -76{,}400$ lbf/in², and $\sigma_y = 0$, the principal stresses are found (from standard methods) to be 1300 lbf/in² and $-77{,}700$ lbf/in². Draw the modified Mohr diagram and plot these principal values. The point is not close to the failure line, so the assumed shear stress of 10,000 lbf/in² is too low.

Repeat the process with shear stresses of 30,000 lbf/in² and 40,000 lbf/in².

τ	σ_1	σ_2
30,000	10,400	−86,800
40,000	17,100	−93,500

The third point (corresponding to a shear stress of 40,000 lbf/in²) is essentially on the failure line, so this is the maximum allowable shear stress. (It is a coincidence that this is also the ultimate tensile strength.)

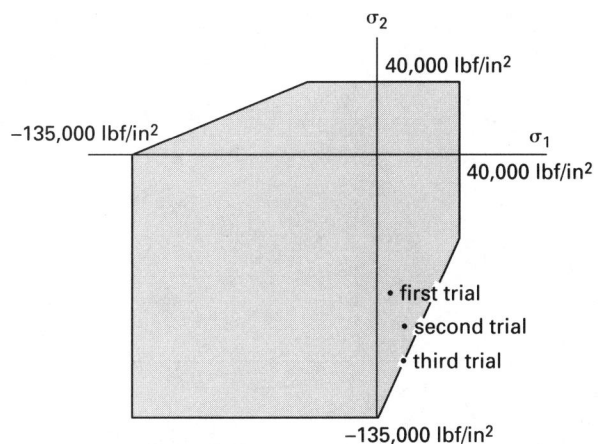

5. STATIC LOADING OF DUCTILE MATERIALS: UNIAXIAL LOADING

Ductile materials fail by yielding, not by fracture. The basic method of designing with ductile materials in uniaxial loading is to keep the maximum stress below the yield strength. This is known as the *maximum stress theory*. Stress concentration factors must be included. The failure criterion is

$$\sigma > S_y \quad \text{[failure criterion]} \qquad 50.6$$

The failure criterion for the equivalent *maximum strain theory* is

$$\epsilon > \frac{S_y}{E} \quad \text{[failure criterion]} \qquad 50.7$$

6. STATIC LOADING OF DUCTILE MATERIALS: MAXIMUM SHEAR STRESS THEORY

With the conservative *maximum shear stress theory*, shear stress is used to indicate yielding (i.e., failure). Loading is not limited to shear and torsion. Loading can include normal stresses as well as shear stresses. According to the maximum shear stress theory, yielding occurs when the maximum shear stress exceeds the yield strength in shear.[2] It is implicit in this theory that the yield strength in shear is half of the tensile yield strength.[3]

$$S_{ys} = \frac{S_{yt}}{2} \qquad 50.8$$

From the combined stress theory, the maximum shear stress, $\tau_{\max}$, for *triaxial loading* is the maximum of the three combined shear stresses. (For biaxial loading, only Eq. 50.9 is used.)

$$\tau_{12} = \frac{\sigma_1 - \sigma_2}{2} \qquad 50.9$$

$$\tau_{23} = \frac{\sigma_2 - \sigma_3}{2} \qquad 50.10$$

$$\tau_{31} = \frac{\sigma_3 - \sigma_1}{2} \qquad 50.11$$

$$\tau_{\max} = \max\left(\tau_{12}, \tau_{23}, \tau_{31}\right) \qquad 50.12$$

The failure criterion is

$$\tau_{\max} > S_{ys} = \frac{S_{yt}}{2} \quad \text{[failure criterion]} \qquad 50.13$$

---

[2]The application of this theory, as well as those that follow, depends on being able to find the maximum shear stress, $\tau_{\max}$. Even in some cases of uniaxial loading, finding the maximum shear stress can be tricky. Two cylindrical rollers or two spheres in contact, for example, represent cases of uniaxial compressive loading, which have well-documented but non-obvious maximum shear stresses.

[3]If the symmetrical shape of the failure envelope is accepted, it is easy to justify the assumption that the yield strength in shear is one-half of the yield strength in tension. For a pure shear loading, the two principal stresses will each be equal and opposite (with magnitudes equal to the applied shear stress). Plotting the locus of points with $\sigma_1 = -\sigma_2$, the failure envelope is encountered at $S_{yt}/2$.

The combinations of allowable principal stresses for the maximum shear stress theory are shown graphically in Fig. 50.4 for the case of biaxial loading (i.e., $\sigma_3 = 0$). The shape of this failure envelope is similar to that of the Coulomb-Mohr theory for brittle materials, but the limits are based on yield strengths, not on ultimate strengths. Since the tensile and compressive yield strengths are assumed equal for ductile materials, the failure envelope is symmetrical. In Fig. 50.4, the limits are not divided by 2 as might be suspected from Eqs. 50.8 through 50.13. This is because the failure envelope is used with the principal stress, not with the shear stress.

The factor of safety with the maximum shear stress theory is

$$\text{FS} = \frac{S_{ys}}{\tau_{\max}} = \frac{S_{yt}}{2\tau_{\max}} \qquad 50.14$$

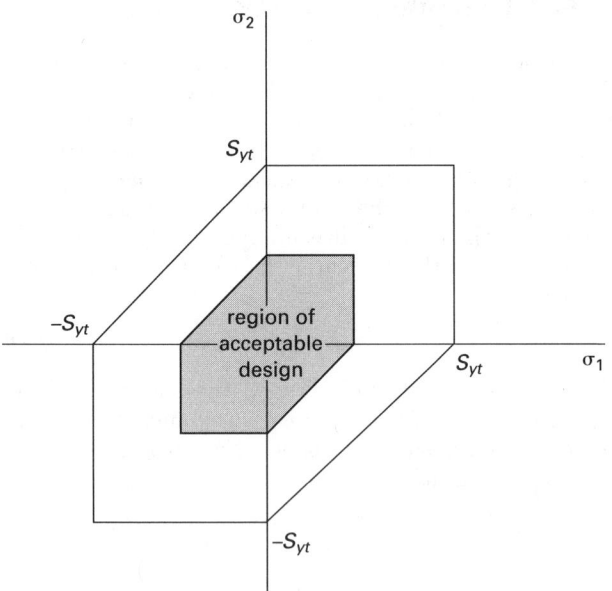

Figure 50.4 *Maximum Shear Stress Theory*

Example 50.3

Strain gages attached to a bearing support show the three principal stresses to be 10,600 lbf/in^2, 2400 lbf/in^2, and -9200 lbf/in^2. The support is cast aluminum with a tensile yield strength of 24,000 lbf/in^2. Use the maximum shear stress theory to determine the factor of safety.

Solution

Calculate the combined shear stresses.

$$\tau_{12} = \frac{\sigma_1 - \sigma_2}{2}$$

$$= \frac{10,600 \, \dfrac{\text{lbf}}{\text{in}^2} - 2400 \, \dfrac{\text{lbf}}{\text{in}^2}}{2}$$

$$= 4100 \, \text{lbf/in}^2$$

$$\tau_{23} = \frac{\sigma_2 - \sigma_3}{2}$$

$$= \frac{2400 \, \dfrac{\text{lbf}}{\text{in}^2} - \left(-9200 \, \dfrac{\text{lbf}}{\text{in}^2}\right)}{2}$$

$$= 5800 \, \text{lbf/in}^2$$

$$\tau_{13} = \frac{\sigma_1 - \sigma_3}{2}$$

$$= \frac{10,600 \, \dfrac{\text{lbf}}{\text{in}^2} - \left(-9200 \, \dfrac{\text{lbf}}{\text{in}^2}\right)}{2}$$

$$= 9900 \, \text{lbf/in}^2$$

The maximum shear stress is 9900 lbf/in^2.

From Eq. 50.14, the factor of safety is

$$\text{FS} = \frac{S_{yt}}{2\tau_{\max}}$$

$$= \frac{24,000 \, \dfrac{\text{lbf}}{\text{in}^2}}{(2)\left(9900 \, \dfrac{\text{lbf}}{\text{in}^2}\right)}$$

$$= 1.21$$

7. STATIC LOADING OF DUCTILE MATERIALS: STRAIN ENERGY THEORY

The *strain energy theory* calculates the energy per unit volume. This energy is compared to the energy that causes yielding, which is known as the *modulus of resilience*. Since the stress-strain curve is essentially a straight line, the strain energy (i.e., the area under the curve) for any uniaxial loading less than the yield strength is

$$U = \frac{\sigma \epsilon}{2} \qquad 50.15$$

For biaxial loading on the principal planes, the strain energy is calculated from superposition.

$$U = \frac{\sigma_1 \epsilon_1 + \sigma_2 \epsilon_2}{2} \qquad 50.16$$

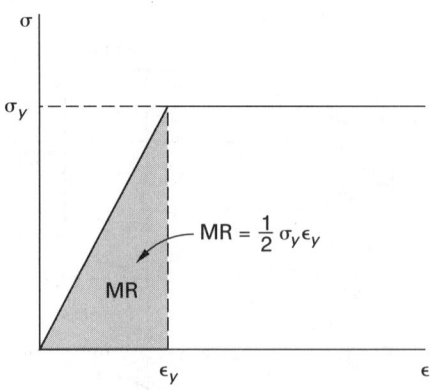

Figure 50.5 *Strain Energy as Area Under the Stress-Strain Curve*

Strain is generally not measured directly, but it can be calculated from the stress and material properties.

$$\epsilon_1 = \left(\frac{1}{E}\right)(\sigma_1 - \nu\sigma_2) \qquad 50.17$$

$$\epsilon_2 = \left(\frac{1}{E}\right)(\sigma_2 - \nu\sigma_1) \qquad 50.18$$

$$U = \left(\frac{1}{2E}\right)(\sigma_1^2 + \sigma_2^2 - 2\nu\sigma_1\sigma_2) \qquad 50.19$$

Failure is assumed to occur when the strain energy, U, exceeds the modulus of resilience. The modulus of resilience can be determined from a simple tensile test and is calculated as the area under the stress-strain curve up to the yield point.[4] At failure in a simple tensile test, $\sigma_1 = S_{yt}$ and $\sigma_2 = 0$. Therefore, from Eq. 50.19, the modulus of resilience is

$$\text{MR} = \left(\frac{1}{2E}\right)[S_{yt}^2 + (0)^2 - (2)(0)] = \frac{S_{yt}^2}{2E} \qquad 50.20$$

Eliminating the $^1/_2E$ terms, the failure criterion is

$$\sigma_1^2 + \sigma_2^2 - 2\nu\sigma_1\sigma_2 > S_{yt}^2 \quad \text{[failure criterion]} \qquad 50.21$$

The factor of safety is

$$\text{FS} = \frac{S_{yt}}{\sqrt{\sigma_1^2 + \sigma_2^2 - 2\nu\sigma_1\sigma_2}} \qquad 50.22$$

8. STATIC LOADING OF DUCTILE MATERIALS: DISTORTION ENERGY THEORY

The *distortion energy theory* (also known as the *theory of constant energy of distortion, von Mises theory, von Mises-Hencky theory*, and *octahedral shear-stress theory*) is similar in development to the strain energy method but is more strict. It is commonly used to predict tensile and shear failure in steel parts. The *von Mises stress* (also known as the *effective stress*), σ', is calculated from the principal stresses. For biaxial loading,

$$\sigma' = \sqrt{\sigma_1^2 + \sigma_2^2 - \sigma_1\sigma_2} \qquad 50.23$$

For *triaxial loading*, the von Mises stress is

$$\sigma' = \sqrt{\left(\tfrac{1}{2}\right)\left[(\sigma_1 - \sigma_2)^2 + (\sigma_2 - \sigma_3)^2 + (\sigma_3 - \sigma_1)^2\right]} \qquad 50.24$$

The failure criterion is

$$\sigma' > S_{yt} \quad \text{[failure criterion]} \qquad 50.25$$

---

[4]The strain energy theory is seldom used to predict failures. However, it is similar in development to the distortion energy theory, which has supplanted it, so the strain energy theory is presented here as part of the logical progression of failure theories.

The factor of safety is

$$\text{FS} = \frac{S_{yt}}{\sigma'} \qquad 50.26$$

If the loading is pure torsion at failure, then $\sigma_1 = \sigma_2 = \pm\tau_{\max}$, and $\sigma_3 = 0$. If $\tau_{\max}$ is substituted for σ in Eq. 50.24 (with $\sigma_3 = 0$), an expression for the yield strength in shear is derived. Equation 50.27 predicts a larger yield strength in shear than did the maximum shear stress theory ($0.5S_{yt}$).

$$S_{ys} = \tau_{\max,\text{failure}} = \frac{S_{yt}}{\sqrt{3}} = 0.577S_{yt} \qquad 50.27$$

Example 50.4

The steel used in a shaft has a tensile yield strength of 110,000 lbf/in² and a Poisson's ratio of 0.3. The shaft is simultaneously acted upon by a longitudinal compressive stress of 60,000 lbf/in² and by a torsional stress of 40,000 lbf/in². Compare the factor of safety using (a) maximum shear stress, (b) strain energy, and (c) distortion energy.

Solution

From combined stress theory, the maximum shear stress is

$$\tau_{\max} = \tfrac{1}{2}\sqrt{(\sigma_x - \sigma_y)^2 + (2\tau)^2}$$

$$= \tfrac{1}{2}\sqrt{\left(60{,}000\ \frac{\text{lbf}}{\text{in}^2}\right)^2 + \left[(2)\left(40{,}000\ \frac{\text{lbf}}{\text{in}^2}\right)\right]^2}$$

$$= 50{,}000\ \text{lbf/in}^2$$

The principal stresses are

$$\sigma_1, \sigma_2 = \left(\tfrac{1}{2}\right)(\sigma_x + \sigma_y) \pm \tau_{\max}$$

$$= \left(\tfrac{1}{2}\right)\left(60{,}000\ \frac{\text{lbf}}{\text{in}^2}\right) \pm 50{,}000\ \frac{\text{lbf}}{\text{in}^2}$$

$$= 80{,}000\ \text{lbf/in}^2,\ -20{,}000\ \text{lbf/in}^2$$

(a) From Eq. 50.14, the factor of safety for the maximum shear stress theory is

$$\text{FS} = \frac{S_{yt}}{2\tau_{\max}}$$

$$= \frac{110{,}000\ \dfrac{\text{lbf}}{\text{in}^2}}{(2)\left(50{,}000\ \dfrac{\text{lbf}}{\text{in}^2}\right)}$$

$$= 1.1$$

(b) From Eq. 50.22, the factor of safety for the strain energy theory is

$$FS = \frac{S_{yt}}{\sqrt{\sigma_1^2 + \sigma_2^2 - 2\nu\sigma_1\sigma_2}}$$

$$= \frac{110,000 \, \frac{lbf}{in^2}}{\sqrt{\begin{array}{c}\left(80,000 \, \frac{lbf}{in^2}\right)^2 + \left(-20,000 \, \frac{lbf}{in^2}\right)^2 \\ - (2)(0.3)\left(80,000 \, \frac{lbf}{in^2}\right)\left(-20,000 \, \frac{lbf}{in^2}\right)\end{array}}}$$

$$= 1.25$$

(c) From Eq. 50.23, the von Mises stress is

$$\sigma' = \sqrt{\sigma_1^2 + \sigma_2^2 - \sigma_1\sigma_2}$$

$$= \sqrt{\begin{array}{c}\left(80,000 \, \frac{lbf}{in^2}\right)^2 + \left(-20,000 \, \frac{lbf}{in^2}\right)^2 \\ - \left(80,000 \, \frac{lbf}{in^2}\right)\left(-20,000 \, \frac{lbf}{in^2}\right)\end{array}}$$

$$= 91,652 \, lbf/in^2$$

From Eq. 50.26, the factor of safety for the distortion energy theory is

$$FS = \frac{S_{yt}}{\sigma'} = \frac{110,000 \, \frac{lbf}{in^2}}{91,652 \, \frac{lbf}{in^2}}$$

$$= 1.20$$

9. ALTERNATING STRESS: SODERBERG LINE

Many parts are subject to a combination of static and reversed loading, as illustrated in Fig. 50.6 for sinusoidal loading. For these parts, failure cannot be determined solely by comparing stresses with the yield strength or endurance limit. The combined effects of the average stress and the amplitude of the reversal must be considered. This is done graphically on a diagram that plots the mean stress versus the alternating stresses.

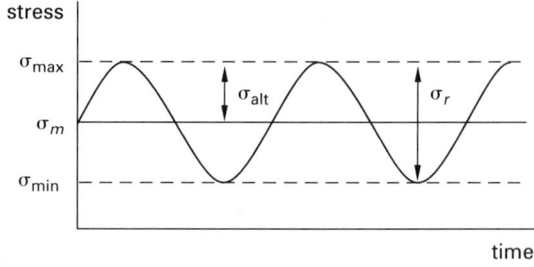

Figure 50.6 *Sinusoidal Fluctuating Stresses*

The *mean stress* is

$$\sigma_m = \frac{\sigma_{max} + \sigma_{min}}{2} \qquad 50.28$$

The *alternating stress* is half of the *range stress*.

$$\sigma_r = \sigma_{max} - \sigma_{min} \qquad 50.29$$
$$\sigma_{alt} = \tfrac{1}{2}\sigma_r = \left(\tfrac{1}{2}\right)(\sigma_{max} - \sigma_{min}) \qquad 50.30$$

A criterion for acceptable design (or for failure) is established by graphically relating the yield strength and the endurance limit. One method of relating this information is a Soderberg line, particularly suited for normal stresses in ductile materials. However, it is the most conservative of the fluctuating stress theories.

Figure 50.7 illustrates how an area of acceptable design is developed by drawing a straight line (the *Soderberg line* or the *failure line*) from the endurance limit, S_e, to the yield strength, S_{yt}. Both of these values should be divided by a suitable factor of safety to define the allowable design area. If the point (σ_m, σ_{alt}) falls below the allowable stress line, the design is acceptable.

Stress concentration factors are applied to the alternating stress only. This is justified with ductile materials, such as steel, which yield around discontinuities, reducing a constant stress.

Figure 50.7 illustrates how the Soderberg *equivalent stress*, σ_{eq}, is determined from the operating point. The factor of safety is

$$FS = \frac{S_e}{\sigma_{eq}} = \frac{S_e}{\sigma_{alt} + \left(\dfrac{S_e}{S_{yt}}\right)\sigma_m} \qquad 50.31$$

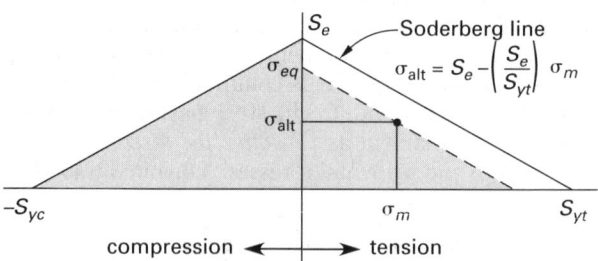

Figure 50.7 *Soderberg Line and Equivalent Stress*

10. ALTERNATING STRESS: GOODMAN LINE

The Soderberg line is a conservative criterion, and it is not often used. Also, the envelope of failures is not truly linear, but follows more of a parabolic line, named the *Gerber line* or *Gerber parabolic relationship*, extending above the Soderberg line from the endurance limit to the ultimate tensile strength.

The *Goodman line* (also known as the *modified Goodman line*) is less conservative than the Soderberg line and is more easily constructed than the Gerber line.[5] It is applicable for steel, aluminum, titanium, and some magnesium alloys.[6] Figure 50.8 illustrates the Goodman line, as well as the method for determining the Goodman equivalent stress.[7] The Goodman factor of safety is

$$\text{FS} = \frac{S_e}{\sigma_{\text{eq}}} = \frac{S_e}{\sigma_{\text{alt}} + \left(\dfrac{S_e}{S_{ut}}\right)\sigma_m} \qquad 50.32$$

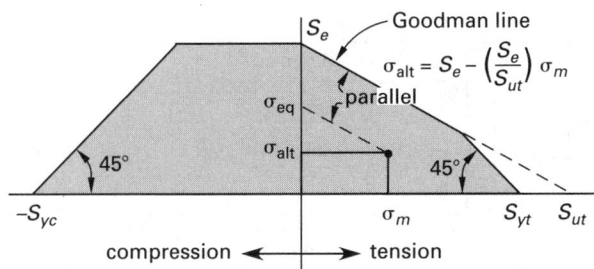

Figure 50.8 *Goodman Line and Equivalent Stress*

For a ductile material experiencing an alternating shear stress, the yield and endurance strengths in shear can be calculated from distortion energy theory for use in drawing the *Goodman shear line*.[8]

$$S_{ys} = (0.577)S_{yt} \qquad 50.33$$
$$S_{es} = (0.577)S_e \qquad 50.34$$

Example 50.5

An aircraft bell crank is made from aluminum with the following properties: yield strength, 40,000 lbf/in²; ultimate strength, 45,000 lbf/in²; and endurance limit

---

[5]There is no significant distinction between a *Goodman line* and a *modified Goodman line*. The modifications were made early in the theory's development, and both names are now used to represent a straight line drawn on the fatigue diagram between the endurance and ultimate tensile strengths.

[6]The Goodman line should not be used with gray cast iron and some types of magnesium. Failures for these materials do not follow the theory well. The envelope of failure for these materials is bounded by the parabolic *Smith line*, which is similar to the Gerber line, except that it runs under the straight line connecting the endurance and ultimate strengths. However, brittle materials such as gray cast iron are not usually considered satisfactory for stresses that fluctuate significantly unless very large factors of safety are included.

[7]The Soderberg, Goodman, Gerber, and Smith lines are drawn on *fatigue diagrams*. Strictly speaking, they are Soderberg *lines*, not Soderberg *diagrams*. The distinction is more critical for Goodman lines, because there is a Goodman diagram for fatigue loading, but it is constructed differently. However, the correct names are not consistently applied, and the terms "Soderberg diagram" and "Goodman diagram" are often heard.

[8]Theoretically, the coefficient 0.5 could be used instead of 0.577 if the maximum shear stress theory was elected. However, this is probably never done in practice.

13,500 lbf/in² reduced to 8500 lbf/in² by various derating factors. The bell crank is subjected to minimum and maximum tensile stresses of 6500 lbf/in² and 9500 lbf/in², respectively. The bell crank is to have an indefinite life, and a factor of safety in excess of 3 is required. Is the design acceptable?

Solution

The mean and alternating stresses are

$$\sigma_m = \left(\tfrac{1}{2}\right)(\sigma_{\max} + \sigma_{\min})$$
$$= \left(\tfrac{1}{2}\right)\left(9500\ \frac{\text{lbf}}{\text{in}^2} + 6500\ \frac{\text{lbf}}{\text{in}^2}\right)$$
$$= 8000\ \text{lbf/in}^2$$
$$\sigma_a = \left(\tfrac{1}{2}\right)(\sigma_{\max} - \sigma_{\min})$$
$$= \left(\tfrac{1}{2}\right)\left(9500\ \frac{\text{lbf}}{\text{in}^2} - 6500\ \frac{\text{lbf}}{\text{in}^2}\right)$$
$$= 1500\ \text{lbf/in}^2$$

The material properties are divided by the safety factor.

$$\frac{S_{yt}}{3} = \frac{40,000\ \dfrac{\text{lbf}}{\text{in}^2}}{3} = 13,333\ \text{lbf/in}^2$$

$$\frac{S_{ut}}{3} = \frac{45,000\ \dfrac{\text{lbf}}{\text{in}^2}}{3} = 15,000\ \text{lbf/in}^2$$

$$\frac{S_e}{3} = \frac{8500\ \dfrac{\text{lbf}}{\text{in}^2}}{3} = 2833\ \text{lbf/in}^2$$

Since the operating point lies above the failure line, the design is not acceptable.

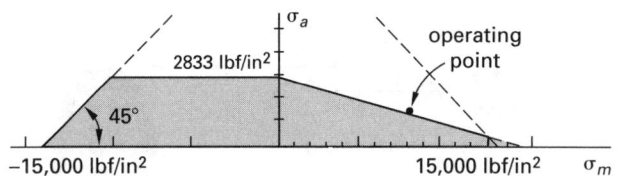

11. CUMULATIVE FATIGUE

If a part is subjected to $\sigma_{\max,1}$ for n_1 cycles, $\sigma_{\max,2}$ for n_2 cycles, and so on, the part will accumulate varying amounts of fatigue damage during each series of cycles. *Miner's rule* (also known as the *Palmgren-Miner cycle ratio summation formula, fatigue interaction formula,* and *cumulative usage factor rule*) can be used to evaluate cumulative damage.[9] In Eq. 50.35, the N_i are the fatigue lives for the corresponding stress levels.

$$\sum \frac{n_i}{N_i} > C \quad \text{[failure criterion]} \qquad 50.35$$

---

[9]Miner's rule does not take into account the increase in the endurance limit that results when virgin material is understressed.

A value of 1.0 is commonly used for C. However, the exact value should be determined from experimentation appropriate to the material and application. Values between approximately 0.7 and 2.2 have been reported.

12. ALTERNATING COMBINED STRESSES

If there are variations in biaxial or triaxial stresses, a conservative approach is to use a combination of the distortion energy and Goodman line.

step 1: Calculate the principal mean stresses from the combined stress theory.

step 2: Calculate the principal alternating stresses.

step 3: Calculate the mean and alternating von Mises stresses.

$$\sigma'_m = \sqrt{\sigma_{m,1}^2 + \sigma_{m,2}^2 - \sigma_{m,1}\sigma_{m,2}} \qquad 50.36$$

$$\sigma'_{\text{alt}} = \sqrt{\sigma_{\text{alt},1}^2 + \sigma_{\text{alt},2}^2 - \sigma_{\text{alt},1}\sigma_{\text{alt},2}} \qquad 50.37$$

step 4: Plot the mean and alternating von Mises stresses in relationship to a standard Goodman line.

PRACTICE PROBLEMS

1. A shaft with a 1.125 in (28.6 mm) diameter receives 400 in-lbf (45 N·m) of torque through a pinned sleeve. The pin is manufactured from steel with a tensile yield strength of 73.9 ksi (510 MPa). Using a factor of safety of 2.5, what pin diameter is needed?

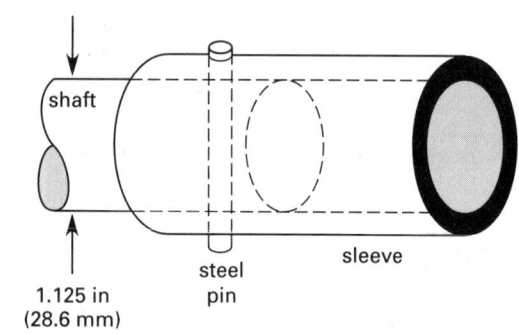

2. A $^5/_8$ in UNC bolt and nut carry a simple tensile load that fluctuates between 1000 lbf and 8000 lbf (4400 N and 35,000 N). Both the bolt and nut are made from a material with a yield strength of 70,000 lbf/in² (480 MPa) and an endurance limit of 30,000 lbf/in² (205 MPa). Including a stress concentration factor for the threads, what is the factor of safety for the bolt?

3. (*Time limit: one hour*) The spool in the spool valve shown is constructed of aluminum with a yield strength of 19,000 lbf/in² (130 MPa). A 500 psig (3.5 MPa) pressure differential exists across the spool shaft. The shaft has a 1.0 in (25 mm) outside diameter and a 0.050 in (1.3 mm) wall thickness. The end disks have a 2.0 in (50 mm) outside diameter and a 0.375 in (9.5 mm) thickness. The end disks are rigidly attached to the tube. Use the distortion energy theory to calculate the factor of safety for the spool.

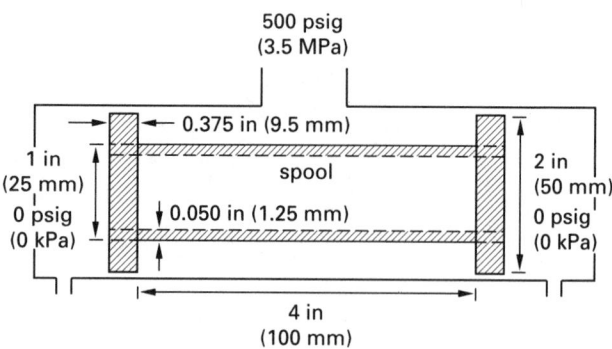

4. (*Time limit: one hour*) The pressure in a small pressure vessel varies continually between the extremes of 50 psig and 350 psig (340 kPa and 2400 kPa). The vessel is closed by a $^1/_2$ in (12 mm) plate. The plate is attached to the vessel flange with six $^3/_8$-24 UNF bolts evenly spaced around a $9^1/_2$ in (240 mm) circle. Each bolt is tightened to an initial preload of 3700 lbf (16.4 kN). The bolts and nuts are constructed of cold-rolled steel with a 90 ksi (620 MPa) yield strength and 110 ksi (760 MPa) ultimate strength. The plate, flange, and vessel are constructed of steel with a 30 ksi (205 MPa) yield strength and a 50 ksi (345 MPa) ultimate strength. The vessel is intended to be used indefinitely. Neglect the effects of the gasket (not shown). Neglect bending of the plate. Using a thread stress concentration factor of 2 and appropriate endurance strength derating factors, what is the factor of safety?

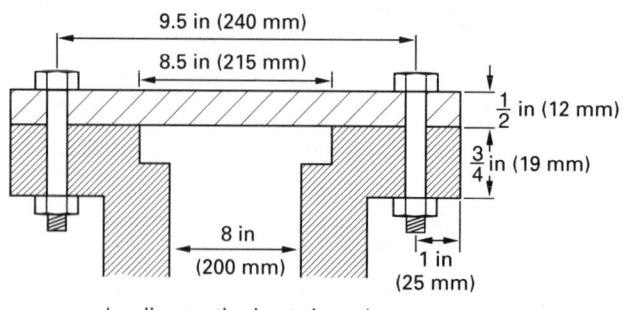

(sealing method not shown)

51

Basic Machine Design

1. Slender Columns 51-1
2. Intermediate Columns 51-3
3. Eccentrically Loaded Columns 51-3
4. Thin-Walled Cylindrical Tanks 51-3
5. Thick-Walled Cylinders 51-4
6. Thin-Walled Spherical Tanks 51-5
7. Interference Fits 51-5
8. Stress Concentrations for
   Press-Fitted Shafts in Flexure 51-7
9. Bolts 51-8
10. Rivet and Bolt Connections 51-10
11. Bolt Preload 51-11
12. Bolt Torque to Obtain Preload 51-12
13. Fillet Welds 51-13
14. Circular Shaft Design 51-13
15. Torsion in Thin-Walled,
    Noncircular Shells 51-14
16. Torsion in Solid, Noncircular Members . . 51-15
17. Eccentrically Loaded Bolted Connections . 51-15
18. Eccentrically Loaded Welded Connections 51-17
19. Flat Plates 51-17
    Practice Problems 51-18

Nomenclature

a	dimension	in	m
A	area	in^2	m^2
b	dimension	in	m
b	width	in	m
c	distance from neutral axis		
	to extreme fiber	in	m
C	circumference	in	m
C	end-restraint coefficient	–	–
d	diameter	in	m
e	eccentricity	in	m
E	modulus of elasticity	lbf/in^2	Pa
f	coefficient of friction	–	–
F	force	lbf	N
FS	factor of safety	–	–
G	shear modulus	lbf/in^2	Pa
I	interference	in	in
I	moment of inertia	in^4	m^4
J	polar moment of inertia	in^4	m^4
k	radius of gyration	in	m
k	stiffness	lbf/in	N/m
L	length	in	m
M	moment	in-lbf	N·m
n	modular ratio	–	–
n	number of connectors	–	–
N	normal force	lbf	N
p	perimeter	in	m
p	pressure	lbf/in^2	Pa

q	shear flow	lbf/in	N/m
r	radius	in	m
t	thickness	in	m
T	torque	in-lbf	N·m
U	energy	in-lbf	J
y	weld size	in	m

Symbols

α	thread half-angle	deg	deg
ϵ	strain	in/in	m/m
θ	lead angle	deg	deg
ν	Poisson's ratio	–	–
σ	normal stress	lbf/in^2	Pa
τ	shear stress	lbf/in^2	Pa
ϕ	angle	rad	rad

Subscripts

a	allowable
c	centroidal, circumferential, or collar
cr	critical
e	Euler or effective
eq	equivalent
h	hoop
i	inside or initial
l	longitudinal
m	mean
o	outside
p	bearing
r	radial
t	tension, thread, or transformed
T	torque
v	vertical
y	yield

1. SLENDER COLUMNS

Very short compression members are known as *piers*. Long compression members are known as *columns*. Failure in piers occurs when the applied stress exceeds the yield strength of the material. However, very long columns fail by sideways *buckling* long before the compressive stress reaches the yield strength. Buckling failure is sudden, often without significant initial sideways bending. The load at which a column fails is known as the *critical load* or *Euler load*.

The *Euler load* is the theoretical maximum load that an initially straight column can support without buckling. For columns with frictionless or pinned ends, this load is given by Eq. 51.1.

$$F_e = \frac{\pi^2 EI}{L^2} = \frac{\pi^2 EA}{\left(\dfrac{L}{k}\right)^2} \qquad 51.1$$

The corresponding column stress is given by Eq. 51.2. This stress cannot exceed half of the compressive yield strength of the column material.

$$\sigma_e = \frac{F_e}{A} = \frac{\pi^2 E}{\left(\dfrac{L}{k}\right)^2} \qquad 51.2$$

The quantity L/k is known as the *slenderness ratio*. Long columns have high slenderness ratios. The smallest slenderness ratio for which Eq. 51.2 is valid is the *critical slenderness ratio*. Typical critical slenderness ratios range from 80 to 120. The critical slenderness ratio becomes smaller as the compressive yield strength increases.

L is the longest unbraced column length. If a column is braced against buckling at some point between its two ends, the column is known as a *braced column*, and L will be less than the full column height. Columns with rectangular cross sections have two radii of gyration, k_x and k_y, and therefore, will have two slenderness ratios. The largest slenderness ratio will govern the design, as this results in the smallest vertical loading.

Columns do not always have frictionless or pinned ends. Often, a column will be fixed ("clamped," "built in," etc.) at its top and base. In such cases, the *effective length*, L', must be used in place of L in Eqs. 51.1 and 51.2.

$$L' = CL \qquad 51.3$$

$$\sigma_e = \frac{F_e}{A} = \frac{\pi^2 E}{\left(\dfrac{CL}{k}\right)^2} \qquad 51.4$$

C is the *end restraint coefficient*, which varies from 0.5 to 2.0 according to Table 51.1. For most real columns, the design values of C should be used since infinite stiffness of the supporting structure is not achievable.

Euler's curve for columns, line BCD in Fig. 51.1, is generated by plotting the *Euler stress* (Eq. 51.2) versus the slenderness ratio. Since the material's compressive yield strength cannot be exceeded, a horizontal line AC is added to limit applications to the region below. Theoretically, members with slenderness ratios less than $(SR)_C$ could be treated as pure compression members. However, this is not done in practice.

Defects in materials, errors in manufacturing, inabilities to achieve theoretical end conditions, and eccentricities frequently combine to cause column failures in the region around point C. Therefore, this region is excluded by designers.

Table 51.1 *Theoretical End Restraint Coefficients*

illus.	end conditions	C ideal	recommended for design
(a)	both ends pinned	1	1.0*
(b)	both ends built in	0.5	0.65*–0.90
(c)	one end pinned, one end built in	0.707	0.80*–0.90
(d)	one end built in, one end free	2	2.0–2.1
(e)	one end built in, one end fixed against rotation but free	1	1.2*
(f)	one end pinned, one end fixed against rotation but free	2	2.0*

*AISC values

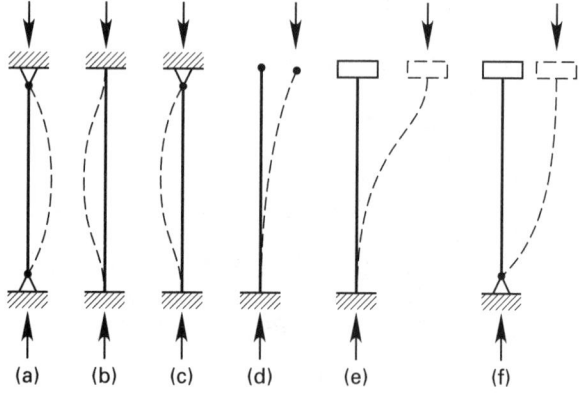

The empirical procedure used to exclude the failure area is to draw a parabolic curve from point A through a tangent point T on the Euler curve at a stress of $\frac{1}{2}S_y$. The corresponding value of the slenderness ratio of any end restraint coefficient, C, is

$$(SR)_T = \frac{1}{C}\sqrt{\frac{2\pi^2 E}{S_y}} \qquad 51.5$$

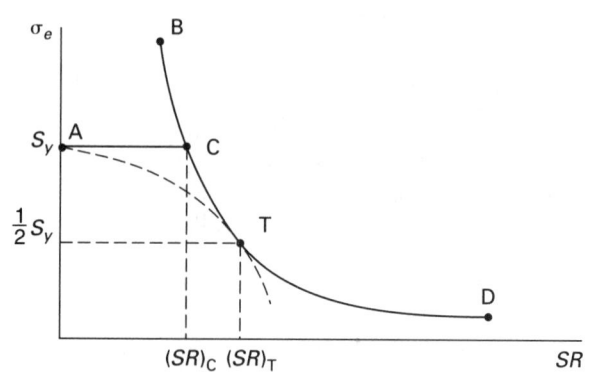

Figure 51.1 *Euler's Curve*

Example 51.1

A steel member is used as an 8.5 ft long column. The ends are pinned. What is the maximum allowable compressive stress in order to have a factor of safety of 3.0? Use the following data for the column.

$$E = 2.9 \times 10^7 \text{ lbf/in}^2$$

$$S_{yt} = 36{,}000 \text{ lbf/in}^2$$

$$k = 0.569 \text{ in}$$

Solution

First, check the slenderness ratio to see if this is a long column.

$$\frac{L}{k} = \frac{(8.5 \text{ ft})\left(12 \; \frac{\text{in}}{\text{ft}}\right)}{0.569 \text{ in}} = 179.3 \quad [\text{ok}]$$

From Eq. 51.2, the Euler stress is

$$\sigma_e = \frac{\pi^2 E}{\left(\dfrac{L}{k}\right)^2} = \frac{\pi^2 \left(2.9 \times 10^7 \; \dfrac{\text{lbf}}{\text{in}^2}\right)}{(179.3)^2}$$

$$= 8903 \text{ lbf/in}^2$$

Since 8903 lbf/in^2 is less than half of the yield strength of $36{,}000 \text{ lbf/in}^2$, the Euler formula is valid. The allowable working stress is

$$\sigma_a = \frac{\sigma_e}{\text{FS}} = \frac{8903 \; \dfrac{\text{lbf}}{\text{in}^2}}{3} = 2968 \text{ lbf/in}^2$$

2. INTERMEDIATE COLUMNS

Columns with slenderness ratios less than the critical slenderness ratio but that are too long to be piers are known as *intermediate columns*. The *parabolic formula* (also known as the *J. B. Johnson formula*) is used to describe the parabolic line between points A and T on Fig. 51.1. The critical stress is given by Eq. 51.6, where a and b are curve-fit constants.

$$\sigma_{\text{cr}} = \frac{P_{\text{cr}}}{A} = a - b\left(\frac{CL}{k}\right)^2 \qquad \textbf{\textit{51.6}}$$

It is commonly assumed that the stress at point A is S_y and the stress at point T is $S_y/2$. In that case, the parabolic formula becomes

$$\sigma_{\text{cr}} = S_y - \left(\frac{1}{E}\right)\left(\frac{S_y}{2\pi}\right)^2\left(\frac{CL}{k}\right)^2 \qquad \textbf{\textit{51.7}}$$

3. ECCENTRICALLY LOADED COLUMNS

Accidental eccentricities are introduced during the course of normal manufacturing, so the load on real columns is rarely axial. The *secant formula* is one of the methods available for determining the critical column stress and critical load with eccentric loading.[1]

$$\sigma_{\max} = \sigma_{\text{ave}}(1 + \text{amplification factor})$$

$$= \left(\frac{F}{A}\right)\left[1 + \left(\frac{ec}{k^2}\right)\sec\left(\frac{\pi}{2}\sqrt{\frac{F}{F_e}}\right)\right]$$

$$= \left(\frac{F}{A}\right)\left[1 + \left(\frac{ec}{k^2}\right)\sec\left(\frac{L}{2k}\sqrt{\frac{F}{AE}}\right)\right]$$

$$= \left(\frac{F}{A}\right)\left[1 + \left(\frac{ec}{k^2}\right)\sec\phi\right] \qquad \textbf{\textit{51.8}}$$

$$\phi = \left(\tfrac{1}{2}\right)\left(\frac{L}{k}\right)\sqrt{\frac{F}{AE}} \qquad \textbf{\textit{51.9}}$$

For a given *eccentricity*, e, or eccentricity ratio, ec/k^2, and an assumed value of the buckling load, F, Eq. 51.9 is solved by trial and error for the slenderness ratio, L/k. Equations 51.8 and 51.9 converge quickly to the known L/k ratio when assumed values of F are substituted. (L/k is smaller when F is larger.)

4. THIN-WALLED CYLINDRICAL TANKS

In general, tanks under internal pressure experience circumferential, longitudinal, and radial stresses. If the wall thickness is small, the radial stress component is negligible and can be disregarded. A cylindrical tank is a *thin-walled tank* if its wall thickness-to-internal diameter ratio is less than approximately 0.1.[2]

$$\frac{t}{d_i} = \frac{t}{2r_i} < 0.1 \qquad [\text{thin-walled}] \qquad \textbf{\textit{51.10}}$$

The *hoop stress*, σ_h, also known as *circumferential stress* and *tangential stress*, for a cylindrical thin-walled tank under internal pressure is derived from the free-body diagram of a cylinder half.[3] Since the cylinder is assumed to be thin-walled, it is not important which radius (e.g., inner, mean, or outer) is used in Eq. 51.11. However, the inner radius is used by common convention.

$$\sigma_h = \frac{pr}{t} \qquad \textbf{\textit{51.11}}$$

---

[1]The design of timber, steel, and reinforced concrete building columns is very code-intensive. None of the theoretical methods presented in this section are acceptable for building design.

[2]There is some confusion about the thin-wall/thick-wall criterion. This is because a cylinder is essentially thin-wall when $t/2r < 0.1$, but Lamé's solution is for thick-walled cylinders with $t/r > 0.1$. The latter (thick-wall) criterion should not be used to predict the applicability of the former (thin-wall) case.

[3]There is no simple way, including the more exact Lamé solutions, of evaluating theoretical stresses in thin-walled cylinders under external pressure, since failure is by collapse, not yielding. However, empirical equations exist for predicting the *collapsing pressure*.

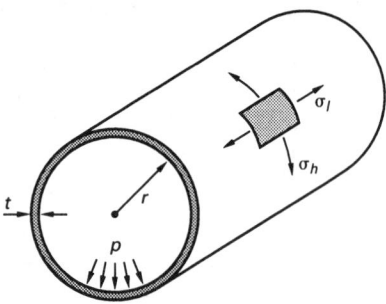

Figure 51.2 *Stresses in a Thin-Walled Tank*

The axial forces on the ends of the cylindrical tank produce a stress, known as the *longitudinal stress* or *long stress*, σ_l, directed along the tank's longitudinal axis.

$$\sigma_l = \frac{pr}{2t} \qquad 51.12$$

Unless the tank is subject to torsion, there is no shear stress. Accordingly, the hoop and long stresses are the principal stresses. They do not combine into larger stresses. Their combined effect should be evaluated according to the appropriate failure theory.

The increase in length due to pressurization is easily determined from the longitudinal strain.

$$\Delta L = L\epsilon_l$$
$$= L\left(\frac{\sigma_l - \nu\sigma_h}{E}\right) \qquad 51.13$$

The increase in circumference (from which the radial increase can also be determined) due to pressurization is

$$\Delta C = C\epsilon_h$$
$$= \pi d_o\left(\frac{\sigma_h - \nu\sigma_l}{E}\right) \qquad 51.14$$

Example 51.2

A thin-walled pressurized tank is supported at both ends, as shown. Points A and B are located midway between the supports and at the upper and lower surfaces, respectively. Evaluate the maximum stresses on the tank.

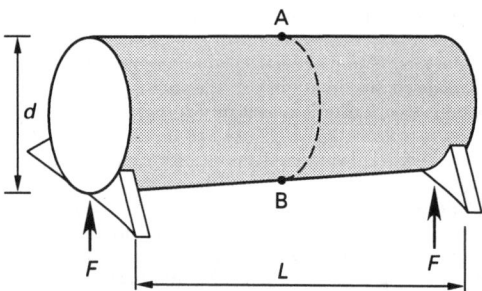

Solution

Since the tank is thin-walled, the radial stress is essentially zero. The hoop and longitudinal stresses are given by Eqs. 51.11 and 51.12, respectively. In addition, point A experiences a bending stress. The bending stress is

$$\sigma_b = \frac{Mc}{I}$$
$$c = \frac{d}{2}$$
$$M = \frac{FL}{2}$$
$$I = k^2 A = \left(\frac{d}{2}\right)^2 \pi dt$$

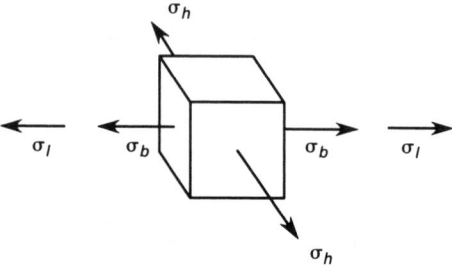

The bending stress is compressive at point A and tensile at point B. At point B, the bending stress has the same sign (tensile) as the longitudinal stress, and these two stresses add to each other. There is no torsional stress, so the resultant normal stresses are the principal stresses. These can be used directly with the failure theories (in particular, the distortion energy theory) presented in Chap. 50.

$$\sigma_1 = \sigma_h$$
$$\sigma_2 = \sigma_l + \sigma_b$$

5. THICK-WALLED CYLINDERS

A thick-walled cylinder has a wall thickness-to-diameter ratio greater than 0.1. Figure 51.3 illustrates a thick-walled tank under both internal and external pressures.

In thick-walled tanks, radial stress is significant and cannot be disregarded. In *Lamé's solution*, a thick-walled cylinder is assumed to be made up of thin laminar rings. This method shows that the radial and circumferential stresses vary with location within the tank wall. (The term *circumferential stress* is preferred over *hoop stress* when dealing with thick-walled cylinders.) Compressive stresses are negative.

$$\sigma_c = \frac{r_i^2 p_i - r_o^2 p_o + \dfrac{(p_i - p_o)\, r_i^2 r_o^2}{r^2}}{r_o^2 - r_i^2} \qquad 51.15$$

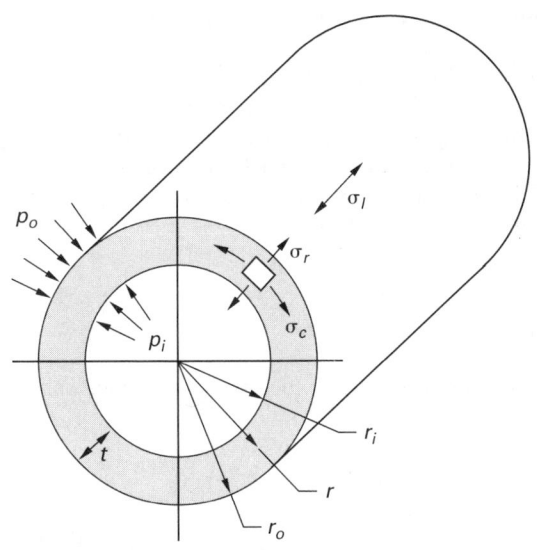

Figure 51.3 *Thick-Walled Cylinder*

The *diametral strain* (which is the same as the *circumferential* and *radial strains*) is given by Eq. 51.18. Radial stresses are always compressive (hence they are negative), and algebraic signs must be observed with Eq. 51.18. Since the circumferential and radial stresses depend on location within the wall thickness, the strain can be evaluated at inner, outer, and any intermediate locations within the wall.

$$\epsilon = \frac{\Delta d}{d} = \frac{\Delta C}{C} = \frac{\Delta r}{r}$$
$$= \frac{\sigma_c - \nu(\sigma_r + \sigma_l)}{E} \qquad 51.18$$

$$\sigma_r = \frac{r_i^2 p_i - r_o^2 p_o - \dfrac{(p_i - p_o)\, r_i^2 r_o^2}{r^2}}{r_o^2 - r_i^2} \qquad 51.16$$

$$\sigma_l = \frac{p_i r_i^2}{r_o^2 - r_i^2} \qquad 51.17$$

At every point in the cylinder, the circumferential, radial, and long stresses are the principal stresses. Unless an external torsional shear stress is added, it is not necessary to use the combined stress equations. Failure theories can be applied directly.

The cases of main interest are those of internal or external pressure only. The stress equations for these cases are summarized in Table 51.2. The maximum shear and normal stresses occur at the inner surface for both internal and external pressure.

6. THIN-WALLED SPHERICAL TANKS

There is no unique axis in a spherical tank or in the spherical ends of a cylindrical tank. Therefore, the hoop and long stresses are identical.

$$\sigma = \frac{pr}{2t} \qquad 51.19$$

7. INTERFERENCE FITS

When assembling two pieces, interference fitting is often more economical than pinning, keying, or splining. The assembly operation can be performed in a hydraulic press, either with both pieces at room temperature or after heating the outer piece and cooling the inner piece. The former case is known as a *press fit* or *interference fit*; the latter as a *shrink fit*.

If two cylinders are pressed together, the pressure acting between them will expand the outer cylinder (placing it into tension) and will compress the inner cylinder. The *interference*, I, is the difference in dimensions between the two cylinders. *Diametral interference* and *radial interference* are both used.[4]

$$I_{\text{diametral}} = 2I_{\text{radial}}$$
$$= d_{o,\text{inner}} - d_{i,\text{outer}}$$
$$= |\Delta d_{o,\text{inner}}| + |\Delta d_{i,\text{outer}}| \qquad 51.20$$

If the two cylinders have the same length, the thick-wall cylinder equations can be used. The materials used for the two cylinders do not need to be the same. Since

Table 51.2 *Stresses in Thick-Walled Cylinders*

stress	external pressure, p	internal pressure, p
$\sigma_{c,o}$	$\dfrac{-\left(r_o^2 + r_i^2\right) p}{r_o^2 - r_i^2}$	$\dfrac{2r_i^2 p}{r_o^2 - r_i^2}$
$\sigma_{r,o}$	$-p$	0
$\sigma_{c,i}$	$\dfrac{-2r_o^2 p}{r_o^2 - r_i^2}$	$\dfrac{\left(r_o^2 + r_i^2\right) p}{r_o^2 - r_i^2}$
$\sigma_{r,i}$	0	$-p$
$\tau_{\max}$	$\frac{1}{2}\sigma_{c,i}$	$\left(\frac{1}{2}\right)(\sigma_{c,i} + p)$

---

[4]Theoretically, the interference can be given to either the inner or outer cylinder, or it can be shared by both cylinders. However, in the case of a surface-hardened shaft with a standard diameter, the interference is usually all given to the disk. Otherwise, it may be necessary to machine the shaft and remove some of the hardened surface.

there is no longitudinal stress from an interference fit and since the radial stress is negative, the strain from Eq. 51.18 is

$$\epsilon = \frac{\Delta d}{d} = \frac{\Delta C}{C} = \frac{\Delta r}{r}$$
$$= \frac{\sigma_c + \nu\sigma_r}{E} \qquad \textbf{51.21}$$

Equation 51.22 applies to the general case where both cylinders are hollow and have different moduli of elasticity and Poisson's ratios. The outer cylinder is designated as the *hub*; the inner cylinder is designated as the *shaft*. If the shaft is solid, use $r_{i,\text{shaft}} = 0$ in Eq. 51.22.

$$I_{\text{diametral}} = 2I_{\text{radial}}$$
$$= \left(\frac{2pr_{o,\text{shaft}}}{E_{\text{hub}}}\right)\left(\frac{r_{o,\text{hub}}^2 + r_{o,\text{shaft}}^2}{r_{o,\text{hub}}^2 - r_{o,\text{shaft}}^2} + \nu_{\text{hub}}\right)$$
$$+ \left(\frac{2pr_{o,\text{shaft}}}{E_{\text{shaft}}}\right)\left(\frac{r_{o,\text{shaft}}^2 + r_{i,\text{shaft}}^2}{r_{o,\text{shaft}}^2 - r_{i,\text{shaft}}^2} - \nu_{\text{shaft}}\right)$$
$$\textbf{51.22}$$

In the special case where the shaft is solid and is made from the same material as the hub, the diametral interference is given by Eq. 51.23.

$$I_{\text{diametral}} = 2I_{\text{radial}}$$
$$= \left(\frac{4pr_{\text{shaft}}}{E}\right)\left[\frac{1}{1 - \left(\dfrac{r_{\text{shaft}}}{r_{o,\text{hub}}}\right)^2}\right] \qquad \textbf{51.23}$$

The maximum assembly force required to overcome friction during a press-fitting operation is given by Eq. 51.24. The coefficient of friction is highly variable. Values in the range of 0.03 to 0.33 have been reported. This relationship is approximate because the coefficient of friction is not known with certainty and the assembly force affects the pressure, p, through Poisson's ratio.

$$F_{\text{max}} = fN = 2\pi fpr_{\text{shaft}}L_{\text{interface}} \qquad \textbf{51.24}$$

The maximum torque that the press-fitted hub can withstand or transmit is given by Eq. 51.25. This can be greater or less than the shaft's torsional shear capacity. Both values should be calculated.

$$T_{\text{max}} = 2\pi fpr_{\text{shaft}}^2 L_{\text{interface}} \qquad \textbf{51.25}$$

Most interference fits are designed to keep the contact pressure or the stress below a given value. Designs of interference fits limited by strength generally use the distortion energy failure criterion. That is, the maximum shear stress is compared with the shear strength determined from the failure theory.

Example 51.3

A steel cylinder has inner and outer diameters of 1.0 in and 2.0 in (25 mm and 50 mm), respectively. The cylinder is pressurized internally to 10,000 lbf/in² (70 MPa). The modulus of elasticity is 2.9×10^7 lbf/in² (200 GPa), and Poisson's ratio is 0.3. What is the radial strain at the inside face?

Solution

The stresses at the inner face are found from Table 51.2.

$$\sigma_{c,i} = \frac{(r_o^2 + r_i^2)p}{r_o^2 - r_i^2}$$
$$= \frac{((1.0 \text{ in})^2 + (0.5 \text{ in})^2)\left(10,000 \dfrac{\text{lbf}}{\text{in}^2}\right)}{(1.0 \text{ in})^2 - (0.5 \text{ in})^2}$$
$$= 16,667 \text{ lbf/in}^2$$
$$\sigma_{r,i} = -p = -10,000 \text{ lbf/in}^2$$
$$\sigma_l = \frac{F}{A} = \frac{p\pi r_i^2}{\pi(r_o^2 - r_i^2)}$$
$$= \frac{\left(10,000 \dfrac{\text{lbf}}{\text{in}^2}\right)(0.5 \text{ in})^2}{(1.0 \text{ in})^2 - (0.5 \text{ in})^2}$$
$$= 3333 \text{ lbf/in}^2$$

The circumferential stress increases the radial strain; the radial and longitudinal stresses decrease the radial strain. The radial strain is

$$\frac{\Delta r}{r} = \frac{\sigma_{c,i} - \nu(\sigma_{r,i} + \sigma_l)}{E}$$
$$= \frac{16,667 \dfrac{\text{lbf}}{\text{in}^2} - (0.3)\left(-10,000 \dfrac{\text{lbf}}{\text{in}^2} + 3333 \dfrac{\text{lbf}}{\text{in}^2}\right)}{2.9 \times 10^7 \dfrac{\text{lbf}}{\text{in}^2}}$$
$$= 6.44 \times 10^{-4}$$

Example 51.4

A hollow aluminum cylinder is pressed over a hollow brass cylinder as shown. Both cylinders are 2 in long. The interference is 0.004 in. The average coefficient of friction during assembly is 0.25. (a) What is the maximum shear stress in the brass? (b) What initial disassembly force is required to separate the two cylinders?

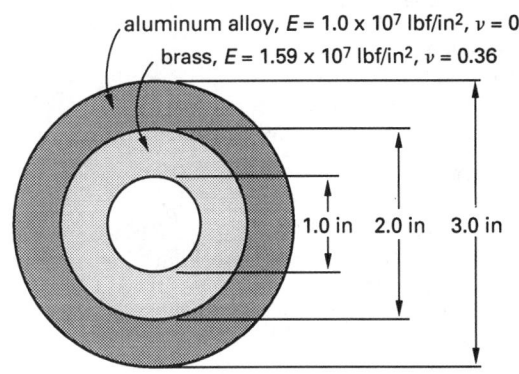

aluminum alloy, $E = 1.0 \times 10^7$ lbf/in^2, $\nu = 0.33$
brass, $E = 1.59 \times 10^7$ lbf/in^2, $\nu = 0.36$

1.0 in   2.0 in   3.0 in

Solution

(a) Work with the aluminum outer cylinder, which is under internal pressure.

$$\sigma_{c,i} = \frac{(r_o^2 + r_i^2)p}{r_o^2 - r_i^2}$$

$$= \frac{\left((1.5 \text{ in})^2 + (1.0 \text{ in})^2\right)p}{(1.5 \text{ in})^2 - (1.0 \text{ in})^2}$$

$$= 2.6p$$

$$\sigma_{r,i} = -p$$

From Eq. 51.18, the diametral strain is

$$\epsilon = \frac{\sigma_{c,i} - \nu(\sigma_{r,i} + \sigma_l)}{E}$$

$$= \frac{2.6p - (0.33)(-p)}{1.0 \times 10^7 \; \frac{\text{lbf}}{\text{in}^2}}$$

$$= 2.93 \times 10^{-7}p$$

$$\Delta d = \epsilon_d = (2.93 \times 10^{-7}p)(2.0 \text{ in})$$

$$= 5.86 \times 10^{-7}p$$

Now work with the brass inner cylinder, which is under external pressure.

$$\sigma_{c,o} = \frac{-(r_o^2 + r_i^2)p}{r_o^2 - r_i^2}$$

$$= \frac{-\left((1.0 \text{ in})^2 + (0.5 \text{ in})^2\right)p}{(1.0 \text{ in})^2 - (0.5 \text{ in})^2}$$

$$= -1.667p$$

$$\sigma_{r,o} = -p$$

From Eq. 51.18, the diametral strain is

$$\epsilon = \frac{\sigma_{c,o} - \nu(\sigma_{r,o} + \sigma_l)}{E}$$

$$= \frac{-1.667p - (0.36)(-p)}{1.59 \times 10^7 \; \frac{\text{lbf}}{\text{in}^2}}$$

$$= -0.822 \times 10^{-7}p$$

$$\Delta d = \epsilon_d = (-0.822 \times 10^{-7}p)(2.0 \text{ in})$$

$$= -1.644 \times 10^{-7}p$$

The diametral interference is known to be 0.004 in. From Eq. 51.20,

$$I_{\text{diametral}} = |\Delta d_{o,\text{inner}}| + |\Delta d_{i,\text{outer}}|$$

$$0.004 \text{ in} = 5.86 \times 10^{-7}p + |-1.644 \times 10^{-7}p|$$

$$p = 5330 \text{ lbf/in}^2$$

From Table 51.2, the circumferential stress at the inner face of the brass (under external pressure) is

$$\sigma_{c,i} = \frac{-2r_o^2 p}{r_o^2 - r_i^2}$$

$$= \frac{(-2)(1.0 \text{ in})^2 \left(5330 \; \frac{\text{lbf}}{\text{in}^2}\right)}{(1.0 \text{ in})^2 - (0.5 \text{ in})^2}$$

$$= -14{,}213 \text{ lbf/in}^2$$

Also from Table 51.2, the maximum shear stress is

$$\tau_{\text{max}} = (0.5)\sigma_{c,i} = (0.5)\left(-14{,}213 \; \frac{\text{lbf}}{\text{in}^2}\right)$$

$$= -7107 \text{ lbf/in}^2$$

(b) The initial force necessary to disassemble the two cylinders is the same as the maximum assembly force.

$$F_{\text{max}} = 2\pi f p r_{\text{shaft}} t_{\text{hub}}$$

$$= (2\pi)(0.25)\left(5330 \; \frac{\text{lbf}}{\text{in}^2}\right)(1 \text{ in})(2 \text{ in})$$

$$= 16{,}745 \text{ lbf}$$

8. STRESS CONCENTRATIONS FOR PRESS-FITTED SHAFTS IN FLEXURE

When a shaft carrying a press-fitted hub (whose thickness is less than the shaft length) is loaded in flexure, there will be an increase in shaft bending stress in the vicinity of the inner hub edge. The fatigue life of the shaft can be seriously affected by this stress increase. The extent of the increase depends on the magnitude of the bending stress and the contact pressure, and can be as high as 2.0 or more.

Some designs attempt to reduce the increase in shaft stress by grooving the disk (to allow the disk to flex). Other designs rely on various treatments to increase the fatigue strength of the shaft. For an unmodified simple press-fit, the stress concentration factor (to be applied to the bending stress calculated from Mc/I) is given by Fig. 51.4.

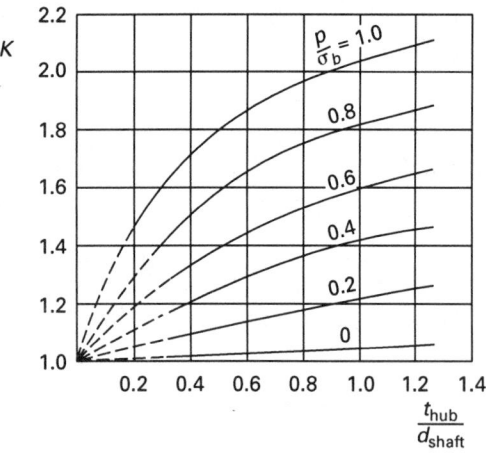

("Fatigue of Shafts at Fitted Members, with a Related Photoelastic Analysis," reproduced from *Transactions of the ASME*, Vol. 57, 1935, and Vol. 58, 1936, with permission of the American Society of Mechanical Engineers.)

Figure 51.4 *Stress Concentration Factor for Press Fit*

9. BOLTS

There are three leading specifications for bolt thread families: ANSI, ISO metric, and DIN metric.[5] ANSI (essentially identical to SAE, ASTM, ISO-inch standard) is widely used in the United States. DIN (Deutches Institute für Normung) fasteners are widely available and broadly accepted.[6] ISO metric (the International Organization for Standardization, which first met in 1961) fasteners are used in large volume by U.S. car manufacturers. The CEN (European Committee for Standardization) standards promulgated by the European Community (EC) have essentially adopted the ISO standards.

An American National (Unified) thread is specified by the sequence of parameters S(×L)-N-F-A-(H-E), where S is the thread outside diameter (nominal size), L is the optional shank length, N is the number of threads per inch, F is the thread pitch family, A is the class (allowance), and H and E are the optional hand and engagement length designations. The letter R can be added to the thread pitch family to indicate the thread roots are radiused (for better fatigue resistance). For example, a $^3/_8 \times$ 1-16UNC-2A bolt is $^3/_8$ in in diameter, 1 in in length, and has 16 Unified Coarse threads per inch rolled with a class 2A accuracy.[7] A UNRC bolt would be identical except for radiused roots. Table 51.3 lists some (but not all) values for these parameters.

---

[5]Other fastener families include the Italian UNI, Swiss VSM, Japanese JIS, and United Kingdom's BS series.
[6]To add to the confusion, many DIN standards are identical to ISO standards, with only slight differences in the tolerance ranges. However, the standards are not interchangeable in every case.
[7]Threads are generally rolled, not cut, into a bolt.

Table 51.3 *Representative American National (Unified) Bolt Thread Designations*[a]

S: Size
  1 through 12
  $^1/_4''$ through $^9/_{16}''$ in $^1/_{16}''$ increments
  $^5/_8''$ through $1^1/_2''$ in $^1/_8''$ increments
  $1^3/_4''$ through $4''$ in $^1/_4''$ increments

F: Thread Family
  UNC and NC—Unified Coarse[b]
  UNF and NF—Unified Fine[b]
  UNEF and NEF—Unified Extra Fine[c]
  8N—8 threads per inch
  12UN and 12N—12 threads per inch
  16UN and 16N—16 threads per inch
  UN, UNS, and NS—special series

A: Allowance (A—external threads, B—internal threads)[d]
  1A and 1B—liberal allowance for ease of assembly with dirty or damaged threads
  2A and 2B—normal production allowance (sufficient for plating)
  3A and 3B—close tolerance work with no allowance

H: Hand
  blank—right-hand thread
  LH—left-hand thread

[a]In addition to fastener thread families, there are other special-use threads such as Acme, stub, square, buttress, and worm series.
[b]Previously known as United States Standard or American Standard.
[c]The UNEF series is the same as the SAE (Society of Automotive Engineers) fine series.
[d]Allowance classes 2 and 3 (without the A and B designation) were used prior to industry transition to the Unified classes.

The *grade* of a bolt indicates the fastener material and is marked on the bolt cap.[8] In this regard, the marking depends on whether an SAE grade or ASTM designation is used. The minimum *proof load* (i.e., the maximum load the bolt can support without acquiring a permanent set) increases with the grade. The *proof strength* is the proof load divided by the tensile stress area. Table 51.4 lists how the caps of the bolts are marked to distinguish among the major grades.[9] If a bolt is manufactured in the United States, its cap must also show the logo or mark of the manufacturer.

A metric thread is specified by an M or MJ and a diameter and a pitch in millimeters, in that order. Thus, M10 × 1.5 is a thread having a nominal major diameter of 10 mm and a pitch of 1.5 mm. The MJ series have rounded root fillets and larger minor diameters.

Head markings on metric bolts indicate their *property class* and correspond to the approximate tensile strength in MPa/100. For example, a bolt marked 8.8 would correspond to a medium carbon, quenched and tempered bolt with an approximate tensile strength of 880 MPa. (The minimum of the tensile strength range for property class 8.8 is 830 MPa.)

---

[8]The *type* of a structural bolt should not be confused with the *grade* of a structural rivet.
[9]Optional markings can also be used.

Machine Design

Table 51.4 *Selected Bolt Grades and Designations*
  LC = low-carbon; MC = medium-carbon; Q&T = quenched and tempered; CD = cold-drawn
  (180°, 120)

standard	head marking[i]	material type	proof strength (ksi)		tensile strength (ksi)	
SAE grades						
grade 1	none	LC or MC	33		55	60
grade 2	none	LC or MC	55[a]	33[a]	74[a]	60[b]
grade 4	none	CD MC	–		115	
grade 5	3 tics, 360°	Q&T MC	85[c]	74[d]	120[c]	105[d]
grade 5.1	3 tics, 180°		85		120	
grade 5.2	3 tics, 120°	Q&T LC martensite	85		120	
grade 7	5 tics, 360°	Q&T MC alloy	105		133	
grade 8	6 tics, 360°	Q&T MC alloy	120		150	
grade 8.1	none		120		150	
grade 8.2	6 tics, 180°	Q&T LC martensite	120		150	
ISO designations						
class 4.6	none	LC or MC	220 MPa		400 MPa	
class 4.8			310 MPa		420 MPa	
class 5.8	none	LC or MC	380 MPa		520 MPa	
class 8.8	8.8 or 88	Q&T MC	600 MPa		830 MPa	
class 9.8	9.8		650 MPa		900 MPa	
class 10.9	10.9 or 109	Q&T alloy steel	830 MPa		1040 MPa	
class 12.9	12.9		970 MPa		1220 MPa	
ASTM designations						
A307 grades A, B	none	LC	–		60	
A325 type 1	3 tics, 360°, A325	Q&T MC	85[e]	74[f]	120[e]	105[f]
A325 type 2	3 tics, 120°, A325	Q&T LC martensite	85[e]	74[f]	120[e]	105[f]
A325 type 3	A325	Q&T weathering steel	85[e]	74[f]	120[e]	105[f]
A354 grade BC	BC	Q&T alloy steel	105[g]		125[g]	
A354 grade BB	BB	Q&T alloy steel	80[g]		105[g]	
A354 grade BD	6 tics, 360°	Q&T alloy steel	120[h]		150[h]	
A449	3 tics, 360°	Q&T MC	85[c]	74[d]	120[c]	105[d]
A490 type 1	A490	Q&T alloy steel	120		150–170	
A490 type 3	A490	Q&T weathering steel	–		–	

[a] $1/4$–$3/4$ in
[b] $3/4$–$1^1/_2$ in
[c] $1/4$–1 in
[d] 1–$1^1/_2$ in
[e] $1/2$–1 in
[f] $1^1/_8$ in–$1^1/_2$ in
[g] $1/4$–$2^1/_2$ in
[h] $1/4$–$1^1/_2$ in
[i] Tics are spread over the arc indicated.

Table 51.5 *Dimensions of American Unified Standard Threaded Bolts*[a]

nominal size	threads per inch	major diameter (in)	minor area (in^2)	tensile stress area (in^2)
coarse series				
1/4	20	0.2500	0.0269	0.0318
5/16	18	0.3125	0.0454	0.0524
3/8	16	0.3750	0.0678	0.0775
7/16	14	0.4375	0.0933	0.1063
1/2	13	0.5000	0.1257	0.1419
9/16	12	0.5625	0.162	0.182
5/8	11	0.6250	0.202	0.226
3/4	10	0.7500	0.302	0.334
7/8	9	0.8750	0.419	0.462
1	8	1.0000	0.551	0.606
fine series				
1/4	28	0.2500	0.0326	0.0364
5/16	24	0.3125	0.0524	0.0580
3/8	24	0.3750	0.0809	0.0878
7/16	20	0.4375	0.1090	0.1187
1/2	20	0.5000	0.1486	0.1599
9/16	18	0.5625	0.189	0.203
5/8	18	0.6250	0.240	0.256
3/4	16	0.7500	0.351	0.373
7/8	14	0.8750	0.480	0.509
1	12	1.0000	0.625	0.663

(Multiply in by 25.4 to obtain mm.)
(Multiply in^2 by 645 to obtain mm^2.)
[a]Based on ANSI B1.1-1974.

10. RIVET AND BOLT CONNECTIONS

Figure 51.5 illustrates a tension *lap joint* connection using rivet or bolt connectors.[10] Unless the plate material is very thick, the effects of eccentricity are disregarded. A connection of this type can fail in shear, tension, or bearing. A common design procedure is to determine the number of connectors based on shear stress and then to check the bearing and tensile stresses.

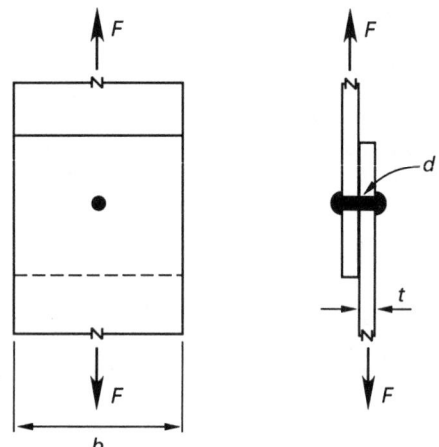

Figure 51.5 *Tension Lap Joint*

[10]Rivets are no longer used in building construction, but they are still extensively used in manufacturing.

One of the failure modes is shearing of the connectors. In the case of *single shear*, each connector supports its proportionate share of the load. In *double shear*, each connector has two shear planes, and the stress per connector is halved.[11] The shear stress in a cylindrical connector is

$$\tau = \frac{F}{\frac{\pi}{4}d^2} \qquad 51.26$$

The number of required connectors, as determined by shear, is

$$n = \frac{\tau}{\text{allowable shear stress}} \qquad 51.27$$

(a) single shear   (b) double shear

Figure 51.6 *Single and Double Shear*

The plate can fail in tension. If there are n connector holes of diameter d in a line across the width, b, of the plate, the cross-sectional area in the plate remaining to resist the tension is

$$A_t = t(b - nd) \qquad 51.28$$

The number of connectors across the plate width must be chosen to keep the tensile stress less than the allowable stress. The maximum tensile stress in the plate will be

$$\sigma_t = \frac{F}{A_t} \qquad 51.29$$

The plate can also fail by *bearing* (i.e., crushing). The number of connectors must be chosen to keep the actual *bearing stress* below the allowable bearing stress. For one connector, the bearing stress in the plate is

$$\sigma_p = \frac{F}{dt} \qquad 51.30$$

$$n = \frac{\sigma_p}{\text{allowable bearing stress}} \qquad 51.31$$

[11]"Double shear" is not the same as "double rivet" or "double butt." *Double shear* means that there are two shear planes in one rivet. *Double rivet* means that there are two rivets along the force path. *Double butt* refers to the use of two backing plates (i.e., "scabs") used on either side to make a tension connection between two plates. Similarly, *single butt* refers to the use of a single backing plate to make a tension connection between two plates.

The plate can also fail by shear tear-out, as illustrated in Fig. 51.7. The shear stress is

$$\tau = \frac{F}{2A} = \frac{F}{2t\left(L - \dfrac{d}{2}\right)} \qquad 51.32$$

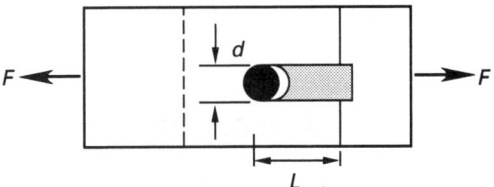

Figure 51.7 *Shear Tear-Out*

The *joint efficiency* is the ratio of the strength of the joint divided by the strength of a solid (i.e., unpunched or undrilled) plate.

11. BOLT PRELOAD

Consider the ungasketed connection shown in Fig. 51.8. The load varies from $F_{\min}$ to $F_{\max}$. If the bolt is initially snug but without initial tension, the force in the bolt also will vary from $F_{\min}$ to $F_{\max}$. If the bolt is tightened so that there is an initial *preload force*, F_i, greater than $F_{\max}$ in addition to the applied load, the bolt will be placed in tension and the parts held together will be in compression.[12] When a load is applied, the bolt tension will increase even more, but the compression in the parts will decrease.

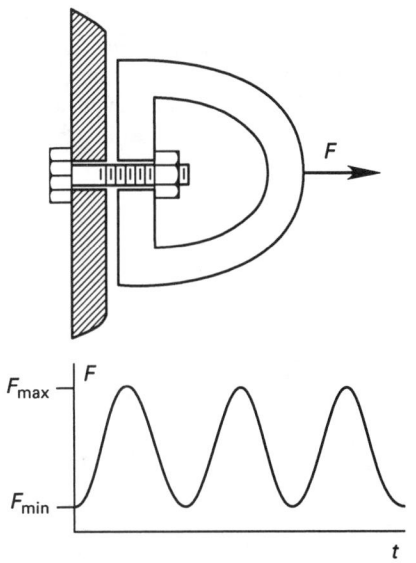

Figure 51.8 *Bolted Tension Joint with Varying Load*

---

[12]If the initial preload force is less than $F_{\max}$, the bolt will once again carry the entire load.

The amount of compression in the parts will vary as the applied load varies. Thus, the clamped members will carry some of the applied load, since this varying load has to "uncompress" the clamped part as well as lengthen the bolt. The net result is the reduction of the variation of the force in the bolt. The initial tension produces a larger mean stress, but the overall result is the reduction of the alternating stress. Thus, preloading is an effective method of reducing the alternating stress in bolted tension connection.

It is convenient to define the *spring constant*, k, of the bolt. The *grip*, L, is the thickness of the parts being connected by the bolt (not the bolt length). It is common to use the nominal diameter of the bolt, disregarding the reduction due to threading.

$$k_{\text{bolt}} = \frac{F}{\Delta L} = \frac{A_{\text{bolt}}E_{\text{bolt}}}{L_{\text{bolt}}} \qquad 51.33$$

The actual spring constant for a bolted part, k_{part}, is difficult to determine if the clamped area is not small and well defined. The only accurate way to determine the stiffness of a part in a bolted joint is through experimentation. If the clamped parts are flat plates, various theories can be used to laboriously calculate the effective load-bearing areas of the flanges.

One simple rule of thumb is that the bolt force spreads out to three times the bolt-hole diameter. Of course, the hole diameter needs to be considered (i.e., needs to be subtracted) in calculating the effective force area. If the modulus of elasticity is the same for the bolt and the clamped parts, using this rule of thumb, the larger area results in the parts being eight times more stiff than the bolts.

$$k_{\text{parts}} = \frac{A_{e,\text{parts}}E_{\text{parts}}}{L_{\text{parts}}} \qquad 51.34$$

If the clamped parts have different moduli of elasticities, including if a gasket constitutes one of the layers compressed by the bolt, the composite spring constant can be found from Eq. 51.35.[13]

$$\frac{1}{k_{\text{parts,composite}}} = \frac{1}{k_1} + \frac{1}{k_2} + \frac{1}{k_3} + \cdots \qquad 51.35$$

The bolt and the clamped parts all carry parts of the applied load.

$$F_{\text{bolt}} = F_i + \frac{k_{\text{bolt}}F_{\text{applied}}}{k_{\text{bolt}} + k_{\text{parts}}} \qquad 51.36$$

$$F_{\text{parts}} = \frac{k_{\text{parts}}F_{\text{applied}}}{k_{\text{bolt}} + k_{\text{parts}}} - F_i \qquad 51.37$$

---

[13]If a soft washer or gasket is used, its spring constant can control Eq. 51.35.

O-ring (metal and elastomeric) seals permit metal-to-metal contact and affect the effective spring constant of the parts very little. However, the seal force tends to separate bolted parts and must be added to the applied force. The seal force can be obtained from the seal deflection and seal stiffness or from manufacturer's literature.

For static loading, recommended amounts of preloading often are specified as a percentage of the *proof strength* (or *proof load* in psi).[14] For bolts, the proof load is slightly less than the yield strength. Traditionally, preload has been specified conservatively as 75% of proof for reusable connectors and 90% of proof for one-use connectors.[15] Connectors with some ductility can safely be used beyond the yield point, and 100% is now in widespread use.[16] When understood, advantages of preloading to 100% of proof load often outweigh the disadvantages.[17]

If the applied load varies, the forces in the bolt and parts will also vary. In that case, the preload must be determined from an analysis of the Goodman line.

Tightening of a tension bolt will induce a torsional stress in the bolt.[18] Where the bolt is to be locked in place, the torsional stress can be removed without greatly affecting the preload by slightly backing off the bolt. If the bolt is subject to cyclic loading, the bolt will probably slip back by itself, and it is reasonable to neglect the effects of torsion in the bolt altogether. (This is the reason that well-designed connections allow for a loss of 5 to 10% of the initial preload during routine use.)

Stress concentrations at the beginning of the threaded section are significant in cyclic loading.[19] To avoid a reduction in fatigue life, the alternating stress used in the Goodman line should be multiplied by an appropriate stress concentration factor, K. For fasteners with rolled threads, an average factor of 2.2 for SAE grades 0 to 2 (metric grades 3.6 to 5.8) is appropriate. For SAE grades 4 to 8 (metric grades 6.6 to 10.9), an average

factor of 3.0 is appropriate. Stress concentration factors for the fillet under the bolt head are different, but lower than these values. Stress concentration factors for cut threads are much higher.

The stress in a bolt depends on its load-carrying area. This area is typically obtained from a table of bolt properties. The effects of threads usually are ignored, so the area is based on the major (nominal) diameter.

$$\sigma_{\text{bolt}} = \frac{KF}{A} \qquad 51.38$$

12. BOLT TORQUE TO OBTAIN PRELOAD

During assembly, the preload tension is not monitored directly. Rather, the torque required to tighten the bolt is used to determine when the proper preload has been reached. Methods of obtaining the required preload include the standard torque wrench, the *run-of-the-nut method* (e.g., turning the bolt some specific angle past snugging torque), and computerized automatic assembly.

The standard manual torque wrench does not provide precise, reliable preloads, since the fraction of the torque going into bolt tension is variable.[20] Torque-, angle-, and time-monitoring equipment, usually part of an automated assembly operation, is essential to obtaining precise preloads on a consistent basis. It automatically applies the snugging torque and specified rotation, then checks the results with torque and rotation sensors. The computer warns of out-of-spec conditions.

The *Maney formula* is a simple relationship between the initial bolt tension, F_i, and the installation torque, T. The *torque coefficient*, K_T (also known as the *bolt torque factor* and the *nut factor*) used in Eq. 51.40 depends mainly on the coefficient of friction, f. The torque coefficient for lubricated bolts generally varies from 0.15 to 0.20, and a value of 0.2 is commonly used.[21] With antiseize lubrication, it can drop as low as 0.12. (The torque coefficient is not the same as the coefficient of friction.)

$$T = K_T d_{\text{bolt}} F_i \qquad 51.39$$

$$K_T = \frac{f_c r_c}{d_{\text{bolt}}} + \left(\frac{r_t}{d_{\text{bolt}}}\right)\left(\frac{\tan\theta + f_t \sec\alpha}{1 - f_t \tan\theta \sec\alpha}\right) \qquad 51.40$$

$$\tan\theta = \frac{\text{lead per revolution}}{2\pi r_t} \qquad 51.41$$

f_c is the coefficient of friction at the collar (fastener bearing face). r_c is the mean collar radius (i.e., the effective radius of action of the friction forces on the

---

[14]This is referred to as a "rule of thumb" specification, because a mathematical analysis is not performed to determine the best preload.

[15]Some U.S. military specifications call for 80% of proof load in tension fasteners and only 30% for shear fasteners. The object of keeping the stresses below yielding is to be able to reuse the bolts.

[16]Even under normal elastic loading of a bolt, local plastic deformation occurs in the bolt-head fillet and thread roots. Since the stress-strain curve is nearly flat at the yield point, a small amount of elongation into the plastic region does not increase the stress or tension in the bolt.

[17]The disadvantages are: (1) Field maintenance probably won't be possible, as manually running up bolts to 100% proof will result in many broken bolts. (2) Bolts should not be reused, as some will have yielded. (3) The highest-strength bolts do not exhibit much plastic elongation and ordinarily should not be run up to 100% proof load.

[18]An argument for the conservative 75% of proof load preload limit is that the residual torsional stress will increase the bolt stress to 90% or higher anyway, and the additional 10% needed to bring the preload up to 100% probably won't improve economic performance much.

[19]Stress concentrations are frequently neglected for static loading.

[20]Even with good lubrication, about 50% of the torque goes into overcoming friction between the head and collar/flange, another 40% is load in thread friction, and only the remaining 10% goes into tensioning the connector.

[21]With a coefficient of friction of 0.15, the torque coefficient is approximately 0.20 for most bolt sizes, regardless of whether the threads are coarse or fine.

bearing face). r_t is the effective radius of action of the frictional forces on the thread surfaces. Similarly, f_t is the coefficient of friction between the thread contact surfaces. θ is the *lead angle*, also known as the *helix thread angle*. α is the thread half angle (30° for UNF threads). $d_{t,m}$ is the mean thread diameter, and N is the number of threads per inch.

13. FILLET WELDS

The common *fillet weld* is shown in Fig. 51.9. Such welds are used to connect one plate to another. The applied load, F, is assumed to be carried in shear by the *effective weld throat*. The *effective throat size*, t_e, is related to the weld size, y, by Eq. 51.42.

$$t_e = 0.707y \qquad 51.42$$

Neglecting any increased stresses due to eccentricity, the shear stress in a fillet lap weld is

$$\tau = \frac{F}{bt_e} \qquad 51.43$$

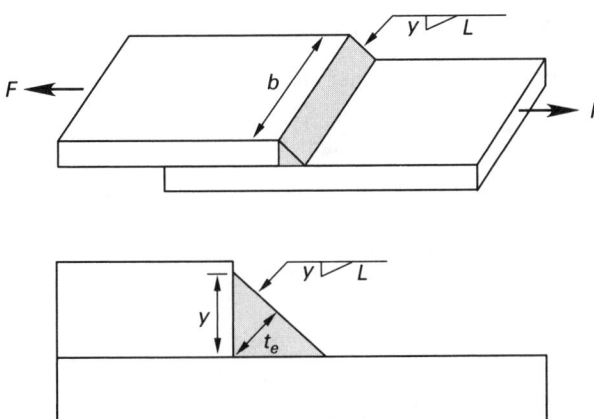

Figure 51.9 *Fillet Lap Weld and Symbol*

Weld (filler) metal should have a strength equal to or greater than the base material. Properties of filler metals are readily available from their manufacturers and, for standard rated welding rod, from engineering handbooks.

14. CIRCULAR SHAFT DESIGN

Shear stress occurs when a shaft is placed in torsion. The shear stress at the outer surface of a bar of radius r, which is torsionally loaded by a torque, T, is

$$\tau = G\phi = \frac{Tr}{J} \qquad 51.44$$

The total strain energy due to torsion is

$$U = \frac{T^2 L}{2GJ} \qquad 51.45$$

J is the shaft's polar moment of inertia. For a solid round shaft,

$$J = \frac{\pi r^4}{2} = \frac{\pi d^4}{32} \qquad 51.46$$

For a hollow round shaft,

$$J = \left(\frac{\pi}{2}\right)\left(r_o^4 - r_i^4\right) \qquad 51.47$$

If a shaft of length L carries a torque T, the angle of twist (in radians) will be

$$\phi = \frac{\tau}{G} = \frac{TL}{GJ} \qquad 51.48$$

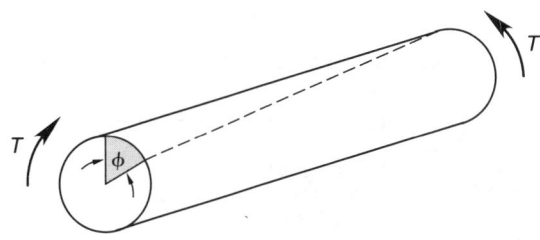

Figure 51.10 *Torsional Deflection of a Circular Shaft*

G is the *shear modulus*. For steel, it is approximately equal to 11.5×10^6 lbf/in² (8.0×10^4 MPa). The shear modulus also can be calculated from the modulus of elasticity.

$$G = \frac{E}{(2)(1+\nu)} \qquad 51.49$$

The torque, T, carried by a shaft spinning at n revolutions per minute is related to the transmitted horsepower.

$$T_{\text{in-lbf}} = \frac{(63{,}025)(\text{horsepower})}{n_{\text{rpm}}} \qquad 51.50$$

If a statically loaded shaft without axial loading experiences a bending stress, $\sigma_x = Mc/I$ (i.e., is loaded in flexure), in addition to torsional shear stress, $\tau = Tr/J$, the maximum shear stress from the combined stress theory is

$$\tau_{\max} = \sqrt{\left(\frac{\sigma_x}{2}\right)^2 + \tau^2} \qquad 51.51$$

$$\tau_{\max} = \frac{16}{\pi d^3}\sqrt{M^2 + T^2} \qquad 51.52$$

The equivalent normal stress from the distortion energy theory is

$$\sigma' = \frac{16}{\pi d^3}\sqrt{4M^2 + 3T^2} \qquad 51.53$$

The diameter can be determined by setting the shear and normal stresses equal to the maximum allowable shear (as calculated from the maximum shear stress theory, $S_y/2(\text{FS})$, or from the distortion energy theory, $\sqrt{3}S_y/2(\text{FS})$, and normal stresses, respectively).

Equations 51.51 and 51.52 should not be used with dynamically loaded shafts (i.e., those that are turning). Fatigue design of shafts should be designed according to a specific code (e.g., ANSI or ASME) or should use a fatigue analysis (e.g., Goodman, Soderberg, or Gerber).

Example 51.5

The press-fitted aluminum alloy-brass cylinder described in Ex. 51.4 is used as a shaft. The press fit is adequate to maintain nonslipping contact between the two materials. The shaft carries a steady torque of 24,000 in-lbf. There is no bending stress. What is the maximum torsional shear stress in the (a) aluminum and (b) brass?

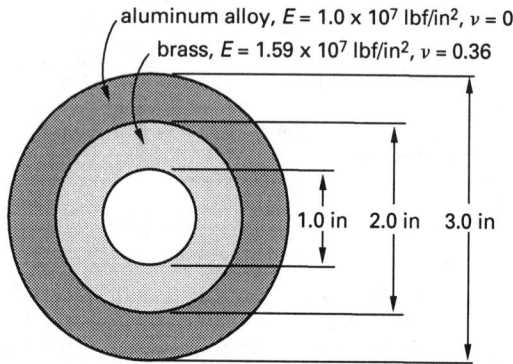

aluminum alloy, $E = 1.0 \times 10^7$ lbf/in², $\nu = 0.33$
brass, $E = 1.59 \times 10^7$ lbf/in², $\nu = 0.36$

1.0 in 2.0 in 3.0 in

Solution

The stronger material (as determined from the shear modulus, G) should be converted to an equivalent area of the weaker material.

For the aluminum,

$$G_{\text{aluminum}} = \frac{E}{(2)(1+\nu)}$$
$$= \frac{1.0 \times 10^7 \ \frac{\text{lbf}}{\text{in}^2}}{(2)(1+0.33)}$$
$$= 3.76 \times 10^6 \ \text{lbf/in}^2$$

For the brass,

$$G_{\text{brass}} = \frac{E}{(2)(1+\nu)}$$
$$= \frac{1.59 \times 10^7 \ \frac{\text{lbf}}{\text{in}^2}}{(2)(1+0.36)}$$
$$= 5.85 \times 10^6 \ \text{lbf/in}^2$$

The brass is the stronger material. The modular shear ratio is

$$n = \frac{G_{\text{brass}}}{G_{\text{aluminum}}}$$
$$= \frac{5.85 \times 10^6 \ \frac{\text{lbf}}{\text{in}^2}}{3.76 \times 10^6 \ \frac{\text{lbf}}{\text{in}^2}}$$
$$= 1.56$$

The polar moment of inertia of the aluminum is

$$J_{\text{aluminum}} = \left(\frac{\pi}{2}\right)(r_o^4 - r_i^4)$$
$$= \left(\frac{\pi}{2}\right)((1.5 \ \text{in})^4 - (1.0 \ \text{in})^4)$$
$$= 6.38 \ \text{in}^4$$

The equivalent polar moment of inertia of the brass is

$$J_{\text{brass}} = n\left(\frac{\pi}{2}\right)(r_o^4 - r_i^4)$$
$$= (1.56)\left(\frac{\pi}{2}\right)((1.0 \ \text{in})^4 - (0.5 \ \text{in})^4) \ x$$
$$= 2.30 \ \text{in}^4$$

The total equivalent polar moment of inertia is

$$J_{\text{total}} = J_{\text{aluminum}} + J_{\text{brass}}$$
$$= 6.38 \ \text{in}^4 + 2.30 \ \text{in}^4 = 8.68 \ \text{in}^4$$

(a) The maximum torsional shear stress in the aluminum occurs at the outer edge.

$$\tau = \frac{Tr}{J} = \frac{(24,000 \ \text{in-lbf})(1.5 \ \text{in})}{8.68 \ \text{in}^4}$$
$$= 4147 \ \text{lbf/in}^2$$

(b) The maximum torsional shear stress in the brass is

$$\tau = \frac{nTr}{J} = \frac{(1.56)(24,000 \ \text{in-lbf})(0.5 \ \text{in})}{8.68 \ \text{in}^4}$$
$$= 2157 \ \text{lbf/in}^2$$

15. TORSION IN THIN-WALLED, NONCIRCULAR SHELLS

Shear stress due to torsion in a thin-walled, noncircular shell (also known as a *closed box*) acts around the perimeter of the shell, as shown in Fig. 51.11. The shear stress, τ, is given by Eq. 51.54. A is the area enclosed by the centerline of the shell.

$$\tau = \frac{T}{2At} \hspace{3cm} 51.54$$

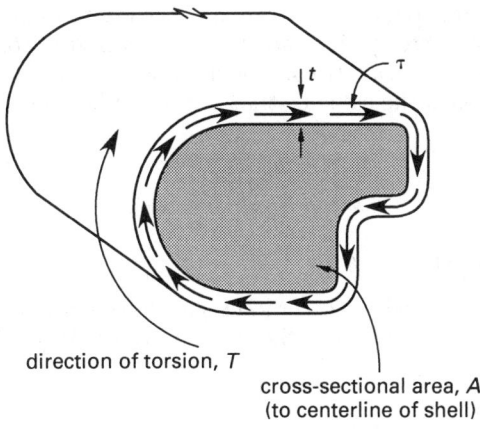

direction of torsion, T

cross-sectional area, A
(to centerline of shell)

Figure 51.11 *Torsion in Thin-Walled Shells*

The shear stress at any point is not proportional to the distance from the centroid of the cross section. Rather, the *shear flow*, q, around the shell is constant, regardless of whether the wall thickness is constant or variable.[22] The shear flow is the shear per-unit length of the centerline path.[23] At any point where the shell thickness is t,

$$q = \tau t = \frac{T}{2A} \quad \text{[constant]} \qquad 51.55$$

When the wall thickness, t, is constant, the angular twist depends on the perimeter, p, of the shell as measured along the centerline of the shell wall.

$$\phi = \frac{TLp}{4A^2 tG} \qquad 51.56$$

16. TORSION IN SOLID, NONCIRCULAR MEMBERS

When a noncircular solid member is placed in torsion, the shear stress is not proportional to the distance from the centroid of the cross section. The maximum shear usually occurs close to the point on the surface that is nearest the centroid.

Shear stress, τ, and angular deflection, ϕ, due to torsion are functions of the cross-sectional shape. They cannot be specified by simple formulas that apply to all sections. Table 51.6 lists the governing equations for several basic cross sections. These formulas have been derived by dividing the member into several concentric thin-walled closed shells and summing the torsional strength, T, provided by each shell.

---

[22]The concept of shear flow can also be applied to a regular beam in bending, although there is little to be gained by doing so. Removing the dimension b in the general beam shear stress equation, $q = VQ/I$.
[23]Shear flow is not analogous to magnetic flux or other similar quantities, because the shear flow path does not need to be complete (i.e., does not need to return to its starting point).

Table 51.6 *Torsion in Solid, Noncircular Shapes*

cross section	K in formula $\phi = TL/KG$	τ (max)
ellipse	$\dfrac{\pi a^3 b^3}{a^2 + b^2}$	$\dfrac{2T}{\pi ab^2}$ (at ends of minor axis)
square	$0.1406a^4$	$\dfrac{T}{0.208a^3}$ (at midpoint of each side)
rectangle	*	$\dfrac{T(3a + 1.8b)}{8a^2b^2}$ (at midpoint of each longer side)
equilateral triangle	$\dfrac{a^4\sqrt{3}}{80}$	$\dfrac{20T}{a^3}$ (at midpoint of each side)
slotted tube	$\dfrac{2\pi rt^3}{3}$	$\dfrac{T(6\pi r + 1.8t)}{4\pi^2 r^2 t^2}$ (along both edges remote from ends)

$$*ab^3 \left[\frac{16}{3} - \left(\frac{3.36b}{a}\right)\left(1 - \frac{b^4}{12a^4}\right)\right]$$

17. ECCENTRICALLY LOADED BOLTED CONNECTIONS

An eccentrically loaded connection is illustrated in Fig. 51.12. The bracket's natural tendency is to rotate about the centroid of the connector group. The shear stress in the connectors includes both the direct vertical shear and the torsional shear stress. The sum of these shear stresses is limited by the shear strength of the critical connector, which in turn determines the capacity of the connection.[24]

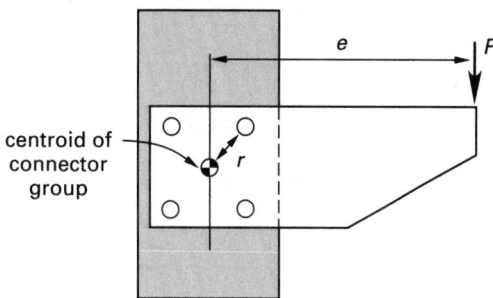

Figure 51.12 *Eccentrically Loaded Connection*

Analysis of an eccentric connection is similar to the analysis of a shaft under torsion. The shaft torque, T, is analogous to the moment, Fe, on the connection. The shaft's radius corresponds to the distance from the centroid of the fastener group to the *critical fastener*. The critical fastener is the one for which the vector sum of the vertical and torsional shear stresses is the greatest.

$$\tau = \frac{Tr}{J} = \frac{Fer}{J} \qquad 51.57$$

The polar moment of intertia, J, is calculated from the parallel axis theorem. Since bolts and rivets have little resistance to twisting in their holes, their individual polar moments of inertia are omitted.[25] Only r^2A terms in the parallel axis theorem are used. r_i is the distance from the fastener group centroid to the centroid (i.e., center) of the ith fastener, which has an area of A_i.

$$J = \sum_i r_i^2 A_i \qquad 51.58$$

The torsional shear stress is directed perpendicularly to a line between each fastener and the connector group centroid. The direction of the shear stress is the same as the rotation of the connection.

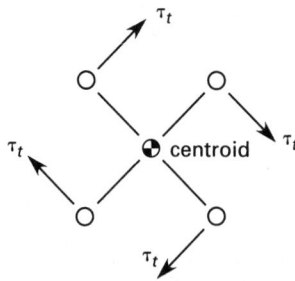

Figure 51.13 *Direction of Torsional Shear Stress*

---

[24]This type of analysis is known as an *elastic analysis* of the connection. Although it is traditional, it tends to understate the capacity of the connection.
[25]In spot-welded and welded stud connections, the torsional resistance of each connector can be considered.

Once the torsional shear stress has been determined in the critical fastener, it is added in a vector sum to the direct vertical shear stress. The direction of the vertical shear stress is the same as that of the applied force.

$$\tau_v = \frac{F}{nA} \qquad 51.59$$

Example 51.6

All fasteners used in the bracket shown have a nominal $1/2$ in diameter. What is the stress in the most critical fastener?

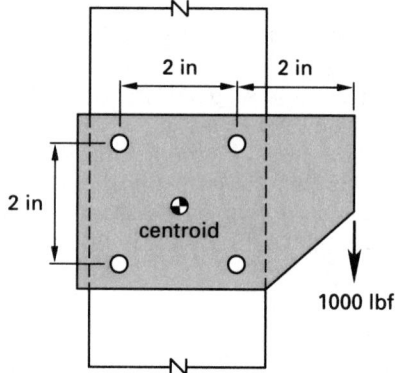

Solution

Since the fastener group is symmetrical, the centroid is centered within the four fasteners. This makes the eccentricity of the load equal to 3 in. Each fastener is located a distance r from the centroid, where

$$r = \sqrt{x^2 + y^2} = \sqrt{(1 \text{ in})^2 + (1 \text{ in})^2} = 1.414 \text{ in}$$

The area of each fastener is

$$A_i = \frac{\pi}{4}d^2 = \left(\frac{\pi}{4}\right)(0.5 \text{ in})^2 = 0.1963 \text{ in}^2$$

Using the parallel axis theorem for polar moments of inertia, and disregarding the individual torsional resistances of the fasteners,

$$J = \sum_i r_i^2 A_i = (4)\left[(1.414 \text{ in})^2(0.1963 \text{ in}^2)\right]$$

$$= 1.570 \text{ in}^4$$

The torsional shear stress on each fastener is

$$\tau_t = \frac{Fer}{J} = \frac{(1000 \text{ lbf})(3 \text{ in})(1.414 \text{ in})}{1.570 \text{ in}^4} = 2702 \text{ lbf/in}^2$$

Each torsional shear stress can be resolved into a horizontal shear stress, τ_{tx}, and a vertical shear stress, τ_{ty}. Both of these components are equal to

$$\tau_{tx} = \tau_{ty} = \frac{(\sqrt{2})\left(2702 \dfrac{\text{lbf}}{\text{in}^2}\right)}{2} = 1911 \text{ lbf/in}^2$$

The direct vertical shear downward is

$$\tau_v = \frac{F}{nA} = \frac{1000 \text{ lbf}}{(4)(0.1963 \text{ in}^2)}$$
$$= 1274 \text{ lbf/in}^2$$

The two right fasteners have vertical downward components of torsional shear stress. The direct vertical shear is also downward. These downward components add, making both right fasteners critical.

The total stress in each of these fasteners is

$$\tau = \sqrt{\tau_{tx}^2 + (\tau_{ty} + \tau_v)^2}$$
$$= \sqrt{\left(1911 \frac{\text{lbf}}{\text{in}^2}\right)^2 + \left(1911 \frac{\text{lbf}}{\text{in}^2} + 1274 \frac{\text{lbf}}{\text{in}^2}\right)^2}$$
$$= 3714 \text{ lbf/in}^2$$

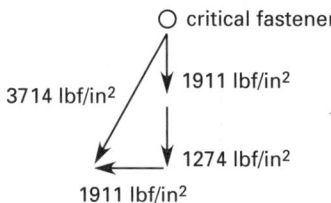

18. ECCENTRICALLY LOADED WELDED CONNECTIONS

The traditional elastic analysis of an eccentrically loaded welded connection is virtually the same as for a bolted connection, with the additional complication of having to determine the polar moment of inertia of the welds.[26] This can be done either by taking the welds as lines or by assuming each weld has an arbitrary thickness, t. After finding the centroid of the weld group, the rectangular moments of inertia of the individual welds are taken about that centroid using the parallel axis theorem. These rectangular moments of inertia are added to determine the polar moment of inertia. This laborious process can be shortened by use of App. 51.A.

The torsional shear stress, calculated from Mr/J (where r is the distance from the centroid of the weld group to the most distant weld point), is added to the direct shear to determine the maximum shear stress at the critical weld point.

[26]Steel building design does not use an elastic analysis to design eccentric brackets, either bolted or welded. The design methodology is highly proceduralized and codified.

19. FLAT PLATES

Flat plates under uniform pressure are separated into two edge-support conditions: simply supported and built-in edges.[27] Commonly accepted working equations are summarized in Table 51.7. It is assumed that (1) the plates are of "medium" thickness (meaning that the thickness is equal to or less than one-fourth of the minimum dimension of the plate), (2) the pressure is no more than will produce a maximum deflection less than or equal to one-half of the thickness, (3) the plates are constructed of isotropic, elastic material, and (4) the stress does not exceed the yield strength.

Example 51.7

A steel pipe with an inside diameter of 10 in (254 mm) is capped by welding round mild steel plates on its ends. The allowable stress is 11,100 lbf/in² (77 MPa). The internal gage pressure in the pipe is maintained at 500 lbf/in² (3.5 MPa). What plate thickness is required?

SI Solution

A fixed edge approximates the welded edges of the plate. From Table 51.7, the maximum bending stress is

$$\sigma_{max} = \frac{3pr^2}{4t^2}$$
$$t = \sqrt{\frac{3pr^2}{4\sigma_{max}}}$$
$$= \sqrt{\frac{(3)(3.5 \text{ MPa})\left(\frac{254 \text{ mm}}{2}\right)^2}{(4)(77 \text{ MPa})}}$$
$$= 23.4 \text{ mm}$$

Customary U.S. Solution

A fixed edge approximates the welded edges of the plate. From Table 51.7, the maximum bending stress is

$$\sigma_{max} = \frac{3pr^2}{4t^2}$$
$$t = \sqrt{\frac{3pr^2}{4\sigma_{max}}}$$
$$= \sqrt{\frac{(3)\left(500 \frac{\text{lbf}}{\text{in}^2}\right)\left(\frac{10 \text{ in}}{2}\right)^2}{(4)\left(11,100 \frac{\text{lbf}}{\text{in}^2}\right)}}$$
$$= 0.919 \text{ in}$$

[27]Fixed-edge conditions are theoretical and are seldom achieved in practice. Considering this fact and other simplifying assumptions that are made to justify the use of Table 51.7, a value of $\nu = 0.3$ can be used without loss of generality.

Table 51.7 *Flat Plates Under Uniform Pressure*

shape	edge condition	maximum stress	deflection at center
circular	simply supported	$\dfrac{\left(\dfrac{3}{8}\right)pr^2(3+\nu)}{t^2}$ (at center)	$\dfrac{\left(\dfrac{3}{16}\right)pr^4(1-\nu)(5+\nu)}{Et^3}$
	built-in	$\dfrac{\left(\dfrac{3}{4}\right)pr^2}{t^2}$ (at edge)	$\dfrac{\left(\dfrac{3}{16}\right)pr^4(1-\nu^2)}{Et^3}$
rectangular	simply supported	$\dfrac{C_1pb^2}{t^2}$ (at center)	$\dfrac{C_2pb^4}{Et^3}$
	built-in	$\dfrac{C_3pb^2}{t^2}$ (at centers of long edges)	$\dfrac{C_4pb^4}{Et^3}$

$\dfrac{a}{b}$	1.0	1.2	1.4	1.6	1.8	2	3	4	5	∞
C_1	0.287	0.376	0.453	0.517	0.569	0.610	0.713	0.741	0.748	0.750
C_2	0.044	0.062	0.077	0.091	0.102	0.111	0.134	0.140	0.142	0.142
C_3	0.308	0.383	0.436	0.487	0.497	0.500	0.500	0.500	0.500	0.500
C_4	0.0138	0.0188	0.023	0.025	0.027	0.028	0.028	0.028	0.028	0.028

PRACTICE PROBLEMS

Note: Unless instructed otherwise in a problem, use the following properties:

steel: $E = 30 \times 10^6$ lbf/in^2 $(20 \times 10^4$ MPa$)$
$G = 11.5 \times 10^6$ lbf/in^2 $(8.0 \times 10^4$ MPa$)$
$\alpha = 6.5 \times 10^{-6}$ 1/°F $(1.2 \times 10^{-5}$ 1/°C$)$
$\nu = 0.3$
aluminum: $E = 10 \times 10^6$ lbf/in^2 $(70 \times 10^3$ MPa$)$
copper: $E = 17.5 \times 10^6$ lbf/in^2 $(12 \times 10^4$ MPa$)$

1. The yield strength of a structural steel member is 36,000 lbf/in^2 (250 MPa). The tensile stress is 8240 lbf/in^2 (57 MPa). What is the factor of safety in tension?

2. A structural steel member 50 ft (15 m) long is used as a long column to support 75,000 lbf (330 kN). Both ends are built-in, and there are no intermediate supports. A factor of safety of 2.5 is used. What is the required moment of inertia?

3. A long steel column with a yield strength of 36,000 lbf/in^2 (250 MPa) has pinned ends. The column is 25 ft (7.5 m) long, has a cross-sectional area of 25.6 in^2 (165 cm^2), a centroidal moment of inertia of 350 in^4 (14,600 cm^2), and a distance from the neutral axis to the extreme fiber of 7 in (180 mm). It carries an axial concentric load of 100,000 lbf (440 kN) and an eccentric load of 150,000 lbf (660 kN) located 3.33 in (80 mm) from the longitudinal axial axis. Use the secant formula to determine the stress factor of safety.

4. A rectangular tank with dimensions of 2 ft by 2 ft by 1 ft (60 cm by 60 cm by 30 cm) is pressurized to 2 lbf/in^2 gage (14 kPa). The steel plate used to construct a tank has a thickness of 0.25 in (6.3 mm). Neglecting stress concentration factors at the corners and edges, what is the factor of safety?

5. A short section of pipe with an inside diameter of 1.750 in (44.5 mm) is to be produced by turning and boring bar stock. The steel has an allowable normal stress of 20,000 lbf/in^2 (140 MPa). The pipe will be pressurized internally to 2000 lbf/in^2 gage (14 MPa). Disregard longitudinal stress and other end effects. Use a thick-walled analysis to determine the required outside diameter.

6. A pressurized cylinder has an inside diameter of 0.742 in (18.8 mm) and an outside diameter of 1.486 in (37.7 mm). The cylinder is subjected to an external pressure of 400 lbf/in^2 gage (2.8 MPa). What is the maximum stress developed in the cylinder?

7. A shell with an outside diameter of 16 in (406 mm) and a wall thickness of 0.10 in (2.54 mm) is subjected to a 40,000 lbf/in² (280 MPa) tensile load and a 400,000 in-lbf torque (45 kN·m). What are the (a) principal stresses and (b) maximum shear stress?

8. (*Time limit: one hour*) A projectile launcher is formed by shrinking a jacket over a tube of the same length. The assembly is made with the maximum allowable diametral interference. The tube inside and outside diameters are 4.7 in and 7.75 in (119 mm and 197 mm), respectively. The outside diameter of the jacket is 12 in (300 mm). The jacket and tube are both steel with a Poisson's ratio of 0.3 and modulus of elasticity of 2.96×10^7 lbf/in² (204 GPa). The maximum allowable stress at the interface is 18,000 lbf/in² (124 MPa). (a) What is the diametral interference? (b) What is the maximum circumferential stress in the jacket? (c) What is the minimum circumferential stress in the tube?

9. (*Time limit: one hour*) Two hollow cylinders of the same length are press-fitted together with a diametral interference of 0.012 in (0.3 mm). The outer cylinder has a wall thickness of 0.079 in (2 mm) and an outside diameter of 4.7 in (120 mm). The inner cylinder has a wall thickness of 0.12 in (3 mm). Both cylinders are the same high-strength material with a Poisson's ratio of 0.3 and modulus of elasticity of 3×10^7 lbf/in² (207 GPa). What is the circumferential stress at the interface between the two cylinders?

10. A ³/₄-16 UNF steel bolt is used without washers to clamp two rigid steel plates, each 2 in (50 mm) thick. 0.75 in (19 mm) of the threaded section of the bolt remains under the nut. The nut has six threads. The nut is tightened until the stress in the bolt body is 40,000 lbf/in² (280 MPa). The bolt's modulus of elasticity is 2.9×10^7 lbf/in² (200 GPa). How much does the bolt stretch?

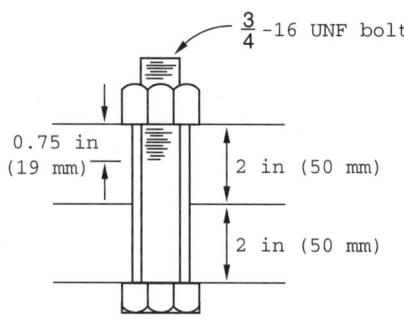

11. A class 30 gray cast-iron hub with an outside diameter of 12 in (300 mm) is pressed onto a soft steel solid shaft with an outside diameter of 6 in (150 mm). Poisson's ratio and modulus of elasticity for the cast iron are 0.27 and 1.45×10^7 lbf/in² (100 GPa), respectively. Poisson's ratio and modulus of elasticity for the steel are 0.30 and 2.9×10^7 lbf/in² (200 GPa), respectively. The stress-strain curve for gray cast iron becomes nonlinear at low stresses. What is the maximum radial interference such that the stress-strain relationship remains linear?

12. The bracket shown is attached to a column with three 0.75 in (19 mm) bolts arranged in an equilateral triangular layout. A force is applied with a moment arm of 20 in (500 mm) measured to the centroid of the bolt group. The maximum shear stress in the bolts is limited to 15,000 lbf/in² (100 MPa). What is the maximum force that the connection can support?

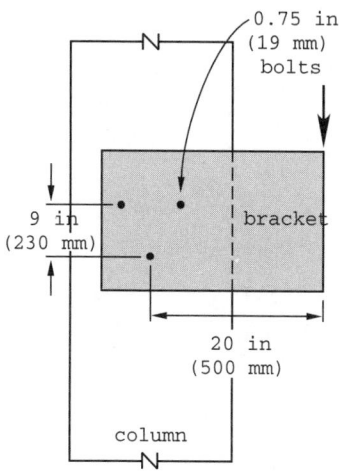

13. A fillet weld is used to secure a steel bracket to a column. The bracket supports a 10,000 lbf (44 kN) force applied 12 in (300 mm) from the face of the column, as shown. The maximum shear stress in the weld material is 8000 lbf/in² (55 MPa). What size fillet weld is required?

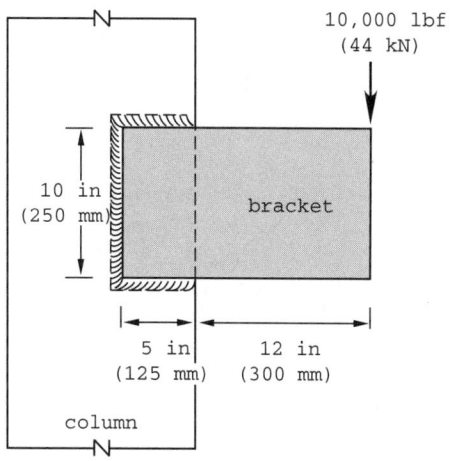

14. A 2 in (50 mm) diameter steel solid shaft rotates at 3500 rpm. A 16 in (406 mm) outside diameter, 2 in (50 mm) thick steel disk flywheel is press fitted to the shaft. The contact pressure between the shaft and flywheel is 1250 lbf/in² (8.6 MPa) when the shaft is turning. Centrifugal expansion of the shaft diameter is negligible. For both the shaft and the flywheel, Poisson's ratio is 0.3, modulus of elasticity is 2.9×10^7 lbf/in² (20 GPa), and mass density is 0.283 lbm/in³ (7830 kg/m³). What interference is required when the shaft is not turning?

52 Advanced Machine Design

1.	Springs	52-2
2.	Spring Materials	52-2
3.	Allowable Spring Stresses: Static Loading	52-3
4.	Safe Spring Stresses: Fatigue Loading	52-4
5.	Helical Compression Springs: Static Loading	52-4
6.	Helical Compression Springs: Design	52-6
7.	Buckling of Helical Compression Springs	52-7
8.	Helical Compression Springs: Dynamic Response	52-8
9.	Helical Springs: High Temperature	52-8
10.	Helical Extension Springs	52-8
11.	Helical Torsion Springs	52-9
12.	Flat and Leaf Springs	52-9
13.	Velocity, Power, and Torque in Rotating Members	52-10
14.	Spur Gear Terminology	52-10
15.	Tooth Thickness	52-11
16.	Gear Sets and Gear Drives	52-11
17.	Mesh Efficiency	52-12
18.	Force Analysis of Spur Gears	52-12
19.	Helical Gears	52-12
20.	Force Analysis of Helical Gears	52-12
21.	Allowable Stresses for Gear Design	52-13
22.	Spur Gear Strength: Lewis Beam Strength	52-13
23.	Design Guidelines for Gears	52-13
24.	AGMA Gear Design Method	52-14
25.	Limit Wear Design	52-14
26.	Bevel Gears	52-14
27.	Worm Gear Sets	52-14
28.	Flat Belt Drives	52-15
29.	V-Belts	52-16
30.	Brakes and Brake Materials	52-16
31.	Energy Dissipation in Brakes and Clutches	52-16
32.	Block Brakes	52-17
33.	Band Brakes	52-18
34.	Disk and Plate Clutches	52-18
35.	Cone Clutches	52-19
36.	Bushings	52-19
37.	Lubricants and Lubrication	52-19
38.	Journal Bearings	52-20
39.	Journal Bearing Frictional Losses	52-21
40.	Journal Bearings: Raimondi and Boyd Method	52-22
41.	Power Screws and Screw Jacks	52-22
	Practice Problems	52-23

Nomenclature

a	dynamic speed constant	ft/min	–
a	brake dimension	in	m
A	allowance	–	–
A	area	in^2	m^2
b	belt width	in	m
b	brake dimension	in	m
c	center-to-center distance	in	m
c	brake dimension	–	–
c	clearance	in	m
c	specific heat	Btu/lbm-°F	kJ/kg·°C
C	belt correction factor	–	–
C	spring index	–	–
d	diameter	in	mm
d	pitch diameter	in	mm
D	diameter	in	mm
e	eccentricity	in	mm
E	energy	in-lbf	J
E	modulus of elasticity	lbf/in^2	Pa
f	coefficient of friction	–	–
f	frequency	Hz	Hz
F	force	lbf	N
FS	factor of safety	–	–
g	acceleration of gravity	ft/sec^2	m/s^2
g_c	gravitational constant	$ft\text{-}lbm/lbf\text{-}sec^2$	–
G	shear modulus	lbf/in^2	Pa
h	film thickness	in	mm
I	rotational mass moment of inertia	$lbm\text{-}in^2$	$kg·m^2$
J	AGMA geometry factor	–	–
k	spring constant	lbf/in	N/m
k	gear speed factor	–	–
k	torsional spring constant	in-lbf/rev	N·m/rev
K	factor	–	–
L	lead	in	m
L	length	in	m
m	mass	lbm	kg
m	belt mass per unit length	lbm/in	kg/m
m	module	–	mm
M	moment	in-lbf	N·m
n	number of coils	–	–
n	rotational speed	rpm	rpm
N	normal force	lbf	N
N	number of teeth on gear	–	–
N	number	–	–
p	circular pitch	in	mm
p	pressure	lbf/in^2	Pa
p	spring pitch	in	m
P	diametral pitch	1/in	1/m
P	power	hp	kW
q	generated heat	Btu/min	kW
r	radius	in	m
R	radius	in	m
s	AGMA stress number	lbf/in^2	Pa

S	Sommerfeld number	–	–
S	strength	lbf/in^2	Pa
t	thickness	in	mm
t	time	sec	s
T	torque	in-lbf	N·m
U	energy	in-lbf	J
v	velocity	ft/sec	m/s
VR	velocity ratio	–	–
w	width	in	mm
W	Wahl correction factor	–	–
W	power jack load	lbf	N
Y	Lewis form factor	–	–

Symbols

α	cone angle	–	–
γ	pitch angle	deg	deg
Γ	pitch angle	deg	deg
δ	deflection	in	m
ϵ	eccentricity ratio	–	–
η	efficiency	–	–
θ	angle	deg	deg
μ	absolute viscosity	lbf-sec/in^2	Pa·s
ν	kinematic viscosity	SSU	m^2/s
ρ	density	lbm/in^3	kg/m^3
σ	normal stress	lbf/in^2	Pa
τ	shear stress	lbf/in^2	Pa
ϕ	angle	rad	rad
ϕ	pressure angle	deg	deg
ψ	helix angle	deg	deg
ω	rotational speed	rad/sec	rad/s

Subscripts

0	minimum
a	active, allowable, application, or axial
alt	alternating
b	bending
c	centrifugal or clash
d	diametral or dynamic
e	endurance
eq	equivalent
f	free or frictional
i	initial or inner
k	kinetic
L	life or load
m	mean, mechanical, or load distribution
n	normal
o	outer
p	pitch, at pitch circle, potential, or pulley
r	radial or at radius r
R	reliability
s	service, shear, size, or solid
t	tangential, tensile, or total
T	temperature
u	ultimate
v	velocity
w	working
y	yield

1. SPRINGS

An *ideal spring* is assumed to be perfectly elastic within its working range. The deflection is assumed to follow *Hooke's law*.[1] The *spring constant*, k, is also known as the *stiffness, spring rate, scale,* and *k-value*.[2]

$$F = k\delta \qquad 52.1$$

$$k = \frac{F_1 - F_2}{\delta_1 - \delta_2} \qquad 52.2$$

A spring stores energy when it is compressed or extended. By the *work-energy principle*, the energy storage is equal to the work to displace the spring. The potential energy of a spring whose ends have been displaced a total distance δ is

$$\Delta E_p = \frac{1}{2}k\delta^2 \qquad 52.3$$

If a mass, m, is dropped from height h onto a spring, the compression, δ, can be found by equating the change in potential energy to the energy storage.

$$mg(h + \delta) = \frac{1}{2}k\delta^2 \qquad \text{[SI]} \qquad 52.4(a)$$

$$m\left(\frac{g}{g_c}\right)(h + \delta) = \frac{1}{2}k\delta^2 \qquad \text{[U.S.]} \qquad 52.4(b)$$

Within the elastic region, this energy can be recovered by restoring the spring to its original unstressed condition. It is assumed that there is no permanent set, and no energy is lost through external friction or *hysteresis* (internal friction) when the spring returns to its original length.[3]

The entire applied load is felt by each spring in a series of springs linked end-to-end. The *equivalent (composite) spring constant* for springs in series is

$$\frac{1}{k_{\text{eq}}} = \frac{1}{k_1} + \frac{1}{k_2} + \frac{1}{k_3} + \cdots \qquad 52.5$$

Springs in parallel (e.g., concentric springs) share the applied load. The equivalent spring constant for springs in parallel is

$$k_{\text{eq}} = k_1 + k_2 + k_3 + \cdots \qquad 52.6$$

2. SPRING MATERIALS

A wide variety of materials are used for springs, including high-carbon steel, stainless steel and various alloys, nickel-based alloys (e.g., inconel), and copper-based

---

[1]A spring can be perfectly elastic even though it does not follow Hooke's law. The deviation from proportionality, if any, occurs at very high loads. The difference in theoretical and actual spring forces is known as the *straight-line error*.

[2]Another unfortunate name for the spring constant, k, that is occasionally encountered is the *spring index*. This is not the same as the spring index, C. The units will determine which meaning is intended.

[3]There is essentially no hysteresis in properly formed compression, extension, and open-wound helical torsion springs.

alloys (e.g., phosphor-bronze and silicon-bronze). "Super-alloys" are used for high-temperature and highly corrosive environments. Spring rate, fatigue strength, temperature range, corrosion resistance, magnetic properties, and cost are all considerations.

Springs manufactured from prehardened materials are generally stress-relieved in a low-temperature process by heating to between 400°F and 800°F (200°C and 430°C) after forming. Springs with intricate shapes must be manufactured from annealed materials and be subsequently strengthened in high-temperature processes. They are first quenched to full hardness and then tempered. Age-hardenable materials (e.g., beryllium copper) can be strengthened simply by heating after forming.

Most springs are cold-wound. Springs with wire diameters much in excess of $1/2$ or $5/8$ in (12 or 16 mm) are wound while red hot. Although the design methods are essentially the same for hot-wound and cold-wound springs, the allowable stresses are reduced approximately 20%, and the modulus is reduced slightly (approximately 9% for the shear modulus and approximately 5% for the elastic modulus). There are other unique issues and special needs associated with the manufacturing of hot-wound springs, as well.

Materials suitable for fatigue service include music wire (ASTM A228), carbon and alloy valve spring wire (ASTM A230), chrome-vanadium (ASTM A232), beryllium copper (ASTM B197), phosphor bronze (ASTM B159), and, to a lesser degree, type-302 stainless steel (ASTM A313). *Shotpeening (stresspeening)* is one of the best methods for increasing a spring's fatigue life.

3. ALLOWABLE SPRING STRESSES: STATIC LOADING

Helical compression and extension springs experience torsional shear stresses. The yield strength in shear is the theoretical maximum stress. There are three common ways of choosing the maximum allowable shear stress for static service.[4]

(a) Selecting the allowable stress based on some percentage of the ultimate tensile strength is the most common method. For ferrous materials except for austenitic stainless, the percentage is approximately 45 to 65%. For nonferrous and austenitic stainless, the percentage is approximately 35 to 55%. The lower limit should be used for unconstrained designs (i.e., the spring diameter can be as large as necessary to keep the stresses low). The higher limit is used when space is limited and higher stresses are unavoidable.[5]

---

[4]These methods apply to helical compression and extension springs. Recommended percentages are different for other types of springs.

[5]For highly precise springs with minimum hysteresis, creep, and drift, the percentages quoted in this section should be reduced.

Table 52.1 *Properties of Typical Spring Materials (room temperature[a])*

material	modulus of elasticity (lbf/in^2)	shear modulus (lbf/in^2)	density (lbm/in^3)
high-carbon wire			
music wire ASTM A228	30×10^6	11.5×10^6	0.284
hard-drawn ASTM A227	30×10^6	11.5×10^6	0.284
oil-tempered ASTM A229	30×10^6	11.5×10^6	0.284
valve spring ASTM A230	30×10^6	11.5×10^6	0.284
alloy-steel wire			
chrome-vanadium SAE 6150, AISI 6150, ASTM A232	30×10^6	11.5×10^6	0.284
chrome-silicon AISI 9254, ASTM A401	30×10^6	11.5×10^6	0.284
silicon manganese AISI 9260	30×10^6	11.5×10^6	0.284
stainless steel wire			
AISI 302, ASTM A313	28×10^6	10.0×10^6	0.280
AISI 410, 420	28×10^6	11.0×10^6	0.286
17-7 PH[b]	29.5×10^6	11.0×10^6	0.286
18-2	29×10^6	9.8×10^6	0.272
nickel-chrome A286	29×10^6	10.4×10^6	0.290
copper alloys			
phosphor bronze ASTM B159	15×10^6	6.3×10^6	0.320
silicon bronze ASTM B99(A)	15×10^6	5.6×10^6	0.308
silicon bronze ASTM B99(A)	17×10^6	6.4×10^6	0.316
beryllium-copper ASTM B197	18.5×10^6	7.0×10^6	0.297
nickel alloys			
inconel 600	31×10^6	11.0×10^6	0.307
inconel X750	31.5×10^6	11.5×10^6	0.298
Ni Span C902	27×10^6	9.7×10^6	0.294

(Multiply lbf/in^2 by 6.89×10^{-6} to obtain GPa.)
(Multiply lbm/in^3 by 27.7×10^3 to obtain kg/m^3.)
Adapted from *Spring Design Handbook*, Barnes Group (Associated Spring Corporation), Bristol, CT.
[a]Properties vary with temperature. Compiled from various sources.
[b]PH—precipitation hardened.

(b) Selecting the allowable stress based on the yield strength in shear is probably the most theoretically rigorous method. The yield strength in shear can be calculated from the tensile yield strength using either the maximum shear stress or the distortion energy theory.[6] If called for, a factor of safety of approximately 1.5 is appropriate for ferrous springs.

---

[6]The tensile yield strength can be estimated for ferrous spring materials as 75% of the ultimate tensile strength.

Table 52.2 Approximate Minimum Ultimate Tensile Strengths[a] of Typical Spring Materials
(ksi, at room temperature[b])

					wire size					
	in:	0.02	0.04	0.06	0.08	0.10	0.15	0.20	0.30	0.40
material	mm:	0.5	1.0	1.5	2.0	2.5	3.8	5.1	7.6	10
high-carbon wire										
music wire ASTM A228		350	315	296	282	271	253	240		
hard-drawn ASTM A227		285	255	235	225	215	205	190	175	16
oil-tempered ASTM A229		295	265	250	235	230	215	195	180	17
valve spring ASTM A230						210	205	195		
alloy-steel wire										
chrome-vanadium SAE 6150,										
AISI 6150, ASTM A232			280	265	260	245	230	220	205	
stainless steel wire										
AISI 302, ASTM A313		300	275	260	245	235	215	190	160	
17-7 PH		275	255	242	235	223	211			
18-2		296	270	265	255	253				
copper alloys										
phosphor bronze										
ASTM B159		150	140	138	135	132	130	128	122	

(Multiply in by 25.4 to obtain mm.)
(Multiply ksi by 6.9 to obtain MPa.)
[a] ASTM specifications provide a range of ultimate strengths. Table values approximately correspond to the minimum of the range.
[b] Properties vary with temperature.
Adapted from *Design Handbook*, 1987 Edition, Barnes Group (Associated Spring), Bristol, CT.

(c) Some specifications limit the torsional shear stress to a percentage of the tensile yield strength.

Some springs (e.g., flat leaf springs and helical torsion springs) experience a bending stress. Such springs are limited by the tensile yield strength of the spring material.

4. SAFE SPRING STRESSES: FATIGUE LOADING

Two methods can be used to design or analyze springs that are repeatedly stressed. The more rigorous method is to use a modified Goodman diagram. (The Wahl factor is applied to both the mean and alternating stress.) A simpler method is to design for static loading using a reduced maximum shear stress. An approximation of the endurance strength in shear can be calculated from the ultimate tensile strength and a factor from Table 52.3. Since the table values are conservative, the lives of most springs designed to it will exceed the numbers of cycles listed. However, a factor of safety may still be used or required.

$$\tau_{\max} = \frac{S_e}{\text{FS}} = \frac{(\text{factor})(S_{ut})}{\text{FS}} \qquad 52.7$$

Table 52.3 Allowable Torsional Stress in Nonreversed Fatigue[a,b]
(Fraction of Ultimate Tensile Strength)

	shear stress		bending stress	
fatigue life (cycles)	not shot-peened spring	shot-peened spring	not shot-peened spring	shot-peened spring
10,000	0.45[c]	0.45	0.80	0.80
100,000	0.35	0.42	0.53	0.62
1,000,000	0.33	0.40	0.50	0.60
10,000,000	0.30	0.36	0.48	0.58

[a] Round wire, room temperature operation, noncorrosive environment, without surging.
[b] For completely reversed service (i.e., when the average stress is zero), use 65% of the values in this table.
[c] 35% for phosphor bronze and type-302 stainless steel.
Source: *Design Handbook*, 1987 Edition, Barnes Group (Associated Spring), Bristol, CT.

5. HELICAL COMPRESSION SPRINGS: STATIC LOADING

The *spring index*, C, is the ratio of the mean coil diameter to the wire diameter.[7] It is difficult to wind springs

---

[7] This section is only for helical compression springs manufactured from round wire. Springs can also be manufactured from wire with a square or rectangular cross section. The design equations are different in that case.

with small (i.e., less than 4) spring indexes, and the operating stresses will be high. The wire cannot be easily bent to the desired small radius. On the other hand, springs with large indexes (i.e., greater than 12) are flimsy and tend to buckle. Most springs have indexes between 8 and 10, although 5 is a typical value for clutch springs.

$$C = \frac{D}{d} \qquad \text{52.8}$$

$$D = \frac{D_i + D_o}{2} \qquad \text{52.9}$$

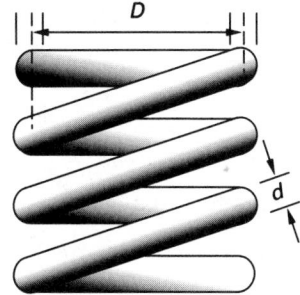

Figure 52.1 *Helical Compression Spring*

The spring deflection can be written in terms of the mean coil diameter or the spring index.

$$\delta = \frac{F}{k} = \frac{8FD^3 n_a}{Gd^4} = \frac{8FC^3 n_a}{Gd} \qquad \text{52.10}$$

The number of *active coils* (i.e., "turns") in a helical spring is less than or equal to the total number of coils, depending on the method of finishing the ends. When a helical compression spring has *plain ends* (i.e., neither squared nor ground), all of the coils contribute to spring force. In that case, the total number of coils, n_t, is equal to the number of active coils, n_a. However, most designs call for squaring and/or grinding the ends in order to obtain better seating for the spring. The number of end coils, n^*, is given in Table 52.4 for various end treatments. Normally, there should be at least two active coils.

$$n_a = n_t - n^* \qquad \text{52.11}$$

The *spring constant* is given by the *load-deflection equation*, also called the *spring rate equation*.

$$k = \frac{F}{\delta} = \frac{Gd^4}{8D^3 n_a} = \frac{Gd}{8C^3 n_a} \qquad \text{52.12}$$

Helical spring stress equations are derived from a superposition of torsional and direct shear stresses.[8] The *Wahl correction factor*, W, corrects for the curvature.[9] It is not a true stress concentration factor, but rather a factor that corrects the average stress to the maximum

---

[8]The *direct shear stress* is simply the force supported by the spring divided by the wire cross-sectional area, $\pi d^2/4$.
[9]Equation 52.13 is approximate, but the error is less than 2%.

stress. The maximum shear stress occurs at the inner face of the wire coil where the torsional and direct shear stresses are additive.

$$W = \frac{4C - 1}{4C - 4} + \frac{0.615}{C} \qquad \text{52.13}$$

$$\tau_{\text{max}} = \frac{8FCW}{\pi d^2} = \frac{8FDW}{\pi d^3} \qquad \text{52.14}$$

The end treatment shown in Fig. 52.2 affects a spring's solid length and pitch. The exact effects are not always obvious. Table 52.4 contains relationships that have become accepted in recent years for design use.

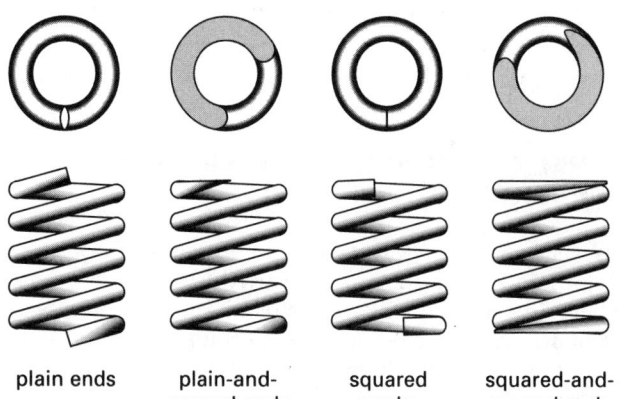

plain ends   plain-and-   squared   squared-and-
           ground ends   ends   ground ends

Figure 52.2 *End Treatment of Helical Compression Springs*

The solid deflection is calculated from the free and solid heights.

$$\delta_s = h_f - h_s \qquad \text{52.15}$$

The *spring pitch (coil pitch)*, p, is the mean coil separation. The *solid height (compressed height)* is given in Table 52.4. The *clash allowance*, A_c, is the percentage difference in solid and working deflections. It should be approximately 20%.

$$A_c = \frac{\delta_s - \delta_w}{\delta_w} \qquad \text{52.16}$$

The total spring *wire length* is the length of wire in the coils. The *active wire length* is calculated from the number of active coils.

$$L_a = \pi D n_a \qquad \text{52.17}$$

The *helix direction* for single springs can be either right hand or left hand. If the spring works over a threaded member, the winding direction should be opposite of the thread direction. With two *nested springs* (i.e., one spring inside the other), the winding directions must be opposite to prevent intermeshing. Also for nested springs, the outer spring should support approximately $2/3$ of the total load. The solid and free heights of both springs should be approximately the same.

Table 52.4 *Effect of End Treatment on Helical Compression Springs*

| | | end treatment | | |
| | | | | |
variable	plain	plain and ground	squared only	squared and ground
end coils, n^*	0	1	2	2
total coils, n_t	n_a	$n_a + 1$	$n_a + 2$	$n_a + 2$
free height, h_f	$pn_a + d$	$p(n_a + 1)$	$pn_a + 3d$	$pn_a + 2d$
solid height^a, h_s	$d(n_t + 1)$	dn_t	$d(n_t + 1)$	dn_t
pitch, p	$\dfrac{h_f - d}{n_a}$	$\dfrac{h_f}{n_a + 1}$	$\dfrac{h_f - 3d}{n_a}$	$\dfrac{h_f - 2d}{n_a}$

^{a}Use the effective wire diameter when calculating the solid height. This includes the nominal wire diameter, the wire tolerance, and the paint or plating thickness.

6. HELICAL COMPRESSION SPRINGS: DESIGN

Conventional spring design is an iterative procedure. One or more parameters are varied until the requirements are satisfied. Often one or more parameters are unknown and must be assumed to complete the design. For example, when the wire diameter is unknown, the allowable stress and Wahl factor can both only be estimated. (For the initial iteration, it is common to assume a Wahl factor of 1.1.) The outside diameter of the spring may also be a limited factor, as when the spring must fit in a hole.

Using the standard commercial wire size (or sheet metal gauge for flat springs) from App. 52.A will result in the most economical design.[10] However, this usually cannot be done without slightly altering one of the other parameters (coil diameter, spring rate, etc.).

Example 52.1

A steel spring is manufactured from W & M No. 4 wire with a spring index of 6. The load on the spring fluctuates between 140.2 lbf and 219.8 lbf (620 N and 970 N). The spring material has a yield strength in shear of 120,000 lbf/in^2 (830 MPa) and an endurance strength in shear of 100,000 lbf/in^2 (690 MPa). A factor of safety of 1.5 is required. Is the spring satisfactory?

SI Solution

From App. 52.A, the wire diameter is 0.2253 in.

$$d = (0.2253 \text{ in}) \left(25.4 \, \frac{\text{mm}}{\text{in}} \right) = 5.723 \text{ mm}$$

The Wahl correction factor is

$$
\begin{aligned}
W &= \frac{4C - 1}{4C - 4} + \frac{0.615}{C} \\
&= \frac{(4)(6) - 1}{(4)(6) - 4} + \frac{0.615}{6} \\
&= 1.2525
\end{aligned}
$$

From Eq. 52.14, the maximum and minimum stresses are

$$
\begin{aligned}
\tau_{\max} &= \frac{8 F_{\max} C W}{\pi d^2} \\
&= \frac{(8)(970 \text{ N})(6)(1.2525)}{\pi \left(\dfrac{5.723 \text{ mm}}{1000 \, \frac{\text{mm}}{\text{m}}} \right)^2} \\
&= 566.75 \text{ MPa}
\end{aligned}
$$

$$
\begin{aligned}
\tau_{\min} &= \frac{8 F_{\min} C W}{\pi d^2} \\
&= \frac{(8)(620 \text{ N})(6)(1.2525)}{\pi \left(\dfrac{5.723 \text{ mm}}{1000 \, \frac{\text{mm}}{\text{m}}} \right)^2} \\
&= 362.25 \text{ MPa}
\end{aligned}
$$

The mean and alternating stresses are

$$
\begin{aligned}
\tau_m &= \left(\tfrac{1}{2} \right) (\tau_{\max} + \tau_{\min}) \\
&= \left(\tfrac{1}{2} \right) (566.75 \text{ MPa} + 362.25 \text{ MPa}) \\
&= 464.5 \text{ MPa} \\
\tau_a &= \left(\tfrac{1}{2} \right) (\tau_{\max} - \tau_{\min}) \\
&= \left(\tfrac{1}{2} \right) (566.75 \text{ MPa} - 362.25 \text{ MPa}) \\
&= 102.25 \text{ MPa}
\end{aligned}
$$

---

[10]To avoid misunderstanding, it is generally better to call out wire (flat stock) by diameter (thickness) rather than gauge number.

The allowable region is defined by the material properties reduced by the factor of safety.

$$\frac{S_{es}}{\text{FS}} = \frac{690 \text{ MPa}}{1.5} = 460 \text{ MPa}$$

$$\frac{S_{ys}}{\text{FS}} = \frac{830 \text{ MPa}}{1.5} = 553.3 \text{ MPa}$$

The point corresponding to the actual stresses is just outside the allowable region.

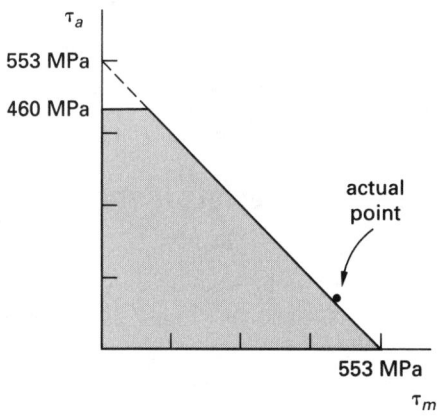

Customary U.S. Solution

From App. 52.A, the wire diameter is 0.2253 in. The Wahl correction factor is

$$
\begin{aligned}
W &= \frac{4C-1}{4C-4} + \frac{0.615}{C} \\
&= \frac{(4)(6)-1}{(4)(6)-4} + \frac{0.615}{6} \\
&= 1.2525
\end{aligned}
$$

From Eq. 52.14, the maximum and minimum stresses are

$$
\begin{aligned}
\tau_{\max} &= \frac{8F_{\max}CW}{\pi d^2} \\
&= \frac{(8)(219.8 \text{ lbf})(6)(1.2525)}{\pi(0.2253 \text{ in})^2} \\
&= 82,866 \text{ lbf/in}^2 \\
\tau_{\min} &= \frac{8F_{\min}CW}{\pi d^2} \\
&= \frac{(8)(140.2 \text{ lbf})(6)(1.2525)}{\pi(0.2253 \text{ in})^2} \\
&= 52,856 \text{ lbf/in}^2
\end{aligned}
$$

The mean and alternating stresses are

$$
\begin{aligned}
\tau_m &= \left(\tfrac{1}{2}\right)\left(\tau_{\max} + \tau_{\min}\right) \\
&= \left(\tfrac{1}{2}\right)\left(82,866 \, \frac{\text{lbf}}{\text{in}^2} + 52,856 \, \frac{\text{lbf}}{\text{in}^2}\right) \\
&= 67,861 \text{ lbf/in}^2 \\
\tau_a &= \left(\tfrac{1}{2}\right)\left(\tau_{\max} - \tau_{\min}\right) \\
&= \left(\tfrac{1}{2}\right)\left(82,866 \, \frac{\text{lbf}}{\text{in}^2} - 52,856 \, \frac{\text{lbf}}{\text{in}^2}\right) \\
&= 15,005 \text{ lbf/in}^2
\end{aligned}
$$

The allowable region is defined by the material properties reduced by the factor of safety.

$$\frac{S_{es}}{\text{FS}} = \frac{100,000 \, \dfrac{\text{lbf}}{\text{in}^2}}{1.5} = 66,667 \text{ lbf/in}^2$$

$$\frac{S_{ys}}{\text{FS}} = \frac{120,000 \, \dfrac{\text{lbf}}{\text{in}^2}}{1.5} = 80,000 \text{ lbf/in}^2$$

The point corresponding to the actual stresses is just outside the allowable region.

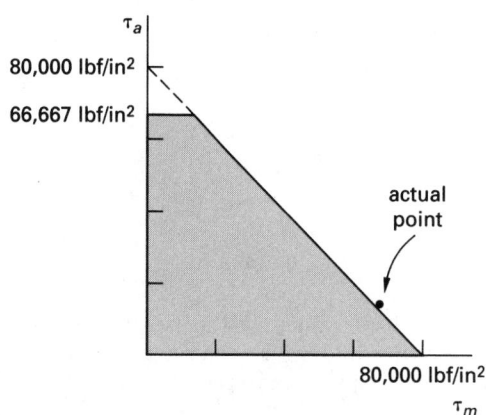

7. BUCKLING OF HELICAL COMPRESSION SPRINGS

Tall springs (i.e., with free heights more than 4 to 5 times their mean coil diameters) and weak springs (i.e., with ratios of mean coil to wire diameters less than 5) can buckle when heavily loaded. A simple method of checking for buckling is to compare the ratio of working deflection to free height against the limiting values given in Table 52.5. Springs that tend to buckle can be guided by placing them over a tube or rod.

Table 52.5 Approximate Maximum δ_w/h_f to Prevent Spring Buckling

h_f/D	both ends pivoting ball	one end pivoting ball, other squared and ground	both ends squared and ground
1	0.72	–	0.72
2	0.63	–	0.71
3	0.38	–	0.71
4	0.20	0.46	0.63
5	0.11	0.25	0.53
6	0.07	0.18	0.38
7	0.05	0.14	0.26
8	0.04	0.10	0.19
9	–	0.08	0.16
10	–	0.07	0.14

Adapted from *Design Handbook*, 1987 Edition, Barnes Group (Associated Spring), Bristol, CT.

8. HELICAL COMPRESSION SPRINGS: DYNAMIC RESPONSE

At certain speeds, resonance between the frequency of the applied force and one of the spring's natural frequencies may occur. The spring may experience excessive vibrations (*surging*) and fail prematurely.

The *fundamental frequency (fundamental harmonic frequency)*, f_0, in Hz of a helical compression spring with both ends fixed can be calculated from Eq. 52.18. To avoid resonance with any of the harmonics, the spring's fundamental frequency should be at least 13 times the exciting frequency.[11]

$$f_0 = (1.12 \times 10^3) \left(\frac{d}{D^2 n_a} \right) \sqrt{\frac{G}{\rho}} \quad \text{[SI]} \quad 52.18(a)$$

$$f_0 = \left(\frac{d}{9D^2 n_a} \right) \sqrt{\frac{G g_c}{\rho}} \quad \text{[U.S.]} \quad 52.18(b)$$

Table 52.1 lists values of the shear modulus, G, for various spring materials. The density, ρ, of virtually all iron-based spring materials is 0.284 lbm/in³ (7870 kg/m³), and $G = 11.5$ lbf/in² (79 MPa). Equation 52.19 can be used directly for steel springs.

$$f_0 \approx (3.5 \times 10^5) \left(\frac{d_{mm}}{D_{mm}^2 n_a} \right) \quad \text{[SI; steel]} \quad 52.19(a)$$

$$f_0 \approx (14{,}000) \left(\frac{d_{in}}{D_{in}^2 n_a} \right) \quad \text{[U.S.; steel]} \quad 52.19(b)$$

Other methods of preventing resonance include winding the spring with a variable pitch (which changes the natural frequency), using a rubbing friction device, and

---

[11]Some authorities recommend 15 to 20 times the fundamental frequency.

using combinations of multiple springs to achieve the desired spring rate.[12]

9. HELICAL SPRINGS: HIGH TEMPERATURE

Operation at high temperature affects both the elastic modulus (shear and tensile) as well as the strength (yield, ultimate, and endurance). Sustained operation at high enough temperatures may also result in creep, set, and relaxation. Evaluating the effects of high temperatures requires knowing the high-temperature material properties.

Recommended maximum spring service temperatures are 250 to 300°F (90 to 150°C) for high-carbon steel, 350°F (175°C) for oil-tempered carbon steel, 425 to 475°F (220 to 245°C) for alloy steel, 500 to 900°F (260 to 480°C) for stainless steel, and 1000 to 1100°F (540 to 590°C) for nickel alloys.

10. HELICAL EXTENSION SPRINGS

Helical *extension springs* can be wound with loose or tight coils. With tight-wound coils, there is an *initial tension*, F_i, that must be applied in order to separate the coils.

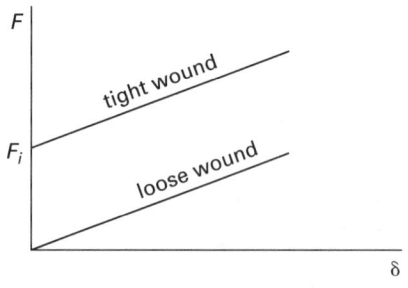

Figure 52.3 Deflection Curves for Extension Springs

The initial stress is given by Eq. 52.20.

$$\tau_i = \frac{8WDF_i}{\pi d^3} \quad 52.20$$

After the applied load has increased to F_i, the spring behavior is predicted by compression spring formulas.

$$\tau = \tau_i + \frac{8FDW}{\pi d^3} \quad 52.21$$

$$\delta = \left(8n_a D^3 \right) \left(\frac{F - F_i}{Gd^4} \right) \quad 52.22$$

Extension springs often fail at their hooks from stress concentrations due to hook curvature. Two critical points are shown in Fig. 52.4. Point A is highly stressed

---

[12]In extreme cases, a special spring can be manufactured from stranded wire.

in bending. Point B is highly stressed in shear. Equations 52.23 and 52.24 are simplified expressions for the maximum bending and shear stresses in the hook. In effect, the ratio of the radii constitute the stress concentration factors. It is recommended that r_4 be greater than twice the wire diameter.

$$\sigma_A = \left(\frac{16FD}{\pi d^3}\right)\left(\frac{r_1}{r_3}\right) \qquad 52.23$$

$$\tau_B = \left(\frac{8FD}{\pi d^3}\right)\left(\frac{r_2}{r_4}\right) \qquad 52.24$$

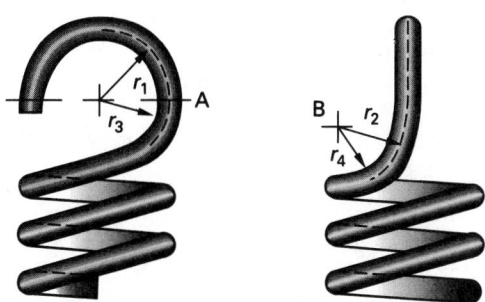

Figure 52.4 *Points of Stress Concentration in Extension Springs*

11. HELICAL TORSION SPRINGS

Helical torsion springs manufactured from round wire are essentially round cantilever beams.[13] Loading produces a bending stress. Most torsion springs operate over an arbor. A clearance of about 10% between the arbor and spring is generally adequate to prevent binding. The bending stress is largest at the inner radius of the spring.

$$\sigma = \frac{32K_b M}{\pi d^3} \qquad 52.25$$

$$M = \mathrm{FL} \qquad 52.26$$

The Wahl stress correction factors for bending at the inner and outer faces are

$$K_{b,i} = \frac{4C^2 - C - 1}{4C(C-1)} \qquad 52.27$$

$$K_{b,o} = \frac{4C^2 + C - 1}{4C(C+1)} \qquad 52.28$$

In the absence of friction, the spring constant, k, for helical torsion springs will be the same for any angular deflection. The angular spring rate in in-lbf/rev

---

(N·mm/rev) is given by Eq. 52.29 but does not include the contribution of the long spring ends, which can be treated as springs as well.[14]

$$k_{\mathrm{angular}} = \frac{M}{\theta_{\mathrm{revolutions}}} = \frac{Ed^4}{10.8Dn_a} \qquad 52.29$$

If the long ends are flexible and are of the same wire diameter, their contributions to the active coils is one-third of their moment arm (measured from the center of the coil).

$$n_a = n_{\mathrm{body}} + n_{\mathrm{ends}} = n_{\mathrm{body}} + \frac{L_1 + L_2}{3\pi D} \qquad 52.30$$

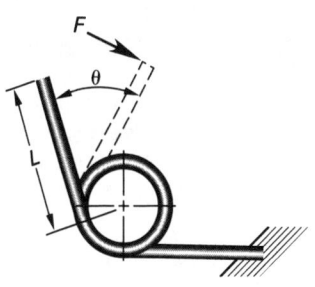

Figure 52.5 *Helical Torsion Spring*

12. FLAT AND LEAF SPRINGS

Flat springs are constructed as simply supported or cantilever beams. They can be flat, curved, or nested (as in a leaf spring). The traditional beam deflection tables can be used with simple flat springs when the deflections are small.[15]

Since beam bending stress is proportional to the thickness of the beam, stress can be kept low by using several low-thickness springs instead of a single thick spring. *Leaf springs* (*leaf set*), as commonly used in cars, consist of several flat springs, one atop the other. The capacity of a leaf set is the sum of the capacities of each spring. The springs slide longitudinally over one another. The effect of this sliding and the resultant friction is difficult to evaluate.

A modification of the elastic modulus of the basic beam equations is required if the spring is very wide. Different methods (typically graphical) are required if the deflection is large (i.e, 30% or more of the spring length).[16]

---

[14]The factor 10.8 is a practical replacement for the theoretical conversion from radians to revolutions of $64/2\pi = 10.18$ that accounts for friction between adjacent spring coils and between the spring and the arbor.

[15]The common beam equations assume that the deflection is small (i.e., less than a few percent of the length) and that the load remains perpendicular to the beam at all times.

[16]When the beam is bent extensively, the maximum stress in a cantilever beam may not even occur at the fixed end. Complexities such as this call for a different analysis method.

---

13. VELOCITY, POWER, AND TORQUE IN ROTATING MEMBERS

Equation 52.31 gives the tangential velocity for a rotating circular member with diameter d. If the pitch circle diameter is used with gears, the tangential velocity is referred to as the *pitch circle velocity*.

$$v_t = \pi d n_{\text{rpm}} \quad \text{[consistent units]} \qquad 52.31$$

The relationship between transmitted horsepower, torque, and rotational speed is

$$P_{\text{kW}} = \frac{T_{\text{N·m}} n_{\text{rpm}}}{9549} \quad \text{[SI]} \qquad 52.32(a)$$

$$P_{\text{hp}} = \frac{T_{\text{in-lbf}} n_{\text{rpm}}}{63{,}025} \quad \text{[U.S.]} \qquad 52.32(b)$$

14. SPUR GEAR TERMINOLOGY

Spur gears have the simplest type of teeth. The teeth faces of a spur gear are parallel to the axis of rotation. Figure 52.6 illustrates a typical spur gear and some of the terminology used to describe gear and tooth geometry.

The *pitch circle* is an imaginary circle on which the gear lever arm is based. The *pitch point* is an imaginary point of tangency between the pitch circles of two meshing gears.

The *addendum* is the radial distance from the pitch circle to the top of the tooth. For full-depth gears, it is equal to the reciprocal of the diametral pitch (i.e., $1/P$). The *base circle* is the circle that is tangent to the line of action. The *clearance* is the difference between the dedendum and the addendum. The *clearance circle* is the circle that is tangent to the addendum of the meshing gear. The *dedendum* is the radial distance from the pitch circle to the root circle. For full-depth gears, it is equal to either $1.25/P$ or $1.35/P$, depending on the clearance wanted. The *tooth face* is the tooth area

between the pitch circle and the addendum circle. The *face width* is the axial width of the tooth. The *flank* is the tooth area between the pitch circle and the dedendum. The *land* is the flat surface at the top of each tooth.

Whole depth is the distance from the addendum circle to the dedendum circle. It is equal to the working depth plus the clearance. The *working depth* is the distance that a tooth from a meshing gear extends into the space between two teeth.

Three meanings for the term "pitch" are used for spur gears. The *diametral pitch*, P, is the number of teeth per inch of pitch circle (i.e., number of teeth per inch around the gear). The diametral pitch is the same for all meshing teeth. The *circular pitch*, p, is the distance between corresponding tooth points along the pitch circle. It is equal to the tooth thickness plus the curved separation distance between teeth. The circular pitch is the same for all meshing teeth.

The *normal pitch* (also known as the *base pitch*) is the distance from a point on one gear to the corresponding point measured along the base circle. It is also the distance from a point to the same corresponding point on the meshing gear tooth.

The *module*, m, of a gear is the ratio of the pitch diameter to the number of teeth. Thus, the module is the reciprocal of the diametral pitch. The module (with units of mm per tooth) is the common SI index of tooth size.

$$m = \frac{d}{N} = \frac{p}{\pi} = \frac{1}{P} \quad \text{[module]} \qquad 52.33$$

$$P = \frac{N}{d} = \frac{\pi}{p} = \frac{1}{m} \quad \text{[diametral pitch]} \qquad 52.34$$

$$p = \frac{\pi d}{N} = \frac{\pi}{P} = \pi m \quad \text{[circular pitch]} \qquad 52.35$$

$$d = \frac{N}{P} = mN \quad \text{[pitch diameter]} \qquad 52.36$$

Figure 52.6 *Spur Gear Terminology*

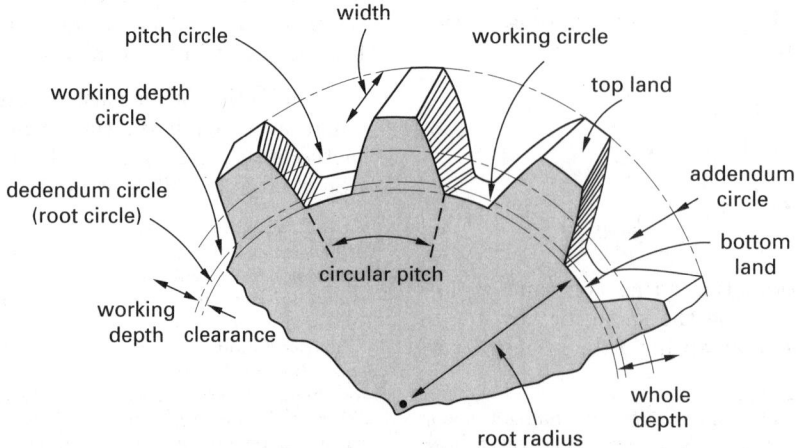

Not every diametral pitch is available. To be economical, designs should make use of the standard diametral pitches. Common "coarse" series diametral pitches include 1, $1\frac{1}{4}$, $1\frac{1}{2}$, $1\frac{3}{4}$, 2, $2\frac{1}{4}$, $2\frac{1}{2}$, 3, 4, 6, 8, 10, 12, and 16 teeth/in. Common "fine" series diametral pitches include 20, 24, 32, 40, 48, 80, 96, 120, 150, and 200 teeth/in.

Most gears in use today are *involute gears* (i.e., have involute-cut teeth). An involute of a circle is the curve traced by the end of a taut string that is unwound from the circumference. The base circle is defined as the circle from which the involute is generated.

The *line of action* (also known as the *pressure line* and *generating line*) is a line passing through the pitch point that is tangent to both base circles. It is the distance between the intersections of the pressure line and the addendum circles. In Fig. 52.7, line A-A is tangent to the base circles and is the line of action. Angle ϕ, determined by the points of tangency, is the *pressure angle* (also known as the *angle of obliquity*). This involute is called a "$\phi°$ involute." In the United States, $20°$, $22\frac{1}{2}°$, and $25°$ pressure angles are in common use. The once-used older $14\frac{1}{2}°$ pressure angle has essentially become obsolete as it produces larger gears.

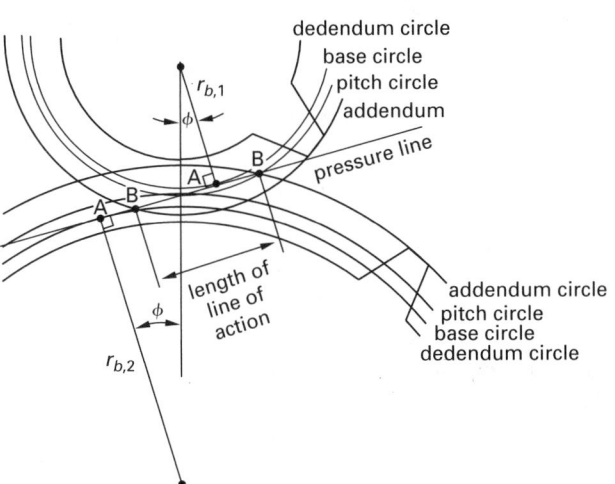

Figure 52.7 *Meshing Gear Terminology*

The pressure angle is proportional to the center-to-center distance of the gears, but a small deviation (i.e., error) in center-to-center distance will change the pressure angle slightly. However, changes in center-to-center spacing and backlash don't change the velocity ratio or the general performance of gear sets with involute gears. This is the main reason that involute gearing is widely used.

Figure 52.7 also illustrates the length of the line of contact between teeth on meshing gears. Line B-B is the section of the line of action between the points where it crosses the two addenda circles. This is sometimes called the *length of the line of action*.

15. TOOTH THICKNESS

Ideally, the tooth thickness will be one-half of the circular pitch. However, to avoid binding from gears whose teeth are inadvertently manufactured with thicker teeth, a clearance must be included in the design. The clearance is known as the *backlash*. The amount of backlash is the distance along the pitch circle, but it is generally measured with feeler gauges. Backlash generally varies between $0.03/P$ and $0.05/P$. Backlash has no effect on gear tooth action for involute-cut gears.

16. GEAR SETS AND GEAR DRIVES

In a simple set of two external gears in contact (referred to as a *mesh* or a *gear set*), one gear drives the other. Often, the smaller gear drives the larger, in which case the smaller gear is referred to as the driving *pinion*. The larger gear in the set is referred to as the driven *gear*. The *center distance* is the distance between the centers of the pinion and gear.

The force between two meshing teeth will be the same. Since the same power ideally will be transmitted by each gear, the product of torque and rotational speed are the same.

$$v_{\text{pinion}} = v_{\text{gear}} \qquad 52.37$$
$$F_{\text{pinion}} = F_{\text{gear}} \qquad 52.38$$

The angular *velocity ratio*, VR, for a pair of gears can be calculated in a number a ways.

$$\text{VR} = \frac{n_{\text{pinion}}}{n_{\text{gear}}} = \frac{\omega_{\text{pinion}}}{\omega_{\text{gear}}} = \frac{r_{\text{gear}}}{r_{\text{pinion}}} = \frac{N_{\text{gear}}}{N_{\text{pinion}}} \qquad 52.39$$

At times, one pair of teeth will carry all of the force and will transmit all of the power. At other times, two (or more) pairs may be in contact. The average number of tooth pairs in contact is the *contact ratio*. The contact ratio is usually between 1.2:1 and 1.6:1. (1.2:1 means that one pair of teeth is in contact at all times, and a second pair is in contact 20% of the time.) For a good design, the contact ratio should be approximately 1.5:1.

$$\begin{aligned} \text{contact ratio} &= \frac{\text{length of line of action}}{p\cos\phi} \\ &= \frac{(\text{length of line of action})\,P}{\pi\cos\phi} \qquad 52.40 \end{aligned}$$

A gear set is "prime" when the numbers of teeth on each meshing gear have no common factor except 1.0. This is a desirable condition, as all teeth tend to wear evenly.

The *service factor* is a single measure that combines the external load dynamics of an application with a gear drive's reliability and operating life. The service factor is applied to the motor's rated performance. When selecting components, either the actual load must be multiplied by the service factor, or all catalog ratings (horsepower, torque, and overhung load) must be divided by the service factor. Acceptable service factor values are determined from experience or are specified by various authorities. The American Gear Manufacturer's Association (AGMA) publishes service factors that depend on the application service class number and type of prime mover. Values range from 0.8 to 2.25.

17. MESH EFFICIENCY

Ideally, the input power is passed through each gear to the next in line.

$$P_{\text{gear}} = P_{\text{pinion}} \qquad 52.41$$

$$T_{\text{gear}} n_{\text{gear}} = T_{\text{pinion}} n_{\text{pinion}} \qquad 52.42$$

In reality, each gear set will dissipate some of the input power. This is accounted for by the *efficiency of the gear train* (i.e., the *mesh efficiency*), η_{mesh}.

$$\eta_{\text{mesh}} = \frac{P_{\text{output}}}{P_{\text{input}}} \qquad 52.43$$

In the case of a pinion driving a gear,

$$P_{\text{gear}} = \eta_{\text{mesh}} P_{\text{pinion}} \qquad 52.44$$

$$T_{\text{gear}} n_{\text{gear}} = \eta_{\text{mesh}} T_{\text{pinion}} n_{\text{pinion}} \qquad 52.45$$

18. FORCE ANALYSIS OF SPUR GEARS

The *tangential force* (also known as the *transmitted load*) at the pitch circle is the useful force, and it is the only component that transmits power. It can be found from the transmitted torque. Other force components (e.g., radial and axial) can be determined from the tangential component.

$$F_t = \frac{2T}{d} = \frac{P_{\text{kW}}\left(1000\ \frac{\text{W}}{\text{kW}}\right)}{v_{t,\text{m/s}}} \qquad [\text{SI}] \quad 52.46(a)$$

$$F_t = \frac{2T}{d} = \frac{P_{\text{hp}}\left(33{,}000\ \frac{\text{ft-lbf}}{\text{hp-min}}\right)}{v_{t,\text{ft/min}}} \qquad [\text{U.S.}] \quad 52.46(b)$$

The radial force on a spur gear tooth is

$$F_r = F_t \tan \phi \qquad 52.47$$

The total force on a spur gear tooth is

$$F_{\text{total}} = \sqrt{F_t^2 + F_r^2}$$

$$= \frac{F_t}{\cos \phi} = \frac{F_r}{\sin \phi} \qquad 52.48$$

19. HELICAL GEARS

Figure 52.8 illustrates the three angles associated with a *helical gear*. In addition to the pressure angle (*normal pressure angle*), ϕ_n, there is a *tangential pressure angle*, ϕ_t, and a *helix angle*, ψ. The 20° normal pressure angle is the most common. Helix angles commonly range from 0° to 30° and seldom approach 45°.

$$\cos \psi = \frac{\tan \phi_n}{\tan \phi_t} \qquad 52.49$$

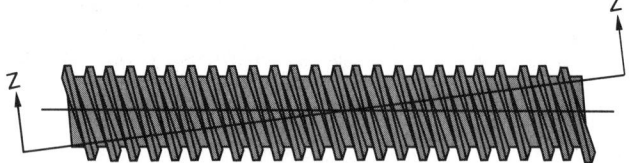

Figure 52.8 *Helical Gear*

The relationship between the *transverse circular pitch* (usually called *circular pitch*), p_t, and the *normal circular pitch*, p_n, is

$$p_t = \frac{p_n}{\cos \psi} \qquad 52.50$$

The *axial pitch* is

$$p_a = \frac{p_t}{\tan \psi} \qquad 52.51$$

The *normal diametral pitch* is

$$P_n = \frac{P_t}{\cos \psi} \qquad 52.52$$

20. FORCE ANALYSIS OF HELICAL GEARS

There are three force components on a helical gear: radial, axial, and tangential. As with spur gears, the tangential force is the useful force, and it is the only component that transmits power. It is calculated from Eq. 52.46. The radial and axial components are calculated from the tangential force.

$$F_r = F_t \tan \phi_t \qquad 52.53$$

$$F_a = F_t \tan \psi \qquad 52.54$$

The three force components combine into the total force, F, on the gear. Although total force is of academic interest, the transmitted tangential force, F_t, is used to design the gear.

$$F = \sqrt{F_r^2 + F_a^2 + F_t^2} \qquad 52.55$$

The radial, axial, and tangential force components in terms of the total force are

$$F_r = F \sin \phi_n \qquad 52.56$$
$$F_a = F \cos \phi_n \sin \psi \qquad 52.57$$
$$F_t = F \cos \phi_n \cos \psi \qquad 52.58$$

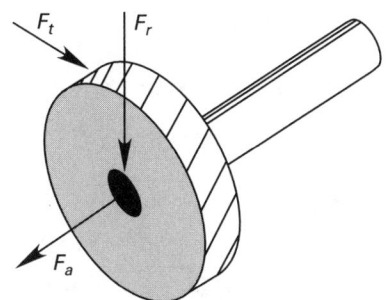

Figure 52.9 *Forces on a Helical Gear*

21. ALLOWABLE STRESSES FOR GEAR DESIGN

The simplest method to calculate an allowable stress for gear tooth design is to divide the tensile yield strength by a substantial factor (e.g., 3 to 5). This method requires judgment and experience.

Another method is to calculate the allowable stress as some percentage (e.g., 75%) of the endurance strength in reversed bending. Since most gear stresses vary repeatedly only from zero to a maximum value (i.e., are not completely reversed), this recommendation builds in a considerable factor of safety.

The AGMA *bending strength* should only be used with the AGMA procedures. It is not an allowable stress in the general sense.

22. SPUR GEAR STRENGTH: LEWIS BEAM STRENGTH

The *Lewis beam strength theory* has been largely superseded by AGMA procedures, but it is of historical interest.[17] The theory assumes that one tooth carries the entire tangential load as a cantilever beam.[18] The force is applied at the tip of the tooth, parallel to the top

[17]The Lewis method can be used for quick estimates.
[18]The Lewis beam strength theory was first proposed in 1892.

land. The radial component of force is disregarded. The maximum stress occurs at the tooth root. The maximum allowable tangential force is given by Eq. 52.59. The *form factor*, Y, varies approximately between 0.2 and 0.5 and has been widely tabulated in gear and machine-design textbooks.[19]

$$F_{t,\max} = \frac{\sigma_a w Y}{P} \qquad 52.59$$

Alternatively, the actual bending stress can be determined.

$$\sigma = \frac{F_t P}{w Y} \qquad 52.60$$

The original Lewis beam strength theory assumed a static loading of the tooth. This is appropriate for low-speed operation only. To account for speed and dynamic effects, the allowable Lewis beam strength force is reduced by a *speed factor*. In the *Barth speed factor*, given by Eq. 52.62, the constant a is 600 for ordinary industrial gears and gears with cast teeth and 1200 for accurately cut gears running as high as 6000 ft/min.[20]

$$F_{t,\max} = \frac{k_d \sigma_a w Y}{P} \qquad 52.61$$

$$k_d = \frac{a}{a + v_{t,\text{fpm}}} \qquad 52.62$$

Additional extensions have been made to the Lewis theory to account for the stress concentration at the root of the tooth.

23. DESIGN GUIDELINES FOR GEARS

The following general rules can be used to simplify designs of gears.

- The face width, w, of spur gears should be 3 to 5 times the circular pitch, π/P.

- Increasing the face width will decrease wear and fatigue, but the dynamic load will increase.

- A coarser tooth (smaller diametral pitch) will improve the fatigue performance but not the wear performance. For a given tooth design, only a harder material will improve the wear.

- Larger pitch diameters have lower tangential forces and lower dynamic loadings.

- Decreasing the tooth error decreases the dynamic loading.

[19]There are two factors named "form factor." The *Lewis form factor* is typically written with a lower case y. The *form factor* is written with an upper case Y. Since $Y = \pi y$, one can be derived from the other.
[20]Over the years, more sophisticated methods for incorporating the speed effects have been proposed. Since this section covers a methodology that has been essentially superseded by the AGMA procedures, only the historical Barth speed factor is presented.

24. AGMA GEAR DESIGN METHOD

The AGMA formula for bending stress in the tooth (referred to as a *bending stress number*, s_b) is given by Eq. 52.63. K_a is an application factor, K_v is the dynamic (velocity) factor, K_s is the size factor, K_m is the load distribution factor, and J is the geometry factor. P is the diametral pitch in teeth per inch, and m is the module in mm per tooth.

$$s_{b,\text{MPa}} = \left(\frac{F_t K_a}{K_v}\right)\left(\frac{1}{mw}\right)\left(\frac{K_s K_m}{J}\right) \quad \text{[SI]} \quad \textbf{52.63(a)}$$

$$s_{b,\text{lbf/in}^2} = \left(\frac{F_t K_a}{K_v}\right)\left(\frac{P}{w}\right)\left(\frac{K_s K_m}{J}\right) \quad \text{[U.S.]} \quad \textbf{52.63(b)}$$

The allowable design stress (referred to as the *allowable bending stress number*) is calculated from the AGMA *allowable tensile bending strength number*, S_t, obtained from AGMA tables.[21] K_L is the life factor, K_T is the temperature factor, and K_R is the reliability factor.

$$\sigma_a = \frac{S_t K_L}{K_T K_R} \qquad \textbf{52.64}$$

AGMA publishes a similar procedure for comparing the surface stress to the allowable contact stress. This is sometimes referred to as the AGMA *pitting resistance* or AGMA *surface durability*.

All AGMA methods are highly dependent on having the AGMA charts and tables.[22] Since different charts and tables are needed for different tooth depths and pressure angles, these charts and tables are extensive. Some are included in engineering handbooks and machine design textbooks, but the material is too extensive to include here.

25. LIMIT WEAR DESIGN

Gear teeth can fail by compressive surface fatigue (excessive gear tooth contact stress) as well as by tooth breakage. Either factor can be the controlling factor in determining tooth width. *Limit wear design* is the procedure that accounts for this aspect of gear design. The general procedure is to keep the limit wear load (capacity) greater than the dynamic load. The dynamic load is calculated from the transmitted tangential force, F_t, with corrections for *errors in action* (e.g., inaccuracies in tooth form, cutting, and spacing, mounting misalignments, and rotational inertia).[23]

---

[21] Some authorities have suggested that some percentage (e.g., 75%) of the endurance strength in reversed bending can be used. However, the pure AGMA procedure requires that the specific strengths from its tables be used. Conversely, the AGMA table values should not be used for any other purpose than gear tooth design.

[22] Refer to the latest edition of *AGMA Standard for Rating Pitting Resistance and Bending Strength of Spur and Helical Involute Gear Teeth*.

[23] The *Buckingham equation* was widely used for many years to calculate dynamic effects until the AGMA equations were developed.

26. BEVEL GEARS

Bevel gearsets are used with intersecting shafts. Gears can be straight or spiral.[24] Figure 52.10 illustrates a straight bevel gearset. The most common shaft angle is 90°, although any angle can be accommodated. The circular pitch and pitch diameter are as defined for spur gears. The *pitch angles*, γ and Γ, are the angles of the forward *pitch cones* projected back to the apex.

$$\tan \gamma = \frac{N_{\text{pinion}}}{N_{\text{gear}}} \qquad \textbf{52.65}$$

$$\tan \Gamma = \frac{N_{\text{gear}}}{N_{\text{pinion}}} \qquad \textbf{52.66}$$

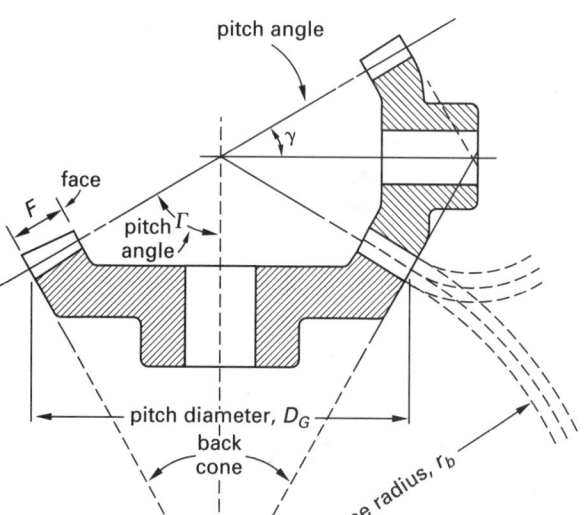

Figure 52.10 *Straight Bevel Gearset*

27. WORM GEAR SETS

Worm drives, consisting of a *worm* and a *worm gear*, are used to turn noncoplanar shafts oriented at right angles with high speed ratios (e.g., 2:1 or 3:1). Worm drives are ordinarily irreversible: A worm will turn a gear, but a gear will not turn a worm. Worms are usually doubly or triply threaded (i.e., have *lead ratios* of 2:1 or 3:1). The relationship between the lead angle, θ, and the worm diameter is

$$\tan \theta = \frac{\text{lead}}{\pi d_{\text{worm}}}$$

$$= \frac{p_{\text{gear}}(\text{lead ratio})}{\pi d_{\text{worm}}} \qquad \textbf{52.67}$$

---

[24] *Hypoid gears* are essentially spiral bevel gears for shafts that do not intersect.

It is customary to calculate the velocity ratio using the number of threads on the worm and number of teeth on the gear.

$$\text{VR} = \frac{n_{\text{worm}}}{n_{\text{gear}}} = \frac{N_{\text{gear}}}{N_{\text{worm}}}$$

$$= \frac{d_{\text{gear}}}{d_{\text{worm}} \tan \theta} \qquad 52.68$$

The power and torque transmission by a worm-gear train is the same as for any two gears in mesh.

$$T_{\text{gear}} n_{\text{gear}} = T_{\text{worm}} n_{\text{worm}} \qquad 52.69$$

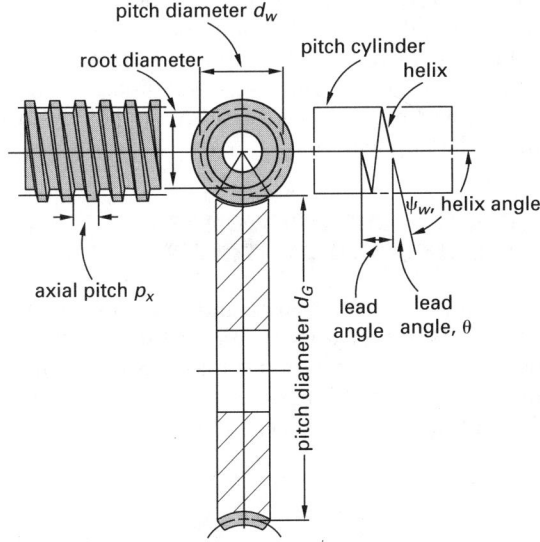

Figure 52.11 Worm and Worm Gear Terminology

28. FLAT BELT DRIVES

There are two types of *belt drives*, as shown in Fig. 52.12. With the *open drive*, the shafts turn in the same direction. With the *crossed drive*, they turn in opposite directions.

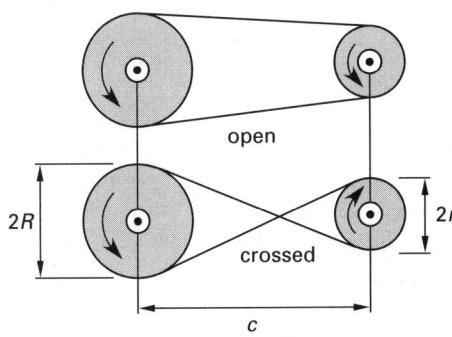

Figure 52.12 Open and Crossed Belt Drives

Analysis of flat belts is essentially an extension of pulley friction. The relationship between the tight and slack side tensions, F_1 and F_2, respectively, is given by Eq. 52.70 for both open and crossed belts. ϕ must be in radians. The coefficient of friction for belts is approximately 0.4 for leather, 0.8 for urethane, and 0.5 to 0.8 for polyamide.

$$\frac{F_1 - F_c}{F_2 - F_c} = e^{f\phi} \qquad 52.70$$

Equation 52.71 gives the centrifugal loading per unit length due to the belt. The density of leather belting is approximately 0.035 to 0.045 lbm/in^3 (970 to 1250 kg/m^3), and the density of modern polyamide and urethane belting is approximately 0.038 to 0.045 lbm/in^3 (1050 to 1250 kg/m^3). m is the mass of the belt per unit length.

$$F_c = m\text{v}^2 = \rho bt\text{v}^2 \qquad \text{[SI]} \qquad 52.71(a)$$

$$F_c = \frac{m\text{v}^2}{g_c} = \frac{\rho bt\text{v}^2}{g_c} \qquad \text{[U.S.]} \qquad 52.71(b)$$

The familiar power-torque relationships from Eq. 52.32 apply. The net tension responsible for the transmission of horsepower is

$$P_{\text{kW}} = \frac{(F_1 - F_2)\text{v}_{\text{m/s}}}{1000 \, \dfrac{\text{W}}{\text{kW}}} \qquad \text{[SI]} \qquad 52.72(a)$$

$$P_{\text{hp}} = \frac{(F_1 - F_2)\text{v}_{\text{ft/min}}}{33,000 \, \dfrac{\text{ft-lbf}}{\text{hp-min}}} \qquad \text{[U.S.]} \qquad 52.72(b)$$

Equation 52.73 relates the maximum and minimum tensions to the initial tension, F_i, in the belt at the time of assembly.

$$F_i = \frac{F_1 + F_2}{2} \qquad 52.73$$

The angle of contact, ϕ, for open belts depends on the radii of the two pulleys and the center-to-center distance, c.

$$\phi = \pi + 2\sin^{-1}\left(\frac{R - r}{c}\right) \qquad \text{[large pulley]} \qquad 52.74$$

$$\phi = \pi - 2\sin^{-1}\left(\frac{R - r}{c}\right) \qquad \text{[small pulley]} \qquad 52.75$$

Equations 52.76 and 52.77 approximate the belt lengths.

$$L_{\text{open drive}} = 2c + \left(\frac{\pi}{2}\right)(2R + 2r) + \frac{(2R - 2r)^2}{4c} \qquad 52.76$$

$$L_{\text{crossed drive}} = 2c + \left(\frac{\pi}{2}\right)(2R + 2r) + \frac{(2R + 2r)^2}{4c} \qquad 52.77$$

Ultimately, a belt is limited by the maximum tension it can support. From Eq. 52.72, with $F_2 = 0$ and $F_1 = 2F_i$, the maximum power that can be transmitted is approximately

$$P_{\text{kW}} = \frac{F_i \text{v}_{\text{m/s}}}{500} \quad \text{[SI]} \quad \textit{52.78(a)}$$

$$P_{\text{hp}} = \frac{F_i \text{v}_{\text{ft/min}}}{16{,}500} \quad \text{[U.S.]} \quad \textit{52.78(b)}$$

Although leather belts were used extensively in the past, they have been largely replaced by polyamide and urethane belts. Manufacturers rate belt materials in terms of the horsepower (per unit width) that can be transmitted at various speeds. The rated power is derated by various factors. In Eq. 52.79, C_p is the pulley correction factor, C_{v} is the velocity correction factor, and K is the service factor. F_a is the allowable belt tension as specified by the belt manufacturer. Tables of allowable belt tension derating factors are included in most engineering handbooks.

$$P_{\text{kW}} = \frac{C_p C_{\text{v}} F_a \text{v}_{\text{m/s}}}{(500) K_s} \quad \text{[SI]} \quad \textit{52.79(a)}$$

$$P_{hp} = \frac{C_p C_{\text{v}} F_a \text{v}_{\text{ft/min}}}{(16{,}500) K_s} \quad \text{[U.S.]} \quad \textit{52.79(a)}$$

29. V-BELTS

Design of V-belt drives is largely dependent on use of manufacturers' literature and tables from engineering handbooks. Sheave diameters are understood to be the diameter of the pitch diameters, as needed in determining the velocity ratios. V-belt cross sections have been standardized and are designated as A, B, C, D, or E and the inside diameter. Thus, a B100 V-belt would have a standard B cross section and an inside diameter of 100 in.

As with flat belts, the power that can be transmitted is derated by various factors, including a *service factor* that accounts for the type of prime mover and the nature of the load. In some cases, the arc of contact may be considered in derating the belt.

The number of belts is calculated from the transmitted power, the rated power per belt, and a contact arc factor (needed when the contact angle is less than 180°).

The center-to-center distance should be less than three times the sum of the two sheave diameters or less than the diameter of the larger sheave. Belt length (referred to as *pitch length* and *effective length*) can be calculated in the same manner as for open flat belts. The length of a known belt can also be calculated by adding a small correction to the inside diameter of the belt. Depending on the belt type, those corrections are: A, 1.3 in; B, 1.8 in; C, 2.9 in; D, 3.3 in; and E, 4.5 in.

30. BRAKES AND BRAKE MATERIALS

A *brake* is a device for bringing motion to a halt by converting kinetic energy into heat. Brakes can be designed to be self-energizing, self-locking, or neither.[25] With a *self-energizing brake*, the moment created by the friction forces acts to reduce the actuating force needed. With a *self-locking brake*, the frictional force is so large that no actuating force is needed at all. A self-locking brake is seldom desirable, as a negative force is required to disengage it.

Chrysotile asbestos-based friction brake products (*pads* and *linings*) have been highly refined since the material was introduced in the early 1900s. Four classes of non-asbestos products are used in response to environmental and health concerns about asbestos: non-asbestos organic (NAO), resin-bonded metallic (semimetallic), sintered metallic, and carbon-carbon. Only the semimetallic and NAO varieties show promise as alternatives for common automotive brake linings. However, semimetallic linings are not readily applicable to most drum brake applications.

31. ENERGY DISSIPATION IN BRAKES AND CLUTCHES

Energy dissipated in brakes and clutches can be calculated as the change in kinetic (rotational or translational) energy of the braking mass. For example, the change in rotational kinetic energy of a spinning mass with rotational mass moment of inertia, I, is

$$\Delta E_k = \frac{I \left(\omega_1^2 - \omega_2^2 \right)}{2} \quad \text{[SI]} \quad \textit{52.80(a)}$$

$$\Delta E_k = \frac{I \left(\omega_1^2 - \omega_2^2 \right)}{2 g_c} \quad \text{[U.S.]} \quad \textit{52.80(b)}$$

The maximum theoretical temperature increase occurs if all of the kinetic energy is absorbed by the brake or clutch material. This is an extreme assumption, as some of the energy will be removed, particularly for braking devices that are run "wet."[26] For steel brake and clutch parts, c_p is approximately 0.12 Btu/lbm-°F (0.5 kJ/kg·K).

$$\Delta T = \frac{\Delta E_k}{m_{\text{brake material}} c_p} \quad \textit{52.81}$$

Various limits are imposed on clutches and brakes, including maximum energy dissipation rate per unit area, maximum temperature, and maximum contact pressure. Table 52.6 lists general limits based on application. Table 52.7 lists limits based on brake material.

---

[25] *Hydraulic braking* or *hydraulically actuated braking* refers to the transmission of the applied braking force (e.g., from a foot pedal) to the brakes via a hydraulic brake line. The actuating braking force is reconstituted in the brake cylinder.
[26] Frictional heat can be removed by an oil spray or bath.

Machine Design

Table 52.6 *General Limits on Energy Dissipation Rates for Clutches and Brakes*

condition	maximum energy dissipation rate	
	hp/in^2	kW/m^2
continuous load, poor heat removal	0.85	980
intermittent load, poor heat removal, long recovery period	1.7	2000
continuous load, good heat removal (e.g., oil bath)	2.5	2900
vehicular brakes	5.0	5800

(Multiply hp/in^2 by 1156 to obtain kW/m^2.)
(Compiled from various sources.)

The energy dissipation rate (i.e., power) can be calculated from Eq. 52.82.

$$P_{\text{braking}} = \left(\frac{F_f}{A}\right)\text{v} = \left(\frac{fN}{A}\right)\text{v} = fp\text{v} \qquad 52.82$$

During braking, the number of revolutions (or distance traveled) required to bring a spinning (moving) mass to a standstill has little relationship to the original speed. With a uniform deceleration, the average speed will be one-half of the original speed. Then, the braking time to come to a complete standstill can be calculated from the original kinetic energy and the average rate of frictional energy dissipated.

$$t_{\text{braking}} = \frac{E_{k,\text{original}}}{F_f \text{v}_{\text{ave}}} \qquad 52.83$$

32. BLOCK BRAKES

A *block brake* (i.e., an *external shoe brake*) is illustrated in Fig. 52.13. The pressure will be distributed approximately uniformly if the block *contact angle*, ϕ, is less than 60°. The coefficient of sliding friction, f, is used. The frictional retarding force is calculated from the normal force.

$$F_f = fN \qquad 52.84$$

Regardless of the location of the pivot point, the normal force can be calculated from the torque capacity, T, of the brake.

$$N = \frac{F_f}{f} = \frac{T}{rf} \qquad 52.85$$

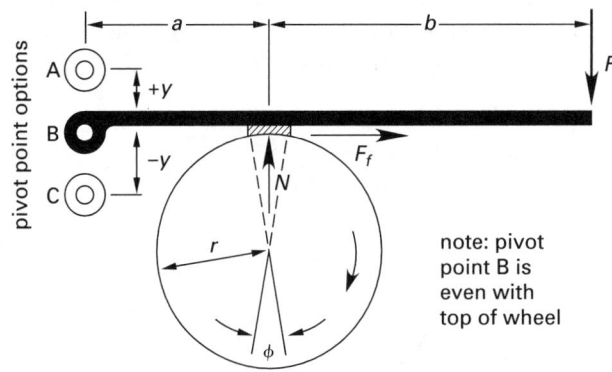

Figure 52.13 *Block Brake*

If the contact angle is greater than 60°, an equivalent coefficient of friction, f', should be used in place of the normal coefficient of sliding friction. In Eq. 52.86, ϕ must be in radians.

$$f' = \frac{f\left[4\sin\left(\dfrac{\phi}{2}\right)\right]}{\phi + \sin\phi} \qquad 52.86$$

Table 52.7 *Typical Clutch and Brake Design Parameters*[a]

application	coefficient of friction		max. temp.		max. pressure	
	oily	dry	°F	°C	psi	kPa
CI/CI	0.05	0.15–0.20	600	320	150–250	1000–1750
PM/CI	0.05–0.1	0.1–0.4	1000	540	150	1000
PM/HS	0.05–0.1	0.1–0.3	1000	540	300	2100
WA/CI-ST	0.1–0.2	0.3–0.6	350–500	175–260	50–100	350–700
MA/CI-ST	0.08–0.12	0.2–0.5	500	260	50–150	350–1000
IA/CI-ST	0.12	0.32	500–750	260–400	150	1000

(Multiply psi by 6.9 to obtain kPa.)
[a]CI—cast iron; PM—powdered metal; HS—hardened steel; CI-ST—cast iron or steel; WA—woven asbestos; MA—molded asbestos; IA—impregnated asbestos.
(Compiled from various sources.)

The required actuating force, F, at the end of the brake arm can be determined by taking moments about the pivot point. Depending on the pivot point, moments will result from the normal force, N, and the frictional force, F_f, as well as from the actuating force, F.

If the wheel turns clockwise, pivot point C results in a self-energizing brake. Similarly, pivot point A is self-energizing for counterclockwise rotation. If $a = fy$, the brake will be completely self-energizing, and the actuating force will be zero. If $a < fy$, the brake will be self-locking.

33. BAND BRAKES

The band brake shown in Fig. 52.14 is evaluated from belt friction equations. The *braking torque* (*braking moment*) is given by Eq. 52.87, in which F_1 is the tight-side tension. The angle ϕ is in radians.

$$T = r(F_1 - F_2) \qquad 52.87$$

$$F_1 - F_2 = F_2(e^{f\phi} - 1) \qquad 52.88$$

$$\frac{F_1}{F_2} = e^{f\phi} \qquad 52.89$$

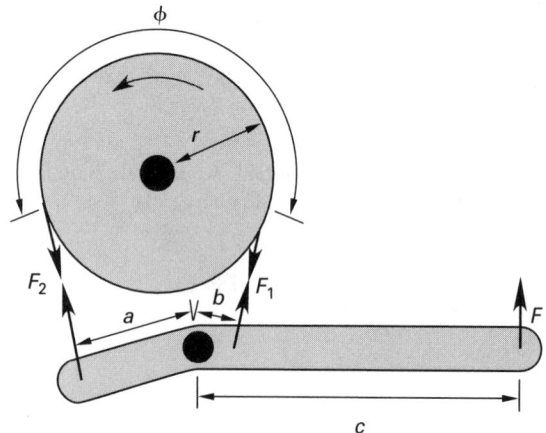

Figure 52.14 *Band Brake*

The applied force, F, is found by taking moments about the pivot point.

$$F = \frac{aF_2 - bF_1}{c}$$
$$= \left(\frac{T}{cr}\right)\left(\frac{a - be^{f\phi}}{e^{f\phi} - 1}\right) \qquad 52.90$$

If $be^{f\phi} > a$, the applied force F will be negative and the brake will be self-locking.

The contact pressure is maximum at the toe of the brake. (The *toe* is where the anchored end of the band first contacts the drum.)

$$p = \frac{F_1}{A} = \frac{F_1}{rw} \qquad 52.91$$

The required contact surface width, w, is calculated from the allowable contact pressure of the band material.

34. DISK AND PLATE CLUTCHES

A clutch is a device for the connection of an initially stationary shaft to a rotating shaft. The torque and power transmitted through a clutch are related by Eq. 52.32.

The torque capacity of a clutch is proportional to the number of friction planes. In the case of a single two-sided automobile clutch disk between two plates, there are two friction planes. The maximum transmitted torque will be twice the torque value per surface calculated from Eq. 52.32.

If a clutch assembly is considered to be rigid, the initial wear will be in the outer areas, where the frictional work is greater. After a period of time (break-in), the pressure distribution will change and the wear will become uniform. *Uniform wear* is equivalent to assuming that all work occurs at the average frictional radius. Uniform wear is the conservative assumption in terms of allowable torque and transmitted horsepower.

With uniform wear, the maximum pressure, p_{max}, will occur at the inner radius. The pressure at any other radius, r, is

$$p_r = \frac{p_{max}r_i}{r} \quad \text{[uniform wear]} \qquad 52.92$$

The axial application force is

$$F = 2\pi p_{max}r_i(r_o - r_i) \quad \text{[uniform wear]} \qquad 52.93$$

The torque per contact surface is carried at the mean radius.

$$T = \tfrac{1}{2}fF(r_o + r_i)$$
$$= \pi f p_{max}r_i(r_o^2 - r_i^2) \quad \text{[uniform wear]} \qquad 52.94$$

If the clutch assembly is known to be semiflexible, a *uniform pressure* distribution can be assumed. The axial application force, F, is also the normal force. Since the pressure is uniform, the maximum pressure, p_{max}, occurs everywhere on the clutch.

$$F = N = \frac{F_f}{f}$$
$$= \pi p_{max}(r_o^2 - r_i^2) \quad \text{[uniform pressure]} \qquad 52.95$$

The frictional radius, r_f, is the mean radius for frictional purposes. It can be used to determine the limiting torque per contact plane.

$$T = F_f r_f$$
$$= \tfrac{2}{3}\pi f p_{max}(r_o^3 - r_i^3)$$
$$\text{[uniform pressure]} \qquad 52.96$$

$$r_f = \frac{(2)(r_o^3 - r_i^3)}{(3)(r_o^2 - r_i^2)} \quad \text{[uniform pressure]} \qquad 52.97$$

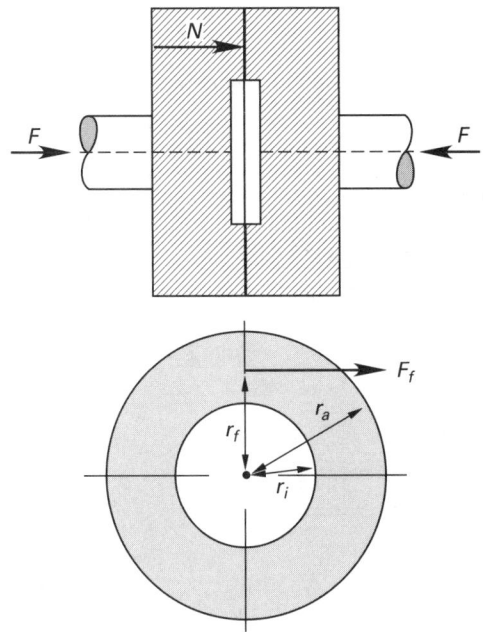

Figure 52.15 *Generalized Plate Clutch*

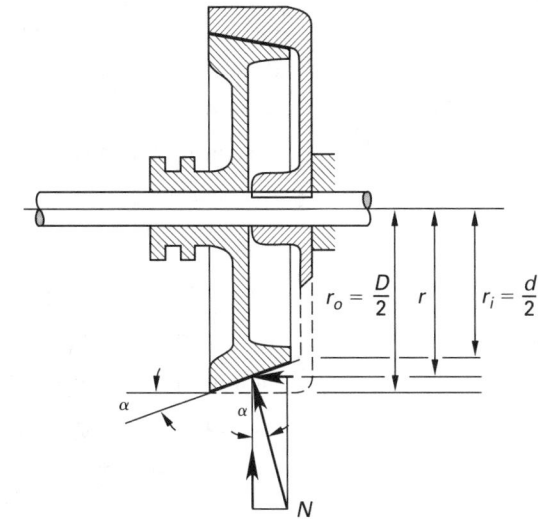

Figure 52.16 *Cone Clutch*

35. CONE CLUTCHES

Figure 52.16 illustrates a *cone clutch*, with its *cone angle*, α. Cone clutches, like disk clutches, can be evaluated for uniform wear or uniform pressure. For uniform wear, the axial force, F, necessary to hold the cone in the cup and generate a force, N, normal to the contact surface is

$$F = N \sin \alpha = \left(\frac{\pi p_{\max} d}{2}\right)(D - d) \qquad \textit{52.98}$$

The frictional force is

$$F_f = fN = \frac{fF}{\sin \alpha} \qquad \textit{52.99}$$

The frictional torque is

$$\begin{aligned} T_f &= \frac{F_f d_m}{2} = \frac{f F_{\text{axial}} d_m}{2 \sin \alpha} \\ &= \left(\frac{\pi f p_{\max} d}{8 \sin \alpha}\right)(D^2 - d^2) \\ &= \left(\frac{F f}{4 \sin \alpha}\right)(D + d) \end{aligned} \qquad \textit{52.100}$$

For uniform pressure,

$$F = \left(\frac{\pi p}{4}\right)(D^2 - d^2) \qquad \textit{52.101}$$

$$\begin{aligned} T &= \left(\frac{\pi f p}{12 \sin \alpha}\right)(D^3 - d^3) \\ &= \left(\frac{F f}{3 \sin \alpha}\right)\left(\frac{D^3 - d^3}{D^2 - d^2}\right) \end{aligned} \qquad \textit{52.102}$$

36. BUSHINGS

Low-load, low-speed applications may be able to use bushings without lubrication. This option is particularly attractive with low-temperature applications where keeping cold oil thin enough to circulate would be difficult. Bushings can be constructed from plastic (e.g., nylon), brass, carbon graphite, or oil-impregnated powdered metal. Bushings are generally limited by lateral load and heat. Hence, the bushing design and selection procedure is to make checks of the contact pressure (based on the projected bearing area), the maximum tangential velocity, and the product of pressure and velocity against upper limits established for the material.

37. LUBRICANTS AND LUBRICATION

In the United States, *crankcase (engine) oils* and *gear oils* are typically called out by *SAE grades* (10–70), although the AGMA also has a grading method. In SI countries, the ISO *viscosity grade*, VG, is used.[27] Viscosity of a lubricating oil should be evaluated at its bulk temperature.

In Fig. 52.17, the viscosity of lubricants is expressed in *reyns*, which is equivalent to lbf-sec/in^2. Viscosities in centipoise (cP), centistokes (cSt), and Saybolt seconds (SSU) (also known as *Saybolt universal viscosity*, SUV) are also widely encountered.

$$\mu_{\text{reyns}} = \frac{\mu_{cP}}{6.89 \times 10^6} \qquad \textit{52.103}$$

$$\mu_{\text{Pa}\cdot\text{s}} = \frac{\mu_{cP}}{1000} \qquad \textit{52.104}$$

$$\mu_{\text{Pa}\cdot\text{s}} = \rho_{\text{kg/m}^3} \nu_{\text{m}^2/\text{s}} \qquad \textit{52.105}$$

$$\nu_{\text{m}^2/\text{s}} = \nu_{\text{cSt}} 10^{-6} \qquad \textit{52.106}$$

$$\nu_{\text{cSt}} = 0.22\,\text{SSU} - \frac{180}{\text{SSU}} \qquad \textit{52.107}$$

---

[27]The VG rating is the nominal kinematic viscosity in centistokes (cSt) at 40°C (104°F).

Figure 52.17 *Viscosity of Common Lubricating Oils*

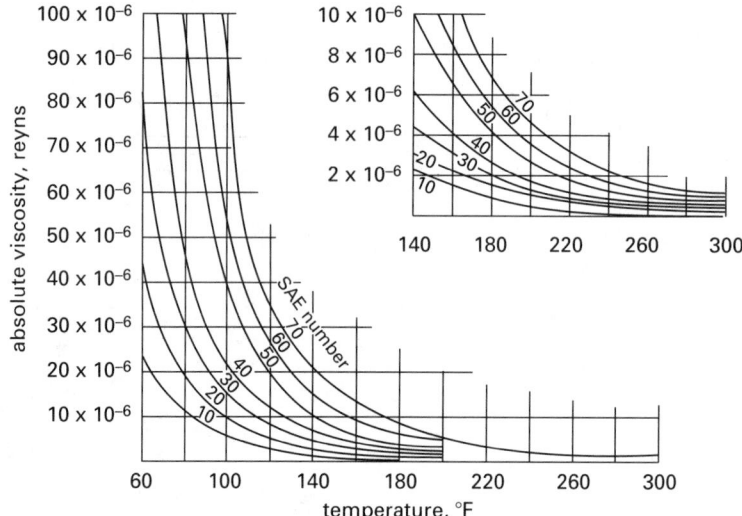

The *pour point* is the lowest temperature at which the oil will flow when chilled under standard test procedures.[28] For example, the approximate pour point for SAE 10W is 15°F (−9°C); for SAE 30, it is −5°F (−20°C); and for SAE 10W-40, it is 30°F (−1°C). Additives can be used to depress the pour points (i.e., make them usable at lower temperatures). Since naphthenic petroleum oils and synthetics contain no wax, they can be cooled below the pour points of most oils.

Although the pour point is of academic interest, the viscosity limit is a more practical temperature cutoff. The *viscosity limit* is the temperature below which the oil simply will not flow fast enough to return to the pan or pass through suction and drain pipes, pumps, and filters. The most restrictive viscosity requirements are found in large circulating oil systems, such as those used in industrial turbomachinery. For these systems, the viscosity limit may be close to room temperature.

38. JOURNAL BEARINGS

Figure 52.18 illustrates a *journal bearing*. The *eccentricity ratio*, ϵ (also known as the *attitude*), of the shaft in the bearing will vary with loading. The *diametral clearance*, c_d, is twice the *radial clearance*, c_r. However, the eccentricity ratios c_d/d and c_r/r are the same.[29]

---

[28]ASTM Method D-97.
[29]When the ratio is written as "c/r" it understood that c is the radial clearance. A radial clearance would not be combined with a diameter.

$$\epsilon = \frac{2e}{c_d} \qquad \textit{52.108}$$

$$c_d = D - d \qquad \textit{52.109}$$

$$c_d = 2c_r \qquad \textit{52.110}$$

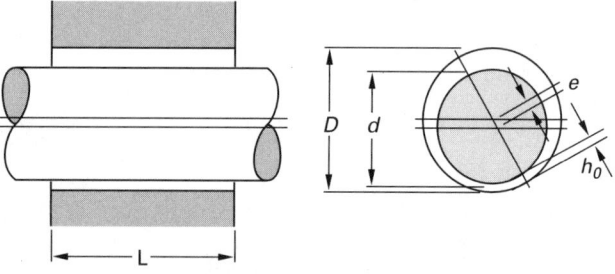

Figure 52.18 *Journal Bearing Nomenclature*

The bearing "pressure" is based on the projected area of the shaft.

$$p = \frac{\text{lateral shaft load}}{\text{projected area}}$$

$$= \frac{\text{lateral shaft load}}{Ld} \qquad \textit{52.111}$$

The relationship between eccentricity, diametral clearance, and minimum film thickness, h_0, is

$$h_0 = \tfrac{1}{2}c_d(1 - \epsilon) \qquad \textit{52.112}$$

Table 52.8 *Typical Journal Bearing Parameters*

application	L/d	average pressure[a]	
		psi	kPa
steam turbines	–	2–10	14–70
pumps	2	80–100	550–700
machine tools	2–4	80–100	550–700
electric motors	2	100–200	700–1400
automotive mains	0.5–0.8	500–600	3500–4100
automotive wristpins	–	2000–3000	14 000–21 000

(Multiply psi by 6.9 to obtain kPa.)
[a]Maximum pressure can be twice as high as the average pressure.
(Compiled from various sources.)

39. JOURNAL BEARING FRICTIONAL LOSSES

The viscosity of journal bearings will cause a frictional torque that opposes rotation. If the bearing is unloaded (i.e., the eccentricity is zero), or if the eccentricity is low and the rotational speed is high, the *Petroff equation* can be used to find the frictional torque. In Eq. 52.113, d is the bearing diameter and L is its length.

$$T_{f,\text{N·m}} = \frac{\pi^2 \mu_{\text{Pa·s}} d^3 L n_{\text{rps}}}{c_d} \quad \text{[SI]} \quad 52.113(a)$$

$$T_{f,\text{in-lbf}} = \left(\frac{d}{2}\right) f(\text{lateral bearing load})$$
$$= \frac{\pi^2 \mu_{\text{reyns}} d^3 L n_{\text{rps}}}{c_d} \quad \text{[U.S.]} \quad 52.113(b)$$

The effective coefficient of friction is

$$f = \frac{2\pi^2 \mu_{\text{reyns}} n_{\text{rps}} d}{p c_d} \quad 52.114$$

The frictional horsepower, $P_{f,\text{hp}}$, can be found from frictional torque by using Eq. 52.32. Frictional heating varies with the square of the rotational speed. The usual range of operating temperatures is 140 to 160°F (60 to 70°C). Most lubricants start to deteriorate above 200°F (90°C). Frictional heating, q_f, is dissipated in an oil cooler or through contact with a cooler surface. The specific heat for petroleum oils is approximately 0.42 to 0.49 Btu/lbm-°F (1.8 to 2.0 kJ/kg·K).

$$q_f = P_{f,\text{hp}} = \dot{m} c_p \Delta T \quad 52.115$$

The ratio of loaded to unloaded friction torque can be correlated to the load number. The *load number* is calculated from Eq. 52.116.

$$N_L = \left(\frac{p}{\mu_{\text{reyns}} n_{\text{rps}}}\right) \left(\frac{c_d}{L}\right)^2 \quad 52.116$$

Example 52.2

SAE 20 oil at 150°F (65°C) bulk temperature is used in a lightly loaded journal bearing. The bearing has a length of 5 in (12.7 cm) and a diameter of 3 in (7.62 cm). The shaft turns at 1800 rpm. The diametral clearance is 0.005 in (0.127 mm). What is the frictional power?

SI Solution

From Fig. 52.17, the viscosity of SAE 20 oil at 65°C is approximately 2.5×10^{-6} reyns. Combining Eqs. 52.103 and 52.104,

$$\mu_{\text{Pa·s}} = \frac{(6.89 \times 10^6) \mu_{\text{reyns}}}{1000}$$
$$= \frac{(6.89 \times 10^6)(2.5 \times 10^{-6} \text{ reyns})}{1000}$$
$$= 0.017225 \text{ Pa·s}$$

From Eq. 52.113, the frictional torque is

$$T_f = \frac{\pi^2 \mu d^3 L n_{\text{rps}}}{c_d}$$
$$= \frac{\pi^2 (0.017225 \text{ Pa·s})(0.0762 \text{ m})^3 \times (0.127 \text{ m})\left(1800 \, \frac{\text{rev}}{\text{min}}\right)}{\left(60 \, \frac{\text{sec}}{\text{min}}\right)\left(\frac{0.127 \text{ mm}}{1000 \, \frac{\text{mm}}{\text{m}}}\right)}$$
$$= 2.257 \text{ N·m}$$

Equation 52.32 gives the frictional power.

$$P_{\text{kW}} = \frac{T_{\text{N·m}} n_{\text{rpm}}}{9549}$$
$$= \frac{(2.257 \text{ N·m})\left(1800 \, \frac{\text{rev}}{\text{min}}\right)}{9549}$$
$$= 0.425 \text{ kW}$$

Customary U.S. Solution

From Fig. 52.17, the viscosity of SAE 20 oil at 150°F is approximately 2.5×10^{-6} reyns. From Eq. 52.113, the frictional torque is

$$T_f = \frac{\pi^2 \mu d^3 L n_{\text{rps}}}{cd}$$
$$= \frac{\pi^2 \left(2.5 \times 10^{-6} \, \frac{\text{lbf-sec}}{\text{in}^2}\right)(3 \text{ in})^3(5 \text{ in})\left(1800 \, \frac{\text{rev}}{\text{min}}\right)}{\left(60 \, \frac{\text{sec}}{\text{min}}\right)(0.005)}$$
$$= 19.99 \text{ in-lbf}$$

Equation 52.32 gives the frictional power.

$$P_{f,\text{hp}} = \frac{T_{\text{in-lbf}}n_{\text{rpm}}}{63,025}$$

$$= \frac{(19.99 \text{ in-lbf})\left(1800 \dfrac{\text{rev}}{\text{min}}\right)}{63,025}$$

$$= 0.571 \text{ hp}$$

40. JOURNAL BEARINGS: RAIMONDI AND BOYD METHOD

With the *Raimondi and Boyd method*, journal bearing performance is correlated with the *bearing characteristic number*, S, also known as the *Sommerfeld number*.[30]

$$S = \left(\frac{d}{c_d}\right)^2 \left(\frac{\mu n_{\text{rps}}}{p}\right) \qquad 52.117$$

The coefficient of friction, f, is needed when calculating the frictional torque and power. It is calculated from Eq. 52.118 using the coefficient of friction variable from App. 52.B.

$$f = \left(\frac{c_d}{d}\right) \text{ (coefficient of friction variable)} \qquad 52.118$$

The minimum film thickness, h_0, is calculated similarly from Eq. 52.119 using the minimum thickness variable from App. 52.B. (Appendix 52.B has two vertical scales and presents the eccentricity ratio, ϵ, directly.)

$$h_0 = r\left(\frac{c_d}{d}\right) \text{ (minimum film thickness variable)} \qquad 52.119$$

41. POWER SCREWS AND SCREW JACKS

A *power screw* changes angular position into linear position (i.e., changes rotary motion into traversing motion). The linear positioning can be horizontal (as in vices and lathes) or vertical, as in jacks. Cross sections of square and Acme threads, both commonly used in power screws, are shown in Fig. 52.19.[31] For square threads, the *thread angle*, ϕ, is zero.

Square power screws are designated by the mean diameter (d_m), pitch (p), and *lead angle* (θ). The *pitch*, p, is the distance between corresponding points on a thread. The *lead*, L, is the distance the screw advances each revolution. Often, double- and triple-threaded screws are used. The lead, L, is one, two, or three times the pitch for single-, double-, and triple-threaded screws, respectively.

$$L = \pi d_m \tan \theta \qquad 52.120$$

---

[30]Correlations have also been developed for various lubricant flows, maximum film pressure, and location of the minimum film thickness in addition to the correlations for the coefficient of friction and minimum film thickness presented in App. 52.B.

[31]The 10° modified thread is essentially equivalent to the square thread, but it is more economical to manufacture than the square thread.

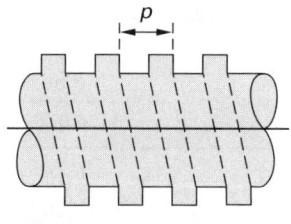

square thread

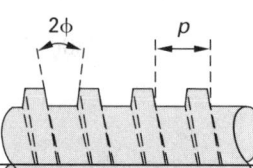

Acme thread

Figure 52.19 *Power Screw Threads*

The torque required to turn a square screw in motion against an axial force F (i.e., "raise" the load) is[32]

$$T = \left(\frac{Fd_m}{2}\right)\left(\frac{\tan \theta + f}{1 - f\tan \theta}\right) \qquad 52.121$$

The torque required to turn the screw in motion in the direction of the applied axial force (i.e., "lower" the load) is given by Eq. 52.122. If the torque is zero or negative (as it would be if the lead was large or friction was low), then the screw is not self-locking and the load will lower by itself causing the screw to spin (i.e, will "overhaul"). The screw will be self-locking when $\tan \theta \leq f$.

$$T = \left(\frac{Fd_m}{2}\right)\left(\frac{\tan \theta - f}{1 + f\tan \theta}\right) \qquad 52.122$$

The torque calculated in Eqs. 52.121 and 52.122 is required to overcome thread friction and to raise the load (i.e., axially compress the screw). Typically, only 10 to 15% of the torque goes into axial compression of the screw. The remainder is used to overcome friction. The mechanical efficiency of the screw is the ratio of torque without friction to the torque with friction. The torque without friction can be calculated from Eq. 52.121 or 52.122 (depending on the travel direction) using $f = 0$.

$$\eta_m = \frac{T_{f=0}}{T} \qquad 52.123$$

In the absence of an antifriction ring, an additional torque will be required to overcome friction in the collar. Since the collar is generally flat, the normal force is the jack load, W, for the purpose of calculating the frictional force.

$$T_{\text{collar}} = Wf_{\text{collar}}r_{\text{collar}} \qquad 52.124$$

---

[32]The relationship is different for Acme and other threads.

PRACTICE PROBLEMS

1. A spring with 12 active coils and a spring index of 9 supports a static load of 50 lbf (220 N) with a deflection of 0.5 in (12 mm). The shear modulus of the spring material is 1.2×10^7 lbf/in^2 (83 GPa). What are the (a) theoretical wire diameter and (b) mean spring diameter?

2. A severe service valve spring is to be manufactured from unpeened ASTM A230 steel wire in standard W & M sizes operating continuously between 20 lbf and 30 lbf (100 N and 150 N). The valve lift is 0.3 in (8 mm). The spring index is 10. The factor of safety is 1.5. What are the (a) wire diameter, (b) spring constant, (c) number of active coils, (d) total number of coils, (e) solid height, (f) spring force at solid height, (g) deflection at solid height, and (h) minimum free height?

3. The material in a spring wire has a shear modulus of 1.2×10^7 lbf/in^2 (83 GPa). The maximum allowable stress is 50,000 lbf/in^2 (350 MPa). The spring index is 7. A 700 lbm (320 kg) object falls from a height of 46 in (120 cm) above the tip of the spring, impacts squarely on the spring, and deflects the spring 10 in (26 cm). What are the (a) wire diameter, (b) mean coil diameter, and (c) number of active coils?

4. A 6 in (150 mm) wide, 24 in (610 mm) long cantilever steel spring supports an 800 lbf (3.5 kN) load at its tip. The deflection is to be less than 1 in (25 mm). The bending stress is limited to 50,000 lbf/in^2 (345 MPa). (a) What is the minimum thickness as limited by deflection alone? (b) What is the minimum thickness as limited by bending stress?

5. A gear train is to have a speed reduction of 600:1. The gears used can have no fewer than 12 teeth and no more than 96 teeth. (a) How many stages are needed? (b) How many teeth should be in each gear in the gear train?

6. A mechanism is driven by a 550 hp (410 kW) motor. The motor turns at 1200 rpm, but a pair of old $14\frac{1}{2}°$ gears on 15 in (380 mm) centers is to reduce the speed of the mechanism to 270 rpm. The pinion is to be SAE 1045 steel with an endurance strength of 90,000 lbf/in^2 (620 MPa). The gear is to be cast steel with an endurance strength of 50,000 lbf/in^2 (345 MPa). A safety factor of 3 is used in the design. Use the Lewis beam strength theory. (a) What are the pitch diameters? (b) Given a gear face width of 6 in (150 mm), what is the diametral pitch? (c) Given a diametral pitch of 1.5 (a module of 17 mm), what should be the face width for the gear?

7. Two 20° involute spur gears are mounted such that their centers are 15 in (380 mm) apart. The pinion is untreated steel with an allowable stress of 30,000 lbf/in^2 (210 MPa). The gear is cast steel with a maximum strength of 50,000 lbf/in^2 (345 MPa). The gear set reduces the speed of a 250 hp motor (190 kW) from 250 rpm to $83\frac{1}{3}$ rpm. A factor of safety of 3 is used. Use the Lewis beam strength theory. (a) What are the pitch diameters? (b) What is the diametral pitch? (c) How many teeth are on each gear? (d) What is the minimum face width of the pinion? (e) What is the minimum face width of the gear?

8. A wet, steel-backed asbestos clutch with hardened steel plates is being designed to transmit 300 in-lbf (33 N·m). The coefficient of friction is 0.12. Slip will occur at 300% of the rated torque. The maximum and minimum friction surface diameters are 4.5 in and 2.5 in (115 mm and 65 mm), respectively. The contact pressure is 100 psi (700 kPa). (a) How many plates are needed? (b) How many disks are needed?

9. (*Time limit: one hour*) A 3 in (76 mm) diameter shaft turns at 1200 rpm in a journal bearing. The bearing has an axial length of 3.5 in (90 mm) and a radial clearance ratio (c_r/r) of 0.001. The transverse load allocated to the bearing is 880 lbf (4 kN). The lubricating oil has a temperature of 165°F (75°C) and a viscosity of 1.184×10^{-6} lbf-sec/in^2 (8.16 cP). (a) What is the minimum film thickness? (b) What power is lost to friction? (c) Is this bearing operating under, at, or above its capacity?

10. (*Time limit: one hour*) Two concentric springs are constructed with squared-and-ground ends from oil-hardened steel. The ultimate tensile strength for the steel is 204,000 lbf/in^2 (1.4 GPa). The shear modulus for the steel is 11.5×10^6 lbf/in^2 (79 GPa). The springs support a static force of 150 lbf (660 N). The spring dimensions and properties are as follows.

inner spring
wire diameter:	0.177 in	(4.5 mm)
mean coil diameter:	1.5 in	(38 mm)
free length:	4.5 in	(115 mm)
total number of coils:	12.75	

outer spring
wire diameter:	0.2253 in	(5.723 mm)
mean coil diameter:	2.0 in	(51 mm)
free length:	3.75 in	(95.3 mm)
total number of coils:	10.25	

(a) What is the deflection of the inner spring? (b) What is the maximum force exerted by the inner spring? (c) What is the maximum shear stress in the inner spring? (d) What is the factor of safety in shear for the inner spring? (e) Specify the winding helix direction for each spring.

11. (*Time limit: one hour*) The power transmission system shown consists of helical gears and AISI 1045 cold-drawn steel shafting. The gears have a 25° helix angle, a 20° normal pressure angle, and a diametral pitch of 5 (module of 5 mm). The yield strength of the 1045 steel is 69,000 lbf/in^2 (480 MPa). The mesh efficiency of each gear set is 98%. Loading is slow and steady. Use a factor of safety of 2. (a) What is the speed of the output shaft? (b) What is the torque output? (c) What is the minimum shaft diameter at section A-A assuming static loading?

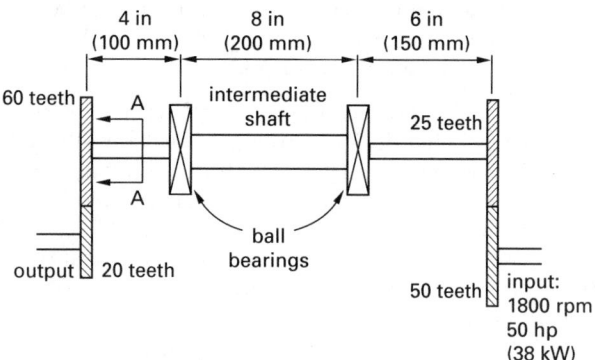

12. A piece of equipment is driven by a 15 hp (11.2 kW) motor through a standard B90 V-belt. The motor runs at 1750 rpm. The nominal speed of the equipment must be 800 rpm. The motor's sheave diameter is 10 in (254 mm). (a) What is the pitch length of the belt? (b) What is the pitch diameter of the sheave on the equipment drive?

53 Pressure Vessels

1. Introduction 53-1
2. Compliance 53-2
3. Exemptions 53-2
4. Stamping Requirements 53-2
5. Design Elements 53-3
6. Service Application 53-3
7. Materials 53-4
8. Heads 53-4
9. Loading Type 53-5
10. Allowable Stresses 53-5
11. Temperature Effects 53-5
12. Impact Testing 53-5
13. Design Pressure and Temperature 53-6
14. Maximum Allowable Working Pressure . . . 53-6
15. Actual Stresses 53-6
16. Corrosion 53-6
17. Welding Specifications 53-6
18. Weld Types 53-7
19. Weld Applications 53-7
20. Joint Efficiency 53-8
21. Weld Examination 53-9
22. Wall Thickness 53-9
23. Nozzle Necks 53-11
24. Reinforcement of Openings 53-11
25. Flanged Joints 53-14
26. Bolt Strengths 53-14
27. Pipes 53-15
28. Pressure Testing 53-16
29. Vessel Volumes 53-16

Nomenclature

A	area	in²	mm²
c	corrosion allowance	in	mm
D	inside diameter	in	mm
E	efficiency	–	–
h	depth of head	in	mm
K	factor	–	–
L	crown radius	in	mm
M	factor	–	–
p	pressure[1,2]	lbf/in²	Pa
r	knuckle radius	in	mm
R	inside radius	in	mm
s	actual stress	lbf/in²	Pa
S	maximum allowable stress	lbf/in²	Pa
t	thickness	in	mm
V	volume	in³	mm³
w	width	in	mm
W	weld size	in	mm
x	ratio	–	–

Symbols

α	one-sided taper angle (half of apex angle)	deg	deg

Subscripts

a	allowable
c	corroded
h	head
n	nozzle
o	outside
p	pad
r	required

[1]The variable for pressure in the *ASME Boiler and Pressure Vessel Code* is the uppercase P. A lowercase p is used in this chapter for consistency with the rest of this book.
[2]All pressures expressed in this chapter are gage pressures.

1. INTRODUCTION

The American Society of Mechanical Engineers (ASME) established its Boiler and Pressure Vessel Committee in 1911. The Committee establishes rules governing the design, fabrication, inspection, and repair of boilers and pressure vessels and interprets these rules when questions arise. The rules constitute the *ASME Boiler and Pressure Vessel Code*, BPVC, which consists of the 11 sections listed in Table 53.1.

Table 53.1 *Sections in the ASME Boiler and Pressure Vessel Code*

Section	Title
I	Power Boilers
II	Material Specifications
III (Div. 1)	Nuclear Power Plant Components
III (Div. 2)	Concrete Reactor Vessel Containments
IV	Heating Boilers
V	Nondestructive Examination
VI	Recommended Rules for Care and Operation of Heating Boilers
VII	Recommended Rules for Care of Power Boilers
VIII (Div. 1)	Pressure Vessels
VIII (Div. 2)	Pressure Vessels—Alternate Rules
IX	Welding and Brazing Qualifications
X	Fiber-Reinforced Plastic Pressure Vessels
XI	Rules for In-Service Inspection of Nuclear Power Plant Components

This chapter covers only pressure vessels with curved shells that are under internal pressure and are designed in accordance with Sec. VIII, "Pressure Vessels," Div. 1.[3,4,5] Division 1 of Sec. VIII covers pressure vessels operating between 15 and 3000 psig (103 kPa and 20.7 MPa).[6,7] Section VIII covers non-nuclear applications. Pressure vessels intended for use in commercial nuclear power plants are covered in Sec. III of the BPVC.

BPVC jurisdiction over pipes and fittings extends from the pressure vessel only up to the first connection (i.e., pipe joint) to piping and added equipment, regardless of whether the connection is a welded, bolted, or threaded flange or a proprietary sealing pipe joint.[8,9]

2. COMPLIANCE

Compliance with the BPVC by manufacturers is voluntary. However, most customers prefer knowing that the pressure vessels they order are designed and built to a known standard, and most states require use of ASME certified pressure vessels as part of their state law. Manufacturers who build unfired pressure vessels in compliance with the code are able to mark the nameplate of their products with the U stamp (mark) shown in Fig. 53.1.[10]

Figure 53.1 *Pressure Vessel Stamp*

3. EXEMPTIONS

Not all pressure vessels fall within the jurisdiction of the BPVC. Some classes of vessels are outside the scope of the BPVC, including vessels of any size operating below 15 psig (103 kPa), vessels intended for domestic water service and with a nominal capacity not in excess of 120 gal (454 L), and vessels less than 6 in (152 mm) in diameter (without limitation to pressure or length). Any vessel exempt from inspection but nevertheless built in accordance with the BPVC may be stamped with the U symbol.

4. STAMPING REQUIREMENTS

Pressure vessels must be permanently marked with information about their construction and type of service. This information may be stamped on the vessel in a conspicuous location (e.g., near an opening or manway) or may be on an attached nameplate.

Nameplates describing a pressure vessel are attached directly to the shell.[11] The nameplate of a pressure vessel designed and constructed in accordance with the BPVC will contain the official "U" stamp and all of the following: manufacturer's name (listed after the words "certified by"), vessel serial number, year built, maximum allowable working pressure and corresponding temperature, and minimum design metal temperature and corresponding pressure.[12] For pressure vessels that are intended for service below −20°F (−29°C), the minimum allowable temperature is also listed.

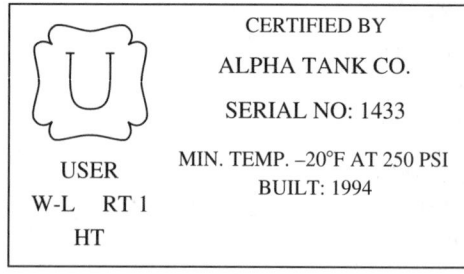

Figure 53.2 *Typical Name Plate*

---

[3]The BPVC also covers pressure vessels under external pressure.
[4]The alternate rules in BPVC Sec. VIII, Div. 2 place greater constraints on the materials and on the design, manufacturing, and testing processes. However, Div. 2 permits the pressure vessel to be designed to a lower factor of safety (i.e., a factor of safety of 3 compared to the theoretical values of 4 or 5 used in Div. 1). The shell and heads of a pressure vessel designed in accordance with Div. 2 will be thinner than a comparable pressure vessel designed to Div. 1. The end result is a reduction in material cost. Most of the rules in Div. 2 pertain only to stationary vessels installed in fixed locations.
[5]Even so, this chapter is only an introduction to the full BPVC.
[6]Division 2, being analysis-oriented, has no pressure limits.
[7]Large low-pressure vessels operating above 0.5 psig (3.4 kPa) but below 15 psig (103 kPa) are designed and built in accordance with API standards. Atmospheric vessels containing flammable and combustible liquids are designed and built in accordance with codes developed by Underwriters' Laboratories (UL) and the American Petroleum Institute (API) and with other acceptable good standards.
[8]Except for threaded plug closures used for inspection, BPVC jurisdiction ends at the first thread joint.
[9]The term *joint* is used to refer to a *welded seam*. The words "joint," "seam," and (sometimes) "weld" are synonymous.
[10]The mark shown in Fig. 53.1 is one of many specified by ASME: A, field assembly of power boilers; E, electric boilers; H, heating boilers (steel plate or cast iron sectional); HLW, lined potable water heaters; M, miniature boilers; N, nuclear components; NPT, nuclear component partials; NA, nuclear installation/assembly; NV, nuclear safety valves; PP, pressure piping; RP, reinforced plastic pressure vessels; U, U2, pressure vessels; UM, miniature pressure vessels; UV, pressure vessel safety valves; V, boiler safety valves.

[11]Duplicate nameplates on supports or at other locations must be marked "Duplicate."
[12]It has been common in some metric countries to specify pressure in either bars or kilograms per square cm (kg/cm^2). Multiply psi by 0.06895 to obtain bars. Multiply psi by 0.07031 to obtain kg/cm^2.

Figure 53.3 *Parts of a Pressure Vessel*

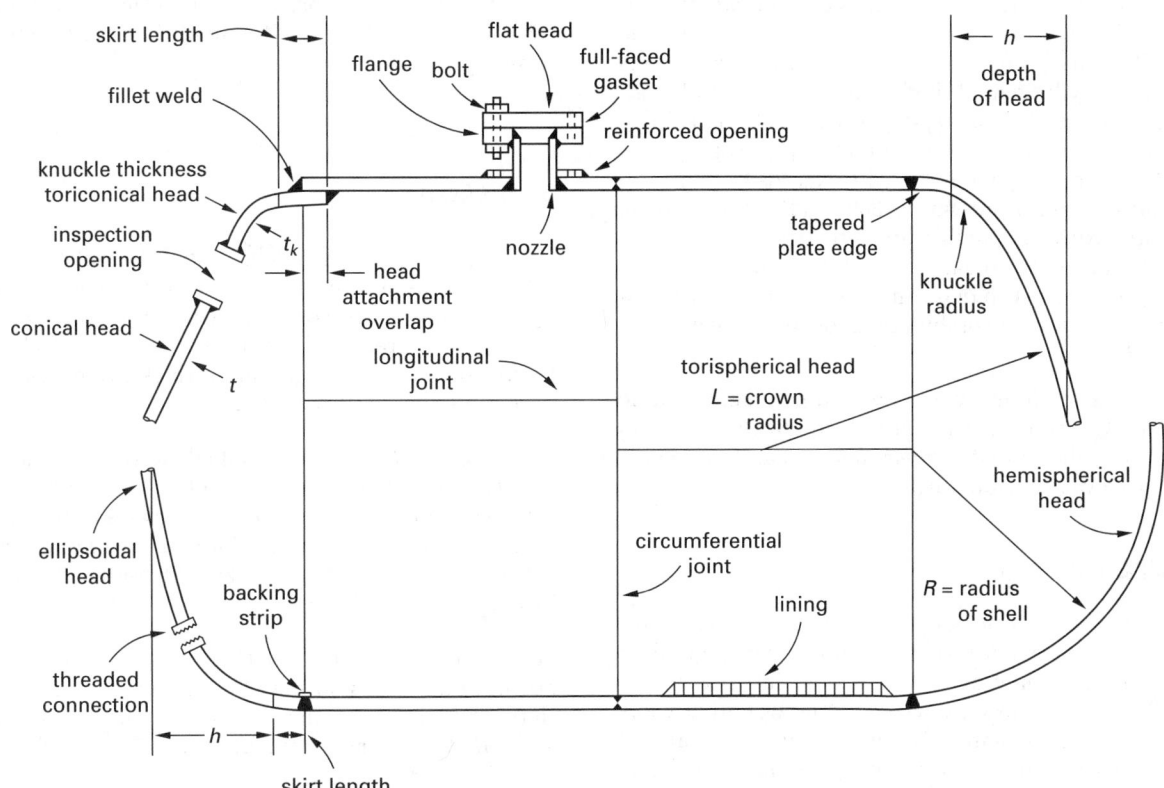

The method of construction and type of service must also be listed. One or more of the following abbreviations will be used: USER, when inspected by a user's inspector; W, arc or gas welded; RES, resistance welded; S, seamless; F, forge welded; B, brazed; IT, impact test performed; L, lethal service; UB, unfired steam boiler; D or DF, direct firing; RT-1, fully radiographed; RT-2, some joints partially radiographed; RT-3, spot radiographed; RT-4, radiographed but other categories not applicable; HT, postweld heat treated; PHT, parts of vessel heat treated.

5. DESIGN ELEMENTS

Figure 53.3 illustrates the various parts of a pressure vessel. The pressure vessel can be divided into shell-type and plate-type elements. A *shell-type element* resists internal pressure through tension (i.e., "membrane action"). A *plate-type element* resists internal pressure through bending. Shell-type elements can be cylindrical, spherical, ellipsoidal, torispherical, or toriconical.

The main body of a pressure vessel is known as the *shell*. A shell can be seamless or seamed. External pipes and equipment are connected to a pressure vessel at *nozzles*. Seamless pipe used for a nozzle is an example of a *seamless shell*.

6. SERVICE APPLICATION

Special restrictions are placed on vessels containing lethal substances, operating below $-20°F$ $(-29°C)$, used for steam generation, or subject to direct firing.

The minimum wall thickness of shells and heads is $1/16$ in (1.6 mm), exclusive of any corrosion allowance. However, for pressure vessels intended for compressed air, steam, and water service, the minimum wall thickness of shells and heads is $3/32$ in (2.4 mm).[13] If the calculated wall thickness is less than $1/4$ in (6.4 mm), the minimum corrosion allowance is $1/16$ in (1.6 mm) or one-sixth of the calculated thickness, whichever is greater, except that the sum of the calculated thickness and corrosion allowance need not exceed $1/4$ in (6.4 mm).[14] Pressure vessels intended for compressed air service and those subject to internal corrosion or erosion must be designed with a suitable inspection opening.

A *lethal substance* is a poisonous gas or liquid that is toxic or lethal in even small quantities.[15] Vessels intended for use with lethal substances may not be constructed from SA-283 carbon steel or cast iron. Expanded connections may not be used. All butt-welded joints must be fully radiographed. Postweld heat treatment is required when vessels carrying lethal substances are constructed of carbon or low-alloy steel.

---

[13] A good rule of thumb for determining the minimum thickness of a pressure vessel is $t_{min,in} = (d_{in} + 100)/1000$. This thickness will result in a vessel that is stiff enough for handling during fabrication and erection.

[14] A corrosion allowance is not required for seamless vessel parts designed with 0.85 joint efficiency or for vessels that are not radiographed.

[15] Several lethal substances regularly encountered in the chemical processing industry include hydrocyanic acid, carbonyl chloride, cyanogen, mustard gas, and xylyl bromide.

The BPVC does not specify material selection based on vessel content. However, hydrogen and caustic embrittlement are important noncode issues that must be dealt with during the design process. Pressure vessels exposed to hydrogen at high temperatures and high pressures are subject to *hydrogen embrittlement*. The vessel can be considered to be subject to hydrogen damage if the hydrogen partial pressure exceeds 87 psia (600 kPa) and the vessel pressure is 300°F (150°C) or more, or if the hydrogen partial pressure exceeds 145 psia (1000 kPa absolute) regardless of temperature. A *Nelson diagram* is used to determine materials that are appropriate for various combinations of temperature and pressure.[16]

Caustic soda (sodium hydroxide) is a common alkaline material. Carbon steel is subject to *caustic cracking* (sometimes called *caustic embrittlement*) and rapid corrosion at elevated temperatures.

7. MATERIALS

Pressure vessels can be constructed from various materials including carbon steel, low-alloy steel, high-alloy steel, nonferrous metal, cast iron, and integrally clad plate. The BPVC specifies which materials and at what temperatures these materials can be used. Material categories are designated by P-numbers, as shown in Table 53.2.

Table 53.2 Material Designations

designation	material
P-1	carbon steels with tensile strengths between 40,000 and 75,000 psi (276 and 517 MPa)
P-3	alloy steel with up to $^3/_4$% chromium; alloy steels with up to 2% total alloying ingredients
P-4	alloy steel with chromium between $^3/_4$% and 2%; alloy steels with more than 2% total alloying ingredients
P-5	alloy steels with up to 10% total alloying ingredients
P-8	austenitic stainless steels
P-9	nickel alloy steels
P-10	other steel alloys

Carbon steels are the most common material chosen for noncorrosive environments between −20 and 800°F (−29 and 426°C). However, carbon steel weakens with

---

[16]A *Nelson diagram* is a *p-T* diagram indicating the pressure-temperature combinations below which failure of a partial alloy have not been reported. Since the Nelson diagram is based on a historical record, it is possible that failures will still occur in the regions previously indicated as eventless. Refer to API publication 941.

long exposure to temperatures higher than 785°F (418°C) through a process known as *graphitization*. For service above 800°F (426°C), materials must be selected carefully.

8. HEADS

Heads may be spherical, ellipsoidal, torispherical, conical, or flat, as shown in Fig. 53.4. Ellipsoidal and hemispherical heads are common, while torispherical heads appear more frequently in thin vessels. BPVC formulas for calculating the minimum thickness for each head type are given in Table 53.9.

Torispherical heads are specified by their inside diameter (D), crown (dish) radius (L), knuckle radius (r), and head thickness (t_h). These variables are shown in Fig. 53.4. An *ASME flanged and dished head* is a torispherical head for which the knuckle radius is 6% of the inside crown radius.

When a head is no thicker than its shell, the head does not need a flange and may be butt-welded to the shell. In practice, however, most nonhemispherical heads have *straight flanges* (i.e., straight longitudinal necks).

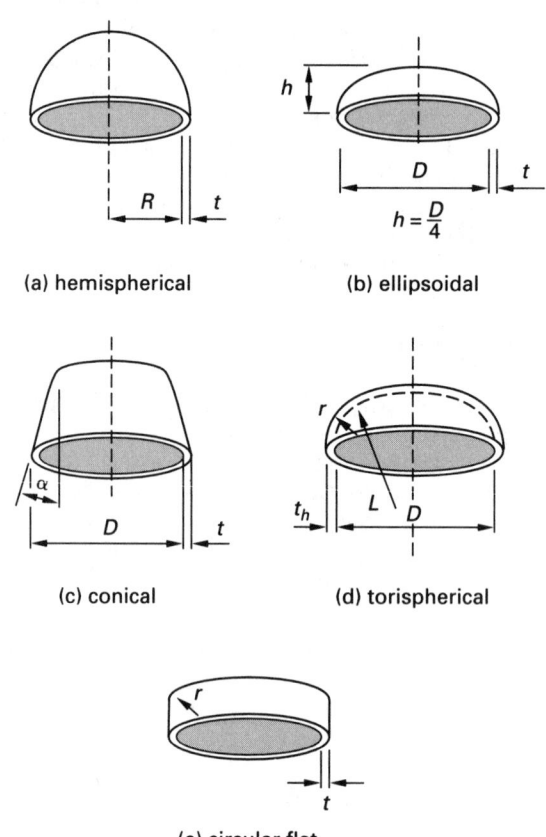

(a) hemispherical

(b) ellipsoidal

(c) conical

(d) torispherical

(e) circular flat

Figure 53.4 Head Shapes

Table 53.3 Maximum Allowable Stress for Carbon and Low-Alloy Steels (thousands of psi)

specification		for metal temperature not exceeding °F							
number	grade	−20 to 650	700	750	800	850	900	950	1050
SA-283	C	12.7	–	–	–	–	–	–	–
SA-285	C	13.8	13.3	12.1	10.2	8.4	6.5	–	–
SA-515	55	13.8	13.3	12.1	10.2	8.4	6.5	4.5	2.5
SA-515	60	15.0	14.4	13.0	10.8	8.7	6.5	4.5	2.5
SA-515	65	16.3	15.5	13.9	11.4	9.0	6.5	4.5	2.5
SA-515	70	17.5	16.6	14.8	12.0	9.3	6.5	4.5	2.5
SA-516	55	13.8	13.3	12.1	10.2	8.4	6.5	4.5	2.5
SA-516	60	15.0	14.4	13.0	10.8	8.7	6.5	4.5	2.5
SA-516	65	16.3	15.5	13.9	11.4	9.0	6.5	4.5	2.5
SA-516	70	17.5	16.6	14.8	12.0	9.3	6.5	4.5	2.5
SA-105		17.5	16.6	14.8	12.0	9.3	6.5	4.5	2.5
SA-181	I	15.0	14.4	13.0	10.8	8.7	6.5	4.5	2.5
SA-350	LF1	15.0	14.4	13.0	10.8	7.8	5.0	3.0	1.5
SA-350	LF2	17.5	16.6	14.8	12.0	7.8	5.0	3.0	1.5
SA-53	B	15.0	14.4	13.0	10.8	8.7	6.5	–	–
SA-106	B	15.0	14.4	13.0	10.8	8.7	6.5	4.5	2.5
SA-193	B7 ≤ $2^1/_2$ in	25.0	25.0	23.6	21.0	17.0	12.5	8.5	4.5

(Multiply psi by 6.895 to obtain kPa.)

9. LOADING TYPE

In addition to loads from internal pressurization, pressure vessels must be designed for the weight of the vessel and its contents, static reactions from attachments, wind pressure, seismic forces, temperature gradients and thermal expansion, and dynamic forces from cyclic variations in pressure or temperature.

10. ALLOWABLE STRESSES

The base allowable stress in tension, S_a, is specified in the BPVC (see Table 53.3) and depends on the material and the temperature. (The allowable stress for parts in compression also depends on the slenderness ratio.) The BPVC specifies that the *maximum allowable stress* used for design purposes is S_a for tensile stresses and for general primary membrane stress induced by any combination of loading. For most internal-pressure configurations, allowable stress in compression is not the controlling factor.[17]

The maximum allowable stress is $1.5S_a$ for the sum of primary membrane stress and bending stress due to any combination of loadings except those for wind and earthquake. Wind and earthquake loadings are not considered to act simultaneously. When wind or earthquake loads are present, the maximum allowable stress for carbon and low-alloy steels below 700°F (371°C) is $1.2S_a$.

[17]The *critical height* of a vertical vessel under internal pressure and subject to wind loading is $pR/16t$. Above this height, compressive stress governs.

Table 53.3 lists the maximum allowable stress for common carbon and low-alloy steels. Values in Table 53.3 may be interpolated.

11. TEMPERATURE EFFECTS

Table 53.4 lists the modulus of elasticity for ferrous materials as a function of temperature.

Table 53.4 Modulus of Elasticity of Ferrous Materials (millions of psi)

temperature		carbon steels		austenitic stainless steel
(°F)	(°C)	$C \leq 0.3\%$	$C > 0.3\%$	
70	21	29.5	29.3	28.3
200	93	28.8	28.6	27.6
300	149	28.3	28.6	27.6
400	204	27.7	27.5	26.5
500	260	27.3	27.1	25.8
600	316	26.7	26.5	25.3
700	371	25.5	25.3	24.8
800	427	24.2	24.0	24.1
900	482	22.4	22.3	23.5
1000	538	20.4	20.2	22.8
1100	593	18.0	17.9	22.1

(Multiply psi by 6.895 to obtain kPa.)

12. IMPACT TESTING

Brittle fracture is a concern when a pressure vessel is designed to handle cryogenic fluids or is exposed to severe winter temperatures. Only *notch-tough materials* should be used when the design temperature is very low. Notch toughness is determined by *impact testing* the pressure vessel material. The cutoff temperature below which impact testing is required depends on the material and the shell thickness. In general, most ferrous metals require special attention whenever the design temperature is less than $-20°F$ ($-29°C$).[18]

Because there is no marked decrease in ductility at low temperature, vessels manufactured from wrought aluminum do not require impact tests until the design temperature is below $-452°F$ ($-269°C$). Other non-ferrous metals, including copper, copper alloys, nickel, and nickel alloys, can be used down to $-325°F$ ($-198°C$) before impact testing is required.

Impact testing is not required if all of the four following conditions are met: (1) the wall thickness is not greater than $1/2$ in (12.7 mm), (2) the vessel is hydrostatically tested, (3) the design temperature is between -20 and $650°F$ (-29 and $343°C$), and (4) thermal or mechanical shock loadings are not a controlling design element.

13. DESIGN PRESSURE AND TEMPERATURE

The BPVC requires that pressure vessels be designed for the most severe combination of pressure and temperature that will be encountered during normal operation, regardless of whether the combination is short term or infrequent.

The *operating pressure*, p, is the pressure at the top of the vessel at which the vessel normally operates. The maximum difference in pressures between the inside and the outside of a vessel or between any two chambers of a combination vessel should also be evaluated. The *design pressure* is the operating pressure plus a reasonable safety margin.[19] The design pressure usually coincides with the relief valve setting.

The *design temperature* is the pressure vessel's metal temperature corresponding to the design pressure. Since the temperature can vary with location and through the thickness, the mean temperature is specified on the nameplate.

14. MAXIMUM ALLOWABLE WORKING PRESSURE

The *maximum allowable working pressure*, MAWP, is specified by the manufacturer. It is the maximum pressure permissible at the top of the vessel in its normal operating position and temperature, in corroded condition, and while under the effects of other expected loadings (e.g., wind and external pressure).[20] MAWP is calculated for different parts of the pressure vessel based on BPVC equations adjusted for static head. The overall MAWP is the smallest of the adjusted values. The maximum allowable working pressure is the design pressure in the thickness equations.

The term *maximum allowable pressure new and cold*, MAWP N & C, is specified by the manufacturer. It is the maximum pressure for the vessel when new (not corroded) and at room temperature.

15. ACTUAL STRESSES

Standard thin-wall equations are used to evaluate the actual tensile stresses in cylindrical shells. For a cylinder with mean radius R, the longitudinal stress and circumferential stresses are

$$s_{\text{longitudinal}} = \frac{pR}{2t} \qquad 53.1$$

$$s_{\text{circumferential}} = \frac{pR}{t} \qquad 53.2$$

Equations 53.1 and 53.2 can also be used to calculate the stresses in longitudinal and circumferential joints.[21]

Equations 53.1 and 53.2 may not be used to calculate wall thickness.

16. CORROSION

The wall thickness of pressure vessels subject to corrosion, erosion, or abrasion must be increased over what is calculated during the design. A corrosion allowance of 0.005 in per year (approximately $1/16$ in (1.6 mm) in 12 years) during the useful life of the pressure vessel is typical.[22] Determining the useful life is a matter of economics. In the absence of other information, useful lives of 10 years (minor vessels) and 15 to 20 years (major vessels) are appropriate. The wall thickness calculated from the equations in Tables 53.8 and 53.9 are exclusive of the corrosion allowance.

17. WELDING SPECIFICATIONS

Specifications for all of the variables in the welding process are contained in a *welding procedure specification*, WPS. The WPS details such items as the welding method, type of joint, filler metal, preheating, and postweld heat treatment. The validity of the WPS is supported by a *procedure qualification record*, PQR,

---

[18]Austenitic steels do not require impact testing if welding is performed according to specific procedures.
[19]A reasonable safety margin is 10% or 25 psi (170 kPa), whichever is greater.
[20]A pressure vessel can have more than one operating temperature and hence, more than one MAWP.

[21]The term *girth seam* is sometimes used when referring to a circumferential joint.
[22]A corrosion allowance is not specified by the BPVC for thick-walled pressure vessels. Thin-walled pressure vessels whose calculated design thickness are less than 0.25 in (6.4 mm) and who are intended for steam, water, or compressed air service must use a corrosion allowance of at least one-sixth of the calculated design thickness, except that the sum of the design thickness and corrosion allowance need not exceed 0.25 in. Other exceptions apply.

which includes test results on weldability and other factors. Only *welders* and *welder operators* qualified under Sec. IX of the BPVC may construct or repair a pressure vessel.[23]

Welding may be automatic, semiautomatic, or manual, though not all methods may be useful in all cases.[24] Electrodes used in inert gas metal-arc welding may be consumable or nonconsumable. Single or multiple electrodes may be used. Necessary information to completely specify the joint type includes the number of weld passes, the type of pass (i.e., string or weaving bead), the size of weld wire or electrode, and (for automatic welding) the rate of travel and wire feed.

Backing strips may be used to ensure that butt welding from one side penetrates the entire plate thickness. The backing strip becomes fused to the weld.

The position (horizontal, vertical, overhead, etc.) of the welding needs to be specified because not all welders are certified in all positions. Categories are 1G (flat), 2G (horizontal), 3G (vertical), 4G (overhead), 5G (horizontal, fixed), and 6G (inclined, fixed). While it is best to restrict welding to a particular position, welding in multiple positions is permitted. If vertical welding (position 3G) is used, the direction should be specified as uphill or downhill.

Filler metals are specified by filler metal group number (F). For submerged arc welding, the flux must be specified. For inert arc welding, the shielding gas type (e.g., carbon dioxide or argon), composition (e.g., 75% carbon dioxide and 25% argon), and flow rate must be specified.

The BPVC specifies when preheating and postweld heat treatment are required. Some (e.g., P-4 and P-5) materials must always be preheated. Above certain minimum thicknesses, the BPVC requires postweld heat treatment.[25]

18. WELD TYPES

The BPVC specified six types of weld joints. Type 1 weld joints are double-welded butt joints. The quality of weld is the same inside and outside of the vessel with double-welded butt joints. Backing strips, if used, are removed after welding. After the weld is made on one side, the second side of the joint is cleaned and

rewelded. The weld quality is identical on both sides of the joint. Type 2 welds are single-welded butt joints with backing strips that remain in place after welding. Type 3 welds are single-welded butt joints without backing strips. Type 4 joints are double full-fillet lap joints. Type 5 joints are single full-fillet lap joints with plug welds. Type 6 joints are single full-fillet lap joints without plug welds. The six weld types are shown in Fig. 53.5 with their typical welding symbols.[26]

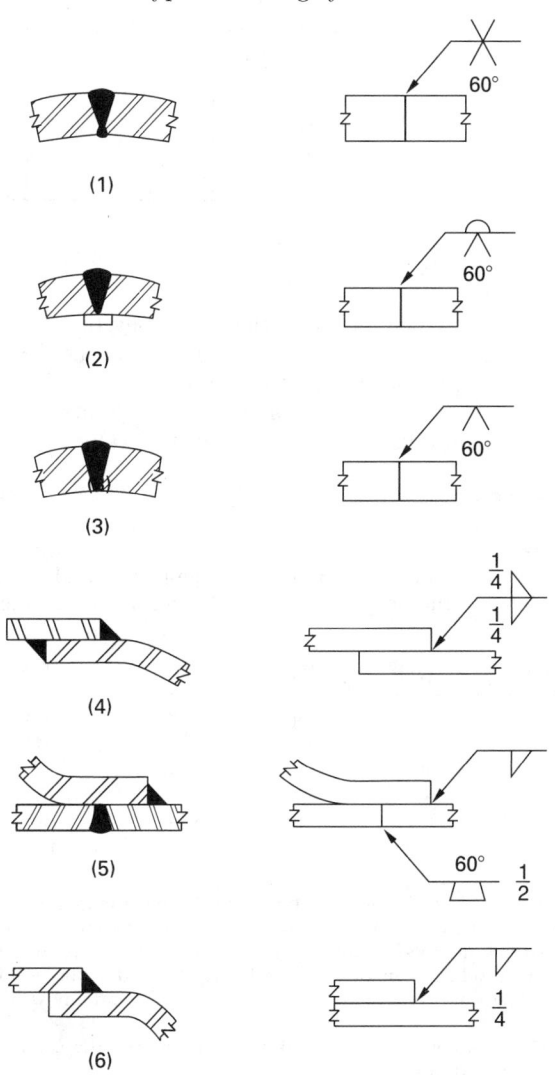

[a]Designations: 1, double-weld butt joint; 2, single-weld butt joint with integral backing strip; 3, single-weld butt joint without backing strip; 4, double-full fillet lap joint; 5, single-full fillet lap joint with plug welds; 6, single-full fillet lap joint without plug welds.

Figure 53.5 *Types of Welds and Symbols[a]*

19. WELD APPLICATIONS

Certain weld types are more appropriate for certain locations on the pressure vessel. Table 53.5 lists the types of welds by pressure vessel location permitted by the BPVC for gas and arc welding methods.

---

[23]A *welder* is qualified in manual and semiautomatic welding. A *welding operator* is qualified to use automatic welding equipment.
[24]In addition to BPVC restrictions, the circumstances of welding limit certain operations. For example, in large vessels without manways, the closing joint must be welded from the outside. Also, small vessels less than approximately 18 to 24 in (460 to 610 mm) cannot be welded from the inside.
[25]For P-1 carbon steel vessels welded without preheating, postweld heat treatment is required for thicknesses greater than $1\frac{1}{4}$ in (31.8 mm). The limit increases to $1\frac{1}{2}$ in (38.1 mm) thickness if the piece has been preheated to 200°F (93°C). Vessels intended for lethal service must be postweld heat treated. Unless exempt from impact testing, carbon steel vessels intended for service below −20°F (−29°C) must be postweld heat treated.

[26]Standard American Welding Society (AWS) symbols are used.

Figure 53.6 *Four Types of Welds by Location*

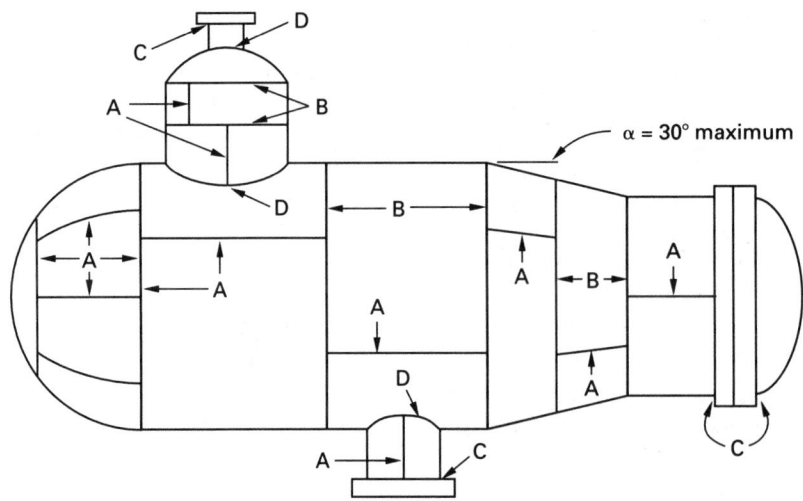

Type 1 joints have no limitations and may be used for all applications. Type 2 joints can also be used anywhere, except that a joint to a plate with a preformed offset can only be used for circumferential joints. Other restrictions are listed in the footnotes of Table 53.5.

Table 53.5 *Permitted Weld Types by Location*

general joint description joint type[a]	longitudinal[b] (A)[g]	circumferential (B)[g]	flange-plate attachment (C)[g]	openings and nozzle attachment (D)[g]
1	yes	yes	yes	yes
2	yes	yes	yes	yes
3	no	yes[c]	yes	no
4	yes[d]	yes[e]	yes	no
5	no	yes[f]	yes[f]	no
6	yes[f]	yes[f]	no	no

[a]Designations: 1, double-weld butt joint; 2, single-weld butt joint with integral backing strip; 3, single-weld butt joint without backing strip; 4, double-full fillet lap joint; 5, single-full fillet lap joint with plug welds; 6, single-full fillet lap joint without plug welds.
[b]Including welds in and on hemispherical heads, and any weld in a sphere.
[c]Thicknesses not over $\frac{5}{8}$ in (16 mm) and outside diameter not over 24 in (610 mm).
[d]Thicknesses not over $\frac{3}{8}$ in (9.5 mm).
[e]Thicknesses not over $\frac{5}{8}$ in (16 mm).
[f]Restrictions apply.
[g]Refer to Fig. 53.6.

Category A welds include (1) longitudinal joints in the main shell, connecting chambers, diameter transitions, and nozzles; (2) any joint in sphere, formed or flat head, or the side plates of a flat-sided vessel; and (3) circumferential joints connecting hemispherical heads to main shells, to transitions in diameter, to nozzles, or to connecting chambers.

Category B welds include (1) circumferential welded joints in the main shell, connecting chambers, nozzles, and diameter transitions; and (2) circumferential welds connecting formed heads (other than hemispherical) to main shells, diameter transitions, nozzles, and connecting chambers.

Category C welds include (1) welded joints connecting flanges, tubesheets or flat heads to main shells, formed heads, diameter transitions, nozzles, and connecting chambers; and (2) any welded joint connecting one side plate to another side plate of a flat-sided vessel.

Category D welds connect chambers of nozzles to main shells, spheres, diameter transitions, heads, and flat sided vessels.

20. JOINT EFFICIENCY

The joint efficiency value, E, used in calculations for a seamless shell is 1.0. For welded joints subject to tension, the values depend on the type of weld and testing processes.[27] The strongest joints are double-welded butt joints (joint type 1). Joint efficiencies for the butt joints and other weld types are summarized in Tables 53.6 and 53.7.

Table 53.6 *Efficiencies of Welded Joints in Shells*

joint type[a]	radiography		
	full	spot	none
1	1.00	0.85	0.70
2	0.90	0.80	0.65
3	–	–	0.60
4	–	–	0.55
5	–	–	0.50
6	–	–	0.45

[a]Designations: 1, double-weld butt joint; 2, single-weld butt joint with integral backing strip; 3, single-weld butt joint without backing strip; 4, double-full fillet lap joint; 5, single-full fillet lap joint with plug welds; 6, single-full fillet lap joint without plug welds.

---

[27]The efficiency of a butt weld joint in compression is 100%.

Table 53.7 *Efficiencies of Welded Joints in Heads*

| head type | joint type[a] | radiography | | |
		full	spot	none
hemispherical	1	1.00	0.85	0.70
	2	0.90	0.80	0.65
others	any	1.00	1.00	0.85

[a]1, double-weld butt joint; 2, single-weld butt joint with integral backing strip.

21. WELD EXAMINATION

Common methods of *nondestructive examination*, NDE, used on welded joints of pressure vessels are radiography and ultrasonic examination.[28,29] Full radiographic inspection of joints is mandatory for (1) all longitudinal welds and welds at openings when the joint efficiency is taken as 1.0 or 0.90; (2) all butt welds joined by electroslag and electrogas methods; (3) with some exceptions, all butt welds where the material thickness exceeds $1^1/_2$ in (38 mm); (4) all butt welds in unfired steam boilers operating at 50 psig (345 kPa) or above; and (5) all butt welds in vessels containing lethal substances.[30]

Spot radiographic inspection can be used to obtain an economical spot check of welding quality.

Radiographic inspection is optional for butt-welded joints that are not required to be fully radiographed, and is not required when the vessel is designed for external pressure or when the joint has been designed (i.e., the joint efficiency value has been chosen) for no inspection. Radiographic inspection is not required for circumferential butt joints 10 in (254 mm) in diameter or smaller in nozzles and connecting chambers.

Full ultrasonic inspection is required for electroslag and electrogas welds in ferritic material.

22. WALL THICKNESS

Pressure vessels can be designed as thin- and thick-shells, though most are thin-walled. The minimum thickness of a thin-walled, cylindrical shell of inside radius R is usually governed by the circumferential stress

and is calculated from Eq. 53.3.[31] The *effective allowable stress*, SE, is the product of the BPVC's tabulated allowable stress value and the joint efficiency.

$$t = \frac{pR}{SE - 0.6p} \qquad [p < 0.385SE; t \le 0.5R] \qquad \text{53.3}$$

The minimum thickness of a thin-walled, spherical shell or a hemispherical head of inside radius R is calculated from Eq. 53.4. For heads without a flange, use the efficiency of the head-to-shell joint if it is less than the efficiency of the joint in the head.

$$t = \frac{pR}{2SE - 0.2p} \qquad [p < 0.665SE; t \le 0.356R] \qquad \text{53.4}$$

Table 53.8 contains the code formulas for designing thin-walled cylindrical shells.

Table 53.8 *Required Wall Thickness of Cylindrical Shells[a]*

member	minimum thickness[a], t	maximum pressure, p	limitation
thin shell (in terms of inside radius)			
longitudinal joint	$\dfrac{pR}{SE - 0.6p}$	$\dfrac{SEt}{R + 0.6t}$	$p \le 0.385SE$[b] $t \le 0.5R$
circumferential joint	$\dfrac{pR}{2SE + 0.4q}$	$\dfrac{2SEt}{R - 0.4t}$	$p \le 1.25SE$ $t \le 0.5R$
thin shell (in terms of outside radius R_o)	$\dfrac{pR_o}{SE + 0.4p}$	$\dfrac{SEt}{R_o - 0.4t}$	$p \le 0.385SE$ $t \le 0.5R$

[a]The minimum thickness of shells and heads used for compressed air, water, or steam is $^3/_{32}$ in (2.4 mm). The minimum thickness for plates of welded construction is $^1/_{16}$ in (1.6 mm).
[b]Equations based on stress across a circumferential joint governs only if the efficiency of the circumferential joint is less than one-half of the longitudinal joint's efficiency.

The minimum thickness of a head depends on its shape, as indicated in Table 53.9.

---

[28]Most other methods of NDE, including eddy current, acoustic emission, liquid penetrant, and magnetic particle testing, are also used with pressure vessels.
[29]Radiography is sometimes referred to as "X raying." However, radiography can use either X rays generated from high electrical voltages or gamma rays generated from a radioactive isotope capsule.
[30]Other requirements based on thickness are: P-1 materials, 1.25 in (32 mm); P-3 materials (groups 1, 2, and 3), 0.75 in (19.1 mm); P-4 materials (groups 1 and 2), 0.625 in (15.9 mm); P-5 materials (groups 1 and 2), 0.0 in (0 mm); P-9A and P-9B materials (group 1), 0.625 in (15.9 mm); P-10A and P-10F materials (groups 1 and 6), 0.75 in (19.1 mm); P-10B and P-10C materials (groups 2 and 3), 0.625 in (15.9 mm).

[31]In the absence of additional loads from wind, earthquake or attachments, the stress in a circumferential joint will govern only when the efficiency of the circumferential joint is less than one-half of the longitudinal joint efficiency.

Machine Design

Table 53.9 *Required Thickness of Heads*

head type	minimum thickness, t	maximum internal pressure, p	limitation
thin hemispherical (inside radius)	$\dfrac{pR}{2SE - 0.2p}$	$\dfrac{2SEt}{R + 0.2t}$	$p \leq 0.665SE$
(outside radius)	$\dfrac{pR_o}{2SE + 0.8p}$	$\dfrac{2SEt}{R_o - 0.8t}$	$t \leq 0.356R$
ellipsoidal (inside diameter)	$\dfrac{pDK}{2SE - 0.2p}$	$\dfrac{2SEt}{KD + 0.2t}$	$K = \left(\tfrac{1}{6}\right)\left[2 + \left(\dfrac{D}{2h}\right)^2\right]$
(outside diameter)	$\dfrac{pD_oK}{2SE + 2p(K - 0.1)}$	$\dfrac{2SEt}{KD_o - 2t(K - 0.1)}$	$K \leq 1.0$
torispherical (inside length)	$\dfrac{pLM}{2SE - 0.2p}$	$\dfrac{2SEt}{LM + 0.2t}$	$\dfrac{L}{r} \leq 16\tfrac{2}{3}$
(outside length)	$\dfrac{pL_oM}{2SE + p(M - 0.2)}$	$\dfrac{2SEt}{ML_o - t(M - 0.2)}$	$M = \left(\tfrac{1}{4}\right)\left(3 + \sqrt{\dfrac{L}{r}}\right)$
conical (inside diameter)	$\dfrac{pD}{2\cos\alpha(SE - 0.6p)}$	$\dfrac{2SEt\cos\alpha}{D + 1.2t\cos\alpha}$	$\alpha \leq 30°$
(outside diameter)	$\dfrac{pD_o}{2\cos\alpha(SE + 0.4p)}$	$\dfrac{SEt\cos\alpha}{D_o - 0.8t\cos\alpha}$	
circular flat (inside diameter)	$D\sqrt{\dfrac{0.13p}{SE}}$	$\dfrac{7.69SEt^2}{D^2}$	$D \leq 24$ in $0.05 \leq \dfrac{t}{D} \leq 0.25$ $t \geq t_{\text{shell}}$

Although plates of any thickness can be produced upon order and heads can be "spun" to any thickness, maximum economy will be achieved when standard-thickness plates are used. Table 53.10 lists the thicknesses of plates commonly available in the United States.

Table 53.10 *Standard Plate Thicknesses*

in		mm
1/16	(0.062)	1.59
5/64	(0.078)	1.98
3/32	(0.090)	2.38
1/8	(0.125)	3.18
3/16	(0.188)	4.76
1/4	(0.250)	6.35
5/16	(0.313)	7.94
3/8	(0.375)	9.53
7/16	(0.437)	11.1
1/2	(0.500)	12.7
every 1/16 to 1 in		
every 1/8 to 2 in		

Example 53.1

A cylindrical pressure vessel is being designed to operate at 125 psig (862 kPa) design pressure and 700°F (371°C) design temperature. When new, the vessel will be 90 in (2286 mm) inside diameter. It will be constructed of SA-515 grade-70 plate with spot-examined, double-welded butt joints throughout. A 0.125 in (3.175 mm) corrosion allowance is to be included. (a) What wall thickness is required? (b) What standard wall thickness plate should be used?

SI Solution

(a) The joint efficiency of a spot-examined double-welded butt joint is 0.85. From Table 53.4, the allowable stress for SA-515-70 plate at 700°F is 16,600 lbf/in^2.

$$S = \left(16{,}600\ \frac{\text{lbf}}{\text{in}^2}\right)\left(6.895\ \frac{\text{kPa·in}^2}{\text{lbf}}\right)$$
$$= 114\,457\ \text{kPa}$$

In its corroded condition, the vessel's inside radius will be

$$R_c = \frac{2286\ \text{mm}}{2} + 3.175\ \text{mm} = 1146.175\ \text{mm}$$

Check that Eq. 53.3 can be used.

$$0.385SE = (0.385)(114\,457 \text{ kPa})(0.85)$$
$$= 37\,456 \text{ kPa} > 862 \text{ kPa} \quad [\text{ok}]$$

From Eq. 53.3, the wall thickness is

$$t = \frac{pR}{SE - 0.6p} + c$$
$$= \frac{(862 \text{ kPa})(1146.175 \text{ mm})}{(114\,457 \text{ kPa})(0.85) - (0.6)(862 \text{ kPa})} + 3.175 \text{ mm}$$
$$= 10.21 \text{ mm} + 3.175 \text{ mm} = 13.385 \text{ mm}$$

By inspection, this is less than $0.5R$.

(b) From Table 53.10, select a 9/16 in plate with a thickness of 11.1 mm.

Customary U.S. Solution

(a) The joint efficiency of a spot-examined, double-welded butt joint is 0.85. From Table 53.4, the allowable stress for SA-515-70 plate at 700°F is 16,600 psi. In its corroded condition, the vessel's inside radius will be

$$R_c = \frac{90 \text{ in}}{2} + 0.125 \text{ in} = 45.125 \text{ in}$$

Check that Eq. 53.3 can be used.

$$0.385SE = (0.385)\left(16{,}600 \, \frac{\text{lbf}}{\text{in}^2}\right)(0.85)$$
$$= 5432 \text{ lbf/in}^2 > 125 \text{ lbf/in}^2 \quad [\text{ok}]$$

From Eq. 53.3, the wall thickness is

$$t = \frac{pR}{SE - 0.6p} + c$$
$$= \frac{\left(125 \, \dfrac{\text{lbf}}{\text{in}^2}\right)(45.125 \text{ in})}{\left(16{,}600 \, \dfrac{\text{lbf}}{\text{in}^2}\right)(0.85) - (0.6)\left(125 \, \dfrac{\text{lbf}}{\text{in}^2}\right)}$$
$$\quad + 0.125 \text{ in}$$
$$= 0.402 \text{ in} + 0.125 \text{ in} = 0.527 \text{ in}$$

By inspection, this is less than $0.5R$.

(b) From Table 53.10, select a 9/16 in plate with a thickness of 0.5625 in.

23. NOZZLE NECKS

Nozzles are openings in the pressure vessel that connect the vessel to subsequent parts of the processing system. The cylindrical section between a pressure vessel and a flange is the *nozzle neck*. The required thickness of the nozzle neck on a pressure vessel under internal pressure is the minimum of the following four values: (1) the thickness computed from the loadings (internal pressurization, reactions, etc.) plus an allowance for corrosion, (2) the required vessel wall thickness (based on $E = 1$) plus an allowance for corrosion, (3) the minimum thickness of standard pipe (considering a 12.5% manufacturing tolerance) plus an allowance for corrosion, and (4) $^1/_{16}$ in (1.6 mm).

Vessels subject to internal corrosion, erosion, or mechanical abrasion must have inspection openings. *Manways* are large openings used for inspection, cleaning, and repair.[32] The required thickness for manways and *handholes* is the thickness computed from the loading (inclusive of an allowance for corrosion).

24. REINFORCEMENT OF OPENINGS

Depending on the shell wall thickness and the size of the opening, reinforcement at the shell-neck connection point may be required.[33] The basic rule is that the cross-sectional area of the reinforcement must equal the cross-sectional area removed less any "excess" area not needed to resist the internal pressure.

Reinforcement is most easily obtained by adding a *pad*, an annular flat plate exterior to and surrounding the nozzle.[34] Some manufacturers replace all of the removed area with pad area as a general practice. This requires little or no calculations, but results in over-reinforced openings. The cross-sectional pad area is

$$A_p = 2w_p t_p \qquad \qquad 53.5$$

For vessels under internal pressure, the cross-sectional area of the required reinforcement is given by Eq. 53.6, in which D is the finished opening diameter in its corroded condition and t is the theoretical thickness of the shell or head as calculated from the code equations.[35,36]

$$A_r = D_c t_r \qquad \qquad 53.6$$

---

[32] A *davit* (or *davit arm*) is a swinging support constructed as part of the vessel and that supports the manway cover when it is unbolted and moved aside.

[33] Reinforcement of pressure vessels not subject to rapid pressure fluctuations is not required when the vessel thickness is $^3/_8$ in (9.5 mm) or less and the opening diameter is 3.5 in (88.9 mm) or less, or when the vessel thickness is more than $^3/_8$ in (9.5 mm) and the opening diameter is 2.375 in (60.3 mm) or less.

[34] Other methods of reinforcement are to use a heavier nozzle neck or longer extension of the nozzle inside the vessel.

[35] Reinforcement of extraordinarily large openings requires a different procedure. For pressure vessels with diameters of 60 in (1.52 m) or less, large openings are those that are more than one-half the vessel diameter or larger than 20 in (508 mm). For pressure vessels with diameters of more than 60 in (1524 mm), large openings are those that are more than one-third the vessel diameter or larger than 40 in (1016 mm).

[36] If the opening and its reinforcement are entirely within the spherical portion of a flanged and dished head, t_r is the thickness required using $M = 1$. Refer to the BPVC for calculating the required thickness for reinforcement purposed for openings in conical and ellipsoidal heads.

Figure 53.7 *Reinforcement of an Opening*

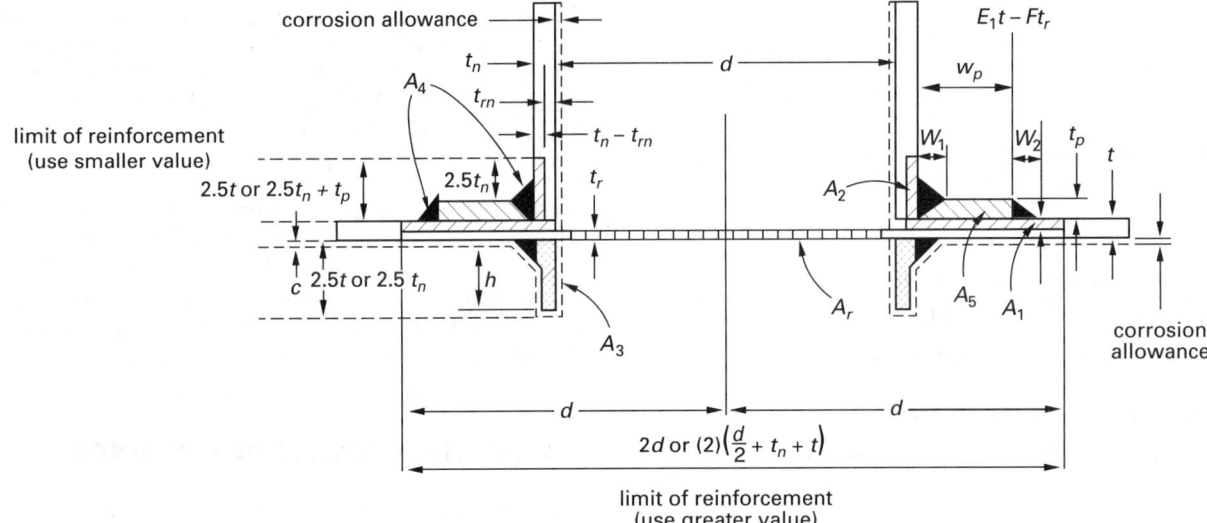

Since corrosion ultimately will reduce the available area, corrosion is accounted for by deducting the *corrosion allowance, c*, from the actual shell and nozzle thicknesses and by adding twice the corrosion allowance to the opening diameter.

The actual shell thickness may be larger than the theoretical thickness, and the difference contributes to the area available. This difference is known as an *excess* and is based on the actual thickness less the corrosion allowance (i.e., is based on the corroded thickness). Excess area comes from excess thickness in the shell or head (A_1), excess area in nozzle wall (A_2), inward projecting nozzle area (A_3), and cross-sectional area of the welds (A_4). (Refer to Fig. 53.7 for some of the dimensions used.)

$$A_1 = \text{greater of} \begin{Bmatrix} (Et - t_r)D \\ (2)(Et - t_r)(t + t_n) \end{Bmatrix} \quad 53.7$$

The weld efficiency, E, is 1.00 when the opening is in a solid plate or when it passes through a type B (circumferential) joint. If the opening passes through any other welded joint, the efficiency of the welded joint is used. When the welded joint has not been radiographically inspected, $E = 0.85$ for a type 1 joint and $E = 0.80$ for a type 2 joint.

$$A_2 = \text{smaller of} \begin{Bmatrix} (2)(2.5t_n)(t_n - t_{rn}) \\ (2)(2.5t)(t_n - t_{rn}) \end{Bmatrix} \quad 53.8$$

$$A_3 = (2h)(t_n - c) \quad 53.9$$

$$A_4 = W_1^2 + W_2^2 \quad 53.10$$

The criterion for a pad is

$$A_r > A_1 + A_2 + A_3 + A_4 \quad \text{[pad needed]} \quad 53.11$$

The pad area required is

$$A_{p,r} = A_r - (A_1 + A_2 + A_3 + A_4) \quad 53.12$$

Not all material around an opening is effective reinforcement. To be considered effective, the material must be located within the *limits of reinforcement*. Material located around a 2*d*-diameter (approximately) circle concentric with the opening axis is considered effective, as is material located up to (approximately) $2.5t$ or $2.5t_n$ above and below the vessel surface. This assumption is built into the calculation of A_2. (Refer to Fig. 53.7 for exact limits of reinforcement.)

If the shell material is stronger than the pad, nozzle, weld, or other reinforcement material, then the areas are modified in proportion to the ratio of their allowable stresses.[37] The modification is two-fold: The required area is increased, and the area considered available as reinforcement is reduced.

No reduction in the required area is made when the pad, nozzle, weld, or other reinforcement material is stronger than the shell material.

Some openings are considered to be *integrally reinforced*. This occurs when the reinforcement is inherent in the shell or nozzle, or when the opening is built up by welding.[38] Integral reinforcement requires use of a code correction factor, F, in Eqs. 53.7 through 53.10 to compensate for variations in stress.

Example 53.2

A pressure vessel has an inside diameter of 46 in. Vessel walls are 0.5625 in SA-516-60 plate. The vessel is intended for service at 350 psig and 200°F. An 8 in (nominal) diameter nozzle with 0.594 in wall thickness

---

[37]The scaling factor is not the ratio of moduli of elasticity as it is with beam reinforcement subject to flexure.
[38]An installed pad is not considered to be integral reinforcement.

is constructed of SA-53-B material. The nozzle extends 1.50 in into the pressure vessel and does not interrupt any of the main vessel seams. $^3/_8$ in fillet welds are used throughout. All seams are fully radiographed. Corrosion is not a factor in the design. (a) Verify that additional reinforcement is required. (b) Design the reinforcement pad using 0.250 in thick SA-516-60 plate.

Solution

(a) Determine the required thicknesses of the shell and nozzle. From Table 53.4, the maximum allowable stress is 15,000 lbf/in^2 for both the shell and nozzle materials. The efficiency of fully radiographed welds is $E = 1.00$.

The shell inside radius is

$$R = \frac{46 \text{ in}}{2} = 23 \text{ in}$$

From Eq. 53.3, the required wall thickness is

$$
\begin{aligned}
t &= \frac{pR}{SE - 0.6p} \\
&= \frac{\left(350 \dfrac{\text{lbf}}{\text{in}^2}\right)(23.0 \text{ in})}{\left(15{,}000 \dfrac{\text{lbf}}{\text{in}^2}\right)(1.00) - (0.6)\left(350 \dfrac{\text{lbf}}{\text{in}^2}\right)} \\
&= 0.544 \text{ in}
\end{aligned}
$$

The 0.594 in wall thickness and 8 in nominal diameter indicates that the nozzle is constructed from schedule-100 pipe. The actual inside diameter of schedule-100 pipe is 7.437 in. The nozzle inside radius is

$$R = \frac{7.437 \text{ in}}{2} = 3.7185 \text{ in}$$

From Eq. 53.3, the required wall thickness of the nozzle is

$$
\begin{aligned}
t &= \frac{pR}{SE - 0.6p} \\
&= \frac{\left(350 \dfrac{\text{lbf}}{\text{in}^2}\right)(3.7185 \text{ in})}{\left(15{,}000 \dfrac{\text{lbf}}{\text{in}^2}\right)(1.00) - (0.6)\left(350 \dfrac{\text{lbf}}{\text{in}^2}\right)} \\
&= 0.088 \text{ in}
\end{aligned}
$$

From Eq. 53.6, the total reinforcement required is

$$
\begin{aligned}
A_r &= D t_r = (7.437 \text{ in})(0.544 \text{ in}) \\
&= 4.046 \text{ in}^2
\end{aligned}
$$

The reinforcement available is calculated in four parts. (Areas are rounded down to understate the available reinforcement.)

$$
\begin{aligned}
(Et - t_r)D &= [(1.00)(0.5625 \text{ in} - 0.544 \text{ in})] \\
&\quad \times (7.437 \text{ in}) \\
&= 0.137 \text{ in}^2
\end{aligned}
$$

$$
\begin{aligned}
(2)(Et - t_r)(t + t_n) &= (2)[(1.00)(0.5625 \text{ in} - 0.544)] \\
&\quad \times (0.5625 \text{ in} + 0.594 \text{ in}) \\
&= 0.042 \text{ in}^2
\end{aligned}
$$

$$
\begin{aligned}
A_1 &= \text{greater of } \{0.137 \text{ in}^2, \ 0.042 \text{ in}^2\} \\
&= 0.137 \text{ in}^2
\end{aligned}
$$

$$
\begin{aligned}
(2)(2.5 t_n)(t_n - t_{rn}) &= (2)[(2.5)(0.594 \text{ in})] \\
&\quad \times (0.594 \text{ in} - 0.088 \text{ in}) \\
&= 1.502 \text{ in}^2
\end{aligned}
$$

$$
\begin{aligned}
(2)(2.5 t)(t_n - t_{rn}) &= (2)[(2.5)(0.5625 \text{ in})] \\
&\quad \times (0.594 \text{ in} - 0.088 \text{ in}) \\
&= 1.423 \text{ in}^2
\end{aligned}
$$

$$
\begin{aligned}
A2 &= \text{smaller of } \{1.502 \text{ in}^2, \ 1.423 \text{ in}^2\} \\
&= 1.423 \text{ in}^2
\end{aligned}
$$

The nozzle projects 1.50 in into the vessel. The limit of reinforcement parallel to the nozzle wall is the smaller of $2.5t$ or $2.5t_n$. In this case, $t < t_n$. The limit of reinforcement is

$$2.5t = (2.5)(0.5625 \text{ in}) = 1.406 \text{ in}$$

$$
\begin{aligned}
A_3 &= (2)(t_n - c)h = (2)(0.594 \text{ in} - 0)(1.406 \text{ in}) \\
&= 1.670 \text{ in}^2
\end{aligned}
$$

Reinforcement is provided by the nozzle-to-shell weld. (If a reinforcement pad is used, reinforcement is also provided by the pad-to-shell weld if that weld is within the limits of reinforcement. The pad-to-shell area is disregarded.)

$$
\begin{aligned}
A_4 &= W^2 = (0.375 \text{ in})^2 \\
&= 0.140 \text{ in}^2
\end{aligned}
$$

The total reinforcement available is

$$
\begin{aligned}
A_1 + A_2 + A_3 + A_4 &= 0.137 \text{ in}^2 + 1.423 \text{ in}^2 \\
&\quad + 1.670 \text{ in}^2 + 0.140 \text{ in}^2 \\
&= 3.370 \text{ in}^2
\end{aligned}
$$

Since $A_{\text{available}} < A_r$, reinforcing is needed.

(b) The pad area required is

$$
\begin{aligned}
A_{pr} &= A_r - (A_1 + A_2 + A_3 + A_4) \\
&= 4.046 \text{ in}^2 - 3.370 \text{ in}^2 = 0.676 \text{ in}^2
\end{aligned}
$$

Since the pad thickness is 0.250 in, the width of the pad is

$$w_p = \frac{0.676 \text{ in}^2}{0.250 \text{ in}} = 2.704 \text{ in}$$

The outside diameter of the 8 in nozzle is 8.625 in. The outside diameter of the circular pad is

$$8.625 \text{ in} + 2.704 \text{ in} = 11.329 \text{ in}$$

25. FLANGED JOINTS

Flanged joints (with flat cover plates) are needed to disassemble, inspect, and clean pressure vessels. Joints may be bolted or boltless pressure-actuated. *Bolted joints*, where sealing gaskets are compressed by bolt forces, are more common. In boltless joints (which may be of the *axially locked joint* and *pressure-actuated joint* varieties), the internal pressure compresses and seals the gasket. Because of the relative size advantages, a boltless joint may be superior at pressures over 2000 psig (14 MPa) and when the shell diameter is roughly 20 in (510 mm) or when the flange thickness exceeds $1^1/_2$ in (38 mm).

The three main types of bolted flanges are ring flange, tapered hub (also known as *welding neck*) flange, and lap-joint flange. The *lap-joint flange* is used for low-pressure, low-cost pressure vessels. Joints may be hubbed or hubless. Advantages of this flange type are low cost and ease of bolt hole alignment. The backing ring can be constructed of a different material from the shell and lap ring, an important consideration when expensive alloys are used.

The *ring flange* is suitable for low and moderate pressure. It consists of an annular plate welded to the end of a cylindrical nozzle (shell). Bolts are spaced equidistantly around the bolt circle. The number of bolts is commonly a multiple of four. An unconfined gasket is used between the annular plate and the closure plate. The gasket usually extends to the inner edge of the bolt line so that the bolts can help center the gasket. A full-face gasket may cover the entire flange area and extend beyond the bolt circle, but is typically used for pressure less than 100 psig (700 kPa).[39]

For reliable and safe operation up to a pressure of approximately 5000 psig (35 MPa), the *tapered-hub flange* can be used. This flange is (roughly) L-shaped and is butt-welded to the shell opening.

Depending on the design, the mating area of the flange surfaces and/or cover plate may or may not compress the gasket. An unconfined and prestressed gasket is commonly used with flat-faced ring flanges. Such a gasket can expand inward and outward when tightened. Since the gasket is unconfined, there is no protection against *gasket blowout*.

Semiconfined gaskets are confined in single-step male-female types of joints. Gaskets are completely confined in tongue-and-groove, ring, double-step male-female joints. Fully confined gaskets are appropriately chosen when there are significant fluctuations in pressure and temperature.

---

[39]In full-face gaskets, the material outside of the bolt ring is not effective in sealing.

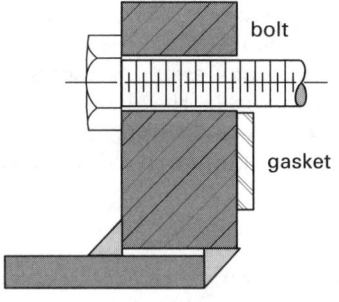

(a) ring flange with flat face

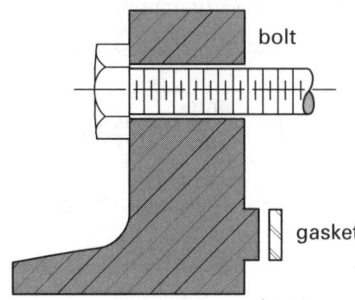

(b) welding neck flange with tongue facing

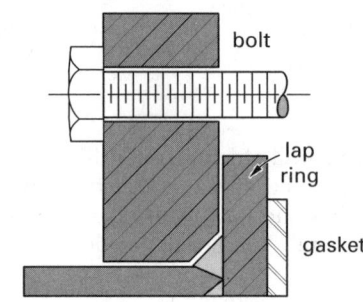

(c) lap joint with raised-face flange

Figure 53.8 *Types of Flanges*

The major concern in regard to flange and gasket choice is flange leakage. Gaskets chosen for operation under internal pressure are much less effective under vacuum. Sheet gaskets may be "sucked in," though this is countered by specifying a spiral-wound gasket. Leakage at a large nozzle can sometimes be eliminated by eliminating the flange (i.e., by welding piping directly to the nozzle).

26. BOLT STRENGTHS

The maximum allowable stress per bolt for standard thread and 8-thread series bolts can be found from Table 53.11. The allowable load per bolt is calculated as the product of the allowable stress and the bolt area.

Table 53.11 Maximum Allowable Working Stress per Bolt

bolt diameter (in)	number of threads per inch	area at bottom of thread (in²)	allowable material stress (psi)			
			7000[a]	16,250	18,750	20,000
			stress per bolt (psi)			
standard thread						
$^1/_2$	13	0.126	882	2047	2362	2520
$^5/_8$	11	0.202	1414	3282	3787	4040
$^3/_4$	10	0.302	2114	4907	5662	6040
$^7/_8$	9	0.419	2933	6808	7856	8380
1	8	0.551	3857	8953	10,331	11,020
$1^1/_8$	7	0.693	4851	11,261	12,993	13,860
$1^1/_4$	7	0.890	6230	14,462	16,687	17,800
$1^3/_8$	6	1.054	7378	17,127	19,762	21,080
$1^1/_2$	6	1.294	9058	21,027	24,262	25,880
$1^5/_8$	$5^1/_2$	1.515	10,605	24,618	28,406	30,300
$1^3/_4$	5	1.744	12,208	28,340	32,700	34,880
$1^7/_8$	5	2.049	14,343	33,296	38,418	40,980
2	$4^1/_2$	2.300	16,100	37,375	43,125	46,000
$2^1/_4$	$4^1/_2$	3.020	21,140	49,075	56,625	60,400
$2^1/_2$	4	3.715	26,005	60,368	69,656	74,300
$2^3/_4$	4	4.618	32,326	75,042	86,587	92,360
3	4	5.620	39,340	91,325	105,375	112,400
8-thread series						
$1^1/_8$	8	0.728	5096	11,830	13,650	14,560
$1^1/_4$	8	0.929	6503	15,096	17,418	18,580
$1^3/_8$	8	1.155	8085	18,768	21,656	23,100
$1^1/_2$	8	1.405	9835	22,831	26,343	28,100
$1^5/_8$	8	1.680	11,760	27,300	31,500	33,600
$1^3/_4$	8	1.980	13,860	32,175	37,125	39,600
$1^7/_8$	8	2.304	16,128	37,440	43,200	46,080
2	8	2.652	18,564	43,095	49,725	53,040
$2^1/_4$	8	3.423	23,961	55,623	64,181	68,460
$2^1/_2$	8	4.292	30,044	69,745	80,475	85,840
$2^3/_4$	8	5.259	36,813	85,458	98,606	105,180
3	8	6.324	44,268	102,765	118,575	126,480

[a]Maximum temperature, 450°F.

27. PIPES

The required thickness and maximum allowable pressure of pipes under internal pressure are given by Eqs. 53.13 and 53.14. The allowable stress, S, for A53B and A106B steel pipe is usually 15,000 psi (103 MPa) within the temperature range of $-20°F$ to $650°F$ ($-29°C$ to $343°C$). The joint efficiency, E, of seamless pipe is 1.0. The diameter, D, and radius, R, are inside values.

$$t = \frac{pR}{SE - 0.6p} \qquad \text{53.13}$$

$$p = \frac{SEt}{R + 0.6t} \qquad \text{53.14}$$

When selecting pipe, a manufacturing tolerance of $\pm 12.5\%$ must be considered. That is, it must be assumed that the pipe selected will have an actual wall thickness of 0.875 times the nominal thickness. This reduction is in addition to the corrosion allowance. Additional thickness is required if the pipe is to be threaded.

Derated strength (S) values for operation at higher temperatures can be calculated by multiplying the room-temperature strength by the factors in Table 53.12.

Table 53.12 *Pipe Strength Temperature Derating Factors*[a]

temperature (°F)	(°C)	multiplicative derating factor
650	343	1.000
700	371	0.9566
750	399	0.8633
800	427	0.7200
850	454	0.5766
900	482	0.4333
950	510	0.3000
1000	538	0.1666

[a]A53B pipe may not be used above 900°F (482°C). A106B pipe may not be used above 1000°F (538°C).

Example 53.3

An 8 in (nominal) diameter, schedule-80 seamless pipe with welded flanges is constructed of A106B steel. The pipe carries 700°F steam. The corrosion allowance is 0.050 in. What is the maximum allowable pressure for this configuration?

Solution

For a schedule-80, 8 in pipe, the wall thickness is 0.500 in and the outside diameter is 8.625 in.

Taking the manufacturing tolerance into account, the minimum wall thickness is

$$(0.875)t = (0.875)(0.500 \text{ in}) = 0.4375 \text{ in}$$

Subtracting the corrosion allowance, the design wall thickness is

$$t = 0.4375 \text{ in} - 0.05 \text{ in} = 0.3875 \text{ in}$$

The inside diameter is

$$D = D_o - 2t = 8.625 \text{ in} - (2)(0.3875 \text{ in})$$
$$= 7.850 \text{ in}$$

From Table 53.12 for 700°F operation, the maximum allowable stress derating factor is 0.9566.

$$S = (0.9566)\left(15{,}000 \; \frac{\text{lbf}}{\text{in}^2}\right) = 14{,}349 \text{ lbf/in}^2$$

From Eq. 53.14, the maximum allowable pressure is

$$p = \frac{2SEt}{D + 1.2t}$$
$$= \frac{(2)\left(14{,}349 \; \frac{\text{lbf}}{\text{in}^2}\right)(1.00)(0.3875 \text{ in})}{7.850 \text{ in} + (1.2)(0.3875 \text{ in})}$$
$$= 1337 \text{ lbf/in}^2$$

28. PRESSURE TESTING

Pressure vessels under internal pressure are normally tested hydrostatically with water.[40] However, vessels that cannot safely be filled with water, that cannot be dried, or that cannot tolerate traces of the test liquid can be tested pneumatically with air.

The hydrostatic *test pressure* is 150% of the maximum allowable pressure new and cold multiplied by the ratio of the allowable stress at the test temperature to the allowable stress at the design temperature. For pneumatic tests, the test pressure is 125% of the maximum allowable pressure new and cold multiplied by the ratio of the allowable stress at the test temperature to the allowable stress at the design temperature. When calculations are not performed to determine the maximum allowable pressure, the hydrostatic test pressure for pressure vessels is 150% of the design pressure.

For cast-iron pressure vessels, the test pressure is 200% of the design pressure unless the design pressure is less than 30 psig (207 kPa), in which case, the test pressure is 60 psig (414 kPa) or 250% of the design pressure, whichever is less. A corrosion allowance is included when calculating all test pressures.

Most testing is not carried out at the operating temperature, however. Since material strengths decrease at higher temperatures, the test pressure is increased according to the ratio of the allowable stress at the test temperature to the allowable stress at the design temperature.

With some exceptions, magnetic particle or liquid penetrant testing is required (1) around all openings and attachments when the pressure vessel is to be pneumatically tested, (2) for all welds on 36% nickel steel, (3) for welds on austenitic chromium-nickel alloy-steel more than $^3/_4$ in (33 mm), and (4) on welds to plates greater than $^1/_2$ in (13 mm).

Special rules apply to pressure vessels whose operating pressure is limited by flange strength or that have multiple chambers, are subject to external pressure, or operate at below-atmospheric pressures.

29. VESSEL VOLUMES

Volumes of spherical and cylindrical vessels are easily calculated. However, head volumes are more difficult to determine. The following equations give the approximate volume for hemispherical, torispherical, and elliptical heads, exclusive of their straight flanges, if any.

$$V = 1.958D^3 \quad \text{[hemispherical]} \qquad \text{53.15}$$
$$V = 0.582D^3 \quad \text{[torispherical]} \qquad \text{53.16}$$
$$V = 0.954D^3 \quad \text{[elliptical]} \qquad \text{53.17}$$

---

[40]Even pressure vessels that are normally under internal pressure may sometimes draw a vacuum, as during a steamout. Other less predictable instances of vacuum failures occur when the contents of a vessel are being drained while the vent line is closed or blocked, or when a filter element is clogged or under-sized.

54

Properties of Solid Bodies

1. Center of Gravity 54-1
2. Mass and Weight 54-1
3. Inertia 54-1
4. Mass Moment of Inertia 54-2
5. Parallel Axis Theorem 54-2
6. Radius of Gyration 54-2
7. Principal Axes 54-2
   Practice Problems 54-2

Nomenclature

a	acceleration	ft/sec^2	m/s^2
d	distance	ft	m
g	gravitational acceleration	ft/sec^2	m/s^2
g_c	gravitational constant	lbm-ft/lbf-sec^2	n.a.
h	height	ft	m
I	mass moment of inertia	lbm-ft^2	kg·m^2
k	radius of gyration	ft	m
L	length	ft	m
m	mass	lbm	kg
r	radius	ft	m
V	volume	ft^3	m^3
w	weight	lbf	N

Symbols

ρ	density	lbm/ft^3	kg/m^3

Subscripts

c	centroidal
i	inner
o	outer

1. CENTER OF GRAVITY

A solid body will have both a center of gravity and a centroid, but the locations of these two points will not necessarily coincide. The earth's attractive force, called *weight*, can be assumed to act through the *center of gravity* (also known as the *center of mass*). Only when the body is homogeneous will the *centroid of the volume* coincide with the center of gravity.[1]

For simple objects and regular polyhedrons, the location of the center of gravity can be determined by inspection. It will always be located on an axis of symmetry. The location of the center of gravity can also be determined mathematically if the object can be described mathematically.

$$x_c = \frac{\int x\,dm}{m} \qquad 54.1$$

$$y_c = \frac{\int y\,dm}{m} \qquad 54.2$$

$$z_c = \frac{\int z\,dm}{m} \qquad 54.3$$

If the object can be divided into several smaller constituent objects, the location of the composite center of gravity can be calculated from the centers of gravity of each of the constituent objects.

$$x_c = \frac{\sum m_i x_{ci}}{\sum m_i} \qquad 54.4$$

$$y_c = \frac{\sum m_i y_{ci}}{\sum m_i} \qquad 54.5$$

$$z_c = \frac{\sum m_i z_{ci}}{\sum m_i} \qquad 54.6$$

2. MASS AND WEIGHT

The mass, m, of a homogeneous solid object is calculated from its mass density and volume. Mass is independent of the strength of the gravitational field.

$$m = \rho V \qquad 54.7$$

The weight, w, of an object depends on the strength of the gravitational field, g.

$$w = mg \qquad \text{[SI]} \qquad 54.8(a)$$

$$w = \frac{mg}{g_c} \qquad \text{[U.S.]} \qquad 54.8(b)$$

3. INERTIA

Inertia (the *inertial force* or *inertia vector*), $m\mathbf{a}$, is the resistance the object offers to attempts to accelerate it (i.e., change its velocity) in a linear direction. Although the mass, m, is a scalar quantity, the acceleration, $\mathbf{a}$, is a vector.

---

[1] The study of nonhomogeneous bodies is beyond the scope of this book. Homogeneity is assumed for all solid objects.

4. MASS MOMENT OF INERTIA

The *mass moment of inertia* measures a solid object's resistance to changes in rotational speed about a specific axis. I_x, I_y, and I_z are the mass moments of inertia with respect to the x-, y-, and z-axes. They are not components of a resultant value.[2]

The *centroidal mass moment of inertia*, I_c, is obtained when the origin of the axes coincides with the object's center of gravity. Although it can be found mathematically from Eqs. 54.9 through 54.11, it is easier to use App. 54.A for simple objects.

$$I_x = \int (y^2 + z^2)\, dm \qquad 54.9$$

$$I_y = \int (x^2 + z^2)\, dm \qquad 54.10$$

$$I_z = \int (x^2 + y^2)\, dm \qquad 54.11$$

5. PARALLEL AXIS THEOREM

Once the centroidal mass moment of inertia is known, the *parallel axis theorem* is used to find the mass moment of inertia about any parallel axis.

$$I_{\text{any parallel axis}} = I_c + md^2 \qquad 54.12$$

For a composite object, the parallel axis theorem must be applied for each of the constituent objects.

$$I = I_{c,1} + m_1 d_1^2 + I_{c,2} + m_2 d_2^2 + \cdots \qquad 54.13$$

6. RADIUS OF GYRATION

The *radius of gyration*, k, of a solid object represents the distance from the rotational axis at which the object's entire mass could be located without changing the mass moment of inertia.

$$k = \sqrt{\frac{I}{m}} \qquad 54.14$$

$$I = k^2 m \qquad 54.15$$

7. PRINCIPAL AXES

An object's mass moment of inertia depends on the orientation of axes chosen. The *principal axes* are the axes for which the *products of inertia* are zero. Equipment rotating about a principal axis will draw minimum power during speed changes.

Finding the principal axes through calculation is too difficult and time consuming to be used with most rotating equipment. Furthermore, the rotating axis is generally fixed. Therefore, *balancing operations* are used to change the distribution of mass about the rotational axis. A device, such as a rotating shaft, flywheel, or crank, is said to be *statically balanced* if its center of mass lies on the axis of rotation. It is said to be *dynamically balanced* if the center of mass lies on the axis of rotation and the products of inertia are zero.

PRACTICE PROBLEMS

1. A spoked flywheel has an outside diameter of 60 in (1500 mm). The rim thickness is 6 in (150 mm), and the width is 12 in (300 mm). The cylindrical hub has an outside diameter of 12 in (300 mm), a thickness of 3 in (75 mm), and a width of 12 in (300 mm). The rim and hub are connected by six equally spaced cylindrical radial spokes, each with a diameter of 4.25 in (110 mm). All parts of the flywheel are ductile cast iron with a density of 0.256 lbm/in³ (7080 kg/m³). What is the rotational mass moment of inertia of the entire flywheel?

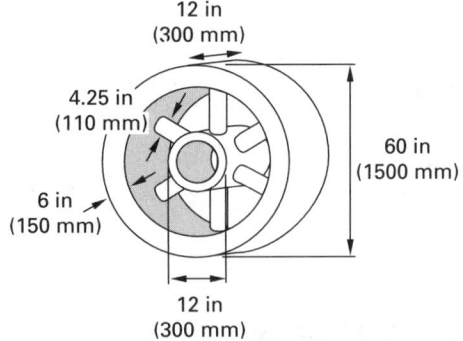

---

[2]At first, it may be confusing to use the same symbol, I, for area and mass moments of inertia. However, the problem types are distinctly dissimilar, and both moments of inertia are seldom used simultaneously.

55

Kinematics

1.	Introduction to Kinematics	55-1
2.	Particles and Rigid Bodies	55-1
3.	Coordinate Systems	55-1
4.	Conventions of Representation	55-2
5.	Linear Particle Motion	55-2
6.	Distance and Speed	55-2
7.	Uniform Motion	55-3
8.	Uniform Acceleration	55-3
9.	Linear Acceleration	55-4
10.	Projectile Motion	55-4
11.	Rotational Particle Motion	55-6
12.	Relationship Between Linear and Rotational Variables	55-7
13.	Normal Acceleration	55-7
14.	Coriolis Acceleration	55-8
15.	Particle Motion in Polar Coordinates	55-9
16.	Relative Motion	55-9
17.	Dependent Motion	55-11
18.	General Plane Motion	55-11
19.	Rotation About a Fixed Axis	55-12
20.	Instantaneous Center of Acceleration	55-13
21.	Slider Rods	55-13
22.	Slider-Crank Assemblies	55-13
	Practice Problems	55-14

Nomenclature

a	acceleration	ft/sec^2	m/s^2
d	distance	ft	m
g	gravitational acceleration	ft/sec^2	m/s^2
H	height	ft	m
l	length	ft	m
n	rotating speed	rpm	rpm
r	radius	ft	m
R	range	ft	m
R	earth's radius	ft	m
s	distance	ft	m
t	time	sec	s
T	flight time	sec	s
v	velocity	ft/sec	m/s
z	elevation	ft	m

Symbols

α	angular acceleration	rad/sec^2	rad/s^2
β	angle	deg	deg
γ	angle	deg	deg
θ	angular position	rad	rad
ϕ	angle or latitude	deg	deg
ω	angular velocity	rad/sec	rad/s

Subscripts

0	initial
ϕ	transverse
a	acceleration
c	Coriolis
H	to maximum altitude
n	normal
O	center
r	radial
t	tangential

1. INTRODUCTION TO KINEMATICS

Dynamics is the study of moving objects. The subject is divided into kinematics and kinetics. *Kinematics* is the study of a body's motion independent of the forces on the body. It is a study of the geometry of motion without consideration of the causes of motion. Kinematics deals only with relationships among position, velocity, acceleration, and time. (Kinetics is covered in Chap. 56.)

2. PARTICLES AND RIGID BODIES

Bodies in motion can be considered *particles* if rotation is absent or insignificant. Particles do not possess rotational kinetic energy. All parts of a particle have the same instantaneous displacement, velocity, and acceleration.

A *rigid body* does not deform when loaded and can be considered a combination of two or more particles that remain at a fixed, finite distance from each other. At any given instant, the parts (particles) of a rigid body can have different displacements, velocities, and accelerations.

3. COORDINATE SYSTEMS

The position of a particle is specified with reference to a *coordinate system*. The description takes the form of an ordered sequence $(q_1, q_2, q_3, \ldots)$ of numbers called *coordinates*. A coordinate can represent a position along an axis, as in the rectangular coordinate system, or it can represent an angle, as in the polar, cylindrical, and spherical coordinate systems.

In general, the number of *degrees of freedom* is equal to the number of coordinates required to completely specify the state of an object. If each of the coordinates is independent of the others, the coordinates are known as *holonomic coordinates*.

The state of a particle is completely determined by the particle's location. In three-dimensional space, the locations of particles in a system of m particles must be specified by $3m$ coordinates. However, the number of required coordinates can be reduced in certain cases. The position of each particle constrained to motion on a surface (i.e., on a two-dimensional system) can be specified by only two coordinates. A particle constrained to moving on a curved path requires only one coordinate.[1]

The state of a rigid body is a function of orientation as well as position. Six coordinates are required to specify the state: three for orientation and three for location.

4. CONVENTIONS OF REPRESENTATION

Consider the particle shown in Fig. 55.1. Its position (as well as its velocity and acceleration) can be specified in three primary forms: vector form, rectangular coordinate form, and unit vector form.

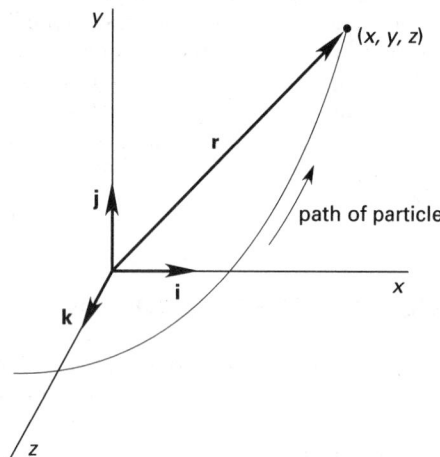

Figure 55.1 *Position of a Particle*

The vector form of the particle's position is $\mathbf{r}$, where the vector $\mathbf{r}$ has both magnitude and direction. The rectangular coordinate form is (x, y, z). The unit vector form is

$$\mathbf{r} = x\mathbf{i} + y\mathbf{j} + z\mathbf{k} \qquad 55.1$$

5. LINEAR PARTICLE MOTION

A *linear system* is one in which particles move only in straight lines. (Another name is *rectilinear system*.) The relationships among position, velocity, and acceleration for a linear system are given by Eqs. 55.2 through

---

[1]The curve can be a straight line, as in the case of a mass hanging on a spring and oscillating up and down. In this case, the coordinate will be a linear coordinate.

55.4. When values of t are substituted into these equations, the position, velocity, and acceleration are known as *instantaneous values*.

$$s(t) = \int \mathrm{v}(t)\, dt = \int \left(\int a(t)\, dt \right) dt \qquad 55.2$$

$$\mathrm{v}(t) = \frac{ds(t)}{dt} = \int a(t)\, dt \qquad 55.3$$

$$a(t) = \frac{d\mathrm{v}(t)}{dt} = \frac{d^2 s(t)}{dt^2} \qquad 55.4$$

The average velocity and acceleration over a period from t_1 to t_2 are

$$\mathrm{v}_{\text{ave}} = \frac{\int\limits_1^2 \mathrm{v}(t)\, dt}{t_2 - t_1} = \frac{s_2 - s_1}{t_2 - t_1} \qquad 55.5$$

$$a_{\text{ave}} = \frac{\int\limits_1^2 a(t)\, dt}{t_2 - t_1} = \frac{\mathrm{v}_2 - \mathrm{v}_1}{t_2 - t_1} \qquad 55.6$$

Example 55.1

A particle is constrained to move along a straight line. The velocity and location are both zero at $t = 0$. The particle's velocity as a function of time is

$$\mathrm{v}(t) = 8t - 6t^2$$

(a) What are the acceleration and position functions?
(b) What is the instantaneous velocity at $t = 5$?

Solution

(a)
$$a(t) = \frac{d\,\mathrm{v}(t)}{dt} = \frac{d(8t - 6t^2)}{dt}$$
$$= 8 - 12t$$

$$s(t) = \int \mathrm{v}(t)\, dt = \int (8t - 6t^2)\, dt$$
$$= 4t^2 - 2t^3$$

(b) Substituting $t = 5$ into the $\mathrm{v}(t)$ function,

$$\mathrm{v}(5) = (8)(5) - (6)(5)^2$$
$$= -110 \quad [\text{backward}]$$

6. DISTANCE AND SPEED

The terms "displacement" and "distance" have different meanings in kinematics. *Displacement* (or *linear displacement*) is the net change in a particle's position as determined from the position function, $s(t)$. *Distance traveled* is the accumulated length of the path traveled during all direction reversals, and can be found by adding the path lengths covered during periods in which the velocity sign does not change. Thus, distance is always greater than or equal to displacement.

$$\text{displacement} = s(t_2) - s(t_1) \qquad 55.7$$

Similarly, "velocity" and "speed" have different meanings: *velocity* is a vector, having both magnitude and direction; *speed* is a scalar quantity, equal to the magnitude of velocity. When specifying speed, direction is not considered.

Example 55.2

What distance is traveled during the period $t = 0$ to $t = 6$ by the particle described in Ex. 55.1?

Solution

Start by determining when, if ever, the velocity becomes negative. (This can be done by inspection, graphically, or algebraically.) Solving for the roots of the velocity equation, the velocity changes from positive to negative at

$$t = \frac{4}{3}$$

The initial displacement is zero. From the position function, the position at $t = 4/3$ is

$$s\left(\frac{4}{3}\right) = (4)\left(\frac{4}{3}\right)^2 - (2)\left(\frac{4}{3}\right)^3 = 2.37$$

The displacement while the velocity is positive is

$$\Delta s = s\left(\frac{4}{3}\right) - s(0) = 2.37 - 0$$

$$= 2.37$$

The position at $t = 6$ is

$$s(6) = (4)(6)^2 - (2)(6)^3 = -288$$

The displacement while the velocity is negative is

$$\Delta s = s(6) - s\left(\frac{4}{3}\right) = -288 - 2.37$$

$$= -290.37$$

The total distance traveled is

$$2.37 + 290.37 = 292.74$$

7. UNIFORM MOTION

The term *uniform motion* means uniform velocity. The velocity is constant and the acceleration is zero. For a constant velocity system, the position function varies linearly with time.

$$s(t) = s_0 + vt \qquad 55.8$$
$$v(t) = v \qquad 55.9$$
$$a(t) = 0 \qquad 55.10$$

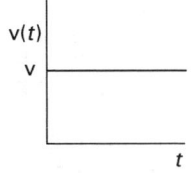
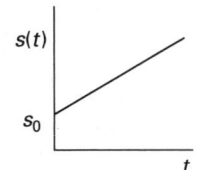

Figure 55.2 *Constant Velocity System*

8. UNIFORM ACCELERATION

The acceleration is constant in many cases. (Gravitational acceleration, where $a = g$, is a notable example.) If the acceleration is constant, the a term can be taken out of the integrals in Eqs. 55.12 and 55.13.

$$a(t) = a \qquad 55.11$$
$$v(t) = a \int dt = v_0 + at \qquad 55.12$$
$$s(t) = a \iint dt^2 = s_0 + v_0 t + \frac{1}{2}at^2 \qquad 55.13$$

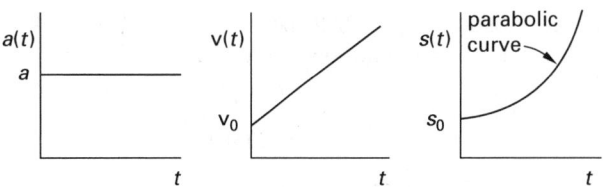

Figure 55.3 *Uniform Acceleration*

Table 55.1 summarizes the equations required to solve most uniform acceleration problems.

Table 55.1 *Uniform Acceleration Formulas*[a]

to find	given these	use this equation
a	t, v_0, v	$a = \dfrac{v - v_0}{t}$
a	t, v_0, s	$a = \dfrac{2s - 2v_0 t}{t^2}$
a	v_0, v, s	$a = \dfrac{v^2 - v_0^2}{2s}$
s	t, a, v_0	$s = v_0 t + \frac{1}{2}at^2$
s	a, v_0, v	$s = \dfrac{v^2 - v_0^2}{2a}$
s	t, v_0, v	$s = \frac{1}{2}t(v_0 + v)$
t	a, v_0, v	$t = \dfrac{v - v_0}{a}$
t	a, v_0, s	$t = \dfrac{\sqrt{v_0^2 + 2as} - v_0}{a}$
t	v_0, v, s	$t = \dfrac{2s}{v_0 + v}$
v_0	t, a, v	$v_0 = v - at$
v_0	t, a, s	$v_0 = \dfrac{s}{t} - \frac{1}{2}at$
v_0	a, v, s	$v_0 = \sqrt{v^2 - 2as}$
v	t, a, v_0	$v = v_0 + at$
v	a, v_0, s	$v = \sqrt{v_0^2 + 2as}$

[a]The table can be used for rotational problems by substituting α, ω, and θ for a, v, and s, respectively.

Example 55.3

A locomotive traveling at 80 kph locks its wheels and skids 95 m before coming to a complete stop. If the deceleration is constant, how many seconds will it take for the locomotive to come to a standstill?

Solution

First, convert the 80 kph to meters per second.

$$v_0 = \frac{\left(80 \ \frac{\text{km}}{\text{h}}\right)\left(1000 \ \frac{\text{m}}{\text{km}}\right)}{3600 \ \frac{\text{s}}{\text{h}}} = 22.22 \ \text{m/s}$$

In this problem, $v_0 = 22.2$ m/s, $v = 0$, and $s = 95$ m are known. t is the unknown. From Table 55.1,

$$t = \frac{2s}{v_0 + v} = \frac{(2)(95 \ \text{m})}{22.22 + 0 \ \frac{\text{m}}{\text{s}}}$$

$$= 8.55 \ \text{s}$$

9. LINEAR ACCELERATION

Linear acceleration means that the acceleration increases uniformly with time. Figure 55.4 shows how the velocity and position vary with time.[2]

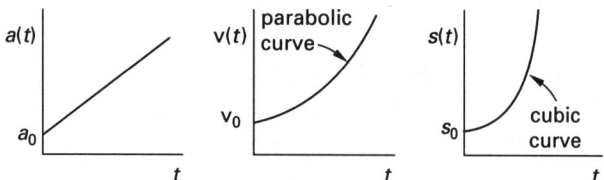

Figure 55.4 *Linear Acceleration*

10. PROJECTILE MOTION

A *projectile* is placed into motion by an initial impulse. (Kinematics deals only with dynamics during the flight. The force acting on the projectile during the launch phase is covered in kinetics.) Neglecting air drag, once the projectile is in motion it is acted upon only by the downward gravitational acceleration (i.e., its own weight). Thus, projectile motion is a special case of motion under constant acceleration.

Consider a general projectile set into motion at an angle of ϕ (from the horizontal plane) and initial velocity v_0. Its range is R, the maximum altitude attained is H, and the total flight time is T. In the absence of air drag, the following rules apply to the case of a level target.[3]

---

[2]Because of the successive integrations, if the acceleration function is a polynomial of degree n, the velocity function will be a polynomial of degree $n + 1$. Similarly, the position function will be a polynomial of degree $n + 2$.

[3]The case of projectile motion with air friction cannot be handled in kinematics, since a retarding force acts continuously on the projectile. In kinetics, various assumptions (e.g., friction varies linearly with the velocity or with the square of the velocity) can be made to include the effect of air friction.

- The trajectory is parabolic.

- The impact velocity is equal to initial velocity, v_0.

- The impact angle is equal to the initial launch angle, ϕ.

- The range is maximum when $\phi = 45°$.

- The time for the projectile to travel from the launch point to the apex is equal to the time to travel from apex to impact point.

- The time for the projectile to travel from the apex of its flight path to impact is the same time an initially stationary object would take to fall a distance H.

Table 55.2 contains the solutions to most common projectile problems. These equations are derived from the laws of uniform acceleration and conservation of energy.

Example 55.4

A projectile is launched at 600 ft/sec (180 m/s) with a 30° inclination from the horizontal. The launch point is on a plateau 500 ft (150 m) above the plane of impact. Neglecting friction, find the maximum altitude, H, above the plane of impact, the total flight time, T, and the range, R.

SI Solution

The maximum altitude above the impact plane includes the height of the plateau and the elevation achieved by the projectile.

$$H = z + \frac{v_0^2 \sin^2 \phi}{2g}$$

$$= 150 \ \text{m} + \frac{\left(180 \ \frac{\text{m}}{\text{s}}\right)^2 (\sin^2 30°)}{(2)\left(9.81 \ \frac{\text{m}}{\text{s}^2}\right)} = 563 \ \text{m}$$

The total flight time includes the time to reach the maximum altitude and the time to fall from the maximum altitude to the impact plane below.

$$T = t_H + t_{\text{fall}}$$

$$= \frac{v_0 \sin \phi}{g} + \sqrt{\frac{2H}{g}}$$

$$= \frac{\left(180 \ \frac{\text{m}}{\text{s}}\right)(\sin 30°)}{9.81 \ \frac{\text{m}}{\text{s}^2}} + \sqrt{\frac{(2)(563 \ \text{m})}{9.81 \ \frac{\text{m}}{\text{s}^2}}}$$

$$= 19.9 \ \text{s}$$

Table 55.2 *Projectile Motion Equations*
 (ϕ may be negative for projection downward)

	level target	target above	target below	horizontal projection
				$v_0 = v_{x'}\ \phi = 0°$
$x(t)$	$(v_0 \cos \phi)t$			$v_0 t$
$y(t)$	$(v_0 \sin \phi)t - \frac{1}{2}gt^2$			$H - \frac{1}{2}gt^2$
$v_x(t)$	$v_0 \cos \phi$			v_0
$v_y(t)$	$v_0 \sin \phi - gt$			$-gt$
$v(t)$	$\sqrt{v_0^2 - 2gy} = \sqrt{v_0^2 - 2gtv_0 \sin \phi + g^2t^2}$			$\sqrt{v_0^2 + g^2t^2}$
$v(y)$	$\sqrt{v_0^2 - 2gy}$			$\sqrt{v_0^2 + 2g(H-y)}$
H	$\dfrac{v_0^2 \sin^2 \phi}{2g}$	$\dfrac{v_0^2 \sin^2 \phi}{2g}$	$z + \dfrac{v_0^2 \sin^2 \phi}{2g}$	$\frac{1}{2}gt^2$
R	$v_0 T \cos \phi$			$v_0 T$
	$\dfrac{v_0^2 \sin 2\phi}{g}$	$\left(\dfrac{v_0 \cos \phi}{g}\right)\left[v_0 \sin \phi + \sqrt{v_0^2 \sin^2\phi - 2gz}\right]$	$\left(\dfrac{v_0 \cos \phi}{g}\right)\left[v_0 \sin \phi + \sqrt{2gz + v_0^2 \sin^2\phi}\right]$	
T	$\dfrac{R}{v_0 \cos \phi}$			$\sqrt{\dfrac{2H}{g}}$
	$\dfrac{2v_0 \sin \phi}{g}$	$\dfrac{v_0 \sin \phi}{g} + \sqrt{\dfrac{(2)(H-z)}{g}}$	$\dfrac{v_0 \sin \phi}{g} + \sqrt{\dfrac{2H}{g}}$	
t_H	$\dfrac{v_0 \sin \phi}{g} = \dfrac{T}{2}$	$\dfrac{v_0 \sin \phi}{g}$		

The x-component of velocity is

$$v_x = v_0 \cos \phi = \left(180 \; \frac{m}{s}\right)(\cos 30°)$$

$$= 156 \; m/s$$

The range is

$$R = v_x T = \left(156 \; \frac{m}{s}\right)(19.9 \; s)$$

$$= 3100 \; m$$

Customary U.S. Solution

The maximum altitude above the impact plane is given by Table 55.2.

$$H = 500 \; ft + \frac{\left(600 \; \frac{ft}{sec}\right)^2 (\sin^2 30°)}{(2)\left(32.2 \; \frac{ft}{sec^2}\right)}$$

$$= 1898 \; ft$$

The total flight time is

$$T = \frac{\left(600 \; \frac{ft}{sec}\right)(\sin 30°)}{32.2 \; \frac{ft}{sec^2}} + \sqrt{\frac{(2)(1898 \; ft)}{32.2 \; \frac{ft}{sec^2}}}$$

$$= 20.2 \; sec$$

The maximum range is

$$R = (v_0 \cos \phi) T = \left(600 \; \frac{ft}{sec}\right)(\cos 30°)(20.2 \; sec)$$

$$= 10{,}500 \; ft \; (1.99 \; mi)$$

Example 55.5

A bomber flies horizontally at 275 mph at an altitude of 9000 ft. At what viewing angle, ϕ, from the bomber to the target should the bombs be dropped?

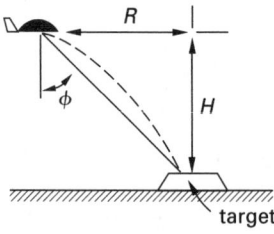

Solution

This is a case of horizontal projection. The falling time depends only on the altitude of the bomber. From Table 55.2,

$$T = \sqrt{\frac{2H}{g}} = \sqrt{\frac{(2)(9000 \; ft)}{32.2 \; \frac{ft}{sec^2}}}$$

$$= 23.64 \; sec$$

If air friction is neglected, the bomb has the same horizontal velocity as the bomber. Since the time of flight, T, is known, the distance traveled during that time can be calculated.

$$R = v_0 T = \frac{\left(275 \; \frac{mi}{hr}\right)\left(5280 \; \frac{ft}{mi}\right)(23.64 \; sec)}{3600 \; \frac{sec}{hr}}$$

$$= 9535 \; ft$$

The viewing angle is found from trigonometry.

$$\phi = \arctan\left(\frac{R}{H}\right) = \arctan\left(\frac{9535 \; ft}{9000 \; ft}\right)$$

$$= 46.7°$$

11. ROTATIONAL PARTICLE MOTION

Rotational particle motion (also known as *angular motion* and *circular motion*) is motion of a particle around a circular path.

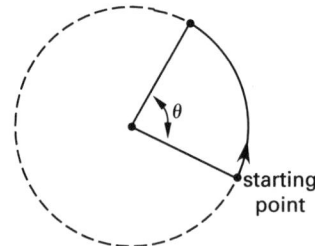

Figure 55.5 *Rotational Particle Motion*

The behavior of a rotating particle is defined by its *angular position*, θ, *angular velocity*, ω, and *angular acceleration*, α, functions. These variables are analogous to the $s(t)$, $v(t)$, and $a(t)$ functions for linear systems. Angular variables can be substituted one-for-one in place of linear variables in most equations.

The relationships among angular position, velocity, and acceleration for a rotational system are given by Eqs. 55.14 through 55.16. When values of t are substituted into these equations, the position, velocity, and acceleration are known as *instantaneous values*.

$$\theta(t) = \int \omega(t) \, dt = \iint \alpha(t) \, dt^2 \qquad 55.14$$

$$\omega(t) = \frac{d\theta(t)}{dt} = \int \alpha(t) \, dt \qquad 55.15$$

$$\alpha(t) = \frac{d\omega(t)}{dt} = \frac{d^2\theta(t)}{dt^2} \qquad 55.16$$

The average velocity and acceleration are

$$\omega_{ave} = \frac{\int_1^2 \omega(t) \, dt}{t_2 - t_1} = \frac{\theta_2 - \theta_1}{t_2 - t_1} \qquad 55.17$$

$$\alpha_{ave} = \frac{\int_1^2 \alpha(t) \, dt}{t_2 - t_1} = \frac{\omega_2 - \omega_1}{t_2 - t_1} \qquad 55.18$$

Example 55.6

A turntable starts from rest and accelerates uniformly at 1.5 rad/sec². How many revolutions will it take before a rotational speed of $33^1/_3$ rpm is attained?

Solution

First, convert $33^1/_3$ rpm into radians per second. Since there are 2π radians per complete revolution,

$$\omega = \frac{\left(33\frac{1}{3}\ \frac{\text{rev}}{\text{min}}\right)\left(2\pi\ \frac{\text{rad}}{\text{rev}}\right)}{60\ \frac{\text{sec}}{\text{min}}} = 3.49\ \text{rad/sec}$$

α, ω_0, and ω are known. θ is unknown. (This is analogous to knowing a, v_0, and v, and not knowing s.) From Table 55.1,

$$\theta = \frac{\omega^2 - \omega_0^2}{2\alpha} = \frac{\left(3.49\ \frac{\text{rad}}{\text{sec}}\right)^2 - \left(0\ \frac{\text{rad}}{\text{sec}}\right)^2}{(2)\left(1.5\ \frac{\text{rad}}{\text{sec}^2}\right)}$$

$$= 4.06\ \text{rad}$$

Converting from radians to revolutions,

$$n = \frac{4.06\ \text{rad}}{2\pi\ \frac{\text{rad}}{\text{rev}}} = 0.65\ \text{rev}$$

Example 55.7

A flywheel is brought to a standstill from 400 rpm in 8 s. (a) What was its average angular acceleration in rad/s during that period? (b) How far (in radians) did the flywheel travel?

Solution

(a) The initial rotational speed must be expressed in radians per second. Since there are 2π radians per revolution,

$$\omega_0 = \frac{\left(400\ \frac{\text{rev}}{\text{min}}\right)\left(2\pi\ \frac{\text{rad}}{\text{rev}}\right)}{60\ \frac{\text{s}}{\text{min}}}$$

$$= 41.9\ \text{rad/s}$$

t, ω_0, and ω are known, and α is unknown. These variables are analogous to t, v_0, v, and a in Table 55.1.

$$\alpha = \frac{\omega - \omega_0}{t} = \frac{0 - 41.9\ \frac{\text{rad}}{\text{s}}}{8\ \text{s}}$$

$$= -5.24\ \text{rad/s}^2$$

(b) t, ω_0, and ω are known, and θ is unknown. Again, from Table 55.1,

$$\theta = \tfrac{1}{2}t(\omega + \omega_0) = \left(\tfrac{1}{2}\right)(8\ \text{s})\left(41.9\ \frac{\text{rad}}{\text{s}} + 0\right)$$

$$= 168\ \text{rad}\quad (26.7\ \text{rev})$$

12. RELATIONSHIP BETWEEN LINEAR AND ROTATIONAL VARIABLES

A particle moving in a curvilinear path will also have instantaneous linear velocity and linear acceleration. These linear variables will be directed tangentially to the path and, therefore, are known as *tangential velocity* and *tangential acceleration*, respectively. In general, the linear variables can be obtained by multiplying the rotational variables by the path radius, r.

$$v_t = \omega r \qquad\qquad 55.19$$
$$v_{t,x} = v_t \cos\phi = \omega r \cos\phi \qquad\qquad 55.20$$
$$v_{t,y} = v_t \sin\phi = \omega r \sin\phi \qquad\qquad 55.21$$
$$a_t = \frac{dv_t}{dt} = \alpha r \qquad\qquad 55.22$$

If the path radius is constant, as it would be in rotational motion, the linear distance (i.e., the *arc length*) traveled is

$$s = \theta r \qquad\qquad 55.23$$

13. NORMAL ACCELERATION

A moving particle will continue tangentially to its path unless constrained otherwise. For example, a rock twirled on a string will move in a circular path only as long as there is tension in the string. When the string is released, the rock will move off tangentially.

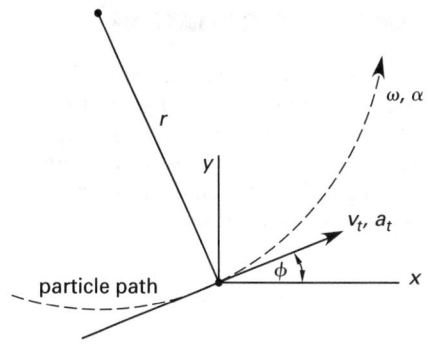

Figure 55.6 *Tangential Variables*

The twirled rock is acted upon by the tension in the string. In general, a restraining force will be directed toward the center of rotation. Whenever a mass experiences a force, an acceleration is acting.[4] The acceleration has the same sense as the applied force (i.e., is

---

[4]This is a direct result of Newton's second law of motion, covered in Chap. 56.

directed toward the center of rotation). Since the inward acceleration is perpendicular to the tangential velocity and acceleration, it is known as *normal acceleration*, a_n.

$$a_n = \frac{v_t^2}{r} = r\omega^2 = v_t\omega \qquad 55.24$$

The *resultant acceleration*, a, is the vector sum of the tangential and normal accelerations. The magnitude of the resultant acceleration is

$$a = \sqrt{a_t^2 + a_n^2} \qquad 55.25$$

The x- and y-components of the resultant acceleration are

$$a_x = a_n \sin\phi \pm a_t \cos\phi \qquad 55.26$$
$$a_y = a_n \cos\phi \mp a_t \sin\phi \qquad 55.27$$

The normal and tangential accelerations can be expressed in terms of the x- and y-components of the resultant acceleration (not shown in Fig. 55.7).

$$a_n = a_x \sin\phi \pm a_y \cos\phi \qquad 55.28$$
$$a_t = a_x \cos\phi \mp a_y \sin\phi \qquad 55.29$$

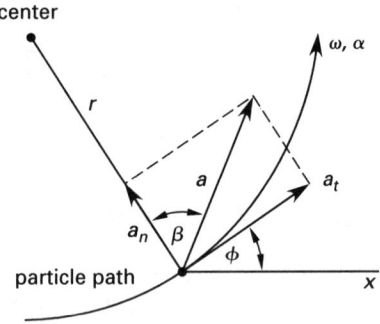

Figure 55.7 *Normal Acceleration*

14. CORIOLIS ACCELERATION

Consider a particle moving with linear radial velocity v_r away from the center of a flat disk rotating with constant velocity ω. Since $v_t = r\omega$, the particle's tangential velocity will increase as it moves away from the center of rotation. This increase is said to be produced by the tangential *Coriolis acceleration*, a_c.

$$a_c = 2v_r\omega \qquad 55.30$$

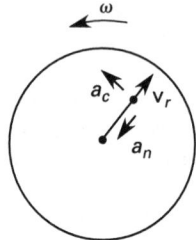

Figure 55.8 *Coriolis Acceleration on a Rotating Disk*

Coriolis acceleration also acts on particles moving on rotating spheres. Consider an aircraft flying with constant air speed v from the equator to the north pole while the earth rotates below it. Three accelerations act on the aircraft: normal, radial, and Coriolis accelerations, shown in Fig. 55.9. The Coriolis acceleration depends on the latitude, ϕ, because the earth's tangential velocity is less near the poles than at the equator.

$$a_n = r\omega^2 = R\omega^2 \cos\phi \qquad 55.31$$
$$a_r = \frac{v^2}{R} \qquad 55.32$$
$$a_c = 2\omega v_x = 2\omega v \sin\phi \qquad 55.33$$

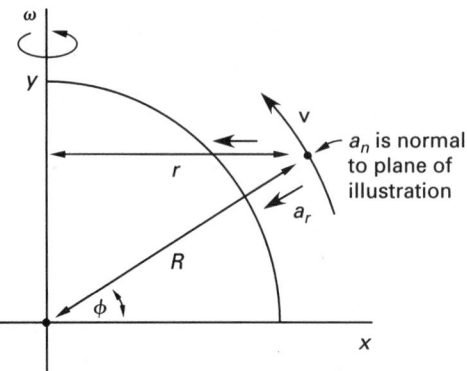

Figure 55.9 *Coriolis Acceleration on a Rotating Sphere*

Example 55.8

A slider moves with a constant velocity of 20 ft/sec along a rod rotating at 5 rad/sec. What is the magnitude of the slider's total acceleration when the slider is 4 ft from the center of rotation?

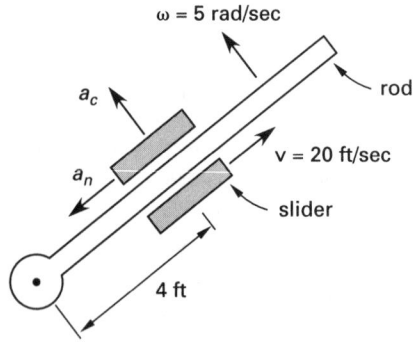

Solution

The normal acceleration is

$$a_n = r\omega^2 = (4 \text{ ft})\left(5\ \frac{\text{rad}}{\text{sec}}\right)^2$$
$$= 100 \text{ ft/sec}^2$$

The Coriolis acceleration is

$$a_c = 2\text{v}\omega = (2)\left(20\ \frac{\text{ft}}{\text{sec}}\right)\left(5\ \frac{\text{rad}}{\text{sec}}\right)$$

$$= 200\ \text{ft/sec}^2$$

The total acceleration is

$$a = \sqrt{a_n^2 + a_c^2} = \sqrt{\left(100\ \frac{\text{ft}}{\text{sec}^2}\right)^2 + \left(200\ \frac{\text{ft}}{\text{sec}^2}\right)^2}$$

$$= 223.6\ \text{ft/sec}^2$$

15. PARTICLE MOTION IN POLAR COORDINATES

In polar coordinates, the path of a particle is described by a radius vector, **r**, and an angle, ϕ. Since the velocity of a particle is not usually directed radially out from the center of the coordinate system, it can be divided into two perpendicular components. The terms *normal* and *tangential* are not used with polar coordinates. Rather, the terms *radial* and *transverse* are used. Figure 55.10 illustrates the *radial* and *transverse components* of velocity in a polar coordinate system.

Figure 55.10 also illustrates the unit radial and unit transverse vectors, $\mathbf{e}_r$ and $\mathbf{e}_\phi$, used in the vector forms of the motion equations.

$$\text{position: } \mathbf{r} = r\mathbf{e}_r \qquad 55.34$$

$$\text{velocity: } \mathbf{v} = \text{v}_r\mathbf{e}_r + \text{v}_\phi\mathbf{e}_\phi = \frac{dr}{dt}\mathbf{e}_r + r\frac{d\phi}{dt}\mathbf{e}_\phi \quad 55.35$$

$$\text{acceleration: } \mathbf{a} = a_r\mathbf{e}_r + a_\phi\mathbf{e}_\phi$$
$$= \left(\frac{d^2r}{dt^2} - r\left(\frac{d\phi}{dt}\right)^2\right)\mathbf{e}_r$$
$$+ \left(r\frac{d^2\phi}{dt^2} + 2\frac{dr}{dt}\frac{d\phi}{dt}\right)\mathbf{e}_\phi \qquad 55.36$$

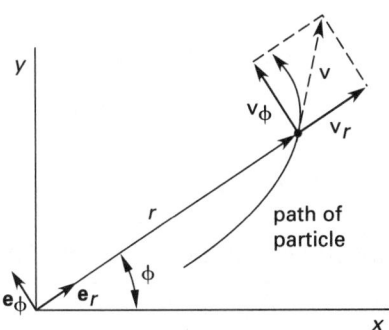

Figure 55.10 Radial and Transverse Components

The magnitudes of the radial and transverse components of velocity and acceleration are given by Eqs. 55.37 through 55.40.

$$\text{v}_r = \frac{dr}{dt} \qquad 55.37$$

$$\text{v}_\phi = r\frac{d\phi}{dt} \qquad 55.38$$

$$a_r = \frac{d^2r}{dt^2} - r\left(\frac{d\phi}{dt}\right)^2 \qquad 55.39$$

$$a_\phi = r\frac{d^2\phi}{dt^2} + 2\frac{dr}{dt}\frac{d\phi}{dt} \qquad 55.40$$

If the radial and transverse components of acceleration and velocity are known, they can be used to calculate the tangential and normal accelerations in a rectangular coordinate system.

$$a_t = \frac{a_r\text{v}_r + a_\phi\text{v}_\phi}{\text{v}_t} \qquad 55.41$$

$$a_n = \frac{a_\phi\text{v}_r - a_r\text{v}_\phi}{\text{v}_t} \qquad 55.42$$

16. RELATIVE MOTION

The term *relative motion* is used when motion of a particle is described with respect to something else in motion. The particle's position, velocity, and acceleration may be specified with respect to another moving particle or with respect to a moving frame of reference, known as a *Newtonian* or *inertial frame of reference*.

In Fig. 55.11, two particles, A and B, are moving with different velocities along a straight line. The separation between the two particles at any specific instant is the *relative position*, $s_{\text{B/A}}$, of B with respect to A, calculated as the difference between their two *absolute positions*.

$$s_{\text{B/A}} = s_\text{B} - s_\text{A} \qquad 55.43$$

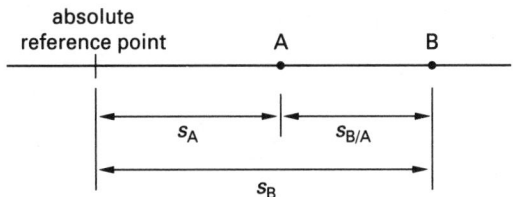

Figure 55.11 Relative Motion of Two Particles

Similarly, the *relative velocity* and *relative acceleration* of B with respect to A are the differences between the two *absolute velocities* and *absolute accelerations*, respectively.

$$\text{v}_{\text{B/A}} = \text{v}_\text{B} - \text{v}_\text{A} \qquad 55.44$$

$$a_{\text{B/A}} = a_\text{B} - a_\text{A} \qquad 55.45$$

Particles A and B are not constrained to move along a straight line. However, the subtraction must be done in vector or graphical form in all but the simplest cases.

$$\mathbf{s}_{B/A} = \mathbf{s}_B - \mathbf{s}_A \qquad 55.46$$

$$\mathbf{v}_{B/A} = \mathbf{v}_B - \mathbf{v}_A \qquad 55.47$$

$$\mathbf{a}_{B/A} = \mathbf{a}_B - \mathbf{a}_A \qquad 55.48$$

Since vector subtraction and addition operations can be performed graphically, many relative motion problems can be solved by a simplified graphical process.

Example 55.9

A stream flows at 5 kph. At what upstream angle, ϕ, should a 10 kph boat be piloted in order to reach the shore directly opposite the initial point?

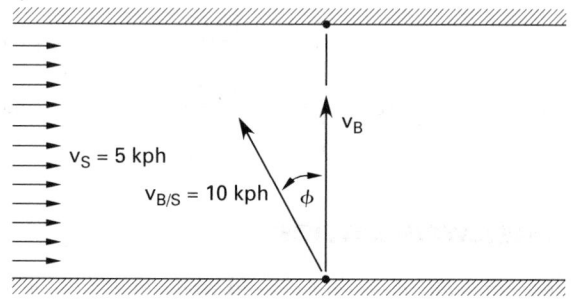

Solution

From Eq. 55.47, the absolute velocity of the boat, v_B, with respect to the shore is equal to the vector sum of the absolute velocity of the stream, v_S, and the relative velocity of the boat with respect to the stream, $v_{B/S}$. The magnitudes of these two velocities are known.

$$v_B = v_S + v_{B/S}$$

Since vector addition is accomplished graphically by placing the two vectors head to tail, the angle can be determined from trigonometry.

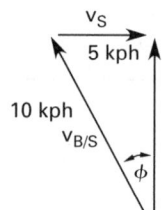

$$\sin\phi = \frac{v_S}{v_{B/S}} = \frac{5 \text{ kph}}{10 \text{ kph}} = 0.5$$

$$\phi = \arcsin(0.5) = 30°$$

Example 55.10

A stationary member of a marching band tosses a 2.0 ft long balanced baton straight up into the air and then begins walking forward at 4 mph. At a particular moment, the baton is 20 ft in the air and is falling back toward the earth with a velocity of 30 ft/sec. The tip of the baton is rotating at 140 rpm in the orientation shown.

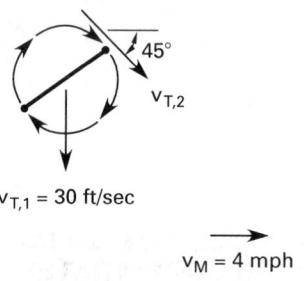

(a) What is the absolute velocity of the baton tip with respect to the ground? (b) What is the relative velocity of the baton tip with respect to the band member?

Solution

(a) The baton tip has two absolute velocity components. The first, with a magnitude of $v_{T,1} = 30$ ft/sec, is directed vertically downward. The second, with a magnitude of $v_{T,2}$, is directed as shown in the figure. The baton's radius, r, is 1 ft. From Eq. 55.19,

$$v_{T,2} = r\omega = \frac{(1 \text{ ft})\left(140 \dfrac{\text{rev}}{\text{min}}\right)\left(2\pi \dfrac{\text{rad}}{\text{rev}}\right)}{60 \dfrac{\text{sec}}{\text{min}}}$$

$$= 14.7 \text{ ft/sec}$$

The vector sum of these two absolute velocities is the velocity of the tip, $\mathbf{v}_T$, with respect to the earth.

$$\mathbf{v}_T = \mathbf{v}_{T,1} + \mathbf{v}_{T,2}$$

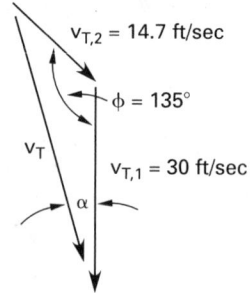

The velocity of the tip is found from the law of cosines.

$$v_T = \sqrt{v_{T,1}^2 + v_{T,2}^2 - 2v_{T,1}v_{T,2}\cos\phi}$$

$$= \sqrt{(30)^2 + (14.7)^2 - (2)(30)(14.7)(\cos 135°)}$$

$$= 41.7 \text{ ft/sec}$$

(b) Angle α is found from the law of sines.

$$\frac{\sin\alpha}{14.7\,\frac{ft}{sec}} = \frac{\sin135°}{41.7\,\frac{ft}{sec}}$$

$$\alpha = 14.4°$$

The band member's absolute velocity, v_M, is

$$v_M = \frac{\left(4\,\frac{mi}{hr}\right)\left(5280\,\frac{ft}{mi}\right)}{3600\,\frac{sec}{hr}} = 5.87\ ft/sec$$

From Eq. 55.44, the velocity of the tip with respect to the band member is

$$v_{T/M} = v_T - v_M$$

Subtracting a vector is equivalent to adding its negative. The velocity triangle is as shown. The law of cosines is used again to determine the relative velocity.

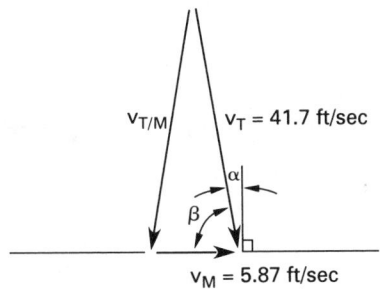

$$\beta = 90° - \alpha = 90° - 14.4° = 75.6°$$

$$v_{T/M} = \sqrt{v_T^2 + v_M^2 - 2v_Tv_M\cos\beta}$$
$$= \sqrt{(41.7)^2 + (5.87)^2 - (2)(41.7)(5.9)(\cos75.6°)}$$
$$= 40.6\ ft/sec\ (27.7\ mph)$$

17. DEPENDENT MOTION

When the position of one particle in a multiple-particle system depends on the position of one or more other particles, the motions are said to be "dependent." A block-and-pulley system with one fixed rope end, as illustrated by Fig. 55.12, is a *dependent system*.

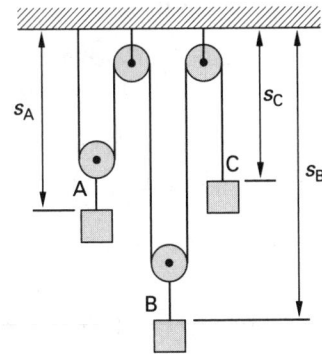

Figure 55.12 *Dependent System*

The following statements define the behavior of a dependent block-and-pulley system.

- Since the length of the rope is constant, the sum of the rope segments representing distances between the blocks and pulleys is constant. By convention, the distances are measured from the top of the block to the support point.[5] Since in Fig. 55.12 there are two ropes supporting block A, two ropes supporting block B, and one rope supporting block C,

$$2s_A + 2s_B + s_C = \text{constant} \qquad 55.49$$

- Since the position of the nth block in an n-block system is determined when the remaining $n - 1$ positions are known, the number of *degrees of freedom* is one less than the number of blocks.

- The movement, velocity, and acceleration of a block supported by two ropes are half the same quantities of a block supported by one rope.

- The relative relationships between the blocks' velocities or accelerations are the same as the relationships between the blocks' positions. For Fig. 55.12,

$$2v_A + 2v_B + v_C = 0 \qquad 55.50$$
$$2a_A + 2a_B + a_C = 0 \qquad 55.51$$

18. GENERAL PLANE MOTION

Rigid body *plane motion* can be described in two dimensions. Examples include rolling wheels, gear sets, and linkages. Plane motion can be considered as the sum of a translational component and a rotation about a fixed axis, as illustrated by Fig. 55.13.

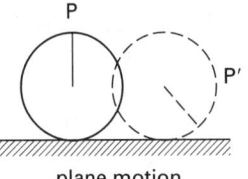

plane motion

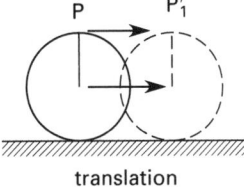

translation

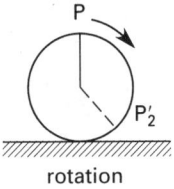

rotation

Figure 55.13 *Components of Plane Motion*

[5]In measuring distances, the finite diameters of the pulleys and the lengths of rope wrapped around the pulleys are disregarded.

19. ROTATION ABOUT A FIXED AXIS

Analysis of the rotational component of a rigid body's plane motion can sometimes be simplified if the location of the body's instantaneous center is known. Using the instantaneous center reduces many relative motion problems to simple geometry. The *instantaneous center* (also known as the *instant center* and IC) is a point at which the body could be fixed (pinned) without changing the instantaneous angular velocities of any point on the body. Thus, with the angular velocities, the body seems to rotate about a fixed instantaneous center.

The instantaneous center is located by finding two points for which the absolute velocity directions are known. Lines drawn perpendicular to these two velocities will intersect at the instantaneous center. (This graphical procedure is slightly different if the two velocities are parallel, as Fig. 55.14 shows. In that case, use is made of the fact that the tangential velocity is proportional to the distance from the instantaneous center.) For a rolling wheel, the instantaneous center is the point of contact with the supporting surface.

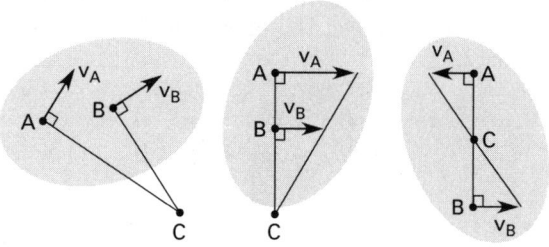

Figure 55.14 *Graphical Method of Finding the Instantaneous Center*

The absolute velocity of any point P on a wheel rolling (Fig. 55.15) with translational velocity, v_O, can be found by geometry. Assume that the wheel is pinned at C and rotates with its actual angular velocity, $\omega = v_O/r$. The direction of the point's velocity will be perpendicular to the line of length l between the instantaneous center and the point.

$$v = l\omega = \frac{lv_O}{r} \qquad 55.52$$

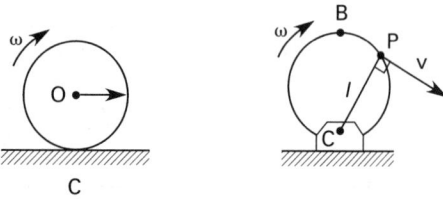

Figure 55.15 *Instantaneous Center of a Rolling Wheel*

Equation 55.52 is valid only for a velocity referenced to the instantaneous center, point C. Table 55.3 can be used to find the velocities with respect to other points.

Table 55.3 *Relative Velocities of a Rolling Wheel*

	reference point		
point	O	C	B
v_O	0	$v_O\rightarrow$	$\leftarrow v_O$
v_C	$\leftarrow v_O$	0	$\leftarrow 2v_O$
v_B	$v_O\rightarrow$	$2v_O\rightarrow$	0

Example 55.11

A truck with 35 in diameter tires travels at a constant 35 mph. What is the absolute velocity of point P on the circumference of the tire?

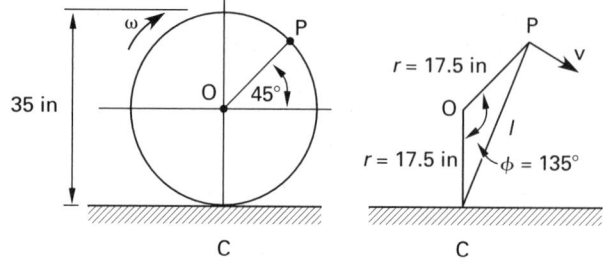

Solution

The translational velocity of the center of the wheel is

$$v_O = \frac{\left(35\ \dfrac{\text{mi}}{\text{hr}}\right)\left(5280\ \dfrac{\text{ft}}{\text{mi}}\right)}{3600\ \dfrac{\text{sec}}{\text{hr}}} = 51.33\ \text{ft/sec}$$

The tire radius is

$$r = \frac{35\ \text{in}}{(2)\left(12\ \dfrac{\text{in}}{\text{ft}}\right)} = 1.458\ \text{ft}$$

The angular velocity of the wheel is

$$\omega = \frac{v_O}{r} = \frac{51.33\ \dfrac{\text{ft}}{\text{sec}}}{1.458\ \text{ft}} = 35.21\ \text{rad/sec}$$

The instantaneous center is the contact point, C. The law of cosines is used to find the distance l.

$$l^2 = r^2 + r^2 - 2r^2\cos\phi = 2r^2(1 - \cos\phi)$$
$$l = \sqrt{(2)(1.458\ \text{ft})^2(1 - \cos 135°)} = 2.694\ \text{ft}$$

From Eq. 55.52, the absolute velocity of point P is

$$v_P = l\omega = (2.694\ \text{ft})\left(35.21\ \frac{\text{rad}}{\text{sec}}\right)$$
$$= 94.9\ \text{ft/sec}$$

20. INSTANTANEOUS CENTER OF ACCELERATION

The *instantaneous center of acceleration* is used to compute the absolute acceleration of a point as if a body is in pure rotation about that point. It is the same as the instantaneous center of rotation only for a body starting from rest and accelerating uniformly with angular acceleration, α.

$$a = l\alpha = \frac{la_O}{r} \qquad 55.53$$

In general, the instantaneous center of acceleration, C_a, will be deflected an angle β from the absolute acceleration vectors, as shown in Fig. 55.16. The relationship among the angle β, the instantaneous acceleration, α, and the instantaneous velocity, ω, is

$$\tan \beta = \frac{\alpha}{\omega^2} \qquad 55.54$$

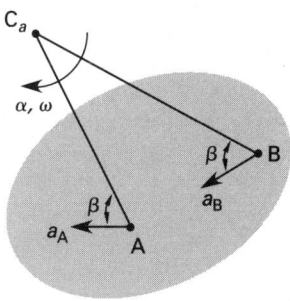

Figure 55.16 *Instantaneous Center of Acceleration*

The absolute acceleration, a, determined from Eq. 55.53 is the same as the *resultant acceleration* in Fig. 55.7.

21. SLIDER RODS

The absolute velocity of any point, P, on a slider rod assembly can be found from the instantaneous center concept. The instantaneous center, C, is located by extending perpendiculars from the velocity vectors.

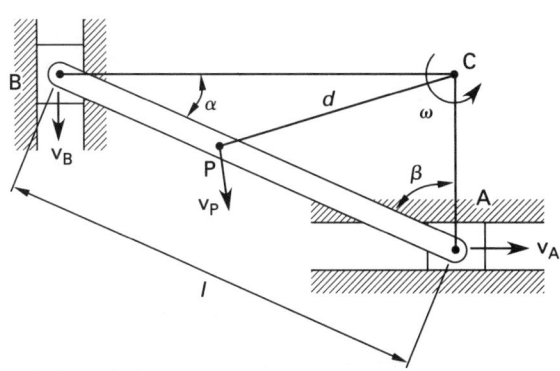

Figure 55.17 *Instantaneous Center of Slider Rod Assembly*

If the velocity with respect to point C of one end of the slider is known, say v_A, then v_B can be found from geometry. Since the slider can be assumed to rotate about point C with angular velocity ω,

$$\omega = \frac{v_A}{AC} = \frac{v_A}{l\cos\beta} = \frac{v_B}{BC} = \frac{v_B}{l\cos\alpha} \qquad 55.55$$

Since $\cos\alpha = \sin\beta$,

$$v_B = v_A(\tan\beta) \qquad 55.56$$

If the velocity with respect to point C of any other point P is required, it can be found from

$$v_P = d\omega \qquad 55.57$$

22. SLIDER-CRANK ASSEMBLIES

Figure 55.18 illustrates a slider-crank assembly for which points A and D are in the same plane and at the same elevation. The velocity of any point, P, on the rod can be found if the distance to the instantaneous center is known.

$$v_P = d\omega \qquad 55.58$$

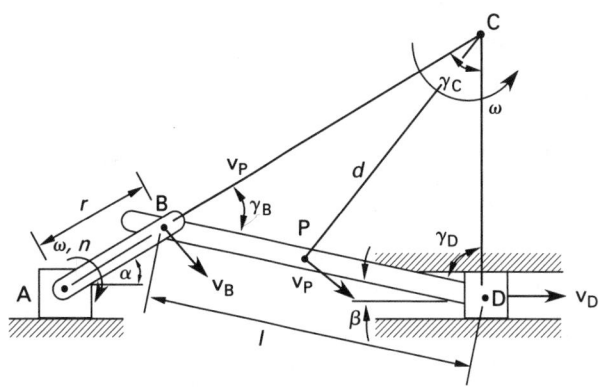

Figure 55.18 *Slider-Crank Assembly*

The tangential velocity of point B on the crank with respect to point A is perpendicular to the end of the crank. Slider D moves with a horizontal velocity. The intersection of lines drawn perpendicular to these velocity vectors locates the instantaneous center, point C. At any given instant, the connecting rod seems to rotate about point C with angular velocity ω.

$$\omega = \frac{2\pi n}{60} \qquad 55.59$$

The velocity of point D, v_D, is

$$v_D = (CD)\omega \qquad 55.60$$

Similarly, the velocity of point B, v_B, is

$$v_B = (BC)\omega \qquad 55.61$$

Dynamics and Vibrations

The following geometric relationships exist between the various points.

$$\frac{\sin \alpha}{l} = \frac{\sin \beta}{r} \qquad 55.62$$

$$\frac{\sin \gamma_D}{BC} = \frac{\sin \gamma_B}{CD} = \frac{\sin \gamma_C}{l} \qquad 55.63$$

$$\gamma_B = \alpha + \beta \qquad 55.64$$

$$\gamma_D = 90 - \beta \qquad 55.65$$

$$\gamma_C = 90 - \alpha \qquad 55.66$$

PRACTICE PROBLEMS

1. The center of a wheel with an outer diameter of 24 in (610 mm) is moving at 28 mi/hr (12.5 m/s). There is no slippage between the wheel and surface. A valve stem is mounted 6 in (150 mm) from the center. What are the velocity and direction of the valve stem when it is 45° from the horizontal?

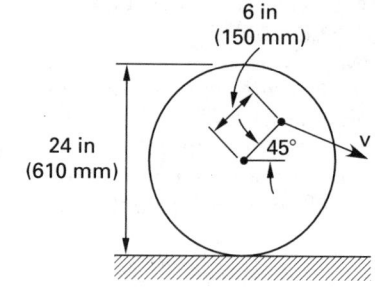

2. A projectile is launched with an initial velocity of 900 ft/sec (270 m/s). The target is 12,000 ft (3600 m) away and 2000 ft (600 m) higher than the launch point. Air friction is to be neglected. At what angle should the projectile be launched?

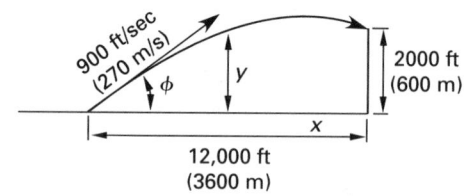

56 Kinetics

Dynamics and Vibrations

1. Introduction to Kinetics 56-2
2. Rigid Body Motion 56-2
3. Stability of Equilibrium Positions 56-2
4. Constant Forces 56-2
5. Linear Momentum 56-2
6. The Ballistic Pendulum 56-3
7. Angular Momentum 56-3
8. Newton's First Law of Motion 56-4
9. Newton's Second Law of Motion 56-4
10. Centripetal Force 56-5
11. Newton's Third Law of Motion 56-5
12. Dynamic Equilibrium 56-5
13. Flat Friction 56-6
14. Wedges 56-7
15. Belt Friction 56-7
16. Rolling Resistance 56-7
17. Motion of Rigid Bodies 56-8
18. Constrained Motion 56-9
19. Cable Tension from a Suspended Mass . . 56-11
20. Impulse 56-12
21. Impulse-Momentum Principle 56-13
22. Impulse-Momentum Principle in
    Open Systems 56-14
23. Impacts 56-15
24. Coefficient of Restitution 56-15
25. Rebound from Stationary Planes 56-15
26. Complex Impacts 56-16
27. Velocity-Dependent Force 56-17
28. Varying Mass 56-17
29. Central Force Fields 56-17
30. Newton's Law of Gravitation 56-18
31. Kepler's Laws of Planetary Motion 56-18
32. Space Mechanics 56-19
33. Vehicle Dynamics 56-19
34. Skidding Distance 56-21
35. Roadway Banking 56-21
    Practice Problems 56-22

Nomenclature

a	acceleration	ft/sec^2	m/s^2
a	coefficient of rolling resistance	ft	m
a	semimajor axis length	ft	m
A	area	ft^2	m^2
b	semiminor axis length	ft	m
BSFC	brake specific fuel consumption	lbm/hp-hr	kg/kW·h
C	coefficient	–	–
C	coefficient of viscous damping (linear)	lbf-sec/ft	N·s/m
C	coefficient of viscous damping (quadratic)	lbf-sec^2/ft^2	N·s^2/m^2
C	constant used in space mechanics	1/ft	1/m
d	diameter	ft	m
e	coefficient of restitution	–	–
e	superelevation	ft	m
E	energy	ft-lbf	J
f	coefficient of friction	–	–
F	force	lbf	N
g	acceleration due to gravity	ft/sec^2	m/s^2
g_c	gravitational constant	lbm-ft/ lbf-sec^2	n.a.
G	grade	ft/ft	m/m
G	universal gravitational constant	lbf-ft^2/lbm^2	N·m^2/kg^2
h	height	ft	m
h	angular momentum	ft^2-lbm/sec	m^2·kg/s
i	slippage fraction	–	–
I	mass moment of inertia	lbm-ft^2	kg·m^2
Imp	linear impulse	lbf-sec	N·s
Imp	angular impulse	lbf-ft-sec	N·m·s
k	spring constant	lbf/ft	N/m
m	mass	lbm	kg
$\dot{m}$	mass flow rate	lbm/sec	kg/s
m	slope	–	–
M	mass of the earth	lbm	kg
n	rotational speed	rev/sec	rev/s
N	normal force	lbf	N
p	momentum	lbm-ft/sec	kg·m/s
P	power	ft-lbf/sec	W
Q	flow rate	gal/hr	L/h
r	radius	ft	m
R	gear ratio	–	–
s	distance	ft	m
s'	fuel economy	mi/gal	km/L
t	time	sec	s
T	torque	ft-lbf	N·m
v	velocity	ft/sec	m/s
w	weight	lbf	–
W	work	ft-lbf	J

Symbols

α	angular acceleration	rad/sec^2	rad/s^2
δ	deflection	ft	m
ϵ	eccentricity	–	–
η	efficiency	–	–
θ	angular position	rad	rad
ρ	density	lbm/ft^3	kg/m^3
ϕ	angle	deg or rad	deg or rad
ω	angular velocity	rad/sec	rad/s

Subscripts

0	initial
b	braking
c	centripetal

PROFESSIONAL PUBLICATIONS, INC. BELMONT, CA

C instant center
D drag
f final or frictional
g gravitational
i inertial
k kinetic (dynamic)
m mechanical
n normal
O center or centroidal
p periodic
r rolling
s static
t tangential or terminal
w wedge

1. INTRODUCTION TO KINETICS

Kinetics is the study of motion and the forces that cause motion. Kinetics includes an analysis of the relationship between the force and mass for translational motion and between torque and moment of inertia for rotational motion. Newton's laws form the basis of the governing theory in the subject of kinetics.

2. RIGID BODY MOTION

The most general type of motion is *rigid body motion*. There are five types.

- *pure translation*: The orientation of the object is unchanged as its position changes. (Motion can be in straight or curved paths.)

- *rotation about a fixed axis*: All particles within the body move in concentric circles about the *axis of rotation*.

- *general plane motion*: The motion can be represented in two dimensions (i.e., the *plane of motion*).

- *motion about a fixed point*: This describes any three-dimensional motion with one fixed point, such as a spinning top or a truck-mounted crane. The distance from a fixed point to any particle in the body is constant.

- *general motion*: This is any motion not falling into one of the other four categories.

Figure 56.1 illustrates the terms yaw, pitch, and roll as they relate to general motion. *Yaw* is a left or right swinging motion of the leading edge. *Pitch* is an up or down swinging motion of the leading edge. *Roll* is rotation about the leading edge's longitudinal axis.

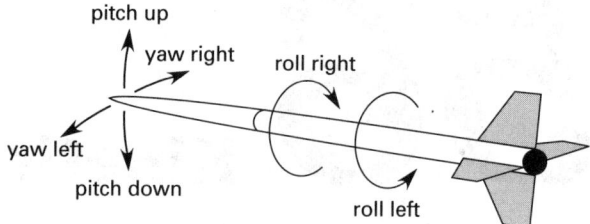

Figure 56.1 *Yaw, Pitch, and Roll*

3. STABILITY OF EQUILIBRIUM POSITIONS

Stability is defined in terms of a body's relationship with an equilibrium position. *Neutral equilibrium* exists if a body, when displaced from its equilibrium position, remains in its displaced state. *Stable equilibrium* exists if the body returns to the original equilibrium position after experiencing a displacement. *Unstable equilibrium* exists if the body moves away from the equilibrium position. These terms are illustrated by Fig. 56.2.

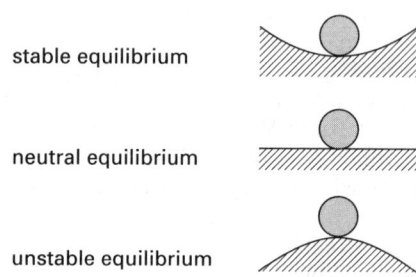

Figure 56.2 *Types of Equilibrium Positions*

4. CONSTANT FORCES

Force is a push or a pull that one body exerts on another, including gravitational, electrostatic, magnetic, and contact influences. Forces that do not vary with time are *constant forces*.

Strictly speaking, actions of other bodies on a rigid body are known as *external forces*. External forces are responsible for external motion of a body. *Internal forces* hold together parts of a rigid body.

5. LINEAR MOMENTUM

The vector *linear momentum* (usually just *momentum*) is defined by Eq. 56.1.[1] It has the same direction as the velocity vector. Momentum has units of force $\times$ time (e.g., lbf-sec or N·s).

$$\mathbf{p} = m\mathbf{v} \qquad \text{[SI]} \qquad 56.1(a)$$

$$\mathbf{p} = \frac{m\mathbf{v}}{g_c} \qquad \text{[U.S.]} \qquad 56.1(b)$$

---

[1] The symbols **P**, **mom**, $m\mathbf{v}$, and others are also used for momentum. Some authorities assign no symbol and just use the word momentum.

Momentum is conserved when no external forces act on a particle. If no forces act on the particle, the velocity and direction of the particle are unchanged. The *law of conservation of momentum* states that the linear momentum is unchanged if no unbalanced forces act on the particle. This does not prohibit the mass and velocity from changing, however. Only the product of mass or velocity is constant. Depending on the nature of the problem, momentum can be conserved in any or all of the three coordinate directions.

$$\sum m_0 \mathbf{v}_0 = \sum m_f \mathbf{v}_f \qquad \textit{56.2}$$

6. THE BALLISTIC PENDULUM

Figure 56.3 illustrates a *ballistic pendulum*. A projectile of known mass but unknown velocity is fired into a hanging target (the *pendulum*). The projectile is captured by the pendulum, which moves forward and upward. Kinetic energy is not conserved during impact because some of the projectile's kinetic energy is transformed into heat. However, momentum is conserved during impact, and the movement of the pendulum can be used to calculate the impact velocity of the projectile.[2]

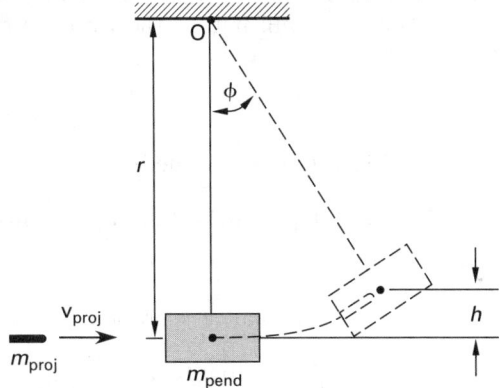

Figure 56.3 *Ballistic Pendulum*

Since no external forces act on the block during impact, the momentum of the system is conserved.

$$\mathbf{P}_{\text{before impact}} = \mathbf{P}_{\text{after impact}} \qquad \textit{56.3}$$

$$m_{\text{proj}} \mathbf{v}_{\text{proj}} = (m_{\text{proj}} + m_{\text{pend}}) \mathbf{v}_{\text{pend}} \qquad \textit{56.4}$$

Although kinetic energy before impact is not conserved, the total remaining energy after impact is conserved. That is, once the projectile has been captured by the pendulum, the kinetic energy of the pendulum-projectile combination is converted totally to potential energy as the pendulum swings upward.

---

[2]In this type of problem, it is important to be specific about when the energy and momentum are evaluated. During impact, kinetic energy is not conserved, but momentum is conserved. After impact, as the pendulum swings, energy is conserved but momentum is not conserved because gravity (an external force) acts on the pendulum during its swing.

$$\left(\tfrac{1}{2}\right)(m_{\text{proj}} + m_{\text{pend}}) \mathbf{v}_{\text{pend}}^2 = (m_{\text{proj}} + m_{\text{pend}}) gh \qquad \textit{56.5}$$

$$\mathbf{v}_{\text{pend}} = \sqrt{2gh} \qquad \textit{56.6}$$

The relationship between the rise of the pendulum, h, and the swing angle, ϕ, is

$$h = r(1 - \cos\phi) \qquad \textit{56.7}$$

Since the time during which the force acts is not well defined, there is no single equivalent force that can be assumed to initiate the motion. Any force that produces the same impulse over a given contact time will be applicable.

7. ANGULAR MOMENTUM

The vector *angular momentum* (also known as *moment of momentum*) taken about a point O is the moment of the linear momentum vector. Angular momentum has units of distance × force × time (e.g., ft-lbf-sec or N·m·s). It has the same direction as the rotation vector and can be determined by use of the right-hand rule. (That is, it acts in a direction perpendicular to the plane containing the position and linear momentum vectors.)

$$\mathbf{h}_O = \mathbf{r} \times m\mathbf{v} \qquad \text{[SI]} \qquad \textit{56.8(a)}$$

$$\mathbf{h}_O = \frac{\mathbf{r} \times m\mathbf{v}}{g_c} \qquad \text{[U.S.]} \qquad \textit{56.8(b)}$$

Any of the methods normally used to evaluate cross-products can be used with angular momentum. The scalar form of Eq. 56.8 is

$$h_O = rmv \sin\phi \qquad \text{[SI]} \qquad \textit{56.9(a)}$$

$$h_O = \frac{rmv \sin\phi}{g_c} \qquad \text{[U.S.]} \qquad \textit{56.9(b)}$$

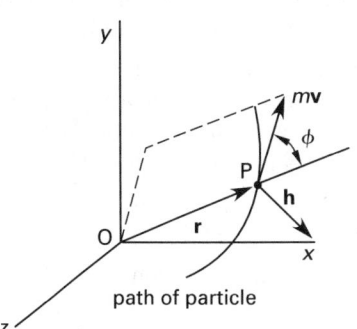

Figure 56.4 *Angular Momentum*

For a rigid body rotating about an axis passing through its center of gravity located at point O, the scalar value of angular momentum is given by Eq. 56.10.

$$h_O = I\omega \qquad \text{[SI]} \qquad \textit{56.10(a)}$$

$$h_O = \frac{I\omega}{g_c} \qquad \text{[U.S.]} \qquad \textit{56.10(b)}$$

Dynamics and Vibrations

8. NEWTON'S FIRST LAW OF MOTION

Much of this chapter is based on Newton's laws of motion. *Newton's first law of motion* can be stated in several forms.

common form: A particle will remain in a state of rest or will continue to move with constant velocity unless an unbalanced external force acts on it.

law of conservation of momentum form: If the resultant external force acting on a particle is zero, then the linear momentum of the particle is constant.

9. NEWTON'S SECOND LAW OF MOTION

Newton's second law of motion is stated as follows.

second law: The acceleration of a particle is directly proportional to the force acting on it and inversely proportional to the particle mass. The direction of acceleration is the same as the force of direction.

This law can be stated in terms of the force vector required to cause a change in momentum. The resultant force is equal to the rate of change of linear momentum.

$$\mathbf{F} = \frac{d\mathbf{p}}{dt} \qquad 56.11$$

If the mass is constant with respect to time, the scalar form of Eq. 56.11 is[3]

$$F = m\left(\frac{d\mathrm{v}}{dt}\right) = ma \qquad \text{[SI]} \quad 56.12(a)$$

$$F = \left(\frac{m}{g_c}\right)\left(\frac{d\mathrm{v}}{dt}\right) = \frac{ma}{g_c} \quad \text{[U.S.]} \quad 56.12(b)$$

Equation 56.12 can be written in rectangular coordinates form (i.e., in terms of x- and y- component forces), in polar coordinates form (i.e, tangential and normal components), and in cylindrical coordinates form (i.e., radial and transverse components).

Although Newton's laws do not specifically deal with rotation, there is an analogous relationship between torque and change in angular momentum. For a rotating body, the torque, $\mathbf{T}$, required to change the angular momentum is

$$\mathbf{T} = \frac{d\mathbf{h}_0}{dt} \qquad 56.13$$

If the moment of inertia is constant, the scalar form of Eq. 56.13 is

$$T = I\left(\frac{d\omega}{dt}\right) = I\alpha \qquad \text{[SI]} \quad 56.14(a)$$

$$T = \left(\frac{I}{g_c}\right)\left(\frac{d\omega}{dt}\right) = \frac{I\alpha}{g_c} \quad \text{[U.S.]} \quad 56.14(b)$$

---

[3]Equation 56.12 shows that force is a scalar multiple of acceleration. Any consistent set of units can be used. For example, if both sides are divided by the acceleration of gravity (i.e., so that acceleration in Eq. 56.12 is in gravities), the force will have units of *g-forces* or *gees* (i.e., multiples of the gravitational force).

Example 56.1

The acceleration in m/s^2 of a 40 kg body is specified by the equation

$$a(t) = 8 - 12t$$

What is the instantaneous force acting on the body at $t = 6$ s?

Solution

The acceleration is

$$a(6) = 8\,\frac{\mathrm{m}}{\mathrm{s}^2} - \left(12\,\frac{\mathrm{m}}{\mathrm{s}^3}\right)(6\ \mathrm{s}) = -64\ \mathrm{m/s}^2$$

From Newton's second law, the instantaneous force is

$$F = ma = (40\ \mathrm{kg})\left(-64\,\frac{\mathrm{m}}{\mathrm{s}^2}\right)$$
$$= -2560\ \mathrm{N}$$

Example 56.2

During start-up, a 4.0 ft diameter wheel with centroidal moment of inertia of 1610 lbm-ft^2 is subjected to tight-side and loose-side belt tensions of 200 lbf and 100 lbf, respectively. A frictional torque of 15 ft-lbf is acting to resist pulley rotation. (a) What is the angular acceleration? (b) How long will it take the wheel to reach a speed of 120 rpm?

Solution

(a) From Eq. 56.24, the net torque is

$$T = rF_{\mathrm{net}} = (2\ \mathrm{ft})(200\ \mathrm{lbf} - 100\ \mathrm{lbf}) - 15\ \mathrm{ft\text{-}lbf}$$
$$= 185\ \mathrm{ft\text{-}lbf}$$

From Eq. 56.14, the angular acceleration is

$$\alpha = \frac{g_c T}{I} = \frac{\left(32.2\,\dfrac{\mathrm{lbm\text{-}ft}}{\mathrm{lbf\text{-}sec}^2}\right)(185\ \mathrm{ft\text{-}lbf})}{1610\ \mathrm{lbm\text{-}ft}^2}$$
$$= 3.7\ \mathrm{rad/sec}^2$$

(b) The rotational speed is

$$\omega = \frac{\left(120\,\dfrac{\mathrm{rev}}{\mathrm{min}}\right)\left(2\pi\,\dfrac{\mathrm{rad}}{\mathrm{rev}}\right)}{60\,\dfrac{\mathrm{sec}}{\mathrm{min}}}$$
$$= 12.6\ \mathrm{rad/sec}$$

This is a case of constant angular acceleration starting from rest.

$$t = \frac{\omega}{\alpha} = \frac{12.6\,\dfrac{\mathrm{rad}}{\mathrm{sec}}}{3.7\,\dfrac{\mathrm{rad}}{\mathrm{sec}^2}}$$
$$= 3.4\ \mathrm{sec}$$

10. CENTRIPETAL FORCE

Newton's second law says there is a force for every acceleration a body experiences. For a body moving around a curved path, the total acceleration can be separated into tangential and normal components. By Newton's second law, there are corresponding forces in the tangential and normal directions. The force associated with the normal acceleration is known as the *centripetal force*.[4]

$$F_c = ma_n = \frac{mv_t^2}{r} \quad \text{[SI]} \quad \textit{56.15(a)}$$

$$F_c = \frac{mv_t^2}{g_c r} \quad \text{[U.S.]} \quad \textit{56.15(b)}$$

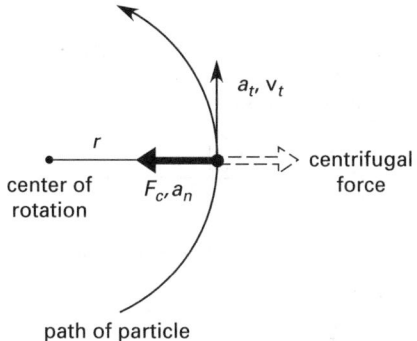

Figure 56.5 *Centripetal Force*

The centripetal force is a real force on the body toward the center of rotation. The so-called *centrifugal force* is an apparent force on the body directed away from the center of rotation. The centripetal and centrifugal forces are equal in magnitude but opposite in sign.

Example 56.3

A 4500 lbm (2000 kg) car travels at 40 mph (65 kph) around a curve with a radius of 200 ft (60 m). What is the centripetal force?

SI Solution

The tangential velocity is

$$v_t = \frac{\left(65 \ \frac{\text{km}}{\text{h}}\right)\left(1000 \ \frac{\text{m}}{\text{km}}\right)}{3600 \ \frac{\text{s}}{\text{h}}} = 18.06 \ \text{m/s}$$

From Eq. 56.15(a), the centripetal force is

$$F_c = \frac{mv_t^2}{r} = \frac{(2000 \ \text{kg})\left(18.06 \ \frac{\text{m}}{\text{s}}\right)^2}{60 \ \text{m}}$$
$$= 10\,900 \ \text{N}$$

[4]The term *normal force* is reserved for the plane reaction in friction calculations.

Customary U.S. Solution

The tangential velocity is

$$v_t = \frac{\left(40 \ \frac{\text{mi}}{\text{hr}}\right)\left(5280 \ \frac{\text{ft}}{\text{mi}}\right)}{3600 \ \frac{\text{sec}}{\text{hr}}} = 58.7 \ \text{ft/sec}$$

From Eq. 56.15(b), the centripetal force is

$$F_c = \frac{mv_t^2}{g_c r}$$
$$= \frac{(4500 \ \text{lbm})\left(58.7 \ \frac{\text{ft}}{\text{sec}}\right)^2}{\left(32.2 \ \frac{\text{lbm-ft}}{\text{lbf-sec}^2}\right)(200 \ \text{ft})}$$
$$= 2400 \ \text{lbf}$$

An unbalanced rotating body (vehicle wheel, clutch disk, rotor of an electrical motor, etc.) will experience a dynamic *unbalanced force*. Though the force is essentially centripetal in nature and is given by Eq. 56.15, it is generally difficult to assign a value to the radius. For that reason, the force is often determined directly on the rotating body or from the deflection of its supports.

Since the body is rotating, the force will be experienced in all directions perpendicular to the axis of rotation. If the supports are flexible, the force will cause the body to vibrate, and the frequency of vibration will essentially be the rotational speed. If the supports are rigid, the bearings will carry the unbalanced force and transmit it to other parts of the frame.

11. NEWTON'S THIRD LAW OF MOTION

Newton's third law of motion is

> *third law:* For every acting force between two bodies, there is an equal but opposite reacting force on the same line of action.

$$\mathbf{F}_{\text{reacting}} = -\mathbf{F}_{\text{acting}} \quad \textit{56.16}$$

12. DYNAMIC EQUILIBRIUM

An accelerating body is not in static equilibrium. Accordingly, the familiar equations of statics ($\sum F = 0$ and $\sum M = 0$) do not apply. However, if the so-called *inertial force*, $m\mathbf{a}$, is included in the static equilibrium equation, the body is said to be in *dynamic equilibrium*.[5,6] This is known as *D'Alembert's principle*. Since the inertial force acts to oppose changes in motion, it is negative in the summation.

[5]Other names for the inertial force are *inertia vector* (when written as $m\mathbf{a}$), *dynamic reaction*, and *reversed effective force*. The term $\sum \mathbf{F}$ is known as the *effective force*.

[6]*Dynamic* and *equilibrium* are contradictory terms. A better term is *simulated equilibrium*, but this form has not caught on.

$$\sum \mathbf{F} - m\mathbf{a} = 0 \quad \text{[SI]} \quad \textit{56.17(a)}$$

$$\sum \mathbf{F} - \frac{m\mathbf{a}}{g_c} = 0 \quad \text{[U.S.]} \quad \textit{56.17(b)}$$

It should be clear that D'Alembert's principle is just a different form of Newton's second law, with the ma term transposed to the left-hand side.

The analogous rotational form of the dynamic equilibrium principle is

$$\sum \mathbf{T} - I\alpha = 0 \quad \text{[SI]} \quad \textit{56.18(a)}$$

$$\sum \mathbf{T} - \frac{I\alpha}{g_c} = 0 \quad \text{[U.S.]} \quad \textit{56.18(b)}$$

13. FLAT FRICTION

Friction is a force that always resists motion or impending motion. It always acts parallel to the contacting surfaces. The frictional force, F_f, exerted on a stationary body is known as *static friction*, *Coulomb friction*, and *fluid friction*. If the body is moving, the friction is known as *dynamic friction* and is less than the static friction.

The actual magnitude of the frictional force depends on the *normal force*, N, and the *coefficient of friction*, f, between the body and the surface.[7] For a body resting on a horizontal surface, the normal force is the weight of the body.

$$N = mg \quad \text{[SI]} \quad \textit{56.19(a)}$$

$$N = \frac{mg}{g_c} \quad \text{[U.S.]} \quad \textit{56.19(b)}$$

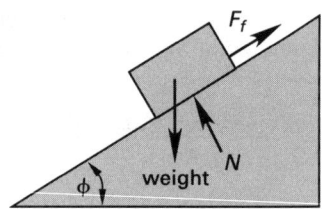

Figure 56.6 *Frictional and Normal Forces*

If the body rests on an inclined surface, the normal force is calculated from the weight.

$$N = mg \cos\phi \quad \text{[SI]} \quad \textit{56.20(a)}$$

$$N = \frac{mg \cos\phi}{g_c} \quad \text{[U.S.]} \quad \textit{56.20(b)}$$

The maximum static frictional force, F_f, is the product of the coefficient of friction, f, and the normal force, N.

---

[7]The symbol μ is also widely used by engineers to represent the coefficient of friction.

(The subscripts s and k are used to distinguish between the static and dynamic (kinetic) coefficients of friction.)

$$F_{f,\text{max}} = f_s N \qquad \textit{56.21}$$

The frictional force acts only in response to a disturbing force. If a small disturbing force (i.e., a force less than $F_{f,\text{max}}$) acts on a body, then the frictional force will equal the disturbing force, and the maximum frictional force will not develop. This is known as the *equilibrium phase*. The *motion impending phase* is where the disturbing force equals the maximum frictional force, $F_{f,\text{max}}$. Once motion begins, however, the coefficient of friction drops slightly, and a lower frictional force opposes movement. These cases are illustrated in Fig. 56.7.

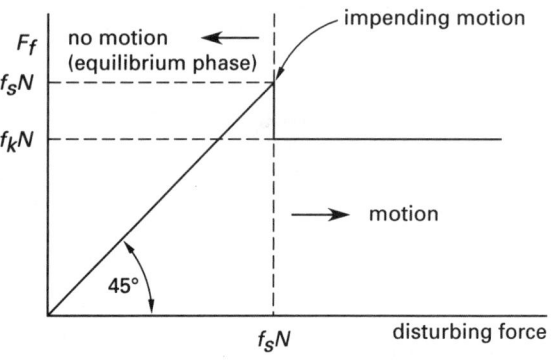

Figure 56.7 *Frictional Force versus Disturbing Force*

A body on an inclined plane will not begin to slip down the plane until the component of weight parallel to the plane exceeds the frictional force. If the plane's inclination angle can be varied, the body will not slip until the angle reaches a critical angle known as the *angle of repose* or *angle of static friction*, ϕ. Equation 56.22 relates this angle to the coefficient of static friction.

$$\tan\phi = f_s \qquad \textit{56.22}$$

Tabulations of coefficients of friction distinguish between types of surfaces and between static and dynamic cases. They might also list values for dry conditions and oiled conditions. The term *dry* is synonymous with *nonlubricated*. The ambiguous term *wet*, although a natural antonym for *dry*, is sometimes used to mean *oily*. However, it usually means wet with water, as in tires on a wet roadway after a rain. Typical values of the coefficient of friction are given in Table 56.1.[8]

---

[8]Experimental and reported values of the coefficient of friction vary greatly from researcher to researcher and experiment to experiment. The values in Table 56.1 are more for use in solving practice problems than serving as the last word in available data.

Table 56.1 Typical Coefficients of Friction

material	condition	dynamic	static
cast iron on cast iron	dry	0.15	1.00
plastic on steel	dry	0.35	0.45
grooved rubber on pavement	dry	0.40	0.55
bronze on steel	oiled	0.07	0.09
steel on graphite	dry	0.16	0.21
steel on steel	dry	0.42	0.78
steel on steel	oiled	0.08	0.10
steel on asbestos-faced steel	dry	0.11	0.15
steel on asbestos-faced steel	oiled	0.09	0.12
press fits (shaft in hole)	oiled	–	0.10–0.15

A special case of the angle of repose is the *angle of internal friction*, ϕ, of soil, grain, or other granular material. The angle made by a pile of granular material depends on how much friction there is between the granular particles. Liquids have angles of internal friction of zero, because they do not form piles.

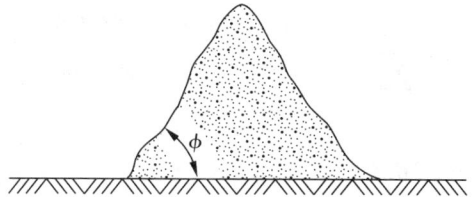

Figure 56.8 Angle of Internal Friction

14. WEDGES

Wedges are machines that are able to raise heavy loads. The wedge angles are chosen so that friction will keep the wedge in place once it is driven between the load and support. As with any situation where friction is present, the frictional force is parallel to the contacting surfaces.

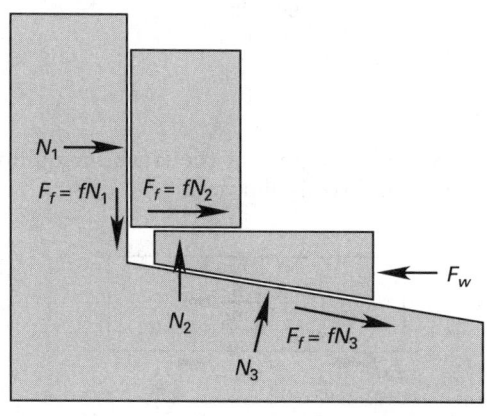

Figure 56.9 Typical Wedge Problem

15. BELT FRICTION

Friction from a belt, rope, or band wrapped around a pulley or sheave is responsible for the transfer of torque. Except at start-up, one side of the belt (the tight side) will have a higher tension than the other (the slack side). The basic relationship between these belt tensions and the coefficient of friction neglects centrifugal effects and is given by Eq. 56.23.[9] (The angle of wrap, ϕ, must be expressed in radians.)

$$\frac{F_{\max}}{F_{\min}} = e^{f\phi} \qquad 56.23$$

The net transmitted torque is

$$T = (F_{\max} - F_{\min})\, r \qquad 56.24$$

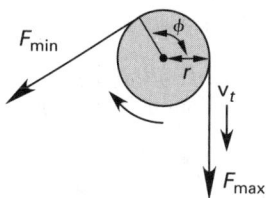

Figure 56.10 Belt Friction

The power transmitted by the belt running at tangential velocity v_t is given by Eq. 56.25.[10]

$$P = (F_{\max} - F_{\min})\, v_t \qquad 56.25$$

The centrifugal force experienced by the belt should be considered when the velocity or belt mass is very large. Equation 56.26 can be used, where m is the mass per unit length of belt.

$$\frac{F_{\max} - F_c}{F_{\min} - F_c} = \frac{F_{\max} - m v_t^2}{F_{\min} - m v_t^2} = e^{f\phi} \quad \text{[SI]} \qquad 56.26$$

16. ROLLING RESISTANCE

Rolling resistance opposes motion, but it is not friction. Rather, it is caused by the deformation of the rolling body and the supporting surface. Rolling resistance is characterized by a *coefficient of rolling resistance,*

---

[9]This equation does not apply to V-belts. V-belt design and analysis is dependent on the cross-sectional geometry of the belt.
[10]When designing a belt system, the horsepower to be transmitted should be multiplied by a *service factor* to obtain the *design power*. Service factors range from 1.0 to 1.5 and depend on the nature of the power source, the load, and the starting characteristics.

Dynamics and Vibrations

a, which has units of length.[11] Since this deformation is very small, the rolling resistance in the direction of motion is

$$F_r = \frac{mga}{r} \qquad \text{[SI]} \qquad 56.27(a)$$

$$F_r = \frac{mga}{rg_c} = \frac{wa}{r} \qquad \text{[U.S.]} \qquad 56.27(b)$$

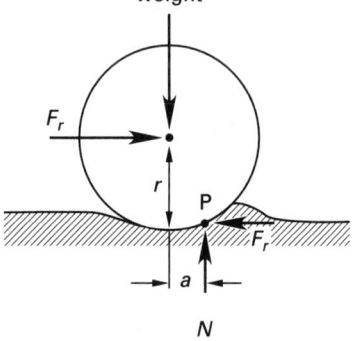

Figure 56.11 *Rolling Resistance*

The term *coefficient of rolling friction*, f_r, is occasionally encountered, although friction is not the cause of rolling resistance.

$$f_r = \frac{F_r}{w} = \frac{a}{r} \qquad\qquad 56.28$$

17. MOTION OF RIGID BODIES

When a rigid body experiences pure translation, its position changes without any change in orientation. At any instant, all points on the body have the same displacement, velocity, and acceleration. The behavior of a rigid body in translation is given by Eqs. 56.29 and 56.30. All equations are written for the center of mass. (These equations represent Newton's second law written in component form.)

$$\sum F_x = ma_x \qquad \text{[SI]} \qquad 56.29$$

$$\sum F_y = ma_y \qquad \text{[SI]} \qquad 56.30$$

When a torque acts on a rigid body, the rotation will be about the center of gravity unless the body is constrained otherwise. In the case of rotation, the torque and angular acceleration are related by Eq. 56.31.

$$T = I\alpha \qquad \text{[SI]} \qquad 56.31(a)$$

$$T = \frac{I\alpha}{g_c} \qquad \text{[U.S.]} \qquad 56.31(b)$$

[11] Rolling resistance is traditionally derived by assuming the roller encounters a small step in its path a distance a in front of the center of gravity. The forces acting on the roller are the weight and driving force acting through the centroid and the normal force and rolling resistance acting at the contact point. Equation 56.27 is derived by taking moments about the contact point, P.

Euler's equations of motion are used to analyze the motion of a rigid body about a fixed point, O. This class of problem is particularly difficult because the mass moments of products of inertia change with time if a fixed set of axes is used. Therefore, it is more convenient to define the x-, y-, and z-axes with respect to the body. Such an action is acceptable because the angular momentum about the origin, $\mathbf{h}_O$, corresponding to a given angular velocity, ω, is independent of the choice of coordinate axes.

An infinite number of axes can be chosen. (A general relationship between moments and angular momentum is given in most dynamics textbooks.) However, if the origin is at the mass center and the x-, y-, and z-axes coincide with the principal axes of inertia of the body (such that the product of inertia is zero), the angular momentum of the body about the origin (i.e., point O at $(0,0,0)$) is given by the simplified relationship

$$\mathbf{h}_O = I_x\omega_x\mathbf{i} + I_y\omega_y\mathbf{j} + I_z\omega_z\mathbf{k} \qquad 56.32$$

The three scalar Euler equations of motion can be derived from this simplified relationship.

$$\sum M_x = I_x\alpha_x - (I_y - I_z)\omega_y\omega_z \qquad 56.33$$

$$\sum M_y = I_y\alpha_y - (I_z - I_x)\omega_z\omega_x \qquad 56.34$$

$$\sum M_z = I_z\alpha_z - (I_x - I_y)\omega_x\omega_y \qquad 56.35$$

Example 56.4

A 5000 lbm truck skids with a deceleration of 15 ft/sec². (a) What is the coefficient of sliding friction? (b) What are the frictional forces and normal reactions (per axle) at the tires?

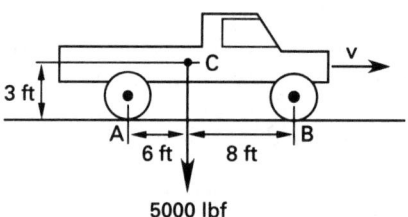

Solution

(a) The free-body diagram of the truck in equilibrium with the inertial force is shown.

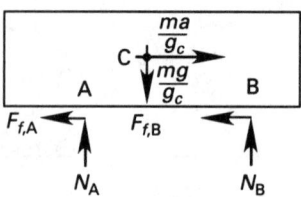

The equation of dynamic equilibrium in the horizontal direction is

$$\sum F_x = \frac{ma}{g_c} - F_{f,\text{A}} - F_{f,\text{B}} = 0$$

$$= \frac{ma}{g_c} - Nf = 0$$

$$\frac{(5000\ \text{lbm})\left(15\ \dfrac{\text{ft}}{\text{sec}^2}\right)}{32.2\ \dfrac{\text{lbm-ft}}{\text{lbf-sec}^2}} - (5000\ \text{lbf})f = 0$$

The coefficient of friction is

$$f = 0.466$$

(b) The vertical reactions at the tires can be found by taking moments about one of the contact points.

$$\sum M_{\text{A}}: \quad 14N_{\text{B}} - (6\ \text{ft})(5000\ \text{lbf})$$

$$- (3\ \text{ft})\left(\frac{5000\ \text{lbm}}{32.2\ \dfrac{\text{lbm-ft}}{\text{lbf-sec}^2}}\right)\left(15\ \frac{\text{ft}}{\text{sec}^2}\right) = 0$$

$$N_{\text{B}} = 2642\ \text{lbf}$$

The remaining vertical reaction is found by summing vertical forces.

$$\sum F_y: \quad N_{\text{A}} + N_{\text{B}} - \frac{mg}{g_c} = 0$$

$$N_{\text{A}} + 2642\ \text{lbf} - 5000\ \text{lbf} = 0$$

$$N_{\text{A}} = 2358\ \text{lbf}$$

The horizontal frictional forces at the front and rear axles are

$$F_{f,\text{A}} = (0.466)(2358\ \text{lbf}) = 1099\ \text{lbf}$$

$$F_{f,\text{B}} = (0.466)(2642\ \text{lbf}) = 1231\ \text{lbf}$$

18. CONSTRAINED MOTION

Figure 56.12 shows a cylinder (or sphere) on an inclined plane. If there is no friction, there will be no torque to start the cylinder rolling. Regardless of the angle, the cylinder will slide down the incline in *unconstrained motion*.

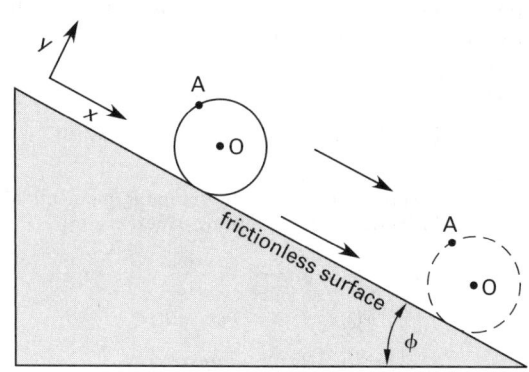

Figure 56.12 *Unconstrained Motion*

The acceleration sliding down the incline can be calculated by writing Newton's second law for an axis parallel to the plane.[12] Once the acceleration is known, the velocity can be found from the constant-acceleration equations.

$$ma_{\text{O},x} = mg\sin\phi \qquad\qquad \textit{56.36}$$

If friction is sufficiently large, or if the inclination is sufficiently small, there will be no slipping. This condition occurs if

$$\phi < \arctan(f_s) \qquad\qquad \textit{56.37}$$

The frictional force acting at the cylinder's radius r supplies a torque that starts and keeps the cylinder rolling. The frictional force is

$$F_f = fN \qquad\qquad \textit{56.38}$$

$$F_f = fmg\cos\phi \qquad [\text{SI}] \qquad \textit{56.39(a)}$$

$$F_f = \frac{fmg\cos\phi}{g_c} \qquad [\text{U.S.}] \qquad \textit{56.39(b)}$$

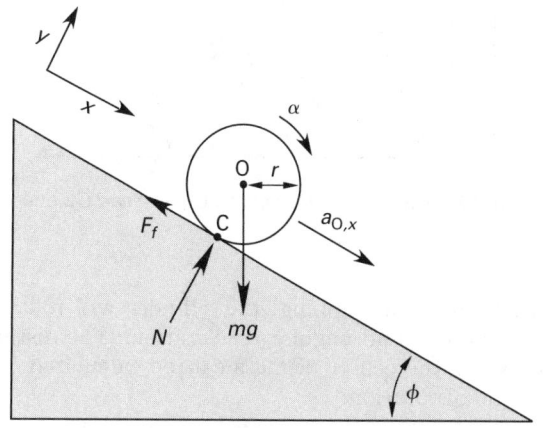

Figure 56.13 *Constrained Motion*

Thus, with no slipping, the cylinder has two degrees of freedom (the x-directional and rotation), and motion of the center of mass must simultaneously satisfy (i.e., is constrained by) two equations. (This excludes motion perpendicular to the plane.) This is called *constrained motion*.

$$mg\sin\phi - F_f = ma_{\text{O},x} \qquad [\text{SI}] \qquad \textit{56.40}$$

$$F_f r = I_{\text{O}}\alpha \qquad [\text{SI}] \qquad \textit{56.41}$$

The mass moment of inertia used in calculating angular acceleration can be either the centroidal moment of inertia, I_{O}, or the moment of inertia taken about the contact point, I_{C}, depending on whether torques (moments) are evaluated with respect to point O or point C, respectively.

---

[12]Most inclined plane problems are conveniently solved by resolving all forces into components parallel and perpendicular to the plane.

If moments are evaluated with respect to point O, the coefficient of friction, f, must be known, and the centroidal moment of inertia (found from a table) can be used. If moments are evaluated with respect to the contact point, the frictional and normal forces drop out of the torque summation. The cylinder instantaneously rotates as though it were pinned at point C. If the centroidal moment of inertia, I_O, is known, the parallel axis theorem can be used to find the required moment of inertia.

$$I_C = I_O + mr^2 \qquad\qquad 56.42$$

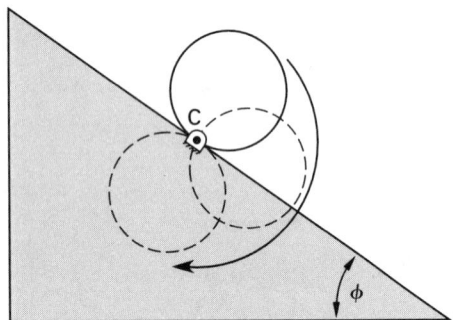

Figure 56.14 *Instantaneous Center of a Constrained Cylinder*

When there is no slipping, the cylinder will roll with constant linear and angular acceleration. The distance traveled by the center of mass can be calculated from the angle of rotation.

$$s_O = r\theta \qquad\qquad 56.43$$

If $\phi \geq \arctan f_s$, the cylinder will simultaneously roll and slide down the incline. The analysis is similar to the no-sliding case, except that the coefficient of sliding friction is used. Once sliding has started, the inclination angle can be reduced to $\arctan f_k$, and rolling with sliding will continue.

Example 56.5

A 150 kg cylinder with radius 0.3 m is pulled up a plane inclined at 30° as fast as possible without the cylinder slipping. The coefficient of friction is 0.236. There is a groove in the cylinder at radius = 0.2 m. A rope in the groove applies a force of 500 N up the ramp. What is the linear acceleration of the cylinder?

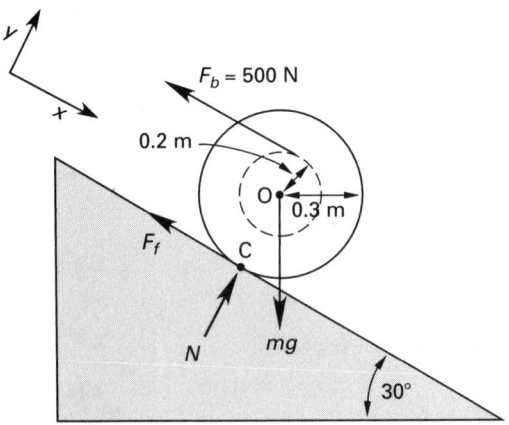

Solution

To solve by summing forces in the x-direction:

The normal force is

$$N = mg\cos\phi = (150 \text{ kg})\left(9.81 \ \frac{\text{m}}{\text{s}^2}\right)(\cos 30°)$$
$$= 1274.4 \text{ N}$$

The frictional maximum (friction impending) force is

$$F_f = fN = (0.236)(1274.4 \text{ N}) = 300.8 \text{ N}$$

The summation of forces in the x-direction is

$$ma_{O,x} = mg\sin\phi - F_f - F_b$$
$$a_{O,x} = \frac{(150 \text{ kg})\left(9.81 \ \frac{\text{m}}{\text{s}^2}\right)(\sin 30°) - 300.8 \text{ N} - 500 \text{ N}}{150 \text{ kg}}$$
$$= -0.433 \text{ m/s}^2 \quad [\text{up the incline}]$$

To solve by taking moment about the contact point:[13]

From App. 54.A, the centroidal mass moment of inertia of the cylinder is

$$I_O = \tfrac{1}{2}mr^2 = (0.5)(150 \text{ kg})(0.3 \text{ m})^2$$
$$= 6.75 \text{ kg·m}^2$$

The mass moment of inertia with respect to the contact point, C, is given by the parallel axis theorem.

$$I_C = I_O + mr^2 = \tfrac{1}{2}mr^2 + mr^2 = \tfrac{3}{2}mr^2$$
$$= \left(\tfrac{3}{2}\right)(150 \text{ kg})(0.3 \text{ m})^2 = 20.25 \text{ kg·m}^2$$

---

[13]This example can also be solved by summing moments about the center. If this is done, the governing equations are

$$I_O = \tfrac{1}{2}mr^2$$
$$M_O = I_O\alpha = F_b r' - F_f r$$
$$\tfrac{1}{2}mr^2\alpha = F_b r' - \mu mg\cos\phi \, r$$
$$a_{O,x} = r\alpha$$

The x-component of the weight acts through the center of gravity. (This term dropped out when moments were taken with respect to the center of gravity.)

$$(mg)_x = (150 \text{ kg}) \left(9.81 \ \frac{\text{m}}{\text{s}^2}\right) (\sin 30°)$$
$$= 735.8 \text{ N}$$

The summation of torques about point C gives the angular acceleration with respect to point C.

$$(735.8 \text{ N})(0.3 \text{ m})$$
$$- (500 \text{ N})(0.3 \text{ m} + 0.2 \text{ m}) = (20.25 \text{ kg·m}^2)\alpha$$
$$\alpha = -1.445 \text{ rad/s}^2$$

The linear acceleration can be calculated from the angular acceleration and the distance between points C and O.

$$a_{O,x} = r\alpha = (0.3 \text{ m}) \left(-1.445 \ \frac{\text{rad}}{\text{s}^2}\right)$$
$$= -0.433 \text{ m/s}^2 \quad \text{[up the incline]}$$

19. CABLE TENSION FROM A SUSPENDED MASS

When a mass hangs motionless from a cable, or when the mass is moving with a uniform velocity, the cable tension will equal the weight of the mass. However, when the mass is accelerating, the weight must be reduced by the inertial force. If the mass experiences a downward acceleration equal to the gravitational acceleration, there is no tension in the cable. Thus, the two cases shown in Fig. 56.15 are not the same.

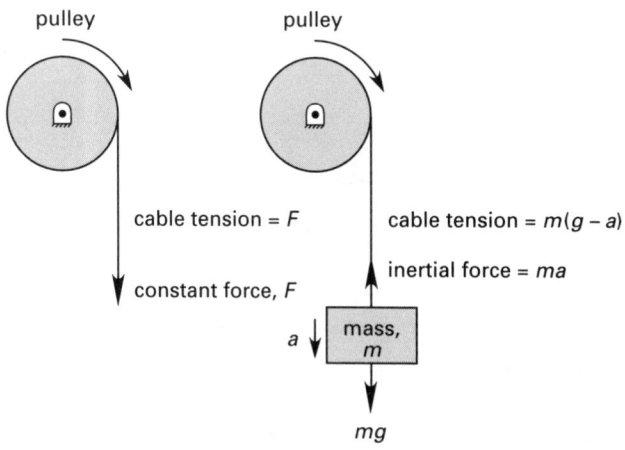

Figure 56.15 *Cable Tension from a Suspended Mass*

Example 56.6

A 10.0 lbm (4.6 kg) mass hangs from a rope wrapped around a 2.0 ft (0.6 m) diameter pulley (centroidal moment of inertia of 70 lbm-ft^2 (2.9 kg·m^2)). (a) What is the angular acceleration of the pulley? (b) What is the linear acceleration of the mass?

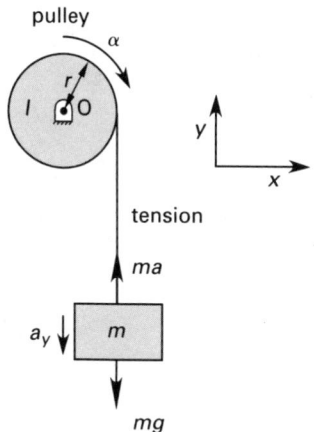

SI Solution

(a) The two equations of motion are

$$\sum F_{y,\text{mass}}: \quad \text{tension} + ma - mg = 0$$
$$\text{tension} - (4.6 \text{ kg}) \left(9.81 \ \frac{\text{m}}{\text{s}^2}\right) = -(4.6 \text{ kg})a_y$$
$$\sum M_{O,\text{pulley}}: \quad (\text{tension})r = I\alpha$$
$$(\text{tension})(0.3 \text{ m}) = (2.9 \text{ kg·m}^2)\alpha$$

Both a_y and α are unknown but are related by $a_y = r\alpha$. Substituting into the $\sum F$ equation and eliminating the tension,

$$\frac{(2.9 \text{ kg·m}^2)\alpha}{0.3 \text{ m}} - 45.1 \text{ N} = -(4.6 \text{ kg})(0.3 \text{ m})\alpha$$
$$9.67\alpha - 45.1 = -1.38\alpha$$
$$\alpha = 4.08 \text{ rad/s}^2$$

(b) The linear acceleration of the mass is

$$a_y = r\alpha = (0.3 \text{ m}) \left(4.08 \ \frac{\text{rad}}{\text{s}^2}\right)$$
$$= 1.22 \text{ m/s}^2$$

Customary U.S. Solution

(a) The two equations of motion are

$$\sum F_{y,\text{mass}}: \quad \text{tension} + \frac{ma_y}{g_c} - \frac{mg}{g_c} = 0$$
$$\text{tension} - \frac{(10 \text{ lbm}) \left(32.2 \ \frac{\text{ft}}{\text{sec}^2}\right)}{32.2 \ \frac{\text{lbm-ft}}{\text{lbf-sec}^2}} = -\frac{(10 \text{ lbm})a_y}{32.2 \ \frac{\text{lbm-ft}}{\text{lbf-sec}^2}}$$

$$\sum M_{O,\text{pulley}}: \quad (\text{tension})r = I\alpha$$

$$(\text{tension})(1.0 \text{ ft}) = (70 \text{ lbm-ft}^2)\alpha$$

$$a_y = r\alpha$$

(b) Substituting into the $\sum F$ equation and eliminating the tension,

$$\frac{(70 \text{ lbm-ft}^2)\,\alpha}{\left(32.2 \dfrac{\text{lbm-ft}}{\text{lbf-sec}^2}\right)(1.0 \text{ ft})} - 10 \text{ lbf} = -\frac{(10 \text{ lbm})(1.0 \text{ ft})\alpha}{32.2 \dfrac{\text{lbm-ft}}{\text{lbf-sec}^2}}$$

$$2.174\,\alpha - 10 = -0.311\,\alpha$$

$$\alpha = 4.02 \text{ rad/sec}^2$$

$$a_y = r\alpha = (1.0 \text{ ft})\left(4.02 \frac{\text{rad}}{\text{sec}^2}\right)$$

$$= 4.02 \text{ ft/sec}^2$$

Example 56.7

A 300 lbm cylinder ($I_O = 710$ lbm-ft^2) has a narrow groove cut in it as shown. One end of the cable is wrapped around the cylinder in the groove, while the other end supports a 200 lbm mass. The pulley is massless and frictionless, and there is no slipping. What are the linear accelerations of the 200 lbm mass and the cylinder?

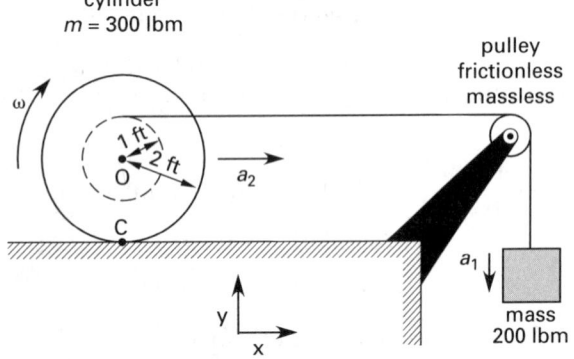

Solution

Since there is no slipping, there is friction between the cylinder and the plane. However, the coefficient of friction is not given. Therefore, moments must be taken about the contact point (the instantaneous center). The moment of inertia about the contact point is

$$I_C = I_O + mr^2$$
$$= 710 \text{ lbm-ft}^2 + (300 \text{ lbm})(2 \text{ ft})^2 = 1910 \text{ lbm-ft}^2$$

The first equation is a summation of forces on the mass.

$$\sum F_y: \quad \text{tension} + \frac{ma_1}{g_c} - \frac{mg}{g_c} = 0$$

The second equation is a summation of moments about the instantaneous center. The frictional force passes through the instantaneous center and is disregarded.

$$\sum M_C: \quad (\text{tension})(2.0 + 1.0 \text{ ft}) = \frac{I_C\alpha}{g_c}$$

Since there are two unknowns, a third equation is needed. This is the relationship between the linear and angular accelerations. a_2 is the acceleration of point O, located 2 ft from point C. a_1 is the acceleration of the cable, whose groove is located 3 ft from point C.

$$\alpha = \frac{a_2}{2.0} = \frac{a_1}{3.0}$$

$$a_1 = \tfrac{3}{2}a_2$$

Solving the three equations simultaneously yields

$$\text{cylinder: } a_2 = 10.4 \text{ ft/sec}^2$$

$$\text{mass: } a_1 = 15.6 \text{ ft/sec}^2$$

$$\text{tension} = 103 \text{ lbf}$$

$$\alpha = 5.2 \text{ rad/sec}^2$$

20. IMPULSE

Impulse, **Imp**, is a vector quantity equal to the change in momentum.[14] Units of linear impulse are the same as for linear momentum: lbf-sec and N·s. Units of lbf-ft-sec and N·m·s are used for angular impulse. Equations 56.44 and 56.45 define the scalar magnitudes of *linear impulse* and *angular impulse*. Figure 56.16 illustrates that impulse is represented by the area under the F-t (or T-t) curve.

$$\text{Imp} = \int_{t_1}^{t_2} F\, dt \quad [\text{linear}] \qquad\qquad 56.44$$

$$\text{Imp} = \int_{t_1}^{t_2} T\, dt \quad [\text{angular}] \qquad\qquad 56.45$$

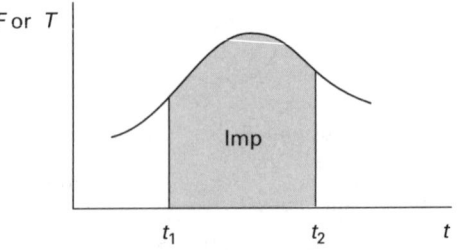

Figure 56.16 *Impulse*

---

[14]Although **Imp** is the most common notation, engineers have no universal symbol for impulse. Some authors use I, $\mathbf{I}$, and i, but these symbols can be mistaken for moment of inertia. Other authors merely use the word *impulse* in their equations.

If the applied force or torque is constant, impulse is easily calculated. A large force acting for a very short period of time is known as an *impulsive force*.

$$\text{Imp} = F(t_2 - t_1) \quad \text{[linear]} \qquad 56.46$$

$$\text{Imp} = T(t_2 - t_1) \quad \text{[angular]} \qquad 56.47$$

If the impulse is known, the average force acting over the duration of the impulse is

$$F_{\text{ave}} = \frac{\text{Imp}}{\Delta t} \qquad 56.48$$

21. IMPULSE-MOMENTUM PRINCIPLE

The change in momentum is equal to the applied *impulse*. This is known as the *impulse-momentum* principle. For a linear system with constant force and mass, the scalar magnitude form of this principle is

$$\text{Imp} = \Delta p \qquad 56.49$$

$$F(t_2 - t_1) = m(v_2 - v_1) \quad \text{[SI]} \qquad 56.50(a)$$

$$F(t_2 - t_1) = \frac{m(v_2 - v_1)}{g_c} \quad \text{[U.S.]} \qquad 56.50(b)$$

For an angular system with constant torque and moment of inertia, the analogous equations are

$$T(t_2 - t_1) = I(\omega_2 - \omega_1) \quad \text{[SI]} \qquad 56.51(a)$$

$$T(t_2 - t_1) = \frac{I(\omega_2 - \omega_1)}{g_c} \quad \text{[U.S.]} \qquad 56.51(b)$$

Example 56.8

A 1.62 oz (0.046 kg) marble attains a velocity of 170 mph (76 m/s) in a hunting slingshot. Contact with the sling is 1/25th of a second. What is the average force on the marble during contact?

SI Solution

From Eq. 56.50(a), the average force is

$$F = \frac{m\Delta v}{\Delta t} = \frac{(0.046 \text{ kg})\left(76 \dfrac{\text{m}}{\text{s}}\right)}{\dfrac{1}{25} \text{ s}}$$

$$= 87.4 \text{ N}$$

Customary U.S. Solution

The mass of the marble is

$$m = \frac{1.62 \text{ oz}}{16 \dfrac{\text{oz}}{\text{lbm}}} = 0.101 \text{ lbm}$$

The velocity of the marble is

$$v = \frac{\left(170 \dfrac{\text{mi}}{\text{hr}}\right)\left(5280 \dfrac{\text{ft}}{\text{mi}}\right)}{3600 \dfrac{\text{sec}}{\text{hr}}} = 249.3 \text{ ft/sec}$$

From Eq. 56.50(b), the average force is

$$F = \frac{m\Delta v}{g_c \Delta t} = \frac{(0.101 \text{ lbm})\left(249.3 \dfrac{\text{ft}}{\text{sec}}\right)}{\left(32.2 \dfrac{\text{lbm-ft}}{\text{lbf-sec}^2}\right)\left(\dfrac{1}{25} \text{ sec}\right)}$$

$$= 19.5 \text{ lbf}$$

Example 56.9

A 2000 kg cannon fires a 10 kg projectile horizontally at 600 m/s. It takes 0.007 s for the projectile to pass through the barrel. The cannon has a spring mechanism to absorb the recoil. (a) What is the cannon's initial recoil velocity? (b) What force is exerted on the recoil spring?

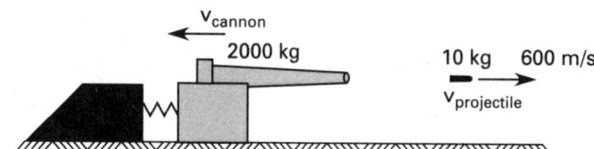

Solution

(a) The accelerating force is applied to the projectile quickly, and external forces such as gravity and friction are not a significant factor. Therefore, momentum is conserved.

$$\sum p: \quad m_{\text{projectile}} \Delta v_{\text{projectile}} = m_{\text{cannon}} \Delta v_{\text{cannon}}$$

$$(10 \text{ kg})\left(600 \dfrac{\text{m}}{\text{s}}\right) = (2000 \text{ kg})(v_{\text{cannon}})$$

$$v_{\text{cannon}} = 3 \text{ m/s}$$

(b) From Eq. 56.50(a), the recoil force is

$$F = \frac{m\Delta v}{\Delta t} = \frac{(10 \text{ kg})\left(600 \dfrac{\text{m}}{\text{s}}\right)}{0.007 \text{ s}}$$

$$= 8.57 \times 10^5 \text{ N}$$

Dynamics and Vibrations

22. IMPULSE-MOMENTUM PRINCIPLE IN OPEN SYSTEMS

The impulse-momentum principle can be used to determine the forces acting on flowing fluids (i.e., in open systems). This is the method used to calculate forces in jet engines and on pipe bends, and forces due to other changes in flow geometry. Equation 56.52 is rearranged in terms of a mass flow rate.

$$F = \frac{m\Delta \mathrm{v}}{\Delta t} = \dot{m}\Delta \mathrm{v} \qquad \text{[SI]} \qquad 56.52(a)$$

$$F = \frac{m\Delta \mathrm{v}}{g_c \Delta t} = \frac{\dot{m}\Delta \mathrm{v}}{g_c} \qquad \text{[U.S.]} \qquad 56.52(b)$$

Example 56.10

Air enters a jet engine at 1500 ft/sec (450 m/s) and leaves at 3000 ft/sec (900 m/s). The thrust produced is 10,000 lbf (44 500 N). Disregarding the small amount of fuel added during combustion, what is the mass flow rate?

SI Solution

From Eq. 56.52(a),

$$\dot{m} = \frac{F}{\Delta \mathrm{v}} = \frac{44\,500 \text{ N}}{900 \dfrac{\text{m}}{\text{s}} - 450 \dfrac{\text{m}}{\text{s}}}$$

$$= 98.9 \text{ kg/s}$$

Customary U.S. Solution

From Eq. 56.52(b),

$$\dot{m} = \frac{Fg_c}{\Delta \mathrm{v}} = \frac{(10{,}000 \text{ lbf})\left(32.2 \dfrac{\text{lbm-ft}}{\text{lbf-sec}^2}\right)}{3000 \dfrac{\text{ft}}{\text{sec}} - 1500 \dfrac{\text{ft}}{\text{sec}}}$$

$$= 215 \text{ lbm/sec}$$

Example 56.11

20 kg of sand fall continuously each second on a conveyor belt moving horizontally at 0.6 m/s. What power is required to keep the belt moving?

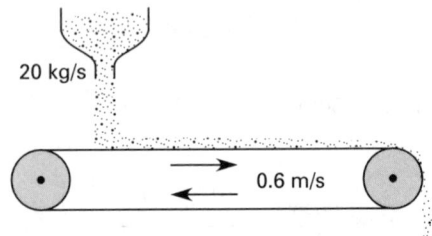

Solution

From Eq. 56.52(a), the force on the sand is

$$F = \dot{m}\Delta \mathrm{v} = \left(20 \frac{\text{kg}}{\text{s}}\right)\left(0.6 \frac{\text{m}}{\text{s}}\right)$$

$$= 12 \text{ N}$$

The power required is

$$P = F\mathrm{v} = (12 \text{ N})\left(0.6 \frac{\text{m}}{\text{s}}\right)$$

$$= 7.2 \text{ W}$$

Example 56.12

A 6×9, $^5/_8$ in diameter hoisting cable (area of 0.158 in^2, modulus of elasticity of 12×10^6 lbf/in^2) carries a 1000 lbm load at its end. The load is being lowered vertically at the rate of 4 ft/sec. When 200 ft of cable has been reeled out, the take-up reel suddenly locks. Neglect the cable mass. What are the (a) cable stretch, (b) maximum force in the cable, (c) maximum stress in the cable, and (d) approximate time for the load to come to a stop vertically?

Solution

(a) The stiffness of the cable is

$$k = \frac{F}{x} = \frac{AE}{L}$$

$$= \frac{(0.158 \text{ in}^2)\left(12 \times 10^6 \dfrac{\text{lbf}}{\text{in}^2}\right)}{(200 \text{ ft})\left(12 \dfrac{\text{in}}{\text{ft}}\right)}$$

$$= 790 \text{ lbf/in}$$

Neglecting the cable mass, the kinetic energy of the moving load is

$$KE = \frac{m\mathrm{v}^2}{2g_c}$$

$$= \frac{(1000 \text{ lbm})\left(4 \dfrac{\text{ft}}{\text{sec}}\right)^2\left(12 \dfrac{\text{in}}{\text{ft}}\right)}{(2)\left(32.2 \dfrac{\text{lbm-ft}}{\text{lbf-sec}^2}\right)}$$

$$= 2981 \text{ in-lbf}$$

By the work-energy principle, the decrease in kinetic energy is equal to the work of lengthening the cable (i.e., the energy stored in the spring).

$$\Delta KE = \tfrac{1}{2}k\delta^2$$

$$2981 \text{ in-lbf} = \left(\tfrac{1}{2}\right)\left(790 \frac{\text{lbf}}{\text{in}}\right)\delta^2$$

$$\delta = 2.75 \text{ in}$$

(b) The maximum force in the cable is

$$F = k\delta = \left(790 \; \frac{\text{lbf}}{\text{in}}\right)(2.75 \; \text{in})$$

$$= 2173 \; \text{lbf}$$

(c) The maximum tensile stress in the cable is

$$\sigma = \frac{F}{A} = \frac{2173 \; \text{lbf}}{0.158 \; \text{in}^2}$$

$$= 13{,}750 \; \text{lbf/in}^2$$

(d) Since the tensile force in the cable increases from zero to the maximum while the load decelerates, the average decelerating force is half of the maximum force. From the impulse momentum principle,

$$F\Delta t = \frac{m\Delta \text{v}}{g_c}$$

$$\left(\tfrac{1}{2}\right)(2173 \; \text{lbf})\Delta t = \frac{(1000 \; \text{lbm})\left(4 \; \dfrac{\text{ft}}{\text{sec}}\right)}{\left(32.2 \; \dfrac{\text{lbm-ft}}{\text{lbf-sec}^2}\right)}$$

$$\Delta t = 0.114 \; \text{sec}$$

23. IMPACTS

According to Newton's second law, momentum is conserved unless a body is acted upon by an external force such as gravity or friction from another object. In an *impact* or *collision*, contact is very brief, and the effect of external forces is insignificant. Therefore, momentum is conserved, even though energy may be lost through heat generation and deformation of the bodies.

Consider two particles, initially moving with velocities v_1 and v_2 on a collision path, as shown in Fig. 56.17. The conservation of momentum equation can be used to find the velocities after impact, v_1' and v_2'. (Observe algebraic signs with velocities.)

$$m_1\text{v}_1 + m_2\text{v}_2 = m_1\text{v}_1' + m_2\text{v}_2' \qquad 56.53$$

Figure 56.17 *Direct Central Impact*

The impact is said to be an *inelastic impact* if kinetic energy is lost. (Other names for an inelastic impact are *plastic impact* and *endoergic impact*.[15]) The impact

---

[15]Theoretically, there is an *exoergic impact* also (i.e., one in which kinetic energy is gained during the impact). However, this can occur only in special cases, such as in nuclear reactions.

is said to be *perfectly inelastic* or *perfectly plastic* if the two particles stick together and move on with the same final velocity.[16] The impact is said to be an *elastic impact* only if kinetic energy is conserved.

$$m_1\text{v}_1^2 + m_2\text{v}_2^2 = m_1\text{v}_1'^2 + m_2\text{v}_2'^2 \Big|_{\text{elastic impact}} \qquad 56.54$$

24. COEFFICIENT OF RESTITUTION

A simple way to determine whether the impact is elastic or inelastic is by calculating the *coefficient of restitution*, e. The collision is inelastic if $e < 1.0$, perfectly inelastic if $e = 0$, and elastic if $e = 1.0$. The coefficient of restitution is the ratio of relative velocity differences along a mutual straight line. (When both impact velocities are not directed along the same straight line, the coefficient of restitution should be calculated separately for each velocity component.)

$$e = \frac{\text{relative separation velocity}}{\text{relative approach velocity}}$$

$$= \frac{\text{v}_1' - \text{v}_2'}{\text{v}_2 - \text{v}_1} \qquad 56.55$$

25. REBOUND FROM STATIONARY PLANES

Figure 56.18 illustrates the case of an object rebounding from a massive, stationary plane.[17] This is an impact where $m_2 = \infty$ and $\text{v}_2 = 0$. The impact force acts perpendicular to the plane, regardless of whether the impact is elastic or inelastic. Therefore, the x-component of velocity is unchanged. Only the y-component of velocity is affected, and then only if the impact is inelastic.

$$\text{v}_x = \text{v}_x' \qquad 56.56$$

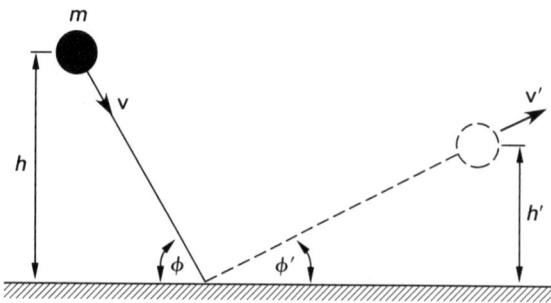

Figure 56.18 *Rebound from a Stationary Plane*

---

[16]In traditional textbook problems, clay balls should be considered perfectly inelastic.

[17]The particle path is shown as a straight line in Fig. 56.18 for convenience. The path will be a straight line only when the particle is dropped straight down. Otherwise, the path will be parabolic.

The coefficient of restitution can be used to calculate the *rebound angle*, *rebound height*, and *rebound velocity*.

$$e = \frac{\sin \phi'}{\sin \phi} = \sqrt{\frac{h'}{h}} = \frac{-v'_y}{v_y} \qquad 56.57$$

Example 56.13

A golf ball dropped vertically from a height of 8.0 ft (2.4 m) onto a hard surface rebounds to a height of 6.0 ft (1.8 m). What are the (a) impact velocity, (b) rebound velocity, and (c) coefficient of restitution?

SI Solution

The impact velocity of the golf ball is found by equating the initial potential energy to the incident kinetic energy.

$$\tfrac{1}{2}mv^2 = mgh$$

$$v = \sqrt{2gh} = \sqrt{(2)\left(9.81\ \frac{m}{s^2}\right)(2.4\ m)}$$

$$= -6.86\ m/s \quad \text{[negative because down]}$$

The rebound velocity can be found from the rebound height.

$$v' = \sqrt{2gh'} = \sqrt{(2)\left(9.81\ \frac{m}{s^2}\right)(1.8\ m)}$$

$$= 5.94\ m/s$$

From Eq. 56.55 with $v_2 = v'_2 = 0$ (or Eq. 56.57),

$$e = \frac{5.94\ \dfrac{m}{s} - 0}{0 - \left(-6.86\ \dfrac{m}{s}\right)} = 0.87$$

Customary U.S. Solution

The impact velocity is

$$v = \sqrt{2gh} = \sqrt{(2)\left(32.2\ \frac{ft}{sec^2}\right)(8.0\ ft)}$$

$$= 22.7\ ft/sec$$

Similarly, the rebound velocity is

$$v' = \sqrt{(2)\left(32.2\ \frac{ft}{sec^2}\right)(6.0\ ft)} = 19.7\ ft/sec$$

From Eq. 56.55 with $v_2 = v'_2 = 0$ (or from Eq. 56.57),

$$e = \frac{19.7\ \dfrac{ft}{sec} - 0}{0 - \left(-22.7\ \dfrac{ft}{sec}\right)} = 0.87$$

26. COMPLEX IMPACTS

The simplest type of impact problem is the direct central impact, shown in Fig. 56.17. An impact is said to be a *direct impact* when the velocities of the two bodies are perpendicular to the contacting surfaces. *Central impact* occurs when the force of the impact is along the line of connecting centers of gravity. Round bodies (i.e., spheres) always experience central impact, whether or not the impact is direct.

When the velocities of the bodies are not along the same line, the impact is said to be an *oblique impact*, as illustrated in Fig. 56.19. The coefficient of restitution can be used to find the x-components of the resultant velocities. Since impact is central, the y-components of velocities will be unaffected by the collision.

$$v_{1y} = v'_{1y} \qquad 56.58$$

$$v_{2y} = v'_{2y} \qquad 56.59$$

$$e = \frac{v'_{1x} - v'_{2x}}{v_{2x} - v_{1x}} \qquad 56.60$$

$$m_1 v_{1x} + m_2 v_{2x} = m_1 v'_{1x} + m_2 v'_{2x} \qquad 56.61$$

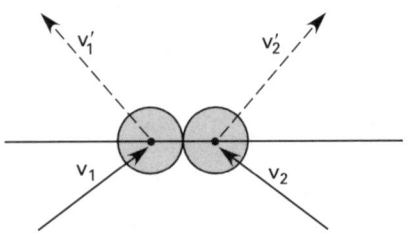

Figure 56.19 *Central Oblique Impact*

Eccentric impacts are neither direct nor central. The coefficient of restitution can be used to calculate the linear velocities immediately after impact along a line normal to the contact surfaces. Since the impact is not central, the bodies will rotate. Other methods must be used to calculate the rate of rotation.

$$e = \frac{v'_{1n} - v'_{2n}}{v_{2n} - v_{1n}} \qquad 56.62$$

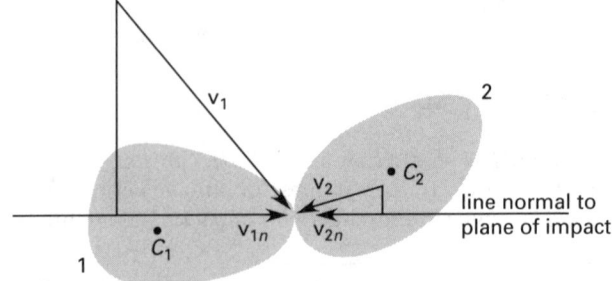

Figure 56.20 *Eccentric Impact*

27. VELOCITY-DEPENDENT FORCE

A force that is a function of velocity is known as a *velocity-dependent force*. A common example of a velocity-dependent force is the *viscous drag* a particle experiences when falling through a fluid. There are two main cases of viscous drag: linear and quadratic.

A *linear velocity-dependent force* is proportional to the first power of the velocity. A linear relationship is typical of a particle falling slowly through a fluid (i.e., viscous drag in laminar flow). In Eq. 56.63, C is a constant of proportionality known as the *viscous coefficient* or *coefficient of viscous damping*.

$$F_b = C\text{v} \qquad \qquad 56.63$$

In the case of a particle falling slowly through a viscous liquid, the differential equation of motion and its solution are derived from Newton's second law.

$$mg - C\text{v} = ma \qquad \text{[SI]} \qquad 56.64(a)$$

$$\frac{mg}{g_c} - C\text{v} = \frac{ma}{g_c} \qquad \text{[U.S.]} \qquad 56.64(b)$$

$$\text{v}(t) = \text{v}_t\left(1 - e^{-\frac{Ct}{m}}\right) \qquad \text{[SI]} \qquad 56.65(a)$$

$$\text{v}(t) = \text{v}_t\left(1 - e^{-\frac{Cg_ct}{m}}\right) \qquad \text{[U.S.]} \qquad 56.65(b)$$

Equation 56.65 shows that the velocity asymptotically approaches a final value known as the *terminal velocity*, v_t. For laminar flow, the terminal velocity is

$$\text{v}_t = \frac{mg}{C} \qquad \text{[SI]} \qquad 56.66(a)$$

$$\text{v}_t = \frac{mg}{Cg_c} \qquad \text{[U.S.]} \qquad 56.66(b)$$

A *quadratic velocity-dependent force* is proportional to the second power of the velocity. A quadratic relationship is typical of a particle falling quickly through a fluid (i.e., turbulent flow).

$$F_b = C\text{v}^2 \qquad \qquad 56.67$$

In the case of a particle falling quickly through a liquid under the influence of gravity, the differential equation of motion is

$$mg - C\text{v}^2 = ma \qquad \text{[SI]} \qquad 56.68(a)$$

$$\frac{mg}{g_c} - C\text{v}^2 = \frac{ma}{g_c} \qquad \text{[U.S.]} \qquad 56.68(b)$$

For turbulent flow, the terminal velocity is

$$\text{v}_t = \sqrt{\frac{mg}{C}} \qquad \text{[SI]} \qquad 56.69(a)$$

$$\text{v}_t = \sqrt{\frac{mg}{Cg_c}} \qquad \text{[U.S.]} \qquad 56.69(b)$$

If a skydiver falls far enough, a turbulent terminal velocity of approximately 125 mph (200 kph) will be achieved. With a parachute (but still in turbulent flow), the terminal velocity is reduced to approximately 25 mph (40 kph).

28. VARYING MASS

Integral momentum equations must be used when the mass of an object varies with time. Most varying mass problems are complex, but the simplified case of an ideal rocket can be evaluated. This discussion assumes constant gravitational force, constant fuel usage, and constant exhaust velocity. (For brevity of presentation, all of the following equations are presented in consistent form.)

The forces acting on the rocket are its thrust, F, and gravity. Newton's second law is

$$F(t) - F_{\text{gravity}} = \frac{d}{dt}(m(t)\text{v}(t)) \qquad 56.70$$

If $\dot{m}$ is the constant fuel usage, the thrust and gravitational forces are

$$F(t) = \dot{m}\text{v}_{\text{exhaust, absolute}} = \dot{m}(\text{v}_{\text{exhaust}} - \text{v}(t)) \qquad 56.71$$

$$F_{\text{gravity}} = m(t)g \qquad 56.72$$

The velocity as a function of time is found by solving the following differential equation.

$$\dot{m}\text{v}_{\text{exhaust,absolute}} - m(t)g = \dot{m}\text{v}(t) + m(t)\left(\frac{d\text{v}(t)}{dt}\right) \qquad 56.73$$

$$\text{v}(t) = \text{v}_0 - gt + \text{v}_{\text{exhaust}}\ln\left(\frac{m_0}{m_0 - \dot{m}t}\right) \qquad 56.74$$

The final *burnout velocity*, v_f, depends on the initial mass, m_0, and the final mass, m_f.

$$\text{v}_f = \text{v}_0 - g\left(\frac{m_0}{\dot{m}}\right)\left(1 - \frac{m_f}{m_0}\right) + \text{v}_{\text{exhaust}}\ln\left(\frac{m_0}{m_f}\right) \qquad 56.75$$

A simple relationship exists for a rocket starting from standstill in a gravity-free environment (i.e., $\text{v}_0 = 0$ and $g = 0$).

$$\frac{m_f}{m_0} = e^{-\text{v}_f/\text{v}_{\text{exhaust}}} \qquad 56.76$$

29. CENTRAL FORCE FIELDS

When the force on a particle is always directed toward or away from a fixed point, the particle is moving in a *central force field*. Examples of central force fields are gravitational fields (*inverse-square attractive fields*) and electrostatic fields (*inverse-square repulsive fields*). Particles traveling in inverse-square attractive fields can

Dynamics and Vibrations

have circular, elliptical, parabolic, or hyperbolic paths. Particles traveling in inverse-square repulsive fields always travel in hyperbolic paths.

The fixed point, O in Fig. 56.21, is known as the *center of force*. The magnitude of the force depends on the distance between the particle and the center of force.

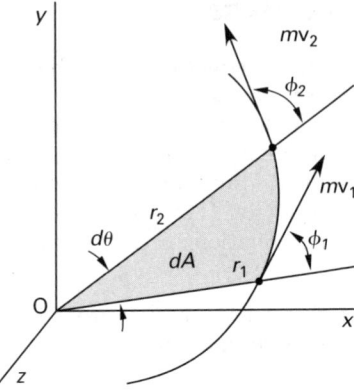

Figure 56.21 *Motion in a Central Force Field*

The angular momentum of a particle moving in a central force field is constant.

$$\mathbf{h}_O = \mathbf{r} \times m\mathbf{v} = \text{constant} \qquad 56.77$$

Equation 56.77 can be written in scalar form.

$$r_1 m v_1 \sin \phi_1 = r_2 m v_2 \sin \phi_2 \qquad 56.78$$

For a particle moving in a central force field, the *areal velocity* is constant.

$$\text{areal velocity} = \frac{dA}{dt} = \tfrac{1}{2} r^2 \frac{d\theta}{dt} = \frac{h_O}{2} \qquad 56.79$$

30. NEWTON'S LAW OF GRAVITATION

For a particle far enough away from a large body, gravity can be considered to be a central force field. *Newton's law of gravitation*, also known as *Newton's law of universal gravitation*, describes the force of attraction between the two masses. The law states that the attractive gravitational force between the two masses is directly proportional to the product of masses, is inversely proportional to the square of the distance between their centers of mass, and is directed along a line passing through the centers of gravity of both masses.

$$F = \frac{G m_1 m_2}{r^2} \qquad 56.80$$

G is *Newton's gravitational constant* (*Newton's universal constant*). Approximate values of G for the earth are given in Table 56.2 for different sets of units. For an earth-particle combination, the product Gm_{earth} has the value of 4.39×10^{14} lbf-ft^2/lbm (4.00×10^{14} N·m^2/kg).

Table 56.2 *Approximate Values of the Gravitational Constant, G*

6.673×10^{-11}	N·m^2/kg^2
6.673×10^{-8}	cm^3/g·s^2
3.436×10^{-8}	lbf-ft^2/slug^2
3.320×10^{-11}	lbf-ft^2/lbm^2
3.436×10^{-8}	ft^4/lbf-sec^4

31. KEPLER'S LAWS OF PLANETARY MOTION

Kepler's three *laws of planetary motion* are as follows.

- *the law of orbits*: The path of each planet is an ellipse with the sun at one focus.

- *the law of areas*: The radius vector drawn from the sun to a planet sweeps equal areas in equal times. (The areal velocity is constant.)

$$\frac{dA}{dt} = \frac{h_0}{2} = \text{constant} \qquad 56.81$$

- *the law of periods*: The square of a planet's periodic time is proportional to the cube of the semi-major axis of its orbit.

$$t_p^2 \propto a^3 \qquad 56.82$$

The *periodic time* referenced in Kepler's third law is the time required for a satellite to travel once around the parent body. If the parent body rotates in the same plane and with the same periodic time as the satellite, the satellite will always be above the same point on the parent body. This condition defines a *geostationary orbit*. Referring to Fig. 56.22, the periodic time is

$$t_p = \frac{2\pi a b}{h_0} = \frac{\pi a b}{\dfrac{dA}{dt}} = \frac{\pi a b}{\text{areal velocity}} \qquad 56.83$$

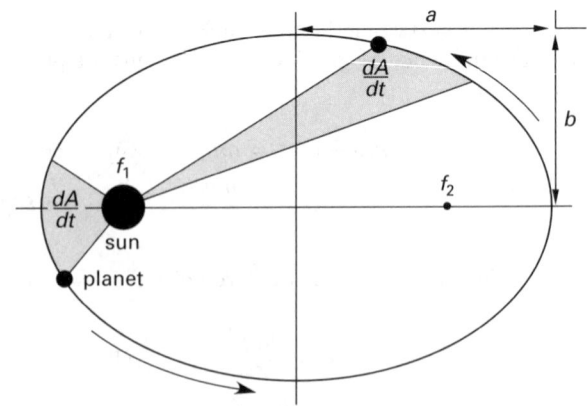

Figure 56.22 *Planetary Motion*

32. SPACE MECHANICS

Figure 56.23 illustrates the motion of a satellite that is released in a path parallel to the earth's surface. (The dotted line represents the launch phase and is not relevant to the analysis.)

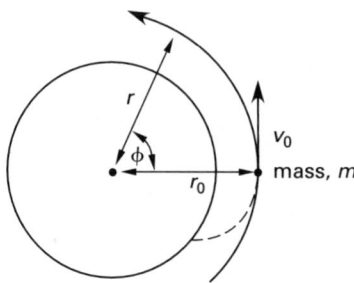

Figure 56.23 *Space Mechanics Nomenclature*

At the instant of release, the magnitude of the angular momentum, **h**, is

$$\mathbf{h} = h_0 = r_0 m v_0 \qquad 56.84$$

The force exerted on the satellite by the earth is given by Newton's law of gravitation. For any angle, ϕ, swept out by the satellite, the separation distance can be found from Eq. 56.85. C is a constant.

$$\frac{1}{r} = \frac{GM}{h^2} + C\cos\phi \qquad 56.85$$

$$C = \frac{1}{r_0} - \frac{GM}{h^2} \qquad 56.86$$

The *orbit eccentricity*, ϵ, can be calculated and used to determine the type of orbit, as listed in Table 56.3.

$$\epsilon = \frac{Ch^2}{GM} \qquad 56.87$$

Table 56.3 *Orbit Eccentricities*

value of ϵ	type of orbit
> 1.0	nonreturning hyperbola
= 1.0	nonreturning parabola
< 1.0	ellipse
= 0.0	circle

The limiting value of the initial release velocity, $v_{0,\max}$, to prevent a nonreturning orbit ($\epsilon = 1.0$) is known as the *escape velocity*. For the earth, the escape velocity is approximately 7.0 mi/sec (25,000 mph, or 11.2 km/s).

$$v_{\text{escape}} = v_{0,\max} = \sqrt{\frac{2GM}{r_0}} \qquad 56.88$$

The release velocity, $v_{0,\text{circular}}$, that results in a circular orbit is

$$v_{0,\text{circular}} = \sqrt{\frac{GM}{r_0}} \qquad 56.89$$

For an elliptical orbit, the minimum separation is known as the *perigee distance*. The maximum separation distance is known as the *apogee*. The terms perigee and apogee are traditionally used for earth satellites. The terms *perihelion* (closest) and *aphelion* (farthest) are used to describe distances between the sun and earth.

33. VEHICLE DYNAMICS

A moving surface vehicle (e.g., car, truck, or bus) is retarded by five major forms of resistance: inertia, grade, rolling, curve, and air resistance. The *inertia resistance* (*inertial resisting force*) is calculated from the vehicle mass and acceleration.

$$F_i = ma \qquad \text{[SI]} \qquad 56.90(a)$$

$$F_i = \frac{ma}{g_c} = \frac{wa}{g} \qquad \text{[U.S.]} \qquad 56.90(b)$$

For a vehicle ascending a constant incline ϕ, the *grade resistance* is the component of the vehicle weight, w, acting parallel and down a frictionless inclined surface. The *grade* or *gradient*, G, is the slope in the direction of roadway. Grade is usually specified in percent. For example, a roadway with a 6% grade would increase in elevation 6 ft for every 100 ft of travel. For grades commonly encountered in road and highway design, there is essentially no difference in the road length and the horizontal measurement.

$$F_g = w\sin\phi$$
$$\approx w\tan\phi = \frac{wG\%}{100} \qquad 56.91$$

The *rolling resistance* can be calculated theoretically as presented in Sec. 16. For analysis of vehicle dynamics, the coefficient of rolling friction, f_r, used in Eq. 56.92 is generally considered to be constant.[18] For modern vehicles, it is reported as approximately 0.010 to 0.015, and is approximately constant at 13.5 lbf per 1000 lbm of mass (130 N per kg) for all speeds up to 60 mph (97 kph). For higher speeds, it should be increased 10% for every 10% increase in speed above 60 mph (97 kph). Rolling resistance is even higher for gravel roads and roads in poor condition.

$$F_r = f_r w\cos\phi$$
$$\approx f_r w \qquad 56.92$$

Curve resistance is the force exerted on the vehicle by its front wheels when the direction of travel is changing. Components of the curve resistance also impede

---

[18]An approximate relationship between speed and coefficient of rolling resistance for vehicles operating on typical paved roads was reported in 1957 by Taborek as

$$f_r = (0.01)\left(1 + \frac{v_{\text{ft/sec}}}{147}\right)$$

the vehicle's forward motion, decreasing the fuel economy. As with centrifugal force, it is a function of the curve radius and vehicle speed. Curve resistance is generally obtained experimentally or from a table of typical values. When the vehicle is traveling in a straight direction, the curve resistance is zero.

Air resistance is the drag force given by Eq. 56.93, where A is the frontal cross-sectional area.

$$F_D = \frac{C_D A \rho v^2}{2} \qquad \text{[SI]} \qquad 56.93(a)$$

$$F_D = \frac{C_D A \rho v^2}{2g_c} \qquad \text{[U.S.]} \qquad 56.93(b)$$

In the absence of any information about the drag coefficient and/or air density, the average air resistance for modern cars can be approximated by

$$F_D \approx 0.0011 A_{\text{m}^2} v^2_{\text{km/h}} \qquad \text{[SI]} \qquad 56.94(a)$$

$$F_D \approx 0.0006 A_{\text{ft}^2} v^2_{\text{mi/hr}} \qquad \text{[U.S.]} \qquad 56.94(b)$$

The *propulsion power* (excluding the power used by power steering, air conditioner, and other accessories) is given by Eq. 56.95, where the sum of all resistances may be referred to as the *tractive force* or *tractive effort*. (Appropriate conversions can be made to horsepower and kilowatts.) For a typical vehicle, only about 50% of the nominal engine horsepower (brake power) rating is available at the flywheel for propulsion (i.e., after accessories). If the power is known, Eq. 56.95 can be solved for the maximum acceleration or velocity. Since the velocity also appears in the expression for drag force, a trial and error solution may be needed.

$$P = (F_i + F_g + F_r + F_c + F_D)v \qquad 56.95$$

Regardless of how much power a vehicle's engine may be capable of developing, there is a point beyond which the use of additional power will merely result in spinning the driving tires. (The maximum tractive force is $f_s \times w$.) Beyond that point, no additional tractive effort can be generated to overcome resistance. The maximum *tractive force* as limited by wheel spinning is found by summing moments about some point on the vehicle, typically the tire-roadway contact point for the nondriven wheels. The moment from the unknown tractive force balances the moments from all of the resistance forces.[19]

Analysis of performance under acceleration and deceleration is the same as for any body under uniform acceleration. A vehicle's instantaneous forward velocity and tangential velocity of the tires are the same.

$$v_{\text{vehicle}} = v_{t,\text{tire}} = \pi d_{\text{tire}} n_{\text{tire}} \qquad 56.96$$

---

[19]Unless specific information is known about the location of the vehicle's center of gravity and other properties, assumptions may need to be made about the point of application of the resistance forces.

The engine-wheel speed ratio depends on the transmission and differential gear ratios. The *transmission gear ratio* is the ratio of the engine speed to the driveshaft speed. Unless there is an *overdrive gear*, the typical transmission gear ratio used for highest speeds is approximately 1:1. The differential gear ratio (typically 2.5:1 to 4:1) is the ratio of driveshaft to tire speeds.

$$n_{\text{wheel}} = \frac{n_{\text{engine}}}{R_{\text{transmission}} R_{\text{differential}}} \qquad 56.97$$

$$R_{\text{transmission}} = \frac{n_{\text{engine}}}{n_{\text{driveshaft}}} \qquad 56.98$$

$$R_{\text{differential}} = \frac{n_{\text{driveshaft}}}{n_{\text{wheel}}} \qquad 56.99$$

The relationship between brake power and engine-generated torque is

$$P_{\text{kW}} = \frac{T_{\text{N·m}} n_{\text{rpm}}}{9549} \qquad \text{[SI]} \qquad 56.100(a)$$

$$P_{\text{hp}} = \frac{T_{\text{in-lbf}} n_{\text{rpm}}}{63{,}025} \qquad \text{[U.S.]} \qquad 56.100(b)$$

Only 75 to 90% of the engine power or torque reaches the rear wheels. This represents the mechanical efficiency, η_m. The remainder is lost in the transmission, differential, and other gear-reduction devices. The engine-generated tractive effort reaching the driving wheels is

$$F_{\text{tractive}} = \frac{\eta_m T R_{\text{transmission}} R_{\text{differential}}}{r_{\text{tire}}} \qquad 56.101$$

Equation 56.102 gives the relationship between vehicle speed and engine speed. The loss fraction, i, is typically 0.02 to 0.05 and primarily accounts for slippage in the clutch and automatic transmission.

$$v = \frac{2\pi n_{\text{rev/sec}}(1 - i)}{R_{\text{transmission}} R_{\text{differential}}} \qquad 56.102$$

The mass and volumetric rates of fuel consumption can be found from the horsepower and brake specific fuel consumption.

$$\dot{m}_{\text{fuel,kg/h}} = (P_{\text{brake,kW}})(\text{BSFC}_{\text{kg/kW·h}})$$
$$\text{[SI]} \qquad 56.103(a)$$

$$\dot{m}_{\text{fuel,lbm/hr}} = (P_{\text{brake,hp}})(\text{BSFC}_{\text{lbm/hp-hr}})$$
$$\text{[U.S.]} \qquad 56.103(b)$$

$$\dot{Q}_{\text{fuel,L/h}} = \frac{\dot{m}_{\text{fuel,kg/h}}\left(1000 \dfrac{\text{L}}{\text{m}^3}\right)}{\rho_{\text{kg/m}^3}} \qquad \text{[SI]} \qquad 56.104(a)$$

$$\dot{Q}_{\text{fuel,gal/hr}} = \frac{\dot{m}_{\text{fuel,lbm/hr}}\left(7.48 \dfrac{\text{gal}}{\text{ft}^3}\right)}{\rho_{\text{lbm/ft}^3}} \qquad \text{[U.S.]} \qquad 56.104(b)$$

The distance traveled per volume of fuel (also known as *fuel economy*), s', is

$$s'_{\text{fuel,km/L}} = \frac{\text{v}_{\text{km/h}}}{\dot{Q}_{\text{fuel,L/h}}} \quad \text{[SI]} \quad \textit{56.105(a)}$$

$$s'_{\text{fuel,mi/gal}} = \frac{\text{v}_{\text{mi/hr}}}{\dot{Q}_{\text{fuel,gal/hr}}} \quad \text{[U.S.]} \quad \textit{56.105(b)}$$

34. SKIDDING DISTANCE

When a driver applies the brakes to a moving vehicle, the deceleration will be dependent on the driver. The harder the driver applies the brakes, the greater the deceleration will be. However, if the vehicle begins to skid, the deceleration no longer will be affected by how hard the brakes are applied. Rather, the deceleration will depend only on the coefficient of friction between the tires and roadway.

$$a = fg \qquad\qquad \textit{56.106}$$

The skidding distance to come to a complete stop on a level surface is

$$s_{\text{skidding}} = \frac{\text{v}^2}{2a} \qquad\qquad \textit{56.107}$$

Equation 56.107 is good for calculating skidding distances on a level surface only. When skidding is uphill, gravity will help slow the vehicle, and the distance traveled to a standstill will be less. Similarly, the distance will be greater when the vehicle skids downhill. If the vehicle skids on a slope inclined at an angle θ from the horizontal (positive if uphill, negative if downhill), the *skidding distance* is

$$s_{\text{skidding}} = \frac{\text{v}^2}{2g(f\cos\theta + \sin\theta)} \qquad \textit{56.108}$$

The *stopping distance* is different from the skidding distance. The difference is the distance traveled during the driver's *reaction time*, known as *PIEV time*. PIEV is an acronym for perception, identification, emotion, and volition.

$$s_{\text{stopping}} = \text{v}t_{\text{PIEV}} + s_{\text{skidding}} \qquad \textit{56.109}$$

35. ROADWAY BANKING

If a vehicle travels in a circular path with instantaneous radius r and tangential velocity v_t, it will experience an apparent centrifugal force. The centrifugal force is resisted by a combination of roadway banking (*superelevation*) and *sideways friction*.[20] If the roadway is banked

so that friction is not required to resist the centrifugal force, the superelevation angle, ϕ, can be calculated from Eq. 56.110.[21]

$$\tan\phi = \frac{\text{v}_t^2}{gr} \qquad\qquad \textit{56.110}$$

Equation 56.110 can be solved for the *normal speed* corresponding to the geometry of a curve.

$$\text{v}_t = \sqrt{gr\tan\phi} \qquad\qquad \textit{56.111}$$

When friction is used to counteract some of the centrifugal force, the *side friction factor*, f, between the tires and roadway is incorporated into the calculation of the superelevation angle.

$$\tan\phi = \frac{\text{v}_t^2}{gr} - f \qquad\qquad \textit{56.112}$$

If the banking angle, ϕ, is set to zero, Eq. 56.112 can be used to calculate the maximum velocity of a vehicle making a turn when there is no banking.

The *superelevation*, e, is the rise in the outer edge of the roadway. This is illustrated in Fig. 56.24.

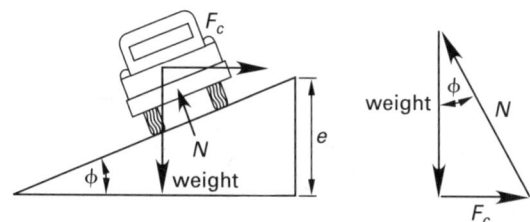

Figure 56.24 *Roadway Banking*

Example 56.14

A 4000 lbm car travels at 40 mph around a banked curve with a radius of 500 ft. What should be the superelevation angle so that tire friction is not needed to prevent the car from sliding?

Solution

The tangential velocity of the car is

$$\text{v}_t = \frac{\left(40\ \dfrac{\text{mi}}{\text{hr}}\right)\left(5280\ \dfrac{\text{ft}}{\text{mi}}\right)}{3600\ \dfrac{\text{sec}}{\text{hr}}} = 58.7\ \text{ft/sec}$$

From Eq. 56.110,

$$\phi = \arctan\left(\frac{\text{v}_t^2}{gr}\right) = \arctan\left(\frac{\left(58.7\ \dfrac{\text{ft}}{\text{sec}}\right)^2}{\left(32.2\ \dfrac{\text{ft}}{\text{sec}^2}\right)(500\ \text{ft})}\right)$$

$$= 12.1°$$

---

[20]The *superelevation* is the slope (in ft/ft or m/m) in the transverse direction (i.e., across the roadway).

[21]Generally it is not desirable to rely on roadway banking alone, since a particular superelevation angle would correspond to only a single speed.

Example 56.15

A vehicle is traveling at 70 mph when it enters a circular curve. The curve radius is 240 ft. The sideways sliding coefficient of friction between the tires and the roadway is 0.57. (a) At what minimum angle from the horizontal must the curve be banked in order to prevent the vehicle from sliding off the top of the curve? (b) If the roadway is banked at 20° from the horizontal, what is the maximum vehicle speed such that no sliding occurs?

Solution

(a) The speed of the vehicle is

$$v_t = \frac{\left(70 \ \frac{\text{mi}}{\text{hr}}\right)\left(5280 \ \frac{\text{ft}}{\text{mi}}\right)}{3600 \ \frac{\text{sec}}{\text{hr}}}$$

$$= 102.7 \text{ ft/sec}$$

From Eq. 56.112, the required banking angle is

$$\phi = \arctan\left(\frac{v_t^2}{gr} - f\right)$$

$$= \arctan\left(\frac{\left(102.7 \ \frac{\text{ft}}{\text{sec}}\right)^2}{\left(32.2 \ \frac{\text{ft}}{\text{sec}^2}\right)(240 \text{ ft})} - 0.57\right)$$

$$= 38.5°$$

(b) Solve Eq. 56.112 for the velocity.

$$v = \sqrt{gr(\tan\phi + f)}$$

$$= \sqrt{\left(32.2 \ \frac{\text{ft}}{\text{sec}^2}\right)(240 \text{ ft})(\tan 20° + 0.57)}$$

$$= 85 \text{ ft/sec} \quad (58 \text{ mi/hr})$$

PRACTICE PROBLEMS

1. The rim of a spoked cast-iron flywheel in a punch press has a mean diameter of 30 in (76.2 cm) and a width of 12 in (30.5 cm). The flywheel rotates at 200 rpm. It supplies 1500 ft-lbf (2 kJ) of energy and slows to 175 rpm each time a punch press is cycled. The density of the cast iron is 0.26 lbm/in^3 (7200 kg/m^3). The hub and arms increase the rotational moment of inertia by 10%. What is the rim thickness?

57

Mechanisms and Power Transmission Systems

1. Energy Storage in Flywheels 57-1
2. Rotating Solid Disks 57-2
3. Rotating Rings 57-2
4. Rotating Hubs 57-2
5. Rotating Fluid Masses 57-3
6. Rim Flywheels 57-3
7. Advanced Flywheel Systems 57-3
8. Gear Train Terminology 57-4
9. Design of Gear Trains 57-5
10. Simple Epicyclic Gear Sets 57-5
11. Analysis of Simple Epicyclic Gear Sets . . . 57-6
12. Design of Simple Epicyclic Gear Trains . . . 57-6
13. Graphic Representation of
    Epicyclic Gear Sets 57-7
14. Analysis of Complex Epicyclic
    Gear Trains 57-8
15. Analysis of Epicyclic Bevel Gear Sets 57-9
16. Cams 57-10
17. Analysis of Cam Performance 57-11
    Practice Problems 57-13

Nomenclature

a	acceleration	ft/sec^2	m/s^2
C	center-to-center distance	ft	m
C	coefficient	–	–
d	diameter	ft	m
d	pitch circle diameter	ft	m
E	energy	ft-lbf	J
E	modulus of elasticity	lbf/ft^2	Pa
F	force	lbf	N
g_c	gravitational constant	ft-lbm/ lbf^2-sec^2	–
I	mass moment of inertia	lbm-ft^2	kg·m^2
k	radius of gyration	ft	m
k	stiffness	lbf/ft	N/m
L	full lifter rise	ft	m
m	mass	lbm	kg
M	whole number	–	–
n	rotational speed	rev/sec	rev/s
N	number (quantity)	–	–
P	diametral pitch	1/ft	1/m
P	power	hp	kW
r	radial distance	ft	m
r	radius	ft	m
R	instantaneous lifter rise	ft	m
S	strength	lbf/ft^2	Pa
t	thickness	ft	m
t	time	sec	s
T	torque	ft-lbf	N·m
TV	train value	–	–
v	velocity	ft/sec	m/s
VR	velocity ratio	–	–
W	work	ft-lbf	J
x	vertical lifter rise	ft	m
x	offset distance	ft	m
y	distance	ft	m

Symbols

α	rotational acceleration	rad/sec^2	rad/s^2
ϵ	strain	–	–
θ	angle of rotation	deg	deg
ν	Poisson's ratio	–	–
ρ	density	lbm/ft^3	kg/m^3
σ	normal stress	lbf/ft^2	Pa
ϕ	pressure angle	deg	deg
ω	rotational velocity	rad/sec	rad/s

Subscripts

a	acceleration
b	base circle
c	centroidal
f	fluctuation
i	inner
k	kinetic
m	mean
n	natural
o	outer
p	pitch
r	radial, rotational, or at radius r
s	shear
t	tangential
u	ultimate
v	velocity
y	y- (vertical) component

1. ENERGY STORAGE IN FLYWHEELS

A *flywheel* is a device that stores energy, reduces speed fluctuations in machinery, and provides large amounts of energy for external work. The rotational kinetic energy of a flywheel is

$$E_{k,r} = \frac{I\omega^2}{2} = \frac{I(2\pi n_{\text{rps}})^2}{2} \qquad \text{[SI]} \qquad 57.1(a)$$

$$E_{k,r} = \frac{I\omega^2}{2g_c} = \frac{I(2\pi n_{\text{rps}})^2}{2g_c} \qquad \text{[U.S.]} \qquad 57.1(b)$$

The mass moment of inertia of a flywheel can be calculated from the total flywheel mass and the radius of gyration, k. The quantity mk^2 is sometimes referred to as the *flywheel effect*.

$$I = mk^2 \qquad \text{[SI]} \qquad 57.2(a)$$

$$I = \frac{mk^2}{g_c} \qquad \text{[U.S.]} \qquad 57.2(b)$$

Speed fluctuations occur when work is performed at the expense of flywheel energy. By the work-energy principle, the change in kinetic energy is equal to the work performed. For example, the work performed by a punch press in punching N holes of diameter d through a sheet of thickness t is

$$W = F_{\text{ave}}t = \frac{F_{\text{max}}t}{2} = \frac{\pi N d t S_{us}}{2} \qquad 57.3$$

The change in rotational kinetic energy is

$$\Delta E_{k,r} = \frac{I(\omega_2^2 - \omega_1^2)}{2} = C_f I \omega_m^2 \qquad \text{[SI]} \qquad 57.4(a)$$

$$\Delta E_{k,r} = \frac{I(\omega_2^2 - \omega_1^2)}{2g_c} = \frac{C_f I \omega_m^2}{g_c} \qquad \text{[U.S.]} \qquad 57.4(b)$$

The *coefficient of fluctuation*, C_f, can be as low as 0.02 to 0.05 for geared drives, generators, and alternators. For punching, shearing, and pressing operations, it is approximately 0.05 to 0.10. For stamping mills and crushing operations, it can be as high as 0.20.

$$C_f = \left| \frac{\omega_2 - \omega_1}{\omega_m} \right| \qquad 57.5$$

The required mass to keep the fluctuation within the desired limits is given by Eq. 57.6. v_c is the tangential velocity at the centroid of the rim cross-sectional area.

$$m = \frac{\Delta E_k}{C_f (v_c)^2} \qquad \text{[SI]} \qquad 57.6(a)$$

$$m = \frac{g_c \Delta E_k}{C_f (v_c)^2} \qquad \text{[U.S.]} \qquad 57.6(b)$$

The relationship between the horsepower and torque of the prime mover that recharges the flywheel is

$$P_{\text{kW}} = \frac{T_{\text{N·m}} n_{\text{rpm}}}{9549} \qquad \text{[SI]} \qquad 57.7(a)$$

$$P_{\text{hp}} = \frac{T_{\text{in-lbf}} n_{\text{rpm}}}{63{,}025} \qquad \text{[U.S.]} \qquad 57.7(b)$$

Given a constant (or average) torque, the time to accelerate the flywheel from ω_1 to ω_2 is given by

$$T = I\alpha = \frac{I(\omega_2 - \omega_1)}{t} \qquad \text{[SI]} \qquad 57.8(a)$$

$$T = \frac{I\alpha}{g_c} = \frac{I(\omega_2 - \omega_1)}{g_c t} \qquad \text{[U.S.]} \qquad 57.8(b)$$

Flywheels are characterized by their *specific power* (i.e., the power output divided by the mass of the flywheel

system) and their *specific energy* (the delivered energy divided by the total mass). These two parameters are relatively independent because the specific power is governed by the discharging device, while the energy is governed by the flywheel design (geometry and materials).

2. ROTATING SOLID DISKS

For a solid disk (i.e., with no hole for a shaft) whose thickness is small compared with its outer radius, r_o, the radial and tangential stresses at a radius r are given by Eqs. 57.9 and 57.10. Both stresses are maximum and equal at the center, where $r = 0$.

$$\sigma_r = \left(\frac{3+\nu}{8} \right) \rho \omega^2 (r_o^2 - r^2) \qquad \text{[SI]} \qquad 57.9(a)$$

$$\sigma_r = \left(\frac{3+\nu}{8} \right) \frac{\rho \omega^2}{g_c} (r_o^2 - r^2) \qquad \text{[U.S.]} \qquad 57.9(b)$$

$$\sigma_t = (\rho \omega^2) \left[\left(\frac{3+\nu}{8} \right) r_o^2 - \left(\frac{1+3\nu}{8} \right) r^2 \right]$$
$$\text{[SI]} \qquad 57.10(a)$$

$$\sigma_t = \left(\frac{\rho \omega^2}{g_c} \right) \left[\left(\frac{3+\nu}{8} \right) r_o^2 - \left(\frac{1+3\nu}{8} \right) r^2 \right]$$
$$\text{[U.S.]} \qquad 57.10(b)$$

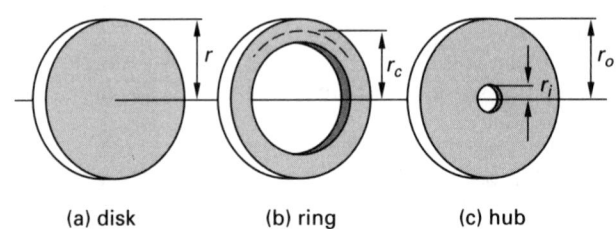

(a) disk   (b) ring   (c) hub

Figure 57.1 *Rotating Bodies*

3. ROTATING RINGS

The tangential (hoop) stress in a rotating uniform circular ring with mean (centroidal) radius r_c and essentially any cross section is given by Eq. 57.11. This stress is not fundamentally different from the hoop stress that exists in a cylinder under internal pressurization, except that the stress is caused by inertial forces.

$$\sigma_t = \rho r_c^2 \omega^2 \qquad \text{[SI]} \qquad 57.11(a)$$

$$\sigma_t = \frac{\rho r_c^2 \omega^2}{g_c} \qquad \text{[U.S.]} \qquad 57.11(b)$$

4. ROTATING HUBS

A rotating annular disk, also known as a *hub*, is a solid disk with a circular hole in the center. For a hub with constant thickness, the radial and tangential stresses at a radius r are

$$\sigma_r = \left(\frac{3+\nu}{8}\right)\rho\omega^2\left(r_o^2 + r_i^2 - \left(\frac{r_o r_i}{r}\right)^2 - r^2\right)$$
$$\text{[SI]} \quad 57.12(a)$$

$$\sigma_r = \left(\frac{3+\nu}{8}\right)\frac{\rho\omega^2}{g_c}\left(r_o^2 + r_i^2 - \left(\frac{r_o r_i}{r}\right)^2 - r^2\right)$$
$$\text{[U.S.]} \quad 57.12(b)$$

$$\sigma_t = \left(\frac{3+\nu}{8}\right)\rho\omega^2\left(r_o^2 + r_i^2 + \left(\frac{r_o r_i}{r}\right)^2 + \frac{(1+3\nu)r^2}{3+\nu}\right)$$
$$\text{[SI]} \quad 57.13(a)$$

$$\sigma_t = \left(\frac{3+\nu}{8}\right)\frac{\rho\omega^2}{g_c}\left(r_o^2 + r_i^2 + \left(\frac{r_o r_i}{r}\right)^2 + \frac{(1+3\nu)r^2}{3+\nu}\right)$$
$$\text{[U.S.]} \quad 57.13(b)$$

The tangential stress will be maximum at the inner boundary.[1]

$$\sigma_{t,\max} = \left(\frac{\rho\omega^2}{4}\right)\left[(3+\nu)r_o^2 + (1-\nu)r_i^2\right] \quad \text{[SI]} \quad 57.14(a)$$

$$\sigma_{t,\max} = \left(\frac{\rho\omega^2}{4g_c}\right)\left[(3+\nu)r_o^2 + (1-\nu)r_i^2\right] \quad \text{[U.S.]} \quad 57.14(b)$$

The maximum tangential stress will usually be larger than the maximum radial stress. However, the maximum radial stress should also be checked. The maximum radial stress occurs at the *geometric mean radius*, $\sqrt{r_o r_i}$.

$$\sigma_{r,\max} = \left(\frac{3+\nu}{8}\right)\rho\omega^2(r_o - r_i)^2 \quad \text{[SI]} \quad 57.15(a)$$

$$\sigma_{r,\max} = \left(\frac{3+\nu}{8g_c}\right)\rho\omega^2(r_o - r_i)^2 \quad \text{[U.S.]} \quad 57.15(b)$$

The change in the inner radius (i.e., the change in radial interference), Δr_i, at the inside radius of the hub due to rotation is found from the tangential stress.

$$\sigma_{t,\max} = E\epsilon_i = E\left(\frac{\Delta r_i}{r}\right) \quad 57.16$$

5. ROTATING FLUID MASSES

When a full cylindrical tank containing liquid is spun around a vertical axis, the centripetal force of the liquid against the tank wall causes the same stresses (e.g., hoop stresses) as would an equivalent internal pressurization. Equation 57.17 gives the hydrostatic pressure in the fluid at any radius r from the axis of rotation. The pressure is the same in all directions.

---

[1]It is interesting that the maximum tangential stress approaches a value that is twice as large as that of a solid disk. This is an example of a *geometric stress concentration factor* caused by a hole.

$$p_r = \frac{\rho\omega^2 r^2}{2} \quad \text{[SI]} \quad 57.17(a)$$

$$p_r = \frac{\rho\omega^2 r^2}{2g_c} \quad \text{[U.S.]} \quad 57.17(b)$$

Equation 57.17 is for a tank that is completely full. It is more difficult to determine pressure when the tank is partially full. The free liquid surface will take on a parabolic shape. The pressure at any point is found from the density and the depth (beneath the parabolic surface) at that point. If the tank is closed and initially pressurized, that pressurization will increase the pressure at a depth accordingly.

6. RIM FLYWHEELS

Rim flywheels consist of a heavy rotating ring connected by spokes to a hub. The contributions of the hub and spokes to the moment of inertia are generally small (i.e., less than 10%).

Assuming that the arms do not restrain the rim from expanding, and the rim thickness is small compared to the mean radius, the tangential (hoop) stress in a rim flywheel is given by Eq. 57.18. v_c is the tangential velocity at the centroid of the rim cross-sectional area.[2] The *burst velocity* can be calculated by setting the tangential stress equal to the ultimate tensile strength with no factor of safety. A large factor of safety (e.g., 10 or larger) has traditionally been used with the cast-iron flywheels that were common in the past.

$$\sigma_t = \rho v_c^2 = \rho(\omega r_c)^2 = \rho(2\pi r_c n_{\text{rps}})^2 \quad \text{[SI]} \quad 57.18(a)$$

$$\sigma_t = \frac{\rho v_c^2}{g_c} = \frac{\rho(\omega r_c)^2}{g_c} = \frac{\rho(2\pi r_c n_{\text{rps}})^2}{g_c} \quad \text{[U.S.]} \quad 57.18(b)$$

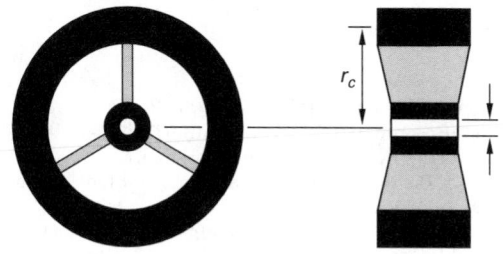

Figure 57.2 *Rim Flywheel*

7. ADVANCED FLYWHEEL SYSTEMS

Advanced flywheel systems (high-performance *flywheel rotors*) are used in innovative energy conservation, transportation, and space applications, but particularly in low-emission electric and hybrid vehicles. They consist of a high-strength flywheel, low-loss bearings, and a charging/discharging system. Advanced flywheels are

---

[2]A general rule of thumb for cast-iron rim flywheels of average size is that the mean velocity should not exceed 6000 ft/min.

constructed of high-strength composite materials to enable them to resist the high speeds (35,000 to 50,000 rpm and above), high tangential velocities (1200 m/s and above), and high temperatures needed for efficient operation.[3]

Rotor suspension systems have historically used ball bearings. However, modern designs focus on noncontact magnetic bearings.

Charging and discharging systems can be mechanical or electrical. (For maximum efficiency in transportation applications, a *continuously variable transmission* (CVT) is needed in mechanically coupled systems.) Charging does not have to be by the same methods as discharging. For example, with a *tandem flywheel system*, one end of the rotor shaft is used for charging the system through mechanical or electrical means, while the discharging equipment is mounted on the other end. In space, where volumetric efficiency is more critical, *concentric flywheel systems* combine electrical charging and discharging methods. Magnetic energy transfer occurs between the rotating rotor and the stationary stator in the frame.

Table 57.1 Typical Characteristics of Materials
Used in Advanced Flywheels

material	ultimate strength[a] (GPa)	density (kg/m³)	specific energy[b] (kJ/kg)
high-strength steel	1.5–2.1	7700–8000	250
E-glass/epoxy	1.4	1900	700
S-glass/epoxy	2.1	1900	1100
Kevlar™/epoxy	1.9	1400	1400
graphite/epoxy	1.6–3.2	1500	1100

(Multiply GPa by 1.45×10^5 to obtain lbf/in².)
(Multiply kg/m³ by 0.0624 to obtain lbm/in³.)
(Multiply kJ/kg by 334 to obtain ft-lbf/lbm.)
[a]Strength of the matrix. Individual fibers will have much higher strengths.
[b]Values given merely indicate the approximate range. Actual values would depend on the design.
Kevlar™ is a trademark of the DuPont Company.

8. GEAR TRAIN TERMINOLOGY

An *external gear* is any gear whose teeth "point" away from the axis of rotation, compared to an *internal gear*, whose teeth point in toward the axis of rotation.

[3]Since rotation is in a vacuum, rotors cannot be convectively cooled.

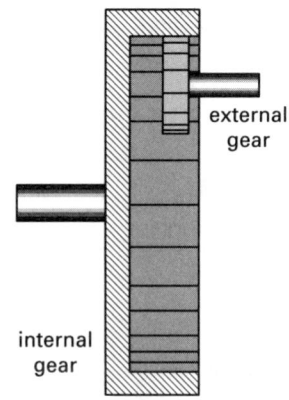

Figure 57.3 External and Internal Gears

A *simple gear set* (*gear train*) consists of two gears with fixed centers in mesh. The gears turn in opposite directions if both are external gears. If one of the gears is an internal gear, the gears turn in the same direction. Each pair of gears constitutes a *stage*. For multiple gears in mesh, each pair of gears in succession constitute a stage, so that one gear can be part of two stages. An *idler gear*, also known as an *intermediate gear*, is a gear that provides a change in shaft rotation direction without a change in rotational speed.

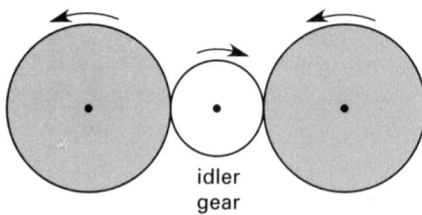

Figure 57.4 Idler Gear

A *compound gear* consists of two gears mounted on a single (usually short) shaft. *Reverted gear sets* are those whose input and output shafts are in-line, usually using one or more compound gears. A reverted gear set always has an even number of stages. Only the gears in mesh need to have the same diametral pitches.

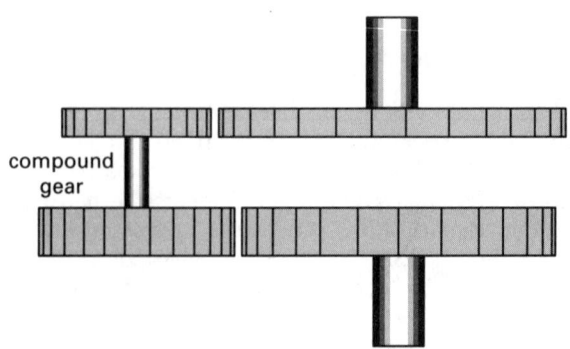

Figure 57.5 Reverted Gear Set and Compound Gears

The *velocity ratio* (*mesh ratio* or *gear ratio*), VR, is the ratio of the input to output speeds. In Eq. 57.19, d is the pitch circle diameter.

$$\text{VR} = \frac{n_{\text{in}}}{n_{\text{out}}} = \frac{N_{\text{out}}}{N_{\text{in}}} = \frac{d_{p,\text{out}}}{d_{p,\text{in}}} \qquad \textbf{57.19}$$

If all of the gears have fixed centers (i.e., fixed axes of rotation), the velocity ratio is the product of all of the numbers of teeth on the driven gears divided by the product of all of the numbers of teeth on the driving gears.

The *train value*, TV, is the reciprocal of the velocity ratio. (Figure 57.8 can be used to determine the train value for common configurations.)

$$\text{TV} = \frac{n_{\text{out}}}{n_{\text{in}}} = \frac{N_{\text{in}}}{N_{\text{out}}} = \frac{1}{\text{VR}} \qquad \textbf{57.20}$$

9. DESIGN OF GEAR TRAINS

Finding the number of teeth that each gear should have in order to achieve a particular train ratio is time consuming, as a trial-and-error solution may be needed. A particularly desired gear ratio may not exactly be achievable, since each gear must contain an integral number of teeth. Other constraints may be placed on the design, including the minimum and maximum numbers of teeth and the maximum number of stages.

All gears in mesh must have the same diametral pitch. Therefore, for any two gears in mesh, the relationship between the diametral pitch, pitch circle diameters, and number of teeth is

$$P_1 = P_2 \qquad \textbf{57.21}$$

$$\frac{N_1}{\pi d_1} = \frac{N_2}{\pi d_2} \qquad \textbf{57.22}$$

The center-to-center distance, C, is the sum of the two pitch radii.

$$C = \frac{d_1 + d_2}{2} = \frac{\dfrac{N_1}{\pi P} + \dfrac{N_2}{\pi P}}{2} \qquad \textbf{57.23}$$

Equations 57.22 and 57.23 can be written for all pairs of meshing gears and solved simultaneously for unknown quantities. Knowledge of various gear ratios between the meshing gears can be used to simplify the simultaneous equations.

For reverted gear trains, the total train ratio should ideally be shared equally between all stages. Since this ratio may not be feasible, the total train ratio should be factored into stage ratios that numerically are not too far apart.

10. SIMPLE EPICYCLIC GEAR SETS

Epicyclic gear sets (also known as *planetary gear sets*) are characterized by one or more gears that do not have fixed axes of rotation. They have two inputs (one of which may be fixed or stationary) and one output.

Compared with gear sets where all gears have fixed centers, epicyclic gear sets can have much higher gear ratios, are more compact, have lower tooth loadings and pitch-line velocities, offer in-line input and output shafts, may be easier to lubricate, and generally are less expensive.

The simplest type of epicyclic gear set is shown in Fig. 57.6. It consists of a *sun gear, ring gear* (also known as an *annulus gear*), and one or more *planet gears* (also referred to as *planets, planet pinions*, and *spider gears*). The rotating bent yoke that connects the planets to their shaft is known as the *planet carrier, arm*, and *spider*.

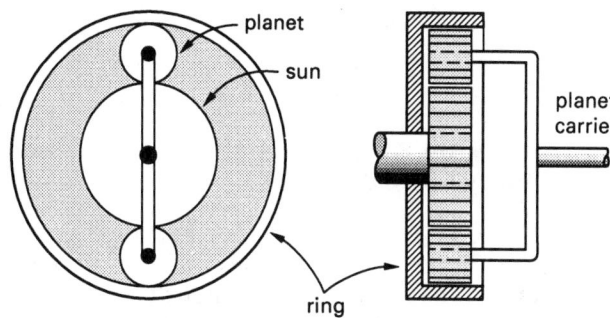

Figure 57.6 *Simple Epicyclic Gear Set*

The planets rotate about their own axes and revolve around the sun gear. During rotation, a point on a planet gear traces out epicyclic curves, hence the name. There are generally one to four planets. The number of planets does not affect the output speed, but the maximum power transmission is essentially proportional to the number of planets. The number of planets is limited by space so that the planets do not "overlap." Either the ring gear or the carrier can be fixed. If one is fixed, then the other must rotate.

When the carrier is fixed and the ring gear is the output, the arrangement is known as a *star gear set*. When the sun gear is fixed and the input is the ring gear, the system is called a *solar gear set*.

Simple epicyclic gear trains operate in one of three different modes:

mode 1: The driving and driven members turn in the same direction, and the overall gear ratio is between 2 and ∞. The overall gear ratio for this mode is

$$\text{VR} = \frac{N_{\text{sun}} + N_{\text{ring}}}{N_{\text{sun}}} \qquad \textbf{57.24}$$

mode 2: The driving and driven members turn in the same direction, and the overall gear ratio is between 1 and 2. The overall gear ratio for this mode is

$$\text{VR} = \frac{N_{\text{sun}} + N_{\text{ring}}}{N_{\text{ring}}} \qquad \textbf{57.25}$$

mode 3: The driving and driven members turn in opposite directions, and the overall gear ratio is between 1 and ∞.

$$\text{VR} = \frac{N_{\text{ring}}}{N_{\text{sun}}} \qquad 57.26$$

The fundamental relationships between the number of teeth on the gears is derived from the fact that all gears are in mesh and must have the same diametral pitch. In Eq. 57.28, M is some whole number.

$$N_{\text{sun}} + 2N_{\text{planet}} = N_{\text{ring}} \qquad 57.27$$

$$N_{\text{sun}} + N_{\text{ring}} = (M)(\text{no. of planets}) \qquad 57.28$$

11. ANALYSIS OF SIMPLE EPICYCLIC GEAR SETS

Analysis of a simple epicyclic gear set starts with calculating the train value, TV. This value is the same regardless of whether the gear set is reducing or augmenting. However, it is negative if the sun and ring gears rotate in different directions when the arm is locked. (TV is always negative when the ring gear is an internal gear.) Although the symbol ω is used for convenience, the relationship is valid for rotational speed in rad/sec, rev/min, and rev/sec. A consistent sign convention should be chosen for directions of rotation (e.g., positive for clockwise and negative for counterclockwise).

$$\text{TV} = \frac{N_{\text{ring}}}{N_{\text{sun}}} = \frac{\omega_{\text{sun}} - \omega_{\text{carrier}}}{\omega_{\text{ring}} - \omega_{\text{carrier}}} \qquad 57.29$$

Equation 57.30 is Eq. 57.29 solved for the speed of the sun gear. It gives the relationship between the known train value and the unknown rotational speeds.

$$\omega_{\text{sun}} = (\text{TV})\omega_{\text{ring}} + (1 - \text{TV})\omega_{\text{carrier}} \qquad 57.30$$

Planet speed is not part of Eq. 57.30. The rotational speed (with respect to their own axis) and direction of the planets can be found from Eq. 57.31. The quantity is negative because the planets and sun gear turn in opposite directions. A consistent sign convention must be used.

$$\frac{N_{\text{planets}}}{N_{\text{sun}}} = \frac{-(\omega_{\text{sun}} - \omega_{\text{carrier}})}{\omega_{\text{planets}} - \omega_{\text{carrier}}} \qquad 57.31$$

Example 57.1

A simple epicyclic gear set with two planets is similar to Fig. 57.6. The sun gear has 32 teeth, the planets have 16 teeth, and the ring has 64 teeth. The sun gear turns clockwise at 100 rev/min. The ring is fixed. Find the (a) speed and (b) direction of the carrier.

Solution

(a) If the arm was locked and the ring was free to rotate, the sun and ring gears would rotate in different directions. Therefore, the train value is negative.[4]

$$\text{TV} = \frac{-64 \text{ teeth}}{32 \text{ teeth}} = -2$$

Since the ring gear is fixed, $\omega_{\text{ring}} = 0$. For this example, decide that clockwise rotation is positive. Use Eq. 57.30 with the speeds given in rev/min.

$$100 \text{ rev/min} = (0)(-2) + \omega_{\text{carrier}}\big(1 - (-2)\big)$$

$$\omega_{\text{carrier}} = 33.3 \text{ rev/min}$$

(b) Since ω_{carrier} is positive, the carrier rotation is clockwise.

12. DESIGN OF SIMPLE EPICYCLIC GEAR TRAINS

As with fixed-center gear trains, analysis of epicyclic gear trains is easier than the design process. However, the following procedure can used.

step 1: Determine the mode of operation (Sec. 10) and gear ratio.

step 2: Use Table 57.2 to determine which gear in the set must be the smallest.

Table 57.2 *Smallest Epicyclic Gear*

mode	gear ratio	smallest gear
1	$2 < R \le 4$	planet
1	$4 \le R < \infty$	sun
2	$1 < R \le \frac{4}{3}$	sun
2	$\frac{4}{3} \le R < 2$	planet
3	$1 < R \le 3$	planet
3	$3 \le R < \infty$	sun

step 3: Depending on which gear is the smallest, use either Table 57.3 or Table 57.4 to determine the number of teeth on the other gears in terms of the gear ratio and the number of teeth on the smallest gear.[5]

Table 57.3 *Number of Teeth when the Sun Gear is the Smallest (N_{sun} and R known)*

no. of teeth	mode 1	mode 2	mode 3
N_{planet}	$\dfrac{(R-2)N_{\text{sun}}}{2}$	$\dfrac{(2-R)N_{\text{sun}}}{2R-2}$	$\dfrac{(R-1)N_{\text{sun}}}{2}$
N_{ring}	$(R-1)N_{\text{sun}}$	$\dfrac{N_{\text{sun}}}{R-1}$	RN_{sun}

---

[4]It may be easier to remember that the train value is always negative whenever the ring gear is in an internal gear.
[5]Generally, no gear should have fewer than 15 to 18 teeth for proper operation.

Table 57.4 *Number of Teeth when the Planet Gear is the Smallest*
(N_{planet} *and R known*)

no. of teeth	mode 1	mode 2	mode 3
N_{sun}	$\dfrac{2N_{\text{planet}}}{R-2}$	$\dfrac{(2R-2)N_{\text{planet}}}{2-R}$	$\dfrac{2N_{\text{planet}}}{R-1}$
N_{ring}	$\dfrac{(2R-2)N_{\text{planet}}}{R-2}$	$\dfrac{2N_{\text{planet}}}{2-R}$	$\dfrac{2RN_{\text{planet}}}{R-1}$

Example 57.2

A simple epicyclic gear drive has a speed ratio of 1.7:1. The driving and driven members turn in the same direction. Each planet is to have 18 teeth. (a) How many teeth are on each gear? (b) How many planets should there be?

Solution

(a) From Sec. 10, the gear set operates in mode 2. From Table 57.2, the planets are the smallest gears. There are two relationships from Table 57.4.

$$N_{\text{sun}} = \frac{(2R-2)N_{\text{planet}}}{2-R} = \frac{((2)(1.7)-2)(18)}{2-1.7}$$
$$= 84 \text{ teeth}$$
$$N_{\text{ring}} = \frac{2N_{\text{planet}}}{2-R} = \frac{(2)(18)}{2-1.7}$$
$$= 120$$

(b) Use Eq. 57.28.

$$N_{\text{sun}} + N_{\text{ring}} = (M)(\text{no. of planets})$$
$$84 + 120 = 204 = (M)(\text{no. of planets})$$

Both M and the number of planets are whole numbers, and both must be factors of 204—that is, 2, 3, 4, or 6. There could be 2, 3, or 4 planets. Six planets would overlap and would be in excess.

13. GRAPHIC REPRESENTATION OF EPICYCLIC GEAR SETS

A standard graphic method (with variations) is used for showing more complex epicyclic gear trains in two dimensions. Figure 57.7 illustrates how a simple gear set with three planets would be diagrammed. The gears are shown as solid blocks or lines. The shafts are shown as lines. Notice that in Fig. 57.7(a), even though the actual carrier output shaft is on the same side as the input shaft, the output shaft is diagrammed on the opposite side. In both Fig. 57.7(a) and (b), notice that only one planet is shown in the two-dimensional representation.

Figure 57.7 *Graphical Representation of Epicyclic Gear Sets*

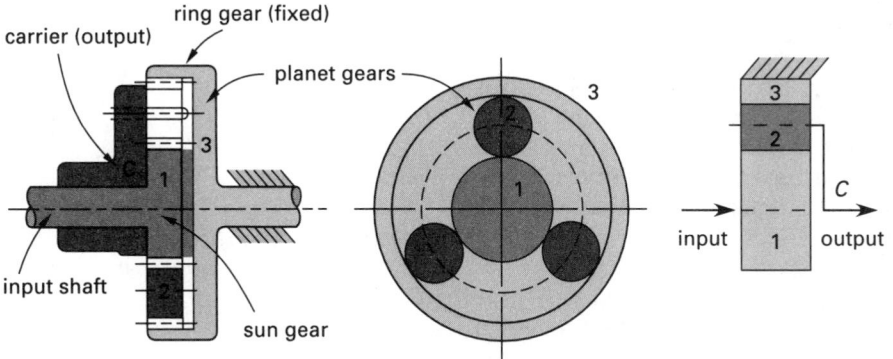

(a) simple epicyclic gear set

(b) complex epicyclic gear set

14. ANALYSIS OF COMPLEX EPICYCLIC GEAR TRAINS

There are several ways of analyzing more complex epicyclic gear trains. Some methods are intuitive, some are procedural, some are graphical, and some are tabular. The procedure presented here is tabular and does not require any significant visualization or intuitive reasoning.

step 1: Identify all of the gears that have the same center of rotation as the arm (i.e., have the same planetary axis). Include stationary gears (such as fixed ring gears), but do not include the arm, any gear that is attached to it, or any fixed gear that is not centered on the planetary axis. Prepare a three-row matrix with a column for each of these gears.

step 2: Write "ω_{carrier}" in the first row for each gear in the table. (Items in quotation marks are to be written exactly as indicated.) Do not substitute numerical values until step 7.[6]

step 3: The speed of one of the gears in the table may be unknown and desired. Refer to this gear as gear z. If all of the gear speeds except the carrier speed are known, choose gear z arbitrarily. For that gear, put "ω_z" into the third row for column z.

step 4: Put "$\omega_z - \omega_{\text{carrier}}$" into the second row for column z.

step 5: Use Fig. 57.8 to determine the velocity relationships, ω/ω_z, between gear z and all other gears in the table. Use a consistent sign convention and consider the directions of rotation if other configurations are used in the gear set.

step 6: Multiply the velocity relationships from step 5 by "$\omega_z - \omega_{\text{carrier}}$" and put the product into row 2.

step 7: Substitute all known values of ω into the third row for all empty columns and into ω_{carrier} wherever it appears. If the velocity of any gear i is unknown, insert "ω_i."

step 8: For each gear whose speed is known, write the *characteristic equation*.

$$\text{row } 1 + \text{row } 2 = \text{row } 3 \qquad 57.32$$

---

[6]The method presented is well founded in rigorous theory. However, it is somewhat "magical" in its manner of deriving the answer from innocuous steps and is sometimes referred to as a "hocus-pocus" solution method.

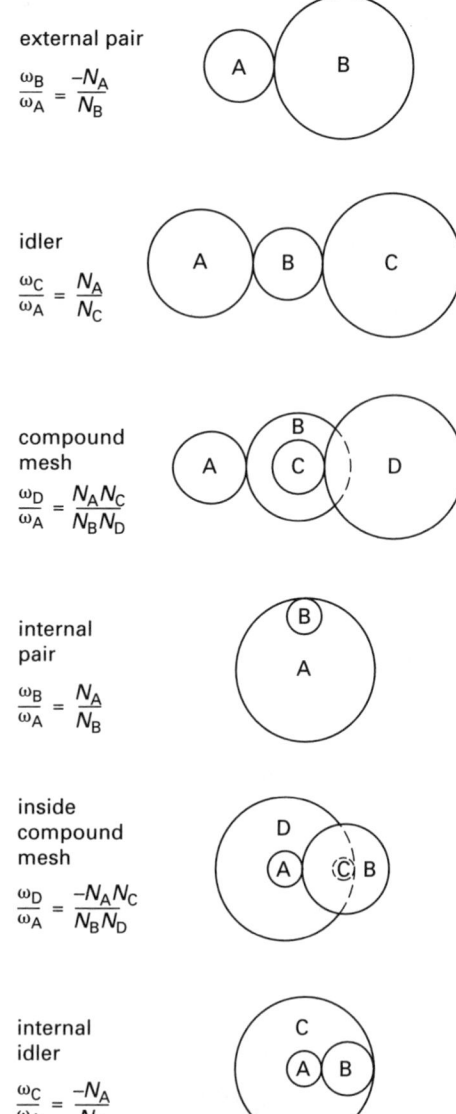

external pair

$$\frac{\omega_B}{\omega_A} = \frac{-N_A}{N_B}$$

idler

$$\frac{\omega_C}{\omega_A} = \frac{N_A}{N_C}$$

compound mesh

$$\frac{\omega_D}{\omega_A} = \frac{N_A N_C}{N_B N_D}$$

internal pair

$$\frac{\omega_B}{\omega_A} = \frac{N_A}{N_B}$$

inside compound mesh

$$\frac{\omega_D}{\omega_A} = \frac{-N_A N_C}{N_B N_D}$$

internal idler

$$\frac{\omega_C}{\omega_A} = \frac{-N_A}{N_C}$$

Figure 57.8 *Speed Relationships Between Gears with Fixed Centers (gear A is always the input)*

Example 57.3

An epicyclic gear train is constructed as shown. The ring gear is fixed. The planet carrier is the output and turns at 20 rev/min. Gear A has 16 teeth, gear B has 32 teeth, gear C has 24 teeth, and gear D has 72 teeth. What are the (a) speed and (b) direction of gear A?

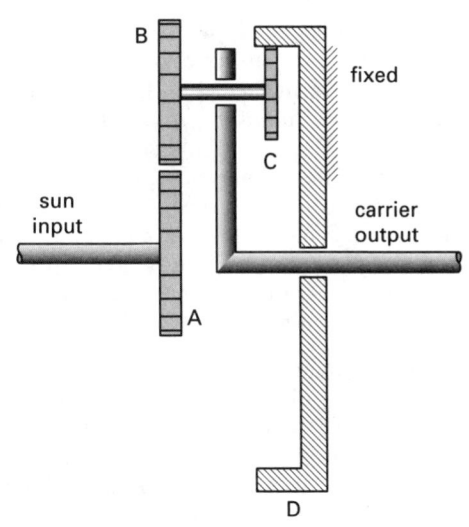

Solution

step 1: Gears A and D are concentric with the arm.

	A	D
row 1		
row 2		
row 3		

step 2:

	A	D
row 1	ω_{carrier}	ω_{carrier}
row 2		
row 3		

step 3: The speed of gear A is unknown.

	A	D
row 1	ω_{carrier}	ω_{carrier}
row 2		
row 3	ω_A	

step 4:

	A	D
row 1	ω_{carrier}	ω_{carrier}
row 2	$\omega_A - \omega_{\text{carrier}}$	
row 3	ω_A	

step 5: From Fig. 57.8, the path from gear A to gear D is an inside compound mesh. The speed ratio is

$$\frac{\omega_D}{\omega_A} = \frac{-N_A N_C}{N_B N_D}$$
$$= \frac{-(16)(24)}{(32)(72)} = -\frac{1}{6}$$

step 6:

	A	D
row 1	ω_{carrier}	ω_{carrier}
row 2	$\omega_A - \omega_{\text{carrier}}$	$\left(-\frac{1}{6}\right)\left(\omega_A - \omega_{\text{carrier}}\right)$
row 3	ω_A	

step 7: Choose the direction of rotation of the carrier as the positive direction. Since gear D is fixed, $\omega_D = 0$. Also, $\omega_{\text{carrier}} = +20$.

	A	D
row 1	20	20
row 2	$\omega_A - 20$	$\left(-\frac{1}{6}\right)\left(\omega_A - 20\right)$
row 3	ω_A	0

step 8: The characteristic equation for column D is

$$\text{row } 1 + \text{row } 2 = \text{row } 3$$
$$20\,\frac{\text{rev}}{\text{min}} + \left(-\frac{1}{6}\right)\left(\omega_A - 20\,\frac{\text{rev}}{\text{min}}\right) = 0$$
$$\omega_A = 140 \text{ rev/min}$$

$$\left[\begin{array}{l}\text{positive, so in the same}\\ \text{direction as the carrier}\end{array}\right]$$

15. ANALYSIS OF EPICYCLIC BEVEL GEAR SETS

Many epicyclic gear sets make use of bevel gears. This is the standard design for traditional automotive differentials, as shown in Fig. 57.9.

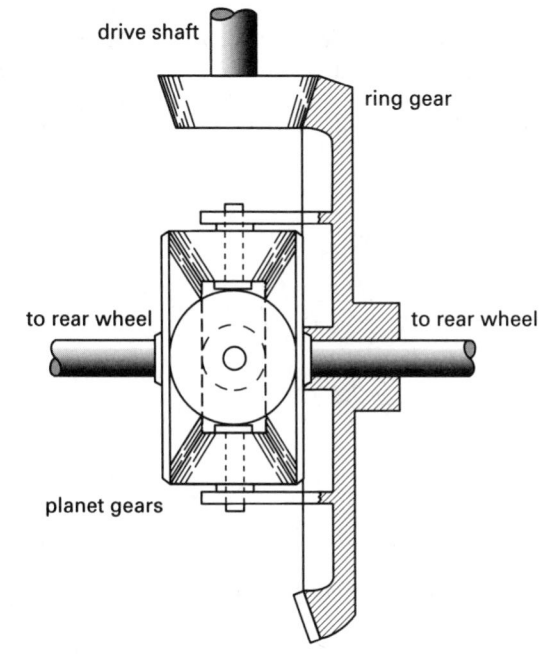

Figure 57.9 *Automotive Differential*

Dynamics and Vibrations

In the simple epicyclic gear set shown in Fig. 57.6, the power path is from the planet carrier, through the planets, and on to the sun, or the reverse. With a bevel gear set shown in Fig. 57.9, the power path is from the drive shaft to the rotating ring gear, through the bevel gears, and on to the two axle shafts.

The simple formula used in Sec. 11 can be used with bevel gear sets if an analogy is made to the traditional (simple) epicyclic gear set. The two identical bevel gears are essentially idlers and are analogous to the planets. The two identical gears driving the axle shafts are analogous to the ring and sun gears that are normally separated by planets. The actual automotive ring gear turns with the bevel gears (i.e., the "planets") and is analogous to the carrier.

Since the analogous sun-planet-ring combination is "rectangular," and since the analogous sun and ring gears would turn at the same speeds but in opposite directions if the carrier was locked, the train value, TV, is always -1.

Equation 57.30 is written in terms of the actual automotive gear names using a train value of -1.[7] The speed of the ring gear is found from the number of teeth on it and the drive shaft gear.

$$\omega_{\text{left rear wheel}} + \omega_{\text{right rear wheel}} = 2\omega_{\text{ring gear}} \qquad 57.33$$

When a vehicle is traveling in a straight line, both wheels turn at the same speed, equal to the rotational speed of the ring gear. When the vehicle travels around a bend, the inner wheel turns slower than the outer wheel.

Example 57.4

A car is jacked up on one side such that the left wheel is stationary on the road surface while the right wheel is free to turn in the air. The engine is run such that the drive shaft turns at 1000 rev/min. An automotive differential similar to Fig. 57.9 is used. The drive shaft gear has 17 teeth, and the ring gear has 54 teeth. What is the speed of the right wheel?

Solution

The speed of the ring gear is

$$\omega_{\text{ring gear}} = \left(\frac{N_{\text{drive shaft}}}{N_{\text{ring gear}}}\right)\omega_{\text{drive shaft}}$$

$$= \left(\frac{17}{54}\right)\left(1000\ \frac{\text{rev}}{\text{min}}\right) = 314.8\ \text{rev/min}$$

Use Eq. 57.33.

$$\omega_{\text{left rear wheel}} = 2\omega_{\text{ring gear}} - \omega_{\text{right rear wheel}}$$

$$0 = (2)\left(314.8\ \frac{\text{rev}}{\text{min}}\right) - \omega_{\text{right rear wheel}}$$

$$\omega_{\text{right rear wheel}} = 629.6\ \text{rev/min}$$

---

[7]The ring gear referenced in Eq. 57.33 is the actual ring gear shown in Fig. 57.9.

16. CAMS

Cams are devices that convert regular rotating motion into irregular translating motion. Cams drive another element, known as a *follower*, *lifter*, or *tappet*. There are various types of cams. This section is concerned only with *plate cams*, the type traditionally used in internal combustion engines to open and close valves. There are various types of followers for plate cams, although the *flat-faced follower* and *roller lifter* are the most common.[8] Where cost is not a factor, the roller lifter is preferred because of its lower friction.

Cam lobes can be shaped in a variety of ways, depending on what type of displacement is desired. Theoretically, any curve can be used to design the cam profile including circular arcs, parabolas, cubics, quartics, and sinusoids.[9] Common standard-motion lobe shapes (i.e., *profiles*) are designated as uniform, parabolic (constant acceleration), harmonic, and cycloidal.[10] Figure 57.10 illustrates the displacement, velocity, and acceleration curves for some of these lobe designs over the rotation necessary to cause full rise.

Some of the curves have advantages over the others. *Jerk* (*geometric jerk*) is a term used to describe the *third derivative of motion*, sometimes called the *second acceleration*. The parabolic cam has three infinite jerks per rise. It is the worst of the three cam types for high-speed use. The harmonic cam also has one infinite jerk. The cycloidal cam has a higher acceleration than the other two cam profiles, but the jerk is finite. Therefore, the cycloidal cam is among the best of the three for high-speed operation.

Figure 57.11 illustrates a generalized cam with a roller follower on its base circle. (The lobe is not shown.) The *base circle* is the smallest circle that can be drawn tangent to the cam surface. x is the *follower offset* (which is zero for *radial cams*), and the follower's movement is radial with respect to the cam profile. r is the separation of the cam and follower centers. R is the vertical follower rise (displacement) at a particular angle of cam rotation (not necessarily the maximum rise or *lift*, L). The *dwell* is the duration, usually measured in cam angle degrees, for which the rise is constant (i.e., the lifter position is unchanged at the top or bottom position). The subscript b refers to when the follower is on the base circle (i.e., is at its lowest point). From the Pythagorean theorem,

$$\begin{aligned}
r^2 &= y^2 + x^2 \\
&= (y_b + R)^2 + x^2 \\
&= y_b^2 + 2y_b d + d^2 + R^2 \\
&= r_b^2 + 2y_b R + R^2 \qquad 57.34
\end{aligned}$$

---

[8]Pointed and spherical-faced followers can be used for special applications.
[9]A *trapezoidal profile*, for example, is an alternating sequence of cubic and parabolic sections.
[10]Though they are easy to manufacture, cam lobes constructed from circular arcs are not covered here.

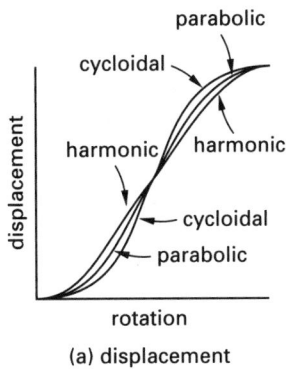

(a) displacement

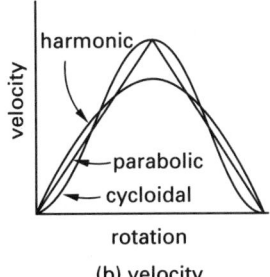

(b) velocity

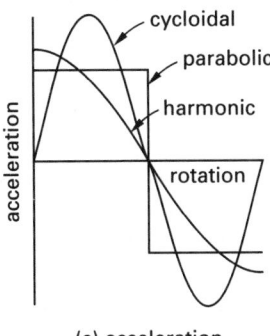

(c) acceleration

Figure 57.10 *Cam Performance vs. Lobe Profile*

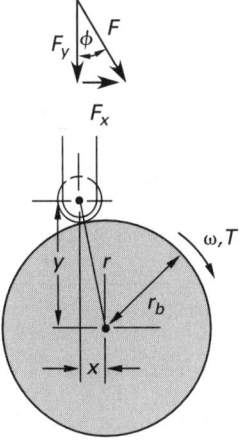

Figure 57.11 *Cam and Follower*

The torque on the camshaft due to the follower force depends on the rotational angle and the vertical force component.

$$T = \frac{F_y \mathrm{v}}{\omega} \qquad 57.37$$

In high-speed operation, a cam follower may bounce out of contact with the cam. This is referred to as *jump*. One theory proposes that jump will not occur if at least two full cycles of vibration can occur in the follower train during the positive acceleration time interval of motion. (Jump may not occur even with fewer than two cycles, but this must be investigated.) In Eq. 57.38, θ_1 is the angle through which the cam rotates during the positive acceleration period. ω_n is the natural frequency of the spring system, and ω is the cam's speed of rotation.

$$\left(\frac{\theta_1}{360°} \right) \left(\frac{\omega_n}{\omega} \right) \geq 2 \qquad 57.38$$

Representing the stiffness of the spring used to hold the follower against the cam as k_1 and the stiffness of the follower train as k_2, the natural frequency of the follower system is given by Eq. 57.39. m is the lumped follower mass.

$$\omega_n = \sqrt{\frac{k_1 + k_2}{m}} \qquad 57.39$$

17. ANALYSIS OF CAM PERFORMANCE

Table 57.5 contains theoretical expressions for the displacement, velocity, and acceleration for three different lobe profiles. In some cases, simplified expressions may be available. For example, the maximum velocity, acceleration, and displacement experienced by a lifter moving a maximum distance L during cam rotation θ_0 are given by Eqs. 57.40 and 57.41. Values of velocity and acceleration coefficients are given in Table 57.6. Maximum deceleration is the same as maximum acceleration.

For flat-faced lifters, the force on a cam and follower is in line with the vertical axis of the follower. For roller lifters, the force may be offset by a *pressure angle* (ϕ) as shown in Fig. 57.11. The pressure angle is usually kept below 30° to 35° to prevent the lifter from binding or locking the cam. The pressure angle is lower for larger base-circle diameters. Referring to Eq. 57.35, it can be seen that the larger the offset, the smaller the pressure angle. For this reason, pressure angle considerations are less important when an offset follower is used.

$$\tan \phi = \frac{\dfrac{\mathrm{v}}{\omega} - x}{r + R} \qquad 57.35$$

For a radial cam, $x = 0$ and $r = y$.

$$\tan \phi = \frac{\mathrm{v}}{\omega(y + R)} \qquad \text{[radial cams]} \qquad 57.36$$

Table 57.5 Cam Lobe Dynamics[a]
(θ and θ_0 are in radians)

type of lobe	displacement (rise, R) maximum at $\theta/\theta_0 = 1$	velocity maximum at $\theta/\theta_0 = 0.5$	acceleration
parabolic	accelerating, $R = 2L\left(\dfrac{\theta}{\theta_0}\right)^2$ $\left(\dfrac{\theta}{\theta_0} \le 0.5\right)$	$\dfrac{dR}{dt} = \dfrac{4L\omega\theta}{\theta_0^2}$	$\dfrac{d^2R}{dt^2} = \dfrac{4L\omega^2}{\theta_0^2}$ (constant)
	decelerating, $R = L\left[1 - (2)\left(1 - \dfrac{\theta}{\theta_0}\right)^2\right]$ $\left(\dfrac{\theta}{\theta_0} \ge 0.5\right)$	$\dfrac{dR}{dt} = \left(\dfrac{4L\omega}{\theta_0}\right)\left(1 - \dfrac{\theta}{\theta_0}\right)$	$\dfrac{d^2R}{dt^2} = -\dfrac{4L\omega^2}{\theta_0^2}$ (constant)
harmonic	$R = \left(\dfrac{L}{2}\right)\left[1 - \cos\left(\dfrac{\pi\theta}{\theta_0}\right)\right]$	$\dfrac{dR}{dt} = \left(\dfrac{\pi L\omega}{2\theta_0}\right)\sin\left(\dfrac{\pi\theta}{\theta_0}\right)$	$\dfrac{d^2R}{dt^2} = \left(\dfrac{\pi^2 L\omega^2}{2\theta_0^2}\right)\cos\left(\dfrac{\pi\theta}{\theta_0}\right)$ (maximum at $\theta/\theta_0 = 0$)
cycloidal	$R = \left(\dfrac{L}{\pi}\right)\left[\dfrac{\pi\theta}{\theta_0} - \dfrac{1}{2}\sin\left(\dfrac{2\pi\theta}{\theta_0}\right)\right]$	$\dfrac{dR}{dt} = \left(\dfrac{L\omega}{\theta_0}\right)\left[1 - \cos\left(\dfrac{2\pi\theta}{\theta_0}\right)\right]$	$\dfrac{d^2R}{dt^2} = \left(\dfrac{2\pi L\omega^2}{\theta_0^2}\right)\sin\left(\dfrac{2\pi\theta}{\theta_0}\right)$ (maximum at $\theta/\theta_0 = 0.25$)

[a] θ_0 is the cam rotation for full lifter rise, L. This table can be used for nonsymmetrical lobes if the analysis is performed in two phases, with two values of θ_0. In each phase, θ_0 is taken as twice the angular range for that phase.

Table 57.6 Maximum Cam Performance Coefficients

lobe profile	C_v	C_a	C_{jerk}
parabolic	2	4	∞
cycloidal	2	2π	$4\pi^2$
simple harmonic	$\dfrac{\pi}{2}$	$\dfrac{\pi^2}{2}$	$\dfrac{\pi^3}{2}$

To use Eqs. 57.40 and 57.41, ω and θ_0 must both use the same units (e.g., radians).[11]

$$v_{max} = \frac{C_v \omega L}{\theta_0} \qquad 57.40$$

$$a_{max} = \frac{C_a \omega^2 L}{\theta_0^2} \qquad 57.41$$

Example 57.5

A cycloidal cam produces a full lift of 1.5 in during an acceleration stroke occurring over 50° of cam rotation. Determine the (a) maximum velocity and (b) maximum acceleration.

---

[11] As written, Eqs. 57.40 and 57.41 have units of distance/sec and distance/sec², respectively. In some problems, units of distance/deg and distance/deg² are needed. Results in these forms can be calculated by omitting ω from the equations.

Solution

The rotational speed of the cam is unknown. The velocity must be determined in units per degree rather than in units per second.

(a) Use Eq. 57.40 and Table 57.6. The maximum velocity is

$$v_{max} = \frac{C_v L}{\theta_0} = \frac{(2)(1.5 \text{ in})}{50°}$$
$$= 0.06 \text{ in/deg}$$

(b) Use Eq. 57.41 and Table 57.6. The maximum acceleration is

$$a_{max} = \frac{C_a L}{\theta_0^2} = \frac{(2\pi)(1.5 \text{ in})}{(50°)^2}$$
$$= 0.00377 \text{ in/deg}^2$$

Example 57.6

A radial cam with cycloidal lobes turns at 1200 rpm. The rise stroke occurs over 120° of cam rotation. The total lift is 25 mm. The follower is held against the cam by a spring with spring constant of 5 N/mm. The spring compression is 5 mm when the follower is on its base circle. The lumped mass of the follower is 0.8 kg. Stiffness of the follower train is to be disregarded. When the cam turns 50° from the start of the lifter movement, what are the (a) angular velocity, (b) rise, (c) inertial force on the lifter, (d) total force on the spring, (e) net force on the cam, and (f) shaft torque on the cam due to one lifter?

Solution

(a) The angular velocity is

$$\omega = 2\pi n_{\text{rps}} = \frac{(2\pi)\left(1200\ \dfrac{\text{rev}}{\text{min}}\right)}{60\ \dfrac{\text{s}}{\text{min}}}$$

$$= 125.66\ \text{rad/s}$$

(b) The angular rotation for rise is

$$\theta_0 = \frac{(120°)(2\pi)}{360°}$$

$$= 2.094\ \text{rad}$$

A cam angle of 50° is

$$\theta = \frac{(50°)(2\pi)}{360°}$$

$$= 0.873\ \text{rad}$$

From Table 57.5, the lifter displacement is

$$R = \left(\frac{L}{\pi}\right)\left[\frac{\pi\theta}{\theta_0} - \frac{1}{2}\sin\left(\frac{2\pi\theta}{\theta_0}\right)\right]$$

$$= \left(\frac{25\ \text{mm}}{\pi}\right)\left[\pi\left(\frac{0.873\ \text{rad}}{2.094\ \text{rad}}\right)\right.$$

$$\left. - \frac{1}{2}\sin\left(\frac{2\pi \times 0.873\ \text{rad}}{2.094\ \text{rad}}\right)\right]$$

$$= 8.44\ \text{mm}$$

(c) From Table 57.5, the lifter acceleration is

$$a = \left(\frac{2\pi L\omega^2}{\theta_0^2}\right)\sin\left(\frac{2\pi\theta}{\theta_0}\right)$$

$$= \left[\frac{(2\pi)(25\ \text{mm})\left(125.66\ \dfrac{\text{rad}}{\text{s}}\right)^2}{(2.094\ \text{rad})^2}\right]$$

$$\times \sin\left(\frac{2\pi \times 0.875\ \text{rad}}{2.094\ \text{rad}}\right)$$

$$= 2.792 \times 10^5\ \text{mm/s}^2$$

The inertial force is

$$F = ma = \frac{(0.8\ \text{kg})\left(2.792 \times 10^5\ \dfrac{\text{mm}}{\text{s}^2}\right)}{1000\ \dfrac{\text{mm}}{\text{m}}}$$

$$= 223.4\ \text{N}$$

(d) The force on the spring is

$$F = k\delta = \left(5\ \frac{\text{N}}{\text{mm}}\right)(5\ \text{mm} + 8.44\ \text{mm})$$

$$= 67.2\ \text{N}$$

(e) Since the spring forces the lifter toward the cam and the inertial force resists movement toward the cam, the net force on the cam is

$$F = 223.4\ \text{N} - 67.2\ \text{N} = 156.2\ \text{N}$$

(f) From Table 57.5, the lifter velocity is

$$\text{v} = \left(\frac{L\omega}{\theta_0}\right)\left(1 - \cos\left(\frac{2\pi\theta}{\theta_0}\right)\right)$$

$$= \left((25\ \text{mm})\left(\frac{125.66\ \dfrac{\text{rad}}{\text{s}}}{2.094\ \text{rad}}\right)\right)$$

$$\times \left(1 - \cos\left(\frac{2\pi \times 0.873\ \text{rad}}{2.094\ \text{rad}}\right)\right)$$

$$= 2801\ \text{mm/s}$$

$$T = \frac{F_y \text{v}}{\omega}$$

$$= \frac{(156.2\ \text{N})\left(2801\ \dfrac{\text{mm}}{\text{s}}\right)}{\left(1000\ \dfrac{\text{mm}}{\text{m}}\right)\left(125.66\ \dfrac{\text{rad}}{\text{s}}\right)}$$

$$= 3.48\ \text{N·m}$$

PRACTICE PROBLEMS

1. A simple epicyclic gearbox with one planet has gears with 24, 40, and 104 teeth on the sun, planet, and internal ring gears, respectively. The sun rotates clockwise at 50 rpm. The ring gear is fixed. What is the rotational velocity of the planet carrier?

2. Refer to the epicyclic gear set illustrated. Gear A rotates counterclockwise on a fixed center at 100 rpm. The ring gear rotates. The planet carrier rotates clockwise at 60 rpm. Each gear has the number of teeth indicated. What is the rotational speed of the sun gear?

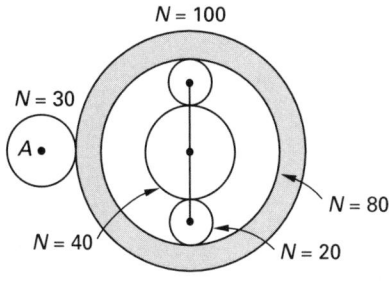

3. An epicyclic gear train consists of a ring gear, three planets, and a fixed sun gear. 15 hp (11 kW) are transmitted through the input ring gear, which turns clockwise at 1500 rpm. The diametral pitch is 10 per inch with a 20° pressure angle. The pitch diameters are at 5 in, 2½ in, and 10 in (127 mm, 63.5 mm, 254 mm) for

the sun, planets, and ring gear, respectively. (a) How many teeth are on each gear? (b) What are the speeds of the sun, ring, and carrier gears? (c) In what directions do the sun, ring, and carrier gears turn? (d) What torques are on the input and output shafts?

4. A flywheel is designed as a solid disk, 2 in (51 mm) thick, 20 in (510 mm) in diameter, and with a concentric 4 in (102 mm) diameter mounting hole. It is manufactured from cast iron with an ultimate tensile strength of 30,000 lbf/in^2 (207 MPa) and a Poisson's ratio of 0.27. The density of the cast iron is 0.26 lbm/in^3 (7200 kg/m^3). Using a factor of safety of 10, what is the maximum safe speed for the flywheel?

5. *(Time limit: one hour)* An automobile differential gear set is arranged as shown. Gear C is turned by the automobile driveshaft. The axle shafts are driven by gears A and B. What are the (a) speed and (b) direction of rotation for gear B?

gear A: 30 teeth, 50 rpm counterclockwise as viewed
             from the gear looking up the shaft
gear B: 30 teeth
gear C: 18 teeth, 600 rpm counterclockwise as
             viewed from the gear looking up the shaft
gear D: 54 teeth
gear E: 15 teeth
gear F: 15 teeth

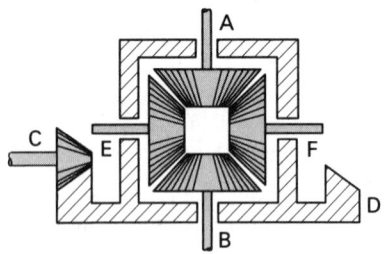

6. *(Time limit: one hour)* The epicyclic gear set shown has an overall speed reduction of 3:1. The planet carrier is the driven element. The sun gear turns clockwise at 1000 rpm. The planet has 20 teeth and a diametral pitch of 10. (a) What is the ratio of the numbers of teeth on the ring and sun gears? (b) What is the angular velocity (in rpm) of the planet with respect to its own axis?

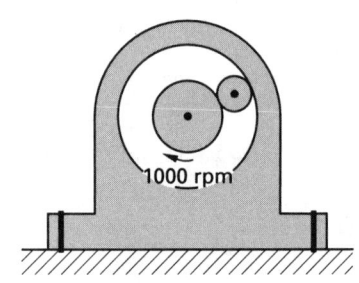

7. *(Time limit: one hour)* A gear train is constructed as shown. The output turns at 250 rpm counterclockwise. How many teeth does gear C have?

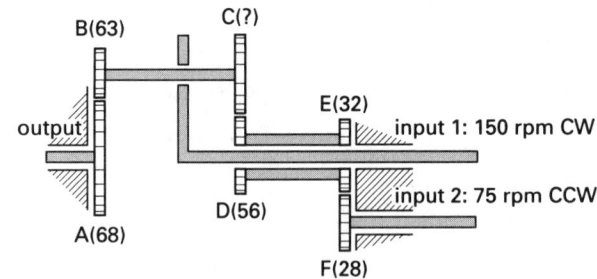

8. *(Time limit: one hour)* A cam is turning at a constant speed of 120 rpm. A radial follower rises 0.5 in (12 mm) with constant acceleration as the cam turns 60°. The follower returns to rest with constant acceleration during the next 90° of cam movement. (a) How far does the follower move during the 150° of rotation? (b) What is the magnitude of the acceleration during the first 60°? (c) What is the magnitude of the deceleration during the last 90°?

58 Vibrating Systems

1. Types of Vibrations 58-2
2. Ideal Components 58-2
3. Static Deflection 58-2
4. Free Vibration 58-2
5. Initial Conditions 58-4
6. Vertical versus Horizontal Oscillation 58-4
7. Conservation of Energy 58-5
8. Free Rotation 58-5
9. Summary of Free Vibration
       Performance Equations 58-5
10. Rayleigh's Method 58-5
11. Transformed Linear Stiffness 58-8
12. Transformed Angular Stiffness 58-8
13. Damped Free Vibrations 58-9
14. Undamped Forced Vibrations 58-10
15. Magnification Factor 58-10
16. Damped Forced Vibrations 58-11
17. Vibration Isolation and Control 58-12
18. Isolation from Active Base 58-14
19. Vibrations in Shafts 58-15
20. Second Critical Shaft Speed 58-17
21. Critical Speed of Stepped Shafts 58-17
22. Vibrations in Thin Plates 58-17
23. Balancing 58-18
24. Modal Vibrations 58-18
25. Multiple Degree of Freedom Systems . . . 58-19
26. Shock 58-19
27. Vibration and Shock Testing 58-20
28. Vibration Instrumentation 58-21
       Practice Problems 58-21

Nomenclature

a	acceleration	ft/sec²	m/s²
a	plate short side dimension	ft	m
A	amplitude	ft	m
A	amplitude	rad	rad
A	maximum acceleration amplitude	g's	g's
AR	amplitude ratio	–	–
b	plate long side dimension	ft	m
C	coefficient of viscous damping (linear)	lbf-sec/ft	N·s/m
d	diameter or distance	ft	m
D	vibration parameter	ft-lbf	N·m
e	eccentricity	ft	m
E	energy	ft-lbf	J
E	modulus of elasticity	psf	Pa
f	frequency	Hz	Hz
F	force	lbf	N
g	gravitational acceleration	ft/sec²	m/s²
g_c	gravitational constant	lbm-ft/lbf-sec²	n.a.
G	shear modulus	lbf/ft²	Pa
G	shock transmission	g's	g's
h	height	ft	m
I	area moment of inertia	–	–
I	mass moment of inertia	lbm-ft²	kg·m²
J	polar area moment of inertia	ft⁴	m⁴
k	spring constant	lbf/ft	N/m
k_r	torsional spring constant	ft-lbf/rad	N·m/ rad
K	vibration constant	–	–
L	length	ft	m
m	mass	lbm	kg
MF	magnification factor	–	–
r	radius	ft	m
r	ratio of forcing to natural frequency	–	–
r	root	sec⁻¹	s⁻¹
SG	specific gravity	–	–
t	time	sec	s
T	period	sec	s
TR	transmissibility	–	–
v	velocity	ft/sec	m/s
w	load per unit area	lbf/ft²	N/m²
w	load per unit length	lbf/ft	N/m
x	position	ft	m

Symbols

α	angular acceleration	rad/sec²	rad/s²
β	magnification factor	–	–
δ	deflection	ft	m
δ	logarithmic decrement	–	–
ζ	damping ratio	–	–
η	isolation efficiency	–	–
ν	Poisson's ratio	–	–
ϕ	angle	rad	rad
ω	natural frequency	rad/sec	rad/s

Subscripts

0	initial
b	block
c	complementary
d	damped
eq	equivalent
f	forced
l	liquid
n	nth mode
p	particular
r	rotational
st	static

PROFESSIONAL PUBLICATIONS, INC. BELMONT, CA

1. TYPES OF VIBRATIONS

Vibration is an oscillatory motion about an equilibrium point.[1] If the motion is the result of a disturbing force that is applied once and then removed, the motion is known as *natural* (or *free*) *vibration*. If a force of impulse is applied repeatedly to a system, the motion is known as *forced vibration*.

Within both of the categories of natural and forced vibrations are the subcategories of damped and undamped vibrations. If there is no damping (i.e., no friction), a system will experience free vibrations indefinitely. This is known as *free vibration* and *simple harmonic motion*.

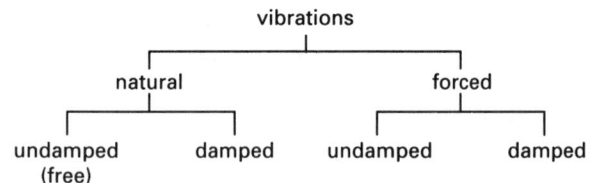

Figure 58.1 *Types of Vibrations*

The performance (behavior) of some mechanical systems can be defined by a single variable. Such systems are given the name *single degree of freedom (SDOF) systems*. For example, the position of a mass hanging from a spring is defined by the one variable $x(t)$.[2] Systems requiring two or more variables to define the positions of all parts are known as *multiple degree of freedom (MDOF) systems*.

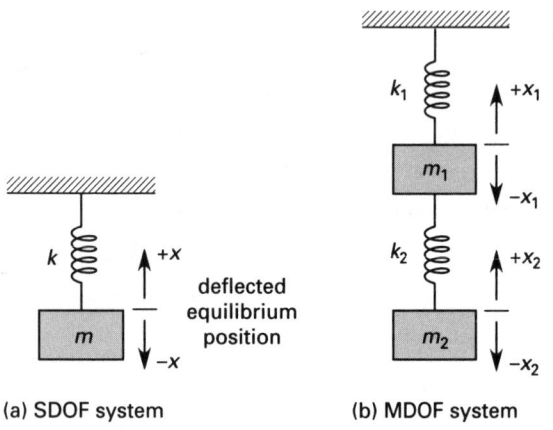

Figure 58.2 *Single and Multiple Degree of Freedom Systems*

---

[1]Although this chapter is presented in terms of mechanical vibrations, the concepts are equally applicable to electrical, fluid, and other types of systems.

[2]Although the convention is by no means universal, the variable x is commonly used as the position variable in oscillatory systems, even when the motion is in the vertical (y) direction.

2. IDEAL COMPONENTS

When used to describe components in a vibrating system, the adjectives *perfect* and *ideal* generally imply *linearity* and the absence of friction and damping. The behavior of a *linear component* can be described by a linear equation. For example, the linear equation $F = kx$ describes a linear spring; however, the quadratic equation $F = Cv^2$ describes a nonlinear dashpot. Similarly, $F = ma$ and $F = Cv$ are linear inertial and viscous forces, respectively.

3. STATIC DEFLECTION

An important concept used in calculating the behavior of a vibrating system is the *static deflection*, δ_{st}. This is the deflection of a mechanical system due to gravitational force alone.[3] (The disturbing force is not considered.) In calculating the static deflection, it is extremely important to distinguish between mass and weight. Figure 58.3 illustrates several cases of static deflection.

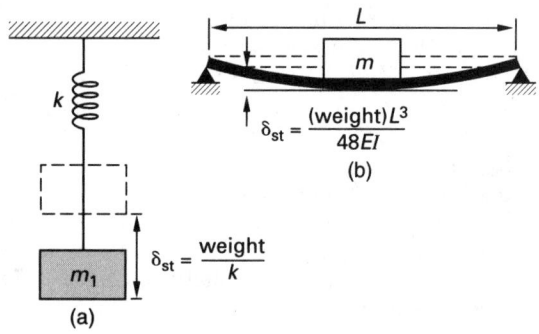

Figure 58.3 *Examples of Static Deflection*

4. FREE VIBRATION

The simple mass and ideal spring illustrated in Fig. 58.3 is an example of free vibration. After the mass is displaced and released, it will oscillate up and down. Since there is no friction (i.e., the vibration is undamped), the oscillations will continue forever.

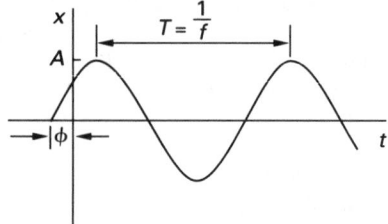

Figure 58.4 *Free Vibration*

The system is initially at rest. The mass is hanging on the spring, and the *equilibrium position* is the static deflection. The system is disturbed by a downward force (i.e., the mass is pulled downward from its static deflection and released).

---

[3]The term *deformation* is used synonymously with *deflection*.

After the initial disturbing force is removed, the object will be acted upon by the restoring force ($-kx$) and the inertial force ($-ma$). Both of these forces are proportional to the displacement from the equilibrium point, and they are opposite in sign from the displacement. From D'Alembert's principle,

$$\sum F = 0: -ma - kx = 0$$

$$m \frac{d^2x}{dt^2} = -kx \quad \text{[SI]} \quad 58.1$$

The solution to this second-order differential equation is easily derived.

$$x(t) = x_0 \cos \omega t + \left(\frac{v_0}{\omega}\right) \sin \omega t \quad 58.2$$

ω is known as the *natural frequency of vibration* or *angular frequency*. It has units of radians per second. It is not the same as the *linear frequency*, f, which has units of hertz (formerly known as cycles per second). The *period of oscillation*, T, is the reciprocal of the linear frequency.

$$\omega = \sqrt{\frac{k}{m}} \quad \text{[SI]} \quad 58.3(a)$$

$$\omega = \sqrt{\frac{kg_c}{m}} \quad \text{[U.S.]} \quad 58.3(b)$$

$$f = \frac{\omega}{2\pi} = \frac{1}{T} \quad 58.4$$

$$T = \frac{1}{f} = \frac{2\pi}{\omega} \quad 58.5$$

Equation 58.3 can be written in terms of the weight of the object suspended from the spring or the static deflection.

$$\text{weight} = mg \quad \text{[SI]} \quad 58.6(a)$$

$$\text{weight} = \frac{mg}{g_c} \quad \text{[U.S.]} \quad 58.6(b)$$

$$\omega = \sqrt{\frac{kg}{\text{weight}}} = \sqrt{\frac{g}{\delta_{\text{st}}}} \quad 58.7$$

Equation 58.7 is extremely useful. It is not limited to the simple mass-on-a-spring arrangement shown in Fig. 58.3(a). It can be used with a variety of systems, including those involving beams, shafts, and plates. Example 58.2 illustrates how this is done.

Equation 58.8 is an alternate form of the solution to Eq. 58.1. A is the *amplitude* and ϕ is the phase angle.

$$x(t) = A \cos(\omega t - \phi) \quad 58.8$$

$$A = \sqrt{(x_0)^2 + \left(\frac{v_0}{\omega}\right)^2} \quad 58.9$$

$$\phi = \arctan\left(\frac{v_0}{\omega x_0}\right) \quad 58.10$$

The position, velocity, and acceleration are all sinusoidal with time. The maximum values are

$$x_{\max} = A \quad 58.11$$

$$v_{\max} = A\omega \quad 58.12$$

$$a_{\max} = A\omega^2 \quad 58.13$$

Example 58.1

A 120 lbm (54 kg) mass is supported by three springs as shown. The initial displacement is 2.0 in (5.0 cm) downward from the equilibrium position. No external forces act on the mass after it is released. What are the maximum velocity and acceleration?

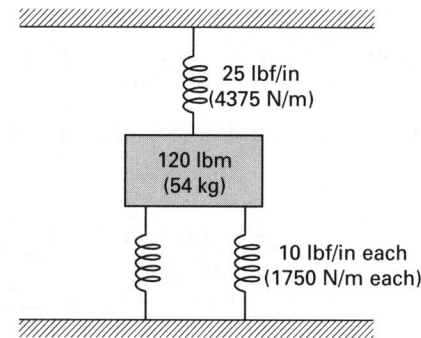

SI Solution

Since the springs are in parallel, they all share the applied load. The equivalent spring constant is

$$k_{\text{eq}} = k_1 + k_2 + k_3$$

$$= 4375 \frac{\text{N}}{\text{m}} + 1750 \frac{\text{N}}{\text{m}} + 1750 \frac{\text{N}}{\text{m}} = 7875 \text{ N/m}$$

The static deflection is

$$\delta_{\text{st}} = \frac{\text{weight}}{k} = \frac{mg}{k} = \frac{(54 \text{ kg})\left(9.81 \frac{\text{m}}{\text{s}^2}\right)}{7875 \frac{\text{N}}{\text{m}}}$$

$$= 0.0673 \text{ m}$$

The natural frequency is given by Eq. 58.7. (Compare this to the value calculated from Eq. 58.3.)

$$\omega = \sqrt{\frac{g}{\delta_{\text{st}}}} = \sqrt{\frac{9.81 \frac{\text{m}}{\text{s}^2}}{0.0673 \text{ m}}}$$

$$= 12.07 \text{ rad/s}$$

Since the mass is pulled down and released, the initial conditions are

$$v_0 = 0$$

$$x_0 = 5.0 \text{ cm} \quad (0.05 \text{ m})$$

From Eqs. 58.12 and 58.13, the maximum velocity and acceleration are

$$v_{max} = A\omega = (0.05 \text{ m}) \left(12.07 \frac{\text{rad}}{\text{s}} \right)$$

$$= 0.604 \text{ m/s}$$

$$a_{max} = A\omega^2 = (0.05 \text{ m}) \left(12.07 \frac{\text{rad}}{\text{s}} \right)^2$$

$$= 7.28 \text{ m/s}^2$$

(Radians are dimensionless.)

Customary U.S. Solution

The equivalent spring constant is

$$k_{eq} = 25 \frac{\text{lbf}}{\text{in}} + 10 \frac{\text{lbf}}{\text{in}} + 10 \frac{\text{lbf}}{\text{in}} = 45 \text{ lbf/in}$$

Referring to Fig. 58.3 and Eq. 58.6, the static deflection is

$$\delta_{st} = \frac{\text{weight}}{k} = \frac{mg}{kg_c} = \frac{(120 \text{ lbm}) \left(32.2 \frac{\text{ft}}{\text{sec}^2} \right)}{\left(45 \frac{\text{lbf}}{\text{in}} \right) \left(32.2 \frac{\text{lbm-ft}}{\text{lbf-sec}^2} \right)}$$

$$= 2.67 \text{ in}$$

The natural frequency is given by Eq. 58.7.

$$\omega = \sqrt{\frac{g}{\delta_{st}}} = \sqrt{\frac{\left(32.2 \frac{\text{ft}}{\text{sec}^2} \right) \left(12 \frac{\text{in}}{\text{ft}} \right)}{2.67 \text{ in}}}$$

$$= 12.03 \text{ rad/sec}$$

Since the mass is pulled down and released, the initial conditions are

$$v_0 = 0$$

$$x_0 = \frac{2 \text{ in}}{12 \frac{\text{in}}{\text{ft}}} = 0.167 \text{ ft}$$

From Eqs. 58.12 and 58.13, the maximum velocity and acceleration are

$$v_{max} = A\omega = (0.167 \text{ ft}) \left(12.03 \frac{\text{rad}}{\text{sec}} \right)$$

$$= 2.0 \text{ ft/sec}$$

$$a_{max} = A\omega^2 = (0.167 \text{ ft}) \left(12.03 \frac{\text{rad}}{\text{sec}} \right)^2$$

$$= 24.2 \text{ ft/sec}^2$$

(Radians are dimensionless.)

Example 58.2

A diving board is supported by a frictionless pivot at one end and by an unyielding, frictionless fulcrum, as indicated. A diver of mass m stands at the free end and bounces up and down. What is the frequency of oscillation?

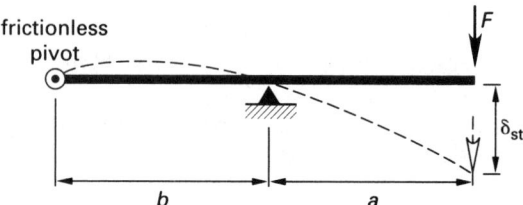

Solution

The deflection curve of the beam is shown by the dotted line. The tip force is

$$F = mg \qquad \text{[SI]}$$

$$F = \frac{mg}{g_c} \qquad \text{[U.S.]}$$

Use standard beam tables to determine the deflection. If the diver were to stand perfectly still, the static deflection at the tip would be

$$\delta_{st} = \frac{Fa^2(a + b)}{3EI}$$

From Eq. 58.7, the linear natural frequency is

$$f = \left(\frac{1}{2\pi} \right) \sqrt{\frac{g}{\delta_{st}}} = \left(\frac{1}{2\pi} \right) \sqrt{\frac{3EIg}{Fa^2(a + b)}}$$

5. INITIAL CONDITIONS

With natural, undamped vibrations, the initial conditions (i.e., initial position and velocity) do not affect the natural period of oscillation. However, the amplitude of the oscillations will be affected, as indicated by Eq. 58.9.

6. VERTICAL VERSUS HORIZONTAL OSCILLATION

As long as friction is absent, the two cases of oscillation shown in Fig. 58.5 are equivalent (i.e., will have the same frequency and amplitude). Although it may seem that there is an extra gravitational force with vertical motion, the weight of the body is completely canceled by the opposite spring force when the system is in equilibrium. Therefore, vertical oscillations about an equilibrium point are equivalent to horizontal oscillations about the unstressed point.

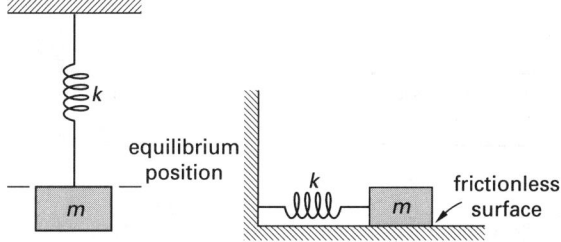

Figure 58.5 *Vertical and Horizontal Oscillations*

7. CONSERVATION OF ENERGY

The conservation of energy in vibrating systems requires the kinetic energy at the static equilibrium position to equal the stored elastic energy at the position of maximum displacement. For the mass-spring system shown in Fig. 58.5, the energy conservation equation is

$$\frac{1}{2}kx_{max}^2 = \frac{1}{2}mv_{max}^2 \qquad \text{[SI]} \qquad \textbf{58.14(a)}$$

$$\frac{1}{2}kx_{max}^2 = \frac{mv_{max}^2}{2g_c} \qquad \text{[U.S.]} \qquad \textbf{58.14(b)}$$

The velocity function is derived by taking the derivative of the position function.

$$x(t) = x_{max}\sin\omega t \qquad \textbf{58.15}$$

$$v(t) = \frac{dx(t)}{dt} = \omega x_{max}\cos\omega t \qquad \textbf{58.16}$$

Equation 58.16 shows that $v_{max} = \omega x_{max}$. Substituting this into Eq. 58.14 derives the natural frequency of vibration.

$$\omega^2 = \frac{k}{m} \qquad \text{[SI]} \qquad \textbf{58.17(a)}$$

$$\omega^2 = \frac{kg_c}{m} \qquad \text{[U.S.]} \qquad \textbf{58.17(b)}$$

8. FREE ROTATION

The so-called *torsional pendulum* in Fig. 58.6 can be analyzed in a manner similar to the spring-mass combination. Ignoring the mass and moment of inertia of the shaft, the differential equation is

$$-k_r\phi = I\,\frac{d^2\phi}{dt^2} \qquad \text{[SI]} \qquad \textbf{58.18}$$

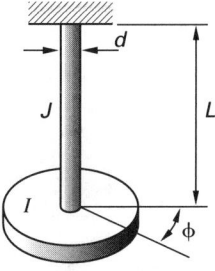

Figure 58.6 *Torsional Pendulum*

The *torsional spring constant*, k_r, used in Eq. 58.18 is

$$k_r = \frac{GJ}{L} = \frac{\pi d^4 G}{32L} \qquad \textbf{58.19}$$

If the shaft is connected to the support through a torsional spring, or if the shaft consists of several sections of different diameters (i.e., a *stepped shaft*), the equivalent torsional spring constant can be calculated in the same manner as for springs in series.

$$\frac{1}{k_{eq}} = \frac{1}{k_{r1}} + \frac{1}{k_{r2}} + \cdots \qquad \textbf{58.20}$$

The solution to Eq. 58.18 is directly analogous to the solution for the spring-mass system. Equations 58.21 through 58.27 summarize the governing equations.

$$\phi(t) = \phi_0\cos\omega t + \left(\frac{\omega_0}{\omega}\right)\sin\omega t = A\cos\left(\omega t - \phi\right) \qquad \textbf{58.21}$$

$$\omega = \sqrt{\frac{k_r}{I}} \qquad \text{[SI]} \qquad \textbf{58.22(a)}$$

$$\omega = \sqrt{\frac{k_r g_c}{I}} \qquad \text{[U.S.]} \qquad \textbf{58.22(b)}$$

$$A = \sqrt{(\phi_0)^2 + \left(\frac{\omega_0}{\omega}\right)^2} \qquad \textbf{58.23}$$

$$\phi = \arctan\left(\frac{\omega_0}{\omega\phi_0}\right) \qquad \textbf{58.24}$$

$$\phi_{max} = A \qquad \textbf{58.25}$$

$$\omega_{max} = A\omega \qquad \textbf{58.26}$$

$$\alpha_{max} = A\omega^2 \qquad \textbf{58.27}$$

9. SUMMARY OF FREE VIBRATION PERFORMANCE EQUATIONS

Most equations of motion can be easily derived for free vibrations without damping. However, Table 58.1 provides a convenient summary of several common cases.

10. RAYLEIGH'S METHOD

Usually, the mass of the spring (beam, bar, shaft, etc.) is disregarded when calculating the frequency or period of vibration of a simple system. This is done to simplify the solution, although the mass of the spring element actually does affect the frequency. The exact solution is generally too complex, but *Rayleigh's method* can be used to derive answers that will usually be less than 5% in error.

Rayleigh's method is to increase the oscillating object's mass by a fraction of the spring mass.

- For spring-mass systems, add $\frac{1}{3}$ of the spring mass to the oscillating object mass.

- For simply supported beams loaded at the center, add 17/35 of the beam mass to the carried mass.

Table 58.1 *Performance of Simple Oscillatory Systems*
(small deflections; consistent units)

mechanism	natural frequency (ω)	linear frequency (f)	period (T)
mass and spring	$\sqrt{\dfrac{k}{m}}$	$\left(\dfrac{1}{2\pi}\right)\sqrt{\dfrac{k}{m}}$	$2\pi\sqrt{\dfrac{m}{k}}$
mass on massless beam (I = area moment of inertia of cross section)	$\sqrt{\dfrac{48EI}{mL^3}}$	$\left(\dfrac{1}{2\pi}\right)\sqrt{\dfrac{48EI}{mL^3}}$	$2\pi\sqrt{\dfrac{mL^3}{48EI}}$
constrained compound pendulum	$\sqrt{\dfrac{mgL+kd^2}{mL^2}}$	$\left(\dfrac{1}{2\pi}\right)\sqrt{\dfrac{mgL+kd^2}{mL^2}}$	$2\pi\sqrt{\dfrac{mL^2}{mgL+kd^2}}$
simple pendulum	$\sqrt{\dfrac{g}{L}}$	$\left(\dfrac{1}{2\pi}\right)\sqrt{\dfrac{g}{L}}$	$2\pi\sqrt{\dfrac{L}{g}}$
compound pendulum	$\sqrt{\dfrac{mgd}{I_0}}$	$\left(\dfrac{1}{2\pi}\right)\sqrt{\dfrac{mgd}{I_0}}$	$2\pi\sqrt{\dfrac{I_0}{mgd}}$
conical pendulum	$\sqrt{\dfrac{g}{h}}$	$\left(\dfrac{1}{2\pi}\right)\sqrt{\dfrac{g}{h}}$	$2\pi\sqrt{\dfrac{h}{g}}$

Table 58.1 (continued)

mechanism	natural frequency (ω)	linear frequency (f)	period (T)
two masses and spring	$\sqrt{\dfrac{k(m_1 + m_2)}{m_1 m_2}}$	$\left(\dfrac{1}{2\pi}\right)\sqrt{\dfrac{k(m_1 + m_2)}{m_1 m_2}}$	$2\pi\sqrt{\dfrac{m_1 m_2}{k(m_1 + m_2)}}$
torsional mass and spring	$\sqrt{\dfrac{JG}{I_0 L}}$	$\left(\dfrac{1}{2\pi}\right)\sqrt{\dfrac{JG}{I_0 L}}$	$2\pi\sqrt{\dfrac{I_0 L}{JG}}$
two torsional masses	$\sqrt{\dfrac{JG(I_1 + I_2)}{I_1 I_2 L}}$	$\left(\dfrac{1}{2\pi}\right)\sqrt{\dfrac{JG(I_1 + I_2)}{I_1 I_2 L}}$	$2\pi\sqrt{\dfrac{I_1 I_2 L}{JG(I_1 + I_2)}}$
(pulley, m_1, m_2, k)	$\sqrt{\dfrac{k}{m_1 + \dfrac{m_2}{2}}}$	$\left(\dfrac{1}{2\pi}\right)\sqrt{\dfrac{k}{m_1 + \dfrac{m_2}{2}}}$	$2\pi\sqrt{\dfrac{m_1 + \dfrac{m_2}{2}}{k}}$
floating block	$\sqrt{\dfrac{g(\mathrm{SG}_l)}{L(\mathrm{SG}_b)}}$	$\left(\dfrac{1}{2\pi}\right)\sqrt{\dfrac{g(\mathrm{SG}_l)}{L(\mathrm{SG}_b)}}$	$2\pi\sqrt{\dfrac{L(\mathrm{SG}_b)}{g(\mathrm{SG}_l)}}$
U-tube and liquid	$\sqrt{\dfrac{2g}{L}}$	$\left(\dfrac{1}{2\pi}\right)\sqrt{\dfrac{2g}{L}}$	$2\pi\sqrt{\dfrac{L}{2g}}$
cantilever spring	$\sqrt{\dfrac{3EI}{mL^3}}$	$\left(\dfrac{1}{2\pi}\right)\sqrt{\dfrac{3EI}{mL^3}}$	$2\pi\sqrt{\dfrac{mL^3}{3EI}}$

- For cantilever beams loaded at the free end, add $33/140$ of the beam mass to the carried mass.

- For circular shafts in torsion, add $1/3$ of the shaft mass moment of inertia to the mass moment of inertia of the rotating load.

11. TRANSFORMED LINEAR STIFFNESS

Transformers are used in electrical and electronic circuits to convert one voltage to another. The analogous mechanical device is the lever. Figure 58.7 illustrates a simple lever-spring system. The pivot point is frictionless; the springs are ideal; the lever is infinitely stiff and has negligible mass.

The *lever ratio* is

$$\frac{F_1}{F_2} = \frac{\delta_2}{\delta_1} = \frac{L_2}{L_1} \qquad 58.28$$

The transformed mass is the mass that, if hung from spring k_1, would oscillate at the same speed as the end of the spring.[4]

$$m' = \left(\frac{L_2}{L_1}\right)^2 m \qquad 58.29$$

The total equivalent spring stiffness (referred to spring 1) is

$$k' = k_1 + \left(\frac{a}{L_1}\right) k_2 \qquad 58.30$$

The natural frequency is

$$\omega = \sqrt{\frac{k'}{m'}} \qquad 58.31$$

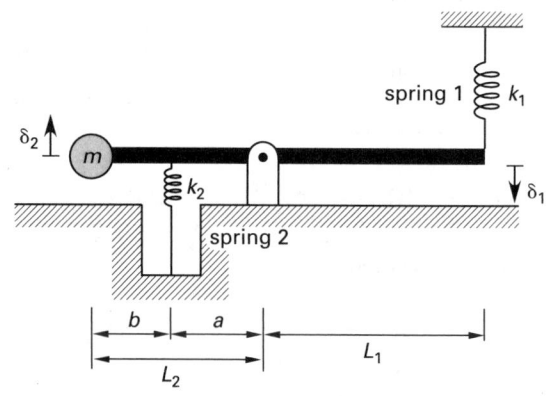

(a) original system

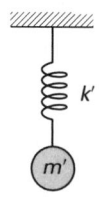

(b) equivalent system

Figure 58.7 *Lever-Coupled Linear System*

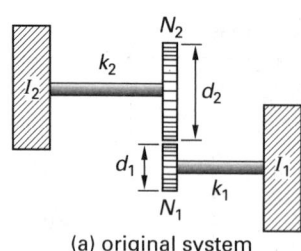

(a) original system

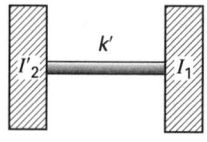

(b) equivalent system

Figure 58.8 *Gear-Coupled Torsional System*

12. TRANSFORMED ANGULAR STIFFNESS

In a meshing gear set, the transmitted force and power are the same for both meshing gears. However, a gear set changes the torque. Therefore, a meshing gear set is a torsional analog to an electrical transformer.

Consider the loaded gear set shown in Fig. 58.8. The masses of shafts 1 and 2 are insignificant compared with the torsional loads, I_1 and I_2. Also, the tooth stiffness is typically disregarded in analyzing this type of system. Power transfer across the gear set is without losses.

The original system can be converted to the equivalent simple, torsional system shown in Fig. 58.8(b). The equivalent load I_2' of shaft 2 (referred to shaft 1) is

$$I_2' = \left(\frac{\omega_2}{\omega_1}\right)^2 I_2 = \left(\frac{N_1}{N_2}\right)^2 I_2$$
$$= \left(\frac{d_1}{d_2}\right)^2 I_2 \qquad 58.32$$

The equivalent torsional stiffness is

$$k_2' = \left(\frac{N_1}{N_2}\right)^2 k_2 = \left(\frac{d_1}{d_2}\right)^2 k_2 \qquad 58.33$$

---

[4]A different equivalent system could be formed by keeping the mass the same and by multiplying all of the spring constants by $(L_1/L_2)^2$. This would have the effect of producing a system where the spring end moves at the speed of the mass.

The total equivalent torsional stiffness, k', is

$$\frac{1}{k'} = \frac{1}{k_1} + \frac{1}{k_2} \qquad 58.34$$

The natural frequency is

$$\omega = \sqrt{k'\left(\frac{I_1 + I_2'}{I_1 I_2'}\right)} \qquad 58.35$$

13. DAMPED FREE VIBRATIONS

When friction resists the oscillatory motion, the system is said to be damped. Friction can occur internally, between two surfaces, or due to motion through a liquid. (Air friction can be disregarded in most problems.) The third type of friction is known as *viscous damping*. Figure 58.9 illustrates the *dashpot* symbol used to represent a source of viscous damping.

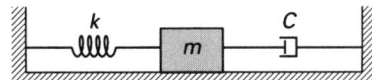

Figure 58.9 *Spring-Mass System with Dashpot*

The viscous damping force can be a function of v or v^2. If velocity is high through the liquid, the viscous damping force will be a function of v^2. Only low-velocity, *linear damping* is covered in this chapter. With linear damping, the damping force is proportional to velocity. C is known as the *coefficient of viscous damping*.

$$F = Cv = C\frac{dx}{dt} \qquad 58.36$$

The differential equation of motion is

$$m\frac{d^2x}{dt^2} = -kx - C\frac{dx}{dt} \qquad \text{[SI]} \quad 58.37$$

The general solution is given by Eq. 58.38. The constants A and B must be determined from initial conditions.

$$x(t) = Ae^{r_1 t} + Be^{r_2 t} \qquad 58.38$$

The roots of Eq. 58.38 are

$$r_1, r_2 = \frac{-C}{2m} \pm \sqrt{\left(\frac{C}{2m}\right)^2 - \frac{k}{m}} \qquad \text{[SI]} \quad 58.39(a)$$

$$r_1, r_2 = g_c\left(\frac{-C}{2m} \pm \sqrt{\left(\frac{C}{2m}\right)^2 - \frac{k}{mg_c}}\right) \quad \text{[U.S.]} \quad 58.39(b)$$

The *damping ratio* (*damping factor*), ζ, is defined as

$$\zeta = \frac{C}{2m\omega} = \frac{C}{2\sqrt{mk}} = \frac{C}{C_{\text{critical}}} \qquad \text{[SI]} \quad 58.40(a)$$

$$\zeta = \frac{Cg_c}{2m\omega} = \frac{C}{2\sqrt{\dfrac{mk}{g_c}}} = \frac{C}{C_{\text{critical}}} \quad \text{[U.S.]} \quad 58.40(b)$$

If $\zeta < 1.0$ (i.e., $C < C_{\text{critical}}$), the radical in Eq. 58.39 will be negative, and both roots will be imaginary. The oscillations are said to be *underdamped*. This case is also known as *light damping*. Motion will be oscillatory with diminishing magnitude, as illustrated in Fig. 58.10.

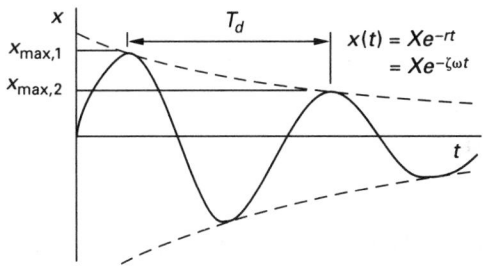

Figure 58.10 *Underdamped Oscillation*

ω_d is the *damped frequency*. It is not the same as the natural frequency, ω, which is calculated assuming $C = 0$.

$$\omega_d = \omega\sqrt{1 - \zeta^2} \qquad 58.41$$

The *logarithmic decrement*, δ, is the natural logarithm of the ratio of two successive amplitudes.

$$\delta = \ln\frac{x_n}{x_{n+1}} = \frac{2\pi\zeta}{\sqrt{1 - \zeta^2}}$$
$$= \zeta\omega T_d \qquad 58.42$$

If $\zeta > 1.0$ (i.e., $C > C_{\text{critical}}$), the radical is positive and both roots are real. The motion is said to be *overdamped*. This case is also known as *heavy damping*. There will be a gradual return to the equilibrium, but no oscillation.

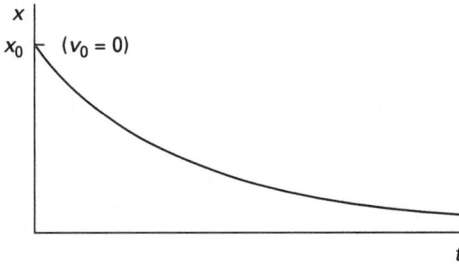

Figure 58.11 *Overdamped Oscillation*

If $\zeta = 1.0$ (i.e., $C = C_{\text{critical}}$), the radical is zero, and the motion is said to be *critically damped*. Such motion is also known as *dead-beat motion*. There is no overshoot, and the return is the fastest of the three types of motion. The *critical damping coefficient* is

$$C_{\text{critical}} = 2m\omega = 2\sqrt{km} = \frac{C}{\zeta} \qquad \text{[SI]} \quad 58.43(a)$$

$$C_{\text{critical}} = \frac{2m\omega}{g_c} = 2\sqrt{\frac{km}{g_c}} = \frac{C}{\zeta} \quad \text{[U.S.]} \quad 58.43(b)$$

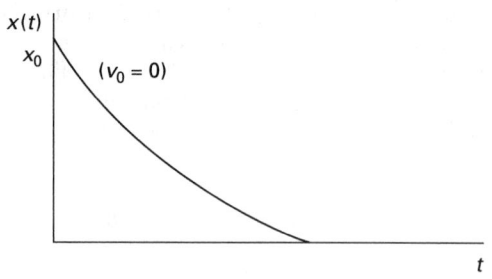

Figure 58.12 *Critically Damped Oscillation*

14. UNDAMPED FORCED VIBRATIONS

When an external disturbing force, $F(t)$, acts on the system, the system is said to be forced. Although the *forcing function* is usually considered to be periodic, it need not be (as in the case of impulse, step, and random functions).[5] However, an initial disturbance (i.e., when a mass is displaced and released to oscillate freely) is not an example of forced vibration.

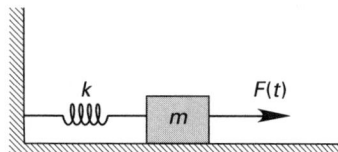

Figure 58.13 *Forced Vibrations*

Consider a sinusoidal periodic force with a *forcing frequency* ω_f and maximum value F_0.

$$F(t) = F_0 \cos \omega_f t \qquad 58.44$$

The differential equation of motion is

$$m \frac{d^2 x}{dt^2} = -kx + F_0 \cos \omega_f t \quad \text{[SI]} \qquad 58.45$$

The solution to Eq. 58.45 consists of the sum of two parts: a complementary solution and a particular solution. The *complementary solution* is obtained by setting $F_0 = 0$ (i.e., solving the homogeneous differential equation). The solution is

$$x_c(t) = A \cos \omega t + B \sin \omega t \qquad 58.46$$

The *particular solution* is found by assuming its form and substituting that function into Eq. 58.47.

$$x_p(t) = D \cos \omega_f t \qquad 58.47$$

$$D = \frac{F_0}{m(\omega^2 - \omega_f^2)} \quad \text{[SI]} \qquad 58.48$$

---

[5]The sinusoidal case is very important, since Fourier transforms can be used to model any forcing function in terms of sinusoids.

Therefore, the solution of Eq. 58.45 is

$$x(t) = A \cos \omega t + B \sin \omega t$$
$$+ \left(\frac{F_0}{m(\omega^2 - \omega_f^2)} \right) \cos \omega_f t \quad \text{[SI]} \qquad 58.49(a)$$

$$x(t) = A \cos \omega t + B \sin \omega t$$
$$+ \left(\frac{F_0 g_c}{m(\omega^2 - \omega_f^2)} \right) \cos \omega_f t \quad \text{[U.S.]} \qquad 58.49(b)$$

15. MAGNIFICATION FACTOR

The *magnification factor*, β (also known as the *amplitude ratio* and *amplification factor*), is defined as the ratio of the steady-state vibration amplitude, D, to the *pseudo-static deflection*, F_0/k.

$$\beta = \frac{D}{\dfrac{F_0}{k}} = \left| \frac{1}{1 - \left(\dfrac{\omega_f}{\omega} \right)^2} \right| \qquad 58.50$$

Figure 58.14 illustrates the magnification factor for various values of ω_f/ω. *Resonance* occurs when ω_f equals or nearly equals ω. Oscillations are theoretically infinite.[6] Resonance leads to rapid failure in structures and mechanical equipment.

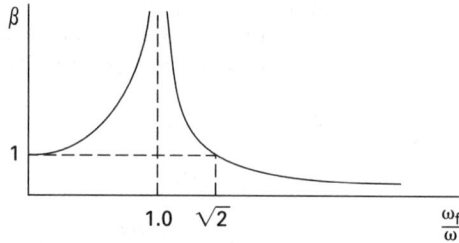

Figure 58.14 *Magnification Factor (no damping)*

If the ratio $\omega_f/\omega = \sqrt{2}$, the magnification factor is

$$\beta = \left| \frac{1}{1 - (\sqrt{2})^2} \right| = 1 \qquad 58.51$$

Therefore, ω_f/ω must be greater than $\sqrt{2}$ for the system to have an oscillation magnitude smaller than the static deflection alone.

When ω_f is significantly greater than ω, the magnification factor is close to zero, and the system will be nearly stationary.

Example 58.3

A 250 lbm (113.6 kg) motor turns at 1000 rpm (16.66 rps). It is mounted on a resilient pad having a stiffness of 3000 lbf/in (525 kN/m). Due to an unbalanced

---

[6]Damping is always present in real systems and keeps the excursions finite.

condition, a periodic force of 20 lbf (89 N) is applied in the vertical direction once each revolution. If the motor is constrained to move vertically and damping is negligible, what is the amplitude of vibration?

SI Solution

The natural frequency of the system is

$$\omega = \sqrt{\frac{k}{m}} = \sqrt{\frac{525\,000\ \dfrac{\text{N}}{\text{m}}}{113.6\ \text{kg}}}$$

$$= 68.0\ \text{rad/s}$$

The forcing frequency is

$$\omega_f = \left(16.66\ \frac{\text{rev}}{\text{s}}\right)\left(2\pi\ \frac{\text{rad}}{\text{rev}}\right) = 104.7\ \text{rad/s}$$

The pseudo-static deflection is

$$\frac{F_0}{k} = \frac{89\ \text{N}}{525\,000\ \dfrac{\text{N}}{\text{m}}} = 1.70 \times 10^{-4}\ \text{m}$$

The magnification factor is

$$\beta = \left|\frac{1}{1 - \left(\dfrac{\omega_f}{\omega}\right)^2}\right| = \left|\frac{1}{1 - \left(\dfrac{104.7\ \dfrac{\text{rad}}{\text{s}}}{68.0\ \dfrac{\text{rad}}{\text{s}}}\right)^2}\right|$$

$$= 0.730$$

The amplitude of oscillation is calculated from Eq. 58.50.

$$D = \beta\left(\frac{F_0}{k}\right) = (0.730)(1.70 \times 10^{-4}\ \text{m})$$

$$= 1.24 \times 10^{-4}\ \text{m}$$

Customary U.S. Solution

The natural frequency of the system is

$$\omega = \sqrt{\frac{kg_c}{m}} = \sqrt{\frac{\left(3000\ \dfrac{\text{lbf}}{\text{in}}\right)\left(32.2\ \dfrac{\text{lbm-ft}}{\text{lbf-sec}^2}\right)\left(12\ \dfrac{\text{in}}{\text{ft}}\right)}{250\ \text{lbm}}}$$

$$= 68.1\ \text{rad/sec}$$

The forcing frequency is

$$\omega_f = \frac{\left(1000\ \dfrac{\text{rev}}{\text{min}}\right)\left(2\pi\ \dfrac{\text{rad}}{\text{rev}}\right)}{60\ \dfrac{\text{sec}}{\text{min}}} = 104.7\ \text{rad/sec}$$

The pseudo-static deflection is

$$\frac{F_0}{k} = \frac{20\ \text{lbf}}{3000\ \dfrac{\text{lbf}}{\text{in}}} = 0.00667\ \text{in}$$

The magnification factor is

$$\beta = \left|\frac{1}{1 - \left(\dfrac{\omega_f}{\omega}\right)^2}\right| = \left|\frac{1}{1 - \left(\dfrac{104.7\ \dfrac{\text{rad}}{\text{sec}}}{68.1\ \dfrac{\text{rad}}{\text{sec}}}\right)^2}\right|$$

$$= 0.733$$

The amplitude of oscillation is calculated from Eq. 58.50.

$$D = \beta\left(\frac{F_0}{k}\right) = (0.733)(0.00667\ \text{in})$$

$$= 0.0049\ \text{in}$$

16. DAMPED FORCED VIBRATIONS

If a viscous damping force is added to a sinusoidally forced system, as in Fig. 58.15, the differential equation of motion is

$$m\,\frac{d^2x}{dt^2} = -kx - C\frac{dx}{dt} + F_0\cos\omega_f t \qquad \text{[SI]} \qquad \textbf{58.52}$$

The solution to Eq. 58.52 has several terms. As a result of the damping force, the complementary solution has decaying exponentials. Therefore, the complementary solution is also known as the *transient component* because its contribution to the system performance decreases rapidly. However, the transient terms do contribute to the initial performance. For this reason, initial cycles may experience displacements greater than the steady-state values. The particular solution is known as the *steady-state component*.

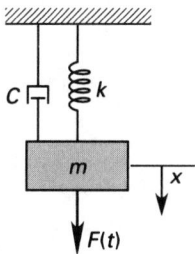

Figure 58.15 Damped Forced Oscillations

Equation 58.53 defines the magnification factor, β, for steady-state damped forced vibrations. The magnification factor for the undamped case (Eq. 58.50) can be derived by setting $\zeta = 0_1$.

$$\beta = \frac{D}{\frac{F_0}{k}} = \left| \frac{1}{\sqrt{\left[1 - \left(\frac{\omega_f}{\omega}\right)^2\right]^2 + \left[\frac{C\omega_f}{m\omega^2}\right]^2}} \right|$$

$$= \left| \frac{1}{\sqrt{\left[1 - \left(\frac{\omega_f}{\omega}\right)^2\right]^2 + \left[\frac{2C\omega_f}{C_{\text{critical}}\omega}\right]^2}} \right|$$

$$= \left| \frac{1}{\sqrt{(1 - r^2)^2 + (2\zeta r)^2}} \right| \qquad \text{[SI]} \qquad 58.53$$

$$r = \frac{\omega_f}{\omega} \qquad\qquad 58.54$$

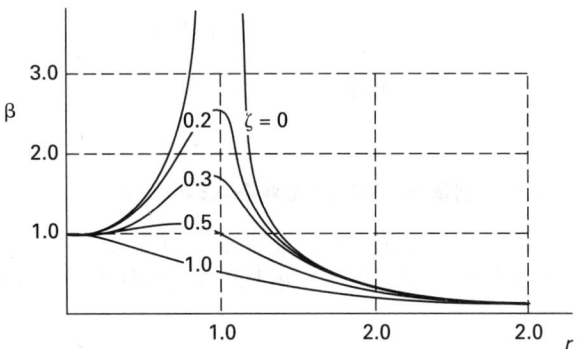

Figure 58.16 Magnification Factor

17. VIBRATION ISOLATION AND CONTROL

It is often desired to isolate a rotating machine from its surroundings, to limit the vibrations that are transmitted to the supports, and to reduce the amplitude of the machine's vibrations.

The *transmissibility* is the ratio of the transmitted force (i.e., the force transmitted to the supports) to the applied force (i.e., the force from the imbalance).

$$\text{TR} = \frac{|F_{\text{transmitted}}|}{F_{\text{applied}}}$$

$$= \beta_d \sqrt{1 + (2r\zeta)^2} \qquad\qquad 58.55$$

$$\beta_d = \frac{1}{\sqrt{(1 - r^2)^2 + (2\zeta r)^2}} \qquad\qquad 58.56$$

In some cases, the transmissibility may be reported in units of *decibels*. Unlike linear transmissibility, which is always positive, logarithmic transmissibility can be negative.

$$\text{TR}_{\text{db}} = 20 \log (\text{TR}_{\text{linear}}) \qquad\qquad 58.57$$

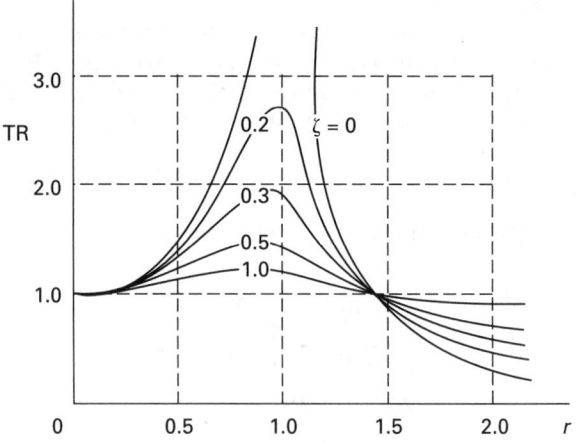

Figure 58.17 Linear Transmissibility

The magnitude of oscillations in vibrating equipment can be reduced and the equipment isolated from the surroundings by mounting on resilient pads or springs. The isolated system must have a natural frequency less than $1/\sqrt{2} = 0.707$ times the disturbing (forcing) frequency. That is, the transmissibility will be reduced below 1.0 only if $\omega_f/\omega > \sqrt{2}$. Otherwise, the attempted isolation will actually increase the transmitted force.

The natural frequency is found from Eq. 58.7, where the static deflection is calculated from the mass and stiffness of the pad or spring.

The amount of isolation is characterized by the *isolation efficiency*, also known as the *percent of isolation* and *degree of isolation*. Suggested isolation efficiencies for satisfactory operation are listed in Table 58.2.

$$\eta = 1 - \left| \frac{1}{\left(\frac{f_f}{f}\right)^2 - 1} \right| \qquad\qquad 58.58$$

Table 58.2 Typical Isolation Efficiencies

equipment	isolation efficiency
centrifugal compressor	0.98
reciprocating compressor	
0 to15 hp	0.85
15 to 150 hp	0.90
centrifugal fans	
800 rpm and higher	0.90 to 0.95
centrifugal pumps	0.95
pipe mounts	0.95

Isolation materials and isolator devices have specific deflection characteristics. If the isolation efficiency is known, it can be used to determine the type of isolator or isolation device used based on the static deflection. Table 58.3 lists typical ranges of isolation materials and devices. Table 58.4 lists typical damping factors.

A *tuned system* is one for which the natural frequency of the vibration absorber is equal to the frequency that is to be eliminated (i.e., the forcing frequency). In theory, this is easy to accomplish: the mass and spring constant of the absorber are varied until the desired natural frequency is achieved. This is known as "tuning" the system.

Table 58.3 *Typical Deflection of Isolation Materials and Devices*

approximate deflection		materials
in	mm	
$0-\frac{1}{16}$	0–2	cork, natural rubber, felt, lead/asbestos, fiberglass
$\frac{1}{16}-\frac{1}{4}$	2–6	neoprene pads, neoprene mounts, multiple layers of felt or cork
$\frac{1}{4}-1\frac{1}{2}$	6–40	steel coil springs, multiple layers of natural rubber or neoprene pads
$1\frac{1}{2}-15$	40–380	steel coil or leaf springs

(Multiply in by 25.4 to obtain mm.)

Table 58.4 *Typical Damping Factors of Isolators*

material	ζ
steel spring	0.005
natural rubber	0.05
neoprene	0.05
cork	0.06
felt	0.06
metal mesh	0.12
air damper	0.17
friction-damped spring	0.33

Example 58.4

A 4000 lbm (1800 kg) machine is rotating at 1000 rpm. The rotating component has an imbalance of 100 lbm (45 kg) acting with an eccentricity of 2 in (50 mm). The machine is already supported by isolation mounts having a combined stiffness of 30,000 lbf/in (5.3 MN/m), but vibration is still excessive. In order to reduce the amplitude of vibration, a viscous damper is connected between the machine and a rigid support. The damping ratio is 0.2. What are the (a) amplitude of oscillation and (b) transmitted force?

SI Solution

(a) The static deflection of the mounts is

$$\delta_{st} = \frac{mg}{k}$$

$$= \frac{(1800 \text{ kg}) \left(9.81 \frac{\text{m}}{\text{s}^2}\right)}{\left(5.3 \frac{\text{MN}}{\text{m}}\right) \left(10^6 \frac{\text{N}}{\text{MN}}\right)}$$

$$= 0.00333 \text{ m}$$

The natural frequency is

$$f = \left(\frac{1}{2\pi}\right) \sqrt{\frac{g}{\delta_{st}}}$$

$$= \left(\frac{1}{2\pi}\right) \sqrt{\frac{9.81 \frac{\text{m}}{\text{s}^2}}{0.00333 \text{ m}}}$$

$$= 8.638 \text{ Hz}$$

The damped frequency is

$$f_d = f\sqrt{1 - \zeta^2}$$

$$= (8.638 \text{ Hz})\sqrt{1 - (0.2)^2} = 8.463 \text{ Hz}$$

The forcing frequency is

$$f_f = \frac{1000 \frac{\text{rev}}{\text{min}}}{60 \frac{\text{s}}{\text{min}}} = 16.667 \text{ Hz}$$

The angular forcing frequency is

$$\omega_f = 2\pi f_f = 2\pi(16.667 \text{ Hz})$$

$$= 104.7 \text{ rad/sec}$$

The out-of-balance force caused by the rotating eccentric mass is

$$F_f = m\omega^2 r$$

$$= \frac{(45 \text{ kg}) \left(104.7 \frac{\text{rad}}{\text{s}}\right)^2 (50 \text{ mm})}{1000 \frac{\text{mm}}{\text{m}}}$$

$$= 24\,665 \text{ N}$$

The ratio of frequencies is

$$r = \frac{\omega_f}{\omega} = \frac{16.667 \text{ Hz}}{8.569 \text{ Hz}}$$

$$= 1.945$$

From Eq. 58.53, the magnification factor is

$$\text{MF} = \left| \frac{1}{\sqrt{(1 - r^2)^2 + (2\zeta r)^2}} \right|$$

$$= \frac{1}{\sqrt{[1 - (1.945)^2]^2 + [(2)(0.2)(1.945)]^2}}$$

$$= 0.346$$

The amplitude of oscillation is

$$x = (\text{MF})\left(\frac{F_f}{k}\right) = \frac{(0.346)(24\,665\text{ N})}{\left(5.3\,\dfrac{\text{MN}}{\text{m}}\right)\left(10^6\,\dfrac{\text{N}}{\text{MN}}\right)}$$

$$= 0.00161\text{ m}$$

(b) The transmissibility is

$$\text{TR} = (\text{MF})\sqrt{1+(2r\zeta)^2}$$
$$= (0.346)\sqrt{1+[(2)(1.945)(0.2)]^2}$$
$$= 0.438$$

The transmitted force is

$$F = (\text{TR})F_f = (0.438)(24\,665\text{ N})$$
$$= 10\,803\text{ N}$$

Customary U.S. Solution

(a) The static deflection of the mounts is

$$\delta_{\text{st}} = \frac{\text{weight}}{k} = \left(\frac{m}{k}\right)\left(\frac{g}{g_c}\right)$$

$$= \left(\frac{4000\text{ lbm}}{30{,}000\,\dfrac{\text{lbf}}{\text{in}}}\right)\left(\frac{32.2\,\dfrac{\text{ft}}{\text{sec}^2}}{32.2\,\dfrac{\text{ft-lbm}}{\text{lbf-sec}^2}}\right)$$

$$= 0.1333\text{ in}$$

The natural frequency is

$$f = \left(\frac{1}{2\pi}\right)\sqrt{\frac{g}{\delta_{\text{st}}}}$$

$$= \left(\frac{1}{2\pi}\right)\sqrt{\frac{\left(32.2\,\dfrac{\text{ft}}{\text{sec}^2}\right)\left(12\,\dfrac{\text{in}}{\text{ft}}\right)}{0.1333\text{ in}}}$$

$$= 8.569\text{ Hz}$$

The damped frequency is

$$f_d = f\sqrt{1-\zeta^2}$$
$$= (8.569\text{ Hz})\sqrt{1-(0.2)^2} = 8.396\text{ Hz}$$

The forcing frequency is

$$f_f = \frac{1000\,\dfrac{\text{rev}}{\text{min}}}{60\,\dfrac{\text{sec}}{\text{min}}} = 16.667\text{ Hz}$$

The angular forcing frequency is

$$\omega_f = 2\pi f_f = 2\pi(16.667\text{ Hz})$$
$$= 104.7\text{ rad/sec}$$

The out-of-balance force caused by the rotating eccentric mass is

$$F_f = \frac{m\omega^2 r}{g_c}$$

$$= \frac{(100\text{ lbm})\left(104.7\,\dfrac{\text{rad}}{\text{sec}}\right)^2(2\text{ in})}{\left(32.2\,\dfrac{\text{ft-lbm}}{\text{lbf-sec}^2}\right)\left(12\,\dfrac{\text{in}}{\text{ft}}\right)}$$

$$= 5674\text{ lbf}$$

The ratio of frequencies is

$$r = \frac{\omega_f}{\omega} = \frac{16.667\text{ Hz}}{8.569\text{ Hz}} = 1.945$$

From Eq. 58.53, the magnification factor is

$$\text{MF} = \left|\frac{1}{\sqrt{(1-r^2)^2+(2\zeta r)^2}}\right|$$

$$= \frac{1}{\sqrt{[1-(1.945)^2]^2+[(2)(0.2)(1.945)]^2}}$$

$$= 0.346$$

The amplitude of oscillation is

$$x = (\text{MF})\left(\frac{F_f}{k}\right) = \frac{(0.346)(5674\text{ lbf})}{30{,}000\,\dfrac{\text{lbf}}{\text{in}}}$$

$$= 0.0654\text{ in}$$

(b) The transmissibility is

$$\text{TR} = (\text{MF})\sqrt{1+(2r\zeta)^2}$$
$$= (0.346)\sqrt{1+[(2)(1.945)(0.2)]^2}$$
$$= 0.438$$

The transmitted force is

$$F = (\text{TR})F_f = (0.438)(5674\text{ lbf})$$
$$= 2485\text{ lbf}$$

18. ISOLATION FROM ACTIVE BASE

In some cases, a machine is to be isolated from an active base. The base (floor, supports, etc.) vibrates, and the magnitude of the vibration seen by the machine is to be limited or reduced. This case is not fundamentally different from the case of a vibrating machine being isolated from a stationary base.

The concept of transmissibility is replaced by the amplitude ratio (magnification factor or amplification factor). This is the ratio of the transmitted displacement (deflection, excursion, motion, etc.) to the applied displacement. That is, it is the ratio of the maximum mass motion to the maximum base motion. The amplification ratio is numerically identical to the transmissibility calculated in Eq. 58.55.

$$
\begin{aligned}
\text{AR} &= \frac{\delta_{\text{dynamic}}}{\delta_{\text{static}}} \\
&= \frac{\delta_{\text{dynamic}}}{\dfrac{F}{k}} = \text{TR}
\end{aligned}
\qquad 58.59
$$

Example 58.5

The suspension system of a car is modeled as a perfect spring and dashpot connecting a massless wheel and a supported mass, m, as shown. The car enters a bumpy area with forward velocity v and initial mass position y_0. The profile of the road surface is modeled as a perfect sinusoid with a peak-to-peak distance of L. The sinusoid is described mathematically by the equation $y(x) = y_{\text{max}} \sin 2\pi(x - x_0)/L$.

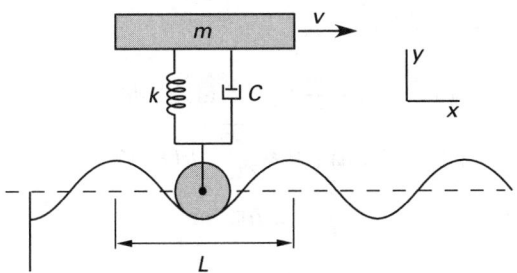

What are the (a) forcing frequency, (b) undamped natural frequency of the system, and (c) horizontal velocity at resonance? (d) For a particular speed, how would the steady-state amplitude of the mass be determined?

Solution

(a) Since the horizontal car velocity is v, the frequency of the forcing function will be

$$
f_f = \frac{\text{v}}{L}
$$

(b) The undamped natural frequency is found from Eq. 58.3.

$$
f = \left(\frac{1}{2\pi}\right)\sqrt{\frac{k}{m}}
$$

(c) Resonance occurs when the forcing frequency equals the natural frequency.

$$
\begin{aligned}
f_f &= f \\
\frac{\text{v}}{L} &= \left(\frac{1}{2\pi}\right)\sqrt{\frac{k}{m}} \\
\text{v} &= \left(\frac{L}{2\pi}\right)\sqrt{\frac{k}{m}}
\end{aligned}
$$

(d) The damping ratio is

$$
\zeta = \frac{C}{C_{\text{critical}}}
$$

The amplitude ratio can be found from either Fig. 58.17 or Eq. 58.53. The steady-state amplitude will be $(\text{AR})y_{\text{max}}$.

19. VIBRATIONS IN SHAFTS

A shaft's natural frequency of vibration is referred to as the *critical speed*. This is the rotational speed in revolutions per second that just equals the lateral natural frequency of vibration. Therefore, vibration in shafts is basically an extension of lateral vibrations (e.g., whipping "up and down") in beams. Rotation is disregarded, and the shaft is considered only from the standpoint of lateral vibrations.

General practice is to keep the operating speed well below the first critical speed. For shafts with distributed or multiple loadings, it may be important to know the second critical speed. However, higher critical speeds are usually well out of the range of operation.

For shafts with constant cross-sectional areas and simple loading configurations, the deflection can be found from beam formulas. Shafts with single antifriction (i.e., ball and roller) bearings at each end can be considered to be simply supported, while shafts with sleeve bearings or two side-by-side antifriction bearings at each shaft end can be considered to have fixed built-in supports.[7]

Once the deflection is known, Eq. 58.7 can be used to find the critical speed.

A shaft carrying no load other than its own weight can be considered as a uniformly loaded beam. The maximum deflection at midspan can be found from beam tables.

The classical analysis method of a shaft carrying single or multiple inertial loads (flywheel, pulley, etc.) assumes that the shaft itself is weightless.[8]

Equation 58.60 gives the critical speed (i.e., the fundamental frequency of vibration) of a shaft carrying multiple masses in terms of the static deflection at the mass.[9] In theory, all that is necessary is to stop the rotation and measure the deflection (from the horizontal) at each mass. The assumption that the static and rotating deflection curves are identical is not exactly true.

---

[7]Sleeve bearings are assumed to be fixed supports, not because they have the mechanical strength to prevent binding, but because sleeve bearings cannot operate and would not be operating with an angled shaft.

[8]The mass of the shaft can be included with Rayleigh's method and Dunkerley's approximation.

[9]This equation is sometimes known as the *Rayleigh equation* or the *Rayleigh-Ritz equation*, as it is based on the Rayleigh method of equating the maximum kinetic energy to the maximum potential energy.

However, the speed error is less than approximately 5%, generally on the high side.

$$f = \left(\frac{1}{2\pi}\right)\sqrt{\frac{g\sum m_i\delta_{st,i}}{\sum m_i\delta_{st,i}^2}} \qquad 58.60$$

When there are only two rotating masses on the shaft, the total deflections can be calculated by superposition. However, Eq. 58.60 is difficult to use when there are more than two masses, as the number of deflection calculations is the square of the number of masses. In that case, the *Dunkerley approximation* is used.[10] In Eq. 58.61, the f_i are the natural frequencies of vibration when mass m_i alone is on the shaft. The Dunkerley equation generally underestimates the critical speed.

$$\left(\frac{1}{f}\right)^2 = \sum\left(\frac{1}{f_i}\right)^2 \qquad 58.61$$

If a shaft's critical speed is unsuitable, Eq. 58.62 can be used to calculate the approximate diameter of a shaft that will be acceptable.

$$d_{new} = d_{old}\sqrt{\frac{f_{new}}{f_{old}}} \qquad 58.62$$

Typical values used in the analysis of steel shafts are: density (ρ), 0.28 lbm/in³ (7750 kg/m³); modulus of elasticity (E), 2.9×10^7 lbf/in² (200 GPa); gravitational constant (g_c), 386.4 in-lbm/lbf-sec².

Example 58.6

A 30 lbm (14 kg) flywheel is supported on an overhung steel shaft as shown. The shaft diameter is 1.0 in (25 mm), and the shaft mass is negligible. What is the critical speed of the shaft?

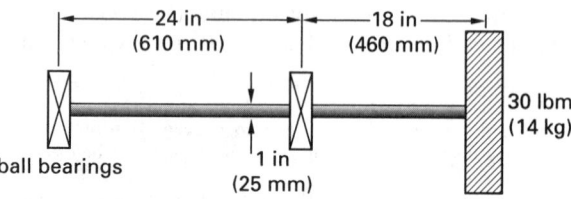

24 in (610 mm) — 18 in (460 mm)

30 lbm (14 kg)

ball bearings   1 in (25 mm)

---

[10]The spelling "Dunkerly" is also found in the literature.

SI Solution

The radius of the shaft is

$$r = \frac{d}{2} = \frac{25\text{ mm}}{2} = 12.5\text{ mm}$$

The moment of inertia of the circular cross section is

$$I = \frac{\pi r^4}{4} = \frac{\pi\left(\dfrac{12.5\text{ mm}}{1000\,\frac{\text{mm}}{\text{m}}}\right)^4}{4}$$
$$= 1.917\times10^{-8}\text{ m}^4$$

The ball bearings prevent vertical but not angular deflection. The deflection at the flywheel is

$$\delta_{st} = \frac{Fa^2(a+b)}{3EI}$$
$$= \frac{mga^2(a+b)}{3EI}$$
$$= \frac{(14\text{ kg})\left(9.81\,\frac{\text{m}}{\text{s}^2}\right)(0.46\text{ m})^2(0.46\text{ m}+0.61\text{ m})}{(3)(200\text{ GPa})\left(10^9\,\frac{\text{Pa}}{\text{GPa}}\right)(1.917\times10^{-8}\text{ m}^4)}$$
$$= 2.703\times10^{-3}\text{ m} \quad (2.703\text{ mm})$$

From Eq. 58.7, the natural frequency is

$$f = \left(\frac{1}{2\pi}\right)\sqrt{\frac{g}{\delta_{st}}}$$
$$= \left(\frac{1}{2\pi}\right)\sqrt{\frac{9.81\,\frac{\text{m}}{\text{s}^2}}{2.703\times10^{-3}\text{ m}}}$$
$$= 9.59\text{ Hz}$$

Customary U.S. Solution

The radius of the shaft is

$$r = \frac{d}{2} = \frac{1\text{ in}}{2} = 0.5\text{ in}$$

The moment of inertia of the circular cross section is

$$I = \frac{\pi r^4}{4} = \frac{\pi(0.5\text{ in})^4}{4}$$
$$= 0.0491\text{ in}^4$$

The ball bearings prevent vertical but not angular deflection. The deflection at the flywheel is

$$\delta_{st} = \frac{Fa^2(a+b)}{3EI}$$

$$= \left(\frac{ma^2(a+b)}{3EI}\right)\left(\frac{g}{g_c}\right)$$

$$= \left(\frac{(30 \text{ lbm})(18 \text{ in})^2(18 \text{ in} + 24 \text{ in})}{(3)\left(2.9 \times 10^7 \frac{\text{lbf}}{\text{in}^2}\right)(0.0491 \text{ in}^4)}\right)$$

$$\times \left(\frac{32.2 \frac{\text{ft}}{\text{sec}^2}}{32.2 \frac{\text{ft-lbm}}{\text{lbf-sec}^2}}\right)$$

$$= 0.0956 \text{ in}$$

From Eq. 58.7, the natural frequency is

$$f = \left(\frac{1}{2\pi}\right)\sqrt{\frac{g}{\delta_{st}}}$$

$$= \left(\frac{1}{2\pi}\right)\sqrt{\frac{\left(32.2 \frac{\text{ft}}{\text{sec}^2}\right)\left(12 \frac{\text{in}}{\text{ft}}\right)}{0.0956 \text{ in}}}$$

$$= 10.1 \text{ Hz}$$

20. SECOND CRITICAL SHAFT SPEED

The second critical speed for a simply supported shaft uniformly loaded along its length is four times the fundamental critical speed.

The second critical speed for a massless shaft carrying two concentrated masses, m_1 and m_2, can be derived from Eq. 58.63. The positive roots ω_1 and ω_2 in this bi-quadratic equation are the first and second natural frequencies (critical speeds), respectively. The a_{ij} constants are the deflections at the location of mass i due to a unit mass at the location of mass j. Only three deflection calculations are needed because $a_{12} = a_{21}$.

$$\frac{1}{\omega^4} - \left(\frac{1}{\omega^2}\right)(a_{11}m_1 + a_{22}m_2)$$

$$+ (a_{11}a_{22} - a_{12}a_{21})m_1 m_2 = 0 \qquad \textit{58.63}$$

21. CRITICAL SPEED OF STEPPED SHAFTS

The Dunkerley equation is one of the few simple methods for evaluating the critical speed of a stepped shaft. The mass of each shaft section is assumed to be concentrated at the midpoints of their respective lengths. The "shaft" itself is considered to be massless. This rather crude approximation provides surprisingly good results. As a further simplification, the shaft section with the smallest diameter can be disregarded.

22. VIBRATIONS IN THIN PLATES

The natural frequency of thin plates and diaphragms depends on the shape (e.g., circular, square, or rectangular) and the method of mounting (e.g., fixed or free edges). Completely fixed edges are difficult to achieve in practice, so the formulas used to calculate the natural frequencies are derived from a blend of heuristic and theoretical methods.

Equation 58.64 is an approximate equation based on a modification of Eq. 58.7. As an approximation, this equation can be used with any uniformly loaded round, square, rectangular, elliptical, or triangular plates with any edge conditions.[11] The maximum error is small, generally less than 3%. δ_{st} is the maximum static deflection produced by the self-mass of the plate and any uniformly distributed mass attached to the plate and vibrating with it.

$$f = \left(\frac{1.277}{2\pi}\right)\sqrt{\frac{g}{\delta_{st}}} \qquad \textit{58.64}$$

If certain assumptions are made, the fundamental natural frequency of a plate can be derived. These assumptions are that the plate material is elastic, homogeneous, and of uniform thickness. Also, the plate thickness and deflections are assumed to be small in comparison to the size of the plate. w is the uniform load per unit area, including the self-weight per unit area (calculated as the specific weight times the thickness). In Eq. 58.66, a is the shorter edge distance. Values of K are given in Table 58.5.

$$f = \left(\frac{K}{2\pi}\right)\sqrt{\frac{Dg}{wr^4}} \qquad \text{[circular plates]} \qquad \textit{58.65}$$

$$f = \left(\frac{K}{2\pi}\right)\sqrt{\frac{Dg}{wa^4}} \qquad \text{[rectangular plates]} \qquad \textit{58.66}$$

$$D = \frac{Et^3}{(12)(1-\nu^2)} \qquad \textit{58.67}$$

Table 58.5 Vibration Coefficients for Plates (fundamental mode)

case	K					
circular plate						
fixed edges	10.2					
simply supported edges	4.99					
free edges[a]	5.25					
rectangular, $a \times b$ sides	$a/b = 1.0$	0.8	0.6	0.4	0.2	0.0
fixed edges	36.0	29.9	25.9	23.6	22.6	22.4
simply supported edges	19.7	16.2	13.4	11.5	10.3	9.87

[a]Supported at inner surface; edges free to flutter.

---

[11]The deflection calculated must correspond to the actual edge conditions.

23. BALANCING

To be "balanced," a rotating component must be statically and dynamically in balance. For two masses m_i rotating at radii r_i and mounted 180° out of phase, the requirement for *static balance* is given by Eq. 58.68. Static balancing is usually sufficient for rotating disks, thin wheels, and gears.

$$\sum m_i r_i = 0 \quad \text{[static balance]} \qquad \textbf{58.68}$$

Dynamic balance requires that there also be no unbalanced couples. When all of the masses are in the same plane of rotation, static balance and dynamic balance will be achieved simultaneously. When rotation is not in the same plane, all of the couples must also balance. In general, two balancing masses are required to provide static and dynamic balance where there is an unbalanced couple.

$$\sum m_i r_i x_i \quad \text{[dynamic balance]} \qquad \textbf{58.69}$$

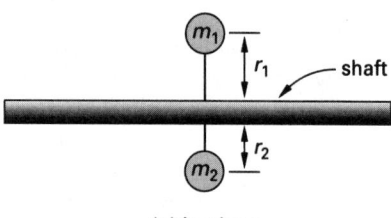

(a) in plane

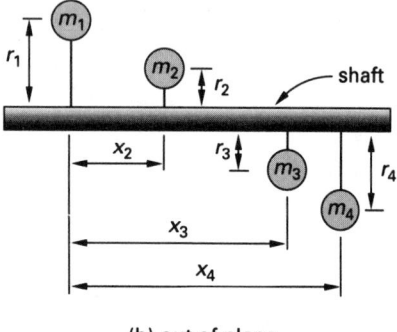

(b) out of plane

Figure 58.18 *Static and Dynamic Balance*

Example 58.7

A 2 ft (610 mm) diameter steel disk is mounted with a 0.33 ft (100 mm) eccentricity on a shaft. The mass of the disk is 570 lbm (260 kg). The assembly is balanced by two counterweights located 1.5 ft (460 mm) and 3.0 ft (910 mm) from the disk, respectively, each mounted at a radius of 1.0 ft (300 mm) from the shaft's centerline. What are the masses of the two counterweights?

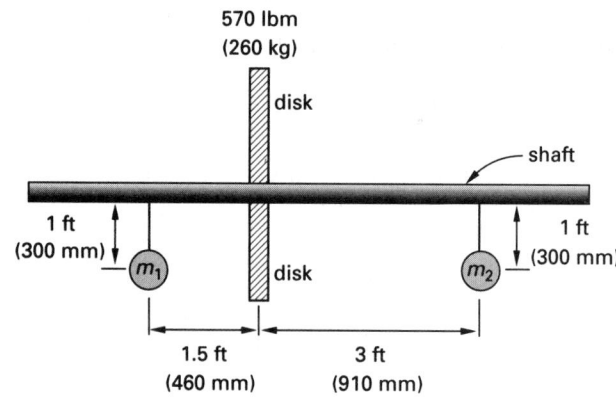

SI Solution

Take moments about the position of the left-hand counterweight. The couple produced by the eccentric disk mass must be balanced by the right counterweight. Use Eq. 58.69.

$$m_{\text{disk}} x_{\text{disk}} e = m_{\text{right}} x_{\text{right}} r_{\text{right}}$$
$$(260 \text{ kg})(460 \text{ mm})(100 \text{ mm}) = (m_{\text{right}})(460 \text{ mm}$$
$$+ 910 \text{ mm})(300 \text{ mm})$$
$$m_{\text{right}} = 29.1 \text{ kg}$$

Use Eq. 58.68 to satisfy the criterion for static balance.

$$m_{\text{left}} r_{\text{left}} + m_{\text{right}} r_{\text{right}} - m_{\text{disk}} e = 0$$
$$m_{\text{left}}(300 \text{ mm}) + (29.1 \text{ kg})(300 \text{ mm})$$
$$- (260 \text{ kg})(100 \text{ mm}) = 0$$
$$m_{\text{left}} = 57.6 \text{ kg}$$

Customary U.S. Solution

Take moments about the position of the left-hand counterweight. The couple produced by the eccentric disk mass must be balanced by the right counterweight. Use Eq. 58.69.

$$m_{\text{disk}} x_{\text{disk}} e = m_{\text{right}} x_{\text{right}} r_{\text{right}}$$
$$(570 \text{ lbm})(1.5 \text{ ft})(0.33 \text{ ft}) = (m_{\text{right}})(1.5 \text{ ft} + 3 \text{ ft})(1 \text{ ft})$$
$$m_{\text{right}} = 62.7 \text{ lbm}$$

Use Eq. 58.68 to satisfy the criterion for static balance.

$$m_{\text{left}} r_{\text{left}} + m_{\text{right}} r_{\text{right}} - m_{\text{disk}} e = 0$$
$$m_{\text{left}}(1 \text{ ft}) + (62.7 \text{ lbm})(1 \text{ ft})$$
$$- (570 \text{ lbm})(0.33 \text{ ft}) = 0$$
$$m_{\text{left}} = 125.4 \text{ lbm}$$

24. MODAL VIBRATIONS

In addition to the fundamental (first) frequency calculated up to this point, there can be higher-order frequencies (*harmonics* or higher *modes*). Generally, if a system is protected against resonance at its fundamental frequency, it will be protected against resonance at the even higher harmonic frequencies.

When a beam, shaft, or plate vibrates laterally at its fundamental frequency, all of it will be on one side of the equilibrium position at any given moment. This is not true with higher modes. There will be one or more positions (i.e., *nodes*) where the deflection curve will pass through the equilibrium position. Except for the simplest cases, the location of the nodes and the *modal shape* are found in handbooks.

The nth harmonic frequency for beams and shafts with uniformly distributed masses on simple supports is n^2 times the fundamental frequency. If both ends are fixed, then the higher modal frequencies are very nearly n^2 times the fundamental frequency, but not exactly. Cantilever beams with distributed masses are not handled so easily. Beams with lumped masses do not have higher modes at all.

Equation 58.70 calculates the nth natural frequency for a cantilever beam with uniformly distributed load w (including its own weight) per unit length over its entire length, L. Table 58.6 gives the values of the vibration constant, K_n, and the locations of the nodes.

$$f_n = \left(\frac{K_n}{2\pi}\right) \sqrt{\frac{EIg}{wL^4}} \qquad 58.70$$

Table 58.6 Vibration Constants for Cantilever Beams[a]

mode	K_n	position of nodes (x/L, from fixed end)			
1	3.52	0.0			
2	22.0	0.0	0.783		
3	61.7	0.0	0.504	0.868	
4	121	0.0	0.358	0.644	0.905
5	200	0.0	0.279	0.500	0.723	0.926

[a] Any uniform cross section; uniformly distributed load w (including own weight) per unit length; length, L.

25. MULTIPLE DEGREE OF FREEDOM SYSTEMS

A system is designed as a *multiple degree of freedom (MDOF) system* when it takes two or more independent variables to define the position of independent parts of the system.

MDOF systems may oscillate at a single natural frequency (e.g., when all of the masses are moving in the same direction), or different parts of the system may oscillate independently at their own frequencies. The *amplitude ratio* is the ratio of the maximum deflection (excursion, movement, etc.) of one part of the system to the maximum deflection of another.

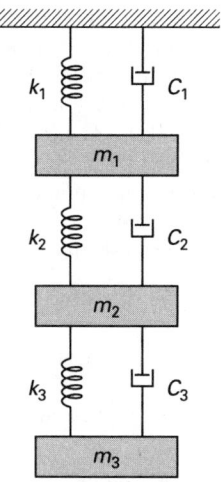

Figure 58.19 *System with Three Degrees of Freedom*

Closed-form solutions for various simple systems can be calculated, but real-world situations are more difficult. For example, consider a circuit board mounted on a bulkhead (plate or diaphragm). The circuit board has its own vibration characteristics; the bulkhead also has its own. When they are assembled together, a third vibration characteristic is produced. The circuit board and the bulkhead are said to be "coupled" because each affects the other.

Even when it is difficult to determine the actual performance characteristics of an MDOF system, damage from resonance can still be prevented by proper design. The modes are "uncoupled" as much as possible. The *octave rule* is a simple rule of thumb for uncoupling the modes. The octave rule requires the uncoupled natural frequency of each added component to be at least twice the natural frequency of the element to which it is attached. In other words, the natural frequency should be doubled every time an additional degree of freedom is added to the system.

When space, weight, or other constraints make it impossible to design modal interaction out of the system, testing and other (usually heuristic or graphical) methods must be used to determine the maximum dynamic forces on the assembly.

26. SHOCK

Vibration is a steady-state, regular phenomenon. *Shock* is a transient phenomenon. Shock results in a sharp, nearly sudden change in velocity. A *shock pulse* (*shock impulse*) is a disturbing force characterized by a rise and subsequent decay of acceleration in a very short period of time. A shock pulse is described by its peak amplitude (usually in gravities), its duration (in milliseconds), and an overall shape (triangular, rectangular, half-sine, etc.). Shock pulses, particularly those that are complex, are often depicted graphically as an acceleration-time curve.

The change in velocity, Δv, can be found as area under an acceleration-time curve or as the area under a force-time curve divided by the mass. For complex curves, the area can be found by integration or by breaking it into simpler sections. Equations 58.71 through 58.76 give the change in velocity for several regular shocks of maximum acceleration amplitude A_0 and duration t_0.

$$\Delta v = \sqrt{2gh} \quad \text{[inelastic drop from height } h\text{]} \qquad 58.71$$

$$\Delta v = 2\sqrt{2gh} \quad \text{[elastic drop from height } h\text{]} \qquad 58.72$$

$$\Delta v = \frac{2A_0 g t_0}{\pi} \quad \text{[half-sine acceleration]} \qquad 58.73$$

$$\Delta v = A_0 g t_0 \quad \text{[rectangular acceleration]} \qquad 58.74$$

$$\Delta v = \frac{A_0 g t_0}{2} \quad \text{[triangular acceleration]} \qquad 58.75$$

$$\Delta v = \frac{A_0 g t_0}{2} \quad \text{[versed sine acceleration]} \qquad 58.76$$

The *shock transmission (transmitted shock)*, G, is an acceleration parameter (with units of gravities) that depends on the natural frequency, f, and the change in velocity, Δv, due to the shock.

$$G = \frac{2\pi f \Delta v}{g} = \frac{\omega \Delta v}{g} \qquad 58.77$$

The *dynamic deflection* of a linear isolator that experiences a shock pulse is

$$\delta = \frac{\Delta v}{2\pi f} = \frac{\Delta v}{\omega} \qquad 58.78$$

Isolating a system against shocks is very different than isolating the system against vibration. Isolators must be capable of absorbing shock energy instantly. The energy may be dissipated in an inelastic isolator (e.g., crush insulation), or the energy may be released at the damped natural frequency of the system later in the cycle.

Example 58.8

A sensitive piece of electronic equipment is mounted on an isolation system with a natural frequency of 15 Hz. The equipment and mount are subjected to a standard 15 g, 11 msec half-sine shock test. What are the (a) maximum shock transmission and (b) isolation deflection?

SI Solution

(a) From Eq. 58.73, the change in velocity is

$$\Delta v = \frac{2A_0 g t_0}{\pi}$$

$$= \frac{(2)(15 \text{ g}) \left(9.81 \dfrac{\text{m}}{\text{s}^2 \cdot \text{g}} \right) (11 \text{ ms})}{\pi \left(1000 \dfrac{\text{ms}}{\text{s}} \right)}$$

$$= 1.03 \text{ m/s}$$

From Eq. 58.77, the shock transmission is

$$G = \frac{2\pi f \Delta v}{g}$$

$$= \frac{(2\pi)(15 \text{ Hz}) \left(1.03 \dfrac{\text{m}}{\text{s}} \right)}{9.81 \dfrac{\text{m}}{\text{s}^2 \cdot \text{g}}}$$

$$= 9.9 \text{ g's} \quad (9.9 \text{ gravities})$$

(b) Use Eq. 58.78 to find the dynamic linear deflection.

$$\delta = \frac{\Delta v}{2\pi f}$$

$$= \frac{1.03 \dfrac{\text{m}}{\text{s}}}{(2\pi)(15 \text{ Hz})}$$

$$= 0.0109 \text{ m}$$

Customary U.S. Solution

(a) From Eq. 58.73, the change in velocity is

$$\Delta v = \frac{2A_0 g t_0}{\pi}$$

$$= \frac{(2)(15 \text{ g}) \left(32.2 \dfrac{\text{ft}}{\text{sec}^2 \cdot \text{g}} \right) \left(12 \dfrac{\text{in}}{\text{ft}} \right) (11 \text{ msec})}{\pi \left(1000 \dfrac{\text{msec}}{\text{sec}} \right)}$$

$$= 40.59 \text{ in/sec}$$

From Eq. 58.77, the shock transmission is

$$G = \frac{2\pi f \Delta v}{g}$$

$$= \frac{(2\pi)(15 \text{ Hz}) \left(40.59 \dfrac{\text{in}}{\text{sec}} \right)}{\left(32.2 \dfrac{\text{ft}}{\text{sec}^2 \cdot \text{g}} \right) \left(12 \dfrac{\text{in}}{\text{ft}} \right)}$$

$$= 9.9 \text{ g's} \quad (9.9 \text{ gravities})$$

(b) Use Eq. 58.78 to find the dynamic linear deflection.

$$\delta = \frac{\Delta v}{2\pi f}$$

$$= \frac{40.59 \dfrac{\text{in}}{\text{sec}}}{(2\pi)(15 \text{ Hz})}$$

$$= 0.431 \text{ in}$$

27. VIBRATION AND SHOCK TESTING

There are two basic types of vibration testing: constant- and random-frequency tests. Constant-frequency *sinusoidal tests* (also known as *harmonic tests*) were the earliest types of tests used. They detect, one at a time, the resonant frequencies of an item. Random tests excite all of the resonant frequencies simultaneously and duplicate the buffeting that equipment will typically experience.

Most vibration tests are performed on a *shaker table* (*exciter*). Tables can be electrodynamic, hydraulic, or mechanical. Electrodynamic shakers produce the highest frequencies, but they are limited to shaking smaller pieces of equipment with low forces and displacements. Hydraulic shakers (also known as *hydrashakers*) use hydraulic fluid to drive a piston. Test frequencies are limited to approximately 500 Hz. Forces can be quite large. Mechanical shakers use eccentric cams to produce the traditional sinusoidal test. They cannot produce random vibrations. Mechanical shakers are limited to approximately 55 Hz.

Since there are many types of shock, there are many types of shock tests used. Hammers, spring-loaded rigs, air guns, and drops into sand pits are in use. Shipping containers are often subjected to a *drop test*, where the equipment is mounted on a table and then dropped onto a concrete floor. The *swing test* is similar—the test specimen swings into a concrete wall. A common test for aircraft-mounted equipment is a *machine drop test* producing a half-sine shock of 15 g's lasting 11 msec. (The nature of the shock is controlled by the resilience of the contact surface.) Missile components subjected to explosive impulses (e.g., explosive separation of stages) may be tested with a 6 msec, 100 g shock.

28. VIBRATION INSTRUMENTATION

There are two main types of *vibration sensors*: proximity and casing devices. *Proximity displacement transducers* (*proximity probes*) do not touch the vibrating item. They sense vibrations by establishing an electric/magnetic field between the probe tip and the object. Changes in the field produced by the minute movement of the conductive surface are detected by the instrument.

Casing transducers that touch or are mounted on vibrating equipment come in two varieties: accelerometers and velocity transducers. The output of an *accelerometer* is proportional to acceleration. Accelerometers generally use a piezoelectric crystal with an internally mounted reference mass. The voltage produced by the crystal varies as the attached mass is vibrated. Accelerometers can also be constructed as strain gages mounted on small flexible members. A *velocity transducer* is an electromagnetic device with a coil and core (or magnet), one of which is stationary and the other that is moving. The voltage produced is proportional to the relative velocity of the core through the coil. The terms *vibration pick-up* and *vibrometer* refer to devices that produce velocity-dependent voltages.

PRACTICE PROBLEMS

1. A 2 in (50 mm) steel shaft 40 in (1020 mm) long is supported on frictionless bearings at its two ends. The shaft carries a 100 lbm disk (45 kg) 15 in (380 mm) from the left bearing, and a 75 lbm (34 kg) disk 25 in (640 mm) from the left bearing. The shaft weight is negligible. There is no damping. What is the critical speed of the shaft?

2. A 300 lbm (140 kg) electromagnet at the end of a cable holds 200 lbm (90 kg) of scrap metal. The total equivalent stiffness of the cable and crane boom is 1000 lbf/in (175 kN/m). The current to the electromagnet is cut off suddenly, and the scrap falls away. Neglect damping. (a) What is the frequency of oscillation of the electromagnet? (b) What will be the minimum cable tension?

3. An 800 lbm (360 kg), single-cylinder vertical compressor operates at 1200 rpm. The compressor is mounted to the floor on four identical, equally loaded springs at its corners. It is desired to reduce the maximum unbalanced force from 25 lbf to 3 lbf (110 N to 13 N). Damping is negligible. What will be the new maximum oscillation?

4. A uniform bar with a mass of 5 lbm (2.3 kg) carries a concentrated mass of 3 lbm (1.4 kg) at its free end. The bar is hinged at one end and supported by an outboard spring as shown. The deflection of the spring from its unstressed position is 0.55 in (1.4 cm). There is no damping. What is the natural frequency of vibration?

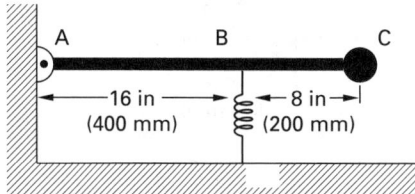

5. When an 8 lbm (3.6 kg) mass is attached on the end of a spring, the spring stretches 5.9 in (150 mm). A dashpot with a damping coefficient of 0.50 lbf-sec/ft (7.3 N·s/m) opposes movement of the mass. A forcing function of $4\cos 2t$ lbf ($18\cos 2t$ N) is applied to the mass. The mass is initially at rest. What are the (a) natural frequency, (b) damping ratio, and (c) maximum excursion? (d) What is the response of the system?

6. A 50 lbm (23 kg) motor is supported by four identical, equally loaded springs, each with a spring constant of 1000 lbf/in (175 kN/m). When the motor is turning at 800 rpm, the rotor imbalance is equivalent to a 1 oz (30 g) mass located 5 in (130 mm) from the shaft's longitudinal axis. The damping factor is $1/8$ of critical damping. What is the maximum vertical vibration amplitude?

7. (*Time limit: one hour*) A 175 lbm (80 kg), single-cylinder air compressor is mounted on four identical, equally loaded corner springs. The motor turns at 1200 rpm. During each cycle, a disturbing force is generated by a 3.6 lbm (1.6 kg) imbalance acting at a radius of 3 in (75 mm). Damping is insignificant. (a) What individual spring stiffness is required so that only 5% of the dynamic force is transmitted to the base? (b) What is the amplitude of vibration?

8. *(Time limit: one hour)* Eight horizontal high-strength steel plates are supported on rigid rollers and support a 20,000 lbm (9100 kg) load, as shown. Each of the plates is 40 in × 30 in × $\frac{1}{2}$ in (1020 mm × 760 mm × 12 mm). The modulus of elasticity of the steel is 2.9 × 10^7 lbf/in^2 (200 GPa). The masses of the plates and rollers are insignificant compared to the supported load. The yield point of the steel is not exceeded. (a) What is the total vertical static deflection of the load? (b) What is the maximum stress in the plates? (c) What is the natural frequency of oscillation in the vertical direction?

9. A centrifugal fan has 8 driving blades and 64 fan blades. The fan turns at 600 rpm and is driven by a 1725 rpm, 60 Hz, 4 pole motor. The fan and motor pulley are $11\frac{1}{2}$ in (290 mm) and 4 in (100 mm) in diameter, respectively. The drive belt has a total length of 72 in (1.83 m). What frequencies of sound and vibration are produced?

10. When not running, a hydraulic oil pump compresses a cork mounting pad 0.02 in (0.5 mm). The pump is turned at 1725 rpm. What is the transmissibility of the pad?

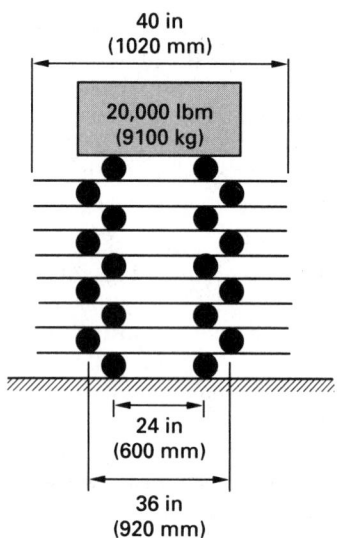

59 Modeling of Engineering Systems

1. Introduction to Engineering Systems
    Modeling 59-1
2. Elements 59-2
3. Energy Sources and Forcing Functions . . . 59-2
4. Through- and Across-Variables 59-3
5. System Diagrams 59-3
6. System Equations 59-3
7. Mechanical Systems 59-4
8. Mechanical Energy Transformations 59-5
9. Rotational Systems 59-5
10. Fluid Systems 59-7
    Practice Problems 59-8

Nomenclature

a	acceleration	ft/sec^2	m/s^2
a	ratio of transformation	–	–
A	area	ft^2	m^2
C	damping coefficient	lbf-sec/ft	N·s/m
C	capacitance	F	F
C	capacitance (fluid)	ft^5/lbf	m^5/N
C_r	rotational damping coefficient	ft-lbf-sec/rad	N·m·s/rad
f	forcing function	various	various
F	force	lbf	N
g_c	gravitational constant	lbm-ft/ lbf-sec^2	n.a.
h	depth	ft	m
I	current	A	A
I	inertance (fluid)	lbf/ft^4	N/m^4
I	mass moment of inertia	lbm-ft^2	kg·m^2
k	spring constant	lbf/ft	N/m
k_r	rotational spring constant	ft-lbf/rad	N·m/rad
l	length	ft	m
L	inductance	H	H
m	mass	lbm	kg
n	number of teeth	–	–
p	pressure	lbf/ft^2	N/m^2
Q	flow rate	ft^3/sec	m^3/s
r	response function	various	various
R	resistance	Ω	Ω
t	time	sec	s
T	torque	ft-lbf	N·m
v	velocity	ft/sec	m/s
V	voltage	V	V
x	position	ft	m

Symbols

α	angular acceleration	rad/sec^2	rad/s^2
γ	specific weight	lbf/ft^3	N/m^3
θ	angular position	rad	rad
ω	angular velocity	rad/sec	rad/s

Subscripts

m	mass
C	dashpot
f	fluid
k	spring
r	rotational

1. INTRODUCTION TO ENGINEERING SYSTEMS MODELING

The ultimate benefit derived from a model is the ability to predict the behavior (known as the *response*) of a real-world system. Some models (scale models, working models, mock-ups, etc.) are physical, but others (such as the ones in this chapter) are mathematical. The goal of modeling is to develop a differential equation or other mathematical function, known as the *response function*, that predicts how the system will behave (i.e., what position, acceleration, or velocity it will have). The response function is commonly transformed into the Laplace s-variable domain. (See Sec. 60.3.)

It is rarely possible to write the response function by observation, and several methods of developing the response function are available. This chapter takes the *two-port black box* approach—"two-port" because there are two pairs of inputs to the model, and "black box" because the inner workings of the model are irrelevant once the response function is known.[1]

Figure 59.1 Two-Port Black Box Model

Many types of real-world systems (e.g., long-term weather prediction) are too complex to be modeled mathematically. Others lend themselves to special types of modeling theory, such as those in Chap. 61. This chapter is limited to mechanical and fluid systems that can be modeled by *idealized (linear) elements* (components, devices, etc.) (see Sec. 45.2).[2] In some cases,

---

[1]It is important to recognize that the mathematical response function is the model. Drawing system diagrams and taking other steps to derive the response function are merely developmental aids.
[2]Almost any linear flow process can be modeled in this manner. Fluid flow and heat transfer systems are other common applications.

nonlinear elements can be considered linear in limited operation ranges. Elements such as mechanical springs and electrical resistors that absorb and dissipate energy are known as *passive elements*. Energy sources are *active elements*.

The response function for a model is derived from a *system equation*, which is usually a differential equation. The *order of the system (model)* is the highest order derivative in the system equation.

It is not necessary to work separate problems to find position, velocity, and acceleration response functions for a system. If the position function $x(t)$ is known, it can be differentiated to give $v(t)$ and $a(t)$. If any one of the three response functions is known, the other two can be derived.

A *single degree of freedom* (SDOF) *system* can be completely defined by one response variable. A single mass on a spring, a swinging pendulum, and a rotating pulley are examples of SDOF systems. In each case, one variable (x or θ) defines the position of the major system element.

Systems in which multiple components have their own values of the dependent variable are known as *multiple degree of freedom* (MDOF) *systems*. The degree of freedom of the system is the number of unrelated dependent variables needed to specify the behavior of all major system elements.

2. ELEMENTS

Each physical device in the system is an *element*. All of the ideal elements are considered to be two-port devices. Springs and dashpots have two ends, each of which can have a different value of the response variable. For example, the velocity of both ends of a shock absorber need not be the same. Even though masses do not have ends in the traditional sense, they are considered to be two-port devices as well.

A dependent variable that describes the performance of an element is a *response variable* (also known as a *state variable*). Each element in the model will have its own response variable, such as position, velocity, and acceleration in mechanical systems and voltage in electrical systems. Time is the independent variable.

3. ENERGY SOURCES AND FORCING FUNCTIONS

Some systems start with and gradually use up stored energy; other systems receive energy on a one-time basis. Still others receive energy on a continuous basis. The equation describing the amount of energy introduced as a function of time is the *forcing function*. Although a wide variety of forcing functions are possible, engineering systems easily become too complicated for manual solutions unless limited to simple types.

An energy source need not be an actual physical component such as a wound spring, battery, or fuel cell. Anything that produces motion in the system, including potential energy or an applied force, can be an energy source.

In modeling, the source or method of energy generation may not be known. For example, a velocity source may produce a specific velocity regardless of the system mass, or a current source may produce a specific current regardless of circuit elements. How this is accomplished need not be explicitly known.

The *homogeneous forcing function* is the zero function (i.e., no energy at all).[3] The homogeneous case does not preclude an initial disturbance or a previous amount of potential energy. However, the forcing function is homogeneous if it ceases to act as soon as the system begins to move. For example, a spring-mass system that is displaced, released, and allowed to oscillate freely experiences a zero force after the release.

A *unit step* has zero magnitude up to a particular instant (say t_1) and a magnitude of 1 thereafter. It can be multiplied by a scalar if the magnitude of the actual forcing function has any other value.

$$F(t) = \begin{cases} 0 & t < t_1 \\ 1 & t \geq t_1 \end{cases} \qquad 59.1$$

Hitting a bell with a hammer is an example of an impulse. A *unit impulse* (also known as a *pulse*) is a limited-duration force whose total impulse (that is, $F\Delta t$) is 1. The unit pulse can be multiplied by a scalar if the actual pulse has an impulse different than 1. Usually the time, Δt, is extremely short.

$$F(t) = \begin{cases} 0 & t < t_1 \\ \dfrac{1}{\Delta t} & t_1 \leq t < t_1 + \Delta t \\ 0 & t \geq t + \Delta t \end{cases} \qquad 59.2$$

Some forces increase with time. The slope of a *unit ramp* forcing function is 1. If a forcing function changes at any other rate, the unit ramp can be multiplied by a scalar.

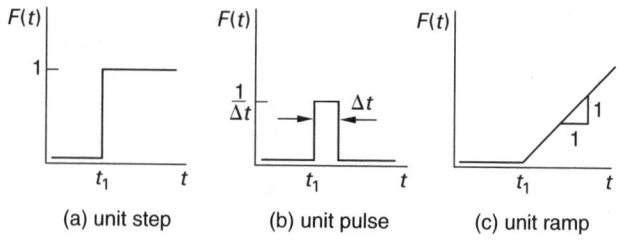

| (a) unit step | (b) unit pulse | (c) unit ramp |

Figure 59.2 *Common Forcing Function Profiles*

---

[3]This is consistent with homogeneous differential equations defined in Sec. 10.1.

The unit step, ramp, impulse, parabola, and so on, are known as *singularity functions* because of the singularity that exists at the point of discontinuity in each function. The effects of integrating and differentiating these functions are listed in Table 59.1.

Table 59.1 *Operations on Forcing Functions*

function	function when differentiated	function when integrated
unit impulse	–	unit step
unit step	unit impulse	unit ramp
unit ramp	unit step	unit parabola
unit parabola	unit ramp	(third degree)
unit exponential	unit exponential	unit exponential
unit sinusoid	unit sinusoid	unit sinusoid

The most common forcing functions used in the analysis of engineering systems are sinusoids. When combined with Fourier series analysis, sinusoids can be used to approximate all other forcing functions.

4. THROUGH- AND ACROSS-VARIABLES

Through-variables have different values at the two ends of an element; *across-variables* have the same value. For example, force is a through-variable since it is passed through objects. Consider a mass hanging on a spring. The gravitational force on the mass is passed through the spring to the support.

The velocity of a mass, however, is measured with respect to an inertial (stationary) frame of reference. Since the "ground" is connected to a mass (see Rule 59.1, Sec. 5), velocity is the across-variable.

5. SYSTEM DIAGRAMS

The system elements are interconnected to produce a system diagram with the following procedure. The diagram is then used to derive the system equation. System diagrams for mechanical systems do not always resemble the topology of the systems they represent. Example 59.1(a) illustrates a system of two elements (a mass and dashpot) connected in series that actually has a parallel system diagram.

step 1: Decide on a dependent response variable. Velocity (v or ω) is preferred in mechanical systems, but position (x or θ) and acceleration (a or α) can be used if desired.

step 2: Identify all parts of the system that have different values of the response variable. It is not necessary to know the actual values, only to recognize where the variable changes. One of the values will always be zero (corresponding to zero velocity) and is known as the *ground level*.

step 3: Start the system diagram by drawing a horizontal line for each different value of the dependent variable identified in step 2.

step 4: Insert and connect the passive elements (masses, springs, etc.) to the appropriate horizontal lines.

Rule 59.1: One end of a mass always connects to the lowest (ground) level.

step 5: Insert and connect the active (energy) sources to the appropriate horizontal lines.

Rule 59.2: One end of an energy source always connects to the lowest (ground) level.

A *line diagram* is a variation of the system diagram in which the levels associated with different values of the response variable (the horizontal lines from step 3) are replaced with nodes and the element symbols are replaced with their values.

6. SYSTEM EQUATIONS

The *system equation* is derived from a system or line diagram. The system equation is a differential equation containing the dependent variable, but it does not explicitly give the response variable. Traditional methods and Laplace transforms (both in Chap. 10) can be used to derive the response variable.

The principle used to obtain a system equation from a system diagram is a conservation law analogous to Kirchhoff's current law that says "...what goes in must come out...."[4] For mechanical systems, force is the conserved quantity. Specifically, the total force supplied by the source equals the forces leaving through all parallel branches (known as *legs*) in the system diagram. Use of the following two rules is illustrated in Ex. 59.1.

Rule 59.3: The force passing through a leg consisting of elements in series can be determined from the conditions across any of the elements in that leg.

Corollary to Rule 59.3: The force passing through a leg consisting of elements in series can be determined from the conditions across the first element in that leg.

Rule 59.4: When writing a difference in response variable values (e.g., $v_2 - v_1$ or $x_2 - x_1$), the first subscript is the same as the node number for which the equation is being written.

---

[4]Electrical current is a through-variable. Kirchhoff's current law applies to any through-variable.

7. MECHANICAL SYSTEMS

A mechanical system consists of interconnected lumped masses, linear springs, linear dashpots, and energy sources. The response variable for a mechanical system is usually the position of one of the masses, although velocity and acceleration can also be chosen.

A *lumped mass* is a rigid body that acts like a particle. All parts of the mass experience identical displacements, velocities, and accelerations. An *ideal lumped mass* is one for which Newtonian (i.e., nonrelativistic) physics applies. The governing equation for a lumped mass is Newton's second law.[5]

$$F = ma = mx'' \qquad \text{[ideal lumped mass]} \qquad \text{[SI]} \qquad \textbf{59.3(a)}$$

$$F = \frac{ma}{g_c} = \frac{mx''}{g_c} \qquad \text{[ideal lumped mass]} \qquad \text{[U.S.]} \qquad \textbf{59.3(b)}$$

An ideal spring is massless, has a constant stiffness (k), and is immune to set, creep, and fatigue. Hooke's law predicts its performance but is rewritten to explicitly include the positions of both spring ends. (If both ends are displaced the same distance, there will be no change in extension, compression, or force.)

$$F = k(x_2 - x_1) \qquad \text{[ideal spring]} \qquad \textbf{59.4}$$

A *dashpot (damper)*, of which an automobile shock absorber is an example, is a device that opposes motion and dissipates energy. An ideal linear dashpot is massless and applies a force opposing motion that is proportional to velocity. Since both ends of the dashpot move, the governing equation is

$$F = C(v_2 - v_1)$$
$$= C(x_2' - x_1') \qquad \text{[ideal dashpot]} \qquad \textbf{59.5}$$

Example 59.1

For the three systems shown, draw the system diagram, draw the line diagram, and write the system equation. Use consistent units.

(a) A mass is connected through a damper to a solid wall as shown. The mass slides without friction on its support and is acted upon by a force.

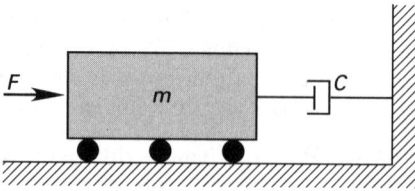

(b) A force is applied through a damper to a mass. The mass rolls on frictionless bearings.

---

[5]It is understood that force, F, and acceleration, a, are functions of time.

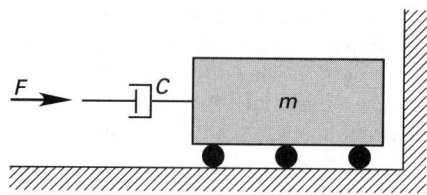

(c) A shock absorber with an integral coil spring is acted upon by a force applied to one end.

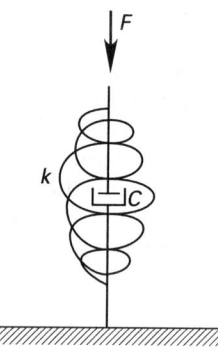

Solution

(a) Choose velocity as the response variable. All parts of the mass move with the same velocity: Call it v_1. The plunger also moves with velocity v_1 since it is attached to the mass. The body of the damper is attached to the stationary wall. Call this velocity $v = 0$.

Two horizontal lines are drawn: the top line for v_1 and the bottom line for $v = 0$. One end of the damper travels at v_1; the other end travels at $v = 0$. Therefore, connect the dashpot to these lines. The mass moves at v_1, so one of its lines connects to v_1. By Rule 59.1 (Sec. 5), the other end connects to $v = 0$. The force contacts the mass, so one end of the force connects to v_1. By Rule 59.2, the other end connects to $v = 0$.

The system and line diagrams are

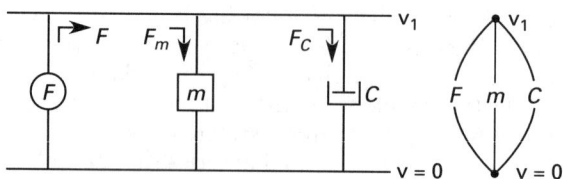

The force leaving the "source" splits—some of it going through the mass and some of it going through the dashpot. The conservation law is written to conserve the force in the v_1 line.

$$F = F_m + F_C$$

Using Rule 59.4 and expanding with Eqs. 59.3 and 59.5,

$$F = ma_1 + C(v_1 - 0)$$

The differential equation is

$$F = mx_1'' + Cx_1'$$

(b) There are three velocities in this example: the velocity of the plunger (call this v_1), the velocity of the dashpot body and mass (call this v_2), and $v = 0$. The system line diagrams are

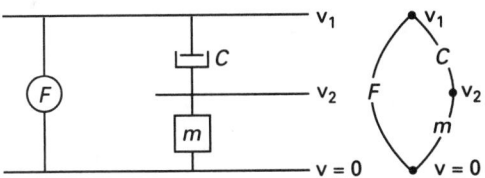

The system equation is written to conserve force in the v_1 line. The force passing through the dashpot is the same force experienced by the mass and is not additive. (This is the essence of Rule 59.3.) Rule 59.4 is used to write the dashpot force as $C(v_1 - v_2)$.

$$F = F_C$$
$$F = C(v_1 - v_2) = C(x_1' - x_2')$$

Since the same force is experienced by both the dashpot and mass, the system equation could be written in terms of the governing equation of the mass. Whether or not this is a better choice will depend on the initial conditions and other information that is available.

$$F = F_m = ma_2 = mx_2''$$

(c) There are only two velocities here. The system and line diagrams are

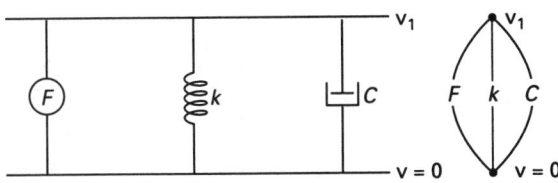

Force in the v_1 line is conserved. The system equation is

$$F = F_C + F_k$$
$$= Cv_1 + kx_1 = Cx_1' + kx_1$$

8. MECHANICAL ENERGY TRANSFORMATIONS

Levers transform one force into another at the expense of the distance the ends travel. The ratio of transformation depends on the lengths of the lever on both sides of the fulcrum. Whether the ratio is less than or greater than one is a matter of preference as long as the transformed force and velocity are correct. The ratio will be negative because the direction of the force and velocity is reversed by a lever. The transformation is represented in system diagrams by the symbol for an electrical transformer.

Example 59.2

What are the system equations for the system shown?

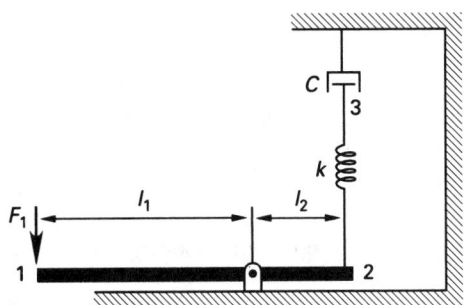

Solution

The lever transforms the force and displacement at point 1 into force and displacement at point 2. The system diagram is

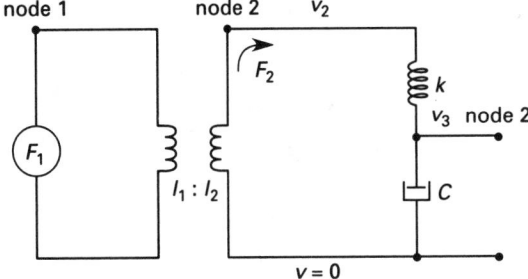

One of the system equations is based on node 2.

$$F_2 = k(x_2 - x_3)$$

The following additional equations are based on the lever transformation ratio. The minus signs indicate that the displacement and force at opposite ends of the lever are in opposite directions.

$$x_2 = \left(-\frac{l_2}{l_1}\right) x_1$$

$$F_2 = \left(-\frac{l_1}{l_2}\right) F_1$$

9. ROTATIONAL SYSTEMS

Rotational systems are directly analogous to mechanical systems. Flywheels have rotational mass moments of inertia (I), torsion springs have torsional stiffness (k_r), and fluid couplings have rotational viscous damping (C_r). The governing equations for these elements are

$$T = I\alpha \quad \text{[ideal flywheel]} \quad \text{[SI]} \qquad \textbf{59.6(a)}$$

$$T = \frac{I\alpha}{g_c} \quad \text{[ideal flywheel]} \quad \text{[U.S.]} \qquad \textbf{59.6(b)}$$

$$T = k_r(\theta_2 - \theta_1) \quad \text{[ideal torsion spring]} \quad 59.7$$

$$T = C_r(\omega_2 - \omega_1) \quad \text{[ideal fluid coupling]} \quad 59.8$$

Angular velocity (ω in rad/s) is usually chosen as the response variable, although angular position (θ in radians) or acceleration (α in rad/s^2) can be used if desired. As with the translational mechanical systems, any one of these three response variables can be used to write the system equation, after which the others can be found by integration or differentiation.

Energy can be provided to rotational systems by constant-torque or constant-velocity sources. Gear sets transform torque and velocity. The ratio of transformation (a) is the ratio of numbers of teeth or diameters. The ratio is negative because each set of gears reverses the direction of rotation. Whether or not the ratio n_1/n_2 or n_2/n_1 is used is not important as long as the transformed dependent variable is correct. When a gear set increases the rotational speed, the torque decreases, and vice versa.

$$a = -\frac{T_1}{T_2} = -\frac{\omega_2}{\omega_1} = \frac{n_1}{n_2} \quad 59.9$$

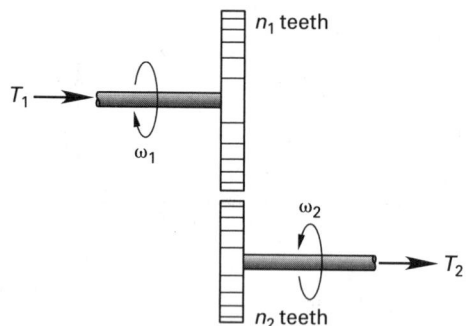

Figure 59.3 *Rotational Transformer*

Example 59.3

Two flywheels are connected by a flexible shaft. The second flywheel is acted upon by a linear viscous force. Draw the system diagram and write the system equation.

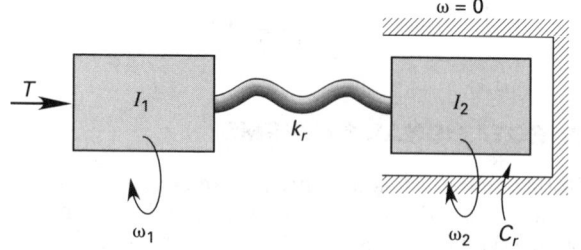

Solution

Choose angular velocity as the response variable. There are three different angular velocities—two for the flywheels and one for the stationary reference. The system diagram is

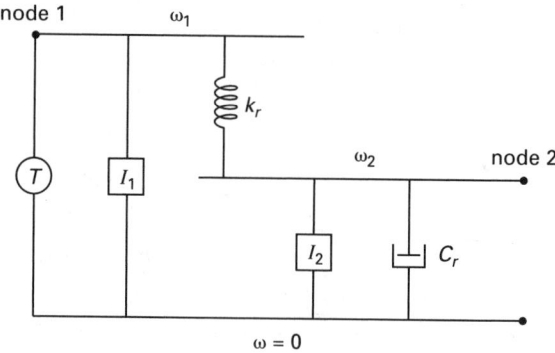

The system equations are

$$\text{node 1:} \quad T = I_1\alpha_1 + k_r(\theta_1 - \theta_2)$$
$$= I_1\theta_1'' + k_r(\theta_1 - \theta_2)$$
$$\text{node 2:} \quad 0 = C_r\omega_2 + I_2\alpha_2 + k_r(\theta_2 - \theta_1)$$
$$= C_r\theta_2' + I_2\theta_2'' + k_r(\theta_2 - \theta_1)$$

Example 59.4

A motor drives a flywheel through a set of reduction gears. The flywheel is connected to the driven gear by a flexible shaft. All other gears and shafts have infinite stiffness. What is the system equation?

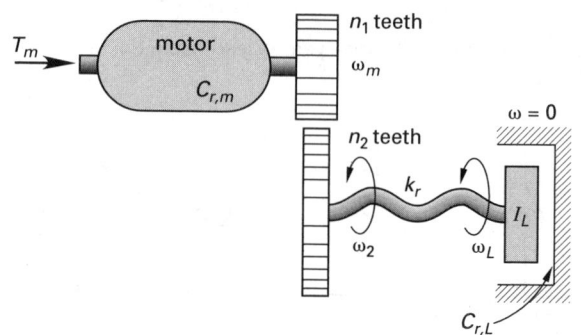

Solution

Angular velocity, ω, is chosen as the response variable. The gear set is represented by the symbol for an electrical transformer. The ratio of transformation is less than 1. This means that the motor's torque will be decreased while the rotational speed is increased. The system equations are

$$\text{node } m: \quad T_m = I_m\alpha_m + C_{r,m}\omega_m + aT_2$$
$$\text{node 2:} \quad T_2 = k_r(\theta_2 - \theta_L)$$
$$\text{node } L: \quad 0 = C_{r,L}\omega_L + I_L\alpha_L + k_r(\theta_L - \theta_2)$$

Since ω_m, ω_2, ω_L, and T_2 are all unknown, a fourth equation is needed.

$$a\omega_m = -\omega_2$$

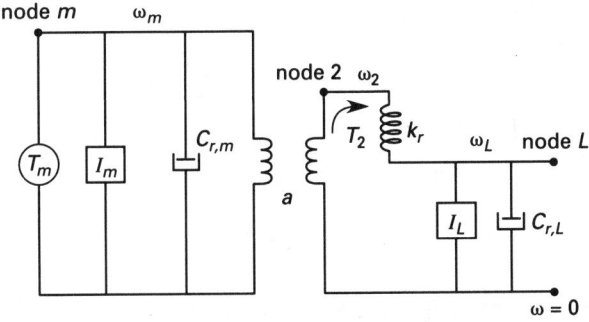

10. FLUID SYSTEMS

Figure 59.4 illustrates an open fluid *reservoir* with a constant vertical cross section. The flow rate out of the reservoir can be calculated from the cross-sectional area and the rate of change in the surface elevation. In Eq. 59.10, C_f is known as the *fluid capacitance*. In system diagrams, open reservoirs always connect to the lowest (ground) level.

$$Q = A\left(\frac{dh}{dt}\right) = \frac{A\left(\frac{dp}{dt}\right)}{\gamma}$$

$$= C_f\left(\frac{dp}{dt}\right) \qquad 59.10$$

$$C_f = \frac{A}{\gamma} \qquad 59.11$$

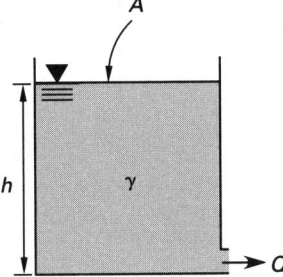

Figure 59.4 *Fluid Reservoir*

The Darcy equation indicates that *flow resistance* is proportional to the square of the flow quantity. As a simplification over a narrow range of flows, the flow resistance in simple systems analysis problems is assumed to be proportional to the flow quantity. R_f is the *fluid resistance coefficient* and is often found by experimentation.

$$p_2 - p_1 = R_f Q \qquad 59.12$$

Newton's second law (or the impulse-momentum principle) is the basis for defining the *fluid inertance (fluid inductance)*, I, which accounts for the inertia of the fluid flow.

$$F = m\left(\frac{dv}{dt}\right) \qquad 59.13$$

$$A(p_2 - p_1) = (\gamma A l)\left(\frac{\frac{dQ}{dt}}{A}\right) \qquad 59.14$$

$$p_2 - p_1 = \left(\frac{\gamma l}{A}\right)\left(\frac{dQ}{dt}\right) = I\left(\frac{dQ}{dt}\right) \qquad 59.15$$

$$I = \frac{\gamma l}{A} \qquad 59.16$$

Equation 59.15 is integrated to obtain an expression for the flow quantity.

$$Q = \left(\frac{1}{I}\right)\int (p_2 - p_1)dt \qquad 59.17$$

Example 59.5

A pump is used to keep liquid flowing through a filter pack. Pipe friction and potential head are insignificant. What is the system equation?

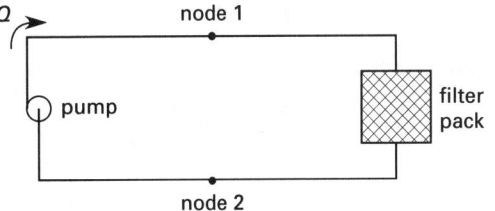

Solution

Pressure varies along the flow path and is the response variable. There are two different pressures: before and after the filter pack. This is analogous to a series electrical circuit of a battery and resistor. The system diagram is

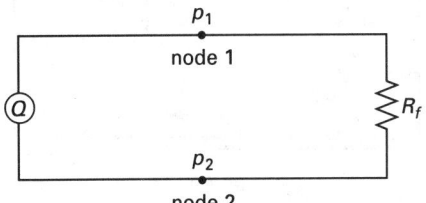

The system equation is

$$Q = \left(\frac{1}{R_f}\right)(p_2 - p_1)$$

Example 59.6

A reservoir discharges through a long pipe and is not refilled. What is the system equation that defines the pressure at the tank bottom as a function of time?

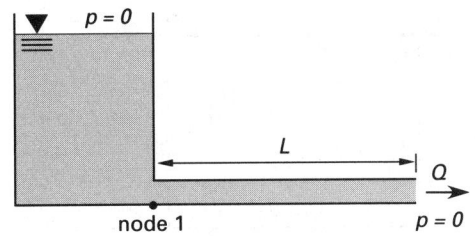

Control Systems

Solution

Although there is flow, there is no external energy source (i.e., there is no pump) in this system. This is analogous to an electrical capacitor discharging through a resistor. The system diagram is

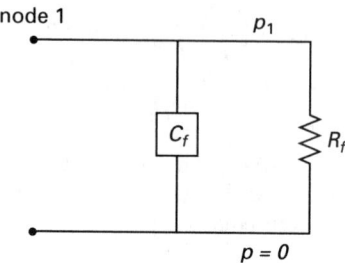

The system equation is

$$0 = \frac{p_1}{R_f} + C_f \left(\frac{dp_1}{dt} \right)$$

Example 59.7

A pump transfers liquid from one reservoir to another. Pipe friction is insignificant. What are the system equations that define the pressure at the bottom of the reservoirs?

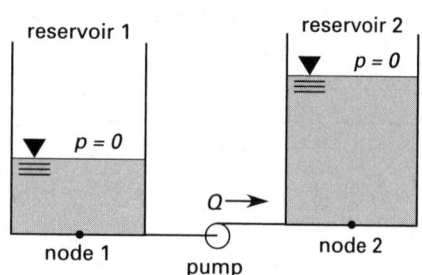

Solution

Both open reservoirs connect to the lowest level. The system diagram is

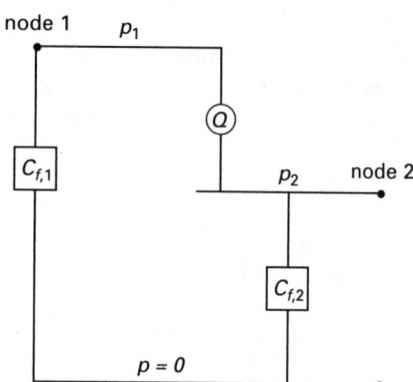

The system equations are

$$node\ 1: \quad Q = C_{f,1} \left(\frac{dp_1}{dt} \right)$$

$$node\ 2: \quad Q = C_{f,2} \left(\frac{dp_2}{dt} \right)$$

PRACTICE PROBLEMS

For each of the systems of ideal elements shown, (a) draw the system diagram and (b) write the differential equations.

1.

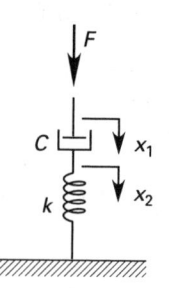

2.

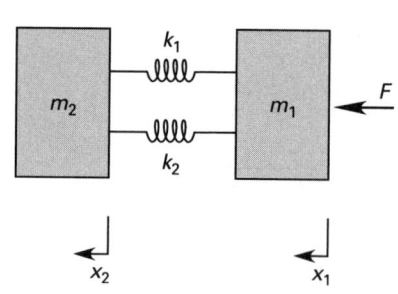

3.

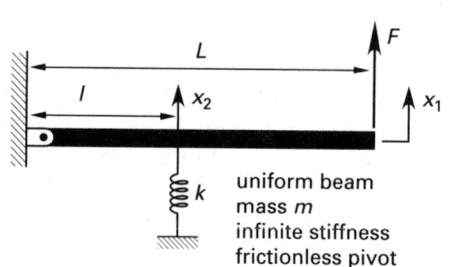

4.

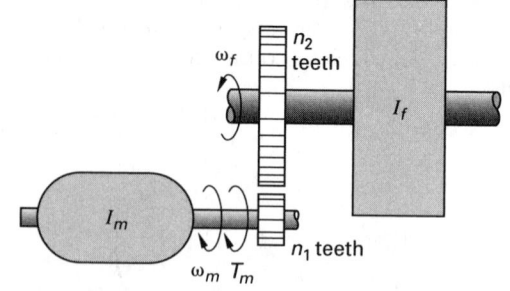

5.

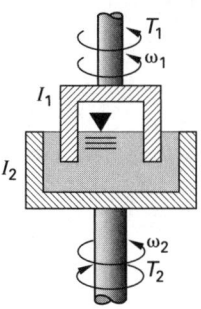

6. The coupling of a railroad car is modeled as the mechanical system shown. Assume all elements are linear. What are the system equations that describe the positions x_1 and x_2 as functions of time?

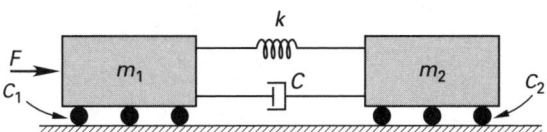

7. Water is discharged freely at a constant rate into an open tank. Water flows out of the tank through a drain with a resistance to flow. (a) Draw the system diagram using idealized elements, and (b) write the differential equations that describe the response of the system.

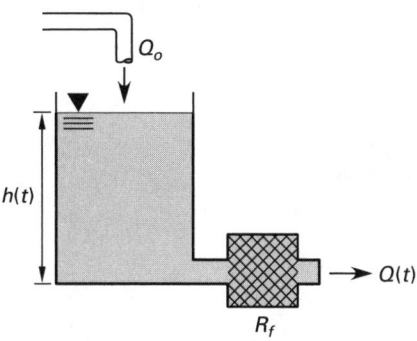

8. Water is pumped into the bottom of an open tank. (a) Draw the system diagram using idealized elements, and (b) write the differential equations that describe the response of the system.

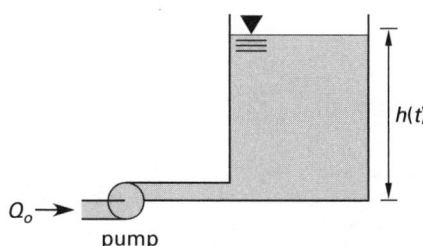

60 Analysis of Engineering Systems

1.	Types of Response	60-1
2.	Graphical Solution	60-2
3.	Classical Solution Method	60-3
4.	Feedback Theory	60-4
5.	Sensitivity	60-5
6.	Block Diagram Algebra	60-5
7.	Predicting System Time Response	60-7
8.	Predicting Time Response from a Related Response	60-7
9.	Initial and Final Values	60-7
10.	Special Cases of Steady-State Response	60-7
11.	Poles and Zeros	60-8
12.	Predicting System Time Response from Response Pole-Zero Diagrams	60-9
13.	Frequency Response	60-10
14.	Gain Characteristic	60-10
15.	Phase Characteristic	60-11
16.	Stability	60-11
17.	Bode Plots	60-11
18.	Root-Locus Diagrams	60-12
19.	Hurwitz Test	60-12
20.	Routh Criterion	60-12
21.	Nyquist Analysis	60-13
22.	Application to Control Systems	60-13
23.	Application to Tachometer Control	60-13
24.	State Model Representation	60-14
	Practice Problems	60-15

Nomenclature

A	steady-state response
B	damping coefficient
BW	bandwidth
C	capacitance or constant
$e(t)$	error
$E(s)$	error, $\mathcal{L}[e(t)]$
$f(t)$	forcing function
$F(s)$	forcing function, $\mathcal{L}[f(t)]$
$G(s)$	forward transfer function
h	step height
$H(s)$	reverse transfer function
$i(t)$	current
$I(s)$	current, $\mathcal{L}[i(t)]$
j	$\sqrt{-1}$
k	spring stiffness
K	gain
L	inductance
L	length of line
M	fraction overshoot
n	degrees of freedom
n	order of the system
n	system type

N	Nyquist's number
$p(t)$	arbitrary function
$P(s)$	arbitrary function, $\mathcal{L}[p(t)]$
P	number of poles
Q	quality factor
r	real value (root)
$r(t)$	time response
$R(s)$	response function, $\mathcal{L}(r(t))$
S	sensitivity
t	time
$T(t)$	transfer function
$T(s)$	transfer function, $\mathcal{L}(T(t))$
u	input variable
v	velocity
V	voltage
x	position
x	state variable
y	output variable
Z	number of zeros

Symbols

α	pole angle
β	zero angle
ϵ	a small number
ζ	damping ratio
τ	time constant
ω	natural frequency

Subscripts

d	damped
f	forced or feedback
i	in
n	natural
o	out
p	peak or pole
r	rise
s	settling
t	total
z	zero

1. TYPES OF RESPONSE

Natural response (also known as *initial condition response*, *homogeneous response*, and *unforced response*) is how a system behaves when energy is applied and is subsequently removed. The system is left alone and is allowed to do what it would naturally, without the application of further disturbing forces. In the absence of friction, natural response is characterized by a sinusoidal response function. Otherwise, the response function will contain exponentially decaying sinusoids.

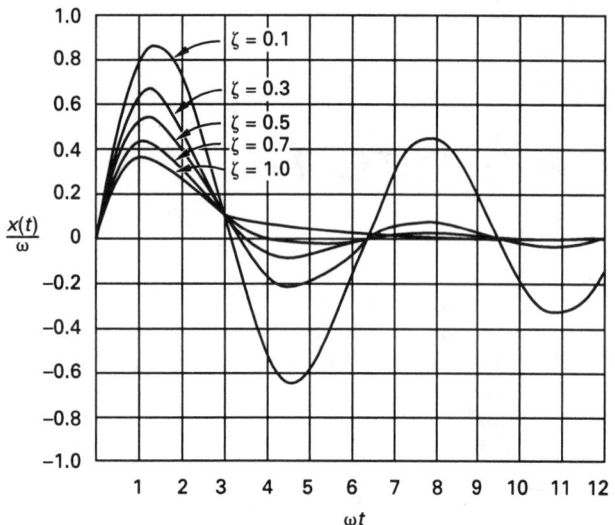

Figure 60.1 *Natural Response*

Forced response is the behavior of a system that is acted on by a force that is applied periodically. Forced response in the absence of friction is characterized by sinusoidal terms having the same frequency as the forcing function.

Natural and forced responses are present simultaneously in forced systems. The sum of the two responses is the *total response*.[1] This is the reason that differential equations are solved by adding a particular solution to the homogeneous solution. The homogeneous solution corresponds to the natural response; the particular solution corresponds to the forced response.

$$\frac{\text{total}}{\text{response}} = \frac{\text{natural}}{\text{response}} + \frac{\text{forced}}{\text{response}} \qquad 60.1$$

Since the influence of decaying functions disappears after a few cycles, natural response is sometimes referred to as *transient response*. Once the transient response effects have died out, the total response will consist entirely of forced terms. This is the *steady-state response*.

2. GRAPHICAL SOLUTION

Graphical solutions are available in limited cases, particularly those with homogeneous, step, and sinusoidal inputs. When the system equation is a homogeneous second-order linear differential equation with constant coefficients (Sec. 10.6) in the form of Eq. 60.2, the natural time response can be determined from Fig. 60.1. ω is the natural frequency and ζ is the damping ratio.

$$x'' + 2\zeta\omega x' + \omega^2 x = 0 \qquad 60.2$$

---

[1]This response is a function of time, and, therefore, is referred to as *time response* to distinguish it from *frequency response* (Sec. 13).

When the system equation is a second-order linear differential equation with constant coefficients and the forcing function is a step of height h (as in Eq. 60.3), the time response can be determined from Fig. 60.2.

$$x'' + 2\zeta\omega x' + \omega^2 x = \omega^2 h \qquad 60.3$$

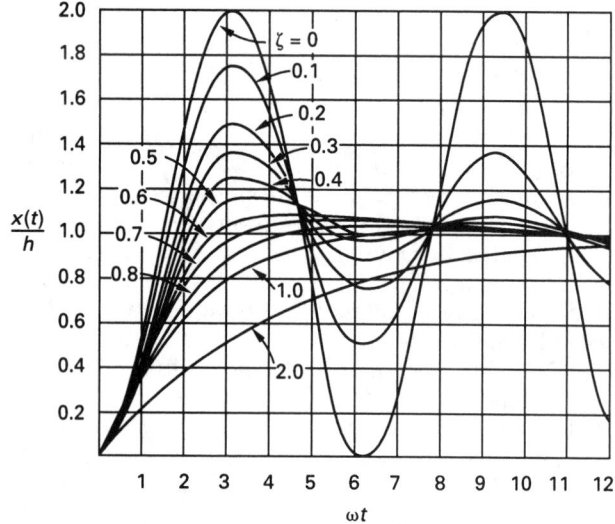

Figure 60.2 *Response to a Unit Step*

Figure 60.2 illustrates that a system responding to a step will eventually settle to the steady-state position of the step, but that when damping is low ($\zeta < 1$), there will be *overshoot*. Figure 60.3 illustrates this and other parameters of second-order response to a step input. The settling time depends on the *tolerance* (i.e., the separation of the actual and steady-state responses). The *time delay*, t_d, in Fig. 60.3 is the time to reach 50% of the steady-state value. The *time constant*, τ, is the time to reach approximately 63% of the steady-state value.

$$\omega_d = \text{damped frequency} = \omega\sqrt{1-\zeta^2} \qquad 60.4$$

$$t_r = \text{rise time} = \frac{\pi - \arccos\zeta}{\omega_d} \qquad 60.5$$

$$t_p = \text{peak time} = \frac{\pi}{\omega_d} \qquad 60.6$$

$$M_p = \text{peak gain} \quad [\text{fraction overshoot}]$$

$$= \exp\left(\frac{-\pi\zeta}{\sqrt{1-\zeta^2}}\right) \qquad 60.7$$

$$t_s = \text{settling time} = \frac{3.91}{\zeta\omega} \quad [2\% \text{ criterion}] \qquad 60.8$$

$$t_s = \frac{3.00}{\zeta\omega} \quad [5\% \text{ criterion}] \qquad 60.9$$

$$\tau = \text{time constant} = \frac{1}{\zeta\omega} \qquad 60.10$$

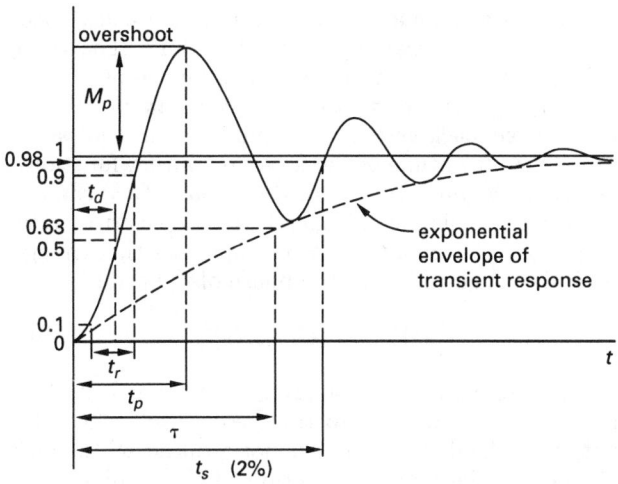

Figure 60.3 *Second-Order Step Time Response Parameters*

3. CLASSICAL SOLUTION METHOD

As described in Sec. 59.1, a system model can be thought of as a block where the *input signal* is the forcing function, $f(t)$, and the *output signal* is the response function, $r(t)$. The standard analytical method of determining the response of a system from its system equation uses Laplace transforms. Accordingly, the classical solution approach does not determine the output signal or response function directly. Rather, it derives the *transfer function*, $T(s)$.[2] The transfer function is also known as the *rational function*.

$$T(s) = \mathcal{L}\left(\frac{r(t)}{f(t)}\right) \qquad 60.11$$

Transfer functions in the s-domain are Laplace transformations of the corresponding time-domain functions. Transforming $r(t)/f(t)$ into $T(s)$ is more than a simple change of variables. The s symbol can be thought of as a derivative operator; similarly, the integration operator is represented as $1/s$. By convention, Laplace transforms are represented by uppercase letters, while operand functions are represented by lowercase letters.

Example 60.1

A system equation in differential equation form has been derived for the output force from a mechanical network. Convert the system equation to the s-domain.

$$f(t) = k \int (v_1 - v_2)dt + m\frac{d(v_2 - v_1)}{dt}$$

---

[2]Strictly speaking, $r(t)/f(t)$ is the *transfer function* and $\mathcal{L}(r(t)/f(t))$ is the *transform of the transfer function*. However, the distinction is seldom made.

Solution

Replace all derivative operators by s; replace all integration operators by $1/s$. Replace time velocity, $v(t)$, with s-domain velocity, $v(s)$.

$$F(s) = \frac{kv_1}{s} - \frac{kv_2}{s} + smv_2 - smv_1$$

Example 60.2

Determine the transfer function for the system shown and draw its black box representation.

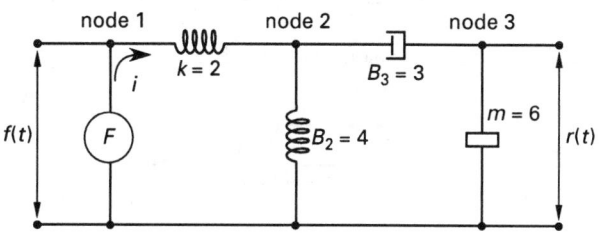

Solution

First, write the system equations. There are two loop "currents," so two simultaneous equations will be needed. One system equation can be written for each of the three nodes, so there will be one redundant system equation. (At this point, it is not obvious which of the two system equations can be used to derive the transfer function.)

node 1:

$$f(t) = k \int (v_1 - v_2)dt = 2 \int (v_1 - v_2)dt$$

node 2:

$$0 = k \int (v_2 - v_1)dt + B_2v_2 + B_3(v_2 - v_3)$$
$$= 2 \int (v_2 - v_1)dt + 4v_2 + (3)(v_2 - v_3)$$

node 3:

$$0 = m\frac{dv_3}{dt} + B_3(v_3 - v_2)$$
$$= 6\frac{dv_3}{dt} + (3)(v_3 - v_2)$$

Next, convert the system equations to the s-domain by substituting s for derivative operations and $1/s$ for integration operations.

node 1:

$$F(s) = \frac{(2)(v_1 - v_2)}{s}$$

node 2:

$$0 = \frac{(2)(v_2 - v_1)}{s} + 4v_2 + (3)(v_2 - v_3)$$

node 3:

$$0 = 6sv_3 + (3)(v_3 - v_2)$$

The transfer function is the ratio of the output to input velocities, v_3/v_1. It does not depend on v_2. The equation for node 1 cannot be used unless $i(t)$ is known, which it (generally) is not. v_2 is eliminated from the equations for nodes 2 and 3. From the second and third nodes,

$$v_2 = \frac{2v_1 + 3sv_3}{2 + 7s}$$

$$v_2 = v_3(1 + 2s)$$

The transfer function, $T(s)$, is found by equating these two expressions and solving for v_3/v_1.

$$T(s) = \frac{v_3}{v_1} = \frac{1}{7s^2 + 4s + 1}$$

The black box representation of this system is

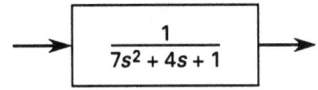

4. FEEDBACK THEORY

The output signal is returned as input in a feedback loop (feedback system). A basic feedback system consists of two black box units (a *dynamic unit* and a *feedback unit*), a pick-off point (take-off point), and a summing point (*comparator* or *summer*). The summing point is assumed to perform positive addition unless a minus sign is present. The incoming signal, V_i, is combined with the feedback signal, V_f, to give the *error (error signal)*, e. Whether addition or subtraction is used in Eq. 60.12 depends on whether the summing point is additive (i.e., a positive feedback system) or subtractive (i.e., a negative feedback system), respectively. $E(s)$ is the *error transfer function (error gain)*.

$$\begin{aligned} E(s) &= \mathcal{L}(e(t)) = V_i(s) \pm V_f(s) \\ &= V_i(s) \pm H(s)V_o(s) \end{aligned} \qquad 60.12$$

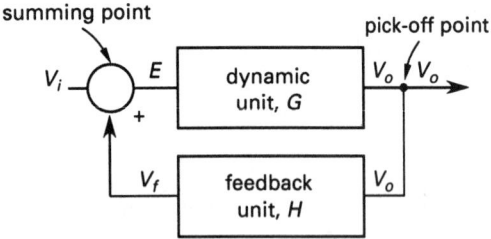

Figure 60.4 *Feedback System*

The ratio $E(s)/V_i(s)$ is the *error ratio (actuating signal ratio)*.

$$\frac{E(s)}{V_i(s)} = \frac{1}{1 + G(s)H(s)} \quad \text{[negative feedback]} \quad 60.13$$

$$= \frac{1}{1 - G(s)H(s)} \quad \text{[positive feedback]} \quad 60.14$$

Since the dynamic and feedback units are black boxes, each has an associated transfer function. The transfer function of the dynamic unit is known as the *forward transfer function (direct transfer function)*, $G(s)$. In most feedback systems—amplifier circuits in particular—the magnitude of the forward transfer function is known as the *forward gain* or *direct gain*. $G(s)$ can be a scalar if the dynamic unit merely scales the error. However, $G(s)$ is normally a complex operator that changes both the magnitude and the phase of the error.

$$V_o(s) = G(s)E(s) \qquad 60.15$$

The pick-off point transmits the output signal, V_o, from the dynamic unit back to the feedback element. The output of the dynamic unit is not reduced by the pick-off point. The transfer function of the feedback unit is the *reverse transfer function (feedback transfer function, feedback gain, etc.)*, $H(s)$, which can be a simple magnitude-changing scalar or a phase-shifting function.

$$V_f(s) = H(s)V_o(s) \qquad 60.16$$

The ratio $V_f(s)/V_i(s)$ is the *feedback ratio (primary feedback ratio)*.

$$\frac{V_f(s)}{V_i(s)} = \frac{G(s)H(s)}{1 + G(s)H(s)} \quad \text{[negative feedback]} \quad 60.17$$

$$= \frac{G(s)H(s)}{1 - G(s)H(s)} \quad \text{[positive feedback]} \quad 60.18$$

The *loop transfer function (loop gain, open-loop gain, or open-loop transfer function)* is the gain after going around the loop one time, $\pm G(s)H(s)$.

The *overall transfer function (closed-loop transfer function, control ratio, system function, closed-loop gain, etc.)*, $G_{\text{loop}}(s)$, is the overall transfer function of the feedback system. The quantity $1 + G(s)H(s) = 0$ is the *characteristic equation*. The *order of the system* is the largest exponent of s in the characteristic equation. (This corresponds to the highest-order derivative in the system equation.)

$$G_{\text{loop}}(s) = \frac{V_o(s)}{V_i(s)} = \frac{G(s)}{1 + G(s)H(s)} \quad \text{[negative feedback]}$$
$$60.19$$

$$= \frac{G(s)}{1 - G(s)H(s)} \quad \text{[positive feedback]}$$
$$60.20$$

With positive feedback and $G(s)H(s)$ less than 1.0, G_{loop} will be larger than $G(s)$. This increase in gain is a characteristic of positive feedback systems. As $G(s)H(s)$ approaches 1.0, the closed-loop transfer function increases without bound, usually an undesirable effect.

In a negative feedback system, the denominator of Eq. 60.19 will be greater than 1.0. Although the closed-loop transfer function will be less than $G(s)$, there may be other desirable effects. Generally, a system with negative feedback will be less sensitive to variations in temperature, circuit component values, input signal frequency, and signal noise. Other benefits include distortion reduction, increased stability, and impedance matching.[3]

Example 60.3

A high-gain non-inverting operational amplifier has a gain of 10^6. Feedback is provided by a resistor voltage divider. What is the closed-loop gain?

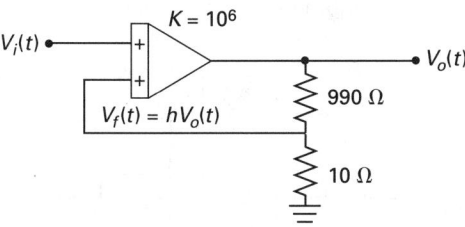

Solution

The fraction of the output signal appearing at the summing point depends on the resistances in the divider circuit.

$$h = \frac{10\ \Omega}{10\ \Omega + 990\ \Omega} = 0.01$$

Since the feedback path only scales the feedback signal, the feedback will be positive. From Eq. 60.19, the closed-loop gain is

$$K_{\text{loop}} = \frac{K}{1 - Kh} = \frac{10^6}{1 - (10^6)(0.01)} \approx -100$$

5. SENSITIVITY

For large loop gains ($G(s)H(s) \gg 1$) in negative feedback systems, the overall gain will be approximately $1/H(s)$. Thus, the forward gain will not be a factor, and by choice of $H(s)$, the output can be made insensitive to variations in $G(s)$.

In general, the *sensitivity* of any variable, A, with respect to changes in another parameter, B, is

$$S_B^A = \frac{d \ln A}{d \ln B} = \frac{\dfrac{dA}{A}}{\dfrac{dB}{B}} \approx \left(\frac{\Delta A}{\Delta B}\right)\left(\frac{B}{A}\right) \qquad \textbf{60.21}$$

[3] For circuits to be directly connected in series without affecting their performance, all input impedances must be infinite and all output impedances must be zero.

The sensitivity of the loop transfer function with respect to the forward transfer function is

$$
\begin{aligned}
S_{G(s)}^{G_{\text{loop}}(s)} &= \left(\frac{\Delta G_{\text{loop}}(s)}{\Delta G(s)}\right)\left(\frac{G(s)}{G_{\text{loop}}(s)}\right) \\
&= \frac{1}{1 + G(s)H(s)} \quad \text{[negative feedback]} \qquad \textbf{60.22} \\
&= \frac{1}{1 - G(s)H(s)} \quad \text{[positive feedback]} \qquad \textbf{60.23}
\end{aligned}
$$

Example 60.4

A closed-loop gain of -100 is required from a circuit, and the output signal must not vary by more than $\pm 1\%$. An amplifier is available, but its output varies by $\pm 20\%$. How can this amplifier be used?

Solution

A closed-loop sensitivity of 0.01 is required. This means

$$\frac{\Delta V_o}{V_o} = \frac{\Delta G_{\text{loop}} V_i}{G_{\text{loop}} V_i} = \frac{\Delta G_{\text{loop}}}{G_{\text{loop}}} = 0.01$$

Similarly, the existing amplifier has a sensitivity of 20%. This means

$$\frac{\Delta G}{G} = 0.20$$

The ratio of these variations corresponds to the definition of sensitivity (Eq. 60.21). For positive feedback,

$$\frac{\Delta G_{\text{loop}} G}{\Delta G\, G_{\text{loop}}} = \frac{0.01}{0.20} = \frac{1}{1 - GH}$$

Solving, $GH = -19$.

Solving for G from Eq. 60.20,

$$G_{\text{loop}} = -100 = \frac{G}{1 - GH} = \frac{G}{1 - (-19)}$$

Solving, $G = -2000$. Finally, solve for H.

$$H = \frac{GH}{G} = \frac{-19}{-2000} = 0.0095$$

6. BLOCK DIAGRAM ALGEBRA

The functions represented by several interconnected black boxes (*cascaded blocks*) can be simplified into a single block operation. Some of the most important simplification rules of block diagram algebra are shown in Fig. 60.5. Case 3 represents the standard feedback model.

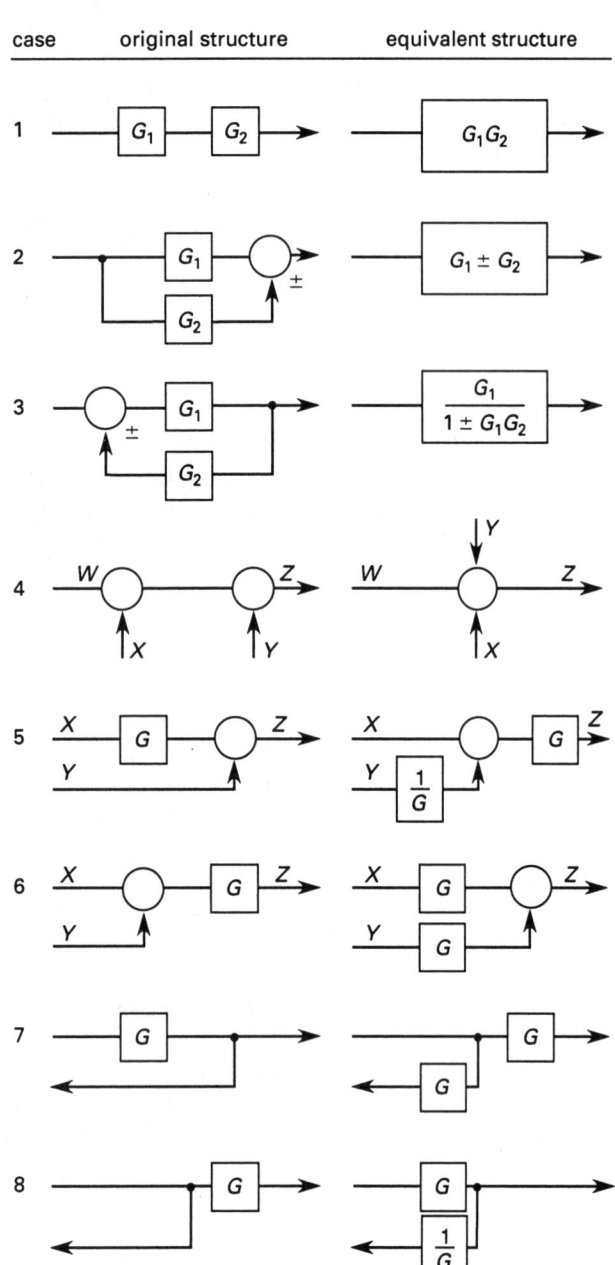

Figure 60.5 *Rules of Simplifying Block Diagrams*

Solution

Use case 5 to move the second summing point back to the first summing point.

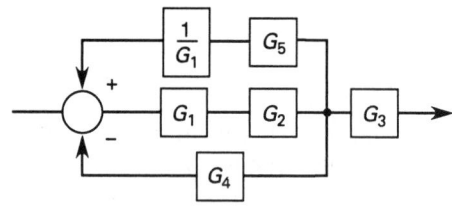

Use case 1 to combine boxes in series.

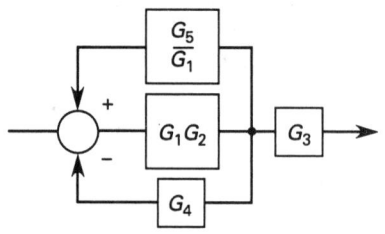

Use case 2 to combine the two feedback loops.

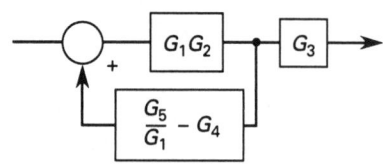

Use case 8 to move the pick-off point outside the G_3 box.

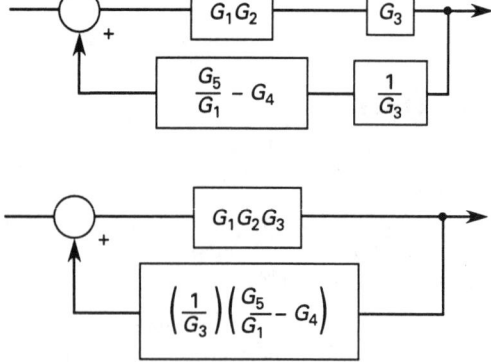

Example 60.5

A complex block system is constructed from five blocks and two summing points. What is the overall transfer function?

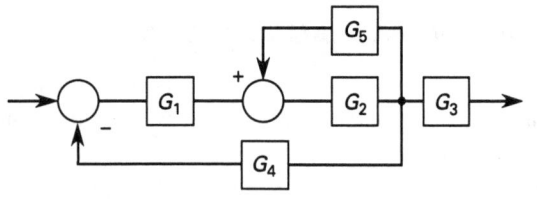

Use case 3 to determine the system gain.

$$G_{\text{loop}} = \frac{G_1 G_2 G_3}{1 - (G_1 G_2 G_3)\left(\dfrac{1}{G_3}\right)\left[\left(\dfrac{G_5}{G_1}\right) - G_4\right]}$$

$$= \frac{G_1 G_2 G_3}{1 - G_2 G_5 + G_1 G_2 G_4}$$

7. PREDICTING SYSTEM TIME RESPONSE

The transfer function is derived without knowledge of the input and is insufficient to predict the time response of the system. The system time response will depend on the form of the input function. Since the transfer function is expressed in the s-domain, the forcing and response functions must be also. (Laplace transforms of step, pulses, sinusoids, and other functions were evaluated in Sec. 10.10.)

$$R(s) = T(s)F(s) \qquad 60.24$$

The time-based response function, $r(t)$, is found by performing the inverse Laplace transform.

$$r(t) = \mathcal{L}^{-1}(R(s)) \qquad 60.25$$

Example 60.6

A mechanical system is acted upon by a constant force of eight units starting at $t = 0$. What is the time-based response function, $r(t)$, if the transfer function is

$$T(s) = \frac{6}{(s+2)(s+4)}$$

Solution

The forcing function is a step of height 8 at $t = 0$. The Laplace transform of a unit step (Sec. 10.10) is $1/s$. Therefore, $F(s) = 8/s$.

From Eq. 60.24,

$$R(s) = T(s)F(s) = \left(\frac{6}{(s+2)(s+4)}\right)\left(\frac{8}{s}\right)$$

$$= \frac{48}{s(s+2)(s+4)}$$

The response function is found from Eq. 60.25 and a table of Laplace transforms (App. 10.A). (Remember that a product of linear terms in the denominator of $R(s)$ is equivalent to a sum of terms in $r(t)$.)

$$r(t)\mathcal{L}^{-1} = \frac{48}{s(s+2)(s+4)}$$

$$= 6 - 12e^{-2t} + 6e^{-4t}$$

The last two terms are decaying exponentials, which represent the transient natural response. The first term does not vary with time; it is the steady-state response.

8. PREDICTING TIME RESPONSE FROM A RELATED RESPONSE

In some cases, it may be possible to use a known response to one input to determine the response to another input. For example, the impulse function is the derivative of the step function (see Sec. 59.3), so the response to an impulse is the derivative of the response to a step function.

9. INITIAL AND FINAL VALUES

The initial and final (steady-state) values of any function, $P(s)$, can be found from the *initial* and *final value theorems*, respectively, providing the limits exist. Equations 60.26 and 60.27 are particularly valuable in determining the steady-state response (substitute $R(s)$ for $P(s)$) and the steady-state error (substitute $E(s)$ for $P(s)$).

$$\lim_{t \to 0+} p(t) = \lim_{s \to \infty}(sP(s)) \quad \text{[initial value]} \qquad 60.26$$

$$\lim_{t \to \infty} p(t) = \lim_{s \to 0}(sP(s)) \quad \text{[final value]} \qquad 60.27$$

Example 60.7

What is the final value of the response function $r(t)$ if

$$R(s) = \frac{1}{s(s+1)}$$

Solution

From Eq. 60.27, $R(s)$ is multiplied by s and the limit taken as s tends to zero.

$$R(\infty) = \lim_{s \to 0}\left(\frac{s}{s(s+1)}\right) = \lim_{s \to 0}\left(\frac{1}{s+1}\right) = 1$$

10. SPECIAL CASES OF STEADY-STATE RESPONSE

In addition to determining the steady-state response from the final value theorem (Sec. 9), the steady-state response to a specific input can be easily derived from the transfer function, $T(s)$, in a few specialized cases. For example, the steady-state response function for a system acted upon by an impulse is simply the transfer function. That is, a pulse has no long-term effect on a system.

$$R(\infty) = T(s) \quad \text{[pulse input]} \qquad 60.28$$

The steady-state response for a *step input* (often referred to as a *d-c input*) is obtained by substituting 0 for s everywhere in the transfer function. (If the step has magnitude h, the steady-state response is multiplied by h.)

$$R(\infty) = T(0) \quad \text{[unit step input]} \qquad 60.29$$

The steady-state response for a sinusoidal input is obtained by substituting $j\omega_f$ for s everywhere in the transfer function, $T(s)$. The output will have the same frequency as the input. It is particularly convenient to perform sinusoidal calculations using phasor notation (as illustrated in Ex. 60.9).

$$R(\infty) = T(j\omega_f) \qquad 60.30$$

Example 60.8

What is the steady-state response of the system in Ex. 60.6 when acted upon by a step of height 8?

Solution

Substitute 0 for s in $T(s)$ and multiply by 8.

$$R(\infty) = (8)(T(0)) = (8)\left(\frac{6}{(0+2)(0+4)}\right) = 6$$

Example 60.9

What is the steady-state response when a sinusoidal forcing function of $4\sin(2t+45°)$ is applied to a system whose transfer function is

$$T(s) = \frac{-1}{7s^2 + 7s + 1}$$

Solution

The angular frequency of the forcing function is $\omega_f = 2$ rad/s. Substitute $j2$ for s in $T(s)$, and simplify the expression by recognizing that $j^2 = -1$.

$$T(j2) = \frac{-1}{(7)(j2)^2 + (7)(j2) + 1}$$

$$= \frac{-1}{-28 + 14j + 1} = \frac{-1}{-27 + 14j}$$

Next, convert $T(j2)$ to phasor (polar) form. The magnitude and angle of the denominator are

$$\text{magnitude} = \sqrt{(14)^2 + (-27)^2} = 30.4138$$

$$\text{angle} = 180° - \arctan\left(\frac{14}{27}\right) = 152.59°$$

However, this is the negative reciprocal of $T(j2)$.

$$T(j2) = \frac{-1}{30.4138\underline{/152.59°}} = -0.03288\underline{/-152.59°}$$

The forcing function expressed in phasor form is $4\underline{/45°}$. From Eq. 60.11, the steady-state response is

$$v(t) = T(t)f(t) = (-0.03288\underline{/-152.59°})(4\underline{/45°})$$

$$= -0.1315\underline{/-107.6°}$$

11. POLES AND ZEROS

A *pole* is a value of s that makes a function, $P(s)$, infinite. Specifically, a pole makes the denominator of $P(s)$ zero.[4] A *zero* of the function makes the numerator of $P(s)$ (and hence $P(s)$ itself) zero. Poles and zeros need not be real or unique; they can be imaginary and repeated within a function.

A *pole-zero diagram* is a plot of poles and zeros in the *s-plane*—a rectangular coordinate system with real and imaginary axes. Zeros are represented by $\bigcirc$'s; poles are represented as $\times$'s. Poles off the real axis always occur in conjugate pairs known as *pole pairs*.

Sometimes it is necessary to derive the function $P(s)$ from its pole-zero diagram. This will be only partially successful since repeating identical poles and zeros are not usually indicated on the diagram. Also, scale factors (scalar constants) are not shown.

Example 60.10

Draw the pole-zero diagram for the following transfer function.

$$T(s) = \frac{(5)(s+3)}{(s+2)(s^2+2s+2)}$$

Solution

The numerator is zero when $s = -3$. This is the only zero of the transfer function.

The denominator is zero when $s = -2$ and $s = -1 \pm j$. These three values are the poles of the transfer function.

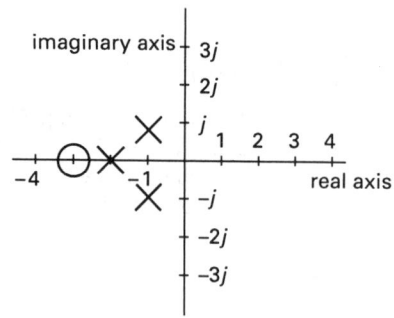

Example 60.11

A pole-zero diagram for a transfer function $T(s)$ has a single pole at $s = -2$ and a single zero at $s = -7$. What is the corresponding function?

Solution

$$T(s) = \frac{K(s+7)}{s+2}$$

The scale factor K must be determined by some other means.

---

[4]Pole values are the system *eigenvalues*.

12. PREDICTING SYSTEM TIME RESPONSE FROM RESPONSE POLE-ZERO DIAGRAMS

A response pole-zero diagram based on $R(s)$ can be used to predict how the system responds to a specific input. (Note that this pole-zero diagram must be based on the product $T(s)F(s)$ since that is how $R(s)$ is calculated. Plotting the product $T(s)F(s)$ is equivalent to plotting $T(s)$ and $F(s)$ separately on the same diagram.)

The system will experience an *exponential decay* when a single pole falls on the real axis. A pole with a value of $-r$, corresponding to the linear term $(s+r)$, will decay at the rate of $e^{-t/r}$. The quantity $1/r$ is the decay *time constant*, the time for the response to achieve approximately 63% of its steady-state value. Thus, the farther left the point is located from the vertical imaginary axis, the faster the motion will die out.

Undamped sinusoidal oscillation will occur if a pole pair falls on the imaginary axis. A conjugate pole pair with the value of $\pm j\omega$ indicates oscillation with a natural frequency of ω rad/s.

Pole pairs to the left of the imaginary axis represent *decaying sinusoidal* response. The closer the poles are to the real (horizontal) axis, the slower will be the oscillations. The closer the poles are to the imaginary (vertical) axis, the slower will be the decay. The *natural frequency*, ω, of undamped oscillation can be determined from a *conjugate pole pair* having values of $r \pm \omega_f$.

$$\omega = \sqrt{r^2 + \omega_f^2} \qquad 60.31$$

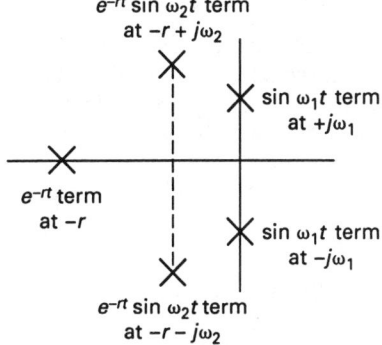

Figure 60.6 *Types of Response Determined by Pole Location*

The magnitude and phase shift can be determined for any input frequency from the pole-zero diagram with the following procedure: Locate the angular frequency, ω_f, on the imaginary axis. Draw a line from each pole (i.e., a pole-line) and from zero (i.e., a zero-line) of $T(s)$ to this point. The angle of each of these lines is the angle between it and the horizontal real axis. The overall magnitude is the product of the lengths of the zero-lines divided by the product of the lengths of the pole-lines. (The scale factor must also be included because it is not shown on the pole-zero diagram.) The phase is the sum of the pole-angles less the sum of the zero-angles.

$$|R| = \frac{K \prod_z |L_z|}{\prod_p |L_p|} = \frac{K \prod_z \text{length}}{\prod_p \text{length}} \qquad 60.32$$

$$\underline{/R} = \sum_p \alpha - \sum_z \beta \qquad 60.33$$

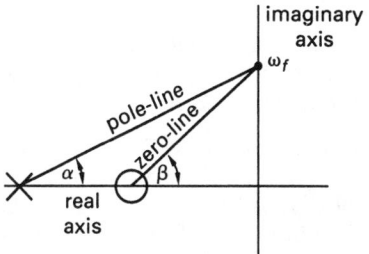

Figure 60.7 *Calculating Magnitude and Phase from a Pole-Zero Diagram*

Example 60.12

What is the response of a system with the response pole-zero diagram representing $R(s) = T(s)F(s)$ shown?

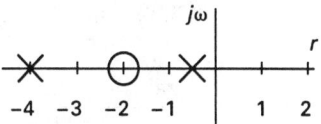

Solution

The poles are at $r = -\frac{1}{2}$ and $r = -4$. The response is

$$r(t) = C_1 e^{-\frac{1}{2}t} + C_2 e^{-4t}$$

Constants C_1 and C_2 must be found from other data.

Example 60.13

What is the system response if $T(s) = (s+2)/(s+3)$ and the input is a unit step?

Solution

The transform of a unit step is $1/s$. The response is

$$R(s) = T(s)F(s) = \frac{s+2}{s(s+3)}$$

The response pole-zero diagram is

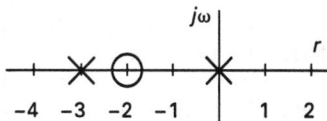

The pole at $r = 0$ contributes the exponential $C_1 e^{-0t}$ (or just C_1) to the total response. The pole at $r = -3$ contributes the term $C_2 e^{-3t}$. The total response is

$$r(t) = C_1 + C_2 e^{-3t}$$

13. FREQUENCY RESPONSE

The gain and phase angle frequency response of a system will change as the forcing frequency is varied. (The dependence of $r(t)$ on ω_f was illustrated in Ex. 60.9.) The *frequency response* is the variation in these parameters, always with a sinusoidal input. *Gain* and *phase characteristics* are plots of the steady-state gain and phase angle responses with a sinusoidal input versus frequency. While a linear frequency scale can be used, frequency response is almost always presented against a logarithmic frequency scale.

The steady-state gain response is expressed in decibels, while the steady-state phase angle response is expressed in degrees. The gain is calculated from Eq. 60.34 where $|T(s)|$ is the absolute value of the steady-state response.

$$\text{gain} = 20\log|T(j\omega)| \quad [\text{in dB}] \qquad 60.34$$

A doubling of $|T(j\omega)|$ is referred to as an *octave* and corresponds to a 6.02 dB increase. A tenfold increase in $|T(j\omega)|$ is a *decade* and corresponds to a 20 dB increase.

$$\begin{aligned}\text{number of octaves} &= \frac{\text{gain}_2 - \text{gain}_1 \quad [\text{in dB}]}{6.02} \\ &= 3.32 \times \text{number of decades} \end{aligned}$$
$$60.35$$

$$\begin{aligned}\text{number of decades} &= \frac{\text{gain}_2 - \text{gain}_1 \quad [\text{in dB}]}{20} \\ &= 0.301 \times \text{number of octaves}\end{aligned}$$
$$60.36$$

14. GAIN CHARACTERISTIC

The *gain characteristic* (*M*-curve for magnitude) is a plot of the gain as ω_f is varied. It is possible to make a rough sketch of the gain characteristic by calculating the gain at a few points (pole frequencies, $\omega = 0$, $\omega = \infty$, etc.). The curve will usually be asymptotic to several lines. The frequencies at which these asymptotes intersect are *corner frequencies*. The peak gain, M_p, coincides with the natural (resonant) frequency of the system.[5] Large peak gains indicate lowered stability and large overshoots. The *gain crossover point*, if any, is the frequency at which $\log(\text{gain}) = 0$.

The *half-power points (cut-off frequencies)* are the frequencies for which the gain is 0.707 (i.e., $\sqrt{2}/2$ times the peak value. (This is equivalent to saying the gain is 3 dB less than the peak gain.) The *cut-off rate* is the slope of the gain characteristic in dB/octave at a half-power point. The frequency difference between the half-power points is the *bandwidth*, BW. The *closed-loop*

---

bandwidth is the frequency range over which the closed-loop gain falls 3 dB below its value at $\omega = 0$. (The term "bandwidth" often means closed-loop bandwidth.) The *quality factor*, Q, is

$$Q = \frac{\omega_n}{\text{BW}} \qquad 60.37$$

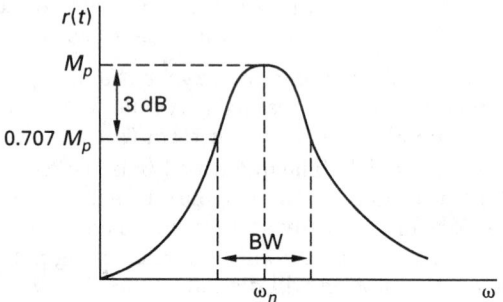

Figure 60.8 *Bandwidth*

Since a low or negative gain (compared to higher parts of the curve) effectively represents attenuation, the gain characteristic can be used to distinguish between low- and high-pass filters. A low-pass filter will have a large gain at low frequencies and a small gain at high frequencies. Conversely, a high-pass filter will have a high gain at high frequencies and a low gain at low frequencies.

It may be possible to determine certain parameters (e.g., the natural frequency and bandwidth) from the transfer function directly. For example, when the denominator of $T(s)$ is a single linear term of the form $s + r$, the bandwidth will be equal to r. Thus, the bandwidth and time constant are reciprocals.

Another important case is when $T(s)$ has the form of Eq. 60.38. (Compare the form of $T(s)$ here to its form in Eq. 60.11 in Sec. 3.) The coefficient of the s^2 term must be 1, and ω_n must be much larger than BW so that the pole is close to the imaginary axis. The zero defined by constants a and b in the numerator is not significant.

$$T(s) = \frac{as + b}{s^2 + (\text{BW})s + \omega_n^2} \qquad 60.38$$

Example 60.14

What are the maximum gain, bandwidth, upper half-power frequency, and half-power gain of a system whose transfer function is

$$T(s) = \frac{1}{s + 5}$$

Solution

The steady-state response to sinusoidal input is determined by substituting $j\omega$ for s in $T(s)$. When $\omega = 0$, $T(s)$ will have a value of $1/5 = 0.2$, and $T(s)$ decreases thereafter. Thus, the maximum gain is 0.2. The

---

[5]The gain characteristic peaks when the forcing frequency equals the natural frequency. It is also said that this peak corresponds to the resonant frequency. Strictly speaking, this is true, although the gain may not actually be resonant (i.e., be infinite).

bandwidth is 5 rad/s. Since the lower half-plane frequency is implicitly 0 rad/s, the upper half-power frequency is 5 rad/s, at which point the gain will be $(0.707)(0.2) = 0.141$.

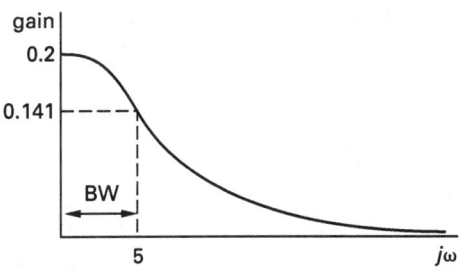

Example 60.15

Predict the natural frequency and bandwidth for the following transfer function.

$$T(s) = \frac{s + 19}{s^2 + 7s + 1000}$$

Solution

The form of this equation is the same as Eq. 60.38. The bandwidth is BW = 7, and the natural frequency is $\sqrt{1000} = 31.62$ rad/s.

15. PHASE CHARACTERISTIC

The phase angle response will also change as the forcing frequency is varied. The *phase characteristic* (α *curve*) is a plot of the phase angle as ω_f is varied.

16. STABILITY

A stable system will remain at rest unless disturbed by external influence and will return to a rest position once the disturbance is removed. A pole with a value of $-r$ on the real axis corresponds to an exponential response of e^{-rt}. Since e^{-rt} is a decaying signal, the system is stable. Similarly, a pole of $+r$ on the real axis corresponds to an exponential response of e^{rt}. Since e^{rt} increases without limit, the system is unstable.

Since any pole to the right of the imaginary axis corresponds to a positive exponential, a *stable system* will have poles only in the left half of the s-plane. If there is an isolated pole on the imaginary axis, the response is stable. However, a conjugate pole pair on the imaginary axis corresponds to a sinusoid that does not decay with time. Such a system is considered to be unstable.

Passive systems (i.e., the homogeneous case) are not acted upon by a forcing function and are always stable. In the absence of an energy source, exponential growth

cannot occur. *Active systems* contain one or more energy sources and may be stable or unstable.

There are several *frequency response (domain) analysis* techniques for determining the stability of a system, including Bode plot, root-locus diagram, Routh stability criterion, Hurwitz test, and Nichols chart. The term *frequency response* almost always means the steady-state response to a sinusoidal input.

The value of the denominator of $T(s)$ is the primary factor affecting stability. When the denominator approaches zero, the system increases without bound. In the typical feedback loop, the denominator is $1 \pm GH$, which can be zero only if $|GH| = 1$. It is logical, then, that most of the methods for investigating stability (e.g., Bode plots, root-locus, Nyquist analysis, and the Nichols chart) investigate the value of the open-loop transfer function, GH. Since $\log(1) = 0$, the requirement for stability is that $\log(GH)$ must not equal 0 dB.

A negative feedback system will also become unstable if it changes to a positive feedback system, which can occur when the feedback signal is changed in phase more than 180°. Therefore, another requirement for stability is that the phase angle change must not exceed 180°.

17. BODE PLOTS

Bode plots are gain and phase characteristics for the open-loop $G(s)H(s)$ transfer function that are used to determine the *relative stability* of a system. The gain characteristic is a plot of $20 \log(|G(s)H(s)|)$ versus ω for a sinusoidal input. (It is important to recognize that the Bode plots, though similar in appearance to the gain and phase frequency response charts, are used to evaluate stability and do not describe the closed-loop system response.)

The *gain margin* is the number of decibels that the open-loop transfer function, $G(s)H(s)$, is below 0 dB at the *phase crossover frequency* (i.e., where the phase angle is −180°). (If the gain happens to be plotted on a linear scale, the gain margin is the reciprocal of the gain at the phase crossover point.) The gain margin must be positive for a stable system, and the larger it is, the more stable the system will be.

The *phase margin* is the number of degrees the phase angle is above −180° at the *gain crossover point* (i.e., where the logarithmic gain is 0 dB or the actual gain is 1).

In most cases, large positive gain and phase margins will ensure a stable system. However, the margins could have been measured at other than the crossover frequencies. Therefore, a Nyquist stability plot is needed to verify the absolute stability of a system.

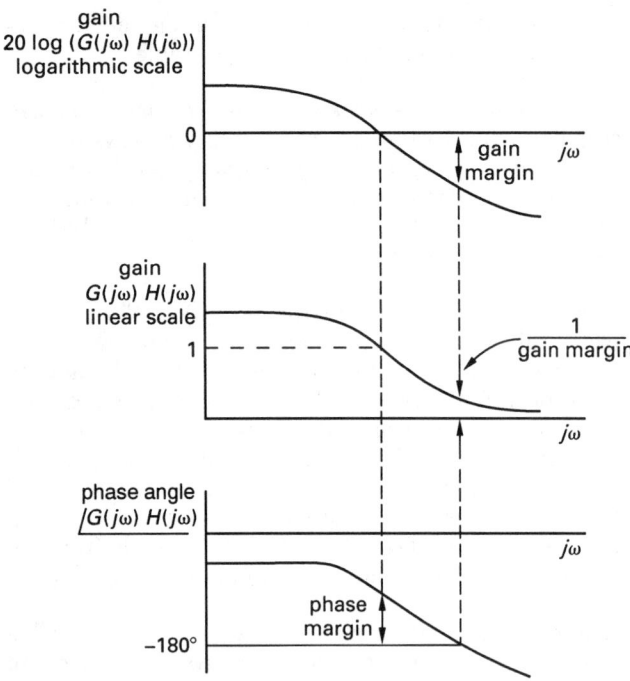

Figure 60.9 *Gain and Phase Margin Bode Plots*

18. ROOT-LOCUS DIAGRAMS

A *root-locus diagram* is a pole-zero diagram showing how the poles of $G(s)H(s)$ move when one of the system parameters (e.g., the gain factor) in the transfer function is varied. The diagram gets its name from the need to find the roots of the denominator (i.e., the poles). The locus of points defined by the various poles is a line or curve that can be used to predict *points of instability* or other critical operating points. A point of instability is reached when the line crosses the imaginary axis into the right-hand side of the pole-zero diagram.

A root-locus curve may not be contiguous, and multiple curves will exist for different sets of roots. Sometimes the curve splits into two branches. In other cases, the curve leaves the real axis at *breakaway points* and continues on with constant or varying slopes approaching asymptotes. One branch of the curve will start at each open-loop pole and end at an open-loop zero.

Example 60.16

Draw the root-locus diagram for a feedback system with open-loop transfer function $G(s)H(s)$. K is a scalar constant that can be varied.

$$G(s)H(s) = \frac{Ks(s+1)(s+2)}{s(s+2) + K(s+1)}$$

Solution

The poles are the zeros of the denominator.

$$s_1, s_2 = -\left(\tfrac{1}{2}\right)(2+K) \pm \sqrt{1 + \tfrac{1}{4}K^2}$$

Since the second term can be either added or subtracted, there are two roots for each value of K. Allowing K to vary from zero to infinity produces a root-locus diagram with two distinct branches. The first branch extends from the pole at the origin to the zero at $s = -1$. The second branch extends from the pole at $s = -2$ to $-\infty$. All poles and zeros are not shown. Since neither branch crosses into the right half, the system is stable for all values of K.

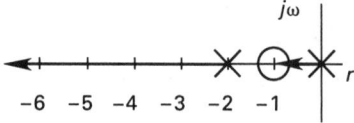

19. HURWITZ TEST

A stable system has poles only in the left half of the s-plane. These poles correspond to roots of the *characteristic equation* (Sec. 4). The characteristic equation can be expanded into a polynomial of the form

$$a_0 s^n + a_1 s^{n-1} + \cdots + a_{n-1}s + a_n = 0 \qquad 60.39$$

The *Hurwitz stability criterion* requires that all coefficients be present and have the same sign (which is equivalent to requiring all coefficients to be positive). If the coefficients differ in sign, the system is unstable. If the coefficients are all alike in sign, the system may or may not be stable. The Routh criterion (test) should be used in that case.

20. ROUTH CRITERION

The *Routh criterion*, like the Hurwitz test, uses the coefficients of the polynomial characteristic equation. A table (the *Routh table*) of these coefficients is formed. The Routh-Hurwitz criterion states that the number of sign changes in the first column of the table equals the number of positive (unstable) roots. Therefore, the system will be stable if all entries in the first column have the same sign.

The table is organized in the following manner.

a_0	a_2	a_4	$a_6 \cdots$
a_1	a_3	a_5	$a_7 \cdots$
b_1	b_2	b_3	$b_4 \cdots$
c_1	c_2	c_3	$c_4 \cdots$
$\vdots$	$\vdots$	$\vdots$	$\vdots$

The remaining coefficients are calculated in the following pattern until all values are zero.

$$b_1 = \frac{a_1 a_2 - a_0 a_3}{a_1} \qquad 60.40$$

$$b_2 = \frac{a_1 a_4 - a_0 a_5}{a_1} \qquad 60.41$$

$$b_3 = \frac{a_1 a_6 - a_0 a_7}{a_1} \qquad 60.42$$

$$c_1 = \frac{b_1 a_3 - a_1 b_2}{b_1} \qquad 60.43$$

Special methods are used if there is a zero in the first column but nowhere else in that row. One of the methods is to substitute a small number, represented by ϵ or δ, for the zero and calculate the remaining coefficients as usual.

Example 60.17

Evaluate the stability of a system that has a characteristic equation of

$$s^3 + 5s^2 + 6s + C = 0$$

Solution

All of the polynomial terms are present (Hurwitz criterion), so the Routh table is

s^3	1	6
s^2	5	C
s^1	$\dfrac{30 - C}{5}$	0
s^0	C	

To be stable, all of the entries in the first column must be positive, which requires $0 < C < 30$.

21. NYQUIST ANALYSIS

The Nyquist Analysis is a particularly useful graphical method when time delays are present in a system or when frequency response data are available. *Nyquist's stability criterion* is $N = P - Z$, where P is the number of poles in the right half of the s-plane, Z is the number of zeros in the right half of the s-plane, and N is the number of encirclements (revolutions) of $1 + G(s)H(s)$ around the critical point. N may be positive, negative, or zero.

22. APPLICATION TO CONTROL SYSTEMS

A control system monitors a process and makes adjustments to maintain performance within certain acceptable limits. Feedback is implicitly a part of all control systems.[6] The *controller (control element)* is the part of the control system that establishes the acceptable limits of performance, usually by setting its own reference inputs. The controller transfer function for a proportional controller is a constant: $G_1(s) = K$.[7] The *plant (controlled system)* is the part of the system that responds to the controller. Both of these are in the forward loop. The input signal, $R(s)$, in Fig. 60.10 is known in a control system as the *command* or *reference value*. Figure 60.10 is known as a *control logic diagram* or *control logic block diagram*.

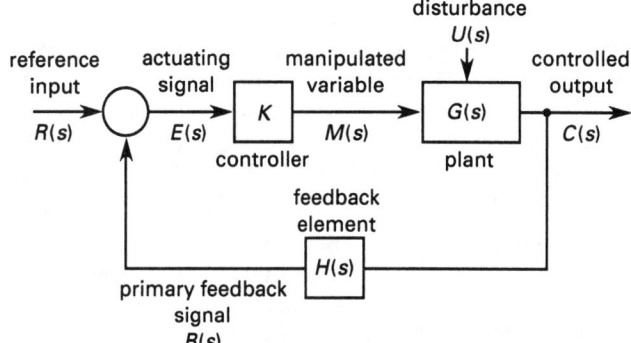

Figure 60.10 *Typical Feedback Control System*

A *servomechanism* is a special type of control system in which the controlled variable is mechanical position, velocity, or acceleration. In many servomechanisms, $H(s) = 1$ (i.e., unity feedback) and it is desired to keep the output equal to the reference input (i.e., maintain a zero error function). If the input, $R(s)$, is constant, the descriptive terms *regulator* and *regulating system* are used.

23. APPLICATION TO TACHOMETER CONTROL

Tachometers are used to measure rotational speeds. Each rotation of the tachometer shaft produces one or more inductive or photoelectric pulses. The rotational speed is determined by counting and scaling the number of pulses per period.

To maintain a particular rotational speed, the output of the tachometer is fed back to a control circuit. The control logic block diagram of a *constant-speed control system* is shown in Fig. 60.11. Control circuits may contain various electrical and electronic devices.[8]

---

[6]Not all controlled systems are feedback systems. The positions of many precision devices (e.g., print heads in dot matrix printers, or cutting heads on some numerically controlled machines) are controlled by precision *stepper motors*. However, unless the device has feedback (e.g., a position sensor), it will have no way of knowing if it gets out of control.

[7]Sometimes the notation K_n is used for K, where n is the type of the system. In a *type 0 system*, a constant error signal results in a constant value of the output signal. In a *type 1 system*, a constant error signal results in a constant rate of change of the output signal. In a *type 2 system*, a constant error signal produces a constant second derivative of the output variable.

[8]The electronic circuit may contain low-pass filters, differential amplifiers, booster amplifiers, an adjustable voltage source, and so on. The actual integration of these devices is not generally part of mechanical engineering.

Figure 60.11 *Block Diagram for Constant-Speed Motor Control*

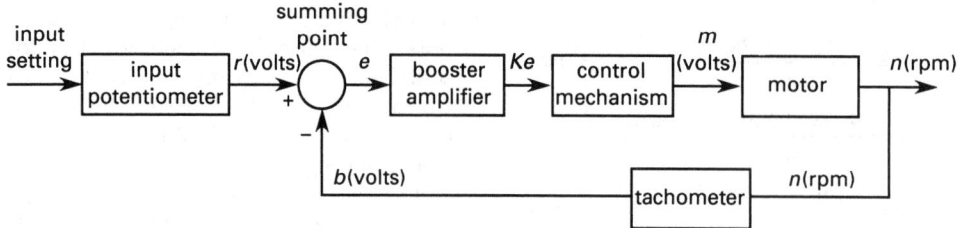

The desired speed is usually set by adjusting a voltage level. This is illustrated as an input potentiometer, which feeds the desired speed into a *summing point*. The tachometer feeds the actual speed into the summing point as well, which actually computes the difference (known as the *error*, e) between the two signals.

The error is amplified from e to Ke, and the amplified signal is fed into an appropriate *control mechanism*. There are many methods, both analog and digital, by which the signal can be translated by the control mechanism into a change in motor input. Figure 60.11 only requires that the transfer function (i.e., from Ke to m) be known. The changed motor speed, n, is converted to a tachometer signal, b, which is fed back to the summing point to complete the loop.

24. STATE MODEL REPRESENTATION

While the classical methods of designing and analyzing control systems are adequate for most situations, state model representations are preferred for more complex cases, particularly those with multiple inputs and outputs or when behavior is nonlinear or varies with time. This evaluation method almost always is carried out on a digital or analog computer.

The state variables completely define the dynamic state (position, voltage, pressure, etc.), $x_i(t)$, of the system at time t. (In simple problems, the number of state variables corresponds to the number of *degrees of freedom*, n, of the system.) The n state variables are written in matrix form as a state vector, $\mathbf{X}$.

$$\mathbf{X} = \begin{pmatrix} x_1 \\ x_2 \\ x_3 \\ \vdots \\ x_n \end{pmatrix} \qquad 60.44$$

It is a characteristic of state models that the state vector is acted upon by a first-degree derivative operator, d/dt, to produce a differential term, $\mathbf{X}'$, of order 1.

$$\mathbf{X}' = \frac{d\mathbf{X}}{dt} \qquad 60.45$$

Equations 60.46 and 60.47 show the general form of a state model representation: $\mathbf{U}$ is an r-dimensional (i.e., an $r \times 1$ matrix) *control vector*; $\mathbf{Y}$ is an m-dimensional (i.e., an $m \times 1$ matrix) *output vector*; $\mathbf{A}$ is an $n \times n$ *system matrix*; $\mathbf{B}$ is an $n \times r$ *control matrix*; and $\mathbf{C}$ is an $m \times n$ *output matrix*. The actual unknowns are the x_i's. The y_i's, which may not be needed in all problems, are only linear combinations of the x_i's. (For example, the x's might represent spring end positions; the y's might represent stresses in the spring. Then, $y = k\Delta x$.) Equation 60.46 is the *state equation*, and Eq. 60.47 is the *response equation*.

$$\mathbf{X}' = \mathbf{A}\mathbf{X} + \mathbf{B}\mathbf{U} \qquad 60.46$$
$$\mathbf{Y} = \mathbf{C}\mathbf{X} \qquad 60.47$$

A conventional block diagram can be modified to show the multiplicity of signals in a state model, as shown in Fig. 60.12.[9] The actual physical system does not need to be a feedback system. The form of Eqs. 60.46 and 60.47 is the sole reason that a feedback diagram is appropriate.

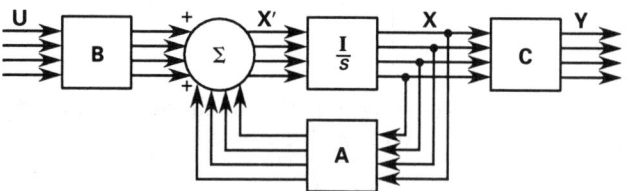

Figure 60.12 *State Variable Diagram*

A state variable model permits only first-degree derivatives, so additional x_i state variables are used for higher-order terms (e.g., acceleration).

System controllability exists if all of the system states can be controlled by the inputs, $\mathbf{U}$. In state model language, system controllability means that an arbitrary initial state can be steered to an arbitrary target state in a finite amount of time. *System observability* exists if the initial system states can be predicted from knowing the inputs, $\mathbf{U}$, and observing the outputs, $\mathbf{Y}$.[10]

---

[9]The block $\mathbf{I}/s$ is a diagonal identity matrix with elements of $1/s$. This effectively is an integration operator.
[10]*Kalman's theorem* based on matrix rank is used to determine system controllability and observability.

Example 60.18

Write the state variable formulation of a mechanical system's transfer function $T(s)$.

$$T(s) = \frac{7s^2 + 3s + 1}{s^3 + 4s^2 + 6s + 2}$$

Solution

Recognize the transfer function as the quotient of two terms, and multiply $T(s)$ by the dimensionless quantity x/x.

$$T(s) = \frac{Y(s)}{U(s)} = \left(\frac{7s^2 + 3s + 1}{s^3 + 4s^2 + 6s + 2}\right)\left(\frac{x}{x}\right)$$
$$Y(s) = 7s^2x + 3sx + x$$
$$U(s) = s^3x + 4s^2x + 6sx + 2x$$

$Y(s)$ and $U(s)$ represent the following differential equations.

$$y(t) = 7x''(t) + 3x'(t) + x(t)$$
$$u(t) = x'''(t) + 4x''(t) + 6x'(t) + 2x(t)$$

Make the following substitutions. ($x_4(t)$ is not needed because one level of differentiation is built into the state model.)

$$x_1(t) = x(t)$$
$$x_2(t) = x'(t) = x_1'(t)$$
$$x_3(t) = x''(t) = x_2'(t)$$

Write the first derivative variables in terms of the $x_i(t)$ to get the **A** matrix entries.

$$x_1' = \quad 0x_1(t) + 1x_2(t) + 0x_3(t) + 0$$
$$x_2' = \quad 0x_1(t) + 0x_2(t) + 1x_3(t) + 0$$
$$x_3' = -2x_1(t) - 6x_2(t) - 4x_3(t) + u(t)$$

Determine the coefficients of the **C** matrix by rewriting $y(t)$ in the same variable order.

$$y(t) = 1x_1(t) + 3x_2(t) + 7x_3(t)$$

From Eqs. 60.46 and 60.47, the state variable representation of $T(s)$ is

$$\begin{bmatrix} x_1' \\ x_2' \\ x_3' \end{bmatrix} = \begin{bmatrix} 0 & 1 & 0 \\ 0 & 0 & 1 \\ -2 & -6 & -4 \end{bmatrix} \begin{bmatrix} x_1 \\ x_2 \\ x_3 \end{bmatrix} + \begin{bmatrix} 0 \\ 0 \\ 1 \end{bmatrix} [r(t)]$$

$$[y(t)] = [1 \ 3 \ 7] \begin{bmatrix} x_1 \\ x_2 \\ x_3 \end{bmatrix}$$

PRACTICE PROBLEMS

1. Simplify the following block diagrams and determine the overall system gain.

(a)

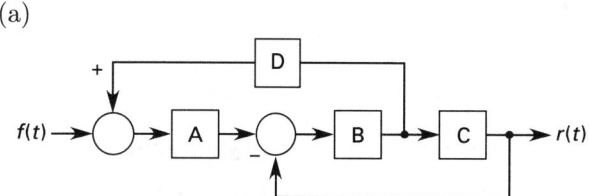

(b)

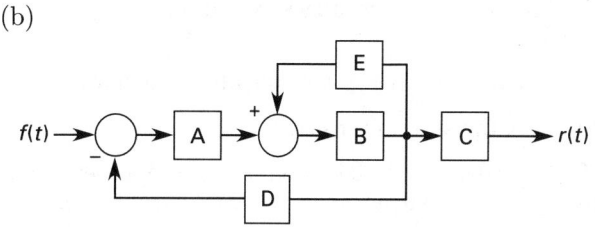

2. *(Time limit: one hour)* A mass of 100 lbm (45 kg) is supported uniformly by a spring system. The spring system has a combined stiffness of 1200 lbf/ft (17.5 kN/m). A dashpot with a damping coefficient of 60 lbf-sec/ft (880 N·s/m) has been installed. (a) What is the undamped natural frequency? (b) What is the damping ratio? (c) Sketch the magnitude and phase characteristics of the frequency response. (d) Sketch the response to a unit step input.

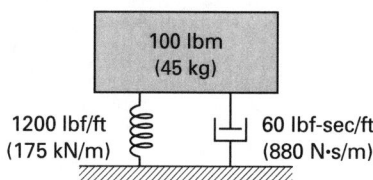

3. *(Time limit: one hour)* A constant-speed motor/ magnetic clutch drive train is monitored and controlled by a speed-sensing tachometer. The entire system is modeled as a control system block diagram, as shown. (The lowercase letters represent small-signal increments from the reference values.) When the control system is operating, the desired motor speed, n (in rpm), is set with a speed-setting potentiometer. The setting is compared to the tachometer output. The comparator output error (in volts), controls the clutch. A current, i (in amps), passes through the clutch coil. The external load torque, $t_L m$ (in in-lbf), is seen by the clutch and is countered by the clutch output torque, t (in in-lbf).

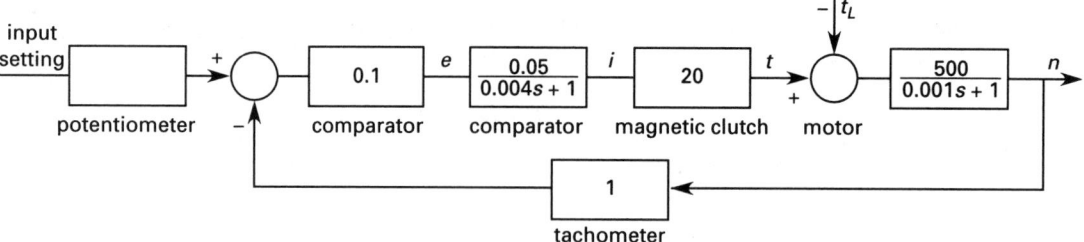

(a) Plot the open-loop frequency response.

(b) What is the open-loop steady-state gain?

(c) Plot the unity feedback closed-loop frequency response.

(d) What is the closed-loop steady-state gain?

(e) Plot the system sensitivity.

(f) Describe the closed-loop response to a step change in the desired output angular velocity. Is it damped or oscillatory? Is there a steady-state error? Why or why not?

(g) Describe the closed-loop response to a step change in the load torque. Is the response damped or oscillatory? Is there a steady-state error? Why or why not?

(h) Assume that you have to select the comparator gain and that it doesn't have to be 0.1. Using the root-locus method or either the Routh or Nyquist stability criterion, find the limits of the comparator gain that cause the closed-loop system to be unstable.

(i) How can you improve the steady-state response of the closed-loop system to constant disturbances in the load torque, t_L?

61 Management Science

1. Introduction 61-1
2. Critical Path Techniques 61-3
3. Queuing Theory 61-6
4. Reliability 61-9
5. Preventative Maintenance 61-10
6. Replacement 61-11
7. Linear Programming 61-12
8. Plant Location 61-13
9. Facilities Layout 61-14
10. Work Measurement 61-15
11. Work Sampling 61-17
12. Assembly Line Balancing 61-18
13. Wage Incentive Plans 61-20
14. Quality Control Charts 61-22
15. Quality Acceptance Sampling 61-25
16. Risk Assessment 61-27
17. Work Methods 61-27
18. Behavioral Science 61-27
19. Human Factors Engineering 61-29
20. Machine Safeguarding 61-30
    Practice Problems 61-30

Nomenclature

c	cycle time	various
c	number of failures (defects)	–
c_2	ratio of $\overline{\sigma}/\sigma'$	–
C	confidence level	–
C	cost	various
d_2	ratio of $\overline{R}/\sigma'$	–
D	duration	various
k	number	–
L	expected line length	–
L_q	expected queue length	–
M	number of consecutive samples	–
MTBF	mean time between failures	various
MTBFO	mean time before failure outage	various
MTTF	mean time to failure	various
MTTR	mean time to repair	various
n	number	–
n	actual number of observations	–
N	sample size	–
N	theoretical number of observations	–
p	observed fraction (n/N)	–
p	true fraction defective	–
$p\{x\}$	probability of event x	–
P	precision	–
Q	quantity	–
R	range	various
R	reliability	–
RB	relative bonus	–
RE	relative earnings	–

s	number of servers in the system	–
s	sample standard deviation	various
t	time	various
t_C	Student's t-distribution factor	–
W	waiting time in the system	various
W_q	waiting time in the queue	various
x	general variable	–
z	standard normal variable	–
Z	objective function	–

Symbols

α	producer's risk	–
β	consumer's risk	–
ϵ	absolute error	various
λ	mean of the Poisson distribution	–
λ	mean arrival rate	1/time
λ	mean failure rate	1/time
μ	distribution mean	various
μ	mean service (repair) rate	1/time
ν	number of degrees of freedom	–
ρ	utilization factor, $\lambda/s\mu$	–
σ	standard deviation (of the sample)	various
σ'	standard deviation of the population	various

Subscripts

ave	average
C	at confidence level C
q	queue
std	standard
t	total, or at time t

1. INTRODUCTION

Management science, also known as *quantitative business analysis, operations research,* and *management systems modeling*, is used to develop mathematical models of real-world situations. This chapter presents various quantitative business analysis techniques used to analyze manufacturing and industrial models. Accordingly, this chapter is more concerned with solutions to problems than with explaining why the problems need to be solved or with listing advantages and disadvantages of solutions. Though they may seem to be obscure, all of the techniques presented in this chapter are commonly taught in operations research (O.R.), industrial engineering, and MBA curricula.[1]

---

[1] *Operations research* developed as a field of its own during World War II when optimizing modeling techniques were used to determine the best way for submarines to patrol a specific region.

PROFESSIONAL PUBLICATIONS, INC. BELMONT, CA

A *deterministic model* is a mathematical model that is built around a set of fixed rules such that any given input always results in a specific output. If an input can produce a variety of outputs determined by rules of probability, the model is known as a *probabilistic* or *stochastic model*.

A common aspect of most management science techniques is the goal of arriving at an optimum solution (regardless of whether the goal is actually realized). The process of optimizing is unique to each type of problem. Calculus is not generally used in optimizing.[2] Optimizing real-world problems always requires a computer, though optimization by hand is possible with simple problems.[3]

Some management science problems attempt to optimize a specific mathematical function known as the *objective function*, Z. If Z is a profit function, it is optimized by maximization; if Z is a cost or time function, it is optimized by minimization.[4] Some management science techniques can maximize only, so in cases requiring minimization, the negative of the objective function is maximized.

Objective functions are restricted from increasing without being bound by *constraints* placed on one or more of the function's variables. These constraints are typically mathematical representations of how resources are limited or combined. Non-negativity constraints are common in mathematical programming problems.

If the objective function and its constraints are linear combinations of the independent variables, the model is said to be a *linear model*. Otherwise, the model is nonlinear.

Not all manufacturing management problems need to be solved by complex or obscure procedures. Some problems (e.g., facilities layout) do not have a general solution procedure and must be solved by exhaustive enumeration. Many problems, such as Exs. 61.1 and 61.2, can be solved simply by using common sense and logical thinking to minimize the total cost.

---

[2]One obvious exception is how the economic order quantity is calculated. (See Ex. 61.1 and Chap. 72.) The EOQ formula is derived by taking the derivative of the total cost function.

[3]Some management science techniques, though interesting, are too obtuse, time consuming, or complex for solving by hand. Subjects that have been omitted from or given only a mere mention in this edition include nonlinear programming, dynamic programming, integer programming, simulation (including the Monte Carlo technique), decision theory, and Markov processes. Furthermore, most management science subjects have many complicated variations that are omitted from this chapter. Simple forecasting and the economic order quantity (EOQ) model, also traditional management science subjects, are covered in Chap. 72.

[4]There is an important difference between *cost* and *price*. Both represent an amount paid, but the distinction depends on who makes the payment and when the payment is made. To one party, the cost of materials incorporated into a manufactured item is the price paid by that party for those materials. That is, there is no difference. However, the cost to one party to acquire or produce an item is much lower than the price at which the item is later sold to a second party.

Example 61.1

A particular part is used by a company at a uniform rate of 120 units per month. The part is obtained from the supplier at a cost of \$20 per unit. The company's cost of stocking the product is \$0.06 per unit per month. The prorated cost of placing an order and putting shipments into inventory is \$0.07 per unit over all normal order quantities. The company's effective monthly interest rate on borrowed money is 0.5% per month. The inventory is initially full. Orders arrive instantaneously when needed, and shortages do not occur. What is the optimum stocking quantity of the product?

Solution

This is essentially an economic order quantity (EOQ) problem. (See Chap. 69.) However, the ordering cost is expressed per unit ordered, and therefore, the total ordering cost is initially unknown. An iterative approach is necessary.

The interest expense on money tied up in a unit of inventory is

$$\left(\frac{0.005}{\text{month}}\right)\left(\frac{\$20}{\text{unit}}\right) = \frac{\$0.10}{\text{unit-month}}$$

Let Q be the quantity ordered. The total cost per month is

$$C_t = \text{cost of ordering} + \text{cost of stocking}$$
$$= \left(\frac{\text{no. of orders}}{\text{month}}\right)\left(\frac{\text{cost}}{\text{order}}\right)$$
$$+ (\text{average monthly inventory})$$
$$\times (\text{stocking} + \text{interest costs})$$

Initially assume that the order quantity, Q, is 100. Then, the cost of placing an order will be

$$\left(\frac{100 \text{ units}}{\text{order}}\right)\left(\frac{\$0.07}{\text{unit}}\right) = \$7.00/\text{order}$$

The inventory drops linearly from Q to zero over time, so the average inventory at any moment is $Q/2$.

$$C_t = \left(\frac{\dfrac{120}{\text{month}}}{Q}\right)(\$7.00) + \left(\frac{Q}{2}\right)(\$0.06 + \$0.10)$$
$$= \frac{\$840}{Q} + \$0.08Q$$

This is minimized by setting the derivative of the cost function equal to zero.

$$\frac{dC_t}{dQ} = \frac{-\$840}{Q^2} + 0.08 = 0$$
$$Q = 102.5 \text{ units}$$

This is close to the initial estimate of Q. (Further iterations refine the value to exactly 105 units.)

Example 61.2

Three different semi-automatic machines are being evaluated as a replacement to a completely manual operation. Each machine is mutually exclusive, and only one machine can be selected. Each machine has a different level of automation, cost of operation, and fraction of generated defects. Which machine should be selected in each production quantity range?

machine	set-up cost	per unit material cost	per unit labor cost	fraction defective
A	$200	$0.47	0.56	0.06
B	$700	$0.52	0.35	0.03
C	$1200	$0.54	0.27	0.02

Solution

When a process has a *scrap rate* greater than zero or a *yield* less than 100% (i.e., a fraction of the items produced are defective), the total cost per saleable item produced should be minimized.

$$\frac{cost}{saleable\ item} = \frac{\dfrac{cost}{aggregate\ item}}{1 - fraction\ defective}$$

Solving this example requires determining each machine's cost per saleable item as a function of production quantity. The setup cost is a *fixed cost*, allocated over all units produced. The material and labor costs are *variable costs*. The fraction defective is a scale factor that increases the cost of all saleable items produced. Let x represent the total number of all items (saleable and nonsaleable) produced. The unit cost per saleable item is

$$C = \frac{material\ cost + labor\ cost + \dfrac{set\text{-}up\ cost}{x}}{1 - fraction\ defective}$$

$$C_A = \frac{\$0.47 + \$0.56 + \dfrac{\$200}{x}}{1 - 0.06}$$

$$= \$1.10 + \frac{\$212.77}{x}$$

$$C_B = \frac{\$0.52 + \$0.35 + \dfrac{\$700}{x}}{1 - 0.03}$$

$$= \$0.90 + \frac{\$721.65}{x}$$

$$C_C = \frac{\$0.54 + \$0.27 + \dfrac{\$1200}{x}}{1 - 0.02}$$

$$= \$0.83 + \frac{\$1224.49}{x}$$

The three cost equations are not linear functions, and straight lines cannot be drawn between two points on the curves. Graphical or algebraic operations show that C_A is minimum for $0 < x < 2544$, C_B is minimum for $2544 < x < 7183$, and C_C is minimum for $x > 7183$.

2. CRITICAL PATH TECHNIQUES

Definitions

activity: any subdivision of a project whose execution requires time and other resources.

critical path: a path connecting all activities that have minimum or zero slack times. The critical path is the longest path through the network.

duration: the time required to perform an activity. All durations are *normal durations* unless otherwise referred to as *crash durations*.

event: the beginning or completion of an activity.

event time: actual time at which an event occurs.

float: same as slack time.

slack time: the minimum time that an activity can be delayed without causing the project to fall behind schedule. Slack time is always minimum or zero along the critical path.

Introduction

Critical path techniques are used to graphically represent the multiple relationships between stages in a complicated project. The graphical network shows the *precedence relationships* between the various activities. The graphical network can be used to control and monitor the progress, cost, and resources of a project. A critical path technique will also identify the most critical activities in the project.

Critical path techniques use *directed graphs* to represent a project. These graphs are made up of *arcs* (arrows) and *nodes* (junctions). The placement of the arcs and nodes completely specifies the precedences of the project. Durations and precedences are usually given in a *precedence table (matrix)*.

Activity-on-Node Networks: The Critical Path Method

One technique is known as the *critical path method*, CPM. This deterministic method is applicable when all activity durations are known in advance. CPM is usually represented as an *activity-on-node* model since arcs are used to specify precedence and the nodes actually represent the activities. Events are not represented on the graph, other than as the heads and tails of the arcs. Two *dummy nodes* taking zero time may be used to specify the start and finish of the project.

Solving a CPM Problem

The solution to a critical path method problem reveals the earliest and latest times that an activity can be started and finished. It also identifies the *critical path* and generates the *slack times* for each activity.

The following procedure may be used to solve a CPM problem. To facilitate the solution, each node should be replaced by a square that has been quartered. The compartments have the meanings indicated by the key.

ES	EF
LS	LF

key

ES: Earliest Start

EF: Earliest Finish

LS: Latest Start

LF: Latest Finish

step 1: Place the project start time or date in the **ES** and **EF** positions of the start activity. The start time is zero for relative calculations.

step 2: Consider any unmarked activity, all of whose predecessors have been marked in the **EF** and **ES** positions. (Go to step 4 if there are none.) Mark in its **ES** position the largest number marked in the **EF** position of those predecessors.

step 3: Add the activity time to the **ES** time and write this in the **EF** box. Go to step 2.

step 4: Place the value of the latest finish date in the **LS** and **LF** boxes of the finish mode.

step 5: Consider unmarked predecessors whose successors have all been marked. Their **LF** is the smallest **LS** of the successors. Go to step 7 if there are no unmarked predecessors.

step 6: The **LS** for the new node is **LF** minus its activity time. Go to step 5.

step 7: The slack time for each node is **LS−ES** or **LF−EF**.

step 8: The critical path encompasses nodes for which the slack time equals **LS−ES** from the start node. There may be more than one critical path.

Example 61.3

Using the precedence table given, construct the precedence matrix and draw an activity-on-node network.

activity	duration (days)	predecessors
A, start	0	–
B	7	A
C	6	A
D	3	B
E	9	B, C
F	1	D, E
G	4	C
H, finish	0	F, G

Solution

The precedence matrix is

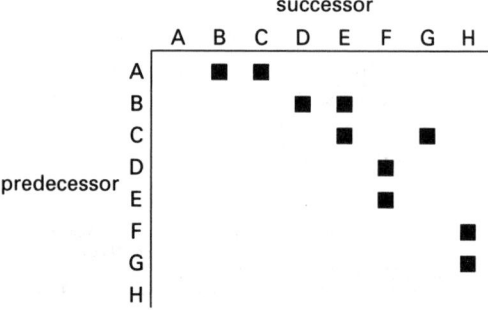

The activity-on-node network is

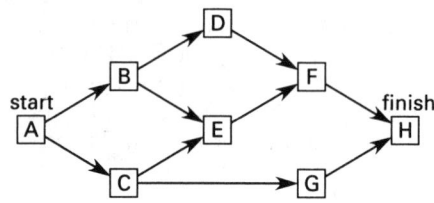

Example 61.4

Complete the network for the previous example and find the critical path. Assume the desired completion date is in 19 days.

Solution

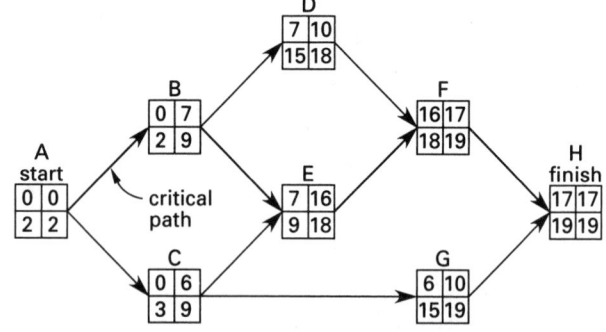

Activity-on-Branch Networks

In an *activity-on-branch* network, the arcs represent the activities, which are labeled with letters of the alphabet, and the nodes represent events, which are numbered. The activity durations may appear in parentheses near the activity letter.

The activity-on-branch method is complicated by the frequent requirement for *dummy activities* and nodes to maintain precedence. Consider the following part of a precedence table:

activity	predecessors
L	–
M	–
N	L, M
P	M

Note that activity P depends on the completion of only M. Figure 61.1(a) is an activity-on-branch representation of this precedence. However, N depends on the completion of both L and M. It would be incorrect to draw the network as Fig. 61.1(b) since the activity N appears twice. To represent the project, the dummy activity X must be used, as shown in Fig. 61.1(c).

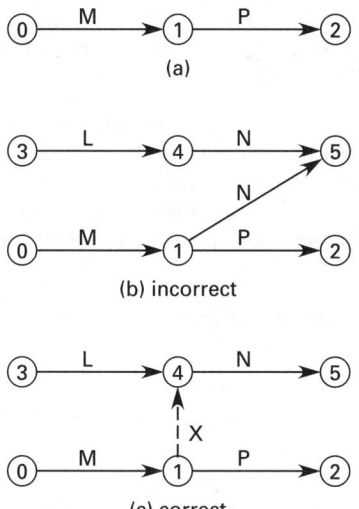

Figure 61.1 Activity-on-Branch Networks

If two activities have the same starting and ending events, a dummy node is required to give one activity a uniquely identifiable completion event. This is illustrated in Fig. 61.2(b).

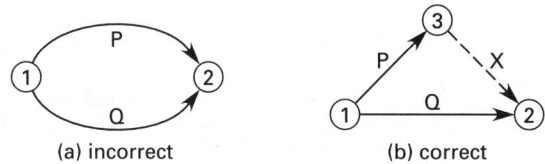

Figure 61.2 Use of a Dummy Node

The solution method for an activity-on-branch problem is essentially the same as for the activity-on-node problem, requiring forward and reverse passes to determine earliest and latest dates.

Example 61.5

Represent the project in Ex. 61.3 as an activity-on-branch network.

Solution

event	event description
0	start project
1	finish B, start D
2	finish C, start G
3	finish B and C, start E
4	finish D and E, start F
5	finish F and G

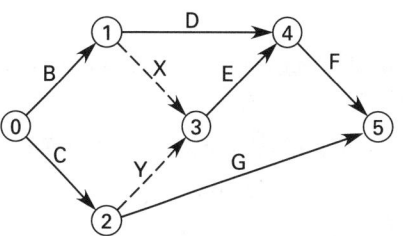

Stochastic Critical Path Models

Stochastic models differ from deterministic models only in the way in which the activity durations are found. Whereas durations are known explicitly for the deterministic model, the time for a stochastic activity is distributed as a random variable.

This stochastic nature complicates the problem greatly since the actual distribution is often unknown. Such problems are solved as a deterministic model using the mean of an assumed duration distribution as the activity duration.

The most common stochastic critical path model is PERT, which stands for *program evaluation and review technique*. In PERT, all duration variables are assumed to come from a *beta distribution*, with mean and standard deviation given by Eqs. 61.1 and 61.2, respectively.

$$\mu = \left(\tfrac{1}{6}\right)\left(t_{\text{minimum}} + 4t_{\text{most likely}} + t_{\text{maximum}}\right) \quad \textit{61.1}$$

$$\sigma = \left(\tfrac{1}{6}\right)\left(t_{\text{maximum}} - t_{\text{minimum}}\right) \quad \textit{61.2}$$

The project *completion time* for large projects is assumed to be normally distributed with mean equal to the critical path length, and overall variance equal to the sum of the variances along the critical path.

The probability that a project duration will exceed some length (D) can be found from Eq. 61.3. z is the standard normal variable.

$$p\{\text{duration} > D\} = p\{x > z\} \qquad \textit{61.3}$$

$$z = \frac{D - \mu_{\text{critical path}}}{\sigma_{\text{critical path}}} \qquad \textit{61.4}$$

Example 61.6

The mean times and variances for activities along a PERT critical path are given. What is the probability that the project's completion time will be (a) less than 14 days, (b) more than 14 days, (c) more than 23 days, and (d) between 14 and 23 days?

activity mean time (days)	activity standard deviation (days)
9	1.3
4	0.5
7	2.6

Solution

The most likely completion time is the sum of the mean activity times.

$$\mu_{\text{critical path}} = 9 \text{ days} + 4 \text{ days} + 7 \text{ days}$$
$$= 20 \text{ days}$$

The variance of the project's completion times is the sum of the variances along the critical path. Variance, σ^2, is the square of the standard deviation, σ.

$$(\sigma_{\text{critical path}})^2 = (1.3 \text{ days})^2 + (0.5 \text{ days})^2$$
$$+ (2.6 \text{ days})^2$$
$$= 8.7 \text{ days}^2$$

$$\sigma_{\text{critical path}} = \sqrt{8.7 \text{ days}^2}$$
$$= 2.95 \text{ days} \quad [\text{use 3 days}]$$

The standard normal variable corresponding to 14 days is

$$z = \frac{D - \mu_{\text{critical path}}}{\sigma_{\text{critical path}}}$$
$$= \frac{14 \text{ days} - 20 \text{ days}}{3 \text{ days}} = -2.0$$

The area under the standard normal curve is 0.4772 for $0.0 < z < -2.0$. Since the normal curve is symmetrical, the negative sign is irrelevant in determining the area.

The standard normal variable corresponding to 23 days is

$$z = \frac{D - \mu_{\text{critical path}}}{\sigma_{\text{critical path}}}$$
$$= \frac{23 \text{ days} - 20 \text{ days}}{3 \text{ days}} = 1.0$$

The area under the standard normal curve is 0.3413 for $0.0 < z < 1.0$.

(a) $p\{\text{duration} < 14\} = p\{x < -2.0\} = 0.5 - 0.4772$
$$= 0.0228 \ (2.28\%)$$

(b) $p\{\text{duration} > 14\} = p\{x > -2.0\} = 0.4772 + 0.5$
$$= 0.9772 \ (97.72\%)$$

(c) $p\{\text{duration} > 23\} = p\{x > 1.0\} = 0.5 - 0.3413$
$$= 0.1587 \ (15.87\%)$$

(d) $p\{14 < \text{duration} < 23\} = p\{-2.0 < x < 1.0\}$
$$= 0.4772 + 0.3413$$
$$= 0.8185 \ (81.85\%)$$

3. QUEUING THEORY

Introduction

A *queue* is a waiting line. *Queuing theory* can predict the length of a waiting line, the average time a customer can expect to spend in the queue, and the probability that n customers will be in the queue.

Different queuing models have been developed to account for differences in the distribution of arrivals, distribution of service times, number of servers, size of the calling population, and order in which the customers are served.

Most of the models are too complicated to be presented here. However, two models are given after a brief listing of the general relationships. The relationships given here are for *steady-state operation*. This means that the service facility has been open and in operation for some time.

General Relationships

$$L = \lambda W \qquad \textit{61.5}$$
$$L_q = \lambda W_q \qquad \textit{61.6}$$
$$W = W_q + \frac{1}{\mu} \qquad \textit{61.7}$$
$$\lambda < \mu s \qquad \textit{61.8}$$
$$\text{average service time} = 1/\mu \qquad \textit{61.9}$$
$$\text{average time between arrivals} = 1/\lambda \qquad \textit{61.10}$$

The M/M/1 System

It is assumed that the following points are true in the $M/M/1$ model.[5]

- There is only one server $(s = 1)$.

- The calling population is infinite.

---

[5]In the nomenclature of queuing theory, an exponentially distributed interarrival or service time is given the symbol M (for Markovian). Other options include D (deterministic, constant, regular, etc.), E or K (for Erlangian or gamma distribution), and G (general distribution).

- The service times are exponentially distributed with mean μ. That is, the probability of a customer's remaining service time exceeding h (after already spending time with the server) is

$$p\{t > h\} = e^{-\mu h} \qquad 61.11$$

Notice that Eq. 61.11 is independent of the time already spent with the server.

- The arrival rate is distributed as Poisson with mean λ. The probability of x customers arriving in the next period is

$$p\{x\} = \frac{e^{-\lambda} \lambda^x}{x!} \qquad 61.12$$

The following relationships are valid for the $M/M/1$ system.

$$p\{0\} = 1 - \rho = \frac{\mu - \lambda}{\mu} \qquad 61.13$$

$$p\{n\} = p\{0\}\rho^n \qquad 61.14$$

$$W = \frac{1}{\mu - \lambda} = W_q - \frac{1}{\mu} = \frac{L}{\lambda} \qquad 61.15$$

$$W_q = \frac{\rho}{\mu - \lambda} = \frac{L_q}{\lambda} \qquad 61.16$$

$$L = \frac{\lambda}{\mu - \lambda} = L_q + \rho \qquad 61.17$$

$$L_q = \frac{\rho \lambda}{\mu - \lambda} \qquad 61.18$$

Example 61.7

Given an $M/M/1$ system with $\mu = 20$ customers per hr and $\lambda = 12$ per hr, find the steady-state values of W, W_q, L, and L_q. What is the probability that there will be five customers in the system?

Solution

$$\rho = \frac{\lambda}{\mu} = \frac{12}{20} = 0.6$$

$$W = \frac{1}{\mu - \lambda} = \frac{1}{20 - 12} = 0.125 \text{ hr}$$

$$W_q = \frac{\rho}{\mu - \lambda} = \frac{0.6}{20 - 12} = 0.075 \text{ hr}$$

$$L = \frac{\lambda}{\mu - \lambda} = \frac{12}{20 - 12} = 1.5 \text{ customers}$$

$$L_q = \frac{\rho \lambda}{\mu - \lambda} = \frac{(0.6)(12)}{20 - 12} = 0.9 \text{ customers}$$

$$p\{0\} = 1 - \rho = 1 - 0.6 = 0.4$$

$$p\{5\} = p\{0\}\rho^n = (0.4)(0.6)^5 = 0.031$$

The *M/M/s* System

The same assumptions are used for the $M/M/s$ system as were used for the $M/M/1$ system, except that there are s servers instead of only one. Each server has a mean service rate μ. All servers draw from a single line so that the first person in line goes to the first available server. Each server does not have its own line. However, if customers are allowed to change the lines they are in so that they go to any available server, this model may also be used to predict the performance of a multiple server system where each server has its own line.

$$\rho = \frac{\lambda}{s\mu} \qquad 61.19$$

$$W = W_q + \frac{1}{\mu} \qquad 61.20$$

$$W_q = \frac{L_q}{\lambda} \qquad 61.21$$

$$L_q = \frac{p\{0\}\rho^{s+1}}{s!(1 - \rho)^2} \qquad 61.22$$

$$L = L_q + \rho \qquad 61.23$$

$$p\{0\} = \frac{1}{\dfrac{\rho^s}{s!\left[1 - \left(\dfrac{\rho}{s}\right)\right]} + \displaystyle\sum_{j=0}^{s-1} \dfrac{\rho^j}{j!}} \qquad 61.24$$

$$p\{n\} = \frac{p\{0\}\rho^n}{n!} \quad [n \le s] \qquad 61.25$$

$$p\{n\} = \frac{p\{0\}\rho^n}{s!s^{n-s}} \quad [n > s] \qquad 61.26$$

Figure 61.3 is a graphical solution to the $M/M/s$ multiple server model.

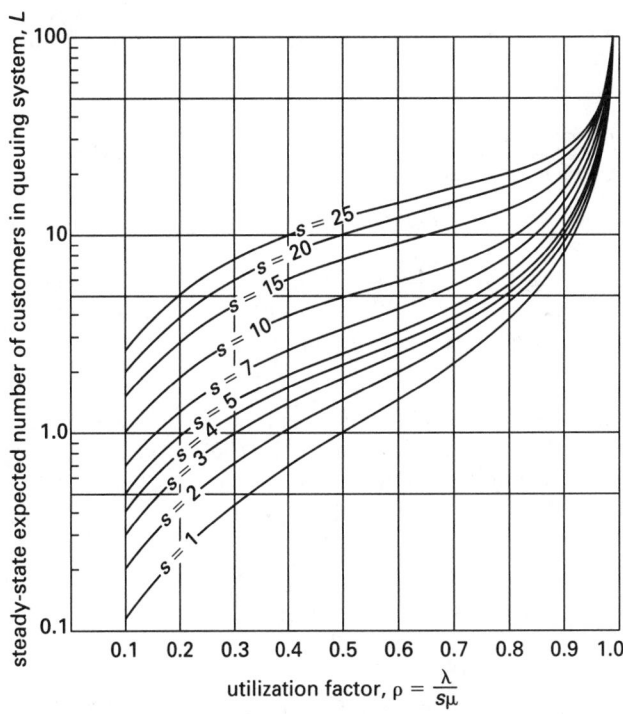

Reprinted from *Operations Research*, 6th Ed., by Frederick S. Hillier and Gerald J. Lieberman, Holden-Day, Inc., with permission of McGraw-Hill, Inc., © 1974.

Figure 61.3 *Mean Number in System (L) for M/M/s System*

Example 61.8

A company has several identical machines operating in parallel. The average breakdown rate is 0.7 machines per week. There is one repair station for the entire company. It takes a maintenance worker one entire week to repair a machine, although the time is reduced in proportion to the number of maintenance workers assigned to the repair. Each maintenance worker is paid $400 per month. Machine downtime is valued at $800 per week. Other costs (additional tools, etc.) are to be disregarded. What is the optimum number of maintenance workers at the repair station?

Solution

Assume the number of breakdowns per week is a Poisson distribution. Then, this example can be solved with queuing theory. Using two or more maintenance workers only decreases the repair time, so this is a single-server model, even with multiple maintenance workers.

The average number of machines breaking down each week is the mean arrival rate, $\lambda = 0.7$ per week. With one worker, the repair rate, μ, is 1.0 per week.

The average time a machine is out of service (waiting for its turn to be repaired and during the repair) is W, the "time in the system."

$$W = \frac{1}{\mu - \lambda} = \frac{1}{1 - 0.7}$$
$$= 3.33 \text{ weeks}$$

With one maintenance worker, the average downtime cost in a week is

$$\left(\frac{\text{downtime cost}}{\text{machine-week}}\right)(\text{no. of machines})(\text{no. of weeks})$$

$$= (\text{downtime cost})\lambda W$$

$$= \left(\frac{\$800}{\text{machine-week}}\right)(0.7 \text{ machines})(3.33 \text{ weeks})$$

$$= (\$800)(2.33) = \$1864$$

However, the product of λ and W is the same as the average number of machines in the system, L. The average number of machines in the system, L (i.e., being repaired or waiting for repair), is

$$L_1 = \frac{\lambda}{\mu - \lambda} = \frac{0.7}{1 - 0.7}$$
$$= 2.33$$

The total average weekly cost with one worker is the sum of the costs of the worker and the downtime.

$$C_{t,1} = (1 \text{ worker})\left(\frac{\$400}{\text{week}}\right) + (2.33)(\$800) = \$2264$$

With two workers, the values are

$$\mu = (2 \text{ workers})\left(\frac{1}{\text{worker-week}}\right) = 2/\text{week}$$

$$L_2 = \frac{0.7}{2 - 0.7} = 0.538$$

$$C_{t,2} = (2)\left(\frac{\$400}{\text{week}}\right) + (0.538)(\$800) = \$1230$$

With three workers, the values are

$$\mu = (3 \text{ workers})\left(\frac{1}{\text{worker-week}}\right) = 3/\text{week}$$

$$L_3 = \frac{0.7}{3 - 0.7} = 0.304$$

$$C_{t,3} = (3)\left(\frac{\$400}{\text{week}}\right) + (0.304)(\$800) = \$1443$$

Adding workers will increase the costs above C_3. Two maintenance workers should staff the maintenance station.

Example 61.9

Twenty identical machines are in operation. The hourly reliability for any one machine is 90%. (That is, the probability of a machine breaking down in any given hour is 10%.) The cost of downtime is $5 per hour. Each broken machine requires one technician for repair, and the average repair time is one hour. If all technicians are busy, broken machines wait idle. Each technician costs $2.5 per hour. How many separate technicians should be used?

Solution

This is a multiple-server model. It is assumed that the $M/M/s$ assumptions are satisfied. The mean arrival rate, λ, is $(0.10)(20) = \$2$ per hour. The repair rate, μ, is $1 per hour. Clearly, one technician cannot handle the workload, nor can two technicians. The average number of machines in the system, L (i.e., being repaired or waiting for repair), is calculated for two, three, four, and five servers. The total cost per hour is calculated as

$$C = 2.5s + 5L$$

s	p_0	L_q	L	cost per hour
2	–	–	–	infinite
3	0.11	0.91	2.91	22.0
4	0.13	0.17	2.17	20.9
5	0.13	0.04	2.04	22.7

The minimum hourly cost is achieved with four technicians.

4. RELIABILITY

A *fault* in a machine or other system is a known cause of breakdown. An *error* is an undesired state within the machine that might lead to improper operation. A *failure* occurs when the machine fails to operate as expected or intended. A *fault-tolerant system* contains provisions to avoid failures after faults occur and cause errors in the machine.

In the most common case, units fail permanently and are neither repaired nor replaced. Reliability of a single item (machine, unit, piece of equipment, etc.) is characterized by its *mean time to failure*, MTTF. The term *mean time between forced outages*, MTBFO, is used with redundant systems in place of MTTF. *Coverage* is the probability of the system reconfiguring itself when a fault occurs. *Redundancy* is the primary tool used to increase reliability and coverage. Systems with two units in parallel are known as *duplex systems*. Systems with three units are known as *triple modular redundancy*, TMR, systems.[6]

The exponential distribution is most frequently used in reliability calculations.[7] The *failure rate*, λ, is the expected number of failures per unit time. Then, the *reliability* is

$$R_t = e^{-\lambda t} \qquad \textit{61.27}$$

The MTTF is found by integrating the reliability function. Therefore, the MTTF is the reciprocal of the failure (arrival) rate.

$$\text{MTTF} = \int_0^\infty R_t dt = \int_0^\infty e^{-\lambda t} dt$$
$$= \frac{1}{\lambda} \qquad \textit{61.28}$$

The probability of exactly c failures in time t is given by the Poisson distribution. Appendix 61.D can be used to find the probability of c or fewer failures in time t.

$$p\{c\} = \frac{\left(\dfrac{t}{\text{MTTF}}\right)^c e^{-\frac{t}{\text{MTTF}}}}{c!} \qquad \textit{61.29}$$

For a 1-out-of-n *fully redundant system* (a parallel system with n identical redundant units, only one of which needs to be operational for the system to operate), the reliability is

$$R_{\text{1-out-of-}n \text{ system}} = 1 - (1 - R_t)^n \qquad \textit{61.30}$$

---

[6]With TMR systems, only one unit is required for successful operation. When one unit fails, the system becomes a duplex system until the failed unit is repaired. With logic, software, electronic, and computer systems, failure can be determined by comparing the output of each of the three units. In effect, the two good units "vote" to determine which unit is faulty and should be shut down.

[7]The three-parameter *Weibull distribution* is more descriptive, flexible, and powerful than the negative exponential distribution. It has gained acceptance primarily in the aerospace industry because of its ability to model the failure distribution more exactly. Its complexity, however, makes application to noncritical applications cumbersome.

For a 1-out-of-2 redundant system (also known as a *2-1-0 system*),

$$R_{\text{1-out-of-2}} = 2e^{-\lambda t} - e^{-2\lambda t} \qquad \textit{61.31}$$
$$\text{MTTF}_{\text{1-out-of-2}} = \int_0^\infty R_t dt$$
$$= \int_0^\infty 2e^{-\lambda t} - e^{-2\lambda t}$$
$$= \frac{1.5}{\lambda} \qquad \textit{61.32}$$

For a 1-out-of-3 fully redundant system (also known as a *3-2-1-0 system*),

$$R_{\text{1-out-of-3}} = 3e^{-\lambda t} - 3e^{-2\lambda t} + e^{-3\lambda t} \qquad \textit{61.33}$$
$$\text{MTTF}_{\text{1-out-of-3}} = \frac{11}{6\lambda} \qquad \textit{61.34}$$

In some cases, failed units are repaired on-line. The average repair time is the *mean time to repair*, MTTR, and is the reciprocal of the repair rate, μ. The *mean time to failure*, MTTF, is the average time a unit operates before failing. The *mean time between failures*, MTBF, is the length of time between when the original and repaired units start.

$$\text{MTBF} = \text{MTTF} + \text{MTTR} \qquad \textit{61.35}$$
$$\text{MTTR} = \frac{1}{\mu} \qquad \textit{61.36}$$

Availability is the probability that a system will be operating at any given time. The system *uptime* is calculated by multiplying the availability by the theoretically maximum number of operational hours (e.g., 8760 hours per year).[8] *Unavailability* and *downtime* are similarly calculated.

$$\text{availability} = \frac{\text{MTTF}}{\text{MTBF}} \qquad \textit{61.37}$$
$$\text{unavailability} = 1 - \text{availability} \qquad \textit{61.38}$$
$$\text{uptime} = (\text{availability})\left(8760\,\frac{\text{hr}}{\text{yr}}\right) \qquad \textit{61.39}$$
$$\text{downtime} = (\text{unavailability})\left(8760\,\frac{\text{hr}}{\text{yr}}\right) \qquad \textit{61.40}$$

The MTBF values for fully redundant systems can be calculated from a Markov model of the system.[9]

$$\text{MTBF}_{\text{1-out-of-2}} = \frac{\mu}{2\lambda^2} \qquad \textit{61.41}$$
$$\text{MTBF}_{\text{1-out-of-3}} = \frac{\mu^2}{3\lambda^3} \qquad \textit{61.42}$$

---

[8]If the machine does not operate 24 hours per day or 365 days per year, the number of hours will be accordingly reduced.

[9]Markov chains is a management science subject not discussed in this chapter.

Repairing a machine as soon as it breaks down is always the preferred course of action. In some cases, however, a faulty machine can be repaired only at regular intervals. This is particularly true for unattended equipment that is inspected only at periodic intervals, often called the *proof test interval*, PTI. A complete failure will occur if the system's redundancy is not adequate to sustain multiple faults during the PTI.

It is not unexpected that the system's MTBF is a function of the PTI. A small decrease in the PTI can greatly increase the MTBF. The MTBFs, availabilities, and *average downtime*, ADT, for periodically repaired equipment operating with full redundancy is

$$\text{ADT}_{1\text{-out-of-2}} = \frac{\text{PTI}}{2} \qquad 61.43$$

$$\text{MTBF}_{1\text{-out-of-2}} = \frac{1}{\lambda^2(\text{PTI})} \qquad 61.44$$

$$\text{availability}_{1\text{-out-of-2}} = \frac{1 - \lambda^2(\text{PTI})^2}{3} \qquad 61.45$$

$$\text{ADT}_{1\text{-out-of-3}} = \frac{\text{PTI}}{\sqrt{3}} \qquad 61.46$$

$$\text{MTBF}_{1\text{-out-of-3}} = \frac{1}{\lambda^3(\text{PTI})^2} \qquad 61.47$$

$$\text{availability}_{1\text{-out-of-3}} = \frac{1 - \lambda^3(\text{PTI})^3}{4} \qquad 61.48$$

The *mean down time*, MDT, for a k-out-of-n system with periodic repair is

$$\text{MDT} = \frac{\text{PTI}}{n - k + 2} \qquad 61.49$$

Example 61.10

One hundred items are tested to failure. Two failed at $t = 1$, 5 at $t = 2$, 7 at $t = 3$, 20 at $t = 4$, 35 at $t = 5$, and 31 at $t = 6$. Find the probability of failure in any period, the conditional probability of failure, and the mean time to failure.

Solution

elapsed time t	failures $F(t)$	survivors $S(t)$	probability of failure $0.01F(t)$	conditional probability of failure $F(t)/S(t-1)$
0	0	100	0	0
1	2	98	0.02	0.02
2	5	93	0.05	0.051
3	7	86	0.07	0.075
4	20	66	0.20	0.233
5	35	31	0.35	0.530
6	31	0	0.31	1.00

The mean time to failure is

$$\text{MTTF} = \frac{\begin{array}{c}(2)(1) + (5)(2) + (7)(3) \\ + (20)(4) + (35)(5) + (31)(6)\end{array}}{100}$$
$$= 4.74$$

Example 61.11

The overall reliability of a fully redundant system must be at least 0.99 over a year's time. The system will be designed to operate as long as any one of multiple identical, parallel units is operational. The MTTF of each unit is 0.8 years. What level of redundancy is required?

Solution

For each unit,

$$\lambda = \frac{1}{\text{MTTF}} = \frac{1}{0.8 \text{ yr}} = 1.25 \text{ 1/yr}$$
$$R_{1 \text{ yr}} = e^{-\lambda t} = e^{(-1.25 \text{ 1/yr})(1 \text{ yr})}$$
$$= 0.2865$$

From Eq. 61.30,

$$R_{1\text{-out-of-}n \text{ system}} = 1 - (1 - R_t)^n$$
$$0.99 = 1 - (1 - 0.2865)^n$$
$$n = 13.6 \quad [\text{use } 14]$$

5. PREVENTATIVE MAINTENANCE

The value of *preventative maintenance*, PM, to prevent breakdowns is undisputed.[10] However, it is not as easy to decide on the frequency and timing of PM, the size of maintenance facilities and number of staff, location and centralization issues, and the quantity of spares to be carried. Quantitative business analysis techniques can be used to formulate some of these PM policies.[11]

The general concept in optimizing PM policies is to minimize the total cost of operation, taking into consideration the costs of preventative maintenance, downtime, and repair. Sometimes the costs are fixed, as when specific penalties must be paid when output is not achieved. At other times, the costs are related to hourly rates and the duration of downtime. The time to failure of a machine and the times for both repair and preventative maintenance are generally not fixed, and they are not always normally distributed either. However, unless simulation is used, it is almost always necessary to work with the average times (e.g., *mean times to failure*, MTTF).

The following guidelines should be considered when establishing PM policies, particularly for single machines. When there are several identical machines operating in parallel, the problem more closely resembles a waiting-line (queuing) problem. Breakdowns are comparable to

---

[10]Maintenance to correct disrepair is known as *remedial maintenance*.

[11]Queuing theories are covered in Sec. 3. Replacement policies are covered in Sec. 6. Even when there are no specific quantitative techniques for solving a particular problem type, simulation can be used to determine the outcome of most preventative maintenance policies.

arrivals in the line, and repair stations (repair crews) are the stations. The optimum solution takes into consideration the costs of idle maintenance crews.

- PM is more applicable when the time-to-breakdown distribution has low variability because the time before a breakdown can be more accurately predicted.

- PM is only useful when its cost is less than the cost of the breakdown. In the absence of cost information, PM is useful when the average PM time is less than the average repair time.

- PM is more applicable when there is little or no inventory of the item produced by the broken machine.

6. REPLACEMENT

Introduction

Replacement and *renewal models* determine the most economical time to replace existing equipment. Replacement processes fall into two categories depending on the life pattern of the equipment, which either deteriorates gradually (becomes obsolete or less efficient) or fails suddenly.

In the case of gradual deterioration, the solution consists of balancing the cost of new equipment against the cost of maintenance or decreased efficiency of the old equipment. Several models are available for cases with specialized assumptions, but no general solution methods exist.

In the case of sudden failure (e.g., light bulbs), the solution method consists of finding a replacement frequency that minimizes the costs of the required new items, the labor for replacement, and the expected cost of failure. The solution is made difficult by the probabilistic nature of the life spans.

Deterioration Models

The replacement decision criterion with deterioration models is the present worth of all future costs associated with each policy. Solution is by trial and error, calculating the present worth of each policy and incrementing the replacement period by one time period for each iteration.

Example 61.12

Item A is currently in use. Its maintenance cost is $400 this year and is increasing each year by $30. Item A can be replaced by item B at a current cost of $3500. However, the purchase cost of B is increasing by $50 each year. Item B has no maintenance costs. Disregarding income taxes, find the optimum replacement year. Use 10% as the interest rate.

Solution

Calculate the present worth, P, of the various policies.

policy 1: Replacement at $t = 5$ (starting the 6th year)

$$P(A) = (-\$400)(P/A, 10\%, 5) - (\$30)(P/G, 10\%, 5)$$
$$= -\$1722$$
$$P(B) = \big(\$3500 + (5)(\$50)\big)(P/F, 10\%, 5) = -\$2328$$

policy 2: Replacement at $t = 6$

$$P(A) = (-\$400)(P/A, 10\%, 6) - (\$30)(P/G, 10\%, 6)$$
$$= -\$2033$$
$$P(B) = -\big(\$3500 + (6)(\$50)\big)(P/F, 10\%, 6) = -\$2145$$

policy 3: Replacement at $t = 7$

$$P(A) = (-\$400)(P/A, 10\%, 7) - (\$30)(P/G, 10\%, 7)$$
$$= -\$2330$$
$$P(B) = -\big(\$3500 + (7)(\$50)\big)(P/F, 10\%, 7) = -\$1976$$

The present worth of A drops below the present worth of B sometime between $t = 6$ and $t = 7$. Replacement should take place at that time.

Sudden Failure Models

The time between installation and failure is not constant for members in the general equipment population. Therefore, in order to solve a failure model, it is necessary to have the distribution of individual item lives (*mortality curve*). The conditional probability of failure in a small time interval, say from t to $t + \delta t$, is calculated from the mortality curve. This probability is *conditional* since it is conditioned on nonfailure up to time t.

The conditional probability of failure may decrease with time (as with *infant mortality*), remain constant (as with an exponential reliability distribution and failure from random causes), or increase with time (as with items that deteriorate with use). If the conditional probability of failure decreases or remains constant over time, operating items should never be replaced prior to failure.

It is usually assumed that all failures occur at the end of a period. The problem is to find the period that minimizes the total cost. The expression for the number of units failing in time t is

$$F(t) = n\left[p\{t\} + \sum_{i=1}^{t-1} p\{i\}p\{t-i\}\right.$$
$$\left. + \sum_{j=2}^{t-1}\left(\sum_{i=1}^{j-1} p\{i\}p\{j-1\}p\{t-j\} + \cdots\right)\right] \quad \textbf{61.50}$$

The term $np\{t\}$ gives the number of failures in time t from the original group.

The term $n\Sigma p\{i\}p\{t-i\}$ gives the number of failures in time t from the set of items that replaced the original items.

The third probability term times n gives the number of failures in time t from the set of items that replaced the first replacement set.

It can be shown that $F(t)$ with replacement will converge to a steady state limiting rate of

$$\overline{F(t)} = \frac{n}{\text{MTBF}} \qquad \textbf{61.51}$$

The optimum policy is to replace all items in the group, including items just recently installed, when the total cost per period is minimized. That is, try to find T such that $K(T)/T$ is minimized.

$$K(T) = nC_1 + C_2 \sum_{t=0}^{T-1} F(t) \qquad \textbf{61.52}$$

Discounting (i.e., taking into consideration the time value of money) is usually not included in the total cost formula since the time periods are considered short. If the equipment has an unusually long life, discounting would be required.

There are some cases where group replacement is always more expensive than just replacing the failures as they occur. Group replacement will be the most economical policy if Eq. 61.53 is valid.

$$C_2[\overline{F(t)}] > \frac{K(T)}{T}\bigg|_{\text{minimum}} \qquad \textbf{61.53}$$

If the opposite inequality holds, group replacement may still be the optimum policy. Further analysis is then required.

7. LINEAR PROGRAMMING

Introduction

Mathematical programming is a modeling procedure applicable to problems for which the objective and resource limitations can be described mathematically.

If the objective function and all resource constraints are linear (polynomials of degree 1 only), the procedure is known as *linear programming*.

If the variables can take on only integer values, a procedure known as *integer programming* is required. If the objective or constraint functions include higher-order polynomials or other functions, then nonlinear optimization techniques such as *gradient search* or *dynamic programming* must be used.

Formulation of a Linear Programming Problem

All linear programming problems have a similar format. Each has an *objective function* that is to be optimized. Usually the objective function is to be maximized, as in the case of a profit function. If the objective is to minimize some function, such as cost, the problem may be turned into a maximization problem by maximizing the negative of the original function.

Each linear programming problem also has a set of functions called *constraints* that limit the independent variables.

Solving Linear Programming Problems

Linear programming problems are generally solved by computer, although some simple problems may be solved by hand with a procedure known as the *simplex method*. Specialized methods allowing easy manual solutions are available for certain classes of problems, primarily the *transportation* and *assignment problems*.

The *simplex method* is an iterative procedure for solving linear programming problems with any number of variables.[12] It works with a matrix (called a *tableau*) of values formed from the resource limitations. Since the method is iterative and systematic, it is easily computerized. While it is possible to solve even multidimensional linear programming problems manually using the simplex method, the computational effort is burdensome, the time requirement is high, and the possibility for accidental errors is high.[13] Therefore, using the simplex method for more than familiarization with the theory is not recommended.

If a linear programming problem can be formulated in terms of only two variables, x_1 and x_2, it can be solved graphically by the following procedure.

step 1: Graph all of the constraints and determine the *feasible region*. Usually this will result in a *convex polygon*.

step 2: Evaluate the objective function, Z, at each corner of the polygon.

step 3: The values of x_1 and x_2 that optimize Z are the coordinates of the corner at which Z is optimized.

Once a solution is found, it is possible to determine the effect on the objective function of changing one of the program parameters. This is known as *sensitivity analysis* and is very important in instances where the accuracy of collected data is unknown.

---

[12]Linear programming problems with two variables are easily solved graphically. See Sec. 7.
[13]Also, the learning time is high.

Example 61.13

A cattle rancher feeds a mixture of three types of cattle food and wants to minimize the cost of feeding his cattle. The costs per pound of the three food types are

food type	cost per pound
1	1.5
2	2.5
3	3.5

The rancher is also concerned with meeting published nutritional information on minimum daily requirements (MDR) given in milligrams per animal. The composition of each food type is known and the contributions for each vitamin in milligrams per pound are

vitamin	MDR (mg)	food type 1	food type 2	food type 3
A	100	1	7	13
B	200	3	9	15
C	300	5	11	17

It is also physically impossible for an animal to eat more than the following amounts per day.

food type	maximum feeding (pounds)
1	50
2	40
3	30

Formulate this problem as a linear programming model to determine how much of each food the rancher should use.

Solution

Let x_i be the number of pounds of food i purchased per animal. Then, the objective function to be minimized is the total cost per animal.

$$Z = 1.5x_1 + 2.5x_2 + 3.5x_3$$

The constraints on this problem are[14]

$$
\begin{aligned}
x_1 + 7x_2 + 13x_3 &\geq 100 \\
3x_1 + 9x_2 + 15x_3 &\geq 200 \\
5x_1 + 11x_2 + 17x_3 &\geq 300 \\
x_1 &\leq 50 \\
x_2 &\leq 40 \\
x_3 &\leq 30 \\
x_1 &\geq 0 \\
x_2 &\geq 0 \\
x_3 &\geq 0
\end{aligned}
$$

---

[14]Depending on the reason for the maximum feeding limits, it may be more appropriate to replace the three single constraints with a single *interaction equation*.

$$\frac{x_1}{50} + \frac{x_2}{40} + \frac{x_3}{30} \leq 1$$

Example 61.14

Solve the following linear programming problem.

$$\max Z = 2x_1 + x_2$$

Constraints:

(a)	$x_1 + 4x_2 \leq 24$	
(b)	$x_1 + 2x_2 \leq 14$	
(c)	$2x_1 - x_2 \leq 8$	
(d)	$x_1 - x_2 \leq 3$	
(e)	$x_1 \geq 0$	
(f)	$x_2 \geq 0$	

Solution

Since there are only two variables, this problem can be solved graphically.

The region enclosed by the constraints is shown.

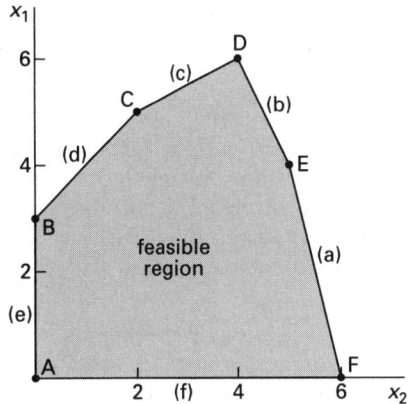

The coordinates and Z-values for each corner are

corner	coordinates (x_2, x_1)	Z
A	(0,0)	0
B	(0,3)	6
C	(2,5)	12
D	(4,6)	16
E	(5,4)	13
F	(6,0)	6

Z is maximized when $x_1 = 6$ and $x_2 = 4$.

8. PLANT LOCATION

The reasons for locating a single new facility at a particular location are as varied as the industries represented. Some reasons are quantitative; others are not. Some reasons look to immediate gains; others look to the future. However, the traditional management science goal in locating a single plant is to minimize the sum of all the quantifiable costs affected by a choice of location. These costs include, among other things, facilities acquisition, labor, taxes, utilities, shipping, and transportation.

The traditional *"plant location problem"* is much more narrowly defined. In fact, the name for this problem type is a misnomer, with the name "multiplant" *fixed-charge problem* being more appropriate. The plant location problem determines where to locate one or more new plants within the framework of existing *plants (sources)* and existing demand locations or *distribution centers (destinations, demands, sinks,* etc.). All locations produce the same product; all distribution centers accept the same product.

Typically, production and distribution expenses are the major cost elements. All other costs, including initial fixed costs, are omitted unless they can be represented as per-unit costs associated with each unique combination of sources and destinations. The object is to minimize the total cost of production and distribution by specifying which new plant(s) should open and how many units each plant should ship to each distribution center.

Constraints ensure that the total demands are satisfied and that the total production capacities are not exceeded. In real-world problems, individual shipments may also be limited, as when the path between one plant and its destination has a limited capacity. These limitations are known as *arc capacities.*

As defined, the plant location problem is a specific case of linear programming known as the *transportation problem*. The transportation problem is concerned with allocating shipments from sources to destinations so that the total transportation cost is minimized. Specific algorithms, manual and computerized, exist for solving this problem.

Example 61.15

Existing plants A and B have insufficient combined capacity to meet product demand. Plants C and D are being evaluated as possible new locations. Either or both can be selected. Either or both of the existing two plants can be closed. The capacities of each plant location are: A, 8; B, 6; C, 5; D, 7.

Warehouses 1, 2, and 3 have demands for 9, 4, and 5, respectively.

The per-unit cost, C_{ij}, associated with each combination of plant (i) and warehouse (j) are given in the following table.

warehouse	plant location			
	A	B	C	D
1	17	15	19	12
2	22	24	10	8
3	39	32	45	16

Formulate this problem as a linear programming program.

Solution

Let the variable Q_{ij} represent the quantity shipped from plant location i to warehouse j. Then, the objective function is

$$\text{minimize } Z = \sum_{i=A}^{D} \sum_{j=1}^{3} C_{ij} Q_{ij}$$

The production capacity constraints are

$$Q_{A1} + Q_{A2} + Q_{A3} \leq 8$$
$$Q_{B1} + Q_{B2} + Q_{B3} \leq 6$$
$$Q_{C1} + Q_{C2} + Q_{C3} \leq 5$$
$$Q_{D1} + Q_{D2} + Q_{D3} \leq 7$$

The demand constraints are

$$Q_{A1} + Q_{B1} + Q_{C1} \geq 9$$
$$Q_{A2} + Q_{B2} + Q_{C2} \geq 4$$
$$Q_{A3} + Q_{B3} + Q_{C3} \geq 5$$

The non-negativity constraints are

$$\text{all } Q_{ij} \geq 0$$

(This solution is the classical linear programming formulation of the *transportation problem*, not a fixed-charge problem.)

9. FACILITIES LAYOUT

Facilities layout (plant layout) problems are numerous in variety and complexity. Laying out facilities involves locating departments and/or operations with respect to one another. In traditional *process layout*, machines with the same function are grouped together. In *product layout* (product-oriented layout), the layout depends on the sequencing of production operations. If the same equipment is used at two different times, it is duplicated in a product layout.

Some computerized methods exist for exhaustively evaluating alternatives. Manual layout techniques are even more limited. Often, paper-cutting is combined with intuition to come up with a layout. Departments are sized to a particular scale and are cut out of paper. The pieces of paper are slid around until a layout "works."

Except for the artificial case of a small number of equally sized, equally shaped departments or operations whose locations are limited to a rectangular grid, it is unlikely that all possible layouts will be considered.[15] The "optimum" layout may actually be merely the best that could be found given the amount of time available.

An alternate manual method is to construct a graph whose "vertices" (nodes) are the departments or operations. The "edges" (line segments) are drawn between two vertices if adjacent associated departments are desired. The edges may be weighted to indicate the level of traffic between the departments. The goal is to rearrange the vertices so that no edges cross. If this can be done, then the layout can be planar. If the departments are somewhat flexible in terms of size and shape, it is possible to have the desired adjacencies.

Certain simplifying assumptions are usually made with both computerized and manual methods. For example, all layouts may be required to be two dimensional. Departments may be assumed to be square or rectangular. When the locations of specific pieces of equipment within the department are unknown, it is assumed that all movement into and out of the department originates and terminates at the centroid of the departmental area. Also, only highly repetitive movements between departments are considered. Once-in-a-while travel is excluded from the analysis.

Almost all facility layout procedures—manual and computerized, exact, trial-and-error, and heuristic— attempt to minimize the transportation cost, sometimes referred to as *movement*.[16] In simple cases, this may mean minimizing the product of trips between departments and the distances between their centroids. In more complex cases, the product of trips and distances may also be multiplied by volumes, weights, and labor rates.

Nonquantitative factors also need to be considered. Sometimes, as when equipment, records, or personnel are shared, it is absolutely necessary that departments be located next to each other. In other cases, as when safety is compromised, it may be absolutely essential to separate departments. In most cases, the *nearness priorities* are in between being absolutely necessary and being absolutely undesirable.[17] The ways that nonquantitative factors are presented and incorporated into the solution vary from case to case.

---

[15]The number of layout variations, including mirror images, with n equally sized square departments is $n!$.

[16]A *trial-and-error method* depends on insight, intuition, and ingenuity to come up with a solution. A *heuristic method* follows a procedure and/or uses rules of thumb to derive an answer. Neither is an optimizing technique.

[17]The *Muther nearness priorities* (developed by Richard Muther in the 1950s) are (1) absolutely necessary, (2) very important, (3) important, (4) ok (ordinary importance), (5) unimportant, and (6) undesirable.

10. WORK MEASUREMENT

Work measurement determines how long it takes to complete an operation. The *standard time* (also known as the *product standard* or *production standard*) for an operation is the time required by a qualified and trained worker working at a normal pace. Standard times can be determined in a variety of ways, including by statistical analysis of actual measurements (*time studies*), analysis of historical data, or by analysis of the physical motions associated with the process (i.e., *predetermined systems*). *Learning curves* can be incorporated where appropriate.

Standard times are used to develop production schedules (*factory loading*), estimate costs and output, determine worker effectiveness, and establish the basis for incentive labor plans. Standard times are combined with actual labor rates to develop *standard costs* of products.

Time standards are usually not developed directly for large operations (e.g., building a television). For various reasons (including reduction of variance), standards are developed for each significant *element* (also known as a *task*) of the operation (e.g., tightening a cover screw or soldering a resistor to the printed circuit board). Such task data can then be used to determine standard times for other products with identical tasks.

In determining the tasks requiring time standards, it is normal to separate fixed and variable tasks. For example, setting up the workplace at the beginning of each shift is a fixed task that would be studied separately. Applying adhesive strips to the product on an assembly line is a variable task. It is also common to separate handling and material transport tasks from processing tasks.

Actual time measurements during normal operation can be taken by hand with a stopwatch or derived from the analysis of video recordings.[18] Operations can be continuously observed, or measurements can be taken on a random basis.[19] To establish a truly representative average, measurements should be taken from as many individuals as possible.

The raw average time for an operation is not used as a production standard for two reasons. First, different people work at different levels of effort, and workers do not always work at a normal pace when being observed. This is taken into consideration by the observer who

---

[18]The stopwatch is used in two different ways. In *snapback recording*, the stopwatch is zeroed out after each measurement. In *continuous recording*, the elapsed times are recorded. The interval times are determined later by subtraction.

[19]Though the sampling is random, the random schedule is established in advance with the aid of random numbers. The investigator follows a predetermined observation schedule that is unknown to the worker being observed.

applies a subjective *performance factor (performance rating)* to the raw data.[20] Expressing the performance factor as a decimal, the *normal time* is

$$t_{\text{normal}} = (t_{\text{ave}})(\text{performance factor}) \qquad 61.54$$

Second, *time allowances* must be added for personal time, normal delays, defective material, and fatigue.[21] Such allowances, usually expressed as percentages, depend greatly on the nature of the operation. Allowances for hard manual labor are much larger than allowances for sitting workers. Expressing the performance factor and allowance as decimals, the standard time is[22]

$$
\begin{aligned}
t_{\text{std}} &= \frac{(t_{\text{ave}})(\text{performance factor})}{1 - \text{allowance}} \\
&= \frac{t_{\text{normal}}}{1 - \text{allowance}} \qquad 61.55
\end{aligned}
$$

The variability of time study data is commonly expressed in terms of a *coefficient of variation*. This is simply the percentage variation calculated from Eq. 61.56. Notice that the sample standard deviation, s_t, of the time measurements is used.

$$\text{coefficient of variation} = \left(\frac{s_t}{t_{\text{ave}}}\right) \times 100\% \qquad 61.56$$

$$s_t = \sqrt{\frac{\sum t^2 - \dfrac{(\sum t)^2}{N}}{N-1}} \qquad 61.57$$

The number of observations is determined from the desired precision of the final answer and the observed variability of the times measured. The more variable the times and the greater the required *precision* ($\pm P\%$), the more observations will be required to achieve a desired *confidence level* ($C\%$).[23,24] The most common confidence level is 95%, though 90% and 99% are also used.[25] A 95% confidence level (i.e., a 0.95 *confidence coefficient*) means that the probability is 95% that the desired precision will be met.

---

[20]In theory, either the standard time or the performance rating must be known. However, the investigator traditionally observes the time and rates the performance simultaneously.

[21]A minimum of 5% should be added for personal time. Allowances for normal delays are determined by observation. Allowances for fatigue are highly subjective and depend on the nature of the operation.

[22]It is also common to simply add the allowance percentage, though the two methods are not equivalent. The method used is usually a matter of company policy.

$$t_{\text{std}} = t_{\text{normal}}(1 + \text{allowance})$$

[23]The *confidence interval* and *confidence level* are different. The confidence interval is twice the precision, expressed in absolute terms. If a $\pm 2\%$ precision is required for a 1.0 minute operation, the confidence interval will be 0.04 minute.

[24]The precision is sometimes called the *required accuracy*, even though a required accuracy of $\pm 5\%$ is obviously inappropriate. "Required maximum inaccuracy" is more descriptive of the actual intent of the term.

[25]The actual value used will be a matter of company policy.

Equation 61.58 can be used for estimating the sample size required. If N is already greater than n, no further sampling is required.

$$N = \left(\frac{t_C s_t}{P t_{\text{ave}}}\right)^2 \qquad 61.58$$

The actual maximum inaccuracy (error), ϵ, is calculated from Eq. 61.59, which uses a two-tail distribution factor, t_C, from Table 61.1 or App. 61.A.[26]

$$\epsilon = P t_{\text{ave}} = t_C s_t \qquad 61.59$$

$$P = \frac{\epsilon}{t_{\text{ave}}} = \frac{t_C s_t}{t_{\text{ave}}} \qquad 61.60$$

Table 61.1 *Two-Tail Distribution Factors*[a]

confidence level, C	z_C or $t_C(\infty)$
90%	1.645
95%	1.960
97.5%	2.240
98%	2.326
99%	2.476
99.9%	3.270

[a]Standard normal distribution is equivalent to a Student's *t*-distribution with infinite degrees of freedom.

Time standards can also be developed from systems of *standard data*, also known as *universal standard data* and *synthetic standard data*.[27] Standard data systems are particularly valuable when there is no operation to observe, as when the product is still at the blueprint stage. An analyst (who is familiar with the steps and equipment necessary to complete an operation) then subdivides the operation into tasks and the tasks into even finer micromotions.

For example, the task of moving a small object from a conveyor belt to a workstation might be subdivided into the micromotions of eye focus, reach, pickup and grasp, move, position, and release micromotions. Times for these micromotions are taken from extensive tables of standard data. Such tables may list time in minutes or in TMUs (time measurement units) equal to 0.0006 minutes.

---

[26]When the standard deviation of the underlying population is not known and must be estimated from the sample, as is usually the case, *Student's t-distribution* should be used in place of the normal distribution. This level of sophistication is particularly necessary when the number of observations is small. Values of t_C are determined from App. 61.A. When the *actual* error is being calculated from n actual observations, t_C should be determined with $n - 1$ degrees of freedom. When the required number of observations, N, is being calculated, and the degrees of freedom is unknown (i.e., $N - 1$ is unknown) or expected to be large, the degrees of freedom is assumed to be infinite. Values from Student's *t*-distribution for infinite degrees of freedom are identical to *z*-values from the normal table. (See Table 61.1.)

[27]*Methods-Time-Measurement* (MTM), *Work Factor, Basic Motion Time Study*, and MODAPS are commercial systems of standard data systems.

Example 61.16

A highly experienced worker was repeatedly observed in a time study. The average time taken by the worker to complete a task was 0.80 min. The observing analyst gave the worker a performance rating of 120%. The standard allowance is 10%. What is the standard time for the task?

Solution

From Eq. 61.55, the standard time is

$$t_{\text{std}} = \frac{(t_{\text{ave}})(\text{performance factor})}{1 - \text{allowance}}$$

$$= \frac{(0.80 \text{ min})(1.2)}{1 - 0.10} = 1.07 \text{ min}$$

Example 61.17

The estimate of an operation's time is to be obtained from 10 measurements. (a) Given a required confidence level of 90%, what is the precision of the estimate? (b) Given a required precision of 5%, what is the confidence level? (c) How many additional observations are required to achieve the 5% precision with a 90% confidence level?

times: 0.32, 0.35, 0.34, 0.36, 0.38, 0.40, 0.40, 0.36, 0.31, 0.38

Solution

(a) The average time and sample standard deviation are

$$t_{\text{ave}} = \frac{\Sigma t}{N} = \frac{3.60}{10} = 0.36$$

From Eq. 61.57, the sample standard deviation is

$$s_t = \sqrt{\frac{\Sigma(t^2) - \dfrac{(\Sigma t)^2}{N}}{N - 1}}$$

$$= \sqrt{\frac{1.3046 - \dfrac{(3.6)^2}{10}}{10 - 1}}$$

$$= 0.03091$$

Since the number of samples is less than 20 (or 25 or 30), the normal curve should not be used. From App. 61.A, the t_C factor for 90% two-tailed confidence limits with $n - 1 = 9$ degrees of freedom is 1.83. The maximum error (with 90% confidence) is

$$\epsilon = t_C s_t = (1.83)(0.03091)$$

$$= 0.0566$$

The precision is

$$P = \frac{\epsilon}{t_{\text{ave}}} = \frac{0.0566}{0.36}$$

$$= 0.157 \quad (\pm 15.7\%)$$

(b) With a precision of $\pm P\%$, the maximum error is

$$\epsilon = P t_{\text{ave}} = (0.05)(0.36) = 0.018$$

The value of t_C is

$$t_C = \frac{\epsilon}{s_t} = \frac{0.018}{0.03091}$$

$$= 0.582$$

From App. 61.A with $n - 1 = 9$ degrees of freedom, the value of 0.582 can be interpolated between the 40% and 50% (two-tail) columns. Thus, the confidence level is between 40% and 50%.

(c) From Eq. 61.58, the total number of observations required is

$$N = \left(\frac{t_C s_t}{P t_{\text{ave}}} \right)^2$$

$$= \left[\frac{(1.83)(0.03091)}{(0.05)(0.36)} \right]^2$$

$$= 9.9 \quad [\text{use 10 observations}]$$

In this case, no additional observations are required.

11. WORK SAMPLING

Work sampling is a method of achieving the results of a stopwatch study without using a stopwatch.[28] It directly determines the fraction of time spent in a particular operation. Work sampling can be used to indirectly establish time standards and to rate performance.

The basic procedure is to observe a worker at random times over a specific time period (e.g., 8 or 40 hours) and to simply note whether or not the worker was performing a given task. The fraction of observations is identical to the fraction of time the worker spends on the task. This fraction is combined with the time period to determine the average time for the task.

$$t_{\text{std}} = \frac{\begin{array}{c}(\text{fixed time period})(\text{fraction observed}) \\ \times (\text{performance factor})\end{array}}{(1 - \text{allowance})(\text{no. of pieces produced})}$$

61.61

The accuracy of the average depends on the total number of observations, N. It is a fundamental principle in work sampling studies that the total required number of observations is proportional to the amount of time spent on the given task. Since the time is initially unknown, an estimate is made and the number of observations is iteratively refined as observations are taken. The observed fraction of observations of the task is

$$\bar{p} = \frac{n}{N} = \frac{\text{no. of observations of task}}{\text{total no. of observations}} \qquad 61.62$$

---

[28]Work sampling is particularly useful in determining the fraction of time that a worker or group of workers is idle.

The sample standard deviation of the observed fraction is the standard deviation of the binomial distribution.

$$s_p = \sqrt{\frac{\overline{p}(1-\overline{p})}{N}} \qquad 61.63$$

It is important not to confuse the maximum fractional error, commonly referred to as the *precision*, with the maximum absolute error. Since $\overline{p}$ varies from 0 to 1, the fractional and absolute errors are both less than 1.00. The maximum actual inaccuracy is

$$\epsilon = P\overline{p} = t_C s_p \qquad 61.64$$

The total number of observations required to achieve the desired precision and confidence is

$$N = \left(\frac{t_C \overline{p}(1-\overline{p})}{\epsilon}\right)^2$$
$$= (t_C)^2 \left(\frac{1-\overline{p}}{P^2 \overline{p}}\right) \qquad 61.65$$

Example 61.18

Work sampling is to be used to determine the fraction of time that a worker is idle. The initial estimate of idleness is 12%. The required precision is ±10%. The required confidence level is 95%. (a) Determine the number of observations that should be initially scheduled. (b) Assuming the fraction of time idle is 12%, with the given confidence level, what is the range of values of the fraction of idleness?

Solution

(a) The initial estimate of the fraction is $\overline{p} = 0.12$, and the required precision, P, is 0.10. Assuming a large number of samples, the two-tail normal curve parameter for a 95% confidence level is found in Table 61.1 to be 1.960.

From Eq. 61.65, the number of observations is

$$N = \frac{(t_C)^2(1-\overline{p})}{P^2\overline{p}}$$
$$= \frac{(1.960)^2(1.00 - 0.12)}{(0.10)^2(0.12)}$$
$$= 2817$$

(b) With a precision (i.e., fractional error) of 10%, the absolute error is

$$\epsilon = P\overline{p} = (0.10)(0.12) = 0.012$$

The range of values will be

$$\overline{p} + \epsilon = 0.12 \pm 0.012$$
$$= 0.132,\ 0.108$$

12. ASSEMBLY LINE BALANCING

Line balancing determines which tasks will be performed progressively at multiple assembly stations. Some tasks must precede others; some tasks can be performed at any point in the assembly; some tasks (e.g., installation of fasteners and final tightening) can be split between stations. Line balancing can also determine how many stations are needed and which tasks will be performed in parallel to increase throughput.

The cumulative durations of all the tasks at a particular station cannot exceed the cycle time, which is the reciprocal of the *production rate*. Generally, the *cycle time* is known.[29]

$$\text{cycle time} = \frac{1}{\text{production rate}} \qquad 61.66$$

Line balancing establishes the capacity of the line. Work cannot pass down the line any faster than it can pass through the *bottleneck* of the line (i.e., the station with the longest set of tasks). Therefore, the longest set of tasks establishes the minimum cycle time. The *target cycle time* is the ideal case and can be calculated if the number of stations is known.

$$\text{target cycle time} = \frac{\Sigma(\text{all task times})}{\text{no. of stations}} \qquad 61.67$$

As a management science problem, the simultaneous goals are to maximize the throughput of the line, minimize the cumulative idle time of all the stations, minimize the total labor cost, and minimize the initial investment in station equipment. In most initial studies, only the goal of minimizing idle time is considered. The total idle time is

$$\begin{aligned}\text{idle time} = &\ (\text{no. of stations})(\text{cycle time}) \\ &- \Sigma(\text{all task times}) \qquad 61.68\end{aligned}$$

The percentage *delay* (also known as *balance delay*) is defined by Eq. 61.69. The best arrangement will have the lowest balance delay. However, an optimal solution does not require a zero balance delay. In fact, a zero balance delay may not be possible, as when the sum of task times is not an integer multiple of the cycle time. Optimality is still assured if the number of stations is the smallest integer greater than the sum of task times divided by the cycle time.

$$\text{balance delay} = \frac{\text{idle time} \times 100\%}{(\text{no. of stations})(\text{cycle time})} \qquad 61.69$$

Trial and error methods of balancing lines by hand are of the shuffle-and-reshuffle variety. Heuristic methods also require considerable personal ingenuity. The number of possibilities is often very large, and therefore

---

[29]The problem is much more difficult when the cycle time is not known. Not only do all arrangements of tasks need to be evaluated, but also the arrangements need to be evaluated over all reasonable cycle times.

modern linear and dynamic programming models, as well as techniques based on exhaustive enumeration, are solved with computers. Regardless of the method used, solutions are often not unique.

The following three different heuristic methods are used to balance lines.

- The task with the largest time (that will fit in the station's available time) is assigned to that station.

- The task with the most successors (that will fit in the station's available time) is assigned to that station.

- The task with the greatest sum of task times of its successor tasks (that will fit in the station's available time) is assigned to that station.

Standard times (Sec. 10) to perform the various station tasks are used in the balancing process. However, except for robot operators and machine-controlled lines, task times are actually random. Sometimes it takes longer; sometimes it takes less time. This implies that almost every line will be unavoidably unbalanced. If the work is rigidly paced to the cycle time (as it would be if the work was permanently attached to the conveyance system), some work pieces might be left unfinished if the previous part required more time than usual. This scenario introduces what is probably one of the most important requirements for maximizing the line throughput: station inventory.

Line throughput will be maximized if each station has a backlog of unfinished work.[30] In this case, the conveyance system is used merely to bring work to stations rather than to pace the stations. If a station is busy when a new piece of work arrives, the station operator merely places that piece into his or her inventory of unfinished work. If all tasks are finished early, before a new piece of work arrives, the operator begins on a piece from the inventory. The station is never idle.

Example 61.19

(a) Design a heuristically balanced line for a cycle time of 10. (b) Calculate the idle time. (c) Calculate the balance delay. (d) What would be the effect of decreasing the cycle time to 9?

task	predecessors	time
start	none	0
A	start	7
B	start	4
C	B	5
D	A,C	2
E	C	8
F	D,E	4
finish	F	0

---

[30]Measurable increases in line output have been reported by selectively unbalancing the line and ensuring that each station has a backlog of work.

Solution

(a) First, draw a simple precedence diagram to visualize the problem.

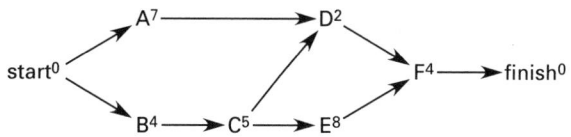

Station 1: The task with the largest time without predecessors will be assigned. (The "start" task, used merely for consistency in drawing directed graphs, is disregarded.) Task A is the largest and is assigned to station 1. That takes up 7 of the cycle time, leaving $10 - 7 = 3$ still available. B is too long, and C's predecessor is incomplete. However, D has a duration of 2, which will fit, leaving 1. No tasks can be added to fill the remaining time.

Station 2: Task A is taken; task B is the only possibility. Task B, with a time of 4, is assigned to station 2. Continuing down the list, only task C can be performed. Assigning task C to station 2 increases the station time to a total of 9. No further tasks can be added.

Station 3: Tasks A, B, C, and D have been taken. Task E is the only possibility and is assigned to station 3. The time of 8 does not leave enough time to combine task F.

Station 4: Tasks A, B, C, D, and E have been taken. Task F is assigned to station 4.

The heuristic solution is

> Station 1: A, D (busy 9; idle 1)
> Station 2: B, C (busy 9; idle 1)
> Station 3: E (busy 8; idle 2)
> Station 4: F (busy 4; idle 6)

(b) The sum of task times is

$$7 + 4 + 5 + 2 + 8 + 4 = 30$$

The idle time is

$$\text{idle time} = (\text{no. of stations})(\text{cycle time})$$
$$- \Sigma(\text{all task times})$$
$$= (4 \text{ stations})(10) - 30 = 10$$

(c) The balance delay is

$$\text{balance delay} = \frac{\text{idle time} \times 100\%}{(\text{no. of stations}) \, (\text{cycle time})}$$

$$= \frac{(10)(100\%)}{(4)(10)} = 25\%$$

Plant Engineering

(d) Each station has at least one unit of idle time, so decreasing the cycle time to 9 would not change the assignment of tasks. However, the throughput would be increased and the percentage balance delay would be decreased.

$$\text{balance delay} = \frac{(36 - 30)(100\%)}{(4)(9)} = 16.7\%$$

13. WAGE INCENTIVE PLANS

The goal of *wage incentive plans* is to motivate workers to produce more by allowing them to earn more.[31] *Individual incentive plans* are tied to each worker's output; *group incentive plans* are tied to the output of a department or other group. Individual plans can usually be modified for use as group plans.

The simplest wage incentive plan is *piecework* or *piecerate*, in which workers are paid in proportion to the number of pieces completed. Piecework payment can be very effective, but is difficult to administer under minimum-wage laws.

Standard-hour plans are similar to piecework, but they guarantee workers a *base wage* regardless of output. Above the *performance standard* (i.e., the output corresponding to the standard payment), workers are paid in proportion to the output. The difference between the base and actual wages is the *bonus*. *Productivity* (also called *efficiency*) is the percentage of the standard output achieved. Output may be expressed in units of standard hours.

The bonus can be 100% of the increase in output above the standard or it can be less (though usually no less than 50% of the increase).[32] Plans with a 100% bonus are known as *one-for-one plans* or *100% premium plans*, while plans with bonuses less than 100% are known as *gain-sharing plans*. The percentage of the earned bonus kept by the worker is known as *labor's participation* in the plan.

Effective plans allow workers to earn bonuses of at least 25 to 30% of their base wages (i.e., the *relative earnings* are 125 to 130%). However, depending on the standard, some workers may be unable to consistently achieve even the standard output. Rather than encouraging higher output, such plans might discourage workers, resulting in even lower output than without the incentive program. For this reason, many incentive programs begin paying a bonus at output less than 100% of standard. Such reduced-standard plans can be one-for-one or gain sharing.

---

[31]As a very general rule, output before wage incentives are implemented will be 60 to 90% of the output based on standard times. With wage incentives, productivity will increase to 130% or more. Although workers may not work significantly faster with wage incentives, they reduce their idle time.
[32]The lower bonus may be justified on the basis that the remainder is needed to pay for the administrative costs of running the incentive program.

Wage incentive plans can be identified in one of three ways: by name, by specifying the participation and productivity at which the bonus begins, or by specifying the standard bonus. The *standard bonus* is bonus earned at a productivity of 100%. Obviously, the standard bonus is zero for any 100% standard plan. For any linear plan, the standard bonus depends on the *bonus factor* defined by Eq. 61.70.

$$\text{standard bonus} = (\text{bonus factor})(\text{participation})$$
$$\text{[linear plans]} \qquad 61.70$$

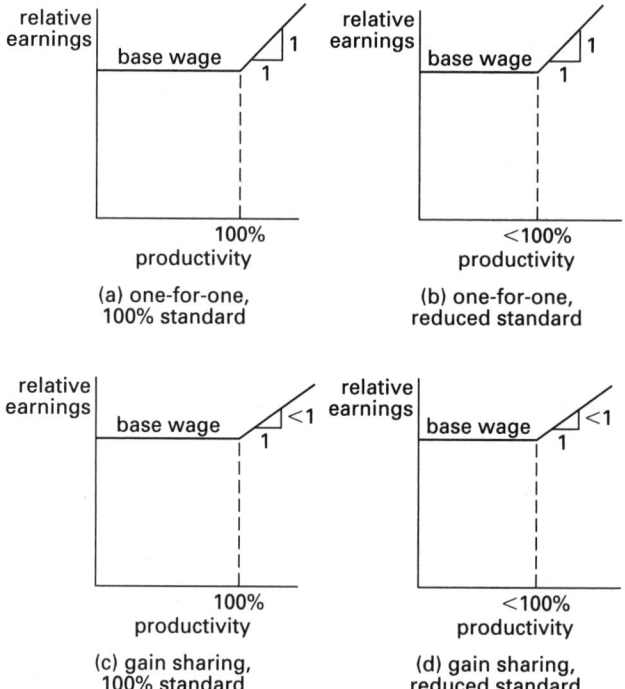

Figure 61.4 *Four Types of Linear Incentive Plans*

The *relative earnings*, RE, are

$$\text{RE} = 1 - \text{participation}$$
$$+ (\text{productivity})(\text{participation})(1+\text{bonus factor})$$
$$61.71$$

The *relative bonus*, RB, is

$$\text{RB} = \text{RE} - 1 \qquad 61.72$$

Figure 61.4 illustrates four types of *linear incentive plans*. *Curvilinear plans*, also known as *self-regulating plans*, are also used, particularly when there is the possibility of runaway production or cheating. Payment under curvilinear plans asymptotically approaches a maximum "ceiling" wage. The maximum relative earnings is one plus the fractional participation. The nature of the curve is specified by giving the productivity level at which the bonus begins and at least one other point on the curve.

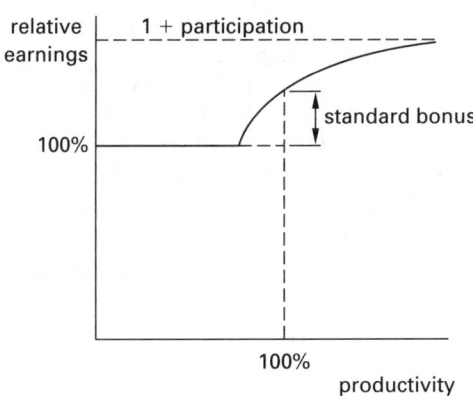

Figure 61.5 *Curvilinear Plan*

For curvilinear plans, the standard bonus and relative earnings are

$$\frac{\text{standard}}{\text{bonus}} = \frac{(\text{bonus factor})(\text{participation})}{1 + \text{bonus factor}}$$
$$\text{[curvilinear plans]} \quad 61.73$$

$$\text{RE} = 1 + \frac{(\text{participation})[\text{productivity} + (\text{productivity})(\text{bonus factor}) - 1]}{\text{productivity} + (\text{productivity}) \times (\text{bonus factor})}$$
$$\text{[curvilinear plans]} \quad 61.74$$

For both linear and curvilinear reduced-standard plans, the productivity at which the bonus begins can be found from Eq. 61.75.

$$\frac{\text{minimum bonus}}{\text{productivity}} = \frac{1}{1 + \text{bonus factor}}$$
$$\begin{bmatrix} \text{linear and} \\ \text{curvilinear plans} \end{bmatrix} \quad 61.75$$

Example 61.20

A linear wage incentive program provides for 50% participation and a 25% standard bonus. A worker produces 1900 units in 8 hr. The standard time for each unit is 0.004 hr. What are (a) the bonus factor, (b) the worker's relative earnings, and (c) the worker's relative bonus?

Solution

(a) From Eq. 61.70, the bonus factor is

$$\text{bonus factor} = \frac{\text{standard bonus}}{\text{participation}} = \frac{0.25}{0.50}$$
$$= 0.50 \quad (50\%)$$

(b) The worker's productivity is

$$\text{productivity} = \frac{\text{standard time}}{\text{actual time}}$$
$$= \frac{(1900 \text{ units})\left(0.004 \, \dfrac{\text{hr}}{\text{unit}}\right)}{8 \text{ hr}}$$
$$= 0.95 \quad (95\%)$$

From Eq. 61.71, the relative earnings are

$$\text{RE} = 1 - \text{participation}$$
$$\qquad + (\text{productivity})(\text{participation})(1 + \text{bonus factor})$$
$$= 1 + (0.95)(0.50)(1 + 0.50) - 0.50$$
$$= 1.2125 \quad (121.25\%)$$

(c) The relative bonus is

$$\text{RB} = \text{RE} - 1 = 1.2125 - 1$$
$$= 0.2125 \quad (21.25\%)$$

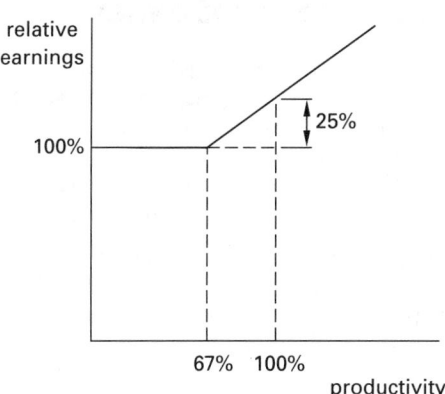

Example 61.21

A curvilinear wage incentive program provides for 75% participation and a 37.5% standard bonus. A worker produces 1900 units in 8 hr. The standard time for each unit is 0.004 hr. What is the worker's relative earnings?

Solution

As in Ex. 61.20, the productivity is 95%. The bonus factor is calculated from Eq. 61.73.

$$\text{standard bonus} = \frac{(\text{bonus factor})(\text{participation})}{1 + \text{bonus factor}}$$
$$0.375 = \frac{(\text{bonus factor})(0.75)}{1 + \text{bonus factor}}$$
$$\text{bonus factor} = 1.00$$

From Eq. 61.74, the relative earnings are

$$\text{RE} = 1 + \frac{(\text{participation})[\text{productivity} + (\text{productivity})(\text{bonus factor}) - 1]}{\text{productivity} + (\text{productivity})(\text{bonus factor})}$$
$$= 1 + (0.75)\left(\frac{0.95 + (0.95)(1.0) - 1.0}{0.95 + (0.95)(1.00)}\right)$$
$$= 1.355 \quad (135.5\%)$$

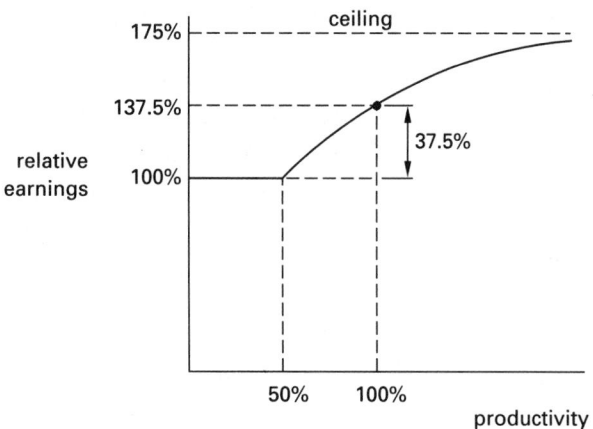

14. QUALITY CONTROL CHARTS

Statistical quality control, SQC, also known as *statistical process control*, SPC, uses several techniques to ensure that a minimum quality level is consistently obtained from production processes. Typical SQC tasks include routine monitoring of process output, sampling incoming raw materials, and testing finished work.

Monitoring process output and charting the results are often the most visible aspects of SQC. Small samples of work are tested at regular or random intervals, and the results are shown graphically.[33] The graphs are known as *control charts* or *Shewhart control charts* because they show, in addition to the measured values, the *control limits* (i.e., what the limits of acceptable values are).[34]

Control charts can be prepared for the average value of some process variable (the *x-bar chart*), the range or other measures of dispersion (*R-chart*, *s-chart*, or *σ-chart*), the fraction defective (*p-chart*), number of defects per unit (*c-chart*), or any combination thereof. Charts that require a measurement of a process variable are known as *variable charts*. The *p*-chart and *c*-chart are examples of *attribute charts*, where only the condition of an item needs to be determined.

It is a basic assumption that random effects are present in every process. Variation within certain limits is inevitable, and if the magnitudes of the limits or the variation are unacceptable, they must be reduced by change in manufacturing or product design. If the magnitudes are acceptable, then correction is required only when the results exceed the magnitude expected on the basis of random effects.

A process is "in control" as long as the process variable is within the control limits. For a process in control, the variable is assumed to be normally distributed with population mean μ and population standard deviation

σ'. When the process changes by more than $\pm 2\sigma'$ (the *warning limits*), a change might be occurring, and the process should be watched. When the process changes by more than $\pm 3\sigma'$ (the *action limits*), the process is considered to have gone out of control.[35] In that case, the process is evaluated and, if necessary, changed.[36]

The *average run length*, ARL, is the average number of samples from an in-control process that will be tested before a point outside the control limits will be encountered.

$$\text{ARL} = \frac{1}{1 - C} \qquad \text{61.76}$$

A *suspicious event* (also known as a *shift* in central tendency) indicates that the process average has shifted. A suspicious event occurs when more than seven consecutive points are on one side of the average, even though the points are within the control limits. Other suspicious events occur when 10 out of 11, 12 out of 14, 14 out of 17, and 16 out of 20 points are on the same side of the average. These criteria are known as *zone tests*.

Stratification occurs when almost all of the points on the chart are near the centerline. This is usually too good to be true, as samples from normal populations do not react in this manner. Usually, stratification indicates that the control limits have been incorrectly calculated.

A *mixture pattern* of points is essentially the opposite of stratification. A mixture pattern occurs when all of the points are near the two control limits. Normal populations do not behave in this manner either, but bimodal distributions do. The usual interpretation is that the product is coming from two different statistical populations. Differences in equipment, raw materials, adjustment, or operators can explain the bimodality.

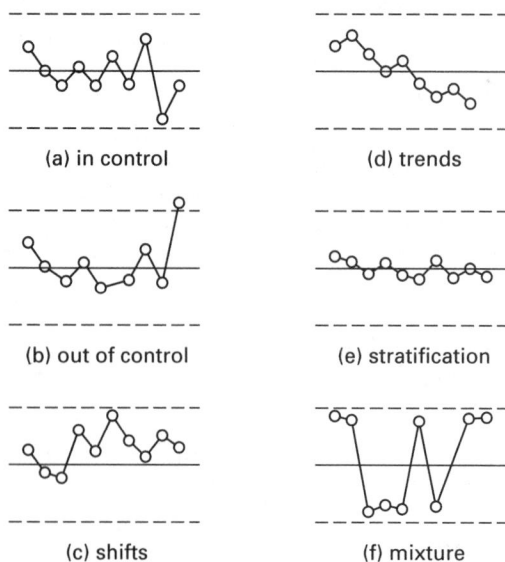

Figure 61.6 *Interpretation of SPC Charts*

---

[33]The graphs are often conspicuously posted at the entrances of departments.
[34]Control limits have nothing to do with *specification limits*. Specification limits determine if the product is acceptable to the customer. Control limits determine if the process is statistically in control.

[35]With 3σ control limits, no more than 0.27% (i.e., 27 times out of 1000) of the points are expected to be outside of the control limits.
[36]Although three standard deviations is the usual value, any multiple of standard deviations can be chosen for the action limits.

In a typical control chart application, a sample of N objects is taken and the quality parameter, x, is measured for each of the N items. The average of the sample, $\overline{x}$, is calculated. This process is repeated M times (over M days, etc.).[37] The grand average, $\overline{\overline{x}}$, and (in theory) the standard deviation, $\sigma_{\overline{x}}$, of the process averages are calculated from the M values of $\overline{x}$.[38]

The "x-bar chart" ($\overline{x}$ chart) for process averages is drawn with a process average of $\overline{\overline{x}}$ and horizontal control limits of $\pm 3\sigma_{\overline{x}}$. As long as the process is in control, the sample averages will be within the *upper control limit*, UCL, and *lower control limit*, LCL, defined by Eq. 61.77.

$$\text{LCL}_{x\text{-bar}}, \ \text{UCL}_{x\text{-bar}} = \overline{\overline{x}} \pm 3\sigma_{\overline{x}} \qquad \textit{61.77}$$

If the lower control limit is negative, that limit is set to zero. If one or more of the M averages is found to be outside of the control limits, those values are excluded and the procedure is repeated to find the control chart values for a process in control.

The unbiased estimators of the population average, μ, and standard deviation, σ', of the $N \times M$ individual values (not the M averages) are $\overline{\overline{x}}$ and $\sqrt{N}\sigma_{\overline{x}}$. Therefore, the upper and lower control limits could also be written as Eq. 61.78, if these population parameters are known.

$$\text{LCL}_{x\text{-bar}}, \ \text{UCL}_{x\text{-bar}} = \overline{\overline{x}} \pm \frac{3\sigma'}{\sqrt{N}} \qquad \textit{61.78}$$

For various reasons, including timely response and job enrichment, the steps and calculations necessary to maintain x-bar (and other) charts are often performed by workers who are not trained in higher mathematics or statistical analysis. Therefore, a simplified method based on the *range* of values (i.e., the maximum measurement less the minimum measurement) can be used to calculate the standard deviation.[39]

$$R = \text{maximum sample value} - \text{minimum sample value} \qquad \textit{61.79}$$

The range is calculated for each of the M samples. Then, the average range, $\overline{R}$, is calculated and converted to the population standard deviation, σ'. A similar calculation can be done if the individual standard

---

[37]For the assumption of a normal distribution to be valid, M must be greater than approximately 20.

[38]Though s^2 is an unbiased estimator of the population variance, σ'^2, s is a biased estimator of σ' and is not used. However, if s is calculated for each sample, it can be used if corrected according to

$$\sigma' = \left(\frac{1}{c_2}\right)\sqrt{\left(\frac{N-1}{N}\right)}(\overline{s})$$

[39]Most quality monitoring today, even on the "factory floor," uses the computer. The R-chart is still an option on most quality software, but there is little to justify its use. The s-chart and σ-chart are more powerful and detect changes in variability faster.

Table 61.2 d_2 and c_2 Factors for Estimating σ' from $\overline{R}$ and $\overline{\sigma}$ [a]

N	d_2	c_2
2	1.128	0.5642
3	1.693	0.7236
4	2.059	0.7979
5	2.326	0.8407
6	2.534	0.8686
7	2.704	0.8882
8	2.847	0.9027
9	2.970	0.9139
10	3.078	0.9227
11	3.173	0.9300
12	3.258	0.9359
13	3.336	0.9410
14	3.407	0.9453
15	3.472	0.9490
16	3.532	0.9523
17	3.588	0.9551
18	3.640	0.9576
19	3.689	0.9599
20	3.735	0.9619
21	3.778	0.9638
22	3.819	0.9655
23	3.858	0.9670
24	3.895	0.9684
25	3.931	0.9696
30	4.086	0.9748
35	4.213	0.9784
40	4.322	0.9811
45	4.415	0.9832
50	4.498	0.9849
55	4.572	0.9863
60	4.639	0.9874
65	4.699	0.9884
70	4.755	0.9892
75	4.806	0.9900
80	4.854	0.9906
85	4.898	0.9912
90	4.939	0.9916
95	4.978	0.9921
100	5.015	0.9925

[a]Sampling from a normally distributed population assumed.

deviation, σ, is calculated for each of the M samples. The factors, c_2 and d_2, used in Eq. 61.80 depend on the sample size, N.

$$\sigma' = \frac{\overline{R}}{d_2} = \frac{\overline{\sigma}}{c_2} \qquad \textit{61.80}$$

The lower and upper control limits for the x-bar chart can be written as

$$\text{LCL}_{x\text{-bar}}, \ \text{UCL}_{x\text{-bar}} = \overline{\overline{x}} \pm \frac{3\overline{R}}{d_2\sqrt{N}} \qquad \textit{61.81}$$

Plant Engineering

The variability of the process can also be tracked by charting the range values in an *R-chart*.[40] The average range is assumed to be zero, and Eqs. 61.82 and 61.83 give the lower and upper control limits. The values depend on the confidence level (i.e., the percentage of points that are expected to fall outside the limits when the process is in control). Since control limits on the *R*-chart are not symmetrical, values of D_{LCL} and D_{UCL} are tabulated in Table 61.3.[41]

$$\text{UCL}_p = D_{UCL}\overline{R} \qquad 61.82$$

$$\text{LCL}_p = D_{LCL}\overline{R} \qquad 61.83$$

Table 61.3 *Control Limit D-Factors for R-Charts*
(for percentage outside limits)

	D_{LCL}			D_{UCL}		
N	0.1%	0.5%	2.5%	2.5%	0.5%	0.1%
2	0.00	0.01	0.04	3.17	3.97	4.65
3	0.06	0.13	0.30	3.68	4.42	5.06
4	0.20	0.34	0.59	3.98	4.69	5.31
5	0.37	0.55	0.85	4.20	4.89	5.48
6	0.53	0.75	1.07	4.36	5.03	5.62
7	0.69	0.92	1.25	4.49	5.15	5.73
8	0.83	1.08	1.41	4.60	5.25	5.82
9	0.97	1.21	1.55	4.70	5.34	5.90
10	1.08	1.33	1.67	4.78	5.42	5.97

The *p-chart* is particularly useful in monitoring the fraction of nonquantitative rejects (i.e., *nonconformances*— the quantity of units that don't work or "just aren't right" for any reason). The *np-chart* monitors the number of rejects in a standard sample size. The variable p is the fraction (or percentage) defective in each sample of N items. $\overline{p}$ is the average fraction defective taken over M samples, equal to the total number of defects found in M samples divided by the total number of samples, $N \times M$.

The average and control limits both depend on $\overline{p}$, which is determined historically. The lower and upper control limits for the p-chart are given by Eq. 61.84. N_i is the number of samples in period i, usually a constant. The second term in Eq. 61.84 is three times the standard deviation of a binomial distribution.[42]

$$\text{LCL}_p, \text{ UCL}_p = \overline{p} \pm 3\sqrt{\frac{\overline{p}(1-\overline{p})}{N_i}} \qquad 61.84$$

The *c-chart* tracks the number of defects (i.e., *nonconformities*) per sample of constant size. (The sample size may be one complete assembly or single unit.) The defects may be of any variety, anywhere in the assembly. Over a period of M samples, the average number of defects, $\overline{c}$, is determined. The probability of any number of defects per sample is assumed to be a Poisson distribution, and since the mean and variance of a Poisson distribution are the same, the upper and lower control limits are given by Eq. 61.85.[43]

$$\text{LCL}_c, \text{ UCL}_c = \overline{c} \pm 3\sqrt{\overline{c}} \qquad 61.85$$

For most other charts, the average and limits can be allowed to change and shift over time. However, for an in-control process monitored by a *c*-chart, the average and limits are fixed. If quality deterioration occurs and $\overline{c}$ increases, changing the average and limits would imply that the process is still in control when it is not.

There is an important difference between nonconformances and nonconformities. A *nonconformance* is a *defective*—an item that does not meet specification. A *nonconformity* is a *defect*. For example, an acceptance specification may say that a painted object may not have more than three paint defects. An object with two bubbles in its paint will have two nonconformities but will not be a nonconformance.

There are three common variations of control charts: (1) When the sample size changes from time to time, the horizontal control limits will move inward or outward with the changes. (2) The average may show a steady, shifting trend represented by an inclined average line, as when there is uniform tool wear. The parallel control limits will be similarly inclined. (3) In continuous processes, such as liquid product manufacturing or refining, an argument can be made for using a moving average (i.e., recalculating the average over the most recent M samples). The smoothing effect of a moving average accurately represents the effect of product blending.

Example 61.22

An *x*-bar chart with $\pm 2.5\sigma$ limits is maintained for a process. If the process is in control, how many samples would you expect to be taken before a point outside the control limits is seen?

Solution

From a standard normal table, the fractional confidence level is the fraction inside $z = \pm 2.5$.

$$C = (2)(0.4938) = 0.9876$$

From Eq. 61.76, the average run length is

$$\text{ARL} = \frac{1}{1-C} = \frac{1}{1-0.9876}$$

$$= \frac{1}{0.0124} = 80.65$$

---

[40] σ charts are also used. For more information on σ charts, refer to a textbook on statistical quality control.

[41] Though the factors are not symmetrical, the probabilities are. For example, with the 95% factors, 2.5% will be outside of each control limit. However, asymmetrical probabilities could be chosen if desired.

[42] The p-chart and np-chart use the normal approximation to the binomial distribution. For this approximation to be reliable, the sample size should be greater than approximately 25, $\overline{np}$ should be greater than or equal to approximately 4, and np should be greater than or equal to approximately 1.

[43] Unlike the normal distribution, the Poisson distribution is not symmetrical. Since the control limits in the c-chart are symmetrical, they do not represent equal probabilities of exceedance.

15. QUALITY ACCEPTANCE SAMPLING

Acceptance sampling is the testing of a sample taken from a *lot* (batch or process) in order to determine if the entire lot should be accepted or rejected. Acceptance sampling is appropriate when testing is destructive or when 100% testing would be too expensive. To design an *acceptance plan* (also known as a *Dodge-Romig plan*), the number in the sample and the *acceptance number* (i.e, the maximum allowable number of defects in the sample) must be specified.

In *single acceptance plans*, a sample of size n out of a total lot size of N is tested. If the number of defects is equal to or less than c, the lot is accepted. The plan can be described graphically by an *operating characteristic (OC) curve*, which plots the *probability of acceptance* (also known as the *producer's acceptance risk*) versus the *lot quality* (i.e., the true fraction or percentage defective).[44] Points on the OC curve are determined from the binomial or, more preferably, from the Poisson approximation to the binomial. In practice, however, acceptance plans are generally designed by referring to tables of predetermined plans.

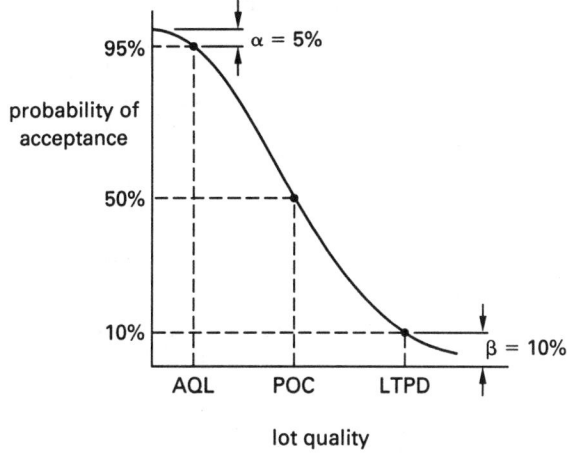

Figure 61.7 *Acceptance Plan Operation Characteristic Curve*

The *producer's risk*, α, also known as the *supplier's risk*, is the probability of a *Type I error* (i.e., rejecting a good lot—that is, of finding trouble when none exists). The *consumer's risk* is the probability of a *Type II error* (i.e., accepting a bad lot—that is, of not finding trouble when it exists). Once a sampling plan has been determined (i.e., the OC curve is established) and the actual incoming fraction defective is known, Eq. 61.86 can be applied.

$$\alpha + \beta = 100\% \qquad 61.86$$

The only way to decrease the consumer's risk without increasing the producer's risk is to increase the sample size (i.e., to change the sampling plan). Sampling plans based on large samples have steeper OC curves.

---

[44]Operating characteristic curves are actually a series of discontinuous points since lot items are finite and discrete. However, they are never drawn in that manner.

(The ideal curve, corresponding to 100% sampling), is a vertical line at the lot quality.) Plans with smaller samples have more gradually inclined curves. When the sample history shows unsatisfactory quality, a *tightened plan* with more samples can be implemented. When the part or supplier has a history of extremely high quality, a *reduced plan* with fewer samples can be used.

Three points, usually those corresponding to 95%, 50%, and 10% probabilities, are given special consideration on the OC curve. The lot quality corresponding to a 95% probability of acceptance (i.e., a 95% consumer's risk) is known as the *acceptable quality level*, AQL.[45] The AQL is sometimes written as the "$p_{95\%}$ quality" or the "$p_{0.95}$ point." The AQL is the percentage defective that is "satisfactory" (i.e., a lot with percentage defective equal to $p_{95\%}$ or less is a good lot). The probability will be $1 - \alpha = 5\%$ that the lot will be rejected if the fraction defective is $p_{95\%}$ or less.

The lot quality corresponding to a 50% probability of acceptance (i.e., producer's and consumer's risks both equal to 50%) is known as the "$p_{50\%}$ quality" or $p_{0.50}$ point, sometimes referred to as the "*point of control*" (POC) or "*indifference quality level.*"

The lot quality corresponding to a 10% probability of acceptance (i.e., a 10% consumer's risk) is known as the *lot tolerance percent (fraction) defective*, LTPD, also known as the "$p_{10\%}$ quality" or, less frequently, the *rejectable quality level*, RQL.

It is generally easier to design a sampling plan that has specific values of AQL and LTPD than to find the AQL and LTPD for a known plan. There are few published tables for the latter case. The traditional method is to plot the OC curve and read the values of fraction defective corresponding to probabilities of acceptance of 95% and 10%. (See Ex. 61.24.)

A Poisson distribution is used to describe the probability of any item being defective. The average number of defects in a sample is

$$\lambda = (\text{fraction defective})(\text{sample size}) \qquad 61.87$$

$$p\{m \text{ defects}\} = \frac{\lambda^m e^{-\lambda}}{m!} \qquad 61.88$$

The probability that a lot will be accepted is

$$p\{\text{acceptance}\} = \sum_{m=0}^{c} \{m \text{ defects}\} \qquad 61.89$$

In practice, most acceptance plans are "designed" by using published tables or, at the very least, by using *cumulative summation ("cumsum") tables* for a particular distribution. However, if the $p_{50\%}$ quality is known, a single-sampling plan can be easily designed. Once the number of defects, c, has been (arbitrarily) chosen, Eq. 61.90 is used to calculate the sample size. There will

---

[45]Any probability could be used. However, the 95% consumer's risk (5% producer's risk) is traditional.

be a family of plans (with different values of c) that satisfy Eq. 61.90. Plans with large values of N and c are better than plans with small values, as the slope of the OC curve is steeper near the $p_{50\%}$ point. In Eq. 61.90, $p_{50\%}$ is expressed in decimal form.

$$N = \frac{c + 0.67}{p_{50\%}}$$ *61.90*

If all items in rejected lots are subsequently screened (i.e., 100% tested) with only good items being passed, the screened lots will contain no defectives.[46] When the various lots are combined, some lots will have been acceptance-sample passed and some will have been 100% screened. The *average outgoing quality limit*, AOQL, is the maximum fraction (percentage) defective of accepted items in the combined lots. In the long run, regardless of incoming quality, item quality will be better than or equal to the AOQL.[47]

The AOQL cannot be read directly from the OC chart. Given a specific OC curve, the AOQL is the maximum value of the product of the lot quality and the probability of acceptance (i.e., the abscissa and ordinate of points on the OC curve). The AOQL can be rapidly found from the curve by trial and error.

With a *double-acceptance plan*, a sample of n_1 is taken out of the original lot size of N. The lot is accepted if the number of defects is c_1 or fewer and rejected if greater than c_2. Otherwise, a second sample of n_2 is taken. The lot is accepted if the total number of defectives from both samples is c_2 or fewer.

Example 61.23

A sample of 100 items is taken from a large lot known to have a 3% defect rate. The acceptance number is 2. What is the probability of acceptance?

Solution

The lot will be accepted if 0, 1, or 2 defects are found. The Poisson probability distributed is assumed. The average number of defects in the sample is

$$\lambda = (\text{fraction defective})(\text{sample size})$$
$$= (0.03)(100)$$
$$= 3$$
$$p\{m \text{ defects}\} = \frac{\lambda^m e^{-\lambda}}{m!}$$

---

[46]Theoretically, each defective item should be replaced with a good item. However, if the lot size is very large, it doesn't really make any difference if a defective item is merely discarded.
[47]Exceptions are possible in the short run.

$$p\{\text{acceptance}\} = p\{0 \text{ defects}\} + p\{1 \text{ defect}\}$$
$$+ p\{2 \text{ defects}\}$$
$$= \frac{3^0 e^{-3}}{0!} + \frac{3^1 e^{-3}}{1!} + \frac{3^2 e^{-3}}{2!}$$
$$= 0.0498 + 0.1494 + 0.2240$$
$$= 0.4232 \quad (42.32\%)$$

(This is the same value that is obtained from App. 61.C with $\lambda = 3$ and $c = 2$.)

Example 61.24

A sampling plan is designed to accept the entire lot if there are 2 or fewer defective in a sample of 100 items. The actual fraction defective of the lot is unknown. (a) Draw the operating characteristic curve. (b) What is the approximate AQL for a producer's risk of 5%? (c) What is the approximate LTPD for a consumer's risk of 10? (d) What is the approximate AOQL?

Solution

(a) Points on the curve can be generated by repeating Ex. 61.23 for different values of the fraction defective, p.

fraction defective, p	Np	$p\{\text{acceptance}\}$
0.00	0.00	1.000
0.01	1.00	0.920
0.02	2.00	0.677
0.03	3.00	0.423
0.04	4.00	0.238
0.05	5.00	0.125
0.06	6.00	0.062

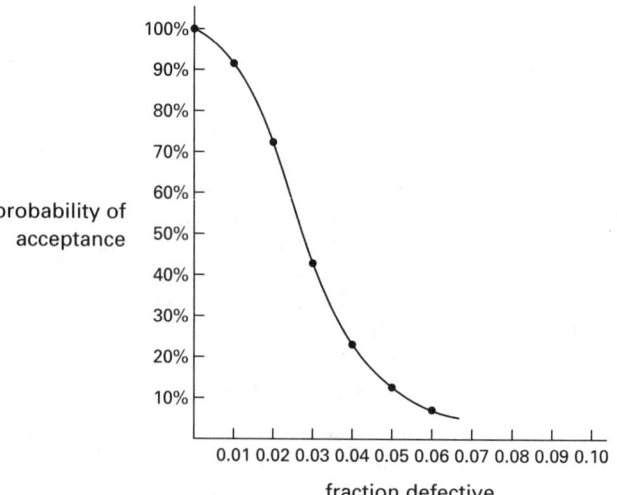

(b) The producer's risk is 5%, so the probability of acceptance is 95%. The AQL is between 0 and 1%. (The actual value is 0.82%.)

(c) The probability of acceptance is the same as the consumer's risk: 10%. The LTPD is between 5% and 6%. (The actual value is 5.3%.)

(d) The AOQL is the maximum product of the fraction defective and the probability of acceptance. This is approximately $(0.02)(0.677) = 0.01354$ (1.35%).

16. RISK ASSESSMENT

Risk assessment is an analytical process that determines the probability that some mishap will occur. It is applied primarily to human health risks and safety hazards or accidents from industrial practices.

Health risk assessment is based on toxicological studies. Data are usually gathered from exposure of laboratory animals to a chemical or process, although human exposure data can be used if available.[48] Tests are run to determine if a chemical is a *carcinogen* (i.e., causes cancer), a *mutagen* (i.e., causes mutations), a *teratogen* (i.e., causes birth defects), a *nephrotoxin* (i.e., harms the kidneys), a *neurotoxin* (i.e., damages the brain or nervous system), or a *genotoxin* (i.e., harms the genes).

Safety risk assessment of potentially unsafe equipment or practices is traditionally called *hazard analysis*.[49] In the past, attention has primarily been focused on processes and equipment. However, hazard analysis is also applicable to material storage, shipping, transportation, and waste disposal. An analysis of each point in the process is performed using a fault-tree or event-tree.

Fault-tree analysis, FTA, is a deductive logic model. Unwanted events (i.e., failures or accidents) are first assumed. Then, the conditions are identified that could bring about the event. All possible contributors to the unwanted event are considered. The event is shown graphically at the top of a network. It is linked through various event statements, logic gates, and probabilities to more basic fault events located laterally and below, producing a graphical tree having branches of sequences and system conditions.

Failure modes and effect analysis, FMEA, is essentially the reverse of fault-tree analysis. It starts with the components of the system and, by focusing on the weaknesses and failure susceptibilities, evaluates how the components can contribute to the unwanted event. The basis for FMEA is, essentially, a parts list showing how each assembly is broken down into subassemblies and basic components. The appearance of an FMEA analysis is tabular, with the columns used for failure modes, causes, symptoms, redundancies, consequences of failure, frequencies, probabilities, and so on.

Life-cycle analysis is used to prevent failures of equipment that have limited lives or that accumulate damage. Though the probability of failure may ideally be low, the history of the equipment may also be unknown. Life cycle analysis commonly focuses on time, history, and condition to determine the *degree of damage* (i.e., any reduction in useful life) present in a unit. One approach is to use reporting systems (i.e., operating logs and historical data). Another approach uses ongoing *material conditions monitoring*, MCM, and testing of the actual equipment.

17. WORK METHODS

The goal of *work methods (methods engineering)* is the reduction of fabrication and assembly time, worker effort, and manufacturing cost. This is accomplished through many activities, including initial *design for manufacture and assembly*, DFMA, selection of methods to be used, human factors engineering, work measurement, plant layout, assembly line balancing, and administration of the manufacturing process. Work methods, which is more worker- and workplace-oriented, goes beyond traditional *value engineering*, which is product-design oriented.

18. BEHAVIORAL SCIENCE

Behavioral science deals with the psychological aspects of the job to increase job productivity. The subject draws from at least three major disciplines: psychology of the individual, sociology of the group, and anthropology of the culture. It is an outgrowth of the "human relations" thinking of the 1930s, whose goal was to increase the happiness of each worker.[50] Current behavioral science, however, seeks to minimize the tensions that limit productivity.[51] The following material briefly summarizes the most popular behavioral science theories.

Improvements in worker satisfaction and decreases in worker dissatisfaction are often accomplished by modifying the structure of the job.[52] *Job flexibility* is a technique giving a worker the ability to move from job task to job task. *Job enlargement* is a technique extending the number of tasks a worker performs within a job.[53] *Horizontal job enlargement* adds new production

---

[48]The best-known exposure test is the *Ames test*, whereby microbes, cells, or test animals are exposed to a chemical in order to determine its carcinogenicity.

[49]The name *hazard assessment by risk analysis*, HARA, refers to a specific method developed by Phillips Petroleum Co. (Bartleville, OK) in the 1980s, and is not a redundant description.

[50]"A happy employee is a productive employee" was the thinking.

[51]There are many who believe that motivational programs are not "honest," since management tries to convince employees to do what management wants. Wage incentive programs encourage employee dishonesty and errors by emphasizing quantity, not quality. Theory X management, with its implied punitive action if goals are not achieved, has never been effective.

[52]There is little or no evidence that workers want a social aspect to their jobs, job enlargement, or more autonomy. Though the relationship between satisfaction, absenteeism, and turnover has been established, the relationship between satisfaction and productivity is questionable.

[53]There are advantages to keeping a job small in scope. Learning time is low, mental effort is reduced, the pay rate can be lower for untrained workers, and supervision is reduced. However, such simple jobs also result in high turnover, increased

Plant Engineering

activities to a job. *Vertical job enlargement* adds planning, inspection, and other nonproduction tasks to the job. *Job enrichment* is a subjective *result* felt by the worker when some technique (perhaps flexibility or enlargement) is used.

Dr. Abraham Maslow's *need hierarchy theory* holds that certain needs become dominant only when lesser needs are satisfied. Although some needs can be sublimated and others overlapped, the theory requires the low-level needs to be satisfied before the higher-level needs can be realized.[54] The need hierarchy theory explains why money is a poor motivator of an affluent individual.

Table 61.4 Maslow's Need Hierarchy
(in order of lower to higher needs)

- *physiological needs:* air, food, water

- *safety needs:* protection against danger, threat, deprivation, arbitrary decisions; need for security in a dependent relationship

- *social needs:* belonging, association, acceptance, giving and receiving love, friendship

- *ego needs:* self-respect, confidence, achievement, self-image, group image, reputation, status, recognition, appreciation

- *self-fulfillment needs:* realizing self-potential, self-development, creativity

The *theory of influence* attempts to explain the effectiveness of supervisors. The most effective supervisors are those who help their workers benefit. For example, supervisors who are close (socially) to their workers and side with them in disputes are effective only if they have enough influence to help the workers. Consequently, knowledge and training are useless unless supervisors have the power to implement what they have learned.

The theory of influence includes five main tenets: (1) Employees think well of supervisors who help them reach their goals and meet their needs. (2) An influential supervisor will be able to help workers. (3) An influential supervisor who is also a disciplinarian will cause dissatisfaction. (4) A supervisor with no influence will not be able to affect worker satisfaction in any way. (5) Increases in supervisor influence are necessary to increase worker satisfaction.

Frederick Herzberg's *motivation-maintenance theory* explains worker satisfaction in terms of satisfiers and dissatisfiers. The *dissatisfiers* (also called *maintenance/*

motivation factors) do not motivate employees; they can only dissatisfy them. Dissatisfiers include salary, fringe benefits, company policy, administration, supervision, working conditions, and interpersonal relations. *Satisfiers* (also known as *motivators*) determine job satisfaction. Common satisfiers are achievement, recognition, the type and nature of the work, responsibility, and advancement. Dissatisfiers must be eliminated before the satisfiers can work.

Two ways of thinking, named "theory X" and "theory Y," have been proposed to describe the extremes of management style. The largely pessimistic *theory X* is based on the assumption that workers inherently dislike and avoid work. Therefore, workers must be coerced into work by threats of punishment. Rewards are not sufficient.

Theory X assumes that the average employee wants to be directed, avoids responsibility, and seeks the security of an employee-employer relationship.[55] Theory X is pessimistic about the effectiveness of employers to satisfy or motivate their employees. According to the theory, by satisfying the physiological and safety (lower level) needs, employees shift the emphasis to higher needs, which cannot be satisfied. Employees, unable to derive satisfaction from their work, behave according to theory X.

Theory Y assumes that workers find expenditure of effort to be natural and not inherently distasteful. It assumes that the average worker learns to accept and enjoy responsibility. It assumes that creativity is widely distributed among employees and that the potentials of average employees are only partially realized. Theory Y places the blame for worker laziness, indifference, and lack of cooperation in the lap of management.

The philosophy of a *zero-defect program* is to expect perfect work from everybody. This counteracts the thinking of workers who have been conditioned to believe that everybody is imperfect and that errors are natural.[56] In a true zero-defect program, standards of performance are set for each worker. Workers are periodically checked against these performance requirements, and recognition is given when the goals are met.

Zero-defect programs develop constant, conscious desires and effort to do the job correctly the first time. This is accomplished by emphasizing what employees have for their own: pride and desire. Employees are continually reminded that their jobs are important, that the product is important, and that management thinks their efforts are important. The challenge of perfection is presented, and the importance of perfection is explained. Management sets an example by expecting zero defects from itself.

---

absenteeism, and low pride in the job (and, subsequently, low quality). Job enlargement generally increases pride and quality, reduces inspection and material handling, and decreases turnover and absenteeism. However, training time is greater, tooling costs are higher, and record-keeping is more complex.
[54]The ego and self-fulfillment needs are rarely satisfied.

---

[55]Theory X is supported by much evidence. Workers exist in a continuum of wants, needs, and desires. Many of the needs are satisfied only off the job. Therefore, work is considered by the worker as a punishment or a price paid for off-the-job satisfaction.
[56]Indeed, perfection is the standard for some professions (e.g., doctors, lawyers, and engineers).

In *quality circle programs*, also known as *quality control circles* (QCC) and *TEAM programs*, worker groups actively participate in measuring and improving departmental operations. Regular (weekly or monthly) meetings are held, with attendance usually being voluntary. Workers who attend are encouraged to suggest topics and ideas for improvement. Topics discussed include effective ways to improve quality, safety, satisfaction, and productivity. Because the lowest-level worker is involved in goal-establishment, quality circles are known as a *bottom-up approach* to goal setting.[57]

Management by objectives, MBO, is a *top-down approach* to goal setting. In theory, the senior company officer establishes realistic, attainable goals for company performance and for his or her immediate managers (who accept the goals). These managers establish realistic goals for lower-level supervisors (who accept their goals). The supervisors, in turn, work out obtainable goals for their subordinates, and so on.

19. HUMAN FACTORS ENGINEERING

Unlike behavioral science (Sec. 18) that deals with the psychological effects of work on the individual, *human factors engineering*, also known as *human engineering* and *ergonomics*, is concerned with the physical effects of work. Human factors attempts to increase or optimize the efficiency and reliability of the *machine-worker system* (also known as the *man-machine system*). The "optimization" process used is not mathematical but is based on thoughtful design technique.[58]

The three most common goals of human factors engineering are protecting the operator, minimizing the required conscious thought and effort, and assuring that the machine-worker system will operate successfully. These goals are usually accomplished by foolproofing the *machine-worker interface* (i.e., the controls and physical devices used) and designing the work environment for long-term health and safety.

Typical foolproofing techniques include locating controls in visible and accessible locations, varying the control handle and knob shapes for different functions, sizing and locating levers and wheels so that adequate operating force can be applied, mounting switches with uniform on-off directions, and arranging gauges so that their dials point in a uniform direction during proper operation.

Anthropometric data may be used to correctly size the equipment and work environment. Such items as range of motion, sitting height, leg room, and speed of motion are extensively tabulated by age, sex, nationality,

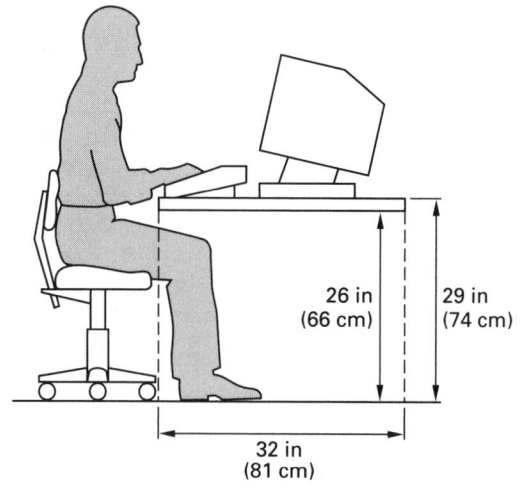

(a) side view

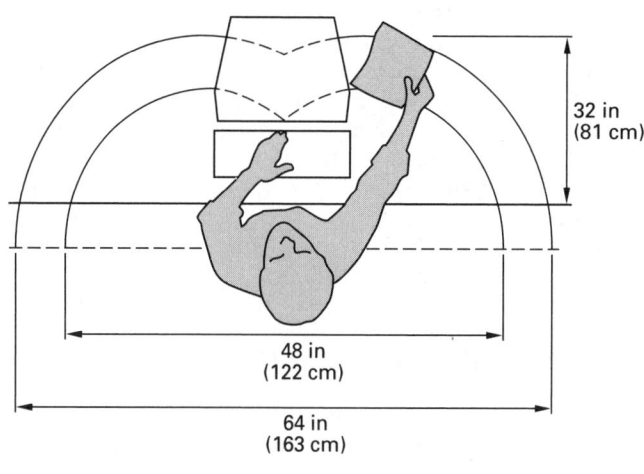

(b) top view

Figure 61.8 *Typical Design of Computer Workstations*

and population percentile. Since it is impossible to design the machine-worker interface for every individual, the reach (as one example) required to operate a control should be no greater than the shortest reach of all workers expected to operate it. Designs frequently intentionally exclude the upper and lower 5% of the population sizes.

Static anthropometry deals with human dimensions such as height, length of forearms, foot size, and so on. *Dynamic anthropometry* deals with limits of motion, such as how far a person can reach in front of them. Both types of data are used in workplace design. Table 61.5 lists typical dimensions developed from anthropometric data for video display terminals, VDTs.[59]

---

[57]Subordinate workers do not dictate to supervisors in *bottom-up* methods. Workers develop recommendations that are passed up, but the decisions are made by supervisors.

[58]It is sometimes difficult to decide when improving the system involves human factors engineering and when it does not. Human factors engineering encompasses the subtopics of safety, industrial hygiene, design of training, and management and supervision, as well as system design.

---

[59]Local laws, regulations, and other standards may apply.

Table 61.5 Recommended Dimensions for VDT Workstations

	inches	centimeters
workstation surface		
minimum height	29	74
width	48–64	122–163
depth	32	81
legroom		
minimum height	26	66
minimum width	26	66
eye-to-screen distance	20–30	51–76
minimum height of home-row		
key from floor	29	74
keyboard thickness	1–2	2.5–5

Attention is also given to the work environment, including temperature, humidity, illumination, and noise level. The "comfort range" for temperature and humidity depend on the nature of the work, dress, and the ventilation. Special consideration is required when the working temperature is not in the range of 65°F to 80°F (26.4°C to 38.4°C). Ear protection should be provided, used, and required when the noise level is above 85 to 90 dB.[60] The level of illumination provided depends on the nature of the work. Typically, storage areas require 5 to 20 fc (54 to 229 lux), office work requires 100 to 200 fc (1100 to 2200 lux), and fine assembly requires 500 to 1000 fc (5400 to 1100 lux).[61]

20. MACHINE SAFEGUARDING

Most machines must be safeguarded to prevent injury to workers. For example, some machines are belt-driven. The belt must be enclosed by a safeguard (i.e., a *guard* or *shield*) to prevent a worker's clothing, hair, body, tools, or workpieces from being drawn into the drive mechanism. In general, safeguards should (1) prevent contact, (2) be secure, (3) protect from entering objects, (4) create no new hazards, (5) create no interference with work, and (6) allow safe repair and maintenance.

A worker should not be able to easily remove or tamper with a safeguard. A safeguard that can be made ineffective accidentally or in order to speed up operation is little better than no safeguard. Safeguards and safety devices should be made of durable materials that will withstand the conditions of normal use. They must be firmly secured to the machine.

The safeguard should ensure that no objects can fall into moving parts. A small tool that is dropped into a rotating machine can become a dangerous projectile.

---

[60]The U.S. Occupational Safety and Health Act (OSHA) of 1970 established 90 dB as the maximum noise level for a 9 hour exposure. Higher levels, up to a maximum of 130 dB, are permitted for shorter intervals.

[61]The *foot-candle* is equal to a lumen per square foot. The unit of illumination in SI units is the *lux*, equal to a lumen per square meter.

A safeguard defeats its own purpose if it creates a hazard of its own. Safeguards should not contain jagged holes or unfinished, sharp edges. Edges should be rolled or bolted in such a way that sharp edges are eliminated.

A safeguard that impedes a worker from performing the job quickly and comfortably might be overridden or disregarded. Proper safeguarding should enhance efficiency as it relieves worker apprehension about injury.

It should be possible to lubricate, maintain, and repair equipment without removing any safeguards. Locating oil reservoirs outside the guard with a line leading to the lubrication point will eliminate the need for a worker or maintenance worker to enter hazardous areas.

There are many ways to safeguard machinery. The type of operation, nature of the raw material, method of handling, physical layout of the work area, and the production requirements or limitations all have to be taken into consideration. In general, power transmission apparatus is best protected by fixed guards that enclose the danger area. For hazards at the point of operation, where moving equipment performs work on stock, Table 61.6 lists options for safeguarding. Miscellaneous methods, such as signs and awareness barriers, should also be considered.

Table 61.6 Machine Safeguarding Options

- *guard types*
  - fixed
  - interlocked
  - adjustable
  - self-adjusting
- *presence-sensing devices*
  - electromechanical (pressure-sensitive body bar, floor mat, etc.)
  - foot switch
  - photoelectric (optical)
  - infrared
  - capacitive (radio-frequency)
- *actuation control*
  - proper presence
  - proper sequence
  - two-switch (two-hand) actuation
  - remote actuation
- *feeding and ejection options*
  - automatic
  - semiautomatic
  - remote (robotic)

PRACTICE PROBLEMS

1. *(Time limit: one hour)* Printed circuit boards are manufactured in four consecutive departmental operations. Each operation occurs on a different machine. Employees in all departments work from 8:00 a.m. to 5:00 p.m. and have one hour total for lunch and personal breaks. The target rate is 900,000 completed units per year. Units found to be defective are discarded. No units are produced during set-up, downtime, maintenance, or record-keeping periods.

	department			
	1	2	3	4
production time (sec/unit)	6	10	11	45
set-up time (min/day)	16	8	20	5
downtime (min/day)	12	10	15	0
maintenance time (min/day)	8	12	8	0
record-keeping (min/day)	6	6	6	30
percentage defects	4%	6%	3%	2%

(a) What is the maximum number of completed circuit boards that can be produced in one year if there is one machine per department? (b) If additional machines can be added for any or all operations, what is the most efficient method of meeting the target production rate? (There are no changes in the defect rates.) (c) What is the efficiency of each department if the capacity is increased per part (b)?

2. *(Time limit: one hour)* Four workers perform operations 1, 2, 3, and 4 in sequence on a manual assembly line. Each station performs its operation only once on the product before sending the product on to the next operation. Operation times at the stations are as given. (Travel times are included in the operation times.)

station	time (min)
1	0.6
2	0.6
3	0.9
4	0.8

A fifth "floating" station has the ability to assist any of the four stations. The fifth station works with the same efficiencies and times as the four stations. There is no fixed assignment for this fifth station. The fifth station is allowed to help any station that needs it.

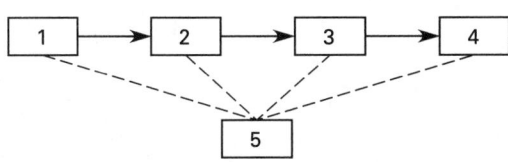

The operators of all five stations are permitted a 10 minute break each hour. (a) What is the maximum number of products that can be produced assuming that the fifth station is assigned to work optimally? Neglect the initial (transient) performance. (b) What is the fraction of station 5's time allocated to each operation?

3. The activities that constitute a project are listed below. The project starts at $t = 0$. (a) Draw the critical path network. (b) Indicate the critical path. (c) What is the earliest finish? (d) What is the latest finish? (e) What is the slack along the critical path?

(f) What is the float along the critical path?

activity	predecessors	successors	duration
start	–	A	0
A	start	B,C,D	7
B	A	G	6
C	A	E,F	5
D	A	G	2
E	C	H	13
F	C	H,I	4
G	D,B	I	18
H	E,F	finish	7
I	F,G	finish	5
finish	H,I	–	0

4. PERT activities constituting a short project are listed with their characteristic completion times. If the project starts at $t = 15$, what is the probability that the project will be completed on or before $t = 42$?

activity	predecessors	successors	t_{min}	t_{likely}	t_{max}
start	–	A	0	0	0
A	start	B,D	1	2	5
B	A	C	7	9	20
C	B	D	5	12	18
D	A,C	finish	2	4	7
finish	D	–	0	0	0

5. *(Time limit: one hour)* Your manufacturing facility produces two models of municipal transit buses, designated B-1 and B-2. You can sell as many of either model as you produce. The per-bus profits are $800,000 for model B-1 and $650,000 for model B-2. You would like to maximize your company's profit by determining the number of each model buses to produce. However, it is not a a simple matter of producing only B-1 models because certain common parts are in limited supply.

part	number available
gage sending units	2000
wheel housing flares	1800
intake grilles	3600

Due to differences in design, the quantity of each common part varies between the two models.

part	number in model	
	B-1	B-2
gage sending units	8	10
wheel housing flares	6	4
intake grilles	3	2

6. A small company makes two chemicals, designated C-1 and C-2. The process for both includes fermentation and purification. Labor limits all fermentation operations to 300 hr per month, and purification is limited to 120 hr per month. Each unit of product C-1 requires

10 hr for fermentation and 8 hr for purification. Each unit of product C-2 requires 20 hr for fermentation and 3 hr for purification. The profit per unit of product C-1 is $3000; the profit per unit of product C-2 is $5000. How much should the company make of each chemical per month?

7. A linear wage incentive program provides for 50% participation and a bonus that begins at a productivity level of 66.7% of standard. A worker produces 1900 units in 8 hr. The standard time for each unit is 0.004 hr. What is the worker's relative earnings?

62 Instrumentation and Measurements[1]

1.	Accuracy	62-1
2.	Precision	62-2
3.	Stability	62-2
4.	Calibration	62-2
5.	Error Types	62-2
6.	Error Magnitudes	62-3
7.	Potentiometers	62-3
8.	Transducers	62-3
9.	Sensors	62-3
10.	Variable-Inductance Transducers	62-4
11.	Variable-Reluctance Transducers	62-4
12.	Variable-Capacitance Transducers	62-4
13.	Other Electrical Transducers	62-4
14.	Photosensitive Conductors	62-5
15.	Resistance Temperature Detectors	62-5
16.	Thermistors	62-5
17.	Thermocouples	62-6
18.	Strain Gages	62-8
19.	Wheatstone Bridges	62-9
20.	Strain Gage Detection Circuits	62-9
21.	Strain Gage in Unbalanced Resistance Bridge	62-10
22.	Bridge Constant	62-10
23.	Stress Measurements in Known Directions	62-10
24.	Stress Measurements in Unknown Directions	62-12
25.	Load Cells	62-12
26.	Dynamometers	62-12
27.	Indicator Diagrams	62-14

Nomenclature

A	area	in^2	m^2
b	base length	in	m
BC	bridge constant	–	–
c	distance from neutral axis	in	m
C	concentration	various	various
d	diameter	in	m
E	modulus of elasticity	psi	Pa
F	force	lbf	N
G	shear modulus	psi	Pa
GF	gage factor	–	–
h	height	in	m
I	current	A	A
I	moment of inertia	in^4	m^4
J	polar moment of inertia	in^4	m^4
k	constant	various	various
k	deflection constant	lbf/in	N/m
K	factor	–	–
L	shaft length	in	m

M	moment	in-lbf	N·m
n	number	–	–
n	rotational speed	rpm	rpm
p	pressure	psi	Pa
P	permeability	various	various
P	power	hp	kW
Q	statical moment	in^3	m^3
r	radius	in	m
R	resistance	Ω	Ω
t	thickness	in	m
T	temperature	1/°R	K
T	torque	in-lbf	N·m
V	voltage	V	V
VR	voltage ratio	–	–
y	deflection	in	m

Symbols

α	temperature coefficient	1/°R	1/K
β	temperature coefficient	$1/°R^2$	$1/K^2$
β	constant	°R	K
γ	shear strain	–	–
ϵ	strain	–	–
η	efficiency	–	–
θ	angle of twist	deg	deg
ν	Poisson's ratio	–	–
ρ	resistivity	Ω-in	Ω·cm
σ	stress	psi	Pa
τ	shear stress	psi	Pa
ϕ	angle of twist	rad	rad

Subscripts

b	battery
g	gage
o	original
r	ratio
ref	reference
t	transverse or total
T	at temperate T
x	in x-direction
y	in y-direction

---

[1] Measurement of fluid pressure is covered in Chap. 15; measurement of fluid flow is covered in Chap. 17.

1. ACCURACY

A measurement is said to be *accurate* if it is substantially unaffected by (i.e., is insensitive to) all variation outside of the measurer's control.

For example, suppose a rifle is aimed at a point on a distant target and several shots are fired. The target point represents the "true value" of a measurement—the value that should be obtained. The impact points represent the measured values—what is obtained. The distance from the centroid of the points of impact to the target point is a measure of the alignment accuracy between the barrel and the sights. This difference between the true and measured values is known as the measurement *bias*.

2. PRECISION

Precision is not synonymous with *accuracy*. Precision is concerned with the repeatability of the measured results. If a measurement is repeated with identical results, the experiment is said to be precise. The average distance of each impact from the centroid of the impact group is a measure of precision. Thus, it is possible to take highly precise measurements and still have a large bias.

Most measurement techniques (e.g., taking multiple measurements and refining the measurement methods or procedures) that are intended to improve accuracy actually increase the precision.

Sometimes, the term *reliability* is used with regard to the precision of a measurement. A *reliable measurement* is the same as a *precise estimate*.

3. STABILITY

Stability and *insensitivity* are synonymous terms. (Conversely, *instability* and *sensitivity* are synonymous.) A stable measurement is insensitive to minor changes in the measurement process.

Example 62.1

At 65°F (18°C), the centroid of an impact group on a rifle target is 2.1 in (5.3 cm) from the sight-in point. At 80°F (27°C), the distance is 2.3 in (5.8 cm). What is the sensitivity to temperature?

SI Solution

$$\begin{aligned} \frac{\text{sensitivity to}}{\text{temperature}} &= \frac{\Delta \text{ measurement}}{\Delta \text{ temperature}} \\ &= \frac{5.8 \text{ cm} - 5.3 \text{ cm}}{27°\text{C} - 18°\text{C}} \\ &= 0.0556 \text{ cm/°C} \end{aligned}$$

Customary U.S. Solution

$$\begin{aligned} \frac{\text{sensitivity to}}{\text{temperature}} &= \frac{\Delta \text{ measurement}}{\Delta \text{ temperature}} \\ &= \frac{2.3 \text{ in} - 2.1 \text{ in}}{80°\text{F} - 65°\text{F}} \\ &= 0.0133 \text{ in/°F} \end{aligned}$$

4. CALIBRATION

Calibration is used to determine or verify the scale of the measurement device. In order to calibrate a measurement device, one or more known values of the quantity to be measured (temperature, force, torque, etc.) are applied to the device and the behavior of the device is noted. (If the measurement device is linear, it may be adequate to use just a single calibration value. This is known as *single-point calibration*.)

Once a measurement device has been calibrated, the calibration signal should be reapplied as often as necessary to prove the reliability of the measurements. In some electronic measurement equipment, the calibration signal is applied continuously.

5. ERROR TYPES

Measurement errors can be categorized as *systematic (fixed) errors*, *random (accidental) errors*, *illegitimate errors*, and *chaotic errors*.

Systematic errors, such as improper calibration, use of the wrong scale, and incorrect (though consistent) technique, are essentially constant or similar in nature over time. *Loading error* is a systematic error and occurs when the act of measuring alters the true value.[2] Some *human errors*, if present in each repetition of the measurement, are also systematic. Systematic errors can be reduced or eliminated by refinement of the experimental method.

Random errors are caused by random and irregular influences generally outside the control of the measurer. Such errors are introduced by fluctuations in the environment, changes in the experimental method, and variations in materials and equipment operation. Since the occurrence of these errors is irregular, their effects can be reduced or eliminated by multiple repetitions of the experiment.

There is no reason to expect or tolerate *illegitimate errors* (e.g., errors in computations and other blunders). These are essentially mistakes that can be avoided through proper care and attention.

Chaotic errors, such as resonance, vibration, or experimental "noise," essentially mask or entirely invalidate the experimental results. Unlike the random errors previously mentioned, chaotic disturbances are sufficiently large to reduce the experimental results to meaninglessness.[3] Chaotic errors must be eliminated.

---

[2]For example, inserting an air probe into a duct will change the flow pattern and velocity around the probe.

[3]Much has been written in recent years about *chaos theory*. This theory holds that, for many processes, the ending state is dependent on imperceptible differences in the starting state. Future weather conditions and the landing orientation of a finely balanced spinning top are often used as examples of states that are greatly affected by their starting conditions.

6. ERROR MAGNITUDES

If a single measurement is taken of some quantity whose true value is known, the *error* is simply the difference between the true and measured values. However, the true value is never known in an experiment, and measurements are usually taken several times, not just once. Therefore, many conventions exist for estimating the unknown error.[4]

When most experimental quantities are measured, the measurements tend to cluster around some "average value." The measurements will be distributed according to some distribution, such as linear, normal, Poisson, and so on. The measurements can be graphed in a *histogram* and the distribution inferred. Usually, the data will be normally distributed.[5]

Certain error terms used with normally distributed data have been standardized. These are listed in Table 62.1.

Table 62.1 *Normal Distribution Error Terms*

term	number of standard deviations	percent certainty	approximate odds of being incorrect
probable error	0.6745	50	1 in 2
mean deviation	0.6745	50	1 in 2
standard deviation	1.000	68.3	1 in 3
one-sigma error	1.000	68.3	1 in 3
90% error	1.6449	90	1 in 10
two-sigma error	2.000	95	1 in 20
three-sigma error	3.000	99.7	1 in 370
maximum error[a]	3.29	99.9+	1 in 1000

[a]The true maximum error is theoretically infinite.

7. POTENTIOMETERS

A *potentiometer (potentiometer transducer, variable resistor)* is a resistor with a sliding third contact. It converts linear or rotary motion into a variable resistance (voltage).[6] It consists of a resistance oriented in a linear or angular manner and a variable-position contact point known as the *tap*. A voltage is applied across the entire resistance, causing current to flow through the resistance. The voltage at the tap will vary with tap position.

---

[4]This may be a good time to review the material in Chap. 11.
[5]The results of all numerical experiments are not automatically normally distributed. The throw of a die (one of two dice) is linearly distributed. Emissive power of a heated radiator is skewed with respect to wavelength. However, the means of groups of experimental data generally will be normally distributed, even if the raw measurements are not.
[6]There is a voltage-balancing device that shares the name *potentiometer (potentiometer circuit)*. An unknown voltage source can be measured by adjusting a calibrated voltage until a null reading is obtained on a voltage meter. The applications are sufficiently different that no confusion occurs when the "pot is adjusted."

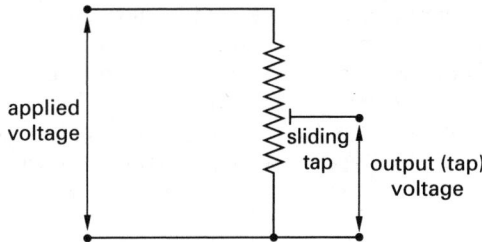

Figure 62.1 *Potentiometer*

8. TRANSDUCERS

Physical quantities are often measured with transducers (*detector-transducers*). A *transducer* converts the measured quantity to a second quantity that is measured. For example, a Bourdon tube pressure gage converts pressure to angular displacement; a strain gage converts stress to resistance change. Transducers are primarily mechanical in nature (e.g., pitot tube, spring devices, Bourdon tube pressure gage) or electrical in nature (e.g., thermocouple, strain gage, moving-core transformer).

9. SENSORS

While the term "transducer" is commonly used for devices that respond to mechanical input (force, pressure, torque, etc.), the term *sensor* is commonly applied to devices that respond to varying chemical conditions.[7] For example, an electrochemical sensor might respond to a specific gas, compound, or ion (known as a *target substance* or *species*). Two types of electrochemical sensors are in use today: potentiometric and amperometric.

Potentiometric sensors generate a measurable voltage at their terminals. In electrochemical sensors taking advantage of half-cell reactions at electrodes, the generated voltage is proportional to the absolute temperature, T, and is inversely proportional to the number of electrons, n, taking part in the chemical reaction at the half-cell. In Eq. 62.1, p_1 is the partial pressure of the target substance at the measurement electrode; p_2 is the partial pressure of the target substance at the reference electrode.

$$V \propto \left(\frac{T_{\text{absolute}}}{n} \right) \ln \left(\frac{p_1}{p_2} \right) \qquad 62.1$$

Amperometric sensors (also known as *voltammetric sensors*) generate a measurable current at their terminals. In the conventional electrochemical sensors known as *diffusion-controlled cells*, a high-conductivity acid or alkaline liquid electrolyte is used with a gas-permeable membrane that transmits ions from the outside to the inside of the sensor. A reference voltage is applied to two terminals within the electrolyte, and the current generated at a (third) sensing electrode is measured.

---

[7]The categorization is common but not universal. The terms "transducer," "sensor," and "pickup" are often used loosely.

Plant Engineering

The maximum current generated is known as the *limiting current*. Current is proportional to the concentration (C) of the target substance, the permeability (P), and the exposed sensor (membrane) area (A), and the number of electrons transferred per molecule detected (n). The current is inversely proportional to the membrane thickness (t).

$$I \propto \frac{nPCA}{t} \qquad 62.2$$

10. VARIABLE-INDUCTANCE TRANSDUCERS

Inductive transducers contain a wire coil and a moving permeable *core*.[8] As the core moves, the flux linkage through the coil changes. The change in inductance affects the overall impedance of the detector circuit.

The *differential transformer* or *linear variable differential transformer* (LVDT) is an important type of *variable-inductance transducer*. It converts linear motion into a change in voltage. The transformer is supplied with a low AC voltage. When the core is centered between the two secondary windings, the LVDT is said to be in its *null position*.

Movement of the core changes the magnetic flux linkage between the primary and secondary windings. Over a reasonable displacement range, the output voltage is proportional to the displacement of the core from the null position, hence the description "linear." The voltage changes phase (by 180°) as the core passes through the null position.

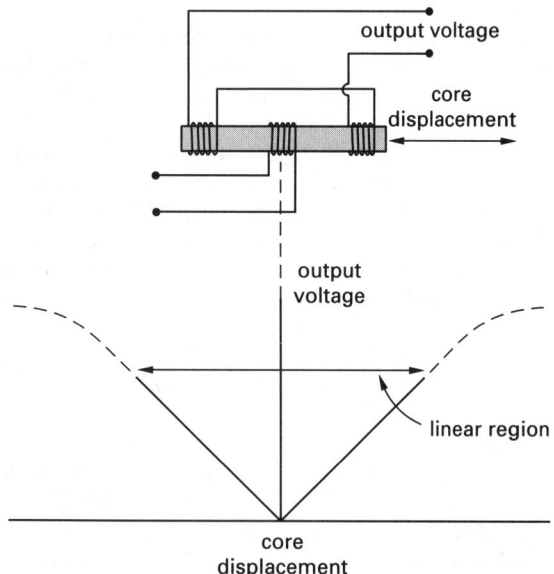

Figure 62.2 *Linear Variable Differential Transformer Schematic and Performance Characteristic*

---

[8]The term *core* is used even if the cross section of the coil and core are not circular.

Sensitivity of a LVDT is measured in mV/in (mV/cm) of core movement. The sensitivity and output voltage depend on the frequency of the applied voltage (i.e., the *carrier frequency*) and are directly proportional to the magnitude of the applied voltage.

11. VARIABLE-RELUCTANCE TRANSDUCERS

A *variable-reluctance transducer (pickup)* is essentially a permanent magnet and a coil in the vicinity of the process being monitored.[9] There are no moving parts in this type of transducer. However, some of the magnet's magnetic flux passes through the surroundings, and the presence or absence of the process changes the coil voltage. Two typical applications of variable-reluctance pickups are measuring liquid levels and determining the rotational speed of a gear.

12. VARIABLE-CAPACITANCE TRANSDUCERS

In *variable-capacitance transducers*, the capacitance of a device can be modified by changing the plate separation, plate area, or dielectric constant of the medium separating the plates.

13. OTHER ELECTRICAL TRANSDUCERS

The *piezoelectric effect* is the name given to the generation of an electrical voltage when placed under stress.[10] *Piezoelectric transducers* generate a small voltage when stressed (strained). Since voltage is developed during the application of changing strain but not while strain is constant, piezoelectric transducers are limited to dynamic applications. Piezoelectric transducers may suffer from low voltage output, instability, and limited ranges in operating temperature and humidity.

The *photoelectric effect* is the generation of an electrical voltage when a material is exposed to light.[11] Devices using this effect are known as *photocells, photovoltaic cells, photosensors,* or *light-sensitive detectors,* depending on the applications. The sensitivity need not be to light in the visible spectrum. Photoelectric detectors can be made that respond to infrared and ultraviolet radiation. The magnitude of the voltage (or of the

---

[9]*Reluctance* depends on the area, length, and permeability of the medium through which the magnetic flux passes.
[10]Quartz, table sugar, potassium sodium tartarate (Rochelle salt), and barium titanate are examples of piezoelectric materials. Quartz is commonly used to provide a stable frequency in electronic oscillators. Barium titanate is used in some ultrasonic cleaners and sonar-like equipment.
[11]While the *photogenerative (photovoltaic)* definition is the most common definition, the term "photoelectric" can also be used with *photoconductive devices* (those whose resistance changes with light) and *photoemissive devices* (those that emit light when a voltage is applied).

current in an attached circuit) depends on the amount of illumination. If the cell is reverse-biased by an external battery, its operation is similar to a constant-current source.[12]

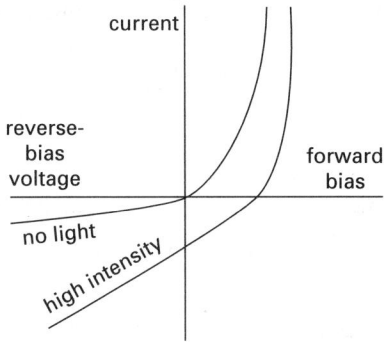

Figure 62.3 *Photovoltaic Device Characteristic Curves*

14. PHOTOSENSITIVE CONDUCTORS

Cadmium sulfide and *cadmium selenide* are two compounds that decrease in resistance when exposed to light. Cadmium sulfide is most sensitive to light in the 5000Å to 6000Å (0.5 to 0.6 μm) range, while cadmium selenide shows peak sensitivities in the 7000Å to 8000Å range (0.7 to 0.8 μm). These compounds are used in *photosensitive conductors*. Due to hysteresis, photosensitive conductors do not react instantaneously to changes in intensity.[13] High-speed operation requires high light intensities and careful design.

15. RESISTANCE TEMPERATURE DETECTORS

Resistance temperature detectors (RTDs), also known as *resistance thermometers*, make use of changes in their resistance to determine changes in temperature. A fine wire is wrapped around a form and protected with glass or a ceramic coating. Nickel and copper are commonly used for industrial RTDs. Platinum is used when precision resistance thermometry is required. RTDs are connected through resistance bridges (Sec. 19) to compensate for lead resistance.

Resistance in most conductors increases with temperature. The resistance at a given temperature can be calculated from the *coefficients of thermal resistance*, α and β.[14] The variation of resistance with temperature

---

is nonlinear, though β is small and is often insignificant over short temperature ranges. Therefore, a linear relationship is often assumed and only α is used. In Eq. 62.3, R_{ref} is the resistance at the reference temperature, T_{ref}, usually 100 Ω at 32°F (0°C).

$$R_T \approx R_{ref}(1 + \alpha\Delta T + \beta\Delta T^2) \qquad 62.3$$

$$\Delta T = T - T_{ref} \qquad 62.4$$

In commercial RTDs, α is referred to by the literal term *alpha-value*. There are two applicable alpha values for platinum, depending on the purity. Commercial platinum RTDs produced in the United States generally have alpha values of 0.00392 1/°C, while RTDs produced in Europe and other countries generally have alpha values of 0.00385 1/°C.

16. THERMISTORS

Thermistors are temperature-sensitive semiconductors constructed from oxides of manganese, nickel, and cobalt and from sulfides of iron, aluminum, and copper. Thermistor materials are encapsulated in glass or ceramic materials to prevent penetration of moisture. Unlike RTDs, the resistance of thermistors decreases as the temperature increases.

Thermistor temperature-resistance characteristics are exponential. Depending on the brand, material, and construction, β typically varies between 3400 K and 3900 K.

$$R = R_o e^k \qquad 62.5$$

$$k = \beta\left(\frac{1}{T} - \frac{1}{T_o}\right) \quad [T \text{ in K}] \qquad 62.6$$

Thermistors can be connected to measurement circuits with copper wire and soldered connections. Compensation of lead wire effects is not required because resistance of thermistors is very large, far greater than the resistance of the leads. Since the negative temperature characteristic makes it difficult to design customized detection circuits, some thermistor and instrumentation standardization has occurred. The most common thermistors have resistances of 2252 Ω at 77°F (25°C), and most instrumentation is compatible with them. Other standardized resistances are 5000 Ω and 10,000 Ω at 25°C.

Thermistors typically are less precise and more unpredictable than metallic resistors. Since resistance varies exponentially, most thermistors are suitable for use only up to approximately 550°F (290°C).

Table 62.2 *Typical Resistivities and Coefficients of Thermal Resistance*[a]

conductor	resistivity[b] ($\Omega \cdot$cm)	$\alpha^{c,d}$ (1/°C)
alumel[e]	28.1×10^{-6}	0.0024 @ 212°F (100°C)
aluminum	2.82×10^{-6}	0.0039 @ 68°F (20°C)
		0.0040 @ 70°F (21°C)
brass	7×10^{-6}	0.002 @ 68°F (20°C)
constantan[f,g]	49×10^{-6}	0.00001 @ 68°F (20°C)
chromel[h]	–	
copper, annealed	1.724×10^{-6}	0.0043 @ 32°F (0°C)
		0.0039 @ 70°F (21°C)
		0.0037 @ 100°F (38°C)
		0.0031 @ 200°F (93°C)
gold	2.44×10^{-6}	0.0034 @ 68°F (20°C)
iron (99.98% pure)	10×10^{-6}	0.005 @ 68°F (20°C)
isoelastic[i]	112×10^{-6}	0.00047
lead	22×10^{-6}	0.0039
magnesium	4.6×10^{-6}	0.004 @ 68°F (20°C)
manganin[j]	44×10^{-6}	0.0000 @ 68°F (20°C)
monel[k]	42×10^{-6}	0.002 @ 68°F (20°C)
nichrome[l]	100×10^{-6}	0.0004 @ 68°F (20°C)
nickel	7.8×10^{-6}	0.006 @ 68°F (20°C)
platinum	10×10^{-6}	0.0039 @ 32°F (0°C)
		0.0036 @ 70°F (21°C)
platinum-iridium[m]	24×10^{-6}	0.0013
platinum-rhodium[n]	18×10^{-6}	0.0017 @ 212°F (100°C)
silver	1.59×10^{-6}	0.004 @ 68°F (20°C)
tin	11.5×10^{-6}	0.0042 @ 68°F (20°C)
tungsten (drawn)	5.8×10^{-6}	0.0045 @ 70°F (21°C)

[a]Compiled from various sources. Data is not to be taken too literally, as values depend on composition and cold working.
[b]At 20°C (68°F)
[c]Values vary with temperature. Common values given when no temperature is specified.
[d]Multiply 1/°C by 5/9 to obtain 1/°F. Multiply ppm/°F by 1.8×10^{-6} to obtain 1/°C.
[e]Trade name for 94% Ni, 2.5% Mn, 2% Al, 1% Si, 0.5% Fe (TM of Hoskins Manufacturing Co.)
[f]60% Cu, 40% Ni, also known by trade names *Advance, Eureka,* and *Ideal.*
[g]Constantan is also the name given to the composition 55% Cu and 45% Ni, an alloy with slightly different properties.
[h]Trade name for 90% Ni, 10% Cr (TM of Hoskins Manufacturing Co.)
[i]36% Ni, 8% Cr, 0.5% Mo, remainder Fe
[j]9 to 18% Mn, 11 to 4% Ni, remainder Cu
[k]33% Cu, 67% Ni
[l]75% Ni, 12% Fe, 11% Cr, 2% Mn
[m]95% Pt, 5% Ir
[n]90% platinum, 10% rhodium

17. THERMOCOUPLES

A *thermocouple* consists of two wires of dissimilar metals joined at both ends.[15] One set of ends, typically called a *junction*, is kept at a known *reference*

[15]The joint may be made by simply twisting the ends together. However, to achieve a higher mechanical strength and a better electrical connection, the ends should be soldered, brazed, or welded.

temperature while the other junction is exposed to the unknown temperature.[16] In a laboratory, the reference junction is often maintained at the *ice point*, 32°F (0°C), in an ice/water bath for convenience in later analysis. In commercial applications, the reference temperature can be any value, with appropriate compensation being made.

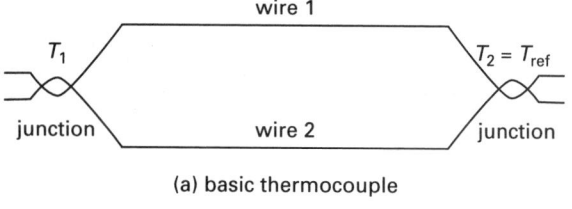

(a) basic thermocouple

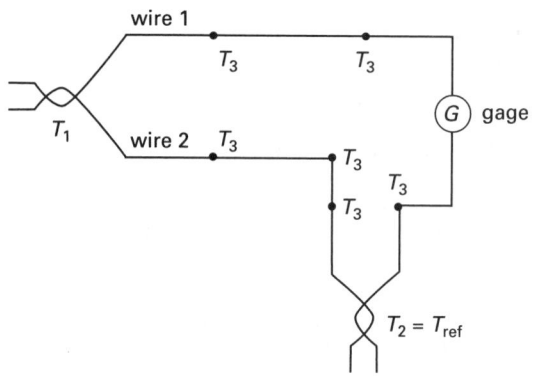

(b) thermocouple in measurement circuit

Figure 62.4 *Thermocouple*

Thermocouple materials, standard ANSI designations, and approximate useful temperature ranges are given in Table 62.3.[17] The "usable" temperature range can be much larger than the useful range. The most significant factors limiting the useful temperature range, sometimes referred to as the *error limits range*, are linearity, the rate at which the material will erode due to oxidation at higher temperatures, irreversible magnetic effects above magnetic critical points, and longer stabilization periods at higher temperatures.

A voltage is generated when the temperatures of the two junctions are different. This phenomenon is known as the *Seebeck effect*.[18] Referring to the polarities of

[16]The *ice point* is the temperature at which liquid water and ice are in equilibrium. Other standardized temperature references are: *oxygen point*, −297.346°F (90.19K); *steam point*, 212.0°F (373.16K); *sulfur point*, 832.28°F (717.76K); *silver point*, 1761.4°F (1233.96K); *gold point*, 1945.4°F (1336.16K).
[17]It is not uncommon to list the two thermocouple materials with "vs." (as in *versus*). For example, a copper/constantan thermocouple might be designated as "copper vs. constantan."
[18]The inverse of the Seebeck effect, that current flowing through a junction of dissimilar metals will cause either heating or cooling, is the *Peltier effect*, though the term is generally used in regard to cooling applications. An extension of the Peltier effect, known as the *Thompson effect*, is that heat will be carried along the conductor. Both the Peltier and Thompson effects occur simultaneously with the Seebeck effect. However, the Peltier and Thompson effects are so minuscule that they can be disregarded.

Table 62.3 *Typical Temperature Ranges of Thermocouple Materials*[a]

materials	ANSI designation	useful range, °F (°C)
copper-constantan	T	−300 to 700 (−180 to 370)
chromel-constantan	E	32 to 1600 (0 to 870)
iron-constantan[b]	J	32 to 1400 (0 to 760)
chromel-alumel	K	32 to 2300 (0 to 1260)
platinum-10% rhodium	S	32 to 2700 (0 to 1480)
platinum-13% rhodium	R	32 to 2700 (0 to 1480)
Pt-6% Rh-Pt-30% Rh	B	1600 to 3100 (870 to 1700)
tungsten-Tu-25% rhenium	–	to 4200[c] (2320)
Tu-5% rhenium-Tu-26% rhenium	–	to 4200[c] (2320)
Tu-3% rhenium-Tu-25% rhenium	–	to 4200[c] (2320)
iridium-rhodium	–	to 3500[c] (1930)
nichrome-constantan	–	to 1600[c] (870)
nichrome-alumel	–	to 2200[c] (1200)

[a]Actual values will depend on wire gage, atmosphere (oxidizing or reducing), use (continuous or intermittent), and manufacturer.
[b]Nonoxidizing atmospheres only.
[c]Approximate usable temperature range. Error limit range is less.

the voltage generated, one metal is known as the *positive element* while the other is the *negative element*. The generated voltage is small, and thermocouples are calibrated in $\mu V/°F$ or $\mu V/°C$. An amplifier may be required to provide usable signal levels, although thermocouples can be connected in series (a *thermopile*) to increase the value.[19] The accuracy (referred to as the *calibration*) of thermocouples is approximately $1/2$ to $3/4$%, though manufacturers produce thermocouples with various guaranteed accuracies.

The voltage generated by a thermocouple is given by Eq. 62.7. Since the *thermoelectric constant*, k_T, varies with temperature, thermocouple problems are solved with published tables of total generated voltage versus temperature. (See App. 62.A.)

$$V = k_T(T - T_{\text{ref}}) \qquad 62.7$$

Generation of thermocouple voltage in a measurement circuit is governed by three laws. The *law of homogeneous circuits* states that the temperature distribution along one or both of the thermocouple leads is irrelevant. Only the junction temperatures contribute to the generated voltage.

The *law of intermediate metals* states that an intermediate length of wire placed within one leg or at the junction of the thermocouple circuit will not affect the voltage generated as long as the two new junctions are

[19]There is no special name for a combination of thermocouples connected in parallel.

at the same temperature. This law permits the use of a measuring device, soldered connections, and extension leads.

The *law of intermediate temperatures* states that if a thermocouple generates voltage V_1 when its junctions are at T_1 and T_2, and it generates voltage V_2 when its junctions are at T_2 and T_3, then it will generate voltage $V_1 + V_2$ when its junctions are at T_1 and T_3.

Example 62.2

A type-K (chromel-alumel) thermocouple produces a voltage of 10.79 mV. The "cold" junction is kept at 32°F (0°C) by an ice bath. What is the temperature of the hot junction?

Solution

Since the cold junction temperature corresponds to the reference temperature, the hot junction temperature is read directly from App. 62.A as 510°F.

Example 62.3

A type-K (chromel-alumel) thermocouple produces a voltage of 10.87 mV. The "cold" junction is at 70°F. What is the temperature of the hot junction?

Solution

Use the law of intermediate temperatures. From App. 62.A, the thermoelectric constant for 70°F is 0.84 mV. If the cold junction had been at 32°C, the generated voltage would have been higher. The corrected reading is

$$10.87 \text{ mV} + 0.84 \text{ mV} = 11.71 \text{ mV}$$

The temperature corresponding to this voltage is 550°F.

Example 62.4

A type-T (copper-constantan) thermocouple is connected directly to a voltage meter. The temperature of the meter's screw-terminals are measured by a nearby thermometer as 70°F. The thermocouple generates 5.262 mV. What is the temperature of its hot junction?

Solution

There are two connections at the meter. However, both connections are at the same temperature, so the meter can be considered to be a length of different wire, and the law of intermediate metals applies. Since the meter connections are not at 32°F, the law of intermediate temperatures applies. The corrected voltage is

$$5.262 \text{ mV} + 0.832 \text{ mV} = 6.094 \text{ mV}$$

From App. 62.A, 6.094 mV corresponds to 280°F.

18. STRAIN GAGES

A *bonded strain gage* is a metallic resistance device that is cemented to the surface of the unstressed member.[20] The gage consists of a metallic conductor (known as the *grid*) on a backing (known as the *substrate*).[21] The substrate and grid experience the same strain as the surface of the member. The resistance of the gage changes as the member is stressed due to changes in conductor cross section and intrinsic changes in resistivity with strain. Temperature effects must be compensated by the circuitry or by using a second unstrained gage as part of the bridge measurement system. (See Sec. 19.)

When simultaneous strain measurements in two or more directions are needed, it is convenient to use a commercial *rosette strain gage*. A rosette consists of two or more *grids* properly oriented for application as a single unit.

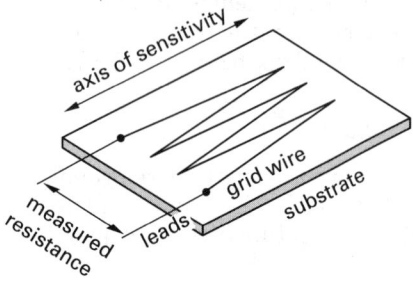

(a) folded-wire strain gage

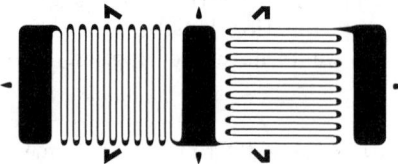

(b) commercial two-element rosette

Figure 62.5 *Strain Gage*

The *gage factor (strain sensitivity factor)*, GF, is the ratio of the fractional change in resistance to the fractional change in length (strain) along the detecting axis of the gage. The gage factor is a function of the gage material. It can be calculated from the grid material's properties and configuration. The higher the gage factor, the greater the sensitivity of the gage. From a

---

[20]A *bonded strain gage* is constructed by bonding the conductor to the surface of the member. An *unbonded strain gage* is constructed by wrapping the conductor tightly around the member or between two points on the member.

Strain gages on rotating shafts are usually connected through *slip rings* to the measurement circuitry.

[21]The grids of strain gages were originally of the folded-wire variety. For example, nichrome wire with a total resistance under 1000 Ω was commonly used. Modern strain gages are generally of the foil type manufactured by printed circuit techniques. Semiconductor gages are also used when extreme sensitivity (i.e., gage factors in excess of 100) is required. However, semiconductor gages are extremely temperature-sensitive.

practical standpoint, however, the gage factor and gage resistance are provided by the gage manufacturer. Only the change in resistance is measured.

$$\text{GF} = 1 + 2\nu + \frac{\dfrac{\Delta\rho}{\rho_o}}{\dfrac{\Delta L}{L_o}}$$

$$= \frac{\dfrac{\Delta R_g}{R_g}}{\dfrac{\Delta L}{L_o}} = \frac{\dfrac{\Delta R_g}{R_g}}{\epsilon} \qquad 62.8$$

Table 62.4 *Approximate Gage Factors*[a]

material	GF
constantan	2.0
iron, soft	4.2
isoelastic	3.5
manganin	0.47
monel	1.9
nichrome	2.0
nickel	-12[b]
platinum	4.8
platinum-iridium	5.1

[a]Other properties of strain gage materials are listed in Table 62.2.
[b]Value depends on amount of preprocessing and cold working.

Constantan and isoelastic wires and metal foil with gage factors of approximately 2 and initial resistances of less than 1000 Ω (typically 120 Ω, 350 Ω, 600 Ω, and 700 Ω) are commonly used. In practice, the gage factor and initial gage resistance, R_g, are specified by the manufacturer of the gage. Once the strain sensitivity factor is known, the strain, ϵ, can be determined from the change resistance. Strain is often reported in units of μin/in (μm/m) and is given the name *microstrain*.

$$\epsilon = \frac{\Delta R_g}{(\text{GF})R_g} \qquad 62.9$$

Theoretically, a strain gage should not respond to strain in its transverse direction. However, the turn-around end-loops are also made of strain-sensitive material, and the end-loop material contributes to a nonzero sensitivity to strain in the transverse direction. Equation 62.10 defines the *transverse sensitivity factor*, K_t, which is of academic interest in most problems. The transverse sensitivity factor is seldom greater than 2%.

$$K_t = \frac{(\text{GF})_{\text{transverse}}}{(\text{GF})_{\text{longitudinal}}} \qquad 62.10$$

Example 62.5

A strain gage with a nominal resistance of 120 Ω and gage factor of 2.0 is used to measure a strain of 1 μin/in. What is the change in resistance?

Solution

From Eq. 62.9,

$$\Delta R_g = (\text{GF})R_g\epsilon$$
$$= (2.0)(120 \ \Omega)\left(1 \times 10^{-6} \ \frac{\text{in}}{\text{in}}\right) = 2.4 \times 10^{-4} \ \Omega$$

19. WHEATSTONE BRIDGES

The *Wheatstone bridge* shown in Fig. 62.6 is one type of *resistance bridge*.[22] The bridge can be used to determine the unknown resistance of a resistance transducer (e.g., thermistor or resistance-type strain gage), say R_1 in Fig. 62.6. The potentiometer is adjusted (i.e., the bridge is "balanced") until no current flows through the meter or until there is no voltage across the meter (hence the name *null indicator*).[23,24] When the bridge is balanced and no current flows through the meter leg, Eqs. 62.11 through 62.14 are applicable.

$$I_2 = I_4 \quad \text{[balanced]} \qquad \textit{62.11}$$
$$I_1 = I_3 \quad \text{[balanced]} \qquad \textit{62.12}$$
$$V_1 + V_3 = V_2 + V_4 \quad \text{[balanced]} \qquad \textit{62.13}$$
$$\frac{R_1}{R_2} = \frac{R_3}{R_4} \quad \text{[balanced]} \qquad \textit{62.14}$$

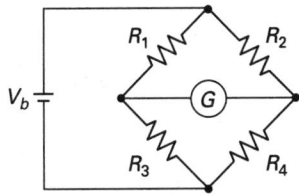

Figure 62.6 *Series-Balance Wheatstone Bridge*

Since any one of the four resistances can be the unknown, up to three of the remaining resistances can be fixed or adjustable, and the battery and meter can be connected to either of two diagonal corners, it is sometimes confusing to apply Eq. 62.14 literally. However, the following bridge law statement can be used to help formulate the proper relationship: *When a series*

Wheatstone bridge is null-balanced, the ratio of resistance of any two adjacent arms equals the ratio of resistance of the remaining two arms, taken in the same sense. In this statement, "taken in the same sense" means that both ratios must be formed reading either left to right, right to left, top to bottom, or bottom to top.

20. STRAIN GAGE DETECTION CIRCUITS

The resistance of a strain gage can be measured by placing the gage in either a ballast circuit or bridge circuit. A *ballast circuit* consists of a voltage source (V_b) of less than 10V (typical), a current-limiting ballast resistance (R_b), and the strain gage of known resistance (R_g) in series. This is essentially a voltage-divider circuit. (See Chap. 68.) The change in voltage (ΔV_g) across the strain gage is measured. The strain (ϵ) can be determined from Eq. 62.15.

$$\Delta V_g = \frac{(\text{GF})\epsilon V_b R_b R_g}{(R_b + R_g)^2} \qquad \textit{62.15}$$

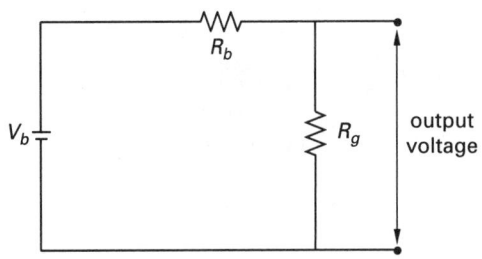

Figure 62.7 *Ballast Circuit*

Ballast circuits do not provide temperature compensation, nor is their sensitivity adequate for measuring static strain. Ballast circuits, where used, are often limited to measurement of transient strains. A bridge detection circuit overcomes these limitations.

Figure 62.8 illustrates how a strain gage can be used with a resistance bridge. Gage 1 measures the strain, while *dummy gage* 2 provides temperature compensation.[25] The meter voltage is a function of the input (battery) voltage and the resistors. (As with bridge circuits, the input voltage is typically less than 10 V.) The variable resistance is used for balancing the bridge prior to the strain. When the bridge is balanced, V_{meter} is zero.

When the gage is strained, the bridge becomes unbalanced. Assuming the bridge is initially balanced, the voltage at the meter (known as the *voltage deflection* from the null condition) will be[26]

$$V_{\text{meter}} = V_b\left(\frac{R_1}{R_1 + R_3} - \frac{R_2}{R_2 + R_4}\right) \quad \left[\tfrac{1}{4} \text{ bridge}\right] \quad \textit{62.16}$$

---

[22]Other types of resistance bridges are the *differential series balance bridge, shunt balance bridge,* and *differential shunt balance bridge.* These differ in the manner in which the adjustable resistor is incorporated into the circuit.

[23]This gives rise to the alternate names of *zero-indicating bridge* and *null-indicating bridge.*

[24]The unknown resistance can also be determined from the amount of voltage unbalance shown by the meter reading, in which case, the bridge is known as a *deflection bridge* rather than a *null-indicating bridge.* Deflection bridges are described in Sec. 20.

[25]This is a "quarter-bridge" or "1/4-bridge" configuration, as described in Sec. 22. The strain gage used for temperature compensation is not active.

[26]Equation 62.16 applies to the unstrained condition as well. However, if the gage is unstrained and the bridge is balanced, the bracketed resistance term is zero.

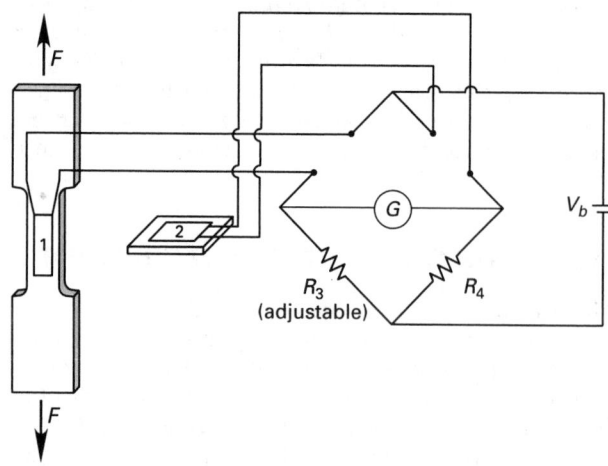

Figure 62.8 *Strain Gage in Resistance Bridge*

For a single strain gage in a resistance bridge and neglecting lead resistance, the voltage deflection is related to the strain by Eq. 62.17.

$$V_{\text{meter}} = \frac{(\text{GF})\epsilon V_b}{4 + (2)(\text{GF})\epsilon}$$

$$\approx \left(\tfrac{1}{4}\right)(\text{GF})\epsilon V_b \quad \left[\tfrac{1}{4} \text{ bridge}\right] \qquad 62.17$$

21. STRAIN GAGE IN UNBALANCED RESISTANCE BRIDGE

A resistance bridge does not need to be balanced prior to use as long as an accurate digital voltmeter is used in the detection circuit. The voltage ratio difference, ΔVR, is defined as the fractional change in the output voltage from the unstrained to the strained condition.

$$\Delta\text{VR} = \left(\frac{V_{\text{meter}}}{V_b}\right)_{\text{strained}} - \left(\frac{V_{\text{meter}}}{V_b}\right)_{\text{unstrained}} \qquad 62.18$$

If the only resistance change between the strained and unstrained conditions is in the strain gage and lead resistance is disregarded, the fractional change in gage resistance for a single strain gage in a resistance bridge is

$$\frac{\Delta R_g}{R_g} = \frac{-4\Delta(\text{VR})}{1 + 2\Delta(\text{VR})} \quad \left[\tfrac{1}{4} \text{ bridge}\right] \qquad 62.19$$

Since the fractional change in gage resistance also occurs in the definition of the gage factor (Eq. 62.9), the strain is

$$\epsilon = \frac{-4\Delta\text{VR}}{(\text{GF})(1 + 2\Delta\text{VR})} \quad \left[\tfrac{1}{4} \text{ bridge}\right] \qquad 62.20$$

22. BRIDGE CONSTANT

The voltage deflection can be doubled (or quadrupled) by using two (or four) strain gages in the bridge circuit. The larger voltage deflection is more easily detected, resulting in more accurate measurements.

Use of multiple strain gages is generally limited to configurations where symmetrical strain is available on the member. For example, a beam in bending experiences the same strain on the top and bottom faces. Therefore, if the temperature-compensation strain gage shown in Fig. 62.8 is bonded to the bottom of the beam, the resistance change would double.

The *bridge constant* (BC) is the ratio of the actual voltage deflection to the voltage deflection from a single gage. Depending on the number and orientation of the gages used, bridge constants of 1.0, 1.3, 2.0, 2.6, and 4.0 may be encountered (for materials with a Poisson's ratio of 0.3).

Figure 62.9 illustrates how (up to) four strain gages can be connected in a Wheatstone bridge circuit. The total strain indicated will be the algebraic sum of the four strains detected. For example, if all four strains are equal in magnitude, ϵ_1 and ϵ_4 are tensile (i.e., positive), and ϵ_2 and ϵ_3 are compressive (i.e., negative), then the bridge constant would be 4.

$$\epsilon_t = \epsilon_1 - \epsilon_2 - \epsilon_3 + \epsilon_4 \qquad 62.21$$

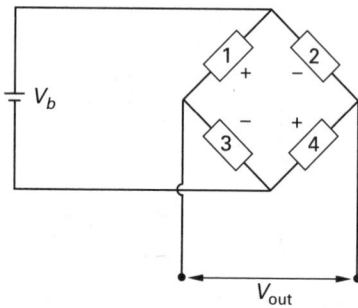

[a]The 45° orientations shown are figurative. Actual gage orientation can be in any direction.

Figure 62.9 *Wheatstone Bridge Strain Gage Circuit*[a]

23. STRESS MEASUREMENTS IN KNOWN DIRECTIONS

Strain gages are the most frequently used method of determining stress in a member. Stress can be calculated from strain, or the measurement circuitry can be calibrated to give the stress directly.

For stress in only one direction (i.e., the *uniaxial stress* case), such as a simple bar in tension, only one strain gage is required. The stress can be calculated from *Hooke's law*.

$$\sigma = E\epsilon \qquad 62.22$$

When a surface, such as that of a pressure vessel, experiences simultaneous stresses in two directions (the *biaxial stress* case), the strain in one direction affects the strain in the other direction.[27] Therefore, two strain gages are needed, even if the stress in only one direction is needed. The strains actually measured by the gages are known as the *net strains*.

$$\epsilon_x = \frac{\sigma_x - \nu\sigma_y}{E} \qquad 62.23$$

$$\epsilon_y = \frac{\sigma_y - \nu\sigma_x}{E} \qquad 62.24$$

The stresses are determined by solving Eqs. 62.23 and 62.24 simultaneously.

$$\sigma_x = \frac{E(\epsilon_x + \nu\epsilon_y)}{1 - \nu^2} \qquad 62.25$$

$$\sigma_y = \frac{E(\epsilon_y + \nu\epsilon_x)}{1 - \nu^2} \qquad 62.26$$

Figure 62.9 shows how four strain gages can be interconnected in a bridge circuit. Figure 62.10 shows how (up to) four strain gages would be physically oriented on a test specimen to measure different types of stress. Not all four gages are needed in all cases. If four gages are used, the arrangement is said to be a *full bridge*. If only one or two gages are used, the terms *quarter-bridge* ($^1/_4$-bridge) and *half-bridge* ($^1/_2$-bridge), respectively, apply.

In the case of up to four gages applied to detect bending strain (Fig. 62.10(a)), the bridge constant (BC) can be 1.0 (one gage in position 1), 2.0 (two gages in positions 1 and 2), or 4.0 (all four gages). The relationships between the stress, strain, and applied force are

$$\sigma = E\epsilon = \frac{E\epsilon_t}{\text{BC}} \qquad 62.27$$

$$\sigma = \frac{Mc}{I} = \frac{Mh}{2I} \qquad 62.28$$

$$I = \frac{bh^3}{12} \quad \text{[rectanglar section]} \qquad 62.29$$

For axial strain (Fig. 62.10(b)) and a material with a Poisson's ratio of 0.3, the bridge constant can be 1.0 (one gage in position 1), 1.3 (two gages in positions 1 and 2), 2.0 (two gages in positions 1 and 3), or 2.6 (all four gages).

$$\sigma = E\epsilon = \frac{E\epsilon_t}{\text{BC}} \qquad 62.30$$

$$\sigma = \frac{F}{A} \qquad 62.31$$

$$A = bh \quad \text{[rectanglular section]} \qquad 62.32$$

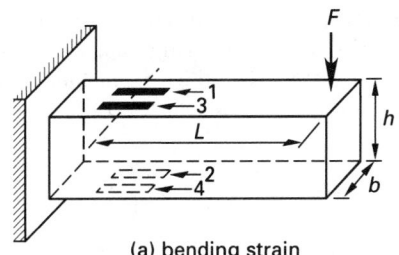

(a) bending strain

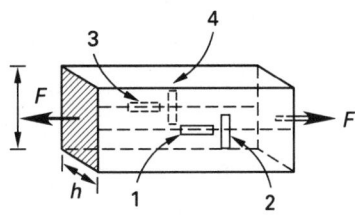

(b) axial strain

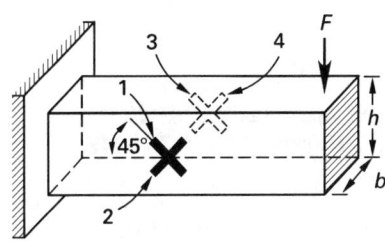

(c) shear strain

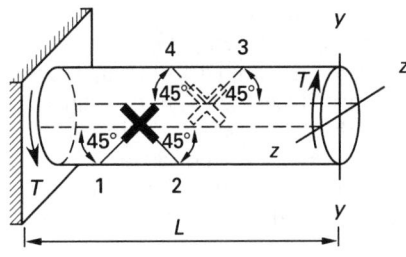

(d) torsional strain

Figure 62.10 *Orientation of Strain Gages*

For shear strain (Fig. 62.10(c)) and a material with a Poisson's ratio of 0.3, the bridge constant can be 2.0 (two gages in positions 1 and 2) or 4.0 (all four gages). The shear strain is twice the axial strain at 45°.

$$\tau = G\gamma = 2G\epsilon = \frac{2G\epsilon_t}{\text{BC}} \qquad 62.33$$

$$\tau = \frac{FQ}{bI} \qquad 62.34$$

$$\gamma = 2\epsilon \quad \text{[at 45°]} \qquad 62.35$$

$$Q_{\text{max}} = \frac{bh^2}{8} \quad \text{[rectanglar section]} \qquad 62.36$$

$$G = \frac{E}{(2)(1 + \nu)} \qquad 62.37$$

For torsional strain (Fig. 62.10(d)) and a material with a Poisson's ratio of 0.3, the bridge constant can be 2.0 (two gages in positions 1 and 2) or 4.0 (all four gages). The shear strain is twice the axial strain at 45°.

$$\tau = G\gamma = 2G\epsilon = \frac{2G\epsilon_t}{BC} \qquad 62.38$$

$$\tau = \frac{Tr}{J} = \frac{Td}{2J} \qquad 62.39$$

$$\gamma = 2\epsilon \quad [\text{at } 45°] \qquad 62.40$$

$$J = \frac{\pi d^4}{32} \quad [\text{solid circular}] \qquad 62.41$$

$$\phi = \frac{TL}{JG} \qquad 62.42$$

$$G = \frac{E}{(2)(1+\nu)} \qquad 62.43$$

24. STRESS MEASUREMENTS IN UNKNOWN DIRECTIONS

In order to calculate the maximum stresses (i.e., the principal stresses) on the surface shown in Sec. 23, the gages were oriented in the known directions of the principal stresses.

In most cases, however, the directions of the principal stresses are not known. Therefore, rosettes of at least three gages are used to obtain information in a third direction. Rosettes of three gages (*rectangular* and *equiangular (delta) rosettes*) are used for this purpose. *T-delta rosettes* include a fourth strain gage to refine and validate the results of the three primary gages. Table 62.5 can be used to calculate the principal stresses.

25. LOAD CELLS

Load cells are used to measure force. A load cell is a transducer that converts a tensile or compressive force into an electrical signal. Though the details of the load cell vary with the application, the basic elements are (1) a member that is strained by the force and (2) a strain detection system (e.g., strain gage). The force is calculated from the observed deflection, y. In Eq. 62.44, the spring constant, k, is known as the load cell's *deflection constant*.

$$F = ky \qquad 62.44$$

Because of their low cost and simple construction, *bending beam load cells* are the most common variety of load cell. Two strain gages, one on the top and the other mounted on the bottom of a cantilever bar, are used. *Shear beam load cells* (which detect force by measuring the shear stress) can be used where the shear does not vary considerably with location, as in the web of an I-beam cross section.[28] The common S-shaped load cell constructed from a machined steel block can be instrumented as either a bending beam or shear beam load cell.

Load cell applications are categorized into classes, with class III (using a single load cell) being the most common. Commercial load cells meet standardized limits on errors due to temperature, nonlinearity, and hysteresis. The *temperature effect on output* (TEO) is typically stated in percentage change per 100°F (55.5°C) change in temperature.

Nonlinearity errors are reduced in proportion to the load cell's derating (i.e., using the load cell to measure forces less than its rated force). For example, a 2:1 derating will reduce the nonlinearity errors by 50%. Hysteresis is not normally reduced by derating.

The overall error of force measurement can be reduced by a factor of $1/\sqrt{n}$ (where n is the number of load cells that share the load equally) by using more than one load cell. Conversely, the applied force can vary by $\sqrt{n}$ times the known accuracy of a single load cell without decreasing the error.

26. DYNAMOMETERS

Torque from large motors and engines is measured by a *dynamometer*. *Absorption dynamometers* (e.g., the simple *friction brake, Prony brake, water brake,* and *fan brake*) dissipate energy as the torque is measured. Opposing torque in pumps and compressors must be supplied by a *driving dynamometer*, which has its own power input. *Transmission dynamometers* (e.g., *torque meters, torsion dynamometers*) use strain gages to sense torque. They do not absorb or provide energy.

Using a brake dynamometer involves measuring a force, a moment arm, and the angular speed of rotation. The familiar torque-power-speed relationships are used with absorption dynamometers.

$$T = Fr \qquad 62.45$$

$$P_{\text{ft-lbf/min}} = 2\pi T_{\text{ft-lbf}} n_{\text{rpm}} \qquad 62.46$$

$$P_{\text{kW}} = \frac{T_{\text{N·m}} n_{\text{rpm}}}{9549} \quad [\text{SI}] \qquad 62.47(a)$$

$$P_{\text{hp}} = \frac{2\pi F_{\text{lbf}} r_{\text{ft}} n_{\text{rpm}}}{33,000}$$

$$= \frac{2\pi T_{\text{ft-lbf}} n_{\text{rpm}}}{33,000} \quad [\text{U.S.}] \qquad 62.47(b)$$

---

[28]While shear in a rectangular beam varies parabolically with distance from the neutral axis, shear in the web of an I-beam is essentially constant at F/A. The flanges carry very little of the shear load.

    Other advantages of the shear beam load cell include protection from the load and environment, high side load rejection, lower creep, faster RTZ (return to zero) after load removal, and higher tolerance of vibration, dynamic forces, and noise.

Table 62.5 *Stress-Strain Relationships for Strain Gage Rosettes*[a]

type of rosette	rectangular	equiangular (delta)	T-delta
principal strains, ϵ_p, ϵ_q	$\left(\frac{1}{2}\right)\left(\epsilon_a + \epsilon_c \pm \sqrt{(2)(\epsilon_a - \epsilon_b)^2 + (2)(\epsilon_b - \epsilon_c)^2}\right)$	$\left(\frac{1}{3}\right)\left(\epsilon_a + \epsilon_b + \epsilon_c \pm \sqrt{\begin{array}{l}(2)(\epsilon_a - \epsilon_b)^2 + (2)(\epsilon_b - \epsilon_c)^2 \\ \quad + (2)(\epsilon_c - \epsilon_a)^2\end{array}}\right)$	$\left(\frac{1}{2}\right)\left(\epsilon_a + \epsilon_d \pm \sqrt{(\epsilon_a - \epsilon_d)^2 + \left(\frac{4}{3}\right)(\epsilon_b - \epsilon_c)^2}\right)$
principal stresses, σ_1, σ_2	$\left(\frac{E}{2}\right)\left(\frac{\epsilon_a + \epsilon_c}{1 - \nu} \pm \frac{1}{1 + \nu} \times \sqrt{(2)(\epsilon_a - \epsilon_b)^2 + (2)(\epsilon_b - \epsilon_c)^2}\right)$	$\left(\frac{E}{3}\right)\left(\frac{\epsilon_a + \epsilon_b + \epsilon_c}{1 - \nu} \pm \frac{1}{1 + \nu} \times \sqrt{\begin{array}{l}(2)(\epsilon_a - \epsilon_b)^2 + (2)(\epsilon_b - \epsilon_c)^2 \\ \quad + (2)(\epsilon_c - \epsilon_a)^2\end{array}}\right)$	$\left(\frac{E}{2}\right)\left(\frac{\epsilon_a + \epsilon_d}{1 - \nu} \pm \frac{1}{1 + \nu} \times \sqrt{(\epsilon_a - \epsilon_d)^2 + \left(\frac{4}{3}\right)(\epsilon_b - \epsilon_c)^2}\right)$
maximum shear, τ_{max}	$\left(\frac{E}{(2)(1 + \nu)}\right) \times \sqrt{(2)(\epsilon_a - \epsilon_b)^2 + (2)(\epsilon_b - \epsilon_c)^2}$	$\left(\frac{E}{(3)(1 + \nu)}\right) \times \sqrt{\begin{array}{l}(2)(\epsilon_a - \epsilon_b)^2 + (2)(\epsilon_b - \epsilon_c)^2 \\ \quad + (2)(\epsilon_c - \epsilon_a)^2\end{array}}$	$\left(\frac{E}{(2)(1 + \nu)}\right) \times \sqrt{(\epsilon_a - \epsilon_d)^2 + \left(\frac{4}{3}\right)(\epsilon_b - \epsilon_c)^2}$
$\tan 2\theta$[b]	$\dfrac{2\epsilon_b - \epsilon_a - \epsilon_c}{\epsilon_a - \epsilon_c}$	$\dfrac{\sqrt{3}(\epsilon_c - \epsilon_b)}{2\epsilon_a - \epsilon_b - \epsilon_c}$	$\left(\dfrac{2}{\sqrt{3}}\right)\left(\dfrac{\epsilon_c - \epsilon_b}{\epsilon_a - \epsilon_d}\right)$
$0 < \theta < +90°$	$\epsilon_b > \dfrac{\epsilon_a + \epsilon_c}{2}$	$\epsilon_c > \epsilon_b$	$\epsilon_c > \epsilon_b$

[a] θ is measured in the counterclockwise direction from the a-axis of the rosette to the axis of the algebraically larger stress.
[b] θ is the angle from gage A axis to axis of maximum normal stress.

Plant Engineering

Some brakes and dynamometers are constructed with a "standard" brake arm whose length is 5.252 ft. In that case, the horsepower calculation conveniently reduces to

$$P_{\text{hp}} = \frac{F_{\text{lbf}} n_{\text{rpm}}}{1000} \quad \text{["standard arm" brake]} \qquad 62.48$$

If an absorption dynamometer uses a DC generator to dissipate energy, the generated voltage (V in volts) and line current (I in amps) are used to determine the power. Equations 62.47 and 62.48 are used to determine the torque.

$$P_{\text{hp}} = \frac{IV}{\eta \left(1000 \, \dfrac{\text{W}}{\text{kW}}\right) \left(0.7457 \, \dfrac{\text{W}}{\text{hp}}\right)} \quad \text{[absorption]}$$

$$62.49$$

For a driving dynamometer using a DC motor,

$$P_{\text{hp}} = \frac{IV\eta}{\left(1000 \, \dfrac{\text{W}}{\text{kW}}\right) \left(0.7457 \, \dfrac{\text{W}}{\text{hp}}\right)} \quad \text{[driving]} \qquad 62.50$$

Torque can be measured directly by a *torque meter* mounted to the power shaft. Either the angle of twist (ϕ) or the shear strain (τ/G) are measured. The torque in a solid shaft of diameter d and length L is

$$T = \frac{JG\phi}{L} = \left(\frac{\pi}{32}\right) d^4 \left(\frac{G\phi}{L}\right)$$

$$= \left(\frac{\pi}{16}\right) d^3 \tau \quad \text{[solid round]} \qquad 62.51$$

27. INDICATOR DIAGRAMS

Indicator diagrams are plots of pressure versus volume and are encountered in the testing of reciprocating engines. In the past, indicator diagrams were actually drawn on an *indicator card* wrapped around a drum through a mechanical linkage of arms and springs. This method is mechanically complex and is not suitable for rotational speeds above 2000 rpm. Modern records of pressure and volume are produced by signals from electronic transducers recorded in real time by computers.

Analysis of indicator diagrams produced by mechanical devices requires knowing the spring constant, also known as the *spring scale*. The *mean effective pressure* (MEP) is calculated by dividing the area of the diagram by the width of the plot and then multiplying by the spring constant.

63 Manufacturing Processes

1. Chip Formation	63-1
2. Cutting Tool Speeds and Forces	63-2
3. Tool Materials	63-4
4. Temperature and Cooling Fluids	63-4
5. Tool Life	63-5
6. Abrasives and Grinding	63-5
7. Chipless (Nontraditional) Machining	63-6
8. Cold- and Hot-Working Operations	63-6
9. Presswork	63-6
10. Forging	63-7
11. Sand Molding	63-8
12. Gravity Molding	63-9
13. Die Casting	63-9
14. Centrifugal Casting	63-9
15. Investment Casting	63-9
16. Continuous Casting	63-9
17. Plastic Molding	63-9
18. Powder Metallurgy	63-10
19. High Energy Rate Forming	63-11
20. Gas Welding	63-11
21. Arc Welding	63-12
22. Soldering and Brazing	63-12
23. Adhesive Bonding	63-13
24. Manufacture of Metal Pipe	63-13
25. Surface Finishing and Coatings	63-13
Practice Problems	63-14

Nomenclature

A	area	ft^2	m^2
b	chip width	ft	m
c	specific heat	Btu/lbm-°F	J/kg·°C
d	depth of cut	ft	m
D	diameter	ft	m
E	energy	ft-lbf	J
f	feed rate	ft/rev	m/rev
F	force	lbf	N
g	gravitational acceleration	ft/sec^2	m/s^2
g_c	gravitational constant	lbm-ft/lbf-sec^2	n.a.
h	height	ft	m
J	Joule's constant (778.17)	ft-lbf/Btu	n.a.
J	rotational moment of inertia	lbm-ft^2	kg·m^2
L	length	ft	m
m	mass	slug	kg
n	constant	–	–
n	rotational speed	rpm	rpm
N	number of items	–	–
p	pressure	lbf/ft^2	Pa
P	power	ft-lbf/min	W
r	chip thickness ratio	–	–
r	radius	ft	m
S	strength	lbf/ft^2	Pa
t	thickness	ft	m
t	time	sec	s
T	temperature	°F	°C
T	time	sec	s
U	specific cutting energy	ft-lbf/ft^3	J/m^3
v	cutting speed	ft/min	m/s
V	volume	ft^3	m^3
Z_w	metal removal rate	ft^3/min	m^3/s

Symbols

α	true rake angle	rad	rad
β	rake face resultant angle	rad	rad
θ	clearance angle	rad	rad
ρ	density	lbm/ft^3	kg/m^3
σ	normal stress	lbf/ft^2	Pa
τ	shear stress	lbf/ft^2	Pa
ϕ	shear angle	rad	rad
ω	wedge angle	rad	rad
ω	rotational speed	rad/sec	rad/s

Subscripts

c	chip
f	final
h	horizontal
n	normal
o	original (undeformed)
p	constant pressure
R	resultant
s	shear
t	tangent
u	ultimate
v	vertical

1. CHIP FORMATION

One of the most common ways that a workpiece can be shaped is by removing material through chip-forming operations such as turning, drilling, multitooth operations (e.g., milling, broaching, sawing, and filing), and specialized processes, such as thread and gear cutting.

The toughness of a workpiece can be determined from the nature of the chips produced. Brittle materials produce discrete fragments, known as *discontinuous chips*, *segmented chips*, or *type-one chips*. Ductile materials form long, helix-coiled string chips, known as *continuous chips* or *type-two chips*.[1] *Chip-breaker* grooves are often ground in the cutting tool face to cause long chips to break into shorter, more manageable pieces. Chip formation is optimum when chips are produced in the shapes of sixes and nines.

2. CUTTING TOOL SPEEDS AND FORCES

Figure 63.1 shows a chip produced during *orthogonal cutting* (i.e., two-dimensional cutting). Cutting involves both compressive and shear stresses. The chip expands from its undeformed thickness, t_o, to t_c (due to the release of compressive stress) as it slides over the cutting tool. The *chip thickness ratio*, $r = t_o/t_c$, is typically around 0.5.

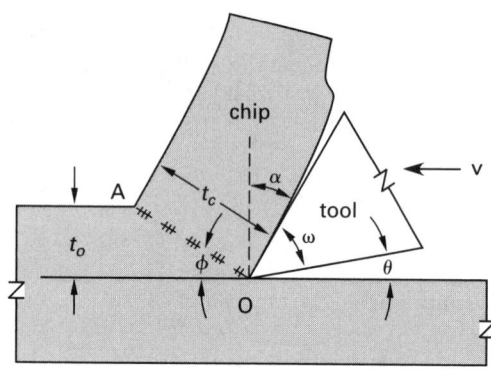

Figure 63.1 *Tool-Workpiece-Chip Geometry*

The cutting energy is minimum when the *shear angle*, ϕ, is approximately given by Eq. 63.1. (The angle β is shown in Fig. 63.2. It can be determined from Eq. 63.8 and Eq. 63.9, using F_n and F_v from toolpost dynamometer data.)

$$\phi_{\text{minimum energy}} = \frac{\pi}{4} + \frac{\alpha}{2} - \frac{\beta}{2} \qquad \textit{63.1}$$

The angle at which the tool meets the workpiece is characterized by *true rake angle*, α. This angle has a major impact on chip formation, and it depends on the tool shape and set-up orientation of the tool. Small rake angles result in excessive compression, high tool forces, and excessive friction. Chips produced are thick, hot, and highly deformed. Conventional cutting tools are oriented with positive rake angles; sintered carbide and ceramic tools used for cutting steel are frequently designed to be oriented with negative rake angles to provide additional support of the cutting edge.

Figure 63.1 also shows the *clearance angle* (*relief angle*), θ, and the *wedge angle*, ω. The sum of the rake, clearance, and wedge angles is $\pi/2$ (90°).

$$\alpha + \theta + \omega = \frac{\pi}{2} \qquad \textit{63.2}$$

---

[1] There is also a *type-three chip*, a continuous chip with a *built-up edge* (BUE), which is not discussed here.

The rake angle, shear angle, and chip thickness ratio are related by Eq. 63.3 and Eq. 63.4.

From Fig. 63.1,

$$r = \frac{t_o}{t_c} = \frac{\sin \phi}{\cos (\phi - \alpha)} \qquad \textit{63.3}$$

Rearranging Eq. 63.3 and applying trigonometric identities,

$$\tan \phi = \frac{r \cos \alpha}{1 - r \sin \alpha} \qquad \textit{63.4}$$

The relative velocity difference between the tool and the workpiece is the *cutting speed*, v. The cutting speed can be calculated from the diameter, D, of a rotating workpiece and the rotational speed, n. If D is in feet and n is in revolutions per minute, the cutting speed will be in traditional (in the United States) units of feet per minute.[2]

$$\text{v} = \pi D n \qquad \textit{63.5}$$

The velocity of the chip relative to the tool face is the *chip velocity*, v_c. The *shear velocity*, v_s, is the velocity of the chip relative to the workpiece.

$$\text{v}_c = r\text{v} = \frac{\text{v} \sin \phi}{\cos (\phi - \alpha)} \qquad \textit{63.6}$$

$$\text{v}_s = \frac{\text{v} \cos \alpha}{\cos (\phi - \alpha)} \qquad \textit{63.7}$$

Figure 63.2 illustrates that the resultant force, F_R, between the tool and the chip can be resolved into tangential force, F_t, and normal force, F_n, relative to the rake face of the tool, or alternatively, into a horizontal cutting force, F_h, and a vertical thrust force, F_v, relative to the cutting surface. The horizontal and vertical forces are commonly measured by a strain-gage toolpost dynamometer. Equations 63.8 and 63.9 relate these two sets of forces.

$$F_t = F_h \sin \alpha + F_v \cos \alpha \qquad \textit{63.8}$$
$$F_n = F_h \cos \alpha - F_v \sin \alpha \qquad \textit{63.9}$$

A third set of axes can be chosen relative to the shear plane. The forces parallel, F_s, and normal, F_{ns}, to the shear plane can be derived from the horizontal and vertical forces. These are also shown in Fig. 63.2.

$$F_s = F_h \cos \phi - F_v \sin \phi \qquad \textit{63.10}$$
$$F_{ns} = F_h \sin \phi + F_v \cos \phi \qquad \textit{63.11}$$

Shear stress is the primary parameter affecting the cutting energy requirement. The average shear stress, τ, is F_s divided by the area of the shear plane, A_s, which depends on the chip width, b.

---

[2] The symbol CS is also used for cutting speed.

Table 63.1 Typical Cutting Speeds (ft/min)

material	high-speed steel		carbide	
	rough	finish	rough	finish
cast iron	50–60	80–110	120–200	350–400
semisteel[a]	40–50	65–90	140–160	250–300
malleable iron[a]	80–110	110–130	250–300	300–400
steel casting[a] (0.35C)	45–60	70–90	150–180	200–250
brass (85-5-5)	200–300	200–300	600–1000	600–1000
bronze (80-10-10)[a]	110–150	150–180	600	1000
aluminum	400	700	800	1000
SAE 1020[a]	80–100	100–120	300–400	300–400
SAE 1050[a]	60–80	100	200	200
stainless steel[a]	100–120	100–120	240–300	240–300

(Multiply ft/min by 5.08×10^{-3} to obtain m/s.)
[a] Appropriate lubricants used to achieve listed speeds.
Used with permission from Myron L. Begeman and B. H. Amstead, *Manufacturing Processes*, 5th Edition, copyright © 1963, by John Wiley & Sons, Inc.

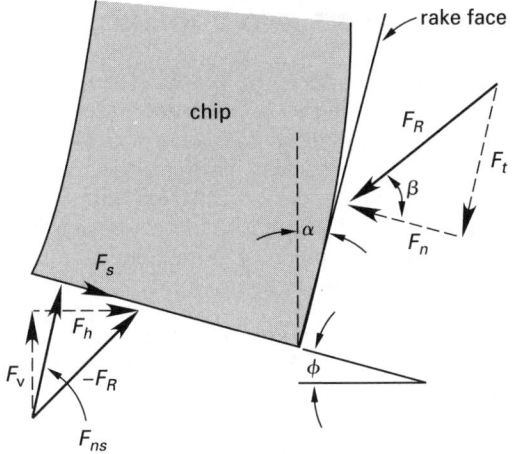

Figure 63.2 Force Components in Orthogonal Cutting

$$A_s = \frac{bt_o}{\sin\phi} \qquad 63.12$$

$$\tau = \frac{F_s}{A_s} = \frac{F_s \sin\phi}{bt_o} \qquad 63.13$$

The average normal stress is

$$\sigma = \frac{F_{ns}}{A_s} = \frac{F_{ns} \sin\phi}{bt_o} \qquad 63.14$$

The energy required per unit cutting time is the cutting power, P.

$$P = F_h \mathrm{v} \qquad 63.15$$

If F_h is in pounds, and if v is in feet per minute, the horsepower requirement is

$$\text{cutting horsepower} = \frac{F_h \mathrm{v}}{33{,}000} \qquad 63.16$$

The *metal removal rate*, Z_w, is

$$Z_w = bt_o \mathrm{v} \qquad 63.17$$

The energy expended per unit volume removed, known as the *specific cutting energy*, is

$$U = \frac{P}{Z_w} = \frac{F_h \mathrm{v}}{Z_w} = \frac{F_h}{bt_o} \qquad 63.18$$

For a simple lathe (turning) operation on a cylindrical work piece of diameter D, the cutting time to make a cut of length L in a single pass is

$$t = \frac{L}{fn_{\mathrm{rpm}}} = \frac{LD}{f\mathrm{v}} \qquad 63.19$$

Example 63.1

A cylindrical cast steel cylinder with a diameter of 6.00 in is faced in a lathe. The procedure removes the outer 0.40 in of the bar. The lathe develops a maximum power of 20 hp. The unit power limit for this material and process is 2.0 hp-min/in³. The cutting pressure is maintained at its maximum limit throughout. What is the approximate minimum time to face the bar?

Solution

The bar radius is

$$r = \frac{6.00 \text{ in}}{2} = 3.00 \text{ in}$$

The volume of material to be removed is

$$V = AL = \pi r^2 L = \pi (3.00 \text{ in})^2 (0.4 \text{ in})$$
$$= 11.31 \text{ in}^3$$

If the lathe develops the full 20 hp power throughout the operation, the minimum cutting time is

$$t = V \left(\frac{\text{unit power}}{\text{lathe power}} \right)$$

$$= (11.31 \text{ in}^3) \left(\frac{2.0 \frac{\text{hp-min}}{\text{in}^3}}{20 \text{ hp}} \right)$$

$$= 1.13 \text{ min}$$

Example 63.2

A cylindrical workpiece with a diameter of 4.00 in is turned on a lathe. The cutting speed is 200 ft/min. The depth of cut is 0.20 in. The feed is 0.010 ipr (inches per revolution). How long will it take to make a 14 in cut?

Solution

The cutting time is

$$t = \frac{LD}{f\text{v}}$$

$$= \frac{(14 \text{ in})(4.00 \text{ in})}{\left(12 \frac{\text{in}}{\text{ft}} \right) \left(0.010 \frac{\text{in}}{\text{rev}} \right) \left(200 \frac{\text{ft}}{\text{min}} \right)}$$

$$= 2.33 \text{ min}$$

3. TOOL MATERIALS

Carbon tool steel is plain carbon steel with approximately 0.9 to 1.3% carbon, which has been hardened and tempered. It can be given a good edge, but is restricted to use below 400 to 600°F (200 to 300°C) to prevent further tempering.

High-speed steel (HSS) contains tungsten or chromium and retains its hardness up to approximately 1100°F (600°C), a property known as *red hardness*. The common 18-4-1 formulation contains 18% tungsten, 4% chromium, and 1% vanadium. Other categories include *molybdenum high-speed steels* and *superhigh-speed steels*. Tools made with these steels can be run approximately twice as fast as carbon steel tools.

Cast nonferrous cutting tools have similar characteristics to carbides and are used in an as-cast condition. A common composition contains 45% cobalt, 34% chromium, 18% tungsten, and 2% carbon. Cast nonferrous tools are brittle but can be used up to approximately 1700°F (925°C) and operate at speeds twice that of HSS tools.

Sintered carbides are produced through powder metallurgy from nonferrous metals (e.g., tungsten carbide and titanium carbide with some cobalt). Carbide tools are commonly of the throw-away type. They are very hard, can be used up to 2200°F (1200°C), and operate at cutting speeds two to five times as fast as HSS tools. However, they are less tough and cannot be used where impact forces are significant.

Ceramic tools manufactured from aluminum oxide have the same expected life as carbide tools but can operate at speeds from two to three times higher. They operate below 2000°F (1100°C).

Diamonds and diamond dust are used in specific cases, usually in finishing operations.

In addition to speed and temperature considerations, there should be no possibility of welding between the chip and tool material. Diamonds, for example, are soluble in the presence of high-temperature iron. Also, aluminum oxide tools are not satisfactory for machining aluminum.

4. TEMPERATURE AND COOLING FLUIDS

Friction is greatly reduced in *free-machining steels* that have had sulfur added as an alloying ingredient. However, only approximately 25% of the heat developed in cutting is due to friction between the tool and the workpiece. The remainder results from compression and shear stresses. Only 20 to 40% of this heat is removed by the tool and workpiece. The remainder must be removed by the chips and cooling fluids.

If all the heat generated goes into the chip (which does not actually occur), the *adiabatic chip temperature* is given by Eq. 63.20.

$$T_c = T_o + \frac{U}{\rho c_p} \qquad \text{[SI]} \qquad 63.20(a)$$

$$T_c = T_o + \frac{U}{\rho c_p J} \qquad \text{[U.S.]} \qquad 63.20(b)$$

Cutting fluids are used to reduce friction, remove heat, remove chips, and protect against corrosion. Gases, such as air, carbon dioxide, and water vapor, can be used, but they do not remove heat well, cannot be reused, and may require an exhaust system. Water is a good heat remover, but it promotes rust. (Addition of *sal soda* to water produces an efficient, inexpensive cutting fluid that does not promote rusting.)

Straight-cutting oils (i.e., petroleum-based nonsoluble oils) reduce friction and do not cause rust but are less efficient at heat removal than water. Therefore, emulsions of water and oil or water-miscible fluids (soluble oils) are often used with steel. Kerosene lubricants are commonly used with aluminum.

Chlorinated or sulfurized oils are used to decrease friction.[3] Other additives are used to inhibit rust, clean the workpiece, soften water, promote film formation, and inhibit bacterial growth.

5. TOOL LIFE

Tools wear and fail through abrasion, loss of hardness, and fracture. Three common types of failure are *flank wear*, *crater wear*, and *nose failure*. The life of a tool, T (expressed in minutes), is the length of time it will cut satisfactorily before requiring grinding, and depends on the conditions of use. The *tool life equation*, also known as *Taylor's equation*, relates cutting speed, v, and tool life, T, for a particular combination of tool and workpiece.

$$vT^n = \text{constant} \qquad 63.21$$

The exponent n is an empirical constant that must be determined for each tool-workpiece setup. Typical values are 0.1 for high-speed steel, 0.2 for carbides, and 0.4 for ceramics.

Since the tool feed rate, f, and depth of cut, d, are also important parameters affecting tool life, Taylor's equation has been expanded into Eq. 63.22. (The depth of cut, d, is the same as the chip thickness, t_o, shown in Fig. 63.1.)

$$vT^n d^x f^y = \text{constant} \qquad 63.22$$

6. ABRASIVES AND GRINDING

Grinding (i.e., *abrasive machining*) is used as a finishing operation since very fine and dimensionally accurate surface finishes can be produced. However, grinding is also used for gross material removal. In fact, grinding is the only economical way to cut hardened steel.

Most modern grinding wheels are produced from aluminum oxide. However, grinding wheels can be produced from either *natural abrasives* or *synthetic abrasives*, as described in Table 63.2.

Table 63.2 *Types of Grinding Wheel Abrasives*

natural abrasives

    sandstone
    solid quartz
    emery (50 to 60% Al_2O_3 plus iron oxide)
    corundum (75 to 90% Al_2O_3 plus iron oxide)
    garnet
    diamond

synthetic abrasives

    silicon carbide, SiC
    aluminum oxide, Al_2O_3
    boron carbide

---

[3]Chlorine and sulfur form metallic chlorides and sulfides at cutting temperatures. These compounds have low shear strength, and therefore friction is reduced. Chlorinated oils work better at low speeds, whereas sulfurized oils work better under severe conditions.

Abrasive grit size is measured by the smallest standard-size screen through which the grains will pass. *Coarse grits*, for example, will pass through #6 (i.e., having 6 uniform openings per inch) to #24 screens, inclusive, but will be retained on any finer screen. Table 63.3 summarizes the abrasive size designations.

Table 63.3 *Abrasive Grit Sizes*

designations	screen sizes	
	English	metric (mm)
coarse	#6–#24	4.23–1.06
medium	#30–#60	0.847–0.423
fine[a]	#70–#600	0.363–0.042

(Multiply in by 25.4 to obtain mm.)
[a]Sizes #240 through #600 are also known as *flour grit*.

Snagging describes very rough grinding, such as that performed in foundries to remove gates, fins, and risers from castings.

Honing is grinding in which very little material, 0.001 to 0.005 in (0.025 to 0.13 mm) is removed. Its purpose is to size the workpiece, to remove tool marks from a prior operation, and to produce very smooth surfaces. Coolants, such as sulfurized mineral-base oils and kerosene, are used to cool the workpiece and to flush away small chips. Because the stones are moved with an oscillatory pattern, honing leaves a characteristic cross-hatch pattern.

Lapping is used to produce dimensionally accurate surfaces by removing less than 0.001 in (0.025 mm). Parts are lapped to produce a close fit and to correct minor surface imperfections.

After any cutting or standard grinding operation, the surface of a workplace will consist of *smear metal* (a fragmented, noncrystalline surface). *Superfinishing* or *ultrafinishing* using light pressure, short but fast oscillations of the stone, and copious amounts of lubricant-coolant, removes the smear metal and leaves a solid crystalline metal surface. The operation is similar to honing, but the stone moves with a different motion. There is essentially no dimensional change in the workpiece.

Other nonprecision methods of abrasion can be used to improve the surface finish and to remove burrs, scale, and oxides. Such methods include buffing, wire brushing, tumbling (i.e., barrel finishing), polishing, and vibratory finishing.

Centerless grinding is a method of grinding that does not require clamping, chucking, or holding round workpieces. The workpiece is supported between two abrasive wheels by a work-rest blade. One wheel rotates at the normal speed and does the actual grinding. The other wheel, the *regulating wheel*, is mounted at a slight offset angle and turns more slowly. Its purpose is to rotate and position the workpiece.

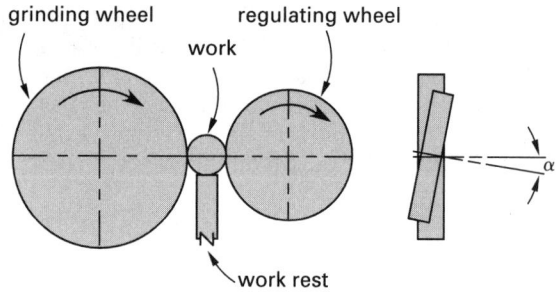

Figure 63.3 *Centerless Grinding*

7. CHIPLESS (NONTRADITIONAL) MACHINING

Electrical discharge machining (EDM), also known as *electrodischarge machining, electrospark machining*, and *electronic erosion*, uses high-energy electrical discharges (i.e., sparks) to shape an electrically conducting workpiece. Thousands of controlled sparks are generated per second between a cutting head and the workpiece, while a servomechanism controls the separating gap. EDM requires the cutting to be performed in a dielectric liquid. The final cut surface consists of small craters melted by the arcs.

EDM can be used with all conductive metals, regardless of melting point, toughness, and hardness. Since there is no contact between the tool and the workpiece, delicate and intricate cutting is possible. However, the metal removal rate is low. Also, the tool material is lost much faster than the workpiece material. Wear ratios for the tool and workpiece vary between 20:1 for common brass tools to 4:1 for expensive tool materials.

Electrochemical machining (ECM) removes metal by electrolysis in a high-current deplating operation. Current densities of 1500 to 2000 A/in^2 (230 to 310 A/cm^2) are common. A tool electrode (the cathode) with the approximate profile desired to be given to the workpiece (the anode) is brought close to the workpiece. The separation is maintained by a servomechanism. A water-based electrolyte (e.g., sodium chlorate solution) is forced between the tool and workpiece. The electrolyte completes the circuit and removes the free ions.

ECM shapes and cuts metal of any hardness or toughness. Relatively high (compared with EDM) metal removal rates are possible. Unlike EDM, the tool is not consumed or changed in shape. The *current efficiency* is defined as the volume of metal removed per unit energy used. Typical units are cubic inches per 1000 ampere-minutes.

Electrochemical grinding (ECG), also known as *electrolytic grinding*, a variant of electrochemical machining, is used to shape and sharpen carbide cutting tools. It uses a rotating metal disk electrode with diamond dust (typically) bonded on the surface. Less than 1% of the workpiece material is removed by conventional grinding. The remainder is removed by electrolysis.

Chemical milling (*chem-milling*), typically used in the manufacture of printed circuit boards, is the selective removal of material not protected by a mask. Some masks are scribed and removed by hand, but most are *photosensitive resists*. When photosensitive resists are used, the workpiece is coated with a light-sensitive emulsion. The emulsion is then exposed through a negative and developed, which removes the unexposed emulsion. Finally, the workpiece is placed in a *reagent* (the *etchant*), which removes only unmasked workpiece metal.

Chemical milling works with almost any metal, such as copper, aluminum, magnesium, and steel. Although the removal rate is low, very large areas can be processed. For highly accurate work, the tendency of the etchant to undercut the mask must be known and compensated for. The *etching radius* (*etch factor*) is one method of quantifying this tendency.

Ultrasonic machining (USM) or *ultrasonic impact machining* works with metallic and nonmetallic materials of any hardness. USM can be used to shape hard and brittle materials such as glass, ceramics, crystals, and gem stones, as well as tool steel and other metals. Ultrasonic energy with a frequency between 15,000 and 30,000 Hz is generated in a *transducer* through magnetostrictive and piezoelectric effects. Wear of the transducer is minimal.

The transducer is separated from the workpiece by a slurry of abrasive particles. The ultrasonic energy generated is used to hurl fine abrasive particles against the workpiece at ultrasonic velocities. The same abrasives used for grinding wheels are used with USM: aluminum oxide, silicon carbide, and boron carbide. Grit sizes of 280 mesh or finer are common.

Laser machining is used to cut or burn very small holes in the workpiece with high dimensional accuracy.

8. COLD- AND HOT-WORKING OPERATIONS

Whether a workpiece is considered cold worked or hot worked depends on whether the working temperature is below or above the recrystallization temperature, respectively. Table 63.4 categorizes most common forming operations.

9. PRESSWORK

Presswork is a general term used to denote the blanking, bending and forming, and shearing of thin-gage metals. Presses (also known as *brakes*) are used with dies and punches to form the workpieces. Press forces are very high, and press capacities (known as *tonnage*) are often quoted in tons.

With *progressive dies*, the workpiece advances through a sequence of operations. Each of the press operations is performed at a *station*. Progressive dies can be of the *strip die* or *transfer die* varieties.

Table 63.4 Cold- and Hot-Working Operations

cold working
    bending
    coining
    cold forging
    cold rolling
    cutting
    drawing
    drilling
    extruding
    grinding
    hobbing
    peening and burnishing
    riveting and staking
    rolling
    shearing, trimming, blanking, and piercing
    sizing
    spinning
    squeezing (e.g., swaging)
    thread rolling

hot working
    bending
    extruding and drawing (bar and wire)
    hot forging
    hot rolling
    piercing
    pipe welding
    spinning and shear forming
    swaging

Shearing operations (blanking, punching, notching, etc.) cut pieces from flat plates, strips, and coil stock. Since the cutting is a shearing operation, the press force required to blank N items at one time is given by Eq. 63.23.

$$F = NS_{us}Lt \qquad \text{63.23}$$

The distinction between blanking and punching is relative. *Blanking* produces usable pieces (i.e., *blanks*), leaving the source piece behind as scrap. *Punching* is the operation of removing scrap blanks from the workpiece, leaving the source piece as the final product.

Bending and forming operations are often considered in the same category. *Bending dies* are used in press brakes to bend along a straight axis. *Forming dies* bend and form the blank along a curved axis and may incorporate other operations (e.g., notching, piercing, lancing, and cut-offs). There is little or no metal flow in a die-forming operation. The tension and compression on either surface of the blank are approximately equal.

Spring-back, *bend allowance*, and bending pressure can be calculated for bending and forming operations. However, it is generally necessary to make test runs to determine these values under realistic conditions.

Drawing is a cold-forming process that converts a flat blank into a hollow vessel (e.g., beverage cans). Drawing sheet metal blanks results in plastic metal flow along a curved axis. Double-acting presses may be required to accomplish deep draws.

Coining, as used in the production of coins, is a severe operation requiring high tonnage, due to the fact that the metal flow is completely confined within the die cavity. Because of this, coining is used mainly to form small parts.

Embossing forms shallow raised letters or other designs in relief on the surface of sheet metal blanks. It differs from coining in that the workpiece is not confined.

Swaging operations reduce the workpiece area by cold flowing the metal into a die cavity by a high compressive force or impact. It is applicable to small parts requiring close finishes. *Sizing* and *cold heading* of bolts and rivets are related operations.

10. FORGING

Forging is the repeated hammering of a workpiece to obtain the desired shape. Forging can be a cold-work process but is commonly considered to be a hot-work process when the term forging is used. Hot-work forging is carried out above the recrystallization temperature to produce a strain-free product. Table 63.5 lists the approximate temperatures for hot forging.

Table 63.5 Approximate Hot-Forging Temperatures

material	temperature	
	°F	°C
steel	2000–2300	1100–1250
copper alloys	1400–1700	750–925
magnesium alloys	600	300
aluminum alloys	700–850	375–450

The oldest form of forging is similar to what is done by blacksmiths. Commercial *hammer forging*, *smith forging*, or *open die forging* consists of repeatedly hammering the workpiece (known as the *stock*) in a powered forge. Accuracy is low since the shape is not defined by dies, and considerable operator skill is required.

Drop forging (*closed-die forming*) relies on closed-impression dies to produce the desired shape. One-half of the die set is stationary; the other half is attached to the hammer. The metal flows plastically into the die upon impact by the forge hammer. The forging blows are repeated at the rate of several times a minute (for *gravity drop hammers*, also known as *board hammers*) to more than 300 times a minute (for *powered hammers*).[4] Progressive forging operations are used to significantly change the shape of a part over several steps.

---

[4]Steam or compressed air can be used to lift the gravity drop hammer back into place, but the hammer is not powered during its downward travel.

The total forming energy (work) supplied by a gravity drop hammer is the potential energy of the hammer.

$$E = mgh \qquad \text{[SI]} \qquad 63.24(a)$$

$$E = \frac{mgh}{g_c} \qquad \text{[U.S.]} \qquad 63.24(b)$$

The total forming energy supplied by a powered drop hammer falling from height h is

$$E = (mg + pA)h \qquad \text{[SI]} \qquad 63.25(a)$$

$$E = \left(\frac{mg}{g_c} + pA\right) h \qquad \text{[U.S.]} \qquad 63.25(b)$$

The total forming energy supplied by an eccentric-crank press can be calculated from its rotational speeds before and after the impact.

$$E = \tfrac{1}{2} J \left(\omega_o^2 - \omega_f^2\right)$$
$$= \tfrac{1}{2} J \left(\frac{\pi}{30}\right)^2 \left(n_o^2 - n_f^2\right) \qquad \text{[SI]} \qquad 63.26(a)$$

$$E = \tfrac{1}{2} \left(\frac{J}{g_c}\right) \left(\omega_o^2 - \omega_f^2\right)$$
$$= \tfrac{1}{2} \left(\frac{J}{g_c}\right) \left(\frac{\pi}{30}\right)^2 \left(n_o^2 - n_f^2\right) \qquad \text{[U.S.]} \qquad 63.26(b)$$

In *impactor forging* or *counterblow forging*, the workpiece is held in position while the dies are hammered horizontally into it from both sides. *Upset forming* involves holding and applying pressure to round heated blanks. The part is progressively formed from one end to the other and, characteristic of upset forming, becomes shorter in length but larger in diameter.

With *press forging*, the part is shaped by a slow squeezing action, rather than rapid impacts. This allows more forging energy to be used in shaping rather than in transmission to the machine and foundation. The pressing action can be obtained through screw or hydraulic action.

Following forging, the part will have a thin projection of excess metal known as *flash* at the *parting line*. The flash is trimmed off by *trimmer dies* in a subsequent operation. Also, since the hot-worked part will be covered with scale, the part is cleaned in acid (an operation known as *pickling*). Additional processing may include shotpeening, tumbling, and heat treatment.

11. SAND MOLDING

With *sand molding*, a mold is produced by packing sand around a pattern. After the pattern is removed, the remaining cavity has the desired shape. To facilitate the removal of the pattern, all surfaces parallel to the direction of withdrawal are slightly tapered. This taper is called *draft*. After molten metal is poured into the mold, the sand is removed, exposing the completed cast part.

Various types of foundry sand are used, including green (moist) sand, dry (baked) sand, and carbon dioxide process sands. Carbon dioxide process sands contain approximately 4% silicate of soda (Na_2SiO_3), which hardens upon exposure to carbon dioxide. Pure, dry silica sand has no binding capacity and is not suited for molding. Various types of clay can be added to dry sand to improve its bonding characteristics (*cohesiveness*).

Other additives permit gases to escape during casting (*permeability*) and enhance the sand's heat resistance (*refractoriness*). Small amounts of organic matter can be added to the sand to enhance its *collapsibility*. The organic matter burns out when exposed to the hot metal, permitting the mold to be easily removed.

The *gating system* shown in Fig. 63.4 is the set of openings and passages that brings molten metal to the mold cavity in a controlled manner. The metal is poured into a *sprue hole* and enters a vertical passage known as a *downgate*. The metal then passes directly into the distribution passageways known as *runners* before entering the cavity. An entrance to the cavity may be constricted to control the rate of fill, and such constrictions are known as *gates*. *Risers* serve as accumulators to feed molten metal into the cavity during initial shrinkage.

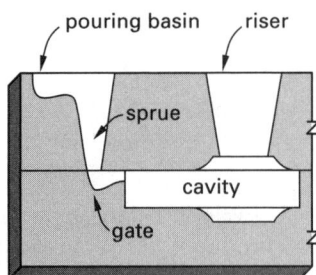

Figure 63.4 *Sand-Casting Gating System*

After hardening, the gates and risers can be broken off from iron castings, but a torch or cutting wheel is necessary with steel castings. Raw castings will be covered with sand and scale that must be removed by tumbling or sand blasting.

If a part has a recess or hole, a *core* must be placed in the mold. Cores come in *green sand core* and *dry sand core* varieties. Green sand cores are formed by the pattern itself. Dry sand cores are formed and baked in a separate operation and inserted into the sand mold.

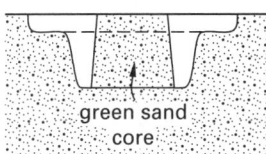

Figure 63.5 *Green and Dry Sand Cores*

12. GRAVITY MOLDING

With *gravity molding* (*gravity die casting* or *permanent molding*), molten metal is poured into a metal or graphite mold. Pressure is not used to fill the mold.[5] The mold may be coated prior to filling to prevent the casting from sticking to the mold's interior. Both ferrous and nonferrous metals (including magnesium, aluminum, and copper alloys) can be gravity molded. This method has the advantage (over sand casting) that a new mold is not required for each casting.

13. DIE CASTING

Die casting (*pressure die casting*) is suitable for creating parts of zinc, aluminum, copper, magnesium, and lead/tin alloys. (More than 75% of all die casting uses zinc alloys.) Molten metal is forced under pressure into a permanent metallic mold known as a *die*. Dies that produce one part per injection are known as *single cavity dies*. Dies that produce more than one part per injection are known as *multiple cavity dies*. The metal can be introduced by a plunger or compressed air but never by gravity alone. The casting pressure is maintained until solidification is complete.

Dies for zinc, tin, and lead alloys are usually made from high-carbon and alloy steels, although low-carbon steel can be used for zinc casting dies. Dies for aluminum, magnesium, and copper alloys (which melt at higher temperatures) are made out of heat-resisting alloy steel.

Hot-chamber die casting and *cold-chamber die casting* are the two main variations of die casting. The hot-chamber method is limited to alloys (e.g., zinc, tin, and lead) that have melting temperatures below 1000°F (550°C) and that do not attack the injection apparatus. The injection apparatus (the *gooseneck*) is submerged in the molten metal, and low temperatures limit corrosion.

Brass (and other copper alloys), aluminum, and magnesium have high melting temperatures and require higher injection pressures. They also corrode ferrous machine parts and become contaminated by the iron they pick up. Brass and bronze, with their 1600 to 1900°F (875 to

1050°C) melting temperatures, particularly attack the steel in die casting machines. These alloys are usually melted in a separate furnace and ladled into the plunger cavity. This is the principle of the *cold-chamber method*.

After solidification, the sprues, gates, runners, and overflows are cut off in *trimming dies*. Die castings will typically be harder on the outside than on the inside due to the chilling action of the die. Also, gases may have been trapped inside the part, making the interior of the casting porous. Porous castings are brittle and subject to fracture.

14. CENTRIFUGAL CASTING

If the mold is rapidly rotated, the molten metal will be forced into the mold by centripetal action while the metal solidifies. This process is known as *centrifugal casting* or *centrifuging*. This method is particularly useful in producing objects with round and symmetrical (e.g., hexagonal) outer surfaces such as gun barrels and brake drums.

15. INVESTMENT CASTING

Casting methods that produce a molding cavity from a wax pattern are known as *investment casting*, *precision casting*, or the *lost-wax process*. These methods are suited to small, complex shapes and casting of precious metals.

A positive image of the part to be cast is created from wax.[6] The image is coated with the *investment material*, which can be finely ground refractory, plaster of paris or another ceramic material, or rubber, which becomes the mold. The mold is heated, and the liquid wax is poured out. The mold is then filled with molten metal. Centrifuging may be used to ensure complete filling. After the metal solidifies, the mold is broken off.

If more than one image is to be cast, a *master pattern* is made out of wood, steel, or plastic. The master pattern is used to make a *master die*. The wax patterns are then made in the master die.

16. CONTINUOUS CASTING

Any process in which molten material is continuously poured into a mold is known as *continuous casting*. This method can be used to produce sheets of glass, copper slabs, and brass or bronze bars. Continuous casting of bars is similar to extrusion, except that a cooling apparatus is included as part of the extrusion head.

17. PLASTIC MOLDING

Thermosetting compounds are purchased in liquid form, which makes them easy to combine with additives. Thermoplastic materials are commonly purchased in

---

[5] There is limited application (e.g., the production of railway wheels and some steel ingots) to a process known as *pressure pouring*, in which the molten metal is forced into the mold by air pressure. Pressure pouring differs from die casting in that ferrous alloys are used.

[6] Other variations of this process use plastics with low melting points, lead, and even frozen mercury.

granular form. They are mixed with additives in a *muller* (i.e., a bulk mixer) before transfer to the feed hoppers. Thermoplastic materials can also be molded into small pellets called *preforms* for easier handling in subsequent melting operations. Common additives for plastics are color pigments and tints, stabilizers, plasticizers, fillers, and resins.

The *hot compression molding* process is the oldest plastic forming process but is used extensively only for thermosetting polymers. (Compression molding is similar to the coining process for metals.) A measured amount of plastic material is placed in the open cavity of a heated mold. The mold is closed, and pressure is applied. The plastic flows into the mold, taking on its shape. Compression molding of thermoplastic resins is not very practical.[7]

Some plastic parts can be produced by *cold molding*. In this process, the part is simply cold pressed in a mold and then heated outside the press to fuse the particles.

The main method of forming thermoplastic resins is *injection molding*. The plastic molding compound is gravity fed into a heating chamber, where it is plasticized. Heating and metering is done by the *torpedo*, illustrated in Fig. 63.6. The molten plastic is injected under pressure into a water-cooled mold, where it solidifies almost immediately. (The mold temperature is constant.)

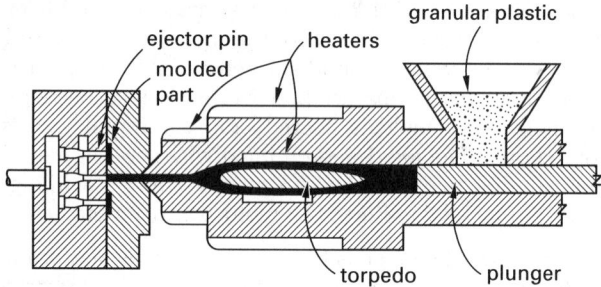

Figure 63.6 *Injection Molding Equipment*

Transfer molding involves the heating of thermosetting plastic powder or preforms under pressure outside the mold cavity. The molten plastic is then forced from the transfer chamber through the gate and runner system into the mold cavity. The plastic is cured in the mold by maintaining the pressure and temperature. This method differs from injection molding in that the mold is kept heated and the plastic part is ejected while it is still hot.

Blow molding and *vacuum forming* both rely on an air-pressure differential to draw a heated thermoplastic sheet around a pattern or into a mold. The plastic retains the shape of the mold after cooling.

---

[7]Unless a mold is cooled before the part is removed, distortion of thermoplastics can result.

Most thermoplastics can be *extrusion-formed* into shapes (including sheets) of any length.[8] Solid plastic in granulated or powdered form is fed by a screw-feed mechanism into a heating chamber and then extruded through a die. The plastic is cooled by contact with air or water after extrusion.

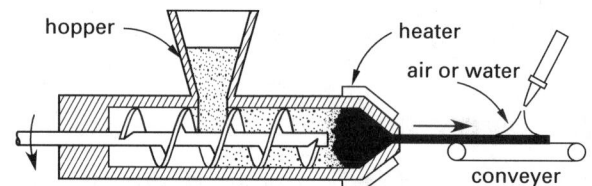

Figure 63.7 *Extrusion Process*

18. POWDER METALLURGY

Useful parts can be made by compressing a metal powder into shape and bonding the particles with heat—the principle of *powder metallurgy*. Typical powder metallurgy products include tungsten carbide cutting tools, copper motor brushes, bronze porous bearings, auto connecting rods and transmission parts, iron magnets, and filters.

Compared with machined products, parts made by powder metallurgy are relatively weak and expensive. However, the process is used for parts that cannot be easily produced in other manners. These are porous products, parts with complex shapes, items made from materials that are difficult to machine (e.g., tungsten carbide), and products that require the characteristics of two materials (e.g., copper/graphite electrical contacts).

The two most common types of powders are *iron-based powders* and *copper-based powders*. However, aluminum, nickel, tungsten, and other metals are also used. Bronze powders are regularly used to produce porous bearings; brass and iron are applicable to small machine parts where strength is important.

Metallic powders must be very fine. Most metal powders are produced by *atomizing* (i.e., using a jet of air to break up a fine stream of molten metal). Other methods include reduction of oxides, electrolytic deposition, and precipitation from a liquid or gas.

Pure metal powders are often mixed to improve manufacturing or performance characteristics. Graphite improves lubricating qualities and is added to powders used for bearings and electrical contacts. Cobalt and other metals are added to tungsten carbide to improve bonding.

Powders can be pressed into their final shape with a punch and die. The ejected shape is known as a *briquette* or *green compact*. With *isostatic molding*, hydraulic pressure is applied in all directions to the powder. Other methods of forming the green compact

---

[8]Thermosetting plastics harden too quickly to be extruded.

are centrifuging, *slip casting* (i.e., slurry casting), extrusion, and rolling. Due to internal friction, the density of a powder metallurgy part will not be consistent throughout but will be higher at the surface.

Sintering is heating to 70 to 90% of the melting point of the metal. The temperature is maintained for up to three hours, although the duration is commonly less than an hour. Typical sintering temperatures are 1600°F (875°C) for copper, 2050°F (1120°C) for iron, 2150°F (1175°C) for stainless steel, and 2700°F (1475°C) for tungsten carbide. To prevent the formation of metallic oxides, sintering must be performed in an inert or reducing gas (e.g., nitrogen) atmosphere.

19. HIGH ENERGY RATE FORMING

High energy rate forming (HERF) is the name given to several processes that plastically deform metals with blasts of high-pressure shock waves. Although these processes have traditionally been used to form thin metals, they are also applicable to other manufacturing needs such as powder metallurgy, forging, and welding.

The most common HERF method is *explosive forming*, as illustrated in Fig. 63.8. A small amount of low- or high-explosive is detonated. The resulting shock waves travel through the surrounding medium and force the metal into a shape determined by the dies. The medium can be either gas or liquid.

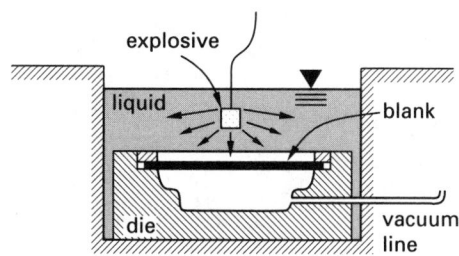

Figure 63.8 *Explosive Forming*

Another process using explosives is *explosive bonding*. A detonation is used to drive two similar metals together. When used with explosives in sheet form, this method has been successful in producing combinations of two metallic sheets (i.e., *cladding*). The resulting bond is almost metallurgically complete.

With *electro-hydraulic forming* (also known as *electro-spark forming*), the forming pressure is obtained from the discharge of massive amounts of stored electricity. The electrical energy is built up in a capacitor bank. The energy used can be changed by adding or removing capacitors from the circuit. The discharge is across a spark gap between two electrodes in a nonconducting medium. Upon discharge, the electrical energy is converted directly into work. The usual medium is liquid.

Magnetic forming (*magnetic pulse forming*) is another example of the direct conversion of electrical energy into work. As with electrospark forming, a large amount of electrical energy is built up in a capacitor bank. A special expendable forming coil is placed around a part to be compressed or within a part to be expanded. (Magnetic forming can also be used for embossing if the workpiece is placed between the forming coil and the embossing die.) When the capacitor bank discharges, the current in the forming coil induces a current and a force in the workpiece. This force stresses the workpiece beyond its elastic limit.

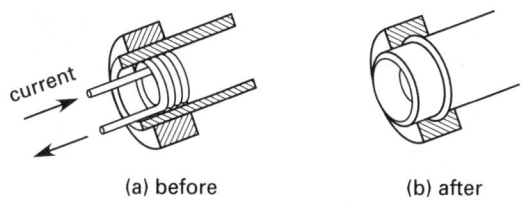

Figure 63.9 *Magnetic Forming*

20. GAS WELDING

With *welding*, two metals are fused (i.e., melted) together by localized heat or pressure. This is known as *fusion* or *coalescence*. A welding rod of similar metal can be used to fill large voids between the two pieces but is not always necessary. The main types of welds are the *bead*, *groove*, *fillet*, and *plug (spot) welds* shown in Fig. 63.10.

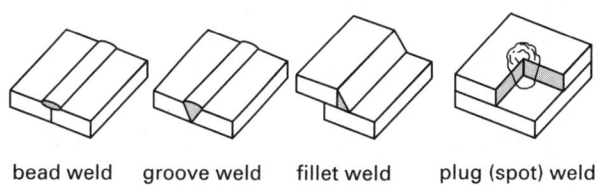

Figure 63.10 *Types of Fusion Welds*

In *gas welding processes*, a combustible fuel and oxidizing gas are combined in a *torch* (*blowpipe*). Although natural gas and hydrogen can be used as fuels, *acetylene gas* (C_2H_2) is most common. Oxygen gas is the oxidizer, hence the names *oxyhydrogen welding* and *oxyacetylene welding*. The maximum welding temperature with oxyhydrogen welding is approximately 3600°F (2000°C), and with oxyacetylene welding, it is approximately 6300°F (3500°C).

MAPP gas (*methyacetylene propadiene*) is also extensively used. It is safer to store, is more dense, and provides more energy per unit volume than acetylene.

The proportions of oxygen and acetylene can be adjusted to obtain three different welding conditions: reducing, neutral, and oxidizing flames. Figure 63.11 illustrates the reactions that occur with a *neutral flame*, which is obtained when the oxygen:acetylene proportions are approximately 1:1 by volume. The inner luminous cone of the flame is distinctly blue in color and is the hottest part of the flame. The outer envelope is only slightly luminous and may be difficult to see. Oxygen for the combustion of the outer envelope comes from the atmosphere.

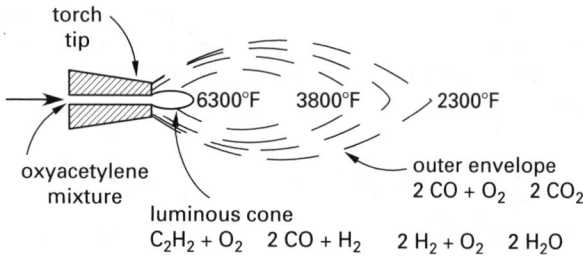

Figure 63.11 *Neutral Oxyacetylene Flame*

If the gas proportions are adjusted to an excess of acetylene, the flame is known as a *reducing flame* or *carburizing flame*. This process is used to weld many nonferrous metals (including Monel metal and nickel alloys as well as some alloy steels) and in applying several types of hard surfacing materials.

An *oxidizing flame* requires an excess of oxygen. Oxidizing fusion has some application to brass and bronze but is generally undesirable.[9]

21. ARC WELDING

Temperatures of up to 10,000°F (5550°C) can be obtained from *arc welding* using either DC or AC electrical current. The arc is created by first touching the electrode to the workpiece, establishing the current flow, and then moving the electrode slightly away. The current flow is maintained by the arc, and the electrical energy is converted to heat, which melts the metal.

If a *carbon electrode* is used to create an arc, a welding rod must be used to supply the filler material. Alternatively, in *metal electrode welding*, the electrode is itself melted by the arc and becomes the filler material.

Electrodes can be bare, but most have coatings known as *flux* that melt into slag and improve other welding characteristics.[10] The molten slag floats on and covers the molten metal, inhibiting the high temperature formation of oxides that weaken most welds. Welding with

coated electrodes is known as *shielded metal arc welding* (SMAW). Most coatings have a significant amount of SiO_2 and/or TiO_2, plus small amounts of oxides of other metals. Hardened slag is chipped away after the weld has cooled.

In the United States, welding rods are classified according to the tensile strength (in ksi—thousands of lbf/in^2) of the deposited material. Thus, an E70 welding rod will produce a bead with a minimum nonstress relieved tensile strength of 70,000 lbf/in^2.

With *submerged-arc welding*, the flux is granular and is dispensed from a feed tube ahead of the welding process. The tip of the electrode and the arc are buried in the granular flux. The arc melts the granular flux, forming a coating that protects against oxidation.

Another method of protecting the molten weld metal from oxidation is by shielding with an inert gas. This is done when welding magnesium, aluminum, stainless steels, and some steels. Argon gas is commonly used, although helium and argon-helium mixtures are also used. (Carbon dioxide gas can be used to shield the weld when working with plain-carbon and low-alloy steels.) This is the principle of *inert gas shielded arc welding*. With *TIG welding* (tungsten inert gas), the arc issues from a water-cooled, nonconsumable tungsten electrode.

MIG welding (metal inert gas) is similar to TIG welding except that a consumable wire is used as the electrode. This method is also known as the *GMAW (gas metal arc welding) process*.

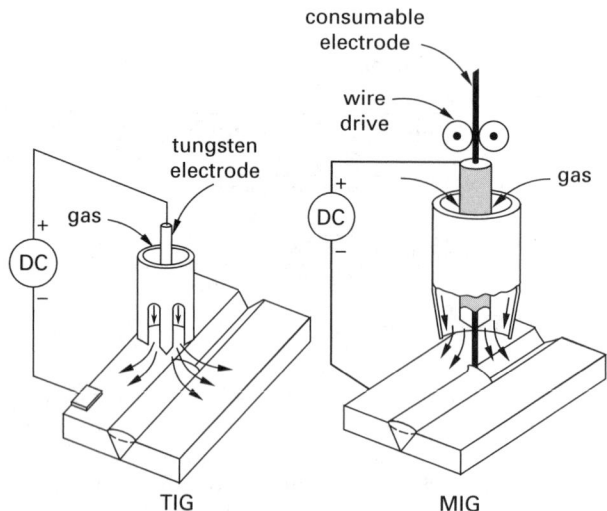

Figure 63.12 *TIG and MIG Welding Processes*

22. SOLDERING AND BRAZING

Soldering and *brazing* both use a molten dissimilar metal as glue between the two pieces. Generally, soldering uses a lead-tin filler with melting points below 800°F (425°C), whereas brazing uses copper-zinc or silver-based alloys with melting points above 800°F (425°C).

---

[9]An oxyacetylene *cutting torch* uses oxygen to cut steel, but the process is different from regular welding. A cutting torch uses a neutral flame to heat the steel. The oxygen used to oxidize (cut) the metal issues from a separate orifice in the torch and does not participate in the combustion of the acetylene.

[10]In addition to providing a protection from oxidation, the next most important function of the coating is to stabilize the arc (i.e., reduce the effect of variations in the separation of electrode and workpiece). Other functions include reducing weld metal splatter, adding alloying ingredients, changing the weld bead shape, improving overhead and vertical weldability, and providing additional filler material.

Soldering is not a fusion process, since parts do not melt. Since some alloying occurs, brazing is considered to be a fusion process.

Usually, the solder or brazing material is drawn into the space separating the pieces by capillary action. However, a chemical flux can be used to remove oxides from the surfaces, inhibit additional oxidation, and improve cohesion of the filler material.

23. ADHESIVE BONDING

Both thermoplastic and thermosetting polymers are used as structural adhesives. Polymers of both types are sometimes combined to obtain the performance characteristics of both. A low-viscosity primer may be used to prepare the bonding surface for the adhesive. Almost all adhesive bonds are of the *lap-joint variety*.

Structural adhesives can be used to join any similar or dissimilar materials. However, even structural adhesives are limited to low-strength and low-temperature (e.g., less than 500°F or 250°C) applications.[11]

Thermoplastic adhesives, such as polyamides, vinyls, and nonvulcanized neoprene rubbers, soften when heated and cannot be used for elevated temperatures. They are generally used for nonstructural applications.

Thermosetting adhesives (e.g., epoxies, isocyanate, phenolic rubbers and vinyls, vulcanized rubbers, and neoprene) are used as structural adhesives. They must be used with elevated temperatures and where creep is unacceptable. Heat, pressure, ultraviolet radiation, and/or chemical hardeners must be used to activate the curing process.

The performance of an adhesive is determined by its toughness, tensile strength, peel strength, and temperature resistance. Adhesives with high tensile and shear strengths are usually hard and brittle and have low *peel strengths*. Adhesives with high peel strengths are usually more ductile and have fair tensile and shear strengths.

24. MANUFACTURE OF METAL PIPE

Pipes and tubes can be either seamed or seamless. Seamless pipe is made by piercing, whereas seamed pipe is made by forming and butt- or lap-welding the joined edges.

Thin-wall pipe can be formed by drawing a heated flat skelp through a welding bell. Prior to forming, the edges of the skelp are heated by flame or induction to the forging temperature, and joining occurs spontaneously in the welding bell, as shown in Fig. 63.13.

---

[11]Certain *ceramic adhesives* have useful ranges up to 1000°F (550°C).

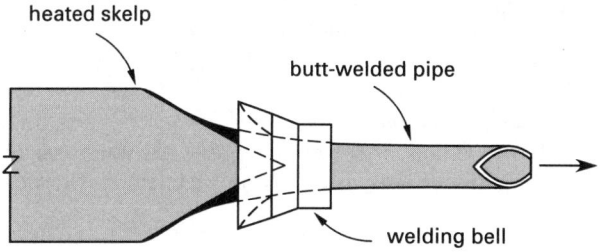

Figure 63.13 *Butt-Welding in a Welding Bell*

Pipes with larger diameters and wall thicknesses are formed with *roll forming* and *electric butt-welding*, in which the skelp is cold-rolled into circular shape by a series of roller pairs. One of the last rolling operations incorporates induction heating to bring the pipe edges up to forging temperature before they are pressed together. The flash is subsequently removed from the inside and outside of the pipe.

The skelp of *lap-welded pipe* passes through rollers that overlap the edges. Unlike roll forming, however, a fixed mandrel is placed inside the pipe. The heated skelp is rolled between the roller and mandrel at high pressure, which welds the heated edges together.

Seamless pipe is manufactured by heating a solid round bar known as a *billet* to forging temperature and piercing it with a mandrel. Subsequent operations with rollers strengthen, size, and finish the pipe.

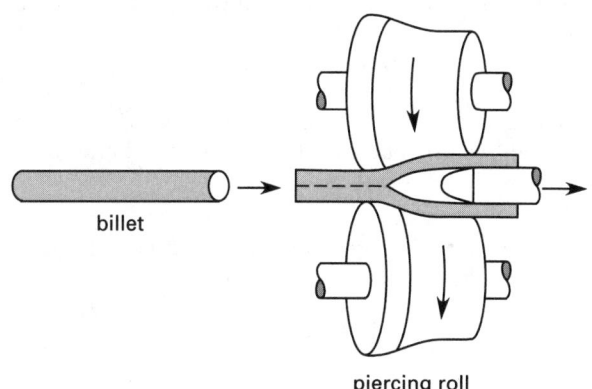

Figure 63.14 *Production of Seamless Pipe*

25. SURFACE FINISHING AND COATINGS

Finishes protect and improve the appearance of surfaces. Some processes involve material removal, and others involve material addition. Some of the common finishing methods and coating systems are listed as follows.

- *abrasive cleaning*: shooting sand (i.e., *sand blasting*), steel grit, or steel shot against workpieces to remove casting sand, scale, and oxidation.

- *anodizing*: an electroplating-acid bath oxidation process for aluminum and magnesium. The workpiece is the anode in the electrical circuit.

- *barrel finishing (tumbling)*: rotating parts in a barrel filled with an abrasive or non-abrasive medium. Widely used to remove burrs, flash, scale, and oxides.

- *buffing*: a fine finishing operation, similar to polishing, using a very fine polishing compound (e.g., *rouge*).

- *burnishing*: a fine grinding or peening operation designed to leave a characteristic pattern on the surface of the workpiece.

- *calorizing*: the diffusing of aluminum into a steel surface, producing an aluminum oxide that protects the steel from high-temperature corrosion.

- *electroplating*: the electro-deposition of a coating onto the workpiece. Electrical current is used to drive ions in solution to the part. The workpiece is the cathode in the electrical circuit.

- *galvanizing*: a zinc coating applied to low-carbon steel to improve corrosion resistance. The coating can be applied in a hot dip bath, by electroplating or by dry tumbling (sheradizing).

- *hard surfacing*: the creation (by spraying, plating, fusion welding, or heat treatment) of a hard metal surface in a softer product.

- *honing*: a grinding operation using stones moving in a reciprocating pattern. Leaves a characteristic cross-hatch pattern.

- *lapping*: a fine grinding operation used to obtain exact fit and dimensional accuracy.

- *metal spraying*: the spraying of molten metal onto a product. Methods include *metallizing, metal powder spraying,* and *plasma flame spraying*.

- *organic finishes*: the covering of surfaces with an organic film of paint, enamel, or lacquer.

- *painting*: see *organic finishes*.

- *parkerizing*: application of a thin phosphate coating on steel to improve corrosion resistance. This process is known as *bonderizing* when used as a primer for paints.

- *pickling*: a process in which metal is dipped in dilute acid solutions to remove dirt, grease, and oxides.

- *polishing*: abrasion of parts against wheels or belts coated with polishing compounds.

- *sheradizing*: a specific method of zinc galvanizing in which parts are tumbled in zinc dust at high temperatures.

- *superfinishing*: a super-fine grinding operation used to expose nonfragmented, crystalline base metal.

- *tin-plating*: a hot-dip or electroplate application of tin to steel.

PRACTICE PROBLEMS

1. A 10 mil (250 μm) adhesive with a shear strength of 1500 lbf/in^2 (10 MPa) is used in a lap joint between two 0.20 in (5 mm) aluminum sheets. The aluminum has a yield strength of 15,000 lbf/in^2 (100 MPa). Assume the adhesive is loaded in pure shear, but use a stress concentration factor of 2. If the joint is to be as strong as the aluminum, what width (overlap) of adhesive joint is required?

64 Material Handling and Processing

1. Wire Rope 64-1
2. Elevators 64-3
3. Bulk Storage Piles 64-4
4. Bins and Hoppers 64-5
5. Feeding Dry Bulk Solids 64-5
6. Pneumatic Conveyors 64-6
7. Briquetting Bulk Solids 64-6
8. Size Reduction Processes 64-7
9. Belt Conveyors 64-7
10. Bucket Elevators 64-9
11. Cyclone Separators 64-10
12. Hydrocyclones 64-10
13. Sieves and Screens 64-10
14. Air Classifiers 64-10
15. Suspensions of Solids 64-12

Nomenclature

d	diameter or particle size	in	cm
d	diameter or particle size	ft	m
e	number of persons per trip	–	–
E	modulus of elasticity	lbf/in^2	Pa
f	coefficient of friction	–	–
F	force	lbf	N
g	acceleration of gravity	ft/sec^2	m/s^2
g_c	gravitational constant	lbm-ft/lbf-sec^2	–
h	fluid head	ft	m
h	height	ft	m
I	rotational mass moment of inertia	lbm-ft^2	kg·m^2
L	length	ft	m
m	mass	lbm	kg
$\dot{m}$	mass flow rate	lbm/sec	kg/s
M	moment	ft-lbf	N·m
n	rotational speed	rpm	rpm
n	rotational speed	rps	rps
n	number of (quantity of)	–	–
N	dimensionless number	–	–
p	pressure	lbf/in^2	Pa
P	peak 5 min population	–	–
P	power	ft-lbf/sec	W
Q	flow rate	ft^3/sec	m^3/s
r	radius	ft	m
r	belt rating	lbf/ft	N/m
Re	Reynolds number	–	–
S	strength	lbf/in^2	Pa
SG	specific gravity	–	–
t	time	various	various
T	tension	lbf	N
T	torque	ft-lbf	N·m
v	velocity	ft/sec	m/s
V	volume	ft^3	m^3
w	width	ft	m
W	weight	lbf	N

Symbols

θ	angle of wrap	rad	rad
μ	viscosity	lbf/ft-sec	Pa·s
ρ	density	lbm/ft^3	kg/m^3
σ	stress	lbf/in^2	Pa
ϕ	angle of internal friction	deg	deg
ω	rotational velocity	rad/sec	rad/s

Subscripts

e	effective
F	force
P	power
Q	quantity
r	rope
sh	sheave
t	tank, tensile, or total
u	ultimate
v	velocity
w	wire

1. WIRE ROPE

Wire rope is constructed by first winding individual *wires* into *strands*, and then winding the strands into rope. Wire rope is specified by its diameter and numbers of strands and wires. The most common *hoisting cable* is 6×19, consisting of six strands of 19 wires each, wound around a core. This configuration is sometimes referred to as "standard wire rope." Other common configurations are 6×7 (stiff *transmission* or *haulage rope*), 8×19 (extra-flexible hoisting rope), and the abrasion-resistant 6×37. The diameter and area of a wire rope are based on the circle that just encloses the rope.

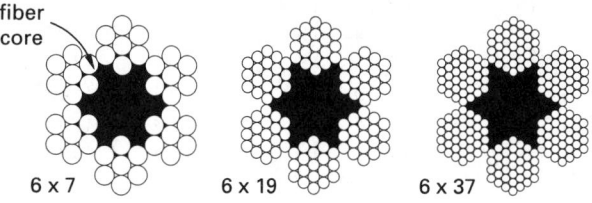

Figure 64.1 *Wire Rope Cross Sections*

Wire rope can be obtained in a variety of materials and cross sections. In the past, wire ropes were available in iron, cast steel, traction steel (TS), mild plow steel

(MPS), and plow steel (PS) grades. Modern wire ropes are generally available only in improved plow steel (IPS) and extra-improved plow steel (EIP) grades.[1] *Monitor* and *blue center steels* are essentially the same as improved plow steel.

Table 64.1 *Minimum Strengths of Wire Materials*

material	ultimate strength, $S_{ut,w}$ (ksi)
iron	65
cast steel	140
extra-strong cast steel	160
plow steel	175–210
improved plow steel	200–240
extra-improved plow steel	240–280

(Multiply ksi by 0.00689 to obtain MPa.)

In Table 64.2, the ultimate strength, $S_{ut,r}$, is the ultimate tensile load that the rope can carry without breaking. This is different from the ultimate strength, $S_{ut,w}$, of each wire given in Table 64.1. The rope's tensile strength will be only 80 to 95% of the combined tensile strengths of the individual wires. The modulus of elasticity for steel ropes is more a function of how the rope is constructed than the type of steel used. (See Table 64.3.)

While manufacturer's data should be relied on whenever possible, general properties of 6×19 wire rope are given in Table 64.2.

Table 64.2 *Properties of 6×19 Steel Wire Rope (improved plow steel, fiber core)*

diameter		mass		tensile strength[a,b], $S_{ut,r}$	
in	(mm)	lbm/ft	(kg/m)	tons[c]	(tonnes)
$1/4$	(6.4)	0.11	(0.16)	2.74	(2.49)
$3/8$	(9.5)	0.24	(0.35)	6.10	(5.53)
$1/2$	(13)	0.42	(0.63)	10.7	(9.71)
$5/8$	(16)	0.66	(0.98)	16.7	(15.1)
$7/8$	(22)	1.29	(1.92)	32.2	(29.2)
$1 1/8$	(29)	2.13	(3.17)	52.6	(47.7)
$1 3/8$	(35)	3.18	(4.73)	77.7	(70.5)
$1 5/8$	(42)	4.44	(6.61)	107	(97.1)
$1 7/8$	(48)	5.91	(8.80)	141	(128)
$2 1/8$	(54)	7.59	(11.3)	179	(162)
$2 3/8$	(60)	9.48	(14.1)	222	(201)
$2 5/8$	(67)	11.6	(17.3)	268	(243)

[a] Add $7 1/2$% for wire ropes with steel cores.
[b] Deduct 10% for galvanized wire ropes.
[c] tons of 2000 pounds

[1] The term "plow steel" is somewhat traditional, as hard-drawn AISI 1070 or AISI 1080 might actually be used.

The central core can be of natural (e.g., hemp) or synthetic fibers or, for higher-temperature use, steel strands or cable. Core designations are FC for *fiber core*, IWRC for *independent wire rope core*, and WSC for *wire-strand core*. Wire rope is protected against corrosion by lubrication carried in the saturated fiber core. Steel-cored ropes are approximately 7.5% stronger than fiber-cored ropes.

Structural rope, structural strand, and *aircraft cabling* are similar in design to wire rope but are intended for permanent installation in bridges and aircraft, respectively. Structural rope and strand are galvanized to prevent corrosion, while aircraft cable is usually manufactured from corrosion-resistant steel.[2,3] Galvanized ropes should not be used for hoisting, as the galvanized coating will be worn off. Structural rope and strand have a nominal tensile strength of 220 ksi (1.5 GPa) and a modulus of elasticity of approximately 20,000 ksi (140 GPa) for diameters between $3/8$ and 4 in (0.95 and 10.2 cm).

The most common winding is *regular lay* in which the wires are wound in one direction and the strands are wound around the core in the opposite direction. Regular lay ropes do not readily kink or unwind. Wires and strands in *lang lay* ropes are wound in the same direction, resulting in a wear-resistant rope that is more prone to unwinding. Lang lay ropes should not be used to support loads that are held in free suspension.

In addition to considering the primary tensile dead load, when selecting wire rope, the significant effects of bending and sheave-bearing pressure must be considered. Self-weight may also be a factor for long cables. Appropriate dynamic factors should be applied to allow for acceleration, deceleration, and impacts, or the acceleration forces. In general for hoisting and hauling, the working load should not exceed 20% of the breaking strength (i.e., a minimum factor of safety of 5 should be used).[4]

If d_w is the nominal wire diameter in inches,[5] d_r is the nominal wire rope diameter in inches, and d_{sh} is the sheave diameter in inches, the stress from bending around a drum or sheave is given by Eq. 64.1. E_w is the modulus of elasticity of the wire material (approximately 3×10^7 psi (207 GPa for steel), not the rope's modulus of elasticity, though the latter is widely used in this calculation.[6]

$$\sigma_{\text{bending}} = \frac{d_w E_w}{d_{\text{sh}}} \qquad 64.1$$

[2] Manufacture of structural rope and strand in the United States are in accordance with ASTM A603 and ASTM A586, respectively.
[3] Galvanizing usually reduces the strength of wire rope by approximately 10%.
[4] Factors of safety are much higher and may be as high as 8 to 12 for elevators and hoists carrying passengers.
[5] For 6×19 standard wire rope, the outer wires are typically 1/13 to 1/16 of the rope diameter. For 6×7 haulage rope, the ratio is approximately 1/9.
[6] While E in Eq. 64.1 is often referred to as "the modulus of elasticity of the wire rope," it is understood that the true meaning is the modulus of elasticity of the wire rope material.

To reduce stress and eliminate permanent set in wire ropes, the diameter of the sheave should be kept as large as is practical, ideally 45 to 90 times the rope diameter. Alternatively, the minimum diameter of the sheave or drum may be stated as 400 times the diameter of the individual outer wires in the rope.[7] Table 64.3 lists minimum diameters for specific rope types.

Table 64.3 *Typical Characteristics of Steel Wire Rope*

configuration	mass (lbm/ft)	area (in²)	minimum sheave diameter	modulus of elasticity (psi)
6×7	$1.50d_r^2$	$0.380d_r^2$	42–$72d_r$	14×10^6
6×19	$1.60d_r^2$	$0.404d_r^2$	30–$45d_r$	12×10^6
6×37	$1.55d_r^2$	$0.404d_r^2$	18–$27d_r$	11×10^6
8×19	$1.45d_r^2$	$0.352d_r^2$	21–$31d_r$	10×10^6

(Multiply lbm/ft by 1.49 to obtain kg/m.)
(Multiply in² by 6.45 to obtain cm².)
(Multiply psi by 6.9×10^{-6} to obtain GPa.)

To prevent wear and fatigue of the sheave or drum, the radial bearing pressure should be kept as low as possible. Actual maximum bearing pressures are highly dependent on the sheave material, type of rope, and application. For 6×19 wire ropes, the acceptable bearing pressure can be as low as 500 psi (3.5 MPa) for cast-iron sheaves and as high as 2500 psi (17 MPa) for alloy steel sheaves. The approximate bearing pressure of the wire rope on the sheave or drum depends on the tensile force in the rope and is given by Eq. 64.2.

$$p_{\text{bearing}} = \frac{2F_t}{d_r d_{\text{sh}}} \qquad 64.2$$

Fatigue failure in wire rope can be avoided by keeping the ratio $p_{\text{bearing}}/S_{ut,w}$ below approximately 0.014 for 6×19 wire rope.[8] ($S_{ut,w}$ is the ultimate tensile strength of the wire material, not of the rope.)

2. ELEVATORS

Elevators classified for freight or passenger use are supported by counterweighted cables.[9] Both geared and gearless traction machines are used for power. Geared machines are suitable for elevator speeds up to approximately 300 ft/min (1.5 m/s); gearless machines (where the driving sheave is mounted directly to the motor shaft) should be used for higher speeds.

[7]For elevators and mine hoists, the sheave-to-wire diameter ratio may be as high as 1000.
[8]A maximum ratio of 0.001 is often quoted for wire rope regardless of configuration.
[9]Design, selection, and use of elevators are governed by local and national codes.

Elevator cable is a special category of wire rope with its own material grading categories. For satisfactory rope life, the sheave diameter is typically 40 to 50 times the rope diameter.

Elevator cabling passes over and is turned by a hoisting sheave. The required frictional force between cable and sheave is obtained by use of multiple cables and wraps. Roping is said to be *half-wrapped* if it passes over the sheave only once (or less) and *full-wrapped* if there are two successive half-wraps with an idler sheave in between.

The *roping* represents the ratios of load-to-cable tension and rope-to-car speeds. With 1:1 roping, the two ends of the rope are connected to the car and counterweight, and car and rope speeds are identical. 1:1 roping is suitable for light cars with speeds in excess of 600 ft/min (3 m/s). With 2:1 roping, the cables are attached at the penthouse level, and the car speed is half the rope speed. 3:1 roping, wherein the rope passes through deflector (idler) sheaves at the penthouse level before connecting to the car and counterweight, also is used occasionally for very heavy loads.

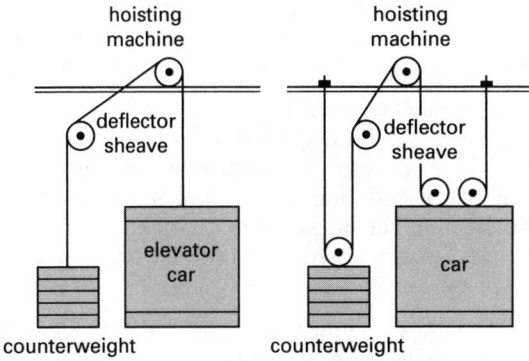

(a) 1:1 roping, single-wrap   (b) 2:1 roping, single-wrap

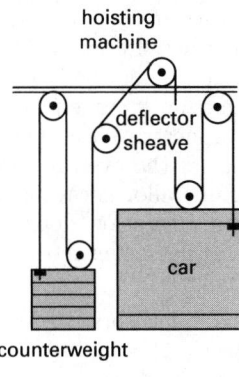

(c) 3:1 roping, single-wrap

Figure 64.2 *Variations on Elevator Roping*

Since the elevator weight is balanced by a counterweight, the motor only has to accelerate the unbalanced load. Passenger elevators are typically overbalanced (i.e., balanced to 40% or more of the fully loaded elevator weight). With rises in excess of approximately 100 ft, the weight of the cable between the traction device and counterweight should be counteracted by the weight of static cables or chains connected to the car bottom and shaft ceiling.

There are no height or speed limitations for the counterweighted design, and overtravel is inherently limited. If either the elevator or counterweight goes too high, one or the other bottoms out, eliminating the driving friction.[10]

The maximum travel rate (elevator speed) depends on the number of floors served. Commercial traction machines operate routinely at 1000 ft/min (5 m/s) and are capable of speeds between approximately 25 and 1600 ft/min (0.13 and 8 m/s), with the highest speeds applicable only to the tallest buildings. Although local building codes may apply, a speed of 150 ft/min (0.75 m/s) is reasonable for 3 or fewer floors, 200 ft/min (1 m/s) for 4 floors, 300 ft/min (1.5 m/s) for 5 or 6 floors, 400 ft/min (2 m/s) for 7 to 9 floors, and 500 ft/min (2.5 m/s) for 10 to 13 floors.

For maximum comfort, acceleration and deceleration should be limited to $\frac{1}{8}$ g. For travel rates of 300 ft/min (1.5 m/s) or higher, the acceleration should be 3 to 5 ft/sec^2 (0.9 to 1.5 m/s^2). Acceleration should taper off after reaching 80% of the desired travel speed. Acceleration of slower elevators is proportionally lower.

Elevator capacity should be based on an average passenger weight of 150 lbm. Office buildings often have total capacities of between 2500 and 4000 passenger pounds. The service population and *peak load service* (i.e., the maximum number of passengers served in a 5 min period) may be estimated by rules of thumb or may be specified by contract or code.[11]

The average wait, known as the *interval*, for an elevator generally should not exceed 20 to 30 sec, although elevator costs must be balanced against the costs of waiting and other factors (crowding, safety, etc.). The theoretical number of elevators required should be calculated using queuing theory.[12] Equation 64.3 can be used if the 5 min peak service population is P, the average round-trip time is t, and the average peak period car loading (in persons per round trip) is e. The average round-trip time will depend on the average distance traveled per trip and will include allowances for elevator movement, acceleration and deceleration, door opening and closing, reaction time, and passenger movement.

$$\text{number of elevators} = \frac{P_{\text{peak,5 min}} t_{\text{min}}}{5e} \qquad 64.3$$

The *dispatch interval* is the average round trip time divided by the number of elevators.

---

[10]Since there is no inherent safety against overtravel, drum-wound cabling is prohibited by code for passenger use. Similarly, almost all hydraulic elevator designs are obsolete.

[11]For example, the peak 5 min service load may be estimated as 10 to 15% of the building population.

[12]For continuity of service, at least two elevators should be installed in critical applications (e.g., in hospitals).

3. BULK STORAGE PILES

Large quantities of free-flowing bulk solid materials are typically stored as mass piles that are conical or rectangular in shape. (Bins and hoppers are discussed in Sec. 4.) The angle formed between the surface of the pile and the horizontal, ϕ, is the *angle of repose* (also known as the *angle of natural slope, angle of natural friction,* and *angle of internal friction*).[13] Typical values for common industrial bulk materials are given in Table 64.4.[14]

Table 64.4 *Typical Properties of Bulk Materials*

material	angle of repose (degrees)	bulk density (lbm/ft^3)	(kg/m^3)
ash, anthracitic	45	40–45	640–720
ash, soft coal	40–45	35–40	560–640
cement, portland	40	90	1400
cinders, coal	25–50	40	640
clay, compact	20–25	110	1800
coal, anthracitic[a]	30–45	52–57	830–910
coal, bituminous[a]	35	44–52	700–830
coke, loose	30–45	23–35	370–560
earth, common	30–45	70–80	1100–1300
gravel	30–40	100	1600
lime, ground	43	60	960
sand, moist	15–30	110–130	1800–2100
slag, blast furnace	25	80–90	1300–1400
soda ash, light	37	20–35	320–560
sugar, granulated	30–45	50–55	800–880

(Multiply lbm/ft^3 by 16.0 to obtain kg/m^3.)
[a]Exercise care in storing coal. Spontaneous combustion is a problem when most coals, with the exception of anthracite, are stored for long periods. Accumulated coal gas represents an explosion hazard. Freshly pulverized coal is hygroscopic in nature.

Though the actual pile peak will be slightly rounded, the volume and height can be calculated assuming a conical shape. In Eqs. 64.4 and 64.5, r is the radius at the base of the pile and h is the vertical height.

$$V = \tfrac{1}{3}\pi r^2 h = \tfrac{1}{3}\pi r^3 \tan \phi \qquad 64.4$$

$$h = \frac{3V}{\pi r^2} = \left(\frac{3V \tan^2 \phi}{\pi} \right)^{\frac{1}{3}} \qquad 64.5$$

Rectangular piles can be assumed to have straight-sloped sides (inclined at the angle of repose) and four quarter-conical corner sections. Volumes of rectangular piles can be calculated from the dimensions of these individual sections.

---

[13]The angle of a conical pile will be slightly less than the true angle of repose of the material due to the effect of impact.

[14]The actual angle of repose will depend on the moisture content and grading. Fines carry most of the moisture.

4. BINS AND HOPPERS

Smaller quantities of dry bulk solids, including powders, pellets, flakes, and granules, are generally inventoried in conical bins, or wedge-shaped hoppers. Bins and hoppers must be carefully chosen to avoid the problems of arching, rat-holing, and segregation.

Arching or *bridging* occurs when the opening at the base of the bin is too small. The strength of the compressed or interlocked bulk solid is enough to form an arched cap above the bin opening. This stops flow. Arching is eliminated by either reducing cohesion in the material or increasing the opening size.

Rat-holing or *piping* occurs when the slope of the bin walls is not great enough for gravity to overcome the friction between the material and the walls. Material near the walls is trapped as dead (stagnant) storage, and a hole through the material appears from the opening to the top of the bin. *Segregation* occurs naturally during filling, when coarse particles fall to the sides and fine particles concentrate in the center.

Storage hoppers and bins are categorized as funnel-flow, mass-flow, and expanded-flow types, depending on the type of flow they promote. *Funnel-flow bins* have long, upright walls that narrow only near the bottom. They are commonly used for coarse (i.e., $1/4$ in or larger), dry materials that don't pack or deteriorate. Though the storage capacity of a funnel-flow bin is the largest of the three types, the effective capacity may be reduced by dead storage. Arching may also be a problem unless the opening is very wide. Since funnel-flow bins are particularly prone to segregation and formation of dead storage, they should not be used for materials that spoil or are damaged by long-term residence.

Mass-flow bins have steeper conical sections than funnel-flow bins. Though their theoretical storage capacity is the smallest of the three types of bins, mass-flow bins are less prone to arching and rat-holing. Material leaves the bin on a first-in, first-out basis, and segregated particles are efficiently remixed during discharge. Therefore, mass-flow bins are suitable for fine powders, cohesive bulk materials, and materials that degrade when stored for extended periods of time. *Expanded-flow* bins combine the designs of both funnel- and mass-flow bins. They have performance characteristics intermediate of the two types.

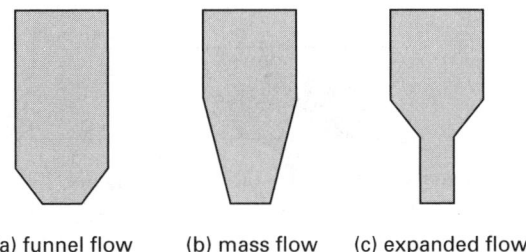

(a) funnel flow   (b) mass flow   (c) expanded flow

Figure 64.3 Conical Bin Geometries

Wedge-shaped rectangular hoppers may offer advantages over conical bins. Typically, a wedge-shaped hopper can be 10° to 12° less steep without experiencing funnel flow. Since the diameter of a conical opening must be twice the width of a slot opening, hoppers can get by with smaller-width openings or provide higher flow rates. Rectangular hoppers use floor space more efficiently, and, for a given volume, hoppers require less headroom than conical bins.

In the past, the design of bins and hoppers was largely based on the desired volume and the angle of repose. Wall angles greater than the angle of repose were chosen. Modern bin and hopper design is based on the *angle of kinematic friction* (i.e., the angle of friction between the material and the storage unit wall), the stress in the material forming an arch, and numerous other factors. This constitutes an engineering economic storage analysis.[15]

5. FEEDING DRY BULK SOLIDS

Depending on their characteristics, dry bulk powders, pellets, and flakes are categorized as being floodable, easy-flowing, difficult-flowing, or cohesive. *Floodable materials* (e.g., flour, gypsum, and diatomaceous earth) behave like liquids and require a positive sealing device such as intermeshed twin screws. *Easy-flowing materials* (e.g., sand and plastic pellets) can be transported by almost any system. *Difficult-flowing* materials (e.g., fiberglass and rubber particles) and *cohesive materials* (e.g., pigments and titanium dioxide) often require custom feeders and flow-aid devices.

Fully enclosed *single-screw feeders (screw conveyors)* are suitable for free-flowing powders and granular, pelletized, and flaked products (e.g., salt, sugar and plastic pellets). The path should be reasonably level and less than approximately 200 ft (60 m) long.[16] The screw may be of constant pitch or, in order to reduce the power required to shear material out of the overlying hopper load, of variable pitch with a tighter helix at the hopper.

Twin-screw feeders use two screws, either intermeshed or non-intermeshed. Intermeshed screws must turn in the same direction. Twin-screw feeders are suitable for almost all materials, including floodable, cohesive, and sticky materials. However, pelletized materials may jam or be crushed between intermeshed screws.

Volumetric flow rates and power requirements for screw feeders are proportional to length and depend on both the empty volume (for frictional effects) and the mass of material carried (for inertial effects). Power calculations are well documented by screw feeder manufacturers, each of which has its own unique set of coefficients and constants.

[15]This method of analysis was perfected by Jenike and Johansen in the early 1960s.
[16]There is little loss in capacity for inclinations below 15°. However, an inclination of 20° can reduce the capacity of open screw conveyors by up to 50%.

Rotary feeders are used for very fine, free-flowing powders. A rotor with radial vanes (blades) turns within a housing. A close tolerance between the vanes and housing keeps the material in the feed hopper from the discharge stream. Top-fed units are preferred for maximum throughput, as each cavity is completely filled during rotation of the rotor. Side-fed units are preferred for pelletized products since each cavity will be only partially filled, limiting shearing and tearing between the vanes and housing.

Belt feeders are used to supply high volumes of fragile materials, ores, and powders. Capacity is a function of the belt speed and width. Belt speeds up to 120 ft/min (0.6 m/s) are used, though 60 ft/min (0.3 m/s) is an upper limit for smaller feeders. *Vibratory feeders* are used to move abrasive or friable solids that could damage or be damaged by screw or rotary feeders.

For most noncyclical feeding devices, the feed rate is a function of the material bulk density, velocity, and flow cross section. The *turndown* of a feeder type refers to the ease and linearity of reducing the feed rate. Screw conveyors have large and linear turndowns; vibratory feeders have limited and nonlinear turndowns.

6. PNEUMATIC CONVEYORS

Pneumatic conveyors are used for dusts, coal, flyash, grains, granular material that is dry and free-flowing, and pellets less than approximately $1/4$ in (6 mm) in size. Material may be moved by positive pressure, vacuum, or a combination of the two. Positive-pressure systems are generally not used for multipoint feeding (i.e., from multiple bins or hoppers). Vacuum systems are generally not used for multipoint delivery.

Most pneumatic conveying systems operate with air speeds of 3000 to 7500 ft/min (15 to 38 m/s). 3000 ft/min (15 m/s), the lower practical limit, is required for light materials such as flour and dusts, and 4000 to 6000 ft/min (20 to 30 m/s) is required for heavier materials such as grains. Higher velocities, up to 20,000 ft/min (100 m/s), also may be used if economical.

The primary factor determining air velocity is the bulk density of the material. Equation 64.6 gives the theoretical air velocity to lift a smooth spherical particle of size d. Since most particles are neither spherical nor uniform, Eq. 64.6 should be used as a general check. Actual conveying velocities will be approximately 50% greater than calculated.

$$v_{\text{ft/min}} = (15{,}000) \left(\frac{\text{SG}}{\text{SG}+1} \right) \sqrt{d_{\text{inches}}} \qquad \textbf{64.6}$$

The required air volume must be determined by trial and error. Values range from 35 to 40 ft^3 per pound for heavy, compact materials, to 90 ft^3 per pound for light, fluffy materials (e.g., cotton).

Rotary piston-type blowers are commonly used when the conveying distance is long and the flow rate is high. Their volumetric flow rates are largely insensitive to changes in the system pressure. The discharge:inlet pressure ratio for positive-displacement blowers is limited to approximately 2:1. This means that positive-pressure systems can operate up to two atmospheres; vacuum systems are limited to half an atmosphere.

Fans can be used as the motive power when the pressure differential required is not too great. Fans are ideal if the product can be conveyed through the fan itself, as is the case with shredded products (e.g., paper, sawdust, and polyethylene trimmings). Fan-driven systems generally operate at lower velocities.

Blow-tank systems are used for conveying materials long distances. A blast of air accelerates a plug of material through the pipeline. Delivery is cyclical, and a large volume of pressurized air is released with each plug. Lines must be sized to accommodate the instantaneous plug feed rate, not the average feed rate.

7. BRIQUETTING BULK SOLIDS

Briquetting and *pressure compaction* use pressure with or without a binder to form loose bulk solids into briquettes or pellets. Briquetting can improve the handling, storage, salability, and environmental acceptance (through dust reduction) of the product. Products commonly briquetted include charcoal, scrap metal, fertilizers, animal food, and salt for water softeners.

The process of producing enlarged particles is known as *agglomeration*. Confined compaction and extrusion are the two primary methods used. Roll briquetters, tabletizers, pelletizers, presses, screw extruders, and roll-extrusion pellet mills are typical agglomeration devices. A rotary knife device cuts extruded product into pellets of the desired length.

In roll machines, the *nip* is the gap between the rolling surfaces. The *angle of nip* defines the effective area of consolidation. Above the angle of nip, slip occurs within the material and between the material and the rollers, and the material is not captured by the rollers. Below the angle of nip, the material is caught and drawn into the rollers. The angle of nip is a primary factor in the compaction pressure. By using a screw conveyor to precompress the material, smaller-diameter rolls can be used to achieve the same briquetting pressure (compared to a gravity-fed system).

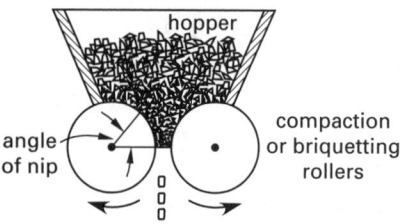

Figure 64.4 *Angle of Nip*

Briquetting pressures depend on the product. Coal, charcoal, coke, fertilizers, and animal feed can be briquetted at low pressures of 5 to 20 ksi (35 to 140 MPa). Medium pressures of 20 to 50 ksi (140 to 350 MPa) are needed for soft metal (e.g., brass and aluminum) turnings and borings and glass-making mixtures. High pressures of 50 to 80 ksi (350 to 550 MPa) are used for ductile metals, steel-chips and turnings, mill scale, and electric- and blast-furnace dust. Very high pressures, 80 ksi to more than 100 ksi (550 to 700 MPa), are needed for metal powers and for titanium and stainless steel turnings.

Some materials are briquetted with a binder. *Matrix binders*, including wax, paraffin, clay, dry starch, and asphalt, fill the majority of the voids in the material. *Film binders* are added in liquid form.[17] As the solvent evaporates or cures, the binder remains to form solid bridges between particles. *Chemical-reaction binding* may be used in certain cases where a chemical reaction occurs between the binder and the briquetted material.

8. SIZE REDUCTION PROCESSES

Size reduction, also known as *comminution* and *pulverizing*, produces multiple pieces of smaller material from larger pieces. Large pieces are commonly crushed to size in primary operations, while smaller pieces are cut or ground in secondary operations. For example, freshly lined coal and friable ores are typically reduced in size in *roll crushers*, the rolls of which can be smooth, corrugated, or toothed. *Jaw* and *gyratory crushers*, which are more efficient and require less maintenance than roll crushers, are now generally used for harder ores.

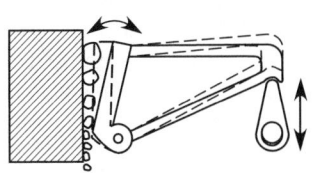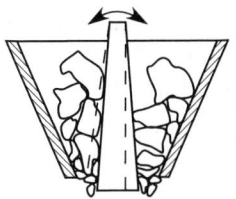

Figure 64.5 *Jaw and Gyratory Crushers*

Rotary cutters, common in the recycling industry, are used for tough or fibrous materials such as pulp, wood, paper, fiberglass, plastics, and rags. Multiple cutting blades (known as *fly knives*) on a rotating rotor are juxtaposed to stationary blades (*bed knives*). Though there may be up to 16 fly knives per cutting head, typically only 2 to 4 bed knives are used. The softest, stickiest, and toughest materials require the fewest bed knives.

Disk mills using steel disks are similar in operation to old flour mills using millstones. They are used for grinding tough organic materials such as wood pulp.

Ring roller mills are used to size-reduce nonmetallic minerals (e.g., coal) and dry chemicals (limestone, starch, pigments, etc.) requiring a finished size of approximately 35 to 800 μm. Ring roller mills have rollers forced inside a cylindrical ring. Either the rollers or the ring can rotate. Operation of bowl- and ball-bearing mills is similar.

Agitator ball mills, also known as *tube-mills*, are commonly used for wet grinding. They use a cylindrical grinding bin partially filled with grinding balls. The grinding efficiency is low, and the ground product typically has a wide distribution of finished sizes.

In *mechanical impact mills*, including *hammer mills*, free-moving particles strike a single beating surface at high speed. Either the particles can be thrown outward by centrifugal force to a stationary beating surface, or the beaters can revolve at high speed.

Fine impact mills are mechanical impact devices that accept material smaller than approximately 4 in (10 mm). Only nonabrasive, brittle, and free-flowing materials are suitable for fine impact mills. Coal for direct firing is commonly pulverized in high-speed (225 rpm and above) impact pulverizers. *Jet mills* accelerate the material by gas or steam jets, with impact between particles or solid surfaces reducing the particle size.

In *autogenous* and *semiautogenous* (SAG) grinding, the material being ground serves as the grinding media.[18] These methods have largely replaced conventional grinding methods for secondary grinding (i.e., following the primary gyratory crusher) in mineral processing plants for materials less than approximately 16 in (0.4 m) in initial size.

9. BELT CONVEYORS

A *belt conveyor* consists of an endless belt and is used for carrying material from one place to another. The belt is driven by a pulley and is supported on rollers (idlers) or (less often) on a flat runway. The belt may be flat or troughed. Carried material is placed on and removed from the belt along its run. Complete design of a belt conveyor will encompass selection of the belt material, the method of support, the driving method, and accessories to maintain belt tension and to load, unload, and clean the belt.

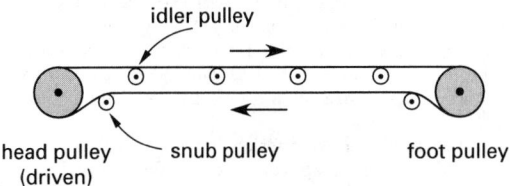

Figure 64.6 *Typical Belt Conveyor*

---

[17]Plain water is the most common film binder.

[18]In SAG grinding, the material is supplemented by steel balls.

Power is required to (1) overcome friction in all of the moving parts, (2) move the load horizontally, (3) move the load vertically, and (4) drive any belt-powered accessories such as trippers.

Equation 64.7 calculates the power required to drive the belt in terms of the belt velocity, v, and net tension, T_{net}. The net tension, also known as the *effective tension*, is the difference between the tight- and loose-side belt tensions.[19] The required horsepower (kilowatts) can be calculated by dividing by the appropriate factor (e.g., 33,000 ft-lbf/hp-min).

$$P = (T_1 - T_2)\text{v} = T_{net}\text{v} \qquad \textbf{64.7}$$

The belt tensions can be found from consideration of the belt wrap angle and belt friction.[20] The coefficient of friction, f, between the belt and pulley is approximately 0.25 for dry, clean rubber belts on cast-iron pulleys and 0.35 when the pulleys are lagged or rubber-coated.[21] For dusty environments, the corresponding coefficients of friction are approximately 0.20 and 0.27. For wet work, the coefficient is approximately 0.20 for bare or lagged pulleys. Centrifugal force is usually not a factor in conveyor belt design.

$$\frac{T_1}{T_2} = e^{f\theta} \qquad [\theta \text{ in radians}] \qquad \textbf{64.8}$$

The belt tension in Eq. 64.7 results primarily from friction in the rotating idlers. Other frictional factors (i.e., friction of the carried material with the skirt boards) are considered to be minor in comparison and are accounted for by including an allowance with the idler friction. Modern designs use sealed antifriction bearings that do not require periodic lubrication.

Belts are seldom inclined more than 30° from the horizontal, and material slippage typically limits the slope to the 15 to 20° range.[22] If the conveyor is inclined, the power required to lift a mass of material per unit time, $\dot{m}$, through a height, h, is given by Eq. 64.9. Theoretically, an inclination of the belt does not affect the power required to drive it empty.[23]

$$P = \dot{m}gh \qquad [\text{SI}] \qquad \textbf{64.9(a)}$$

$$P = \frac{\dot{m}gh}{g_c} \qquad [\text{U.S.}] \qquad \textbf{64.9(b)}$$

---

[19]There are cases—primarily when the conveyor is inclined, uses a tandem drive, or has a holdback—where the maximum tension is greater than T_1.

[20]It is seldom possible to get more than 180° with a plain pulley or more than 260° with a snub-nose pulley. However, with a *tandem-pulley drive* (i.e., a second drive pulley with a reverse bend in the belt), the combined angle of wrap can be up to approximately 410°. Tandem drives are seldom used in practice because of their expense.

[21]Ideally, the coefficient of friction between a rubber-coated pulley and a rubber belt is approximately 0.55. However, the value of 0.35 should be used to account for dirty conditions.

[22]Dry silica sand is limited to approximately 15°. Paper-wrapped packages are limited to 16°. Fine coal may be inclined up to about 22° if evenly loaded and 20° if lumpy. Run-of-mine coal and coke are limited to inclines of approximately 18°.

[23]There is some additional stress in the belt.

In the absence of measurements of actual tension or specific manufacturer's data, little can be done to accurately predict the required driving power, though several quick approximations are widely used. A simple method for first approximations of horsepower for conveyors with modern antifriction idler bearings carrying materials with densities of 50 to 100 lbm/ft^3 (800 to 1600 kg/m^3) is given by Eq. 64.10.[24]

$$P_{hp} = 0.4 + \left(\frac{\dot{m}_{\text{lbm/hr}}}{2000}\right)\left(\frac{0.00325 L_{ft}}{100} + \frac{0.01 h_{ft}}{10}\right) \qquad \textbf{64.10}$$

Belt speed should be no faster than necessary to provide the capacity under conditions of full loading. However, it may be uneconomical (compared to other material handling methods) to run conveyor belts at much less than 150 ft/min (0.75 m/s). The main factor in selecting conveyor speed is belt wear due to contact with the carried material. It is more desirable to load a belt deep than shallow, as more material is carried for the same amount of belt damage. Because of this, wider belts can run faster than narrow belts because the loading can be deeper.

Approximate maximum speeds for common-design conveyors carrying coal, ore, and gravel vary linearly from 300 ft/min (1.5 m/s) for 12 in wide (0.3 m) belts to 600 ft/min (3 m/s) for 48 in wide (1.2 m) and wider belts. Approximate maximum speeds for conveyors carrying heavy grain (e.g., wheat and corn) vary linearly from 400 ft/min (2 m/s) for 12 in wide belts (0.3 m) to 700 ft/min (3.5 m/s) for 48 in wide (1.2 m) and wider belts. Lighter grains (e.g., bran, oats, and flour) can be carried at up to 500 ft/min (2.5 m/s) without being blown off the belt by air resistance.

Driving pulley diameter is not a factor in theoretical power calculation unless the belt is too stiff to bend around the pulley. Large pulleys reduce fatigue wear in the belt but cost more and take up greater space. The driving pulley size is traditionally selected on the basis of 5 in (13 cm) of diameter for every belt ply. For foot and snub pulleys, use 4 in (10 cm); for tandem drives, use 6 in (15 cm).

Fabric belt materials are rated according to their maximum tension per ply per unit width. Calling this rating r and the width w, the required number of plies depends on the tight-side tension.

$$n = \frac{T_1}{rw} \qquad \textbf{64.11}$$

---

[24]This is a modification of the old rule of thumb calculating the horsepower as "2% of the tons per hour for every 100 feet or horizontal distance plus 1% of the tons per hour for every 10 feet of vertical distance." The 2% figure, applicable to early greased fittings, has been substantially reduced by use of low-friction bearings.

10. BUCKET ELEVATORS

A *bucket elevator (belt elevator)* for carrying bulk materials consists of buckets, a belt, a method of driving the belt, and accessories for loading and unloading the material, maintaining belt tension, and so on. Material may be carried at low speed and discharged by gravity, as in a *continuous bucket elevator*, or materials may be carried at high speed and discharged by centrifugal force, as in a *centrifugal discharge belt elevator*.

Figure 64.7 illustrates the *head pulley* of a typical belt elevator. The *foot pulley* is located in the *loading boot* at the bottom of the elevator. Loading is usually by the passage of the buckets through the material contained in the loading boot.

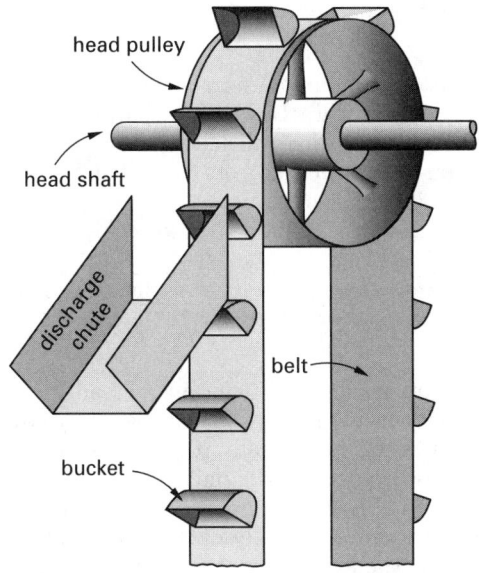

Figure 64.7 *Belt Elevator*

Since the carried material cascades over the previously emptied descending bucket, most continuous bucket elevators are inclined, and bucket spacing is closer than in centrifugal elevators. Buckets in continuous bucket elevators are generally angular, while buckets for centrifugal elevators have rounded bottoms.

Belt elevators are driven similarly to belt conveyors—using friction contact between a pulley and the belt. However, since snub and tandem pulleys cannot be used with bucketed belts, the contact angle cannot exceed 180°. Increased tension is used to obtain the desired speed and capacity. Coefficients of friction for belt elevators are essentially the same as for belt conveyors. Where the belt, buckets, and load are light, intrinsic belt tension may be sufficient. For heavier loads or wet or dusty environments, artificial tension may be needed.

Since bucket loading and delivery of material to the boot can vary over time, belt elevators and their motors should be sized for their expected peak minute loading, not for their average hourly loading. The theoretical capacity (mass per unit time) of a belt elevator is given

by Eq. 64.12. However, in many situations, conditions for obtaining full bucket loads do not exist. Grain elevators typically run at 85 to 90% of their capacity; coal and ore elevators run at 75% or less.

$$\text{capacity} = \frac{(\text{bucket volume})(\text{belt speed})}{\text{bucket spacing}} \quad \textit{64.12}$$

Tension in the belt is the result of pulley and shaft friction, drag of the buckets through the loading boot, and weight of the material in the lifting buckets. The first two items are not easily estimated and must be derived indirectly from power tests of the elevator. The power required only to lift the material is given by Eq. 64.13. Appropriate conversion factors are required to convert to horsepower or kilowatts.[25]

$$P = (\text{total material weight})(\text{elevator speed})$$
$$= \left(\frac{\text{weight}}{\text{unit time}}\right)(\text{elevator height}) \quad \textit{64.13}$$

Location of discharge chutes for centrifugal elevators is somewhat heuristic. Some material is always scattered and spilled, but the amount will be small in good designs. The angle and location of the discharge chute are determined from an analysis of the centrifugal force on the carried material assuming frictionless discharge.[26] Discharge chutes for centrifugal elevators carrying liquids and dry, free-flowing materials like grain are commonly located near the top of the wheel where material weight and centrifugal force are equal. In that case, where the head pulley has a radius r,

$$W = mg = \frac{mv^2}{r} \qquad \text{[SI]} \quad \textit{64.14(a)}$$

$$W = \frac{mg}{g_c} = \frac{mv^2}{g_c r} \qquad \text{[U.S.]} \quad \textit{64.14(b)}$$

$$v = \sqrt{gr} = \frac{2\pi r n_{\text{rpm}}}{60} \quad \textit{64.15}$$

Slower running speeds (than calculated by Eq. 64.15) are required when the material is hard, lumpy, dusty, or moist enough to stick to the buckets. Belt speeds for coal, ashes and cinders, coke, stone, ores, salt, and fertilizers are typically 80% or less of that calculated from Eq. 64.15.[27] Maximum speeds are typically 60 ft/min (0.3 m/s) when buckets are loaded by separate systems and less than 30 ft/min (0.15 m/s) when the buckets fill in a submerged boot.

---

[25] A rule of thumb is that the horsepower required to drive a boot-loaded bucket elevator is equal to the tons per hour multiplied by the lift in feet and divided by 500.
[26] It is possible to proceed by assuming a coefficient of friction between the bucket and the contents, but the results are highly unreliable and generally do not lead to any practical result.
[27] Other factors, such as method of loading and bucket wear, are also considered when determining belt speed for hard-to-handle materials.

11. CYCLONE SEPARATORS

Separation is the process of sorting different materials from each other.[28] Solid particles are easily removed from gas streams by reverse flow *cyclone separators (vortex separators)* with tangential or involute inlets.

Cyclone specification and selection are mainly accomplished by reference to manufacturers' data and known designs. The two primary performance parameters are the pressure drop and the fractional-efficiency curve. The pressure difference between the inlet and outlet, typically $1/2$ to 1 in of water (0.13 to 0.25 kPa), is a measure of the energy consumption.

Cyclones collect larger particles more easily than smaller ones. The *efficiency* of the separation is the fraction by weight of particles collected compared to the total amount in the feed. The fractional-efficiency curves give the percent of particles collected versus particle size. Particles with a nominal diameter equal to the *separation mesh*, d_{50}, will be separated from the air flow with a collection efficiency of 50%.

Efficiency is proportional to the rotational velocity (approximately 3000 ft/min (15 m/s) for dust-collection applications), which is itself proportional to the pressure drop. Both are proportional to the power put into the air stream.

12. HYDROCYCLONES

Hydrocyclones (also known as *hydraulic cyclones* and *hydroclones*) are used to separate solids suspended in liquid streams. They are common in ore processing plants because of their compactness and economy of operation.[29]

The primary cyclone design characteristic is the cyclone diameter, and the primary input variable is the *separation mesh* (or *cut point*), d_{50}—the particle diameter that is captured with an efficiency of 50%. Other factors affecting cyclone design and selection are feed flow rate, operating pressure, and pressure drop.

Like air cyclones, hydrocyclones consist of a cylindrical section joined to a conical section. The suspension is pumped tangentially into the upper part of the cylindrical section. Centrifugal force acts on solid particles, and the most coarse particles exit in the *underflow* at the bottom with a small amount of liquid. The *overflow* consists of most of the liquid and the finer particles. Half of the particles with diameters equal to d_{50} will appear in the overflow and half will appear in the underflow. The overflow is removed through a cylindrical pipe known as the *vortex finder*.

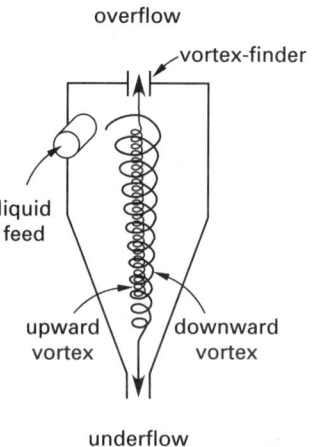

Figure 64.8 *Hydrocyclone*

It is common for manufacturers of hydrocyclones to supply performance data based on separating sand from water. The separation efficiency for other solids and liquids can be calculated from Eq. 64.16. The density terms can be replaced by specific gravity terms (1.00 for water, 2.65 for sand, and 1.4 for coal).

$$d_{50,\text{actual}} = d_{50,\text{sand-water}} \sqrt{\frac{\rho_{\text{sand}} - \rho_{\text{water}}}{\rho_{\text{solid}} - \rho_{\text{liquid}}}} \qquad 64.16$$

The pressure needed to maintain a steady circulation (i.e., velocity head) inside the cyclone and to overcome the friction losses is normally provided by a centrifugal pump. Pressure drop is the only major operating cost, and it is probably not economical to exceed a loss of 5 atm. Smaller units typically operate with a loss of 4 to 5 atm, while the loss in larger units may be as low as 1 atm.

13. SIEVES AND SCREENS

Sieving is the process of separating different sizes of the same material. Vibratory shakers and air-jet fluidization are used to increase the performance of sieving. Sieving is practical for particles down to approximately 30 to 100 microns.

In the past, wet classification was accomplished with stationary and vibrating screens. This method has high operating costs, however, and has been essentially abandoned in favor of hydrocyclones.

14. AIR CLASSIFIERS

Classification is the process of separating different grades or sizes of the same material, and often follows a grinding process.[30] *Air classifier mills* combine the processes of classification and grinding, recycling the coarse

---

[28] *Cyclone steam separators* separate different water phases (e.g., steam and liquid droplets).

[29] Other methods for separating solids from liquids include settling with and without floc (sedimentation) and centrifuging. Centrifuging is more appropriate for particles in the micrometer size range.

[30] *Fractioning* is the process of separating a material into a variety of specifically sized groups.

material for further grinding. Particles and minerals most suitable for air classification are uniform, homogeneous, spherical, smooth, and dry.[31]

Air classifiers are primarily used to grade bulk powders. The *selectivity* or *efficiency* is the probability of a particle entering the coarse stream. The *selectivity curve* is a curve of particle size versus selectivity. The *cut point*, designated d_{50}, is the midpoint of the selectivity curve. This is the size of particles that appear in equal numbers in the fine and coarse streams. The *bandwidth* is the range of particle sizes captured.[32] The cutoff values are the top-size and bottom-size limits.[33]

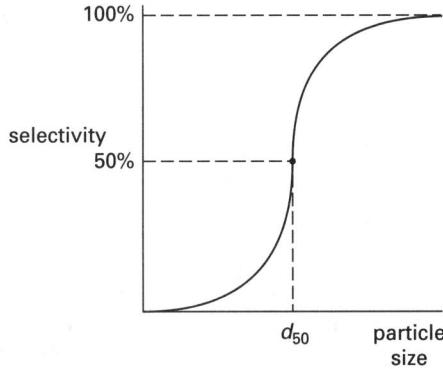

Figure 64.9 *Typical Selectivity Curve*

There are many ways to describe the *sharpness*, a measure of how tight the particle-size distribution is around the cut point. The *yield* or *recovery* is the fraction by weight of the end product in the desired size range, compared to the total amount of material in the feed.

Air classification encompasses elutriation, free vortex classification, and forced vortex classification. *Elutriation* is a crude method of separating most of the fine and coarse particles. Feed is introduced into the airflow, which raises the fine particles but allows the coarse particles to fall. Elutriation is effective in the same range of sizes as sieving but has a greater throughput and is more flexible. The cut point depends on the air velocity.

Free vortex classification is the process that occurs in a traditional cyclone. *Forced vortex classification*, also known as *centrifugal classification* and *mechanical classification*, is used to obtain the most precise particle size control. The cut points of centrifugal classifiers are in the 1 to 100 micron range.

Depending on the classifier design, a rotating rotor (i.e., the *rejector* or *classifier wheel*) forces the feed stream into circular motion. The rotor is constructed from blades, rods, or spokes, or as a hollow perforated disk to allow smaller particles closer to the rotational axis to escape to the central outlet. Coarse particles are thrown to the outer housing and collected. *Sidedraft classifiers* perform better in all respects (i.e., yield, product recovery, and efficiency) than conventional gravity-fed *updraft classifiers*.

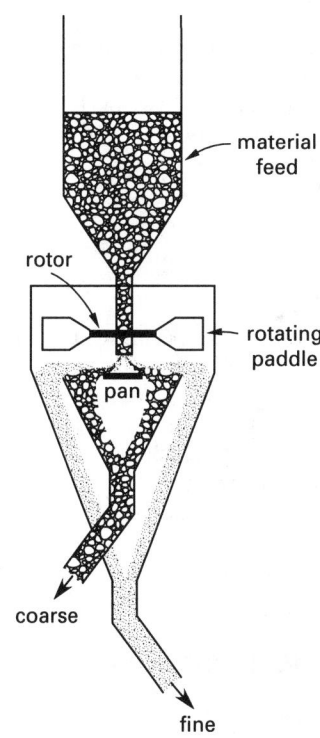

Figure 64.10 *Mechanical Air Classifier*

The separation can be adjusted by varying the rotor speed and air flow. Rejection increases with rotor speed. Rejection also can be increased by increasing the number of blades or rods in the rejector. Therefore, an increase in rejection increases fines but decreases the yield. Yield, in turn, can be increased by increasing the air velocity, but this makes the product more coarse and increases the pressure drop.[34]

Operating power requirements of disk rotor-type of classifiers are not easily modeled. For steady-state operation, the power calculations have been based on the rotational mass moment of inertia ($I = mr^2$) of the rotor and the start-up time. Using a typical starting time (t_{start}) of 10 sec, the calculated steady-state power is conservative.

$$P = \frac{I\omega^2}{2t_{\text{start}}} \qquad \text{[SI]} \qquad 64.17(a)$$

$$P = \frac{I\omega^2}{2g_c t_{\text{start}}} \qquad \text{[U.S.]} \qquad 64.17(b)$$

Equation 64.17 is generally not applicable to higher-efficiency classifiers that draw more power while operating at higher speeds in more dense particle clouds.

---

[31]Grinding, drying, solvent removal, extraction, homogenizing, dispersion, or deagglomeration processes may be necessary before air classification.

[32]Some materials, such as chromatographic material and toners, have bandwidths of 1 to 5 microns.

[33]The *cut-off values* and *cut-size* are different quantities.

[34]Pressure drop is proportional to the square of the air velocity.

For these classifiers, the power requirement consists of two components: the power to distribute the feed and the power to overcome the drag force (i.e., friction between the rotor and the particle cloud).

$$P_t = P_{\text{feed}} + P_{\text{drag}} \qquad \textit{64.18}$$

The power to distribute the feed is approximated by calculating the kinetic energy imparted to the feed. Particle velocities cannot exceed the rotor's peripheral velocity, so that value can be used to conservatively estimate power requirements if the distribution of particle velocities is not known. $\dot{m}$ is the material feed rate.

$$P_{\text{feed}} = \frac{\dot{m}\text{v}^2}{2} \qquad [\text{SI}] \qquad \textit{64.19(a)}$$

$$P_{\text{feed}} = \frac{\dot{m}\text{v}^2}{2g_c} \qquad [\text{U.S.}] \qquad \textit{64.19(b)}$$

The power dissipated in overcoming the drag force depends on the number of vanes or rods in the rotor, n, and the effective peripheral velocity, v_e. The effective peripheral velocity is the resultant of the rotor's peripheral velocity and the air velocity of the feed. The drag force is calculated by traditional means using the drag coefficient of each vane or rod.

$$P_{\text{drag}} = nF_{\text{drag}}\text{v}_e \qquad \textit{64.20}$$

15. SUSPENSIONS OF SOLIDS

Many industries use rotating impellers to keep solid material in suspension. The blades of *radial-flow impellers* (*paddle-type impellers, turbine impellers*, etc.) are parallel to the drive shaft. *Axial-flow impellers* (*propellers, pitched-blade impellers*, etc.) have blades inclined with respect to the drive shaft. Though radial-flow impellers were used extensively in the past, axial-flow impellers are almost always used today.

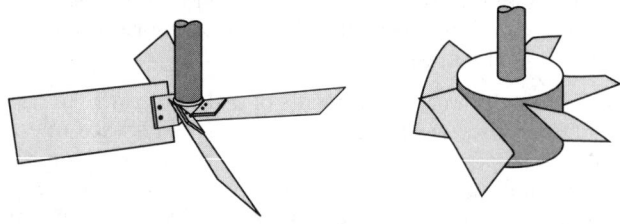

Figure 64.11 *Typical Axial Flow Mixing Impellers*

The *mixing Reynolds number* depends on the impeller diameter, d, suspension specific gravity, SG, liquid viscosity, μ, and rotational speed, n (in rps). Flow is in transition for $10 < \text{Re} < 10{,}000$, laminar below 10, and turbulent above 10,000.

$$\text{Re} = \frac{d^2 n \rho}{\mu} \qquad [\text{SI}] \qquad \textit{64.21(a)}$$

$$\text{Re} = \frac{d^2 n \rho}{g_c \mu} \qquad [\text{U.S.}] \qquad \textit{64.21(b)}$$

The dimensionless *power number*, N_P, of the impeller is defined by the power required to drive the impeller. (For any given level of suspension, Eqs. 64.22 through 64.31 apply to geometrically similar impellers and turbulent flow.[35])

$$P = N_P n^3 d^5 \rho \qquad [\text{SI}] \qquad \textit{64.22(a)}$$

$$P = \frac{N_P n^3 d^5 \rho}{g_c} \qquad [\text{U.S.}] \qquad \textit{64.22(b)}$$

$$P = \rho g Q h_{\text{v}} \qquad [\text{SI}] \qquad \textit{64.23(a)}$$

$$P = \frac{\rho g Q h_{\text{v}}}{g_c} \qquad [\text{U.S.}] \qquad \textit{64.23(b)}$$

The impeller's dimensionless *flow number*, N_Q, is defined by the flow rate equation.

$$Q = N_Q n d^3 \qquad \textit{64.24}$$

The velocity head can be calculated from a combination of the power and flow numbers.

$$h_{\text{v}} = \frac{N_P n^2 d^2}{N_Q g} \qquad \textit{64.25}$$

The *force number*, N_F, is defined by the force equation. F is the peak transverse force in normal operation. The upper shaft bending moment, M, is calculated by multiplying the peak force by the shaft length. (Units of area may be incorporated into N_F.)

$$F = \frac{N_F N_P \rho n^2 d^2}{g_c} \qquad \textit{64.26}$$

$$M = FL \qquad \textit{64.27}$$

A key equation for evaluating impeller design is the ratio of power to flow rate, obtained by dividing Eq. 64.23 by Eq. 64.24.

$$\frac{P}{Q} = \left(\frac{N_P}{N_Q}\right) n^2 d^2 \rho \qquad [\text{SI}] \qquad \textit{64.28(a)}$$

$$\frac{P}{Q} = \frac{\left(\dfrac{N_P}{N_Q}\right) n^2 d^2 \rho}{g_c} \qquad [\text{U.S.}] \qquad \textit{64.28(b)}$$

Equations 64.23 and 64.24 can be combined in other ways to express the P/Q ratio in terms of the fixed variables. If Q and d are fixed, rotational speed and power are variable. The ratio N_P/N_Q^3 characterizes the solids suspension performance of the impeller. Power is proportional to N_P/N_Q^3.

$$\frac{P}{Q} = \frac{\left(\dfrac{N_P}{N_Q^3}\right) Q^2 \rho}{d^4} \qquad [\text{SI}] \qquad \textit{64.29(a)}$$

$$\frac{P}{Q} = \frac{\left(\dfrac{N_P}{N_Q^3}\right) Q^2 \rho}{d^4} \qquad [\text{U.S.}] \qquad \textit{64.29(b)}$$

[35]Different levels of suspension range from being merely in motion (but not off the bottom) to being uniformly dispersed throughout the liquid.

For suspensions produced in small tanks, the ratio of the impeller diameter to tank diameter, d/d_t, is a characteristic in many relationships. For common d/d_t ratios (e.g., 0.25 to 0.50) and a specific impeller design, Eqs. 64.30 and 64.31 apply.[36]

$$P \propto \left(\frac{d}{d_t}\right)^{-\frac{1}{5}} \qquad \text{64.30}$$

$$T \propto \left(\frac{d}{d_t}\right)^{\frac{2}{3}} \qquad \text{64.31}$$

Vibration near the critical speed can be a major problem with low d/d_t ratios and with modern, high-efficiency, high-speed impellers. Mixing speed should be well below the first critical speed of the shaft. Other important design factors include tip speed and shaft bending moment.

---

[36]There is a risk of the impeller being "sanded in" if power to the mixer is lost. A considerably higher torque than predicted by Eq. 64.31 will be required to get the impeller moving again. The motor, shaft, and impeller design must take this into consideration.

65 Fire Protection Systems[1,2]

1. Schedule-Designed versus Hydraulically Designed Sprinkler Systems 65-1
2. Tree versus Loop Sprinkler Systems 65-1
3. Sprinkler System Diagrams 65-2
4. Supply Pumps for Fire Fighting 65-3
5. Sprinkler Head Characteristics 65-4
6. Sprinkler Discharge Characteristics 65-5
7. Sprinkler Protection Area 65-6
8. Discharge Density 65-6
9. Distance Between Sprinklers 65-6
10. Friction Losses 65-7
11. Minor Losses 65-7
12. Velocity Pressure 65-8
13. Normal Pressure 65-8
14. Hydraulic Design Concepts 65-8
15. Hydraulic Design Procedure: Pressure Along Branches 65-9
16. Hydraulic Design Procedure: Pressure at Cross Mains 65-12
17. Hydraulic Design Procedure: Pressure in Risers 65-13

Nomenclature[3]

A	area	ft^2	m^2
C	Hazen-Williams coefficient	–	–
d	internal pipe diameter	in	mm
g	acceleration due to gravity	ft/sec^2	m/s^2
h	height of fluid column	ft	m
K	orifice constant	gpm-$\sqrt{\text{psi}}$	L/s$\sqrt{\text{kPa}}$
L	length or distance	ft	m
m	mass	lbm	kg
$\dot{m}$	mass leak rate	lbm/sec	kg/s
n	number of sprinklers	–	–
p	absolute pressure	psfa	kPa
p	gauge pressure	psig	kPa
q	water flow rate increment	gpm	L/s
Q	flow rate	gpm	L/s
t	time	sec	s
v	velocity	ft/sec	m/s
V	volume	gal	L
$\dot{V}$	volumetric flow rate	gal/sec	L/s

Symbols

γ	specific weight	lbf/ft^3	n.a.
ρ	mass density	lbm/ft^3	kg/m^3

Subscripts

d	discharge	n	normal
e	elevation difference	t	total
f	friction	v	velocity

1. SCHEDULE-DESIGNED VERSUS HYDRAULICALLY DESIGNED SPRINKLER SYSTEMS

Pipes used in sprinkler systems can be sized in one of two ways. Before calculators and computers made sizing calculations simpler, the traditional manner was to use a *pipe schedule*—a table dictating the maximum number of fire sprinklers that can be served by any size of pipe. Different occupancy hazards call for different schedules. This method of design has been in use for more than a century.

Most contemporary sprinkler system designs (and all designs for deluge and water spray systems) are based on hydraulic calculations. The total water supply requirements for hydraulically designed systems are lower than those for schedule-designed systems. Also, pipe sizes can be reduced (down the run) in hydraulically designed systems. For these two reasons, hydraulically designed systems are generally more economical to install.

2. TREE VERSUS LOOP SPRINKLER SYSTEMS

Pipe networks can be designed as tree, gridded, or loop systems. With common *tree systems*, the first sprinkler to operate discharges at a greater-than-design rate due to the nature of the flow's declining pressure. *Loop and gridded systems* cannot eliminate the flow-related friction loss, but by providing multiple (parallel) paths to an open sprinkler, the friction loss can be reduced or minimized. Therefore, sprinkler pipe sizes can be smaller (or sprinkler spacing can be larger) with loop

---

[1]This chapter is an abridgement of Part 4, "Design of Water Sprinkler Systems," *Fire and Explosion Protection Systems* (Lindeburg), published by Professional Publications, Inc. This chapter only covers the basics of designing the piping system, a very small part of relevant fire protection engineering. *Fire and Explosion Protection Systems* includes chapters covering classification of buildings and equipment, detection and warning systems, fire protection methods, explosion protection systems, numerous additional worked examples, annotated references to the various code sections, and additional appendices.

[2]The authoritative references for subjects in this chapter are NFPA 13 ("Installation of Sprinkler Systems") and NFPA 20 ("Centrifugal Fire Pumps"), both of which are recommended.

[3]Most modern fire protection design is done in SI units. However, the major reference (*Standard for the Installation of Sprinkler Systems*, NFPA 13) uses bars as the unit of pressure, so this non-SI metric unit is included in this chapter.

systems than with tree systems. Gridded systems are commonly used when the protected area is large and roughly rectangular.

Although loop and gridded designs have a slight economic benefit because they need fewer sprinklers or smaller piping, they require more piping and more complex design calculations. The cost of materials may be slightly greater for loop systems than for tree systems. With modern computer design methods, design costs are probably not an important factor.

A *circulating closed-loop system*, which is a permitted application, is a wet pipe-looped sprinkler system in which the pipes have a second purpose—usually to carry heating or cooling water. Such systems are also known as *automatic sprinkler systems with nonfire protection connections*. Water is only circulated through the system; it is not used or removed from the system.

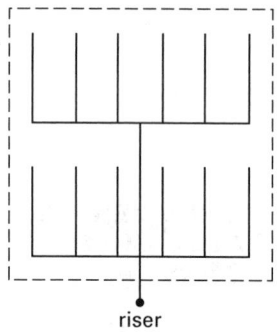

(a) tree system (side central feed shown)

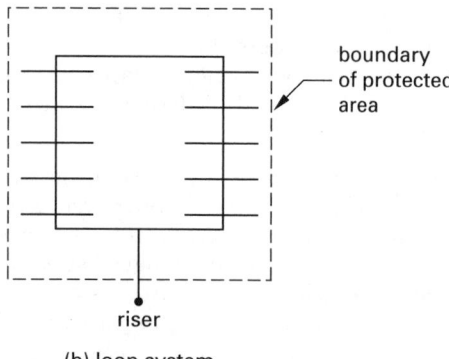

(b) loop system

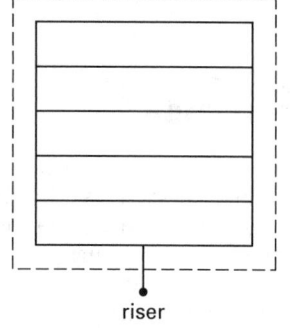

(c) gridded system

Figure 65.1 *Tree, Loop, and Gridded Pipe Systems*

3. SPRINKLER SYSTEM DIAGRAMS

Descriptions of sprinkler systems make use of the following component terms. Other terms are illustrated in Fig. 65.2. Diagrams of sprinkler systems make use of the abbreviations listed in Table 65.1.

- *branch line*: a pipe containing sprinklers crossing a cross main

- *cross main*: a pipe directly supplying cross lines containing sprinklers

- *feed main*: a supply riser or a cross main

- *riser*: a pipe, approximately vertical, usually extending one full story

- *system riser*: an above-ground pipe, approximately vertical, bringing water from the supply source

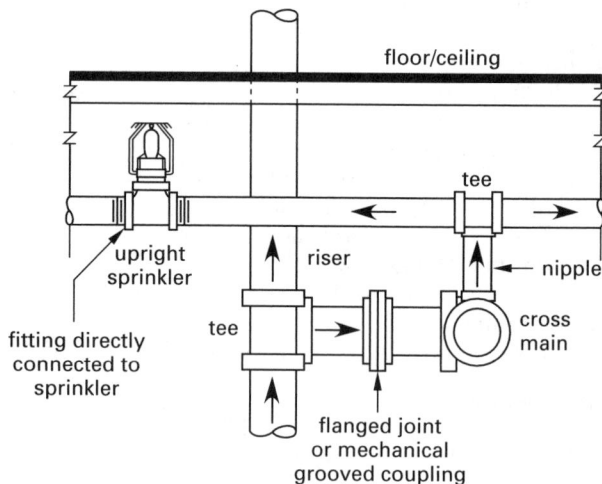

Figure 65.2 *Typical Sprinkler System Components*

Table 65.1 *Standard Abbreviations for Sprinkler Diagrams*

ALV	alarm valve
BV	butterfly (wafer) check valve
Cr	cross
CV	swing check valve
Del V	deluge valve
DPV	dry-pipe valve
EP	90° elbow
EE	45° elbow
GV	gate valve
Lt.E	long-turn elbow
OS&Y	outside stem (screw) and yoke manual control valve
PIV	post indicator valve (control valve)
St	strainer
TP	tee, flow turned 90°
WCV	butterfly (wafer) check valve

4. SUPPLY PUMPS FOR FIRE FIGHTING[4]

Water from tanks and reservoirs usually flows under gravity action to its fire protection systems. However, depending on the elevations, supply pumps may be required. In some cases, the municipal system can supply adequate volume but at too low a pressure. In such cases, booster pumps can be used to boost the pressure of the municipal water supply. All pumps used in fire protection systems must be approved and/or listed.

Approved fire pumps are almost always horizontal (centrifugal) pumps. However, vertical (centrifugal or turbine) pumps are also used depending on which is more economical and appropriate. Pumps may be single- or multiple-stage. Horizontal pumps must operate with positive suction head. This is particularly important with remote starting. Horizontal pumps must be split-case, end-suction, or in-line types. Single-stage, end-suction, and in-line pumps are limited to under 500 gpm.

To avoid loss of prime and other suction lift problems, *vertical submersible turbine pumps* (i.e., "sump pumps") with submerged impellers can be used when drawing water from a deep well or sump. Vertical pumps can operate without priming, and they are used in ponds, streams, and pits. However, the elevation difference between the source (when pumping at 150% of rated capacity) and the ground surface is limited to 200 ft.

Pumps may be driven by diesel engines, but electric drives are preferred because of their simplicity. Electric power must be available from a reliable source (i.e., one experiencing fewer than two outages per year) or from two independent sources. Spark-ignited gasoline-powered engines should not be used.

Pumps should start automatically unless other sources are capable of simultaneously supplying water for all firefighting and industrial demands. Each pump should have a manual shutdown switch.

A *pressure-maintenance pump* (also known as a *jockey pump* or a *makeup pump*) is usually a low-volume, electrically driven centrifugal pump. The impeller discharge is directly into the water sprinkler line. Operation is triggered by the initial water pressure drop resulting from the opening of a few sprinklers or other leakage. The automatic controller starts the primary fire pump when the capacity of the pressure-maintenance pump is exceeded.

There are three limiting points on the pump head-discharge curve:

- *churn* or *shutoff*, where the pump operates at rated speed with the discharge valve closed

- *rated (100%) capacity* and *rated (100%) head*

- *overload*, 65% of the rated head

[4]NFPA 20, *Centrifugal Fire Pumps*, is the accepted authority on this subject.

In selecting horizontal and vertical shaft turbine-type pumps for fire supply use, the churn (shutoff) head must not be greater than 140% of the rated head. This requirement eliminates pumps with high shutoff pressures and steep curves. The capacity at overload should not be less than 150% of the rated capacity. Figure 65.3 illustrates the pump characteristics that are described by these requirements.

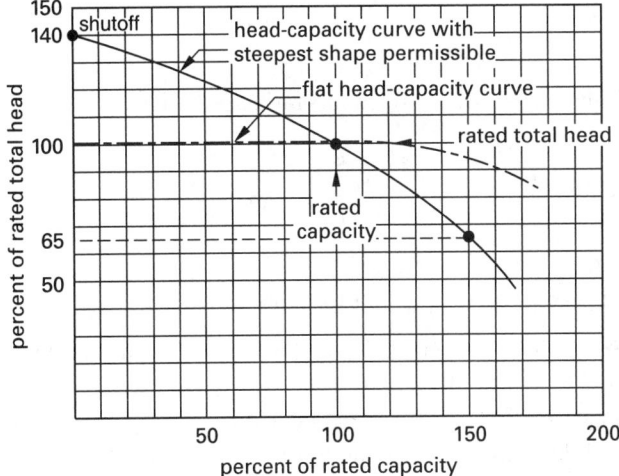

Reprinted with permission from NFPA 20, *Installation of Centrifugal Fire Pumps*, Copyright © 1990, National Fire Protection Association, Quincy, MA 02269. This reprinted material is not the complete and official position of the National Fire Protection Association, on the referenced subject which is represented only by the standard in its entirety.

Figure 65.3 *Standard Fire Protection Pump Curve*

Standard rated capacities of approved horizontal and vertical fire pumps are 25, 50, 100, 150, 200, 250, 300, 400, 450, 500, 750, 1000, 1250, 1500, 2000, 2500, 3000, 3500, 4000, 4500, and 5000 gpm (95, 189, 379, 568, 757, 946, 1136, 1514, 1703, 1892, 2839, 3785, 4731, 5677, 7570, 9462, 11 355, 13 247, 15 140, 17 032 and 18 925 L/min). Pumps are rated at net pressures in excess of 40 psi (280 kPa; 2.8 bars).

Pressure ratings range from (approximately) 40 to 200 psi (280 kPa to 1.40 MPa; 2.8 to 14 bars) for horizontal pumps and from 75 to 280 psi (520 kPa to 1.9 MPa; 5.2 to 19 bars) for vertical pumps. Pressure ratings include the pressure boost of all stages in multistage pumps.

It is general practice to choose a pump for fire protection systems based on overload capacity. This is illustrated in Ex. 65.1. However, some local authorities may require a pump to be sized below overload (i.e., with a higher capacity).

Example 65.1

A fire protection system requires 1300 gpm (82 L/s), including hose streams of water at 80 psi (550 kPa). A suction tank is the source of the water. The equivalent pressure corresponding to the suction lift is 4 psi (28 kPa; 0.28 bars). What are the capacity and pressure ratings for an appropriate pump?

SI Solution

Meet the 82 L/s demand with the pump's overload (i.e., 150%) capacity. A trial rated capacity is

$$Q = \frac{82 \frac{L}{s}}{1.50} = 54.7 \text{ L/s}$$

The nearest standard pump rating is 63 L/s.

$$\frac{82 \frac{L}{s}}{63 \frac{L}{s}} = 1.30$$

Therefore, 82 L/s would be 130% of capacity.

From Fig. 65.3, at 130% of capacity the total pressure is 80% of the rated pressure.

The total pump pressure is divided between the 550 kPa sprinkler and 28 kPa suction lift pressures. The total net pressure is

$$p_{net} = 550 \text{ kPa} + 28 \text{ kPa} = 578 \text{ kPa}$$

The rated pressure at 63 L/s is

$$p_{rated} = \frac{578 \text{ kPa}}{0.80} = 723 \text{ kPa}$$

The pump rating should be 63 L/s at 723 kPa.

Customary U.S. Solution

Meet the 1300 gpm demand with the pump's overload (i.e., 150%) capacity. A trial rated capacity is

$$Q = \frac{1300 \text{ gpm}}{1.50} = 867 \text{ gpm}$$

The nearest standard pump rating is 1000 gpm. Therefore, 1300 gpm would be 130% of capacity.

From Fig. 65.3, at 130% of capacity the total pressure is 80% of the rated pressure.

The total pump pressure is divided between the 80 psi sprinkler and 4 psi suction lift pressures. The total net pressure is

$$p_{net} = 80 \text{ psi} + 4 \text{ psi} = 84 \text{ psi}$$

The rated pressure at 1000 gpm is

$$p_{rated} = \frac{84 \text{ psi}}{0.80} = 105 \text{ psi}$$

The pump rating should be 1000 gpm at 105 psi.

5. SPRINKLER HEAD CHARACTERISTICS

The most common sprinkler types are the *upright sprinkler* (which discharges water upward against the deflector), the *pendant sprinkler* (which discharges water downward against the deflector), and the *sidewall sprinkler* (which emits water in a one-quarter sphere spray away from the wall, with a small amount directed at the wall).[5] Figure 65.4 illustrates the uniform water distribution pattern that is characteristic of sprinklers in use since 1953.

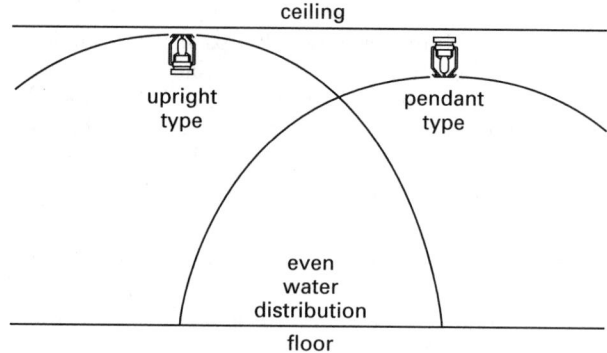

Figure 65.4 *Distribution Pattern from Ceiling Sprinklers*

Several types of sprinklers have been used in the past including fusible link, glass bulb, and soldier puck sprinklers. Modern sprinklers come in a variety of designs, but all sprinklers contain several important elements: a nozzle, a deflector, and a release mechanism. The release mechanism for the old-type *fusible-link* (also known as soldered *link-and-lever) sprinkler* is a soldered fusible link that separates at a specific temperature. Figure 65.5 illustrates a link-and-lever sprinkler.

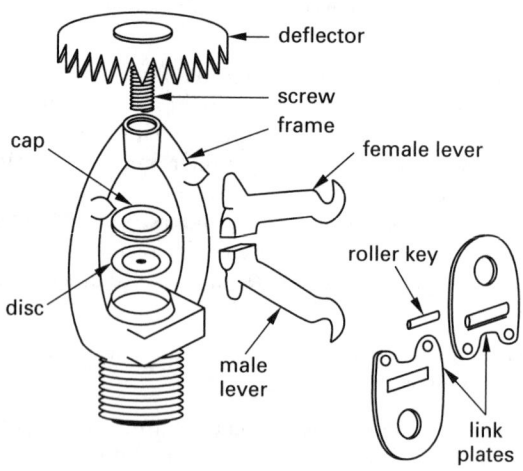

Figure 65.5 *Soldered Link-and-Lever Automatic Sprinkler*

---

[5]Older (i.e., pre-1953) sprinklers that discharge a significant fraction (i.e., 40%) of the water against the ceiling are no longer used except for special cases, such as wharves and other wood construction.

6. SPRINKLER DISCHARGE CHARACTERISTICS

The discharge from a sprinkler depends on the normal (i.e., static) pressure at the nozzle entrance. The minimum design pressure depends on the sprinkler head design and required coverage. Minimum pressure for any sprinkler is 7 psi (48 kPa; 0.48 bars), the pressure required to produce a 15-gpm flow (0.95 L/s; 57 L/min) in a standard $\frac{1}{2}$ in (12.7 mm) orifice sprinkler.

If the *orifice constant* (also known as *nozzle constant*, *orifice discharge coefficient*, or *K-factor*), K, is known, Eq. 65.1 determines the relationship between the flow rate, Q, in gpm (L/s), and normal pressure, p, in psi (kPa).

$$Q_{\text{L/s}} = K_{\text{SI}}\sqrt{p} \qquad \text{[SI]} \qquad \textbf{\textit{65.1(a)}}$$

$$Q = K\sqrt{p} \qquad \text{[U.S.]} \qquad \textbf{\textit{65.1(b)}}$$

$$Q_{\text{L/min}} = K_{\text{bars}}\sqrt{p_{\text{bars}}} \qquad [p \text{ in bars}] \qquad \textbf{\textit{65.1(c)}}$$

$$K_{\text{SI}} = 0.024K \qquad \text{[SI]} \qquad \textbf{\textit{65.2(a)}}$$

$$K_{\text{bars}} = 14K \qquad [\text{in bars}] \qquad \textbf{\textit{65.2(b)}}$$

Although there are many sprinkler designs, most have similar operating characteristics. For example, the standard heads have a 0.5 in (12.7 mm) orifice, and they discharge 15 gpm (0.95 L/s; 57 L/min) at 7 psi (48 kPa; 0.48 bars). This is equivalent to a discharge constant, K, of approximately 5.6 to 5.8 (0.134 to 0.140). The coefficient of discharge, C_d, is taken as 0.75 to 0.78. Figure 65.6 illustrates the discharge characteristics for standard 0.5 in (12.7 mm) and $^{17}/_{32}$ in (13.5 mm) orifices.

Table 65.2 *Sprinkler Discharge Characteristics*

nominal orifice size (in)	orifice type	K-factor	percent of nominal $\frac{1}{2}$ in discharge
$^1/_4$	small	1.3–1.5	25
$^5/_{16}$	small	1.8–2.0	33.3
$^3/_8$	small	2.6–2.9	50
$^7/_{16}$	small	4.0–4.4	75
$^1/_2$	standard	5.3–5.8	100
$^{17}/_{32}$	large	7.4–8.2	140
$^5/_8$	extra large	11.0–11.5	200
$^5/_8$	large drop	11.0–11.5	200
$^3/_4$	ESFR	13.5–14.5	250

Reprinted with permission from NFPA 13, *Installation of Sprinkler Systems*, Copyright © 1991, National Fire Protection Association, Quincy, MA 02269. This reprinted material is not the complete and official position of the National Fire Protection Association, on the referenced subject which is represented only by the standard in its entirety.

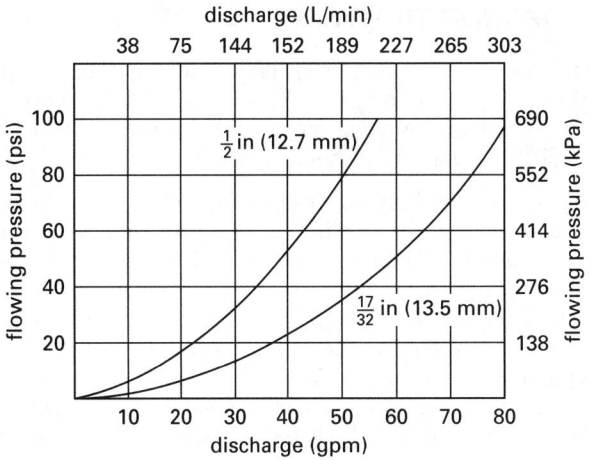

pressure at sprinkler psi (kPa)	approximate discharge gpm (L/min)	
	$\frac{1}{2}$ in (12.7 mm)	$\frac{17}{32}$ in (13.5 mm)
10 (69)	18 (68)	25 (95)
15 (103)	22 (83)	31 (117)
20 (138)	25 (95)	36 (136)
25 (172)	28 (106)	40 (151)
35 (241)	33 (125)	47 (178)
50 (345)	40 (151)	57 (216)
75 (517)	48 (182)	69 (261)
100 (690)	56 (212)	80 (303)

Reprinted with permission from *Fire Protection Handbook*, 17th Edition, Copyright © 1991, National Fire Protection Association, Quincy, MA 02269.

Figure 65.6 *Orifice Discharge versus Static Pressure ($^1/_2$ in and $^{17}/_{32}$ in diameter orifices)*

Heads with smaller orifices can be used for light hazard occupancies when hydraulically designed. Strainers are required for orifices less than $^3/_8$ in (9.5 mm) in diameter. Strainers are also a good idea with high-velocity flows. Strainers are often installed in risers.

Example 65.2

A sprinkler has a discharge constant of 5.6 (0.13 for liters and seconds). What is the discharge from the sprinkler if the normal pressure is 23 psi (160 kPa)?

SI Solution

$$Q = K\sqrt{p} = 0.13\sqrt{160 \text{ kPa}}$$
$$= 1.64 \text{ L/s}$$

Customary U.S. Solution

$$Q = K\sqrt{p} = 5.6\sqrt{23 \text{ psi}}$$
$$= 26.9 \text{ gpm}$$

7. SPRINKLER PROTECTION AREA

Theoretical coverage per sprinkler can vary from 60 to 400 ft^2 (5.4 to 36 m^2) of floor area. However, the common operating range is more in the order of 80 to 120 ft^2 (7.2 to 10.8 m^2) of floor area.

The spacing and layout design of sprinkler systems will be governed by the maximum floor area that one sprinkler can be expected to protect (i.e., the *protection area*). This, according to *Standard for the Installation of Sprinkler Systems* (NFPA 13), is in turn determined by two factors: the hazard classification and the type of ceiling construction. (Sidewall spray sprinklers have their own different area and spacing requirements.) Other sprinklers may be listed for larger coverage areas.

For example, light hazard areas as listed in Table 65.3 require one sprinkler for every 200 ft^2 (18 m^2) of floor area if the design is by schedule. The requirement is relaxed to one sprinkler for every 225 ft^2 (20.3 m^2) if the system is hydraulically calculated. For open-wood joist ceilings, one sprinkler is required for every 130 ft^2 (12 m^2).

When calculating numbers of sprinklers required for areas by dividing the floor area by the sprinkler coverage, fractional numbers are rounded up to the next higher integer. Similarly, when calculating numbers of sprinklers required for branches by dividing the design length by the sprinkler spacing, fractional numbers are rounded up to the next higher integer.

Table 65.3 Maximum Sprinkler Protection Areas

	light hazard	ordinary hazard	extra hazard[e]	high-piled storage[f]
unobstructed ceiling[a]	225 ft^2 [b]	130 ft^2	100 ft^2	100 ft^2
noncombustible obstructed ceiling	225 ft^2 [b]	130 ft^2	100 ft^2	100 ft^2
combustible obstructed ceiling	168 ft^2 [c,d]	130 ft^2	100 ft^2	100 ft^2

(Multiply ft^2 by 0.0929 to obtain m^2.)
(Multiply gpm/ft^2 by 40.75 to obtain L/min·m^2.)
[a] Unobstructed construction excludes wood truss construction.
[b] 200 ft^2 (18 m^2) when designed by schedule.
[c] 130 ft^2 (11.7 m^2) for light combustible framing members spaced less than 3 ft on center.
[d] 225 ft^2 (22.3 m^2) for heavy framing members spaced 3 ft or more on center.
[e] 90 ft^2 when designed by schedule; 130 ft^2 (11.7 m^2) when hydraulically designed and density is less than 0.25 gpm/ft^2 (10.2 L/min·m^2).
[f] 130 ft^2 (11.7 m^2) when hydraulically designed according to NFPA 231 and NFPA 231C and density is less than 0.25 gpm/ft^2 (10.2 L/min·m^2).

Reprinted with permission from NFPA 13, *Installation of Sprinkler Systems*, Copyright © 1991, National Fire Protection Association, Quincy, MA 02269. This reprinted material is not the complete and official position of the National Fire Protection Association, on the referenced subject which is represented only by the standard in its entirety.

Even though the number of sprinklers is determined assuming that all sprinklers will discharge at the same rate, this does not usually occur in practice. The discharge rate depends on the pressure at the nozzle, which decreases toward the last open sprinkler. This introduces a conservative element in the design. The first (by time) sprinkler to open will discharge at a higher-than-average design rate until other sprinklers open.

8. DISCHARGE DENSITY

Another design criterion, increasingly used in severe hazard areas (including high-piled storage and combustible liquid storage), calls for a discharge density in gpm/ft^2 (L/m^2·s). Values between 0.15 and 0.6 gpm/ft^2 (0.10 and 0.40 L/m^2·s) are typical.

Sizing supply lines using discharge density design criteria is known as the *area/density method*. When specified by code or building officials, the area/density method takes precedence over other design criteria.

In order to determine the discharge density, the design area must first be known. This can be specified by local ordinances, the local fire or building officials, insurance requirements, or the fire protection engineer. The design area is not the same as the coverage per sprinkler or the total room/building area. The total discharge from sprinklers in the design area is added to the hose allowance to determine the required water supply.

9. DISTANCE BETWEEN SPRINKLERS

The distance between sprinklers on lines should be approximately the same as the distance between lines. The maximum spacing between adjacent sprinkler heads and branch lines is generally 15 ft (4.5 m) for light and ordinary hazard occupancies and 12 ft (3.6 m) for extra hazard and high-piled storage occupancies. There are two rare exceptions to these spacing requirements: bays and low discharge densities. (Sidewall spray sprinklers have their own, different area and spacing requirements.)

The maximum distance to the nearest wall is generally one-half of the required spacing but not less than 4 in (10.2 cm).

Sprinklers that are spaced too close together may interfere with other sprinklers' activations. For example, the water discharge from one sprinkler may cool and delay or prevent an adjacent sprinkler from opening. If sprinklers are spaced less than every 6 ft (1.8 m), baffles are required between them. This limitation might be increased to 8 ft by the local authority or the insurance underwriter. NFPA 13 should be consulted for baffle requirements for other types of sprinklers, such as large drop and early suppression fast response (ESFR) sprinklers.

Structural members (including joists and trusses) and nonflat ceiling construction greatly complicate spacing and placement standards.

10. FRICTION LOSSES

The Hazen-Williams equation (Eq. 65.3) should be used to calculate the friction loss in hydraulically designed sprinkler systems. C is the Hazen-Williams coefficient, commonly known as the C-value. The actual (not nominal) pipe diameter should be used with Eq. 65.3.

$$p_{f,\text{kPa}} = \frac{(1.18 \times 10^8)L_\text{m}Q_{\text{L/s}}^{1.85}}{C^{1.85}d_{\text{mm}}^{4.87}} \quad \text{[SI]} \quad \textbf{65.3(a)}$$

$$p_{f,\text{psi}} = \frac{4.52L_\text{ft}Q_{\text{gpm}}^{1.85}}{C^{1.85}d_{\text{in}}^{4.87}} \quad \text{[U.S.]} \quad \textbf{65.3(b)}$$

Use Eq. 65.3(c) to calculate a pressure drop in bars from a flow rate in L/min.

$$p_{f,\text{bars}} = \frac{(6.05 \times 10^5)L_\text{m}Q_{\text{L/min}}^{1.85}}{C^{1.85}d_{\text{mm}}^{4.87}} \quad \text{[in bars]} \quad \textbf{65.3(c)}$$

Table 65.4 gives the Hazen-Williams C-values to be used. The local building or fire officials, however, may specify other values. The C-value is assumed to be constant for a specific pipe roughness and is independent of velocity.

Table 65.4 *Hazen-Williams C-Values for Sprinkler System Design*

type of pipe	C
unlined cast or ductile iron	100
black steel (for dry and preaction systems)	100
black steel (for wet and deluge systems)	120
galvanized (all uses)	120
plastic (listed)	150
cement-lined cast or ductile iron	140
copper tube or stainless steel	150

New unlined steel pipe has a C-value of 140. However, as the pipe ages, the friction will increase. For that reason, 120 is the value used for wet systems. Dry systems have an even greater tendency to develop deposits and corrosion. Therefore, the C-value for dry systems is taken as 100. Both of these assumptions are conservative. When the pipe is new, greater-than-design flows will be achieved.

Graphical solutions to the Hazen-Williams equation are permitted and are commonly used. Appendices 65.A and 65.B are graphical solutions for $C = 120$. Friction losses for other C-values can be calculated by multiplying the graphical friction loss by the appropriate factor from Table 65.5.

Table 65.5 *Multipliers for Other C-Values*

actual C-value	basis of graphical value	
	$C = 100$	$C = 120$
150	0.472	0.662
140	0.537	0.752
130	0.615	0.862
120	0.714	1.00
110	0.838	1.18
100	1.00	1.40
90	1.22	1.70
80	1.51	2.12
70	1.93	2.71
60	2.57	3.61

11. MINOR LOSSES

Ideally, all sources of friction, bends, valves, meters, and strainers should be recognized. However, in hydraulic calculations for sprinkler systems, only fittings involving changes in flow direction are included. Minor losses from straight-through run-of-tees and straight-across crosses are disregarded. Friction losses due to tapered reducers and reducers directly adjacent (attached) to spray nozzles are also disregarded.

Minor losses from tees at the top of risers are included with the cross mains, losses for tees at the bases of risers are included with the risers, and losses from crosses or tees at cross-main feed junctions are included with the cross mains.

For tees and reducing elbows, the equivalent length is based on the velocity and/or diameter of the smaller outlet.

Values for standard elbows are typically used for abrupt 90° turns constructed with threaded fittings. The long-turn elbow values are used with flanged, welded, or other mechanical connections.

Friction losses for fittings connected directly to sprinklers are omitted.

The equivalent lengths of valves and fittings for $C = 120$ are given in App. 65.C. Loss data for specialized elements (such as pressure-reducing valves, deluge valves, alarm valves, dry pipe valves, and strainers) must be obtained from the manufacturers.

Friction losses for other C-values can be calculated by multiplying the equivalent lengths in App. 65.C by the appropriate factor in Table 65.6. (These factors assume that the friction loss through the fitting is independent of the piping C-value.)

Plant Engineering

Table 65.6 *Multiplying Factors for Equivalent Lengths*

actual C-value	multiplying factor
100	0.713
120	1.00
130	1.16
140	1.33
150	1.51

12. VELOCITY PRESSURE

Velocity pressure, since it is small, may be omitted at the discretion of the designer. This normally introduces a conservative error. However, if the velocity pressure is much more than 5% of the total pressure, it should be considered. If considered, velocity pressure must be included in calculations for both nonlooped branch lines and cross mains.

$$p_{v,kPa} = \frac{9.81v^2}{2g} \quad \text{[SI]} \qquad \text{65.4(a)}$$

$$p_{v,psi} = \frac{0.433v^2}{2g} \quad \text{[U.S.]} \qquad \text{65.4(b)}$$

$$p_{v,bars} = \frac{0.0981v^2}{2g} \quad \text{[in bars]} \qquad \text{65.4(c)}$$

Even when velocity pressure is considered, however, the method used to include it is peculiar to sprinkler design. Specifically, the velocity pressure downstream of a sprinkler is used for end outlets, while velocity pressure upstream (on the supply side) is used for other outlets.

End outlets include the last sprinkler on a dead-end branch, the final flowing branch on a dead-end cross main, any sprinkler with a flow split on a gridded branch line, and any branch line with a flow split on a loop system.

When velocity pressure is based on the upstream flow rate (which is initially unknown), the velocity pressure is determined by trial and error. This procedure starts with an estimated flow rate for the upstream side of the nozzle. This flow rate is used to determine the trial velocity pressure, and in turn, the normal pressure and a new flow rate. The procedure is repeated until the estimated and calculated flow rates converge sufficiently.

In analysis problems that consider velocity pressure, the discharge from the next-to-last sprinkler in a line may be lower than the last sprinkler. This condition may or may not require attention. If the discharge density from the next-to-last sprinkler exceeds the minimum and is less than approximately 3% of the total design flow, the design probably should be kept. Otherwise, the pipe supplying the end sprinkler should be increased in size.

Example 65.3

The flow through a 1 in (nominal) schedule-40 branch line is 36 gpm (2.3 L/s). What is the velocity pressure?

SI Solution

The internal cross-sectional area of a 1 in line is 5.574×10^{-4} m^2.

The velocity is

$$v = \frac{\dot{V}}{A} = \frac{\left(2.3 \frac{\text{L}}{\text{s}}\right)\left(0.001 \frac{\text{m}^3}{\text{L}}\right)}{5.574 \times 10^{-4} \text{ m}^2}$$

$$= 4.13 \text{ m/s}$$

From Eq. 65.4, the velocity pressure is

$$p_v = \frac{9.81v^2}{2g}$$

$$= \frac{\left(9.81 \frac{\text{kPa}}{\text{m}}\right)\left(4.13 \frac{\text{m}}{\text{s}}\right)^2}{(2)\left(9.81 \frac{\text{m}}{\text{s}^2}\right)}$$

$$= 8.53 \text{ kPa}$$

Customary U.S. Solution

The internal cross-sectional area of a 1 in line is 0.0060 ft^2.

The velocity is

$$v = \frac{\dot{V}}{A} = \frac{(36 \text{ gpm})\left(0.002228 \frac{\text{ft}^3}{\text{sec-gpm}}\right)}{0.0060 \text{ ft}^2}$$

$$= 13.37 \text{ ft/sec}$$

From Eq. 65.4, the velocity pressure is

$$p_v = \frac{0.433v^2}{2g}$$

$$= \frac{\left(0.433 \frac{\text{psi}}{\text{ft}}\right)\left(13.37 \frac{\text{ft}}{\text{sec}}\right)^2}{(2)\left(32.2 \frac{\text{ft}}{\text{sec}^2}\right)}$$

$$= 1.20 \text{ psi}$$

13. NORMAL PRESSURE

The *normal pressure*, usually referred to in other subjects as the *static pressure*, is the difference between the total pressure and the velocity pressure.

$$p_n = p_t - p_v \qquad \text{65.5}$$

14. HYDRAULIC DESIGN CONCEPTS

Hydraulic calculations, whether for analysis or design, can be performed by hand or by computer. However, to standardize designs, certain conventions (which follow) have been established. This includes standard-ized terminology (i.e., *normal pressure* instead of *static pressure*) and nomenclature.[6] Alternatively, more so-phisticated methods may be used. However, it may be more difficult to justify or explain these methods to the local officials.

Analysis calculations of sprinkler systems start at the hydraulically most-remote nozzle using the minimum nozzle pressure (e.g., 20 psig). A worksheet similar to App. 65.D must be used. This most-remote nozzle is not necessarily the most-distant nozzle. Rather, it is the nozzle whose supply-to-nozzle path experiences the greatest friction loss. It may take several trial sets of calculations to verify which sprinkler is the most remote hydraulically.

To begin the design, either the pressure or the discharge quantity must be known for the most-remote nozzle. If the pressure is known, the discharge quantity is found from the pressure-volume curve for the sprinkler. If the density requirement is known, it can be used to calculate the discharge quantity. The pressure at the nozzle is then found from the pressure-volume curve. Calculations of normal pressure then proceed, fitting by fitting, back to the point of water supply.

When there is only one sprinkler, as there is for typ-ical protection of a closet or washroom, a reasonable assumption is that the sprinkler will be independent of the other sprinklers in the main room. As long as the sole sprinkler can discharge at the necessary rate, it can be omitted from the hydraulic calculations for main ar-eas greater than 1500 ft^2 (135 m^2).

Branch calculations should be performed separately (i.e., on their own sheets). Calculations are normally omitted for identical and symmetrical branches.

Sprinkler systems supplied by loops are more difficult to analyze and design. Although looped systems can be designed by hand using the Hardy-Cross method, it is more expedient to use computerized methods.

15. HYDRAULIC DESIGN PROCEDURE: PRESSURE ALONG BRANCHES

The following analysis procedure starts at the hydrauli-cally most-distant sprinkler. Note the difference in how velocity pressure is calculated for the third-to-last sprin-kler in a branch and beyond.

step 1: Calculate the discharge rate (in gpm) for the last sprinkler from the area and density or from the normal pressure and discharge characteristics, whichever is known. Either Eq. 65.1 or graphical characteristics can be used. (No sprinkler may discharge at less than the minimum design rate.)

step 2: Using the flow rate, the pipe length to the next sprinkler, and the *C*-value, determine the Hazen-Williams friction pressure loss (in psi). Either Eq. 65.3, App. 65.A, or App. 65.B may be used. If the Hazen-Williams equation is solved graphically, convert the loss to the value corresponding to the actual *C*-value.

step 3: Add the friction pressure loss from step 2 to the normal pressure from step 1. This is the total pressure available at the next-to-last sprinkler.

step 4: If it is to be included in the calculations, calculate the velocity pressure (in psi) in the section of pipe used in step 2 from Eq. 65.4.

step 5: Subtract the velocity pressure from the to-tal pressure. This is the normal pressure available at the next-to-last sprinkler. No-tice that the downstream, not upstream, ve-locity pressure is used with the next-to-last sprinkler.

step 6: Calculate the discharge from the next-to-last sprinkler from the normal pressure and sprin-kler discharge characteristics. (No sprinkler may discharge at less than the minimum de-sign rate.)

step 7: Using the cumulative flow rate and pipe length to the next sprinkler, determine the Hazen-Williams friction pressure loss (in psi). Convert pressure loss to proper *C*-value if necessary.

step 8: If velocity pressure is to be considered, cal-culate it from the flow rate downstream of the next sprinkler.

step 9: Calculate the total pressure at the next up-stream sprinkler by adding the normal pres-sure at the previous sprinkler, the friction pressure from step 7, and the velocity pres-sure from step 8.

step 10: Estimate the sprinkler discharge for the next upstream sprinkler. (Finding the dis-charge from the third-to-last and all subse-quent sprinklers is an iterative process. The upstream velocity is used with all sprinklers except the next-to-last. Since the upstream velocity is unknown, it must be estimated. While the estimated variable could be either the upstream velocity or the upstream flow rate, it is common (and easier) to estimate the sprinkler discharge.)

---

[6]The conventions and standardized nomenclature in this chapter are taken from *Standard for the Installation of Sprinkler Systems* (NFPA 13).

step 11: Calculate the flow rate in the upstream pipeline by adding the estimated sprinkler discharge and the cumulative discharge from subsequent sprinklers.

step 12: Calculate the upstream velocity pressure from the flow rate determined in step 11.

step 13: Calculate the normal pressure by subtracting the velocity pressure calculated in step 12 from the total pressure calculated in step 9.

step 14: Determine a corrected discharge from the sprinkler using the normal pressure calculated in step 13. (No sprinkler may discharge at less than the minimum design rate.)

step 15: Compare the discharge assumed in step 9 with the corrected discharge calculated in step 14. If they are sufficiently the same, repeat steps 7 through 15 for all remaining sprinklers. If the two values are different, estimate a new discharge and repeat steps 10 through 15.

Example 65.4

The hydraulically most-remote branch of a sprinkler system consists of a 1 in (nominal) schedule-40 steel pipe and three sprinklers with standard $1/2$ in diameter orifices. The last sprinkler is at the end of the branch, and the distance between sprinklers is 10 ft. The minimum pressure to any sprinkler is 10 psig. The discharge constant for the sprinklers is 5.6. What is the volume of water discharged from each of the three open sprinklers?

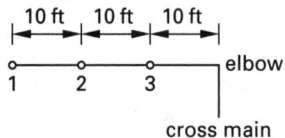

Solution

Follow the procedure in Sec. 15.

step 1: The normal (static) pressure at sprinkler 1 is 10 psig. From Eq. 65.1, the discharge from sprinkler 1 is

$$Q_1 = K\sqrt{p_{n,1}} = \left(5.6\frac{\text{gpm}}{\sqrt{\text{psig}}}\right)\sqrt{10 \text{ psig}}$$

$$= 17.71 \text{ gpm}$$

step 2: The Hazen-Williams coefficient for steel pipe is $C = 120$. From App. 65.A, the friction loss in 1 in of pipe with a flow rate of 17.7 gpm is approximately 0.10 psig per foot. The pressure loss in 10 ft of pipe is 1.0 psig.

Alternatively, Eq. 65.3 can be used. The inside diameter of a 1 in pipe is 1.049 in. The friction loss between sprinklers 1 and 2 is

$$p_{f,1\text{-}2} = \frac{(4.52)LQ^{1.85}}{C^{1.85}d^{4.87}}$$

$$= \frac{(4.52)(10 \text{ ft})(17.71 \text{ gpm})^{1.85}}{(120)^{1.85}(1.049 \text{ in})^{4.87}}$$

$$= (0.0051)(17.71 \text{ gpm})^{1.85}$$

$$= 1.04 \text{ psig}$$

step 3: The total pressure at sprinkler 2 (as illustrated) is

$$p_{t,2} = p_{n,1} + p_{f,1\text{-}2} = 10 \text{ psig} + 1.04 \text{ psig}$$

$$= 11.04 \text{ psig}$$

step 4: The cross-sectional area of a 1 in diameter schedule-40 steel pipe is 0.0060 ft^2. The velocity in the pipe between sprinklers 1 and 2 is

$$\text{v} = \frac{\dot{V}}{A} = \frac{(17.7 \text{ gpm})\left(0.002228\dfrac{\text{ft}^3}{\text{sec-gpm}}\right)}{0.0060 \text{ ft}^2}$$

$$= (17.7 \text{ gpm})\left(0.3713\frac{\text{ft}}{\text{sec-gpm}}\right)$$

$$= 6.57 \text{ ft/sec}$$

From Eq. 65.4, the velocity pressure of the flow between sprinklers 1 and 2 is

$$p_{v,1\text{-}2} = \frac{0.433\text{v}^2}{2g}$$

$$= \frac{\left(0.433\dfrac{\text{psi}}{\text{ft}}\right)\left(6.57\dfrac{\text{ft}}{\text{sec}}\right)^2}{(2)\left(32.2\dfrac{\text{ft}}{\text{sec}^2}\right)}$$

$$= (6.724 \times 10^{-3})\left(6.57\frac{\text{ft}}{\text{sec}}\right)^2$$

$$= 0.29 \text{ psig}$$

step 5: The normal pressure to be used with sprinkler 2 is

$$p_{n,2} = p_{t,2} - p_{v,1\text{-}2} = 11.04 \text{ psig} - 0.29 \text{ psig}$$

$$= 10.75 \text{ psig} \quad [>10 \text{ psig, so OK}]$$

step 6: The discharge from sprinkler 2 is

$$Q_2 = K\sqrt{p_{n,2}} = \left(5.6\frac{\text{gpm}}{\sqrt{\text{psig}}}\right)\sqrt{10.75 \text{ psig}}$$

$$= 18.36 \text{ gpm}$$

step 7: The quantity flowing between sprinklers 2 and 3 is

$$Q_{2\text{-}3} = Q_1 + Q_2 = 17.71 \text{ gpm} + 18.36 \text{ gpm}$$

$$= 36.07 \text{ gpm}$$

The friction loss between sprinklers 2 and 3 is

$$p_{f,2\text{-}3} = (0.0051)(36.07 \text{ gpm})^{1.85}$$

$$= 3.88 \text{ psig}$$

step 8: The velocity between sprinklers 2 and 3 is

$$\text{v} = (36.07 \text{ gpm})\left(0.3713 \frac{\text{ft}}{\text{sec-gpm}}\right)$$

$$= 13.4 \text{ ft/sec}$$

The velocity pressure between sprinklers 2 and 3 is

$$p_{v,2\text{-}3} = (6.724 \times 10^{-3})\left(13.4 \frac{\text{ft}}{\text{sec}}\right)^2$$

$$= 1.21 \text{ psig}$$

step 9: The total pressure at sprinkler 3 is

$$p_{t,3} = p_{n,2} + p_{f,2\text{-}3} + p_{v,2\text{-}3}$$

$$= 10.75 \text{ psig} + 3.88 \text{ psig} + 1.21 \text{ psig}$$

$$= 15.84 \text{ psig}$$

step 10: Estimate the discharge from sprinkler 3 as 20 gpm. The normal pressure that would cause this discharge is not calculated.

step 11: The estimated flow rate from the cross-main elbow to sprinkler 3 is

$$Q_{3,\text{estimated}} = Q_{2\text{-}3} + Q_{3,\text{estimated}}$$

$$= 36.07 \text{ gpm} + 20 \text{ gpm} = 56.07 \text{ gpm}$$

step 12: The estimated velocity in the pipe from the cross-main elbow to sprinkler 3 is

$$\text{v}_{\text{elbow-3, estimated}} = (56.07 \text{ gpm})\left(0.3713 \frac{\text{ft}}{\text{sec-gpm}}\right)$$

$$= 20.82 \text{ ft/sec}$$

The estimated velocity pressure between the cross-main elbow and sprinkler 3 is

$$p_{v,\text{elbow-3,estimated}} = (6.724 \times 10^{-3})\left(20.82 \frac{\text{ft}}{\text{sec}}\right)^2$$

$$= 2.91 \text{ psig}$$

step 13: The normal pressure at sprinkler 3 is

$$p_{n,3} = p_{t,3} - p_{v,\text{elbow-3,estimated}}$$

$$= 15.84 \text{ psig} - 2.91 \text{ psig} = 12.93 \text{ psig}$$

step 14: The corrected discharge from sprinkler 3 is

$$Q_{3,\text{corrected}} = K\sqrt{p_{n,3}} = \left(5.6 \frac{\text{gpm}}{\sqrt{\text{psig}}}\right)\sqrt{12.93 \text{ psig}}$$

$$= 20.13 \text{ gpm}$$

step 15: The discharge from sprinkler 3 was assumed in step 10 to be 20 gpm. It was calculated as 20.13 gpm in step 14. Assuming this is sufficiently close, 20.13 gpm would be used in subsequent steps.

The standardized sprinkler systems worksheet (also seen in App. 65.D), if used, would appear as follows.

Contract No. _____   Sheet No. ___1___ of ___1___

Name and Location _Example 65.4_

nozzle type and location	flow gpm (L/min)	pipe size in	fitting and devices	pipe equivalent length	friction loss psi/ft (bars/m)	required pressure psi (bars)		normal pressure	K = 5.6 notes
			length	10	C = 120	p_t 10	p_t		$q = 5.6\sqrt{10} = 17.71$
1	q 17.71		fittings	0		p_f 1.04	p_v		steps 1–2
	Q 17.71		total	10		p_e 0	p_n		
			length	10		p_t 11.04	p_t 11.04		$q = 5.6\sqrt{10.75} = 18.36$
2	q 18.36		fittings	0		p_f 3.88	p_v 0.29		steps 3–7
	Q 36.07		total	10		p_e 0	p_n 10.75		
			length	10		p_t 15.84	p_t 15.84		$p_t = 10.75 + 3.88 + 1.21$
3	q 20.13		fittings	0		p_f	p_v 2.91		$= 15.84$
	Q 56.20		total	10		p_e	p_n 12.93		$q = 5.6\sqrt{12.93} = 20.13$ steps 8–14

16. HYDRAULIC DESIGN PROCEDURE: PRESSURE AT CROSS MAINS

The total pressure at the cross-main connection to a branch line is the normal pressure at the nearest open sprinkler plus the friction loss and the velocity pressure in the pipe between the sprinkler and the cross-main connection. Thus, the connection is assumed to be an end outlet, and the velocity downstream of the cross-main connection is used to calculate the total pressure. Minor losses, such as tee and nipple losses, must be included in the friction loss.

The pressure at each subsequent upstream cross-main-to-branch connection is calculated similarly to that for branch lines, except that the velocity head (already included from the most-remote cross-main connection) is assumed to be unchanged. The normal pressure at each successive cross-main-to-branch line connection is assumed to be the normal pressure of the last cross-main connection plus the friction pressure loss between the two branch connections.[7]

The analysis procedure works back up the cross main from branch to subsequent upstream branch. While the normal pressure is known, the flow through each subsequent branch (not the hydraulically most-remote) is not. If the subsequent branch is identical to a previous branch, the subsequent branch can be treated as an orifice with the orifice constant, K, calculated from the flow rate and normal pressure at the cross main for the previous branch. All identical branches will have the same orifice constant. The discharge for the subsequent branch is simply calculated from Eq. 65.1.

If the subsequent branch is different from all previous branches, the new branch flow is initially calculated based on any conveniently assumed pressure at the end sprinkler in that branch, working up the branch back to the cross main. The final calculated normal pressure will not coincide with the normal pressure calculated by working up the cross main. This means that the calculated flow will also be incorrect.

Pressures at hydraulic junction points must balance within 0.5 psi (3.5 kPa; 0.035 bars). Pressure differences and their corresponding flows greater than this tolerance must be corrected. Orifice plates and sprinklers with mixed orifice sizes generally cannot be used to balance the system, although there are exceptions for small rooms.

The corrected flow quantity in the branch is calculated from Eq. 65.6.

$$\frac{Q_{\text{corrected}}}{Q_{\text{calculated}}} = \sqrt{\frac{p_{\text{cross main}}}{p_{\text{calculated}}}} \qquad 65.6$$

---

[7]This is normally a valid assumption, since the velocities in the cross mains will be low. However, if necessary, a rigorous calculation based on actual velocities can be performed. This is seldom necessary.

When two branches on opposite sides of a cross main have the same configuration, the flow calculated from one can be doubled. When the two opposite branches are different in configuration, the hydraulically shorter flow will have to be adjusted by using Eq. 65.6. The effect of this adjustment is to increase the flow in the shorter branch.

The corrected flow quantity is added to the cumulative flow in the cross main. Then, the next upstream branch is handled similarly until all branches in the design area are calculated. Once the flows from the design area have been calculated, the flow in the cross main is assumed constant all the way back to the supply valve. The distance from the design area to the supply valve is used to calculate the pressure friction loss based on the discharge from all sprinklers in the design area. However, branches outside the design area do not influence the flow rate in the cross main.

Example 65.5

The sprinkler branch from Ex. 65.4 is fed by a $1\frac{1}{2}$ in (nominal) schedule-40 steel pipe cross main. What are the total and normal pressures in the cross main at the entrance to the elbow?

```
        branch A
        o────────┐elbow
        3         │
                  │
```

Solution

From Ex. 65.4, the flow rate between the elbow and sprinkler 3 is

$$Q_{\text{elbow-3}} = 36.07 \text{ gpm} + 20.13 \text{ gpm} = 56.20 \text{ gpm}$$

The velocity between sprinkler 3 and the elbow is

$$v = (56.20 \text{ gpm})\left(0.3713 \frac{\text{ft}}{\text{sec-gpm}}\right)$$
$$= 20.87 \text{ ft/sec}$$

The velocity pressure between sprinkler 3 and the elbow is

$$p_{v,\text{elbow-3}} = (6.724 \times 10^{-3})\left(20.87 \frac{\text{ft}}{\text{sec}}\right)^2$$
$$= 2.93 \text{ psig}$$

Since the flow is turned $90°$, the minor loss of the elbow (based on the smaller diameter) must be included with the friction loss. The equivalent length of a 1 in elbow is given in App. 65.C as 2 ft. The total friction loss between sprinkler 3 and the elbow is

$$p_{f,\text{elbow-3}} = \frac{(4.52)LQ^{1.85}}{C^{1.85}d^{4.87}}$$
$$= \frac{(4.52)(10 \text{ ft} + 2 \text{ ft})(56.20 \text{ gpm})^{1.85}}{(120)^{1.85}(1.049 \text{ in})^{4.87}}$$
$$= 10.56 \text{ psig}$$

The total pressure at the end of the cross main at the entrance to the elbow at branch A is

$$p_{t,A} = p_{n,3} + p_{v,\text{elbow-3}} + p_{f,\text{elbow-3}}$$
$$= 12.93 \text{ psig} + 2.93 \text{ psig} + 10.56 \text{ psig}$$
$$= 26.42 \text{ psig}$$

The normal pressure at the end of the cross main at the entrance to the elbow at branch A excludes the velocity pressure.

$$p_{n,A} = p_{n,3} + p_{f,\text{elbow-3}}$$
$$= 12.93 \text{ psig} + 10.56 \text{ psig}$$
$$= 23.49 \text{ psig}$$

Example 65.6

The $1^{1}/_{2}$ in (nominal) cross main evaluated in Exs. 65.4 and 65.5 contains two additional opposing branches. Both branches use 1 in (nominal) schedule-40 steel pipe. The two opposing branches have different configurations. The branch flow rates and normal pressures at the cross connection based on an assumed minimum pressure of $p_{\text{minimum}} = 10$ psig are given. The distance along the cross main between branches A and B/C is 10 ft. Branch A, evaluated in Ex. 65.5, is the hydraulically most-remote. What is the total flow rate in the cross main?

```
        branch A
        ┌──────┐ elbow
        │
        │
        │  cross
   ─────┼─────────┬─────────
   branch B        branch C
   Q_B = 40 gpm    Q_C = 35 gpm
   p_{n,B} = 15 psig  p_{n,C} = 13 psig
```

Solution

The diameter of $1^{1}/_{2}$ in diameter pipe is 1.610 in.

The minor loss of the elbow has already been included. The cross is a straight-through connection, and its minor loss is disregarded. The friction pressure loss between the elbow and the cross is

$$p_{f,\text{elbow-cross}} = \frac{(4.52)LQ^{1.85}}{C^{1.85}d^{4.87}}$$
$$= \frac{(4.52)(10 \text{ ft})(56.20 \text{ gpm})^{1.85}}{(120)^{1.85}(1.610 \text{ in})^{4.87}}$$
$$= 1.09 \text{ psig}$$

The normal pressure at the cross is

$$p_{n,\text{cross}} = p_{n,\text{elbow}} + p_{f,\text{elbow-cross}}$$
$$= 23.49 + 1.09 \text{ psig} = 24.58 \text{ psig}$$

The calculation for the normal pressure for branches B and C was not 24.58 psig. Therefore, the residual pressure at the ends of branches B and C will not be 10 psig—it will be higher. The flow rates are also incorrect. Equation 65.6 is used to correct the flows.

$$Q_{B,\text{corrected}} = (Q_{B,\text{calculated}})\sqrt{\frac{p_{\text{cross main}}}{p_{\text{calculated}}}}$$
$$= (40 \text{ gpm})\sqrt{\frac{24.58 \text{ psig}}{15 \text{ psig}}} = 51.2 \text{ gpm}$$

$$Q_{C,\text{corrected}} = (35 \text{ gpm})\sqrt{\frac{24.58 \text{ psig}}{13 \text{ psig}}} = 48.1 \text{ gpm}$$

When branches A, B, and C are all open, the total flow into the cross main will be

$$Q_{\text{total}} = 56.2 \text{ gpm} + 51.2 \text{ gpm} + 48.1 \text{ gpm} = 155.5 \text{ gpm}$$

17. HYDRAULIC DESIGN PROCEDURE: PRESSURE IN RISERS

The total pressure at the top of a riser is calculated by adding the normal pressure at the nearest (downstream) flowing branch, the total friction loss between the branch and the top of the riser (including the minor loss for the riser-to-cross-main fitting), and the velocity pressure in the cross main at the riser connection. Thus, the total pressure is based on the velocity downstream of the riser-to-cross-main connection.

The pressure at the bottom of a riser is the pressure at the top of the riser plus the friction loss in the riser plus the elevation (i.e., static) pressure corresponding to the change in elevation. (Each foot of water height corresponds to approximately 0.434 psi.)

Example 65.7

The $1^{1}/_{2}$ in diameter cross main of Exs. 65.4 and 65.5 continues 10 ft and then connects to a $1^{1}/_{2}$ in diameter, 12 ft high riser. The riser is fed by a 3 in feed main at the lower level. The riser-to-cross-main connection at the top of the riser is through a standard 90° elbow. The connection to the feed main at the base of the riser is through a tee. What are the normal pressures at the top and base of the riser?

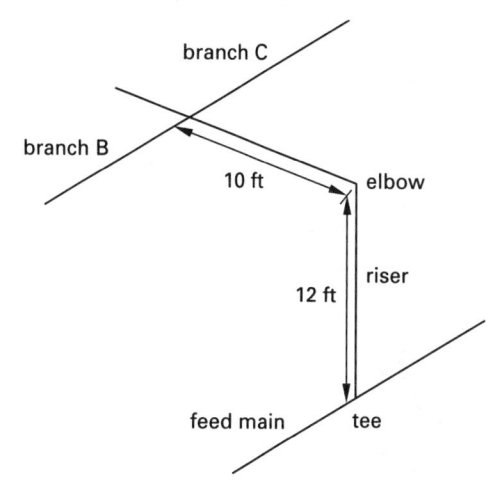

Solution

The equivalent length of a standard $1^1/_2$ in 90° elbow is given in App. 65.C as 4 ft. The friction loss between branches B and C and the top of the riser is

$$p_{f,\text{cross-top}} = \frac{(4.52)LQ^{1.85}}{C^{1.85}d^{4.87}}$$
$$= \frac{(4.52)(10 \text{ ft} + 4 \text{ ft})(155.5 \text{ gpm})^{1.85}}{(120)^{1.85}(1.610 \text{ in})^{4.87}}$$
$$= 10.05 \text{ psig}$$

The normal pressure at the top of the riser is

$$p_{n,\text{top}} = p_{n,\text{cross}} + p_{f,\text{cross-top}}$$
$$= 24.58 \text{ psig} + 10.05 \text{ psig} = 34.63 \text{ psig}$$

The friction loss in the 12 ft of riser is

$$p_{f,\text{riser}} = \frac{(4.52)LQ^{1.85}}{C^{1.85}d^{4.87}}$$
$$= \frac{(4.52)(12 \text{ ft})(155.5 \text{ gpm})^{1.85}}{(120)^{1.85}(1.610 \text{ in})^{4.87}}$$
$$= 8.62 \text{ psig}$$

The term *base of the riser* is somewhat ambiguous and probably means the feed main. Thus, the pressure at the base of the riser is the required feed main pressure. From App. 65.C, the equivalent length of a $1^1/_2$ in tee is 8 ft. The friction loss in the tee is

$$p_{f,\text{tee}} = \frac{(4.52)LQ^{1.85}}{C^{1.85}d^{4.87}}$$
$$= \frac{(4.52)(8 \text{ ft})(155.5 \text{ gpm})^{1.85}}{(120)^{1.85}(1.610 \text{ in})^{4.87}}$$
$$= 5.74 \text{ psig}$$

In addition, the elevation pressure head from 12 ft of water is

$$p_{\text{elevation}} = \gamma h$$
$$= \frac{\left(62.4 \dfrac{\text{lbf}}{\text{ft}^3}\right)(12 \text{ ft})}{144 \dfrac{\text{in}^2}{\text{ft}^2}}$$
$$= 5.20 \text{ psig}$$

The pressure at the base of the riser is

$$p_{n,\text{base}} = p_{n,\text{top}} + p_{f,\text{riser}} + p_{f,\text{tee}} + p_{\text{elevation}}$$
$$= 34.63 \text{ psig} + 8.62 \text{ psig} + 5.74 \text{ psig} + 5.20 \text{ psig}$$
$$= 54.19 \text{ psig}$$

66 Environmental Engineering

Part 1: General Concepts 66-2

Part 2: Types and Sources of Pollution
2. The Environment 66-3
3. Pollutants 66-3
4. Wastes 66-3
5. Pollution Sources 66-4

Part 3: Environmental Issues
    (alphabetical)
6. Introduction 66-4
7. Acid Gas 66-4
8. Acid Rain 66-5
9. Allergens and Microorganisms . . . 66-5
10. Asbestos 66-5
11. Bottom Ash 66-6
12. Disposal of Ash 66-6
13. Carbon Dioxide 66-6
14. Carbon Monoxide 66-8
15. Chlorofluorocarbons 66-8
16. Cooling Tower Blowdown 66-9
17. Condenser Cooling Water 66-9
18. Dioxins 66-9
19. Dust, General 66-9
20. Dust, Coal 66-10
21. Fugitive Emissions 66-10
22. Gasoline-Related Pollution 66-10
23. Global Warming 66-11
24. Landfill Gas 66-11
25. Leachate 66-12
26. Lead 66-12
27. Municipal Solid Waste 66-12
28. Nitrogen Oxides 66-13
29. Odors 66-15
30. Oil Spills in Navigable Waters . . . 66-15
31. Ozone 66-15
32. Particulate Matter 66-15
33. PCBs 66-15
34. Plastics 66-15
35. Radon 66-16
36. Rainwater Runoff 66-16
37. Smog 66-17
38. Smoke 66-17
39. Spills, Hazardous 66-17
40. Sulfur Oxides 66-17
41. Tires 66-18
42. Trihalomethanes 66-18
43. Volatile Inorganic Compounds . . . 66-19
44. Volatile Organic Compounds . . . 66-19
45. Water Vapor 66-19

Part 4: Storage, Handling, and Transfer
    of Hazardous Materials
46. General Storage 66-19
47. Storage Tanks 66-19
48. Disposition of Hazardous Wastes . . . 66-20

Part 5: Testing and Sampling
49. Emissions Sampling 66-20
50. Emissions Monitoring Systems 66-21

Part 6: Remediation Processes and Equipment
    (alphabetical)
51. Introduction 66-21
52. Absorption, Gas (General) 66-21
53. Absorption, Gas (Spray Towers) 66-22
54. Absorption, Gas (in Packed Towers) . . 66-22
55. Adsorption (Activated Carbon) 66-23
56. Adsorption (Solvent Recovery) 66-23
57. Adsorption, Hazardous Waste 66-24
58. Advanced Flue Gas Cleanup 66-24
59. Advanced Oxidation 66-24
60. Baghouses 66-24
61. Bioremediation 66-25
62. Biofiltration 66-25
63. Bioreaction 66-25
64. Bioventing 66-25
65. Coal Conditioning 66-26
66. Cyclone Separators 66-26
67. Dechlorination 66-26
68. Electrostatic Precipitators 66-26
69. Flue Gas Recirculation 66-27
70. Fluidized-Bed Combustors 66-27
71. Injection Wells 66-28
72. Incinerators, Fluidized-Bed Combustion 66-29
73. Incineration, General 66-30
74. Incineration, Hazardous Waste 66-31
75. Incineration, Infrared 66-31
76. Incineration, Liquids 66-31
77. Incineration, Municipal Solid Waste . . 66-31
78. Incineration, Oxygen-Enriched 66-34
79. Incineration, Plasma 66-34
80. Incineration, Soil 66-34
81. Incineration, Solids and Sludges 66-34
82. Landfilling 66-34
83. Incineration, Vapors from 66-34
84. Low Excess-Air Burners 66-34
85. Low-NOx Burners 66-35
86. Mechanical Seals 66-35
87. Multiple Pipe Containment 66-36
88. Ozonation 66-36
89. Scrubbing, General 66-37
90. Scrubbing, Chemical 66-37

Plant
Engineering

91. Scrubbing, Dry 66-37
92. Scrubbing, Wet 66-37
93. Selective Catalytic Reduction 66-37
94. Selective Noncatalytic Reduction 66-38
95. Soil Washing 66-38
96. Sorbent Injection 66-38
97. Sparging 66-38
98. Staged Combustion 66-38
99. Stripping, Air 66-38
100. Stripping, Steam 66-40
101. Thermal Desorption 66-40
102. Vacuum Extraction 66-40
103. Vapor Condensing 66-40
104. Vitrification 66-40
105. Wastewater Treatment Processes 66-41

Nomenclature

a	interfacial area per volume of packing	1/ft	1/m
A	area	ft^2	m^2
B	cyclone inlet width	ft	m
B	volumetric fraction	–	–
C	concentration	ppm	ppm
C	concentration	various	various
C	ESP constant	–	–
D	diameter	ft	m
e	excess air fraction by volume	–	–
E	mass emission rate	lbm/ MMBtu	kg/MJ
g	gravitational acceleration	ft/sec^2	m/s^2
G	gas loading rate	ft^3/min-ft^2	m^3/s·m^2
G	MSW generation rate, per capita	lbm/day	kg/day
h	head	ft	m
H	cyclone inlet height	ft	m
H	Henry's law constant	atm	kPa
HHV	higher heating value	MMBtu/lbm	kJ/kg
HRR	heat release rate	Btu/hr-ft^2	W/m^2
HTU	height of a transfer unit	ft	m
k	reaction rate constant	1/sec	1/s
K	ESP constant	–	–
K_G	coefficient of gas mass transfer	ft/sec	m/s
$K_G a$	gas mass transfer coefficient	1/sec	1/s
$K_L a$	liquid mass transfer coefficient	1/sec	1/s
L	length	ft	m
L	liquid loading rate	ft^3/min-ft^2	m^3/s·m^2
LF	loading factor	–	–
m	mass	lbm	kg
$\dot{m}$	mass flow rate	lbm/hr	kg/h
MW	molecular weight	lbm/ lbmol	kg/ kmol
N	population size	–	–
N	number of transfer units (alt.)	–	–

NTU	number of transfer units	–	–
p	vapor pressure	lbf/in^2	kPa
p	partial pressure	atm	atm
q	heat generation rate	MMBtu/hr	kW
Q	volumetric flow rate	ft^3/sec	m^3/s
r	cyclone radius	ft	m
R	stripping factor	–	–
S	separation factor	–	–
SFC	specific feed characteristics	Btu/lbm	J/kg
t	time	sec	s
T	temperature	°R	K
v	velocity	ft/sec	m/s
V	volume	ft^3	m^3
$\dot{V}$	pollution emission rate	ft^3/hr	m^3/h
$\dot{V}$	volumetric flow rate	ft^3/min	L/s
w	drift velocity	ft/sec	m/s
x	fraction by weight	–	–
z	packing height	ft	m

Symbols

α	coefficient of linear thermal expansion	ft/ft-°F	m/m·°C
γ	specific weight	lbf/ft^3	–
η	efficiency	%	%

Superscripts
* equilibrium

Subscripts

a	air
dp	dew point
e	equivalent
G	gas
L	liquid
O	overall
p	plate
s	sulfur
t	total
th	theoretical
w	water

Part 1: General Concepts

Pollution prevention (sometimes referred to as "P2") can be achieved in a number of ways. The most desirable method is reduction at the source. Reduction is accomplished through process modifications, raw material quantity reduction, substitution of materials, improvements in material quality, and increased efficiencies. However, traditional *end-of-pipe* treatment and disposal processes after the pollution is generated remain the main focus. *Pollution control* is the limiting of pollutants in a planned and systematic manner.

Table 66.1 *Pollution Prevention Hierarchy*
(from most to least desirable)

source reduction
recycling
waste separation and concentration
waste exchange
energy/material recovery
waste treatment
disposal

Pollution control, hazardous waste, and other environmental regulations vary from nation to nation and are constantly changing. Therefore, regulation-specific issues, including timetables, permit application processes, enforcement, and penalties for violations, are either omitted from this chapter or discussed in general terms. For a similar reason, few limits on pollutant emission rates and concentrations are given in this chapter.[1] Those that are given should be considered merely representative and typical of the general range of values.

Part 2: Types and Sources of Pollution

2. THE ENVIRONMENT

Specific regulations often deal with parts of the *environment*, such as the atmosphere (i.e., the "air"), oceans and other surface water, subsurface water, and the soil. *Nonattainment areas* are geographical areas identified by regulation that already do not currently meet *national ambient air quality standards* (NAAQS). Nonattainment is usually the result of geography, concentrations of industrial facilities, and excessive vehicular travel, and can apply to any of the regulated substances (e.g., ozone, oxides of sulfur and nitrogen, and lead).

3. POLLUTANTS

A *pollutant* is a material or substance that is accidentally or intentionally introduced to the environment in a quantity that exceeds what occurs naturally. Not all pollutants are toxic or hazardous, but the issue is moot when regulations limiting permissible concentrations are specific. As defined by regulations in the United States, *hazardous air pollutants* (HAPs), also known as *air toxics*, consist of trace metals (e.g., lead, beryllium, mercury, cadmium, nickel, and arsenic) and other substances for a total of approximately 200 "listed" substances.[2]

Another categorization defined by regulation in the United States separates pollutants into *criteria pollutants* (e.g., sulfur dioxide, nitrogen oxides, carbon monoxide, volatile organic compounds, particulate matter, and lead) and *noncriteria pollutants* (e.g., fluorides, sulfuric acid mist, reduced sulfur compounds, vinyl chloride, asbestos, beryllium, mercury, and other heavy metals).

4. WASTES

Process wastes are generated during manufacturing. *Intrinsic wastes* are part of the design of the product and manufacturing process. Examples of intrinsic wastes are impurities in the reactants and raw materials, by-products, residues, and spent materials. Reduction of intrinsic wastes usually means redesigning the product or manufacturing process. *Extrinsic wastes*, on the other hand, are usually reduced by administrative controls, maintenance, training, or recycling. Examples of extrinsic wastes are fugitive leaks and discharges during material handling, testing, or process shutdown.

Solid wastes are garbage, refuse, biosolids, and containerized solid, liquid, and gaseous wastes.[3] *Hazardous waste* is defined as solid waste, alone or in combination with other solids, that because of its quantity, concentration, or physical, chemical, or infectious characteristics may either (a) cause an increase in mortality or serious (irreversible or incapacitating) illness, or (b) pose a present or future hazard to health or the environment when improperly treated, stored, transported, disposed, or managed.

A substance is *reactive* if it reacts violently with water; if the pH is less than 2 or greater than 12.5, it is *corrosive*; if its flash point is less than 140°F, it is *ignitable*. The *toxicity characteristic leaching procedure* (TCLP)[4] test is used to determine if the substance is *toxic*. The TCLP tests for the presence of 8 metals (e.g., chromium, lead, and mercury) and 25 organic compounds (e.g., benzene and chlorinated hydrocarbons).

Hazardous wastes can be categorized by the nature of their sources. *F-wastes* (e.g., spent solvents and distillation residues) originate from nonspecific sources; *K-wastes* (e.g., separator sludge from petroleum refining) are generated from industry-specific sources; *P-wastes* (e.g., hydrogen cyanide) are acutely hazardous discarded commercial products, off-specification products, and spill residues; *U-wastes* (e.g., benzene and hydrogen sulfide) are other discarded commercial products, off-specification products, and spill residues.

---

[1]Another factor complicating the publication of specific regulations is that the maximum-permitted concentrations and emissions depend on the size and nature of the source.
[2]The most dangerous air toxics include asbestos, benzene, cadmium, carbon tetrachloride, chlorinated dioxins and dibenzofurans, chromium, ethylene dibromide, ethylene dichloride, ethylene oxide and methylene chloride, radionuclides, vinyl chloride, and emissions from coke ovens. Most of these substances are carcinogenic.

---

[3]The term *biosolids* is replacing the term *sludge* as it refers to organic waste produced from biological wastewater treatment processes. Sludge from industrial processes and flue gas cleanup (FGC) devices retains its name.
[4]Probably no subject in this book has more acronyms than environmental engineering. All of the acronyms used in this chapter are in actual use; none were invented for the benefit of the chapter.

Plant Engineering

Two notable "rules" pertain to hazardous wastes. The *mixture rule* states that any solid waste mixed with hazardous waste becomes hazardous. The *derived from rule* states that any waste derived from the treatment of a hazardous waste (e.g., ash from the incineration of hazardous waste) remains a hazardous waste.[5]

5. POLLUTION SOURCES

A *pollution source* is any facility that produces pollution. A *generator* is any facility that generates hazardous waste. The term "major" (i.e., a *major source*) is defined differently for each class of nonattainment areas.[6]

The combustion of fossil fuels (e.g., coal, fuel oil, and natural gas) to produce steam in electrical generating plants is the most significant source of air-borne pollution. As such, this industry is among the most highly regulated.

Specific regulations pertain to generators of hazardous waste. Generators of more than certain amounts (e.g., 100 kg/month) must be registered (i.e., with the Environmental Protection Agency, EPA). Restrictions are placed on generators in the areas of storage, personnel training, shipping, treatment, and disposal.

Part 3: Environmental Issues

6. INTRODUCTION

Mechanical engineers face many environmental issues. This part of the chapter discusses (in alphabetical order) some of them. In some cases, "listed substances" (e.g., those that are specifically regulated) are discussed. In other cases, environmental issues are discussed in general terms.

Environmental engineering covers an immense subject area, and this chapter is merely an introduction to some of the topics. Some subjects "belong" to other engineering disciplines. For example, municipal wastewater plants are typically designed by civil engineers. Other subjects, such as the storage and destruction of nuclear wastes, are specialized topics subject to changing politics, complex legislation, and sometimes-untested technologies. Finally, some wastes are considered nonhazardous industrial wastes and are virtually unregulated. They, too, are not discussed in this chapter.

Processing of medical wastes, personal safety, and the physiological effects of exposure, as important as they are, are beyond the scope of this chapter.

Other philosphical and political issues, such as nuclear fuel versus fossil fuel, plastic bags versus paper bags, and disposable diapers versus cloth diapers, are similarly not covered.

7. ACID GAS

Acid gas generally refers to sulfur trioxide, SO_3, in the flue gas.[7,8] *Sulfuric acid*, H_2SO_4, formed when sulfur trioxide combines with water, has a low vapor pressure and, consequently, a high boiling point. Hydrochloric (HCl), hydrofluoric (HF), and nitric (HNO_3) acids also form in smaller quantities. However, unlike sulfuric acid, they do not lower the vapor pressure of water significantly. Therefore, any sulfuric acid present will control the dew point.

Sulfur trioxide has a large affinity for water, forming a strong acid even at very low concentrations. At the elevated temperatures in a stack, sulfuric acid attacks steel, almost all plastics, cement, and mortar. Sulfuric acid can be prevented from forming by keeping the temperature of the flue gas above the dew-point temperature. This may require preheating equipment prior to start-up and postheating during shutdown.

Stack dew points have been reported by various researchers to be in the range of 225 to 300°F. However, the actual value is dependent on the amount of SO_3 in the flue gas, and therefore, on the amount of sulfur in the fuel. The theoretical equilibrium relationships are too complex to be useful, and empirical correlations or graphical methods are used. Equation 66.1 can be used to determine the dew point based on the partial pressures, p_w and p_s, of the water vapor and sulfur trioxide, respectively, in units of atmospheres.[9,10]

$$\frac{1000}{T_{dp,K}} = 1.7842 + 0.0269\log_{10}p_w - 0.1029\log_{10}p_s$$
$$+ 0.0329\log_{10}p_w \log_{10}p_s \qquad \textit{66.1}$$

Hydrochloric acid does not normally occur unless the fuel has a high chlorine content, as do chlorinated solvents, municipal solid wastes (MSW), and refuse-derived fuels (RDF). Hydrochloric acid formed during the combustion of MSW and RDF can be removed by semidry scrubbing. HCl removal efficiencies of 90 to 99% are common. (See also Acid Rain and Sulfur Oxides.)

---

[5]Hazardous waste should not be incinerated with nonhazardous waste, as all of the ash would be considered hazardous by these rules.
[6]A nonattainment area is classified as marginal, moderate, serious, severe, or extreme based on the average pollution (e.g., ozone) level measured in the area.

[7]The term *stack gas* is used interchangeably with *flue gas*.
[8]Sulfur dioxide normally is not a source of acidity in the flue gas.
[9]This correlation was reported by F. H. Verhoff and J. T. Banchero in *Chemical Engineering Progress*, 1974, Vol. 70, p. 71.
[10]*Dalton's law* states that the partial pressure of a component is volumetrically weighted. Therefore, knowing the volumetric flue gas concentration, C_A, in ppmv is equivalent to knowing the partial pressures.

$$p_A = p_t\left(\frac{C_A}{10^6}\right)$$

Example 66.1

At a particular point in a stack, the flue gas has a total pressure of 30.2 in Hg. The flue gas is 8% water vapor by volume, and the sulfur trioxide concentration is 100 ppm. What is the approximate dew-point temperature at that point?

Solution

From Dalton's law, the partial pressures are volumetrically weighted.

$$p_w = B_w p_t$$
$$= \frac{(0.08)(30.2 \text{ in Hg})}{29.92 \frac{\text{in Hg}}{\text{atm}}} = 0.0807 \text{ atm}$$

$$p_s = B_s p_t$$
$$= \frac{(100 \text{ ppm})(30.2 \text{ in Hg})}{\left(29.92 \frac{\text{in Hg}}{\text{atm}}\right)(10^6 \text{ ppm})}$$
$$= 1.01 \times 10^{-4} \text{ atm}$$

Use Eq. 66.1.

$$\frac{1000}{T_{dp,K}} = 1.7842 + 0.0269 \log_{10}(0.0807 \text{ atm})$$
$$- 0.1029 \log_{10}(1.01 \times 10^{-4} \text{ atm})$$
$$+ 0.0329 \log_{10}(0.0807 \text{ atm})$$
$$\times \log_{10}(1.01 \times 10^{-4} \text{ atm})$$
$$= 2.3097$$

$$T_{dp} = \frac{1000 \text{K}}{2.3097} = 433\text{K} \quad (160°\text{C}, \ 320°\text{F})$$

8. ACID RAIN

Acid rain consists of weak solutions of sulfuric, hydrochloric, and to a lesser extent, nitric acids. These acids are formed when emissions of sulfur oxides (SOx), hydrogen chloride (HCl), and nitrogen oxides (NOx) return to the ground in rain, fog, or snow, or as dry particles and gases. Acid rain affects lakes and streams, damages buildings and monuments, contributes to reduced visibility, and affects certain forest tree species. Acid rain may also represent a health hazard. (See also Acid Gas and Sulfur Oxides.)

9. ALLERGENS AND MICROORGANISMS

Allergens such as molds, viruses, bacteria, animal droppings, mites, cockroaches, and pollen can cause allergic reactions in humans. One form of bacteria, *legionella*, causes the potentially fatal *Legionnaire's disease*. Inside buildings, allergens and microorganisms become particularly concentrated in standing water, carpets, HVAC filters and humidifier pads, and locations where birds and rodents have taken up residence.

Some buildings cause large numbers of people to simultaneously become sick, particularly after a major renovation or change has been made. This is known as *sick building syndrome* or *building-related illness*. This phenomenon can be averted by using building materials that do not release vapor over time (e.g., as does plywood impregnated with formaldehyde) or that harbor other irritants. Carpets can accumulate dusts. Repainting, wallpapering, and installing of new flooring can release new airborne chemicals. Areas must be flushed with fresh air until all noticeable effects have been eliminated.

Care must also be taken to ensure that filters in the HVAC system do not harbor microorganisms and are not contaminated by bird or rodent droppings. Air intakes must not be located near areas of chemical storage or parking garages.

10. ASBESTOS

Asbestos is a fibrous silicate mineral material that is inert, strong, and incombustible. Once released into the air, its fibers are light enough to stay airborne for a long time.

Asbestos has typically been used in woven and compressed forms in furnace insulation, gaskets, pipe coverings, boards, roofing felt, and shingles, and has been used as a filler and reinforcement in paint, asphalt, cement, and plastic. Asbestos is no longer banned outright in industrial products. However, regulations, well-publicized health risks associated with cancer and *asbestosis*, and potential liabilities have driven producers to investigate alternatives.

No single product has emerged as a suitable replacement for all asbestos applications. (Almost all replacements are more costly.) *Fiberglass* is an insulator with superior tensile strength but low heat resistance. Fiberglass has a melting temperature of approximately 1000°F (538°C). However, fiberglass treated with hydrochloric acid to leach out most of the silica (SiO_2) can withstand 2000 to 3500°F (1090 to 1930°C).

In typical static sealing applications, *aramid fibers* (known by the trade names Kevlar™ and Twaron™) are particularly useful up to approximately 800°F (427°C). However, aramid fibers cannot withstand the caustic, high-temperature environment encountered in curing concrete.

Table 66.2 lists some of the appropriate alternatives by application.

Table 66.2 Asbestos Substitutes

application	substitute material
insulation boards	mineral wool, fiberglass, foams (polyurethane, cellulose, styrene, and polyimide)
braided packing seals	polytetrafluoroethylene (PTFE), carbon, graphite
flange and furnace door gaskets	*reinforcing fibers*: cellulose, carbon, glass, polyvinyl alcohol, polyamide, polyester, polyacrylonitrile, aramid, and polyolefin; *binders*: nitrile butadiene rubber (NBR)
concrete filler; flame retardant	fiberglass, graphite, polypropylene, acrylics
thixotropic[a] agent in paints, coatings, adhesives, and sealants	polyethylene
additive to plastics and fiberglass to improve heat resistance	vermiculite

[a]A *thixotropic substance* is thick and gel-like when stationary but becomes free-flowing when stirred or agitated.

11. BOTTOM ASH

Ash is the residue left after combustion of a fossil fuel. *Bottom ash* (*bottoms ash* or *bottoms*) is the ash that is removed from the combustor after a fuel is burned. (The rest of the ash is flyash.) The ash falls through the combustion grates into quenching troughs below. It may be continuously removed by *submerged scraper conveyors* (SSCs), screw-type devices, or ram dischargers. The ash can be dewatered to approximately 15% moisture content by compression or by being drawn up a dewatering slope. Bottom ash is combined with conditioned flyash on the way to the ash storage bunker.[11] Most combined ash is eventually landfilled.

12. DISPOSAL OF ASH

Combined ash (bottom ash and flyash) is usually landfilled. Other occasional uses for high-quality combustion ash (not ash from the incineration of municipal solid waste) include roadbed subgrades, road surfaces ("ashphalt"), and building blocks.

---

[11]Approximately half of the electrical generating plants in the United States use wet flyash handling.

13. CARBON DIOXIDE

Carbon dioxide, though an environmental issue, is not a hazardous material and is not regulated as a pollutant.[12] Carbon dioxide is not an environmental or human toxin, is not flammable, and is not explosive. Skin contact with solid or liquid carbon dioxide presents a freezing hazard. Other than the remote potential for causing frostbite, carbon dioxide has no long-term health effects. Its major hazard is that of asphyxiation, by excluding oxygen from the lungs.

As with other products of combustion, the fraction of carbon dioxide in a flue gas can be arbitrarily reduced by the introduction of dilution air into the flue stream. Therefore, carbon dioxide is reported on a standardized basis—typically as a dry volumetric fraction at some percentage (e.g., 3%) of oxygen.[13] The standardized value can be calculated from stoichiometric relationships. Since there is essentially no carbon dioxide in air (approximately 0.03% by volume), carbon dioxide in a flue gas has a unique source—carbon in the fuel. Knowing the theoretical flue gas composition is sufficient to calculate the standardized value. The analysis is independent of stack temperature.

For a standardized value of 3% oxygen, Table 66.3 gives the dry carbon dioxide volumetric fraction directly, based on the volumetric ratio of hydrogen to carbon in the fuel. For any other percentage of oxygen, the following procedure can be used.

step 1: Obtain the volumetric fuel composition. Gaseous fuel compositions are normally reported on a volumetric basis. Solid and liquid fuels are reported on a weight (gravimetric) basis. Convert weight basis analyses to a volumetric basis by dividing the gravimetric percentages by their respective atomic (or, molecular) weights. Combustion products are gaseous. (For gases, molar and volumetric ratios are the same.)

step 2: Write and balance the stoichiometric combustion equation using the volumetric fractions as coefficients for the fuel elements. Disregard trace emissions (NO, SO_2, CO, etc.) that contribute less than 1% to the flue gas volume, and disregard oxygen contributed by the fuel. Include a variable amount of excess air. Include nitrogen for the combustion and excess air at the ratio of 3.773 volumes of nitrogen for each volume of oxygen.

---

[12]Industrial exposure is regulated by the U.S. Occupational Safety and Health Administration (OSHA).

[13]The phrases "at 3% O_2" and "3% excess air" are not equivalent. The former means that oxygen comprises 3% of the gaseous reaction products by volume. The latter means that 3% more air (3% more oxygen) is provided in reactants than is needed.

step 3: Divide the carbon dioxide volume (i.e., the balanced reaction coefficient) by the sum of all flue gas volumes, excluding water vapor, and set this ratio equal to the standardized volumetric fraction. (The fraction can be multiplied by 10^6 to obtain the volumetric fraction in ppm, though this is seldom done for carbon dioxide.)

The carbon dioxide concentration (mass emission per dry standard volume, C) can be calculated from the carbon dioxide's molecular weight (MW = 44) and Eq. 66.8.

Table 66.3 *Theoretical CO_2 Fraction at 3% Oxygen (dry, volumetric basis)*

H/C ratio	CO_2	H/C ratio	CO_2
0	0.18	2.1	0.12723
0.1	0.17651	2.2	0.12548
0.2	0.17816	2.3	0.12378
0.3	0.16993	2.4	0.12212
0.4	0.16682	2.5	0.12050
0.5	0.16382	2.6	0.11893
0.6	0.16093	2.7	0.11740
0.7	0.15814	2.8	0.11590
0.8	0.15544	2.9	0.11445
0.9	0.15283	3.0	0.11303
1.0	0.15031	3.1	0.11165
1.1	0.14787	3.2	0.11029
1.2	0.14551	3.3	0.10898
1.3	0.14323	3.4	0.10768
1.4	0.14101	3.5	0.10643
1.5	0.13886	3.6	0.10520
1.6	0.13678	3.7	0.10400
1.7	0.13476	3.8	0.10283
1.8	0.13279	3.9	0.10168
1.9	0.13089	4.0	0.10056
2.0	0.12903		

Example 66.2

Coal has the following gravimetric composition: carbon, 86.5%; hydrogen, 11.75%; nitrogen (N_2), 0.39%; sulfur, 0.40%; ash, 0.01%; and oxygen (O_2), 0.96%. Determine the theoretical carbon dioxide concentration in parts per million on a dry volumetric (ppmvd) basis at 3% oxygen.

Solution

step 1: Disregarding the combustion of sulfur to SO_2 and other elements present in small quantities, flue gases will be products of carbon and hydrogen combustion and the excess air. The atomic weight of carbon is 12 lbm/lbmol. Since the hydrogen is present in elemental form, not as H_2 gas, the atomic weight is 1 lbm/lbmol. Consider 100 lbm of fuel. The

carbon content will be $(0.865)(100 \text{ lbm}) = 86.5$ lbm. The volumetric ratios (number of moles) are

$$\text{C:} \quad \frac{86.5 \text{ lbm}}{12 \frac{\text{lbm}}{\text{lbmol}}} = 7.208 \text{ lbmol}$$

$$\text{H:} \quad \frac{11.75 \text{ lbm}}{1 \frac{\text{lbm}}{\text{lbmol}}} = 11.75 \text{ lbmol}$$

step 2: The unbalanced combustion reaction is

$$7.208\text{C} + 11.75\text{H} + n_1\text{O}_2 + 3.773n_1\text{N}_2$$
$$\longrightarrow n_2\text{CO}_2 + n_3\text{H}_2\text{O} + 3.773n_1\text{N}_2$$

The balanced combustion reaction is

$$7.208\text{C} + 11.75\text{H} + 10.146\text{O}_2 + 38.281\text{N}_2$$
$$\longrightarrow 7.208\text{CO}_2 + 5.875\text{H}_2\text{O} + 38.281\text{N}_2$$

Let e represent the excess air fraction. All of the excess oxygen and nitrogen will appear in the flue gas.

$$7.208\text{C} + 11.75\text{H} + (1+e)10.146\text{O}_2$$
$$+ (1+e)38.281\text{N}_2$$
$$\longrightarrow 7.208\text{CO}_2 + 5.875\text{H}_2\text{O}$$
$$+ (1+e)38.281\text{N}_2 + 10.146e\text{O}_2$$

At 3% O_2,

$$0.03 = \frac{10.146e}{7.208 + (1+e)(38.281) + 10.146e}$$
$$e \approx 0.157 \quad (15.7\% \text{ excess air})$$

The balanced combustion reaction at 3% oxygen is

$$7.208\text{C} + 11.75\text{H} + 11.739\text{O}_2 + 44.291\text{N}_2$$
$$\longrightarrow 7.208\text{CO}_2 + 5.875\text{H}_2\text{O} + 44.291\text{N}_2 + 1.593\text{O}_2$$

step 3: The theoretical carbon dioxide fraction, on a dry volumetric basis at 3% oxygen, is

$$\frac{7.208}{7.208 + 44.291 + 1.593} = 0.1358 \quad (13.58\%)$$

(This is the same value as obtained from Table 66.3 for a volumetric fuel ratio of H/C = 11.75/7.208 = 1.63.)

Example 66.3

When the fuel described in Ex. 66.2 is burned, the carbon dioxide in the stack gas is measured on a wet, volumetric basis to be 10.4%. What is this value corrected to 3% O_2?

Solution

Since the theoretical carbon dioxide volumetric fraction can be calculated from the fuel composition, the actual carbon dioxide fraction in the stack is irrelevant. (The amount of excess air could be found from this value, however.) The theoretical carbon dioxide fraction, on a dry volumetric basis at 3% oxygen, is still 13.55%.

14. CARBON MONOXIDE

Carbon monoxide, CO, is formed during incomplete combustion of carbon in fuels. This is usually the result of an oxygen deficiency at lower temperatures. Carbon monoxide displaces oxygen in the bloodstream and represents an asphyxiation hazard. Carbon monoxide does not contribute to smog.

Generation of carbon monoxide can be minimized by furnace monitoring and control. For industrial sources, the American Boiler Manufacturer's Association (ABMA) recommends limiting carbon monoxide to 400 ppm (corrected to 3% O_2) in oil- and gas-fired industrial boilers. This value can usually be met with reasonable ease. Local ordinances may be more limiting, however.

Most carbon monoxide released in highly populated areas comes from vehicles. Vehicular traffic may cause the CO concentration to exceed regulatory limits. For this reason, *oxygenated fuels* are required to be sold in those areas during certain parts of the year. Oxygenated gasoline has a minimum oxygen content of approximately 2.0%. Oxygen is added to gasoline by adding additives such as ethanol (ethyl alcohol) or methyl tertiary butyl ether (MTBE).

Minimization of carbon monoxide is compromised by efforts to minimize nitrogen oxides. Control of these pollutants is inversely related.

15. CHLOROFLUOROCARBONS

Most atmospheric oxygen is in the form of two-atom molecules, O_2. However, there is a thin layer in the stratosphere about 12 miles up where *ozone* molecules, O_3, are found in large quantities. Ozone filters out ultraviolet radiation that damages crops and causes skin cancer.

Chlorofluorocarbons (i.e., chlorinated fluorocarbons, such as Freon™) contribute to the deterioration of the Earth's ozone layer. Ozone in the atmosphere is depleted in a complex process involving pollutants, wind patterns, and atmospheric ice. As chlorofluorocarbon molecules rise through the atmosphere, solar energy breaks the chlorine free. The chlorine molecules attach themselves to ozone molecules, and the new structure eventually decomposes into chlorine oxide and normal oxygen, O_2. The depletion process is particularly pronounced in the Antarctic because that continent's dry, cold air is filled with ice crystals on whose surfaces the chlorine and ozone can combine. Also, the prevailing winter wind isolates and concentrates the chlorofluorocarbons.

Table 66.4 *Typical Replacement Compounds for Chlorofluorocarbons*

designation	applications
HCFC 22	low- and medium-temperature refrigerant; blowing agent; propellant
HCFC 123	replacement for CFC-11; industrial chillers and applications where potential for exposure is low; somewhat toxic; blowing agent; replacement for perchloroethylene (dry cleaning fluid)
HCFC 124	industrial chillers; blowing agent
HFC 134a	replacement for CFC-12; medium-temperature refrigeration systems; centrifugal and reciprocating chillers; propellant
HCFC 141b	replacement for CFC-11 as a blowing agent; solvent
HCFC 142b	replacement for CFC-12 as a blowing agent; propellant
IPC (isopropyl chloride)	replacement for CFC-11 as a blowing agent

The depletion is not limited to the Antarctic, but occurs throughout the northern hemisphere, including virtually all of the United States. At various points during the year in the Northern Hemisphere, the ozone deficit is as much as 6%.

The 1987 Montreal Protocol (conference) resulted in an international agreement to phase out world-wide production of chemical compounds that have ozone-depletion characteristics.

The Clean Air Act (Title VI) follows the Montreal Protocol and prohibits production of chlorofluorcarbons (and Halon) in the United States. The 1990 Clean Air Act required that class I chemicals, including chlorofluorcarbons (CFCs), Halons, and carbon tetrachloride, be phased out by the year 2000. The fumigant methyl chloride, also a class I chemical, was given a phase-out deadline of 2002. The 1990 Clean Air Act required that class II chemicals, including hydrochlorofluorcarbons (HCFCs), be phased out by the year 2030.

Special allowances are made for aviation safety, national security, and fire suppression and explosion prevention if safe or effective substitutes are not available for those purposes. Excise taxes are used as interim disincentives to produce the compounds. Large reserves and recycling, however, will probably ensure that chlorofluorocarbons and Halons are in use for many years after the deadlines have passes.

Possible replacements for chlorofluorocarbons (CFCs) include hydrochlorofluorcarbons (HCFCs) and hydrofluorocarbons (HFCs), both of which are environmentally more benign than CFCs, and blends of HCFCs and HFCs. The additional hydrogen atoms in the molecules make them less stable, allowing nearly all chlorine to dissipate in the lower atmosphere before reaching the ozone layer. The lifetime of HCFC molecules is 2 to 25 years, compared with 100 years or longer for CFCs. The net result is that HCFCs have only 2 to 10% of the ozone-depletion ability of CFCs. HFCs have no chlorine and thus cannot deplete the ozone layer.

Most chemicals intended to replace CFCs still have chlorine, but at reduced levels. Additional studies are determining if HCFCs and HFCs accumulate in the atmosphere, how they decompose, and whether any by-products could damage the environment.

16. COOLING TOWER BLOWDOWN

State-of-the art reuse programs in *cooling towers* (CTs) may recirculate water 15 to 20 times before it is removed through blowdown. Pollutants such as metals, herbicides, and pesticides originally in the makeup water are concentrated to five or six times their incoming concentrations. Most CTs are constructed with copper alloy condenser tubes, so the recirculated water becomes contaminated with copper ions as well. CT water also usually contains chlorine compounds or other biocides added to inhibit biofouling. Ideally, no water should leave the plant (i.e., a *zero-discharge facility*). If discharged, CT blowdown must be treated prior to disposal.

17. CONDENSER COOLING WATER

Approximately half of the electrical generation plants in the United States use *once-through (OT) cooling water*. The discharged cooling water may be a chemical or thermal pollutant. Since fouling in the main steam condenser significantly reduces performance, water can be treated by the intermittent addition of chlorine, chlorine dioxide, bromine, or ozone, and these chemicals may be present in residual form. *Total residual chlorine* (TRC) is regulated. Methods of chlorine control include *targeted chlorination* (the frequent application of small amounts of chlorine where needed) and *dechlorination*.

18. DIOXINS

Dioxins are a family of chlorinated dibenzo-*p*-dioxins (CDDs). The term *dioxin*, however, is commonly used to refer to the specific congener 2,3,7,8-tetrachlorodibenzo-*p*-dioxin (TCDD). Primary sources of dioxin include herbicides containing 2,4-T, 2,4,5-trichlorophenol, and hexachlorophene. Other potential sources include incinerated municipal and industrial waste, leaded gasoline exhaust, chlorinated chemical wastes, incinerated polychlorinated biphenyls (PCBs), and any combustion in the presence of chlorine.

The exact mechanism of dioxin formation during incineration is complex but probably requires free chlorine (in the form of HCl vapor), heavy metal concentrations (often found in the ash), and a critical temperature window of 570 to 840°F (300 to 450°C). Dioxins in incinerators probably form near waste heat boilers, which operate in this temperature range.

Dioxin destruction is difficult because it is a large organic molecule with a high boiling point. Most destruction methods rely on high temperature since temperatures of 1550°F (850°C) denature the dioxins. Other methods include physical immobilization (i.e., vitrification), dehalogenation, oxidation, and catalytic cracking using catalysts such as platinum.

Dioxins liberated during the combustion of municipal solid waste (MSW) and refuse-derived fuel (RDF) can be controlled by the proper design and operation of the furnace combustion system. Once formed, they can be removed by end-of-pipe processes, including activated charcoal (AC) injection. Success has also been reported using the vanadium oxide catalyst used for NOx removal.

19. DUST, GENERAL

Dusts or *fugitive dusts* are any solid particulate matter (PM) that becomes airborne, with the exception of PM emitted from the exhaust stack of a combustion process. Nonhazardous fugitive dusts are commonly generated when a material (e.g., coal) is unloaded from trucks and railcars. Dusts are also generated by manufacturing, construction, earth-moving, sand blasting, demolition, and vehicle movement.

Dust hazards can be categorized into three types: (1) Breathing airborne dust or vapors, particularly those that carry hazardous compounds is the major concern. Even without toxic compounds, odors can be objectionable. Dusts are easily observed and can cover cars and other objects left outside. (2) Dusts can transport hazardous materials, contaminating the environment far from the original source. (3) In closed environments, even nontoxic dusts can represent an explosion hazard.

Dust emission from *spot sources* (e.g., manufacturing processes such as grinders) is accomplished by *inertial separators* such as cyclone separators. Potential dust sources (e.g., truck loads and loose piles) can be covered, and dust generation can be reduced by spraying water mixed with other compounds.

There are three mechanisms of dust control by spraying: (1) In *particle capture* (as occurs in a spray curtain at a railcar unloading station), suspended particles are knocked down, wetted, and captured by liquid droplets. (2) In *bulk agglomeration* (as when a material being carried on a screw conveyor is sprayed), the moisture keeps the dust with the material being transported. (3) Spraying roads and coal piles to inhibit wind-blown dust is an example of *surface stabilization*.

Plant Engineering

Wetting agents are *surfactant* formulations added to water to improve water's ability to wet and agglomerate fine particles.[14] The resulting solution can be applied as liquid spray or as a foam.[15] Humectant binders (e.g., magnesium chloride, and calcium chloride) and adhesive binders (e.g., waste oil) may also be added to the mixture to make the dust adhere to the contact surface if other water-based methods are ineffective.[16]

Surface stabilization of materials stored outside and exposed to wind, rain, freeze-thaw cycles, and ultraviolet radiation is enhanced by the addition of *crusting agents*.

20. DUST, COAL

Clean coals, western low-sulfur coal, eastern low-sulfur coal, eastern high-sulfur coal, low-rank lignite coal, and varieties in between have their own peculiar handling characteristics.[17] *Dry ultra-fine coal* (DUC) and coal slurries have their own special needs.

Western coals have a lower sulfur content, but because they are easily fractured, they generate more dust. Western coals also pose higher fire and explosion hazards than eastern coals. Water misting or foam must be applied to coal cars, storage piles, and conveyer transfer points. Adequate ventilation in storage silos and bunkers is also required, with the added benefit of reducing methane accumulation.

DUC is as fine as talcum (10 μm with less than 10% moisture).[18] It must be containerized for transport, because traditional railcars are not sufficiently airtight. Also, since the minimum oxygen content for combustion is 14 to 15%, DUC should be maintained in a pressurized, oxygen-depleted environment until used. Pneumatic flow systems are used to transport DUC through the combustion plant.

21. FUGITIVE EMISSIONS

Equipment leaks from plant equipment are known as *fugitive emissions* (FEs). Leaks from pump and valve seals are common sources, though compressors, pressure-relief devices, connectors, sampling systems, closed vents, and storage vessels are also potential sources. FEs are reduced administratively by *leak detection and repair programs* (LDARs).

Common causes of fugitive emissions are (1) equipment in poor condition, (2) off-design pump operation, (3) inadequate seal characteristics, and (4) inadequate

boiling-point margin in the seal chamber. Other pump/shaft/seal problems that can increase emissions include improper seal-face material, excessive seal-face loading and seal-face deflections, and improper pressure balance ratio.

Inadequate *boiling-point margin* (BPM) results in poor seal performance and face damage. BPM is the difference between the seal chamber temperature and the initial boiling point temperature of the pumped product at the seal chamber pressure. When seals operate close to the BPM, seal-generated heat can cause the pumped product between the seal faces to flash and the seal to run dry. A minimum BPM of 15°F (9°C) or a 25 psig (172 kPa) pressure margin is recommended to avoid flashing. Even greater margins will result in longer seal life and reduced emissions.

22. GASOLINE-RELATED POLLUTION

Gasoline-related pollution is primarily in the form of unburned hydrocarbons, nitrogen oxides, lead, and carbon monoxide. (The requirement for lead-free gasoline has severely curtailed gasoline-related lead pollution.) Though a small reduction in gasoline-related pollution can be achieved by blending detergents into gasoline, reformulation is required for significant improvements. Table 66.5 lists typical characteristics for traditional and reformulated gasolines.

The potential for smog formation can be reduced by reformulating gasoline and/or reducing the summer *Reid vapor pressure* (RVP) of the blend. Regulatory requirements to reduce the RVP can be met, in the short run, by using less butane in the gasoline. However, reformulating to remove pentanes is required when the RVP is required to achieve 7 psig (48 kPa) or below. Sulfur content can be reduced by pretreating the refinery feed in hydrodesulfurization (HDS) units. Heavier gasoline components must be removed to lower the 50% and 90% *distillation temperatures* (T50 and T90).

---

[14] A *surface-acting agent (surfactant)* is a soluble compound that reduces a liquid's surface tension or reduces the interfacial tension between a liquid and a solid.

[15] Foam is an increasingly popular means of reducing the potential for explosions in secondary coal crushers.

[16] A *humectant* is a substance that absorbs or retains moisture.

[17] A valuable resource for this subject is NFPA 850, *Fire Protection for Fossil Fueled Steam and Combustion Turbine Electrical Generating Plants*, National Fire Protection Association, Quincy, MA.

[18] A μm is a *micrometer*, 10^6 m, and commonly referred to as a *micron*.

Table 66.5 *Typical Gasoline Characteristics*

fuel parameter	traditional	reformulated
sulfur	150 ppmw[a]	40 ppmw
benzene	2% by vol	1% by vol
olefins	9.9% by vol	4% by vol
aromatic hydrocarbons	32% by vol	25% by vol
oxygen	0	1.8 to 2.2% by wt
T90	330°F (166°C)	300°F (149°C)
T50	220°F (104°C)	210°F (99°C)
RVP	8.5 psig (59 kPa)	7 psig (48 kPa)

[a]ppmw = parts per million by weight

23. GLOBAL WARMING

The *global warming* hypothesis is that increased levels of atmospheric carbon dioxide, CO_2, from the combustion of carbon-rich fossil fuels and other *greenhouse gases* (e.g., water vapor, methane, nitrous oxide, and chlorofluorocarbons) trap an increasing amount of solar radiation in a *greenhouse effect*, gradually increasing the Earth's temperature. It is claimed that the most recent cycle of increases in carbon dioxide began with the industrial revolution.

Recent studies have shown that atmospheric carbon dioxide is increasing at the rate of about 1% per year. (For example, in one year carbon dioxide might increase from 350 ppmv to 354 ppmv. By comparison, oxygen is approximately 209,500 ppmv.) There has also been a 100% increase in atmospheric methane since the beginning of the industrial revolution. According to some researchers, there has been a corresponding global temperature increase of approximately 0.9°F (0.5°C) since the year 1890.[19]

In addition to a temperature increase, other evidence cited by supporters of the global warming hypothesis are several record-breaking hot summers during the 1980s, widespread aberrations in the traditional seasonal weather patterns (e.g., hurricane-like storms in England), and a 10 to 30 cm rise in sea level over the last century.[20]

The global warming hypothesis is disputed by many scientists and has not been proved to be an absolute truth. Arguments against the hypothesis center around the fact that manufactured carbon dioxide is a small fraction of what is naturally released (e.g., by wetlands, in rain forest fires, and during volcanic eruptions). It is argued that, in the face of such massive contributors, and since the Earth's temperature has remained essentially constant for millenia, the Earth already possesses some built-in mechanism, currently not perfectly understood, that reduces the Earth's temperature swings.

The environmentalists' claim that there will be a temperature increase of 3 to 9°F by the year 2100 and that this increase will be catastrophic is also disputed by many scientists. On the contrary, most scientists agree that the deprivation and human suffering that would result from a reduction or cessation in the burning of fossil fuels will be significantly higher than the effects of a slightly increased temperature.

Regardless of the validity of the global warming hypothesis, some major power generation industries have adopted goals of reducing carbon dioxide emissions. However, efforts to reduce carbon dioxide emissions by converting fossil fuel from one form to another are questioned by many engineers. Natural gas produces the

least amount of carbon dioxide of any fossil fuel. Therefore, conversion of coal to a gas or liquid fuel in order to lower the carbon-to-hydrogen ratio would appear to lessen carbon dioxide emissions at the point of final use. However, the conversion processes consume energy derived from carbon-containing fuel. This additional consumption, taken over all sites, results in a net increase in carbon dioxide emission of 10 to 200%, depending on the process.

The use of ethanol as an alternative for gasoline is also problematic. Manufacturing processes that produce ethanol give off (at least) twice as much carbon dioxide as the gasoline being replaced produces during combustion.

Most synthetic fuels are intrinsically less efficient (based on their actual heating values compared with those theoretically obtainable from the fuels' components in elemental form). This results in an increase in the amount of fuel consumed. Thus, fossil fuels should be used primarily in their raw forms until cleaner sources of energy are available.

24. LANDFILL GAS

Most closed landfills are covered by a thick soil layer. As the covered refuse decays beneath the soil, the natural anaerobic biological reaction generates a low-Btu *landfill gas* (LFG) consisting of approximately 50% methane, carbon dioxide, and trace amounts of other gases. If uncontrolled, LFG migrates to the surface. If the LFG accumulates, an explosion hazard results. If it escapes, other environmental problems (including objectionable odors) occur. Therefore, synthetic and compacted clay liners and clay trenches are used to prevent gas from spreading laterally. Wells and collection pipes are used to collect and incinerate LFG in flares. However, emissions from flaring are also problematic.

Alternatives to flaring include using the LFG to produce hot water or steam for heating or electricity generation. During the 1980s, reciprocating engines and combustion turbines powered by LFG were tried. However, such engines generated relatively high emissions of their own due to impurities and composition variations in the fuel. True Rankine-cycle power plants (generally without reheat) avoid this problem, since boilers are less sensitive to impurities.

One problem with using LFG commercially is that LFG is withdrawn from landfills at less than atmospheric pressure. Conventional furnace burners need approximately 5 psig at the boiler front. Low-pressure burners requiring 2 psig are available, though expensive. Therefore, some of the plant power must be used to pressurize the LFG.

Although production is limited, LFG is produced for a long period after a landfill site is closed. Production slowly drops 3 to 5% annually to approximately 30% of its original value after about 20 to 25 years, which is

---

[19]Other researchers can detect no discernible upward trend, and some offer a counter-argument. Based on the retreat of the northern-most lines capable of growing oranges since 1850, some believe that the weather is generally becoming colder, not hotter.
[20]The rise in the level of the oceans is disputed.

considered to be the economical life of a gas-reclamation system. Approximately 40 ft^3 of gas will be produced per cubic yard (1.5 m^3 of gas per cubic meter) of landfill.

25. LEACHATE

Leachates are highly concentrated liquid wastes produced in landfills.[21] The leachate becomes more concentrated as the landfill ages. Leachate forms from liquids brought into the landfill, water run-off, and precipitation. Leachate can contaminate surrounding soil and groundwater since it percolates downward through the refuse and collects in a drainage layer at the lower landfill boundary. Leachate must be removed to reduce hydraulic head and to reduce unacceptable concentrations of hazardous substances.

Leachate from *municipal solid waste landfills* (MSWLs) can generally be treated with biological (i.e., trickling filter and activated sludge) processes.

26. LEAD

Lead, even in low concentrations, is toxic. Inhaled lead accumulates in the blood, bones, and vital organs. It can produce stomach cramps, fatigue, aches, and nausea. It causes irreparable damage to the brain, particularly in young and unborn children, and high blood pressure in adults. At high concentrations, lead damages the nervous system and can be fatal.

Lead was outlawed in paint in the late 1970s.[22] Lead has also been removed from most gasoline blends. Lead continues to be used in large quantities in automobile batteries and plating and metal-finishing operations. However, these manufacturing operations are tightly regulated. Lead enters the atmosphere during the combustion of fossil fuels and the smelting of sulfide ore. Lead enters lakes and streams primarily from acid mine drainage.

For lead in industrial and municipal wastewater, current remediation methods include pH adjustment with lime or alkali hydroxides, coagulation-sedimentation, reverse osmosis, and zeolite ion exchange.

27. MUNICIPAL SOLID WASTE

Municipal solid waste (MSW, previously known as *"garbage"*) has traditionally been disposed of in *municipal solid waste landfills* (MSWLs, previously referred to as *"dumps"*).[23] Waste is placed in layers, typically 2 to 3 ft thick (0.6 to 0.9 m), and is compacted to the

cell height before soil is added as a cover. Two to five passes by a tracked bulldozer are sufficient to compact the MSW to 800 to 1500 lbm/yd^3 (470 to 890 kg/m^3). (A density of 1000 lbm/ft^3 or 590 kg/m^3 is used in design studies.) The *lift* is the height of the covered layer, as shown in Fig. 66.1. When the landfill layer is full, the ratio of solid waste volume to soil cover volume will be approximately between 4:1 and 3:1.

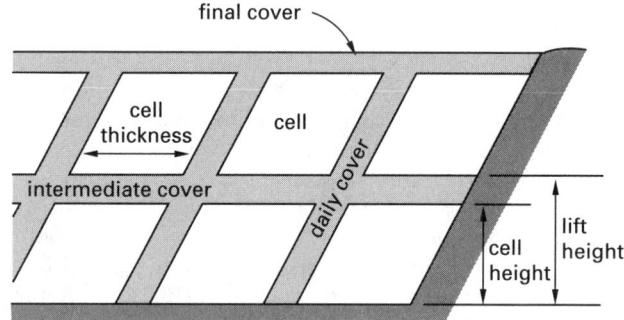

Figure 66.1 *Landfill Cells*

The *cell height* is typically taken as 8 ft (2.4 m) for design studies, although it can actually be much higher. The height should be chosen to minimize the cover material requirement consistent with the regulatory requirements. Cell slopes will be less than 40°, and typically less than 20 to 30°.

MSW is generated at the average rate of approximately 5 to 8 lbm/capita-day (2.3 to 3.6 kg/capita-day).[24] The daily increase in landfill volume is predicted by Eq. 66.2.[25] N is the population size, G is the per-capita MSW generation rate, and γ is the landfill overall specific weight, calculated as a weighted average of soil and compacted MSW specific weights.[26]

$$\Delta V_{\text{day}} = \frac{NG(\text{LF})}{\gamma} \qquad \text{66.2}$$

The *loading factor*, LF, is 1.25 for a 4:1 volumetric ratio, and is calculated from Eq. 66.3 for other ratios.

$$\text{LF} = \frac{V_{\text{MSW}} + V_{\text{cover soil}}}{V_{\text{MSW}}} \qquad \text{66.3}$$

Suitable (economical and safe) landfill sites are becoming difficult to find, and once identified, are objected to by residents near the site and in the MSW transport corridor. This is referred to as the *NIMBY* (not in my backyard) *syndrome*.[27]

---

[21]The leachate from the ash of incinerated MSW is largely salt water.
[22]Workers can be exposed to lead if they strip away lead-based paint, demolish old buildings, or weld or cut metals that have been coated with lead-based paint.
[23]In 1988, 73% of the nation's MSW was landfilled, 14% was incinerated, and 13% was recycled.

[24]5 lbm/capita-day is a conservative estimate commonly used in design studies.
[25]Multiply cubic yards by 27/43,560 to get acre-feet (ac-ft).
[26]Soils have specific weights between 70 and 130 lbf/ft^3 (densities between 1100 and 2100 kg/m^3). A value of 100 lbf/ft^3 (1600 kg/m^3) is commonly used in design studies.
[27]In fact, NIMBY applies to landfills, incinerators, and any commercial or processing plant dealing with hazardous wastes.

Table 66.6 Typical Nationwide Characteristics of MSW

component	percentage by weight
paper	40
yard waste	18
glass	8
plastic	7
steel	7
food waste	6
aluminum	2
other	12

28. NITROGEN OXIDES

Nitrogen oxides (NOx) are one of the primary causes of smog formation.[28] NOx from the combustion of coal is primarily *nitric oxide* (NO) with small quantities of *nitrogen dioxide* (NO_2).[29] NO_2 can be a primary or a secondary pollutant. Although some NO_2 is emitted directly from combustion sources, it is also produced from the oxidation of nitric oxide, NO, in the atmosphere.

NOx is produced in two ways: (1) *thermal NOx* produced at high temperatures from free nitrogen and oxygen, and (2) *fuel NOx* (or *fuel-bound NOx*) formed from the decomposition/combustion of fuel-bound nitrogen.[30] When natural gas and light distillate oil is burned, almost all of the NOx produced is thermal. Residual fuel oil, however, can have a nitrogen content as high as 0.3%, and 50 to 60% of this can be converted to NOx. Coal has an even higher nitrogen content.[31]

Thermal NOx is usually produced in small (but significant) quantities when excess oxygen is present at the highest temperature point in a furnace, such that nitrogen (N_2) can dissociate.[32] Dissociation of N_2 and O_2 is negligible, and little or no thermal NOx is produced below approximately 3000°F (1650°C).

Formation of thermal NOx is approximately exponential with temperature. For this reason, many NOx-reduction techniques attempt to reduce the *peak flame temperature* (PFT). NOx formation can be reduced also by injecting urea or ammonia reagents directly into the furnace.[33] The relationship between NOx production and excess oxygen is somewhat inconclusive, but NOx production appears to vary directly with the square root of the oxygen concentration.

In existing plants, retrofit NOx-reduction techniques include fuel-rich combustion (i.e., staged air burners), flue gas recirculation, changing to a low-nitrogen fuel, reduced air preheat, installing low-NOx burners, and using overfire air.[34] Low-NOx burners using controlled flow/split flame or internal fuel staging technology are essentially direct replacement (i.e., "plug in") units, differing from the original burners primarily in nozzle configuration. However, some fuel supply and air modifications may also be needed. Use of overfire air requires windbox modifications and separate ducts.

Lime spray-dryer scrubbers of the type found in electrical generating plants do not remove all of the NOx. Further reduction requires that the remaining NOx be destroyed. Reburn, selective catalytic reduction (SCR), and selective noncatalytic methods are required.

Scrubbing, incineration, and other end-of-pipe methods typically have been used to reduce NOx emissions from stationary sources such as gas turbine-generators, although these methods are unwieldy for the smallest units. Water/steam injection, catalytic combustion, and *selective catalytic reduction* (SCR) are particularly suited for gas-turbine and combined-cycle installations. SCR is also effective in NOx reduction in all heater and boiler applications.

The volumetric fraction (in ppm) of NOx in the flue gas is calculated from the molecular weight and mass flow rates. The molecular weight of NOx is assumed to be 46.[35] The ratio of $\dot{m}_{NOx}/\dot{m}_{flue\ gas}$, percentage of water vapor, and flue gas molecular weight are derived from flue gas analysis.

$$B_{NOx} = \left(\frac{\dot{m}_{NOx}}{\dot{m}_{flue\ gas}}\right)(10^6)\left(\frac{MW_{flue\ gas}}{MW_{NOx}}\right)$$
$$\times \left(\frac{100\%}{100\% - \%H_2O}\right) \qquad 66.4$$

Since the apparent concentration of NOx could be arbitrarily decreased without reducing the NOx production rate by simply diluting the combustion gas with excess air, it is common to correct NOx readings to 3% O_2 by volume, dry basis. (This is indicated by the units *ppmvd*.) Standardizing is accomplished by multiplying the measured NOx reading (in ppmvd) by the O_2 correction factor from Eq. 66.5. (Corrections to 7%, 12%,

---

[28]In Los Angeles County, approximately 75% of the NOx is produced by vehicles, with the remainder being produced by combustion operations, such as boilers and heaters.

[29]Other oxides are produced in insignificant amounts. These include *nitrous oxide* (N_2O), N_2O_4, N_2O_5, and NO_3, all of which are eventually oxidized to (and reported) as NO_2.

[30]Some engineers further divide the production of fuel-bound NOx into low- and high-temperature processes and declare a third fuel-related NOx-production method known as *prompt-NOx*. Prompt-NOx is the generation of the first 15 to 20 ppm of NOx from partial combustion of the fuel at lower temperatures.

[31]In order of increasing fuel-bound NOx production potential, common boiler fuels are: methanol, ethanol, natural gas, propane, butane, ultralow-nitrogen fuel oil, fuel oil No. 2, fuel oil No. 6, and coal.

[32]The *mean residence time* at the high temperature points of the combustion gases is also an important parameter. At temperatures below 2500°F (1370°C), several minutes of exposure may be required to generate any significant quantities of NOx. At temperatures above 3000°F (1650°C), dissociation can occur in less than a second. At the highest temperatures—3600°F (1980°C) and above—dissociation takes less than a tenth of a second.

[33]*Urea* (NH_2CONH_2), also known as *carbamide urea*, is a water-soluble organic compound prepared from ammonia. Urea has significant biological and industrial usefulness.

[34]Air is injected into the furnace at high velocity over the combustion bed to create turbulence and to provide oxygen.

[35]46 is the molecular weight of NO_2. The predominant oxide in the flue gas is NO, and NO_2 may be only 10 to 15% of the total NOx. Furthermore, NO is measured, not NO_2. However, NO has a short half-life and is quickly oxidized to NO_2 in the atmosphere.

and 15% oxygen content are also used. For a correction to any other percentage O_2, replace the 3 in Eq. 66.5 with the new value.)

$$
\begin{aligned}
O_2 \text{ correction factor} &= \frac{21\% - 3\%}{21\% - \%O_2} \\
&= \frac{18\%}{21\% - \%O_2} \qquad 66.5
\end{aligned}
$$

If the flue gas analysis is on a wet basis, Eq. 66.6 can be used to calculate the multiplicative factor.

$$
\begin{aligned}
O_2 \text{ correction factor} &= \frac{21\% - 3\%}{21\% - \left(\dfrac{100\%}{100\% - \%H_2O}\right)(\%O_2)} \\
&= \frac{18\%}{21\% - \left(\dfrac{100\%}{100\% - \%H_2O}\right)(\%O_2)}
\end{aligned}
$$

$$66.6$$

Some regulations specify NOx limitations in terms of pounds per hour or in terms of pounds of NOx per million Btus (MMBtu) of gross heat released. The mass emission rate, E, in lbm/MMBtu, is calculated from the concentration in pounds per dry, standard cubic foot (dscf).

$$
\begin{aligned}
E_{\text{lbm/MMBtu}} &= \frac{C_{\text{lbm/dscf}}\dot{V}_{\text{NOx,dscf/hr}}}{q_{\text{MMBtuh}}} \\
&= \frac{C_{\text{lbm/dscf}}(100)(V_{\text{th,CO}_2})}{(C_{\text{m,CO}_2,\%})(HHV_{\text{MMBtu/lbm}})} \qquad 66.7
\end{aligned}
$$

The relationship between the NOx concentrations, C, expressed in pounds per dry standard volume and ppm can be calculated from Eq. 66.8.[36] Conversions between ppm and pounds are made assuming NOx has a molecular weight of 46.

$$C_{\text{g/m}^3} = (4.15 \times 10^{-5})C_{\text{ppm}}(MW) \quad \text{[SI]} \qquad 66.8(a)$$

$$C_{\text{lbm/ft}^3} = (2.59 \times 10^{-9})C_{\text{ppm}}(MW) \quad \text{[U.S.]} \qquad 66.8(b)$$

The theoretical volume of carbon dioxide, $V_{\text{th,CO}_2}$, produced can be determined stoichiometrically. However, Table 66.7 can be used for quick estimates with an accuracy of approximately $\pm 5.9\%$.

Table 66.7 Approximate CO_2 Production for Various Fuels (with 0% excess air)

fuel	standard[a] ft^3/10^6 Btu	standard[a] m^3/10^6 cal
coal, anthracite	1980	0.222
coal, bituminous	1810	0.203
coal, lignite	1810	0.203
gas, butane	1260	0.142
gas, natural	1040	0.117
gas, propane	1200	0.135
oil	1430	0.161

[a]Standard conditions are 70°F (21°C) and 1 atm pressure.

[36]The constants in Eq. 66.8 are the same and can be used for any gas.

Equation 66.7 is based on the theoretical volume of carbon dioxide gas produced per pound of fuel. Other approximations and correlations are based on the total dry volume of the flue gas in standard cubic feet per million Btu at 3% oxygen, $V_{t,\text{dry}}$. For quick estimates on furnaces burning natural gas, propane, and butane, V_t is approximately 10,130 ft^3/MMBtu; for fuel oil, V_t is approximately 10,680 ft^3/MMBtu.

$$
E_{\text{lbm/MMBtu}} = (C_{\text{ppmv at 3\% O}_2})(V_{t,\text{dry}})
$$
$$
\times \left[\frac{46\ \dfrac{\text{lbm}}{\text{lbmol}}}{\left(379.3\ \dfrac{\text{ft}^3}{\text{lbmol}}\right)(10^6)}\right] \qquad 66.9
$$

As with all pollutants, the maximum allowable concentration or discharge of NOx is subject to continuous review and revision. Actual limits may depend on the type of geographical location, type of fuel, size of the facility, and so on. For steam/electrical (gas turbine) plants, the general target is 25 to 40 ppm. The lower values apply to combustion of natural gas, and the higher values apply to combustion of distillate oil. Even lower values (down to 9 to 10 ppm) are imposed in some areas.

Example 66.4

A combustion turbine produces 25 lbm/hr of NOx while generating 550,000 lbm/hr of exhaust gases. The exhaust gas contains 10% water vapor and 11% oxygen by volume. What is the NOx concentration in ppm, dry volume basis, corrected to 3% oxygen?

Solution

Assume the molecular weight of NOx is 46 lbm/lbmol. Some of the gas analysis is missing, so the flue gas is assumed to be mostly nitrogen (the largest component of air) with a molecular weight of 28 lbm/lbmol.

The uncorrected, dry NOx volumetric fraction is

$$
\begin{aligned}
B_{\text{NOx}} &= \left(\frac{\dot{m}_{\text{NOx}}}{\dot{m}_{\text{flue gas}}}\right) \\
&\quad \times \left(\frac{MW_{\text{flue gas}}}{MW_{\text{NOx}}}\right)(10^6)\left(\frac{100\%}{100\% - \%H_2O}\right) \\
&= \left(\frac{25\ \dfrac{\text{lbm}}{\text{hr}}}{550,000\ \dfrac{\text{lbm}}{\text{hr}}}\right)\left(\frac{46\ \dfrac{\text{lbm}}{\text{lbmol}}}{28\ \dfrac{\text{lbm}}{\text{lbmol}}}\right) \\
&\quad \times (10^6)\left(\frac{100\%}{100\% - 10\%}\right) \\
&= 83.0\ \text{ppmvd} \quad \text{[uncorrected]}
\end{aligned}
$$

Equation 66.5 corrects the value to 3% oxygen.

O_2 correction factor

$$= \frac{21\% - 3\%}{21\% - \left(\dfrac{100\%}{100\% - \%H_2O} \right) (\%O_2)}$$

$$= \frac{18\%}{21\% - \left(\dfrac{100\%}{(100\% - 10\%)} \right) (11\%)}$$

$$= 2.05$$

The corrected value is

$$C = (2.05)(83.0 \text{ ppmvd}) = 170 \text{ ppmvd}$$

29. ODORS

Odors of unregulated substances can be eliminated at their source, contained by sealing and covering, diluted to unnoticeable levels with clean air, or removed by simple water washing, chemical scrubbing (using acid, alkali, or sodium hypochlorite), bioremediation, and activated carbon adsorption.

30. OIL SPILLS IN NAVIGABLE WATERS

Intentional and accidental releases of oil in navigable waters are prohibited. Deleterious effects of such spills include large-scale biological (i.e., sea life and wildlife) damage and destruction of scenic and recreational sites. Long-term toxicity can be harmful for microorganisms that normally live in the sediment.[37]

31. OZONE

Ground-level ozone is a secondary pollutant. Ozone is not usually emitted directly, but is formed from hydrocarbons and nitrogen oxides (NO and NO_2) in the presence of sunlight. *Oxidants* are byproducts of reactions between combustion products. Nitrogen oxides react with other organic substances (e.g., hydrocarbons) to form the oxidants ozone and peroxyacyl nitrates (PAN) in complex *photochemical reactions*. Ozone and PAN are usually considered to be the major components of smog.[38] (See also Smog.)

32. PARTICULATE MATTER

Particulate matter (PM), also known as *aerosols*, is defined as all particles that are emitted by a combustion source. Particulate matter with aerodynamic diameters of less than or equal to a nominal 10 μm is known as *PM-10*. Particulate matter is generally inorganic in nature. It can be categorized into metals (or heavy metals), acids, bases, salts, and nonmetallic inorganics.

Metallic inorganic PM from incinerators is controlled with baghouses or electrostatic precipitators (ESPs), while nonmetallic inorganics are removed by scrubbing (wet absorption). Flue gas PM, such as flyash and lime particles from desulfurization processes, can be removed by fabric baghouses and electrostatic precipitators. These processes must be used with other processes that remove NOx and SO_2.

High temperatures cause the average flue gas particle to decrease in size toward or below 10 μm. With incineration, emission of trace metals into the atmosphere increases significantly. Because of this, incinerators should not operate above 1650°F (900°C.)[39]

33. PCBs

Polychlorinated biphenyls (PCBs) are organic compounds (i.e., *chlorinated organics*) manufactured in oily liquid and solid forms through the late 1970s, and subsequently prohibited. PCBs are carcinogenic and can cause skin lesions and reproductive problems. PCBs build up, rather than dissipate, in the food chain, accumulating in fatty tissues. Most PCBs were used as dielectric and insulating liquids in large electrical transformers and capacitors, and in ballasts for fluorescent lights (which contain capacitors). PCBs were also used as heat transfer and hydraulic fluids, as dye carriers in carbonless copy paper, and as plasticizers in paints, adhesives, and caulking compounds.

Incineration of PCB liquids and PCB-contaminated materials (usually soil) has long been used as an effective mediation technique. However, PCB-contaminated soil apparently can be effectively treated in situ by adding high-calcium flyash or (more expensive) aluminum powder. The reduction mechanism is unclear, but catalysis with an unidentified catalyst is most likely. The destruction of PCB is total. (The addition of quicklime alone has not been supported as a treatment method.)

Specialized PCB processes targeted at cleaning PCB from spent oil are also available. Final PCB concentrations are below detectable levels, and the cleaned oil can be recycled or used as fuel.

34. PLASTICS

Plastics, of which there are six main chemical polymers, generally do not degrade once disposed of, and therefore, are considered a disposal issue. Disposal is not a problem per se, however, since plastics are lightweight, inert, and do not harm the environment when discarded.

A distinction is made between *biodegrading* and *recycling*. Most plastics, such as the polyethylene bags used to protect pressed shirts from the dry cleaners and to mail some magazines, are not biodegradable but are recyclable. Also, all plastics can be burned for their fuel value.

---

[37]This has given rise to a new *sediment testing technology*.
[38]The term *oxidant* sometimes is used to mean the original emission products of NOx and hydrocarbons.

[39]The upper temperature limit for biosolids (sewage sludge) incineration may be regulated.

The collection and sorting problems often render low-volume plastic recycling efforts uneconomical. Complicating the drive toward recycling is the fact that many of the six different types cannot be distinguished visually, and they cannot be recycled successfully when intermixed. Also, some plastic products consist of layers of different polymers that cannot be separated mechanically.

Table 66.8 *Polymers*

polymer	common use
low-density polyethylene (LDPE)	grocery bags; food wrap
high-density polyethylene (HDPE)	detergent; milk bottles; oil containers; toys
polyethylene terephthalate (PET)	clear beverage containers
polypropylene (PP)	labels; bottles; housewares;
polyvinyl chloride (PVC)	clear bottles
polystyrene (PS)	styrofoam cups; "clam shell" food containers

Sorting in low-volume applications is performed visually and manually. Commercial high-volume methods include hydrocycloning, flotation with flocculation (for all polymers), x-ray fluorescence (primarily for PVC detection), and near-infrared spectroscopy (primarily for separating PVC, PET, PP, PE, and PS). Mass spectroscopy is also promising, but has yet to be commercialized.

Unsorted plastics can be melted and reformed into some low-value products. This operation is known as *downcycling*, since each successful cycle further degrades the material. However, this method is suitable only for a small fraction of the overall recyclable plastic.

Other operations that can reuse the compounds found in plastic products are hydrogenation, pyrolysis, and gasification. *Hydrogenation* is the conversion of mixed plastic scrap to "syncrude," synthetic crude oil, in a high-temperature (i.e., 750 to 880°F (400 to 470°C)), high-pressure (i.e., 2200 to 4400 psig (15 to 30 MPa)), hydrogen-rich atmosphere. Since the end product is a crude oil substitute, hydrogenation operations must be integrated into refinery or petrochemical operations.

Gasification and pyrolysis are stand-alone operations that do not require integration with a refinery. *Pyrolysis* takes place in a fluidized bed between 750°F and 1475°F (400°C and 800°C). Cracked polymer gas or other inert gas fluidizes the sand bed, which promotes good mixing and heat transfer, resulting in liquid and gaseous petroleum products.

Gasification operates at higher temperatures, 1650 to 3600°F (900 to 2000°C), and lower pressures, around 870 psig (6 MPa). The waste stream is pyrolyzed at lower temperatures before being processed by the gasifier. The gas can be used on-site to generate steam.

Gasification has the added advantage of being able to treat the entire municipal solid waste stream, avoiding the need for sorting plastics.

The development and use of a biodegradable plastic is an alternative to recycling. Original research in degradable plastics focused on nondegradable polymers mixed with small amounts of materials that reacted with sunlight and soil microbes. Such products, however, were criticized by environmentalists as being waste that broke down into smaller nondegradable pieces. Polymers that are derived from agricultural sources, rather than from petroleum, are now being investigated in the search for totally biodegradable plastics.

Success has been reported with polymers based on blackstrap molasses (a sugar production waste residue) and potato or corn starches. Another promising degradable polymer is polyhydroxybutyrate-valerate (PBHV), derived from the bacterium *alcaligenes eutrophus*. The bacteria, when deprived of phosphorus, feed on organic acids and sugars, converting more than 80% of their body mass into PBHV.

Some engineers point out that biodegrading is not even a desirable characteristic for plastics and that being nonbiodegradable is not harmful. Biodegrading converts materials (such as the paper bags often preferred over plastic bags) to water and carbon dioxide, contributing to the greenhouse effect without even receiving the energy benefit of incineration. Biodegrading of most substances also results in gases and leachates that can be more harmful to the environment than the original substance. In a landfill, biodegrading serves no useful purpose, since the space occupied by the degraded plastic does not create additional useful space (volume).

35. RADON

Radon gas is a radioactive gas produced from the natural decay of radium within the rocks beneath a building.[40] Radon accumulates in unvented areas (e.g., basements), in stagnant water, and in air pockets formed when the ground settles beneath building slabs. Radon also can be brought into the home by radon-saturated well water used in baths and showers. The maximum acceptable concentration is approximately 3 pCi/L.

Radon mitigation methods include (1) pressurizing to prevent the infiltration of radon, (2) installing of depressurization systems to intercept radon below grade and vent it safely, (3) removing radon-producing soil, and (4) abandoning radon-producing sites.

36. RAINWATER RUNOFF

Rainwater percolating through piles of coal, flyash, mine tailings, and other stored substances can absorb toxic compounds and eventually make its way into the earth, possibly contaminating soil and underground aquifers.

---

[40]Radon also can be generated in concrete made from slag derived from uranium mines. Such mines are primarily located in Colorado, Oregon, Florida, and South Dakota.

37. SMOG

Photochemical smog (usually, just *smog*) consists of ground-level ozone and peroxyacyl nitrates (PAN). Smog is produced by the sunlight-induced reaction of ozone *precursors*, primarily nitrogen dioxide (NOx), hydrocarbons, and volatile organic compounds (VOCs). NOx and hydrocarbons are emitted by combustion sources such as automobiles, refineries, and industrial boilers. VOCs are emitted by manufacturing processes, dry cleaners, gasoline stations, print shops, painting operations, and municipal wastewater treatment plants. (See also Ozone.)

38. SMOKE

Smoke results from incomplete combustion and indicates unacceptable combustion conditions. In addition to being a nuisance problem, smoke contributes to air pollution and reduced visibility. Smoke generation can be minimized by proper furnace monitoring and control.

Opacity can be measured by a variety of informal and formal methods, including transmissometers mounted on the stack. The sum of the *opacity* (the fraction of light blocked) and the *transmittance* (the fraction of light transmitted) is 1.0.

Optical density is calculated from Eq. 66.10. The *smoke spot number* (SSN) can also be used to quantify smoke levels.

$$\text{optical density} = \log_{10}\left(\frac{1}{1 - \text{opacity}}\right) \qquad 66.10$$

Visible moisture plumes with opacities of 40% are common at large steam-generators even when there are no unburned hydrocarbons emitted. High-sulfur fuels and the presence of ammonium chloride (a byproduct of some ammonia-injection processes) seem to increase formation of visible plumes. Moisture plumes from saturated gas streams can be avoided by reheating prior to discharge to the atmosphere.

39. SPILLS, HAZARDOUS

Contamination by a hazardous material can occur accidentally (e.g., from a spill) or intentionally (e.g., a previously used chemical-holding lagoon). Soil contaminated with hazardous materials from leaks of *underground storage tanks* (commonly known as *UST wastes*) and spills is itself a hazardous waste.

The type of waste determines what laws are applicable, what permits are required, and what remediation methods are used.[41] With contaminated soil, spilled substances can be (1) solid and nonhazardous or (2) nonhazardous liquid petroleum products (e.g., "UST

nonhazardous"), and Resource Conservation and Recovery Act- (RCRA-) listed (3) hazardous substances, and (4) toxic substances.

Cleaning up a hazardous waste requires removing the waste from whatever air, soil, and water (lakes, rivers, and oceans) that have been contaminated. The term *remediation* is sometimes used to mean the corrective steps taken to return the environment to its original condition. *Stabilization* refers to the act of reducing the waste concentrations to lower levels so that the waste can be transported, stored, or landfilled.

Remediation methods are classified as available or innovative. *Available methods* can be implemented immediately without being further tested. *Innovative methods* are new, unproven methods in various stages of study.

The remediation method used depends on the waste type. The two most common available methods are landfilling after stabilization and incineration.[42] Incineration can occur in rotary kilns, injection incinerators, infrared incineration, and fluidized-bed combustors. Landfilling requires the contaminated soil to be stabilized chemically or by other means prior to disposal. Innovative technologies for general VOC-contaminated soil include vacuum extraction, bioremediation, thermal desorption, and soil washing. Innovative methods for dioxin removal include oxygen-enriched incineration, plasma incineration, chemical dehalogenation, and ultraviolet radiation.

40. SULFUR OXIDES

Sulfur oxides (SOx), consisting of *sulfur dioxide* (SO2) and *sulfur trioxide* (SO3), are the primary cause of acid rain. *Sulfurous acid* (H2SO3) and *sulfuric acid* (H2SO4) are produced when oxides of sulfur react with moisture in the flue gas. Both of these acids are corrosive.

$$SO_2 + H_2O \longrightarrow H_2SO_3$$
$$SO_3 + H_2O \longrightarrow H_2SO_4$$

Sulfur dioxide (like carbon dioxide and nitrogen oxides) emissions must be reported on a standardized basis. Equation 66.11 can be used to calculate the volumetric, dry basis concentration from a wet stack gas sample.

$$\frac{SO_{2\,[\text{dry,at 3\% O}_2]}}{SO_{2\,[\text{wet,at stack O}_2]}} = \frac{CO_{2\,[\text{th,dry,at 3\% O}_2]}}{CO_{2\,[\text{wet,at stack O}_2]}} \qquad 66.11$$

The sulfur dioxide concentration (mass per dry standard volume, C) can be calculated from sulfur dioxide's molecular weight (MW = 64.07) and Eq. 66.8.[43]

---

[41] Permits must be obtained from the U.S. Environmental Protection Agency (EPA) whenever hazardous wastes (including hazardous-waste contaminated soil) are incinerated.

[42] Other innovative technologies include in situ and ex situ bioremediation, chemical treatment, in situ flushing, in situ vitrification, soil vapor extraction, soil washing, solvent extraction, and thermal desorption.

[43] See Ftn. 36.

Plant Engineering

Fuel switching (coal substitution/blending (CS/B)) is the burning of low-sulfur fuel. (Low-sulfur fuels are approximately 0.25 to 0.65% sulfur by weight, compared to high-sulfur coals with 2.4 to 3.5% sulfur.) However, unlike nitrogen oxides, which can be prevented during combustion, formation of sulfur oxides cannot be avoided when low-cost, high-sulfur fuels are burned.

Some air quality regulations regarding SOx production may be met by a combination of options. These options include fuel switching, flue gas scrubbing, derating, and allowance trading. The most economical blend of these options will vary from location to location.

In addition to fuel switching, available technology options for retrofitting existing coal-fired plants include wet scrubbing, dry scrubbing, sorbent injection, repowering with clean coal technology (CCT), and co-firing with natural gas.

Example 66.5

A wet stack gas analysis shows 230 ppm of SO_2 and 10.5% CO_2. Based on stoichiometric combustion, the theoretical carbon dioxide percentage at 3% oxygen should have been 13.4%. What is the SO_2 concentration in ppm on a dry basis, standardized to 3% oxygen?

Solution

Use Eq. 66.11.

$$SO_2 \text{ [dry,at 3\% O}_2\text{]} = SO_2 \text{ [wet,at stack O}_2\text{]}$$
$$\times \frac{CO_2 \text{ [th,dry,at 3\% O}_2\text{]}}{CO_2 \text{ [wet,at stack O}_2\text{]}}$$
$$= \frac{(230 \text{ ppm})(13.4\%)}{10.5\%} = 294 \text{ ppm}$$

41. TIRES

Discarded tires are more than a disposal problem. At 15,000 Btu/lbm (35 MJ/kg), their energy content is 80% of crude oil. Discarded tires are wasted energy resources. While tires can be incinerated, other processes can be used to gasify them to produce clean synthetic gas (i.e., *syngas*). The low-sulfur, hydrogen-rich syngas can subsequently be burned in combined-cycle plants or used as a feedstock for ammonia and methanol production, or the hydrogen can be recovered for separate use. Tires can also be converted in a nonchemical process into a strong asphalt-rubber pavement. In one process, asphalt binder is mixed with ground tire rubber and held at 375°F (190°C) for 30 to 60 minutes. Pavement is then produced in the normal fashion.

42. TRIHALOMETHANES

Trihalomethanes (THMs) are halogenated disinfection byproducts (DBPs). These organic chemicals are produced during the disinfection of water and, therefore, constitute a drinking water quality problem. The chemically active elements of chlorine, iodine, and bromine react with various *organic precursors* to produce THMs.

Chlorine in various forms is widely used to disinfect water. Bromine can be present in gaseous chlorine as an impurity. Bromine also results from reacting chlorine with the bromide present in high-salinity water. Iodine is seldom used in disinfection. Therefore, only four THMs are found in significant quantities: $CHCl_3$ (trichloromethane, also known as chloroform); $CHBr_3$ (tribromomethane, also known as bromoform); $CHBrCl_2$ (bromodichloromethane); and $CHBr_2Cl$ (dibromochloromethane).

The *organic precursors* that react with chlorine to produce THMs are primarily naturally occurring (e.g., humic and fulvic acids from decaying vegetation). The precursors are not themselves harmful, but the THMs produced from them are presumed to be carcinogenic.[44]

Two main options exist for reducing THMs: formation avoidance and removal after formation. The first option includes (1) using ozone, chlorine dioxide, or potassium permanganate (60 to 90% THM formation reduction), (2) dechlorination after chlorination to prevent reaction of chlorine, and (3) adding ammonia to water prior to discharge to enhance chloramine formation, since chloramines suppress THM formation. The second option includes (1) using activated carbon to remove THMs (25 to 60% THM reduction), (2) using activated carbon, adsorbents, or zeolite filters to remove precursors, (3) moving the chlorination point to the end of the treatment process so that most precursors are removed in earlier processes prior to disinfection (70 to 75% THM reduction), (4) optimizing coagulation and settling water treatment processes to improve precursor removal, and (5) selecting water with fewer precursors.

Follow-up studies are necessary when changing to alternate disinfectants. An alternate disinfectant and its byproducts should be evaluated to determine disinfecting power, residual power, toxicity, and other health effects. Costs of operation will increase when alternate disinfectants are used, although moving the point of application may not result in any significant increase in operating costs after a modest capital expenditure is made.

Ozone, frequently considered as an alternative disinfectant, is less expensive than chlorine dioxide, but it costs more than using chloramines or changing the point of chlorine application. An emerging consensus holds that ozonation alone is relatively ineffective (5 to 20% reduction at typical ozone doses) in controlling THM. Also, ozonation converts the precursors to *biodegradable organic matter* (BOM). Unless a biological process subsequently removes the BOM, aftergrowths in the distribution system may become a problem. Finally, ozonation does not leave any residual disinfectant in the water for use throughout the distribution system.

---

[44]Tests have shown that chloroform in high doses is carcinogenic to rats. Other THMs are considered carcinogenic by association.

Therefore, emphasis is on ozonation in combination with biological treatment to destroy precursors, followed by chlorine application for residual formation.

43. VOLATILE INORGANIC COMPOUNDS

Volatile inorganic compounds (VICs) include H_2S, NOx (except N_2O), SO_2, HCl, NH_3, and many other less common compounds.

44. VOLATILE ORGANIC COMPOUNDS

Volatile organic compounds (VOCs) (e.g., benzene, chloroform, formaldehyde, methylene chloride, naphtalene, phenol, toluene, and trichloroethylene) are highly soluble in water. (VOCs are also listed in Tables 66.10 and 66.16.) VOCs that have leaked from storage tanks or have been discharged often end up in groundwater and drinking supplies.

There are a large number of methods for removing VOCs, including incineration, chemical scrubbing with oxidants, water washing, air or steam stripping, activated charcoal adsorption processes, SCR (selective catalytic reduction), and bioremediation. Incineration of VOCs is fast and 99%+ effective, but incineration requires large amounts of fuel and produces NOx. Using SCR with heat recovery after incineration reduces the energy input but adds to the expense.

45. WATER VAPOR

Water vapor emitted from sources is not generally considered to be a pollutant.

Part 4: Storage, Handling, and Transfer of Hazardous Materials

46. GENERAL STORAGE

Storage of *hazardous materials (hazmats)* is often governed by local building codes in addition to state and federal regulations. Types of construction, maximum floor areas, and building layout may all be restricted.[45]

Good engineering judgment is called for in areas not specifically governed by the building code. Engineering consideration will need to be given to the following aspects of storage facility design: (1) spill containment provisions, (2) chemical resistance of construction and storage materials, (3) likelihood of and resistance to explosions, (4) exiting, (5) ventilation, (6) electrical design, (7) storage method, (8) personnel emergency equipment, (9) security, and (10) spill cleanup provisions.

47. STORAGE TANKS

Underground storage tanks (USTs) have traditionally been used to store bulk chemicals and petroleum products. Fire and explosion risks are low with USTs, but subsurface pollution is common since inspection is limited. Since 1988, the U.S. Environmental Protection Agency (EPA) has required USTs to have secondary containment, corrosion protection, and leak detection. UST operators also must carry insurance in an amount sufficient to clean up a tank failure.

Because of the cost of complying with UST legislation, above-ground storage tanks (ASTs) are becoming more popular. AST strengths and weaknesses are the reverse of USTs: ASTs reduce pollution caused by leaks, but the expected damage due to fire and explosion is greatly increased. Because of this, some local ordinances prohibit all ASTs for petroleum products.[46]

The following factors should be considered when deciding between USTs and ASTs: (1) space available, (2) zoning ordinances, (3) secondary containment, (4) leak detection equipment, (5) operating limitations, and (6) economics.

Most ASTs are constructed of carbon or stainless steel. These provide better structural integrity and fire resistance than fiberglass-reinforced plastic and other composite tanks. Tanks can be either field-erected or factory-fabricated (capacities greater than approximately 50,000 gal (190 kL). Factory-fabricated ASTs are usually designed according to UL-142 (Underwriters Laboratories *Standard for Safety*), which dictates steel type, wall thickness, and characteristics of compartments, bulkheads, and fittings. Most ASTs are not pressurized, but those that are must be designed in accordance with the ASME *Boiler and Pressure Vessel Code*, Section VIII.

NFPA 30 ("Flammable and Combustible Liquids," National Fire Protection Association, Quincy, MA) specifies the minimum separation distances between ASTs, other tanks, structures, and public right-of-ways. The separation is a function of tank type, size, and contents. NFPA 30 also specifies installation, spill control, venting, and testing.

ASTs must be double-walled, concrete-encased, or contained in a dike or vault to prevent leaks and spills, and they must meet fire codes. Dikes should have a capacity in excess (e.g., 110 to 125%) of the tank volume. ASTs (as do USTs) must be equipped with overfill

---

[45]For example, flammable materials stored in rack systems are typically limited to heights of 25 ft (8.3 m).

[46]The American Society of Petroleum Operations Engineers (AS-POE) policy statement states, "Above-ground storage of liquid hydrocarbon motor fuels is inherently less safe than underground storage. Above-ground storage of Class 1 liquids (gasoline) should be prohibited at facilities open to the public."

prevention systems. Piping should be above-ground wherever possible. Reasonable protection against vandalism and hunters' bullets is also necessary.[47]

Though it's a good idea, ASTs are not typically required to have leak-detection systems. Methodology for leak detection is evolving, but currently includes vacuum or pressure monitoring, electronic gauging, and optical and sniffing sensors. Double-walled tanks may also be fitted with sensors within the interstitial space.

Operationally, ASTs present special problems. In hot weather, volatile substances vaporize and represent an additional leak hazard. In cold weather, viscous contents may need to be heated (often by steam tracing).

ASTs are not necessarily less expensive than USTs, but they are generally thought to be so. Additional hidden costs of regulatory compliance, secondary containment, fire protection, and land acquisition must also be considered.

48. DISPOSITION OF HAZARDOUS WASTES

When a hazardous waste is disposed of, it must be taken to a registered *treatment, storage, or disposal facility* (TSDF). The EPA's *land ban* specifically prohibits the disposal of hazardous wastes on land prior to treatment. Incineration at sea is also prohibited. Wastes must be treated to specific maximum concentration limits by specific technology prior to disposal in landfills.

Once treated to specific regulated concentrations, hazardous waste residues can be disposed of by incineration or landfilling, or less frequently, by deep-well injection. All disposal facilities must meet detailed design and operational standards.

Part 5: Testing and Sampling

49. EMISSIONS SAMPLING

There are numerous sampling procedures, as most pollutants have their own specific characteristics, nuances, and idiosyncrasies.

Sampling of emissions in flue gases can either be continuous or by spot sampling using wet chemistry or whole air methods. With *wet chemistry sampling methods*, a sample is collected in a container, solid adsorbent, or liquid-impinger train. The sample is then moved from the field to the laboratory for analysis.

With *whole air sampling*, a volume of flue gas is collected in a bag-like container (i.e., a *Tedlar bag*), and the container is transported to the laboratory for analysis by gas chromatography (GC). Flame ionization detection (GC-FID) and a mass spectrophotometer (GC-MS) are both used. For some pollutants, residence time in the container is critical. Volatile organic compounds (VOCs) with short half-lives can be captured and tested in a sampling train (VOST), which allows for longer sample-holding times.

Particulates are detected and measured by EPA method 5, which though complex, has become a de facto standard, or EPA method 17 (in-stack filtration). With method 5, the sampling device includes a series of absorbers connected in tandem. The absorption train is followed by a gas-drying tube and a vacuum pump. Several samples are commonly taken. The amount of particulates is determined from the filter's weight increase. Particles in the sampling path are washed out, dried, and included in the weight. Stack gas moisture content is determined from the impinger train. Oxygen, carbon dioxide, and (by difference) nitrogen volumes are determined by analyzing a sample collected separately in a sample bag.

Tests of gas streams should be unbiased, which requires sampling to be isokinetic. The gas velocity and ratio of dry gas mass to pollutant mass are not disturbed by *isokinetic testing*.[48,49]

The number of samples taken depends on the proximity of the sampling point to upstream and downstream disturbances. The number of samples can be minimized by locating the sampling point such that flow is undisturbed by upstream and downstream changes of direction, changes in diameter, and equipment in the duct (i.e., after a "good straight run"). Sampling should be at least eight equivalent diameters downstream and two equivalent diameters upstream from disturbances in flow that would cause the flow pattern to be unsymmetrical (i.e., would cause swirling). For rectangular ducts, the equivalent diameter is

$$D_e = \frac{2 \times \text{height} \times \text{width}}{\text{height} + \text{width}} \qquad 66.12$$

Since a pollutant may not be dispersed evenly across the duct, the average concentration must be determined by taking multiple samples across the flow cross section. This process is known as *traversing* the area, and the sampling points are known as *traverse points*. For circular ducts, the cross section is divided into annular

---

[47]Approximately 20% of all spills from ASTs result from vandalism.

[48]With *super-isokinetic* sampling, too little pollution is indicated. This can occur if the vacuum draws a sample such that sampling velocity is higher than the bulk flow velocity. Since particles are subject to inertial forces and gases are not, the particle count will be essentially unaffected, but the metered volume will be higher, resulting in a lower apparent concentration. Conversely, if the sample is drawn too slowly (i.e., *sub-isokinetic* sampling), the metered volume would be lower, and the apparent pollution concentration will be too high.

[49]Solid and liquid substances in flue gases require isokinetic sampling. Gaseous substances do not.

areas according to a standardized procedure. A minimum of six measurements should be taken along each perpendicular diameter. Twenty-four is the usual maximum number of traverse points. (Figure 66.2 shows the locations for a ten-point traverse.) For rectangular ducts, the cross section is divided into a minimum of nine equal rectangular areas of the same shape and with an aspect ratio between 1:1 and 1:2. The traverse points are located at the centroids of these areas.

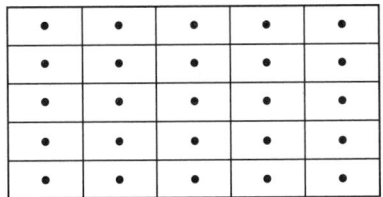

(a) rectangular stack
(measure at center of at least 9 equal areas)

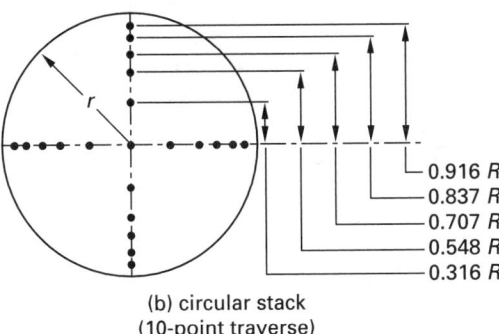

0.916 R
0.837 R
0.707 R
0.548 R
0.316 R

(b) circular stack
(10-point traverse)

Figure 66.2 Traverse Sampling Points

50. EMISSIONS MONITORING SYSTEMS

Continuous emissions monitoring systems (CEMs) are used in large installations, though smaller installations may also be subject to CEM requirements. CEM not only ensures compliance with regulatory limits, but also permits the installation to participate in any allowance trading components of the regulations. CEMs are meant as compliance monitors, not merely operational indicators. The name "continuous emissions monitor system" has come to mean combined equipment for determining and recording SO_2 and NOx concentrations, volumetric flow, opacity, *diluent gases* (O_2 or CO_2), total hydrocarbons, and HCl.

CEMs consist of components that (1) obtain the sample, (2) analyze the sample, and (3) acquire and record analyses in real time.[50] With *straight extractive* CEMs, a sample is drawn from the stack by a vacuum pump and sent down a heated line to an analyzer at the base of

the stack. This method must keep the stack gas heated above its 250 to 300°F (120 to 150°C) dew point to keep the moisture from condensing. Since samples are drawn by vacuum, leaks in the tube can lead to errors.

With *dilution extraction* CEMs, the gas is diluted with clean air or other inert carrier gas and passed to the analyzer located at a lower level. Dilution ratios are generally high—between 50:1 and 300:1, although they may be as low as approximately 12:1 at the outlet of a FGD unit. Since the dew point is lowered to below $-10°F$ ($-23°C$) by the dilution, samples passed to the analyzer do not need to be heated. Samples are passed under positive pressure, ensuring that outside air and other gases cannot enter the tube. Dilution extraction analyses are on a wet-basis.

With in situ CEM methods, direct measurements are taken by advanced electro-optical techniques. All instrumentation is located on the stack or duct. Data is sent by wire to remote monitors and the *data acquisition/reporting system* (DAS). The term *in situ* also implies that the flue gas analysis will be "wet" (i.e., the volumetric fraction will include water vapor).

There are three common methods of determining volumetric flow rates: (1) thermal sensing using a hotwire anemometer or thermal dispersion, (2) differential pressure using a single-point pitot tube or multipoint annubar, and (3) acoustic velocimetry using ultrasonic transducers.

With the advent of mass emission measurement requirements, pollutant (e.g., SO_2 and CO_2) concentration and velocity monitors are being combined.

Part 6: Remediation Processes and Equipment

51. INTRODUCTION

This part of the chapter discusses (in alphabetical order) the methods and equipment that can be used to reduce or eliminate pollution. Legislation often requires use of the *best available control technology* (BACT—also known as the *best available technology*—BAT) and the *maximum achievable control technology* (MACT), *lowest achievable emission rate* (LAER), and *reasonably available control technology* (RACT) in the design of pollution-prevention systems.

52. ABSORPTION, GAS (GENERAL)

Gas *absorption processes* remove a gaseous substance (the *target substance*) from a gas stream by dissolving

---

[50]The term "continuous" is somewhat of a misnomer. Though sampling is "automatic and frequent," data are rarely collected more frequently than once every 15 minutes.

it in a liquid solvent.[51] Absorption can be used for flue gas cleanup (FGC) to remove sulfur dioxide, hydrogen sulfide, hydrogen chloride, chlorine, ammonia, nitrogen oxides, and light hydrocarbons. Gas absorption equipment includes packed towers, spray towers and chambers, and venturi absorbers.[52]

53. ABSORPTION, GAS (SPRAY TOWERS)

In a general spray tower or spray chamber (i.e., a *wet scrubber*), liquid and gas flow countercurrently or crosscurrently. The gas moves through a liquid spray that is carried downward by gravity. A mist eliminator removes entrained liquid from the gas flow. The liquid can be recirculated. The removal efficiency is moderate.

Spray towers are characterized by their low pressure drops—typically 1 to 2 iwg[53] (0.25 to 0.5 kPa), and their liquid-to-gas ratios—typically 20 to 100 gal/1000 ft³ (3 to 14 L/m³). For self-contained units, fan power is also low—approximately 3×10^{-4} kW/ft³ of gas moved.

Flooding of spray towers is an operational difficulty where the liquid spray is carried up the column by the gas stream. Flooding occurs when the gas stream velocity approaches the *flooding velocity*. To prevent this, the tower diameter is chosen as approximately 50 to 75% of the flooding velocity.

Flue gas desulfurization (FGD) wet scrubbers can remove approximately 90 to 95% of the sulfur with efficiencies of 98% claimed by some installations. Stack effluent leaves at approximately 150°F (65°C) and is saturated (or nearly saturated) with moisture. Dense steam plumes may be present unless the scrubbed stream is reheated.

Collected sludge waste and flyash are removed within the scrubber by purely inertial means. After being dewatered, the sludge is landfilled.[54]

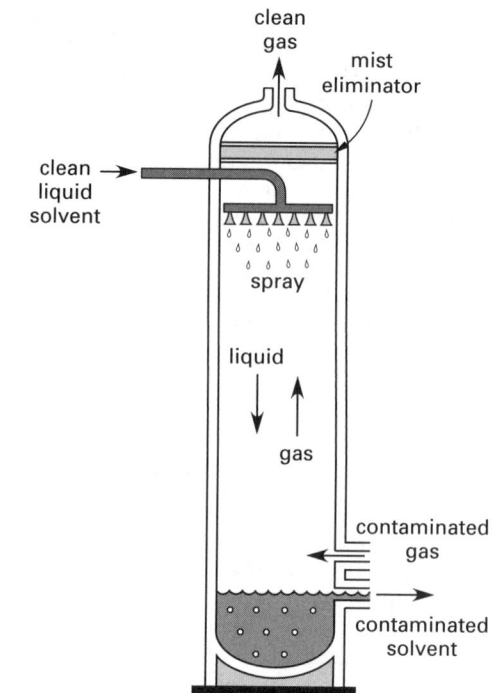

Figure 66.3 *Spray Tower Absorber*

Spray towers, though simple in concept and requiring little energy to operate, are not simple to operate and maintain. Other disadvantages of spray scrubbers include high water usage, generation of wastewater or sludge, and low efficiency at removing particles smaller than 5 μm. Relative to dry scrubbing, the production of wet sludge and the requirement for a sludge-handling system is the major disadvantage of wet scrubbing.

54. ABSORPTION, GAS (IN PACKED TOWERS)

In a packed tower, clean liquid flows from top to bottom over the tower packing media, usually consisting of synthetic engineered shapes designed to maximize liquid-surface contact area. (See Fig. 66.17.) The contaminated gas flows countercurrently, from bottom to top, although crosscurrent designs also exist. As in spray towers, the liquid can be recirculated, and a mist-eliminator is used.

Pressure drops are in the 1 to 8 iwg (0.25 to 2.0 kPa) range. Typical liquid-to-gas ratios are 10 to 20 gal/1000 ft³ (1 to 3 L/m³). Although the pressure drop is greater than in spray towers, the removal efficiency is much higher.

---

[51]Scrubbing, gas absorption, and stripping are distinguished by their target substances, the carrier flow phase, and the directions of flow. *Scrubbing* is the removal of particulate matter from a gas flow by exposing the flow to a liquid or slurry spray. *Gas absorption* is a countercurrent operation for the removal of a target gas from a gas mixture by exposing the mixture to a liquid bath or spray. *Stripping*, also known as *gas desorption* (also a countercurrent operation), is the removal of a dissolved gas or other volatile component from liquid by exposing the liquid to air or steam. Stripping is the reverse operation of gas absorption.
   Packed towers can be used for both gas absorption and stripping processes, and the processes look similar. The fundamental difference is that in gas absorption processes, the target substance (i.e., the substance to be removed) is in the gas flow moving up the tower, and in stripping processes, the target substance is in the liquid flow moving down the tower.
[52]Spray and packed towers, though they are capable, are not generally used for desulfurization of furnace or incineration combustion gases, as scrubbers are better suited for this task.
[53]"iwg" is the abbreviation for "inches water gage," also referred to as iwc for "inches water column" and "inches of water."
[54]A typical 500 MW plant burning high-sulfur fuel can produce as much as 10^7 ft³ (300,000 m³) of dewatered sludge per year.

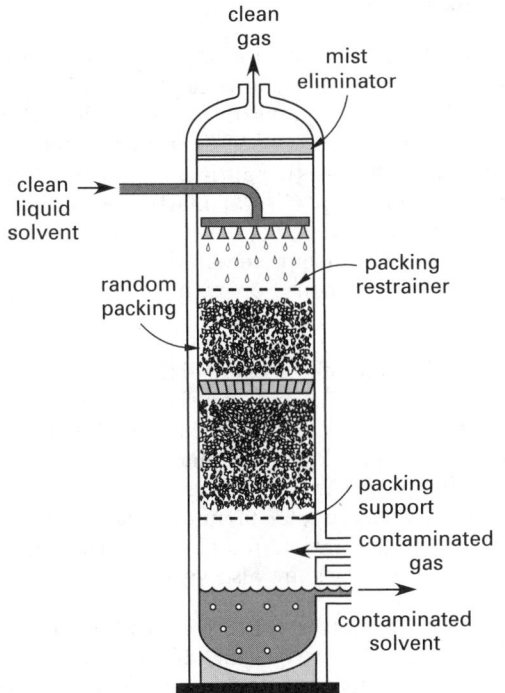

Figure 66.4 *Packed Bed Spray Tower*

55. ADSORPTION (ACTIVATED CARBON)

Granular activated carbon (GAC), also known as *activated carbon* and *activated charcoal* (AC), processes are effective in removing a number of compounds, including volatile organic compounds (VOCs), heavy metals (e.g., lead and mercury), and dioxins.[55] AC is an effective adsorbent[56] for both air and water streams, and a VOC removal efficiency of 99% can be achieved. AC can be manufactured from almost any carbonaceous raw material, but wood, coal, and coconut shell are widely used.

Pollution control processes use AC in both solvent capture-recovery (i.e., recycling) and capture-destruction processes. AC is available in powder and granular form. Granules are preferred for use in recovery systems since the AC can be regenerated when *breakthrough*, also known as *breakpoint*, occurs. This is when the AC has become saturated with the solvent and traces of the solvent begin to appear in the exit air. Until breakpoint, removal efficiency is essentially constant. The *retentivity* of the AC is the ratio of adsorbed solvent mass to carbon mass.

---

[55]The term *activated* refers to the high-temperature removal of tarry substances from the interior of the carbon granule, leaving a highly porous structure.

[56]An *adsorbent* is a substance with high surface area per unit weight, an intricate pore structure, and a hydrophobic surface. An *adsorbent material* traps substances in fluid (liquid and gaseous) form on its exposed surfaces. In addition to activated carbon, other common industrial adsorbents include alumina, bauxite, bone char, Fuller's earth, magnesia, and silica gel.

56. ADSORPTION (SOLVENT RECOVERY)

Activated carbon (AC) for solvent recovery is used in a cyclic process where it is alternately exposed to the target substance and then regenerated by the removal of the target substance. For solvent recovery to be effective, the VOC inlet concentration should be at least 700 ppm. Regeneration of the AC is accomplished by heating, usually by passing low-pressure (e.g., 5 psig) steam over the AC to raise the temperature above the solvent-capture temperature.

There are three main processes for solvent recovery. In a traditional *open-loop recovery system*, a countercurrent air stripper separates VOC from the incoming stream. The resulting VOC air/vapor stream passes through an AC bed. Periodically, the air/vapor stream is switched to an alternate bed, and the first bed is regenerated by passing steam through it. The VOC is recovered from the steam. The VOC-free air stream is freely discharged. Operation of a *closed-loop recovery system* is the same as an open loop system, except that the VOC-free air stream is returned to the air stripper.

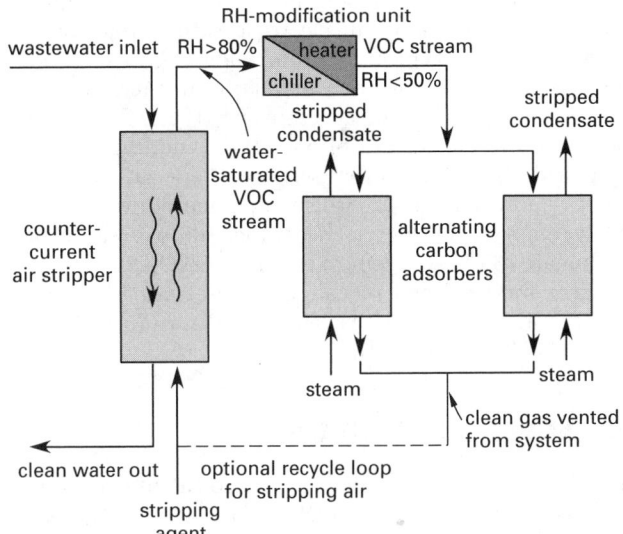

Figure 66.5 *Stripping-AC Process with Recovery*

For VOC-laden gas flows from 5000 to 100,000 SCFM (2.3 to 47 kL/s), a third type of solvent recovery system involves a *rotor concentrator*. VOCs are continuously adsorbed onto a multilayer, corrugated wheel whose honeycomb structure has been coated with powdered AC. The wheel area is divided into three zones: one for adsorption, one for desorption (i.e., regeneration), and one for cooling. Each of the zones is isolated from the other by tight sealing. The wheel rotates at low speed, continuously exposing new portions of the streams to each of the three zones. Though the equipment is expensive, operational efficiencies are high—95 to 98%

57. ADSORPTION, HAZARDOUS WASTE

AC is particularly attractive for flue gas cleanup (FGC) at installations that burn spent oil, electrical cable, biosolids (i.e., sewage sludge), waste solvents, or tires as supplemental fuels. As with liquids, flue gases can be cleaned by passing them through fixed AC. Heavy metal dioxins are quickly adsorbed by the AC—in the first 20 cm or so. AC injection (direct or spray) can remove 60 to 90% of the target substance present.

The spent AC creates its own waste disposal problem. For some substances (e.g., dioxins), AC can be incinerated. Heavy metals in AC must be removed by a wash process. Another process in use is vitrifying the AC and flyash into a glassy, unleachable substance. *Vitrification* is a high-temperature process that turns incinerator ash into a safe, glass-like material. In some processes, heavy-metal salts are recovered separately. No gases and no hazardous wastes are formed.

58. ADVANCED FLUE GAS CLEANUP

Most air pollution control systems reduce NOx in the burner and remove SOx in the stack. *Advanced flue gas cleanup* (AFGC) methods combine processes to remove both NOx and SOx in the stack. Particulate matter is removed by an electrostatic precipitator or baghouse as is typical. Promising AFGC methods include wet scrubbing with metal chelates such as ferrous ethylenediaminetetraacetate (Fe(II)-EDTA) (NOx/SOx removal efficiencies of 60%/90%), adding sodium hydroxide injection to dry scrubbing operations (35% NOx removal), in-duct sorbent injection of urea (80%/NOx removal), in-duct sorbent injection of sodium bicarbonate (35% NOx removal), and the NOXSO process using a fluidized-bed of sodium-impregnated alumina sorbent at approximately 250°F (120°C) (70 to 90%/90% NOx/SOx removal).

59. ADVANCED OXIDATION

The term *advanced oxidation* refers to the use of ozone, hydrogen peroxide, ultraviolet radiation, and other exotic methods that produce free hydroxyl radicals (OH^-). (See also Scrubbing, Chemical.)

Table 66.9 *Relative Oxidation Powers of Common Oxidants*

oxidant	oxidation power (relative to Cl)
fluorine	2.25
hydroxyl radical (OH)	2.05
ozone (O_3)	1.52
permanganate radical (MnO_4)	1.23
chlorine dioxide (ClO_2)	1.10
hypochlorous acid	1.10
chlorine	1.00
bromine	0.80
iodine	0.40
oxygen	0.29

60. BAGHOUSES

Baghouses have a reputation for excellent particulate removal, down to 0.005 grains/ft³ (dry), with particulate emissions of 0.01 grains/ft³ being routine. Removal efficiencies are in excess of 99% and are often as high as 99.99% (weight basis). Fabric filters have a high efficiency for removing particular matter less than 10 μm in size. Because of this, baghouses are effective at collecting air toxics, which preferentially condense on these particles at the baghouse operating temperature—less than 300°F (150°C).

Baghouse filter fabric has micro-sized holes but may be felted or woven, with the material depending on the nature of the flue gas and particulates. Woven fabrics are better suited for lower filtering velocities in the 1 to 2 ft/min (0.3 to 0.7 m/min) range, while felted fabrics are better at 5 ft/min (1.7 m/min). The fabric is often used in tube configuration, but envelopes (i.e., flat bags) and pleated cartridges are also available.

The particulate matter collects on the outside of the bag, forming a cake-like coating. If lime has been introduced into the stream in a previous step, some of the cake will consist of unreacted lime. As the gases pass through this cake, additional neutralizing takes place.

When the pressure drop across the filter reaches a preset limit (usually 6 to 8 iwg (1.5 to 2 kPa)), the cake is dislodged by mechanical shaking, reverse-air cleaning, or pulse-jetting. The dislodged flyash cake falls into collection hoppers below the bags. Flyash is transported by pressure- or vacuum-conveying systems to the conditioning system (consisting of surge bins, rotary feeders, and pug mills).

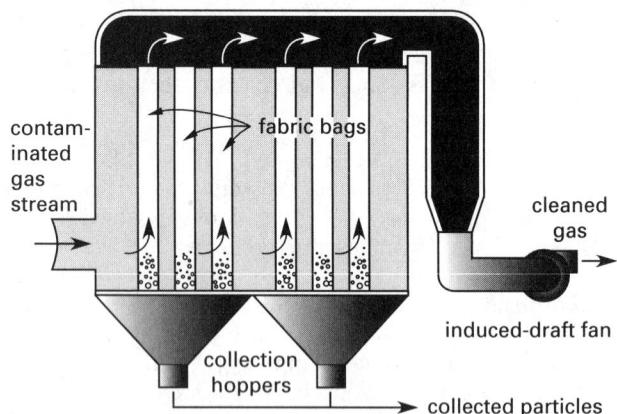

Figure 66.6 *Typical Baghouse*

Baghouses are characterized by their air-to-cloth ratios and pressure drop. The *air-to-cloth ratio*, also known as *filter ratio, superficial face velocity*, and *filtering velocity*, is the ratio of the air volumetric flow rate in ft³/min (m³/s) to the exposed surface area in ft² (m²). After canceling, the units are ft/min (m/s), hence the name *filtering velocity*. The higher the ratio, the smaller the baghouse and the higher the pressure drop. Shaker- and reverse-air baghouses with woven fabrics have typical air-to-cloth ratios ranging from 2.0 to 2.5 ft/min (0.01 to 0.0125 m/s). Pulse-jet collectors with felted fabrics have higher ratios, ranging from 3.5 to 5.0 ft/min (0.0175 to 0.025 m/s).

Advantages of baghouses include high efficiency and performance that is essentially independent of flow rate, particle size, and particle (electrical) resistivity. Also, baghouses produce the lowest opacity (generally less than 10, which is virtually invisible). Disadvantages include clogging, difficult cleaning, and bag breakage.

61. BIOREMEDIATION

Bioremediation encompasses the methods of biofiltration, bioreaction, bioreclamation, activated sludge, trickle filtration, fixed-film biological treatment, landfilling, and injection wells (for in situ treatment of soils and groundwater). Bioremediation relies on microorganisms in a moist, oxygen-rich environment to oxidize solid, liquid, or gaseous organic compounds, producing carbon dioxide and water. Bioremediation is effective for removing volatile organic compounds (VOCs) and easy-to-degrade organic compounds such as BTEXs (benzene, toluene, ethylbenzene, and xylene).[57] Wood-preserving wastes, such as creosote and other polynuclear aromatic hydrocarbons (PAHs), can also be treated.

Bioremediation can be carried out in open tanks, packed columns, beds of porous synthetic materials, composting piles, or soil. The effectiveness of bioremediation depends on the nature of the process, the time and physical space available, and the degradability of the substance. Table 66.10 categorizes gases according to their general degradabilities.

Table 66.10 Degradability of Volatile Organic and Inorganic Gases

rapidly degradable VOCs	rapidly reactive VICs	slowly degradable VOCs	very slowly degradable VOCs
alcohols	H_2S	hydrocarbons[a]	halogenated
aldehydes	NOx (but	phenols	hydrocarbons[b]
ketones	not N_2O)	methylene	polyaromatic
ethers	SO_2	chloride	hydrocarbons
esters	HCl		CS_2
organic acids	NH_3		
amines	PH_3		
thiols	SiH_4		
other	HF		
molecules			
containing O,			
N, or S			
functional groups			

[a]Aliphatics degrade faster than aromatics, such as xylene, toluene, benzene, and styrene.
[b]such as trichloroethylene, trichloroethane, carbon tetrachloride, and pentachlorophenol

[57]BTEXs are common ingredients in gasoline.

Though slow, limited by microorganisms with the specific affinity for the chemicals present, and susceptible to compounds that are toxic to the microorganisms, bioremediation has the advantage of destroying substances rather than merely concentrating them. Bioremediation is less effective when a variety of different compounds are present simultaneously.

62. BIOFILTRATION

The term *biofiltration* refers to the use of composting and soil beds. A *biofilter* is a bed of soil or compost through which runs a distribution system of perforated pipe. Contaminated air or liquid flows through the pipes and into the bed. Volatile organic compounds (VOCs) are oxidized to CO_2 by microorganisms. Volatile inorganic compounds (VICs) are oxidized to acids (e.g., HNO_3 and H_2SO_4) and salts.

Biofilters require no fuel or chemicals when processing VOCs, and the operational lifetime is essentially infinite. For VICs, the lifetime depends on the soil's capacity to neutralize acids. Though reaction times are long and absorption capacities are low (hence the large areas required), the oxidation continually regenerates (rather than depletes) the treatment capacity. Once operational, biofiltration is probably the least expensive method of eliminating VOCs and VICs.

The *removal efficiency* of a biofilter is given by Eq. 66.13. k is an empirical *reaction rate constant*, and t is the *bed residence time* of the carrier fluid (water or air) in the bed. The reaction rate constant depends on the temperature but is primarily a function of the biodegradability of the target substance. Biofiltration typically removes 80 to 99% of volatile organic and inorganic compounds (VOCs and VICs).

$$\eta = 1 - \frac{C_{\text{out}}}{C_{\text{in}}} = 1 - e^{-kt} \qquad 66.13$$

63. BIOREACTION

Bioreactors (reactor tanks) are open or closed tanks containing dozens or hundreds of slowly rotating disks covered with a biological film of microorganisms (i.e., colonies). Closed tanks can be used to maintain anaerobic or other design atmosphere conditions. For example, methanotrophic bacteria are useful in breaking down chlorinated hydrocarbons (e.g., trichlorethylene (TCE), dichloroethylene (DCE), and vinyl chloride (VC)) that would otherwise be considered nonbiodegradable. Methanotrophic bacteria derive their food from methane gas that is added to the bioreactor. An enzyme, known as MMO, secreted by the bacteria breaks down the chlorinated hydrocarbons.

64. BIOVENTING

Bioventing is the treatment of contaminated soil in a large plastic-covered tank. Clean air, water, and nutrients are continuously supplied to the tank while off-gases are suctioned off. The off-gas is cleaned with activated carbon (AC) adsorption or with thermal or catalytic oxidation prior to discharge. Bioventing has been used successfully to remove volatile hydrocarbon compounds (e.g., gasoline and BTEX compounds) from soil.

65. COAL CONDITIONING

Coal intended for electrical generating plants can be modified into "self-scrubbing" coal by conditioning prior to combustion. Conditioning consists of physical separation by size, by cleaning (e.g., cycloning to remove noncombustible material, including up to 90% of the pyritic sulfur), and by the optional addition of limestone and other additives to capture SO_2 during combustion. Small-sized material is pelletized to reduce loss of fines and particulate emissions (dust).

66. CYCLONE SEPARATORS

Inertial separators of the *double-vortex, single cyclone* variety are suitable for collecting medium- and large-sized (i.e., greater than 15 μm) particles from spot sources. During operation, particulate matter in the incoming gas stream spirals downward at the outside and upward at the inside.[58] The particles, because of their greater mass, move toward the outside wall, where they drop into a collection bin.

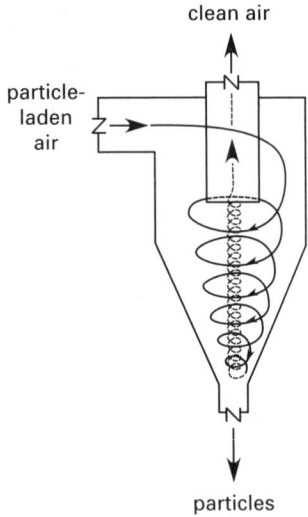

Figure 66.7 *Double-Vortex, Single Cyclone*

The *cut size* is the diameter of particles collected with a 50% efficiency. Separation efficiency varies directly with (1) particle diameter, (2) particle density, (3) inlet velocity, (4) cyclone body length (ratio of cyclone body diameter to outlet diameter), (5) smoothness of inner wall, (6) number of gas revolutions, and (7) amount of particles in the flow. Collection efficiency decreases with increases in (1) gas viscosity, (2) gas density, (3) cyclone diameter, (4) gas outlet diameter, (5) gas inlet duct width, and (6) inlet area.

Collection efficiencies are not particularly high, and dusts (5 to 10 μm) are too fine for most cyclones. For geometrically similar cyclones, the collection efficiency varies directly with the dimensionless *separation factor*, S.

$$S = \frac{v_{inlet}^2}{rg} \qquad 66.14$$

Pressure drop, h, in feet of air across a cyclone can be roughly (i.e., with an accuracy of only approximately $\pm 30\%$) estimated by Eq. 66.15. H and B are the height and width of the rectangular cyclone inlet duct, respectively; D is the gas exit duct diameter; and K is an empirical constant that varies from approximately 7.5 to 18.4, with $K = 13$ being a common design value.

$$h = \frac{KBHv_{inlet}^2}{2gD^2} \qquad 66.15$$

67. DECHLORINATION

Industrial and municipal wastewaters containing excessive amounts of *total residual chlorine* (TRC) must be dechlorinated prior to discharge. Sulfur dioxide, sodium metabisulfate, and sulfite salts are effective dechlorinating agents. For sulfur dioxide, the dose is approximately 0.9 lbm per pound (kg per kg) of chlorine to be removed. Reaction is essentially instantaneous, being completed within ten pipe diameters at turbulent flow. For open channels, a submerged weir may be necessary to obtain the necessary turbulence. TRC is reduced to less than detectable levels.

68. ELECTROSTATIC PRECIPITATORS

Electrostatic precipitators (ESPs) are used to remove particulate matter from gas streams. Collection efficiency for particulate matter is usually in the 95 to 99% range. ESPs are preferred over scrubbers because they are more economical to operate, dependable, and predictable, and because they don't produce a moisture plume.

ESPs used on steam generator/electrical utility units treat gas that is approximately 280 to 300°F (140 to 150°C) with moisture being superheated.[59] This high temperature enhances buoyancy and plume dissipation. However, ESPs generally cannot be used with moist

---

[58]The number of gas revolutions varies approximately between 0.5 and 10.0, with averages of 1.5 revolutions for simple cyclones and 5 revolutions for high-efficiency cyclones.

[59]The flue gas, at approximately 1400°F (760°C), is cooled to this temperature range in a conditioning tower. Injected water increases the moisture content of the flue gas to approximately 25% by volume.

flows, mists, or sticky or hygroscopic particles. Scrubbers should be used in those cases. Relatively humid flows can be treated, although entrained water droplets can insulate particles, lowering their resistivities.

In operation, the gas passes over negatively charged tungsten *corona wires* or grids. Particles are attracted to positively charged collection plates.[60] The speed at which the particles move toward the plate is known as the *drift velocity*, w. Drift velocity is approximately 0.20 to 0.30 ft/sec (0.06 to 0.09 m/s), with 0.25 ft/sec (0.075 m/s) being a reasonable design value. Periodically, *rappers* vibrate the collection plates and dislodge the particles, which drop into collection hoppers.

One of the most important factors affecting the collection efficiency is the *specific collection area* (SCA), the area (in ft^2) of the collection plates divided by the volumetric flow rate (in actual ft^3/min) of the air. The higher the SCA, the higher will be the collection efficiency, with 900 ft^2/1000 ft^3/min (3000 m^2/1000 m^3/min) being at the high end of the scale. Pressure drop across ESPs is low (e.g., less than approximately 0.5 iwg (0.13 kPa)).

Use of ESPs is most efficient when particles have resistivities in the 10^{10} to 10^{14} $\Omega \cdot$cm range. If the particle resistivity is much lower, particles may give up their charge upon contacting the collection plate and become reentrained in (i.e., reintroduced into) the gas. If the resistivity is too high, the particles may be difficult to dislodge and may insulate the collection plates. Very small particles (0.1 to 1.0 μm) reduce the corona efficiency, and particles less than 2 to 3 μm are more easily reentrained.

The theoretical efficiency of a plate precipitator is calculated from Eq. 66.16.[61] K is a measure of the ease with which the particles can be captured, and C depends on the precipitator size, gas velocity and volume, and voltage. K and C are constants for a particular installation. t is the time the particle remains in the electrical field. w is the drift velocity, A_p is the collection plate area, and Q is the volumetric air flow rate.

$$\eta = 1 - K^{Ct} = 1 - e^{\frac{-wA_p}{Q}} \qquad \text{66.16}$$

Advantages of ESPs include high reliability, low maintenance, low power requirements, and low pressure drop. Disadvantages are sensitivity to particle size and resistivity, and the need to heat ESPs during start-up and shut-down to avoid corrosion from acid gas condensation.

---

[60]The *tubular ESP*, used for collecting moist or sticky particles, is a variation on this design.

[61]Design of ESPs is almost entirely empirical, being based on previous experience with similar processes or pilot studies.

Table 66.11 *Typical Electrostatic Precipitator Design Parameters*

parameter	typical range U.S.	SI
efficiency	90 to 98%	
gas velocity	2 to 4 ft/sec	0.6 to 1.2 m/s
gas temperature		
standard	$\leq 700°$F	$\leq 370°$C
high-temperature	$\leq 1000°$F	$\leq 540°$C
special	$\leq 1300°$F	$\leq 700°$C
drift velocity	0.1 to 0.7 ft/sec	0.03 to 0.21 m/s
treatment/		
residence time	2 to 10 sec	
draft pressure loss	0.1 to 0.5 iwg	0.025 to 0.125 kPa
plate spacing	12 to 16 in	30 to 41 cm
plate height	30 to 50 ft	9 to 15 m
plate length/		
height ratio	1.0 to 2.0	
applied voltage	30 to 75 kV	

69. FLUE GAS RECIRCULATION

NOx emissions can be reduced when thermal dissociation is the primary NOx source (as it is when low-nitrogen fuels such as natural gas are burned) by recirculating a portion (15 to 25%) of the flue gas back into the furnace. This process is known as *flue gas recirculation* (FGR). The recirculated gas absorbs heat energy from the flame and lowers the peak temperature. Thermal NOx formation can be reduced by up to 50%. The recirculated gas should not be more than 600°F (315°C).

70. FLUIDIZED-BED COMBUSTORS

Fluidized-bed combustors (FBCs) are increasingly being used in steam/electric generation systems and for destruction of hazardous wastes. A *bubbling bed FBC*, as shown in Fig. 66.8, consists of four major components: (1) a windbox (plenum) that receives the fluidizing/combustion air, (2) an air distribution plate that transmits the air at 10 to 30 ft/sec (3 to 9 m/s) from the windbox to the bed and prevents the bed material from sifting into the windbox, (3) the fluid bed of inert material (usually sand or product ash), and (4) the freeboard area above the bed.

During the operation of a bubbling bed fluidized-bed combustor, the inert bed is levitated by the upcoming air, taking on many characteristics of a fluid (hence, the name FBC). The bed "boils" vigorously, offering an extremely large heat-transfer area, resulting in the thorough mixing of feed combustibles and air. Combustible material usually represents less than 1% of the bed mass, so the rest of the bed acts as a large thermal flywheel.

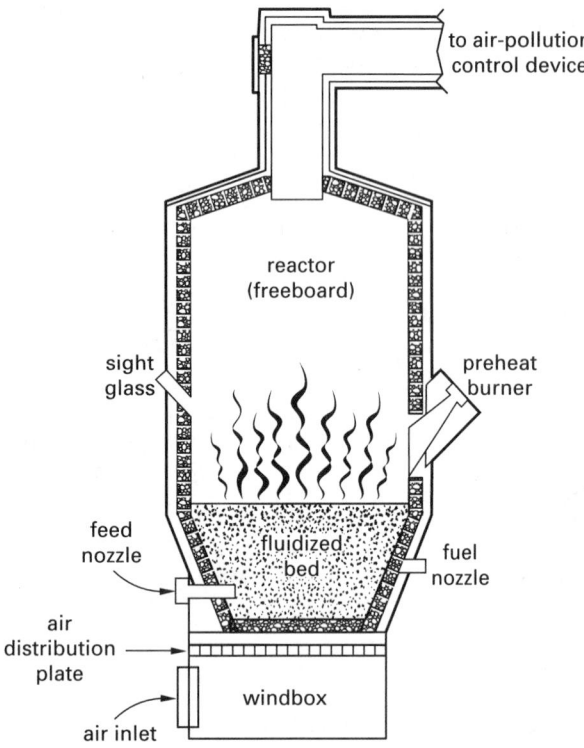

Figure 66.8 *Fluidized-Bed Combustor*

Combustion of volatile materials is completed in the freeboard area. Ash is generally reduced to small size so that it exits with the flue gas. In some cases (e.g., deliberate pelletization or wastes with high ash content), ash can accumulate in the bed. The fluid nature of the bed allows the ash to float on its surface, where it is removed through an overflow drain.

Most FBC systems use forced-air with a single blower at the front end. If there are significant losses due to heat recovery or pollution control systems, an exhaust fan may also be used. In that case, the *null (balanced draft) point* should be in the freeboard area.

Large variations in the composition of the flue gases (known as *puffing*) is minimized by the long residence time and the large heat reservoir of the bed. Air pollution control equipment common to most boilers and incinerators are used with fluidized-bed combustors. Either wet scrubbers or baghouses can be used.

The temperature of the bed can be as high as approximately 1900°F (1040°C), though in most applications, temperatures this high are neither required nor desirable.[62] Most systems operate in the 1400 to 1650°F (760 to 900°C) range.

Three main options exist for reducing temperatures with overautogeneous fuels:[63] (1) Water can be injected

---

[62]It is a common misconception that extremely high temperatures are required to destroy hazardous waste.

[63]A *subautogenous waste* has a heating value too low to sustain combustion and requires a supplemental fuel. Conversely, an *overautogeneous waste* has a heating value in excess of what is required to sustain combustion and requires temperature control.

into the bed. This has the disadvantage of reducing downstream heat recovery. (2) Excess air can be injected. This requires the entire system to be sized for the excess air, increasing its cost. (3) Heat-exchange coils can be placed within the bed itself, in which case, the name *fluidized-bed boiler* is applicable.

Air can be preheated to approximately 1000 to 1250°F (540 to 680°C) for use with subautogenous fuels, or auxiliary fuel can be used.

Contempory FBC boilers for steam/electricity generation are typically of the *circulating fluidized-bed boiler design*. An important aspect of FBC boiler operation is the in-bed gas desulfurization and dechlorination that occurs when limestone and other solid reagents are injected into the combustion area. In addition, circulating fluidized-bed boilers have very low NOx emissions (i.e., less than 200 ppm).

71. INJECTION WELLS

Properly treated and stabilized liquid and low-viscosity wastes can be injected under high pressure into appropriate strata 2000 to 6000 ft (600 to 1800 m) below the surface. The wastes displace natural fluids, and the injection well is capped to maintain the pressure. Injection wells fail primarily by waste plumes through fractures, cracks, fault slips, and seepage around the well casing.

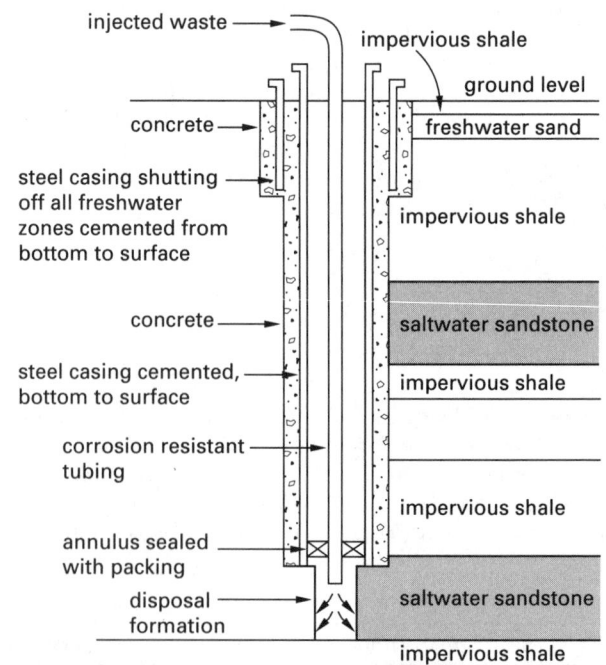

Figure 66.9 *Typical Injection Well Installation*

72. INCINERATORS, FLUIDIZED-BED COMBUSTION

In an FBC, combustion is efficient and excess air required is low (e.g., 25 to 50%). Destruction is essentially complete due to the long residence time (5 to 8 sec for gases, and even longer for solids), high turbulence, and exposure to oxygen. Combustion control is simple, and combustion is often maintained within a 15°F (8°C) band. Both overautogeneous and subautogenous feeds can be handled. Due to the thermal mass of the bed, the FBC temperature drops slowly—at the rate of about 10°F/hr (5°C/h) after shutdown. Start-up is fast and operation can be intermittent.

Relative to other hazardous waste disposal methods, NOx and metal emissions are low, and the organic content of the ash is very low (e.g., below 1%). Because of the long residence time, FBC systems do not usually require afterburners or secondary combustion chambers (SCCs). Due to the turbulent combustion process, the combustion temperature can be 200 to 300°F (110 to 170°C) lower than in rotary kilns. These factors translate into fuel savings, fewer or no NOx emissions, and lower metal emissions.

Most limitations of FBC systems relate to the feed. The feed must be of a form (small size and roughly regular in shape) that can be fluidized. This makes FBCs ideal for non-atomizable liquids, slurries, sludges, tars, and granular solids. It is more appropriate (and economical) to destroy atomizable liquid wastes in a boiler or special injection furnace and large bulky wastes in a rotary kiln or incinerator.

FBCs are usually inapplicable when the feed material or its ash melts below the bed temperature.[64] If the feed melts, it will agglomerate and defluidize the bed. When it is necessary to burn a feed with low melting temperature, two methods can be used. The bed can be operated as a *chemically active bed*, where operation is close to but below the ash melting point of 1350 to 1450°F (730 to 790°C). A small amount of controlled melting is permitted to occur, and the agglomerated ash is removed.

When a higher temperature is desirable in order to destroy hazardous materials, the melting point of feed can be increased by injecting low-cost additives (e.g., calcium hydroxide or kaolin clay). These combine with salts of alkali metals to form refractory-like materials with melting points in the 1950 to 2350°F (1070 to 1290°C) range.

The design of a fluidizing-bed incinerator starts with a heat balance. Energy enters the FBC during combustion of the feed and auxiliary fuel and from sensible heat contributions of the air and fuel. Some of the heat is recovered and reused. The remainder of the heat is lost through sensible heating of the combustion products, water vapor, excess air, and ash, and through radiation from the combustor vessel and associated equipment. Since these items may not all be known in advance, the following assumptions are reasonable.

- Radiation losses are approximately 5%.

- Sensible heat losses from the ash are minimal, particularly if the feed contains significant amounts of moisture.

- Excess air will be approximately 40%.

- Combustion temperature will be approximately 1400°F (760°C).

- If air is preheated, the preheat temperature will be approximately 1000°F (540°C).

The fuel's *specific feed characteristic* (SFC) is the ratio of the higher (gross) heating value to the moisture content. Typical values range from a low of approximately 1000 Btu/lbm (2300 kJ/kg) for wastewater treatment plant biosolids to more than 60,000 Btu/lbm (140 MJ/kg) for barks, sawdust, and RDF.[65] Making the previous assumptions and using the SFC as an indicator, the following generalizations can be made:

- Fuels with SFCs less than 2600 Btu/lbm (6060 kJ/kg) require a 1000°F (540°C) hot windbox and are autogenous at that value and subautogenous below that value. The SFC drops to 2400 Btu/lbm (5600 kJ/kg) with air preheated to 1200°F (650°C). Below the subautogenous SFC, water evaporation is the controlling design factor, and auxiliary fuel is required.

- Fuels with SFC of 4000 Btu/lbm (9300 kJ/kg) are autogenous with a cold windbox and are overautogenous above that value. Combustion is the controlling design factor for overautogenous SFCs.

Example 66.6

A biosolid sludge is dewatered to 75% water by weight. The solids have a higher heating value of 6500 Btu/lbm (15,000 kJ/kg). The dewatered sludge enters a fluidized-bed combustor with a 1000°F (540°C) windbox at the rate of 15,000 lbm/hr (1.9 kg/s). (a) What is the specific feed characteristic? (b) Approximately what energy (in Btu/hr) must the auxiliary fuel supply?

Solution

(a) The dewatered sludge consists of 25% combustible solids and 75% moisture. The total mass of sludge required to obtain 1.0 lbm of water is

$$m_{\text{sludge}} = \frac{m_w}{x_w}$$

$$= \frac{1 \text{ lbm}}{0.75} = 1.333 \text{ lbm}$$

---

[64]The melting temperature of a eutectic mixture of two components may be lower than the individual melting temperatures.

[65]Although the units are the same, the specific feed characteristic is not the same as the heating value.

The mass of combustible solids is

$$m_{\text{solids}} = 0.25 m_{\text{sludge}} = (0.25)(1.333 \text{ lbm})$$
$$= 0.3333 \text{ lbm}$$

The specific feed characteristic is

$$\text{SFC} = \frac{\text{total heat of combustion}}{m_w} = \frac{(m_{\text{solid}})(\text{HHV})}{m_w}$$

$$= \frac{(0.3333 \text{ lbm})\left(6500 \, \dfrac{\text{Btu}}{\text{lbm}}\right)}{1 \text{ lbm}}$$

$$= 2166 \text{ Btu/lbm}$$

(b) Making the assumptions discussed, operation with a 1000°F hot windbox requires a SFC of approximately 2600 Btu/lbm. Therefore, the energy per pound of water in the fuel that an auxiliary fuel must provide is

$$2600 \text{ Btu} - 2166 \text{ Btu} = 434 \text{ Btu}$$

The energy supplied by the auxiliary fuel is

$$\dot{m}_{\text{fuel}} x_w (\text{SFC deficit}) = \left(15{,}000 \, \frac{\text{lbm}}{\text{hr}}\right)(0.75)$$
$$\times \left(434 \, \frac{\text{Btu}}{\text{lbm}}\right)$$
$$= 4.88 \times 10^6 \text{ Btu/hr}$$

73. INCINERATION, GENERAL

Most rotary kiln and liquid injection incinerators have primary and *secondary combustion chambers* (SCCs). Kiln temperatures are approximately 1200 to 1400°F (650 to 760°C) for soil incineration and up to 1700°F (930°C) for other waste types.[66] SCC temperatures are higher—1800°F (980°C) for most hazardous wastes and 2200°F (1200°C) for liquid PCBs. The waste heat may be recovered in a boiler, but the combustion gas must be cooled prior to further processing. *Thermal ballast* can be accomplished by injecting large amounts of excess air (typical when liquid fuels are burned) or by quenching with a water-spray (typical in rotary kilns).

SCCs are necessary to destroy toxics in the offgases. SCCs are vertical units with high-swirl vortex-type burners. These produce high *destruction removal efficiencies* (DREs) with low retention times (e.g., 0.5 sec) and moderate-to-high temperatures, even for chlorinated compounds. When soil with fine clay is incinerated, fines can build up in the SCC, causing slagging and other problems. A refractory-lined cyclone located after the primary combustion chamber can be used to reduce particle carryover to the SCC.

[66]Temperatures higher than 1400°F (760°C) may cause incinerated soil to vitrify and clog the incinerator.

Prior to full operation, incinerators must be tested in a trial burn with a *principal organic hazardous constituent* (POHC) that is in or has been added to the waste. The POHC must be destroyed with a DRE of at least 99.99% by weight.

Table 66.12 Representative Incinerator Performance

	type of incinerator/use			
			soil incinerators	
	rotary kiln	liquid injection	hazard-ous	nonhaz-ardous
waste heating value,				
(Btu/lbm)	15,000	20,000	0	0
(MJ/kg)	35	46	0	0
kiln temp				
(°F)	1700	–	1650	850
(°C)	925	–	900	450
SCC temp				
(°F)	1800–2200	2000–2200	1800	1400
(°C)	980–1200	1100–1200	980	750
SCC mean residence time (sec)	2	2	2	1
O_2 in stack gas (%)	10%	12%	9%	6%

For nontoxic organics, a DRE of 95% is a common requirement. Hazardous wastes require a DRE of 99.99%. Certain hazardous wastes, including PCBs and dioxins, require a 99.9999% DRE. This is known as the *six nines rule*.

Emission limitations of some pollutants depend on the incoming concentration and the height of the stack.[67] Thus, in certain circumstances, raising the stack is the most effective method of being in compliance. This is considered justified on the basis that ground-level concentrations will be lower with higher stacks.

Rules of thumb regarding incinerator performance are:

- Stoichiometric combustion requires approximately 725 lbm (330 kg) of air for each million Btu of fuel or waste burned.

- 100% excess air is required.

- Stack gas dew point is approximately 180°F (80°C).

- Water-spray-quenched flue gas is approximately 40% moisture by weight.

[67]Some of the metallic pollutants treated this way include antimony, arsenic, barium, beryllium, cadmium, chromium, lead, mercury, silver, and thallium.

Example 66.7

What is the mass (in lbm/hr) of water required to spray-quench combustion gases from the incineration of hazardous waste with a heating rate of 50 MMBtu/hr? No additional fuel is added to the incinerator.

Solution

Use the rules of thumb. Assume 100% excess air is required. The total dry air required is

$$m_a = \frac{(2)\left(50 \times 10^6 \dfrac{\text{Btu}}{\text{hr}}\right)(725 \text{ lbm})}{10^6 \dfrac{\text{Btu}}{\text{hr}}}$$

$$= 72{,}500 \text{ lbm/hr} \quad [\text{dry}]$$

Since the quenched combustion gas is 40% water by weight, it is 60% dry air by weight. The total mass of wet combustion gas produced per hour is

$$m_t = \frac{m_a}{x_a} = \frac{m_a}{0.6} = \frac{72{,}500 \dfrac{\text{lbm}}{\text{hr}}}{0.6}$$

$$= 1.21 \times 10^5 \text{ lbm/hr}$$

The required mass of quenching water is

$$m_w = x_w m_t = 0.4 m_t = (0.4)\left(1.21 \times 10^5 \dfrac{\text{lbm}}{\text{hr}}\right)$$

$$= 4.84 \times 10^4 \text{ lbm/hr}$$

74. INCINERATION, HAZARDOUS WASTES

Most hazardous waste incinerators use rotary kilns. Waste in solid and paste form enter a rotating drum where it is burned at 1850 to 2200°F (1000 to 1200°C). (See Table 66.12 for other representative performance characteristics.) Slag is removed at the bottom, and toxic gases exit to a tall, vortex secondary combustion chamber (SCC). Gases remain in the SCC for 2 to 4 sec where they are completely burned at approximately 1850°F (1000°C). Liquid wastes are introduced into and destroyed by the SCC as well. Heat from off-gases may be recovered in a boiler. Typical flue gas cleaning processes include electrostatic precipitation, two-stage scrubbing, and NOx removal.

Common problems with hazardous waste incinerators include (1) inadequate combustion efficiency (easily caused by air leakage in the drum and uneven fuel loading) resulting in incomplete combustion of the primary organic hazardous component (POHC), emission of CO, NOx, and *products of incomplete combustion* (also known as *partially incinerated compounds* or PICS) and metals, (2) meeting low dioxin limits, and (3) minimizing the toxicity of slag and flyash.

These problems are addressed by (1) reducing air leaks in the drum and (2) introducing air to the SCC through multiple sets of ports at specific levels. Gas is burned in the SCC in substoichiometric conditions, with the vortex ensuring adequate mixing to obtain complete combustion. Dioxin formation can be reduced by eliminating the waste-heat recovery process, since the lower temperatures present near waste-heat boilers are ideal for dioxin formation. Once formed, dioxin is removed by traditional end-of-pipe methods.

75. INCINERATION, INFRARED

Infrared incineration (II) is effective for reducing dioxins to undetectable levels. The basic II system consists of a waste feed conveyor, an electrical-heated primary chamber, a gas-fired afterburner, and a typical flue gas cleanup (FGC) system (i.e., scrubber, electrostatic precipitator, and/or baghouse). Electrical heating elements heat organic wastes to their combustion temperatures. Off-gas is burned in a secondary combustion chamber (SCC).

76. INCINERATION, LIQUIDS

Liquid-injection incinerators can be used for atomizable liquids. Such incinerators have burners that fire directly into a refractory-lined chamber. If the liquid waste contains salts or metals, a downfired liquid-injection incinerator is used with a submerged quench to capture the molten material. A typical flue gas cleanup (FGC) system (i.e., scrubber, electrostatic precipitator, and/or baghouse) completes the system.

Incineration of organic liquid wastes usually requires little external fuel, since the wastes are overautogenous and have good heating values. The heating value is approximately 20,000 Btu/lbm (47 MJ/kg) for solvents and approximately 8000 to 18,000 Btu/lbm (19 to 42 MJ/kg) for chlorinated compounds.

77. INCINERATION, MUNICIPAL SOLID WASTE

Incinerator/generator facilities with separation capability are known as *resource-recovery plants*, commonly referred to as *waste-to-energy facilities*. Incineration of *municipal solid waste* (MSF) results in a 90% reduction in waste disposal volume and a mass reduction of 75%.

In *no-boiler incinerators*, MSW is efficiently burned without steam generation. However, incineration is often accompanied by electrical power generation. In *waste-to-energy* (WTE) plants, the combustion heat is used to generate steam and electrical power. *Mass burning* is the incineration of unprocessed MSW, typically on stoker grates or in rotary and waterwall combustors to generate steam. With mass burning, MSW is unloaded into a pit and then moved by crane to the

Figure 66.10 *Large-Scale Hazardous Waste Incinerator*

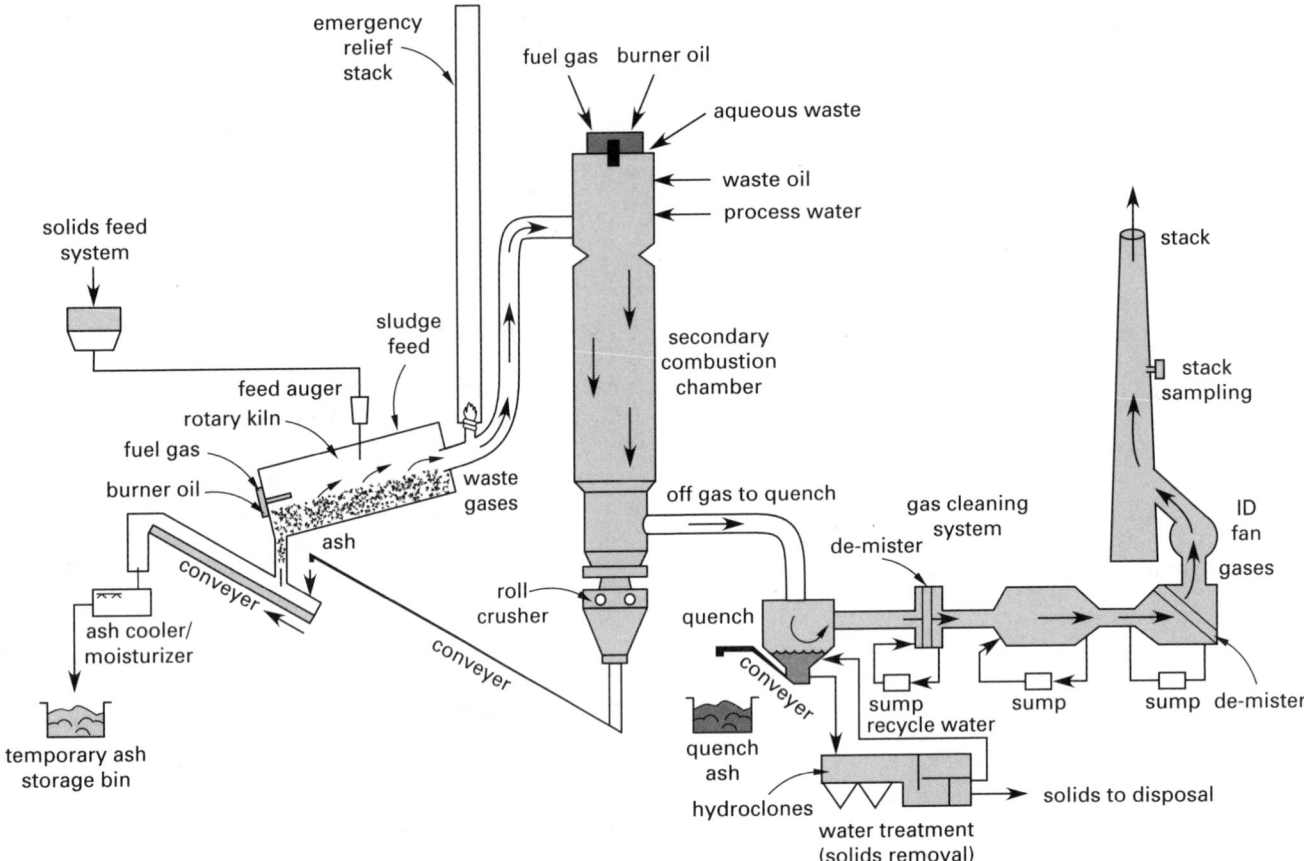

furnace-conveying mechanism. Approximately 27% of the MSW remains as ash, which consists of glass, sand, stones, aluminum, and other noncombustible materials. It and the flyash are usually collected and disposed of in municipal landfills.[68] Capacities of typical mass burn units vary from less than 400 tpd (360 Mg/d) to a high of 3000 tpd (2700 Mg/d), with the majority of units processing 1000 to 2000 tpd (910 to 1800 Mg/d).

MSW, as collected, has a heating value of approximately 4500 Btu/lbm (10 MJ/kg), though higher values have been reported. Per 1000 tpd (1000 Mg/d) of MSW incinerated, the yields are approximately 150,000 to 200,000 lbm/hr (75 to 1000 Mg/h) of 650 to 750 psig (4.5 to 5.2 MPa) steam at 700 to 750°F (370 to 400°C) and 25 to 30 MW (27.6 to 33.1 MW) of gross electrical power. Approximately 10% of the electrical power is used internally, and units generating much less than approximately 10 MW (gross) may use all of their generated electrical power internally.

The *burning rate* varies from approximately 40 to 60 lbm/hr-ft^2 (200 to 300 kg/h·m^2) and is the fueling rate divided by the total effective grate area. Maximum

heat release rates are approximately 300,000 Btu/hr-ft^2 (3.4 GJ/h·m^2). The *heat release rate*, HRR, is defined as

$$\text{HRR} = \frac{(\text{fueling rate})(\text{HHV})}{\text{total effective grate area}} \qquad 66.17$$

Refuse-derived fuel (RDF) is derived from MSW. First generation RDF plants used "crunch and burn" technology. In these plants, the MSW is shredded after ferrous metals are removed magnetically. First-generation RDF plants suffer from the same problems that have plagued early mass burn units, including ash with excessive quantities of noncombustible materials such as glass, grit and sand, and aluminum.

Second-generation plants incorporated screens and air classifiers to reduce noncombustible materials and to increase the recovery of some materials.[69] The ash content is reduced and the energy content of the RDF is increased. The MSW is converted into RDF pellets 2^1/$_2$ to 6 in (6.4 to 15 cm) in size and introduced through feed ports above a traveling grate. Some of the fuel is burned in suspension, with the rest burned on the grate. Grate speed is varied so that the fuel is completely incinerated by the time it reaches the ash rejection ports at the front of the burner.

---

[68]The U.S. Environmental Protection Agency (EPA) has ruled that ash from the incineration of MSW is not a hazardous waste. However, this is hotly contested and is subject to ongoing evaluation.

[69]*Material recovery facilities* (MRFs) specialize in sorting out recyclables from MSF.

Table 66.13 *Typical Ultimate Analyses of MSW and RDF*

| | percentage by weight | |
element	MSW	RDF
carbon	26.65	31.00
water	25.30	27.14
ash	23.65	13.63
oxygen	19.61	22.72
hydrogen	3.61	4.17
chlorine	0.55	0.66
nitrogen	0.46	0.49
sulfur	0.17	0.19
TOTAL	100.00	100.00

Large, 2000 to 3000 tpd (1800 to 2700 Mg/d), third-generation RDF plants, illustrated in Fig. 66.11, retain greater than 95% of the original MSW combustibles while reducing *mass yield* (i.e., the ratio of RDF mass to MSW mass) to below 85%.

RDF has a heating value of approximately 5500 to 5900 Btu/lbm (12 to 14 MJ/kg). The moisture and ash contents of RDF are approximately 24% and 12%, respectively.

Performance of a typical third-generation facility burning RDF is similar to a mass-burn unit. For each 1000 tpd (1000 Mg/d) of MSW collected, approximately 150,000 to 250,000 lbm/hr (75 to 125 Mg/h) of 750 to 850 psig (5.2 to 5.9 MPa) steam at 750 to 825°F (400 to 440°C) can be generated, resulting in approximately 30 to 40 MW (33.1 to 44.1 MW) of electrical power.

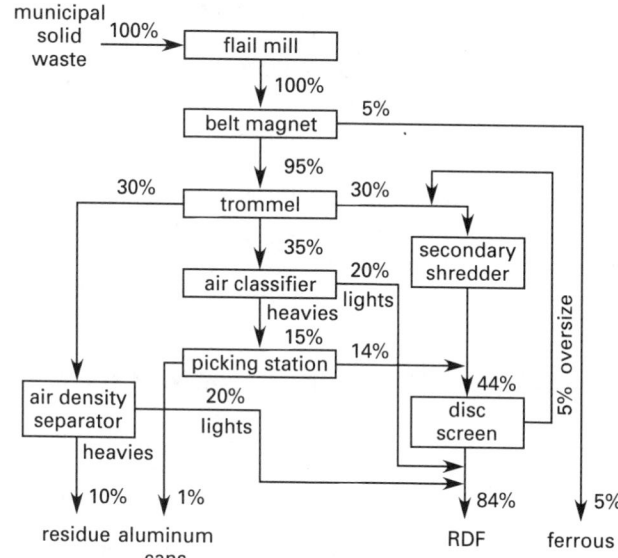

[a]Evaporative water losses not shown.
[b]Special lines to process bulky wastes and shred tires not shown.

Figure 66.12 *Typical RDF Processing*[a,b]

Natural gas is introduced at startup to bring the furnace up to the required 1800°F (1000°C) operating temperature. Natural gas can also be used upon demand, as when a load of particularly wet RDF lowers the furnace temperature.

Coal, oil, or natural gas can be used as a back-up fuel if RDF is unavailable or as an intended part of a *co-firing* design. Co-firing installations are relatively rare, and

Figure 66.11 *Typical Mass-Burn Waste-to-Energy Plant*

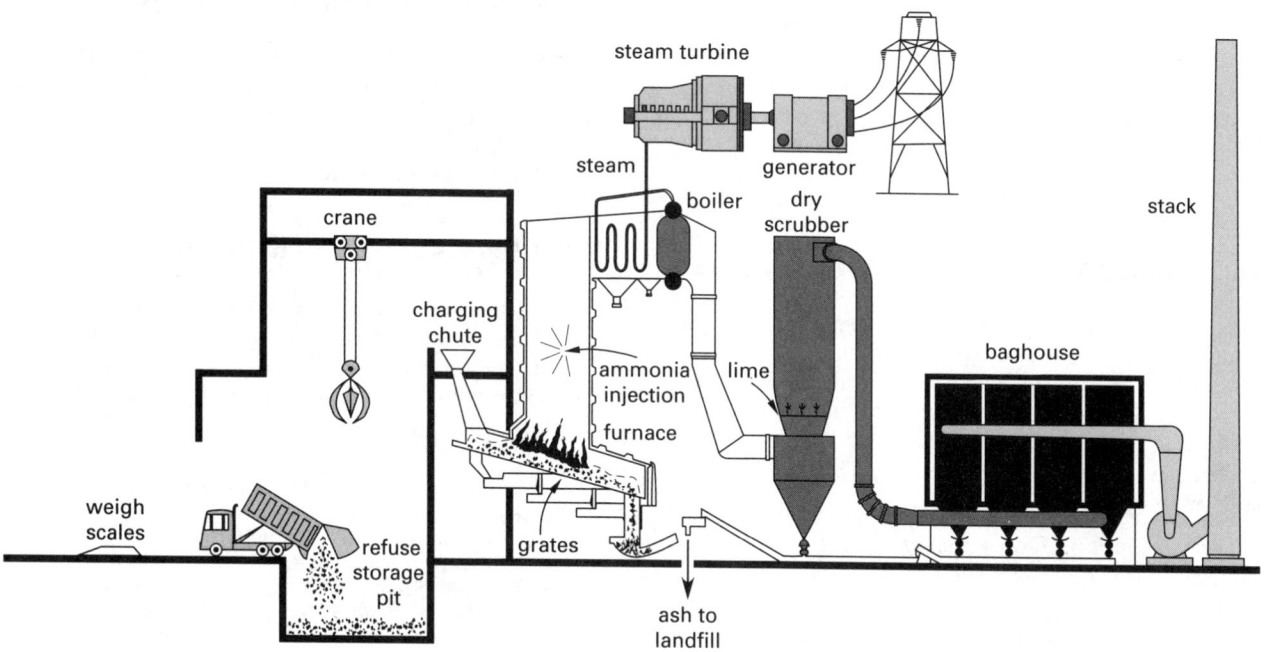

many have discontinued burning RDF due to poor economic performance. Typical problems with co-firing are (1) upper furnace wall slagging, (2) decreased efficiencies in electrostatic precipitators, (3) increased boiler tube corrosion, (4) excessive amounts of bottom ash, and (5) difficulties in receiving and handling RDF.

RDF is a low-sulfur fuel, but like MSW, RDF is high in ash and chlorine. Relative to coal, RDF produces less SO_2 but more hydrogen chloride. Bottom ash can also contain trace organics and lead, cadmium, and other heavy metals.

78. INCINERATION, OXYGEN-ENRICHED

Oxygen-enriched incineration is intended primarily for dioxin removal and is operationally similar to that of a rotary kiln. However, the burner includes oxidant jets. The jets aspirate furnace gases to provide more oxygen for combustion. Apparent advantages are low NOx production and increased incinerator feed rates.

79. INCINERATION, PLASMA

A wide variety of solid, liquid, and gaseous wastes can be treated in a *plasma incinerator*. Wastes are heated by an electric arc to higher than 5000°F (2760°C), dissociating them into component atoms. Upon cooling, atoms recombine into hydrogen gas, nitrogen gas, carbon monoxide, hydrochloric acid, and particulate carbon. The ash cools to a nonleachable, vitrified matrix. Off-gases pass through a normal train of cyclone, baghouse, and scrubbing operations. The process has a very high DRE. However, energy requirements are high.

80. INCINERATION, SOIL

Incineration can completely decontaminate soil. Incinerators are often thought of as being fixed facilities, as are cement kilns and special-use (e.g., Superfund) incinerators. However, mobile incinerators can be brought to sites when the soil quantities are large (e.g., 2000 to 100,000 tons (1800 to 90,000 Mg)) and enough setup space is available.[70] (See also Incineration, Solids and Sludges.)

81. INCINERATION, SOLIDS AND SLUDGES

Rotary kilns and fluidized-bed incinerators are commonly used to incinerate solids and sludges. The feed system depends on the waste's physical characteristics. Ram feeders are used for boxed or drummed solids. Bulk solids are fed via chutes or screw feeders. Sludges are fed via lances or by premixing with solids.

The constant rotation of the shell moves wastes through rotary kilns and promotes incineration. External fuel is required in rotary kilns if the heating value of the waste

is below 1200 Btu/lbm (2800 kJ/mg) (i.e., is subautogenous). Additional fuel is required in the secondary chamber, as well.

Fluidized-bed incinerators work best when the waste is consistent in size and texture. An important benefit is the ability to introduce limestone and other solid reagents to the bed in order to remove HCl and SO_2.

82. LANDFILLING

Since 1992, new and expanded municipal landfills in the United States must satisfy strict design regulations and are designated *"Subtitle D landfills."*[71,72] The regulations apply to any landfill designed to hold *municipal solid waste* (MSW), biosolids, and ash from MSW combustion. Construction of landfills is prohibited near sensitive areas such as airports, floodplains, wetlands, earthquake zones, and geologically unstable terrain. Air quality control methods are required to control emission of dust, odors, and landfill gas. Runoff from storms must be controlled.

Subtitle D landfills have sophisticated liners and leachate collection systems. While states can specify greater protection, the basic bottom requirements are a 30 mil flexible PVC membrane liner (FML)[73] and at least 2 ft of compacted soil with a maximum hydraulic conductivity of 1.2×10^{-5} ft/hr (1×10^{-7} cm/s). A series of wells are required to detect high hydraulic heads and accumulation of heavy metals and volatile organic compounds (VOCs) in the leachate. The minimum thickness is 20 mils for the top FML, and the maximum hydraulic conductivity of the cover soil is 1.8×10^{-5} ft/hr (1.5×10^{-7} cm/s).

83. INCINERATION, VAPORS FROM

Vapor incinerators, also known as *afterburners* and *flares*, convert combustible materials (gases, vapors, and particulate matter) in the stack gas to carbon dioxide and water. Afterburners can be either direct-flame or catalytic in operation.

84. LOW EXCESS-AIR BURNERS

NOx formation in gas-fired boilers can be reduced by maintaining excess air below 5%.[74] *Low excess-air burners* use a forced-draft and self-recirculating combustion chamber configuration to approximate multistaged combustion.

---

[70]Modified asphalt batch processing plants can be used to incinerate soils contaminated with low-heating value, nonchlorinated hydrocarbons.

[71]The reference is to Subtitle D of the RCRA.
[72]Closures are also regulated.
[73]If the membrane is high-density polyethylene (HDPE), the minimum thickness is 60 mils. Since 30 mil PVC costs less than 60 mil HDPE, bidded public projects may end up using the less expensive product, while contracted private projects will opt for the superior protection of the 60 mil product.
[74]Reducing excess air from 30% to 10%, for example, can reduce NOx emissions by 30%.

85. LOW-NOx BURNERS

Low (or *ultralow*) *NOx burners* (LNB) in gas-fired applications use a combination of staged-fuel burning and internal flue gas recirculation (FGR). Recirculation within a burner is induced by either the pressure of the fuel gas or other agents (e.g., medium-pressure steam or compressed air).

86. MECHANICAL SEALS

Fugitive emissions from pumps and other rotating equipment can be reduced or eliminated using current technology by the proper selection and installation of mechanical seals. The three major classes of mechanical seals are single seals, tandem seals (dual seals placed next to each other), and double seals (dual seals mounted back-to-back or face-to-face).

The most economical and reliable sealing device is a *single seal*. It has a minimum number of parts and requires no support devices. However, since the pumped product is usually the lubricant for the seal face, small amounts of the product escape into the environment. Emissions are generally below 1000 ppmv, and are often below 100 ppmv.

Tandem seals consist of two seal assemblies separated by a buffer fluid at a pressure lower than the seal-chamber liquid. The primary inner seal operates at full pump pressure. The outer seal operates in the buffer fluid. Tandem seals can achieve zero emissions when used with a vapor recovery system, provided that the pumped product's specific gravity is less than that of the buffer fluid and the product is immiscible with the buffer fluid.

If the buffer fluid used in a tandem dual seal is a controlled substance, emissions from the outer seal must also meet the emission limits. Seals using glycol as the buffer fluid typically achieve zero emissions. Seals using diesel oil or kerosene have emissions in the 25 to 100 ppmv range.

Double seals are recommended for hazardous fluids with specific gravities less than 0.4 and are a good choice when a vapor recovery system is not available. (Low specific-gravity liquids do not lubricate the seal well.) Double seals consist of two seal assemblies connected by a common collar; they operate with a barrier fluid kept at a pressure higher than that of the pumped product. Double seals can reduce the emission rate to zero.

Figure 66.14 shows the relationship between typical emissions and specific gravity for the different seal types.

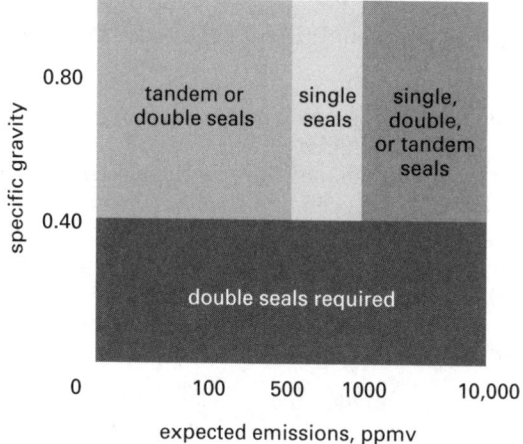

[a] maximum seal size, 6 in (153 mm); maximum pressure, 600 psig (40 bar); maximum speed, 3600 rpm

Figure 66.14 *Representative Emissions to Atmosphere for Mechanical Seals [a] (1 cm from source)*

Figure 66.13 *Subtitle D Liner and Cover Detail*

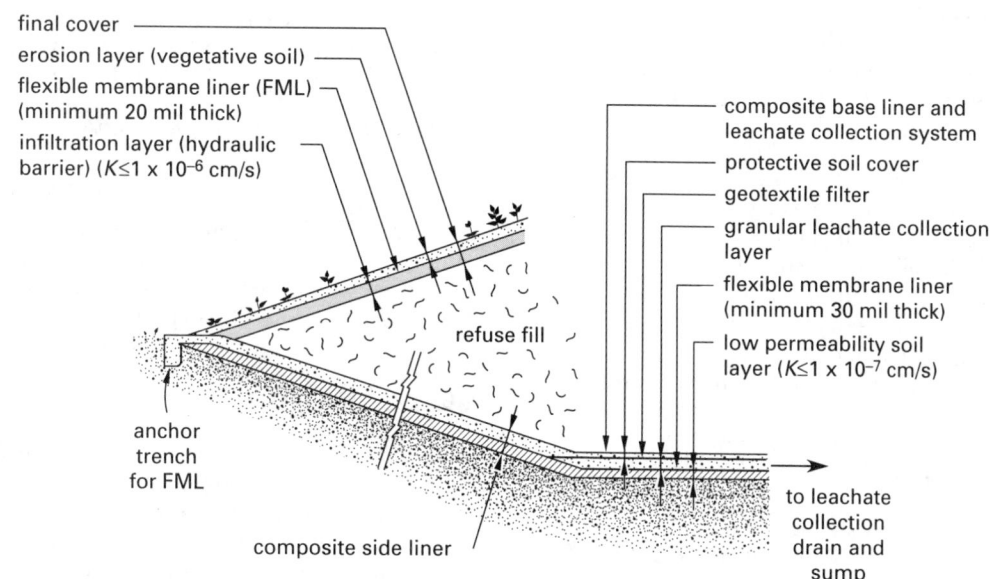

87. MULTIPLE PIPE CONTAINMENT

The U.S. Environmental Protection Agency (EPA) requires *multiple pipe containment* (MPC) in the storage and transmission of hazardous fluids.[75] MPC systems consist of a carrier pipe or bundle of pipes, a common casing, and a leak detector consisting of redundant instrumentation with automatic shutdown features. Casings are generally not pressurized. If one of the pipes within the bundle develops a leak, the fluid will remain in the casing and the detector will signal an alarm and automatically shut off the liquid flow.

Some of the many factors that go into the design of an MPC system are (1) the number of pipes to be grouped together; (2) the weight and strength of the pipes, carrier, and supports; (3) the compatibility of pipe materials and fluids; (4) differential expansion of the container and carrier pipes; and (5) a method of leak detection.

Carrier pipes within the casing should be supported and separated from each other by internal supports (i.e., perforated baffles within the casing). Carrier pipes pass through oval-shaped holes in the internal supports. Holes are oval and oversized to allow flexing of the carrier pipes due to expansion. Other holes through the internal supports allow venting and draining of the casing in the event of a leak. If multiple detectors are used, isolation baffles can be included to separate the casing into smaller sections.

When a pipe bundle changes elevation as well as direction, the carrier pipes may change their positions relative to one another. A pipe on the outside (i.e., at 3 o'clock) may end up being the lower pipe (i.e., at 6 o'clock).[76] This change in relative positioning is known as *rotation*. Rotation complicates the accommodation of pipes that must enter and exit the bundle at specific locations.

Rotation occurs when a change in direction is located at the top or bottom of an elevation change. Therefore, unless rotation can be tolerated, changes in direction should not be combined with changes in elevation. In Fig. 66.15(a), pipe A remains to the left of pipe B. In Fig. 66.15(b), pipe A starts out to the left of pipe B and ends up above pipe B.

Another problem associated with MPC systems is expansion of the containment casing and carrier pipes caused by differences in ambient and fluid temperatures and the use of different pipe materials. For pipes containing direct changes, there must be space at each elbow to permit the pipe expansion. If both ends of a pipe are fixed, expansion and contraction will cause the pipes to flex unless Z-bends or loops are included.

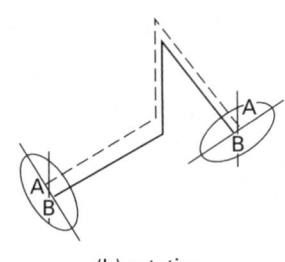

(a) no rotation

(b) rotation

Figure 66.15 *Rotation in Pipe Bundles*

Differential expansion must be calculated from the largest expected temperature difference using Eq. 66.18.

$$\Delta L = \alpha L \Delta T \qquad 66.18$$

Table 66.14 *Approximate Coefficients of Thermal Expansion of Piping Materials (at 70°F (21°C))*

pipe material	U.S. (ft/ft-°F)	SI (m/m·°C)
carbon steel	6.1×10^{-6}	1.1×10^{-5}
chlorinated polyvinyl chloride (CPVC)	3.8×10^{-5}	6.8×10^{-5}
copper	9.5×10^{-6}	1.7×10^{-5}
fiberglass-reinforced polyethylene (FRP)	8.5×10^{-6}	1.5×10^{-5}
polyethylene (PE)	8.3×10^{-5}	1.5×10^{-4}
polyvinyl chloride (PVC)	3.0×10^{-5}	5.4×10^{-5}
stainless steel	9.1×10^{-6}	1.6×10^{-5}

(Multiply ft/ft-°F by 1.8 to get m/m·°C.)

Computer programs are commonly used to design pipe networks. These programs are helpful in locating guides and anchors and in performing a stress analysis on the final design. Drains, purge points, baffles, and joints are generally chosen by experience.

88. OZONATION

Ozonation is one of several advanced oxidation methods capable of reducing pollutants from water. Ozone is a powerful oxidant produced by electric discharge through liquid oxygen. Ozone is routinely used in the water treatment industry for disinfection.

---

[75]In addition to multiple pipe containment, other factors that contribute to reduction of fugitive emissions are (1) the use of tongue-in-groove flanges, and (2) the reduction of as many nozzles as possible from storage tanks and reactor vessels.

[76]It is important to consistently "face" the same way (e.g., in the direction of flow).

Table 66.15 Oxidation of Industrial Wastewater Pollutants[a]

	Cl_22	ClO_2	$KMnO_4$	O_3	H_2O_2	OH^-
amines	C	P	P	P	P	C
ammonia	C	N	P	N	N	N
bacteria	C	C	C	C	P	C
carbohydrates	P	P	P	P	N	C
chlorinated solvents	P	P	P	P	N	C
phenols	P	C	C	C	P	C
sulfides	C	C	C	C	C	C

[a]C = complete reaction; P = partially effective; N = not effective

89. SCRUBBING, GENERAL

Scrubbing is the act of removing particulate matter from a gaseous stream, although substances in gaseous and liquid forms may also be removed.[77]

90. SCRUBBING, CHEMICAL

Chemical scrubbing using oxidizing compounds such as chlorine, ozone, hypochlorite, or permanganate rapidly destroys volatile organic compounds (VOCs). Efficiencies are typically 95% for highly reactive substances, but are lower for hydrocarbons and substances with low reactivities.

91. SCRUBBING, DRY

Dry scrubbing, also known as *dry absorption* and *semi-dry scrubbing*, is one form of sorbent injection. It is commonly used to remove SO_2 from flue gas. In operation, flue gases pass through a scrubbing chamber where a slurry of lime and water is sprayed through them.[78] The slurry is produced by high-speed rotary atomizers. The sorbent (lime, limestone, hydrated lime, sodium bicarbonate, or sodium sesquicarbonate–trona–reagent) is injected into the flue gas in either dry or slurry form. The flue gas heat drives off the water. In either case, a dry powder is carried through the flue gas system. An electrostatic precipitator or baghouse captures the flyash and calcium sulfate particulates.

Dry scrubbing has an SO_2 removal efficiency of approximately 50 to 75%, with some installations reporting 90% efficiencies. NOx removal is not usually intended and is essentially zero. Though the removal efficiency is lower than with wet scrubbing, an advantage of dry scrubbing is that the waste is dry, requiring no sludge-handling equipment. The lower efficiency may be sufficient in older power plants with less stringent regulations. (See also Sorbent Injection.)

---

[77]As it relates to air pollution control, the term *scrubbing* is loosely used. The term may be used in reference to any process that removes any substance from flue gas. In particular, any process that removes SO_2 by passing flue gas through lime or limestone is referred to as *scrubbing*.

[78]In addition to lime, limestone, sodium carbonate, and magnesium oxide can be used. Because they cost less, lime and limestone are the most common.

Particulate removal efficiencies of 90% can be achieved. Since wet scrubbers normally have a lower particulate removal efficiency than baghouses, they are usually combined with electrostatic precipitators.

92. SCRUBBING, WET

See Absorption, Gas (Spray Towers).

93. SELECTIVE CATALYTIC REDUCTION

Selective catalytic reduction (SCR) uses ammonia in combination with a catalyst to reduce nitrogen oxides (NOx) to nitrogen gas and water.[79] Vaporized ammonia is injected into the flue gas; the mixture passes through a catalytic reaction bed approximately 0.5 to 1.0 sec later. The reaction bed can be constructed from honeycomb plates or parallel-ridged plates. Alternatively, the reaction bed may consist of a packed bed of rings, pellets, or other shapes. The catalyst lowers the NOx decomposition activation energy. NOx and NH_3 combine on the catalyst's surface, producing nitrogen and water.

One mole of ammonia is required for every mole of NOx removed. However, in order to maximize the reduction efficiency, approximately 5 to 10 ppm of unreacted ammonia is left behind.[80] The chemical reaction is

$$O_2 + 4NO + 4NH_3 \longrightarrow 4N_2 + 6H_2O$$

The optimum temperature for SCR is 600 to 700°F (315 to 370°C). Gas velocities are typically around 20 ft/sec (6 m/s). The pressure drop is 3 to 4 iwg (0.75 to 1 kPa). NOx removal efficiency is typically 90% but can range from 70 to 95% depending on the application.[81] SCR produces no liquid waste.

Several conditions can cause deactivation of the catalyst. *Poisoning* occurs when trace quantities of specific materials (e.g., arsenic, lead, phosphorous, other heavy metals, silicon, and sulfur from SO_2) react with the catalyst and lower its activity. Poisoning by SO_2 can be reduced by keeping the flue gas temperature above 608°F (320°C) and/or using poison-resisting compounds. *Masking* (or *plugging*) occurs when the catalytic surface becomes covered with fine particle dust, unburned

---

[79]Vanadium oxide, titanium, and platinum are metallic catalysts; zeolites and ceramic catalysts are also used.

[80]Leftover ammonia in the flue gas is referred to as *ammonia slip* and is usually measured in ppm. 50 to 100 ppm would be considered an excessive ammonia slip. Since NOx and NH_3 react on a 1:1 molar basis, slip is easily calculated. In the following equations, the units can be either molar or volumetric (ppmvd, lbm/hr, mol/hr, scfm, etc.), but they must be consistent.

$$NH_{3,slip} = NH_{3,feed} - NH_{3,reacted}$$
$$= NH_{3,feed} - (NOx_{in} - NOx_{out})$$
$$= NH_{3,feed} - (NOx_{in})(\text{removal efficiency})$$

[81]Removal efficiencies of 95 to 99% are possible when SCR is used for VOC removal.

solids, or ammonium salts. Internal cleaning devices remove surface contaminants such as ash deposits and are used to increase the activity of the equipment.

94. SELECTIVE NONCATALYTIC REDUCTION

Selective noncatalytic reduction (SNCR) involves injecting ammonia or urea into the upper parts of the combustion chamber (or into a thermally favorable location downstream of the combustion chamber) to reduce NOx.[82] SNCR is effective when the oxygen content is low (e.g., 1%) and the combustion temperature is controlled. If the temperature is too high, the NH_3 will react more with oxygen than with NOx, forming even more NOx. If the temperature is too low, the reactions slow and unreacted ammonia enters the flue gas. The optimal temperature range is 1600 to 1750°F (870 to 950°C) for NH_3 injection and up to 1900°F (1040°C) for urea. (In general, the temperature ranges for NH_3 and urea are similar. However, various hydrocarbon additives can be used with urea to lower the temperature range.) The reactions are

$$6NO + 4NH_3 \longrightarrow 5N_2 + 6H_2O$$
$$6NO_2 + 8NH_3 \longrightarrow N_2 + 12H_2O$$

The NOx reduction efficiency is approximately 20 to 50% with an ammonia slip of 20 to 30 ppm. The actual efficiency is highly dependent on the injection geometries and interrelations between ammonia slip, ash, and sulfur.

Since the NOx reduction efficiency is relatively low, the SNCR process cannot usually satisfy NOx regulations by itself. The use of urea must be balanced against the potential for ammonia slip and the conversion of NO to nitrous oxide. These and other problems relating to formation of ammonium salts have kept SNCR from gaining widespread popularity in small installations.

95. SOIL WASHING

Soil washing is effective in removing heavy metals, wood preserving wastes (PAHs), and BTEX compounds from contaminated soil. Soil washing is a two-step process. In the first step, soil is mixed with water to dissolve the contaminants. Additives are used as required to improve solubility. In the second step, additional water is used to flush the soil and to separate the fine soil from coarser particles. (Semivolatile materials concentrate in the fines.) Metals are extracted by adding chelating agents to the wash water.[83] The contaminated wash water is subsequently treated.

---

[82] *Urea*, which decomposes to NH_3 and carbon dioxide inside the combustion chamber, is safer and easier to handle.

[83] A *chelate* is a ring-like molecular structure formed by unshared electrons of neighboring atoms. A *chelating agent (chelant)* is an organic compound in which atoms form bonds with metals in solution. By combining with metal ions, chelates control the damaging effects of trace metal contamination. Ethylenediaminetetraacetic acid (EDTA) types are the leading industrial chelants.

96. SORBENT INJECTION

Sorbent injection (FSI) involves injecting a limestone slurry directly into the upper parts of the combustion chamber (or into a thermally favorable location downstream) to reduce SOx. Heat calcines the limestone into reactive lime. Fast drying prevents wet particles from building up in the duct. Lime particles are captured in a scrubber with or without an electrostatic precipitator (ESP). With only an ESP, the SOx removal efficiency is approximately 50%.

97. SPARGING

Sparging is the process of using air injection wells to bubble air through groundwater. The air pushes volatile contaminants into the overlying soil above the aquifer where they can be captured by vacuum extraction.

98. STAGED COMBUSTION

Staged combustion methods are primarily used with gas-fired burners to reduce formation of nitrogen oxides (NOx). Both staged-air burner and staged-fuel burner systems are used. *Staged-air burner systems* reduce NOx production 20 to 35% by admitting the combustion air through primary and secondary paths around each fuel nozzle. The fuel burns partially in the fuel-rich zone. Fuel-borne nitrogen is converted to compounds that are subsequently oxidized to nitrogen gas. Secondary air completes combustion and controls the flame size and shape. Combustion temperature is lowered by recirculation of combustion products within the burner. Staged burners have few disadvantages, the main one being longer flames.

In *staged-fuel burner systems*, a portion of the fuel gas is initially burned in a fuel-lean (air-rich) combustion. The peak flame temperature is reduced, with a corresponding reduction in thermal NOx production. The remainder of the fuel is injected through secondary nozzles. Combustion gases from the first stage dilute the combustion in the second stage, reducing peak temperature and oxygen content. Reductions in NOx formation of 50 to 60% are possible. Flame length is less than with staged-air burners, and the required excess air is lower.

99. STRIPPING, AIR

Air strippers are primarily used to remove volatile organic compounds (VOCs) or other target substances from water. In operation, contaminated water enters a stripping tower at the top, and fresh air enters at the bottom. The effectiveness of the process depends on the volatility of the compound, its temperature and concentration, and the liquid-air contact area. However, removal efficiencies of 80 to 90% (and above) are common for VOCs.

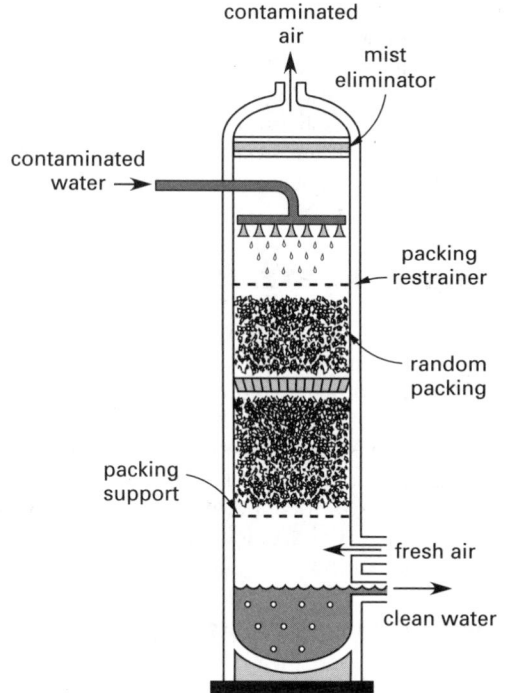

Figure 66.16 *Schematic of Air Stripping Operation*

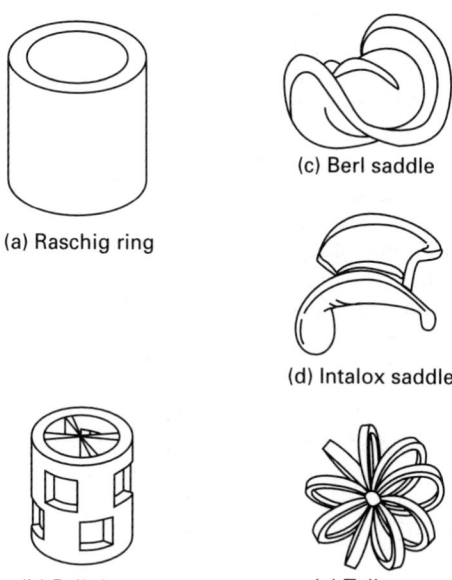

Figure 66.17 *Packing Media Types*

There are three types of stripping towers—*packed towers* filled with synthetic *packing media* (polypropylene balls, rings, saddles, etc.), *spray towers*, and (less frequently for VOC removal) *tray towers* with horizontal trays arranged in layers vertically. *Redistribution rings (wall wipers)* prevent channeling down the inside of the tower. (*Channeling* is the flow of liquid through a few narrow paths rather than an over-the-bed packing.) The stripping air is generated by a small blower at the column base. A mist eliminator at the top eliminates entrained water from the air.

As the contaminated water passes over packing media in a packed tower, the target substance leaves the liquid and enters the air where the concentration is lower. The mole fraction of the target substance in the water, x, decreases; the mole fraction of the target substance in the air, y, increases.

The discharged air, known as *off-gas*, is discharged to a process that destroys or recovers the target substance. (Since the quantities are small, recovery is rarer.) Destruction of the target substance can be accomplished by flaring, carbon absorption, and incineration. Since flaring is dangerous and carbon absorption creates a secondary waste if the AC is not regenerated, incineration is often preferred.

Henry's law, as it applies to water treatment, states that at equilibrium, the vapor pressure of target substance A is directly proportional to the target substance's mole fraction, x_A. (The maximum mole fraction is the substance's *solubility*.) Table 66.16 lists representative values of Henry's law constant, H.[84] When multiple compounds are present in the water, the *key component* is the one that is the most difficult to remove (i.e., the component with the highest concentration and lowest Henry's constant).

$$p_A = H_A x_A \qquad 66.19$$

Table 66.16 *Typical Henry's Law Constants for Selected Volatile Organic Compounds (at low pressures and 25°C)*

VOC	Henry's law constant, H (atm)[a]
1,1,2,2-tetrachloroethane	24.02
1,1,2-trichloroethane	47.0
propylene dichloride	156.8
methylene chloride	177.4
chloroform	188.5
1,1,1-trichloroethane	273.56
1,2-dichloroethene	295.8
1,1-dichloroethane	303.0
hexachloroethane	547.7
hexachlorobutadiene	572.7
trichloroethylene	651.0
1,1-dichloroethene	834.03
perchloroethane	1596.0
carbon tetrachloride	1679.17

(Multiply atm by 101.3 to obtain kPa.)
[a] Henry's law indicates that the units of H are atmospheres per mole fraction. Mole fraction is dimensionless and does not appear in the units for H.

---

[84] Henry's law is fairly accurate for most gases when the partial pressure is less than 1 atm. If the partial pressure is greater than 1 atm, Henry's law constants will vary with partial pressure.

Plant Engineering

The liquid *mass-transfer coefficient*, $K_L a$, is the product of the *coefficient of liquid mass transfer*, K, and the *interfacial area* per volume of packing, a. The mass transfer coefficient is a measure of the efficiency of the air stripper as a whole. The higher the mass transfer coefficient, the higher the efficiency. The $K_L a$ value is largely a function of the size and shape of the packing material, and it is usually obtained from the packing manufacturer or theoretical correlations. For some types of packing, the gas-phase resistance is greater than the liquid-phase resistance, and the gas mass-transfer coefficient, $K_G a$, may be given.

The *stripping factor*, R (the reciprocal of the *absorption factor*), is given by Eq. 66.20. Units for the gas and liquid flow rates, G and L, depend on the correlations used to solve the stripping equations. The air flow rate is limited by the acceptable pressure drop through the tower.

$$R = \frac{HG}{L} \qquad \textit{66.20}$$

The *transfer unit method* is a convenient way of designing and analyzing the performance of stripping towers. A *transfer unit* is a measure of the difficulty of the mass-transfer operation and depends on the solubility and concentrations. The overall number of transfer units is expressed as NTU_{OG} or NTU_{OL} (alternatively, N_{OG} or N_{OL}), depending on whether the gas or liquid resistance dominates.[85] The NTU value depends on the incoming and desired concentrations and the material flow rates.[86]

The *height of the packing*, z, is the effective height of the tower. It is calculated from the NTUs and the *height of a transfer unit*, HTU, using Eq. 66.21. The HTU value depends on the packing media.

$$z = (\text{HTU}_{OG})(\text{NTU}_{OG}) = (\text{HTU}_{OL})(\text{NTU}_{OL}) \quad \textit{66.21}$$

Ten percent additional height should be added to the theoretical value as a safety factor. Packing heights greater than 30 to 40 ft (9.1 to 12.2 m) are not recommended since the packing might be crushed, and greater heights produce little or no increase in removal efficiency.

The blower power depends on the packing shape and size and the air-to-water ratio. Typical pressure drops are 0.5 to 1.0 iwg per vertical foot (0.38 to 0.75 kPa/m) of packing. To minimize the blower power requirement, the ratio of tower diameter to gross packing dimension should be between 8 and 15. Blower fan power increases with increasing air-to-water ratios.

---

[85]The gas film resistance controls when the solubility of the target substance in the liquid is high. Conversely, when the solubility is low, the liquid film resistance controls. In flue gas cleanup (FGC) work, the gas film resistance usually controls.

[86]Stripping is one of many mass-transfer operations studied by chemical engineers. Determining the number of transfer units required and the height of a single transfer unit are typical chemical engineering calculations.

A significant problem for air strippers removing contaminants from water is fouling of the packing media through biological growth and solids (e.g., iron complexes) deposition. Biological growth can be inhibited by continuous or batch use of a *biocide* that does not interfere with the off-gas system.[87] Another problem is *flooding*, which occurs when excess liquid flow rates impede the flow of the air.

100. STRIPPING, STEAM

Steam stripping is more effective than air stripping for removing semi- and non-volatile compounds, such as diesel fuel, oil, and other organic compounds with boiling points up to approximately 400°F (200°C). Operation of a steam stripper is similar to that of an air stripper, except that steam is used in place of the air. Steam strippers can be operated at or below atmospheric pressure. Higher vacuums will remove greater amounts of the compound.

101. THERMAL DESORPTION

Thermal desorption is primarily used to remove volatile organic compounds (VOCs) from contaminated soil. The soil is heated directly or indirectly to approximately 1000°F (540°C) to evaporate the volatiles. This method differs from incineration in that the released gases are not burned but are captured in a subsequent process step (e.g., activated carbon filtration).

102. VACUUM EXTRACTION

Vacuum extraction is used to remove many types of volatile organic compounds (VOCs) from soil. The VOCs are pulled from the soil through a well dug in the contaminated area. Air is withdrawn from the well, vacuuming volatile substances with it.

103. VAPOR CONDENSING

Some vapors can be removed simply by cooling them to below their dew points. Traditional contact- (open) and surface- (closed) condensers can be used.

104. VITRIFICATION

Vitrification melts and forms slag and ash wastes into glass-like pellets. Heavy metals and toxic compounds cannot leach out, and the pellets can be disposed of in

---

[87]A *biocide* is a chemical that kills living things, particularly microorganisms. Biocide categories include chlorinated isocyanurates (used in swimming pool disinfectants and dishwashing detergents), sodium bromide, inorganics (used in wood treatments), quaternaries ("quats") used in hard surface cleaners and sanitizers, and iodophors (used in human skin disinfectants). By comparison, *biostats* do not kill microorganisms already present, but they retard further growth of microorganisms from the moment they are incorporated. Biostats are organic acids and salts (e.g., sodium and potassium benzoate, sorbic acid, and potassium sorbate used "to preserve freshness" in foods).

hazardous waste landfills. Vitrification can occur in the incineration furnace or in a stand-alone process. Stand-alone vitrification occurs in an electrically heated vessel where the temperature is maintained at 2200 to 2370°F (1200 to 1300°C) for up to 20 hr or so. Since the electric heating is nonturbulent, flue gas cleaning systems are not needed.

105. WASTEWATER TREATMENT PROCESSES

Many industrial processes use water for cooling, rinsing, or mixing. Such water must be treated prior to being discharged to *publicly owned treatment works* (POTWs).[88] Table 66.17 lists some of the polluting characteristics of industrial wastewaters.

Table 66.17 Types of Pollution from Industrial Wastewater

ammonia
biochemical oxygen demand (BOD)
carbon, total organic (TOC)
chemical oxygen demand (COD)
chloride
flow rate
metals, soluble
metals, nonsoluble
nitrate
nitrite
organic compounds, acid-extractable
organic compounds, base/neutral extractable
organic compounds, volatile (VOC)
pH
phosphorous
sodium
solids, total suspended (TSS)
sulfate
surfactants
temperature
whole-effluent toxicity (LC_{50})

Most large-volume industrial wastewaters go through the following processes: (1) flow equalization, (2) neutralization, (3) oil and grease removal, (4) suspended solids removal, (5) metals removal, and (6) VOC removal. These processes are similar, in many cases, to processes with the same names used to treat municipal wastewater in a POTW.

Flow equalization reduces the chance of under- and overloading a treatment process. Equalization of both hydraulic and chemical loading is required. *Hydraulic loading* is usually equalized by storing wastewater during high flow periods and discharging it during periods of low flow. *Chemical loading* is equalized by use of an equalization basin with mechanical mixing or air agitation. Mixing with air has the added advantages of oxidizing reducing compounds, stripping away volatiles, and eliminating odors.

There are two reasons for water *neutralization*. First, the pH of water is regulated and must be between 6.0 and 9.0 when discharged. Second, water that is too acidic or alkaline may not be properly processed by subsequent, particularly biological, processes. The optimum pH for biological processes is between 6.5 and 7.5. The effectiveness of the processes is greatly reduced with pHs below 4.0 and above 9.5.

Acidic water is neutralized by adding lime (oxides and hydroxides of calcium and magnesium), limestone, or some other caustic solution. Alkaline waters are treated with sulfuric or hydrochloric acid or carbon dioxide gas.[89] Large volumes of oil and grease float to the surface and are skimmed off. Smaller volumes may require a dissolved-air process. Removing emulsified oil is more complex and may require use of chemical coagulants.

The method used to remove suspended solids depends on the solid size. Most industrial plants do not require strainers, bar screens, and fine screens to remove large solids (i.e., those larger than 1 in (25 mm)). *Grit* (i.e., sand and gravel) is removed in grit chambers by simple sedimentation. Fine solids are categorized into *settleable solids* (diameters more than 1 μm) and *colloids* (diameters between 0.001 μm and 1 μm). Settleable solids and colloids are removed in a sedimentation tank, with chemical coagulants or dissolved-air flotation being used to assist colloidal particles in settling out.

Most metal (e.g., lead) removal occurs by precipitating its hydroxide, although precipitation as carbonates or sulfides is also used.[90] A caustic substance (e.g., lime) is added to the water to raise the pH below the solubility limit of the metal ion. When they come out of solution, the metallic compounds are *flocculated* into larger flakes that ultimately settle out. Floc is mechanically removed as inorganic heavy-metal sludge, which has a 96 to 99% water content by weight. The sludge is dewatered in a drying bed, vacuum filter, or filter press to 65 to 85% water. Depending on its composition, the sludge can be landfilled or treated as a hazardous waste by incineration or other methods.

Volatile organic compounds (VOCs) such as benzene and toluene are removed by air stripping or adsorption in activated charcoal (AC) towers. Nonvolatile organic compounds (NVOCs) are removed by biological processes such as lagooning, trickle filtration, and activated sludge.

Although destruction of biological health hazards is not always needed for industrial wastewater, chemical oxidants (see Table 66.9) are still needed to destroy odors, control bacterial growth downstream, and eliminate sulfur compounds.[91] Chemical oxidants can also reduce heavy metals (e.g., iron, manganese, silver, and lead) that were not removed in previous operations.

---

[88]Rainwater runoff from some industrial plants can also be a hazardous waste.

[89]In water, carbon dioxide gas forms *carbonic acid.*
[90]Ion exchange, activated charcoal, and reverse osmosis methods can be used but may be more expensive.
[91]Meat-packing and dairy (e.g., cheese) plants are examples of industrial processes that require chlorination to destroy bacteria in wastewater.

Plant Engineering

67 Electricity and Electrical Equipment

1. Electric Charge 67-2
2. Current 67-2
3. Voltage Sources 67-2
4. Resistivity and Resistance 67-2
5. Conductivity and Conductance 67-2
6. Ohm's Law 67-3
7. Power in DC Circuits 67-3
8. Electrical Circuit Symbols 67-3
9. Resistors in Combination 67-3
10. Simple Series Circuits 67-3
11. Simple Parallel Circuits 67-4
12. Voltage and Current Dividers 67-4
13. AC Voltage Sources 67-4
14. Maximum, Effective, and Average Values . . 67-5
15. Impedance 67-5
16. Reactance 67-5
17. Admittance 67-5
18. Resistors in AC Circuits 67-5
19. Capacitors 67-5
20. Inductors 67-6
21. Transformers 67-6
22. Ohm's Law for AC Circuits 67-6
23. Phase Angle 67-6
24. Power in AC Circuits 67-6
25. Power Factor 67-7
26. Cost of Electrical Energy 67-7
27. Power Factor Correction 67-8
28. Three-Phase Electricity 67-8
29. Three-Phase Loads 67-9
30. Line and Phase Values 67-9
31. Input Power Three-Phase Systems 67-9
32. Rotating Machines 67-10
33. Regulation 67-10
34. Torque and Power 67-10
35. NEMA Motor Classifications 67-11
36. Nameplate Values 67-11
37. Service Factor 67-11
38. Duty Cycle 67-12
39. Induction Motors 67-12
40. Typical Induction Motor Performance . . . 67-13
41. Synchronous Motors 67-13
42. DC Machines 67-13
43. Choice of Motor Types 67-15
44. Losses in Rotating Machines 67-15
45. Efficiency of Rotating Machines 67-15
46. High-Efficiency Motors and Drives 67-16

Nomenclature

a	turns ratio	–	–
A	area	ft^2	m^2
B	susceptance	S	S
C	capacitance	F	F
d	diameter	ft	m
E	energy usage (demand)	kW-hr	kW·h
f	linear frequency	Hz	Hz
G	conductance	$1/\Omega$	$1/\Omega$
i	varying current	A	A
I	current	A	A
l	length	ft	m
L	inductance	H	H
n	speed	rpm	rpm
N	number of turns	–	–
pf	power factor	–	–
p	number of poles	–	–
P	power	W	W
Q	charge	C	C
Q	reactive power	VAR	VAR
R	resistance	Ω	Ω
s	slip	–	–
S	apparent power	VA	VA
sf	service factor	–	–
SR	speed regulation	–	–
t	time	sec	s
T	period	sec	s
T	torque	ft-lbf	N·m
v	varying voltage	V	V
V	voltage	V	V
VR	voltage regulation	–	–
X	reactance	Ω	Ω
Y	admittance	S	S
Z	impedance	Ω	Ω

Symbols

η	efficiency	–	–
θ	phase angle	rad	rad
ρ	resistivity	Ω-ft	Ω·m
σ	conductivity	$1/\Omega$-ft	$1/\Omega$·m
ϕ	angle	rad	rad
ω	angular frequency	rad/sec	rad/s

Subscripts

a	armature
ave	average
C	capacitive
Cu	copper
e	equivalent
eff	effective

f field
l line
L inductive
m maximum
p phase or primary
R resistive
s secondary
t total

1. ELECTRIC CHARGE

The charge on an electron is one *electrostatic unit* (esu). Since an esu is very small, electrostatic charges are more conveniently measured in *coulombs* (C). One coulomb is approximately 6.24×10^{18} esu. Another unit of charge, the *faraday*, is sometimes encountered in the description of ionic and plating processes. One faraday is equal to one mole of electrons, approximately 96,500 C.

2. CURRENT

Current, I, is the movement of electrons. By historical convention, current moves in a direction opposite to the flow of electrons (i.e., current flows from the positive terminal to the negative terminal). Current is measured in *amperes* (A) and is the time rate change in charge.

$$I = \frac{dQ}{dt} \qquad 67.1$$

3. VOLTAGE SOURCES

A net *voltage*, V, causes electrons to move, hence the common synonym *electromotive force* (emf).[1] With *direct-current* (DC) voltage sources, the voltage may vary in amplitude, but not in polarity. In simple problems where a battery serves as the voltage source, the magnitude is also constant.

With *alternating-current* (AC) voltage sources, the magnitude and polarity both vary with time. Due to the method of generating electrical energy, AC voltages are typically sinusoidal.

4. RESISTIVITY AND RESISTANCE

Resistance is the property of a *resistor* or resistive circuit to impede current flow.[2] A circuit with zero resistance is a *short circuit*, whereas an *open circuit* has infinite resistance. Adjustable resistors are known as *potentiometers* and *rheostats*.

---

[1] The symbol E, (derived from the name electromotive force), has been commonly used to represent voltage induced by electromagnetic induction.
[2] *Resistance* is not the same as *inductance* (Sec. 29), which is the property of a device to impede a *change* in current flow.

Resistance of a circuit element depends on the *resistivity*, ρ (in Ω-in or $\Omega \cdot$cm), of the material and the geometry of the current path through the element. The area, A, of circular conductors is often measured in *circular mils*, abbreviated *cmils*, the area of a 0.001 in diameter circle.

$$R = \frac{\rho l}{A} \qquad 67.2$$

$$A_{\text{cmils}} = \left(\frac{d_{\text{in}}}{0.001} \right)^2 \qquad 67.3$$

$$A_{\text{in}^2} = (7.854 \times 10^{-7})(A_{\text{cmils}}) \qquad 67.4$$

Resistivity depends on temperature. For most conductors, resistivity increases linearly with temperature. The resistivity of standard *IACS (International Annealed Copper Standard)* copper wire at $20°C$ is approximately

$$\rho_{\text{IACS Cu,20°C}} = 1.7241 \times 10^{-6} \ \Omega \cdot \text{cm}$$

$$= 0.67879 \times 10^{-6} \ \Omega \text{-in}$$

$$= 10.371 \ \Omega \text{-cmil/ft} \qquad 67.5$$

Example 67.1

What is the resistance of a parallelepiped (1 cm $\times$ 1 cm $\times$ 1 m long) of IACS copper if current flows between the two smaller faces?

Solution

From Eqs. 67.2 and 67.5, the resistance is

$$R = \frac{\rho l}{A} = \frac{(1.7241 \times 10^{-6} \ \Omega \cdot \text{cm})(100 \ \text{cm})}{(1 \ \text{cm})(1 \ \text{cm})}$$

$$= 1.724 \times 10^{-4} \ \Omega$$

5. CONDUCTIVITY AND CONDUCTANCE

The reciprocals of resistivity and resistance are *conductivity* (σ) and *conductance* (G), respectively. The unit of conductance is the *siemens* (S).[3]

$$\sigma = \frac{1}{\rho} \qquad 67.6$$

$$G = \frac{1}{R} \qquad 67.7$$

Percent conductivity is the ratio of a substance's conductivity to the conductivity of standard IACS copper. (See Sec. 4.)

$$\% \ \text{conductivity} = \frac{\sigma}{\sigma_{\text{Cu}}} \times 100\%$$

$$= \frac{\rho_{\text{Cu}}}{\rho} \times 100\% \qquad 67.8$$

---

[3] The siemens is the inverse of an ohm and is the same as the obsolete unit, the *mho*.

6. OHM'S LAW

The *voltage drop*, also known as the *IR drop*, across a circuit or circuit element with resistance R is given by *Ohm's law*.[4]

$$V = IR \quad \text{[DC circuits]} \qquad 67.9$$

$$v(t) = i(t)R \quad \text{[AC circuits]} \qquad 67.10$$

7. POWER IN DC CIRCUITS

The *power*, P (in watts), dissipated across two terminals with resistance R and voltage drop V is given by *Joule's law*, Eq. 67.11.

$$P = IV = I^2R = \frac{V^2}{R} = V^2G \quad \text{[DC circuits]} \qquad 67.11$$

$$P(t) = i(t)v(t) = i(t)^2R = \frac{v(t)^2}{R} = v(t)^2G$$
$$\text{[AC circuits]} \qquad 67.12$$

8. ELECTRICAL CIRCUIT SYMBOLS

Figure 67.1 illustrates symbols typically used to diagram electrical circuits in this book.

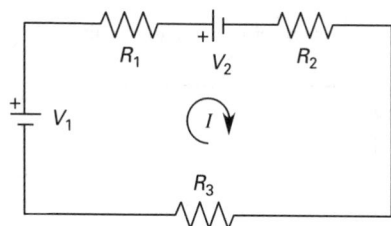

(a) DC voltage source   (b) AC voltage source   (c) current source

(d) resistor   (e) inductor   (f) capacitor

N_1 N_2

(g) transformer

Figure 67.1 *Symbols for Electrical Circuit Elements*

9. RESISTORS IN COMBINATION

Resistors in series are added to obtain the total (equivalent) resistance of a circuit.

$$R_e = R_1 + R_2 + R_3 + \cdots \quad \text{[series]} \qquad 67.13$$

Resistors in parallel are combined by adding their reciprocals. This is a direct result of the fact that conductances in parallel add.

$$G_e = G_1 + G_2 + G_3 + \cdots \quad \text{[parallel]} \qquad 67.14$$

$$\frac{1}{R_e} = \frac{1}{R_1} + \frac{1}{R_2} + \frac{1}{R_3} + \cdots \quad \text{[parallel]} \qquad 67.15$$

---

[4]This book uses the convention that uppercase letters represent fixed, maximum, or effective values, and lowercase letters represent values that change with time.

For two resistors in parallel, the equivalent resistance is

$$R_e = \frac{R_1R_2}{R_1 + R_2} \quad \text{[two parallel resistors]} \qquad 67.16$$

10. SIMPLE SERIES CIRCUITS

Figure 67.2 illustrates a simple series DC circuit and its equivalent circuit.

- The current is the same through all circuit elements.

$$I = I_{R1} = I_{R2} = I_{R3} \qquad 67.17$$

- The equivalent resistance is the sum of the individual resistance.

$$R_e = R_1 + R_2 + R_3 \qquad 67.18$$

- The equivalent applied voltage is the sum of all voltage sources (polarities considered).

$$V_e = \pm V_1 \pm V_2 \qquad 67.19$$

- The sum of all of the voltage drops across the components in the circuit (a *loop*) is equal to the equivalent applied voltage. This fact is known as *Kirchhoff's voltage law*.

$$V_e = \Sigma IR_j = I\Sigma R_j = IR_e \qquad 67.20$$

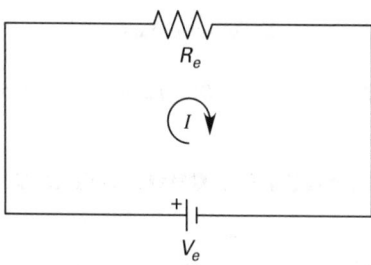

(a) original series circuit

(b) equivalent circuit

Figure 67.2 *Simple Series DC Circuit and Its Equivalent*

Plant Engineering

11. SIMPLE PARALLEL CIRCUITS

Figure 67.3 illustrates a simple parallel DC circuit with one voltage source and its equivalent circuit.

- The voltage drop is the same across all legs.

$$V = V_{R1} = V_{R2} = V_{R3}$$
$$= I_1 R_1 = I_2 R_2 = I_3 R_3 \qquad \text{67.21}$$

- The reciprocal of the equivalent resistance is the sum of the reciprocals of the individual resistances.

$$\frac{1}{R_e} = \frac{1}{R_1} + \frac{1}{R_2} + \frac{1}{R_3} \qquad \text{67.22}$$

$$G_e = G_1 + G_2 + G_3 \qquad \text{67.23}$$

- The sum of all of the leg currents is equal to the total current. This fact is an extension of *Kirchhoff's current law*: The current flowing out of a connection (*node*) is equal to the current flowing into it.

$$I = I_1 + I_2 + I_3$$
$$= \frac{V}{R_1} + \frac{V}{R_2} + \frac{V}{R_3} = V(G_1 + G_2 + G_3) \qquad \text{67.24}$$

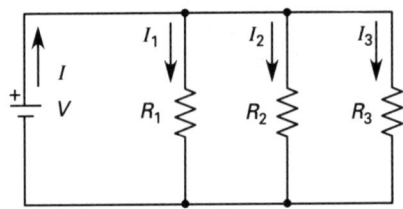

(a) original parallel circuit

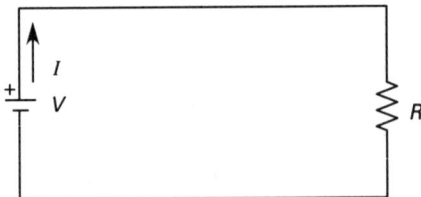

(b) equivalent circuit

Figure 67.3 *Simple Parallel DC Circuit and Its Equivalent*

12. VOLTAGE AND CURRENT DIVIDERS

Figure 67.4(a) illustrates a *voltage divider circuit*. The voltage across resistor 2 is

$$V_2 = V \left(\frac{R_2}{R_1 + R_2} \right) \qquad \text{67.25}$$

Figure 67.4(b) illustrates a *current divider circuit*. The current through resistor 2 is

$$I_2 = I \left(\frac{R_1}{R_1 + R_2} \right) \qquad \text{67.26}$$

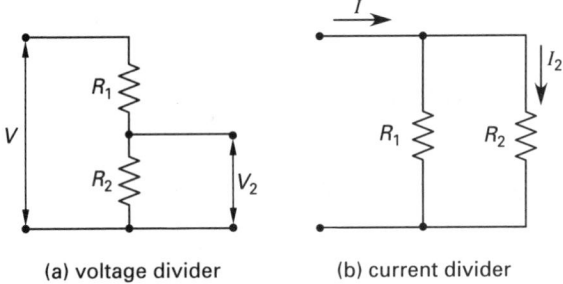

(a) voltage divider   (b) current divider

Figure 67.4 *Divider Circuits*

13. AC VOLTAGE SOURCES

The term *alternating current* (AC) almost always means that the current is produced from the application of a voltage with sinusoidal waveform.[5] Sinusoidal variables can be specified without loss of generality as either sines or cosines. If a sine waveform is used, Eq. 67.27 gives the instantaneous voltage as a function of time. V_m is the *maximum value*, also known as the *amplitude*, of the sinusoid. If $v(t)$ is not zero at $t = 0$, a *phase angle*, θ, must be included.

$$v(t) = V_m \sin(\omega t + \theta) \quad \text{[trigonometric form]} \qquad \text{67.27}$$

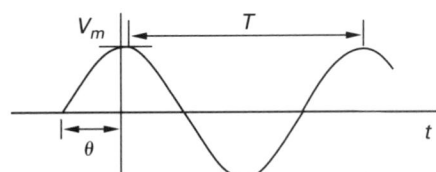

Figure 67.5 *Sinusoidal Waveform*

Figure 67.5 illustrates the *period*, T, of the waveform. The *frequency*, f (also known as *linear frequency*), of the sinusoid is the reciprocal of the period and is expressed in hertz (Hz).[6] *Angular frequency*, ω, in radians per second (rad/s), can also be specified.

$$f = \frac{1}{T} = \frac{\omega}{2\pi} \qquad \text{67.28}$$

$$\omega = 2\pi f = \frac{2\pi}{T} \qquad \text{67.29}$$

---

[5]Other alternating waveforms commonly encountered in commercial applications are the square, triangular, and sawtooth waveforms.
[6]In the United States, the standard frequency is 60 Hz. In Japan, the British Isles and Commonwealth countries, continental Europe, and some Mediterranean, Near Eastern, African, and South American countries, the standard is 50 Hz.

14. MAXIMUM, EFFECTIVE, AND AVERAGE VALUES

The *maximum value* (Fig. 67.5) of a sinusoidal voltage is usually not specified in commercial and residential power systems. The *effective value*, also known as the *root-mean-square (rms) value*, is usually specified when referring to single- and three-phase voltages.[7] A DC current equal in magnitude to the effective value of a sinusoidal AC current produces the same heating effect as the sinusoid. The scale reading of a typical AC current meter is proportional to the effective current.

$$V_{\text{eff}} = \frac{V_m}{\sqrt{2}} \approx 0.707 V_m \qquad 67.30$$

The *average value* of a symmetrical sinusoidal waveform is zero. However, the average value of a rectified sinusoid (or the average value of a sinusoid taken over half of the cycle) is $V_{\text{ave}} = 2V_m/\pi$. A DC current equal to the average value of a *rectified AC current* has the same electrolytic action (e.g., capacitor charging, plating, and ion formation) as the rectified sinusoid.[8]

15. IMPEDANCE

Simple alternating current circuits can be composed of three different types of passive circuit components—resistors, inductors, and capacitors. Each type of component affects both the magnitude of the current flowing as well as the phase angle (Sec. 23) of the current. For individual and combinations of components, these two effects are quantified by the *impedance*, $\mathbf{Z}$. Impedance is a complex quantity with a magnitude (in ohms) and an associated *impedance angle*, ϕ. It is usually written in *phasor (polar) form* (e.g., $\mathbf{Z} \equiv Z/\underline{\phi}$).

Multiple impedances in a circuit combine in the same manner as resistances: Impedances in series add; reciprocals of impedances in parallel add. However, the addition must use complex (i.e., vector) arithmetic.

16. REACTANCE

Impedance, like any complex quantity, can also be written in rectangular form. In this case, impedance is written as the complex sum of the resistive (R) and reactive (X) components, both having units of ohms. The resistive and reactive components combine trigonometrically in the impedance triangle. The reactive component, X, is known as the *reactance*.

$$\mathbf{Z} \equiv R \pm jX \qquad 67.31$$

$$R = Z\cos\phi \quad \text{[resistive part]} \qquad 67.32$$

$$X = Z\sin\phi \quad \text{[reactive part]} \qquad 67.33$$

---

[7]The standard U.S. 110 V household voltage (also commonly referred to as 115 V, 117 V, and 120 V) is an effective value. This is sometimes referred to as the *nominal system voltage* or just *system voltage*. Other standard voltages used around the world, expressed as effective values, are 208 V, 220 V, 230 V, 240 V, 480 V, 550 V, 575 V, and 600 V.

[8]A *rectified waveform* has had all of its negative values converted to positive values of equal absolute value.

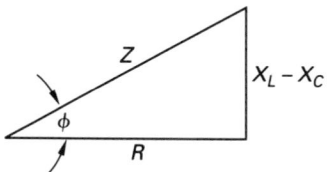

Figure 67.6 *Impedance Triangle of a Complex Circuit*

17. ADMITTANCE

The reciprocal of impedance is the complex quantity *admittance*, $\mathbf{Y}$. Admittance can be used to analyze parallel circuits, since admittance of parallel circuit elements add together.

$$\mathbf{Y} = \frac{1}{\mathbf{Z}} \equiv \frac{1}{Z}/\underline{-\phi} \qquad 67.34$$

The reciprocal of the resistive part of the impedance is the *conductance*, G, with units of siemens, S. The reciprocal of the reactive part of impedance is the *susceptance*, B.

$$G = \frac{1}{R} \qquad 67.35$$

$$B = \frac{1}{X} \qquad 67.36$$

18. RESISTORS IN AC CIRCUITS

An ideal resistor, with resistance R, has no inductance or capacitance. The magnitude of the impedance is equal to the resistance, R, in ohms. The impedance angle is zero. Therefore, current and voltage are in-phase in a purely resistive circuit.

$$\mathbf{Z}_R \equiv R/\underline{0°} \equiv R + j0 \qquad 67.37$$

19. CAPACITORS

A *capacitor* stores electrical charge. The charge on a capacitor is proportional to its *capacitance*, C (in farads, F), and voltage.[9]

$$Q = CV \qquad 67.38$$

An ideal capacitor has no resistance or inductance. The magnitude of the impedance is the *capacitive reactance*, X_C, in ohms. The impedance angle is $-\pi/2$ ($-90°$). Therefore, current leads the voltage by $90°$ in a purely capacitive circuit.

$$\mathbf{Z}_C \equiv X_C/\underline{-90°} \equiv 0 - jX_C \qquad 67.39$$

$$X_C = \frac{1}{\omega C} = \frac{1}{2\pi f C} \qquad 67.40$$

---

[9]Since a farad is a very large unit of capacitance, most capacitors are measured in microfarads, μF.

Plant Engineering

20. INDUCTORS

An ideal *inductor*, with an *inductance* L (in henries, H), has no resistance or capacitance. The magnitude of the impedance is the *inductive reactance*, X_L, in ohms. The impedance angle is $\pi/2$ (90°). Therefore, current lags the voltage by 90° in a purely inductive circuit.

$$\mathbf{Z}_L \equiv X_L \underline{/90°} \equiv 0 + jX_L \qquad 67.41$$

$$X_L = \omega L = 2\pi f L \qquad 67.42$$

21. TRANSFORMERS

Transformers are used to change voltages, isolate circuits, and match impedances. Transformers usually consist of two coils of wire wound on magnetically permeable cores. One coil, designated as the *primary coil*, serves as the input; the other coil, the *secondary coil*, is the output. The primary current produces a magnetic flux in the core; the magnetic flux, in turn, induces a voltage in the secondary coil. In an *ideal transformer* (*loss-less transformer* or *100% efficient transformer*), the coils have no electrical resistance, and all magnetic flux lines pass through both coils.

The ratio of the numbers of primary to secondary coil windings is the *turns ratio (ratio of transformation)*, a. If the turns ratio is greater than 1.0, the transformer decreases voltage and is a *step-down transformer*. If the turns ratio is less than 1.0, the transformer increases voltage and is a *step-up transformer*.

$$a = \frac{N_p}{N_s} \qquad 67.43$$

In an ideal transformer, the power transferred from the primary side equals the power received by the secondary side.

$$I_p V_p = I_s V_s \qquad 67.44$$

$$a = \frac{V_p}{V_s} = \frac{I_s}{I_p} = \sqrt{\frac{Z_p}{Z_s}} \qquad 67.45$$

22. OHM'S LAW FOR AC CIRCUITS

Ohm's law (Sec. 6) can be written in phasor (polar) form. Voltage and current can be represented by their maximum, effective (rms), or average values. However, both must be represented in the same manner.

$$\mathbf{V} = \mathbf{IZ} \qquad 67.46$$

$$V = IZ \quad \text{[magnitudes only]} \qquad 67.47$$

$$\phi_V = \phi_I + \phi_Z \quad \text{[angles only]} \qquad 67.48$$

23. PHASE ANGLE

The current and current phase angle of a circuit are determined by using Ohm's law (Sec. 6) in phasor form. The current *phase angle*, ϕ_I, is the angular difference between when the current and voltage waveforms peak.

$$I = \frac{V}{Z} \qquad 67.49$$

$$I = \frac{V}{Z} \quad \text{[magnitudes only]} \qquad 67.50$$

$$\phi_I = \phi_V - \phi_Z \quad \text{[angles only]} \qquad 67.51$$

Example 67.2

An inventor's black box is connected across standard household voltage (110 V rms). The current drawn is 1.7 A with a lagging phase angle of 14° with respect to the voltage. What is the impedance of the black box?

Solution

From Eq. 67.49,

$$\mathbf{Z} = \frac{\mathbf{V}}{\mathbf{I}} = \frac{110 \text{ V} \underline{/0°}}{1.7 \text{ A} \underline{/14°}}$$

$$= 64.71 \ \Omega \underline{/-14°}$$

24. POWER IN AC CIRCUITS

In a purely resistive circuit, all of the current drawn contributes to dissipated energy. The flow of current causes resistors to increase in temperature, and heat is transferred to the environment.

In a typical AC circuit containing inductors and capacitors as well as resistors, some of the current drawn does not cause heating.[10] Rather, the current charges capacitors and creates magnetic fields in inductors. Since the voltage alternates in polarity, capacitors alternately charge and discharge. Thus, energy is repeatedly drawn and returned by capacitors. Similarly, energy is repeatedly drawn and returned by inductors as their magnetic fields form and collapse.

Current through resistors in AC circuits causes heating, just as in DC circuits. The power dissipated is represented by the *real power* vector, $\mathbf{P}$. Current through capacitors and inductors contributes to reactive power, represented by the *reactive power* vector, $\mathbf{Q}$. Reactive power does not contribute to heating. For convenience, both real and reactive power are considered to be complex (vector) quantities, with magnitudes and associated angles. Real and reactive power combine as vectors into the *complex power* vector, $\mathbf{S}$, as is shown in Fig. 67.7. The angle, ϕ, is known as the overall *impedance angle*.

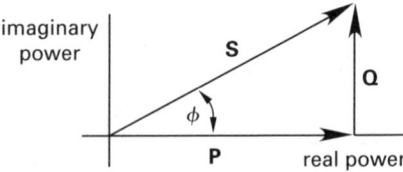

Figure 67.7 *Complex Power Triangle (lagging)*

---

[10]The notable exception is a resonant circuit in which the inductive and capacitive reactances are equal. In that case, the circuit is purely resistive in nature.

The magnitude of real power is known as the *average power*, P, and is measured in watts. The magnitude of the reactive power vector is also known as *reactive power*, Q, and is measured in VARs (volt-amps-reactive). The magnitude of the complex power vector is the *apparent power*, S, measured in VAs (volt-amps). The apparent power is easily calculated from measurements of the line current and line voltage.

$$S = I_{\text{eff}}V_{\text{eff}} = \tfrac{1}{2}I_m V_m \qquad 67.52$$

25. POWER FACTOR

The complex power triangle shown in Fig. 67.7 is congruent to the impedance triangle (Fig. 67.6), and the *power angle*, ϕ, is identical to the overall impedance angle.

$$S^2 = P^2 + Q^2 \qquad 67.53$$
$$P = S\cos\phi \qquad 67.54$$
$$Q = S\sin\phi \qquad 67.55$$
$$\phi = \arctan\left(\frac{Q}{P}\right) \qquad 67.56$$

The *power factor*, pf (also occasionally referred to as the *phase factor*), for a sinusoidal voltage is $\cos\phi$. By convention, the power factor is usually given in percent value rather than by its equivalent decimal value. For a purely resistive circuit, pf = 100%; for a purely reactive circuit, pf = 0.

$$\text{pf} = \frac{P}{S} \qquad 67.57$$

Since the cosine is positive for both positive and negative angles, the terms "leading" and "lagging" are used to describe the nature of the circuit. In a circuit with a *leading power factor* (i.e., in a *leading circuit*), the load is primarily capacitive in nature. In a circuit with a *lagging power factor* (i.e., in a *lagging circuit*), the load is primarily inductive in nature.

26. COST OF ELECTRICAL ENERGY

Except for large industrial users, electrical meters at service locations usually measure and record real power only. Electrical utilities charge on the basis of the total energy used. Energy usage, commonly referred to as the *usage* or *demand*, is measured in kilowatt-hours, abbreviated kWh or kW-hr.

$$\text{cost} = (\text{cost per kW-hr}) \times E_{\text{kW-hr}} \qquad 67.58$$

The cost per kW-hr may not be a single value but may be tiered so that cost varies with cumulative usage. The lowest rate is the *baseline rate*.[11] To encourage conservation, the incremental cost of energy increases with

increases in monthly usage.[12] To encourage cutbacks during the day, the cost may also increase during periods of peak demand.[13] The increase in cost for usage during peak demand may be accomplished by varying the rate, additively, or by use of a *peak demand multiplier*. There may also be different rates for summer and winter usage.

Although only real power is dissipated, reactive power contributes to total current. (Reactive power results from the current drawn in supplying the magnetization energy in motors and charges on capacitors.) Therefore, the distribution system (wires, transformers, etc.) must be sized to carry the total current, not just the current supplying the heating effect. When real power alone is measured at the service location, the power factor is routinely monitored and its effect is built into the charge per kW-hr. This has the equivalent effect of charging the user for apparent power usage.

Example 67.3

A small office normally uses 700 kW-hr per month of electrical energy. The company adds a 1.5 kW heater (to be used at the rate of 1000 kW-hr/month) and a 5 hp motor with a mechanical efficiency of 90% (to be used 240 hours per month). What is the incremental cost of adding the heater and motor? The tiered rate structure is

electrical usage (kW-hr)	rate ($/kW-hr)
less than 350	0.1255
350 to 999	0.1427
1000 to 3999	0.1693

Solution

Motors are rated by their real power output, which is less than their real power demand. The incremental electrical usage per month is

$$1000\text{ kW-hr} + \frac{(5\text{ hp})\left(0.7457\,\dfrac{\text{kW}}{\text{hp}}\right)\times(240\text{ hr})}{0.90} = 1994\text{ kW-hr}$$

The cumulative monthly electrical usage is

$$700\text{ kW-hr} + 1994\text{ kW-hr} = 2694\text{ kW-hr}$$

The company was originally in the second tier. The new usage will be billed partially at the second and third tier rates.

---

[11]There might also be a *lifeline rate* for low-income individuals with very low usage.

[12]There are different rate structures for different categories of users. While increased use within certain categories of users (e.g., residential) results in a higher cost per kW-hr, larger users in another category may pay substantially less per kW-hr due to their volume "buying power."

[13]The day may be divided into *peak periods*, *partial-peak periods*, and *off-peak periods*.

The incremental cost is

$$(999 \text{ kW-hr} - 700 \text{ kW-hr})\left(0.1427 \frac{\$}{\text{kW-hr}}\right)$$

$$+ (2694 \text{ kW-hr} - 999 \text{ kW-hr})\left(0.1693 \frac{\$}{\text{kW-hr}}\right)$$

$$= \$329.63$$

27. POWER FACTOR CORRECTION

Inasmuch as apparent power is paid for but only real power is used to drive motors or provide light and heating, it may be possible to reduce electrical utility charges by reducing the power angle without changing the real power. This strategy, known as *power factor correction*, is routinely accomplished by changing the circuit reactance in order to reduce the reactive power. The change in reactive power needed to change the power angle from ϕ_1 to ϕ_2 is

$$\Delta Q = P(\tan \phi_1 - \tan \phi_2) \qquad 67.59$$

When a circuit is capacitive (i.e., leading), induction motors (Sec. 39) can be connected across the line to improve the power factor. In the more common situation, when a circuit is inductive (i.e., lagging), capacitors or synchronous capacitors (Sec. 41) can be added across the line. The size (in farads) of capacitor required is

$$C = \frac{\Delta Q}{\pi f V_m^2} \quad [V_m \text{ maximum}] \qquad 67.60$$

$$C = \frac{\Delta Q}{2\pi f V_{\text{eff}}^2} \quad [V_{\text{eff}} \text{ effective}] \qquad 67.61$$

Capacitors for power factor correction are generally rated in kVA. Equation 67.59 can be used to find that rating.

Example 67.4

A 60 Hz, 5 hp induction motor draws 53 A (rms) at 117 V (rms) with a 78.5% electrical-to-mechanical energy conversion efficiency. What capacitance should be connected across the line to increase the power factor to 92%?

Solution

The apparent power is found from the observed voltage and the current.

$$S = IV = \frac{(53 \text{ A})(117 \text{ V})}{1000 \frac{\text{VA}}{\text{kVA}}}$$

$$= 6.201 \text{ kVA}$$

The real power drawn from the line is calculated from the real work done by the motor.

$$P_{\text{electrical}} = \frac{P_{\text{out}}}{\eta} = \frac{(5 \text{ hp})(0.7457 \text{ kW/hp})}{0.785}$$

$$= 4.750 \text{ kW}$$

The reactive power and power angle are calculated from the real and apparent powers.

$$Q_1 = \sqrt{S^2 - P^2} = \sqrt{(6.201 \text{ kVA})^2 + (4.750 \text{ kW})^2}$$

$$= 3.986 \text{ kVAR}$$

$$\phi_1 = \arccos\left(\frac{4.750 \text{ kW}}{6.201 \text{ kVA}}\right) = 40.00°$$

The desired power factor angle is

$$\phi_2 = \arccos(0.92) = 23.07°$$

The reactive power after the capacitor is installed is

$$Q_2 = P \tan \phi_2 = (4.750 \text{ kW})(\tan 23.07°)$$

$$= 2.023 \text{ kVAR}$$

The required capacitance is found from Eq. 67.61.

$$C = \frac{\Delta Q}{2\pi f V_{\text{eff}}^2}$$

$$= \frac{(3.986 \text{ kVAR} - 2.023 \text{ kVAR})\left(1000 \frac{\text{VAR}}{\text{kVAR}}\right)}{(2\pi)(60 \text{ Hz})(117 \text{ V})^2}$$

$$= 3.8 \times 10^{-4} \text{ F } (380 \text{ } \mu\text{F})$$

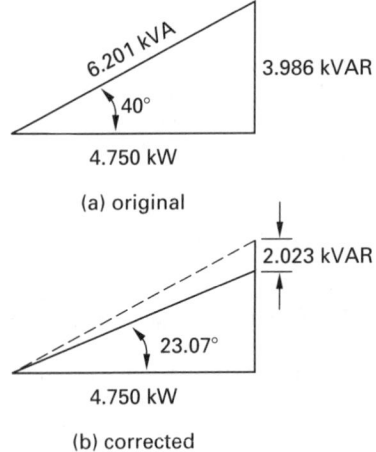

(a) original

(b) corrected

28. THREE-PHASE ELECTRICITY

Smaller electric loads, such as household loads, are served by single-phase power. The power company delivers a sinusoidal voltage of fixed frequency and amplitude connected between two wires—a phase wire and a neutral wire. Large electric loads, large buildings, and industrial plants are served by three-phase power. Three voltage signals are connected between three phase wires and a single neutral wire. The phases have equal frequency and amplitude, but they are out of phase by 120° (electrical) with each other. Such *three-phase systems* use smaller conductors to distribute electricity.[14] Thus, for the same delivered power, three-phase distribution systems have fewer losses and are more efficient.

Three-phase motors provide a more uniform torque than do single-phase motors whose torque production pulsates.[15] Three-phase induction motors require no additional starting windings or associated switches. When rectified, three-phase voltage has a smoother waveform.

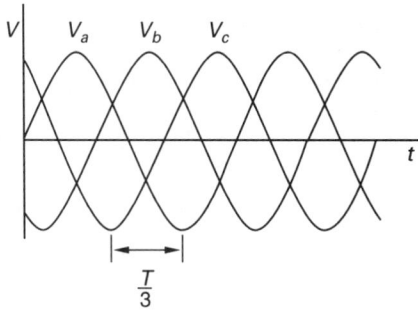

Figure 67.8 *Three-Phase Voltage*

29. THREE-PHASE LOADS

Three impedances are required to fully load a three-phase voltage source. In three-phase motors and other devices, these impedances are three separate motor windings. Similarly, three-phase transformers have three separate sets of primary and secondary windings.

The impedances in a three-phase system are said to be *balanced loads* when they are identical in magnitude and angle. The voltages, line currents, and real, apparent, and reactive powers are all identical in a balanced system. Also, the power factor is the same for each phase. Therefore, only one phase of a balanced system needs to be analyzed (i.e., a *one-line analysis*).

30. LINE AND PHASE VALUES

The *line current*, I_l, is the current carried by the distribution lines (wires). The *phase current*, I_p, is the current flowing through each of the three separate loads (i.e., the phase) in the motor or device. Line and phase currents are both vector quantities.

Depending on how the motor or device is internally wired, the line and phase currents may or may not be the same. Figure 67.9 illustrates delta- and

---

[14]The upper case Greek letter phi is often used as an abbreviation for the word "phase." For example, "3Φ" would be interpreted as "three-phase."

[15]Single-phase motors require auxiliary windings for starting, since one phase alone cannot get the magnetic field rotating.

wye-connected loads. For balanced *wye-connected loads*, the line and phase currents are the same. For balanced *delta-connected loads*, they are not.

$$I_p = I_l \quad \text{[wye]} \qquad 67.62$$

$$I_p = \frac{I_l}{\sqrt{3}} \quad \text{[delta]} \qquad 67.63$$

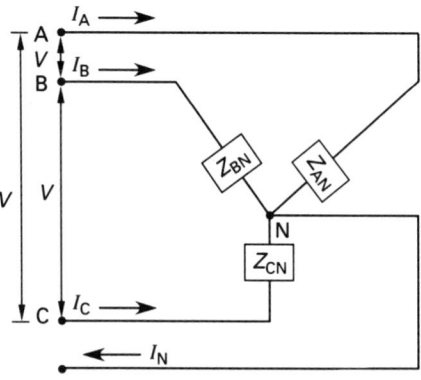

(a) wye-connected loads

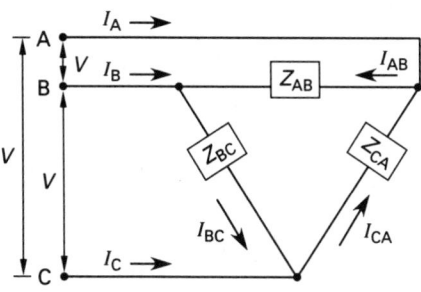

(b) delta-connected loads

Figure 67.9 *Wye- and Delta-Connected Loads*

Similarly, the *line voltage*, V_l (same as *line-to-line voltage*, commonly referred to as the *terminal voltage*), and *phase voltage*, V_p, may not be the same. With balanced delta-connected loads, the full line voltage appears across each phase. With balanced wye-connected loads, the line voltage appears across two loads.

$$V_p = \frac{V_l}{\sqrt{3}} \quad \text{[wye]} \qquad 67.64$$

$$V_p = V_l \quad \text{[delta]} \qquad 67.65$$

31. INPUT POWER THREE-PHASE SYSTEMS

Each impedance in a balanced system dissipates the same real *phase power*, P_p. The power dissipated in a balanced three-phase system is three times the phase power and is calculated in the same manner for both delta- and wye-connected loads.

$$P_t = 3P_p = 3V_p I_p \cos \phi$$
$$= \sqrt{3} V_l I_l \cos \phi \qquad 67.66$$

The real power component is sometimes referred to as "power in kW." Apparent power is sometimes referred to as "kVA value" or "power in kVA."

$$S_t = 3S_p = 3V_pI_p$$
$$= \sqrt{3}V_lI_l \qquad 67.67$$

32. ROTATING MACHINES

Rotating machines are categorized as AC and DC machines. Both categories include machines that use electrical power (i.e., motors) and those that generate electrical power (alternators and generators).[16] Machines can be constructed in either single-phase or polyphase configurations, although single-phase machines may be inferior in terms of economics and efficiency. (See Sec. 28.)

Large AC motors are almost always three-phase. However, since the phases are balanced, it is necessary to analyze one phase only of the motor. Torque and power are divided evenly among the three phases.

33. REGULATION

Rotating machines (motors and alternators), as well as power supplies, are characterized by changes in voltage and speed under load. *Voltage regulation* is defined as

$$VR = \frac{\text{no-load voltage} - \text{full-load voltage}}{\text{full-load voltage}} \times 100\% \qquad 67.68$$

Speed regulation is defined as

$$SR = \frac{\text{no-load speed} - \text{full-load speed}}{\text{full-load speed}} \times 100\% \qquad 67.69$$

34. TORQUE AND POWER

For rotating machines, torque and power are basic operational parameters. It takes mechanical power to turn an alternator or generator. A motor converts electrical power into mechanical power. In SI units, power is given in watts (W) and kilowatts (kW). One horsepower (hp) is equivalent to 0.7457 kilowatts (kW). The relationship between torque and power is

$$T_{\text{ft-lbf}} = (5252)\left(\frac{P_{\text{horsepower}}}{n_{\text{rpm}}}\right) \qquad 67.70$$

$$T_{\text{N·m}} = (9549)\left(\frac{P_{\text{kW}}}{n_{\text{rpm}}}\right) \qquad 67.71$$

There are many important torque parameters for motors. The *starting torque* (also known as *static torque*, *break-away torque*, and *locked-rotor torque*) is the turning effort exerted by the motor when starting from rest.

---

[16]An *alternator* produces AC potential. A *generator* produces DC potential.

Pull-up torque (acceleration torque) is the minimum torque developed during the period of acceleration from rest to full-speed. *Pull-in torque* (as developed in synchronous motors) is the maximum torque that brings the motor back to synchronous speed (Sec. 39). *Nominal pull-in torque* is the torque that is developed at 95% of synchronous speed.

The *full-load torque (steady-state torque)* occurs at the rated speed and horsepower. Full-load torque is supplied to the load on a continuous basis. The full-load torque establishes the temperature increase that the motor must be able to withstand without deterioration. The *rated torque* is developed at rated speed and rated horsepower. The maximum torque that the motor can develop at its synchronous speed is the *pull-out torque*. *Breakdown torque* is the maximum torque that the motor can develop without stalling (i.e., without coming rapidly to a complete stop).

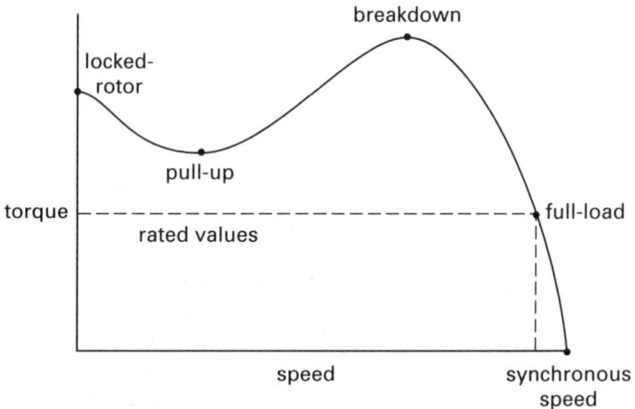

Figure 67.10 *Induction Motor Torque-Speed Characteristic (typical of design B frames)*

Motors are rated according to their output power (*rated power* or *brake power*). Thus, a 5 hp motor will normally deliver a full five horsepower when running at its rated speed. While the rated power is not affected by the motor's energy conversion efficiency, η, the electrical power input to the motor is.

$$P_{\text{electrical}} = \frac{P_{\text{rated}}}{\eta} \qquad 67.72$$

Table 67.1 lists standard motor sizes by rated horsepower.[17] The rated horsepower should be greater than the calculated brake power requirements. Since the rated power is the power actually produced, motors are not selected on the basis of their efficiency or electrical power input. The smaller motors listed in Table 67.1 are generally single-phase motors. The larger motors listed are three-phase motors.

---

[17]For economics, standard motor sizes should be specified.

Table 67.1 Typical Standard Motor Sizes (horsepower)[a]

$\frac{1}{8}$	$\frac{1}{6}$	$\frac{1}{4}$	$\frac{1}{3}$	$\frac{1}{2}$	$\frac{3}{4}$
1	$1\frac{1}{2}$	2	3	5	$7\frac{1}{2}$
10	15	20	25	30	40 50 60 75
100	125	150	200	250	

(Multiply hp by 0.7457 to obtain kW.)
[a] $1/8$ hp and $1/6$ hp motors are less common.

35. NEMA MOTOR CLASSIFICATIONS

Motors are classified by their NEMA design type—A, B, C, D, and F.[18,19] These designs are collectively referred to as *NEMA frame motors*. All NEMA motors of a given frame size are interchangeable as to bolt holes, shaft diameter, height, length, and various other dimensions.

- frame design A: Three-phase, squirrel-cage motor with high locked-rotor (starting) current but also higher breakdown torques; capable of handling intermittent overloads without stalling; 1 to 3% slip at full load.

- frame design B: Three-phase, squirrel-cage motor capable of withstanding full-voltage starting; locked-rotor and breakdown torques suitable for most applications; 1 to 3% slip at full load; often designated for "normal" usage; the most common design.[20]

- frame design C: Three-phase, squirrel-cage motor capable of withstanding full-voltage starting; high locked-rotor torque for special applications (e.g., conveyors and compressors); 1 to 3% slip at full load.

- frame design D: Three-phase, squirrel-cage motor capable of withstanding full-voltage starting; develops 275% locked-rotor torque for special applications; high breakdown torque; depending on design, 5 to 13% slip at full load; commonly known as a *high-slip motor*; used for cranes, hoists, oil well pump jacks, and valve actuators.

- frame design F: Three-phase, squirrel-cage motors with low starting current and low starting torque; used to meet applicable regulations regarding starting current; limited availability.

36. NAMEPLATE VALUES

A *nameplate* is permanently affixed to each motor's housing or frame. This nameplate is embossed or engraved with the *rated values* of the motor (*nameplate*

values or *full-load values*). Nameplate information may include some or all of the following: voltage,[21] frequency,[22] number of phases, rated power (output power), running speed, duty cycle (Sec. 38), locked-rotor and breakdown torques, starting current, current drawn at rated load (in kVA/hp), ambient temperature, temperature rise at rated load, temperature rise at service factor (Sec. 37), insulation rating, power factor, efficiency, frame size, and enclosure type.

It is important to recognize that the rated power of the motor is the actual output power (i.e., power delivered to the load), not the input power. Only the electrical power input is affected by the motor efficiency. (See Eq. 67.72.)

Nameplates are also provided on transformers. Transformer nameplate information includes the two winding voltages (either of which can be the primary winding), frequency, and kVA rating. Apparent power in kVA, not real power, is used in the rating because heating is proportional to the square of the supply current. As with motors, continuous operation at the rated values will not result in excessive heat build-up.

37. SERVICE FACTOR

The horsepower and torque ratings listed on the nameplate of an AC motor can be maintained on a continuous basis without overheating. However, motors can be operated at slightly higher loads without exceeding a safe temperature rise.[23] The ratio of the safe to rated loads (horsepower or torque) is the *service factor*, sf, usually expressed as a decimal.

$$\text{sf} = \frac{\text{maximum safe load}}{\text{nameplate load}} \quad\quad 67.73$$

Service factors vary from 1.15 to 1.4, with the lower values being applicable to the larger, more efficient motors. Typical values of service factor are 1.4 (up to $1/8$ hp motors), 1.35 ($1/6$ to $1/3$ hp motors), 1.25 ($1/2$ to 1 hp motors), and 1.15 (1 or $1^1/2$ to 200 hp).

When running above the rated load, the motor speed, temperature, power factor, full-load current, and efficiency will differ from the nameplate values. However, the locked-rotor and breakdown torques will remain the same, as will the starting current.

Active current is proportional to torque, and hence, is proportional to the horsepower developed (Eq. 67.75).

---

[18] NEMA stands for the National Electrical Manufacturers Association.
[19] There is no type E.
[20] Design B is estimated to be used in 90% of all applications.

[21] Standard NEMA nameplate voltages (effective) are 200 V, 230 V, 460 V, and 575 V. NEMA motors are capable of operating in a range of only ±10% of their rated voltages. Thus, 230 V motors should not be used on 208 V systems.
[22] While some 60 Hz motors (notably those intended for 230 V operation) can be used at 50 Hz, most others (e.g., those intended for 200 V operation) are generally not suitable for use at 50 Hz.
[23] Higher temperatures have a deteriorating effect on the winding insulation. A general rule of thumb is that a motor loses two or three hours of useful life for each hour run at the service factor load.

The active current (line or phase) drawn is also proportional to the service factor. The current drawn per phase is given by Eq. 67.75.[24]

$$I_{\text{active}} = I_l(\text{pf}) \qquad 67.74$$

$$I_{\text{actual},p} = \frac{(\text{sf})(P_{\text{rated},p})}{V\eta(\text{pf})}$$

$$= \frac{P_{\text{actual},p}}{V\eta(\text{pf})} \quad [\text{per phase}; P \text{ in watts}] \quad 67.75$$

Equation 67.75 can also be used when a motor is developing less than its rated power. In this case, the service factor can be considered as the fraction of the rated power being developed.

Example 67.5

What is the approximate phase current drawn by a three-phase, 75 hp, 230 V (rms) motor running at 88% power factor and its rated load?

Solution

No information is given about the efficiency, which is taken as 100%. The service factor is 1.00 because the motor is running at its rated load. From Eq. 67.75, the approximate phase current is

$$I_p = \frac{(\text{sf})(P_p)}{V_p\eta(\text{pf})}$$

$$= \frac{(1.00)\left(\dfrac{75 \text{ hp}}{3 \text{ phases}}\right)\left(0.7457 \, \dfrac{\text{kW}}{\text{hp}}\right)\left(1000 \, \dfrac{\text{W}}{\text{kW}}\right)}{(230 \text{ V})(1.00)(0.88)}$$

$$= 92.1 \text{ A} \quad [\text{per phase}]$$

38. DUTY CYCLE

Motors are categorized according to their *duty cycle: continuous-duty* (24 hr/day); *short-time duty* (15 to 30 min); and *special-duty* (application specific).

39. INDUCTION MOTORS

The three-phase induction motor is, by far, the most frequently used motor in industry. In an induction motor, the magnetic field rotates at the synchronous speed. The *synchronous speed* can be calculated from the number of poles and frequency. The frequency, f, is either 60 Hz (in the United States) or 50 Hz (in

Europe and other locations). The number of poles, p, must be even.[25] The most common motors have 2, 4, and 6 poles.

$$n_{\text{synchronous}} = \frac{120f}{p} \quad [\text{rpm}] \qquad 67.76$$

Due to friction and other factors, rotors (and hence, the motor shafts) in induction motors run slightly slower than their synchronous speeds. The percentage difference is known as the *slip*, s. Slip is seldom greater than 10%, and it is usually much less than that. 4% is a typical value.

$$s = \frac{n_{\text{synchronous}} - n_{\text{actual}}}{n_{\text{synchronous}}} \qquad 67.77$$

The rotor's actual speed is[26]

$$n_{\text{actual}} = (1 - s)n_{\text{synchronous}} \qquad 67.78$$

Induction motors are usually specified in terms of the *kVA ratings*. The kVA rating is not the same as the motor power in kilowatts, although one can be calculated from the other if the motor's power factor is known. The power factor generally varies from 0.8 to 0.9 depending on the motor size.

$$\text{kVA rating} = \frac{P_{\text{kW}}}{\text{pf}} \qquad 67.79$$

$$P_{\text{kW}} = 0.7457 P_{\text{mechanical,hp}} \qquad 67.80$$

Induction motors can differ in the manner in which their rotors are constructed. A *wound rotor* is similar to an armature winding in a dynamo. Wound rotors have high-torque and soft-starting capabilities. There are no wire windings at all in a *squirrel-cage* rotor. Most motors use squirrel-cage rotors. Typical torque-speed characteristics of a design B induction motor are shown in Fig. 67.10.

Example 67.6

A pump is driven by a three-phase induction motor running at its rated values. The motor's nameplate lists the following rated values: 50 hp, 440 V, 92% lagging power factor, 90% efficiency, 60 Hz, 4 poles. The motor's windings are delta-connected. The pump efficiency is 80%. When running under the pump load, the slip is 4%. What are the (a) total torque developed, (b) torque developed per phase, and (c) line current?

---

[24]It is important to recognize the difference between the rated (i.e., nameplate) power and the actual power developed. The actual power should not be combined with the service factor, since the actual power developed is the product of the rated power and the service factor.

[25]There are various forms of Eq. 67.76. As written, the speed is given in rpm, and the number of poles, p, is twice the number of *pole pairs*. When the synchronous speed is specified as f/p, it is understood that the speed is in revolutions per second (rps) and p is the number of pole *pairs*.

[26]Some motors (i.e., *integral gear motors*) are manufactured with integral speed reducers. Common standard output speeds are 37, 45, 56, 68, 84, 100, 125, 155, 180, 230, 280, 350, 420, 520, and 640 rpm. While the integral gear motor is more compact, lower in initial cost, and easier to install than a separate motor with belt drive, coupling, and guard, the separate motor and reducer combination may nevertheless be preferred for its flexibility, especially in replacing and maintaining the motor.

Solution

(a) The synchronous speed is

$$n_{\text{synchronous}} = \frac{120f}{p} = \frac{(120)(60 \text{ Hz})}{4}$$
$$= 1800 \text{ rpm}$$

From Eq. 67.78, the rotor speed is

$$n_{\text{actual}} = (1 - s)n_{\text{synchronous}} = (1 - 0.04)(1800 \text{ rpm})$$
$$= 1728 \text{ rpm}$$

Since the motor is running at its rated values, the motor delivers 50 hp to the pump. The pump's efficiency is irrelevant. From Eq. 67.70, the total torque developed by all three phases is

$$T_t = (5252)\left(\frac{P_{\text{horsepower}}}{n_{\text{rpm}}}\right) = \frac{(5252)(50 \text{ hp})}{1728 \text{ rpm}}$$
$$= 152.0 \text{ ft-lbf} \quad [\text{total}]$$

(b) The torque developed per phase is one-third of the total torque developed.

$$T_p = \frac{T_t}{3} = \frac{152.0 \text{ ft-lbf}}{3 \text{ phases}}$$
$$= 50.67 \text{ ft-lbf} \quad [\text{per phase}]$$

(c) The total electrical input power is given by Eq. 67.72.

$$P_{\text{electrical}} = \frac{P_{\text{rated}}}{\eta} = \frac{(50 \text{ hp})\left(0.7457 \, \dfrac{\text{kW}}{\text{hp}}\right)}{0.90}$$
$$= 41.43 \text{ kW} \quad [\text{total}]$$

Since the power factor is less than 1.0, more current is being drawn than is being converted into useful work. From Eq. 67.79, the apparent power in kVA per phase is

$$S_{\text{kVA}} = \frac{P_{\text{kW}}}{\text{pf}} = \frac{41.43 \text{ kW}}{(3)(0.92)}$$
$$= 15.01 \text{ kVA} \quad [\text{per phase}]$$

The phase current is

$$I = \frac{S}{V} = \frac{(15.01 \text{ kVA})\left(1000 \, \dfrac{\text{VA}}{\text{kVA}}\right)}{440 \text{ V}}$$
$$= 34.11 \text{ A}$$

Since the motor's windings are delta-connected across the three lines, the line current is

$$I_l = \sqrt{3}I_p = (\sqrt{3})(34.11 \text{ A})$$
$$= 59.08 \text{ A}$$

40. TYPICAL INDUCTION MOTOR PERFORMANCE

The following rules of thumb can be used for initial estimates of induction motor performance.

- At 1800 rpm, a motor will develop a torque of 3 ft-lbf/hp.

- At 1200 rpm, a motor will develop a torque of 4.5 ft-lbf/hp.

- At 550 V, a three-phase motor will draw 1 A/hp.

- At 440 V, a three-phase motor will draw 1.25 A/hp.

- At 220 V, a three-phase motor will draw 2.5 A/hp.

- At 220 V, a single-phase motor will draw 5 A/hp.

- At 110 V, a single-phase motor will draw 10 A/hp.

41. SYNCHRONOUS MOTORS

Synchronous motors are essentially dynamo alternators operating in reverse. The stator field frequency is fixed, so regardless of load, the motor runs only at a single speed—the synchronous speed given by Eq. 67.76. Stalling occurs when the motor's counter torque is exceeded. For some equipment that must be driven at constant speed, such as large air or gas compressors, the additional complexity of synchronous motors is justified.

Power factor can be adjusted manually by varying the field current. With *normal excitation* field current, the power factor is 1.0. With *over-excitation*, the power factor is leading, and the field current is greater than normal. With *under-excitation*, the power factor is lagging, and the field current is less than normal.

Since a synchronous motor can be adjusted to draw leading current, it can be used for power factor correction. A synchronous motor used purely for power factor correction is referred to as a *synchronous capacitor* or *synchronous condenser*. A power factor of 80% is often specified or used with synchronous capacitors.

42. DC MACHINES

DC motors and generators can be wired in one of three ways: series, shunt, and compound. Operational characteristics are listed in Table 67.2. Equations 67.70 and 67.71 can be used to calculate torque and power.

Mechanical Engineering Reference Manual

Table 67.2 Operational Characteristics of DC and AC Machines

	motors			generators		
	shunt	series	compound	shunt	series	compound
equivalent circuit						
line voltage, V	V	V	V	V	V	V
line current, I_l	I_l	$I_l = I_a$	I_l	$I_l = I_a - I_f$	$I_l = I_a$	$I_l = I_a - I_f$
field current, I_f	$I_f = \dfrac{V}{R_f}$	$I_f = I_a$	$I_f = \dfrac{V}{R_{f1}}$	$I_f = \dfrac{V}{R_f}$	$I_f = I_a$	$I_f = \dfrac{V}{R_{f1}}$
armature current, I_a	$I_a = I_l - I_f$	$I_a = I_l$	$I_a = I_l - I_f$	$I_a = I_l + I_f$	$I_a = I_l$	$I_a = I_l + I_f$
armature circuit loss, V_a	$V_a = I_a R_a$	$V_a = I_a(R_a + R_f)$	$V_a = I_a(R_a + R_{f2})$	$V_a = I_a R_a$	$V_a = I_a(R_a + R_f)$	$V_a = I_a(R_a + R_{f2})$
counter emf, E_S	$E_S = V - V_a$ $= V - I_a R_a$	$E_S = V - V_a$ $= V - I_a(R_a + R_f)$	$E_S = V - I_a \times (R_a + R_{f2})$	$E_S = V + V_a$ $= V + I_a R_a$	$E_S = V + V_a$ $= V + I_a(R_a + R_f)$	$E_S = V + V_a$ $= V + I_a(R_a + R_{f2})$
power in kW or hp [DC machines]	$P = VI_l$	$P = VI_l$	$P = VI_l$	$\text{hp} = \dfrac{2\pi nT}{33{,}000}$	$\text{hp} = \dfrac{2\pi nT}{33{,}000}$	$\text{hp} = \dfrac{2\pi nT}{33{,}000}$
power in kVA [AC machines]	$P = VI_l \cos\phi$	$P = VI_l \cos\phi$	$P = VI_l \cos\phi$	$\text{hp} = \dfrac{2\pi nT}{33{,}000}$	$\text{hp} = \dfrac{2\pi nT}{33{,}000}$	$\text{hp} = \dfrac{2\pi nT}{33{,}000}$
power out, hp or kW [DC machines]	$\text{hp} = \dfrac{2\pi nT}{33{,}000}$	$\text{hp} = \dfrac{2\pi nT}{33{,}000}$	$\text{hp} = \dfrac{2\pi nT}{33{,}000}$	$P = VI_l$	$P = VI_l$	$P = VI_l$
power out, kVA [AC machines]	$\text{hp} = \dfrac{2\pi nT}{33{,}000}$	$\text{hp} = \dfrac{2\pi nT}{33{,}000}$	$\text{hp} = \dfrac{2\pi nT}{33{,}000}$	$P = VI_l \cos\phi$	$P = VI_l \cos\phi$	$P = VI_l \cos\phi$

PROFESSIONAL PUBLICATIONS, INC. BELMONT, CA

43. CHOICE OF MOTOR TYPES

Squirrel-cage induction motors are commonly chosen because of their simple construction, low maintenance, and excellent efficiencies.[27] A wound-rotor induction motor should be used only if it is necessary to achieve a low starting kVA, controllable kVA, controllable torque, or variable speed.[28]

While induction motors are commonly used, synchronous motors are suitable for many applications normally handled by a NEMA design B squirrel-cage motor. They have adjustable power factors and higher efficiency. Their initial cost may also be less.

Selecting a motor type is greatly dependent on the power, torque, and speed requirements of the rotating load. Table 67.3, Fig. 67.11, and Apps. 67.A and 67.B can be used as starting points in the selection process.

Table 67.3 *Recommended Motor Voltage and Power Ranges*

voltage	horsepower
direct current	
115	0 to 30 (max)
230	0 to 200 (max)
550 or 600	$\frac{1}{2}$ and upward
alternating current, one-phase	
110, 115, or 120	0 to $1\frac{1}{2}$
220, 230, or 240	0 to 10
440 or 550	5 to 10*
alternating current, two- and three-phase	
110, 115, or 120	0 to 15
208, 220, 230, or 240	0 to 200
440 or 550	0 to 500
2200 or 2300	40 and upward
4000	75 and upward
6600	400 and upward

*not recommended.
(Multiply hp by 0.7457 to obtain kW.)

44. LOSSES IN ROTATING MACHINES

Losses in rotating machines are typically divided into the following categories: armature copper losses, field copper losses, mechanical losses (including friction and windage), core losses (including hysteresis losses, eddy current losses, and brush resistance losses), and stray losses.

Copper losses (also known as I^2R *losses*) are real power losses due to wire and winding resistance.

$$P_{Cu} = \Sigma I^2 R \qquad 67.81$$

---

[27]The larger the motor and the higher the speed, the higher the efficiency. Large 3600 rpm induction motors have excellent performance.
[28]A constant-speed motor with a slip coupling could also be used.

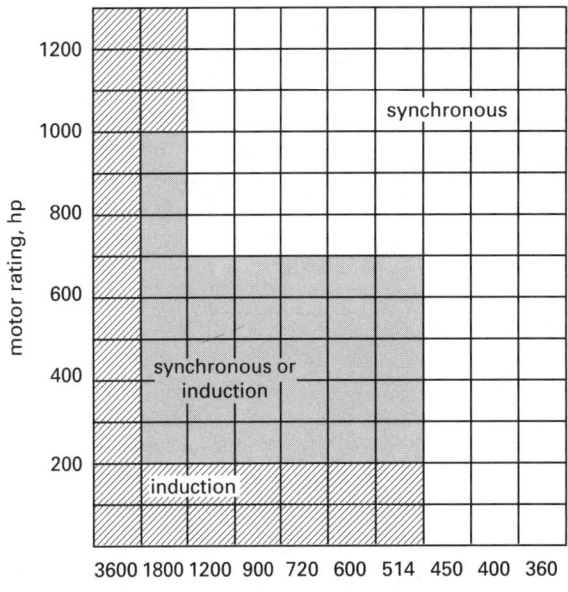

[a]Adapted from *Mechanical Engineering*, Design Manual NAVFAC DM-3, Department of the Navy, © 1972.

Figure 67.11 *Motor Rating According to Speed*[a]
(general guidelines)

Core losses (also known as *iron losses*) are constant losses that are independent of the load and, for that reason, are also known as *open-circuit* and *no-load losses*.

Mechanical losses (also known as *rotational losses*) include brush and bearing friction and *windage* (air friction). (Windage is a no-load loss, but is not an electrical core loss.) Mechanical losses are determined by measuring the power input at the rated speed and with no load.

Stray losses are due to non-uniform current distribution in the conductors. Stray losses are approximately 1% for DC machines and are zero for AC machines.

45. EFFICIENCY OF ROTATING MACHINES

Only real power is used to compute the efficiency of a rotating machine. This efficiency is sometimes referred to as *overall efficiency* and *commercial efficiency*.

$$\begin{aligned}
\eta &= \frac{\text{output power}}{\text{input power}} \\
&= \frac{\text{output power}}{\text{output power} + \text{power losses}} \\
&= \frac{\text{input power} - \text{power losses}}{\text{input power}} \qquad 67.82
\end{aligned}$$

Example 67.7

A DC shunt motor draws 40 A at 112 V when fully loaded. When running without a load at the same speed, it draws only 3 A at 106 V. The field resistance is 100 Ω; and the armature resistance is 0.125 Ω. (a) What is the efficiency of the motor? (b) What power (in hp) does the motor deliver at full load?

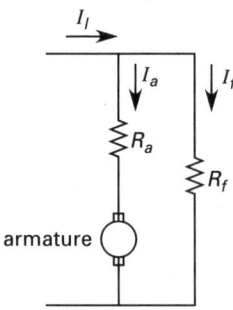

Solution

(a) The field current is

$$I_f = \frac{V}{R_f} = \frac{112 \text{ V}}{100 \text{ } \Omega}$$
$$= 1.12 \text{ A}$$

The total full-load line current is known to be 40 A. The full-load armature current is

$$I_a = I_l - I_f = 40 \text{ A} - 1.12 \text{ A}$$
$$= 38.88 \text{ A}$$

The field copper loss is

$$P_{\text{Cu},f} = I_f^2 R_f = (1.12 \text{ A})^2 (100 \text{ } \Omega)$$
$$= 125.4 \text{ W}$$

The armature copper loss is

$$P_{\text{Cu},a} = I_a^2 R_a = (38.88 \text{ A})^2 (0.125 \text{ } \Omega)$$
$$= 189.0 \text{ W}$$

The total copper loss is

$$P_{\text{Cu},t} = P_{\text{Cu},f} + P_{\text{Cu},a} = 125.4 \text{ W} + 189.0 \text{ W}$$
$$= 314.4 \text{ W}$$

Stray power is determined from the no-load conditions. At no load, the field current is

$$I_f = \frac{V}{R_f} = \frac{106 \text{ V}}{100 \text{ } \Omega}$$
$$= 1.06 \text{ A}$$

At no load, the armature current is

$$I_a = I_l - I_f = 3 \text{ A} - 1.06 \text{ A}$$
$$= 1.94 \text{ A}$$

The stray power loss is

$$P_{\text{stray}} = V_l I_a - I_a^2 R_a$$
$$= (106 \text{ V})(1.94 \text{ A}) - (1.94 \text{ A})^2 (0.125 \text{ } \Omega)$$
$$= 205.2 \text{ W}$$

The stray power loss is assumed to be independent of the load. The total losses at full load are

$$P_{\text{loss}} = P_{\text{Cu},t} + P_{\text{stray}} = 314.4 \text{ W} + 205.2 \text{ W}$$
$$= 519.6 \text{ W}$$

The power input to the motor when fully loaded is

$$P_{\text{input}} = I_l V_l = (40 \text{ A})(112 \text{ V}) = 4480 \text{ W}$$

The efficiency is

$$\eta = \frac{\text{input power} - \text{power losses}}{\text{input power}}$$
$$= \frac{4480 \text{ W} - 519.6 \text{ W}}{4480 \text{ W}} = 0.884 \quad (88.4\%)$$

(b) The real power delivered at full load is

$$P_{\text{real}} = \text{input power} - \text{power losses}$$
$$= \frac{4480 \text{ W} - 519.6 \text{ W}}{\left(1000 \frac{\text{W}}{\text{kW}}\right)\left(0.7457 \frac{\text{kW}}{\text{hp}}\right)}$$
$$= 5.3 \text{ hp}$$

46. HIGH-EFFICIENCY MOTORS AND DRIVES

A premium, energy-efficient motor will have approximately 50% of the losses of a conventional motor. Due to the relatively high overall efficiency enjoyed by all motors, however, this translates into only a 5% increase in overall efficiency.

High-efficiency motors are often combined with *variable-frequency drives* (VFDs) to achieve continuous-variable speed control.[29] VFDs can substantially reduce the power drawn by the process. For example, with a motor-driven pump or fan, the motor can be slowed down instead of closing a valve or damper when the flow requirements decrease. Since the required motor horsepower varies with the cube root of the speed, for processes that do not always run at full-flow (fluid pumping, metering, flow control, etc.), the energy savings can be substantial.

A VFD uses an electronic controller to produce a variable-frequency signal that is not an ideal sine wave. This results in additional heating (e.g., an increase of 20 to 40%) from copper and core losses in the motor. In pumping applications, though, the process load drops off faster than additional heat is produced. Thus, the process power savings dominate.

Most low- and medium-power motors implement VFD with DC voltage intermediate circuits, a technique known as *voltage-source inversion*. Voltage-source inversion is further subdivided into *pulse-width modulation* (PWM) and *pulse-amplitude modulation* (PAM).

Under VFD control, the motor torque is approximately proportional to the applied voltage and drawn current but inversely proportional to the applied frequency.

$$T \propto \frac{VI}{f} \qquad \textit{67.83}$$

---

[29] Prior to VFDs, there were two common ways to change the speed of an induction motor: (1) increasing the slip, and (2) changing the number of pole-pairs. Increasing the slip was accomplished by under-magnetizing the motor so that it received less input voltage than it was built for. Dropping or adding the number of active poles resulted in the speed changing by a factor of 2 (or $^1/_2$).

68 Illumination and Sound

1.	Luminous Flux	68-1
2.	Luminous Intensity	68-2
3.	Illuminance	68-2
4.	Luminance	68-3
5.	Interaction of Light with Matter	68-3
6.	Sound and Noise	68-3
7.	Physiological Basis of Sound	68-4
8.	Longitudinal Sound Waves	68-4
9.	Propagation Velocity of Longitudinal Sound Waves	68-4
10.	Doppler Effect	68-5
11.	Sound Meter Measurements	68-5
12.	Sound Fields	68-5
13.	Sound Pressure Level	68-6
14.	Sound Power Level	68-6
15.	Directivity	68-6
16.	Converting Sound Pressure into Sound Power	68-6
17.	Frequency Content of Noise	68-7
18.	Octave Band Analysis	68-7
19.	Combining Multiple Sources	68-7
20.	Broad Band Measurements	68-7
21.	Loudness	68-8
22.	Noise Criterion Curves	68-8
23.	Acoustic Absorber Materials	68-8
24.	Sound Absorption Coefficient	68-8
25.	Room Constant	68-8
26.	Acoustic Barrier Materials	68-9
27.	Sound Transmission Class	68-9
28.	Noise Reduction	68-9
29.	Insertion Loss	68-9
30.	Transmission Loss	68-9
31.	Attenuation	68-9
32.	Noise Dose	68-9
33.	Allowable Exposure Limits	68-10
34.	Noise Reduction Rating	68-10
	Practice Problems	68-10

Nomenclature

a	speed of sound	ft/sec	m/s
A	area	ft^2	m^2
c	speed of light	ft/sec	m/s
C	permitted exposure	hr	h
D	dose	–	–
E	irradiance	lm/ft^2	lm/m^2
E	modulus of elasticity	lbf/ft^2	Pa
f	frequency	Hz	Hz
g_c	gravitational constant	lbm-ft/ lbf-sec^2	n.a.
I	light intensity	lm/sr	lm/sr
I	sound intensity	dB	dB
IL	insertion loss	dB	dB
k	ratio of specific heats	–	–
L	sound level	dB	dB
M	brightness	lm/ft^2	lm/m^2
MW	molecular weight	lbm/lbmol	kg/kmol
NR	noise reduction	dB	dB
p	pressure	lbf/ft^2	Pa
Q	directivity	–	–
Q	quantity of light	lm	lm
r	distance	ft	m
R	room constant	ft^2	m^2
R	specific gas constant	ft-lbf/ lbm-°R	kJ/kg·K
R^*	universal gas constant	ft-lbf/ lbmol-°R	kJ/kmol·K
S	surface absorption	sabins	sabins
t	time	sec	s
T	period	sec	s
T	temperature	°R	K
TL	transmission loss	dB	dB
v	velocity	ft/sec	m/s
W	power	W	W

Symbols

α	sound absorption coefficient	–	–
η	efficiency	–	–
λ	wavelength	ft	m
ρ	density	lbm/ft^3	kg/m^3
τ	transmission coefficient	–	–
Φ	luminous flux	lm	lm
ω	solid angle	sr	sr

Subscripts

l	luminous
p	pressure
ref	reference
t	total
W	power

1. LUMINOUS FLUX

The amount of visible light emitted from a source is the *luminous flux*, Φ, with units of lumens (lm).[1] Not all of the energy emitted will be in the visible region, though. Therefore, the total source power is not a good indicator

---

[1] A lumen is approximately equal to 1.47×10^{-3} W of visible light with a wavelength of 5.55×10^{-7} m.

of the visible lighting effect. The ratio of luminous flux to total radiant energy (also known as the *quantity of light*), Q, is the *luminous efficiency*, η_l.

Efficiency of a lamp is the ratio of light output in lumens to the lamp's input in lumens. *Efficacy* is the ratio of the lamp's output in lumens to its input power in watts.

$$\eta_l = \frac{\Phi}{Q} \qquad 68.1$$

The *luminous emittance* (also known as *brightness*), M, of a source is the luminous flux per unit area of the source. (Compare this with *illumination* of a receiver covered in Sec. 3, which is the flux per unit of receiving area.) In the United States, the unit of brightness is the lm/ft^2 (foot-lambert).[2]

$$M = \frac{\Phi}{A_{\text{source}}} \qquad 68.2$$

Table 68.1 *Typical Luminous Efficacy*

source		lm/W
	candle	0.1
25 W	tungsten lamp	10
100 W	tungsten lamp	16
1000 W	tungsten lamp	22
40 W	fluorescent lamp	80
400 W	mercury fluorescent lamp	58
1000 W	carbon arc	60
100 W	mercury arc	35
400 W	metal halide lamp	85
1000 W	metal halide lamp	100
400 W	high-pressure sodium lamp	125
1000 W	high-pressure sodium lamp	130
180 W	low-pressure sodium lamp	180

Example 68.1

A 100 W lamp has a luminous flux of 4400 lm. What is the lamp's luminous efficacy?

Solution

From Eq. 68.1,

$$\eta_l = \frac{\Phi}{Q} = \frac{4400 \text{ lm}}{100 \text{ W}} = 44 \text{ lm/W}$$

Example 68.2

A tungsten filament emitting 1600 lm has a surface area of 0.35 in^2 (2.26×10^{-4} m^2). What is its brightness?

SI Solution

$$M = \frac{\Phi}{A_{\text{surface}}} = \frac{1600 \text{ lm}}{2.26 \times 10^{-4} \text{ m}^2} = 7.08 \times 10^6 \text{ lm/m}^2$$

Customary U.S. Solution

$$M = \frac{\Phi}{A_{\text{surface}}} = \frac{1600 \text{ lm}}{\dfrac{0.35 \text{ in}^2}{144 \dfrac{\text{in}^2}{\text{ft}^2}}} = 6.58 \times 10^5 \text{ lm/ft}^2$$

2. LUMINOUS INTENSITY

The *luminous intensity*, I, of a source is a measure of flux per unit solid angle. Its units are lumens per steradian (lm/sr), referred to by the synonymous names of *candle*, *candela*, and *candlepower*.[3] Equation 68.3 relates flux from an omnidirectional point source to intensity measured on a spherical receptor of radius r.

$$I = \frac{\Phi}{\omega} = \frac{\Phi r^2}{A} = \frac{\Phi_t r^2}{4\pi r^2} = \frac{\Phi_t}{4\pi} \qquad 68.3$$

Example 68.3

Two lumens pass through a circular hole (diameter = 0.5 m) in a screen located 6.0 m from an omnidirectional source. (a) What is the luminous intensity of the source? (b) What luminous flux is emitted by the source?

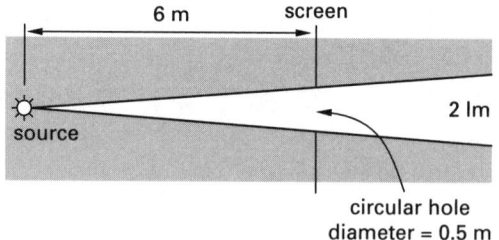

Solution

(a) The solid angle subtended by the hole is

$$\omega = \frac{A}{r^2} = \frac{\left(\dfrac{\pi}{4}\right)(0.5 \text{ m})^2}{(6 \text{ m})^2} = 0.005454 \text{ sr}$$

From Eq. 68.3, the intensity is

$$I = \frac{\Phi}{\omega} = \frac{2 \text{ lm}}{0.005454 \text{ sr}} = 366.7 \text{ lm/sr}$$

(b) From Eq. 68.3, the luminous flux is

$$\Phi_t = 4\pi I = (4\pi \text{ sr})\left(366.7 \frac{\text{lm}}{\text{sr}}\right)$$
$$= 4608 \text{ lm}$$

3. ILLUMINANCE

The *illuminance (illumination)*, E, is a measure of luminous flux per incident area. Typical units are lux (lm/m^2) and foot-candles (lm/ft^2). The name *irradiance* is used when the units are W/m^2.

---

[2]The term *lambert* in the U.S. is a source of confusion. The old lambert unit is the same as lm/cm^2. A foot-lambert is the same as lm/ft^2.

[3]Intensity is not a measure of power, as the term *candlepower* may imply.

One foot-candle is the illumination from a one *candle-power* (cp) source at a distance of one foot.

$$E = \frac{\Phi}{A_{\text{receptor}}} \qquad 68.4$$

For an omnidirectional source and a spherical receptor of radius r,

$$E = \frac{\Phi_t}{4\pi r^2} \qquad 68.5$$

Equation 68.5 shows that illumination follows the *distance squared law*.

$$E_1 r_1^2 = E_2 r_2^2 \qquad 68.6$$

Table 68.2 *Recommended Illuminance*

location	lm/ft² (fc)	lux
roadway	1–2	10–20
living room	5–15	50–150
library (reading)	30–70	300–700
evening sports	30–100	300–1000
factory (assembly)	100–200	1000–2000
office	100–200	1000–2000
factory (fine assembly)	500–1000	5000–10 000
hospital operating room	2000–2500	20 000–25 000

(Multiply foot-candles by 10.764 to obtain lux.)
(Multiply foot-candles by 0.10764 to obtain hectolux.)

Example 68.4

A lamp radiating hemispherically and rated at 2000 lm is positioned 20 ft (6 m) above the ground. What is the illumination on a walkway directly below the lamp?

SI Solution

$$E = \frac{\Phi}{A} = \frac{\Phi}{\frac{1}{2}A_{\text{sphere}}} = \frac{2000 \text{ lm}}{\left(\frac{1}{2}\right)(4\pi)(6 \text{ m})^2} = 8.84 \text{ lux}$$

Customary U.S. Solution

$$E = \frac{\Phi}{A} = \frac{2000 \text{ lm}}{\left(\frac{1}{2}\right)(4\pi)(20 \text{ ft})^2} = 0.796 \text{ lm/ft}^2 \text{ (fc)}$$

4. LUMINANCE

Illuminance of an object is the quantity of light from a source reaching the object. The amount of light reflected from the object's surface is referred to as *luminance* or *brightness* of the object. Luminance depends on the *reflectance* of the surface and is measured in essentially the same units as illuminance.[4] For example, if an object has a reflectance of 40% and is illuminated at 150 fc, its brightness is $(0.40)(150 \text{ fc}) = 60 \text{ fL}$.

5. INTERACTION OF LIGHT WITH MATTER

Light travels through a vacuum as an electromagnetic wave. When the light makes contact with matter, some of the wave energy is absorbed by the matter, causing electrons to jump into higher energy states. (A polished metal surface, for example, will absorb only about 10% of the incident energy, reflecting the remaining 90% away.) Some of this absorbed energy is re-emitted when the electrons drop back to a lower energy level. Generally, the re-emitted light will not be at its original wavelength.

If the reflecting surface is smooth, the *reflection angle* for most of the light will be the same as the *incident angle* and the light is said to be *regularly reflected* (i.e., the case of *specular reflection*). If the surface is rough, however, the light will be scattered and reflected randomly (the case of *diffuse reflection*).

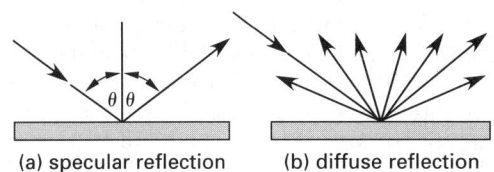

    (a) specular reflection   (b) diffuse reflection

Figure 68.1 *Specular and Diffuse Reflection*

The energy that is absorbed is said to be *refracted*. In the case of an *opaque material*, the refracted energy is absorbed within a very thin layer and converted to heat. Light is able to pass through a *transparent material* without being absorbed. Light is partially absorbed in a *translucent material*.

6. SOUND AND NOISE

Sound is made up of the pressure waves and vibrations that are received from a vibrating source. *Noise* is unwanted sound. Noise can either be continuous or impulsive. For the purpose of classification, *impact noise (impulse noise)* is noise whose peak levels occur at intervals greater than one second.

Background noise is the "normal" noise in the local environment. Background noise consists of environmental noises that are not components of the noise to be measured or studied. It should be measured separately and subtracted from any noise measurements made while equipment or other sources are operating.

---

[4]A *lambert* is equal to 929 foot-lamberts. A *millilambert* is equal to 0.929 foot-lamberts. Other units include the *stilb* (candle/cm²) and *nit* (candle/m²).

A *pure tone* is one that consists essentially of a single frequency. Relatively pure tones can be generated at simple harmonic frequencies in rotating equipment. For example, a five-bladed fan turning at 50 rps would generate harmonics with a fundamental frequency of 250 Hz. Pure tones and harmonics are of interest because they are difficult to attenuate. However, most noise is white noise, not pure tones. *White noise* is relatively evenly distributed over the frequency spectrum.

Structure-borne noise (also known as *mechanical noise*) is noise that is generated by mechanical elements in moving equipment. Panels, covers, housings, and other membrane-like elements are frequent sources of structure-borne noise. These elements produce noise at their own natural frequencies, not just at the cycling frequency of the equipment.

7. PHYSIOLOGICAL BASIS OF SOUND

Longitudinal sound waves in the 20 to 20,000 Hz range can be heard by most people. The average human ear is most sensitive to frequencies around 3000 Hz. Even in this range, however, the intensity must be great enough for a sound to be detected. *Intensity* is the amount of energy passing through a unit area each second. For example, the *threshold of audibility* at 3000 Hz is approximately 10^{-12} W/m². If the sound is too intense (i.e., the intensity is above the *threshold of pain*, approximately 10 W/m²), the sound will be experienced painfully. The thresholds of audibility and pain are both somewhat frequency-dependent.

Loudness is a qualitative sensation that can be quantified approximately by comparing the sound intensity to a reference intensity (usually the assumed threshold of audibility, 10^{-12} W/m²). Loudness is perceived by most individuals approximately logarithmically and is calculated in that manner.

$$\text{loudness} = 10 \log\left(\frac{I}{I_0}\right)$$
$$= 10 \log\left(\frac{I}{10^{-12}\,\frac{W}{m^2}}\right) \qquad 68.7$$

8. LONGITUDINAL SOUND WAVES

Sound, like light and other electromagnetic radiation, is a wave phenomenon. However, sound has two distinct differences from electromagnetic waves. First, sound is transmitted as *longitudinal waves*, also known as *compression waves*. Such waves are alternating compressions and expansions of the medium through which they travel. In comparison, electromagnetic waves are *transverse waves*. Second, longitudinal waves require a medium through which to travel—they cannot travel through a vacuum, as can electromagnetic waves.

Table 68.3 *Approximate Loudness of Various Sources*

source	loudness (dB)	perception
jet engine, thunder	120	painful
jackhammer	110	deafening
sheet metal shop	90	very loud
street noise	70	loud
office noise, normal speech	50	moderate
quiet conversation	30	quiet
whisper	20	faint
anechoic room	10	very faint
none	0	silence

For ease of visualization, longitudinal waves may be represented by transverse waves. The *wavelength* is the distance between successive compressions (or expansions), as shown in Fig. 68.2.

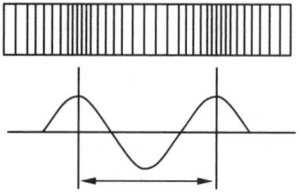

Figure 68.2 *Longitudinal Waves*

The usual relationships among wavelength, propagation velocity, period, and frequency are valid.

$$a = f\lambda \qquad 68.8$$
$$T = \frac{1}{f} \qquad 68.9$$

9. PROPAGATION VELOCITY OF LONGITUDINAL SOUND WAVES

The *propagation velocity*, commonly known as the *speed of sound* and *sonic velocity*, of longitudinal waves depends on the compressibility of the supporting medium. For solids, the compressibility is accounted for by the elastic modulus, E. For liquids, it is customary to refer to the elastic modulus as the *bulk modulus*.

$$a = \sqrt{\frac{E}{\rho}} \qquad \text{[SI]} \qquad 68.10(a)$$

$$a = \sqrt{\frac{Eg_c}{\rho}} \qquad \text{[U.S.]} \qquad 68.10(b)$$

For ideal gases, the propagation velocity is predicted by Eq. 68.11.

$$a = \sqrt{kRT} = \sqrt{\frac{kR^*T}{MW}} \qquad \text{[SI]} \qquad 68.11(a)$$

$$a = \sqrt{kg_cRT} = \sqrt{\frac{kg_cR^*T}{MW}} \qquad \text{[U.S.]} \qquad 68.11(b)$$

Table 68.4 lists approximate values for the speed of sound in various media.

Table 68.4 *Approximate Speeds of Sound
(at one atmospheric pressure)*

material	speed of sound m/s	speed of sound ft/sec
air	330 at 0°C	1130 at 70°F
aluminum	4990	16,400
carbon dioxide	260 at 0°C	870 at 70°F
hydrogen	970 at 0°C	3310 at 70°F
steel	5150	16,900
water	1490 at 20°C	4880 at 70°F

10. DOPPLER EFFECT

If the distance between a sound source and a listener is changing, the frequency heard will differ from the frequency emitted. If the separation distance is decreasing, the frequency will be shifted higher; if the separation distance is increasing, the frequency will be shifted lower. This shifting is known as the *Doppler effect*. The ratio of observed frequency (f') to emitted frequency (f) depends on the local speed of sound (a) and the absolute velocities of the source and observer (v_s and v_o). In Eq. 68.12, v_s is positive if the source moves away from the observer and is negative otherwise; v_o is positive if the observer moves toward the source and negative otherwise.

$$\frac{f'}{f} = \frac{a + v_o}{a + v_s} \qquad \text{68.12}$$

Example 68.5

On a particular day, the local speed of sound is 1130 ft/sec (344 m/s). What frequency is heard by a stationary pedestrian when a 1200 Hz siren on an emergency vehicle moves away at 45 mph (72 km/h)?

SI Solution

The observer's velocity is zero. The source's velocity is positive since the vehicle is moving away. From Eq. 68.12, the heard frequency is

$$f' = f\left(\frac{a + v_o}{a + v_s}\right)$$

$$= (1200 \text{ Hz})\left(\frac{344 \frac{\text{m}}{\text{s}} + 0}{344 \frac{\text{m}}{\text{s}} + \frac{\left(72 \frac{\text{km}}{\text{h}}\right)\left(1000 \frac{\text{m}}{\text{km}}\right)}{3600 \frac{\text{s}}{\text{h}}}}\right)$$

$$= (1200 \text{ Hz})(0.9451) = 1134 \text{ Hz}$$

Customary U.S. Solution

The source velocity is positive and must be converted from mi/hr to ft/sec.

$$v_s = \frac{\left(45 \frac{\text{mi}}{\text{hr}}\right)\left(5280 \frac{\text{ft}}{\text{mi}}\right)}{3600 \frac{\text{sec}}{\text{hr}}} = 66 \text{ ft/sec}$$

From Eq. 68.12,

$$f' = f\left(\frac{a + v_o}{a + v_s}\right) = (1200 \text{ Hz})\left(\frac{1130 \frac{\text{ft}}{\text{sec}} + 0}{1130 \frac{\text{ft}}{\text{sec}} + 66 \frac{\text{ft}}{\text{sec}}}\right)$$

$$= (1200 \text{ Hz})(0.9448) = 1134 \text{ Hz}$$

11. SOUND METER MEASUREMENTS

Rather than attempt to describe sound in terms of subjective loudness, simple *sound meters (noise meters)* measure noise using various response curves. (A *response curve* is basically a set of weighting values that can be applied to each frequency detected.) Meter response curves have been standardized worldwide, and most industrial-quality meters respond accurately to three curves, designated as the A-, B-, and C-scales. The slow-response A-scale approximates human hearing the most accurately, so it is used the most frequently.[5,6]

When sound measurements are reported, a sound "level" always means a logarithmic value expressed in decibels. Sound level units are reported along with the scale used. For example, "15 dBA" (also written as "dB(A)" and "dB-A") indicates that the A-scale was used.

Some meters often also provide a fourth, unweighted reading using the basic frequency response of the microphone and meter circuitry. This reading is different for different meters. The terms *unweighted response, flat response, 20 kHz response,* and *linear response* all refer to this case.

Peak levels from impulse sources are measured with special equipment (i.e., an impact meter or oscilloscope) that captures and holds the maximum instantaneous noise level.

12. SOUND FIELDS

Sound that is received directly from a source without any significant reflection (including that from the floor or ground) is known as *free-field* sound. Similarly, free-field measurements are those that do not include any contribution from reflections. Measurements made near

---

[5] The A-curve discriminates against lower frequencies, the C-curve is nearly flat, and the B-curve falls in between.
[6] Refer to ANSI standard S1.4 for sound level meters.

walls or other objects where reflected sound is significant are called *reverberant field* measurements. Most in-room measurements are reverberant field measurements. In a reverberant field, sound pressure levels do not fall off with increases in distance from the source. Sound levels are essentially the same anywhere in a reverberant field.

Measurements made close to a noise source are called *near-field measurements*. They can be in error due to directionality effects and reflections from the source itself. The near-field effects will be significant if the measurement is taken closer than one wavelength (based on the lowest frequency of interest) from the source. Measurements made at large distances from a noise source are called *far-field measurements*. Far-field measurements can be in error due to background noise. In a true far-field environment, the sound pressure level will decay 6 dB for each doubling of the distance from the source.

Anechoic rooms (free-field throughout) and *reverberant rooms* (reverberant field throughout) represent extremes in measurement environments. Such rooms are routinely used for measuring directionality effects and absorption material performance.

13. SOUND PRESSURE LEVEL

Sound propagates as pressure fluctuates. The actual pressure at a particular point is the *sound pressure level*, L_p. (The symbol SPL is also used in the literature.) Sound pressure level is the quantity that is actually measured in sound meters.

Since the human ear responds to variations in air pressure over a range of more than a million to one, a logarithmic power ratio scale is used to measure sound in decibels. The minimum perceptible pressure amplitude (i.e., the approximate *threshold of hearing*) has been standardized as 20 μPa, and this value is used as a reference pressure.[7] The acoustic energy is proportional to the square of the sound pressure. For an rms (root-mean-squared or "effective") sound pressure, p, in pascals, the sound pressure level is

$$L_p = 10 \log \left(\frac{p}{p_{\text{ref}}} \right)^2 = 20 \log \left(\frac{p}{p_{\text{ref}}} \right)$$
$$= 20 \log \left(\frac{p}{2 \times 10^{-5} \text{ Pa}} \right) \qquad 68.13$$

Usually, the sound pressure level is sampled at several single frequencies over the audio spectrum. These samples are referred to as "narrow band" or "octave band" measurements. The individual measurements can be combined into a *broad band* value by weighting each by standardized values.[8]

---

[7]At one time, this was thought to be the threshold of hearing for an average young person at 1000 Hz.
[8]Weighting and combining are done automatically by the sound meter.

14. SOUND POWER LEVEL

The *sound power level* is used to measure the total acoustic power emitted by a source of sound. Sound power is, therefore, independent of the environment. Sound power is not measured by sound meters; it must be calculated. A standard reference level of 1×10^{-12} W is used.[9] The literature uses L_W and PWL as symbols for sound power level. For a source of sound emitting W watts, the sound power level in decibels is

$$L_W = 10 \log \left(\frac{W}{W_{\text{ref}}} \right) = 10 \log \left(\frac{W}{1 \times 10^{-12} \text{ W}} \right) \quad 68.14$$

15. DIRECTIVITY

The *directivity*, Q (also known as the *directivity factor*), affects the sound power that appears as sound pressure in a near-field measurement. (Directivity is not an issue in far-field measurements.) Directivity depends on the orientation of the source and the physical location of the source in relationship to walls and other solid surfaces. Directivity is 1 for an isotropic source radiating into free space, 2 for a source on the surface of an infinite flat plane, 4 for a source at the intersection of two perpendicular planes, and 8 for a source in the corner of three mutually perpendicular reflecting planes.

16. CONVERTING SOUND PRESSURE INTO SOUND POWER

Sound pressure level and sound power level are both reported in decibels. However, there is no simple conversion between them because sound pressure level is affected by directivity and separation distance from the source, while sound power level is not. However, the two scales have been established such that a decibel change in one quantity will produce the same decibel change in the other quantity. Therefore, a decibel decrease in sound power will produce the same decibel decrease in sound pressure.

The sound pressure at a distance r from a noise source with known sound power, directivity, and room constant (which is covered in Sec. 25) is

$$L_{p,\text{dBA}} = L_W + 10 \log \left(\frac{Q}{4\pi r^2} + \frac{4}{R} \right) \quad \text{[SI]} \quad 68.15(a)$$

$$L_{p,\text{dBA}} = 10.5 + L_W + 10 \log \left(\frac{Q}{4\pi r^2} + \frac{4}{R} \right)$$
$$\text{[U.S.]} \quad 68.15(b)$$

Example 68.6

The combined sound pressure level from four identical machines at a point equally distant from each is 100 dBA. Three machines are shut down. What is the new sound pressure level?

---

[9]Many years ago, a reference of 10^{-13} W was used. There may still be some references to this value in older literature.

Solution

When three of the four machines are shut down, the total radiated sound power will be 25% of the original condition. The reduction in sound power level will be

$$L_{W,1} - L_{W,2} = 10 \log \left(\frac{W_1}{W_2} \right)$$
$$= 10 \log (4) = 6.02 \text{ dB} \quad (\text{use 6 dB})$$

The change in sound pressure level is the same as the change in sound power level. Therefore, the new sound pressure level will be

$$L_p = 100 \text{ dBA} - 6 \text{ dB} = 94 \text{ dBA}$$

17. FREQUENCY CONTENT OF NOISE

Detailed knowledge of the frequency content of noise is required for most noise control problems. *Frequency content* is described by dividing the sound spectrum into a series of frequency ranges called *bands*. Frequency bands are established by dividing the frequency range of interest into bands of either equal bandwidths or equal percentages.[10] The sound level in each frequency band can be reported or plotted against the midpoint of the frequency band.

18. OCTAVE BAND ANALYSIS

Percentage bandwidths and center frequencies have been standardized worldwide. The most common (and widest) standard bandwidth is 50%, which means that the ratio between the frequencies at the two ends of the band is 1:2. For example, the band whose center frequency is 1000 Hz spans the range from 707 Hz to 1414 Hz. Since musical pitch frequency ratios of 1:2 are called *octaves*, the name *octave band* has been adopted.

Center frequencies for octave bands have been standardized at 63, 125, 250, 500, 1000, 2000, 4000, and 8000 Hz.[11,12] These bands are numbered 1, 2, 3, 4, 5, 6, 7, and 8, respectively. Octave bands can also be subdivided for better frequency detail. The most common subdivision is the *one-third octave band*.

---

[10]Equal bandwidths are generally used only when the frequency range of interest is small. There is little standardization of narrow bandwidths, primarily because narrow band analysis is used more for detailed investigation of noise sources than for reporting absolute levels.

[11]A now-obsolete series of octave bands was used until the late 1950s.

[12]Octave bands centered at 31.5 Hz and 16,000 Hz are also occasionally encountered. However, these two end bands are usually omitted, as they are at the extreme limits of most peoples' hearing abilities. For example, the A-weighted correction for the 31.5 Hz is −39.2 dB. This is such a large reduction that the contribution from this band is negligible except for all but the most powerful or pure-tone sources.

19. COMBINING MULTIPLE SOURCES

Multiple sound sources combine to produce more sound. However, multiple sound pressure levels or sound power levels expressed in decibels cannot be directly added. Two 90 dB sources do not produce a 180 dB source. For multiple sound sources, the combined power is given by Eq. 68.16. This is known as an *unweighted sum* because the individual readings are used without modification.

$$L = 10 \log \sum 10^{W_i/10} \qquad \textbf{68.16}$$

Using Eq. 68.16 to add or subtract the sound from two or more sources is not the same as determining the *change* in sound pressure level. The relationship between the old, new, and change in sound levels is that of a simple linear combination (i.e., addition or subtraction).

Equation 68.16 illustrates an important noise reduction principle: The noisiest source must be identified and treated before significant overall noise reduction can be achieved. Sources with sound levels more than a few decibels lower than the noisiest source make a minor contribution to the total sound level.

Example 68.7

What is the unweighted combined sound pressure from two machines, one with a sound pressure of 89 dB and the other with a sound pressure of 94 dB?

Solution

Use Eq. 68.16.

$$L = 10 \log \sum 10^{W_i/10}$$
$$= 10 \log \left(10^{\frac{89}{10}} + 10^{\frac{94}{10}} \right) = 95.2 \text{ dB}$$

20. BROAD BAND MEASUREMENTS

Octave band measurements contain enough information about the frequency content to permit calculation of the equivalent A-weighted level. The correction from Table 68.5 is made to each octave band measurement, and the corresponding levels are added using Eq. 68.16.

Table 68.5 *A-Weighting Corrections from Octave Band Analysis*

center frequency of band (Hz)	correction to be added (dB)[a]
63	− 26.2
125	− 16.1
250	− 8.6
500	− 3.2
1000	0.0
2000	+ 1.2
4000	+ 1.0
8000	− 1.1

[a]referenced to 20 μPa

Example 68.8

The octave band measurements of a sound source are: 63 Hz, 60 dB; 125 Hz, 72 dB; 250 Hz, 85 dB; 500 Hz, 90 dB; 1000 Hz, 90 dB; 2000 Hz, 80 dB; 4000 Hz, 83 dB; 8000 Hz, 70 dB. What is the A-weighted sound level?

Solution

Add the corrections from Table 68.5 to the measurements.

frequency	measurement	correction	corrected value
63	60	−26.2	33.8
125	72	−16.1	55.9
250	85	−8.6	76.4
500	90	−3.2	86.8
1000	90	0	90
2000	80	+1.2	81.2
4000	83	+1.0	84
8000	70	−1.1	68.9

Use Eq. 68.16.

$$L = 10 \log \sum 10^{W_i/10}$$
$$= 10 \log (10^{3.38} + 10^{5.59} + 10^{7.64} + 10^{8.68} + 10^9$$
$$+ 10^{8.12} + 10^{8.4} + 10^{6.89})$$
$$= 92.8 \text{ dBA}$$

21. LOUDNESS

Human hearing is not equally sensitive to pressure at all frequencies. In addition, some types of noise are more annoying than others. Subjective (i.e., perceived) noise level is called *loudness*.

There are two loudness scales: the *phon scale* and the *sone scale*. One *phon* is numerically equal to the sound pressure level in decibels at a frequency of 1000 Hz. At 1000 Hz, the values of loudness and L_p are the same.

A common loudness scale used by fan and duct manufacturers is the sone scale. By definition, one *sone* is the loudness of a 1000 Hz sound with a sound pressure of 0.02 μbar.[13] Sones are calculated from octave band sound pressure level measurements made through frequency filters in much the same way as are dBA measurements.[14]

Sone values are easier to use than decibel values because the sone scale is linear, not logarithmic. A doubling of sones is perceived by most people to be a doubling of the loudness.

---

[13]This is approximately the loudness of a quiet refrigerator in a quiet kitchen.
[14]However, sone values are combined and manipulated differently than decibel (dBA) values.

22. NOISE CRITERION CURVES

One way of specifying limits on loudness is to require all octave band measurements to be below a certain value. However, since the human ear is not as sensitive to low frequencies as it is to high frequencies, the limiting value will depend on the frequency. *Noise criterion* (NC) *curves* (also known as *noise criteria curves*) are curves of acceptable sound pressure level plotted against frequency. There are several such curves, and these are designated NC-15, NC-20, and so on. The NC curve used depends on the type of location. NC-35 is commonly used for such locations as apartment buildings, hotels, executive offices, and laboratories. NC-45 is commonly used for washrooms, kitchens, lobbies, banking areas, and open offices.

23. ACOUSTIC ABSORBER MATERIALS

Acoustic absorbers are materials that prevent reflection of incident sound. Fibrous glass, draperies, and open cell foams are acoustic absorbers. Absorbers generally are used in combination with barriers and dampers since absorbers do not block noise or reduce vibration when used alone. Absorbers work best at higher frequencies unless they are very thick or are tuned to low frequencies.

24. SOUND ABSORPTION COEFFICIENT

The *sound absorption coefficient*, α, is the ratio of sound energy absorbed by the surface compared to the sound energy striking the surface. Hard surfaces absorb essentially no sound energy and have sound absorption coefficients close to zero. Soft surfaces absorb most of the sound energy striking them. A value of 1.0 indicates a perfect absorber.

Since the sound absorption coefficient varies with frequency, the *noise reduction coefficient* rating (NRC) is used to simplify comparisons of materials. The NRC is the arithmetic average of the four absorption coefficients measured at frequencies of 250, 500, 1000, and 2000 Hz. Coefficients of common building materials and furnishings are listed in App. 68.A.

Absorptive products seldom can reduce ambient noise by more than 6 to 8 dB. They reduce noise reflection, but they do not stop sound transmission.

25. ROOM CONSTANT

If a room contains absorbing materials with known absorption coefficients, the average absorption coefficient can be calculated. For this method of determining the average absorption coefficient to be valid, the absorption materials must be well distributed in the room so that the value applies to the majority of locations. In Eq. 68.17, the product αA is known as the *surface absorption* or *sabin area*, S, with units of sabins.[15] One

---

[15]The spelling *sabine* is also encountered.

sabin is one square foot of perfectly absorbing surface. Generally, the surface absorption should be 20 to 50% of the room area. Otherwise, the room will be reverberatory.

$$S = \alpha A \qquad 68.17$$

$$\overline{\alpha} = \frac{\sum A_i \alpha_i}{\sum A_i} = \frac{\sum S_i}{\sum A_i} \qquad 68.18$$

The acoustic absorption characteristics of the room can be described by its *room constant*, *R*. A room with perfectly absorbing surfaces (i.e., an anechoic room) has an infinite room constant. A room with perfectly reflecting surfaces (i.e., a reverberant room) has a room constant of zero.

$$R = \frac{\overline{\alpha}A}{1 - \overline{\alpha}} = \frac{S}{1 - \overline{\alpha}} \qquad 68.19$$

26. ACOUSTIC BARRIER MATERIALS

Acoustic barriers block sound. The best performing *barrier materials* have high mass per unit area. Due to their low cost, gypsum board, plywood, and masonry are often used for this purpose.

27. SOUND TRANSMISSION CLASS

The *sound transmission class* (STC) is a single-number rating that is based on a match of the barrier's transmission loss curve to one of several standard ASTM contours. STC is particularly useful in selecting noise barriers for speech isolation. However, STC ratings are not accurate enough for most industrial noise control work.

28. NOISE REDUCTION

Received sound can be reduced by modifying either the sound source, the transmission path, or the area occupied by the source. The process of reducing the noise is known as *noise reduction, sound abatement,* and *attenuation*. The amount of noise reduction, NR, is a decrease in sound pressure level due to such modifications.

$$NR = L_{p,1} - L_{p,2} \qquad 68.20$$

When adding noise reduction material to a room, the new surface absorption (sabin area) should be 3 to 10 times the original surface absorption. In a room with well-distributed absorption surfaces and at large distances from a single noise source, or for the case of many small noise sources distributed in the room, the noise reduction is

$$NR = 10 \log \left(\frac{\sum S_1}{\sum S_2} \right) = 10 \log \left(\frac{R_1(1 - \overline{\alpha}_1)}{R_2(1 - \overline{\alpha}_2)} \right) \qquad 68.21$$

29. INSERTION LOSS

Insertion loss, IL, is the term used to describe the amount of noise reduction at a particular place after the noise source has been isolated. The source is usually isolated by covering it with an enclosure or by constructing a wall between it and occupants. (The wall or enclosure has been "inserted" into the sound transmission path.) The decrease in sound pressure level is the same as the decrease in the sound power level.

$$IL = L_{p,1} - L_{p,2} = L_{W,1} - L_{W,2}$$
$$= 10 \log \left(\frac{W_1}{W_2} \right) = 20 \log \left(\frac{p_1}{p_2} \right) \qquad 68.22$$

30. TRANSMISSION LOSS

The *transmission loss* is the portion of the sound energy lost when sound passes through a wall or barrier. The fraction of the incident energy transmitted through a barrier is designated as the *transmission coefficient*, τ. Appendix 68.B lists the transmission loss of several common materials used for noise control. The decrease in sound pressure level is the same as the decrease in the sound power level.

$$TL = L_{W,1} - L_{W,2}$$
$$= 10 \log \left(\frac{W_1}{W_2} \right) = 10 \log \left(\frac{1}{\tau} \right) \qquad 68.23$$

Transmission loss depends on the sound frequency, increasing by approximately 4 to 5 dB for each doubling of the frequency or for each doubling of the mass per unit area.

31. ATTENUATION

Attenuation is the amount by which sound is decreased as it travels from the source to the receiver. This general term is used interchangeably with noise reduction, insertion loss, and transmission loss. An *attenuator* is any device (e.g., a muffler) that can be inserted into the sound path to reduce the noise received.

32. NOISE DOSE

The *noise dose*, *D*, is the actual length of exposure, *C*, divided by the permissible (i.e., legal) duration of exposure, *t*. Preventative steps must be taken if the daily noise dose is more than 1.0. When the noise level varies during the measurement period, the noise dose is

$$D = \frac{C_1}{t_1} + \frac{C_2}{t_2} + \frac{C_3}{t_3} + \cdots \qquad 68.24$$

Unlike a traditional sound meter, which measures the instantaneous sound level at one location, a sound *dosimeter* worn by an employee monitors and integrates the sound level over the entire work shift.

PROFESSIONAL PUBLICATIONS, INC. BELMONT, CA

33. ALLOWABLE EXPOSURE LIMITS

In the United States, the Occupational Safety and Health Act (OSHA) sets maximum limits on daily sound exposure. The "all-day" eight-hour noise level limit in the United States is 90 dBA. This is higher than the maximum level permitted in other countries (e.g., 85 dBA in Germany and Japan). The limit may be even lower in some countries in the future.[16] In the United States, employees may not be exposed to steady sound levels above 115 dBA, regardless of the duration. Impact sound levels are limited to 140 dBA.

In the United States, hearing protection, educational programs, periodic examinations, and other actions are required for workers whose eight-hour exposure is more than 85 dBA or whose noise dose exceeds 50% of the action levels.

Table 68.6 *Typical Permissible Noise Exposure Levels*[a,b]

sound level (dBA)	exposure (hours per day)
90	8
92	6
95	4
97	3
100	2
102	$1^1/_2$
105	1
110	$^1/_2$
115	$^1/_4$ or less

[a]without hearing protection
[b]Federal Register, Title 29, Sec. 1910.95, Table G-16

34. NOISE REDUCTION RATING

Hearing protection devices (earplugs, muffs, etc.) for use where noise is continuous (as opposed to impulsive) are rated by a *noise reduction rating* (NRR). The NRR is, essentially, an average insertion loss. Ideally, the level of noise entering a person's protected ear is the difference between the A-weighted noise level and the NRR. For environments dominated by frequencies below 500 Hz, the C-weighted noise level should be used. With impulsive noise (e.g., gunfire), the NRR may not be an accurate indicator of hearing protection.

PRACTICE PROBLEMS

1. What is the total sound pressure level when a 40 dB source is placed adjacent to a 35 dB source?

2. With no machinery operating, the background noise in a room has a sound pressure level of 43 dB. With the machinery operating, the sound pressure level is 45 dB. What is the sound pressure level due to the machinery alone?

3. An unenclosed source produces a sound pressure level of 100 dB. An enclosure is constructed from a material having a transmission loss of 30 dB. The sound pressure level inside the enclosure from the enclosed source increases to 110 dB. What is the reduction in sound pressure level outside the enclosure?

4. 4 ft (1.2 m) from an isotropic sound source, the sound pressure level is 92 dB. What is the sound pressure level 12 ft (3.6 m) from the source?

5. Octave band measurements of a noise source were made. The measurements were 85, 90, 92, 87, 82, 78, 65, and 54 dB at frequencies of 63, 125, 250, 500, 1000, 2000, 4000, and 8000 Hz, respectively. What is the overall A-weighted sound pressure?

6. What is the maximum possible reduction in sound pressure level if the number of sabins is 50% of the total room area?

7. A room has dimensions of 100 ft × 400 ft × 20 ft (30 m × 120 m × 6 m). All surfaces are precast concrete. 40% of the walls are treated acoustically with a material having a sound absorption coefficient of 0.8. What is the reduction in sound pressure level?

8. An office has dimensions of 20 ft × 50 ft × 10 ft (6 m × 15 m × 3 m). The floor is covered with roll vinyl. The walls and ceiling are sheetrock. 20% of the walls are glass windows. There are 15 desks, 15 occupants, and 5 miscellaneous sabins. After complaints from the occupants, the ceiling is treated with a sound absorbing material with a sound absorption coefficient of 0.7. What is the reduction in sound level?

9. A room has dimensions of 15 ft × 20 ft × 10 ft (4.5 m × 6 m × 3 m). The sound absorption coefficients are 0.03, 0.5, and 0.06 for the floor, ceiling, and walls, respectively. A machine with a sound power level of 65 dB is located at the intersection of the floor and wall, 7.5 ft (2.25 m) from the perpendicular walls. The ambient sound pressure level is 50 dB everywhere in the room. What is the sound pressure level 5 ft (1.5 m) from the machine?

---

[16]Levels as low as 75 dBA have been proposed in Europe.

69 Engineering Economic Analysis

1.	Irrelevant Characteristics	69-2
2.	Multiplicity of Solution Methods	69-2
3.	Precision and Significant Digits	69-2
4.	Nonquantifiable Factors	69-2
5.	Year-End Convention	69-3
6.	Cash Flow Diagrams	69-3
7.	Types of Cash Flows	69-3
8.	Typical Problem Types	69-4
9.	Implicit Assumptions	69-4
10.	Equivalence	69-5
11.	Single-Payment Equivalence	69-5
12.	Standard Cash Flow Factors and Symbols	69-6
13.	Calculating Uniform Series Equivalence	69-7
14.	Finding Past Values	69-8
15.	Times to Double and Triple an Investment	69-8
16.	Varied and Nonstandard Cash Flows	69-9
17.	The Meaning of Present Worth and i	69-11
18.	Simple and Compound Interest	69-11
19.	Extracting the Interest Rate: Rate of Return	69-12
20.	Rate of Return versus Return on Investment	69-14
21.	Minimum Attractive Rate of Return	69-14
22.	Typical Alternative-Comparison Problem Formats	69-14
23.	Durations of Investments	69-14
24.	Choice of Alternatives: Comparing One Alternative with Another Alternative	69-14
25.	Choice of Alternatives: Comparing an Alternative with a Standard	69-16
26.	Ranking Mutually Exclusive Multiple Projects	69-17
27.	Alternatives with Different Lives	69-17
28.	Opportunity Costs	69-18
29.	Replacement Studies	69-18
30.	Treatment of Salvage Value in Replacement Studies	69-18
31.	Economic Life: Retirement at Minimum Cost	69-19
32.	Life-Cycle Cost	69-20
33.	Capitalized Assets versus Expenses	69-20
34.	Purpose of Depreciation	69-20
35.	Depreciation Basis of an Asset	69-20
36.	Depreciation Methods	69-21
37.	Accelerated Depreciation Methods	69-23
38.	Book Value	69-23
39.	Amortization	69-24
40.	Depletion	69-24
41.	Basic Income Tax Considerations	69-24
42.	Taxation at the Times of Asset Purchase and Sale	69-25
43.	Depreciation Recovery	69-26
44.	Other Interest Rates	69-28
45.	Rate and Period Changes	69-28
46.	Bonds	69-29
47.	Probabilistic Problems	69-30
48.	Fixed and Variable Costs	69-31
49.	Accounting Costs and Expense Terms	69-32
50.	Accounting Principles	69-33
51.	Cost Accounting	69-36
52.	Cost of Goods Sold	69-37
53.	Break-Even Analysis	69-37
54.	Pay-Back Period	69-38
55.	Management Goals	69-38
56.	Inflation	69-39
57.	Consumer Loans	69-39
58.	Forecasting	69-41
59.	Learning Curves	69-42
60.	Economic Order Quantity	69-42
61.	Sensitivity Analysis	69-43
62.	Value Engineering	69-43
	Practice Problems	69-44

Nomenclature

A	annual amount	$
B	present worth of all benefits	$
BV_j	book value at end of the jth year	$
C	cost, or present worth of all costs	$
d	declining balance depreciation rate	decimal
D	demand	various
D	depreciation	$
DR	present worth of after-tax depreciation recovery	$
e	constant inflation rate	decimal
E_0	initial amount of an exponentially growing cash flow	$
EAA	equivalent annual amount	$
EUAC	equivalent uniform annual cost	$
f	federal income tax rate	decimal
F	forecasted quantity	various
F	future worth	$
g	exponential growth rate	decimal
G	uniform gradient amount	$
i	effective rate per period (usually per year)	decimal per unit time
i'	effective interest rate corrected for inflation	decimal
k	number of compounding periods per year	–
m	an integer	–
n	number of compounding periods, or life of asset	–

Economics

P	present worth	\$
r	nominal rate per year (rate per annum)	decimal per unit time
ROI	return on investment	\$
ROR	rate of return	decimal per unit time
s	state income tax rate	decimal
S_n	expected salvage value in year n	\$
t	composite tax rate	decimal
t	time	years (typical)
T	a quantity equal to $\frac{1}{2}n(n+1)$	–
TC	tax credit	\$
z	a quantity equal to $\frac{1+i}{1-d}$	decimal

Symbols

α	smoothing coefficient for forecasts	–
ϕ	effective rate per period (r/k)	decimal
$\mathcal{E}$	expected value	various

Subscripts

0	initial
j	at time j
n	at time n
t	at time t

1. IRRELEVANT CHARACTERISTICS

In its simplest form, an *engineering economic analysis* is a study of the desirability of making an investment.[1] The decision-making principles in this chapter can be applied by individuals as well as by companies. The nature of the spending opportunity or industry is not important. Farming equipment, personal investments, and multimillion dollar factory improvements can all be evaluated using the same principles.

Similarly, the applicable principles are largely insensitive to the monetary units. Although *dollars* are used in this chapter, it is equally convenient to use British pounds, Japanese yen, or German marks.

Finally, this chapter may give the impression that investment alternatives must be evaluated on a year-by-year basis. Actually, the *effective period* can be defined as a day, month, century, or any other convenient period of time.

2. MULTIPLICITY OF SOLUTION METHODS

Most economic conclusions can be reached in more than one manner. There are usually several different analyses that will eventually result in identical answers.[2] Other

---

[1]This subject is also known as *engineering economics* and *engineering economy*. There is very little, if any, true economics in this subject.
[2]Because of round-off errors, particularly when factors are taken from tables, these different calculations will produce slightly different numerical results (e.g., \$49.49 versus \$49.50). However, this type of divergence is well known and accepted in engineering economic analysis.

than the pursuit of elegant solutions in a timely manner, there is no reason to favor one procedural method over another.[3]

3. PRECISION AND SIGNIFICANT DIGITS

The full potential of electronic calculators will never be realized in engineering economic analyses. Considering that calculations are based on estimates of far-future cash flows, and that unrealistic assumptions (no inflation, identical cost structures of replacement assets, etc.) are routinely made, it makes little sense to carry cents along in calculations.

The calculations in this chapter have been designed to illustrate and review the principles presented. Because of this, greater precision than is normally necessary in everyday problems may be used. Though used, such precision is not warranted.

Unless there is some compelling reason to strive for greater precision, the following rules are presented for use in reporting final answers to engineering economic analysis problems.

- Omit fractional parts of the dollar (i.e., cents).

- Report and record a number to a maximum of four significant digits unless the first digit of that number is 1, in which case, a maximum of five significant digits should be written. For example,

\$49	not	\$49.43
\$93,450	not	\$93,453
\$1,289,700	not	\$1,289,673

4. NONQUANTIFIABLE FACTORS

An engineering economic analysis is a quantitative analysis. Some factors cannot be introduced as numbers into the calculations. Such factors are known as *nonquantitative factors*, *judgment factors*, and *irreducible factors*. Typical nonquantifiable factors are

- preferences
- political ramifications
- urgency
- goodwill
- prestige
- utility
- corporate strategy
- environmental effects
- health and safety rules
- reliability
- political risks

---

[3]This does not imply that approximate methods, simplifications, and rules of thumb are acceptable.

Since these factors are not included in the calculations, the policy is to disregard the issues entirely. Of course, the factors should be discussed in a final report. The factors are particularly useful in breaking ties between competing alternatives that are economically equivalent.

5. YEAR-END CONVENTION

Except in short-term transactions, it is simpler to assume that all receipts and disbursements (cash flows) take place at the end of the year in which they occur.[4] This is known as the *year-end convention*. The exceptions to the year-end convention are initial project cost (purchase cost), trade-in allowance, and other cash flows that are associated with the inception of the project at $t = 0$.

On the surface, such a convention appears grossly inappropriate, since repair expenses, interest payments, corporate taxes, and so on, seldom coincide with the end of a year. However, the convention greatly simplifies engineering economic analysis problems, and it is justifiable on the basis that the increased precision associated with a more rigorous analysis is not warranted (due to the numerous other simplifying assumptions and estimates initially made in the problem).

There are various established procedures, known as *rules* or *conventions*, imposed by the Internal Revenue Service on U.S. taxpayers. An example is the *half-year rule*, which permits only half of the first-year depreciation to be taken in the first year of an asset's life when certain methods of depreciation are used. These rules are subject to constantly changing legislation and are not covered in this book. The implementation of such rules is outside the scope of engineering practice and is best left to accounting professionals.

6. CASH FLOW DIAGRAMS

Although they are not always necessary in simple problems (and they are often unwieldy in very complex problems), *cash flow diagrams* can be drawn to help visualize and simplify problems having diverse receipts and disbursements.

The following conventions are used to standardize cash flow diagrams.

- The horizontal (time) axis is marked off in equal increments, one per period, up to the duration (or *horizon*) of the project.

- Two or more transfers in the same period are placed end-to-end, and these may be combined.

- Expenses incurred before $t = 0$ are called *sunk costs*. Sunk costs are not relevant to the problem unless they have tax consequences in an after-tax analysis.

- *Receipts* are represented by arrows directed upward. *Disbursements* are represented by arrows directed downward. The arrow length is proportional to the magnitude of the cash flow.

Example 69.1

A mechanical device will cost $20,000 when purchased. Maintenance will cost $1000 each year. The device will generate revenues of $5000 each year for five years, after which the salvage value is expected to be $7000. Draw and simplify the cash flow diagram.

Solution

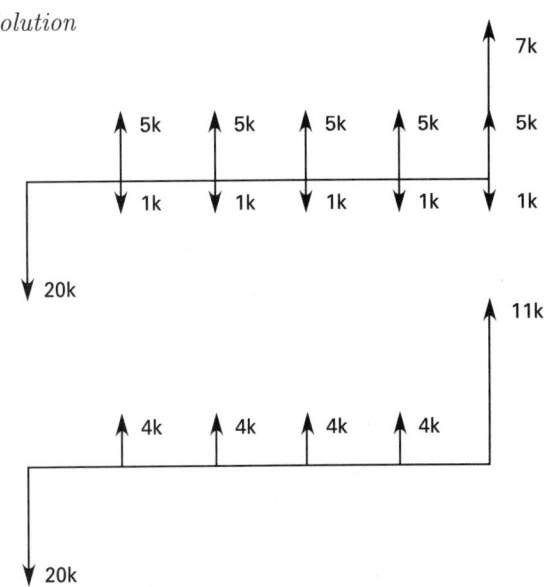

7. TYPES OF CASH FLOWS

To evaluate a real-world project, it is necessary to present the project's cash flows in terms of standard cash flows that can be handled by engineering economic analysis techniques. The standard cash flows are single payment cash flow, uniform series cash flow, gradient series cash flow, and the infrequently encountered exponential gradient series cash flow.

A *single payment cash flow* can occur at the beginning of the time line (designated as $t = 0$), at the end of the time line (designated as $t = n$), or at any time in between.

The *uniform series cash flow* consists of a series of equal transactions starting at $t = 1$ and ending at $t = n$. The symbol A is typically given to the magnitude of each individual cash flow.[5]

---

[4] A *short-term transaction* typically has a lifetime of five years or less and has payments or compounding that are more frequent than once per year.

[5] Notice that the cash flows do not begin at $t = 0$. This is an important concept with all of the series cash flows. This convention has been established to accommodate the timing of annual maintenance (and similar) cash flows for which the year-end convention is applicable.

The *gradient series cash flow* starts with a cash flow (typically given the symbol G) at $t = 2$, and increases by G each year until $t = n$, at which time the final cash flow is $(n-1)G$.

An *exponential gradient series cash flow* is based on a phantom value (typically given the symbol E_0) at $t = 0$, and grows or decays exponentially according to the following relationship.[6]

$$\text{amount at time } t = E_t = E_0(1+g)^t$$
$$[t = 1, 2, 3, \ldots, n] \quad \textbf{69.1}$$

In Eq. 69.1, g is the *exponential growth rate*, which can be either positive or negative. Exponential gradient cash flows are rarely seen in economic justification projects assigned to engineers.[7]

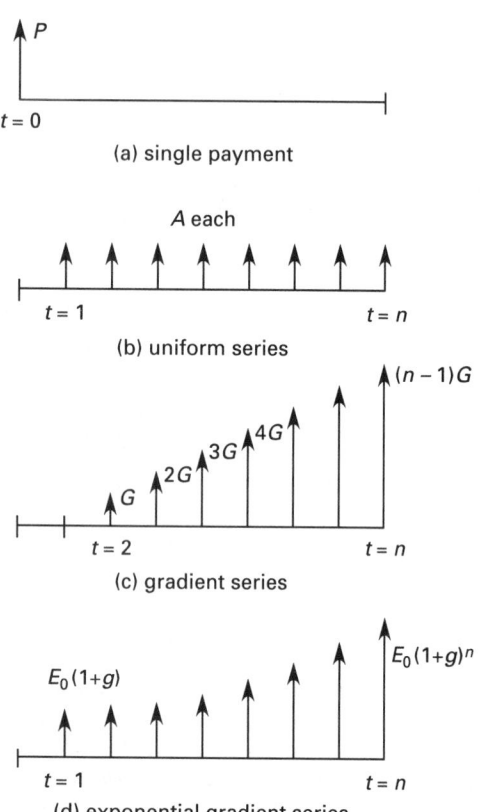

(a) single payment

(b) uniform series

(c) gradient series

(d) exponential gradient series

Figure 69.1 *Standard Cash Flows*

8. TYPICAL PROBLEM TYPES

There is a wide variety of problem types that, collectively, are considered to be engineering economic analysis problems.

---

[6]Notice the convention for an exponential cash flow series: The first cash flow E_0 is at $t = 1$, as in the uniform annual series. However, the first cash flow is $E_0(1+g)$. The cash flow of E_0 at $t = 0$ is absent (i.e., is a *phantom cash flow*).

[7]For one of the few discussions on exponential cash flow, see *Capital Budgeting*, Robert V. Oakford, The Ronald Press Company, New York, 1970.

By far, the majority of engineering economic analysis problems are *alternative comparisons*. In these problems, two or more mutually exclusive investments compete for limited funds. A variation of this is a *replacement/retirement analysis*, which is repeated each year to determine if an existing asset should be replaced. Finding the percentage return on an investment is a *rate of return problem*, one of the alternative comparison solution methods.

Investigating interest and principal amounts in loan payments is a *loan repayment problem*. An *economic life analysis* will determine when an asset should be retired. In addition, there are miscellaneous problems involving economic order quantity, learning curves, break-even points, product costs, and so on.

9. IMPLICIT ASSUMPTIONS

Several assumptions are implicitly made when solving engineering economic analysis problems. Some of these assumptions are made with the knowledge that they are or will be poor approximations of what really will happen. The assumptions are made, regardless, for the benefit of obtaining a solution.

The most common assumptions are the following.

- The year-end convention is applicable.

- There is no inflation now, nor will there be any during the lifetime of the project.

- Unless otherwise specifically called for, a before-tax analysis is needed.

- The effective interest rate used in the problem will be constant during the lifetime of the project.

- Nonquantifiable factors can be disregarded.

- Funds invested in a project are available and are not urgently needed elsewhere.

- Excess funds continue to earn interest at the effective rate used in the analysis.

This last assumption, like most of the assumptions listed, is almost never specifically mentioned in the body of a solution. However, it is a key assumption when comparing two alternatives that have different initial costs.

For example, suppose two investments, one costing $10,000 and the other costing $8000, are to be compared at 10%. It is obvious that $10,000 in funds is available, otherwise the costlier investment would not be under consideration. If the smaller investment is chosen, what is done with the remaining $2000? The last assumption yields the answer: the $2000 is "put to work" in some investment earning (in this case) 10%.

10. EQUIVALENCE

Industrial decision makers using engineering economic analysis are concerned with the magnitude and timing of a project's cash flow as well as with the total profitability of that project. In this situation, a method is required to compare projects involving receipts and disbursements occurring at different times.

By way of illustration, consider $100 placed in a bank account that pays 5% effective annual interest at the end of each year. After the first year, the account will have grown to $105. After the second year, the account will have grown to $110.25.

Assume that you will have no need for money during the next two years, and any money received will immediately go into your 5% bank account. Then, which of the following options would be more desirable?

option a: $100 now

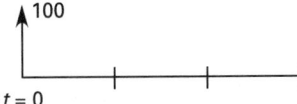

option b: $105 to be delivered in one year.

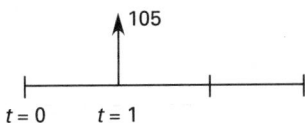

option c: $110.25 to be delivered in two years

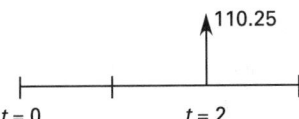

As illustrated, none of the options is superior under the assumptions given. If the first option is chosen, you will immediately place $100 into a 5% account, and in two years the account will have grown to $110.25. In fact, the account will contain $110.25 at the end of two years regardless of the option chosen. Therefore, these alternatives are said to be *equivalent*.

Equivalence may or may not be the case, depending on the interest rate. Thus, an alternative that is acceptable to one decision maker may be unacceptable to another. The interest rate that is used in actual calculations is known as the *effective interest rate*.[8] If compounding is

---

[8]The adjective *effective* distinguishes this interest rate from other interest rates (e.g., nominal interest rates) that are not meant to be used directly in calculating equivalent amounts.

once a year, it is known as the *effective annual interest rate*. However, effective quarterly, monthly, daily, and so on, interest rates are also used.

The fact that $100 today grows to $105 in one year (at 5% annual interest) is an example of what is known as the *time value of money* principle. This principle simply articulates what is obvious: Funds placed in a secure investment will increase to an equivalent future amount. The procedure for determining the present investment from the equivalent future amount is known as *discounting*.

11. SINGLE-PAYMENT EQUIVALENCE

The equivalence of any present amount, P, at $t = 0$, to any future amount, F, at $t = n$, is called the *future worth* and can be calculated from Eq. 69.2.

$$F = P(1 + i)^n \qquad 69.2$$

The factor $(1 + i)^n$ is known as the *single payment compound amount factor* and has been tabulated in App. 69.B for various combinations of i and n.

Similarly, the equivalence of any future amount to any present amount is called the *present worth* and can be calculated from Eq. 69.3.

$$P = F(1 + i)^{-n} = \frac{F}{(1 + i)^n} \qquad 69.3$$

The factor $(1 + i)^{-n}$ is known as the *single payment present worth factor*.[9]

The interest rate used in Eqs. 69.2 and 69.3 must be the effective rate per period. Also, the basis of the rate (annually, monthly, etc.) must agree with the type of period used to count n. Thus, it would be incorrect to use an effective annual interest rate if n was the number of compounding periods in months.

Example 69.2

How much should you put into a 10% (effective annual rate) savings account in order to have $10,000 in five years?

Solution

This problem could also be stated: What is the equivalent present worth of $10,000 five years from now if money is worth 10% per year?

$$P = F(1 + i)^{-n} = (\$10,\!000)(1 + 0.10)^{-5}$$
$$= \$6209$$

The factor 0.6209 would usually be obtained from the tables.

---

[9]The *present worth* is also called the *present value* and *net present value*. These terms are used interchangeably and no significance should be attached to the terms *value*, *worth*, and *net*.

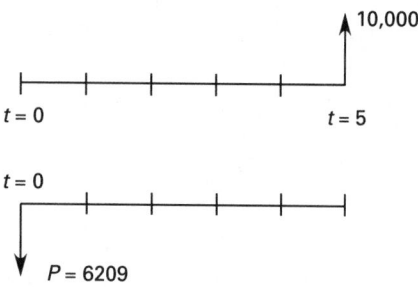

12. STANDARD CASH FLOW FACTORS AND SYMBOLS

Equations 69.2 and 69.3 may give the impression that solving engineering economic analysis problems involves a lot of calculator use, and, in particular, a lot of exponentiation. Such calculations may be necessary from time to time, but most problems are simplified by the use of tabulated values of the factors.

Rather than actually writing the formula for the compound amount factor (which converts a present amount to a future amount), it is common convention to substitute the standard functional notation of $(F/P, i\%, n)$. Thus, the future value in n periods of a present amount would be symbolically written as

$$F = P(F/P, i\%, n) \qquad 69.4$$

Similarly, the present worth factor has a functional notation of $(P/F, i\%, n)$. The present worth of a future amount n periods hence would be symbolically written as

$$P = F(P/F, i\%, n) \qquad 69.5$$

Values of these *cash flow (discounting) factors* are tabulated in App. 69.B. There is often initial confusion about whether the (F/P) or (P/F) column should be used in a particular problem. There are several ways of remembering what the functional notations mean.

One method of remembering which factor should be used is to think of the factors as conditional probabilities. The conditional probability of event **A** given that event **B** has occurred is written as $p\{\mathbf{A}|\mathbf{B}\}$, where the given event comes after the vertical bar. In the standard notational form of discounting factors, the given amount is similarly placed after the slash. What you want comes before the slash. (F/P) would be a factor to find F given P.

Another method of remembering the notation is to interpret the factors algebraically. Thus, the (F/P) factor could be thought of as the fraction F/P. Algebraically, Eq. 69.4 would be

$$F = P(F/P) \qquad 69.6$$

This algebraic approach is actually more than an interpretation. The numerical values of the discounting factors are consistent with this algebraic manipulation.

Thus, the (F/A) factor could be calculated as $(F/P) \times (P/A)$. This consistent relationship can be used to calculate other factors that might be occasionally needed, such as (F/G) or (G/P). For instance, the annual cash flow that would be equivalent to a uniform gradient may be found from

$$A = G(P/G, i\%, n)(A/P, i\%, n) \qquad 69.7$$

Formulas for the compounding and discounting factors are contained in Table 69.1. Normally, it will not be necessary to calculate factors from the formulas. Appendix 69.B is adequate for solving most problems.

Example 69.3

What factor will convert a gradient cash flow ending at $t = 8$ to a future value at $t = 8$? (That is, what is the $(F/G, i\%, 8)$ factor?) The effective annual interest rate is 10%.

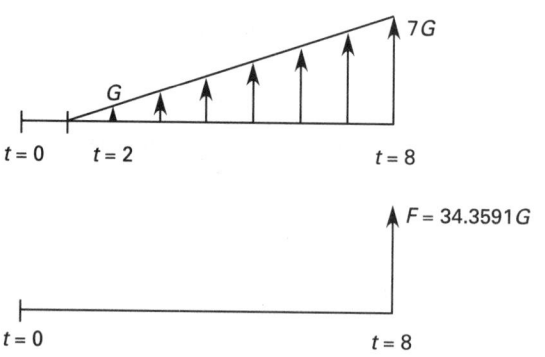

Solution

method 1: From Table 69.1, the $(F/G, 10\%, 8)$ factor is

$$(F/G, 10\%, 8) = \frac{(1+i)^n - 1}{i^2} - \frac{n}{i}$$
$$= \frac{(1.10)^8 - 1}{(0.10)^2} - \frac{8}{0.10} = 34.3589$$

method 2: The tabulated values of (P/G) and (F/P) in App. 69.B can be used to calculate the factor.

$$(F/G, 10\%, 8) = (P/G, 10\%, 8)(F/P, 10\%, 8)$$
$$= (16.0287)(2.1436) = 34.3591$$

The (F/G) factor could also have been calculated as the product of the (A/G) and (F/A) factors.

Table 69.1 Discount Factors for Discrete Compounding

factor name	converts	symbol	formula
single payment compound amount	P to F	$(F/P, i\%, n)$	$(1+i)^n$
single payment present worth	F to P	$(P/F, i\%, n)$	$(1+i)^{-n}$
uniform series sinking fund	F to A	$(A/F, i\%, n)$	$\dfrac{i}{(1+i)^n - 1}$
capital recovery	P to A	$(A/P, i\%, n)$	$\dfrac{i(1+i)^n}{(1+i)^n - 1}$
uniform series compound amount	A to F	$(F/A, i\%, n)$	$\dfrac{(1+i)^n - 1}{i}$
uniform series present worth	A to P	$(P/A, i\%, n)$	$\dfrac{(1+i)^n - 1}{i(1+i)^n}$
uniform gradient present worth	G to P	$(P/G, i\%, n)$	$\dfrac{(1+i)^n - 1}{i^2(1+i)^n} - \dfrac{n}{i(1+i)^n}$
uniform gradient future worth	G to F	$(F/G, i\%, n)$	$\dfrac{(1+i)^n - 1}{i^2} - \dfrac{n}{i}$
uniform gradient uniform series	G to A	$(A/G, i\%, n)$	$\dfrac{1}{i} - \dfrac{n}{(1+i)^n - 1}$

13. CALCULATING UNIFORM SERIES EQUIVALENCE

A cash flow that repeats each year for n years without change in amount is known as an *annual amount* and is given the symbol A. As an example, a piece of equipment may require annual maintenance, and the maintenance cost will be an annual amount. Although the equivalent value for each of the n annual amounts could be calculated and then summed, it is more expedient to use one of the uniform series factors. For example, it is possible to convert from an annual amount to a future amount by use of the (F/A) factor.

$$F = A(F/A, i\%, n) \qquad 69.8$$

A *sinking fund* is a fund or account into which annual deposits of A are made in order to accumulate F at $t = n$ in the future. Since the annual deposit is calculated as $A = F(A/F, i\%, n)$, the (A/F) factor is known as the *sinking fund factor*. An *annuity* is a series of equal payments (A) made over a period of time.[10] Usually, it is necessary to "buy into" an investment (a bond, and

insurance policy, etc.) in order to ensure the annuity. In the simplest case of an annuity that starts at the end of the first year and continues for n years, the purchase price (P) is

$$P = A(P/A, i\%, n) \qquad 69.9$$

The present worth of an *infinite (perpetual) series* of annual amounts is known as a *capitalized cost*. There is no $(P/A, i\%, \infty)$ factor in the tables, but the capitalized cost can be calculated simply as

$$P = \frac{A}{i} \qquad [i \text{ in decimal form}] \qquad 69.10$$

Alternatives with different lives will generally be compared by way of *equivalent uniform annual cost* (EUAC). An EUAC is the annual amount that is equivalent to all of the cash flows in the alternative. The EUAC differs in sign from all of the other cash flows. Costs and expenses expressed as EUACs, which would normally be considered negative, are actually positive. The term *cost* in the designation EUAC serves to make clear the meaning of a positive number.

---

[10]An annuity may also consist of a lump sum payment made at some future time. However, this interpretation is not considered in this chapter.

Example 69.4

Maintenance costs for a machine are $250 each year. What is the present worth of these maintenance costs over a 12-year period if the interest rate is 8%?

Solution

$$P = A(P/A, 8\%, 12) = (-\$250)(7.5361)$$
$$= -\$1884$$

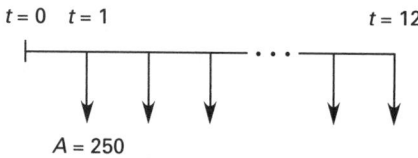

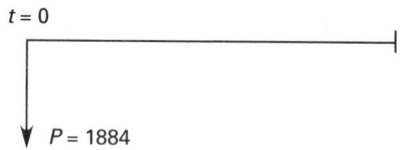

14. FINDING PAST VALUES

From time to time, it will be necessary to determine an amount in the past equivalent to some current (or future) amount. For example, you might have to calculate the original investment made 15 years ago given a current annuity payment.

Such problems are solved by placing the $t = 0$ point at the time of the original investment, and then calculating the past amount as a P value. For example, the original investment, P, can be extracted from the annuity, A, by using the standard cash flow factors.

$$P = A(P/A, i\%, n) \qquad \textbf{69.11}$$

The choice of $t = 0$ is flexible. As a general rule, the $t = 0$ point should be selected for convenience in solving a problem.

Example 69.5

You are currently paying $250 per month to lease your office phone equipment. You have three years (36 months) left on the five-year (60-month) lease. What would have been an equivalent purchase price two years ago? The effective interest rate per month is 1%.

Solution

The solution of this example is not affected by the fact that investigation is being performed in the middle of the horizon. This is a simple calculation of present worth.

$$P = A(P/A, 1\%, 60)$$
$$= (-\$250)(44.9550) = -\$11,239$$

15. TIMES TO DOUBLE AND TRIPLE AN INVESTMENT

If an investment doubles in value (in n compounding periods and with $i\%$ effective interest), the ratio of current value to past investment will be 2.

$$F/P = (1 + i)^n = 2 \qquad \textbf{69.12}$$

Similarly, the ratio of current value to past investment will be 3 if an investment triples in value. This can be written as

$$F/P = (1 + i)^n = 3 \qquad \textbf{69.13}$$

It is a simple matter to extract the number of periods, n, from Eqs. 69.12 and 69.13 to determine the *doubling time* and *tripling time*, respectively. For example, the doubling time is

$$n = \frac{\log 2}{\log (1 + i)} \qquad \textbf{69.14}$$

When a quick estimate of the doubling time is needed, the *rule of 72* can be used. The doubling time is approximately $72/i$.

The tripling time is

$$n = \frac{\log 3}{\log (1 + i)} \qquad \textbf{69.15}$$

Equations 69.14 and 69.15 form the basis of Table 69.2.

Table 69.2 *Doubling and Tripling Times for Various Interest Rates*

interest rate ($i\%$)	doubling time (periods)	tripling time (periods)
1	69.7	110.4
2	35.0	55.5
3	23.4	37.2
4	17.7	28.0
5	14.2	22.5
6	11.9	18.9
7	10.2	16.2
8	9.01	14.3
9	8.04	12.7
10	7.27	11.5
11	6.64	10.5
12	6.12	9.69
13	5.67	8.99
14	5.29	8.38
15	4.96	7.86
16	4.67	7.40
17	4.41	7.00
18	4.19	6.64
19	3.98	6.32
20	3.80	6.03

16. VARIED AND NONSTANDARD CASH FLOWS

Gradient Cash Flow

A common situation involves a uniformly increasing cash flow. If the cash flow has the proper form, its present worth can be determined by using the *uniform gradient factor*, $(P/G, i\%, n)$. The uniform gradient factor finds the present worth of a uniformly increasing cash flow that starts in year two (not in year one).

There are three common difficulties associated with the form of the uniform gradient. The first difficulty is that the initial cash flow occurs at $t = 2$. This convention recognizes that annual costs, if they increase uniformly, begin with some value at $t = 1$ (due to the year-end convention) but do not begin to increase until $t = 2$. The tabulated values of (P/G) have been calculated to find the present worth of only the increasing part of the annual expense. The present worth of the base expense incurred at $t = 1$ must be found separately with the (P/A) factor.

The second difficulty is that, even though the $(P/G, i\%, n)$ factor is used, there are only $n - 1$ actual cash flows. It is clear that n must be interpreted as the *period number* in which the last gradient cash flow occurs, not the number of gradient cash flows.

Finally, the sign convention used with gradient cash flows may seem confusing. If an expense increases each year (as in Ex. 69.6), the gradient will be negative, since it is an expense. If a revenue increases each year, the gradient will be positive. In most cases, the sign of the gradient depends on whether the cash flow is an expense or a revenue.[11]

Example 69.6

Maintenance on an old machine is $100 this year but is expected to increase by $25 each year thereafter. What is the present worth of five years of the costs of maintenance? Use an interest rate of 10%.

Solution

In this problem, the cash flow must be broken down into parts. (Notice that the five-year gradient factor is used even though there are only four nonzero gradient cash flows.)

$$P = A(P/A, 10\%, 5) + G(P/G, 10\%, 5)$$
$$= (-\$100)(3.7908) - (\$25)(6.8618)$$
$$= -\$551$$

---

[11]This is not a universal rule. It is possible to have a uniformly decreasing revenue as in Fig. 69.2(c). In this case, the gradient would be negative.

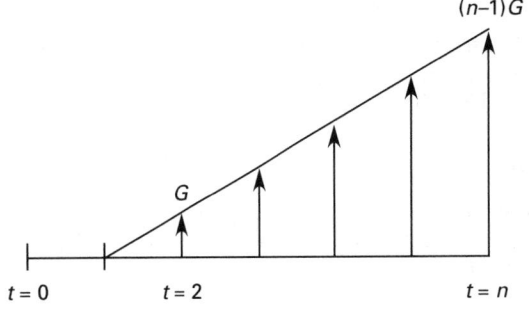

(a) positive gradient cash flow

$$P = G(P/G)$$

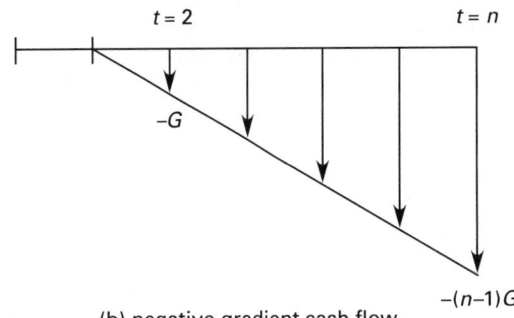

(b) negative gradient cash flow

$$P = -G(P/G)$$

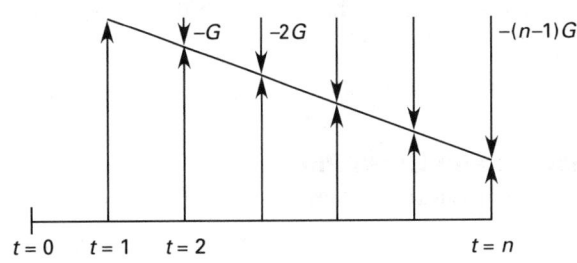

(c) decreasing revenue incorporating a negative gradient

$$P = S(P/A) - G(P/G)$$

Figure 69.2 *Positive and Negative Gradient Cash Flows*

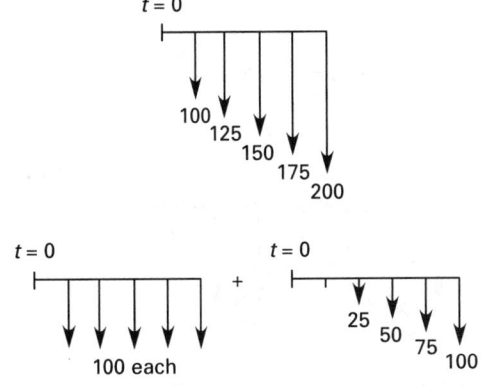

Stepped Cash Flows

Stepped cash flows are easily handled by the technique of *superposition of cash flows*. This technique is illustrated by Ex. 69.7.

Example 69.7

An investment costing $1000 returns $100 for the first five years and $200 for the following five years. How would the present worth of this investment be calculated?

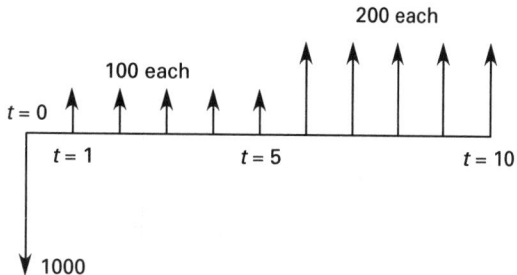

Solution

Using the principle of superposition, the revenue cash flow can be thought of as $200 each year from $t = 1$ to $t = 10$, with a negative revenue of $100 from $t = 1$ to $t = 5$. Superimposed, these two cash flows make up the actual performance cash flow.

$$P = -\$1000 + (\$200)(P/A, i\%, 10) - (\$100)(P/A, i\%, 5)$$

Missing and Extra Parts of Standard Cash Flows

A missing or extra part of a standard cash flow can also be handled by superposition. For example, suppose an annual expense is incurred each year for ten years, except in the ninth year. (The cash flow is illustrated in Figure 69.3.) The present worth could be calculated as a subtractive process.

$$P = A(P/A, i\%, 10) - A(P/F, i\%, 9) \qquad \textbf{69.16}$$

Alternatively, the present worth could be calculated as an additive process.

$$P = A(P/A, i\%, 8) + A(P/F, i\%, 10) \qquad \textbf{69.17}$$

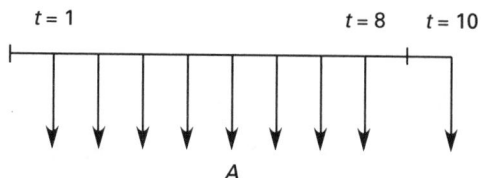

Figure 69.3 *Cash Flow with a Missing Part*

Delayed and Premature Cash Flows

There are cases when a cash flow matches a standard cash flow exactly, except that the cash flow is delayed or starts sooner than it should. Often, such cash flows can be handled with superposition. At other times, it may be more convenient to shift the time axis. This shift is known as the *projection method*. Example 69.8 illustrates the projection method.

Example 69.8

An expense of $75 is incurred starting at $t = 3$ and continues until $t = 9$. There are no expenses or receipts until $t = 3$. Use the projection method to determine the present worth of this stream of expenses.

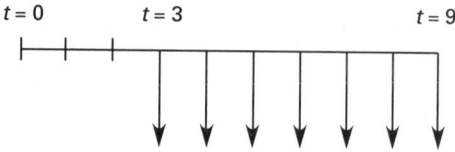

Solution

First, determine a cash flow at $t = 2$ that is equivalent to the entire expense stream. If $t = 0$ was where $t = 2$ actually is, the present worth of the expense stream would be

$$P' = (-\$75)(P/A, i\%, 7)$$

P' is a cash flow at $t = 2$. It is now simple to find the present worth (at $t = 0$) of this future amount.

$$P = P'(P/F, i\%, 2) = (-\$75)(P/A, i\%, 7)(P/F, i\%, 2)$$

Cash Flows at Beginnings of Years: The Christmas Club Problem

This type of problem is characterized by a stream of equal payments (or expenses) starting at $t = 0$ and ending at $t = n - 1$. It differs from the standard annual cash flow in the existence of a cash flow at $t = 0$ and the absence of a cash flow at $t = n$. This problem gets its name from the service provided by some savings institutions whereby money is automatically deposited each week or month (starting immediately, when the savings plan is opened) in order to accumulate money to purchase Christmas presents at the end of the year.

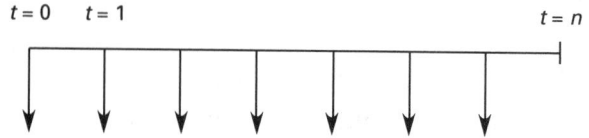

Figure 69.4 *Cash Flows at Beginnings of Years*

It may seem that the present worth of the savings stream can be determined by directly applying the (P/A) factor. However, this is not the case, since the Christmas

Club cash flow and the standard annual cash flow differ. The Christmas Club problem is easily handled by superposition, as illustrated by Ex. 69.9.

Example 69.9

How much can you expect to accumulate by $t = 10$ for a child's college education if you deposit $300 at the beginning of each year for a total of ten payments?

Solution

Notice that the first payment is made at $t = 0$ and that there is no payment at $t = 10$. The future worth of the first payment is calculated with the (F/P) factor. The absence of the payment at $t = 10$ is handled by superposition. Notice that this "correction" is not multiplied by a factor.

$$F = (\$300)(F/P, i\%, 10) + (\$300)(F/A, i\%, 10) - \$300$$

$$= (\$300)(F/A, i\%, 11) - \$300$$

17. THE MEANING OF PRESENT WORTH AND i

If $100 is invested in a 5% bank account (using annual compounding), you can remove $105 one year from now; if this investment is made, you will receive a *return on investment* (ROI) of $5. The cash flow diagram and the present worth of the two transactions are

$$P = -\$100 + (\$105)(P/F, 5\%, 1)$$

$$= -\$100 + (\$105)(0.9524) = 0$$

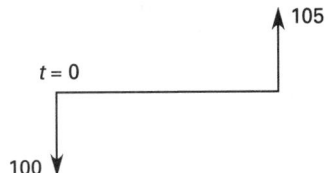

Figure 69.5 *Cash Flow Diagram*

Notice that the present worth is zero even though you will receive a $5 return on your investment.

However, if you are offered $120 for the use of $100 over a one-year period, the cash flow diagram and present worth (at 5%) would be

$$P = -\$100 + (\$120)(P/F, 5\%, 1)$$

$$= -\$100 + (\$120)(0.9524) = \$14.29$$

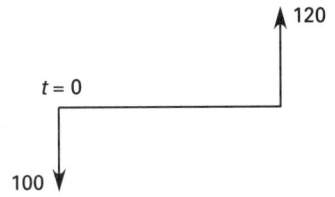

Figure 69.6 *Cash Flow Diagram*

Therefore, the present worth of an alternative is seen to be equal to the equivalent value at $t = 0$ of the increase in return above that which you would be able to earn in an investment offering $i\%$ per period. In the previous case, $14.29 is the present worth of ($20 - $5), the difference in the two ROIs.

The present worth is also the amount that you would have to be given to dissuade you from making an investment, since placing the initial investment amount along with the present worth into a bank account earning $i\%$ will yield the same eventual return on investment. Relating this to the previous paragraphs, you could be dissuaded from investing $100 in an alternative that would return $120 in one year by a $t = 0$ payment of $14.29. Clearly, ($100 + $14.29) invested at $t = 0$ will also yield $120 in one year at 5%.

Income-producing alternatives with negative present worths are undesirable, and alternatives with positive present worths are desirable because they increase the average earning power of invested capital. (In some cases, such as municipal and public works projects, the present worths of all alternatives are negative, in which case, the least negative alternative is best.)

The selection of the interest rate is difficult in engineering economics problems. Usually, it is taken as the average rate of return that an individual or business organization has realized in past investments. Alternatively, the interest rate may be associated with a particular level of risk. Usually, i for individuals is the interest rate that can be earned in relatively *risk-free investments*.

18. SIMPLE AND COMPOUND INTEREST

If $100 is invested at 5%, it will grow to $105 in one year. During the second year, 5% interest continues to be accrued, but on $105, not on $100. This is the principle of *compound interest*: The interest accrues interest.[12]

If only the original principal accrues interest, the interest is said to be *simple interest*. Simple interest is rarely encountered in engineering economic analyses, but the concept may be incorporated into short-term transactions.

19. EXTRACTING THE INTEREST RATE: RATE OF RETURN

An intuitive definition of the *rate of return* (ROR) is the effective annual interest rate at which an investment accrues income. That is, the rate of return of a project is the interest rate that would yield identical profits if all money was invested at that rate. Although this definition is correct, it does not provide a method of determining the rate of return.

---

[12]This assumes, of course, that the interest remains in the account. If the interest is removed and spent, only the remaining funds accumulate interest.

It was previously seen that the present worth of a $100 investment invested at 5% is zero when $i = 5\%$ is used to determine equivalence. Thus, a working definition of rate of return would be the effective annual interest rate that makes the present worth of the investment zero. Alternatively, rate of return could be defined as the effective annual interest rate that will discount all cash flows to a total present worth equal to the required initial investment.

It is tempting, but impractical, to determine a rate of return analytically. It is simply too difficult to extract the interest rate from the equivalence equation. For example, consider a $100 investment that pays back $75 at the end of each of the first two years. The present worth equivalence equation (set equal to zero in order to determine the rate of return) is

$$P = 0 = -\$100 + (\$75)(1+i)^{-1} + (\$75)(1+i)^{-2} \quad \textbf{69.18}$$

Solving Eq. 69.18 requires finding the roots of a quadratic equation. In general, for an investment or project spanning n years, the roots of an nth-order polynomial would have to be found. It should be obvious that an analytical solution would be essentially impossible for more complex cash flows. (The rate of return in this example is 31.87%.)

If the rate of return is needed, it can be found from a trial-and-error solution. To find the rate of return of an investment, proceed as follows.

step 1: Set up the problem as if to calculate the present worth.

step 2: Arbitrarily select a reasonable value for i. Calculate the present worth.

step 3: Choose another value of i (not too close to the original value), and again solve for the present worth.

step 4: Interpolate or extrapolate the value of i that gives a zero present worth.

step 5: For increased accuracy, repeat steps 2 and 3 with two more values that straddle the value found in step 4.

A common, although incorrect, method of calculating the rate of return involves dividing the annual receipts or returns by the initial investment. (See Sec. 54.) However, this technique ignores such items as salvage, depreciation, taxes, and the time value of money. This technique also is inadequate when the annual returns vary.

It is possible that more than one interest rate will satisfy the zero present worth criteria. This confusing situation occurs whenever there is more than one change in sign in the investment's cash flow.[13] Table 69.3 indicates the numbers of possible interest rates as a function of the number of sign reversals in the investment's cash flow.

Table 69.3 Multiplicity of Rates of Return

number of sign reversals	number of distinct rates of return
0	0
1	0 or 1
2	0, 1, or 2
3	0, 1, 2, or 3
4	0, 1, 2, 3, or 4
m	$0, 1, 2, 3, \ldots m-1, m$

Difficulties associated with interpreting the meaning of multiple rates of return can be handled with the concepts of external investment and external rate of return. An *external investment* is an investment that is distinct from the investment being evaluated (which becomes known as the internal investment). The *external rate of return*, which is the rate of return earned by the external investment, does not need to be the same as the rate earned by the internal investment.

Generally, the multiple rates of return indicate that the analysis must proceed as though money will be invested outside of the project. The mechanics of how this is done are not covered here.

Example 69.10

What is the rate of return on invested capital if $1000 is invested now with $500 being returned in year 4 and $1000 being returned in year 8?

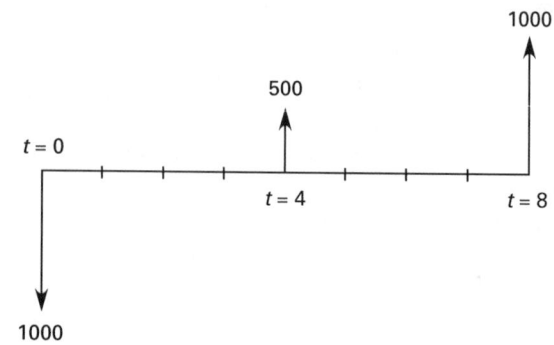

---

[13]There will always be at least one change of sign in the cash flow of a legitimate investment. (This excludes municipal and other tax-supported functions.) At $t = 0$, an investment is made (a negative cash flow). Hopefully, the investment will begin to return money (a positive cash flow) at $t = 1$ or shortly thereafter. Although it is possible to conceive of an investment in which all of the cash flows were negative, such an investment would probably be classified as a *hobby*.

Solution

First, set up the problem as a present worth calculation. Try $i = 5\%$.

$$P = -\$1000 + (\$500)(P/F, 5\%, 4) + (\$1000)$$
$$\times (P/F, 5\%, 8)$$
$$= -\$1000 + (\$500)(0.8227) + (\$1000)(0.6768)$$
$$= \$88$$

Next, try a larger value of i to reduce the present worth. If $i = 10\%$,

$$P = -\$1000 + (\$500)(P/F, 10\%, 4) + (\$1000)$$
$$\times (P/F, 10\%, 8)$$
$$= -\$1000 + (\$500)(0.6830) + (\$1000)(0.4665)$$
$$= -\$192$$

Using simple interpolation, the rate of return is

$$\text{ROR} = 5\% + \left(\frac{\$88}{\$88 + \$192}\right)(10\% - 5\%)$$
$$= 6.6\%$$

A second iteration between 6% and 7% yields 6.39%.

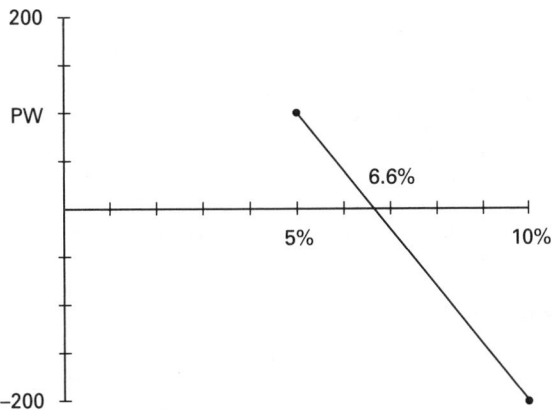

Example 69.11

An existing biomedical company is developing a new drug. A venture capital firm gives the company $25 million initially and $55 million more at the end of the first year. The drug patent will be sold at the end of year 5 to the highest bidder, and the biomedical company will receive $80 million. (The venture capital firm will receive everything in excess of $80 million.) The firm invests unused money in short-term commercial paper earning 10% effective interest per year through its bank. In the mean time, the biomedical company incurs development expenses of $50 million annually for the first

three years. The drug is to be evaluated by a government agency and there will be neither expenses nor revenues during the fourth year. What is the biomedical company's rate of return on this investment?

Solution

Normally, the rate of return is determined by setting up a present worth problem and varying the interest rate until the present worth is zero. Writing the cash flows, though, shows that there are two reversals of sign: one at $t = 2$ (positive to negative) and the other at $t = 5$ (negative to positive). Therefore, there could be two interest rates that produce a zero present worth. (In fact, there actually are two interest rates: 10.7% and 41.4%.)

time	cash flow (millions)
0	+25
1	+55 − 50 = +5
2	−50
3	−50
4	0
5	+80

However, this problem can be reduced to one with only one sign reversal in the cash flow series. The initial $25 million is invested in commercial paper (an *external investment* having nothing to do with the drug development process) during the first year at 10%. The accumulation of interest and principal after one year is

$$(25)(1 + 0.10) = 27.5$$

This 27.5 is combined with the 5 (the money remaining after all expenses are paid at $t = 1$) and invested externally, again at 10%. The accumulation of interest and principal after one year (i.e., at $t = 2$) is

$$(27.5 + 5)(1 + 0.10) = 35.75$$

This 35.75 is combined with the development cost paid at $t = 2$.

The cash flow for the development project (the internal investment) is

time	cash flow (millions)
0	0
1	0
2	35.75 − 50 = −14.25
3	−50
4	0
5	+80

Now, there is only one sign reversal in the cash flow series. The *internal rate of return* on this development project is found by the traditional method to be 10.3%. Notice that this is different from the rate the company can earn from investing externally in commercial paper.

20. RATE OF RETURN VERSUS RETURN ON INVESTMENT

Rate of return (ROR) is an effective annual interest rate, typically stated in percent per year. *Return on investment* (ROI) is a dollar amount. Thus, *rate of return* and *return on investment* are not synonymous.

Return on investment can be calculated in two different ways. The accounting method is to subtract the total of all investment costs from the total of all net profits (i.e., revenues less expenses). The time value of money is not considered.

In engineering economic analysis, the return on investment is calculated from equivalent values. Specifically, the present worth (at $t = 0$) of all investment costs is subtracted from the future worth (at $t = n$) of all net profits.

When there are only two cash flows, a single investment amount and a single payback, the two definitions of return on investment yield the same numerical value. When there are more than two cash flows, the returns on investment will be different depending on which definition is used.

21. MINIMUM ATTRACTIVE RATE OF RETURN

A company may not know what effective interest rate, i, to use in engineering economic analysis. In such a case, the company can establish a minimum level of economic performance that it would like to realize on all investments. This criterion is known as the *minimum attractive rate of return* (MARR). Unlike the effective interest rate, i, the minimum attractive rate of return is not used in numerical calculations.[14] It is used only in comparisons with the rate of return.

Once a rate of return for an investment is known, it can be compared to the minimum attractive rate of return. To be a viable alternative, the rate of return must be greater than the minimum attractive rate of return.

The advantage of using comparisons to the minimum attractive rate of return is that an effective interest rate, i, never needs to be known. The minimum attractive rate of return becomes the correct interest rate for use in present worth and equivalent uniform annual cost calculations.

22. TYPICAL ALTERNATIVE-COMPARISON PROBLEM FORMATS

With the exception of some investment and rate of return problems, the typical problem involving engineering economics will have the following characteristics.

- An interest rate will be given.
- Two or more alternatives will be competing for funding.
- Each alternative will have its own cash flows.
- It is necessary to select the best alternative.

23. DURATIONS OF INVESTMENTS

Because they are handled differently, short-term investments and short-lived assets need to be distinguished from investments and assets that constitute an infinitely lived project. Short-term investments are easily identified: a drill press that is needed for three years or a temporary factory building that is being constructed to last five years.

Investments with perpetual cash flows are also (usually) easily identified: maintenance on a large flood control dam and revenues from a long-span toll bridge. Furthermore, some items with finite lives can expect renewal on a repeated basis.[15] For example, a major freeway with a pavement life of 20 years is unlikely to be abandoned; it will be resurfaced or replaced every 20 years.

Actually, if an investment's finite lifespan is long enough, it can be considered an infinite investment because money 50 or more years from now has little impact on current decisions. The $(P/F, 10\%, 50)$ factor, for example, is 0.0085. Thus, one dollar at $t = 50$ has an equivalent present worth of less than one penny. Since these far-future cash flows are eclipsed by present cash flows, long-term investments can be considered finite or infinite without significant impact on the calculations.

24. CHOICE OF ALTERNATIVES: COMPARING ONE ALTERNATIVE WITH ANOTHER ALTERNATIVE

Several methods exist for selecting a superior alternative from among a group of proposals. Each method has its own merits and applications.

Present Worth Method

When two or more alternatives are capable of performing the same functions, the superior alternative will have the largest present worth. The *present worth method* is restricted to evaluating alternatives that are mutually exclusive and that have the same lives. This method is suitable for ranking the desirability of alternatives.

Example 69.12

Investment A costs $10,000 today and pays back $11,500 two years from now. Investment B costs $8000 today and pays back $4500 each year for two years. If an interest rate of 5% is used, which alternative is superior?

---

[14]Not everyone adheres to this rule. Some people use "minimum attractive rate of return" and "effective interest rate" interchangeably.

[15]The term *renewal* can be interpreted to mean replacement or repair.

Solution

$$P(A) = -\$10,000 + (\$11,500)(P/F, 5\%, 2)$$
$$= -\$10,000 + (\$11,500)(0.9070)$$
$$= \$431$$
$$P(B) = -\$8000 + (\$4500)(P/A, 5\%, 2)$$
$$= -\$8000 + (\$4500)(1.8594)$$
$$= \$367$$

Alternative A is superior and should be chosen.

Capitalized Cost Method

The present worth of a project with an infinite life is known as the *capitalized cost* or *life cycle cost*. Capitalized cost is the amount of money at $t = 0$ needed to perpetually support the project on the earned interest only. Capitalized cost is a positive number when expenses exceed income.

In comparing two alternatives, each of which is infinitely lived, the superior alternative will have the lowest capitalized cost.

Normally, it would be difficult to work with an infinite stream of cash flows since most economics tables do not list factors for periods in excess of 100 years. However, the (A/P) discounting factor approaches the interest rate as n becomes large. Since the (P/A) and (A/P) factors are reciprocals of each other, it is possible to divide an infinite series of equal cash flows by the interest rate in order to calculate the present worth of the infinite series. This is the basis of Eq. 69.19.

$$\text{capitalized cost} = \text{initial cost} + \frac{\text{annual costs}}{i} \qquad \textit{69.19}$$

Equation 69.19 can be used when the annual costs are equal in every year. If the operating and maintenance costs occur irregularly instead of annually, or if the costs vary from year to year, it will be necessary to somehow determine a cash flow of equal annual amounts (EAA) that is equivalent to the stream of original costs.

The equal annual amount may be calculated in the usual manner by first finding the present worth of all the actual costs, and then multiplying the present worth by the interest rate (the (A/P) factor for an infinite series). However, it is not even necessary to convert the present worth to an equal annual amount since Eq. 69.20 will convert the equal amount back to the present worth.

$$\text{capitalized cost} = \text{initial cost} + \frac{\text{EAA}}{i}$$
$$= \text{initial cost} + \frac{\text{present worth}}{\text{of all expenses}} \qquad \textit{69.20}$$

Example 69.13

What is the capitalized cost of a public works project that will cost \$25,000,000 now and will require \$2,000,000 in maintenance annually? The effective annual interest rate is 12%.

Solution

Worked in millions of dollars, from Eq. 69.19, the capitalized cost is

$$\text{capitalized cost} = 25 + (2)(P/A, 12\%, \infty)$$
$$= 25 + \frac{2}{0.12} = 41.67$$

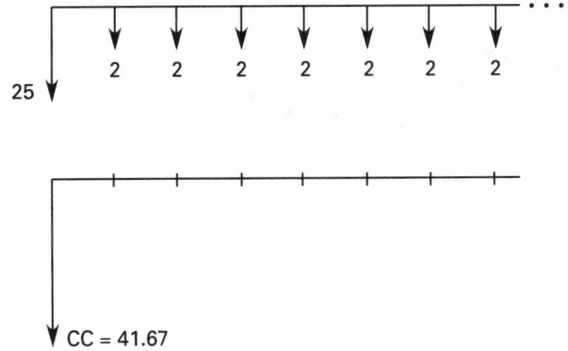

Annual Cost Method

Alternatives that accomplish the same purpose but that have unequal lives must be compared by the *annual cost method*.[16] The annual cost method assumes that each alternative will be replaced by an identical twin at the end of its useful life (infinite renewal). This method, which may also be used to rank alternatives according to their desirability, is also called the *annual return method* or *capital recovery method*.

Restrictions are that the alternatives must be mutually exclusive and repeatedly renewed up to the duration of the longest-lived alternative. The calculated annual cost is known as the *equivalent uniform annual cost* (EUAC) or just *equivalent annual cost*. Cost is a positive number when expenses exceed income.

Example 69.14

Which of the following alternatives is superior over a 30-year period if the interest rate is 7%?

	alternative A	alternative B
type	brick	wood
life	30 years	10 years
initial cost	\$1800	\$450
maintenance	\$5/year	\$20/year

---

[16]Of course, the annual cost method can be used to determine the superiority of assets with identical lives as well.

Solution

$$EUAC(A) = (\$1800)(A/P, 7\%, 30) + \$5$$
$$= (\$1800)(0.0806) + \$5$$
$$= \$150$$
$$EUAC(B) = (\$450)(A/P, 7\%, 10) + \$20$$
$$= (\$450)(0.1424) + \$20$$
$$= \$84$$

Alternative B is superior since its annual cost of operation is the lowest. It is assumed that three wood facilities, each with a life of 10 years and a cost of $450, will be built to span the 30-year period.

25. CHOICE OF ALTERNATIVES: COMPARING AN ALTERNATIVE WITH A STANDARD

With specific economic performance criteria, it is possible to qualify an investment as acceptable or unacceptable without having to compare it with another investment. Two such performance criteria are the benefit-cost ratio and the minimum attractive rate of return.

Benefit-Cost Ratio Method

The *benefit-cost ratio* method is often used in municipal project evaluations where benefits and costs accrue to different segments of the community. With this method, the present worth of all benefits (irrespective of the beneficiaries) is divided by the present worth of all costs. The project is considered acceptable if the ratio equals or exceeds 1.0, that is, if $B/C \geq 1.0$.

When the benefit-cost ratio method is used, disbursements by the initiators or sponsors are *costs*. Disbursements by the users of the project are known as *disbenefits*. It is often difficult to determine whether a cash flow is a cost or a disbenefit (whether to place it in the numerator or denominator of the benefit-cost ratio calculation).

Regardless of where the cash flow is placed, an acceptable project will always have a benefit-cost ratio greater than or equal to 1.0, although the actual numerical result will depend on the placement. For this reason, the benefit-cost ratio method should not be used to rank competing projects.

The benefit-cost ratio method of comparing alternatives is used extensively in transportation engineering where the ratio is often (but not necessarily) written in terms of annual benefits and annual costs instead of present worths. Another characteristic of highway benefit-cost ratios is that the route (road, highway, etc.) is usually already in place and that various alternative upgrades are being considered. There will be existing benefits

and costs associated with the current route. Therefore, the *change* (usually an increase) in benefits and costs is used to calculate the benefit-cost ratio.[17]

$$B/C = \frac{\Delta^{\text{user}}_{\text{benefits}}}{\Delta^{\text{investment}}_{\text{cost}} + \Delta\text{maintenance} - \Delta^{\text{residual}}_{\text{value}}} \qquad 69.21$$

Notice that the change in *residual value (terminal value)* appears in the denominator as a negative item. An increase in the residual value would decrease the denominator.

Example 69.15

By building a bridge over a ravine, a state department of transportation can shorten the time it takes to drive through a mountainous area. Estimates of costs and benefits (due to decreased travel time, fewer accidents, reduced gas usage, etc.) have been prepared. Should the bridge be built? Use the benefit-cost ratio method of comparison.

	millions
initial cost	40
capitalized cost of perpetual annual maintenance	12
capitalized value of annual user benefits	49
residual value	0

Solution

If Eq. 69.21 is used, the benefit-cost ratio is

$$B/C = \frac{49}{40 + 12 + 0} = 0.942$$

Since the benefit-cost ratio is less than 1.00, the bridge should not be built.

If the maintenance costs are placed in the numerator (per Ftn. 17), the benefit-cost ratio value will be different but the conclusion will not change.

$$B/C_{\text{alternate method}} = \frac{49 - 12}{40} = 0.925$$

Rate of Return Method

The minimum attractive rate of return (MARR) has already been introduced as a standard of performance against which an investment's actual *rate of return*

---

[17]This discussion of highway benefit-cost ratios is not meant to imply that everyone agrees with Eq. 69.21. In *Economic Analysis for Highways* (International Textbook Company, Scranton, PA, 1969), author Robley Winfrey takes a strong stand on one aspect of the benefits versus disbenefits issue: highway maintenance. According to Winfrey, regular highway maintenance costs should be placed in the numerator as a subtraction from the user benefits. Some have called this mandate the *Winfrey method*.

(ROR) is compared. If the rate of return is equal to or exceeds the minimum attractive rate of return, the investment is qualified. This is the basis for the *rate of return method* of alternative selection.

Finding the rate of return can be a long, iterative process. Usually, the actual numerical value of rate of return is not needed; it is sufficient to know whether or not the rate of return exceeds the minimum attractive rate of return. This *comparative analysis* can be accomplished without calculating the rate of return simply by finding the present worth of the investment using the minimum attractive rate of return as the effective interest rate (i.e., $i = $ MARR). If the present worth is zero or positive, the investment is qualified. If the present worth is negative, the rate of return is less than the minimum attractive rate of return.

26. RANKING MUTUALLY EXCLUSIVE MULTIPLE PROJECTS

Ranking of multiple investment alternatives is required when there is sufficient funding for more than one investment. Since the best investments should be selected first, it is necessary to be able to place all investments into an ordered list.

Ranking is relatively easy if the present worths, future worths, capitalized costs, or equivalent uniform annual costs have been calculated for all the investments. The highest ranked investment will be the one with the largest present or future worth, or the smallest capitalized or annual cost. Present worth, future worth, capitalized cost, and equivalent uniform annual cost can all be used to rank multiple investment alternatives.

However, neither rates of return nor benefit-cost ratios should be used to rank multiple investment alternatives. Specifically, if two alternatives both have rates of return exceeding the minimum acceptable rate of return, it is not sufficient to select the alternative with the highest rate of return.

An *incremental analysis*, also known as a *rate of return on added investment study*, should be performed if rate of return is used to select between investments. An incremental analysis starts by ranking the alternatives in order of increasing initial investment. Then, the cash flows for the investment with the lower initial cost are subtracted from the cash flows for the higher-priced alternative on a year-by-year basis. This produces, in effect, a third alternative representing the costs and benefits of the added investment. The added expense of the higher priced investment is not warranted unless the rate of return of this third alternative exceeds the minimum attractive rate of return as well. The choice criterion is to select the alternative with the higher initial investment if the incremental rate of return exceeds the minimum attractive rate of return.

An incremental analysis is also required if ranking is to be done by the benefit-cost ratio method. The incremental analysis is accomplished by calculating the ratio of differences in benefits to differences in costs for each possible pair of alternatives. If the ratio exceeds 1.0, alternative 2 is superior to alternative 1. Otherwise, alternative 1 is superior.[18]

$$\frac{B_2 - B_1}{C_2 - C_1} \geq 1 \quad \text{[alternative 2 superior]} \qquad 69.22$$

27. ALTERNATIVES WITH DIFFERENT LIVES

Comparison of two alternatives is relatively simple when both alternatives have the same life. For example, a problem might be stated: "Which would you rather have: car A with a life of three years, or car B with a life of five years?"

However, care must be taken to understand what is going on when the two alternatives have different lives. If car A has a life of three years, and car B has a life of five years, what happens at $t = 3$ if the five-year car is chosen? If a car is needed for five years, what happens at $t = 3$ if the three-year car is chosen?

In this type of situation, it is necessary to distinguish between the length of the need (the *analysis horizon*) and the lives of the alternatives or assets intended to meet that need. The lives do not have to be the same as the horizon.

Finite Horizon with Incomplete Asset Lives

If an asset with a five-year life is chosen for a three-year need, the disposition of the asset at $t = 3$ must be known in order to evaluate the alternative. If the asset is sold at $t = 3$, the salvage value is entered into the analysis (at $t = 3$), and the alternative is evaluated as a three-year investment. The fact that the asset is sold when it has some useful life remaining does not affect the analysis horizon.

Similarly, if a three-year asset is chosen for a five-year need, something about how the need is satisfied during the last two years must be known. Perhaps a rental asset will be used. Or, perhaps the function will be "farmed out" to an outside firm. In any case, the costs of satisfying the need during the last two years enter the analysis, and the alternative is evaluated as a five-year investment.

If both alternatives are "converted" to the same life, any of the alternative selection criteria (present worth method, annual cost method, etc.) can be used to determine which alternative is superior.

---

[18]It goes without saying that the benefit-cost ratios for all investment alternatives by themselves must also be equal to or greater than 1.0.

Finite Horizon with Integer Multiple Asset Lives

It is common to have a long-term horizon (need) that must be met with short-lived assets. In special instances, the horizon will be an integer number of asset lives. For example, a company may be making a 12-year transportation plan, and may be evaluating two cars: one with a three-year life, and another with a four-year life.

In this example, four of the first car or three of the second car are needed to reach the end of the 12-year horizon.

If the horizon is an integer number of asset lives, any of the alternative selection criteria can be used to determine which is superior. If the present worth method is used, all alternatives must be evaluated over the entire horizon. (In this example, the present worth of 12 years of car purchases and use must be determined for both alternatives.)

If the equivalent uniform annual cost method is used, it may be possible to base the calculation of annual cost on one lifespan of each alternative only. It may not be necessary to incorporate all of the cash flows into the analysis. (In the running example, the annual cost over three years would be determined for the first car; the annual cost over four years would be determined for the second car.) This simplification is justified if the subsequent asset replacements (renewals) have the same cost and cash flow structure as the original asset. This assumption is typically made implicitly when the annual cost method of comparison is used.

Infinite Horizon

If the need horizon is infinite, it is not necessary to impose the restriction that asset lives of alternatives be integer multiples of the horizon. The superior alternative will be replaced (renewed) whenever it is necessary to do so, forever.

Infinite horizon problems are almost always solved with either the annual cost or capitalized cost method. It is common to (implicitly) assume that the cost and cash flow structure of the asset replacements (renewals) are the same as the original asset.

28. OPPORTUNITY COSTS

An *opportunity cost* is an imaginary cost representing what will not be received if a particular strategy is rejected. It is what you will lose if you do or do not do something. As an example, consider a growing company with an existing operational computer system. If the company trades in its existing computer as part of an upgrade plan, it will receive a *trade-in allowance*. (In other problems, a *salvage value* may be involved.)

If one of the alternatives being evaluated is not to upgrade the computer system at all, the trade-in allowance (or, salvage value in other problems) will not be realized. The amount of the trade-in allowance is an opportunity cost that must be included in the problem analysis.

Similarly, if one of the alternatives being evaluated is to wait one year before upgrading the computer, the *difference in trade-in allowances* is an opportunity cost that must be included in the problem analysis.

29. REPLACEMENT STUDIES

An investigation into the retirement of an existing process or piece of equipment is known as a *replacement study*. Replacement studies are similar in most respects to other alternative comparison problems: An interest rate is given, two alternatives exist, and one of the previously mentioned methods of comparing alternatives is used to choose the superior alternative. Usually, the annual cost method is used on a year-by-year basis.

In replacement studies, the existing process or piece of equipment is known as the *defender*. The new process or piece of equipment being considered for purchase is known as the *challenger*.

30. TREATMENT OF SALVAGE VALUE IN REPLACEMENT STUDIES

Since most defenders still have some market value when they are retired, the problem of what to do with the salvage arises. It seems logical to use the salvage value of the defender to reduce the initial purchase cost of the challenger. This is consistent with what would actually happen if the defender were to be retired.

By convention, however, the defender's salvage value is subtracted from the defender's present value. This does not seem logical, but it is done to keep all costs and benefits related to the defender with the defender. In this case, the salvage value is treated as an opportunity cost that would be incurred if the defender is not retired.

If the defender and the challenger have the same lives and a present worth study is used to choose the superior alternative, the placement of the salvage value will have no effect on the net difference between present worths for the challenger and defender. Although the values of the two present worths will be different depending on the placement, the difference in present worths will be the same.

If the defender and the challenger have different lives, an annual cost comparison must be made. Since the salvage value would be "spread over" a different number of years depending on its placement, it is important to abide by the conventions listed in this section.

There are a number of ways to handle salvage value in retirement studies. The best way is to calculate the cost of keeping the defender one more year. In addition to the usual operating and maintenance costs, that cost includes an opportunity interest cost incurred by not selling the defender, and also a drop in the salvage value if the defender is kept for one additional year. Specifically,

$$
\begin{aligned}
\text{EUAC (defender)} =\ & \text{next year's maintenance costs} \\
& + i\,(\text{current salvage value}) \\
& + \text{current salvage} \\
& - \text{next year's salvage} \qquad \textit{69.23}
\end{aligned}
$$

It is important in retirement studies not to double count the salvage value. That is, it would be incorrect to add the salvage value to the defender and at the same time subtract it from the challenger.

Equation 69.23 contains the difference in salvage value between two consecutive years. This calculation shows that the defender/challenger decision must be made on a year-by-year basis. One application of Eq. 69.23 will not usually answer the question of whether the defender should remain in service indefinitely. The calculation must be repeatedly made as long as there is a drop in salvage value from one year to the next.

31. ECONOMIC LIFE: RETIREMENT AT MINIMUM COST

As an asset grows older, its operating and maintenance costs typically increase. Eventually, the cost to keep the asset in operation becomes prohibitive, and the asset is retired or replaced. However, it is not always obvious when an asset should be retired or replaced.

As the asset's maintenance cost is increasing each year, the amortized cost of its initial purchase is decreasing. It is the sum of these two costs that should be evaluated to determine the point at which the asset should be retired or replaced. Since an asset's initial purchase price is likely to be high, the amortized cost will be the controlling factor in those years when the maintenance costs are low. Therefore, the EUAC of the asset will decrease in the initial part of its life.

However, as the asset grows older, the change in its amortized cost decreases while maintenance cost increases. Eventually, the sum of the two costs reaches a minimum and then starts to increase. The age of the asset at the minimum cost point is known as the *economic life* of the asset. The economic life generally is less than the length of need and the technological lifetime of the asset.

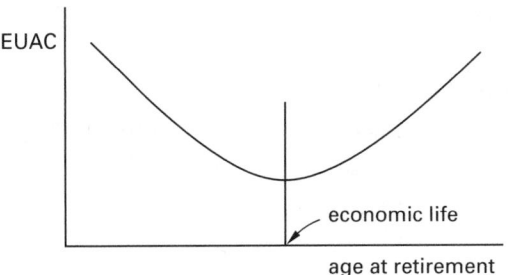

Figure 69.7 *EUAC versus Age at Retirement*

The determination of an asset's economic life is illustrated by Ex. 69.16.

Example 69.16

Buses in a municipal transit system have the characteristics listed. When should the city replace its buses if money can be borrowed at 8%?

initial cost of bus: $120,000

year	maintenance cost	salvage value
1	35,000	60,000
2	38,000	55,000
3	43,000	45,000
4	50,000	25,000
5	65,000	15,000

Solution

The annual maintenance is different each year. Each maintenance cost must be spread over the life of the bus. This is done by first finding the present worth and then amortizing the maintenance costs. If a bus is kept for one year and then sold, the annual cost will be

$$
\begin{aligned}
\text{EUAC}(1) =\ & (\$120{,}000)(A/P, 8\%, 1) \\
& + (\$35{,}000)(A/F, 8\%, 1) \\
& - (\$60{,}000)(A/F, 8\%, 1) \\
=\ & (\$120{,}000)(1.0800) + (\$35{,}000)(1.000) \\
& - (\$60{,}000)(1.000) \\
=\ & \$104{,}600
\end{aligned}
$$

If a bus is kept for two years and then sold, the annual cost will be

$$
\begin{aligned}
\text{EUAC}(2) =\ & [\$120{,}000 + (\$35{,}000)(P/F, 8\%, 1)] \\
& \times (A/P, 8\%, 2) \\
& + (\$38{,}000 - \$55{,}000)(A/F, 8\%, 2) \\
=\ & [\$120{,}000 + (\$35{,}000)(0.9259)](0.5608) \\
& + (\$38{,}000 - \$55{,}000)(0.4808) \\
=\ & \$77{,}296
\end{aligned}
$$

If a bus is kept for three years and then sold, the annual cost will be

$$\begin{aligned}
\text{EUAC}(3) &= [\$120{,}000 + (35{,}000)(P/F, 8\%, 1) \\
&\quad + (\$38{,}000)(P/F, 8\%, 2)](A/P, 8\%, 3) \\
&\quad + (\$43{,}000 - \$45{,}000)(A/F, 8\%, 3) \\
&= [\$120{,}000 + (\$35{,}000)(0.9259) \\
&\quad + (\$38{,}000)(0.8573)](0.3880) \\
&\quad - (\$2000)(0.3080) \\
&= \$71{,}158
\end{aligned}$$

This process is continued until the annual cost begins to increase. In this example, EUAC(4) is \$71,700. Therefore, the buses should be retired after three years.

32. LIFE-CYCLE COST

The *life-cycle cost* of an alternative is the equivalent value (at $t = 0$) of the alternative's cash flow over the alternative's lifespan. Since the present worth is evaluated using an effective interest rate of i (which would be the interest rate used for all engineering economic analyses), the life-cycle cost is the same as the alternative's present worth. If the alternative has an infinite horizon, the life-cycle cost and capitalized cost will be identical.

33. CAPITALIZED ASSETS VERSUS EXPENSES

High expenses reduce profit, which in turn, reduces income tax. It seems logical to label each and every expenditure, even an asset purchase, as an expense. As an alternative to this *expensing the asset*, it may be decided to capitalize the asset. *Capitalizing the asset* means that the cost of the asset is divided into equal or unequal parts, and only one of these parts is taken as an expense each year. Expensing is clearly the more desirable alternative, since the after-tax profit is increased early in the asset's life.

There are long-standing accounting conventions as to what can be expensed and what must be capitalized.[19] Some companies capitalize everything—regardless of cost—with expected lifetimes greater than one year. Most companies, however, expense items whose purchase costs are below a cut-off value. A cut-off value in the range of \$250 to \$500, depending on the size of the company, is chosen as the maximum purchase cost of an expensed asset. Assets costing more than this are capitalized.

It is not necessary for a large corporation to keep track of every lamp, desk, and chair for which the purchase price is greater than the cut-off value. Such assets, all

of which have the same lives and have been purchased in the same year, can be placed into groups or *asset classes*. A group cost, equal to the sum total of the purchase costs of all items in the group, is capitalized as though the group was an identifiable and distinct asset itself.

34. PURPOSE OF DEPRECIATION

Depreciation is an *artificial expense* that spreads the purchase price of an asset or other property over a number of years.[20] Depreciating an asset is an example of capitalization, as previously defined. The inclusion of depreciation in engineering economic analysis problems will increase the after-tax present worth (profitability) of an asset. The larger the depreciation, the greater will be the profitability. Therefore, individuals and companies eligible to utilize depreciation want to maximize and accelerate the depreciation available to them.

Although the entire property purchase price is eventually recognized as an expense, the net recovery from the expense stream never equals the original cost of the asset. That is, depreciation cannot realistically be thought of as a fund (an annuity or sinking fund) that accumulates capital to purchase a replacement at the end of the asset's life. The primary reason for this is that the depreciation expense is reduced significantly by the impact of income taxes, as will be seen in later sections.

35. DEPRECIATION BASIS OF AN ASSET

The *depreciation basis* of an asset is the part of the asset's purchase price that is spread over the *depreciation period*, also known as the *service life*.[21] Usually, the depreciation basis and the purchase price are not the same.

A common depreciation basis is the difference between the purchase price and the expected salvage value at the end of the depreciation period. That is,

$$\text{depreciation basis} = C - S_n \qquad \textit{69.24}$$

There are several methods of calculating the year-by-year depreciation of an asset. Equation 69.24 is not universally compatible with all depreciation methods.

---

[19]For example, purchased vehicles must be capitalized; payments for leased vehicles can be expensed. Repainting a building with paint that will last five years is an expense, but the replacement cost of a leaking roof must be capitalized.

[20]The IRS tax regulations allow depreciation on almost all forms of *business property* except land. The following types of property are distinguished: *real* (e.g., buildings used for business), *residential* (e.g., buildings used as rental property), and *personal* (e.g., equipment used for business). Personal property does *not* include items for personal use (such as a personal residence), despite its name. *Tangible personal property* is distinguished from *intangible property* (goodwill, copyrights, patents, trademarks, franchises, agreements not to compete, etc.).

[21]The *depreciation period* is selected to be as short as possible within recognized limits. This depreciation will not normally coincide with the *economic life* or *useful life* of an asset. For example, a car may be capitalized over a depreciation period of three years. It may become uneconomical to maintain and use at the end of an economic life of nine years. However, the car may be capable of operation over a useful life of 25 years.

Some methods do not consider the salvage value. This is known as an *unadjusted basis*. When the depreciation method is known, the depreciation basis can be rigorously defined.[22]

36. DEPRECIATION METHODS

Generally, tax regulations do not allow the cost of an asset to be treated as a deductible expense in the year of purchase. Rather, portions of the depreciation basis must be allocated to each of the n years of the asset's depreciation period. The amount that is allocated each year is called the *depreciation*.

Various methods exist for calculating an asset's depreciation each year.[23] Although the depreciation calculations may be considered independently (for the purpose of determining book value or as an academic exercise), it is important to recognize that depreciation has no effect on engineering economic analyses unless income taxes are also considered.

Straight Line Method

With the *straight line method*, depreciation is the same each year. The depreciation basis $(C - S_n)$ is allocated uniformly to all of the n years in the depreciation period. Each year, the depreciation will be

$$D = \frac{C - S_n}{n} \qquad 69.25$$

Constant Percentage Method

The *constant percentage method*[24] is similar to the straight line method in that the depreciation is the same each year. If the fraction of the basis used as depreciation is $1/n$, there is no difference between the constant percentage and straight line methods. The two methods differ only in what information is available. (With the straight line method, the life is known. With the constant percentage method, the depreciation fraction is known.)

Each year, the depreciation will be

$$D = \text{(depreciation fraction)(depreciation basis)}$$
$$= \text{(depreciation fraction)}(C - S_n) \qquad 69.26$$

---

[22]For example, with the Accelerated Cost Recovery System (ACRS) the *depreciation basis* is the total purchase cost, regardless of the expected salvage value. With declining balance methods, the depreciation basis is the purchase cost less any previously taken depreciation.

[23]This discussion gives the impression that any form of depreciation may be chosen regardless of the nature and circumstances of the purchase. In reality, the IRS tax regulations place restrictions on the higher-rate (accelerated) methods, such as declining balance and sum-of-the-years' digits methods. Furthermore, the *Economic Recovery Act of 1981* and the *Tax Reform Act of 1986* substantially changed the laws relating to personal and corporate income taxes.

[24]The *constant percentage method* should not be confused with the declining balance method, which used to be known as the *fixed percentage on diminishing balance method*.

Sum-of-the-Years' Digits Method

In *sum-of-the-years' digits* (SOYD) depreciation, the digits from 1 to n inclusive are summed. The total, T, can also be calculated from

$$T = \tfrac{1}{2}n(n + 1) \qquad 69.27$$

The depreciation in year j can be found from Eq. 69.28. Notice that the depreciation in year j, D_j, decreases by a constant amount each year.

$$D_j = \frac{(C - S_n)(n - j + 1)}{T} \qquad 69.28$$

Double Declining Balance Method[25]

Double declining balance[26] (DDB) depreciation is independent of salvage value. Furthermore, the book value never stops decreasing, although the depreciation decreases in magnitude. Usually, any book value in excess of the salvage value is written off in the last year of the asset's depreciation period. Unlike any of the other depreciation methods, double declining balance depends on accumulated depreciation.

$$D_{\text{first year}} = \frac{2C}{n} \qquad 69.29$$

$$D_j = \frac{(2)\left(C - \sum_{m=1}^{j-1} D_m\right)}{n} \qquad 69.30$$

Calculating the depreciation in the middle of an asset's life appears particularly difficult with double declining balance, since all previous years' depreciation amounts seem to be required. It appears that the depreciation in the sixth year, for example, cannot be calculated unless the values of depreciation for the first five years are calculated. However, this is not true.

Depreciation in the middle of an asset's life can be found from the following equations. (d is known as the *depreciation rate*.)

$$d = \frac{2}{n} \qquad 69.31$$

$$D_j = dC(1 - d)^{j-1} \qquad 69.32$$

Statutory Depreciation Systems

In the United States, property placed into service in 1981 and thereafter must use the *Accelerated Cost Recovery System* (ACRS), and after 1986, the *Modified*

---

[25]In the past, the *declining balance method* has also been known as the *fixed percentage of book value* and *fixed percentage on diminishing balance method*.

[26]Double declining balance depreciation is a particular form of *declining balance depreciation,* as defined by the IRS tax regulations. Declining balance depreciation includes 125% declining balance and 150% declining balance depreciations that can be calculated by substituting 1.25 and 1.50, respectively, for the 2 in Eq. 69.29.

Accelerated Cost Recovery System (MACRS) or other statutory method. Other methods (straight line, declining balance, etc.) cannot be used except in special cases.

Property placed into service in 1980 or before must continue to be depreciated according to the method originally chosen (e.g., straight line, declining balance, or sum-of-the-years' digits). ACRS and MACRS cannot be used.

Under ACRS and MACRS, the cost recovery amount in the jth year of an asset's cost recovery period is calculated by multiplying the initial cost by a factor.

$$D_j = C \times \text{factor} \qquad 69.33$$

The initial cost used is not reduced by the asset's salvage value for ACRS and MACRS calculations. The factor used depends on the asset's cost recovery period. Such factors are subject to continuing legislation changes. Current tax publications should be consulted before using this method.

Table 69.4 *Representative ACRS and MACRS Depreciation Factors*[a]

year j	recovery period (n)		
	3 years ACRS/MACRS	5 years ACRS/MACRS	10 years ACRS/MACRS
1	0.25/0.3333	0.15/0.2000	0.08/0.100
2	0.38/0.4445	0.22/0.3200	0.14/0.180
3	0.37/0.1481	0.21/0.1920	0.12/0.1440
4	0/0.0741	0.21/0.1152	0.10/0.1152
5		0.21/0.1152	0.10/0.0922
6		0/0.0576	0.10/0.0737
7			0.09/0.0655
8			0.09/0.0655
9			0.09/0.0656
10			0.09/0.0655
11			0/0.0328

[a]MACRS values are for the "half-year" convention. This table gives typical values only. Since these factors are subject to continuing revision, they should not be used without consulting an accounting professional.

Production or Service Output Method

If an asset has been purchased for a specific task and that task is associated with a specific lifetime amount of output or production, the depreciation may be calculated by the fraction of total production produced during the year. The depreciation is not expected to be the same each year.

$$D_j = (C - S_n)\left(\frac{\text{actual output in year } j}{\text{estimated lifetime output}}\right) \qquad 69.34$$

Sinking Fund Method

The *sinking fund method* is seldom used in industry because the initial depreciation is low. The formula for sinking fund depreciation (which increases each year) is

$$D_j = (C - S_n)(A/F, i\%, n)(F/P, i\%, j - 1) \qquad 69.35$$

Disfavored Methods

Three other depreciation methods are mentioned here, not because they are currently accepted or in widespread use, but because they are still occasionally encountered in the literature.[27]

The *sinking fund plus interest on first cost* depreciation method, like the following two methods, is an attempt to include the *opportunity interest cost* on the purchase price with the depreciation. That is, the purchasing company not only incurs an annual expense due to the drop in book value, but it also loses the interest on the purchase price. The formula for this method is

$$D = (C - S_n)(A/F, i\%, n) + Ci \qquad 69.36$$

The *straight line plus interest on first cost* method is similar. Its formula is

$$D = \left(\frac{1}{n}\right)(C - S_n) + Ci \qquad 69.37$$

The *straight line plus average interest method* assumes that the opportunity interest cost should be based on the book value only, not on the full purchase price. Since the book value changes each year, an average value is used. The depreciation formula is

$$D = \left(\frac{C - S_n}{n}\right)\left(1 + \frac{i(n + 1)}{2}\right) + iS_n \qquad 69.38$$

Example 69.17

An asset is purchased for $9000. Its estimated economic life is ten years, after which it will be sold for $200. Find the depreciation in the first three years using straight line, double declining balance, and sum-of-the-years' digits depreciation methods.

---

[27]These three depreciation methods should not be used in the usual manner (e.g., in conjunction with the income tax rate). These methods are attempts to calculate a more accurate annual cost of an alternative. Sometimes they give misleading answers. Their use cannot be recommended. They are included in this chapter only for the sake of completeness.

Solution

SL: $D = \dfrac{\$9000 - \$200}{10}$ $= \$880$ each year

DDB: $D_1 = \dfrac{(2)(\$9000)}{10}$ $= \$1800$ in year 1

  $D_2 = \dfrac{(2)(\$9000 - \$1800)}{10}$ $= \$1440$ in year 2

  $D_3 = \dfrac{(2)(\$9000 - \$3240)}{10}$ $= \$1152$ in year 3

SOYD: $T = \left(\tfrac{1}{2}\right)(10)(11) = 55$

  $D_1 = \left(\dfrac{10}{55}\right)(\$9000 - \$200) = \$1600$ in year 1

  $D_2 = \left(\dfrac{9}{55}\right)(\$8800)$ $= \$1440$ in year 2

  $D_3 = \left(\dfrac{8}{55}\right)(\$8800)$ $= \$1280$ in year 3

37. ACCELERATED DEPRECIATION METHODS

An *accelerated depreciation method* is one that calculates a depreciation amount greater than a straight line amount. Double declining balance and sum-of-the-years' digits methods are accelerated methods. The ACRS and MACRS methods are explicitly accelerated methods. Straight line and sinking fund methods are not accelerated methods.

Use of an accelerated depreciation method may result in unexpected tax consequences when the depreciated asset or property is disposed of. Professional tax advice should be obtained in this area.

38. BOOK VALUE

The difference between original purchase price and accumulated depreciation is known as *book value*.[28] At the end of each year, the book value (which is initially equal to the purchase price) is reduced by the depreciation in that year.

It is important to properly synchronize depreciation calculations. It is difficult to answer the question, "What is the book value in the fifth year?" unless the timing of the book value change is mutually agreed upon. It is better to be specific about an inquiry by identifying

---

[28]The balance sheet of a corporation usually has two asset accounts: the *equipment account* and the *accumulated depreciation account*. There is no book value account on this financial statement, other than the implicit value obtained from subtracting the accumulated depreciation account from the equipment account. The book values of various assets, as well as their original purchase cost, date of purchase, salvage value, and so on, and accumulated depreciation appear on detail sheets or other peripheral records for each asset.

when the book value change occurs. For example, the following question is unambiguous: "What is the book value at the end of year 5, after subtracting depreciation in the fifth year?" or "What is the book value after five years?"

Unfortunately, this type of care is seldom taken in book value inquiries, and it is up to the respondent to exercise reasonable care in distinguishing between beginning-of-year book value and end-of-year book value. To be consistent, the book value equations in this chapter have been written in such a way that the year subscript (j) has the same meaning in book value and depreciation calculations. That is, BV_5 means the book value at the end of the fifth year, after five years of depreciation, including D_5, has been subtracted from the original purchase price.

There can be a great difference between the book value of an asset and the *market value* of that asset. There is no legal requirement for the two values to coincide, and no intent for book value to be a reasonable measure of market value.[29] Therefore, it is apparent that book value is merely an accounting convention with little practical use. Even when a depreciated asset is disposed of, the book value is used to determine the consequences of disposal, not the price the asset should bring at sale.

The calculation of book value is relatively easy, even for the case of the declining balance depreciation method.

For the straight line depreciation method, the book value at the end of the jth year, after the jth depreciation deduction has been made, is

$$BV_j = C - \frac{j(C - S_n)}{n} = C - jD \qquad 69.39$$

For the sum-of-the-years' digits method, the book value is

$$BV_j = (C - S_n)\left[1 - \frac{j(2n + 1 - j)}{n(n + 1)}\right] + S_n \qquad 69.40$$

For the declining balance method, the book value is

$$BV_j = C(1 - d)^j \qquad 69.41$$

For the sinking fund method, the book value is calculated directly as

$$BV_j = C - (C - S_n)(A/F, i\%, n)(F/A, i\%, j) \qquad 69.42$$

---

[29]Common examples of assets with great divergences of book and market values are buildings (rental houses, apartment complexes, factories, etc.) and company luxury automobiles (Porsches, Mercedes, etc.) during periods of inflation. Book values decrease, but actual values increase.

Of course, the book value at the end of year j can always be calculated for any method by successive subtractions (i.e., subtraction of the accumulated depreciation), as Eq. 69.43 illustrates.

$$\text{BV}_j = C - \sum_{m=1}^{j} D_m \qquad \textbf{69.43}$$

Figure 69.8 illustrates the book value of a hypothetical asset depreciated using several depreciation methods. Notice that the double declining balance method initially produces the fastest write-off, while the sinking fund method produces the slowest write-off. Note, also, that the book value does not automatically equal the salvage value at the end of an asset's depreciation period with the double declining balance method.[30]

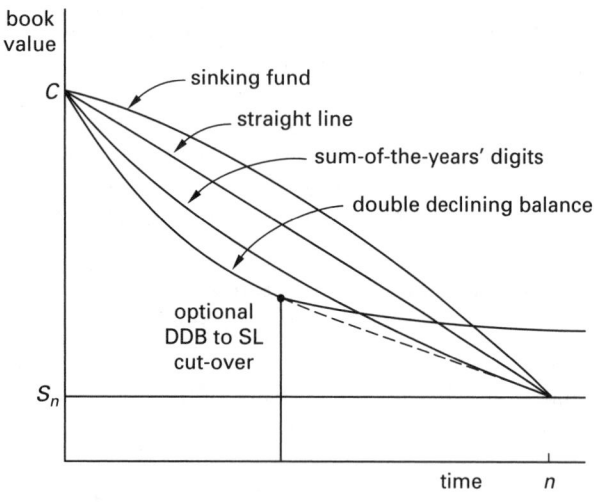

Figure 69.8 *Book Value with Different Depreciation Methods*

Example 69.18

For the asset described in Ex. 69.17, calculate the book value at the end of the first three years if sum-of-the-years' digits depreciation is used. The book value at the beginning of year 1 is $9000.

Solution

From Eq. 69.43,

$$\text{BV}_1 = \$9000 - \$1600 = \$7400$$
$$\text{BV}_2 = \$7400 - \$1440 = \$5960$$
$$\text{BV}_3 = \$5960 - \$1280 = \$4680$$

---

[30]This means that the straight line method of depreciation may result in a lower book value at some point in the depreciation period than if double declining balance is used. A *cut-over* from double declining balance to straight line may be permitted in certain cases. Finding the *cut-over point*, however, is usually done by comparing book values determined by both methods. The analytical method is complicated.

39. AMORTIZATION

Amortization and depreciation are similar in that they both divide up the cost basis or value of an asset. In fact, in certain cases, the term "amortization" may be used in place of the term "depreciation." However, depreciation is a specific form of amortization.

Amortization spreads the cost basis or value of an asset over some base. The base can be time, units of production, number of customers, and so on. The asset can be tangible (e.g., a delivery truck or building) or intangible (e.g., goodwill or a patent).

If the asset is tangible, if the base is time, and if the length of time is consistent with accounting standards and taxation guidelines, then the term "depreciation" is appropriate. However, if the asset is intangible, or if the base is some other variable, or if some length of time other than the customary period is used, then the term "amortization" is more appropriate.[31]

Example 69.19

A company purchases complete and exclusive patent rights to an invention for $1,200,000. It is estimated that, once commercially produced, the invention will have a specific but limited market of 1200 units. For the purpose of allocating the patent right cost to production cost, what is the amortization rate in dollars per unit?

Solution

Clearly, the patent should be amortized at the rate of

$$\frac{\$1,200,000}{1200 \text{ units}} = \$1000 \text{ per unit}$$

40. DEPLETION

Depletion is another artificial deductible operating expense designed to compensate mining organizations for decreasing mineral reserves. Since original and remaining quantities of minerals are seldom known accurately, the *depletion allowance* is calculated as a fixed percentage of the organization's gross income. These percentages are usually in the 10 to 20% range and apply to such mineral deposits as oil, natural gas, coal, uranium, and most metal ores.

41. BASIC INCOME TAX CONSIDERATIONS

The issue of income taxes is often overlooked in academic engineering economic analysis exercises. Such a position is justifiable when an organization (e.g., a nonprofit school, a church, or the government) pays no

---

[31]From time to time, the U.S. Congress has allowed certain types of facilities (e.g., emergency, grain storage, and pollution control) to be written off more rapidly than would otherwise be permitted in order to encourage investment in such facilities. The term "amortization" has been used with such write-off periods.

income taxes. However, if an individual or organization is subject to income taxes, the income taxes must be included in an economic analysis of investment alternatives.

Assume that an organization pays a fraction f of its profits to the federal government as income taxes. If the organization also pays a fraction s of its profits as state income taxes and if state taxes paid are recognized by the federal government as tax-deductible expenses, then the composite tax rate is

$$t = s + f - sf \qquad 69.44$$

The basic principles used to incorporate taxation into engineering economic analyses are the following.

- Initial purchase expenditures are unaffected by income taxes.

- Salvage revenues are unaffected by income taxes.

- Deductible expenses, such as operating costs, maintenance costs, and interest payments, are reduced by the fraction t (e.g., multiplied by the quantity $1 - t$).

- Revenues are reduced by the fraction t (e.g., multiplied by the quantity $1 - t$).

- Since tax regulations allow the depreciation in any year to be handled as if it were an actual operating expense, and since operating expenses are deductible from the income base prior to taxation, the after-tax profits will be increased. If D is the depreciation, the net result to the after-tax cash flow will be the addition of tD. Depreciation is multiplied by t and added to the appropriate year's cash flow, increasing that year's present worth.

For simplicity, most engineering economics practice problems involving income taxes specify a single income tax rate. In practice, however, federal and most state tax rates depend on the income level. Each range of incomes and its associated tax rate are known as *income bracket* and *tax bracket*, respectively. For example, the state income tax rate might be 4% for incomes up to and including \$30,000, and 5% for incomes above \$30,000. The income tax for a taxpaying entity with an income of \$50,000 would have to be calculated in two parts.

$$\text{tax} = (0.04)(\$30{,}000) + (0.05)(\$50{,}000 - \$30{,}000)$$
$$= \$2200$$

Income taxes and depreciation have no bearing on municipal or governmental projects since municipalities, states, and the U.S. government pay no taxes.

Example 69.20

A corporation that pays 53% of its profit in income taxes invests \$10,000 in an asset that will produce a \$3000 annual revenue for eight years. If the annual expenses are \$700, salvage after eight years is \$500, and 9% interest is used, what is the after-tax present worth? Disregard depreciation.

Solution

$$
\begin{aligned}
P =\ & -\$10{,}000 + (\$3000)(P/A, 9\%, 8)(1 - 0.53) \\
& - (\$700)(P/A, 9\%, 8)(1 - 0.53) \\
& + (\$500)(P/F, 9\%, 8) \\
=\ & -\$10{,}000 + (\$3000)(5.5348)(0.47) \\
& - (\$700)(5.5348)(0.47) + (\$500)(0.5019) \\
=\ & -\$3766
\end{aligned}
$$

42. TAXATION AT THE TIMES OF ASSET PURCHASE AND SALE

There are numerous rules and conventions that governmental tax codes and the accounting profession impose on organizations. Engineers are not normally expected to be aware of most of the rules and conventions, but occasionally it may be necessary to incorporate their effects into an engineering economic analysis.

Tax Credit

A *tax credit* (also known as an *investment tax credit* or *investment credit*) is a one-time credit against income taxes.[32] Therefore, it is added to the after-tax present worth as a last step in an engineering economic analysis. Such tax credits may be allowed by the government from time to time for equipment purchases, employment of various classes of workers, rehabilitation of historic landmarks, and so on.

A tax credit is usually calculated as a fraction of the initial purchase price or cost of an asset or activity.

$$\text{TC} = \text{fraction} \times \text{initial cost} \qquad 69.45$$

When the tax credit is applicable, the fraction used is subject to legislation. A professional tax expert or accountant should be consulted prior to applying the tax credit concept to engineering economic analysis problems.

Since the investment tax credit reduces the buyer's tax liability, a tax credit should be included only in after-tax engineering economic analyses. The credit is assumed to be received at the end of the year.

Gain or Loss on the Sale of a Depreciated Asset

If an asset that has been depreciated over a number of prior years is sold for more than its current book value, the difference between the book value and selling price

---

[32]Strictly, *tax credit* is the more general term, and applies to a credit for doing anything creditable. An *investment tax credit* requires an investment in something (usually real property or equipment).

is taxable income in the year of the sale. Alternatively, if the asset is sold for less than its current book value, the difference between the selling price and book value is an expense in the year of the sale.

Example 69.21

One year, a company makes a $5000 investment in an historic building. The investment is not depreciable, but it does qualify for a one-time 20% tax credit. In that same year, revenue is $45,000 and expenses (exclusive of the $5000 investment) are $25,000. The company pays a total of 53% in income taxes. What is the after-tax present worth of this year's activities if the company's interest rate for investment is 10%?

Solution

The tax credit is

$$TC = (0.20)(\$5000) = \$1000$$

This tax credit is assumed to be received at the end of the year. The after-tax present worth is

$$
\begin{aligned}
P &= -\$5000 + (\$45,000 - \$25,000)(1 - 0.53) \\
&\quad \times (P/F, 10\%, 1) + (\$1000)(P/F, 10\%, 1) \\
&= -\$5000 + (\$20,000)(0.47)(0.9091) \\
&\quad + (\$1000)(0.9091) \\
&= \$4455
\end{aligned}
$$

43. DEPRECIATION RECOVERY

The economic effect of depreciation is to reduce the income tax in year j by tD_j. The present worth of the asset is also affected: The present worth is increased by $tD_j(P/F, i\%, j)$. The after-tax present worth of all depreciation effects over the depreciation period of the asset is called the *depreciation recovery* (DR).[33]

$$DR = t \sum_{j=1}^{n} D_j (P/F, i\%, j) \qquad \textbf{69.46}$$

Straight line depreciation recovery from an asset is easily calculated, since the depreciation is the same each year. Assuming the asset has a constant depreciation of D and depreciation period of n years, the depreciation recovery is

$$DR = tD(P/A, i\%, n) \qquad \textbf{69.47}$$

$$D = \frac{C - S_n}{n} \qquad \textbf{69.48}$$

---

[33]Since the depreciation benefit is reduced by taxation, depreciation cannot be thought of as an annuity to fund a replacement asset.

Sum-of-the-years' digits depreciation recovery is also relatively easily calculated, since the depreciation decreases uniformly each year.

$$DR = \left(\frac{t(C - S_n)}{T}\right)(n(P/A, i\%, n) - (P/G, i\%, n)) \qquad$$
$$\textbf{69.49}$$

Finding *declining balance depreciation recovery* is more involved. There are three difficulties. The first (the apparent need to calculate all previous depreciations in order to determine the subsequent depreciation) has already been addressed by Eq. 69.32.

The second difficulty is that there is no way to ensure (that is, to force) the book value to be S_n at $t = n$. Therefore, it is common to write off the remaining book value (down to S_n) at $t = n$ in one lump sum. This assumes $BV_n \geq S_n$.

The third difficulty is that of finding the present worth of an *exponentially decreasing cash flow*. Although the proof is omitted here, such exponential cash flows can be handled with the *exponential gradient factor*, (P/EG).[34]

$$(P/EG, z - 1, n) = \frac{z^n - 1}{z^n(z - 1)} \qquad \textbf{69.50}$$

$$z = \frac{1 + i}{1 - d} \qquad \textbf{69.51}$$

Then, as long as $BV_n > S_n$, the declining balance depreciation recovery is

$$DR = tC\left(\frac{d}{1 - d}\right)(P/EG, z - 1, n) \qquad \textbf{69.52}$$

Example 69.22

For the asset described in Ex. 69.17, calculate the after-tax depreciation recovery with straight line and sum-of-the-years' digits depreciation methods. Use 6% interest with 48% income taxes.

Solution

SL:
$$
\begin{aligned}
DR &= (0.48)(\$880)(P/A, 6\%, 10) \\
&= (0.48)(\$880)(7.3601) \\
&= \$3109
\end{aligned}
$$

SOYD: The depreciation series can be thought of as a constant $1600 term with a negative $160 gradient.

$$
\begin{aligned}
DR &= (0.48)(\$1600)(P/A, 6\%, 10) \\
&\quad - (0.48)(\$160)(P/G, 6\%, 10) \\
&= (0.48)(\$1600)(7.3601) \\
&\quad - (0.48)(\$160)(29.6023) \\
&= \$3379
\end{aligned}
$$

---

[34]The (P/A) columns in App. 69.B can be used for (P/EG) as long as the interest rate is assumed to be $z - 1$.

Table 69.5 *Depreciation Calculation Summary*

method	depreciation basis	depreciation in year j (D_j)	book value after jth depreciation (BV_j)	after-tax depreciation recovery (DR)	supplementary formulas
straight line (SL)	$C - S_n$	$\dfrac{C - S_n}{n}$ (constant)	$C - jD$	$tD(P/A, i\%, n)$	
constant percentage	$C - S_n$	fraction $\times (C - S_n)$ (constant)	$C - jD$	$tD(P/A, i\%, n)$	
sum-of-the-years' digits (SOYD)	$C - S_n$	$\dfrac{(C - S_n) \times (n - j + 1)}{T}$	$(C - S_n) \times \left(1 - \dfrac{j(2n + 1 - j)}{n(n+1)}\right) + S_n$	$\dfrac{t(C - S_n)}{T} \times (n(P/A, i\%, n) - (P/G, i\%, n))$	$T = \frac{1}{2}n(n+1)$
double declining balance (DDB)	C	$dC(1 - d)^{j-1}$	$C(1 - d)^j$	$tC\left(\dfrac{d}{1 - d}\right) \times (P/EG, z - 1, n)$	$d = \dfrac{2}{n}$; $z = \dfrac{1+i}{1-d}$ $(P/EG, z - 1, n) = \dfrac{z^n - 1}{z^n(z - 1)}$
sinking fund (SF)	$C - S_n$	$(C - S_n) \times (A/F, i\%, n) \times (F/P, i\%, j - 1)$	$C - (C - S_n) \times (A/F, i\%, n) \times (F/A, i\%, j)$	$\dfrac{(t)(C - S_n)(A/F, i\%, n)}{1 + i}$	
accelerated cost recovery system (ACRS/MACRS)	C	$C \times$ factor	$C - \displaystyle\sum_{m=1}^{j} D_m$	$t \displaystyle\sum_{j=1}^{n} D_j(P/F, i\%, j)$	
production or service output	$C - S_n$	$(C - S_n) \times \left(\dfrac{\text{actual output in year } j}{\text{lifetime output}}\right)$	$C - \displaystyle\sum_{m=1}^{j} D_m$	$t \displaystyle\sum_{j=1}^{n} D_j(P/F, i\%, j)$	

Notice that the ten-year (P/G) factor is used even though there are only nine years in which the gradient reduces the initial $1600 amount.

Example 69.23

What is the after-tax present worth of the asset described in Ex. 69.20 if straight line, sum-of-the-years' digits, and double declining balance depreciation methods are used?

Solution

Using SL, the depreciation recovery is

$$\text{DR} = (0.53)\left(\frac{\$10{,}000 - \$500}{8}\right)(P/A, 9\%, 8)$$

$$= (0.53)\left(\frac{\$9500}{8}\right)(5.5348)$$

$$= \$3483$$

Using SOYD, the depreciation recovery is calculated as follows.

$$T = \left(\tfrac{1}{2}\right)(8)(9) = 36$$

$$\text{depreciation base} = \$10{,}000 - \$500 = \$9500$$

$$D_1 = \left(\frac{8}{36}\right)(\$9500) = \$2111$$

$$G = \text{gradient} = \left(\frac{1}{36}\right)(\$9500)$$

$$= \$264$$

$$\text{DR} = (0.53)[(\$2111)(P/A, 9\%, 8)$$

$$- (\$264)(P/G, 9\%, 8)]$$

$$= (0.53)[(\$2111)(5.5348)$$

$$- (\$264)(16.8877)]$$

$$= \$3830$$

Using DDB, the depreciation recovery is calculated as follows.[35]

$$d = \frac{2}{8} = 0.25$$

$$z = \frac{1 + 0.09}{1 - 0.25} = 1.4533$$

$$(P/EG, z - 1, n) = \frac{(1.4533)^8 - 1}{(1.4533)^8(0.4533)} = 2.095$$

From Eq. 69.52,

$$\text{DR} = (0.53)\left(\frac{(0.25)(\$10,000)}{0.75}\right)(2.095)$$

$$= \$3701$$

The after-tax present worth, neglecting depreciation, was previously found to be $-\$3766$.

The after-tax present worths, including depreciation recovery, are

SL: $P = -\$3766 + \$3483 = -\$283$
SOYD: $P = -\$3766 + \$3830 = \$64$
DDB: $P = -\$3766 + \$3701 = -\$65$

44. OTHER INTEREST RATES

The *effective interest rate per period*, i (also called *yield* by banks), is the only interest rate that should be used in equivalence equations. The interest rates at the top of the factor tables in App. 69.B are implicitly all effective interest rates. Usually, the period will be one year, hence the name *effective annual interest rate*. However, there are other interest rates in use as well.

The term *nominal interest rate*, r (*rate per annum*), is encountered when compounding is more than once per year. The nominal rate does not include the effect of compounding and is not the same as the effective rate. And, since the effective interest rate can be calculated from the nominal rate only if the number of compounding periods per year is known, nominal rates cannot be compared unless the method of compounding is specified. The only practical use for a nominal rate per year is for calculating the effective rate per period.

45. RATE AND PERIOD CHANGES

If there are k compounding periods during the year (two for semiannual compounding, four for quarterly compounding, twelve for monthly compounding, etc.), and the nominal rate is r, the *effective rate per compounding period* is

$$\phi = \frac{r}{k} \qquad\qquad 69.53$$

---

[35]This method should start by checking that the book value at the end of the depreciation period is greater than the salvage value. In this example, such is the case. However, the step is not shown.

The effective annual rate, i, can be calculated from the effective rate per period, ϕ, by using Eq. 69.54.

$$i = (1 + \phi)^k - 1$$

$$= \left(1 + \frac{r}{k}\right)^k - 1 \qquad\qquad 69.54$$

Sometimes, only the effective rate per period (e.g., per month) is known. However, that will be a simple problem since compounding for n periods at an effective rate per period is not affected by the definition or length of the period.

The following rules may be used to determine which interest rate is given in a problem.

- Unless specifically qualified in the problem, the interest rate given is an annual rate.

- If the compounding is annual, the rate given is the effective rate. If compounding is other than annual, the rate given is the nominal rate.

The effective annual interest rate determined on a *daily compounding basis* will not be significantly different than if *continuous compounding* is assumed.[36] In the case of continuous (or daily) compounding, the discounting factors can be calculated directly from the nominal interest rate and number of years, without having to find the effective interest rate per period. Table 69.6 can be used to determine the discount factors for continuous compounding.

Table 69.6 Discount Factors for Continuous Compounding
(n is the number of years)

symbol	formula
$(F/P, r\%, n)$	e^{rn}
$(P/F, r\%, n)$	e^{-rn}
$(A/F, r\%, n)$	$\dfrac{e^r - 1}{e^{rn} - 1}$
$(F/A, r\%, n)$	$\dfrac{e^{rn} - 1}{e^r - 1}$
$(A/P, r\%, n)$	$\dfrac{e^r - 1}{1 - e^{-rn}}$
$(P/A, r\%, n)$	$\dfrac{1 - e^{-rn}}{e^r - 1}$

---

[36]The number of *banking days in a year* (250, 360, etc.) must be specifically known.

Example 69.24

A savings and loan offers a nominal rate of $5^1/_4\%$ compounded daily over 365 days in a year. What is the effective annual rate?

Solution

method 1: Use Eq. 69.54.

$$r = 0.0525, \ k = 365$$

$$i = \left(1 + \frac{0.0525}{365}\right)^{365} - 1 = 0.0539$$

method 2: Assume daily compounding is the same as continuous compounding.

$$i = (F/P, r\%, 1) - 1$$
$$= e^{0.0525} - 1 = 0.0539$$

Example 69.25

A real estate investment trust pays $7,000,000 for a 100-unit apartment complex. The trust expects to sell the complex in ten years for $15,000,000. In the meantime, it expects to receive an average rent of $900 per month from each apartment. Operating expenses are expected to be $200 per month per occupied apartment. A 95% occupancy rate is predicted. In similar investments, the trust has realized a 15% effective annual return on its investment. Compare the expected present worth of this investment when calculated assuming (a) annual compounding (i.e., the year-end convention), and (b) monthly compounding. Disregard taxes, depreciation, and all other factors.

Solution

(a) The net annual income will be

$$(0.95)(100 \text{ units})\left(\$900 \ \frac{\$}{\text{unit-month}}\right.$$

$$\left. - \$200 \ \frac{\$}{\text{unit-month}}\right)\left(12 \ \frac{\text{months}}{\text{year}}\right)$$

$$= \$798,000/\text{year}$$

The present worth of ten years of operation is

$$P = -\$7,000,000 + (\$798,000)(P/A, 15\%, 10)$$
$$+ (\$15,000,000)(P/F, 15\%, 10)$$
$$= -\$7,000,000 + (\$798,000)(5.0188)$$
$$+ (\$15,000,000)(0.2472)$$
$$= \$713,000$$

(b) The net monthly income is

$$(0.95)(100 \text{ units})\left(\$900 \ \frac{\$}{\text{unit-month}}\right.$$

$$\left. - \$200 \ \frac{\$}{\text{unit-month}}\right) = \$66,500/\text{month}$$

Equation 69.54 is used to calculate the effective monthly rate, ϕ, from the effective annual rate, $i = 15\%$, and the number of compounding periods per year, $k = 12$.

$$\phi = (1 + i)^{1/k} - 1$$
$$= (1 + 0.15)^{1/12} - 1 = 0.011715 \ \ (1.1715\%)$$

The number of compounding periods in ten years is

$$n = (10 \text{ years})\left(12 \ \frac{\text{months}}{\text{year}}\right) = 120 \text{ months}$$

The present worth of 120 months of operation is

$$P = -\$7,000,000 + (\$66,500)(P/A, 1.1715\%, 120)$$
$$+ (\$15,000,000)(P/F, 1.1715\%, 120)$$

Since table values for 1.1715% discounting factors are not available, the factors are calculated from Table 69.1.

$$(P/A, 1.1715\%, 120) = (1 + i)^n - \frac{1}{i(1 + i)^n}$$

$$= \frac{(1 + 0.011715)^{120} - 1}{(0.011715)(1 + 0.011715)^{120}}$$

$$= 64.261$$

$$(P/F, 1.1715\%, 120) = (1 + i)^{-n} = (1 + 0.011715)^{-120}$$
$$= 0.2472$$

The present worth over 120 monthly compounding periods is

$$P = -\$7,000,000 + (\$66,500)(64.261)$$
$$+ (\$15,000,000)(0.2472)$$
$$= \$981,357$$

46. BONDS

A *bond* is a method of long-term financing commonly used by governments, states, municipalities, and very large corporations.[37] The bond represents a contract to pay the bondholder specific amounts of money at specific times. The holder purchases the bond in exchange for specific payments of interest and principal. Typical municipal bonds call for quarterly or semiannual interest payments, and a payment of the *face value of the bond* on the *date of maturity* (end of the bond period).[38] Due to the practice of discounting in the bond market, a bond's face value and its purchase price generally will not coincide.

---

[37]In the past, 30-year bonds were typical. Shorter term 10-year, 15-year, 20-year, and 25-year bonds are also commonly issued.
[38]A *fully amortized bond* pays back interest and principal throughout the life of the bond. There is no balloon payment.

Economics

In the past, a bondholder had to submit a coupon or ticket in order to receive an interim interest payment. This has given rise to the term *coupon rate*, which is the nominal annual interest rate on which the interest payments are made. Coupon books are seldom used with modern bonds, but the term survives. The coupon rate determines the magnitude of the semiannual (or otherwise) interest payments during the life of the bond. The bondholder's own effective interest rate should be used for economic decisions about the bond.

Actual *bond yield* is the bondholder's actual rate of return of the bond, considering the purchase price, interest payments, and face value payment (or, value realized if the bond is sold before it matures). By convention, bond yield is calculated as a nominal rate (rate per annum), not an effective rate per year. The bond yield should be determined by finding the effective rate of return per payment period (e.g., per semiannual interest payment) as a conventional rate of return problem. Then, the nominal rate can be found by multiplying the effective rate per period by the number of payments per year, as in Eq. 69.54.

Example 69.26

What is the maximum amount an investor should pay for a 25-year bond with a $20,000 face value and 8% coupon rate (interest only paid semiannually)? The bond will be kept to maturity. The investor's effective annual interest rate for economic decisions is 10%.

Solution

For this problem, take the compounding period to be six months. Then, there are 50 compounding periods. Since 8% is a nominal rate, the effective bond rate per period is calculated from Eq. 69.53 as

$$\phi_{\text{bond}} = \frac{r}{k} = \frac{8\%}{2} = 4\%$$

The bond payment received semiannually is

$$(0.04)(\$20,000) = \$800$$

10% is the investor's effective rate per year, so Eq. 69.54 is again used to calculate the effective analysis rate per period.

$$0.10 = (1 + \phi)^2 - 1$$
$$\phi = 0.04881 \quad (4.88\%)$$

The maximum amount that the investor should be willing to pay is the present worth of the investment.

$$P = (800)(P/A, 4.88\%, 50) + (20,000)(P/F, 4.88\%, 50)$$

Table 69.1 can be used to calculate the following factors.

$$(P/A, 4.88\%, 50) = \frac{(1.0488)^{50} - 1}{(0.0488)(1.0488)^{50}} = 18.600$$

$$(P/F, 4.88\%, 50) = \frac{1}{(1.0488)^{50}} = 0.09233$$

Then, the present worth is

$$P = (\$800)(18.600) + (\$20,000)(0.09233)$$
$$= \$16,727$$

47. PROBABILISTIC PROBLEMS

If an alternative's cash flows are specified by an implicit or explicit probability distribution rather than being known exactly, the problem is *probabilistic*.

Probabilistic problems typically possess the following characteristics.

- There is a chance of loss that must be minimized (or, rarely, a chance of gain that must be maximized) by selection of one of the alternatives.

- There are multiple alternatives. Each alternative offers a different degree of protection from the loss. Usually, the alternatives with the greatest protection will be the most expensive.

- The magnitude of loss or gain is independent of the alternative selected.

Probabilistic problems are typically solved using annual costs and expected values. An *expected value* is similar to an *average value* since it is calculated as the mean of the given probability distribution. If cost 1 has a probability of occurrence, p_1, cost 2 has a probability of occurrence, p_2, and so on, the expected value is

$$\mathcal{E}\{\text{cost}\} = p_1(\text{cost } 1) + p_2(\text{cost } 2) + \cdots \qquad \textbf{69.55}$$

Example 69.27

Flood damage in any year is given according to the following table. What is the present worth of flood damage for a ten-year period? Use 6% as the effective annual interest rate.

damage	probability
0	0.75
$10,000	0.20
$20,000	0.04
$30,000	0.01

Solution

The expected value of flood damage in any given year is

$$\mathcal{E}\{\text{damage}\} = (0)(0.75) + (\$10,000)(0.20)$$
$$+ (\$20,000)(0.04) + (\$30,000)(0.01)$$
$$= \$3100$$

The present worth of ten years of expected flood damage is

$$\text{present worth} = (\$3100)(P/A, 6\%, 10)$$
$$= (\$3100)(7.3601)$$
$$= \$22{,}816$$

Example 69.28

A dam is being considered on a river that periodically overflows and causes $600,000 damage. (The damage is essentially the same each time the river causes flooding.) The project horizon is 40 years. A 10% interest rate is being used.

Three different designs are available, each with different costs and storage capacities.

design alternative	cost	maximum capacity
A	$500,000	1 unit
B	$625,000	1.5 units
C	$900,000	2.0 units

The national weather service has provided a statistical analysis of annual rainfall runoff from the watershed draining into the river.

units annual rainfall	probability
0	0.10
0.1 to 0.5	0.60
0.6 to 1.0	0.15
1.1 to 1.5	0.10
1.6 to 2.0	0.04
2.1 or more	0.01

Which design alternative would you choose assuming the dam is essentially empty at the start of each rainfall season?

Solution

The sum of the construction cost and the expected damage should be minimized. If alternative A is chosen, it will have a capacity of 1 unit. Its capacity will be exceeded (causing $600,000 damage) when the annual rainfall exceeds 1 unit. Therefore, the expected value of the annual cost of alternative A is

$$\mathcal{E}\{\text{EUAC(A)}\} = (\$500{,}000)(A/P, 10\%, 40)$$
$$+ (\$600{,}000)(0.10 + 0.04 + 0.01)$$
$$= (\$500{,}000)(0.1023) + (\$600{,}000)(0.15)$$
$$= \$141{,}150$$

Similarly,

$$\mathcal{E}\{\text{EUAC(B)}\} = (\$625{,}000)(A/P, 10\%, 40)$$
$$+ (\$600{,}000)(0.04 + 0.01)$$
$$= (\$625{,}000)(0.1023) + (\$600{,}000)(0.05)$$
$$= \$93{,}938$$

$$\mathcal{E}\{\text{EUAC(C)}\} = (\$900{,}000)(A/P, 10\%, 40)$$
$$+ (\$600{,}000)(0.01)$$
$$= (\$900{,}000)(0.1023) + (\$600{,}000)(0.01)$$
$$= \$98{,}070$$

Alternative B should be chosen.

48. FIXED AND VARIABLE COSTS

The distinction between fixed and variable costs depends on how these costs vary when an independent variable changes. For example, factory or machine production is frequently the independent variable. However, it could just as easily be vehicle miles driven, hours of operation, or quantity (mass, volume, etc.).

If a cost is a function of the independent variable, the cost is said to be a *variable cost*. The change in cost per unit variable change (i.e., what is usually called the *slope*) is known as the *incremental cost*. Material and labor costs are examples of variable costs. They increase in proportion to the number of product units manufactured.

If a cost is not a function of the independent variable, the cost is said to be a *fixed cost*. Rent and lease payments are typical fixed costs. These costs will be incurred regardless of production levels.

Some costs have both fixed and variable components, as Fig. 69.9 illustrates. The fixed portion can be determined by calculating the cost at zero production.

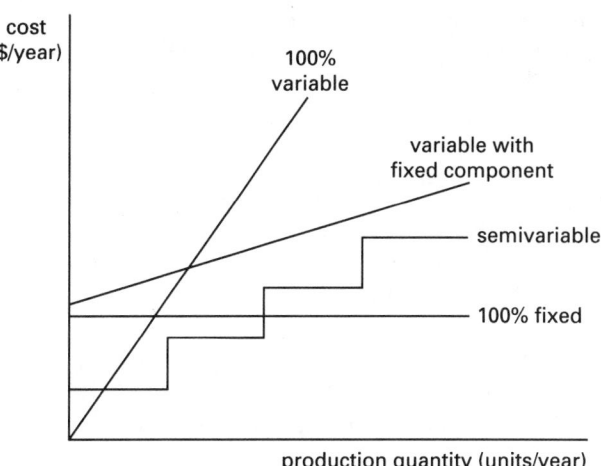

Figure 69.9 *Fixed and Variable Costs*

An additional category of cost is the *semivariable cost.* This type of cost increases step-wise. Semivariable cost structures are typical of situations where *excess capacity* exists. For example, supervisory cost is a step-wise function of the number of production shifts. Also, labor cost for truck drivers is a step-wise function of weight (volume) transported. As long as a truck has room left (i.e., excess capacity), no additional driver is needed. As soon as the truck is filled, labor cost will increase.

Table 69.7 *Summary of Fixed and Variable Costs*

> *fixed costs*
>> rent
>> property taxes
>> interest on loans
>> insurance
>> janitorial service expense
>> tooling expense
>> setup, cleanup, and tear-down expenses
>> depreciation expense
>> marketing and selling costs
>> cost of utilities
>> general burden and overhead expense
>
> *variable costs*
>> direct material costs
>> direct labor costs
>> cost of miscellaneous supplies
>> payroll benefit costs
>> income taxes
>> supervision costs

49. ACCOUNTING COSTS AND EXPENSE TERMS

The accounting profession has developed special terms for certain groups of costs. When annual costs are incurred due to the functioning of a piece of equipment, they are known as *operating and maintenance* (O&M) *costs.* The annual costs associated with operating a business (other than the costs directly attributable to production) are known as *general, selling, and administrative* (GS&A) *expenses.*

Direct labor costs are costs incurred in the factory, such as assembly, machining, and painting labor costs. *Direct material costs* are the costs of all materials that go into production.[39] Typically, both direct labor and direct material costs are given on a per-unit or per-item basis. The sum of the direct labor and direct material costs is known as the *prime cost.*

There are certain additional expenses incurred in the factory, such as the costs of factory supervision, stockpicking, quality control, factory utilities, and miscellaneous supplies (cleaning fluids, assembly lubricants,

routing tags, etc.) that are not incorporated into the final product. Such costs are known as *indirect manufacturing expenses* (IME) or *indirect material and labor costs.*[40] The sum of the per-unit indirect manufacturing expense and prime cost is known as the *factory cost.*

Research and development (R&D) *costs* and *administrative expenses* are added to the factory cost to give the *manufacturing cost* of the product.

Additional costs are incurred in marketing the product. Such costs are known as *selling expenses* or *marketing expenses.* The sum of the selling expenses and manufacturing cost is the *total cost* of the product.

Figure 69.10 illustrates these terms.[41]

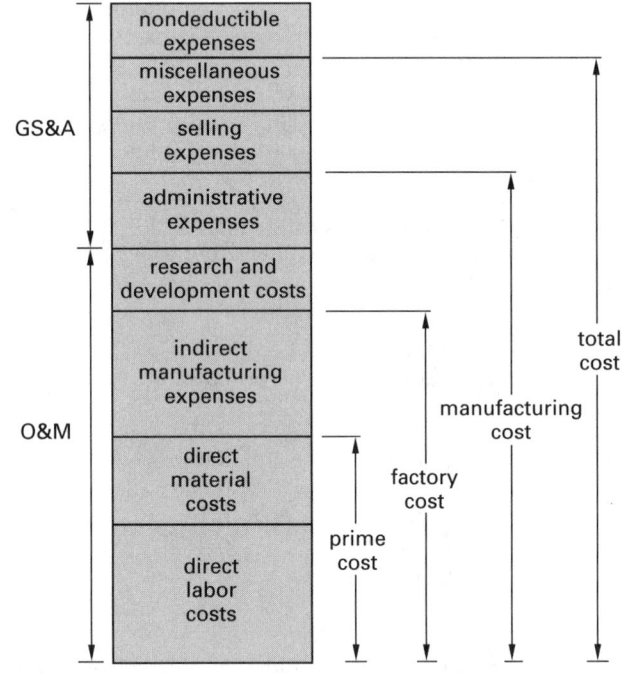

Figure 69.10 *Costs and Expenses Combined*

The distinctions among the various forms of cost (particularly with overhead costs) are not standardized. Each company must develop a classification system to deal with the various cost factors in a consistent manner. There are also other terms in use (e.g., *raw materials, operating supplies, general plant overhead*), but these terms must be interpreted within the framework of each company's classification system. Table 69.8 is typical of such classification systems.

---

[39]There may be problems with pricing the material when it is purchased from an outside vendor and the stock on hand derives from several shipments purchased at different prices.

[40]The *indirect material* and *labor costs* usually exclude costs incurred in the office area.

[41]Notice that *total cost* does not include income taxes.

Table 69.8 *Typical Classification of Expenses*

direct material expenses
    items purchased from other vendors
    manufactured assemblies
direct labor expenses
    machining and forming
    assembly
    finishing
    inspection
    testing
factory overhead expenses
    supervision
    benefits
        pension
        medical insurance
        vacations
    wages overhead
        unemployment compensation taxes
        social security taxes
        disability taxes
    stock-picking
    quality control and inspection
    expediting
    rework
    maintenance
    miscellaneous supplies
        routing tags
        assembly lubricants
        cleaning fluids
        wiping cloths
        janitorial supplies
    packaging (materials and labor)
    factory utilities
    laboratory
    depreciation on factory equipment
research and development expenses
    engineering (labor)
    patents
    testing
    prototypes (material and labor)
    drafting
    O&M of R&D facility
administrative expenses
    corporate officers
    accounting
    secretarial/clerical/reception
    security (protection)
    medical (nurse)
    employment (personnel)
    reproduction
    data processing
    production control
    depreciation on nonfactory equipment
    office supplies
    office utilities
    O&M of offices
selling expenses
    marketing (labor)
    advertising
    transportation (if not paid by customer)
    outside sales force (labor and expenses)
    demonstration units
    commissions
    technical service and support
    order processing
    branch office expenses
miscellaneous expenses
    insurance
    property taxes
    interest on loans
nondeductible expenses
    federal income taxes
    fines and penalties

50. ACCOUNTING PRINCIPLES

Basic Bookkeeping

An accounting or *bookkeeping system* is used to record historical financial transactions. The resultant records are used for product costing, satisfaction of statutory requirements, reporting of profit for income tax purposes, and general company management.

Bookkeeping consists of two main steps: recording the transactions, followed by categorization of the transactions.[42] The transactions (receipts and disbursements) are recorded in a *journal (book of original entry)* to complete the first step. Such a journal is organized in a simple chronological and sequential manner.[43] The transactions are then categorized (into interest income, advertising expense, etc.) and posted (i.e., entered or written) into the appropriate *ledger account.*

The ledger accounts together constitute the *general ledger* or *ledger*. All ledger accounts can be classified into one of three types: *asset accounts, liability accounts,* and *owners' equity accounts*. Strictly speaking, income and expense accounts, kept in a separate journal, are included within the classification of owners' equity accounts.

Together, the journal and ledger are known simply as "the books" of the company.

Balancing the Books

In a business environment, *balancing the books* means more than reconciling the checkbook and bank statements. All accounting entries must be posted in such a way as to maintain the equality of the *basic accounting equation,*

$$\text{assets} = \text{liability} + \text{owners' equity} \qquad 69.56$$

In a *double-entry bookkeeping system*, the equality is maintained within the ledger system by entering each transaction into two balancing ledger accounts. For example, paying a utility bill would decrease the cash account (an asset account) and decrease the utility expense account (a liability account) by the same amount.

Transactions are either *debits* or *credits*, depending on their sign. Increases in asset accounts are debits; decreases are credits. For liability and equity accounts, the opposite is true: Increases are credits, and decreases are debits.[44]

---

[42]These two steps are not to be confused with the *double-entry bookkeeping method*.

[43]The two-step process is more typical of a *manual bookkeeping system* than a computerized *general ledger system*. However, even most computerized systems produce reports in journal entry order, as well as account summaries.

[44]There is a difference in sign between asset and liability accounts. Thus, an increase in an expense account is actually a decrease. The accounting profession, apparently, is comfortable with the common confusion that exists between debits and credits.

Economics

Cash and Accrual Systems[45]

The simplest form of bookkeeping is based on the *cash system*. The only transactions that are entered into the journal are those that represent cash receipts and disbursements. In effect, a checkbook register or bank deposit book could serve as the journal.

During a given period (e.g., month or quarter), expense liabilities may be incurred even though the payments for those expenses have not been made. For example, an invoice (bill) may have been received but not paid. Under the *accrual system*, the obligation is posted into the appropriate expense account before it is paid.[46] Analogous to expenses, under the accrual system, income will be claimed before payment is received. Specifically, a sales transaction can be recorded as income when the customer's order is received, when the outgoing invoice is generated, or when the merchandise is shipped.

Financial Statements

Each period, two types of corporate financial statements are typically generated: the *balance sheet* and *profit and loss* (P&L) *statements*.[47] The profit and loss statement, also known as a *statement of income and retained earnings*, is a summary of sources of *income* or *revenue* (interest, sales, fees charged, etc.) and *expenses* (utilities, advertising, repairs, etc.) for the period. The expenses are subtracted from the revenues to give a *net income* (generally, before taxes).[48] Figure 69.11 illustrates a simple profit and loss statement.

The *balance sheet* presents the *basic accounting equation* in tabular form. The balance sheet lists the major categories of assets and outstanding liabilities. The difference between asset values and liabilities is the *equity*, as defined in Eq. 69.56. This equity represents what would be left over after satisfying all debts by liquidating the company.

---

[45]There is also a distinction made between cash flows that are known and those that are expected. It is a *standard accounting principle* to record losses in full, at the time they are recognized, even before their occurrence. In the construction industry, for example, losses are recognized in full and projected to the end of a project as soon as they are foreseeable. Profits, on the other hand, are recognized only as they are realized (typically, as a percentage of project completion). The difference between cash and accrual systems is a matter of *bookkeeping*. The difference between loss and profit recognition is a matter of *accounting convention*. Engineers seldom need to be concerned with the accounting tradition.
[46]The expense for an item or service might be accrued even *before* the invoice is received. It might be recorded when the purchase order for the item or service is generated, or when the item or service is received.
[47]Other types of financial statements (*statements of changes in financial position, cost of sales statements*, inventory and asset reports, etc.) also will be generated, depending on the needs of the company.
[48]Financial statements also can be prepared with percentages (of total assets and net revenue) instead of dollars, in which case they are known as *common size financial statements*.

revenue

interest	2000	
sales	237,000	
returns	<23,000>	
net revenue		216,000

expenses

salaries	149,000	
utilities	6000	
advertising	28,000	
insurance	4000	
supplies	1000	
net expenses		188,000

period net income	28,000
beginning retained earnings	63,000
net year-to-date earnings	91,000

Figure 69.11 *Simplified Profit and Loss Statement*

There are several terms that appear regularly on balance sheets.

- *current assets:* cash and other assets that can be converted quickly into cash, such as accounts receivable, notes receivable, and merchandise (inventory). Also known as *liquid assets*.

- *fixed assets:* relatively permanent assets used in the operation of the business and relatively difficult to convert into cash. Examples are land, buildings, and equipment. Also known as *non-liquid assets*.

- *current liabilities:* liabilities due within a short period of time (e.g., within one year) and typically paid out of current assets. Examples are accounts payable, notes payable, and other accrued liabilities.

- *long-term liabilities:* obligations that are not totally payable within a short period of time (e.g., within one year).

Figure 69.12 is a simplified balance sheet.

Analysis of Financial Statements

Financial statements are evaluated by management, lenders, stockholders, potential investors, and many other groups for the purpose of determining the *health of the company*. The health can be measured in terms of *liquidity* (ability to convert assets to cash quickly), *solvency* (ability to meet debts as they become due), and *relative risk* (of which one measure is *leverage*—the portion of total capital contributed by owners).

The analysis of financial statements involves several common ratios, usually expressed as percentages. The following are some frequently encountered ratios.

- *current ratio:* an index of short-term paying ability.

$$\text{current ratio} = \frac{\text{current assets}}{\text{current liabilities}}$$

ASSETS

current assets

cash	14,000	
accounts receivable	36,000	
notes receivable	20,000	
inventory	89,000	
prepaid expenses	3000	
total current assets		162,000

plant, property,
and equipment

land and buildings	217,000	
motor vehicles	31,000	
equipment	94,000	
accumulated depreciation	<52,000>	
total fixed assets		290,000
total assets		452,000

LIABILITIES AND OWNERS' EQUITY

current liabilities

accounts payable	66,000	
accrued income taxes	17,000	
accrued expenses	8000	
total current liabilities		91,000

long-term debt

notes payable	117,000	
mortgage	23,000	
total long-term debt		140,000

owners' and stockholders'
equity

stock	130,000	
retained earnings	91,000	
total owners' equity		221,000

total liabilities and owners' equity   452,000

Figure 69.12 Simplified Balance Sheet

- *quick (or acid-test) ratio:* a more stringent measure of short-term debt-paying ability. The *quick assets* are defined to be current assets minus inventories and prepaid expenses.

$$\text{quick ratio} = \frac{\text{quick assets}}{\text{current liabilities}}$$

- *receivable turnover:* a measure of the average speed with which accounts receivable are collected.

$$\text{receivable turnover} = \frac{\text{net credit sales}}{\text{average net receivables}}$$

- *average age of receivables:* number of days, on the average, in which receivables are collected.

$$\text{average age of receivables} = \frac{365}{\text{receivable turnover}}$$

- *inventory turnover:* a measure of the speed with which inventory is sold, on the average.

$$\text{inventory turnover} = \frac{\text{cost of goods sold}}{\text{average cost of inventory on hand}}$$

- *days supply of inventory on hand:* number of days, on the average, that the current inventory would last.

$$\text{days supply of inventory on hand} = \frac{365}{\text{inventory turnover}}$$

- *book value per share of common stock:* number of dollars represented by the balance sheet owners' equity for each share of common stock outstanding.

$$\text{book value per share of common stock} = \frac{\text{common shareholders' equity}}{\text{number of outstanding shares}}$$

- *gross margin:* gross profit as a percentage of sales. (Gross profit is sales less cost of goods sold.)

$$\text{gross margin} = \frac{\text{gross profit}}{\text{net sales}}$$

- *profit margin ratio:* percentage of each dollar of sales that is net income.

$$\text{profit margin} = \frac{\text{net income before taxes}}{\text{net sales}}$$

- *return on investment ratio:* shows the percent return on owners' investment.

$$\text{return on investment} = \frac{\text{net income}}{\text{owners' equity}}$$

- *price-earnings ratio:* indication of relationship between earnings and market price per share of common stock, useful in comparisons between alternative investments.

$$\text{price-earnings} = \frac{\text{market price per share}}{\text{earnings per share}}$$

51. COST ACCOUNTING

Cost accounting is the system that determines the cost of manufactured products. Cost accounting is called *job cost accounting* if costs are accumulated by part number or contract. It is called *process cost accounting* if costs are accumulated by departments or manufacturing processes.

Cost accounting is dependent on historical and recorded data. The unit product cost is determined from actual expenses and numbers of units produced. Allowances (i.e., budgets) for future costs are based on these historical figures. Any deviation from historical figures is called a *variance*. Where adequate records are available, variances can be divided into *labor variance* and *material variance*.

When determining a unit product cost, the direct material and direct labor costs are generally clear-cut and easily determined. Furthermore, these costs are 100% variable costs. However, the indirect cost per unit of product is not as easily determined. Indirect costs (*burden*, *overhead*, etc.) can be fixed or semivariable costs. The amount of indirect cost allocated to a unit will depend on the unknown future overhead expense as well as the unknown future production (*vehicle size*).

A typical method of allocating indirect costs to a product is as follows.

step 1: Estimate the total expected indirect (and overhead) costs for the upcoming year.

step 2: Determine the most appropriate vehicle (basis) for allocating the overhead to production. Usually, this vehicle is either the number of units expected to be produced or the number of direct hours expected to be worked in the upcoming year.

step 3: Estimate the quantity or size of the overhead vehicle.

step 4: Divide expected overhead costs by the expected overhead vehicle to obtain the unit overhead.

step 5: Regardless of the true size of the overhead vehicle during the upcoming year, one unit of overhead cost is allocated per unit of overhead vehicle.

Once the prime cost has been determined and the indirect cost calculated based on projections, the two are combined into a *standard factory cost* or *standard cost*, which remains in effect until the next budgeting period (usually a year).

During the subsequent manufacturing year, the standard cost of a product is not generally changed merely because it is found that an error in projected indirect costs or production quantity (vehicle size) has been made. The allocation of indirect costs to a product is assumed to be independent of errors in forecasts. Rather, the difference between the expected and actual expenses, known as the *burden (overhead) variance*, experienced during the year is posted to one or more *variance accounts*.

Burden (overhead) variance is caused by errors in forecasting both the actual indirect expense for the upcoming year and the overhead vehicle size. In the former case, the variance is called *burden budget variance*; in the latter, it is called *burden capacity variance*.

Example 69.29

A company expects to produce 8000 items in the coming year. The current material cost is $4.54 each. Sixteen minutes of direct labor are required per unit. Workers are paid $7.50 per hour. 2133 direct labor hours are forecasted for the product. Miscellaneous overhead costs are estimated at $45,000.

Find (a) the expected direct material cost, (b) the direct labor cost, (c) the prime cost, (d) the burden as a function of production and direct labor, and (e) the total cost.

Solution

(a) The direct material cost was given as $4.54.

(b) The direct labor cost is

$$\left(\frac{16}{60}\right)(\$7.50) = \$2.00$$

(c) The prime cost is

$$\$4.54 + \$2.00 = \$6.54$$

(d) If the burden vehicle is production, the burden rate is $45,000/8000 = $5.63 per item, making the total cost

$$\$4.54 + \$2.00 + \$5.63 = \$12.17$$

(e) If the burden vehicle is direct labor hours, the burden rate is $45,000/2133 = $21.10 per hour, making the total cost

$$\$4.54 + \$2.00 + \left(\frac{16}{60}\right)(\$21.10) = \$12.17$$

Example 69.30

The actual performance of the company in Ex. 69.29 is given by the following figures.

$$\text{actual production: } 7560$$
$$\text{actual overhead costs: } \$47,000$$

What are the burden budget variance and the burden capacity variance?

Solution

The burden capacity variance is

$$\$45,000 - (7560)(\$5.63) = \$2437$$

The burden budget variance is

$$\$47,000 - \$45,000 = \$2000$$

The overall burden variance is

$$\$47,000 - (7560)(\$5.63) = \$4437$$

The sum of the burden capacity and burden budget variances should equal the overall burden variance.

$$\$2437 + \$2000 = \$4437$$

52. COST OF GOODS SOLD

Cost of goods sold (COGS) is an accounting term that represents an inventory account adjustment.[49] Cost of goods sold is the difference between the starting and ending inventory valuations. That is,

$$\begin{aligned} \text{COGS} &= \text{starting inventory valuation} \\ &\quad - \text{ending inventory valuation} \end{aligned} \quad \text{69.57}$$

Cost of goods sold is subtracted from *gross profit* to determine the *net profit* of a company. Despite the fact that cost of goods sold can be a significant element in the profit equation, the inventory adjustment may not be made each accounting period (e.g., each month) due to the difficulty in obtaining an accurate inventory valuation.

With a *perpetual inventory system*, a company automatically maintains up-to-date inventory records, either through an efficient stocking and stock-releasing system, or through a *point of sale* (POS) *system* integrated with the inventory records. If a company only counts its inventory (i.e., takes a *physical inventory*) at regular intervals (e.g., once a year), it is said to be operating on a *periodic inventory system*.

Inventory accounting is a source of many difficulties. The inventory value is calculated by multiplying the quantity on hand by the standard cost. In the case of completed items actually assembled or manufactured at the company, this standard cost usually is the manufacturing cost, although factory cost also can be used. In the case of purchased items, the standard cost will be the cost per item charged by the supplying vendor. In some cases, delivery and transportation costs will be included in this standard cost.

It is not unusual for the elements in an item's inventory to come from more than one vendor, or from one vendor in more than one order. Inventory valuation is more difficult if the price paid is different for these different purchases. There are four methods of determining the cost of elements in inventory. Any of these methods can be used (if applicable), but the method must be used consistently from year to year. The four methods are as follows.

- *specific identification method:* Each element can be uniquely associated with a cost. Inventory elements with serial numbers fit into this costing scheme. Stock, production, and sales records must include the serial number.

- *average cost method:* The standard cost of an item is the average of (recent or all) purchase costs for that item.

- *first-in, first-out* (FIFO) *method:* This method keeps track of how many of each item were purchased each time and the number remaining out of each purchase, as well as the price paid at each purchase. The inventory system assumes that the oldest elements are issued first.[50] Inventory value is a weighted average dependent on the number of elements from each purchase remaining. Items issued no longer contribute to the inventory value.

- *last-in, first-out* (LIFO) *method:* This method keeps track of how many of each item were purchased each time and the number remaining out of each purchase, as well as the price paid at each purchase.[51] The inventory value is a weighted average dependent on the number of elements from each purchase remaining. Items issued no longer contribute to the inventory value.

53. BREAK-EVEN ANALYSIS

Special Nomenclature

f	fixed cost that does not vary with production
a	*incremental cost* to produce one additional item (also called *marginal cost* or *differential cost*)
Q	quantity sold
p	*incremental value* (price)
R	total revenue
C	total cost

Break-even analysis is a method of determining when the value of one alternative becomes equal to the value of another. A common application is that of determining when costs exactly equal revenue. If the manufactured quantity is less than the break-even quantity, a loss is incurred. If the manufactured quantity is greater than the break-even quantity, a profit is made.

---

[49]The cost of goods sold inventory adjustment is posted to the *COGS expense account*.

[50]If all elements in an item's inventory are identical, and if all shipments of that item are agglomerated, there will be no way to guarantee that the oldest element in inventory is issued first. But, unless *spoilage* is a problem, it really does not matter.

[51]See Ftn. 50.

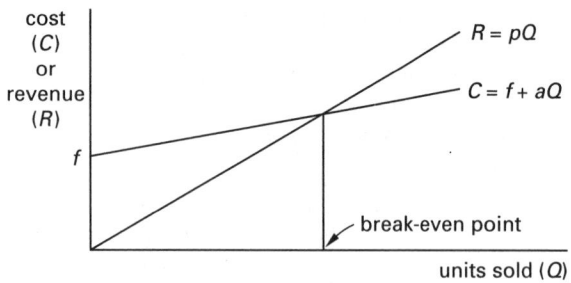

Figure 69.13 *Break-Even Quantity*

Assuming no change in the inventory, the *break-even point* can be found by setting costs equal to revenue $(C = R)$.

$$C = f + aQ \qquad 69.58$$

$$R = pQ \qquad 69.59$$

$$Q^* = \frac{f}{p - a} \qquad 69.60$$

An alternative form of the break-even problem is to find the number of units per period for which two alternatives have the same total costs. Fixed costs are to be spread over a period longer than one year using the equivalent uniform annual cost (EUAC) concept. One of the alternatives will have a lower cost if production is less than the break-even point. The other will have a lower cost for production greater than the break-even point.

Example 69.31

Two plans are available for a company to obtain automobiles for its salesmen. How many miles must the cars be driven each year for the two plans to have the same costs? Use an interest rate of 10%. (Use the year-end convention for all costs.)

plan A: Lease the cars and pay $0.15 per mile.

plan B: Purchase the cars for $5000. Each car has an economic life of three years, after which it can be sold for $1200. Gas and oil cost $0.04 per mile. Insurance is $500 per year.

Solution

Let x be the number of miles driven per year. Then, the EUAC for both alternatives is

$$\text{EUAC(A)} = 0.15x$$

$$\begin{aligned}\text{EUAC(B)} &= 0.04x + \$500 + (\$5000)(A/P, 10\%, 3) \\ &\quad - (\$1200)(A/F, 10\%, 3) \\ &= 0.04x + \$500 + (\$5000)(0.4021) \\ &\quad - (\$1200)(0.3021) \\ &= 0.04x + 2148\end{aligned}$$

Setting EUAC(A) and EUAC(B) equal and solving for x yields 19,527 miles per year as the break-even point.

54. PAY-BACK PERIOD

The *pay-back period* is defined as the length of time, usually in years, for the cumulative net annual profit to equal the initial investment. It is tempting to introduce equivalence into pay-back period calculations, but by convention, this is generally not done.[52]

$$\text{pay-back period} = \frac{\text{initial investment}}{\text{net annual profit}} \qquad 69.61$$

Example 69.32

A ski resort installs two new ski lifts at a total cost of $1,800,000. The resort expects the annual gross revenue to increase by $500,000 while it incurs an annual expense of $50,000 for lift operation and maintenance. What is the pay-back period?

Solution

From Eq. 69.61,

$$\text{pay-back period} = \frac{\$1,800,000}{\$500,000 - \$50,000} = 4\,\text{years}$$

55. MANAGEMENT GOALS

Depending on many factors (market position, age of the company, age of the industry, perceived marketing and sales windows, etc.), a company may select one of many production and marketing strategic goals. Three such strategic goals are

- maximization of product demand

- minimization of cost

- maximization of profit

Such goals require knowledge of how the dependent variable (e.g., demand quantity or quantity sold) varies as a function of the independent variable (e.g., price). Unfortunately, these three goals are not usually satisfied simultaneously. For example, minimization of product cost may require a large production run to realize economies of scale, while the actual demand is too small to take advantage of such economies of scale.

---

[52]Equivalence (i.e., interest and compounding) generally is not considered when calculating the "pay-back period." However, if it is desirable to include equivalence, then the term *pay-back period* should not be used. Other terms, such as *cost recovery period* or *life of an equivalent investment*, should be used. Unfortunately, this convention is not always followed in practice.

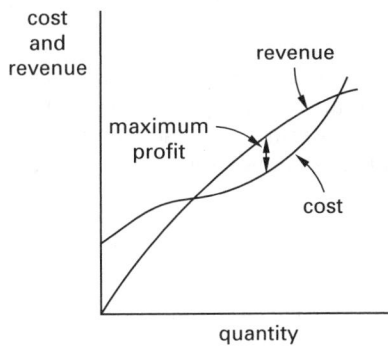

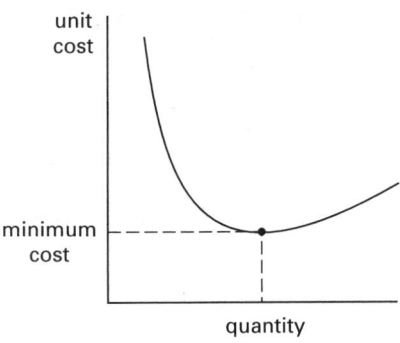

Figure 69.14 *Graphs of Management Goal Functions*

If sufficient data are available to plot the independent and dependent variables, it may be possible to optimize the dependent variable graphically. Of course, if the relationship between independent and dependent variables is known algebraically, the dependent variable can be optimized by taking derivatives or by use of other numerical methods.

56. INFLATION

It is important to perform economic studies in terms of *constant value dollars*. One method of converting all cash flows to constant value dollars is to divide the flows by some annual *economic indicator* or price index.

If indicators are not available, cash flows can be adjusted by assuming that inflation is constant at a decimal rate (e) per year. Then, all cash flows can be converted to $t = 0$ dollars by dividing by $(1 + e)^n$, where n is the year of the cash flow.

An alternative is to replace the effective annual interest rate (i) with a value corrected for inflation. This corrected value (i') is

$$i' = i + e + ie \qquad \textit{69.62}$$

This method has the advantage of simplifying the calculations. However, precalculated factors are not available for the non-integer values of i'. Therefore, Table 69.1 must be used to calculate the factors.

Example 69.33

What is the uninflated present worth of a $2000 future value in two years if the average inflation rate is 6% and i is 10%?

Solution

$$P = \frac{\$2000}{(1.10)^2(1.06)^2} = \$1471$$

Example 69.34

Repeat Ex. 69.33 using i'.

Solution

$$i' = 0.10 + 0.06 + (0.10)(0.06) = 0.166$$

$$P = \frac{\$2000}{(1.166)^2} = \$1471$$

57. CONSUMER LOANS

Special Nomenclature

BAL_j	balance after the jth payment
LV	principal total value loaned (cost minus down payment)
j	payment or period number
N	total number of payments to pay off the loan
PI_j	jth interest payment
PP_j	jth principal payment
PT_j	jth total payment
ϕ	effective rate per period (r/k)

Many different arrangements can be made between a borrower and a lender. With the advent of creative financing concepts, it often seems that there are as many variations of loans as there are loans made. Nevertheless, there are several traditional types of transactions. Real estate or investment texts, or a financial consultant, should be consulted for more complex problems.

Simple Interest

Interest due does not compound with a *simple interest loan*. The interest due is merely proportional to the length of time that the principal is outstanding. Because of this, simple interest loans are seldom made for long periods (e.g., more than one year). (For loans less than one year, it is commonly assumed that a year consists of 12 months of 30 days each.)

Example 69.35

A $12,000 simple interest loan is taken out at 16% per annum interest rate. The loan matures in two years with no intermediate payments. How much will be due at the end of the second year?

PROFESSIONAL PUBLICATIONS, INC. BELMONT, CA

Economics

Solution

The interest each year is

$$PI = (0.16)(\$12{,}000) = \$1920$$

The total amount due in two years is

$$PT = \$12{,}000 + (2)(\$1920) = \$15{,}840$$

Example 69.36

$4000 is borrowed for 75 days at 16% per annum simple interest. How much will be due at the end of 75 days?

Solution

$$\text{amount due} = \$4000 + (0.16)\left(\frac{75}{360}\right)(\$4000) = \$4133$$

Loans with Constant Amount Paid Toward Principal

With this loan type, the payment is not the same each period. The amount paid toward the principal is constant, but the interest varies from period to period. The equations that govern this type of loan are

$$BAL_j = LV - (j)(PP) \qquad \text{69.63}$$

$$PI_j = \phi(BAL)_{j-1} \qquad \text{69.64}$$

$$PT_j = PP + PI_j \qquad \text{69.65}$$

$$PP = \frac{LV}{N} \qquad \text{69.66}$$

$$N = \frac{LV}{PP} \qquad \text{69.67}$$

$$LV = (PP + PI_1)(P/A, \phi, N) \\ - PI_N(P/G, \phi, N) \qquad \text{69.68}$$

$$1 = \left(\frac{1}{N} + \phi\right)(P/A, \phi, N) \\ - \left(\frac{\phi}{N}\right)(P/G, \phi, N) \qquad \text{69.69}$$

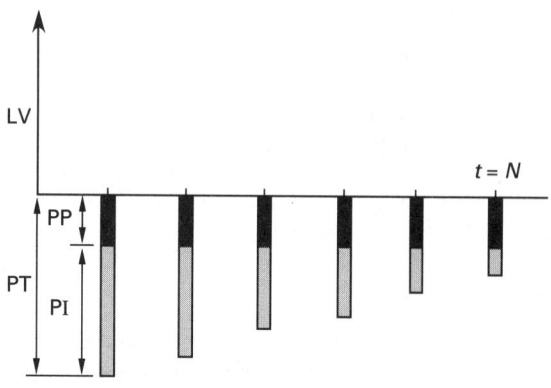

Figure 69.15 *Loan with Constant Amount Paid Toward Principal*

Example 69.37

A $12,000 six-year loan is taken from a bank that charges 15% effective annual interest. Payments toward the principal are uniform, and repayments are made at the end of each year. Tabulate the interest, total payments, and the balance remaining after each payment is made.

Solution

The amount of each principal payment is

$$PP = \frac{\$12{,}000}{6} = \$2000$$

At the end of the first year (before the first payment is made), the principal balance is $12,000 (i.e., $BAL_0 = \$12{,}000$). From Eq. 69.64, the interest payment is

$$PI_1 = (0.15)(\$12{,}000) = \$1800$$

The total first payment is

$$PT_1 = PP + PI = \$2000 + \$1800 \\ = \$3800$$

The following table is similarly constructed.

j	BAL_j	PP_j	PI_j	PT_j
		(in dollars)		
0	12,000	–	–	–
1	10,000	2000	1800	3800
2	8000	2000	1500	3500
3	6000	2000	1200	3200
4	4000	2000	900	2900
5	2000	2000	600	2600
6	0	2000	300	2300

Direct Reduction Loans

This is the typical "interest paid on unpaid balance" loan. The amount of the periodic payment is constant, but the amounts paid toward the principal and interest both vary.

$$BAL_{j-1} = PT\left(\frac{1 - (1+\phi)^{j-1-N}}{\phi}\right) \qquad \text{69.70}$$

$$PI_j = \phi(BAL)_{j-1} \qquad \text{69.71}$$

$$PP_j = PT - PI_j \qquad \text{69.72}$$

$$BAL_j = BAL_{j-1} - PP_j \qquad \text{69.73}$$

$$N = \frac{-\ln\left(1 - \frac{\phi(LV)}{PT}\right)}{\ln(1+\phi)} \qquad \text{69.74}$$

Equation 69.74 calculates the number of payments necessary to pay off a loan. This equation can be solved with effort for the total periodic payment (PT) or the initial value of the loan (LV). It is easier, however, to use the $(A/P, i\%, n)$ factor to find the payment and loan value.

$$PT = LV(A/P, \phi\%, N) \qquad \textit{69.75}$$

If the loan is repaid in yearly installments, then i is the effective annual rate. If the loan is paid off monthly, then i should be replaced by the effective rate per month (ϕ from Eq. 69.54). For monthly payments, N is the number of months in the loan period.

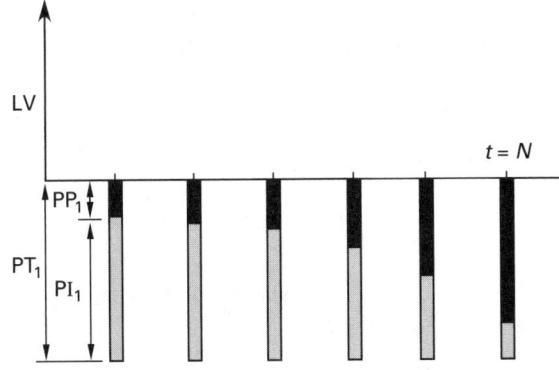

Figure 69.16 Direct Reduction Loan

Example 69.38

A $45,000 loan is financed at 9.25% per annum. The monthly payment is $385. What are the amounts paid toward interest and principal in the 14th period? What is the remaining principal balance after the 14th payment has been made?

Solution

The effective rate per month is

$$\phi = \frac{r}{k} = \frac{0.0925}{12}$$
$$= 0.0077083\ldots \quad [\text{say, } 0.007708]$$

$$N = \frac{-\ln\left(1 - \frac{(0.007708)(45,000)}{385}\right)}{\ln(1 + 0.007708)} = 301$$

$$BAL_{13} = (\$385)\left(\frac{1 - (1 + 0.007708)^{14-1-301}}{0.007708}\right)$$
$$= \$44,476.39$$

$$PI_{14} = (0.007708)(\$44,476.39) = \$342.82$$

$$PP_{14} = \$385 - \$342.82 = \$42.18$$

$$BAL_{14} = \$44,476.39 - \$42.18 = \$44,434.21$$

Direct Reduction Loans with Balloon Payments

This type of loan has a constant periodic payment, but the duration of the loan is insufficient to completely pay back the principal (i.e, the loan is not fully amortized). Therefore, all remaining unpaid principal must be paid back in a lump sum when the loan matures. This large payment is known as a *balloon payment*.[53]

Equations 69.71 through 69.75 also can be used with this type of loan. The remaining balance after the last payment is the balloon payment. This balloon payment must be repaid along with the last regular payment calculated.

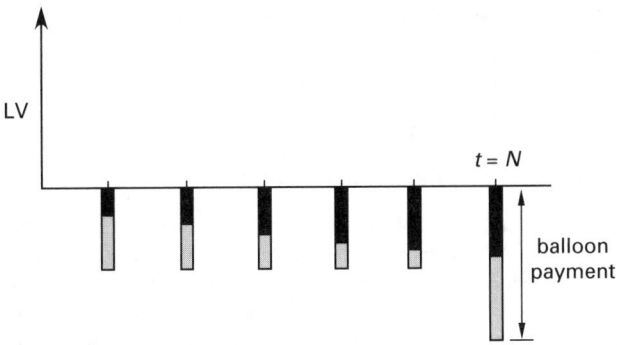

Figure 69.17 Direct Reduction Loan with Balloon Payment

58. FORECASTING

There are many types of forecasting models, although most are variations of the basic types.[54] All models produce a *forecast* (F_{t+1}) of some quantity (*demand* is used in this section) in the next period based on actual measurements (D_j) in current and prior periods. All of the models also try to provide *smoothing* (or *damping*) of extreme data points.

Forecasts by Moving Averages

The method of *moving average forecasting* weights all previous demand data points equally and provides some smoothing of extreme data points. The amount of smoothing increases as the number of data points, n, increases.

$$F_{t+1} = \frac{1}{n} \sum_{m=t+1-n}^{t} D_m \qquad \textit{69.76}$$

---

[53] The term *balloon payment* may include the final interest payment as well. Generally, the problem statement will indicate whether the balloon payment is inclusive or exclusive of the regular payment made at the end of the loan period.

[54] For example, forecasting models that take into consideration steady (linear), cyclical, annual, and seasonal trends are typically variations of the exponentially weighted model. A truly different forecasting tool, however, is *Monte Carlo simulation*.

Forecasts by Exponentially Weighted Averages

With *exponentially weighted forecasts*, the more current (most recent) data points receive more weight. This method uses a *weighting factor* (α) also known as a *smoothing coefficient*, which typically varies between 0.01 and 0.30. An initial forecast is needed to start the method. Forecasts immediately following are sensitive to the accuracy of this first forecast. It is common to choose $F_0 = D_1$ to get started.

$$F_{t+1} = \alpha D_t + (1 - \alpha)F_t \qquad 69.77$$

59. LEARNING CURVES

Special Nomenclature

R	decimal learning curve rate (2^{-b})
T_1	time or cost for the first item
T_n	time or cost for the nth item
n	total number of items produced
b	learning curve constant

The more products that are made, the more efficient the operation becomes due to experience gained. Therefore, direct labor costs decrease.[55] Usually, a *learning curve* is specified by the decrease in cost each time the cumulative quantity produced doubles. If there is a 20% decrease per doubling, the curve is said to be an 80% learning curve (i.e., the *learning curve rate*, R, is 80%).

Then, the time to produce the nth item is

$$T_n = T_1 n^{-b} \qquad 69.78$$

The total time to produce units from quantity n_1 to n_2 inclusive is approximately given by Eq. 69.79. T_1 is a constant, the time for item 1, and does not correspond to n unless $n_1 = 1$.

$$\int_{n_1}^{n_2} T_n \, dn \approx \left(\frac{T_1}{1-b}\right)\left(\left(n_2 + \tfrac{1}{2}\right)^{1-b} - \left(n_1 - \tfrac{1}{2}\right)^{1-b}\right)$$

$$69.79$$

The *average time per unit* over the production from n_1 to n_2 is the above total time from Eq. 69.79 divided by the quantity produced, $(n_2 - n_1 + 1)$.

$$T_{\text{ave}} = \frac{\displaystyle\int_{n_1}^{n_2} T_n \, dn}{n_2 - n_1 + 1} \qquad 69.80$$

Table 69.9 lists representative values of the *learning curve constant* (b). For learning curve rates not listed in the table, Eq. 69.81 can be used to find b.

$$b = \frac{-\log_{10} R}{\log_{10}(2)} = \frac{-\log_{10} R}{0.301} \qquad 69.81$$

---

[55]Remember that learning curve reductions apply only to direct labor costs. They are not applied to indirect labor or direct material costs.

Table 69.9 Learning Curve Constants

learning curve rate (R)	b
0.70 (70%)	0.515
0.75 (75%)	0.415
0.80 (80%)	0.322
0.85 (85%)	0.234
0.90 (90%)	0.152
0.95 (95%)	0.074

Example 69.39

A 70% learning curve is used with an item whose first production time is 1.47 hr. (a) How long will it take to produce the 11th item? (b) How long will it take to produce the 11th through 27th items?

Solution

(a) From Eq. 69.78,

$$T_{11} = (1.47 \text{ hr})(11)^{-0.515} = 0.428 \text{ hr}$$

(b) The time to produce the 11th item through 27th item is given by Eq. 69.79.

$$T \approx \left(\frac{1.47 \text{ hr}}{1 - 0.515}\right)\left[(27.5)^{1-0.515} - (10.5)^{1-0.515}\right]$$

$$= 5.643 \text{ hr}$$

60. ECONOMIC ORDER QUANTITY

Special Nomenclature

a	constant depletion rate (items/unit time)
h	inventory storage cost ($/item-unit time)
H	total inventory storage cost between orders ($)
K	fixed cost of placing an order ($)
Q	order quantity (original quantity on hand)

The *economic order quantity* (EOQ) is the order quantity that minimizes the inventory costs per unit time. Although there are many different EOQ models, the simplest is based on the following assumptions.

- Reordering is instantaneous. The time between order placement and receipt is zero.

- Shortages are not allowed.

- Demand for the inventory item is deterministic (i.e., is not a random variable).

- Demand is constant with respect to time.

- An order is placed when the inventory is zero.

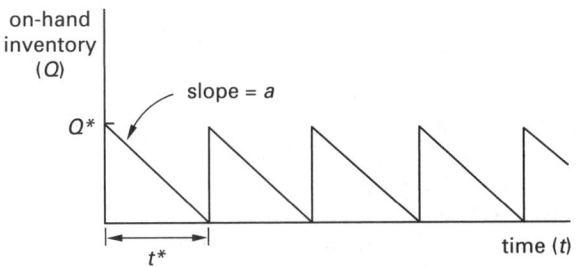

Figure 69.18 *Inventory with Instantaneous Reorder*

If the original quantity on hand is Q, the stock will be depleted at

$$t^* = \frac{Q}{a} \qquad \text{69.82}$$

The total inventory storage cost between t_0 and t^* is

$$H = \tfrac{1}{2}Qht^* = \frac{Q^2 h}{2a} \qquad \text{69.83}$$

The total inventory and ordering cost per unit time is

$$C_t = \frac{aK}{Q} + \frac{hQ}{2} \qquad \text{69.84}$$

C_t can be minimized with respect to Q. The economic order quantity and time between orders are

$$Q^* = \sqrt{\frac{2aK}{h}} \qquad \text{69.85}$$

$$t^* = \frac{Q^*}{a} \qquad \text{69.86}$$

61. SENSITIVITY ANALYSIS

Data analysis and forecasts in economic studies require estimates of costs that will occur in the future. There are always uncertainties about these costs. However, these uncertainties are insufficient reason not to make the best possible estimates of the costs. Nevertheless, a decision between alternatives often can be made more confidently if it is known whether or not the conclusion is sensitive to moderate changes in data forecasts. Sensitivity analysis provides this extra dimension to an economic analysis.

The sensitivity of a decision is determined by inserting a range of estimates for critical cash flows and other parameters. If radical changes can be made to a cash flow without changing the decision, the decision is said to be *insensitive* to uncertainties regarding that cash flow. However, if a small change in the estimate of a cash flow will alter the decision, that decision is said to be very *sensitive* to changes in the estimate. If the decision is sensitive only for a limited range of cash flow values, the term *variable sensitivity* is used. Figure 69.19 illustrates these terms.

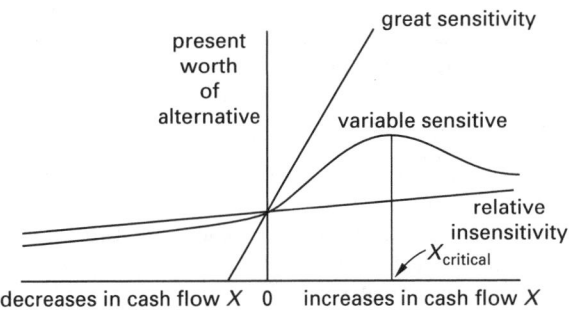

Figure 69.19 *Types of Sensitivity*

An established semantic tradition distinguishes between risk analysis and uncertainty analysis. *Risk analysis* addresses variables that have a known or estimated probability distribution. In this regard, statistics and probability theory can be used to determine the probability of a cash flow varying between given limits. On the other hand, *uncertainty analysis* is concerned with situations in which there is not enough information to determine the probability or frequency distribution for the variables involved.

As a first step, sensitivity analysis should be applied one at a time to the dominant factors. Dominant cost factors are those that have the most significant impact on the present value of the alternative.[56] If warranted, additional investigation can be used to determine the sensitivity to several cash flows varying simultaneously. Significant judgment is needed, however, to successfully determine the proper combinations of cash flows to vary. It is common to plot the dependency of the present value on the cash flow being varied in a two-dimensional graph. Simple linear interpolation is used (within reason) to determine the critical value of the cash flow being varied.

62. VALUE ENGINEERING

The *value* of an investment is defined as the ratio of its return (performance or utility) to its cost (effort or investment). The basic object of *value engineering* (VE, also referred to as *value analysis*) is to obtain the maximum per-unit value.[57]

Value engineering concepts often are used to reduce the cost of mass-produced manufactured products. This is done by eliminating unnecessary, redundant, or superfluous features, by redesigning the product for a less expensive manufacturing method, and by including features for easier assembly without sacrificing utility and

---

[56]In particular, engineering economic analysis problems are sensitive to the choice of effective interest rate (i), and to accuracy in cash flows at or near the beginning of the horizon. The problems will be less sensitive to accuracy in far-future cash flows, such as salvage value and subsequent generation replacement costs.
[57]Value analysis, the methodology that has become today's value engineering, was developed in the early 1950s by Lawrence D. Miles, an analyst at General Electric.

function.[58] However, the concepts are equally applicable to one-time investments, such as buildings, chemical processing plants, and space vehicles. In particular, value engineering has become an important element in all federally funded work.[59]

Typical examples of large-scale value engineering work are using stock-sized bearings and motors (instead of custom manufactured units), replacing rectangular concrete columns with round columns (which are easier to form), and substituting custom buildings with prefabricated structures.

Value engineering is usually a team effort. And, while the original designers may be on the team, usually outside consultants are utilized. The cost of value engineering is usually returned many times over by reduced construction and life-cycle costs.

PRACTICE PROBLEMS

1. At 6% effective annual interest, how much will be accumulated if $1000 is invested for ten years?

2. At 6% effective annual interest, what is the present worth of $2000 that becomes available in four years?

3. At 6% effective annual interest, how much should be invested to accumulate $2000 in 20 years?

4. At 6% effective annual interest, what year-end annual amount deposited over seven years is equivalent to $500 invested now?

5. At 6% effective annual interest, what will be the accumulated amount at the end of ten years if $50 is invested at the end of each year for ten years?

6. At 6% effective annual interest, how much should be deposited at the start of each year for ten years in order to empty the fund by drawing out $200 at the end of each year for ten years?

7. At 6% effective annual interest, how much should be deposited at the start of each year for five years to accumulate $2000 on the date of the last deposit?

8. At 6% effective annual interest, how much will be accumulated in ten years if three payments of $100 are deposited every other year for four years, with the first payment occurring at $t = 0$?

9. $500 is compounded monthly at a 6% effective annual interest rate. How much will have accumulated in five years?

---

[58]Some people say that value engineering is "the act of going over the plans and taking out everything that is interesting."

[59]U.S. Government Office of Management and Budget Circular A-131 outlines value engineering for federally funded construction projects.

10. What is the effective annual rate of return on an $80 investment that pays back $120 in seven years?

11. A new machine will cost $17,000 and will have a resale value of $14,000 after five years. Special tooling will cost $5000. The tooling will have a resale value of $2500 after five years. Maintenance will be $2000 per year. The effective annual interest rate is 6%. What will be the average annual cost of ownership during the next five years?

12. An old covered wooden bridge can be strengthened at a cost of $9000, or it can be replaced for $40,000. The present salvage value of the old bridge is $13,000. It is estimated that the reinforced bridge will last for 20 years, will have an annual cost of $500, and will have a salvage value of $10,000 at the end of 20 years. The estimated salvage value of the new bridge after 25 years is $15,000. Maintenance for the new bridge would cost $100 annually. The effective annual interest rate is 8%. Which is the best alternative?

13. A firm expects to receive $32,000 each year for 15 years from sales of a product. An initial investment of $150,000 will be required to manufacture the product. Expenses will run $7530 per year. Salvage value is zero, and straight-line depreciation is used. The income tax rate is 48%. What is the after-tax rate of return?

14. A public works project has initial costs of $1,000,000, benefits of $1,500,000, and disbenefits of $300,000. (a) What is the benefit/cost ratio? (b) What is the excess of benefits over costs?

15. A speculator in land pays $14,000 for property that he expects to hold for ten years. $1000 is spent in renovation, and a monthly rent of $75 is collected from the tenants. (Use the year-end convention.) Taxes are $150 per year, and maintenance costs are $250 per year. What must be the sale price in ten years to realize a 10% rate of return?

16. What is the effective annual interest rate for a payment plan of 30 equal payments of $89.30 per month when a lump sum payment of $2000 would have been an outright purchase?

17. A depreciable item is purchased for $500,000. The salvage value at the end of 25 years is estimated at $100,000. What is the depreciation in each of the first three years using the (a) straight line, (b) sum-of-the-years' digits, and (c) double-declining balance methods?

18. Equipment that is purchased for $12,000 now is expected to be sold after ten years for $2000. The estimated maintenance is $1000 for the first year, but it is expected to increase $200 each year thereafter. The effective annual interest rate is 10%. What are the (a) present worth and (b) annual cost?

19. A new grain combine with a 20-year life can remove seven pounds of rocks from its harvest per hour. Any rocks left in its output hopper will cause $25,000 damage in subsequent processes. Several investments are available to increase the rock-removal capacity, as listed in the table. The effective annual interest rate is 10%. What should be done?

rock removal rate	probability of exceeding rock removal rate	required investment to achieve removal rate
7	0.15	0
8	0.10	$15,000
9	0.07	$20,000
10	0.03	$30,000

20. (*Time limit: one hour*) A mechanism that costs $10,000 has operating costs and salvage values as given. An effective annual interest rate of 20% is to be used.

year	operating cost	salvage value
1	$2000	$8000
2	$3000	$7000
3	$4000	$6000
4	$5000	$5000
5	$6000	$4000

(a) What is the economic life of the mechanism? (b) Assuming that the mechanism has been owned and operated for four years already, what is the cost of owning and operating the mechanism for one more year?

21. (*Time limit: one hour*) A salesperson intends to purchase a car for $50,000 for personal use, driving 15,000 miles per year. Insurance for personal use costs $2000 per year, and maintenance costs $1500 per year. The car gets 15 miles per gallon, and gasoline costs $1.50 per gallon. The resale value after five years will be $10,000. The salesperson's employer has asked that the car be used for business driving of 50,000 miles per year and has offered a reimbursement of $0.30 per mile. Using the car for business would increase the insurance cost to $3000 per year and maintenance to $2000 per year. The salvage value after five years would be reduced to $5000. If the employer purchased a car for the salesperson to use, the initial cost would be the same, but insurance, maintenance, and salvage would be $2500, $2000, and $8000, respectively. The salesperson's effective annual interest rate is 10%. (a) Is the reimbursement offer adequate? (b) With a reimbursement of $0.30 per mile, how many miles must the car be driven per year to justify the employer buying the car for the salesperson to use?

22. (*Time limit: one hour*) Alternatives A and B are being evaluated. The effective annual interest rate is 10%. What alternative is economically superior?

	alternative A	alternative B
first cost	$80,000	$35,000
life	20 years	10 years
salvage value	$7000	0
annual costs		
years 1–5	$1000	$3000
years 6–10	$1500	$4000
years 11–20	$2000	0
additional cost		
year 10	$5000	0

23. (*Time limit: one hour*) A car is needed for three years. Plans A and B for acquiring the car are being evaluated. An effective annual interest rate of 10% is to be used. Which plan is economically superior?

plan A: lease the car for $0.25/mile (all inclusive)

plan B: purchase the car for $30,000
keep the car for three years
sell the car after three years for $7200
pay $0.14 per mile for oil and gas
pay other costs of $500 per year

24. (*Time limit: one hour*) Two methods are being considered to meet strict air pollution control requirements over the next ten years. Method A uses equipment with a life of ten years. Method B uses equipment with a life of five years that will be replaced with new equipment with an additional life of five years. Capacities of the two methods are different, but operating costs do not depend on the throughput. Operation is 24 hours per day, 365 days per year. The effective annual interest rate for this evaluation is 7%.

	method A	method B	
	years 1–10	years 1–5	years 6–10
installation cost	$13,000	$6000	$7000
equipment cost	$10,000	$2000	$2200
operating cost			
per hour	$10.50	$8.00	$8.00
salvage value	$5000	$2000	$2000
capacity (tons/yr)	50	20	20
life	10 years	5 years	5 years

(a) What is the uniform annual cost per ton for each method? (b) Over what range of throughput (in tons/yr) does each method have the minimum cost?

25. (*Time limit: one hour*) A transit district has asked for your assistance in determining the proper fare for its bus system. An effective annual interest rate of 7% is to be used. The following additional information was compiled for your study.

cost per bus	$60,000
bus life	20 years
salvage value	$10,000
miles driven per year	37,440
number of passengers per year	80,000
operating cost	$1.00 per mile in the first year, increasing $0.10 per mile each year thereafter

(a) If the fare is to remain constant for the next 20 years, what is the break-even fare per passenger? (b) If the transit district decides to set the per-passenger fare at $0.35 for the first year, by what amount should the per-passenger fare go up each year thereafter such that the district can break even in 20 years? (c) If the transit district decides to set the per-passenger fare at $0.35 for the first year, and the per-passenger fare goes up $0.05 each year thereafter, what additional governmental subsidy (per passenger) is needed for the district to break even in 20 years?

70 Engineering Law[1]

1. Forms of Company Ownership 70-1
2. Sole Proprietorships 70-1
3. Partnerships 70-2
4. Corporations 70-2
5. Agency 70-2
6. General Contracts 70-2
7. Standard Boilerplate Clauses 70-3
8. Subcontracts 70-4
9. Parties to a Construction Contract 70-4
10. Standard Contracts for
    Design Professionals 70-4
11. Consulting Fee Structure 70-4
12. Discharge of a Contract 70-5
13. Torts 70-5
14. Breach of Contract, Negligence,
    Misrepresentation, and Fraud 70-5
15. Strict Liability in Tort 70-6
16. Manufacturing and Design Liability 70-6
17. Damages 70-7
18. Insurance 70-7

1. FORMS OF COMPANY OWNERSHIP

There are three basic forms of company ownership in the United States: (1) sole proprietorship, (2) partnership, and (3) corporation.[2] Each of these forms of ownership has advantages and disadvantages.

2. SOLE PROPRIETORSHIPS

A *sole proprietorship (single proprietorship)* is the easiest form of ownership to establish. Other than the necessary licenses (which apply to all forms of ownership), no legal formalities are required to start business operations. A sole proprietor (the owner) has virtually total control of the business and makes all supervisory and management decisions.

Legally, there is no distinction between the sole proprietor and the sole proprietorship (the business). This is the greatest disadvantage of this form of business. The owner is solely responsible for the operation of the business, even if the owner hires others for assistance. The owner assumes personal, legal, and financial liability for all acts and debts of the company. If the company debts remain unpaid, or in the event there is a legal judgment against the company, the owner's personal assets (home, car, savings, etc.) can be seized or attached.

Another disadvantage of the sole proprietorship is the lack of significant organizational structure. In times of business crisis or trouble, there may be no one to share the responsibility or to help make decisions. When the owner is sick or dies, there may be no way to continue the business.

There is also no distinction between the incomes of the business and the owner. Therefore, the business income is taxed at the owner's income tax rate. Depending on the owner's financial position, the success of the business, and the tax structure, this can be an advantage or a disadvantage.[3]

3. PARTNERSHIPS

A *partnership* (also known as a *general partnership*) is ownership by two or more persons known as *general partners*. Legally, this form is very similar to a sole proprietorship, and the two forms of business have many of the same advantages and disadvantages. For example, with the exception of an optional *partnership agreement*, there are a minimum of formalities to setting up business. The partners make all business and management decisions themselves according to an agreed-upon process. The business income is split among the partners and taxed at the partners' individual tax rates.[4] Continuity of the business is still a problem since most partnerships are automatically dissolved upon the withdrawal or death of one of the partners.[5]

One advantage of a partnership over a sole proprietorship is the increase in available funding. Not only do more partners bring in more start-up capital, but the resource pool may make business credit easier to obtain. Also, the partners bring a diversity of skills and talents.

---

[1]The author is not giving legal advice in this chapter, nor is this chapter intended to be a substitute for professional advice. Law is not always black and white. For every rule there are exceptions. For every legal principle, there are variations. For every type of injury, there are numerous legal precedents. This chapter covers the superficial basics of a small subset of U.S. law affecting engineers.

[2]The discussion of forms of company ownership in Sections 2, 3, and 4 applies equally to service-oriented companies (e.g., consulting engineering firms) and product-oriented companies.

[3]To use a simplistic example, if the corporate tax rates are higher than the individual tax rates, it would be *financially* better to be a sole proprietor because the company income would be taxed at a lower rate.

[4]The percentage split is specified in the partnership agreement.

[5]Some or all of the remaining partners may want to form a new partnership, but this is not always possible.

Law and Ethics

Unless the partnership agreement states otherwise, each partner can individually obligate (i.e., *bind*) the partnership without the consent of the other partners. Similarly, each partner has personal responsibility and liability for the acts and debts of the partnership company, just as sole proprietors do. In fact, each partner assumes the *sole* responsibility, not just a proportionate share. If one or more partners are unable to pay, the remaining partners shoulder the entire debt. The possibility of one partner having to pay for the actions of another partner must be considered when choosing this form of business ownership.

A *limited partnership* differs from a general partnership in that one or more of the partners is silent. The *limited partners* make a financial contribution to the business and receive a share of the profit but do not participate in the management and cannot bind the partnership. While *general partners* have unlimited personal liabilities, limited partners are generally liable only to the extent of their investment.[6] A written partnership agreement is required, and the agreement must be filed with the proper authorities.

4. CORPORATIONS

A corporation is a legal entity (i.e., a legal person) distinct from the founders and owners. The separation of ownership and management makes the corporation a fundamentally different kind of business form than a sole proprietorship or partnership, with very different advantages and disadvantages.

A corporation becomes legally distinct from its founders upon formation and proper registration. Ownership of the corporation is through shares of stock, which are distributed to the founders and investors according to some agreed-upon investment and distribution rule. Thus, the founders and investors become the stockholders (i.e., owners) of the corporation. A *closely held (private) corporation* is one in which all stock is owned by a family or small group of coinvestors. A *public corporation* is one whose stock is available for the public-at-large to purchase.

There is no mandatory connection between ownership and management functions. The decision-making power is vested in the executive officers and a *board of directors* that governs by majority vote. The stockholders elect the board of directors which, in turn, hires the executive officers, management, and other employees. Employees of the corporation may or may not be stockholders.

Disadvantages (at least for a person or persons who could form a partnership or sole proprietorship instead) include the higher corporate tax rate, difficulty in forming (some states require a minimum number of persons on the board of directors), and additional legal and accounting paperwork.

However, since a corporation is distinctly separate from its founders and investors, those individuals are not liable for the acts and debts of the corporation. Debts are paid from the corporate assets. Income to the corporation is not taxable income to the owners. (Only the salaries, if any, paid to the employees by the corporation are taxable to the employees.) Even if the corporation were to go bankrupt, the assets of the owners would not ordinarily be subject to seizure or attachment.

A corporation offers the best guarantee of continuity of operation in the event of the death, incapacity, or retirement of the founders since, as a legal entity, it is distinct from the founders and owners.

5. AGENCY

In some contracts, decision-making authority and right of action are transferred from one party (the owner, or *principal*) who would normally have that authority to another person (the *agent*). For example, in construction contracts, the engineer is ordinarily the agent of the owner. Agents are limited in what they can do by the scope of the agency agreement. Within that scope, however, an agent acts on behalf of the principal, and the principal is liable for the acts of the agent and is bound by contracts made in the principal's name by the agent.

Agents are required to execute their work with care, skill, and diligence. Specifically, agents have *fiduciary responsibility* toward their principal, meaning that agent must be honest and loyal. Agents will be liable for damages resulting from a lack of diligence, loyalty, and/or honesty. If the agents misrepresented their skills when obtaining the agency, they can be liable for breach of contract or fraud.

6. GENERAL CONTRACTS

A *contract* is a legally binding agreement or promise to exchange goods or services.[7] A written contract is merely a documentation of the agreement. Some agreements must be in writing, but most agreements for engineering services can be verbal, particularly if the parties to the agreement know each other well.[8] Written contract documents do not need to contain intimidating legal language, but all agreements must satisfy three basic requirements to be enforceable (binding).

---

[6]That is, if the partnership fails or is liquidated to pay debts, the limited partners lose no more than their initial investments.

[7]Not all agreements are legally binding (i.e., enforceable). Two parties may agree on something, but unless the agreement meets all of the requirements and conditions of a contract, the parties cannot hold each other to the agreement.

[8]All states have a *statute of frauds* that, among other things, specifies what types of contracts must be in writing to be enforceable. These include contracts for the sale of land, contracts requiring more than one year for performance, contracts for the sale of goods over $500 in value, contracts to satisfy the debts of another, and marriage contracts. Contracts to provide engineering services do not fall under the statute of frauds.

- There must be a clear, specific, and definite *offer* with no room for ambiguity or misunderstanding.

- There must be some form of conditional future *consideration* (i.e., payment).[9]

- There must be an *acceptance* of the offer.

There are other conditions that the agreement must meet to be enforceable. These conditions are not normally part of the explicit agreement but represent the conditions under which the agreement was made.

- The agreement must be *voluntary* for all parties.

- All parties must have *legal capacity* (i.e., be mentally competent, of legal age, and uninfluenced by drugs).

- The purpose of the agreement must be *legal.*

For small projects, a simple *letter of agreement* on one party's stationery may suffice. For larger, complex projects, a more formal document may be required. Some clients prefer to use a *purchase order*, which can function as a contract if all basic requirements are met.

Regardless of the format of the written document—letter of agreement, purchase order, or standard form—a contract should include the following features.[10]

- introduction, preamble, or preface indicating the purpose of the contract

- name, address, and business forms of both contracting parties

- signature date of the agreement

- effective date of the agreement (if different from the signature date)

- duties and obligations of both parties

- deadlines and required service dates

- fee amount

- fee schedule and payment terms

- agreement expiration date

- standard boilerplate clauses

- signatures of parties or their agents

- declaration of authority of the signatories to bind the contracting parties

- supporting documents

7. STANDARD BOILERPLATE CLAUSES

It is common for full-length contract documents to include important *boilerplate clauses.* These clauses have specific wordings that should not normally be changed, hence the name "boilerplate." Some of the most common boilerplate clauses are paraphrased here.

- Delays and inadequate performance due to war, strikes, and acts of God and nature are forgiven (*force majeure*).

- The contract document is the complete agreement, superseding all previous verbal and written agreements.

- The contract can be modified or canceled only in writing.

- Parts of the contract that are determined to be void or unenforceable shall not affect the enforceability of the remainder of the contract (*severability*). Alternatively, parts of the contract that are determined to be void or unenforceable shall be rewritten to accomplish their intended purpose without affecting the remainder of the contract.

- None (or one, or both) of the parties can (or cannot) assign its (or their) rights and responsibilities under the contract (*assignment*).

- All notices provided for in the agreement must be in writing and sent to the address in the agreement.

- Time is of the essence.[11]

- The subject headings of the agreement paragraphs are for convenience only and do not control the meaning of the paragraphs.

- The laws of the state in which the contract is signed must be used to interpret and govern the contract.

- Disagreements shall be arbitrated according to the rules of the American Arbitration Association.

- Any lawsuits related to the contract must be filed in the county and state in which the contract is signed.

---

[9]Actions taken or payments made prior to the agreement are irrelevant. Also, it does not matter to the courts whether the exchange is based on equal value or not.

[10]*Construction contracts* are unique unto themselves. Items that might also be included as part of the *contract documents* are the agreement form, the general conditions, drawings, specifications, and addenda.

[11]Without this clause in writing, damages for delay cannot be claimed.

- Obligations under the agreement are unique, and in the event of a breach, the defaulting party waives the defense that the loss can be adequately compensated by monetary damages (*specific performance*).

- In the event of a lawsuit, the prevailing party is entitled to an award of reasonable attorneys' and court fees.[12]

- Consequential damages are not recoverable in a lawsuit.

8. SUBCONTRACTS

When a party to a contract engages a third party to perform the work in the original contract, the contract with the third party is known as a *subcontract*. Whether or not responsibilities can be subcontracted under the original contract depends on the content of the *assignment clause* in the original contract.

9. PARTIES TO A CONSTRUCTION CONTRACT

A specific set of terms has developed for referring to parties in consulting and construction contracts. The *owner* of a construction project is the person, partnership, or corporation that actually owns the land, assumes the financial risk, and ends up with the completed project. The *developer* contracts with the architect and/or engineer for the design and with the contractors for the construction of the project. In some cases, the owner and developer are the same, in which case the term *owner-developer* can be used.

The *architect* designs the project according to established codes and guidelines but leaves most stress and capacity calculations to the *engineer*.[13] Depending on the construction contract, the engineer may work for the architect, or vice versa, or both may work for the developer.

Once there are approved plans, the developer hires *contractors* to do the construction. Usually, the entire construction project is awarded to a *general contractor*. Due to the nature of the construction industry, separate *subcontracts* are used for different tasks (electrical, plumbing, mechanical, framing, fire sprinkler installation, finishing, etc.). The general contractor who hires all of these different *subcontractors* is known as the *prime contractor* (or *prime*). (The subcontractors can also work directly for the owner-developer, although

this is less common.) The prime contractor is responsible for all of the acts of the subcontractors and is liable for any damage suffered by the owner-developer due to those acts.

Construction is managed by an agent of the owner-developer known as the *construction manager*, who may be the engineer, the architect, or someone else.

10. STANDARD CONTRACTS FOR DESIGN PROFESSIONALS

Several of the design engineering societies have produced standard agreement forms and other standard documents for design professionals.[14] Among other standard forms, notices, and agreements, the following standard contracts are available.[15]

- standard contract between engineer and client

- standard contract between engineer and architect

- standard contract between engineer and contractor

- standard contract between owner and construction manager

The major advantage of the standard contracts is that the meanings of the clauses are well established, not only among the design professionals and their clients but also in the courts. The clauses in these contracts have already been litigated many times. Where a clause has been found to be unclear or ambiguous, it has been rewritten to accomplish its intended purpose.

11. CONSULTING FEE STRUCTURE

Compensation for consulting engineering services can incorporate one or more of the following concepts.

- *lump-sum fee*: This is a predetermined fee agreed upon by client and engineer. This payment can be used for small projects where the scope of work is clearly defined.

- *cost plus fixed fee*: All costs (labor, material, travel, etc.) incurred by the engineer are paid by the client. The client also pays a predetermined fee as profit. This method has an advantage when the scope of services cannot be determined accurately in advance. Detailed records must be kept by the engineer in order to allocate costs among different clients.

---

[12]Without this clause in writing, attorneys' fees and court costs are rarely recoverable.

[13]On simple small projects, such as wood-framed residential units, the design may be developed by a *building designer*. The legal capacities of building designers vary from state to state.

[14]There are two main sources of standard forms. The American Consulting Engineers' Council (ACEC), National Society of Professional Engineers (NSPE), and American Society of Civil Engineers (ASCE) have produced one set. Working independently, the American Institute of Architects (AIA) and the Associated General Contractors of America (AGC) have produced another.

[15]The Construction Specifications Institute (CSI) has produced standard specifications for materials.

- *per diem fee*: The engineer is paid a specific sum for each day spent on the job. Usually, certain direct expenses (e.g., travel and reproduction) are billed in addition to the per diem rate.

- *salary plus*: The client pays for the employees on an engineer's payroll (the salary) plus an additional percentage to cover indirect overhead and profit plus certain direct expenses.

- *retainer*: This is a minimum amount paid by the client, usually in total and in advance, for a normal amount of work expected during an agreed-upon period. None of the retainer is returned, regardless of how little work the engineer performs. The engineer can be paid for additional work beyond what is normal, however. Some direct costs, such as travel and reproduction expenses, may be billed directly to the client.

- *percentage of construction cost*: This method, which is widely used in construction design contracts, pays the architect and/or the engineer a percentage of the final total cost of the project. Cost of land, financing, and legal fees are generally not included in the construction cost, and other costs (plan revisions, project management labor, value engineering, etc.) are billed separately.

12. DISCHARGE OF A CONTRACT

A contract is normally discharged when all parties have satisfied their obligations. However, a contract can also be terminated for the following reasons.

- mutual agreement of all parties to the contract

- impossibility of performance (e.g., death of a party to the contract)

- illegality of the contract

- material breach by one or more parties to the contract

- fraud on the part of one or more parties

- failure (i.e., loss or destruction) of consideration (e.g., the burning of a building one party expected to own or occupy upon satisfaction of the obligations)

Some contracts may be dissolved by actions of the court (e.g., bankruptcy), passage of new laws and public acts, or a declaration of war.

Extreme difficulty (including economic hardship) in satisfying the contract does not discharge it, even if it becomes more costly or less profitable than originally anticipated.

13. TORTS

A *tort* is a civil wrong committed by one person causing damage to another person or person's property, emotional well-being, or reputation.[16] It is a breach of the rights of an individual to be secure in person or property. In order to correct the wrong, a civil lawsuit (*tort action* or *civil complaint*) is brought by the alleged injured party (the *plaintiff*) against the *defendant*. To be a valid *tort action* (i.e., lawsuit), there must have been injury (i.e., damage). Generally, there will be no contract between the two parties, so the tort action cannot claim a breach of contract.[17]

Tort law is concerned with compensation for the injury, not punishment. Therefore, tort awards usually consist of general, compensatory, and special damages and rarely include punitive and exemplary damages. (See Sec. 17 for definitions of these damages.)

14. BREACH OF CONTRACT, NEGLIGENCE, MISREPRESENTATION, AND FRAUD

A *breach of contract* occurs when one of the parties fails to satisfy all of its obligations under a contract. The breach can be *willful* (as in a contractor walking off a construction job) or *unintentional* (as in providing less than adequate quality work or materials). A *material breach* is defined as nonperformance that results in the injured party receiving something substantially less than or different from what the contract intended.

Normally, the only redress that an *injured party* has through the courts in the event of a breach of contract is to force the breaching party to provide *specific performance*—that is, to satisfy all remaining contract provisions and to pay for any damage caused. Normally, *punitive damages* (to punish the breaching party) are unavailable.

Negligence is an action, willful or unwillful, taken without proper care or consideration for safety, resulting in damages to property or injury to persons. "Proper care" is a subjective term, but in general it is the diligence that would be exercised by a reasonably prudent person.[18] Damages sustained by a negligent act are recoverable in a tort action. (See Sec. 14.) If the plaintiff

---

[16]The difference between a *civil tort (lawsuit)* and a *criminal lawsuit* is the alleged injured party. A *crime* is a wrong against society. A criminal lawsuit is brought by the state against a defendant.

[17]It is possible for an injury to be both a breach of contract and a tort. Suppose an owner has an agreement with a contractor to construct a building, and the contract requires the contractor to comply with all state and federal safety regulations. If the owner is subsequently injured on a stairway because there was no guardrail, the injury could be recoverable both as a tort and as a breach of contract. If a third party unrelated to the contract was injured, however, that party could recover only through a tort action.

[18]Negligence of a design professional (e.g., an engineer or architect) is the absence of a *standard of care* (i.e., customary and normal care and attention) that would have been provided by other engineers. It is highly subjective.

was partially at fault (as in the case of *comparative negligence*), the defendant will be liable only for the portion of the damage caused by the defendant.

Punitive damages are available, however, if the breaching party was fraudulent in obtaining the contract. In addition, the injured party has the right to void (nullify) the contract entirely. A *fraudulent act* is basically a special case of misrepresentation (i.e., an intentionally false statement known to be false at the time it is made). Misrepresentation that does not result in a contract is a tort. When a contract is involved, misrepresentation can be a breach of that contract (i.e., *fraud*).

Unfortunately, it is extremely difficult to prove *compensatory fraud* (i.e., fraud for which damages are available). Proving fraud requires showing *beyond a reasonable doubt* (1) a reckless or intentional misstatement of a material fact (2) meant to deceive, (3) resulting in misleading the innocent party to contract (4) to the innocent party's detriment.

For example, if an engineer claims to have experience in designing steel buildings but actually has none, the court might consider the misrepresentation a fraudulent action. If, however, the engineer has some experience, but an insufficient amount to do an adequate job, the engineer probably will not be considered to have acted fraudulently.

15. STRICT LIABILITY IN TORT

Strict liability in tort means that the injured party wins if the injury can be proven. It is not necessary to prove negligence, breach of explicit or implicit warranty, or the existence of a contract (*privity of contract*). Strict liability in tort is most commonly encountered in product liability cases. A defect in a product, regardless of how the defect got there, is sufficient to create strict liability in tort.

Case law surrounding defective products has developed and refined the following requirements for winning a strict liability in tort case. The following points must be proved.

- The product was defective in manufacture, design, labeling, and so on.

- The product was defective when used.

- The defect rendered the product unreasonably dangerous.

- The defect caused the injury.

- The specific use of the product that caused the damage was reasonably foreseeable.

16. MANUFACTURING AND DESIGN LIABILITY

Case law makes a distinction between *design professionals* (architects, structural engineers, building designers, etc.) and manufacturers of consumer products. Design professionals are generally consultants whose primary product is a design service sold to sophisticated clients. Consumer product manufacturers produce a specific product line sold through wholesalers and retailers to the unsophisticated public.

The law treats design professionals favorably. Such professionals are expected to meet a *standard of care* and skill that can be measured by comparison with the conduct of other professionals. However, professionals are not expected to be infallible. In the absence of a contract provision to the contrary, design professionals are not held to be guarantors of their work in the strict sense of legal liability. Damages incurred due to design errors are recoverable through tort actions, but proving a breach of contract requires showing negligence (i.e., not meeting the standard of care).

On the other hand, the law is much stricter with consumer product manufacturers, and perfection is (essentially) expected of them. They are held to the standard of strict liability in tort without regard to negligence. A manufacturer is held liable for all phases of the design and manufacturing of a product being marketed to the public.[19]

Prior to 1916, the court's position toward product defects was exemplified by the expression *caveat emptor* ("let the buyer beware").[20] Subsequent court rulings have clarified that "... a manufacturer is strictly liable in tort when an article [it] places on the market, knowing that it will be used without inspection, proves to have a defect that causes injury to a human being."[21]

Although all defectively designed products can be traced back to a design engineer or team, only the manufacturing company is usually held liable for injury caused by the product. This is more a matter of economics than justice. The company has liability insurance; the product design engineer (who is merely an employee of the company) probably does not. Unless

---

[19]The reason for this is that the public is not considered to be as sophisticated as a client who contracts with a design professional for building plans.

[20]1916, *McPherson vs. Buick*. McPherson bought a Buick from a car dealer. The car had a defective wooden steering wheel, and there was evidence that reasonable inspection would have uncovered the defect. The steering wheel injured McPherson, who then sued Buick. Buick defended itself under the ancient *prerequisite of privity* (i.e., the requirement of a face-to-face contractual relationship in order for liability to exist), since the dealer, not Buick, had sold the car to McPherson, and no contract between Buick and McPherson existed. The judge disagreed, thus establishing the concept of *third party liability* (i.e., manufacturers are responsible to consumers even though consumers do not buy directly from manufacturers).

[21]1963, *Greenman vs. Yuba Power Products*. Greenman purchased and was injured by an electric power tool.

the product design or manufacturing process is intentionally defective, or unless the defect is known in advance and covered up, the product design engineer will rarely be punished by the courts.[22]

17. DAMAGES

An injured party can sue for *damages* as well as for specific performance. Damages are the award made by the court for losses incurred by the injured party.

- *General* or *compensatory damages* are awarded to make up for the injury that was sustained.

- *Special damages* are awarded for the direct financial loss due to the breach of contract.

- *Nominal damages* are awarded when responsibility has been established but the injury is so slight as to be inconsequential.

- *Liquidated damages* are amounts that are specified in the contract document itself for nonperformance.

- *Punitive* or *exemplary damages* are awarded, usually in tort and fraud cases, to punish and make an example of the defendant (i.e., to deter others from doing the same thing).

- *Consequential damages* provide compensation for indirect losses that are incurred by the injured party but are not directly related to the contract.

18. INSURANCE

Most design firms and many independent design professionals carry *errors and omissions insurance* to protect them from claims due to their mistakes. Such policies are costly, and for that reason, some professionals choose to "go bare."[23] Policies protect against inadvertent mistakes only, not against willful, knowing, or conscious efforts to defraud or deceive.

---

[22]Of course, the engineer can expect to be discharged from the company. However, for strategic reasons, this discharge probably will not occur until after the company loses the case.

[23]Going bare appears foolish at first glance, but there is a perverted logic behind the strategy. One-person consulting firms (and perhaps, firms that are not profitable) are "judgment-proof." Without insurance or other assets, these firms would be unable to pay any large judgments against them. When damage victims (and their lawyers) find this out in advance, they know that judgments will be uncollectable. So, often the lawsuit never makes its way to trial.

71 Engineering Ethics

1. Creeds, Codes, Canons, Statutes,
   and Rules 71-1
2. Purpose of a Code of Ethics 71-1
3. Ethical Priorities 71-2
4. Dealing with Clients and Employers 71-2
5. Dealing with Suppliers 71-2
6. Dealing with Other Engineers 71-3
7. Dealing with (and Affecting) the Public . . 71-3
8. Competitive Bidding 71-3

1. CREEDS, CODES, CANONS, STATUTES, AND RULES

It is generally conceded that an individual acting on his or her own cannot be counted on to always act in a proper and moral manner. Creeds, statutes, rules, and codes all attempt to complete the guidance needed for an engineer to do "...the correct thing."

A *creed* is a statement or oath, often religious in nature, taken or assented to by an individual in ceremonies. For example, the *Engineers' Creed* adopted by the National Society of Professional Engineers is[1]

> I pledge...
>
> ... to give the utmost of performance;
>
> ... to participate in none but honest enterprise;
>
> ... to live and work according to the laws of man and the highest standards of professional conduct;
>
> ... to place service before profit, the honor and standing of the profession before personal advantage, and the public welfare above all other considerations.
>
> In humility and with need for Divine Guidance, I make this pledge.

A *code* is a system of nonstatutory, nonmandatory canons of personal conduct. A *canon* is a fundamental belief that usually encompasses several rules. For example, the code of ethics of the American Society of Civil Engineers (ASCE) contains the following seven canons.

1. Engineers shall hold paramount the safety, health, and welfare of the public in the performance of their professional duties.

2. Engineers shall perform services only in areas of their competence.

3. Engineers shall issue public statements only in an objective and truthful manner.

4. Engineers shall act in professional matters for each employer or client as faithful agents or trustees and shall avoid conflicts of interest.

5. Engineers shall build their professional reputation on the merit of their service and shall not compete unfairly with others.

6. Engineers shall act in such a manner as to uphold and enhance the honor, integrity, and dignity of the engineering profession.

7. Engineers shall continue their professional development throughout their careers and shall provide opportunities for the professional development of those engineers under their supervision.

A *rule* is a guide (principle, standard, or norm) for conduct and action in a certain situation. A *statutory rule* is enacted by the legislative branch of state or federal government and carries the weight of law. Some U.S. engineering registration boards have statutory *rules of professional conduct*.

2. PURPOSE OF A CODE OF ETHICS

Many different sets of *codes of ethics* (*canons of ethics, rules of professional conduct*, etc.) have been produced by various engineering societies, registration boards, and other organizations.[2] The purpose of these ethical

---

[1] The *Faith of an Engineer* adopted by the Accreditation Board for Engineering and Technology (ABET) is a similar but more detailed creed.

[2] All of the major engineering technical and professional societies in the United States (ASCE, IEEE, ASME, AIChE, NSPE, etc.) and throughout the world have adopted codes of ethics. Most U.S. societies have endorsed the *Code of Ethics of Engineers* developed by the Accreditation Board for Engineering and Technology (ABET), formerly the Engineers' Council for Professional Development (ECPD). The National Council of Examiners for Engineering and Surveying (NCEES) has developed its *Model Rules of Professional Conduct* as a guide for state registration boards in developing guidelines for the professional engineers in those states.

Law and Ethics

guidelines is to guide the conduct and decision making of engineers. Most codes are primarily educational. Nevertheless, from time to time they have been used by the societies and regulatory agencies as the basis for disciplinary actions.

Fundamental to ethical codes is the requirement that engineers render faithful, honest, professional service. In providing such service, engineers must represent the interests of their employers or clients and, at the same time, protect public health, safety, and welfare.

There is an important distinction between what is legal and what is ethical. Many legal actions can be violations of codes of ethical or professional behavior.[3] For example, an engineer's contract with a client may give the engineer the right to assign the engineer's responsibilities, but doing so without informing the client would be unethical.

Ethical guidelines can be categorized on the basis of who is affected by the engineer's actions—the client, vendors and suppliers, other engineers, or the public at large.[4]

3. ETHICAL PRIORITIES

There are frequently conflicting demands on engineers. While it is impossible to use a single decision-making process to solve every ethical dilemma, it is clear that ethical considerations will force engineers to subjugate their own self-interests. Specifically, the ethics of engineers dealing with others need to be considered in the following order from highest to lowest priority.

- society and the public
- the law
- the engineering profession
- the engineer's client
- the engineer's firm
- other involved engineers
- the engineer personally

4. DEALING WITH CLIENTS AND EMPLOYERS

The most common ethical guidelines affecting engineers' interactions with their employer (the *client*) can be summarized as follows.[5]

---

[3]Whether the guidelines emphasize ethical behavior or professional conduct is a matter of wording. The intention is the same: to provide guidelines that transcend the requirements of the law.
[4]Some authorities also include ethical guidelines for dealing with the employees of an engineer. However, these guidelines are no different for an engineering employer than they are for a supermarket, automobile assembly line, or airline employer. Ethics is not a unique issue when it comes to employees.
[5]These general guidelines contain references to contractors, plans, specifications, and contract documents. This language is common, though not unique, to the situation of an engineer supplying design services to an owner-developer or architect. However, most of the ethical guidelines are general enough to apply to engineers in industry as well.

- Engineers should not accept assignments for which they do not have the skill, knowledge, or time.

- Engineers must recognize their own limitations. They should use associates and other experts when the design requirements exceed their abilities.

- The client's interests must be protected. The extent of this protection exceeds normal business relationships and transcends the legal requirements of the engineer-client contract.

- Engineers must not be bound by what the client wants in instances where such desires would be unsuccessful, dishonest, unethical, unhealthy, or unsafe.

- Confidential client information remains the property of the client and must be kept confidential.

- Engineers must avoid conflicts of interest and should inform the client of any business connections or interests that might influence their judgment. Engineers should also avoid the *appearance* of a conflict of interest when such an appearance would be detrimental to the profession, their client, or themselves.

- The engineers' sole source of income for a particular project should be the fee paid by their client. Engineers should not accept compensation in any form from more than one party for the same services.

- If the client rejects the engineer's recommendations, the engineer should fully explain the consequences to the client.

- Engineers must freely and openly admit to the client any errors made.

All courts of law have required an engineer to perform in a manner consistent with normal professional standards. This is not the same as saying an engineer's work must be error-free. If an engineer completes a design, has the design and calculations checked by another competent engineer, and an error is subsequently shown to have been made, the engineer may be held responsible, but will probably not be considered negligent.

5. DEALING WITH SUPPLIERS

Engineers routinely deal with manufacturers, contractors, and vendors (*suppliers*). In this regard, engineers have great responsibility and influence. Such a relationship requires that engineers deal justly with both clients and suppliers.

An engineer will often have an interest in maintaining good relationships with suppliers since this often leads to future work. Nevertheless, relationships with suppliers must remain highly ethical. Suppliers should not be encouraged to feel that they have any special favors coming to them because of a long-standing relationship with the engineer.

The ethical responsibilities relating to suppliers are listed as follows.

- The engineer must not accept or solicit gifts or other valuable considerations from a supplier during, prior to, or after any job. An engineer should not accept discounts, allowances, commissions, or any other indirect compensation from suppliers, contractors, or other engineers in connection with any work or recommendations.

- The engineer must enforce the plans and specifications (i.e., the *contract documents*) but must also interpret the contract documents fairly.

- Plans and specifications developed by the engineer on behalf of the client must be complete, definite, and specific.

- Suppliers should not be required to spend time or furnish materials that are not called for in the plans and contract documents.

- The engineer should not unduly delay the performance of suppliers.

6. DEALING WITH OTHER ENGINEERS

Engineers should try to protect the engineering profession as a whole, to strengthen it, and to enhance its public stature. The following ethical guidelines apply.

- An engineer should not attempt to maliciously injure the professional reputation, business practice, or employment position of another engineer. However, if there is proof that another engineer has acted unethically or illegally, the engineer should advise the proper authority.

- An engineer should not review someone else's work while the other engineer is still employed unless the other engineer is made aware of the review.

- An engineer should not try to replace another engineer once the other engineer has received employment.

- An engineer should not use the advantages of a salaried position to compete unfairly (i.e., moonlight) with other engineers who have to charge more for the same consulting services.

- Subject to legal and proprietary restraints, an engineer should freely report, publish, and distribute information that would be useful to other engineers.

7. DEALING WITH (AND AFFECTING) THE PUBLIC

In regard to the social consequences of engineering, the relationship between an engineer and the public is essentially straightforward. Responsibilities to the public demand that the engineer place service to humankind above personal gain. Furthermore, proper ethical behavior requires that an engineer avoid association with projects that are contrary to public health and welfare or that are of questionable legal character.

- Engineers must consider the safety, health, and welfare of the public in all work performed.

- Engineers must uphold the honor and dignity of their profession by refraining from self-laudatory advertising, by explaining (when required) their work to the public, and by expressing opinions only in areas of knowledge.

- When engineers issue a public statement, they must clearly indicate if the statement is being made on anyone's behalf (i.e., if anyone is benefitting from their position).

- Engineers must keep their skills at a state-of-the-art level.

- Engineers should develop public knowledge and appreciation of the engineering profession and its achievements.

- Engineers must notify the proper authorities when decisions adversely affecting public safety and welfare are made.[6]

8. COMPETITIVE BIDDING

The ethical guidelines for dealing with other engineers presented here and in more detailed codes of ethics no longer include a prohibition on *competitive bidding*. Until 1971, most codes of ethics for engineers considered competitive bidding detrimental to public welfare, since cost cutting normally results in a lower quality design.

However, in a 1971 case against the National Society of Professional Engineers that went all the way to the U.S. Supreme Court, the prohibition against competitive bidding was determined to be a violation of the Sherman Antitrust Act (i.e., it was an unreasonable restraint of trade).

---

[6]This practice has come to be known as *whistle-blowing*.

The opinion of the Supreme Court does not *require* competitive bidding—it merely forbids a prohibition against competitive bidding in NSPE's code of ethics. The following points must be considered.

- Engineers and design firms may individually continue to refuse to bid competitively on engineering services.

- Clients are not required to seek competitive bids for design services.

- Federal, state, and local statutes governing the procedures for procuring engineering design services, even those statutes that prohibit competitive bidding, are not affected.

- Any prohibitions against competitive bidding in individual state engineering registration laws remain unaffected.

- Engineers and their societies may actively and aggressively lobby for legislation that would prohibit competitive bidding for design services by public agencies.

72 Engineering Registration in the United States

1. What Registration Is 72-1
2. The U.S. Registration Procedure 72-1
3. National Council of Examiners
   for Engineering and Surveying 72-1
4. Reciprocity Among States 72-2
5. Uniform Examinations 72-2
6. Applying for the Examination 72-2
7. Examination Dates 72-3
8. FE Examination Format 72-3
9. PE Examination Format 72-3

1. WHAT REGISTRATION IS

Engineering registration (also known as *engineering licensing*) in the United States is an examination process by which a state's *board of engineering licensing* (typically referred to as the "engineers' board" or "board of registration") determines and certifies that you have achieved a minimum level of competence.[1] This process is intended to protect the public by preventing unqualified individuals from offering engineering services.

Most engineers in the United States do not need to be registered.[2] In particular, most engineers who work for companies that design and manufacture products are exempt from the licensing requirement. This is known as the *industrial exemption*, something that is built into the laws of most states.[3]

Nevertheless, there are many good reasons for wanting to become a registered engineer. For example, you cannot offer consulting engineering services in any state unless you are registered in that state. Even within a product-oriented corporation, you may find that employment, advancement, and managerial positions are limited to registered engineers.

Once you have met the registration requirements, you will be allowed to use the titles *Professional Engineer* (PE), *Registered Engineer* (RE), or *Consulting Engineer* (CE) as permitted by your state.

---

[1]Registration of engineers is not unique to the United States. However, the practice of requiring a degreed engineer to take an examination is not common in other countries. Registration in many countries requires a degree and may also require experience, references, and demonstrated knowledge of ethics and law, but no technical examination.

[2]Less than one-third of the degreed engineers in the United States are registered.

[3]Only one or two states have abolished the industrial exemption. There has always been a lot of "talk" among engineers about abolishing it, but there has been little success in actually trying to do so. One of the reasons is that manufacturers' lobbies are very strong.

Although the registration process is similar in each of the 50 states, each has its own registration law. Unless you offer consulting engineering services in more than one state, however, you will not need to be registered in the other states.

2. THE U.S. REGISTRATION PROCEDURE

The registration procedure is similar in all states. You will take two eight-hour written examinations. The full process requires you to complete two applications, one for each of the two examinations. The first examination is the *Fundamentals of Engineering* (FE) *examination*, formerly known (and still commonly referred to) as the *Engineer-In-Training* (E-I-T) *examination*.[4] This examination covers basic subjects from all of the mathematics, physics, chemistry, and engineering courses you took during your university years.

The second examination is the *Professional Engineering* (PE) *examination*, also known as the *Principles and Practices* (P&P) *examination*. This examination covers only the subjects in your engineering discipline (e.g., civil, mechanical, electrical, and others).

The actual details of registration qualifications, experience requirements, minimum education levels, fees, oral interviews, and examination schedules vary from state to state. Contact your state's registration board for more information.

3. NATIONAL COUNCIL OF EXAMINERS FOR ENGINEERING AND SURVEYING

The *National Council of Examiners for Engineering and Surveying* (NCEES) in Clemson, South Carolina, writes, prints, distributes, and scores the national FE and PE examination.[5] The individual states purchase the examinations from NCEES and administer them. NCEES does not distribute applications to take the examinations, administer the examinations or appeals, or notify you of the results. These tasks are all performed by the individual states.

---

[4]The terms *engineering intern* (EI) and *intern engineer* (IE) have also been used in the past to designate the status of an engineer who has passed the first exam. These uses are rarer but may still be encountered in some states.

[5]National Council of Examiners for Engineering and Surveying, P.O. Box 1686, Clemson, SC 29633, (803) 654-6824.

Law and Ethics

4. RECIPROCITY AMONG STATES

With minor exceptions, having a license from one state will not permit you to practice engineering in another state. You must have a professional engineering license from each state in which you work. For most engineers, this is not a problem, but for some it is. Luckily, it is not too difficult to get a license from every state you work in once you have a license from one of them.

All states use the NCEES examinations. If you take and pass the FE or PE examination in one state, your certificate or license will be honored by all of the other states. Upon proper application, payment of fees, and proof of your license, you will be issued a license by the new state. Although there may be other special requirements imposed by a state, it will not be necessary to retake the FE or PE examinations.[6] The issuance of an engineering license based on another state's licensing is known as *reciprocity* or *comity*.

5. UNIFORM EXAMINATIONS

Although each state has its own licensing law and is, theoretically, free to administer its own exams, none does so for the major disciplines. All states have chosen to use the NCEES exams. Each state administers the exams on the same days. The exams from all the states are sent to NCEES and are graded by the same graders. Each state adopts the cut-off passing scores recommended by NCEES. These practices have led to the term *uniform examination*.

6. APPLYING FOR THE EXAMINATION

While the exam administration process is essentially standardized among the states, there is a lot of variation in the application process. Each state has its own application forms, charges different fees, and has different age, education, and experience requirements. Therefore, you will have to request an application from the state in which you want to become registered in order to find out what is required. It is generally sufficient for you to phone for this application; a written request is unnecessary. Telephone numbers for all U.S. boards of registration (states and territories) are given in Table 72.1. Some states have a fee for the application.

As with any other important document, it is a good idea to keep a copy of your examination application and send the original application by certified mail, requesting a receipt of delivery. Keep your proof of mailing and delivery receipt with your copy of the application.

All states make special accommodations for persons who are physically challenged or who have other special needs. Be sure to communicate your need to the state board well in advance of the examination day.

---

[6]For example, California requires all civil engineering applicants to pass special examinations in seismic design and surveying in addition to their regular eight-hour PE exams. Licensed engineers from other states only have to pass these two special exams. They do not need to retake the PE exam.

Table 72.1 *Phone Numbers of Boards of Registration (States and Territories)*

Alabama	(334) 242-5568
Alaska	(907) 465-2540
Arizona	(602) 255-4053
Arkansas	(501) 324-9085
California	(916) 263-2222
Colorado	(303) 894-7788
Connecticut	(860) 566-3290
Delaware	(302) 577-6500
District of Columbia	(202) 727-7454
Florida	(904) 488-9912
Georgia	(404) 656-3926
Guam	(671) 646-9386
Hawaii	(808) 586-3000
Idaho	(208) 334-3860
Illinois	(217) 782-8556
Indiana	(317) 232-2980
Iowa	(515) 281-5602
Kansas	(913) 296-3053
Kentucky	(502) 573-2680
Louisiana	(504) 295-8522
Maine	(207) 287-3236
Maryland	(410) 333-6322
Massachusetts	(617) 727-9957
Michigan	(517) 335-1669
Minnesota	(612) 296-2388
Mississippi	(601) 359-6160
Missouri	(573) 751-0047
Montana	(406) 444-4285
Nebraska	(402) 471-2407
Nevada	(702) 688-1231
New Hampshire	(603) 271-2219
New Jersey	(201) 504-6460
New Mexico	(505) 827-7561
New York	(518) 474-3846
North Carolina	(919) 781-9499
North Dakota	(701) 258-0786
Ohio	(614) 466-3650
Oklahoma	(405) 521-2874
Oregon	(503) 378-4180
Pennsylvania	(717) 783-7049
Puerto Rico	(809) 722-2122
Rhode Island	(401) 277-2565
South Carolina	(803) 737-9260
South Dakota	(605) 394-2510
Tennessee	(615) 741-3221
Texas	(512) 440-7723
Utah	(801) 530-6551
Vermont	(802) 828-2363
Virginia	(804) 367-8512
Virgin Islands	(809) 774-3130
Washington	(360) 753-6966
West Virginia	(304) 558-3554
Wisconsin	(608) 266-1397
Wyoming	(307) 777-6155

7. EXAMINATION DATES

The national FE and PE examinations are administered twice a year, on the same weekends in all states. Table 72.2 contains the dates of upcoming examination periods.

Table 72.2 *Dates of U.S. Engineering Licensing Exams*[a]

year	Spring exam	Fall exam
1999	April 23–24	October 29–30
2000	April 14–15	October 27–28

[a]Subject to change without notice.

8. FE EXAMINATION FORMAT

The FE examination consists of two four-hour sessions, separated by a one-hour lunch period. There are 120 multiple-choice problems in the morning and 60 multiple-choice problems in the afternoon. All problems have four answer choices, from which you are to choose the best single answer. Afternoon problems are slightly more difficult and have double weight.

Questions from all undergraduate technical courses appear in the examination, including mathematics, chemistry, physics, and engineering.[7]

The FE exam is essentially all in SI units. There are a few non-SI problems and some opportunities to choose between identical SI and non-SI problems.

NCEES provides its own reference booklet for use in the examination. You may bring your own calculator and pencils, but you are not allowed to use any of your own books or notes.

9. PE EXAMINATION FORMAT

The NCEES PE examination consists of two four-hour sessions, separated by a one-hour lunch period. It only covers subjects in your major field of study (e.g., mechanical engineering). There are several variations in PE exam format, but the most common type has ten complex problems in the morning session and ten complex problems in the afternoon session.[8] You must choose four of the morning problems and four of the afternoon problems. There are no required problems, and you are free to select which problems to work.

All problems in the morning session are of the "essay" (also known as "free response") variety, which means that the solutions are worked in the answer booklets provided. Solutions are reviewed by graders and scored according to their degree of completion and correctness.

All problems in the afternoon session are *objectively scored*. This means that each problem consists of ten multiple-choice questions, which are graded by computer. There is no penalty for guessing or wrong answers, but there is also no partial credit for correct methods and assumptions.

Unlike the FE exam where SI units are used extensively, virtually all of the problems on the PE exam are in customary U.S. units. There are a few exceptions (e.g., occasional problems in chemical engineering), but most problems use pounds, feet, seconds, gallons, and British thermal units.

The PE examination is open book. While some states have a few restrictions, generally you may bring in the calculators and books of your choice.

Law and Ethics

---

[7]The format of the FE exam, other valuable information, and tips on passing are given in greater detail in *Engineer-in-Training Review Manual*, (Lindeburg), and *Engineer-In-Training Reference Manual* (Lindeburg), both published by Professional Publications.

[8]Other PE exam variations include twelve problems in each session (used in the civil and electrical engineering disciplines), and four problems (i.e., a *no choice exam*) in each session (as used in many of the smaller disciplines such as industrial and environmental engineering).

Appendices Table of Contents

1.A   Conversion Factors A-1
1.B   Common SI Unit Conversion Factors A-3
7.A   Mensuration of Two-Dimensional Areas A-7
7.B   Mensuration of Three-Dimensional Volumes A-9
9.A   Abbreviated Table of Indefinite Integrals A-10
10.A Laplace Transforms A-11
11.A Areas Under the Standard Normal Curve A-12
14.A Properties of Water at Atmospheric Pressure
        (English units) A-13
14.B Properties of Water at Atmospheric Pressure (SI units) A-13
14.C Viscosity of Water in Other Units A-14
14.D Properties of Air at Atmospheric Pressure
        (English units) A-14
14.E Properties of Air at Atmospheric Pressure (SI units) . A-14
14.F Properties of Common Liquids A-15
14.G Properties of Uncommon Fluids A-16
14.H Vapor Pressures of Various Hydrocarbons and Water . A-17
14.I Specific Gravity of Hydrocarbons A-18
14.J Viscosity Conversion Chart A-19
14.K Viscosity Index Chart: 0–100 V.I. A-20
16.A Area, Wetted Perimeter, and Hydraulic Radius
        of Partially Filled Circular Pipes A-21
16.B Dimensions of Welded and Seamless Steel
        Pipe (selected sizes) (English units) A-22
16.C Dimensions of Welded and Seamless Steel Pipe
        (schedules 40 and 80) (SI units) A-25
16.D Dimensions of Copper Water Tubing (English units) . A-26
16.E Dimensions of Brass and Copper Water Tubing
        (English units) A-27
16.F Dimensions of Seamless Steel Boiler (BWG)
        Tubing (English units) A-28
16.G Dimensions of PVC Pipe (English units) A-29
16.H Standard ANSI Piping Symbols A-30
17.A Specific Roughness and Hazen-Williams
        Constants for Various Pipe Materials A-31
17.B Darcy Friction Factors A-32
17.C Flow of Water Through Schedule-40 Steel Pipe . . . A-36
17.D Equivalent Length of Straight Pipe for Various
        (Generic) Fittings A-37
19.A Symbols for Fluid Power Equipment A-38
21.A Atomic Numbers and Weights of the Elements A-39
21.B Periodic Table of the Elements A-40
21.C Water Chemistry $CaCO_3$ Equivalents A-41
22.A Heats of Combustion for Common Compounds A-43
24.A Properties of Saturated Steam by Temperature . . . A-44
24.B Properties of Saturated Steam by Pressure A-46
24.C Properties of Superheated Steam A-47
24.D Properties of Compressed Water A-49
24.E Enthalpy-Entropy (Mollier) Diagram for Steam . . . A-50
24.F Properties of Low-Pressure Air A-51
24.G Properties of Saturated Refrigerant-12
        by Temperature A-53
24.H Properties of Saturated Refrigerant-12 by Pressure . A-54
24.I Properties of Superheated Refrigerant-12 . . . . . . A-55
24.J Pressure-Enthalpy Diagram for Refrigerant-12 . . . A-57
24.K Properties of Saturated Refrigerant-22
        by Temperature A-58
24.L Pressure-Enthalpy Diagram for Refrigerant-22 . . . A-60
24.M Pressure-Enthalpy Diagram for Refrigerant HFC-134a A-61
24.N Properties of Saturated Steam by Temperature . . . A-62
24.O Properties of Saturated Steam by Pressure A-64
24.P Properties of Superheated Steam A-65
24.Q Properties of Compressed Water A-66
24.R Enthalpy-Entropy (Mollier) Diagram for Steam . . . A-67
24.S Properties of Low-Pressure Air A-68
24.T Properties of Saturated Refrigerant-12 by
        Temperature A-70
24.U Pressure-Enthalpy Diagram for Refrigerant-12 . . . A-72
24.V Properties of Saturated Refrigerant-22
        by Temperature A-73
24.W Pressure-Enthalpy Diagram for Refrigerant-22 . . . A-75
24.X Pressure-Enthalpy Diagram for Refrigerant HFC-134a A-76
24.Y Physical Properties of Selected Solids A-77
24.Z Generalized Compressibility Charts A-78
26.A Isentropic Flow Factors A-79
26.B Isentropic Flow and Normal Shock Parameters . . . A-80
26.C Fanno Flow Factors A-81
26.D Rayleigh Flow Factors A-82

26.E International Standard Atmosphere A-83
34.A Representative Thermal Conductivity A-84
34.B Properties of Metals and Alloys A-85
34.C Properties of Nonmetals A-86
34.D Transient Heat Flow Charts
        (solid spheres of radius r_o) A-87
34.E Transient Heat Flow Charts
        (infinite solid circular cylinders of r_o) A-88
34.F Transient Heat Flow Charts
        (infinite flat slabs of thickness $2L$) A-89
34.G Heisler Transient Heat Flow Chart (temperature
        at center of a sphere of radius r_o) A-90
34.H Heisler Transient Heat Flow Chart (temperature
        at center of infinite cylinder of radius r_o) . . . A-91
34.I Heisler Transient Heat Flow Chart (temperature
        at center of infinite slab of thickness $2L$) A-92
35.A Properties of Saturated Water (English units) A-93
35.B Properties of Saturated Water (SI units) A-93
35.C Properties of Atmospheric Air (English units) A-94
35.D Properties of Atmospheric Air (SI units) A-94
35.E Properties of Saturated Steam at
        One Atmosphere (English units) A-95
35.F Properties of Saturated Steam at
        One Atmosphere (SI units) A-95
36.A Correction Factor, F_c, for the Logarithmic Mean
        Temperature Difference (one shell pass, even
        number of tube passes) A-96
36.B Correction Factor, F_c, for the Logarithmic Mean
        Temperature Difference (two shell passes,
        multiple of four tube passes) A-96
36.C Characteristics of Birmingham Wire Gage
        (BWG) Size Tubing A-97
36.D Heat Exchanger Effectiveness A-98
37.A Emissivities of Various Surfaces A-100
38.A ASHRAE Psychrometric Chart No. 1, Normal
        Temperature—Sea Level (32–120°)
        (customary U.S. units) A-101
38.B ASHRAE Psychrometric Chart No. 1, Normal
        Temperature—Sea Level (0–50°) (SI units) A-102
38.C ASHRAE Psychrometric Chart No. 2,
        Low Temperature—Sea Level (–40–50°)
        (customary U.S. units) A-103
38.D ASHRAE Psychrometric Chart No. 3, High
        Temperature—Sea Level (60–250°)
        (customary U.S. units) A-104
40.A Representative Insulating Properties of
        Selected Building Materials A-105
46.A Typical Mechanical Properties of
        Representative Metals A-107
46.B Typical Mechanical Properties of Thermoplastic
        Resins and Composites A-110
48.A Centroids and Area Moments of Inertia for
        Basic Shapes A-111
49.A Elastic Beam Deflection Equations A-112
49.B Stress Concentration Factors A-115
51.A Properties of Weld Groups A-116
52.A Spring Wire Diameters and Sheet Metal Gauges . . A-117
52.B Journal Bearing Correlations A-118
54.A Mass Moments of Inertia A-119
57.A Standard Epicyclic Gear Train Ratios A-120
61.A Percentile Values of Student's t-Distribution A-121
61.B Single Sampling Plan Table for Various
        Producer's and Consumer's Risks A-122
61.C Cumulative Probability Chart A-123
62.A Thermoelectric Constants for Thermocouples A-124
65.A Friction Loss in Schedule-40 Steel Pipe
        (customary U.S. units) A-125
65.B Friction Loss in Schedule-40 Steel Pipe (SI units) . A-126
65.C Equivalent Lengths of Valves and Fittings for
        Fire Protection Systems A-127
65.D Standardized Sprinkler System Worksheet A-128
67.A Polyphase Motor Classifications and Characteristics . A-129
67.B DC and Single-Phase Motor Classifications and
        Characteristics A-130
68.A Noise Reduction and Absorption Coefficients A-131
68.B Transmission Loss Through Common Materials . . . A-132
69.A Standard Cash Flow Factors A-133
69.B Factor Tables A-134

Appendices

APPENDIX 1.A
Conversion Factors

multiply	by	to obtain	multiply	by	to obtain
acres	0.4047	hectares	feet	30.48	centimeters
	43,560.0	square feet		0.3048	meters
	1.5625×10^{-3}	square miles		1.645×10^{-4}	miles (nautical)
ampere-hours	3600.0	coulombs		1.894×10^{-4}	miles (statute)
angstrom units	3.937×10^{-9}	inches	feet/min	0.5080	centimeters/sec
	1×10^{-4}	microns	feet/sec	0.592	knots
astronomical units	1.496×10^{8}	kilometers		0.6818	miles/hour
atmospheres	76.0	centimeters of mercury	foot-pounds	1.285×10^{-3}	Btu
				5.051×10^{-7}	horsepower-hr
atomic mass unit	9.316×10^{8}	electron-volts		3.766×10^{-7}	kilowatt-hours
	1.492×10^{-10}	joules	foot-pound/sec	4.6272	Btu/hr
	1.66×10^{-27}	kilograms		1.818×10^{-3}	horsepower
BeV (also GeV)	10^{9}	electron-volts		1.356×10^{-3}	kilowatts
Btu	3.93×10^{-4}	horsepower-hours	furlongs	660.0	feet
	778.3	foot-pounds		0.125	miles (U.S.)
	2.93×10^{-4}	kilowatt hours	gallons	0.1337	cubic feet
	1.0×10^{-5}	therms		3.785	liters
Btu/hr	0.2161	foot-pounds/sec	gallons H_2O	8.3453	pounds H_2O
	3.929×10^{-4}	horsepower	gallons/min	8.0208	cubic feet/hr
	0.293	watts		0.002228	cubic feet/sec
bushels	2150.4	cubic inches	GeV (also BeV)	10^{9}	electron-volts
calories, gram (mean)	3.9683×10^{-3}	Btu (mean)	grams	10^{-3}	kilograms
centares	1.0	square meters		3.527×10^{-2}	ounces (avoirdupois)
centimeters	1×10^{-5}	kilometers		3.215×10^{-2}	ounces (troy)
	1×10^{-2}	meters		2.205×10^{-3}	pounds
	10.0	millimeters	hectares	2.471	acres
	3.281×10^{-2}	feet		1.076×10^{5}	square feet
	0.3937	inches	horsepower	2545.0	Btu/hr
chains	792.0	inches		42.44	Btu/min
coulombs	1.036×10^{-5}	faradays		550	foot-pounds/sec
cubic centimeters	0.06102	cubic inches		0.7457	kilowatts
	2.113×10^{-3}	pints (U.S. liquid)		745.7	watts
cubic feet	0.02832	cubic meters	horsepower-hour	2545.0	Btu
	7.4805	gallons		1.976×10^{-6}	foot-pounds
cubic feet/min	62.43	pounds H_2O/min		0.7457	kilowatt hours
cubic feet/sec	448.831	gallons/min	hours	4.167×10^{-2}	days
	0.64632	millions of gallons per day		5.952×10^{-3}	weeks
			inches	2.540	centimeters
cubits	18.0	inches		1.578×10^{-5}	miles
days	86,400.0	seconds	inches, H_2O	5.199	lbf/ft^2
degrees (angle)	1.745×10^{-2}	radians		0.0361	psi
degrees/sec	0.1667	revolutions/min		0.0735	inches, mercury
dynes	1×10^{-5}	newtons	inches, mercury	70.7	lbf/ft^2
electron volts	1.074×10^{-9}	atomic mass units		0.491	psi
	10^{-9}	BeV (also GeV)		13.60	inches, H_2O
	1.602×10^{-19}	joules	joules	6.705×10^{9}	atomic mass units
	1.783×10^{-36}	kilograms		9.480×10^{-4}	Btu
	10^{-6}	MeV		1×10^{7}	ergs
faradays/sec	96,500	amperes (absolute)		6.242×10^{18}	electron-volts
fathoms	6.0	feet		1.113×10^{-17}	kilograms

(continued)

Support Material

APPENDIX 1.A *(continued)*
Conversion Factors

multiply	by	to obtain	multiply	by	to obtain
kilograms	6.025×10^{26}	atomic mass units	parsecs	3.086×10^{13}	kilometers
	5.610×10^{35}	electron-volts		1.9×10^{13}	miles
	8.987×10^{16}	joules	pascal-sec	1000	centipoise
	2.205	pounds		10	poise
kilometers	3281.0	feet		0.02089	pound force-sec/ft^2
	1000.0	meters		0.6720	pound mass/ft-sec
	0.6214	miles		0.02089	slug/ft-sec
kilometers/hr	0.5396	knots	pints (liquid)	473.2	cubic centimeters
kilowatts	3412.9	Btu/hr		28.87	cubic inches
	737.6	foot-pound/sec		0.125	gallons
	1.341	horsepower		0.5	quarts (liquid)
kilowatt-hours	3413.0	Btu	poise	0.002089	pound-sec/ft^2
knots	6076.0	feet/hr	pounds	0.4536	kilograms
	1.0	nautical miles/hr		16.0	ounces
	1.151	statute miles/hr		14.5833	ounces (troy)
light years	5.9×10^{12}	miles		1.21528	pounds (troy)
links (surveyor)	7.92	inches	pounds/ft^2	0.006944	pounds/in^2
liters	1000.0	cubic centimeters	pounds/in^2	2.308	feet, H$_2$O
	61.02	cubic inches		27.7	inches, H$_2$O
	0.2642	gallons (U.S. liquid)		2.037	inches, mercury
	1000.0	milliliters		144	pounds/ft^2
	2.113	pints	quarts (dry)	67.20	cubic inches
MeV	10^6	electron-volts	quarts (liquid)	57.75	cubic inches
meters	100.0	centimeters		0.25	gallons
	3.281	feet		0.9463	liters
	1×10^{-3}	kilometers	radians	57.30	degrees
	5.396×10^{-4}	miles (nautical)		3438.0	minutes
	6.214×10^{-4}	miles (statute)	revolutions	360.0	degrees
	1000.0	millimeters	revolutions/min	6.0	degrees/sec
microns	1×10^{-6}	meters	rods	16.5	feet
miles (nautical)	6076	feet		5.029	meters
	1.853	kilometers	rods (surveyor)	5.5	yards
	1.1516	miles (statute)	seconds	1.667×10^{-2}	minutes
miles (statute)	5280.0	feet	square meters/sec	10^6	centistokes
	1.609	kilometers		10.76	square feet/sec
	0.8684	miles (nautical)		10^4	stokes
miles/hr	88.0	feet/min	slugs	32.174	pounds
milligrams/liter	1.0	parts/million	stokes	0.0010764	square feet/sec
milliliters	1×10^{-3}	liters	tons (long)	1016.0	kilograms
millimeters	3.937×10^{-2}	inches		2240.0	pounds
newtons	1×10^5	dynes		1.120	tons (short)
ohms (international)	1.0005	ohms (absolute)	tons (short)	907.1848	kilograms
ounces	28.349527	grams		2000.0	pounds
	6.25×10^{-2}	pounds		0.89287	tons (long)
ounces (troy)	1.09714	ounces (avoirdupois)	volts (absolute)	3.336×10^{-3}	statvolts
			watts	3.4121	Btu/hr
				1.341×10^{-3}	horsepower
			yards	0.9144	meters
				4.934×10^{-4}	miles (nautical)
				5.682×10^{-4}	miles (statute)

APPENDIX 1.B
Common SI Unit Conversion Factors

multiply	by	to obtain
AREA		
circular mil	506.7	square micrometer
square feet	0.0929	square meter
square kilometer	0.3861	square mile
square meter	10.764	square feet
	1.196	square yard
square micrometer	0.001974	circular mil
square mile	2.590	square kilometer
square yard	0.8361	square meter
ENERGY		
Btu (international)	1.0551	kilojoule
erg	0.1	microjoule
foot-pound	1.3558	joule
horsepower-hour	2.6485	megajoule
joule	0.7376	foot-pound
	0.10197	meter-kilogram force
kilogram-calorie (international)	4.1868	kilojoule
kilojoule	0.9478	Btu
	0.2388	kilogram-calorie
kilowatt-hour	3.6	megajoule
megajoule	0.3725	horsepower-hour
	0.2778	kilowatt-hour
	0.009478	therm
meter-kilogram force	9.8067	joule
microjoule	10.0	erg
therm	105.506	megajoule
FORCE		
dyne	10.0	micronewton
kilogram force	9.8067	newton
kip	4448.2	newton
micronewton	0.1	dyne
newton	0.10197	kilogram force
	0.0002248	kip
	3.597	ounce force
	0.2248	pound force
ounce force	0.2780	newton
pound force	4.4482	newton
HEAT		
Btu/ft^2-hr	3.1546	$watt/m^2$
Btu/ft^2-°F	5.6783	$watt/m^2 \cdot °C$
Btu/ft^3	0.0373	$megajoule/m^3$
Btu/ft^3-°F	0.06707	$megajoule/m^3 \cdot °C$
Btu/hr	0.2931	watt
Btu/lbm	2326	joule/kg
Btu/lbm-°F	4186.8	joule/kg·°C
Btu-in/ft^2-hr-°F	0.1442	watt/meter·°C
joule/kg	0.000430	Btu/lbm
joule/kg·°C	0.0002388	Btu/lbm-°F
$megajoule/m^3$	26.839	Btu/ft^3
$megajoule/m^3 \cdot °C$	14.911	Btu/ft^3-°F
watt	3.4121	Btu/hr
watt/m·°C	6.933	Btu-in/ft^2-hr-°F
$watt/m^2$	0.3170	Btu/ft^2-hr
$watt/m^2 \cdot °C$	0.1761	Btu/ft^2-hr

(continued)

APPENDIX 1.B *(continued)*
Common SI Unit Conversion Factors

multiply	by	to obtain
LENGTH		
Angstrom	0.1	nanometer
foot	0.3048	meter
inch	25.4	millimeter
kilometer	0.6214	mile
	0.540	mile (international nautical)
meter	3.2808	foot
	1.0936	yard
micrometer	1.0	micron
micron	1.0	micrometer
mil	0.0254	millimeter
mile	1.6093	kilometer
mile (international nautical)	1.852	kilometer
millimeter	0.0394	inch
	39.370	mil
nanometer	10.0	Angstrom
yard	0.9144	meter
MASS (weight)		
grain	64.799	milligram
gram	0.0353	ounce (avoirdupois)
	0.03215	ounce (troy)
kilogram	2.2046	pounds-mass
	0.068522	slug
	0.0009842	ton (long—2240 lbm)
	0.001102	ton (short—2000 lbm)
milligram	0.0154	grain
ounce (avoirdupois)	28.350	gram
ounce (troy)	31.1035	gram
pounds-mass	0.4536	kilogram
slug	14.5939	kilogram
ton (long—2240 lbm)	1016.047	kilogram
ton (short—2000 lbm)	907.185	kilogram
PRESSURE		
bar	100.0	kilopascal
inch, H_2O (20°C)	0.2486	kilopascal
inch, Hg (20°C)	3.3741	kilopascal
kilogram force/cm^2	98.067	kilopascal
kilopascal	0.01	bar
	4.0219	inch, H_2O (20°C)
	0.2964	inch, Hg (20°C)
	0.0102	kilogram force/cm^2
	7.528	millimeter Hg (20°C)
	0.1450	psi
	0.009869	standard atmosphere (760 torr)
	7.5006	torr
millimeter Hg (20°C)	0.13284	kilopascal
psi	6.8948	kilopascal
standard atmosphere (760 torr)	101.325	kilopascal
torr	0.13332	kilopascal

(continued)

APPENDIX 1.B *(continued)*
Common SI Unit Conversion Factors

multiply	by	to obtain
POWER		
Btu (international)/hr	0.2931	watt
foot-pound/sec	1.3558	watt
horsepower	0.7457	kilowatt
kilowatt	1.341	horsepower
	0.2843	tons of refrigeration
meter-kilogram force/second	9.8067	watt
tons of refrigeration	3.517	kilowatt
watt	3.4122	Btu (international)/hour
	0.7376	foot-pound/second
	0.10197	meter-kilogram force/second
TEMPERATURE		
Celsius	$\frac{9}{5}°C + 32$	Fahrenheit
Fahrenheit	$\frac{5}{9}(°F - 32)$	Celsius
Kelvin	$\frac{9}{5}$	Rankine
Rankine	$\frac{5}{9}$	Kelvin
TORQUE		
gram force centimeter	0.098067	millinewton·meter
kilogram force meter	9.8067	newton-meter
millinewton	10.197	gram force centimeter
newton-meter	0.10197	kilogram force meter
	0.7376	foot-pound
	8.8495	inch-pound
foot-pound	1.3558	newton-meter
inch-pound	0.1130	newton-meter
VELOCITY		
feet/second	0.3048	meters/second
kilometers/hour	0.6214	miles/hour
meters/second	3.2808	feet/second
	2.2369	miles/hour
miles/hour	1.60934	kilometers/hour
	0.44704	meters/second
VISCOSITY		
centipoise	0.001	pascal-seconds
centistoke	1×10^{-6}	meter^2/second
pascal-second	1000	centipoise
meter^2/second	1×10^6	centistoke
VOLUME (capacity)		
cubic centimeter	0.06102	cubic inch
cubic foot	28.3168	liter
cubic inch	16.3871	cubic centimeter
cubic meter	1.308	cubic yard
cubic yard	0.7646	cubic meter

(continued)

Support
Material

APPENDIX 1.B *(continued)*
Common SI Unit Conversion Factors

multiply	by	to obtain
gallon (U.S.)	3.785	liter
liter	0.2642	gallon (U.S.)
	2.113	pint (U.S. fluid)
	1.0567	quart (U.S. fluid)
	0.03531	cubic foot
milliliter	0.0338	ounce (U.S. fluid)
ounce (U.S. fluid)	29.574	milliliter
pint (U.S. fluid)	0.4732	liter
quart (U.S. fluid)	0.9464	liter

VOLUME FLOW (gas–air)

cubic meter/second	2119	standard cubic foot/minute
liter/second	2.119	standard cubic foot/minute
microliter/second	0.000127	standard cubic foot/hour
milliliter/second	0.002119	standard cubic foot/minute
	0.127133	standard cubic foot/hour
standard cubic foot/minute	0.0004719	cubic meter/second
	0.4719	liter/second
	471.947	milliliter/second
standard cubic foot/hour	7866	microliter/second
	7.8658	milliliter/second

VOLUME FLOW (liquid)

gallon/hour (U.S.)	0.001052	liter/second
gallon/minute (U.S.)	0.06309	liter/second
liter/second	951.02	gallon/hour (U.S.)
	15.850	gallon/min (U.S.)

APPENDIX 7.A
Mensuration of Two-Dimensional Areas

Nomenclature
A total surface area
b base
c chord length
d distance
h height
L length
p perimeter
r radius
s side (edge) length, arc length
θ vertex angle, in radians
ϕ central angle, in radians

Circular Sector

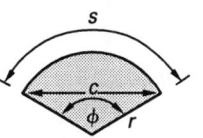

$$A = \tfrac{1}{2}\phi r^2 = \tfrac{1}{2}sr$$

$$\phi = \frac{s}{r}$$

$$s = r\phi$$

$$c = 2r\sin\left(\frac{\phi}{2}\right)$$

Triangle

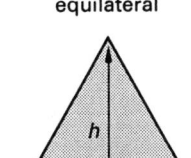 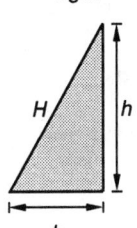 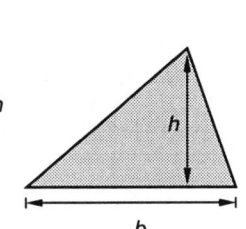

equilateral   right   oblique

$$A = \tfrac{1}{2}bh = \frac{\sqrt{3}}{4}b^2 \qquad A = \tfrac{1}{2}bh \qquad A = \tfrac{1}{2}bh$$

$$h = \frac{\sqrt{3}}{2}b \qquad\qquad H^2 = b^2 + h^2$$

Parabola

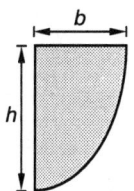

$$A = \tfrac{2}{3}bh$$

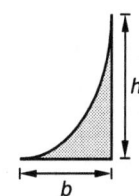

$$A = \tfrac{1}{3}bh$$

Circle

$$p = 2\pi r$$

$$A = \pi r^2 = \frac{p^2}{4\pi}$$

Ellipse

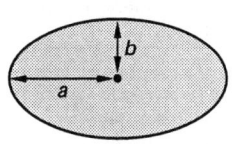

$$A = \pi ab$$

$$p = 2\pi\sqrt{\tfrac{1}{2}(a^2 + b^2)}$$

Circular Segment

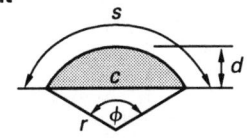

$$A = \tfrac{1}{2}r^2(\phi - \sin\phi)$$

$$\phi = \frac{s}{r} = (2)\left(\arccos\frac{r-d}{r}\right)$$

$$c = 2r\sin\left(\frac{\phi}{2}\right)$$

(continued)

PROFESSIONAL PUBLICATIONS, INC. BELMONT, CA

APPENDIX 7.A *(continued)*
Mensuration of Two-Dimensional Areas

Trapezoid

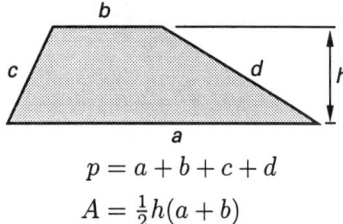

$$p = a + b + c + d$$
$$A = \tfrac{1}{2}h(a + b)$$

The trapezoid is isosceles if $c = d$.

Parallelogram

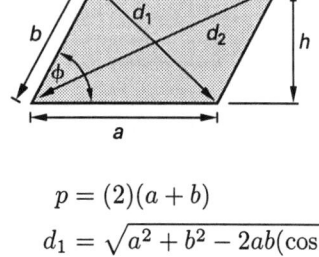

$$p = (2)(a + b)$$
$$d_1 = \sqrt{a^2 + b^2 - 2ab(\cos \phi)}$$
$$d_2 = \sqrt{a^2 + b^2 + 2ab(\cos \phi)}$$
$$d_1^2 + d_2^2 = (2)(a^2 + b^2)$$
$$A = ah = ab(\sin \phi)$$

If $a = b$, the parallelogram is a rhombus.

Regular Polygon (*n* equal sides)

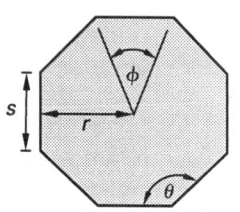

$$\phi = \frac{2\pi}{n}$$
$$\theta = \frac{\pi(n - 2)}{n} = \pi - \phi$$
$$p = ns$$
$$s = 2r\left(\tan\left(\frac{\theta}{2}\right)\right)$$
$$A = \tfrac{1}{2}nsr$$

sides	name	area (A) when diameter of inscribed circle = 1	area (A) when side = 1	radius (r) of circumscribed circle when side = 1	length (L) of side when radius (r) of circumscribed circle = 1	length (L) of side when perpendicular to center = 1	perpendicular (p) to center when side = 1
3	triangle	1.299	0.433	0.577	1.732	3.464	0.289
4	square	1.000	1.000	0.707	1.414	2.000	0.500
5	pentagon	0.908	1.720	0.851	1.176	1.453	0.688
6	hexagon	0.866	2.598	1.000	1.000	1.155	0.866
7	heptagon	0.843	3.634	1.152	0.868	0.963	1.038
8	octagon	0.828	4.828	1.307	0.765	0.828	1.207
9	nonagon	0.819	6.182	1.462	0.684	0.728	1.374
10	decagon	0.812	7.694	1.618	0.618	0.650	1.539
11	undecagon	0.807	9.366	1.775	0.563	0.587	1.703
12	dodecagon	0.804	11.196	1.932	0.518	0.536	1.866

APPENDIX 7.B
Mensuration of Three-Dimensional Volumes

Nomenclature
A area
b base
h height
r radius
R radius
s side (edge) length
V volume

Sphere

$$V = \frac{4\pi r^3}{3}$$
$$A = 4\pi r^2$$

Right Circular Cone

$$V = \frac{\pi r^2 h}{3}$$
$$A = \pi r \sqrt{r^2 + h^2}$$

(does not include base area)

Right Circular Cylinder

$$V = \pi r^2 h$$
$$A = 2\pi r h$$

(does not include end area)

Spherical Segment

Surface area of a spherical segment of radius r cut out by an angle θ_0 rotated from the center about a radius, r, is

$$A = 2\pi r^2 \left(1 - \cos\theta_0\right)$$
$$\phi = \frac{A}{r^2} = 2\pi \left(1 - \cos\theta_0\right)$$

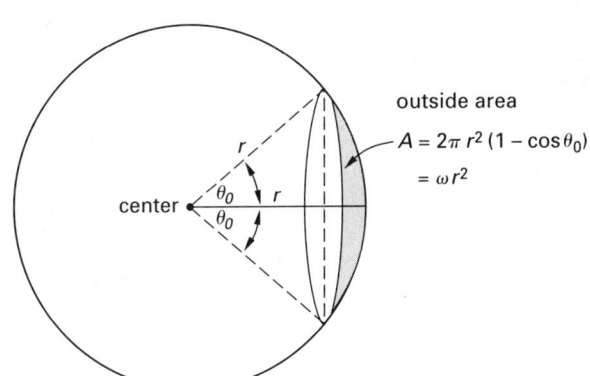

Paraboloid of Revolution

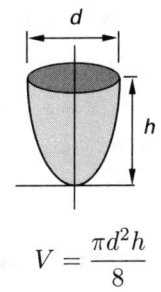

$$V = \frac{\pi d^2 h}{8}$$

Torus

$$A = 4\pi^2 R r$$
$$V = 2\pi^2 R r^2$$

Regular Polyhedra (identical faces)

name	number of faces	form of faces	total surface area	volume
tetrahedron	4	equilateral triangle	$1.7321s^2$	$0.1179s^3$
cube	6	square	$6.0000s^2$	$1.0000s^3$
octahedron	8	equilateral triangle	$3.4641s^2$	$0.4714s^3$
dodecahedron	12	regular pentagon	$20.6457s^2$	$7.6631s^3$
isosahedron	20	equilateral triangle	$8.6603s^2$	$2.1817s^3$

The radius of a sphere inscribed within a regular polyhedron is

$$r = \frac{3V_{\text{polyhedron}}}{A_{\text{polyhedron}}}$$

APPENDIX 9.A
Abbreviated Table of Indefinite Integrals
(In each case, add a constant of integration.
All angles are measured in radians.)

General Formulas

1. $\int dx = x$
2. $\int c\,dx = c\int dx$
3. $\int (dx + dy) = \int dx + \int dy$
4. $\int u\,dv = uv - \int v\,du$ (integration by parts)

Algebraic Forms

5. $\displaystyle \int x^n\,dx = \frac{x^{n+1}}{n+1} \qquad n \neq -1$

6. $\displaystyle \int x^{-1}dx = \int \frac{dx}{x} = \ln|x|$

7. $\displaystyle \int (ax+b)^n\,dx = \frac{(ax+b)^{n+1}}{a(n+1)} \qquad n \neq -1$

8. $\displaystyle \int \frac{dx}{ax+b} = \frac{1}{a}\ln(ax+b)$

9. $\displaystyle \int \frac{x\,dx}{ax+b} = \frac{1}{a^2}[ax+b - b\ln(ax+b)]$

10. $\displaystyle \int \frac{x\,dx}{(ax+b)^2} = \frac{1}{a^2}\left[\frac{b}{ax+b} + \ln(ax+b)\right]$

11. $\displaystyle \int \frac{dx}{x(ax+b)} = \frac{1}{b}\ln\left(\frac{x}{ax+b}\right)$

12. $\displaystyle \int \frac{dx}{x(ax+b)^2} = \frac{1}{b(ax+b)} + \frac{1}{b^2}\ln\left(\frac{x}{ax+b}\right)$

13. $\displaystyle \int \frac{dx}{x^2+a^2} = \frac{1}{a}\tan^{-1}\left(\frac{x}{a}\right)$

14. $\displaystyle \int \frac{dx}{a^2-x^2} = \frac{1}{a}\tanh^{-1}\left(\frac{x}{a}\right)$

15. $\displaystyle \int \frac{x\,dx}{ax^2+b} = \frac{1}{2a}\ln(ax^2+b)$

16. $\displaystyle \int \frac{dx}{x(ax^n+b)} = \frac{1}{bn}\ln\left(\frac{x^n}{ax^n+b}\right)$

17. $\displaystyle \int \frac{dx}{ax^2+bx+c} = \frac{1}{\sqrt{b^2-4ac}}\ln\left(\frac{2ax+b-\sqrt{b^2-4ac}}{2ax+b+\sqrt{b^2-4ac}}\right) \quad b^2 > 4ac$

18. $\displaystyle \int \frac{dx}{ax^2+bx+c} = \frac{2}{\sqrt{4ac-b^2}}\tan^{-1}\left(\frac{2ax+b}{\sqrt{4ac-b^2}}\right) \quad b^2 < 4ac$

19. $\displaystyle \int \sqrt{a^2-x^2}\,dx = \frac{x}{2}\sqrt{a^2-x^2} + \frac{a^2}{2}\sin^{-1}\left(\frac{x}{a}\right)$

20. $\displaystyle \int x\sqrt{a^2-x^2}\,dx = -\tfrac{1}{3}(a^2-x^2)^{\frac{3}{2}}$

21. $\displaystyle \int \frac{dx}{\sqrt{a^2-x^2}} = \sin^{-1}\left(\frac{x}{a}\right)$

22. $\displaystyle \int \frac{x\,dx}{\sqrt{a^2-x^2}} = -\sqrt{a^2-x^2}$

APPENDIX 10.A
Laplace Transforms

$f(t)$	$\mathcal{L}(f(t))$	$f(t)$	$\mathcal{L}(f(t))$
$\delta(t)$ (unit impulse at $t = 0$)	1	e^{at}	$\dfrac{1}{s - a}$
$\delta(t - c)$ (unit impulse at $t = c$)	e^{-cs}	$e^{at}\sin bt$	$\dfrac{b}{(s - a)^2 + b^2}$
1 or u_0 (unit step at $t = 0$)	$\dfrac{1}{s}$	$e^{at}\cos bt$	$\dfrac{s - a}{(s - a)^2 + b^2}$
u_c (unit step at $t = c$)	$\dfrac{e^{-cs}}{s}$	$e^{at}t^n$ (n is a positive integer)	$\dfrac{n!}{(s - a)^{n+1}}$
t (unit ramp at $t = 0$)	$\dfrac{1}{s^2}$	$1 - e^{-at}$	$\dfrac{a}{s(s + a)}$
$\dfrac{t^{n-1}}{(n - 1)!}$	$\dfrac{1}{s^n}$	$e^{-at} + at - 1$	$\dfrac{a^2}{s^2(s + a)}$
$\sin at$	$\dfrac{a}{s^2 + a^2}$	$\dfrac{e^{-at} - e^{-bt}}{b - a}$	$\dfrac{1}{(s + a)(s + b)}$
$at - \sin at$	$\dfrac{a^3}{s^2(s^2 + a^2)}$	$\dfrac{(c - a)e^{-at} - (c - b)e^{-bt}}{b - a}$	$\dfrac{s + c}{(s + a)(s + b)}$
$\sinh at$	$\dfrac{a}{s^2 - a^2}$	$\dfrac{1}{ab} + \dfrac{be^{-at} - ae^{-bt}}{ab(a - b)}$	$\dfrac{1}{s(s + a)(s + b)}$
$t\sin at$	$\dfrac{2as}{(s^2 + a^2)^2}$	$t\sinh at$	$\dfrac{2as}{(s^2 - a^2)^2}$
$\cos at$	$\dfrac{s}{s^2 + a^2}$	$t\cosh at$	$\dfrac{s^2 + a^2}{(s^2 - a^2)^2}$
$1 - \cos at$	$\dfrac{a^2}{s(s^2 + a^2)}$	rectangular pulse, magnitude M, duration a	$\left(\dfrac{M}{s}\right)(1 - e^{-as})$
$\cosh at$	$\dfrac{s}{s^2 - a^2}$	triangular pulse, magnitude M, duration $2a$	$\left(\dfrac{M}{as^2}\right)(1 - e^{-as})^2$
$t\cos at$	$\dfrac{s^2 - a^2}{(s^2 + a^2)^2}$	sawtooth pulse, magnitude M, duration a	$\left(\dfrac{M}{as^2}\right)[1 - (as + 1)e^{-as}]$
t^n (n is a positive integer)	$\dfrac{n!}{s^{n+1}}$	sinusoidal pulse, magnitude M, duration π/a	$\left(\dfrac{Ma}{s^2 + a^2}\right)(1 + e^{-\pi s/a})$

Support
Material

APPENDIX 11.A
Areas Under the Standard Normal Curve
(0 to z)

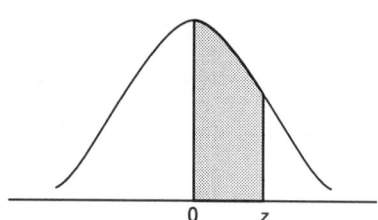

z	0	1	2	3	4	5	6	7	8	9
0.0	0.0000	0.0040	0.0080	0.0120	0.0160	0.0199	0.0239	0.0279	0.0319	0.0359
0.1	0.0398	0.0438	0.0478	0.0517	0.0557	0.0596	0.0636	0.0675	0.0714	0.0754
0.2	0.0793	0.0832	0.0871	0.0910	0.0948	0.0987	0.1026	0.1064	0.1103	0.1141
0.3	0.1179	0.1217	0.1255	0.1293	0.1331	0.1368	0.1406	0.1443	0.1480	0.1517
0.4	0.1554	0.1591	0.1628	0.1664	0.1700	0.1736	0.1772	0.1808	0.1844	0.1879
0.5	0.1915	0.1950	0.1985	0.2019	0.2054	0.2088	0.2123	0.2157	0.2190	0.2224
0.6	0.2258	0.2291	0.2324	0.2357	0.2389	0.2422	0.2454	0.2486	0.2518	0.2549
0.7	0.2580	0.2612	0.2642	0.2673	0.2704	0.2734	0.2764	0.2794	0.2823	0.2852
0.8	0.2881	0.2910	0.2939	0.2967	0.2996	0.3023	0.3051	0.3078	0.3106	0.3133
0.9	0.3159	0.3186	0.3212	0.3238	0.3264	0.3289	0.3315	0.3340	0.3365	0.3389
1.0	0.3413	0.3438	0.3461	0.3485	0.3508	0.3531	0.3554	0.3577	0.3599	0.3621
1.1	0.3643	0.3665	0.3686	0.3708	0.3729	0.3749	0.3770	0.3790	0.3810	0.3830
1.2	0.3849	0.3869	0.3888	0.3907	0.3925	0.3944	0.3962	0.3980	0.3997	0.4015
1.3	0.4032	0.4049	0.4066	0.4082	0.4099	0.4115	0.4131	0.4147	0.4162	0.4177
1.4	0.4192	0.4207	0.4222	0.4236	0.4251	0.4265	0.4279	0.4292	0.4306	0.4319
1.5	0.4332	0.4345	0.4357	0.4370	0.4382	0.4394	0.4406	0.4418	0.4429	0.4441
1.6	0.4452	0.4463	0.4474	0.4484	0.4495	0.4505	0.4515	0.4525	0.4535	0.4545
1.7	0.4554	0.4564	0.4573	0.4582	0.4591	0.4599	0.4608	0.4616	0.4625	0.4633
1.8	0.4641	0.4649	0.4656	0.4664	0.4671	0.4678	0.4686	0.4693	0.4699	0.4706
1.9	0.4713	0.4719	0.4726	0.4732	0.4738	0.4744	0.4750	0.4756	0.4761	0.4767
2.0	0.4772	0.4778	0.4783	0.4788	0.4793	0.4798	0.4803	0.4808	0.4812	0.4817
2.1	0.4821	0.4826	0.4830	0.4834	0.4838	0.4842	0.4846	0.4850	0.4854	0.4857
2.2	0.4861	0.4864	0.4868	0.4871	0.4875	0.4878	0.4881	0.4884	0.4887	0.4890
2.3	0.4893	0.4896	0.4898	0.4901	0.4904	0.4906	0.4909	0.4911	0.4913	0.4916
2.4	0.4918	0.4920	0.4922	0.4925	0.4927	0.4929	0.4931	0.4932	0.4934	0.4936
2.5	0.4938	0.4940	0.4941	0.4943	0.4945	0.4946	0.4948	0.4949	0.4951	0.4952
2.6	0.4953	0.4955	0.4956	0.4957	0.4959	0.4960	0.4961	0.4962	0.4963	0.4964
2.7	0.4965	0.4966	0.4967	0.4968	0.4969	0.4970	0.4971	0.4972	0.4973	0.4974
2.8	0.4974	0.4975	0.4976	0.4977	0.4977	0.4978	0.4979	0.4979	0.4980	0.4981
2.9	0.4981	0.4982	0.4982	0.4983	0.4984	0.4984	0.4985	0.4985	0.4986	0.4986
3.0	0.4987	0.4987	0.4987	0.4988	0.4988	0.4989	0.4989	0.4989	0.4990	0.4990
3.1	0.4990	0.4991	0.4991	0.4991	0.4992	0.4992	0.4992	0.4992	0.4993	0.4993
3.2	0.4993	0.4993	0.4994	0.4994	0.4994	0.4994	0.4994	0.4995	0.4995	0.4995
3.3	0.4995	0.4995	0.4996	0.4996	0.4996	0.4996	0.4996	0.4996	0.4996	0.4997
3.4	0.4997	0.4997	0.4997	0.4997	0.4997	0.4997	0.4997	0.4997	0.4997	0.4998
3.5	0.4998	0.4998	0.4998	0.4998	0.4998	0.4998	0.4998	0.4998	0.4998	0.4998
3.6	0.4998	0.4998	0.4999	0.4999	0.4999	0.4999	0.4999	0.4999	0.4999	0.4999
3.7	0.4999	0.4999	0.4999	0.4999	0.4999	0.4999	0.4999	0.4999	0.4999	0.4999
3.8	0.4999	0.4999	0.4999	0.4999	0.4999	0.4999	0.4999	0.4999	0.4999	0.4999
3.9	0.5000	0.5000	0.5000	0.5000	0.5000	0.5000	0.5000	0.5000	0.5000	0.5000

APPENDIX 14.A
Properties of Water at Atmospheric Pressure
(English units)

temperature (°F)	density (lbm/ft^3)	density (slug/ft^3)	absolute viscosity (lbf-sec/ft^2)	kinematic viscosity (ft^2/sec)	surface tension (lbf/ft)	vapor pressure head[a,b] (ft)	bulk modulus (lbf/in^2)
32	62.42	1.940	3.746×10^{-5}	1.931×10^{-5}	0.518×10^{-2}	0.20	293×10^3
40	62.43	1.940	3.229×10^{-5}	1.664×10^{-5}	0.514×10^{-2}	0.28	294×10^3
50	62.41	1.940	2.735×10^{-5}	1.410×10^{-5}	0.509×10^{-2}	0.41	305×10^3
60	62.37	1.938	2.359×10^{-5}	1.217×10^{-5}	0.504×10^{-2}	0.59	311×10^3
70	62.30	1.936	2.050×10^{-5}	1.059×10^{-5}	0.500×10^{-2}	0.84	320×10^3
80	62.22	1.934	1.799×10^{-5}	0.930×10^{-5}	0.492×10^{-2}	1.17	322×10^3
90	62.11	1.931	1.595×10^{-5}	0.826×10^{-5}	0.486×10^{-2}	1.62	323×10^3
100	62.00	1.927	1.424×10^{-5}	0.739×10^{-5}	0.480×10^{-2}	2.21	327×10^3
110	61.86	1.923	1.284×10^{-5}	0.667×10^{-5}	0.473×10^{-2}	2.97	331×10^3
120	61.71	1.918	1.168×10^{-5}	0.609×10^{-5}	0.465×10^{-2}	3.96	333×10^3
130	61.55	1.913	1.069×10^{-5}	0.558×10^{-5}	0.460×10^{-2}	5.21	334×10^3
140	61.38	1.908	0.981×10^{-5}	0.514×10^{-5}	0.454×10^{-2}	6.78	330×10^3
150	61.20	1.902	0.905×10^{-5}	0.476×10^{-5}	0.447×10^{-2}	8.76	328×10^3
160	61.00	1.896	0.838×10^{-5}	0.442×10^{-5}	0.441×10^{-2}	11.21	326×10^3
170	60.80	1.890	0.780×10^{-5}	0.413×10^{-5}	0.433×10^{-2}	14.20	322×10^3
180	60.58	1.883	0.726×10^{-5}	0.385×10^{-5}	0.426×10^{-2}	17.87	313×10^3
190	60.36	1.876	0.678×10^{-5}	0.362×10^{-5}	0.419×10^{-2}	22.29	313×10^3
200	60.12	1.868	0.637×10^{-5}	0.341×10^{-5}	0.412×10^{-2}	27.61	308×10^3
212	59.83	1.860	0.593×10^{-5}	0.319×10^{-5}	0.404×10^{-2}	35.38	300×10^3

[a] Based on actual densities, not on standard "cold, clear water."
[b] Can also be calculated from steam tables as $(p_{\text{saturation}})$ $(144 \text{ in}^2/\text{ft}^2)$ (v) (g/g_c).

APPENDIX 14.B
Properties of Water at Atmospheric Pressure[a]
(SI units)

temperature (°C)	density (kg/m^3)	absolute viscosity (Pa·s)	kinematic viscosity (m^2/s)	vapor pressure (kPa)	bulk modulus (kPa)
0	999.87	1.7921×10^{-3}	1.792×10^{-6}	0.611	204×10^4
4	1000.00	1.5674×10^{-3}	1.567×10^{-6}	0.813	206×10^4
10	999.73	1.3077×10^{-3}	1.371×10^{-6}	1.228	211×10^4
20	998.23	1.0050×10^{-3}	1.007×10^{-6}	2.338	220×10^4
25	997.08	0.8937×10^{-3}	8.963×10^{-7}	3.168	221×10^4
30	995.68	0.8007×10^{-3}	8.042×10^{-7}	4.242	223×10^4
40	992.25	0.6560×10^{-3}	6.611×10^{-7}	7.375	227×10^4
50	988.07	0.5494×10^{-3}	5.560×10^{-7}	12.333	230×10^4
60	983.24	0.4688×10^{-3}	4.768×10^{-7}	19.92	228×10^4
70	977.81	0.4061×10^{-3}	4.153×10^{-7}	31.16	225×10^4
80	971.83	0.3565×10^{-3}	3.668×10^{-7}	47.34	221×10^4
90	965.34	0.3165×10^{-3}	3.279×10^{-7}	70.10	216×10^4
100	958.38	0.2838×10^{-3}	2.961×10^{-7}	101.325	207×10^4

[a] Compiled from various sources.

Support Material

APPENDIX 14.C
Viscosity of Water in Other Units
(English units)

temperature	absolute viscosity	kinematic viscosity	
(°F)	(cP)	(cSt)	(SSU)
32	1.79	1.79	33.0
50	1.31	1.31	31.6
60	1.12	1.12	31.2
70	0.98	0.98	30.9
80	0.86	0.86	30.6
85	0.81	0.81	30.4
100	0.68	0.69	30.2
120	0.56	0.57	30.0
140	0.47	0.48	29.7
160	0.40	0.41	29.6
180	0.35	0.36	29.5
212	0.28	0.29	29.3

Reprinted with permission from the *Hydraulic Handbook*, copyright © 1988, by Fairbanks Morse Pump, Kansas City, Kansas.

APPENDIX 14.D
Properties of Air at Atmospheric Pressure
(English units)

temperature	density		kinematic viscosity	absolute viscosity
(°F)	(slug/ft^3)	(lbm/ft^3)	(ft^2/sec)	(lbf-sec/ft^2)
0	0.00268	0.0862	12.6×10^{-5}	3.28×10^{-7}
20	0.00257	0.0827	13.6×10^{-5}	3.50×10^{-7}
40	0.00247	0.0794	14.6×10^{-5}	3.62×10^{-7}
60	0.00237	0.0763	15.8×10^{-5}	3.74×10^{-7}
68	0.00233	0.0752	16.0×10^{-5}	3.75×10^{-7}
80	0.00228	0.0735	16.9×10^{-5}	3.85×10^{-7}
100	0.00220	0.0709	18.0×10^{-5}	3.96×10^{-7}
120	0.00215	0.0684	18.9×10^{-5}	4.07×10^{-7}
250	0.00174	0.0559	27.3×10^{-5}	4.74×10^{-7}

APPENDIX 14.E
Properties of Air at Atmospheric Pressure
(SI units)

temperature (°C)	density (kg/m^3)	absolute viscosity (Pa·s)	kinematic viscosity (m^2/s)
0	1.293	1.709×10^{-5}	1.322×10^{-5}
20	1.205	1.80×10^{-5}	1.49×10^{-5}
50	1.093	1.951×10^{-5}	1.785×10^{-5}
100	0.946	2.175×10^{-5}	2.474×10^{-5}
150	0.834	2.385×10^{-5}	3.077×10^{-5}
200	0.746	2.582×10^{-5}	3.724×10^{-5}
250	0.675	2.770×10^{-5}	4.416×10^{-5}
300	0.616	2.946×10^{-5}	5.145×10^{-5}
350	0.567	3.113×10^{-5}	5.907×10^{-5}
400	0.525	3.277×10^{-5}	6.721×10^{-5}
450	0.488	3.433×10^{-5}	7.750×10^{-5}
500	0.457	3.583×10^{-5}	8.436×10^{-5}

APPENDIX 14.F
Properties of Common Liquids

liquid	temp, °F	specific gravity[a]	kinematic viscosity		
			centistokes	SSU	ft²/sec
acetone	68	0.792	0.41		
alcohol, ethyl (ethanol)	68	0.789	1.52	31.7	1.65×10^{-5}
(C_2H_5OH)	104	0.772	1.2	31.5	
alcohol, methyl (methanol)	68	0.79			
(CH_3OH)	59		0.74		
	0	0.810			
ammonia	0	0.662	0.30		
benzene	60	0.885			
	32	0.899			
butane	−50		0.52		
	30		0.35		
	60	0.584			
carbon tetrachloride	68	1.595			
castor oil	68	0.96			1110×10^{-5}
	104	0.95	259–325	1200–1500	
	130		98–130	450–600	
ethylene glycol	60	1.125			
	70		17.8	88.4	
Freon-11	70	1.49	21.1	0.21	
Freon-12	70	1.33	21.1	0.27	
fuel oils, no. 1 to no. 6	60	0.82–0.95			
fuel oil no. 1	70		2.39–4.28	34–40	
	100		−2.69	32–35	
fuel oil no. 2	70		3.0–7.4	36–50	
	100		2.11–4.28	33–40	
fuel oil no. 3	70		2.69–5.84	35–45	
	100		2.06–3.97	32.8–39	
fuel oil no. 5A	70		7.4–26.4	50–125	
	100		4.91–13.7	42–72	
fuel oil no. 5B	70		26.4–	125–	
	100		13.6–67.1	72–310	
fuel oil no. 6	122		97.4–660	450–3000	
	160		37.5–172	175–780	
gasoline	60	0.728			0.73×10^{-5}
	80	0.719			0.66×10^{-5}
	100	0.710			0.60×10^{-5}
glycerine	68	1.261			
kerosene	60	0.78–0.82			
	68		2.17	35	
jet fuel (JPI, 3, 4, 5, 6)	−30		7.9	52	
	60	0.78–0.82			
mercury	70	13.55	21.1	0.118	
	100	13.55	37.8	0.11	
oils, SAE 5 to 150	60	0.88–0.94			
SAE-5W	0		1295 max	6000 max	
SAE-10W	0		1295–2590	6000–12,000	
SAE-20W	0		2590–10,350	12,000–48,000	
SAE-20	210		5.7–9.6	45–58	
SAE-30	210		9.6–12.9	58–70	
SAE-40	210		12.9–16.8	70–85	
SAE-50	210		16.8–22.7	85–110	
saltwater (5%)	39	1.037			
	68		1.097	31.1	
saltwater (25%)	39	1.196			
	60	1.19	2.4	34	
seawater	59	1.025			

[a]Measured with respect to 60°F water.

PROFESSIONAL PUBLICATIONS, INC. BELMONT, CA

Support Material

APPENDIX 14.G
Properties of Uncommon Fluids

Typical fluid viscosities are listed in the following table. The values given for thixotropic fluids are effective viscosities at normal pumping shear rates. Effective viscosity can vary greatly with changes in solids content, concentration, and so on.

Fluid Type:　　N—Newtonian　　　　T—Thixotropic　　　D—Dilatant

Fluid	Specific Gravity	Viscosity CPS	Fluid Type
Adhesives			
"box" adhesives	1	3000	T
PVA	1.3	100	T
rubber and solvents	1.0	15,000	N
Bakery			
batter	1	2200	T
butter, melted	0.98	18 @ 140°F	N
egg, whole	0.5	60 @ 50°F	N
emulsifier		20	T
frosting	1	10,000	T
lecithin		3250 @ 125°F	T
77% sweetened condensed milk	1.3	10,000 @ 77°F	N
yeast slurry, 15%	1	180	T
Beer, wine			
beer	1.0	1.1 @ 40°F	N
brewers concentrated yeast, 80% solids		16,000 @ 40°F	T
wine	1.0		
Confectionary			
caramel	1.2	400 @ 140°F	
chocolate	1.1	17,000 @ 120°F	T
fudge, hot	1.1	36,000	T
toffee	1.2	87,000	T
Cosmetics, Soaps			
face cream		10,000	T
gel, hair	1.4	5000	T
shampoo		5000	T
toothpaste		20,000	T
hand cleaner		2000	T
Dairy			
cottage cheese	1.08	225	T
cream	1.02	20 @ 40°F	N
milk	1.03	1.2 @ 60°F	N
cheese, process		30,000 @ 160°F	T
yogurt		1100	T
Detergents			
detergent concentrate		10	N
Dyes and Inks			
ink, printers	1 to 1.38	10,000	T
dye	1.1	10	N
gum		5000	T
Fats and Oils			
corn oil	0.92	30	N
lard	0.96	60 @ 100°F	N
linseed oil	0.93	30 @ 100°F	N
peanut oil	0.92	42 @ 100°F	N
soybean oil	0.95	36 @ 100°F	N
vegetable oil	0.92	3 @ 300°F	N
Foods, Miscellaneous			
black bean paste		10,000	T
cream style corn		130 @ 190°F	T
catsup	1.11	560 @ 145°F	T
pablum		4500	T
pear pulp		4000 @ 160°F	T
potato, mashed	1.0	20,000	T
potato skins and caustic		20,000 @ 100°F	T
prune juice	1.0	60 @ 120°F	T
orange juice concentrate	1.1	5000 @ 38°F	T
tapioca pudding	0.7	1000 @ 235°F	T
mayonnaise	1.0	5000 @ 75°F	T
tomato paste, 33%	1.14	7000	T
honey	1.5	1500 @ 100°F	

Fluid	Specific Gravity	Viscosity CPS	Fluid Type
Meat Products			
animal fat, melted	0.9	43 @ 110°F	N
ground beef fat	0.9	11,000 @ 60°F	T
meat emulsion	1.0	22,000 @ 40°F	T
pet food	1.0	11,000 @ 40°F	T
pork fat slurry	1.0	650 @ 40°F	T
Paint			
auto paint, metallic		200	T
solvents	0.8–0.9	0.5 to 10	N
titanium dioxide slurry		10,000	T
varnish	1.06	140 @ 100°F	
turpentine	0.86	2 @ 60°F	
Paper and Textile			
black liquor tar		2000 @ 300°F	
paper coating, 35%		400	
sulfide, 6%		1600	
black liquor	1.3	1100 @ 122°F	
black liquor soap		7000 @ 122°F	
Petroleum and Petroleum Products			
asphalt, unblended	1.3	500 to 2500	
gasoline	0.7	0.8 @ 60°F	N
kerosene	0.8	3 @ 68°F	N
fuel oil no. 6	0.9	660 @ 122°F	N
auto lube oil SAE 40	0.9	200 @ 100°F	N
auto trans oil SAE 90	0.9	320 @ 100°F	N
propane	0.46	0.2 @ 100°F	N
tars	1.2	wide range	
Pharmaceuticals			
castor oil	0.96	350	N
cough syrup	1.0	190	N
"stomach" remedy		1500	T
pill pastes		5000	T
Plastics, Resins			
butadiene	0.94	0.17 @ 40°F	
polyster resin (typ)	1.4	3000	
PVA resin (typ)	1.3	65,000	T
Starches, Gums			
corn starch sol 22°B	1.18	32	T
corn starch sol 25°B	1.21	300	T
Sugar, Syrups, Molasses			
corn syrup 41°Be	1.39	15,000 @ 60°F	N
corn syrup 45°Be	1.45	12,000 @ 130°F	N
glucose	1.42	10,000 @ 100°F	
molasses			
A	1.4 to 1.47	280 to 5000 @ 100°F	
B	1.43 to 1.48	1400 to 13,000 @ 100°F	
C	1.46 to 1.49	2600 to 5500 @ 100°F	
sugar syrups			
60 brix	1.29	75 at 60°F	N
68 brix	1.34	360 @ 60°F	N
76 brix	1.39	4000 @ 60°F	N
Water and Waste Treatment			
clarified sewage sludge	1.1	2000 range	

Used with permission from *Waukesha Pump Engineering Manual*, © 1976 by Waukesha Cherry-Burrell. A United Dominion Company.

APPENDIX 14.H
Vapor Pressures of Various Hydrocarbons and Water
(Cox chart)

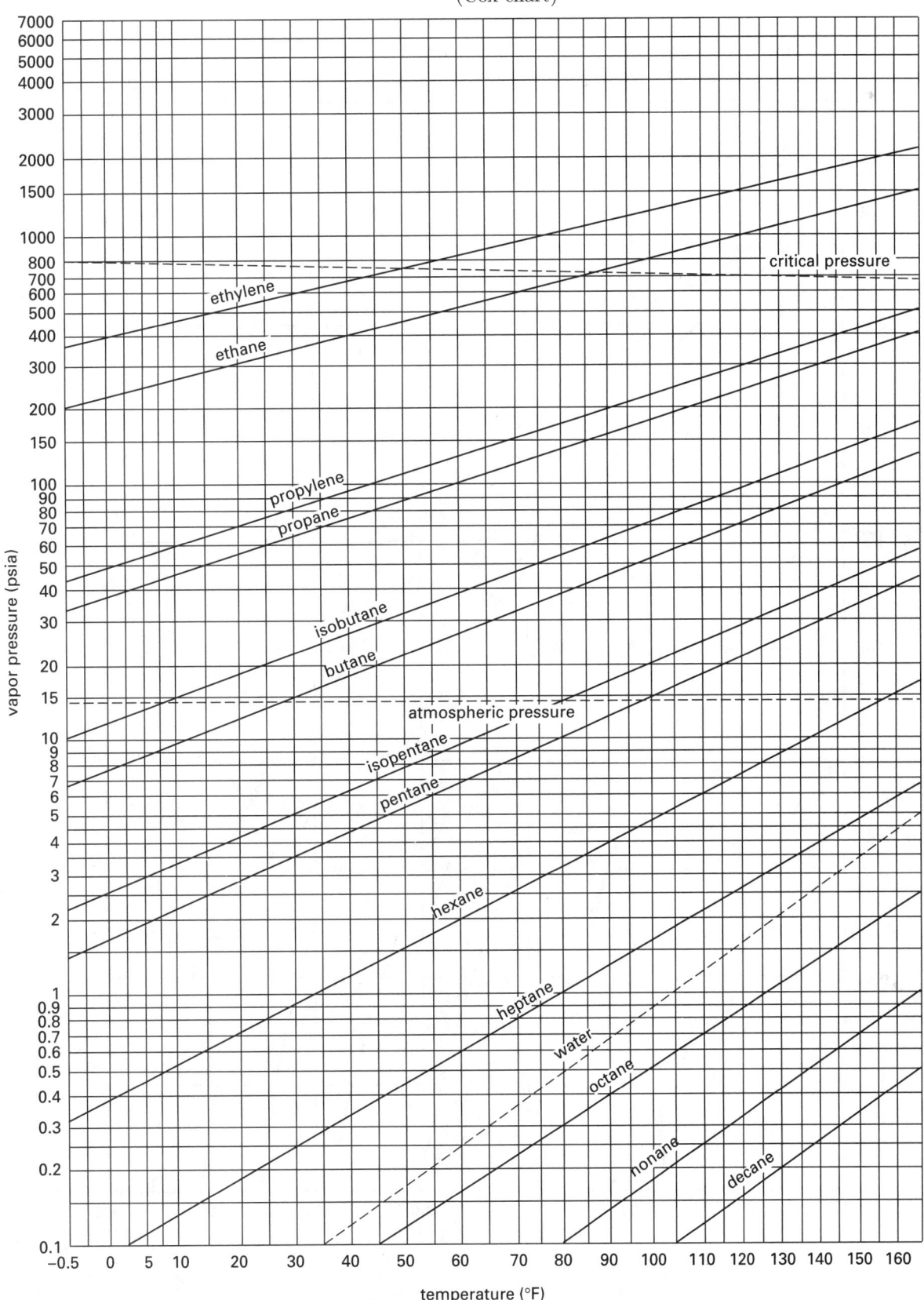

Used with permission from *Hydraulic Handbook*, 10th ed., by Colt Industries, Fairbanks Morse Pump Division, Kansas City, Kansas, 1977.

APPENDIX 14.I
Specific Gravity of Hydrocarbons

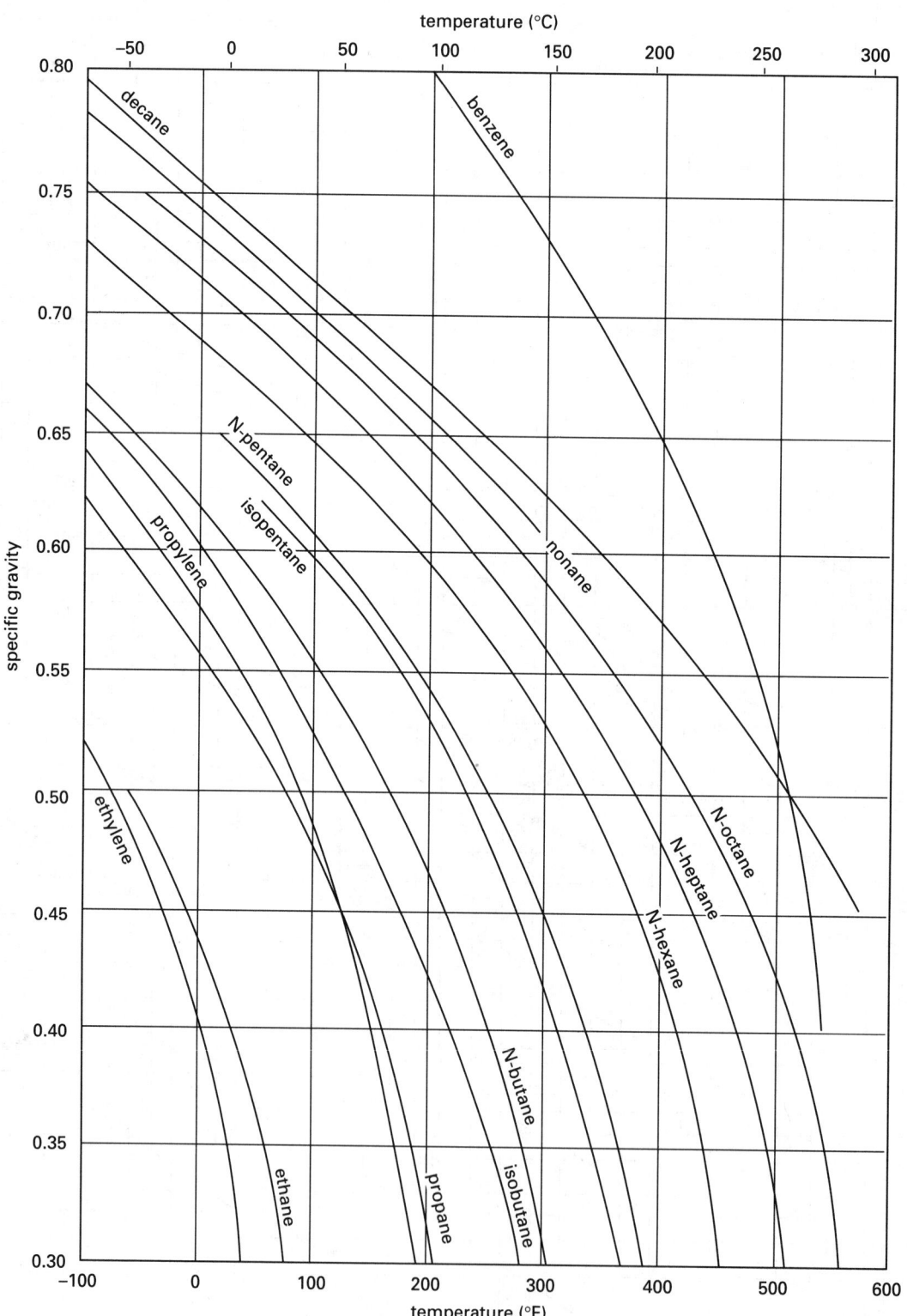

Reproduced with permission of Ingersoll-Dresser Pump Company from *Cameron Hydraulic Data: A handy reference on the subject of hydraulics and steam*, 18th Edition, edited by C. C. Heald, Ingersoll-Dresser Pumps, 1995.

APPENDIX 14.J
Viscosity Conversion Chart
(Approximate Conversions for Newtonian Fluids)

APPENDIX 14.K
Viscosity Index Chart: 0–100 V.I.
(per ASTM D2270)

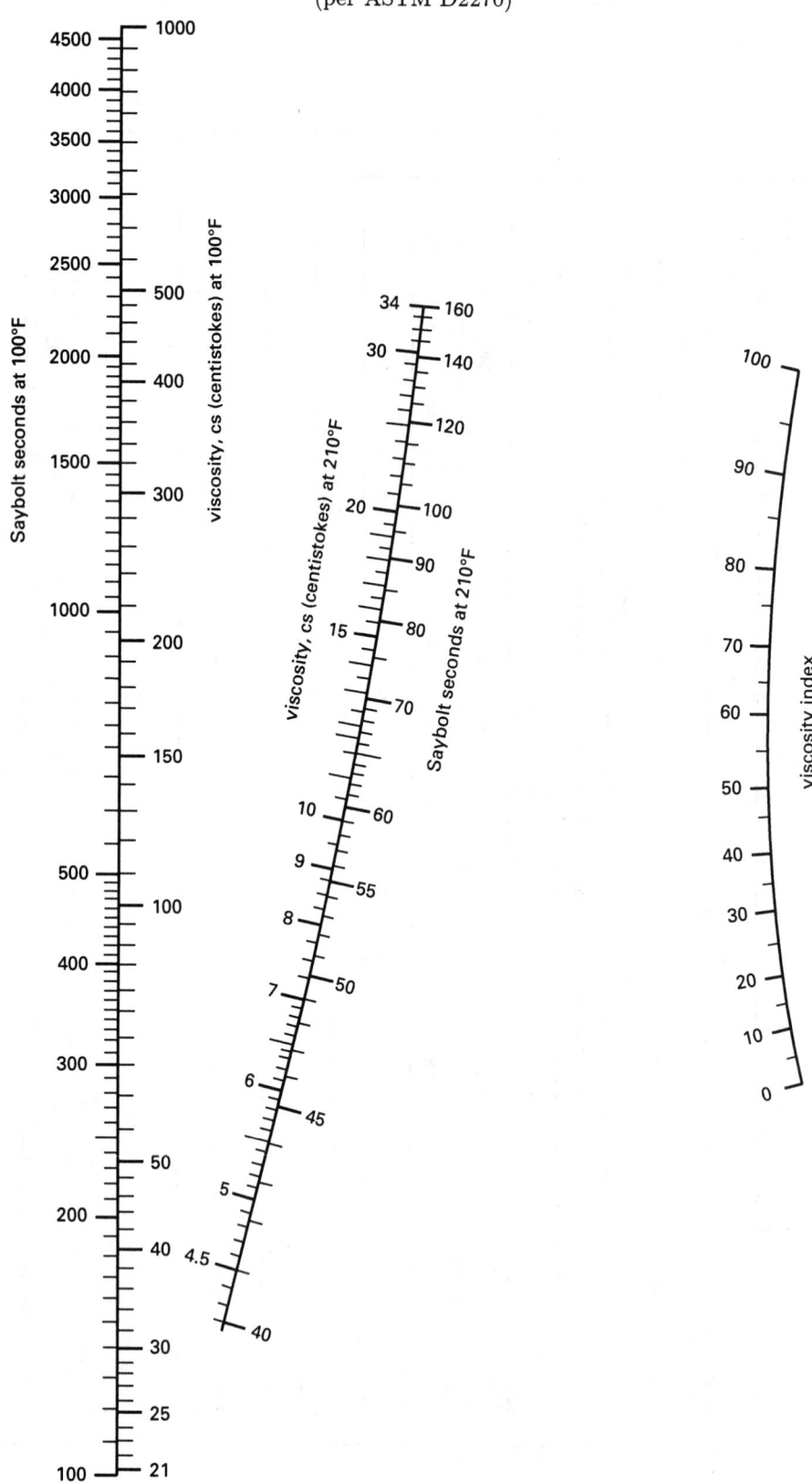

Copyright ASTM. Reprinted with permission.

APPENDIX 16.A
Area, Wetted Perimeter, and Hydraulic Radius
of Partially Filled Circular Pipes

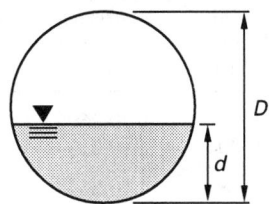

$\dfrac{d}{D}$	$\dfrac{\text{area}}{D^2}$	$\dfrac{\text{wetted perimeter}}{D}$	$\dfrac{r_h}{D}$	$\dfrac{d}{D}$	$\dfrac{\text{area}}{D^2}$	$\dfrac{\text{wetted perimeter}}{D}$	$\dfrac{r_h}{D}$
0.01	0.0013	0.2003	0.0066	0.51	0.4027	1.5908	0.2531
0.02	0.0037	0.2838	0.0132	0.52	0.4127	1.6108	0.2561
0.03	0.0069	0.3482	0.0197	0.53	0.4227	1.6308	0.2591
0.04	0.0105	0.4027	0.0262	0.54	0.4327	1.6509	0.2620
0.05	0.0147	0.4510	0.0326	0.55	0.4426	1.6710	0.2649
0.06	0.0192	0.4949	0.0389	0.56	0.4526	1.6911	0.2676
0.07	0.0242	0.5355	0.0451	0.57	0.4625	1.7113	0.2703
0.08	0.0294	0.5735	0.0513	0.58	0.4723	1.7315	0.2728
0.09	0.0350	0.6094	0.0574	0.59	0.4822	1.7518	0.2753
0.10	0.0409	0.6435	0.0635	0.60	0.4920	1.7722	0.2776
0.11	0.0470	0.6761	0.0695	0.61	0.5018	1.7926	0.2797
0.12	0.0534	0.7075	0.0754	0.62	0.5115	1.8132	0.2818
0.13	0.0600	0.7377	0.0813	0.63	0.5212	1.8338	0.2839
0.14	0.0688	0.7670	0.0871	0.64	0.5308	1.8546	0.2860
0.15	0.0739	0.7954	0.0929	0.65	0.5404	1.8755	0.2881
0.16	0.0811	0.8230	0.0986	0.66	0.5499	1.8965	0.2899
0.17	0.0885	0.8500	0.1042	0.67	0.5594	1.9177	0.2917
0.18	0.0961	0.8763	0.1097	0.68	0.5687	1.9391	0.2935
0.19	0.1039	0.9020	0.1152	0.69	0.5780	1.9606	0.2950
0.20	0.1118	0.9273	0.1206	0.70	0.5872	1.9823	0.2962
0.21	0.1199	0.9521	0.1259	0.71	0.5964	2.0042	0.2973
0.22	0.1281	0.9764	0.1312	0.72	0.6054	2.0264	0.2984
0.23	0.1365	1.0003	0.1364	0.73	0.6143	2.0488	0.2995
0.24	0.1449	1.0239	0.1416	0.74	0.6231	2.0714	0.3006
0.25	0.1535	1.0472	0.1466	0.75	0.6318	2.0944	0.3017
0.26	0.1623	1.0701	0.1516	0.76	0.6404	2.1176	0.3025
0.27	0.1711	1.0928	0.1566	0.77	0.6489	2.1412	0.3032
0.28	0.1800	1.1152	0.1614	0.78	0.6573	2.1652	0.3037
0.29	0.1890	1.1373	0.1662	0.79	0.6655	2.1895	0.3040
0.30	0.1982	1.1593	0.1709	0.80	0.6736	2.2143	0.3042
0.31	0.2074	1.1810	0.1755	0.81	0.6815	2.2395	0.3044
0.32	0.2167	1.2025	0.1801	0.82	0.6893	2.2653	0.3043
0.33	0.2260	1.2239	0.1848	0.83	0.6969	2.2916	0.3041
0.34	0.2355	1.2451	0.1891	0.84	0.7043	2.3186	0.3038
0.35	0.2450	1.2661	0.1935	0.85	0.7115	2.3462	0.3033
0.36	0.2546	1.2870	0.1978	0.86	0.7186	2.3746	0.3026
0.37	0.2642	1.3078	0.2020	0.87	0.7254	2.4038	0.3017
0.38	0.2739	1.3284	0.2061	0.88	0.7320	2.4341	0.3008
0.39	0.2836	1.3490	0.2102	0.89	0.7384	2.4655	0.2995
0.40	0.2934	1.3694	0.2142	0.90	0.7445	2.4981	0.2980
0.41	0.3032	1.3898	0.2181	0.91	0.7504	2.5322	0.2963
0.42	0.3130	1.4101	0.2220	0.92	0.7560	2.5681	0.2944
0.43	0.3229	1.4303	0.2257	0.93	0.7612	2.6061	0.2922
0.44	0.3328	1.4505	0.2294	0.94	0.7662	2.6467	0.2896
0.45	0.3428	1.4706	0.2331	0.95	0.7707	2.6906	0.2864
0.46	0.3527	1.4907	0.2366	0.96	0.7749	2.7389	0.2830
0.47	0.3627	1.5108	0.2400	0.97	0.7785	2.7934	0.2787
0.48	0.3727	1.5308	0.2434	0.98	0.7816	2.8578	0.2735
0.49	0.3827	1.5508	0.2467	0.99	0.7841	2.9412	0.2665
0.50	0.3927	1.5708	0.2500	1.00	0.7854	3.1416	0.2500

APPENDIX 16.B
Dimensions of Welded and Seamless Steel Pipe[a,b]
(selected sizes)[c]
(English units)

nominal diameter (in)	schedule	outside diameter (in)	wall thickness (in)	internal diameter (in)	internal area (in^2)	internal diameter (ft)	internal area (ft^2)
$^1/_8$	40 (S)	0.405	0.068	0.269	0.0568	0.0224	0.00039
	80 (X)		0.095	0.215	0.0363	0.0179	0.00025
$^1/_4$	40 (S)	0.540	0.088	0.364	0.1041	0.0303	0.00072
	80 (X)		0.119	0.302	0.0716	0.0252	0.00050
$^3/_8$	40 (S)	0.675	0.091	0.493	0.1909	0.0411	0.00133
	80 (X)		0.126	0.423	0.1405	0.0353	0.00098
$^1/_2$	40 (S)	0.840	0.109	0.622	0.3039	0.0518	0.00211
	80 (X)		0.147	0.546	0.2341	0.0455	0.00163
	160		0.187	0.466	0.1706	0.0388	0.00118
	(XX)		0.294	0.252	0.0499	0.0210	0.00035
$^3/_4$	40 (S)	1.050	0.113	0.824	0.5333	0.0687	0.00370
	80 (X)		0.154	0.742	0.4324	0.0618	0.00300
	160		0.218	0.614	0.2961	0.0512	0.00206
	(XX)		0.308	0.434	0.1479	0.0362	0.00103
1	40 (S)	1.315	0.133	1.049	0.8643	0.0874	0.00600
	80 (X)		0.179	0.957	0.7193	0.0798	0.00500
	160		0.250	0.815	0.5217	0.0679	0.00362
	(XX)		0.358	0.599	0.2818	0.0499	0.00196
$1^1/_4$	40 (S)	1.660	0.140	1.380	1.496	0.1150	0.01039
	80 (X)		0.191	1.278	1.283	0.1065	0.00890
	160		0.250	1.160	1.057	0.0967	0.00734
	(XX)		0.382	0.896	0.6305	0.0747	0.00438
$1^1/_2$	40 (S)	1.900	0.145	1.610	2.036	0.1342	0.01414
	80 (X)		0.200	1.500	1.767	0.1250	0.01227
	160		0.281	1.338	1.406	0.1115	0.00976
	(XX)		0.400	1.100	0.9503	0.0917	0.00660
2	40 (S)	2.375	0.154	2.067	3.356	0.1723	0.02330
	80 (X)		0.218	1.939	2.953	0.1616	0.02051
	160		0.343	1.689	2.240	0.1408	0.01556
	(XX)		0.436	1.503	1.774	0.1253	0.01232
$2^1/_2$	40 (S)	2.875	0.203	2.469	4.788	0.2058	0.03325
	80 (X)		0.276	2.323	4.238	0.1936	0.02943
	160		0.375	2.125	3.547	0.1771	0.02463
	(XX)		0.552	1.771	2.464	0.1476	0.01711
3	40 (S)	3.500	0.216	3.068	7.393	0.2557	0.05134
	80 (X)		0.300	2.900	6.605	0.2417	0.04587
	160		0.437	2.626	5.416	0.2188	0.03761
	(XX)		0.600	2.300	4.155	0.1917	0.02885
$3^1/_2$	40 (S)	4.000	0.226	3.548	9.887	0.2957	0.06866
	80 (X)		0.318	3.364	8.888	0.2803	0.06172
4	40 (S)	4.500	0.237	4.026	12.73	0.3355	0.08841
	80 (X)		0.337	3.826	11.50	0.3188	0.07984
	120		0.437	3.626	10.33	0.3022	0.07171
	160		0.531	3.438	9.283	0.2865	0.06447
	(XX)		0.674	3.152	7.803	0.2627	0.05419
5	40 (S)	5.563	0.258	5.047	20.01	0.4206	0.1389
	80 (X)		0.375	4.813	18.19	0.4011	0.1263
	120		0.500	4.563	16.35	0.3803	0.1136
	160		0.625	4.313	14.61	0.3594	0.1015
	(XX)		0.750	4.063	12.97	0.3386	0.09004

(continued)

APPENDIX 16.B *(continued)*
Dimensions of Welded and Seamless Steel Pipe[a,b]
(selected sizes)[c]
(English units)

nominal diameter (in)	schedule	outside diameter (in)	wall thickness (in)	internal diameter (in)	internal area (in^2)	internal diameter (ft)	internal area (ft^2)
6	40 (S)	6.625	0.280	6.065	28.89	0.5054	0.2006
	80 (X)		0.432	5.761	26.07	0.4801	0.1810
	120		0.562	5.501	23.77	0.4584	0.1650
	160		0.718	5.189	21.15	0.4324	0.1469
	(XX)		0.864	4.897	18.83	0.4081	0.1308
8	20	8.625	0.250	8.125	51.85	0.6771	0.3601
	30		0.277	8.071	51.16	0.6726	0.3553
	40 (S)		0.322	7.981	50.03	0.6651	0.3474
	60		0.406	7.813	47.94	0.6511	0.3329
	80 (X)		0.500	7.625	45.66	0.6354	0.3171
	100		0.593	7.439	43.46	0.6199	0.3018
	120		0.718	7.189	40.59	0.5990	0.2819
	140		0.812	7.001	38.50	0.5834	0.2673
	(XX)		0.875	6.875	37.12	0.5729	0.2578
	160		0.906	6.813	36.46	0.5678	0.2532
10	20	10.75	0.250	10.250	82.52	0.85417	0.5730
	30		0.307	10.136	80.69	0.84467	0.5604
	40 (S)		0.365	10.020	78.85	0.83500	0.5476
	60 (X)		0.500	9.750	74.66	0.8125	0.5185
	80		0.593	9.564	71.84	0.7970	0.4989
	100		0.718	9.314	68.13	0.7762	0.4732
	120		0.843	9.064	64.53	0.7553	0.4481
	140 (XX)		1.000	8.750	60.13	0.7292	0.4176
	160		1.125	8.500	56.75	0.7083	0.3941
12	20	12.75	0.250	12.250	117.86	1.0208	0.8185
	30		0.330	12.090	114.80	1.0075	0.7972
	(S)		0.375	12.000	113.10	1.0000	0.7854
	40		0.406	11.938	111.93	0.99483	0.7773
	(X)		0.500	11.750	108.43	0.97917	0.7530
	60		0.562	11.626	106.16	0.96883	0.7372
	80		0.687	11.376	101.64	0.94800	0.7058
	100		0.843	11.064	96.14	0.92200	0.6677
	120 (XX)		1.000	10.750	90.76	0.89583	0.6303
	140		1.125	10.500	86.59	0.87500	0.6013
	160		1.312	10.126	80.53	0.84383	0.5592
14 O.D.	10	14.00	0.250	13.500	143.14	1.1250	0.9940
	20		0.312	13.376	140.52	1.1147	0.9758
	30 (S)		0.375	13.250	137.89	1.1042	0.9575
	40		0.437	13.126	135.32	1.0938	0.9397
	(X)		0.500	13.000	132.67	1.0833	0.9213
	60		0.593	12.814	128.96	1.0679	0.8956
	80		0.750	12.500	122.72	1.0417	0.8522
	100		0.937	12.126	115.48	1.0105	0.8020
	120		1.093	11.814	109.62	0.98450	0.7612
	140		1.250	11.500	103.87	0.95833	0.7213
	160		1.406	11.188	98.31	0.93233	0.6827
16 O.D.	10	16.00	0.250	15.500	188.69	1.2917	1.3104
	20		0.312	15.376	185.69	1.2813	1.2895
	30 (S)		0.375	15.250	182.65	1.2708	1.2684
	40 (X)		0.500	15.000	176.72	1.2500	1.2272
	60		0.656	14.688	169.44	1.2240	1.1767
	80		0.843	14.314	160.92	1.1928	1.1175
	100		1.031	13.938	152.58	1.1615	1.0596
	120		1.218	13.564	144.50	1.1303	1.0035
	140		1.437	13.126	135.32	1.0938	0.9397
	160		1.593	12.814	128.96	1.0678	0.8956

(continued)

APPENDIX 16.B *(continued)*
Dimensions of Welded and Seamless Steel Pipe[a,b]
(selected sizes)[c]
(English units)

nominal diameter (in)	schedule	outside diameter (in)	wall thickness (in)	internal diameter (in)	internal area (in²)	internal diameter (ft)	internal area (ft²)
18 O.D.	10	18.00	0.250	17.500	240.53	1.4583	1.6703
	20		0.312	17.376	237.13	1.4480	1.6467
	(S)		0.375	17.250	233.71	1.4375	1.6230
	30		0.437	17.126	230.36	1.4272	1.5997
	(X)		0.500	17.000	226.98	1.4167	1.5762
	40		0.562	16.876	223.68	1.4063	1.5533
	60		0.750	16.500	213.83	1.3750	1.4849
	80		0.937	16.126	204.24	1.3438	1.4183
	100		1.156	15.688	193.30	1.3073	1.3423
	120		1.375	15.250	182.65	1.2708	1.2684
	140		1.562	14.876	173.81	1.2397	1.2070
	160		1.781	14.438	163.72	1.2032	1.1370
20 O.D.	10	20.00	0.250	19.500	298.65	1.6250	2.0739
	20 (S)		0.375	19.250	291.04	1.6042	2.0211
	30 (X)		0.500	19.000	283.53	1.5833	1.9689
	40		0.593	18.814	278.00	1.5678	1.9306
	60		0.812	18.376	265.21	1.5313	1.8417
	80		1.031	17.938	252.72	1.4948	1.7550
	100		1.281	17.438	238.83	1.4532	1.6585
	120		1.500	17.000	226.98	1.4167	1.5762
	140		1.750	16.500	213.83	1.3750	1.4849
	160		1.968	16.064	202.67	1.3387	1.4075
24 O.D.	10	24.00	0.250	23.500	433.74	1.9583	3.0121
	20 (S)		0.375	23.250	424.56	1.9375	2.9483
	(X)		0.500	23.000	415.48	1.9167	2.8852
	30		0.562	22.876	411.01	1.9063	2.8542
	40		0.687	22.626	402.07	1.8855	2.7922
	60		0.968	22.060	382.20	1.8383	2.6542
	80		1.218	21.564	365.21	1.7970	2.5362
	100		1.531	20.938	344.32	1.7448	2.3911
	120		1.812	20.376	326.92	1.6980	2.2645
	140		2.062	19.876	310.28	1.6563	2.1547
	160		2.343	19.310	292.87	1.6092	2.0337
30 O.D.	10	30.00	0.312	29.376	677.76	2.4480	4.7067
	(S)		0.375	29.250	671.62	2.4375	4.6640
	20 (X)		0.500	29.000	660.52	2.4167	4.5869
	30		0.625	28.750	649.18	2.3958	4.5082

(Multiply in by 25.4 to obtain mm.)
(Multiply in² by 645 to obtain mm².)
[a]Designations per ANSI B36.10.
[b]The "S" wall thickness was formerly designated as "standard weight." Standard weight and schedule-40 are the same for all diameters through 10 in. For diameters between 12 in and 24 in, standard weight pipe has a wall thickness of 0.75 in. The "X" wall thickness was formerly designated as "extra strong." Extra strong weight and schedule-80 are the same for all diameters through 8 in. For diameters between 10 in and 24 in, extra strong weight pipe has a wall thickness of 0.50 in. The "XX" wall thickness was formerly designed as "double extra strong." Double extra strong weight pipe does not have a corresponding schedule number.
[c]Pipe sizes and weights in most common usage listed. Other weights and sizes exist.

APPENDIX 16.C
Dimensions of Welded and Seamless Steel Pipe
Schedules 40 and 80
(SI units)

nominal pipe size (in)	schedule number	wall thickness (mm)	inside diameter (mm)	inside cross-sectional area (m²)
$1/8$	40	1.73	6.83	0.3664×10^{-4}
	80	2.41	5.46	0.2341×10^{-4}
$1/4$	40	2.24	9.25	0.6720×10^{-4}
	80	3.02	7.67	0.4620×10^{-4}
$3/8$	40	2.31	12.52	1.231×10^{-4}
	80	3.20	10.74	0.9059×10^{-4}
$1/2$	40	2.77	15.80	1.961×10^{-4}
	80	3.73	13.87	1.511×10^{-4}
$3/4$	40	2.87	20.93	3.441×10^{-4}
	80	3.91	18.85	2.791×10^{-4}
1	40	3.38	26.64	5.574×10^{-4}
	80	4.45	24.31	4.641×10^{-4}
$1 1/4$	40	3.56	35.05	9.648×10^{-4}
	80	4.85	32.46	8.275×10^{-4}
$1 1/2$	40	3.68	40.89	13.13×10^{-4}
	80	5.08	38.10	11.40×10^{-4}
2	40	3.91	52.50	21.65×10^{-4}
	80	5.54	49.25	19.05×10^{-4}
$2 1/2$	40	5.16	62.71	30.89×10^{-4}
	80	7.01	59.00	27.30×10^{-4}
3	40	5.49	77.92	47.69×10^{-4}
	80	7.62	73.66	42.61×10^{-4}
$3 1/2$	40	5.74	90.12	63.79×10^{-4}
	80	8.08	85.45	57.35×10^{-4}
4	40	6.02	102.3	82.19×10^{-4}
	80	8.56	97.18	74.17×10^{-4}
5	40	6.55	128.2	129.1×10^{-4}
	80	9.53	122.3	117.5×10^{-4}
6	40	7.11	154.1	186.5×10^{-4}
	80	10.97	146.3	168.1×10^{-4}
8	40	8.18	202.7	322.7×10^{-4}
	80	12.70	193.7	294.7×10^{-4}
10	40	9.27	254.5	508.6×10^{-4}
	80	15.06	242.9	463.4×10^{-4}
12	40	10.31	303.2	721.9×10^{-4}
	80	17.45	289.0	655.6×10^{-4}
14	40	11.10	333.4	872.8×10^{-4}
	80	19.05	317.5	791.5×10^{-4}
16	40	12.70	381.0	1140×10^{-4}
	80	21.41	363.6	1038×10^{-4}
18	40	14.05	428.7	1443×10^{-4}
	80	23.80	409.6	1317×10^{-4}
20	40	15.06	477.9	1793×10^{-4}
	80	26.19	455.6	1630×10^{-4}
24	40	17.45	574.7	2593×10^{-4}
	80	30.94	547.7	2356×10^{-4}

APPENDIX 16.D
Dimensions of Copper Water Tubing
(English units)

classification	nominal tube size (in)	outside diameter (in)	wall thickness (in)	inside diameter (in)	transverse area (in²)	safe working pressure (psi)
hard	1/4	3/8	0.025	0.325	0.083	1000
	3/8	1/2	0.025	0.450	0.159	1000
	1/2	5/8	0.028	0.569	0.254	890
	3/4	7/8	0.032	0.811	0.516	710
	1	1 1/8	0.035	1.055	0.874	600
	1 1/4	1 3/8	0.042	1.291	1.309	590
type "M" 250 psi working pressure	1 1/2	1 5/8	0.049	1.527	1.831	580
	2	2 1/8	0.058	2.009	3.17	520
	2 1/2	2 5/8	0.065	2.495	4.89	470
	3	3 1/8	0.072	2.981	6.98	440
	3 1/2	3 5/8	0.083	3.459	9.40	430
	4	4 1/8	0.095	3.935	12.16	430
	5	5 1/8	0.109	4.907	18.91	400
	6	6 1/8	0.122	5.881	27.16	375
	8	8 1/8	0.170	7.785	47.6	375
hard	3/8	1/2	0.035	0.430	0.146	1000
	1/2	5/8	0.040	0.545	0.233	1000
	3/4	7/8	0.045	0.785	0.484	1000
	1	1 1/8	0.050	1.025	0.825	880
type "L" 250 psi working pressure	1 1/4	1 3/8	0.055	1.265	1.256	780
	1 1/2	1 5/8	0.060	1.505	1.78	720
	2	2 1/8	0.070	1.985	3.094	640
	2 1/2	2 5/8	0.080	2.465	4.77	580
	3	3 1/8	0.090	2.945	6.812	550
	3 1/2	3 5/8	0.100	3.425	9.213	530
	4	4 1/8	0.110	3.905	11.97	510
	5	5 1/8	0.125	4.875	18.67	460
	6	6 1/8	0.140	5.845	26.83	430
hard	1/4	3/8	0.032	0.311	0.076	1000
	3/8	1/2	0.049	0.402	0.127	1000
	1/2	5/8	0.049	0.527	0.218	1000
	3/4	7/8	0.065	0.745	0.436	1000
	1	1 1/8	0.065	0.995	0.778	780
type "K" 400 psi working pressure	1 1/4	1 3/8	0.065	1.245	1.217	630
	1 1/2	1 5/8	0.072	1.481	1.722	580
	2	2 1/8	0.083	1.959	3.014	510
	2 1/2	2 5/8	0.095	2.435	4.656	470
	3	3 1/8	0.109	2.907	6.637	450
	3 1/2	3 5/8	0.120	3.385	8.999	430
	4	4 1/8	0.134	3.857	11.68	420
	5	5 1/8	0.160	4.805	18.13	400
	6	6 1/8	0.192	5.741	25.88	400
soft	1/4	3/8	0.032	0.311	0.076	1000
	3/8	1/2	0.049	0.402	0.127	1000
	1/2	5/8	0.049	0.527	0.218	1000
	3/4	7/8	0.065	0.745	0.436	1000
	1	1 1/8	0.065	0.995	0.778	780
type "K" 250 psi working pressure	1 1/4	1 3/8	0.065	1.245	1.217	630
	1 1/2	1 5/8	0.072	1.481	1.722	580
	2	2 1/8	0.083	1.959	3.014	510
	2 1/2	2 5/8	0.095	2.435	4.656	470
	3	3 1/8	0.109	2.907	6.637	450
	1 1/2	2 5/8	0.120	3.385	8.999	430
	4	4 1/8	0.134	3.857	11.68	420
	5	5 2/8	0.160	4.805	18.13	400
	6	6 1/8	0.192	5.741	25.88	400

(Multiply in by 25.4 to obtain mm. Multiply in² by 645 to obtain mm².)

APPENDIX 16.E
Dimensions of Brass and Copper Tubing
(English units)

regular

pipe size (in)	nominal dimensions (in)			cross-sectional area of bore (in²)	lbm/ft	
	O.D.	I.D.	wall		red brass	copper
1/8	0.405	0.281	0.062	0.062	0.253	0.259
1/4	0.540	0.376	0.082	0.110	0.447	0.457
3/8	0.675	0.495	0.090	0.192	0.627	0.641
1/2	0.840	0.626	0.107	0.307	0.934	0.955
3/4	1.050	0.822	0.114	0.531	1.270	1.300
1	1.315	1.063	0.126	0.887	1.780	1.820
1 1/4	1.660	1.368	0.146	1.470	2.630	2.690
1 1/2	1.900	1.600	0.150	2.010	3.130	3.200
2	2.375	2.063	0.156	3.340	4.120	4.220
2 1/2	2.875	2.501	0.187	4.910	5.990	6.120
3	3.500	3.062	0.219	7.370	8.560	8.750
3 1/2	4.000	3.500	0.250	9.620	11.200	11.400
4	4.500	4.000	0.250	12.600	12.700	12.900
5	5.562	5.062	0.250	20.100	15.800	16.200
6	6.625	6.125	0.250	29.500	19.000	19.400
8	8.625	8.001	0.312	50.300	30.900	31.600
10	10.750	10.020	0.365	78.800	45.200	46.200
12	12.750	12.000	0.375	113.000	55.300	56.500

extra strong

pipe size (in)	nominal dimensions (in)			cross-sectional area of bore (in²)	lbm/ft	
	O.D.	I.D.	wall		red brass	copper
1/8	0.405	0.205	0.100	0.033	0.363	0.371
1/4	0.540	0.294	0.123	0.068	0.611	0.625
3/8	0.675	0.421	0.127	0.139	0.829	0.847
1/2	0.840	0.542	0.149	0.231	1.230	1.250
3/4	1.050	0.736	0.157	0.425	1.670	1.710
1	1.315	0.951	0.182	0.710	2.460	2.510
1 1/4	1.660	1.272	0.194	1.270	3.390	3.460
1 1/2	1.990	1.494	0.203	1.750	4.100	4.190
2	2.375	1.933	0.221	2.94	5.670	5.800
2 1/2	2.875	2.315	0.280	4.21	8.660	8.850
3	3.500	2.892	0.304	6.57	11.600	11.800
3 1/2	4.000	3.358	0.321	8.86	14.100	14.400
4	4.500	3.818	0.341	11.50	16.900	17.300
5	5.562	4.812	0.375	18.20	23.200	23.700
6	6.625	5.751	0.437	26.00	32.200	32.900
8	8.625	7.625	0.500	45.70	48.400	49.500
10	10.750	9.750	0.500	74.70	61.100	62.400

(Multiply in by 25.4 to obtain mm.)
(Multiply in² by 645 to obtain mm².)

Support Material

APPENDIX 16.F
Dimensions of Seamless Steel Boiler (BWG) Tubing[a,b,c]
(English units)

O.D. (in)	BWG	wall thickness (in)	O.D. (in)	BWG	wall thickness (in)
1	13	0.095	3	12	0.109
	12	0.109		11	0.120
	11	0.120		10	0.134
	10	0.134		9	0.148
1¼	13	0.095	3¼	11	0.120
	12	0.109		10	0.134
	11	0.120		9	0.148
	10	0.134		8	0.165
1½	13	0.095	3½	11	0.120
	12	0.109		10	0.134
	11	0.120		9	0.148
	10	0.134		8	0.065
1¾	13	0.095	4	10	0.134
	12	0.109		9	0.148
	11	0.120		8	0.165
	10	0.134		7	0.180
2	13	0.095	4½	10	0.134
	12	0.109		9	0.148
	11	0.120		8	0.165
	10	0.034		7	0.180
2¼	13	0.095	5	9	0.148
	12	0.109		8	0.165
	11	0.120		7	0.180
	10	0.134		6	0.203
2½	12	0.109	5½	9	0.148
	11	0.120		8	0.165
	10	0.134		7	0.180
	9	0.148		6	0.203
2¾	12	0.109	6	7	0.180
	11	0.120		6	0.203
	10	0.134		5	0.220
	9	0.148		4	0.238

(Multiply in by 25.4 to obtain mm.)
(Multiply in² by 645 to obtain mm².)
[a] Abstracted from information provided by United States Steel Corporation.
[b] Values in this table are not to be used for tubes in condensers and heat-exchangers unless those tubes are specified by BWG.
[c] Birmingham wire gauge, commonly used for ferrous tubing, is identical with Stubs iron-wire gauge.

APPENDIX 16.G
Dimensions of PVC Pipe[a,b]
(English units)

nominal size (in)	schedule	wall thickness[c] (in)	O.D. (in)	I.D. (in)
1/4	40	0.088	0.540	0.364
	80	0.119	0.540	0.302
1/2	40	0.109	0.840	0.622
	80	0.147	0.840	0.546
3/4	40	0.113	1.050	0.824
	80	0.154	1.050	0.742
1	40	0.133	1.315	1.049
	80	0.179	1.315	0.957
1 1/4	40	0.140	1.660	1.380
	80	0.191	1.660	1.278
1 1/2	40	0.145	1.900	1.610
	80	0.200	1.900	1.500
2	40	0.154	2.375	2.067
	80	0.218	2.375	1.939
3	40	0.216	3.500	3.068
	80	0.300	3.500	2.900
4	40	0.237	4.500	4.026
	80	0.337	4.500	3.826

(Multiply in by 25.4 to obtain mm.)
(Multiply in^2 by 645 to obtain mm^2.)
[a] Abstracted from ASTM Specification D1785-85.
[b] Two strengths of PVC are in use. A maximum (bursting) stress of 6400 psig (44 MPa) applies to PVC types 1120, 1220, and 4116. A maximum (bursting) stress of 5000 psig (35 MPa) applies to PVC types 2112, 2116, and 2120.
[c] Minimum wall thickness, with tolerances of −0%, +10%.

Support Material

APPENDIX 16.H
Standard ANSI Piping Symbols

	flanged	screwed	bell and spigot	welded	soldered
joint					
elbow—90°					
elbow—45°					
elbow—turned up					
elbow—turned down					
elbow—long radius					
reducing elbow					
tee					
tee—outlet up					
tee—outlet down					
side outlet tee—outlet up					
cross					
reducer—concentric					
reducer—eccentric					
lateral					
gate valve					
globe valve					
check valve					
stop cock					
safety valve					
expansion joint					
union					
sleeve					
bushing					

APPENDIX 17.A
Specific Roughness and Hazen-Williams Constants
for Various Pipe Materials

type of pipe or surface	ϵ (ft) range	design	C range	clean	design
steel					
welded and seamless	0.0001–0.0003	0.0002	150–80	140	100
interior riveted, no projecting rivets				139	100
projecting girth rivets				130	100
projecting girth and horizontal rivets				115	100
vitrified, spiral-riveted, flow with lap				110	100
vitrified, spiral-riveted, flow against lap				100	90
corrugated				60	60
mineral					
concrete	0.001–0.01	0.004	152–85	120	100
cement-asbestos			160–140	150	140
vitrified clays					110
brick sewer					100
iron					
cast, plain	0.0004–0.002	0.0008	150–80	130	100
cast, tar (asphalt) coated	0.0002–0.0006	0.0004	145–50	130	100
cast, cement lined	0.000008	0.000008		150	140
cast, bituminous lined	0.000008	0.000008	160–130	148	140
cast, centrifugally spun	0.00001	0.00001			
galvanized, plain	0.0002–0.0008	0.0005			
wrought, plain	0.0001–0.0003	0.0002	150–80	130	100
miscellaneous					
copper and brass	0.000005	0.000005	150–120	140	130
wood stave	0.0006–0.003	0.002	145–110	120	110
transite	0.000008	0.000008			
lead, tin, glass		0.000005	150–120	140	130
plastic		0.000005	150–120	140	130
fiberglass	0.000017	0.000017	160–150	150	150

Support Material

APPENDIX 17.B
Darcy Friction Factors
(turbulent flow)

relative roughness (ϵ/D)

Reynolds no.	0.00000	0.000001	0.0000015	0.00001	0.00002	0.00004	0.00005	0.00006	0.00008
2×10^3	0.0495	0.0495	0.0495	0.0495	0.0495	0.0495	0.0495	0.0495	0.0495
2.5×10^3	0.0461	0.0461	0.0461	0.0461	0.0461	0.0461	0.0461	0.0461	0.0461
3×10^3	0.0435	0.0435	0.0435	0.0435	0.0435	0.0436	0.0436	0.0436	0.0436
4×10^3	0.0399	0.0399	0.0399	0.0399	0.0399	0.0399	0.0400	0.0400	0.0400
5×10^3	0.0374	0.0374	0.0374	0.0374	0.0374	0.0374	0.0374	0.0375	0.0375
6×10^3	0.0355	0.0355	0.0355	0.0355	0.0355	0.0356	0.0356	0.0356	0.0356
7×10^3	0.0340	0.0340	0.0340	0.0340	0.0340	0.0341	0.0341	0.0341	0.0341
8×10^3	0.0328	0.0328	0.0328	0.0328	0.0328	0.0328	0.0329	0.0329	0.0329
9×10^3	0.0318	0.0318	0.0318	0.0318	0.0318	0.0318	0.0318	0.0319	0.0319
1×10^4	0.0309	0.0309	0.0309	0.0309	0.0309	0.0309	0.0310	0.0310	0.0310
1.5×10^4	0.0278	0.0278	0.0278	0.0278	0.0278	0.0279	0.0279	0.0279	0.0280
2×10^4	0.0259	0.0259	0.0259	0.0259	0.0259	0.0260	0.0260	0.0260	0.0261
2.5×10^4	0.0245	0.0245	0.0245	0.0245	0.0246	0.0246	0.0246	0.0247	0.0247
3×10^4	0.0235	0.0235	0.0235	0.0235	0.0235	0.0236	0.0236	0.0236	0.0237
4×10^4	0.0220	0.0220	0.0220	0.0220	0.0220	0.0221	0.0221	0.0222	0.0222
5×10^4	0.0209	0.0209	0.0209	0.0209	0.0210	0.0210	0.0211	0.0211	0.0212
6×10^4	0.0201	0.0201	0.0201	0.0201	0.0201	0.0202	0.0203	0.0203	0.0204
7×10^4	0.0194	0.0194	0.0194	0.0194	0.0195	0.0196	0.0196	0.0197	0.0197
8×10^4	0.0189	0.0189	0.0189	0.0189	0.0190	0.0190	0.0191	0.0191	0.0192
9×10^4	0.0184	0.0184	0.0184	0.0184	0.0185	0.0186	0.0186	0.0187	0.0188
1×10^5	0.0180	0.0180	0.0180	0.0180	0.0181	0.0182	0.0183	0.0183	0.0184
1.5×10^5	0.0166	0.0166	0.0166	0.0166	0.0167	0.0168	0.0169	0.0170	0.0171
2×10^5	0.0156	0.0156	0.0156	0.0157	0.0158	0.0160	0.0160	0.0161	0.0163
2.5×10^5	0.0150	0.0150	0.0150	0.0151	0.0152	0.0153	0.0154	0.0155	0.0157
3×10^5	0.0145	0.0145	0.0145	0.0146	0.0147	0.0149	0.0150	0.0151	0.0153
4×10^5	0.0137	0.0137	0.0137	0.0138	0.0140	0.0142	0.0143	0.0144	0.0146
5×10^5	0.0132	0.0132	0.0132	0.0133	0.0134	0.0137	0.0138	0.0140	0.0142
6×10^5	0.0127	0.0128	0.0128	0.0129	0.0131	0.0133	0.0135	0.0136	0.0139
7×10^5	0.0124	0.0124	0.0124	0.0126	0.0127	0.0131	0.0132	0.0134	0.0136
8×10^5	0.0121	0.0121	0.0121	0.0123	0.0125	0.0128	0.0130	0.0131	0.0134
9×10^5	0.0119	0.0119	0.0119	0.0121	0.0123	0.0126	0.0128	0.0130	0.0133
1×10^6	0.0116	0.0117	0.0117	0.0119	0.0121	0.0125	0.0126	0.0128	0.0131
1.5×10^6	0.0109	0.0109	0.0109	0.0112	0.0114	0.0119	0.0121	0.0123	0.0127
2×10^6	0.0104	0.0104	0.0104	0.0107	0.0110	0.0116	0.0118	0.0120	0.0124
2.5×10^6	0.0100	0.0100	0.0101	0.0104	0.0108	0.0113	0.0116	0.0118	0.0123
3×10^6	0.0097	0.0098	0.0098	0.0102	0.0105	0.0112	0.0115	0.0117	0.0122
4×10^6	0.0093	0.0094	0.0094	0.0098	0.0103	0.0110	0.0113	0.0115	0.0120
5×10^6	0.0090	0.0091	0.0091	0.0096	0.0101	0.0108	0.0111	0.0114	0.0119
6×10^6	0.0087	0.0088	0.0089	0.0094	0.0099	0.0107	0.0110	0.0113	0.0118
7×10^6	0.0085	0.0086	0.0087	0.0093	0.0098	0.0106	0.0110	0.0113	0.0118
8×10^6	0.0084	0.0085	0.0085	0.0092	0.0097	0.0106	0.0109	0.0112	0.0118
9×10^6	0.0082	0.0083	0.0084	0.0091	0.0097	0.0105	0.0109	0.0112	0.0117
1×10^7	0.0081	0.0082	0.0083	0.0090	0.0096	0.0105	0.0109	0.0112	0.0117
1.5×10^7	0.0076	0.0078	0.0079	0.0087	0.0094	0.0104	0.0108	0.0111	0.0116
2×10^7	0.0073	0.0075	0.0076	0.0086	0.0093	0.0103	0.0107	0.0110	0.0116
2.5×10^7	0.0071	0.0073	0.0074	0.0085	0.0093	0.0103	0.0107	0.0110	0.0116
3×10^7	0.0069	0.0072	0.0073	0.0084	0.0092	0.0103	0.0107	0.0110	0.0116
4×10^7	0.0067	0.0070	0.0071	0.0084	0.0092	0.0102	0.0106	0.0110	0.0115
5×10^7	0.0065	0.0068	0.0070	0.0083	0.0092	0.0102	0.0106	0.0110	0.0115

(continued)

APPENDIX 17.B *(continued)*
Darcy Friction Factors
(turbulent flow)

relative roughness (ϵ/D)

Reynolds no.	0.0001	0.0001501	0.00020	0.0003	0.00030	0.00035	0.0004	0.0006	0.0008
2×10^3	0.0495	0.0496	0.0496	0.0496	0.0497	0.0497	0.0498	0.0499	0.0501
2.5×10^3	0.0461	0.0462	0.0462	0.0463	0.0463	0.0463	0.0464	0.0466	0.0467
3×10^3	0.0436	0.0437	0.0437	0.0437	0.0438	0.0438	0.0439	0.0441	0.0442
4×10^3	0.0400	0.0401	0.0401	0.0402	0.0402	0.0403	0.0403	0.0405	0.0407
5×10^3	0.0375	0.0376	0.0376	0.0377	0.0377	0.0378	0.0378	0.0381	0.0383
6×10^3	0.0356	0.0357	0.0357	0.0358	0.0359	0.0359	0.0360	0.0362	0.0365
7×10^3	0.0341	0.0342	0.0343	0.0343	0.0344	0.0345	0.0345	0.0348	0.0350
8×10^3	0.0329	0.0330	0.0331	0.0331	0.0332	0.0333	0.0333	0.0336	0.0339
9×10^3	0.0319	0.0320	0.0321	0.0321	0.0322	0.0323	0.0323	0.0326	0.0329
1×10^4	0.0310	0.0311	0.0312	0.0313	0.0313	0.0314	0.0315	0.0318	0.0321
1.5×10^4	0.0280	0.0281	0.0282	0.0283	0.0284	0.0285	0.0285	0.0289	0.0293
2×10^4	0.0261	0.0262	0.0263	0.0264	0.0265	0.0266	0.0267	0.0272	0.0276
2.5×10^4	0.0248	0.0249	0.0250	0.0251	0.0252	0.0254	0.0255	0.0259	0.0264
3×10^4	0.0238	0.0239	0.0240	0.0241	0.0243	0.0244	0.0245	0.0250	0.0255
4×10^4	0.0223	0.0224	0.0226	0.0227	0.0229	0.0230	0.0232	0.0237	0.0243
5×10^4	0.0212	0.0214	0.0216	0.0218	0.0219	0.0221	0.0223	0.0229	0.0235
6×10^4	0.0205	0.0207	0.0208	0.0210	0.0212	0.0214	0.0216	0.0222	0.0229
7×10^4	0.0198	0.0200	0.0202	0.0204	0.0206	0.0208	0.0210	0.0217	0.0224
8×10^4	0.0193	0.0195	0.0198	0.0200	0.0202	0.0204	0.0206	0.0213	0.0220
9×10^4	0.0189	0.0191	0.0194	0.0196	0.0198	0.0200	0.0202	0.0210	0.0217
1×10^5	0.0185	0.0188	0.0190	0.0192	0.0195	0.0197	0.0199	0.0207	0.0215
1.5×10^5	0.0172	0.0175	0.0178	0.0181	0.0184	0.0186	0.0189	0.0198	0.0207
2×10^5	0.0164	0.0168	0.0171	0.0174	0.0177	0.0180	0.0183	0.0193	0.0202
2.5×10^5	0.0158	0.0162	0.0166	0.0170	0.0173	0.0176	0.0179	0.0190	0.0199
3×10^5	0.0154	0.0159	0.0163	0.0166	0.0170	0.0173	0.0176	0.0188	0.0197
4×10^5	0.0148	0.0153	0.0158	0.0162	0.0166	0.0169	0.0172	0.0184	0.0195
5×10^5	0.0144	0.0150	0.0154	0.0159	0.0163	0.0167	0.0170	0.0183	0.0193
6×10^5	0.0141	0.0147	0.0152	0.0157	0.0161	0.0165	0.0168	0.0181	0.0192
7×10^5	0.0139	0.0145	0.0150	0.0155	0.0159	0.0163	0.0167	0.0180	0.0191
8×10^5	0.0137	0.0143	0.0149	0.0154	0.0158	0.0162	0.0166	0.0180	0.0191
9×10^5	0.0136	0.0142	0.0148	0.0153	0.0157	0.0162	0.0165	0.0179	0.0190
1×10^6	0.0134	0.0141	0.0147	0.0152	0.0157	0.0161	0.0165	0.0178	0.0190
1.5×10^6	0.0130	0.0138	0.0144	0.0149	0.0154	0.0159	0.0163	0.0177	0.0189
2×10^6	0.0128	0.0136	0.0142	0.0148	0.0153	0.0158	0.0162	0.0176	0.0188
2.5×10^6	0.0127	0.0135	0.0141	0.0147	0.0152	0.0157	0.0161	0.0176	0.0188
3×10^6	0.0126	0.0134	0.0141	0.0147	0.0152	0.0157	0.0161	0.0176	0.0187
4×10^6	0.0124	0.0133	0.0140	0.0146	0.0151	0.0156	0.0161	0.0175	0.0187
5×10^6	0.0123	0.0132	0.0139	0.0146	0.0151	0.0156	0.0160	0.0175	0.0187
6×10^6	0.0123	0.0132	0.0139	0.0145	0.0151	0.0156	0.0160	0.0175	0.0187
7×10^6	0.0122	0.0132	0.0139	0.0145	0.0151	0.0155	0.0160	0.0175	0.0187
8×10^6	0.0122	0.0131	0.0139	0.0145	0.0150	0.0155	0.0160	0.0175	0.0187
9×10^6	0.0122	0.0131	0.0139	0.0145	0.0150	0.0155	0.0160	0.0175	0.0187
1×10^7	0.0122	0.0131	0.0138	0.0145	0.0150	0.0155	0.0160	0.0175	0.0186
1.5×10^7	0.0121	0.0131	0.0138	0.0144	0.0150	0.0155	0.0159	0.0174	0.0186
2×10^7	0.0121	0.0130	0.0138	0.0144	0.0150	0.0155	0.0159	0.0174	0.0186
2.5×10^7	0.0121	0.0130	0.0138	0.0144	0.0150	0.0155	0.0159	0.0174	0.0186
3×10^7	0.0120	0.0130	0.0138	0.0144	0.0150	0.0155	0.0159	0.0174	0.0186
4×10^7	0.0120	0.0130	0.0138	0.0144	0.0150	0.0155	0.0159	0.0174	0.0186
5×10^7	0.0120	0.0130	0.0138	0.0144	0.0150	0.0155	0.0159	0.0174	0.0186

(continued)

Support Material

APPENDIX 17.B *(continued)*
Darcy Friction Factors
(turbulent flow)

relative roughness (ϵ/D)

Reynolds no.	0.001	0.0015	0.002	0.0025	0.003	0.0035	0.004	0.006	0.008
2×10^3	0.0502	0.0506	0.0510	0.0513	0.0517	0.0521	0.0525	0.0539	0.0554
2.5×10^3	0.0469	0.0473	0.0477	0.0481	0.0485	0.0489	0.0493	0.0509	0.0524
3×10^3	0.0444	0.0449	0.0453	0.0457	0.0462	0.0466	0.0470	0.0487	0.0503
4×10^3	0.0409	0.0414	0.0419	0.0424	0.0429	0.0433	0.0438	0.0456	0.0474
5×10^3	0.0385	0.0390	0.0396	0.0401	0.0406	0.0411	0.0416	0.0436	0.0455
6×10^3	0.0367	0.0373	0.0378	0.0384	0.0390	0.0395	0.0400	0.0421	0.0441
7×10^3	0.0353	0.0359	0.0365	0.0371	0.0377	0.0383	0.0388	0.0410	0.0430
8×10^3	0.0341	0.0348	0.0354	0.0361	0.0367	0.0373	0.0379	0.0401	0.0422
9×10^3	0.0332	0.0339	0.0345	0.0352	0.0358	0.0365	0.0371	0.0394	0.0416
1×10^4	0.0324	0.0331	0.0338	0.0345	0.0351	0.0358	0.0364	0.0388	0.0410
1.5×10^4	0.0296	0.0305	0.0313	0.0320	0.0328	0.0335	0.0342	0.0369	0.0393
2×10^4	0.0279	0.0289	0.0298	0.0306	0.0315	0.0323	0.0330	0.0358	0.0384
2.5×10^4	0.0268	0.0278	0.0288	0.0297	0.0306	0.0314	0.0322	0.0352	0.0378
3×10^4	0.0260	0.0271	0.0281	0.0291	0.0300	0.0308	0.0317	0.0347	0.0374
4×10^4	0.0248	0.0260	0.0271	0.0282	0.0291	0.0301	0.0309	0.0341	0.0369
5×10^4	0.0240	0.0253	0.0265	0.0276	0.0286	0.0296	0.0305	0.0337	0.0365
6×10^4	0.0235	0.0248	0.0261	0.0272	0.0283	0.0292	0.0302	0.0335	0.0363
7×10^4	0.0230	0.0245	0.0257	0.0269	0.0280	0.0290	0.0299	0.0333	0.0362
8×10^4	0.0227	0.0242	0.0255	0.0267	0.0278	0.0288	0.0298	0.0331	0.0361
9×10^4	0.0224	0.0239	0.0253	0.0265	0.0276	0.0286	0.0296	0.0330	0.0360
1×10^5	0.0222	0.0237	0.0251	0.0263	0.0275	0.0285	0.0295	0.0329	0.0359
1.5×10^5	0.0214	0.0231	0.0246	0.0259	0.0271	0.0281	0.0292	0.0327	0.0357
2×10^5	0.0210	0.0228	0.0243	0.0256	0.0268	0.0279	0.0290	0.0325	0.0355
2.5×10^5	0.0208	0.0226	0.0241	0.0255	0.0267	0.0278	0.0289	0.0325	0.0355
3×10^5	0.0206	0.0225	0.0240	0.0254	0.0266	0.0277	0.0288	0.0324	0.0354
4×10^5	0.0204	0.0223	0.0239	0.0253	0.0265	0.0276	0.0287	0.0323	0.0354
5×10^5	0.0202	0.0222	0.0238	0.0252	0.0264	0.0276	0.0286	0.0323	0.0353
6×10^5	0.0201	0.0221	0.0237	0.0251	0.0264	0.0275	0.0286	0.0323	0.0353
7×10^5	0.0201	0.0221	0.0237	0.0251	0.0264	0.0275	0.0286	0.0322	0.0353
8×10^5	0.0200	0.0220	0.0237	0.0251	0.0263	0.0275	0.0286	0.0322	0.0353
9×10^5	0.0200	0.0220	0.0236	0.0251	0.0263	0.0275	0.0285	0.0322	0.0353
1×10^6	0.0199	0.0220	0.0236	0.0250	0.0263	0.0275	0.0285	0.0322	0.0353
1.5×10^6	0.0198	0.0219	0.0235	0.0250	0.0263	0.0274	0.0285	0.0322	0.0352
2×10^6	0.0198	0.0218	0.0235	0.0250	0.0262	0.0274	0.0285	0.0322	0.0352
2.5×10^6	0.0198	0.0218	0.0235	0.0249	0.0262	0.0274	0.0285	0.0322	0.0352
3×10^6	0.0197	0.0218	0.0235	0.0249	0.0262	0.0274	0.0285	0.0321	0.0352
4×10^6	0.0197	0.0218	0.0235	0.0249	0.0262	0.0274	0.0284	0.0321	0.0352
5×10^6	0.0197	0.0218	0.0235	0.0249	0.0262	0.0274	0.0284	0.0321	0.0352
6×10^6	0.0197	0.0218	0.0235	0.0249	0.0262	0.0274	0.0284	0.0321	0.0352
7×10^6	0.0197	0.0218	0.0234	0.0249	0.0262	0.0274	0.0284	0.0321	0.0352
8×10^6	0.0197	0.0218	0.0234	0.0249	0.0262	0.0274	0.0284	0.0321	0.0352
9×10^6	0.0197	0.0218	0.0234	0.0249	0.0262	0.0274	0.0284	0.0321	0.0352
1×10^7	0.0197	0.0218	0.0234	0.0249	0.0262	0.0273	0.0284	0.0321	0.0352
1.5×10^7	0.0197	0.0217	0.0234	0.0249	0.0262	0.0273	0.0284	0.0321	0.0352
2×10^7	0.0197	0.0217	0.0234	0.0249	0.0262	0.0273	0.0284	0.0321	0.0352
2.5×10^7	0.0196	0.0217	0.0234	0.0249	0.0262	0.0273	0.0284	0.0321	0.0352
3×10^7	0.0196	0.0217	0.0234	0.0249	0.0262	0.0273	0.0284	0.0321	0.0352
4×10^7	0.0196	0.0217	0.0234	0.0249	0.0262	0.0273	0.0284	0.0321	0.0352
5×10^7	0.0196	0.0217	0.0234	0.0249	0.0262	0.0273	0.0284	0.0321	0.0352

(continued)

APPENDIX 17.B *(continued)*
Darcy Friction Factors
(turbulent flow)

relative roughness (ϵ/D)

Reynolds no.	0.01	0.015	0.02	0.025	0.03	0.035	0.04	0.045	0.05
2×10^3	0.0568	0.0602	0.0635	0.0668	0.0699	0.0730	0.0760	0.0790	0.0819
2.5×10^3	0.0539	0.0576	0.0610	0.0644	0.0677	0.0709	0.0740	0.0770	0.0800
3×10^3	0.0519	0.0557	0.0593	0.0628	0.0661	0.0694	0.0725	0.0756	0.0787
4×10^3	0.0491	0.0531	0.0570	0.0606	0.0641	0.0674	0.0707	0.0739	0.0770
5×10^3	0.0473	0.0515	0.0555	0.0592	0.0628	0.0662	0.0696	0.0728	0.0759
6×10^3	0.0460	0.0504	0.0544	0.0583	0.0619	0.0654	0.0688	0.0721	0.0752
7×10^3	0.0450	0.0495	0.0537	0.0576	0.0613	0.0648	0.0682	0.0715	0.0747
8×10^3	0.0442	0.0489	0.0531	0.0571	0.0608	0.0644	0.0678	0.0711	0.0743
9×10^3	0.0436	0.0484	0.0526	0.0566	0.0604	0.0640	0.0675	0.0708	0.0740
1×10^4	0.0431	0.0479	0.0523	0.0563	0.0601	0.0637	0.0672	0.0705	0.0738
1.5×10^4	0.0415	0.0466	0.0511	0.0553	0.0592	0.0628	0.0664	0.0698	0.0731
2×10^4	0.0407	0.0459	0.0505	0.0547	0.0587	0.0624	0.0660	0.0694	0.0727
2.5×10^4	0.0402	0.0455	0.0502	0.0544	0.0584	0.0621	0.0657	0.0691	0.0725
3×10^4	0.0398	0.0452	0.0499	0.0542	0.0582	0.0619	0.0655	0.0690	0.0723
4×10^4	0.0394	0.0448	0.0496	0.0539	0.0579	0.0617	0.0653	0.0688	0.0721
5×10^4	0.0391	0.0446	0.0494	0.0538	0.0578	0.0616	0.0652	0.0687	0.0720
6×10^4	0.0389	0.0445	0.0493	0.0536	0.0577	0.0615	0.0651	0.0686	0.0719
7×10^4	0.0388	0.0443	0.0492	0.0536	0.0576	0.0614	0.0650	0.0685	0.0719
8×10^4	0.0387	0.0443	0.0491	0.0535	0.0576	0.0614	0.0650	0.0685	0.0718
9×10^4	0.0386	0.0442	0.0491	0.0535	0.0575	0.0613	0.0650	0.0684	0.0718
1×10^5	0.0385	0.0442	0.0490	0.0534	0.0575	0.0613	0.0649	0.0684	0.0718
1.5×10^5	0.0383	0.0440	0.0489	0.0533	0.0574	0.0612	0.0648	0.0683	0.0717
2×10^5	0.0382	0.0439	0.0488	0.0532	0.0573	0.0612	0.0648	0.0683	0.0717
2.5×10^5	0.0381	0.0439	0.0488	0.0532	0.0573	0.0611	0.0648	0.0683	0.0716
3×10^5	0.0381	0.0438	0.0488	0.0532	0.0573	0.0611	0.0648	0.0683	0.0716
4×10^5	0.0381	0.0438	0.0487	0.0532	0.0573	0.0611	0.0647	0.0682	0.0716
5×10^5	0.0380	0.0438	0.0487	0.0531	0.0572	0.0611	0.0647	0.0682	0.0716
6×10^5	0.0380	0.0438	0.0487	0.0531	0.0572	0.0611	0.0647	0.0682	0.0716
7×10^5	0.0380	0.0438	0.0487	0.0531	0.0572	0.0611	0.0647	0.0682	0.0716
8×10^5	0.0380	0.0437	0.0487	0.0531	0.0572	0.0611	0.0647	0.0682	0.0716
9×10^5	0.0380	0.0437	0.0487	0.0531	0.0572	0.0610	0.0647	0.0682	0.0716
1×10^6	0.0380	0.0437	0.0487	0.0531	0.0572	0.0610	0.0647	0.0682	0.0716
1.5×10^6	0.0379	0.0437	0.0487	0.0531	0.0572	0.0610	0.0647	0.0682	0.0716
2×10^6	0.0379	0.0437	0.0487	0.0531	0.0572	0.0610	0.0647	0.0682	0.0716
2.5×10^6	0.0379	0.0437	0.0487	0.0531	0.0572	0.0610	0.0647	0.0682	0.0716
3×10^6	0.0379	0.0437	0.0487	0.0531	0.0572	0.0610	0.0647	0.0682	0.0716
4×10^6	0.0379	0.0437	0.0486	0.0531	0.0572	0.0610	0.0647	0.0682	0.0716
5×10^6	0.0379	0.0437	0.0486	0.0531	0.0572	0.0610	0.0647	0.0682	0.0716
6×10^6	0.0379	0.0437	0.0486	0.0531	0.0572	0.0610	0.0647	0.0682	0.0716
7×10^6	0.0379	0.0437	0.0486	0.0531	0.0572	0.0610	0.0647	0.0682	0.0716
8×10^6	0.0379	0.0437	0.0486	0.0531	0.0572	0.0610	0.0647	0.0682	0.0716
9×10^6	0.0379	0.0437	0.0486	0.0531	0.0572	0.0610	0.0647	0.0682	0.0716
1×10^7	0.0379	0.0437	0.0486	0.0531	0.0572	0.0610	0.0647	0.0682	0.0716
1.5×10^7	0.0379	0.0437	0.0486	0.0531	0.0572	0.0610	0.0647	0.0682	0.0716
2×10^7	0.0379	0.0437	0.0486	0.0531	0.0572	0.0610	0.0647	0.0682	0.0716
2.5×10^7	0.0379	0.0437	0.0486	0.0531	0.0572	0.0610	0.0647	0.0682	0.0716
3×10^7	0.0379	0.0437	0.0486	0.0531	0.0572	0.0610	0.0647	0.0682	0.0716
4×10^7	0.0379	0.0437	0.0486	0.0531	0.0572	0.0610	0.0647	0.0682	0.0716
5×10^7	0.0379	0.0437	0.0486	0.0531	0.0572	0.0610	0.0647	0.0682	0.0716

Support Material

APPENDIX 17.C
Flow of Water Through Schedule-40 Steel Pipe

pressure drop per 1000 ft of schedule-40 steel pipe, in lbf/in²

discharge (gal/min)	velocity (ft/sec)	pressure drop	velocity (ft/sec)	pressure drop	velocity (ft/sec)	pressure drop	velocity (ft/sec)	pressure drop	velocity (ft/sec)	pressure drop	velocity (ft/sec)	pressure drop	velocity (ft/sec)	pressure drop	velocity (ft/sec)	pressure drop	velocity (ft/sec)	pressure drop
	1 in		**1¼ in**		**1½ in**		**2 in**		**2½ in**		**3 in**		**3½ in**		**4 in**		**5 in**	
1	0.37	0.49																
2	0.74	1.70	0.43	0.45														
3	1.12	3.53	0.64	0.94	0.47	0.44												
4	1.49	5.94	0.86	1.55	0.63	0.74												
5	1.86	9.02	1.07	2.36	0.79	1.12												
6	2.24	12.25	1.28	3.30	0.95	1.53	0.57	0.46										
8	2.98	21.1	1.72	5.52	1.26	2.63	0.76	0.75										
10	3.72	30.8	2.14	8.34	1.57	3.86	0.96	1.14	0.67	0.48								
15	5.60	64.6	3.21	17.6	2.36	8.13	1.43	2.33	1.00	0.99								
20	7.44	110.5	4.29	29.1	3.15	13.5	1.91	3.86	1.34	1.64	0.87	0.59						
25			5.36	43.7	3.94	20.2	2.39	5.81	1.68	2.48	1.08	0.67	0.81	0.42				
30			6.43	62.9	4.72	29.1	2.87	8.04	2.01	3.43	1.30	1.21	0.97	0.60				
35			7.51	82.5	5.51	38.2	3.35	10.95	2.35	4.49	1.52	1.58	1.14	.079	0.88	0.42		
40					6.30	47.8	3.82	13.7	2.68	5.88	1.74	2.06	1.30	1.00	1.01	0.53		
45					7.08	60.6	4.30	17.4	3.00	7.14	1.95	2.51	1.46	1.21	1.13	0.67		
50					7.87	74.7	4.78	20.6	3.35	8.82	2.17	3.10	1.62	1.44	1.26	0.80		
60							5.74	29.6	4.02	12.2	2.60	4.29	1.95	2.07	1.51	1.10	**5 in**	
70							6.69	38.6	4.69	15.3	3.04	5.84	2.27	2.71	1.76	1.50	1.12	0.48
80							7.65	50.3	5.37	21.7	3.48	7.62	2.59	3.53	2.01	1.87	1.28	0.63
90	**6 in**						8.60	63.6	6.04	26.1	3.91	9.22	2.92	4.46	2.26	2.37	1.44	0.80
100	1.11	0.39					9.56	75.1	6.71	32.3	4.34	11.4	3.24	5.27	2.52	2.81	1.60	0.95
125	1.39	0.56							8.38	48.2	5.42	17.1	4.05	7.86	3.15	4.38	2.00	1.48
150	1.67	0.78							10.06	60.4	6.51	23.5	4.86	11.3	3.78	6.02	2.41	2.04
175	1.94	1.06							11.73	90.0	7.59	32.0	5.67	14.7	4.41	8.20	2.81	2.78
200	2.22	1.32	**8 in**								8.68	39.7	6.48	19.2	5.04	10.2	3.21	3.46
225	2.50	1.66	1.44	0.44							9.77	50.2	7.29	23.1	5.67	12.9	3.61	4.37
250	2.78	2.05	1.60	0.55							10.85	61.9	8.10	28.5	6.30	15.9	4.01	5.14
275	3.06	2.36	1.76	0.63							11.94	75.0	8.91	34.4	6.93	18.3	4.41	6.22
300	3.33	2.80	1.92	0.75							13.02	84.7	9.72	40.9	7.56	21.8	4.81	7.41
325	3.61	3.29	2.08	0.88									10.53	45.5	8.18	25.5	5.21	8.25
350	3.89	3.62	2.24	0.97									11.35	52.7	8.82	29.7	5.61	9.57
375	4.16	4.16	2.40	1.11									12.17	60.7	9.45	32.3	6.01	11.0
400	4.44	4.72	2.56	1.27									12.97	68.9	10.08	36.7	6.41	12.5
425	4.72	5.34	2.72	1.43									13.78	77.8	10.70	41.5	6.82	14.1
450	5.00	5.96	2.88	1.60	**10 in**								14.59	87.3	11.33	46.5	7.22	15.0
475	5.27	6.66	3.04	1.69	1.93	0.30									11.96	51.7	7.62	16.7
500	5.55	7.39	3.20	1.87	2.04	0.63									12.59	57.3	8.02	18.5
550	6.11	8.94	3.53	2.26	2.24	0.70									13.84	69.3	8.82	22.4
600	6.66	10.6	3.85	2.70	2.44	0.86									15.10	82.5	9.62	26.7
650	7.21	11.8	4.17	3.16	2.65	1.01	**12 in**										10.42	31.3
700	7.77	13.7	4.49	3.69	2.85	1.18	2.01	0.48									11.22	36.3
750	8.32	15.7	4.81	4.21	3.05	1.35	2.15	0.55									12.02	41.6
800	8.88	17.8	5.13	4.79	3.26	1.54	2.29	0.62	**14 in**								12.82	44.7
850	9.44	20.2	5.45	5.11	3.46	1.74	2.44	0.70	2.02	0.43							13.62	50.5
900	10.00	22.6	5.77	5.73	3.66	1.94	2.58	0.79	2.14	0.48							14.42	56.6
950	10.55	23.7	6.09	6.38	3.87	2.23	2.72	0.88	2.25	0.53							15.22	63.1
1000	11.10	26.3	6.41	7.08	4.07	2.40	2.87	0.97	2.38	0.59							16.02	70.0
1100	12.22	31.8	7.05	8.56	4.48	2.74	3.16	1.18	2.61	0.68	**16 in**						17.63	84.6
1200	13.32	37.8	7.69	10.2	4.88	3.27	3.45	1.40	2.85	0.81	2.18	0.40						
1300	14.43	44.4	8.33	11.3	5.29	3.86	3.73	1.56	3.09	0.95	2.36	0.47						
1400	15.54	51.5	8.97	13.0	5.70	4.44	4.02	1.80	3.32	1.10	2.54	0.54						
1500	16.65	55.5	9.62	15.0	6.10	5.11	4.30	2.07	3.55	1.19	2.73	0.62						
1600	17.76	63.1	10.26	17.0	6.51	5.46	4.59	2.36	3.80	1.35	2.91	0.71	**18 in**					
1800	19.98	79.8	11.54	21.6	7.32	6.91	5.16	2.98	4.27	1.71	3.27	0.85	2.58	0.48				
2000	22.20	98.5	12.83	25.0	8.13	8.54	5.73	3.47	4.74	2.11	3.63	1.05	2.88	0.56				
2500			16.03	39.0	10.18	12.5	7.17	5.41	5.93	3.09	4.54	1.63	3.59	0.88	**20 in**			
3000			19.24	52.4	12.21	18.0	8.60	7.31	7.12	4.45	5.45	2.21	4.31	1.27	3.45	0.73		
3500			22.43	71.4	14.25	22.9	10.03	9.95	8.32	6.18	6.35	3.00	5.03	1.52	4.03	0.94	**24 in**	
4000			25.65	93.3	16.28	29.9	11.48	13.0	9.49	7.92	7.25	3.92	5.74	2.12	4.61	1.22	3.19	0.51
4500					18.31	37.8	12.90	15.4	10.67	9.36	8.17	4.97	6.47	2.50	5.19	1.55	3.59	0.60
5000					20.35	46.7	14.34	18.9	11.84	11.6	9.08	5.72	7.17	3.08	5.76	1.78	3.99	0.74
6000					24.42	67.2	17.21	27.3	14.32	15.4	10.88	8.24	8.62	4.45	6.92	2.57	4.80	1.00
7000					28.50	85.1	20.08	37.2	16.60	21.0	12.69	12.2	10.04	6.06	8.06	3.50	5.68	1.36
8000							22.95	45.1	18.98	27.4	14.52	13.6	11.48	7.34	9.23	4.57	6.38	1.78
9000							25.80	57.0	21.35	34.7	16.32	17.2	12.92	9.27	10.37	5.36	7.19	2.25
10,000							28.63	70.4	23.75	42.9	18.16	21.2	14.37	11.5	11.53	6.63	7.96	2.78
12,000							34.38	93.6	28.50	61.8	21.80	30.9	17.23	16.5	13.83	9.54	9.57	3.71
14,000									33.20	84.0	25.42	41.6	20.10	20.7	16.14	12.0	11.18	5.05
16,000											29.05	54.4	22.96	27.1	18.43	15.7	12.77	6.60

(Multiply in wg/ft by 36.1 to obtain lbf/in²-1000 ft.)

Reproduced with permission from *Design of Fluid Systems Hook-Ups*, published by Spirax Sarco®, Inc., © 1992.

APPENDIX 17.D
Equivalent Length of Straight Pipe for Various (Generic) Fittings
(in feet, turbulent flow only, for any fluid)

fittings			¼	⅜	½	¾	1	1¼	1½	2	2½	3	4	5	6	8	10	12	14	16	18	20	24
regular 90°ell	screwed	steel	2.3	3.1	3.6	4.4	5.2	6.6	7.4	8.5	9.3	11.0	13.0											
		c.i.										9.0	11.0											
	flanged	steel			0.92	1.2	1.6	2.1	2.4	3.1	3.6	4.4	5.9	7.3	8.9	12.0	14.0	17.0	18.0	21.0	23.0	25.0	30.0
		c.i.										3.6	4.8		7.2	9.8	12.0	15.0	17.0	19.0	22.0	24.0	28.0
long radius 90°ell	screwed	steel	1.5	2.0	2.2	2.3	2.7	3.2	3.4	3.6	3.6	4.0	4.6											
		c.i.										3.3	3.7											
	flanged	steel			1.1	1.3	1.6	2.0	2.3	2.7	2.9	3.4	4.2	5.0	5.7	7.0	8.0	9.0	9.4	10.0	11.0	12.0	14.0
		c.i.										2.8	3.4		4.7	5.7	6.8	7.8	8.6	9.6	11.0	11.0	13.0
regular 45°ell	screwed	steel	0.34	0.52	0.71	0.92	1.3	1.7	2.1	2.7	3.2	4.0	5.5											
		c.i.										3.3	4.5											
	flanged	steel			0.45	0.59	0.81	1.1	1.3	1.7	2.0	2.6	3.5	4.5	5.6	7.7	9.0	11.0	13.0	15.0	16.0	18.0	22.0
		c.i.										2.1	2.9		4.5	6.3	8.1	9.7	12.0	13.0	15.0	17.0	20.0
tee-line flow	screwed	steel	0.79	1.2	1.7	2.4	3.2	4.6	5.6	7.7	9.3	12.0	17.0											
		c.i.										9.9	14.0											
	flanged	steel			0.69	0.82	1.0	1.3	1.5	1.8	1.9	2.2	2.8	3.3	3.8	4.7	5.2	6.0	6.4	7.2	7.6	8.2	9.6
		c.i.										1.9	2.2		3.1	3.9	4.6	5.2	5.9	6.5	7.2	7.7	8.8
tee-branch flow	screwed	steel	2.4	3.5	4.2	5.3	6.6	8.7	9.9	12.0	13.0	17.0	21.0											
		c.i.										14.0	17.0											
	flanged	steel			2.0	2.6	3.3	4.4	5.2	6.6	7.5	9.4	12.0	15.0	18.0	24.0	30.0	34.0	37.0	43.0	47.0	52.0	62.0
		c.i.										7.7	10.0		15.0	20.0	25.0	30.0	35.0	39.0	44.0	49.0	57.0
180° return bend (regular)	screwed	steel	2.3	3.1	3.6	4.4	5.2	6.6	7.4	8.5	9.3	11.0	13.0											
		c.i.										9.0	11.0											
	flanged	steel			0.92	1.2	1.6	2.1	2.4	3.1	3.6	4.4	5.9	7.3	8.9	12.0	14.0	17.0	18.0	21.0	23.0	25.0	30.0
		c.i.										3.6	4.8		7.2	9.8	12.0	15.0	17.0	19.0	22.0	24.0	28.0
180° return bend (long radius)	flanged	steel			1.1	1.3	1.6	2.0	2.3	2.7	2.9	3.4	4.2	5.0	5.7	7.0	8.0	9.0	9.4	10.0	11.0	12.0	14.0
		c.i.										2.8	3.4		4.7	5.7	6.8	7.8	8.6	9.6	11.0	11.0	13.0
globe valve	screwed	steel	21.0	22.0	22.0	24.0	29.0	37.0	42.0	54.0	62.0	79.0	110.0											
		c.i.										65.0	86.0											
	flanged	steel			38.0	40.0	45.0	54.0	59.0	70.0	77.0	94.0	120.0	150.0	190.0	260.0	310.0	390.0						
		c.i.										77.0	99.0		150.0	210.0	270.0	330.0						
gate valve	screwed	steel	0.32	0.45	0.56	0.67	0.84	1.1	1.2	1.5	1.7	1.9	2.5											
		c.i.										1.6	2.0											
	flanged	steel								2.6	2.7	2.8	2.9	3.1	3.2	3.2	3.2	3.2	3.2	3.2	3.2	3.2	3.2
		c.i.										2.3	2.4		2.6	2.7	2.8	2.9	2.9	3.0	3.0	3.0	3.0
angle valve	screwed	steel	12.8	15.0	15.0	15.0	17.0	18.0	18.0	18.0	18.0	18.0	18.0											
		c.i.										15.0	15.0											
	flanged	steel			15.0	15.0	17.0	18.0	18.0	21.0	22.0	28.0	38.0	50.0	63.0	90.0	120.0	140.0	160.0	190.0	210.0	240.0	300.0
		c.i.										23.0	31.0		52.0	74.0	98.0	120.0	150.0	170.0	200.0	230.0	280.0
swing check valve	screwed	steel	7.2	7.3	8.0	8.8	11.0	13.0	15.0	19.0	22.0	27.0	38.0											
		c.i.										22.0	31.0											
	flanged	steel			3.8	5.3	7.2	10.0	12.0	17.0	21.0	27.0	38.0	50.0	63.0	90.0	120.0	140.0						
		c.i.										22.0	31.0		52.0	74.0	98.0	120.0						
coupling or union	screwed	steel	0.14	0.18	0.21	0.24	0.29	0.36	0.39	0.45	0.47	0.53	0.65											
		c.i.										0.44	0.52											
bell mouth inlet		steel	0.04	0.07	0.10	0.13	0.18	0.26	0.31	0.43	0.52	0.67	0.95	1.3	1.6	2.3	2.9	3.5	4.0	4.7	5.3	6.1	7.6
		c.i.										0.55	0.77		1.3	1.9	2.4	3.0	3.6	4.3	5.0	5.7	7.0
square mouth inlet		steel	0.44	0.68	0.96	1.3	1.8	2.6	3.1	4.3	5.2	6.7	9.5	13.0	16.0	23.0	29.0	35.0	40.0	47.0	53.0	61.0	76.0
		c.i.										5.5	7.7		13.0	19.0	24.0	30.0	36.0	43.0	50.0	57.0	70.0
inlet — re-entrant pipe		steel	0.88	1.4	1.9	2.6	3.6	5.1	6.2	8.5	10.0	13.0	19.0	25.0	32.0	45.0	58.0	70.0	80.0	95.0	110.0	120.0	150.0
		c.i.										11.0	15.0		26.0	37.0	49.0	61.0	73.0	86.0	100.0	110.0	140.0

Support Material

APPENDIX 19.A
Symbols for Fluid Power Equipment[a]

(a) spring (spring-loaded)	(m) hydraulic motor, fixed capacity (two directions of flow)	(y) flow control valve
two winding or one winding (b) solenoid	(n) hydraulic motor, variable capacity (one direction of flow)	(z) shut-off valve
(c) adjustable symbol	(o) actuating cylinder (single acting)	(aa) electric motor
(d) directional arrow (oil)	(p) actuating cylinder (double acting)	(bb) internal combustion engine
(e) directional arrow (air or gas)	(q) two-way, two-position control valve (normally closed)	(cc) coupling
(f) fluid flow line	(r) two-way, two-position control valve (normally open)	(dd) accumulator
(g) shaft or lever	(s) three-way, two-position control valve (normally open)	(ee) cooler
(h) reservoir (open)	(t) four-way, two-position control valve	(ff) heater
(i) reservoir (closed)	(u) check (nonreturn) valve	(gg) pressure gage
(j) filter or strainer	(v) shuttle valve	(hh) temperature gage
(k) pump, fixed capacity (one direction of flow)	(w) pressure control valve	(ii) flow meter
(l) pump, variable capacity (two directions of flow)	(x) pressure relief valve	

[a]These symbols are consistent with ANSI standard Y32.10 and ISO 1219.

APPENDIX 21.A
Atomic Numbers and Weights of the Elements
(referred to Carbon-12)

name	symbol	atomic number	atomic weight	name	symbol	atomic number	atomic weight
actinium	Ac	89	–	mercury	Hg	80	200.59
aluminum	Al	13	26.9815	molybdenum	Mo	42	95.94
americium	Am	95	–	neodymium	Nd	60	144.24
antimony	Sb	51	121.75	neon	Ne	10	20.183
argon	Ar	18	39.948	neptunium	Np	93	237.048
arsenic	As	33	74.9216	nickel	Ni	28	58.71
astatine	At	85	–	niobium	Nb	41	92.906
barium	Ba	56	137.34	nitrogen	N	7	14.0067
berkelium	Bk	97	–	nobelium	No	102	–
beryllium	Be	4	9.0122	osmium	Os	76	190.2
bismuth	Bi	83	208.980	oxygen	O	8	15.9994
boron	B	5	10.811	palladium	Pd	46	106.4
bromine	Br	35	79.904	phosphorus	P	15	30.9738
cadmium	Cd	48	112.40	platinum	Pt	78	195.09
calcium	Ca	20	40.08	plutonium	Pu	94	–
californium	Cf	98	–	polonium	Po	84	–
carbon	C	6	12.01115	potassium	K	19	39.102
cerium	Ce	58	140.12	praseodymium	Pr	59	140.907
cesium	Cs	55	132.905	promethium	Pm	61	–
chlorine	Cl	17	35.453	protactinium	Pa	91	231.036
chromium	Cr	24	51.996	radium	Ra	88	–
cobalt	Co	27	58.9332	radon	Rn	86	226.025
copper	Cu	29	63.546	rhenium	Re	75	186.2
curium	Cm	96	–	rhodium	Rh	45	102.905
dysprosium	Dy	66	162.50	rubidium	Rb	37	85.47
einsteinium	Es	99	–	ruthenium	Ru	44	101.07
erbium	Er	68	167.26	samarium	Sm	62	150.35
europium	Eu	63	151.96	scandium	Sc	21	44.956
fermium	Fm	100	–	selenium	Se	34	78.96
fluorine	F	9	18.9984	silicon	Si	14	28.086
francium	Fr	87	–	silver	Ag	47	107.868
gadolinium	Gd	64	157.25	sodium	Na	11	22.9898
gallium	Ga	31	69.72	strontium	Sr	38	87.62
germanium	Ge	32	72.59	sulfur	S	16	32.064
gold	Au	79	196.967	tantalum	Ta	73	180.948
hafnium	Hf	72	178.49	technetium	Tc	43	–
helium	He	2	4.0026	tellurium	Te	52	127.60
holmium	Ho	67	164.930	terbium	Tb	65	158.924
hydrogen	H	1	1.00797	thallium	Tl	81	204.37
indium	In	49	114.82	thorium	Th	90	232.038
iodine	I	53	126.9044	thulium	Tm	69	168.934
iridium	Ir	77	192.2	tin	Sn	50	118.69
iron	Fe	26	55.847	titanium	Ti	22	47.90
krypton	Kr	36	83.80	tungsten	W	74	183.85
lanthanum	La	57	138.91	uranium	U	92	238.03
lead	Pb	82	207.19	vanadium	V	23	50.942
lithium	Li	3	6.939	xenon	Xe	54	131.30
lutetium	Lu	71	174.97	ytterbium	Yb	70	173.04
magnesium	Mg	12	24.312	yttrium	Y	39	88.905
manganese	Mn	25	54.9380	zinc	Zn	30	65.37
mendelevium	Md	101	–	zirconium	Zr	40	91.22

Support
Material

APPENDIX 21.B
Periodic Table of the Elements
(referred to Carbon-12)

The Periodic Table of Elements (Long Form)

The number of electrons in filled shells is shown in the column at the extreme left; the remaining electrons for each element are shown immediately below the symbol for each element. Atomic numbers are enclosed in brackets. Atomic weights (rounded, based on carbon-12) are shown above the symbols. Atomic weight values in parentheses are those of the isotopes of longest half-life for certain radioactive elements whose atomic weights cannot be precisely quoted without knowledge of origin of the element.

metals — transition metals — nonmetals

period	IA	IIA	IIIB	IVB	VB	VIB	VIIB	VIII	VIII	VIII	IB	IIB	IIIA	IVA	VA	VIA	VIIA	0
1 (0)	1.0079 H[1] 1																	4.0026 He[2] 2
2	6.939 Li[3] 1	9.0122 Be[4] 2											10.81 B[5] 3	12.01115 C[6] 4	14.0067 N[7] 5	15.9994 O[8] 6	18.994 F[9] 7	20.183 Ne[10] 8
3 (2,8)	22.9898 Na[11] 1	24.312 Mg[12] 2											26.9815 Al[13] 3	28.086 Si[14] 4	30.9738 P[15] 5	32.064 S[16] 6	35.453 Cl[17] 7	39.948 Ar[18] 8
4 (2,8)	39.098 K[19] 8,1	40.08 Ca[20] 8,2	44.956 Sc[21] 9,2	47.90 Ti[22] 10,2	50.942 V[23] 11,2	51.996 Cr[24] 13,1	54.938 Mn[25] 13,2	55.847 Fe[26] 14,2	58.933 Co[27] 15,2	58.71 Ni[28] 16,2	63.546 Cu[29] 18,1	65.38 Zn[30] 18,2	69.72 Ga[31] 18,3	72.59 Ge[32] 18,4	74.922 As[33] 18,5	78.96 Se[34] 18,6	79.904 Br[35] 18,7	83.80 Kr[36] 18,8
5 (2,8,18)	85.47 Rb[37] 8,1	87.62 Sr[38] 8,2	88.905 Y[39] 9,2	91.22 Zr[40] 10,2	92.906 Nb[41] 12,1	95.94 Mo[42] 13,1	(98) Tc[43] 14,1	101.07 Ru[44] 15,1	102.905 Rh[45] 16,1	106.4 Pd[46] 18	107.868 Ag[47] 18,1	112.40 Cd[48] 18,2	114.82 In[49] 18,3	118.69 Sn[50] 18,4	121.75 Sb[51] 18,5	127.60 Te[52] 18,6	126.904 I[53] 18,7	131.30 Xe[54] 18,8
6 (2,8,18)	132.905 Cs[55] 18,8,1	137.34 Ba[56] 18,8,2	* (57-71)	178.49 Hf[72] 32,10,2	180.948 Ta[73] 32,11,2	183.85 W[74] 32,12,2	186.2 Re[75] 32,13,2	190.2 Os[76] 32,14,2	192.2 Ir[77] 32,15,2	195.09 Pt[78] 32,17,1	196.967 Au[79] 32,18,1	200.59 Hg[80] 32,18,2	204.37 Tl[81] 32,18,3	207.19 Pb[82] 32,18,4	208.980 Bi[83] 32,18,5	(210) Po[84] 32,18,6	(210) At[85] 32,18,7	(222) Rn[86] 32,18,8
7 (2,8,18,32)	(223) Fr[87] 18,8,1	226.025 Ra[88] 18,8,2	† (89-103)	Rf[104] 32,10,2	Ha[105] 32,11,2	[106] 32,12,2	[107] 32,12,2	[108]										

*lanthanide series

138.91 La[57] 18,9,2	140.12 Ce[58] 20,8,2	140.907 Pr[59] 21,8,2	144.24 Nd[60] 22,8,2	(147) Pm[61] 23,8,2	150.35 Sm[62] 24,8,2	151.96 Eu[63] 25,8,2	157.25 Gd[64] 25,9,2	158.924 Tb[65] 27,8,2	162.50 Dy[66] 28,8,2	164.930 Ho[67] 29,8,2	167.26 Er[68] 30,8,2	168.934 Tm[69] 31,8,2	173.04 Yb[70] 32,8,2	174.97 Lu[71] 32,9,2

*actinide series

(227) Ac[89] 18,9,2	232.038 Th[90] 18,10,2	231.036 Pa[91] 20,9,2	238.03 U[92] 21,9,2	237.048 Np[93] 23,8,2	(242) Pu[94] 24,8,2	(243) Am[95] 25,8,2	(247) Cm[96] 25,9,2	(247) Bk[97] 26,9,2	(249) Cf[98] 28,8,2	(254) Es[99] 29,8,2	(253) Fm[100] 30,8,2	(256) Md[101] 31,8,2	(254) No[102] 32,8,2	(257) Lr[103] 32,9,2

APPENDIX 21.C
Water Chemistry CaCO₃ Equivalents

cations	formula	ionic weight	equivalent weight	factor
aluminum	Al^{+3}	27.0	9.0	5.56
ammonium	NH_4^+	18.0	18.0	2.78
calcium	Ca^{+2}	40.1	20.0	2.50
hydrogen	H^+	1.0	1.0	50.00
ferrous iron	Fe^{+2}	55.8	27.9	1.79
ferric iron	Fe^{+3}	55.8	18.6	2.69
magnesium	Mg^{+2}	24.3	12.2	4.10
manganese	Mn^{+2}	54.9	27.5	1.82
potassium	K^+	39.1	39.1	1.28
sodium	Na^+	23.0	23.0	2.18
cupric copper	Cu^{+2}	63.6	31.8	1.57
cuprous copper	Cu^{+3}	63.6	21.2	2.36

anions	formula	ionic weight	equivalent weight	factor
bicarbonate	HCO_3^-	61.0	61.0	0.82
carbonate	CO_3^{-2}	60.0	30.0	1.67
chloride	Cl^-	35.5	35.5	1.41
fluoride	F^-	19.0	19.0	2.66
nitrate	NO_3^-	62.0	62.0	0.81
hydroxide	OH^-	17.0	17.0	2.94
phosphate (tribasic)	PO_4^{-3}	95.0	31.7	1.58
phosphate (dibasic)	HPO_4^{-2}	96.0	48.0	1.04
phosphate (monobasic)	$H_2PO_4^-$	97.0	97.0	0.52
sulfate	SO_4^{-2}	96.1	48.0	1.04
sulfite	SO_3^{-2}	80.1	40.0	1.25

compounds	formula	molecular weight	equivalent weight	factor
aluminum hydroxide	$Al(OH)_3$	78.0	26.0	1.92
aluminum sulfate	$Al_2(SO_4)_3$	342.1	57.0	0.88
alumina	Al_2O_3	102.0	17.0	2.94
sodium aluminate	$Na_2Al_2O_4$	164.0	27.3	1.83
calcium bicarbonate	$Ca(HCO_3)_2$	162.1	81.1	0.62
calcium carbonate	$CaCO_3$	100.1	50.1	1.00
calcium chloride	$CaCl_2$	111.0	55.5	0.90
calcium hydroxide (pure)	$Ca(OH)_2$	74.1	37.1	1.35
calcium hydroxide (90%)	$Ca(OH)_2$	—	41.1	1.22
calcium sulfate (anhydrous)	$CaSO_4$	136.2	68.1	0.74
calcium sulfate (gypsum)	$CaSO_4{\cdot}2H_2O$	172.2	86.1	0.58
calcium phosphate	$Ca_3(PO_4)_2$	310.3	51.7	0.97
disodium phosphate	$Na_2HPO_4{\cdot}12H_2O$	358.2	119.4	0.42
disodium phosphate (anhydrous)	Na_2HPO_4	142.0	47.3	1.06

(continued)

APPENDIX 21.C (*continued*)
Water Chemistry CaCO$_3$ Equivalents

compounds	formula	molecular weight	equivalent weight	factor
ferric oxide	Fe$_2$O$_3$	159.6	26.6	1.88
iron oxide (magnetic)	Fe$_3$O$_4$	321.4	–	–
ferrous sulfate (copperas)	FeSO$_4 \cdot$ 7H$_2$O	278.0	139.0	0.36
magnesium oxide	MgO	40.3	20.2	2.48
magnesium bicarbonate	Mg(HCO$_3$)$_2$	146.3	73.2	0.68
magnesium carbonate	MgCO$_3$	84.3	42.2	1.19
magnesium chloride	MgCl$_2$	95.2	47.6	1.05
magnesium hydroxide	Mg(OH)$_2$	58.3	29.2	1.71
magnesium phosphate	Mg$_3$(PO$_4$)$_2$	263.0	43.8	1.14
magnesium sulfate	MgSO$_4$	120.4	60.2	0.83
monosodium phosphate	NaH$_2$PO$_4 \cdot$ H$_2$O	138.1	46.0	1.09
monosodium phosphate (anhydrous)	NaH$_2$PO$_4$	120.1	40.0	1.25
metaphosphate	NaPO$_3$	102.0	34.0	1.47
silica	SiO$_2$	60.1	30.0	1.67
sodium bicarbonate	NaHCO$_3$	84.0	84.0	0.60
sodium carbonate	Na$_2$CO$_3$	106.0	53.0	0.94
sodium chloride	NaCl	58.5	58.5	0.85
sodium hydroxide	NaOH	40.0	40.0	1.25
sodium nitrate	NaNO$_3$	85.0	85.0	0.59
sodium sulfate	Na$_2$SO$_4$	142.0	71.0	0.70
sodium sulfite	Na$_2$SO$_3$	126.1	63.0	0.79
tetrasodium EDTA	(CH$_2$)$_2$N$_2$(CH$_2$COONa)$_4$	380.2	95.1	0.53
trisodium phosphate	Na$_3$PO$_4 \cdot$ 12H$_2$O	380.2	126.7	0.40
trisodium phosphate (anhydrous)	Na$_3$PO$_4$	164.0	54.7	0.91
trisodium NTA	(CH$_2$)$_3$N(COONa)$_3$	257.1	85.7	0.58

gases

ammonia	NH$_3$	17	17	2.94
carbon dioxide	CO$_2$	44	22	2.27
hydrogen	H$_2$	2	1	50.00
oxygen	O$_2$	32	8	6.25
hydrogen sulfide	H$_2$S	34	17	2.94

acids

carbonic	H$_2$CO$_3$	62.0	31.0	1.61
hydrochloric	HCl	36.5	36.5	1.37
phosphoric	H$_3$PO$_4$	98.0	32.7	1.53
sulfuric	H$_2$SO$_4$	98.1	49.1	1.02

(Multiply the concentration (in mg/L) of the substance by the corresponding factors to obtain the equivalent concentration in mg/L as CaCO$_3$. For example, 70 mg/L of Mg^{++} as substance would be (70 mg/L)(4.1) = 287 mg/L as CaCO$_3$.)

APPENDIX 22.A
Heats of Combustion for Common Compounds[a]

| substance | formula | molecular weight | specific volume (ft^3/lbm) | heat of combustion | | | |
| | | | | Btu/ft^3 | | Btu/lbm | |
				gross (high)	net (low)	gross (high)	net (low)
carbon	C	12.01				14,093	14,093
carbon dioxide	CO$_2$	44.01	8.548				
carbon monoxide	CO	28.01	13.506	322	322	4347	4347
hydrogen	H$_2$	2.016	187.723	325	275	60,958	51,623
nitrogen	N$_2$	28.016	13.443				
oxygen	O$_2$	32.000	11.819				
paraffin series (alkanes)							
methane	CH$_4$	16.041	23.565	1013	913	23,879	21,520
ethane	C$_2$H$_6$	30.067	12.455	1792	1641	22,320	20,432
propane	C$_3$H$_8$	44.092	8.365	2590	2385	21,661	19,944
n-butane	C$_4$H$_{10}$	58.118	6.321	3370	3113	21,308	19,680
isobutane	C$_4$H$_{10}$	58.118	6.321	3363	3105	21,257	19,629
n-pentane	C$_5$H$_{12}$	72.144	5.252	4016	3709	21,091	19,517
isopentane	C$_5$H$_{12}$	72.144	5.252	4008	3716	21,052	19,478
neopentane	C$_5$H$_{12}$	72.144	5.252	3993	3693	20,970	19,396
n-hexane	C$_6$H$_{14}$	86.169	4.398	4762	4412	20,940	19,403
olefin series (alkenes and alkynes)							
ethylene	C$_2$H$_4$	28.051	13.412	1614	1513	21,644	20,295
propylene	C$_3$H$_6$	42.077	9.007	2336	2186	21,041	19,691
n-butene	C$_4$H$_8$	56.102	6.756	3084	2885	20,840	19,496
isobutene	C$_4$H$_8$	56.102	6.756	3068	2869	20,730	19,382
n-pentene	C$_5$H$_{10}$	70.128	5.400	3836	3586	20,712	19,363
aromatic series							
benzene	C$_6$H$_6$	78.107	4.852	3751	3601	18,210	17,480
toluene	C$_7$H$_8$	92.132	4.113	4484	4284	18,440	17,620
xylene	C$_8$H$_{10}$	106.158	3.567	5230	4980	18,650	17,760
miscellaneous gases							
acetylene	C$_2$H$_2$	26.036	14.344	1499	1448	21,500	20,776
air		28.9	13.063				
ammonia	NH$_3$	17.031	21.914	441	365	9668	8001
digester gas[b]	–	–					
ethyl alcohol	C$_2$H$_5$OH	46.067	8.221	1600	1451	13,161	11,929
hydrogen sulfide	H$_2$S	34.076	10.979	647	596	7100	6545
iso-octane	C$_8$H$_{18}$	114.2				20,590	19,160
methyl alcohol	CH$_3$OH	32.041	11.820	868	768	10,259	9078
naphthalene	C$_{10}$H$_8$	128.162	2.955	5854	5654	17,298	16,708
sulfur	S	32.06				3983	3983
sulfur dioxide	SO$_2$	64.06	5.770				
water vapor	H$_2$O	18.016	21.017				

(Multiply Btu/lbm by 2.326 to obtain kJ/kg.)
(Multiply Btu/ft^3 by 37.25 to obtain kJ/m^3.)
[a]Gas volumes listed are at 60°F (16°C) and 1 atm.
[b]Digester gas from wastewater treatment plants is approximately 65% methane and 35% carbon dioxide by volume. Use composite properties of these two gases.

APPENDIX 24.A
Properties of Saturated Steam by Temperature
(customary U.S. units)

| temp. (°F) | press. (lbf/in²) | specific volume (ft³/lbm) | | internal energy (Btu/lbm) | | enthalpy (Btu/lbm) | | | entropy (Btu/lbm-°R) | | temp. (°F) |
		sat. liquid v_f	sat. vapor v_g	sat. liquid u_f	sat. vapor u_g	sat. liquid h_f	evap. h_{fg}	sat. vapor h_g	sat. liquid s_f	sat. vapor s_g	
32	0.0886	0.1602	3305	−0.01	1021.2	−0.01	1075.4	1075.4	−0.00003	2.1870	32
35	0.0999	0.01602	2948	2.99	1022.2	3.00	1073.7	1076.7	0.00607	2.1764	35
40	0.1217	0.01602	2445	8.02	1023.9	8.02	1070.9	1078.9	0.01617	2.1592	40
45	0.1475	0.01602	2037	13.04	1025.5	13.04	1068.1	1081.1	0.02618	2.1423	45
50	0.1780	0.01602	1704	18.06	1027.2	18.06	1065.2	1083.3	0.03607	2.1259	50
52	0.1917	0.01603	1589	20.06	1027.8	20.07	1064.1	1084.2	0.04000	2.1195	52
54	0.2064	0.01603	1482	22.07	1028.5	22.07	1063.0	1085.1	0.04391	2.1131	54
56	0.2219	0.01603	1383	24.08	1029.1	24.08	1061.9	1085.9	0.04781	2.1068	56
58	0.2386	0.01603	1292	26.08	1029.8	26.08	1060.7	1086.8	0.05159	2.1005	58
60	0.2563	0.01604	1207	28.08	1030.4	28.08	1059.6	1087.7	0.05555	2.0943	60
62	0.2751	0.01604	1129	30.09	1031.1	30.09	1058.5	1088.6	0.05940	2.0882	62
64	0.2952	0.01604	1056	32.09	1031.8	32.09	1057.3	1089.4	0.06323	2.0821	64
66	0.3165	0.01604	988.4	34.09	1032.4	34.09	1056.2	1090.3	0.06704	2.0761	66
68	0.3391	0.01605	925.8	36.09	1033.1	36.09	1055.1	1091.2	0.07084	2.0701	68
70	0.3632	0.01605	867.7	38.09	1033.7	38.09	1054.0	1092.0	0.07463	2.0642	70
72	0.3887	0.01606	813.7	40.09	1034.4	40.09	1052.8	1092.9	0.07839	2.0584	72
74	0.4158	0.01606	763.5	42.09	1035.0	42.09	1051.7	1093.8	0.08215	2.0526	74
76	0.4446	0.01606	716.8	44.09	1035.7	44.09	1050.6	1094.7	0.08589	2.0469	76
78	0.4750	0.01607	673.3	46.09	1036.3	46.09	1049.4	1095.5	0.08961	2.0412	78
80	0.5073	0.01607	632.8	48.08	1037.0	48.09	1048.3	1096.4	0.09332	2.0356	80
82	0.5414	0.01608	595.0	50.08	1037.6	50.08	1047.2	1097.3	0.09701	2.0300	82
84	0.5776	0.01608	559.8	52.08	1038.3	52.08	1046.0	1098.1	0.1007	2.0245	84
86	0.6158	0.01609	527.0	54.08	1038.9	54.08	1044.9	1099.0	0.1044	2.0190	86
88	0.6562	0.01609	496.3	56.07	1039.6	56.07	1043.8	1099.9	0.1080	2.0136	88
90	0.6988	0.01610	467.7	58.07	1040.2	58.07	1042.7	1100.7	0.1117	2.0083	90
92	0.7439	0.01611	440.9	60.06	1040.9	60.06	1041.5	1101.6	0.1153	2.0030	92
94	0.7914	0.01611	415.9	62.06	1041.5	62.06	1040.4	1102.4	0.1189	1.9977	94
96	0.8416	0.01612	392.4	64.05	1042.2	64.06	1039.2	1103.3	0.1225	1.9925	96
98	0.8945	0.01612	370.5	66.05	1042.8	66.05	1038.1	1104.2	0.1261	1.9874	98
100	0.9503	0.01613	350.0	68.04	1043.5	68.05	1037.0	1105.0	0.1296	1.9822	100
110	1.276	0.01617	265.1	78.02	1046.7	78.02	1031.3	1109.3	0.1473	1.9574	110
120	1.695	0.01621	203.0	87.99	1049.9	88.00	1025.5	1113.5	0.1647	1.9336	120
130	2.225	0.01625	157.2	97.97	1053.0	97.98	1019.8	1117.8	0.1817	1.9109	130
140	2.892	0.01629	122.9	107.95	1056.2	107.96	1014.0	1121.9	0.1985	1.8892	140
150	3.722	0.01634	97.0	117.95	1059.3	117.96	1008.1	1126.1	0.2150	1.8684	150
160	4.745	0.01640	77.2	127.94	1062.3	127.96	1002.2	1130.1	0.2313	1.8484	160
170	5.996	0.01645	62.0	137.95	1065.4	137.97	996.2	1134.2	0.2473	1.8293	170
180	7.515	0.01651	50.2	147.97	1068.3	147.99	990.2	1138.2	0.2631	1.8109	180
190	9.343	0.01657	41.0	158.00	1071.3	158.03	984.1	1142.1	0.2787	1.7932	190
200	11.529	0.01663	33.6	168.04	1074.2	168.07	977.9	1145.9	0.2940	1.7762	200
210	14.13	0.01670	27.82	178.1	1077.0	178.1	971.6	1149.7	0.3091	1.7599	210
212	14.70	0.01672	26.80	180.1	1077.6	180.2	970.3	1150.5	0.3121	1.7567	212
220	17.19	0.01677	23.15	188.2	1079.8	188.2	965.3	1153.5	0.3241	1.7441	220
230	20.78	0.01685	19.39	198.3	1082.6	198.3	958.8	1157.1	0.3388	1.7289	230
240	24.97	0.01692	16.33	208.4	1085.3	208.4	952.3	1160.7	0.3534	1.7143	240
250	29.82	0.01700	13.83	218.5	1087.9	218.6	945.6	1164.2	0.3677	1.7001	250
260	35.42	0.01708	11.77	228.6	1090.5	228.8	938.8	1167.6	0.3819	1.6864	260
270	41.83	0.01717	10.07	238.8	1093.0	239.0	932.0	1170.9	0.3960	1.6731	270
280	49.18	0.01726	8.65	249.0	1095.4	249.2	924.9	1174.1	0.4099	1.6602	280
290	57.53	0.01735	7.47	259.3	1097.7	259.4	917.8	1177.2	0.4236	1.6477	290
300	66.98	0.01745	6.472	269.5	1100.0	269.7	910.4	1180.2	0.4372	1.6356	300
310	77.64	0.01755	5.632	279.8	1102.1	280.1	903.0	1183.0	0.4507	1.6238	310
320	89.60	0.01765	4.919	290.1	1104.2	290.4	895.3	1185.8	0.4640	1.6123	320
330	103.00	0.01776	4.312	300.5	1106.2	300.8	887.5	1188.4	0.4772	1.6010	330
340	117.93	0.01787	3.792	310.9	1108.0	311.3	879.5	1190.8	0.4903	1.5901	340

(continued)

APPENDIX 24.A *(continued)*
Properties of Saturated Steam by Temperature
(customary U.S. units)

temp. (°F)	press. (lbf/in²)	specific volume (ft³/lbm)		internal energy (Btu/lbm)		enthalpy (Btu/lbm)			entropy (Btu/lbm-°R)		temp. (°F)
		sat. liquid v_f	sat. vapor v_g	sat. liquid u_f	sat. vapor u_g	sat. liquid h_f	evap. h_{fg}	sat. vapor h_g	sat. liquid s_f	sat. vapor s_g	
350	134.53	0.01799	3.346	321.4	1109.8	321.8	871.3	1193.1	0.5033	1.5793	350
360	152.92	0.01811	2.961	331.8	1111.4	332.4	862.9	1195.2	0.5162	1.5688	360
370	173.23	0.01823	2.628	342.4	1112.9	343.0	854.2	1197.2	0.5289	1.5585	370
380	195.60	0.01836	2.339	353.0	1114.3	353.6	845.4	1199.0	0.5416	1.5483	380
390	220.2	0.01850	2.087	363.6	1115.6	364.3	836.2	1200.6	0.5542	1.5383	390
400	247.1	0.01864	1.866	374.3	1116.6	375.1	826.8	1202.0	0.5667	1.5284	400
410	276.5	0.01878	1.673	385.0	1117.6	386.0	817.2	1203.1	0.5792	1.5187	410
420	308.5	0.01894	1.502	395.8	1118.3	396.9	807.2	1204.1	0.5915	1.5091	420
430	343.3	0.01909	1.352	406.7	1118.9	407.9	796.9	1204.8	0.6038	1.4995	430
440	381.2	0.01926	1.219	417.6	1119.3	419.0	786.3	1205.3	0.6161	1.4900	440
450	422.1	0.01943	1.1011	428.6	1119.5	430.2	775.4	1205.6	0.6282	1.4806	450
460	466.3	0.01961	0.9961	439.7	1119.6	441.4	764.1	1205.5	0.6404	1.4712	460
470	514.1	0.01980	0.9025	450.9	1119.4	452.8	752.4	1205.2	0.6525	1.4618	470
480	565.5	0.02000	0.8187	462.2	1118.9	464.3	740.3	1204.6	0.6646	1.4524	480
490	620.7	0.02021	0.7436	473.6	1118.3	475.9	727.8	1203.7	0.6767	1.4430	490
500	680.0	0.02043	0.6761	485.1	1117.4	487.7	714.8	1202.5	0.6888	1.4335	500
520	811.4	0.02091	0.5605	508.5	1114.8	511.7	687.3	1198.9	0.7130	1.4145	520
540	961.5	0.02145	0.4658	532.6	1111.0	536.4	657.5	1193.8	0.7374	1.3950	540
560	1131.8	0.02207	0.3877	548.4	1105.8	562.0	625.0	1187.0	0.7620	1.3749	560
580	1324.3	0.02278	0.3225	583.1	1098.9	588.6	589.3	1178.0	0.7872	1.3540	580
600	1541.0	0.02363	0.2677	609.9	1090.0	616.7	549.7	1164.4	0.8130	1.3317	600
620	1784.4	0.02465	0.2209	638.3	1078.5	646.4	505.0	1151.4	0.8398	1.3075	620
640	2057.1	0.02593	0.1805	668.7	1063.2	678.6	453.4	1131.9	0.8681	1.2803	640
660	2362	0.02767	0.1446	702.3	1042.3	714.4	391.1	1105.5	0.8990	1.2483	660
680	2705	0.03032	0.1113	741.7	1011.0	756.9	309.8	1066.7	0.9350	1.2068	680
700	3090	0.03666	0.0744	801.7	947.7	822.7	167.5	990.2	0.9902	1.1346	700
705.4	3204	0.05053	0.05053	872.6	872.6	902.5	0	902.5	1.0580	1.0580	705.4

Steam Tables: Thermodynamic Properties of Water Including Vapor, Liquid, and Solid Phases. Joseph H. Keenan, Frederick G. Keyes, Philip G. Hill, and Joan G. Moore. Copyright © 1978. Adapted with permission of John Wiley & Sons, Inc.

Support Material

APPENDIX 24.B
Properties of Saturated Steam by Pressure
(customary U.S. units)

absolute press. (lbf/in²)	temp. (°F)	specific volume (ft³/lbm)		internal energy (Btu/lbm)		enthalpy (Btu/lbm)			entropy (Btu/lbm-°R)			absolute press. (lbf/in²)
		sat. liquid v_f	sat. vapor v_g	sat. liquid u_f	sat. vapor u_g	sat. liquid h_f	evap. h_{fg}	sat. vapor h_g	sat. liquid s_f	evap. s_{fg}	sat. vapor s_g	
0.4	72.84	0.01606	792.0	40.94	1034.7	40.94	1052.3	1093.3	0.0800	1.9760	2.0559	0.4
0.6	85.19	0.01609	540.0	53.26	1038.7	53.27	1045.4	1098.6	0.1029	1.9184	2.0213	0.6
0.8	94.35	0.01611	411.7	62.41	1041.7	62.41	1040.2	1102.6	0.1195	1.8773	1.9968	0.8
1.0	101.70	0.01614	333.6	69.74	1044.0	69.74	1036.0	1105.8	0.1327	1.8453	1.9779	1.0
1.2	107.88	0.01616	280.9	75.90	1046.0	75.90	1032.5	1108.4	0.1436	1.8190	1.9626	1.2
1.5	115.65	0.01619	227.7	83.65	1048.5	83.65	1028.0	1111.7	0.1571	1.7867	1.9438	1.5
2.0	126.04	0.01623	173.75	94.02	1051.8	94.02	1022.1	1116.1	0.1750	1.7448	1.9198	2.0
3.0	141.43	0.01630	118.72	109.38	1056.6	109.39	1013.1	1122.5	0.2009	1.6852	1.8861	3.0
4.0	152.93	0.01636	90.64	120.88	1060.2	120.89	1006.4	1127.3	0.2198	1.6426	1.8624	4.0
5.0	162.21	0.01641	73.53	130.15	1063.0	130.17	1000.9	1131.0	0.2349	1.6093	1.8441	5.0
6.0	170.03	0.01645	61.98	137.98	1065.4	138.00	996.2	1134.2	0.2474	1.5819	1.8292	6.0
7.0	176.82	0.01649	53.65	144.78	1067.4	144.80	992.1	1136.9	0.2581	1.5585	1.8167	7.0
8.0	182.84	0.01653	47.35	150.81	1069.2	150.84	988.4	1139.3	0.2675	1.5383	1.8058	8.0
9.0	188.26	0.01656	42.41	156.25	1070.8	156.27	985.1	1141.4	0.2760	1.5203	1.7963	9.0
10	193.19	0.01659	38.42	161.20	1072.2	161.23	982.1	1143.3	0.2836	1.5041	1.7877	10
14.696	211.99	0.01672	26.80	180.10	1077.6	180.15	970.4	1150.5	0.3121	1.4446	1.7567	14.696
15	213.03	0.01672	26.29	181.14	1077.9	181.19	969.7	1150.9	0.3137	1.4414	1.7551	15
20	227.96	0.01683	20.09	196.19	1082.0	196.26	960.1	1156.4	0.3358	1.3962	1.7320	20
25	240.08	0.01692	16.31	208.44	1085.3	208.52	952.2	1160.7	0.3535	1.3607	1.7142	25
30	250.34	0.01700	13.75	218.84	1088.0	218.93	945.4	1164.3	0.3682	1.3314	1.6996	30
35	259.30	0.01708	11.90	227.93	1090.3	228.04	939.3	1167.4	0.3809	1.3064	1.6873	35
40	267.26	0.01715	10.50	236.03	1092.3	236.16	933.8	1170.0	0.3921	1.2845	1.6767	40
45	274.46	0.01721	9.40	243.37	1094.0	243.51	928.8	1172.3	0.4022	1.2651	1.6673	45
50	281.03	0.01727	8.52	250.08	1095.6	250.24	924.2	1174.4	0.4113	1.2476	1.6589	50
55	287.10	0.01733	7.79	256.28	1097.0	256.46	919.9	1176.3	0.4196	1.2317	1.6513	55
60	292.73	0.01738	7.177	262.1	1098.3	262.2	915.8	1178.0	0.4273	1.2170	1.6443	60
65	298.00	0.01743	6.647	267.5	1099.5	267.7	911.9	1179.6	0.4345	1.2035	1.6380	65
70	302.96	0.01748	6.209	272.6	1100.6	272.8	908.3	1181.0	0.4412	1.1909	1.6321	70
75	307.63	0.01752	5.818	277.4	1101.6	277.6	904.8	1182.4	0.4475	1.1790	1.6265	75
80	312.07	0.01757	5.474	282.0	1102.6	282.2	901.4	1183.6	0.4534	1.1679	1.6213	80
85	316.29	0.01761	5.170	286.3	1103.5	286.6	898.2	1184.8	0.4591	1.1574	1.6165	85
90	320.31	0.01766	4.898	290.5	1104.3	290.8	895.1	1185.9	0.4644	1.1475	1.6119	90
95	324.16	0.01770	4.654	294.5	1105.0	294.8	892.1	1186.9	0.4695	1.1380	1.6075	95
100	327.86	0.01774	4.434	298.3	1105.8	298.6	889.2	1187.8	0.4744	1.1290	1.6034	100
110	334.82	0.01781	4.051	305.5	1107.1	305.9	883.7	1189.6	0.4836	1.1122	1.5958	110
120	341.30	0.01789	3.730	312.3	1108.3	312.7	878.5	1191.1	0.4920	1.0966	1.5886	120
130	347.37	0.01796	3.457	318.6	1109.4	319.0	873.5	1192.5	0.4999	1.0822	1.5821	130
140	353.08	0.01802	3.221	324.6	1110.3	325.1	868.7	1193.8	0.5073	1.0688	1.5761	140
150	358.48	0.01809	3.016	330.2	1111.2	330.8	864.2	1194.9	0.5142	1.0562	1.5704	150
160	363.60	0.01815	2.836	335.6	1112.0	336.2	859.8	1196.0	0.5208	1.0443	1.5651	160
170	368.47	0.01821	2.676	340.8	1112.7	341.3	855.6	1196.9	0.5270	1.0330	1.5600	170
180	373.13	0.01827	2.553	345.7	1113.4	346.3	851.5	1197.8	0.5329	1.0223	1.5552	180
190	377.59	0.01833	2.405	350.4	1114.0	351.0	847.5	1198.6	0.5386	1.0122	1.5508	190
200	381.86	0.01839	2.289	354.9	1114.6	355.6	843.7	1199.3	0.5440	1.0025	1.5465	200
250	401.04	0.01865	1.845	375.4	1116.7	376.2	825.8	1202.1	0.5680	0.9594	1.5274	250
300	417.43	0.01890	1.544	393.0	1118.2	394.1	809.8	1203.9	0.5883	0.9232	1.5115	300
350	431.82	0.01912	1.327	408.7	1119.0	409.9	795.0	1204.9	0.6060	0.8917	1.4977	350
400	444.70	0.01934	1.162	422.8	1119.5	424.2	781.2	1205.5	0.6218	0.8638	1.4856	400
450	456.39	0.01955	1.033	435.7	1119.6	437.4	768.2	1205.6	0.6360	0.8385	1.4745	450
500	467.13	0.01975	0.928	447.7	1119.4	449.5	755.8	1205.3	0.6490	0.8154	1.4644	500
550	477.07	0.01994	0.842	458.9	1119.1	460.9	743.9	1204.8	0.6611	0.7941	1.4551	550
600	486.33	0.02013	0.770	469.4	1118.6	471.7	732.4	1204.1	0.6723	0.7742	1.4464	600
700	503.23	0.02051	0.656	488.9	1117.0	491.5	710.5	1202.0	0.6927	0.7378	1.4305	700
800	518.36	0.02087	0.569	506.6	1115.0	509.7	689.6	1199.3	0.7110	0.7050	1.4160	800
900	532.12	0.02123	0.501	523.0	1112.6	526.6	669.5	1196.0	0.7277	0.6750	1.4027	900
1000	544.75	0.02159	0.446	538.4	1109.9	542.4	650.0	1192.4	0.7432	0.6471	1.3903	1000
1100	556.45	0.02195	0.401	552.9	1106.8	557.4	631.0	1188.3	0.7576	0.6209	1.3786	1100
1200	567.37	0.02232	0.362	566.7	1103.5	571.7	612.3	1183.9	0.7712	0.5961	1.3673	1200
1300	577.60	0.02269	0.330	579.9	1099.8	585.4	593.8	1179.2	0.7841	0.5724	1.3565	1300
1400	587.25	0.02307	0.302	592.7	1096.0	598.6	575.5	1174.1	0.7964	0.5497	1.3461	1400
1500	596.39	0.02346	0.277	605.0	1091.8	611.5	557.2	1168.7	0.8082	0.5276	1.3359	1500
1600	605.06	0.02386	0.255	616.9	1087.4	624.0	538.9	1162.9	0.8196	0.5062	1.3258	1600
1700	613.32	0.02428	0.236	628.6	1082.7	636.2	520.6	1156.9	0.8307	0.4852	1.3159	1700
1800	621.21	0.02472	0.218	640.0	1077.7	648.3	502.1	1150.4	0.8414	0.4645	1.3060	1800
1900	628.76	0.02517	0.203	651.3	1072.3	660.1	483.4	1143.5	0.8519	0.4441	1.2961	1900
2000	636.00	0.02565	0.188	662.4	1066.6	671.9	464.4	1136.3	0.8623	0.4238	1.2861	2000
2250	652.90	0.02698	0.157	689.9	1050.6	701.1	414.8	1115.9	0.8876	0.3728	1.2604	2250
2500	668.31	0.02860	0.131	717.7	1031.0	730.9	360.5	1091.4	0.9131	0.3196	1.2327	2500
2750	682.46	0.03077	0.107	747.3	1005.9	763.0	297.4	1060.4	0.9401	0.2604	1.2005	2750
3000	695.52	0.03431	0.084	783.4	968.8	802.5	213.0	1015.5	0.9732	0.1843	1.1575	3000
3203.6	705.44	0.05053	0.0505	872.6	872.6	902.5	0	902.5	1.0580	0	1.0580	3203.6

Steam Tables: Thermodynamic Properties of Water Including Vapor, Liquid, and Solid Phases (English Units). Joseph H. Keenan, Frederick G. Keyes, Philip G. Hill, and Joan G. Moore. Copyright © 1969. Adapted with permission of John Wiley & Sons, Inc.

APPENDIX 24.C
Properties of Superheated Steam
(customary U.S. units)

specific volume (v) in ft³/lbm; enthalpy (h) in Btu/lbm; entropy (s) in Btu/lbm-°R

absolute pressure (psia) (sat. temp., °F)		temperature (°F)								
		200	300	400	500	600	700	800	900	1000
1.0 (101.70)	v	392.5	452.3	511.9	571.5	631.1	690.7	750.3	809.9	869.5
	h	1150.1	1195.7	1241.8	1288.5	1336.1	1384.5	1433.7	1483.8	1534.8
	s	2.0508	2.1150	2.1720	2.2235	2.2706	2.3142	2.3550	2.3932	2.4294
5.0 (162.21)	v	78.15	90.24	102.24	114.20	126.15	138.08	150.01	161.94	173.86
	h	1148.6	1194.8	1241.2	1288.2	1335.8	1384.3	1433.5	1483.7	1534.7
	s	1.8715	1.9367	1.9941	2.0458	2.0930	2.1367	2.1775	2.2158	2.2520
10.0 (193.19)	v	38.85	44.99	51.03	57.04	63.03	69.01	74.98	80.95	86.91
	h	1146.6	1193.7	1240.5	1287.7	1335.5	1384.0	1433.3	1483.5	1534.6
	s	1.7927	1.8592	1.9171	1.9690	2.0164	2.0601	2.1009	2.1393	2.1755
14.696 (211.99)	v	....	30.52	34.67	38.77	42.86	46.93	51.00	55.07	59.13
	h	....	1192.6	1239.9	1287.3	1335.2	1383.8	1433.1	1483.4	1534.5
	s	....	1.8157	1.8741	1.9263	1.9737	2.0175	2.0584	2.0967	2.1330
20.0 (227.96)	v	....	22.36	25.43	28.46	31.47	34.77	37.46	40.45	43.44
	h	....	1191.5	1239.2	1286.8	1334.8	1383.5	1432.9	1483.2	1534.3
	s	....	1.7805	1.8395	1.8919	1.9395	1.9834	2.0243	2.0627	2.0989
60.0 (292.73)	v	....	7.260	8.353	9.399	10.425	11.440	12.448	13.452	14.454
	h	....	1181.9	1233.5	1283.0	1332.1	1381.4	1431.2	1481.8	1533.2
	s	....	1.6496	1.7134	1.7678	1.8165	1.8609	1.9022	1.9408	1.9773
100.0 (327.86)	v	....	....	4.934	5.587	6.216	6.834	7.445	8.053	8.657
	h	....	....	1227.5	1279.1	1329.3	1379.2	1429.6	1480.5	1532.1
	s	....	....	1.6517	1.7085	1.7582	1.8033	1.8449	1.8838	1.9204
150.0 (358.48)	v	....	....	3.221	3.679	4.111	4.531	4.944	5.353	5.759
	h	....	....	1219.5	1274.1	1325.7	1376.6	1427.5	1478.8	1530.7
	s	....	....	1.5997	1.6598	1.7110	1.7568	1.7989	1.8381	1.8750
200.0 (381.86)	v	....	....	2.361	2.724	3.058	3.379	3.693	4.003	4.310
	h	....	....	1210.8	1268.8	1322.1	1373.8	1425.3	1477.1	1529.3
	s	....	....	1.5600	1.6239	1.6767	1.7234	1.7660	1.8055	1.8425
250.0 (401.04)	v	....	....	....	2.150	2.426	2.688	2.943	3.193	3.440
	h	....	....	....	1263.3	1318.3	1371.1	1423.2	1475.3	1527.9
	s	....	....	....	1.5948	1.6494	1.6970	1.7401	1.7799	1.8172
300.0 (417.43)	v	....	....	....	1.766	2.004	2.227	2.442	2.653	2.860
	h	....	....	....	1257.5	1314.5	1368.3	1421.0	1473.6	1526.5
	s	....	....	....	1.5701	1.6266	1.6751	1.7187	1.7589	1.7964
400.0 (444.70)	v	....	....	....	1.2843	1.4760	1.6503	1.8163	1.9776	2.136
	h	....	....	....	1245.2	1306.6	1362.5	1416.6	1470.1	1523.6
	s	....	....	....	1.5282	1.5892	1.6397	1.6884	1.7252	1.7632

(continued)

Support Material

APPENDIX 24.C *(continued)*
Properties of Superheated Steam
(customary U.S. units)

specific volume (v) in ft^3/lbm; enthalpy (h) in Btu/lbm; entropy (s) in Btu/lbm-°R

absolute pressure (psia) (sat. temp., °F)		temperature (°F) 500	600	700	800	900	1000	1100	1200	1400	1600
450	v	1.123	1.300	1.458	1.608	1.752	1.894	2.034	2.172	2.444	
	h	1238.5	1302.5	1359.6	1414.4	1468.3	1522.2	1576.3	1630.8	1741.7	
(456.4)	s	1.5097	1.5732	1.6248	1.6701	1.7113	1.7495	1.7853	1.8192	1.8823	
500	v	0.992	1.158	1.304	1.441	1.572	1.701	1.827	1.952	2.198	
	h	1231.5	1298.3	1356.7	1412.1	1466.5	1520.7	1575.1	1629.8	1741.0	
(467.1)	s	1.4923	1.5585	1.6112	1.6571	1.6987	1.7471	1.7731	1.8072	1.8704	
600	v	0.795	0.946	1.073	1.190	1.302	1.411	1.517	1.622	1.829	
	h	1216.2	1289.5	1350.6	1407.6	1462.9	1517.8	1572.7	1627.8	1739.5	
(486.3)	s	1.4592	1.5320	1.5872	1.6343	1.6766	1.7155	1.7519	1.7861	1.8497	
700	v	....	0.793	0.907	1.011	1.109	1.204	1.296	1.387	1.565	
	h	....	1280.2	1344.4	1402.9	1459.3	1514.9	1570.2	1625.8	1738.1	
(503.2)	s	....	1.5081	1.5661	1.6145	1.6576	1.6970	1.7337	1.7682	1.8321	
800	v	....	0.677	0.783	0.876	0.964	1.048	1.130	1.210	1.367	
	h	....	1270.4	1338.0	1398.2	1455.6	1511.9	1567.8	1623.8	1736.6	
(518.3)	s	....	1.4861	1.5471	1.5969	1.6408	1.6807	1.7178	1.7526	1.8167	
900	v	....	0.587	0.686	0.772	0.851	0.927	1.001	1.073	1.214	
	h	....	1260.0	1331.4	1393.4	1451.9	1508.9	1565.4	1621.7	1735.1	
(532.1)	s	....	1.4652	1.5297	1.5810	1.6257	1.6662	1.7036	1.7386	1.8031	
1000	v	....	0.514	0.608	0.688	0.761	0.831	0.898	0.963	1.091	1.215
	h	....	1248.8	1324.6	1388.5	1448.1	1505.9	1562.9	1619.7	1733.7	1849.3
(544.7)	s	....	1.4450	1.5135	1.5665	1.6120	1.6530	1.6908	1.7261	1.7909	1.8499
1200	v	....	0.402	0.491	0.562	0.626	0.685	0.743	0.798	0.906	1.011
	h	....	1223.6	1310.2	1378.4	1440.4	1499.7	1557.9	1615.5	1730.7	1847.1
(567.4)	s	....	1.4054	1.4837	1.5402	1.5876	1.6297	1.6682	1.7040	1.7696	1.8290
1400	v	....	0.318	0.406	0.471	0.529	0.582	0.632	0.681	0.774	0.865
	h	....	1193.1	1294.8	1367.9	1432.5	1493.5	1552.8	1611.4	1727.8	1844.8
(587.2)	s	....	1.3641	1.4562	1.5168	1.5661	1.6094	1.6487	1.6851	1.7513	1.8111
1600	v	....	....	0.342	0.403	0.466	0.504	0.549	0.592	0.675	0.755
	h	....	....	1278.1	1357.0	1424.4	1487.1	1547.7	1607.1	1724.8	1842.6
(605.1)	s	....	....	1.4299	1.4953	1.5468	1.5913	1.6315	1.6684	1.7354	1.7955
1800	v	....	....	0.291	0.350	0.399	0.443	0.484	0.524	0.598	0.670
	h	....	....	1259.9	1345.7	1416.1	1480.7	1542.5	1602.9	1721.8	1840.4
(621.2)	s	....	....	1.4042	1.4753	1.5291	1.5749	1.6159	1.6534	1.7211	1.7817
2000	v	....	....	0.249	0.307	0.353	0.395	0.433	0.469	0.537	0.602
	h	....	....	1239.8	1333.8	1407.6	1474.1	1537.2	1598.6	1718.8	1838.2
(636.0)	s	....	....	1.3782	1.4562	1.5126	1.5598	1.6017	1.6398	1.7082	1.7692

Steam Tables: Thermodynamic Properties of Water Including Vapor, Liquid, and Solid Phases (English Units). Joseph H. Keenan, Frederick G. Keyes, Philip G. Hill, and Joan G. Moore. Copyright © 1969. Adapted with permission of John Wiley & Sons, Inc.

APPENDIX 24.D
Compressed Water

			Temperature, °F							
			32	100	200	300	400	500	600	700
Absolute Pressure, psia (sat. temp.)	saturated liquid	P v_f h_f s_f	0.08854 0.016022 0 0	0.9492 0.016132 67.97 0.12948	11.526 0.016634 167.99 0.29382	67.013 0.017449 269.59 0.43694	247.31 0.018639 374.97 0.56638	680.8 0.020432 487.82 0.68871	1542.9 0.023629 617.0 0.8131	3093.7 0.03692 823.3 0.9905
200 (381.79)		$(v - v_f) \times 10^{-5 a}$ $(h - h_f)$ $(s - s_f) \times 10^{-3 b}$	−1.1 +0.61 +0.03	−1.1 +0.54 −0.05	−1.1 +0.41 −0.21	−1.1 +0.23 −0.21				
400 (444.59)		$(v - v_f) \times 10^{-5 a}$ $(h - h_f)$ $(s - s_f) \times 10^{-3 b}$	−2.3 +1.21 +0.04	−2.1 +1.09 −0.16	−2.2 +0.88 −0.47	−2.8 +0.61 −0.56	−2.1 +0.16 −0.40			
800 (518.23)		$(v - v_f) \times 10^{-5 a}$ $(h - h_f)$ $(s - s_f) \times 10^{-3 b}$	−4.6 +2.39 +0.10	−4.0 +2.17 −0.40	−4.4 +1.78 −0.97	−5.6 +1.35 −1.27	−6.5 +0.61 −1.48	−1.7 −0.05 −0.53		
1000 (544.61)		$(v - v_f) \times 10^{-5 a}$ $(h - h_f)$ $(s - s_f) \times 10^{-3 b}$	−5.7 +2.99 +0.15	−5.1 +2.70 −0.53	−5.4 +2.21 −1.20	−6.9 +1.75 −1.64	−8.7 +0.84 −2.00	−6.4 −0.14 −1.41		
1500 (596.23)		$(v - v_f) \times 10^{-5 a}$ $(h - h_f)$ $(s - s_f) \times 10^{-3 b}$	−8.4 +4.48 +0.20	−7.5 +3.99 −0.86	−8.1 +3.36 −1.79	−10.4 +2.70 −2.53	−14.1 +1.44 −3.32	−17.3 −0.29 −3.56		
2000 (635.82)		$(v - v_f) \times 10^{-5 a}$ $(h - h_f)$ $(s - s_f) \times 10^{-3 b}$	−11.0 +5.97 +0.22	−9.9 +5.31 −1.18	−10.8 +4.51 −2.39	−13.8 +3.64 −3.42	−19.5 +2.03 −4.57	−27.8 −0.38 −5.58	−32.6 −2.5 −4.3	
3000 (695.36)		$(v - v_f) \times 10^{-5 a}$ $(h - h_f)$ $(s - s_f) \times 10^{-3 b}$	−16.3 +9.00 +0.28	−14.7 +7.88 −1.79	−16.0 +6.76 −3.56	−20.7 +5.49 −5.12	−30.0 +3.33 −7.03	−47.1 −0.41 −9.42	−87.9 −6.9 −12.4	
4000		$(v - v_f) \times 10^{-5 a}$ $(h - h_f)$ $(s - s_f) \times 10^{-3 b}$	−21.5 +11.88 +0.29	−19.2 +10.49 −2.42	−21.0 +9.03 −4.74	−27.5 +7.41 −6.77	−40.0 +4.71 −9.40	−64.5 −0.16 −13.03	−132.2 −10.0 −19.3	−821 −59.5 −55.8
5000		$(v - v_f) \times 10^{-5 a}$ $(h - h_f)$ $(s - s_f) \times 10^{-3 b}$	−26.7 +14.75 +0.22	−23.6 +13.08 −3.07	−26.0 +11.30 −5.92	−34.0 +9.36 −8.40	−49.6 +6.08 −11.74	−80.5 +0.25 −16.47	−169.3 −12.1 −25.3	−1017 −76.9 −75.3

[a] Multiply table values of Δv by 10^5.
[b] Multiply table values of ΔS by 10^3.

Steam Tables, Thermodynamic Properties of Water Including Vapor, Liquid, and Solid Phases (English Units), by J. H. Keenan, F. G. Keyes, P. G. Hill, and J. G. Moore, copyright © 1969. Adapted by permission of John Wiley & Sons, Inc.

Support Material

APPENDIX 24.E
Enthalpy-Entropy (Mollier) Diagram for Steam
(customary U.S. units)

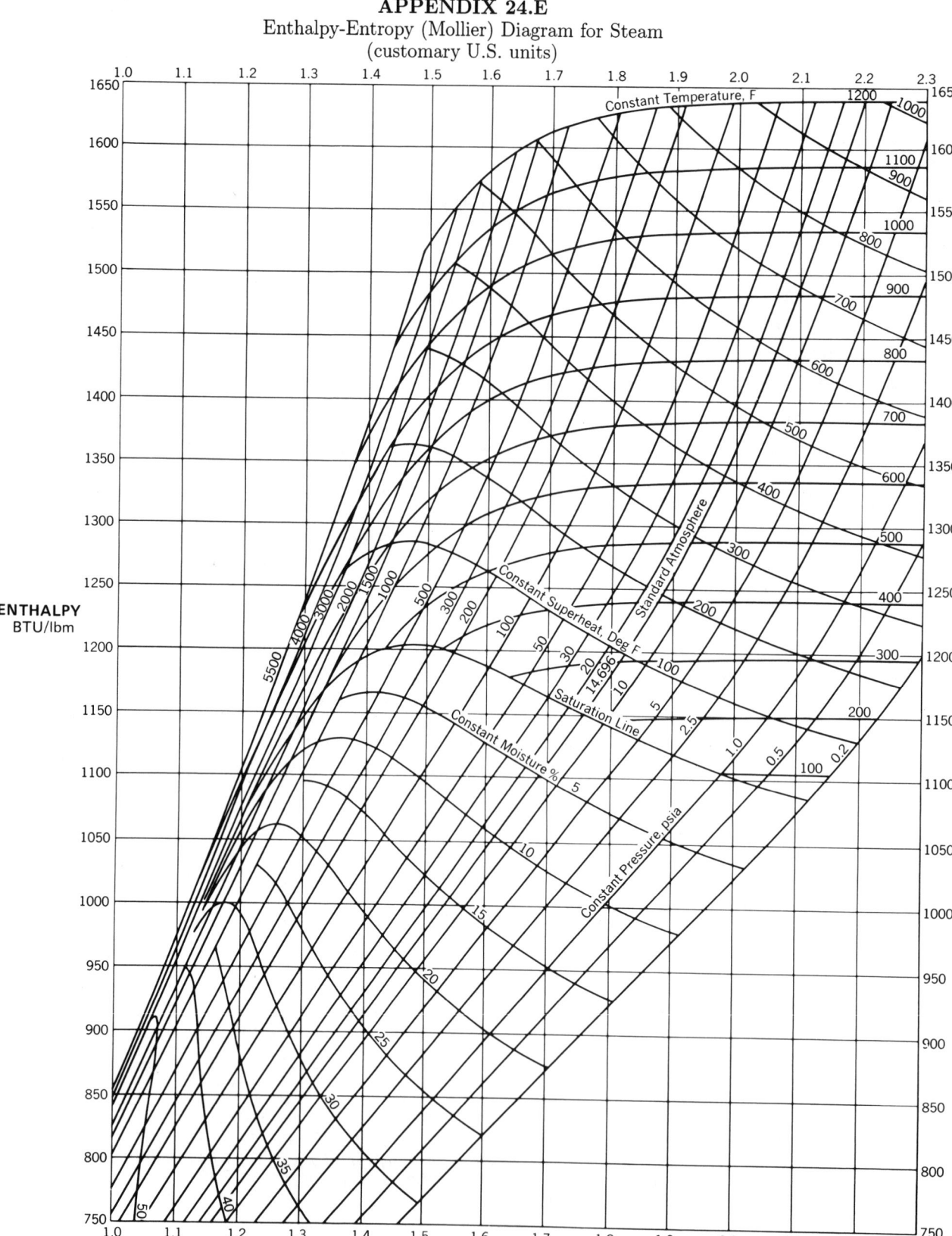

Used with permission from *Steam: Its generation and use*, 40th Edition, ed. by S. C. Stultz and J. B. Kitto, copyright © 1992, by The Babcock & Wilcox Company.

APPENDIX 24.F
Properties of Low-Pressure Air
(customary U.S. units)

T in °R; h and u in Btu/lbm; ϕ in Btu/lbm-°R

T	h	p_r	u	v_r	ϕ	T	h	p_r	u	v_r	ϕ
360	85.97	0.3363	61.29	396.6	0.50369	1700	422.59	90.95	306.06	6.924	0.88758
380	90.75	0.4061	64.70	346.6	0.51663	1750	436.12	101.98	316.16	6.357	0.89542
400	95.53	0.4858	68.11	305.0	0.52890	1800	449.71	114.0	326.32	5.847	0.90308
420	100.32	0.5760	71.52	270.1	0.54058	1850	463.37	127.2	336.55	5.388	0.91056
440	105.11	0.6776	74.93	240.6	0.55172	1900	477.09	141.5	346.85	4.974	0.91788
460	109.90	0.7913	78.36	215.33	0.56235	1950	490.88	157.1	357.20	4.598	0.92504
480	114.69	0.9182	81.77	193.65	0.57255	2000	504.71	174.0	367.61	4.258	0.93205
500	119.48	1.0590	85.20	174.90	0.58233	2050	518.61	192.3	378.08	3.949	0.93891
520	124.27	1.2147	88.62	158.58	0.59172	2100	532.55	212.1	388.60	3.667	0.94564
537	128.34	1.3593	91.53	146.34	0.59945	2150	546.54	233.5	399.17	3.410	0.95222
540	129.06	1.3860	92.04	144.32	0.60078	2200	560.59	256.6	409.78	3.176	0.95868
560	133.86	1.5742	95.47	131.78	0.60950	2250	574.69	281.4	420.46	2.961	0.96501
580	138.66	1.7800	98.90	120.70	0.61793	2300	588.82	308.1	431.16	2.765	0.97123
600	143.47	2.005	102.34	110.88	0.62607	2350	603.00	336.8	441.91	2.585	0.97732
620	148.28	2.249	105.78	102.12	0.63395	2400	617.22	367.6	452.70	2.419	0.98331
640	153.09	2.514	109.21	94.30	0.64159	2450	631.48	400.5	463.54	2.266	0.98919
660	157.92	2.801	112.67	87.27	0.64902	2500	645.78	435.7	474.40	2.125	0.99497
680	162.74	3.111	116.12	80.96	0.65621	2550	660.12	473.3	485.31	1.996	1.00064
700	167.56	3.446	119.58	75.25	0.66321	2600	674.49	513.5	496.26	1.876	1.00623
720	172.39	3.806	123.04	70.07	0.67002	2650	688.90	556.3	507.25	1.765	1.01172
740	177.23	4.193	126.51	65.38	0.67665	2700	703.35	601.9	518.26	1.662	1.01712
760	182.08	4.607	129.99	61.10	0.68312	2750	717.83	650.4	529.31	1.566	1.02244
780	186.94	5.051	133.47	57.20	0.68942	2800	732.33	702.0	540.40	1.478	1.02767
800	191.81	5.526	136.97	53.63	0.69558	2850	746.88	756.7	551.52	1.395	1.03282
820	196.69	6.033	140.47	50.35	0.70160	2900	761.45	814.8	562.66	1.318	1.03788
840	201.56	6.573	143.98	47.34	0.70747	2950	776.05	876.4	573.84	1.247	1.04288
860	206.46	7.149	147.50	44.57	0.71323	3000	790.68	941.4	585.04	1.180	1.04779
880	211.35	7.761	151.02	42.01	0.71886	3050	805.34	1011	596.28	1.118	1.05264
900	216.26	8.411	154.57	39.64	0.72438	3100	820.03	1083	607.53	1.060	1.05741
920	221.18	9.102	158.12	37.44	0.72979	3150	834.75	1161	618.82	1.006	1.06212
940	226.11	9.834	161.68	35.41	0.73509	3200	849.48	1242	630.12	0.9546	1.06676
960	231.06	10.61	165.26	33.52	0.74030	3250	864.24	1328	641.46	0.9069	1.07134
980	236.02	11.43	168.83	31.76	0.74540	3300	879.02	1418	652.81	0.8621	1.07585
1000	240.98	12.30	172.43	30.12	0.75042	3350	893.83	1513	664.20	0.8202	1.08031
1040	250.95	14.18	179.66	27.17	0.76019	3400	908.66	1613	675.60	0.7807	1.08470
1080	260.97	16.28	186.93	24.58	0.76964	3450	923.52	1719	687.04	0.7436	1.08904
1120	271.03	18.60	194.25	22.30	0.77880	3500	938.40	1829	698.48	0.7087	1.09332
1160	281.14	21.18	201.63	20.29	0.78767	3550	953.30	1946	709.95	0.6759	1.09755
1200	291.30	24.01	209.05	18.51	0.79628	3600	968.21	2068	721.44	0.6449	1.10172
1240	301.52	27.13	216.53	16.93	0.80466	3650	983.15	2196	732.95	0.6157	1.10584
1280	311.79	30.55	224.05	15.52	0.81280	3700	998.11	2330	744.48	0.5882	1.10991
1320	322.11	34.31	231.63	14.25	0.82075	3750	1013.1	2471	756.04	0.5621	1.11393
1360	332.48	38.41	239.25	13.12	0.82848	3800	1028.1	2618	767.60	0.5376	1.11791
1400	342.90	42.88	246.93	12.10	0.83604	3850	1043.1	2773	779.19	0.5143	1.12183
1440	353.37	47.75	254.66	11.17	0.84341	3900	1058.1	2934	790.80	0.4923	1.12571
1480	363.89	53.04	262.44	10.34	0.85062	3950	1073.2	3103	802.43	0.4715	1.12955
1520	374.47	58.78	270.26	9.578	0.85767	4000	1088.3	3280	814.06	0.4518	1.13334
1560	385.08	65.00	278.13	8.890	0.86456	4050	1103.4	3464	825.72	0.4331	1.13709
1600	395.74	71.73	286.06	8.263	0.87130	4100	1118.5	3656	837.40	0.4154	1.14079
1650	409.13	80.89	296.03	7.556	0.87954	4150	1133.6	3858	849.09	0.3985	1.14446

(continued)

APPENDIX 24.F *(continued)*
Properties of Low-Pressure Air
(customary U.S. units)

T in °R; h and u in Btu/lbm; ϕ in Btu/lbm-°R

T	h	p_r	u	v_r	ϕ
4200	1148.7	4067	860.81	0.3826	1.14809
4300	1179.0	4513	884.28	0.3529	1.15522
4400	1209.4	4997	907.81	0.3262	1.16221
4500	1239.9	5521	931.39	0.3019	1.16905
4600	1270.4	6089	955.04	0.2799	1.17575
4700	1300.9	6701	978.73	0.2598	1.18232
4800	1331.5	7362	1002.5	0.2415	1.18876
4900	1362.2	8073	1026.3	0.2248	1.19508
5000	1392.9	8837	1050.1	0.2096	1.20129
5100	1423.6	9658	1074.0	0.1956	1.20738
5200	1454.4	10539	1098.0	0.1828	1.21336
5300	1485.3	11481	1122.0	0.1710	1.21923

Gas Tables: Thermodynamic Properties of Air, Products of Combustion and Component Gases, Compressible Flow Functions, 2nd Edition (English Units). Joseph H. Keenan, Jing Chao, and Joseph Kaye. Copyright © 1980. Adapted with permission from John Wiley & Sons, Inc.

APPENDIX 24.G
Properties of Saturated Refrigerant-12 (R-12) by Temperature
(customary U.S. units)

| temp. (°F) | pressure (psia) | specific volume (ft³/lbm) | | enthalpy (Btu/lbm) | | | entropy (Btu/lbm-°R) | | | temp. (°F) |
| | | sat. liquid | sat. vapor | sat. liquid | evap. | sat. vapor | sat. liquid | evap. | sat. vapor | |
T	p	v_f	v_g	h_f	h_{fg}	h_g	s_f	s_{fg}	s_g	T
−60	5.37	0.01036	6.516	−4.20	75.33	71.13	−0.0102	0.1681	0.1783	−60
−50	7.13	0.01047	5.012	−2.11	74.42	72.31	−0.0050	0.1717	0.1767	−50
−40	9.32	0.0106	3.911	0.00	73.50	73.50	0.00000	0.17517	0.17517	−40
−30	12.02	0.0107	3.088	2.03	72.67	74.70	0.00471	0.16916	0.17387	−30
−20	15.28	0.0108	2.474	4.07	71.80	75.87	0.00940	0.16335	0.17275	−20
−10	19.20	0.0109	2.003	6.14	70.91	77.05	0.01403	0.15772	0.17175	−10
0	23.87	0.0110	1.637	8.25	69.96	78.21	0.01869	0.15222	0.17091	0
5	26.51	0.0111	1.485	9.32	69.47	78.79	0.02097	1.14955	0.17052	5
10	29.35	0.0112	1.351	10.39	68.97	79.36	0.02328	0.14687	0.17015	10
20	35.75	0.0113	1.121	12.55	67.94	80.49	0.02783	0.14166	0.16949	20
30	43.16	0.0115	0.939	14.76	66.85	81.61	0.03233	0.13654	0.16887	30
40	51.68	0.0116	0.792	17.00	65.71	82.71	0.03680	0.13153	0.15833	40
50	61.39	0.0118	0.673	19.27	64.51	83.78	0.04126	0.12659	0.16785	50
60	72.41	0.0119	0.575	21.57	63.25	84.82	0.04568	0.12173	0.16741	60
70	84.82	0.0121	0.493	23.90	61.92	85.82	0.05009	0.11692	0.16701	70
80	98.76	0.0123	0.425	26.28	60.52	86.80	0.05446	0.11215	0.16662	80
86	107.9	0.0124	0.389	27.72	59.65	87.37	0.05708	0.10932	0.16640	86
90	114.3	0.0125	0.368	28.70	59.04	87.74	0.05882	0.10742	0.16624	90
100	131.6	0.0127	0.319	31.16	57.46	88.62	0.06316	0.10268	0.16584	100
110	150.7	0.0129	0.277	33.65	55.78	89.43	0.06749	0.09793	0.16542	110
120	171.8	0.0132	0.240	36.16	53.99	90.15	0.07180	0.09315	0.16495	120
233	596.9	0.02870	0.02870	78.86	0	78.86	0.1359	0	0.1359	233

Reproduced by permission of the DuPont Company.

APPENDIX 24.H
Properties of Saturated Refrigerant-12 (R-12) by Pressure
(customary U.S. units)

| pressure (psia) | temp. (°F) | specific volume (ft³/lbm) | | enthalpy (Btu/lbm) | | | entropy (Btu/lbm-°R) | | | pressure (psia) |
| | | sat. liquid | sat. vapor | sat. liquid | evap. | sat. vapor | sat. liquid | evap. | sat. vapor | |
p	T	v_f	v_g	h_f	h_{fg}	h_g	s_f	s_{fg}	s_g	p
5	−62.5	0.01034	6.953	−4.73	75.56	70.83	−0.0115	0.1943	0.1788	5
10	−37.3	0.0106	3.662	0.54	73.28	73.82	0.00127	0.17360	0.17487	10
15	−20.8	0.0108	2.518	3.91	71.87	75.78	0.00902	0.16381	0.17283	15
20	−8.2	0.0109	1.925	6.53	70.74	77.27	0.01488	0.15672	0.17160	20
30	11.1	0.0112	1.324	10.62	68.86	79.48	0.02410	0.14597	0.17007	30
40	25.9	0.0114	1.009	13.86	67.30	81.16	0.03049	0.13865	0.16914	40
50	38.3	0.0116	0.817	16.58	65.94	82.52	0.03597	0.13244	0.16841	50
60	48.7	0.0117	0.688	18.96	64.69	83.65	0.04065	0.12726	0.16791	60
80	66.3	0.0120	0.521	23.01	62.44	85.45	0.04844	0.11872	0.16716	80
100	80.9	0.0123	0.419	26.49	60.40	86.89	0.05483	0.11176	0.16659	100
120	93.4	0.0126	0.419	29.53	58.52	88.05	0.06030	0.10580	0.16610	120
140	104.5	0.0128	0.298	32.28	56.71	88.99	0.06513	0.10053	0.16566	140
160	114.5	0.0130	0.260	34.78	54.99	89.77	0.06958	0.09564	0.16522	160
180	123.7	0.0133	0.228	37.07	53.31	90.38	0.07337	0.09139	0.16476	180
200	132.1	0.0135	0.202	39.21	51.65	90.86	0.07694	0.08730	0.16424	200
220	139.9	0.0138	0.181	41.22	50.28	91.50	0.08021	0.08354	0.16375	220
596	233.6	0.02870	0.02870	78.86	0	78.86	0.1359	0	0.1359	596

Reproduced by permission of the DuPont Company.

APPENDIX 24.I
Properties of Superheated Refrigerant-12 (R-12)
(customary U.S. units)

specific volume (v) in ft^3/lbm; enthalpy (h) in Btu/lbm; entropy (s) in Btu/lbm-°R

pressure (psia) (sat. temp.)		temperature (°F)											
		−40	−20	0	20	40	60	80	100	150	200	250	300
5 (−62.5)	v	7.363	7.726	8.088	8.450	8.812	9.173	9.533	9.893	10.79	11.69	....	....
	h	73.72	76.36	79.05	81.78	84.56	87.41	90.30	93.25	100.84	108.75	....	....
	s	0.1859	0.1920	0.1979	0.2038	0.2095	0.2150	0.2205	0.2258	0.2388	0.2518	....	....
10 (−37.3)	v	....	3.821	4.006	4.189	4.371	4.556	4.740	4.923	5.379	5.831	6.281	....
	h	....	76.11	78.81	81.56	84.35	87.19	90.11	93.05	100.66	108.63	116.88	....
	s	....	0.1801	0.1861	0.1919	0.1977	0.2033	0.2087	0.2141	0.2271	0.2396	0.2517	....
15 (−20.8)	v	....	2.521	2.646	2.771	2.895	3.019	3.143	3.266	3.571	3.877	4.191	....
	h	....	75.89	78.59	81.37	84.18	87.03	89.94	92.91	100.53	108.49	116.78	....
	s	....	0.17307	0.17913	0.18499	0.19074	0.19635	0.20185	0.20723	0.22028	0.23282	0.24491	....
20 (−8.2)	v	....	....	1.965	2.060	2.155	2.250	2.343	2.437	2.669	2.901	3.130	....
	h	....	....	78.39	81.14	83.97	86.85	89.78	92.75	100.40	108.38	116.67	....
	s	....	....	0.17407	0.17996	0.18573	0.19138	0.19688	0.20229	0.21537	0.22794	0.24005	....
25 (2.2)	v	....	....	....	1.712	1.793	1.873	1.952	2.031	2.227	2.422	2.615	....
	h	....	....	....	80.95	83.78	86.67	89.61	92.56	100.26	108.26	116.56	....
	s	....	....	....	0.17637	0.18216	0.18783	0.19336	0.19748	0.21190	0.22450	0.23665	....
30 (11.1)	v	....	....	....	1.364	1.430	1.495	1.560	1.624	1.784	1.943	2.099	....
	h	....	....	....	80.75	83.59	86.49	89.43	92.42	100.12	108.13	116.45	....
	s	....	....	....	0.17278	0.17859	0.18429	0.18983	0.19527	0.20843	0.22105	0.23325	....
35 (18.9)	v	....	....	....	1.109	1.237	1.295	1.352	1.409	1.550	1.689	1.827	....
	h	....	....	....	80.49	83.40	86.30	89.26	92.26	99.98	108.01	116.33	....
	s	....	....	....	0.16963	0.17591	0.18162	0.18719	0.19266	0.20584	0.21849	0.23069	....
40 (25.9)	v	....	....	....	....	1.044	1.095	1.144	1.194	1.315	1.435	1.554	....
	h	....	....	....	....	83.20	86.11	89.09	92.09	99.83	107.88	116.21	....
	s	....	....	....	....	0.17322	0.17896	0.18455	0.19004	0.20325	0.21592	0.22813	....
50 (38.3)	v	....	....	....	....	0.821	0.863	0.904	0.944	1.044	1.142	1.239	1.332
	h	....	....	....	....	82.76	85.72	88.72	91.75	99.54	107.62	116.00	124.69
	s	....	....	....	....	0.16895	0.17475	0.18040	0.18591	0.19923	0.21196	0.22419	0.23600
60 (48.7)	v	....	....	....	....	....	0.708	0.743	0.778	0.863	0.946	1.028	1.108
	h	....	....	....	....	....	85.33	88.35	91.41	99.24	107.36	115.54	124.29
	s	....	....	....	....	....	0.17120	0.17689	0.18246	0.19585	0.20865	0.22094	0.23280
70 (57.9)	v	....	....	....	....	....	0.553	0.642	0.673	0.750	0.824	0.896	0.967
	h	....	....	....	....	....	84.94	87.96	91.05	98.94	107.10	115.54	124.29
	s	....	....	....	....	....	0.16765	0.17399	0.17961	0.19310	0.20597	0.21830	0.23020
80 (66.3)	v	....	....	....	....	....	....	0.540	0.568	0.636	0.701	0.764	0.826
	h	....	....	....	....	....	....	87.56	90.68	98.64	106.84	115.30	124.08
	s	....	....	....	....	....	....	0.17108	0.17675	0.19035	0.20328	0.21566	0.22760
90 (73.6)	v	....	....	....	....	....	....	....	0.505	0.568	0.627	0.685	0.742
	h	....	....	....	....	....	....	....	90.31	98.32	106.56	115.07	123.88
	s	....	....	....	....	....	....	....	0.17443	0.18813	0.20111	0.21356	0.22554
100 (80.9)	v	....	....	....	....	....	....	....	0.442	0.499	0.553	0.606	0.657
	h	....	....	....	....	....	....	....	89.93	97.99	106.29	114.84	123.67
	s	....	....	....	....	....	....	....	0.17210	0.18590	0.19894	0.21145	0.22347
120 (93.4)	v	....	....	....	....	....	....	....	0.357	0.407	0.454	0.500	0.543
	h	....	....	....	....	....	....	....	89.13	97.30	105.70	114.35	123.25
	s	....	....	....	....	....	....	....	0.16803	0.18207	0.19529	0.20792	0.22000

(continued)

APPENDIX 24.I *(continued)*
Properties of Superheated Refrigerant-12 (R-12)
(customary U.S. units)

pressure (psia) (sat. temp.)		temperature (°F)											
		−40	−20	0	20	40	60	80	100	150	200	250	300
140 (104.5)	v	....	....	....	....	....	....	....	....	0.341	0.383	0.423	0.462
	h	....	....	....	....	....	....	....	....	96.65	105.14	113.85	122.85
	s	....	....	....	....	....	....	....	....	0.17868	0.19205	0.20479	0.21701
160 (114.5)	v	....	....	....	....	....	....	....	....	0.318	0.335	0.372	0.408
	h	....	....	....	....	....	....	....	....	95.82	104.50	113.33	122.39
	s	....	....	....	....	....	....	....	....	0.17561	0.18927	0.20213	0.21444
180 (123.7)	v	....	....	....	....	....	....	....	....	0.294	0.287	0.321	0.353
	h	....	....	....	....	....	....	....	....	94.99	103.85	112.81	121.92
	s	....	....	....	....	....	....	....	....	0.17254	0.18648	0.19947	0.21187
200 (132.1)	v	....	....	....	....	....	....	....	....	0.241	0.255	0.288	0.317
	h	....	....	....	....	....	....	....	....	94.16	103.12	112.20	121.42
	s	....	....	....	....	....	....	....	....	0.16970	0.18395	0.19717	0.20970
220 (139.9)	v	....	....	....	....	....	....	....	....	0.188	0.232	0.254	0.282
	h	....	....	....	....	....	....	....	....	93.32	102.39	111.59	120.91
	s	....	....	....	....	....	....	....	....	0.16685	0.18142	0.19387	0.20753

Reproduced by permission of the DuPont Company.

APPENDIX 24.J
Pressure-Enthalpy Diagram for Refrigerant-12 (R-12)
(customary U.S. units)

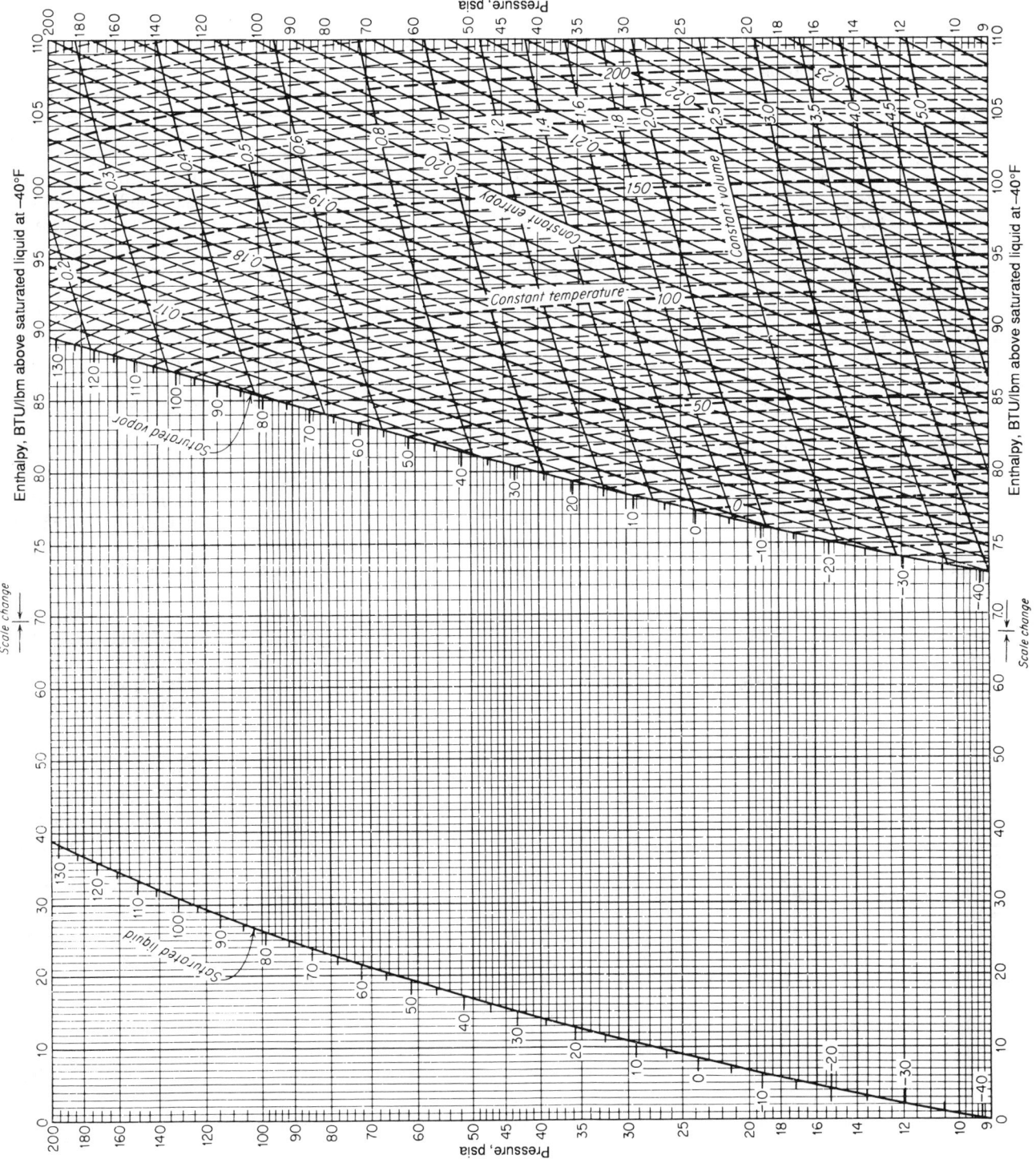

Reproduced with permission from the DuPont Company.

APPENDIX 24.K
Properties of Saturated Refrigerant-22 (R-22) by Temperature
(customary U.S. units)

temp. (°F)	saturation pressure		volume (ft³/lbm)	density (lbm/ft³)	enthalpy (Btu/lbm)		entropy (Btu/lbm-°R)	
	(psia)	(psig)	v_g	ρ_g	h_f	h_g	s_f	s_g
−130	0.68858	28.519*	59.170	96.313	−24.388	89.888	−0.065456	0.28118
−120	1.0725	27.738*	39.078	95.416	−21.538	91.049	−0.056942	0.27452
−110	1.6199	26.623*	26.578	94.509	−18.738	92.211	−0.048818	0.26848
−100	2.3802	25.075*	18.558	93.590	−15.980	93.371	−0.041046	0.26298
−90	3.4111	22.976*	13.268	92.660	−13.259	94.527	−0.033590	0.25798
−80	4.7793	20.191*	9.6902	91.717	−10.570	95.676	−0.026418	0.25342
−75	5.6131	18.493*	8.3419	91.241	−9.2346	96.247	−0.022929	0.25128
−70	6.5603	16.564*	7.2139	90.761	−7.9050	96.815	−0.019500	0.24924
−65	7.6317	14.383*	6.2655	90.278	−6.5802	97.380	−0.016128	0.24728
−60	8.8386	11.926*	5.4641	89.791	−5.2593	97.942	−0.012808	0.24541
−50	11.707	6.0851*	4.2039	88.807	−2.6263	99.055	−0.006316	0.24189
−48	12.361	4.7548*	3.9962	88.608	−2.1007	99.275	−0.005039	0.24122
−46	13.042	3.3666*	3.8007	88.408	−1.5753	99.495	−0.003770	0.24056
−44	13.754	1.9186*	3.6168	88.208	−1.0501	99.714	−0.002507	0.23991
−42	14.495	0.4090*	3.4437	88.007	−0.5250	99.932	−0.001250	0.23927
−41.47	14.696	0.0	3.3997	87.954	−0.3865	99.990	−0.000920	0.23910
−40	15.268	0.5717	3.2805	87.806	0.0	100.15	0.0	0.23864
−38	16.072	1.3763	3.1267	87.604	0.5250	100.37	0.001244	0.23802
−36	16.910	2.2138	2.9816	87.401	1.0500	100.58	0.002482	0.23741
−34	17.781	3.0852	2.8446	87.197	1.5751	100.80	0.003714	0.23681
−32	18.687	3.9914	2.7152	86.993	2.1003	101.01	0.004940	0.23622
−30	19.629	4.9333	2.5930	86.788	2.6257	101.22	0.006161	0.23564
−28	20.608	5.9119	2.4774	86.582	3.1512	101.44	0.007377	0.23506
−26	21.624	6.9283	2.3680	86.375	3.6771	101.65	0.008587	0.23450
−24	22.679	7.9832	2.2645	86.168	4.2032	101.86	0.009792	0.23394
−22	23.774	9.0778	2.1664	85.960	4.7297	102.07	0.010993	0.23340
−20	24.909	10.213	2.0735	85.751	5.2566	102.28	0.012188	0.23285
−18	26.086	11.390	1.9854	85.542	5.7840	102.48	0.013379	0.23232
−16	27.306	12.610	1.9018	85.331	6.3119	102.69	0.014566	0.23180
−14	28.569	13.873	1.8225	85.120	6.8403	102.90	0.015748	0.23128
−12	29.877	15.181	1.7472	84.908	7.3693	103.10	0.016926	0.23077
−10	31.231	16.535	1.6757	84.695	7.8989	103.30	0.018100	0.23027
−8	32.632	17.936	1.6077	84.481	8.4292	103.51	0.019270	0.22977
−6	34.081	19.385	1.5430	84.266	8.9603	103.71	0.020436	0.22928
−4	35.579	20.883	1.4815	84.051	9.4921	103.91	0.021598	0.22880
−2	37.127	22.431	1.4230	83.834	10.025	104.10	0.022757	0.22832

(continued)

APPENDIX 24.K *(continued)*
Properties of Saturated Refrigerant-22 (R-22) by Temperature
(customary U.S. units)

temp. (°F)	saturation pressure		volume (ft³/lbm)	density (lbm/ft³)	enthalpy (Btu/lbm)		entropy (Btu/lbm-°R)	
	(psia)	(psig)	v_g	ρ_f	h_f	h_g	s_f	s_g
0	38.726	24.030	1.3672	83.617	10.558	104.30	0.023912	0.022785
2	40.378	25.682	1.3141	83.399	11.093	104.50	0.025064	0.22738
4	42.083	27.387	1.2635	83.179	11.628	104.69	0.026213	0.22693
6	43.843	29.147	1.2152	82.959	12.164	104.89	0.027359	0.22647
8	45.658	30.962	1.1692	82.738	12.702	105.08	0.028502	0.22602
10	47.530	32.834	1.1253	82.516	13.240	105.27	0.029642	0.22558
12	49.461	34.765	1.0833	82.292	13.779	105.46	0.030779	0.22515
14	51.450	36.754	1.0433	82.068	14.320	105.64	0.031913	0.22471
16	53.501	38.805	1.0050	81.843	14.862	105.83	0.033045	0.22429
18	55.612	40.916	0.96841	81.616	15.405	106.02	0.034175	0.22387
20	57.786	43.090	0.93343	81.389	15.950	106.20	0.035302	0.22345
25	63.505	48.809	0.85246	80.815	17.317	106.65	0.038110	0.22243
30	69.641	54.945	0.77984	80.234	18.693	107.09	0.040905	0.22143
35	76.215	61.519	0.71454	79.645	20.078	107.52	0.043689	0.22046
40	83.246	68.550	0.65571	79.049	21.474	107.94	0.046464	0.21951
45	90.754	76.058	0.60258	78.443	22.880	108.35	0.049229	0.21858
50	98.758	84.062	0.55451	77.829	24.298	108.74	0.051987	0.21767
55	107.28	92.583	0.51093	77.206	25.728	109.12	0.054739	0.21677
60	116.34	101.64	0.47134	76.572	27.170	109.49	0.057486	0.21589
65	125.95	111.26	0.43531	75.928	28.626	109.84	0.060228	0.21502
70	136.15	121.45	0.40245	75.273	30.095	110.18	0.062968	0.21416
80	158.36	143.66	0.34497	73.926	33.077	110.80	0.068441	0.21246
90	183.14	168.44	0.29668	72.525	36.121	111.35	0.073911	0.21077
100	210.67	195.97	0.25582	71.061	39.233	111.81	0.079400	0.20907
110	241.13	226.44	0.22102	69.524	42.422	112.17	0.084906	0.20734
120	274.73	260.03	0.19118	67.901	45.694	112.42	0.090444	0.20554
130	311.66	296.96	0.16542	66.174	49.064	112.52	0.096033	0.20365
140	352.14	337.45	0.14300	64.319	52.550	112.47	0.10170	0.20161
150	396.42	381.72	0.12334	62.301	56.177	112.20	0.10749	0.19938
160	444.75	430.06	0.10590	60.068	59.989	111.67	0.11345	0.19684
170	497.46	482.76	0.090228	57.532	64.055	110.76	0.11970	0.19386
180	554.89	540.19	0.075819	54.533	68.504	109.30	0.12640	0.19018
190	617.52	602.82	0.061991	50.703	73.617	106.88	0.13399	0.18518
200	686.02	671.32	0.046923	44.671	80.406	101.99	0.14394	0.17666
**205.07	723.4	708.7	0.03123	32.03	91.58	91.58	0.1605	0.1605

*in Hg vacuum
**critical point

Reprinted by permission of the American Society of Heating, Refrigerating, and Air Conditioning Engineers from the *1993 ASHRAE Handbook: Fundamentals.*

Support Material

APPENDIX 24.L
Pressure-Enthalpy Diagram for Refrigerant-22 (R-22)
(customary U.S. units)

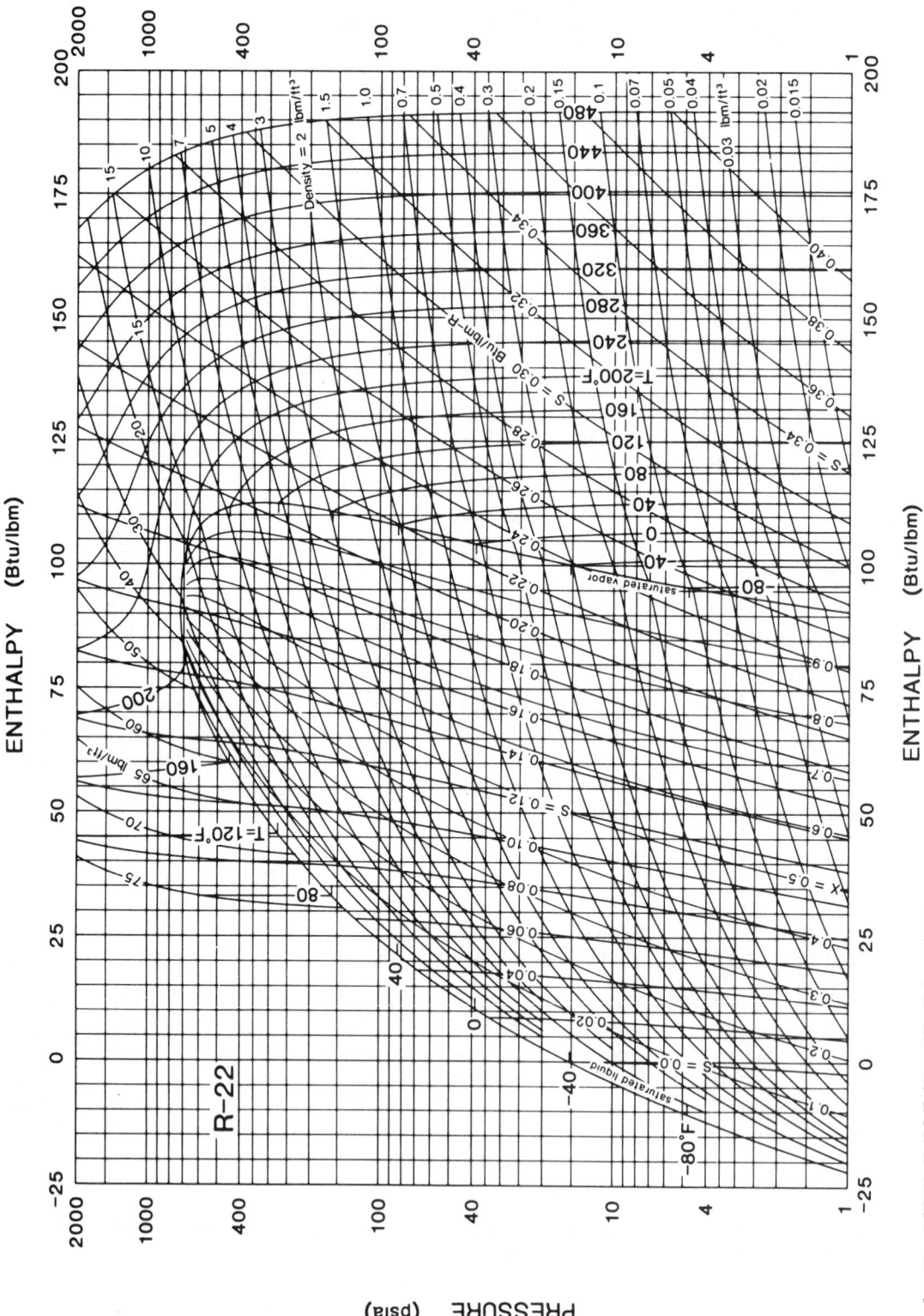

APPENDIX 24.M
Pressure-Enthalpy Diagram for Refrigerant HFC-134a
(customary U.S. units)

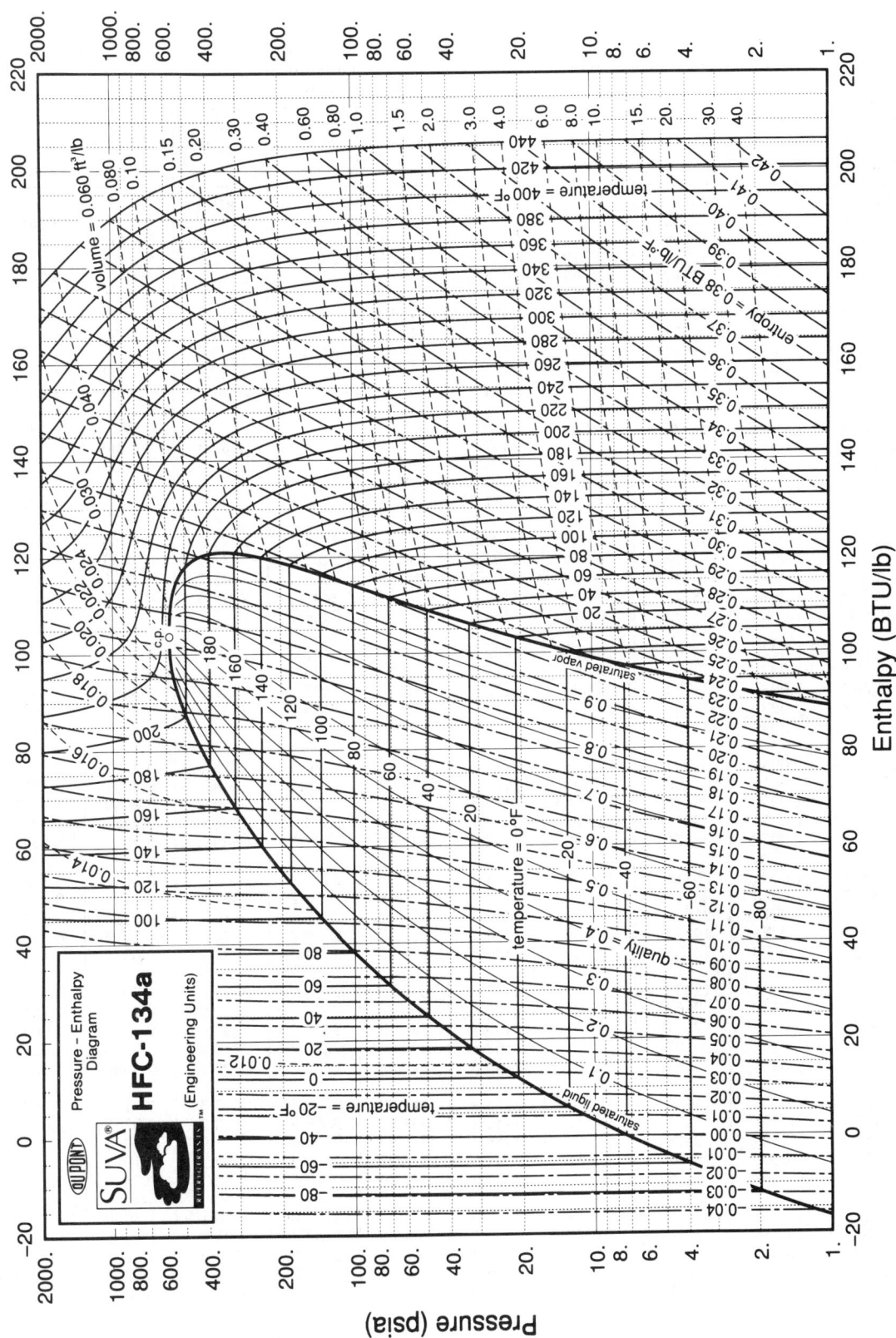

Reproduced by permission of the DuPont Company.

APPENDIX 24.N
Properties of Saturated Steam by Temperature
(SI units)

temp. (°C)	absolute pressure (bars)	specific volume (cm³/g)		internal energy (kJ/kg)		enthalpy (kJ/kg)			entropy (kJ/kg·K)		temp. (°C)
		sat. liquid v_f	sat. vapor v_g	sat. liquid u_f	sat. vapor u_g	sat. liquid h_f	evap. h_{fg}	sat. vapor h_g	sat. liquid s_f	sat. vapor s_g	
0.01	0.00611	1.0002	206136	0.00	2375.3	0.01	2501.3	2501.4	0.0000	9.1562	0.01
4	0.00813	1.0001	157232	16.77	2380.9	16.78	2491.9	2508.7	0.0610	9.0514	4
5	0.00872	1.0001	147120	20.97	2382.3	20.98	2489.6	2510.6	0.0761	9.0257	5
6	0.00935	1.0001	137734	25.19	2383.6	25.20	2487.2	2512.4	0.0912	9.0003	6
8	0.01072	1.0002	120917	33.59	2386.4	33.60	2482.5	2516.1	0.1212	8.9501	8
10	0.01228	1.0004	106379	42.00	2389.2	42.01	2477.7	2519.8	0.1510	8.9008	10
11	0.01312	1.0004	99857	46.20	2390.5	46.20	2475.4	2521.6	0.1658	8.8765	11
12	0.01402	1.0005	93784	50.41	2391.9	50.41	2473.0	2523.4	0.1806	8.8524	12
13	0.01497	1.0007	88124	54.60	2393.3	54.60	2470.7	2525.3	0.1953	8.8285	13
14	0.01598	1.0008	82848	58.79	2394.7	58.80	2468.3	2527.1	0.2099	8.8048	14
15	0.01705	1.0009	77926	62.99	2396.1	62.99	2465.9	2528.9	0.2245	8.7814	15
16	0.01818	1.0011	73333	67.18	2397.4	67.19	2463.6	2530.8	0.2390	8.7582	16
17	0.01938	1.0012	69044	71.38	2398.8	71.38	2461.2	2532.6	0.2535	8.7351	17
18	0.02064	1.0014	65038	75.57	2400.2	75.58	2458.8	2534.4	0.2679	8.7123	18
19	0.02198	1.0016	61293	79.76	2401.6	79.77	2456.5	2536.2	0.2823	8.6897	19
20	0.02339	1.0018	57791	83.95	2402.9	83.96	2454.1	2538.1	0.2966	8.6672	20
21	0.02487	1.0020	54514	88.14	2404.3	88.14	2451.8	2539.9	0.3109	8.6450	21
22	0.02645	1.0022	51447	92.32	2405.7	92.33	2449.4	2541.7	0.3251	8.6229	22
23	0.02810	1.0024	48574	96.51	2407.0	96.52	2447.0	2543.5	0.3393	8.6011	23
24	0.02985	1.0027	45883	100.70	2408.4	100.70	2444.7	2545.4	0.3534	8.5794	24
25	0.03169	1.0029	43360	104.88	2409.8	104.89	2442.3	2547.2	0.3674	8.5580	25
26	0.03363	1.0032	40994	109.06	2411.1	109.07	2439.9	2549.0	0.3814	8.5367	26
27	0.03567	1.0035	38774	113.25	2412.5	113.25	2437.6	2550.8	0.3954	8.5156	27
28	0.03782	1.0037	36690	117.42	2413.9	117.43	2435.2	2552.6	0.4093	8.4946	28
29	0.04008	1.0040	34733	121.60	2415.2	121.61	2432.8	2554.5	0.4231	8.4739	29
30	0.04246	1.0043	32894	125.78	2416.6	125.79	2430.5	2556.3	0.4369	8.4533	30
31	0.04496	1.0046	31165	129.96	2418.0	129.97	2428.1	2558.1	0.4507	8.4329	31
32	0.04759	1.0050	29540	134.14	2419.3	134.15	2425.7	2559.9	0.4644	8.4127	32
33	0.05034	1.0053	28011	138.32	2420.7	138.33	2423.4	2561.7	0.4781	8.3927	33
34	0.05324	1.0056	26571	142.50	2422.0	142.50	2421.0	2563.5	0.4917	8.3728	34
35	0.05628	1.0060	25216	146.67	2423.4	146.68	2418.6	2565.3	0.5053	8.3531	35
36	0.05947	1.0063	23940	150.85	2424.7	150.86	2416.2	2567.1	0.5188	8.3336	36
38	0.06632	1.0071	21602	159.20	2427.4	159.21	2411.5	2570.7	0.5458	8.2950	38
40	0.07384	1.0078	19523	167.56	2430.1	167.57	2406.7	2574.3	0.5725	8.2570	40
45	0.09593	1.0099	15258	188.44	2436.8	188.45	2394.8	2583.2	0.6387	8.1648	45
50	0.1235	1.0121	12032	209.32	2443.5	209.33	2382.7	2592.1	0.7038	8.0763	50
55	0.1576	1.0146	9568	230.21	2450.1	230.23	2370.7	2600.9	0.7679	7.9913	55
60	0.1994	1.0172	7671	251.11	2456.6	251.13	2358.5	2609.6	0.8312	7.9096	60
65	0.2503	1.0199	6197	272.02	2463.1	272.06	2346.2	2618.3	0.8935	7.8310	65
70	0.3119	1.0228	5042	292.95	2469.6	292.98	2333.8	2626.8	0.9549	7.7553	70
75	0.3858	1.0259	4131	313.90	2475.9	313.93	2321.4	2635.3	1.0155	7.6824	75
80	0.4739	1.0291	3407	334.86	2482.2	334.91	2308.8	2643.7	1.0753	7.6122	80
85	0.5783	1.0325	2828	355.84	2488.4	355.90	2296.0	2651.9	1.1343	7.5445	85
90	0.7014	1.0360	2361	376.85	2494.5	376.92	2283.2	2660.1	1.1925	7.4791	90
95	0.8455	1.0397	1982	397.88	2500.6	397.96	2270.2	2668.1	1.2500	7.4159	95
100	1.014	1.0435	1673	418.94	2506.5	419.04	2257.0	2676.1	1.3069	7.3549	100
110	1.433	1.0516	1210	461.14	2518.1	461.30	2230.2	2691.5	1.4185	7.2387	110
120	1.985	1.0603	891.9	503.50	2529.3	503.71	2202.6	2706.3	1.5276	7.1296	120
130	2.701	1.0697	668.5	546.02	2539.9	546.31	2174.2	2720.5	1.6344	7.0269	130
140	3.613	1.0797	508.9	588.74	2550.0	589.13	2144.7	2733.9	1.7391	6.9299	140

(continued)

APPENDIX 24.N *(continued)*
Properties of Saturated Steam by Temperature
(SI units)

temp. (°C)	absolute pressure (bars)	specific volume (cm³/g)		internal energy (kJ/kg)		enthalpy (kJ/kg)			entropy (kJ/kg·K)		temp. (°C)
		sat. liquid v_f	sat. vapor v_g	sat. liquid u_f	sat. vapor u_g	sat. liquid h_f	evap. h_{fg}	sat. vapor h_g	sat. liquid s_f	sat. vapor s_g	
150	4.758	1.0905	392.8	631.68	2559.5	632.20	2114.3	2746.5	1.8418	6.8379	150
160	6.178	1.1020	307.1	674.86	2568.4	675.55	2082.6	2758.1	1.9427	6.7502	160
170	7.917	1.1143	242.8	718.33	2576.5	719.21	2049.5	2768.7	2.0419	6.6663	170
180	10.02	1.1274	194.1	762.09	2583.7	763.22	2015.0	2778.2	2.1396	6.5857	180
190	12.54	1.1414	156.5	806.19	2590.0	807.62	1978.8	2786.4	2.2359	6.5079	190
200	15.54	1.1565	127.4	850.65	2595.3	852.45	1940.7	2793.2	2.3309	6.4323	200
210	19.06	1.1726	104.4	895.53	2599.5	897.76	1900.7	2798.5	2.4248	6.3585	210
220	23.18	1.1900	86.19	940.87	2602.4	943.62	1858.5	2802.1	2.5178	6.2861	220
230	27.95	1.2088	71.58	986.74	2603.9	990.12	1813.8	2804.0	2.6099	6.2146	230
240	33.44	1.2291	59.76	1033.2	2604.0	1037.3	1766.5	2803.8	2.7015	6.1437	240
250	39.73	1.2512	50.13	1080.4	2602.4	1085.4	1716.2	2801.5	2.7927	6.0730	250
260	46.88	1.2755	42.21	1128.4	2599.0	1134.4	1662.5	2796.6	2.8838	6.0019	260
270	54.99	1.3023	35.64	1177.4	2593.7	1184.5	1605.2	2789.7	2.9751	5.9301	270
280	64.12	1.3321	30.17	1227.5	2586.1	1236.0	1543.6	2779.6	3.0668	5.8571	280
290	74.36	1.3656	25.57	1278.9	2576.0	1289.1	1477.1	2766.2	3.1594	5.7821	290
300	85.81	1.4036	21.67	1332.0	2563.0	1344.0	1404.9	2749.0	3.2534	5.7045	300
320	112.7	1.4988	15.49	1444.6	2525.5	1461.5	1238.6	2700.1	3.4480	5.5362	320
340	145.9	1.6379	10.80	1570.3	2464.6	1594.2	1027.9	2622.0	3.6594	5.3357	340
360	186.5	1.8925	6.945	1725.2	2351.5	1760.5	720.5	2481.0	3.9147	5.0526	360
374.14	220.9	3.155	3.155	2029.6	2029.6	2099.3	0	2099.3	4.4298	4.4298	374.14

Steam Tables: Thermodynamic Properties of Water Including Vapor, Liquid, and Solid Phases (English Units).
Joseph H. Keenan, Frederick G. Keyes, Philip G. Hill, and Joan G. Moore. Copyright © 1969. Adapted by
permission of John Wiley & Sons, Inc.

APPENDIX 24.O
Properties of Saturated Steam by Pressure
(SI units)

absolute pressure (bars)	temp. (°C)	specific volume (cm³/g)		internal energy (kJ/kg)		enthalpy (kJ/kg)			entropy (kJ/kg·K)		absolute pressure (bars)
		sat. liquid v_f	sat. vapor v_g	sat. liquid u_f	sat. vapor u_g	sat. liquid h_f	evap. h_{fg}	sat. vapor h_g	sat. liquid s_f	sat. vapor s_g	
0.04	28.96	1.0040	34800	121.45	2415.2	121.46	2432.9	2554.4	0.4226	8.4746	0.04
0.06	36.16	1.0064	23739	151.53	2425.0	151.53	2415.9	2567.4	0.5210	8.3304	0.06
0.08	41.51	1.0084	18103	173.87	2432.2	173.88	2403.1	2577.0	0.5926	8.2287	0.08
0.10	45.81	1.0102	14674	191.82	2437.9	191.82	2392.8	2584.7	0.6493	8.1502	0.10
0.20	60.06	1.0172	7649	251.38	2456.7	251.40	2358.3	2609.7	0.8320	7.9085	0.20
0.30	69.10	1.0223	5229	289.20	2468.4	289.23	2336.1	2625.3	0.9439	7.7686	0.30
0.40	75.87	1.0265	3993	317.53	2477.0	317.58	2319.2	2636.8	1.0259	7.6700	0.40
0.50	81.33	1.0300	3240	340.44	2483.9	340.49	2305.4	2645.9	1.0910	7.5939	0.50
0.60	85.94	1.0331	2732	359.79	2489.6	359.86	2293.6	2653.5	1.1453	7.5320	0.60
0.70	89.95	1.0360	2365	376.63	2494.5	376.70	2283.3	2660.0	1.1919	7.4797	0.70
0.80	93.50	1.0380	2087	391.58	2498.8	391.66	2274.1	2665.8	1.2329	7.4346	0.80
0.90	96.71	1.0410	1869	405.06	2502.6	405.15	2265.7	2670.9	1.2695	7.3949	0.90
1.00	99.63	1.0432	1694	417.36	2506.1	417.46	2258.0	2675.5	1.3026	7.3594	1.00
1.50	111.4	1.0528	1159	466.94	2519.7	467.11	2226.5	2693.6	1.4336	7.2233	1.50
2.00	120.2	1.0605	885.7	504.49	2529.5	504.70	2201.9	2706.7	1.5301	7.1271	2.00
2.50	127.4	1.0672	718.7	535.10	2537.2	535.37	2181.5	2716.9	1.6072	7.0527	2.50
3.00	133.6	1.0732	605.8	561.15	2543.6	561.47	2163.8	2725.3	1.6718	6.9919	3.00
3.50	138.9	1.0786	524.3	583.95	2546.9	584.33	2148.1	2732.4	1.7275	6.9405	3.50
4.00	143.6	1.0836	462.5	604.31	2553.6	604.74	2133.8	2738.6	1.7766	6.8959	4.00
4.50	147.9	1.0882	414.0	622.25	2557.6	623.25	2120.7	2743.9	1.8207	6.8565	4.50
5.00	151.9	1.0926	374.9	639.68	2561.2	640.23	2108.5	2748.7	1.8607	6.8212	5.00
6.00	158.9	1.1006	315.7	669.90	2567.4	670.56	2086.3	2756.8	1.9312	6.7600	6.00
7.00	165.0	1.1080	272.9	696.44	2572.5	697.22	2066.3	2763.5	1.9922	6.7080	7.00
8.00	170.4	1.1148	240.4	720.22	2576.8	721.11	2048.0	2769.1	2.0462	6.6628	8.00
9.00	175.4	1.1212	215.0	741.83	2580.5	742.83	2031.1	2773.9	2.0946	6.6226	9.00
10.0	179.9	1.1273	194.4	761.68	2583.6	762.81	2015.3	2778.1	2.1387	6.5863	10.0
15.0	198.3	1.1539	131.8	843.16	2594.5	844.84	1947.3	2792.2	2.3150	6.4448	15.0
20.0	212.4	1.1767	99.63	906.44	2600.3	908.79	1890.7	2799.5	2.4474	6.3409	20.0
25.0	224.0	1.1973	79.98	959.11	2603.1	962.11	1841.0	2803.1	2.5547	6.2575	25.0
30.0	233.9	1.2165	66.68	1004.8	2604.1	1008.4	1795.7	2804.2	2.6457	6.1869	30.0
35.0	242.6	1.2347	57.07	1045.4	2603.7	1049.8	1753.7	2803.4	2.7253	6.1253	35.0
40.0	250.4	1.2522	49.78	1082.3	2602.3	1087.3	1714.1	2801.4	2.7964	6.0701	40.0
45.0	257.5	1.2692	44.06	1116.2	2600.1	1121.9	1676.4	2798.3	2.8610	6.0199	45.0
50.0	264.0	1.2859	39.44	1147.8	2597.1	1154.2	1640.1	2794.3	2.9202	5.9734	50.0
60.0	275.6	1.3187	32.44	1205.4	2589.7	1213.4	1571.0	2784.3	3.0267	5.8892	60.0
70.0	285.9	1.3513	27.37	1257.6	2580.5	1267.0	1505.1	2772.1	3.1211	5.8133	70.0
80.0	295.1	1.3842	23.52	1305.6	2569.8	1316.6	1441.3	2758.0	3.2068	5.7432	80.0
90.0	303.4	1.4178	20.48	1350.5	2557.8	1363.3	1378.9	2742.1	3.2858	5.6772	90.0
100	311.1	1.4524	18.03	1393.0	2544.4	1407.6	1317.1	2724.7	3.3596	5.6141	100
110	318.2	1.4886	15.99	1433.7	2529.8	1450.1	1255.5	2705.6	3.4295	5.5527	110
120	324.8	1.5267	14.26	1473.0	2513.7	1491.3	1193.6	2684.9	3.4962	5.4924	120
130	330.9	1.5671	12.78	1511.1	2496.1	1531.5	1130.7	2662.2	3.5606	5.4323	130
140	336.8	1.6107	11.49	1548.6	2476.8	1571.1	1066.5	2637.6	3.6232	5.3717	140
150	342.2	1.6581	10.34	1585.6	2455.5	1610.5	1000.0	2610.5	3.6848	5.3098	150
160	347.4	1.7107	9.306	1622.7	2431.7	1650.1	930.6	2580.6	3.7461	5.2455	160
170	352.4	1.7702	8.364	1660.2	2405.0	1690.3	856.9	2547.2	3.8079	5.1777	170
180	357.1	1.8397	7.489	1698.9	2374.3	1732.0	777.1	2509.1	3.8715	5.1044	180
190	361.5	1.9243	6.657	1739.9	2338.1	1776.5	688.0	2464.5	3.9388	5.0228	190
200	365.8	2.036	5.834	1785.6	2293.0	1826.3	583.4	2409.7	4.0139	4.9269	200
220.9	374.1	3.155	3.155	2029.6	2029.6	2099.3	0	2099.3	4.4298	4.4298	220.9

Steam Tables: Thermodynamic Properties of Water Including Vapor, Liquid, and Solid Phases (English Units). Joseph H. Keenan, Frederick G. Keyes, Philip G. Hill, and Joan G. Moore. Copyright © 1969. Adapted by permission of John Wiley & Sons, Inc.

APPENDIX 24.P
Properties of Superheated Steam
(SI units)

specific volume (v) in m³/kg; enthalpy (h) in kJ/kg; entropy (s) in kJ/kg·K

absolute pressure (kPa) (sat. temp. °C)		temperature (°C)							
		100	150	200	250	300	360	420	500
10 (45.81)	v	17.196	19.512	21.825	24.136	26.445	29.216	31.986	35.679
	h	2687.5	2783.0	2879.5	2977.3	3076.5	3197.6	3320.9	3489.1
	s	8.4479	8.6882	8.9038	9.1002	9.2813	9.4821	9.6682	9.8978
50 (81.33)	v	3.418	3.889	4.356	4.820	5.284	5.839	6.394	7.134
	h	2682.5	2780.1	2877.7	2976.0	3075.5	3196.8	3320.4	3488.7
	s	7.6947	7.9401	8.1580	8.3556	8.5373	8.7385	8.9249	9.1546
75 (91.78)	v	2.270	2.587	2.900	3.211	3.520	3.891	4.262	4.755
	h	2679.4	2778.2	2876.5	2975.2	3074.9	3196.4	3320.0	3488.4
	s	7.5009	7.7496	7.9690	8.1673	8.3493	8.5508	8.7374	8.9672
100 (99.63)	v	1.6958	1.9364	2.172	2.406	2.639	2.917	3.195	3.565
	h	2672.2	2776.4	2875.3	2974.3	3074.3	3195.9	3319.6	3488.1
	s	7.3614	7.6134	7.8343	8.0333	8.2158	8.4175	8.6042	8.8342
150 (111.37)	v	....	1.2853	1.4443	1.6012	1.7570	1.9432	2.129	2.376
	h	....	2772.6	2872.9	2972.7	3073.1	3195.0	3318.9	3487.6
	s	....	7.4193	7.6433	7.8438	8.0720	8.2293	8.4163	8.6466
400 (143.63)	v	....	0.4708	0.5342	0.5951	0.6548	0.7257	0.7960	0.8893
	h	....	2752.8	2860.5	2964.2	3066.8	3190.3	3315.3	3484.9
	s	....	6.9299	7.1706	7.3789	7.5662	7.7712	7.9598	8.1913
700 (164.97)	v	....	....	0.2999	0.3363	0.3714	0.4126	0.4533	0.5070
	h	....	....	2884.8	2953.6	3059.1	3184.7	3310.9	3481.7
	s	....	....	6.8865	7.1053	7.2979	7.5063	7.6968	7.9299
1000 (179.91)	v	....	....	0.2060	0.2327	0.2579	0.2873	0.3162	0.3541
	h	....	....	2827.9	2942.6	3051.2	3178.9	3306.5	3478.5
	s	....	....	6.6940	6.9247	7.1229	7.3349	7.5275	7.7622
1500 (198.32)	v	....	....	0.13248	0.15195	0.16966	0.18988	0.2095	0.2352
	h	....	....	2796.8	2923.3	3037.6	3169.2	3299.1	3473.1
	s	....	....	6.4546	6.7090	6.9179	7.1363	7.3323	7.5698
2000 (212.42)	v	....	....	....	0.11144	0.12547	0.14113	0.15616	0.17568
	h	....	....	....	2902.5	3023.5	3159.3	3291.6	3467.6
	s	....	....	....	6.5453	6.7664	6.9917	7.1915	7.4317
2500 (223.99)	v	....	....	....	0.08700	0.09890	0.11186	0.12414	0.13998
	h	....	....	....	2880.1	3008.8	3149.1	3284.0	3462.1
	s	....	....	....	6.4085	6.6438	6.8767	7.0803	7.3234
3000 (233.90)	v	....	....	....	0.07058	0.08114	0.09233	0.10279	0.11619
	h	....	....	....	2855.8	2993.5	3138.7	3276.3	3456.5
	s	....	....	....	6.2872	6.5390	6.7801	6.9878	7.2338

Steam Tables: Thermodynamic Properties of Water Including Vapor, Liquid, and Solid Phases (SI Edition). Joseph H. Keenan, Frederick G. Keyes, Philip G. Hill, and Joan G. Moore. Copyright © 1978. Adapted with permission of John Wiley & Sons, Inc.

APPENDIX 24.Q
Properties of Compressed Water
(SI units)

T (°C)	v (cm³/g)	u (kJ/kg)	h (kJ/kg)	s (kJ/kg·K)	v (cm³/g)	u (kJ/kg)	h (kJ/kg)	s (kJ/kg·K)
	$p = 25$ bars $= 2.5$ MPa ($T_{sat} = 223.99$°C)				$p = 50$ bars $= 5.0$ MPa ($T_{sat} = 263.99$°C)			
20	1.0006	83.80	86.30	0.2961	0.9995	83.65	88.65	0.2956
40	1.0067	167.25	169.77	0.5715	1.0056	166.95	171.97	0.5705
80	1.0280	334.29	336.86	1.0737	1.0268	333.72	338.85	1.0720
100	1.0423	418.24	420.85	1.3050	1.0410	417.52	422.72	1.3030
140	1.0784	587.82	590.52	1.7369	1.0768	586.76	592.15	1.7343
180	1.1261	761.16	763.97	2.1375	1.1240	759.63	765.25	2.1341
200	1.1555	849.9	852.8	2.3294	1.1530	848.1	853.9	2.3255
220	1.1898	940.7	943.7	2.5174	1.1866	938.4	944.4	2.5128
Sat.	1.1973	959.1	962.1	2.5546	1.2859	1147.8	1154.2	2.9202
	$p = 75$ bars $= 7.5$ MPa ($T_{sat} = 290.59$°C)				$p = 100$ bars $= 10.0$ MPa ($T_{sat} = 311.06$°C)			
20	0.9984	83.50	90.99	0.2950	0.9972	83.36	93.33	0.2945
40	1.0045	166.64	174.18	0.5696	1.0034	166.35	176.38	0.5686
80	1.0256	333.15	340.84	1.0704	1.0245	332.59	342.83	1.0688
100	1.0397	416.81	424.62	1.3011	1.0385	416.12	426.50	1.2992
140	1.0752	585.72	593.78	1.7317	1.0737	584.68	595.42	1.7292
180	1.1219	758.13	766.55	2.1308	1.1199	756.65	767.84	2.1275
220	1.1835	936.2	945.1	2.5083	1.1805	934.1	945.9	2.5039
260	1.2696	1124.4	1134.0	2.8763	1.2645	1121.1	1133.7	2.8699
Sat.	1.3677	1282.0	1292.2	3.1649	1.4524	1393.0	1407.6	3.3596
	$p = 150$ bars $= 15.0$ MPa ($T_{sat} = 342.24$°C)				$p = 200$ bars $= 20.0$ MPa ($T_{sat} = 365.81$°C)			
20	0.9950	83.06	97.77	0.2934	0.9928	82.77	102.62	0.2923
40	1.0013	165.76	180.78	0.5666	0.9992	165.17	185.16	0.5646
80	1.0222	331.48	346.81	1.0656	1.0199	330.40	350.80	1.0624
100	1.0361	414.74	430.28	1.2955	1.0337	413.39	434.06	1.2917
140	1.0707	582.66	598.72	1.7242	1.0678	580.69	602.04	1.7193
180	1.1159	753.76	770.50	2.1210	1.1120	750.95	773.20	2.1147
220	1.1748	929.9	947.5	2.4953	1.1693	925.9	949.3	2.4870
260	1.2550	1114.6	1133.4	2.8576	1.2462	1108.6	1133.5	2.8459
300	1.3770	1316.6	1337.3	3.2260	1.3596	1306.1	1333.3	3.2071
Sat.	1.6581	1585.6	1610.5	3.6848	2.036	1785.6	1826.3	4.0139
	$p = 250$ bars $= 25$ MPa (supercritical)				$p = 300$ bars $= 30$ MPa (supercritical)			
20	0.9907	82.47	107.24	0.2911	0.9886	82.17	111.84	0.2899
40	0.9971	164.60	189.52	0.5626	0.9951	164.04	193.89	0.5607
100	1.0313	412.08	437.85	1.2881	1.0290	410.78	441.66	1.2844
200	1.1344	834.5	862.8	2.2961	1.1302	831.4	865.3	2.2893
300	1.3442	1296.6	1330.2	3.1900	1.3304	1287.9	1327.8	3.1741

Steam Tables: Thermodynamic Properties of Water Including Vapor, Liquid, and Solid Phases (SI Edition). Joseph H. Keenan, Frederick G. Keyes, Philip G. Hill, and Joan G. Moore. Copyright © 1978. Adapted with permission of John Wiley & Sons, Inc.

APPENDIX 24.R
Enthalpy-Entropy (Mollier) Diagram for Steam
(SI units)

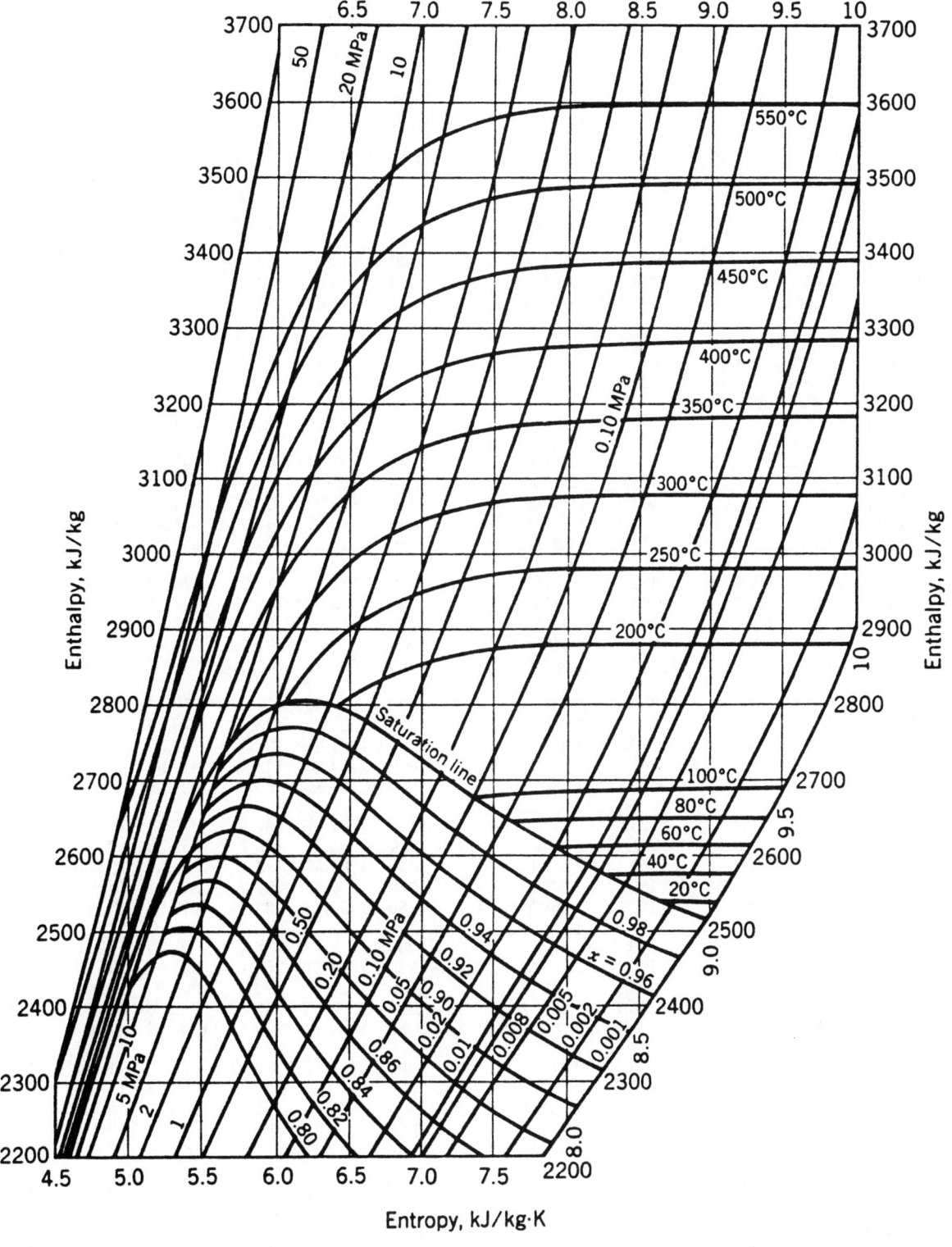

Steam Tables: Thermodynamic Properties of Water Including Vapor, Liquid, and Solid Phases (SI Edition). Joseph H. Keenan, Frederick G. Keyes, Philip G. Hill, and Joan G. Moore. Copyright © 1978. Adapted with permission of John Wiley & Sons, Inc.

APPENDIX 24.S
Properties of Low-Pressure Air
(SI units)

T in K; h and u in kJ/kg; ϕ in kJ/kg·K

T	h	p_r	u	v_r	ϕ	T	h	p_r	u	v_r	ϕ
200	199.97	0.3363	142.56	1707	1.29559	650	659.84	21.86	473.25	85.34	2.49364
210	209.97	0.3987	149.69	1512	1.34444	660	670.47	23.13	481.01	81.89	2.50985
220	219.97	0.4690	156.82	1346	1.39105	670	681.14	24.46	488.81	78.61	2.52589
230	230.02	0.5477	164.00	1205	1.43557	680	691.82	25.85	496.62	75.50	2.54175
240	240.02	0.6355	171.13	1084	1.47824	690	702.52	27.29	504.45	72.56	2.55731
250	250.05	0.7329	178.28	979	1.51917	700	713.27	28.80	512.33	69.76	2.57277
260	260.09	0.8405	185.45	887.8	1.55848	710	724.04	30.38	520.23	67.07	2.58810
270	270.11	0.9590	192.60	808.0	1.59634	720	734.82	32.02	528.14	64.53	2.60319
280	280.13	1.0889	199.75	738.0	1.63279	730	745.62	33.72	536.07	62.13	2.61803
285	285.14	1.1584	203.33	706.1	1.65055	740	756.44	35.50	544.02	59.82	2.63280
290	290.16	1.2311	206.91	676.1	1.66802	750	767.29	37.35	551.99	57.63	2.64737
295	295.17	1.3068	210.49	647.9	1.68515	760	778.18	39.27	560.01	55.54	2.66176
300	300.19	1.3860	214.07	621.2	1.70203	770	789.11	41.31	568.07	53.39	2.67595
305	305.22	1.4686	217.67	596.0	1.71865	780	800.03	43.35	576.12	51.64	2.69013
310	310.24	1.5546	221.25	572.3	1.73498	790	810.99	45.55	584.21	49.86	2.70400
315	315.27	1.6442	224.85	549.8	1.75106	800	821.95	47.75	592.30	48.08	2.71787
320	320.29	1.7375	228.42	528.6	1.76690	820	843.98	52.59	608.59	44.84	2.74504
325	325.31	1.8345	232.02	508.4	1.78249	840	866.08	57.60	624.95	41.85	2.77170
330	330.34	1.9352	235.61	489.4	1.79783	860	888.27	63.09	641.40	39.12	2.79783
340	340.42	2.149	242.82	454.1	1.82790	880	910.56	68.98	657.95	36.61	2.82344
350	350.49	2.379	250.02	422.2	1.85708	900	732.93	75.29	674.58	34.31	2.84856
360	360.58	2.626	257.24	393.4	1.88543	920	955.38	82.05	691.28	32.18	2.87324
370	370.67	2.892	264.46	367.2	1.91313	940	977.92	89.28	708.08	30.22	2.89748
380	380.77	3.176	271.69	343.4	1.94001	960	1000.55	97.00	725.02	28.40	2.92128
390	390.88	3.481	278.93	321.5	1.96633	980	1023.25	105.2	741.98	26.73	2.94468
400	400.98	3.806	286.16	301.6	1.99194	1000	1046.04	114.0	758.94	25.17	2.96770
410	411.12	4.153	293.43	283.3	2.01699	1020	1068.89	123.4	776.10	23.72	2.99034
420	421.26	4.522	300.69	266.6	2.04142	1040	1091.85	133.3	793.36	22.39	3.01260
430	431.43	4.915	307.99	251.1	2.06533	1060	1114.86	143.9	810.62	21.14	3.03449
440	441.61	5.332	315.30	236.8	2.08870	1080	1137.89	155.2	827.88	19.98	3.05608
450	451.80	5.775	322.62	223.6	2.11161	1100	1161.07	167.1	845.33	18.896	3.07732
460	462.02	6.245	329.97	211.4	2.13407	1120	1184.28	179.7	862.79	17.886	3.09825
470	472.24	6.742	337.32	200.1	2.15604	1140	1207.57	193.1	880.35	16.946	3.11883
480	482.49	7.268	344.70	189.5	2.17760	1160	1230.92	207.2	897.91	16.064	3.13916
490	492.74	7.824	352.08	179.7	2.19876	1180	1254.34	222.2	915.57	15.241	3.15916
500	503.02	8.411	359.49	170.6	2.21952	1200	1277.79	238.0	933.33	14.470	3.17888
510	513.32	9.031	366.92	162.1	2.23993	1220	1301.31	254.7	951.09	13.747	3.19834
520	523.63	9.684	374.36	154.1	2.25997	1240	1324.93	272.3	968.95	13.069	3.21751
530	533.98	10.37	381.84	146.7	2.27967	1260	1348.55	290.8	986.90	12.435	3.23638
540	544.35	11.10	389.34	139.7	2.29906	1280	1372.24	310.4	1004.76	11.835	3.25510
550	554.74	11.86	396.86	133.1	2.31809	1300	1395.97	330.9	1022.82	11.275	3.27345
560	565.17	12.66	404.42	127.0	2.33685	1320	1419.76	352.5	1040.88	10.747	3.29160
570	575.59	13.50	411.97	121.2	2.35531	1340	1443.60	375.3	1058.94	10.247	3.30959
580	586.04	14.38	419.55	115.7	2.37348	1360	1467.49	399.1	1077.10	9.780	3.32724
590	596.52	15.31	427.15	110.6	2.39140	1380	1491.44	424.2	1095.26	9.337	3.34474
600	607.02	16.28	434.78	105.8	2.40902	1400	1515.42	450.5	1113.52	8.919	3.36200
610	617.53	17.30	442.42	101.2	2.42644	1420	1539.44	478.0	1131.77	8.526	3.37901
620	628.07	18.36	450.09	96.92	2.44356	1440	1563.51	506.9	1150.13	8.153	3.39586
630	638.63	19.84	457.78	92.84	2.46048	1460	1587.63	537.1	1168.49	7.801	3.41247
640	649.22	20.64	465.50	88.99	2.47716	1480	1611.79	568.8	1186.95	7.468	3.42892

(continued)

APPENDIX 24.S *(continued)*
Properties of Low-Pressure Air
(SI units)

T in K; h and u in kJ/kg; ϕ in kJ/kg·K

T	h	p_r	u	v_r	ϕ	T	h	p_r	u	v_r	ϕ
1500	1635.97	601.9	1205.41	7.152	3.44516	1700	1880.1	1025	1392.7	4.761	3.5979
1520	1660.23	636.5	1223.87	6.854	3.46120	1750	1941.6	1161	1439.8	4.328	3.6336
1540	1684.51	672.8	1242.43	6.569	3.47712	1800	2003.3	1310	1487.2	3.944	3.6684
1560	1708.82	710.5	1260.99	6.301	3.49276	1850	2065.3	1475	1534.9	3.601	3.7023
1580	1733.17	750.0	1279.65	6.046	3.50829	1900	2127.4	1655	1582.6	3.295	3.7354
1600	1757.57	791.2	1298.30	5.804	3.52364	1950	2189.7	1852	1630.6	3.022	3.7677
1620	1782.00	834.1	1316.96	5.574	3.53879	2000	2252.1	2068	1678.7	2.776	3.7994
1640	1806.46	878.9	1335.72	5.355	3.55381	2050	2314.6	2303	1726.8	2.555	3.8303
1660	1830.96	925.6	1354.48	5.147	3.56867	2100	2377.4	2559	1775.3	2.356	3.8605
1680	1855.50	974.2	1373.24	4.949	3.58335	2150	2440.3	2837	1823.8	2.175	3.8901
						2200	2503.2	3138	1872.4	2.012	3.9191
						2250	2566.4	3464	1921.3	1.864	3.9474

Gas Tables: Thermodynamic Properties of Air, Products of Combustion and Component Gases, Compressible Flow Functions, 2nd Edition (English Units). Joseph H. Keenan, Jing Chao, and Joseph Kaye. Copyright © 1980. Adapted with permission from John Wiley & Sons, Inc.

Support Material

APPENDIX 24.T
Properties of Saturated
Refrigerant-12 (R-12) by Temperature
(SI units)

temp. (°C)	pressure (MPa)	volume vapor (m³/kg)	density liquid (kg/m³)	enthalpy		entropy	
				liquid (kJ/kg)	vapor (kJ/kg)	liquid (kJ/kg·K)	vapor (kJ/kg·K)
−100	0.001174	10.122	1678.0	112.69	306.46	0.60441	1.7235
−95	0.001851	6.6005	1665.0	116.92	308.67	0.62845	1.7048
−90	0.002836	4.4264	1651.9	121.14	310.90	0.65182	1.6879
−85	0.004230	3.0449	1638.7	125.36	313.16	0.67457	1.6727
−80	0.006160	2.1438	1625.5	129.59	315.44	0.69673	1.6589
−75	0.008774	1.5416	1612.1	133.81	317.74	0.71835	1.6465
−70	0.012246	1.1301	1598.7	138.06	320.05	0.73948	1.6353
−65	0.016776	0.84332	1585.2	142.32	322.38	0.76015	1.6252
−60	0.022591	0.63956	1571.5	146.58	324.71	0.78040	1.6161
−55	0.029944	0.49230	1557.8	150.87	327.05	0.80025	1.6079
−50	0.039115	0.38415	1543.9	155.18	329.40	0.81974	1.6005
−45	0.050408	0.30355	1529.9	159.51	331.74	0.83890	1.5938
−40	0.064152	0.24264	1515.7	163.86	334.09	0.85775	1.5879
−35	0.080701	0.19603	1501.4	168.25	336.43	0.87632	1.5825
−30	0.10043	0.15993	1486.9	172.67	338.76	0.89462	1.5777
−29.79	0.101325	0.15861	1486.3	172.85	338.86	0.89538	1.5775
−25	0.12373	0.13166	1472.3	177.12	341.08	0.91269	1.5734
−20	0.15101	0.10929	1457.4	181.61	343.39	0.93053	1.5696
−15	0.18272	0.09142	1442.4	186.14	345.69	0.94817	1.5662
−10	0.21928	0.07702	1427.1	190.72	347.96	0.96561	1.5632
−5	0.26117	0.06531	1411.5	195.33	350.22	0.98289	1.5605
0	0.30885	0.05571	1395.6	200.00	352.44	1.0000	1.5581
2	0.32966	0.05236	1389.2	201.88	353.32	1.0068	1.5572
4	0.35150	0.04925	1382.7	203.77	354.20	1.0136	1.5564
6	0.37441	0.04637	1376.2	205.66	355.07	1.0203	1.5556
8	0.39842	0.04369	1369.6	207.57	355.93	1.0271	1.5548
10	0.42356	0.04119	1363.0	209.48	356.79	1.0338	1.5541
12	0.44986	0.03887	1356.2	211.40	357.65	1.0405	1.5533
14	0.47737	0.03670	1349.5	213.33	358.49	1.0472	1.5527
16	0.50610	0.03468	1342.6	215.27	359.33	1.0538	1.5520
18	0.53610	0.03279	1335.7	217.22	360.16	1.0605	1.5514
20	0.56740	0.03102	1328.7	219.18	360.98	1.0671	1.5508
22	0.60003	0.02937	1321.6	221.14	361.80	1.0737	1.5502
24	0.63403	0.02782	1314.5	223.12	362.60	1.0803	1.5497
26	0.66943	0.02637	1307.3	225.11	363.40	1.0868	1.5491
28	0.70626	0.02500	1299.9	227.10	364.19	1.0934	1.5486
30	0.74457	0.02372	1292.5	229.11	364.96	1.0999	1.5481
32	0.78439	0.02252	1285.0	231.12	365.73	1.1064	1.5476
34	0.82574	0.02138	1277.4	233.15	366.48	1.1130	1.5471
36	0.86868	0.02032	1269.7	235.18	367.22	1.1195	1.5466
38	0.91324	0.01931	1261.9	237.23	367.95	1.1259	1.5461

(continued)

APPENDIX 24.T *(continued)*
Properties of Saturated
Refrigerant-12 (R-12) by Temperature
(SI units)

temp. (°C)	pressure (MPa)	volume vapor (m³/kg)	density liquid (kg/m³)	enthalpy		entropy	
				liquid (kJ/kg)	vapor (kJ/kg)	liquid (kJ/kg·K)	vapor (kJ/kg·K)
40	0.95944	0.01836	1253.9	239.29	368.67	1.1324	1.5456
42	1.0073	0.01746	1245.9	241.36	369.37	1.1389	1.5451
44	1.0570	0.01662	1237.7	243.44	370.06	1.1453	1.5446
46	1.1084	0.01581	1299.3	245.54	370.73	1.1518	1.5441
48	1.1616	0.01506	1220.9	247.64	371.38	1.1582	1.5435
50	1.2167	0.01434	1212.2	249.76	372.02	1.1647	1.5430
52	1.2736	0.01366	1203.5	251.90	372.64	1.1711	1.5425
54	1.3325	0.01301	1194.5	254.04	373.24	1.1776	1.5419
56	1.3934	0.01239	1185.4	256.21	373.82	1.1840	1.5413
58	1.4562	0.01181	1176.1	258.38	374.38	1.1904	1.5407
60	1.5212	0.01126	1166.6	260.58	374.91	1.1969	1.5401
62	1.5883	0.01073	1156.9	262.79	375.42	1.2033	1.5394
64	1.6575	0.01023	1146.9	265.02	375.90	1.2098	1.5387
66	1.7289	0.009746	1136.7	267.27	376.36	1.2162	1.5379
68	1.8026	0.009289	1126.3	269.54	376.78	1.2227	1.5371
70	1.8786	0.008852	1115.6	271.83	377.17	1.2292	1.5362
72	1.9570	0.008434	1104.6	274.15	377.53	1.2357	1.5353
74	2.0378	0.008034	1093.3	276.49	377.85	1.2423	1.5343
76	2.1210	0.007651	1081.6	278.86	378.13	1.2489	1.5332
78	2.2069	0.007283	1069.6	281.27	378.36	1.2555	1.5320
80	2.2953	0.006931	1057.2	283.70	378.54	1.2622	1.5308
85	2.5282	0.006106	1024.1	289.97	378.75	1.2792	1.5271
90	2.7790	0.005350	987.60	296.56	378.52	1.2968	1.5225
95	3.0490	0.004648	946.44	303.58	377.67	1.3152	1.5165
100	3.3399	0.003980	898.55	311.26	375.88	1.3351	1.5083
105	3.6538	0.003311	839.10	320.08	372.41	1.3576	1.4960
110	3.9943	0.002517	746.58	331.91	364.02	1.3875	1.4713
*111.80	4.125	0.00179	558.	348.4	348.4	1.430	1.430

*critical point
Reprinted with permission from *1993 ASHRAE Handbook: Fundamentals*, SI Edition, American Society of
Heating, Refrigerating, and Air Conditioning Engineers, Inc., Atlanta, GA, ©1993.

APPENDIX 24.U
Pressure-Enthalpy Diagram for Refrigerant-12 (R-12)
(SI units)

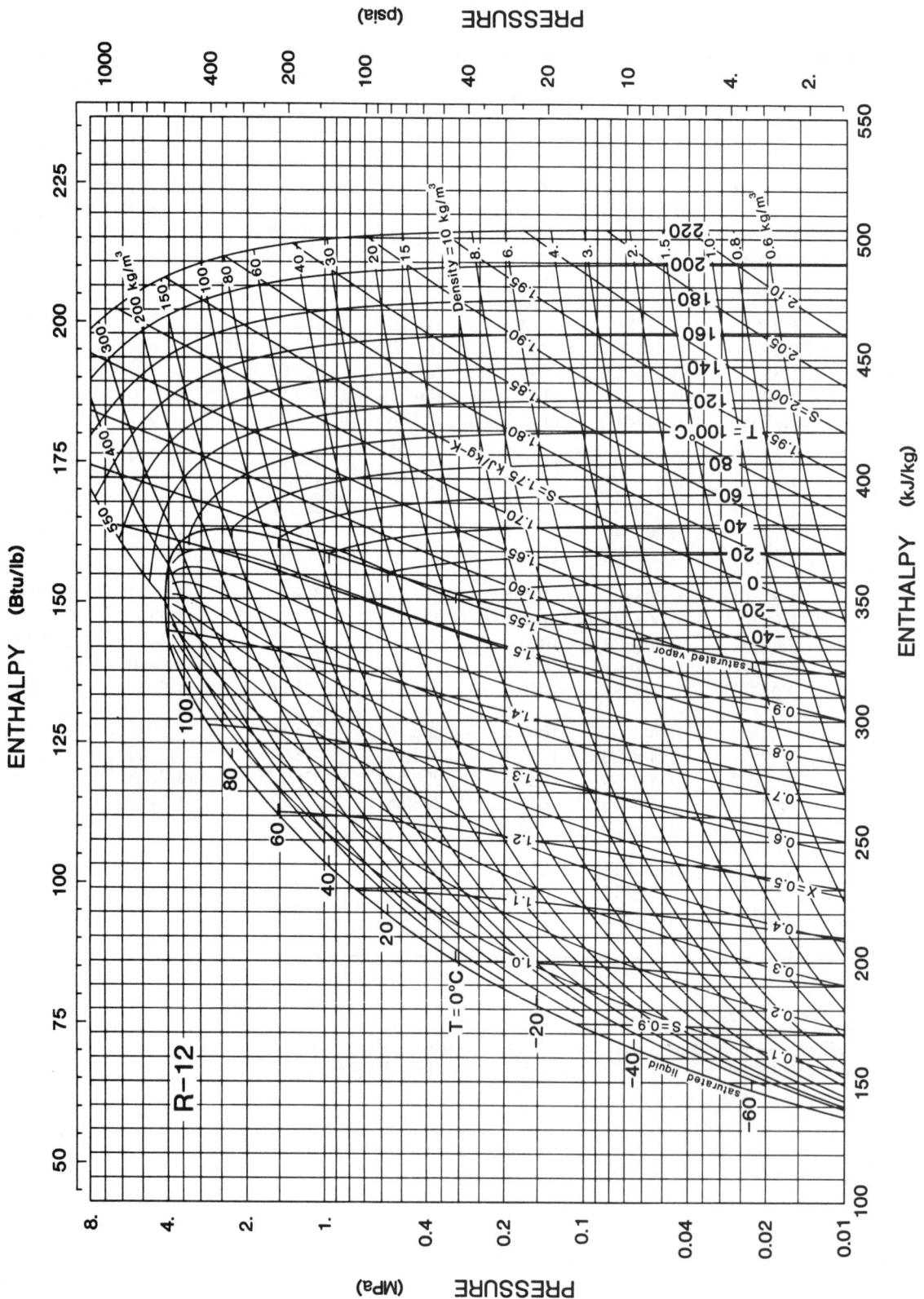

Reprinted with permission from *1993 ASHRAE Handbook: Fundamentals*, SI Edition, American Society of Heating, Refrigerating, and Air Conditioning Engineers, Inc., Atlanta, GA, ©1993.

APPENDIX 24.V
Properties of Saturated
Refrigerant-22 (R-22) by Temperature
(SI units)

temp. (°C)	pressure (MPa)	volume vapor (m³/kg)	density liquid (kg/m³)	enthalpy liquid (kJ/kg)	enthalpy vapor (kJ/kg)	entropy liquid (kJ/kg·K)	entropy vapor (kJ/kg·K)
−90	0.004748	3.6939	1542.8	98.575	364.20	0.55032	2.006
−85	0.007084	2.5394	1529.9	104.54	366.63	0.58246	1.9754
−80	0.010308	1.7883	1516.8	110.42	369.06	0.61326	1.9524
−75	0.014662	1.2870	1503.6	116.21	371.49	0.64284	1.9312
−70	0.020424	0.94477	1490.3	121.92	373.91	0.67132	1.9117
−65	0.027914	0.70609	1476.7	127.58	376.32	0.69879	1.8938
−60	0.037491	0.53641	1463.1	133.18	378.72	0.72535	1.8773
−55	0.049556	0.41362	1449.2	138.74	381.10	0.75109	1.8621
−50	0.064549	0.32330	1435.2	144.27	383.45	0.77610	1.8479
−45	0.082947	0.25586	1421.0	149.77	385.77	0.80044	1.8348
−40.82	0.101325	0.21223	1408.9	154.37	387.69	0.82034	1.8246
−40	0.10527	0.20480	1406.5	155.26	388.06	0.82419	1.8227
−38	0.11542	0.18790	1400.7	157.46	388.96	0.83354	1.8180
−36	0.12632	0.17268	1394.8	159.66	389.86	0.84281	1.8135
−34	0.13801	0.15894	1388.9	161.86	390.75	0.85200	1.8091
−32	0.15053	0.14651	1382.9	164.06	391.64	0.86113	1.8049
−30	0.16391	0.13524	1376.9	166.26	392.52	0.87018	1.8007
−28	0.17821	0.12502	1370.9	168.46	393.39	0.87917	1.7967
−26	0.19346	0.11573	1364.8	170.67	394.25	0.88810	1.7927
−24	0.20969	0.10726	1358.7	172.89	395.10	0.89697	1.7889
−22	0.22696	0.09954	1352.6	175.10	395.95	0.90579	1.7851
−20	0.24531	0.09249	1346.4	177.33	396.79	0.91455	1.7815
−18	0.26477	0.08603	1340.1	179.56	397.62	0.92327	1.7779
−16	0.28540	0.08012	1333.8	181.79	398.43	0.93194	1.7744
−14	0.30724	0.07470	1327.5	184.04	399.24	0.94057	1.7710
−12	0.33034	0.06971	1321.1	186.29	400.04	0.94916	1.7677
−10	0.35474	0.06513	1314.6	188.55	400.83	0.95771	1.7644
−8	0.38049	0.06090	1308.1	190.82	401.61	0.96623	1.7612
−6	0.40763	0.05701	1301.5	193.10	402.37	0.97471	1.7581
−4	0.43622	0.05341	1294.9	195.39	403.12	0.98317	1.7550
−2	0.46630	0.05008	1288.2	197.69	403.87	0.99160	1.7520
0	0.49792	0.04700	1281.5	200.00	404.59	1.0000	1.7490
2	0.53113	0.04415	1274.7	202.32	405.31	1.0084	1.7461
4	0.56599	0.04150	1267.8	204.66	406.01	1.0167	1.7432
6	0.60254	0.03904	1260.8	207.01	406.70	1.0251	1.7404
8	0.64083	0.03675	1253.8	209.37	407.37	1.0334	1.7376
10	0.68091	0.03462	1246.7	211.74	408.03	1.0417	1.7349
12	0.72285	0.03263	1239.5	214.13	408.67	1.0500	1.7322
14	0.76668	0.03078	1232.3	216.54	409.29	1.0583	1.7295
16	0.81246	0.02905	1224.9	218.96	409.90	1.0665	1.7269
18	0.86025	0.02743	1217.5	221.40	410.49	1.0748	1.7243

(continued)

APPENDIX 24.V *(continued)*
Properties of Saturated
Refrigerant-22 (R-22) by Temperature
(SI units)

temp. (°C)	pressure (MPa)	volume vapor (m³/kg)	density liquid (kg/m³)	enthalpy liquid (kJ/kg)	enthalpy vapor (kJ/kg)	entropy liquid (kJ/kg·K)	entropy vapor (kJ/kg·K)
20	0.91009	0.02592	1210.0	223.85	411.06	1.0831	1.7217
22	0.96205	0.02451	1202.4	226.32	411.61	1.0913	1.7191
24	1.0162	0.02318	1194.6	228.80	412.14	1.0996	1.7165
26	1.0725	0.02193	1186.8	231.31	412.65	1.1078	1.7140
28	1.1312	0.02076	1178.9	233.83	413.13	1.1160	1.7114
30	1.1921	0.01967	1170.8	236.38	413.60	1.1243	1.7089
32	1.2555	0.01863	1162.6	238.94	414.03	1.1325	1.7063
34	1.3213	0.01766	1154.3	241.52	414.45	1.1408	1.7038
36	1.3896	0.01674	1145.9	244.13	414.83	1.1490	1.7012
38	1.4605	0.01588	1137.3	246.75	415.19	1.1573	1.6987
40	1.5340	0.01506	1128.6	249.40	415.52	1.1656	1.6961
42	1.6102	0.01429	1119.7	252.07	415.82	1.1739	1.6934
44	1.6892	0.01356	1110.6	254.77	416.08	1.1822	1.6908
46	1.7710	0.01287	1101.4	257.49	416.31	1.1905	1.6881
48	1.8556	0.01221	1091.9	260.24	416.50	1.1989	1.6854
50	1.9432	0.01159	1082.3	263.02	416.65	1.2072	1.6826
52	2.0339	0.01101	1072.4	265.83	416.75	1.2156	1.6798
54	2.1276	0.01045	1062.3	268.67	416.81	1.2241	1.6769
56	2.2244	0.009915	1051.9	271.55	416.83	1.2326	1.6739
58	2.3245	0.009409	1041.3	274.46	416.79	1.2411	1.6709
60	2.4279	0.008927	1030.3	277.41	416.69	1.2497	1.6677
65	2.7015	0.007816	1001.3	284.98	416.16	1.2714	1.6594
70	2.9975	0.006819	969.68	292.88	416.14	1.2937	1.6500
75	3.3175	0.005917	934.38	301.22	413.46	1.3169	1.6393
80	3.6633	0.005086	893.89	310.18	410.88	1.3414	1.6265
85	4.0370	0.004301	845.17	320.13	406.90	1.3681	1.6104
90	4.4413	0.003517	780.60	331.96	400.28	1.3996	1.5877
95	4.8808	0.002547	660.94	350.67	384.95	1.4490	1.5421
*96.15	4.988	0.00195	513	368.1	368.1	1.496	1.496

*critical point
Reprinted with permission from *1993 ASHRAE Handbook: Fundamentals*, SI Edition, American Society of Heating, Refrigerating, and Air Conditioning Engineers, Inc., Atlanta, GA, ©1993.

APPENDIX 24.W
Pressure-Enthalpy Diagram for Refrigerant-22 (R-22)
(SI units)

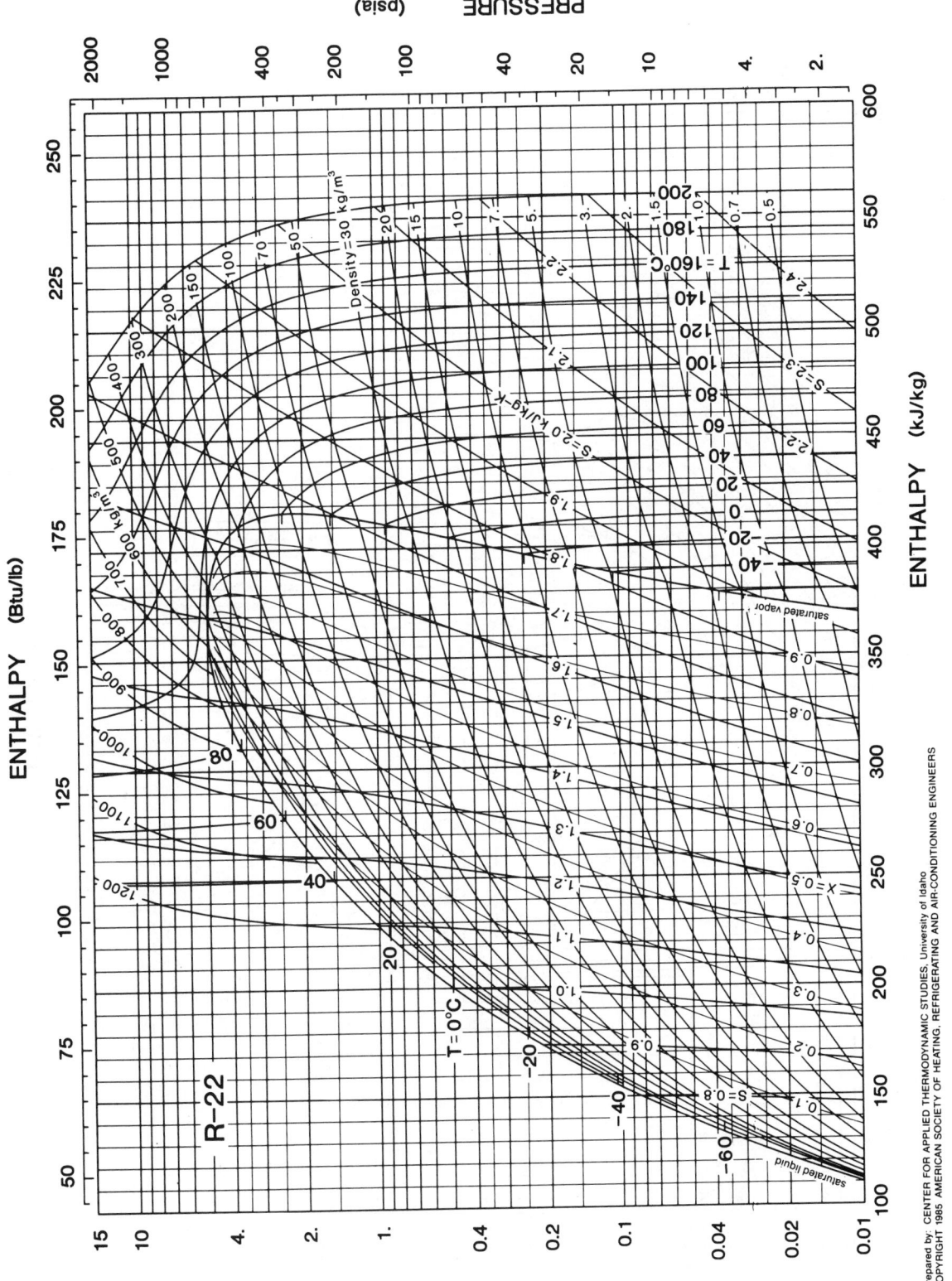

Prepared by: CENTER FOR APPLIED THERMODYNAMIC STUDIES, University of Idaho
COPYRIGHT 1985 AMERICAN SOCIETY OF HEATING, REFRIGERATING AND AIR-CONDITIONING ENGINEERS

Support
Material

APPENDIX 24.X
Pressure-Enthalpy Diagram for Refrigerant HFC-134a
(SI units)

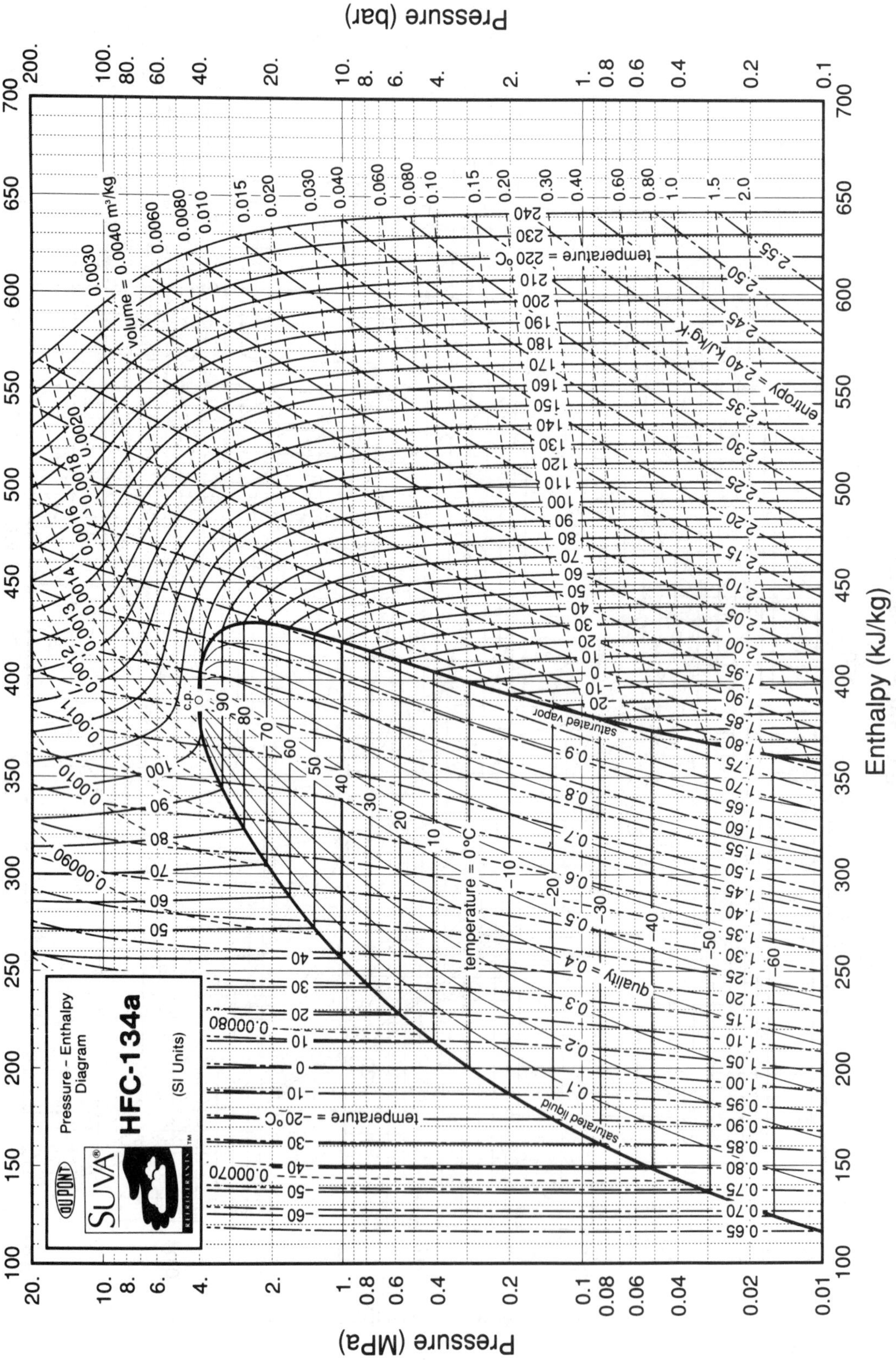

Reproduced by permission of the DuPont Company.

APPENDIX 24.Y
Physical Properties of Selected Solids
(customary U.S. units)

material	ρ (lbm/ft³) (68°F) (20°C)	c_p (Btu/lbm-°F) (68°F) (20°C)	α (ft²/hr) (68°F) (20°C)	k (Btu/hr-ft-°F) (68°F) (20°C)	(212°F) (100°C)	(572°) (300°C)
metals						
aluminum	168.6	0.224	3.55	132	132	133
copper	555	0.092	3.98	223	219	213
gold	1206	0.031	4.52	169	170	172
iron	492	0.122	0.83	42.3	39.0	31.6
lead	708	0.030	0.80	20.3	19.3	17.2
magnesium	109	0.248	3.68	99.5	96.8	91.4
nickel	556	0.111	0.87	53.7	47.7	36.9
platinum	1340	0.032	0.09	40.5	41.9	43.5
silver	656	0.057	6.42	240	237	209
tin	450	0.051	1.57	36	36	
tungsten	1206	0.032	2.44	94	87	77
uranium α	1167	0.027	0.53	16.9	17.2	19.6
zinc	446	0.094	1.55	65	63	58
alloys						
aluminum 2024	173	0.23	1.76	70.2		
brass (70% Cu, 30% Zn)	532	0.091	1.27	61.8	73.9	85.3
constantan (60% Cu, 40% Ni)	557	0.098	0.24	13.1	15.4	
iron, cast	455	0.100	0.65	29.6	26.8	
nichrome V	530	0.106	0.12	7.06	7.99	9.94
stainless steel	488	0.110	0.17	9.4	10.0	13
steel, mild (1%C)	488	0.113	0.45	24.8	24.8	22.9
nonmetals						
asbestos	36	0.25		0.092	0.11	0.125
brick (fire clay)	144	0.22			0.65	
brick (masonry)	106	0.20		0.38		
brick (chrome)	188	0.20			0.67	
concrete	150	0.21		0.70		
corkboard	10	0.4		0.025		
diatomaceous earth, powdered	14	0.2		0.03		
glass, window	170	0.2		0.45		
glass, Pyrex™	140	0.2		0.63	0.67	0.84
Kaolin firebrick	19					0.052
85% magnesia	17			0.038	0.041	
sandy loam, 4% H$_2$O	104	~0.4		0.54		
sandy loam, 10% H$_2$O	121			1.08		
rock wool	~10	0.2		0.023	0.033	
wood, oak, $\perp$ to grain	51	0.57		0.12		
wood, oak, $\parallel$ to grain	51	0.57		0.23		

(Multiply lbm/ft³ by 16.018 to obtain kg/m³.)
(Multiply Btu/lbm-°F by 4186.8 to obtain J/kg·K.)
(Multiply ft²/hr by 0.0929 to obtain m²/h.)
(Multiply Btu/hr-ft-°F by 1.7307 to obtain W/m·K.)

Engineering Heat Transfer, James R. Welty, copyright © 1974. Adapted by permission of John Wiley & Sons, Inc.

APPENDIX 24.Z
Generalized Compressibility Charts

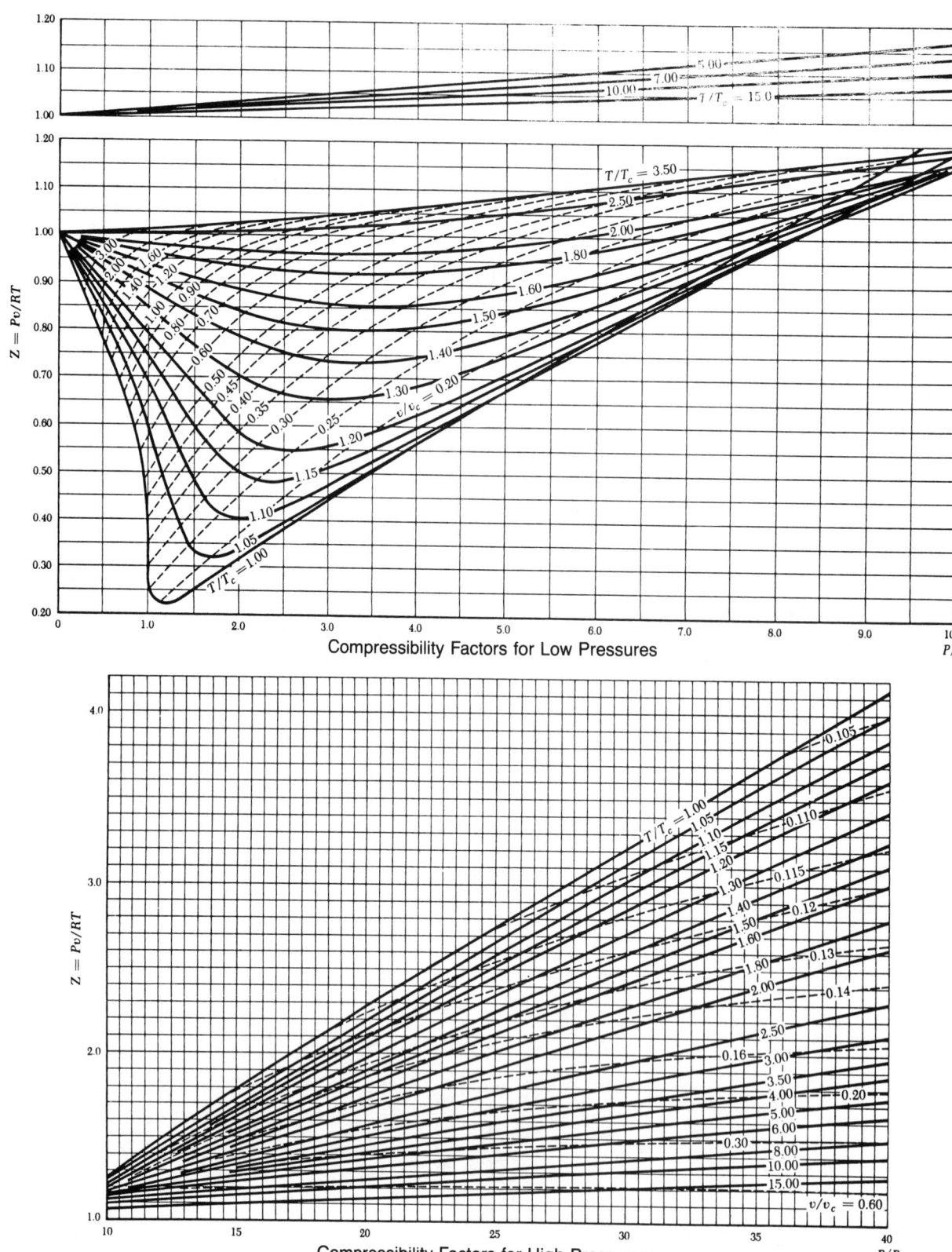

Reprinted with permission from *ASME Transactions*, Vol. 76, Edward F. Obert and L. C. Nelson, American Society of Mechanical Engineers, 1954.

APPENDIX 26.A
Isentropic Flow Factors
($k = 1.4$)

M = Mach number, p = pressure, p_0 = total pressure, ρ = density, ρ_0 = total density, T = temperature, T_0 = total temperature, A = area at a point, A^* = throat area for M = 1, v = velocity, and a^* = speed of sound at throat.

M	p/p_0	ρ/ρ_0	T/T_0	A/A^*	v/a^*	M	p/p_0	ρ/ρ_0	T/T_0	A/A^*	v/a^*
0.00	1.0000	1.0000	1.0000	—	0.0000	0.51	0.8374	0.8809	0.9506	1.3212	0.5447
0.01	0.9999	1.0000	1.0000	57.8737	0.0110	0.52	0.8317	0.8766	0.9487	1.3034	0.5548
0.02	0.9997	0.9998	0.9999	28.9420	0.0219	0.53	0.8259	0.8723	0.9468	1.2865	0.5649
0.03	0.9994	0.9996	0.9998	19.3005	0.0329	0.54	0.8201	0.8679	0.9449	1.2703	0.5750
0.04	0.9989	0.9992	0.9997	14.4815	0.0438	0.55	0.8142	0.8634	0.9430	1.2549	0.5851
0.05	0.9983	0.9988	0.9995	11.5914	0.0548	0.56	0.8082	0.8589	0.9410	1.2403	0.5951
0.06	0.9975	0.9982	0.9993	9.6659	0.0657	0.57	0.8022	0.8544	0.9390	1.2263	0.6051
0.07	0.9966	0.9976	0.9990	8.2915	0.0766	0.58	0.7962	0.8498	0.9370	1.2130	0.6150
0.08	0.9955	0.9968	0.9987	7.2616	0.0876	0.59	0.7901	0.8451	0.9349	1.2003	0.6249
0.09	0.9944	0.9960	0.9984	6.4613	0.0985	0.60	0.7840	0.8405	0.9328	1.1882	0.6348
0.10	0.9930	0.9950	0.9980	5.8218	0.1094	0.61	0.7778	0.8357	0.9307	1.1767	0.6447
0.11	0.9916	0.9940	0.9976	5.2992	0.1204	0.62	0.7716	0.8310	0.9286	1.1656	0.6545
0.12	0.9900	0.9928	0.9971	4.8643	0.1313	0.63	0.7654	0.8262	0.9265	1.1551	0.6643
0.13	0.9883	0.9916	0.9966	4.4969	0.1422	0.64	0.7591	0.8213	0.9243	1.1451	0.6740
0.14	0.9864	0.9903	0.9961	4.1824	0.1531	0.65	0.7528	0.8164	0.9221	1.1356	0.6837
0.15	0.9844	0.9888	0.9955	3.9103	0.1639	0.66	0.7465	0.8115	0.9199	1.1265	0.6934
0.16	0.9823	0.9873	0.9949	3.6727	0.1748	0.67	0.7401	0.8066	0.9176	1.1179	0.7031
0.17	0.9800	0.9857	0.9943	3.4635	0.1857	0.68	0.7338	0.8016	0.9153	1.1097	0.7127
0.18	0.9776	0.9840	0.9936	3.2779	0.1965	0.69	0.7274	0.7966	0.9131	1.1018	0.7223
0.19	0.9751	0.9822	0.9928	3.1123	0.2074	0.70	0.7209	0.7916	0.9107	1.0944	0.7318
0.20	0.9725	0.9803	0.9921	2.9635	0.2182	0.71	0.7145	0.7865	0.9084	1.0873	0.7413
0.21	0.9697	0.9783	0.9913	2.8293	0.2290	0.72	0.7080	0.7814	0.9061	1.0806	0.7508
0.22	0.9668	0.9762	0.9904	2.7076	0.2398	0.73	0.7016	0.7763	0.9037	1.0742	0.7602
0.23	0.9638	0.9740	0.9895	2.5968	0.2506	0.74	0.6951	0.7712	0.9013	1.0681	0.7696
0.24	0.9607	0.9718	0.9886	2.4956	0.2614	0.75	0.6886	0.7660	0.8989	1.0624	0.7789
0.25	0.9575	0.9694	0.9877	2.4027	0.2722	0.76	0.6821	0.7609	0.8964	1.0570	0.7883
0.26	0.9541	0.9670	0.9867	2.3173	0.2829	0.77	0.6756	0.7557	0.8940	1.0519	0.7975
0.27	0.9506	0.9645	0.9856	2.2385	0.2936	0.78	0.6691	0.7505	0.8915	1.0471	0.8068
0.28	0.9470	0.9619	0.9846	2.1656	0.3043	0.79	0.6625	0.7452	0.8890	1.0425	0.8160
0.29	0.9433	0.9592	0.9835	2.0979	0.3150	0.80	0.6560	0.7400	0.8865	1.0382	0.8251
0.30	0.9395	0.9564	0.9823	2.0351	0.3257	0.81	0.6495	0.7347	0.8840	1.0342	0.8343
0.31	0.9355	0.9535	0.9811	1.9765	0.3364	0.82	0.6430	0.7295	0.8815	1.0305	0.8433
0.32	0.9315	0.9506	0.9799	1.9218	0.3470	0.83	0.6365	0.7242	0.8789	1.0270	0.8524
0.33	0.9274	0.9476	0.9787	1.8707	0.3576	0.84	0.6300	0.7189	0.8763	1.0237	0.8614
0.34	0.9231	0.9445	0.9774	1.8229	0.3682	0.85	0.6235	0.7136	0.8737	1.0207	0.8704
0.35	0.9188	0.9413	0.9761	1.7780	0.3788	0.86	0.6170	0.7083	0.8711	1.0179	0.8793
0.36	0.9143	0.9380	0.9747	1.7358	0.3893	0.87	0.6106	0.7030	0.8685	1.0153	0.8882
0.37	0.9098	0.9347	0.9734	1.6961	0.3999	0.88	0.6041	0.6977	0.8659	1.0129	0.8970
0.38	0.9052	0.9313	0.9719	1.6587	0.4104	0.89	0.5977	0.6924	0.8632	1.0108	0.9058
0.39	0.9004	0.9278	0.9705	1.6234	0.4209	0.90	0.5913	0.6870	0.8606	1.0089	0.9146
0.40	0.8956	0.9243	0.9690	1.5901	0.4313	0.91	0.5849	0.6817	0.8579	1.0071	0.9233
0.41	0.8907	0.9207	0.9675	1.5587	0.4418	0.92	0.5785	0.6764	0.8552	1.0056	0.9320
0.42	0.8857	0.9170	0.9659	1.5289	0.4522	0.93	0.5721	0.6711	0.8525	1.0043	0.9406
0.43	0.8807	0.9132	0.9643	1.5007	0.4626	0.94	0.5658	0.6658	0.8498	1.0031	0.9493
0.44	0.8755	0.9094	0.9627	1.4740	0.4729	0.95	0.5595	0.6604	0.8471	1.0021	0.9578
0.45	0.8703	0.9055	0.9611	1.4487	0.4833	0.96	0.5532	0.6551	0.8444	1.0014	0.9663
0.46	0.8650	0.9016	0.9594	1.4246	0.4936	0.97	0.5469	0.6498	0.8416	1.0008	0.9748
0.47	0.8596	0.8976	0.9577	1.4018	0.5038	0.98	0.5407	0.6445	0.8389	1.0003	0.9832
0.48	0.8541	0.8935	0.9560	1.3801	0.5141	0.99	0.5345	0.6392	0.8361	1.0001	0.9916
0.49	0.8486	0.8894	0.9542	1.3595	0.5243	1.00	0.5283	0.6339	0.8333	1.0000	1.0000
0.50	0.8430	0.8852	0.9524	1.3398	0.5345						

APPENDIX 26.B
Isentropic Flow and Normal Shock Parameters
($k = 1.4$)

(x refers to upstream conditions. y refers to downstream conditions. For example, $p_x/p_{0,y}$ is the ratio of static pressure before the shock wave to total pressure after the shock wave.)

M	p/p_0	ρ/ρ_0	T/T_0	A/A^*	v/a^*	M_y	p_y/p_x	ρ_x/ρ_y	T_y/T_x	$p_{0,y}/p_{0,x}$	$p_x/p_{0,y}$
1.00	0.5283	0.6339	0.8333	1.0000	1.0000	1.0000	0.1000×10^1	0.1000×10^1	0.1000×10^1	1.0000	0.5283
1.10	0.4684	0.5817	0.8052	1.0079	1.0812	0.9118	0.1245×10^1	0.1169×10^1	0.1065×10^1	0.9989	0.4689
1.20	0.4124	0.5311	0.7764	1.0304	1.1583	0.8422	0.1513×10^1	0.1342×10^1	0.1128×10^1	0.9928	0.4154
1.30	0.3609	0.4829	0.7474	1.0663	1.2311	0.7860	0.1805×10^1	0.1516×10^1	0.1191×10^1	0.9794	0.3685
1.40	0.3142	0.4374	0.7184	1.1149	1.2999	0.7397	0.2120×10^1	0.1690×10^1	0.1255×10^1	0.9582	0.3280
1.50	0.2724	0.3950	0.6897	1.1762	1.3646	0.7011	0.2458×10^1	0.1862×10^1	0.1320×10^1	0.9298	0.2930
1.60	0.2353	0.3557	0.6614	1.2502	1.4254	0.6684	0.2820×10^1	0.2032×10^1	0.1388×10^1	0.8952	0.2628
1.70	0.2026	0.3197	0.6337	1.3376	1.4825	0.6405	0.3205×10^1	0.2198×10^1	0.1458×10^1	0.8557	0.2368
1.80	0.1740	0.2868	0.6068	1.4390	1.5360	0.6165	0.3613×10^1	0.2359×10^1	0.1532×10^1	0.8127	0.2142
1.90	0.1492	0.2570	0.5807	1.5553	1.5861	0.5956	0.4045×10^1	0.2516×10^1	0.1608×10^1	0.7674	0.1945
2.00	0.1278	0.2300	0.5556	1.6875	1.6330	0.5774	0.4500×10^1	0.2667×10^1	0.1687×10^1	0.7209	0.1773
2.10	0.1094	0.2058	0.5313	1.8369	1.6769	0.5613	0.4978×10^1	0.2812×10^1	0.1770×10^1	0.6742	0.1622
2.20	0.9352×10^{-1}	0.1841	0.5081	2.0050	1.7179	0.5471	0.5480×10^1	0.2951×10^1	0.1857×10^1	0.6281	0.1489
2.30	0.7997×10^{-1}	0.1646	0.4859	2.1931	1.7563	0.5344	0.6005×10^1	0.3085×10^1	0.1947×10^1	0.5833	0.1371
2.40	0.6840×10^{-1}	0.1472	0.4647	2.4031	1.7922	0.5231	0.6553×10^1	0.3212×10^1	0.2040×10^1	0.5401	0.1266
2.50	0.5853×10^{-1}	0.1317	0.4444	2.6367	1.8257	0.5130	0.7125×10^1	0.3333×10^1	0.2137×10^1	0.4990	0.1173
2.60	0.5012×10^{-1}	0.1179	0.4252	2.8960	1.8571	0.5039	0.7720×10^1	0.3449×10^1	0.2238×10^1	0.4601	0.1089
2.70	0.4295×10^{-1}	0.1056	0.4068	3.1830	1.8865	0.4956	0.8338×10^1	0.3559×10^1	0.2343×10^1	0.4236	0.1014
2.80	0.3685×10^{-1}	0.9463×10^{-1}	0.3894	3.5001	1.9140	0.4882	0.8980×10^1	0.3664×10^1	0.2451×10^1	0.3895	0.9461×10^{-1}
2.90	0.3165×10^{-1}	0.8489×10^{-1}	0.3729	3.8498	1.9398	0.4814	0.9645×10^1	0.3763×10^1	0.2563×10^1	0.3577	0.8848×10^{-1}
3.00	0.2722×10^{-1}	0.7623×10^{-1}	0.3571	4.2346	1.9640	0.4752	0.1033×10^2	0.3857×10^1	0.2679×10^1	0.3283	0.8291×10^{-1}
3.10	0.2345×10^{-1}	0.6852×10^{-1}	0.3422	4.6573	1.9866	0.4695	0.1104×10^2	0.3947×10^1	0.2799×10^1	0.3012	0.7785×10^{-1}
3.20	0.2023×10^{-1}	0.6165×10^{-1}	0.3281	5.1210	2.0079	0.4643	0.1178×10^2	0.4031×10^1	0.2922×10^1	0.2762	0.7323×10^{-1}
3.30	0.1748×10^{-1}	0.5554×10^{-1}	0.3147	5.6286	2.0278	0.4596	0.1254×10^2	0.4112×10^1	0.3049×10^1	0.2533	0.6900×10^{-1}
3.40	0.1512×10^{-1}	0.5009×10^{-1}	0.3019	6.1837	2.0466	0.4552	0.1332×10^2	0.4188×10^1	0.3180×10^1	0.2322	0.6513×10^{-1}
3.50	0.1311×10^{-1}	0.4523×10^{-1}	0.2899	6.7896	2.0642	0.4512	0.1412×10^2	0.4261×10^1	0.3315×10^1	0.2129	0.6157×10^{-1}
3.60	0.1138×10^{-1}	0.4089×10^{-1}	0.2784	7.4501	2.0808	0.4474	0.1495×10^2	0.4330×10^1	0.3454×10^1	0.1953	0.5829×10^{-1}
3.70	0.9903×10^{-2}	0.3702×10^{-1}	0.2675	8.1691	2.0964	0.4439	0.1580×10^2	0.4395×10^1	0.3596×10^1	0.1792	0.5526×10^{-1}
3.80	0.8629×10^{-2}	0.3355×10^{-1}	0.2572	8.9506	2.1111	0.4407	0.1668×10^2	0.4457×10^1	0.3743×10^1	0.1645	0.5247×10^{-1}
3.90	0.7532×10^{-2}	0.3044×10^{-1}	0.2474	9.7990	2.1250	0.4377	0.1758×10^2	0.4516×10^1	0.3893×10^1	0.1510	0.4987×10^{-1}
4.00	0.6586×10^{-2}	0.2766×10^{-1}	0.2381	10.7187	2.1381	0.4350	0.1850×10^2	0.4571×10^1	0.4047×10^1	0.1388	0.4747×10^{-1}
4.10	0.5769×10^{-2}	0.2516×10^{-1}	0.2293	11.7147	2.1505	0.4324	0.1944×10^2	0.4624×10^1	0.4205×10^1	0.1276	0.4523×10^{-1}
4.20	0.5062×10^{-2}	0.2292×10^{-1}	0.2208	12.7916	2.1622	0.4299	0.2041×10^2	0.4675×10^1	0.4367×10^1	0.1173	0.4314×10^{-1}
4.30	0.4449×10^{-2}	0.2090×10^{-1}	0.2129	13.9549	2.1732	0.4277	0.2140×10^2	0.4723×10^1	0.4532×10^1	0.1080	0.4120×10^{-1}
4.40	0.3918×10^{-2}	0.1909×10^{-1}	0.2053	15.2099	2.1837	0.4255	0.2242×10^2	0.4768×10^1	0.4702×10^1	0.9948×10^{-1}	0.3938×10^{-1}
4.50	0.3455×10^{-2}	0.1745×10^{-1}	0.1980	16.5622	2.1936	0.4236	0.2346×10^2	0.4812×10^1	0.4875×10^1	0.9170×10^{-1}	0.3768×10^{-1}
4.60	0.3053×10^{-2}	0.1597×10^{-1}	0.1911	18.0178	2.2030	0.4217	0.2452×10^2	0.4853×10^1	0.5052×10^1	0.8459×10^{-1}	0.3609×10^{-1}
4.70	0.2701×10^{-2}	0.1464×10^{-1}	0.1846	19.5828	2.2119	0.4199	0.2560×10^2	0.4893×10^1	0.5233×10^1	0.7809×10^{-1}	0.3459×10^{-1}
4.80	0.2394×10^{-2}	0.1343×10^{-1}	0.1783	21.2637	2.2204	0.4183	0.2671×10^2	0.4930×10^1	0.5418×10^1	0.7214×10^{-1}	0.3319×10^{-1}
4.90	0.2126×10^{-2}	0.1233×10^{-1}	0.1724	23.0671	2.2284	0.4167	0.2784×10^2	0.4966×10^1	0.5607×10^1	0.6670×10^{-1}	0.3187×10^{-1}
5.00	0.1890×10^{-2}	0.1134×10^{-1}	0.1667	25.0000	2.2361	0.4152	0.2900×10^2	0.5000×10^1	0.5800×10^1	0.6172×10^{-1}	0.3062×10^{-1}
5.10	0.1683×10^{-2}	0.1044×10^{-1}	0.1612	27.0696	2.2433	0.4138	0.3018×10^2	0.5033×10^1	0.5997×10^1	0.5715×10^{-1}	0.2945×10^{-1}
5.20	0.1501×10^{-2}	0.9620×10^{-2}	0.1561	29.2833	2.2503	0.4125	0.3138×10^2	0.5064×10^1	0.6197×10^1	0.5297×10^{-1}	0.2834×10^{-1}
5.30	0.1341×10^{-2}	0.8875×10^{-2}	0.1511	31.6491	2.2569	0.4113	0.3260×10^2	0.5093×10^1	0.6401×10^1	0.4913×10^{-1}	0.2730×10^{-1}
5.40	0.1200×10^{-2}	0.8197×10^{-2}	0.1464	34.1748	2.2631	0.4101	0.3385×10^2	0.5122×10^1	0.6610×10^1	0.4560×10^{-1}	0.2631×10^{-1}
5.50	0.1075×10^{-2}	0.7578×10^{-2}	0.1418	36.8690	2.2691	0.4090	0.3512×10^2	0.5149×10^1	0.6822×10^1	0.4236×10^{-1}	0.2537×10^{-1}
5.60	0.9643×10^{-3}	0.7012×10^{-2}	0.1375	39.7402	2.2748	0.4079	0.3642×10^2	0.5175×10^1	0.7038×10^1	0.3938×10^{-1}	0.2448×10^{-1}
5.70	0.8663×10^{-3}	0.6496×10^{-2}	0.1334	42.7974	2.2803	0.4069	0.3774×10^2	0.5200×10^1	0.7258×10^1	0.3664×10^{-1}	0.2364×10^{-1}
5.80	0.7794×10^{-3}	0.6023×10^{-2}	0.1294	46.0500	2.2855	0.4059	0.3908×10^2	0.5224×10^1	0.7481×10^1	0.3412×10^{-1}	0.2284×10^{-1}
5.90	0.7021×10^{-3}	0.5590×10^{-2}	0.1256	49.5075	2.2905	0.4050	0.4044×10^2	0.5246×10^1	0.7709×10^1	0.3179×10^{-1}	0.2208×10^{-1}
6.00	0.6334×10^{-3}	0.5194×10^{-2}	0.1220	53.1798	2.2953	0.4042	0.4183×10^2	0.5268×10^1	0.7941×10^1	0.2965×10^{-1}	0.2136×10^{-1}
6.10	0.5721×10^{-3}	0.4829×10^{-2}	0.1185	57.0772	2.2998	0.4033	0.4324×10^2	0.5289×10^1	0.8176×10^1	0.2767×10^{-1}	0.2067×10^{-1}
6.20	0.5173×10^{-3}	0.4495×10^{-2}	0.1151	61.2102	2.3042	0.4025	0.4468×10^2	0.5309×10^1	0.8415×10^1	0.2584×10^{-1}	0.2002×10^{-1}
6.30	0.4684×10^{-3}	0.4187×10^{-2}	0.1119	65.5899	2.3084	0.4018	0.4614×10^2	0.5329×10^1	0.8658×10^1	0.2416×10^{-1}	0.1939×10^{-1}
6.40	0.4247×10^{-3}	0.3904×10^{-2}	0.1088	70.2274	2.3124	0.4011	0.4762×10^2	0.5347×10^1	0.8905×10^1	0.2259×10^{-1}	0.1880×10^{-1}
6.50	0.3855×10^{-3}	0.3643×10^{-2}	0.1058	75.1343	2.3163	0.4004	0.4912×10^2	0.5365×10^1	0.9156×10^1	0.2115×10^{-1}	0.1823×10^{-1}
6.60	0.3503×10^{-3}	0.3402×10^{-2}	0.1030	80.3227	2.3200	0.3997	0.5065×10^2	0.5382×10^1	0.9411×10^1	0.1981×10^{-1}	0.1768×10^{-1}
6.70	0.3187×10^{-3}	0.3180×10^{-2}	0.1002	85.8049	2.3235	0.3991	0.5220×10^2	0.5399×10^1	0.9670×10^1	0.1857×10^{-1}	0.1716×10^{-1}
6.80	0.2902×10^{-3}	0.2974×10^{-2}	0.9758×10^{-1}	91.5935	2.3269	0.3985	0.5378×10^2	0.5415×10^1	0.9933×10^1	0.1741×10^{-1}	0.1667×10^{-1}
6.90	0.2646×10^{-3}	0.2785×10^{-2}	0.9504×10^{-1}	97.9017	2.3302	0.3979	0.5538×10^2	0.5430×10^1	0.1020×10^2	0.1634×10^{-1}	0.1619×10^{-1}
7.00	0.2416×10^{-3}	0.2609×10^{-2}	0.9259×10^{-1}	104.1429	2.3333	0.3974	0.5700×10^2	0.5444×10^1	0.1047×10^2	0.1535×10^{-1}	0.1573×10^{-1}
7.10	0.2207×10^{-3}	0.2446×10^{-2}	0.9024×10^{-1}	110.9309	2.3364	0.3968	0.5864×10^2	0.5459×10^1	0.1074×10^2	0.1443×10^{-1}	0.1530×10^{-1}
7.20	0.2019×10^{-3}	0.2295×10^{-2}	0.8797×10^{-1}	118.0799	2.3393	0.3963	0.6031×10^2	0.5472×10^1	0.1102×10^2	0.1357×10^{-1}	0.1488×10^{-1}
7.30	0.1848×10^{-3}	0.2155×10^{-2}	0.8578×10^{-1}	125.6046	2.3421	0.3958	0.6200×10^2	0.5485×10^1	0.1130×10^2	0.1277×10^{-1}	0.1448×10^{-1}
7.40	0.1694×10^{-3}	0.2025×10^{-2}	0.8367×10^{-1}	133.5200	2.3448	0.3954	0.6372×10^2	0.5498×10^1	0.1159×10^2	0.1202×10^{-1}	0.1409×10^{-1}
7.50	0.1554×10^{-3}	0.1904×10^{-2}	0.8163×10^{-1}	141.8415	2.3474	0.3949	0.6546×10^2	0.5510×10^1	0.1188×10^2	0.1133×10^{-1}	0.1372×10^{-1}
7.60	0.1427×10^{-3}	0.1792×10^{-2}	0.7967×10^{-1}	150.5849	2.3499	0.3945	0.6722×10^2	0.5522×10^1	0.1217×10^2	0.1068×10^{-1}	0.1336×10^{-1}
7.70	0.1312×10^{-3}	0.1687×10^{-2}	0.7777×10^{-1}	159.7665	2.3523	0.3941	0.6900×10^2	0.5533×10^1	0.1247×10^2	0.1008×10^{-1}	0.1302×10^{-1}
7.80	0.1207×10^{-3}	0.1589×10^{-2}	0.7594×10^{-1}	169.4030	2.3546	0.3937	0.7081×10^2	0.5544×10^1	0.1277×10^2	0.9510×10^{-2}	0.1269×10^{-1}
7.90	0.1111×10^{-3}	0.1498×10^{-2}	0.7417×10^{-1}	179.5114	2.3569	0.3933	0.7264×10^2	0.5555×10^1	0.1308×10^2	0.8982×10^{-2}	0.1237×10^{-1}
8.00	0.1024×10^{-3}	0.1414×10^{-2}	0.7246×10^{-1}	190.1094	2.3591	0.3929	0.7450×10^2	0.5565×10^1	0.1339×10^2	0.8488×10^{-2}	0.1207×10^{-1}
8.10	0.9449×10^{-4}	0.1334×10^{-2}	0.7081×10^{-1}	201.2148	2.3612	0.3925	0.7638×10^2	0.5575×10^1	0.1370×10^2	0.8025×10^{-2}	0.1177×10^{-1}
8.20	0.8723×10^{-4}	0.1260×10^{-2}	0.6921×10^{-1}	212.8461	2.3632	0.3922	0.7828×10^2	0.5585×10^1	0.1402×10^2	0.7592×10^{-2}	0.1149×10^{-1}
8.30	0.8060×10^{-4}	0.1191×10^{-2}	0.6767×10^{-1}	225.0221	2.3652	0.3918	0.8020×10^2	0.5594×10^1	0.1434×10^2	0.7187×10^{-2}	0.1122×10^{-1}
8.40	0.7454×10^{-4}	0.1126×10^{-2}	0.6617×10^{-1}	237.7622	2.3671	0.3915	0.8215×10^2	0.5603×10^1	0.1466×10^2	0.6806×10^{-2}	0.1095×10^{-1}
8.50	0.6898×10^{-4}	0.1066×10^{-2}	0.6472×10^{-1}	251.0862	2.3689	0.3912	0.8412×10^2	0.5612×10^1	0.1499×10^2	0.6449×10^{-2}	0.1070×10^{-1}

APPENDIX 26.C
Fanno Flow Factors
$(k = 1.4)$

M	p/p^*	$a/a^* = \rho^*/\rho$	T/T^*	$p_0/p_0{}^*$	$4fL/D$
0.00	∞	0.	1.200	∞	∞
0.05	21.903	0.0547	1.199	11.592	280.02
0.10	10.944	0.1094	1.197	5.822	66.922
0.12	9.116	0.131	1.1965	4.864	45.408
0.14	7.809	0.153	1.195	4.182	32.511
0.16	6.829	0.175	1.194	3.673	24.198
0.18	6.066	0.196	1.192	3.278	18.543
0.20	5.455	0.218	1.1905	2.963	14.533
0.25	4.355	0.272	1.185	2.403	8.483
0.30	3.619	0.3257	1.178	2.035	5.299
0.35	3.092	0.379	1.171	1.778	3.453
0.40	2.696	0.431	1.163	1.590	2.308
0.45	2.386	0.483	1.153	1.448	1.566
0.50	2.138	0.534	1.143	1.340	1.069
0.52	2.052	0.555	1.138	1.303	0.917
0.54	1.972	0.575	1.134	1.270	0.787
0.56	1.897	0.595	1.129	1.240	0.673
0.58	1.828	0.615	1.124	1.213	0.576
0.60	1.763	0.635	1.119	1.188	0.491
0.65	1.618	0.684	1.106	1.135	0.325
0.70	1.493	0.732	1.093	1.094	0.208
0.75	1.385	0.779	1.078	1.062	0.127
0.80	1.289	0.825	1.064	1.038	0.072
0.85	1.205	0.870	1.048	1.020	0.0363
0.90	1.129	0.914	1.0327	1.009	0.0145
0.95	1.061	0.958	1.0165	1.002	0.0033
1.00	1.000	1.000	1.000	1.000	0.000
1.20	0.804	1.158	0.932	1.030	0.0336
1.50	0.606	1.365	0.827	1.176	0.136
1.60	0.557	1.425	0.794	1.250	0.172
1.70	0.513	1.483	0.760	1.338	0.208
1.80	0.474	1.536	0.728	1.439	0.242
1.90	0.439	1.586	0.697	1.555	0.274
2.00	0.408	1.633	0.667	1.687	0.305
2.50	0.292	1.826	0.533	2.637	0.432
3.00	0.218	1.964	0.428	4.235	0.522
3.50	0.1685	2.064	0.348	6.789	0.586
4.00	0.134	2.138	0.286	10.719	0.633
4.50	0.108	2.194	0.237	16.562	0.667
5.00	0.0894	2.236	0.200	25.000	0.694

Support
Material

APPENDIX 26.D
Rayleigh Flow Factors
$(k = 1.4)$

M	p/p^*	$p_0/p_0{}^*$	T/T^*	$T_0/T_0{}^*$	$a/a^* = \rho^*/\rho$
0.00	2.400	1.268	0.000	0.000	0.000
0.05	2.392	1.266	0.0143	0.0119	0.00598
0.10	2.367	1.259	0.056	0.0468	0.0237
0.12	2.353	1.255	0.079	0.0667	0.0339
0.14	2.336	1.251	0.107	0.089	0.0458
0.16	2.317	1.246	0.137	0.115	0.0593
0.18	2.296	1.241	0.1708	0.143	0.0744
0.20	2.273	1.235	0.2066	0.1735	0.091
0.25	2.207	1.218	0.304	0.257	0.138
0.30	2.131	1.198	0.409	0.3468	0.192
0.35	2.048	1.178	0.514	0.439	0.251
0.40	1.961	1.157	0.615	0.529	0.314
0.45	1.870	1.135	0.708	0.614	0.378
0.50	1.778	1.114	0.790	0.691	0.444
0.52	1.741	1.106	0.819	0.720	0.470
0.54	1.704	1.098	0.847	0.747	0.497
0.56	1.668	1.090	0.872	0.772	0.523
0.58	1.632	1.083	0.896	0.796	0.549
0.60	1.596	1.075	0.917	0.819	0.574
0.65	1.508	1.058	0.961	0.868	0.637
0.70	1.424	1.043	0.993	0.908	0.697
0.75	1.343	1.030	1.014	0.940	0.755
0.80	1.266	1.019	1.025	0.964	0.810
0.85	1.193	1.011	1.028	0.981	0.862
0.90	1.125	1.045	1.0245	0.992	0.911
0.95	1.060	1.001	1.0146	0.998	0.957
1.00	1.000	1.000	1.000	1.000	1.000
1.20	0.796	1.0194	0.912	0.978	1.146
1.50	0.578	1.122	0.753	0.909	1.301
1.60	0.523	1.176	0.702	0.884	1.340
1.70	0.475	1.240	0.654	0.859	1.375
1.80	0.433	1.316	0.609	0.836	1.405
1.90	0.396	1.403	0.567	0.814	1.431
2.00	0.363	1.503	0.529	0.794	1.455
2.50	0.246	2.222	0.378	0.710	1.538
3.00	0.176	3.424	0.280	0.654	1.588
3.50	0.132	5.328	0.214	0.616	1.619
4.00	0.1025	8.227	0.168	0.589	1.641
4.50	0.0818	12.502	0.135	0.569	1.656
5.00	0.0667	18.634	0.111	0.555	1.667

APPENDIX 26.E
International Standard Atmosphere

customary U.S. units				SI units		
altitude (ft)	temperature (°R)	pressure (psia)		altitude (m)	temperature (K)	pressure (bar)
0	518.7	14.696		0	288.15	1.01325
1000	515.1	14.175				
2000	511.6	13.664		500	284.9	0.9546
3000	508.0	13.168		1000	281.7	0.8988
4000	504.4	12.692		1500	278.4	0.8456
				2000	275.2	0.7950
5000	500.9	12.225		2500	271.9	0.7469
6000	497.3	11.778				
7000	493.7	11.341		3000	268.7	0.7012
8000	490.2	10.914		3500	265.4	0.6578
9000	486.6	10.501		4000	262.2	0.6166
				4500	258.9	0.5775
10,000	483.0	10.108		5000	255.7	0.5405
11,000	479.5	9.720				
12,000	475.9	9.347		5500	252.4	0.5054
13,000	472.3	8.983		6000	249.2	0.4722
14,000	468.8	8.630		6500	245.9	0.4408
				7000	242.7	0.4111
15,000	465.2	8.291		7500	239.5	0.3830
16,000	461.6	7.962				
17,000	458.1	7.642		8000	236.2	0.3565
18,000	454.5	7.338		8500	233.0	0.3315
19,000	450.9	7.038		9000	229.7	0.3080
				9500	226.5	0.2858
20,000	447.4	6.753		10 000	223.3	0.2650
21,000	443.8	6.473				
22,000	440.2	6.203		10 500	220.0	0.2454
23,000	436.7	5.943		11 000	216.8	0.2270
24,000	433.1	5.693		11 500	216.7	0.2098
				12 000	216.7	0.1940
25,000	429.5	5.452		12 500	216.7	0.1793
26,000	426.0	5.216				
27,000	422.4	4.990		13 000	216.7	0.1658
28,000	418.8	4.774		13 500	216.7	0.1533
29,000	415.3	4.563		14 000	216.7	0.1417
				14 500	216.7	0.1310
30,000	411.7	4.362		15 000	216.7	0.1211
31,000	408.1	4.165				
32,000	404.6	3.978		15 500	216.7	0.1120
33,000	401.0	3.797		16 000	216.7	0.1035
34,000	397.5	3.625		16 500	216.7	0.09572
				17 000	216.7	0.08850
35,000	393.9	3.458		17 500	216.7	0.08182
36,000	392.7	3.296				
37,000	392.7	3.143		18 000	216.7	0.07565
38,000	392.7	2.996		18 500	216.7	0.06995
39,000	392.7	2.854		19 000	216.7	0.06467
				19 500	216.7	0.05980
40,000	392.7	2.721		20 000	216.7	0.05529
41,000	392.7	2.593				
42,000	392.7	2.475		22 000	218.6	0.04047
43,000	392.7	2.358		24 000	220.6	0.02972
44,000	392.7	2.250		26 000	222.5	0.02188
				28 000	224.5	0.01616
45,000	392.7	2.141		30 000	226.5	0.01197
46,000	392.7	2.043				
47,000	392.7	1.950		32 000	228.5	0.00889
48,000	392.7	1.857				
49,000	392.7	1.768				
50,000	392.7	1.690				
51,000	392.7	1.611				
52,000	392.7	1.532				
53,000	392.7	1.464				
54,000	392.7	1.395				
55,000	392.7	1.331				
56,000	392.7	1.267				
57,000	392.7	1.208				
58,000	392.7	1.154				
59,000	392.7	1.100				
60,000	392.7	1.046				
61,000	392.7	0.997				
62,000	392.7	0.953				
63,000	392.7	0.909				
64,000	392.7	0.864				
65,000	392.7	0.825				

APPENDIX 34.A
Representative Thermal Conductivity[a,b]
(at 32°F (0°C) unless specified otherwise)

material	$\dfrac{\text{Btu-ft}}{\text{hr-ft}^2\text{-°F}}$	$\dfrac{\text{W}}{\text{m·K}}$
air	0.014	0.024
aluminum	117	202
asbestos	0.087	0.15
brass	56	97
brick, fire clay (400°F)	0.58	1.0
concrete	0.5	0.9
copper	224	388
cork	0.025	0.043
fiberglass	0.03	0.05
glass	0.63	1.1
glass, Pyrex™	0.68	1.2
gold (68°F)	169	292
hydrogen (100°F)	0.11	0.19
ice	1.3	2.2
iron, cast (4% C, 68°F)	30	52
iron, pure	36	62
lead (70°F)	20	35
mercury	4.83	8.36
nickel	34.4	59.5
oxygen	0.016	0.028
rubber, soft	0.10	0.17
silver	242	419
steel (1% C)	27	47
tungsten	92	160
water	0.32	0.55
zinc	65	110

(Multiply Btu-ft/hr-ft²-°F by 12 to get Btu-in/hr-ft²-°F.)

(Multiply Btu-ft/hr-ft²-°F by 1.73073 to get W/m·K.)

(Multiply Btu-ft/hr-ft²-°F by 4.1365×10^{-3} to get cal·cm/s·cm²·°C.)

[a] Values of thermal conductivity are typically accurate to only ±20%, although in some cases the error may be as small as ±10%.

[b] Values are compiled from a variety of sources.

APPENDIX 34.B
Properties of Metals and Alloys[a,b]

metal	thermal conductivity, k $\left(\dfrac{\text{Btu-ft}}{\text{hr-ft}^2\text{-°F}}\right)$				c_p $\left(\dfrac{\text{Btu}}{\text{lbm-°F}}\right)$	ρ $\left(\dfrac{\text{lbm}}{\text{ft}^3}\right)$
	32°F (0°C)	212°F (100°C)	572°F (300°C)	932°F (500°C)	32°F (0°C)	32°F (0°C)
aluminum alloy	92	104	–	–	–	–
aluminum, pure	117	119	133	155	0.208	169
brass (70% Cu, 30% Zn)	58	60	66	–	0.092	532
bronze (75% Cu, 25% Sn)	15	–	–	–	0.082	540
copper, pure	224	218	212	207	0.091	558
iron, cast, alloy	30	28.3	27.0	–	0.10	455
iron, cast, plain	33	31.8	27.7	24.8	0.11	474
iron, pure	35.8	36.6	–	–	0.104	491
lead	20.1	19	18	–	0.030	705
magnesium	91	92	–	–	0.232	109
nickel/chrome	7.5	9.2	–	–	–	–
silver	242	238	–	–	0.056	655
steel, carbon (1% C)	26.5	26.0	25.0	22.0	0.11	490
steel, stainless	8.0	9.3	11.0	12.8	0.11	488
tin	36	34	–	–	0.054	456
zinc	65	64	59	–	0.091	446

(Multiply Btu-ft/hr-ft^2-°F by 12 to get Btu-in/hr-ft^2-°F.)
(Multiply Btu-ft/hr-ft^2-°F by 1.7307 to get W/m·K.)
(Multiply Btu-ft/hr-ft^2-°F by 4.1365 $\times 10^{-3}$ to get cal·cm/s·cm^2-°C.)
(Multiply Btu/lbm-°F by 4186.8 to obtain J/kg·K.)
(Multiply lbm/ft^3 by 16.0185 to obtain kg/m^3.)

[a]Values of thermal conductivity are typically accurate to only ±20%, although in some cases the error may be as small as ±10%.

[b]Values are compiled from a variety of sources.

APPENDIX 34.C
Properties of Nonmetals[a,b]

material	average temperature (°F)	k (Btu/hr-ft-°F)	c_p (Btu/lbm-°F)	ρ (lbm/ft³)
asbestos	32	0.087	0.25	36
	392	0.12		
brick, building	70	0.38	0.20	106
brick, fire-clay	392	0.58	0.20	144
	1832	0.95		
brick, Kaolin				
insulating	932	0.15		27
	2102	0.26		
firebrick	392	0.05		19
	1400	0.11		
concrete, stone	70	0.54	0.20	144
with 10% moisture	70	0.70		140
diatomaceous earth				
powdered	100	0.030	0.21	14
	300	0.036		
	600	0.046		
glass, window	70	0.45	0.2	170
glass wool (fine)	20	0.022		
	100	0.031		1.5
	200	0.043		
glass wool (packed)	20	0.016		
	100	0.022		6.0
	200	0.029		
ice	32	1.28	0.46	57
magnesia, 85%, molded pipe	32	0.032		17
covering ($T < 600°F$)	200	0.037		
molded pipe covering,	400	0.051		26
diatomaceous silica	1600	0.088		
sand, dry	68	0.20		95
sand, with 10% water	68	0.60		100
soil, dry	70	0.20	0.44	
soil, wet	70	1.5		
wood, oak				
perpendicular to grain	70	0.12	0.57	51
parallel to grain	70	0.20	0.57	
wood, pine				
perpendicular to grain	70	0.06	0.67	31
parallel to grain	70	0.14	0.67	

(Multiply Btu-ft/hr-ft²-°F by 12 to get Btu-in/hr-ft²-°F.)

(Multiply Btu-ft/hr-ft²-°F by 1.7307 to get W/m·K.)

(Multiply Btu-ft/hr-ft²-°F by 4.1365 ×10⁻³ to get cal·cm/s·cm²·°C.)

(Multiply Btu/lbm-°F by 4186.8 to obtain J/kg·K.)

(Multiply lbm/ft³ by 16.0185 to obtain kg/m³.)

[a] Values of thermal conductivity are typically accurate to only ±20%, although in some cases the error may be as small as ±10%.

[b] Values are compiled from a variety of sources.

APPENDIX 34.D
Transient Heat Flow Charts
(solid spheres of radius r_o)

(The variable r_o used in this chart is the outside radius of the sphere,
not the characteristic length.)

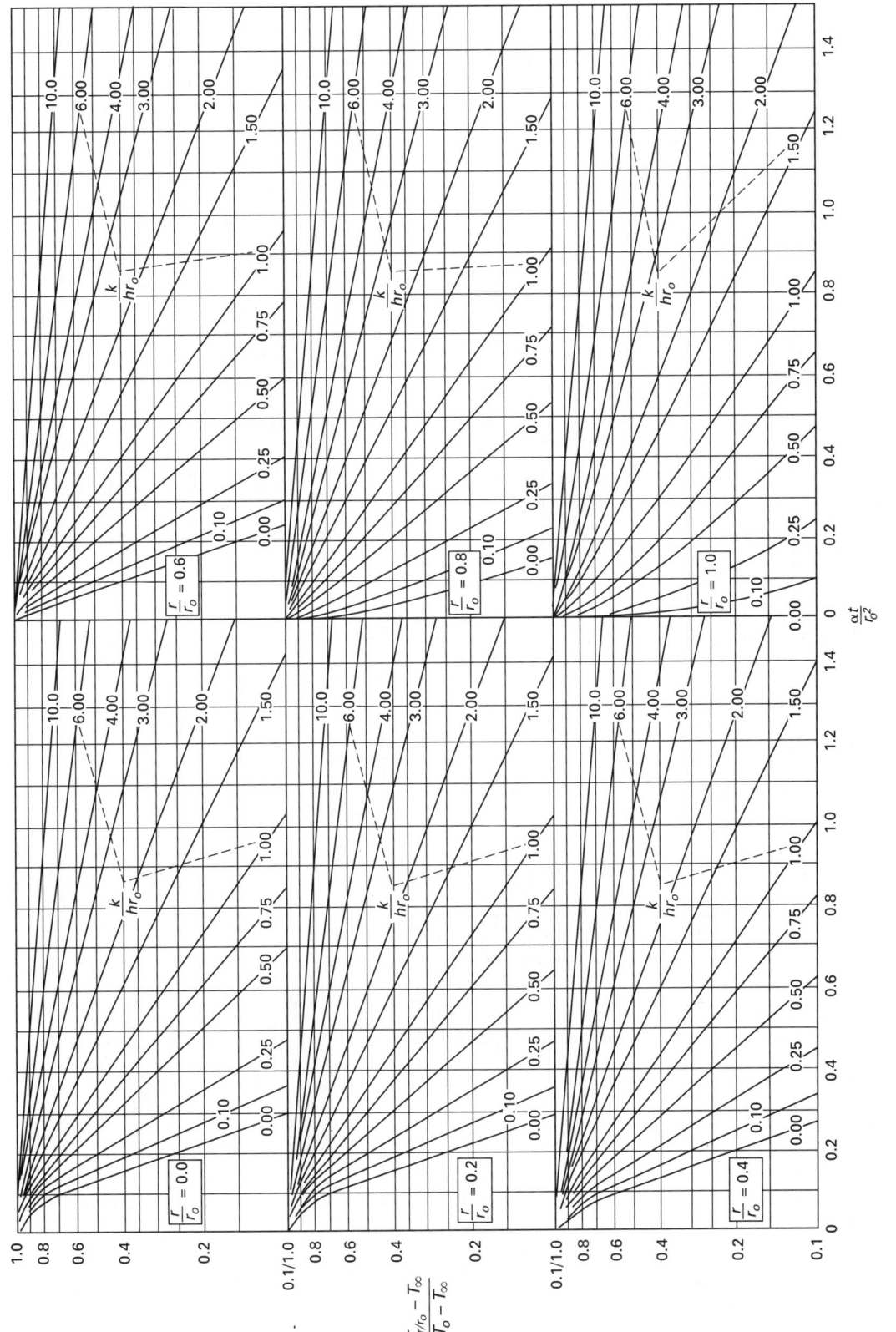

APPENDIX 34.E
Transient Heat Flow Charts
(infinite solid circular cylinders of radius r_o)
(The variable r_o used in this chart is the outside radius of the cylinder, not the characteristic length.)

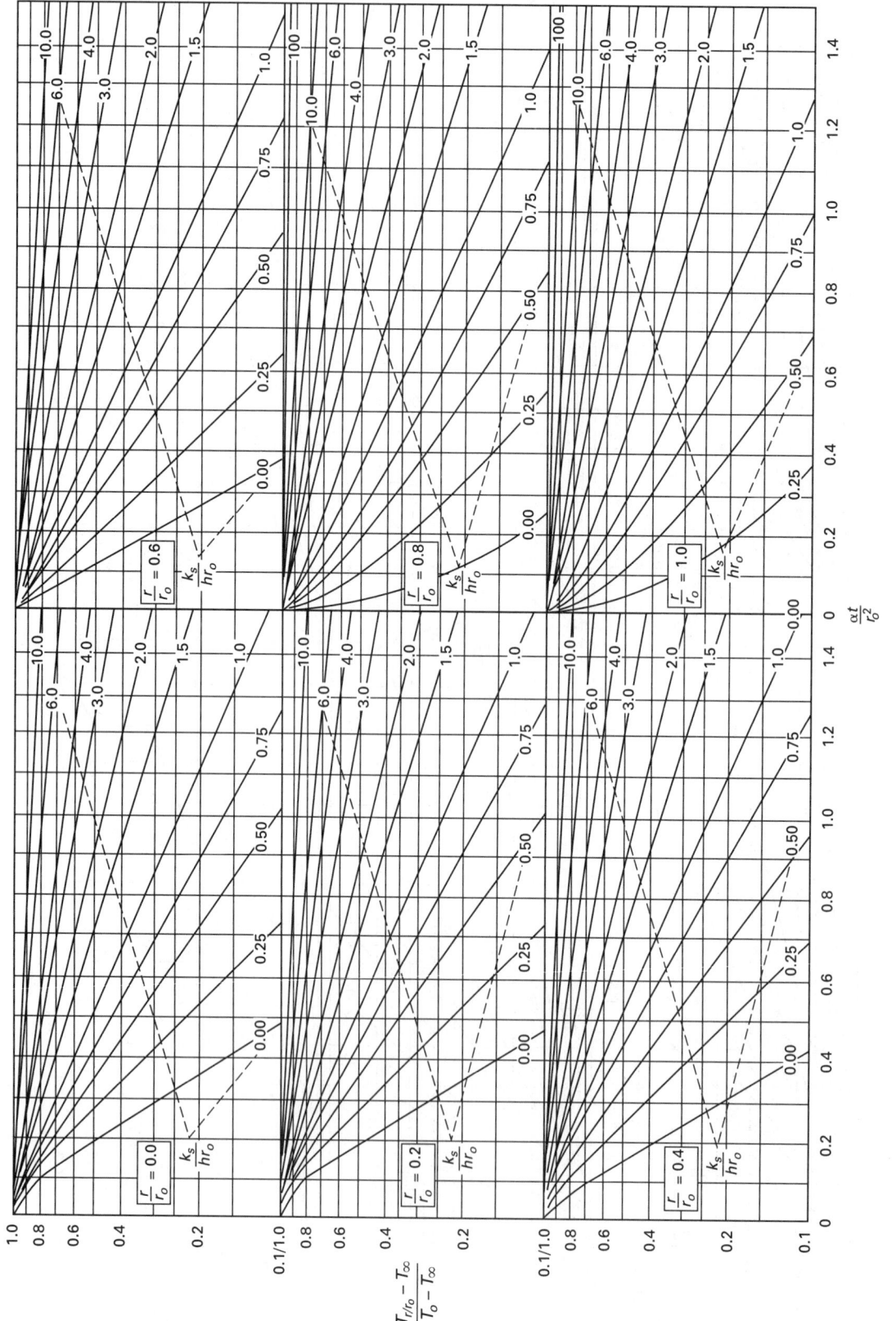

APPENDIX 34.F
Transient Heat Flow Charts
(infinite flat slabs of thickness $2L$)

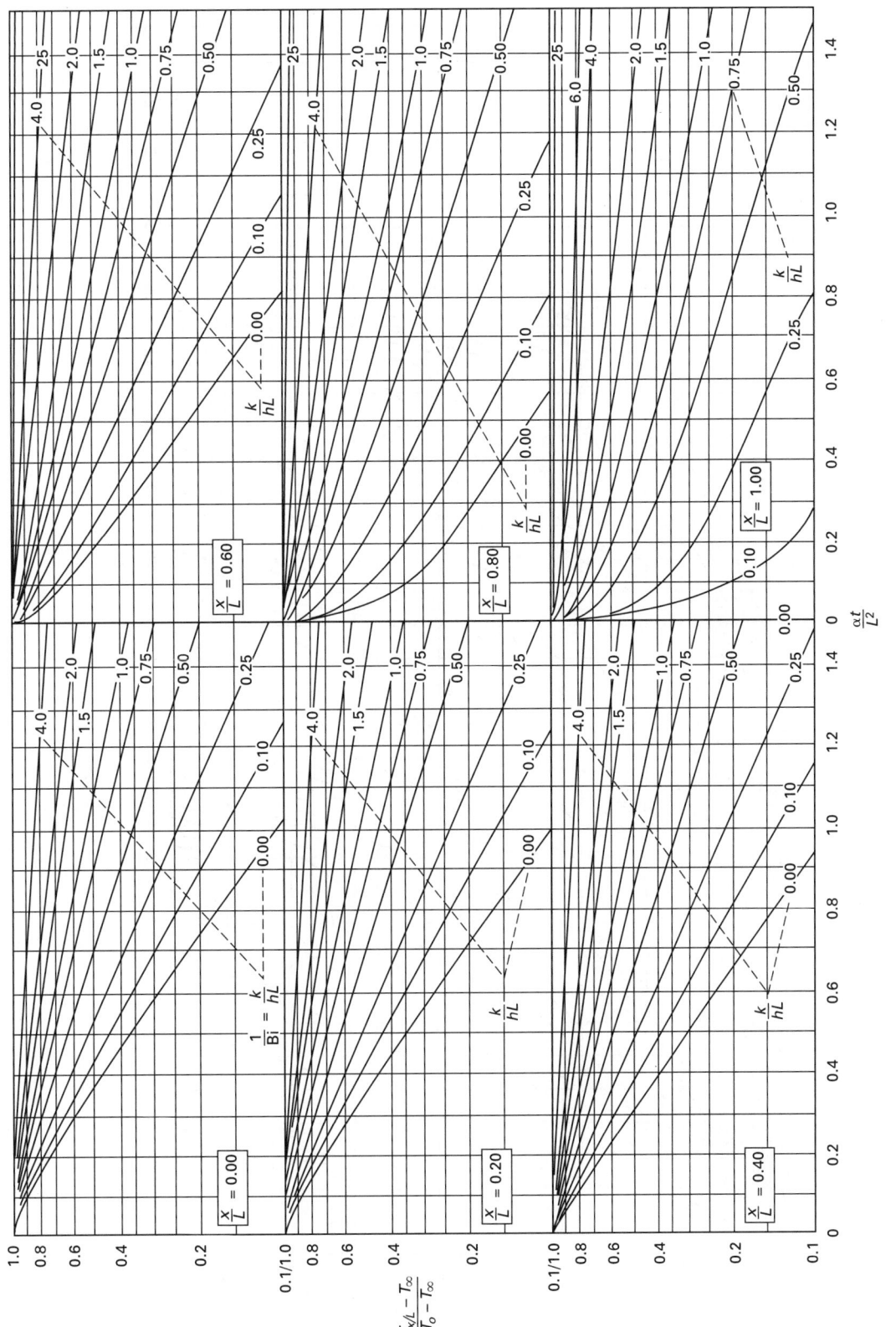

APPENDIX 34.G
Heisler Transient Heat Flow Chart
(temperature at center of a sphere of radius r_o)

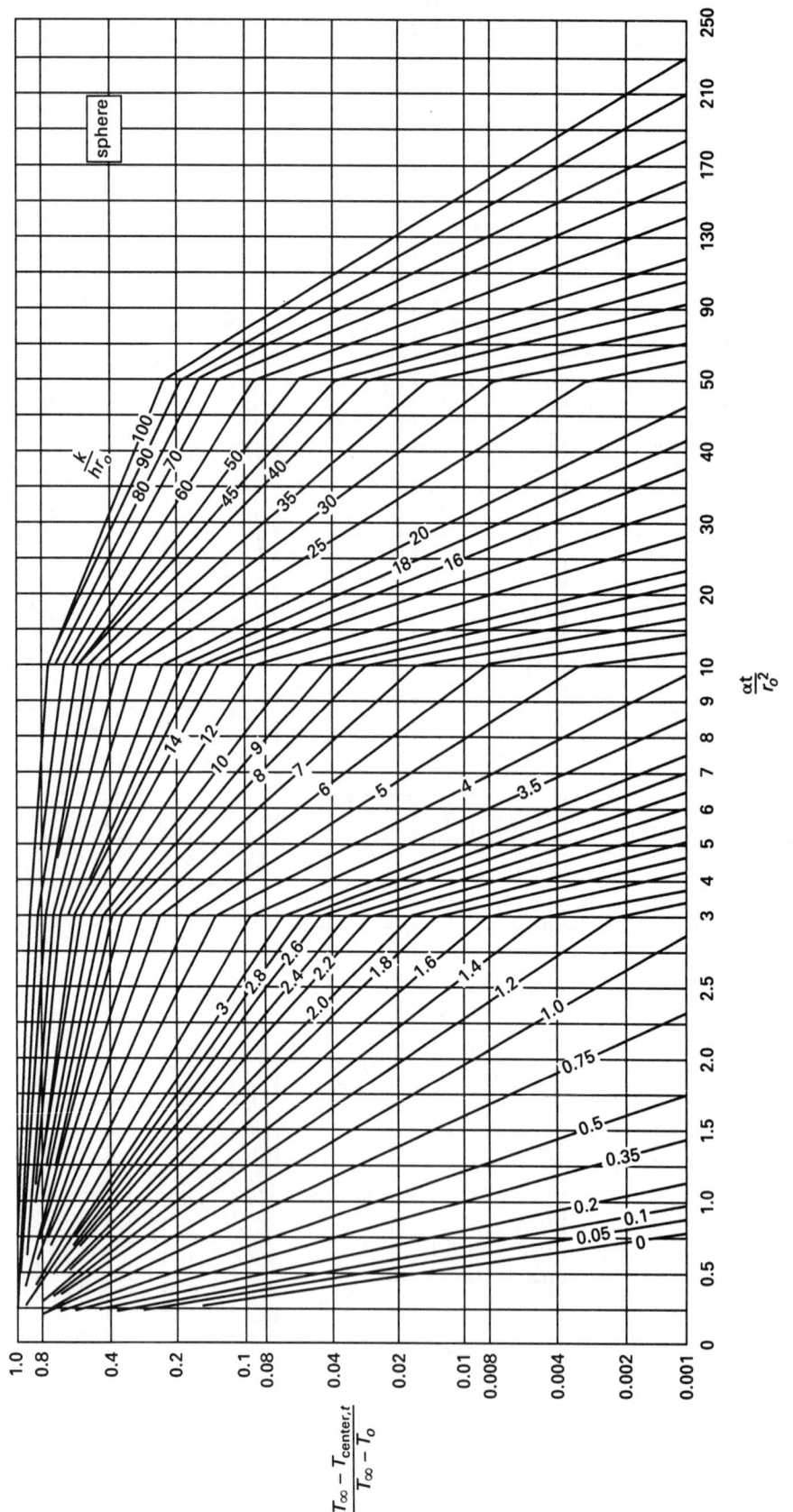

APPENDIX 34.H
Heisler Transient Heat Flow Chart
(temperature at center of infinite cylinder of radius r_o)

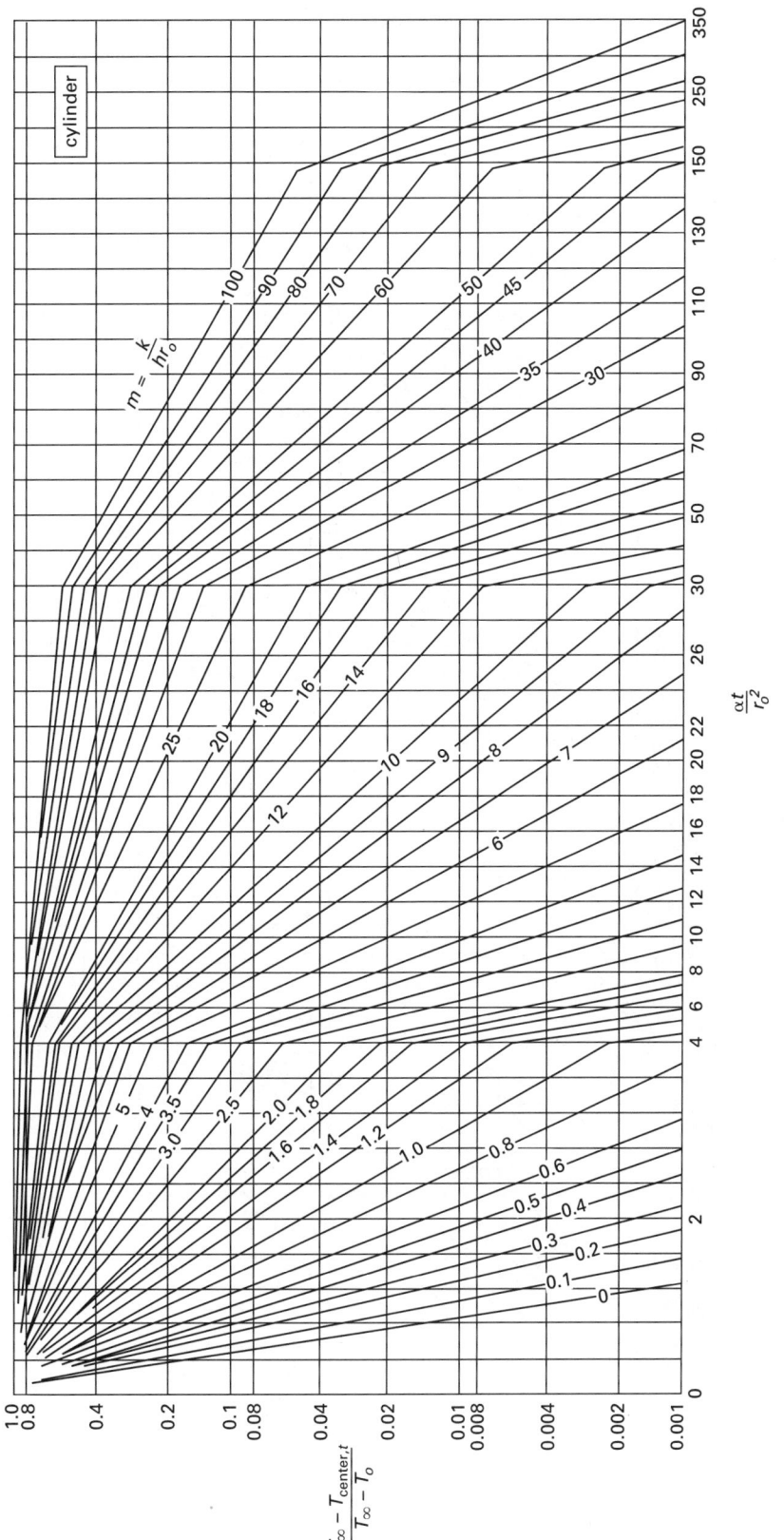

APPENDIX 34.I
Heisler Transient Heat Flow Chart
(temperature at center of infinite slab of thickness 2L)

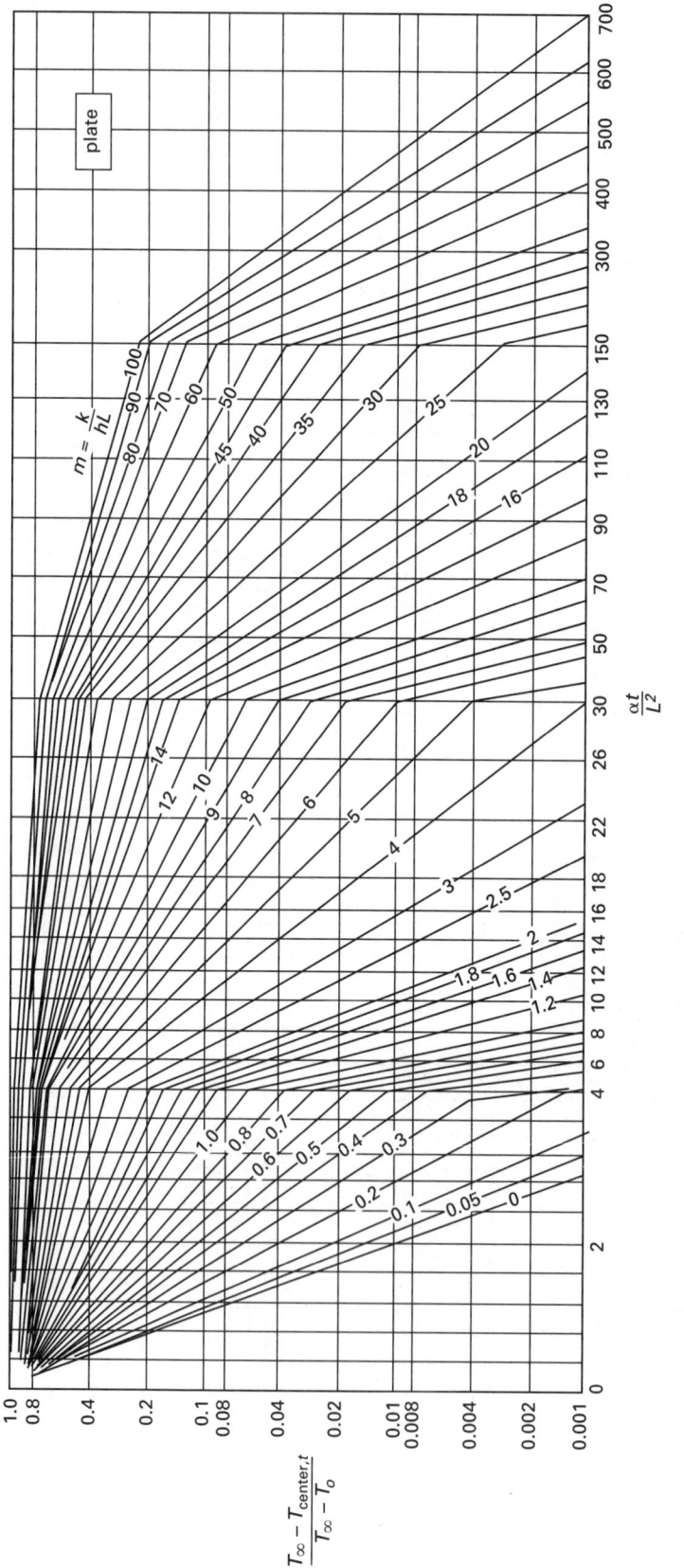

APPENDIX 35.A
Properties of Saturated Water
(English units)

T (°F)	ρ (lbm/ft^3)	c_p (Btu/lbm-°F)	μ (lbm/ft-sec)	ν (ft^2/sec)	k (Btu/hr-ft-°F)	Pr	β (1/°F)	$\dfrac{g\beta\rho^2}{\mu^2}$ (1/ft^3-°F)
32	62.4	1.01	1.20×10^{-3}	1.93×10^{-5}	0.319	13.7	-0.37×10^{-4}	
40	62.4	1.00	1.04×10^{-3}	1.67×10^{-5}	0.325	11.6	0.20×10^{-4}	2.3×10^6
50	62.4	1.00	0.88×10^{-3}	1.40×10^{-5}	0.332	9.55	0.49×10^{-4}	8.0×10^6
60	62.3	0.999	0.76×10^{-3}	1.22×10^{-5}	0.340	8.03	0.85×10^{-4}	18.4×10^6
70	62.3	0.998	0.658×10^{-3}	1.06×10^{-5}	0.347	6.82	1.2×10^{-4}	34.6×10^6
80	62.2	0.998	0.578×10^{-3}	0.93×10^{-5}	0.353	5.89	1.5×10^{-4}	56.0×10^6
90	62.1	0.997	0.514×10^{-3}	0.825×10^{-5}	0.359	5.13	1.8×10^{-4}	85.0×10^6
100	62.0	0.998	0.458×10^{-3}	0.740×10^{-5}	0.364	4.52	2.0×10^{-4}	118×10^6
150	61.2	1.00	0.292×10^{-3}	0.477×10^{-5}	0.384	2.74	3.1×10^{-4}	440×10^6
200	60.1	1.00	0.205×10^{-3}	0.341×10^{-5}	0.394	1.88	4.0×10^{-4}	1.11×10^9
250	58.8	1.01	0.158×10^{-3}	0.269×10^{-5}	0.396	1.45	4.8×10^{-4}	2.14×10^9
300	57.3	1.03	0.126×10^{-3}	0.220×10^{-5}	0.395	1.18	6.0×10^{-4}	4.00×10^9
350	55.6	1.05	0.105×10^{-3}	0.189×10^{-5}	0.391	1.02	6.9×10^{-4}	6.24×10^9
400	53.6	1.08	0.091×10^{-3}	0.170×10^{-5}	0.381	0.927	8.0×10^{-4}	8.95×10^9
450	51.6	1.12	0.080×10^{-3}	0.155×10^{-5}	0.367	0.876	9.0×10^{-4}	12.1×10^9
500	49.0	1.19	0.071×10^{-3}	0.145×10^{-5}	0.349	0.87	10.0×10^{-4}	15.3×10^9
550	45.9	1.31	0.064×10^{-3}	0.139×10^{-5}	0.325	0.93	11.0×10^{-4}	17.8×10^9
600	42.4	1.51	0.058×10^{-3}	0.137×10^{-5}	0.292	1.09	12.0×10^{-4}	20.6×10^9

(Multiply Btu/lbm-°F by 4187 to obtain J/kg·K.)
(Multiply lbm/sec-ft by 3600 to obtain lbm/hr-ft.)
(Multiply lbm/sec-ft by 1.488 to obtain kg/s·m.)
(Multiply ft^2/sec by 0.0929 to obtain m^2/s.)
(Multiply Btu/hr-ft-°F by 1.730 to obtain W/m·K.)
(Multiply 1/°F by 5/9 to obtain 1/K.)
(Multiply 1/ft^3-°F by 19.611 to obtain 1/m^3·K.)
Used with permission from Frank Kreith, *Principles of Heat Transfer*, HarperCollins, © 1965.

APPENDIX 35.B
Properties of Saturated Water
(SI units)

T (°C)	T (K)	ρ (kg/m^3)	c_p (kJ/kg·K)	μ (kg/m·s)	k (W/m·K)	Pr	β (1/K)	$\dfrac{g\beta\rho^2}{\mu^2}$ (1/K·m^3)
0	273.2	999.6	4.229	1.786×10^{-3}	0.5694	13.3	-0.630×10^{-4}	
15.6	288.8	998.0	4.187	1.131×10^{-3}	0.5884	8.07	1.44×10^{-4}	10.93×10^8
26.7	299.9	996.4	4.183	0.860×10^{-3}	0.6109	5.89	2.34×10^{-4}	30.70×10^8
37.8	311.0	994.7	4.183	0.682×10^{-3}	0.6283	4.51	3.24×10^{-4}	68.0×10^8
65.6	338.8	981.9	4.187	0.432×10^{-3}	0.6629	2.72	5.04×10^{-4}	256.2×10^8
93.3	366.5	962.7	4.229	0.3066×10^{-3}	0.6802	1.91	6.66×10^{-4}	642×10^8
121.1	394.3	943.5	4.271	0.2381×10^{-3}	0.6836	1.49	8.46×10^{-4}	1300×10^8
148.9	422.1	917.9	4.312	0.1935×10^{-3}	0.6836	1.22	10.08×10^{-4}	2231×10^8
204.4	477.6	858.6	4.522	0.1384×10^{-3}	0.6611	0.950	14.04×10^{-4}	5308×10^8
260.0	533.2	784.9	4.982	0.1042×10^{-3}	0.6040	0.859	19.8×10^{-4}	$11\,030\times10^8$
315.6	588.8	679.2	6.322	0.0862×10^{-3}	0.5071	1.07	31.5×10^{-4}	$19\,260\times10^8$

(Multiply lbm/ft^3 by 16.0185 to obtain kg/m^3.)
(Multiply Btu/lbm-°F by 4187 to obtain J/kg·K.)
(Multiply lbm/sec-ft by 3600 to obtain lbm/hr-ft.)
(Multiply lbm/sec-ft by 1.488 to obtain kg/s·m.)
(Multiply ft^2/sec ty 0.0929 to obtain m^2/s.)
(Multiply Btu/hr-ft-°F by 1.730 to obtain W/m·K.)
(Multiply 1/°F by 5/9 to obtain 1/K.)
(Multiply 1/ft^3-°F by 19.611 to obtain 1/m^3·K.)
Used with permission from Prentice-Hall, Inc.

APPENDIX 35.C
Properties of Atmospheric Air[a]
(English units)

T (°F)	ρ (lbm/ft^3)	c_p (Btu/lbm-°F)	μ (lbm/ft-sec)	ν (ft^2/sec)	k (Btu/hr-ft-°F)	Pr	β (1/°F)	$\dfrac{g\beta\rho^2}{\mu^2}$ (1/ft^3-°F)
0	0.086	0.239	1.110×10^{-5}	0.130×10^{-3}	0.0133	0.73	2.18×10^{-3}	4.2×10^6
32	0.081	0.240	1.165×10^{-5}	0.145×10^{-3}	0.0140	0.72	2.03×10^{-3}	3.16×10^6
100	0.071	0.240	1.285×10^{-5}	0.180×10^{-3}	0.0154	0.72	1.79×10^{-3}	1.76×10^6
200	0.060	0.241	1.440×10^{-5}	0.239×10^{-3}	0.0174	0.72	1.52×10^{-3}	0.850×10^6
300	0.052	0.243	1.610×10^{-5}	0.306×10^{-3}	0.0193	0.71	1.32×10^{-3}	0.444×10^6
400	0.046	0.245	1.75×10^{-5}	0.378×10^{-3}	0.0212	0.689	1.16×10^{-3}	0.258×10^6
500	0.0412	0.247	1.890×10^{-5}	0.455×10^{-3}	0.0231	0.683	1.04×10^{-3}	0.159×10^6
600	0.0373	0.250	2.000×10^{-5}	0.540×10^{-3}	0.0250	0.685	0.943×10^{-3}	0.106×10^6
700	0.0341	0.253	2.14×10^{-5}	0.625×10^{-3}	0.0268	0.690	0.862×10^{-3}	70.4×10^3
800	0.0314	0.256	2.25×10^{-5}	0.717×10^{-3}	0.0286	0.697	0.794×10^{-3}	49.8×10^3
900	0.0291	0.259	2.36×10^{-5}	0.815×10^{-3}	0.0303	0.705	0.735×10^{-3}	36.0×10^3
1000	0.0271	0.262	2.47×10^{-5}	0.917×10^{-3}	0.0319	0.713	0.685×10^{-3}	26.5×10^3
1500	0.0202	0.276	3.00×10^{-5}	1.47×10^{-3}	0.0400	0.739	0.510×10^{-3}	7.45×10^3
2000	0.0161	0.286	3.54×10^{-5}	2.14×10^{-3}	0.0471	0.753	0.406×10^{-3}	2.84×10^3
2500	0.0133	0.292	3.69×10^{-5}	2.80×10^{-3}	0.051	0.763	0.338×10^{-3}	1.41×10^3
3000	0.0114	0.297	3.86×10^{-5}	3.39×10^{-3}	0.054	0.765	0.289×10^{-3}	0.815×10^3

(Multiply lbm/ft^3 by 16.0185 to obtain kg/m^3.)
(Multiply Btu/lbm-°F by 4187 to obtain J/kg·K.)
(Multiply lbm/sec-ft by 3600 to obtain lbm/hr-ft.)
(Multiply lbm/sec-ft by 1.488 to obtain kg/s·m.)
(Multiply ft^2/sec by 0.0929 to obtain m^2/s.)
(Multiply Btu/hr-ft-°F by 1.730 to obtain W/m·K.)
(Multiply 1/°F by 5/9 to obtain 1/K.)
(Multiply 1/ft^3-°F by 19.611 to obtain 1/m^3·K.)
[a]μ, k, c_p, and Pr do not greatly depend on pressure and may be used over a wide range of pressures.
Used with permission from Frank Kreith, *Principles of Heat Transfer*, HarperCollins, © 1965.

APPENDIX 35.D
Properties of Atmospheric Air[a]
(SI units)

T (°C)	T (K)	ρ (kg/m^3)	c_p (kJ/kg·K)	μ (kg/m·s)	k (W/m·K)	Pr	β (1/K)	$\dfrac{g\beta\rho^2}{\mu^2}$ (1/K·m^3)
−17.8	255.4	1.379	1.0048	1.62×10^{-5}	0.02250	0.720	3.92×10^{-3}	2.79×10^8
0	273.2	1.293	1.0048	1.72×10^{-5}	0.02423	0.715	3.65×10^{-3}	2.04×10^8
10.0	283.2	1.246	1.0048	1.78×10^{-5}	0.02492	0.713	3.53×10^{-3}	1.72×10^8
37.8	311.0	1.137	1.0048	1.90×10^{-5}	0.02700	0.705	3.22×10^{-3}	1.12×10^8
65.6	338.8	1.043	1.0090	2.03×10^{-5}	0.02925	0.702	2.95×10^{-3}	0.775×10^8
93.3	366.5	0.964	1.0090	2.15×10^{-5}	0.03115	0.694	2.74×10^{-3}	0.534×10^8
121.1	394.3	0.895	1.0132	2.27×10^{-5}	0.03323	0.692	2.54×10^{-3}	0.386×10^8
148.9	422.1	0.838	1.0174	2.37×10^{-5}	0.03531	0.689	2.38×10^{-3}	0.289×10^8
176.7	449.9	0.785	1.0216	2.50×10^{-5}	0.03721	0.687	2.21×10^{-3}	0.214×10^8
204.4	477.6	0.740	1.0258	2.60×10^{-5}	0.03894	0.686	2.09×10^{-3}	0.168×10^8
232.2	505.4	0.700	1.0300	2.71×10^{-5}	0.04084	0.684	1.98×10^{-3}	0.130×10^8
260.0	533.2	0.662	1.0341	2.80×10^{-5}	0.04258	0.680	1.87×10^{-3}	0.104×10^8

(Multiply lbm/ft^3 by 16.0185 to obtain kg/m^3.)
(Multiply Btu/lbm-°F by 4187 to obtain J/kg·K.)
(Multiply lbm/sec-ft by 3600 to obtain lbm/hr-ft.)
(Multiply lbm/sec-ft by 1.488 to obtain kg/s·m.)
(Multiply ft^2/sec by 0.0929 to obtain m^2/s.)
(Multiply Btu/hr-ft-°F by 1.730 to obtain W/m·K.)
(Multiply 1/°F by 5/9 to obtain 1/K.)
(Multiply 1/ft^3-°F by 19.611 to obtain 1/m^3·K.)
[a]μ, k, c_p, and Pr do not greatly depend on pressure and may be used over a wide range of pressures.
Used with permission from Prentice-Hall, Inc.

APPENDIX 35.E
Properties of Saturated Steam at One Atmosphere[a]
(English units)

T (°F)	ρ (lbm/ft^3)	c_p (Btu/lbm-°F)	μ (lbm/ft-sec)	ν (ft^2/sec)	k (Btu/hr-ft-°F)	Pr	β (1/°F)	$\dfrac{g\beta\rho^2}{\mu^2}$ (1/ft^3-°F)
212	0.0372	0.451	0.870×10^{-5}	0.234×10^{-3}	0.0145	0.96	1.49×10^{-3}	0.877×10^6
300	0.0328	0.456	1.00×10^{-5}	0.303×10^{-3}	0.0171	0.95	1.32×10^{-3}	0.459×10^6
400	0.0288	0.462	1.13×10^{-5}	0.395×10^{-3}	0.0200	0.94	1.16×10^{-3}	0.243×10^6
500	0.0258	0.470	1.265×10^{-5}	0.490×10^{-3}	0.0228	0.94	1.04×10^{-3}	0.139×10^6
600	0.0233	0.477	1.420×10^{-5}	0.610×10^{-3}	0.0257	0.94	0.943×10^{-3}	82×10^3
700	0.0213	0.485	1.555×10^{-5}	0.725×10^{-3}	0.0288	0.93	0.862×10^{-3}	52.1×10^3
800	0.0196	0.494	1.70×10^{-5}	0.855×10^{-3}	0.0321	0.92	0.794×10^{-3}	34.0×10^3
900	0.0181	0.50	1.810×10^{-5}	0.987×10^{-3}	0.0355	0.91	0.735×10^{-3}	23.6×10^3
1000	0.0169	0.51	1.920×10^{-5}	1.13×10^{-3}	0.0388	0.91	0.685×10^{-3}	17.1×10^3
1200	0.0149	0.53	2.14×10^{-5}	1.44×10^{-3}	0.0457	0.88	0.603×10^{-3}	9.4×10^3
1400	0.0133	0.55	2.36×10^{-5}	1.78×10^{-3}	0.053	0.87	0.537×10^{-3}	5.49×10^3
1600	0.0120	0.56	2.58×10^{-5}	2.14×10^{-3}	0.061	0.87	0.485×10^{-3}	3.38×10^3
1800	0.0109	0.58	2.81×10^{-5}	2.58×10^{-3}	0.068	0.87	0.442×10^{-3}	2.14×10^3
2000	0.0100	0.60	3.03×10^{-5}	3.03×10^{-3}	0.076	0.86	0.406×10^{-3}	1.43×10^3
2500	0.0083	0.64	3.58×10^{-5}	4.30×10^{-3}	0.096	0.86	0.338×10^{-3}	0.603×10^3
3000	0.0071	0.67	4.00×10^{-5}	5.75×10^{-3}	0.114	0.86	0.289×10^{-3}	0.293×10^3

(Multiply lbm/ft^3 by 16.0185 to obtain kg/m^3.)
(Multiply Btu/lbm-°F by 4187 to obtain J/kg·K.)
(Multiply lbm/sec-ft by 3600 to obtain lbm/hr-ft.)
(Multiply lbm/sec-ft by 1.488 to obtain kg/s·m.)
(Multiply ft^2/sec by 0.0929 to obtain m^2/s.)
(Multiply Btu/hr-ft-°F by 1.730 to obtain W/m·K.)
(Multiply 1/°F by 5/9 to obtain 1/K.)
(Multiply 1/ft^3-°F by 19.611 to obtain 1/m^3·K.)
[a]μ, k, c_p, and Pr do not greatly depend on pressure and may be used over a wide range of pressures.
Used with permission from Frank Kreith, *Principles of Heat Transfer*, HarperCollins, © 1965.

APPENDIX 35.F
Properties of Saturated Steam at One Atmosphere[a]
(SI units)

T (°C)	T (K)	ρ (kg/m^3)	c_p (kJ/kg·K)	μ (kg/m·s)	k (W/m·K)	Pr	β (1/K)	$\dfrac{g\beta\rho^2}{\mu^2}$ (1/K·m^3)
100.0	373.2	0.596	1.888	1.295×10^{-5}	0.02510	0.96	2.68×10^{-3}	0.557×10^8
148.9	422.1	0.525	1.909	1.488×10^{-5}	0.02960	0.95	2.38×10^{-3}	0.292×10^8
204.4	477.6	0.461	1.934	1.682×10^{-5}	0.03462	0.94	2.09×10^{-3}	0.154×10^8
260.0	533.2	0.413	1.968	1.883×10^{-5}	0.03946	0.94	1.87×10^{-3}	0.0883×10^8
315.6	588.8	0.373	1.997	2.113×10^{-5}	0.04448	0.94	1.70×10^{-3}	52.1×10^5
371.1	644.3	0.341	2.030	2.314×10^{-5}	0.04985	0.93	1.55×10^{-3}	33.1×10^5
426.7	699.9	0.314	2.068	2.529×10^{-5}	0.05556	0.92	1.43×10^{-3}	21.6×10^5

(Multiply lbm/ft^3 by 16.0185 to obtain kg/m^3.)
(Multiply Btu/lbm-°F by 4187 to obtain J/kg·K.)
(Multiply lbm/sec-ft by 3600 to obtain lbm/hr-ft.)
(Multiply lbm/sec-ft by 1.488 to obtain kg/s·m.)
(Multiply ft^2/sec by 0.0929 to obtain m^2/s.)
(Multiply Btu/hr-ft-°F by 1.730 to obtain W/m·K.)
(Multiply 1/°F by 5/9 to obtain 1/K.)
(Multiply 1/ft^3-°F by 19.611 to obtain 1/m^3·K.)
[a]μ, k, c_p, and Pr do not greatly depend on pressure and may be used over a wide range of pressures.
Used with permission from Prentice-Hall, Inc.

APPENDIX 36.A
Correction Factor, F_c, for the Logarithmic Mean Temperature Difference
(one shell pass, even number of tube passes)

$$R = \frac{T_{shell,in} - T_{shell,out}}{T_{tube,out} - T_{tube,in}}$$

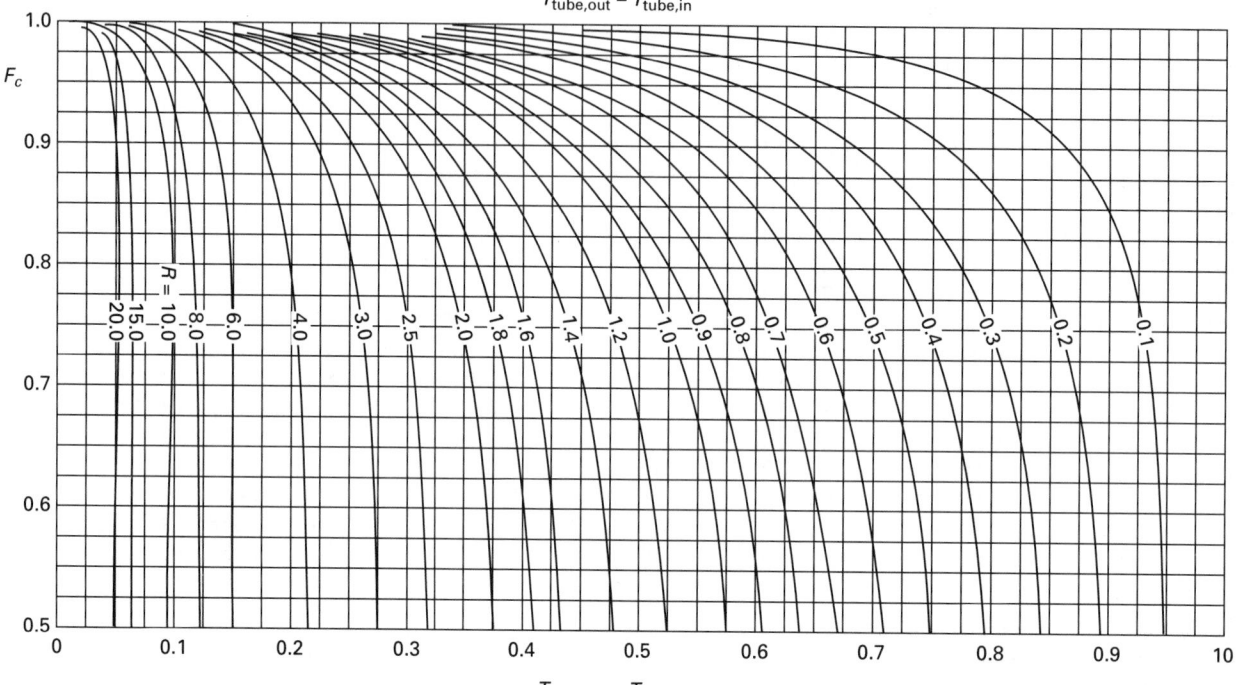

$$\frac{T_{tube,out} - T_{tube,in}}{T_{shell,in} - T_{tube,in}}$$

© 1988 by Tubular Exchanger Manufacturers Association.

APPENDIX 36.B
Correction Factor, F_c, for the Logarithmic Mean Temperature Difference
(two shell passes, multiple of four tube passes)

$$R = \frac{T_{shell,in} - T_{shell,out}}{T_{tube,out} - T_{tube,in}}$$

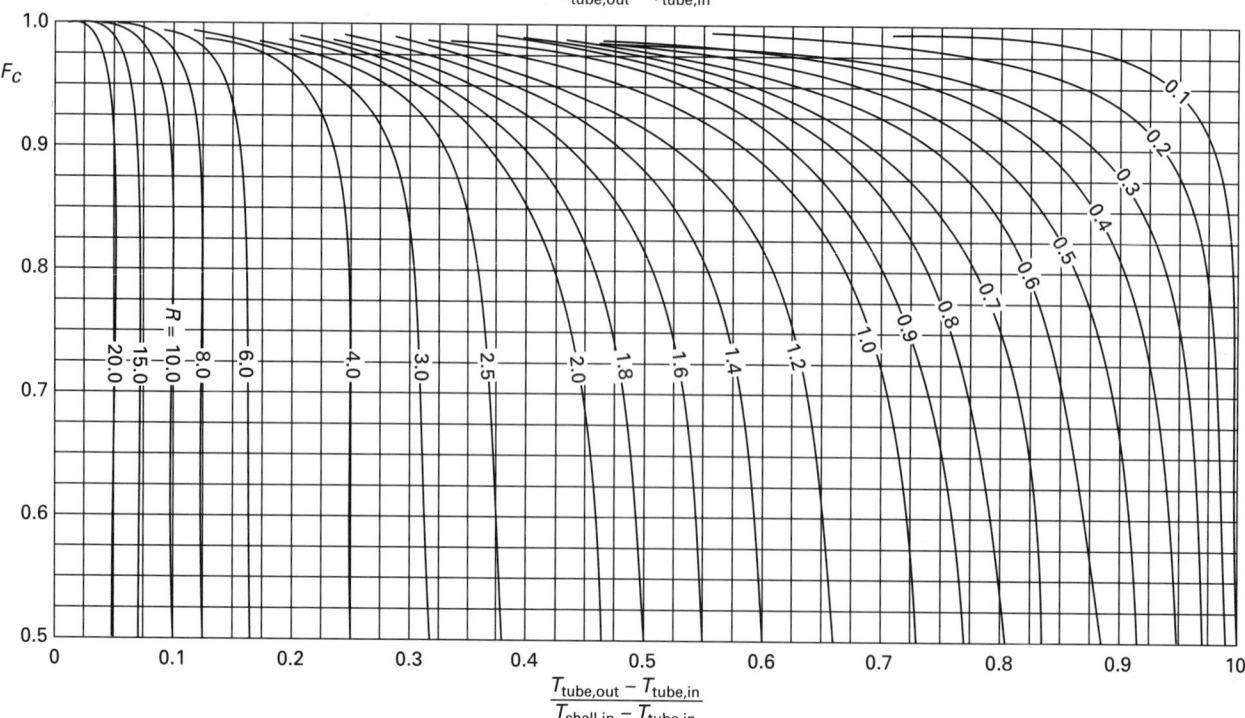

$$\frac{T_{tube,out} - T_{tube,in}}{T_{shell,in} - T_{tube,in}}$$

© 1988 by Tubular Exchanger Manufacturers Association.

APPENDIX 36.C
Characteristics of Birmingham Wire Gage (BWG) Size Tubing

tube OD (in)	BWG (gage)	tube ID (in)	thickness (in)	internal area (in²)	external surface per ft length (ft²/ft)	internal surface per ft length (ft²/ft)
1/4	22	0.194	0.028	0.0296	0.0654	0.0508
1/4	24	0.206	0.022	0.0333	0.0654	0.0539
1/4	26	0.214	0.018	0.0360	0.0654	0.0560
1/4	27	0.218	0.016	0.0373	0.0654	0.0571
3/8	18	0.277	0.049	0.0603	0.0982	0.0725
3/8	20	0.305	0.035	0.0731	0.0982	0.0798
3/8	22	0.319	0.028	0.0799	0.0982	0.0835
3/8	24	0.331	0.022	0.0860	0.0982	0.0867
1/2	16	0.370	0.065	0.1075	0.1309	0.0969
1/2	18	0.402	0.049	0.1269	0.1309	0.1052
1/2	20	0.430	0.035	0.1452	0.1309	0.1126
1/2	22	0.444	0.028	0.1548	0.1309	0.1162
5/8	12	0.407	0.109	0.1301	0.1636	0.1066
5/8	13	0.435	0.095	0.1486	0.1636	0.1139
5/8	14	0.459	0.083	0.1655	0.1636	0.1202
5/8	15	0.481	0.072	0.1817	0.1636	0.1259
5/8	16	0.495	0.065	0.1924	0.1636	0.1296
5/8	17	0.509	0.058	0.2035	0.1636	0.1333
5/8	18	0.527	0.049	0.2181	0.1636	0.1380
5/8	19	0.541	0.042	0.2299	0.1636	0.1416
5/8	20	0.555	0.035	0.2419	0.1636	0.1453
3/4	10	0.482	0.134	0.1825	0.1963	0.1262
3/4	11	0.510	0.120	0.2043	0.1963	0.1335
3/4	12	0.532	0.109	0.2223	0.1963	0.1393
3/4	13	0.560	0.095	0.2463	0.1963	0.1466
3/4	14	0.584	0.083	0.2679	0.1963	0.1529
3/4	15	0.606	0.072	0.2884	0.1963	0.1587
3/4	16	0.620	0.065	0.3019	0.1963	0.1623
3/4	17	0.634	0.058	0.3157	0.1963	0.1660
3/4	18	0.652	0.049	0.3339	0.1963	0.1707
3/4	20	0.680	0.035	0.3632	0.1963	0.1780
1	8	0.670	0.165	0.3526	0.2618	0.1754
1	10	0.732	0.134	0.4208	0.2618	0.1916
1	11	0.760	0.120	0.4536	0.2618	0.1990
1	12	0.782	0.109	0.4803	0.2618	0.2047
1	13	0.810	0.095	0.5153	0.2618	0.2121
1	14	0.834	0.083	0.5463	0.2618	0.2183
1	15	0.856	0.072	0.5755	0.2618	0.2241
1	16	0.870	0.065	0.5945	0.2618	0.2278
1	18	0.902	0.049	0.6390	0.2618	0.2361
1	20	0.930	0.035	0.6793	0.2618	0.2435
1 1/4	7	0.890	0.180	0.6221	0.3272	0.2330
1 1/4	8	0.920	0.165	0.6648	0.3272	0.2409
1 1/4	10	0.982	0.134	0.7574	0.3272	0.2571
1 1/4	11	1.010	0.120	0.8012	0.3272	0.2644
1 1/4	12	1.032	0.109	0.8365	0.3272	0.2702
1 1/4	13	1.060	0.095	0.8825	0.3272	0.2775
1 1/4	14	1.084	0.083	0.9229	0.3272	0.2838
1 1/4	16	1.120	0.065	0.9852	0.3272	0.2932
1 1/4	18	1.152	0.049	1.0423	0.3272	0.3016
1 1/4	20	1.180	0.035	1.0936	0.3272	0.3089
1 1/2	10	1.232	0.134	1.1921	0.3927	0.3225
1 1/2	12	1.282	0.109	1.2908	0.3927	0.3356
1 1/2	14	1.334	0.083	1.3977	0.3927	0.3492
1 1/2	16	1.370	0.065	1.4741	0.3927	0.3587
2	11	1.760	0.120	2.4328	0.5236	0.4608
2	12	1.782	0.109	2.4941	0.5236	0.4665
2	13	1.810	0.095	2.5730	0.5236	0.4739
2	14	1.834	0.083	2.6417	0.5236	0.4801

Support Material

APPENDIX 36.D
Heat Exchanger Effectiveness[a]

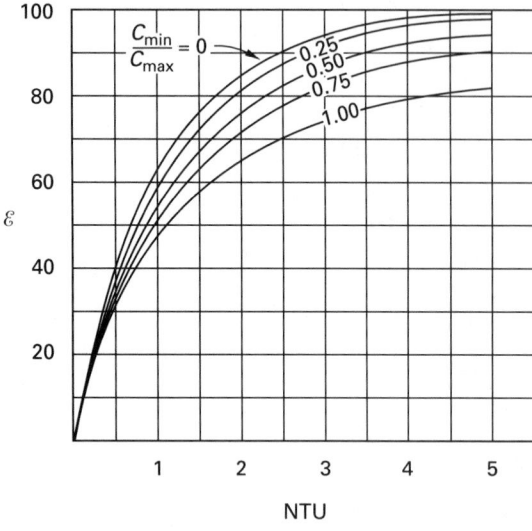

(a) single-pass, shell-and-tube, counterflow

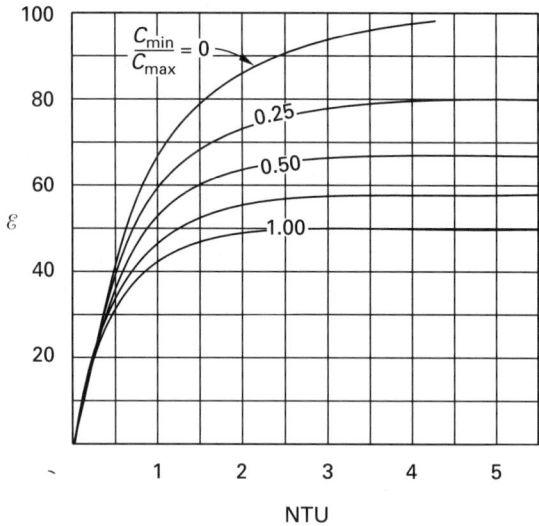

(b) single-pass, shell-and-tube, parallel flow

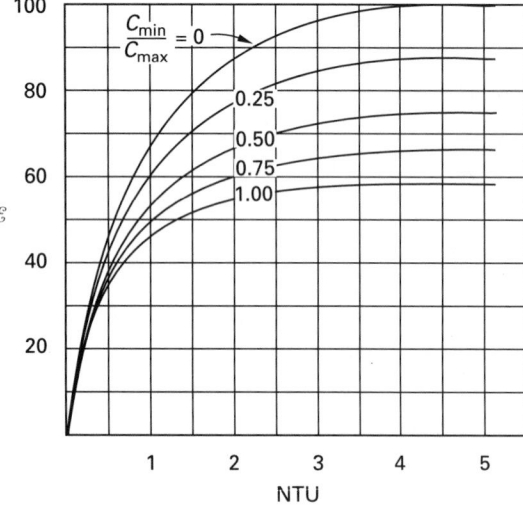

(c) one shell pass, two, four, six, eight . . . tube passes, parallel counterflow[a]

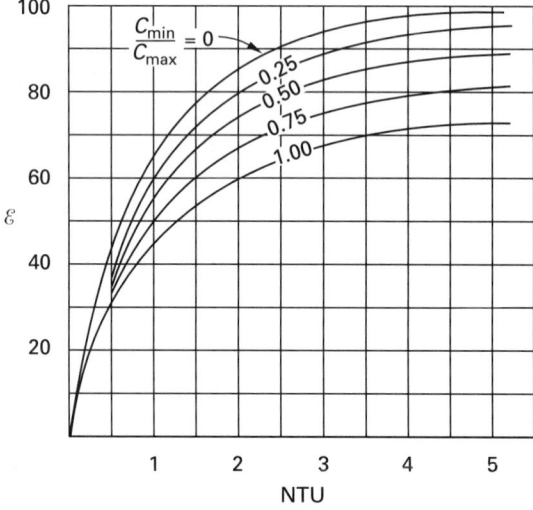

(d) two shell passes, four, eight, or twelve . . . tube passes, multipass counterflow

(continued)

APPENDIX 36.D *(continued)*
Heat Exchanger Effectiveness

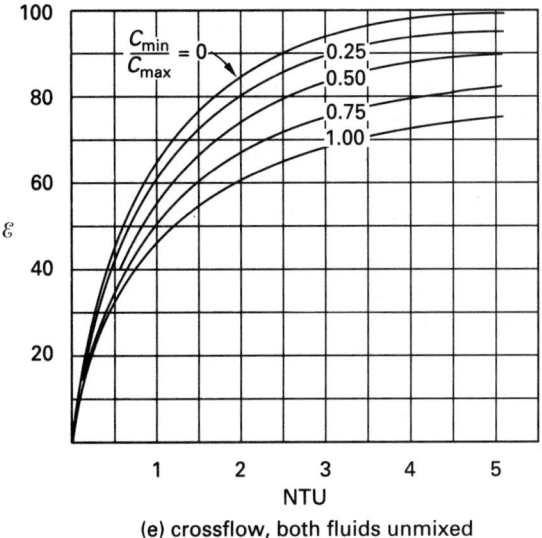

(e) crossflow, both fluids unmixed

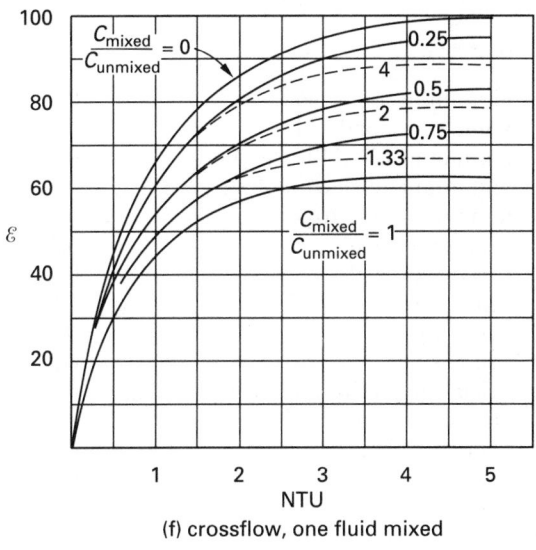

(f) crossflow, one fluid mixed

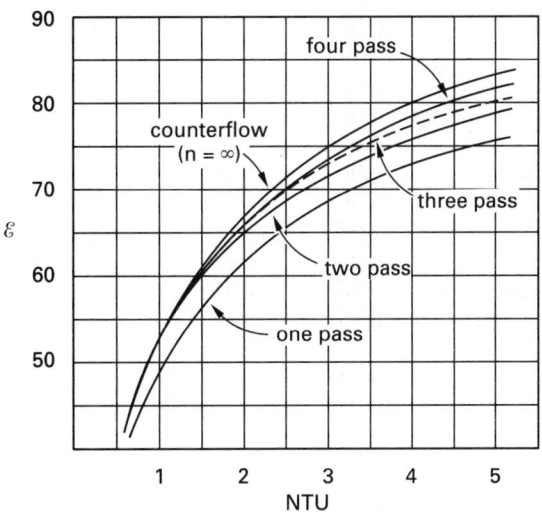

(g) crossflow, multiple passes, both fluids unmixed

[a]Formulas used to generate the curves for one shell pass are exact for two, four, and six tube passes, but produce a small error (approximately 1 to 5%) for higher multiples. The error associated with reading the graphs is generally higher than the error introduced by the formulas.

Support Material

APPENDIX 37.A
Emissivities of Various Surfaces[a]

material	wavelength 9.3 μm average temperature 100°F 38°C	5.4 μm 500°F 260°C	3.6 μm 1000°F 538°C	1.8 μm 2500°F 1371°C	0.6 μm solar
aluminum, polished	0.04	0.05	0.08	0.19	
aluminum, oxidized	0.11	0.12	0.18		
aluminum (24-ST), weathered	0.4	0.32	0.27		
aluminum, surface roofing	0.22				
aluminum, anodized	0.94	0.42	0.60	0.34	
brass, polished	0.10	0.10			
brass, oxidized	0.61				
brick, red	0.93				0.7
brick, fire clay	0.9		0.7	0.75	
brick, silica	0.9		0.75	0.84	
brick, magnesite refractory	0.9			0.4	
chromium, polished	0.08	0.17	0.26	0.40	0.49
copper, polished	0.04	0.05	0.18	0.17	
copper, oxidized	0.87	0.83	0.77		
enamel, white	0.9				
glass	0.9				
ice (at 32°F)	0.97				
iron, polished	0.06	0.08	0.13	0.25	0.45
iron, cast, oxidized	0.63	0.66	0.76		
iron, galvanized, new	0.23			0.42	0.66
iron, galvanized, dirty	0.28			0.90	0.89
iron, oxide	0.96		0.85		0.74
iron, molten				0.3–0.4	
magnesium	0.07	0.13	0.18	0.24	0.30
paper, white	0.95		0.82	0.25	0.28
paint, aluminized lacquer	0.65	0.65			
paint, lacquer, black or white	0.96	0.98			
paint, lampback	0.96	0.97		0.97	0.97
paint, white (ZnO)	0.95		0.91		0.18
paint, enamel, white	0.9				
stainless steel, 18-8, polished	0.15	0.18	0.22		
stainless steel, 18-8, weathered	0.85	0.85	0.85		
steel tube, oxidized		0.8			
steel plate, rough	0.94	0.97	0.98		
tungsten filament	0.03			0.18	0.35 (6000°F)
water	0.96				
wood	0.93				
zinc, polished	0.02	0.03	0.04	0.06	0.46
zinc, galvanized sheet	0.25				

[a] Compiled from various sources.

APPENDIX 38.A
ASHRAE Psychrometric Chart No. 1, Normal Temperature—
Sea Level (32–120°) (customary U.S. units)

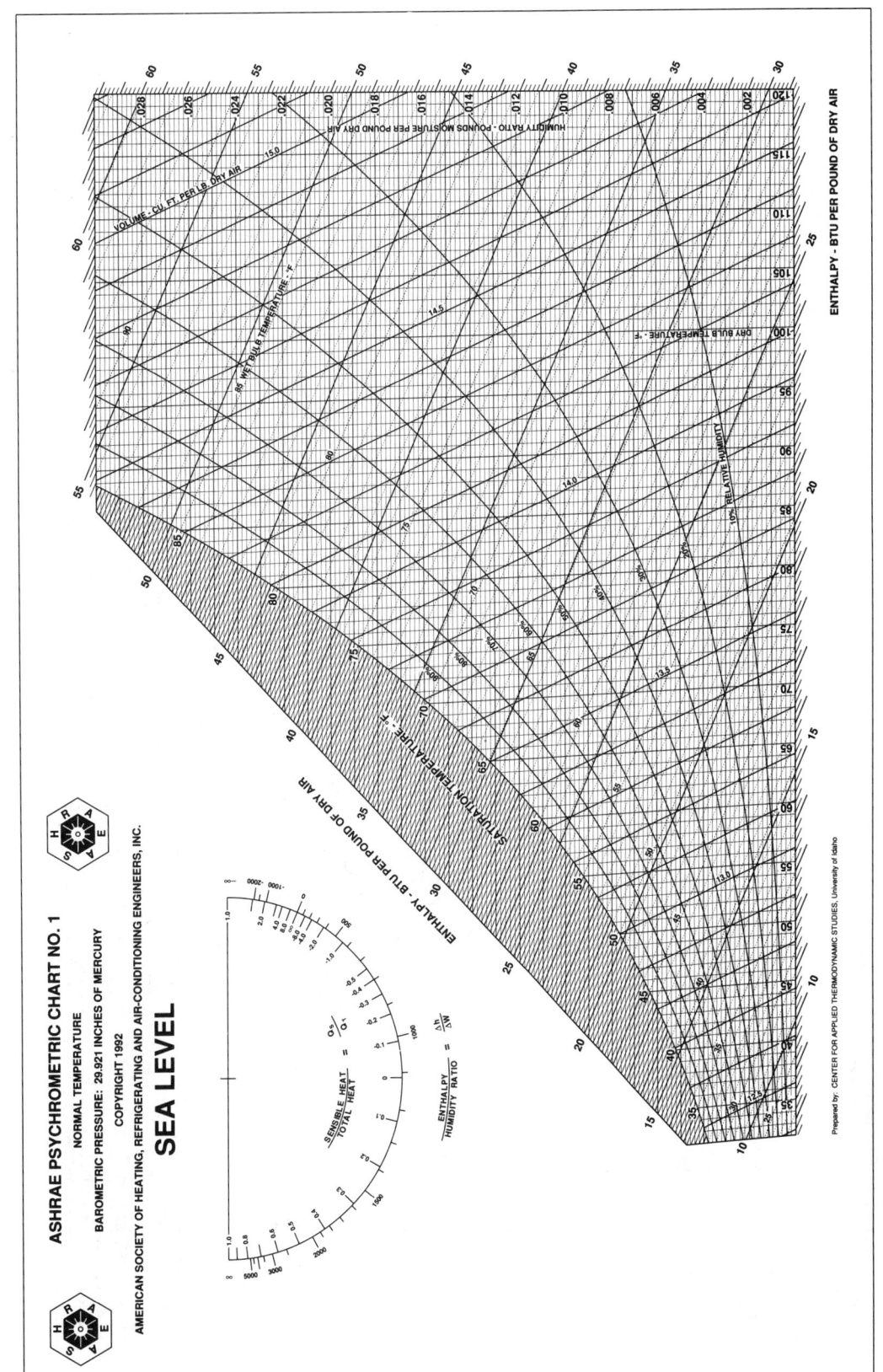

APPENDIX 38.B
ASHRAE Psychrometric Chart No. 1, Normal Temperature—
Sea Level (0–50°) (SI units)

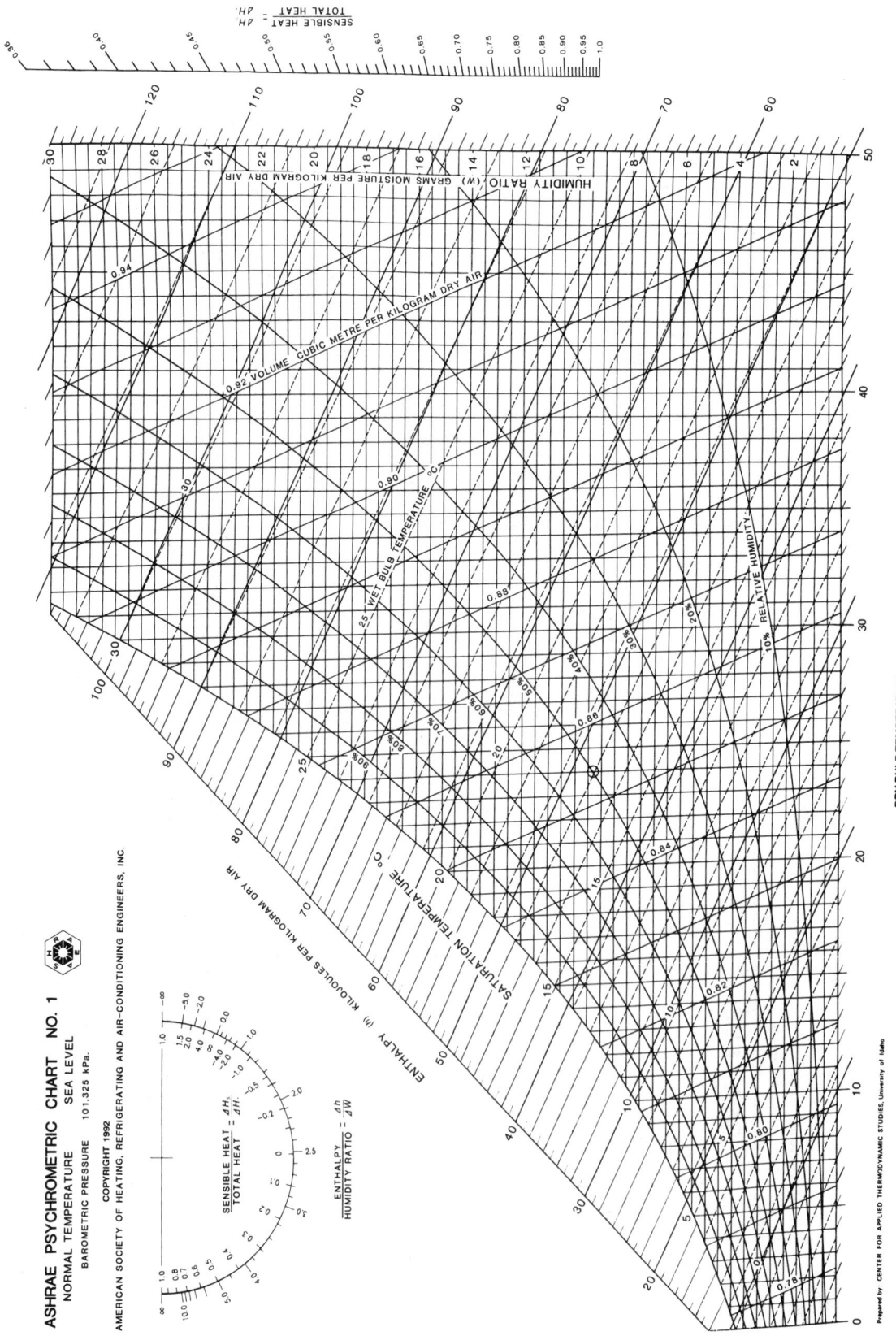

Copyright © 1992 by the American Society of Heating, Refrigerating and Air-Conditioning Engineers, Inc. Used by permission.

APPENDIX 38.C
ASHRAE Psychrometric Chart No. 2, Low Temperature—
Sea Level (−40–50°) (customary U.S. units)

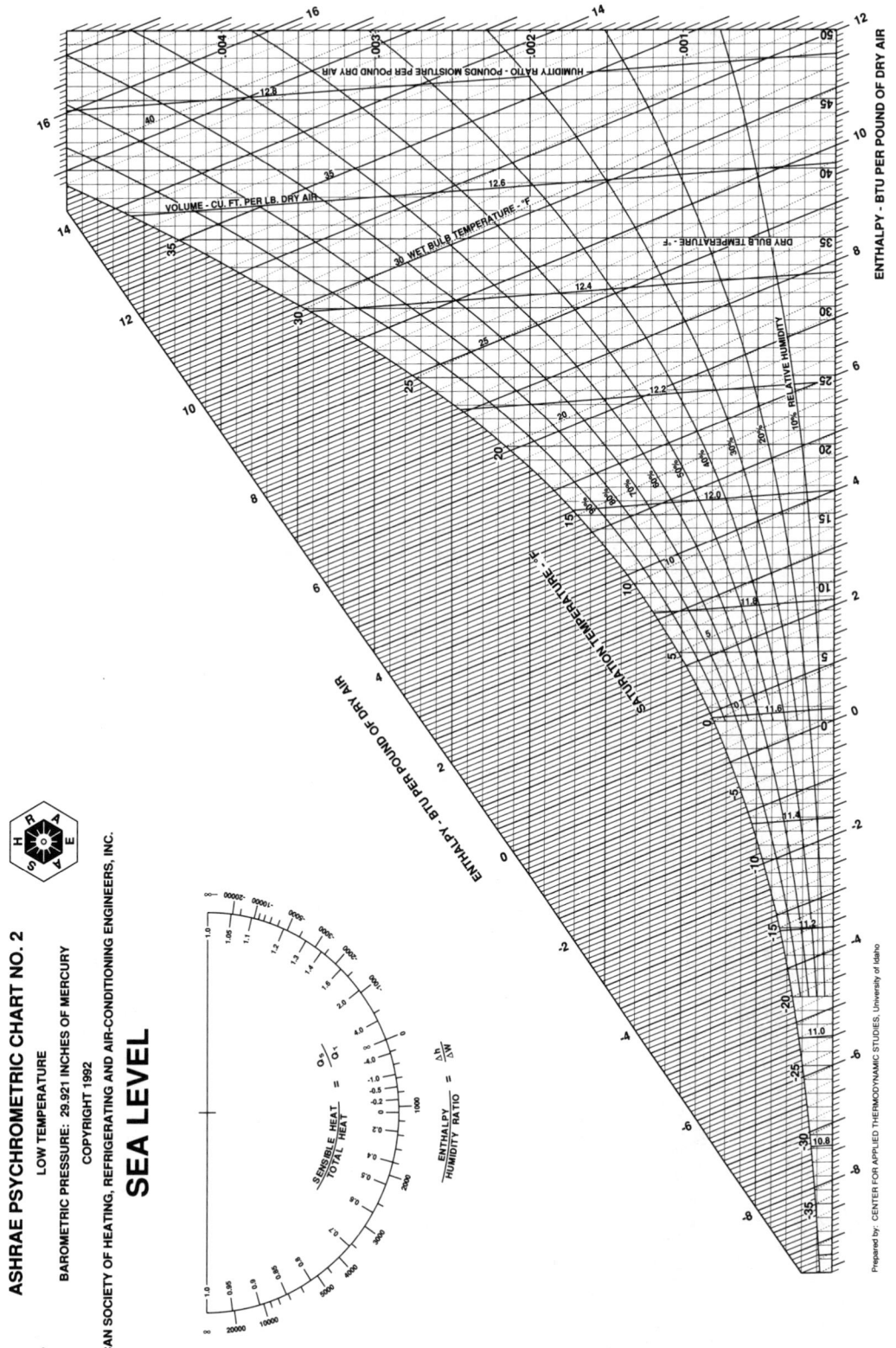

ASHRAE PSYCHROMETRIC CHART NO. 2
LOW TEMPERATURE
BAROMETRIC PRESSURE: 29.921 INCHES OF MERCURY
COPYRIGHT 1992
AMERICAN SOCIETY OF HEATING, REFRIGERATING AND AIR-CONDITIONING ENGINEERS, INC.
SEA LEVEL

Copyright © 1992 by the American Society of Heating, Refrigerating and Air-Conditioning Engineers, Inc. Used by permission.

APPENDIX 38.D
ASHRAE Psychrometric Chart No. 3, High Temperature—
Sea Level (60–250°) (customary U.S. units)

ASHRAE PSYCHROMETRIC CHART NO. 3

HIGH TEMPERATURE

BAROMETRIC PRESSURE: 29.921 INCHES OF MERCURY

COPYRIGHT 1992

AMERICAN SOCIETY OF HEATING, REFRIGERATING AND AIR-CONDITIONING ENGINEERS, INC.

SEA LEVEL

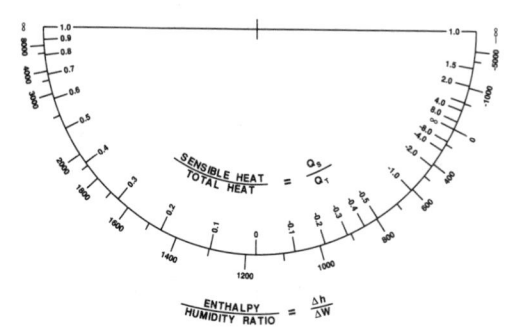

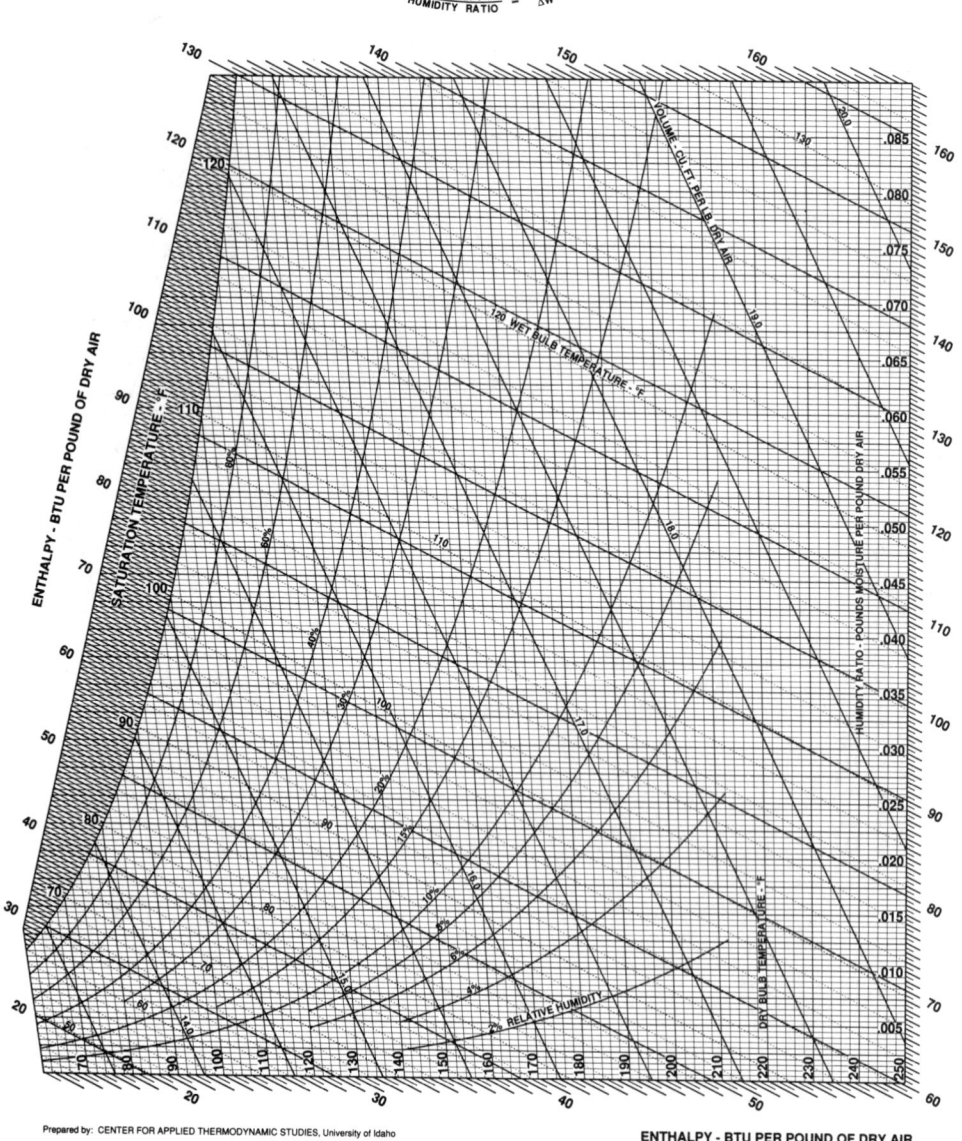

Prepared by: CENTER FOR APPLIED THERMODYNAMIC STUDIES, University of Idaho

ENTHALPY - BTU PER POUND OF DRY AIR

APPENDIX 40.A
Representative Insulating Properties of
Selected Building Materials[a]

usage	description	density lbm/ft³	density lbm/ft²	resistance per inch ft²-°F-hr Btu-in	resistance for thickness listed ft²-°F-hr Btu
building board	asbestos-cement, $^1/_8$ in	120	1.25	0.25	0.03
	gypsum or plaster board, $^1/_2$ in	50	5.00		0.45
	gypsum or plaster board, $^5/_8$ in	50	6.25		0.56
	plywood, $^1/_2$ in	34	1.42	1.25	0.63
	wood fiber board	26		2.38	
	fir/pine sheathing, 0.72 in	32	2.08		0.98
flooring	asphalt tile, $^1/_8$ in	120	1.25		0.04
	carpet and fiber pad				2.08
	carpet and rubber pad				1.23
	ceramic tile, 1 in				0.08
	cork tile, $^1/_8$ in	25	0.26	2.22	0.28
	linoleum, $^1/_8$ in	80	0.83		0.08
	plywood subfloor, $^5/_8$ in	34	1.77		0.78
	rubber/plastic tile, $^1/_8$ in	110	1.15		0.02
	hardwood, $^3/_4$ in	45	2.81		0.68
insulation (blanket/batt)	fibrous mineral wool, including fiberglass	0.8–2.0		3.85	
	wood fiber	3.2–3.6		4.00	
insulation (board/slab)	glass fiber	9.5		4.00	
	acoustical tile, $^1/_2$ in	22.4	0.93		1.19
	acoustical tile, $^3/_4$ in	22.4	1.4		1.78
	cellular glass	9.0		2.5	
	foamed plastic	1.62		3.45	
insulation (loose)	paper/pulp products	2.5–3.5		3.57	
	mineral wool (glass, slag, or rock)	2.0–5.0		3.33	
	expanded vermiculite	7.0		2.08	
roof insulation	all types for use above deck	15.6	1.3 per inch thickness	2.78	
masonry materials (concrete)	cement mortar	116		0.20	
	lightweight aggregates, depending on density	120		0.19	
		80		0.40	
		40		0.86	
		20		1.43	
	oven-dried sand, gravel, or stone aggregate	140		0.11	
	stucco	116		0.20	
plastering materials	cement (sand), $^1/_2$ in	116	4.8	0.20	0.10
	cement (sand), $^3/_4$ in	116	7.2	0.20	0.15
	gypsum plaster, lightweight aggregate, $^1/_2$ in	45	1.88		0.32
	lightweight aggregate on metal lath	45		0.59	
	sand aggregate, $^1/_2$ in	105	4.4	0.18	0.09
	sand aggregate on metal lath, $^3/_4$ in	105	6.6		0.13
roofing	asbestos-cement shingle	120			0.21
	asphalt roll roofing	70			0.15
	asphalt shingles	70			0.44
	built-up roofing, $^3/_8$ in	70	2.2		0.33
	wood shingles	40			0.94

(continued)

APPENDIX 40.A (*continued*)
Representative Insulating Properties of
Selected Building Materials[a]

usage	description	density lbm/ft^3	density lbm/ft^2	resistance per inch ft^2-°F-hr Btu-in	resistance for thickness listed ft^2-°F-hr Btu
siding material (on flat face)	16 in shingles				
	with 7^1/$_2$ in exposure				0.87
	with 12 in exposure				1.19
	siding				
	asbestos-cement, ^{1}/$_4$ in lap				0.21
	asphalt roll siding				0.15
	wood, bevel (^{1}/$_2$ in × 8 in lap)				0.81
	wood, bevel (^{3}/$_4$ in × 10 in lap)				1.05
	structural glass				0.10
woods	hardwoods	45		0.91	
	softwoods	32		1.25	
masonry	brick, common, 4 in	120	40	0.20	0.80
	brick, face, 4 in	130	43	0.11	0.44
	hollow clay tile, 4 in	48			1.11
	stone (lime or sand)	150		0.08	
	concrete blocks (3 oval core)				
	with sand/gravel				
	aggregate, 3 in	76	19		0.40
	6 in	64	32		0.91
	12 in	63	63		1.28
	lightweight aggregate, 3 in	60	15		0.56
	6 in	46	23		1.50
	12 in	43	43		1.89
poured concrete	8 in thick	30	20		10.0[b]
		80	53		4.00[b]
		140	93		1.72[b]
	12 in thick	30	30		14.28[b]
		80	80		5.55[b]
		140	140		2.17[b]
miscellaneous	adobe 8 in		26		2.94[b]
	glass (single sheet, ^{1}/$_4$ in, vertical)				0.88[b]
	glass (double, vertical)				
	^{1}/$_4$ in air space				1.63[b]
	^{1}/$_2$ in air space				1.82[b]
	glass (single, horizontal)				summer: 1.16[b]
					winter: 0.71[b]
	glass (double, horizontal)				summer: 2.00[b]
	^{1}/$_4$ in air space				winter: 1.43[b]
	door, 1 in nominal, wood				1.45[b]
	1^1/$_2$ in nominal, wood				1.92[b]
	2 in nominal, wood				2.18[b]
	door, glass (herculite), ^{3}/$_4$ in thick				0.95[b]

(Multiply lbm/ft^3 by 16.0 to obtain kg/m^3.)
(Multiply lbm/ft^2 by 4.88 to obtain kg/m^2.)
(Multiply ft^2-hr-°F/Btu-in by 6.93 to obtain m·°C/W.)
(Multiply ft^2-hr-°F/Btu by 0.176 to obtain m^2·°C/W.)
[a]Compiled from a variety of sources to support practice problems in this text.
[b]Includes appropriate inside and outside film coefficients.

APPENDIX 46.A
Typical Mechanical Properties of Representative Metals
(room temperature)

The following mechanical properties are not guaranteed since they are averages for various sizes, product forms, and methods of manufacture. Thus, this data is not for design use, but is intended only as a basis for comparing alloys and tempers.

material designation, composition, typical use, and source if applicable	condition, heat treatment	S_{ut} (ksi)	S_{yt} (ksi)
IRON BASED			
Armco ingot iron, for fresh and saltwater piping	normalized	44	24
AISI 1020, plain carbon steel, for general machine parts and screws and carburized parts	hot rolled cold worked	65 78	43 66
AISI 1030, plain carbon steel, for gears, shafts, levers, seamless tubing, and carburized parts	cold drawn	87	74
AISI 1040, plain carbon steel, for high-strength parts, shafts, gears, studs, connecting rods, axles, and crane hooks	hot rolled cold worked hardened	91 100 113	58 88 86
AISI 1095, plain carbon steel, for handtools, music wire springs, leaf springs, knives, saws, and agricultural tools such as plows and disks	annealed hot rolled hardened	100 142 180	53 84 118
AISI 1330, manganese steel, for axles and drive shafts	annealed cold drawn hardened	97 113 122	83 93 100
AISI 4130, chromium-molybdenum steel, for high-strength aircraft structures	annealed hardened	81 161	52 137
AISI 4340, nickel-chromium-molybdenum steel, for large-scale, heavy-duty, high-strength structures	annealed as rolled hardened	119 192 220	99 147 200
AISI 2315, nickel steel, for carburized parts	as rolled cold drawn	85 95	56 75
AISI 2330, nickel steel	as rolled cold drawn annealed normalized	98 110 80 95	65 90 50 61
AISI 3115, nickel-chromium steel for carburized parts	cold drawn as rolled annealed	95 75 71	70 60 62
STAINLESS STEELS			
AISI 302, stainless steel, most widely used, same as 18-8	annealed cold drawn	90 105	35 60
AISI 303, austenitic stainless steel, good machineability	annealed cold worked	90 110	35 75
AISI 304, austenitic stainless steel, good machineability and weldability	annealed cold worked	85 110	30 75
AISI 309, stainless steel, good weldability, high strength at high temperatures, used in furnaces and ovens	annealed cold drawn	90 110	35 65
AISI 316, stainless steel, excellent corrosion resistance	annealed cold drawn	85 105	35 60
AISI 410, magnetic, martensitic, can be quenched and tempered to give varying strength	annealed cold drawn oil quenched and drawn	60 180 110	32 150 91
AISI 430, magnetic, ferritic, used for auto and architectural trim and for equipment in food and chemical industries	annealed cold drawn	60 100	35
AISI 502, magnetic, ferritic, low cost, widely used in oil refineries	annealed	60	25

(continued)

APPENDIX 46.A *(continued)*
Typical Mechanical Properties of Representative Metals
(room temperature)

material designation, composition, typical use, and source if applicable	condition, heat treatment	S_{ut} (ksi)	S_{yt} (ksi)
ALUMINUM BASED			
2011, for screw machine parts, excellent machineability, but not weldable, and corrosion sensitive	T3	55	43
	T8	59	45
2014, for aircraft structures, weldable	T3	63	40
	T4, T451	61	37
	T6, T651	68	60
2017, for screw machine parts	T4, T451	62	40
2018, for engine cylinders, heads, and pistons	T61	61	46
2024, for truck wheels, screw machine parts, and aircraft structures	T3	65	45
	T4, T351	64	42
	T361	72	57
2025, for forgings	T6	58	37
2117, for rivets	T4	43	24
2219, high-temperature applications (up to 600°F), excellent weldability and machineabilty	T31, T351	52	36
	T37	57	46
	T42	52	27
3003, for pressure vessels and storage tanks, poor machineability but good weldability, excellent corrosion resistance	0	16	6
	H12	19	18
	H14	22	21
	H16	26	25
3004, same characteristics as 3003	0	26	10
	H32	31	25
	H34	35	29
	H36	38	33
4032, pistons	T6	55	46
5083, unfired pressure vessels, cryogenics, towers, and drilling rigs	0	42	21
	H116, H117, H321	46	33
5154, saltwater services, welded structures, and storage tanks	0	35	17
	H32	39	30
	H34	42	33
5454, same characteristics as 5154	0	36	17
	H32	40	30
	H34	44	35
5456, same characteristics as 5154	0	45	23
	H111	47	33
	H321, H116, H117	51	37
6061, corrosion resistant and good weldability, used in railroad cars	T4	33	19
	T6	42	37
7178, Alclad, corrosion-resistant	0	33	15
	T6	88	78

CAST IRON (note redefinition of columns)		S_{ut} (ksi)	S_{us} (ksi)	S_{uc} (ksi)
gray cast iron	class 20	30	32.5	30
	class 25	25	34	100
	class 30	30	41	110
	class 35	35	49	125
	class 40	40	52	135
	class 50	50	64	160
	class 60	60	60	150

(continued)

APPENDIX 46.A *(continued)*
Typical Mechanical Properties of Representative Metals
(room temperature)

material designation, composition, typical use, and source if applicable	condition, heat treatment	S_{ut} (ksi)	S_{yt} (ksi)
COPPER BASED			
copper, commercial purity	annealed (furnace cool from 400°C)	32	10
	cold drawn	45	40
cartridge brass: 70% Cu, 30% Zn	cold rolled (annealed 400°C, furnace cool)	76	63
copper-beryllium (1.9% Be, 0.25% Co)	annealed, wqf 1450°F	70	
	cold rolled	200	
	hardened after annealing	200	150
phosphor-bronze, for springs	wire, 0.025 in and under	145	
	0.025 in to 0.0625 in	135	
	0.125 in to 0.250 in	125	
monel metal	cold-drawn bars, annealed	70	30
red brass	sheet and strip half-hard	51	
	hard	63	
	spring	78	
yellow brass	sheet and strip half-hard	55	
	hard	68	
	spring	86	
NICKEL BASED			
pure nickel, magnetic, high corrosion resistance	annealed (ht 1400°F, acrt)	46	8.5
	annealed at 2050°F	125	75
Inconel X, type 550, excellent high temperature properties	annealed and age hardened	175	110
	annealed (wqf 1600°F)	100	45
K-monel, excellent high temperature properties and corrosion resistance	age hardened spring stock	185	160
Invar, 36% Ni, 64% Fe, low coefficient of expansion (0.9×10^{-6} %/°C)	annealed (wqf 800°C)	71	40
REFRACTORY METALS (properties at room temperature)			
molybdenum	as rolled	100	75
tantalum	annealed at 1050°C in vacuum	60	45
	as rolled	110	100
titanium, commercial purity	annealed at 1200°F	95	80
titanium, 6% Al, 4% V	annealed at 1400°F, acrt	135	130
	heat treated (wqf 1750°F, ht 1000°F, acrt)	170	150
titanium, 4% Al, 4% Mn OR 5% Al, 2.75% Cr, 1.25% Fe OR 5% Al, 1.5% Fe, 1.4% Cr, 1.2% Mo	wqf 1450°F, ht 900°F, acrt	185	170
tungsten, commercial purity	hard wire	600	540

MAGNESIUM		S_{ut} (ksi)	S_{yt} (ksi)	S_{us} (ksi)
AZ92, for sand and permanent-mold casting	as cast	24	14	
	solution treated	39	14	
	aged	39	21	
AZ91, for die casting	as cast	33	21	
AZ31X (sheet)	annealed	35	20	
	hard	40	31	
AZ80X, for structural shapes	extruded	48	32	
	extruded and aged	52	37	
ZK60A, for structural shapes	extruded	49	38	
	extruded and aged	51	42	
AZ31B (sheet and plate), for structural shapes in use below 300°F	temper 0	32	15	17
	temper 1124	34	18	18
	temper 1126	35	21	18
	temper F	32	16	17

Abbreviations:
   wqf: water-quench from
   acrt: air-cooled to room temperature
   ht: heated to

APPENDIX 46.B
Typical Mechanical Properties of
Thermoplastic Resins and Composites
(room temperature, after post-mold annealing)

base resin and glass content (% by wt)	specific gravity	tensile yield strength (psi)	flexural modulus (ksi)	flexural strength (psi)	impact strength, Izod notched/unnotched (ft-lbf/in)	deflection temperature, at 264 psi (°F)	coefficient of thermal expansion (10^{-5} in/in-°F)
ASTM test⟶	D792	D638	D790	D790	D256	D648	D696
polyimide (30)	1.48	14,000	950	22,000	1.8/9	530	2.3
ethylene tetrafluoro-ethylene (ETFE) (20)	1.82	11,000	750	15,000	7/13	440	2.0
fluorinated ethylene-propylene (FEP) (20)	2.21	5000	800	10,500	8/17	350	2.4
polyphenylene sulfide (40)	1.56	23,000	1800	32,000	1.5/11	505	1.1
polyethersulfone (40)	1.68	22,000	1600	30,000	1.6/12	420	1.6
nylon 6/6 (50)	1.57	32,000	2200	46,500	3.3/20	500	1.0
polyester (40)	1.62	22,000	1600	32,000	2/12	475	1.05
polysulfone (40)	1.55	20,000	1600	27,000	2/16	370	1.2
polyarylsulfone (0)	1.36	13,000	395	17,000	5/40	525	2.6
poly-p-oxybenzoate (0)	1.40	14,000	700	17,000	1/3	560	1.6
polyamide-imide (0)	1.40	27,400	665	30,500	2.5/14	525	1.8

(Multiply ksi by 6.89 to obtain MPa.)
(Multiply psi by 0.006895 to obtain MPa.)
(Multiply ft-lbf/in by 53.38 to obtain N·m/m.)
(Multiply in/in-°F by 1.8 to obtain m/m·K.)

APPENDIX 48.A
Centroids and Area Moments of Inertia
for Basic Shapes

shape		x_c	y_c	A	I, J	k
rectangle		$\dfrac{b}{2}$	$\dfrac{h}{2}$	bh	$I_x = \dfrac{bh^3}{3}$ $I_{c,x} = \dfrac{bh^3}{12}$ $J_c = \left(\dfrac{1}{12}\right)bh(b^2 + h^2)$	$k_x = \dfrac{h}{\sqrt{3}}$ $k_{c,x} = \dfrac{h}{2\sqrt{3}}$
triangular area		$\dfrac{2b}{3}$	$\dfrac{h}{3}$	$\dfrac{bh}{2}$	$I_x = \dfrac{bh^3}{12}$ $I_{c,x} = \dfrac{bh^3}{36}$	$k_x = \dfrac{h}{\sqrt{6}}$ $k_{c,x} = \dfrac{h}{3\sqrt{2}}$
trapezoid			$h\left(\dfrac{b+2t}{3b+3t}\right)$	$\dfrac{(b+t)h}{2}$	$I_x = \dfrac{(b+3t)h^3}{12}$ $I_{c,x} = \dfrac{(b^2 + 4bt + t^2)h^3}{(36)(b+t)}$	$k_x = \left(\dfrac{h}{\sqrt{6}}\right)\sqrt{\dfrac{b+3t}{b+t}}$ $k_{c,x} = \dfrac{h\sqrt{2(b^2+4bt+t^2)}}{(6)(b+t)}$
circle		0	0	πr^2	$I_x = I_y = \dfrac{\pi r^4}{4}$ $J_c = \dfrac{\pi r^4}{2}$	$k_x = \dfrac{r}{2}$
quarter-circular area		$\dfrac{4r}{3\pi}$	$\dfrac{4r}{3\pi}$	$\dfrac{\pi r^2}{4}$	$I_x = I_y = \dfrac{\pi r^4}{16}$ $J_o = \dfrac{\pi r^4}{8}$	
semicircular area		0	$\dfrac{4r}{3\pi}$	$\dfrac{\pi r^2}{2}$	$I_x = I_y = \dfrac{\pi r^4}{8}$ $I_{c,x} = 0.1098r^4$ $J_o = \dfrac{\pi r^4}{4}$ $J_c = 0.5025r^4$	$k_x = \dfrac{r}{2}$ $k_{c,x} = 0.264r$
quarter-elliptical area		$\dfrac{4a}{3\pi}$	$\dfrac{4b}{3\pi}$	$\dfrac{\pi ab}{4}$	$I_x = \dfrac{\pi ab^3}{8}$ $I_y = \dfrac{\pi a^3 b}{8}$	
semielliptical area		0	$\dfrac{4b}{3\pi}$	$\dfrac{\pi ab}{2}$	$J_o = \dfrac{\pi ab(a^2 + b^2)}{8}$	
semiparabolic area		$\dfrac{3a}{8}$	$\dfrac{3h}{5}$	$\dfrac{2ah}{3}$		
parabolic area		0	$\dfrac{3h}{5}$	$\dfrac{4ah}{3}$	$I_x = \dfrac{4ah^3}{7}$ $I_y = \dfrac{4ha^3}{15}$ $I_{c,x} = \dfrac{16ah^3}{175}$	$k_x = h\sqrt{\dfrac{3}{7}}$ $k_y = \dfrac{a}{\sqrt{5}}$
parabolic spandrel		$\dfrac{3a}{4}$	$\dfrac{3h}{10}$	$\dfrac{ah}{3}$	$I_x = \dfrac{ah^3}{21}$ $I_y = \dfrac{3ha^3}{15}$	
general spandrel		$\left(\dfrac{n+1}{n+2}\right)a$	$\left(\dfrac{n+1}{4n+2}\right)h$	$\dfrac{ah}{n+1}$		
circular sector		$\dfrac{2r\sin\alpha}{3\alpha}$	0	αr^2		

APPENDIX 49.A
Elastic Beam Deflection Equations
(w is the load per unit length.)

Case 1: Cantilever with End Load

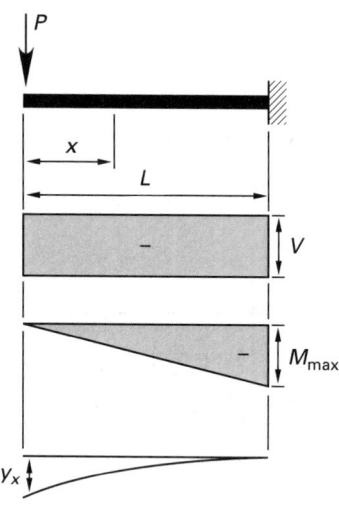

reactions:
$$R_l = 0$$
$$R_r = P$$
shear:
$$V = -P \text{ (constant)}$$
moments:
$$M_x = -Px$$
$$M_{max} = -PL$$
end slope:
$$\phi_l = +\frac{PL^2}{2EI}$$
$$\phi_r = 0$$
deflection:
$$y_x = \left(-\frac{P}{6EI}\right)(2L^3 - 3L^2x + x^3)$$
$$y_{max} = -\frac{PL^3}{3EI} \text{ at } x = 0$$

Case 2: Cantilever with Uniform Load

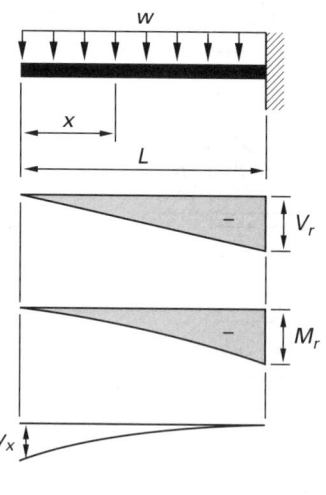

reactions:
$$R_l = 0$$
$$R_r = wL$$
shear:
$$V_x = -wx$$
$$V_{max} = -wL = V_r$$
moments:
$$M_x = -\frac{wx^2}{2}$$
$$M_{max} = -\frac{wL^2}{2} = M_r$$
end slope:
$$\phi_l = +\frac{wL^3}{6EI}$$
$$\phi_r = 0$$
deflection:
$$y_x = \left(\frac{w}{24EI}\right)(3L^4 - 4L^3x + x^4)$$
$$y_{max} = -\frac{wL^4}{8EI} \text{ at } x = 0$$

Case 3: Cantilever with Triangular Load

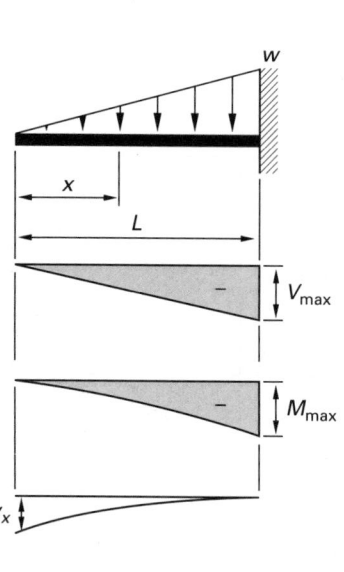

reactions:
$$R_l = 0$$
$$R_r = \frac{wL}{2}$$
shear:
$$V_x = -\frac{wx^2}{2L}$$
$$V_{max} = \frac{wL}{2} \text{ at } x = L$$
moments:
$$M_x = -\frac{wx^3}{6L}$$
$$M_{max} = \frac{wL^2}{6} \text{ at } x = L$$
end slope:
$$\phi_l = +\frac{wL^3}{24EI}$$
$$\phi_r = 0$$
deflection:
$$y_x = \left(-\frac{w}{120EIL}\right)(4L^5 - 5L^4x + x^5)$$
$$y_{max} = -\frac{wL^4}{30EI} \text{ at } x = 0$$

Case 4: Propped Cantilever with Uniform Load

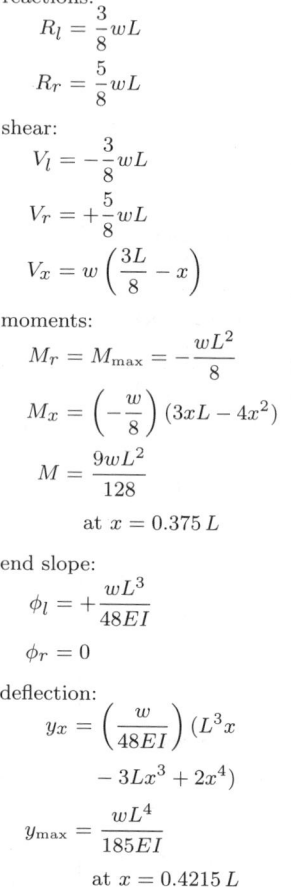

reactions:
$$R_l = \frac{3}{8}wL$$
$$R_r = \frac{5}{8}wL$$
shear:
$$V_l = -\frac{3}{8}wL$$
$$V_r = +\frac{5}{8}wL$$
$$V_x = w\left(\frac{3L}{8} - x\right)$$
moments:
$$M_r = M_{max} = -\frac{wL^2}{8}$$
$$M_x = \left(-\frac{w}{8}\right)(3xL - 4x^2)$$
$$M = \frac{9wL^2}{128}$$
$$\text{at } x = 0.375\,L$$
end slope:
$$\phi_l = +\frac{wL^3}{48EI}$$
$$\phi_r = 0$$
deflection:
$$y_x = \left(\frac{w}{48EI}\right)(L^3x - 3Lx^3 + 2x^4)$$
$$y_{max} = \frac{wL^4}{185EI}$$
$$\text{at } x = 0.4215\,L$$

(continued)

APPENDIX 49.A *(continued)*
Elastic Beam Deflection Equations
(w is the load per unit length.)

Case 5: Cantilever with End Moment

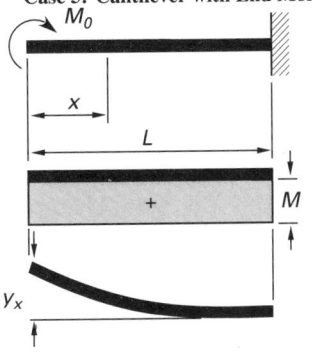

reactions:
$$R_l = 0$$
$$R_r = 0$$
shear:
$$V = 0$$
moments:
$$M = M_0 = M_{max}$$
end slope:
$$\phi_l = -\frac{M_0 L}{EI}$$
$$\phi_r = 0$$
deflection:
$$y_x = \left(+\frac{M_0}{2EI}\right)$$
$$\times (L^2 - 2xL + x^2)$$
$$y_{max} = +\frac{M_0 L^2}{2EI} \text{ at } x = 0$$

Case 6: Simple Beam with Center Load

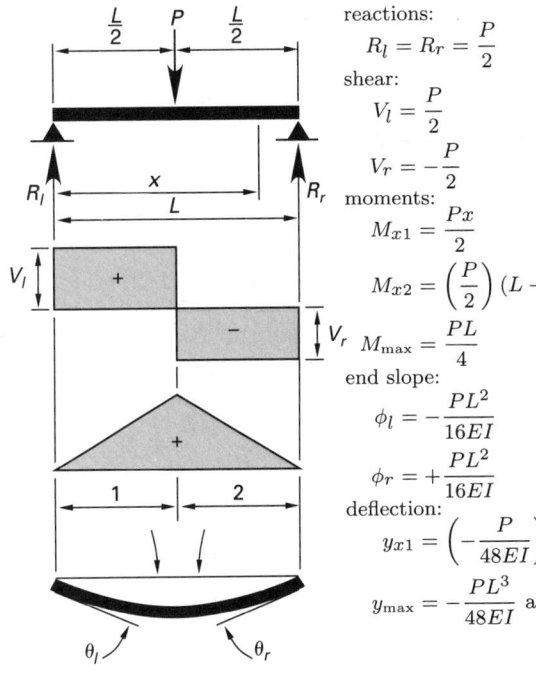

reactions:
$$R_l = R_r = \frac{P}{2}$$
shear:
$$V_l = \frac{P}{2}$$
$$V_r = -\frac{P}{2}$$
moments:
$$M_{x1} = \frac{Px}{2}$$
$$M_{x2} = \left(\frac{P}{2}\right)(L - x)$$
$$M_{max} = \frac{PL}{4}$$
end slope:
$$\phi_l = -\frac{PL^2}{16EI}$$
$$\phi_r = +\frac{PL^2}{16EI}$$
deflection:
$$y_{x1} = \left(-\frac{P}{48EI}\right)(3xL^2 - 4x^3)$$
$$y_{max} = -\frac{PL^3}{48EI} \text{ at } x = \frac{L}{2}$$

Case 7: Simple Beam with Intermediate Load

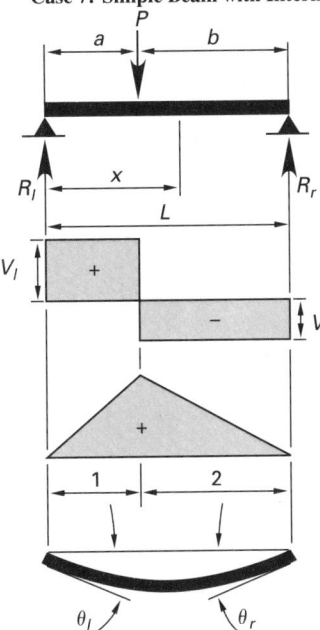

reactions:
$$R_l = \frac{Pb}{L}$$
$$R_r = \frac{Pa}{L}$$
shear:
$$V_l = +\frac{Pb}{L}$$
$$V_r = -\frac{Pa}{L}$$
moments:
$$M_{x1} = \frac{Pbx}{L}$$
$$M_{x2} = \frac{Pa(L - x)}{L}$$
$$M_{max} = \frac{Pab}{L} \text{ at } x = a$$
end slope:
$$\phi_l = -\frac{Pab\left(1 + \frac{b}{L}\right)}{6EI}$$
$$\phi_r = \frac{Pab\left(1 + \frac{a}{L}\right)}{6EI}$$
deflection:
$$y_{x1} = \left(\frac{Pb}{6EIL}\right)(L^2 x - b^2 x - x^3)$$
$$y_{x2} = \left(\frac{Pb}{6EIL}\right)\left[\left(\frac{L}{b}\right)(x - a)^3\right.$$
$$\left. + (L^2 - b^2)x - x^3\right]$$
$$y = \frac{Pa^2 b^2}{3EIL} \text{ at } x = a$$
$$y_{max} = \left(\frac{0.06415\,Pb}{EIL}\right)(L^2 - b^2)^{\frac{3}{2}}$$
$$\text{at } x = \sqrt{\frac{a(L + b)}{3}}$$

Case 8: Simple Beam with Two Loads

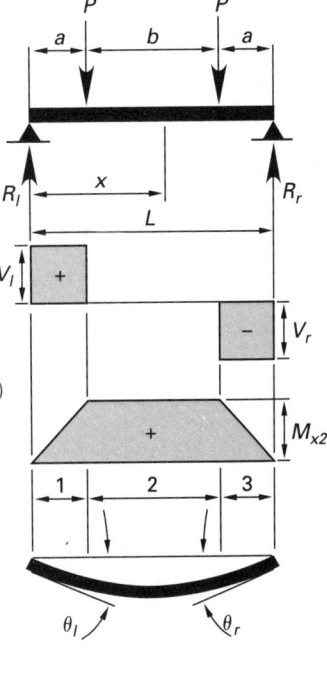

reactions:
$$R_l = R_r = P$$
shear:
$$V_l = +P$$
$$V_r = -P$$
moments:
$$M_{x1} = Px$$
$$M_{x2} = Pa$$
$$M_{x3} = P(L - x)$$
end slope:
$$\phi_l = -\frac{Pa(a + b)}{2EI}$$
$$\phi_r = +\frac{Pa(a + b)}{2EI}$$
deflection:
$$y_{x1} = \left(\frac{P}{6EI}\right)(3Lax$$
$$- 3a^2 x - x^3)$$
$$y_{x2} = \left(\frac{P}{6EI}\right)(3Lax$$
$$- 3ax^2 - a^3)$$
$$y_{max} = \left(\frac{P}{24EI}\right)(3L^2 a - 4a^3)$$
$$\text{at } x = \frac{L}{2}$$

(continued)

Support Material

APPENDIX 49.A *(continued)*
Elastic Beam Deflection Equations
(w is the load per unit length.)

Case 9: Simple Beam with Uniform Load

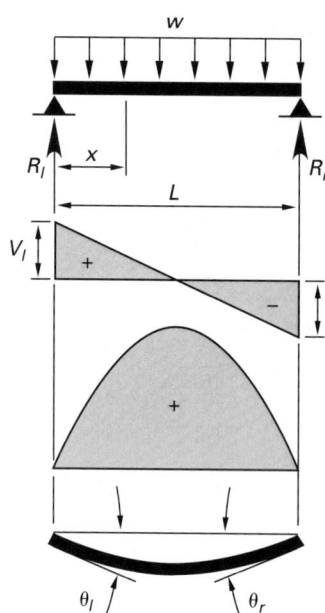

reactions:
$$R_l = R_r = \frac{wL}{2}$$
shear:
$$V_l = +\frac{wL}{2}$$
$$V_r = -\frac{wL}{2}$$
moments:
$$M = \left(\frac{w}{2}\right)(x^2 - Lx)$$
$$M_{max} = \frac{wL^2}{8}$$
end slope:
$$\phi_l = \frac{wL^3}{24EI}$$
$$\phi_r = +\frac{wL^3}{24EI}$$
deflection:
$$y_x = \left(-\frac{w}{24EI}\right)(L^3x - 2Lx^3 + x^4)$$
$$y_{max} = \frac{5wL^4}{384EI} \text{ at } x = \frac{L}{2}$$

Case 10: Simple Beam with Triangular Load

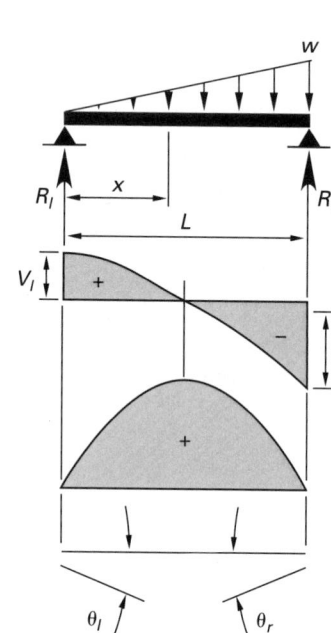

reactions:
$$R_l = \frac{wL}{6}$$
$$R_r = \frac{wL}{3}$$
shear:
$$V_l = +\frac{wL}{6}$$
$$V_r = -\frac{wL}{3}$$
$$V_x = \left(\frac{wL}{6}\right)\left[1 - 3\left(\frac{x}{L}\right)^2\right]$$
moments:
$$M_x = \left(\frac{w}{6}\right)\left(Lx - \frac{x^3}{L}\right)$$
$$M_{max} = \frac{wL^2}{9\sqrt{3}} = 0.0642wL^2$$
$$\text{at } x = 0.577L$$
end slope:
$$\phi_l = \frac{7wL^3}{360EI}$$
$$\phi_r = +\frac{wL^3}{45EI}$$
deflection:
$$y_x = \left(-\frac{w}{360EI}\right)\left(7L^3x - 10Lx^3 + \frac{3x^5}{L}\right)$$
$$y_{max} = -(0.00652)\left(\frac{wL^4}{EI}\right)$$
$$\text{at } x = 0.519L$$

Case 11: Simple Beam with Overhung Load

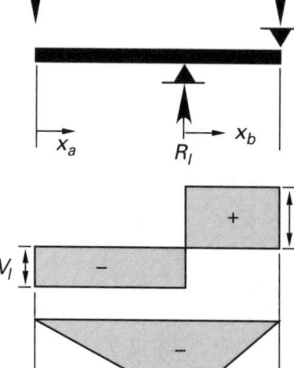

reactions:
$$R_l = \left(\frac{P}{b}\right)(b + a)$$
$$R_r = \frac{-Pa}{b}$$
shear:
$$V_l = -P$$
$$V_r = \frac{Pa}{b}$$
moments:
$$M_a = Px_a$$
$$M_b = \left(\frac{Pa}{b}\right)(b - x_b)$$
$$M_{max} = Fa \text{ at } x_a = a$$
deflections:
$$y_a = \left(\frac{F}{3EI}\right)\left[(a^2 + ab)(a - x_a) + \left(\frac{x_a}{2}\right)(x_a^2 - a^2)\right]$$
$$y_b = \left(\frac{Fax_b}{6EI}\right)\left[3x_b - \left(\frac{x_b^2}{b}\right) - 2b\right]$$
$$y_{tip} = \left(\frac{Fa^2}{3EI}\right)(a + b) \text{ [max down]}$$
$$y_{max} = (0.06415)\left(\frac{Fab^2}{EI}\right) \text{ at } x_b = 0.4226b \text{ [max up]}$$

APPENDIX 49.B
Stress Concentration Factors

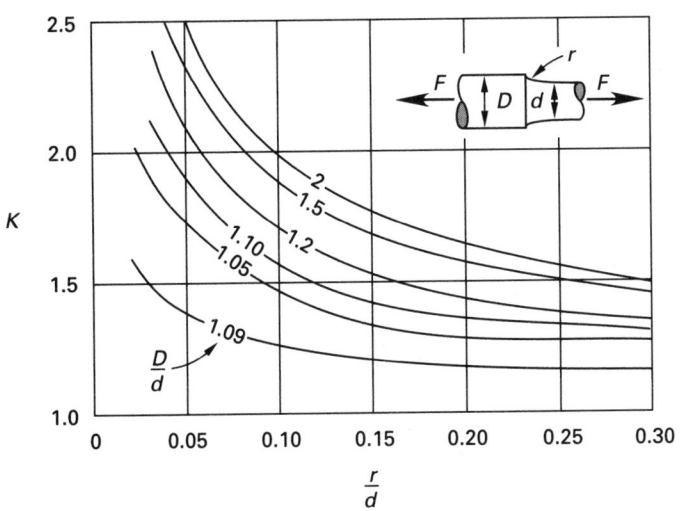

(a) stress concentration factor, K, for filleted shaft
in tension (basis: smaller section)

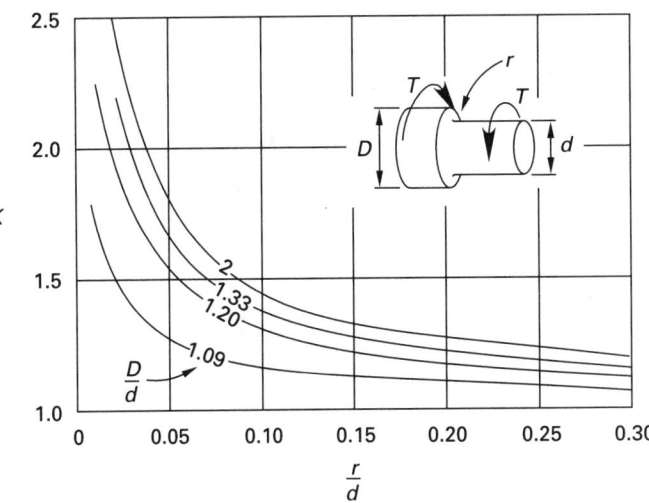

(b) stress concentration factor, K, for filleted shaft
in torsion (basis: smaller section)

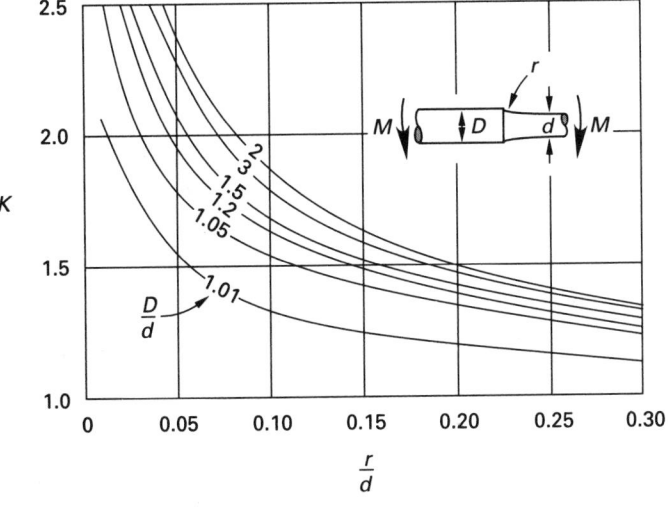

(c) stress concentration factor, K, for a shaft with shoulder
fillet in bending (basis: smaller section)

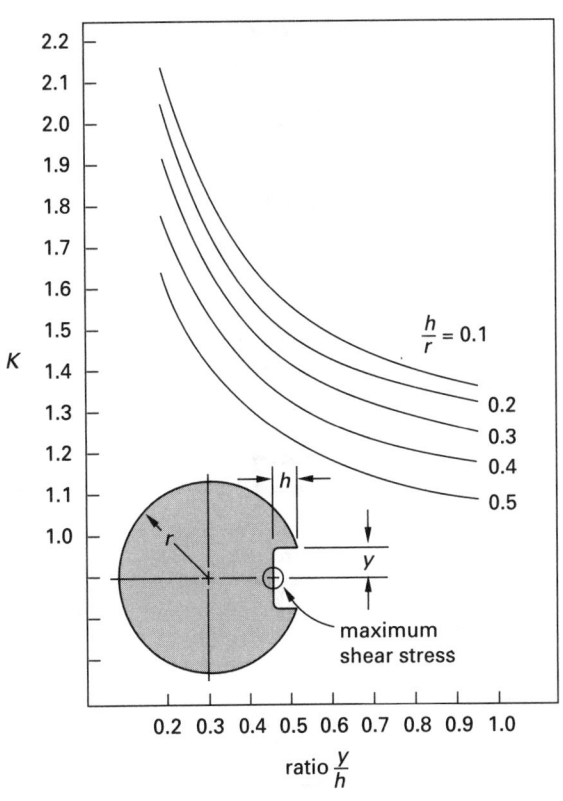

(d) stress concentration factor, K, for slotted shaft
in torsion (based on unslotted section)

Figures (a), (b), and (c) reprinted with permission from *Machine Design Magazine*, Vol. 23. © 1951, A Penton Publication.

Figure (d) reprinted with permission from *Machine Design*, September 22, 1977. © 1977, A Penton Publication.

APPENDIX 51.A
Properties of Weld Groups
(treated as lines)

weld configuration	centroid location	section modulus $S = I_{c,x}/\overline{y}$	polar moment of inertia $J = I_{c,x} + I_{c,y}$
	$\overline{y} = \dfrac{d}{2}$	$\dfrac{d^2}{6}$	$\dfrac{d^3}{12}$
	$\overline{y} = \dfrac{d}{2}$	$\dfrac{d^2}{3}$	$\dfrac{d(3b^2 + d^2)}{6}$
	$\overline{y} = \dfrac{d}{2}$	bd	$\dfrac{b(3d^2 + b^2)}{6}$
	$\overline{y} = \dfrac{d^2}{(2)(b+d)}$ $\overline{x} = \dfrac{b^2}{(2)(b+d)}$	$\dfrac{4bd + d^2}{6}$	$\dfrac{(b+d)^4 - 6b^2d^2}{(12)(b+d)}$
	$\overline{x} = \dfrac{b^2}{2b+d}$	$bd + \dfrac{d^2}{6}$	$\dfrac{8b^3 + 6bd^2 + d^3}{12} - \dfrac{b^4}{2b+d}$
	$\overline{y} = \dfrac{d^2}{b+2d}$	$\dfrac{2bd + d^2}{3}$	$\dfrac{b^3 + 6b^2d + 8d^3}{12} - \dfrac{d^4}{2d+b}$
	$\overline{y} = \dfrac{d}{2}$	$bd + \dfrac{d^2}{3}$	$\dfrac{(b+d)^3}{6}$
	$\overline{y} = \dfrac{d^2}{b+2d}$	$\dfrac{2bd + d^2}{3}$	$\dfrac{b^3 + 8d^3}{12} - \dfrac{d^4}{b+2d}$
	$\overline{y} = \dfrac{d}{2}$	$bd + \dfrac{d^2}{3}$	$\dfrac{b^3 + 3bd^2 + d^3}{6}$
	$\overline{y} = r$	πr^2	$2\pi r^3$

APPENDIX 52.A
Spring Wire Diameters and Sheet Metal Gauges[a]

no. of gauge	Washburn and Moen	Brown and Sharpe	Birmingham or Stubbs	U.S. standard for plate (iron and steel)	Stubbs steel wire	imperial wire gauge	Morse twist drill and steel wire	wood and machine screws	Amer. S. & W. piano and music wire
6–0	.....	.....	.....	.....	.....	0.464	.....	.....	.....
5–0	.....	.....	.....	.....	.....	0.432	.....	.....	0.005
4–0	0.394	0.460	0.454	.....	.....	0.400	.....	.....	0.006
3–0	0.362	0.410	0.425	.....	.....	0.372	.....	0.032	0.007
2–0	0.331	0.365	0.380	.....	.....	0.348	.....	0.045	0.008
0	0.307	0.325	0.340	0.313	.....	0.324	.....	0.058	0.009
1	0.283	0.289	0.300	0.281	0.227	0.300	0.228	0.071	0.010
2	0.263	0.258	0.284	0.265	0.219	0.276	0.221	0.084	0.011
3	0.244	0.229	0.259	0.250	0.212	0.252	0.213	0.097	0.012
4	0.225	0.204	0.238	0.234	0.207	0.232	0.209	0.110	0.013
5	0.207	0.182	0.220	0.219	0.204	0.212	0.206	0.124	0.014
6	0.192	0.162	0.203	0.203	0.201	0.192	0.204	0.137	0.016
7	0.177	0.144	0.180	0.188	0.199	0.176	0.201	0.150	0.018
8	0.162	0.128	0.165	0.172	0.197	0.160	0.199	0.163	0.020
9	0.148	0.114	0.148	0.156	0.194	0.144	0.196	0.176	0.022
10	0.135	0.102	0.134	0.141	0.191	0.128	0.194	0.189	0.024
11	0.120	0.091	0.120	0.125	0.188	0.116	0.191	0.203	0.026
12	0.105	0.081	0.109	0.109	0.185	0.104	0.189	0.216	0.029
13	0.092	0.072	0.095	0.094	0.182	0.092	0.185	0.229	0.031
14	0.080	0.064	0.083	0.078	0.180	0.080	0.182	0.242	0.033
15	0.072	0.057	0.072	0.070	0.178	0.072	0.180	0.255	0.035
16	0.063	0.051	0.065	0.063	0.175	0.064	0.177	0.268	0.037
17	0.054	0.045	0.058	0.056	0.172	0.056	0.173	0.282	0.039
18	0.047	0.040	0.049	0.050	0.168	0.048	0.170	0.295	0.041
19	0.041	0.036	0.042	0.044	0.164	0.040	0.166	0.308	0.043
20	0.035	0.032	0.035	0.038	0.161	0.036	0.161	0.321	0.045
21	0.032	0.028	0.032	0.034	0.157	0.032	0.159	0.334	0.047
22	0.028	0.025	0.028	0.031	0.155	0.028	0.157	0.347	0.049
23	0.025	0.023	0.025	0.028	0.153	0.024	0.154	0.360	0.051
24	0.023	0.020	0.022	0.025	0.151	0.022	0.152	0.374	0.055
25	0.020	0.018	0.020	0.022	0.148	0.020	0.150	0.387	0.059
26	0.018	0.016	0.018	0.019	0.146	0.018	0.147	0.400	0.063
27	0.0173	0.0141	0.016	0.017	0.143	0.0164	0.144	0.413	0.067
28	0.0162	0.0126	0.014	0.016	0.139	0.0149	0.141	0.426	0.071
29	0.015	0.0112	0.013	0.014	0.134	0.0136	0.136	0.439	0.075
30	0.014	0.010	0.012	0.013	0.127	0.0124	0.129	0.453	0.080
31	0.0132	0.0089	0.010	0.011	0.120	0.0116	0.120	0.466	0.085
32	0.0128	0.0079	0.009	0.010	0.115	0.0108	0.116	0.479	0.090
33	0.0118	0.007	0.008	0.009	0.112	0.010	0.113	0.492	0.095
34	0.0104	0.0063	0.007	0.0086	0.110	0.0092	0.111	0.505	0.100
35	0.0095	0.0056	0.005	0.0078	0.108	0.0084	0.110	0.518	0.106
36	0.009	0.005	0.004	0.007	0.106	0.0076	0.1065	0.532	0.112
37	.....	0.0044	.....	0.0066	0.103	0.0068	0.104	0.545	0.118
38	.....	0.0039	.....	0.0062	0.101	0.006	0.1015	0.558	0.124
39	.....	0.0035	.....	.....	0.099	0.0052	0.0995	0.571	0.130
40	.....	0.0031	.....	.....	0.097	0.0048	0.098	0.584	0.138
41	.....	.....	.....	.....	0.095	.....	0.096	0.597	0.146
42	.....	.....	.....	.....	0.092	.....	0.094	0.611	0.154
43	.....	.....	.....	.....	0.088	.....	0.089	0.624	0.162
44	.....	.....	.....	.....	0.085	.....	0.086	0.637	0.170
45	.....	.....	.....	.....	0.081	.....	0.082	0.650	0.180
46	.....	.....	.....	.....	0.079	.....	0.081	0.663	.....
47	.....	.....	.....	.....	0.077	.....	0.079	0.676	.....
48	.....	.....	.....	.....	0.075	.....	0.076	0.690	.....
49	.....	.....	.....	.....	0.072	.....	0.073	0.703	.....
50	.....	.....	.....	.....	0.069	.....	0.070	0.716	.....

(Multiply in by 25.4 to obtain mm.)
[a]Birmingham Wire Gauge (BWG), commonly used for ferrous tubing, is identical with Stubs iron-wire gauge.

Support Material

APPENDIX 52.B
Journal Bearing Correlations

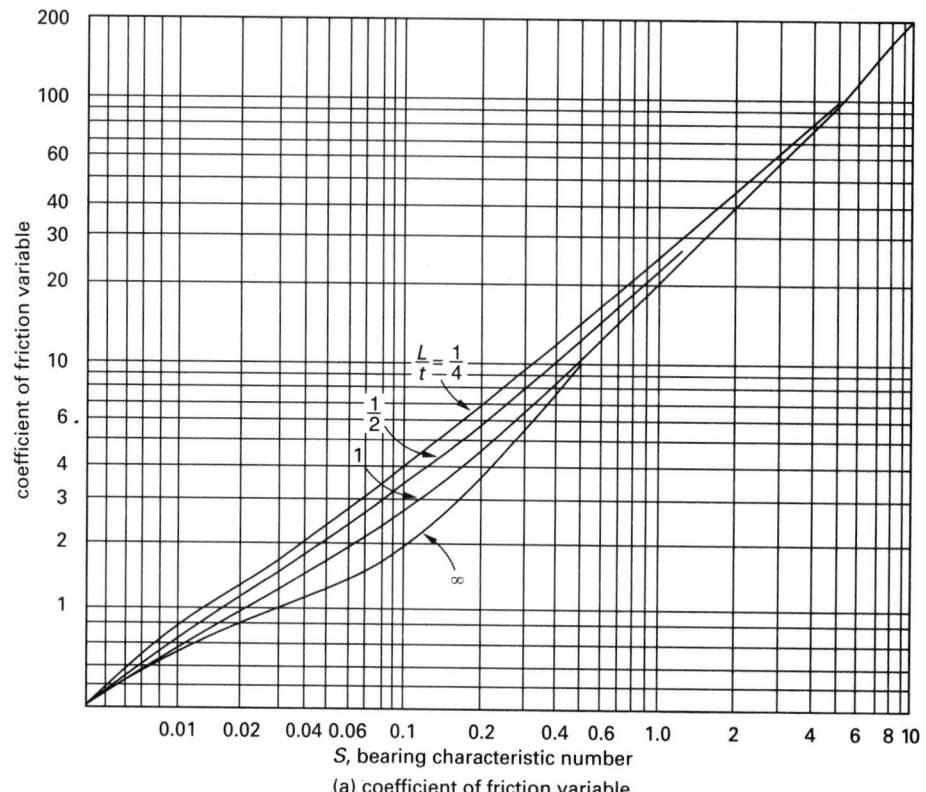

(a) coefficient of friction variable

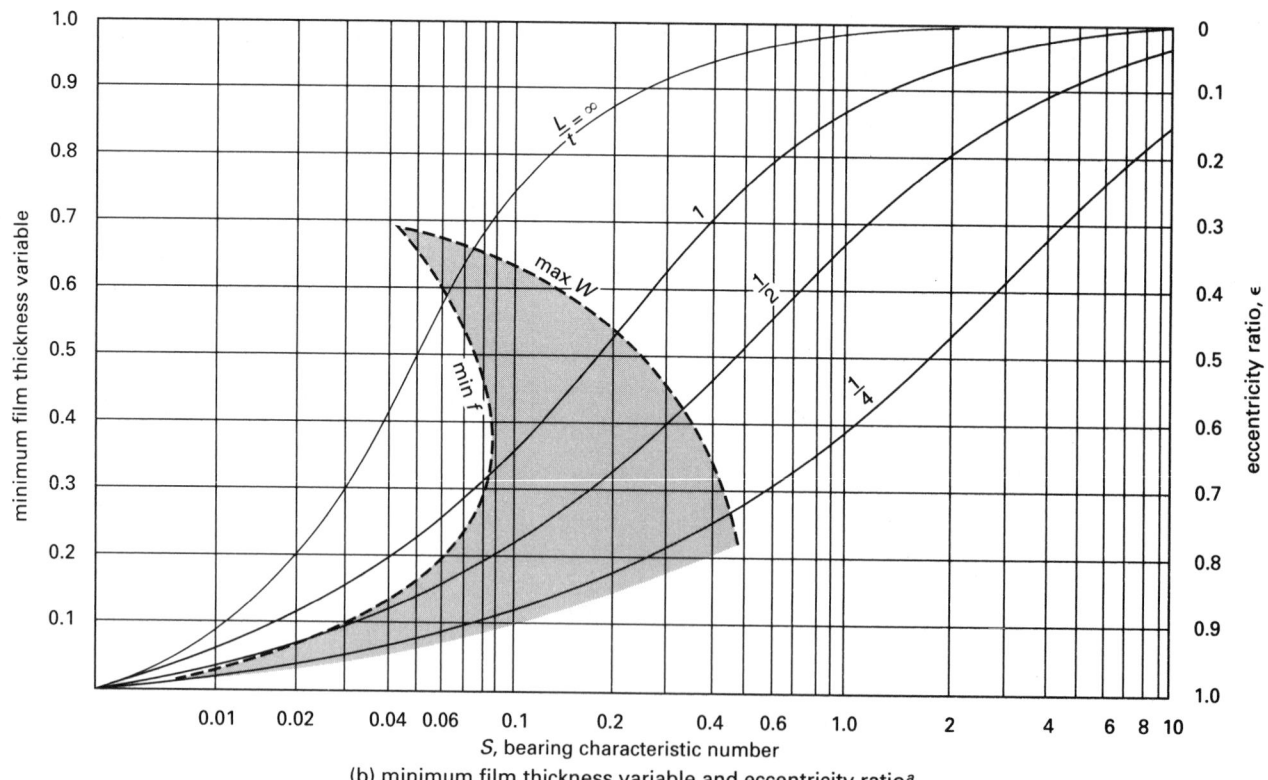

(b) minimum film thickness variable and eccentricity ratio[a]

[a]The left edge of the shaded area represents the optimum minimum thickness, resulting in the minimum friction. The right edge of the shaded area represents the optimum minimum thickness for maximum load-carrying ability.

APPENDIX 54.A
Mass Moments of Inertia (centroids at points labeled C)

slender rod		$I_y = I_z = \dfrac{mL^2}{12}$ $I_y' = I_z' = \dfrac{mL^2}{3}$
thin rectangular plate		$I_x = \dfrac{m(b^2 + c^2)}{12}$ $I_y = \dfrac{mc^2}{12}$ $I_z = \dfrac{mb^2}{12}$
rectangular parallelepiped		$I_x = \dfrac{m(b^2 + c^2)}{12}$ $I_y = \dfrac{m(c^2 + a^2)}{12}$ $I_z = \dfrac{m(a^2 + b^2)}{12}$ $I_{x'} = \dfrac{m(4b^2 + c^2)}{12}$
thin disk, radius r		$I_x = \dfrac{mr^2}{2}$ $I_y = I_z = \dfrac{mr^2}{4}$
solid circular cylinder, radius r		$I_x = \dfrac{mr^2}{2}$ $I_y = I_z = \dfrac{m(3r^2 + L^2)}{12}$
solid circular cone, base radius r		$I_x = \dfrac{3mr^2}{10}$ $I_y = I_z = \left(\dfrac{3m}{5}\right)\left(\dfrac{r^2}{4} + h^2\right)$
sphere, radius r		$I_x = I_y = I_z = \dfrac{2mr^2}{5}$
hollow circular cylinder, inner radius r_i, outer radius r_o		$I_x = \dfrac{m\left(r_o^2 + r_i^2\right)}{2}$ $= \left(\dfrac{\pi \rho L}{2}\right)\left(r_o^4 - r_i^4\right)$

PROFESSIONAL PUBLICATIONS, INC. BELMONT, CA

APPENDIX 57.A
Standard Epicyclic Gear Train Ratios
(for fixed gears, set $n = 0$)

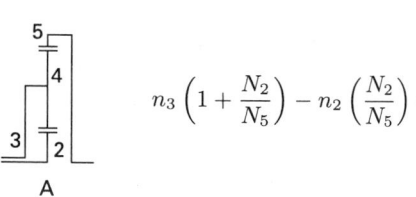

$$n_3\left(1 + \frac{N_2}{N_5}\right) - n_2\left(\frac{N_2}{N_5}\right) = n_5$$

A

$$n_3\left(1 + \frac{N_2 N_5}{N_4 N_6}\right) - n_2\left(\frac{N_2 N_5}{N_4 N_6}\right) = n_6$$

B

$$n_3\left(1 + \frac{N_2 N_5 N_7}{N_4 N_6 N_8}\right) - n_2\left(\frac{N_2 N_5 N_7}{N_4 N_6 N_8}\right) = n_8$$

C

$$n_3\left(1 + \frac{N_2 N_5 N_7}{N_4 N_6 N_8}\right) - n_2\left(\frac{N_2 N_5 N_7}{N_4 N_6 N_8}\right) = n_8$$

D

$$n_3\left(1 + \frac{N_2 N_5}{N_4 N_7}\right) - n_2\left(\frac{N_2 N_5}{N_4 N_7}\right) = n_7$$

E

$$n_3\left(1 + \frac{N_2 N_6}{N_5 N_7}\right) - n_2\left(\frac{N_2 N_6}{N_5 N_7}\right) = n_7$$

F

$$n_3\left(1 - \frac{N_2 N_5}{N_4 N_6}\right) + n_2\left(\frac{N_2 N_5}{N_4 N_6}\right) = n_6$$

G

$$n_3\left(1 - \frac{N_2 N_5}{N_4 N_6}\right) + n_2\left(\frac{N_2 N_5}{N_4 N_6}\right) = n_6$$

H

$$n_3\left(1 - \frac{N_2}{N_6}\right) + n_2\left(\frac{N_2}{N_6}\right) = n_6$$

I

$$n_3\left(1 - \frac{N_2 N_5 N_7}{N_4 N_6 N_8}\right) + n_2\left(\frac{N_2 N_5 N_7}{N_4 N_6 N_8}\right) = n_8$$

J

$$n_3\left(1 - \frac{N_2 N_5}{N_4 N_7}\right) + n_2\left(\frac{N_2 N_5}{N_4 N_7}\right) = n_7$$

K

$$n_3\left(1 - \frac{N_2 N_6}{N_5 N_7}\right) + n_2\left(\frac{N_2 N_6}{N_5 N_7}\right) = n_7$$

L

APPENDIX 61.A
Percentile Values for Student's t-Distribution
(ν degrees of freedom; confidence level C;
shaded area $= p$)

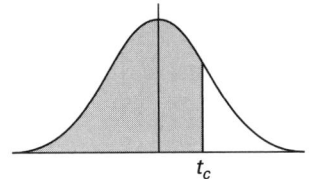

	two-tail									
	$t_{99\%}$	$t_{98\%}$	$t_{95\%}$	$t_{90\%}$	$t_{80\%}$	$t_{60\%}$	$t_{50\%}$	$t_{40\%}$	$t_{20\%}$	$t_{10\%}$
	one-tail									
ν	$t_{99.5\%}$	$t_{99\%}$	$t_{97.5\%}$	$t_{95\%}$	$t_{90\%}$	$t_{80\%}$	$t_{75\%}$	$t_{70\%}$	$t_{60\%}$	$t_{55\%}$
1	63.66	31.82	12.71	6.31	3.08	1.376	1.000	0.727	0.325	0.158
2	9.92	6.96	4.30	2.92	1.89	1.061	0.816	0.617	0.289	0.142
3	5.84	4.54	3.18	2.35	1.64	0.978	0.765	0.584	0.277	0.137
4	4.60	3.75	2.78	2.13	1.53	0.941	0.741	0.569	0.271	0.134
5	4.03	3.36	2.57	2.02	1.48	0.920	0.727	0.559	0.267	0.132
6	3.71	3.14	2.45	1.94	1.44	0.906	0.718	0.553	0.265	0.131
7	3.50	3.00	2.36	1.90	1.42	0.896	0.711	0.549	0.263	0.130
8	3.36	2.90	2.31	1.86	1.40	0.889	0.706	0.546	0.262	0.130
9	3.25	2.82	2.26	1.83	1.38	0.883	0.703	0.543	0.261	0.129
10	3.17	2.76	2.23	1.81	1.37	0.879	0.700	0.542	0.260	0.129
11	3.11	2.72	2.20	1.80	1.36	0.876	0.697	0.540	0.260	0.129
12	3.06	2.68	2.18	1.78	1.36	0.873	0.695	0.539	0.259	0.128
13	3.01	2.65	2.16	1.77	1.35	0.870	0.694	0.538	0.259	0.128
14	2.98	2.62	2.14	1.76	1.34	0.868	0.692	0.537	0.258	0.128
15	2.95	2.60	2.13	1.75	1.34	0.866	0.691	0.536	0.258	0.128
16	2.92	2.58	2.12	1.75	1.34	0.865	0.690	0.535	0.258	0.128
17	2.90	2.57	2.11	1.74	1.33	0.863	0.689	0.534	0.257	0.128
18	2.88	2.55	2.10	1.73	1.33	0.862	0.688	0.534	0.257	0.127
19	2.86	2.54	2.09	1.73	1.33	0.861	0.688	0.533	0.257	0.127
20	2.84	2.53	2.09	1.72	1.32	0.860	0.687	0.533	0.257	0.127
21	2.83	2.52	2.08	1.72	1.32	0.859	0.686	0.532	0.257	0.127
22	2.82	2.51	2.07	1.72	1.32	0.858	0.686	0.532	0.256	0.127
23	2.81	2.50	2.07	1.71	1.32	0.858	0.685	0.532	0.256	0.127
24	2.80	2.49	2.06	1.71	1.32	0.857	0.685	0.531	0.256	0.127
25	2.79	2.48	2.06	1.71	1.32	0.856	0.684	0.531	0.256	0.127
26	2.78	2.48	2.06	1.71	1.32	0.856	0.684	0.531	0.256	0.127
27	2.77	2.47	2.05	1.70	1.31	0.855	0.684	0.531	0.256	0.127
28	2.76	2.47	2.05	1.70	1.31	0.855	0.683	0.530	0.256	0.127
29	2.76	2.46	2.04	1.70	1.31	0.854	0.683	0.530	0.256	0.127
30	2.75	2.46	2.04	1.70	1.31	0.854	0.683	0.530	0.256	0.127
40	2.70	2.42	2.02	1.68	1.30	0.851	0.681	0.529	0.255	0.126
60	2.66	2.39	2.00	1.67	1.30	0.848	0.679	0.527	0.254	0.126
120	2.62	2.36	1.98	1.66	1.29	0.845	0.677	0.526	0.254	0.126
∞	2.58	2.33	1.96	1.645	1.28	0.842	0.674	0.524	0.253	0.126

Support Material

APPENDIX 61.B
Single Sampling Plan Table
for Various Producer's and Consumer's Risks
(values of np)

Choose the column for the appropriate consumer's risk, β. If the producer's risk, α, is known, calculate β as $1 - \alpha$. Choose the row corresponding to the number of defects, c. Read the table value. Calculate the sample size, N, as the table value divided by the percent of defectives, p, in the lot.

c	β 99%	95%	90%	50%	10%	5%	1%
0	0.010	0.051	0.105	0.693	2.303	2.996	4.605
1	0.149	0.355	0.532	1.678	3.890	4.744	6.638
2	0.436	0.818	1.102	2.674	5.322	6.296	8.406
3	0.823	1.366	1.745	3.672	6.681	7.754	10.045
4	1.279	1.970	2.433	4.671	7.994	9.154	11.605
5	1.785	2.613	3.152	5.670	9.275	10.513	13.108
6	2.330	3.286	3.895	6.670	10.532	11.842	14.571
7	2.906	3.981	4.656	7.669	11.771	13.148	16.000
8	3.507	4.695	5.432	8.669	12.995	14.434	17.403
9	4.130	5.426	6.221	9.669	14.206	15.705	18.783
10	4.771	6.169	7.021	10.668	15.407	16.962	20.145
11	5.428	6.924	7.829	11.668	16.598	18.208	21.490
12	6.099	7.690	8.646	12.668	17.782	19.442	22.821
13	6.782	8.464	9.470	13.668	18.958	20.668	24.139
14	7.477	9.246	10.300	14.668	20.128	21.886	25.446
15	8.181	10.035	11.135	15.668	21.292	23.098	26.743
16	8.895	10.831	11.976	16.668	22.452	24.302	28.031
17	9.616	11.633	12.822	17.668	23.606	25.500	29.310
18	10.346	12.442	13.672	18.668	24.756	26.692	30.581
19	11.082	13.254	14.525	19.668	25.902	27.879	31.845
20	11.825	14.072	15.383	20.668	27.045	29.062	33.103
21	12.574	14.894	16.244	21.668	28.184	30.241	34.355
22	13.329	15.719	17.108	22.668	29.320	31.416	35.601
23	14.088	16.548	17.975	23.668	30.453	32.586	36.841
24	14.853	17.382	18.844	24.668	31.584	33.752	38.077
25	15.623	18.218	19.717	25.667	32.711	34.916	39.308
30	19.532	22.444	24.113	30.667	38.315	40.690	45.401
35	23.525	26.731	28.556	35.667	43.872	46.404	51.409
40	27.587	31.066	33.038	40.667	49.390	52.069	57.347
45	31.704	35.441	37.550	45.667	54.878	57.695	63.231
50	35.867	39.849	42.089	50.667	60.339	63.287	69.066

Adapted with permission from *Statistical Quality Control*, 3rd Ed., by Eugene L. Grant, McGraw-Hill Book Company, copyright © 1964.

APPENDIX 61.C
Cumulative Probability Chart[a,b]

Calculate the average number of defectives or failures as $\lambda = Np$ (i.e., sample size times lot fraction defective).

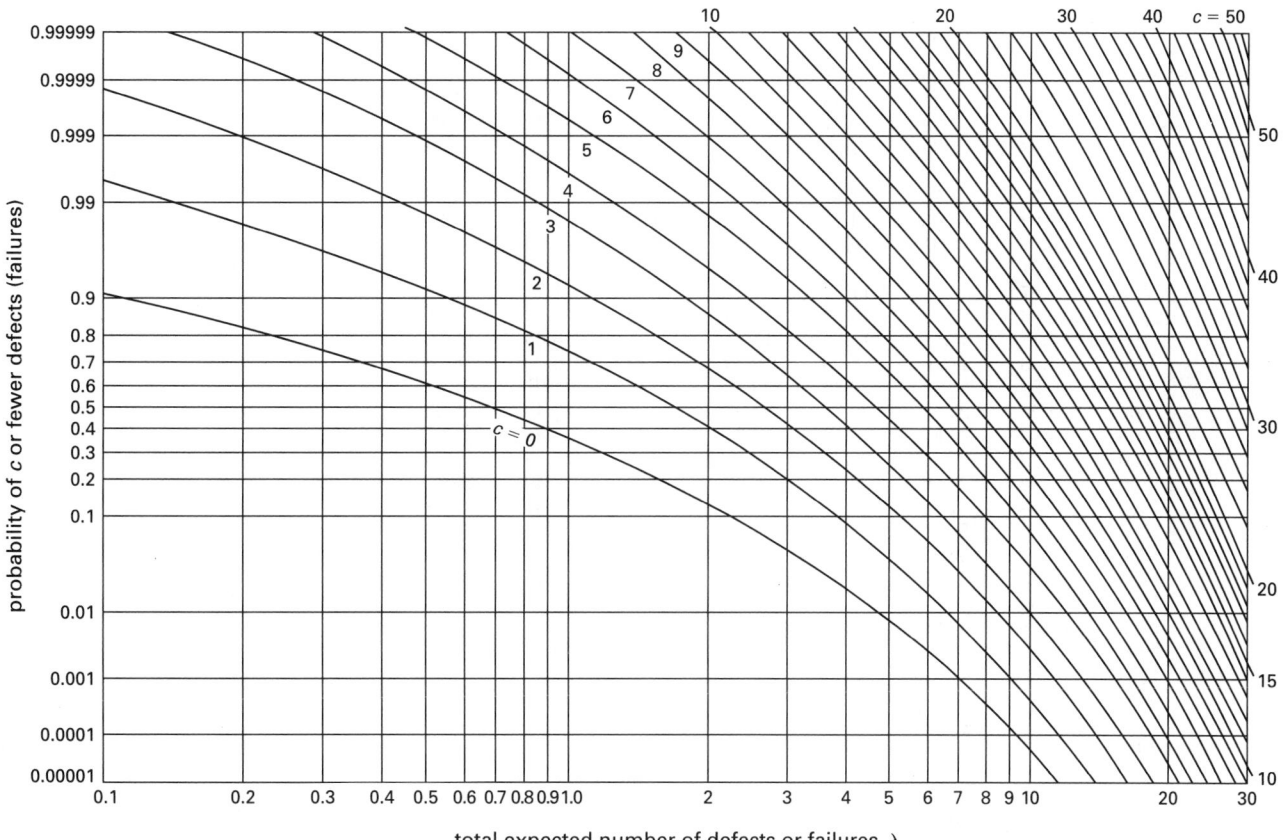

total expected number of defects or failures, λ

[a] Use this appendix to find the probability of c or fewer defects in a sample of size N from a lot with fraction defective p, or to find the probability of c or fewer failures when N units are operated for duration t (total test time $T = Nt$) with an individual failure rate (i.e., probability of failing in time T) of r.

[b] This appendix is based on an approximation to the binomial distribution calculated as a summation of the first c terms of the Poisson distribution with $\lambda = pN$, sometimes known as the *Poisson exponential binomial limit*. This approximation is good when $p < 0.10$.

Used with permission from *Sampling Inspection Tables: Single and Double Sampling*, by Harold F. Dodge and Harry G. Romig, John Wiley & Sons, Inc. © 1944.

APPENDIX 62.A
Thermoelectric Constants for Thermocouples
(mV, reference 32°F (0°C))

(a) chromel-alumel (type K)

°F	0	10	20	30	40	50	60	70	80	90
					millivolts					
−300	−5.51	−5.60								
−200	−4.29	−4.44	−4.58	−4.71	−4.84	−4.96	−5.08	−5.20	−5.30	−5.41
−100	−2.65	−2.83	−3.01	−3.19	−3.36	−3.52	−3.69	−3.84	−4.00	−4.15
−0	−0.68	−0.89	−1.10	−1.30	−1.50	−1.70	−1.90	−2.09	−2.28	−2.47
+0	−0.68	−0.49	−0.26	−0.04	0.18	0.40	0.62	0.84	1.06	1.29
100	1.52	1.74	1.97	2.20	2.43	2.66	2.89	3.12	3.36	3.59
200	3.82	4.05	4.28	4.51	4.74	4.97	5.20	5.42	5.65	5.87
300	6.09	6.31	6.53	6.76	6.98	7.20	7.42	7.64	7.87	8.09
400	8.31	8.54	8.76	8.98	9.21	9.43	9.66	9.88	10.11	10.34
500	10.57	10.79	11.02	11.25	11.48	11.71	11.94	12.17	12.40	12.63
600	12.86	13.09	13.32	13.55	13.78	14.02	14.25	14.48	14.71	14.95
700	15.18	15.41	15.65	15.88	16.12	16.35	16.59	16.82	17.06	17.29
800	17.53	17.76	18.00	18.23	18.47	18.70	18.94	19.18	19.41	19.65
900	19.89	20.13	20.36	20.60	20.84	21.07	21.31	21.54	21.78	22.02
1000	22.26	22.49	22.73	22.97	23.20	23.44	23.68	23.91	24.15	24.39
1100	24.63	24.86	25.10	25.34	25.57	25.81	26.05	26.28	26.52	26.75
1200	26.98	27.22	27.45	27.69	27.92	28.15	28.39	28.62	28.86	29.09
1300	29.32	29.56	29.79	30.02	30.25	30.49	30.72	30.95	31.18	31.42
1400	31.65	31.88	32.11	32.34	32.57	32.80	33.02	33.25	33.48	33.71
1500	33.93	34.16	34.39	34.62	34.84	35.07	35.29	35.52	35.75	35.97
1600	36.19	36.42	36.64	36.87	37.09	37.31	37.54	37.76	37.98	38.20
1700	38.43	38.65	38.87	39.09	39.31	39.53	39.75	39.96	40.18	40.40
1800	40.62	40.84	41.05	41.27	41.49	41.70	41.92	42.14	42.35	42.57
1900	42.78	42.99	43.21	43.42	43.63	43.85	44.06	44.27	44.49	44.70
2000	44.91	45.12	45.33	45.54	45.75	45.96	46.17	46.38	46.58	46.79

(b) iron-constantan (type J)

°F	0	10	20	30	40	50	60	70	80	90
					millivolts					
−300	−7.52	−7.66								
−200	−5.76	−5.96	−6.16	−6.35	−6.53	−6.71	−6.89	−7.06	−7.22	−7.38
−100	−3.49	−3.73	−3.97	−4.21	−4.44	−4.68	−4.90	−5.12	−5.34	−5.55
−0	−0.89	−1.16	−1.43	−1.70	−1.96	−2.22	−2.48	−2.74	−2.99	−3.24
+0	−0.89	−0.61	−0.34	−0.06	0.22	0.50	0.79	1.07	1.36	1.65
100	1.94	2.23	2.52	2.82	3.11	3.41	3.71	4.01	4.31	4.61
200	4.91	5.21	5.51	5.81	6.11	6.42	6.72	7.03	7.33	7.64
300	7.94	8.25	8.56	8.87	9.17	9.48	9.79	10.10	10.41	10.72
400	11.03	11.34	11.65	11.96	12.26	12.57	12.88	13.19	13.50	13.81
500	14.12	14.42	14.73	15.04	15.34	15.65	15.96	16.26	16.57	16.88
600	17.18	17.49	17.80	18.11	18.41	18.72	19.03	19.34	19.64	19.95
700	20.26	20.56	20.87	21.18	21.48	21.79	22.10	22.40	22.71	23.01
800	23.32	23.63	23.93	24.24	24.55	24.85	25.16	25.47	25.78	26.09
900	26.40	26.70	27.02	27.33	27.64	27.95	28.26	28.58	28.89	29.21
1000	29.52	29.84	30.16	30.48	30.80	31.12	31.44	31.76	32.08	32.40
1100	32.72	33.05	33.37	33.70	34.03	34.36	34.68	35.01	35.35	35.68
1200	36.01	36.35	36.69	37.02	37.36	37.71	38.05	38.39	38.74	39.08
1300	39.43	39.78	40.13	40.48	40.83	41.19	41.54	41.90	42.25	42.61

(c) copper-constantan (type T)

°F	0	10	20	30	40	50	60	70	80	90
					millivolts					
−300	−5.284	−5.379								
−200	−4.111	−4.246	−4.377	−4.504	−4.627	−4.747	−4.863	−4.974	−5.081	−5.185
−100	−2.559	−2.730	−2.897	−3.062	−3.223	−3.380	−3.533	−3.684	−3.829	−3.972
−0	−0.670	−0.872	−1.072	−1.270	−1.463	−1.654	−1.842	−2.026	−2.207	−2.385
+0	−0.670	−0.463	−0.254	−0.042	0.171	0.389	0.609	0.832	1.057	1.286
100	1.517	1.751	1.987	2.226	2.467	2.711	2.958	3.207	3.458	3.712
200	3.967	4.225	4.486	4.749	5.014	5.280	5.550	5.821	6.094	6.370
300	6.647	6.926	7.208	7.491	7.776	8.064	8.352	8.642	8.935	9.229
400	9.525	9.823	10.123	10.423	10.726	11.030	11.336	11.643	11.953	12.263
500	12.575	12.888	13.203	13.520	13.838	14.157	14.477	14.799	15.122	15.447

Based on National Bureau of Standards Circular No. 561, April 1955.

APPENDIX 65.A
Friction Loss in Schedule-40 Steel Pipe
(customary U.S. units)
(Hazen-Williams $C = 120$)[a]

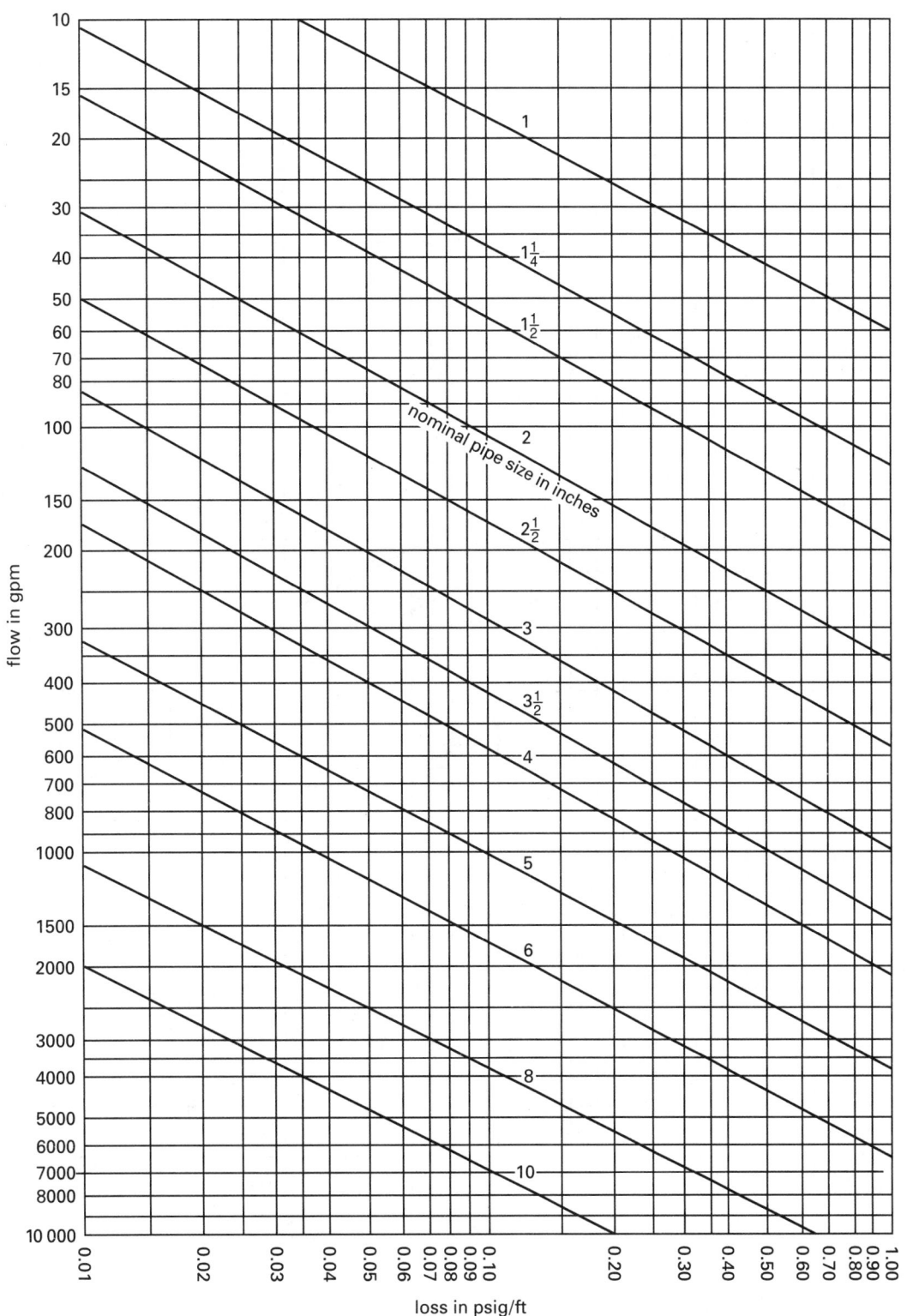

[a] Use Table 65.5 to convert to other C-values.

Reprinted with permission from NFPA 15, *Water Spray Fixed Systems for Fire Protection*, Copyright © 1990, National Fire Protection Association, Quincy, MA 02269. This reprinted material is not the complete and official position of the National Fire Protection Association, on the referenced subject which is represented only by the standard in its entirety.

APPENDIX 65.B
Friction Loss in Schedule-40 Steel Pipe
(SI units)
(Hazen-Williams $C = 120$)[a]

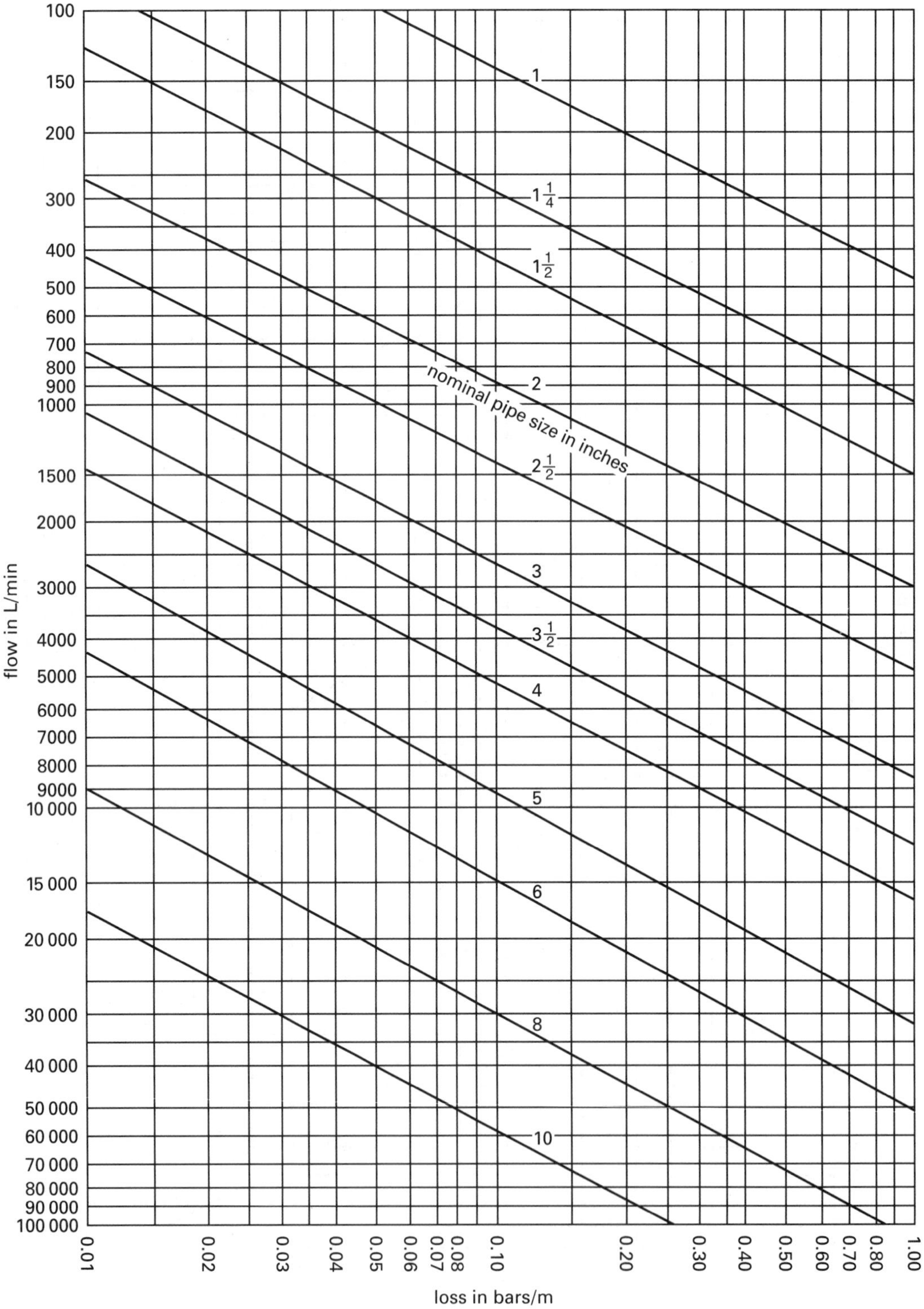

(Multiply bars by 100 to get kPa.)
[a]Use Table 65.5 to convert to other C-values.

Reprinted with permission from NFPA 15, *Water Spray Fixed Systems for Fire Protection*, Copyright
© 1990, National Fire Protection Association, Quincy, MA 02269. This reprinted material is not the
complete and official position of the National Fire Protection Association, on the referenced subject
which is represented only by the standard in its entirety.

APPENDIX 65.C
Equivalent Lengths of Valves and Fittings
for Fire Protection Systems
(Hazen-Williams $C = 120$)
(all types of approved pipe)
(fittings and valves expressed in equivalent ft (m) of pipe)

	$^3/_4$ in		1 in		$1^1/_4$ in		$1^1/_2$ in		2 in		$2^1/_2$ in		3 in	
45° elbow	1	(0.3)	1	(0.3)	1	(0.3)	2	(0.6)	2	(0.6)	3	(0.9)	3	(0.9)
90° standard elbow	2	(0.6)	2	(0.6)	3	(0.9)	4	(1.2)	5	(1.5)	6	(1.8)	7	(2.1)
90° long turn elbow	1	(0.3)	2	(0.6)	2	(0.6)	2	(0.6)	3	(0.9)	4	(1.2)	5	(1.5)
tee or cross (flow turned 90°)	4	(1.2)	5	(1.5)	6	(1.8)	8	(2.4)	10	(3.1)	12	(3.7)	15	(4.6)
gate valve	–		–		–		–		1	(0.3)	1	(0.3)	1	(0.3)
butterfly valve	–		–		–		–		6	(1.8)	7	(2.1)	10	(3.1)
swing check valve[a]	4	(1.2)	5	(1.5)	7	(2.1)	9	(2.7)	11	(3.4)	14	(4.3)	16	(4.9)

	$3^1/_2$ in		4 in		5 in		6 in		8 in		10 in		12 in	
45° elbow	3	(0.9)	4	(1.2)	5	(1.5)	7	(2.1)	9	(2.7)	11	(3.4)	13	(4.0)
90° standard elbow	8	(2.4)	10	(3.1)	12	(3.7)	14	(4.3)	18	(5.5)	22	(6.7)	27	(8.2)
90° long turn elbow	5	(1.5)	6	(1.8)	8	(2.4)	9	(2.7)	13	(4.0)	16	(4.9)	18	(5.5)
tee or cross (flow turned 90°)	17	(5.2)	20	(6.1)	25	(7.6)	30	(9.2)	35	(10.7)	50	(15.3)	60	(18.3)
gate valve	1	(0.3)	2	(0.6)	2	(0.6)	3	(0.9)	4	(1.2)	5	(1.5)	6	(1.8)
butterfly valve	–		12	(3.7)	9	(2.7)	10	(3.1)	12	(3.7)	19	(5.8)	21	(6.4)
swing check valve[a]	19	(5.8)	22	(6.7)	27	(8.2)	32	(9.8)	45	(13.7)	55	(16.8)	65	(19.8)

[a]average values

Support Material

APPENDIX 65.D
Standardized Sprinkler System Worksheet

Contract No. _____ Sheet No. _____ of _____

Name and Location _____

nozzle type and location	flow gpm (L/min)	pipe size (in)	fitting and devices	pipe equivalent length	friction loss (psi/ft) (bars/m)	required pressure (psi) (bars)	normal pressure	notes
___ q ___ Q			length ___ fittings ___ total			p_t ___ p_f ___ p_e	p_t ___ p_v ___ p_n	
___ q ___ Q			length ___ fittings ___ total			p_t ___ p_f ___ p_e	p_t ___ p_v ___ p_n	
___ q ___ Q			length ___ fittings ___ total			p_t ___ p_f ___ p_e	p_t ___ p_v ___ p_n	
___ q ___ Q			length ___ fittings ___ total			p_t ___ p_f ___ p_e	p_t ___ p_v ___ p_n	
___ q ___ Q			length ___ fittings ___ total			p_t ___ p_f ___ p_e	p_t ___ p_v ___ p_n	
___ q ___ Q			length ___ fittings ___ total			p_t ___ p_f ___ p_e	p_t ___ p_v ___ p_n	
___ q ___ Q			length ___ fittings ___ total			p_t ___ p_f ___ p_e	p_t ___ p_v ___ p_n	
___ q ___ Q			length ___ fittings ___ total			p_t ___ p_f ___ p_e	p_t ___ p_v ___ p_n	
___ q ___ Q			length ___ fittings ___ total			p_t ___ p_f ___ p_e	p_t ___ p_v ___ p_n	

APPENDIX 67.A
Polyphase Motor Classifications and Characteristics

speed regulations	speed control	starting torque	breakdown torque	application
general-purpose squirrel cage (NEMA design B)				
Drops about 3% for large to 5% for small sizes.	None, except multi-speed types, designed for two to four fixed speeds.	100% for large; 275% for 1 hp 4 pole unit.	200% of full load.	Constant-speed service where starting is not excessive. Fans, blowers, rotary compressors, and centrifugal pumps.
high-torque squirrel cage (NEMA design C)				
Drops about 3% for large to 6% for small sizes.	None, except multi-speed types, designed for two and four fixed speeds.	250% of full load for high-speed to 200% for low-speed designs.	200% of full load.	Constant-speed where fairly high starting torque is required infrequently with starting current about 550% of full load. Reciprocating pumps and compressors, crushers, etc.
high-slip squirrel cage (NEMA design D)				
Drops about 10 to 15% from no load to full load.	None, except multi-speed types, designed for two to four fixed speeds.	225 to 300% full load, depending on speed with rotor resistance.	200%. Will usually not stall until loaded to max torque, which occurs at standstill.	Constant-speed and high starting torque, if starting is not too frequent, and for high-peak loads with or without flywheels. Punch presses, shears, elevators, etc.
low-torque squirrel cage (NEMA design F)				
Drops about 3% for large to 5% for small sizes.	None, except multi-speed types, designed for two to four fixed speeds.	50% of full load for high-speed to 90% for low-speed designs.	135 to 170% of full load.	Constant speed service where starting duty is light. Fans, blowers, centrifugal pumps and similar loads.
wound rotor				
With rotor rings short circuited, drops about 3% for large to 5% for small sizes.	Speed can be reduced to 50% by rotor resistance. Speed varies inversely as load.	Up to 300% depending on external resistance in rotor circuit and how distributed.	300% when rotor slip rings are short circuited.	Where high starting torque with low starting current or where limited speed control is required. Fans, centrifugal and plunger pumps, compressors, conveyors, hoists, cranes, etc.
synchronous				
Constant.	None, except special motors designed for two fixed speeds.	40% for slow to 160% for medium speed 80% pf. Specials develop higher.	Unity-pf motors 170%, 80%-pf motors 225%. Specials, up to 300%	For constant speed service, direct connection to slow-speed machines and where pf correction is required.

Adapted from *Mechanical Engineering*, Design Manual NAVFAC DM-3, Department of the Navy, © 1972.

APPENDIX 67.B
DC and Single-Phase Motor Classifications and Characteristics

speed regulations	speed control	starting torque	breakdown torque	application
series				
Varies inversely as load. Races on light loads and full voltage.	Zero to maximum depending on control and load.	High. Varies as square of voltage. Limited by commutation, heating, and line capacity.	High. Limited by commutation, heating, and line capacity.	Where high starting torque is required and speed can be regulated. Traction, bridges, hoists, gates, car dumpers, car retarders.
shunt				
Drops 3 to 5% from no load to full load.	Any desired range depending on design, type of system.	Good. With constant field, varies directly as voltage applied to armature.	High. Limited by commutation, heating, and line capacity.	Where constant or adjustable speed is required and starting conditions are not severe. Fans, blowers, centrifugal pumps, conveyors, wood and metal-working machines, elevators.
compound				
Drops 7 to 20% from no load to full load depending on amount of compounding.	Any desired range, depending on design, type of control.	Higher than for shunt, depending on amount of compounding.	High. Limited by commutation, heating and line capacity.	Where high starting torque and fairly constant speed is required. Plunger pumps, punch presses, shears, bending rolls, geared elevators, conveyors, hoists.
split-phase				
Drops about 10% from no load to full load.	None.	75% for large to 175% for small sizes.	150% for large to 200% for small sizes.	Constant-speed service where starting is easy. Small fans, centrifugal pumps and light-running machines, where polyphase is not available.
capacitors				
Drops about 5% for large to 10% for small sizes.	None.	150 to 350% of full load depending on design and size.	150% for large to 200% for small sizes.	Constant-speed service for any starting duty and quiet operation, where polyphase current cannot be used.
commutator type				
Drops about 5% for large to 10% for small sizes.	Repulsion induction, none. Brush-shifting types, four to one at full load.	250% for large to 350% for small sizes.	150% for large to 250%.	Constant-speed service for any starting duty where speed control is required and polyphase current cannot be used.

Adapted from *Mechanical Engineering*, Design Manual NAVFAC DM-3, Department of the Navy, © 1972.

APPENDIX 68.A
Noise Reduction and Absorption Coefficients

material	noise absorption coefficients						NRC
	125 Hz	250 Hz	500 Hz	1000 Hz	2000 Hz	4000 Hz	
brick, unglazed	0.03	0.03	0.03	0.04	0.05	0.07	0.04
brick, unglazed, painted	0.01	0.01	0.02	0.02	0.02	0.03	0.02
carpet							
$^1/_8$ in pile height	0.05	0.05	0.10	0.20	0.30	0.40	0.16
$^1/_4$ in pile height	0.05	0.10	0.15	0.30	0.50	0.55	0.26
$^3/_{16}$ in combined pile and foam	0.05	0.10	0.10	0.30	0.40	0.50	0.23
$^5/_{16}$ in combined pile and foam	0.05	0.15	0.30	0.40	0.50	0.60	0.34
concrete block, painted	0.10	0.05	0.06	0.07	0.09	0.08	0.07
fabrics							
light velour, 10 oz per sq yd, hung straight, in contact with wall	0.03	0.04	0.11	0.17	0.24	0.35	0.14
medium velour, 14 oz per sq yd, draped to half area	0.07	0.31	0.49	0.75	0.70	0.60	0.56
heavy velour, 18 oz per sq yd, draped to half area	0.14	0.35	0.55	0.72	0.70	0.65	0.62
floors							
concrete or terrazzo	0.01	0.01	0.01	0.02	0.02	0.02	0.02
linoleum, asphalt, rubber or cork tile on concrete	0.02	0.03	0.03	0.03	0.03	0.02	0.03
wood	0.15	0.11	0.10	0.07	0.06	0.07	0.09
wood parquet in asphalt on concrete	0.04	0.04	0.07	0.06	0.06	0.07	0.06
glass							
$^1/_4$ in, sealed, large panes	0.05	0.03	0.02	0.02	0.03	0.02	0.03
24 oz, operable windows (in closed condition)	0.10	0.05	0.04	0.03	0.03	0.03	0.04
gypsum board, $^1/_2$ in nailed to 2 × 4's 16 in o.c., painted	0.10	0.08	0.05	0.03	0.03	0.03	0.05
marble or glazed tile	0.01	0.01	0.01	0.01	0.02	0.02	0.01
plaster, gypsum or lime,							
rough finish or lath	0.02	0.03	0.04	0.05	0.04	0.03	0.04
same, with smooth finish	0.02	0.02	0.03	0.04	0.04	0.03	0.03
hardwood plywood paneling							
$^1/_4$ in thick, wood frame	0.58	0.22	0.07	0.04	0.03	0.07	0.09
water surface, as in a swimming pool	0.01	0.01	0.01	0.01	0.02	0.03	0.01
wood roof decking, tougue-and-groove cedar	0.24	0.19	0.14	0.08	0.13	0.10	0.14
air, sabins per 1000 cubic ft at 50% RH				0.9	2.3	7.2	
audience, seated, depending on spacing and upholstery of seats*	2.5–4.0	3.5–5.0	4.0–5.5	4.5–6.5	5.0–7.0	4.5–7.0	
seats, heavily upholstered with fabric*	1.5–3.5	3.5–4.5	4.0–5.0	4.0–5.5	3.5–5.5	3.5–4.5	
seats, heavily upholstered with leather, plastic, etc.*	2.5–3.5	3.0–4.5	3.0–4.0	2.0–4.0	1.5–3.5	1.0–3.0	
seats, lightly upholstered with leather, plastic, etc.*			1.5–2.0				
seats, wood veneer, no upholstery*	0.15	0.20	0.25	0.30	0.50	0.50	

*Values given are in sabins per person or unit of seating at the indicated frequency.
Derived from *Performance Data for Acoustical Materials Bulletin*, Acoustical and Board Products Association, 1975.

APPENDIX 68.B
Transmission Loss Through Common Materials
(decibels)

material	density per unit area (lbm/ft^2)	125 Hz	250 Hz	500 Hz	1000 Hz	2000 Hz	4000 Hz	8000 Hz
lead								
$^1/_{32}$ in thick	2	22	24	29	33	40	43	49
$^1/_{64}$ in thick	1	19	20	24	27	33	39	43
plywood								
$^3/_4$ in thick	2	24	22	27	28	25	27	35
$^1/_4$ in thick	0.7	17	15	20	24	28	27	25
lead vinyl	0.5	11	12	15	20	26	32	37
lead vinyl	1.0	15	17	21	28	33	37	43
steel								
18-gauge	2.0	15	19	31	32	35	48	53
16-gauge	2.5	21	30	34	37	40	47	52
sheet metal (viscoelastic laminate core)	2	15	25	28	32	39	42	47
plexiglass								
$^1/_4$ in thick	1.45	16	17	22	28	33	35	35
$^1/_2$ in thick	2.9	21	23	26	32	32	37	37
1 in thick	5.8	25	28	32	32	34	46	46
glass								
$^1/_8$ in thick	1.5	11	17	23	25	26	27	28
$^1/_4$ in thick	3	17	23	25	27	28	29	30
double glass								
$^1/_4 \times ^1/_2 \times ^1/_4$ in		23	24	24	27	28	30	36
$^1/_4 \times 6 \times ^1/_4$ in		25	28	31	37	40	43	47
$^5/_8$ in gypsum								
on 2×2 in stud		23	28	33	43	50	49	50
on staggered stud		26	35	42	52	57	55	57
concrete, 4 in thick	48	29	35	37	43	44	50	55
concrete block, 6 in	36	33	34	35	38	46	52	65
panels of 16 gauge steel, 4 in absorbent, 20 gauge steel		25	35	43	48	52	55	56

(Multiply lbm/ft^2 by 4.882 to obtain kg/m^2.)

Reprinted from *Industrial Noise Control Manual*, HEW Publication NIOSH 75-183, 1975.

APPENDIX 69.A
Standard Cash Flow Factors

multiply	by	to obtain

 F $(P/F, i\%, n)$ P 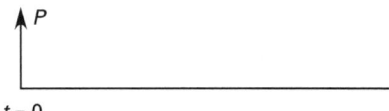

 P $(F/P, i\%, n)$ F 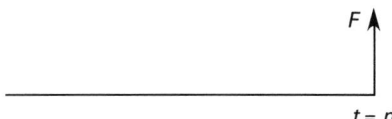

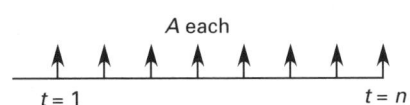 A $(P/A, i\%, n)$ P 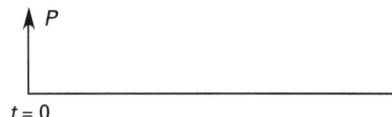

 P $(A/P, i\%, n)$ A 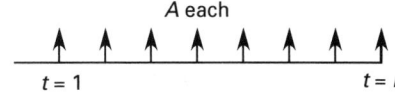

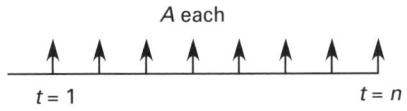 A $(F/A, i\%, n)$ F 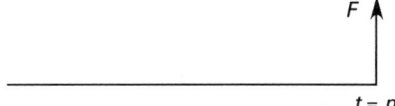

 F $(A/F, i\%, n)$ A 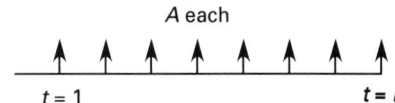

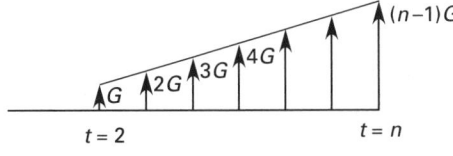 G $(P/G, i\%, n)$ P 

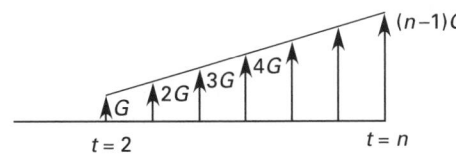 G $(A/G, i\%, n)$ A 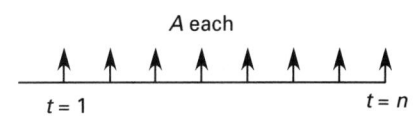

Support Material

PROFESSIONAL PUBLICATIONS, INC. BELMONT, CA

APPENDIX 69.B
Factor Tables

$$I = 0.50\%$$

n	P/F	P/A	P/G	F/P	F/A	A/P	A/F	A/G	n
1	0.9950	0.9950	0.0000	1.0050	1.0000	1.0050	1.0000	0.0000	1
2	0.9901	0.9851	0.9901	1.0100	2.0050	0.5038	0.4988	0.4988	2
3	0.9851	2.9702	2.9604	1.0151	3.0150	0.3367	0.3317	0.9967	3
4	0.9802	3.9505	5.9011	1.0202	4.0301	0.2531	0.2481	1.4938	4
5	0.9754	4.9259	9.8026	1.0253	5.0503	0.2030	0.1980	1.9900	5
6	0.9705	5.8964	14.6552	1.0304	6.0755	0.1696	0.1646	2.4855	6
7	0.9657	6.8621	20.4493	1.0355	7.1059	0.1457	0.1407	2.9801	7
8	0.9609	7.8230	27.1755	1.0407	8.1414	0.1278	0.1228	3.4738	8
9	0.9561	8.7791	34.8244	1.0459	9.1821	0.1139	0.1089	3.9668	9
10	0.9513	9.7304	43.3865	1.0511	10.2280	0.1028	0.0978	4.4589	10
11	0.9466	10.6770	52.8526	1.0564	11.2792	0.0937	0.0887	4.9501	11
12	0.9419	11.6189	63.2136	1.0617	12.3356	0.0861	0.0811	5.4406	12
13	0.9372	12.5562	74.4602	1.0670	13.3972	0.0796	0.0746	5.9302	13
14	0.9326	13.4887	86.5835	1.0723	14.4642	0.0741	0.0691	6.4190	14
15	0.9279	14.4166	99.5743	1.0777	15.5365	0.0694	0.0644	6.9069	15
16	0.9233	15.3399	113.4238	1.0831	16.6142	0.0652	0.0602	7.3940	16
17	0.9187	16.2586	128.1231	1.0885	17.6973	0.0615	0.0565	7.8803	17
18	0.9141	17.1728	143.6634	1.0939	18.7858	0.0582	0.0532	8.3658	18
19	0.9096	18.0824	160.0360	1.0994	19.8797	0.0553	0.0503	8.8504	19
20	0.9051	18.9874	177.2322	1.1049	20.9791	0.0527	0.0477	9.3342	20
21	0.9006	19.8880	195.2434	1.1104	22.0840	0.0503	0.0453	9.8172	21
22	0.8961	20.7841	214.0611	1.1160	23.1944	0.0481	0.0431	10.2993	22
23	0.8916	21.6757	233.6768	1.1216	24.3104	0.0461	0.0411	10.7806	23
24	0.8872	22.5629	254.0820	1.1272	25.4320	0.0443	0.0393	11.2611	24
25	0.8828	23.4456	275.2686	1.1328	26.5591	0.0427	0.0377	11.7407	25
26	0.8784	24.3240	297.2281	1.1385	27.6919	0.0411	0.0361	12.2195	26
27	0.8740	25.1980	319.9523	1.1442	28.8304	0.0397	0.0347	12.6975	27
28	0.8697	26.0677	343.4332	1.1499	29.9745	0.0384	0.0334	13.1747	28
29	0.8653	26.9330	367.6625	1.1556	31.1244	0.0371	0.0321	13.6510	29
30	0.8610	27.7941	392.6324	1.1614	32.2800	0.0360	0.0310	14.1265	30
31	0.8567	28.6508	418.3348	1.1672	33.4414	0.0349	0.0299	14.6012	31
32	0.8525	29.5033	444.7618	1.1730	34.6086	0.0339	0.0289	15.0750	32
33	0.8482	30.3515	471.9055	1.1789	35.7817	0.0329	0.0279	15.5480	33
34	0.8440	31.1955	499.7583	1.1848	36.9606	0.0321	0.0271	16.0202	34
35	0.8398	32.0354	528.3123	1.1907	38.1454	0.0312	0.0262	16.4915	35
36	0.8356	32.8710	557.5598	1.1967	39.3361	0.0304	0.0254	16.9621	36
37	0.8315	33.7025	587.4934	1.2027	40.5328	0.0297	0.0247	17.4317	37
38	0.8274	34.5299	618.1054	1.2087	41.7354	0.0290	0.0240	17.9006	38
39	0.8232	35.3531	649.3883	1.2147	42.9441	0.0283	0.0233	18.3686	39
40	0.8191	36.1722	681.3347	1.2208	44.1588	0.0276	0.0226	18.8359	40
41	0.8151	36.9873	713.9372	1.2269	45.3796	0.0270	0.0220	19.3022	41
42	0.8110	37.7983	747.1886	1.2330	46.6065	0.0265	0.0215	19.7678	42
43	0.8070	38.6053	781.0815	1.2392	47.8396	0.0259	0.0209	20.2325	43
44	0.8030	39.4082	815.6087	1.2454	49.0788	0.0254	0.0204	20.6964	44
45	0.7990	40.2072	850.7631	1.2516	50.3242	0.0249	0.0199	21.1595	45
46	0.7950	41.0022	886.5376	1.2579	51.5758	0.0244	0.0194	21.6217	46
47	0.7910	41.7932	922.9252	1.2642	52.8337	0.0239	0.0189	22.0831	47
48	0.7871	42.5803	959.9188	1.2705	54.0978	0.0235	0.0185	22.5437	48
49	0.7832	43.3635	997.5116	1.2768	55.3683	0.0231	0.0181	23.0035	49
50	0.7793	44.1428	1035.6966	1.2832	56.6452	0.0227	0.0177	23.4624	50
51	0.7754	44.9182	1074.4670	1.2896	57.9284	0.0223	0.0173	23.9205	51
52	0.7716	45.6897	1113.8162	1.2961	59.2180	0.0219	0.0169	24.3778	52
53	0.7677	46.4575	1153.7372	1.3026	60.5141	0.0215	0.0165	24.8343	53
54	0.7639	47.2214	1194.2236	1.3091	61.8167	0.0212	0.0162	25.2899	54
55	0.7601	47.9814	1235.2686	1.3156	63.1258	0.0208	0.0158	25.7447	55
60	0.7414	51.7256	1448.6458	1.3489	69.7700	0.0193	0.0143	28.0064	60
65	0.7231	55.3775	1675.0272	1.3829	76.5821	0.0181	0.0131	30.2475	65
70	0.7053	58.9394	1913.6427	1.4178	83.5661	0.0170	0.0120	32.4680	70
75	0.6879	62.4136	2163.7525	1.4536	90.7265	0.0160	0.0110	34.6679	75
80	0.6710	65.8023	2424.6455	1.4903	98.0677	0.0152	0.0102	36.8474	80
85	0.6545	69.1075	2695.6389	1.5280	105.5943	0.0145	0.0095	39.0065	85
90	0.6383	72.3313	2976.0769	1.5666	113.3109	0.0138	0.0088	41.1451	90
95	0.6226	75.4757	3265.3298	1.6061	121.2224	0.0132	0.0082	43.2633	95
100	0.6073	78.5426	3562.7934	1.6467	129.3337	0.0127	0.0077	45.3613	100

(continued)

APPENDIX 69.B *(continued)*
Factor Tables

$$I = 0.75\%$$

n	P/F	P/A	P/G	F/P	F/A	A/P	A/F	A/G	n
1	0.9926	0.9926	0.0000	1.0075	1.0000	1.0075	1.0000	0.0000	1
2	0.9852	1.9777	0.9852	1.0151	2.0075	0.5056	0.4981	0.4981	2
3	0.9778	2.9556	2.9408	1.0227	3.0226	0.3383	0.3308	0.9950	3
4	0.9706	3.9261	5.8525	1.0303	4.0452	0.2547	0.2472	1.4907	4
5	0.9633	4.8894	9.7058	1.0381	5.0756	0.2045	0.1970	1.9851	5
6	0.9562	5.8456	14.4866	1.0459	6.1136	0.1711	0.1636	2.4782	6
7	0.9490	6.7946	20.1808	1.0537	7.1595	0.1472	0.1397	2.9701	7
8	0.9420	7.7366	26.7747	1.0616	8.2132	0.1293	0.1218	3.4608	8
9	0.9350	8.6716	34.2544	1.0696	9.2748	0.1153	0.1078	3.9502	9
10	0.9280	9.5996	42.6064	1.0776	10.3443	0.1042	0.0967	4.4384	10
11	0.9211	10.5207	51.8174	1.0857	11.4219	0.0951	0.0876	4.9253	11
12	0.9142	11.4349	61.8740	1.0938	12.5076	0.0875	0.0800	5.4110	12
13	0.9074	12.3423	72.7632	1.1020	13.6014	0.0810	0.0735	5.8954	13
14	0.9007	13.2430	84.4720	1.1103	14.7034	0.0755	0.0680	6.3786	14
15	0.8940	14.1370	96.9876	1.1186	15.8137	0.0707	0.0632	6.8606	15
16	0.8873	15.0243	110.2973	1.1270	16.9323	0.0666	0.0591	7.3413	16
17	0.8807	15.9050	124.3887	1.1354	18.0593	0.0629	0.0554	7.8207	17
18	0.8742	16.7792	139.2494	1.1440	19.1947	0.0596	0.0521	8.2989	18
19	0.8676	17.6468	154.8671	1.1525	20.3387	0.0567	0.0492	8.7759	19
20	0.8612	18.5080	171.2297	1.1612	21.4912	0.0540	0.0465	9.2516	20
21	0.8548	19.3628	188.3253	1.1699	22.6524	0.0516	0.0441	9.7261	21
22	0.8484	20.2112	206.1420	1.1787	23.8223	0.0495	0.0420	10.1994	22
23	0.8421	21.0533	224.6682	1.1875	25.0010	0.0475	0.0400	10.6714	23
24	0.8358	21.8891	243.8923	1.1964	26.1885	0.0457	0.0382	11.1422	24
25	0.8296	22.7188	263.8029	1.2054	27.3849	0.0440	0.0365	11.6117	25
26	0.8234	23.5422	284.3888	1.2144	28.5903	0.0425	0.0350	12.0800	26
27	0.8173	24.3595	305.6387	1.2235	29.8047	0.0411	0.0336	12.5470	27
28	0.8112	25.1707	327.5416	1.2327	31.0282	0.0397	0.0322	13.0128	28
29	0.8052	25.9759	350.0867	1.2420	32.2609	0.0385	0.0310	13.4774	29
30	0.7992	26.7751	373.2631	1.2513	33.5029	0.0373	0.0298	13.9407	30
31	0.7932	27.5683	397.0602	1.2607	34.7542	0.0363	0.0288	14.4028	31
32	0.7873	28.3557	421.4675	1.2701	36.0148	0.0353	0.0278	14.8636	32
33	0.7815	29.1371	446.4746	1.2796	37.2849	0.0343	0.0268	15.3232	33
34	0.7757	29.9128	472.0712	1.2892	38.5646	0.0334	0.0259	15.7816	34
35	0.7699	30.6827	498.2471	1.2989	39.8538	0.0326	0.0251	16.2387	35
36	0.7641	31.4468	524.9924	1.3086	41.1527	0.0318	0.0243	16.6946	36
37	0.7585	32.2053	552.2969	1.3185	42.4614	0.0311	0.0236	17.1493	37
38	0.7528	32.9581	580.1511	1.3283	43.7798	0.0303	0.0228	17.6027	38
39	0.7472	33.7053	608.5451	1.3383	45.1082	0.0297	0.0222	18.0549	39
40	0.7416	34.4469	637.4693	1.3483	46.4465	0.0290	0.0215	18.5058	40
41	0.7361	35.1831	666.9144	1.3585	47.7948	0.0284	0.0209	18.9556	41
42	0.7306	35.9137	696.8709	1.3686	49.1533	0.0278	0.0203	19.4040	42
43	0.7252	36.6389	727.3297	1.3789	50.5219	0.0273	0.0198	19.8513	43
44	0.7198	37.3587	758.2815	1.3893	51.9009	0.0268	0.0193	20.2973	44
45	0.7145	38.0732	789.7173	1.3997	53.2901	0.0263	0.0188	20.7421	45
46	0.7091	38.7823	821.6283	1.4102	54.6898	0.0258	0.0183	21.1856	46
47	0.7039	39.4862	854.0056	1.4207	56.1000	0.0253	0.0178	21.6280	47
48	0.6986	40.1848	886.8404	1.4314	57.5207	0.0249	0.0174	22.0691	48
49	0.6934	40.8782	920.1243	1.4421	58.9521	0.0245	0.0170	22.5089	49
50	0.6883	41.5664	953.8486	1.4530	60.3943	0.0241	0.0166	22.9476	50
51	0.6831	42.2496	988.0050	1.4639	61.8472	0.0237	0.0162	23.3850	51
52	0.6780	42.9276	1022.5852	1.4748	63.3111	0.0233	0.0158	23.8211	52
53	0.6730	43.6006	1057.5810	1.4859	64.7859	0.0229	0.0154	24.2561	53
54	0.6680	44.2686	1092.9842	1.4970	66.2718	0.0226	0.0151	24.6898	54
55	0.6630	44.9316	1128.7869	1.5083	67.7688	0.0223	0.0148	25.1223	55
60	0.6387	48.1734	1313.5189	1.5657	75.4241	0.0208	0.0133	27.2665	60
65	0.6153	51.2963	1507.0910	1.6253	83.3709	0.0195	0.0120	29.3801	65
70	0.5927	54.3046	1708.6065	1.6872	91.6201	0.0184	0.0109	31.4634	70
75	0.5710	57.2027	1917.2225	1.7514	100.1833	0.0175	0.0100	33.5163	75
80	0.5500	59.9944	2132.1472	1.8180	109.0725	0.0167	0.0092	35.5391	80
85	0.5299	62.6838	2352.6375	1.8873	118.3001	0.0160	0.0085	37.5318	85
90	0.5104	65.2746	2577.9961	1.9591	127.8790	0.0153	0.0078	39.4946	90
95	0.4917	67.7704	2807.5694	2.0337	137.8225	0.0148	0.0073	41.4277	95
100	0.4737	70.1746	3040.7453	2.1111	148.1445	0.0143	0.0068	43.3311	100

(continued)

APPENDIX 69.B *(continued)*
Factor Tables

$I = 1.00\%$

n	P/F	P/A	P/G	F/P	F/A	A/P	A/F	A/G	n
1	0.9901	0.9901	0.0000	1.0100	1.0000	1.0100	1.0000	0.0000	1
2	0.9803	1.9704	0.9803	1.0201	2.0100	0.5075	0.4975	0.4975	2
3	0.9706	2.9410	2.9215	1.0303	3.0301	0.3400	0.3300	0.9934	3
4	0.9610	3.9020	5.8044	1.0406	4.0604	0.2563	0.2463	1.4876	4
5	0.9515	4.8534	9.6103	1.0510	5.1010	0.2060	0.1960	1.9801	5
6	0.9420	5.7955	14.3205	1.0615	6.1520	0.1725	0.1625	2.4710	6
7	0.9327	6.7282	19.9168	1.0721	7.2135	0.1486	0.1386	2.9602	7
8	0.9235	7.6517	26.3812	1.0829	8.2857	0.1307	0.1207	3.4478	8
9	0.9143	8.5660	33.6959	1.0937	9.3685	0.1167	0.1067	3.9337	9
10	0.9053	9.4713	41.8435	1.1046	10.4622	0.1056	0.0956	4.4179	10
11	0.8963	10.3676	50.8067	1.1157	11.5668	0.0965	0.0865	4.9005	11
12	0.8874	11.2551	60.5687	1.1268	12.6825	0.0888	0.0788	5.3815	12
13	0.8787	12.1337	71.1126	1.1381	13.8093	0.0824	0.0724	5.8607	13
14	0.8700	13.0037	82.4221	1.1495	14.9474	0.0769	0.0669	6.3384	14
15	0.8613	13.8651	94.4810	1.1610	16.0969	0.0721	0.0621	6.8143	15
16	0.8528	14.7179	107.2734	1.1726	17.2579	0.0679	0.0579	7.2886	16
17	0.8444	15.5623	120.7834	1.1843	18.4304	0.0643	0.0543	7.7613	17
18	0.8360	16.3983	134.9957	1.1961	19.6147	0.0610	0.0510	8.2323	18
19	0.8277	17.2260	149.8950	1.2081	20.8109	0.0581	0.0481	8.7017	19
20	0.8195	18.0456	165.4664	1.2202	22.0190	0.0554	0.0454	9.1694	20
21	0.8114	18.8570	181.6950	1.2324	23.2392	0.0530	0.0430	9.6354	21
22	0.8034	19.6604	198.5663	1.2447	24.4716	0.0509	0.0409	10.0998	22
23	0.7954	20.4558	216.0660	1.2572	25.7163	0.0489	0.0389	10.5626	23
24	0.7876	21.2434	234.1800	1.2697	26.9735	0.0471	0.0371	11.0237	24
25	0.7798	22.0232	252.8945	1.2824	28.2432	0.0454	0.0354	11.4831	25
26	0.7720	22.7952	272.1957	1.2953	29.5256	0.0439	0.0339	11.9409	26
27	0.7644	23.5596	292.0702	1.3082	30.8209	0.0424	0.0324	12.3971	27
28	0.7568	24.3164	312.5047	1.3213	32.1291	0.0411	0.0311	12.8516	28
29	0.7493	25.0658	333.4863	1.3345	33.4504	0.0399	0.0299	13.3044	29
30	0.7419	25.8077	355.0021	1.3478	34.7849	0.0387	0.0287	13.7557	30
31	0.7346	26.5423	377.0394	1.3613	36.1327	0.0377	0.0277	14.2052	31
32	0.7273	27.2696	399.5858	1.3749	37.4941	0.0367	0.0267	14.6532	32
33	0.7201	27.9897	422.6291	1.3887	38.8690	0.0357	0.0257	15.0995	33
34	0.7130	28.7027	446.1572	1.4026	40.2577	0.0348	0.0248	15.5441	34
35	0.7059	29.4086	470.1583	1.4166	41.6603	0.0340	0.0240	15.9871	35
36	0.6989	30.1075	494.6207	1.4308	43.0769	0.0332	0.0232	16.4285	36
37	0.6920	30.7995	519.5329	1.4451	44.5076	0.0325	0.0225	16.8682	37
38	0.6852	31.4847	544.8835	1.4595	45.9527	0.0318	0.0218	17.3063	38
39	0.6784	32.1630	570.6616	1.4741	47.4123	0.0311	0.0211	17.7428	39
40	0.6717	32.8347	596.8561	1.4889	48.8864	0.0305	0.0205	18.1776	40
41	0.6650	33.4997	623.4562	1.5038	50.3752	0.0299	0.0199	18.6108	41
42	0.6584	34.1581	650.4514	1.5188	51.8790	0.0293	0.0193	19.0424	42
43	0.6519	34.8100	677.8312	1.5340	53.3978	0.0287	0.0187	19.4723	43
44	0.6454	35.4555	705.5853	1.5493	54.9318	0.0282	0.0182	19.9006	44
45	0.6391	36.0945	733.7037	1.5648	56.4811	0.0277	0.0177	20.3273	45
46	0.6327	36.7272	762.1765	1.5805	58.0459	0.0272	0.0172	20.7524	46
47	0.6265	37.3537	790.9938	1.5963	59.6263	0.0268	0.0168	21.1758	47
48	0.6203	37.9740	820.1460	1.6122	61.2226	0.0263	0.0163	21.5976	48
49	0.6141	38.5881	849.6237	1.6283	62.8348	0.0259	0.0159	22.0178	49
50	0.6080	39.1961	879.4176	1.6446	64.4632	0.0255	0.0155	22.4363	50
51	0.6020	39.7981	909.5186	1.6611	66.1078	0.0251	0.0151	22.8533	51
52	0.5961	40.3942	939.9175	1.6777	67.7689	0.0248	0.0148	23.2686	52
53	0.5902	40.9844	970.6057	1.6945	69.4466	0.0244	0.0144	23.6823	53
54	0.5843	41.5687	1001.5743	1.7114	71.1410	0.0241	0.0141	24.0945	54
55	0.5785	42.1472	1032.8148	1.7285	72.8525	0.0237	0.0137	24.5049	55
60	0.5504	44.9550	1192.8061	1.8167	81.6697	0.0222	0.0122	26.5333	60
65	0.5237	47.6266	1358.3903	1.9094	90.9366	0.0210	0.0110	28.5217	65
70	0.4983	50.1685	1528.6474	2.0068	100.6763	0.0199	0.0099	30.4703	70
75	0.4741	52.5871	1702.7340	2.1091	110.9128	0.0190	0.0090	32.3793	75
80	0.4511	54.8882	1879.8771	2.2167	121.6715	0.0182	0.0082	34.2492	80
85	0.4292	57.0777	2059.3701	2.3298	132.9790	0.0175	0.0075	36.0801	85
90	0.4084	59.1609	2240.5675	2.4486	144.8633	0.0169	0.0069	37.8724	90
95	0.3886	61.1430	2422.8811	2.5735	157.3538	0.0164	0.0064	39.6265	95
100	0.3697	63.0289	2605.7758	2.7048	170.4814	0.0159	0.0059	41.3426	100

(continued)

APPENDIX 69.B *(continued)*
Factor Tables

$$I = 1.50\%$$

n	P/F	P/A	P/G	F/P	F/A	A/P	A/F	A/G	n
1	0.9852	0.9852	0.0000	1.0150	1.0000	1.0150	1.0000	0.0000	1
2	0.9707	1.9559	0.9707	1.0302	2.0150	0.5113	0.4963	0.4963	2
3	0.9563	2.9122	2.8833	1.0457	3.0452	0.3434	0.3284	0.9901	3
4	0.9422	3.8544	5.7098	1.0614	4.0909	0.2594	0.2444	1.4814	4
5	0.9283	4.7826	9.4229	1.0773	5.1523	0.2091	0.1941	1.9702	5
6	0.9145	5.6972	13.9956	1.0934	6.2296	0.1755	0.1605	2.4566	6
7	0.9010	6.5982	19.4018	1.1098	7.3230	0.1516	0.1366	2.9405	7
8	0.8877	7.4859	25.6157	1.1265	8.4328	0.1336	0.1186	3.4219	8
9	0.8746	8.3605	32.6125	1.1434	9.5593	0.1196	0.1046	3.9008	9
10	0.8617	9.2222	40.3675	1.1605	10.7027	0.1084	0.0934	4.3772	10
11	0.8489	10.0711	48.8568	1.1779	11.8633	0.0993	0.0843	4.8512	11
12	0.8364	10.9075	58.0571	1.1956	13.0412	0.0917	0.0767	5.3227	12
13	0.8240	11.7315	67.9454	1.2136	14.2368	0.0852	0.0702	5.7917	13
14	0.8118	12.5434	78.4994	1.2318	15.4504	0.0797	0.0647	6.2582	14
15	0.7999	13.3432	89.6974	1.2502	16.6821	0.0749	0.0599	6.7223	15
16	0.7880	14.1313	101.5178	1.2690	17.9324	0.0708	0.0558	7.1839	16
17	0.7764	14.9076	113.9400	1.2880	19.2014	0.0671	0.0521	7.6431	17
18	0.7649	15.6726	126.9435	1.3073	20.4894	0.0638	0.0488	8.0997	18
19	0.7536	16.4262	140.5084	1.3270	21.7967	0.0609	0.0459	8.5539	19
20	0.7425	17.1686	154.6154	1.3469	23.1237	0.0582	0.0432	9.0057	20
21	0.7315	17.9001	169.2453	1.3671	24.4705	0.0559	0.0409	9.4550	21
22	0.7207	18.6208	184.3798	1.3876	25.8376	0.0537	0.0387	9.9018	22
23	0.7100	19.3309	200.0006	1.4084	27.2251	0.0517	0.0367	10.3462	23
24	0.6995	20.0304	216.0901	1.4295	28.6335	0.0499	0.0349	10.7881	24
25	0.6892	20.7196	232.6310	1.4509	30.0630	0.0483	0.0333	11.2276	25
26	0.6790	21.3986	249.6065	1.4727	31.5140	0.0467	0.0317	11.6646	26
27	0.6690	22.0676	267.0002	1.4948	32.9867	0.0453	0.0303	12.0992	27
28	0.6591	22.7267	284.7958	1.5172	34.4815	0.0440	0.0290	12.5313	28
29	0.6494	23.3761	302.9779	1.5400	35.9987	0.0428	0.0278	12.9610	29
30	0.6398	24.0158	321.5310	1.5631	37.5387	0.0416	0.0266	13.3883	30
31	0.6303	24.6461	340.4402	1.5865	39.1018	0.0406	0.0256	13.8131	31
32	0.6210	25.2671	359.6910	1.6103	40.6883	0.0396	0.0246	14.2355	32
33	0.6118	25.8790	379.2691	1.6345	42.2986	0.0386	0.0236	14.6555	33
34	0.6028	26.4817	399.1607	1.6590	43.9331	0.0378	0.0228	15.0731	34
35	0.5939	27.0756	419.3521	1.6839	45.5921	0.0369	0.0219	15.4882	35
36	0.5851	27.6607	439.8303	1.7091	47.2760	0.0362	0.0212	15.9009	36
37	0.5764	28.2371	460.5822	1.7348	48.9851	0.0354	0.0204	16.3112	37
38	0.5679	28.8051	481.5954	1.7608	50.7199	0.0347	0.0197	16.7191	38
39	0.5595	29.3646	502.8576	1.7872	52.4807	0.0341	0.0191	17.1246	39
40	0.5513	29.9158	524.3568	1.8140	54.2679	0.0334	0.0184	17.5277	40
41	0.5431	30.4590	546.0814	1.8412	56.0819	0.0328	0.0178	17.9284	41
42	0.5351	30.9941	568.0201	1.8688	57.9231	0.0323	0.0173	18.3267	42
43	0.5272	31.5212	590.1617	1.8969	59.7920	0.0317	0.0167	18.7227	43
44	0.5194	32.0406	612.4955	1.9253	61.6889	0.0312	0.0162	19.1162	44
45	0.5117	32.5523	635.0110	1.9542	63.6142	0.0307	0.0157	19.5074	45
46	0.5042	33.0565	657.6979	1.9835	65.5684	0.0303	0.0153	19.8962	46
47	0.4967	33.5532	680.5462	2.0133	67.5519	0.0298	0.0148	20.2826	47
48	0.4894	34.0426	703.5462	2.0435	69.5652	0.0294	0.0144	20.6667	48
49	0.4821	34.5247	726.6884	2.0741	71.6087	0.0290	0.0140	21.0484	49
50	0.4750	34.9997	749.9636	2.1052	73.6828	0.0286	0.0136	21.4277	50
51	0.4680	35.4677	773.3629	2.1368	75.7881	0.0282	0.0132	21.8047	51
52	0.4611	35.9287	796.8774	2.1689	77.9249	0.0278	0.0128	22.1794	52
53	0.4543	36.3830	820.4986	2.2014	80.0938	0.0275	0.0125	22.5517	53
54	0.4475	36.8305	844.2184	2.2344	82.2952	0.0272	0.0122	22.9217	54
55	0.4409	37.2715	868.0285	2.2679	84.5296	0.0268	0.0118	23.2894	55
60	0.4093	39.3803	988.1674	2.4432	96.2147	0.0254	0.0104	25.0930	60
65	0.3799	41.3378	1109.4752	2.6320	108.8028	0.0242	0.0092	26.8393	65
70	0.3527	43.1549	1231.1658	2.8355	122.3638	0.0232	0.0082	28.5290	70
75	0.3274	44.8416	1352.5600	3.0546	136.9728	0.0223	0.0073	30.1631	75
80	0.3039	46.4073	1473.0741	3.2907	152.7109	0.0215	0.0065	31.7423	80
85	0.2821	47.8607	1592.2095	3.5450	169.6652	0.0209	0.0059	33.2676	85
90	0.2619	49.2099	1709.5439	3.8189	187.9299	0.0203	0.0053	34.7399	90
95	0.2431	50.4622	1824.7224	4.1141	207.6061	0.0198	0.0048	36.1602	95
100	0.2256	51.6247	1937.4506	4.4320	228.8030	0.0194	0.0044	37.5295	100

(continued)

APPENDIX 69.B *(continued)*
Factor Tables

$$I = 2.00\%$$

n	P/F	P/A	P/G	F/P	F/A	A/P	A/F	A/G	n
1	0.9804	0.9804	0.0000	1.0200	1.0000	1.0200	1.0000	0.0000	1
2	0.9612	1.9416	0.9612	1.0404	2.0200	0.5150	0.4950	0.4950	2
3	0.9423	2.8839	2.8458	1.0612	3.0604	0.3468	0.3268	0.9868	3
4	0.9238	3.8077	5.6173	1.0824	4.1216	0.2626	0.2426	1.4752	4
5	0.9057	4.7135	9.2403	1.1041	5.2040	0.2122	0.1922	1.9604	5
6	0.8880	5.6014	13.6801	1.1262	6.3081	0.1785	0.1585	2.4423	6
7	0.8706	6.4720	18.9035	1.1487	7.4343	0.1545	0.1345	2.9208	7
8	0.8535	7.3255	24.8779	1.1717	8.5830	0.1365	0.1165	3.3961	8
9	0.8368	8.1622	31.5720	1.1951	9.7546	0.1225	0.1025	3.8681	9
10	0.8203	8.9826	38.9551	1.2190	10.9497	0.1113	0.0913	4.3367	10
11	0.8043	9.7868	46.9977	1.2434	12.1687	0.1022	0.0822	4.8021	11
12	0.7885	10.5753	55.6712	1.2682	13.4121	0.0946	0.0746	5.2642	12
13	0.7730	11.3484	64.9475	1.2936	14.6803	0.0881	0.0681	5.7231	13
14	0.7579	12.1062	74.7999	1.3195	15.9739	0.0826	0.0626	6.1786	14
15	0.7430	12.8493	85.2021	1.3459	17.2934	0.0778	0.0578	6.6309	15
16	0.7284	13.5777	96.1288	1.3728	18.6393	0.0737	0.0537	7.0799	16
17	0.7142	14.2919	107.5554	1.4002	20.0121	0.0700	0.0500	7.5256	17
18	0.7002	14.9920	119.4581	1.4282	21.4123	0.0667	0.0467	7.9681	18
19	0.6864	15.6785	131.8139	1.4568	22.8406	0.0638	0.0438	8.4073	19
20	0.6730	16.3514	144.6003	1.4859	24.2974	0.0612	0.0412	8.8433	20
21	0.6598	17.0112	157.7959	1.5157	25.7833	0.0588	0.0388	9.2760	21
22	0.6468	17.6580	171.3795	1.5460	27.2990	0.0566	0.0366	9.7055	22
23	0.6342	18.2922	185.3309	1.5769	28.8450	0.0547	0.0347	10.1317	23
24	0.6217	18.9139	199.6305	1.6084	30.4219	0.0529	0.0329	10.5547	24
25	0.6095	19.5235	214.2592	1.6406	32.0303	0.0512	0.0312	10.9745	25
26	0.5976	20.1210	229.1987	1.6734	33.6709	0.0497	0.0297	11.3910	26
27	0.5859	20.7069	244.4311	1.7069	35.3443	0.0483	0.0283	11.8043	27
28	0.5744	21.2813	259.9392	1.7410	37.0512	0.0470	0.0270	12.2145	28
29	0.5631	21.8444	275.7064	1.7758	38.7922	0.0458	0.0258	12.6214	29
30	0.5521	22.3965	291.7164	1.8114	40.5681	0.0446	0.0246	13.0251	30
31	0.5412	22.9377	307.9538	1.8476	42.3794	0.0436	0.0236	13.4257	31
32	0.5306	23.4683	324.4035	1.8845	44.2270	0.0426	0.0226	13.8230	32
33	0.5202	23.9886	341.0508	1.9222	46.1116	0.0417	0.0217	14.2172	33
34	0.5100	24.4986	357.8817	1.9607	48.0338	0.0408	0.0208	14.6083	34
35	0.5000	24.9986	374.8826	1.9999	49.9945	0.0400	0.0200	14.9961	35
36	0.4902	25.4888	392.0405	2.0399	51.9944	0.0392	0.0192	15.3809	36
37	0.4806	25.9695	409.3424	2.0807	54.0343	0.0385	0.0185	15.7625	37
38	0.4712	26.4406	426.7764	2.1223	56.1149	0.0378	0.0178	16.1409	38
39	0.4619	26.9026	444.3304	2.1647	58.2372	0.0372	0.0172	16.5163	39
40	0.4529	27.3555	461.9931	2.2080	60.4020	0.0366	0.0166	16.8885	40
41	0.4440	27.7995	479.7535	2.2522	62.6100	0.0360	0.0160	17.2576	41
42	0.4353	28.2348	497.6010	2.2972	64.8622	0.0354	0.0154	17.6237	42
43	0.4268	28.6616	515.5253	2.3432	67.1595	0.0349	0.0149	17.9866	43
44	0.4184	29.0800	533.5165	2.3901	69.5027	0.0344	0.0144	18.3465	44
45	0.4102	29.4902	551.5652	2.4379	71.8927	0.0339	0.0139	18.7034	45
46	0.4022	29.8923	569.6621	2.4866	74.3306	0.0335	0.0135	19.0571	46
47	0.3943	30.2866	587.7985	2.5363	76.8172	0.0330	0.0130	19.4079	47
48	0.3865	30.6731	605.9657	2.5871	79.3535	0.0326	0.0126	19.7556	48
49	0.3790	31.0521	624.1557	2.6388	81.9406	0.0322	0.0122	20.1003	49
50	0.3715	31.4236	642.3606	2.6916	84.5794	0.0318	0.0118	20.4420	50
51	0.3642	31.7878	660.5727	2.7454	87.2710	0.0315	0.0115	20.7807	51
52	0.3571	32.1449	678.7849	2.8003	90.0164	0.0311	0.0111	21.1164	52
53	0.3501	32.4950	696.9900	2.8563	92.8167	0.0308	0.0108	21.4491	53
54	0.3432	32.8383	715.1815	2.9135	95.6731	0.0305	0.0105	21.7789	54
55	0.3365	33.1748	733.3527	2.9717	98.5865	0.0301	0.0101	22.1057	55
60	0.3048	34.7609	823.6975	3.2810	114.0515	0.0288	0.0088	23.6961	60
65	0.2761	36.1975	912.7085	3.6225	131.1262	0.0276	0.0076	25.2147	65
70	0.2500	37.4986	999.8343	3.9996	149.9779	0.0267	0.0067	26.6632	70
75	0.2265	38.6771	1084.6393	4.4158	170.7918	0.0259	0.0059	28.0434	75
80	0.2051	39.7445	1166.7868	4.8754	193.7720	0.0252	0.0052	29.3572	80
85	0.1858	40.7113	1246.0241	5.3829	219.1439	0.0246	0.0046	30.6064	85
90	0.1683	41.5869	1322.1701	5.9431	247.1567	0.0240	0.0040	31.7929	90
95	0.1524	42.3800	1395.1033	6.5617	278.0850	0.0236	0.0036	32.9189	95
100	0.1380	43.0984	1464.7527	7.2446	312.2323	0.0232	0.0032	33.9863	100

(continued)

APPENDIX 69.B (continued)
Factor Tables

$$I = 3.00\%$$

n	P/F	P/A	P/G	F/P	F/A	A/P	A/F	A/G	n
1	0.9709	0.9709	0.0000	1.0300	1.0000	1.0300	1.0000	0.0000	1
2	0.9426	1.9135	0.9426	1.0609	2.0300	0.5226	0.4926	0.4926	2
3	0.9151	2.8286	2.7729	1.0927	3.0909	0.3535	0.3235	0.9803	3
4	0.8885	3.7171	5.4383	1.1255	4.1836	0.2690	0.2390	1.4631	4
5	0.8626	4.5797	8.8888	1.1593	5.3091	0.2184	0.1884	1.9409	5
6	0.8375	5.4172	13.0762	1.1941	6.4684	0.1846	0.1546	2.4138	6
7	0.8131	6.2303	17.9547	1.2299	7.6625	0.1605	0.1305	2.8819	7
8	0.7894	7.0197	23.4806	1.2668	8.8923	0.1425	0.1125	3.3450	8
9	0.7664	7.7861	29.6119	1.3048	10.1591	0.1284	0.0984	3.8032	9
10	0.7441	8.5302	36.3088	1.3439	11.4639	0.1172	0.0872	4.2565	10
11	0.7224	9.2526	43.5330	1.3842	12.8078	0.1081	0.0781	4.7049	11
12	0.7014	9.9540	51.2482	1.4258	14.1920	0.1005	0.0705	5.1485	12
13	0.6810	10.6350	59.4196	1.4685	15.6178	0.0940	0.0640	5.5872	13
14	0.6611	11.2961	68.0141	1.5126	17.0863	0.0885	0.0585	6.0210	14
15	0.6419	11.9379	77.0002	1.5580	18.5989	0.0838	0.0538	6.4500	15
16	0.6232	12.5611	86.3477	1.6047	20.1569	0.0796	0.0496	6.8742	16
17	0.6050	13.1661	96.0280	1.6528	21.7616	0.0760	0.0460	7.2936	17
18	0.5874	13.7535	106.0137	1.7024	23.4144	0.0727	0.0427	7.7081	18
19	0.5703	14.3238	116.2788	1.7535	25.1169	0.0698	0.0398	8.1179	19
20	0.5537	14.8775	126.7987	1.8061	26.8704	0.0672	0.0372	8.5229	20
21	0.5375	15.4150	137.5496	1.8603	28.6765	0.0649	0.0349	8.9231	21
22	0.5219	15.9369	148.5094	1.9161	30.5368	0.0627	0.0327	9.3186	22
23	0.5067	16.4436	159.6566	1.9736	32.4529	0.0608	0.0308	9.7093	23
24	0.4919	16.9355	170.9711	2.0328	34.4265	0.0590	0.0290	10.0954	24
25	0.4776	17.4131	182.4336	2.0938	36.4593	0.0574	0.0274	10.4768	25
26	0.4637	17.8768	194.0260	2.1566	38.5530	0.0559	0.0259	10.8535	26
27	0.4502	18.3270	205.7309	2.2213	40.7096	0.0546	0.0246	11.2255	27
28	0.4371	18.7641	217.5320	2.2879	42.9309	0.0533	0.0233	11.5930	28
29	0.4243	19.1885	229.4137	2.3566	45.2189	0.0521	0.0221	11.9558	29
30	0.4120	19.6004	241.3613	2.4273	47.5754	0.0510	0.0210	12.3141	30
31	0.4000	20.0004	253.3609	2.5001	50.0027	0.0500	0.0200	12.6678	31
32	0.3883	20.3888	265.3993	2.5751	52.5028	0.0490	0.0190	13.0169	32
33	0.3770	20.7658	277.4642	2.6523	55.0778	0.0482	0.0182	13.3616	33
34	0.3660	21.1318	289.5437	2.7319	57.7302	0.0473	0.0173	13.7018	34
35	0.3554	21.4872	301.6267	2.8139	60.4621	0.0465	0.0165	14.0375	35
36	0.3450	21.8323	313.7028	2.8983	63.2759	0.0458	0.0158	14.3688	36
37	0.3350	22.1672	325.7622	2.9852	66.1742	0.0451	0.0151	14.6957	37
38	0.3252	22.4925	337.7956	3.0748	69.1594	0.0445	0.0145	15.0182	38
39	0.3158	22.8082	349.7942	3.1670	72.2342	0.0438	0.0138	15.3363	39
40	0.3066	23.1148	361.7499	3.2620	75.4013	0.0433	0.0133	15.6502	40
41	0.2976	23.4124	373.6551	3.3599	78.6633	0.0427	0.0127	15.9597	41
42	0.2890	23.7014	385.5024	3.4607	82.0232	0.0422	0.0122	16.2650	42
43	0.2805	23.9819	397.2852	3.5645	85.4839	0.0417	0.0117	16.5660	43
44	0.2724	24.2543	408.9972	3.6715	89.0484	0.0412	0.0112	16.8629	44
45	0.2644	24.5187	420.6325	3.7816	92.7199	0.0408	0.0108	17.1556	45
46	0.2567	24.7754	432.1856	3.8950	96.5015	0.0404	0.0104	17.4441	46
47	0.2493	25.0247	443.6515	4.0119	100.3965	0.0400	0.0100	17.7285	47
48	0.2420	25.2667	455.0255	4.1323	104.4084	0.0396	0.0096	18.0089	48
49	0.2350	25.5017	466.3031	4.2562	108.5406	0.0392	0.0092	18.2852	49
50	0.2281	25.7298	477.4803	4.3839	112.7969	0.0389	0.0089	18.5575	50
51	0.2215	25.9512	488.5535	4.5154	117.1808	0.0385	0.0085	18.8258	51
52	0.2150	26.1662	499.5191	4.6509	121.6962	0.0382	0.0082	19.0902	52
53	0.2088	26.3750	510.3742	4.7904	126.3471	0.0379	0.0079	19.3507	53
54	0.2027	26.5777	521.1157	4.9341	131.1375	0.0376	0.0076	19.6073	54
55	0.1968	26.7744	531.7411	5.0821	136.0716	0.0373	0.0073	19.8600	55
60	0.1697	27.6756	583.0526	5.8916	163.0534	0.0361	0.0061	21.0674	60
65	0.1464	28.4529	631.2010	6.8300	194.3328	0.0351	0.0051	22.1841	65
70	0.1263	29.1234	676.0869	7.9178	230.5941	0.0343	0.0043	23.2145	70
75	0.1089	29.7018	717.6978	9.1789	272.6309	0.0337	0.0037	24.1634	75
80	0.0940	30.2008	756.0865	10.6409	321.3630	0.0331	0.0031	25.0353	80
85	0.0811	30.6312	791.3529	12.3357	377.8570	0.0326	0.0026	25.8349	85
90	0.0699	31.0024	823.6302	14.3005	443.3489	0.0323	0.0023	26.5667	90
95	0.0603	31.3227	853.0742	16.5782	519.2720	0.0319	0.0019	27.2351	95
100	0.0520	31.5989	879.8540	19.2186	607.2877	0.0316	0.0016	27.8444	100

(continued)

APPENDIX 69.B *(continued)*
Factor Tables

$$I = 4.00\%$$

n	P/F	P/A	P/G	F/P	F/A	A/P	A/F	A/G	n
1	0.9615	0.9615	0.0000	1.0400	1.0000	1.0400	1.0000	0.0000	1
2	0.9246	1.8861	0.9246	1.0816	2.0400	0.5302	0.4902	0.4902	2
3	0.8890	2.7751	2.7025	1.1249	3.1216	0.3603	0.3203	0.9739	3
4	0.8548	3.6299	5.2670	1.1699	4.2465	0.2755	0.2355	1.4510	4
5	0.8219	4.4518	8.5547	1.2167	5.4163	0.2246	0.1846	1.9216	5
6	0.7903	5.2421	12.5062	1.2653	6.6330	0.1908	0.1508	2.3857	6
7	0.7599	6.0021	17.0657	1.3159	7.8983	0.1666	0.1266	2.8433	7
8	0.7307	6.7327	22.1806	1.3686	9.2142	0.1485	0.1085	3.2944	8
9	0.7026	7.4353	27.8013	1.4233	10.5828	0.1345	0.0945	3.7391	9
10	0.6756	8.1109	33.8814	1.4802	12.0061	0.1233	0.0833	4.1773	10
11	0.6496	8.7605	40.3772	1.5395	13.4864	0.1141	0.0741	4.6090	11
12	0.6246	9.3851	47.2477	1.6010	15.0258	0.1066	0.0666	5.0343	12
13	0.6006	9.9856	54.4546	1.6651	16.6268	0.1001	0.0601	5.4533	13
14	0.5775	10.5631	61.9618	1.7317	18.2919	0.0947	0.0547	5.8659	14
15	0.5553	11.1184	69.7355	1.8009	20.0236	0.0899	0.0499	6.2721	15
16	0.5339	11.6523	77.7441	1.8730	21.8245	0.0858	0.0458	6.6720	16
17	0.5134	12.1657	85.9581	1.9479	23.6975	0.0822	0.0422	7.0656	17
18	0.4936	12.6593	94.3498	2.0258	25.6454	0.0790	0.0390	7.4530	18
19	0.4746	13.1339	102.8933	2.1068	27.6712	0.0761	0.0361	7.8342	19
20	0.4564	13.5903	111.5647	2.1911	29.7781	0.0736	0.0336	8.2091	20
21	0.4388	14.0292	120.3414	2.2788	31.9692	0.0713	0.0313	8.5779	21
22	0.4220	14.4511	129.2024	2.3699	34.2480	0.0692	0.0292	8.9407	22
23	0.4057	14.8568	138.1284	2.4647	36.6179	0.0673	0.0273	9.2973	23
24	0.3901	15.2470	147.1012	2.5633	39.0826	0.0656	0.0256	9.6479	24
25	0.3751	15.6221	156.1040	2.6658	41.6459	0.0640	0.0240	9.9925	25
26	0.3607	15.9828	165.1212	2.7725	44.3117	0.0626	0.0226	10.3312	26
27	0.3468	16.3296	174.1385	1.8834	47.0842	0.0612	0.0212	10.6640	27
28	0.3335	16.6631	183.1424	2.9987	49.9676	0.0600	0.0200	10.9909	28
29	0.3207	16.9837	192.1206	3.1187	52.9663	0.0589	0.0189	11.3120	29
30	0.3083	17.2920	201.0618	3.2434	56.0849	0.0578	0.0178	11.6274	30
31	0.2965	17.5885	209.9556	3.3731	59.3283	0.0569	0.0169	11.9371	31
32	0.2851	17.8736	218.7924	3.5081	62.7015	0.0559	0.0159	12.2411	32
33	0.2741	18.1476	227.5634	3.6484	66.2095	0.0551	0.0151	12.5396	33
34	0.2636	18.4112	236.2607	3.7943	69.8579	0.0543	0.0143	12.8324	34
35	0.2534	18.6646	244.8768	3.9461	73.6522	0.0536	0.0136	13.1198	35
36	0.2437	18.9083	253.4052	4.1039	77.5983	0.0529	0.0129	13.4018	36
37	0.2343	19.1426	261.8399	4.2681	81.7022	0.0522	0.0122	13.6784	37
38	0.2253	19.3679	270.1754	4.4388	85.9703	0.0516	0.0116	13.9497	38
39	0.2166	19.5845	278.4070	4.6164	90.4091	0.0511	0.0111	14.2157	39
40	0.2083	19.7928	286.5303	4.8010	95.0255	0.0505	0.0105	14.4765	40
41	0.2003	19.9931	294.5414	4.9931	99.8265	0.0500	0.0100	14.7322	41
42	0.1926	20.1856	302.4370	5.1928	104.8196	0.0495	0.0095	14.9828	42
43	0.1852	20.3708	310.2141	5.4005	110.0124	0.0491	0.0091	15.2284	43
44	0.1780	20.5488	317.8700	5.6165	115.4129	0.0487	0.0087	15.4690	44
45	0.1712	20.7200	325.4028	5.8412	121.0294	0.0483	0.0083	15.7047	45
46	0.1646	20.8847	332.8104	6.0748	126.8706	0.0479	0.0079	15.9356	46
47	0.1583	21.0429	340.0914	6.3178	132.9454	0.0475	0.0075	16.1618	47
48	0.1522	21.1951	347.2446	6.5705	139.2632	0.0472	0.0072	16.3832	48
49	0.1463	21.3415	354.2689	6.8333	145.8337	0.0469	0.0069	16.6000	49
50	0.1407	21.4822	361.1638	7.1067	152.6671	0.0466	0.0066	16.8122	50
51	0.1353	21.6175	367.9289	7.3910	159.7738	0.0463	0.0063	17.0200	51
52	0.1301	21.7476	374.5638	7.6866	167.1647	0.0460	0.0060	17.2232	52
53	0.1251	21.8727	381.0686	7.9941	174.8513	0.0457	0.0057	17.4221	53
54	0.1203	21.9930	387.4436	8.3138	182.8454	0.0455	0.0055	17.6167	54
55	0.1157	22.1086	393.6890	8.6464	191.1592	0.0452	0.0052	17.8070	55
60	0.0951	22.6235	422.9966	10.5196	237.9907	0.0442	0.0042	18.6972	60
65	0.0781	23.0467	449.2014	12.7987	294.9684	0.0434	0.0034	19.4909	65
70	0.0642	23.3945	472.4789	15.5716	364.2905	0.0427	0.0027	20.1961	70
75	0.0528	23.6804	493.0408	18.9453	448.6314	0.0422	0.0022	20.8206	75
80	0.0434	23.9154	511.1161	23.0498	551.2450	0.0418	0.0018	21.3718	80
85	0.0357	24.1085	526.9384	28.0436	676.0901	0.0415	0.0015	21.8569	85
90	0.0293	24.2673	540.7369	34.1193	827.9833	0.0412	0.0012	22.2826	90
95	0.0241	24.3978	552.7307	41.5114	1012.7846	0.0410	0.0010	22.6550	95
100	0.0198	24.5050	563.1249	50.5049	1237.6237	0.0408	0.0008	22.9800	100

(continued)

APPENDIX 69.B *(continued)*
Factor Tables

$I = 5.00\%$

n	P/F	P/A	P/G	F/P	F/A	A/P	A/F	A/G	n
1	0.9524	0.9524	0.0000	1.0500	1.0000	1.0500	1.0000	0.0000	1
2	0.9070	1.8594	0.9070	1.1025	2.0500	0.5378	0.4878	0.4878	2
3	0.8638	2.7232	2.6347	1.1576	3.1525	0.3672	0.3172	0.9675	3
4	0.8227	3.5460	5.1028	1.2155	4.3101	0.2820	0.2320	1.4391	4
5	0.7835	4.3295	8.2369	1.2763	5.5256	0.2310	0.1810	1.9025	5
6	0.7462	5.0757	11.9680	1.3401	6.8019	0.1970	0.1470	2.3579	6
7	0.7107	5.7864	16.2321	1.4071	8.1420	0.1728	0.1228	2.8052	7
8	0.6768	6.4632	20.9700	1.4775	9.5491	0.1547	0.1047	3.2445	8
9	0.6446	7.1078	26.1268	1.5513	11.0266	0.1407	0.0907	3.6758	9
10	0.6139	7.7217	31.6520	1.6289	12.5779	0.1295	0.0795	4.0991	10
11	0.5847	8.3064	37.4988	1.7103	14.2068	0.1204	0.0704	4.5144	11
12	0.5568	8.8633	43.6241	1.7959	15.9171	0.1128	0.0628	4.9219	12
13	0.5303	9.3936	49.9879	1.8856	17.7130	0.1065	0.0565	5.3215	13
14	0.5051	9.8986	56.5538	1.9799	19.5986	0.1010	0.0510	5.7133	14
15	0.4810	10.3797	63.2880	2.0789	21.5786	0.0963	0.0463	6.0973	15
16	0.4581	10.8378	70.1597	2.1829	23.6575	0.0923	0.0423	6.4736	16
17	0.4363	11.2741	77.1405	2.2920	25.8404	0.0887	0.0387	6.8423	17
18	0.4155	11.6896	84.2043	2.4066	28.1324	0.0855	0.0355	7.2034	18
19	0.3957	12.0853	91.3275	2.5270	30.5390	0.0827	0.0327	7.5569	19
20	0.3769	12.4622	98.4884	2.6533	33.0660	0.0802	0.0302	7.9030	20
21	0.3589	12.8212	105.6673	2.7860	35.7193	0.0780	0.0280	8.2416	21
22	0.3418	13.1630	112.8461	2.9253	38.5052	0.0760	0.0260	8.5730	22
23	0.3256	13.4886	120.0087	3.0715	41.4305	0.0741	0.0241	8.8971	23
24	0.3101	13.7986	127.1402	3.2251	44.5020	0.0725	0.0225	9.2140	24
25	0.2953	14.0939	134.2275	3.3864	47.7271	0.0710	0.0210	9.5238	25
26	0.2812	14.3752	141.2585	3.5557	51.1135	0.0696	0.0196	9.8266	26
27	0.2678	14.6430	148.2226	3.7335	54.6691	0.0683	0.0183	10.1224	27
28	0.2551	14.8981	155.1101	3.9201	58.4026	0.0671	0.0171	10.4114	28
29	0.2429	15.1411	161.9126	4.1161	62.3227	0.0660	0.0160	10.6936	29
30	0.2314	15.3725	168.6226	4.3219	66.4388	0.0651	0.0151	10.9691	30
31	0.2204	15.5928	175.2333	4.5380	70.7608	0.0641	0.0141	11.2381	31
32	0.2099	15.8027	181.7392	4.7649	75.2988	0.0633	0.0133	11.5005	32
33	0.1999	16.0025	188.1351	5.0032	80.0638	0.0625	0.0125	11.7566	33
34	0.1904	16.1929	194.4168	5.2533	85.0670	0.0618	0.0118	12.0063	34
35	0.1813	16.3742	200.5807	5.5160	90.3203	0.0611	0.0111	12.2498	35
36	0.1727	16.5469	206.6237	5.7918	95.8363	0.0604	0.0104	12.4872	36
37	0.1644	16.7113	212.5434	6.0814	101.6281	0.0598	0.0098	12.7186	37
38	0.1566	16.8679	218.3378	6.3855	107.7095	0.0593	0.0093	12.9440	38
39	0.1491	17.0170	224.0054	6.7048	114.0950	0.0588	0.0088	13.1636	39
40	0.1420	17.1591	229.5452	7.0400	120.7998	0.0583	0.0083	13.3775	40
41	0.1353	17.2944	234.9564	7.3920	127.8398	0.0578	0.0078	13.5857	41
42	0.1288	17.4232	240.2389	7.7616	135.2318	0.0574	0.0074	13.7884	42
43	0.1227	17.5459	245.3925	8.1497	142.9933	0.0570	0.0070	13.9857	43
44	0.1169	17.6628	250.4175	8.5572	151.1430	0.0566	0.0066	14.1777	44
45	0.1113	17.7741	255.3145	8.9850	159.7002	0.0563	0.0063	14.3644	45
46	0.1060	17.8801	260.0844	9.4343	168.6852	0.0559	0.0059	14.5461	46
47	0.1009	17.9810	264.7281	9.9060	178.1194	0.0556	0.0056	14.7226	47
48	0.0961	18.0772	269.2467	10.4013	188.0254	0.0553	0.0053	14.8943	48
49	0.0916	18.1687	273.6418	10.9213	198.4267	0.0550	0.0050	15.0611	49
50	0.0872	18.2559	277.9148	11.4674	209.3480	0.0548	0.0048	15.2233	50
51	0.0831	18.3390	282.0673	12.0408	220.8154	0.0545	0.0045	15.3808	51
52	0.0791	18.4181	286.1013	12.6428	232.8562	0.0543	0.0043	15.5337	52
53	0.0753	18.4934	290.0184	13.2749	245.4990	0.0541	0.0041	15.6823	53
54	0.0717	18.5651	293.8208	13.9387	258.7739	0.0539	0.0039	15.8265	54
55	0.0683	18.6335	297.5104	14.6356	272.7126	0.0537	0.0037	15.9664	55
60	0.0535	18.9293	314.3432	18.6792	353.5837	0.0528	0.0028	16.6062	60
65	0.0419	19.1611	328.6910	23.8399	456.7980	0.0522	0.0022	17.1541	65
70	0.0329	19.3427	340.8409	30.4264	588.5285	0.0517	0.0017	17.6212	70
75	0.0258	19.4850	351.0721	38.8327	756.6537	0.0513	0.0013	18.0176	75
80	0.0202	19.5965	359.6460	49.5614	971.2288	0.0510	0.0010	18.3526	80
85	0.0158	19.6838	366.8007	63.2544	1245.0871	0.0508	0.0008	18.6346	85
90	0.0124	19.7523	372.7488	80.7304	1597.6073	0.0506	0.0006	18.8712	90
95	0.0097	19.8059	377.6774	103.0347	2040.6935	0.0505	0.0005	19.0689	95
100	0.0076	19.8479	381.7492	131.5013	2610.0252	0.0504	0.0004	19.2337	100

(continued)

APPENDIX 69.B *(continued)*
Factor Tables

$I = 6.00\%$

n	P/F	P/A	P/G	F/P	F/A	A/P	A/F	A/G	n
1	0.9434	0.9434	0.0000	1.0600	1.0000	1.0600	1.0000	0.0000	1
2	0.8900	1.8334	0.8900	1.1236	2.0600	0.5454	0.4854	0.4854	2
3	0.8396	2.6730	2.5692	1.1910	3.1836	0.3741	0.3141	0.9612	3
4	0.7921	3.4651	4.9455	1.2625	4.3746	0.2886	0.2286	1.4272	4
5	0.7473	4.2124	7.9345	1.3382	5.6371	0.2374	0.1774	1.8836	5
6	0.7050	4.9173	11.4594	1.4185	6.9753	0.2034	0.1434	2.3304	6
7	0.6651	5.5824	15.4497	1.5036	8.3938	0.1791	0.1191	2.7676	7
8	0.6274	6.2098	19.8416	1.5938	9.8975	0.1610	0.1010	3.1952	8
9	0.5919	6.8017	24.5768	1.6895	11.4913	0.1470	0.0870	3.6133	9
10	0.5584	7.3601	29.6023	1.7908	13.1808	0.1359	0.0759	4.0220	10
11	0.5268	7.8869	34.8702	1.8983	14.9716	0.1268	0.0668	4.4213	11
12	0.4970	8.3838	40.3369	2.0122	16.8699	0.1193	0.0593	4.8113	12
13	0.4688	8.8527	45.9629	2.1329	18.8821	0.1130	0.0530	5.1920	13
14	0.4423	9.2950	51.7128	2.2609	21.0151	0.1076	0.0476	5.5635	14
15	0.4173	9.7122	57.5546	2.3966	23.2760	0.1030	0.0430	5.9260	15
16	0.3936	10.1059	63.4592	2.5404	25.6725	0.0990	0.0390	6.2794	16
17	0.3714	10.4773	69.4011	2.6928	28.2129	0.0954	0.0354	6.6240	17
18	0.3503	10.8276	75.3569	2.8543	30.9057	0.0924	0.0324	6.9597	18
19	0.3305	11.1581	81.3062	3.0256	33.7600	0.0896	0.0296	7.2867	19
20	0.3118	11.4699	87.2304	3.2071	36.7856	0.0872	0.0272	7.6051	20
21	0.2942	11.7641	93.1136	3.3996	39.9927	0.0850	0.0250	7.9151	21
22	0.2775	12.0416	98.9412	3.6035	43.3923	0.0830	0.0230	8.2166	22
23	0.2618	12.3034	104.7007	3.8197	46.9958	0.0813	0.0213	8.5099	23
24	0.2470	12.5504	110.3812	4.0489	50.8156	0.0797	0.0197	8.7951	24
25	0.2330	12.7834	115.9732	4.2919	54.8645	0.0782	0.0182	9.0722	25
26	0.2198	13.0032	121.4684	4.5494	59.1564	0.0769	0.0169	9.3414	26
27	0.2074	13.2105	126.8600	4.8223	63.7058	0.0757	0.0157	9.6029	27
28	0.1956	13.4062	132.1420	5.1117	68.5281	0.0746	0.0146	9.8568	28
29	0.1846	13.5907	137.3096	5.4184	73.6398	0.0736	0.0136	10.1032	29
30	0.1741	13.7648	142.3588	5.7435	79.0582	0.0726	0.0126	10.3422	30
31	0.1643	13.9291	147.2864	6.0881	84.8017	0.0718	0.0118	10.5740	31
32	0.1550	14.0840	152.0901	6.4534	90.8898	0.0710	0.0110	10.7988	32
33	0.1462	14.2302	156.7681	6.8406	97.3432	0.0703	0.0103	11.0166	33
34	0.1379	14.3681	161.3192	7.2510	104.1838	0.0696	0.0096	11.2276	34
35	0.1301	14.4982	165.7427	7.6861	111.4348	0.0690	0.0090	11.4319	35
36	0.1227	14.6210	170.0387	8.1473	119.1209	0.0684	0.0084	11.6298	36
37	0.1158	14.7368	174.2072	8.6361	127.2681	0.0679	0.0079	11.8213	37
38	0.1092	14.8460	178.2490	9.1543	135.9042	0.0674	0.0074	12.0065	38
39	0.1031	14.9491	182.1652	9.7035	145.0585	0.0669	0.0069	12.1857	39
40	0.0972	15.0463	185.9568	10.2857	154.7620	0.0665	0.0065	12.3590	40
41	0.0917	15.1380	189.6256	10.9029	165.0477	0.0661	0.0061	12.5264	41
42	0.0865	15.2245	193.1732	11.5570	175.9505	0.0657	0.0057	12.6883	42
43	0.0816	15.3062	196.6017	12.2505	187.5076	0.0653	0.0053	12.8446	43
44	0.0770	15.3832	199.9130	12.9855	199.7580	0.0650	0.0050	12.9956	44
45	0.0727	15.4558	203.1096	13.7646	212.7435	0.0647	0.0047	13.1413	45
46	0.0685	15.5244	206.1938	14.5905	226.5081	0.0644	0.0044	13.2819	46
47	0.0647	15.5890	209.1681	15.4659	241.0986	0.0641	0.0041	13.4177	47
48	0.0610	15.6500	212.0351	16.3939	256.5645	0.0639	0.0039	13.5485	48
49	0.0575	15.7076	214.7972	17.3775	272.9584	0.0637	0.0037	13.6748	49
50	0.0543	15.7619	217.4574	18.4202	290.3359	0.0634	0.0034	13.7964	50
51	0.0512	15.8131	220.0181	19.5254	308.7561	0.0632	0.0032	13.9137	51
52	0.0483	15.8614	222.4823	20.6969	328.2814	0.0630	0.0030	14.0267	52
53	0.0456	15.9070	224.8525	21.9387	348.9783	0.0629	0.0029	14.1355	53
54	0.0430	15.9500	227.1316	23.2550	370.9170	0.0627	0.0027	14.2402	54
55	0.0406	15.9905	229.3222	24.6503	394.1720	0.0625	0.0025	14.3411	55
60	0.0303	16.1614	239.0428	32.9877	533.1282	0.0619	0.0019	14.7909	60
65	0.0227	16.2891	246.9450	44.1450	719.0829	0.0614	0.0014	15.1601	65
70	0.0169	16.3845	253.3271	59.0759	967.9322	0.0610	0.0010	15.4613	70
75	0.0126	16.4558	258.4527	79.0569	1300.9487	0.0608	0.0008	15.7058	75
80	0.0095	16.5091	262.5493	105.7960	1746.5999	0.0606	0.0006	15.9033	80
85	0.0071	16.5489	265.8096	141.5789	2342.9817	0.0604	0.0004	16.0620	85
90	0.0053	16.5787	268.3946	189.4645	3141.0752	0.0603	0.0003	16.1891	90
95	0.0039	16.6009	270.4375	253.5463	4209.1042	0.0602	0.0002	16.2905	95
100	0.0029	16.6175	272.0471	339.3021	5638.3681	0.0602	0.0002	16.3711	100

(continued)

APPENDIX 69.B *(continued)*
Factor Tables

$I = 7.00\%$

n	P/F	P/A	P/G	F/P	F/A	A/P	A/F	A/G	n
1	0.9346	0.9346	0.0000	1.0700	1.0000	1.0700	1.0000	0.0000	1
2	0.8734	1.8080	0.8734	1.1449	2.0700	0.5531	0.4831	0.4831	2
3	0.8163	2.6243	2.5060	1.2250	3.2149	0.3811	0.3111	0.9549	3
4	0.7629	3.3872	4.7947	1.3108	4.4399	0.2952	0.2252	1.4155	4
5	0.7130	4.1002	7.6467	1.4026	5.7507	0.2439	0.1739	1.8650	5
6	0.6663	4.7665	10.9784	1.5007	7.1533	0.2098	0.1398	2.3032	6
7	0.6227	5.3893	14.7149	1.6058	8.6540	0.1856	0.1156	2.7304	7
8	0.5820	5.9713	18.7889	1.7182	10.2598	0.1675	0.0975	3.1465	8
9	0.5439	6.5152	23.1404	1.8385	11.9780	0.1535	0.0835	3.5517	9
10	0.5083	7.0236	27.7156	1.9672	13.8164	0.1424	0.0724	3.9461	10
11	0.4751	7.4987	32.4665	2.1049	15.7836	0.1334	0.0634	4.3296	11
12	0.4440	7.9427	37.3506	2.2522	17.8885	0.1259	0.0559	4.7025	12
13	0.4150	8.3577	42.3302	2.4098	20.1406	0.1197	0.0497	5.0648	13
14	0.3878	8.7455	47.3718	2.5785	22.5505	0.1143	0.0443	5.4167	14
15	0.3624	9.1079	52.4461	2.7590	25.1290	0.1098	0.0398	5.7583	15
16	0.3387	9.4466	57.5271	2.9522	27.8881	0.1059	0.0359	6.0897	16
17	0.3166	9.7632	62.5923	3.1588	30.8402	0.1024	0.0324	6.4110	17
18	0.2959	10.0591	67.6219	3.3799	33.9990	0.0994	0.0294	6.7225	18
19	0.2765	10.3356	72.5991	3.6165	37.3790	0.0968	0.0268	7.0242	19
20	0.2584	10.5940	77.5091	3.8697	40.9955	0.0944	0.0244	7.3163	20
21	0.2415	10.8355	82.3393	4.1406	44.8652	0.0923	0.0223	7.5990	21
22	0.2257	11.0612	87.0793	4.4304	49.0057	0.0904	0.0204	7.8725	22
23	0.2109	11.2722	91.7201	4.7405	53.4361	0.0887	0.0187	8.1369	23
24	0.1971	11.4693	96.2545	5.0724	58.1767	0.0872	0.0172	8.3923	24
25	0.1842	11.6536	100.6765	5.4274	63.2490	0.0858	0.0158	8.6391	25
26	0.1722	11.8258	104.9814	5.8074	68.6765	0.0846	0.0146	8.8773	26
27	0.1609	11.9867	109.1656	6.2139	74.4838	0.0834	0.0134	9.1072	27
28	0.1504	12.1371	113.2264	6.6488	80.6977	0.0824	0.0124	9.3289	28
29	0.1406	12.2777	117.1622	7.1143	87.3465	0.0814	0.0114	9.5427	29
30	0.1314	12.4090	120.9718	7.6123	94.4608	0.0806	0.0106	9.7487	30
31	0.1228	12.5318	124.6550	8.1451	102.0730	0.0798	0.0098	9.9471	31
32	0.1147	12.6466	128.2120	8.7153	110.2182	0.0791	0.0091	10.1381	32
33	0.1072	12.7538	131.6435	9.3253	118.9334	0.0784	0.0084	10.3219	33
34	0.1002	12.8540	134.9507	9.9781	128.2588	0.0778	0.0078	10.4987	34
35	0.0937	12.9477	138.1353	10.6766	138.2369	0.0772	0.0072	10.6687	35
36	0.0875	13.0352	141.1990	11.4239	148.9135	0.0767	0.0067	10.8321	36
37	0.0818	13.1170	144.1441	12.2236	160.3374	0.0762	0.0062	10.9891	37
38	0.0765	13.1935	146.9730	13.0793	172.5610	0.0758	0.0058	11.1398	38
39	0.0715	13.2649	149.6883	13.9948	185.6403	0.0754	0.0054	11.2845	39
40	0.0668	13.3317	152.2928	14.9745	199.6351	0.0750	0.0050	11.4233	40
41	0.0624	13.3941	154.7892	16.0227	214.6096	0.0747	0.0047	11.5565	41
42	0.0583	13.4524	157.1807	17.1443	230.6322	0.0743	0.0043	11.6842	42
43	0.0545	13.5070	159.4702	18.3444	247.7765	0.0740	0.0040	11.8065	43
44	0.0509	13.5579	161.6609	19.6285	266.1209	0.0738	0.0038	11.9237	44
45	0.0476	13.6055	163.7559	21.0025	285.7493	0.0735	0.0035	12.0360	45
46	0.0445	13.6500	165.7584	22.4726	306.7518	0.0733	0.0033	12.1435	46
47	0.0416	13.6916	167.6714	24.0457	329.2244	0.0730	0.0030	12.2463	47
48	0.0389	13.7305	169.4981	25.7289	353.2701	0.0728	0.0028	12.3447	48
49	0.0363	13.7668	171.2417	27.5299	378.9990	0.0726	0.0026	12.4387	49
50	0.0339	13.8007	172.9051	29.4570	406.5289	0.0725	0.0025	12.5287	50
51	0.0317	13.8325	174.4915	31.5190	435.9860	0.0723	0.0023	12.6146	51
52	0.0297	13.8621	176.0037	33.7253	467.5050	0.0721	0.0021	12.6967	52
53	0.0277	13.8898	177.4447	36.0861	501.2303	0.0720	0.0020	12.7751	53
54	0.0259	13.9157	178.8173	38.6122	537.3164	0.0719	0.0019	12.8500	54
55	0.0242	13.9399	180.1243	41.3150	575.9286	0.0717	0.0017	12.9215	55
60	0.0173	14.0392	185.7677	57.9464	813.5204	0.0712	0.0012	13.2321	60
65	0.0123	14.1099	190.1452	81.2729	1146.7552	0.0709	0.0009	13.4760	65
70	0.0088	14.1604	193.5185	113.9894	1614.1342	0.0706	0.0006	13.6662	70
75	0.0063	14.1964	196.1035	159.8760	2269.6574	0.0704	0.0004	13.8136	75
80	0.0045	14.2220	198.0748	224.2344	3189.0627	0.0703	0.0003	13.9273	80
85	0.0032	14.2403	199.5717	314.5003	4478.5761	0.0702	0.0002	14.0146	85
90	0.0023	14.2533	200.7042	441.1030	6287.1854	0.0702	0.0002	14.0812	90
95	0.0016	14.2626	201.5581	618.6697	8823.8535	0.0701	0.0001	14.1319	95
100	0.0012	14.2693	202.2001	867.7163	12381.6618	0.0701	0.0001	14.1703	100

(continued)

APPENDIX 69.B *(continued)*
Factor Tables

$$I = 8.00\%$$

n	P/F	P/A	P/G	F/P	F/A	A/P	A/F	A/G	n
1	0.9259	0.9259	0.0000	1.0800	1.0000	1.0800	1.0000	0.0000	1
2	0.8573	1.7833	0.8573	1.1664	2.0800	0.5608	0.4808	0.4808	2
3	0.7938	2.5771	2.4450	1.2597	3.2464	0.3880	0.3080	0.9487	3
4	0.7350	3.3121	4.6501	1.3605	4.5061	0.3019	0.2219	1.4040	4
5	0.6806	3.9927	7.3724	1.4693	5.8666	0.2505	0.1705	1.8465	5
6	0.6302	4.6229	10.5233	1.5869	7.3359	0.2163	0.1363	2.2763	6
7	0.5835	5.2064	14.0242	1.7138	8.9228	0.1921	0.1121	2.6937	7
8	0.5403	5.7466	17.8061	1.8509	10.6366	0.1740	0.0940	3.0985	8
9	0.5002	6.2469	21.8081	1.9990	12.4876	0.1601	0.0801	3.4910	9
10	0.4632	6.7101	25.9768	2.1589	14.4866	0.1490	0.0690	3.8713	10
11	0.4289	7.1390	30.2657	2.3316	16.6455	0.1401	0.0601	4.2395	11
12	0.3971	7.5361	34.6339	2.5182	18.9771	0.1327	0.0527	4.5957	12
13	0.3677	7.9038	39.0463	2.7196	21.4953	0.1265	0.0465	4.9402	13
14	0.3405	8.2442	43.4723	2.9372	24.2149	0.1213	0.0413	5.2731	14
15	0.3152	8.5595	47.8857	3.1722	27.1521	0.1168	0.0368	5.5945	15
16	0.2919	8.8514	52.2640	3.4259	30.3243	0.1130	0.0330	5.9046	16
17	0.2703	9.1216	56.5883	3.7000	33.7502	0.1096	0.0296	6.2037	17
18	0.2502	9.3719	60.8426	3.9960	37.4502	0.1067	0.0267	6.4920	18
19	0.2317	9.6036	65.0134	4.3157	41.4463	0.1041	0.0241	6.7697	19
20	0.2145	9.8181	69.0898	4.6610	45.7620	0.1019	0.0219	7.0369	20
21	0.1987	10.0168	73.0629	5.0338	50.4229	0.0998	0.0198	7.2940	21
22	0.1839	10.2007	76.9257	5.4365	55.4568	0.0980	0.0180	7.5412	22
23	0.1703	10.3711	80.6726	5.8715	60.8933	0.0964	0.0164	7.7786	23
24	0.1577	10.5288	84.2997	6.3412	66.7648	0.0950	0.0150	8.0066	24
25	0.1460	10.6748	87.8041	6.8485	73.1059	0.0937	0.0137	8.2254	25
26	0.1352	10.8100	91.1842	7.3964	79.9544	0.0925	0.0125	8.4352	26
27	0.1252	10.9352	94.4390	7.9881	87.3508	0.0914	0.0114	8.6363	27
28	0.1159	11.0511	97.5687	8.6271	95.3388	0.0905	0.0105	8.8289	28
29	0.1073	11.1584	100.5738	9.3173	103.9659	0.0896	0.0096	9.0133	29
30	0.0994	11.2578	103.4558	10.0627	113.2832	0.0888	0.0088	9.1897	30
31	0.0920	11.3498	106.2163	10.8677	123.3459	0.0881	0.0081	9.3584	31
32	0.0852	11.4350	108.8575	11.7371	134.2135	0.0875	0.0075	9.5197	32
33	0.0789	11.5139	111.3819	12.6760	145.9506	0.0869	0.0069	9.6737	33
34	0.0730	11.5869	113.7924	13.6901	158.6267	0.0863	0.0063	9.8208	34
35	0.0676	11.6546	116.0920	14.7853	172.3168	0.0858	0.0058	9.9611	35
36	0.0626	11.7172	118.2839	15.9682	187.1021	0.0853	0.0053	10.0949	36
37	0.0580	11.7752	120.3713	17.2456	203.0703	0.0849	0.0049	10.2225	37
38	0.0537	11.8289	122.3579	18.6253	220.3159	0.0845	0.0045	10.3440	38
39	0.0497	11.8786	124.2470	20.1153	238.9412	0.0842	0.0042	10.4597	39
40	0.0460	11.9246	126.0422	21.7245	259.0565	0.0839	0.0039	10.5699	40
41	0.0426	11.9672	127.7470	23.4625	280.7810	0.0836	0.0036	10.6747	41
42	0.0395	12.0067	129.3651	25.3395	304.2435	0.0833	0.0033	10.7744	42
43	0.0365	12.0432	130.8998	27.3666	329.5830	0.0830	0.0030	10.8692	43
44	0.0338	12.0771	132.3547	29.5560	356.9496	0.0828	0.0028	10.9592	44
45	0.0313	12.1084	133.7331	31.9204	386.5056	0.0826	0.0026	11.0447	45
46	0.0290	12.1374	135.0384	34.4741	418.4261	0.0824	0.0024	11.1258	46
47	0.0269	12.1643	136.2739	37.2320	452.9002	0.0822	0.0022	11.2028	47
48	0.0249	12.1891	137.4428	40.2106	490.1322	0.0820	0.0020	11.2758	48
49	0.0230	12.2122	138.5480	43.4274	530.3427	0.0819	0.0019	11.3451	49
50	0.0213	12.2335	139.5928	46.9016	573.7702	0.0817	0.0017	11.4107	50
51	0.0197	12.2532	140.5799	50.6537	620.6718	0.0816	0.0016	11.4729	51
52	0.0183	12.2715	141.5121	54.7060	671.3255	0.0815	0.0015	11.5318	52
53	0.0169	12.2884	142.3923	59.0825	726.0316	0.0814	0.0014	11.5875	53
54	0.0157	12.3041	143.2229	63.8091	785.1141	0.0813	0.0013	11.6403	54
55	0.0145	12.3186	144.0065	68.9139	848.9232	0.0812	0.0012	11.6902	55
60	0.0099	12.3766	147.3000	101.2571	1253.2133	0.0808	0.0008	11.9015	60
65	0.0067	12.4160	149.7387	148.7798	1847.2481	0.0805	0.0005	12.0602	65
70	0.0046	12.4428	151.5326	218.6064	2720.0801	0.0804	0.0004	12.1783	70
75	0.0031	12.4611	152.8448	321.2045	4002.5566	0.0802	0.0002	12.2658	75
80	0.0021	12.4735	153.8001	471.9548	5886.9354	0.0802	0.0002	12.3301	80
85	0.0014	12.4820	154.4925	693.4565	8655.7061	0.0801	0.0001	12.3772	85
90	0.0010	12.4877	154.9925	1018.9151	12723.9386	0.0801	0.0001	12.4116	90
95	0.0007	12.4917	155.3524	1497.1205	18701.5069	0.0801	0.0001	12.4365	95
100	0.0005	12.4943	155.6107	2199.7613	27484.5157	0.0800	0.0000	12.4545	100

(continued)

APPENDIX 69.B *(continued)*
Factor Tables

$$I = 9.00\%$$

n	P/F	P/A	P/G	F/P	F/A	A/P	A/F	A/G	n
1	0.9174	0.9174	0.0000	1.0900	1.0000	1.0900	1.0000	0.0000	1
2	0.8417	1.7591	0.8417	1.1881	2.0900	0.5685	0.4785	0.4785	2
3	0.7722	2.5313	2.3860	1.2950	3.2781	0.3951	0.3051	0.9426	3
4	0.7084	3.2397	4.5113	1.4116	4.5731	0.3087	0.2187	1.3925	4
5	0.6499	3.8897	7.1110	1.5386	5.9847	0.2571	0.1671	1.8282	5
6	0.5963	4.4859	10.0924	1.6771	7.5233	0.2229	0.1329	2.2498	6
7	0.5470	5.0330	13.3746	1.8280	9.2004	0.1987	0.1087	2.6574	7
8	0.5019	5.5348	16.8877	1.9926	11.0285	0.1807	0.0907	3.0512	8
9	0.4604	5.9952	20.5711	2.1719	13.0210	0.1668	0.0768	3.4312	9
10	0.4224	6.4177	24.3728	2.3674	15.1929	0.1558	0.0658	3.7978	10
11	0.3875	6.8052	28.2481	2.5804	17.5603	0.1469	0.0569	4.1510	11
12	0.3555	7.1607	32.1590	2.8127	20.1407	0.1397	0.0497	4.4910	12
13	0.3262	7.4869	36.0731	3.0658	22.9534	0.1336	0.0436	4.8182	13
14	0.2992	7.7862	39.9633	3.3417	26.0192	0.1284	0.0384	5.1326	14
15	0.2745	8.0607	43.8069	3.6425	29.3609	0.1241	0.0341	5.4346	15
16	0.2519	8.3126	47.5849	3.9703	33.0034	0.1203	0.0303	5.7245	16
17	0.2311	8.5436	51.2821	4.3276	36.9737	0.1170	0.0270	6.0024	17
18	0.2120	8.7556	54.8860	4.7171	41.3013	0.1142	0.0242	6.2687	18
19	0.1945	8.9501	58.3868	5.1417	46.0185	0.1117	0.0217	6.5236	19
20	0.1784	9.1285	61.7770	5.6044	51.1601	0.1095	0.0195	6.7674	20
21	0.1637	9.2922	65.0509	6.1088	56.7645	0.1076	0.0176	7.0006	21
22	0.1502	9.4424	68.2048	6.6586	62.8733	0.1059	0.0159	7.2232	22
23	0.1378	9.5802	71.2359	7.2579	69.5319	0.1044	0.0144	7.4357	23
24	0.1264	9.7066	74.1433	7.9111	76.7898	0.1030	0.0130	7.6384	24
25	0.1160	9.8226	76.9265	8.6231	84.7009	0.1018	0.0118	7.8316	25
26	0.1064	9.9290	79.5863	9.3992	93.3240	0.1007	0.0107	8.0156	26
27	0.0976	10.0266	82.1241	10.2451	102.7231	0.0997	0.0097	8.1906	27
28	0.0895	10.1161	84.5419	11.1671	112.9682	0.0989	0.0089	8.3571	28
29	0.0822	10.1983	86.8422	12.1722	124.1354	0.0981	0.0081	8.5154	29
30	0.0754	10.2737	89.0280	13.2677	136.3076	0.0973	0.0073	8.6657	30
31	0.0691	10.3428	91.1024	14.4618	149.5752	0.0967	0.0067	8.8083	31
32	0.0634	10.4062	93.0690	15.7633	164.0370	0.0961	0.0061	8.9436	32
33	0.0582	10.4644	94.9314	17.1820	179.8003	0.0956	0.0056	9.0718	33
34	0.0534	10.5178	96.6935	18.7284	196.9823	0.0951	0.0051	9.1933	34
35	0.0490	10.5668	98.3590	20.4140	215.7108	0.0946	0.0046	9.3083	35
36	0.0449	10.6118	99.9319	22.2512	236.1247	0.0942	0.0042	9.4171	36
37	0.0412	10.6530	101.4162	24.2538	258.3759	0.0939	0.0039	9.5200	37
38	0.0378	10.6908	102.8158	26.4367	282.6298	0.0935	0.0035	9.6172	38
39	0.0347	10.7255	104.1345	28.8160	309.0665	0.0932	0.0032	9.7090	39
40	0.0318	10.7574	105.3762	31.4094	337.8824	0.0930	0.0030	9.7957	40
41	0.0292	10.7866	106.5445	34.2363	369.2919	0.0927	0.0027	9.8775	41
42	0.0268	10.8134	107.6432	37.3175	403.5281	0.0925	0.0025	9.9546	42
43	0.0246	10.8380	108.6758	40.6761	440.8457	0.0923	0.0023	10.0273	43
44	0.0226	10.8605	109.6456	44.3370	481.5218	0.0921	0.0021	10.0958	44
45	0.0207	10.8812	110.5561	48.3273	525.8587	0.0919	0.0019	10.1603	45
46	0.0190	10.9002	111.4103	52.6767	574.1860	0.0917	0.0017	10.2210	46
47	0.0174	10.9176	112.2115	57.4176	626.8628	0.0916	0.0016	10.2780	47
48	0.0160	10.9336	112.9625	62.5852	684.2804	0.0915	0.0015	10.3317	48
49	0.0147	10.9482	113.6661	68.2179	746.8656	0.0913	0.0013	10.3821	49
50	0.0134	10.9617	114.3251	74.3575	815.0836	0.0912	0.0012	10.4295	50
51	0.0123	10.9740	114.9420	81.0497	889.4411	0.0911	0.0011	10.4740	51
52	0.0113	10.9853	115.5193	88.3442	970.4908	0.0910	0.0010	10.5158	52
53	0.0104	10.9957	116.0593	96.2951	1058.8349	0.0909	0.0009	10.5549	53
54	0.0095	11.0053	116.5642	104.9617	1155.1301	0.0909	0.0009	10.5917	54
55	0.0087	11.0140	117.0362	114.4083	1260.0918	0.0908	0.0008	10.6261	55
60	0.0057	11.0480	118.9683	176.0313	1944.7921	0.0905	0.0005	10.7683	60
65	0.0037	11.0701	120.3344	270.8460	2998.2885	0.0903	0.0003	10.8702	65
70	0.0024	11.0844	121.2942	416.7301	4619.2232	0.0902	0.0002	10.9427	70
75	0.0016	11.0938	121.9646	641.1909	7113.2321	0.0901	0.0001	10.9940	75
80	0.0010	11.0998	122.4306	986.5517	10950.5741	0.0901	0.0001	11.0299	80
85	0.0007	11.1038	122.7533	1517.9320	16854.8003	0.0901	0.0001	11.0551	85
90	0.0004	11.1064	122.9758	2335.5266	25939.1842	0.0900	0.0000	11.0726	90
95	0.0003	11.1080	123.1287	3593.4971	39916.6350	0.0900	0.0000	11.0847	95
100	0.0002	11.1091	123.2335	5529.0408	61422.6755	0.0900	0.0000	11.0930	100

(continued)

APPENDIX 69.B *(continued)*
Factor Tables

$I = 10.00\%$

n	P/F	P/A	P/G	F/P	F/A	A/P	A/F	A/G	n
1	0.9091	0.9091	0.0000	1.1000	1.0000	1.1000	1.0000	0.0000	1
2	0.8264	1.7355	0.8264	1.2100	2.1000	0.5762	0.4762	0.4762	2
3	0.7513	2.4869	2.3291	1.3310	3.3100	0.4021	0.3021	0.9366	3
4	0.6830	3.1699	4.3781	1.4641	4.6410	0.3155	0.2155	1.3812	4
5	0.6209	3.7908	6.8618	1.6105	6.1051	0.2638	0.1638	1.8101	5
6	0.5645	4.3553	9.6842	1.7716	7.7156	0.2296	0.1296	2.2236	6
7	0.5132	4.8684	12.7631	1.9487	9.4872	0.2054	0.1054	2.6216	7
8	0.4665	5.3349	16.0287	2.1436	11.4359	0.1874	0.0874	3.0045	8
9	0.4241	5.7590	19.4215	2.3579	13.5795	0.1736	0.0736	3.3724	9
10	0.3855	6.1446	22.8913	2.5937	15.9374	0.1627	0.0627	3.7255	10
11	0.3505	6.4951	26.3963	2.8531	18.5312	0.1540	0.0540	4.0641	11
12	0.3186	6.8137	29.9012	3.1384	21.3843	0.1468	0.0468	4.3884	12
13	0.2897	7.1034	33.3772	3.4523	24.5227	0.1408	0.0408	4.6988	13
14	0.2633	7.3667	36.8005	3.7975	27.9750	0.1357	0.0357	4.9955	14
15	0.2394	7.6061	40.1520	4.1772	31.7725	0.1315	0.0315	5.2789	15
16	0.2176	7.8237	43.4164	4.5950	35.9497	0.1278	0.0278	5.5493	16
17	0.1978	8.0216	46.5819	5.0545	40.5447	0.1247	0.0247	5.8071	17
18	0.1799	8.2014	49.6395	5.5599	45.5992	0.1219	0.0219	6.0526	18
19	0.1635	8.3649	52.5827	6.1159	51.1591	0.1195	0.0195	6.2861	19
20	0.1486	8.5136	55.4069	6.7275	57.2750	0.1175	0.0175	6.5081	20
21	0.1351	8.6487	58.1095	7.4002	64.0025	0.1156	0.0156	6.7189	21
22	0.1228	8.7715	60.6893	8.1403	71.4027	0.1140	0.0140	6.9189	22
23	0.1117	8.8832	63.1462	8.9543	79.5430	0.1126	0.0126	7.1085	23
24	0.1015	8.9847	65.4813	9.8497	88.4973	0.1113	0.0113	7.2881	24
25	0.0923	9.0770	67.6964	10.8347	98.3471	0.1102	0.0102	7.4580	25
26	0.0839	9.1609	69.7940	11.9182	109.1818	0.1092	0.0092	7.6186	26
27	0.0763	9.2372	71.7773	13.1100	121.0999	0.1083	0.0083	7.7704	27
28	0.0693	9.3066	73.6495	14.4210	134.2099	0.1075	0.0075	7.9137	28
29	0.0630	9.3696	75.4146	15.8631	148.6309	0.1067	0.0067	8.0489	29
30	0.0573	9.4269	77.0766	17.4494	164.4940	0.1061	0.0061	8.1762	30
31	0.0521	9.4790	78.6395	19.1943	181.9434	0.1055	0.0055	8.2962	31
32	0.0474	9.5264	80.1078	21.1138	201.1378	0.1050	0.0050	8.4091	32
33	0.0431	9.5694	81.4856	23.2252	222.2515	0.1045	0.0045	8.5152	33
34	0.0391	9.6086	82.7773	25.5477	245.4767	0.1041	0.0041	8.6149	34
35	0.0356	9.6442	83.9872	28.1024	271.0244	0.1037	0.0037	8.7086	35
36	0.0323	9.6765	85.1194	30.9127	299.1268	0.1033	0.0033	8.7965	36
37	0.0294	9.7059	86.1781	34.0039	330.0395	0.1030	0.0030	8.8789	37
38	0.0267	9.7327	87.1673	37.4043	364.0434	0.1027	0.0027	8.9562	38
39	0.0243	9.7570	88.0908	41.1448	401.4478	0.0125	0.0025	9.0285	39
40	0.0221	9.7791	88.9525	45.2593	442.5926	0.1023	0.0023	9.0962	40
41	0.0201	9.7991	89.7560	49.7852	487.8518	0.1020	0.0020	9.1596	41
42	0.0183	9.8174	90.5047	54.7637	537.6370	0.1019	0.0019	9.2188	42
43	0.0166	9.8340	91.2019	60.2401	592.4007	0.1017	0.0017	9.2741	43
44	0.0151	9.8491	91.8508	66.2641	652.6408	0.1015	0.0015	9.3258	44
45	0.0137	9.8628	92.4544	72.8905	718.9048	0.1014	0.0014	9.3740	45
46	0.0125	9.8753	93.0157	80.1795	791.7953	0.1013	0.0013	9.4190	46
47	0.0113	9.8866	93.5372	88.1975	871.9749	0.1011	0.0011	9.4610	47
48	0.0103	9.8969	94.0217	97.0172	960.1723	0.1010	0.0010	9.5001	48
49	0.0094	9.9063	94.4715	106.7190	1057.1896	0.1009	0.0009	9.5365	49
50	0.0085	9.9148	94.8889	117.3909	1163.9085	0.1009	0.0009	9.5704	50
51	0.0077	9.9226	95.2761	129.1299	1281.2994	0.1008	0.0008	9.6020	51
52	0.0070	9.9296	95.6351	142.0429	1410.4293	0.1007	0.0007	9.6313	52
53	0.0064	9.9360	95.9679	156.2472	1552.4723	0.1006	0.0006	9.6586	53
54	0.0058	9.9418	96.2763	171.8719	1708.7195	0.1006	0.0006	9.6840	54
55	0.0053	9.9471	96.5619	189.0591	1880.5914	0.1005	0.0005	9.7075	55
60	0.0033	9.9672	97.7010	304.4816	3034.8164	0.1003	0.0003	9.8023	60
65	0.0020	9.9796	98.4705	490.3707	4893.7073	0.1002	0.0002	9.8672	65
70	0.0013	9.9873	98.9870	789.7470	7887.4696	0.1001	0.0001	9.9113	70
75	0.0008	9.9921	99.3317	1271.8954	12708.9537	0.1001	0.0001	9.9410	75
80	0.0005	9.9951	99.5606	2048.4002	20474.0021	0.1000	0.0000	9.9609	80
85	0.0003	9.9970	99.7120	3298.9690	32979.6903	0.1000	0.0000	9.9742	85
90	0.0002	9.9981	99.8118	5313.0226	53120.2261	0.1000	0.0000	9.9831	90
95	0.0001	9.9988	99.8773	8556.6760	85556.7605	0.1000	0.0000	9.9889	95
100	0.0001	9.9993	99.9202	13780.6123	137796.1234	0.1000	0.0000	9.9927	100

(continued)

APPENDIX 69.B (continued)
Factor Tables

$I = 12.00\%$

n	P/F	P/A	P/G	F/P	F/A	A/P	A/F	A/G	n
1	0.8929	0.8929	0.0000	1.1200	1.0000	1.1200	1.0000	0.0000	1
2	0.7972	1.6901	0.7972	1.2544	2.1200	0.5917	0.4717	0.4717	2
3	0.7118	2.4018	2.2208	1.4049	3.3744	0.4163	0.2963	0.9246	3
4	0.6355	3.0373	4.1273	1.5735	4.7793	0.3292	0.2092	1.3589	4
5	0.5674	3.6048	6.3970	1.7623	6.3528	0.2774	0.1574	1.7746	5
6	0.5066	4.1114	8.9302	1.9738	8.1152	0.2432	0.1232	2.1720	6
7	0.4523	4.5638	11.6443	2.2107	10.0890	0.2191	0.0991	2.5515	7
8	0.4039	4.9676	14.4714	2.4760	12.2997	0.2013	0.0813	2.9131	8
9	0.3606	5.3282	17.3563	2.7731	14.7757	0.1877	0.0677	3.2574	9
10	0.3220	5.6502	20.2541	3.1058	17.5487	0.1770	0.0570	3.5847	10
11	0.2875	5.9377	23.1288	3.4785	20.6546	0.1684	0.0484	3.8953	11
12	0.2567	6.1944	25.9523	3.8960	24.1331	0.1614	0.0414	4.1897	12
13	0.2292	6.4235	28.7024	4.3635	28.0291	0.1557	0.0357	4.4683	13
14	0.2046	6.6282	31.3624	4.8871	32.3926	0.1509	0.0309	4.7317	14
15	0.1827	6.8109	33.9202	5.4736	37.2797	0.1468	0.0268	4.9803	15
16	0.1631	6.9740	36.3670	6.1304	42.7533	0.1434	0.0234	5.2147	16
17	0.1456	7.1196	38.6973	6.8660	48.8837	0.1405	0.0205	5.4353	17
18	0.1300	7.2497	40.9080	7.6900	55.7497	0.1379	0.0179	5.6427	18
19	0.1161	7.3658	42.9979	8.6128	63.4397	0.1358	0.0158	6.8375	19
20	0.1037	7.4694	44.9676	9.6463	72.0524	0.1339	0.0139	6.0202	20
21	0.0926	7.5620	46.8188	10.8038	81.6987	0.1322	0.0122	6.1913	21
22	0.0826	7.6446	48.5543	12.1003	92.5026	0.1308	0.0108	6.3514	22
23	0.0738	7.7184	50.1776	13.5523	104.6029	0.1296	0.0096	6.5010	23
24	0.0659	7.7843	51.6929	15.1786	118.1552	0.1285	0.0085	6.6406	24
25	0.0588	7.8431	53.1046	17.0001	133.3339	0.1275	0.0075	6.7708	25
26	0.0525	7.8957	54.4177	19.0401	150.3339	0.1267	0.0067	6.8921	26
27	0.0469	7.9426	55.6369	21.3249	169.3740	0.1259	0.0059	7.0049	27
28	0.0419	7.9844	56.7674	23.8839	190.6989	0.1252	0.0052	7.1098	28
29	0.0374	8.0218	57.8141	26.7499	214.5828	0.1247	0.0047	7.2071	29
30	0.0334	8.0552	58.7821	29.9599	241.3327	0.1241	0.0041	7.2974	30
31	0.0298	8.0850	59.6761	33.5551	271.2926	0.1237	0.0037	7.3811	31
32	0.0266	8.1116	60.5010	37.5817	304.8477	0.1233	0.0033	7.4586	32
33	0.0238	8.1354	61.2612	42.0915	342.4294	0.1229	0.0029	7.5302	33
34	0.0212	8.1566	61.9612	47.1425	384.5210	0.1226	0.0026	7.5965	34
35	0.0189	8.1755	62.6052	52.7996	431.6635	0.1223	0.0023	7.6577	35
36	0.0169	8.1924	63.1970	59.1356	484.4631	0.1221	0.0021	7.7141	36
37	0.0151	8.2075	63.7406	66.2318	543.5987	0.1218	0.0018	7.7661	37
38	0.0135	8.2210	64.2394	74.1797	609.8305	0.1216	0.0016	7.8141	38
39	0.0120	8.2330	64.6967	83.0812	684.0102	0.1215	0.0015	7.8582	39
40	0.0107	8.2438	65.1159	93.0510	767.0914	0.1213	0.0013	7.8988	40
41	0.0096	8.2534	65.4997	104.2171	860.1424	0.1212	0.0012	7.9361	41
42	0.0086	8.2619	65.8509	116.7231	964.3595	0.1210	0.0010	7.9704	42
43	0.0076	8.2696	66.1722	130.7299	1081.0826	0.1209	0.0009	8.0019	43
44	0.0068	8.2764	66.4659	146.4175	1211.8125	0.1208	0.0008	8.0308	44
45	0.0061	8.2825	66.7342	163.9876	1358.2300	0.1207	0.0007	8.0572	45
46	0.0054	8.2880	66.9792	183.6661	1522.2176	0.1207	0.0007	8.0815	46
47	0.0049	8.2928	67.2028	205.7061	1705.8838	0.1206	0.0006	8.1037	47
48	0.0043	8.2972	67.4068	230.3908	1911.5898	0.1205	0.0005	8.1241	48
49	0.0039	8.3010	67.5929	258.0377	2141.9806	0.1205	0.0005	8.1427	49
50	0.0035	8.3045	67.7624	289.0022	2400.0182	0.1204	0.0004	8.1597	50
51	0.0031	8.3076	67.9169	323.6825	2689.0204	0.1204	0.0004	8.1753	51
52	0.0028	8.3103	68.0576	362.5243	3012.7029	0.1203	0.0003	8.1895	52
53	0.0025	8.3128	68.1856	406.0273	3375.2272	0.1203	0.0003	8.2025	53
54	0.0022	8.3150	68.3022	454.7505	3781.2545	0.1203	0.0003	8.2143	54
55	0.0020	8.3170	68.4082	509.3206	4236.0050	0.1202	0.0002	8.2251	55
60	0.0011	8.3240	68.8100	897.5969	7471.6411	0.1201	0.0001	8.2664	60
65	0.0006	8.3281	69.0581	1581.8725	13173.9374	0.1201	0.0001	8.2922	65
70	0.0004	8.3303	69.2103	2787.7998	23223.3319	0.1200	0.0000	8.3082	70
75	0.0002	8.3316	69.3031	4913.0558	40933.7987	0.1200	0.0000	8.3181	75
80	0.0001	8.3324	69.3594	8658.4831	72145.6925	0.1200	0.0000	8.3241	80
85	0.0001	8.3328	69.3935	15259.2057	127151.7140	0.1200	0.0000	8.3278	85
90	0.0000	8.3330	69.4140	26891.9342	224091.1185	0.1200	0.0000	8.3300	90
95	0.0000	8.3332	69.4263	47392.7766	394931.4719	0.1200	0.0000	8.3313	95
100	0.0000	8.3332	69.4336	83522.2657	696010.5477	0.1200	0.0000	8.3321	100

(continued)

Support Material

APPENDIX 69.B *(continued)*
Factor Tables

$I = 15.00\%$

n	P/F	P/A	P/G	F/P	F/A	A/P	A/F	A/G	n
1	0.8696	0.8696	0.0000	1.1500	1.0000	1.1500	1.0000	0.0000	1
2	0.7561	1.6257	0.7561	1.3225	2.1500	0.6151	0.4651	0.4651	2
3	0.6575	2.2832	2.0712	1.5209	3.4725	0.4380	0.2880	0.9071	3
4	0.5718	2.8550	3.7864	1.7490	4.9934	0.3503	0.2003	1.3263	4
5	0.4972	3.3522	5.7751	2.0114	6.7424	0.2983	0.1483	1.7228	5
6	0.4323	3.7845	7.9368	2.3131	8.7537	0.2642	0.1142	2.0972	6
7	0.3759	4.1604	10.1924	2.6600	11.0668	0.2404	0.0904	2.4498	7
8	0.3269	4.4873	12.4807	3.0590	13.7268	0.2229	0.0729	2.7813	8
9	0.2843	4.7716	14.7548	3.5179	16.7858	0.2096	0.0596	3.0922	9
10	0.2472	5.0188	16.9795	4.0456	20.3037	0.1993	0.0493	3.3832	10
11	0.2149	5.2337	19.1289	4.6524	24.3493	0.1911	0.0411	3.6549	11
12	0.1869	5.4206	21.1849	5.3503	29.0017	0.1845	0.0345	3.9082	12
13	0.1625	5.5831	23.1352	6.1528	34.3519	0.1791	0.0291	4.1438	13
14	0.1413	5.7245	24.9725	7.0757	40.5047	0.1747	0.0247	4.3624	14
15	0.1229	5.8474	26.9630	8.1371	47.5804	0.1710	0.0210	4.5650	15
16	0.1069	5.9542	28.2960	9.3576	55.7175	0.1679	0.0179	4.7522	16
17	0.0929	6.0472	29.7828	10.7613	65.0751	0.1654	0.0154	4.9251	17
18	0.0808	6.1280	31.1565	12.3755	75.8364	0.1632	0.0132	5.0843	18
19	0.0703	6.1982	32.4213	14.2318	88.2118	0.1613	0.0113	5.2307	19
20	0.0611	6.2593	33.5822	16.3665	102.4436	0.1598	0.0098	5.3651	20
21	0.0531	6.3125	34.6448	18.8215	118.8101	0.1584	0.0084	5.4883	21
22	0.0462	6.3587	35.6150	21.6447	137.6316	0.1573	0.0073	5.6010	22
23	0.0402	6.3988	36.4988	24.8915	159.2764	0.1563	0.0063	5.7040	23
24	0.0349	6.4338	37.3023	28.6252	184.1678	0.1554	0.0054	5.7979	24
25	0.0304	6.4641	38.0314	32.9190	212.7930	0.1547	0.0047	5.8834	25
26	0.0264	6.4906	38.6918	37.8568	245.7120	0.1541	0.0041	5.9612	26
27	0.0230	6.5135	39.2890	43.5353	283.5688	0.1535	0.0035	6.0319	27
28	0.0200	6.5335	39.8283	50.0656	327.1041	0.1531	0.0031	6.0960	28
29	0.0174	6.5509	40.3146	57.5755	377.1697	0.1527	0.0027	6.1541	29
30	0.0151	6.5660	40.7526	66.2118	434.7451	0.1523	0.0023	6.2066	30
31	0.0131	6.5791	41.1466	76.1435	500.9569	0.1520	0.0020	6.2541	31
32	0.0114	6.5905	41.5006	87.5651	577.1005	0.1517	0.0017	6.2970	32
33	0.0099	6.6005	41.8184	100.6998	664.6655	0.1515	0.0015	6.3357	33
34	0.0086	6.6091	42.1033	115.8048	765.3654	0.1513	0.0013	6.3705	34
35	0.0075	6.6166	42.3586	133.1755	881.1702	0.1511	0.0011	6.4019	35
36	0.0065	6.6231	42.5872	153.1519	1014.3457	0.1510	0.0010	6.4301	36
37	0.0057	6.6288	42.7916	176.1246	1167.4975	0.1509	0.0009	6.4554	37
38	0.0049	6.6338	42.9743	202.5433	1343.6222	0.1507	0.0007	6.4781	38
39	0.0043	6.6380	43.1374	232.9248	1546.1655	0.1506	0.0006	6.4985	39
40	0.0037	6.6418	43.2830	267.8635	1779.0903	0.1506	0.0006	6.5168	40
41	0.0032	6.6450	43.4128	308.0431	2046.9539	0.1505	0.0005	6.5331	41
42	0.0028	6.6478	43.5286	354.2495	2354.9969	0.1504	0.0004	6.5478	42
43	0.0025	6.6503	43.6317	407.3870	2709.2465	0.1504	0.0004	6.5609	43
44	0.0021	6.6524	43.7235	468.4950	3116.6334	0.1503	0.0003	6.5725	44
45	0.0019	6.6543	43.8051	538.7693	3585.1285	0.1503	0.0003	6.5830	45
46	0.0016	6.6559	43.8778	619.5847	4123.8977	0.1502	0.0002	6.5923	46
47	0.0014	6.6573	43.9423	712.5224	4743.4824	0.1502	0.0002	6.6006	47
48	0.0012	6.6585	43.9997	819.4007	5456.0047	0.1502	0.0002	6.6080	48
49	0.0011	6.6596	44.0506	942.3108	6275.4055	0.1502	0.0002	6.6146	49
50	0.0009	6.6605	44.0958	1083.6574	7217.7163	0.1501	0.0001	6.6205	50
51	0.0008	6.6613	44.1360	1246.2061	8301.3737	0.1501	0.0001	6.6257	51
52	0.0007	6.6620	44.1715	1433.1370	9547.5798	0.1501	0.0001	6.6304	52
53	0.0006	6.6626	44.2031	1648.1075	10980.7167	0.1501	0.0001	6.6345	53
54	0.0005	6.6631	44.2311	1895.3236	12628.8243	0.1501	0.0001	6.6382	54
55	0.0005	6.6636	44.2558	2179.6222	14524.1479	0.1501	0.0001	6.6414	55
60	0.0002	6.6651	44.3431	4383.9987	29219.9916	0.1500	0.0000	6.6530	60
65	0.0001	6.6659	44.3903	8817.7874	58778.5826	0.1500	0.0000	6.6593	65
70	0.0001	6.6663	44.4156	17735.7200	118231.4669	0.1500	0.0000	6.6627	70
75	0.0000	6.6665	44.4292	35672.8680	237812.4532	0.1500	0.0000	6.6646	75
80	0.0000	6.6666	44.4364	71750.8794	478332.5293	0.1500	0.0000	6.6656	80
85	0.0000	6.6666	44.4402	144316.6470	962104.3133	0.1500	0.0000	6.6661	85
90	0.0000	6.6666	44.4422	290272.3252	1935142.1680	0.1500	0.0000	6.6664	90
95	0.0000	6.6667	44.4433	583841.3276	3892268.8509	0.1500	0.0000	6.6665	95
100	0.0000	6.6667	44.4438	1174313.4507	7828749.6713	0.1500	0.0000	6.6666	100

(continued)

APPENDIX 69.B *(continued)*
Factor Tables

$I = 20.00\%$

n	P/F	P/A	P/G	F/P	F/A	A/P	A/F	A/G	n
1	0.8333	0.8333	0.0000	1.2000	1.0000	1.2000	1.0000	0.0000	1
2	0.6944	1.5278	0.6944	1.4400	2.2000	0.6545	0.4545	0.4545	2
3	0.5787	2.1065	1.8519	1.7280	3.6400	0.4747	0.2747	0.8791	3
4	0.4823	2.5887	3.2986	2.0736	5.3680	0.3863	0.1863	1.2742	4
5	0.4019	2.9906	4.9061	2.4883	7.4416	0.3344	0.1344	1.6405	5
6	0.3349	3.3255	6.5806	2.9860	9.9299	0.3007	0.1007	1.9788	6
7	0.2791	3.6046	8.2551	3.5832	12.9159	0.2774	0.0774	2.2902	7
8	0.2326	3.8372	9.8831	4.2998	16.4991	0.2606	0.0606	2.5756	8
9	0.1938	4.0310	11.4335	5.1598	20.7989	0.2481	0.0481	2.8364	9
10	0.1615	4.1925	12.8871	6.1917	25.9587	0.2385	0.0385	3.0739	10
11	0.1346	4.3271	14.2330	7.4301	32.1504	0.2311	0.0311	3.2893	11
12	0.1122	4.4392	15.4667	8.9161	39.5805	0.2253	0.0253	3.4841	12
13	0.0935	4.5327	16.5883	10.6993	48.4966	0.2206	0.0206	3.6597	13
14	0.0779	4.6106	17.6008	12.8392	59.1959	0.2169	0.0169	3.8175	14
15	0.0649	4.6755	18.5095	15.4070	72.0351	0.2139	0.0139	3.9588	15
16	0.0541	4.7296	19.3208	18.4884	87.4421	0.2114	0.0114	4.0851	16
17	0.0451	4.7746	20.0419	22.1861	105.9306	0.2094	0.0094	4.1976	17
18	0.0376	4.8122	20.6805	26.6233	128.1167	0.2078	0.0078	4.2975	18
19	0.0313	4.8435	21.2439	31.9480	154.7400	0.2065	0.0065	4.3861	19
20	0.0261	4.8696	21.7395	38.3376	186.6880	0.2054	0.0054	4.4643	20
21	0.0217	4.8913	22.1742	46.0051	225.0256	0.2044	0.0044	4.5334	21
22	0.0181	4.9094	22.5546	55.2061	271.0307	0.2037	0.0037	4.5941	22
23	0.0151	4.9245	22.8867	66.2474	326.2369	0.2031	0.0031	4.6475	23
24	0.0126	4.9371	23.1760	79.4968	392.4842	0.2025	0.0025	4.6943	24
25	0.0105	4.9476	23.4276	95.3962	471.9811	0.2021	0.0021	4.7352	25
26	0.0087	4.9563	23.6460	114.4755	567.3773	0.2018	0.0018	4.7709	26
27	0.0073	4.9636	23.8353	137.3706	681.8528	0.2015	0.0015	4.8020	27
28	0.0061	4.9697	23.9991	164.8447	819.2233	0.2012	0.0012	4.8291	28
29	0.0051	4.9747	24.1406	197.8136	984.0680	0.2010	0.0010	4.8527	29
30	0.0042	4.9789	24.2628	237.3763	1181.8816	0.2008	0.0008	4.8731	30
31	0.0035	4.9824	24.3681	284.8516	1419.2579	0.2007	0.0007	4.8908	31
32	0.0029	4.9854	24.4588	341.8219	1704.1095	0.2006	0.0006	4.9061	32
33	0.0024	4.9878	24.5368	410.1863	2045.9314	0.2005	0.0005	4.9194	33
34	0.0020	4.9898	24.6038	492.2235	2456.1176	0.2004	0.0004	4.9308	34
35	0.0017	4.9915	24.6614	590.6682	2948.3411	0.2003	0.0003	4.9406	35
36	0.0014	4.9929	24.7108	708.8019	3539.0094	0.2003	0.0003	4.9491	36
37	0.0012	4.9941	24.7531	850.5622	4247.8112	0.2002	0.0002	4.9564	37
38	0.0010	4.9951	24.7894	1020.6747	5098.3735	0.2002	0.0002	4.9627	38
39	0.0008	4.9959	24.8204	1224.8096	6119.0482	0.2002	0.0002	4.9681	39
40	0.0007	4.9966	24.8469	1469.7716	7343.8578	0.2001	0.0001	4.9728	40
41	0.0006	4.9972	24.8696	1763.7259	8813.6294	0.2001	0.0001	4.9767	41
42	0.0005	4.9976	24.8890	2116.4711	10577.3553	0.2001	0.0001	4.9801	42
43	0.0004	4.9980	24.9055	2539.7653	12693.8263	0.2001	0.0001	4.9831	43
44	0.0003	4.9984	24.9196	3047.7183	15233.5916	0.2001	0.0001	4.9856	44
45	0.0003	4.9986	24.9316	3657.2620	18281.3099	0.2001	0.0001	4.9877	45
46	0.0002	4.9989	24.9419	4388.7144	21938.5719	0.2000	0.0000	4.9895	46
47	0.0002	4.9991	24.9506	5266.4573	26327.2863	0.2000	0.0000	4.9911	47
48	0.0002	4.9992	24.9581	6319.7487	31593.7436	0.2000	0.0000	4.9924	48
49	0.0001	4.9993	24.9644	7583.6985	37913.4923	0.2000	0.0000	4.9935	49
50	0.0001	4.9995	24.9698	9100.4382	45497.1908	0.2000	0.0000	4.9945	50
51	0.0001	4.9995	24.9744	10920.5258	54597.6289	0.2000	0.0000	4.9953	51
52	0.0001	4.9996	24.9783	13104.6309	65518.1547	0.2000	0.0000	4.9960	52
53	0.0001	4.9997	24.9816	15725.5571	78622.7856	0.2000	0.0000	4.9966	53
54	0.0001	4.9997	24.9844	18870.6685	94348.3427	0.2000	0.0000	4.9971	54
55	0.0000	4.9998	24.9868	22644.8023	113219.0113	0.2000	0.0000	4.9976	55
60	0.0000	4.9999	24.9942	56347.5144	281732.5718	0.2000	0.0000	4.9989	60
65	0.0000	5.0000	24.9975	140210.6469	701048.2346	0.2000	0.0000	4.9995	65
70	0.0000	5.0000	24.9989	348888.9569	1744439.7847	0.2000	0.0000	4.9998	70
75	0.0000	5.0000	24.9995	868147.3693	4340731.8466	0.2000	0.0000	4.9999	75

(continued)

APPENDIX 69.B *(continued)*
Factor Tables

$$I = 25.00\%$$

n	P/F	P/A	P/G	F/P	F/A	A/P	A/F	A/G	n
1	0.8000	0.8000	0.0000	1.2500	1.0000	1.2500	1.0000	0.0000	1
2	0.6400	1.4400	0.6400	1.5625	2.2500	0.6944	0.0444	0.4444	2
3	0.5120	1.9520	1.6640	1.9531	3.8125	0.5123	0.2623	0.8525	3
4	0.4096	2.3616	2.8928	2.4414	5.7656	0.4234	0.1734	1.2249	4
5	0.3277	2.6893	4.2035	3.0518	8.2070	0.3718	0.1218	1.5631	5
6	0.2621	2.9514	5.5142	3.8147	11.2588	0.3383	0.0888	1.8683	6
7	0.2097	3.1611	6.7725	4.7684	15.0735	0.3163	0.0663	2.1424	7
8	0.1678	3.3289	7.9469	5.9605	19.8419	0.3004	0.0504	2.3872	8
9	0.1342	3.4631	9.0207	7.4506	25.8023	0.2888	0.0388	2.6048	9
10	0.1074	3.5705	9.9870	9.3132	33.2529	0.2801	0.0301	2.7971	10
11	0.0859	3.6564	10.8460	11.6415	42.5661	0.2735	0.0235	2.9663	11
12	0.0687	3.7251	11.6020	14.5519	54.2077	0.2684	0.0184	3.1145	12
13	0.0550	3.7801	12.2617	18.1899	68.7596	0.2645	0.0145	3.2437	13
14	0.0440	3.8241	12.8334	22.7374	86.9495	0.2615	0.0115	3.3559	14
15	0.0352	3.8593	13.3260	28.4217	109.6868	0.2591	0.0091	3.4530	15
16	0.0281	3.8874	13.7482	35.5271	138.1085	0.2572	0.0072	3.5366	16
17	0.0225	3.9099	14.1085	44.4089	173.6357	0.2558	0.0058	3.6084	17
18	0.0180	3.9279	14.4147	55.5112	218.0446	0.2546	0.0046	3.6698	18
19	0.0144	3.9424	14.6741	69.3889	273.5558	0.2537	0.0037	3.7222	19
20	0.0115	3.9539	14.8932	86.7362	342.9447	0.2529	0.0029	3.7667	20
21	0.0092	3.9631	15.0777	108.4202	429.6809	0.2523	0.0023	3.8045	21
22	0.0074	3.9705	15.2326	135.5253	538.1011	0.2519	0.0019	3.8365	22
23	0.0059	3.9764	15.3625	169.4066	673.6264	0.2515	0.0015	3.8634	23
24	0.0047	3.9811	15.4711	211.7582	843.0329	0.2512	0.0012	3.8861	24
25	0.0038	3.9849	15.5618	264.6978	1054.7912	0.2509	0.0009	3.9052	25
26	0.0030	3.9879	15.6373	330.8722	1319.4890	0.2508	0.0008	3.9212	26
27	0.0024	3.9903	15.7002	413.5903	1650.3612	0.2506	0.0006	3.9346	27
28	0.0019	3.9923	15.7524	516.9879	2063.9515	0.2505	0.0005	3.9457	28
29	0.0015	3.9938	15.7957	646.2349	2580.9394	0.2504	0.0004	3.9551	29
30	0.0012	3.9950	15.8316	807.7936	3227.1743	0.2503	0.0003	3.9628	30
31	0.0010	3.9960	15.8614	1009.7420	4034.9678	0.2502	0.0002	3.9693	31
32	0.0008	3.9968	15.8859	1262.1774	5044.7098	0.2502	0.0002	3.9746	32
33	0.0006	3.9975	15.9062	1577.7218	6306.8872	0.2502	0.0002	3.9791	33
34	0.0005	3.9980	15.9229	1972.1523	7884.6091	0.2501	0.0001	3.9828	34
35	0.0004	3.9984	15.9367	2465.1903	9856.7613	0.2501	0.0001	3.9858	35
36	0.0003	3.9987	15.9481	3081.4879	12321.9516	0.2501	0.0001	3.9883	36
37	0.0003	3.9990	15.9574	3851.8599	15403.4396	0.2501	0.0001	3.9904	37
38	0.0002	3.9992	15.9651	4814.8249	19255.2994	0.2501	0.0001	3.9921	38
39	0.0002	3.9993	15.9714	6018.5311	24070.1243	0.2500	0.0000	3.9935	39
40	0.0001	3.9995	15.9766	7523.1638	30088.6554	0.2500	0.0000	3.9947	40
41	0.0001	3.9996	15.9809	9403.9548	37611.8192	0.2500	0.0000	3.9956	41
42	0.0001	3.9997	15.9843	11754.9435	47015.7740	0.2500	0.0000	3.9964	42
43	0.0001	3.9997	15.9872	14693.6794	58770.7175	0.2500	0.0000	3.9971	43
44	0.0001	3.9998	15.9895	18367.0992	73464.3969	0.2500	0.0000	3.9976	44
45	0.0000	3.9998	15.9915	22958.8740	91831.4962	0.2500	0.0000	3.9980	45
46	0.0000	3.9999	15.9930	28698.5925	114790.3702	0.2500	0.0000	3.9984	46
47	0.0000	3.9999	15.9943	35873.2407	143488.9627	0.2500	0.0000	3.9987	47
48	0.0000	3.9999	15.9954	44841.5509	179362.2034	0.2500	0.0000	3.9989	48
49	0.0000	3.9999	15.9962	56051.9386	224203.7543	0.2500	0.0000	3.9991	49
50	0.0000	3.9999	15.9969	70064.9232	280255.6929	0.2500	0.0000	3.9993	50
51	0.0000	4.0000	15.9975	87581.1540	350320.6161	0.2500	0.0000	3.9994	51
52	0.0000	4.0000	15.9980	109476.4425	437901.7701	0.2500	0.0000	3.9995	52
53	0.0000	4.0000	15.9983	136845.5532	547378.2126	0.2500	0.0000	3.9996	53
54	0.0000	4.0000	15.9986	171056.9414	684223.7658	0.2500	0.0000	3.9997	54
55	0.0000	4.0000	15.9989	213821.1768	855280.7072	0.2500	0.0000	3.9997	55
60	0.0000	4.0000	15.9996	652530.4468	2610117.7872	0.2500	0.0000	3.9999	60

(continued)

APPENDIX 69.B *(continued)*
Factor Tables

$$I = 30.00\%$$

n	P/F	P/A	P/G	F/P	F/A	A/P	A/F	A/G	n
1	0.7692	0.7692	0.0000	1.3000	1.0000	1.3000	1.0000	0.000	1
2	0.5917	1.3609	0.5917	1.6900	2.3000	0.7348	0.4348	0.434	2
3	0.4552	1.8161	1.5020	2.1970	3.9900	0.5506	0.2506	0.827	3
4	0.3501	2.1662	2.5524	2.8561	6.1870	0.4616	0.1616	1.178	4
5	0.2693	2.4356	3.6297	3.7129	9.0431	0.4106	0.1106	1.490	5
6	0.2072	2.6427	4.6656	4.8268	12.7560	0.3784	0.0784	1.765	6
7	0.1594	2.8021	5.6218	6.2749	17.5828	0.3569	0.0569	2.006	7
8	0.1226	2.9247	6.4800	8.1573	23.8577	0.3419	0.0419	2.215	8
9	0.0943	3.0190	7.2343	10.6045	32.0150	0.3312	0.0312	2.396	9
10	0.0725	3.0915	7.8872	13.7858	42.6195	0.3235	0.0235	2.551	10
11	0.0558	3.1473	8.4452	17.9216	56.4053	0.3177	0.0177	2.683	11
12	0.0429	3.1903	8.9173	23.2981	74.3270	0.3135	0.0135	2.795	12
13	0.0330	3.2233	9.3135	30.2875	97.6250	0.3102	0.0102	2.889	13
14	0.0254	3.2487	9.6437	39.3738	127.9125	0.3078	0.0078	2.968	14
15	0.0195	3.2682	9.9172	51.1859	167.2863	0.3060	0.0060	3.034	15
16	0.0150	3.2832	10.1426	66.5417	218.4722	0.3046	.0046	3.089	16
17	0.0116	3.2948	10.3276	86.5042	285.0139	0.3035	0.0035	3.134	17
18	0.0089	3.3037	10.4788	112.4554	371.5180	0.3027	0.0027	3.171	18
19	0.0068	3.3105	10.6019	146.1920	483.9734	0.3021	0.0021	3.202	19
20	0.0053	3.3158	10.7019	190.0496	630.1655	0.3016	0.0016	3.227	20
21	0.0040	3.3198	10.7828	247.0645	820.2151	0.3012	0.0012	3.248	21
22	0.0031	3.3230	10.8482	321.1839	1067.2796	0.3009	0.0009	3.264	22
23	0.0024	3.3254	10.9009	417.5391	1388.4635	0.3007	0.0007	3.278	23
24	0.0018	3.3272	10.9433	542.8008	1806.0026	0.3006	0.0006	3.289	24
25	0.0014	3.3286	10.9773	705.6410	2348.8033	0.3004	0.0004	3.297	25
26	0.0011	3.3297	11.0045	917.3333	3054.4443	0.3003	0.0003	3.305	26
27	0.0008	3.3305	11.0263	1192.5333	3971.7776	0.3003	0.0003	3.310	27
28	0.0006	3.3312	11.0437	1550.2933	5164.3109	0.3002	0.0002	3.315	28
29	0.0005	3.3317	11.0576	2015.3813	6714.6042	0.3001	0.0001	3.318	29
30	0.0004	3.3321	11.0687	2619.9956	8729.9855	0.3001	0.0001	3.321	30
31	0.0003	3.3324	11.0775	3405.9943	11349.9811	0.3001	0.0001	3.324	31
32	0.0002	3.3326	11.0845	4427.7926	14755.9755	0.3001	0.0001	3.326	32
33	0.0002	3.3328	11.0901	5756.1304	19183.7681	0.3001	0.0001	3.327	33
34	0.0001	3.3329	11.0945	7482.9696	24939.8985	0.3000	0.0000	3.328	34
35	0.0001	3.3330	11.0980	9727.8604	32422.8681	0.3000	0.0000	3.329	35
36	0.0001	3.3331	11.1007	12646.2186	42150.7285	0.3000	0.0000	3.330	36
37	0.0001	3.3331	11.1029	16440.0841	54796.9471	0.3000	0.0000	3.331	37
38	0.0000	3.3332	11.1047	21372.1094	71237.0312	0.3000	0.0000	3.331	38
39	0.0000	3.3332	11.1060	27783.7422	92609.1405	0.3000	0.0000	3.331	39
40	0.0000	3.3332	11.1071	36118.8648	120392.8827	0.3000	0.0000	3.332	40
41	0.0000	3.3333	11.1080	46954.5243	156511.7475	0.3000	0.0000	3.332	41
42	0.0000	3.3333	11.1086	61040.8815	203466.2718	0.3000	0.0000	3.332	42
43	0.0000	3.3333	11.1092	79353.1460	264507.1533	0.3000	0.0000	3.332	43
44	0.0000	3.3333	11.1096	103159.0898	343860.2993	0.3000	0.0000	3.332	44
45	0.0000	3.3333	11.1099	134106.8167	447019.3890	0.3000	0.0000	3.333	45
46	0.0000	3.3333	11.1102	174338.8617	581126.2058	0.3000	0.0000	3.333	46
47	0.0000	3.3333	11.1104	226640.5202	755465.0675	0.3000	0.0000	3.333	47
48	0.0000	3.3333	11.1105	294632.6763	982105.5877	0.3000	0.0000	3.333	48
49	0.0000	3.3333	11.1107	383022.4792	1276738.2640	0.3000	0.0000	3.333	49
50	0.0000	3.3333	11.1108	497929.2230	1659760.7433	0.3000	0.0000	3.333	50

(continued)

APPENDIX 69.B *(continued)*
Factor Tables

$I = 40.00\%$

n	P/F	P/A	P/G	F/P	F/A	A/P	A/F	A/G	n
1	0.7143	0.7143	0.0000	1.4000	1.0000	1.4000	1.0000	0.000	1
2	0.5102	1.2245	0.5102	1.9600	2.4000	0.8167	0.4167	0.416	2
3	0.3644	1.5889	1.2391	2.7440	4.3600	0.6294	0.2294	0.779	3
4	0.2603	1.8492	2.0200	3.8416	7.1040	0.5408	0.1408	1.092	4
5	0.1859	2.0352	2.7637	5.3782	10.9456	0.4914	0.0914	1.358	5
6	0.1328	2.1680	3.4278	7.5295	16.3238	0.4613	0.0613	1.581	6
7	0.0949	2.2628	3.9970	10.5414	23.8534	0.4419	0.0419	1.766	7
8	0.0678	2.3306	4.4713	14.7579	34.3947	0.4291	0.0291	1.918	8
9	0.0484	2.3790	4.8585	20.6610	49.1526	0.4203	0.0203	2.042	9
10	0.0346	2.4136	5.1696	28.9255	69.8137	0.4143	0.0143	2.141	10
11	0.0247	2.4383	5.4166	40.4957	98.7391	0.4101	0.0101	2.221	11
12	0.0176	2.4559	5.6106	56.6939	139.2348	0.4072	0.0072	2.284	12
13	0.0126	2.4685	5.7618	79.3715	195.9287	0.4051	0.0051	2.334	13
14	0.0090	2.4775	5.8788	111.1201	275.3002	0.4036	0.0036	2.372	14
15	0.0064	2.4839	5.9688	155.5681	386.4202	0.4026	0.0026	2.403	15
16	0.0046	2.4885	6.0376	217.7953	541.9883	0.4018	0.0018	2.426	16
17	0.0033	2.4918	6.0901	304.9135	759.7837	0.4013	0.0013	2.444	17
18	0.0023	2.4941	6.1299	426.8789	1064.6971	0.4009	0.0009	2.457	18
19	0.0017	2.4958	6.1601	597.6304	1491.5760	0.4007	0.0007	2.468	19
20	0.0012	2.4970	6.1828	836.6826	2089.2064	0.4005	0.0005	2.476	20
21	0.0009	2.4979	6.1998	1171.3556	2925.8889	0.4003	0.0003	2.482	21
22	0.0006	2.4985	6.2127	1639.8978	4097.2445	0.4002	0.0002	2.486	22
23	0.0004	2.4989	6.2222	2295.8569	5737.1423	0.4002	0.0002	2.490	23
24	0.0003	2.4992	6.2294	3214.1997	8032.9993	0.4001	0.0001	2.492	24
25	0.0002	2.4994	6.2347	4499.8796	11247.1990	0.4001	0.0001	2.494	25
26	0.0002	2.4996	6.2387	6299.8314	15747.0785	0.4001	0.0001	2.495	26
27	0.0001	2.4997	6.2416	8819.7640	22046.9099	0.4000	0.0000	2.496	27
28	0.0001	2.4998	6.2438	12347.6696	30866.6739	0.4000	0.0000	2.497	28
29	0.0001	2.4999	6.2454	17286.7374	43214.3435	0.4000	0.0000	2.498	29
30	0.0000	2.4999	6.2466	24201.4324	60501.0809	0.4000	0.0000	2.498	30
31	0.0000	2.4999	6.2475	33882.0053	84702.5132	0.4000	0.0000	2.499	31
32	0.0000	2.4999	6.2482	47434.8074	118584.5185	0.4000	0.0000	2.499	32
33	0.0000	2.5000	6.2487	66408.7304	166019.3260	0.4000	0.0000	2.499	33
34	0.0000	2.5000	6.2490	92972.2225	232428.0563	0.4000	0.0000	2.499	34
35	0.0000	2.5000	6.2493	130161.1116	325400.2789	0.4000	0.0000	2.499	35
36	0.0000	2.5000	6.2495	182225.5562	455561.3904	0.4000	0.0000	2.499	36
37	0.0000	2.5000	6.2496	255115.7786	637786.9466	0.4000	0.0000	2.499	37
38	0.0000	2.5000	6.2497	357162.0901	892902.7252	0.4000	0.0000	2.499	38
39	0.0000	2.5000	6.2498	500026.9261	1250064.8153	0.4000	0.0000	2.499	39
40	0.0000	2.5000	6.2498	700037.6966	1750091.7415	0.4000	0.0000	2.499	40
41	0.0000	2.5000	6.2499	980052.7752	2450129.4381	0.4000	0.0000	2.500	41
42	0.0000	2.5000	6.2499	1372073.8853	3430182.2133	0.4000	0.0000	2.500	42
43	0.0000	2.5000	6.2499	1920903.4394	4802256.0986	0.4000	0.0000	2.500	43
44	0.0000	2.5000	6.2500	2689264.8152	6723159.5381	0.4000	0.0000	2.500	44
45	0.0000	2.5000	6.2500	3764970.7413	9412424.3533	0.4000	0.0000	2.500	45

Index

A

ABET (*see* Accreditation Board for
    Engineering and Technology)
Ablation, 34-2 (ftn)
Abrasive, 45-19
    cleaning, 63-13
    machining, 63-5
    type, 63-5 (tbl)
Abscissa, 7-2
Absolute
    convergence, 3-12
    English system, 1-4
    position, 55-9
    pressure, 14-2
    temperature, 24-4, 24-5
Absorbent, 33-7
Absorber, 31-2, 33-7
Absorptance, 31-2
Absorption
    coefficient, 21-7
    coefficient, sound, 68-8
    cycle, 33-6
    dry, 66-37
    dynamometer, 62-12
    factor, 66-40
    process, 66-21
ABS pipe, 16-10
Abundance, relative, 21-2
Accelerated
    Cost Recovery System, 69-21
    depreciation, 69-23
Acceleration
    centripetal, 15-19 (ftn)
    Coriolis, 55-8
    head, 18-2
    linear, 55-4
    normal, 55-8
    of fluid mass, 15-19
    of gravity, 1-2
    resultant, 55-8
    second, 57-10
    tangential, 55-7
    uniform, 55-3
Accelerometer, 58-21
Acceptance, 61-25, 70-3
    plan, 61-25
    quality level, 61-25
Accounting, 69-36
Accreditation Board for Engineering and
    Technology, 71-1 (ftn)
Accrual system, 69-34
Accumulation, 8-7
Accumulator, 17-32, 19-5
Accuracy, 11-13, 61-16 (ftn), 62-1
Acetylene gas, 63-11
Acid, 21-8
    carbonic, 21-13, 21-15, 66-41 (ftn)
    gas, 66-4
    rain, 66-5
    -test ratio, 69-35

Acidity, 21-13
Acme thread, 52-22
Acoustic
    absorber, 68-8
    barrier, 68-9
    emission testing, 46-11
    holography, 46-11
Acoustical velocity, 14-14
Across-variable, 59-3
Actinide, 21-2
Actinon, 21-2
Action
    capillary, 14-11
    chimney, 22-16
    error in, 52-14
    galvanic, 21-11
    limit, 61-22
    line of, 5-1, 43-2, 52-11
Activated carbon, 66-23
Active
    base, isolation, 58-14
    coil, 52-5
    element, 59-2
    solar system, 31-2
    system, 60-11
    wire length, 52-5
Activity, 61-3
    -on-branch, 61-5
    -on-node, 61-3
Actuator, 19-6, 42-1
Acute angle, 6-1
Addendum, 52-10
Addition
    of vectors, 5-3
    polymerization, 45-15
Adhesion, 14-11 (ftn)
Adhesive bonding, 63-13
Adiabatic
    compressibility, 14-13
    compression, 15-13
    compression efficiency, 32-3
    efficiency, 27-8
    flame temperature, 22-15
    flow with friction, 26-15
    head, 32-5
    process, 25-2
    saturation process, 38-7, 38-10
    tip, 34-17
Adjacent
    angle, 6-2
    side, 6-2
Adjoint, classical, 4-5
Admiralty bronze, 45-12
Admittance, 67-5
Adsorbent, 38-11, 38-12 (ftn), 66-23
Adsorption, 66-23, 66-24
Advanced
    flue gas cleanup, 66-24
    flywheel, 57-3
    oxidation, 66-24

reactor, 30-2
    scrubbing, 22-5
    turbine system, 29-14 (ftn), 29-17
Advantage
    mechanical, 43-10
    pulley, 43-11
Aeroderivative turbine, 29-14
Aero horsepower, 17-34
Affinity
    law, 18-17
    law, fan, 20-8
Afterburner, 66-34
Aftercooler, 29-10, 32-7
Age hardening, 45-8 (ftn), 45-11, 47-9
Agency, 70-2
Agent, 66-38 (ftn), 70-2
Agglomeration, 64-6
AGMA gear design, 52-14
Agreement, letter of, 70-3
Air
    -and-water system, 42-1
    atmospheric, 22-8
    bypass, recirculating, 41-3
    change method, 39-1
    classification, 64-10
    conditioner, 33-2
    conditioning load, 41-1
    conditioning process, 38-6
    -cooled exchanger, 36-12
    density ratio, 20-2 (ftn)
    excess, 22-12
    film coefficient, 35-4
    filter, 39-5
    -fuel ratio, 22-9, 22-12, 29-3
    handling unit, 39-6, 42-2
    heater, 27-15
    heater ash, 22-3
    horsepower, 20-4
    moist, 38-2, 38-3
    pollutant, 66-3
    preheater, 27-15
    primary, 27-3
    refrigeration cycle, 33-6
    resistance, 56-20
    sample, whole, 66-20
    saturated, 38-2
    space, 35-7, 40-3
    -standard Carnot cycle, 29-4
    -standard cycle, 29-2
    -standard diesel cycle, 29-10
    -standard dual cycle, 29-11
    -standard Otto cycle, 29-5
    stripping, 66-38
    table, 24-10
    -to-cloth ratio, 66-24
    toxic, 66-3
    unsaturated, 38-2
    volume control, 42-1
    washer, 38-2 (ftn), 38-10, 38-11

Aircraft cable, 64-2
Airflow, cooling, 36-24
Airfoil, 17-33
Alclad, 45-11 (ftn)
Alcohol, 22-7
Algebra, 3-1, 4-3
Algebraic equation, 3-2
Alias component, 9-7 (ftn)
Alkali metal, 21-2
Alkaline
    earth metal, 21-2
    fuel cell, 31-8
Alkalinity, 21-13
    methyl orange, 21-14
    phenolphthalein, 21-14
Alkane, 22-1
Alkyne, 22-1
All-air system, 42-1
Allergen, 66-5
Allotrope, 47-4
Allowable stress, 50-1
    effective, of pressure vessel, 53-9
    gear, 52-13
    pressure vessel, 53-4
    spring, 52-3
    working pressure, 19-3
Allowance
    clash, 52-5
    exposure, 40-1
    time, 61-16
Alloy
    aluminum, 45-11
    binary, 47-1
    copper, 45-12
    miscible, 47-1
    SAP, 45-19
    unified numbering system, 45-1
Alloying ingredient, 47-1
All-volatile treatment, 21-16
Alnico, 45-13
Alphabet, Greek, 3-1
Alpha-iron, 47-4
Alternating
    combined stress, 50-8
    current, 67-2, 67-4
    stress, 50-6
Alternator, 67-10 (ftn)
Altitude
    effect on power, 29-13
    pressure at, 20-2
Alumina, 45-11
Aluminum
    alloy, 45-11
    bronze, 45-12
    designations, 45-11
    hardening of, 47-9
    production of, 45-11
    properties of, 45-11
    tempers, 47-10
Amagat-Leduc's rule, 24-17
Amagat's law, 24-17
American Society of Civil Engineers,
    71-1
Amine, 21-15
Ammonia slip, 66-37 (ftn)
Amorphous film, 31-4
Amortization, 69-24
Ampere, 67-2
Amperometric sensor, 62-3
Amplifier, pressure, 32-2
Amplitude, 58-3, 58-19

Analogy, electrical/thermal, 34-5, 34-12
Analysis
    chemical, 21-6
    fault-tree, 61-27
    Fourier, 9-6
    frequency, 9-7 (ftn)
    gravimetric, 22-2
    hazard, 61-27
    proximate, 22-2
    sensitivity, 69-43
    signature, 9-7 (ftn)
    time-series, 9-7 (ftn)
    ultimate, 21-4, 21-5, 22-2
    volumetric, 22-2
Analytic function, 8-1
Analyzer
    FFT, 9-7
    signal, 9-7
    spectrum, 9-7
Andrade's equation, 46-15
Anechoic room, 68-6
Anemometer, 17-21
Angle, 6-2 (see also type)
    between lines, 6-1
    clearance, 63-2
    convergent, 26-5
    deviation, 26-11
    direction, 5-2, 7-5
    factor, 37-2
    function of, 6-2
    helix, 52-12
    helix thread, 51-13
    lead, 52-22
    lift valve, 16-12
    Mach, 26-11 (ftn)
    of contact, 14-12
    of depression, 6-2
    of elevation, 6-2
    of internal friction, 46-7, 56-7, 64-4
    of kinetic friction, 64-5
    of natural friction, 64-4
    of natural slope, 64-4
    of nip, 64-6
    of obliquity, 52-11
    of repose, 56-6, 64-4
    of rupture, 46-7
    of static friction, 56-6
    pitch, 52-14
    plane, 6-1
    power, 67-7
    pressure, 52-11, 52-12, 57-11
    rake, 63-2
    relief, 63-2
    semivertex, 26-11
    shock, 26-11
    solid, 6-6
    thread, 52-22
    trihedral, 6-5
    valve, 16-11
    wedge, 63-2
Angles between figures, 7-8
Angular
    acceleration, 55-6
    frequency, 58-3, 67-4
    impulse, 56-12
    momentum, 56-3
    motion, 55-6
    orientation, 5-1
    perspective, 2-3
    position, 55-6
    velocity, 55-6
Anisotropic material, 5-1, 34-2 (ftn),
    45-19

Annealing, 47-9
Annual
    amount, 69-7
    cost, 69-7
    cost method, 69-15
    return method, 69-15
Annuity, 69-7
Annular flow, 36-8
Annulus gear, 57-5
Anode, sacrificial, 21-12
Anodizing, 63-14
Anthracite, 22-4
Anthropometric data, 61-29
Anticipation control, 42-2
Antiderivative, 9-1
Antoine equation, 14-9 (ftn)
Apex, 43-22
Aphelion, 56-19
API scale, 14-4
Apogee, 56-19
Apparatus
    dew point, 38-8
    Orsat, 22-11
    sensible heat ratio, 38-6
Apparent modulus, 46-10
Application
    for exam, 72-2
    point of, 5-1
Approach, 27-14, 36-13, 38-12
    point, 29-20
    temperature, 36-13
    velocity of, 16-4, 17-16
Approximation
    Dunkerley, 58-16
    small angle, 6-3
Approximations, series, 8-8
Aquation, 21-7
Aramid fiber, 66-5
Arc, 61-3
    furnace, 45-5
    length (by integration), 9-5
    welding, 63-12
Archimedes' principle, 15-14
Arching, 64-5
Architect, 70-4
Area
    between two curves, 9-4
    by integration, 9-4
    centroid, 48-1
    daylight, 20-24
    density method, 65-6
    first moment of, 44-5, 48-2
    free, 20-24
    irregular, 7-1
    logarithmic mean, 34-8
    moment of inertia, 48-3
    reduction in, 46-6
    second moment of, 48-3
    sprinkler protection, 65-6
    transformation method, 49-19
Areas, law of, 56-18
Argon-oxygen-decarburization, 45-2 (ftn)
Argument, 3-8
Arithmetic
    sequence, 3-11
    series, 3-11
Arm, 43-2, 57-5
Aromatic liquid, 14-9 (ftn)
Arrangement factor, 37-2, 37-3
Array, 4-1
Arrest, thermal, 47-2

Artificial
  abrasive, 45-19
  aging, 47-9
  expense, 69-20
Asbestos, 66-5
  brake lining, 52-16
  cement (pipe), 16-10
  substitutes, 66-5
Asbestosis, 66-5
ASCE (see American Society of Civil
  Engineers)
"As fired," 22-3
Ash, 22-3, 66-6
ASME long-radius nozzle, 17-26 (ftn)
Aspect ratio, 17-33
  duct, 20-14
  ellipse, 7-11
Assembly line balancing, 61-18
Asset, 69-33, 69-34
Assignment, 61-12, 70-3
Associative law
  matrix, 4-4
  of addition, 3-3
  of multiplication, 3-3
  sets, 11-2
Aston process, 45-10
Asymmetrical function, 7-4
Asymptote, 7-2
Asymptotic, 7-2
Atactic polymer, 45-16
Atmosphere
  of earth, 15-12
  standard, 24-5
Atmospheric
  air, 22-8
  fluidized-bed combustion, 22-5
  head, 18-5 (ftn)
Atom, 21-1
Atomic
  fraction, 47-1 (ftn)
  mass unit, 21-1
  number, 21-2
  percent, 47-1 (ftn)
  structure, 21-1
  weight, 21-1
Attack, intergranular, 21-11
Attemperator, 27-13
Attenuation, sound, 68-9
Attitude, bearing, 52-20
Attribute chart, 61-22
Augmented matrix, 4-1
Austempering, 47-7
Austenite, 47-5
Austenitic stainless steel, 21-11 (ftn),
  45-8
Austenitizing, 47-7
Autogenous
  grinding, 64-7
  waste, 66-28 (ftn)
Autoignition temperature, 22-8
Auxiliary
  equation, 10-1
  view, 2-2
Availability, 25-11, 61-9
Available
  draft, 22-16
  hydrogen, 22-3, 22-14
  methods, 66-17
Average
  outgoing quality limit, 61-26
  power, 67-7
  pressure, 15-6
  run length, 61-22

  speed, 24-17
  time-weighted, 39-5
  value, 9-4, 67-5
  velocity, 16-8
Avogadro's
  hypothesis, 21-4
  law, 24-13
  number, 21-4
Axial
  -centrifugal compressor, 32-1
  exhaust, 29-14
  -flow compressor, 32-1
  -flow fan, 20-3
  -flow impeller, 18-4, 64-12
  -flow turbine, 18-20, 18-23
  loading, 49-11
  member, 43-11
  pitch, 52-12
  strain, 46-4
Axis
  conjugate, 7-11
  neutral, 49-10
  oblique, 2-2
  parabolic, 7-10
  principal, 48-7, 54-2
  rotation, 48-7
Axonometric view, 2-3

B
Babcock formula, 17-9
Backfitting, 27-16
Backflush, thermal, 27-12
Background noise, 68-3
Backing strip, 53-7
Back
  pressure, 27-11
  pressure critical ratio, 26-12, 26-13
  work ratio, 29-17
Backlash, 52-11
Backpressure turbine, 27-8
Backward-curved fan, 20-4
Baffle, 36-13
Baghouse, 66-24
Bainite, 47-7
Bakelite, 45-15 (ftn)
Balance
  delay, 61-18
  sheet, 69-34
Balanced
  draft, 22-16
  draft point, 66-28
  electrical load, 67-9
Balancing, 54-2, 58-18
  chemical equations, 21-6
  duct, 20-18
  line, 61-18
Ballast circuit, 62-9
Balling scale, 14-4 (ftn)
Ballistic pendulum, 56-3
Balloon payment, 69-41
Ball valve, 16-11
Band
  brake, 52-18
  dead, 42-5
  sound, 68-6, 68-7
Bandwidth, 60-10
  classification, 64-11
  of frequency analysis, 9-8
Banking, roadway, 56-21
Barlow formula, 19-3
Barometer, 15-2, 15-5

Barometric
  height relationship, 15-12
  pressure, 14-3
Barrel finishing, 63-14
Barrier, sound, 68-9
Barth speed factor, 52-13
Base, 12-1, 21-8
  -8 system, 12-2
  -10 number, 12-1
  -16 system, 12-2
  -b number, 12-1
  circle, 52-10, 57-10
  exchange, 21-14
  metal, 47-1
  of riser, 65-14
  pitch, 52-10
  unit, 1-5
  wage, 61-20
Baseline rate, 67-7
Base-load unit, 27-16
Basic Motion Time Study, 61-16 (ftn)
Basis, depreciation, 69-20
Bathtub distribution, 11-7
Battery, 31-10 (see also type)
Baumé scale, 14-5
Bauxite, 45-11
Bayer process, 45-11
Beam
  -column, 49-11
  conjugate, 49-15
  continuous, 44-5
  curved, 49-19
  deflection, 49-13 to 49-16
  fixed-end, 44-1, 44-6
  indeterminate, 44-1
  shear stress, 49-9
  wide, 49-16, 52-9
Bearing, 51-10
  characteristic number, 52-22
  journal, 52-20
  stiffener, 49-18
Bed
  chemically active, 66-29
  moisture level, 22-3
  packed, 36-10
  residence time, 66-25
Behavior science, 61-27
Bell-shaped curve, 11-6
Belt
  conveyor, 64-7
  elevator, 64-9
  feeder, 64-6
  flat, 52-15
  friction, 56-7
  tension, effective, 64-8
  V-, 52-16
Beltline, reactor, 30-2
Bending
  moment diagram, 49-8
  stress, beam, 49-10
  stress number, 52-14
Benefit-cost ratio, 69-16
Bent-tube boiler, 27-4
Benzoyl peroxide, 45-15 (ftn)
Bernoulli equation, 16-2, 17-14
Beryllium, 45-12
Bessel equation, 10-5
Bessemer process, 45-3
Best available technology, 66-21
Beta
  distribution, 61-5
  -iron, 47-4
  ratio, 17-23
Betz coefficient, 31-5

Bevel gear, 52-14, 57-9
Bias, 11-13, 62-2
Biaxial stress, 49-5, 50-1
Bidding, competitive, 71-3
Bifunctionality, 45-15
Billet, 45-10 (ftn), 63-13
Billion, 1-7 (tbl)
Bin, 27-3, 64-5 (see also type)
Binary
    alloy, 47-1
    cycle, 28-7, 31-7
    digit, 12-1
    number system, 12-1
Binder, 64-7
Binding, 27-14
Bingham
    fluid, 14-7
    plastic model, 17-11
Binomial
    coefficient, 11-2
    distribution, 11-4
    theorem, 3-3
Biocide, 27-12, 66-40 (ftn)
Biofilter, 66-25
Bioreactor, 66-25
Bioremediation, 66-25
Biosolid, 66-3 (ftn)
Biostat, 66-40 (ftn)
Biot number, 1-9 (tbl), 34-4, 34-13
Bioventing, 66-25
Bisection method, 3-4, 13-1
Bit, 12-1, 12-4
Bituminous coal, 22-4
Black body, 37-1
Black steel pipe, 16-10
Blade
    jet force on, 17-29
    pitch control, 20-3
Blanking, 63-7
Blasius
    equation, 17-5
    solution, 17-38, 36-4
Blast
    furnace, 45-1
    gas, 22-7 (ftn)
    gate, 20-18
    governing, 27-16
Bleed, 27-8
    fraction, 27-14
    -off, 38-13
    rate, 27-8
Block
    and tackle, 43-11
    brake, 52-17
    diagram, 60-5
Bloom, 45-10 (ftn)
Blow, 20-24
    molding, 63-10
    -off, flame, 27-3
    -out, gasket, 53-14
    -tank system, 64-6
Blowdown, 21-15, 27-5, 38-13, 66-9
Blower, 32-1
Blower power, 20-4
Blowing the whistle, 71-3 (ftn)
Blowoff, 27-5
Blue
    center steel, 64-2
    embrittlement, 47-9
Board
    hammer, 63-7
    of directors, 70-2
    of engineering licensing, 72-1

Boardman formula, 19-3
Bode plot, 60-11
Body
    black, 37-1
    -centered tetragonal, 47-7 (ftn)
    gray, 37-2, 37-3
Boiler, 27-4 (see also type)
    and Pressure Vessel Code, 53-1
    efficiency, 22-16, 27-4
    feedwater characteristics, 21-15
    feedwater quality, 21-11
    horsepower, 22-16, 27-4
    throttle valve, 27-16
    once-through, 27-1
    waste-heat, 29-20
Boilerplate clause, 70-3
Boiling, 14-9
    film, 35-10
    nucleate, 35-10
    -point margin, 66-10
    pool, 35-10
    water reactor, 30-2
Bolt, 51-8
    family, 51-8
    grade, 51-8
    marking, 51-9
    preload, 51-11
    standards, 51-8
    strength, pressure vessel, 53-14
    stress concentration, 51-12
    torque, 51-12
Boltzmann constant, 24-17
Bomb calorimeter, 22-14
Bond, 69-29
    fully amortized, 69-29 (ftn)
    yield, 69-30
Bonded strain gage, 62-8
Bonding, adhesive, 63-13
Bonus, 61-20
Bookkeeping, 69-33
Book value, 69-23
Boom, sonic, 14-14
Booster, 32-2
Booster humidification, 39-3
Boost, fan, 22-16
Bore, 29-2
Bottleneck, 61-18
Bottom
    ash, 22-3, 66-6
    -dead-center, 29-2, 32-3 (ftn)
Bottoming cycle, 29-21
Bound
    lower, 9-4
    upper, 9-4
    vector, 5-1 (ftn)
Boundary
    irregular, 7-1
    layer, 16-8, 17-37, 36-2
    system, 25-1
    value, 10-2
Bourdon pressure gauge, 15-2
Box, closed, 51-14
Boyle's law, 25-7
Braced column, 51-2
Brake, 52-16, 63-6
    band, 52-18
    block, 52-17
    dynamometer, 62-12
    energy dissipation, 52-16
    external shoe, 52-17
    lining, 52-16
    pad, 52-16
    power, 20-4, 29-3, 29-4

    properties, 29-3
    pump power, 18-7
    specific fuel consumption, 29-3
    torque, 29-4
    prony, 29-3
Braking torque, 52-18
Brale indentor, 46-12
Branch
    line, 65-2
    loss coefficient, 20-16
    takeoff, 20-16
Branching, 45-16
Brass, 16-10, 45-12 (ftn)
Brayton
    cooling cycle, 33-6
    cycle, 29-15, 29-18, 29-19
Brazing, 63-12
Breach of contract, 70-5
Breakaway point, 60-12
Break-even point, 69-37
Breaking strength, 46-3
Breakpoint, 66-23
Breakthrough, 66-23
Breeder reactor, 30-1 (ftn)
Breeze, 22-5
Brickwork, 27-4
Bridge
    constant, 62-10
    truss, 43-12
    Wheatstone, 62-9
Bridging, 64-5
Brightness, 68-2, 68-3
Brinell hardness, 46-12
Briquetting, 64-6
British thermal unit, 23-1, 24-5
Brittle
    failure, 46-13
    material, 46-16, 50-1, 50-2
Brix scale, 14-4 (ftn)
Broad band measurements, 68-7
Bromley equation, 35-11
Bronze
    government, 45-12
    type, 45-12
Btu (see British thermal unit)
Bubbling fluidized-bed combustor, 66-27
Bucket elevator, 64-9
Buckingham equation, 52-14 (ftn)
Buckling, 51-1
    beam, 49-18
    spring, 52-7
Buffer, 21-9
Buffing, 63-14
Building
    designer, 70-4
    -related illness, 66-5
Built
    -in end beam, 44-6
    -up edge, 63-2 (ftn)
Bulk
    modulus, 14-13, 68-4
    solid, 64-5
    storage pile, 64-4
    temperature, 34-4, 35-3, 36-2, 36-3
    velocity, 16-8
Bundle, tube, 36-9
Bunker C oil, 22-5
Buoyancy, 15-14, 15-16
Burden, 69-36
Burning rate, municipal solid waste, 66-32
Burnishing, 63-14
Burnout velocity, 56-17
Burst
    pressure, 19-3
    velocity, 57-3

Bushing, 52-19
Business property, 69-20 (ftn)
Butterfly valve, 16-11
Byers-Aston process, 45-10
Bypass
    factor, 38-7, 41-4
    governing, 27-16
    recirculating air, 41-3

C
Cabinet projection, 2-3
Cable, 43-17 to 43-20
    catenary, 43-19
    hoisting, 64-1
    parabolic, 43-17
    tension, 56-11
CaCO₃ equivalent, 21-13
Calculus, fundamental theorem of, 9-4
Calibration, 62-2
Calorimeter, bomb, 22-14
Calorizing, 63-14
Cal sil, 34-9
Cam, 57-10
Candlepower, 68-3
Canon, 71-1
Canonical form, row, 4-2
Capacitance, 67-5
    fluid, 59-7
    thermal, 34-12
Capacitive reactance, 67-5
Capacitor, 67-5
Capacity
    compressor, 32-2
    control, 20-3
    factor, 27-15
    heat, 24-7
    rate, thermal, 36-20
    refrigeration, 33-3
    specific, battery, 31-10
    turndown, 27-13
Capillary
    action, 14-11
    -action heat pipe, 34-22
Capitalized cost, 69-7, 69-15
Capitalizing an asset, 69-20
Capital recovery method, 69-15
Carbide, 45-19, 47-5, 63-4
Carbon
    dioxide buildup, 39-3
    dioxide emission, 66-6
    fixed, 22-4
    granular activated, 66-23
    monoxide emission, 66-8
    steel, 45-5
    temper, 45-10
Carbonate
    fuel cell, 31-8
    hardness, 21-14
Carbonic acid, 21-13, 21-15, 66-41 (ftn)
Carborundum, 45-19
Carburetor, 29-2
Carburizing, 47-10, 63-12
Carcinogen, 61-27
Card, indicator, 29-8
Carnot cycle, 25-10 (ftn), 25-11, 28-2
    air-standard, 29-4
    refrigeration, 33-3
Carrier
    arm, 57-5
    equation, 38-2
    slurry, 17-11

Cartesian
    coordinate system, 7-3
    equation form, 3-3 (ftn)
    triad, 5-2
    unit vector, 5-2
Cash
    flow, 69-3, 69-6
    system, 69-34
Casing transducer, 58-21
Casting
    die, 63-9
    type, 63-8, 63-9
Cast iron, 45-9, 47-4, 47-5 (see also type)
    heat treating, 47-8
    pipe, 16-10
Catalytic
    converter, 29-2
    cracking, 22-3
    reduction, selective, 66-37
Catenary cable, 43-19
Cauchy
    equation, 10-5
    number, 1-9 (tbl)
    -Schwartz theorem, 5-3
Caustic
    embrittlement, 21-11
    phosphate treatment, 21-16
Cavalier projection, 2-3
Caveat emptor, 70-6
Cavitation, 18-13, 18-14
    coefficient, 18-14
    corrosion, 21-12
c-chart, 61-22, 61-24
Ceiling, heat transfer, 40-2
Cell
    fuel, 31-7
    landfill, 66-12
    photovoltaic, 31-4
    solar, 31-4
Cellular fill, 38-11 (ftn)
Cement, asbestos (pipe), 16-10
Cementation, 47-10
Cemented carbide, 45-19
Cementite, 47-5
Center, 7-2
    distance, 52-11
    instant, 55-12, 55-13
    of buoyancy, 15-16
    of force, 56-17
    of gravity, 9-6, 54-1
    of mass, 54-1
    of pressure, 15-4, 15-9, 15-10 (fig), 43-5
    of vision, 2-3
    -radius form, 7-9
Centerless grinding, 63-5
Centerline velocity, 16-8
Centigrade, 24-5 (ftn)
Centipoise, 14-7
Centistoke, 14-8
Central
    force field, 56-17
    impact, 56-16
    limit theorem, 11-11
    receiver system, 31-3
    system, 27-3
    tendency, 11-10
    view, 2-2
Centrifugal
    casting, 63-9
    fan, 20-3
    force, 56-5
    pump, 18-2, 18-4

Centripetal
    acceleration, 15-19 (ftn)
    force, 56-5
Centroid, 9-6
    area, 48-1
    line, 48-2
    of volume, 54-1
Centroidal
    mass moment of inertia, 54-2
    moment of inertia, 48-3
Ceramic, 45-17,
    adhesive, 63-13 (ftn)
    tool, 63-4
    toughened, 45-19
Cermet, 45-19
Cetane number, 22-7
cgs system, 1-4
Challenger, 69-18
Chamber
    checker, 45-4
    property, 26-2
Change, allotropic, 47-4
Channeling, 66-39
Chaos theory, 62-2 (ftn)
Chaotic error, 62-2
Char, 22-5
Characteristic
    curve
        compressor, 32-8
        fan, 20-5
        system, 20-7
    dimension, 16-6, 34-3
    equation, 4-7, 10-1, 60-4, 60-12
    exhaust velocity, 26-9
    length, 26-9, 34-3, 35-2, 36-8
    log, 3-5
    number, bearing, 52-22
    polynomial, 4-7
    specific feed, 66-29
Characteristics
    of metals, 45-1
    phase, 60-11
Charcoal, activated, 66-23
Charge, electric, 67-2
Charles' law, 25-7
Charpy test, 46-13
Chart (see also type)
    control, 11-13, 61-22
    Cox, 14-9
    Heisler, 34-13
    pressure-enthalpy, 24-11
    psychrometric, 38-3
    time-temperature, 34-13
Checker chamber, 45-4
Check valve, 16-12
Chelant, 21-16, 66-38 (ftn)
Chelate, 66-38 (ftn)
Chelating agent, 66-38 (ftn)
Chemical
    dehumidification, 38-11
    dehydration, 38-11
    equations, balancing, 21-6
    loading, 66-41
    milling, 63-6
    reaction, 21-5
    scrubbing, 66-37
Chemically active bed, 66-29
Chiller, 33-2, 33-9
Chimney, 22-16
Chip
    -breaker groove, 63-2
    forming, 63-1
    temperature, 63-4

thickness ratio, 63-2
type, 63-1
velocity, 63-2
Chloride stress corrosion, 45-9
Chlorination, 66-9
Chlorine, 27-12, 66-26
Chlorofluorocarbon, 66-8
Choked flow, 26-4
Choke point, 32-8
Chord, 17-33, 43-12
Chrysotile-asbestos brake lining, 52-16
Churn, pump, 65-3
Circle, 7-9
  base, 52-10
  clearance, 52-10
  Mohr's, 48-7, 49-6
  pitch, 52-10
  unit, 6-1
Circular
  fin, 34-18
  mil, 67-2
  motion, 55-6
  permutation, 5-5
  pitch, 52-10
  shaft, design, 51-13
  transcendental function, 6-2
Circulating
  closed-loop system, 65-2
  fluidized-bed boiler, 66-28
Circulation, 8-7 (ftn), 17-33, 29-20
Circumferential
  strain, 51-5
  stress, 51-3, 51-4
Cis form, 3-8
Civil
  complaint, 70-5
  tort, 70-5 (ftn)
Cladding, 63-11
Clash allowance, 52-5
Class
  limits, 11-9
  property, bolt, 51-8
  sound transmission, 68-9
Classical adjoint, 4-5
Claude cycle, 31-6 (ftn)
Clausius-Clapeyron equation, 14-9 (ftn)
Clay, vitrified (pipe), 16-10
Cleanliness factor, 27-11
Clean room, 39-5
Clearance, 52-10
  angle, 63-2
  circle, 52-10
  diametral, 52-20
  percent, 32-3
  radial, 52-20
  volume, 29-2, 32-3
Client (dealing with), 71-2
Clinker, 22-3
Clinographic projection, 2-3
Clock spring, 52-9 (ftn)
Closed
  box, 51-14
  combustor, 29-16
  -cycle OTEC, 31-6
  die forging, 63-7
  feedwater heater, 27-14, 36-14 (ftn)
  -loop gain, 60-4
  -loop recovery, 66-23
  system, 25-1
  turbine cycle, 29-15
Closely held corporation, 70-2
Cloud point, 22-7
Clutch, 52-16, 52-19

Coal, 22-4
  ash, 22-3
  blending, 66-18
  cleaning, 22-5
  dry ultrafine, 66-10
  -fueled turbine, 29-15
  micronized, 22-5, 27-3 (ftn)
  substitution, 66-18
  upgrading, 22-5
  -water mixture, 29-15
Coalescence, 63-9
Coarsening temperature, 47-8
Cocurrent
  -crossflow, 36-11
  flow, 36-10
Code, 71-1
Coefficient
  absorption, 21-7
  Betz, 31-5
  binomial, 11-2
  branch loss, 20-16
  cavitation, 18-14
  confidence, 61-16
  correlation, 11-13
  drag, 17-34, 36-4 (ftn)
  end restraint, 51-2
  film, 34-6, 35-2
  fluid resistance, 59-7
  Joule-Kelvin, 25-3
  Joule-Thompson, 25-3
  loss, 17-12, 20-14
  mass-transfer, 66-40
  matrix, 4-6
  noise reduction, 68-8
  nozzle, 26-8
  of compressibility, 14-12
  of contraction, 17-16, 17-24
  of discharge, 17-17, 17-24
  of entry, 20-15
  of expansion, thermal, 14-12 (ftn)
  of flow, 17-24, 26-15
  of fluctuation, 57-2
  of forced convection, 36-2
  of friction, 15-12, 56-6
  of lift, 17-32
  of linear expansion, 49-3
  of performance, 33-2
  of restitution, 56-15
  of rolling friction, 56-8
  of rolling resistance, 56-7, 56-19
  of the instrument, 17-22
  of thrust, 26-8
  of transformation, 5-2
  of variation, 11-11, 61-16
  of velocity, 17-16, 17-23
  of viscosity, 14-6
  of viscous damping, 56-17, 58-9
  of volumetric expansion, 49-3
  overall heat transfer, 34-6, 36-15
  power, 31-5
  profile drag, 36-4 (ftn)
  radiant heat transfer, 37-4
  roughness, 17-7
  Seebeck, 31-9
  shading, 41-2
  skin friction, 17-38, 36-4
  slab edge, 40-4
  smoothing, 69-42
  sound absorption, 68-8
  static regain, 20-16
  strength, 46-5
  torque, 51-12
  transmission, 68-9

valve flow, 17-13
virial, 24-19
Coefficients, method of undetermined,
  10-4
Cofactor
  matrix, 4-1
  of entry, 4-2
Co-firing, 66-33
Cogeneration cycle, 29-21
Cogeners, 21-2
Coherent unit system, 1-2
Cohesion, 14-11 (ftn), 46-7
Coil, 36-6
  active, 52-5
  efficiency, 38-7
  of film coefficient, 35-7
  load, 41-1 (ftn)
  pitch, 52-5
  sensible heat ratio, 38-6, 41-4
Coining, 63-7
Coke, 22-5, 22-7 (ftn), 45-2 (ftn)
Colburn equation, 36-9
Cold
  air-standard cycle, 29-2 (ftn)
  -end drive, 29-14
  flow, 46-15
  heading, 63-7
  reserve, 27-15
  -working, 47-9, 63-6 (tbl)
Colebrook equation, 17-5
Collapse, duct, 20-18
Collapsing pressure, 51-3 (ftn)
Collector
  concentrating, 31-2
  dust, 39-5
  efficiency, 31-2
  evacuated tube, 31-2
  flat plate, 31-2
  focusing, 31-2
  solar, 31-2
Collinear, 7-2, 43-6
Collision, 56-15
Colloid, 66-41
Column
  braced, 51-2
  intermediate, 51-3
  matrix, 4-1
  rank, 4-5 (ftn)
  slender, 51-1
Combination, 11-2
  direct, 21-6
  method, 20-21
Combined
  ash, 66-6
  cycle, 29-22
  diesel or gas turbine system,
    29-10 (ftn)
  film coefficient, 34-17
  heat transfer coefficient, 37-4
  stress, 49-5
Combining weight, 21-5
Combustible loss, 22-15
Combustion
  complete, 22-12
  efficiency, 22-16
  fluidized-bed, 22-5
  heat of, 22-14
  incomplete, 22-11
  loss, 22-15
  reaction, 22-8
  staged, 66-38
  stoichiometric, 22-9
  stratified charge, 29-2

temperature, 22-15
turbine, 29-14
Combustor, 29-12, 29-14, 29-16, 66-27
Comfort range, 39-2
Comity, 72-2
Command, 60-13
Commercial bronze, 45-12 (ftn)
Comminution, 64-7
Common
    glass, 45-17
    ratio, 3-11
Commutative law
    matrix, 4-4
    of addition, 3-3
    of multiplication, 3-3
    sets of, 11-2
Compaction, 64-6
Comparative negligence, 70-6
Comparator, 60-4
Comparison test, 3-12
Compatibility method, 44-2
Compensatory
    damages, 70-7
    fraud, 70-6
Competitive bidding, 71-3
Complaint, 70-5
Complement
    nines, 12-3
    number, 12-3
    ones, 12-3
    set, 11-1
    tens, 12-3
    twos, 12-3
Complementary
    angle, 6-2
    equation, 10-1
    probability, 11-4
    solution, 10-4, 58-10
Complete combustion, 22-12
Completion time, 61-5
Complex
    conjugate, 3-8
    matrix, 4-1
    number, 3-1, 3-7
    number operations, 3-8
    plane, 3-7
    power, 67-6
Component
    alias, 9-7 (ftn)
    linear, 58-2
    moment, 43-3
    steady-state, 58-11
    transient, 58-11
Components of a vector, 5-2
Composite
    complex wall, 34-6, 34-7
    cylinder, 34-8
    material, 45-19
    material, flywheel, 57-4
    spring constant, 52-2
    structure, 49-19
Composition, eutectic, 47-3
Compound, 21-3
    amount factor, 69-5
    gear, 57-4
    interest, 69-11
Compounding period, 69-28
Compressed
    -air energy storage, 31-9 (ftn)
    -air storage tank, 32-2
    height, 52-5
Compressibility, 14-1, 14-12, 24-9
    adiabatic, 14-13

coefficient of, 14-12
factor, 14-12 (ftn), 24-19
isentropic, 14-13
isothermal, 14-13
Compressible
    flow through orifice, 26-15
    fluid dynamics, 26-1
    fluid, flow of, 17-26
Compression
    adiabatic, 15-13
    cycle, vapor, 33-3
    efficiency, adiabatic, 32-3
    ignition, 29-2
    isothermal, 15-12
    polytropic, 15-13
    ratio, 29-3, 29-5, 29-11, 32-4 (ftn)
    ratio, isentropic, 29-4
    spring, helical, 52-4, 52-6
    stage, 32-1
    stroke, 29-2
    wave, 68-4
Compressive strength, 46-7
Compressor, 32-1 (see also type)
    air/gas, 32-1
    capacity, 32-2
    characteristic curve, 32-8
    control, 32-2
    scroll, 33-7
    work, 32-4
Concave, 7-2
Concavity, 7-2
Concentrated force, 43-2
Concentrating collector, 31-2
Concentration
    -cell corrosion, 21-11
    ionic, 21-8
    stress, 49-4, 51-7
    tower buildup, 38-13
Concentric
    cylinder viscometer, 14-6 (ftn)
    flywheel system, 57-4
    loading, 49-11
Concrete
    density, 15-11 (ftn)
    pipe, 16-10
Concurrent force, 43-6
Condensate
    depression, 27-11
    polishing, 21-16
    well, 27-11
Condensation
    dropwise, 35-8
    filmwise, 35-8
    polymerization, 45-15
    Reynolds number, 35-9
    vapor, 35-7 to 35-9
Condenser, 27-11, 33-1 (see also type)
    cooling water, 66-9
    duty, 27-11
    hot well subcooling, 27-11
Condensing
    cycle, 28-2
    vapor, film, 35-7
Condition
    initial, 58-4
    line, turbine, 27-8
    standard, 22-2
Conditional
    convergence, 3-12
    probability, 11-4
    probability of failure, 11-7

Conditioner, sludge, 21-16
Conditions of equilibrium, 43-6
Conductance, 34-2, 67-2, 67-5
    overall, 36-15
    thermal, 40-2
Conduction, 34-2
Conductive, unit, 34-6
Conductivity, 67-2
    overall, 34-6
    thermal, 34-2
Cone
    clutch, 52-19
    pitch, 52-14
Confidence
    coefficient, 61-16
    interval, 61-16 (ftn)
    level, 11-11, 61-16
    limit, 11-11
Configuration factor, 37-2
Congruency, 7-3
Conical pile, 64-4
Conic section, 7-8
Conjugate
    axis, 7-11
    beam, 49-15
    complex, 3-8
Connection
    bolt, 51-10
    eccentric, 51-15, 51-17
Consequential damages, 70-7
Conservation
    law, radiation, 37-1
    methods, heat, 40-6
    of energy, 58-5
    of momentum, 17-27, 56-3
Consideration, 70-3
Consistent
    deformation method, 44-2, 49-19
    system, 3-7
    unit system, 1-2
Constant
    Boltzmann, 24-17
    bridge, 62-10
    coefficients, 10-1 to 10-3
    elastic relationships, 46-8
    Euler's, 9-8 (ftn)
    gravitational, 1-2
    Henry's law, 21-7
    Joule's, 23-1, 24-6
    matrix, 4-6
    Newton's gravitation, 56-18
    of integration, 9-1, 10-1
    outlet, 20-25
    percentage method, 69-21 (ftn)
    pressure process, 25-2, 25-7
    reaction rate, 66-25
    room, 68-8
    solar, 31-2, 37-7
    -speed control, 60-13
    spring, 49-2, 52-2
    sprinkler orifice, 65-5
    Stefan-Boltzmann, 37-2
    temperature process, 25-2, 25-8
    thermoelectric, 62-7
    time, 34-12
    torsional spring, 58-5
    volume process, 25-2, 25-8
Constrained motion, 56-9
Constraint, 61-2, 61-12
Construction contract, 70-3 (ftn)
Construction manager, 70-4
Consulting engineer, 72-1

Contact
  angle, 14-12
  ratio, 52-11
  resistance, 34-6 (ftn)
Content, moisture, 24-9
Continuity equation, 17-2
Continuous
  beam, 44-1, 44-5
  bucket elevator, 64-9
  compounding, 69-28
  distribution function, 11-5
  duty rating, 29-4
  emissions monitoring, 66-21
Contract, 70-2
  breach of, 70-5
  documents, 70-3 (ftn)
  discharge of, 70-5
  privity, 70-6
  requirement for, 70-2
  standard, 70-4
Contraction, 17-12, 17-13, 17-16, 17-24
Contractor, 70-4
Contraflexure, point of, 7-2, 8-2, 49-16
Control (see also type)
  blade pitch, 20-3
  capacity, 20-3
  chart, 11-13, 61-22
  compressor, 32-2
  differential, 42-2
  element, 60-13
  limit, 61-22, 61-23
  mass, 25-1
  mechanism, 60-14
  of heat exchange operations, 36-23
  of turbine, 27-16
  point, 42-2, 61-25
  ratio, 60-4
  surface, 25-1
  system, 60-13
  tachometer, 60-13
  valve, hydraulic, 19-2
  velocity, 39-4
  volume, 25-1
Controllability, 60-14
Controllable pitch, 20-3
Controlled
  -circulation boiler, 27-4
  -cooling transformation curve, 47-6
Controller, 60-13
Convection
  forced, 36-2
  natural/free, 35-2
  Newton's law of, 36-2
  preheater, 27-15
  superheater, 27-12
Convective heat transfer coefficient, 34-6
Convention
  half-year, 69-3
  sign, thermal, 25-4
  year-end, 69-3
Conventional duct system, 20-17
Convergence, 3-11, 3-12
Convergent
  angle, 26-5
  -divergent nozzle, 26-5
  sequence, 3-10
  series, 3-11
Converging-diverging nozzle, 17-23
Conversion
  base, 12-3
  power, 23-5 (tbl)
Converter, 29-2, 45-3

Convex, 7-2
  hull, 7-2, 7-3
  polygon, 61-12
Convexity, 7-3
Conveyor
  belt, 64-7
  pneumatic, 64-6
  screw, 64-5
  submerged scraper, 66-6
Convolution integral, 10-7
Cooler, 33-9
Cooling
  coil, 33-8
  compressor, 32-7
  degree day, 41-2
  duty, 38-12
  efficiency, tower, 38-12
  electronic enclosures, 36-23
  evaporative, 38-7, 38-10
  load, 41-1
  load factor, 41-2
  Newton's law of, 10-10
  sensible, 38-8
  tower, 38-11, 38-13
  tower blowdown, 66-9
  water, 36-14
  with dehumidification, 38-8
  with humidification, 38-8
Coordinate
  holonomic, 55-2
  system, 7-3
Coordinates of fluid steam, 17-17
Coplanar, 7-2, 43-6
Copolymer, 45-13
Copper
  alloy, 45-12
  -beryllium, 45-12
  loss, 67-15
  pipe, 16-10
  production, 45-12
  tubing, 16-10
Core, 49-12, 62-4 (ftn), 63-8
  loss, 67-15
  wire rope, 64-2
Coriolis acceleration, 55-8
Corner frequency, 60-10
Corona wire, 66-27
Corporation, 70-2
Corrected fin length, 34-18
Correlation
  coefficient, 11-13
  film, 35-3
  Othmer, 14-11
  Seider-Tate, 36-5, 36-6
Corresponding state, 24-19
Corrosion, 21-10
  cavitation, 21-12
  chloride, 45-9
  concentration cell, 21-11
  cracking, intergranular stress, 30-2
  cracking, stress, 45-9
  crevice, 21-11
  erosion, 21-11
  fatigue, 21-12
  fretting, 21-12
  galvanic, 21-11
  intergranular, 21-10
  microbiologically influenced, 27-12
  pressure vessel, 53-6
  -resistant steel, 45-5
  stress, 21-12
  two-metal, 21-11
Corrosive substance, 66-3

Cosine
  direction, 5-2, 7-5, 43-2
  function, integral, 9-8
Cosines, law of, 6-5, 6-6
Cost, 61-2 (ftn), 69-31 (see also type)
  accounting, 69-36
  capitalized, 69-7
  electricity, 18-8
  of goods sold, 69-37
  opportunity, 69-18
  plus fee, 70-4
  standard, 61-15
  sunk, 69-3
Couette flow, 36-3 (ftn)
Coulomb, 67-2
  friction, 56-5
  -Mohr theory, 50-2
Counterblow forging, 63-8
Counter-crossflow, 36-11
Countercurrent flow, 36-10
Counterflow, 36-10
Couple, 43-4, 43-9
Coupon rate, 69-30
Cox chart, 14-9
CPM (see Critical path method)
Crack
  coefficient, 39-2
  length method, 39-2
  stress corrosion, 45-9
  weld bead, 45-9
Cracking, hydrocarbon, 22-2
Cramer's rule, 3-7, 4-6
Crankcase oil, 52-19
Crank end, 29-2
Crank-slider, 55-13
Credit, 69-33
  investment, 69-25
  tax, 69-25
Creed, 71-1
Creep, 46-10, 46-15
Crest (dam), 15-11
Crevice, 21-11
Cristobalite, 45-18
Criteria pollutant, 66-3
Critical
  back-pressure ratio, 26-12, 26-13
  damping, 58-9
  damping coefficient, 58-9
  fastener, 51-16
  flow, 16-8
  gas constant, 26-4
  height, pressure vessel, 53-5 (ftn)
  insulation thickness, 34-10
  isobar, 24-3
  line, 47-4
  load, column, 51-1
  path method, 61-3, 61-4
  point, 8-2, 24-3, 24-8, 26-4, 47-4
  pressure ratio, 26-4
  property, 24-8, 26-5
  radius, 34-10
  ratio, 26-4
  Reynolds number, 36-3
  slenderness ratio, 51-2
  speed, shaft, 58-15, 58-17
  temperature, 47-4, 47-7
  zone, 16-8
Cross
  -drum boiler, 27-4
  -linking, 45-13, 45-16
  main, 65-2, 65-12
  product, 43-2
  product, vector, 5-4
  temperature, 36-14

Crossed drive, 52-15
Crossflow
  cylinder, 36-8
  heat exchanger, 36-11
Crossover
  frequency, 60-11
  gain, 60-11
  point, 60-10
Crud, 21-16
Crushing, 64-7
Crusting agent, 66-10
Cryogen, 24-2 (ftn)
Cryogenic fluids, 24-2 (ftn)
Cryolite, 45-11
Crystallinity, polymer, 45-16
Crystallization, water of, 21-4
Cube, conduction, 34-11
Culvert, 17-19
Cumulative
  fatigue, 50-7
  frequency, 11-9
  usage factor rule, 50-7
Cunife, 45-13
Cup
  -and-bob viscometer, 14-6 (ftn)
  -and-cone failure, 46-4
Cupola, 45-10 (ftn)
Curl, 8-7
Current
  asset, 69-34
  divider, 67-4
  electrical, 67-2
  liability, 69-34
  meter, 17-21
  ratio, 69-34
Curve, 7-3
  bell-shaped, 11-6
  degree, 7-3
  Euler's, 51-2
  fan characteristic, 20-5
  flow, 46-5
  inversion, 25-3
  learning, 69-42
  load, 27-15
  load-elongation, 46-1
  mortality, 61-11
  operating characteristic, 61-25
  performance, 18-15
  reloading, 46-6
  resistance, 56-19
  stress-strain, 46-1, 46-3
  symmetry, 7-4
  system, 18-15, 20-7
  time-temperature-transformation, 47-6
  unloading, 46-6
Curved beam, 49-19
Cut-and-sum method, 43-15
Cut-in/out pressure, 32-2
Cut
  -off
    frequency, 60-10
    governing, 27-16
    rate, 60-10
    ratio, 29-11
    volume, 29-11
  -point, 64-10, 64-11
  size, 66-26
Cutting, 64-7
  energy, 63-3
  force, 63-2
  hardness, 46-12
  oil, 63-4
  power, 63-3

  speed, 63-2, 63-3 (tbl)
  tool material, 63-4
  torch, 63-12 (ftn)
Cyaniding, 47-10
Cycle, 25-10, 28-1 (*see also type*)
  absorption, 33-6
  air refrigeration, 33-6
  air-standard, 29-2
  air-standard diesel, 29-10
  air-standard dual, 29-11
  air-standard Otto, 29-5
  binary, 28-7, 31-7
  bottoming, 29-21
  Brayton, 29-15, 29-18, 29-19, 33-6
  Carnot, 25-10 (ftn), 25-11, 28-2, 29-4,
    33-3
  Claude, 31-6 (ftn)
  cogeneration, 29-21
  combined, 29-22
  direct fuel reforming, 31-8
  dual-flash, 31-7
  Ericsson, 29-13
  flash steam, 31-7
  four-stroke, 29-2
  internal fuel reforming, 31-8
  Joule, 29-15
  nuclear power, 30-1
  power, 25-10
  Rankine, 28-3 to 28-5
  redesign, 27-16
  regenerative, 28-7
  solar power, 31-3
  Stirling, 29-12
  supercritical, 28-7
  time, 61-18
  topping, 29-21
  two-stroke, 29-2
  vapor compression, 33-3
Cycles of concentration, 38-13
Cyclic
  strain, 46-9 (ftn)
  stress, 46-9 (ftn)
Cycloid, 7-3
Cycloidal cam profile, 57-10
Cyclone, 66-26
Cyclone separator, 64-10
Cylinder
  composite, 34-8
  conduction through, 34-8
  crossflow, 36-8
  hydraulic, 19-6
  -operated pump, 18-2
  pressure, peak, 29-8 (ftn)
  thick-walled, 51-4
Cylindrical coordinate system, 7-3 (tbl),
  7-4 (fig)

D
D'Alembert's
  paradox, 17-2 (ftn)
  principle, 56-5
Dalton's law, 22-13, 24-17, 66-4 (ftn)
Dam, 15-11
Damage
  hydrogen, 21-11
  punitive, 70-5
Damages, 70-7
Damped
  forced vibration, 58-11
  free vibration, 58-9
  frequency, 58-9, 60-2

Damper, 59-4
  control, face-and-bypass, 42-1
  duct, 20-18
  face, 42-2 (ftn)
  pulsation (pump), 18-2
Damping
  coefficient, critical, 58-9
  critical, 58-9
  factor, 58-9
  heavy, 58-9
  light, 58-5
  linear, 58-9
  ratio, 58-9
  viscous, 58-9
Dangerous section, 49-11
Darcy
  equation, 17-7
  friction factor, 17-5
  -Weisbach equation, 17-7
Darrieus rotor, 31-4
Dashpot, 58-9, 59-4
Davit, 53-11 (ftn)
Davonius rotor, 31-4 (ftn)
Daylight area, 20-24
Dead
  band, 42-5
  -beat motion, 58-9
  load, 43-4
Deaerating heater, 27-14
Deaerator, 27-14
Debit, 69-33
Decade, 60-10
Decay
  exponential, 10-9
  weld, 21-10, 45-9
Dechlorination, 66-26
Decibel, 58-12
Decile, 11-10
Decimal number, 12-1
Declining balance method, 69-21, 69-26
Decomposition, 21-6
Decrement, logarithmic, 58-9
Dedendum, 52-10
Defect, 6-6, 61-24
Defendant, 70-5
Defender, 69-18
Definite integral, 9-1 (ftn), 9-4
Deflection, 58-2 (ftn)
  beam, 49-13 to 49-16
  bridge, 62-9
  dynamic, 58-20
  pseudo-static, 58-10
  static, 58-2
  temperature, 46-10
  truss, 49-17
Deformation, 58-2 (ftn)
  consistent, method of, 49-19
  elastic, 44-1, 49-2
  thermal, 49-3
Degenerate
  circle, 7-10
  ellipse, 7-11
Degree, 6-1
  day, 40-5, 41-2
  of a curve, 7-3
  of a polynomial, 3-3
  of damage, 61-27
  of freedom, 27-11, 47-3, 55-2, 58-2,
    58-19, 60-14
  of indeterminacy, 43-7, 44-1
  of isolation, 58-12
  of redundancy, 44-1
  of saturation, 38-2

Dehalogenation agent, 27-12
Deionized water, 21-16
Delay, 61-18
Del operator, 8-5
Delta
  -connection, 67-9
  -iron, 47-4
  rosette, 62-12
Demand
  electrical, 67-7
  factor, 27-15
Demineralized water, 21-16
de Moivre's theorem, 3-8
de Morgan's law, sets, 11-2
Density, 24-5
  concrete, 15-11 (ftn)
  factor, fan, 20-20
  fin, 29-20
  function, 11-4
  mass, 1-3
  of a fluid, 14-3
  optical, 66-17
  ratio, air, 20-2 (ftn)
  sprinkler discharge, 65-6
  watt, 40-4 (ftn)
  weight, 1-3
Denting, 21-16 (ftn)
Deoxidizer, 45-4
Dependent
  event, 11-3
  motion, 55-11
  variable, 3-2
Depletion, 69-24
Depreciation, 69-20, 69-21
  accelerated, 69-23
  basis, 69-20
  rate, 69-21
  recovery, 69-26
Depression (see also type)
  angle of, 6-2
  wet-bulb, 38-2
Depress of polymerization, 45-14
Depth, 2-2, 52-10
Derivative, 8-1
  directional, 8-6
  Laplace transform of, 10-6
  of motion, 57-10
Derived
  from rule, 66-4
  unit, 1-5, 1-6 (tbl)
Descartes' rule of signs, 3-4
Design
  condition, indoor, 39-2
  condition, outside, 40-2
  factor, 19-3
  for manufacture and assembly, 61-27
  power, 56-7 (ftn)
  pressure, 26-5, 26-13
  professional, 70-6
  temperature, inside, 41-1
Designations, aluminum, 45-11
Desorption
  gas, 66-22 (ftn)
  thermal, 66-40
Destruction removal efficiency, 66-30
Destructive test, 46-13
Desulfurization, flue gas, 22-5, 66-22
Desuperheater, 27-13
Detector-transducer, 62-3
Deterioration model, 61-11
Determinacy, 43-7
Determinant, 4-3
Deterministic model, 61-2
Deuterium, 21-2

Developer, 70-4
Deviation
  angle, 26-11
  standard, 11-10
Device, pressure measurement, 15-2
Dew point
  apparatus, 38-8
  flue gas, 22-13
  stack, 66-4
  temperature, 38-2
Dezincification, 21-11, 45-12
Diabatic flow, frictionless, 26-19
Diagonal matrix, 4-1
Diagram
  block, 60-5
  cash flow, 69-3
  equilibrium, 24-3, 47-1
  fatigue, 50-7 (ftn)
  free-body, 43-8
  iron-carbon, 47-4
  isothermal transformation, 47-6
  Mollier, 24-9
  moment, 49-8
  Nelson, 53-4
  phase, 47-1
  pole-zero, 60-8, 60-9
  pressure-enthalpy, 24-11
  root-locus, 60-12
  shear, 49-8
  time-temperature, 47-2
Diameter
  equivalent, 16-6, 17-9, 17-36, 20-14
  hydraulic, 16-6, 17-9, 20-14 (ftn), 36-8
Diametral
  clearance, 52-20
  interference, 51-5
  pitch, 52-10
  strain, 51-5
Diamond, 63-4
Diaphragm
  gauge, 15-2
  pump, 18-3
  vibration, 58-17
Die
  casting, 63-9
  types, 63-6
Diesel
  cycle, air-standard, 29-10
  engine, 29-10
  fuel, 22-7
Difference
  divided, 13-3
  Rth-order, 3-10
  table, divided, 13-3 (ftn)
Differential
  automotive, 57-9
  control, 42-2
  equation, 10-1, 10-8
  gear ratio, 56-20
  manometer, 15-3
  term, 9-1
  transformer, 62-4
Differentiation, 8-4
Diffuser, 20-24
Diffusion-controlled cell, 62-3
Diffusivity, thermal, 34-4
Digit, 3-1, 12-1
Digital
  control, 42-5
  -to-analog, 42-5
  -to-proportional, 42-5
Dilatant fluid, 14-6, 14-7
Dilation, 49-2
Diluent gas, 66-21

Dilution
  exhaust gas, 29-2
  extraction, 66-21
  toxicity, 39-5
  ventilation, 39-5
Dimension
  characteristic, 16-6, 34-3
  nominal, 16-10
  primary, 1-7
Dimensionless
  group, 1-7, 1-9 (tbl)
  number, 1-7, 1-9 (tbl)
Dimensions of variables, 1-8 (tbl)
Dimetric view, 2-3
Dioxin, 66-9
Direct
  -acting pump, 18-2
  coal-fueled turbine, 29-15
  combination, 21-6
  -condensing tower, 38-13
  -contact heater, 27-13
  current, 67-2
  digital control, 42-5
  -firing system, 27-3
  fuel reforming cycle, 31-8
  impact, 56-16
  injection, 29-3
  iron-making process, 45-7
  labor, 69-32
  transfer function, 60-4
Directed graph, 61-3
Direction
  angle, 5-2, 7-4
  cosine, 5-2, 7-5, 43-2
  helix, 52-5
  number, 7-5
Directional derivative, 8-6
Directivity, sound, 68-7
Directrix, 7-10
Disbenefit, 69-16
Disbursement, 69-3
Discharge, 18-4
  coefficient, 17-17, 17-24
  coefficient, nozzle, 26-8
  coefficient, sprinkler, 65-5
  density, 65-6
  from tank, 17-16, 17-17
  of contract, 70-5
  side, 18-4
Discounting, 69-5
Discriminant, 3-3, 7-9
Dish
  engine system, 31-3
  Stirling system, 31-3
Disjoint set, 11-1
Disk
  clutch, 52-19
  rotating, 57-2
Dispatch interval, 64-4
Dispersant, 21-16
Dispersion, 11-10, 45-19
Dispersoid, 45-19
Displacement, 21-6, 55-2
  meter, 17-21
  piston, 32-3
Displacer, 29-12
Dissatisfier, 61-28
Dissolved oxygen process, 45-3
Distance, 55-2
  between figures, 7-7
  skidding, 56-21

-squared law, 68-3
stopping, 56-21
Distillate oil, 22-6
Distillation temperature, 66-10
Distortion energy theory, 46-8, 50-5
Distributed
  collector system, 31-3
  load, 43-4, 43-6
Distribution (see also type)
  bathtub, 11-7
  beta, 61-5
  binomial, 11-4
  continuous, 11-6
  discrete, 11-4
  exponential, 11-6, 61-9
  frequency, 11-9
  Gaussian, 11-6
  hypergeometric, 11-5
  multiple hypergeometric, 11-5
  normal, 11-6
  point, 20-24
  Poisson, 11-5
  Student's t-, 61-16 (ftn)
  velocity, 16-8
  Weibull, 61-9
Distributive law, 3-3
  matrix, 4-4
  sets, 11-2
District heating, 29-21
Dittus-Boelter equation, 36-6
Divergence, 8-7
Divergent sequence, 3-10
Divided
  difference, 13-3 (ftn)
  -flow fitting, 20-16
Division (matrix), 4-4
Dodge-Romig plan, 61-25
Domestic heat source, 40-4 (ftn)
Dominant load, 49-18
Doppler effect, 68-4
Dosimeter, sound, 68-9
Dot product, 5-3
Double
  -acceptance plan, 61-26
  -acting engine, 29-2
  -acting pump, 18-2
  -angle formula, 6-4
  butt, 51-10 (ftn)
  -dabble method, 12-1 (ftn)
  declining balance, 69-21, 69-26
  -entry bookkeeping, 69-33
  integral, 9-3
  integration method, 49-13
  rivet, 51-10 (ftn)
  root, 3-3
  seal, 66-35
  shear, 51-10
  -suction pump, 18-4
  -tempered steel, 45-7 (ftn)
Doubling time, 69-8
Downcomer, 27-4
Downcycling, 66-16
Downtime, 61-9, 61-10
Downturn, 36-23
Draft, 22-16, 63-8
  balanced, 22-16
  cooling tower, 38-11
  forced, 22-16
  gauge, 15-3 (ftn)
  induced, 22-16
  loss, 22-16
  natural, 22-16
  tube, 18-20

Drag, 17-34 (see also type)
  coefficient, 17-34, 36-4 (ftn)
  disk, 17-34
  sphere, 17-34
  vehicular, 56-20
Draught, 22-16 (ftn)
Drawing, 2-2 (ftn), 47-9, 63-7
Drift, 38-13, 66-27
Drip, 27-14
Drive
  belt, 52-15
  flat belt, 52-15
  train efficiency, 56-20
  worm, 52-14
Driving dynamometer, 62-12
Droop, compressor, 32-9
Drop
  forging, 63-7
  spherical, 10-10
  test, 58-21
Dropwise condensation, 35-8
Drum
  mud, 27-4 (ftn)
  steam, 27-4
Dry
  absorption, 66-37
  -bulb temperature, 38-1
  bulk solid, 64-5
  combustor, 29-14
  compression, 33-4
  cooling tower, 38-13
  flue gas loss, 22-15
  sand, 63-8
  scrubbing, 66-37
  ultrafine coal, 66-10
Dual
  cycle, air-standard, 29-11
  -duct system, 42-1
  -flash cycle, 31-7
Duct, 36-8 (ftn)
  collapse, 20-18
  exhaust, 20-18, 20-25
  fan, 20-3
  friction loss, 20-11
  heat gain to, 41-2
  leakage, 20-18
  rectangular, 20-14
  supply, 20-18
Ductile
  cast iron, 16-10, 45-9
  material, 46-5, 50-3, 50-4
Ductility, 46-5
Ductility transition temperature, 46-14
Dulong and Petit law, 24-7
Dulong's formula, 22-14
Dummy
  activity, 61-5
  node, 61-3
  unit load, 44-6
Dump, 27-11 (ftn), 66-12
Dunkerley approximation, 58-16
Du Novy torsion balance, 14-11 (fig)
Duplex
  pump, 18-2
  stainless steel, 45-9
Duplexed steel, 45-4
Duralumin, 45-12 (ftn)
Dust, 66-9
  coal, 66-10
  collector, 39-5
  control, 66-9

Duty, 29-21
  condenser, 27-11
  cooling, 38-12
  cycle, 67-12
  heat, 36-14
  point, fan, 20-7 (ftn)
  rating, continuous, 29-4
Dwell, 57-10
Dynamic
  analysis, 49-7
  balance, 54-2, 58-18
  compressor, 32-1, 32-2, 32-8
  deflection, 58-20
  energy, 16-2 (ftn)
  equilibrium, 56-5
  friction, 56-6
  head, total, 18-6
  programming, 61-12
  reaction, 56-5 (ftn)
  response, 49-7
  response, spring, 52-8
  similarity, 17-38, 18-19, 20-9
  unit, 60-4
  viscosity, 14-6 (ftn)
Dynamics, 26-1, 55-1
Dynamometer, 29-3, 62-12
Dyne, 1-4

E
Earnings, relative, 61-20
Earth's atmosphere, 15-12, 15-13
Earths, rare, 21-2
Eastern coal, 22-4
Eccentric
  impact, 56-16
  load, column, 51-3
  load, connection, 51-15, 51-17
  loading, 49-11
Eccentricity, 7-8, 7-11, 7-12, 49-11, 51-3
  dam, 15-12
  orbit, 56-19
  ratio, bearing, 52-20
EC grade aluminum, 45-11 (ftn)
Echelon form, row-reduced, 4-2
Echelon matrix, 4-1
Eckert number, 1-9 (tbl)
Economic
  indicator, 69-39
  insulation thickness, 34-10
  life, 69-19
  order quantity, 61-2 (ftn), 69-42
Economizer, 22-3, 27-15
Eddy current testing, 46-11
Eductor, 27-3 (ftn)
Effect
  Doppler, 68-5
  evaporator, 27-14
  flywheel, 57-1
  greenhouse, 31-2
  Hall, 31-8 (ftn)
  isotope, 21-2 (ftn)
  Magnus, 17-34
  Peltier, 62-6 (ftn)
  photoelectric, 62-4
  piezoelectric, 15-2, 62-4
  refrigeration, 33-3
  Seebeck, 62-6
  stack, 22-16
  system, 20-7
  Thompson, 62-6 (ftn)
Effective
  allowable stress, pressure vessel, 53-9
  exhaust velocity, 2-8
  force, 56-5 (ftn)

head, 17-16
interest rate, 69-5, 69-28
length, column, 51-2
length, V-belt, 52-16
period, 69-2
pressure, 29-3, 29-8
sensible heat ratio, 38-7
tension, belt, 64-8
time, 26-9
value, 67-5
weld throat, 51-13
wind velocity, 39-2
Effectiveness
fin, 34-16
heat exchanger, 36-20
lever, 15-14
method, 36-20
regenerator, 29-19
Efficiency, 23-6
adiabatic, 27-8
adiabatic compression, 32-3
boiler, 22-16, 27-4
coil, 38-7
collector, 31-2
combustion, 22-16
cooling tower, 38-12
destruction removal, 66-30
drive train, 56-20
factor, heat removal, 31-2
fin, 34-16
furnace, 22-16
gear train, 52-12
humidification, 38-7
isentropic, turbine, 27-8
isolation, 58-12
joint, 16-10, 51-11, 53-8
lever, 15-14
luminous, 68-2
mechanical, 29-3
mesh, 52-12
method, 36-20
motor, 67-15
nozzle, 26-8
overall, pump, 18-8
polytropic, 32-5
production, 61-20
pulley, 43-11
pumping (temperature), 18-19
pump isentropic, 27-6
radiator, 34-16
regenerator, 29-19
relative, 29-3
removal, 66-25
saturation, 38-7, 38-11
static, 20-4
thermal, 22-16, 25-10, 28-1, 29-3
volumetric, 29-3, 32-4
Efflux, speed of, 17-16
Effort, 43-10
Eigenvalue, 4-7, 60-8 (ftn)
Eigenvector, 4-7
Ejector pump, 18-2
Elastic
analysis, 51-16
constants, relationships, 46-8
deformation, 44-1, 49-2
failure, 49-18
impact, 56-15
limit, 46-2
line, 49-14
region, 46-2
strain, 46-2
toughness, 46-6

Elasticity, modulus of, 46-2, 49-2
Elastomer, 45-13, 45-15
Electric
arc furnace, 45-5
charge, 67-2
motor, hydraulic power, 19-5
vehicle, 31-10
Electrical
discharge machining, 63-6
generator, 27-15
output, 27-1
rating, power, 27-1
thermal analogy, 34-5, 34-12
Electricity, cost, 18-8
Electrochemical
grinding, 63-6
machining, 63-6
Electrolytic grinding, 63-6
Electromotive force, 67-2
Electronic enclosures, 36-23
Electroplating, 63-14
Electropneumatic switch, 42-1
Electrospark machining, 63-6
Electrostatic precipitator, 66-26
Electrostatic unit, 67-2
Elementary row operation, 4-2
Element, transition, 21-2
Elevation
angle of, 6-2
drawing, 2-2
Elevator, 64-3, 64-9
Elimination, Gauss-Jordan, 4-2
Elinvar, 45-13
Ellipse, 7-11
Elliptic integral function, 9-8 (ftn)
Elongation at failure, 46-5
Elutriation, 64-11
Embossing, 63-7
Embrittlement
blue, 47-9
caustic, 21-11
hydrogen, 21-11, 53-4
neutron, 30-2
Emery, 45-19
Emission, fugitive, 66-10
Emissions
monitoring, 66-21
sampling, 66-20
Emissive power, 37-1
Emissivity, 37-1, 37-3
Emulsion, 19-2 (ftn), 21-7 (ftn)
Enclosed air space, 35-7
Enclosure hood, 39-4
Enclosures, cooling electronic, 36-23
End
-around carry, 12-4
post, 43-12
-quench test, 47-7
restraint coefficient, 51-2
Endoergic impact, 56-15
Endothermic reaction, 21-9
Endpoint, pressure-drop, 21-16
Endurance
limit, 46-9
ratio, 46-9
strength, 46-9
Energy, 23-1
conservation, 58-5
conservation law, 23-1
content of moist air, 38-3
cutting, 63-3
dynamic, 16-2 (ftn)
-efficiency ratio, 23-6 (ftn), 33-2

falling mass, 23-3 (ex)
flow, 16-2, 23-3, 24-6, 25-6
flywheel, 57-1
grade line, 16-9, 17-15
gravitational, 23-2
impact, 16-4
internal, 23-4, 24-5
kinetic, 16-2, 23-3
line, 16-9
potential, 16-2, 23-2
pressure, 16-2, 23-3
-production efficiency, 23-6
sink, 17-15, 25-10
solar, 31-2
source, 17-15, 25-10, 31-1
specific, 16-1, 16-9, 23-1
specific, battery, 31-10
spring, 23-3
stagnation, 16-4
static, 16-2, 23-3
storage, 31-9
strain, 46-6, 49-2, 49-11
total, 16-2, 16-4
-use efficiency, 23-6
velocity, 16-2 (ftn)
-work principle, 23-3, 23-4
Engine
diesel, 29-10
gas-diesel, 29-10
oil, 52-19
Stirling, 29-12
Engineering
dimensional system, 1-7
intern, 72-1 (ftn)
registration, 72-1
strain, 46-2
stress, 46-2
Engineer-In-Training exam, 72-1
Engineers' Creed, 71-1
English
engineering system, 1-2
gravitational system, 1-3
Enlargement, 17-12, 17-13
Enthalpy, 24-5
of formation, 21-9
of moist air, 38-3
of reaction, 21-9
Entrance region, 36-5
Entropy, 24-6
function, 24-11
production, 24-6
Entry, coefficient of, 20-15
Entry length, thermal, 36-5
Environment, 25-1, 66-3
Environmental engineering, 66-4
Eötvös number, 17-38
Epicyclic gears, 57-9
Epicycloid, 7-3
Epidemic, 10-10
Equal
-friction method, 20-19
vectors, 5-1
Equality of matrices, 4-3
Equation, 3-2
Andrade's, 46-15
Antoine, 14-9 (ftn)
auxiliary, 10-1
Bernoulli, 16-2, 17-14
Bessel, 10-5
Blasius, 17-5
Bromley, 35-11
Buckingham, 52-14 (ftn)
Cauchy, 10-5

characteristic, 10-1
Clausius-Clapeyron, 14-9 (ftn)
Colebrook, 17-5
complementary, 10-1
Darcy, 17-7
Darcy-Weisbach, 17-7
differential, 10-1
Euler's, 3-8, 10-5
Gauss' hypergeometric, 10-5
general flow, 26-2
Hagen-Poiseuille, 17-7
Hazen-Williams, 17-7, 65-7
Legendre, 10-5
Nikuradse, 17-5
nonparametric, 3-3
Nusselt, 35-3, 36-5
of state, 24-12, 24-13
    real gas, 24-19
    van der Waals', 24-19
    virial, 24-19
parametric, 3-2
Petroff, 52-21
quadratic, 3-3
reduced, 10-1
Rohsenow, 14-10
steady-flow energy, 26-2
Taylor's, 63-5
three-moment, 44-5
tool life, 63-5
Torricelli's, 17-16
Weisbach, 17-7
Weymouth, 26-17
Zuber, 14-10
Equations
   of motion, Euler's, 56-8
   simultaneous linear, 3-7
Equiangular rosette, 62-12
Equilateral hyperbola, 7-12
Equilibrium, 43-20, 56-2
   conditions, 43-6
   diagram, 24-3, 47-1
   dynamic, 56-5
   isopiestic, 14-9
   moisture content, 45-17
   phase, 56-6
   position, 58-2
   solid, 24-3
   static, 15-16
   thermal, 24-4, 25-5, 37-1, 37-6
Equipment
   high-side, 33-7
   low-side, 33-8
Equipotential line, 17-3
Equity, 69-34
Equivalence, economic, 69-5
Equivalent
   $CaCO_3$, 21-13
   chemical, 21-4
   diameter, 16-7, 17-9, 17-36
   diameter, duct, 20-14
   evaporization, 27-5
   fluid height, 15-5 (tbl)
   Joule, 23-1
   length, 17-11
   length method, duct, 20-15
   resultant, 43-4
   spring constant, 52-2
   stress, 50-6
   temperature difference, 41-2
   uniform annual cost, 69-7
   vector, 5-1 (ftn)
   weight, 21-4
   weight, milligram, 21-8
Erf, 9-8

Ergonomics, 61-29
Ericsson cycle, 29-13
Erosion corrosion, 21-11
Error, 11-13, 60-4, 60-14, 61-9, 62-2 (*see also type*)
   function, 9-8, 11-6
   gain, 60-4
   in action, 52-14
   magnitude, 62-3
   ratio, 60-4
   signal, 60-4
   transfer function, 60-4
   type I/II, 61-25
Errors and omission insurance, 70-7
Escape velocity, 56-19
Estimator, unbiased, 11-10
Etching radius, 63-6
Ethanol, 22-7
Ethic, 71-1
Ethical priority, 71-2
Ethyl alcohol, 22-7
Euler
   constant, 9-8 (ftn)
   curve, 51-2
   equation, 3-8
   equations of motion, 56-8
   load, 51-1
   number, 1-9 (tbl)
Eutectic
   alloy, 47-3
   cast iron, 47-5
   composition, 47-3
   line, 47-3
   reaction, 47-3
   temperature, 47-3
Eutectoid
   reaction, 47-3
   steel, 47-5
   temperature, 47-4
Evacuated-tube collector, 31-2
Evaluation, nondestructive, 46-11
Evaporation, 10-10, 38-13
Evaporative cooling, 38-7, 38-10
Evaporative pan humidification, 39-3
Evaporator, 27-14 (ftn), 33-1, 33-8
Evaporization, 35-10, 35-11
   equivalent, 27-5
   factor of, 27-4
Even symmetry, 7-4, 9-7 (tbl)
Event, 11-2, 61-3
Exact equations, first-order, 10-3
Exams, licensing, 72-1, 72-2
Exam date, 72-2
Excess air, 22-12
Exchange, ion, 21-14
Exchanger, heat, 36-10
Excitation, motor, 67-13
Exciter, 58-21
Exemplary damages, 70-7
Exfoliation, 21-11
Exhaust,
   duct, 20-18, 20-25
   gas dilution, 29-2
   hood, fume, 39-4
   stroke, 29-2
   velocity, characteristic, 26-9
   velocity, effective, 26-8
Exit property, 26-5
Exoergic impact, 56-15 (ftn)
Exothermic reaction, 21-9
Expanded-flow bin, 64-5
Expander, 29-14
Expansion
   by cofactor, 4-3

coefficient of, 14-12 (ftn)
   factor, 17-26
   linear, coefficient, 49-3
   over-, 26-13
   series, 8-8
   thermal, 49-3
   two-stage, 27-8
   under-, 26-13
   valve, 33-1
   volumetric, coefficient, 49-3
Expected value, 69-30
Expense, 69-32
Expensing an asset, 69-20
Explosive
   forming, 63-11
   limit, lower, 39-5
Exponent, 3-5
   polytropic, 15-13, 25-2
   strain-hardening, 46-5
Exponential
   decay, 10-9, 60-9
   distribution, 11-6, 11-7, 61-9
   form, 3-8
   function, integral, 9-8
   gradient cash flow, 69-4
   growth, 10-9, 69-4
   reliability, 11-7
   weighted forecast, 69-42
Exposed fraction, 15-15
Exposure
   allowance, 40-1
   limit, short-term, 39-5
Extended surfaces, 34-16
Extension spring, 52-8
Extensive property, 24-4
External
   force, 43-2, 56-2
   gear, 57-4
   pressure, 15-14
   rate of return, 69-12
   shoe brake, 52-17
   work, 23-1
Externality, nuclear 30-2
Externally fired combined cycle, 29-15
Extraction
   cycle, 28-2
   rate, 27-8
   turbine, 27-8, 27-13
Extractive metallurgy, 45-1
Extreme
   fiber, 49-11
   point, 8-2
   stress, 49-5
   temperature difference, 36-13
Extrinsic waste, 66-3
Extrusion, 63-10

F
Face
   -and-bypass damper control, 42-1
   damper, 42-2 (ftn)
   value, 69-29
   velocity, 66-24
   width, tooth, 52-10
Facilities layout, 61-14
Factor (*see also type*)
   absorption, 66-40
   Barth speed, 52-13
   bolt torque, 51-12
   bonus, 61-20
   bypass, 41-4
   capacity, 27-15

cleanliness, 27-11
coil, 38-7
compressibility, 14-12 (ftn), 24-19
cooling load, 41-2
damping, 58-9
Darcy friction, 17-5
demand, 27-15
density, fan, 20-2
design, 19-3
directivity, 68-7
emissivity, 37-3
expansion, 17-26
Fanning friction, 17-5 (ftn), 26-15, 26-18
fatigue notch, 46-10
fatigue stress concentration, 46-9, 46-10
form, gear, 52-13
fouling, 36-17
gage, 62-8
heat removal efficiency, 31-2
heat transfer, 36-22
impact, 16-4
integrating, 10-2
interference, 31-5
joint quality, 16-10
load, 27-15
loading, 31-9
magnification, 58-10
normal shock, 26-9
nut, 51-12
of evaporization, 27-4
of safety, 15-11, 15-12, 50-1
output, 27-16
performance, 38-10
performance rating, 61-16
phase, 67-7
piping geometry, 17-13
plant, 27-15
power, 67-7, 67-8
reheat, 28-6
sensible heat, 38-6
service, 40-4, 56-7 (ftn), 67-11
service, gear drive, 52-11
service, V-belt, 52-16
shading, 31-2
shape, 34-20, 37-2
side friction, 56-21
stress concentration, 49-4
stripping, 66-40
transmission, 26-18
transverse sensitivity, 62-8
use, 27-16
velocity of approach, 17-23
Wahl, 52-5
water hammer, 19-3
Factorial, 3-2 (tbl, ftn)
Factors versus strength, 46-16
Factory
    cost, 69-32, 69-36
    loading, 61-15
Failure, 11-3
    brittle, 46-13
    cup-and-cone, 46-4
    line, 50-6
    mean time to, 11-7, 61-9
    mode and effect analysis, 61-27
    sudden, 61-11
    theory, 50-1
Failures, mean time between, 61-9
Faith of an engineer, 71-1
Falling mass energy, 23-3 (ex)
Family, chemical, 21-2

Fan
    axial flow, 20-3
    boost, 22-16
    centrifugal, 20-3
    curve, 20-5
    duct, 20-3
    homologous, 20-9
    law, 20-8
    non-overloading, 20-4
    paddle wheel, 20-4
    power, 20-4
    propeller, 20-3
    radial, 20-4
    radial tip, 20-4
    rating table, 20-5
    shaving wheel, 20-4
    similarity, 20-9
    specific speed, 20-4
    squirrel cage, 20-3
    straight-blade, 20-4
    tubeaxial, 20-3
    vaneaxial, 20-3
Fanning friction factor, 17-5 (ftn), 26-15, 26-18, 36-4 (ftn)
Fanno flow, 26-15
Faraday, 67-2
Far-field, 68-6
Fastener, 51-8, 51-16
Fast Fourier transform, 9-7
Fatigue
    corrosion, 21-12
    cumulative, 50-7
    diagram, 50-7 (ftn)
    failure, 46-8
    interaction formula, 50-7
    life, 46-9
    loading, spring, 52-4
    notch factor, 46-10
    notch sensitivity, 46-10
    ratio, 46-9
    strength, 46-9
    stress concentration, 49-5
    stress concentration factor, 46-10
    test, 46-8
Fault, 61-9
    -tolerant system, 61-9
    -tree analysis, 61-27
Feasible region, 61-12
Feedback, 60-4
    gain, 60-4
    negative, 60-4
    positive, 60-4
    ratio, 60-4
    transfer function, 60-4
Feeder, dry bulk solid, 64-5, 64-6
Feed main, 65-2
Feedwater
    boiler, 21-15, 21-16
    heater, 27-13
    heater, closed, 36-14 (ftn)
Fee structure, 70-4
Fenestration, 41-2 (ftn)
Ferrimagnetic material, 45-18
Ferrispinel, 45-18
Ferrite, 45-18, 47-4, 47-6
Ferritic stainless steel, 21-11 (ftn), 45-8
Ferroelectric, 45-18
FFT (see Fast Fourier transform)
Fiber
    -nickel-cadmium battery, 31-10
    -strengthened, 45-19, 45-20
Fiber core, wire rope, 64-2
Fiberglass, 45-19, 66-5

Fictive temperature, 45-17
Fiduciary responsibility, 70-2
Figure of merit, 31-9
File hardness, 46-13
Filler, 45-13
Fillet weld, 51-13
Fill, tower, 38-11
Film
    amorphous, 31-4
    boiling, 35-10
    coefficient, 34-6
        combined, 34-17
        natural, 35-2
        total, 35-11
    Reynolds number, 35-9
    temperature, 35-3, 36-3
Filming amine, 21-15
Filmwise condensation, 35-8
Filter, 60-10
    air, 39-5
    hydraulic, 19-5
    ratio, 66-24
Filtering velocity, 66-24
Fin, 34-16 to 34-18 (see also type)
    density, 29-20
    effectiveness, 34-16
    efficiency, 34-16
    length, corrected, 34-18
    -tube exchanger, 36-13
Final value, 60-7
Fines (coal), 22-4
Finish, surface, 2-4, 46-9
Finite series, 3-11
Fire
    damper, 20-18
    protection system, 65-1
    -resistant fluid, 19-2
    -tube boiler, 27-4 (ftn)
Firm capacity, 27-15
First
    critical back-pressure ratio, 26-13
    derivative, 8-1
    -in, first-out, 69-37
    law of thermodynamics, 23-2 (ftn), 25-4, 25-5
    moment of a function, 9-6
    moment of area, 44-5, 48-2
    -order, 10-3
    -order, linear, 10-2
Fit
    interference, 51-5
    press, 51-5
    shrink, 51-5
Fitting
    divided flow, 20-16
    friction loss, duct, 20-14
    screwed, 17-12 (ftn)
    threaded, 17-12 (ftn)
Fixed
    asset, 69-34
    carbon, 22-4
    -charge problem, 61-13
    cost, 69-31
    -end beam, 44-1, 44-6
    -end moment, 44-6
    -orifice desuperheater, 27-13
    -point iteration, 13-2
    pulley, 43-11
    vector, 5-1 (ftn)

Flame
  blow-off, 27-3
  hardening, 47-10
  temperature, 22-15, 66-13
  welding, 63-12
Flange, 48-4
  stiffener, 49-18
  straight, 53-4
Flanged
  hood, 39-4
  joint, 53-14
Flank, tooth, 52-10
Flap, 17-33
Flare, 66-11, 66-34
Flash, 63-8
  -back, 27-3
  evaporator, 27-14 (ftn)
  point, 22-5 (ftn)
  steam cycle, 31-7
  tank, 27-14 (ftn), 31-7
Flashing, 27-13, 27-14 (ftn)
Flat
  belt drive, 52-15
  coil spring, 52-9 (ftn)
  -faced follower, 57-10
  friction, 56-6
  plate, 51-17
    collector, 31-2
    evaporation, 35-11
    film coefficient, 35-6
    flow over, 36-3, 36-4
  spring, 52-10
Flexible hydraulic hose, 19-3
Flexural stress, 49-10
Float, 61-3
Floating
  action control, 42-5
  -head heat exchanger, 36-11
  object, 15-15, 15-16
Flocculation, 66-41
Floodable material, 64-5
Flooding, 66-22, 66-40
  velocity, 66-22
Flow
  adiabatic, with friction, 26-15
  annular, 36-8
  around cylinder, 17-36
  choked, 26-4
  coefficient of, 17-24
  coefficient, valve, 17-13
  Couette, 36-3 (ftn)
  critical, 16-8
  curve, 46-5
  energy, 16-2, 23-3, 24-6, 25-6
  equalization, 66-41
  Fanno, 26-15
  frictionless diabatic, 26-19
  isentropic, 26-3
  isothermal, with friction, 26-17
  laminar, 16-7
  measurement, 17-20
  nozzle, 17-25
  number, 64-12
  over flat plate, 17-37
  rate, 20-3
  Rayleigh, 26-19
  shear, 51-15
  streamline, 16-7
  turbulent, 16-8, 17-5
  uniform, 16-9
  viscous, 16-7
  work, 23-3, 25-6
Fluctuation, coefficient of, 57-2

Flue gas, 22-11, 66-4 (ftn)
  calculations, 22-12
  cleanup, 66-24
  desulfurization, 22-5, 66-22
  dew point, 22-13
  recirculation, 66-27
  temperature, 22-13
Fluid
  capacitance, 59-7
  compressible, 26-1
  cryogenic, 24-2 (ftn)
  heat transfer, 31-2, 36-22
  height equivalent, 15-5 (tbl)
  hydraulic, 19-2
  hydraulic fire-resistant, 19-2
  ideal, 14-2
  inductance, 59-7
  inertance, 59-7
  mass, 15-19
  Newtonian, 14-2, 14-6
  power, 19-1
    pump, 19-4
    symbols, 19-1
  pressure, 14-2, 15-14
  real, 14-2
  resistance coefficient, 59-7
  rotating, 57-3
  shear stress, 14-6
  system, modeling, 59-7
  type, 14-2
  viscosity type, 14-6
Fluidized-bed
  boiler, 66-28
  combustion, 22-5, 27-3, 27-4
  combustor, 66-27
Fluorinated hydrocarbon, 33-4
Fluoroplastic, 45-15
Fluoropolymer, 45-15
Flux
  geometric, 37-4
  heat, 35-3, 36-5
  luminous, 68-1
  welding, 63-12
Fly ash, 22-3
Flywheel, 57-1, 57-3
  effect, 57-1
  effect, thermal, 40-5
  rotor, 57-3
F-method, 36-20
Focus, 7-10
Focusing collector, 31-2
Follower, 57-10
Footcandle, 61-30 (ftn)
Foot pulley, 64-9
Force, 56-2 (see also type)
  amplification, 43-10
  central field, 56-17
  centrifugal, 56-5
  centripetal, 56-5
  concentrated, 43-2
  -couple system, 43-4
  cutting, 63-2
  external, 43-2
  gear, 52-12, 52-13
  inertial, 56-5
  internal, 43-2
  majeure, 70-3
  normal, 56-6
  normal, on a dam, 15-12
  number, 64-12
  on a dam, 15-11
  on a blade, jet, 17-29
  on a pipe bend, 17-30
  on-plate, jet, 17-28

  tractive, 56-20
  van der Waals', 24-19
  velocity-dependent, 56-17
  work performed by, 9-4
Forced
  convection, 36-2
  draft, 22-16
  draft tower, 38-11
  outages, mean time between, 61-9
  response, 60-2
  vibration, 58-2, 58-10
Forcing
  frequency, 58-10
  function, 10-1, 10-4, 58-10, 59-2
Forebay, 18-20
Forecasting, 69-41
Forging, 63-7
Form
  cis, 3-8
  drag, 17-34
  exponential, 3-8
  factor, gear, 52-13
  Newton, 13-3 (ftn)
  operational (log), 3-5
  phasor, 3-8
  polar, 3-8
  rectangular, 3-7
  trigonometric, 3-7
Formality, 21-7
Formation
  enthalpy of, 21-9
  heat of, 21-9
Forming, explosive, 63-11
Formula
  Babcock, 17-9
  Barlow, 19-3
  Boardman, 19-3
  double-angle, 6-4
  Dulong's, 22-14
  half-angle, 6-4
  J. B. Johnson, 51-3
  Lamé, 19-3
  Maney, 51-12
  Panhandle, 17-9
  parabolic column, 51-3
  PLAN, 29-8
  secant, 51-3
  Spitzglass, 17-9
  two-angle, 6-4
  weight, 21-4
  Weymouth, 27-9
Forward
  -curved fan, 20-3
  gain, 60-4
  transfer function, 60-4
Fouling, 17-12, 36-17
Fourier
  analysis, 9-6
  inversion, 9-6
  number, 1-9 (tbl), 34-4
  series, 9-6
  transform, fast, 9-7
Fourier's
  law, 34-5
  theorem, 9-6
Four-stroke
  cycle, 29-2
  diesel, 29-10
Fraction
  bleed, 27-14
  gravimetric, 21-5, 22-2, 24-17
  mass, 24-17

mole, 14-6, 22-2, 24-17
partial, 3-6, 9-2
partial pressure, 22-2
submerged, 15-15
volumetric, 22-2, 24-17
Fracture
    appearance transition temperature, 46-14
    strength, 46-3
    transition plastic temperature, 46-14
Frame, 43-12
Frame, inertial reference, 55-9
Francis turbine, 18-22
Fraud, 70-2 (ftn), 70-6
Fraudulent act, 70-6
Free
    area, 20-24
    -body diagram, 43-8, 43-20
    convection, 35-2
    field, 68-5
    -machining steel, 45-5, 63-4
    moment, 43-4, 43-9
    -piston Stirling engine, 29-12
    protection, 42-2
    pulley, 43-11
    rotation, 58-5
    -stream temperature, 36-2
    surface, 16-6
    vibration, 58-2
Freedom, degree of, 47-3, 55-2, 58-19
Freeze stat, 42-2
Freezing point, 47-1 (ftn), 47-2
Freon, 66-8
Frequency
    analysis, 9-7 (ftn), 60-11
    angular, 58-3
    crossover, 60-11
    cut-off, 60-10
    damped, 58-9, 60-2
    distribution, 11-9
    electrical, 67-4
    forcing, 58-10
    fundamental, spring, 52-8
    linear, 58-3
    natural, 9-6, 58-3
    polygon, 11-9
    response, 60-2 (ftn), 60-10, 60-11
Fresnel integral function, 9-8 (ftn)
Fretting corrosion, 21-12
Friction, 56-6 (see also type)
    belt, 56-7
    brake, 62-12
    coefficient of, 15-12, 56-6
    factor
        chart, Moody, 17-6
        Darcy, 17-5
        Fanning, 17-5 (ftn), 26-15, 26-18, 36-4 (ftn)
    flat, 56-6
    head, 18-5
    head loss, 17-4
    internal, angle, 46-7
    loss
        duct, 20-11, 20-14
        gas, 17-9
        hydraulic power, 19-4
        slurry, 17-11
        sprinkler, 65-7
        stack, 22-18
        steam, 17-9
    power, 18-8, 29-3
    power, fan, 20-5
    pressure, 20-11

sideways, 56-21
turbulent flow, 17-7
Frictionless
    diabatic flow, 26-19
    surface, 43-7
Fried heat transfer efficiency factor, 36-22
Froude number, 1-9 (tbl), 17-40
Fuel
    -air ratio, 29-3
    -bound NOx, 66-13
    cell, 31-7 (see also type)
    consumption
        furnace, 40-6
        specific, 29-3
        vehicular, 56-20
    diesel, 22-7
    economy, 56-21
    -injection, 29-2, 29-3
    oil, 22-5
    oxygenated, 66-8
    refuse-derived, 66-32
    switching, 66-18
    utilization, 29-22
    waste, 22-4
Fugitive
    dust, 66-9
    emission, 66-10, 66-35
Full
    -arc admission, 27-16
    -bridge, 62-11
    -load value, motor, 67-11
    -wave symmetry, 9-7 (tbl)
    -wrap, 64-3
Fully
    amortized bond, 69-29 (ftn)
    developed, 36-5, 36-7 (ftn)
    turbulent flow, 16-8
Fume exhaust hood, 39-4
Function
    circular transcendental, 6-2
    error, 9-8, 11-6
    first moment of, 9-6
    forcing, 10-1, 58-10
    hazard, 11-7
    holomorphic, 8-1
    hyperbolic, 6-4
    integral, 9-8
        cosine, 9-8
        elliptic, 9-8 (ftn)
        exponential, 9-8
        Fresnel, 9-8 (ftn)
        gamma, 9-8 (ftn)
        sine, 9-8
    objective, 61-12
    of an angle, 6-2
    of related angles, 6-3
    path, 25-4
    performance, 11-8
    point, 25-4, 25-7
    regular, 8-1
    response, 59-1
    second moment of a, 9-6
    singularity, 59-3
    symmetrical, 7-4
    transfer, 60-3
    trigonometric, 6-2
    unit impulse, 10-6
    unit step, 10-6
    vector, 5-5
Functionality, 45-15

Functions, integral of combination, 9-2
Fundamental
    frequency, spring, 52-8
    speed, shaft, 58-15
    theorem of calculus, 9-4
Fundamentals of Engineering exam, 72-1, 72-3
Funnel-flow bin, 64-5
Furnace, 27-3, 40-5, 40-6
    arc, 45-5
    blast, 45-2
    efficiency, 22-16
    fluidized-bed, 27-3
    induction, 45-5
    pulverized-coal, 27-3
    sorbent-injection, 22-5
    reverberatory, 45-10
Fusion
    heat of, 47-2
    metals, 63-11
Future worth, 69-5
F-waste, 66-3

G
Gage
    factor, 62-8
    pressure, 14-2, 15-2, 15-6 (ftn)
    strain, 62-8, 62-9
Gain
    characteristic, 60-10
    crossover, 60-11
    crossover point, 60-10
    feedback, 60-4
    forward, 60-4
    heat, 41-1
    margin, 60-11
    overall, 60-4
    -sharing plan, 61-20
Galvanic
    action, 21-11
    series, 21-11
Galvanizing, 63-14
Gamma
    integral function, 9-8 (ftn)
    -iron, 47-4
Gangue, 45-2
Garbage, 22-4, 66-12
Gas
    absorption, 66-21, 66-22
    compressor, 32-1
    constant, specific, 24-14
    constant, universal, 24-13
    desorption, 66-22 (ftn)
    -diesel engine, 29-10
    dynamics, 26-1
    friction loss, 17-9
    greenhouse, 66-11
    high velocity, 26-1
    ideal, 24-2
    incompressible, 16-3 (ftn)
    landfill, 66-11
    liquefied petroleum, 22-7
    manufactured, 22-7 (ftn)
    mixture, 24-2, 24-17
    noble, 21-2
    noncondensing, 27-11
    process, metal arc welding, 63-12
    properties, 24-15 (tbl)
    real, 24-2
    synthetic, 22-5, 22-8
    table, 24-10
    turbine, 29-14
    welding, 63-11

Gasification process, 29-15, 66-16
Gasifier, 29-14
Gasket blow-out, 53-14
Gasohol, 22-7
Gasoline, 22-6, 66-10
Gate, 63-8
  blast, 20-18
  valve, 16-11
Gauge
  diaphragm, 15-2
  draft, 15-3 (ftn)
  pressure, 15-2 (*see also* Gage pressure)
  strain, 15-2
Gauss' hypergeometric equation, 10-5
Gaussian distribution, 11-6
Gauss-Jordan elimination, 4-2
Gear (*see also* type)
  bevel, 52-14
  helical, 52-12
  hypoid, 52-14
  involute, 52-11
  motor, 18-10, 67-12 (ftn)
  oil, 52-19
  ratio, 57-5
  set, 52-11, 57-4
  spur, 52-10
  train, 57-4
  versus pinion, 52-11
  worm, 52-14
gees, 56-4 (ftn)
General
  contractor, 70-4
  damages, 70-7
  expense, 69-32
  flow equation, 26-2
  ledger, 69-33
  partner, 70-1
  term, 3-10
  triangle, 6-5
Generating line, 52-11
Generation, internal heat, 34-14
Generator, 67-10 (ftn)
  -absorber-heat-exchanger cycle, 33-7
  electrical, 27-15
  magnetohydrodynamics, 31-8
  pollution, 66-4
  steam, 27-4
  thermionic, 31-9
  thermoelectric, 31-9
Genotoxin, 61-27
Geometric
  factor, 37-2
  flux, 37-4
  jerk, 57-10
  mean, 11-10
  mean radius, 57-3
  sequence, 3-11
  series, 3-11
  similarity, 17-38
Geopressured source, 31-7 (ftn)
Geostationary orbit, 56-18
Geothermal energy, 31-7
Gerber line, 50-6
Geysers, 31-7
g-force, 56-4 (ftn)
Gibbs
  phase rule, 47-3
  theorem, 24-18
Girth seam, 53-6 (ftn)
Glass, 45-17
  common, 45-17
  fiber types 45-19
  transition temperature, 45-17

Glauconite, 21-14
Global warming, 66-11
Globe valve, 16-11
Glow plug, 29-10 (ftn)
Golf ball, 17-35
Goodman line, 50-6
Governing, 27-16
Government bronze, 45-12
Grade
  bolt, 51-8
  line, 16-9, 17-15
  resistance, 56-19
  roadway, 56-19
  steel, 19-3 (ftn)
  viscosity, 52-19
Gradient
  cash flow, 69-9
  roadway, 56-19
  search, 61-12
  series cash flow, 69-4
  temperature, 34-5
  thermal, 34-5
  vector, 8-5
  velocity, 14-6
Graetz number, 1-9 (tbl), 36-2
Grain
  growth stage, 47-9
  properties, 47-8
  size, 47-8
Grand sensible heat ratio, 38-6, 41-4
Granular activated carbon, 66-23
Graphite fiber, 45-20
Graphitization, 45-9
Grashof number, 1-9 (tbl), 35-2
Gravimetric
  analysis, 22-2
  fraction 21-5, 22-2, 24-17
Gravitational
  constant, 1-2
  energy, 23-2
Gravitation, Newton's law of, 56-18
Gravity
  acceleration of, 1-2
  center of, 54-1
  chimneys, 22-16 (ftn)
  dam, 15-11 (ftn)
  damper, 20-18
  die casting, 63-9
  drop hammers, 63-7
  molding, 63-9
  on earth, 1-2
  specific, 14-4
Gray
  body, 37-3
  cast iron, 45-9, 47-8
Greek alphabet, 3-1
Greenhouse
  effect, 31-2
  gas, 66-11
Green sand, 21-14, 63-8
Gridded system, 65-1
Grille, 20-18, 20-24
Grinding, 63-5, 64-7
Grip, 51-11
Grit, 63-5 (tbl), 66-41
Gross
  electrical output, 27-1
  heating value, 22-14
Ground slab, 40-3
Group
  chemical, 21-2
  dimensionless, 1-7, 1-9 (tbl)
  X steel, 45-7

Growth, exponential, 10-9
Guard, 61-30
Guide vane control, compressor, 32-2
Gyration, radius of, 48-6, 54-2

H
Hagen-Poiseuille equation, 17-7
Half
  -angle formula, 6-4
  -bridge, 62-11
  -power point, 60-10
  -wave symmetry, 7-4, 9-7 (tbl)
  -wrap, 64-3
  -year convention, 69-3
Hall effect, 31-8 (ftn)
Hammer
  forging, 63-7
  water, 17-31
Hard
  material, 46-16
  steel, 47-7
  surfacing, 63-14
Hardenability test, 47-7
Hardening
  age, 45-8 (ftn), 45-11, 47-9
  of non-allotropic alloys, 47-9
  precipitation, 45-11, 47-9
  strain, 46-4
  surface, 47-10
Hardness
  carbonate, 21-14
  noncarbonate, 21-14
  number, 46-12
  permanent, 21-14
  red, 63-4
  red/hot, 45-7 (ftn)
  steel maximum, 47-7
  temporary, 21-14
  test, 46-12
  total, 21-14
  water, 21-13
Hardwood, 45-16
Hardy-Cross method, 17-20, 49-17
Harmonic, 58-18
  cam profile, 57-10
  frequency, fundamental, 52-8
  motion, simple, 58-2
  sequence, 3-11
  series, 3-11
  terms, 9-6
  test, vibration, 58-20
Hastelloy, 45-13
Haulage rope, 64-1
Hazard
  analysis, 61-27
  assessment by risk analysis, 61-27 (ftn)
  function, 11-7
Hazardous
  air pollutant, 66-3
  material, 66-19
  waste, 66-3, 66-20, 66-31
Hazen-Williams equation, 17-7, 65-7
Hazmat, 66-19
Head, 15-4 (ftn), 18-5
  absorption (boiler), 27-4
  acceleration, 18-2
  added, 17-14
  adiabatic, 32-5
  atmospheric, 18-5 (ftn)
  end, 29-2
  extracted, 17-14
  friction, 18-5
  impact, 16-5

loss, 17-4
    effect of viscosity, 17-10
    slurry, 17-11
polytropic, 32-5
pressure, 18-5 (ftn)
pressure vessel, 53-4
pulley, 64-9
stack effect, 22-17
static discharge, 18-5
static suction, 18-5
total, 16-3
total dynamic, 18-6
vapor pressure, 18-5 (ftn)
velocity, 18-5
volume of pressure vessel, 53-16
Header-type feedwater heater, 27-14
Head-on machine, 31-4
Health risk assessment, 61-27
Heat
    capacity, 24-7
    content, 24-5
    -driven refrigeration cycle, 33-6
    duty, 36-14
    exchanger, 27-10, 36-10 (*see also type*)
    exchanger effectiveness, 36-20
    flux, 35-3, 36-5
    gain, 40-4, 41-1
    graphically finding, 25-4
    latent, 24-8, 24-9
    load, 27-11, 36-14, 38-12
    metabolic, 39-3
    of combustion, 22-14
    of formation, 21-9
    of fusion, 47-2
    of steel, 45-4
    pipe, 34-20
    pump, 33-2
    rate, 27-5, 28-2
    ratio, sensible, 41-3
    recovery steam generator, 29-20
    rejector, 29-12
    release rate, 66-32
    removal efficiency factor, 31-2
    reservoir, 25-10
    sensible, 24-8
    specific, 24-7, 34-3
    tape, 34-20
    total, 24-5
    tracing, 34-20
    transfer
        coefficient, 35-2
        coefficient, convective, 34-6
        coefficient, overall, 36-15
        coefficient, radiant, 37-4
        factor, 36-22
        fluid, 31-2, 36-22
        unit, 36-20 (ftn)
    transmittance, overall, 34-6
    treatment, cast iron, 47-8
Heater (*see also type*)
    air, 27-15
    feedwater, 27-13
Heating
    district, 29-21
    load, 40-1
    sensible, 38-8
    space, 29-21, 33-2
    value, 22-14
    with dehumidification, 38-11
    with humidification, 38-10
Heats, ratio of specific, 26-3
Heavy
    damping, 58-9
    -duty coil, 36-13

hydrogen, 21-2
    metal, 21-2
Heel (dam), 15-11
Height, 2-2
    compressed, 52-5
    critical, pressure vessel, 53-5 (ftn)
    of packing, 66-40
    of transfer unit, 66-40
    solid, 52-5
    waviness, 2-4
Heisler chart, 34-13
Helical
    compression spring, 52-4
    extension spring, 52-8
    gear, 52-12
    torsion spring, 52-9
Heliostat, 31-3
Helix, 7-12
    angle, 52-12
    direction, 52-5
    pitch, 7-12
    thread angle, 51-13
Hematite, 45-2
Henry's law, 21-7, 66-39
HEPA (*see* High-efficiency particulate
    arrester)
Hess' law, 21-9
Hexadecimal system, 12-2
Hideout, 21-11
Hierarchy, need, 61-28
High
    -alloy austenitic stainless steel, 45-9
    -alloy steel, 45-5
    -carbon steel, 45-5
    -efficiency particulate arrester, 39-6
    energy rate forming, 63-11
    -pass filter, 60-10
    -performance power system, 29-22
    -pressure turbine, 27-8
    -side equipment, 33-7
    -speed steel, 63-4
    -strength steel, 45-5
    -temperature advanced furnace, 29-22
    -temperature reservoir, 25-10
    velocity gas, 26-1
    -velocity system, 20-17
Higher heating value, 22-14
Hilbert-Morgan equation, 36-8
Hinge, 43-10, 49-18
Histogram, 11-8
Hobby, 69-12 (ftn)
Hoisting cable, 64-1
Hole, tap, 15-2
Holographic testing, 46-11
Holomorphic function, 8-1
Holonomic coordinate, 55-2
Homogeneous
    circuits, law of, 62-7
    differential equation, 10-1, 10-2
    function, 59-2
    linear equation, 3-7
    unit system, 1-2
Homologous
    fan, 20-9
    pump, 18-12, 18-18
Honing, 63-5, 63-14
Hood, fume exhaust, 39-4
Hooke's law, 46-2, 49-2, 52-2
Hoop stress, 51-3, 51-4
Hopper, 64-5
Horizon, 69-3

Horizontal
    -axis machine, 31-4
    shear, 49-10
Horsepower, 23-5
    aero, 17-34
    air, 20-4
    boiler, 22-16, 27-4
    hydraulic, 17-14
    shaft, 51-13
Hose, hydraulic, 19-2
Host ingredient, 47-1
Hot
    dry rock source, 31-7 (ftn)
    -end drive, 29-14
    hardness, 45-7 (ftn)
    reserve, 27-15
    well, 27-11, 27-14
    windbox, 29-23
    -wire anemometer, 17-22
    working, 47-9, 63-6 (tbl)
Hourglass failure, 46-7 (ftn)
Hub, 51-6, 57-2
Human
    error, 62-2
    factors, 61-29
Humectant, 66-10 (ftn)
Humidification, 38-8, 39-2
    efficiency, 38-7
    load, 38-8
    steam, 38-10
Humidistat, 42-1
Humidity, 38-2
Hurwitz
    criterion, 60-12
    test, 60-12
HVAC, 42-1
Hybrid throttle pressure operation,
    27-16 (ftn)
Hydrashaker, 58-21
Hydrated molecule, 21-4
Hydration, 21-4, 21-7
Hydraulic
    brake, 52-16 (ftn)
    control valve, 19-2
    cyclone, 64-10
    cylinder, 19-6
    diameter, 16-6, 17-9, 20-14 (ftn), 36-8
    fluid, 19-2
    grade line, 16-9, 17-15
    horsepower, 17-14
    jack, 15-14
    line materials, 19-2
    loading, 66-41
    power, 18-6, 19-1
    press, 15-14
    pump, 19-4
    radius, 16-6
    ram, 15-14, 19-6
Hydraulics, 17-2
Hydrocarbon, 22-1
Hydrochloric acid, 66-4
Hydroclone, 64-10
Hydrocyclone, 64-10
Hydrodynamics, 17-2
Hydrogen
    available, 22-3
    damage, 21-11
    embrittlement, 21-11, 53-4
    heavy, 21-2
    peroxide, 45-13 (ftn)
Hydrogenation, 66-16
Hydrometer, 14-4
Hydro-pneumatic accumulator, 19-5

Hydrostatic
    moment, 15-11
    paradox, 15-4
    pressure (see Pressure)
    pressure test, 53-16
    resultant, 15-6
    torque, 15-11
Hydrostatics, 17-2
Hydrothermal source, 31-7 (ftn)
Hygroscopic, 39-2
Hygrostat, 42-1
Hyperbola, 7-11, 7-12
Hyperbolic
    function, 6-4
        derivative, 8-2
        integral, 9-2
    identity, 6-5
    radian, 6-5
Hypereutectoid steel, 47-5
Hypergeometric
    distribution, 11-5
    equation, Gauss', 10-5
Hypersonic, 14-14
Hypocycloid, 7-3
Hypoeutectic cast iron, 47-5
Hypoeutectoid steel, 47-5
Hypoid gear, 52-14
Hypothesis, 21-4, 52-2
Hypothesis test, 11-12
Hysteresis, 62-5 (ftn)

I
Ice point, 62-6
Ideal
    combustion, 22-9
    fluid, 14-2
    gas, 24-2
    gas, property, 24-15
    machine, 23-6
    radiation, 37-1
    reaction, 21-6
    spring, 52-2
    yield, 21-7
Idempotent law, 11-1
Identity
    hyperbolic, 6-5
    laws, sets of, 11-1
    logarithm, 3-5
    matrix, 4-1
    trigonometric, 6-3
Idler gear, 57-4
Ignitable substance, 66-3
Ignition, 29-2
Ignition temperature, 22-8
Illegitimate error, 62-2
Illuminance, 68-2
Illumination, 68-2
Imaginary number, 3-1
Immiscible liquid, 15-12
Impact, 56-15
    central, 56-16
    direct, 56-16
    eccentric, 56-16
    elastic, 56-15
    endoergic, 56-15
    energy, 16-4
    energy test, 46-13
    factor, 16-4
    head, 16-5
    inelastic, 56-15
    loading, 49-7
    noise, 68-3
    oblique, 56-16

plastic, 56-15
    test, pressure vessel, 53-5, 53-6
    tube, 16-4
Impactor forging, 63-8
Impedance, 67-5
Impeller, 18-4, 18-11, 64-12
Impending motion, 56-6
Implicit differentiation, 8-4
Impulse, 56-12
    fluid, 17-27
    function, 10-6
    -momentum principle, 17-26, 56-13,
        56-14
    noise, 68-3
    shock, 58-19
    specific, 26-8
    total, 26-8
    turbine, 17-29, 18-20, 18-21, 27-8
    unit, 59-2
Impulsive force, 56-13
Incentive plan, wage, 61-20
Inches of water gage, 20-2
Incineration, 22-4
    liquid, 66-31
    MSW, 66-31
    sludge, 66-34
    soil, 66-34
Incinerator, 66-28 to 66-31
Inclined plane, 56-9
Income tax, 69-24
Incomplete combustion, 22-11
Incompressible gas, 16-3 (ftn)
Inconel, 45-13
Inconsistent system, 3-7
Incremental
    analysis, 69-17
    cost, 69-31
Indefinite integral, 9-1 (ftn), 9-4
Independent
    event, 11-3
    variable, 3-2
Indeterminacy, degree of, 44-1
Indeterminate
    statics, 44-1
    truss, 44-6
Index
    roughness height, 2-4
    spring, 52-2 (ftn), 52-4
    viscosity, 14-9, 19-2 (ftn)
Indicated
    power, 29-3
    properties, 29-3
    specific fuel consumption, 29-3
Indicator
    card, 29-8
    drawing, 29-8
"Indifference quality level," 61-25
Indirect
    coal-fired turbine, 29-15
    -condensing tower, 38-13
    injection, 29-3
    manufacturing expense, 69-32
Indoor design condition, 39-2
Induced
    draft, 22-16
    drag, 17-34
Inductance, 59-7, 67-6
Induction
    furnace, 45-5
    hardening, 47-10
    motor, 18-9, 67-12

ratio, 20-24
    reactance, 67-6
    transducer, 62-4
Inductor, 67-6
Industrial exemption, 72-1
Inelastic
    impact, 56-15
    strain, 46-2
Inertance, fluid, 59-7
Inert gas welding, 63-12
Inertia, 54-1
    area moment of, 48-3
    mass moment of, 54-2
    polar moment of, 48-5
    principal moment of, 48-7
    product of, 48-6, 54-2
    resistance, vehicle, 56-19
    thermal, 40-5
    vector, 54-1, 56-5 (ftn)
Inertial
    force, 54-1, 56-5
    frame, 55-9
    resisting force, vehicle, 56-19
    separator, 66-26
Infant mortality, 11-7, 61-11
Infiltration, 39-1, 39-2
    energy to warm, 40-4
    loss, 40-1
Infinite series, 3-11 (ftn)
Inflation, 69-39
Inflection point, 7-2, 8-2, 49-16
Influence diagram, 43-10
Infrared
    incineration, 66-31
    testing, 46-11
Initial
    condition, 58-4
    tension, spring, 52-8
    value, 9-3, 10-1, 60-7
Initiator, 45-15
Injection
    molding, 63-10
    sorbent, 66-38
    steam, 29-14
    well, 66-28
Injured party, 70-5
Inlet
    pressure, net positive, 18-2 (ftn)
    vane, 20-3
Inline sampling, 21-11
Innovative method, 66-17
Insensitivity, 11-13, 62-2
Insertion loss, 68-9
Inside design temperature, 41-1
Insolation, 31-2
Instability, 60-12, 62-2
Instantaneous
    center, 55-12, 55-13
    cooling load, 41-2
    heat gain, 41-1
Instrument, coefficient of, 17-22
Insulated
    pipe, 34-8
    tip, 34-17
Insulation
    pipe, 34-9
    thickness, critical, 34-10
Insurance, 70-7
Intake stroke, 29-2
Intangible property, 69-20 (ftn)
Integer programming, 61-12
Integral
    convolution, 10-7
    cosine function, 9-8

definite, 9-1 (ftn), 9-4
double, 9-3
elliptic function, 9-8 (ftn)
exponential function, 9-8
Fresnel function, 9-8 (ftn)
gamma function, 9-8 (ftn)
indefinite, 9-1, 9-4
of combination of functions, 9-2
of hyperbolic function, 9-2
of transcendental function, 9-1
reinforcement, 53-12
sine function, 9-8
triple, 9-3
Integrand, 9-1
Integrated cycle, 29-22
Integrating factor, 10-2
Integration, 9-1
  by parts, 9-2
  by separation of terms, 9-3
  constant of, 9-1, 10-1
Intensifier, 32-2
Intensity
  luminous, 68-2
  sound, 68-4
Intensive property, 24-4
Interaction factor, 37-2
Intercept form, 7-5
Intercooling, 29-19
Intercooling, compressor, 32-7
Interest
  compound, 69-11
  rate, effective, 69-5, 69-28
  rate, nominal, 69-28
  simple, 69-11
Interfacial area, 66-40
Interference, 51-5
  factor, 31-5
  fit, 51-5
Intergranular, 21-12
  attack, 21-11
  corrosion, 21-10
  stress corrosion cracking, 30-2
Intermediate
  column, 51-3
  gear, 57-4
  metals, law of, 62-7
  -pressure turbine, 27-8
  temperatures, law of, 62-7
Internal
  energy, 23-4, 24-5
  force, 43-2, 43-13, 56-2
  friction, angle, 46-7, 56-7
  fuel reforming cycle, 31-8
  gear, 57-4
  heat gain, 40-4, 41-2
  heat generation, 34-14
  rate of return, 69-12
  work, 23-1, 46-6
Intern engineer, 72-1 (ftn)
Interpolating polynomial, Newton's, 13-3
Interpolation, nonlinear, 13-2, 13-3
Interpolymer, 45-13
Intersecting lines, 2-1
Intersection
  line, 7-6
  set, 11-1
Interval
  confidence, 61-16 (ftn)
  elevator, 64-4
Intrinsic waste, 66-3
Invar, 45-13

Inventory, 69-37
Inverse
  function, 6-4
  Laplace transform, 10-6
  lever rule, 38-5
  matrix, 4-5
  -square field, 56-17
Inversion
  curve, 25-3
  Fourier, 9-6
  point, 25-3
  temperature, 25-3
Invert emulsion, 19-2 (ftn)
Investment
  casting, 63-9
  credit, 69-25
  return on, 69-11, 69-14
Involute gear, 52-11
Ion exchange, 21-14, 21-15
Ionic concentration, 21-8
Iron
  allotropes of, 47-4
  alpha, 47-4
  beta, 47-4
  carbide, 47-5
  -carbon diagram, 47-4
  cast, 47-5
  delta, 47-4
  gamma, 47-4
  loss, 67-15
  wrought, 45-10
Irradiance, 68-2
Irrational number, 3-1
Irreducible factor, 69-2
Irreversibility, 25-12
Irreversible process, 25-3
ISA nozzle, 17-26 (ftn)
Isenthalpic curve, 25-3
Isentropic
  compressibility, 14-13, 24-9
  compression ratio, 29-4
  efficiency, pump, 27-6
  efficiency, turbine, 27-8
  flow, 26-3
  pressure, 26-13
  pressure ratio, 29-4
  process, 25-2, 25-8, 25-9
Isobar, 24-3
Isobaric
  compressibility, 24-9
  process, 25-2
Isochoric process, 25-2
Isokinetic test, 66-20
Isolated system, 25-2
Isolation
  degree of, 58-12
  efficiency, 58-12
  from active base, 58-14
  vibration, 58-12
Isometric
  process, 25-2
  view, 2-3
Isometry, 7-3
Isopiestic equilibrium, 14-9
Isoprene latex, 45-13
Isostatic
  molding, 63-10
  polymer, 45-16
Isotherm, 24-3
Isothermal
  compressibility, 14-13, 24-9
  compression, 15-12
  flat plate, 36-3

flow with friction, 26-17
process, 25-2
transformation diagram, 47-6
Isotope, 21-2 (ftn)
Iteration, fixed-point, 13-2
Izod test, 46-15

J
Jack
  hydraulic, 15-14
  power, 52-22
Jacketed pipe heat exchanger, 36-10
J. B. Johnson formula, 51-3
Jerk, 57-10
Jet
  condenser, 27-11 (ftn)
  force on blade, 17-29
  force on plate, 17-28
  propulsion, 17-27
  pump, 18-2
Job
  cost accounting, 69-36
  enlargement, 61-28
  enrichment, 61-28
  flexibility, 61-28
Jockey pump, 65-3
Joint, 53-2 (ftn)
  efficiency, 16-10, 51-11, 53-8
  flanged, 53-14
  lap, bolted, 51-10
  probability, 11-3
  quality factor, 16-10
Joints, method of, 43-14
Jominy end-quench test, 47-7
Joule, 1-5
  cycle, 29-15
  equivalent, 23-1
  -Kelvin coefficient, 25-3
  -Thompson coefficient, 25-3
Joule's
  constant, 23-1, 24-6
  law, 23-4
Journal, 52-20, 69-33
Judgment factor, 69-2
Jump, follower, 57-11

K
Kalman's theorem, 60-14 (ftn)
Kelvin
  day, 40-5
  -Planck statement, 25-11
  scale, 24-5
Kepler's law, 56-18
Kern, 49-12
Kernal, 49-12
Killed steel, 45-5
Kinematic
  Stirling engine, 29-12
  viscosity, 14-8, 16-7
Kinetic
  energy, 16-2, 23-3
  friction, angle of, 64-5
  gas theory, 24-15
  pump, 18-2
Kinetics, 56-2
Kirchhoff's
  radiation law, 37-1
  voltage law, 67-3
K-monel metal, 45-13
Knives, cutting, 64-7
Knoop hardness, 46-12
Knudsen number, 1-9 (tbl)
Kutta-Joukowsky theorem, 17-34

kVA rating, 67-12
K-waste, 66-3

L
Labor
    direct, 69-32
    variance, 69-36
Lagging circuit, 67-7
Lagrange stream function, 17-3
Lagrangian interpolating polynomial, 13-2
Lag, thermal, 40-5
Lamé formula, 19-3
Lamé's solution, 51-4
Laminae, 16-7
Laminar flow, 16-7, 17-5
Lance, oxygen, 45-3
Land
    ban, 66-20
    tooth, 52-10
Landfill, 66-12
    gas, 66-11
    liner, 66-34
Lang lay, 64-2
Langley, 31-2 (ftn)
Lanthanide, 21-2
Lanthanon, 21-2
Lanthanum barium copper oxide, 45-18
Lap
    joint, 63-13
    joint, bolted, 51-10
Laplace transform, 10-5, 10-6
Lapping, 63-5, 63-14
Laser machining, 63-6
Last-in, first-out, 69-37
Latent heat, 39-3
    fusion, 24-9
    sublimation, 24-9
    vaporization, 24-9
Lateral
    buckling, 49-18
    strain, 46-4
Latex, isoprene, 45-13
Latus rectum, 7-10, 7-11
Law
    affinity, 18-17, 20-8
    Amagat's, 24-17
    Archimedes', 15-14
    associative, addition, 3-3
    associative, multiplication, 3-3
    Avogadro's, 24-13
    Boyle's, 25-7
    Charles', 25-7
    commutative, addition, 3-3
    commutative, multiplication, 3-3
    corresponding state, 24-19
    Dalton's, 22-13, 24-17
    distance-squared, 68-3
    distributive, 3-3
    Dulong and Petit, 24-7
    fan, 20-8
    Fourier's, 34-5
    Henry's, 21-7, 66-39
    Hess', 21-9
    Hooke's, 46-2, 49-2, 52-2
    Joule's, 23-4
    Kepler's, 56-18
    Kirchhoff's radiation, 37-1
    Newton's first, 56-4
    Newton's, of gravitation, 56-18
    Newton's second, 56-4
    Newton's third, 56-5
    of areas, 56-18

of conservation of energy, 23-1
of convection, Newton's, 36-2
of cooling, Newton's, 10-10, 34-11
of cosines, 6-5, 6-6
of definite proportions, 21-3
of homogeneous circuits, 62-7
of intermediate metals, 62-7
of intermediate temperatures, 62-7
of multiple proportions, 21-3
of orbits, 56-18
of periods, 56-18
of sines, 6-5, 6-6
of tangents, 6-5
of thermodynamics, first, closed system, 25-4
of thermodynamics, first, open system, 25-5
of thermodynamics, second, 25-11
Pascal's, 15-4
radiation conservation, 37-1
second, thermodynamics, 24-7 (ftn)
similarity, 18-18
Stefan-Boltzmann, 37-2
Stoke's, 17-35
third, thermodynamics, 24-6
zeroth, thermodynamics, 24-4
Layer, boundary, 16-8 (ftn), 36-2
Layout, facilities, 61-14
Lay, wire rope, 64-2
L-D process, 45-3
Leachate, 66-12
Leaching, selective, 21-11
Lead, 66-12
    -acid battery, 31-10
    angle, 52-22
    ratio, 52-14
    screw, 52-22
Leading circuit, 67-7
Leaf, 52-10
Leak
    detection, 66-10
    points, 27-11 (ftn)
Leakage, duct, 20-18
Lean-burn engine, 29-10
Learning curve, 69-42
Least
    radius of gyration, 48-6
    significant bit, 12-1, 12-4
    significant digit, 12-1
    squares method, 11-13
Ledger account, 69-33
Legendre equation, 10-5
Legionnaire's disease, 66-5
Length, 2-2
    arc, by integration, 9-5
    characteristic, 26-9, 34-3, 35-2, 36-8
    effective, column, 51-2
    equivalent, 17-11
    of the line of action, 52-11
    of tube, 36-15
    thermal entry, 36-5
Lethal substance, 53-3
Letter of agreement, 70-3
Level
    acceptance quality, 61-25
    confidence, 11-11, 61-16
    indifference quality, 61-25
    sound power, 68-7
    sound pressure, 68-6

Lever, 43-10, 59-5
    effectiveness, 15-14
    efficiency, 15-14
    ratio, 58-8
    rule, 38-5, 47-2
Leverage, 69-34
Lewis
    beam strength theory, 52-13
    form factor, 52-13 (ftn)
    number, 1-9 (tbl)
L'Hôpital's Rule, 3-9
Liability
    account, 69-33
    current, 69-34
    in tort, 70-5
Licensing, 72-1
Life
    -cycle analysis, 61-27
    -cycle cost, 69-20
    economic, 69-19
    extension, 29-23
    fatigue, 46-9
    pump, 19-5
    service, 69-20
    tool, 63-5
Lifeline rate, 67-7 (ftn)
Lifer, 57-10
Lift, 17-32
    coefficient of, 17-32
    landfill, 66-12
    rotating cylinders, 17-33
    valve, 16-12
Light
    damping, 58-9
    heat gain from, 40-5
    metal, 21-2
    quantity of, 68-2
    -sensitive detector, 62-4
    -water reactor, 30-1
Lignin, 45-16
Lignite, 22-4
Limit, 3-9 (see also type)
    average outgoing quality, 61-26
    confidence, 11-11
    control, 61-22
    elastic, 46-2
    endurance, 46-9
    lower explosive, 39-5
    of reinforcement, 53-12
    proportionality, 46-2
    simplifying, 3-9
    surge, 32-8
    value, threshold, 39-5
    viscosity, 52-20
    wear design of gears, 52-14
Limited partner, 70-2
Limiting
    current, sensor, 62-4
    reactant, 21-7
    value, 3-9
Line, 6-1 (ftn), 7-3
    balancing, 61-18
    centroid, 48-2
    condition, turbine, 27-8
    critical, 47-4
    current, 67-9
    diagram, 59-3
    energy grade, 16-9, 17-15
    equipotential, 17-3
    eutectic, 47-3
    failure, 50-6
    generating, 52-11
    Gerber, 50-6

Goodman, 50-6
hydraulic grade, 16-9, 17-15
intersection, 7-6
liquidus, 47-1
normal vector, 8-6
of action, 5-1, 43-2, 52-11
pressure, 52-11
Rayleigh, 26-19
refrigerant, 33-9
saturated, 24-3
Smith, 50-7 (ftn)
Soderberg, 50-6
solidus, 47-1
straight, 7-4
voltage, 19-5 (ftn)
Lineal distance, 40-4 (ftn)
Linear
    acceleration, 55-4
    actuator, 19-6
    component, 58-2, 59-1
    damping, 58-9
    differential equation, 10-1
    equations, 3-7
    expansion, coefficient of, 49-3
    first-order, 10-2
    force system, 43-4
    frequency, 58-3, 67-4
    impulse, 56-12
    model, 61-2
    momentum, 56-2
    motion, 55-2
    motor, 19-6 (ftn)
    programming, 61-12
    regression, 11-13
    second-order, 10-3
    spring, 23-2, 23-3 (ftn)
    system, 55-2
    variable differential transformer, 62-4
    watt density, 40-4 (ftn)
Linearity theorem, 10-6
Liner, 45-20, 66-34
Lines
    angle between, 6-1
    intersecting, 2-1
    perpendicular, 2-1
Lining, brake, 52-16
Linz-Donawitz process, 45-3
Liquefied petroleum gas, 22-7
Liquid
    aromatic, 14-9 (ftn)
    -dominated reservoir, 31-7
    fuel, 22-5
    immiscible, 15-12
    metal, 36-7
    metal fast breeder reactor, 30-1 (ftn)
    penetrant testing, 46-11
    saturated, 24-2
    subcooled, 24-2
    -vapor mixture, 24-2, 24-12
    volatile, 14-9, 21-7 (ftn)
Liquidated damages, 70-7
Liquidus line, 47-1
Liter, 1-6
LMTD method, 36-20
Load
    air conditioning, 41-1
    cell, 62-12
    coil, 41-1 (ftn)
    cooling, 41-1
    curve, 27-15
    dead, 43-4
    -deflection equation, 52-5
    distributed, 43-4, 43-6

dominant moving, 49-18
duration curve, 27-15
-elongation curve, 46-1
factor, 27-15
heat, 27-11, 36-14, 38-12
heating, 40-1
humidification, 38-8
moving, 49-18
number, bearing, 52-21
overhung, 18-10
pickup, 40-5
proof, bolt, 51-8, 51-12
refrigeration, 41-1 (ftn)
tower, 38-12
Loading
    axial, 49-11
    boot, 64-9
    chemical, 66-41
    concentric, 49-11
    error, 62-2
    factor, 31-9, 66-12
    hydraulic, 66-41
    impact, 49-7
    pump shaft, 18-10
Loan, 69-39
Local
    buckling, 49-18
    control computer, 42-5
Locus of points, 7-2
Logarithm, 3-5
Logarithmic
    decrement, 58-9
    mean area, 34-8
    mean temperature difference, 36-14
Longitudinal
    fin, 34-18
    stress, 51-4
    wave, 68-4
Long stress, 51-4
Long-term liability, 69-34
Loop
    gain, 60-4
    pipe, 17-20
    primary/secondary, 30-1
    system, 65-1
    transfer function, 60-4
Loss
    coefficient, 17-12, 20-14
    combustion, 22-15
    duct friction, 20-11
    electrical machine, 67-15
    factor, pulley, 43-11
    friction, sprinkler, 65-7
    infiltration, 40-1
    insertion, 68-9
    journal bearing, 52-21
    method, 22-16
    of coolant accident, 30-2
    transmission, 40-1, 68-9
    -wax process, 63-9
    zero-length, 20-15
Lot
    quality, 61-25
    tolerance percent defective, 61-25
Loudness, 68-4, 68-8
Low
    -alloy steel, 45-5
    -carbon steel, 45-5
    excess-air burner, 66-34
    NOx burner, 66-35
    -pass filter, 60-10
    -pressure turbine, 27-8
    -side equipment, 33-8

-sulfur coal, 22-4
-temperature reservoir, 25-10
-velocity system, 20-17
Lower
    bound, 9-4
    control limit, 61-23
    critical point, 47-4
    explosive limit, 39-5
    heating value, 22-14
Lubarsky-Kaufman correlation, 36-7
Lubricant, 52-19
Lumber, oven-dry, 45-17 (ftn)
Lumen, 68-1
Luminance, 68-3
Luminous
    efficiency, 68-2
    emittance, 68-2
    flux, 68-1
    intensity, 68-2
Lump coal, 22-4
Lumped
    mass, 59-4
    parameter method, 34-12
Lump-sum fee, 70-4
Lux, 61-30 (ftn)

M
Mach
    angle, 26-11 (ftn)
    number, 1-9 (tbl), 14-14
    wave, 26-11 (ftn)
Machine, 43-20
    ideal, 23-6
    real, 23-6
    safeguard, 61-30
    -worker system, 61-29
Maclaurin series, 8-7 (ftn)
Macrofouling, 27-12
Macroreticular synthetic resin, 21-15
Magnaflux, 46-11
Magnaglow, 46-11
Magnesia, 22-3 (ftn), 34-9 (ftn)
Magnetic
    flowmeter, 17-22
    particle testing, 46-11
    pulse forming, 63-11
Magnetite, 45-2
Magnetohydrodynamics, 31-8
Magnification factor, 58-10
Magnus effect, 17-34
Maintenance
    motivation factor, 61-28
    preventative, 61-10
Major
    axis, 7-11
    source, pollution, 66-4
Make-up
    pump, 65-3
    water, 21-15
Malleable cast iron, 45-10, 47-8
Management
    -by-objectives, 61-29
    science, 61-1
    systems modeling, 61-1
Maney formula, 51-12
Manganese bronze, 45-12 (ftn)
Man-machine system, 61-29
Manometer, 15-2, 15-3
Mantissa, 3-5
Manufactured gas, 22-7 (ftn)
Manufacturing cost, 69-32
Manway, 53-11
MAPP gas, 63-11

Mapping function, 7-3
Maraging steel, 45-5 (ftn)
Margin
  boiling-point, 66-10
  gain, 60-11
  of safety, 50-1
  phase, 60-11
Market value, 69-23
Martempering, 47-7
Martensite, 47-7
Martensitic stainless steel, 21-11 (ftn), 45-8
Masking, 66-37
Masonry density, 15-12
Mass, 1-1, 24-4, 54-1
  burn, 66-31
  center of, 54-1
  control, 25-1
  -flow bin, 64-5
  fraction, 24-17
  lumped, 59-4
  moment of inertia, 54-2
  -transfer coefficient, 66-40
  yield, 66-33
Material
  anisotropic, 5-1, 34-2
  breach, 70-5
  conditions monitoring, 61-27
  handling, 64-1
  pipe, 16-9
  recovery facility, 66-32 (ftn)
  refractory, 37-4
  spring, 52-3, 52-4
  variance, 69-36
Mathematical programming, 61-12
Matrix, 4-1, 4-2 (see also type)
  algebra, 4-3
  inverse, 4-5
  multiplication, 4-4
  order, 4-1
  transformation, 5-3
Matter
  mineral, 22-3
  volatile, 22-4
Maximum
  flame temperature, 22-15
  normal stress theory, 50-1
  point, 8-2
  shear stress theory, 46-8, 50-3
  strain theory, 50-3
  value, sinusoid, 67-5
  velocity in pipe, 17-3
Maxwell-Boltzmann distribution, 24-17
MBh, 27-4 (ftn), 34-5 (ftn)
Mean, 11-10
  area, logarithmic, 34-8
  arithmetic, 11-10
  bulk temperature, 36-3
  effective pressure, 29-8
  geometric, 11-10
  radius, geometric, 57-3
  stress, 50-6
  time between failures, 61-9
  time between forced outages, 61-9
  time to failure, 11-7, 61-9
  time to repair, 61-9
Measurement
  far-field, 68-6
  flow, 17-20
  near-field, 68-6
Mechanical
  advantage, 43-10
  efficiency, 29-3
  seal, 66-35

similarity, 17-38
system, modeling, 59-4
Mechanism, 43-20
Median, 11-10
Media, packing, 66-39
Medium-carbon steel, 45-5
Melting point, 47-1 (ftn)
Member
  axial, 43-11
  redundant, 43-7, 43-13
  three-force, 43-6
  two-force, 43-6
  zero-force, 43-14
Meniscus, 14-11
Mensuration, 7-1
Mer, 45-13
Merit, figure of, 31-9
Mesh, 52-11
  efficiency, 52-12
  ratio, 57-5
  separation, 64-10
Metabolic, 39-3
Metacenter, 15-16
Metacentric height, 15-17
Metal, 21-2
  alkali, 21-2
  alkaline earth, 21-2
  base, 47-1
  characteristics, 45-1
  heavy, 21-2
  light, 21-2
  removal rate, 63-3
  spraying, 63-14
  transition, 21-3
Metallic properties, 21-2
Metallizing, 63-14
Metalloid, 21-2
Metallurgy, 45-1, 63-10
Metal-metal system, 45-19
Metastable structure, 45-9
Metathesis, 21-6 (ftn)
Meter
  current, 17-21
  displacement, 17-21
  obstruction, 17-21
  orifice, 17-24
  pressure, 14-3
  sound, 68-5
  turbine, 17-21
  variable-area, 17-21
  venturi, 17-23
Metering pump, 18-4
Methane
  landfill, 66-11
  series, 22-2
Methanol, 22-7
Method (see also type)
  area transformation, 49-19
  bisection, 3-4, 13-1
  combination, 20-21
  compatibility, 44-2
  consistent deformation, 44-2, 49-19
  cut-and-sum, 43-15
  double-dabble, 12-1 (ftn)
  engineering, 61-27
  equal-friction, 20-19
  equivalent length, duct, 20-15
  Hardy-Cross, 17-20
  loss, 22-16
  lumped parameter, 34-12
  M-factor, 40-5
  Newton's, 13-2, 34-11, 34-12 (ftn)
  NTU, 36-20
  numerical, 3-4, 13-1

of joints, 43-14
of least squares, 11-13
of sections, 43-15
of undetermined coefficients, 3-6, 10-4
parallelogram, 5-3
polygon, 5-3
Rayleigh's, 58-5
Raymondi and Boyd, 52-22
regula falsi, 13-2 (ftn)
remainder, 12-1
simplex, 61-12
static regain, 20-21
superposition, 44-4
-Time-Measurement, 61-16 (ftn)
total pressure, 20-24
transfer function, 41-2
velocity reduction, 20-19
Winfrey, 69-16 (ftn)
Methyl
  alcohol, 22-7
  orange alkalinity, 21-14
Metric system, 1-4
Meyer hardness, 46-12
M-factor method, 40-5
Mho, 67-2 (ftn)
Microbiologically influenced corrosion, 27-12
Microfin, 33-9 (ftn)
Microfouling, 27-12
Micronized coal, 22-5 (ftn), 27-3 (ftn)
Microstrain, 62-8
MIG welding, 63-12
Miles per gallon, 56-21
Milligram
  equivalent weight, 21-8
  per liter, 21-8, 21-11 (ftn)
Milling, 64-7
Million, parts per, 21-8, 21-11 (ftn)
Mineral matter, 22-3
Miner's rule, 50-7
Minimum
  attractive rate of return, 69-14, 69-16
  point, 8-2
Minor
  axis, 7-11
  loss, 17-11
  loss, sprinkler, 65-7
  of entry, 4-2
Miscible alloy, 47-1
Mixed
  -bed unit, naked, 21-16
  flow pump, 18-4 (ftn)
  triple product, 5-4
Mixer, 64-12
Mixing
  heater, 27-13
  problem, 10-8
  Reynolds number, 64-12
  two air streams, 38-5
Mixture, 47-3
  gas, 24-2, 24-17
  liquid-vapor, 24-2
  pattern, 61-22
  rule, 66-4
mks system, 1-5
MMBtu, 66-14
Modal
  shape, 58-19
  vibration, 58-18
MODAPS, 61-16 (ftn)
Mode, 11-10, 58-18
Model, 17-11, 17-38 (see also type)

Modified
  Goodman line, 50-7
  Mohr theory, 50-2
Modular ratio, 49-19
Modulation, flow rate, 20-3
Module, gear, 52-10
Modulus, 3-8
  apparent, 46-10
  Biot, 34-4
  bulk, 14-13, 68-4
  creep, plastics, 46-10
  of elasticity, 46-2, 49-2
  of elasticity, temperature effect on, 53-5
  of resilience, 46-6, 50-4
  of rigidity, 46-7
  of toughness, 46-6
  secant, 14-13, 46-4
  section, 49-11
  shear, 46-7, 49-2
  tangent bulk, 14-13
  Young's, 46-2, 49-2
Mohr's
  circle, 48-7, 49-6
  theory, modified, 50-2
  theory of rupture, 46-7
Mohs hardness test, 46-12
Moist air, 38-2, 38-3
Moisture
  bed, 22-3
  content, 24-9, 45-16
  separator, 30-1
Molality, 21-7
Molar
  properties, 24-4
  volume, 21-4
Molarity, 21-8
Molding, 63-8 to 63-10
Mole, 21-4
  fraction, 14-6, 21-8, 22-2, 24-17
  percent, 14-6
Molecular weight, 21-4
Molecule, 21-1
  hydrated, 21-4
  spacing, 14-2
  speed, 24-17
Mollier diagram, 24-9
Molluscide, 27-12
Molten carbonate fuel cell, 31-8
Moment, 49-8
  about a line, 43-3
  about a point, 43-2
  arm, 43-2
  component, 43-3
  diagram, 49-8
  first, area, 48-2
  fixed-end, 44-6
  free, 43-4, 43-9
  hydrostatic, 15-11
  of a function, 9-6
  of area, first, 44-5
  of inertia
    area, 48-3
    mass, 54-2
    polar, 48-5
    principal, 48-7
  of momentum, 56-3
  one-way, 49-8 (ftn)
  overturning (dam), 15-11
  righting, 15-16
  second, 48-3
  statical, 48-2, 49-10
Moment area method, 49-14

Momentum
  angular, 56-3
  conservation, 56-3
  fluid, 17-27
  linear, 56-2
Monel metal, 45-13
Monitoring, emissions, 66-21
Monitor steel, 64-2
Monomer, 45-13
Monomial, 3-3
Monotectic reaction, 47-3
Montreal Protocol, 33-2, 66-8
Moody friction factor chart, 17-6
Mortality
  curve, 61-11
  infant, 11-7, 61-11
Most
  probable speed, 24-17
  significant bit, 12-1, 12-4
  significant digit, 12-1
Motion
  constrained, 56-9
  dependent, 55-11
  impending, 56-6
  linear, 55-2
  plane, 55-11
  projectile, 55-4
  relative, 55-9
  simple harmonic, 58-2
  uniform, 55-3
Motivation-maintenance theory, 61-28
Motivator, 61-28
Motor
  electrical, 67-10
  electric, hydraulic power, 19-5
  gear, 18-10
  induction, 67-12
  linear, 19-6 (ftn)
  octane number, 22-6
  size, 18-8
  speed control, 67-15
  stepper, 60-13 (ftn)
  synchronous, 67-13
Mottled cast iron, 45-10
Movement, 61-15
Moving average forecast, 69-41
Moving load, beam, 49-18
MSW (see Municipal solid waste)
MTTF (see Mean time to failure)
MTBE, 22-7
Mud drum, 27-4 (ftn)
Muller, 63-10
Multicouple, 31-9
Multiple
  degree of freedom, 58-2, 58-19, 59-2
  hypergeometric distribution, 11-5
  -pass heat exchanger, 36-10
  pipe containment, 66-36
  reheat, 27-13
  -stage pump, 18-4
Multiplication
  matrix, 4-4
  vector, 5-3
Multirating table, 20-5
Municipal solid waste, 66-12, 66-31, 66-34
Mutagen, 61-27
Muther nearness priority, 61-15 (ftn)

N
Nacelle, 31-4
Naked mixed-bed unit, 21-16

Nameplate
  motor, 67-11
  rating, 18-9
  pressure vessel, 53-2
Napierian logarithm, 3-5
Napthalene series, 22-2
National
  Council of Examiners for Engineering and Surveying, 72-1
  Society of Professional Engineers, 71-1, 71-3
Native copper, 45-12
Natural
  abrasive, 45-19
  -circulation boiler, 27-4
  convection, 35-2
  draft, 22-16
  frequency, 9-6, 58-3, 60-9
  logarithm, 8-5
  response, 60-1
  vibration, 58-2
Near-field, 68-6
Nearness priority, 61-15
Necking down, 46-4
Neck, nozzle, 53-3, 53-11
Need hierarchy, 61-28
Needle valve, 16-11
Negative
  exponential distribution, 11-7
  feedback, 60-4
  number (computer), 12-4
Negligence, 70-5
Nelson diagram, 53-4
Nernst theorem, 24-6
Nested spring, 52-5
Net
  electrical output, 27-1
  heating value, 22-14
  inlet pressure required, 18-13 (ftn)
  positive inlet pressure, 18-2 (ftn)
  positive suction head, 18-13
  radiation transfer, 37-4
  rating, 22-16
  stack temperature, 22-13
  transport theory, 34-2
Network polymer, 45-15
Neurotoxin, 61-27
Neutral
  axis, 49-10
  equilibrium, 56-2
  flame, 63-11
  position, 42-5
  solution, 21-8
Neutralization, 21-9, 66-41
Neutralizing amine, 21-15
Neutron
  embrittlement, 30-2
  gaging, 46-12
  radiography, 46-12
Newton form, 13-3 (ftn)
Newtonian fluid, 14-2, 14-6
Newtonian frame of reference, 55-9
Newton's
  first law, 56-4
  gravitational constant, 56-18
  interpolating polynomial, 13-3
  law of convection, 36-2
  law of cooling, 10-10, 34-11
  law of gravitation, 56-18
  method, 13-2, 34-11, 34-12 (ftn)
  second law, 56-4
  third law, 56-5

Nichrome, 45-13
Nickel, 45-13
    -metal hydride battery, 31-10
    silver, 45-12 (ftn)
Nikuradse equation, 17-5
Nilvar, 45-13
NIMBY, 66-12
Nines complement, 12-3
Nip, angle of, 64-6
Nitric oxide, 66-13
Nitriding, 47-10
Nitrogen
    dioxide, 66-13
    oxides, 66-13
    -oxygen ratio, 22-8
Nitrous oxide, 66-13 (ftn)
Noble gas, 21-2
No-boiler incinerator, 66-31
No choice exam, 72-3 (ftn)
Nocturnal radiation, 37-7
Node, 7-2, 58-19, 61-3
Nodular cast iron, 45-9, 47-8
Noise, 68-3
    criteria curve, 68-8
    dose, 68-9
    meter, 68-5
    reduction, 68-8 to 68-10
Nominal
    damages, 70-7
    dimension (pipe), 16-10
    interest rate, 69-28
    system voltage, 67-5 (ftn)
Nonattainment area, 66-3
Noncarbonate hardness, 21-14
Noncatalytic reduction, selective, 66-38
Noncircular duct, 17-9
Noncondensing cycle, 28-2
Noncondensing gas, 27-11
Nonconformance, 61-24
Noncriteria pollutant, 66-3
Nondestructive testing, 46-11, 46-13
Nonenclosure hood, 39-4
Nonhomogeneous differential equation,
    10-1, 10-4
Nonhomogeneous linear equation, 3-7
Nonlinear interpolation, 13-2, 13-3
Nonmetal, 21-2
Non-overloading fan, 20-4
Non-parametric equation, 3-3
Non-quantifiable factor, 69-2
Nonreverse-flow valve, 16-12
Nonsingular matrix, 4-5
Normal
    acceleration, 55-8
    distribution, 11-6
    force, 15-12, 56-6
    form, 7-5
    line vector, 8-6
    pitch, 52-10, 52-12
    pressure angle, 52-12
    pressure, sprinkler, 65-7, 65-8
    shock wave, 26-9
    speed, 56-21
    stress, 49-2
    stress theory, maximum, 50-1
    temperature and pressure, 22-3
    vector, 7-6
    view, 2-1
Normality, 21-8
Normalizing, 47-7, 47-9

Notch
    -brittle, 46-6 (ftn)
    factor, fatigue, 46-10
    sensitivity, 46-10
    stress concentration, 49-5
    toughness, 46-13, 47-9, 53-5
NOx, 22-5, 66-13
Nozzle, 17-23
    ASME long radius, 17-26 (ftn)
    coefficient, 18-21
    control, 27-16
    convergent-divergent, 26-5
    discharge coefficient, 26-8
    efficiency, 26-8
    flow, 17-25
    ISA, 17-26 (ftn)
    loss, turbine, 18-21
    pressure vessel, 53-3, 53-11
    steam flow through, 26-20
    velocity coefficient, 26-8
    venturi, 26-5
np-chart, 61-24
NSPE (see National Society of
    Professional Engineers)
NTU method, 36-20
Nuclear
    power cycle, 30-1
    sensing, 46-12
Nucleate boiling, 35-10
Nucleon, 21-1
Null
    event, 11-3
    indicator, 62-9
    matrix, 4-1
    point, 39-4, 66-28
    position, 62-4
    region, 7-3
    set, 11-1
Number
    AGMA bending stress, 52-14
    atomic, 21-2
    Avogadro's, 21-4
    base-b, 12-1
    bearing load, 52-21
    binary, 12-1
    Biot, 34-4
    cavitation, 18-14
    cetane, 22-7
    complement, 12-3
    complex, 3-1, 3-7
    condensation Reynolds, 35-9
    dimensionless, 1-7, 1-9 (tbl)
    direction, 7-5
    Eötvös, 17-38
    film Reynolds, 35-9
    flow, 64-12
    force, 64-12
    Fourier, 34-4
    Froude, 17-40
    Graetz, 36-2
    Grashof, 35-2
    imaginary, 3-1
    irrational, 3-1
    Mach, 14-14
    mixing Reynolds, 64-12
    negative (computer), 12-4
    Nusselt, 35-2, 36-2
    octane, 22-6
    of transfer units, 36-20
    oxidation, 21-3
    Peclet, 36-2
    performance, 22-6
    power, 64-12

Prandtl, 35-2, 36-2
rational, 3-1
Rayleigh, 35-3
real, 3-1
Reynolds, 16-7, 17-38, 35-3, 36-2
smoke spot, 22-13, 66-17
Sommerfeld, 52-22
Stanton, 36-2
system, 12-1
type, 3-1
Weber, similarity, 17-40
Numbering system, 3-1
Numerical method, 3-3, 13-1
Nusselt
    equation, 35-3, 36-5
    number, 1-9 (tbl), 35-2, 36-2
    -type correlation, 35-3 (ftn)
Nut
    coal, 22-4
    factor, 51-12
Nyquist analysis, 60-13

O
Objective function, 61-2, 61-12
Objectively scored problems, 72-3
Oblique
    axis, 2-2
    compression shock wave, 26-13
    impact, 56-16
    perspective, 2-3
    shock wave, 26-11
    triangle, 6-5
    view, 2-2
Obliquity, angle of, 52-11
Observability, 60-14
Obstruction meter, 17-21
Obtuse angle, 6-1
Ocean thermal energy conversion, 31-6
Octahedral shear-stress theory, 50-5
Octane number, 22-6
Octal number system, 12-2
Octave, 60-10, 68-7
    band, 68-7
    rule, 58-19
Odd
    symmetrical, 7-4
    symmetry, 9-7 (tbl)
Odor, 66-15
Odor removal, 39-3
Offer, 70-3
Off-gas, 66-39
Offset, parallel, 46-3
Ohm's law, 67-3, 67-6
Oil
    Bunker C, 22-5
    crankcase, 52-19
    cutting, 63-4
    distillate, 22-6
    fuel, 22-5
    residual fuel, 22-6
    spill, 66-15
Olefin series, 22-2
Oligomer, 45-14
Once-through
    boiler, 27-1, 27-4
    cooling water, 66-9
One-for-one plan, 61-20
Ones complement, 12-3
One-tail limit, 11-12
One-way
    moment, 49-8 (ftn)
    shear, 49-7 (ftn)
Opacity, smoke, 66-17

Opaque material, 68-3
Open
  circuit, electrical, 67-2
  combustor, 29-16
  -cycle OTEC, 31-6
  die forging, 63-7
  drive, 52-15
  hearth process, 45-4
  heater, 27-13
  -loop recovery, 66-23
  manometer, 15-3
  system, 25-1
Operating
  characteristic curve, 61-25
  differential, 42-2
  expense, 69-32
  point
    compressor, 32-8
    fan, 20-7
    pump, 18-16
  pressure, pressure vessel, 53-6
Operation
  factor, 27-16
  unit, 27-1
Operational form (log), 3-5
Operations
  on complex numbers, 3-8
  research, 61-1
Opportunity cost, 69-18
Opposite side, 6-2
Optical
  density, 66-17
  holography, 46-11
Optimum staging, 32-7
Orbit
  eccentricity, 56-19
  geostationary, 56-18
  law of, 56-18
Order
  differential equation, 10-1
  matrix, 4-1
  of the system, 59-2, 60-4
Ordinate, 7-2
Organic
  peroxide, 45-15 (ftn)
  precursor, 66-18
  sulfur, 22-3
Orientation, angular, 5-1
Orifice, 17-16
  compressible flow, 26-15
  constant, sprinkler, 65-5
  large, 17-18
  meter, 17-24
  plate, 17-24
  pressure loss, 17-25 (ftn)
Orsat apparatus, 22-11
Orthogonal, 7-2
  vectors, 5-3
  view, 2-2
Orthographic
  oblique view, 2-3
  view, 2-2
Oscillation
  amplitude of, 58-3
  period of, 58-3
OSHA, sound exposure limits, 68-10
Osmotic pressure, 14-9
Othmer correlation, 14-11
Otto cycle, air-standard, 29-5
Outlet
  constant, 20-25
  duct, 20-24
Output factor, 27-16

Outside design condition, 40-2
Overall
  coefficient of heat transfer, 34-6, 40-2
  conductance, 36-15
  conductivity, 34-2, 34-6
  efficiency, pump, 18-8
  gain, 60-4
  heat transfer coefficient, 36-15
  transfer function, 60-4
Overautogenous waste, 66-28 (ftn)
Overdamped, 58-9
Overdrive gear, 56-20
Over-expansion, 26-13
Overfeed stoking, 27-3
Overflow, 64-10
Overflow bit, 12-4
Overhauling screw, 52-22
Overhead, 69-36
Overhung load, 18-10
Overload, pump, 65-3
Overshoot, 42-2, 60-2
Overturning
  factor of safety, 15-11
  moment (dam), 15-11
Owner, 70-4
Owners' equity, 69-33
Ownership, 70-1
Oxidant, 66-15
Oxidation, 21-3
Oxides, nitrogen, 66-13
Oxidizing flame, 63-12
Oxyacetylene welding, 63-11
Oxygen
  -enriched incineration, 66-34
  metabolic, 39-3
  process, 45-3
  scavenging, 21-15
Oxygenate, 22-6
Oxygenated
  fuel, 66-8
  gasoline, 22-6
Oxyhydrogen welding, 63-11
Ozonation, 66-36
Ozone, 66-8, 66-15

P
Pace value, 30-2
Packed
  bed, 36-10
  tower, 66-22, 66-39
Packing
  height of, 66-40
  media, 66-39
  tower, 38-11
Pad, brake, 52-16
Paddle
  impeller, 64-12
  wheel fan, 20-4
Palmgren-Miner cycle ratio, 50-7
Panel
  coil heat exchanger, 36-12
  truss, 43-12
Panhandle formula, 17-9
Pappus-Guldinus, theorem, 48-3
Pappus' theorems, 9-5
Parabola, 7-10
Parabolic
  axis, 7-10
  cable, 43-17
  cam profile, 57-10
  column formula, 51-3
Paradox, hydrostatic, 15-4
Paraffin series, 22-2

Parallel
  axis theorem, 48-4, 54-2
  circuit, electrical, 67-4
  flow, 36-10
  force, 43-6
  line, 7-8
  offset, 46-3
  perspective, 2-3
  pipe, 17-20
  pumps, 18-17
  reliability, 11-8
Parallelogram method, 5-3
Parameters
  catenary, 43-19
  variation of, 10-4
Parametric
  equation, 3-2
  equation, derivative, 8-3
  equation, plane, 7-6
Parent ingredient, 47-1
Parkerizing, 63-14
Partial
  -arc admission, 27-16
  differentiation, 8-4
  emission, forced vortex pump,
    18-11 (ftn)
  fraction, 3-6, 9-2
  pressure, 24-17
  pressure fraction, 22-2
  similarity, 17-40
  solubility, 47-3
  volume, 24-17
Particle
  arrester, 39-6
  -strengthened, 45-19
Particular solution, 10-4 (tbl)
Particulate matter, 66-15
Parting line, 63-8
Partner
  general, 70-1
  limited, 70-2
Partnership, 70-1
Parts, integration by, 9-2
Parts per million, 21-8, 21-11 (ftn)
Pascal's
  law, 15-4, 15-14 (ftn)
  triangle, 3-3
Passivated steel, 45-8
Passive
  element, 59-2
  solar system, 31-2 (ftn)
  system, 60-11
Path
  critical, 61-3, 61-4
  function, 25-4
Payback period, 69-38
PCB (see Polychlorinated biphenyls)
p-chart, 61-22, 61-24
Peak
  cylinder pressure, 29-8 (ftn)
  demand multiplier, 67-7
  flame temperature, 66-13
  gain, 60-2
  load service, elevator, 64-4
  time, 60-2
Pea coal screening, 22-4
Peaking unit, 27-16
Pearlite, 47-6
Peclet number, 1-9 (tbl), 36-2
Peel strength, 63-13
Peltier effect, 62-6 (ftn)

Pendant sprinkler, 65-4
Pendulum
    ballistic, 56-3
    torsional, 58-5
Penstock, 18-20
Percent
    clearance, 32-3
    isolation, 58-12
    mole, 14-6
Percentage
    humidity, 38-2
    yield, 21-7
Percentile rank, 11-10
Per diem fee, 70-5
Perfect
    reaction, 21-6
    spring, 52-2
Performance
    coefficient of, 33-2
    curve, pump, 18-15
    factor, 38-10
    function, 11-8
    number, 22-6
    rating factor, 61-16
    standard, 61-20
Perigee, 56-19
Perihelion, 56-19
Perimeter, wetted, 16-6
Period, 69-2
    chemical, 21-2
    electrical, 67-4
    of oscillation, 58-3
    of waveform, 9-6
Periodic
    table, 21-2
    time, 56-18
    waveform, 9-6
Periods, law of, 56-18
Peritectic reaction, 47-3
Peritectoid reaction, 47-3
Permalloy, 45-13
Permanent
    hardness, 21-14
    molding, 63-9
    set, 46-2
Permivar, 45-13
Permutation, 5-5, 11-2
Peroxide, organic, 45-15 (ftn)
Perpendicular
    axis theorem, 48-5
    line, 7-8
    line principle, 2-1
    lines, 2-1
Perpetual inventory system, 69-37
Personal property, 69-20 (ftn)
Perspective, 2-2, 2-3
PERT (see Program evaluation and
    review technique)
Petroff equation, 52-21
pH, 21-8
Phase, 47-1, 67-9
    angle, 67-4, 67-6
    characteristic, 60-10, 60-11
    crossover, 60-11
    current, 67-9
    diagram, 24-3, 47-1
    factor, 67-7
    margin, 60-11
    rule, Gibbs, 47-3
    voltage, 19-5 (ftn)
Phasor form, 3-8, 67-5
Phenolic plastic, 45-15
Phenolphthalein alkalinity, 21-14

Phon scale, 68-8
Phosphoric acid fuel cell, 31-8
Phosphorus bronze, 45-12
Photochemical
    reaction, 66-15
    smog, 66-17
Photoelectric effect, 62-4
Photosensitive
    conductor, 62-5
    resist, 63-6
Photovoltaic cell, 31-4, 62-4
Physical
    inventory, 69-37
    strain, 46-5
    stress, 46-4
Pickling, 63-8, 63-14
Pickup, 62-3 (ftn)
Pickup load, 40-5
PI control, 42-5
Pictorial drawing, 2-2 (ftn)
PID control, 42-5
Piece-rate, 61-20
Piecework, 61-20
Pier, 49-12, 51-1
PIEV time, 56-21
Piezoelectric, 45-18
    effect, 15-2, 62-4
    transducer, 62-4
Piezometer, 15-2 (ftn)
Pig, 45-3
Pile, bulk storage, 64-4
Pinch point, 29-20
Pinion, 52-11
Pinned support, 43-7
Pipe
    bend, force on, 17-30
    black steel, 16-10
    entrance, 17-13
    exit, 17-12
    heat, 34-20
    hydraulic, 19-2
    insulated, 34-8
    insulation, 34-9
    manufacture, 63-13
    material, 16-9
    maximum velocity in, 17-3
    parallel, 17-20
    pressure vessel, 53-14
    series, 17-19
    wall thickness, 19-3
Piping, 64-5
Piping geometry factor, 17-13
Piston displacement, 32-3
Pit, ash, 22-3
Pitch, 36-9, 56-2
    angle, 52-14
    axial, 52-12
    base, 52-10
    circle, 52-10
    circle velocity, 52-10
    circular, 52-10
    coil, 52-5
    cone, 52-14
    control, blade, 20-3
    diametral, 52-10
    helix, 7-12
    length, V-belt, 52-16
    -limiting, 31-5
    normal, 52-10
        circular, 52-12
        diametral, 52-12
    point, 52-10
    spring, 52-5

transverse circular, 52-12
Pitot-static gage, 17-22
Pitot tube, 16-4, 26-21
Pitting, 21-11
Pitting resistance, gear, 52-14
Plain carbon steel, 45-5
Plain end, 52-5
Plaintiff, 70-5
Plan, 61-20 (see also type)
Planar view, 2-2
Plane, 7-6
    angle, 6-1
    complex, 3-7
    motion, 55-11, 56-2
    sandwiched, 34-6
    tangent, 8-5
Planet
    carrier, 57-5
    gear, 57-5
    pinion, 57-5
Planetary gear, 57-5
PLAN formula, 29-8
Plant, 60-13
    factor, 27-15
    layout, 61-14
    location problem, 61-13
Plasma
    diode, 31-9
    incineration, 66-34
Plastic
    behavior, 49-2 (ftn)
    hinge, 49-18
    impact, 56-15
    molding, 63-9
    pipe, 16-10
    strain, 46-2
    types, 66-16
    waste, 66-15
Plastics, tests on, 46-10
Plate
    absorber, 31-2
    -and-frame heat exchanger, 36-12
    cam, 57-10
    clutch, 52-19
    exchanger, 36-12
    flat, 36-3, 36-4, 51-17
    heat exchanger, 36-12
    heat rate, 27-5
    jet force on, 17-28
    -type element, 53-3
    vibration, 58-17
Plateau, 47-2
Plow steel, 64-2
Plug
    flow, 29-10 (ftn)
    valve, 16-11
Plugging, 66-37
Plume, 38-12
Plunger pump, 18-2
Pneumatic
    actuator, 42-1
    conveyor, 64-6
pOH, 21-8
Pohlhausen solution, 36-4
Point (see also type)
    approach, 29-20
    breakaway, 60-12
    choke, 32-8
    cloud, 22-7
    contraflexure, 8-2
    critical, 8-2, 24-3, 24-8, 26-4
    distribution, 20-24
    extreme, 8-2

force, 43-2
freezing, 47-2
function, 25-4, 25-7
inflection, 7-2, 8-2, 49-16
inversion, 25-3
null, 39-4
of application, 5-1
of contraflexure, 7-2, 49-16
of control, 61-25
of duty, fan, 20-7 (ftn)
of instability, 60-12
of rating, fan, 20-7 (ftn)
of sale, 69-37
operating, 18-16
operating, compressor, 32-8
operating, fan, 20-7
pinch, 29-20
pitch, 52-10
pour, 22-7, 52-20
set, 21-16
singular, 8-1, 8-3
-slope form, 7-5
stone wall, 32-8
terminal, 5-1
triple, 24-8, 47-4
yield, 46-2
Poise, 14-7
Poisoning, 66-37
Poisson distribution, 11-5
Poisson's ratio, 46-4
Polar
   amine, 21-15
   coordinate system, 7-3 (tbl), 7-4 (fig)
   form, 3-8, 7-5
   moment of inertia, 48-5
Pole, 60-8
Pole-zero diagram, 60-8, 60-9
Polishing, 21-16, 63-14
Pollutant, 66-3
Pollution, 66-2 to 66-4
Polychlorinated biphenyls, 66-15
Polygon method, 5-3
Polymer, 45-13
   atactic, 45-16
   crystallinity, 45-16
   isotactic, 45-16
   network, 45-15
   syndiotactic, 45-16
   synthetic, 45-14
Polymerization
   addition, 45-15
   condensation, 45-15
   degree, 45-14
   step reaction, 45-15
Polymorph, 45-18
Polynomial, 3-3
   characteristic, 4-7
   degree, 3-3
   Lagrangian interpolating, 13-2
   Newton's interpolating, 13-3
   standard form, 3-3
Polystyrene resin zeolite, 21-14
Polytropic
   compression, 15-13
   efficiency, 32-5
   equation of state, 25-2
   exponent, 15-13, 19-6, 25-2
   exponent, compression, 32-2
   head, 32-5
   process, 25-2, 25-9
   specific heat, 25-2
Pool boiling, 35-10
Poppet valve, 29-2

Population, 11-3
Position
   equilibrium, 58-2
   valve, 19-2
Positional number system, 12-1
Positive
   displacement pump, 18-2
   feedback, 60-4
Potential
   energy, 16-2, 23-2
   stream, 17-3
Potentiometer, 62-3, 67-2
Potentiometric sensor, 62-3
Poundal, 1-4
Pour point, 22-7, 52-20
Powder metallurgy, 63-10
Power, 23-5
   angle, 67-7
   blower, 20-4
   boosting, 29-14
   brake, 29-3, 29-4
   brake, fan, 20-4
   coefficient, 31-5
   cycle, 25-10
      nuclear, 30-1
      solar, 31-3
      vapor, 28-1
   design, 56-7 (ftn)
   electrical, 27-1, 67-3, 67-6
   emissive, 37-1
   factor, 67-7, 67-8
   fan, 20-4
   fluid, 19-1
   friction, 18-8, 20-5, 29-3
   hydraulic, 18-6, 19-1
   indicated, 29-3
   law (viscosity), 17-11
   number, 64-12
   phase, 67-9
   propulsion, 56-20
   pump, 18-2, 18-6
   screw, 52-22
   screw jacks, 52-22
   spring, 52-9 (ftn)
   specific battery, 31-10
   specific flywheel, 57-2
   stroke, 29-2
   thermal, 27-1
   tidal/wave, 31-7
   -to-heat ratio, 29-22
   -tower system, 31-3
   turbine, 17-29, 18-21
   versus torque, 52-10
   water, 18-6
   wind, 31-4
Powerhouse, 18-20
ppm (see parts per million)
Prandtl number, 1-9 (tbl), 35-2, 36-2
Prandtl's layer theory, 17-37
Precedence table, 61-3
Prechamber, 29-3
Precharge gas pressure, 19-6
Precipitation, 21-7
   -hardened stainless steel, 45-8 (ftn)
   hardening, 45-11, 47-9
   softening, 21-14
Precipitator, electrostatic, 66-26
Precision, 11-13, 61-16, 61-18, 62-2
Precision casting, 63-9
Precombustion chamber, 29-3
Precursor
   smog, 66-17
   trihalomethane, 66-18

Predetermined system, 61-15
Prefixes, SI 1-7
Preheater, 27-15
Preload, 51-11
Prerequisite of privity, 70-6 (ftn)
Present worth, 69-5, 69-11, 69-14
Press
   fit, 51-5
   forging, 63-8
   hydraulic, 15-14
Pressure, 14-2, 15-4, 24-5
   absolute, 14-3
   amplifier, 32-2
   angle, 52-11, 52-12, 57-11
   at a depth, 15-6 (ftn)
   at altitude, 20-2
   average, 15-6
   back, 27-11
   barometric, 14-3
   booster, 32-2
   burst, 19-3
   center of, 15-4, 15-8, 43-5
   collapsing, 51-3 (ftn)
   cut-in/out, 32-2
   design, 26-13
   die casting, 63-9
   drag, 17-34
   -drop endpoint, 21-16
   drop
      friction, 17-4 (ftn)
      hydraulic power, 19-4
      tube bundle, 36-9
   energy, 16-2, 23-3
   -enthalpy diagram, 24-11
   external, 15-14
   flow, 16-6
   fluid, 14-2
   friction, 20-11
   from gas, 15-12
   gage, 14-3
   gauge, 15-2
   head, 18-5 (ftn)
   hydrostatic, 15-4
   isentropic, 26-13
   line, 52-11
   loss, orifice, 17-25 (ftn)
   -maintenance pump, 65-3
   maximum allowable, 53-6
   mean effective, 29-8
   measurement, 15-1
   multiple liquid, 15-12
   on a dam, 15-11
   on curved surface, 15-10
   on plane surface, 15-6 to 15-8
   operating, pressure vessel, 53-6
   osmotic, 14-9
   partial, 24-17
   peak cylinder, 29-8
   pouring, 63-9 (ftn)
   rating, 19-3
   ratio, 24-10, 29-11, 32-1
      design, 26-5
      isentropic, 29-4
   reduction valve, 19-2
   -relief valve, 17-32
   saturation, 14-9
   static, 20-1
   test, hydrostatic, 53-15
   total, 16-3, 20-2, 24-17
   transverse, 17-22
   unit, 14-2
   vapor, 14-9
   velocity, 20-2
   vessel, 53-1, 53-4
Pressurestat, 42-1

Pressurized
    fluidized-bed combustion, 22-5
    liquid, 15-14, 27-7
    tank, 17-17
    water reactor, 30-1
Presswork, 63-6
Preventative maintenance, 61-10
Price, 61-2 (ftn)
Price-earnings ratio, 69-35
Primary
    air, 27-3
    creep, 46-15
    dimension, 1-7
    loop, 30-1
    unit, 1-7
Prime
    contractor, 70-4
    cost, 69-32
    gear set, 52-11
    mover, 18-8 (ftn)
Principal
    Archimedes', 15-14
    axis, 48-7, 54-2
    D'Alembert's, 56-5
    impulse-momentum, 17-26, 56-13, 56-14
    moment of inertia, 48-7
    organic hazardous constituent, 66-30
    stress, 49-5
    view, 2-2
    work-energy, 23-3, 23-4, 52-2
Priority, ethical, 71-2
Privity of contract, 70-6
Probabilistic model, 61-2
Probability, 11-2
    complementary, 11-4
    conditional, 11-4
    density function, 11-4
    joint, 11-3
    of acceptance, 61-25
Problem mixing, 10-8 (see also type)
Process
    adiabatic saturation, 38-7, 38-10
    air, 39-1 (ftn)
    air conditioning, 38-6
    annealing, 47-9
    Aston, 45-10
    base exchange, 21-14
    Bayer, 45-11
    Bessemer, 45-3
    cost accounting, 69-36
    dissolved oxygen, 45-3
    gasification, 29-15
    ion exchange, 21-14
    irreversibility, 25-12
    layout, 61-14
    L-D, 45-3
    open hearth, 45-4
    oxygen, 45-3
    puddling, 45-10
    reversible, 25-3
    steam, 29-21
    thermodynamic, 25-2 (see also type)
    waste, 66-3
    zeolite, 21-14
Producer gas, 22-7 (ftn)
Producer's risk, 61-25
Product, 21-5
    cross, 43-2
    layout, 61-14
    mixed triple, 5-4
    of inertia, 48-6, 54-2
    standard, 61-15

triple scalar, 5-4
    vector, 5-4 to 5-5
Production
    rate, 61-18
    standard, 61-15
Productivity, 61-20
Professional
    engineer, 72-1
    engineering exam, 72-3
Profile
    auxiliary view, 2-2
    cam, 57-10
    drag, 17-34
    drag coefficient, 36-4 (ftn)
    temperature, 34-5
Profit, 69-37
Profit and loss statement, 69-34
Program evaluation and review technique, 61-5
Programming
    dynamic, 61-12
    integer, 61-12
    linear, 61-12
    mathematical, 61-12
Progressive die, 63-6
Projectile motion, 55-4
Projection, 2-3
Projector, 2-2
Prony brake, 29-3, 62-12
Proof
    alcohol, 22-7
    load, bolt, 51-8, 51-12
    strength, 51-8, 51-12
    stress, 46-3 (ftn)
    test interval, 61-10
Propagation, 45-15
Propagation velocity, 68-4
Propeller
    fan, 20-3
    turbine, 18-23
Proper polynomial fraction, 3-6 (ftn)
Proper subset, 11-1
Properties
    gases, 24-15 (tbl)
    metallic, 21-2
    solutions, 14-15
Property
    business, 69-20 (ftn)
    chamber, 26-2
    class, bolt, 51-8
    critical, 26-5
    exit, 26-5
    extensive, 24-4
    intensive, 24-4
    stagnation, 26-2
    static, 26-3
    throat, 26-5
    total, 26-2
Property changes
    in ideal gases, 25-7
    in liquids, 25-10
    in solids, 25-10
Proportional
    action, 42-5
    plus integral and derivative control, 42-5
    plus integral control, 42-5
    region, 46-2
Proportionality limit, 46-2
Proportions, 21-3
Propped cantilever beam, 44-1

Proprietorship, sole, 70-1
Propulsion
    jet, 17-27
    power, 56-20
Protection area, sprinkler, 65-6
Protocol, Montreal, 33-2
Proton exchange membrane fuel cell, 31-8
Prototype, 17-38
Proximate analysis, 22-2
Proximity probe, 58-21
p-sequence, 3-11
p-series, 3-11
Pseudoplastic fluid, 14-7
Pseudo-static deflection, 58-10
Psychrometric chart, 38-3
Psychrometrics, 38-1
Public
    corporation, 70-2
    dealing with, 71-3
Puddling process, 45-10
Puffing, 66-28
Pulley, 43-11
    advantage, 43-11
    efficiency, 43-11
    foot, 64-9
    head, 64-9
    variable-pitch, 20-3
Pulsation
    damper, 18-2
    stabilizer, 18-2
Pulse, shock, 58-19
Pulverized-coal furnace, 27-3
Pulverizing, 64-7
Pump, 17-14, 18-2, 18-7 (see also type)
    boilerfeed, 27-6
    centrifugal, 18-2, 18-4
    drip, 27-14
    efficiency, isentropic, 27-6
    fire fighting, 65-3
    fluid power, 19-4
    homologous, 18-12, 18-18
    jockey, 65-3
    life, 19-5
    metering, 18-4
    mixed flow, 18-4 (ftn)
    overall efficiency, 18-8
    performance curve, 18-15
    power, 18-6
    shaft loading, 18-10
    similarity, 18-8
Pumping
    hydrocarbons, 18-19
    other liquids, 18-19
Punching, 63-7
Punch press, 57-2
Punitive damage, 70-5, 70-7
Purchase order, 70-3
PV cell, 31-4
PVC pipe, 16-10
p-V work, 23-3, 25-6
P-waste, 66-3
Pyritic sulfur, 22-3
Pyrolysis, 66-16
Pythagorean theorem, 6-2

Q
Quadrant, 6-3
Quadratic equation, 3-3
Quality, 24-9
    acceptance plan, 61-25
    circle, 61-29
    control, 61-22

factor, 60-10
  lot, 61-25
  of boiler feedwater, 21-11
Quantitative business analysis, 61-1
Quantity, economic order, 69-42
Quarter
  -bridge, 62-11
  -wave symmetry, 9-7 (tbl)
Quartile, 11-10
Quartz, 45-18
Quartz-crystal transducer, 15-2
Quasiequilibrium process, 25-2
Quasistatic process, 25-2
Quenching, 47-6
Queue, 61-6
Quick ratio, 69-35

R
Rad, 6-1 (ftn)
Radial
  camp, 57-10
  clearance, 52-20
  fan, 20-4
  -flow impeller, 18-4, 64-12
  -flow turbine, 18-20, 18-22
  interference, 51-5
  strain, 51-5
  tip fan, 20-4
  velocity, 55-9
Radian, 6-1, 6-5
Radiant
  heat transfer coefficient, 37-4
  superheater, 27-12
Radiation
  law, 37-1
  loss, 22-15
  nocturnal, 37-7
  solar, 37-7
  thermal, 37-1
Radiator
  efficiency, 34-16
  ideal, 37-1
Radical, 21-3
Radiographic inspection, pressure vessel,
  53-9
Radiography, 46-12
Radioisotope generator, 31-9
Radius
  critical, 34-10
  geometric mean, 57-3
  hydraulic, 16-6
  of gyration, 48-6, 54-2
Radix, 12-1
Radon, 66-16
Raimondi and Boyd method, 52-22
Rainwater runoff, 66-16
Rake angle, 63-2
Ram, hydraulic, 15-14, 19-6
Ramp, unit, 59-2
Random error, 62-2
Range, 11-10
  stress, 50-6
  water, 38-12
Rankine
  cycle, 28-3, 28-4
  scale, 24-5
Rank (matrix), 4-5
Ranking alternatives, 69-17
Rapper, 66-27
Rare earths, 21-2
Rate
  bleed, 27-8
  creep, 46-15
  heat, 27-5, 28-2
  heat release, 66-32
  metal removal, 63-3
  of change, 8-1

of quench, 47-6
  of return, 69-11, 69-14, 69-16
  of shear formation, 14-6
  of strain, 14-6
  production, 61-18
  spring, 52-2
  station, 27-5, 28-2
  steam, 27-15
  thermal capacity, 36-20
  water, 27-15
Rated
  capacity, pump, 65-3
  head, pump, 65-3
  value, motor, 67-11
Rat-holing, 64-5
Rating
  continuous duty, 29-4
  factor, tower, 38-12
  net, 22-16
  point, fan, 20-7 (ftn)
  table, fan, 20-5
Ratio
  air
    density, 20-2 (ftn)
    -fuel, 22-9, 22-12, 29-3
    -to-cloth, 66-24
  amplitude, 58-19
  aspect, 17-33, 20-14
  back work, 29-17
  beta, 17-23
  chip thickness, 63-2
  circulation, 29-20
  common, 3-11
  compression, 29-3, 29-5, 29-11,
    32-4 (ftn)
  contact, 52-11
  critical, 26-4
  critical pressure, 26-4, 26-12, 26-13
  current, 69-34
  cut-off, 29-11
  damping, 58-9
  design pressure, 26-5
  eccentricity, bearing, 52-20
  endurance, 46-9
  energy-efficiency, 23-6 (ftn), 33-2
  error, 60-4
  fatigue, 46-9
  feedback, 60-4
  filter, 66-24
  fuel-air, 29-3
  gear, 57-5
  gear velocity, 52-11
  humidity, 38-2
  induction, 20-24
  isentropic, 29-4
  lead, 52-14
  lever, 58-8
  mesh, 57-5
  modular, 49-19
  nitrogen/oxygen, 22-8
  of concentration, 38-13
  of transformation, 67-6
  Palmgren-Miner cycle, 50-7
  Poisson's, 46-4
  power-to-heat, 29-22
  pressure, 24-10, 29-11, 32-1
  price-earnings, 69-35
  saturation, 38-2
  sensible heat, 38-6, 41-3
  slenderness, 51-2
  specific heat, 15-13, 24-8, 26-3
  speed, 57-5
  spray, 38-11

test, 3-12
  train, 57-5
  turndown, 27-3, 27-13
  velocity, 52-11, 52-15, 57-5
  volume, 24-10
Rational
  function, 60-3
  number, 3-1
Rationalization, 3-8
Ray, 6-1 (ftn)
Rayleigh
  equation, 58-15 (ftn)
  flow, 26-19
  line, 26-19
  number, 35-3
  -Ritz equation, 58-15 (ftn)
Rayleigh's method, 58-5
R-chart, 61-22, 61-24
Reactance, 67-5
Reactant, 21-5, 21-7
Reaction, 43-6, 43-8
  chemical, 21-5
  endothermic, 21-9
  enthalpy, 21-9
  eutectic, 47-3
  exothermic, 21-9
  rate constant, 66-25
  site, 45-15
  stoichiometric, 21-6, 21-7
  time, 56-21
  turbine, 18-20, 18-22, 27-8
Reactive
  power, 67-6, 67-7
  substance, 66-3
Reactor (see also type)
  boiling water, 30-2
  nuclear, 30-1
  pressurized water, 30-1
  reformer, 31-8
  tank, 66-25
Real
  fluid, 14-2
  gas, 24-2, 24-19, 24-20
  machine, 23-6
  number, 3-1
  power, 67-6
Rebound, 56-15
Rebound test, 46-13
Receipt, 69-3
Receivable turnover, 69-35
Receiver tank, 32-2
Reciprocating
  compressor, 32-1, 32-2
  pump, 18-2
Reciprocity, 72-2
Reciprocity theorem for radiation, 37-4
Recirculating air bypass, 41-3
Recirculation, flue gas, 66-27
Recovery, 46-2
  depreciation, 69-26
  stage, 47-9
Recrystallization, 47-8, 47-9
Rectangular
  coordinate system, 7-3
  duct, 20-14
  form, 3-7
  hyperbola, 7-12
  rosette, 62-12
Rectified sinusoid, 67-5
Rectifier, 33-7

Rectilinear system, 55-2
Rectum, latus, 7-10, 7-11
Recuperative heater, 27-15
Recuperative heat exchanger, 36-10 (ftn)
Recuperator, 33-7
Recycle air, 38-11
Red
    hardness, 45-7 (ftn), 63-4
    mud, 45-11
Reduced
    -coke process, 45-7
    equation, 10-1
    plan, 61-25
    variable, 24-19
Reducing flame, 63-12
Reduction in area, 46-6
Redundancy, 61-9
    degree of, 44-1
    reliability, 11-8
Redundant member, 43-7, 43-13
Reference value, 60-13
Reflectance, 68-3
Reflect angle, 6-1
Reflection, light, 68-3
Reflux pool-boiling receiver, 31-3
Reformer, 31-8
Reformulated gasoline, 22-6
Refracted light, 68-3
Refractory
    ash, 22-3
    material, 37-4
Refrigerant, 33-2, 33-9
Refrigeration, 33-1
    capacity, 33-3
    cycle, air, 33-6
    cycle, Carnot, 33-3
    effect, 33-3
    load, 41-1 (ftn)
    tons, 33-3
Refuse, 22-4
Refuse-derived fuel, 22-4, 66-32
Regain, static, 20-15
Regeneration, 21-17
Regenerative
    air preheater, 27-15
    cycle, 28-7
    heat exchanger, 36-10 (ftn)
Regenerator, 29-12, 29-18
    effectiveness, 29-19
    efficiency, 29-19
Region
    elastic, 46-2
    entrance, 36-5
    feasible, 61-12
    proportional, 46-2
    transition, 16-8
Register, 20-24
Registered engineer, 72-1
Registration, 72-1
Regression, linear, 11-13
Regula falsi method, 13-2 (ftn)
Regular
    function, 8-1
    lay, 64-2
Regulating system, 60-13
Regulation
    speed, 67-10
    voltage, 67-10
Regulator, 60-13
Reheat, 41-4
    control, 42-1
    factor, 28-6

temperature, 27-13
    Rankine cycle, 28-5
Reheater, 27-13, 30-1
Reheating turbine, 27-8
Reid vapor pressure, 66-10
Reinforcement, opening, 53-11
Related angle, 6-2
Relative
    abundance, 21-2
    atomic weight, 21-1
    bonus, 61-20
    earnings, 61-20
    efficiency, 29-3
    humidity, 38-2
    motion, 55-9
    position, 55-9
    risk, 69-34
    roughness, 17-4
    stability, 60-11
    time, 34-4
    velocity difference, 17-29
Reliability, 11-7, 11-8, 11-13, 61-9
Reliability (measurement), 62-2
Relief angle, 63-2
Reloading curve, 46-6
Reluctance, 62-4 (ftn)
Remainder method, 12-1
Remedial maintenance, 61-10 (ftn)
Remediation, 66-17
Removal
    efficiency, 66-25
    of heat, ventilation, 39-3
Renewable energy source, 31-1
Renewal, 61-11
Repair, mean time to, 61-9
Rephosphorized steel, 45-5
Replacement, 21-6, 61-11
Replacement study, 69-18
Repose, angle of, 56-6, 64-4
Repowering, 27-8 (ftn), 29-22
Research
    and development cost, 69-32
    octane number, 22-6
Reserve, 27-15 (*see also type*)
Reservoir
    heat, 25-10
    modeling, 59-7
    three-, 17-20
Residual
    fuel oil, 22-6
    gas, 32-3
    value, 69-16
Resilience, 46-6, 50-4
Resin, ion exchange, 21-15
Resistance, 34-3
    bridge, 62-9
    contact, 34-6 (ftn)
    electrical, 67-2
    motion, 14-2
    rolling, 56-7
    shear, 14-1
    temperature detector, 62-5
    thermal, 40-2
    vehicular, 56-19
Resistivity, electrical, 67-2
Resistor, 67-2, 67-3, 67-5
Resolution, 3-6, 9-8
Resonance, 58-10
Resource-recovery plant, 66-31
Response, 59-1 (*see also* Steady-state
    response)
    dynamic, 49-7
    equation, 60-14

frequency, 60-10, 60-11
    function, 59-1
    sinusoidal, 60-2
    static, 49-7
    time, 60-7, 60-9
    types, 60-1, 60-2
    unit step, 60-2
    variable, 59-2
Restitution, coefficient, 56-15
Resulfurized steel, 45-5
Resultant
    acceleration, 55-8
    equivalent, 43-4
    vector, 5-2, 5-3
Resuperheater, 27-13
Retainer fee, 70-5
Retentivity, 66-23
Retort stoking, 27-3
Retrofitting, 27-16
Return on investment, 69-11, 69-14,
    69-15
Reverberant
    field, 68-6
    room, 68-6
Reverberatory furnace, 45-10
Reverse
    -bias, 62-5 (ftn)
    transfer function, 60-4
Reversible
    adiabatic process, 25-8
    process, 25-3
Reverted gear set, 57-4
Revolution
    surface of, 9-5, 48-3
    volume of, 9-5, 48-3
Reyn, 52-19
Reynolds number, 1-9 (tbl), 16-7, 35-3,
    36-2
    condensation, 35-9
    critical, 36-3
    film, 35-9
    mixing, 64-12
    similarity, 17-38
Rheopectic fluid, 14-7
Rheostat, 67-2
Rich-quench-lean combustion, 29-15
Right
    angle, 6-2
    -hand rule, 5-4, 43-3
    triangle, 6-2
Righting moment, 15-16
Rigid
    body, 55-1, 56-8
    body motion, 56-2
    truss, 43-12
Rigidity, 49-2, 49-3
Rigidity, modulus, 46-7
Rim flywheel, 57-3
Rimmed steel, 45-4
Ring
    flange, 53-14
    gear, 57-5
    permutation, 11-2
    rotating, 57-2
Ringelman scale, 22-13
Riser, 27-4 (fig), 63-8, 65-2, 65-13, 65-14
Rise time, 60-2
Risk 61-25 (*see also type*)
    analysis, 69-43
    assessment, 61-27
    -free investment, 69-11
Rivet, 51-10
Rms value, 67-5

Roadway banking, 56-21
Rocket, 56-17
Rockwell hardness, 46-12
Rod, slider, 55-13
Rohsenow equation, 14-10
Roll, 56-2
Roller
    lifter, 57-10
    support, 43-7
Roll forming, 63-13
Rolling
    friction, coefficient of, 56-8
    resistance, 56-7, 56-19
Room
    clean, 39-5
    constant, 68-8
    sensible heat ratio, 38-6
Root, 3-3, 13-1
    double, 3-3
    finding, 13-1
    -locus diagram, 60-12
    -mean-square speed, 24-16
    -mean-square value, 11-10, 67-5
Rope, 43-11 (ftn), 64-1
Roping, 64-3
Rosette strain gate, 62-8
Rotameter, 17-21
Rotary
    action pump, 18-2, 18-3
    feeder, 64-6
    positive-displacement compressor, 32-1
        (ftn)
    turboblower, 32-1 (ftn)
Rotating
    compressor, 32-1
    disk, 57-2
    fluid mass, 57-3
    hub, 57-2
    machine, electrical, 67-10, 67-13
    ring, 57-2
Rotation, 55-6, 56-2
    free, 58-5
    of axis, 48-7
    of fluid mass, 15-19
    pipe bundle, 66-36
Rotational
    symmetry, 7-4
    system, modeling, 59-5
Rotor
    flywheel, 57-3
    motor, 67-12
    wind machine, 31-4
Rotor concentrator, 66-23
Roughness
    coefficient, 17-7
    height index, 2-4
    relative, 17-4
    specific, 17-4
    weight, 2-4
Rough-pipe flow, 17-5
Routh criterion, 60-12
Roving, 45-19 (ftn)
Row
    canonical form, 4-2
    equivalent matrix, 4-2
    matrix, 4-1
    operation, 4-2
    rank, 4-5 (ftn)
    -reduced echelon form, 4-2
    -reduced echelon matrix, 4-1
Rth-order difference, 3-10

Rubbish, 22-4
Rule, 71-1
    Amagat-Leduc's, 24-17
    Cramer's, 3-7, 4-6
    derived from, 66-4
    Gibbs, 24-18
    Gibbs phase, 47-3
    lever, 38-5, 47-2
    L'Hôpital's, 3-9
    Miner's, 50-7
    mixture, 66-4
    octave, 58-19
    of professional conduct, 71-1
    of signs, Descartes', 3-4
    right-hand, 5-4, 43-3
    Simpson's, 7-2
    trapezoidal, 7-1
Rules, SI system, 1-5
Runaway speed, 17-30
Runner, turbine, 18-21
Runoff, 66-16
Run-of-mine coal, 22-4
Run-of-the-nut method, 51-12
Rupture
    angle of, 46-7
    Mohr's theory, 46-7
    strength, 46-15
R-value, 34-3

S

Sabin, 68-9
Sacrificial anode, 21-12
Saddle point, 7-2
Safeguard, machine, 61-30
Safety
    factor of, 50-1
    margin of, 50-1
    risk assessment, 61-27
Salary plus fee, 70-5
Salometer scale, 14-4 (ftn)
Sal soda, 63-4
Salt, crevice, 21-11
Salvage value, 69-18
Sample, 11-2
    space, 11-3
    standard deviation, 11-10
    variance, 11-11
Sampling
    emissions, 66-20
    with replacement, 11-3
    work, 61-17
Sand
    blasting, 63-13
    green, 63-8
    molding, 63-8
Sandwiched plane, 34-6
SAP alloy, 45-19
Satisfier, 61-28
Saturated
    air, 38-2
    hydrocarbon, 22-1
    liquid, 24-2, 24-12
    liquid line, 24-3
    polymer, 45-15
    solution, 21-7
    temperature, 24-4
    vapor, 24-2, 24-12
    vapor line, 24-3
Saturation
    degree of, 38-2
    efficiency, 38-7, 38-11
    pressure, 14-9
    process, adiabatic, 38-7, 38-10

ratio, 38-2
    table, 24-10
Saybolt
    second, 14-6 (ftn)
    seconds
        Furol, 14-6 (ftn)
        universal, 14-6 (ftn)
        universal, conversions, 18-2 (ftn),
            19-1 (ftn)
    universal viscosity, 52-19
    viscometer, 14-6 (ftn)
Scalar, 5-1
    matrix, 4-1
    product, triple, 5-4
Scale
    API, 14-4
    Baumé, 14-5
    model, 17-38
    Ringelman, 22-13
Scaling, 27-12, 36-17
Scavenging, 21-15, 29-2
s-chart, 61-22
Schedule (pipe), 16-10
Schmidt number, 1-9 (tbl)
Scleroscopic test, 46-13
Scratch hardness, 46-12
Screw
    conveyor, 64-5
    -feeder, 64-5
    power, 52-22
Screwed fitting, 17-12 (ftn)
Scroll compressor, 33-7
Scrubber, 66-22
Scrubbing, 66-37
Seal, 66-35
Seal material, 19-2 (ftn)
Seam, 53-2 (ftn), 53-6 (ftn)
Seamless
    pipe, 63-13
    shell, 53-3
Secant
    formula, 51-3
    modulus, 46-4
    modulus (bulk), 14-13
Second
    acceleration, 57-10
    critical back-pressure ratio, 26-13
    critical shaft speed, 58-15
    derivative, 8-1
    law of thermodynamics, 24-7 (ftn),
        25-11
    moment of a function, 9-6
    moment of the area, 48-3
Secondary
    air, 27-3
    combustion chamber, 66-30
    creep, 46-15
Second-order, linear, 10-3
Section
    conic, 7-8
    dangerous, 49-11
    drawing, 2-3
    modulus, 49-11
Sections, method of, 43-15
Seebeck
    coefficient, 31-9
    effect, 62-6
Seider-Tate correlation, 36-5, 36-6
Selective
    catalytic reduction, 66-13, 66-37
    leaching, 21-11
    noncatalytic reduction, 66-38
    surface, 31-2

Selectivity, classification, 64-11
Self-energizing brake, 52-16
Self-locking brake, 52-16
Self-locking screw, 52-22
Self-regulating plan, 61-20
Semidry scrubbing, 66-37
Semi-infinite solid, 34-13
Semi-killed steel, 45-5
Semimajor distance, 7-11
Semiminor distance, 7-11
Semivariable cost, 69-31
Semivertex angle, 26-11
Sense, 5-1
Sensible heat, 24-8, 39-3
    factor, 38-6
    ratio, 38-6, 41-3
Sensible heating/cooling, 38-8
Sensitivity, 11-13, 60-5, 62-2
    analysis, 61-12, 69-43
    fatigue notch, 46-10
Sensitization, 45-9
Sensor, 42-1, 62-3
Separable, first-order, 10-3
Separation
    cyclone, 64-10
    factor, 66-26
    mesh, 64-10
    of terms, 9-2
Separator, moisture, 30-1
Sequence, 3-10
    arithmetic, 3-11
    convergent, 3-10
    divergent, 3-10
    geometric, 3-11
    harmonic, 3-11
    p-, 3-11
    standard, 3-11
Sequential combustion, 29-14
Serial reliability, 11-8
Series, 3-11
    arithmetic, 3-11
    circuit, electrical, 67-3
    expansion, 8-8
    finite, 3-11
    Fourier, 9-6
    galvanic, 21-11
    geometric, 3-11
    harmonic, 3-11
    hydrocarbon, 22-2
    infinite, 3-11
    Maclaurin, 8-7 (ftn)
    of alternating sign, 3-12
    p-sequence, 3-11
    pipe, 17-19
    pumps, 18-17
    Taylor's, 8-7
Server, 61-6, 61-7
Service
    factor, 40-4, 56-7 (ftn), 61-6, 61-7,
        67-11
        gear drive, 52-11
        V-belt, 52-16
    life, 69-20
    pressure vessel, 53-3
Servomechanism, 60-13
Set, 11-1
    laws, 11-1
    null, 11-1
    permanent, 46-2
    point, 21-16, 42-2
    theory, 11-1
    universal, 11-1
Setback, thermostat, 40-6
Setting, 27-4

Settleable solid, 66-41
Settling time, 60-2
Severability, 70-3
Shading
    coefficient, 41-2
    factor, 31-2
Shaft
    critical speed, 58-15, 58-17
    design, 51-13
    loading, pump, 18-10
    stepped, 58-5
    vibration, 58-15
    work, 25-6, 27-8
Shaker table, 58-21
Shape
    factor, 34-20, 37-2
    mode, 58-19
    of a fluid, 14-1
Shaving wheel fan, 20-4
Shear, 49-7
    diagram, 49-8
    flow, 51-15
    formation, rate of, 14-6
    horizontal, 49-10
    modulus, 46-7, 49-2, 51-13
    one-way, 49-7 (ftn)
    rate, 14-6, 17-13
    resistance, 14-1
    single/double, 51-10
    strain, 46-7
    strength, 46-8
    stress, 49-2
        beam, 49-9
        fluid, 14-6, 17-13
        theory, maximum, 50-3
    wind, 31-5
Shearing, 63-7
Sheave, 43-11, 64-2
Shelf, 47-2
Shell, 53-3
    -and-tube heat exchanger, 36-10
    torsion, 51-14
    -type element, 53-3
Sheradizing, 63-14
Sherwood number, 1-9 (tbl)
Shewhart control chart, 61-22
Shield, 61-30
Shielded metal arc welding, 63-12
Shift, 61-22
Shock, 58-19
    angle, 26-11
    pulse, 58-19
    transmission, 58-20
    treatment, 27-12
    wave, 26-9, 26-11, 26-13
Shoe brake, 52-17
Shore hardness, 46-13
Short
    circuit, electrical, 67-2
    -term exposure limit, 39-5
Shotpeen, 52-3
Shotting, 45-10
Shrink fit, 51-5
Shutoff
    pump, 65-3
    service valve, 16-11
Sick building, 66-5
Side friction factor, 56-21
Sidewall sprinkler, 65-4
Sideways friction, 56-21
Siemens, 67-2
Sieving, 64-10

Sigma
    -chart, 61-22
    phase (steel), 45-8 (ftn)
Sign
    alternating, series of, 3-12
    convention (thermal), 25-4
    convention, thermodynamic, 28-1
    of the function, 6-3
Signal analyzer, 9-7
Signature analysis, 9-7 (ftn)
Significant digit, 3-1, 12-1
Silica, 45-18
Siliceous-gel zeolite, 21-14
Silicon
    bronze, 45-12
    carbide, 45-19
Similarity, 17-38
    fan, 20-9
    law (pump), 18-18
    partial, 17-40
Similar triangle, 6-2
Simple
    cycle, 29-17
    harmonic motion, 58-2
    interest, 69-11, 69-39
    support, 43-7
Simplex
    method, 61-12
    pump, 18-2
Simplified boiling water reactor, 30-2
Simpson's rule, 7-2
Simultaneous
    equations, matrix, 4-6
    linear equations, 3-7
Sine function, integral, 9-8
Sines, law of, 6-5, 6-6
Single
    acceptance plan, 61-25
    -acting engine, 29-2
    -acting pump, 18-2
    butt, 51-10 (ftn)
    degree of freedom, 58-2, 59-2
    -duct system, 42-1
    -pass heat exchanger, 36-10
    payment cash flow, 69-3
    payment present worth factor, 69-5
    seal, 66-35
    shear, 51-10
    stage pump, 18-4
    -suction pump, 18-4
Singularity, 4-5
Singularity function, 59-3
Singular matrix, 4-1, 4-5
Singular point, 8-1, 8-3 (ftn)
Sink, 33-1 (ftn)
    energy, 17-15
    reservoir, 25-10
Sinking fund, 69-7
    factor, 69-7
    method, 69-22
    plus interest, 69-22
Sintered carbide, 45-19, 63-4
Sintering, 45-19 (ftn), 63-10, 63-11
Sinusoidal
    response, 60-2
    test, vibration, 58-20
Siphon, 17-19
SI system, 1-5, 1-7
Six nines rule, 66-30
Size, motor, 18-8
Sizing, 63-7
Skelp, 63-13
Skewness, 11-11

Skew symmetric matrix, 4-2
Skidding distance, 56-21
Skin friction, 17-5 (ftn), 17-34
Skin friction coefficient, 17-38, 36-4
Slab, 40-3, 45-10 (ftn)
Slab edge coefficient, 40-4
Slack time, 61-3, 61-4
Slag, 22-3, 29-15, 45-3
Slagging, 29-20 (ftn)
Slender column, 51-1
Slenderness ratio, 51-2
Slider
    crank, 55-13
    rod, 55-13
Sliding
    factor of safety, 15-12
    load unit, 27-16
    plate viscometer test, 14-6
    pressure operation, 27-16
    vector, 5-1 (ftn)
Sling psychrometer, 38-7
Slip, 18-9
    ammonia, 66-37 (ftn)
    electrical, 67-12
    pump, 18-3
Slipping, 56-9
Slope, 8-1
Slope form, 7-5
Slow-closing valve, 17-32
Sludge, 66-3 (ftn)
    conditioner, 21-16
    incineration, 66-34
Slug, 1-3
Slurry, friction loss, 17-11
Small angle approximation, 6-3
Smear metal, 63-5
Smelting, 45-11
Smith
    forging, 63-7
    line, 50-7 (ftn)
Smog, 66-17
Smoke, 22-13, 66-17
Smoke spot number, 22-13, 66-17
Smoothing coefficient, 69-42
Snagging, 63-5
Soderberg line, 50-6
Sodium-sulfur battery, 31-10
Softening
    point, Vicat, 46-11
    precipitation, 21-14
Soft material, 46-16
Softwood, 45-16
Soil
    incineration, 66-34
    washing, 66-38
Solar
    cell, 31-4
    collector, 31-2
    constant, 31-2, 37-7
    cooling load, 41-2
    energy, 31-2
    energy generating system, 31-3
    gear set, 57-5
    power cycle, 31-3
    radiation, 37-7
    system, active, 31-2 (ftn)
Soldering, 63-12
Sole proprietorship, 70-1
Solid, 24-2
    angle, 6-6
    height, 52-5
    oxide fuel cell, 31-8
    polymer fuel cell, 31-8

property, 24-11
-solution alloy, 47-1
suspension, 64-12
waste, 66-3
Solidus line, 47-1
Solubility, 47-3, 66-39
Solute, 14-9, 21-7
Solution, 14-9, 47-3
    buffer, 21-9
    complementary, 10-4, 58-10
    heat treatment, 47-9
    neutral, 21-8
    particular, 10-4
    saturated, 21-7
    trivial, 3-7 (ftn)
Solutions, 14-15, 21-7
Solvency, 69-34
Solvent, 14-9, 21-7
Solvent recovery, 66-23
Sommerfeld number, 52-22
Sone scale, 68-8
Sonic
    boom, 14-14
    velocity, 14-14, 68-4
Sorbent injection, 66-38
Sound, 68-3
    abatement, 68-9
    absorption coefficient, 68-8
    field, 68-5
    intensity, 68-4
    meter, 68-5
    power level, 68-7
    pressure level, 68-6
    speed of, 14-14, 68-4
    transmission class, 68-9
    wave, 68-4
Source, 33-1 (ftn)
    energy, 17-15
    reservoir, 25-10
Space
    heating, 29-21, 33-2
    mechanics, 56-19
Spacing, molecule, 14-2
Span, 17-33
Sparging, 66-38
Spark ignition, 29-2
Special
    damages, 70-7
    triangle, 6-2
Specific
    capacity, battery, 31-10
    collection area, 66-27
    cutting energy, 63-3
    energy, 16-1, 16-9, 23-1
        battery, 31-10
        flywheel, 57-2
    feed characteristic, 66-29
    fuel consumption, 29-3
    gas constant, 24-14
    gravity, 14-4
    heat, 24-7, 34-3
        gas, 24-15
        load, 36-14
        polytropic, 25-2
        ratio of, 15-13, 24-8
        real gas, 24-20
    humidity, 38-2
    impulse, 26-8
    performance, 70-4, 70-5
    power, 31-10
    power, flywheel, 57-2
    properties, 24-4
    roughness, 17-4

speed, 18-11
    fan, 20-4
    suction, 18-15
    turbine, 18-20
    volume, 14-4, 24-5
    weight, 14-5
Specification limit, 61-22 (ftn)
Specimen, 46-1
Spectrum analyzer, 9-7
Specular reflection, 68-3
Speed, 55-3
    control
        compressor, 32-2
        constant, 60-13
        motor, 67-15
        turbine, 27-16
    cutting, 63-2, 63-3
    factor, 52-13
    of efflux, 17-16
    of molecule, 24-17
    of sound, 14-14, 68-4
    ratio, 57-5
    regulation, 67-10
    runaway, 17-30
    specific, 18-11
        fan, 20-4
        turbine, 18-20
    synchronous, 18-9, 67-12
Sphere, 7-12
    flow over, 36-10
    unit, 6-6
Spherical
    coordinate system, 7-3 (tbl), 7-4 (fig)
    defect, 6-6
    drop, 10-10
    excess, 6-6
    shell, conduction, 34-10
    tank, 51-5
    triangle, 6-5
    trigonometry, 6-5
Sphericity, 17-36
Spheroidize-annealing, 47-9
Spheroidizing, 47-9
Spider gear, 57-5
Spill, hazardous, 66-17
Spinel, 45-18
Spinning reserve, 27-15
Spitzglass formula, 17-9
Splitter, damper, 20-18
Spool valve, 19-2
Spot humidification, 39-3
Spray
    chamber, 38-2 (ftn), 38-10 (ftn)
    ratio, 38-11
    tower, 66-22, 66-39
Spring
    allowable stress, 52-3
    -back, 63-7
    buckling, 52-7
    clock, 52-9 (ftn)
    compression, 52-6
    constant, 49-2, 52-2
    constant, torsional, 58-5
    design, 52-2
    dynamic response of, 52-8
    energy, 23-3
    extension, 52-8
    fatigue loading, 52-4
    flat, 52-9
    flat coil, 52-9 (ftn)
    index, 52-2 (ftn), 52-4
    leaf, 52-10
    linear, 23-2, 23-3 (ftn)

material, 52-3, 52-4
nested, 52-5
pitch, 52-5
power, 52-9 (ftn)
rate, 52-2
rate equation, 52-5
torsion, 52-9
work, 23-2
Sprinkler, 65-4
  characteristics, 65-5
  head, 65-4
  protection area, 65-6
  separation, 65-6
  system, 65-2
  system design, 65-9
Sprue hole, 63-8
Spur gear, 52-10
Square matrix, 4-2
Squirrel
  cage fan, 20-3
  case rotor, 67-12
SSF (see Saybolt seconds Furol)
SSU (see Saybolt seconds universal)
Stability, 11-13, 15-16, 56-2, 60-11, 62-2
Stabilization, 66-17
Stable equilibrium, 56-2
Stack
  dew point, 66-4
  effect, 22-16
  effect head, 22-17
  friction loss, 22-18
  gas (see Flue gas)
Stage
  compression, 32-1
  gear, 57-4
  pump, 18-4
Staged
  -air burner, 66-38
  combustion, 66-38
  lean combustion, 29-14
Staging, optimum, 32-7
Stagnation
  energy, 6-4
  point, 16-4
  property, 26-2
  tube, 16-4
Stagnation energy, 16-4
Stainless steel, 21-11 (ftn), 45-5, 45-7,
  45-9
Stall
  angle, 17-33
  -limiting, 31-5
Standard
  atmosphere, 24-5
  bonus, 61-20
  condition, 14-4 (ftn), 22-2
  contract, 70-4
  cost, 61-15, 69-36
  cubic feet per minute, 20-2
  data, 61-16
  deviation, 11-10
  error of the mean, 11-11
  gravity, 1-2
  -hour plan, 61-20
  normal value, 11-6
  of care, 70-5 (ftn), 70-6
  orthographic view, 2-2
  polynomial form, 3-3
  sequence, 3-11
  state, chemical, 21-9
  temperature and pressure, 14-4, 22-2
  time, 61-15, 61-19
Standardized reactor, 30-2
Stanton number, 1-9 (tbl), 36-2

Star gear set, 57-5
Stat, 42-1
State, 24-4
  corresponding, 24-19
  equation, 60-14
  equation of, 24-13
  model, 60-14
  oxidation, 21-3
  standard, chemical, 21-9
  thermodynamic, 24-2
  variable, 59-2
Static
  balance, 54-2, 58-18
  deflection, 58-2
  discharge head, 18-5
  efficiency, 20-4
  energy, 16-2, 23-3
  equilibrium, 15-16, 43-6 (ftn)
  friction, 56-6
  pressure, 20-1
    probe, 17-21
    tube, 15-2
  property, 26-3
  regain, 20-15
  regain method, 20-21
  response, 49-7
  suction head, 18-5
Statically indeterminate, 43-7, 44-1
Statical moment, 48-2, 49-10
Statics, 43-1
Station, 61-18
Station rate, 27-5, 28-2
Statistical
  process control, 11-12, 61-22
  quality control, 61-22
Stator (pump), 18-3
Statute, 71-1
Statute of frauds, 70-2 (ftn)
Steady-
  flow availability function, 25-12
  flow energy equation, 25-5, 26-2
  flow open system, 25-1
  flow system, 25-5
  state component, 58-11
  state response, 60-2, 60-7
Steam
  admission valve, 27-16
  bleed, 27-8
  drum, 27-4
  flow through nozzle, 26-20
  friction loss, 17-9
  generator, 27-4
    heat recovery, 29-20
    nuclear, 30-1
  humidification, 38-10
  injection, 29-14
  process, 29-21
  pump, 18-2
  rate, 27-15
  stripping, 66-40
  table, 24-10
  tracing, 34-20
  turbine, 27-8
Steel, 45-5 47-5 (see also type)
  grate, 19-3 (ftn)
  pipe, 16-10
  pipe, black, 16-10
  pipe, water flow in, 17-8
  carbon, 45-5
  duplexed, 45-4
  free-machining, 63-4
  group, 45-7
  hardened, 47-7
  hardness, maximum, 47-7

killed, 45-5
rimmed, 45-5
semi-killed, 45-5
stainless, 45-5, 45-7
tool, 45-7, 63-4 (see also type)
Stefan-Boltzmann law, 37-2
Step
  function, 10-6
  reaction polymerization, 45-15
  unit, 59-2
Stepped
  cash flow, 69-10
  shaft, 58-5
Stepper motor, 60-13 (ftn)
Steradian, 6-6
Stiffener, beam, 49-18
Stiffness, 49-2
  spring, 52-2
  transformed, 58-8
Stirling
  cycle, 29-12
  engine, 29-12
Stochastic
  critical path model, 61-5
  model, 61-2
Stoichiometric
  combustion, 22-9
  reaction, 21-6, 21-7
Stoke, 14-8
Stoker, 27-3
Stoke's law, 17-35
Stokes number, 1-9 (tbl)
Stone wall point, 32-8
Stopping distance, 56-21
Storage
  energy, 31-9
  pile, bulk, 64-4
  tank, 66-17, 66-19
Stouhal number, 1-9 (tbl)
Stove, 45-2
STP (see Standard temperature
  and pressure)
Straddles, 13-1
Straight
  angle, 6-2
  -blade fan, 20-4
  condensing, 27-11 (ftn), 28-2
  extraction, 66-21
  fin, 34-18
  line
    forms, 7-5
    method, 69-21, 69-26
    plus interest, 69-22
  -tube boiler, 27-4
Strain, 49-2
  axial, 46-4
  circumferential, 51-5
  creep, 46-15
  diametral, 51-5
  elastic, 46-2
  energy, 46-6, 49-2, 49-11, 49-15,
    49-17
  energy theory, 50-4
  engineering, 46-2
  gage, 62-8, 62-9
  gauge, 15-2
  hardening, 46-4
  -hardening exponent, 46-5
  inelastic, 46-2
  lateral, 46-4
  physical, 46-5
  plastic, 46-2
  radial, 51-5

rate of, 14-6
sensitivity factor, 62-8
shear, 46-7
thermal, 49-4
true, 46-5
volumetric, 14-13
Strainer, hydraulic, 19-5
Stratification, 61-22
Stratified charge combustion, 29-2
Stratosphere, 15-12, 15-13 (ftn)
Stream
function, 17-3
potential, 17-3
Streamline, 16-7
Streamline flow, 16-7
Streamtube, 16-7
Strength
breaking, 46-3
coefficient, 46-5
compressive, 46-7
creep, 46-15
design, 49-2 (ftn)
endurance, 46-9
fatigue, 46-9
proof, 51-8, 51-12
rupture, 46-15
shear, 46-8
tensile, 46-3
ultimate, 46-3
versus factors, 46-16
yield, 46-2
Stress, 49-2
allowable, 50-1
gear, 52-13
pressure vessel, 53-5
alternating, 50-6
alternating combined, 50-8
bending, 49-10
biaxial, 49-5, 50-1
circumferential, 51-3, 51-4
combined, 49-5
concentration, 49-4
concentration factor, bolt, 51-12
concentration factor, fatigue, 46-9, 46-10
concentration, press-fit, 51-7
corrosion, 21-12
corrosion cracking, 45-9
effective allowable, pressure vessel, 53-9
engineering, 46-2
equivalent, 50-6
extreme, 49-5
flexural, 49-10
longitudinal, 51-4
mean, 46-8, 50-6
normal, 49-2
physical, 46-4
principal, 49-5
proof, 46-3 (ftn)
range, 50-6
-relaxation, 46-10
relief, 47-8
riser, 49-4
-rupture test, 46-15
shear, 49-2
shear, fluid, 14-6
-strain curve, 46-1, 46-3
tangential, 51-3
triaxial, 50-3, 50-5
true, 46-4
yield, 46-2
Stresspeen, 52-3
Stripping, 66-22 (ftn)
air, 66-38
factor, 66-40
steam, 66-40
Stroke, 29-2
Strong material, 46-16

Structural
cell, 43-12
laminate, 45-20
rope, 64-2
steel, 45-5
strand, 64-2
Structure
atomic, 21-1
composite, 49-19
Student's t-distribution, 61-16 (ftn)
Subautogenous waste, 66-28 (ftn)
Subcontract, 70-4
Subcooled liquid, 24-2, 24-11
Sub-isokinetic, 66-20
Sublimation, 24-9
Submatrix, 4-1
Submerged
arc welding, 63-12
fraction, 15-15
scraper conveyor, 66-6
Subset, 11-1
Subsonic, 14-14
Substance, lethal, 53-3
Success, 11-3
Suction, 18-4
gas, 33-1
head, net positive, 18-13
line, 33-9
specific speed, 18-15
throttling, compressor, 32-2
Sudden failure model, 61-11
Sulfate sulfur, 22-3
Sulfur, 22-3
dioxide, 66-17
in stack gas, 66-4
low-, coal, 22-4
oxides, 66-17
trioxide, 22-3, 66-4, 66-17
Sulfurous acid, 66-17
Summer, 60-4
Summer degree day, 41-2
Summing point, 60-14
Sum-of-the-years' digits method, 69-21, 69-26
Sun gear, 57-5
Sunk cost, 69-3
Superaustenitic stainless steel, 45-8
Supercharger, 29-10
Superconductivity, 45-18
Supercooled liquid, 45-17
Supercritical
cycle, 28-7
-pressure boiler, 27-4
-pressure turbine, 27-8
Superelevation, 56-21
Superferritic stainless steel, 45-8
Superfinishing, 63-5, 63-14
Superheat
table, 24-10
with Rankine cycle, 28-4
Superheated
steam, cooling, 36-20
vapor, 24-2, 24-3, 24-13
Superheater, 27-12 (see also type)
Superhigh-speed steel, 63-4
Super-isokinetic, 66-20
Superposed turbine, 27-8
Superposition, 49-16
loads, 43-16
method, 44-4
theorem, 10-6
Supersaturated solution, 21-7 (ftn)

Supersonic, 14-14
Supplementary
angle, 6-2
firing, 29-20
unit, 1-5
Supplier, dealing with, 71-2
Supplier's risk, 61-25
Supply
duct, 20-18
pump, fire fighting, 65-3
Support type, 43-7
Surcharge, 15-12
Surface
-acting agent, 66-10 (ftn)
condenser, 27-11
control, 25-1
durability, gear, 52-14
finish, 2-4, 46-9
free, 16-6
frictionless, 43-7
hardening, 47-10
of revolution, 9-5, 48-3
runoff, 66-16
selective, 31-2
temperature, 10-10
tension, 14-10
Surfactant, 66-10
Surge
chamber, 18-20
limit, 32-8
tank, 17-32
Surging, 52-8
Surroundings, 25-1
Susceptance, 67-5
Suspension, 21-7
Suspension, solids, 64-12
Suspicious event, 61-22
Swaging, 63-7
Sweating, 34-10
Swept volume, 29-2, 32-3
Swing
temperature, 39-2
test, 58-21
valve, 16-12
Symbols, 3-1, 53-7
Symbols, fluid power, 19-1
Symmetrical
function, 7-4
matrix, 4-2
waveform, 9-7 (tbl)
Symmetry
curve, types, 7-4
even, 9-7 (tbl)
full-wave, 9-7 (tbl)
half-wave, 9-7 (tbl)
odd, 9-7 (tbl)
quarter-wave, 9-7 (tbl)
Synchronous
capacitor, 67-13
motor, 67-13
speed, 18-9, 67-12
Syndiotactic polymer, 45-16
Syngas, 29-14, 66-18
Synthesis, 21-6
Synthetic
gas, 22-5, 22-8 (ftn), 29-14
polymer, 45-14
resin, 21-15
standard data, 61-16
System (see also type)
bin, 27-3
binary, 12-1
boundary, 25-1

bypass factor, 41-4
central, 27-3
closed, 25-1
consistent, 3-7
conventional duct, 20-17
curve, 20-7
diagram, 59-3
direct-firing, 27-3
effect, 20-7
equation, 59-3
force-couple, 43-4
function, 60-4
hexadecimal, 12-2
high-velocity, 20-17
inconsistent, 3-7
isolated, 25-2
linear force, 43-4
low-velocity, 20-17
modeling, 59-1
numbering, 3-1
octal, 12-2
of forces, 43-6
of units, 1-2
open, 25-1
performance curve, 18-15
riser, 65-2
SI, 1-5
steady-flow, 25-5
thermodynamic, 25-1
type, 60-13 (ftn)
unified numbering, 45-1
unit, 27-3
voltage, 67-5 (ftn)
wind energy conversion, 31-4
Systematic error, 62-2

T
T_{250} temperature, 22-3
Table
    divided difference, 13-3
    fan rating, 20-5
    multirating, 20-5
    normal shock, 26-9
    of integrals, 9-1 (ftn)
    periodic, 21-2
Tableau, 61-12
Tachometer control, 60-13
Tail
    curve, 11-6
    race, 18-20
    -water, 18-20
Take-off, branch, 20-16
Tall building, infiltration, 39-2
Tandem
    flywheel system, 57-4
    -pulley drive, 64-8 (ftn)
    seal, 66-35
Tangent, 7-2
    modulus (bulk), 14-13
    plane, 8-5
Tangential
    acceleration, 55-7
    force, gear, 52-12
    pressure angle, 52-12
    stress, 51-3
    turbine, 18-21 (ftn)
    velocity, 18-21, 55-7
Tangents, law of, 6-5
Tangible property, 69-20 (ftn)
Tank
    above-ground, 66-19
    compressed air, 32-2
    discharge, 17-16, 17-17

pressurized, 17-17
reactor, 66-25
spherical, 51-5
thick-walled (*see* Cylinder,
    thick-walled)
thin-walled, 51-3 (*see also* Pressure
    vessel)
time to empty, 17-18
underground, 66-17
Tapered-hub flange, 53-14
Tap hole, 15-2
Tappet, 57-10
Target substance, 62-3
Task, 61-15
Tax credit, 69-25
Taylor's
    equation, 63-5
    formula, 8-7
    series, 8-7
T-delta rosette, 62-12
Tedlar bag, 66-20
Teetering two-blade, 31-6
Teflon, 45-15
Telonomer, 45-14
TEMA (*see* Tubular Exchangers
    Manufacturers Association)
Temper
    carbon, 45-10
    designations, aluminum, 47-10
Temperature, 24-4, 24-17
    absolute, 24-4, 24-5
    approach, 36-13
    at a point, 34-7
    autoignition, 22-8
    bulk, 34-4, 35-3, 36-2, 36-3
    change, airflow, 36-24
    chip, 63-4
    coarsening, 47-8
    combustion, 22-15
    critical, 47-7
    cross, 36-14
    deflection, 46-10
    dew-point, 38-2
    difference, 36-13
        equivalent, 41-2
        logarithmic mean, 36-14
        terminal, 27-11, 27-14
    dry-bulb, 38-1
    effect on output, 62-12
    eutectic, 47-3
    fictive, 45-17
    film, 35-3, 36-3
    flame, 22-15
    flue gas, 22-13
    free-stream, 36-2
    glass transition, 45-17
    gradient, 34-5
    ignition, 22-8
    increase in pressurized liquid, 27-7
    inversion, 25-3
    net stack, 22-13
    profile, 34-5
    ratio, 36-20 (ftn)
    recrystallization, 47-8
    reheat, 27-13
    saturation, 24-4
    surface, 10-10
    swing, 39-2
    -time diagram, 47-2
    transition, 46-14
    wet-bulb, 38-2
Tempered martensite, 47-7
Tempering, 47-7, 47-9

Temporary hardness, 21-14
Tendency, central, 11-10
Tens complement, 12-3
Tensile
    strength, 46-3
    test, 46-1
Tension
    cable, 56-11
    effective belt, 64-8
    surface, 14-10
Tensor, 5-1
Teratogen, 61-27
Term
    general, 3-10
    in a sequence, 3-10
Terminal
    point, 5-1
    temperature difference, 27-11, 27-14
    value, 69-16
    velocity, 17-35, 20-24, 56-17
Terminator, 45-15
Terms
    harmonic, 9-6
    separation of, 9-2
Tertiary
    air, 27-3
    creep, 46-15
Test (*see also* type)
    comparison, 3-12
    fatigue, 46-8
    financial, 69-34, 69-35
    for convergence, 3-11
    hardness, 46-12
    Hurwitz, 60-12
    hydrostatic pressure, 53-15
    Jominy, 47-7
    on plastic, 46-10
    ratio, 3-12
    sliding plate viscometer, 14-6
    tensile, 46-1
    torsion, 46-7
Testing (*see also* type)
    boiler feedwater, 21-11
    isokinetic, 66-20
    nondestructive, 46-11
Tetrafunctionality, 45-15
Theorem
    binomial, 3-3
    buoyancy, 15-14
    Cauchy-Schwartz, 5-3
    central limit, 11-11
    de Moivre's, 3-8
    final value, 60-7
    Gibbs, 24-18
    initial value, 60-7
    Kalman's, 60-14 (ftn)
    Kutta-Joukowsky, 17-34
    linearity, 10-6
    Nernst, 24-6
    of calculus, fundamental, 9-4
    of Fourier, 9-6
    of Pappus, 9-5
    of Pappus-Guldinus, 48-3
    parallel axis, 48-4, 54-2
    perpendicular axis, 48-5
    Pythagorean, 6-2
    reciprocity, for radiation, 37-4
    superposition, 10-6
    time-shifting, 10-6
    transfer axis, 48-4
    Varignon's, 43-3

Theoretical
  maximum velocity, 26-2
  yield, 21-7
Theory
  Coulomb-Mohr, 50-2
  distortion energy, 46-8, 50-5
  failure, 50-1
  maximum
    normal stress, 50-1
    shear stress, 46-8, 50-3
    strain, 50-3
  net transport, 34-2
  octahedral shear-stress, 50-5
  of constant energy distortion, 50-5
  of influence, 61-28
  Prandtl's layer, 17-37
  queuing, 61-6
  set, 11-1
  strain energy, 50-4
  von Mises, 46-8, 50-5
  X, 61-28
  Y, 61-28
Therm, 40-1 (ftn)
Thermal
  arrest, 47-2
  backflushing, 27-12
  ballast, 66-30
  capacitance, 34-12
  capacity rate, 36-20
  coefficient of expansion, 14-12 (ftn)
  conductance, 34-2
  conductivity, 34-2
  cracking, 22-2
  deflection temperature, 46-11
  deformation, 49-3
  desorption, 66-40
  diffusivity, 34-4
  efficiency, 22-16, 25-10, 28-1, 29-3
  energy, 23-3 (ftn)
  entry length, 36-5
  equilibrium, 24-4, 25-5, 37-1, 37-6
  expansion, 49-3
  flywheel effect, 40-5
  gradient, 34-5
  inertia, 40-5
  lag, 40-5
  NOx, 66-13
  radiation, 37-1
  rating, power, 27-1
  resistance, 34-3, 40-2
  strain, 49-4
Thermionic generator, 31-9
Thermistor, 62-5
Thermocouple, 62-6
Thermodynamic
  relation, 25-6
  state, 24-2
  system, 25-1
Thermodynamics
  first law of, closed system, 25-4
  first law of, open system, 25-5
  second law of, 25-11
  third law of, 24-6
  zeroth law of, 24-4
Thermoelectric
  constant, 62-7
  generator, 31-9
Thermopile, 62-7
Thermoplastic, 45-16, 63-13
Thermosetting, 45-16, 63-13
Thermosiphon, 33-6 (ftn), 34-20
Thermostat, 42-1
  mixing, 42-2
  setback, 40-6

Thickness
  insulation, 34-10
  pressure vessel, 53-3, 53-9, 53-15
  tooth, 52-11
Thick-walled cylinder, 51-4
Thin-walled tank, 51-3
Third
  critical back-pressure ratio, 26-13
  derivative of motion, 57-10
  law, thermodynamics, 24-6
Thixotropic fluid, 14-7
Thompson effect, 62-6 (ftn)
Thread
  Acme, 52-22
  angle, 52-22
  angle, helix, 51-13
  stress concentration, 51-12
Threaded fitting, 17-12 (ftn)
Three-
  force member, 43-6
  moment equation, 44-5
  phase electricity, 67-8
  reservoir problem, 17-20
Threshold
  audibility, 68-4
  hearing, 68-6
  limit value, 39-5
  pain, 68-4
Throat
  effective weld, 51-13
  of nozzle, 26-5
Throttle
  control, 27-16
  temperature, 27-13
  valve, 27-16 (ftn)
Throttling
  process, 25-2, 25-3, 25-9
  range, 42-5
  service valve, 16-11
  valve, 27-12, 27-16 (ftn), 33-1
Through-variable, 59-3
Throw, 20-24
Thrust, coefficient, 26-8
Tidal power, 31-7
Tightened plan, 61-25
TIG welding, 63-12
Time (see also type)
  allowance, 61-16
  bed residence, 66-25
  constant, 34-12, 60-2, 60-9
  cycle, 61-18
  delay, 60-2
  effective, 26-9
  relative, 34-4
  response, 60-2 (ftn), 60-7, 60-9
  -series analysis, 9-7 (ftn)
  -shifting theorem, 10-6
  study, 61-15
  -temperature chart, 34-13
  -temperature diagram, 47-2
  -temperature-transformation curve,
    47-6
  to double, 69-8
  to empty tank, 17-18
  to triple, 69-8
  value of money, 69-5
  -weighted average, 39-5
Timed two-position control, 42-2
Tip, adiabatic, 34-17
Tire, 45-13 (ftn)
  -derived fuel, 22-4
  discarded, 66-18
Titania, 22-3 (ftn)
Toe (dam), 15-11
Tolerance, 2-4
Tone, pure, 68-3

Tonnage, 63-6
Tonne, 1-7
Tons, refrigeration, 33-3
Tool
  life, 63-5
  steel, 45-7
    group, 45-7
    type, 63-4
  wear, 63-5
Tooth thickness, 52-11
Top-dead-center, 29-2, 32-3
Topping
  cycle, 29-21
  turbine, 27-8
Torispherical head, 53-4
Torque, 43-2
  bolt, 51-12
  brake, 29-4
  braking, 52-19
  coefficient, 51-12
  electrical motor, 67-10
  hydrostatic, 15-11
  meter, 62-14
  on a gate, 15-11
  pump shaft, 18-10
  shaft, 51-13
  versus power, 52-10
Torricelli's equation, 17-16
Torsion
  balance, 14-10
  shell, 51-14
  solid member, 51-15
  spring, helical, 52-9
  test, 46-7
Torsional
  pendulum, 58-5
  spring constant, 58-5
Tort, 70-5
Tortuous-path valve, 27-12
Total
  dynamic head, 18-6
  energy, 16-2, 16-4
  energy line, 17-15
  equivalent temperature difference,
    41-2
  film coefficient, 35-11
  hardness, 21-14
  head, 16-3
  heat, 24-5
  impulse, 26-8
  pressure, 16-3, 20-2, 24-17
  pressure method, 20-24
  property, 26-2
  residual chlorine, 66-26
  response, 60-2
Toughened ceramic, 45-19
Tough material, 46-6, 46-16
Toughness, 46-6, 46-13
  modulus, 46-6
  notch, 46-13, 47-9, 53-5
  test, 46-13
Tow, 45-20 (ftn)
Tower
  cooling, 38-11
  forced draft, 38-11
  load, 38-12
  losses, 38-13
  packed, 66-22, 66-39
  shadow, 31-4
  spray, 66-39
  tray, 66-39
  unit, 38-12
Town gas, 22-7 (ftn)

Toxicity
dilution, 39-5
leaching characteristic, 66-3
Tracheid, 45-16
Tracing, heat, 34-20
Traction effort, 56-20
Tractive force, 56-20
Trade-in allowance, 69-18
Train
ratio, 57-5
value, 57-5
Transcendental function
circular, 6-2
derivative, 8-2
integral of, 9-1
Transducer, 15-2, 42-1, 58-21, 62-3
Transfer
axis theorem, 48-4
function, 60-3, 60-4
function method, 41-2
molding, 63-10
unit, 66-40
units, number of, 36-20
Transferred duty, 29-21
Transform
fast Fourier, 9-7
Laplace, 10-5
Transformation
coefficient of, 5-2
matrix, 5-3
Transformed stiffness, 58-8
Transformer, 67-6
Transgranular, 21-12
Transient
component, 58-11
heat flow, 34-11
modulus, 34-4
response, 60-2
Transition
element, 21-2, 21-3
metal, 21-3
region, 16-8
temperature, 46-14
temperature, glass, 45-17
Translation, 56-2
Translucent material, 68-3
Trans mer, 45-16
Transmissibility, 58-12
Transmissible vector, 5-1 (ftn)
Transmission
coefficient, 68-9
dynamometer, 62-12
factor, 26-18
gear ratio, 56-20
loss, 40-1, 68-9
rope, 64-1
Transmittance, 31-2, 66-17
Transmitted
load, gear, 52-12
power versus torque, 52-10
shock, 58-20
Transonic, 14-14
Transparent material, 68-3
Transportation problem, 61-12, 61-14
Transpose, 4-4
Transverse
axis, 7-11
circular pitch, 52-12
fin, 34-18
load, truss, 43-16
pressure, 17-22
sensitivity factor, 62-8
velocity, 55-9
wave, 68-4

Trapezoidal
cam profile, 57-10 (ftn)
rule, 7-1
Trash, 22-4
Traverse point, 66-20
Tray tower, 66-39
Treatment
all-volatile, 21-16
caustic phosphate, 21-16
zero-solids, 21-16
Tree system, 65-1
Trestle, 43-12
Triad, Cartesian, 5-2
Trial, 11-2
Triangle, 6-2
general, 6-5
oblique, 6-5
Pascal's, 3-3
right, 6-2
similar, 6-2
special, 6-2
spherical, 6-5
Triangular matrix, 4-2
Triaxial stress, 50-3, 50-5
Tridymite, 45-18
Trifunctionality, 45-15
Trigonometric
identity, 6-3
form, 3-7
function, 6-2
Trigonometry, spherical, 6-5
Trihalomethane, 66-18
Trihedral angle, 6-5
Trimetric view, 2-3
Trimmer die, 63-8, 63-9
Triple
integral, 9-3
junction cell, 31-4
point, 24-8, 47-4
product, mixed, 5-4
product, vector, 5-5
scalar product, 5-4
Triplex pump, 18-2
Tripling time, 69-8
Tripod, 43-22
Tritium, 21-2
Trivial solution, 3-7 (ftn)
Tropopause, 15-13 (ftn)
Troposphere, 15-13
Trough-electric system, 31-3
True
strain, 46-5
stress, 46-4
Truss, 43-12
deflection, 49-17
indeterminate, 44-6
transverse load, 43-16
TTT curve, 47-6
Tube
bundle, 36-9
condensation, 35-9, 35-10
impact, 16-4
-in-tube heat exchanger, 36-10
length required, 36-15
piezometer, 15-2
pitot, 16-4, 26-21
stagnation, 16-4
static pressure, 15-2
Tubeaxial fan, 20-3
Tubesheet feedwater heater, 27-14
Tubing
copper, 16-10
hydraulic, 19-2
Tubular Exchangers Manufacturers
Association, 36-11

Tumbling, 63-14
Tuned system, 58-13
Turbine, 17-14
aeroderivative, 29-14
axial-flow, 18-20, 18-23
combustion, 29-14
control, 27-16
efficiency, 27-8, 27-9
gas, 29-14
high-head, 18-21
impulse, 17-29, 18-21
meter, 17-21
power, 17-29, 18-21
reaction, 18-22
runner, 18-21
specific speed, 18-20
steam, 27-8 (see also type)
throttle valve, 27-16
type, 18-20, 18-22
Turboblower, 32-1 (ftn)
Turbocharger, 29-10, 32-1 (ftn)
Turbocompressor, 32-1 (ftn)
Turbulent flow, 16-8, 17-5
Turndown ratio, 27-3, 27-13
Turns ratio, 67-6
Tuyères, 27-3, 45-2
Twaddel scale, 14-4 (ftn)
Two
-angle formula, 6-4
-force member, 43-6, 43-12
-metal corrosion, 21-11
-point form, 7-5
points, distance, 7-7
-position control, 42-2
-stage expansion, 27-8
-stroke cycle, 29-2
-stroke, diesel, 29-10
tail limit, 11-12
Twos complement, 12-3
Type
I error, 61-25
II error, 61-25
n system, 60-13 (ftn)
number, 3-1
Types of views, 2-1

U
Ultimate
analysis, 21-4, 21-5, 22-2
CO_2, 22-12
strength, 46-3
strength design, 49-2 (ftn)
Ultrafinishing, 63-5
Ultrahigh-strength steel, 45-5
Ultralow
NOx burner, 66-35
particle arrester, 39-6
Ultrasonic
flowmeter, 17-22
machining, 63-6
test, 46-13
Ultrasound testing, 46-11
Unavailability, 61-9
Unbalanced force, 43-1
Unbiased estimator, 11-10
Unburned fuel loss, 22-15
Uncertainty analysis, 69-43
Unconstrained motion, 56-9
Undamped, forced vibration, 58-10
Underdamped, 58-9
Under-expansion, 26-13
Underfeed stoking, 27-3
Underflow, 64-10
Underground storage tank, 66-17

Undetermined coefficients, method of, 3-6, 10-4
Unicouple, 31-9
Unified numbering system, 45-1
Uniform
  acceleration, 55-3
  attack corrosion, 21-10
  cost, annual, 69-7
  exam, 72-2
  flow, 16-9
  gradient factor, 69-9
  motion, 55-3
  pressure, clutch, 52-19
  series cash flow, 69-3
  wear, clutch, 52-19
Union of sets, 11-1
Unit
  air conditioner, 33-2
  circle, 6-1
  conductance, 34-6
  impulse, 59-2
  impulse function, 10-6
  load method, 49-17
  matrix, 4-2
  operation, 27-1
  primary, 1-7
  ramp, 59-2
  sphere, 6-6
  step, 59-2
    function, 10-6
    response, 60-2
  system, 27-3
  tower, 38-12
  transfer, 66-40
  vector, 5-2, 43-2
  vector, Cartesian, 5-2
Unitary
  air conditioner, 33-2
  equipment, 42-1
Units
  consistent system of, 1-2
  of pressure, 14-2
Universal
  gas constant, 24-13
  gravitation, Newton's, 56-18
  sets, 11-1
  standard data, 61-16
Universe, 11-3
Unloading curve, 46-6
Unsaturated
  air, 38-2
  hydrocarbon, 22-1
  position, 45-16
Unstable equilibrium, 56-2
Unsteady heat transfer, 34-11
Upgrade, coal, 22-5
Upper
  bound, 9-4
  control limit, 61-23
  critical point, 47-4
Upright sprinkler, 65-4
Upset forming, 63-8
Uptime, 61-9
Upwind machine, 31-4
Urea, 66-13 (ftn), 66-38 (ftn)
Usage, electrical, 67-7
Use factor, 27-16
Utilization, fuel, 29-22
U-tube
  heat exchanger, 36-12
  manometer (see manometer)
U-waste, 66-3

V
Vacuum, 14-3, 15-2
  extraction, 66-40
  forming, 63-10
Value
  analysis, 69-43
  average (by integration), 9-4
  book, 69-23
  boundary, 10-2
  effective, 67-5
  engineering, 61-27, 69-43
  expected, 69-30
  face, 69-29
  heating, 22-14
  initial, 9-3, 10-1
  limiting, 3-9
  market, 69-23
  Pace, 30-2
  R-, 34-3
  residual, 69-16
  salvage, 69-18
  standard normal, 11-6
  threshold limit, 39-5
Valve (see also type)
  angle, 16-11
  ball, 16-11
  butterfly, 16-11
  check, 16-12
  flow coefficient, 17-13
  gate, 16-11
  globe, 16-11
  hydraulic control, 19-2
  lift, 16-12
  needle, 16-11
  nonreverse-flow, 16-12
  plug, 16-11
  poppet, 29-2
  pressure-relief, 17-32
  spool, 19-2
  steam throttle, 27-16
  swing, 16-12
  throttling, 27-12
  types, 16-11
Vaneaxial fan, 20-3
van der Waal's
  equation, 24-19
  force, 24-19
Vane, inlet, 20-3
Vapor
  compression cycle, 33-3
  condensing, 66-40
  condensing, film, 35-7
  -dominated reservoir, 31-7
  incinerator, 66-34
  power cycle, 28-1
  pressure, 14-9
    head, 18-5 (ftn)
    moist air, 38-2
    Reid, 66-10
  saturated, 24-2
  superheated, 24-2, 24-3
Variable, 3-2
  -area meter, 17-21
  -capacitance transducer, 62-4
  chart, 61-22
  cost, 69-31
  dimension of, 1-8 (tbl)
  flow rate, 20-3
  -inductance transducer, 62-4
  matrix, 4-6
  -pitch pulley, 20-3
  -reluctance transducer, 62-4

  resistor, 62-3
  throttle pressure operation, 27-16
  reduced, 24-19
Variance, 11-11, 69-36
Variation
  coefficient of, 11-1, 61-16
  of parameters, 10-4
Varignon's theorem, 43-3
Varying mass, 56-17
V-belt, 52-16, 56-7 (ftn)
Vector, 5-1
  addition, 5-3
  bound, 5-1 (ftn)
  Cartesian unit, 5-2
  cross product, 5-4
  dot product, 5-3
  equivalent, 5-1 (ftn)
  field, curl, 8-7
  field, divergence, 8-6
  fixed, 5-1 (ftn)
  function, 5-5
  gradient, 8-5
  multiplication, 5-3
  normal line, 8-6
  normal to plane, 7-6
  resultant, 5-3
  sliding, 5-1 (ftn)
  transmissible, 5-1 (ftn)
  triple product, 5-5
  unit, 5-2, 43-2
Vectors
  equal, 5-1
  orthogonal, 5-3
Vehicle
  dynamics, 56-19
  electric, 31-10
Veil, 45-20
Velocity, 55-3
  acoustical, 14-14
  average, 16-8
  bulk, 16-8
  burnout, 56-17
  burst, 57-3
  centerline, 16-8
  characteristic exhaust, 26-9
  coefficient, 17-16, 17-23
  coefficient, nozzle, 26-8
  control, 39-4
  -dependent force, 56-17
  distribution, 16-8
  drift, 66-27
  energy, 16-2 (ftn)
  escape, 56-19
  flooding, 66-22
  gradient, 14-6, 17-13
  head, 18-5
  of approach, 16-4, 17-16, 17-23
  of whirl, 18-10 (ftn)
  pitch circle, 52-10
  pressure, 20-2
  pressure, sprinkler, 65-7
  propagation, 68-4
  ratio, 57-5
  ratio, gear, 52-11, 52-15, 57-5
  reduction method, 20-19
  sonic, 14-14, 68-4
  tangential, 18-21, 55-7
  terminal, 17-35, 56-17
  terminal duct, 20-24
  theoretical maximum, 26-2
  transducer, 58-21
Vena contracta, 17-16, 17-24

Ventilation, 39-1
  dilution, 39-5
  energy to warm, 40-4
  for heat removal, 39-3
Venturi
  meter, 17-23
  nozzle, 26-5
Vertex, parabola, 7-10
Vertical
  angle, 6-2
  -axis machine, 31-4
  buckling, 49-18
  submersible turbine pump, 65-3
Vessel, pressure, 53-1
Vibration, 58-2
  damped, forced, 58-11
  damped, free, 58-9
  forced, 58-2
  free, 58-2
  isolation, 58-12
  natural, 58-2
  pick-up, 58-21
  sensor, 58-21
  shaft, 58-15
  thin plate, 58-17
Vibratory feeder, 64-6
Vibrometer, 58-21
Vicat softening point, 46-11
Vickers hardness, 46-12
View
  auxiliary, 2-2
  axonometric, 2-3
  central, 2-2
  dimetric, 2-3
  factor, 37-2
  isometric, 2-3
  normal, 2-1
  oblique, 2-2
  orthogonal, 2-2
  orthographic, 2-2
  orthographic oblique, 2-3
  perspective, 2-2, 2-3
  planar, 2-2
  principal, 2-2
  profile auxiliary, 2-2
  trimetric, 2-3
  types, 2-1
Virial
  coefficient, 24-19
  equation of state, 24-19
Virtual work, 49-17
Viscometer
  concentric cylinder, 14-6 (ftn)
  cup-and-bob, 14-6 (ftn)
  Saybolt, 14-6 (ftn)
  test, 14-6
Viscosity, 14-6
  coefficient of, 14-6
  conversion, 14-8 (tbl)
  dynamic, 14-6 (ftn)
  effect on head loss, 17-10
  grade, 52-19
  index, 14-9, 19-2 (ftn)
  kinematic, 14-8, 16-7
  limit, 52-20
  power law, 17-11
  Saybolt universal, 52-19
Viscous
  coefficient, 56-17
  damping, 58-9
  flow, 16-7
Vision, center of, 2-3
Vitrification, 45-17, 66-24, 66-40

Vitrified clay (pipe), 16-10
VOC (*see* Volatile organic compounds)
Void region, 7-3
Volatile
  inorganic compounds, 66-19
  liquid, 14-9, 21-7 (ftn)
  matter, 22-4
  organic compounds, 66-19
Volatility, 22-6
Voltage, 67-2
  divider, 67-4
  drop, 67-3
  line, 19-5 (ftn)
  phase, 19-5 (ftn)
  regulation, 67-10
Voltammetric sensor, 62-3
Volume
  clearance, 32-3
  control, 25-1
  cut-off, 29-11
  molar, 21-4
  of a fluid, 14-1
  of pressure vessel head, 53-16
  of revolution, 9-5, 48-3
  partial, 24-17
  ratio, 24-10
  specific, 14-4, 24-5
  splitter, 20-18
  swept, 32-3
Volumetric
  analysis, 22-2
  efficiency, 29-3, 32-4
  efficiency pump, 18-3
  expansion, coefficient, 49-3
  fraction, 22-2, 24-17
  strain, 14-13
von Mises-Hencky theory, 50-5
von Mises theory, 46-8, 50-5
Vortex, 64-10
Vorticity, 8-7
Vulcanization, 45-13

W

Wage incentive plan, 61-20
Wahl correction factor, 52-5
Wake drag, 17-34
Wall
  complex, 34-7
  composite, 34-6
  heat transfer, 40-2
  shear stress factor, 17-5 (ftn)
  thickness, pipe, 19-3
  thickness, pressure vessel, 53-3, 53-9,
    53-15
Warming, global, 66-11
Warning limit, 61-22
Wash, 31-4
Washer, air, 38-10, 38-11
Washing, soil, 66-38
Waste, 66-3 (*see also type*)
  fuel, 22-4
  -heat boiler, 29-20
  -to-energy facility, 66-31
Wastewater, 66-41
Wasting, 21-16
Water
  cooling, 36-14
  deionized, 21-16
  demineralized, 21-16
  film coefficient, 35-3
  flow, steel pipes, 17-8
  gas, 22-7 (ftn)
  hammer, 17-31

hammer factor, 19-3
hardness, 21-13
horsepower, 17-14
make-up, 21-15
-moderated reactor, 30-1
of crystallization, 21-4
of hydration, 21-4
power, 18-6
range, 38-12
rate, 27-15
reactor, boiling, 30-2
reactor, pressure, 30-1
spray humidification, 39-3
supply chemistry, 21-13
-tube boiler, 27-4
tubing, copper, 16-10
vapor, 66-19
Watt, 1-5
Watt density, 40-4 (ftn)
Wave
  Mach, 26-11 (ftn)
  normal shock, 26-9
  oblique shock, 26-11
  power, 31-7
  sound, 68-4
  weak shock, 26-11
Waveform
  frequency, 9-6
  period, 9-6
  symmetrical, 9-7 (tbl)
Wavelength, sound, 68-4
Waviness height, 2-4
Weak
  material, 46-16
  shock wave, 26-11
Wear, tool, 63-5
Web, 48-4
  crippling, 49-18
  stiffener, 49-18
Weber number, 1-9 (tbl), 17-40
Wedge, 56-7
Wedge angle, 63-2
Weibull distribution, 61-9 (ftn)
Weight, 1-1, 1-3, 54-1
  and proportion problem, 21-6
  atomic, 21-1
  combining, 21-5
  density, 1-3
  equivalent, 21-4
  formula, 21-4
  molecular, 21-4
  roughness, 2-4
  specific, 14-5
Weisbach equation, 17-7
Weld
  bead crack, 45-9
  decay, 21-10, 45-9
  examination, 53-9
  fillet, 51-13
  symbols, 53-7
  throat, effective, 51-13
  type, 63-11 (fig)
Welded
  -plate exchanger, 36-12
  seam, 53-2 (ftn)
Welding, 63-11
  bell, 63-13
  neck flange, 53-14
  pressure vessel, 53-6

Well, injection, 66-28
Western coal, 22-4
Wet
  -bulb depression, 38-2
  -bulb temperature, 38-2
  compression, 33-4
  scrubber, 66-22
  scrubbing, 66-37
Wetted perimeter, 16-6
Weymouth
  equation, 26-17
  formula, 17-9
Wheatstone bridge, 62-9
Whirl, velocity of, 18-10 (ftn)
Whisker, 45-19
Whistle-blowing, 71-3 (ftn)
White
  cast iron, 45-9, 47-8
  noise, 68-4
Whole
  air sample, 66-20
  depth, 52-10
Width, 2-2
Wide beams, 49-16, 52-9
Wind
  -axis machine, 31-4
  energy conversion system, 31-4
  farm, 31-6
  power, 31-4
  shear, 31-5
  velocity, effective, 39-2
Windage loss, 38-13, 67-15
Windbox, 29-23
Winfrey method, 69-16 (ftn)
Winter degree day, 40-5

Wire
  film coefficient, 35-7
  length, 52-5
  rope, 64-1
Wood, 22-3, 45-16
Work, 23-1
  compressor, 32-4
  done by force, 23-2
  done by linear spring, 23-2
  done by torque, 23-2
  -energy principle, 23-3, 23-4, 52-2
  external, 23-1
  factor, 61-16 (ftn)
  flow, 23-3, 25-6
  graphically finding, 25-4
  internal, 23-1, 46-6
  measurement, 61-15
  methods, 61-27
  performed by a force, 9-4
  p-V, 23-3, 25-6
  sampling, 61-17
  shaft, 25-6, 27-8
  virtual, 49-17
Working
  cold, 47-9
  depth, 52-10
  hot, 47-9
Worm
  drive, 52-14
  gear, 52-14
Worth
  future, 69-5
  present, 69-5
Wrap, cable, 64-3
Wrought iron, 45-10
Wye-connection, 67-9

X
x-bar chart, 61-22, 61-23
X-ray testing, 46-12

Y
Yaw, 56-2
Year-end convention, 69-3
Yield, 21-7, 69-28
  bond, 69-30
  classification, 64-11
  mass, 66-33
  point, 46-2
  stress, 46-2
Young's modulus, 46-2, 49-2

Z
Zebra mussel, 27-12
Zeolite, 21-14
Zeolite process, 21-14
Zero, 3-3, 13-1, 60-8
  -defect program, 61-28
  -discharge facility, 66-9
  -force member, 43-14
  -length loss, 20-15
  matrix, 4-2
  -solids treatment, 21-16
Zeroth law of thermodynamics, 24-4
Zone test, 61-22
Zuber equation, 14-10

Index of Selected Figures, Tables, Appendices, and Compiled Data

A

Absorption coefficient, A-131
Acceleration head, 18-3
Acceleration, uniform, 55-3
AC machine, 67-14
Accumulator exponents, 19-6
Acoustic velocity, 68-5
Advantage, mechanical, pulley block, 43-11
Air
    atmospheric, A-94
    composition, 22-8
    film coefficient, 35-5, 40-3
    properties, A-14, A-51, A-68
Air space, thermal conductance, 40-3
Alkalinity, 21-14
Allowable stress
    bolt, 53-15
    spring, 52-4
Allowable stress vs. temperature, steel, 53-5
Alloying ingredients
    cast iron, 45-10
    steel, 45-6
Alloys, properties, A-77, A-85
Alphabet, Greek 3-1
Altitude
    pressure at, 20-2
    properties at, A-83
Aluminum
    designations, 45-11
    heat treatment, 45-12
    temper, 47-9
Angle
    contact, 14-12
    shock, wedge, 26-11
Angle of repose, 64-4
ANSI piping symbols, A-30
Area, A-111
Areas, A-7
Arrangement factor, 37-3
Asbestos substitute, 66-6
Atmosphere
    composition, 22-8
    standard, A-83
Atmospheric air properties, A-14, A-94
Atmospheric pressure, 14-3
Atomic number, A-39
Atomic weight, A-39
A-Weighting, octave band, 68-7

B

Band, octave, 68-7
Base units, SI, 1-5
Beam deflection, A-112–A-114
Beam
    curved, 49-19
    vibration, 58-19

Bearing

Bearing
    journal, 52-21
    journal, correlation, A-118
Birmingham water gage tubing, A-28
Birmingham wire gage tubing, A-97
Block diagram, 60-6
Boiler tubing, steel, A-28
Bolt
    allowable stress, 53-15
    dimension, 51-10
    grade and type, 51-9
Brake, material, 52-17
Branch loss coefficient, 20-16
Brass tubing, dimensions, A-27
Buckling, spring, 52-8
Building material, properties, A-105
Bulk material, 64-4
Bulk modulus, water, 14-13
BWG, tubing, A-97
BWG tubing, dimensions, A-28

C

$CaCO_3$ equivalent, A-41, A-42
Cam dynamics, 57-12
Carbon dioxide fraction, 66-7
Carbon dioxide production, 66-14
Cash flow diagram, A-133
Cast iron, 45-9, 45-10
    alloying ingredients, 45-10
Ceiling, thermal coefficient, 40-2
Center of pressure, 15-10
Centroid, A-111
Ceramic, 45-18, 45-19
CFC replacement, 66-8
Characteristics, polyphase motor, A-129
Charge number, 21-3
Chart
    Heisler, A-90–A-92
    Mollier, A-50, A-67
Circular pipe, A-21
Clutch, material, 52-17
CO_2 fraction, 66-7
CO_2 production, 66-14
Coal properties, 22-5
Coefficient of drag, 17-35, 17-36
Coefficient of entry, air, 20-15
Coefficient of heat transfer, 36-15, 40-2
Coefficient of thermal expansion, 49-4, 66-36
Coefficient of thermal resistance, 62-6
Coefficient
    absorption, A-131
    branch loss, 20-16
    discharge, venturi, 17-24
    end restraint, 51-2
    film, 35-2
        air, 35-5, 40-3
        evaporization, water, 35-11
    fitting loss, air, 20-15
    flow, orifice plate, 17-25

friction, 56-7
    loss, 17-12
    noise reduction, A-131
    orifice, 17-17
    power, wind, 31-6
Cold working, 63-7
Column, end coefficient, 51-2
Combustion data, 22-10
Combustion, heat of, A-43
Combustion reaction, 22-9
Composite material, 45-20
Compounding factor, 69-7
Compound, volatile organic, 66-25
Compressed water, properties, A-49, A-66
Compressibility, 14-13
Compressibility factor, A-78
Concentration factor, stress, A-115
Conditions, standard, 14-4
Conductance, thermal, air space, 40-3
Conductivity, thermal, 34-3, 36-16, A-84
Confidence level z-values, 11-12
Conic sections, 7-9
Constant
    gravitational, 56-18
    Hazen-Williams, A-31
        sprinkler, 65-7
    Henry's law, 66-39
    thermoelectric, A-124
    universal gas, 24-14
Constants
    elastic, 46-8
    fundamental, inside front cover
Contact angle, 14-12
Control chart factor, quality, 61-23, 61-24
Control velocity, hood, 39-4
Conversion factors, A-1–A-6, inside front cover
Conversions
    power, 23-5
    viscosity, 14-8, A-19
Copper tubing, dimensions, A-26, A-27
Correction factor, LMTD, A-96
Correlation
    hardness, 46-14
    journal bearing, A-118
    tube-wire, 35-7
Critical point, 24-8
Crossflow, tube, 36-9
Culvert entrance loss, 17-19
Cumulative probability, A-123
Current, electric motor, 19-5
Curve
    learning, 69-42
    normal, A-12
Curved beam, 49-19
Cutting speed, 63-3
Cylinder, thick-wall, 51-5

D

Damping factor, 58-13
Darcy chart, 17-6
Darcy friction factor, A-32–A-35
DC machine, 67-14
Deflection, 49-3
  beam, A-112–A-114
Density, fluid, 14-3
Depreciation calculation, 69-27
Derived units, SI, 1-5, 1-6
Designations, 45-6
  aluminum, 45-11
  TEMA, 36-11
Diagram
  block, 60-6
  Mollier, A-50, A-67
Diameter, equivalent, 16-7
Diffusivity, thermal, 34-4
Dimension, bolt, 51-10
Dimensionless groups, 1-9
Dimensions
  brass tubing, A-27
  BWG, tubing, A-97
  copper tubing, A-26, A-27
  of variables, 1-8
  PVC pipe, A-29
  steel boiler tubing, A-28
  steel pipe, 16-11, A-22–A-25
Discharge coefficient, venturi, 17-24
Discharge, sprinkler, 65-5
Discounting factor, A-134–A-152
Distribution, t-, Student's, A-121
Divided flow fittings, 20-17
Doubling time, 69-8
Drag coefficient, 17-35, 17-36
Dry air, composition, 22-8
Duct friction loss, 20-12, 20-13
Ductile transition temperature, 46-14
Duct velocity, 20-19
Dynamics, cam lobe, 57-12

E

Eccentricity, orbit, 56-19
Economic analysis factor, A-134–A-152
Effectiveness, heat exchanger, A-98, A-99
Efficacy, luminous, 68-2
Efficiency
  fin, 34-17
  pump, 18-8
  weld, joint, 53-9
Elastic constants, 46-8
Elastic deflection, beam, A-112–A-114
Elasticity, modulus, 46-2
Electric machine, 67-14
Electric motor formulas, 19-5
Electric motor size, 18-9
Electrostatic precipitator, 66-27
Elements, A-39, A-40
  types, 21-2
Ellipse, 7-11
Emissivity, A-100
Emissivity factor, 37-3
End restraint coefficient, 51-2
End treatment, spring, 52-6
Engineering economy factor,
  A-134–A-152
Enthalpy of formation, 21-10
Entrance loss, culvert, 17-19
Entry, coefficient of, air, 20-13
Epicyclic gear, 57-6, A-120
Equilibrium conditions, static, 43-6
Equilibrium diagram, iron-carbon, 47-5
Equivalent diameter, 16-7

Equivalent length, 17-12, A-127
  air, 20-13
  fitting, A-37, A-127
Equivalents, fluid height, 15-5
Error, terms, 62-3
Evaporization, film coefficient, water, 35-11
Expansion factor, 17-26
Expansion, thermal, coefficient of, 49-4, 66-36
Exponential growth factor, A-134–A-152
Exponents, rules, 3-5

F

Factor
  arrangement, 37-3
  compounding, 69-7
  compressibility, A-78
  correction, LMTD, A-95
  damping, 58-13
  discounting, A-134–A-152
  economic analysis, A-13–A-152
  emissivity, 37-3
  expansion, 17-26
  fouling, 36-17
  friction, Darcy, A-32–A-35
  gage, 62-8
  magnification, 58-12
  shape, 34-21, 37-2, 37-4
  stress concentration, A-115
    press fit, 51-8
  surface finish, 46-10
  utilization, queue, 61-7
Factors, conversion, A-1–A-6, inside front cover
Fanno flow, A-81
Fan similarity equations, 20-9
Film coefficient, 35-2
  air, 35-5, 40-3
  evaporization, water, 35-11
Fin efficiency, 34-17
Finish, surface, 2-4
  factor, 46-10
Fitting, equivalent length, A-37, A-127
Fitting loss coefficient, air, 20-15
Fittings, divided flow, 20-17
Flat plate, uniform pressure, 51-18
Flow
  factor, isentropic, A-79, A-80
  Fanno, A-81
  heat, transient, A-87–A-92
  Rayleigh, A-82
Flow coefficient, orifice plate, 17-25
Flow diagram, cash, A-133
Flow of water through schedule-40 pipe, A-36, A-125, A-126
Fluid density, 14-3
Fluid height equivalents, 15-5
Fluid power, symbols, A-38
Fluid, transfer, 31-3
Fluid velocity, maximum, 17-3
Fluoropolymer, 45-16
Flywheel, advanced, material, 57-4
Forging temperature, 63-7
Formation, enthalpy of, 21-10
Fouling factor, 36-17
Fourier waveforms, 9-7
Freon replacement, 66-8

Friction, coefficient, 56-7
Friction factor chart, Moody, 17-6
Friction factor, Darcy, A-32–A-35
Friction loss
  air duct, 20-12, 20-13
  water, schedule-40 pipe, A-36, A-125, A-126
Fuel, heat of combustion, A-43
Fuel oil, 22-6
Fuel properties, 22-6
Functions of related angles, 6-3
Fundamental constants, inside front cover

G

Gage factor, 62-8
Gage, strain, 62-13
Gauge
  sheet metal, A-117
  wire, A-117
Galvanic series, 21-12
Gas
  natural, 22-11
  specific heat, 22-2
Gas constant, universal, 24-14
Gasoline, 66-10
Gas properties, 24-15
Gas specific heat, 24-20
Gear, epicyclic, 57-6, A-120
Generator, 67-14
Glass, 45-18
Gravitational constant, 56-18
Greek alphabet, 3-1
Groups, dimensionless, 1-9
Gyration, radius of, A-111

H

Hardening, aluminum, 45-12
Hardness
  correlations, 46-14
  steel, 47-7
Hardness scale, Moh, 46-12
Hardness test, 46-13
Harmonic motion, simple, 58-6
Hazen-Williams constant, A-31
  sprinkler, 65-7
Head, acceleration, 18-3
Heat
  latent, water, 24-9
  specific, 24-8
    gas, 24-20
Heat exchanger designations, TEMA, 36-11
Heat exchanger effectiveness, A-98, A-99
Heat flow, transient, A-87–A-92
Heat generation, occupant, 39-3
Heat of combustion, A-43
Heat transfer, coefficient, 36-15, 40-2
Heat treatment, aluminum, 45-12
Heisler chart, A-90–A-92
Helix, 7-12
Henry's law constant, 66-39
HFC-134a, A-61, A-76
Hood control velocity, 39-4
Horsepower, hydraulic, 18-6, 18-7
Hot working, 63-7
Hydraulic accumulator exponents, 19-6
Hydraulic horsepower, 18-6, 18-7
Hydraulic radius, A-21

Hydrocarbon
    specific gravity, A-18
    vapor pressure, A-17
Hydrocarbon NPSHR, 18-20
Hyperbola, 7-11

I

Identities
    logarithm, 3-5
    trigonometric, 6-2–6-5
Illuminance, recommended, 68-3
Indefinite integrals, 9-1, 9-2, A-10
Index, viscosity, A-20
Insulating properties, building materials,
    A-105
Insulation, thermal properties, 34-10
Integrals, 9-1, 9-2
    indefinite, A-10
International standard atmosphere, A-83
Inversion temperature, 25-3
Iron
    cast, 45-9, 45-10
    refining, 45-3
Iron-carbon diagram, 47-5
Isentropic flow factor, A-79, A-80
Isolation material, 58-13

J

Joint, weld, efficiency, 53-9
Journal bearing, 52-21
Journal bearing correlation, A-118

L

Laminar flow, 36-5, 36-10
Laplace transform, A-11
Latent heat, water, 24-9
Learning curve, 69-42
Length, equivalent, 17-12, A-127
    air, 20-13
    fitting, A-37, A-127
Liquid, properties, A-15, A-16
LMTD correction factor, A-96
Lobe, cam, dynamics, 57-12
Logarithm identities, 3-5
Loss
    entrance, culvert, 17-19
    friction, air duct, 20-12, 20-13
    transmission, A-132
Loss coefficient, 17-12, 17-13
    air, 20-13
    branch, 20-16
Loudness, 68-4
Lubricating oil, 52-20
Luminous efficacy, 68-2

M

Machine, electric, 67-14
Magnification factor, 58-12
Masonry, M-factor, 40-5
Mass moment of inertia, A-119
Material, spring, 52-3, 52-4
Maximum fluid velocity, 17-3
Mechanical advantage, pulley block,
    43-11
Mechanical properties, metal,
    A-107–A-109
Mensuration, A-7–A-9
Mer, 45-15
Metal, properties, mechanical,
    A-107–A-109
Metals, properties, A-77, A-85
M-factor, masonry, 40-5

Modulus
    bulk, water, 14-13
    shear, 46-7
Modulus of elasticity, 46-2
    steel, 53-5
Moh hardness scale, 46-12
Mollier diagram, A-50, A-67
Moment of inertia, A-111
    mass, A-119
Moody friction factor chart, 17-6
Motion, simple harmonic, 58-6
Motor, 67-14
    polyphase, characteristics, A-129
Motor formulas, 19-5
Motor size, 18-9, 67-11
MSW, 66-33
Municipal solid waste, 66-33

N

Natural gas, 22-11
Natural response, 60-2
NEMA motor size, 18-9
Newton's gravitational constant, 56-18
Noise reduction coefficient, A-131
Nonlinear regression curves, 11-14
Nonmetals, properties, A-77, A-86
Non-SI units, 1-6
Normal curve, A-12
Normal shock, A-80
NPSHR, hydrocarbon, 18-20
Number
    atomic, A-39
    oxidation, 21-3
Nusselt equation parameters, 35-4
Nusselt number, 36-5, 36-8, 36-10

O

Occupant heat generation, 39-3
Octave band, 68-7
Oil
    fuel, 22-6
    lubricating, 52-20
Orbit eccentricity, 56-19
Orifice coefficient, 17-17
Orifice plate flow coefficient, 17-25
Oscillatory system, 58-6
Overall coefficient of heat transfer,
    36-15, 40-2
Oxidation number, 21-3
Oxygenates, 22-7

P

Parabola, 7-10
Partially filled pipe, A-21
Periodic table, A-40
Phase diagram, iron-carbon, 47-5
Physical constants, inside front cover
Pile, bulk, 64-4
Pipe
    circular, A-21
    PVC, dimensions, A-29
    schedule-40, flow of water through,
        A-36, A-125, A-126
    steel, dimensions, A-22–A-25
    temperature effect, 53-16
Piping symbols, A-30
Planetary gear, 57-6, A-120
Plan, single-sampling, A-122
Plastic, 66-16
Plate
    flat, uniform pressure, 51-18
    vibration, 58-17

Point
    critical, 24-8
    triple, 24-8
Poisson's ratio, 46-4
Polymer, 45-15–45-17, 66-16
    thermoplastic, 45-17
    thermosetting, 45-17
Polyphase motor, characteristics, A-129
Power coefficient, wind, 31-6
Power conversions, 23-5
Precipitator, 66-27
Press fit, stress concentration factor,
    51-8
Pressure
    atmospheric, 14-3
    center of, 15-10
    vapor, 14-9, A-17
Pressure at altitude, 20-2
Pressure measuring devices, 15-2
Pressure vessel, 53-3
    thickness, 53-9, 53-10
    weld, 53-8
Probability, cumulative, A-123
Projectile, 55-5
Properties
    air, A-14, A-51, A-68
    alloys, A-77, A-85
    atmospheric air, A-94
    cast iron, 45-9, 45-10
    ceramic, 45-18
    clutch/brake material, 52-17
    coal, 22-5
    composite material, 45-20
    flywheel, advanced, 57-4
    fuel, 22-6
    fuel oil, 22-6
    gas, 24-15
    gasoline, 66-10
    glass, 45-18
    insulating, building materials, A-105
    isolation material, 58-13
    liquid, A-15, A-16
    mechanical, metal, A-107–A-109
    metals, A-77, A-85
    nonmetals, A-77, A-86
    oxygenates, 22-7
    refrigerant-12, A-53–A-57, A-70–A-72
    refrigerant-22, A-58–A-60, A-73–A-75
    refrigerant HFC-134a, A-61, A-76
    saturated steam, A-44, A-46, A-62,
        A-64
    saturated water, A-93
    solids, A-77
    spring material, 52-3, 52-4
    standard atmosphere, A-83
    steam, saturated, A-95
    superheated steam, A-47, A-65
    thermal, insulation, 34-10
    thermoplastic, A-110
    transfer fluid, 31-3
    water, A-13, A-14
        compressed, A-49, A-66
    weld group, A-116
    wire, 64-2
    wire rope, 64-2
    wood, 45-17
Property, critical, 24-8
Psychrometric chart, A-101–A-104
Pulley block, mechanical advantage,
    43-11
Pump efficiency, 18-8
Pump velocity ratio, 18-3
PVC pipe, dimensions, A-29

Q

Quadrants, signs of functions, 6-3
Quality control chart, factor, 61-23, 61-24
Queue, utilization factor, 61-7

R

R-12, A-53–A-57, A-70–A-72
R-22, A-58–A-60, A-73–A-75
Radiation, emissivity, A-100
Radius of gyration, 48-6
Ratio, Poisson's, 46-4
Rayleigh flow, A-82
Reaction, combustion, 22-9
Recrystallization temperature, 47-8
Reduction coefficient, noise, A-131
Refining, iron, 45-3
Refrigerant-12, properties, A-53–A-57, A-70–A-72
Refrigerant-22, properties, A-58–A-60, A-73–A-75
Refrigerant HFC-134a, properties, A-61, A-76
Regain, static, 20-23
Regression, nonlinear curves, 11-14
Related angles, functions of, 6-3
Repose, angle, 64-4
Resistance, thermal, coefficient, 62-6
Resistivity, 62-6
Response
    natural, 60-2
    unit step, 60-2
Rope, wire, 64-2, 64-3
Roughness, specific, 17-4, A-31

S

Sample, plan, single-, A-122
Saturated steam, A-95
    properties, A-44, A-46, A-62, A-64
Saturated water, A-93
Schedule-40 pipe, 16-11
Sections, conic, 7-9
Series, galvanic, 21-12
Shape factor, 34-21, 37-2, 37-4
Shear modulus, 46-7
Sheet metal gauge, A-117
SHM, 58-6
Shock angle, wedge, 26-11
Shock, normal, A-80
SI base units, 1-5
SI-derived units, 1-5, 1-6
Signs of functions, 6-3
Similarity, fan, 20-9
Simple harmonic motion, 58-6
Single-sampling plan, A-122
Solids, properties, A-77
Solid waste, municipal, 66-33
Sound, speed, 68-5
Specific gravity, hydrocarbon, A-18
Specific heat, 24-8
    gas, 22-2, 24-20
Specific roughness, 17-4, A-31
Specific speed, 18-8, 18-11
Speed
    cutting, 63-3
    specific, 18-8, 18-11
    synchronous, 18-9
Speed of sound, 68-5
Sphericity, 17-36
Spring
    allowable stress, 52-4
    buckling, 52-8

end treatment, 52-6
    material, 52-3, 52-4
Spring wire gauge, A-117
Sprinkler
    discharge, 65-5
    protection area, 65-6
Sprinkler worksheet, A-128
SQC, 61-22
Stainless steel, characteristics, 45-8
Standard atmosphere, A-83
Standard atmospheric pressure, 14-3
Standard conditions, 14-4
Standard normal curve, A-12
Standard normal values, 11-12
Standard temperature and pressure, 14-4
Static regain, 20-23
Statistical error, terms, 62-3
Steam
    saturated, A-95
        properties, A-44, A-46, A-62, A-64
    superheated, properties, A-47, A-65
Steel
    AISI-SAE designations, 45-6
    allowable stress vs. temperature, 53-5
    alloying ingredients, 45-6
    hardness, 47-7
    stainless, characteristics, 45-8
    temperature effect, 53-5
Steel boiler tubing, A-28
Steel pipe dimensions, 16-11, A-22–A-25
Step, unit, 60-2
Stiffness, 49-3
STP, 14-4
Strain gauge, 62-13
Strength
    ultimate, spring, 52-4
    yield, 46-3
Stress
    allowable
        bolt, 53-15
        spring, 52-4
    cylinder, thick-wall, 51-5
Stress concentration factor, A-115
    press fit, 51-8
Student's t-distribution, A-121
Superheated steam, properties, A-47, A-65
Surface finish, 2-4
Surface finish factor, 46-10
Surface tension, 14-11
Symbols
    fluid power, A-38
    piping, A-30
Synchronous speed, 18-9

T

Table, periodic, A-40
t-distribution, Student's, A-121
TEMA designations, 36-11
Temperature
    forging, 63-7
    inversion, 25-3
    recrystallization, 47-8
    transition, ductile, 46-14
Temperature effect
    pipe, 53-16
    steel, 53-5
Temperature scales, 24-5
Tension, surface, 14-11
Test, hardness, 46-13
Thermal conductance, air space, 40-3
Thermal conductivity, 34-3, 36-16, A-84

Thermal diffusivity, 34-4
Thermal expansion, coefficient of, 49-4, 66-36
Thermal properties, insulation, 34-10
Thermal resistance, coefficient of, 62-6
Thermal treatment, aluminum, 45-12
Thermocouple, constant, A-124
Thermocouple material, 62-7
Thermoelectric constant, A-124
Thermoplastic polymer, 45-17
Thermoplastic, properties, A-110
Thermosetting polymer, 45-17
Thickness, pressure vessel, 53-9, 53-10
Thick-walled cylinder, 51-5
Three-dimensional areas, A-9
Time
    doubling, 69-8
    tripling, 69-8
Torsion, noncircular shape, 51-15
Transfer fluid, 31-3
Transform, Laplace, A-11
Transient heat flow, A-87–A-92
Transmissibility, 58-12
Transmission loss, A-132
Trigonometric identities, 6-2–6-5
Triple point, 24-8
Tripling time, 69-8
Truss, types, 43-11
Tube, crossflow, 36-9
Tube-wire correlation, 35-7
Tubing
    brass, dimensions, A-27
    copper, dimensions, A-26, A-27
    steel boiler, A-28
Two-dimensional areas, A-7

U

Ultimate strength, spring, 52-4
Uniform acceleration, 55-3
Units
    non-SI, 1-6
    SI base, 1-5
    SI-derived, 1-5, 1-6
Unit step, 60-2
Universal gas constant, 24-14
Utilization factor, queue, 61-7
U-value, thermal, 40-2

V

Valence, 21-3
Valves, 16-12
Vapor pressure, 14-9, A-17
Variables, dimensions of, 1-8
Velocity
    control, hood, 39-4
    duct, 20-19
    maximum fluid, 17-3
Velocity ratio, pump, 18-3
Venturi discharge coefficient, 17-24
Vessel, pressure, 53-3
    thickness, 53-9, 53-10
Vibration, 58-6
    beam, 58-19
    plate, 58-17
Viscosity
    conversions, A-19
    oil, 52-20
Viscosity conversions, 14-8
Viscosity index, A-20
VOC, 66-25
Volatile organic compound, 66-25
Voltage, motor, 67-15

W

Wall, thermal coefficient, 40-2
Water
  bulk modulus, 14-13
  compressed, A-49, A-66
  evaporization, film coefficient, 35-11
  flow through schedule-40 pipe, A-36,
    A-125, A-126
  latent heat, 24-9
  properties, A-13, A-14
  saturated, A-93
  vapor pressure, A-17
Water chemistry equivalent, A-41, A-42
Water horsepower, 18-6, 18-7
Water softening equivalent, A-41, A-42
Waveforms, Fourier, 9-7
Wedge, shock angle, 26-11
Weight, atomic, A-39
Weld
  joint, efficiency, 53-9
  pressure vessel, 53-8
Weld group, properties, A-116
Wind, power coefficient, 31-6
Wire, 64-2
Wire gauge, A-117
Wire rope, 64-2, 64-3
Wire-tube correlation, 35-7
Wood, 45-17
Working
  cold-, 63-7
  hot-, 63-7
Worksheet, sprinkler, A-128

Y

Yield strength, 46-3
Young's modulus, 46-2

Z

z-values for confidence levels, 11-12

PROFESSIONAL PUBLICATIONS, INC. BELMONT, CA

Resources to Help You on the PE Exam...and Beyond

Engineering Unit Conversions

Michael R. Lindeburg, PE   168 pages, 6 × 9, hardcover

If you have ever struggled with converting grams to slugs, centistokes to square feet per second, or pounds per million gallon (lbm/MG) to milligrams per liter (mg/L), you will immediately appreciate the time-saving value of this book. With more than 4500 conversions, this is the most complete reference of its kind. Covering traditional English, conventional metric, and SI units in the fields of civil, mechanical, electrical, and chemical engineering, this book puts virtually every engineering conversion at your fingertips.

ASME Mollier Chart of Steam Properties

24″ × 36″ poster

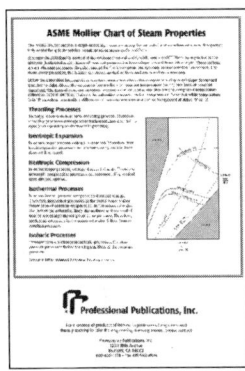

The ASME Mollier Chart is a graph of enthalpy versus entropy for saturated and superheated steam. For the PE candidate and the practicing engineer, this chart offers a quick way to solve certain types of steam cycle problems, including those involving isentropic expansion, isentropic compression, throttling processes, isothermal processes, and isobaric processes. The chart is small enough to use on your desk during the PE exam but large enough to be read to the nearest whole BTU/lbm.

Getting Started as a Consulting Engineer

Michael R. Lindeburg, PE   92 pages, 6 × 9, paperback

Have you ever considered being your own boss as a consulting engineer? The ability to work as a consultant is one of the main reasons that engineers obtain a PE license. If you are interested in consulting—and most engineers do consider this at some point in their careers—you will find all the information you need to get you started in this handy, no-nonsense guide. First, you'll see how consultants typically operate: how they organize their time and how they make their money. Next, you'll learn each of the necessary steps in setting up your own business, from getting appropriate business licenses and becoming incorporated to developing a referral base and negotiating contracts. Reading this short book before you launch your consulting career is essential if you want to avoid the pitfalls and surprises that can slow you down.

To order, contact
Professional Publications, Inc.
1250 Fifth Avenue • Belmont, CA 94002
800-426-1178 • fax 650-592-4519
Order on-line at www.ppi2pass.com